Useful relations

At 298.15 K

RT	$2.4790 \text{ kJ mol}^{-1}$
RT/F	25.693 mV
$RT \ln 10/F$	59.160 mV
kT/hc	207.23 cm^{-1}
kT/e	25.693 meV
$V_m^{\ominus}$	$2.4790 \times 10^{-2} \text{ m}^3 \text{ mol}^{-1}$ $24.790 \text{ L mol}^{-1}$

Conversion factors

1 eV	$1.602\,18 \times 10^{-19} \text{ J}$
	$96.485 \text{ kJ mol}^{-1}$
	8065.5 cm^{-1}
1 cal	4.184^* J
1 atm	101.325^* kPa
	760^* Torr
1 cm^{-1}	$1.9864 \times 10^{-23} \text{ J}$
1 D	$3.335\,64 \times 10^{-30} \text{ C m}$
1 Å	10^{-10} m^*

* Exact value

Mathematical relations

$\pi = 3.141\,592\,653\,59\ldots$

$e = 2.718\,281\,828\,46\ldots$

$\ln x = (\ln 10) \log x = (2.302\,585\ldots) \log x$

$$\frac{dx^n}{dx} = nx^{n-1}$$

$$\int x^n \, dx = \frac{x^{n+1}}{n+1} + \text{constant}$$

$$\int \frac{1}{x} \, dx = \ln x + \text{constant}$$

$$\int_0^\infty x^n e^{-ax} dx = \frac{n!}{a^{n+1}}$$

Prefixes

z	a	f	p	n	μ	m	c	d	Da	k	M	G	T
zepto	atto	femto	pico	nano	micro	milli	centi	deci	deka	kilo	mega	giga	tera
10^{-21}	10^{-18}	10^{-15}	10^{-12}	10^{-9}	10^{-6}	10^{-3}	10^{-2}	10^{-1}	10^{1}	10^{3}	10^{6}	10^{9}	10^{12}

Physical Chemistry

Sixth edition

P.W. Atkins

Professor of Chemistry, University of Oxford, and Fellow of Lincoln College

OXFORD

UNIVERSITY PRESS

Oxford Melbourne Tokyo

OXFORD

UNIVERSITY PRESS

Great Clarendon Street, Oxford OX2 6DP

Oxford University Press is a department of the University of Oxford
and furthers the University's aim of excellence in research, scholarship,
and education by publishing worldwide in

Oxford New York

Athens Auckland Bangkok Bogota Bombay Buenos Aires Calcutta
Cape Town Dar es Salaam Delhi Florence Hong Kong Istanbul
Karachi Kuala Lumpur Madras Madrid Melbourne Mexico City
Nairobi Paris São Paulo Singapore Taipei Tokyo Toronto Warsaw
and associated companies in Berlin Ibadan

Oxford is a registered trade mark of Oxford University Press

First edition 1978
Second edition 1982
Third edition 1986
Fourth edition 1990
Fifth edition 1994
Sixth edition 1998
Reprinted (with corrections) 1999

A catalogue record for this book is available from the British Library

Library of Congress Cataloging in Publication Data
(Data available for)

ISBN 0 19 850102 1 (Hbk)
0 19 850101 3 (Pbk)
0 19 269068 X (ISE)

Printed in the United States of America

General data and fundamental constants

Quantity	Symbol	Value	Power of ten	Units
Speed of light	c	2.997 924 58*	10^8	$\mathrm{m\,s^{-1}}$
Elementary charge	e	1.602 177	10^{-19}	C
Faraday constant	$F = N_A e$	9.648 53	10^4	$\mathrm{C\,mol^{-1}}$
Boltzmann constant	k	1.380 66	10^{-23}	$\mathrm{J\,K^{-1}}$
Gas constant	$R = N_A k$	8.314 51		$\mathrm{J\,K^{-1}\,mol^{-1}}$
		8.314 51	10^{-2}	$\mathrm{L\,bar\,K^{-1}\,mol^{-1}}$
		8.205 78	10^{-2}	$\mathrm{L\,atm\,K^{-1}\,mol^{-1}}$
		6.236 40	10	$\mathrm{L\,Torr\,K^{-1}\,mol^{-1}}$
Planck constant	h	6.626 08	10^{-34}	$\mathrm{J\,s}$
	$\hbar = h/2\pi$	1.054 57	10^{-34}	$\mathrm{J\,s}$
Avogadro constant	N_A	6.022 14	10^{23}	$\mathrm{mol^{-1}}$
Atomic mass unit	u	1.660 54	10^{-27}	kg
Mass				
electron	m_e	9.109 39	10^{-31}	kg
proton	m_p	1.672 62	10^{-27}	kg
neutron	m_n	1.674 93	10^{-27}	kg
Vacuum permittivity	$\varepsilon_0 = 1/c^2\mu_0$	8.854 19	10^{-12}	$\mathrm{J^{-1}\,C^2\,m^{-1}}$
	$4\pi\varepsilon_0$	1.112 65	10^{-10}	$\mathrm{J^{-1}\,C^2\,m^{-1}}$
Vacuum permeability	μ_0	4π*	10^{-7}	$\mathrm{J\,s^2\,C^{-2}\,m^{-1}}$
				$(= \mathrm{T^2\,J^{-1}\,m^3})$
Magneton				
Bohr	$\mu_B = e\hbar/2m_e$	9.274 02	10^{-24}	$\mathrm{J\,T^{-1}}$
nuclear	$\mu_N = e\hbar/2m_p$	5.050 79	10^{-27}	$\mathrm{J\,T^{-1}}$
g value	g_e	2.002 32		
Bohr radius	$a_0 = 4\pi\varepsilon_0\hbar^2/m_e e^2$	5.291 77	10^{-11}	m
Fine-structure constant	$\alpha = \mu_0 e^2 c/2h$	7.297 35	10^{-3}	
	α^{-1}	1.370 36	10^2	
Second radiation constant	$c_2 = hc/k$	1.438 77	10^{-2}	$\mathrm{m\,K}$
Stefan–Boltzmann constant	$\sigma = 2\pi^5 k^4/15h^3 c^2$	5.670 51	10^{-8}	$\mathrm{W\,m^{-2}\,K^{-4}}$
Rydberg constant	$\mathcal{R} = m_e e^4/8h^3 c\varepsilon_0^2$	1.097 37	10^5	$\mathrm{cm^{-1}}$
Standard acceleration of free fall	g	9.806 65*		$\mathrm{m\,s^{-2}}$
Gravitational constant	G	6.672 59	10^{-11}	$\mathrm{N\,m^2\,kg^{-2}}$

* Exact value.

Preface

In preparing this edition, I have been conscious of the need to retain rigour but to make the text more accessible. I have also aimed to keep track of the ever-evolving subject of physical chemistry without producing an over-bloated text or failing to deliver an authoritative account of the subject's largely unchanging core. As usual, I have taken the opportunity that a new edition provides to rework the text at all levels of presentation. Whatever laurels may have accrued from earlier editions, I have not found them soft enough to rest on.

The pedagogical devices have been thoroughly overhauled. Each chapter retains the *Synopsis* that started the chapters in the fifth edition, but now it is more succinct. The conceptual framework of each chapter is now summarized in the *Checklist of key ideas* at the end of the chapter. The actual check boxes can be used to record mastery of, or at least familiarity with, a topic. The *Further reading* sections now follow the chapters they aim to enrich, and I give references to recent accessible articles as well as to more authoritative texts and sources of data. As well as the numerous *Worked examples*, each with their mind-focusing *Method* section and accompanying *Self-test* (note the new name), there are now a number of *Illustrations*. These succinct components provide a no-fuss demonstration of how an equation is used (units always seem to give students trouble), and a brief *Illustration* should help to show how a calculation is done without all the fuss and pomp of a full *Worked example*. Some of the *Illustrations* are also accompanied by *Self-tests*.

There are major innovations in the end-of-chapter *Exercise* and *Problem* sections. The idea behind the two categories remains: an *Exercise* is a straightforward, direct application of an item in the text. A *Problem* is more complex and may draw on the literature. In addition to the scattering of literature-based problems, there is now an additional section provided by Carmen Giunta and Charles Trapp that draws explicitly on the literature. There has always been a problem about how to apportion solutions in the *Solutions Manual* for this text: some users welcome solutions to all *Exercises* and *Problems*; others consider that only half should be answered. In an attempt to please both camps, I have almost doubled the number of *Exercises* by providing a second linked companion *Exercise* in each case. All the 'a' set are answered, as before, in the *Student's Solutions Manual*; the solutions to the 'b' set, however, are given only in the (new) *Instructor's Solutions Manual*. Answers to the *Problems* are now divided approximately equally between the *Student's Manual* and the *Instructor's Manual*.

There is a further point concerning the *Problems*. As well as suggesting deletions of tired problems and providing new replacements, Charles Trapp has collaborated with Marshall Cady to develop a series of over-arching problems, which will be found at the end of each Part. These *MicroProjects* are designed to draw on knowledge from all the chapters in each Part, and to make use of literature data. The *MicroProjects* are intended to be helpful when reviewing the material of each Part of the text, and also provide some interesting applications. Some of them require quite challenging numerical techniques, such as non-linear regression, iterative solutions of sets of coupled equations, numerical integration and differentiation, and a variety of graphing procedures. It is therefore strongly recommended that they are solved by using mathematical software, such as MathCad, Mathematica, or similar programs.

Another obvious change, apart from the design, is the complete refurbishment of the artwork. The producers of the drawing software I use (Corel) produce new versions at more than twice the rate that I produce new editions of this text, so by the time that a revision is due, I am sorely tempted to use the new opportunities that their new version provides. My taste also changes as the years go by. So, I have redrawn all the line art, and have added many new pieces. The second colour has been used more extensively and rationally. Broadly speaking, colour denotes a more abstract component of the illustration; black is closer to reality.

The contents of the chapters themselves have undergone considerable revision. The lowest level of subheadings are now numbered to make assignments easier to specify precisely. The *Introduction and orientation* ('Chapter 0') has been completely rewritten with a change in philosophy. Now I use it to introduce some of the principal background concepts, such as the Boltzmann distribution. In that way the *Molecular interpretations* can become more meaningful. Those *Interpretations*, which were introduced in the fifth edition, have been extended in this edition. They enrich the presentation of thermodynamics and go some way towards helping users who wish to emphasize quantum concepts early in the course.

There has been a number of changes in the content of the chapters. To some extent, these changes are a consequence of incorporating what was *Further information* material into the body of the text. The *Further information* sections now provide accounts of globally important background material (such as classical mechanics and partial differentiation) rather than acting as appendages for individual chapters. Thus, Chapter 1 (gases) now includes a fuller discussion of kinetic theory and collisions and Chapter 10 (electrochemistry) contains an account of the Debye–Hückel theory. I have also redistributed material over the chapters: liquid surfaces have been transferred to Chapter 6 (pure substances) and colloids to Chapter 23 (macromolecules), both from their logical but pedagogically rather awkward home among solid surfaces (Chapter 28) in the fifth edition.

The reorganization of other material (such as the relocation of adiabatic changes into Chapter 4) has avoided a certain amount of repetition, and thus has saved space too. I have managed to find more space by adopting a more succinct style of presentation where I thought it would be acceptable. I hope my readers will distinguish the length of the actual text from the wealth of pedagogical aids and end-of-chapter material that is intended to help the student in all manner of different ways.

Finally, I would like to emphasize that the central layout of the text, its division into three parts and the order of chapters, which has remained unchanged over all its editions, is there more as a trademark than as a rigidly imposed structure. I am well aware that different instructors have different views about the order in which the subject is best presented. I have always taken care to present the material in a flexible way, and know from experience that instructors have no difficulty in adapting the text to their inclinations. This edition should suit them even more than previous editions through the reorganization of material and the greater number of subheadings.

There are two supplements for this text. The *Student's Solutions Manual* has been fully revised and contains full solutions to the 'a' *Exercises* and half the *Problems*. The *Instructor's Solution Manual* is new to this edition. As indicated above, it contains full solutions to all the 'b' *Exercises*, and the other half of the *Problems*.

So much for the description of this new edition. It would not have come about without the input of so many well-wishers, both commissioned and non-commissioned. I try to acknowledge the suggestions from individuals as they arrive, and apologize for not having space to thank them more publicly here. I hope they realize that they are part of the lifeblood of the text. Many were specifically consulted in the course of the preparation of this edition, and I would like to thank all those listed on p. vii.

As to the third component of the production of a text, the first two being the author and the advisors, I would like to thank my publishers for their advice and support throughout the planning, execution, and production phases of this vast and demanding project.

Oxford, September 1997   P.W.A.

Acknowledgements

Ms Susmita Acharya Cardinal Stritch College
Professor David Andrews University of East Anglia
Professor Russell G. Baughman Truman State University
Dr A.J. Blake University of Nottingham
Professor Michael Blandamer University of Leicester
Dr Gary Bond University of Central Lancashire
Dr Colin Boxall University of Central Lancashire
Dr Wendy Brown University of Cambridge
Dr Priscilla C. LeBrun Ashland University
Dr J.N. Chácon University of Paisley
Professor M.A. Chesters University of Nottingham
Professor A.L. Cooksy University of Mississippi
Dr Alan Cooper University of Glasgow
Dr Terence Cosgrove University of Bristol
Dr George Davidson University of Nottingham
Professor Ronald J. Duchovic Indiana University Purdue University Fort Wayne
Dr Andrew Fischer University of Glasgow
Dr Andrew A. Freer University of Glasgow
Professor Ronald S. Friedman Indiana University Purdue University Fort Wayne
Dr Amy E. Frost Mercer University
Dr Chris Gilmore University of Glasgow
Dr Carmen Giunta Le Moyne College
Dr Andrew Glidle University of Glasgow
Professor L. Peter Gold Pennsylvania State University
Professor Dixie J. Goss Hunter College, City University of New York
Dr Heike Gross University of Cambridge
Dr Marjorie Harding University of Liverpool

Professor Hal H. Harris University of Missouri, St Louis
Dr Michael Hey University of Nottingham
Professor Judith Howard University of Durham
Professor M. Lynn James University of Northern Colorado
Dr Robert G. Jones University of Nottingham
Professor Neil R. Kestner Louisiana State University
Professor Kathleen D. Knierim University of Southwestern Louisiana
Mr Ping Li Cleveland State University
Dr Li Liu Brookhaven National Laboratory, New York
Dr Roy Lowry University of Plymouth
Dr P.J. MacDougall Middle Tennessee State University
Dr Martin McCoustra University of Nottingham
Dr I.C. McNeil University of Glasgow
Professor Randy M. Miller California State University, Chico
Dr K.W. Muir University of Glasgow
Professor Lily Ng Cleveland State University
Professor Robert Pecora Stanford University
Dr Katharine Reid University of Nottingham
Professor Robert W. Ricci College of the Holy Cross
Professor Richard Schwenz University of Northern Colorado
Professor R.S. Sinclair University of Paisley
Dr S. Sotiropolous University of Nottingham
Dr Diane Stirling University of Glasgow
Dr Jeremy Titman University of Nottingham
Dr J.K. Tyler University of Glasgow
Dr Adrian Wander University of Cambridge
Dr Brian Webster University of Glasgow
Dr Richard Wheatley University of Nottingham
Professor Charles Wright University of Utah

Copyright acknowledgements

The permission of the respective copyright holders to reproduce tables of data and extracts from them is gratefully acknowledged as follows: Butterworths (28.1, 28.2 [© 1964]), Professor J. G. Calvert (17.1, 17.2 [© 1966], Dr J. Emsley (13.4, 13.5 [© 1989]), Chapman and Hall (16.3 [© 1975], 2.5, 9.1 [© 1986]), CRC Press (1.4, 9.1, 10.2, 10.7, 14.3, 22.1 [© 1979]), Dover Publications Inc. (12.2 [© 1965]), Elsevier Scientific Publishing Company (22.1 [© 1978]), Longman (1.3, 1.5, 2.2, 2.6, 3.1, 7.2, 9.1, 21.3, 22.5, 24.2, 24.4, 24.5 [© 1973]), McGraw Hill Book Company (1.4, 1.5, 2.3, 3.1, 3.2, 4.1, 13.4, 16.2, 22.2, 24.3, 24.7, [© 1975], 5.2 [© 1961], 25.1, 25.2 [© 1965], 3.2 [© 1968]), National Bureau of Standards (2.5, 2.6, [© 1982]), Dr J. Nicholas (25.1, 25.2, 25.4 [© 1976]), Oxford University Press (25.1, 25.2, 25.4 [© 1975], 28.3, 28.4, 28.5 [© 1974]), Pergamon Press Ltd (22.3 [© 1961]), Prentice Hall Inc. (23.1, 23.3 [© 1971]), John Wiley and Sons Inc. (7.1 [© 1975], 22.5 [© 1954], 23.2 23.4 [© 1961]). The sources of the material are quoted at the foot of each table.

The permission of the following individuals, institutions, and journals for reproduction of illustrations is gratefully acknowledged: NMR spectroscopy, John Wiley and Sons Ltd, Professor H. Günther (18.5), Professor G. Erlich (28.33), Dr J. Evans (21.21), Dr A. J. Forty (28.15), Dr J. Foster (28.21), Professor H. Ibach (28.10), Professor M. Karplus (27.24, 27.25), Dr W. Kiefer and Dr H. J. Bernstein (16.53), Professor D. A. King (28.30), Dr G. Morris (18.32), Dr A. H. Narten (24.9), *Proc. Roy. Soc.* (18.6), Oxford University Press (28.5, 28.6), Professor C. F. Quate (28.20), Professor M. W. Roberts (28.9, 28.12), Dr H. M. Rosenberg (28.6), Professor G. A. Somorjai (28.15, 29.18), *Scientific American* (23.21), Dr A. Stevens (21.21), Professor A. H. Zewail (27.9).

Summary of contents

Contents

Conventions

SI units and IUPAC conventions are used throughout, except in a small number of cases.

The default numbering of equations is (n); however, $[n]$ is used to denote a definition and $\{n\}$ is used to indicate that a variable x should be interpreted as $x/x^{\ominus}$ (for instance, $p/p^{\ominus}$), where $x^{\ominus}$ is a standard value. The $\{n\}$ convention simplifies the appearance of many expressions.

A subscript r attached to an equation number indicates that the equation applies only to a reversible change.

A superscript $^{\circ}$ attached to an equation number indicates that the equation applies only to an ideal system, such as a perfect gas or an ideal solution.

Cross-references of the form eqn n are to equations within the current chapter; those of the form eqn $N.n$ are to equations in Chapter N.

The symbol $p^{\ominus}$ denotes 1 bar (10^5 Pa) exactly and $b^{\ominus}$ denotes 1 mol kg^{-1} exactly.

When referring to temperature, T denotes a thermodynamic temperature (for example, on the Kelvin scale) and θ a temperature on the Celsius scale.

For numerical calculations, unless otherwise specified, assume that zeros in data like 10, 100, 1000, etc. are significant (that is, interpret such data as 10., 100., 1000., etc.).

Introduction: orientation and background

This chapter introduces some basic ideas which it will be useful to know about even before they are introduced formally. All the new material will be dealt with in greater detail later in the text.

Physical chemistry is the branch of chemistry that establishes and develops the principles of the subject. Its concepts are used to explain and interpret observations on the physical and chemical properties of matter. Physical chemistry is also essential for developing and interpreting the modern techniques used to determine the structure and properties of matter, such as new synthetic materials and biological macromolecules.

The structure of science

The observations that physical chemistry organizes and explains are summarized by scientific laws. A **law** is a summary of experience. Thus, we shall encounter the laws of thermodynamics, which are summaries of the relations between bulk properties, and particularly observations on the transformations of energy. We shall also encounter the laws of quantum mechanics, which summarize observations on the behaviour of individual particles, such as molecules, atoms, and subatomic particles. The first step in accounting for a law is to propose a **hypothesis**, which is essentially a guess at an explanation in terms of more fundamental concepts. Dalton's atomic hypothesis, which was proposed to account for the laws of chemical composition, is an example. When a hypothesis has become established, perhaps as a result of the success of further experiments it has inspired or by a more elaborate formulation (often in terms of mathematics) that puts it into the context of broader aspects of science, it is promoted to the status of a **theory**. We shall encounter a number of theories in this text: among them will be the theories of chemical equilibrium, atomic structure, and the rates of reactions.

A characteristic of physical chemistry (like other branches of science) is that, to develop theories, it adopts models of the system it is seeking to describe. A **model** is a simplified version of the system that focuses on the essentials of the problem. Once a successful model has been constructed and tested against known observations and any experiments the model inspires, it can be made more sophisticated and incorporate some of the complications that the original model ignored. Thus, models provide the initial framework for discussions, and reality is captured rather like a building is completed, decorated, and furnished. We shall encounter a number of such models. One example is the **kinetic model** of gases, in which a gas is regarded as a collection of particles in ceaseless, random motion. Another example is the **nuclear model** of an atom, and in particular a hydrogen atom, which is used as a basis for the discussion of the structures of all atoms. A third very important type of example is that of a **perfect gas**, which is an idealized model of the gaseous state of matter. This model, which is also the starting point of the discussion of **real gases**, or actual gases, is the basis of many thermodynamic expressions.

It is often convenient to preserve the form of equations developed on the basis of a simple model in any elaboration of the model. The advantage of such a procedure is that the appearance of many equations is then preserved and they remain familiar. An example of such a modification is the replacement of a concentration term in certain thermodynamic expressions (such as an equilibrium constant) by an effective concentration called an **activity**. Physical chemistry helps to make this a practically useful procedure by establishing a relation between the effective concentration and the true concentration.

Matter

A **substance** is a distinct, pure form of matter. The **amount of substance**, n (more colloquially 'number of moles' or 'chemical amount'), in a sample is reported in terms of a unit called a **mole** (mol). The formal definition of 1 mol is that it is the amount of substance that contains as many objects (atoms, molecules, ions, or other specified entities) as there are atoms in exactly 12 g of carbon-12. This number is found experimentally to be approximately 6.02×10^{23}.[1] If a sample contains N entities, the amount of substance it contains is $n = N/N_A$, where N_A is the **Avogadro constant**: $N_A = 6.02 \times 10^{23}\ \text{mol}^{-1}$. Note that N_A is a quantity with units, not a pure number. Conversely, if the amount of substance is n (for example, 2.0 mol O_2), then the number of entities present is nN_A (in this example, 1.2×10^{24} O_2 molecules).

A distinction is made in chemistry between extensive properties and intensive properties. An **extensive property** is a property that depends on the amount of substance in the sample. An **intensive property** is a property that is independent of the amount of substance in the sample. Two examples of extensive properties are mass and volume. Examples of intensive properties are temperature, mass density (mass divided by volume), and pressure. A **molar property**, X_m, is the value of an extensive property, X, of the sample divided by the amount of substance present in the sample. A molar property is intensive because the value of an extensive property X is proportional to the amount of substance, n, so $X_m = X/n$ is independent of the amount of substance in the sample. An example is the molar volume, V_m, the volume of a sample divided by the amount of substance in the sample (the volume per mole). The one exception to the notation X_m is the **molar mass**, which is denoted simply M. The molar mass of an element is the mass per mole of its atoms. The molar mass of a molecular compound is the mass per mole of molecules, and the molar mass of an ionic compound is the mass per mole of formula units.[2] The names 'atomic weight' and 'molecular weight' are still widely used in place of molar mass, but we shall not use them in this text.

---

1   More precise values of the fundamental quantities and conversion factors introduced in this chapter are given inside the front cover.

2   A formula unit is an assembly of ions corresponding to the chemical formula of the compound; so the formula unit NaCl consists of one Na^+ ion and one Cl^- ion.

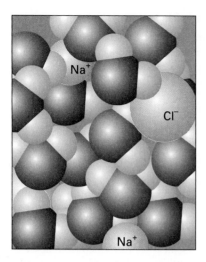

0.1 A schematic indication of the relative sizes of ions and molecules and the average separation of ions in a 1 M NaCl aqueous solution. There are typically about three H_2O molecules between ions. Cations tend to be found near anions, and vice versa. Cations are hydrated by weak bonding with the O atoms of neighbouring H_2O molecules; anions are hydrated by weak bonding through the H atoms.

The **molar concentration** ('molarity') of a solute in a solution refers to the amount of substance of the solute divided by the volume of the solution. Molar concentration is usually expressed in moles per litre ($mol\,L^{-1}$ or $mol\,dm^{-3}$; 1 L is identical to $1\,dm^3$). A solution in which the molar concentration of the solute is $1\,mol\,L^{-1}$ is prepared by dissolving 1 mol of the solute in sufficient solvent to prepare 1 L of solution. Such a solution is widely called a '1 molar' solution and denoted 1 M. The term **molality** refers to the amount of substance of solute divided by the mass of solvent used to prepare the solution. Its units are typically moles of solute per kilogram of solvent ($mol\,kg^{-1}$).

The development of an appreciation of events on an atomic scale is a valuable talent in physical chemistry. In a 1 M NaCl(aq) solution, the average separation between oppositely charged ions is about 1 nm, which is enough to accommodate about three H_2O molecules (Fig. 0.1). A dilute solution typically means a solution of molar concentration of no greater than about $0.01\,mol\,L^{-1}$. In such solutions, the ions are separated by about 10 H_2O molecules.

Energy

The central concept of all explanations in physical chemistry, as in so many other branches of physical science, is that of **energy**. A formal definition of this quantity will be given in Part 1; here we shall make use of the somewhat bald definition: *energy is the capacity to do work*. We shall often make use of the apparently universal law of nature that *energy is conserved*; that is, energy can be neither created nor destroyed. Therefore, although energy can be transferred from one location to another (as when water in a beaker is heated by electricity generated in a power station), the total energy available is constant.

0.1 Contributions to the energy

There are two contributions to the total energy of a system from the matter it contains. The **kinetic energy**, E_K, of a body is the energy it possesses as a result of its motion. For a body of mass m travelling at a speed v, the kinetic energy is $\frac{1}{2}mv^2$, so a heavy body travelling rapidly has a high kinetic energy. A stationary body has zero kinetic energy. The **potential energy**, V, of a body is the energy it possesses as a result of its position. The zero of potential energy is arbitrary. For example, the gravitational potential energy of a body is often set to zero at the surface of the Earth; the electrical potential energy of two charged particles is set to zero when their separation is infinite.

No universal expression for the potential energy can be given because it depends on the type of interaction the body experiences. However, there are two common types of interaction that give rise to simple expressions for the potential energy. One is the potential energy of a body of mass m in the gravitational field close to the surface of the Earth (a gravitational field acts on the mass of a body). If the body is at a height h above the surface of the Earth, then its potential energy is mgh, where g is a constant called the **acceleration of free fall**, $g = 9.81\,m\,s^{-2}$, and $V = 0$ at $h = 0$ (the arbitrary zero mentioned previously). Of greater importance in chemistry is the potential energy of a charged body in the vicinity of another charged body (an electric field acts on the charge carried by a body). If a particle (a point-like body) of charge q_1 is at a distance r in a vacuum from another particle of charge q_2, then their potential energy is given by the expression

$$V = \frac{q_1 q_2}{4\pi\varepsilon_0 r} \tag{1}$$

The constant ε_0 is the **vacuum permittivity**, a fundamental constant with the value $8.85 \times 10^{-12}\,C^2\,J^{-1}\,m^{-1}$. Note that, as remarked previously, $V = 0$ at infinite separation. This very important relation is called the **Coulomb potential energy** and the interaction it

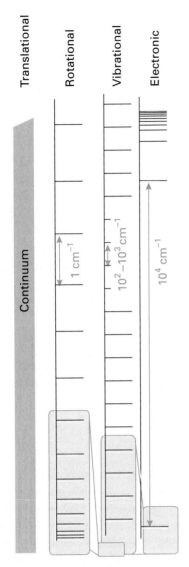

0.2 A representation of the quantization of the energy of different types of motion. Free translational motion in an infinite region is not quantized, and the permitted energy levels form a continuum. Rotation is quantized, and the separation increases as the state of excitation increases. The separation between levels depends on the moment of inertia of the molecule. Vibrational motion is quantized, but note the change in scale between the ladders. The separation of levels depends on the masses of atoms in the molecule and the rigidities of the bonds linking them. Electronic energy levels are quantized, and the separations are typically very large (of the order of 1 eV).

describes is called the **Coulomb interaction** of two charges. The Coulomb interaction is important in chemistry because we deal frequently with the interactions between the charges of electrons, nuclei, and ions.

0.2 Energy units

The SI[3] unit of energy is the **joule** (J), which is defined as[4]

$$1 \text{ J} = 1 \text{ kg m}^2 \text{ s}^{-2} \qquad \qquad [2]$$

A joule is quite a small unit of energy: for instance, each beat of the human heart consumes about 1 J. The unit is named after the nineteenth-century scientist J.P. Joule, who helped to establish the role of energy in science. The molar energy is the energy of a sample divided by the amount of substance; it is normally expressed in joules per mole ($J \text{ mol}^{-1}$) or a multiple of this unit, most commonly kilojoules per mole ($kJ \text{ mol}^{-1}$).

Although the joule is the SI unit of energy, it is sometimes convenient to employ other units. One of the most useful alternative energy units in chemistry is the **electronvolt** (eV): 1 eV is defined as the kinetic energy acquired when an electron is accelerated through a potential difference of 1 V. The relation between electronvolts and joules is $1 \text{ eV} \approx 1.6 \times 10^{-19}$ J. Many processes in chemistry involve energies of a few electronvolts. For example, to remove an electron from a sodium atom requires about 5 eV. Calories (cal) and kilocalories (kcal) are still encountered in the chemical literature: by definition, $1 \text{ cal} = 4.184$ J. An energy of 1 cal is enough to raise the temperature of 1 g of water by 1 °C.

0.3 Equipartition

A molecule has a certain number of degrees of freedom, such as its ability to translate (the motion of its centre of mass through space), rotate around its centre of mass, or vibrate (as its bond lengths and angles change). Many physical and chemical properties depend on the energy associated with each of these modes of motion. For instance, a chemical bond might break if a lot of energy becomes concentrated in it.

The **equipartition theorem** is a useful guide to the average energy associated with each degree of freedom when the sample is at a temperature T.[5] There are two parts to the theorem, one qualitative and the other quantitative. The qualitative part of the theorem tells us that all degrees of freedom have the same average energy. That means that the average kinetic energy of motion parallel to the x-axis is the same as the average kinetic energy of motion parallel to the y-axis and to the z-axis, and each rotational degree of freedom also has the same energy. That is, in a normal sample, the total energy is equally 'partitioned' over all the available modes of motion. One mode of motion is not especially rich in energy at the expense of another.

To introduce the quantitative part of the theorem we have to be more precise about what we mean by 'degree of freedom'. From now on, we shall refer to a *quadratic term* in the energy, a term for the kinetic or potential energy that appears as the square of a coordinate or a velocity (or momentum). For example, because the kinetic energy of a body of mass m free to move in three dimensions is $\frac{1}{2}mv_x^2 + \frac{1}{2}mv_y^2 + \frac{1}{2}mv_z^2$, there are three quadratic terms. The equipartition theorem then goes on to say that the average energy associated with each quadratic term is equal to $\frac{1}{2}kT$, where T is the temperature and k is a fundamental constant called the **Boltzmann constant**. This constant has the value $1.38 \times 10^{-23} \text{ J K}^{-1}$. The

3   We use SI units throughout this text; SI stands for *Système international*, a systematic, coherent set of units based on the metric system. Conversions to alternative units will be given where appropriate.

4   Equation numbers in square brackets indicate a definition.

5   Here and throughout this text, the symbol T will denote temperature on a scale that begins at 0 for the lowest attainable temperature, which is found empirically to lie at -273.15 °C. Later we shall see that this scale corresponds to the Kelvin scale.

Boltzmann constant is related to the gas constant by $R = N_A k$. In passing, it should be noted that one reason why R occurs in so many formulas, including those apparently wholly unrelated to gases, is that it is really the fundamental constant k appearing in disguise. The quantity kT, which has the value 4×10^{-21} J (more briefly 4 zJ, where z, zepto, is the uncommon but useful SI prefix for 10^{-21}) at 25 °C, or 26 meV, is an indication of the average energy carried by a mode of motion at a temperature T.

An important point is that the equipartition theorem is derived from classical physics, and when quantization is important (see below) the theorem is inapplicable. Broadly speaking, we can be confident about using it for average translational motion in gases, guardedly confident about applying it to the rotation of most molecules, and not apply it to their vibrational motion.

0.4 The quantization of energy

The great revolution in physics that occurred in the opening decades of the twentieth century and which introduced **quantum mechanics** is of crucial importance to chemistry. Chemistry is concerned with the behaviour of subatomic particles, particularly electrons, and it is essential to use quantum mechanics when dealing with such small particles. The feature of quantum mechanics that distinguishes it from the **classical mechanics** of Newton and his immediate successors is that matter has a wave-like character. That is, instead of particles and waves being distinct entities, particles have some of the properties of waves and waves have some of the properties of particles. For instance, if a particle has a linear momentum p (the product of mass and velocity, $p = mv$), then according to quantum mechanics it also has (in some sense) a wavelength, λ, given by the **de Broglie relation**:

$$\lambda = \frac{h}{p} \tag{3}$$

where h is the **Planck constant**, a fundamental constant with the value 6.6×10^{-34} J s.

Another feature of quantum mechanics is that energy is **quantized**, or confined to certain discrete values. These permitted energies are called **energy levels** and their values depend on the species. The quantization of energy is most important—in the sense that the allowed energies are widest apart—for particles of small mass confined to small regions of space. Consequently, quantization is very important for electrons in atoms and molecules, but usually unimportant for macroscopic bodies.

(a) The energies of material objects

For particles in containers of macroscopic dimensions the separation of translational energy levels is so small that for all practical purposes their translational motion is unquantized (Fig. 0.2). The separation between energy levels is small for molecular rotational motion, larger for molecular vibrational motion, and greatest for the energies of electrons in atoms and molecules. The separations of energy levels for a small molecule are about 10^{-23} J (0.01 zJ) for rotational motion (which corresponds to 0.01 kJ mol^{-1}), 10^{-20} J (10 zJ) for vibrational motion (10 kJ mol^{-1}), and 10^{-18} J (1 aJ, where a is another uncommon but useful SI prefix, standing for atto, and denoting 10^{-18}) for electronic excitation (10^3 kJ mol^{-1}). The relative values are consistent with the validity of the equipartition theorem for translational and rotational motion, but not for the other modes.

A final point in this connection is that we need to be aware that more than one **state** can correspond to a given energy level. For example, a molecule can rotate in one plane at a certain energy, but it may also be able to rotate in a different plane with the same energy; each different orientation of rotational motion corresponds to a distinct rotational state of the molecule. The number of individual states that belong to one energy level is called the **degeneracy** of that level (Fig. 0.3). If there is only one state of motion corresponding to a

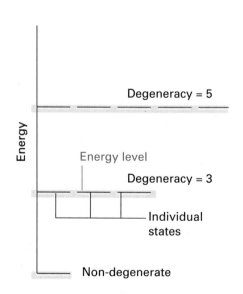

0.3 Several distinct states may correspond to the same energy. That is, each energy level may be degenerate. Three energy levels are shown here, possessing one, three, and five distinct states.

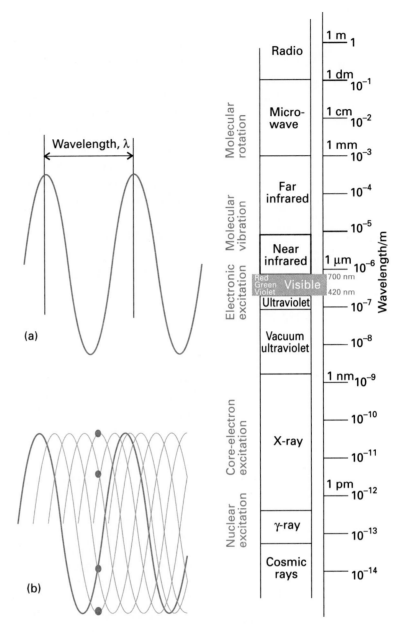

0.4 (a) The wavelength, λ, of a wave is the peak-to-peak distance. (b) The wave is shown travelling to the right at a speed c; at a given location, the instantaneous amplitude of the wave changes through a complete cycle (the four dots show half a cycle) as it passes a given point, and the frequency, ν, is the number of cycles per second that occur at a given point. Wavelength and frequency are related by $\lambda\nu = c$.

0.5 The regions of the electromagnetic spectrum and the types of excitation that give rise to each region.

particular energy, then we say that the level is non-degenerate. Be very careful to distinguish the energy levels (the ladder of possible energies) from the states that correspond to each 'rung' of the ladder.

(b) The energy of the electromagnetic field

An **electromagnetic field** is an oscillating electric and magnetic disturbance that spreads as a wave through empty space, the vacuum. The wave travels at a constant speed called the

speed of light, c, which is about 3×10^8 m s^{-1}. As its name suggests, an electromagnetic field has two components, an **electric field** that acts on charged particles (whether stationary or moving) and a **magnetic field** that acts only on moving charged particles. The electromagnetic field is characterized by a **wavelength**, λ, the distance between the neighbouring peaks of the wave, and its **frequency**, ν, the number of times per second at which its displacement at a fixed point returns to its original value (Fig. 0.4; the frequency is measured in hertz, where 1 Hz $= 1$ s^{-1}). The wavelength and frequency of a wave are related by

$$\lambda \nu = c \tag{4}$$

Therefore, the shorter the wavelength, the higher the frequency. The characteristics of a wave are also reported by giving the **wavenumber**, $\tilde{\nu}$, of the radiation, where

$$\tilde{\nu} = \frac{\nu}{c} = \frac{1}{\lambda} \tag{5}$$

The wavenumber can be interpreted as the number of complete wavelengths in a given length. Wavenumbers are normally reported in reciprocal centimetres (cm^{-1}), so a wavenumber of 5 cm^{-1} indicates that there are 5 complete wavelengths in 1 cm. It is useful to note that the relations $E = h\nu$ and $\nu = c\tilde{\nu}$ can be combined to convert energies to wavenumbers; it turns out, for instance, that 1 eV ≈ 8066 cm^{-1}. The classification of the electromagnetic field according to its frequency and wavelength is summarized in Fig. 0.5.

Quantum mechanics adds to this wave-like description of electromagnetic radiation by introducing the concept of particle-like packets of electromagnetic energy called **photons**. The **intensity** of the radiation is determined by the number of photons in the ray: an intense ray consists of a large number of photons; a feeble ray consists of only a few photons. A human eye can respond to a single photon; a lamp rated at 100 W (where 1 W $= 1$ J s^{-1}; W denotes the watt) generates about 10^{19} photons each second, but even so takes several hours to generate 1 mol of photons. The energy of each photon is determined by its frequency, ν, by

$$E = h\nu \tag{6}$$

This relation implies that photons of microwave radiation have lower energy than photons of visible light (which consists of shorter wavelength, higher frequency radiation). It also implies that the energies of photons of visible light increase as the light is changed from red (longer wavelength) to violet (shorter wavelength).

0.5 The populations of states

The continuous thermal agitation that the molecules experience in a sample at $T > 0$ ensures that they are distributed over the available energy levels. One particular molecule may be in one low energy state at one instant, and then be excited into a high energy state a moment later. Although we cannot keep track of the energy state of a single molecule, we can speak of the average numbers of molecules in each state; these average numbers are constant in time provided the temperature remains the same. The average number of molecules in a state is called the **population** of the state.

Only the lowest energy state is occupied at $T = 0$. Raising the temperature excites some molecules into higher energy states, and more and more states become accessible as the temperature is raised further (Fig. 0.6). Nevertheless, whatever the temperature, there is always a higher population in a state of low energy than one of high energy. The only exception occurs when the temperature is infinite: then all states of the system are equally populated.

The formula for calculating the populations of states of various energies is called the **Boltzmann distribution** and was derived by the Austrian scientist Ludwig Boltzmann

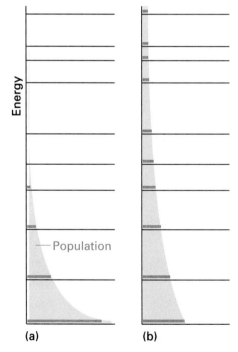

(a)   **(b)**

0.6 The Boltzmann distribution predicts that the population of a state decreases exponentially with the energy of the state. (a) At low temperatures, only the lowest states are significantly populated; (b) at high temperatures, there is significant population in high-energy states as well as in low-energy states. At infinite temperature (not shown), all states are equally populated.

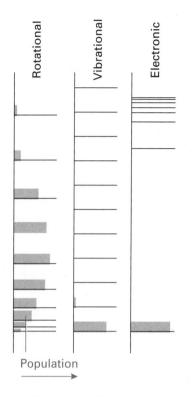

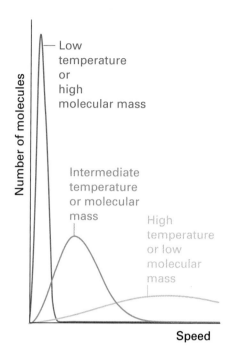

0.7 The Boltzmann distribution for three types of motion at a single temperature. There is a change in scale between the three stacks of levels (recall Fig. 0.2). Only the ground electronic state is populated at room temperature in most systems, and the bulk of the molecules are also in their ground vibrational state. Many rotational states are populated at room temperature as the energy levels are so close. The peculiar shape of the distribution over rotational states arises from the fact that each energy level actually corresponds to a number of degenerate states in which the molecule is rotating at the same speed but in different orientations. Each of these states is populated according to the Boltzmann distribution, and the shape of the distribution reflects the total population of each level.

0.8 The essential content of the Maxwell distribution of molecular speeds is summarized by this diagram. Note how the maximum in the distribution moves to higher speeds as the temperature is increased or, at constant temperature, we consider species of decreasing mass. The distribution also becomes wider as its peak moves to higher speeds.

towards the end of the nineteenth century. This formula gives the ratio of the numbers of particles, N_i/N_j, in states with energies E_i and E_j as

$$\frac{N_i}{N_j} = e^{-(E_i - E_j)/kT} \tag{7}$$

An important point concerning the interpretation of the Boltzmann distribution is that it refers to the populations of *states*, not levels. Because the Boltzmann distribution refers to states, all the members of a degenerate set of states belonging to the same energy level will have the same population. We shall see a consequence of this feature shortly.

A typical energy separation between the ground state and the first electronically excited state of an atom or molecule is about 3 eV, which corresponds to 300 kJ mol^{-1}. In a sample at 25 °C (298 K) the ratio of the populations of the two states is about e^{-121}, or about 10^{-53}. Therefore, essentially every atom or molecule in a sample is in its electronic ground state. The population of the upper state rises to about 1 per cent of the population of the ground state only when the temperature reaches 10^4 °C. The separation of vibrational energy levels is very much less than that of electronic energy levels (about 0.1 eV, corresponding to 10 kJ mol^{-1}), but nevertheless at room temperature only the lowest energy level is significantly populated. Only about 1 in e$^4 \approx 60$ molecules is not in its ground state. Rotational energy levels are much more closely spaced than vibrational energy levels (typically, about 100 to 1000 times closer), and even at room temperature we can expect many rotational states to be occupied. Therefore, when considering the contribution of rotational motion to the properties of a sample, we need to take into account the fact that molecules occupy a wide range of different states, with some rotating rapidly and others slowly.

The important features of the Boltzmann distribution to bear in mind are that the distribution of populations is an exponential function of energy and temperature, and that more states are significantly populated if they are close together in comparison with kT (like rotational and translational states), than if they are far apart (like vibrational and electronic states). Moreover, more states are occupied at high temperatures than at low temperatures. The illustration (Fig. 0.7) summarizes the form of the Boltzmann distribution for some typical sets of energy levels.

The Boltzmann distribution takes a special form when we consider the free translational motion of noninteracting gas molecules. Different energies now correspond to different speeds (because the kinetic energy is equal to $\frac{1}{2}mv^2$), so the Boltzmann formula can be used to predict the proportions of molecules having a specific speed at a particular temperature. The expression giving the proportion of molecules that have a particular speed is called the **Maxwell distribution**, and has the features summarized in Fig. 0.8. The bulge in the distribution represents the fact that the kinetic energy of a molecule depends on its speed, and there are many ways of obtaining a given value of the speed v with different values of the components v_x, v_y, and v_z relative to the three axes, particularly when the speed is high. In other words, translational energy levels are highly degenerate, and the degeneracy increases with energy. Therefore, although the populations of individual states decrease with increasing energy (and hence speed), there are many more states of a given energy at high energies and the product of this rising degeneracy and the falling exponential function has a bulge at an intermediate energy.

Notice how the tail towards high speeds is longer at high temperatures than at low, which indicates that at high temperatures more molecules in a sample have speeds much higher than average. The speed corresponding to the maximum in the graph is the **most probable speed**, the speed most likely to be found for a molecule selected at random. The illustration also shows how the distribution varies with mass for some components of air at $25\,°C$. The lighter molecules move, on average, much faster than the heavier ones.

Further reading

Articles of general interest

I.M. Mills, The choice of names and symbols for quantities in chemistry. *J. Chem. Educ.* **66**, 887 (1989).

G.B. Kauffman, The centenary of physical chemistry. *Educ. in Chem.* **24**, 168 (1987).

Texts and sources of data and information

J.W. Servos, *Physical chemistry from Ostwald to Pauling.* Princeton University Press (1990).

I.M. Mills (ed.), *Quantities, units, and symbols in physical chemistry.* Blackwell Scientific, Oxford (1993).

K.J. Laidler, *The world of physical chemistry.* Oxford University Press (1993).

V. Gold, K.L. Loening, A.D. McNaight, and P. Sehmi, *Compendium of chemical terminology.* Blackwell Scientific, Oxford (1987).

Part 1 Equilibrium

Part 1 of the text develops the concepts that are needed for the discussion of equilibria in chemistry. Equilibria include physical change, such as fusion and vaporization, and chemical change, including electrochemistry. The discussion is in terms of thermodynamics, and particularly in terms of enthalpy and entropy. We shall see that a unified view of equilibrium and the direction of spontaneous change can be obtained in terms of the chemical potential of substances. The chapters of Part 1 deal with the properties of matter in bulk; those of Part 2 will show how these properties stem from the behaviour of individual atoms.

1

The properties of gases

This chapter establishes the properties of gases that will be used throughout the text. It begins with an account of an idealized version of a gas, a perfect gas, and shows how its equation of state may be assembled experimentally. We then see how this relation between the properties of the gas can be explained in terms of the kinetic model, in which the gas is represented by a collection of point masses in continuous random motion. Finally, we see how the properties of real gases differ from those of a perfect gas, and construct an equation of state that describes their properties.

The simplest state of matter is a **gas**, a form of matter that fills any container it occupies. Initially we shall consider only pure gases, but later in the chapter we shall see that the same ideas and equations apply to mixtures of gases too.

The perfect gas

We shall find it helpful to picture a gas as a collection of molecules (or atoms) in continuous random motion, with speeds that increase as the temperature is raised. A gas differs from a liquid in that, except during collisions, the molecules of a gas are widely separated from one another and move in paths that are largely unaffected by intermolecular forces.

1.1 The states of gases

The physical **state** of a sample of a substance is defined by its physical properties, and two samples of a substance that have the same physical properties are in the same state. The state of a pure gas, for example, is specified by giving the values of its volume, V, amount of substance (number of moles), n, pressure, p, and temperature, T. However, it has been established experimentally that it is sufficient to specify only three of these variables, for then the fourth variable is fixed. That is, it is an experimental fact that each substance is described by an **equation of state**, an equation that interrelates these four variables.

The general form of an equation of state is

$$p = f(T, V, n) \tag{1}$$

This equation tells us that, if we know the values of n, T, and V for a particular substance, then we can find its pressure. Each substance is described by its own equation of state, but we know the explicit form of the equation in only a few special cases. One very important example is the equation of state for a **perfect gas**, which has the form

$$p = \frac{nRT}{V} \tag{2}$$

where R is a constant. Much of the rest of this chapter will examine the origin and implications of this equation of state.

(a) Pressure

Pressure is defined as force divided by the area to which the force is applied. The greater the force acting on a given area, the greater the pressure. The origin of the force exerted by a gas is the incessant battering of the molecules on the walls of its container. The collisions are so numerous that they exert an effectively steady force, which is experienced as a steady pressure.

The SI unit of pressure, the pascal (Pa), is defined as 1 newton per metre squared:

$$1\ \text{Pa} = 1\ \text{N m}^{-2} \tag{3}$$

However, several other units are still widely used; they are summarized in Table 1.1. A pressure of 10^5 Pa (1 bar) is the **standard pressure** for reporting data, and we shall denote it $p^{\ominus}$.

Table 1.1 Pressure units

Name	Symbol	Value
pascal	1 Pa	$1\ \text{N m}^{-2}$, $1\ \text{kg m}^{-1}\text{s}^{-2}$
bar	1 bar	10^5 Pa
atmosphere	1 atm	101 325 Pa
torr	1 Torr	$(101\,325/760)\ \text{Pa} = 133.32 \ldots \text{Pa}$
millimetre of mercury	1 mmHg	$133.322 \ldots \text{Pa}$
pound per square inch	1 psi	$6.894\,757 \ldots \text{kPa}$

Example 1.1 Calculating pressure

Suppose Isaac Newton weighed 65 kg. Calculate the pressure he exerted on the ground when wearing (a) boots with soles of total area 250 cm^2 in contact with the ground, (b) ice skates, of total area 2.0 cm^2.

Method Pressure is force divided by area ($p = F/A$), so the calculation depends on being able to calculate the force, F, that Newton exerts on the ground, and then to divide it by the area, A, over which the force is exerted. To calculate the force, we need to know (from elementary physics) that the downward force an object of mass m exerts at the surface of the Earth is $F = mg$, where g is the acceleration of free fall, 9.81 m s^{-2}. Note that

$1 \text{ N} = 1 \text{ kg m s}^{-2}$ and $1 \text{ Pa} = 1 \text{ N m}^{-2}$. Convert the areas to metre squared by using $1 \text{ cm}^2 = 10^{-4} \text{ m}^2$.

Answer The force exerted by Newton is

$$F = (65 \text{ kg}) \times (9.81 \text{ m s}^{-2}) = 6.4 \times 10^2 \text{ N}$$

The force is the same whatever his footwear. The areas 250 cm^2 and 2.0 cm^2 correspond to $2.50 \times 10^{-2} \text{ m}^2$ and $2.0 \times 10^{-4} \text{ m}^2$, respectively. The pressure he exerts in each case is:

$$\text{(a) } p = \frac{6.4 \times 10^2 \text{ N}}{2.50 \times 10^{-2} \text{ m}^2} = 2.6 \times 10^4 \text{ Pa } (26 \text{ kPa})$$

$$\text{(b) } p = \frac{6.4 \times 10^2 \text{ N}}{2.0 \times 10^{-4} \text{ m}^2} = 3.2 \times 10^6 \text{ Pa } (3.2 \text{ MPa})$$

Comment One Newton exerts a force of much more than 1 newton. A pressure of 26 kPa corresponds to 0.26 atm and a pressure of 3.2 MPa corresponds to 32 atm.

- -

Self-test 1.1 Calculate the pressure exerted by a mass of 1.0 kg pressing through the point of a pin of area 1.0×10^{-2} mm^2 at the surface of the Earth.

[0.98 GPa, 9.7×10^3 atm]

If two gases are in separate containers that share a common movable wall (Fig. 1.1), the gas that has the higher pressure will tend to compress (reduce the volume of) the lower pressure gas. The pressure of the former gas will fall as it expands and that of the latter gas will rise as it is compressed. There will come a stage when the two pressures are equal and the wall has no further tendency to move. This condition of equality of pressure on either side of a movable wall (a 'piston') is a state of **mechanical equilibrium** between the two gases. The numerical value of the pressure of a gas is therefore an indication of whether a container that contains the gas will be in mechanical equilibrium with another gas with which it shares a movable wall.

(b) The measurement of pressure

The pressure exerted by the atmosphere is measured with a **barometer**. The original version of a barometer (which was invented by Torricelli, a student of Galileo) was an inverted tube of mercury sealed at the upper end. When the column of mercury is in mechanical equilibrium with the atmosphere, the pressure at its base is equal to that exerted by the atmosphere. It follows that the height of the mercury column is proportional to the external pressure.

Example 1.2 Calculating the pressure exerted by a column of liquid

Derive an equation for the pressure at the base of a column of liquid of mass density ρ and height h at the surface of the Earth.

Method As in Example 1.1, $p = F/A$ and $F = mg$, so we need to know the mass of a column of liquid. The mass of a column of liquid is its mass density, ρ, multiplied by its volume, V. So, the first step is to calculate the volume of a cylindrical column of liquid.

Answer Let the column have cross-sectional area A; then its volume is Ah and its mass is $m = \rho Ah$. The force the column of this mass exerts at its base is

$$F = mg = \rho Ahg$$

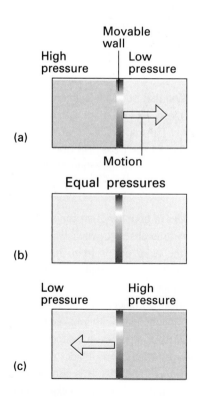

Movable wall

High pressure | Low pressure

(a)

Motion

Equal pressures

(b)

Low pressure | High pressure

(c)

1.1 When a region of high pressure is separated from a region of low pressure by a movable wall, the wall will be pushed into one region or the other, as in (a) and (c). However, if the two pressures are identical, the wall will not move (b). The latter condition is one of mechanical equilibrium between the two regions.

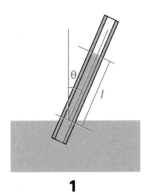

1

The pressure at the base of the column is therefore

$$p = \frac{F}{A} = \frac{\rho Ahg}{A} = \rho gh$$

Comment Note that the pressure is independent of the shape and cross-sectional area of the column. The mass of the column increases as the area, but so does the area on which the force acts, so the two cancel.

- -

Self-test 1.2 Derive an expression for the pressure at the base of a column of liquid of length l held at an angle θ to the vertical (**1**).

$$[p = \rho gl \cos \theta]$$

The pressure of a sample of gas inside a container is measured with a **manometer** (Fig. 1.2). In its simplest form, a manometer is a U-tube filled with a liquid of low volatility (such as silicone oil). The pressure, p, of the gaseous sample balances the pressure exerted by the column of liquid, which is equal to ρgh if the column is of height h (see Example 1.2), plus the external pressure, p_{ex}, if one tube is open to the atmosphere:

$$p = p_{ex} + \rho gh \qquad (4)$$

It follows that the pressure of the sample can be obtained by measuring the height of the column and noting the external pressure. More sophisticated techniques are used at lower pressures. Methods that avoid the complication of having to account for the vapour from the manometer fluid are also available. These superior methods include monitoring the deflection of a diaphragm, either mechanically or electrically, or monitoring the change in a pressure-sensitive electrical property.

(c) Temperature

The concept of temperature springs from the observation that a change in physical state (for example, a change of volume) can occur when two objects are in contact with one another (as when a red-hot metal is plunged into water). Later (Section 2.1) we shall see that the change in state can be interpreted as arising from a flow of energy as heat from one object to another. The **temperature**, T, is the property that tells us the direction of the flow of energy. If energy flows from A to B when they are in contact, then we say that A has a higher temperature than B (Fig. 1.3).

It will prove useful to distinguish between two types of boundary that can separate the objects. A boundary is **diathermic** if a change of state is observed when two objects at different temperatures are brought into contact (the word *dia* is from the Greek for 'through'). A metal container has diathermic walls. A boundary is **adiabatic** if no change occurs even though the two objects have different temperatures.

The temperature is a property that tells us whether two objects would be in **thermal equilibrium** if they were in contact through a diathermic boundary. Thermal equilibrium is established if no change of state occurs when two objects A and B are in contact through a diathermic boundary. Suppose an object A (which we can think of as a block of iron) is in thermal equilibrium with an object B (a block of copper), and that B is also in thermal equilibrium with another object C (a flask of water). Then it has been found experimentally that A and C will also be in thermal equilibrium when they are put in contact (Fig. 1.4). This observation is summarized by the following statement:

The Zeroth Law of thermodynamics: If A is in thermal equilibrium with B, and B is in thermal equilibrium with C, then C is also in thermal equilibrium with A.

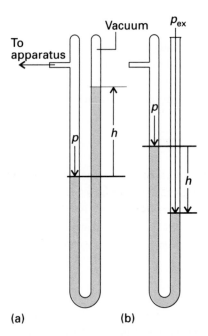

(a)   **(b)**

1.2 Two versions of a manometer used to measure the pressure of a sample of gas. (a) The height difference, h, of the two columns in the sealed-tube manometer is directly proportional to the pressure of the sample, and $p = \rho gh$, where ρ is the density of the liquid. (b) The difference in heights of the columns in the open-tube manometer is proportional to the difference in pressure between the sample and the atmosphere. In the example shown, the pressure of the sample is lower than that of the atmosphere.

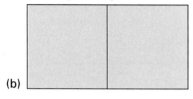

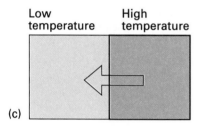

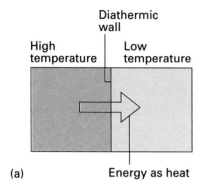

1.3 Energy flows as heat from a region at a higher temperature to one at a lower temperature if the two are in contact through a diathermic wall, as in (a) and (c). However, if the two regions have identical temperatures, there is no net transfer of energy as heat even though the two regions are separated by a diathermic wall (b). The latter condition corresponds to the two regions being at thermal equilibrium.

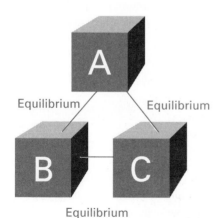

1.4 The experience summarized by the Zeroth Law of thermodynamics is that, if an object A is in thermal equilibrium with B and if B is in thermal equilibrium with C, then B is in thermal equilibrium with A.

The Zeroth Law justifies the concept of temperature and the use of a **thermometer**, a device for measuring the temperature. Thus, suppose that B is a glass capillary containing mercury. Then, when A is in contact with B, the mercury column in the latter has a certain length. According to the Zeroth Law, if the mercury column in B has the same length when it is placed in contact with another object C, then we can predict that no change of state of A and C will occur when they are in contact. Moreover, we can use the length of the mercury column as a measure of the temperatures of A and C.

In the early days of thermometry (and still in laboratory practice today), temperatures were related to the length of a column of liquid, and the difference in lengths shown when the thermometer was first in contact with melting ice and then with boiling water was divided into 100 steps called 'degrees', the lower point being labelled 0. This procedure led to the **Celsius scale** of temperature. In this text, temperatures on the Celsius scale are denoted θ and expressed in degrees Celsius (°C). However, because different liquids expand to different extents, and do not always expand uniformly over a given range, thermometers constructed from different materials showed different numerical values of the temperature between their fixed points. The pressure of a gas, however, can be used to construct a **perfect-gas temperature scale** that is almost independent of the identity of the gas, as we shall explain shortly. The perfect-gas scale turns out to be identical to the **thermodynamic temperature scale** to be introduced in Section 4.2c, so we shall use the latter term from now on to avoid a proliferation of names. On the thermodynamic temperature scale, temperatures are denoted T and are normally reported in kelvins, K. Thermodynamic and Celsius temperatures are related by the exact expression

$$T/\mathrm{K} = \theta/°\mathrm{C} + 273.15 \tag{5}$$

This relation is the current definition of the Celsius scale in terms of the more fundamental Kelvin scale.

Illustration

To express 25.00°C as a temperature in kelvins, we write

$$T/\mathrm{K} = 25.00°\mathrm{C}/°\mathrm{C} + 273.15 = 25.00 + 273.15 = 298.15$$

Note how the units are cancelled like numbers. This is the procedure called 'quantity calculus' in which a physical quantity (such as the temperature) is the product of a numerical value (25.00) and a unit (1°C). Multiplication of both sides by the unit K then gives $T = 298.15$ K.

1.2 The gas laws

The equation of state of a low-pressure gas was established by combining a series of empirical laws. We shall introduce these laws below, and then show how they can be combined into the single equation of state $pV = nRT$.

(a) The individual gas laws

Robert Boyle, acting on the suggestion of a correspondent John Townley, showed in 1661 that, to a good approximation, the pressure and volume of a fixed amount of gas at constant temperature are related by

$$pV = \text{constant} \tag{6}°$$

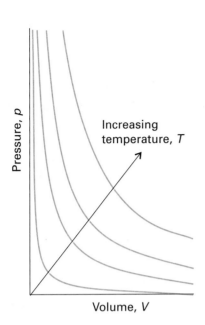

1.5 The pressure–volume dependence of a fixed amount of perfect gas at different temperatures. Each curve is a hyperbola ($pV = $ constant) and is called an isotherm.

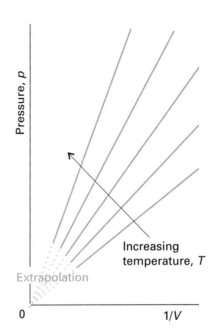

1.6 Straight lines are obtained when the pressure is plotted against $1/V$.

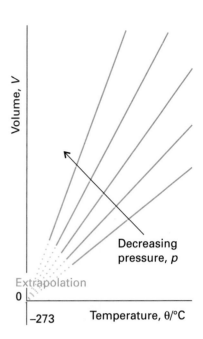

1.7 The variation of the volume of a fixed amount of gas with the temperature constant. Note that in each case they extrapolate to zero volume at $-273.15\,°C$.

This relation is known as **Boyle's law**. That is, at constant temperature, the pressure of a sample of gas is inversely proportional to its volume, and the volume it occupies is inversely proportional to its pressure:

$$p \propto \frac{1}{V} \qquad V \propto \frac{1}{p} \tag{7}°$$

The variation of the pressure of a sample of gas as the volume is changed is depicted in Fig. 1.5. Each of the curves in the graph corresponds to a single temperature and hence is called an **isotherm**. According to Boyle's law, the isotherms of gases are hyperbolas.[1] An alternative depiction, a plot of pressure against 1/volume, is shown in Fig. 1.6.

More modern experiments have shown that Boyle's law is valid only at low pressures, and that real gases obey it only in the limit of the pressure approaching zero (which we write $p \to 0$). Boyle's law is therefore an example of a **limiting law**, a law that is strictly true only in a certain limit, in this case $p \to 0$. Equations that are valid in this limiting sense will be signalled by a ° on the equation number, as in eqn 6.

The molecular explanation of Boyle's law is that, if a sample of gas is compressed to half its volume, then twice as many molecules strike the walls in a given period of time than before it was compressed. As a result, the average force exerted on the walls is doubled. Hence, when the volume is halved the pressure of the gas is doubled, and $p \times V$ is a constant. The reason why Boyle's law applies to all gases regardless of their chemical identity (provided the pressure is low) is that at low pressures the molecules are so far apart on average that they exert no influence on one another, and hence travel independently.

The French scientist Jacques Charles established the next important property of gases. He studied the effect of temperature on the volume of a sample of gas that was subjected to constant pressure. He found that the volume increased linearly with the temperature,

1 A hyperbola is a curve obtained by plotting y against x with $xy = $ **constant**.

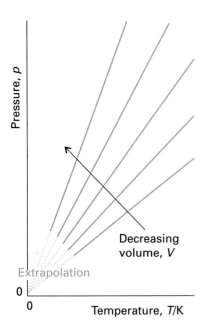

1.8 The pressure also varies linearly with the temperature, and extrapolates to zero at $T = 0$ ($-273.15\,°C$).

whatever the identity of the gas, provided it was at low pressure. Specifically, he established that

$$V = \text{constant} \times (\theta + 273\,°C) \qquad \text{(at constant pressure)} \qquad (8)°$$

(Recall that we use θ to denote temperatures on the Celsius scale.) The linear variation of volume with temperature summarized by this expression is illustrated in Fig. 1.7. The lines are examples of **isobars**, or lines showing the variation of properties at constant pressure.

Equation 8 suggests that the volume of any gas should extrapolate to zero at $\theta = -273\,°C$ and therefore that $-273\,°C$ is a natural zero of a fundamental temperature scale. As we have already indicated, a scale with 0 set at $-273.15\,°C$ is equivalent to the thermodynamic temperature scale devised by Kelvin, so **Charles's law** can be written

$$V = \text{constant} \times T \qquad \text{(at constant pressure)} \qquad (9)°$$

An alternative version of Charles's law, in which the pressure of a sample of gas is monitored under conditions of constant volume is

$$p = \text{constant} \times T \qquad \text{(at constant volume)} \qquad (10)°$$

This version of the law indicates that the pressure of a gas falls to zero as the temperature is reduced to zero (Fig. 1.8).

The molecular explanation of Charles's law lies in the fact that raising the temperature of a gas increases the average speed of its molecules. The molecules collide with the walls more frequently, and do so with greater impact. Therefore they exert a greater pressure on the walls of the container. A problem we leave until later is why the pressure depends so simply on the temperature.

The final piece of experimental information we need is that, at a given pressure and temperature, the molar volume, $V_m = V/n$, the volume per mole of molecules, of a gas is approximately the same regardless of the identity of the gas. This observation implies that the volume of a sample of gas is proportional to the amount of molecules (in moles) present, and that the constant of proportionality is independent of the identity of the gas:

$$V = \text{constant} \times n \qquad \text{(at constant pressure and temperature)} \qquad (11)°$$

This conclusion is a modern form of **Avogadro's principle**, that equal volumes of gases at the same pressure and temperature contain the same numbers of molecules.

(b) The combined gas law

The empirical observations summarized by eqns 6–11 can be combined into a single expression:

$$pV = \text{constant} \times nT$$

This expression is consistent with Boyle's law ($pV = \text{constant}$) when n and T are constant, with both forms of Charles's law ($p \propto T$, $V \propto T$) when n and either V or p are held constant, and with Avogadro's principle ($V \propto n$) when p and T are constant. The constant of proportionality, which experimentally is found to be the same for all gases, is denoted R and is called the **gas constant**. The resulting expression

$$pV = nRT \qquad (12)°$$

is called the **perfect gas equation**. It is the approximate equation of state of any gas, and becomes increasingly exact as the pressure of the gas approaches zero. A gas that obeys eqn 12 exactly under all conditions is called a **perfect gas** (or ideal gas). A **real gas**, an actual gas, behaves more like a perfect gas the lower the pressure, and is described exactly by eqn 12 in the limit of $p \to 0$. The gas constant R can be determined by evaluating $R = pV/nT$ for a gas in the limit of zero pressure (to guarantee that it is behaving perfectly).

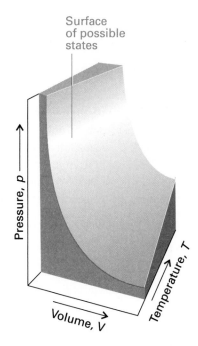

1.9 A region of the p, V, T surface of a fixed amount of perfect gas. The points forming the surface represent the only states of the gas that can exist.

1.10 Sections through the surface shown in Fig. 1.9 at constant temperature give the isotherms shown in Fig. 1.5 and the isobars shown in Fig. 1.7.

Table 1.2 The gas constant in various units

R
$8.31451 \, \mathrm{J \, K^{-1} \, mol^{-1}}$
$8.20578 \times 10^{-2} \, \mathrm{L \, atm \, K^{-1} \, mol^{-1}}$
$8.31451 \times 10^{-2} \, \mathrm{L \, bar \, K^{-1} \, mol^{-1}}$
$8.31451 \, \mathrm{Pa \, m^{3} \, K^{-1} \, mol^{-1}}$
$62.364 \, \mathrm{L \, Torr \, K^{-1} \, mol^{-1}}$
$1.98722 \, \mathrm{cal \, K^{-1} \, mol^{-1}}$

However, a more accurate value can be obtained by measuring the speed of sound in a low-pressure gas and extrapolating its value to zero pressure. The values of R in a variety of units are given in Table 1.2.

The surface in Fig. 1.9 is a plot of the pressure of a fixed amount of perfect gas against its volume and thermodynamic temperature as given by eqn 12. The surface depicts the only possible states of a perfect gas: the gas cannot exist in states that do not correspond to points on the surface. The graphs in Figs 1.5 and 1.8 correspond to the sections through the surface (Fig. 1.10).

The perfect gas equation is of the greatest importance in physical chemistry because it is used to derive a wide range of relations that are used throughout thermodynamics. However, it is also of considerable practical utility for calculating the properties of a gas under a variety of conditions. For instance, the molar volume of a perfect gas under the conditions called **standard ambient temperature and pressure** (SATP), which means 298.15 K and 1 bar (that is, exactly 10^5 Pa), is easily calculated from $V_{\mathrm{m}} = RT/p$ to be $24.790 \, \mathrm{L \, mol^{-1}}$. An earlier definition, **standard temperature and pressure** (STP), was 0 °C and 1 atm; at STP, the molar volume of a perfect gas is $22.414 \, \mathrm{L \, mol^{-1}}$.

Example 1.3 Using the perfect gas equation

In an industrial process, nitrogen is heated to 500 K in a vessel of constant volume. If it enters the vessel at a pressure of 100 atm and a temperature of 300 K, what pressure would it exert at the working temperature if it behaved as a perfect gas?

Method The known and unknown data are summarized in (2). We expect the pressure to be greater on account of the increase in temperature. Moreover, because pressure is proportional to temperature, we can anticipate that the pressure will increase by a factor of T_2/T_1, where T_1 is the initial temperature and T_2 is the final temperature. To proceed more formally, we note that, when a problem involves a change in conditions for a constant

	n	p	V	T
Initial	Same	100	Same	300
Final	Same	?	Same	500

2

amount of gas, the perfect gas equation can be developed as follows. First, write eqn 12 for the initial and final sets of conditions:

$$\frac{p_1 V_1}{T_1} = nR \qquad \frac{p_2 V_2}{T_2} = nR$$

Because n (in this problem) and R are both constant and equal, it follows that

$$\frac{p_1 V_1}{T_1} = \frac{p_2 V_2}{T_2}$$

Any constant quantities (volume in this example) cancel, and the data may then be substituted into the resulting expression.

Answer Cancellation of the volumes (because $V_1 = V_2$) on each side of this expression results in

$$\frac{p_1}{T_1} = \frac{p_2}{T_2}$$

which can be rearranged into

$$p_2 = \frac{T_2}{T_1} \times p_1$$

Substitution of the data then gives

$$p_2 = \frac{500 \text{ K}}{300 \text{ K}} \times (100 \text{ atm}) = 167 \text{ atm}$$

Comment Experiment shows that the pressure is actually 183 atm under these conditions, so the assumption that the gas is perfect leads to a 10 per cent error. The expression $p_1 V_1 / T_1 = p_2 V_2 / T_2$ is commonly called the *combined gas law*.

- -

Self-test 1.3 What temperature would result in the same sample exerting a pressure of 300 atm?

[900 K]

(c) Mixtures of gases

The kind of question we need to answer when dealing with gaseous mixtures is the contribution that each component gas makes to the total pressure of the sample. In the nineteenth century John Dalton made observations that provide the answer and summarized them in a law:

Dalton's law: The pressure exerted by a mixture of perfect gases is the sum of the partial pressures of the gases.

The **partial pressure** of a perfect gas is the pressure that it would exert if it occupied the container alone. That is, if a certain amount of H_2 exerts 25 kPa when present alone in a container, and an amount of N_2 exerts 80 kPa when present alone in the same container at the same temperature, then the total pressure when both are present is the sum of these two partial pressures, or 105 kPa (provided the individual gases and the mixture behave perfectly). More generally, if the partial pressure of a perfect gas A is p_A, that of a perfect gas B is p_B, and so on, then the total pressure when all the gases occupy the same container at the same temperature is

$$p = p_A + p_B + \cdots \qquad (13)°$$

where, for each substance J,

$$p_J = \frac{n_J RT}{V}$$

(14)°

Example 1.4 Using Dalton's law

A container of volume 10.0 L holds 1.00 mol N_2 and 3.00 mol H_2 at 298 K. What is the total pressure in atmospheres if each component behaves as a perfect gas?

Method It follows from eqns 13 and 14 that the total pressure when two gases, A and B, occupy a container is

$$p = p_A + p_B = (n_A + n_B)\frac{RT}{V}$$

To obtain the answer expressed in atmospheres, use $R = 8.206 \times 10^{-2} \text{ L atm K}^{-1} \text{ mol}^{-1}$.

Answer Under the stated conditions

$$p = (1.00 \text{ mol} + 3.00 \text{ mol}) \times \frac{(8.206 \times 10^{-2} \text{ L atm K}^{-1} \text{ mol}^{-1}) \times (298 \text{ K})}{10.0 \text{ L}}$$

$$= 9.78 \text{ atm}$$

Self-test 1.4 Calculate the total pressure when 1.00 mol N_2 and 2.00 mol O_2 are added to the same container (with the nitrogen and hydrogen still inside) at 298 K.

[17.1 atm]

(d) Mole fractions and partial pressures

We can take a step closer to the discussion of mixtures of real gases by introducing the mole fraction of each component J. The **mole fraction**, x_J, is the amount of J expressed as a fraction of the total amount of molecules, n, in the sample:

$$x_J = \frac{n_J} {n} \qquad n = n_A + n_B + \cdots$$

[15]

When no J molecules are present, $x_J = 0$; when only J molecules are present, $x_J = 1$. A mixture of 1.0 mol N_2 and 3.0 mol H_2, and therefore of 4.0 mol molecules in all, consists of mole fractions 0.25 of N_2 and 0.75 of H_2. It follows from the definition of x_J that, whatever the composition of the mixture,

$$x_A + x_B + \cdots = 1$$

(16)

Next, we *define* the **partial pressure**, p_J, of a gas J in a mixture (any gas, not just a perfect gas) as

$$p_J = x_J p$$

[17]

where p is the total pressure of the mixture (Fig. 1.11). Only for a mixture of perfect gases can we identify p_J with the pressure that the gas J would exert if it were alone in the container and calculate its value from eqn 14.

It follows from eqns 16 and 17 that the sum of the partial pressures is equal to the total pressure:

$$p_A + p_B + \cdots = (x_A + x_B + \cdots)p = p$$

(18)

This relation is true for both real and perfect gases.

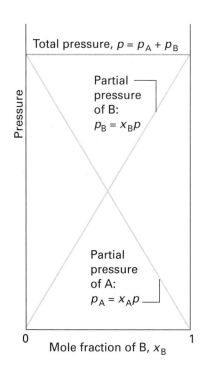

1.11 The partial pressures p_A and p_B of a binary mixture of (real or perfect) gases of total pressure p as the composition changes from pure A to pure B. The sum of the partial pressures is equal to the total pressure. If the gases are perfect, then the partial pressure is also the pressure that each gas would exert if it were present alone in the container.

Figure content: Total pressure, $p = p_A + p_B$ / Partial pressure of B: $p_B = x_B p$ / Partial pressure of A: $p_A = x_A p$ / Pressure / Mole fraction of B, x_B / 0 ... 1

Example 1.5 Calculating partial pressures

The mass percentage composition of dry air at sea level is approximately $N_2 : 75.5; O_2 : 23.2;$ Ar : 1.3. What is the partial pressure of each component when the total pressure is 1.00 atm?

Method We expect species with a high mole fraction to have a proportionally high partial pressure. Partial pressures are defined by eqn 17. To use the equation, we need the mole fractions of the components. To calculate mole fractions, which are defined by eqn 15, we use the fact that the amount of molecules J of molar mass M_J in a sample of mass m_J is $n_J = m_J/M_J$. The mole fractions are independent of the total mass of the sample, so we can choose the latter to be 100 g (which makes the conversion from mass percentages very easy).

Answer The amounts of each type of molecule present in 100 g of air are

$$n(N_2) = \frac{(100 \text{ g}) \times 0.755}{28.02 \text{ g mol}^{-1}} = 2.69 \text{ mol}$$

$$n(O_2) = \frac{(100 \text{ g}) \times 0.232}{32.00 \text{ g mol}^{-1}} = 0.725 \text{ mol}$$

$$n(Ar) = \frac{(100 \text{ g}) \times 0.013}{39.95 \text{ g mol}^{-1}} = 0.033 \text{ mol}$$

Because overall $n = 3.45$ mol, the mole fractions and partial pressures (obtained by multiplying the mole fraction by the total pressure, 1.00 atm) are as follows:

	N_2	O_2	Ar
Mole fraction:	0.780	0.210	0.0096
Partial pressure/atm:	0.780	0.210	0.0096

Comment We have not had to assume that the gases are perfect: partial pressures are defined as $p_J = x_J p$ for any gas.

- -

Self-test 1.5 When carbon dioxide is taken into account, the mass percentages are 75.52 (N_2), 23.15 (O_2), 1.28 (Ar), and 0.046 (CO_2). What are the partial pressures when the total pressure is 0.900 atm?

$$[0.703, 0.189, 0.0084, 0.00027 \text{ atm}]$$

1.3 The kinetic model of gases

We have seen that properties of a perfect gas can be rationalized qualitatively in terms of a model in which the molecules of the gas are in ceaseless random motion. We shall now see how this interpretation can be expressed quantitatively in terms of the **kinetic model** of gases, in which it is assumed that the only contribution to the energy of the gas is from the kinetic energies of the molecules (that is, the potential energy of the interactions between molecules makes a negligible contribution to the total energy of the gas). The kinetic model is one of the most remarkable—and arguably most beautiful—models in physical chemistry for, from a set of very slender assumptions, powerful quantitative conclusions can be deduced.

The kinetic model of gases is based on three assumptions:

1. The gas consists of molecules of mass m in ceaseless random motion.
2. The size of the molecules is negligible, in the sense that their diameters are much smaller than the average distance travelled between collisions.

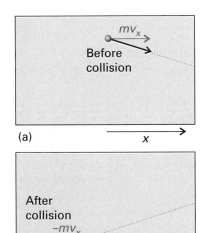

(a)

(b)

1.12 The pressure of a gas arises from the impact of its molecules on the walls. In an elastic collision of a molecule with a wall perpendicular to the x-axis, the x-component of velocity is reversed but the y- and z-components are unchanged.

3.   The molecules do not interact, except that they make perfectly elastic collisions when they are in contact.

An **elastic collision** is one in which no internal modes of motion are excited; that is, the translational energy of the molecules is conserved (remains constant) in a collision.

From these very economical assumptions, it follows that the pressure and volume of the gas are related by the following expression:

$$pV = \tfrac{1}{3}nMc^2 \qquad\qquad (19)°$$

where $M = mN_A$, the molar mass of the molecules, and c is the **root mean square speed** of the molecules, the square root of the mean of the squares of the speeds, v, of the molecules:

$$c = \langle v^2 \rangle^{1/2} \qquad\qquad [20]$$

The derivation of eqn 19 is set out in the following *Justification*.

Justification 1.1

Consider the arrangement in Fig. 1.12. When a particle of mass m that is travelling with a component of velocity v_x parallel to the x-axis collides with the wall on the right and is reflected, its linear momentum (the product of its mass and its velocity) changes from mv_x before the collision to $-mv_x$ after the collision (when it is travelling in the opposite direction). The momentum therefore changes by $2mv_x$ on each collision (the y- and z-components are unchanged). Many molecules collide with the wall in an interval Δt, and the total change of momentum is the product of the change in momentum of each molecule multiplied by the number of molecules that reach the wall during the interval.

Next we calculate the number of molecules that collide with the wall in the interval Δt. Because a molecule with velocity component v_x can travel a distance $v_x\Delta t$ along the x-axis in an interval Δt, all the molecules within a distance $v_x\Delta t$ of the wall will strike it if they are travelling towards it (Fig. 1.13). It follows that, if the wall has area A, then all the particles in a volume $A \times v_x\Delta t$ will reach the wall (if they are travelling towards it). The number density of particles is nN_A/V, where n is the total amount of molecules in the container of volume V and N_A is the Avogadro constant, so the number in the volume $Av_x\Delta t$ is $(nN_A/V) \times Av_x\Delta t$.

On average, at any instant half the particles are moving to the right and half are moving to the left. Therefore, the average number of collisions with the wall during the interval Δt is $\tfrac{1}{2}nN_AAv_x\Delta t/V$. The total momentum change in that interval is the product of this number and the change $2mv_x$:

$$\text{Momentum change} = \frac{nN_AAv_x\Delta t}{2V} \times 2mv_x = \frac{nmN_AAv_x^2\Delta t}{V}$$
$$= \frac{nMAv_x^2\Delta t}{V}$$

where $M = mN_A$.

Next, to identify the force, we calculate the rate of change of momentum, which is this change of momentum divided by the interval Δt during which it occurs:

$$\text{Rate of change of momentum} = \frac{nMAv_x^2}{V}$$

This rate of change of momentum is equal to the force (by Newton's second law of motion). It follows that the pressure, the force divided by the area, is

$$\text{Pressure} = \frac{nMv_x^2}{V}$$

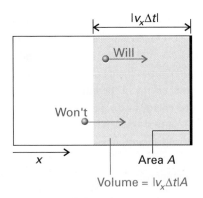

$|v_x\Delta t|$

Will

Won't

Area A

Volume = $|v_x\Delta t|A$

1.13 A molecule will reach the wall on the right within an interval Δt if it is within a distance $v_x\Delta t$ of the wall and travelling to the right.

Not all the molecules travel with the same velocity, so the detected pressure, p, is the average (denoted $\langle \cdots \rangle$) of the quantity just calculated:

$$p = \frac{nM \langle v_x^2 \rangle}{V}$$

This expression already resembles the perfect gas equation of state.

Because the molecules are moving randomly (and there is no net flow in a particular direction), the average speed along x is the same as that in the y- and z-directions. It follows that

$$c^2 = \langle v_x^2 \rangle + \langle v_y^2 \rangle + \langle v_z^2 \rangle = 3 \langle v_x^2 \rangle, \qquad \text{implying that } \langle v_x^2 \rangle = \tfrac{1}{3} c^2$$

Equation 19 now follows immediately.

Equation 19 is one of the key results of the kinetic model. We see that, if the root mean square speed of the molecules depends only on the temperature, then at constant temperature

$$pV = \text{constant}$$

which is the content of Boyle's law.

(a) Molecular speeds

If eqn 19 is to be precisely the equation of state of a perfect gas, its right-hand side must be equal to nRT. It follows that the root mean square speed of the molecules in a gas at a temperature T must be given by the expression

$$c = \left(\frac{3RT}{M} \right)^{1/2} \tag{21}°$$

We can conclude that the root mean square speed of the molecules of a gas is proportional to the square root of the temperature and inversely proportional to the square root of the molar mass. That is, the higher the temperature, the faster the molecules travel on average, and, at a given temperature, heavy molecules travel more slowly on average than light molecules. Sound waves are pressure waves, and for them to propagate the molecules of the gas must move to form regions of high and low pressure. Therefore, we should expect the root mean square speeds of molecules to be comparable to the speed of sound in air (340 m s^{-1}).

Illustration
....

The molar mass of CO_2 is 44.01 g mol^{-1}. Therefore, it follows from eqn 21 that at 298 K

$$c = \left(\frac{3 \times (8.3145 \text{ J K}^{-1} \text{ mol}^{-1}) \times (298 \text{ K})}{44.01 \times 10^{-3} \text{ kg mol}^{-1}} \right)^{1/2} = 411 \text{ m s}^{-1}$$

To obtain this result, we have used R in SI units and $1 \text{ J} = 1 \text{ kg m}^2 \text{ s}^{-2}$.
....

Equation 21 is an expression for the root mean square speed of molecules. However, in an actual gas the speeds of individual molecules span a wide range, and the collisions in the gas continually redistribute the speeds among the molecules. Before a collision, a particular molecule may be travelling rapidly, but after a collision it may be accelerated to a very high speed, only to be slowed again by the next collision. The fraction of molecules that have speeds in the range v to $v + \mathrm{d}v$ is proportional to the width of the range, and is written

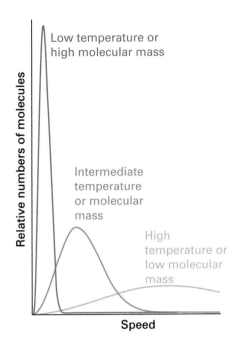

1.14 The distribution of molecular speeds with temperature and molar mass. Note that the most probable speed (corresponding to the peak of the distribution) increases with temperature and with decreasing molar mass and, simultaneously, the distribution becomes broader.

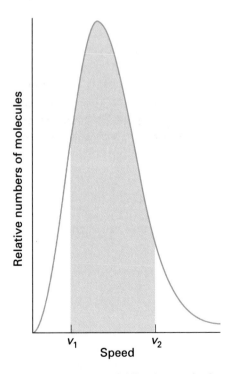

1.15 To calculate the probability that a molecule will have a speed in the range v_1 to v_2, we integrate the distribution between those two limits; the integral is equal to the area of the curve between the limits, as shown shaded here.

$f(v)\mathrm{d}v$, where $f(v)$, which varies with the speed, v, is called the **distribution** of speeds. The precise form of f was derived by J.C. Maxwell, and is

$$f(v) = 4\pi \left(\frac{M}{2\pi RT}\right)^{3/2} v^2 \mathrm{e}^{-Mv^2/2RT} \qquad (22)$$

This expression is called the **Maxwell distribution** of speeds, and is derived in the following *Justification*. Its main features are summarized in Fig. 1.14. We see that the range of speeds broadens as the temperature increases. Lighter molecules also have a broader distribution of speeds than heavy molecules. To use eqn 22 quantitatively to calculate the fraction of molecules in a given narrow range of speeds, Δv, we evaluate $f(v)$ at the speed of interest, then multiply it by the width of the range of speeds of interest (that is, we form $f(v)\Delta v$). To use the distribution to calculate the fraction in a range of speeds that is so wide that it cannot be regarded as almost infinitesimal, we evaluate the integral:

$$\text{Fraction in the range } v_1 \text{ to } v_2 = \int_{v_1}^{v_2} f(v)\,\mathrm{d}v \qquad (23)$$

This integral is the area under the graph of f as a function of v, and except in special cases has to be evaluated numerically using mathematical software (Fig. 1.15).

Justification 1.2

The probability that a molecule has a velocity in the range v_x to $v_x + \mathrm{d}v_x$, v_y to $v_y + \mathrm{d}v_y$, v_z to $v_z + \mathrm{d}v_z$ is proportional to the widths of the ranges and depends on the velocity components, so we write it $f(v_x, v_y, v_z)\,\mathrm{d}v_x\mathrm{d}v_y\mathrm{d}v_z$. The probability that a molecule has a particular velocity parallel to x is independent of the probability that it has a particular velocity parallel to y or z, so $f(v_x, v_y, v_z) = f(v_x)f(v_y)f(v_z)$. Moreover, the probability of finding a molecule with a velocity in the range $+|v_x|$ to $+|v_x| + \mathrm{d}v_x$ is the same as finding it in the range $-|v_x|$ to $-|v_x| - \mathrm{d}v_x$, so $f(v_x)$ must depend on the square of v_x. Hence, $f(v_x) = f(v_x^2)$, and likewise for v_y and v_z. It follows that we can also write $f(v_x, v_y, v_z) = f(v_x^2)f(v_y^2)f(v_z^2)$.

Next, we assume that the probability of a molecule lying in a particular range is independent of its direction of flight. That is, we assume that $f(v_x, v_y, v_z)$ depends only on the speed v, with $v^2 = v_x^2 + v_y^2 + v_z^2$, but not on the individual components. For example, we assume that the probability that a molecule has a velocity in an infinitesimal range at the velocity $(1.0, 2.0, 3.0)$ km s^{-1}, and speed 3.7 km s^{-1}, is the same as the probability that it has a velocity in the same infinitesimal range at the velocity $(2.0, -1.0, -3.0)$ km s^{-1}, or any other set of components corresponding to the speed 3.7 km s^{-1}. It follows that $f(v_x, v_y, v_z) = f(v_x^2 + v_y^2 + v_z^2)$.

By combining the conclusions of these two paragraphs, we can write

$$f(v_x^2 + v_y^2 + v_z^2) = f(v_x^2)f(v_y^2)f(v_z^2)$$

Only an exponential function satisfies such a relation (because $\mathrm{e}^a\mathrm{e}^b = \mathrm{e}^{a+b}$), so we can conclude that

$$f(v_x) = K\mathrm{e}^{\pm\zeta v_x^2}$$

where K and ζ (zeta) are constants and $f(v_x) = f(v_x^2)$. The two constants are the same for $f(v_y)$ and $f(v_z)$ because the distributions are independent of direction. Moreover, because the probability is very small that a molecule will have a very high velocity, we can discard the solution with $+\zeta v_x^2$ in the exponent.

To determine the constant K, we note that a molecule must have a velocity somewhere in the range $-\infty < v_x < \infty$, so

$$\int_{-\infty}^{\infty} f(v_x)\,dv_x = 1 \tag{24}$$

Substitution of the expression for $f(v_x)$ then gives

$$1 = K \int_{-\infty}^{\infty} e^{-\zeta v_x^2}\,dv_x = K\left(\frac{\pi}{\zeta}\right)^{1/2}$$

Therefore, $K = (\zeta/\pi)^{1/2}$. To determine ζ we calculate the root mean square speed. The first step is to write

$$\langle v_x^2 \rangle = \int_{-\infty}^{\infty} v_x^2 f(v_x)\,dv_x = \left(\frac{\zeta}{\pi}\right)^{1/2} \int_{-\infty}^{\infty} v_x^2 e^{-\zeta v_x^2}\,dv_x$$

$$= \left(\frac{\zeta}{\pi}\right)^{1/2} \times \tfrac{1}{2}\left(\frac{\pi}{\zeta^3}\right)^{1/2} = \frac{1}{2\zeta}$$

Therefore,

$$c = (\langle v_x^2 \rangle + \langle v_x^2 \rangle + \langle v_x^2 \rangle)^{1/2} = \left(\frac{3}{2\zeta}\right)^{1/2}$$

However, we already know that $c = (3RT/M)^{1/2}$; therefore $\zeta = M/2RT$. At this stage, therefore, we know that

$$f(v_x) = \left(\frac{M}{2\pi RT}\right)^{1/2} e^{-Mv_x^2/2RT} \tag{25}$$

The probability that a molecule has a velocity in the range v_x to $v_x + dv_x$, v_y to $v_y + dv_y$, v_z to $v_z + dv_z$ is

$$f(v_x, v_y, v_z)\,dv_x dv_y dv_z = f(v_x)f(v_y)f(v_z)\,dv_x dv_y dv_z$$

$$= \left(\frac{M}{2\pi RT}\right)^{3/2} e^{-Mv^2/2RT}\,dv_x dv_y dv_z$$

The probability $f(v)dv$ that the molecules have a speed in the range v to $v + dv$ regardless of direction is the sum of the probabilities that the velocity lies in any of the volume elements $dv_x dv_y dv_z$ forming a spherical shell of radius v (Fig. 1.16). The sum of the volume elements on the right-hand side of the last equation is the volume of this shell, $4\pi v^2\,dv$. Therefore,

$$f(v) = 4\pi \left(\frac{M}{2\pi RT}\right)^{3/2} v^2 e^{-Mv^2/2RT}$$

as given in eqn 22.

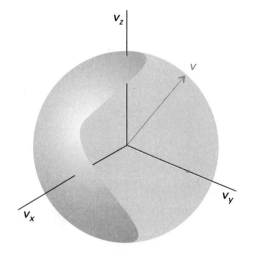

1.16 To evaluate the probability that a molecule has a speed in the range v to $v + dv$, we evaluate the total probability that the molecule will have a speed that is anywhere on the surface of a sphere of radius $v = (v_x^2 + v_y^2 + v_z^2)^{1/2}$ by summing the probabilities that it is in a volume element $dv_x dv_y dv_z$ at a distance v from the origin.

Example 1.6 Calculating the mean speed of molecules in a gas

What is the mean speed, $\bar{c}$, of N_2 molecules in air at 25°C?

Method We are asked to calculate the mean speed, not the root mean square speed. A mean value is calculated by multiplying each speed by the fraction of molecules that have that speed, and then adding all the products together. When the speed varies over a continuous range, the sum is replaced by an integral. To employ this approach here, we note that the

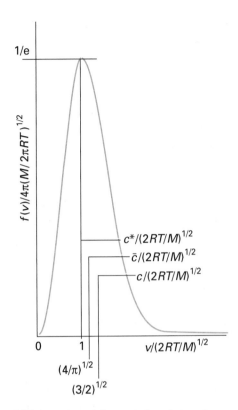

1.17 A summary of the conclusions that can be deduced from the Maxwell distribution for molecules of molar mass M at a temperature T: c^* is the most probable speed, $\bar{c}$ is the mean speed, and c is the root mean square speed.

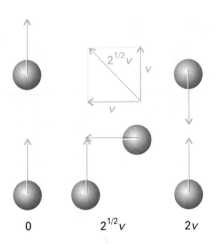

1.18 A simplified version of the argument to show that the relative mean speed of molecules in a gas is related to the mean speed. When the molecules are moving in the same direction, the relative mean speed is zero; it is $2v$ when the molecules are approaching each other. A typical mean direction of approach is from the side, and the mean speed of approach is then $2^{1/2}v$. The last direction of approach is the most characteristic, so the mean speed of approach can be expected to be about $2^{1/2}v$. This value is confirmed by more detailed calculation.

fraction of molecules with a speed in the range v to $v + dv$ is $f(v)\,dv$, so the product of this fraction and the speed is $vf(v)\,dv$. The mean speed, $\bar{c}$, is obtained by evaluating the integral

$$\bar{c} = \int_0^\infty vf(v)\,dv$$

with f given in eqn 22.

Answer The integral required is

$$\bar{c} = 4\pi \left(\frac{M}{2\pi RT}\right)^{3/2} \int_0^\infty v^3 e^{-Mv^2/2RT}\,dv$$

$$= 4\pi \left(\frac{M}{2\pi RT}\right)^{3/2} \times \tfrac{1}{2}\left(\frac{2RT}{M}\right)^2 = \left(\frac{8RT}{\pi M}\right)^{1/2}$$

Substitution of the data then gives

$$\bar{c} = \left(\frac{8 \times (8.3145\ \text{J K}^{-1}\,\text{mol}^{-1}) \times (298\ \text{K})}{\pi \times (28.02 \times 10^{-3}\ \text{kg mol}^{-1})}\right)^{1/2} = 475\ \text{m s}^{-1}$$

To evaluate the integral we have used the standard result from tables of integrals (or software) that

$$\int_0^\infty x^3 e^{-ax^2}\,dx = \frac{1}{2a^2}$$

- -

Self-test 1.6 Evaluate the root mean square speed of the molecules by integration. You will need the integral

$$\int_0^\infty x^4 e^{-ax^4}\,dx = \tfrac{3}{8}\left(\frac{\pi}{a^5}\right)^{1/2}$$

$$[c = (3RT/M)^{1/2}, 515\ \text{m s}^{-1}]$$

As shown in Example 1.6, the Maxwell distribution may be used to evaluate the mean speed, $\bar{c}$, of the molecules in a gas:

$$\bar{c} = \left(\frac{8RT}{\pi M}\right)^{1/2} \tag{26}$$

The most probable speed, c^*, may also be identified from the location of the peak of the distribution:

$$c^* = \left(\frac{2RT}{M}\right)^{1/2} \tag{27}$$

These results are summarized in Fig. 1.17. The **relative mean speed**, $\bar{c}_{\text{rel}}$, the mean speed with which one molecule approaches another, can also be calculated from the distribution:

$$\bar{c}_{\text{rel}} = 2^{1/2}\bar{c} \tag{28}$$

This result is much harder to derive, but the diagram in Fig. 1.18 should help to show that it is plausible. The last result can also be generalized to the relative mean speed of two dissimilar molecules of masses m_A and m_B:

$$\bar{c}_{\text{rel}} = \left(\frac{8kT}{\pi\mu}\right)^{1/2} \qquad \mu = \frac{m_A m_B}{m_A + m_B} \tag{29}$$

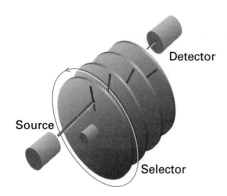

1.19 A velocity selector. The molecules are produced in the source (which may be an oven with a small hole in one wall), and travel in a beam towards the rotating disks. Only if the speed of a molecule is such as to carry it through each slot that rotates into its path will it reach the detector. Thus, the number of slow molecules can be counted by rotating the disks slowly, and the number of fast molecules counted by rotating the disks rapidly.

Table 1.3* Collision cross-sections

	σ/nm^2
Benzene, C_6H_6	0.88
Carbon dioxide, CO_2	0.52
Helium, He	0.21
Nitrogen, N_2	0.43

*More values are given in the *Data section* at the end of this volume.

Note that the molecular masses (not the molar masses) and the **Boltzmann constant**, $k = R/N_A$, appear in this expression; the quantity μ is called the **reduced mass** of the molecules. Equation 29 turns into eqn 28 when the molecules are identical (that is, $m_A = m_B = m$, so $\mu = \frac{1}{2}m$).

The Maxwell distribution has been verified experimentally. For example, molecular speeds can be measured directly with a velocity selector (Fig. 1.19). The spinning disks have slots that permit the passage of only those molecules moving through them at the appropriate speed, and the number of molecules can be determined by collecting them at a detector.

(b) The collision frequency

The kinetic model enables us to make more quantitative our picture of the events that take place in a gas. In particular, it enables us to calculate the frequency with which molecular collisions occur and the distance a molecule travels on average between collisions.[2]

We count a 'hit' whenever the centres of two molecules come within a distance d of each other, where d, the **collision diameter**, is of the order of the actual diameters of the molecules (for impenetrable hard spheres d is the diameter). As we show in the following *Justification*, kinetic theory can be used to deduce that the **collision frequency**, z, the number of collisions made by one molecule divided by the time interval during which the collisions are counted, when there are N molecules in a volume V is

$$z = \sigma \bar{c}_{rel} \mathcal{N} \qquad \sigma = \pi d^2 \tag{30}°$$

with $\mathcal{N} = N/V$ and $\bar{c}_{rel}$ given in eqn 28. The area σ is called the **collision cross-section** of the molecules. Some typical collision cross-sections are given in Table 1.3 (they are obtained by the techniques described in Section 22.5). In terms of the pressure,

$$z = \frac{\sigma \bar{c}_{rel} p}{kT} \tag{31}°$$

Justification 1.3

We consider the positions of all the molecules except one to be frozen. Then we note what happens as one mobile molecule travels through the gas with a mean relative speed $\bar{c}_{rel}$ for a time Δt. In doing so it sweeps out a 'collision tube' of cross-sectional area $\sigma = \pi d^2$ and length $\bar{c}_{rel}\Delta t$, and therefore of volume $\sigma \bar{c}_{rel}\Delta t$ (Fig. 1.20). The number of stationary molecules with centres inside the collision tube is given by the volume of the tube multiplied by the number density $\mathcal{N} = N/V$, and is $\mathcal{N}\sigma \bar{c}_{rel}\Delta t$. The number of hits scored in the interval Δt is equal to this number, so the number of collisions divided by the time interval, is $\mathcal{N}\sigma \bar{c}_{rel}$. The expression in terms of the pressure of the gas is obtained by using the perfect gas equation to write

$$\frac{N}{V} = \frac{nN_A}{V} = \frac{pN_A}{RT} = \frac{p}{kT}$$

Equation 30 tells us that the collision frequency increases with increasing temperature in a sample held at constant volume. The reason is that the mean relative speed increases with temperature. Equation 31 tells us that, at constant temperature, the collision frequency is proportional to the pressure. Such a proportionality is plausible for, the greater the pressure, the greater the number density of molecules in the sample, and the rate at which they encounter one another is greater even though their average speed remains the same. For an N_2 molecule in a sample at 1 atm and 25°C, $z \approx 7 \times 10^9$ s^{-1}, so a given molecule collides

---

2   The kinetic model also provides a basis for calculating how rapidly physical properties are transported through a gas, Chapter 24, and the rates of chemical reactions, Chapter 27.

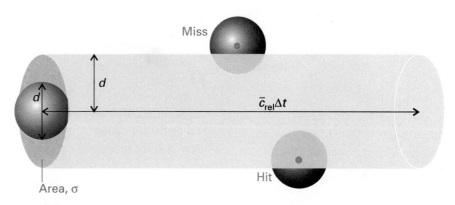

1.20 In an interval Δt, a molecule of diameter d sweeps out a tube of diameter $2d$ and length $\bar{c}_{rel}\Delta t$. As it does so it encounters other molecules with centres that lie within the tube, and each such encounter counts as one collision. In practice, the tube is not straight, but changes direction at each collision. Nevertheless, the volume swept out is the same, and this straightened version of the tube can be used as a basis of the calculation.

about 7×10^9 times each second. We are beginning to appreciate the timescale of events in gases.

(c) The mean free path

Once we have the collision frequency, we can calculate the **mean free path**, λ, the average distance a molecule travels between collisions. If a molecule collides with a frequency z, it spends a time $1/z$ in free flight between collisions, and therefore travels a distance $(1/z)\bar{c}$. Therefore, the mean free path is

$$\lambda = \frac{\bar{c}}{z} \tag{32}$$

Substitution of the expression for z in eqn 31 gives

$$\lambda = \frac{kT}{2^{1/2}\sigma p} \tag{33}$$

Doubling the pressure reduces the mean free path by half. A typical mean free path in nitrogen gas at 1 atm is 70 nm, or about 10^3 molecular diameters. Although the temperature appears in eqn 33, in a sample of constant volume, the pressure is proportional to T, so T/p remains constant when the temperature is increased. Therefore, the mean free path is independent of the temperature in a sample of gas in a container of fixed volume. The distance between collisions is determined by the number of molecules present in the given volume, not by the speed at which they travel.

In summary, a typical gas (N_2 or O_2) at 1 atm and 25 °C can be thought of as a collection of molecules travelling with a mean speed of about 350 m s^{-1}. Each molecule makes a collision within about 1 ns, and between collisions it travels about 10^2 to 10^3 molecular diameters. The kinetic model of gases is valid (and the gas behaves nearly perfectly) if the diameter of the molecules is much smaller than the mean free path ($d \ll \lambda$), for then the molecules spend most of their time far from one another.

Real gases

Real gases do not obey the perfect gas law exactly. Deviations from the law are particularly important at high pressures and low temperatures, especially when a gas is on the point of condensing to liquid.

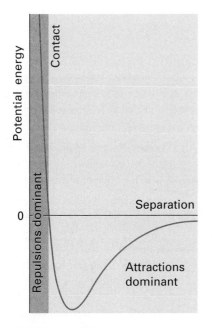

1.21 The variation of the potential energy of two molecules on their separation. High positive potential energy (at very small separations) indicates that the interactions between them are strongly repulsive at these distances. At intermediate separations, where the potential energy is negative, the attractive interactions dominate. At large separations (on the right) the potential energy is zero and there is no interaction between the molecules.

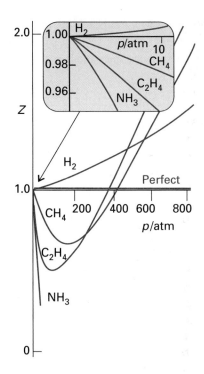

1.22 The variation of the compression factor $Z = pV_m/RT$ with pressure for several gases at $0\,°C$. A perfect gas has $Z = 1$ at all pressures. Notice that, although the curves approach 1 as $p \to 0$, they do so with different slopes.

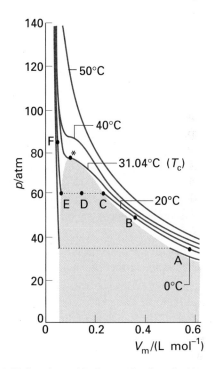

1.23 Experimental isotherms of carbon dioxide at several temperatures. The 'critical isotherm', the isotherm at the critical temperature, is at $31.04\,°C$. The critical point is marked with a star.

1.4 Molecular interactions

Real gases show deviations from the perfect gas law because molecules interact with each other. Repulsive forces between molecules assist expansion and attractive forces assist compression.

Repulsive forces are significant only when molecules are almost in contact: they are short-range interactions, even on a scale measured in molecular diameters (Fig. 1.21). Because they are short-range interactions, repulsions can be expected to be important only when the molecules are close together on average. This is the case at high pressure, when a large number of molecules occupy a small volume. On the other hand, attractive intermolecular forces have a relatively long range and are effective over several molecular diameters. They are important when the molecules are fairly close together but not necessarily touching (at the intermediate separations in Fig. 1.21). Attractive forces are ineffective when the molecules are far apart (well to the right in Fig. 1.21). Intermolecular forces may also be important when the temperature is so low that the molecules travel with such low mean speeds that they can be captured by one another. At low pressure, when the sample occupies a large volume, the molecules are so far apart for most of the time that the intermolecular forces play no significant role, and the gas behaves perfectly. At moderate pressure, when the molecules are on average only a few molecular diameters apart, the attractive forces dominate the repulsive forces. In this case, the gas can be expected to be more compressible than a perfect gas because the forces are helping to draw the molecules together. At high pressure, when the molecules are on average close together, the repulsive forces dominate and the gas can be expected to be less compressible because now the forces help to drive the molecules apart.

(a) The compression factor

That real gases reflect the distance dependence of the forces we have described can be demonstrated by plotting the **compression factor**, Z, against pressure, where Z is defined as

$$Z = \frac{pV_m}{RT} \tag{34}$$

Because, for a perfect gas, $Z = 1$ under all conditions, deviation of Z from 1 is a measure of departure from perfect behaviour.

Some experimental values of Z are plotted in Fig. 1.22. At very low pressures, all the gases shown have $Z \approx 1$ and behave nearly perfectly. At high pressures, all the gases have $Z > 1$, signifying that they are more difficult to compress than a perfect gas (for a given molar volume, the product pV_m is greater than RT). Repulsive forces are now dominant. At intermediate pressures, most gases have $Z < 1$, indicating that the attractive forces are dominant and favour compression.

(b) Virial coefficients

Some experimental isotherms for carbon dioxide are shown in Fig. 1.23. At large molar volumes and high temperatures the real isotherms do not differ greatly from perfect isotherms. The small differences suggest that the perfect gas law is in fact the first term in an expression of the form

$$pV_m = RT\left(1 + B'p + C'p^2 + \cdots\right) \tag{35}$$

Table 1.4* Second virial coefficients, $B/(\text{cm}^3\ \text{mol}^{-1})$

	Temperature	
	273 K	**600 K**
Ar	−21.7	11.9
CO_2	−149.7	−12.4
N_2	−10.5	21.7
Xe	−153.7	−19.6

*More values are given in the *Data section*.

This expression is an example of a common procedure in physical chemistry, in which a simple law (in this case $pV = nRT$) is treated as the first term in a series in powers of a variable (in this case p). A more convenient expansion for many applications is

$$pV_m = RT\left(1 + \frac{B}{V_m} + \frac{C}{V_m^2} + \cdots\right) \tag{36}$$

These two expressions are two versions of the **virial equation of state**.[3] The coefficients $B, C, \ldots$, which depend on the temperature, are the second, third,... **virial coefficients** (Table 1.4); the first virial coefficient is 1. The third virial coefficient, C, is usually less important than the second coefficient, B, in the sense that, at typical molar volumes, $C/V_m^2 \ll B/V_m$.

The virial equation can be used to demonstrate the important point that, although the equation of state of a real gas may coincide with the perfect gas law as $p \to 0$, not all its properties necessarily coincide with those of a perfect gas in that limit. Consider, for example, the value of dZ/dp, the slope of the graph of compression factor against pressure. For a perfect gas $dZ/dp = 0$ (because $Z = 1$ at all pressures), but for a real gas

$$\frac{dZ}{dp} = B' + 2pC' + \cdots \to B' \quad \text{as } p \to 0 \tag{37}$$

However, B' is not necessarily zero, so the slope of Z with respect to p does not necessarily approach 0 (the perfect gas value). Because several properties depend on derivatives (as we shall see), the properties of real gases do not always coincide with the perfect gas values at low pressures. By a similar argument,

$$\frac{dZ}{d(1/V_m)} \to B \quad \text{as } V_m \to \infty \text{ corresponding to } p \to 0 \tag{38}$$

Because the virial coefficients depend on the temperature, there may be a temperature at which $Z \to 1$ with zero slope at low pressure or high molar volume (Fig. 1.24). At this temperature, which is called the **Boyle temperature**, T_B, the properties of the real gas do coincide with those of a perfect gas as $p \to 0$. According to the relation above, Z has zero slope as $p \to 0$ if $B = 0$, so we can conclude that $B = 0$ at the Boyle temperature. It then follows from eqn 36 that $pV_m \approx RT_B$ over a more extended range of pressures than at other temperatures because the first term after 1 (that is, B/V_m) in the virial equation is zero and C/V_m^2 and higher terms are negligibly small. For helium $T_B = 22.64\ \text{K}$; for air $T_B = 346.8\ \text{K}$; more values are given in Table 1.5.

(c) Condensation

Now consider what happens when a sample of gas initially in the state marked A in Fig. 1.23 is compressed at constant temperature (by pushing in a piston). Near A, the pressure of the gas rises in approximate agreement with Boyle's law. Serious deviations from that law begin to appear when the volume has been reduced to B.

At C (which corresponds to about 60 atm for carbon dioxide), all similarity to perfect behaviour is lost, for suddenly the piston slides in without any further rise in pressure: this stage is represented by the horizontal line CDE. Examination of the contents of the vessel shows that just to the left of C a liquid appears, and there are two phases separated by a sharply defined surface. As the volume is decreased from C through D to E, the amount of liquid increases. There is no additional resistance to the piston because the gas can respond by condensing. The pressure corresponding to the line CDE, when both liquid and vapour are present in equilibrium, is called the **vapour pressure** of the liquid at the temperature of the experiment.

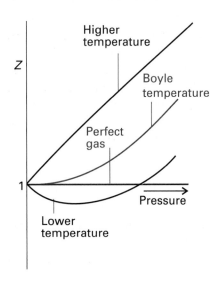

1.24 The compression factor approaches 1 at low pressures, but does so with different slopes. For a perfect gas, the slope is zero, but real gases may have either positive or negative slopes, and the slope may vary with temperature. At the Boyle temperature, the slope is zero and the gas behaves perfectly over a wider range of conditions than at other temperatures.

---

3 The name comes from the Latin word for force.

Table 1.5* Critical constants of gases

	p_c/atm	$V_c/(\text{cm}^3\,\text{mol}^{-1})$	T_c/K	Z_c	T_B/K
Ar	48.0	75.3	150.7	0.292	411.5
CO_2	72.9	94.0	304.2	0.274	714.8
He	2.26	57.8	5.2	0.305	22.6
O_2	50.14	78.0	154.8	0.308	405.9

*More values are given in the *Data section*.

At E, the sample is entirely liquid and the piston rests on its surface. Any further reduction of volume requires the exertion of considerable pressure, as is indicated by the sharply rising line to the left of E. Even a small reduction of volume from E to F requires a great increase in pressure.

(d) Critical constants

The isotherm at the temperature T_c (304.19 K, or 31.04°C for CO_2) plays a special role in the theory of the states of matter. An isotherm slightly below T_c behaves as we have already described: at a certain pressure, a liquid condenses from the gas and is distinguishable from it by the presence of a visible surface. If, however, the compression takes place at T_c itself, then a surface separating two phases does not appear and the volumes at each end of the horizontal part of the isotherm have merged to a single point, the **critical point** of the gas. The temperature, pressure, and molar volume at the critical point are called the **critical temperature**, T_c, **critical pressure**, p_c, and **critical molar volume**, V_c, of the substance. Collectively, p_c, V_c, and T_c are the **critical constants** of a substance (Table 1.5).

At and above T_c, the sample has a single phase which occupies the entire volume of the container. Such a phase is, by definition, a gas. Hence, *the liquid phase of a substance does not form above the critical temperature*. At the critical temperature, a surface does not form and the horizontal part of the isotherm has merged to a single point. The critical temperature of oxygen, for instance, signifies that it is impossible to produce liquid oxygen by compression alone if its temperature is greater than 154.8 K: to liquefy it—to obtain a fluid phase that does not occupy the entire volume—the temperature must first be lowered to below 154.8 K, and then the gas compressed isothermally. The single phase that fills the entire volume at $T > T_c$ may be much denser that we normally consider typical of gases, and the name **supercritical fluid** is preferred.

1.5 The van der Waals equation

Conclusions can be drawn from the virial equations of state only by inserting specific values of the coefficients. It is often useful to have a broader, if less precise, view of all gases. Therefore, we introduce the approximate equation of state suggested by J.H. van der Waals in 1873. This equation is an excellent example of an expression that can be obtained by thinking scientifically about a mathematically complicated but physically simple problem, that is, it is a good example of 'model building'. Van der Waals himself proposed his equation on the basis of experimental evidence available to him in conjunction with rigorous thermodynamic arguments. The **van der Waals equation** is

$$p = \frac{nRT}{V - nb} - a\left(\frac{n}{V}\right)^2 \tag{39a}$$

Table 1.6* Van der Waals coefficients

	$a/(\text{atm L}^2\,\text{mol}^{-2})$	$b/(10^{-2}\,\text{L mol}^{-1})$
Ar	1.363	3.219
CO_2	3.640	4.267
He	0.057	2.370
N_2	1.408	3.913

*More values are given in the *Data section*.

and a derivation is given in the *Justification* below. The equation is often written in terms of the molar volume $V_m = V/n$ as

$$p = \frac{RT}{V_m - b} - \frac{a}{V_m^2} \tag{39b}$$

The constants a and b are called the **van der Waals coefficients**. They are characteristic of each gas but independent of the temperature; some values are listed in Table 1.6.

Justification 1.4

The repulsive interactions between molecules are taken into account by supposing that they cause the molecules to behave as small but impenetrable spheres. The nonzero volume of the molecules implies that instead of moving in a volume V they are restricted to a smaller volume $V - nb$, where nb is approximately the total volume taken up by the molecules themselves. This argument suggests that the perfect gas law $p = nRT/V$ should be replaced by

$$p = \frac{nRT}{V - nb}$$

when repulsions are significant.

The pressure depends on both the frequency of collisions with the walls and the force of each collision. Both the frequency of the collisions and their force are reduced by the attractive forces, which act with a strength proportional to the molar concentration, n/V, of molecules in the sample. Therefore, because both the frequency and the force of the collisions are reduced by the attractive forces, the pressure is reduced in proportion to the square of this concentration. If the reduction of pressure is written as $-a(n/V)^2$, where a is a positive constant characteristic of each gas, the combined effect of the repulsive and attractive forces is the van der Waals equation of state as expressed in eqn 39.

In this *Justification* we have built the van der Waals equation using vague arguments about the volumes of molecules and the effects of forces. It can be derived in other ways, but the present method has the advantage that it shows how to derive the form of an equation out of general ideas. The derivation also has the advantage of keeping imprecise the significance of the coefficients a and b: they are much better regarded as empirical parameters than as precisely defined molecular properties.

Example 1.7 Using the van der Waals equation to estimate a molar volume

Estimate the molar volume of CO_2 at 500 K and 100 atm by treating it as a van der Waals gas.

Method To express eqn 39b as an equation for the molar volume, we need to rearrange it into

$$V_m^3 - \left(b + \frac{RT}{p}\right)V_m^2 + \left(\frac{a}{p}\right)V_m - \frac{ab}{p} = 0$$

Although closed expressions for the roots of a cubic equation can be given, they are very complicated. Unless analytical solutions are essential, it is usually more expedient to solve such equations either with a programmable calculator or with a commercial software package.[4]

4   A procedure is included in the CD that accompanies this text.

Table 1.7 Selected equations of state

	Equation	Reduced form	Critical constants		
			p_c	V_c	T_c
Perfect gas	$p = \dfrac{RT}{V_m}$				
Van der Waals	$p = \dfrac{RT}{V_m - b} - \dfrac{a}{V_m^2}$	$p_r = \dfrac{8T_r}{3V_r - 1} - \dfrac{3}{V_r^2}$	$\dfrac{a}{27b^2}$	$3b$	$\dfrac{8a}{27bR}$
Berthelot	$p = \dfrac{RT}{V_m - b} - \dfrac{a}{TV_m^2}$	$p_r = \dfrac{8T_r}{3V_r - 1} - \dfrac{3}{T_r V_r^2}$	$\dfrac{1}{12}\left(\dfrac{2aR}{3b^3}\right)^{1/2}$	$3b$	$\dfrac{2}{3}\left(\dfrac{2a}{3bR}\right)^{1/2}$
Dieterici	$p = \dfrac{RTe^{-a/RTV_m}}{V_m - b}$	$p_r = \dfrac{e^2 T_r e^{-2/T_r V_r}}{2V_r - 1}$	$\dfrac{a}{4e^2 b^2}$	$2b$	$\dfrac{a}{4Rb}$
Beattie–Bridgman	$p = \dfrac{(1 - \gamma)RT(V_m + \beta) - \alpha}{V_m^2}$ with $\alpha = a_0\left(1 + \dfrac{a}{V_m}\right)$				
	$\beta = b_0\left(1 - \dfrac{b}{V_m}\right)$				
	$\gamma = \dfrac{c_0}{V_m T^3}$				
Virial (Kammerlingh Onnes)	$p = \dfrac{RT}{V_m}\left\{1 + \dfrac{B(T)}{V_m} + \dfrac{C(T)}{V_m^2} + \cdots\right\}$				

Answer According to Table 1.6, $a = 3.640 \text{ atm L}^2 \text{ mol}^{-2}$ and $b = 4.267 \times 10^{-2} \text{ L mol}^{-1}$. Under the stated conditions, $RT/p = 0.410 \text{ L mol}^{-1}$. The coefficients in the equation for V_m are therefore

$$b + RT/p = 0.453 \text{ L mol}^{-1}$$
$$a/p = 3.64 \times 10^{-2} \text{ (L mol}^{-1})^2$$
$$ab/p = 1.55 \times 10^{-3} \text{ (L mol}^{-1})^3$$

Therefore, on writing $x = V_m/(\text{L mol}^{-1})$, the equation to solve is

$$x^3 - 0.453x^2 + (3.64 \times 10^{-2})x - (1.55 \times 10^{-3}) = 0$$

The acceptable root is $x = 0.366$, which implies that $V_m = 0.370 \text{ L mol}^{-1}$.

Comment For a perfect gas under these conditions, the molar volume is 0.410 L mol^{-1}.

- -

Self-test 1.7 Calculate the molar volume of argon at 100°C and 100 atm on the assumption that it is a van der Waals gas.

[0.298 L mol^{-1}]

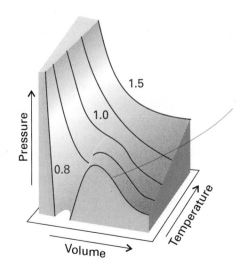

1.25 The surface of possible states allowed by the van der Waals equation. Compare this surface with that shown in Fig. 1.9.

(a) The reliability of the equation

We now examine to what extent the van der Waals equation predicts the behaviour of real gases. It is too optimistic to expect a single, simple expression to be the true equation of state of all substances, and accurate work on gases must resort to the virial equation, use tabulated values of the coefficients at various temperatures, and analyse the systems numerically. The advantage of the van der Waals equation is that it is analytical and allows us to draw some general conclusions about real gases. When the equation fails we must use

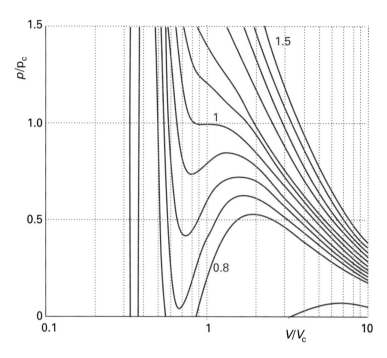

1.26 Van der Waals isotherms at several values of T/T_c. Compare these curves with those in Fig.1.23. The van der Waals loops are normally replaced by horizontal straight lines. The critical isotherm is the isotherm for $T/T_c = 1$.

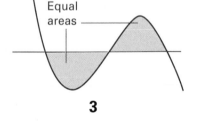

Equal areas

3

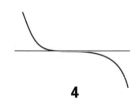

4

one of the other equations of state that have been proposed (some are listed in Table 1.7), invent a new one, or go back to the virial equation.

That having been said, we can begin to judge the reliability of the equation by comparing the isotherms it predicts with the experimental isotherms in Fig. 1.23. Some calculated isotherms are shown in Figs 1.25 and 1.26. Apart from the oscillations below the critical temperature, they do resemble experimental isotherms quite well. The oscillations, the **van der Waals loops**, are unrealistic because they suggest that under some conditions an increase of pressure results in an increase of volume. Therefore they are replaced by horizontal lines drawn so the loops define equal areas above and below the lines: this procedure is called the **Maxwell construction** (3). The van der Waals coefficients, such as those in Table 1.6, are found by fitting the calculated curves to the experimental curves.

(b) The features of the equation

The principal features of the van der Waals equation can be summarized as follows.

(1) *Perfect gas isotherms are obtained at high temperatures and large molar volumes.*

When the temperature is high, RT may be so large that the first term in eqn 39b greatly exceeds the second. Furthermore, if the molar volume is large (in the sense $V_m \gg b$), then the denominator $V_m - b \approx V_m$. Under these conditions, the equation reduces to $p = RT/V_m$, the perfect gas equation.

(2) *Liquids and gases coexist when cohesive and dispersing effects are in balance.*

The van der Waals loops occur when both terms in eqn 39b have similar magnitudes. The first term arises from the kinetic energy of the molecules and their repulsive interactions; the second represents the effect of the attractive interactions.

(3) *The critical constants are related to the van der Waals coefficients.*

For $T < T_c$, the calculated isotherms oscillate, and each one passes through a minimum followed by a maximum. These extrema converge as $T \to T_c$ and coincide at $T = T_c$; at the critical point the curve has a flat inflexion (4). From the properties of curves, we know that an inflexion of this type occurs when both the first and second derivatives are zero. Hence, we can find the critical constants by calculating these derivatives and setting them equal to zero:

$$\frac{dp}{dV_m} = -\frac{RT}{(V_m - b)^2} + \frac{2a}{V_m^3} = 0$$

$$\frac{d^2p}{dV_m^2} = \frac{2RT}{(V_m - b)^3} - \frac{6a}{V_m^4} = 0$$

at the critical point. The solutions of these two equations are

$$V_c = 3b \qquad p_c = \frac{a}{27b^2} \qquad T_c = \frac{8a}{27Rb} \tag{40}$$

These relations can be tested by noting that the **critical compression factor**, Z_c, is predicted to be equal to

$$Z_c = \frac{p_c V_c}{RT_c} = \tfrac{3}{8} \tag{41}$$

for all gases. We see from Table 1.5 that, although $Z_c < \tfrac{3}{8}$ (or 0.375), it is approximately constant (at 0.3) and the discrepancy is reasonably small.

1.6 The principle of corresponding states

An important general technique in science for comparing the properties of objects is to choose a related fundamental property of the same kind and to set up a relative scale on that basis. We have seen that the critical constants are characteristic properties of gases, so it may be that a scale can be set up by using them as yardsticks. We therefore introduce the **reduced variables** of a gas by dividing the actual variable by the corresponding critical constant:

$$p_r = \frac{p}{p_c} \qquad V_r = \frac{V_m}{V_c} \qquad T_r = \frac{T}{T_c} \tag{42}$$

If the reduced pressure of a gas is given, then we can easily calculate its actual pressure by using $p = p_r p_c$, and likewise for the volume and temperature. Van der Waals, who first tried this procedure, hoped that gases confined to the same reduced volume, V_r, at the same reduced temperature, T_r, would exert the same reduced pressure, p_r. The hope was largely fulfilled (Fig. 1.27). The illustration shows the dependence of the compression factor on the reduced pressure for a variety of gases at various reduced temperatures. The success of the procedure is strikingly clear: compare this graph with Fig. 1.22, where similar data are plotted without using reduced variables. The observation that real gases at the same volume and temperature exert the same reduced pressure is called the **principle of corresponding states**. It is only an approximation, and works best for gases composed of spherical molecules; it fails, sometimes badly, when the molecules are non-spherical or polar.

The van der Waals equation sheds some light on the principle. First, we express eqn 39b in terms of the reduced variables, which gives

$$p_r p_c = \frac{RT_r T_c}{V_r V_c - b} - \frac{a}{V_r^2 V_c^2}$$

Then we express the critical constants in terms of a and b by using eqn 40:

$$\frac{a p_r}{27b^2} = \frac{8a T_r}{27b(3bV_r - b)} - \frac{a}{9b^2 V_r^2}$$

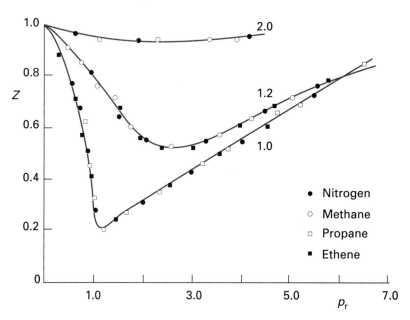

1.27 The compression factors of four gases, including two of those shown in Fig. 1.22, plotted using reduced variables. The use of reduced variables organizes the data on to single curves.

which can be reorganized into

$$p_r = \frac{8T_r}{3V_r - 1} - \frac{3}{V_r^2}$$
(43)

This equation has the same form as the original, but the coefficients a and b, which differ from gas to gas, have disappeared. It follows that, if the isotherms are plotted in terms of the reduced variables (as we did in fact in Fig. 1.26 without drawing attention to the fact), then the same curves are obtained whatever the gas. This is precisely the content of the principle of corresponding states, and so the van der Waals equation is compatible with it.

Looking for too much significance in this apparent triumph is mistaken, because other equations of state also accommodate the principle (Table 1.7). In fact, all we need are two parameters playing the roles of a and b, for then the equation can always be manipulated into reduced form. The observation that real gases obey the principle approximately amounts to saying that the effects of the attractive and repulsive interactions can each be approximated in terms of a single parameter. The importance of the principle is then not so much its theoretical interpretation but the way that it enables the properties of a range of gases to be coordinated on to a single diagram (for example, Fig. 1.27 instead of Fig. 1.22).

Checklist of key ideas

□ gas

The perfect gas

1.1 The states of gases
□ state
□ equation of state
□ perfect gas
□ pressure
□ standard pressure
□ mechanical equilibrium

□ barometer
□ manometer
□ temperature
□ diathermic
□ adiabatic
□ thermal equilibrium
□ Zeroth Law of thermodynamics
□ thermometer
□ Celsius scale
□ perfect-gas temperature scale

□ thermodynamic temperature scale

1.2 The gas laws
□ Boyle's law (6)
□ isotherm
□ limiting law
□ isobar
□ Charles's law (9)
□ Avogadro's principle (11)
□ gas constant

□ perfect gas equation (12)
□ perfect gas
□ real gas
□ standard ambient temperature and pressure (SATP)
□ standard temperature and pressure (STP)
□ Dalton's law
□ partial pressure
□ mole fraction (15)
□ partial pressure defined (17)

1.3 The kinetic model of gases
- [ ] kinetic model of gases
- [ ] elastic collision
- [ ] root mean square speed
- [ ] distribution of speeds
- [ ] Maxwell distribution of speeds (22)
- [ ] relative mean speed (29)
- [ ] Boltzmann constant
- [ ] reduced mass (29)
- [ ] collision diameter
- [ ] collision frequency (30)
- [ ] collision cross-section
- [ ] mean free path (33)

Real gases

1.4 Molecular interactions
- [ ] attractive and repulsive intermolecular forces
- [ ] compression factor (34)
- [ ] virial equation of state (36)
- [ ] virial coefficient
- [ ] Boyle temperature
- [ ] condensation
- [ ] vapour pressure
- [ ] critical point
- [ ] critical temperature
- [ ] critical pressure
- [ ] critical molar volume
- [ ] critical constants

1.5 The van der Waals equation
- [ ] van der Waals equation (39)
- [ ] van der Waals coefficients
- [ ] van der Waals loops
- [ ] Maxwell construction
- [ ] critical compression factor (41)

1.6 The principle of corresponding states
- [ ] reduced variables
- [ ] principle of corresponding states

Further reading

Articles of general interest

D.B. Clark, The ideal gas law at the center of the sun. *J. Chem. Educ.* **66**, 826 (1989).

J.G. Eberhart, The many faces of van der Waals's equation of state. *J. Chem. Educ.* **66**, 906 (1989).

J.M. Alvaiño, J. Veguillas, and S. Valasco, Equations of state, collisional energy transfer, and chemical equilibrium in gases. *J. Chem. Educ.* **66**, 157 (1989).

G. Rhodes, Does a one-molecule gas obey Boyle's law? *J. Chem. Educ.* **69**, 16 (1992).

J.G. Eberhart, A least-squares technique for determining the van der Waals parameters from the critical constants. *J. Chem. Educ.* **69**, 220 (1992).

J.B. Ott, J.R. Goates, and H.T. Hall, Comparison of equations of state. *J. Chem. Educ.* **48**, 515 (1971).

E.F. Meyer and T.P. Meyer, Supercritical fluid: liquid, gas, both, or neither? A different approach. *J. Chem. Educ.* **63**, 463 (1986).

J.L. Pauley and E.H. Davis, *P-V-T* isotherms of real gases: experimental versus calculated values. *J. Chem. Educ.* **63**, 466 (1986).

M. Ross, Equations of state. In *Encyclopedia of applied physics* (ed. G.L. Trigg), 6, 291. VCH, New York (1993).

Texts and sources of data and information

D. Tabor, *Gases, liquids, and solids.* Cambridge University Press, (1979).

A.J. Walton, *Three phases of matter.* Oxford University Press (1983).

J.H. Dymond and E.B. Smith, *The virial coefficients of pure gases and mixtures.* Oxford University Press (1980).

J.O. Hirschfelder, C.F. Curtiss, and R.B. Bird, *The molecular theory of gases and liquids.* Wiley, New York (1954).

B.W. Rossiter and R.C. Baetzgold (ed.), *Physical methods of chemistry*, **VI**. Wiley–Interscience, New York (1992).

Exercises

1.1 (a) A sample of air occupies 1.0 L at 25 °C and 1.00 atm. What pressure is needed to compress it to 100 cm^3 at this temperature?

1.1 (b) A sample of carbon dioxide gas occupies 350 cm^3 at 20 °C and 104 kPa. What pressure is needed to compress it to 250 cm^3 at this temperature?

1.2 (a) (a) Could 131 g of xenon gas in a vessel of volume 1.0 L exert a pressure of 20 atm at 25 °C if it behaved as a perfect gas? If not, what pressure would it exert? (b) What pressure would it exert if it behaved as a van der Waals gas?

1.2 (b) (a) Could 25 g of argon gas in a vessel of volume 1.5 L exert a pressure of 2.0 bar at 30 °C if it behaved as a perfect gas? If not, what pressure would it exert? (b) What pressure would it exert if it behaved as a van der Waals gas?

1.3 (a) A perfect gas undergoes isothermal compression, which reduces its volume by 2.20 L. The final pressure and volume of the gas are 3.78×10^3 Torr and 4.65 L, respectively. Calculate the original pressure of the gas in (a) Torr, (b) atm.

1.3 (b) A perfect gas undergoes isothermal compression, which reduces its volume by 1.80 dm^3. The final pressure and volume of the gas are 1.48×10^3 Torr and 2.14 dm^3, respectively. Calculate the original pressure of the gas in (a) Torr, (b) bar.

1.4 (a) To what temperature must a 1.0 L sample of a perfect gas be cooled from 25 °C in order to reduce its volume to 100 cm³?

1.4 (b) To what temperature must a sample of a perfect gas of volume 500 mL be cooled from 35 °C in order to reduce its volume to 150 cm³?

1.5 (a) A car tyre (that is, an automobile tire) was inflated to a pressure of 24 lb in⁻² (1.00 atm = 14.7 lb in⁻²) on a winter's day when the temperature was −5 °C. What pressure will be found, assuming no leaks have occurred and that the volume is constant, on a subsequent summer's day when the temperature is 35 °C? What complications should be taken into account in practice?

1.5 (b) A sample of hydrogen gas was found to have a pressure of 125 kPa when the temperature was 23 °C. What can its pressure be expected to be when the temperature is 11 °C?

1.6 (a) A sample of 255 mg of neon occupies 3.00 L at 122 K. Use the perfect gas law to calculate the pressure of the gas.

1.6 (b) A homeowner uses 4.00×10^3 m³ of natural gas in a year to heat a home. Assume that natural gas is all methane, CH_4, and that methane is a perfect gas for the conditions of this problem, which are 1.00 atm and 20 °C. What is the mass of gas used?

1.7 (a) In an attempt to determine an accurate value of the gas constant, R, a student heated a 20.000 L container filled with 0.25132 g of helium gas to 500 °C and measured the pressure as 206.402 cm of water in a manometer at 25 °C. Calculate the value of R from these data. (The density of water at 25 °C is 0.99707 g cm⁻³.)

1.7 (b) The following data have been obtained for oxygen gas at 273.15 K. Calculate the best value of the gas constant R from them and the best value of the molar mass of O_2.

p/atm	0.750 000	0.500 000	0.250 000
$V_m/\text{L mol}^{-1}$	29.8649	44.8090	89.6384
$\rho/(\text{g L}^{-1})$	1.07144	0.714 110	0.356 975

1.8 (a) At 500 °C and 699 Torr, the mass density of sulfur vapour is 3.71 g L⁻¹. What is the molecular formula of sulfur under these conditions?

1.8 (b) At 100 °C and 120 Torr, the mass density of phosphorus vapour is 0.6388 kg m⁻³. What is the molecular formula of phosphorus under these conditions?

1.9 (a) Calculate the mass of water vapour present in a room of volume 400 m³ that contains air at 27 °C on a day when the relative humidity is 60 per cent.

1.9 (b) Calculate the mass of water vapour present in a room of volume 250 m³ that contains air at 23 °C on a day when the relative humidity is 53 per cent.

1.10 (a) Given that the density of air at 740 Torr and 27 °C is 1.146 g L⁻¹, calculate the mole fraction and partial pressure of nitrogen and oxygen assuming that (a) air consists only of these two gases, (b) air also contains 1.0 mole per cent Ar.

1.10 (b) A gas mixture consists of 320 mg of methane, 175 mg of argon, and 225 mg of neon. The partial pressure of neon at 300 K is 66.5 Torr. Calculate (a) the volume and (b) the total pressure of the mixture.

1.11 (a) The density of a gaseous compound was found to be 1.23 g L⁻¹ at 330 K and 150 Torr. What is the molar mass of the compound?

1.11 (b) In an experiment to measure the molar mass of a gas, 250 cm³ of the gas was confined in a glass vessel. The pressure was 152 Torr at 298 K and, after correcting for buoyancy effects, the mass of the gas was 33.5 mg. What is the molar mass of the gas?

1.12 (a) The density of air at −85 °C, 0 °C, and 100 °C is 1.877 g L⁻¹, 1.294 g L⁻¹, and 0.946 g L⁻¹, respectively. From these data, and assuming that air obeys Charles's law, determine a value for the absolute zero of temperature in degrees Celsius.

1.12 (b) A certain sample of a gas has a volume of 20.00 L at 0 °C and 1.000 atm. A plot of the experimental data of its volume against the Celsius temperature, θ, at constant p, gives a straight line of slope 0.0741 L (°C)⁻¹. From these data alone (without making use of the perfect gas law), determine the absolute zero of temperature in degrees Celsius.

1.13 (a) Determine the ratios of (a) the mean speeds, (b) the mean kinetic energies of gaseous H_2 molecules and Hg atoms at 20 °C.

1.13 (b) Determine the ratios of (a) the mean speeds, (b) the mean kinetic energies of He atoms and Hg atoms at 25 °C.

1.14 (a) A 1.0 L glass bulb contains 1.0×10^{23} H_2 molecules. If the pressure exerted by the gas is 100 kPa, what are (a) the temperature of the gas, (b) the root mean square speeds of the molecules? (c) Would the temperature be different if they were O_2 molecules?

1.14 (b) The best laboratory vacuum pump can generate a vacuum of about 1 nTorr. At 25 °C and assuming that air consists of N_2 molecules with a collision diameter of 395 pm, calculate (a) the mean speed of the molecules, (b) the mean free path, (c) the collision frequency in the gas.

1.15 (a) At what pressure does the mean free path of argon at 25 °C become comparable to the size of a 1 L vessel that contains it? Take $\sigma = 0.36$ nm².

1.15 (b) At what pressure does the mean free path of argon at 25 °C become comparable to the diameters of the atoms themselves?

1.16 (a) At an altitude of 20 km the temperature is 217 K and the pressure 0.050 atm. What is the mean free path of N_2 molecules? ($\sigma = 0.43$ nm².)

1.16 (b) At an altitude of 15 km the temperature is 217 K and the pressure 12.1 kPa. What is the mean free path of N_2 molecules? ($\sigma = 0.43$ nm².)

1.17 (a) How many collisions does a single Ar atom make in 1.0 s when the temperature is 25 °C and the pressure is (a) 10 atm, (b) 1.0 atm, (c) 1.0 μatm?

1.17 (b) How many collisions per second does an N_2 molecule make at an altitude of 15 km? (See Exercise 1.16b for data.)

1.18 (a) Calculate the mean free path of molecules in air using $\sigma = 0.43 \text{ nm}^2$ at 25 °C and (a) 10 atm, (b) 1.0 atm, (c) 1.0×10^{-6} atm.

1.18 (b) Calculate the mean free path of carbon dioxide molecules using $\sigma = 0.52 \text{ nm}^2$ at 25 °C and (a) 15 atm, (b) 1.0 bar, (c) 1.0 Torr.

1.19 (a) Use the Maxwell distribution of speeds to estimate the fraction of N_2 molecules at 500 K that have speeds in the range 290 to 300 m s^{-1}.

1.19 (b) Use the Maxwell distribution of speeds to estimate the fraction of CO_2 molecules at 300 K that have speeds in the range 200 to 250 m s^{-1}.

1.20 (a) Calculate the pressure exerted by 1.0 mol C_2H_6 behaving as (a) a perfect gas, (b) a van der Waals gas when it is confined under the following conditions: (i) at 273.15 K in 22.414 L, (ii) at 1000 K in 100 cm^3. Use the data in Table 1.6.

1.20 (b) Calculate the pressure exerted by 1.0 mol H_2S behaving as (a) a perfect gas, (b) a van der Waals gas when it is confined under the following conditions: (i) at 273.15 K in 22.414 L, (ii) at 500 K in 150 cm^3. Use the data in Table 1.6.

1.21 (a) Estimate the critical constants of a gas with van der Waals parameters $a = 0.751 \text{ atm L}^2 \text{ mol}^{-2}$ and $b = 0.0226 \text{ L mol}^{-1}$.

1.21 (b) Estimate the critical constants of a gas with van der Waals parameters $a = 1.32 \text{ atm L}^2 \text{ mol}^{-2}$ and $b = 0.0436 \text{ L mol}^{-1}$.

1.22 (a) A gas at 250 K and 15 atm has a molar volume 12 per cent smaller than that calculated from the perfect gas law. Calculate (a) the compression factor under these conditions and (b) the molar volume of the gas. Which are dominating in the sample, the attractive or the repulsive forces?

1.22 (b) A gas at 350 K and 12 atm has a molar volume 12 per cent larger than that calculated from the perfect gas law. Calculate (a) the compression factor under these conditions and (b) the molar volume of the gas. Which are dominating in the sample, the attractive or the repulsive forces?

1.23 (a) In an industrial process, nitrogen is heated to 500 K at a constant volume of 1.000 m^3. The gas enters the container at 300 K and 100 atm. The mass of the gas is 92.4 kg. Use the van der Waals equation to determine the approximate pressure of the gas at its working temperature of 500 K. For nitrogen, $a = 1.408 \text{ L}^2 \text{ atm mol}^{-2}$, $b = 0.0391 \text{ L mol}^{-1}$.

1.23 (b) Cylinders of compressed gas are typically filled to a pressure of 200 bar. For oxygen, what would be the molar volume at this pressure and 25 °C based on (a) the perfect gas equation, (b) the van der Waals equation? For oxygen, $a = 1.378 \text{ L}^2 \text{ atm mol}^{-2}$, $b = 3.183 \times 10^{-2} \text{ L mol}^{-1}$.

1.24 (a) The density of water vapour at 327.6 atm and 776.4 K is 133.2 g dm^{-3}. (a) Determine the molar volume, V_m, of water and the compression factor, Z, from these data. (b) Calculate Z from the van der Waals equation with $a = 5.536 \text{ L}^2 \text{ atm mol}^{-2}$ and $b = 0.03049 \text{ L mol}^{-1}$.

1.24 (b) The density of water vapour at 1.00 bar and 383 K is 0.5678 kg m^{-3}. (a) Determine the molar volume, V_m, of water and the compression factor, Z, from these data. (b) Calculate Z from the van der Waals equation with $a = 5.536 \text{ L}^2 \text{ atm mol}^{-2}$ and $b = 0.03049 \text{ L mol}^{-1}$.

1.25 (a) Suppose that 10.0 mol $C_2H_6(g)$ is confined to 4.860 L at 27 °C. Predict the pressure exerted by the ethane from (a) the perfect gas and (b) the van der Waals equations of state. Calculate the compression factor based on these calculations. For ethane, $a = 5.562 \text{ L}^2 \text{ atm mol}^{-2}$, $b = 0.06380 \text{ L mol}^{-1}$.

1.25 (b) At 300 K and 20 atm, the compression factor of a gas is 0.86. Calculate (a) the volume occupied by 8.2 mmol of the gas under these conditions and (b) an approximate value of the second virial coefficient B at 300 K.

1.26 (a) A vessel of volume 22.4 L contains 2.0 mol H_2 and 1.0 mol N_2 at 273.15 K. Calculate (a) the mole fractions of each component, (b) their partial pressures, and (c) their total pressure.

1.26 (b) A vessel of volume 22.4 L contains 1.5 mol H_2 and 2.5 mol N_2 at 273.15 K. Calculate (a) the mole fractions of each component, (b) their partial pressures, and (c) their total pressure.

1.27 (a) The critical constants of methane are $p_c = 45.6$ atm, $V_c = 98.7 \text{ cm}^3 \text{ mol}^{-1}$, and $T_c = 190.6$ K. Calculate the van der Waals parameters of the gas and estimate the radius of the molecules.

1.27 (b) The critical constants of ethane are $p_c = 48.20$ atm, $V_c = 148 \text{ cm}^3 \text{ mol}^{-1}$, and $T_c = 305.4$ K. Calculate the van der Waals parameters of the gas and estimate the radius of the molecules.

1.28 (a) Use the van der Waals parameters for chlorine to calculate approximate values of (a) the Boyle temperature of chlorine and (b) the radius of a Cl_2 molecule regarded as a sphere.

1.28 (b) Use the van der Waals parameters for hydrogen sulfide to calculate approximate values of (a) the Boyle temperature of the gas and (b) the radius of a H_2S molecule regarded as a sphere.

1.29 (a) Suggest the pressure and temperature at which 1.0 mol of (a) NH_3, (b) Xe, (c) He will be in states that correspond to 1.0 mol H_2 at 1.0 atm and 25 °C.

1.29 (b) Suggest the pressure and temperature at which 1.0 mol of (a) H_2S, (b) CO_2, (c) Ar will be in states that correspond to 1.0 mol N_2 at 1.0 atm and 25 °C.

1.30 (a) A certain gas obeys the van der Waals equation with $a = 0.50 \text{ m}^6 \text{ Pa mol}^{-2}$. Its volume is found to be $5.00 \times 10^{-4} \text{ m}^3 \text{ mol}^{-1}$ at 273 K and 3.0 MPa. From this information calculate the van der Waals constant b. What is the compression factor for this gas at the prevailing temperature and pressure?

1.30 (b) A certain gas obeys the van der Waals equation with $a = 0.76 \text{ m}^6 \text{ Pa mol}^{-2}$. Its volume is found to be $4.00 \times 10^{-4} \text{ m}^3 \text{ mol}^{-1}$ at 288 K and 4.0 MPa. From this information calculate the van der Waals constant b. What is the compression factor for this gas at the prevailing temperature and pressure?

Problems

Numerical problems

1.1 A diving bell has an air space of 3.0 m³ when on the deck of a boat. What is the volume of the air space when the bell has been lowered to a depth of 50 m? Take the mean density of sea water to be 1.025 g cm^{-3} and assume that the temperature is the same as on the surface.

1.2 What pressure difference must be generated across the length of a 15 cm vertical drinking straw in order to drink a water-like liquid of density 1.0 g cm^{-3}?

1.3 Recent communications with the inhabitants of Neptune have revealed that they have a Celsius-type temperature scale, but based on the melting point (0°N) and boiling point (100°N) of their most common substance, hydrogen. Further communications have revealed that the Neptunians know about perfect gas behaviour and they find that, in the limit of zero pressure, the value of pV is 28 L atm at 0°N and 40 L atm at 100°N. What is the value of the absolute zero of temperature on their temperature scale?

1.4 A meterological balloon had a radius of 1.0 m when released at sea level at 20°C and expanded to a radius of 3.0 m when it had risen to its maximum altitude where the temperature was −20°C. What is the pressure inside the balloon at that altitude?

1.5 Deduce the relation between the pressure and mass density, ρ, of a perfect gas of molar mass M. Confirm graphically, using the following data on dimethyl ether at 25°C, that perfect behaviour is reached at low pressures and find the molar mass of the gas.

p/Torr	91.74	188.98	277.3	452.8	639.3	760.0
ρ/(g L^{-1})	0.232	0.489	0.733	1.25	1.87	2.30

1.6 Charles's law is sometimes expressed in the form $V = V_0(1 + \alpha\theta)$, where θ is the Celsius temperature, α is a constant, and V_0 is the volume of the sample at 0°C. The following values for α have been reported for nitrogen at 0°C:

p/Torr	749.7	599.6	333.1	98.6
$10^3\alpha/(°C)^{-1}$	3.6717	3.6697	3.6665	3.6643

From these data calculate the best value for the absolute zero of temperature on the Celsius scale.

1.7 Investigate some of the technicalities of ballooning using the perfect gas law. Suppose your balloon has a radius of 3.0 m and that it is spherical. (a) What amount of H_2 (in moles) is needed to inflate it to 1.0 atm in an ambient temperature of 25°C at sea level? (b) What mass can the balloon lift at sea level, where the density of air is 1.22 kg m^{-3}? (c) What would be the payload if He were used instead of H_2?

1.8 The molar mass of a newly synthesized fluorocarbon was measured in a gas microbalance. This device consists of a glass bulb forming one end of a beam, the whole surrounded by a closed container. The beam is pivoted, and the balance point is attained by raising the pressure of gas in the container, so increasing the buoyancy of the enclosed bulb. In one experiment, the balance point was reached when the fluorocarbon pressure was 327.10 Torr; for the same setting of the pivot, a balance was reached when CHF$_3$ ($M = 70.014$ g mol^{-1}) was introduced at 423.22 Torr. A repeat of the experiment with a different setting of the pivot required a pressure of 293.22 Torr of the fluorocarbon and 427.22 Torr of the CHF$_3$. What is the molar mass of the fluorocarbon? Suggest a molecular formula.

1.9 A constant-volume perfect gas thermometer indicates a pressure of 50.2 Torr at the triple point temperature of water (273.16 K). (a) What change of pressure indicates a change of 1.00 K at this temperature? (b) What pressure indicates a temperature of 100.00°C? (c) What change of pressure indicates a change of 1.00 K at the latter temperature?

1.10 A vessel of volume 22.4 L contains 2.0 mol H_2 and 1.0 mol N_2 at 273.15 K initially. All the H_2 reacted with sufficient N_2 to form NH$_3$. Calculate the partial pressures and the total pressure of the final mixture.

1.11 In an experiment to measure the speed of molecules by a rotating slotted-disk experiment, the apparatus consisted of five coaxial 5.0 cm diameter disks separated by 1.0 cm, the slots in their rims being displaced by 2.0° between neighbours. The relative intensities, $\mathcal{I}$, of the detected beam of Kr atoms for two different temperatures and at a series of rotation rates were as follows:

ν/Hz	20	40	80	100	120
$\mathcal{I}(40\ \text{K})$	0.846	0.513	0.069	0.015	0.002
$\mathcal{I}(100\ \text{K})$	0.592	0.485	0.217	0.119	0.057

Find the distributions of molecular velocities, $f(v_x)$, at these temperatures, and check that they conform to the theoretical prediction for a one-dimensional system.

1.12 Cars were timed by police radar as they passed in both directions below a bridge. Their velocities (kilometres per hour, numbers of cars in parentheses) to the east and west were as follows: 80 E (40), 85 E (62), 90 E (53), 95 E (12), 100 E (2); 80 W (38), 85 W (59), 90 W (60), 100 W (2). What are (a) the mean velocity, (b) the mean speed, (c) the root mean square speed?

1.13 A population consists of people of the following heights (in metres, numbers of individuals in parentheses): 1.80 (1), 1.82 (2), 1.84 (4), 1.86 (7), 1.88 (10), 1.90 (15), 1.92 (9), 1.94 (4), 1.96 (0), 1.98 (1). What are (a) the mean height, (b) the root mean square height of the population?

1.14 Calculate the escape velocity (the minimum initial velocity that will take an object to infinity) from the surface of a planet of radius R. What are the values for (a) the Earth, $R = 6.37 \times 10^6$ m, $g = 9.81$ m s^{-2}, (b) Mars, $R = 3.38 \times 10^6$ m, $m_{\text{Mars}}/m_{\text{Earth}} = 0.108$? At what temperatures do H_2, He, and O_2 molecules have mean speeds equal to their escape speeds? What proportion of the molecules have enough speed to escape when the temperature is (a) 240 K, (b) 1500 K? Calculations of this kind are very important in considering the composition of planetary atmospheres.

1.15 Calculate the molar volume of chlorine gas at 350 K and 2.30 atm using (a) the perfect gas law and (b) the van der Waals equation. Use the answer to (a) to calculate a first approximation to the correction term for attraction and then use successive approximations to obtain a numerical answer for part (b).

1.16 At 273 K, measurements on argon gave $B = -21.7$ cm^3 mol^{-1} and $C = 1200$ cm^6 mol^{-2}, where B and C are the second and third virial coefficients in the expansion of Z in powers of $1/V_m$. Assuming that the perfect gas law holds sufficiently well for the estimation of the second and third terms of the expansion, calculate the compression factor of argon at 100 atm and 273 K. From your result, estimate the molar volume of argon under these conditions.

1.17 Calculate the volume occupied by 1.00 mol N_2 using the van der Waals equation in the form of a virial expansion at (a) its critical temperature, (b) its Boyle temperature. Assume that the pressure is 10 atm throughout. At what temperature is the gas most perfect? Use the following data: $T_c = 126.3$ K, $a = 1.408$ atm L^2 mol^{-2}, $b = 0.0391$ L mol^{-1}.

1.18 The mass density of water vapour at 327.6 atm and 776.4 K is 1.332×10^2 g L^{-1}. Given that for water $T_c = 647.4$ K, $p_c = 218.3$ atm, $a = 5.536$ atm L^2 mol^{-2}, $b = 0.03049$ L mol^{-1}, and $M = 18.02$ g mol^{-1}, calculate (a) the molar volume. Then calculate the compression factor (b) from the data, (c) from the virial expansion of the van der Waals equation.

1.19 The critical volume and critical pressure of a certain gas are 160 cm^3 mol^{-1} and 40 atm, respectively. Estimate the critical temperature by assuming that the gas obeys the Berthelot equation of state. Estimate the radii of the gas molecules on the assumption that they are spheres.

1.20 Estimate the coefficients a and b in the Dieterici equation of state from the critical constants of xenon. Calculate the pressure exerted by 1.0 mol Xe when it is confined to 1.0 L at 25°C.

Theoretical problems

1.21 The Maxwell distribution of speeds was derived from arguments about probability, but it can also be derived from the Boltzmann distribution; see the *Introduction*. Do so.

1.22 Start from the Maxwell–Boltzmann distribution and derive an expression for the most probable speed of a gas of molecules at a temperature T. Go on to demonstrate the validity of the equipartition conclusion (see the *Introduction*) that the average translational kinetic energy of molecules free to move in three dimensions is $\frac{3}{2}kT$.

1.23 Consider molecules that are confined to move in a plane (a two-dimensional gas). Calculate the distribution of speeds and determine the mean speed of the molecules at a temperature T.

1.24 A specially constructed velocity-selector accepts a beam of molecules from an oven at a temperature T but blocks the passage of molecules with a speed greater than the mean. What is the mean speed of the emerging beam, relative to the initial value, treated as a one-dimensional problem?

1.25 What is the proportion of gas molecules having (a) more than, (b) less than the root mean square speed? (c) What are the proportions having speeds greater and smaller than the mean speed?

1.26 Calculate the fractions of molecules in a gas that have a speed in a range Δv at the speed nc^* relative to those in the same range at c^* itself. This calculation can be used to estimate the fraction of very energetic molecules (which is important for reactions). Evaluate the ratio for $n = 3$ and $n = 4$.

1.27 Show that the van der Waals equation leads to values of $Z < 1$ and $Z > 1$, and identify the conditions for which these values are obtained.

1.28 Express the van der Waals equation of state as a virial expansion in powers of $1/V_m$ and obtain expressions for B and C in terms of the parameters a and b. The expansion you will need is $(1 - x)^{-1} = 1 + x + x^2 + \cdots$. Measurements on argon gave $B = -21.7$ cm^3 mol^{-1} and $C = 1200$ cm^6 mol^{-2} for the virial coefficients at 273 K. What are the values of a and b in the corresponding van der Waals equation of state?

1.29 A scientist proposed the following equation of state:

$$p = \frac{RT}{V_m} - \frac{B}{V_m^2} + \frac{C}{V_m^3}$$

Show that the equation leads to critical behaviour. Find the critical constants of the gas in terms of B and C and an expression for the critical compression factor.

1.30 Equations 35 and 36 are expansions in p and $1/V_m$, respectively. Find the relation between B, C and B', C'.

1.31 The second virial coefficient B' can be obtained from measurements of the density ρ of a gas at a series of pressures. Show that the graph of p/ρ against p should be a straight line with slope proportional to B'. Use the data on dimethyl ether in Problem 1.5 to find the values of B' and B at 25°C.

1.32 The equation of state of a certain gas is given by $p = RT/V_m + (a + bT)/V_m^2$, where a and b are constants. Find $(\partial V/\partial T)_p$.

1.33 The following equations of state are occasionally used for approximate calculations on gases: (gas A) $pV_m = RT(1 + b/V_m)$, (gas B) $p(V_m - b) = RT$. Assuming that there were gases that actually obeyed these equations of state, would it be possible to liquefy either gas A or B? Would they have a critical temperature? Explain your answer.

1.34 Derive an expression for the compression factor of a gas that obeys the equation of state $p(V - nb) = nRT$, where b and R are constants. If the pressure and temperature are such that $V_m = 10b$, what is the numerical value of the compression factor?

1.35 The barometric formula

$$p = p_0 e^{-Mgh/RT}$$

relates the pressure of a gas of molar mass M at an altitude h to its pressure p_0 at sea level. Derive this relation by showing that the change in pressure dp for an infinitesimal change in altitude dh where the density is ρ is $dp = -\rho g \, dh$. Remember that ρ depends on the pressure. Evaluate the pressure difference between the top and bottom of (a) a laboratory vessel of height 15 cm, and (b) the World Trade Center, 1350 ft. Ignore temperature variations.

Additional problems supplied by Carmen Giunta and Charles Trapp

1.36 Amedeo Avogadro (*Journal de Physique* (1811)) noted that two volumes of hydrogen combine with one volume of oxygen to form two volumes of water vapour. He also gave the density of water vapour relative to that of air as 0.625 and that of hydrogen as 0.0732. Use this information and Avogadro's principle to compute the molar masses of water vapour and oxygen relative to that of hydrogen.

1.37 The discovery of the element argon by Lord Rayleigh and Sir William Ramsay had its origins in Rayleigh's measurements of the density of nitrogen with an eye toward accurate determination of its molar mass. Rayleigh prepared some samples of nitrogen by chemical reaction of nitrogen-containing compounds; under his standard conditions, a glass globe filled with this 'chemical nitrogen' had a mass of 2.2990 g. He prepared other samples by removing oxygen, carbon dioxide, and water vapour from atmospheric air; under the same conditions, this 'atmospheric nitrogen' had a mass of 2.3102 g (Lord Rayleigh, *Royal Institution Proceedings* **14**, 524 (1895)). With the hindsight of knowing accurate values for the molar masses of nitrogen and argon, compute the mole fraction of argon in the latter sample on the assumption that the former was pure nitrogen and the latter a mixture of nitrogen and argon.

1.38 A substance as elementary and well known as argon still receives research attention. Stewart and Jacobsen have published a review of thermodynamic properties of argon (R.B. Stewart and R.T. Jacobsen, *J. Phys. Chem. Ref. Data* **18**, 639 (1989)) which included the following 300 K isotherm.

p/MPa	0.4000	0.5000	0.6000	0.8000	1.000
$V_m/(\text{L mol}^{-1})$	6.2208	4.9736	4.1423	3.1031	2.4795
p/MPa	1.500	2.000	2.500	3.000	4.000
$V_m/(\text{L mol}^{-1})$	1.6483	1.2328	0.98357	0.81746	0.60998

(a) Compute the second virial coefficient, B, at this temperature. (b) If you have access to nonlinear curve-fitting software, compute the third virial coefficient, C, at this temperature.

1.39 Ozone is a trace atmospheric gas which plays an important role in screening the Earth from harmful ultraviolet light, and the abundance of ozone is commonly reported in Dobson units. One Dobson unit is the thickness, in thousandths of a centimetre, of a column of gas if it were collected as a pure gas at 1.00 atm and 0°C. What amount of O_3 (in moles) is found in a column of atmosphere with a cross-sectional area of 1.00 dm^2 if the abundance is 250 Dobson units (a typical mid-latitude value)? In the seasonal Antarctic ozone hole, the column abundance drops below 100 Dobson units; how many moles of ozone are found in such a column of air above a 1.00 dm^2 area? Most atmospheric ozone is found between 10 and 50 km above the surface of the Earth. If that ozone is spread uniformly through this portion of the atmosphere, what is the average molar concentration corresponding to (a) 250 Dobson units, (b) 100 Dobson units?

1.40 Chlorofluorocarbons such as CCl_3F and CCl_2F_2 have been linked to ozone depletion in Antarctica. As of 1994, these gases were found in quantities of 261 and 509 parts per trillion (10^{12}) by volume (World Resources Institute, *World Resources* (1996–97)). Compute the molar concentration of these gases under conditions typical of (a) the mid-latitude troposphere (10°C and 1.0 atm) and (b) the Antarctic stratosphere (200 K and 0.050 atm).

1.41 In the *standard model* of stellar structure (I. Nicholson, *The sun*, Rand McNally, New York (1982)), the interior of the Sun is thought to consist of 36 per cent H and 64 per cent He by mass, at a density of 158 g cm^{-3}. Both atoms are completely ionized. The approximate dimensions of the nuclei can be calculated from the formula $r_{\text{nucleus}} = 1.4 \times 10^{-15}A^{1/3}$ m, where A is the mass number. The size of the free electron, $r_e \approx 10^{-18}$ m, is negligible compared to the size of the nuclei. (a) Calculate the excluded volume in 1.0 cm^3 of the stellar interior and on that basis decide upon the applicability of the perfect gas law to this system. (b) The standard model suggests that the pressure in the stellar interior is 2.5×10^{11} atm. Calculate the temperature of the Sun's interior based on the perfect gas model. The generally accepted standard model value is 1.6×10^7 K. (c) Would a van der Waals type of equation (with $a = 0$) give a better value for T?

1.42 Problem 1.7 on ballooning is most readily solved (see the *Solutions manual*) with the use of Archimedes' principle, which states that the lifting force is equal to the difference between the weight of the displaced air and the weight of the balloon. Prove Archimedes' principle for the atmosphere from the barometric formula (see Problem 1.35). *Hint*. Assume a simple shape for the balloon, perhaps a right circular cylinder of cross-sectional area A and height h.

1.43 The composition of the atmosphere is roughly 80 per cent nitrogen and 20 per cent oxygen by mass. At what height above the surface of the Earth would the atmosphere become 90 per cent nitrogen and 10 per cent oxygen by mass? Assume that the temperature of the atmosphere is constant at 25°C. What is the pressure of the atmosphere at that height?

1.44 Show that the compression factor, Z, of a van der Waals gas can be expressed as $Z = V_r'/(V_r' - \frac{1}{8}) - 27/64T_rV_r'$, where $V_r' = p_cV_m/RT_c$ is the 'pseudoreduced volume', or alternatively as the solution of the cubic equation $Z^3 - \{(p_r/8T_r) + 1\}Z^2 + (27p_r/64T_r^2)Z - 27p_r^2/512T_r^3 = 0$. Solve this equation for Z for nitrogen, methane, propane, and ethene at $T_r = 1.2$ and $p_r = 3.0$ and compare to the value given in Fig. 1.27.

2

The First Law:
the concepts

This chapter introduces some of the basic concepts of thermodynamics. It concentrates on the conservation of energy—the experimental observation that energy can be neither created nor destroyed—and shows how the principle of conservation of energy can be used to assess the energy changes that accompany physical and chemical processes. Much of this chapter examines the means by which a system can exchange energy with its surroundings in terms of the work it may do or the heat that it may produce. The target concept of the chapter is enthalpy, which is a very useful book-keeping property for keeping track of the heat output (or requirements) of physical processes and chemical reactions at constant pressure.

The release of energy can be used to provide heat when a fuel burns in a furnace, to produce mechanical work when a fuel burns in an engine, and to produce electrical work when a chemical reaction pumps electrons through a circuit. In chemistry, we encounter reactions that can be harnessed to provide heat and work, reactions that liberate energy which is squandered (often to the detriment of the environment) but which give products we require, and reactions that constitute the processes of life. **Thermodynamics**, the study of the transformations of energy, enables us to discuss all these matters quantitatively and to make useful predictions.

The basic concepts

For the purposes of physical chemistry, the universe is divided into two parts, the system and its surroundings. The **system** is the part of the world in which we have a special interest. It may be a reaction vessel, an engine, an electrochemical cell, a biological cell, and so on. The **surroundings** are where we make our measurements. The type of system depends on the characteristics of the boundary that divides it from the surroundings (Fig. 2.1). If matter can be transferred through the boundary between the system and its surroundings the system is classified as **open**. If matter cannot pass through the boundary the system is classified as

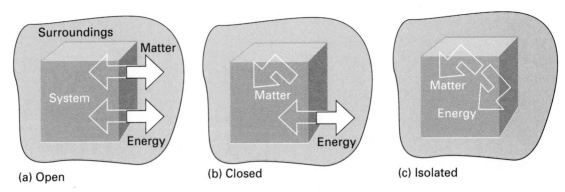

(a) Open   (b) Closed   (c) Isolated

2.1 (a) An open system can exchange matter and energy with its surroundings. (b) A closed system can exchange energy with its surroundings, but it cannot exchange matter. (c) An isolated system can exchange neither energy nor matter with its surroundings.

closed. Both open and closed systems can exchange energy with their surroundings. For example, a closed system can expand and thereby raise a weight in the surroundings, and it may also transfer energy to them if they are at a lower temperature. An **isolated system** is a closed system that has neither mechanical nor thermal contact with its surroundings.

2.1 Work, heat, and energy

The fundamental physical property in thermodynamics is work: **work** is done when an object is moved against an opposing force. It is equivalent to a change in the height of a weight somewhere in the surroundings. An example of doing work is the expansion of a gas that pushes out a piston and raises a weight. A chemical reaction that drives an electric current through a resistance also does work, because the same current could be driven through a motor and used to raise a weight.

The **energy** of a system is its capacity to do work. When work is done on an otherwise isolated system (for instance, by compressing a gas or winding a spring), its capacity to do work is increased, so the energy of the system is increased. When the system does work (when the piston moves out or the spring unwinds), its energy is reduced because it can do less work than before.

Experiments have shown that the energy of a system (its capacity to do work) may be changed by means other than work itself. When the energy of a system changes as a result of a temperature difference between it and its surroundings we say that energy has been transferred as **heat**. When a heater is immersed in a beaker of water (the system), the capacity of the system to do work increases because hot water can be used to do more work than cold water. Not all boundaries permit the transfer of energy even though there is a temperature difference between the system and its surroundings. A boundary that does permit energy transfer as heat (such as steel and glass) is called **diathermic**. A boundary that does not permit energy transfer as heat is called **adiabatic** (Fig. 2.2).

A process that releases energy as heat is called **exothermic**. All combustion reactions are exothermic. Processes that absorb energy as heat are called **endothermic**. An example of an endothermic process is the vaporization of water. An endothermic process in a diathermic container results in energy flowing into the system as heat. An exothermic process in a similar diathermic container results in a release of energy as heat into the surroundings. When an endothermic process takes place in an adiabatic container, it results in a lowering of temperature of the system; an exothermic process results in a rise of temperature. These features are summarized in Fig. 2.3.

2.2 (a) A diathermic system is one that allows energy to escape as heat through its boundary if there is a difference in temperature between the system and its surroundings. (b) An adiabatic system is one that does not permit the passage of energy as heat through its boundary even if there is a temperature difference between the system and its surroundings.

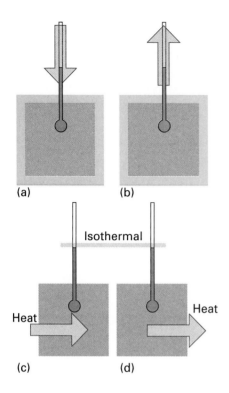

2.3 (a) When an endothermic process occurs in an adiabatic system, the temperature falls; (b) if the process is exothermic, then the temperature rises. (c) When an endothermic process occurs in a diathermic container, energy enters as heat from the surroundings, and the system remains at the same temperature; (d) if the process is exothermic, then energy leaves as heat, and the process is isothermal.

Molecular interpretation 2.1

In molecular terms, *heat is the transfer of energy that makes use of chaotic molecular motion*. The chaotic motion of molecules is called **thermal motion**. The thermal motion of the molecules in the hot surroundings stimulates the molecules in the cooler system to move more vigorously and, as a result, the energy of the system is increased. When a system heats its surroundings, molecules of the system stimulate the thermal motion of the molecules in the surroundings (Fig. 2.4).

In contrast, *work is the transfer of energy that makes use of organized motion* (Fig. 2.5). When a weight is raised or lowered, its atoms move in an organized way. The atoms in a spring move in an orderly way when it is wound; the electrons in an electric current move in an orderly direction when it flows. When a system does work it causes atoms or electrons in its surroundings to move in an organized way. Likewise, when work is done on a system, molecules in the surroundings are used to transfer energy to it in an organized way, as the atoms in a weight are lowered or a current of electrons is passed.

The distinction between work and heat is made in the surroundings. The fact that a falling weight may stimulate thermal motion in the system is irrelevant to the distinction between heat and work: work is identified as energy transfer making use of the organized motion of atoms *in the surroundings*, and heat is identified as energy transfer making use of thermal motion *in the surroundings*. In the compression of a gas, for instance, work is done as the particles of the compressing weight descend in an orderly way, but the effect of the incoming piston is to accelerate the gas molecules to higher average speeds. Because collisions between molecules quickly randomize their directions, the orderly motion of the atoms of the weight is in effect stimulating thermal motion in the gas. We observe the falling weight, the orderly descent of its atoms, and report that work is being done even though it is stimulating thermal motion.

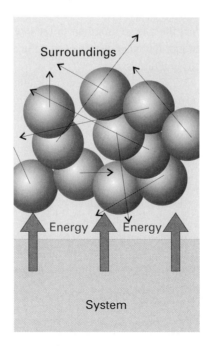

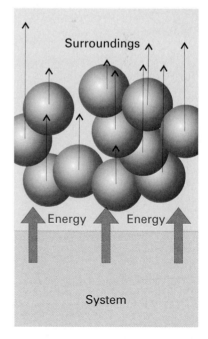

2.4 When energy is transferred to the surroundings as heat, the transfer stimulates disordered motion of the atoms in the surroundings. Transfer of energy from the surroundings to the system makes use of disordered motion (thermal motion) in the surroundings.

2.5 When a system does work, it stimulates orderly motion in the surroundings. For instance, the atoms shown here may be part of a weight that is being raised. The ordered motion of the atoms in a falling weight does work on the system.

2.2 The First Law

In thermodynamics, the total energy of a system is called its **internal energy**, U. The internal energy is the total kinetic and potential energy of the molecules composing the system. We denote by ΔU the change in internal energy when a system changes from an initial state i with internal energy U_i to a final state f of internal energy U_f:

$$\Delta U = U_f - U_i \tag{1}$$

The internal energy is a **state function** in the sense that its value depends only on the current state of the system and is independent of how that state has been prepared. In other words, it is a function of the properties that determine the current state of the system. Changing any one of the state variables (such as the pressure) results in a change in internal energy. The internal energy is an extensive property.

Internal energy, heat, and work are all measured in the same units, the joule (J). Changes in molar internal energy are typically expressed in kilojoules per mole (kJ mol^{-1}).

Molecular interpretation 2.2

Consider the case of a monatomic perfect gas at a temperature T. We know that the kinetic energy of one atom, of mass m, is

$$E_K = \tfrac{1}{2}mv_x^2 + \tfrac{1}{2}mv_y^2 + \tfrac{1}{2}mv_z^2$$

According to the equipartition theorem (see the *Introduction*), the average energy of each term is $\tfrac{1}{2}kT$, where k is the Boltzmann constant. Therefore, the mean energy of the atoms is $\tfrac{3}{2}kT$ and the total energy of the gas (there being no potential energy contribution) is $\tfrac{3}{2}NkT$, or $\tfrac{3}{2}nRT$. We can therefore write

$$U_m = U_m(0) + \tfrac{3}{2}RT$$

where $U_m(0)$ is the molar internal energy at $T = 0$, when all translational motion has ceased and the sole contribution to the internal energy arises from the internal structure of the atoms. This equation shows that the internal energy of a perfect gas increases linearly with temperature.

When the gas consists of polyatomic molecules that can rotate around three axes as well as translate in three dimensions, there is an additional contribution of $\tfrac{3}{2}RT$ arising from the kinetic energy of rotation. In this case, therefore,

$$U_m = U_m(0) + 3RT$$

The internal energy now increases twice as rapidly with temperature compared with the monatomic gas.

Energy is also taken up by the vibrations of molecules. However, these modes cannot be treated classically because the separations between their energy levels are so wide. The expression for the mean energy of an oscillator of frequency ν is worked out by using the quantum mechanical expression for the energy levels (which form a uniform ladder like that illustrated in Fig. 0.7) and the Boltzmann distribution. The resulting expression (which is derived in Section 19.1b) is

$$U_m = U_m(0) + \frac{N_A h\nu}{e^{h\nu/kT} - 1}$$

It may be verified that the second term on the right increases with temperature and approaches RT, the classical expression, when $kT \gg h\nu$.

The internal energy of interacting molecules in condensed phases also has a contribution from the potential energy of their interaction. However, no simple expressions can be written down in general. Nevertheless, the crucial molecular point is that, as the temperature of a system is raised, the internal energy increases as the various modes of motion become more highly excited.

(a) The conservation of energy

It has been found experimentally that the internal energy of a system may be changed either by doing work on the system or by heating it. Whereas we may know how the energy transfer has occurred (because we can see if a weight has been raised or lowered in the surroundings, indicating transfer of energy by doing work, or if ice has melted in the surroundings, indicating transfer of energy as heat), the system is blind to the mode employed. Heat and work are equivalent ways of changing a system's internal energy. A system is like a bank: it accepts deposits in either currency, but stores its reserves as internal energy. It is also found experimentally that, if a system is isolated from its surroundings, then no change in internal energy takes place. We cannot use a system to do work, leave it isolated for a month, and then come back expecting to finding it restored to its original state and ready to do the same work again. The evidence for this property is that no perpetual motion machine of the first kind (a machine that does work without consuming fuel or some other source of energy) has ever been built.

These remarks may be summarized as follows. If we write w for the work done on a system, q for the energy transferred as heat to a system, and ΔU for the resulting change in internal energy, then it follows that

$$\Delta U = q + w \tag{2}$$

Equation 2 is the mathematical statement of the **First Law of thermodynamics**, for it summarizes the equivalence of heat and work and the fact that the internal energy is constant in an isolated system (for which $q = 0$ and $w = 0$). The equation states that the change in internal energy of a closed system is equal to the energy that passes through its boundary as heat or work. It employs the 'acquisitive convention', in which $w > 0$ or $q > 0$ if energy is transferred to the system as work or heat and $w < 0$ or $q < 0$ if energy is lost from the system as work or heat.

Illustration
· ·
If an electric motor produced 15 kJ of energy each second as mechanical work and lost 2 kJ as heat to the surroundings, then the change in the internal energy of the motor each second is

$$\Delta U = -2 \text{ kJ} - 15 \text{ kJ} = -17 \text{ kJ}$$

Suppose that, when a spring was wound, 100 J of work was done on it but 15 J escaped to the surroundings as heat. The change in internal energy of the spring was

$$\Delta U = +100 \text{ kJ} - 15 \text{ kJ} = +85 \text{ kJ}$$

· ·

(b) The formal statement of the First Law

The expression of the First Law that we have given is adequate for most purposes in thermodynamics. However, there are several unsatisfactory features about it, such as how we define and measure 'heat'. This section gives a more sophisticated version of the law, and shows how eqn 2 can be put on a firmer foundation. As the remainder of the text does not depend on this material, it is possible to omit it and go immediately to the following section ('Work and heat').

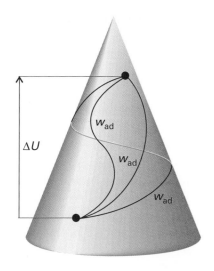

2.6 It is found that the same quantity of work must be done on an adiabatic system to achieve the same change of state even though different means of achieving that work may be used. This path independence implies the existence of a state function, the internal energy. The change in internal energy is like the change in altitude when climbing a mountain: its value is independent of path.

We begin by pretending that we do not know what we mean by 'energy'. We pretend that we know only what is meant by work, because we can observe a weight being raised or lowered in the surroundings. We also know how to measure work by noting the height through which the weight is raised. Throughout this section, work will be the fundamental, measurable quantity, and we define energy, heat, and the First Law in terms of work alone. We shall employ terms that have been established by the Zeroth Law of thermodynamics (Section 1.1), namely, state and temperature and the concepts of adiabatic and diathermic walls.

In an adiabatic system of a particular composition, it is known experimentally that the same increase in temperature is brought about by the same quantity of any kind of work we do on the system. Thus, if 1 kJ of mechanical work is done on the system (by stirring it with rotating paddles, for instance), or 1 kJ of electrical work is done (by passing an electric current through a heater), and so on, then the same rise in temperature is produced. The following statement of the First Law of thermodynamics is a summary of a large number of observations of this kind:

The work needed to change an adiabatic system from one specified state to another specified state is the same however the work is done.

This form of the law looks completely different from the form we gave before, but we shall now see how it implies eqn 2.

Suppose we do work w_{ad} on an adiabatic system to change it from an initial state i to a final state f. The work may be of any kind (mechanical or electrical) and may take the system through different intermediate states (different temperatures and pressures, for instance). We might (in ignorance of the First Law) think that we need to label w_{ad} with the path and to write w_{ad}(mechanical) or w_{ad}(electrical). However, the First Law tells us that w_{ad} is the same for all paths and depends only on the initial and final states. This conclusion is analogous to climbing a mountain: the height we must climb between any two points is independent of the path we take (Fig. 2.6). In mountain climbing we can attach a number, the altitude, A, to each point on the mountain and express the height, h, of the climb as a difference in altitudes:

$$h = A_f - A_i = \Delta A$$

That is, in mountain climbing, the observation that h is independent of the path taken implies the existence of the state function A. The First Law has exactly the same implication. The fact that w_{ad} is independent of the path implies that to each state of the system we can attach a value of a quantity—we call it the 'internal energy', U—and express the work as a difference in internal energies:

$$w_{ad} = U_f - U_i = \Delta U \qquad [3]$$

This equation also shows that we can measure the change in the internal energy of a system by measuring the work needed to bring about the change in an adiabatic system.

(c) The mechanical definition of heat

Suppose we strip away the thermal insulation around the system and make it diathermic. The system is now in thermal contact with its surroundings as we drive it from the same initial state to the same final state. The change in internal energy is the same as before, because U is a state function, but we might find that the work we must do is not the same as before. Thus, whereas we might have needed to do 42 kJ of work when the system was in an adiabatic container, to achieve the same change of state we might now have to do 50 kJ of work. The difference between the work done in the two cases is *defined* as the heat absorbed by the system in the process:

$$q = w_{ad} - w \qquad [4]$$

In the present case, we would conclude that $q = 42$ kJ $- 50$ kJ $= -8$ kJ, and report that 8 kJ of energy had left the system as heat. We see that we now have a purely mechanical definition of heat in terms of work. We know how to measure work in terms of the height through which a weight falls, so we now also have a method for measuring heat in terms of work.

Finally, we can express eqn 4 in a more familiar way. Because we already know that ΔU is (by definition) equal to w_{ad}, the expression for the energy transferred to the system as heat is $q = \Delta U - w$. However, this expression is equivalent to eqn 2, the mathematical form of the First Law that we saw earlier.

Work and heat

The way can now be opened to powerful methods of calculation by switching attention to infinitesimal changes of state (such as infinitesimal change in temperature) and infinitesimal changes in the internal energy dU. Then, if the work done on a system is dw and the energy supplied to it as heat is dq, in place of eqn 2 we have

$$dU = dq + dw \qquad (5)$$

To use eqn 5 we must be able to relate dq and dw to events taking place in the surroundings.

2.3 Expansion work

We begin by discussing **expansion work**, the work arising from a change in volume. This type of work includes the work done by a gas as it expands and drives back the atmosphere. Many chemical reactions result in the generation or consumption of gases (for instance, the thermal decomposition of calcium carbonate or the combustion of octane), and the thermodynamic characteristics of a reaction depend on the work it can do.

(a) The general expression for work

The calculation of expansion work starts from the definition used in physics, which states that the work required to move an object a distance dz against an opposing force of magnitude F is

$$dw = -F\,dz \qquad [6]$$

The negative sign tells us that, when the system moves an object against an opposing force, the internal energy of the system doing the work will decrease. Now consider the arrangement shown in Fig. 2.7, in which one wall of a system is a massless, frictionless, rigid, perfectly fitting piston of area A. If the external pressure is p_{ex}, the force on the outer face of the piston is $F = p_{ex}A$. When the system expands through a distance dz against an external pressure p_{ex}, it follows that the work done is $dw = -p_{ex}A\,dz$. But $A\,dz$ is the change in volume, dV, in the course of the expansion. Therefore, the work done when the system expands by dV against a pressure p_{ex} is

$$dw = -p_{ex}\,dV \qquad (7)$$

To obtain the total work done when the volume changes from V_i to V_f we integrate this expression between the initial and final volumes:

$$w = -\int_{V_i}^{V_f} p_{ex}\,dV \qquad (8)$$

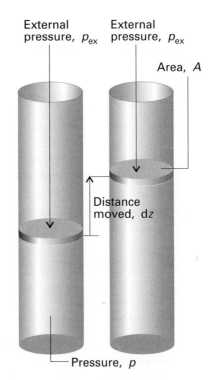

External pressure, p_{ex} External pressure, p_{ex}

Area, A

Distance moved, dz

Pressure, p

2.7 When a piston of area A moves out through a distance dz, it sweeps out a volume $dV = A\,dz$. The external pressure, p_{ex}, is equivalent to a weight pressing on the piston, and the force opposing expansion is $F = p_{ex}A$.

Table 2.1 Varieties of work*

Type of work	dw	Comments	Units[†]
Expansion	$-p_{ex}dV$	p_{ex} is the external pressure	Pa
		dV is the change in volume	m^3
Surface expansion	$\gamma\,d\sigma$	γ is the surface tension	$N\,m^{-1}$
		dσ is the change in area	m^2
Extension	$f\,dl$	f is the tension	N
		dl is the change of length	m
Electrical	$\phi\,dq$	ϕ is the electric potential	V
		dq is the change in charge	C

* In general, the work done on a system can be expressed in the form d$w = -F\,dz$, where F is a 'generalized force' and dz is a 'generalized displacement'.

† For work in joules (J). Note that $1\,N\,m = 1\,J$ and $1\,V\,C = 1\,J$.

The force acting on the piston, $p_{ex}A$, is equivalent to a weight that is raised as the system expands. If the system is compressed instead, then the same weight is lowered in the surroundings and eqn 8 can still be used, but now $V_f < V_i$. It is important to note that it is still the external pressure that determines the magnitude of the work.

Other types of work (for example, electrical work) have analogous expressions, with each one the product of an intensive factor (the pressure, for instance) and an extensive factor (the change in volume). Some are collected in Table 2.1. For the present we continue with the work associated with changing the volume, the expansion work, and see what we can extract from eqn 7.

(b) Free expansion

By **free expansion** we mean expansion against zero opposing force. It occurs when $p_{ex} = 0$. According to eqn 7, d$w = 0$ for each stage of the expansion. Hence, overall:

$$w = 0 \tag{9}$$

That is, no work is done when a system expands freely. Expansion of this kind occurs when a system expands into a vacuum.

(c) Expansion against constant pressure

Now suppose that the external pressure is constant throughout the expansion. For example, the piston may be pressed on by the atmosphere, which exerts the same pressure throughout the expansion. A chemical example of this condition is the expansion of a gas formed in a chemical reaction. Equation 8 may now be evaluated by taking the constant p_{ex} outside the integral:

$$w = -p_{ex}\int_{V_i}^{V_f} dV = -p_{ex}(V_f - V_i)$$

Therefore, if we write the change in volume as $\Delta V = V_f - V_i$,

$$w = -p_{ex}\Delta V \tag{10}$$

This result is illustrated graphically in Fig. 2.8, which makes use of the fact that an integral can be interpreted as an area.[1] The magnitude of w, which is denoted $|w|$, is equal to the area beneath the horizontal line at $p = p_{ex}$ lying between the initial and final volumes. A p,V-graph used to compute expansion work is called an **indicator diagram**; James Watt first used one to indicate aspects of the operation of his steam engine.

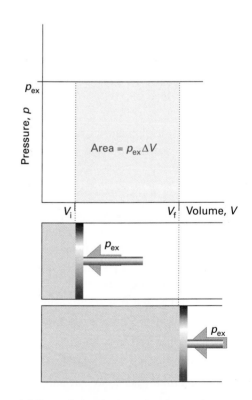

2.8 The work done by a gas when it expands against a constant external pressure, p_{ex}, is equal to the shaded area in this example of an indicator diagram.

1   Specifically, the value of the integral $\int_a^b f(x)\,dx$ is equal to the area under the graph of $f(x)$ between $x = a$ and $x = b$.

(d) Reversible expansion

A **reversible change** in thermodynamics is a change that can be reversed by an infinitesimal modification of a variable. The key word 'infinitesimal' sharpens the everyday meaning of the word 'reversible' as something that can change direction. We say that a system is in **equilibrium** with its surroundings if an infinitesimal change in the conditions in opposite directions results in opposite changes in its state. One example of reversibility that we have encountered already is the thermal equilibrium of two systems with the same temperature. The transfer of energy as heat between the two is reversible because, if the temperature of either system is lowered infinitesimally, then energy flows into the system with the lower temperature. If the temperature of either system at thermal equilibrium is raised infinitesimally, then energy flows out of the hotter system.

Suppose a gas is confined by a piston and that the external pressure, p_{ex}, is set equal to the pressure, p, of the confined gas. Such a system is in mechanical equilibrium with its surroundings (as illustrated in Section 1.1) because an infinitesimal change in the external pressure in either direction causes changes in volume in opposite directions. If the external pressure is reduced infinitesimally, then the gas expands slightly. If the external pressure is increased infinitesimally, then the gas contracts slightly. In either case the change is reversible in the thermodynamic sense. If, on the other hand, the external pressure differs measurably from the internal pressure, then changing p_{ex} infinitesimally will not decrease it below the pressure of the gas and so will not change the direction of the process. Such a system is not in mechanical equilibrium with its surroundings and the expansion is thermodynamically irreversible.

To achieve reversible expansion we set p_{ex} equal to p at each stage of the expansion. In practice, this equalization could be achieved by gradually removing weights from the piston so that the downward force due to the weights always matched the changing upward force due to the pressure of the gas. When we set $p_{ex} = p$, eqn 7 becomes

$$\mathrm{d}w = -p_{ex}\,\mathrm{d}V = -p\,\mathrm{d}V \qquad\qquad (11)_{rev}$$

(Equations valid only for reversible processes are labelled with a subscript 'rev'.) Although the pressure inside the system appears in this expression for the work, it does so only because p_{ex} has been set equal to p to ensure reversibility. The total work of reversible expansion is therefore

$$w = -\int_{V_i}^{V_f} p\,\mathrm{d}V \qquad\qquad (12)_{rev}$$

The integral can be evaluated once we know how the pressure of the confined gas depends on its volume. Equation 12 is the link with the material covered in Chapter 1 for, if we know the equation of state of the gas, then we can express p in terms of V and evaluate the integral.

(e) Isothermal reversible expansion

Consider the isothermal, reversible expansion of a perfect gas. The expansion is made isothermal by keeping the system in thermal contact with its surroundings (which may be a constant-temperature bath). Because the equation of state is $pV = nRT$, we know that at each stage $p = nRT/V$, with V the volume at that stage of the expansion. The temperature T is constant in an isothermal expansion, so (together with n and R) it may be taken outside the integral. It follows that the work of reversible isothermal expansion of a perfect gas from V_i to V_f at a temperature T is

$$w = -nRT\int_{V_i}^{V_f}\frac{\mathrm{d}V}{V} = -nRT\ln\left(\frac{V_f}{V_i}\right) \qquad\qquad (13)_{rev}^{\circ}$$

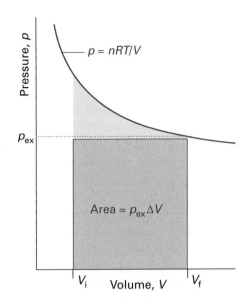

2.9 The work done by a perfect gas when it expands reversibly and isothermally is equal to the area under the isotherm $p = nRT/V$. The work done during the irreversible expansion against the same final pressure is equal to the rectangular area shown slightly darker. Note that the reversible work is greater than the irreversible work.

When the final volume is greater than the initial volume, as in an expansion, the logarithm in eqn 13 is positive and hence $w < 0$. In this case, the system has done work on the surroundings and the internal energy of the system has decreased as a result of the work it has done. The equation also shows that more work is done for a given change of volume when the temperature is increased. The greater pressure of the confined gas then needs a higher opposing pressure to ensure reversibility.

The result of the calculation can be expressed as an indicator diagram, for the magnitude of the work done is equal to the area under the isotherm $p = nRT/V$ (Fig. 2.9). Superimposed on the diagram is the rectangular area obtained for irreversible expansion against constant external pressure fixed at the same final value as that reached in the reversible expansion. More work is obtained when the expansion is reversible (the area is greater) because matching the external pressure to the internal pressure at each stage of the process ensures that none of the system's pushing power is wasted. We cannot obtain more work than for the reversible process because increasing the external pressure even infinitesimally at any stage results in compression. We may infer from this discussion that, because some pushing power is wasted when $p > p_{ex}$, *the maximum work available from a system operating between specified initial and final states and passing along a specified path is obtained when the change takes place reversibly.*

We have introduced the connection between reversibility and maximum work for the special case of a perfect gas undergoing expansion. Later (in Section 4.6b) we shall see that it applies to all substances and to all kinds of work.

Example 2.1 Calculating the work of gas production

Calculate the work done when 50 g of iron reacts with hydrochloric acid in (a) a closed vessel of fixed volume, (b) an open beaker at 25 °C.

Method We need to judge the magnitude of the volume change, and then to decide how the process occurs. If there is no change in volume, there is no expansion work however the process takes place. If the system expands against a constant external pressure, the work can be calculated from eqn 10. A general feature of processes in which a condensed phase changes into a gas is that the volume of the former may usually be neglected relative to that of the gas it forms.

Answer In (a) the volume cannot change, so no work is done and $w = 0$. In (b) the gas drives back the atmosphere and therefore $w = -p_{ex}\Delta V$. We can neglect the initial volume because the final volume (after the production of gas) is so much larger and $\Delta V = V_f - V_i \approx V_f = nRT/p_{ex}$, where n is the amount of H_2 produced. Therefore,

$$w = -p_{ex}\Delta V \approx -p_{ex} \times \frac{nRT}{p_{ex}} = -nRT$$

Because the reaction is

$$Fe(s) + 2HCl(aq) \longrightarrow FeCl_2(aq) + H_2(g)$$

we know that 1 mol H_2 is generated when 1 mol Fe is consumed, and n can be taken as the amount of Fe atoms that react. Because the molar mass of Fe is 55.85 g mol^{-1}, it follows that

$$w \approx -\frac{50 \text{ g}}{55.85 \text{ g mol}^{-1}} \times (8.3145 \text{ J K}^{-1} \text{ mol}^{-1}) \times (298.15 \text{ K})$$

$$\approx -2.2 \text{ kJ}$$

The system (the reaction mixture) does 2.2 kJ of work driving back the atmosphere.

Comment Note that (for this perfect gas system) the external pressure does not affect the final result: the lower the pressure, the larger the volume occupied by the gas, so the effects cancel.

- -

Self-test 2.1 Calculate the expansion work done when 50 g of water is electrolysed under constant pressure at 25 °C.

$$[-10 \text{ kJ}]$$

2.4 Heat transactions

In general, the change in internal energy of a system is

$$dU = dq + dw_{exp} + dw_e \tag{14}$$

where dw_e is work in addition (e for 'extra') to the expansion work, dw_{exp}. For instance, dw_e might be the electrical work of driving a current through a circuit. A system kept at constant volume can do no expansion work, so $dw_{exp} = 0$. If the system is also incapable of doing any other kind of work (if it is not, for instance, an electrochemical cell connected to an electric motor), then $dw_e = 0$ too. Under these circumstances:

$$dU = dq \qquad \text{(at constant volume, no additional work)} \tag{15}$$

We express this relation by writing $dU = dq_V$, where the subscript implies a change at constant volume. For a measurable change,

$$\Delta U = q_V \tag{16}$$

It follows that, by measuring the energy supplied to a constant-volume system as heat ($q > 0$) or obtained from it as heat ($q < 0$) when it undergoes a change of state, we are in fact measuring the change in its internal energy.

(a) Calorimetry

The most common device for measuring ΔU is the **adiabatic bomb calorimeter** (Fig. 2.10). The process we wish to study—which may be a chemical reaction—is initiated inside a constant-volume container, the bomb. The bomb is immersed in a stirred water bath, and the whole device is the calorimeter. The calorimeter is also immersed in an outer water bath. The water in the calorimeter and that of the outer bath are both monitored and adjusted to the same temperature. This arrangement ensures that there is no net loss of heat from the calorimeter to the surroundings (the bath) and hence that the calorimeter is adiabatic.

The change in temperature, ΔT, of the calorimeter is proportional to the heat that the reaction releases or absorbs. Therefore, by measuring ΔT we can determine q_V and hence find ΔU. The conversion of ΔT to q_V is best achieved by calibrating the calorimeter using a process of known energy output and determining the **calorimeter constant**, the constant C in the relation

$$q = C\Delta T \tag{17}$$

The calorimeter constant may be measured electrically by passing a current, I, from a source of known potential, $\mathcal{V}$, through a heater for a known period of time, t:

$$q = I\mathcal{V}t \tag{18}$$

Alternatively, C may be determined by burning a known mass of substance (benzoic acid is often used) that has a known heat output. With C known, it is simple to interpret an observed temperature rise as a release of heat.

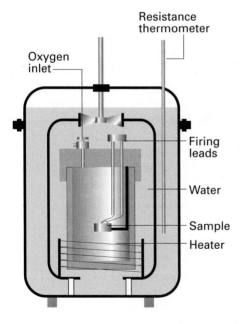

2.10 A constant-volume bomb calorimeter. The 'bomb' is the central vessel, which is massive enough to withstand high pressures. The calorimeter (for which the heat capacity must be known) is the entire assembly shown here. To ensure adiabaticity, the calorimeter is immersed in a water bath with a temperature continuously readjusted to that of the calorimeter at each stage of the combustion.

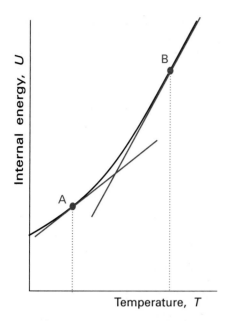

2.11 The internal energy of a system increases as the temperature is raised; this graph shows its variation as the system is heated at constant volume. The slope of the graph at any temperature (as shown by the tangents at A and B) is the heat capacity at constant volume at that temperature. Note that, for the system illustrated, the heat capacity is greater at B than at A.

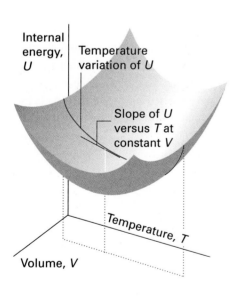

2.12 The internal energy of a system varies with volume and temperature, perhaps as shown here by the surface. The variation of the internal energy with temperature at one particular constant volume is illustrated by the curve drawn parallel to T. The slope of this curve at any point is the partial derivative $(\partial U/\partial T)_V$.

Illustration

If we pass a current of 10.0 A from a 12 V supply for 300 s, then from eqn 18 the energy supplied as heat is

$$q = (10.0\ \text{A}) \times (12\ \text{V}) \times (300\ \text{s}) = 3.6 \times 10^4\ \text{A V s} = 36\ \text{kJ}$$

because 1 A V s = 1 J. If the observed rise in temperature is 5.5 K, then the calorimeter constant is $C = (36\ \text{kJ})/(5.5\ \text{K}) = 6.5\ \text{kJ K}^{-1}$.

(b) Heat capacity

The internal energy of a substance increases when its temperature is raised. The increase depends on the conditions under which the heating takes place, and for the present we shall suppose that the sample is confined to a constant volume. For example, it may be a gas in a container of fixed volume. If the internal energy is plotted against temperature, then a curve like that in Fig. 2.11 may be obtained. The slope of the curve at any temperature is called the **heat capacity** of the system at that temperature. The **heat capacity at constant volume** is denoted C_V and is defined formally as[2]

$$C_V = \left(\frac{\partial U}{\partial T}\right)_V \qquad\qquad [19]$$

The notation is that of a 'partial derivative'. A partial derivative is a slope calculated with all except one variables held constant.[3] In this case, the internal energy varies with the temperature and the volume of the sample, but we are interested only in its variation with the temperature, the volume being held constant (Fig. 2.12).

Heat capacities are extensive properties: 100 g of water, for instance, has 100 times the heat capacity of 1 g of water (and therefore requires 100 times the heat to bring about the same rise in temperature). The **molar heat capacity at constant volume**, $C_{V,m}$, is the heat capacity per mole of material, and is an intensive property (all molar quantities are intensive). Typical values of $C_{V,m}$ for polyatomic gases are close to 25 J K^{-1} mol^{-1}. For certain applications it is useful to know the **specific heat capacity** (more informally, the 'specific heat') of a substance, which is the heat capacity of the sample divided by the mass, usually in grams. The specific heat capacity of water at room temperature is close to 4 J K^{-1} g^{-1}. In general, heat capacities depend on the temperature and decrease at low temperatures. However, over small ranges of temperature at and above room temperature, the variation is quite small and for approximate calculations heat capacities can be treated as almost independent of temperature.

Molecular interpretation 2.3

The heat capacity of a monatomic perfect gas can be calculated by inserting the expression for the internal energy derived in *Molecular interpretation* 2.2. There we saw that $U_m = U_m(0) + \frac{3}{2}RT$, so from eqn 19

$$C_{V,m} = \left(\frac{\partial U_m}{\partial T}\right)_V = \frac{3}{2}R$$

---

2 If the system can change its composition, it is necessary to distinguish between equilibrium and fixed-composition values of C_V. All applications in this chapter refer to a single substance, so this complication can be ignored.

3 Partial derivatives are reviewed in *Further information 1*.

The numerical value is $12.47 \, \mathrm{J\,K^{-1}\,mol^{-1}}$. Similarly, for a gas composed of nonlinear polyatomic molecules

$$C_{V,m} = 3R \qquad (20)$$

or $24.94 \, \mathrm{J\,K^{-1}\,mol^{-1}}$. These values are both independent of temperature.

The heat capacity arising from vibration is obtained by evaluating the derivative of the expression derived using quantum mechanics, and for a vibrational frequency ν is

$$C_{V,m} = Rf^2 \qquad f = \left(\frac{h\nu}{kT}\right) \frac{\mathrm{e}^{-h\nu/2kT}}{1 - \mathrm{e}^{-h\nu/kT}} \qquad (21)$$

The factor $f = 0$ at $T = 0$ and $f \approx 1$ when $kT \gg h\nu$. That is, the contribution of the vibration of a molecule to the molar heat capacity is zero at $T = 0$ and climbs towards its classical value (R) as the temperature is raised. Physically, at low temperatures the gap between the energy levels is so great that the vibrations cannot be excited, so energy cannot be absorbed. As the temperature is raised, more and more energy levels become accessible, and the molecule begins to behave as though its vibrations were not quantized.

The heat capacity can be used to relate a change in internal energy to a change in temperature of a constant-volume system. It follows from eqn 19 that

$$\mathrm{d}U = C_V \, \mathrm{d}T \qquad \text{(at constant volume)} \qquad (22a)$$

That is, an infinitesimal change in temperature brings about an infinitesimal change in internal energy, and the constant of proportionality is the heat capacity at constant volume. If the heat capacity is independent of temperature over the range of temperatures of interest, a measurable change of temperature, ΔT, brings about a measurable increase in internal energy, ΔU, where

$$\Delta U = C_V \Delta T \qquad \text{(at constant volume)} \qquad (22b)$$

Because a change in internal energy can be identified with the heat supplied at constant volume (eqn 16), the last equation can be written

$$q_V = C_V \Delta T \qquad (22c)$$

This relation provides a simple way of measuring the heat capacity of a sample: a measured quantity of heat is supplied to the sample (electrically, for example), and the resulting increase in temperature is monitored. The ratio of the heat supplied to the temperature rise it causes is the heat capacity of the sample.

A large heat capacity implies that, for a given quantity of heat, there will be only a small increase in temperature (the sample has a large capacity for heat). An infinite heat capacity implies that there will be no increase in temperature however much heat is supplied. At a phase transition (such as at the boiling point of water), the temperature of a substance does not rise as heat is supplied (the energy is used to drive the endothermic phase transition, in this case to vaporize the water, rather than to increase its temperature), so, at the temperature of a phase transition, the heat capacity of a sample is infinite. The properties of heat capacities close to phase transitions are treated more fully in Section 6.7.

2.5 Enthalpy

The change in internal energy is not equal to the heat supplied when the system is free to change its volume. Under these circumstances some of the energy supplied as heat to the system is returned to the surroundings as expansion work (Fig. 2.13), so $\mathrm{d}U$ is less than $\mathrm{d}q$. However, we shall now show that in this case the heat supplied at constant pressure is equal to the change in another thermodynamic property of the system, the **enthalpy**, H.

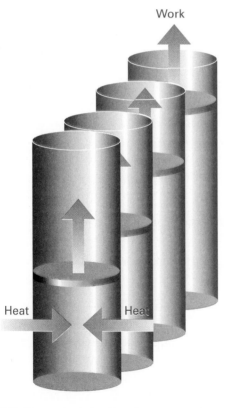

Work

Heat   Heat

2.13 When a system is subjected to constant pressure and is free to change its volume, some of the energy supplied as heat may escape back into the surroundings as work. In such a case, the change in internal energy is smaller than the energy supplied as heat.

(a) The definition of enthalpy

The enthalpy is defined as

$$H = U + pV \qquad [23]$$

where p is the pressure of the system and V is its volume. Because U, p, and V are all state functions, the enthalpy is a state function. As is true of any state function, the change in enthalpy, ΔH, between any pair of initial and final states is independent of the path between them.

The change in enthalpy is equal to the heat supplied at constant pressure to a system (so long as the system does no additional work):

$$dH = dq \qquad \text{(at constant pressure, no additional work)} \qquad (24a)$$

For a measurable change,

$$\Delta H = q_p \qquad (24b)$$

Justification 2.1

For a general infinitesimal change in the state of the system, U changes to $U + dU$, p changes to $p + dp$, and V changes to $V + dV$, so H changes from $U + pV$ to

$$H + dH = (U + dU) + (p + dp)(V + dV)$$
$$= U + dU + pV + p\,dV + V\,dp + dp\,dV$$

The last term is the product of two infinitesimally small quantities, and can be neglected. As a result, after recognizing $U + pV = H$ on the right, we find that H changes to

$$H + dH = H + dU + p\,dV + V\,dp$$

and hence that

$$dH = dU + p\,dV + V\,dp$$

If we now substitute $dU = dq + dw$ into this expression, we get

$$dH = dq + dw + p\,dV + V\,dp$$

If the system is in mechanical equilibrium with its surroundings at a pressure p and does only expansion work, we can write $dw = -p\,dV$ and obtain

$$dH = dq + V\,dp$$

Now we impose the condition that the heating occurs at constant pressure by writing $dp = 0$. Then

$$dH = dq \qquad \text{(at constant pressure, no additional work)}$$

as in eqn 24a.

The result expressed in eqn 24 states that, when a system is subjected to a constant pressure and only expansion work can occur, the change in enthalpy is equal to the energy supplied as heat. For example, if we supply 36 kJ of energy through an electric heater immersed in an open beaker of water, then the enthalpy of the water increases by 36 kJ and we write $\Delta H = +36$ kJ.

(b) The measurement of an enthalpy change

An enthalpy change can be measured calorimetrically by monitoring the temperature change that accompanies a physical or chemical change occurring at constant pressure. For a combustion reaction an adiabatic flame calorimeter may be used to measure ΔT when a

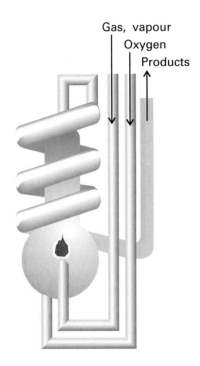

Gas, vapour
Oxygen
Products

2.14 A constant-pressure flame calorimeter consists of this element immersed in a stirred water bath. Combustion occurs as a known amount of reactant is passed through to fuel the flame, and the rise in temperature is monitored.

given amount of substance burns in a supply of oxygen (Fig. 2.14). Another route to ΔH is to measure the internal energy change by using a bomb calorimeter, and then to convert ΔU to ΔH. Because solids and liquids have small molar volumes, for them pV_m is so small that the molar enthalpy and molar internal energy are almost identical ($H_m = U_m + pV_m \approx U_m$). Consequently, if a process involves only solids or liquids, the values of ΔH and ΔU are almost identical. Physically, such processes are accompanied by a very small change in volume, the system does negligible work on the surroundings when the process occurs, so the energy supplied as heat stays entirely within the system.

Example 2.2 Relating ΔH and ΔU

The internal energy change when 1.0 mol $CaCO_3$ in the form of calcite converts to aragonite is $+0.21$ kJ. Calculate the difference between the enthalpy change and the change in internal energy when the pressure is 1.0 bar given that the densities of the solids are 2.71 g cm^{-3} and 2.93 g cm^{-3}, respectively.

Method The starting point for the calculation is the relation between the enthalpy of a substance and its internal energy (eqn 23). The difference between the two quantities can be expressed in terms of the pressure and the difference of their molar volumes, and the latter can be calculated from their molar masses, M, and their mass densities, ρ, by using $\rho = M/V_m$.

Answer The change in enthalpy when the transformation occurs is

$$\Delta H = H(\text{aragonite}) - H(\text{calcite})$$
$$= \{U(a) + pV(a)\} - \{U(c) + pV(c)\}$$
$$= \Delta U + p\{V(a) - V(c)\} = \Delta U + p\Delta V$$

The volume of 1.0 mol $CaCO_3$ (100 g) as aragonite is 34 cm^3, and that of 1.0 mol $CaCO_3$ as calcite is 37 cm^3. Therefore,

$$p\Delta V = (1.0 \times 10^5 \text{ Pa}) \times (34 - 37) \times 10^{-6} \text{ m}^3 = -0.3 \text{ J}$$

(because 1 Pa m^3 = 1 J). Hence,

$$\Delta H - \Delta U = -0.3 \text{ J}$$

which is only 0.1 per cent of the value of ΔU.

Comment It is usually justifiable to ignore the difference between the enthalpy and internal energy of condensed phases, except at very high pressures, when pV is no longer negligible.

Self-test 2.2 Calculate the difference between ΔH and ΔU when 1.0 mol of grey tin (density 5.75 g cm^{-3}) changes to white tin (density 7.31 g cm^{-3}) at 10.0 bar. At 298 K, $\Delta H = +2.1$ kJ.

$$[\Delta H - \Delta U = -4.4 \text{ J}]$$

The enthalpy of a perfect gas is related to its internal energy by using $pV = nRT$ in the definition of H:

$$H = U + pV = U + nRT \tag{25}°$$

This relation implies that the change of enthalpy in a reaction that produces or consumes gas is

$$\Delta H = \Delta U + \Delta n_g RT \qquad (26)°$$

where Δn_g is the change in the amount of gas molecules in the reaction. For example, in the reaction

$$2H_2(g) + O_2(g) \longrightarrow 2H_2O(l) \qquad \Delta n_g = -3 \text{ mol}$$

because 3 mol of gas-phase molecules is replaced by 2 mol of liquid-phase molecules, and at 298 K the enthalpy and internal energy changes taking place in the system are related by

$$\Delta H - \Delta U = (-3 \text{ mol}) \times RT \approx -7.5 \text{ kJ}$$

Note that the difference is expressed in kilojoules, not joules as in Example 2.2. The enthalpy change is smaller than the change in internal energy because, although heat escapes from the system when the reaction occurs, the system contracts when the liquid is formed, so energy is restored to it from the surroundings.

Example 2.3 Calculating a change in enthalpy

Water is heated to boiling under a pressure of 1.0 atm. When an electric current of 0.50 A from a 12 V supply is passed for 300 s through a resistance in thermal contact with it, it is found that 0.798 g of water is vaporized. Calculate the molar internal energy and enthalpy changes at the boiling point (373.15 K).

Method Because the vaporization occurs at constant pressure, the enthalpy change is equal to the heat supplied by the heater. Therefore, the strategy is to calculate the heat supplied (from $q = IVt$), express that as an enthalpy change, and then convert the result to a molar enthalpy change by division by the amount of H_2O molecules vaporized. To convert from enthalpy change to internal energy change, we assume that the vapour is a perfect gas and use eqn 26.

Answer The enthalpy change is

$$\Delta H = q_p = (0.50 \text{ A}) \times (12 \text{ V}) \times (300 \text{ s}) = +1.8 \text{ kJ}$$

Because 0.798 g of water is 0.0443 mol H_2O, the enthalpy of vaporization per mole of H_2O is

$$\Delta H_m = +\frac{1.8 \text{ kJ}}{0.0443 \text{ mol}} = +41 \text{ kJ mol}^{-1}$$

In the process $H_2O(l) \rightarrow H_2O(g)$ the change in the amount of gas molecules is $\Delta n_g = +1$ mol, so

$$\Delta U_m = \Delta H_m - RT = +38 \text{ kJ mol}^{-1}$$

Comment The plus sign is added to positive quantities to emphasize that they represent an increase in internal energy or enthalpy. Notice that the internal energy change is smaller than the enthalpy change because energy has been used to drive back the surrounding atmosphere to make room for the vapour.

- -

Self-test 2.3 The molar enthalpy of vaporization of benzene at its boiling point (353.25 K) is 30.8 kJ mol^{-1}. What is the molar internal energy change? For how long

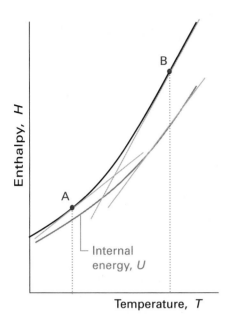

2.15 The slope of a graph of the enthalpy of a system subjected to a constant pressure plotted against temperature is the constant-pressure heat capacity. The slope of the graph may change with temperature, in which case the heat capacity varies with temperature. Thus, the heat capacities at A and B are different. For gases, the slope of the graph of enthalpy versus temperature is steeper than that of the graph of internal energy versus temperature, and $C_{p,\mathrm{m}}$ is larger than $C_{V,\mathrm{m}}$.

Table 2.2* Temperature variation of molar heat capacities, $C_{p,\mathrm{m}}/(\mathrm{J\,K^{-1}\,mol^{-1}}) = a + bT + c/T^2$

	a	$b/(10^{-3}\,\mathrm{K^{-1}})$	$c/(10^5\,\mathrm{K^2})$
C(s, graphite)	16.86	4.77	−8.54
CO_2(g)	44.22	8.79	−8.62
H_2O(l)	75.29	0	0
N_2(g)	28.58	3.77	−0.50

* More values are given in the *Data section* at the end

would the same 12 V source need to supply a 0.50 A current in order to vaporize a 10 g sample?

$$[+27.9\ \mathrm{kJ\,mol^{-1}},\ 6.6 \times 10^2\ \mathrm{s}]$$

Enthalpy and internal energy changes may also be measured by noncalorimetric methods (see Chapters 9 and 10).

(c) The variation of enthalpy with temperature

The enthalpy of a substance increases as its temperature is raised. The relation between the increase in enthalpy and the increase in temperature depends on the conditions (for example, constant pressure or constant volume). The most important condition is constant pressure, and the slope of a graph of enthalpy against temperature at constant pressure is called the **heat capacity at constant pressure**, C_p (Fig. 2.15). More formally:

$$C_p = \left(\frac{\partial H}{\partial T}\right)_p \qquad [27]$$

The heat capacity at constant pressure is the analogue of the heat capacity at constant volume, and is an extensive property.[4] The **molar heat capacity at constant pressure**, $C_{p,\mathrm{m}}$, is the heat capacity per mole of material; it is an intensive property.

The heat capacity at constant pressure is used to relate the change in enthalpy to a change in temperature. For infinitesimal changes of temperature,

$$\mathrm{d}H = C_p\,\mathrm{d}T \qquad \text{(at constant pressure)} \qquad (28a)$$

If the heat capacity is constant over the range of temperatures of interest, then for a measurable increase in temperature

$$\Delta H = C_p\Delta T \qquad \text{(at constant pressure)} \qquad (28b)$$

Because an increase in enthalpy can be equated with the heat supplied at constant pressure, the practical form of the latter equation is

$$q_p = C_p\Delta T \qquad (29)$$

This expression shows us how to measure the heat capacity of a sample: a measured quantity of heat is supplied under conditions of constant pressure (as in a sample exposed to the atmosphere and free to expand), and the temperature rise is monitored.

The variation of heat capacity with temperature can sometimes be ignored if the temperature range is small; this approximation is highly accurate for a monatomic perfect gas (one of the noble gases). However, when it is necessary to take the variation into account, a convenient approximate empirical expression is

$$C_{p,\mathrm{m}} = a + bT + \frac{c}{T^2} \qquad (30)$$

The empirical parameters a, b, and c are independent of temperature. Some typical values are given in Table 2.2.

Example 2.4 Evaluating an increase in enthalpy with temperature

What is the change in molar enthalpy of N_2 when it is heated from 25 °C to 100 °C? Use the heat capacity information in Table 2.2.

---

4   As in the case of C_V, if the system can change its composition it is necessary to distinguish between equilibrium and fixed-composition values. All applications in this chapter refer to pure substances, so this complication can be ignored.

Method The heat capacity of N_2 changes with temperature, so we cannot use eqn 28b (which assumes that the heat capacity of the substance is constant). Therefore, we must use eqn 28a, substitute eqn 30 for the temperature dependence of the heat capacity, and integrate the resulting expression from 25°C to 100°C.

Answer For convenience, we denote the two temperatures T_1 (298 K) and T_2 (373 K). The integrals we require are

$$\int_{H(T_1)}^{H(T_2)} dH = \int_{T_1}^{T_2} \left(a + bT + \frac{c}{T^2} \right) dT$$

Notice how the limits of integration correspond on each side of the equation: the integration over H on the left ranges from $H(T_1)$, the value of H at T_1, up to $H(T_2)$, the value of H at T_2, while on the right the integration over the temperature ranges from T_1 to T_2. Now we use the integrals

$$\int dx = x \qquad \int x \, dx = \tfrac{1}{2}x^2 \qquad \int \frac{dx}{x^2} = -\frac{1}{x}$$

to obtain

$$H(T_2) - H(T_1) = a(T_2 - T_1) + \tfrac{1}{2}b(T_2^2 - T_1^2) - c\left(\frac{1}{T_2} - \frac{1}{T_1} \right)$$

Substitution of the numerical data results in

$$H(373 \text{ K}) = H(298 \text{ K}) + 2.20 \text{ kJ mol}^{-1}$$

If we had assumed a constant heat capacity of 29.14 J K^{-1} mol^{-1} (the value given by eqn 30 at 25°C), we would have found that the two enthalpies differed by 2.19 kJ mol^{-1}.

- -

Self-test 2.4 At very low temperatures the heat capacity of a solid is proportional to T^3, and we can write $C_p = aT^3$. What is the change in enthalpy of such a substance when it is heated from 0 to a temperature T (with T close to 0)?

$$[\Delta H = \tfrac{1}{4}aT^4]$$

(d) The relation between heat capacities

Most systems expand when heated at constant pressure. Such systems do work on the surroundings and some of the energy supplied to them as heat escapes back to the surroundings. As a result, the temperature of the system rises less than when the heating occurs at constant volume. A smaller increase in temperature implies a larger heat capacity, so we conclude that in most cases *the heat capacity at constant pressure of a system is larger than its heat capacity at constant volume.*

There is a simple relation between the two heat capacities of a perfect gas:

$$C_p - C_V = nR \tag{31}°$$

(This relation is derived in Section 3.3a.) It follows that the molar heat capacity of a perfect gas is about 8 J K^{-1} mol^{-1} larger at constant pressure than at constant volume. Because the heat capacity at constant volume of a nonlinear polyatomic gas is about 25 J K^{-1} mol^{-1}, the difference is highly significant and must be taken into account.

2.6 Adiabatic changes

We are now equipped to deal with the changes that occur when a perfect gas expands adiabatically. A decrease in temperature should be expected: because work is done, the

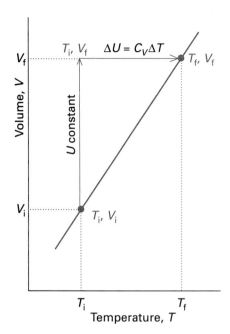

2.16 To achieve a change of state from one temperature and volume to another temperature and volume, we may consider the overall change as composed of two steps. In the first step, the system expands at constant temperature; there is no change in internal energy if the system consists of a perfect gas. In the second step, the temperature of the system is increased at constant volume. The overall change in internal energy is the sum of the changes for the two steps.

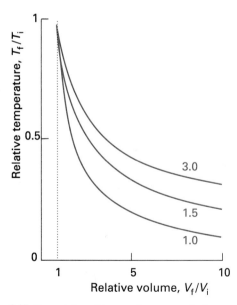

2.17 The variation of temperature as a perfect gas is expanded reversibly and adiabatically. The curves are labelled with different values of $c = C_{V,m}/R$. Note that the temperature falls most steeply for gases with low molar heat capacity.

internal energy falls, and therefore the temperature of the working gas also falls. In molecular terms, the kinetic energy of the molecules falls as work is done, so their average speed decreases, and hence the temperature falls.

(a) The work of adiabatic change

The change in internal energy of a perfect gas when the temperature is changed from T_i to T_f and the volume is changed from V_i to V_f can be expressed as the sum of two steps (Fig. 2.16). In the first step, only the volume changes and the temperature is held constant at its initial value. However, because the internal energy of a perfect gas arises solely from the kinetic energies of the molecules, the overall change in internal energy arises solely from the second step, the change in temperature at constant volume. Provided the heat capacity is independent of temperature, this change is

$$\Delta U = C_V(T_f - T_i) = C_V\Delta T \tag{32}$$

Because the expansion is adiabatic, we know that $q = 0$; because $\Delta U = q + w$, it then follows that $\Delta U = w_{ad}$. Therefore, by equating the two values we have obtained for ΔU, we obtain

$$w_{ad} = C_V\Delta T \tag{33}°$$

That is, the work done during an adiabatic expansion of a perfect gas is proportional to the temperature difference between the initial and final states. That is exactly what we expect on molecular grounds, because the mean kinetic energy is proportional to T, so a change in internal energy arising from temperature alone is also expected to be proportional to ΔT.

To use eqn 33, we need to relate the change in temperature to the change in volume (which we know). The most important type of adiabatic expansion (and the only kind that we need for later) is *reversible* adiabatic expansion, in which the external pressure is matched to the internal pressure throughout. For adiabatic, reversible expansion the initial and final temperatures are related by

$$V_f T_f^c = V_i T_i^c \qquad c = \frac{C_{V,m}}{R} \tag{34}°_{rev}$$

This expression is derived in the following *Justification* and the temperature dependence it implies is shown in Fig. 2.17. All we need do at this stage to find the work done by the system is to solve this relation for T_f,

$$T_f = T_i\left(\frac{V_i}{V_f}\right)^{1/c} \tag{35}°_{rev}$$

calculate ΔT, and substitute into eqn 33.

Justification 2.2

Consider a stage in the expansion when the pressure inside and out is p (the expansion is reversible, so the internal and external pressures are equal at all stages). The work done when the gas expands by dV is $-p\,dV$. However, for a perfect gas, $dU = C_V\,dT$ (by the same argument as in the text for a macroscopic change). Therefore, because $dU = dw$ for an adiabatic change, we can write

$$C_V\,dT = -p\,dV$$

We are dealing with a perfect gas, so we can replace p by nRT/V and obtain

$$C_V\frac{dT}{T} = -nR\frac{dV}{V}$$

To integrate this expression we note that T is equal to T_i when V is equal to V_i, and is equal to T_f when V is equal to V_f at the end of the expansion. Therefore,

$$C_V \int_{T_i}^{T_f} \frac{dT}{T} = -nR \int_{V_i}^{V_f} \frac{dV}{V}$$

(We are taking C_V to be independent of temperature.) Then, because $\int dx/x = \ln x$, we obtain

$$C_V \ln \left(\frac{T_f}{T_i}\right) = -nR \ln \left(\frac{V_f}{V_i}\right)$$

With $c = C_V/nR$ and by using $a \ln x = \ln x^a$ and $-\ln(x/y) = \ln(y/x)$ we obtain

$$\ln \left(\frac{T_f}{T_i}\right)^c = \ln \left(\frac{V_i}{V_f}\right)$$

which implies eqn 34.

Illustration

Consider the adiabatic, reversible expansion of 0.020 mol Ar, initially at 25 °C, from 0.50 L to 1.00 L. The molar heat capacity of argon at constant volume is 12.48 J K^{-1} mol^{-1}, so $c = 1.501$. Therefore, from eqn 35,

$$T_f = (298 \text{ K}) \times \left(\frac{0.50 \text{ L}}{1.00 \text{ L}}\right)^{1/1.501} = 188 \text{ K}$$

It follows that $\Delta T = -110$ K, and therefore, from eqn 33, that

$$w = C_V \Delta T = (0.020 \text{ mol}) \times (12.48 \text{ J K}^{-1} \text{ mol}^{-1}) \times (-110 \text{ K})$$
$$= -27 \text{ J}$$

Note that the temperature change is independent of the mass of gas but the work is not.

Self-test 2.5 Calculate the final temperature, the work done, and the change of internal energy when ammonia is used in a reversible adiabatic expansion from 0.50 L to 2.00 L, the other initial conditions being the same.

[194 K, −56 J, −56 J]

(b) Heat capacity ratio and adiabats

Now we consider the change in pressure that results from an adiabatic, reversible expansion of a perfect gas. We show in the *Justification* below that

$$pV^\gamma = \text{constant} \tag{36}_{rev}^{\circ}$$

where the **heat capacity ratio**, γ, of a substance is defined as

$$\gamma = \frac{C_{p,m}}{C_{V,m}} \tag{37}$$

Because the heat capacity at constant pressure is greater than the heat capacity at constant volume, $\gamma > 1$. For a perfect gas, it follows from eqn 31 that

$$\gamma = \frac{C_{V,m} + R}{C_{V,m}} \tag{38}^{\circ}$$

For a monatomic perfect gas, $C_{V,m} = \frac{3}{2}R$ (see *Molecular interpretation* 2.3), so $\gamma = \frac{5}{3}$. For a gas of nonlinear polyatomic molecules (which can rotate as well as translate), $C_{V,m} = 3R$, so $\gamma = \frac{4}{3}$.

Justification 2.3

The initial and final states of a perfect gas satisfy the perfect gas law regardless of how the change of state takes place, so we can use $pV = nRT$ to write

$$\frac{p_i V_i}{p_f V_f} = \frac{T_i}{T_f}$$

However, we have established (eqn 35) that for a reversible adiabatic change the temperature changes so as to satisfy

$$\frac{T_i}{T_f} = \left(\frac{V_f}{V_i}\right)^{1/c}$$

Then we combine the two expressions, to obtain

$$p_i V_i^{\gamma} = p_f V_f^{\gamma}$$

It follows that $pV^{\gamma} = $ constant as used in the text.

The curves of pressure versus volume for adiabatic, reversible change are known as **adiabats**, and one is illustrated in Fig. 2.18. Because $\gamma > 1$, an adiabat falls more steeply ($p \propto 1/V^{\gamma}$) than the corresponding isotherm ($p \propto 1/V$). The physical reason for the difference is that, in an isothermal expansion, energy flows into the system as heat and maintains the temperature; as a result, the pressure does not fall as much as in an adiabatic expansion.

Illustration

When a sample of argon (for which $\gamma = \frac{5}{3}$) at 100 kPa expands reversibly and adiabatically to twice its initial volume, the final pressure will be

$$p_f = p_i \left(\frac{V_i}{V_f}\right)^{\gamma} = (100 \text{ kPa}) \times \left(\tfrac{1}{2}\right)^{5/3} = 31.5 \text{ kPa}$$

For an isothermal doubling of volume, the final pressure would be 50 kPa.

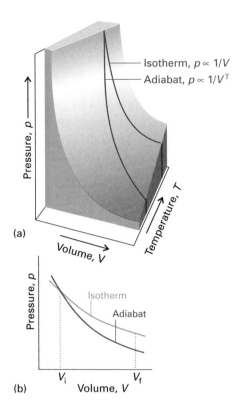

(a)

(b)

Isotherm, $p \propto 1/V$
Adiabat, $p \propto 1/V^{\gamma}$

Isotherm
Adiabat

2.18 An adiabat depicts the variation of pressure with volume when a gas expands reversibly and adiabatically. (a) An adiabat for a perfect gas. (b) Note that the pressure declines more steeply for an adiabat than it does for an isotherm because the temperature decreases in the former.

Thermochemistry

The study of the heat produced or required by chemical reactions is called **thermochemistry**. Thermochemistry is a branch of thermodynamics because a reaction vessel and its contents form a system, and chemical reactions result in the exchange of energy between the system and the surroundings. Thus we can use calorimetry to measure the heat produced or absorbed by a reaction, and can identify q with a change in internal energy (if the reaction occurs at constant volume) or a change in enthalpy (if the reaction occurs at constant pressure). Conversely, if we know the ΔU or ΔH for a reaction, we can predict the heat the reaction can produce.

We have already remarked that a process that releases heat is classified as exothermic and one that absorbs heat is classified as endothermic. Because the release of heat signifies a decrease in the enthalpy of a system (at constant pressure), we can now see that an exothermic process at constant pressure is one for which $\Delta H < 0$. Conversely, because the absorption of heat results in an increase in enthalpy, an endothermic process at constant pressure has $\Delta H > 0$.

2.7 Standard enthalpy changes

Changes in enthalpy are normally reported for processes taking place under a set of standard conditions. In most of our discussions we shall consider the **standard enthalpy change**, $\Delta H^{\ominus}$, the change in enthalpy for a process in which the initial and final substances are in their standard states:

> **The standard state of a substance at a specified temperature is its pure form at 1 bar.**

For example, the standard state of liquid ethanol at 298 K is pure liquid ethanol at 298 K and 1 bar; the standard state of solid iron at 500 K is pure iron at 500 K and 1 bar. The standard enthalpy change for a reaction or a physical process is the difference between the products in their standard states and enthalpy of the reactants in their standard states, all at the same specified temperature.

As an example of a standard enthalpy change, the **standard enthalpy of vaporization**, $\Delta_{vap}H^{\ominus}$, is the enthalpy change per mole when a pure liquid at 1 bar vaporizes to a gas at 1 bar, as in

$$H_2O(l) \longrightarrow H_2O(g) \qquad \Delta_{vap}H^{\ominus}(373 \text{ K}) = +40.66 \text{ kJ mol}^{-1}$$

As implied by the examples, standard enthalpies may be reported for any temperature. However, the conventional temperature for reporting thermodynamic data is 298.15 K (corresponding to 25.00 °C). Unless otherwise mentioned, all thermodynamic data in this text will refer to this conventional temperature.

(a) Enthalpies of physical change

The standard enthalpy change that accompanies a change of physical state is called the **standard enthalpy of transition** and is denoted $\Delta_{trs}H^{\ominus}$ (Table 2.3). The standard enthalpy of vaporization, $\Delta_{vap}H^{\ominus}$, is one example. Another is the standard enthalpy of fusion $\Delta_{fus}H^{\ominus}$, the enthalpy change accompanying the conversion of a solid to a liquid, as in

$$H_2O(s) \longrightarrow H_2O(l) \qquad \Delta_{fus}H^{\ominus}(273 \text{ K}) = +6.01 \text{ kJ mol}^{-1}$$

As in this case, it is sometimes convenient to know the standard enthalpy change at the transition temperature as well as at the conventional temperature.

Because enthalpy is a state function, a change in enthalpy is independent of the path between the two states. This feature is of great importance in thermochemistry, for it implies that the same value of $\Delta H^{\ominus}$ will be obtained however the change is brought about (so long as the initial and final states are the same). For example, we can picture the conversion of a

Table 2.3* Standard enthalpies of fusion and vaporization at the transition temperature, $\Delta_{trs}H^{\ominus}/(\text{kJ mol}^{-1})$

	T_f/K	Fusion	T_b/K	Vaporization
Ar	83.81	1.188	87.29	6.506
C_6H_6	278.61	10.59	353.2	30.8
H_2O	273.15	6.008	373.15	40.656
				44.016 at 298 K
He	3.5	0.021	4.22	0.084

* More values are given in the *Data section*.

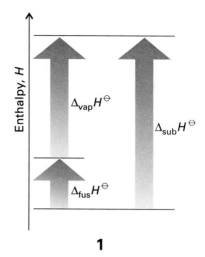

1

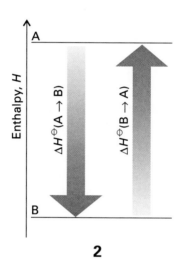

2

solid to a vapour either as occurring by sublimation (the direct conversion from solid to vapour),

$$H_2O(s) \longrightarrow H_2O(g) \qquad \Delta_{sub}H^\ominus$$

or as occurring in two steps, first fusion (melting) and then vaporization of the resulting liquid:

$$H_2O(s) \longrightarrow H_2O(l) \qquad \Delta_{fus}H^\ominus$$
$$H_2O(l) \longrightarrow H_2O(g) \qquad \Delta_{vap}H^\ominus$$
Overall: $H_2O(s) \longrightarrow H_2O(g) \qquad \Delta_{fus}H^\ominus + \Delta_{vap}H^\ominus$

Because the overall result of the indirect path is the same as that of the direct path, the overall enthalpy change is the same in each case (**1**), and we can conclude that (for processes occurring at the same temperature)

$$\Delta_{sub}H^\ominus = \Delta_{fus}H^\ominus + \Delta_{vap}H^\ominus$$

An immediate conclusion is that, because all enthalpies of fusion are positive, the enthalpy of sublimation of a substance is greater than its enthalpy of vaporization (at a given temperature).

Another consequence of H being a state function is that the standard enthalpy changes of a forward process and its reverse must differ only in sign (**2**):

$$\Delta H^\ominus(A \rightarrow B) = -\Delta H^\ominus(A \leftarrow B)$$

For instance, because the enthalpy of vaporization of water is $+44 \text{ kJ mol}^{-1}$ at 298 K, its enthalpy of condensation at that temperature is -44 kJ mol^{-1}.

The different types of enthalpies encountered in thermochemistry are summarized in Table 2.4. We shall meet them again in various locations throughout the text.

(b) Enthalpies of chemical change

Now we consider enthalpy changes that accompany chemical reactions. Broadly speaking, the **standard reaction enthalpy**, $\Delta_r H^\ominus$, is the change in enthalpy when reactants in their standard states change to products in their standard states, as in

$$CH_4(g) + 2O_2(g) \longrightarrow CO_2(g) + 2H_2O(l) \qquad \Delta_r H^\ominus = -890 \text{ kJ mol}^{-1}$$

Table 2.4 Enthalpies of transition

Transition	Process	Symbol*
Transition	Phase $\alpha \longrightarrow$ phase β	$\Delta_{trs}H$
Fusion	$s \longrightarrow l$	$\Delta_{fus}H$
Vaporization	$l \longrightarrow g$	$\Delta_{vap}H$
Sublimation	$s \longrightarrow g$	$\Delta_{sub}H$
Mixing of fluids	Pure $\longrightarrow$ mixture	$\Delta_{mix}H$
Solution	Solute $\longrightarrow$ solution	$\Delta_{sol}H$
Hydration	$X^\pm(g) \longrightarrow X^\pm(aq)$	$\Delta_{hyd}H$
Atomization	Species(s, l, g) $\longrightarrow$ atoms(g)	$\Delta_{at}H$
Ionization	$X(g) \longrightarrow X^+(g) + e^-(g)$	$\Delta_{ion}H$
Electron gain	$X(g) + e^-(g) \longrightarrow X^-(g)$	$\Delta_{eg}H$
Reaction	Reactants $\longrightarrow$ products	$\Delta_r H$
Combustion	Compound(s, l, g) $+ O_2(g) \longrightarrow CO_2(g)$, $H_2O(l, g)$	$\Delta_c H$
Formation	Elements $\longrightarrow$ compound	$\Delta_f H$
Activation	Reactants $\longrightarrow$ activated complex	$\Delta^\ddagger H$

* IUPAC recommendations. In common usage, the transition subscript is often attached to ΔH, as in ΔH_{trs}.

This standard value refers to the reaction in which 1 mol CH_4 in the form of pure methane gas at 1 bar reacts completely with 2 mol O_2 in the form of pure oxygen gas to produce 1 mol CO_2 as pure carbon dioxide at 1 bar and 2 mol H_2O as pure liquid water at 1 bar; the numerical value is for the reaction at 298 K. The combination of a chemical equation and a standard reaction enthalpy is called a **thermochemical equation**. A standard reaction enthalpy refers to the overall process

Pure, unmixed reactants in their standard states

$\longrightarrow$ pure, separated products in their standard states

Except in the case of ionic reactions in solution, the enthalpy changes accompanying mixing and separation are insignificant in comparison with the contribution from the reaction itself.

We said 'broadly speaking' above, because the precise specification of the standard reaction enthalpy is more specific about the significance of the 'per mole' that appears in the value of $\Delta_r H^{\ominus}$. To establish this precise definition, consider the reaction

$$2A + B \longrightarrow 3C + D$$

The standard enthalpy of this reaction is based on the expression

$$\Delta_r H^{\ominus} = \sum_{\text{Products}} \nu H_m^{\ominus} - \sum_{\text{Reactants}} \nu H_m^{\ominus} \tag{39}$$

where the terms on the right are the standard molar enthalpies of the products and reactants weighted by the stoichiometric coefficients, ν, in the chemical equation. For our reaction,

$$\Delta_r H^{\ominus} = \{3H_m^{\ominus}(C) + H_m^{\ominus}(D)\} - \{2H_m^{\ominus}(A) + H_m^{\ominus}(B)\}$$

where $H_m^{\ominus}(J)$ is the standard molar enthalpy of species J at the temperature of interest.

A somewhat more sophisticated way of expressing the definition, which is useful for some of the formal expressions we shall derive (but less so in practice), is obtained by writing the reaction in the symbolic form

$$0 = 3C + D - 2A - B$$

by subtracting the reactants from both sides (and replacing the arrow by an equals sign). This equation has the form

$$0 = \sum_J \nu_J J \tag{40}$$

where J denotes the substances and the ν_J are the corresponding **stoichiometric numbers** in the chemical equation. These numbers have the values

$$\nu_A = -2 \qquad \nu_B = -1 \qquad \nu_C = +3 \qquad \nu_D = +1$$

Note that, in the convention we shall adopt, the stoichiometric numbers of the products are positive and those of the reactants are negative. The standard reaction enthalpy is then defined as

$$\Delta_r H^{\ominus} = \sum_J \nu_J H_m^{\ominus}(J) \tag{41}$$

as may be verified by substituting the values of the stoichiometric numbers given above.

Illustration

For the reaction

$$N_2(g) + 3H_2(g) \longrightarrow 2NH_3(g)$$

Table 2.5* Standard enthalpies of formation and combustion of organic compounds at 298 K

	$\Delta_f H^{\ominus}/(\text{kJ mol}^{-1})$	$\Delta_c H^{\ominus}/(\text{kJ mol}^{-1})$
Benzene, $C_6H_6(l)$	+49.0	−3268
Ethane, $C_2H_6(g)$	−84.7	−1560
Glucose, $C_6H_{12}O_6(s)$	−1274	−2808
Methane, $CH_4(g)$	−74.8	−890
Methanol, $CH_3OH(l)$	−238.7	−726

*More values are given in the *Data section*.

the stoichiometric numbers are

$$\nu(N_2) = -1 \qquad \nu(H_2) = -3 \qquad \nu(NH_3) = +2$$

so the standard reaction enthalpy is

$$\Delta_r H^{\ominus} = 2H_m^{\ominus}(NH_3) - \{H_m^{\ominus}(N_2) + 3H_m^{\ominus}(H_2)\}$$

· ·

Some standard reaction enthalpies have special names and a particular significance. The **standard enthalpy of combustion**, $\Delta_c H^{\ominus}$, is the standard reaction enthalpy for the complete oxidation of an organic compound to CO_2 and H_2O if the compound contains C, H, and O, and to N_2 if N is also present. An example is the combustion of glucose:

$$C_6H_{12}O_6(s) + 6O_2(g) \longrightarrow 6CO_2(g) + 6H_2O(l) \qquad \Delta_c H^{\ominus} = -2808 \text{ kJ mol}^{-1}$$

The value quoted shows that 2808 kJ of heat is released when 1 mol $C_6H_{12}O_6$ burns under standard conditions (at 298 K). Some further values are listed in Table 2.5.

(c) Hess's law

Standard enthalpies of individual reactions can be combined to obtain the enthalpy of another reaction. This application of the First Law is called **Hess's law**:

> **The standard enthalpy of an overall reaction is the sum of the standard enthalpies of the individual reactions into which a reaction may be divided.**

The individual steps need not be realizable in practice: they may be hypothetical reactions, the only requirement being that their chemical equations should balance. The thermodynamic basis of the law is the path-independence of the value of $\Delta_r H^{\ominus}$ and the implication that we may take the specified reactants, pass through any (possibly hypothetical) set of reactions to the specified products, and overall obtain the same change of enthalpy. The importance of Hess's law is that information about a reaction of interest, which may be difficult to determine directly, can be assembled from information on other reactions.

Example 2.5 Using Hess's law

The standard reaction enthalpy for the hydrogenation of propene,

$$CH_2{=}CHCH_3(g) + H_2(g) \longrightarrow CH_3CH_2CH_3(g)$$

is −124 kJ mol^{-1}. The standard reaction enthalpy for the combustion of propane,

$$CH_3CH_2CH_3(g) + 5O_2(g) \longrightarrow 3CO_2(g) + 4H_2O(l)$$

is −2220 kJ mol^{-1}. Calculate the standard enthalpy of combustion of propene.

Method Add and subtract the reactions given, together with any others needed, so as to reproduce the reaction required. Then add and subtract the reaction enthalpies in the same way. Additional data are in Table 2.5.

Answer The combustion reaction we require is

$$C_3H_6(g) + \tfrac{9}{2}O_2(g) \longrightarrow 3CO_2(g) + 3H_2O(l)$$

This reaction can be recreated from the following sum:

	$\Delta_r H^{\ominus}/\text{kJ mol}^{-1}$
$C_3H_6(g) + H_2(g) \longrightarrow C_3H_8(g)$	-124
$C_3H_8(g) + 5O_2(g) \longrightarrow 3CO_2(g) + 4H_2O(l)$	-2220
$H_2O(l) \longrightarrow H_2(g) + \tfrac{1}{2}O_2(l)$	$+286$
$C_3H_6(g) + \tfrac{9}{2}O_2(g) \longrightarrow 3CO_2(g) + 3H_2O(l)$	-2058

Comment The skill to develop is the ability to assemble a given thermochemical equation from others.

- -

Self-test 2.6 Calculate the enthalpy of hydrogenation of benzene from its enthalpy of combustion and the enthalpy of combustion of cyclohexane.

$$[-205 \text{ kJ mol}^{-1}]$$

2.8 Standard enthalpies of formation

The **standard enthalpy of formation**, $\Delta_f H^{\ominus}$, of a substance is the standard reaction enthalpy for the formation of the compound from its elements in their reference states. The **reference state** of an element is its most stable state at the specified temperature and 1 bar. For example, at 298 K the reference state of nitrogen is a gas of N_2 molecules, that of mercury is liquid mercury, that of carbon is graphite, and that of tin is the white (metallic) form. There is one exception to this general prescription of reference states: the reference state of phosphorus is taken to be white phosphorus despite this allotrope not being the most stable form but simply the most reproducible form of the element. Standard enthalpies of formation are expressed as enthalpies per mole of the compound. The standard enthalpy of formation of liquid benzene at 298 K, for example, refers to the reaction

$$6C(s, \text{graphite}) + 3H_2(g) \longrightarrow C_6H_6(l)$$

and is $+49.0 \text{ kJ mol}^{-1}$. The standard enthalpies of formation of elements in their reference states are zero at all temperatures because they are the enthalpies of such 'null' reactions as

$$N_2(g) \longrightarrow N_2(g)$$

Some enthalpies of formation are listed in Tables 2.5 and 2.6.

Table 2.6* Standard enthalpies of formation of inorganic compounds, $\Delta_f H^{\ominus}/(\text{kJ mol}^{-1})$, at 298 K

$H_2O(l)$	-285.83	$H_2O_2(l)$	-187.78
$NH_3(g)$	-46.11	$N_2H_4(l)$	$+50.63$
$NO_2(g)$	$+33.18$	$N_2O_4(g)$	$+9.16$
$NaCl(s)$	-411.15	$KCl(s)$	-436.75

* More values are given in the *Data section*.

(a) The reaction enthalpy in terms of enthalpies of formation

Conceptually, we can regard a reaction as proceeding by decomposing the reactants into their elements and then forming those elements into the products. The value of $\Delta_r H^{\ominus}$ for the overall reaction is the sum of these 'unforming' and forming enthalpies. Because 'unforming' is the reverse of forming, the enthalpy of an unforming step is the negative of the enthalpy of formation (3). Hence, in the enthalpies of formation of substances, we have enough information to calculate the enthalpy of any reaction by using

$$\Delta_r H^{\ominus} = \sum_{\text{Products}} \nu \Delta_f H^{\ominus} - \sum_{\text{Reactants}} \nu \Delta_f H^{\ominus} \tag{42}$$

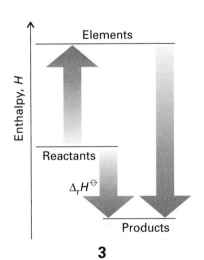

3

where in each case the enthalpies of formation of the species that occur are multiplied by their stoichiometric coefficients.

Illustration

The standard reaction enthalpy of

$$2HN_3(l) + 2NO(g) \longrightarrow H_2O_2(l) + 4N_2(g)$$

is calculated as follows:

$$\begin{aligned} \Delta_r H^{\ominus} &= \{\Delta_f H^{\ominus}(H_2O_2, l) + 4\Delta_f H^{\ominus}(N_2, g)\} \\ &\quad - \{2\Delta_f H^{\ominus}(HN_3, l) + 2\Delta_f H^{\ominus}(NO, g)\} \\ &= \{-187.78 + 4(0)\} - \{2(264.0) + 2(90.25)\} \text{ kJ mol}^{-1} \\ &= -892.3 \text{ kJ mol}^{-1} \end{aligned}$$

The formal expression for eqn 42 in terms of a reaction written as in eqn 40 is

$$\Delta_r H^{\ominus} = \sum_J \nu_J \Delta_f H^{\ominus}(J) \qquad [43]$$

It may be verified that this expression reproduces the value calculated in the *Illustration*.

(b) Group contributions

We have seen that standard reaction enthalpies may be constructed by combining standard enthalpies of formation. The question that now arises is whether we can construct standard enthalpies of formation from a knowledge of the composition of the species. The short answer is that there is no *thermodynamically* exact way of breaking enthalpies of formation down into contributions from individual atoms and bonds. In the past, approximate procedures based on **mean bond enthalpies**, $\Delta H(\text{A—B})$, the enthalpy change associated with the breaking of a specific A—B bond,

$$\text{A—B}(g) \longrightarrow \text{A}(g) + \text{B}(g) \qquad \Delta H(\text{A—B})$$

have been used. However, this procedure is notoriously unreliable, in part because the $\Delta H(\text{A—B})$ are average values for a series of related compounds. Nor does the approach distinguish between geometrical isomers, where the same atoms and bonds may be present, but experimentally the enthalpies of formation might be significantly different.

Somewhat more reliable is an approach in which a molecule is regarded as being built up of **thermochemical groups**, an atom or physical group of atoms bound to at least two other atoms: two examples are shown as (**4**) and (**5**). The enthalpy of formation of the compound is then expressed (approximately, at least) as the sum of the contributions associated with all the thermochemical groups into which the molecule can be divided. A list of the appropriate values is given in Table 2.7 and the method is illustrated in the following example. Table 2.7 also contains information on heat capacity, which is approximately additive in a similar sense.

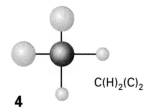

C(H)$_2$(C)$_2$

4

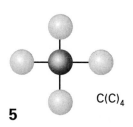

C(C)$_4$

5

Table 2.7* Benson thermochemical groups

Group	$\Delta_f H^{\ominus}$ / (kJ mol^{-1})	$C_{p,m}^{\ominus}$ / $(\text{J K}^{-1}\,\text{mol}^{-1})$
C(H)$_3$(C)	−42.17	25.9
C(H)$_2$(C)$_2$	−20.7	22.8
C(H)(C)$_3$	−6.19	18.7
C(C)$_4$	+8.16	18.2

* More values are given in the *Data section*.

Example 2.6 Using the thermochemical group approach

Estimate the standard enthalpy of formation at 298 K of hexane in (a) the gas phase, (b) the liquid phase.

Method First, identify the thermochemical groups present in the molecule. Then add together the appropriate values in Table 2.7. To obtain the enthalpy of formation of the liquid, include the enthalpy of condensation of the vapour, that is the negative of the enthalpy of vaporization of the liquid (Table 2.3).

Answer The decomposition of the molecule into groups is depicted in (**6**). There are two $C(H)_3(C)$ groups and four $C(H)_2(C)_2$ groups; therefore

$$\Delta_f H^{\ominus}(C_6H_{14}, g) = 2(-42.17 \text{ kJ mol}^{-1}) + 4(-20.7 \text{ kJ mol}^{-1})$$
$$= -167.1 \text{ kJ mol}^{-1}$$

The enthalpy of vaporization of hexane is 28.9 kJ mol^{-1} (Table 2.3); therefore (**7**),

$$\Delta_f H^{\ominus}(C_6H_{14}, l) = (-167.1 \text{ kJ mol}^{-1}) - (28.9 \text{ kJ mol}^{-1}) = -196.0 \text{ kJ mol}^{-1}$$

The experimental value is −198.7 kJ mol^{-1}.

- -

Self-test 2.7 Estimate the standard enthalpy of formation of gaseous 2,2-dimethyl-propane.

$$[-160.52 \text{ kJ mol}^{-1}]$$

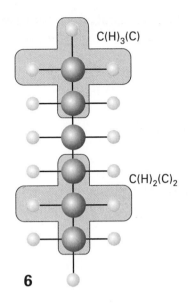

6

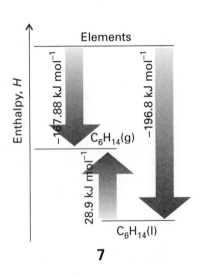

7

2.9 The temperature dependence of reaction enthalpies

The standard enthalpies of many important reactions have been measured at different temperatures, and for serious work these accurate data must be used. However, in the absence of this information, standard reaction enthalpies at different temperatures may be estimated from heat capacities and the reaction enthalpy at some other temperature.

It follows from eqn 27 that, when a substance is heated from T_1 to T_2, its enthalpy changes from $H(T_1)$ to

$$H(T_2) = H(T_1) + \int_{T_1}^{T_2} C_p \, dT \tag{44}$$

(We have assumed that no phase transition takes place in the temperature range of interest.) Because this equation applies to each substance in the reaction, the standard reaction enthalpy changes from $\Delta_r H^{\ominus}(T_1)$ to

$$\Delta_r H^{\ominus}(T_2) = \Delta_r H^{\ominus}(T_1) + \int_{T_1}^{T_2} \Delta_r C_p^{\ominus} \, dT \tag{45}$$

where $\Delta_r C_p^{\ominus}$ is the difference of the molar heat capacities of products and reactants under standard conditions weighted by the stoichiometric coefficients that appear in the chemical equation:

$$\Delta_r C_p^{\ominus} = \sum_{\text{Products}} \nu C_{p,m}^{\ominus} - \sum_{\text{Reactants}} \nu C_{p,m}^{\ominus} \tag{46}$$

More formally,

$$\Delta_r C_p^{\ominus} = \sum_J \nu_J C_{p,m}^{\ominus}(J) \tag{47}$$

Equation 45 is known as **Kirchhoff's law** (Fig. 2.19). It is normally a good approximation to assume that $\Delta_r C_p^{\ominus}$ is independent of the temperature, at least over reasonably limited ranges, as illustrated in the following example. Although the individual heat capacities may vary, their difference varies less significantly. In some cases the temperature dependence of heat capacities is taken into account by using eqn 30.

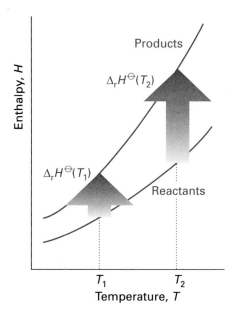

2.19 An illustration of the content of Kirchhoff's law. When the temperature is increased, the enthalpies of the products and the reactants both increase, but may do so to different extents. In each case, the change in enthalpy depends on the heat capacities of the substances. The change in reaction enthalpy reflects the difference in the changes of the enthalpies.

Example 2.7 Using Kirchhoff's law

The standard enthalpy of formation of gaseous H_2O at 298 K is -241.82 kJ mol^{-1}. Estimate its value at $100\,°C$ given the following values of the molar heat capacities at constant pressure: $H_2O(g)$: 33.58 J K^{-1} mol^{-1}; $H_2(g)$: 28.84 J K^{-1} mol^{-1}; $O_2(g)$: 29.37 J K^{-1} mol^{-1}. Assume that the heat capacities are independent of temperature.

Method When $\Delta C_p^{\ominus}$ is independent of temperature in the range T_1 to T_2, the integral in eqn 45 evaluates to $(T_2 - T_1)\Delta_r C_p^{\ominus}$. Therefore,

$$\Delta_r H^{\ominus}(T_2) = \Delta_r H^{\ominus}(T_1) + (T_2 - T_1)\Delta_r C_p^{\ominus}$$

To proceed, write the chemical equation, identify the stoichiometric coefficients, and calculate $\Delta_r C_p^{\ominus}$ from the data.

Answer The reaction is

$$H_2(g) + \tfrac{1}{2}O_2(g) \longrightarrow H_2O(g)$$

so

$$\Delta_r C_p^{\ominus} = C_{p,m}^{\ominus}(H_2O, g) - \{C_{p,m}^{\ominus}(H_2, g) + \tfrac{1}{2}C_{p,m}^{\ominus}(O_2, g)\}$$
$$= -9.94 \text{ J K}^{-1} \text{ mol}^{-1}$$

It then follows that

$$\Delta_f H^{\ominus}(373 \text{ K}) = -241.82 \text{ kJ mol}^{-1} + (75 \text{ K}) \times (-9.94 \text{ J K}^{-1} \text{ mol}^{-1})$$
$$= -242.6 \text{ kJ mol}^{-1}$$

- -

Self-test 2.8 Estimate the standard enthalpy of formation of liquid cyclohexane at 400 K from the data in Table 2.5 in the *Data section* at the end of this volume.

$$[-163 \text{ kJ mol}^{-1}]$$

Checklist of key ideas

☐ thermodynamics

The basic concepts

☐ system
☐ surroundings
☐ open system
☐ closed system
☐ isolated system

2.1 Work, heat, and energy
☐ work
☐ energy
☐ heat
☐ diathermic boundary
☐ adiabatic boundary
☐ exothermic processes
☐ endothermic processes

☐ the molecular interpretation of heat and work
☐ thermal motion

2.2 The First Law
☐ internal energy
☐ state function
☐ First Law of thermodynamics
☐ alternative statement of First Law
☐ definition of heat

Work and heat

2.3 Expansion work
☐ expansion work
☐ work defined

☐ free expansion
☐ expansion against constant pressure
☐ indicator diagram
☐ reversible change
☐ equilibrium
☐ work of reversible expansion (12)
☐ work of isothermal reversible expansion (13)
☐ maximum work and reversible change

2.4 Heat transactions
☐ internal energy change and heat transfer at constant volume
☐ adiabatic bomb calorimeter

☐ calorimeter constant
☐ heat capacity
☐ heat capacity at constant volume
☐ molar heat capacity at constant volume
☐ specific heat capacity
☐ internal energy change and temperature rise (22)

2.5 Enthalpy
☐ enthalpy
☐ enthalpy change and heat transfer at constant pressure
☐ relation between ΔU and ΔH (26)

- ☐ heat capacity at constant pressure
- ☐ molar heat capacity at constant pressure
- ☐ empirical temperature variation (30)
- ☐ relation between heat capacities (31)

2.6 Adiabatic changes
- ☐ work of adiabatic change
- ☐ temperature change accompanying adiabatic change
- ☐ heat capacity ratio

- ☐ pressure–volume relation for adiabatic change
- ☐ adiabat

Thermochemistry
- ☐ thermochemistry

2.7 Standard enthalpy changes
- ☐ standard enthalpy change
- ☐ standard state
- ☐ standard enthalpy of vaporization
- ☐ standard enthalpy of transition ($\Delta_{trs}H^{\ominus}$)

- ☐ addition of enthalpy changes
- ☐ enthalpy change for reverse process
- ☐ standard reaction enthalpy ($\Delta_r H^{\ominus}$)
- ☐ thermochemical equation
- ☐ writing a chemical equation formally
- ☐ stoichiometric numbers
- ☐ standard enthalpy of combustion
- ☐ Hess's law

2.8 Standard enthalpies of formation
- ☐ standard enthalpy of formation ($\Delta_f H^{\ominus}$)
- ☐ reference state
- ☐ combining enthalpies of formation
- ☐ mean bond enthalpy
- ☐ thermochemical group

2.9 The temperature dependence of reaction enthalpies
- ☐ Kirchhoff's law for the temperature dependence of reaction enthalpies

Further reading

Articles of general interest

E.A. Gislason and N.C. Craig, General definitions of work and heat in thermodynamic processes. *J. Chem. Educ.* **64**, 670 (1987).

E.R. Boyko and J.F. Belliveau, Simplification of some thermochemical equations. *J. Chem. Educ.* **67**, 743 (1990).

T. Solomon, Standard enthalpies of formation of ions in solution. *J. Chem. Educ.* **68**, 41 (1991).

R.S. Treptow, Bond energies and enthalpies: an often neglected difference. *J. Chem. Educ.* **72**, 497 (1995).

J. Güèmez, S. Velasco, and M.A. Matías, Thermal coefficients and heat capacities in systems with chemical reaction: the Le Chatelier–Braun principle. *J. Chem. Educ.* **72**, 199 (1995).

R.D. Freeman, Conversion of standard (1 atm) thermodynamic data to the new standard-state pressure, 1 bar (10^5 Pa). *Bull. Chem. Thermodynamics* **25**, 523 (1982).

I.H. Segel and L.D. Segel, Energetics of biological processes. In *Encyclopedia of applied physics* (ed. G.L. Trigg) **6**, 207. VCH, New York (1993).

Texts and sources of data and information

B.W. Rossiter and R.C. Baetzgold (ed.), *Physical methods of chemistry*, **VI**. Wiley–Interscience, New York (1992).

M.L. McGlashan, *Chemical thermodynamics*. Academic Press, London (1979).

W.E. Dasent, *Inorganic energetics*. Cambridge University Press (1982).

D.A. Johnson, *Some thermodynamic aspects of inorganic chemistry*. Cambridge University Press (1982).

S.W. Benson, *Thermochemical kinetics*. Wiley, New York (1976).

J.D. Cox and G. Pilcher, *Thermochemistry of organic and organometallic compounds*. Academic Press, New York (1970).

D.R Stull, E.F. Westrum, and G.C. Sinke, *The chemical thermodynamics of organic compounds*. Wiley–Interscience, New York (1969).

D.B. Wagman, W.H. Evans, V.B. Parker, R.H. Schumm, I. Halow, S.M. Bailey, K.L. Churney, and R.L. Nuttall, *The NBS tables of chemical thermodynamic properties*. Published as *J. Phys. Chem. Reference Data* **11** (Supplement 2), 1982.

M.W. Chase, Jr, C.A. Davies, J.R. Downey, Jr, D.J. Frurip, R.A. McDonald, and A.N. Syverud, *JANAF thermochemical tables. The NBS tables of chemical thermodynamic properties*. Published as *J. Phys. Chem. Reference Data* **14** (Supplement 1), 1985.

J.B. Pedley, R.D. Naylor, and S.P. Kirby, *Thermochemical data of organic compounds*. Chapman & Hall, London (1986).

J.A. Dean (ed.), *Handbook of organic chemistry*. McGraw–Hill, New York (1987).

R.C. Weast (ed.), *Handbook of chemistry and physics*, Vol. 78. CRC Press, Boca Raton (1997).

A.M. James and M.P. Lord (ed.), *Macmillan's chemical and physical data*. Macmillan, London (1992).

J. Emsley, *The elements*. Oxford University Press (1996).

E.F.G. Harrington, Thermodynamic data and their uses. *Chemical thermodynamics: Specialist Periodical Reports*. **1**, 31 (1973).

D.R. Stull and G.C. Sinke, *Thermodynamic properties of the elements*. American Chemical Society, Washington (1956).

CODATA Bulletin 28: Recommended key values of thermodynamics. *J. Chem. Thermo.* **10**, 903 (1978).

Exercises

Assume all gases are perfect unless stated otherwise. Note that 1 bar = 1.01325 atm exactly. Unless otherwise stated, thermochemical data are for 298.15 K.

2.1 (a) Calculate the work done to raise a mass of 1.0 kg through 10 m on the surface of (a) the Earth ($g = 9.81$ m s^{-2}) and (b) the moon ($g = 1.60$ m s^{-2}).

2.1 (b) Calculate the work done to raise a mass of 5.0 kg through 100 m on the surface of (a) the Earth ($g = 9.81$ m s^{-2}) and (b) Mars ($g = 3.73$ m s^{-2}).

2.2 (a) Calculate the work needed for a 65 kg person to climb through 4.0 m on the surface of the Earth.

2.2 (b) Calculate the work needed for a bird of mass 120 g to fly to a height of 50 m from the surface of the Earth.

2.3 (a) A chemical reaction takes place in a container of cross-sectional area 100 cm^2. As a result of the reaction, a piston is pushed out through 10 cm against an external pressure of 1.0 atm. Calculate the work done by the system.

2.3 (b) A chemical reaction takes place in a container of cross-sectional area 50.0 cm^2. As a result of the reaction, a piston is pushed out through 15 cm against an external pressure of 121 kPa. Calculate the work done by the system.

2.4 (a) A sample consisting of 1.00 mol Ar is expanded isothermally at 0°C from 22.4 L to 44.8 L (a) reversibly, (b) against a constant external pressure equal to the final pressure of the gas, and (c) freely (against zero external pressure). For the three processes calculate q, w, ΔU, and ΔH.

2.4 (b) A sample consisting of 2.00 mol He is expanded isothermally at 22°C from 22.8 L to 31.7 L (a) reversibly, (b) against a constant external pressure equal to the final pressure of the gas, and (c) freely (against zero external pressure). For the three processes calculate q, w, ΔU, and ΔH.

2.5 (a) A sample consisting of 1.00 mol of monatomic perfect gas, for which $C_{V,m} = \frac{3}{2}R$, initially at $p_1 = 1.00$ atm and $T_1 = 300$ K, is heated reversibly to 400 K at constant volume. Calculate the final pressure, ΔU, q, and w.

2.5 (b) A sample consisting of 2.00 mol of a perfect gas, for which $C_{V,m} = \frac{5}{2}R$, initially at $p_1 = 111$ kPa and $T_1 = 277$ K, is heated reversibly to 356 K at constant volume. Calculate the final pressure, ΔU, q, and w.

2.6 (a) A sample of 4.50 g of methane gas occupies 12.7 L at 310 K. (a) Calculate the work done when the gas expands isothermally against a constant external pressure of 200 Torr until its volume has increased by 3.3 L. (b) Calculate the work that would be done if the same expansion occurred reversibly.

2.6 (b) A sample of argon of mass 6.56 g occupies 18.5 L at 305 K. (a) Calculate the work done when the gas expands isothermally against a constant external pressure of 7.7 kPa until its volume has increased by 2.5 L. (b) Calculate the work that would be done if the same expansion occurred reversibly.

2.7 (a) In the isothermal reversible compression of 52.0 mmol of a perfect gas at 260 K, the volume of the gas is reduced to one-third its initial value. Calculate w for this process.

2.7 (b) In the isothermal reversible compression of 1.77 mmol of a perfect gas at 273 K, the volume of the gas is reduced to 0.224 of its initial value. Calculate w for this process.

2.8 (a) A sample of 1.00 mol $H_2O(g)$ is condensed isothermally and reversibly to liquid water at 100°C. The standard enthalpy of vaporization of water at 100°C is 40.656 kJ mol^{-1}. Find w, q, ΔU, and ΔH for this process.

2.8 (b) A sample of 2.00 mol $CH_3OH(g)$ is condensed isothermally and reversibly to liquid at 64°C. The standard enthalpy of vaporization of methanol at 64°C is 35.3 kJ mol^{-1}. Find w, q, ΔU, and ΔH for this process.

2.9 (a) A strip of magnesium of mass 15 g is dropped into a beaker of dilute hydrochloric acid. Calculate the work done by the system as a result of the reaction. The atmospheric pressure is 1.0 atm and the temperature 25°C.

2.9 (b) A piece of zinc of mass 5.0 g is dropped into a beaker of dilute hydrochloric acid. Calculate the work done by the system as a result of the reaction. The atmospheric pressure is 1.1 atm and the temperature 23°C.

2.10 (a) Calculate the heat required to melt 750 kg of sodium metal at 371 K.

2.10 (b) Calculate the heat required to melt 500 kg of potassium metal at 336 K. The enthalpy of fusion of potassium is 2.40 kJ mol^{-1}.

2.11 (a) The value of $C_{p,m}$ for a sample of a perfect gas was found to vary with temperature according to the expression $C_{p,m}/(\text{J K}^{-1}) = 20.17 + 0.3665(T/\text{K})$. Calculate q, w, ΔU, and ΔH for 1.00 mol when the temperature of 1.00 mol of gas is raised from 25°C to 200°C (a) at constant pressure, (b) at constant volume.

2.11 (b) The constant-pressure heat capacity of a sample of a perfect gas was found to vary with temperature according to the expression $C_{p,m}/(\text{J K}^{-1}) = 20.17 + 0.4001(T/\text{K})$. Calculate q, w, ΔU, and ΔH for 1.00 mol when the temperature of 1.00 mol of gas is raised from 0°C to 100°C (a) at constant pressure, (b) at constant volume.

2.12 (a) Calculate the final temperature of a sample of argon of mass 12.0 g that is expanded reversibly and adiabatically from 1.0 L at 273.15 K to 3.0 L.

2.12 (b) Calculate the final temperature of a sample of carbon dioxide of mass 16.0 g that is expanded reversibly and adiabatically from 500 mL at 298.15 K to 2.00 L.

2.13 (a) A sample of carbon dioxide of mass 2.45 g at 27.0°C is allowed to expand reversibly and adiabatically from 500 mL to 3.00 L. What is the work done by the gas?

2.13 (b) A sample of nitrogen of mass 3.12 g at 23.0°C is allowed to expand reversibly and adiabatically from 400 mL to 2.00 L. What is the work done by the gas?

2.14 (a) Calculate the final pressure of a sample of carbon dioxide that expands reversibly and adiabatically from 57.4 kPa and 1.0 L to a final volume of 2.0 L. Take $\gamma = 1.4$.

2.14 (b) Calculate the final pressure of a sample of water vapour that expands reversibly and adiabatically from 87.3 Torr and 500 mL to a final volume of 3.0 L. Take $\gamma = 1.3$.

2.15 (a) Calculate the final pressure of a sample of carbon dioxide of mass 2.4 g that expands reversibly and adiabatically from an initial temperature of 278 K and volume 1.0 L to a final volume of 2.0 L. Take $\gamma = 1.4$.

2.15 (b) Calculate the final pressure of a sample of water vapour of mass 1.4 g that expands reversibly and adiabatically from an initial temperature of 300 K and volume 1.0 L to a final volume of 3.0 L. Take $\gamma = 1.3$.

2.16 (a) Calculate the standard enthalpy of formation of butane at 25°C from its standard enthalpy of combustion.

2.16 (b) Calculate the standard enthalpy of formation of hexane at 25°C from its standard enthalpy of combustion.

2.17 (a) When 229 J of energy is supplied as heat at constant pressure to 3.0 mol Ar(g), the temperature of the sample increases by 2.55 K. Calculate the molar heat capacities at constant volume and constant pressure of the gas.

2.17 (b) When 178 J of energy is supplied as heat at constant pressure to 1.9 mol of a gas, the temperature of the sample increases by 1.78 K. Calculate the molar heat capacities at constant volume and constant pressure of the gas.

2.18 (a) A sample of a liquid of mass 25 g is cooled from 290 K to 275 K at constant pressure by the extraction of 1.2 kJ of energy as heat. Calculate q and ΔH and estimate the heat capacity of the sample.

2.18 (b) A sample of a liquid of mass 30.5 g is cooled from 288 K to 275 K at constant pressure by the extraction of 2.3 kJ of energy as heat. Calculate q and ΔH and estimate the heat capacity of the sample.

2.19 (a) When 3.0 mol O_2 is heated at a constant pressure of 3.25 atm, its temperature increases from 260 K to 285 K. Given that the molar heat capacity of O_2 at constant pressure is 29.4 J K^{-1} mol^{-1}, calculate q, ΔH, and ΔU.

2.19 (b) When 2.0 mol CO_2 is heated at a constant pressure of 1.25 atm, its temperature increases from 250 K to 277 K. Given that the molar heat capacity of CO_2 at constant pressure is 37.11 J K^{-1} mol^{-1}, calculate q, ΔH, and ΔU.

2.20 (a) A sample of 4.0 mol O_2 is originally confined in 20 L at 270 K and then undergoes adiabatic expansion against a constant pressure of 600 Torr until the volume has increased by a factor of 3.0. Calculate q, w, ΔT, ΔU, and ΔH. (The final pressure of the gas is not necessarily 600 Torr.)

2.20 (b) A sample of 5.0 mol CO_2 is originally confined in 15 L at 280 K and then undergoes adiabatic expansion against a constant pressure of 78.5 kPa until the volume has increased by a factor of 4.0. Calculate q, w, ΔT, ΔU, and ΔH. (The final pressure of the gas is not necessarily 78.5 kPa.)

2.21 (a) A sample consisting of 3.0 mol of perfect gas at 200 K and 2.00 atm is compressed reversibly and adiabatically until the temperature reaches 250 K. Given that its molar constant-volume heat capacity is 27.5 J K^{-1} mol^{-1}, calculate q, w, ΔU, ΔH, and the final pressure and volume.

2.21 (b) A sample consisting of 2.5 mol of perfect gas at 220 K and 200 kPa is compressed reversibly and adiabatically until the temperature reaches 255 K. Given that its molar constant-volume heat capacity is 27.6 J K^{-1} mol^{-1}, calculate q, w, ΔU, ΔH, and the final pressure and volume.

2.22 (a) A sample consisting of 1.0 mol of perfect gas with $C_V = 20.8$ J K^{-1} is initially at 3.25 atm and 310 K. It undergoes reversible adiabatic expansion until its pressure reaches 2.50 atm. Calculate the final volume and temperature and the work done.

2.22 (b) A sample consisting of 1.5 mol of perfect gas with $C_{p,m} = 20.8$ J K^{-1} mol^{-1} is initially at 230 kPa and 315 K. It undergoes reversible adiabatic expansion until its pressure reaches 170 kPa. Calculate the final volume and temperature and the work done.

2.23 (a) Estimate the change in volume that occurs when a sample of mercury of volume 1.0 cm^3 is heated through 5.0 K at room temperature.

2.23 (b) Estimate the change in volume that occurs when a sample of iron of volume 5.0 cm^3 is heated through 10.0 K at room temperature.

2.24 (a) Consider a system consisting of 2.0 mol CO_2 (assumed to be a perfect gas) at 25°C confined to a cylinder of cross-section 10 cm^2 at 10 atm. The gas is allowed to expand adiabatically and irreversibly against a constant pressure of 1.0 atm. Calculate w, q, ΔU, ΔH, and ΔT when the piston has moved 20 cm.

2.24 (b) Consider a system consisting of 3.0 mol O_2 (assumed to be a perfect gas) at 25°C confined to a cylinder of cross-section 22 cm^2 at 820 kPa. The gas is allowed to expand adiabatically and irreversibly against a constant pressure of 110 kPa. Calculate w, q, ΔU, ΔH, and ΔT when the piston has moved 15 cm.

2.25 (a) A sample consisting of 65.0 g of xenon is confined in a container at 2.00 atm and 298 K and then allowed to expand adiabatically (a) reversibly to 1.00 atm, (b) against a constant pressure of 1.00 atm. Calculate the final temperature in each case.

2.25 (b) A sample consisting of 15.0 g of nitrogen is confined in a container at 220 kPa and 200 K and then allowed to expand adiabatically (a) reversibly to 110 kPa, (b) against a constant pressure of 110 kPa. Calculate the final temperature in each case.

2.26 (a) A certain liquid has $\Delta_{vap}H^\ominus = 26.0$ kJ mol^{-1}. Calculate q, w, ΔH, and ΔU when 0.50 mol is vaporized at 250 K and 750 Torr.

2.26 (b) A certain liquid has $\Delta_{vap}H^\ominus = 32.0$ kJ mol^{-1}. Calculate q, w, ΔH, and ΔU when 0.75 mol is vaporized at 260 K and 765 Torr.

2.27 (a) The standard enthalpy of formation of ethylbenzene is -12.5 kJ mol^{-1}. Calculate its standard enthalpy of combustion.

2.27 (b) The standard enthalpy of formation of phenol is -165.0 kJ mol^{-1}. Calculate its standard enthalpy of combustion.

2.28 (a) Calculate the standard enthalpy of hydrogenation of 1-hexene to hexane given that the standard enthalpy of combustion of 1-hexene is $-4003 \, kJ \, mol^{-1}$.

2.28 (b) Calculate the standard enthalpy of hydrogenation of 1-butene to butane given that the standard enthalpy of combustion of 1-butene is $-2717 \, kJ \, mol^{-1}$.

2.29 (a) The standard enthalpy of combustion of cyclopropane is $-2091 \, kJ \, mol^{-1}$ at 25°C. From this information and enthalpy of formation data for $CO_2(g)$ and $H_2O(g)$, calculate the enthalpy of formation of cyclopropane. The enthalpy of formation of propene is $+20.42 \, kJ \, mol^{-1}$. Calculate the enthalpy of isomerization of cyclopropane to propene.

2.29 (b) From the following data, determine $\Delta_f H^{\ominus}$ for diborane, $B_2H_6(g)$, at 298 K:

(1) $B_2H_6(g) + 3O_2(g) \longrightarrow B_2O_3(s) + 3H_2O(g)$
$$\Delta_r H^{\ominus} = -1941 \, kJ \, mol^{-1}$$

(2) $2B(s) + \frac{3}{2}O_2(g) \longrightarrow B_2O_3(s) \qquad \Delta_r H^{\ominus} = -2368 \, kJ \, mol^{-1}$

(3) $H_2(g) + \frac{1}{2}O_2(g) \longrightarrow H_2O(g) \qquad \Delta_r H^{\ominus} = -241.8 \, kJ \, mol^{-1}$

2.30 (a) Calculate the standard internal energy of formation of liquid methyl acetate from its standard enthalpy of formation, which is $-442 \, kJ \, mol^{-1}$.

2.30 (b) Calculate the standard internal energy of formation of urea from its standard enthalpy of formation, which is $-333.51 \, kJ \, mol^{-1}$.

2.31 (a) The temperature of a bomb calorimeter rose by 1.617 K when a current of 3.20 A was passed for 27.0 s from a 12.0 V source. Calculate the calorimeter constant.

2.31 (b) The temperature of a bomb calorimeter rose by 1.712 K when a current of 2.86 A was passed for 22.5 s from a 12.0 V source. Calculate the calorimeter constant.

2.32 (a) When 120 mg of naphthalene, $C_{10}H_8(s)$, was burned in a bomb calorimeter the temperature rose by 3.05 K. Calculate the calorimeter constant. By how much will the temperature rise when 100 mg of phenol, $C_6H_5OH(s)$, is burned in the calorimeter under the same conditions?

2.32 (b) When 2.25 mg of anthracene, $C_{14}H_{10}(s)$, was burned in a bomb calorimeter the temperature rose by 1.35 K. Calculate the calorimeter constant. By how much will the temperature rise when 135 mg of phenol, $C_6H_5OH(s)$, is burned in the calorimeter under the same conditions?

2.33 (a) When 0.3212 g of glucose was burned in a bomb calorimeter of calorimeter constant $641 \, J \, K^{-1}$ the temperature rose by 7.793 K. Calculate (a) the standard molar enthalpy of combustion, (b) the standard internal energy of combustion, and (c) the standard enthalpy of formation of glucose.

2.33 (b) When 0.2715 g of fructose was burned in a bomb calorimeter of calorimeter constant $437 \, J \, K^{-1}$ the temperature rose by 9.69 K. Calculate (a) the standard molar enthalpy of combustion, (b) the standard internal energy of combustion, and (c) the standard enthalpy of formation of fructose.

2.34 (a) Calculate the standard enthalpy of solution of AgCl(s) in water from the enthalpies of formation of the solid and the aqueous ions.

2.34 (b) Calculate the standard enthalpy of solution of AgBr(s) in water from the enthalpies of formation of the solid and the aqueous ions.

2.35 (a) The standard enthalpy of decomposition of the yellow complex H_3NSO_2 into NH_3 and SO_2 is $+40 \, kJ \, mol^{-1}$. Calculate the standard enthalpy of formation of H_3NSO_2.

2.35 (b) Given that the standard enthalpy of combustion of graphite is $-393.51 \, kJ \, mol^{-1}$ and that of diamond is $-395.41 \, kJ \, mol^{-1}$, calculate the enthalpy of the graphite $\longrightarrow$ diamond transition.

2.36 (a) The mass of a typical (sucrose) sugar cube is 1.5 g. Calculate the energy released as heat when a cube is burned in air. To what height could a 65 kg person climb on the energy a cube provides assuming 25 per cent of the energy is available for work?

2.36 (b) The mass of a typical glucose tablet is 2.5 g. Calculate the energy released as heat when a tablet is burned in air. To what height could a 65 kg person climb on the energy a cube provides assuming 25 per cent of the energy is available for work?

2.37 (a) The standard enthalpy of combustion of propane gas is $-2220 \, kJ \, mol^{-1}$ and the standard enthalpy of vaporization of propane liquid is $+15 \, kJ \, mol^{-1}$. Calculate (a) the standard enthalpy and (b) the standard internal energy of combustion of liquid propane.

2.37 (b) The standard enthalpy of combustion of butane gas is $-2878 \, kJ \, mol^{-1}$ and the standard enthalpy of vaporization of butane liquid is $+21.0 \, kJ \, mol^{-1}$. Calculate (a) the standard enthalpy and (b) the standard internal energy of combustion of liquid butane.

2.38 (a) Express the following reactions in the form $0 = \sum_J \nu_J J$, identify the stoichiometric numbers, and classify the reactions as exothermic or endothermic.

(a) $CH_4(g) + 2O_2(g) \longrightarrow CO_2(g) + 2H_2O(l)$
$$\Delta_r H^{\ominus} = -890 \, kJ \, mol^{-1}$$
(b) $2C(s) + H_2(g) \longrightarrow C_2H_2(g) \qquad \Delta_r H^{\ominus} = +227 \, kJ \, mol^{-1}$
(c) $NaCl(s) \longrightarrow NaCl(aq) \qquad \Delta_r H^{\ominus} = +3.9 \, kJ \, mol^{-1}$

2.38 (b) Express the following reactions in the form $0 = \sum_J \nu_J J$, identify the stoichiometric numbers, and classify the reactions as exothermic or endothermic.

(a) $C(s, diamond) \longrightarrow C(s, graphite) \qquad \Delta_r H^{\ominus} = -1.9 \, kJ \, mol^{-1}$
(b) $Fe_3O_4(s) + CO(g) \longrightarrow 3FeO(s) + CO_2(g)$
$$\Delta_r H^{\ominus} = +35.9 \, kJ \, mol^{-1}$$
(c) $3FeO(s) + CO_2(g) \longrightarrow Fe_3O_4(s) + CO(g)$
$$\Delta_r H^{\ominus} = -35.9 \, kJ \, mol^{-1}$$

2.39 (a) Use standard enthalpies of formation to calculate the standard enthalpies of the following reactions:

(a) $2NO_2(g) \longrightarrow N_2O_4(g)$
(b) $NH_3(g) + HCl(g) \longrightarrow NH_4Cl(s)$

2.39 (b) Use standard enthalpies of formation to calculate the standard enthalpies of the following reactions:

(a) Cyclopropane(g) $\longrightarrow$ propene(g)
(b) $HCl(aq) + NaOH(aq) \longrightarrow NaCl(aq) + H_2O(l)$

2.40 (a) Given the reactions (1) and (2) below, determine (a) $\Delta_r H^{\ominus}$ and $\Delta_r U^{\ominus}$ for reaction (3), (b) $\Delta_f H^{\ominus}$ for both HCl(g) and $H_2O(g)$ all at 298 K. Assume all gases are perfect.

(1) $H_2(g) + Cl_2(g) \longrightarrow 2HCl(g)$ $\Delta_r H^\ominus = -184.62 \text{ kJ mol}^{-1}$
(2) $2H_2(g) + O_2(g) \longrightarrow 2H_2O(g)$ $\Delta_r H^\ominus = -483.64 \text{ kJ mol}^{-1}$
(3) $4HCl(g) + O_2(g) \longrightarrow Cl_2(g) + 2H_2O(g)$

2.40 (b) Given the reactions (1) and (2) below, determine (a) $\Delta_r H^\ominus$ and $\Delta_r U^\ominus$ for reaction (3), (b) $\Delta_f H^\ominus$ for both HI(g) and H_2O(g) all at 298 K. Assume all gases are perfect.

(1) $H_2(g) + I_2(s) \longrightarrow 2HI(g)$ $\Delta_r H^\ominus = +52.96 \text{ kJ mol}^{-1}$
(2) $2H_2(g) + O_2(g) \longrightarrow 2H_2O(g)$ $\Delta_r H^\ominus = -483.64 \text{ kJ mol}^{-1}$
(3) $4HI(g) + O_2(g) \longrightarrow I_2(s) + 2H_2O(g)$

2.41 (a) For the reaction $C_2H_5OH(l) + 3O_2(g) \rightarrow 2CO_2(g) + 3H_2O(g)$, $\Delta_r U^\ominus = -1373 \text{ kJ mol}^{-1}$ at 298 K. Calculate $\Delta_r H^\ominus$.

2.41 (b) For the reaction $2C_6H_5COOH(s) + 13O_2(g) \rightarrow 12CO_2(g) + 6H_2O(g)$, $\Delta_r U^\ominus = -772.7 \text{ kJ mol}^{-1}$ at 298 K. Calculate $\Delta_r H^\ominus$.

2.42 (a) Calculate the standard enthalpies of formation of (a) $KClO_3(s)$ from the enthalpy of formation of KCl, (b) $NaHCO_3(s)$ from the enthalpies of formation of CO_2 and NaOH together with the following information:

$2KClO_3(s) \longrightarrow 2KCl(s) + 3O_2(g)$ $\Delta_r H^\ominus = -89.4 \text{ kJ mol}^{-1}$
$NaOH(s) + CO_2(g) \longrightarrow NaCO_3(s)$ $\Delta_r H^\ominus = -127.5 \text{ kJ mol}^{-1}$

2.42 (b) Calculate the standard enthalpy of formation of NOCl(g) from the enthalpy of formation of NO given in Table 2.5, together with the following information:

$2NOCl(g) \longrightarrow 2NO(g) + Cl_2(g)$ $\Delta_r H^\ominus = +75.5 \text{ kJ mol}^{-1}$

2.43 (a) Use the information in Table 2.5 to predict the standard reaction enthalpy of $2NO_2(g) \rightarrow N_2O_4(g)$ at 100°C from its value at 25°C.

2.43 (b) Use the information in Table 2.5 to predict the standard reaction enthalpy of $2H_2(g) + O_2(g) \rightarrow 2H_2O(l)$ at 100°C from its value at 25°C.

2.44 (a) From the data in Table 2.5, calculate $\Delta_r H^\ominus$ and $\Delta_r U^\ominus$ at (a) 298 K, (b) 378 K for the reaction $C(\text{graphite}) + H_2O(g) \rightarrow CO(g) + H_2(g)$. Assume all heat capacities to be constant over the temperature range of interest.

2.44 (b) Calculate $\Delta_r H^\ominus$ and $\Delta_r U^\ominus$ at 298 K and $\Delta_r H^\ominus$ at 348 K for the hydrogenation of ethyne (acetylene) to ethene (ethylene) from the enthalpy of combustion and heat capacity data in Tables 2.5 and 2.6. Assume the heat capacities to be constant over the temperature range involved.

2.45 (a) Set up a thermodynamic cycle for determining the enthalpy of hydration of Mg^{2+} ions using the following data: enthalpy of sublimation of Mg(s), $+167.2 \text{ kJ mol}^{-1}$; first and second ionization enthalpies of Mg(g), 7.646 eV and 15.035 eV; dissociation enthalpy of $Cl_2(g)$, $+241.6 \text{ kJ mol}^{-1}$; electron-gain enthalpy of Cl(g), -3.78 eV; enthalpy of solution of $MgCl_2(s)$, $-150.5 \text{ kJ mol}^{-1}$; enthalpy of hydration of $Cl^-(g)$, $-383.7 \text{ kJ mol}^{-1}$.

2.45 (b) Set up a thermodynamic cycle for determining the enthalpy of hydration of Ca^{2+} ions using the following data: enthalpy of sublimation of Ca(s), $+178.2 \text{ kJ mol}^{-1}$; first and second ionization enthalpies of Ca(g), $589.7 \text{ kJ mol}^{-1}$ and 1145 kJ mol^{-1}; enthalpy of vaporization of bromine, $+30.91 \text{ kJ mol}^{-1}$; dissociation enthalpy of $Br_2(g)$, $+192.9 \text{ kJ mol}^{-1}$; electron-gain enthalpy of Br(g), $-331.0 \text{ kJ mol}^{-1}$; enthalpy of solution of $CaBr_2(s)$, $-103.1 \text{ kJ mol}^{-1}$; enthalpy of hydration of $Br^-(g)$, -337 kJ mol^{-1}.

2.46 (a) Use the thermochemical groups in Table 2.7 to estimate the standard enthalpy of formation in the gas phase of (a) cyclohexane, (b) 2,4-dimethylhexane.

2.46 (b) Use the thermochemical groups in Table 2.7 to estimate the standard enthalpy of formation of gaseous (a) 2,2,4-trimethylpentane, (b) 2,2-dimethylpropane.

Problems

Assume all gases are perfect unless stated otherwise. Note that 1 atm = 1.013 25 bar. Unless otherwise stated, thermochemical data are for 298.15 K.

Numerical problems

2.1 Calculate the heat needed to raise the temperature of the air in a house from 20°C to 25°C. Assume that the house contains 600 m^3 of air, which should be taken to be a perfect diatomic gas. The density of air is 1.21 kg m^{-3} at 20°C. Calculate ΔU and ΔH for the heating of the air.

2.2 An average human produces about 10 MJ of heat each day through metabolic activity. If a human body were an isolated system of mass 65 kg with the heat capacity of water, what temperature rise would the body experience? Human bodies are actually open systems, and the main mechanism of heat loss is through the evaporation of water. What mass of water should be evaporated each day to maintain constant temperature?

2.3 Consider a perfect gas contained in a cylinder and separated by a frictionless adiabatic piston into two sections, A and B; Section B is in contact with a water bath that maintains it at constant temperature. Initially $T_A = T_B = 300$ K, $V_A = V_B = 2.00$ L, and $n_A = n_B = 2.00$ mol. Heat is supplied to Section A and the piston moves to the right reversibly until the final volume of Section B is 1.00 L. Calculate (a) the work done by the gas in Section A, (b) ΔU for the gas in Section B, (c) q for the gas in B, (d) ΔU for the gas in A, and (e) q for the gas in A. Assume $C_{V,m} = 20.0 \text{ J K}^{-1} \text{ mol}^{-1}$.

2.4 A sample consisting of 1 mol of a monatomic perfect gas (for which $C_{V,m} = \frac{3}{2}R$) is taken through the cycle shown in Fig. 2.20. (a) Determine the temperature at 1,2, and 3. (b) Calculate q, w, ΔU, and ΔH for each step and for the overall cycle. If a numerical answer cannot be obtained from the information given, then write in +,-,0, or ? as appropriate.

2.5 A 5.0 g block of solid carbon dioxide is allowed to evaporate in a vessel of volume 100 cm^3 maintained at 25°C. Calculate the work done when the system expands (a) isothermally against a pressure of

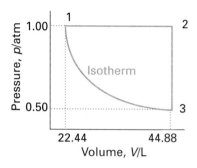

Fig. 2.20

1.0 atm, and (b) isothermally and reversibly to the same volume as in (a).

2.6 A sample consisting of 1.0 mol $CaCO_3(s)$ was heated to 800°C, when it decomposed. The heating was carried out in a container fitted with a piston which was initially resting on the solid. Calculate the work done during complete decomposition at 1.0 atm. What work would be done if instead of having a piston the container were open to the atmosphere?

2.7 A new fluorocarbon of molar mass 102 g mol⁻¹ was placed in an electrically heated vessel. When the pressure was 650 Torr, the liquid boiled at 78°C. After the boiling point had been reached, it was found that a current of 0.232 A from a 12.0 V supply passed for 650 s vaporized 1.871 g of the sample. Calculate the (molar) enthalpy and internal energy of vaporization.

2.8 An object is cooled by the evaporation of liquid methane at its normal boiling point (112 K). What volume of gas at 1.00 atm pressure must be formed from the liquid in order to remove 32.5 kJ of energy as heat from the object?

2.9 The molar heat capacity of ethane is represented in the temperature range 298 K to 400 K by the empirical expression $C_{p,m}/(J\,K^{-1}\,mol^{-1}) = 14.73 + 0.1272(T/K)$. The corresponding expressions for C(s) and $H_2(g)$ are given in Table 2.2. Calculate the standard enthalpy of formation of ethane at 350 K from its value at 298 K.

2.10 A sample of the sugar D-ribose ($C_5H_{10}O_5$) of mass 0.727 g was placed in a calorimeter and then ignited in the presence of excess oxygen. The temperature rose by 0.910 K. In a separate experiment in the same calorimeter, the combustion of 0.825 g of benzoic acid, for which the internal energy of combustion is −3251 kJ mol⁻¹, gave a temperature rise of 1.940 K. Calculate the internal energy of combustion of D-ribose and its enthalpy of formation.

2.11 The standard enthalpy of formation of the metallocene bis-(benzene)chromium was measured in a calorimeter. It was found for the reaction $Cr(C_6H_6)_2(s) \rightarrow Cr(s) + 2C_6H_6(g)$ that $\Delta_r U^\ominus(583\ K) = +8.0\ kJ\,mol^{-1}$. Find the corresponding reaction enthalpy and estimate the standard enthalpy of formation of the compound at 583 K. The constant-pressure molar heat capacity of benzene is 140 J K⁻¹ mol⁻¹ in its liquid range and 28 J K⁻¹ mol⁻¹ as a gas.

2.12 The standard enthalpy of combustion of sucrose is −5645 kJ mol⁻¹. What is the advantage (in kilojoules per mole of energy released as heat) of complete aerobic oxidation compared with anaerobic hydrolysis of sucrose to lactic acid?

Theoretical problems

2.13 With reference to Fig. 2.21, and assuming perfect gas behaviour, calculate: (a) the amount of gas molecules (in moles) in this system and its volume in states B and C, (b) the work done on the gas along the paths ACB and ADB, (c) the work done on the gas along the isotherm AB, (d) q and ΔU for each of the three paths. Take $C_{V,m} = \frac{3}{2}R$ and $T = 313$ K.

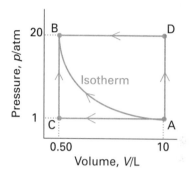

Fig. 2.21

2.14 When a system is taken from state A to state B along the path ACB in Fig. 2.22, 80 J of heat flows into the system and the system does 30 J of work. (a) How much heat flows into the system along path ADB if the work done is 10 J? (b) When the system is returned from state B to A along the curved path, the work done on the system is 20 J. Does the system absorb or liberate heat, and how much? (c) If $U_D - U_A = +40$ J, find the heat absorbed in the processes AD and DB.

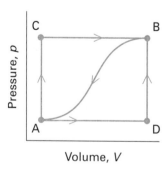

Fig. 2.22

2.15 Show that the value of ΔH for the adiabatic expansion of a perfect gas may be calculated by integration of $dH = V\,dp$, and evaluate the integral for reversible adiabatic expansion.

2.16 Express the work of isothermal reversible expansion of a van der Waals gas in reduced variables and find a definition of reduced work that makes the overall expression independent of the identity of the gas. Calculate the work of isothermal reversible expansion along the critical isotherm from V_c to xV_c.

Additional problems supplied by Carmen Giunta and Charles Trapp

2.17 Since their discovery in 1985, fullerenes have received the attention of many chemical researchers. Kolesov *et al.* have recently reported the standard enthalpy of combustion and of formation of crystalline C_{60} based on calorimetric measurements (V.P. Kolesov, S.M. Pimenova, V.K. Pavlovich, N.B. Tamm, and A.A. Kurskaya, *J. Chem. Thermodynamics* **28**, 1121 (1996)). In one of their runs, they found the standard specific internal energy of combustion to be $-36.0334 \text{ kJ g}^{-1}$ at 298.15 K. Compute $\Delta_c H^{\ominus}$ and $\Delta_f H^{\ominus}$ of C_{60}.

2.18 A thermodynamic study of $DyCl_3$ (E.H.P. Cordfunke, A.S Booji, and M. Yu. Furkaliouk, *J. Chem. Thermodynamics* **28**, 1387 (1996)) determined its standard enthalpy of formation from the following information

(1) $DyCl_3(s) \longrightarrow DyCl_3(aq, \text{in } 4.0 \text{ M HCl})$
$\Delta_r H^{\ominus} = -180.06 \text{ kJ mol}^{-1}$

(2-
) $Dy(s) + 3HCl(aq, 4.0 \text{ M}) \longrightarrow DyCl_3(aq, \text{in } 4.0 \text{ M HCl}) + \frac{3}{2}H_2(g)$
$\Delta_r H^{\ominus} = -699.43 \text{ kJ mol}^{-1}$

(3) $\frac{1}{2}H_2(g) + \frac{1}{2}Cl_2(g) \longrightarrow HCl(aq, 4.0 \text{ M})$
$\Delta_r H^{\ominus} = -158.31 \text{ kJ mol}^{-1}$

Determine $\Delta_f H^{\ominus}(DyCl_3, s)$ from these data.

2.19 Seakins *et al.* (P.W. Seakins, M.J. Pilling, J.T. Niiranen, D. Gutman, and L.N. Krasnoperov, *J. Phys. Chem.* **96**, 9847 (1992)) report $\Delta_f H^{\ominus}$ for a variety of alkyl radicals in the gas phase— information which is applicable to studies of pyrolysis and the oxidation reactions of hydrocarbons. This information can be combined with thermodynamic data on alkenes to determine the reaction enthalpy for possible fragmentation of a large alkyl radical into smaller radicals and alkenes. Use the following set of data to compute the standard reaction enthalpies for three possible fates of the *tert*-butyl radical, namely (a) *tert*-$C_4H_9 \longrightarrow$ *sec*-C_4H_9, (b) *tert*-$C_4H_9 \longrightarrow C_3H_6 + CH_3$, (c) *tert*-$C_4H_9 \longrightarrow C_2H_4 + C_2H_5$.

Species:	C_2H_5	*sec*-C_4H_9	*tert*-C_4H_9
$\Delta_f H^{\ominus}/(\text{kJ mol}^{-1})$	+121.0	+67.5	+51.3

2.20 Alkyl radicals are important intermediates in the combustion and atmospheric chemistry of hydrocarbons. N. Cohen has reported group additivity tables for the thermochemistry of alkyl radicals in the gas phase (N. Cohen, *J. Phys. Chem.* **96**, 9052 (1992)). He suggests computing enthalpies of formation based on the following bond-dissociation energies $(\Delta U^{\ominus})$ for C–H bonds: primary (–(H)C(H)–H), $420.5 \text{ kJ mol}^{-1}$; secondary (–(C)C(H)–H), $410.5 \text{ kJ mol}^{-1}$; tertiary (–(C)C(C)–H), $398.3 \text{ kJ mol}^{-1}$. Estimate $\Delta_f H^{\ominus}$ of (a) C_2H_5, (b) *sec*-C_4H_9, and (c) *tert*-C_4H_9. ($\Delta_f H^{\ominus}$(2-methylpropane, g) $= -134.2 \text{ kJ mol}^{-1}$.)

2.21 Silylene (SiH_2) is a key intermediate in the thermal decomposition of silicon hydrides such as silane (SiH_4) and disilane (Si_2H_6). Moffat *et al.* (H.K. Moffat, K.F. Jensen, and R.W. Carr, *J. Phys. Chem.* **95**, 145 (1991)) report $\Delta_f H^{\ominus}(SiH_2) = +274 \text{ kJ mol}^{-1}$. If $\Delta_f H^{\ominus}(SiH_4) = +34.3 \text{ kJ mol}^{-1}$ and $\Delta_f H^{\ominus}(Si_2H_6) = +80.3 \text{ kJ mol}^{-1}$ (*CRC Handbook* (1995)), compute the standard enthalpies of the following reactions:

(a) $SiH_4(g) \longrightarrow SiH_2(g) + H_2(g)$

(b) $Si_2H_6(g) \longrightarrow SiH_2(g) + SiH_4(g)$

2.22 Silanone (SiH_2O) and silanol (SiH_3OH) are species believed to be important in the oxidation of silane (SiH_4). These species are much more elusive than their carbon counterparts. Darling and Schlegel (C.L. Darling and H.B. Schlegel, *J. Phys. Chem.* **97**, 8207 (1993)) report the following values (converted from calories) from a computational study: $\Delta_f H^{\ominus}(SiH_2O) = -98.3 \text{ kJ mol}^{-1}$ and $\Delta_f H^{\ominus}(SiH_3OH) = -282 \text{ kJ mol}^{-1}$. Compute the standard enthalpies of the following reactions:

(a) $SiH_4(g) + \frac{1}{2}O_2(g) \longrightarrow SiH_3OH(g)$

(b) $SiH_4(g) + O_2(g) \longrightarrow SiH_2O(g) + H_2O(l)$

(c) $SiH_3OH(g) \longrightarrow SiH_2O(g) + H_2(g)$

Note that $\Delta_f H^{\ominus}(SiH_4, g) = +34.3 \text{ kJ mol}^{-1}$ (*CRC Handbook* (1995)).

2.23 *Polytropic processes* are defined as those that satisfy the condition $pV^n = C$, where C is a constant. In one experiment, 1.00 mol of 'air molecules' is compressed from 1.00 bar to 10.0 bar at 25°C by two different combinations of reversible polytropic processes: (1) heating at constant volume to the final pressure, followed by cooling at constant pressure, (2) adiabatic compression to the final volume, followed by cooling at constant volume. (a) Sketch these processes on a pV diagram and identify the value of n for each step in each process. (b) Calculate q, w, ΔU, and ΔH for each step in the processes and for the overall process. Note that the overall process can be accomplished in one isothermal reversible step. Assume air is a perfect diatomic gas with $C_{p,m} = \frac{7}{2}R$.

2.24 For reversible polytropic processes described by the general relation $pV^n = C$, derive the following expressions for work and heat

$$w = \frac{RT_1}{n-1}\left\{\left(\frac{p_2}{p_1}\right)^{(n-1)/n} - 1\right\}$$

$$q = \frac{(n-\gamma)RT_1}{(n-1)(\gamma-1)}\left\{\left(\frac{p_2}{p_1}\right)^{(n-1)/n} - 1\right\}$$

Show that these expressions reduce to already familiar expressions for $n = 0, 1, \gamma,$ and ∞.

2.25 From the enthalpy of combustion data in Table 2.5 for the alkanes methane through octane, test the extent to which the relation $\Delta_c H^{\ominus} = k(M/\text{g mol}^{-1})^n$ holds and find the numerical values for k and n. Predict $\Delta_c H^{\ominus}$ for decane and compare to the known value.

2.26 Ammonia is compressed in a piston-cylinder apparatus from an initial state of 30°C and 500 kPa to a final pressure of 1400 kPa. The following data were obtained during the process.

p/kPa	500	653	802	945	1100	1248	1400
V/L	1.25	1.08	0.96	0.84	0.72	0.60	0.50

(a) Is this a polytropic process? (See Problem 2.23 for the definition of a polytropic process.) If so, what is n? (b) Calculate the work done on the ammonia. (c) What is the final temperature?

3

The First Law:
the machinery

In this chapter we begin to unfold some of the power of thermodynamics by showing how to establish relations between different properties of a system. The procedure we use is based on the experimental fact that the internal energy and the enthalpy are state functions, and we derive a number of relations between observables by exploring the mathematical consequences of these facts. We shall see that one very useful aspect of thermodynamics is that a property can be measured indirectly by measuring others and then combining their values. The relations we derive also enable us to discuss the liquefaction of gases and to establish a quantitative relation between the heat capacities of a substance at constant pressure and constant volume.

We saw in Section 2.2 that properties that are independent of how a sample is prepared are called **state functions**. Such properties can be regarded as functions of variables, such as pressure and temperature, that define the current state of the system. The internal energy and enthalpy are examples of state functions, for they depend on the current state of the system and are independent of its previous history. Properties that relate to the preparation of the state are called **path functions**. Examples of path functions are the work that is done in preparing a state and the energy transferred as heat. We do not speak of a system in a particular state as possessing work or heat. In each case, the energy transferred as work or heat relates to the path being taken, not to the current state itself.

State functions and exact differentials

We can use the mathematical properties of state functions to draw far-reaching conclusions about the relations between physical properties and establish connections that may be completely unexpected. The practical importance of these results is that we can combine measurements of different properties to obtain the value of a property we require.

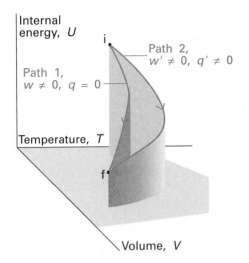

3.1 As the volume and temperature of a system are changed, the internal energy changes. An adiabatic and a non-adiabatic path are shown as Path 1 and Path 2, respectively: they correspond to different values of q and w but to the same value of ΔU.

3.1 State functions

Consider a system undergoing the changes depicted in Fig. 3.1. The initial state of the system is i and in this state the internal energy is U_i. Work is done by the system as it expands adiabatically to a state f. In this state the system has an internal energy U_f and the work done on the system as it changes along Path 1 from i to f is w. Notice our use of language: U is a property of the state; w is a property of the path. Now consider another process, Path 2, in which the initial and final states are the same but in which the expansion is not adiabatic. The internal energies of both the initial and the final states are the same as before (because U is a state function). However, in the second path an energy q' enters the system as heat and the work w' is not the same as w. The work and the heat are path functions. In terms of the mountaineering analogy introduced in Section 2.2, the change in altitude (a state function) is independent of the path, but the distance travelled (a path function) does depend on the path taken between the fixed endpoints.

(a) Exact and inexact differentials

If a system is taken along a path (for example, by heating it), U changes from U_i to U_f, and the overall change is the sum (integral) of all the infinitesimal changes along the path:

$$\Delta U = \int_i^f dU \tag{1}$$

The value of ΔU depends on the initial and final states of the system but is independent of the path between them. This path-independence of the integral is expressed by saying that dU is an **exact differential**. In general, an exact differential is an infinitesimal quantity which, when integrated, gives a result that is independent of the path between the initial and final states.

When a system is heated, the total energy transferred as heat is the sum of all individual contributions at each point of the path:

$$q = \int_{i, \, path}^f dq \tag{2}$$

Notice the difference between this equation and eqn 1. First, we do not write Δq, because q is not a state function and the energy supplied as heat cannot be expressed as $q_f - q_i$. Secondly, it is necessary to specify the path of integration because q depends on the path selected (for example, an adiabatic path has $q = 0$, whereas a nonadiabatic path between the same two states would have $q \neq 0$). This path-dependence is expressed by saying that dq is an **inexact differential**. In general, an inexact differential is an infinitesimal quantity that, when integrated, gives a result that depends on the path between the initial and final states. Often dq is written đq to emphasize that it is inexact.

The work done on a system to change it from one state to another depends on the path taken between the two specified states; for example, it is different if the change takes place adiabatically from if it takes place nonadiabatically. It follows that dw is an inexact differential. It is often written đw.

Example 3.1 Calculating work, heat, and internal energy

Consider a perfect gas inside a cylinder fitted with a piston. Let the initial state be T, V_i and the final state be T, V_f. The change of state can be brought about in many ways, of which the two simplest are the following: Path 1, in which there is free expansion against zero external pressure; Path 2, in which there is reversible, isothermal expansion. Calculate w, q, and ΔU for each process.

Method To find a starting point for a calculation in thermodynamics, it is often a good idea to go back to first principles, and to look for a way of expressing the quantity we are asked to calculate in terms of other quantities that are easier to calculate. Because the internal energy of a perfect gas arises only from the kinetic energy of its molecules, it is independent of volume; therefore, for any isothermal change, $\Delta U = 0$. We also know that in general $\Delta U = q + w$. The question depends on being able to combine the two expressions. We derived a number of expressions for the work done in a variety of processes in Chapter 2, and here we need to select the appropriate ones.

Answer Because $\Delta U = 0$ for both paths and $\Delta U = q + w$, in each case $q = -w$. The work of free expansion is zero (Section 2.3b), so in Path 1, $w = 0$ and $q = 0$. For Path 2, the work is given by eqn 2.13, so $w = -nRT \ln (V_f/V_i)$ and consequently $q = nRT \ln (V_f/V_i)$.

- -

Self-test 3.1 Calculate the values of q, w, and ΔU for an irreversible isothermal expansion of a perfect gas against a constant nonzero external pressure.

$$[q = p_{ex}\Delta V,\ w = -p_{ex}\Delta V,\ \Delta U = 0]$$

(b) Changes in internal energy

We shall now begin to unfold the consequences of dU being an exact differential by noting that, for a closed system of constant composition (the only type of system considered in this chapter), U is a function of volume and temperature.[1] When V changes to $V + dV$ at constant temperature, U changes to

$$U' = U + \left(\frac{\partial U}{\partial V}\right)_T dV$$

The coefficient $(\partial U/\partial V)_T$, the slope of a plot of U against V at constant temperature, is the partial derivative[2] of U with respect to V. If, instead, T changes to $T + dT$ at constant volume, the internal energy changes to

$$U' = U + \left(\frac{\partial U}{\partial T}\right)_V dT$$

Now suppose that V and T both change infinitesimally. The new internal energy, neglecting second-order infinitesimals (those proportional to $dVdT$), is

$$U' = U + \left(\frac{\partial U}{\partial V}\right)_T dV + \left(\frac{\partial U}{\partial T}\right)_V dT$$

As a result of the infinitesimal changes in conditions, the internal energy U' differs from U by the infinitesimal amount dU. Therefore, from the last equation we obtain the very important result that

$$dU = \left(\frac{\partial U}{\partial V}\right)_T dV + \left(\frac{\partial U}{\partial T}\right)_V dT \tag{3}$$

The interpretation of this equation is that, in a closed system of constant composition, any infinitesimal change in the internal energy is proportional to the infinitesimal changes of volume and temperature, the coefficients of proportionality being the partial derivatives.

1　U could be regarded as a function of V, T, and p but, because there is an equation of state, it is possible to express p in terms of V and T, so p is not an independent variable. We could have chosen p, T or p, V as independent variables, but V, T fit our purpose.

2　Partial derivatives are reviewed in *Further information 1*.

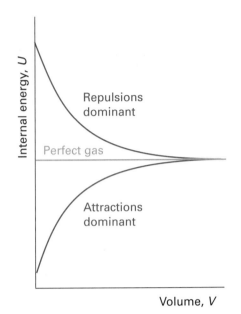

3.2 For a perfect gas, the internal energy is independent of the volume (at constant temperature). If attractions are dominant in a real gas, the internal energy increases with volume because the molecules become further apart on average. If repulsions are dominant, the internal energy decreases as the gas expands.

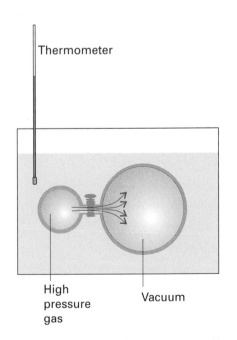

3.3 A schematic diagram of the apparatus used by Joule in an attempt to measure the change in internal energy when a gas expands isothermally. The heat absorbed by the gas is proportional to the change in temperature of the bath.

In every case, a partial derivative is the slope of a graph of the property of interest against *one* of the variables on which it depends (recall Fig. 2.12), all the other variables being held constant. In many cases the slopes have a straightforward physical interpretation, and thermodynamics gets shapeless and difficult only when that meaning is not kept in sight. In the present case, we have already met $(\partial U/\partial T)_V$ in eqn 2.19, where we saw that it is the constant-volume heat capacity, C_V. Therefore, we can write

$$dU = \left(\frac{\partial U}{\partial V}\right)_T dV + C_V\, dT \tag{4}$$

The other coefficient, $(\partial U/\partial V)_T$, plays a major role in thermodynamics because it is a measure of the variation of the internal energy of a substance as the volume it occupies is changed at constant temperature. We shall denote it π_T (because it has the same dimensions as pressure), and call it the **internal pressure**:

$$\pi_T = \left(\frac{\partial U}{\partial V}\right)_T \tag{5}$$

We shall see that the internal pressure is a measure of the strength of the cohesive forces in the sample. Then

$$dU = \pi_T\, dV + C_V\, dT \tag{6}$$

If the internal energy increases ($dU > 0$) as the volume of the sample expands isothermally ($dV > 0$), which is the case when there are attractive forces between the particles, a graph of internal energy against volume slopes upwards and $\pi_T > 0$ (Fig. 3.2). When there are no interactions between the molecules, the internal energy is independent of their separation and hence independent of the volume the sample occupies; hence $\pi_T = 0$ for a perfect gas. The statement $\pi_T = 0$ (that is, the internal energy is independent of the volume occupied by the sample) can be taken to be the definition of a perfect gas, for later we shall see that it implies the equation of state $pV = nRT$.

(c) The Joule experiment

James Joule thought that he could measure π_T by observing the change in temperature of a gas when it is allowed to expand into a vacuum. He used two metal vessels immersed in a water bath (Fig. 3.3). One was filled with air at about 22 atm and the other was evacuated. He then tried to measure the change in temperature of the water of the bath when a stopcock was opened and the air expanded into a vacuum. He observed no change in temperature.

The thermodynamic implications of the experiment are as follows. No work was done in the expansion into a vacuum, so $w = 0$. No heat entered or left the system (the gas) because the temperature of the bath did not change, so $q = 0$. Consequently, within the accuracy of the experiment, $\Delta U = 0$. It follows that U does not change much when a gas expands isothermally and therefore that $\pi_T = 0$.

Joule's experiment was crude. In particular, the heat capacity of the apparatus was so large that the temperature change that gases do in fact cause was too small to measure. His experiment was on a par with Boyle's: he extracted an essential limiting property of a gas, a property of a perfect gas, without detecting the small deviations characteristic of real gases.

Illustration

For ammonia, $\pi_T = 840$ Pa at 300 K and 1.0 bar, and $C_{V,m} = 27.32$ J K^{-1} mol^{-1}. The change in molar internal energy of ammonia when it is heated through 2.0 K and compressed through 100 cm^3 is approximately

$$\Delta U_m \approx (840 \text{ J m}^{-3} \text{ mol}^{-1}) \times (-100 \times 10^{-6} \text{ m}^3)$$
$$+ (27.32 \text{ J K}^{-1} \text{ mol}^{-1}) \times (2.0 \text{ K})$$
$$\approx -0.084 \text{ J mol}^{-1} + 55 \text{ J mol}^{-1} = +55 \text{ J mol}^{-1}$$

Note that the contribution of the heating term greatly dominates that of the compression term for this gas.

(d) Changes in internal energy at constant pressure

Partial derivatives have many useful properties and some that we shall draw on frequently are reviewed in *Further information 1*. Skilful use of them can often turn some unfamiliar quantity into a quantity that can be recognized, interpreted, or measured.

As an example, suppose we want to find out how the internal energy varies with temperature when the pressure of the system is kept constant. If we divide both sides of eqn 6 by dT and impose the condition of constant pressure on the resulting differentials, so that dU/dT on the left becomes $(\partial U/\partial T)_p$, we obtain

$$\left(\frac{\partial U}{\partial T}\right)_p = \pi_T \left(\frac{\partial V}{\partial T}\right)_p + C_V$$

It is usually sensible in thermodynamics to inspect the output of a manipulation like this to see if it contains any recognizable physical quantity. The differential coefficient on the right in this expression is the slope of the plot of volume against temperature (at constant pressure). This property is normally tabulated as the **expansion coefficient**, α, of a substance,[3] which is defined as

$$\alpha = \frac{1}{V}\left(\frac{\partial V}{\partial T}\right)_p \qquad [7]$$

A large value of α means that the volume of the sample responds strongly to changes in temperature. Some experimental values are given in Table 3.1.

Example 3.2 Using the expansion coefficient of a gas

Derive an expression for the expansion coefficient for a perfect gas.

Method The expansion coefficient is defined in eqn 7. To use this expression, we simply substitute the expression for V in terms of T obtained from the equation of state for the gas. As implied by eqn 7, the pressure, p, is treated as a constant.

Answer (a) Because $pV = nRT$, we can write

$$\alpha = \frac{1}{V}\left(\frac{\partial(nRT/p)}{\partial T}\right)_p = \frac{nR}{pV} = \frac{1}{T}$$

The higher the temperature, the less responsive is its volume to a change in temperature.

Table 3.1* Expansion coefficients (α) and isothermal compressibilities (κ_T)

Substance	$\alpha/(10^{-4} \text{ K}^{-1})$	$\kappa_T/(10^{-6} \text{ atm}^{-1})$
Benzene	12.4	92.1
Diamond	0.030	0.187
Lead	0.861	2.21
Water	2.1	49.6

* More values are given in the *Data section* at the end of this volume.

3 As for heat capacities, the expansion coefficients of a mixture depend on whether or not the composition is allowed to change. Throughout this chapter, we deal only with pure substances, so this complication can be disregarded.

Self-test 3.2 Evaluate α for a gas for which the equation of state is $p = nRT/(V - nb)$.

$$[\alpha = (1 - b/V_m)/T]$$

When we introduce the definition of α into the equation for $(\partial U/\partial T)_p$, we obtain

$$\left(\frac{\partial U}{\partial T}\right)_p = \alpha \pi_T V + C_V \tag{8}$$

This equation is entirely general (provided the system is closed and its composition is constant). It expresses the dependence of the internal energy on the temperature at constant pressure in terms of C_V, which can be measured in one experiment, in terms of α, which can be measured in another, and in terms of the quantity π_T. For a perfect gas, $\pi_T = 0$, so

$$\left(\frac{\partial U}{\partial T}\right)_p = C_V \tag{9}°$$

That is, the constant-volume heat capacity of a perfect gas is equal to the slope of a plot of internal energy against temperature at constant pressure as well as (by definition) to the slope at constant volume.

At this stage we know the slope of U with respect to T at constant volume (the constant-volume heat capacity) and the slope of U with respect to T at constant pressure, eqn 8. The fact that the former expression is so simple strongly suggests that it is sensible to treat U as a function of the volume, and to use it in thermodynamics when V is under our control. We saw a hint of that simplicity earlier in the expression $\Delta U = q_V$.

3.2 The temperature dependence of the enthalpy

We can carry out a similar set of operations on the enthalpy, $H = U + pV$. The quantities U, p, and V are all state functions; therefore H is also a state function, and hence dH is an exact differential.

(a) Changes in the enthalpy at constant volume

The variation of enthalpy with temperature at constant pressure is simply the constant-pressure heat capacity, C_p. The simplicity of this relation strongly suggests that H will prove to be a useful thermodynamic function when the pressure is under our control. We saw a sign of that in the relation $\Delta H = q_p$ (eqn 2.24). We shall therefore regard H as a function of p and T, and adapt the argument in Section 3.1 to find an expression for the one temperature variation we currently lack, the variation of H with temperature at constant volume. This relation will prove useful for relating the heat capacities at constant pressure and volume and for a discussion of the liquefaction of gases.

By the same argument as for U (but with p in place of V) we find that, for a closed system of constant composition,

$$dH = \left(\frac{\partial H}{\partial p}\right)_T dp + \left(\frac{\partial H}{\partial T}\right)_p dT \tag{10}$$

We recognize the second coefficient as the definition of the constant-pressure heat capacity, C_p, so

$$dH = \left(\frac{\partial H}{\partial p}\right)_T dp + C_p \, dT \tag{11}$$

The manipulation of this expression is slightly more involved than before, but we show in the *Justification* below that it implies that

$$\left(\frac{\partial H}{\partial T}\right)_V = \left(1 - \frac{\alpha\mu}{\kappa_T}\right)C_p \tag{12}$$

where the **isothermal compressibility**, κ_T, is defined as

$$\kappa_T = -\frac{1}{V}\left(\frac{\partial V}{\partial p}\right)_T \tag{13}$$

and the **Joule–Thomson coefficient**, μ, is defined as

$$\mu = \left(\frac{\partial T}{\partial p}\right)_H \tag{14}$$

Equation 12 applies to any substance. Because all the quantities that appear in it can be measured in suitable experiments, we now know how H varies with T when the volume of the sample is held constant.

Justification 3.1

First, divide eqn 11 through by dT and impose constant volume:

$$\left(\frac{\partial H}{\partial T}\right)_V = \left(\frac{\partial H}{\partial p}\right)_T\left(\frac{\partial p}{\partial T}\right)_V + C_p$$

The third differential coefficient looks like something we ought to recognize, and is perhaps related to $(\partial V/\partial T)_p$, the expansion coefficient. It follows from the chain relation in *Further information 1* that

$$\left(\frac{\partial p}{\partial T}\right)_V\left(\frac{\partial T}{\partial V}\right)_p\left(\frac{\partial V}{\partial p}\right)_T = -1$$

and therefore that

$$\left(\frac{\partial p}{\partial T}\right)_V = -\frac{1}{(\partial T/\partial V)_p(\partial V/\partial p)_T}$$

Unfortunately, $(\partial T/\partial V)_p$ occurs instead of $(\partial V/\partial T)_p$. However, another relation between partial differentials (see *Further information 1*) allows us to invert partial derivatives and to write $(\partial y/\partial x)_z = 1/(\partial x/\partial y)_z$, and leads to

$$\left(\frac{\partial p}{\partial T}\right)_V = -\frac{(\partial V/\partial T)_p}{(\partial V/\partial p)_T} = \frac{\alpha}{\kappa_T}$$

Next, we must change $(\partial H/\partial p)_T$ into something recognizable. The chain relation lets us write this partial derivative as

$$\left(\frac{\partial H}{\partial p}\right)_T = -\frac{1}{(\partial p/\partial T)_H(\partial T/\partial H)_p}$$

Both derivatives may be brought up into the numerator:

$$\left(\frac{\partial H}{\partial p}\right)_T = -\left(\frac{\partial T}{\partial p}\right)_H\left(\frac{\partial H}{\partial T}\right)_p$$

and we can recognize both the constant-pressure heat capacity, C_p, and the Joule–Thomson coefficient μ as defined in the text. Therefore,

$$\left(\frac{\partial H}{\partial p}\right)_T = -\mu C_p \tag{15}$$

When we use this expression in the first equation in this *Justification*, we obtain eqn 12.

(b) The isothermal compressibility

The negative sign in the definition of κ_T, eqn 13, ensures that it is positive, because an increase of pressure, implying a positive dp, brings about a reduction of volume, a negative dV. The isothermal compressibility is obtained from the slope of the plot of volume against pressure at constant temperature (that is, it is proportional to the slope of an isotherm). Some values of κ_T are listed in Table 3.1. Its value for a perfect gas is obtained by substitution of the perfect gas equation of state into eqn 13, which gives

$$\kappa_T = -\frac{1}{V}\left(\frac{\partial(nRT/p)}{\partial p}\right)_T = -\frac{nRT}{V}\left(-\frac{1}{p^2}\right) = \frac{1}{p} \qquad (16)°$$

This expression shows that, the higher the pressure of the gas, the lower its compressibility.

Example 3.3 Using the isothermal compressibility

The isothermal compressibility of water at 20°C and 1 atm is 4.94×10^{-6} atm^{-1}. What change of volume occurs when a sample of volume 50 cm^3 is subjected to an additional 1000 atm at constant temperature?

Method We know from the definition of κ_T that, for an infinitesimal change of pressure at constant temperature, the volume changes by

$$dV = \left(\frac{\partial V}{\partial p}\right)_T dp = -\kappa_T V\, dp$$

Therefore, for a finite change in pressure, we need to integrate both sides. When confronted by an integration, it is often a useful first approximation (for substances other than gases) to suppose that the integrand is constant over the range of integration.

Answer The integral we need to evaluate is

$$\int_{V_i}^{V_f} dV = -\int_{p_i}^{p_f} \kappa_T V\, dp$$

The integral on the left is ΔV. If we suppose that κ_T and V are approximately constant over the range of pressures of interest, we can write

$$\Delta V = -\kappa_T V \int_{p_i}^{p_f} dp = -\kappa_T V \Delta p$$

Substitution of the data into the last expression then gives

$$\Delta V = -(4.94 \times 10^{-6}\ \text{atm}^{-1}) \times (50\ \text{cm}^3) \times (1000\ \text{atm}) = -0.25\ \text{cm}^3$$

Comment Because the compression results in a decrease in volume of only 0.5 per cent, the assumption of constant V and κ_T is probably acceptable as a first approximation. Note that very high pressures are needed to bring about significant changes of volume.

- -

Self-test 3.3 A sample of copper of volume 50 cm^3 is subjected to an additional pressure of 100 atm and a temperature increase of 5.0 K. Estimate the total change in volume.

[8.8 mm^3]

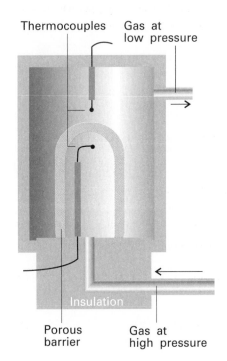

Thermocouples   Gas at low pressure

Insulation

Porous barrier   Gas at high pressure

3.4 A diagram of the apparatus used for measuring the Joule–Thomson effect. The gas expands through the porous barrier, which acts as a throttle, and the whole apparatus is thermally insulated. As explained in the text, this arrangement corresponds to an isenthalpic expansion (expansion at constant enthalpy). Whether the expansion results in a heating or a cooling of the gas depends on the conditions.

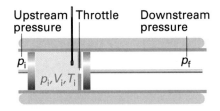

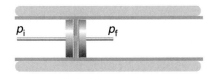

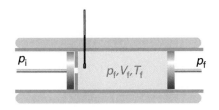

3.5 A diagram representing the thermodynamic basis of Joule-Thomson expansion. The pistons represent the upstream and downstream gases, which maintain constant pressures either side of the throttle. The transition from the top diagram to the bottom diagram, which represents the passage of a given amount of gas through the throttle, occurs without change of enthalpy.

(c) The Joule-Thomson effect

The analysis of the Joule-Thomson coefficient is central to the technological problems associated with the liquefaction of gases. We need to be able to interpret it physically and to measure it.

The cunning required to impose the constraint of constant enthalpy on a change of state was supplied by Joule and William Thomson (later Lord Kelvin). They let a gas expand through a porous barrier from one constant pressure to another, and monitored the difference of temperature that arose from the expansion (Fig. 3.4). The whole apparatus was insulated so that the process was adiabatic. They observed a lower temperature on the low-pressure side, the difference in temperature being proportional to the pressure difference they maintained. This cooling by adiabatic expansion is now called the **Joule-Thomson effect**.

The thermodynamic analysis of the experiment takes as the system a sample of fixed amount of gas. Because all changes to the gas occur adiabatically, $q = 0$. To calculate the work done as the gas passes through the throttle, we consider the passage of a fixed amount of gas from the high-pressure side, where the pressure is p_i, the temperature T_i, and the gas occupies a volume V_i (Fig. 3.5). The gas emerges on the low-pressure side, where the same amount of gas has a pressure p_f, a temperature T_f, and occupies a volume V_f. The gas on the left is compressed isothermally by the upstream gas acting as a piston. The relevant pressure is p_i and the volume changes from V_i to 0; therefore, the work done on the gas is $-p_i(0 - V_i)$, or $p_i V_i$. The gas expands isothermally on the right of the throttle (but possibly at a different constant temperature) against the pressure p_f provided by the downstream gas acting as a piston to be driven out. The volume changes from 0 to V_f, so the work done on the gas in this stage is $-p_f(V_f - 0)$, or $-p_f V_f$. The total work done on the gas is the sum of these two quantities, or $p_i V_i - p_f V_f$. It follows that the change of internal energy of the gas as it moves from one side of the throttle to the other is

$$U_f - U_i = w = p_i V_i - p_f V_f$$

Reorganization of this expression gives

$$U_f + p_f V_f = U_i + p_i V_i, \qquad \text{or } H_f = H_i \qquad (17)$$

Therefore, the expansion occurs without change of enthalpy: it is an **isenthalpic process**, a process at constant enthalpy.

The property measured in the experiment is the ratio of the temperature change to the change of pressure, $\Delta T/\Delta p$. Adding the constraint of constant enthalpy and taking the limit of small Δp implies that the thermodynamic quantity measured is $(\partial T/\partial p)_H$, which is the Joule-Thomson coefficient, μ. In other words, the physical interpretation of μ is that it is the ratio of the change in temperature to the change in pressure when a gas expands under adiabatic conditions.

The modern method of measuring μ is indirect, and involves measuring the **isothermal Joule-Thomson coefficient**, the quantity

$$\mu_T = \left(\frac{\partial H}{\partial p}\right)_T \qquad [18]$$

The two coefficients are related by eqn 15:

$$\mu_T = -C_p \mu \qquad (19)$$

To measure μ_T, the gas is pumped continuously at a steady pressure through a heat exchanger (which brings it to the required temperature), and then through a porous plug inside a thermally insulated container. The steep pressure drop is measured, and the cooling effect is exactly offset by an electric heater placed immediately after the plug (Fig. 3.6). The energy provided by the heater is monitored. Because the heat can be identified with the

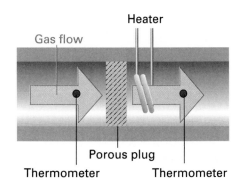

3.6 A schematic diagram of the apparatus used for measuring the isothermal Joule-Thomson coefficient. The electrical heating required to offset the cooling arising from expansion is interpreted as ΔH and used to calculate $(\partial H/\partial p)_T$, which is then converted to μ as explained in the text.

Table 3.2* Inversion temperatures (T_I), normal freezing (T_f) and boiling (T_b) points, and Joule–Thomson coefficients (μ) at 1 atm and 298 K

	T_I/K	T_f/K	T_b/K	$\mu/$ (K atm^{-1})
Ar	723	83.8	87.3	
CO_2	1500		194.7	1.11
He	40		4.2	−0.060
N_2	621	63.3	77.4	0.25

* More values are given in the *Data section*.

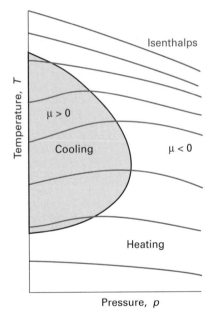

3.7 The sign of the Joule–Thomson coefficient, μ, depends on the conditions. Inside the boundary, the shaded area, it is positive and outside it is negative. The temperature corresponding to the boundary at a given pressure is the 'inversion temperature' of the gas at that pressure. For a given pressure, the temperature must be below a certain value if cooling is required but, if it becomes too low, the boundary is crossed again and heating occurs. Reduction of pressure under adiabatic conditions moves the system along one of the isenthalps, or curves of constant enthalpy. The inversion temperature curve runs through the points of the isenthalps where their slopes change from negative to positive.

value of ΔH for the gas (because $\Delta H = q_p$), and the pressure change Δp is known, the value of μ_T can be obtained from the limiting value of $\Delta H/\Delta p$ as $\Delta p \to 0$, and then converted to μ. Some values obtained in this way are listed in Table 3.2.

Real gases have nonzero Joule–Thomson coefficients and, depending on the identity of the gas, the pressure, the relative magnitudes of the attractive and repulsive intermolecular forces, and the temperature, the sign of the coefficient may be either positive or negative (Fig. 3.7). A positive sign implies that dT is negative when dp is negative, in which case the gas cools on expansion. Gases that show a heating effect ($\mu < 0$) at one temperature show a cooling effect ($\mu > 0$) when the temperature is below their upper **inversion temperature**, T_I (Table 3.2, Fig. 3.8). As indicated in Fig. 3.8, a gas typically has two inversion temperatures, one at high temperature and the other at low.

The 'Linde refrigerator' makes use of Joule–Thomson expansion to liquefy gases (Fig. 3.9). The gas at high pressure is allowed to expand through a throttle; it cools and is circulated past the incoming gas. That gas is cooled, and its subsequent expansion cools it still further. There comes a stage when the circulating gas becomes so cold that it condenses to a liquid.

For a perfect gas, $\mu = 0$; hence, the temperature of a perfect gas is unchanged by Joule–Thomson expansion.[4] This characteristic points clearly to the involvement of intermolecular forces in determining the size of the effect. However, the Joule–Thomson coefficient of a real gas does not necessarily approach zero as the pressure is reduced even though the equation of state of the gas approaches that of a perfect gas. The coefficient is an example of a property mentioned in Section 1.4b that depends on derivatives and not on p, V, and T themselves.

3.3 The relation between C_V and C_p

The constant-pressure heat capacity C_p differs from the constant-volume heat capacity C_V by the work needed to change the volume of the system to maintain constant pressure. This work arises in two ways. One is the work of driving back the atmosphere; the other is the work of stretching the bonds in the material, including any weak intermolecular interactions. In the case of a perfect gas, the second makes no contribution. We shall now derive a general relation between the two heat capacities, and show that it reduces to the perfect gas result in the absence of intermolecular forces.

(a) The relation for a perfect gas

First, we carry through the calculation for a perfect gas. In this special case, we can use eqn 9 to express both heat capacities in terms of derivatives at constant pressure:

$$C_p - C_V = \left(\frac{\partial H}{\partial T}\right)_p - \left(\frac{\partial U}{\partial T}\right)_p \tag{20}$$

Then we introduce

$$H = U + pV = U + nRT$$

into the first term, which results in

$$C_p - C_V = \left(\frac{\partial U}{\partial T}\right)_p + nR - \left(\frac{\partial U}{\partial T}\right)_p = nR \tag{21}°$$

This is the formal derivation of eqn 2.31.

4   Simple adiabatic expansion does cool a perfect gas, because the gas does work; recall Section 2.6.

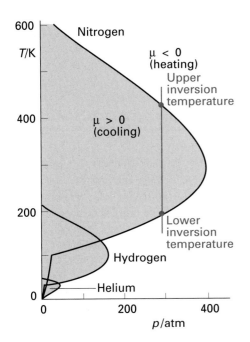

3.8 The inversion temperatures for three real gases, nitrogen, hydrogen, and helium.

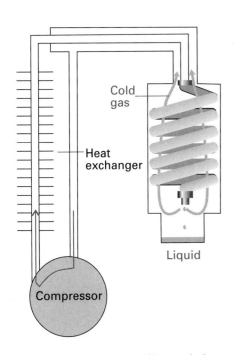

3.9 The principle of the Linde refrigerator is shown in this diagram. The gas is recirculated and, so long as it is beneath its inversion temperature, it cools on expansion through the throttle. The cooled gas cools the high-pressure gas, which cools still further as it expands. Eventually liquefied gas drips from the throttle.

This formula is a thermodynamic expression, which means that it applies to any substance (that is, it is 'universally true'). It reduces to eqn 21 for a perfect gas when we set $\alpha = 1/T$ and $\kappa_T = 1/p$.

Justification 3.2

A useful rule when doing a problem in thermodynamics is to go back to first principles. In the present problem we do this twice, first by expressing C_p and C_V in terms of their definitions and then by inserting the definition $H = U + pV$:

$$C_p - C_V = \left(\frac{\partial H}{\partial T}\right)_p - \left(\frac{\partial U}{\partial T}\right)_V$$

$$= \left(\frac{\partial U}{\partial T}\right)_p + \left(\frac{\partial (pV)}{\partial T}\right)_p - \left(\frac{\partial U}{\partial T}\right)_V$$

We have already calculated the difference of the first and third terms on the right, and eqn 8 lets us write this difference as $\alpha \pi_T V$. The factor αV gives the change in volume when the temperature is raised, and $\pi_T = (\partial U/\partial V)_T$ converts this change in volume into a change in internal energy. We can simplify the remaining term by noting that, because p is constant,

$$\left(\frac{\partial (pV)}{\partial T}\right)_p = p\left(\frac{\partial V}{\partial T}\right)_p = \alpha pV$$

The middle term of this expression identifies it as the contribution to the work of pushing back the atmosphere: $(\partial V/\partial T)_p$ is the change of volume caused by a change of temperature, and multiplication by p converts this expansion into work.

Collecting the two contributions gives

$$C_p - C_V = \alpha(p + \pi_T)V$$

As just remarked, the first term on the right (αpV) is a measure of the work needed to push back the atmosphere; the second term on the right, $\alpha \pi_T V$, is the work required to separate the molecules composing the system.

At this point we can go further by using the result we prove in Section 5.1 that

$$\pi_T = T\left(\frac{\partial p}{\partial T}\right)_V - p$$

When this expression is inserted in the last equation we obtain

$$C_p - C_V = \alpha TV\left(\frac{\partial p}{\partial T}\right)_V$$

The same coefficient as appears here was encountered in *Justification* 3.1, where we saw that it is equal to α/κ_T. Therefore, this expression turns into eqn 22.

Because thermal expansivities, α, of liquids and solids are small, it is tempting to deduce from eqn 22 that for them $C_p \approx C_V$. But this is not always so, because the compressibility κ_T might also be small, so α^2/κ_T might be large. That is, although only a little work need be done to push back the atmosphere, a great deal of work may have to be done to pull atoms apart from one another as the solid expands. As an illustration, for water at $25\,°C$, eqn 22 gives $C_{p,m} = 75.3\ \mathrm{J\,K^{-1}\,mol^{-1}}$ compared with $C_{V,m} = 74.8\ \mathrm{J\,K^{-1}\,mol^{-1}}$. In some cases, the two heat capacities differ by as much as 30 per cent.

Checklist of key ideas

☐ state functions
☐ path functions

State functions and exact differentials

3.1 State functions
☐ exact differential
☐ inexact differential
☐ change in internal energy arising from changes in volume and temperature

☐ internal pressure (π_T, 5)
☐ Joule experiment to show $\pi_T = 0$
☐ expansion coefficient (α, 7)

3.2 The temperature dependence of the enthalpy
☐ variation of enthalpy with temperature at constant volume

☐ isothermal compressibility (κ_T, 13)
☐ Joule–Thomson coefficient (μ, 14)
☐ isothermal compressibility of a perfect gas
☐ Joule–Thomson effect
☐ isenthalpic process
☐ isothermal Joule–Thomson coefficient (μ_T, 18)

☐ inversion temperature
☐ Linde refrigerator

3.3 The relation between C_V and C_p
☐ the relation between C_p and C_V for a perfect gas
☐ the relation between C_p and C_V for a general substance

Further reading

Articles of general interest

S.M. Blinder, Mathematical methods in elementary thermodynamics. *J. Chem. Educ.* **43**, 85 (1966).

E.W. Anacker, S.E. Anacker, and W.J. Swartz, Some comments on partial derivatives in thermodynamics. *J. Chem. Educ.* **64**, 674 (1987).

G.A. Estèvez, K. Yang, and B.B. Dasgupta, Thermodynamic partial derivatives and experimentally measurable quantities. *J. Chem. Educ.* **66**, 890 (1989).

R.A. Alberty, Legendre transforms in chemical thermodynamics. *Chem. Rev.* **94**, 1457 (1994).

Texts and sources of data and information

M.L. McGlashan, *Chemical thermodynamics*. Academic Press, London (1979).

D.M. Hirst, *Mathematics for chemists*. Macmillan, London (1983).

E. Steiner, *The chemistry maths book*. Oxford University Press (1996).

Exercises

Assume that all gases are perfect and that all data refer to 298.15 K unless stated otherwise.

3.1 (a) Show that the following functions have exact differentials: (a) $x^2y + 3y^2$, (b) $x \cos xy$.

3.1 (b) Show that the following functions have exact differentials: (a) x^3y^2, (b) $t(t + e^s) + s$.

3.2 (a) Let $z = ax^2y^3$. Find dz.

3.2 (b) Let $z = x/(1 + y)^2$. Find dz.

3.3 (a) What is the total differential of $z = x^2 + 2y^2 - 2xy + 2x - 4y - 8$? (b) Show that $\partial^2 z/\partial y \partial x = \partial^2 z/\partial x \partial y$ for this function.

3.3 (b) (a) What is the total differential of $z = x^3 - 2xy^2 + 15$? (b) Show that $\partial^2 z/\partial y \partial x = \partial^2 z/\partial x \partial y$ for this function.

3.4 (a) Let $z = xy - y + \ln x + 2$. Find dz and show that it is exact.

3.4 (b) Let $z = x^2y + xy^2$. Find dz and show that it is exact.

3.5 (a) Express $(\partial C_V/\partial V)_T$ as a second-derivative of U and find its relation to $(\partial U/\partial V)_T$. From this relation show that $(\partial C_V/\partial V)_T = 0$ for a perfect gas.

3.5 (b) Express $(\partial C_p/\partial p)_T$ as a second-derivative of H and find its relation to $(\partial H/\partial p)_T$. From this relation show that $(\partial C_p/\partial p)_T = 0$ for a perfect gas.

3.6 (a) By direct differentiation of $H = U + pV$, obtain a relation between $(\partial H/\partial U)_p$ and $(\partial U/\partial V)_p$.

3.6 (b) Confirm that $(\partial H/\partial U)_p = 1 + p(\partial V/\partial U)_p$ by expressing $(\partial H/\partial U)_p$ as the ratio of two derivatives with respect to volume and then using the definition of enthalpy.

3.7 (a) Write an expression for dV given that V is a function of p and T. Deduce an expression for $d \ln V$ in terms of the expansion coefficient and the isothermal compressibility.

3.7 (b) Write an expression for dp given that p is a function of V and T. Deduce an expression for $d \ln p$ in terms of the expansion coefficient and the isothermal compressibility.

3.8 (a) The internal energy of a perfect monatomic gas relative to its value at $T = 0$ is $\frac{3}{2}nRT$. Calculate $(\partial U/\partial V)_T$ and $(\partial H/\partial V)_T$ for the gas.

3.8 (b) The internal energy of a perfect monatomic gas relative to its value at $T = 0$ is $\frac{3}{2}nRT$. Calculate $(\partial H/\partial p)_T$ and $(\partial U/\partial p)_T$ for the gas.

3.9 (a) The coefficient of thermal expansion, α, is defined in eqn 7 and the isothermal compressibility, κ_T, is defined in eqn 13. Starting from the expression for the total differential dV in terms of T and p, show that $(\partial p/\partial T)_V = \alpha/\kappa_T$.

3.9 (b) Evaluate α and κ_T for a perfect gas.

3.10 (a) When a certain freon used in refrigeration was expanded adiabatically from an initial pressure of 32 atm and $0\,°C$ to a final pressure of 1.00 atm, the temperature fell by 22 K. Calculate the Joule–Thomson coefficient, μ, at $0\,°C$, assuming it remains constant over this temperature range.

3.10 (b) A vapour at 22 atm and $5\,°C$ was allowed to expand adiabatically to a final pressure of 1.00 atm; the temperature fell by 10 K. Calculate the Joule–Thomson coefficient, μ, at $5\,°C$, assuming it remains constant over this temperature range.

3.11 (a) For a van der Waals gas, $\pi_T = a/V_m^2$. Calculate ΔU_m for the isothermal reversible expansion of nitrogen gas from an initial volume of 1.00 L to 24.8 L at 298 K. What are the values of q and w?

3.11 (b) For a van der Waals gas, $\pi_T = a/V_m^2$. Calculate ΔU_m for the isothermal reversible expansion of argon from an initial volume of 1.00 L to 22.1 L at 298 K. What are the values of q and w?

3.12 (a) The volume of a certain liquid varies with temperature as

$$V = V'\{0.75 + 3.9 \times 10^{-4}(T/K) + 1.48 \times 10^{-6}(T/K)^2\}$$

where V' is its volume at 300 K. Calculate its expansion coefficient, α, at 320 K.

3.12 (b) The volume of a certain liquid varies with temperature as

$$V = V'\{0.77 + 3.7 \times 10^{-4}(T/K) + 1.52 \times 10^{-6}(T/K)^2\}$$

where V' is its volume at 298 K. Calculate its expansion coefficient, α, at 310 K.

3.13 (a) The isothermal compressibility of copper at 293 K is 7.35×10^{-7} atm^{-1}. Calculate the pressure that must be applied in order to increase its density by 0.08 per cent.

3.13 (b) The isothermal compressibility of lead at 293 K is 2.21×10^{-6} atm^{-1}. Calculate the pressure that must be applied in order to increase its density by 0.08 per cent.

3.14 (a) Given that $\mu = 0.25$ K atm^{-1} for nitrogen, calculate the value of its isothermal Joule–Thomson coefficient. Calculate the energy that must be supplied as heat to maintain constant temperature when 15.0 mol N_2 flows through a throttle in an isothermal Joule–Thomson experiment and the pressure drop is 75 atm.

3.14 (b) Given that $\mu = 1.11$ K atm^{-1} for carbon dioxide, calculate the value of its isothermal Joule–Thomson coefficient. Calculate the energy that must be supplied as heat to maintain constant temperature when 12.0 mol CO_2 flows through a throttle in an isothermal Joule–Thomson experiment and the pressure drop is 55 atm.

3.15 (a) To design a particular kind of refrigerator we need to know the temperature reduction brought about by adiabatic expansion of the refrigerant gas. For one type of freon, $\mu = 1.2$ K atm^{-1}. What pressure difference is needed to produce a temperature drop of 5.0 K?

3.15 (b) For another type of freon (see previous exercise), $\mu = 13.3$ mK kPa^{-1}. What pressure difference is needed to produce a temperature drop of 4.5 K?

Problems

Assume all gases are perfect unless stated otherwise. Unless otherwise stated, thermochemical data are for 298.15 K.

Numerical problems

3.1 The isothermal compressibility of lead is 2.21×10^{-6} atm^{-1}. Express this value in Pa^{-1}. A cube of lead of side 10 cm at $25\,°C$ was to be inserted in the keel of an underwater exploration TV camera, and its designers needed to know the stresses in the equipment. Calculate the change of volume of the cube at a depth of 1.000 km (disregarding the effects of temperature). Take the mean density of sea water as 1.03 g cm^{-3}. Given that the expansion coefficient of lead is 8.61×10^{-5} K^{-1} and that the temperature where the camera operates is $-5\,°C$, calculate the volume of the block taking the temperature into account too.

3.2 Calculate the change in (a) the molar internal energy and (b) the molar enthalpy of water when its temperature is raised by 10 K. Account for the difference between the two quantities.

3.3 The constant-volume heat capacity of a gas can be measured by observing the decrease in temperature when it expands adiabatically and reversibly. If the decrease in pressure is also measured, we can use it to infer the value of γ (the ratio of heat capacities, C_p/C_V) and hence, by combining the two values, deduce the constant-pressure heat capacity. A fluorocarbon gas was allowed to expand reversibly and adiabatically to twice its volume; as a result, the temperature fell from 298.15 K to 248.44 K and its pressure fell from 1522.2 Torr to 613.85 Torr. Evaluate C_p.

3.4 A sample consisting of 1.00 mol of a van der Waals gas is compressed from 20.0 L to 10.0 L at 300 K. In the process, 20.2 kJ of work is done on the gas. Given that $\mu = \{(2a/RT) - b\}/C_{p,\mathrm{m}}$, with $C_{p,\mathrm{m}} = 38.4\ \mathrm{J\,K^{-1}\,mol^{-1}}$, $a = 3.60\ \mathrm{L^2\,atm\,mol^{-2}}$, and $b = 0.44\ \mathrm{L\,mol^{-1}}$, calculate ΔH for the process.

3.5 Estimate γ (the ratio of heat capacities) for xenon at 100 °C and 1.00 atm on the assumption that it is a van der Waals gas (see Problem 3.22).

Theoretical problems

3.6 Determine whether or not $\mathrm{d}z = xy\,\mathrm{d}x + xy\,\mathrm{d}y$ is exact by integrating it around the closed curve formed by the paths $y = x$ and $y = x^2$ between the points $(0,0)$ and $(1,1)$.

3.7 Decide whether $\mathrm{d}q = (RT/p)\,\mathrm{d}p - R\,\mathrm{d}T$ is exact. Then determine whether multiplication of $\mathrm{d}q$ by $1/T$ is exact. Comment on the significance of your result.

3.8 Obtain the total differential of the function $w = xy + yz + xz$. Demonstrate that $\mathrm{d}w$ is exact by integration between the points $(0,0,0)$ and $(1,1,1)$ along the two different paths (a) $z = y = x$ and (b) $z = y = x^2$.

3.9 Derive the relation $C_V = -(\partial U/\partial V)_T(\partial V/\partial T)_U$ from the expression for the total differential of $U(T,V)$.

3.10 Starting from the expression for the total differential of $H(T,p)$, express $(\partial H/\partial p)_T$ in terms of C_p and the Joule–Thomson coefficient, μ.

3.11 Starting from the expression $C_p - C_V = T(\partial p/\partial T)_V(\partial V/\partial T)_p$, use the appropriate relations between partial derivatives to show that

$$C_p - C_V = \frac{T(\partial V/\partial T)_p^2}{(\partial V/\partial p)_T}$$

Evaluate $C_p - C_V$ for a perfect gas.

3.12 From an analysis of Joule's free expansion experiment, demonstrate that it is possible to calculate the change in internal energy of a perfect gas for any process by knowing only C_V and ΔT.

3.13 By the consideration of a suitable cycle involving a perfect gas, demonstrate that $\mathrm{d}q$ is an inexact differential and, therefore, that q is not a state function.

3.14 Use the fact that $(\partial U/\partial V)_T = a/V_\mathrm{m}^2$ for a van der Waals gas to show that $\mu C_{p,\mathrm{m}} \approx (2a/RT) - b$ by using the definition of μ and appropriate relations between partial derivatives. (*Hint*: Use the approximation $pV_\mathrm{m} \approx RT$ when it is justifiable to do so.)

3.15 Obtain the expression for the total differential dp for a van der Waals gas in terms of dT and dV. Also obtain $(\partial V/\partial T)_p$. Demonstrate

that dp is not an exact differential by integrating it from (T_1,V_1) to (T_2,V_2) along the two different paths, namely (a) $(T_1,V_1) \rightarrow (T_2,V_1) \rightarrow (T_2,V_2)$ and (b) $(T_1,V_1) \rightarrow (T_1,V_2) \rightarrow (T_2,V_2)$.

3.16 Take nitrogen to be a van der Waals gas with $a = 1.408\ \mathrm{atm\,L^2\,mol^{-2}}$ and $b = 0.03913\ \mathrm{L\,mol^{-1}}$, and calculate ΔH_m when the pressure on the gas is decreased from 500 atm to 1.00 atm at 300 K. For a van der Waals gas, $\mu = \{(2a/RT) - b\}/C_{p,\mathrm{m}}$. Assume $C_{p,\mathrm{m}} = \frac{7}{2}R$.

3.17 The pressure of a given amount of a van der Waals gas depends on T and V. Find an expression for dp in terms of dT and dV.

3.18 Rearrange the van der Waals equation of state to give an expression for T as a function of p and V (with n constant). Calculate $(\partial T/\partial p)_V$ and confirm that $(\partial T/\partial p)_V = 1/(\partial p/\partial T)_V$. Go on to confirm Euler's chain relation.

3.19 Calculate the isothermal compressibility and the expansion coefficient of a van der Waals gas. Show, using Euler's chain relation, that $\kappa_T R = \alpha(V_\mathrm{m} - b)$.

3.20 Given that $\mu C_p = T(\partial V/\partial T)_p - V$, derive an expression for μ in terms of the van der Waals parameters a and b, and express it in terms of reduced variables. Evaluate μ at 25 °C and 1.0 atm, when the molar volume of the gas is 24.6 $\mathrm{L\,mol^{-1}}$. Use the expression obtained to derive a formula for the inversion temperature of a van der Waals gas in terms of reduced variables, and evaluate it for the xenon sample.

3.21 The thermodynamic equation of state $(\partial U/\partial V)_T = T(\partial p/\partial T)_V - p$ was quoted in the chapter. Derive its partner

$$\left(\frac{\partial H}{\partial p}\right)_T = -T\left(\frac{\partial V}{\partial T}\right)_p + V$$

from it and the general relations between partial differentials.

3.22 Show that for a van der Waals gas,

$$C_{p,\mathrm{m}} - C_{V,\mathrm{m}} = \lambda R \qquad \frac{1}{\lambda} = 1 - \frac{(3V_\mathrm{r} - 1)^2}{4V_\mathrm{r}^3 T_\mathrm{r}}$$

and evaluate the difference for xenon at 25 °C and 10.0 atm.

3.23 The speed of sound, c_s, in a gas of molar mass M is related to the ratio of heat capacities γ by $c_\mathrm{s} = (\gamma RT/M)^{1/2}$. Show that $c_\mathrm{s} = (\gamma p/\rho)^{1/2}$, where ρ is the mass density of the gas. Calculate the speed of sound in argon at 25 °C.

Additional problems supplied by Carmen Giunta and Charles Trapp

3.24 In 1995, the Intergovernmental Panel on Climate Change considered a global average temperature rise of 1.0–3.5 °C likely by the year 2100, with 2.0 °C its best estimate (IPCC Second Assessment Synthesis of Scientific–Technical Information Relevant to Interpreting Article 2 of the UN Framework Convention on Climate Change (1995)). Predict the average rise in sea level due to thermal expansion of sea water based on temperature rises of 1.0 °C, 2.0 °C, and 3.5 °C given that the volume of the Earth's oceans is $1.37 \times 10^9\ \mathrm{km^3}$ and their surface area is $361 \times 10^6\ \mathrm{km^2}$, and state the approximations that go into the estimates.

3.25 Concerns over the harmful effects of chlorofluorocarbons on stratospheric ozone have motivated a search for new refrigerants. One such alternative is 2,2-dichloro-1,1,1-trifluoroethane (refrigerant 123). Younglove and McLinden published a compendium of thermophysical properties of this substance (B.A. Younglove and M. McLinden, *J. Phys. Chem. Ref. Data* **23**, 7 (1994)), from which properties such as the Joule–Thomson coefficient μ can be computed. (a) Compute μ at 1.00 bar and 50°C given that $(\partial H/\partial p)_T = -3.29 \times 10^3$ J MPa^{-1} mol^{-1} and $C_{p,m} = 110.0$ J K^{-1} mol^{-1}.

(b) Compute the temperature change that would accompany adiabatic expansion of 2.0 mol of this refrigerant from 1.5 bar to 0.5 bar at 50°C.

3.26 Another alternative refrigerant (see preceding problem) is 1,1,1,2-tetrafluoroethane (refrigerant HFC-134a). Tillner-Roth and Baehr published a compendium of thermophysical properties of this substance (R. Tillner-Roth and H.D. Baehr, *J. Phys. Chem. Ref. Data* **23**, 657 (1994)), from which properties such as the Joule–Thomson coefficient μ can be computed.

(a) Compute μ at 0.100 MPa and 300 K from the following data (all referring to 300 K).

p/MPa	0.080	0.100	0.12
Specific enthalpy/(kJ kg^{-1})	426.48	426.12	425.76

(The specific constant-pressure heat capacity is 0.7649 kJ K^{-1} kg^{-1}.)

(b) Compute μ at 1.00 MPa and 350 K from the following data (all referring to 350 K).

p/MPa	0.80	1.00	1.2
Specific enthalpy/(kJ kg^{-1})	461.93	459.12	456.15

(The specific constant-pressure heat capacity is 1.0392 kJ K^{-1} kg^{-1}.)

3.27 A cylindrical container of fixed total volume is divided into three sections, S_1, S_2, and S_3. The sections S_1 and S_2 are separated by an adiabatic piston, whereas S_2 and S_3 are separated by a diathermic (heat-conducting) piston. The pistons can slide along the walls of the cylinder without friction. Each section of the cylinder contains 1.00 mol of a perfect diatomic gas. Initially the gas pressure in all three sections is 1.00 bar and the temperature is 298 K. The gas in S_1 is heated slowly until the temperature of the gas in S_3 reaches 348 K. Find the final temperature, pressure, and volume, as well as the change in internal energy for each section. Determine the total energy supplied to the gas in S_1.

3.28 Solve Problem 3.27 for the case where both pistons are (a) adiabatic, (b) diathermic.

3.29 A gas obeying the equation of state $p(V - b) = nRT$ is subjected to a Joule–Thomson expansion. Will the temperature increase, decrease, or remain the same?

3.30 A gas obeys the equation of state $V_m = RT/p + aT^2$ and its constant-pressure heat capacity is given by $C_{p,m} = A + BT + Cp$, where a, A, B, and C are constants independent of T and p. Obtain expressions for (a) the Joule–Thomson coefficient and (b) the constant-volume heat capacity of the gas.

4

The Second Law: the concepts

The purpose of this chapter is to explain the origin of the spontaneity of physical and chemical change. We examine two simple processes and show that a property, the entropy, can be defined, measured, and used to discuss spontaneous changes quantitatively. The chapter also introduces a major subsidiary thermodynamic property, the Gibbs energy. The introduction of the Gibbs energy enables the spontaneity of a process to be expressed solely in terms of the properties of a system (instead of having to consider entropy changes in the system and its surroundings). The Gibbs energy also enables us to predict the maximum non-expansion work that a process can do.

Some things happen naturally; some things don't. A gas expands to fill the available volume, a hot body cools to the temperature of its surroundings, and a chemical reaction runs in one direction rather than another. Some aspect of the world determines the **spontaneous** direction of change, the direction of change that does not require work to be done to bring it about. We *can* confine a gas to a smaller volume, we *can* cool an object with a refrigerator, and we *can* force some reactions to go in reverse (as in the electrolysis of water). However, none of these processes happens spontaneously; each one must be brought about by doing work.

The recognition of two classes of process, spontaneous and non-spontaneous, is summarized by the **Second Law of thermodynamics**. This law may be expressed in a variety of equivalent ways. One statement was formulated by Kelvin:

> **No process is possible in which the sole result is the absorption of heat from a reservoir and its complete conversion into work.**

For example, it has proved impossible to construct an engine like that shown in Fig. 4.1, in which heat is drawn from a hot reservoir and completely converted into work. All real heat engines have both a hot source and a cold sink, and some heat is always discarded into the cold sink and not converted into work. The Kelvin statement is a generalization of another everyday observation, that a ball at rest on a surface has never been observed to leap

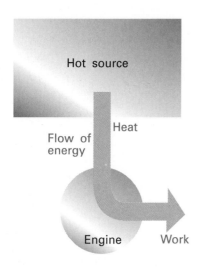

4.1 The Kelvin statement of the Second Law denies the possibility of the process illustrated here, in which heat is changed completely into work, there being no other change. The process is not in conflict with the First Law because energy is conserved.

spontaneously upwards. An upward leap of the ball would be equivalent to the conversion of heat from the surface into work.

The direction of spontaneous change

What determines the direction of spontaneous change? It is not the total energy of the isolated system. The First Law of thermodynamics states that energy is conserved in any process, and we cannot disregard that law now and say that everything tends towards a state of lower energy: the total energy of an isolated system is constant.

Is it perhaps the energy of the system of interest that tends towards a minimum? Two arguments show that this cannot be so. First, a perfect gas expands spontaneously into a vacuum, yet its internal energy remains constant as it does so. Secondly, if the energy of a system does happen to decrease during a spontaneous change, the energy of its surroundings must increase by the same amount (by the First Law). The increase in energy of the surroundings is just as spontaneous a process as the decrease in energy of the system.

When a change occurs, the total energy of an isolated system remains constant but it is parcelled out in different ways. Can it be, therefore, that the direction of change is related to the distribution of energy? We shall see that this idea is the key, and that *spontaneous changes are always accompanied by a dispersal of energy into a more disordered form*.

4.1 The dispersal of energy

The role of the distribution of energy can be illustrated by thinking about a ball (the system of interest) bouncing on a floor (the surroundings). The ball does not rise as high after each bounce because there are inelastic losses in the materials of the ball and floor (that is, the conversion of kinetic energy of the ball's overall motion into the energy of thermal motion). The direction of spontaneous change is towards a state in which the ball is at rest with all its energy degraded into the thermal motion of the atoms of the virtually infinite floor (Fig. 4.2).

A ball resting on a warm floor has never been observed to start bouncing. For bouncing to begin, something rather special would need to happen. In the first place, some of the thermal motion of the atoms in the floor would have to accumulate in a single, small object, the ball. This accumulation requires a spontaneous localization of energy from the myriad vibrations of the atoms of the floor into the much smaller number of atoms that constitute the ball (Fig. 4.3). Furthermore, whereas the thermal motion is disorderly, for the ball to move upwards its atoms must all move in the same direction. The localization of random motion as orderly motion is so unlikely that we can dismiss it as virtually impossible.[1]

We appear to have found the signpost of spontaneous change: we look for the direction of change that leads to the greater chaotic dispersal of the total energy of the isolated system. This principle accounts for the direction of change of the bouncing ball, because its energy is dissipated as thermal motion of the atoms of the floor. The reverse process is not spontaneous because it is highly improbable that the chaotic distribution of energy will become organized into localized, uniform motion. A gas does not spontaneously contract, because to do so the chaotic motion of its molecules would have to take them all into the same region of the container; the opposite change, spontaneous expansion, is a natural consequence of increasing disorder. An object does not spontaneously become warmer than its surroundings because it is highly improbable that the jostling of randomly vibrating atoms in the surroundings will lead to the accumulation of excess thermal motion in the

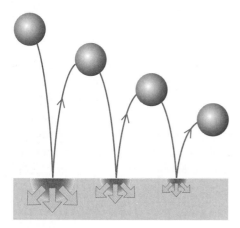

4.2 The direction of spontaneous change for a ball bouncing on a floor. On each bounce some of its energy is degraded into the thermal motion of the atoms of the floor, and that energy disperses. The reverse has never been observed to take place on a macroscopic scale.

---

1   It occurs on a much smaller scale in the form of the fluctuations of position known as Brownian motion.

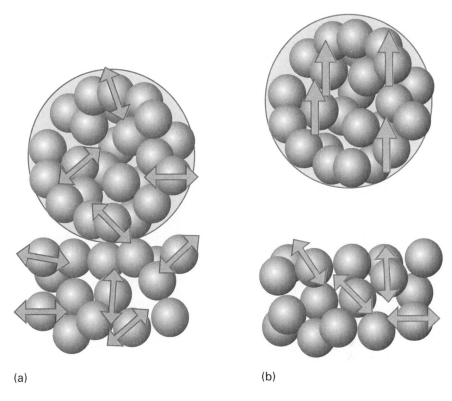

4.3 The molecular interpretation of the irreversibility expressed by the Second Law. (a) A ball resting on a warm surface; the atoms are undergoing thermal motion (chaotic vibration, in this instance), as indicated by the arrows. (b) For the ball to fly upwards, some of the random vibrational motion would have to change into coordinated, directed motion. Such a conversion is highly improbable.

object. The opposite change, the spreading of the object's energy into the surroundings as thermal motion, is a natural consequence of chaos.

It may seem very puzzling that collapse into disorder can account for the formation of such ordered substances as crystals or proteins. Nevertheless, in due course we shall see that organized structures can emerge as energy and matter disperse. We shall see, in fact, that collapse into disorder accounts for change in all its forms.

4.2 Entropy

The First Law of thermodynamics led to the introduction of the internal energy, U. The internal energy is a state function that lets us assess whether a change is permissible: only those changes may occur for which the internal energy of an isolated system remains constant. The law that is used to identify the signpost of spontaneous change, the Second Law of thermodynamics, may also be expressed in terms of another state function, the entropy, S. We shall see that the entropy (which we shall define shortly, but which is a measure of the molecular disorder of a system) lets us assess whether one state is accessible from another by a spontaneous change. The First Law uses the internal energy to identify permissible changes; the Second Law uses the entropy to identify the *spontaneous* changes among those *permissible* changes.

The **Second Law** of thermodynamics can be expressed in terms of the entropy:

The entropy of an isolated system increases in the course of a spontaneous change:

$$\Delta S_{\text{tot}} > 0 \tag{1}$$

where S_{tot} is the total entropy of the system and its surroundings. Thermodynamically irreversible processes (like cooling to the temperature of the surroundings and the free expansion of gases) are spontaneous processes, and hence must be accompanied by an increase in entropy.

(a) The thermodynamic definition of entropy

The thermodynamic definition of entropy concentrates on the change in entropy dS that occurs as a result of a physical or chemical change (in general, as a result of a 'process'). The definition is motivated by the idea that a change in the extent to which energy is dispersed in a disorderly manner depends on the quantity of energy transferred as heat. As we have remarked, heat stimulates disorderly motion in the surroundings. Work, which stimulates uniform motion of atoms in the surroundings, does not change the degree of disorder, and so does not change the entropy.

The thermodynamic definition of entropy is based on the expression

$$dS = \frac{dq_{rev}}{T} \qquad [2]$$

For a measurable change between two states i and f this expression integrates to

$$\Delta S = \int_i^f \frac{dq_{rev}}{T} \qquad (3)$$

That is, to calculate the difference in entropy between any two states of a system, we find a *reversible* path between them, and integrate the heat supplied at each stage of the path divided by the temperature at which the heat is supplied.

Molecular interpretation 4.1 The molecules in a system at high temperature are highly disorganized, either in terms of their locations or in terms of the occupation of their available translational, rotational, and vibrational energy states. A small additional transfer of energy will result in a relatively small additional disorder, much as sneezing in a busy street may be barely noticed. In contrast, the molecules in a system at low temperature have access to far fewer energy states (at $T = 0$, only the lowest state is accessible), and the same quantity of heat will have a pronounced effect on the degree of disorder, much as sneezing in a quiet library can be very disruptive. Hence, the change in entropy when a given quantity of heat is transferred will be greater when it is transferred to a cold body than when it is transferred to a hot body. This argument suggests that the change in entropy should be inversely proportional to the temperature at which the transfer takes place, as in eqn 2.

According to eqn 2, when the heat transferred is expressed in joules and the temperature is in kelvins, the units of entropy are joules per kelvin (J K^{-1}). Molar entropy, the entropy divided by the amount of substance, is expressed in joules per kelvin per mole (J K^{-1} mol^{-1}), the same units as those of the gas constant, R, and molar heat capacities.

Example 4.1 Calculating the entropy change for the isothermal expansion of a perfect gas

Calculate the entropy change of a sample of perfect gas when it expands isothermally from a volume V_i to a volume V_f.

Method The definition of entropy instructs us to find the heat absorbed for a reversible path between the stated initial and final states regardless of the actual manner in which the process takes place. A simplification is that the expansion is isothermal, so the temperature is a constant and may be taken outside the integral in eqn 3. The heat absorbed during a reversible isothermal expansion of a perfect gas can be calculated from $\Delta U = q + w$ and $\Delta U = 0$, which implies that $q = -w$ in general and therefore that $q_{rev} = -w_{rev}$ for a reversible change. The work of reversible isothermal expansion was calculated in Section 2.3 (eqn 2.13).

Answer Because the temperature is constant, eqn 3 becomes

$$\Delta S = \frac{1}{T} \int_i^f dq_{rev} = \frac{q_{rev}}{T}$$

From eqn 2.13, we know that

$$q_{rev} = -w_{rev} = nRT \ln\left(\frac{V_f}{V_i}\right)$$

Therefore, it follows that

$$\Delta S = nR \ln\left(\frac{V_f}{V_i}\right)$$

Comment As an illustration of this formula, when the volume of 1.00 mol of any perfect gas is doubled at any constant temperature,

$$\Delta S = (1.00 \text{ mol}) \times (8.3145 \text{ J K}^{-1} \text{mol}^{-1}) \times \ln 2 = +5.76 \text{ J K}^{-1}$$

Self-test 4.1 Calculate the change in entropy when the pressure of a perfect gas is changed isothermally from p_i to p_f.

$$[\Delta S = nR \ln(p_i/p_f)]$$

The definition in eqn 2 can be used to formulate an expression for the change in entropy of the surroundings, ΔS_{sur}. Consider an infinitesimal transfer of heat dq_{sur} to the surroundings. The surroundings consist of a reservoir of constant volume,[2] so the heat supplied to them can be identified with the change in their internal energy, dU_{sur}. The internal energy is a state function, and dU_{sur} is an exact differential. As we have seen, these properties imply that dU_{sur} is independent of how the change is brought about, and in particular is independent of whether the process is reversible or irreversible. The same remarks therefore apply to dq_{sur}, to which dU_{sur} is equal. Therefore, we can adapt the definition of entropy change in eqn 2 to write

$$dS_{sur} = \frac{dq_{sur,rev}}{T_{sur}} = \frac{dq_{sur}}{T_{sur}} \tag{4}$$

Furthermore, because the temperature of the surroundings is constant whatever the change, for a measurable change

$$\Delta S_{sur} = \frac{q_{sur}}{T_{sur}} \tag{5}$$

That is, regardless of how the change is brought about in the system, the change of entropy of the surroundings can be calculated by dividing the heat transferred by the temperature at which the transfer takes place.

[2] Alternatively, the surroundings can be regarded as being at constant pressure, in which case we could equate dq_{sur} to dH_{sur}.

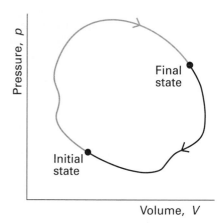

4.4 In a thermodynamic cycle, the overall change in a state function (from the initial state to the final state and then back to the initial state again) is zero.

Equation 5 makes it very simple to calculate the changes in entropy of the surroundings that accompany any process. For instance, for any adiabatic change, $q_{sur} = 0$, so

$$\Delta S_{sur} = 0 \qquad (6)$$

This expression is true however the change takes place, reversibly or irreversibly, provided no local hot spots are formed in the surroundings. That is, it is true so long as the surroundings remain in internal equilibrium. If hot spots do form, then the localized energy may subsequently disperse spontaneously and hence generate more entropy.

Illustration

To calculate the entropy change in the surroundings when 1.00 mol $H_2O(l)$ is formed from its elements under standard conditions at 298 K, we use $\Delta H^{\ominus} = -286$ kJ from Table 2.6. The heat released is supplied to the surroundings, now regarded as being at constant pressure, so $q_{sur} = +286$ kJ. Therefore,

$$\Delta S_{sur} = \frac{2.86 \times 10^5 \text{ J}}{298 \text{ K}} = +959 \text{ J K}^{-1}$$

This strongly exothermic reaction results in an increase in the entropy of the surroundings as heat is released into them.

Self-test 4.2 Calculate the entropy change in the surroundings when 1.00 mol $N_2O_4(g)$ is formed from 2.00 mol $NO_2(g)$ under standard conditions at 298 K.

$$[-192 \text{ J K}^{-1}]$$

(b) The entropy as a state function

The entropy is a state function. To prove this assertion, we need to show that the integral of dS is independent of path. To do so, it is sufficient to prove that the integral of eqn 2 around an arbitrary cycle is zero, for that guarantees that the entropy is the same at the initial and final states of the system regardless of the path taken between them (Fig. 4.4). That is, we need to show that

$$\oint \frac{dq_{rev}}{T} = 0 \qquad (7)$$

where the symbol $\oint$ denotes integration around a closed path.

To prove eqn 7, we first consider the special **Carnot cycle** shown in Fig. 4.5. A Carnot cycle, which is named after the French engineer Sadi Carnot, consists of four reversible stages:

1. Reversible isothermal expansion from A to B at T_h; the entropy change is q_h/T_h, where q_h is the heat supplied to the system from the hot source. In this step, q_h is positive.

2. Reversible adiabatic expansion from B to C. No heat leaves the system, so the change in entropy is zero. In the course of this expansion, the temperature falls from T_h to T_c, the temperature of the cold sink.

3. Reversible isothermal compression from C to D at T_c. Heat is released to the cold sink; the change in entropy of the system is q_c/T_c; in this expression q_c is negative.

4. Reversible adiabatic compression from D to A. No heat enters the system, so the change in entropy is zero. The temperature rises from T_c to T_h.

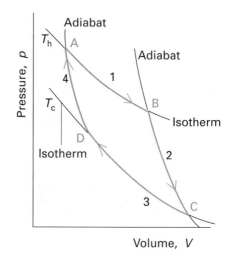

4.5 The basic structure of a Carnot cycle. In step 1, there is isothermal reversible expansion at the temperature T_h. Step 2 is a reversible adiabatic expansion in which the temperature falls from T_h to T_c. In Step 3 there is an isothermal reversible compression at T_c, and that isothermal step is followed by an adiabatic reversible compression, which restores the system to its initial state.

The total change in entropy around the cycle is

$$\oint dS = \frac{q_h}{T_h} + \frac{q_c}{T_c}$$

However, we show in the *Justification* below that

$$\frac{q_h}{q_c} = -\frac{T_h}{T_c} \tag{8}$$

Substitution of this relation into the preceding equation gives zero on the right, which is what we wanted to prove.

Justification 4.1

As explained in Example 4.1, for a perfect gas:

$$q_h = nRT_h \ln\left(\frac{V_B}{V_A}\right) \qquad q_c = nRT_c \ln\left(\frac{V_D}{V_C}\right)$$

From the relations between temperature and volume for reversible adiabatic processes (eqn 2.34):

$$V_A T_h^c = V_D T_c^c \qquad V_C T_c^c = V_B T_h^c$$

Multiplication of the first of these expressions by the second gives

$$V_A V_C T_h^c T_c^c = V_D V_B T_h^c T_c^c$$

which simplifies to

$$\frac{V_A}{V_B} = \frac{V_D}{V_C}$$

Consequently,

$$q_c = nRT_c \ln\left(\frac{V_A}{V_B}\right)$$

and eqn 8 follows.

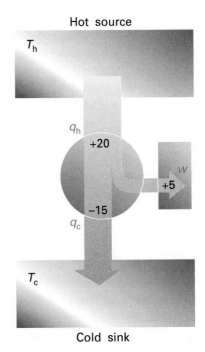

Hot source

T_h

q_h

+20

w

+5

−15

q_c

T_c

Cold sink

4.6 Suppose an energy q_h (for example, 20 kJ) is supplied to the engine and q_c is lost from the engine (for example, $q_c = -15$ kJ) and discarded into the cold reservoir. The work done by the engine is equal to $q_h + q_c$ (for example, 20 kJ + (−15 kJ) = 5 kJ). The efficiency is the work done divided by the heat supplied from the hot source.

Now we need to show that the same conclusion applies to any material, not just a perfect gas. To do so, consider the **efficiency**, ε, of a heat engine:

$$\varepsilon = \frac{\text{work performed}}{\text{heat absorbed}} = \frac{|w|}{q_h} \tag{9}$$

The definition implies that, the greater the work output for a given supply of heat from the hot reservoir, the greater the efficiency of the engine. The definition can be expressed in terms of the heat transactions alone, because (as shown in Fig. 4.6) the work performed by the engine is the difference between the heat supplied by the hot reservoir and that returned to the cold reservoir:

$$\varepsilon = \frac{q_h + q_c}{q_h} = 1 + \frac{q_c}{q_h} \tag{10}$$

(Remember that $q_c < 0$.) It then follows that for a Carnot engine

$$\varepsilon_{rev} = 1 - \frac{T_c}{T_h} \tag{11}$$

Now, the Second Law of thermodynamics implies that *all reversible engines have the same efficiency* regardless of their construction. To see the truth of this statement, suppose two

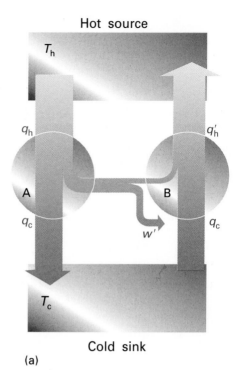

(a)

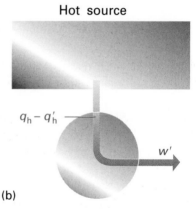

(b)

4.7 (a) The demonstration of the equivalence of the efficiencies of all reversible engines working between the same thermal reservoirs is based on the flow of energy represented in this diagram. (b) The net effect of the processes is the conversion of heat into work without there being a need for a cold sink: this is contrary to the Kelvin statement of the Second Law.

reversible engines are coupled together and run between the same two reservoirs (Fig. 4.7). The working substances and details of construction of the two engines are entirely arbitrary. Initially, suppose that engine A is more efficient than engine B, and that we choose a setting of the controls that causes engine B to acquire the heat q_c from the cold reservoir and to release a certain quantity of heat into the hot reservoir. However, because engine A is more efficient than engine B, not all the work that A produces is needed for this process, and the difference can be used to do work. The net result is that the cold reservoir is unchanged, work has been produced, and the hot reservoir has lost a certain amount of energy. This outcome is contrary to the Kelvin statement of the Second Law, because some heat has been converted directly into work. In molecular terms, the disordered thermal motion of the hot reservoir has been converted into ordered motion characteristic of work. Because the conclusion is contrary to experience, the initial assumption that engines A and B can have different efficiencies must be false. It follows that the relation between the heat transfers and the temperatures (eqn 11) must also be independent of the working material, and therefore that eqn 8 is always true for any substance involved in a Carnot cycle.

To complete the argument, we note that any reversible cycle can be approximated as a collection of Carnot cycles (Fig. 4.8). This approximation becomes exact as the individual cycles are allowed to become infinitesimal. The entropy change around each individual cycle is zero (as demonstrated above), so the sum of entropy changes for all the cycles is zero. However, in the interior of the overall cycle, the entropy change along any path is cancelled by the entropy change along the path it shares with the neighbouring cycle. Therefore, all the entropy changes cancel except for those along the perimeter of the overall cycle. That is,

$$\sum_{\text{all}} \frac{q_{\text{rev}}}{T} = \sum_{\text{perimeter}} \frac{q_{\text{rev}}}{T} = 0$$

In the limit of infinitesimal cycles, the non-cancelling edges of the Carnot cycles match the overall cycle exactly, and the sum becomes an integral. Equation 7 then follows immediately. This result implies that dS is an exact differential and therefore that S is a state function.

(c) The thermodynamic temperature

Suppose we have an engine that is working reversibly between a hot source at a temperature T_h and a cold sink at a temperature T; then we know from eqn 11 that

$$T = (1 - \varepsilon)T_h \qquad (12)$$

This expression enabled Kelvin to define the **thermodynamic temperature scale** in terms of the efficiency of a heat engine. The zero of the scale occurs for a Carnot efficiency of 1. The size of the unit is entirely arbitrary, but on the **Kelvin scale** is defined by setting the temperature of the triple point of water as 273.16 K exactly. Then, if the heat engine has a hot source at the triple point of water, the temperature of the cold sink (the object we want to measure) is found by measuring the efficiency of the engine. This result is independent of the working substance.

An additional point is that, as we saw in Section 2.2c, heat transferred can, in principle, be measured mechanically (in terms of the location of a weight). Therefore, it is possible, in principle at least, to use the distance moved by a weight to measure temperature. Kelvin's definition of the thermodynamic temperature scale puts the measurement of temperature on to a purely mechanical basis.

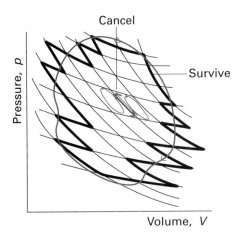

4.8 A general cycle can be divided into small Carnot cycles. The match is exact in the limit of infinitesimally small cycles. Paths cancel in the interior of the collection, and only the perimeter, an increasingly good approximation to the true cycle as the number of cycles increases, survives. Because the entropy change around every individual cycle is zero, the integral of the entropy around the perimeter is zero too.

(d) The Clausius inequality

So far, we have verified that the entropy as defined in eqn 2 is a state function. We now need to verify that the entropy is a signpost of spontaneous change in the sense that $dS_{tot} \geq 0$ for *any* spontaneous change.

Consider a system in thermal and mechanical contact with its surroundings at the same temperature, T. The system and the surroundings are not necessarily in mechanical equilibrium (for instance, a gas might have a greater pressure than that of its surroundings). Any change of state is accompanied by a change in entropy of the system, dS, and of the surroundings, dS_{sur}. Because the process might be irreversible, the total entropy will increase when a process occurs in the system, so we can write

$$dS + dS_{sur} \geq 0, \qquad \text{or } dS \geq -dS_{sur}$$

(The equality applies if the process is reversible.) Because eqn 5 implies that $dS_{sur} = -dq/T$, where dq is the heat supplied to the system during the process (that is, $dq_{sur} = -dq$, because the heat that enters the system comes from the surroundings), it follows that for any change

$$dS \geq \frac{dq}{T} \tag{13}$$

This expression is the **Clausius inequality**.

Suppose the system is isolated from its surroundings. Then $dq = 0$, and the Clausius inequality implies that

$$dS \geq 0 \tag{14}$$

This is exactly the characteristic we need for the entropy to be the signpost of spontaneous change, for it tells us that in an *isolated* system *the entropy of the system alone* cannot decrease when a spontaneous change takes place.

We can illustrate the content of the Clausius inequality in two simple cases. First, suppose a system undergoes an irreversible adiabatic change. Then $dq = 0$ and, by eqn 13, $dS \geq 0$. That is, for this type of spontaneous change, the entropy of the system has increased. Because no heat flows into the surroundings, their entropy remains constant and $dS_{sur} = 0$. Therefore, the total entropy of the system and its surroundings obeys $dS_{total} \geq 0$.

Now consider irreversible isothermal expansion of a perfect gas. As we saw in Example 4.1, for such a change $dq = -dw$ (because $dU = 0$). If the gas expands freely into a vacuum, it does no work and $dw = 0$, which implies that $dq = 0$ too. Therefore, according to the Clausius inequality, $dS \geq 0$. Next, consider the surroundings. No heat is transferred into the surroundings, so $dS_{sur} = 0$. Therefore, in this case too $dS_{tot} \geq 0$.

Another type of irreversible process is spontaneous cooling. Consider a transfer of energy as heat dq from one system—the hot source—at a temperature T_h to another system—the cold sink—at a temperature T_c (Fig. 4.9). When $|dq|$ leaves the hot source, the entropy of the source changes by $-|dq|/T_h$ (a decrease). When $|dq|$ enters the cold sink its entropy changes by $+|dq|/T_c$ (an increase). The overall change in entropy is therefore

$$dS = \frac{|dq|}{T_c} - \frac{|dq|}{T_h} = |dq|\left(\frac{1}{T_c} - \frac{1}{T_h}\right) \tag{15}$$

which is positive (because $T_h \geq T_c$). Hence, cooling (the transfer of heat from hot to cold) is spontaneous, as we know from experience. When the temperatures of the two systems are equal, $dS_{tot} = 0$: the two systems are then at thermal equilibrium.

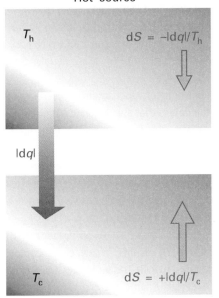

4.9 When energy leaves a hot reservoir as heat, the entropy of the reservoir decreases. When the same quantity of energy enters a cooler reservoir, the entropy increases by a larger amount. Hence, overall there is an increase in entropy and the process is spontaneous. Relative changes in entropy are indicated by the sizes of the arrows.

4.3 Entropy changes accompanying specific processes

We now see how to calculate the entropy changes that accompany a variety of simple processes.

(a) The entropy of phase transition at the transition temperature

Because a change in the degree of molecular order occurs when a substance freezes or boils, we should expect the transition to be accompanied by a change in entropy. For example, when a substance vaporizes, a compact condensed phase changes into a widely dispersed gas, and we can expect the entropy of the substance to increase considerably. The entropy of a solid substance increases when it melts to a liquid, and it also increases when the liquid phase turns into a gas.

Consider a system and its surroundings at the normal transition temperature T_{trs}, the temperature at which two phases are in equilibrium at 1 atm. This temperature is $0\,^\circ\text{C}$ (273 K) for ice in equilibrium with liquid water at 1 atm, and $100\,^\circ\text{C}$ (373 K) for water in equilibrium with its vapour at 1 atm. At the transition temperature, any transfer of heat between the system and its surroundings is reversible because the two phases in the system are in equilibrium. Because at constant pressure $q = \Delta_{trs}H$, the change in molar entropy[3] of the system is

$$\Delta_{trs}S = \frac{\Delta_{trs}H}{T_{trs}} \tag{16}$$

If the phase transition is exothermic ($\Delta_{trs}H < 0$, as in freezing or condensing), then the entropy change is negative. This decrease in entropy is consistent with the system becoming more ordered when a solid forms from a liquid. If the transition is endothermic ($\Delta_{trs}H > 0$, as in melting), then the entropy change is positive, which is consistent with the system becoming more disordered. Melting and vaporizing are endothermic processes, so both are accompanied by an increase in the system's entropy. This increase is consistent with liquids being more disordered than solids, and gases more disordered than liquids. Some experimental entropies of transition are listed in Table 4.1.

In Table 4.2 we list in more detail the standard entropies of vaporization of several liquids at their boiling points. An interesting feature of the data is that a wide range of liquids give approximately the same standard entropy of vaporization (about $85\ \text{J K}^{-1}\,\text{mol}^{-1}$): this empirical observation is called **Trouton's rule**.

Molecular interpretation 4.2 The explanation of Trouton's rule is that a comparable amount of disorder is generated when any liquid evaporates and becomes a gas. Hence, all liquids can be expected to have similar standard entropies of vaporization.

Liquids that show significant deviations from Trouton's rule do so on account of the molecules in the liquid being arranged in a partially orderly manner. In such cases, a greater change of disorder occurs when the liquid evaporates than if the molecules were highly disordered in the liquid. An example is water, where the large entropy of vaporization reflects the presence of structure arising from hydrogen-bonding in the liquid. Hydrogen bonds tend to organize the molecules in the liquid so that they are less random than, for example, the molecules in liquid hydrogen sulfide (which is not hydrogen bonded).

Methane has an unusually low entropy of vaporization. A part of the reason is that the entropy of the gas itself is slightly low (186 J K^{-1} mol^{-1} at 298 K); the entropy of N$_2$ under the same conditions is 192 J K^{-1} mol^{-1}. As we shall see in Chapter 19, light molecules are

---

3 Recall from Section 2.7 that $\Delta_{trs}H$ is an enthalpy change per mole of substance, so $\Delta_{trs}S$ is also a molar quantity.

Table 4.1* Standard entropies (and temperatures) of phase transitions, $\Delta_{trs}S^{\ominus}/(\text{J K}^{-1}\,\text{mol}^{-1})$

	Fusion (at T_f)	Vaporization (at T_b)
Argon, Ar	14.17 (at 83.8 K)	74.53 (at 87.3 K)
Benzene, C_6H_6	38.00 (at 279 K)	87.19 (at 353 K)
Water, H_2O	22.00 (at 273.15 K)	109.0 (at 373.15 K)
Helium, He	4.8 (at 1.8 K and 30 bar)	19.9 (at 4.22 K)

* More values are given in the *Data section* at the end of this volume.

Table 4.2* The standard entropies of vaporization of liquids

	$\Delta_{vap}H^{\ominus}/(\text{kJ mol}^{-1})$	$\theta_b/°C$	$\Delta_{vap}S^{\ominus}/(\text{J K}^{-1}\,\text{mol}^{-1})$
Benzene	+30.8	80.1	+87.2
Carbon tetrachloride	+30.00	76.7	+85.8
Cyclohexane	+30.1	80.7	+85.1
Hydrogen sulfide	+18.7	−60.4	+87.9
Methane	+8.18	−161.5	+73.2
Water	+40.7	100.0	+109.1

* More values are given in the *Data section*.

difficult to excite into rotation; as a result, only a few rotational states are accessible at room temperature, and the disorder associated with the population of rotational states is low.

Example 4.2 Using Trouton's rule

Predict the standard molar enthalpy of vaporization of bromine given that it boils at 59.2 °C.

Method We need to judge whether there is the likelihood of anomalous structural organization in the liquid phase or some anomaly in the gas phase. If there is not, it is permissible to use Trouton's rule in the form

$$\Delta_{vap}H^{\ominus} = T_b \times (85\ \text{J K}^{-1}\,\text{mol}^{-1})$$

Answer There is no hydrogen bonding in liquid bromine and Br_2 is a heavy molecule that is unlikely to display unusual behaviour in the gas phase, so it would seem safe to use Trouton's rule. Substitution of the data then gives

$$\Delta_{vap}H^{\ominus} = (332.4\ \text{K}) \times (85\ \text{J K}^{-1}\,\text{mol}^{-1}) = +28\ \text{kJ mol}^{-1}$$

The experimental value is +29.45 kJ mol^{-1}.

- -

Self-test 4.3 Predict the enthalpy of vaporization of ethane from its boiling point, −88.6 °C.

$$[+16\ \text{kJ mol}^{-1}]$$

(b) The expansion of a perfect gas

We established in Example 4.1 that the change in entropy of a perfect gas that expands isothermally from V_i to V_f is

$$\Delta S = nR \ln\left(\frac{V_f}{V_i}\right) \tag{17}°$$

Because S is a state function, this expression applies whether the change of state occurs reversibly or irreversibly.

If the change is reversible, the entropy change in the surroundings (which are in thermal and mechanical equilibrium with the system) must be such as to give $\Delta S_{tot} = 0$. Therefore, in this case, the change of entropy of the surroundings is the negative of the expression in eqn 17. If the expansion occurs freely ($w = 0$) and irreversibly, and if the temperature remains constant, then $q = 0$. Consequently, $\Delta S_{sur} = 0$, and the total entropy change is given by eqn 17.

(c) The variation of entropy with temperature

Equation 3 can be used to calculate the entropy of a system at a temperature T_f from a knowledge of its entropy at a temperature T_i and the heat supplied to change its temperature from one value to the other:

$$S(T_f) = S(T_i) + \int_i^f \frac{dq_{rev}}{T} \tag{18}$$

We shall be particularly interested in the entropy change when the system is subjected to constant pressure (such as from the atmosphere) during the heating. Then, from the definition of constant-pressure heat capacity (eqn 2.27),

$$dq_{rev} = C_p \, dT$$

so long as the system is doing no non-expansion work. Consequently, at constant pressure:

$$S(T_f) = S(T_i) + \int_i^f \frac{C_p \, dT}{T} \tag{19}$$

The same expression applies at constant volume, but with C_p replaced by C_V. When C_p is independent of temperature in the temperature range of interest, we obtain

$$S(T_f) = S(T_i) + C_p \int_i^f \frac{dT}{T} = S(T_i) + C_p \ln\left(\frac{T_f}{T_i}\right) \tag{20}$$

with a similar expression for heating at constant volume.

Example 4.3 Calculating the entropy change

Calculate the entropy change when argon at 25 °C and 1.00 atm in a container of volume 500 cm^3 is allowed to expand to 1000 cm^3 and is simultaneously heated to 100 °C.

Method Because S is a state function, we are free to choose the most convenient path from the initial state. One such path is reversible isothermal expansion to the final volume, followed by reversible heating at constant volume to the final temperature. The entropy change in the first step is given by eqn 17 and that of the second step, provided C_V is independent of temperature, by eqn 20 (with C_V in place of C_p). In each case we need to know n, the amount of gas, and can calculate it from the perfect gas equation and the data for the initial state. The heat capacity at constant volume can be obtained from the value of $C_{p,m}$ in Table 2.6 and the relation $C_{p,m} - C_{V,m} = R$.

Answer The amount of Ar present (from $n = pV/RT$) is 0.0204 mol. The entropy change in the first step (expansion from 500 cm^3 to 1000 cm^3 at 298 K) is

$$\Delta S = nR \ln 2.00 = +0.118 \text{ J K}^{-1}$$

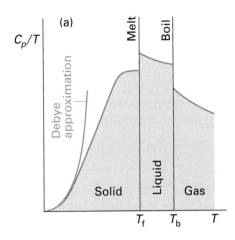

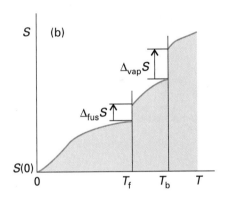

4.10 The determination of entropy from heat capacity data. (a) The variation of C_p/T with the temperature for a sample. (b) The entropy, which is equal to the area beneath the upper curve up to the corresponding temperature, plus the entropy of each phase transition passed.

The entropy change in the second step, from 298 K to 373 K at constant volume, is

$$\Delta S = (0.0204 \text{ mol}) \times (12.48 \text{ J K}^{-1} \text{ mol}^{-1}) \times \ln\left(\frac{373 \text{ K}}{298 \text{ K}}\right) = +0.057 \text{ J K}^{-1}$$

The overall entropy change, the sum of these two changes, is $\Delta S = +0.175 \text{ J K}^{-1}$

Self-test 4.4 Calculate the entropy change when the same initial sample is compressed to 50.0 cm^3 and cooled to $-25\,^\circ$C.

$$[-0.44 \text{ J K}^{-1}]$$

(d) The measurement of entropy

The entropy of a system at a temperature T is related to its entropy at $T = 0$ by measuring its heat capacity C_p at different temperatures and evaluating the integral in eqn 19. The entropy of transition ($\Delta_{\text{trs}}H/T_{\text{trs}}$) must be added for each phase transition between $T = 0$ and the temperature of interest. For example, if a substance melts at T_f and boils at T_b, then its entropy above its boiling temperature is given by

$$S(T) = S(0) + \int_0^{T_f} \frac{C_p(\text{s})\,dT}{T} + \frac{\Delta_{\text{fus}}H}{T_f}$$
$$+ \int_{T_f}^{T_b} \frac{C_p(\text{l})\,dT}{T} + \frac{\Delta_{\text{vap}}H}{T_b} + \int_{T_b}^{T} \frac{C_p(\text{g})\,dT}{T} \tag{21}$$

All the properties required, except $S(0)$, can be measured calorimetrically, and the integrals can be evaluated either graphically or, as is now more usual, by fitting a polynomial to the data and integrating the polynomial analytically. The procedure is illustrated in Fig. 4.10: the area under the curve of C_p/T against T is the integral required. Because $dT/T = d \ln T$, an alternative procedure is to evaluate the area under a plot of C_p against $\ln T$.

One problem with the measurement of entropy is the difficulty of measuring heat capacities near $T = 0$. There are good theoretical grounds for assuming that the heat capacity is proportional to T^3 when T is low (see Section 11.1c), and this dependence is the basis of the **Debye extrapolation**. In this method, C_p is measured down to as low a temperature as possible, and a curve of the form aT^3 is fitted to the data. That fit determines the value of a, and the expression $C_p = aT^3$ is assumed valid down to $T = 0$.

Illustration

The standard molar entropy of nitrogen gas at 25 °C has been calculated from the following data:

	$S_m^{\ominus}/(\text{J K}^{-1} \text{ mol}^{-1})$
Debye extrapolation	1.92
Integration, from 10 K to 35.61 K	25.25
Phase transition at 35.61 K	6.43
Integration, from 35.61 K to 63.14 K	23.38
Fusion at 63.14 K	11.42
Integration, from 63.14 K to 77.32 K	11.41
Vaporization at 77.32 K	72.13
Integration, from 77.32 K to 298.15 K	39.20
Correction for gas imperfection[4]	0.92
Total	192.06

4 See Section 20.5.

Therefore,

$$S_m^{\ominus}(298.15 \text{ K}) = S_m^{\ominus}(0) + 192.1 \text{ J K}^{-1} \text{mol}^{-1}$$

· ·

Example 4.4 Calculating the entropy at low temperatures

The molar constant-pressure heat capacity of a certain solid at 10 K is 0.43 J K^{-1} mol^{-1}. What is its molar entropy at that temperature?

Method Because the temperature is so low, we can assume that the heat capacity varies with temperature as aT^3, in which case we can use eqn 19 to calculate the entropy at a temperature T in terms of the entropy at $T = 0$ and the constant a. When the integration is carried out, it turns out that the result can be expressed in terms of the heat capacity at the temperature T, so the data can be used directly to calculate the entropy.

Answer The integration required is

$$S(T) = S(0) + \int_0^T \frac{aT^3 \, dT}{T} = S(0) + a \int_0^T T^2 \, dT = S(0) + \tfrac{1}{3}aT^3$$

However, because aT^3 is the heat capacity at the temperature T,

$$S(T) = S(0) + \tfrac{1}{3}C_p(T)$$

from which it follows that

$$S_m(10 \text{ K}) = S_m(0) + 0.14 \text{ J K}^{-1} \text{mol}^{-1}$$

- -

Self-test 4.5 For metals, there is also a contribution to the heat capacity from the electrons which is linearly proportional to T when the temperature is low. Find its contribution to the entropy at low temperatures.

$$[S(T) = S(0) + C_p(T)]$$

4.4 The Third Law of thermodynamics

At $T = 0$, all energy of thermal motion has been quenched, and in a perfect crystal all the atoms or ions are in a regular, uniform array. The absence of both spatial disorder and thermal motion suggests that such materials also have zero entropy. This conclusion is consistent with the molecular interpretation of entropy, because $S = 0$ if there is only one way of arranging the molecules.

(a) The Nernst heat theorem

The thermodynamic observation that turns out to be consistent with the view that the entropy of a regular array of molecules is zero at $T = 0$ is known as the **Nernst heat theorem**:

> **The entropy change accompanying any physical or chemical transformation approaches zero as the temperature approaches zero: $\Delta S \rightarrow 0$ as $T \rightarrow 0$.**

As an example of the experimental evidence for this law, consider the entropy of the transition between orthorhombic sulfur, S(α), and monoclinic sulfur, S(β), which can be calculated from the transition enthalpy $(-402 \text{ J mol}^{-1})$ at the transition temperature (369 K):

$$\Delta_{trs}S = S_m(\alpha) - S_m(\beta) = \frac{(-402 \text{ J mol}^{-1})}{369 \text{ K}} = -1.09 \text{ J K}^{-1} \text{ mol}^{-1}$$

The two individual entropies can also be determined by measuring the heat capacities from $T = 0$ up to $T = 369$ K. It is found that

$$S_m(\alpha) = S_m(\alpha, 0) + 37 \text{ J K}^{-1} \text{ mol}^{-1}$$
$$S_m(\beta) = S_m(\beta, 0) + 38 \text{ J K}^{-1} \text{ mol}^{-1}$$

These two values imply that, at the transition temperature,

$$\Delta_{trs}S = S_m(\alpha, 0) - S_m(\beta, 0) - 1 \text{ J K}^{-1} \text{ mol}^{-1}$$

On comparing this value with the one above, we conclude that

$$S_m(\alpha, 0) - S_m(\beta, 0) \approx 0$$

in accord with the theorem.

It follows from the Nernst theorem that, if we arbitrarily ascribe the value zero to the entropies of elements in their perfect crystalline form at $T = 0$, then all perfect crystalline compounds also have zero entropy at $T = 0$ (because the change in entropy that accompanies the formation of the compounds, like the entropy of all transformations at that temperature, is zero). Hence, *all* perfect crystals may be taken to have zero entropy at $T = 0$. This conclusion is summarized by the **Third Law of thermodynamics**:

> **If the entropy of every element in its most stable state at $T = 0$ is taken as zero, then every substance has a positive entropy which at $T = 0$ may become zero, and which does become zero for all perfect crystalline substances, including compounds.**

Note that a non-crystalline perfect state, such as the superfluid state of He (Section 6.3c), is included by the opening phrase.

It should also be noted that the Third Law does not state that entropies *are* zero at $T = 0$: it merely implies that all perfect materials have the same entropy at that temperature. As far as thermodynamics is concerned, choosing this common value as zero is then a matter of convenience. The molecular interpretation of entropy, however, implies that $S = 0$ at $T = 0$.

(b) Third-Law entropies

The choice $S(0) = 0$ for perfect crystals will be made from now on. Entropies reported on the basis of this choice are called **Third-Law entropies** (and often just 'standard entropies'). When the substance is in its standard state at the temperature T, the standard (Third-Law) entropy is denoted $S^{\ominus}(T)$. A list of values at 298 K is given in Table 4.3.

The **standard reaction entropy**, $\Delta_r S^{\ominus}$, is defined, like the standard reaction enthalpy, as the difference between the molar entropies of the pure, separated products and the pure, separated reactants, all substances being in their standard states at the specified temperature:

$$\Delta_r S^{\ominus} = \sum_{\text{Products}} \nu S_m^{\ominus} - \sum_{\text{Reactants}} \nu S_m^{\ominus} \tag{22a}$$

Table 4.3* Standard Third-Law entropies at 298 K

	$S_m^{\ominus}/(\text{J K}^{-1}\text{ mol}^{-1})$
Solids:	
Graphite, C(s)	5.7
Diamond, C(s)	2.4
Sucrose, $C_{12}H_{22}O_{11}$(s)	360.2
Iodine, I_2(s)	116.1
Liquids:	
Benzene, C_6H_6(l)	173.3
Water, H_2O(l)	69.9
Mercury, Hg(l)	76.0
Gases:	
Methane, CH_4(g)	186.3
Carbon dioxide, CO_2(g)	213.7
Hydrogen, H_2(g)	130.7
Helium, He(g)	126.2
Ammonia, NH_3(g)	192.3

* More values are given in the *Data section*.

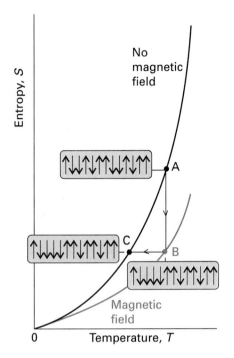

4.11 The technique of adiabatic demagnetization is used to attain very low temperatures. The upper curve shows the variation of the entropy of a paramagnetic system in the absence of an applied field. The lower curve shows the variation in entropy when a field is applied and has made the electron magnets more orderly. The isothermal magnetization step is from A to B; the adiabatic demagnetization step (at constant entropy) is from B to C.

In this expression, each term is weighted by the appropriate stoichiometric coefficient. More formally, in the notation introduced in eqn 2.41,

$$\Delta_r S^{\ominus} = \sum_J \nu_J S_m^{\ominus}(J) \tag{22b}$$

Illustration

. . . .

To calculate the standard reaction entropy of $H_2(g) + \frac{1}{2}O_2(g) \rightarrow H_2O(l)$ at 25 °C, we use the data in Table 2.6 of the *Data section* to write

$$\Delta_r S^{\ominus} = S_m^{\ominus}(H_2O, l) - \{S_m^{\ominus}(H_2, g) + \tfrac{1}{2}S_m^{\ominus}(O_2, g)\}$$

$$= 69.9 - \{130.7 + \tfrac{1}{2}(205.0)\} \ \text{J K}^{-1}\,\text{mol}^{-1} = -163.3 \ \text{J K}^{-1}\,\text{mol}^{-1}$$

The negative value is consistent with the conversion of two gases to a compact liquid.

. . . .

Self-test 4.6 Calculate the standard reaction entropy for the combustion of methane to carbon dioxide and liquid water at 25 °C.

$$[-243 \ \text{J K}^{-1}\,\text{mol}^{-1}]$$

4.5 Reaching very low temperatures

The world record low temperature stands at about 20 nK. Gases may be cooled by Joule–Thomson expansion below their inversion temperature, and temperatures lower than 4 K (the boiling point of helium) can be reached by the evaporation of liquid helium by pumping rapidly through large-diameter pipes. Temperatures as low as about 1 K can be reached in this way, but at lower temperatures helium is too involatile for this procedure to be effective; moreover, the superfluid phase begins to interfere with the cooling process by creeping round the apparatus.

The method used to reach very low temperatures is **adiabatic demagnetization**. In the absence of a magnetic field, the unpaired electrons of a paramagnetic material are orientated at random, but in the presence of a magnetic field there are more β spins ($m_s = -\frac{1}{2}$; these terms are explained in Section 12.8) than α spins ($m_s = +\frac{1}{2}$). In thermodynamic terms, the application of a magnetic field lowers the entropy of a sample (Fig. 4.11) and, at a given temperature, the entropy of a sample is lower when the field is on than when it is off.

A sample of paramagnetic material, such as a d- or f-metal complex, is cooled to about 1 K by using helium. Gadolinium(III) sulfate octahydrate, $Gd_2(SO_4)_3 \cdot 8H_2O$, has been used because each gadolinium ion carries several unpaired electrons but is separated from its neighbours by a coordination sphere of hydrating H_2O molecules. The sample is then exposed to a strong magnetic field while it is surrounded by helium, which provides thermal contact with the cold reservoir. This magnetization step is isothermal, and heat leaves the sample as the electron spins adopt the lower energy state (AB in Fig. 4.11). Thermal contact between the sample and the surroundings is now broken by pumping away the helium and the magnetic field is reduced to zero. This step is adiabatic and effectively reversible, so the state of the sample changes from B to C. At the end of this step the sample is the same as it was at A except that it now has a lower entropy. That lower entropy in the absence of a magnetic field corresponds to a lower temperature. That is, adiabatic demagnetization has cooled the sample.

Even lower temperatures can be reached if nuclear spins (which also behave like small magnets) are used instead of electron spins in the technique of **adiabatic nuclear demagnetization**. This technique was used to reach the current world record (in copper).

Concentrating on the system

Entropy is the basic concept for discussing the direction of natural change, but to use it we have to analyse changes in both the system and its surroundings. We have seen that it is always very simple to calculate the entropy change in the surroundings, and we shall now see that it is possible to devise a simple method for taking that contribution into account automatically. This approach focuses our attention on the system and simplifies discussions. Moreover, it is the foundation of all the applications of chemical thermodynamics that follow.

4.6 The Helmholtz and Gibbs energies

Consider a system in thermal equilibrium with its surroundings at a temperature T. When a change in the system occurs and there is a transfer of energy as heat between the system and the surroundings, the Clausius inequality, eqn 13, reads

$$dS - \frac{dq}{T} \geq 0 \tag{23}$$

This inequality can be developed in two ways according to the conditions (of constant volume or constant pressure) under which the process occurs.

First, consider heat transfer at constant volume. Then, in the absence of non-expansion work, we can write $dq_V = dU$; consequently

$$dS - \frac{dU}{T} \geq 0 \tag{24}$$

The importance of the inequality in this form is that *it expresses the criterion for spontaneous change solely in terms of the state functions of the system*. The inequality is easily rearranged to

$$T\,dS \geq dU \qquad \text{(constant } V, \text{ no non-expansion work)} \tag{25}$$

At either constant internal energy ($dU = 0$) or constant entropy ($dS = 0$), this expression becomes, respectively,

$$dS_{U,V} \geq 0 \qquad dU_{S,V} \leq 0 \tag{26}$$

where the subscripts indicate the constant properties.

Equation 26 expresses the criteria for spontaneous change in terms of properties relating to the system. The first inequality states that, in a system at constant volume and constant internal energy (such as an isolated system), the entropy increases in a spontaneous change. That statement is essentially the content of the Second Law. The second inequality is less obvious, for it says that, if the entropy and volume of the system are constant, then the internal energy must decrease in a spontaneous change. Do not interpret this criterion as a tendency of the system to sink to lower energy. It is a disguised statement about entropy, and should be interpreted as implying that, if the entropy of the system is unchanged, then there must be an increase in entropy of the surroundings, which can be achieved only if the energy of the system decreases as energy flows out as heat.

When heat is transferred at constant pressure, and there is no work other than expansion work, we can write $dq_p = dH$ and obtain

$$T\,dS \geq dH \qquad \text{(constant } p, \text{ no non-expansion work)} \tag{27}$$

At either constant enthalpy or constant entropy this inequality becomes, respectively,

$$dS_{H,p} \geq 0 \qquad dH_{S,p} \leq 0 \tag{28}$$

The interpretations of these inequalities are similar to those of eqn 26. The entropy of the system at constant pressure must increase if its enthalpy remains constant (for there can then be no change in entropy of the surroundings). Alternatively, the enthalpy must decrease if the entropy of the system is constant, for then it is essential to have an increase in entropy of the surroundings.

Because eqns 25 and 27 have the forms $dU - T\,dS \leq 0$ and $dH - T\,dS \leq 0$, respectively, they can be expressed more simply by introducing two more thermodynamic quantities. One is the **Helmholtz energy**, A, which is defined as

$$A = U - TS \tag{29}$$

The other is the **Gibbs energy**, G:

$$G = H - TS \tag{30}$$

All the symbols in these two definitions refer to the system.

When the state of the system changes at constant temperature, the two functions change as follows:

$$\text{(a)} \quad dA = dU - T\,dS \qquad \text{(b)} \quad dG = dH - T\,dS \tag{31}$$

When we introduce eqns 25 and 27, respectively, we obtain the criteria of spontaneous change as

$$dA_{T,V} \leq 0 \qquad dG_{T,p} \leq 0 \tag{32}$$

These inequalities are the most important conclusions from thermodynamics for chemistry. They are developed in subsequent sections and chapters.

(a) Some remarks on the Helmholtz energy

A change in a system at constant temperature and volume is spontaneous if $dA_{T,V} \leq 0$. That is, a change under these conditions is spontaneous if it corresponds to a decrease in the Helmholtz energy. Such systems move spontaneously towards states of lower A if a path is available. The criterion of equilibrium, when neither the forward nor reverse process has a tendency to occur, is

$$dA_{T,V} = 0 \tag{33}$$

The expressions $dA = dU - T\,dS$ and $dA < 0$ are sometimes interpreted as follows. A negative value of dA is favoured by a negative value of dU and a positive value of $T\,dS$. This observation suggests that the tendency of a system to move to lower A is due to its tendency to move towards states of lower internal energy and higher entropy. However, this interpretation is false (even though it is a good rule of thumb for remembering the expression for dA) because the tendency to lower A is solely a tendency towards states of greater overall entropy. *Systems change spontaneously if in doing so the total entropy of the system and its surroundings increases, not because they tend to lower internal energy.* The form of dA may give the impression that systems favour lower energy, but that is misleading: dS is the entropy change of the system, $-dU/T$ is the entropy change of the surroundings (when the volume of the system is constant), and their total tends to a maximum.

(b) Maximum work

It turns out that A carries a greater significance than being simply a signpost of spontaneous change: the change in the Helmholtz energy is equal to the maximum work accompanying a process:

$$dw_{max} = dA \tag{34}$$

As a result, A is sometimes called the 'maximum work function', or the 'work function' (*Arbeit* is the German word for work; hence the symbol A).

Justification 4.2

First we prove that a system does maximum work when it is working reversibly. (This conclusion was demonstrated in Section 2.3e for the expansion of a perfect gas; now we prove its universal validity.) We combine the Clausius inequality $dS \geq dq/T$ in the form $T\,dS \geq dq$ with the First Law, $dU = dq + dw$, and obtain

$$dU \leq T\,dS + dw$$

(dU is smaller than the term on the right because we are replacing dq by $T\,dS$, which in general is larger.) This expression rearranges to

$$dw \geq dU - T\,dS$$

It follows that the most negative value of dw, and therefore the maximum energy that can be obtained from the system as work, is given by

$$dw_{max} = dU - T\,dS$$

and that this work is done only when the path is traversed reversibly (because then the equality applies). Because at constant temperature $dA = dU - T\,dS$, we conclude that $dw_{max} = dA$.

When a macroscopic isothermal change takes place in the system, eqn 34 becomes

$$w_{max} = \Delta A \tag{35}$$

where

$$\Delta A = \Delta U - T\Delta S \tag{36}$$

This expression shows that in some cases, depending on the sign of $T\Delta S$, not all the change in internal energy may be available for doing work. If the change occurs with a decrease in entropy (of the system), in which case $T\Delta S < 0$, then the right-hand side of this equation is not as negative as ΔU itself, and consequently the maximum work is less than $|\Delta U|$. For the change to be spontaneous, some of the energy must escape as heat in order to generate enough entropy in the surroundings to overcome the reduction in entropy in the system (Fig. 4.12). In this case, Nature is demanding a tax on the internal energy as it is converted into work. This is the origin of the alternative name 'Helmholtz free energy' for A, because ΔA is that part of the change in internal energy that we are free to use to do work.

Molecular interpretation 4.3 Further insight into the relation between the work that a system can do and the Helmholtz energy is to recall that work is energy transferred to the surroundings as the uniform motion of atoms. The expression $A = U - TS$ can be interpreted as showing that A is the total internal energy of the system, U, less a contribution that is stored chaotically (the quantity TS). Because chaotically stored energy

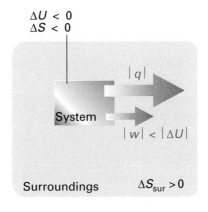

$\Delta U < 0$
$\Delta S < 0$

$|q|$

System

$|w| < |\Delta U|$

Surroundings $\Delta S_{sur} > 0$

4.12 In a system not isolated from its surroundings, the work done may be different from the change in internal energy. Moreover, the process is spontaneous if, overall, the entropy of the global, isolated system increases. In the process depicted here, the entropy of the system decreases, so that of the surroundings must increase in order for the process to be spontaneous, which means that energy must pass from the system to the surroundings as heat. Therefore, less work than $|\Delta U|$ can be obtained.

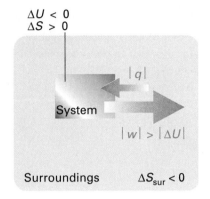

$\Delta U < 0$
$\Delta S > 0$

$|q|$

System

$|w| > |\Delta U|$

Surroundings $\Delta S_{sur} < 0$

4.13 In this process, the entropy of the system increases; hence we can afford to lose some entropy of the surroundings. That is, some of their energy may be lost as heat to the system. This energy can be returned to them as work. Hence the work done can exceed ΔU.

cannot be used to achieve uniform motion in the surroundings, only the part of U that is not stored chaotically, the quantity $U - TS$, is available for conversion into work.

If the change occurs with an increase of entropy of the system (in which case $T\Delta S > 0$), the right-hand side of the eqn 36 is more negative than ΔU. In this case, the maximum work that can be obtained from the system is greater than ΔU. The explanation of this apparent paradox is that the system is not isolated and energy may flow in as heat as work is done. Because the entropy of the system increases, we can afford a reduction of the entropy of the surroundings yet still have, overall, a spontaneous process. Therefore, some heat (no more than the value of $T\Delta S$) may leave the surroundings and contribute to the work the change is generating (Fig. 4.13). Nature is now providing a tax refund.

Example 4.5 Calculating the maximum available work

When 1.000 mol $C_6H_{12}O_6$ (glucose) is oxidized to carbon dioxide and water at 25 °C according to the equation

$$C_6H_{12}O_6(s) + 6O_2(g) \longrightarrow 6CO_2(g) + 6H_2O(l)$$

calorimetric measurements give $\Delta_r U^{\ominus} = -2808$ kJ mol^{-1} and $\Delta_r S = +182.4$ J K^{-1} mol^{-1} at 25 °C. How much of this energy change can be extracted as (a) heat at constant pressure, (b) work?

Method We know that the heat released at constant pressure is equal to the value of ΔH, so we need to relate $\Delta_r H^{\ominus}$ to $\Delta_r U^{\ominus}$, which is given. To do so, we suppose that all the gases involved are perfect, and use eqn 2.26 in the form $\Delta_r H = \Delta_r U + \Delta n_g RT$. For the maximum work available from the process we use eqn 35.

Answer (a) Because $\Delta n_g = 0$, we know that $\Delta_r H^{\ominus} = \Delta_r U^{\ominus} = -2808$ kJ mol^{-1}. Therefore, at constant pressure, the energy available as heat is 2808 kJ mol^{-1}. (b) Because $T = 298$ K, the value of $\Delta_r A^{\ominus}$ is

$$\Delta_r A^{\ominus} = \Delta_r U^{\ominus} - T\Delta_r S^{\ominus} = -2862 \text{ kJ mol}^{-1}$$

Therefore, the combustion of 1.000 mol $C_6H_{12}O_6$ can be used to produce up to 2862 kJ of work.

Comment The maximum work available is greater than the change in internal energy on account of the positive entropy of reaction (which is partly due to the generation of a large number of small molecules from one big one). The system can therefore draw in energy from the surroundings (so reducing their entropy) and make it available for doing work.

- -

Self-test 4.7 Repeat the calculation for the combustion of 1.000 mol $CH_4(g)$ under the same conditions, using data from Table 2.5.

$$[|q_p| = 890 \text{ kJ}, |w_{max}| = 813 \text{ kJ}]$$

(c) Some remarks on the Gibbs energy

The Gibbs energy (the 'free energy') is more common in chemistry than the Helmholtz energy because, at least in laboratory chemistry, we are usually more interested in changes occurring at constant pressure than at constant volume. The criterion $dG_{T,p} \le 0$ carries over into chemistry as the observation that, at constant temperature and pressure, chemical reactions are spontaneous in the direction of decreasing Gibbs energy. Therefore, if we want to know whether a reaction is spontaneous, the pressure and temperature being constant, we assess the change in the Gibbs energy. If G decreases as the reaction proceeds, then the reaction has a spontaneous tendency to convert the reactants into products. If G increases, then the reverse reaction is spontaneous.

The existence of spontaneous endothermic reactions provides an illustration of the role of G. In such reactions, H increases, the system rises spontaneously to states of higher enthalpy, and $dH > 0$. Because the reaction is spontaneous we know that $dG < 0$ despite $dH > 0$; it follows that the entropy of the system increases so much that $T\,dS$ is strongly positive and outweighs dH in $dG = dH - T\,dS$. Endothermic reactions are therefore driven by the increase of entropy of the system, and this entropy change overcomes the reduction of entropy brought about in the surroundings by the inflow of heat into the system ($dS_{sur} = -dH/T$ at constant pressure).

(d) Maximum non-expansion work

The analogue of the maximum work interpretation of ΔA, and the origin of the name free energy, can be found for ΔG. In the *Justification* below, we show that, at constant temperature and pressure, the maximum *non-expansion* work, w_e (with 'e' denoting 'extra'), is given by the change in Gibbs energy:

$$dw_{e,max} = dG \tag{37}$$

The corresponding expression for a measurable change is

$$w_{e,max} = \Delta G \tag{38}$$

This expression is particularly useful for assessing the electrical work that may be produced by fuel cells and electrochemical cells, and we shall see many applications of it.

Justification 4.3

Because $H = U + pV$, in a general change,

$$dH = dq + dw + d(pV)$$

When the change is reversible, $dw = dw_{rev}$ and $dq = dq_{rev} = T\,dS$, so

$$dG = T\,dS + dw_{rev} + d(pV) - T\,dS = dw_{rev} + d(pV)$$

The work consists of expansion work, which for a reversible change is given by $-p\,dV$, and possibly some other kind of work (for instance, the electrical work of pushing electrons through a circuit or of raising a column of liquid); this non-expansion work we denote dw_e. Therefore, with $d(pV) = p\,dV + V\,dp$,

$$dG = (-p\,dV + dw_{e,rev}) + p\,dV + V\,dp = dw_{e,rev} + V\,dp$$

If the change occurs at constant pressure (as well as constant temperature), the last term disappears, and $dG = dw_{e,rev}$. Therefore, at constant temperature and pressure, $dw_{e,rev} = dG$. However, because the process is reversible, the work done must now have its maximum value, so eqn 37 follows.

Example 4.6 Calculating the maximum non-expansion work of a reaction

How much energy is available for sustaining muscular and nervous activity from the combustion of 1.00 mol of glucose molecules under standard conditions at 37°C (blood temperature)? The standard entropy of reaction is $+182.4\ J\,K^{-1}\,mol^{-1}$.

Method The non-expansion work available from the reaction is equal to the change in standard Gibbs energy for the reaction ($\Delta_r G^\ominus$, a quantity defined more fully below). To calculate this quantity, it is legitimate to ignore the temperature dependence of the reaction enthalpy, to obtain $\Delta_r H^\ominus$ from Table 2.5, and to substitute the data into $\Delta_r G^\ominus = \Delta_r H^\ominus - T\Delta_r S^\ominus$.

Answer Because the standard reaction enthalpy is -2808 kJ mol^{-1}, it follows that the standard reaction Gibbs energy is

$$\Delta_r G^\ominus = -2808 \text{ kJ mol}^{-1} - (310 \text{ K}) \times (182.4 \text{ J K}^{-1}\text{ mol}^{-1}) = -2865 \text{ kJ mol}^{-1}$$

Therefore, $w_{e,max} = -2865$ kJ for the combustion of 1 mol glucose molecules, and the reaction can be used to do up to 2865 kJ of non-expansion work.

Comment A person of mass 70 kg would need to do 2.1 kJ of work to climb vertically through 3.0 m; therefore, at least 0.13 g of glucose is needed to complete the task (and in practice significantly more).

- -

Self-test 4.8 How much non-expansion work can be obtained from the combustion of 1.00 mol $CH_4(g)$ under standard conditions at 298 K? Use $\Delta_r S^\ominus = -243$ J K^{-1} mol^{-1}.

[818 kJ]

4.7 Standard molar Gibbs energies

Standard entropies and enthalpies of reaction can be combined to obtain the **standard Gibbs energy of reaction**, $\Delta_r G^\ominus$ (or 'standard reaction Gibbs energy'):

$$\Delta_r G^\ominus = \Delta_r H^\ominus - T\Delta_r S^\ominus \qquad [39]$$

The standard Gibbs energy of reaction is the difference in standard molar Gibbs energies of the products and reactants in their standard states at the temperature specified for the reaction as written. As in the case of standard reaction enthalpies, it is convenient to define the **standard Gibbs energies of formation**, $\Delta_f G^\ominus$:

> The standard Gibbs energy of formation is the standard reaction Gibbs energy for the formation of a compound from its elements in their reference states.

The reference state of an element was defined in Section 2.7. Standard Gibbs energies of formation of the elements in their reference states are zero, because their formation is a 'null' reaction. A selection of values for compounds is given in Table 4.4. From the values there, it is a simple matter to obtain the standard Gibbs energy of reaction by taking the appropriate combination:

$$\Delta_r G^\ominus = \sum_{\text{Products}} \nu \Delta_f G^\ominus - \sum_{\text{Reactants}} \nu \Delta_f G^\ominus \qquad (40a)$$

with each term weighted by the appropriate stoichiometric coefficient. More formally, in the notation introduced in Section 2.7,

$$\Delta_r G^\ominus = \sum_{J} \nu_J \Delta_f G^\ominus(J) \qquad (40b)$$

Table 4.4* Standard Gibbs energies of formation at 298 K

	$\Delta_f G^\ominus/(\text{kJ mol}^{-1})$
Diamond, C(s)	+2.9
Benzene, $C_6H_6(l)$	+124.3
Methane, $CH_4(g)$	−50.7
Carbon dioxide, $CO_2(g)$	−394.4
Water, $H_2O(l)$	−237.1
Ammonia, $NH_3(g)$	−16.5
Sodium chloride, NaCl(s)	−384.1

* More values are given in the *Data section*.

Illustration

To calculate the standard Gibbs energy of the reaction $CO(g) + \frac{1}{2}O_2(g) \rightarrow CO_2(g)$ at 25°C, we write

$$\Delta_r G^\ominus = \Delta_f G^\ominus(CO_2, g) - \{\Delta_f G^\ominus(CO, g) + \frac{1}{2}\Delta_f G^\ominus(O_2, g)\}$$
$$= -394.4 - \{(-137.2) + \frac{1}{2}(0)\} \text{ kJ mol}^{-1} = -257.2 \text{ kJ mol}^{-1}$$

Self-test 4.9 Calculate the standard reaction Gibbs energy for the combustion of $CH_4(g)$ at 298 K.

$$[-818 \text{ kJ mol}^{-1}]$$

Calorimetry (for ΔH directly, and for S via heat capacities) is only one of the ways of determining the values of Gibbs energies. They may also be obtained from equilibrium constants (Chapter 9) and electrochemical measurements (Chapter 10), and they may be calculated using data from spectroscopic observations (Chapter 20). The information in Tables 2.5 and 2.6 of the *Data section*; however, together with the machinery we shall now construct, is all we need in order to draw far-reaching conclusions about reactions and other processes of interest in chemistry.

Checklist of key ideas

☐ Kelvin statement of Second Law of thermodynamics

The direction of spontaneous change

4.1 The dispersal of energy
☐ spontaneity and the rise of disorder
☐ collapse into disorder as the driving force of change

4.2 Entropy
☐ Second Law ($\Delta S_{tot} > 0$)
☐ thermodynamic definition of entropy
☐ entropy change in the surroundings (4,5)
☐ entropy change for an adiabatic process
☐ Carnot cycle
☐ efficiency of a heat engine
☐ Carnot efficiency

☐ proof that entropy is a state function
☐ thermodynamic temperature scale
☐ Kelvin scale
☐ Clausius inequality
☐ entropy as a signpost of spontaneous change
☐ spontaneous cooling

4.3 Entropy changes accompanying specific processes
☐ entropy of phase transition at the transition temperature
☐ Trouton's rule
☐ entropy of expansion of a perfect gas
☐ variation of entropy with temperature (19,20)
☐ measurement of entropy
☐ Debye extrapolation

4.4 The Third Law of thermodynamics
☐ Nernst heat theorem
☐ Third Law of thermodynamics
☐ Third-Law entropy
☐ standard reaction entropy ($\Delta_r S^\ominus$, 22)

4.5 Reaching very low temperatures
☐ adiabatic demagnetization
☐ adiabatic nuclear demagnetization

Concentrating on the system

4.6 The Helmholtz and Gibbs energies
☐ the criteria $dS_{U,V} \geq 0$ and $dU_{S,V} \leq 0$
☐ Helmholtz energy
☐ Gibbs energy

☐ the criteria $dA_{T,V} \leq 0$ and $dG_{T,p} \leq 0$
☐ the criterion of equilibrium at constant temperature and volume
☐ maximum work and the Helmholtz energy (34,35)
☐ the criterion of equilibrium at constant temperature and pressure
☐ maximum non-expansion work and the Gibbs energy (37,38)

4.7 Standard molar Gibbs energies
☐ standard Gibbs energy of reaction ($\Delta_r G^\ominus$, 39)
☐ standard Gibbs energy of formation ($\Delta_f G^\ominus$)
☐ expressing $\Delta_r G^\ominus$ in terms of $\Delta_f G^\ominus$ (40)

Further reading

Articles of general interest

K. Seidman and T.R. Michalik, The efficiency of reversible heat engines: the possible misinterpretation of a corollary to Carnot's theorem. *J. Chem. Educ.* **68**, 208 (1991).

H.B. Hollinger and M.J. Zenzen, Thermodynamic irreversibility: 1. What is it? *J. Chem. Educ.* **68**, 31 (1991).

E.F. Meyer, The Carnot cycle revisited. *J. Chem. Educ.* **65**, 873 (1988).

N.C. Craig, Entropy analyses of four familiar processes. *J. Chem. Educ.* **65**, 760 (1988).

P.G. Nelson, Derivation of the second law of thermodynamics from Boltzmann's distribution law. *J. Chem. Educ.* **65**, 390 (1988).

J. Waser and V. Schomaker, A note on thermodynamic inequalities. *J. Chem. Educ.* **65**, 393 (1988).

P. Djurdjevic and I. Gutman, A simple method for showing that entropy is a function of state. *J. Chem. Educ.* **65**, 399 (1988).

W.H. Cropper, Walther Nernst and the last law. *J. Chem. Educ.* **64**, 3 (1988).

L. Glasser, Order, chaos, and all that! *J. Chem. Educ.* **66**, 997 (1989).

M. Barón, With Clausius from energy to entropy. *J. Chem. Educ.* **66**, 1001 (1989).

D.F.R. Gilson, Order and disorder and entropies of fusion. *J. Chem. Educ.* **69**, 23 (1992).

R.S. Ochs, Thermodynamics and spontaneity. *J. Chem. Educ.* **73**, 952 (1996).

N.C. Craig, Entropy diagrams. *J. Chem. Educ.* **73**, 710 (1996).

H. Frohlich, On the entropy change of surroundings of finite and infinite size. *J. Chem. Educ.* **73**, 716 (1996).

F.J. Hale, Heat engines and refrigerators. In *Encyclopedia of applied physics* (ed. G.L. Trigg), **7**, 383. VCH, New York (1993).

Texts and sources of data and information

P.W. Atkins, *The second law.* Scientific American Books, New York (1994).

J.B. Fenn, Engines, energy, and entropy. W.H. Freeman & Co, New York (1982).

J.M. Smith and H.C. Van Ness, *Introduction to chemical engineering thermodynamics.* McGraw–Hill, New York (1987).

K.E. Bett, J.S. Rowlinson, and G. Saville, *Thermodynamics for chemical engineers.* MIT Press, Cambridge (1975).

Exercises

Assume that all gases are perfect and that data refer to 298.15 K unless otherwise stated.

4.1 (a) Calculate the change in entropy when 25 kJ of energy is transferred reversibly and isothermally as heat to a large block of iron at (a) 0 °C, (b) 100 °C.

4.1 (b) Calculate the change in entropy when 50 kJ of energy is transferred reversibly and isothermally as heat to a large block of copper at (a) 0 °C, (b) 70 °C.

4.2 (a) Calculate the molar entropy of a constant-volume sample of neon at 500 K given that it is 146.22 $J K^{-1} mol^{-1}$ at 298 K.

4.2 (b) Calculate the molar entropy of a constant-volume sample of argon at 250 K given that it is 154.84 $J K^{-1} mol^{-1}$ at 298 K.

4.3 (a) A sample consisting of 1.00 mol of a monatomic perfect gas with $C_{V,m} = \frac{3}{2}R$ is heated from 100 °C to 300 °C at constant pressure. Calculate ΔS (for the system).

4.3 (b) A sample consisting of 1.00 mol of a diatomic perfect gas with $C_{V,m} = \frac{5}{2}R$ is heated from 0 °C to 100 °C at constant pressure. Calculate ΔS (for the system).

4.4 (a) Calculate ΔS (for the system) when the state of 3.00 mol of a monatomic perfect gas, for which $C_{p,m} = \frac{5}{2}R$, is changed from 25 °C and 1.00 atm to 125 °C and 5.00 atm. How do you rationalize the sign of ΔS?

4.4 (b) Calculate ΔS (for the system) when the state of 2.00 mol of a diatomic perfect gas, for which $C_{p,m} = \frac{7}{2}R$, is changed from 25 °C and 1.50 atm to 135 °C and 7.00 atm. How do you rationalize the sign of ΔS?

4.5 (a) A sample consisting of 3.00 mol of a diatomic perfect gas at 200 K is compressed reversibly and adiabatically until its temperature reaches 250 K. Given that $C_{V,m} = 27.5 \, J K^{-1} mol^{-1}$, calculate q, w, ΔU, ΔH, and ΔS.

4.5 (b) A sample consisting of 2.00 mol of a diatomic perfect gas at 250 K is compressed reversibly and adiabatically until its temperature reaches 300 K. Given that $C_{V,m} = 27.5 \, J K^{-1} mol^{-1}$, calculate q, w, ΔU, ΔH, and ΔS.

4.6 (a) Calculate the increase in entropy when 1.00 mol of a monatomic perfect gas with $C_{p,m} = \frac{5}{2}R$, is heated from 300 K to 600 K and simultaneously expanded from 30.0 L to 50.0 L.

4.6 (b) Calculate the increase in entropy when 3.50 mol of a monatomic perfect gas with $C_{p,m} = \frac{5}{2}R$, is heated from 250 K to 700 K and simultaneously expanded from 20.0 L to 60.0 L.

4.7 (a) A system undergoes a process in which the entropy change is $+2.41 \, \mathrm{J\,K^{-1}}$. During the process, 1.00 kJ of heat is added to the system at 500 K. Is the process thermodynamically reversible? Explain your reasoning.

4.7 (b) A system undergoes a process in which the entropy change is $+5.51 \, \mathrm{J\,K^{-1}}$. During the process, 1.50 kJ of heat is added to the system at 350 K. Is the process thermodynamically reversible? Explain your reasoning.

4.8 (a) A sample of aluminium of mass 1.75 kg is cooled at constant pressure from 300 K to 265 K. Calculate (a) the energy that must be removed as heat and (b) the change in entropy of the sample.

4.8 (b) A sample of copper of mass 2.75 kg is cooled at constant pressure from 330 K to 275 K. Calculate (a) the energy that must be removed as heat and (b) the change in entropy of the sample.

4.9 (a) A sample of methane gas of mass 25 g at 250 K and 18.5 atm expands isothermally until its pressure is 2.5 atm. Calculate the change in entropy of the gas.

4.9 (b) A sample of nitrogen gas of mass 35 g at 230 K and 21.1 atm expands isothermally until its pressure is 4.3 atm. Calculate the change in entropy of the gas.

4.10 (a) A sample of perfect gas that initially occupies 15.0 L at 250 K and 1.00 atm is compressed isothermally. To what volume must the gas be compressed to reduce its entropy by $5.0 \, \mathrm{J\,K^{-1}}$?

4.10 (b) A sample of perfect gas that initially occupies 11.0 L at 270 K and 1.20 atm is compressed isothermally. To what volume must the gas be compressed to reduce its entropy by $3.0 \, \mathrm{J\,K^{-1}}$?

4.11 (a) Calculate the change in entropy when 50 g of water at 80°C is poured into 100 g of water at 10°C in an insulated vessel given that $C_{p,m} = 75.5 \, \mathrm{J\,K^{-1}\,mol^{-1}}$.

4.11 (b) Calculate the change in entropy when 25 g of ethanol at 50°C is poured into 70 g of ethanol at 10°C in an insulated vessel, given that $C_{p,m} = 111.5 \, \mathrm{J\,K^{-1}\,mol^{-1}}$.

4.12 (a) Calculate ΔH and ΔS_{tot} when two copper blocks, each of mass 10.0 kg , one at 100°C and the other at 0°C, are placed in contact in an isolated container. The specific heat capacity of copper is $0.385 \, \mathrm{J\,K^{-1}\,g^{-1}}$ and may be assumed constant over the temperature range involved.

4.12 (b) Calculate ΔH and ΔS_{tot} when two iron blocks, each of mass 1.00 kg, one at 200°C and the other at 25°C, are placed in contact in an isolated container. The specific heat capacity of iron is $0.449 \, \mathrm{J\,K^{-1}\,g^{-1}}$ and may be assumed constant over the temperature range involved.

4.13 (a) Consider a system consisting of 2.0 mol $CO_2(g)$, initially at 25°C and 10 atm and confined to a cylinder of cross-section 10.0 cm^2. It is allowed to expand adiabatically against an external pressure of 1.0 atm until the piston has moved outwards through 20 cm. Assume that carbon dioxide may be considered a perfect gas with $C_{V,m} = 28.8 \, \mathrm{J\,K^{-1}\,mol^{-1}}$ and calculate (a) q, (b) w, (c) ΔU, (d) ΔT, (e) ΔS.

4.13 (b) Consider a system consisting of 1.5 mol $CO_2(g)$, initially at 15°C and 9.0 atm and confined to a cylinder of cross-section 100.0 cm^2. The sample is allowed to expand adiabatically against an external pressure of 1.5 atm until the piston has moved outwards through 15 cm. Assume that carbon dioxide may be considered a perfect gas with $C_{V,m} = 28.8 \, \mathrm{J\,K^{-1}\,mol^{-1}}$, and calculate (a) q, (b) w, (c) ΔU, (d) ΔT, (e) ΔS.

4.14 (a) The enthalpy of vaporization of chloroform ($CHCl_3$) is $29.4 \, \mathrm{kJ\,mol^{-1}}$ at its normal boiling point of 334.88 K. Calculate (a) the entropy of vaporization of chloroform at this temperature and (b) the entropy change of the surroundings.

4.14 (b) The enthalpy of vaporization of methanol is $35.27 \, \mathrm{kJ\,mol^{-1}}$ at its normal boiling point of 64.1°C. Calculate (a) the entropy of vaporization of methanol at this temperature and (b) the entropy change of the surroundings.

4.15 (a) Calculate the standard reaction entropy at 298 K of

(a) $2CH_3CHO(g) + O_2(g) \longrightarrow 2CH_3COOH(l)$

(b) $2AgCl(s) + Br_2(l) \longrightarrow 2AgBr(s) + Cl_2(g)$

(c) $Hg(l) + Cl_2(g) \longrightarrow HgCl_2(s)$

4.15 (b) Calculate the standard reaction entropy at 298 K of

(a) $Zn(s) + Cu^{2+}(aq) \longrightarrow Zn^{2+}(aq) + Cu(s)$

(b) $C_{12}H_{22}O_{11}(s) + 12O_2(g) \longrightarrow 12CO_2(g) + 11H_2O(l)$

4.16 (a) Combine the reaction entropies calculated in Exercise 4.15a with the reaction enthalpies, and calculate the standard reaction Gibbs energies at 298 K.

4.16 (b) Combine the reaction entropies calculated in Exercise 4.15b with the reaction enthalpies, and calculate the standard reaction Gibbs energies at 298 K.

4.17 (a) Use standard Gibbs energies of formation to calculate the standard reaction Gibbs energies at 298 K of the reactions in Exercise 4.15a.

4.17 (b) Use standard Gibbs energies of formation to calculate the standard reaction Gibbs energies at 298 K of the reactions in Exercise 4.15b.

4.18 (a) Calculate the standard Gibbs energy of the reaction $4HCl(g) + O_2(g) \rightarrow 2Cl_2(g) + 2H_2O(l)$ at 298 K, from the standard entropies and enthalpies of formation given in Table 2.6.

4.18 (b) Calculate the standard Gibbs energy of the reaction $CO(g) + CH_3OH(l) \rightarrow CH_3COOH(l)$ at 298 K, from the standard entropies and enthalpies of formation given in Tables 2.5 and 2.6.

4.19 (a) The standard enthalpy of combustion of solid phenol (C_6H_5OH) is $-3054 \, \mathrm{kJ\,mol^{-1}}$ at 298 K and its standard molar entropy is $144.0 \, \mathrm{J\,K^{-1}\,mol^{-1}}$. Calculate the standard Gibbs energy of formation of phenol at 298 K.

4.19 (b) The standard enthalpy of combustion of solid urea ($CO(NH_2)_2$) is $-632 \, \mathrm{kJ\,mol^{-1}}$ at 298 K and its standard molar entropy is $104.60 \, \mathrm{J\,K^{-1}\,mol^{-1}}$. Calculate the standard Gibbs energy of formation of urea at 298 K.

4.20 (a) Calculate the change in the entropies of the system and the surroundings, and the total change in entropy, when a sample of nitrogen gas of mass 14 g at 298 K and 1.00 bar doubles its volume in (a) an isothermal reversible expansion, (b) an isothermal irreversible expansion against $p_{ex} = 0$, and (c) an adiabatic reversible expansion.

4.20 (b) Calculate the change in the entropies of the system and the surroundings, and the total change in entropy, when the volume of a sample of argon gas of mass 21 g at 298 K and 1.50 bar increases from 1.20 L to 4.60 L in (a) an isothermal reversible expansion, (b) an isothermal irreversible expansion against $p_{ex} = 0$, and (c) an adiabatic reversible expansion.

4.21 (a) Calculate the change in entropy when a monatomic perfect gas is compressed to half its volume and simultaneously heated to twice its initial temperature.

4.21 (b) Calculate the change in entropy when a diatomic perfect gas is compressed to one-third its volume and simultaneously heated to three times its initial temperature.

4.22 (a) Calculate the maximum non-expansion work per mole that may be obtained from a fuel cell in which the chemical reaction is the combustion of methane at 298 K.

4.22 (b) Calculate the maximum non-expansion work per mole that may be obtained from a fuel cell in which the chemical reaction is the combustion of propane at 298 K.

4.23 (a) (a) Calculate the Carnot efficiency of a primitive steam engine operating on steam at 100°C and discharging at 60°C. (b) Repeat the calculation for a modern steam turbine that operates with steam at 300°C and discharges at 80°C.

4.23 (b) A certain heat engine operates between 1000 K and 500 K. (a) What is the maximum efficiency of the engine? (b) Calculate the maximum work that can be done for each 1.0 kJ of heat supplied by the hot source. (c) How much heat is discharged into the cold sink in a reversible process for each 1.0 kJ supplied by the hot source?

Problems

Assume that all gases are perfect and that data refer to 298 K unless otherwise stated.

Numerical problems

4.1 Calculate the difference in molar entropy (a) between liquid water and ice at −5°C, (b) between liquid water and its vapour at 95°C and 1.00 atm. The differences in heat capacities on melting and on vaporization are $37.3 \, \text{J K}^{-1} \, \text{mol}^{-1}$ and $-41.9 \, \text{J K}^{-1} \, \text{mol}^{-1}$, respectively. Distinguish between the entropy changes of the sample, the surroundings, and the total system, and discuss the spontaneity of the transitions at the two temperatures.

4.2 The heat capacity of chloroform (trichloromethane, $CHCl_3$) in the range 240 K to 330 K is given by $C_{p,m}/(\text{J K}^{-1} \, \text{mol}^{-1}) = 91.47 + 7.5 \times 10^{-2}(T/K)$. In a particular experiment, 1.00 mol $CHCl_3$ is heated from 273 K to 300 K. Calculate the change in molar entropy of the sample.

4.3 A block of copper of mass 2.00 kg $(C_{p,s} = 0.385 \, \text{J K}^{-1} \, \text{g}^{-1})$ and temperature 0°C is introduced into an insulated container in which there is 1.00 mol $H_2O(g)$ at 100°C and 1.00 atm. (a) Assuming all the steam is condensed to water, what will be the final temperature of the system, the heat transferred from water to copper, and the entropy change of the water, copper, and the total system? (b) In fact, some water vapour is present at equilibrium. From the vapour pressure of water at the temperature calculated in (a), and assuming that the heat capacities of both gaseous and liquid water are constant and given by their values at that temperature, obtain an improved value of the final temperature, the heat transferred, and the various entropies. (*Hint*: You will need to make plausible approximations.)

4.4 Consider a perfect gas contained in a cylinder and separated by a frictionless adiabatic piston into two sections A and B. All changes in B are isothermal, that is, a thermostat surrounds B to keep its temperature constant. There is 2.00 mol of the gas in each section. Initially, $T_A = T_B = 300$ K, $V_A = V_B = 2.00$ L. Heat is added to Section A and the piston moves to the right reversibly until the final volume of Section B is 1.00 L. Calculate (a) ΔS_A and ΔS_B, (b) ΔA_A and ΔA_B, (c) ΔG_A and ΔG_B, (d) ΔS of the total system and its surroundings. If numerical values cannot be obtained, indicate whether the values should be positive, negative, or zero or are indeterminate from the information given. (Assume $C_{V,m} = 20 \, \text{J K}^{-1} \, \text{mol}^{-1}$.)

4.5 A Carnot cycle uses 1.00 mol of a monatomic perfect gas as the working substance from an initial state of 10.0 atm and 600 K. It expands isothermally to a pressure of 1.00 atm (step 1), and then adiabatically to a temperature of 300 K (step 2). This expansion is followed by an isothermal compression (step 3), and then an adiabatic compression (step 4) back to the initial state. Determine the values of q, w, ΔU, ΔH, ΔS, and ΔS_{tot}, for each stage of the cycle and for the cycle as a whole. Express your answer as a table of values.

4.6 1.00 mol of a perfect gas at 27°C is expanded isothermally from an initial pressure of 3.00 atm to a final pressure of 1.00 atm in two ways: (a) reversibly, and (b) against a constant external pressure of 1.00 atm. Determine the values of q, w, ΔU, ΔH, ΔS, ΔS_{surr}, and ΔS_{tot} for each path.

4.7 A sample of 1.00 mol of a monatomic perfect gas at 27°C and 1.00 atm is expanded adiabatically in two ways: (a) reversibly to 0.50 atm, and (b) against a constant external pressure of 0.50 atm. Determine the values of q, w, ΔU, ΔH, ΔS, ΔS_{surr}, and ΔS_{tot} for each path where the data permit. Take $C_{V,m} = \frac{3}{2}R$.

4.8 A sample of 1.00 mol of a monatomic perfect gas with $C_{V,m} = \frac{3}{2}R$, initially at 298 K and 10 L, is expanded, with the surroundings maintained at 298 K, to a final volume of 20 L, in three ways: (a) isothermally and reversibly, (b) isothermally against a constant external pressure of 0.50 atm, (c) adiabatically against a constant external pressure of 0.50 atm. Calculate ΔS, ΔS_{surr}, ΔH, ΔT, ΔA, and ΔG for each path. If a numerical answer cannot be obtained from the data, write +, or −, or ? as appropriate.

4.9 The standard molar entropy of $NH_3(g)$ is 192.45 $JK^{-1}mol^{-1}$ at 298 K, and its heat capacity is given by eqn 2.30 with the coefficients given in Table 2.2. Calculate the standard molar entropy at (a) 100°C and (b) 500°C.

4.10 A block of copper of mass 500 g and initially at 293 K is in thermal contact with an electric heater of resistance 1.00 kΩ and negligible mass. A current of 1.00 A is passed for 15.0 s. Calculate the change in entropy of the copper, taking $C_{p,m} = 24.4\ JK^{-1}mol^{-1}$. The experiment is then repeated with the copper immersed in a stream of water that maintains its temperature at 293 K. Calculate the change in entropy of the copper and the water in this case.

4.11 Calculate the standard Helmholtz energy of formation, $\Delta_f A$, of $CH_3OH(l)$ at 298 K from the standard Gibbs energy of formation and the assumption that H_2 and O_2 are perfect gases.

4.12 Calculate the change in entropy when 200 g of (a) water at 0°C, (b) ice at 0°C is added to 200 g of water at 90°C in an insulated container.

4.13 Calculate (a) the maximum work and (b) the maximum non-expansion work that can be obtained from the freezing of supercooled water at −5°C and 1.0 atm. The densities of water and ice are 0.999 $g\,cm^{-3}$ and 0.917 $g\,cm^{-3}$, respectively, at −5°C.

4.14 The molar heat capacity of lead varies with temperature as follows:

T/K	10	15	20	25	30	50
$C_{p,m}/(JK^{-1}mol^{-1})$	2.8	7.0	10.8	14.1	16.5	21.4
T/K	70	100	150	200	250	298
$C_{p,m}/(JK^{-1}mol^{-1})$	23.3	24.5	25.3	25.8	26.2	26.6

Calculate the standard Third-Law entropy of lead at (a) 0°C and (b) 25°C.

4.15 Suppose that an internal combustion engine runs on octane, for which the enthalpy of combustion is −5512 $kJ\,mol^{-1}$, and take the mass of 1 gallon of fuel as 3 kg. What is the maximum height, neglecting all forms of friction, to which a car of mass 1000 kg can be driven on 1.00 gallon of fuel given that the engine cylinder temperature is 2000°C and the exit temperature is 800°C?

4.16 From standard enthalpies of formation, standard entropies, and standard heat capacities available from tables in the *Data section*, calculate the standard enthalpies and entropies at 298 K and 398 K for the reaction $CO_2(g) + H_2(g) \rightarrow CO(g) + H_2O(g)$. Assume that the heat capacities are constant over the temperature range involved.

4.17 The standard reaction Gibbs energy of

$$K_4[Fe(CN)_6] \cdot 3H_2O(s) \longrightarrow$$

$$4K^+(aq) + [Fe(CN)_6]^{4-}(aq) + 3H_2O(l)$$

is +26.120 $kJ\,mol^{-1}$ (I.R. Malcolm, L.A.K. Staveley, and R.D. Worswick, *J. Chem. Soc. Faraday Trans. I*, 1532 (1973). The enthalpy of solution of the trihydrate is +55.000 $kJ\,mol^{-1}$. Calculate (a) the standard molar entropy of the hexacyanoferrate(II) ion in water and (b) the standard reaction entropy given that the standard molar entropy of the solid trihydrate is 599.7 $JK^{-1}mol^{-1}$ and that of the K^+ ion in water is 102.5 $JK^{-1}mol^{-1}$.

4.18 The heat capacity of anhydrous potassium hexacyanoferrate(II) varies with temperature as follows:

T/K	$C_{p,m}/(JK^{-1}mol^{-1})$	T/K	$C_{p,m}/(JK^{-1}mol^{-1})$
10	2.09	100	179.6
20	14.43	110	192.8
30	36.44	150	237.6
40	62.55	160	247.3
50	87.03	170	256.5
60	111.0	180	265.1
70	131.4	190	273.0
80	149.4	200	280.3
90	165.3		

Calculate the molar enthalpy relative to its value at $T = 0$ and the Third-Law entropy at each of these temperatures.

4.19 The compound 1,3,5-trichloro-2,4,6-trifluorobenzene is an intermediate in the conversion of hexachlorobenzene to hexafluorobenzene, and its thermodynamic properties have been examined by measuring its heat capacity over a wide temperature range (R.L. Andon and J.F. Martin, *J. Chem. Soc. Faraday Trans. I*, 871 (1973)). Some of the data are as follows:

T/K	14.14	16.33	20.03	31.15	44.08	64.81
$C_{p,m}/(JK^{-1}mol^{-1})$	9.492	12.70	18.18	32.54	46.86	66.36
T/K	100.90	140.86	183.59	225.10	262.99	298.06
$C_{p,m}/(JK^{-1}mol^{-1})$	95.05	121.3	144.4	163.7	180.2	196.4

Calculate the molar enthalpy relative to its value at $T = 0$ and the Third-Law entropy of the compound at these temperatures.

Theoretical problems

4.20 Show that the integral of dq_{rev}/T round a Carnot cycle is zero. Then show that the integral is negative if the isothermal reversible expansion stage is replaced by an isothermal irreversible expansion.

4.21 Prove that two reversible adiabatic paths can never cross. Assume that the energy of the system under consideration is a function of temperature only. (*Hint*. Suppose that two such paths can intersect, and complete a cycle with the two paths plus one isothermal path. Consider the changes accompanying each stage of the cycle and show that they conflict with the Kelvin statement of the Second Law.)

4.22 Represent the Carnot cycle on a temperature–entropy diagram and show that the area enclosed by the cycle is equal to the work done.

4.23 Find an expression for the change in entropy when two blocks of the same substance and of equal mass, one at the temperature T_h and the other at T_c, are brought into thermal contact and allowed to reach equilibrium. Evaluate the change for two blocks of copper, each of mass 500 g, with $C_{p,m} = 24.4\ JK^{-1}mol^{-1}$, taking $T_h = 500$ K and $T_c = 250$ K.

4.24 A gaseous sample consisting of 1.00 mol molecules is described by the equation of state $pV_m = RT(1 + Bp)$. Initially at 373 K, it undergoes Joule–Thomson expansion from 100 atm to 1.00 atm. Given that $C_{p,m} = \frac{5}{2}R$, $\mu = 0.21$ K atm^{-1}, $B = -0.525(K/T)$ atm^{-1} and that these are constant over the temperature range involved, calculate ΔT and ΔS for the gas.

4.25 The cycle involved in the operation of an internal combustion engine is called the *Otto cycle*. Air can be considered to be the working substance and can be assumed to be a perfect gas. The cycle consists of the following steps: (1) reversible adiabatic compression from A to B, (2) reversible constant volume pressure increase from B to C due to the combustion of a small amount of fuel, (3) reversible adiabatic expansion from C to D, and (4) reversible and constant-volume pressure decrease back to state A. Determine the change in entropy (of the system and of the surroundings) for each step of the cycle and determine an expression for the efficiency of the cycle, assuming that the heat is supplied in Step 2. Evaluate the efficiency for a compression ratio of 10 : 1. Assume that in state A, $V = 4.00$ L, $p = 1.00$ atm, and $T = 300$ K, that $V_A = 10V_B$, $p_C/p_B = 5$, and that $C_{p,m} = \frac{7}{2}R$.

4.26 Prove that the perfect gas temperature scale and the thermodynamic temperature scale based on the Second Law of thermodynamics differ from each other by at most a constant numerical factor.

4.27 The definitions of the enthalpy, Gibbs energy, and Helmholtz energy have all been of the form $g = f + yz$. Show that the addition of the product yz is a general way of converting a function of x and y to a function of x and z in the sense that, if $df = a\,dx - z\,dy$, then $dg = a\,dx + y\,dz$.

Additional problems supplied by Carmen Giunta and Charles Trapp

4.28 Alkyl radicals are important intermediates in the combustion and atmospheric chemistry of hydrocarbons. N. Cohen reports group additivity tables for the thermochemistry of alkyl free radicals (N. Cohen, *J. Phys. Chem.* **96**, 9052 (1992)). A portion of the table follows. Use the table to estimate the standard molar entropies of C_2H_5, *sec*-C_4H_9, and *tert*-C_4H_9. Note that $S_m^\ominus = S_{int}^\ominus - R \ln s$, where $S_{int}^\ominus$ is the so-called intrinsic molar entropy, computed by group additivity and s is a symmetry number. ($s = 6$ for C_2H_5, 3^2 for *sec*-C_4H_9, and 3^4 for *tert*-C_4H_9.)

Group	$S_{int}^\ominus/(\text{J K}^{-1}\text{ mol}^{-1})$
C—(C)(H)$_3$	126.8
·C—(C)(H)$_2$	135.9
·C—(C)$_2$(H)	59.3
·C—(C)$_3$	−29.2
C—(·C)(H)$_3$	126.8
C—(·C)(C)(H)$_2$	42.0

4.29 Use the following enthalpies of formation reported by Seakins *et al.* (P.W. Seakins, M.J. Pilling, J.T. Niiranen, D. Gutman, and L.N. Krasnoperov, *J. Phys. Chem.* **96**, 9847 (1992)) and entropies based on group additivity tables of Cohen (N. Cohen, *J. Phys. Chem.* **96**, 9052 (1992)) to compute $\Delta_r G^\ominus$ for three possible fates of the *tert*-butyl radical at 700 K, namely, (a) *tert*-$C_4H_9 \rightarrow$ *sec*-C_4H_9, (b) *tert*-$C_4H_9 \rightarrow C_3H_6 + CH_3$, (c) *tert*-$C_4H_9 \rightarrow C_2H_4 + C_2H_5$.

Species	$\Delta_f H^\ominus/(\text{kJ mol}^{-1})$	$S_m^\ominus/(\text{J K}^{-1}\text{ mol}^{-1})$
C_2H_5	+121.0	247.8
sec-C_4H_9	+67.5	336.6
tert-C_4H_9	+51.3	314.6

4.30 Given that $S_m^\ominus = 29.79$ J K^{-1} mol^{-1} for bismuth at 100 K and the following tabulated heat capacity data (D.G. Archer, *J. Chem. Eng. Data* **40**, 1015 (1995)), compute the standard molar entropy of bismuth at 200 K.

T/K	100	120	140	150	160	180	200
$C_{p,m}/(\text{J K}^{-1}\text{ mol}^{-1})$	23.00	23.74	24.25	24.44	24.61	24.89	25.11

Compare this value to the value that would be obtained by taking the heat capacity to be constant at 24.44 J K^{-1} mol^{-1} over this range.

4.31 Consider a Carnot engine operating in outer space between the temperatures T_h and T_c. The only way that the engine can discard heat at T_c is by radiation. The power radiated by the engine at T_c follows the Stefan–Boltzmann law (see Section 11.1), which for our purposes here is written $dq_c/dt = kAT_c^4$, where k is a constant related to the Stefan–Boltzmann constant. Find the ratio T_c/T_h that corresponds to a minimum area A of the radiator for a fixed power output and constant T_h.

4.32 Polytropic processes are those that satisfy the relation $pV^n = C$. Make schematic plots of polytropic processes on a pV and TS diagram for $n = 0$, ± 1, γ (the heat capacity ratio), and $\pm\infty$.

5 The Second Law: the machinery

One of the principal applications of thermodynamics is to find relations between properties that might not be thought to be related. Several relations of this kind can be established by making use of the fact that the Gibbs energy is a state function. We also see how to derive expressions for the variation of the Gibbs energy with temperature and pressure. These expressions will prove useful later when we need to discuss the effect of temperature and pressure on equilibrium constants. This chapter also introduces the chemical potential, a property that will be at the centre of discussions in the remaining chapters of this part of the text. We shall also see how to formulate expressions that are valid for real gases.

The Gibbs energy, G, is of central importance to chemistry, and in this chapter it begins to move to the centre of the stage. We shall also meet the 'chemical potential', the quantity on which almost all the most important applications of thermodynamics to chemistry are based.

Combining the First and Second Laws

We have seen that the First Law of thermodynamics may be written

$$dU = dq + dw \tag{1}$$

For a reversible change in a closed system of constant composition, and in the absence of any non-expansion work,

$$dw_{rev} = -p\,dV \qquad dq_{rev} = T\,dS$$

Therefore,

$$dU = T\,dS - p\,dV \tag{2}$$

However, because dU is an exact differential, its value is independent of path. Therefore, the same value of dU is obtained whether the change is brought about irreversibly or reversibly. Consequently, eqn 2 applies to any change—reversible or irreversible—of a closed system that does no non-expansion work. We shall call this combination of the First and Second Laws the **fundamental equation**.

The fact that the fundamental equation applies to both reversible and irreversible changes may be puzzling at first sight. The reason is that only in the case of a reversible change may $T\,dS$ be identified with dq and $-p\,dV$ with dw. When the change is irreversible, $T\,dS > dq$ (the Clausius inequality) and $-p\,dV > dw$. The sum of dw and dq remains equal to the sum of $T\,dS$ and $-p\,dV$, provided the composition is constant.

5.1 Properties of the internal energy

Equation 2 shows that the internal energy of a closed system changes in a simple way when S and V are changed ($dU \propto dS$ and $dU \propto dV$). These simple proportionalities suggest that U should be regarded as a function of S and V. We could regard U as a function of other variables, such as S and p or T and V, because they are all interrelated; but the simplicity of the fundamental equation suggests that $U(S, V)$ is the best choice.

The mathematical consequence of U being a function of S and V is that a change dU can be expressed in terms of the changes dS and dV by[1]

$$dU = \left(\frac{\partial U}{\partial S}\right)_V dS + \left(\frac{\partial U}{\partial V}\right)_S dV \tag{3}$$

This expression states that the change in U is proportional to the change in S and to the change in V, the two coefficients being the slopes of the plots of U against S and V, respectively. When this expression is compared to the thermodynamic relation, eqn 2, we see that, for systems of constant composition,

$$\left(\frac{\partial U}{\partial S}\right)_V = T \qquad \left(\frac{\partial U}{\partial V}\right)_S = -p \tag{4}$$

The first of these two equations is a purely thermodynamic definition of temperature as the ratio of the changes in the internal energy and entropy of a constant-volume, closed, constant-composition system. We are beginning to generate relations between the properties of a system and to discover the power of thermodynamics for establishing unexpected relations.

(a) The Maxwell relations

Because the fundamental equation, eqn 2, is an expression for an exact differential, the coefficients of dS and dV must pass the test for exact differentials (see *Further information 1*). That is,

$$df = g\,dx + h\,dy \text{ is exact if } \left(\frac{\partial g}{\partial y}\right)_x = \left(\frac{\partial h}{\partial x}\right)_y \tag{5}$$

Therefore, because we know that $dU = T\,dS - p\,dV$ is exact, it must be the case that

$$\left(\frac{\partial T}{\partial V}\right)_S = -\left(\frac{\partial p}{\partial S}\right)_V \tag{6}$$

We have generated a relation between quantities which, at first sight, would not seem to be related.

The equation just derived is an example of a **Maxwell relation**. However, apart from being unexpected, it does not look particularly interesting. Nevertheless, it does suggest that there may be other similar relations that are more useful. Indeed, the fact that H, G, and A are all state functions can be used to derive three more Maxwell relations. The argument to obtain

---

1   For a review of partial differential properties, see *Further information 1*.

Table 5.1 The Maxwell relations

$$\left(\frac{\partial T}{\partial V}\right)_S = -\left(\frac{\partial p}{\partial S}\right)_V$$

$$\left(\frac{\partial T}{\partial p}\right)_S = \left(\frac{\partial V}{\partial S}\right)_p$$

$$\left(\frac{\partial p}{\partial T}\right)_V = \left(\frac{\partial S}{\partial V}\right)_T$$

$$\left(\frac{\partial V}{\partial T}\right)_p = -\left(\frac{\partial S}{\partial p}\right)_T$$

them runs in the same way in each case: because H, G, and A are state functions, the expressions for dH, dG, and dA satisfy the relation like eqn 6. All four relations are listed in Table 5.1. In the next section we derive one of them, but as no new principles are involved we shall not derive them all.

(b) The variation of internal energy with volume

The coefficient we have called the internal pressure,

$$\pi_T = \left(\frac{\partial U}{\partial V}\right)_T \tag{7}$$

played a central role in the manipulation of the First Law, and in *Justification* 3.2 we used the relation

$$\pi_T = T\left(\frac{\partial p}{\partial T}\right)_V - p \tag{8}$$

This relation is called a **thermodynamic equation of state** because it expresses a quantity in terms of the two variables T and p and applies to any substance. We are now ready to derive it from the equations we have just established.

We can obtain the coefficient π_T by dividing both sides of eqn 3 by dV, imposing the constraint of constant temperature, and then introducing the two relations in eqn 4:

$$\left(\frac{\partial U}{\partial V}\right)_T = \left(\frac{\partial U}{\partial S}\right)_V\left(\frac{\partial S}{\partial V}\right)_T + \left(\frac{\partial U}{\partial V}\right)_S$$

$$= T\left(\frac{\partial S}{\partial V}\right)_T - p$$

This equation is already beginning to look like the expression we want. One of the Maxwell relations does the job of turning $(\partial S/\partial V)_T$ into something else:

$$\left(\frac{\partial S}{\partial V}\right)_T = \left(\frac{\partial p}{\partial T}\right)_V$$

The substitution of this relation completes the proof of eqn 8.

Example 5.1 Deriving a thermodynamic relation

Show thermodynamically that $\pi_T = 0$ for a perfect gas, and compute its value for a van der Waals gas.

Method Proving a result 'thermodynamically' means basing it entirely on general thermodynamic relations and equations of state, without drawing on molecular arguments (such as the existence of intermolecular forces). We know that, for a perfect gas, $p = nRT/V$, so this relation should be used in eqn 8. Similarly, the van der Waals equation is given in Table 1.7, and for the second part of the question it should be used in eqn 8.

Answer Because $(\partial p/\partial T)_V = nR/V$ for a perfect gas (by differentiation of the equation of state), eqn 8 becomes

$$\pi_T = \frac{nRT}{V} - p = 0$$

The equation of state of a van der Waals gas is

$$p = \frac{nRT}{V - nb} - a\frac{n^2}{V^2}$$

Therefore, because a and b are independent of temperature, we can write

$$\left(\frac{\partial p}{\partial T}\right)_V = \frac{nR}{V - nb}$$

That is,

$$\pi_T = \frac{nRT}{V - nb} - \frac{nRT}{V - nb} + a\frac{n^2}{V^2} = a\frac{n^2}{V^2}$$

Comment This result for π_T implies that the internal energy of a van der Waals gas increases when it expands isothermally (that is, $(\partial U/\partial V)_T > 0$), and that the increase is related to the parameter a, which models the attractive interactions between the particles. A larger molar volume, corresponding to a greater average separation between molecules, implies weaker mean intermolecular attractions.

- -

Self-test 5.1 Calculate π_T for a gas that obeys the virial equation of state.

$$[\pi_T = RT^2(\partial B/\partial T)_V/V_m^2 + \cdots]$$

5.2 Properties of the Gibbs energy

The same arguments that were applied to the fundamental equation for U may be applied to the Gibbs energy $G = H - TS$. When the system undergoes a change of state, G may change because H, T, and S change. For infinitesimal changes in each property,

$$dG = dH - T\,dS - S\,dT$$

Because $H = U + pV$, we know that

$$dH = dU + p\,dV + V\,dp$$

For a closed system doing no non-expansion work, dU can be replaced by the fundamental equation $dU = T\,dS - p\,dV$. The result of these steps is

$$dG = (T\,dS - p\,dV) + p\,dV + V\,dp - T\,dS - S\,dT$$

That is, for a closed system in the absence of non-expansion work and at constant composition

$$dG = V\,dp - S\,dT \tag{9}$$

This expression, which shows that a change in G is proportional to changes in p and T, suggests that G may be best regarded as a function of p and T. It confirms that G is an important quantity in chemistry because the pressure and temperature are usually the variables under our control. In other words, G carries around the combined consequences of the First and Second Laws in a way that makes it particularly suitable for chemical applications.

The same argument that led to eqn 4, when applied to the exact differential dG, now gives

$$\left(\frac{\partial G}{\partial T}\right)_p = -S \qquad \left(\frac{\partial G}{\partial p}\right)_T = V \tag{10}$$

These relations show how the Gibbs energy varies with temperature and pressure. Because S is positive, it follows that G decreases when the temperature is raised at constant pressure and composition. Moreover, the relation shows that G decreases most sharply when the entropy of the system is large. Therefore, the Gibbs energy of the gaseous phase of a substance, which has a high molar entropy, is more sensitive to temperature than its liquid

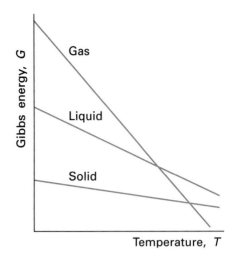

5.1 The variation of the Gibbs energy with the temperature is determined by the entropy. Because the entropy of the gaseous phase of a substance is greater than that of the liquid phase, and the entropy of the solid phase is smallest, the Gibbs energy changes most steeply for the gas phase, followed by the liquid phase, and then the solid phase of the substance.

and solid phases (Fig. 5.1). Because V is positive, G always increases when the pressure of the system is increased at constant temperature (and composition). Because the molar volumes of gases are large, G is more sensitive to pressure for the gaseous phase of a substance than for its liquid and solid phases (Fig. 5.2).

Example 5.2 Calculating the effect of pressure on the Gibbs energy

Calculate the change in the molar Gibbs energy of (a) liquid water treated as an incompressible fluid and (b) water vapour treated as a perfect gas, when the pressure is increased isothermally from 1.0 bar to 2.0 bar at 298 K.

Method In each case, the change in molar Gibbs energy can be obtained by integration of eqn 9 with the temperature held constant (that is, setting $dT = 0$):

$$G_m(p_f) - G_m(p_i) = \int_{p_i}^{p_f} V_m \, dp$$

For an incompressible fluid, the molar volume is independent of the pressure, so V_m can be treated as a constant. For a perfect gas, the molar volume varies with pressure as $V_m = RT/p$, so this expression must be used in the integrand, and the integration performed treating RT as a constant.

Answer For the incompressible liquid, V_m is constant at $18.0 \text{ cm}^3 \text{ mol}^{-1}$, so

$$
\begin{aligned}
G_m(p_f) - G_m(p_i) &= V_m \int_{p_i}^{p_f} dp = V_m \times (p_f - p_i) \\
&= (18.0 \times 10^{-6} \text{ m}^3 \text{ mol}^{-1}) \times (1.0 \times 10^5 \text{ Pa}) \\
&= +1.8 \text{ J mol}^{-1}
\end{aligned}
$$

(because $1 \text{ Pa m}^3 = 1 \text{ N m} = 1 \text{ J}$). For a perfect gas:

$$
\begin{aligned}
G_m(p_f) - G_m(p_i) &= \int_{p_i}^{p_f} \frac{RT}{p} \, dp = RT \ln\left(\frac{p_f}{p_i}\right) \\
&= (2.48 \text{ kJ mol}^{-1}) \times \ln 2.0 = +1.7 \text{ kJ mol}^{-1}
\end{aligned}
$$

Comment Note that G increases in both cases, and that the increase for a gas is about 1000 times greater than for the liquid.

- -

Self-test 5.2 Calculate the change in G_m for ice at $-10°C$ when it has density 0.917 g cm^{-3}, when the pressure is increased from 1.0 bar to 2.0 bar.

$$[+2.0 \text{ J mol}^{-1}]$$

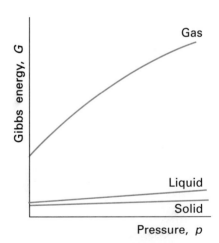

5.2 The variation of the Gibbs energy with the pressure is determined by the volume of the sample. Because the volume of the gaseous phase of a substance is greater than that of the same amount of liquid phase, and the volume of the solid phase is smallest (for most substances), the Gibbs energy changes most steeply for the gas phase, followed by the liquid phase, and then the solid phase of the substance. Because the volumes of the solid and liquid phases of a substance are similar, they vary by similar amounts as the pressure is changed.

(a) The temperature dependence of the Gibbs energy

In due course we shall see that the equilibrium composition of a system depends on the Gibbs energy, and that to discuss the response of the composition to temperature it is necessary to know how G varies with temperature. The first expression in eqn 10 is the starting point; although it expresses the variation of G in terms of the entropy, it can be expressed in terms of the enthalpy by using the definition of G to write $S = (H - G)/T$. Then

$$\left(\frac{\partial G}{\partial T}\right)_p = \frac{G - H}{T} \tag{11}$$

We shall see later that the equilibrium constant of a reaction is related to G/T rather than to G itself,[2] and it turns out the variation of this quantity with temperature is simpler than the temperature variation of G alone. In fact, it is easy to deduce from the last equation (see the *Justification* below) that

$$\left(\frac{\partial}{\partial T} \left(\frac{G}{T} \right) \right)_p = -\frac{H}{T^2}$$ (12)

This expression is called the **Gibbs–Helmholtz equation**. (G–H is a helpful way of remembering what this equation relates.) It shows that, if the enthalpy of the system is known, then the temperature dependence of G/T is also known.

Justification 5.1

First, we write eqn 11 as

$$\left(\frac{\partial G}{\partial T} \right)_p - \frac{G}{T} = -\frac{H}{T}$$

The expression on the left is simplified by noting that

$$\left(\frac{\partial}{\partial T} \left(\frac{G}{T} \right) \right)_p = \frac{1}{T} \left(\frac{\partial G}{\partial T} \right)_p + G \frac{\mathrm{d}}{\mathrm{d}T} \frac{1}{T}$$

$$= \frac{1}{T} \left(\frac{\partial G}{\partial T} \right)_p - \frac{G}{T^2}$$

$$= \frac{1}{T} \left\{ \left(\frac{\partial G}{\partial T} \right)_p - \frac{G}{T} \right\}$$

When we substitute eqn 11 into this expression, we obtain eqn 12.

Example 5.3 Manipulating the Gibbs–Helmholtz equation

Show that

$$\left(\frac{\partial (G/T)}{\partial (1/T)} \right)_p = H$$

Method This example is an exercise in manipulating partial differentials. The desired expression resembles the Gibbs–Helmholtz equation, so eqn 12 is a good starting point. To obtain the desired result, we need to convert the variable of differentiation from T to $1/T$, which can be done by standard techniques of manipulating derivatives.

Answer The left-hand side of eqn 12 can be written

$$\left(\frac{\partial (G/T)}{\partial T} \right)_p = \left(\frac{\partial (G/T)}{\partial (1/T)} \right)_p \frac{\mathrm{d}(1/T)}{\mathrm{d}T} = \left(\frac{\partial (G/T)}{\partial (1/T)} \right)_p \times \left(-\frac{1}{T^2} \right)$$

Substitution of this result into eqn 12 and multiplication of both sides by $-T^2$ gives the expression required.

---

2   In Section 9.1d we derive the result that the equilibrium constant for a reaction is related to its standard reaction free energy by $\Delta_r G^{\ominus}/T = -R \ln K$.

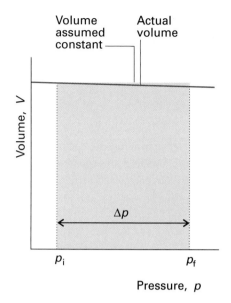

5.3 The difference in Gibbs energy of a solid or liquid at two pressures is equal to the rectangular area shown. We have assumed that the variation of volume with pressure is negligible.

Comment The result shows that, if H is independent of temperature over a range, then a plot of G/T against $1/T$ should be a straight line of slope H. We see the usefulness of this result in Chapter 9.

Self-test 5.3 Find the equation for the temperature dependence of A that corresponds to that just derived for G.

$$[(\partial(A/T)/\partial(1/T))_V = U]$$

The Gibbs–Helmholtz equation is most useful when it is applied to changes, including changes of physical state and chemical reactions at constant pressure. Then, with $\Delta G = G_f - G_i$ for the change of Gibbs energy between the final and initial states, because the equation applies to both G_f and G_i, we can write

$$\left(\frac{\partial(\Delta G/T)}{\partial T}\right)_p = -\frac{\Delta H}{T^2} \tag{13}$$

(b) *The pressure dependence of the Gibbs energy*

To find the Gibbs energy at one pressure in terms of its value at another pressure, the temperature being constant, we set $dT = 0$ in eqn 9 and integrate the remaining expression:

$$G(p_f) = G(p_i) + \int_{p_i}^{p_f} V \, dp \tag{14}$$

For a liquid or solid, the volume changes only slightly as the pressure changes (Fig. 5.3), so V may be treated as a constant and taken outside the integral. Then, for molar quantities,

$$\begin{aligned} G_m(p_f) &= G_m(p_i) + V_m(p_f - p_i) \\ &= G_m(p_i) + V_m\Delta p \end{aligned} \tag{15}$$

where $\Delta p = p_f - p_i$. Under normal laboratory conditions $V_m\Delta p$ is very small, as we saw in Example 5.2, and may be neglected. Hence, we may usually suppose that the Gibbs energies of solids and liquids are independent of pressure. However, if we are interested in geophysical problems, then because pressures in the Earth's interior are huge, their effect on the Gibbs energy cannot be ignored. If the pressures are so great that there are substantial volume changes, we must use the complete expression, eqn 14.

Illustration

Suppose that for a certain phase transition of a solid $\Delta_{trs}V = +1.0$ cm^3 mol^{-1}. Then, for an increase in pressure to 3.0 Mbar, the Gibbs energy of the transition changes from $\Delta_{trs}G(1$ bar$)$ to

$$\begin{aligned} \Delta_{trs}G(3 \text{ Mbar}) &= \Delta_{trs}G(1 \text{ bar}) + (1.0 \times 10^{-6} \text{ m}^3 \text{ mol}^{-1}) \\ &\quad \times (3.0 \times 10^{11} \text{ Pa} - 1.0 \times 10^5 \text{ Pa}) \\ &= \Delta_{trs}G(1 \text{ bar}) + 3.0 \times 10^2 \text{ kJ mol}^{-1} \end{aligned}$$

Self-test 5.4 Calculate the difference in molar Gibbs energy between the top and bottom of a column of mercury in a barometer. The mass density of mercury is 13.6 g cm^{-3}.

$$[+1.5 \text{ J mol}^{-1}]$$

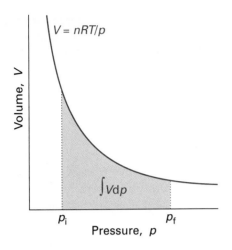

5.4 The difference in Gibbs energy for a perfect gas at two pressures is equal to the area shown below the perfect-gas isotherm.

The molar volumes of gases are large, so the Gibbs energy may depend strongly on the pressure. Furthermore, because the volume also varies markedly with the pressure, we cannot treat it as a constant in the integral in eqn 14 (Fig. 5.4). For a perfect gas we substitute $V = nRT/p$ into the integral, and find

$$G(p_f) = G(p_i) + nRT \int_{p_i}^{p_f} \frac{dp}{p} = G(p_i) + nRT \ln\left(\frac{p_f}{p_i}\right) \qquad (16)°$$

This expression shows that, when the pressure is increased ten-fold at room temperature, the molar Gibbs energy increases by about 6 kJ mol^{-1}. It also follows from this equation that, if we set $p_i = p^{\ominus}$ (the standard pressure of 1 bar), then the Gibbs energy of a perfect gas at a pressure p is related to its standard value by

$$G(p) = G^{\ominus} + nRT \ln\left(\frac{p}{p^{\ominus}}\right) \qquad (17)°$$

5.3 The chemical potential of a pure substance

Now we switch attention from the Gibbs energy itself to a quantity, the chemical potential, that is closely related and which will play a central role in all subsequent discussions of equilibrium. First, we introduce the chemical potential of a pure substance, and in particular the chemical potential of a perfect gas. At this stage its introduction will seem to be no more than a change of notation. However, the definition prepares the ground for the introduction (in Section 7.1b) of the chemical potential of a substance in a mixture (including a reaction mixture), which is a powerful and general concept.

The **chemical potential**, μ, of a pure substance is defined as

$$\mu = \left(\frac{\partial G}{\partial n}\right)_{T,p} \qquad [18]$$

That is, the chemical potential shows how the Gibbs energy of a system changes as a substance is added to it. For a pure substance, the Gibbs energy is simply $G = n \times G_m$, so

$$\mu = \left(\frac{\partial nG_m}{\partial n}\right)_{T,p} = G_m \qquad (19)$$

and the chemical potential is the same as the molar Gibbs energy. For example, the chemical potential of a perfect gas at a pressure p can be written from eqn 17:

$$\mu = \mu^{\ominus} + RT \ln\left(\frac{p}{p^{\ominus}}\right) \qquad (20)°$$

where $\mu^{\ominus}$ is the standard chemical potential, the molar Gibbs energy of the pure gas at 1 bar. The logarithmic variation of the chemical potential with the pressure given by eqn 20 is illustrated in Fig. 5.5.

Real gases: the fugacity

At various stages in the development of physical chemistry it is necessary to switch from a consideration of idealized systems to real systems. In many cases it is desirable to preserve the form of the expressions that have been derived for the idealized system. Then deviations from the idealized behaviour can be expressed most simply. We shall illustrate such a procedure in this section by considering how the expressions that have been derived for perfect gases, particularly eqn 20 for the chemical potential of a perfect gas, are adapted to describe real gases.

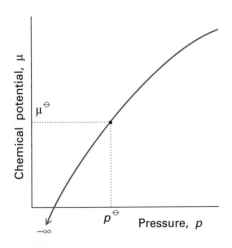

5.5 The chemical potential, μ, of a perfect gas is proportional to ln p, and the standard state is reached at $p^{\ominus}$. Note that, as $p \to 0$, μ becomes negatively infinite.

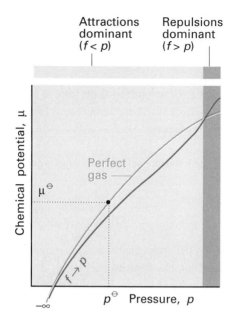

Attractions dominant $(f < p)$ | Repulsions dominant $(f > p)$

Perfect gas

$\mu^{\ominus}$

$f \to p$

$p^{\ominus}$ Pressure, p

$-\infty$

5.6 The chemical potential of a real gas. As $p \to 0$, μ coincides with the value for a perfect gas (shown by the pale line). When attractive forces are dominant (at intermediate pressures), the chemical potential is less than that of a perfect gas and the molecules have a lower 'escaping tendency'. At high pressures, when repulsive forces are dominant, the chemical potential of a real gas is greater than that of a perfect gas. Then the 'escaping tendency' is increased.

5.4 The definition of fugacity

The pressure dependence of the chemical potential of a real gas might resemble that shown in Fig. 5.6. To adapt eqn 20 to this case, we replace the true pressure, p, by an effective pressure, called the **fugacity**, f, and write

$$\mu = \mu^{\ominus} + RT \ln \left(\frac{f}{p^{\ominus}} \right) \qquad [21]$$

The name 'fugacity' comes from the Latin for 'fleetness' in the sense of 'escaping tendency'; fugacity has the same dimensions as pressure. In later chapters we derive thermodynamically exact expressions in terms of chemical potentials, and therefore in terms of fugacities. For example, from elementary chemistry we know that the equilibrium constant for a reaction such as $H_2(g) + Br_2(g) \rightleftharpoons 2HBr(g)$ should be written

$$K_p = \frac{p_{HBr}^2}{p_{H_2} p_{Br_2}}$$

where p_J is the partial pressure of substance J; however, this expression is only an approximation. The thermodynamically exact expression is

$$K = \frac{f_{HBr}^2}{f_{H_2} f_{Br_2}}$$

where f_J is the fugacity of J. Although the latter expression is exact, it is useful only if we know how to interpret the fugacities in terms of the partial pressures. This is the task we deal with in the remainder of this chapter.

5.5 Standard states of real gases

A perfect gas is in its standard state when its pressure is $p^{\ominus}$ (that is, 1 bar): the pressure arises solely from the kinetic energy of the molecules and there are no intermolecular forces to take into account. We aim to recapture this 'kinetic energy only' definition for the standard state of a real gas by picturing it as a hypothetical state in which all the intermolecular forces have been extinguished:

> **The standard state of a real gas is a hypothetical state in which the gas is at a pressure $p^{\ominus}$ and behaving perfectly.**

The advantage of this definition is that it ensures that the standard state of a real gas has the simple properties of a perfect gas. If we had defined the standard state as the one for which $f = p^{\ominus}$, then the standard states of different gases would have had relatively complex properties. The choice of a hypothetical standard state literally standardizes the interactions between the particles by setting them to zero.[3] Then differences of standard chemical potential of different gases arise solely from the internal structure and properties of the molecules, not from the way they interact with each other.

5.6 The relation between fugacity and pressure

We shall write the fugacity as

$$f = \phi p \qquad [22]$$

---

3 The alternative choice, of selecting as the standard state the gas at zero pressure, at which it certainly behaves perfectly, runs into difficulty because $\mu \to -\infty$ as $p \to 0$.

where ϕ is the dimensionless **fugacity coefficient**. In general, ϕ depends on the identity of the gas, the pressure, and the temperature. With the introduction of ϕ, eqn 21 becomes

$$\mu = \mu^{\ominus} + RT \ln (p/p^{\ominus}) + RT \ln \phi \qquad (23)$$

As $\mu^{\ominus}$ refers to a hypothetical 'kinetic energy only' gas, and the term $\ln p$ is the same as for a perfect gas, the term $RT \ln \phi$ must express the entire effect of all the intermolecular forces. Because all gases become perfect as the pressure approaches zero (so $f \rightarrow p$ as $p \rightarrow 0$), we know that $\phi \rightarrow 1$ as $p \rightarrow 0$.

We shall now show that, at a general pressure p, the fugacity coefficient of a gas is given by the expression

$$\ln \phi = \int_0^p \left(\frac{Z-1}{p}\right) dp \qquad (24)$$

where Z is the compression factor of the gas ($Z = pV_m/RT$; this quantity was introduced in Section 1.4a). Equation 24 is an explicit expression for the fugacity coefficient at any pressure p and therefore, through eqn 22, for the fugacity of the gas at that pressure.

Justification 5.2

Equation 14 is true for all gases, whether real or perfect. Expressing it in terms of molar quantities and then using eqn 21 gives

$$\int_{p'}^p V_m \, dp = \mu - \mu' = RT \ln \left(\frac{f}{f'}\right)$$

In this expression, f is the fugacity when the pressure is p and f' is the fugacity when it the pressure is p'. If the gas were perfect we could write

$$\int_{p'}^p V_{\text{perfect,m}} \, dp = \mu_{\text{perfect}} - \mu'_{\text{perfect}} = RT \ln \left(\frac{p}{p'}\right)$$

The difference of the two equations is

$$\int_{p'}^p \left(V_m - V_{\text{perfect,m}}\right) dp = RT\left\{\ln \left(\frac{f}{f'}\right) - \ln \left(\frac{p}{p'}\right)\right\}$$

which can be rearranged to

$$\ln \left(\frac{f}{p} \times \frac{p'}{f'}\right) = \frac{1}{RT} \int_{p'}^p \left(V_m - V_{\text{perfect,m}}\right) dp$$

When $p' \rightarrow 0$, the gas behaves perfectly and f' becomes equal to the pressure p'. Therefore, $p'/f' \rightarrow 1$ as $p' \rightarrow 0$. If we take this limit (which means setting $p'/f' = 1$ on the left and $p' = 0$ on the right), the last equation becomes

$$\ln \left(\frac{f}{p}\right) = \frac{1}{RT} \int_0^p \left(V_m - V_{\text{perfect,m}}\right) dp$$

With $\phi = f/p$,

$$\ln \phi = \frac{1}{RT} \int_0^p \left(V_m - V_{\text{perfect,m}}\right) dp$$

For a perfect gas $V_{\text{perfect,m}} = RT/p$. For a real gas, $V_m = RTZ/p$, where Z is the compression factor. With these two substitutions, we obtain eqn 24.

To evaluate ϕ from eqn 24, we need experimental data on the compression factor from very low pressures up to the pressure of interest. Some information of this kind is available in numerical tables, in which case the integral may be evaluated numerically. Sometimes an algebraic expression is available for Z (for instance, from one of the equations of state,

Table 1.7) and it may be possible to evaluate the integral analytically. Thus, if we know the virial coefficients for the gas we can obtain the fugacity by using

$$\ln \phi = B'p + \tfrac{1}{2}C'p^2 + \cdots \tag{25}$$

This expression was obtained by explicit evaluation of eqn 24.

Example 5.4 Calculating a fugacity

Suppose that the attractive interactions between gas particles can be neglected, and find an expression for the fugacity of a van der Waals gas in terms of the pressure. Estimate its value for ammonia at 10.00 atm and 298.15 K.

Method The starting point for the calculation is eqn 24. To evaluate the integral, we need an analytical expression for Z, which can be obtained from the equation of state. We saw in Section 1.5 that the van der Waals coefficient a represents the attractions between molecules, so it may be set equal to zero in this calculation.

Answer When we neglect a in the van der Waals equation, that equation becomes

$$p = \frac{RT}{V_m - b}$$

and hence

$$Z = 1 + \frac{bp}{RT}$$

The integral in eqn 24 that we require is therefore

$$\int_0^p \left(\frac{Z-1}{p}\right) dp = \int_0^p \left(\frac{b}{RT}\right) dp = \frac{bp}{RT}$$

Consequently, from eqns 24 and 22, the fugacity at the pressure p is

$$f = pe^{bp/RT}$$

From Table 1.6, $b = 3.707 \times 10^{-2}$ L mol^{-1}, so $pb/RT = 1.515 \times 10^{-2}$, giving

$$f = (10.00 \text{ atm}) \times e^{0.01515} = 10.2 \text{ atm}$$

Comment The effect of the repulsive term (as represented by the coefficient b in the van der Waals equation) is to increase the fugacity above the pressure, and so the effective pressure of the gas—its 'escaping tendency'—is greater than if it were perfect.

- -

Self-test 5.5 Find an expression for the fugacity coefficient when the attractive interaction is dominant in a van der Waals gas, and the pressure is low enough to make the approximation $4ap/(RT)^2 \ll 1$. Evaluate the fugacity for ammonia, as above.

[$\ln \phi = -ap/(RT)^2$, 9.32 atm]

It is clear from Fig. 1.27 that for most gases $Z < 1$ up to moderate pressures, but that $Z > 1$ at higher pressures. If $Z < 1$ throughout the range of integration, then the integrand in eqn 24 is negative and $\phi < 1$. This implies that $f < p$ (the molecules tend to stick together) and that the chemical potential of the gas is less than that of a perfect gas. At higher pressures, the range over which $Z > 1$ may dominate the range over which $Z < 1$. The integral is then positive, $\phi > 1$, and $f > p$ (the repulsive interactions are dominant and tend to drive the particles apart). Now the chemical potential of the gas is greater than that of the perfect gas at the same pressure.

Figure 5.7, which has been calculated using the full van der Waals equation of state, shows how the fugacity depends on the pressure in terms of their reduced variables

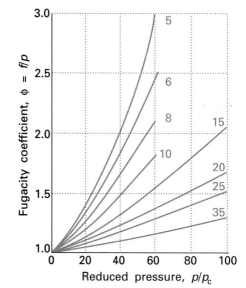

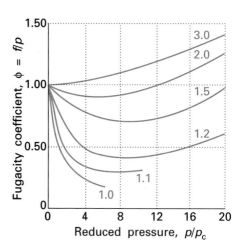

5.7 The fugacity coefficient of a van der Waals gas plotted using the reduced variables of the gas. The curves are labelled with the reduced temperature $T_r = T/T_c$.

Table 5.2* The fugacity of nitrogen at 273 K

p/atm	f/atm
1	0.99955
10	9.9560
100	97.03
1000	1839

*More values are given in the *Data section*.

(Section 1.6). Because the critical constants are available in Table 1.5, the graphs can be used for quick estimates of the fugacities of a wide range of gases. Table 5.2 gives some explicit values for nitrogen.

Illustration

To estimate the fugacity of nitrogen at 500 atm and 0°C, we first convert the reduced pressure ($p_r = p/p_c$) and temperature ($T_r = T/T_c$) of the gas by using the critical pressure and temperature of nitrogen (33.5 atm and 126.2 K); hence $p_r = 14.9$ and $T_r = 2.16$. These values correspond to $\phi = 1.15$ in Fig. 5.7. Therefore, the fugacity of nitrogen is approximately $f = 1.15 \times (500 \text{ atm}) = 575$ atm under the stated conditions. Because $\phi > 1$, the repulsive contributions are dominant in nitrogen at 500 atm and 0°C.

Self-test 5.6 Estimate the fugacity of carbon dioxide at 90°C and 580 atm.

[230 atm]

Checklist of key ideas

Combining the First and Second Laws

☐ fundamental equation

5.1 Properties of the internal energy

☐ the relations $(\partial U/\partial S)_V = T$ and $(\partial U/\partial V)_S = -p$
☐ Maxwell relations (Table 5.1)
☐ thermodynamic equation of state

5.2 Properties of the Gibbs energy

☐ the relations $(\partial G/\partial T)_p = -S$ and $(\partial G/\partial p)_T = V$
☐ the Gibbs–Helmholtz equation
☐ the variation of Gibbs energy with pressure for a condensed phase (15) and a perfect gas (17)

5.3 The chemical potential of a pure substance

☐ chemical potential
☐ chemical potential of a perfect gas (20)

Real gases: the fugacity

5.4 The definition of fugacity

☐ fugacity
☐ the chemical potential of a real gas

☐ equilibrium constants in terms of fugacities

5.5 Standard states of real gases

☐ standard state of a real gas

5.6 The relation between fugacity and pressure

☐ fugacity coefficient
☐ the fugacity coefficient in terms of the compression factor (24) and the virial coefficients

Further reading

Articles of general interest

J.S. Winn, The fugacity of van der Waals gas. *J. Chem. Educ.* **65**, 772 (1988).

R.M. Noyes, Thermodynamics of a process in a rigid container. *J. Chem. Educ.* **69**, 470 (1992).

L.L. Combs, An alternative view of fugacity. *J. Chem. Educ.* **69**, 218 (1992).

L. D'Alessio, On the fugacity of a van der Waals gas: an approximate expression that separates attractive and repulsive forces. *J. Chem. Educ.* **70**, 96 (1993).

R.J. Tykodi, The Gibbs function, spontaneity, and walls. *J. Chem. Educ.* **73**, 398 (1996).

R.M. Noyes, Application of the Gibbs function to chemical systems and subsystems. *J. Chem. Educ.* **73**, 404 (1996).

S.E. Wood and R. Battino, The Gibbs function controversy. *J. Chem. Educ.* **73**, 408, (1996).

R.J. Tykodi, Spontaneity, accessibility, irreversibility, 'useful work': the availability function, the Helmholtz function, and the Gibbs function. *J. Chem. Educ.* **72**, 103 (1995).

Texts and sources of data and information

J.M. Smith and H.C. Van Ness, *Introduction to chemical engineering thermodynamics.* McGraw–Hill, New York (1987).

B.D. Wood, *Applications of thermodynamics.* Addison–Wesley, New York (1982).

G.N. Lewis and M. Randall, *Thermodynamics*, revised by K.S. Pitzer and L. Brewer. McGraw–Hill, New York (1961).

Exercises

Assume all gases are perfect and that the temperature is 298.15 K unless stated otherwise.

5.1 (a) Express $(\partial S/\partial V)_T = (\partial p/\partial T)_V$ in terms of α and κ_T (see eqns 3.7 and 3.13 for definitions).

5.1 (b) Express $(\partial S/\partial p)_T = -(\partial V/\partial T)_p$ in terms of α (see eqns 3.7 and 3.13 for definitions).

5.2 (a) Suppose that 3.0 mmol $N_2(g)$ occupies 36 cm^3 at 300 K and expands to 60 cm^3. Calculate ΔG for the process.

5.2 (b) Suppose that 2.5 mmol $Ar(g)$ occupies 72 dm^3 at 298 K and expands to 100 dm^3. Calculate ΔG for the process.

5.3 (a) The change in the Gibbs energy for a certain constant-pressure process was found to fit the expression $\Delta G/J = -85.40 + 36.5(T/K)$. Calculate the value of ΔS for the process.

5.3 (b) The change in the Gibbs energy for a certain constant-pressure process was found to fit the expression $\Delta G/J = -73.1 + 42.8(T/K)$. Calculate the value of ΔS for the process.

5.4 (a) Calculate the change in Gibbs energy of 35 g of ethanol (mass density 0.789 g cm^{-3}) when the pressure is increased isothermally from 1 atm to 3000 atm.

5.4 (b) Calculate the change in Gibbs energy of 25 g of methanol (mass density 0.791 g cm^{-3}) when the pressure is increased isothermally from 100 kPa to 100 MPa.

5.5 (a) When 2.00 mol of a gas at 330 K and 3.50 atm is subjected to isothermal compression, its entropy decreases by 25.0 J K^{-1}. Calculate (a) the final pressure of the gas and (b) ΔG for the compression.

5.5 (b) When 3.00 mol of a gas at 230 K and 150 kPa is subjected to isothermal compression, its entropy decreases by 15.0 J K^{-1}. Calculate (a) the final pressure of the gas and (b) ΔG for the compression.

5.6 (a) Calculate the change in chemical potential of a perfect gas when its pressure is increased isothermally from 1.8 atm to 29.5 atm at 40 °C.

5.6 (b) Calculate the change in chemical potential of a perfect gas when its pressure is increased isothermally from 92.0 kPa to 252.0 kPa at 50 °C.

5.7 (a) The fugacity coefficient of a certain gas at 200 K and 50 bar is 0.72. Calculate the difference of its chemical potential from that of a perfect gas in the same state.

5.7 (b) The fugacity coefficient of a certain gas at 290 K and 2.1 MPa is 0.68. Calculate the difference of its chemical potential from that of a perfect gas in the same state.

5.8 (a) At 373 K, the second virial coefficient B of xenon is -81.7 cm^3 mol^{-1}. Calculate the value of B' and hence estimate the fugacity coefficient of xenon at 50 atm and 373 K.

5.8 (b) At 100 K, the second virial coefficient B of nitrogen is -160.0 cm^3 mol^{-1}. Calculate the value of B' and hence estimate the fugacity coefficient of nitrogen at 62 MPa and 100 K.

5.9 (a) Estimate the change in the Gibbs energy of 1.0 L of benzene when the pressure acting on it is increased from 1.0 atm to 100 atm.

5.9 (b) Estimate the change in the Gibbs energy of 1.0 L of water when the pressure acting on it is increased from 100 kPa to 300 kPa.

5.10 (a) Calculate the change in the molar Gibbs energy of hydrogen gas when its pressure is increased isothermally from 1.0 atm to 100.0 atm at 298 K.

5.10 (b) Calculate the change in the molar Gibbs energy of oxygen when its pressure is increased isothermally from 50.0 kPa to 100.0 kPa at 500 K.

5.11 (a) The molar Helmholtz energy of a certain gas is given by:

$$A_m = -\frac{a}{V_m} - RT\ln(V_m - b) + f(T)$$

where a and b are constants and $f(T)$ is a function of temperature only. Obtain the equation of state of the gas.

5.11 (b) The molar Gibbs energy of a certain gas is given by:

$$G_m = RT\ln p + A' + B'p + \tfrac{1}{2}C'p^2 + \tfrac{1}{3}D'p^3$$

where A, B, C, and D are constants. Obtain the equation of state of the gas.

5.12 (a) Evaluate $(\partial S/\partial V)_T$ for a van der Waals gas. For an isothermal expansion, will ΔS be greater for a perfect gas or a van der Waals gas? Explain your conclusion.

Problems

Numerical problems

5.1 Calculate $\Delta_r G^{\ominus}(375$ K$)$ for the reaction $2CO(g) + O_2(g) \rightarrow 2CO_2(g)$ from the value of $\Delta_r G^{\ominus}(298$ K$)$, $\Delta_r H^{\ominus}(298$ K$)$, and the Gibbs–Helmholtz equation.

5.2 Estimate the standard reaction Gibbs energy of $N_2(g) + 3H_2(g) \rightarrow 2NH_3(g)$ at (a) 500 K, (b) 1000 K from their values at 298 K.

5.3 At 298 K the standard enthalpy of combustion of sucrose is -5645 kJ mol^{-1} and the standard Gibbs energy of the reaction is -6333 kJ mol^{-1}. Estimate the additional non-expansion work that may be obtained by raising the temperature to blood temperature, 37 °C.

5.4 At 200 K, the compression factor of oxygen varies with pressure as shown below. Evaluate the fugacity of oxygen at this temperature and 100 atm.

p/atm	1.0000	4.00000	7.00000	10.0000	40.00	70.00	100.0
Z	0.9971	0.98796	0.97880	0.96956	0.8734	0.7764	0.6871

Theoretical problems

5.5 Show that, for a perfect gas, $(\partial U/\partial S)_V = T$ and $(\partial U/\partial V)_S = -p$.

5.6 Two of the four Maxwell relations were derived in the text, but two were not. Complete their derivation by showing that $(\partial S/\partial V)_T = (\partial p/\partial T)_V$ and $(\partial T/\partial p)_S = (\partial V/\partial S)_p$.

5.7 Use the Maxwell relations to express the derivatives $(\partial S/\partial V)_T$ and $(\partial V/\partial S)_p$ in terms of the expansion coefficient α and the isothermal compressibility κ_T.

5.8 Use the Maxwell relations and Euler's chain relation to express $(\partial p/\partial S)_V$ in terms of the heat capacities, the expansion coefficient, and the isothermal compressibility.

5.9 Use the Maxwell relations to show that the entropy of a perfect gas depends on the volume as $S \propto R \ln V$.

5.10 Derive the thermodynamic equation of state

$$\left(\frac{\partial H}{\partial p}\right)_T = V - T\left(\frac{\partial V}{\partial T}\right)_p$$

Derive an expression for $(\partial H/\partial p)_T$ for (a) a perfect gas and (b) a van der Waals gas. In the latter case, estimate its value for 1.0 mol Ar(g) at 298 K and 10 atm. By how much does the enthalpy of the argon change when the pressure is increased isothermally to 11 atm?

5.11 Prove the following relation:

$$\left(\frac{\partial H}{\partial V}\right)_T = -V^2\left(\frac{\partial p}{\partial T}\right)_V\left(\frac{\partial(T/V)}{\partial V}\right)_p$$

5.12 Show that, if $B(T)$ is the second virial coefficient of a gas, and $\Delta B = B(T'') - B(T')$, $\Delta T = T'' - T'$, and T is the mean of T'' and T', then

$$\pi_T \approx \frac{RT^2 \Delta B}{V_m^2 \Delta T}$$

Estimate π_T for argon given that $B(250\ \text{K}) = -28.0\ \text{cm}^3\ \text{mol}^{-1}$ and $B(300\ \text{K}) = -15.6\ \text{cm}^3\ \text{mol}^{-1}$ at 275 K at (a) 1.0 atm, (b) 10.0 atm.

5.13 (a) Prove that the heat capacities C_V and C_p of a perfect gas are independent of both volume and pressure. May they depend on the temperature? (b) Deduce an expression for the dependence of C_V on volume of a gas that is described by the equation of state $pV_m/RT = 1 + B/V_m$.

5.14 The Joule coefficient, μ_J, is defined as $\mu_J = (\partial T/\partial V)_U$. Show that $\mu_J C_V = p - (\alpha T/\kappa_T)$.

5.15 Evaluate π_T for a Dieterici gas (Table 1.7). Justify physically the form of the expression obtained.

5.16 Instead of assuming that the volume of a condensed phase is constant when pressure is applied, assume only that the compres-

sibility is constant. Show that, when the pressure is changed isothermally by Δp, G changes to

$$G' = G + V_m \Delta p(1 - \tfrac{1}{2}\kappa_T \Delta p)$$

Assess the error in assuming that a solid is incompressible by applying this expression to the compression of copper when $\Delta p = 500$ atm. (For copper at 25 °C, $\kappa_T = 0.8 \times 10^{-6}\ \text{atm}^{-1}$ and $\rho = 8.93\ \text{g cm}^{-3}$.)

5.17 Derive an expression for the standard reaction Gibbs energy $\Delta_r G^{\ominus}$ (the analogue of the standard reaction enthalpy) at a temperature T' in terms of its value $\Delta_r G^{\ominus}$ at T by using the Gibbs–Helmholtz equation and (a) assuming that $\Delta_r H^{\ominus}$ does not vary with temperature, (b) assuming instead that $\Delta_r C_p^{\ominus}$ does not vary with temperature and using Kirchhoff's law.

5.18 The adiabatic compressibility, κ_S, is defined like κ_T (eqn 3.13) but at constant entropy. Show that for a perfect gas $p\gamma\kappa_S = 1$ (where γ is the ratio of heat capacities).

5.19 Show that, if S is regarded as a function of T and p, then

$$T\,dS = C_V\,dT + T\left(\frac{\partial p}{\partial T}\right)_V dV$$

Calculate the energy that must be transferred as heat to a van der Waals gas that expands reversibly and isothermally from V_i to V_f.

5.20 Suppose that S is regarded as a function of p and T. Show that

$$T\,dS = C_p\,dT - \alpha TV\,dp$$

Hence, show that the energy transferred as heat when the pressure on an incompressible liquid or solid is increased by Δp is equal to $-\alpha TV\Delta p$. Evaluate q when the pressure acting on 100 cm^3 of mercury at 0 °C is increased by 1.0 kbar. ($\alpha = 1.82 \times 10^{-4}\ \text{K}^{-1}$.)

5.21 The volume of a newly synthesized polymer was found to depend exponentially on the pressure as $V = V_0 e^{-p/p^*}$, where p is the excess pressure and p^* is a constant. Deduce an expression for the Gibbs energy of the polymer as a function of excess pressure. What is the natural direction of change of the compressed material when the pressure is relaxed?

5.22 Find an expression for the fugacity coefficient of a gas that obeys the equation of state

$$\frac{pV_m}{RT} = 1 + \frac{B}{V_m} + \frac{C}{V_m^2}$$

Use the resulting expression to estimate the fugacity of argon at 1.00 atm and 100 K using $B = -21.13\ \text{cm}^3\ \text{mol}^{-1}$ and $C = 1054\ \text{cm}^6\ \text{mol}^{-2}$.

5.23 Derive an expression for the fugacity coefficient of a gas that obeys the equation of state

$$\frac{pV_m}{RT} = 1 + \frac{qT}{V_m}$$

where q is a constant, and plot ϕ against $4pq/R$.

Additional problems supplied by Carmen Giunta and Charles Trapp

5.24 In 1995, the Intergovernmental Panel on Climate Change considered a global average temperature rise of 1.0–3.5°C likely by the year 2100, with 2.0°C its best estimate (IPCC Second Assessment Synthesis of Scientific–Technical Information Relevant to Interpreting Article 2 of the UN Framework Convention on Climate Change (1995)). Because water vapour is itself a greenhouse gas, the increase in water vapour content of the atmosphere is of some concern to climate change experts. Predict the relative increase in water vapour in the atmosphere based on a temperature rises of 2.0 K, assuming that the relative humidity remains constant. (The present global mean temperature is 290 K, and the equilibrium vapour pressure of water at that temperature is 0.0189 bar.)

5.25 Nitric acid hydrates have received much attention as possible catalysts for heterogeneous reactions which bring about the Antarctic ozone hole. Worsnop *et al.* investigated the thermodynamic stability of these hydrates under conditions typical of the polar winter stratosphere (D.R. Worsnop, L.E. Fox, M.S. Zahniser, and S.C. Wofsy, *Science* **259**, 71 (1993)). They report thermodynamic data for the sublimation of mono-, di-, and trihydrates to nitric acid and water vapours, $HNO_3 \cdot nH_2O(s) \rightarrow HNO_3(g) + nH_2O(g)$, for $n = 1, 2$, and 3. Given $\Delta_r G^{\ominus}$ and $\Delta_r H^{\ominus}$ for these reactions at 220 K, use the Gibbs–Helmholtz equation to compute $\Delta_r G^{\ominus}$ at 190 K.

n	1	2	3
$\Delta_r G^{\ominus}/(\text{kJ mol}^{-1})$	46.2	69.4	93.2
$\Delta_r H^{\ominus}/(\text{kJ mol}^{-1})$	127	188	237

5.26 In an investigation of thermophysical properties of toluene (R.D. Goodwin, *J. Phys. Chem. Ref. Data* **18**, 1565 (1989)), Goodwin tabulated (among other quantities) the compression factor, Z, at several temperatures and pressures. From the following information, compute the fugacity coefficient of toluene at 600 K and (a) 30.0 bar and (b) 1000 bar.

p/bar	0.500	1.013	2.00	3.00	5.00	
Z	0.994 12	0.988 96	0.979 42	0.969 95	0.951 33	
p/bar	10.00	20.0	30.0	42.4	50.0	
Z	0.905 69	0.812 27	0.701 77	0.471 98	0.223 76	
p/bar	70.0	100.0	200	300	500	1000
Z	0.265 20	0.349 20	0.623 62	0.882 88	1.371 09	2.488 36

5.27 J. Gao and J.H. Weiner in their study of the origin of stress on the atomic level in dense polymer systems (*Science* **266**, 748 (1994)), observe that the tensile force required to maintain the length, l, of a long linear chain of N freely jointed links, each of length a, can be interpreted as arising from an entropic spring. For such a chain, $S(l) = -3kl^2/2Na^2 + C$, where k is the Boltzmann constant and C is a constant. Using thermodynamic relations from this and previous chapters, show that the tensile force obeys Hooke's law, $f = -k_f l$, if we assume that the energy U is independent of l.

5.28 You are told that the differential of pressure consistent with an equation of state is given by *one* of the following two expressions. Determine the equation of state.

$$dp = \frac{2(V - b)\,dV}{RT} + \frac{(V - b)^2\,dT}{RT^2}$$

$$dp = -\frac{RT\,dV}{(V - b)^2} + \frac{R\,dT}{V - b}$$

5.29 At 1 atm, liquid water has its maximum density at 4°C. What can be concluded about the variation of the entropy of liquid water with pressure at constant temperature at 3°C, 4°C, and 5°C?

6 Physical transformations of pure substances

The simplest applications of thermodynamics to chemically significant systems are to the discussion of the phase transitions of pure substances. We shall see that a phase diagram is a map of the pressures and temperatures at which each phase of a substance is the most stable. First, we describe the interpretation of empirically determined phase diagrams for a selection of materials. Then we turn to a consideration of the factors that determine the positions and shapes of the boundaries between the regions on a phase diagram. The practical importance of the expressions we derive is that they show how the vapour pressure of a substance varies with temperature and how the melting point varies with pressure. We shall see that the transitions between phases can be classified by noting how various thermodynamic functions change when the transition occurs. Finally, we consider the thermodynamic discussion of liquid surfaces, such as surface tension and capillary action.

Vaporization, melting, and the conversion of graphite to diamond are all examples of changes of phase without change of chemical composition. In this chapter we describe such processes thermodynamically, using as the guiding principle the tendency of systems at constant temperature and pressure to minimize their Gibbs energy. Because we are dealing with pure substances, the molar Gibbs energy of the system is the same as the chemical potential, μ, so the direction of spontaneous change is in the direction of decreasing chemical potential. Once again we see how the properties of the chemical potential mirror its name: a pure substance with a high chemical potential has a spontaneous tendency to move to a state with lower chemical potential.

Phase diagrams

One of the most succinct ways of presenting the physical changes of state that a substance can undergo is in terms of its phase diagram. We present the concept in this section.

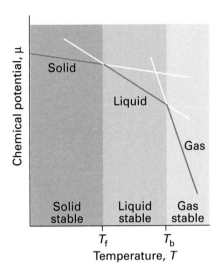

6.1 The schematic temperature dependence of the chemical potential of the solid, liquid, and gas phases of a substance (in practice, the lines are curved). The phase with the lowest chemical potential at a specified temperature is the most stable one at that temperature. The transition temperatures, the melting and boiling temperatures, are the temperatures at which the chemical potentials of two phases are equal.

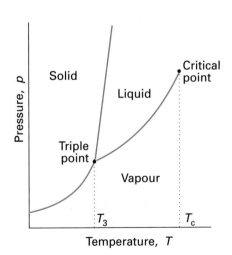

6.2 The general regions of pressure and temperature where solid, liquid, or gas is stable (that is, has lowest chemical potential) are shown on this phase diagram. For example, the solid phase is the most stable phase at low temperatures and high pressures. In the following paragraphs we locate the precise boundaries between the regions.

6.1 The stabilities of phases

A **phase** of a substance is a form of matter that is uniform throughout in chemical composition and physical state. Thus, we speak of solid, liquid, and gas phases of a substance, and of its various solid phases, such as the white and black allotropes of phosphorus. A **phase transition**, the spontaneous conversion of one phase into another phase, occurs at a characteristic temperature for a given pressure. Thus, at 1 atm, ice is the stable phase of water below 0°C, but above 0°C liquid water is more stable. This difference indicates that below 0°C the chemical potential of ice is lower than that of liquid water, and that above 0°C the opposite is true (Fig. 6.1). The **transition temperature**, T_{trs}, is the temperature at which the two chemical potentials are equal and the two phases are in equilibrium at the prevailing pressure.

When considering phase transitions, it is always important to distinguish between the thermodynamic description of the transition and the rate at which the transition actually occurs. A transition that is predicted from thermodynamics to be spontaneous may occur too slowly to be significant in practice. For instance, at normal temperatures and pressures the chemical potential of graphite is lower than that of diamond, so there is a thermodynamic tendency for diamond to change into graphite. However, for this transition to take place, the C atoms must change their locations, which is an immeasurably slow process in a solid except at high temperatures. The rate of attainment of equilibrium is a kinetic problem, and is outside the scope of thermodynamics. In gases and liquids the mobilities of the molecules allow phase transitions to occur rapidly, but in solids thermodynamic instability may be frozen in. Thermodynamically unstable phases that persist because the transition is kinetically hindered are called **metastable phases**. Diamond is a metastable phase of carbon under normal conditions.

6.2 Phase boundaries

The **phase diagram** of a substance shows the regions of pressure and temperature at which its various phases are thermodynamically stable (Fig. 6.2). The lines separating the regions, which are called **phase boundaries**, show the values of p and T at which two phases coexist in equilibrium.

Consider a liquid sample of a pure substance in a closed vessel. The pressure of a vapour in equilibrium with the liquid is called the **vapour pressure** of the substance (Fig. 6.3). Therefore, the liquid–vapour phase boundary in a phase diagram shows how the vapour pressure of the liquid varies with temperature. Similarly, the solid–vapour phase boundary shows the temperature variation of the **sublimation vapour pressure** of the solid. The vapour pressure of a substance increases with temperature because at higher temperatures the Boltzmann distribution populates more heavily the states of higher energy, corresponding to molecules that have escaped from their neighbours.

(a) Critical points and boiling points

When a liquid is heated in an *open* vessel, the liquid vaporizes from its surface. At the temperature at which its vapour pressure would be equal to the external pressure, vaporization can occur throughout the bulk of the liquid and the vapour can expand freely into the surroundings. The condition of free vaporization throughout the liquid is called **boiling**. The temperature at which the vapour pressure of a liquid is equal to the external pressure is called the **boiling temperature** at that pressure. For the special case of an external pressure of 1 atm, the boiling temperature is called the **normal boiling point**, T_b. With the replacement of 1 atm by 1 bar as standard pressure, there is some advantage in using the **standard boiling point** instead: this is the temperature at which the vapour pressure reaches 1 bar. Because 1 bar is slightly less than 1 atm (1.00 bar = 0.987 atm), the

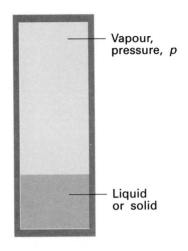

6.3 The vapour pressure of a liquid or solid is the pressure exerted by the vapour in equilibrium with the condensed phase.

Vapour, pressure, p

Liquid or solid

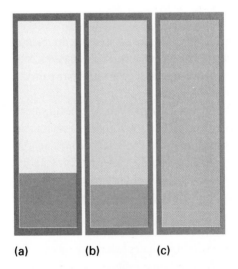

6.4 (a) A liquid in equilibrium with its vapour. (b) When a liquid is heated in a sealed container, the density of the vapour phase increases and that of the liquid decreases slightly. There comes a stage, (c), at which the two densities are equal and the interface between the fluids disappears. This disappearance occurs at the critical temperature. The container needs to be strong: the critical temperature of water is 374 °C and the vapour pressure is then 218 atm.

standard boiling point of a liquid is slightly lower than its normal boiling point. The normal boiling point of water is 100.0 °C; its standard boiling point is 99.6 °C.

Boiling does not occur when a liquid is heated in a *closed* vessel. Instead, the vapour pressure, and hence the density of the vapour, rise continuously as the temperature is raised (Fig. 6.4). At the same time, the density of the liquid decreases slightly as a result of its expansion. There comes a stage when the density of the vapour is equal to that of the remaining liquid and the surface between the two phases disappears. The temperature at which the surface disappears is the **critical temperature**, T_c, of the substance. We first encountered this property in Section 1.4d. The vapour pressure at the critical temperature is called the **critical pressure**, p_c. At and above the critical temperature, a single uniform phase called a **supercritical fluid** fills the container and an interface no longer exists. That is, above the critical temperature, the liquid phase of the substance does not exist.

(b) Melting points and triple points

The temperature at which, under a specified pressure, the liquid and solid phases of a substance coexist in equilibrium is called the **melting temperature**. Because a substance melts at exactly the same temperature as that at which it freezes, the melting temperature of a substance is the same as its freezing temperature. The freezing temperature when the pressure is 1 atm is called the **normal freezing point**, T_f, and its freezing point when the pressure is 1 bar is called the **standard freezing point**. The normal and standard freezing points are negligibly different for most purposes. The normal freezing point is also called the **normal melting point**.

There is a set of conditions under which three different phases of a substance (typically solid, liquid, and vapour) all simultaneously coexist in equilibrium. It is represented by the **triple point**, a point at which the three phase boundaries meet. The temperature at the triple point is denoted T_3. The triple point of a pure substance is outside our control: it occurs at a single definite pressure and temperature characteristic of the substance. The triple point of water lies at 273.16 K and 611 Pa (6.11 mbar, 4.58 Torr), and the three phases of water (ice, liquid water, and water vapour) coexist in equilibrium at no other combination of pressure and temperature. This invariance of the triple point is the basis of its use in the definition of the thermodynamic temperature scale (Section 4.2c).

As can be seen from Fig. 6.2, the triple point marks the lowest pressure at which a liquid phase of a substance can exist. If (as is common) the slope of the solid–liquid phase boundary is as shown in the diagram, then the triple point also marks the lowest temperature at which the liquid can exist; the critical temperature is the upper limit.

6.3 Three typical phase diagrams

We shall now illustrate how these general features appear in the phase diagrams of pure substances.

(a) Water

Figure 6.5 is the phase diagram for water. The liquid–vapour boundary in the phase diagram summarizes how the vapour pressure of liquid water varies with temperature. It also summarizes how the boiling temperature varies with pressure: we simply read off the temperature at which the vapour pressure is equal to the prevailing atmospheric pressure. The solid–liquid boundary shows how the melting temperature varies with the pressure. Its very steep slope indicates that enormous pressures are needed to bring about significant changes. Notice that the line has a negative slope up to 2000 atm, which means that the melting temperature falls as the pressure is raised. The reason for this unusual behaviour can be traced to the decrease in volume that occurs on melting, and hence it being more

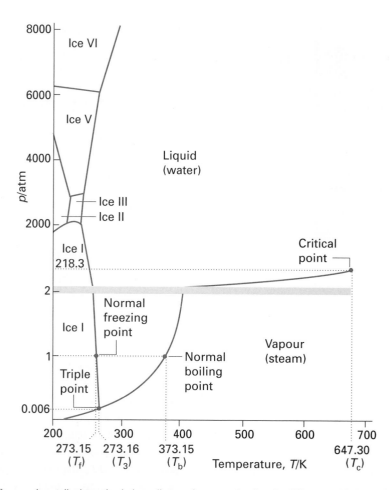

6.5 The experimentally determined phase diagram for water showing the different solid phases. Note the change of vertical scale at 2 atm.

favourable for the solid to transform into the liquid as the pressure is raised. The decrease in volume is a result of the very open molecular structure of ice: the H_2O molecules are held apart (as well as together) by the hydrogen bonds between them, but the structure partially collapses on melting, and the liquid is denser than the solid.

The motion of glaciers may be a consequence of the decrease in melting temperature with pressure: glacial ice melts where it is pressed against the sharp edges of stones and rocks and the glacier inches forwards. However, for many substances surface melting also occurs below the normal melting point, and the explanation of glacier motion (and ice skating) may be more subtle. The reduction in chemical potential of the water below that of the ice may stem from differences in the energy of interaction between water and ice and the rock surface.

At high pressures, different structural forms of ice come into stability as the bonds between molecules are modified by the stress. Some of these phases (which are called ice II, III, V, VI, and VII)[1] melt at high temperatures. Ice VII, for instance, melts at $100°C$ but exists only above 25 kbar. Note that five more triple points occur in the diagram other than the one where vapour, liquid, and ice I coexist. Each one occurs at a definite pressure and temperature that cannot be changed.

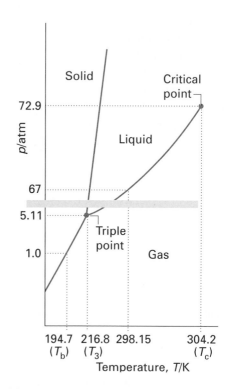

6.6 The experimental phase diagram for carbon dioxide. Note that, as the triple point lies at pressures well above atmospheric, liquid carbon dioxide does not exist under normal conditions (a pressure of at least 5.11 atm must be applied).

---

1   Ice IV was an illusion, or possibly a transient, not thermodynamically stable, state.

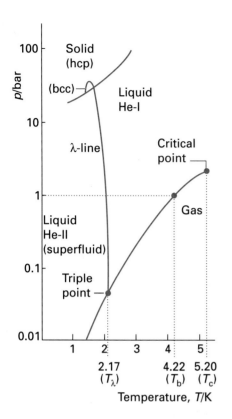

6.7 The phase diagram for helium (^{4}He). The λ-line marks the conditions under which the two liquid phases are in equilibrium. Helium-II is the superfluid phase. Note that a pressure of over 20 bar must be exerted before solid helium can be obtained. The labels hcp and bcc denote different solid phases in which the atoms pack together differently: hcp denotes hexagonal close packing and bcc denotes body-centred cubic (see Section 21.1 for a description of these structures).

(b) Carbon dioxide

The phase diagram for carbon dioxide is shown in Fig. 6.6. The features to notice include the positive slope of the solid–liquid boundary (the direction of this line is characteristic of most substances), which indicates that the melting temperature of solid carbon dioxide rises as the pressure is increased. Notice also that, as the triple point lies above 1 atm, the liquid cannot exist at normal atmospheric pressures whatever the temperature, and the solid sublimes when left in the open (hence the name 'dry ice'). To obtain the liquid, it is necessary to exert a pressure of at least 5.11 atm. Cylinders of carbon dioxide generally contain the liquid or compressed gas; at a temperature of 25 °C that implies a vapour pressure of 67 atm if both gas and liquid are present in equilibrium. When the gas squirts through the throttle it cools by the Joule–Thomson effect, so when it emerges into a region where the pressure is only 1 atm, it condenses into a finely divided snow-like solid.

Supercritical carbon dioxide (that is, highly compressed carbon dioxide above its critical temperature) is used in **supercritical fluid chromatography** (SFC), a form of chromatography in which the supercritical fluid is used as the mobile phase. The technique can be used to separate lipids and phospholipids and to separate fuel oil into alkanes, alkenes, and arenes. A more mundane application of supercritical carbon dioxide is in the decaffeination of coffee, in which caffeine is extracted from green coffee beans.

(c) Helium

The phase diagram of helium is shown in Fig. 6.7. Helium behaves unusually at low temperatures. For instance, the solid and gas phases are never in equilibrium however low the temperature: the He atoms are so light that they vibrate with a large-amplitude motion even at very low temperatures, and the solid simply shakes itself apart. Solid helium can be obtained, but only by holding the atoms together by applying pressure.

When considering helium at low temperatures it is necessary to distinguish between the isotopes ^{3}He and ^{4}He, because quantum mechanical effects become important and these two isotopes differ not only in mass but also in the spin of their nuclei.[2] As a result of quantum mechanical effects, pure helium-4 has a liquid–liquid phase transition at its **λ-line** (lambda line); the reason for this name is explained in Section 6.7. The liquid phase marked He-I behaves like a normal liquid. The other phase, He-II, is a **superfluid**. It is so called because it flows without viscosity. The phase diagram of helium-3 differs from the phase diagram of helium-4, but it also possesses a superfluid phase. Helium-3 is unusual in that the entropy of the liquid is lower than that of the solid, and melting is exothermic.

Phase stability and phase transitions

We shall now see how thermodynamic considerations can account for the features of the phase diagrams we have just described.

6.4 The thermodynamic criterion of equilibrium

We shall base our discussion on the following consequence of the Second Law:

> **At equilibrium, the chemical potential of a substance is the same throughout a sample, regardless of how many phases are present.**

---

2   Nuclear spin is a quantum mechanical property that will be encountered in more detail in Part 2. At this stage it can be visualized as a spinning motion of the nucleus. A ^{4}He nucleus has zero spin; a ^{3}He nucleus has nonzero spin.

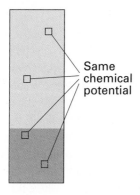

Same chemical potential

6.8 When two or more phases are in equilibrium, the chemical potential of a substance (and, in a mixture, a component) is the same in each phase and is the same at all points in each phase.

When the liquid and solid phases of a substance are in equilibrium, the chemical potential of the substance is the same throughout the liquid and throughout the solid, and is the same in the solid as in the liquid (Fig. 6.8).

To see the validity of this remark, consider a system in which the chemical potential of a substance is μ_1 at one location and μ_2 at another location. The locations may be in the same or in different phases. When an amount dn of the substance is transferred from one location to the other, the Gibbs energy of the system changes by $-\mu_1\,dn$ when material is removed from location 1, and it changes by $+\mu_2\,dn$ when that material is added to location 2. The overall change is therefore $dG = (\mu_2 - \mu_1)\,dn$. If the chemical potential at location 1 is higher than that at location 2, the transfer is accompanied by a decrease in G, and so has a spontaneous tendency to occur. Only if $\mu_1 = \mu_2$ is there no change in G, and only then is the system at equilibrium.

Molecular interpretation 6.1 The tendency towards equality of chemical potential is a disguised form of the tendency to greater entropy as expressed by the Second Law. Suppose that in location 1 the molecules experience less favourable attractive forces than in location 2. Then the entropy of the surroundings will increase if molecules migrate from location 1 to location 2 because heat will be released into the surroundings and increase their disorder. However, to judge whether a process is spontaneous, we need to consider the total entropy change, and there may be a difference in molecular disorder between the two locations. If the disorder of the molecules in location 1 is greater than in location 2, the entropy of the system will decrease when molecules migrate from location 1 to location 2. For example, location 1 might be a gas phase and location 2 might be a liquid phase. The transfer of molecules from location 1 (gas) to location 2 (liquid) is spontaneous if the increase in entropy of the surroundings exceeds the decrease in entropy of the system. The opposite transfer, from location 2 (liquid) to location 1 (gas) is spontaneous if the increase in disorder of the system is greater than the decrease in disorder of the surroundings. The change is spontaneous in neither direction if the changes in entropies of the system and its surroundings cancel, which is expressed by setting the chemical potentials of the two phases equal to each other. The chemical potential takes into account the contributions of the enthalpy change to the surroundings and the change in entropy of the molecules themselves.

6.5 The dependence of stability on the conditions

At low temperatures, the solid phase of a substance has the lowest chemical potential and, provided the pressure is not too low, is usually the most stable at low temperatures. However, the chemical potentials of phases change with temperature in different ways, and as the temperature is raised the chemical potential of another phase (perhaps another solid phase, a liquid, or a gas) will fall below that of the solid. When that happens, a phase transition occurs if it is kinetically feasible to do so.

(a) The temperature dependence of phase stability

The temperature dependence of the Gibbs energy is expressed in terms of the entropy of the system by eqn 5.10 $((\partial G/\partial T)_p = -S)$. Because the chemical potential of a pure substance is equal to the molar Gibbs energy of that substance, it follows that

$$\left(\frac{\partial \mu}{\partial T}\right)_p = -S_m \tag{1}$$

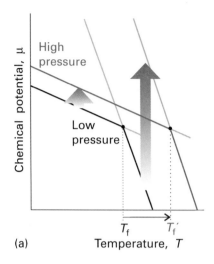

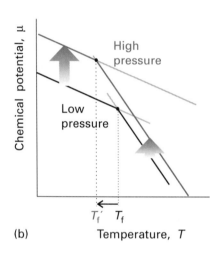

6.9 The pressure dependence of the chemical potential of a substance depends on the molar volume of the phase. The lines show schematically the effect of increasing pressure on the chemical potentials of the solid and liquid phases (in practice, the lines are curved), and the corresponding effects on the freezing temperatures. (a) In this case the molar volume of the solid is less than that of the liquid and $\mu(s)$ increases less than $\mu(l)$. As a result, the freezing temperature rises. (b) Here the molar volume is greater for the solid than the liquid (as for water), $\mu(s)$ increases more strongly than $\mu(l)$, and the freezing temperature is lowered.

This relation shows that, as the temperature is raised, the chemical potential of a pure substance decreases: $S_m > 0$ always, so the slope of a plot of μ against T is negative.

Equation 1 implies that the slope of a plot of μ against temperature is steeper for gases than for liquids, because $S_m(g) > S_m(l)$. The slope is also steeper for a liquid than the corresponding solid, because $S_m(l) > S_m(s)$ almost always, on account of the greater disorder of a liquid. These features were illustrated in Fig. 6.1. The steep negative slope of $\mu(l)$ results in its falling below $\mu(s)$ when the temperature is high enough, and then the liquid becomes the stable phase: the solid melts. The chemical potential of the gas phase plunges steeply downwards as the temperature is raised (because the molar entropy of the vapour is so high), and there comes a temperature at which it lies lowest. Then the gas is the stable phase and the liquid vaporizes. When we bring about a phase transition, what we are actually doing is modifying the relative values of the chemical potentials of the phases. The easiest way of doing that is by changing the temperature of the sample.

(b) The response of melting to applied pressure

Most substances melt at a higher temperature when subjected to pressure. It is as though the pressure is preventing the formation of the less dense liquid phase. Exceptions to this behaviour include water, for which the liquid is denser than the solid. Application of pressure to water encourages the formation of the liquid phase. That is, water freezes at a lower temperature when it is under pressure.

We can rationalize the response of melting temperatures to pressure as follows. The variation of the chemical potential with pressure is expressed (from eqn. 5.10) by

$$\left(\frac{\partial \mu}{\partial p}\right)_T = V_m \qquad (2)$$

This equation shows that the slope of a plot of chemical potential against pressure is equal to the molar volume of the substance. An increase in pressure raises the chemical potential of any pure substance (because $V_m > 0$). In most cases, $V_m(l) > V_m(s)$ and the equation predicts that an increase in pressure increases the chemical potential of the liquid more than that of the solid. As shown in Fig. 6.9a, the effect of pressure in such a case is to raise the melting temperature slightly. For water, however, $V_m(l) < V_m(s)$, and an increase in pressure increases the chemical potential of the solid more than that of the liquid. In this case, the melting temperature is lowered slightly (Fig. 6.9b).

Example 6.1 Assessing the effect of pressure on the chemical potential

Calculate the effect on the chemical potentials of ice and water of increasing the pressure from 1.00 bar to 2.00 bar at $0\,°C$. The density of ice is $0.917\ \mathrm{g\,cm}^{-3}$ and that of liquid water is $0.999\ \mathrm{g\,cm}^{-3}$ under these conditions.

Method From eqn 2, we know that the change in chemical potential of an incompressible substance when the pressure is changed by Δp is $\Delta\mu = V_m \Delta p$. Therefore, to answer the question, we need to know the molar volumes of the two phases of water. These values are obtained from the mass density, ρ, and the molar mass, M, by using $V_m = M/\rho$. We therefore use the expression

$$\Delta\mu = \frac{M\Delta p}{\rho}$$

Convert all data to SI units, and use $1\ \mathrm{Pa\,m}^3 = 1\ \mathrm{J}$.

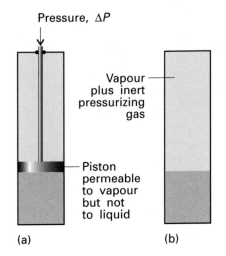

Pressure, ΔP

Vapour plus inert pressurizing gas

Piston permeable to vapour but not to liquid

(a)　　　　(b)

6.10 Pressure may be applied to a condensed phase either (a) by compressing the condensed phase or (b) by subjecting it to an inert pressurizing gas. When pressure is applied, the vapour pressure of the condensed phase increases.

Answer The molar mass of water is 18.02 g mol^{-1} ($1.802 \times 10^{-2} \text{ kg mol}^{-1}$); therefore,

$$\Delta\mu(\text{ice}) = \frac{(1.802 \times 10^{-2} \text{ kg mol}^{-1}) \times (1.00 \times 10^5 \text{ Pa})}{917 \text{ kg m}^{-3}} = +1.97 \text{ J mol}^{-1}$$

$$\Delta\mu(\text{water}) = \frac{(1.802 \times 10^{-2} \text{ kg mol}^{-1}) \times (1.00 \times 10^5 \text{ Pa})}{999 \text{ kg m}^{-3}} = +1.80 \text{ J mol}^{-1}$$

Comment The chemical potential of ice rises more sharply than that of water so, if they are initially in equilibrium at 1 bar, there will be a tendency for the ice to melt at 2 bar.

- -

Self-test 6.1 Calculate the effect of an increase in pressure of 1.00 bar on the liquid and solid phases of carbon dioxide (of molar mass 44.0 g mol^{-1}) in equilibrium with densities 2.35 g cm^{-3} and 2.50 g cm^{-3}, respectively.

$$[\Delta\mu(\text{l}) = +1.87 \text{ J mol}^{-1}, \Delta\mu(\text{s}) = +1.76 \text{ J mol}^{-1}; \text{ solid forms}]$$

(c) The effect of applied pressure on vapour pressure

When pressure is applied to a condensed phase, its vapour pressure rises: in effect, molecules are squeezed out of the phase and escape as a gas. Pressure can be exerted on the condensed phase mechanically or by subjecting it to the applied pressure of an inert gas (Fig. 6.10); in the latter case, the vapour pressure is the partial pressure of the vapour in equilibrium with the condensed phase, and we speak of the **partial vapour pressure** of the substance. One complication (which we ignore here) is that, if the condensed phase is a liquid, then the pressurizing gas might dissolve and change its properties. Another complication is that the gas-phase molecules might attract molecules out of the liquid by the process of **gas solvation**, the attachment of molecules to gas-phase species.

The quantitative relation between the vapour pressure, p, when a pressure ΔP is applied and the vapour pressure, p^*, of the liquid in the absence of an additional pressure is

$$p = p^* e^{V_m \Delta P / RT} \tag{3}°$$

This equation shows how the vapour pressure increases when the pressure acting on the condensed phase is increased.

Justification 6.1

We calculate the vapour pressure of a pressurized liquid by using the fact that at equilibrium the chemical potentials of the liquid and its vapour are equal: $\mu(\text{l}) = \mu(\text{g})$. It follows that, for any change that preserves equilibrium, the resulting change in $\mu(\text{l})$ must be equal to the change in $\mu(\text{g})$; therefore, we can write $d\mu(\text{g}) = d\mu(\text{l})$. When the pressure P on the liquid is increased by dP, the chemical potential of the liquid changes by $d\mu(\text{l}) = V_m(\text{l}) dP$. The chemical potential of the vapour changes by $d\mu(\text{g}) = V_m(\text{g}) dp$ where dp is the change in the vapour pressure we are trying to find. If we treat the vapour as a perfect gas, the molar volume can be replaced by $V_m(\text{g}) = RT/p$, and we obtain

$$d\mu(\text{g}) = \frac{RT \, dp}{p}$$

Next, we equate the changes in chemical potentials of the vapour and the liquid:

$$\frac{RT \, dp}{p} = V_m(\text{l}) \, dP$$

This expression can be integrated once we know the limits of integration.

When there is no additional pressure acting on the liquid, P (the pressure experienced by the liquid) is equal to the normal vapour pressure p^*; so, when $P = p^*$, $p = p^*$ too. When there is an additional pressure ΔP on the liquid, with the result that $P = p + \Delta P$, the vapour pressure is p (the value we want to find). The effect of pressure on the vapour pressure is so small that it is a good approximation to replace the p in $p + \Delta P$ by p^* itself, and to set the upper limit of the integral to $p^* + \Delta P$. The integrations required are therefore as follows:

$$RT \int_{p^*}^{p} \frac{dp}{p} = \int_{p^*}^{p^* + \Delta P} V_m(l)\, dP$$

We now assume that the molar volume of the liquid is the same throughout the small range of pressures involved. Then both integrations are straightforward, and lead to

$$RT \ln\left(\frac{p}{p^*}\right) = V_m(l)\Delta P$$

which rearranges to eqn 3.

Example 6.2 Estimating the effect of pressure on the vapour pressure

Derive an expression from eqn 3 that is valid for small changes in vapour pressure and calculate the fractional increase of the vapour pressure of water for an increase in pressure of 10 bar at 25 °C.

Method The question centres on the approximation of the right-hand side of eqn 3 when the exponent is small. For the approximation, we note that the exponential function e^x is equal to the expansion $1 + x + \frac{1}{2}x^2 + \cdots$; so if $x \ll 1$, a good approximation is $e^x \approx 1 + x$.

Answer If $V_m \Delta p / RT \ll 1$, the exponential function on the right of eqn 3 may be approximated by $1 + V_m \Delta p / RT$:

$$p \approx p^*\left(1 + \frac{V_m \Delta p}{RT}\right)$$

which rearranges to

$$\frac{p - p^*}{p^*} \approx \frac{V_m \Delta P}{RT}$$

For water (which has density 0.997 g cm^{-3} at 25 °C and therefore molar volume $18.1 \text{ cm}^3 \text{ mol}^{-1}$)

$$\frac{V_m \Delta P}{RT} = \frac{(1.81 \times 10^{-5} \text{ m}^3 \text{ mol}^{-1}) \times (1.0 \times 10^6 \text{ Pa})}{(8.3145 \text{ J K}^{-1} \text{ mol}^{-1}) \times (298 \text{ K})} = 7.3 \times 10^{-3}$$

Because $V_m \Delta P / RT \ll 1$, the approximate formula can be used, and we obtain $(p - p^*)/p^* = 7.3 \times 10^{-3}$, an increase of 0.73 per cent.

- -

Self-test 6.2 Calculate the effect of an increase in pressure of 100 bar on the vapour pressure of benzene at 25 °C, which has density 0.879 g cm^{-3}.

[36 per cent; because the change is so large, use eqn 3]

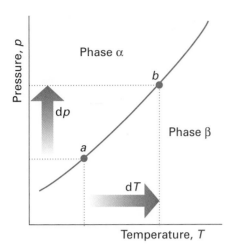

6.11 When pressure is applied to a system in which two phases are in equilibrium (at *a*), the equilibrium is disturbed. It can be restored by changing the temperature, so moving the state of the system to *b*. It follows that there is a relation between d*p* and d*T* that ensures that the system remains in equilibrium as either variable is changed.

6.6 The location of phase boundaries

We can find the precise locations of the phase boundaries—the pressures and temperatures at which two phases can coexist—by making use of the fact that, when two phases are in equilibrium, their chemical potentials must be equal. Therefore, where the phases α and β are in equilibrium,

$$\mu_\alpha(p, T) = \mu_\beta(p, T) \tag{4}$$

By solving this equation for p in terms of T, we get an equation for the phase boundary.

(a) The slopes of the phase boundaries

It turns out to be simplest to discuss the phase boundaries in terms of their slopes, dp/dT.

Let p and T be changed infinitesimally, but in such a way that the two phases α and β remain in equilibrium. The chemical potentials of the phases are initially equal (the two phases are in equilibrium). They remain equal when the conditions are changed to another point on the phase boundary, where the two phases continue to be in equilibrium (Fig. 6.11). Therefore, the changes in the chemical potentials of the two phases must be equal, and we can write $d\mu_\alpha = d\mu_\beta$. Because, from eqn 5.9, we know that

$$d\mu = -S_m\, dT + V_m\, dp$$

for each phase, it follows that

$$-S_{\alpha,m}\, dT + V_{\alpha,m}\, dp = -S_{\beta,m}\, dT + V_{\beta,m}\, dp$$

where $S_{\alpha,m}$ and $S_{\beta,m}$ are the molar entropies of the phases and $V_{\alpha,m}$ and $V_{\beta,m}$ are their molar volumes. Hence

$$(V_{\beta,m} - V_{\alpha,m})\, dp = (S_{\beta,m} - S_{\alpha,m})\, dT \tag{5}$$

which rearranges into the **Clapeyron equation**:

$$\frac{dp}{dT} = \frac{\Delta_{trs}S}{\Delta_{trs}V} \tag{6}$$

In this expression $\Delta_{trs}S = S_{\beta,m} - S_{\alpha,m}$ and $\Delta_{trs}V = V_{\beta,m} - V_{\alpha,m}$ are the entropy and volume of transition. The Clapeyron equation is an exact expression for the slope of the phase boundary and applies to any phase equilibrium of any pure substance.

(b) The solid–liquid boundary

Melting (fusion) is accompanied by a molar enthalpy change $\Delta_{fus}H$ and occurs at a temperature T. The molar entropy of melting at T is therefore $\Delta_{fus}H/T$, and the Clapeyron equation becomes

$$\frac{dp}{dT} = \frac{\Delta_{fus}H}{T\Delta_{fus}V} \tag{7}$$

where $\Delta_{fus}V$ is the change in molar volume that occurs on melting. The enthalpy of melting is positive (the only exception is helium-3) and the volume change is usually positive and always small. Consequently, the slope dp/dT is steep and usually positive (Fig. 6.12).

The formula for the phase boundary can be obtained by integrating dp/dT, assuming that $\Delta_{fus}H$ and $\Delta_{fus}V$ change so little with temperature and pressure that they can be treated as constant. If the melting temperature is T^* when the pressure is p^*, and T when the pressure is p, the integration required is

$$\int_{p^*}^{p} dp = \frac{\Delta_{fus}H}{\Delta_{fus}V} \int_{T^*}^{T} \frac{dT}{T}$$

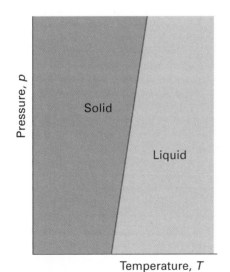

6.12 A typical solid–liquid phase boundary slopes steeply upwards. This slope implies that, as the pressure is raised, the melting temperature rises. Most substances behave in this way.

Therefore, the approximate equation of the solid–liquid boundary is

$$p = p^* + \frac{\Delta_{fus}H}{\Delta_{fus}V} \ln\left(\frac{T}{T^*}\right) \qquad (8)$$

This equation was originally obtained by yet another Thomson–James, the brother of William, Lord Kelvin. When T is close to T^*, the logarithm can be approximated by using

$$\ln\left(\frac{T}{T^*}\right) = \ln\left(1 + \frac{T - T^*}{T^*}\right) \approx \frac{T - T^*}{T^*}$$

because $\ln(1 + x) \approx x$ when $x \ll 1$; therefore,

$$p \approx p^* + \frac{(T - T^*)\Delta_{fus}H}{T^*\Delta_{fus}V} \qquad (9)$$

This expression is the equation of a steep straight line when p is plotted against T (as in Fig. 6.12).

(c) The liquid–vapour boundary

The entropy of vaporization at a temperature T is equal to $\Delta_{vap}H/T$; the Clapeyron equation for the liquid–vapour boundary is therefore

$$\frac{dp}{dT} = \frac{\Delta_{vap}H}{T\Delta_{vap}V} \qquad (10)$$

The enthalpy of vaporization is positive; $\Delta_{vap}V$ is large and positive. Therefore, dp/dT is positive, but it is much smaller than for the solid–liquid boundary. It follows that dT/dp is large, and hence that the boiling temperature is more responsive to pressure than the freezing temperature.

Example 6.3 Estimating the effect of pressure on the boiling point

Estimate the typical size of the effect of increasing pressure on the boiling point of a liquid.

Method To use eqn 10 we need to estimate the right-hand side. At the boiling point, the term $\Delta_{vap}H/T$ is Trouton's constant (Section 4.3a). Because the molar volume of a gas is so much greater than the molar volume of a liquid, we can write

$$\Delta_{vap}V = V_m(g) - V_m(l) \approx V_m(g)$$

and take for $V_m(g)$ the molar volume of a perfect gas (at low pressures, at least).

Answer Trouton's constant has the value $85\ J\,K^{-1}\,mol^{-1}$. The molar volume of a perfect gas is about $25\ L\,mol^{-1}$ at 1 atm and near but above room temperature. Therefore,

$$\frac{dp}{dT} \approx \frac{85\ J\,K^{-1}\,mol^{-1}}{25 \times 10^{-3}\ m^3\,mol^{-1}} = 3.4 \times 10^3\ Pa\,K^{-1}$$

This value corresponds to $0.034\ atm\,K^{-1}$, and hence to $dT/dp = 30\ K\,atm^{-1}$. Therefore, a change of pressure of $+0.1$ atm can be expected to change a boiling temperature by about $+3\ K$.

- -

Self-test 6.3 Estimate dT/dp for water at its normal boiling point using the information in Table 4.2 and $V_m(g) = RT/p$.

$$[28\ K\,atm^{-1}]$$

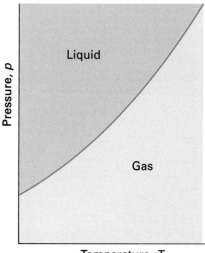

6.13 A typical liquid–vapour phase boundary. The boundary can be regarded as a plot of the vapour pressure against the temperature. Note that, in some depictions of phase diagrams in which a logarithmic pressure scale is used, the phase boundary has the opposite curvature (see Fig. 6.7). That phase boundary terminates at the critical point (not shown).

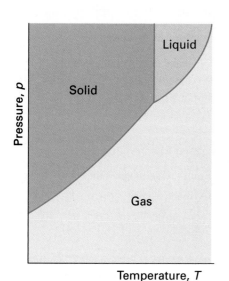

6.14 Near the point where they coincide (at the triple point), the solid–gas boundary has a steeper slope than the liquid–gas boundary because the enthalpy of sublimation is greater than the enthalpy of vaporization and the temperatures that occur in the Clausius–Clapeyron equation for the slope have similar values.

Because the molar volume of a gas is so much greater than the molar volume of a liquid, we can write $\Delta_{vap}V \approx V_m(g)$ (see the example above). Moreover, if the gas behaves perfectly, $V_m(g) = RT/p$. These two approximations turn the exact Clapeyron equation into

$$\frac{dp}{dT} = \frac{\Delta_{vap}H}{T(RT/p)}$$

which rearranges into the **Clausius–Clapeyron equation** for the variation of vapour pressure with temperature:

$$\frac{d \ln p}{dT} = \frac{\Delta_{vap}H}{RT^2} \qquad (11)°$$

(We have used $dx/x = d \ln x$.) If we also assume that the enthalpy of vaporization is independent of temperature, this equation integrates to

$$p = p^* e^{-\chi} \qquad \chi = \frac{\Delta_{vap}H}{R}\left(\frac{1}{T} - \frac{1}{T^*}\right) \qquad (12)°$$

where p^* is the vapour pressure when the temperature is T^* and p the vapour pressure when the temperature is T. Equation 12 is the curve plotted as the liquid–vapour boundary in Fig. 6.13. The line does not extend beyond the critical temperature T_c, because above this temperature the liquid does not exist.

(d) The solid–vapour boundary

The only difference between this case and the last is the replacement of the enthalpy of vaporization by the enthalpy of sublimation, $\Delta_{sub}H$. Because the enthalpy of sublimation is greater than the enthalpy of vaporization, the equation predicts a steeper slope for the sublimation curve than for the vaporization curve at similar temperatures, which is near where they meet (Fig. 6.14).

6.7 The Ehrenfest classification of phase transitions

There are many different types of phase transition, including the common examples of fusion and vaporization and the less common examples of solid–solid, conducting–superconducting, and fluid–superfluid transitions. We shall now see that it is possible to use thermodynamic properties of substances, and in particular the behaviour of the chemical potential, to classify phase transitions into different types. The classification scheme was originally proposed by Paul Ehrenfest, and is known as the **Ehrenfest classification**.

Many familiar phase transitions, like fusion and vaporization, are accompanied by changes of enthalpy and volume. These changes have implications for the slopes of the chemical potentials of the phases at either side of the phase transition. Thus, at the transition from a phase α to another phase β,

$$\left(\frac{\partial \mu_\beta}{\partial p}\right)_T - \left(\frac{\partial \mu_\alpha}{\partial p}\right)_T = V_{\beta,m} - V_{\alpha,m} = \Delta_{trs}V$$

$$\left(\frac{\partial \mu_\beta}{\partial T}\right)_p - \left(\frac{\partial \mu_\alpha}{\partial T}\right)_p = -S_{\beta,m} + S_{\alpha,m} = -\Delta_{trs}S = -\frac{\Delta_{trs}H}{T_{trs}} \qquad (13)$$

Because $\Delta_{trs}V$ and $\Delta_{trs}H$ are nonzero for melting and vaporization, it follows that for such transitions the slopes of the chemical potential plotted against either pressure or temperature are different on either side of the transition (Fig. 6.15a). In other words, the first derivatives of the chemical potentials with respect to pressure and temperature are discontinuous at the transition. A transition for which the first derivative with respect to temperature is discontinuous is classified as a **first-order phase transition**.

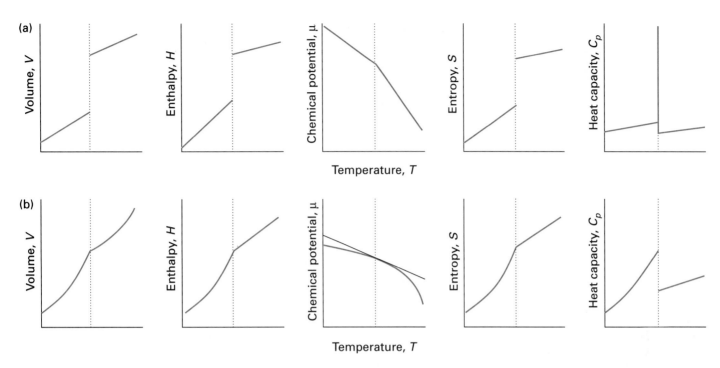

6.15 The changes in thermodynamic properties accompanying (a) first-order and (b) second-order phase transitions.

The constant-pressure heat capacity C_p of a substance is the slope of a plot of the enthalpy with respect to temperature. At a first-order phase transition, H changes by a finite amount for an infinitesimal change of temperature. Therefore, at the transition the heat capacity is infinite. The physical reason is that heating drives the transition rather than raising the temperature. For example, boiling water stays at the same temperature even though heat is being supplied.

A **second-order phase transition** in the Ehrenfest sense is one in which the first derivative of μ with respect to temperature is continuous but its second derivative is discontinuous. A continuous slope of μ (a graph with the same slope on either side of the transition) implies that the volume and entropy (and hence the enthalpy) do not change at the transition (Fig. 6.15b). The heat capacity is discontinuous at the transition but does not become infinite there. An example of a second-order transition is the conducting-superconducting transition in metals at low temperatures.

The term **λ-transition** is applied to a phase transition that is not first-order yet the heat capacity becomes infinite at the transition temperature. Typically, the heat capacity of a system that shows such a transition begins to increase well before the transition (Fig. 6.16), and the shape of the heat capacity curve resembles the Greek letter lambda. This type of transition includes order–disorder transitions in alloys, the onset of ferromagnetism, and the fluid–superfluid transition of liquid helium.

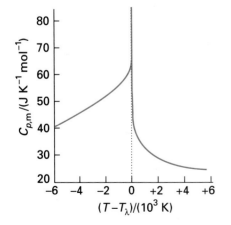

6.16 The λ-curve for helium, where the heat capacity rises to infinity. The shape of this curve is the origin of the name λ-transition.

Molecular interpretation 6.2 One type of second-order transition is associated with a change in symmetry of the crystal structure of a solid. Thus, suppose the arrangement of atoms in a solid is like that represented in Fig. 6.17a, with one dimension (technically, of the unit cell) longer than the other two, which are equal. Such a crystal structure is classified as tetragonal (see Section 21.1). Moreover, suppose the two shorter dimensions increase more than the long dimension when the temperature is raised. There may come a stage when the three dimensions become equal. At that point the crystal has cubic symmetry (Fig. 6.17b), and at higher temperatures it will expand equally in all three

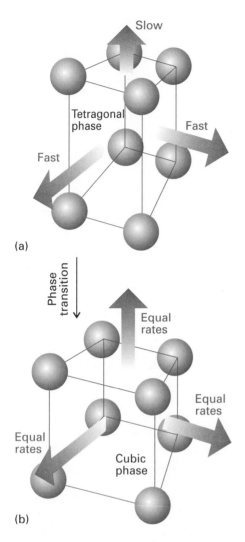

6.17 One version of a second-order phase transition in which (a) a tetragonal phase expands more rapidly in two directions than a third, and hence becomes a cubic phase, which (b) expands uniformly in three directions as the temperature is raised. There is no rearrangement of atoms at the transition temperature, and hence no enthalpy of transition.

Table 6.1* Surface tensions of liquids at 293 K

	$\gamma/(\text{mN m}^{-1})$†
Benzene	28.88
Mercury	472
Methanol	22.6
Water	72.75

* More values are given in the *Data section* at the end of this volume.

† Note that $1\,\text{N m}^{-1} = 1\,\text{J m}^{-2}$.

directions (because there is no longer any distinction between them). The tetragonal → cubic phase transition has occurred, but, as it has not involved a discontinuity in the interaction energy between the atoms or the volume they occupy, the transition is not first-order.

The order–disorder transition in β-brass (CuZn) is an example of a λ-transition. The low-temperature phase is an orderly array of alternating Cu and Zn atoms. The high-temperature phase is a random array of the atoms (Fig. 6.18). At $T = 0$ the order is perfect, but islands of disorder appear as the temperature is raised. The islands form because the transition is cooperative in the sense that, once two atoms have exchanged locations, it is easier for their neighbours to exchange their locations. The islands grow in extent, and merge throughout the crystal at the transition temperature (742 K). The heat capacity increases as the transition temperature is approached because the cooperative nature of the transition means that it is increasingly easy for the heat supplied to drive the phase transition rather than to be stored as thermal motion.

The physical liquid surface

So far, we have concentrated on the properties of the boundaries in the phase diagram of a substance. However, the *physical* boundary between phases, such as the surface where solid is in contact with liquid or liquid is in contact with vapour, has interesting properties. In this section we concentrate on the liquid–vapour interface, which is interesting because it is so mobile. Chapter 28 deals with solid surfaces and their important role in catalysis.

6.8 Surface tension

Liquids tend to adopt shapes that minimize their surface area, for then the maximum number of molecules are in the bulk and hence surrounded by and interacting with neighbours. Droplets of liquids therefore tend to be spherical, because a sphere is the shape with the smallest surface-to-volume ratio. However, there may be other forces present that compete against the tendency to form this ideal shape, and in particular gravity may flatten spheres into puddles or oceans.

Surface effects may be expressed in the language of Helmholtz and Gibbs energies. The link between these quantities and the surface area is the work needed to change the area by a given amount, and the fact that dA and dG are equal (under different conditions) to the work done in changing the energy of a system. The work needed to change the surface area, σ, of a sample by an infinitesimal amount $d\sigma$ is proportional to $d\sigma$, and we write

$$dw = \gamma\, d\sigma \qquad [14]$$

The constant of proportionality, γ, is called the **surface tension**; its dimensions are energy/area and its units are typically joules per metre squared (J m^{-2}). However, as in Table 6.1, values of γ are usually reported in newtons per metre (N m^{-1}, because $1\,\text{J} = 1\,\text{N m}$). The work of surface formation at constant volume and temperature can be identified with the change in the Helmholtz energy, and we can write

$$dA = \gamma\, d\sigma \qquad (15)$$

Because the Helmholtz energy decreases ($dA < 0$) if the surface area decreases ($d\sigma < 0$), surfaces have a natural tendency to contract. This is a more formal way of expressing what we have already described.

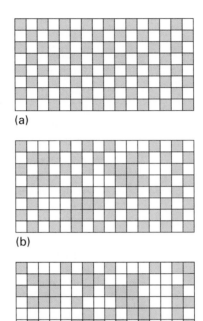

(a)

(b)

(c)

6.18 An order–disorder transition. (a) At $T = 0$, there is perfect order, with different kinds of atoms occupying alternate sites. (b) As the temperature is increased, atoms exchange locations and islands of each kind of atom form in regions of the solid. Some of the original order survives. (c) At and above the transition temperature, the islands occur at random throughout the sample.

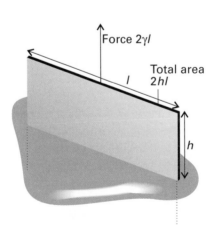

Force $2\gamma l$

Total area $2hl$

l

h

6.19 The model used for calculating the work of forming a liquid film when a wire of length l is raised and pulls the surface with it through a height h.

Example 6.4 Using the surface tension

Calculate the work needed to raise a wire of length l and to stretch the surface of a liquid through a height h in the arrangement shown in Fig. 6.19. Disregard gravitational potential energy.

Method According to eqn 14, the work required to create a surface area given that the surface tension does not vary as the surface is formed is $w = \gamma\sigma$. Therefore, all we need do is to calculate the surface area of the two-sided rectangle formed as the frame is withdrawn from the liquid.

Answer When the wire of length l is raised through a height h it increases the area of the liquid by twice the area of the rectangle (because there is a surface on each side). The total increase is therefore $2lh$ and the work done is $2\gamma lh$.

Comment The work can be expressed as a force × distance by writing it as $2\gamma l \times h$, and identifying γl as the opposing force on the wire of length l. This is why γ is called a tension and why its units are often chosen to be newtons per metre ($\mathrm{N\,m^{-1}}$, so γl is a force in newtons).

- -

Self-test 6.4 Calculate the work of creating a spherical cavity of radius r in a liquid of surface tension γ.

$[4\pi r^2\gamma]$

6.9 Curved surfaces

The minimization of the surface area of a liquid may result in the formation of a curved surface, as in a bubble. We shall now see that there are two consequences of curvature, and hence of the surface tension, that are relevant to the properties of liquids. One is that the vapour pressure of a liquid depends on the curvature of its surface. The other is the capillary rise (or fall) of liquids in narrow tubes.

(a) Bubbles, cavities, and droplets

A **bubble** is a region in which vapour (and possibly air too) is trapped by a thin film; a **cavity** is a vapour-filled hole in a liquid. What are widely called 'bubbles' in liquids are therefore strictly cavities. True bubbles have two surfaces (one on each side of the film); cavities have only one. The treatments of both are similar, but a factor of 2 is required for bubbles to take into account the doubled surface area. A **droplet** is a small volume of liquid at equilibrium surrounded by its vapour (and possibly also air).

The pressure on the concave side of an interface, p_{in}, is always greater than the pressure on the convex side, p_{out}. This relation is expressed by the **Laplace equation**, which is derived in the *Justification* below:

$$p_{in} = p_{out} + \frac{2\gamma}{r} \tag{16}$$

The Laplace equation shows that the difference in pressure decreases to zero as the radius of curvature becomes infinite (when the surface is flat, Fig. 6.20). Small cavities have small radii of curvature, and so the pressure difference across their surface is quite large. For instance, a 'bubble' (actually, a cavity) of radius 0.10 mm in champagne implies a pressure difference of 1.5 kPa, which is enough to sustain a column of water of height 15 cm.

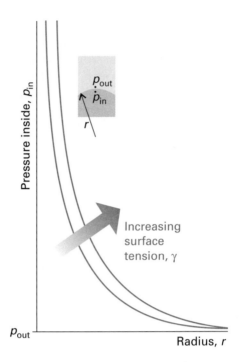

6.20 The dependence of the pressure inside a curved surface on the radius of the surface, for two different values of the surface tension.

Justification 6.2

The cavities in a liquid are at equilibrium when the tendency for their surface area to decrease is balanced by the rise of internal pressure which would then result. When the pressure inside a cavity is p_{in} and its radius is r, the outward force is pressure × area = $4\pi r^2 p_{in}$. The force inwards arises from the external pressure and the surface tension. The former has magnitude $4\pi r^2 p_{out}$. The latter is calculated as follows. The change in surface area when the radius of a sphere changes from r to $r + dr$ is

$$d\sigma = 4\pi(r + dr)^2 - 4\pi r^2 = 8\pi r\, dr$$

(The second-order infinitesimal, $(dr)^2$, is ignored.) The work done when the surface is stretched by this amount is therefore

$$dw = 8\pi\gamma r\, dr$$

As force × distance is work, the force opposing stretching through a distance dr when the radius is r is

$$F = 8\pi\gamma r$$

The total inward force is therefore $4\pi r^2 p_{out} + 8\pi\gamma r$. At equilibrium, the outward and inward forces are balanced, so we can write

$$4\pi r^2 p_{in} = 4\pi r^2 p_{out} + 8\pi\gamma r$$

which rearranges into eqn 16.

We saw in Section 6.5c that the vapour pressure of a liquid depends on the pressure applied to the liquid. Because curving a surface gives rise to a pressure differential of $2\gamma/r$, we can expect the vapour pressure above a curved surface to be different from that above a flat surface. By substituting this value of the pressure difference into eqn 3 we obtain the **Kelvin equation** for the vapour pressure of a liquid when it is dispersed as droplets of radius r:

$$p = p^* e^{2\gamma V_m/rRT} \tag{17}$$

The analogous expression for the vapour pressure inside a cavity can be written at once. The pressure of the liquid outside the cavity is less than the pressure inside, so the only change is in the sign of the exponent in the last expression.

(b) Nucleation

For droplets of water of radius 1 μm and 1 nm, the ratios p/p^* at 25°C are about 1.001 and 3, respectively. The second figure, although quite large, is unreliable because at that radius the droplet is less than about 10 molecules in diameter and the basis of the calculation is suspect. The first figure shows that the effect is usually small; nevertheless it may have important consequences.

Consider, for example, the formation of a cloud. Warm, moist air rises into the cooler regions higher in the atmosphere. At some altitude the temperature is so low that the vapour becomes thermodynamically unstable with respect to the liquid and we expect it to condense into a cloud of liquid droplets. The initial step can be imagined as a swarm of water molecules congregating into a microscopic droplet. Because the initial droplet is so small it has an enhanced vapour pressure. Therefore, instead of growing it evaporates. This effect effectively stabilizes the vapour because an initial tendency to condense is overcome by a heightened tendency to evaporate. The vapour phase is then said to be **supersaturated**. It is thermodynamically unstable with respect to the liquid but not unstable with respect to the small droplets that need to form before the bulk liquid phase can appear, so the formation of the latter by a simple, direct mechanism is hindered.

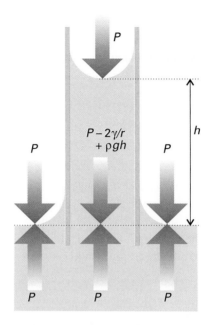

6.21 When a capillary tube is first stood in a liquid, the latter climbs up the walls, so curving the surface. The pressure just under the meniscus is less than that arising from the atmosphere by $2\gamma/r$. The pressure is equal at equal heights throughout the liquid provided the hydrostatic pressure (which is equal to $\rho g h$) cancels the pressure difference arising from the curvature.

Clouds do form, so there must be a mechanism. Two processes are responsible. The first is that a sufficiently large number of molecules might congregate into a droplet so big that the enhanced evaporative effect is unimportant. The chance of one of these **spontaneous nucleation centres** forming is low, and in rain formation it is not a dominant mechanism. The more important process depends on the presence of minute dust particles or other kinds of foreign matter. These particles **nucleate** the condensation (that is, provide centres at which it can occur) by providing surfaces to which the water molecules can attach.

Liquids may be **superheated** above their boiling temperatures and **supercooled** below their freezing temperatures. In each case the thermodynamically stable phase is not achieved on account of the kinetic stabilization that occurs in the absence of nucleation centres. For example, superheating occurs because the vapour pressure inside a cavity is artificially low, so any cavity that does form tends to collapse. This instability is encountered when an unstirred beaker of water is heated, for its temperature may be raised above its boiling point. Violent bumping often ensues as spontaneous nucleation leads to bubbles big enough to survive. To ensure smooth boiling at the true boiling temperature, nucleation centres, such as small pieces of sharp-edged glass or bubbles (cavities) of air, should be introduced. A *bubble chamber*, a device for tracking elementary particles, depends on the nucleation of the evaporation of superheated liquid hydrogen by ionizing radiation.

6.10 Capillary action

The tendency of liquids to rise up capillary tubes (tubes of narrow bore), which is called **capillary action**, is a consequence of surface tension. Consider what happens when a glass capillary tube is first immersed in water or any liquid that has a tendency to adhere to the walls. The energy is lowest when a thin film covers as much of the glass as possible. As this film creeps up the inside wall it has the effect of curving the surface of the liquid inside the tube. This curvature implies that the pressure just beneath the curving meniscus is less than the atmospheric pressure by approximately $2\gamma/r$, where r is the radius of the tube and we assume a hemispherical surface. The pressure immediately under the flat surface outside the tube is p, the atmospheric pressure, but inside the tube under the curved surface it is only $p - 2\gamma/r$. The excess external pressure presses the liquid up the tube until hydrostatic equilibrium (equal pressures at equal depths) has been reached (Fig. 6.21).

(a) Capillary rise

The pressure exerted by a column of liquid of mass density ρ and height h is

$$p = \rho g h \tag{18}$$

This hydrostatic pressure matches the pressure difference $2\gamma/r$ at equilibrium. Therefore, the height of the column at equilibrium is obtained by equating $2\gamma/r$ and $\rho g h$, which gives

$$h = \frac{2\gamma}{\rho g r} \tag{19}$$

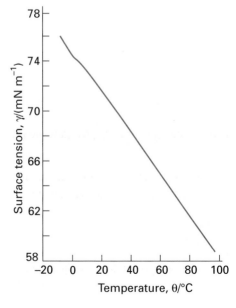

6.22 The variation of the surface tension of water with temperature.

This simple expression provides a reasonably accurate way of measuring the surface tension of liquids. Surface tension decreases with increasing temperature (Fig. 6.22).

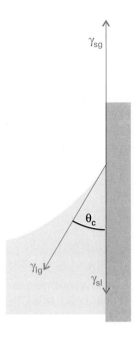

6.23 The balance of forces that results in a contact angle, θ_c.

Illustration

If water at 25 °C rises through 7.36 cm in a capillary of radius 0.20 mm, its surface tension at that temperature is

$$\gamma = \tfrac{1}{2}\rho ghr$$
$$= \tfrac{1}{2} \times (997.1 \text{ kg m}^{-3}) \times (9.81 \text{ m s}^{-2}) \times (7.36 \times 10^{-2} \text{ m})$$
$$\times (2.0 \times 10^{-4} \text{ m}) = 72 \text{ mN m}^{-1}$$

When the adhesive forces between the liquid and the material of the capillary wall are weaker than the cohesive forces within the liquid (as for mercury in glass), the liquid in the tube retracts from the walls. This retraction curves the surface with the concave, high-pressure side downwards. To equalize the pressure at the same depth throughout the liquid the surface must fall to compensate for the heightened pressure arising from its curvature. This compensation results in a capillary depression.

(b) The contact angle

In many cases there is a nonzero angle between the edge of the meniscus and the wall. If this contact angle is θ_c, then eqn 19 should be modified by multiplying the right-hand side by $\cos \theta_c$.

The origin of the contact angle can be traced to the balance of forces at the line of contact between the liquid and the solid (Fig. 6.23). If the solid–gas, solid–liquid, and liquid–gas surface tensions (essentially the energy needed to create a unit area of each of the interfaces) are denoted γ_{sg}, γ_{sl}, and γ_{lg}, respectively, the horizontal forces are in balance if

$$\gamma_{sg} = \gamma_{sl} + \gamma_{lg} \cos \theta_c \tag{20}$$

This expression solves to

$$\cos \theta_c = \frac{\gamma_{sg} - \gamma_{sl}}{\gamma_{lg}} \tag{21}$$

If we note that the work of adhesion of the liquid to the solid (per unit area of contact) is

$$w_{ad} = \gamma_{sg} + \gamma_{lg} - \gamma_{sl} \tag{22}$$

eqn 21 can be written

$$\cos \theta_c = \frac{w_{ad}}{\gamma_{lg}} - 1 \tag{23}$$

We now see that the liquid completely 'wets' (spreads over) the surface fully, corresponding to $\theta_c > 0$, when $w_{ad} > 2\gamma_{lg}$ (Fig. 6.24). The liquid does not wet the surface (corresponding to ($\theta_c > 90°$) when $w_{ad} < \gamma_{lg}$. For mercury in contact with glass, $\theta_c = 140°$, which corresponds to $w_{ad}/\gamma_{lg} = 0.23$, indicating a relatively low work of adhesion of the mercury to glass on account of the strong cohesive forces within mercury.

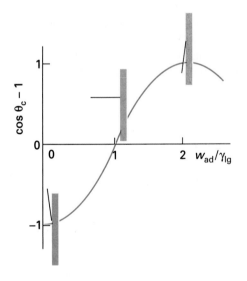

6.24 The variation of contact angle (shown by the semaphore-like object) as the ratio w_{ad}/γ_{lg} changes.

Checklist of key ideas

Phase diagrams

6.1 The stabilities of phases
- [ ] phase
- [ ] phase transition
- [ ] transition temperature
- [ ] metastable phase

6.2 Phase boundaries
- [ ] phase diagram
- [ ] phase boundary
- [ ] vapour pressure
- [ ] sublimation vapour pressure
- [ ] boiling
- [ ] boiling temperature
- [ ] normal boiling point
- [ ] standard boiling point
- [ ] critical temperature
- [ ] critical pressure
- [ ] supercritical fluid
- [ ] melting temperature
- [ ] normal freezing point
- [ ] standard freezing point
- [ ] normal melting point
- [ ] triple point and its significance

6.3 Three typical phase diagrams
- [ ] the interpretation of the phase diagram for water

- [ ] ice polymorphs
- [ ] the interpretation of the phase diagram for carbon dioxide
- [ ] supercritical fluid chromatography
- [ ] the interpretation of the phase diagram for helium
- [ ] λ-line
- [ ] superfluid helium

Phase stability and phase transitions

6.4 The thermodynamic criterion of equilibrium
- [ ] the uniformity of chemical potential throughout a system at equilibrium

6.5 The dependence of stability on the conditions
- [ ] the variation of chemical potential with temperature (1) and its implications for phase stability
- [ ] the variation of chemical potential with pressure (2) and its implications for phase stability

- [ ] partial vapour pressure
- [ ] gas solvation
- [ ] vapour pressure in the presence of applied pressure (3)

6.6 The location of phase boundaries
- [ ] the Clapeyron equation
- [ ] the slope of the solid–liquid boundary (7) and the corresponding formula for the boundary (8,9)
- [ ] the slope of the liquid–vapour boundary (10)
- [ ] the Clausius–Clapeyron equation (11) and the corresponding formula for the boundary (12)
- [ ] the slope of the solid–vapour boundary

6.7 The Ehrenfest classification of phase transitions
- [ ] the Ehrenfest classification of phase transitions
- [ ] first-order phase transition
- [ ] second-order phase transition
- [ ] λ-transition

The physical liquid surface

6.8 Surface tension
- [ ] work of forming a surface (14)
- [ ] surface tension

6.9 Curved surfaces
- [ ] bubble
- [ ] cavity
- [ ] droplet
- [ ] Laplace equation for the vapour pressure at a curved surface (16)
- [ ] Kelvin equation for the vapour pressure of droplets (17)
- [ ] supersaturated phase
- [ ] spontaneous nucleation centres
- [ ] nucleate
- [ ] superheated
- [ ] supercooled

6.10 Capillary action
- [ ] capillary rise and surface tension (19)
- [ ] contact angle and interfacial tension (21)
- [ ] the criteria for surface wetting

Further reading

Articles of general interest

K.M. Scholsky, Supercritical phase transitions at very high pressure. *J. Chem. Educ.* **66**, 989 (1989).

B.L. Earl, The direct relation between altitude and boiling point. *J. Chem. Educ.* **67**, 45 (1990).

C.L. Phelps, N.G. Smart, and C.M. Wai, Past, present, and future applications of supercritical fluid extraction technology. *J. Chem. Educ.* **73**, 1163 (1996).

J.M. Sanchez, Order–disorder transitions. In *Encyclopedia of applied physics* (ed. G.L. Trigg), **13**, 1. VCH, New York (1995).

E.K.H. Salje, Phase transitions, structural. In *Encyclopedia of applied physics* (ed. G.L. Trigg), **13**, 373. VCH, New York (1995).

W.B. Daniels, High-pressure techniques. In *Encyclopedia of applied physics* (ed. G.L. Trigg), **7**, 495. VCH, New York (1995).

Texts and sources of data and information

S.I. Sandler, *Chemical and engineering thermodynamics*. Wiley, New York (1989).

H.E. Stanley, *Introduction to phase transitions and critical phenomena*. Clarendon Press, Oxford (1971).

J.S. Rowlinson and B. Widom, *Molecular theory of capillarity*. Clarendon Press, Oxford (1982).

E.L. Skau and J.C. Arthur, Determination of melting and freezing temperatures. In *Techniques of Chemistry* (ed. A. Weissberger and B.W. Rossiter), **5**, 105. Wiley, New York (1971).

J.R. Anderson, Determination of boiling and condensation temperatures. In *Techniques of Chemistry* (ed. A. Weissberger and B.W. Rossiter), **5**, 199. Wiley, New York (1971).

A.N. Nesmeyanov. *Vapour pressure of the elements*. Pergamon Press, New York (1963).

T. Boublík, V. Fried, and E. Hála, *The vapor pressures of pure substances*. Elsevier, Amsterdam (1984).

J.J. Jasper, The surface tension of pure liquid compounds. *J. Phys. Chem. Ref. Data* **1**, 841 (1972).

Exercises

6.1 (a) The vapour pressure of dichloromethane at 24.1 °C is 400 Torr and its enthalpy of vaporization is 28.7 kJ mol^{-1}. Estimate the temperature at which its vapour pressure is 500 Torr.

6.1 (b) The vapour pressure of a substance at 20.0 °C is 58.0 kPa and its enthalpy of vaporization is 32.7 kJ mol^{-1}. Estimate the temperature at which its vapour pressure is 66.0 kPa.

6.2 (a) The molar volume of a certain solid is 161.0 cm^3 mol^{-1} at 1.00 atm and 350.75 K, its melting temperature. The molar volume of the liquid at this temperature and pressure is 163.3 cm^3 mol^{-1}. At 100 atm the melting temperature changes to 351.26 K. Calculate the enthalpy and entropy of fusion of the solid.

6.2 (b) The molar volume of a certain solid is 142.0 cm^3 mol^{-1} at 1.00 atm and 427.15 K, its melting temperature. The molar volume of the liquid at this temperature and pressure is 152.6 cm^3 mol^{-1}. At 1.2 MPa the melting temperature changes to 429.26 K. Calculate the enthalpy and entropy of fusion of the solid.

6.3 (a) The vapour pressure of a liquid in the temperature range 200 K to 260 K was found to fit the expression $\ln(p/\text{Torr}) = 16.255 - 2501.8/(T/\text{K})$. Calculate the enthalpy of vaporization of the liquid.

6.3 (b) The vapour pressure of a liquid in the temperature range 200 K to 260 K was found to fit the expression $\ln(p/\text{Torr}) = 18.361 - 3036.8/(T/\text{K})$. Calculate the enthalpy of vaporization of the liquid.

6.4 (a) The vapour pressure of benzene between 10 °C and 30 °C fits the expression $\log(p/\text{Torr}) = 7.960 - 1780/(T/\text{K})$. Calculate (a) the enthalpy of vaporization and (b) the normal boiling point of benzene.

6.4 (b) The vapour pressure of a liquid between 15 °C and 35 °C fits the expression $\log(p/\text{Torr}) = 8.750 - 1625/(T/\text{K})$. Calculate (a) the enthalpy of vaporization and (b) the normal boiling point of the liquid.

6.5 (a) When benzene freezes at 5.5 °C its density changes from 0.879 g cm^{-3} to 0.891 g cm^{-3}. Its enthalpy of fusion is 10.59 kJ mol^{-1}. Estimate the freezing point of benzene at 1000 atm.

6.5 (b) When ethanol freezes at −3.65 °C its density changes from 0.789 g cm^{-3} to 0.801 g cm^{-3}. Its enthalpy of fusion is 8.68 kJ mol^{-1}. Estimate the freezing point of the liquid at 100 MPa.

6.6 (a) In July in Los Angeles, the incident sunlight at ground level has a power density of 1.2 kW m^{-2} at noon. A swimming pool of area 50 m^2 is directly exposed to the sun. What is the maximum rate of loss of water assuming that all radiation is absorbed?

6.6 (b) Suppose the incident sunlight at ground level has a power density of 0.87 kW m^{-2} at noon. What is the maximum rate of loss of water from a lake of area 1.0 ha? (1 ha = 10^4 m^2.)

6.7 (a) An open vessel containing (a) water, (b) benzene, (c) mercury stands in a laboratory measuring 5.0 m × 5.0 m × 3.0 m at 25 °C. What mass of each substance will be found in the air if there is no ventilation? (The vapour pressures are (a) 24 Torr, (b) 98 Torr, (c) 1.7 mTorr.)

6.7 (b) On a cold, dry morning after a frost, the temperature was −5 °C and the partial pressure of water in the atmosphere fell to 2.0 Torr. Will the frost sublime? What partial pressure of water would ensure that the frost remained?

6.8 (a) Refer to Fig. 6.5 and describe the changes that would be observed when water vapour at 1.0 atm and 400 K is cooled at constant pressure to 260 K. Suggest the appearance of the cooling curve, a plot of temperature against time.

6.8 (b) Use the phase diagram in Fig. 6.6 to state what would be observed when a sample of carbon dioxide, initially at 1.0 atm and 298 K, is subjected to the following cycle: (a) isobaric (constant-pressure) heating to 320 K, (b) isothermal compression to 100 atm, (c) isobaric cooling to 210 K, (d) isothermal decompression to 1.0 atm, (e) isobaric heating to 298 K.

6.9 (a) Naphthalene, C$_{10}$H$_8$, melts at 80.2 °C. If the vapour pressure of the liquid is 10 Torr at 85.8 °C and 40 Torr at 119.3 °C, use the Clausius–Clapeyron equation to calculate (a) the enthalpy of vaporization, (b) the normal boiling point, and (c) the enthalpy of vaporization at the boiling point.

6.9 (b) The boiling point of hexane is 69.0 °C. Estimate (a) its enthalpy of vaporization and (b) its vapour pressure at 25 °C and 60 °C.

6.10 (a) Calculate the melting point of ice under a pressure of 50 bar. Assume that the density of ice under these conditions is approximately 0.92 g cm^{-3} and that of liquid water is 1.00 g cm^{-3}.

6.10 (b) Calculate the melting point of ice under a pressure of 10 MPa. Assume that the density of ice under these conditions is approximately 0.915 g cm^{-3} and that of liquid water is 0.998 g cm^{-3}.

6.11 (a) What fraction of the enthalpy of vaporization of water is spent on expanding the water vapour?

6.11 (b) What fraction of the enthalpy of vaporization of ethanol is spent on expanding its vapour?

6.12 (a) Calculate the vapour pressure of a spherical droplet of water of radius 10 nm at 20 °C. The vapour pressure of bulk water at that temperature is 2.3 kPa and its density is 0.9982 g cm^{-3}.

6.12 (b) Calculate the vapour pressure of a spherical droplet of water of radius 20.0 nm at 35.0 °C. The vapour pressure of bulk water at that temperature is 5.623 kPa and its density is 994.0 kg m^{-3}.

6.13 (a) The contact angle for water on clean glass is close to zero. Calculate the surface tension of water at 20°C given that at that temperature water climbs to a height of 4.96 cm in a clean glass capillary tube of internal radius 0.300 mm. The density of water at 20°C is 998.2 kg m^{-3}.

6.13 (b) The contact angle for water on clean glass is close to zero. Calculate the surface tension of water at 30°C given that at that temperature water climbs to a height of 9.11 cm in a clean glass capillary tube of internal diameter 0.320 mm. The density of water at 30°C is 0.9956 g cm^{-3}.

6.14 (a) Calculate the pressure differential of water across the surface of a spherical droplet of radius 200 nm at 20°C.

6.14 (b) Calculate the pressure differential of ethanol across the surface of a spherical droplet of radius 220 nm at 20°C. The surface tension of ethanol at that temperature is 22.39 mN m^{-1}.

Problems

Numerical problems

6.1 The temperature dependence of the vapour pressure of solid sulfur dioxide can be approximately represented by the relation $\log(p/\text{Torr}) = 10.5916 - 1871.2/(T/\text{K})$ and that of liquid sulfur dioxide by $\log(p/\text{Torr}) = 8.3186 - 1425.7/(T/\text{K})$. Estimate the temperature and pressure of the triple point of sulfur dioxide.

6.2 Prior to the discovery that freon-12 (CF$_2$Cl$_2$) was harmful to the Earth's ozone layer, it was frequently used as the dispersing agent in spray cans for hair spray, etc. Its enthalpy of vaporization at its normal boiling point of −29.2°C is 20.25 kJ mol^{-1}. Estimate the pressure that a can of hair spray using freon-12 had to withstand at 40°C, the temperature of a can that has been standing in sunlight. Assume that $\Delta_{\text{vap}}H$ is a constant over the temperature range involved and equal to its value at −29.2°C.

6.3 The enthalpy of vaporization of a certain liquid is found to be 14.4 kJ mol^{-1} at 180 K, its normal boiling point. The molar volumes of the liquid and the vapour at the boiling point are 115 cm^3 mol^{-1} and 14.5 dm^3 mol^{-1}, respectively. (a) Estimate dp/dT from the Clapeyron equation and (b) the percentage error in its value if the Clausius–Clapeyron equation is used instead.

6.4 Calculate the difference in slope of the chemical potential against temperature on either side of (a) the normal freezing point of water and (b) the normal boiling point of water. (c) By how much does the chemical potential of water supercooled to −5.0°C exceed that of ice at that temperature?

6.5 Calculate the difference in slope of the chemical potential against pressure on either side of (a) the normal freezing point of water and (b) the normal boiling point of water. The densities of ice and water at 0°C are 0.917 g cm^{-3} and 1.000 g cm^{-3}, and those of water and water vapour at 100°C are 0.958 g cm^{-3} and 0.598 g L^{-1}, respectively. By how much does the chemical potential of water vapour exceed that of liquid water at 1.2 atm and 100°C?

6.6 The enthalpy of fusion of mercury is 2.292 kJ mol^{-1}, and its normal freezing point is 234.3 K with a change in molar volume of +0.517 cm^3 mol^{-1} on melting. At what temperature will the bottom of a column of mercury (density 13.6 g cm^{-3}) of height 10.0 m be expected to freeze?

6.7 50.0 L of dry air was slowly bubbled through a thermally insulated beaker containing 250 g of water initially at 25°C. Calculate the final temperature. (The vapour pressure of water is

approximately constant at 23.8 Torr throughout, and its heat capacity is 75.5 J K^{-1} mol^{-1}. Assume that the air is not heated or cooled and that water vapour is a perfect gas.)

6.8 The vapour pressure, p, of nitric acid varies with temperature as follows:

$\theta/°C$	0	20	40	50	70	80	90	100
p/Torr	14.4	47.9	133	208	467	670	937	1282

What are (a) the normal boiling point and (b) the enthalpy of vaporization of nitric acid?

6.9 The vapour pressure of the ketone carvone ($M = 150.2$ g mol^{-1}), a component of oil of spearmint, is as follows:

$\theta/°C$	57.4	100.4	133.0	157.3	203.5	227.5
p/Torr	1.00	10.0	40.0	100	400	760

What are (a) the normal boiling point and (b) the enthalpy of vaporization of carvone?

6.10 Construct the phase diagram for benzene near its triple point at 36 Torr and 5.50°C using the following data: $\Delta_{\text{fus}}H = 10.6$ kJ mol^{-1}, $\Delta_{\text{vap}}H = 30.8$ kJ mol^{-1}, $\rho(\text{s}) = 0.891$ g cm^{-3}, $\rho(\text{l}) = 0.879$ g cm^{-3}.

Theoretical problems

6.11 Show that, for a transition between two incompressible solid phases, ΔG is independent of the pressure.

6.12 The change in enthalpy is given by $dH = C_p\,dT + V\,dp$. The Clapeyron equation relates dp and dT at equilibrium, and so in combination the two equations can be used to find how the enthalpy changes along a phase boundary as the temperature changes and the two phases remain in equilibrium. Show that $d(\Delta H/T) = \Delta C_p\,d\ln T$.

6.13 In the 'gas saturation method' for the measurement of vapour pressure, a volume V of gas (as measured at a temperature T and a pressure P) is bubbled slowly through the liquid that is maintained at the temperature T and a mass loss m is measured. Show that the vapour pressure, p, of the liquid is related to its molar mass, M, by $p = AmP/(1 + Am)$, where $A = RT/MPV$. The vapour pressure of geraniol ($M = 154.2$ g mol^{-1}), which is a component of oil of roses, was measured at 110°C. It was found that, when 5.00 L of nitrogen at 760 Torr was passed slowly through the heated liquid, the loss of mass was 0.32 g. Calculate the vapour pressure of geraniol.

6.14 Combine the barometric formula (stated in Problem 1.35) for the dependence of the pressure on altitude with the Clausius–Clapeyron equation, and predict how the boiling temperature of a liquid depends on the altitude and the ambient temperature. Take the mean ambient temperature as 20°C and predict the boiling temperature of water at 3000 m.

6.15 Figure 6.1 gives schematic representations of how the chemical potentials of the solid, liquid, and gaseous phases of a substance vary with temperature. All have a negative slope, but it is unlikely that they are truly straight lines as indicated in the illustrations. Derive an expression for the curvatures (specifically, the second derivatives with respect to temperature) of these lines. Is there a restriction on the curvature of these lines? Which state of matter shows the greatest curvature?

6.16 The Clapeyron equation does not apply to second-order phase transitions, but there are two analogous equations, the *Ehrenfest equations*, which do. They are:

$$\frac{dp}{dT} = \frac{\alpha_2 - \alpha_1}{\kappa_{T,2} - \kappa_{T,1}} \qquad \frac{dp}{dT} = \frac{C_{p,m2} - C_{p,m1}}{TV(\alpha_2 - \alpha_1)}$$

where α is the expansion coefficient, κ_T the isothermal compressibility, and the subscripts 1 and 2 refer to two different phases. Derive these two equations. Why does the Clapeyron equation not apply to second-order transitions?

6.17 For a first-order phase transition, to which the Clapeyron equation does apply, prove the relation

$$C_S = C_p - \frac{\alpha V \Delta_{trs} H}{\Delta_{trs} V}$$

where $C_S = (\partial q/\partial T)_S$ is the heat capacity along the coexistence curve of two phases.

Additional problems supplied by Carmen Giunta and Charles Trapp

6.18 A substance as well known as methane still receives research attention. Friend *et al.* have published a review of the thermophysical properties of methane (D.G. Friend, J.F. Ely, and H. Ingham, *J. Phys. Chem. Ref. Data* **18**, 583 (1989)), which included the following data describing the liquid–vapour phase boundary.

T/K	100	108	110	112	114	120
p/MPa	0.034	0.074	0.088	0.104	0.122	0.192
T/K	130	140	150	160	170	190
p/MPa	0.368	0.642	1.041	1.593	2.329	4.521

(a) Plot the liquid–vapour phase boundary. (b) Estimate the standard boiling point of methane. (c) Compute the standard enthalpy of vaporization of methane, given that the molar volumes of the liquid and vapour at the standard boiling point are 3.80×10^{-2} and 8.89 L mol^{-1}, respectively.

6.19 In an investigation of the thermophysical properties of toluene (R.D. Goodwin, *J. Phys. Chem. Ref. Data* **18**, 1565 (1989)), Goodwin presented expressions for two coexistence curves. The solid–liquid coexistence curve is given by:

$$p/\text{bar} = p_3/\text{bar} + 1000 \times (5.60 + 11.727x)x$$

where $x = T/T_3 - 1$ and the triple point pressure and temperature are $p_3 = 0.4362$ μbar and $T_3 = 178.15$ K, respectively. The liquid–vapour curve is given by:

$$\ln(p/\text{bar}) = -10.418/y + 21.157 - 15.996y$$
$$+ 14.015y^2 - 5.0120y^3 + 4.7224(1 - y)^{1.70}$$

where $y = T/T_c = T/(593.95 \text{ K})$. (a) Plot the solid–liquid and liquid–vapour phase boundaries. (b) Estimate the standard melting point of toluene. (c) Estimate the standard boiling point of toluene. (d) Compute the standard enthalpy of vaporization of toluene, given that the molar volumes of the liquid and vapour at the normal boiling point are 0.12 L mol^{-1} and 30.3 L mol^{-1}, respectively.

6.20 In a study of the vapour pressure of chloromethane (A. Bah and N. Dupont-Pavlovsky, *J. Chem. Eng. Data* **40**, 869 (1995)), Bah and Dupont-Pavlovsky presented data for the vapour pressure over solid chloromethane at low temperatures. Some of that data are shown below.

T/K	145.94	147.96	149.93	151.94	153.97	154.94
p/Pa	13.07	18.49	25.99	36.76	50.86	59.56

Estimate the standard enthalpy of sublimation of chloromethane at 150 K. (Take the molar volume of the vapour to be that of a perfect gas, and that of the solid to be negligible.)

6.21 What pressure is required to convert graphite into diamond at 25°C? The following data apply to 25°C and 100 kPa. Assume the specific volume, V_s, and κ_T are constant with respect to pressure changes.

	Graphite	Diamond
$\Delta_f G^{\ominus}/(\text{kJ mol}^{-1})$	0	+2.8678
$V_s/(\text{cm}^3 \text{ g}^{-1})$	0.444	0.284
κ_T/kPa	3.04×10^{-8}	0.187×10^{-8}

7

Simple mixtures

The chapter begins by developing the concept of chemical potential to show that it is a particular case of a class of properties called partial molar quantities. Then it explores how the chemical potential of a substance is used to describe the physical properties of a mixture. The underlying principle to keep in mind is that at equilibrium the chemical potential of a species is the same in every phase. We shall see, by making use of the experimental observations known as Raoult's and Henry's laws, that the chemical potential of a substance can be expressed in terms of its mole fraction in a mixture. With this result established, we can calculate the effect of a solute on certain thermodynamic properties of a solution. These properties include the lowering of vapour pressure of the solvent, the elevation of its boiling point, the depression of its freezing point, and the origin of osmotic pressure. Finally, we see how the chemical potential of a substance in a real mixture can be expressed in terms of a property known as the activity. We see how the activity may be measured, and conclude with a brief discussion of how the standard states of solutes and solvents are defined.

Chemistry deals with mixtures, including mixtures of substances that can react together. Therefore, we need to generalize the foregoing material to deal with substances that are mingled together. As a first step towards dealing with chemical reactions (which are treated in Chapter 9), here we consider mixtures of substances that do not react together. At this stage we shall deal mainly with binary mixtures (mixtures of two components, A and B). We shall therefore often be able to simplify equations by making use of the relation $x_A + x_B = 1$. Another restriction of this chapter is that we shall consider mainly non-electrolyte solutions, in which the solute is not present as ions.

The thermodynamic description of mixtures

We have already seen that the partial pressure, which is the contribution of one component to the total pressure, is used to discuss the properties of mixtures of gases. For a more general description of the thermodynamics of mixtures we need to introduce other analogous 'partial' properties.

7.1 Partial molar quantities

The easiest partial molar property to visualize is the partial molar volume, the contribution that a component of a mixture makes to the total volume of a sample.

(a) Partial molar volume

Imagine a huge volume of pure water at 25 °C. When a further 1 mol H_2O is added, the volume increases by 18 cm^3, and we can report that 18 $cm^3\,mol^{-1}$ is the molar volume of pure water. However, when we add 1 mol H_2O to a huge volume of pure ethanol, the volume increases by only 14 cm^3. The reason for the different increase in volume is that the volume occupied by a given number of water molecules depends on the identity of the molecules that surround them. In the latter case there is so much ethanol present that each H_2O molecule is surrounded by ethanol molecules, and the packing of the molecules results in the H_2O molecules increasing the volume by only 14 cm^3. The quantity 14 $cm^3\,mol^{-1}$ is the partial molar volume of water in pure ethanol. In general, the **partial molar volume** of a substance A in a mixture is the change in volume per mole of A added to a large volume of the mixture.

The partial molar volumes of the components of a mixture vary with composition because the environment of each type of molecule changes as the composition changes from pure A to pure B. It is this changing molecular environment, and the consequential modification of the forces acting between molecules, that results in the variation of the thermodynamic properties of a mixture as its composition is changed. The partial molar volumes of water and ethanol across the full composition range at 25 °C are shown in Fig. 7.1.

The partial molar volume, V_J, of a substance J at some general composition is defined formally as follows:

$$V_J = \left(\frac{\partial V}{\partial n_J}\right)_{p,T,n'} \tag{1}$$

where the subscript n' signifies that the amounts of all other substances present are constant.[1] The partial molar volume is the slope of the plot of the total volume as the amount of J is changed, the pressure, temperature, and amount of the other components being constant (Fig. 7.2). Its value depends on the composition, as we saw for water and ethanol. The definition in eqn 1 implies that, when the composition of the mixture is changed by the addition of dn_A of A and dn_B of B, the total volume of the mixture changes by

$$dV = \left(\frac{\partial V}{\partial n_A}\right)_{p,T,n_B} dn_A + \left(\frac{\partial V}{\partial n_B}\right)_{p,T,n_A} dn_B = V_A\,dn_A + V_B\,dn_B \tag{2}$$

Once the partial molar volumes of the two components of a mixture at the composition (and temperature) of interest are known, we can state the total volume, V, of the mixture by using

$$V = n_A V_A + n_B V_B \tag{3}$$

---

[1] The IUPAC recommendation is to denote a partial molar quantity by $\overline{X}$, but only when there is the possibility of confusion with the quantity X. For instance, the partial molar volume of NaCl in water could be written $\overline{V}(NaCl, aq)$ to distinguish it from the volume of the solution, $V(NaCl, aq)$.

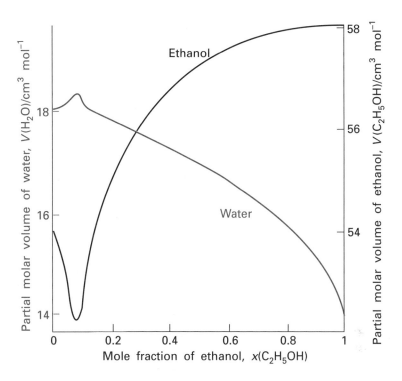

7.1 The partial molar volumes of water and ethanol at 25°C. Note the different scales (water on the left, ethanol on the right).

Justification 7.1

Consider a very large sample of the mixture of the specified composition. Then, when an amount n_A of A is added to the mixture, the composition remains virtually unchanged, the partial molar volume V_A is constant, and the volume of the sample changes by $n_A V_A$. When n_B of B is added, the volume changes by $n_B V_B$, for the same reason. The total change of volume is therefore $n_A V_A + n_B V_B$. The sample now occupies a larger volume, but the proportions of the components are still the same. At this stage, scoop out of the enlarged volume a sample containing n_A of A and n_B of B. Its volume is $n_A V_A + n_B V_B$. Because V is a state function, the same sample could have been prepared simply by mixing the appropriate amounts of A and B. This justifies eqn 3.

Partial molar volumes (and partial molar quantities in general) can be measured in several ways. One method is to measure the dependence of the volume on the composition and to fit the observed volume to a function of the amount of the substance by using a curve-fitting program (that is, by finding the parameters that give a best fit of a particular function to the experimental data). Once the function has been found, its slope can be determined at any composition of interest by differentiation. For instance, if it was found that the volume of a mixture was described by the function

$$V = A + Bn_A + C(n_A^2 - 1)$$

with particular values of the parameters A, B, and C, then the partial molar volume of A at any composition could be obtained from

$$V_A = \left(\frac{\partial V}{\partial n_A}\right)_{p,T,n_B} = B + 2Cn_A$$

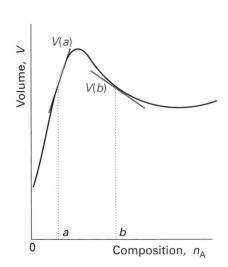

7.2 The partial molar volume of a substance is the slope of the variation of the total volume of the sample plotted against the composition. In general, partial molar quantities vary with the composition, as shown by the different slopes at the compositions a and b. Note that the partial molar volume at b is negative: the overall volume of the sample decreases as A is added.

The partial molar volume of the second component is obtained from eqn 3 arranged into

$$V_B = \frac{V - n_A V_A}{n_B} = \frac{A - (n_A^2 + 1)C}{n_B}$$

Illustration

The total volume of an ethanol solution at 25 °C containing 1.000 kg of water is found to be given by the expression

$$V/mL = 1002.93 + 54.6664b - 0.36394b^2 + 0.028256b^3$$

where b is the numerical value of the molality. Because the amount of CH_3CH_2OH in moles is equal to the numerical value of the molality in moles per kilogram, we can write the partial molar volume of ethanol, V_E, as

$$V_E/(mL\,mol^{-1}) = \left(\frac{\partial(V/mL)}{\partial b}\right)_{p,T,n_W} = 54.6664 - 2(0.36394)b + 3(0.028256)b^2$$

Self-test 7.1 At 25 °C, the density of a 50 per cent by mass ethanol/water solution is 0.914 g cm^{-3}. Given that the partial molar volume of water in the solution is 17.4 cm^3 mol^{-1}, what is the partial molar volume of the ethanol?

[56.3 cm^3 mol^{-1}]

Molar volumes are always positive, but partial molar quantities need not be. For example, the limiting partial molar volume of $MgSO_4$ in water (its partial molar volume in the limit of zero concentration) is -1.4 cm^3 mol^{-1}, which means that the addition of 1 mol $MgSO_4$ to a large volume of water results in a decrease in volume of 1.4 cm^3. The contraction occurs because the salt breaks up the open structure of water as the ions become hydrated, and it collapses slightly.

(b) Partial molar Gibbs energies

The concept of a partial molar quantity can be extended to any extensive state function. For a pure substance, the chemical potential is just another name for the molar Gibbs energy. For a substance in a mixture, the chemical potential is *defined* as being the partial molar Gibbs energy:

$$\mu_J = \left(\frac{\partial G}{\partial n_J}\right)_{p,T,n'} \qquad [4]$$

That is, the chemical potential is the slope of a plot of Gibbs energy against the amount of the component J, with the pressure and temperature (and the amounts of the other substances) held constant (Fig. 7.3). By the same argument that led to eqn 3, it follows that the total Gibbs energy of a binary mixture is

$$G = n_A \mu_A + n_B \mu_B \qquad (5)$$

where μ_A and μ_B are the chemical potentials at the composition of the mixture. That is, the chemical potential of a substance in a mixture is the contribution of that substance to the total Gibbs energy of the mixture.

In an open system of constant composition, the Gibbs energy depends on the composition, the pressure, and the temperature. Thus, G may change when p, T, and the

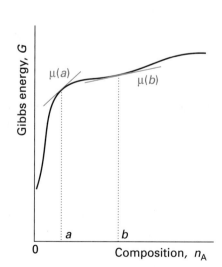

7.3 The chemical potential of a substance is the slope of the total Gibbs energy of a mixture with respect to the amount of substance of interest. In general, the chemical potential varies with composition, as shown for the two values at a and b. In this case, both chemical potentials are positive.

composition change, and, for a system of components A, B..., the equation $dG = V\,dp - S\,dT$ becomes

$$dG = V\,dp - S\,dT + \mu_A\,dn_A + \mu_B\,dn_B + \cdots \tag{6}$$

This expression is the **fundamental equation of chemical thermodynamics**. Its implications and consequences are explored and developed in this and the next three chapters.

At constant pressure and temperature, eqn 6 simplifies to

$$dG = \mu_A\,dn_A + \mu_B\,dn_B + \cdots \tag{7}$$

We saw in Section 4.6c that under the same conditions $dG = dw_{e,max}$. Therefore, at constant temperature and pressure,

$$dw_{e,max} = \mu_A\,dn_A + \mu_B\,dn_B + \cdots \tag{8}$$

That is, non-expansion work can arise from the changing composition of a system. For instance, in an electrochemical cell, the chemical reaction is arranged to take place in two distinct sites (at the two electrodes). The electrical work the cell performs can be traced to its changing composition as products are formed from reactants.

(c) The wider significance of the chemical potential

The chemical potential does more than show how G varies with composition. Because

$$G = U + pV - TS$$

a general infinitesimal change in U for a system of variable composition can be written

$$
\begin{aligned}
dU &= -p\,dV - V\,dp + S\,dT + T\,dS + dG \\
&= -p\,dV - V\,dp + S\,dT + T\,dS + (V\,dp - S\,dT + \mu_A\,dn_A + \mu_B\,dn_B + \cdots) \\
&= -p\,dV + T\,dS + \mu_A\,dn_A + \mu_B\,dn_B + \cdots
\end{aligned}
$$

This expression is the generalization of eqn 5.2 (that $dU = T\,dS - p\,dV$) to systems in which the composition may change. It follows that, at constant volume and entropy,

$$dU = \mu_A\,dn_A + \mu_B\,dn_B + \cdots$$

and hence that

$$\mu_J = \left(\frac{\partial U}{\partial n_J}\right)_{S,V,n'} \tag{9}$$

Therefore, not only does the chemical potential show how G changes when the composition changes, it also shows how the internal energy changes too (but under a different set of conditions). In the same way it is easy to deduce that

$$\text{(a)} \quad \mu_J = \left(\frac{\partial H}{\partial n_J}\right)_{S,p,n'} \qquad \text{(b)} \quad \mu_J = \left(\frac{\partial A}{\partial n_J}\right)_{T,V,n'} \tag{10}$$

Thus we see that the μ_J shows how all the extensive thermodynamic properties U, H, A, and G depend on the composition. This is why the chemical potential is so central to chemistry.

(d) The Gibbs–Duhem equation

Because the total Gibbs energy of a mixture is given by eqn 5 and the chemical potentials depend on the composition, when the compositions are changed infinitesimally we might expect G of a binary system to change by

$$dG = \mu_A\,dn_A + \mu_B\,dn_B + n_A\,d\mu_A + n_B\,d\mu_B$$

However, we have seen that at constant pressure and temperature a change in Gibbs energy is given by eqn 7. Because G is a state function, these two equations must be equal to each other, which implies that at constant temperature and pressure

$$n_A\,d\mu_A + n_B\,d\mu_B = 0 \tag{11}$$

This equation is a special case of the **Gibbs–Duhem equation**:

$$\sum_J n_J\,d\mu_J = 0 \tag{12}$$

The significance of the Gibbs–Duhem equation is that the chemical potential of one component of a mixture cannot change independently of the chemical potentials of the other components: in a binary mixture, if one partial molar quantity increases, the other must decrease:

$$d\mu_B = -\frac{n_A}{n_B}\,d\mu_A \tag{13}$$

The same line of reasoning applies to all partial molar quantities. We can see in Fig. 7.1, for example, that where the partial molar volume of water increases, that of ethanol decreases. Moreover, as eqn 13 shows, and as we can see from Fig 7.1, a small change in the partial molar volume of A corresponds to a large change in the partial molar volume of B if n_A/n_B is large, but the opposite is true when this ratio is small. In practice, the Gibbs–Duhem equation is used to determine the partial molar volume of one component of a binary mixture from measurements of the partial molar volume of the second component.

Example 7.1 Using the Gibbs–Duhem equation

The experimental value of the partial molar volume of $K_2SO_4(aq)$ at 298 K is given by the expression

$$V_{K_2SO_4}/(cm^3\,mol^{-1}) = 32.280 + 18.216b^{1/2}$$

where b is the numerical value of the molality of K_2SO_4. Use the Gibbs–Duhem equation to derive an equation for the molar volume of water in the solution. The molar volume of pure water at 298 K is 18.079 $cm^3\,mol^{-1}$.

Method Let A denote K_2SO_4 and B denote H_2O, the solvent. The Gibbs–Duhem equation for the partial molar volumes of two components is

$$n_A\,dV_A + n_B\,dV_B = 0$$

This relation implies that

$$dV_B = -\frac{n_A}{n_B}\,dV_A$$

Therefore, V_B can be found by integration:

$$V_B = V_B^* - \int \frac{n_A}{n_B}\,dV_A$$

The first step is to change the variable V_A to the molality b, and then to integrate the right-hand side between $b = 0$ (pure B) and the molality of interest.

Answer It follows from the information in the question, that, with A = K_2SO_4,

$$\frac{dV_A}{db} = 9.108b^{-1/2}$$

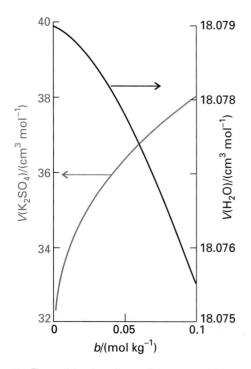

7.4 The partial molar volumes of the components of an aqueous solution of potassium sulfate.

Therefore, the integration required is

$$V_B = V_B^* - 9.108 \int_0^b \frac{n_A}{n_B} b^{-1/2} \, db$$

However, we need to note that the molality is related to the amounts of A and B by

$$b = \frac{n_A}{n_B M_B}$$

where M_B is the molar mass of water (in kilograms per mole). Therefore,

$$V_B = V_B^* - 9.108 M_B \int_0^b b^{1/2} \, db$$
$$= V_B^* - \tfrac{2}{3}(9.108 M_B b^{3/2})$$

It then follows, by substituting the data, that

$$V_B/(cm^3\,mol^{-1}) = 18.079 - 0.1094 b^{3/2}$$

The partial molar volumes are plotted in Fig. 7.4.

Self-test 7.2 Repeat the calculation for a salt A for which $V_A/(cm^3\,mol^{-1}) = 6.218 + 5.146 b - 7.147 b^2$.

$$[V_B/(cm^3\,mol^{-1}) = 18.079 - 0.0464 b^2 + 0.0859 b^3]$$

7.2 The thermodynamics of mixing

The dependence of the Gibbs energy of a mixture on its composition is given by eqn 5, and we know that at constant temperature and pressure systems tend towards a lower Gibbs energy. This is the link we need in order to apply thermodynamics to the discussion of spontaneous changes of composition, as in the mixing of two substances. One simple example of a spontaneous mixing process is that of two gases introduced into the same container. The mixing is spontaneous, so it must correspond to a decrease in G. We shall now see how to express this idea quantitatively.

(a) The Gibbs energy of mixing

Let the amounts of two perfect gases in the two containers be n_A and n_B; both are at a temperature T and a pressure p. At this stage, the chemical potentials of the two gases have their 'pure' values and the Gibbs energy of the total system is given by eqn 5 as

$$G_i = n_A \mu_A + n_B \mu_B$$
$$= n_A \left\{ \mu_A^\ominus + RT \ln \left(\frac{p}{p^\ominus} \right) \right\} + n_B \left\{ \mu_B^\ominus + RT \ln \left(\frac{p}{p^\ominus} \right) \right\}$$

It will be much simpler notationally if we agree to let p denote the pressure *relative* to $p^\ominus$; that is, to replace $p/p^\ominus$ by p, for then we can write

$$G_i = n_A \{ \mu_A^\ominus + RT \ln p \} + n_B \{ \mu_B^\ominus + RT \ln p \} \qquad \{14\}$$

Equations for which this convention is used will be labelled $\{1\}, \{2\}, \ldots$; to use the equations, we have to remember to replace p by $p/p^\ominus$ again. In practice, that simply means using the numerical value of p in bars. After mixing, the partial pressures of the gases are p_A and p_B, with $p_A + p_B = p$. The total Gibbs energy changes to

$$G_f = n_A \{ \mu_A^\ominus + RT \ln p_A \} + n_B \{ \mu_B^\ominus + RT \ln p_A \} \qquad \{15\}$$

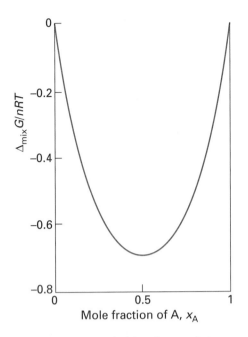

7.5 The Gibbs energy of mixing of two perfect gases and (as discussed later) of two liquids that form an ideal solution. The Gibbs energy of mixing is negative for all compositions and temperatures, so perfect gases mix spontaneously in all proportions.

The difference $G_f - G_i$, the Gibbs energy of mixing, $\Delta_{mix}G$, is therefore

$$\Delta_{mix}G = n_A RT \ln\left(\frac{p_A}{p}\right) + n_B RT \ln\left(\frac{p_B}{p}\right) \qquad (16)$$

At this point we may replace n_J by $x_J n$ and use Dalton's law (Section 1.2c) to write $p_J/p = x_J$ for each component, which gives

$$\Delta_{mix}G = nRT(x_A \ln x_A + x_B \ln x_B) \qquad (17)°$$

Because mole fractions are never greater than 1, the logarithms in this equation are negative, and $\Delta_{mix}G < 0$ (Fig. 7.5). The conclusion that $\Delta_{mix}G$ is negative confirms that perfect gases mix spontaneously in all proportions. However, the equation extends common sense by allowing us to discuss the process quantitatively. We see, for instance, that $\Delta_{mix}G$ is directly proportional to the temperature but is independent of the total pressure.

Example 7.2 Calculating a Gibbs energy of mixing

A container is divided into two equal compartments (Fig. 7.6). One contains 3.0 mol H_2 at 25°C; the other contains 1.0 mol N_2 at 25°C. Calculate the Gibbs energy of mixing when the partition is removed. Assume perfect behaviour.

Method We proceed by calculating the initial Gibbs energy from the chemical potentials. To do so, we need the pressure of each gas. Write the pressure of nitrogen as p; then the pressure of hydrogen as a multiple of p can be found from the gas laws. Next, calculate the Gibbs energy for the system when the partition is removed. The volume of each gas doubles, so its partial pressure falls by a factor of 2.

Answer Given that the pressure of nitrogen is p, the pressure of hydrogen is $3p$; therefore, the initial Gibbs energy is

$$G_i = (3.0\ mol)\{\mu^{\ominus}(H_2) + RT \ln 3p\} + (1.0\ mol)\{\mu^{\ominus}(N_2) + RT \ln p\}$$

The partial pressure of nitrogen falls to $\frac{1}{2}p$ and that of hydrogen falls to $\frac{3}{2}p$. Therefore, the Gibbs energy changes to

$$G_i = (3.0\ mol)\{\mu^{\ominus}(H_2) + RT \ln \tfrac{3}{2}p\} + (1.0\ mol)\{\mu^{\ominus}(N_2) + RT \ln \tfrac{1}{2}p\}$$

The Gibbs energy of mixing is the difference of these two quantities:

$$\Delta_{mix}G = (3.0\ mol)RT \ln\left(\frac{\tfrac{3}{2}p}{3p}\right) + (1.0\ mol)RT \ln\left(\frac{\tfrac{1}{2}p}{p}\right)$$
$$= -(3.0\ mol)RT \ln 2 - (1.0\ mol)RT \ln 2$$
$$= -(4.0\ mol)RT \ln 2 = -6.9\ kJ$$

Comment In this example, the value of $\Delta_{mix}G$ is the sum of two contributions: the mixing itself, and the changes in pressure of the two gases to their final pressure, $2p$. When 3.0 mol H_2 mixes with 1.0 mol N_2 at the same pressure, the change of Gibbs energy is -5.6 kJ independent of the initial common pressure.

- -

Self-test 7.3 Suppose that 2.0 mol H_2 at 2.0 atm and 25°C and 4.0 mol N_2 at 3.0 atm and 25°C are mixed at constant volume. Calculate $\Delta_{mix}G$. What would be the value of $\Delta_{mix}G$ had the pressures been identical initially?

$$[-9.7\ kJ,\ -9.5\ kJ]$$

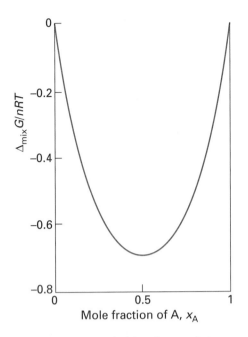

3.0 mol H_2 | 1.0 mol N_2

$3p$ | p

3.0 mol H_2 | 1.0 mol N_2

$p(H_2) = 3p/2$ | $p(N_2) = p/2$

$2p$

7.6 The initial and final states considered in the calculation of the Gibbs energy of mixing of gases at different initial pressures.

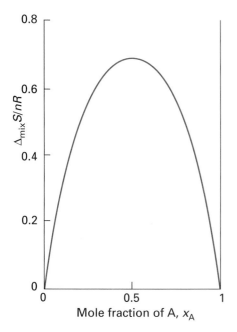

7.7 The entropy of mixing of two perfect gases and (as discussed later) of two liquids that form an ideal solution. The entropy increases for all compositions and temperatures, so perfect gases mix spontaneously in all proportions. Because there is no transfer of heat to the surroundings when perfect gases mix, the entropy of the surroundings is unchanged. Hence, the graph also shows the total entropy of the system plus the surroundings when perfect gases mix.

(b) Other thermodynamic mixing functions

Because $(\partial G/\partial T)_{p,n} = -S$, it follows immediately from eqn 17 that, for a mixture of perfect gases, the **entropy of mixing**, $\Delta_{\mathrm{mix}}S$, is

$$\Delta_{\mathrm{mix}}S = -\left(\frac{\partial \Delta_{\mathrm{mix}}G}{\partial T}\right)_{p,n_A,n_B} = -nR(x_A \ln x_A + x_B \ln x_B) \qquad (18)°$$

Because $\ln x < 0$, it follows that $\Delta_{\mathrm{mix}}S > 0$ for all compositions (Fig. 7.7). This increase in entropy is what we expect when one gas disperses into the other and the system becomes more disordered. The Gibbs energy of mixing in Example 7.2 was calculated as $-(4.0 \text{ mol})RT \ln 2$, so the corresponding entropy of mixing is $+(4.0 \text{ mol})R \ln 2$, or $+23 \text{ J K}^{-1}$.

The isothermal, isobaric (constant pressure) **enthalpy of mixing**, $\Delta_{\mathrm{mix}}H$, of two perfect gases may be found from $\Delta G = \Delta H - T\Delta S$ (because the process is isothermal). From eqns 17 and 18, we find

$$\Delta_{\mathrm{mix}}H = 0 \qquad (19)°$$

The enthalpy of mixing is zero, as we should expect for a system in which there are no interactions between the molecules forming the gaseous mixture. It follows that the whole of the driving force for mixing comes from the increase in entropy of the system, because the entropy of the surroundings is unchanged.

7.3 The chemical potentials of liquids

To discuss the equilibrium properties of liquid mixtures we need to know how the chemical potential of a liquid varies with its composition. To calculate its value, we use the fact that, at equilibrium, the chemical potential of a substance present as a vapour must be equal to its chemical potential in the liquid.

(a) Ideal solutions

We shall denote quantities relating to pure substances by a superscript *, so the chemical potential of pure A is written μ_A^*, and as $\mu_A^*(\mathrm{l})$ when we need to emphasize that A is a liquid. Because the vapour pressure of the pure liquid is p_A^*, it follows from eqn 5.20 that the chemical potential of A in the vapour is $\mu^{\ominus} + RT \ln p_A^*$ (with p_A^* to be interpreted as the relative pressure $p_A^*/p^{\ominus}$). These two chemical potentials are equal at equilibrium (Fig. 7.8), so we can write

$$\mu_A^* = \mu_A^{\ominus} + RT \ln p_A^* \qquad \{20\}$$

If another substance, a solute, is also present in the liquid, the chemical potential of A in the liquid is μ_A and its vapour pressure is p_A. In this case

$$\mu_A = \mu_A^{\ominus} + RT \ln p_A \qquad \{21\}$$

Next, we combine these two equations to eliminate the standard chemical potential of the gas, and obtain

$$\mu_A = \mu_A^* + RT \ln\left(\frac{p_A}{p_A^*}\right) \qquad (22)$$

The final step draws on additional experimental information about the relation between the ratio of vapour pressures and the composition of the liquid. In a series of experiments on mixtures of closely related liquids (such as benzene and methylbenzene), the French chemist François Raoult found that the ratio of the partial vapour pressure of each component to its vapour pressure as a pure liquid, p_A/p_A^*, is approximately equal to the mole fraction of A in the liquid mixture. That is, he established what we now call **Raoult's law**:

$$p_A = x_A p_A^* \qquad (23)°$$

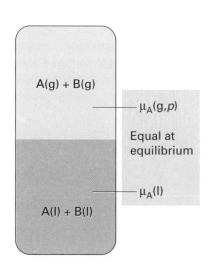

7.8 At equilibrium, the chemical potential of the gaseous form of a substance A is equal to the chemical potential of its condensed phase. The equality is preserved if a solute is also present. Because the chemical potential of A in the vapour depends on its partial vapour pressure, it follows that the chemical potential of liquid A can be related to its partial vapour pressure.

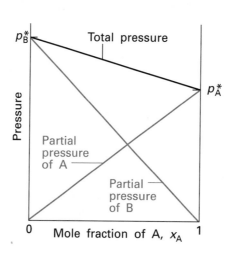

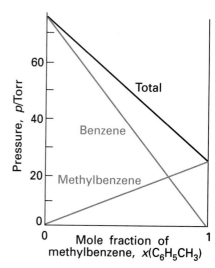

7.9 The total vapour pressure and the two partial vapour pressures of an ideal binary mixture are proportional to the mole fractions of the components.

7.10 Two similar liquids, in this case benzene and methylbenzene (toluene), behave almost ideally, and the variation of their vapour pressures with composition resembles that for an ideal solution.

This law is illustrated in Fig. 7.9. Some mixtures obey Raoult's law very well, especially when the components are structurally similar (Fig. 7.10). Mixtures that obey the law throughout the composition range from pure A to pure B are called **ideal solutions**. When we write equations that are valid only for ideal solutions, we shall label them with a superscript °, as in eqn 23.

For an ideal solution, it follows from eqns 22 and 23 that

$$\mu_A = \mu_A^* + RT \ln x_A \qquad (24)°$$

This important equation can be used as the *definition* of an ideal solution (so that it implies Raoult's law rather than stemming from it). It is in fact a better definition than eqn 23 because it does not assume that the gas is perfect.

Molecular interpretation 7.1 The origin of Raoult's law can be understood in molecular terms by considering the rates at which molecules leave and return to the liquid. The law reflects the fact that the presence of a second component reduces the rate at which A molecules leave the surface of the liquid but does not inhibit the rate at which they return (Fig. 7.11).

The rate at which A molecules leave the surface is proportional to the number of them at the surface, which in turn is proportional to the mole fraction of A:

rate of vaporization $= k x_A$

where k is a constant of proportionality. The rate at which molecules condense is proportional to their concentration in the gas phase, which in turn is proportional to their partial pressure:

rate of condensation $= k' p_A$

At equilibrium, the rates of vaporization and condensation are equal, so

$$k' p_A = k x_A$$

It follows that

$$p_A = \frac{k}{k'} x_A$$

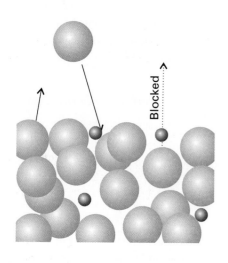

7.11 A pictorial representation of the molecular basis of Raoult's law. The large spheres represent solvent molecules at the surface of a solution (the uppermost line of spheres), and the small spheres are solute molecules. The latter hinder the escape of solvent molecules into the vapour, but do not hinder their return.

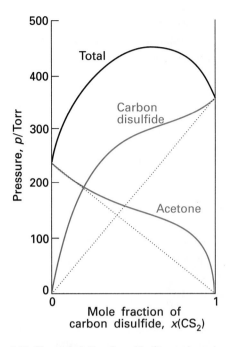

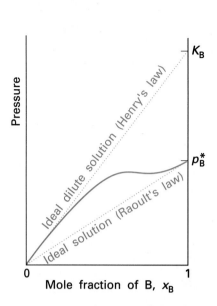

7.12 Strong deviations from ideality are shown by dissimilar liquids (in this case carbon disulfide and acetone (propanone)).

7.13 When a component (the solvent) is nearly pure, it has a vapour pressure that is proportional to the mole fraction with a slope p_B^* (Raoult's law). When it is the minor component (the solute), its vapour pressure is still proportional to the mole fraction, but the constant of proportionality is now K_B (Henry's law).

For the pure liquid, $x_A = 1$; so

$$p_A^* = \frac{k}{k'}$$

Equation 23 then follows by substitution of this relation into the previous one.

Some solutions depart significantly from Raoult's law (Fig. 7.12). Nevertheless, even in these cases the law is obeyed increasingly closely for the component in excess (the solvent) as it approaches purity. The law is therefore a good approximation for the properties of the solvent if the solution is dilute.

(b) Ideal-dilute solutions

In ideal solutions the solute, as well as the solvent, obeys Raoult's law. However, the English chemist William Henry found experimentally that, for real solutions at low concentrations, although the vapour pressure of the solute is proportional to its mole fraction, the constant of proportionality is not the vapour pressure of the pure substance (Fig. 7.13). **Henry's law** is:

$$p_B = x_B K_B \tag{25}°$$

In this expression x_B is the mole fraction of the solute and K_B is an empirical constant (with the dimensions of pressure) chosen so that the plot of the vapour pressure of B against its mole fraction is tangent to the experimental curve at $x_B = 0$.

Mixtures for which the solute obeys Henry's law and the solvent obeys Raoult's law are called **ideal-dilute solutions**. We shall also label equations with a superscript ° when they have been derived from Henry's law.

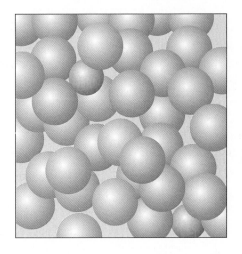

7.14 In a dilute solution, the solvent molecules (the green spheres) are in an environment that differs only slightly from that of the pure solvent. The solute particles, however, are in an environment totally unlike that of the pure solute.

Table 7.1* Henry's law constants for gases in water at 298 K

	K/Torr
CO_2	1.25×10^6
H_2	5.34×10^7
N_2	6.51×10^7
O_2	3.30×10^7

*More values are given in the *Data section* at the end of this volume.

Molecular interpretation 7.2 The difference in behaviour of the solute and solvent at low concentrations (as expressed by Henry's and Raoult's laws, respectively) arises from the fact that in a dilute solution the solvent molecules are in an environment very much like the one they have in the pure liquid (Fig. 7.14). In contrast, the solute molecules are surrounded by solvent molecules, which is entirely different from their environment when pure. Thus, the solvent behaves like a slightly modified pure liquid, but the solute behaves entirely differently from its pure state unless the solvent and solute molecules happen to be very similar. In the latter case, the solute also obeys Raoult's law.

Example 7.3 Investigating the validity of Raoult's and Henry's laws

The vapour pressures of each component in a mixture of propanone (acetone, A) and trichloromethane (chloroform, C) were measured at 35°C with the following results:

x_C	0	0.20	0.40	0.60	0.80	1
p_C/Torr	0	35	82	142	219	293
p_A/Torr	347	270	185	102	37	0

Confirm that the mixture conforms to Raoult's law for the component in large excess and to Henry's law for the minor component. Find the Henry's law constants.

Method Both Raoult's and Henry's laws are statements about the form of the graph of partial vapour pressure against mole fraction. Therefore, plot the partial vapour pressures against mole fraction. Raoult's law is tested by comparing the data with the straight line $p_J = x_J p_J^*$ for each component in the region in which it is in excess (and acting as the solvent). Henry's law is tested by finding a straight line $p_J = x_J K_J$ that is tangent to each partial vapour pressure at low x, where the component can be treated as the solute.

Answer The data are plotted in Fig. 7.15 together with the Raoult's law lines. Henry's law requires $K = 175$ Torr for propanone and $K = 165$ Torr for trichloromethane.

Comment Notice how the system deviates from both Raoult's and Henry's laws even for quite small departures from $x = 1$ and $x = 0$, respectively. We deal with these deviations in Section 7.6.

- -

Self-test 7.4 The vapour pressure of chloromethane at various mole fractions in a mixture at 25°C was found to be as follows:

x	0.005	0.009	0.019	0.024
p/Torr	205	363	756	946

Estimate Henry's law constant.

$$[4 \times 10^4 \text{ Torr}]$$

Some Henry's law data are listed in Table 7.1. As well as providing a link between the mole fraction of solute and its partial pressure, the data in the table may also be used to calculate gas solubilities. The following example illustrates the procedure.

Example 7.4 Using Henry's law

Estimate the molar solubility (the solubility in moles per litre) of oxygen in water at 25°C and a partial pressure of 160 Torr, its partial pressure in the atmosphere at sea level.

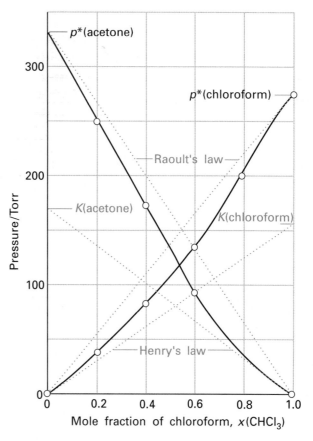

7.15 The experimental partial vapour pressures of a mixture of chloroform (trichloromethane) and acetone (propanone) based on the data in Example 7.3. The values of K are obtained by extrapolating the dilute solution vapour pressures as explained in the Example.

Method The mole fraction of solute is given by Henry's law as $x_J = p_J/K_J$, where p_J is the partial pressure of the gaseous solute J. All we need do is to calculate the mole fraction that corresponds to the stated partial pressure, and then interpret that mole fraction as a molar concentration. For the latter part of the calculation, we calculate the amount of O_2 dissolved in 1.00 kg of water (which corresponds to about 1.00 L water). The solution is dilute, so the expressions for the mole fraction can be simplified.

Answer Because the amount of O_2 dissolved is small, its mole fraction is

$$x(O_2) = \frac{n(O_2)}{n(O_2) + n(H_2O)} \approx \frac{n(O_2)}{n(H_2O)}$$

Hence,

$$n(O_2) \approx x(O_2)n(H_2O) = \frac{p(O_2)n(H_2O)}{K}$$

$$\approx \frac{(160\ \text{Torr}) \times (55.5\ \text{mol})}{3.30 \times 10^7\ \text{Torr}} = 2.69 \times 10^{-4}\ \text{mol}$$

The molality of the saturated solution is therefore $2.69 \times 10^{-4}\ \text{mol kg}^{-1}$, corresponding to a molar concentration of approximately $2.7 \times 10^{-4}\ \text{mol L}^{-1}$.

Comment Knowledge of Henry's law constants for gases in fats and lipids is important for the discussion of respiration, especially when the partial pressure of oxygen is abnormal, as in diving and mountaineering, and for the discussion of the action of gaseous anaesthetics.

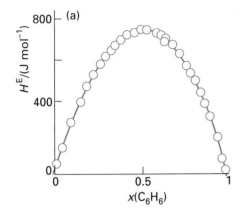

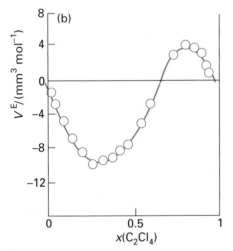

7.16 Experimental excess functions at 25 °C. (a) H^E for benzene/cyclohexane; this graph shows that the mixing is endothermic (because $\Delta_{mix}H = 0$ for an ideal solution). (b) The excess volume, V^E, for tetrachloroethane/cyclopentane; this graph shows that there is a contraction at low tetrachloroethane mole fractions, but an expansion at high mole fractions (because $\Delta_{mix}V = 0$ for an ideal mixture).

Self-test 7.5 Calculate the molar solubility of nitrogen in water exposed to air at 25 °C; partial pressures were calculated in Example 1.5.

[0.505 mmol L^{-1}]

The properties of solutions

In this section we consider the thermodynamics of mixing of liquids. First, we consider the simple case of mixtures of liquids that mix to form an ideal solution. In this way, we identify the thermodynamic consequences of molecules of one species mingling randomly with molecules of the second species. The calculation provides a background for discussing the deviations from ideal behaviour exhibited by real solutions.

7.4 Liquid mixtures

The Gibbs energy of mixing of two liquids to form an ideal solution is calculated in exactly the same way as for two gases (Section 7.2a). The total Gibbs energy before liquids are mixed is

$$G_i = n_A \mu_A^* + n_B \mu_B^*$$

When they are mixed, the individual chemical potentials are given by eqn 24 and the total Gibbs energy is

$$G_f = n_A \{\mu_A^* + RT \ln x_A\} + n_B \{\mu_B^* + RT \ln x_B\}$$

Consequently, the Gibbs energy of mixing is

$$\Delta_{mix}G = nRT \{x_A \ln x_A + x_B \ln x_B\} \tag{26}°$$

where $n = n_A + n_B$. As for gases, it follows that the ideal entropy of mixing of two liquids is

$$\Delta_{mix}S = -nR \{x_A \ln x_A + x_B \ln x_B\} \tag{27}°$$

and the ideal enthalpy of mixing is zero.

Equation 26 is the same as that for two perfect gases, and all the conclusions drawn there are valid here: the driving force for mixing is the increasing entropy of the system as the molecules mingle, and the enthalpy of mixing is zero. It should be noted, however, that solution ideality means something different from gas perfection. In a perfect gas there are no interactions between molecules. In ideal solutions there are interactions, but the average A–B interactions in the mixture are the same as the average A–A and B–B interactions in the pure liquids. The variation of the Gibbs energy of mixing with composition is the same as that already depicted for gases in Fig. 7.5; the same is true of the entropy of mixing, Fig. 7.7.

Real solutions are composed of particles for which A–A, A–B, and B–B interactions are all different. Not only may there be an enthalpy change when liquids mix, but there may also be an additional contribution to the entropy arising from the way in which the molecules of one type might cluster together instead of mingling freely with the others. If the enthalpy change is large and positive or if the entropy change is adverse (because of a reorganization of the molecules that results in an orderly mixture), then the Gibbs energy might be positive for mixing. In that case, separation is spontaneous and the liquids may be immiscible. Alternatively, the liquids might be **partially miscible**, which means that they are miscible only over a certain range of compositions.

The thermodynamic properties of real solutions may be expressed in terms of the **excess functions**, X^E, the difference between the observed thermodynamic function of mixing and the function for an ideal solution. The excess entropy, S^E, for example, is defined as

$$S^E = \Delta_{mix}S - \Delta_{mix}S^{ideal} \tag{28}$$

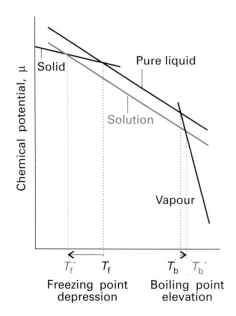

7.17 The chemical potential of a solvent in the presence of a solute. The lowering of the liquid's chemical potential has a greater effect on the freezing point than on the boiling point because of the angles at which the lines intersect (which are determined by entropies; recall Fig. 6.1).

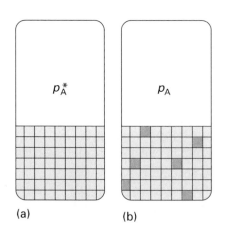

7.18 The vapour pressure of a pure liquid represents a balance between the increased disorder arising from vaporization and the decreased disorder of the surroundings. (a) Here the structure of the liquid is represented highly schematically by the grid of squares. (b) When solute (the dark squares) is present, the disorder of the condensed phase is relatively higher than that of the pure liquid, and there is a decreased tendency to acquire the disorder characteristic of the vapour.

where $\Delta_{\mathrm{mix}}S^{\mathrm{ideal}}$ is given by eqn 27. The excess enthalpy and volume are both equal to the observed enthalpy and volume of mixing, because the ideal values are zero in each case.

Deviations of the excess energies from zero indicate the extent to which the solutions are nonideal. In this connection a useful model system is the **regular solution**, a solution for which $H^{\mathrm{E}} \neq 0$ but $S^{\mathrm{E}} = 0$. A regular solution can be thought of as one in which the two kinds of molecules are distributed randomly (as in an ideal solution) but have different energies of interactions with each other. Two examples of the composition dependence of excess functions are shown in Fig. 7.16.

7.5 Colligative properties

The properties we now consider are the elevation of boiling point, the depression of freezing point, and the osmotic pressure arising from the presence of a solute. In dilute solutions these properties depend only on the number of solute particles present, not their identity. For this reason, they are called **colligative properties** (denoting 'depending on the collection').

We shall assume throughout the following that the solute is not volatile, so it does not contribute to the vapour. We shall also assume that the solute does not dissolve in the solid solvent: that is, the pure solid solvent separates when the solution is frozen. The latter assumption is quite drastic, although it is true of many mixtures; it can be avoided at the expense of more algebra, but that introduces no new principles.

(a) The common features of colligative properties

All the colligative properties stem from the reduction of the chemical potential of the liquid solvent as a result of the presence of solute. The reduction is from μ_{A}^{*} for the pure solvent to $\mu_{\mathrm{A}}^{*} + RT \ln x_{\mathrm{A}}$ when a solute is present ($\ln x_{\mathrm{A}}$ is negative because $x_{\mathrm{A}} < 1$). There is no direct influence of the solute on the chemical potential of the solvent vapour and the solid solvent because the solute appears in neither the vapour nor the solid. As can be seen from Fig. 7.17, the reduction in chemical potential of the solvent implies that the liquid–vapour equilibrium occurs at a higher temperature (the boiling point is raised) and the solid–liquid equilibrium occurs at a lower temperature (the freezing point is lowered).

Molecular interpretation 7.3 The molecular origin of the lowering of the chemical potential is not the energy of interaction of the solute and solvent particles, because the lowering occurs even in ideal solutions (which have zero enthalpy of mixing). If it is not an enthalpy effect, then it must be an entropy effect.

The pure liquid solvent has an entropy that reflects the disorder of its molecules. Its vapour pressure reflects the tendency of the solution towards greater entropy, which can be achieved if the liquid vaporizes to form a more disordered gas. When a solute is present, there is an additional contribution to the entropy of the liquid, even in an ideal solution. Because the entropy of the liquid in the solution is already higher than that of the pure liquid, there is a weaker tendency to form the gas (Fig. 7.18). The effect of the solute appears as a lowered vapour pressure, and hence a higher boiling point.

Similarly, the enhanced molecular randomness of the solution opposes the tendency to freeze. Consequently, a lower temperature must be reached before equilibrium between solid and solution is achieved. Hence, the freezing point is lowered.

The strategy for the quantitative discussion of the elevation of boiling point and the depression of freezing point is to look for the temperature at which, at 1 atm, one phase (the pure solvent vapour or the pure solid solvent) has the same chemical potential as the solvent in the solution. This is the new equilibrium temperature for the phase transition at 1 atm,

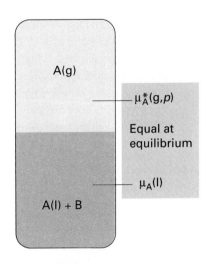

7.19 The heterogeneous equilibrium involved in the calculation of the elevation of boiling point is between A in the pure vapour and A in the mixture, A being the solvent and B an involatile solute.

and hence corresponds to the new boiling point or the new freezing point of the solvent.

(b) The elevation of boiling point

The heterogeneous equilibrium of interest when considering boiling is between the solvent vapour and the solvent in solution at 1 atm (Fig. 7.19). We denote the solvent by A and the solute by B. The equilibrium is established at a temperature for which

$$\mu_A^*(g) = \mu_A^*(l) + RT \ln x_A \tag{29}$$

(The pressure of 1 atm is the same throughout, and will not be written explicitly.) We show in the *Justification* below that this equation implies that the presence of a solute at a mole fraction x_B causes an increase in normal boiling point from T^* to $T^* + \Delta T$, where

$$\Delta T = K x_B \qquad K = \frac{RT^{*2}}{\Delta_{vap}H} \tag{30}°$$

Justification 7.2

Equation 29 rearranges into

$$\ln (1 - x_B) = \frac{\mu_A^*(g) - \mu_A^*(l)}{RT} = \frac{\Delta_{vap}G}{RT}$$

where $\Delta_{vap}G$ is the Gibbs energy of vaporization of the pure solvent (A) and x_B is the mole fraction of the solute; we have used $x_A + x_B = 1$. We now write

$$\Delta_{vap}G = \Delta_{vap}H - T\Delta_{vap}S$$

and ignore the small temperature dependence of $\Delta_{vap}H$ and $\Delta_{vap}S$. Then,

$$\ln (1 - x_B) = \frac{\Delta_{vap}H}{RT} - \frac{\Delta_{vap}S}{R}$$

When $x_B = 0$, the boiling point is that of pure liquid A, T^*, and

$$\ln 1 = \frac{\Delta_{vap}H}{RT^*} - \frac{\Delta_{vap}S}{R}$$

Because $\ln 1 = 0$, the difference of the two equations is

$$\ln (1 - x_B) = \frac{\Delta_{vap}H}{R} \left(\frac{1}{T} - \frac{1}{T^*} \right)$$

We now suppose that the amount of solute present is so small that $x_B \ll 1$. We can then write $\ln (1 - x_B) \approx -x_B$ and hence obtain

$$x_B = \frac{\Delta_{vap}H}{R} \left(\frac{1}{T^*} - \frac{1}{T} \right)$$

Finally, because $T \approx T^*$, it also follows that

$$\frac{1}{T^*} - \frac{1}{T} = \frac{T - T^*}{TT^*} \approx \frac{\Delta T}{T^{*2}}$$

with $\Delta T = T - T^*$. The previous equation then rearranges into eqn 30.

Because eqn 30 makes no reference to the identity of the solute, only to its mole fraction, we conclude that the elevation of boiling point is a colligative property. The value of ΔT does depend on the properties of the solvent, and the biggest changes occur for solvents

Table 7.2* Cryoscopic and ebullioscopic constants

	$K_f/$ $(K/(mol\,kg^{-1}))$	$K_b/$ $(K/(mol\,kg^{-1}))$
Benzene	5.12	2.53
Camphor	40	
Phenol	7.27	3.04
Water	1.86	0.51

* More values are given in the *Data section*.

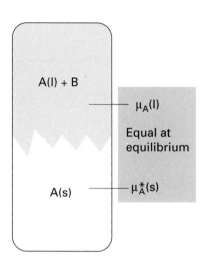

7.20 The heterogeneous equilibrium involved in the calculation of the lowering of freezing point is between A in the pure solid and A in the mixture, A being the solvent and B a solute that is insoluble in solid A.

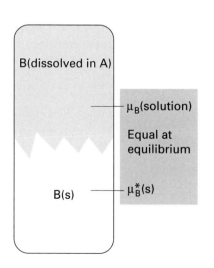

7.21 The heterogeneous equilibrium involved in the calculation of the solubility is between pure solid B and B in the solution.

with high boiling points.[2] For practical applications of eqn 30, we note that the mole fraction of B is proportional to its molality, b, in dilute solutions and write

$$\Delta T = K_b b \tag{31}$$

where K_b is the empirical **ebullioscopic constant** of the solvent (Table 7.2).

(c) The depression of freezing point

The heterogeneous equilibrium now of interest is between pure solid solvent A and the solution with solute present at a mole fraction x_B (Fig. 7.20). At the freezing point, the chemical potentials of A in the two phases are equal:

$$\mu_A^*(s) = \mu_A^*(l) + RT \ln x_A \tag{32}$$

The only difference between this calculation and the last is the appearance of the solid's chemical potential in place of the vapour's. Therefore we can write the result directly from eqn 30:

$$\Delta T = K'x_B \qquad K' = \frac{RT^{*2}}{\Delta_{fus}H} \tag{33}°$$

where ΔT is the freezing point depression, $T^* - T$, and $\Delta_{fus}H$ is the enthalpy of fusion of the solvent. Larger depressions are observed in solvents with low enthalpies of fusion and high melting points. When the solution is dilute, the mole fraction is proportional to the molality of the solute, b, and it is common to write the last equation as

$$\Delta T = K_f b \tag{34}$$

where K_f is the empirical **cryoscopic constant** (Table 7.2). Once the cryoscopic constant of a solvent is known, the depression of freezing point may be used to measure the molar mass of a solute in the method known as **cryoscopy**; however, the technique is of little more than historical interest.

(d) Solubility

Although it is not strictly a colligative property, the solubility of a solute may be estimated by the same techniques as we have been using. When a solid solute is left in contact with a solvent, it dissolves until the solution is saturated. Saturation is a state of equilibrium, with the undissolved solute in equilibrium with the dissolved solute. Therefore, in a saturated solution the chemical potential of the pure solid solute, $\mu_B^*(s)$, and the chemical potential of B in solution, μ_B, are equal (Fig. 7.21). Because the latter is

$$\mu_B = \mu_B^*(l) + RT \ln x_B$$

we can write

$$\mu_B^*(s) = \mu_B^*(l) + RT \ln x_B \tag{35}$$

This expression is the same as the starting equation of the last section, except that the quantities refer to the solute B, not the solvent A.

2 By Trouton's rule (Section 4.3a), $\Delta_{vap}H/T^*$ is a constant; therefore eqn 30 has the form $\Delta T \propto T^*$ and is independent of $\Delta_{vap}H$ itself.

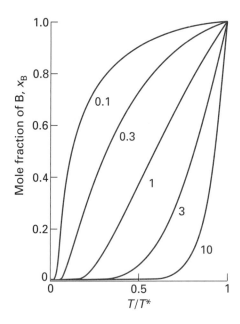

7.22 The variation of solubility (the mole fraction of solute in a saturated solution) with temperature (T^* is the freezing temperature of the solute). Individual curves are labelled with the value of $\Delta_{fus}H/RT^*$.

The starting point is the same but the aim is different. In the present case, we want to find the mole fraction of B in solution at equilibrium when the temperature is T. Therefore, we start by rearranging the last equation into

$$\ln x_B = \frac{\mu_B^*(s) - \mu_B^*(l)}{RT} = -\frac{\Delta_{fus}G}{RT}$$
$$= -\frac{\Delta_{fus}H}{RT} + \frac{\Delta_{fus}S}{R}$$

At the melting point of the solute, T^*, we know that $\Delta_{fus}G = 0$, so $\Delta_{fus}G/RT^* = 0$ too; consequently, this term may be added to the right-hand side to give

$$\ln x_B = -\frac{\Delta_{fus}H}{RT} + \frac{\Delta_{fus}S}{R} + \frac{\Delta_{fus}H}{RT^*} - \frac{\Delta_{fus}S}{R}$$
$$= -\frac{\Delta_{fus}H}{RT} + \frac{\Delta_{fus}H}{RT^*}$$

It then follows that

$$\ln x_B = -\frac{\Delta_{fus}H}{R}\left(\frac{1}{T} - \frac{1}{T^*}\right) \tag{36}°$$

Equation 36 is plotted in Fig. 7.22. It shows that the solubility of B decreases exponentially as the temperature is lowered from its melting point, and that solutes with high melting points and large enthalpies of fusion have low solubilities at normal temperatures. However, the detailed content of eqn 36 should not be treated too seriously because it is based on highly questionable approximations, such as the ideality of the solution. One aspect of its approximate character is that it fails to predict that solutes will have different solubilities in different solvents, for no solvent properties appear in the expression.

(e) Osmosis

The phenomenon of **osmosis** (from the Greek word for 'push') is the spontaneous passage of a pure solvent into a solution separated from it by a **semipermeable membrane**, a membrane permeable to the solvent but not to the solute (Fig. 7.23). The **osmotic pressure**, Π, is the pressure that must be applied to the solution to stop the influx of solvent. One of the most important examples of osmosis is transport of fluids through cell membranes, but it is also the basis of **osmometry**, the determination of molar mass by the measurement of osmotic pressure. Osmometry is widely used to determine the molar masses of macromolecules.

In the simple arrangement shown in Fig. 7.24, the opposing pressure arises from the head of solution that the osmosis itself produces. Equilibrium is reached when the hydrostatic pressure of the column of solution matches the osmotic pressure. The complication in this arrangement is that the entry of solvent into the solution results in its dilution, and so it is more difficult to treat than the arrangement in Fig. 7.23, in which there is no flow and the concentrations remain unchanged.

The thermodynamic treatment of osmosis depends on noting that, at equilibrium, the chemical potential of the solvent must be the same on each side of the membrane. As shown in the *Justification* below, this equality implies that for dilute solutions the osmotic pressure is given by the **van't Hoff equation**:

$$\Pi = [B]RT \tag{37}°$$

where $[B] = n_B/V$ is the molar concentration of the solute.

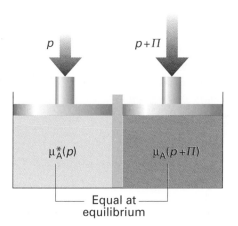

7.23 The equilibrium involved in the calculation of osmotic pressure, Π, is between pure solvent A at a pressure p on one side of the semipermeable membrane and A as a component of the mixture on the other side of the membrane, where the pressure is $p + \Pi$.

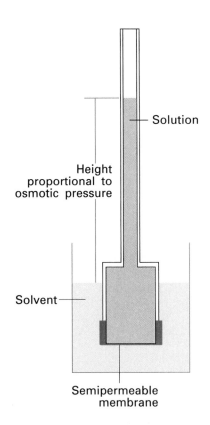

Solution

Height proportional to osmotic pressure

Solvent

Semipermeable membrane

7.24 In a simple version of the osmotic pressure experiment, A is at equilibrium on each side of the membrane when enough has passed into the solution to cause a hydrostatic pressure difference.

Justification 7.3

On the pure solvent side the chemical potential of the solvent, which is at a pressure p, is $\mu_A^*(p)$. On the solution side, the chemical potential is lowered by the presence of the solute, which reduces the mole fraction of the solvent from 1 to x_A. However, the chemical potential of A is raised on account of the greater pressure, $p + \Pi$, that the solution experiences. At equilibrium the chemical potential of A is the same in both compartments, and we can write

$$\mu_A^*(p) = \mu_A(x_A, p + \Pi)$$

The presence of solute is taken into account in the normal way:

$$\mu_A(x_A, p + \Pi) = \mu_A^*(p + \Pi) + RT \ln x_A$$

We saw in Section 5.2b (eqn 5.14) how to take the effect of pressure into account:

$$\mu_A^*(p + \Pi) = \mu_A^*(p) + \int_p^{p+\Pi} V_m \, dp$$

where V_m is the molar volume of the pure solvent A. When these three equations are combined we get

$$-RT \ln x_A = \int_p^{p+\Pi} V_m \, dp$$

For dilute solutions, $\ln x_A$ may be replaced by $\ln(1 - x_B) \approx -x_B$. We may also assume that the pressure range in the integration is so small that the molar volume of the solvent is a constant. That being so, V_m may be taken outside the integral, giving

$$RT x_B = \Pi V_m$$

When the solution is dilute, $x_B \approx n_B/n_A$. Moreover, because $n_A V_m = V$, the total volume of the solvent, the equation simplifies to eqn 37.

Because the effect of osmotic pressure is so readily measurable and large, one of the most common applications of osmometry is to the measurement of molar masses of macromolecules (proteins and synthetic polymers). As these huge molecules dissolve to produce solutions that are far from ideal, it is assumed that the van't Hoff equation is only the first term of a virial-like expansion:

$$\Pi = [B]RT\{1 + B[B] + \cdots\} \tag{38}$$

The additional terms take the nonideality into account; the empirical constant B is called the **osmotic virial coefficient**. The osmotic pressure is measured at a series of mass concentrations, c, and a plot of Π/c against c is used to determine the molar mass of B.

Example 7.5 Using osmometry to determine the molar mass of a macromolecule

The osmotic pressures of solutions of poly(vinyl chloride), PVC, in cyclohexanone at 298 K are given below. The pressures are expressed in terms of the heights of solution (of mass density $\rho = 0.980 \text{ g cm}^{-3}$) in balance with the osmotic pressure. Determine the molar mass of the polymer.

$c/(\text{g L}^{-1})$	1.00	2.00	4.00	7.00	9.00
h/cm	0.28	0.71	2.01	5.10	8.00

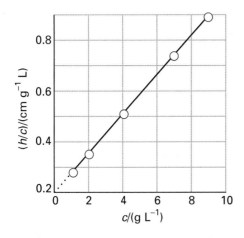

7.25 The plot involved in the determination of molar mass by osmometry. The molar mass is calculated from the intercept at $c = 0$; in Chapter 23 we shall see that additional information comes from the slope.

Method We use eqn 38 with $[B] = c/M$, where c is the mass concentration of the polymer and M is its molar mass. The osmotic pressure is related to the hydrostatic pressure by $\Pi = \rho g h$ (Example 1.2) with $g = 9.81$ m s^{-2}. With these substitutions, eqn 38 becomes

$$\frac{h}{c} = \frac{RT}{\rho g M}\left(1 + \frac{Bc}{M} + \cdots\right) = \frac{RT}{\rho g M} + \left(\frac{RTB}{\rho g M^2}\right)c + \cdots$$

Therefore, to find M, plot h/c against c, and expect a straight line with intercept $RT/\rho g M$ at $c = 0$.

Answer The data give the following values for the quantities to plot:

$c/(\text{g L}^{-1})$	1.00	2.00	4.00	7.00	9.00
$(h/c)/(\text{cm g}^{-1}\text{ L})$	0.28	0.36	0.503	0.729	0.889

The points are plotted in Fig. 7.25. The intercept is at 0.21. Therefore,

$$
\begin{aligned}
M &= \frac{RT}{\rho g} \times \frac{1}{0.21 \text{ cm g}^{-1}\text{ L}} \\
&= \frac{(8.3145 \text{ J K}^{-1}\text{ mol}^{-1}) \times (298 \text{ K})}{(980 \text{ kg m}^{-3}) \times (9.81 \text{ m s}^{-2})} \times \frac{1}{2.1 \times 10^{-3} \text{ m}^4 \text{ kg}^{-1}} \\
&= 1.2 \times 10^2 \text{ kg mol}^{-1}
\end{aligned}
$$

Comment Molar masses of macromolecules are often reported in daltons (Da), with 1 Da = 1 g mol^{-1}. The macromolecule in this example has a molar mass of about 120 kDa.

- -

Self-test 7.6 Estimate the depression of freezing point of the most concentrated of these solutions, taking K_f as about 10 K/(mol kg^{-1}).

[0.8 mK]

Activities

Now we see how to adjust the expressions developed earlier in the chapter to take into account deviations from ideal behaviour. In Section 5.4 we saw how the fugacity was introduced to take into account the effects of gas imperfections in a manner that resulted in the least upset of the form of equations. Here we see how the expressions encountered in the treatment of ideal solutions can also be preserved almost intact by introducing the concept of activity.

7.6 The solvent activity

The general form of the chemical potential of a real or ideal solvent is given by a straightforward modification of eqn 22:

$$\mu_A = \mu_A^* + RT \ln\left(\frac{p_A}{p_A^*}\right) \tag{39}$$

where p_A^* is the vapour pressure of pure A and p_A is the vapour pressure of A when it is a component of a solution. For an ideal solution, the solvent obeys Raoult's law at all concentrations and we write

$$\mu_A = \mu_A^* + RT \ln x_A \tag{40}°$$

The standard state of the solvent is the pure liquid (at 1 bar) and is obtained when $x_A = 1$. The form of the last equation can be preserved when the solution does not obey Raoult's law by writing

$$\mu_A = \mu_A^* + RT \ln a_A \qquad [41]$$

The quantity a_A is the **activity** of A, a kind of 'effective' mole fraction, just as the fugacity is an effective pressure.

Because eqn 39 is true for both real and ideal solutions (the only approximation being the use of pressures rather than fugacities), we can conclude by comparing it with eqn 41 that

$$a_A = \frac{p_A}{p_A^*} \qquad (42)$$

We see there is nothing mysterious about the activity of a solvent: it can be determined experimentally simply by measuring the vapour pressure and then using eqn 42.

Illustration

. . . .

The vapour pressure of 0.500 M $KNO_3(aq)$ at 100°C is 749.7 Torr, so the activity of water in the solution at this temperature is

$$a_A = \frac{749.7 \text{ Torr}}{760.0 \text{ Torr}} = 0.9864$$

. . . .

Because all solvents obey Raoult's law (that $p_A/p_A^* = x_A$) increasingly closely as the concentration of solute approaches zero, the activity of the solvent approaches the mole fraction as $x_A \to 1$:

$$a_A \to x_A \text{ as } x_A \to 1 \qquad (43)$$

As in the case of real gases, a convenient way of expressing this convergence is to introduce the **activity coefficient**, γ, by the definition

$$a_A = \gamma_A x_A \qquad \gamma_A \to 1 \text{ as } x_A \to 1 \qquad [44]$$

at all temperatures and pressures. The chemical potential of the solvent is then

$$\mu_A = \mu_A^* + RT \ln x_A + RT \ln \gamma_A \qquad (45)$$

The standard state of the solvent, the pure liquid solvent at 1 bar, is established when $x_A = 1$.

7.7 The solute activity

The problem with defining activity coefficients and standard states for solutes is that they approach ideal-dilute (Henry's law) behaviour as $x_B \to 0$, not as $x_B \to 1$ (corresponding to pure solute). We shall show how to set up the definitions for a solute that obeys Henry's law exactly, and then show how to allow for deviations.

(a) Ideal-dilute solutions

A solute B that satisfies Henry's law has a vapour pressure given by $p_B = K_B x_B$, where K_B is an empirical constant. In this case, the chemical potential of B is

$$\mu_B = \mu_B^* + RT \ln \left(\frac{p_B}{p_B^*} \right) = \mu_B^* + RT \ln \left(\frac{K_B}{p_B^*} \right) + RT \ln x_B$$

Both K_B and p_B^* are characteristics of the solute, so the second term may be combined with the first to give a new standard chemical potential, which we denote $\mu^\dagger$:

$$\mu_B^\dagger = \mu_B^* + RT \ln \left(\frac{K_B}{p_B^*} \right) \qquad \text{[46]}$$

It then follows that

$$\mu_B = \mu_B^\dagger + RT \ln x_B \qquad (47)^\circ$$

(b) Real solutes

We now permit deviations from ideal-dilute, Henry's law behaviour. For the solute, we introduce a_B in place of x_B in eqn 47, and obtain

$$\mu_B = \mu_B^\dagger + RT \ln a_B \qquad \text{[48]}$$

The standard state remains unchanged in this last stage, and all the deviations from ideality are captured in the activity a_B. The value of the activity at any concentration can be obtained in the same way as for the solvent, but in place of eqn 42 we use

$$a_B = \frac{p_B}{K_B} \qquad (49)$$

As for the solvent, it is sensible to introduce an activity coefficient through

$$a_B = \gamma_B x_B \qquad \text{[50]}$$

Now all the deviations from ideality are captured in the activity coefficient γ_B. Because the solute obeys Henry's law as its concentration goes to zero, it follows that

$$a_B \to x_B \text{ and } \gamma_B \to 1 \text{ as } x_B \to 0 \qquad (51)$$

at all temperatures and pressures. Deviations of the solute from ideality disappear as zero concentration is approached.

Example 7.6 Measuring activity

Use the information in Example 7.3 to calculate the activity and activity coefficient of chloroform in acetone at 25 °C, treating it first as a solvent and then as a solute.

Method For the activity of chloroform as a solvent (the Raoult's law activity), form $a = p/p^*$ and $\gamma = a/x$. For its activity as a solute (the Henry's law activity), form $a = p/K$ and $\gamma = a/x$.

Answer Because $p^* = 293$ Torr and $K = 165$ Torr, we can construct the following table (with x the mole fraction of chloroform):

x	0	0.20	0.40	0.60	0.80	1.00

From Raoult's law (chloroform regarded as the solvent):

a	0	0.12	0.28	0.49	0.75	1.00
γ		0.60	0.70	0.82	0.94	1.00

From Henry's law (chloroform regarded as the solute):

a	0	0.21	0.50	0.86	1.33	1.78
γ	1	1.05	1.25	1.43	1.66	1.78

These values are plotted in Fig. 7.26.

Comment Notice that $\gamma \to 1$ as $x \to 1$ in the Raoult's law case, but that $\gamma \to 1$ as $x \to 0$ in the Henry's law case.

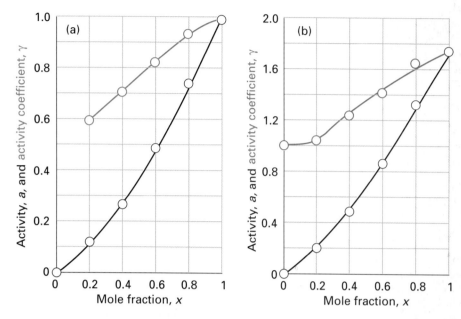

7.26 The variation of activity and activity coefficient of chloroform (trichloromethane) and acetone (propanone) with composition according to (a) Raoult's law, (b) Henry's law.

Self–test 7.7 Calculate the activities and activity coefficients for acetone according to the two conventions.

(c) Activities in terms of molalities

The compositions of mixtures are often expressed as molalities, b, in place of mole fractions. It proves convenient to introduce yet another definition of activity, but one that follows naturally from what we have done so far. First, we note that in dilute solutions the amount of solute is much less than the amount of solvent ($n_B \ll n_A$), so to a good approximation $x_B \approx n_B/n_A$. Because n_B is proportional to the molality b_B, we can write $x_B = \kappa b_B/b^{\ominus}$, where $b^{\ominus} = 1 \text{ mol kg}^{-1}$ and κ is a dimensionless constant. For an ideal-dilute solution it follows that

$$\mu_B = \mu_B^{\dagger} + RT \ln \kappa + RT \ln \left(\frac{b_B}{b^{\ominus}} \right) \tag{52}$$

As before, it will simplify the notation if we adopt the convention of interpreting b as a relative molality (that is, replacing $b/b^{\ominus}$ by b; in practice, using for b its numerical value in moles per kilogram), and signifying this convention by using the label $\{\cdots\}$ for the equation number. Then eqn 52 is written

$$\mu_B = \mu_B^{\dagger} + RT \ln \kappa + RT \ln b_B \tag{53}$$

We now define a new standard chemical potential, $\mu^{\ominus}$, by combining the first two terms on the right, and hence obtain

$$\mu_B = \mu_B^{\ominus} + RT \ln b_B \tag{54}$$

According to this definition, the chemical potential of the solute has its standard value $\mu^{\ominus}$ when the molality of B is equal to $b^{\ominus}$ (that is, at 1 mol kg^{-1}). Note that as $b_B \to 0$, $\mu_B \to -\infty$; that is, as the solution becomes diluted, so the solute becomes increasingly stabilized. The practical consequence of this result is that it is very difficult to remove the last traces of a solute from a solution.

Table 7.3 Standard states

Component	Basis	Standard state	Chemical potential	Limits
Solvent[‡]	Raoult	Pure solvent	$\mu = \mu^* + RT \ln a$	
			$a = p/p^*$ and $a = \gamma x$	$\gamma \to 1$ as $x \to 1$ (pure solvent)
Solute	Henry	(1) A hypothetical state of the pure solute	$\mu = \mu^\dagger + RT \ln a$ $a = p/K$ and $a = \gamma x$	$\gamma \to 1$ as $x \to 0$
		(2) A hypothetical state of the solute at molality $b^\ominus$	$\mu = \mu^\ominus + RT \ln a$ $a = \gamma b/b^\ominus$	$\gamma \to 1$ as $b \to 0$

‡ The implication of this entry is that the activity of a pure solid or liquid is 1. The pressure in each case is 1 bar.

Now, as before, we incorporate deviations from ideality by introducing a dimensionless activity a_B, a dimensionless activity coefficient γ_B, and writing

$$a_B = \gamma_B \frac{b_B}{b^\ominus} \text{ where } \gamma_B \to 1 \text{ as } b_B \to 0 \qquad [55]$$

at all temperatures and pressures. The standard state remains unchanged in this last stage and, as before, all the deviations from ideality are captured in the activity coefficient γ_B. We then arrive at the following succinct expression for the chemical potential of a real solute at any molality:

$$\mu = \mu^\ominus + RT \ln a \qquad (56)$$

It is important to be aware of the different definitions of standard states and activities, and they are summarized in Table 7.3. We shall put them to work in the next few chapters, when we shall see that using them is much easier than defining them.

Checklist of key ideas

The thermodynamic description of mixtures

7.1 Partial molar quantities
- [ ] partial molar volume (1)
- [ ] the total volume of a mixture (3)
- [ ] partial molar Gibbs energy and the definition of the chemical potential (4)
- [ ] the total Gibbs energy of a mixture (5)
- [ ] the fundamental equation of chemical thermodynamics (6)
- [ ] the chemical potential in terms of U (9), H (10a), and A (10b)
- [ ] the Gibbs–Duhem equation (12)

7.2 The thermodynamics of mixing
- [ ] the Gibbs energy of mixing of two perfect gases ($\Delta_{mix}G$, 17)

- [ ] the entropy of mixing of two perfect gases ($\Delta_{mix}S$, 18)
- [ ] the enthalpy of mixing ($\Delta_{mix}H$, 19)

7.3 The chemical potentials of liquids
- [ ] Raoult's law (23)
- [ ] ideal solution
- [ ] the chemical potential of a component of an ideal solution (24)
- [ ] Henry's law (25)
- [ ] ideal-dilute solution
- [ ] Henry's law and gas solubility

The properties of solutions

7.4 Liquid mixtures
- [ ] the Gibbs energy of mixing of two liquids to give an ideal solution (26) and the entropy of mixing (27)

- [ ] partially miscible liquids
- [ ] excess function (X^E, 28)
- [ ] regular solution

7.5 Colligative properties
- [ ] colligative property
- [ ] the origin of colligative properties: the lowering of chemical potential of the solvent
- [ ] the elevation of boiling point (30)
- [ ] ebullioscopic constant (31)
- [ ] the depression of freezing point (33)
- [ ] cryoscopic constant (34)
- [ ] cryoscopy
- [ ] ideal solubility (36)
- [ ] osmosis
- [ ] semipermeable membrane
- [ ] osmotic pressure
- [ ] osmometry
- [ ] van't Hoff equation for the osmotic pressure (37)

Activities
- [ ] osmotic virial coefficient

7.6 The solvent activity
- [ ] solvent activity and chemical potential (41)
- [ ] the determination of activity from vapour pressure (42)
- [ ] the Raoult's law basis of activity and activity coefficient (44)

7.7 The solute activity
- [ ] the chemical potential of a solute in an ideal-dilute solution (47)
- [ ] the activity of a solute (49) and the Henry's law basis of activity coefficient (50)
- [ ] chemical potential of a solute in terms of its molality (54) and the activity on this basis (55)

Further reading

Articles of general interest

E.F. Meyer, Thermodynamics of 'mixing' of ideal gases: a persistent pitfall. *J. Chem. Educ.* **64**, 676 (1987).

M.J. Clugston, A mathematical verification of the second law of thermodynamics from the entropy of mixing. *J. Chem. Educ.* **67**, 203 (1990).

J.J. Carroll, Henry's law: a historical view. *J. Chem. Educ.* **70**, 91 (1993).

M.E. Cardinali and C. Giomini, Boiling temperature vs. composition: an almost-exact explicit equation for a binary mixture following Raoult's law. *J. Chem. Educ.* **66**, 549 (1989).

H.F. Franzen, The freezing point depression law in physical chemistry. *J. Chem. Educ.* **65**, 1077 (1988).

R.L. Scott, Models for phase equilibria in fluid mixtures. *Acc. Chem. Res.* **20**, 97(1987).

A. Vincent, Osmotic pressure and the effect of gravity on solutions. *J. Chem. Educ.* **73**, 998 (1996).

S.F. Ackley, Sea ice. In *Encyclopedia of applied physics* (ed. G.L. Trigg), **17**, 81. VCH, New York (1996).

B. Freeman, Osmosis. In *Encyclopedia of applied physics* (ed. G.L. Trigg), **13**, 59. VCH, New York (1995).

Texts and sources of data and information

S.I. Sandler, *Chemical and engineering thermodynamics.* Wiley, New York (1989).

J.S. Rowlinson and F.L. Swinton, *Liquids and liquid mixtures.* Butterworth, London (1982).

J.N. Murrell and A.D. Jenkins, *Properties of liquids and solutions.* Wiley–Interscience, New York (1994).

M.R.J. Dack (ed.), *Techniques in chemistry. VIII. Solutions and solubilities.* Wiley, New York (1975).

J.R. Overton, Determination of osmotic pressure. In *Techniques of chemistry* (ed. A. Weissberger and B.W. Rossiter), **5**, 309. Wiley, New York (1971).

W.J. Mader and L.T. Brady, Determination of solubility. In *Techniques of chemistry* (ed. A. Weissberger and B.W. Rossiter), **5**, 257. Wiley, New York (1971).

M.R.J. Dack, Solutions and solubilities. In *Techniques of chemistry* (ed. A. Weissberger and B.W. Rossiter), **8**. Wiley, New York (1971).

Exercises

7.1 (a) The partial molar volumes of acetone (propanone) and chloroform (trichloromethane) in a mixture in which the mole fraction of $CHCl_3$ is 0.4693 are 74.166 $cm^3\,mol^{-1}$ and 80.235 $cm^3\,mol^{-1}$, respectively. What is the volume of a solution of mass 1.000 kg?

7.1 (b) The partial molar volumes of two liquids A and B in a mixture in which the mole fraction of A is 0.3713 are 188.2 $cm^3\,mol^{-1}$ and 176.14 $cm^3\,mol^{-1}$, respectively. The molar masses of the A and B are 241.1 $g\,mol^{-1}$ and 198.2 $g\,mol^{-1}$. What is the volume of a solution of mass 1.000 kg?

7.2 (a) At 25 °C, the density of a 50 per cent by mass ethanol–water solution is 0.914 $g\,cm^{-3}$. Given that the partial molar volume of water in the solution is 17.4 $cm^3\,mol^{-1}$, calculate the partial molar volume of the ethanol.

7.2 (b) At 20 °C, the density of a 20 per cent by mass ethanol–water solution is 968.7 $kg\,m^{-3}$. Given that the partial molar volume of ethanol in the solution is 52.2 $cm^3\,mol^{-1}$, calculate the partial molar volume of the water.

7.3 (a) At 300 K, the partial vapour pressures of HCl (that is, the partial pressure of the HCl vapour) in liquid $GeCl_4$ are as follows:

x_{HCl}	0.005	0.012	0.019
p_{HCl}/kPa	32.0	76.9	121.8

Show that the solution obeys Henry's law in this range of mole fractions, and calculate Henry's law constant at 300 K.

7.3 (b) At 310 K, the partial vapour pressures of a substance B dissolved in a liquid A are as follows:

x_B	0.010	0.015	0.020
p_B/kPa	82.0	122.0	166.1

Show that the solution obeys Henry's law in this range of mole fractions, and calculate Henry's law constant at 310 K.

7.4 (a) Predict the partial vapour pressure of HCl above its solution in liquid germanium tetrachloride of molality 0.10 $mol\,kg^{-1}$. For data, see Exercise 7.3a.

7.4 (b) Predict the partial vapour pressure of the component B above its solution in A in Exercise 7.3b when the molality of B is 0.25 $mol\,kg^{-1}$. The molar mass of A is 74.1 $g\,mol^{-1}$.

7.5 (a) Calculate the cryoscopic and ebullioscopic constants of tetrachloromethane.

7.5 (b) Calculate the cryoscopic and ebullioscopic constants of naphthalene.

7.6 (a) The vapour pressure of benzene is 400 Torr at 60.6 °C, but it fell to 386 Torr when 19.0 g of an involatile organic compound was dissolved in 500 g of benzene. Calculate the molar mass of the compound.

7.6 (b) The vapour pressure of 2-propanol is 50.00 kPa at 338.8 °C, but it fell to 49.62 kPa when 8.69 g of an involatile organic compound was dissolved in 250 g of 2-propanol. Calculate the molar mass of the compound.

7.7 (a) The addition of 100 g of a compound to 750 g of CCl_4 lowered the freezing point of the solvent by 10.5 K. Calculate the molar mass of the compound.

7.7 (b) The addition of 5.00 g of a compound to 250 g of naphthalene lowered the freezing point of the solvent by 0.780 K. Calculate the molar mass of the compound.

7.8 (a) The osmotic pressure of an aqueous solution at 300 K is 120 kPa. Calculate the freezing point of the solution.

7.8 (b) The osmotic pressure of an aqueous solution at 288 K is 99.0 kPa. Calculate the freezing point of the solution.

7.9 (a) Consider a container of volume 5.0 L that is divided into two compartments of equal size. In the left compartment there is nitrogen at 1.0 atm and 25 °C; in the right compartment there is hydrogen at the same temperature and pressure. Calculate the entropy and Gibbs energy of mixing when the partition is removed. Assume that the gases are perfect.

7.9 (b) Consider a container of volume 250 mL that is divided into two compartments of equal size. In the left compartment there is argon at 100 kPa and 0 °C; in the right compartment there is neon at the same temperature and pressure. Calculate the entropy and Gibbs energy of mixing when the partition is removed. Assume that the gases are perfect.

7.10 (a) Air is a mixture with a composition given in Self-test 1.5. Calculate the entropy of mixing when it is prepared from the pure (and perfect) gases.

7.10 (b) Calculate the Gibbs energy, entropy, and enthalpy of mixing when 1.00 mol C_6H_{14} (hexane) is mixed with 1.00 mol C_7H_{16} (heptane) at 298 K; treat the solution as ideal.

7.11 (a) What proportions of hexane and heptane should be mixed (a) by mole fraction, (b) by mass in order to achieve the greatest entropy of mixing?

7.11 (b) What proportions of benzene and ethylbenzene should be mixed (a) by mole fraction, (b) by mass in order to achieve the greatest entropy of mixing?

7.12 (a) Use Henry's law and the data in Table 7.1 to calculate the solubility (as a molality) of CO_2 in water at 25 °C when its partial pressure is (a) 0.10 atm, (b) 1.00 atm.

7.12 (b) The mole fractions of N_2 and O_2 in air at sea level are approximately 0.78 and 0.21. Calculate the molalities of the solution formed in an open flask of water at 25 °C.

7.13 (a) A water carbonating plant is available for use in the home and operates by providing carbon dioxide at 5.0 atm. Estimate the molar concentration of the soda water it produces.

7.13 (b) After some weeks of use, the pressure in the water carbonating plant mentioned in the previous exercise has fallen to 2.0 atm. Estimate the molar concentration of the soda water it produces at this stage.

7.14 (a) Calculate the freezing point of a glass of water of volume 250 cm^3 sweetened with 7.5 g of sucrose.

7.14 (b) Calculate the freezing point of a glass of water of volume 200 cm^3 in which 10 g of glucose has been dissolved.

7.15 (a) The enthalpy of fusion of anthracene is 28.8 kJ mol^{-1} and its melting point is 217 °C. Calculate its ideal solubility in benzene at 25 °C.

7.15 (b) Predict the ideal solubility of lead in bismuth at 280 °C given that its melting point is 327 °C and its enthalpy of fusion is 5.2 kJ mol^{-1}.

7.16 (a) The osmotic pressures of solutions of polystyrene in toluene were measured at 25 °C and the pressure was expressed in terms of the height of the solvent of density 1.004 g cm^{-3}:

$c/(\text{g L}^{-1})$	2.042	6.613	9.521	12.602
h/cm	0.592	1.910	2.750	3.600

Calculate the molar mass of the polymer.

7.16 (b) The molar mass of an enzyme was determined by dissolving it in water, measuring the osmotic pressure at 20 °C, and extrapolating the data to zero concentration. The following data were obtained:

$c/(\text{mg cm}^{-3})$	3.221	4.618	5.112	6.722
h/cm	5.746	8.238	9.119	11.990

Calculate the molar mass of the enzyme.

7.17 (a) Substances A and B are both volatile liquids with $p_A^* = 300$ Torr, $p_B^* = 250$ Torr, and $K_B = 9.00$ Torr (concentration expressed in mole fraction). When $x_A = 0.9$, $b_B = 2.22$ mol kg^{-1}, $p_A = 250$ Torr, and $p_B = 25$ Torr. Calculate the activities and activity coefficients of A and B. Use the mole fraction, Raoult's law basis for A and the Henry's law basis (both mole fractions and molalities) for B.

7.17 (b) Given that $p^*(H_2O) = 0.02308$ atm and $p(H_2O) = 0.02239$ atm in a solution in which 0.122 kg of a non-volatile solute ($M = 241$ g mol^{-1}) is dissolved in 0.920 kg water at 293 K, calculate the activity and activity coefficient of water in the solution.

7.18 (a) A dilute solution of bromine in carbon tetrachloride behaves as an ideal-dilute solution. The vapour pressure of pure CCl_4 is 33.85 Torr at 298 K. The Henry's law constant when the concentration of Br_2 is expressed as a mole fraction is 122.36 Torr. Calculate the vapour pressure of each component, the total pressure, and the composition of the vapour phase when the mole fraction of Br_2 is 0.050, on the assumption that the conditions of the ideal-dilute solution are satisfied at this concentration.

7.18 (b) Benzene and toluene form nearly ideal solutions. The boiling point of pure benzene is 80.1 °C. Calculate the chemical potential of benzene relative to that of pure benzene when $x_{\text{benzene}} = 0.30$ at its boiling point. If the activity coefficient of benzene in this solution were actually 0.93 rather than 1.00, what would be its vapour pressure?

7.19 (a) By measuring the equilibrium between liquid and vapour phases of an acetone (A)–methanol (M) solution at 57.2°C at 1.00 atm, it was found that $x_A = 0.400$ when $y_A = 0.516$. Calculate the activities and activity coefficients of both components in this solution on the Raoult's law basis. The vapour pressures of the pure components at this temperature are: $p_A^* = 786$ Torr and

$p_M^* = 551$ Torr. (x_A is the mole fraction in the liquid and y_A the mole fraction in the vapour.)

7.19 (b) By measuring the equilibrium between liquid and vapour phases of a solution at 30°C at 1.00 atm, it was found that $x_A = 0.220$ when $y_A = 0.314$. Calculate the activities and activity coefficients of both components in this solution on the Raoult's law basis. The vapour pressures of the pure components at this temperature are: $p_A^* = 73.0$ kPa and $p_B^* = 92.1$ kPa. (x_A is the mole fraction in the liquid and y_A the mole fraction in the vapour.)

Problems

Numerical problems

7.1 The following table gives the mole fraction of methylbenzene (A) in liquid and gaseous mixtures with butanone at equilibrium at 303.15 K and the total pressure p. Take the vapour to be perfect and calculate the partial pressures of the two components. Plot them against their respective mole fractions in the liquid mixture and find the Henry's law constants for the two components.

x_A	0	0.0898	0.2476	0.3577	0.5194	0.6036	0.7188
y_A	0	0.0410	0.1154	0.1762	0.2772	0.3393	0.4450
p/kPa	36.066	34.121	30.900	28.626	25.239	23.402	20.6984

x_A	0.8019	0.9105	1
y_A	0.5435	0.7284	1
p/kPa	18.592	15.496	12.295

7.2 The volume of an aqueous solution of NaCl at 25°C was measured at a series of molalities b, and it was found that the volume fitted the expression

$$V/\text{cm}^3 = 1003 + 16.62b + 1.77b^{3/2} + 0.12b^2$$

where V is the volume of a solution formed from 1.000 kg of water and b is to be understood as $b/b^{\ominus}$. Calculate the partial molar volume of the components in a solution of molality 0.100 mol kg^{-1}.

7.3 At 18°C the total volume of a solution formed from MgSO$_4$ and 1.000 kg of water fits the expression

$$V/\text{cm}^3 = 1001.21 + 34.69(b - 0.070)^2$$

where b is to be understood as $b/b^{\ominus}$. Calculate the partial molar volumes of the salt and the solvent when in a solution of molality 0.050 mol kg^{-1}.

7.4 The densities of aqueous solutions of copper(II) sulfate at 20°C were measured as set out below. Determine and plot the partial molar volume of CuSO$_4$ in the range of the measurements.

$m(\text{CuSO}_4)$/g	5	10	15	20
ρ/(g cm^{-3})	1.051	1.107	1.167	1.230

where $m(\text{CuSO}_4)$ is the mass of CuSO$_4$ dissolved in 100 g of solution.

7.5 What proportions of ethanol and water should be mixed in order to produce 100 cm^3 of a mixture containing 50 per cent by mass of ethanol? What change in volume is brought about by adding 1.00 cm^3 of ethanol to the mixture? (Use data from Fig. 7.1.)

7.6 The table below lists the vapour pressures of mixtures of iodoethane (I) and ethyl acetate (A) at 50°C. Find the activity coefficients of both components on (a) the Raoult's law basis, (b) the Henry's law basis with I as solute.

x_I	0	0.0579	0.1095	0.1918	0.2353
p_I/Torr	0	20.0	52.7	87.7	105.4
p_A/Torr	280.4	266.1	252.3	231.4	220.8

x_I	0.3718	0.5478	0.6349	0.8253	0.9093	1.0000
p_I/Torr	155.4	213.3	239.1	296.9	322.5	353.4
p_A/Torr	187.9	144.2	122.9	66.6	38.2	0

7.7 The excess Gibbs energy of solutions of methylcyclohexane (MCH) and tetrahydrofuran (THF) at 303.15 K were found to fit the expression

$$G^E = RTx(1 - x)\{0.4857 - 0.1077(2x - 1) + 0.0191(2x - 1)^2\}$$

where x is the mole fraction of the methylcyclohexane. Calculate the Gibbs energy of mixing when a mixture of 1.00 mol MCH and 3.00 mol THF is prepared.

Theoretical problems

7.8 The excess Gibbs energy of a certain binary mixture is equal to $gRTx(1 - x)$ where g is a constant and x is the mole fraction of a solute A. Find an expression for the chemical potential of A in the mixture and sketch its dependence on the composition.

7.9 Use the Gibbs–Duhem equation to derive the Gibbs–Duhem–Margules equation

$$\left(\frac{\partial \ln f_A}{\partial \ln x_A}\right)_{p,T} = \left(\frac{\partial \ln f_B}{\partial \ln x_B}\right)_{p,T}$$

where f is the fugacity. Use the relation to show, when the fugacities are replaced by pressures, that if Raoult's law applies to one component in a mixture, then it must also apply to the other.

7.10 Use the Gibbs–Duhem equation to show that the partial molar volume (or any partial molar property) of a component B can be obtained if the partial molar volume (or other property) of A is known for all compositions up to the one of interest. Do this by proving that

$$V_B = V_B^* + \int_{V_A^*}^{V_A} \frac{x_A}{1 - x_A}\, dV_A$$

7.11 Use the Gibbs–Helmholtz equation to find an expression for $d \ln x_A$ in terms of dT. Integrate $d \ln x_A$ from $x_A = 0$ to the value of interest, and integrate the right-hand side from the transition temperature for the pure liquid A to the value in the solution. Show that, if the enthalpy of transition is constant, then eqns 30 and 33 are obtained.

7.12 The 'osmotic coefficient', ϕ, is defined as $\phi = -(x_A/x_B) \ln a_A$. By writing $r = x_B/x_A$, and using the Gibbs–Duhem equation, show that we can calculate the activity of B from the activities of A over a composition range by using the formula

$$\ln\left(\frac{a_B}{r}\right) = \phi - \phi(0) + \int_0^r \left(\frac{\phi - 1}{r}\right) dr$$

7.13 Show that the osmotic pressure of a real solution is given by

$$\Pi V = -RT \ln a_A$$

Go on to show that, provided the concentration of the solution is low, this expression takes the form

$$\Pi V = \phi RT [B]$$

and hence that the osmotic coefficient, ϕ (which is defined in Problem 7.12), may be determined from osmometry.

Additional problems supplied by Carmen Giunta and Charles Trapp

7.14 Aminabhavi et al. examined mixtures of cyclohexane with various long-chain alkanes (T.M. Aminabhavi, V.B. Patil, M.I. Aralaguppi, J.D. Ortego, and K.C. Hansen, J. Chem. Eng. Data **41**, 526 (1996)). Among their data are the following measurements of the density of a mixture of cyclohexane and pentadecane as a function of the mole fraction of cyclohexane (x_c) at 298.15 K.

x_c	0.6965	0.7988	0.9004
$\rho/(\text{g cm}^{-3})$	0.7661	0.7674	0.7697

Compute the partial molar volume for each component in a mixture which has a mole fraction of cyclohexane of 0.7988.

7.15 Comelli and Francesconi examined mixtures of propionic acid with various other organic liquids at 313.15 K (F. Comelli and

R. Francesconi, J. Chem. Eng. Data **41**, 101 (1996)). They report the excess volume of mixing propionic acid with oxane as $V^E = x_1 x_2 \{a_0 + a_1(x_1 - x_2)\}$, where x_1 is the mole fraction of propionic acid, x_2 that of oxane, $a_0 = -2.4697 \text{ cm}^3 \text{ mol}^{-1}$ and $a_1 = 0.0608 \text{ cm}^3 \text{ mol}^{-1}$. The density of propionic acid at this temperature is $0.97174 \text{ g cm}^{-3}$; that of oxane is $0.86398 \text{ g cm}^{-3}$. (a) Derive an expression for the partial molar volume of each component at this temperature. (b) Compute the partial molar volume for each component in an equimolar mixture.

7.16 Francesconi et al. studied the liquid–vapour equilibria of trichloromethane and 1,2-epoxybutane at several temperatures (R. Francesconi, B. Lunelli, and F. Comelli, J. Chem. Eng. Data **41**, 310 (1996)). Among their data are the following measurements of the mole fractions of trichloromethane in the liquid phase (x_T) and the vapour phase (y_T) at 298.15 K as a function of pressure.

p/kPa	23.40	21.75	20.25	18.75	18.15	20.25	22.50	26.30
x	0	0.129	0.228	0.353	0.511	0.700	0.810	1
y	0	0.065	0.145	0.285	0.535	0.805	0.915	1

Compute the activity coefficients of both components on the basis of Raoult's law.

7.17 Chen and Lee studied the liquid–vapour equilibria of cyclohexanol with several gases at elevated pressures (J.-T. Chen and M.-J. Lee, J. Chem. Eng. Data **41**, 339 (1996)). Among their data are the following measurements of the mole fractions of cyclohexanol in the vapour phase (y_{cyc}) and the liquid phase (x_{cyc}) at 393.15 K as a function of pressure.

p/bar	10.0	20.0	30.0	40.0	60.0	80.0
y_{cyc}	0.0267	0.0149	0.0112	0.00947	0.00835	0.00921
x_{cyc}	0.9741	0.9464	0.9204	0.892	0.836	0.773

Determine the Henry's law constant of CO_2 in cyclohexanol, and compute the activity coefficient of CO_2.

7.18 Equation 36 indicates that solubility is an exponential function of temperature. The data in the table below give the solubility, S, of calcium acetate in water as a function of temperature.

$\theta/^\circ\text{C}$	0	20	40	60	80
$S/(\text{mol L}^{-1})$	36.4	34.9	33.7	32.7	31.7

Determine the extent to which the data fit the exponential $S = S_0 e^{\tau/T}$ and obtain values for S_0 and τ. Express these constants in terms of properties of the solute.

8 Phase diagrams

Phase diagrams for pure substances were introduced in Chapter 6. Now we develop their use systematically and show how they are rich summaries of empirical information about a wide range of systems. To set the stage, we introduce the famous phase rule of Gibbs, which shows the extent to which various parameters can be varied yet the equilibrium between phases preserved. With the rule established, we see how it can be used to discuss the phase diagrams that we met in the two preceding chapters. The chapter then introduces systems of gradually increasing complexity. In each case we shall see how the phase diagram for the system summarizes empirical observations on the conditions under which the various phases of the system are stable.

In this chapter we describe a systematic way of discussing the physical changes mixtures undergo when they are heated or cooled and when their compositions are changed. In particular, we shall see how to use phase diagrams to judge whether two substances are mutually miscible, whether an equilibrium can exist over a range of conditions, or whether the system must be brought to a definite pressure, temperature, and composition before equilibrium is established. Phase diagrams are of considerable commercial and industrial significance, particularly for semiconductors, ceramics, steels, and alloys. They are also the basis of separation procedures in the petroleum industry and of the formulation of foods and cosmetic preparations.

Phases, components, and degrees of freedom

All phase diagrams can be discussed in terms of a relationship, the phase rule, derived by J.W. Gibbs. We shall derive this rule first, and then apply it to a wide variety of systems. The phase rule requires a careful use of terms, so we begin by presenting a number of definitions.

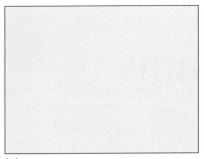

(a)

(b)

8.1 The difference between (a) a single-phase solution, in which the composition is uniform on a microscopic scale, and (b) a dispersion, in which regions of one component are embedded in a matrix of a second component.

8.1 Definitions

The term **phase** was introduced at the start of Chapter 6, where we saw that it signifies a state of matter that is uniform throughout, not only in chemical composition but also in physical state. (The words are Gibbs's.) Thus we speak of the solid, liquid, and gas phases of a substance, and of its various solid phases (as for black phosphorus and white phosphorus). The number of phases in a system is denoted P. A gas, or a gaseous mixture, is a single phase, a crystal is a single phase, and two totally miscible liquids form a single phase. Ice is a single phase ($P = 1$) even though it might be chipped into small fragments. A slurry of ice and water is a two-phase system ($P = 2$) even though it is difficult to map the boundaries between the phases. A system in which calcium carbonate undergoes thermal decomposition consists of two solid phases (one consisting of calcium carbonate and the other of calcium oxide) and one gaseous phase (consisting of carbon dioxide).

An alloy of two metals is a two-phase system ($P = 2$) if the metals are immiscible, but a single-phase system ($P = 1$) if they are miscible. This example shows that it is not always easy to decide whether a system consists of one phase or of two. A solution of solid A in solid B—a homogeneous mixture of the two substances—is uniform on a molecular scale. In a solution, atoms of A are surrounded by atoms of A and B, and any sample cut from the sample, however small, is representative of the composition of the whole.

A dispersion is uniform on a macroscopic scale but not on a microscopic scale, for it consists of grains or droplets of one substance in a matrix of the other. A small sample could come entirely from one of the minute grains of pure A and would not be representative of the whole (Fig. 8.1). Such dispersions are important because, in many advanced materials (including steels), heat treatment cycles are used to achieve the precipitation of a fine dispersion of particles of one phase (such as a carbide phase) within a matrix formed by a saturated solid solution phase. The ability to control this microstructure resulting from phase equilibria makes it possible to tailor the mechanical properties of the materials to a particular application.

By a **constituent** of a system we mean a chemical species (an ion or a molecule) that is present. Thus, a mixture of ethanol and water has two constituents. The term constituent should be carefully distinguished from 'component', which has a more technical meaning. A **component** is a chemically independent constituent of a system. The number of components, C, in a system is the minimum number of independent species necessary to define the composition of *all* the phases present in the system.

When no reaction takes place, the number of components is equal to the number of constituents. Thus, pure water is a one-component system ($C = 1$), because we need only the species H_2O to specify its composition. Similarly, a mixture of ethanol and water is a two-component system ($C = 2$): we need the species H_2O and C_2H_5OH to specify its composition. When a reaction can occur under the conditions prevailing in the system, we need to decide the minimum number of species that, after allowing for reactions in which one species is synthesized from others, can be used to specify the composition of all the phases. Consider, for example, the equilibrium

$$CaCO_3(s) \rightleftharpoons CaO(s) + CO_2(g)$$
Phase 1   Phase 2   Phase 3

in which there are three phases. To specify the composition of the gas phase we need the species CO_2, and to specify the composition of Phase 2 we need the species CaO. However, we do not need an additional species to specify the composition of the Phase 3 because its identity ($CaCO_3$) can be expressed in terms of the other two constituents by making use of the stoichiometry of the reaction. Hence, the system has three constituents but only two components ($C = 2$).

Example 8.1 Counting components

How many components are present in a system in which ammonium chloride undergoes thermal decomposition?

Method Begin by writing down the chemical equation for the reaction and identifying the constituents of the system (all the species present) and the phases. Then decide whether, under the conditions prevailing in the system, any of the constituents can be prepared from any of the other constituents. The removal of these constituents leaves the number of independent constituents. Finally, identify the minimum number of these independent constituents that are needed to specify the composition of all the phases.

Answer The chemical reaction is

$$NH_4Cl(s) \rightleftharpoons NH_3(g) + HCl(g)$$

There are three constituents and two phases (one solid, one gas). However, NH_3 and HCl are formed in fixed stoichiometric proportions by the reaction. Therefore, the compositions of both phases can be expressed in terms of the single species NH_4Cl. It follows that there is only one component in the system ($C = 1$).

Comment If additional HCl (or NH_3) were supplied to the system, the decomposition of NH_4Cl would not give the correct composition of the gas phase and HCl (or NH_3) would have to be invoked as a second component. A system that consists of hydrogen, oxygen, and water at room temperature has three components, despite it being possible to form H_2O from H_2 and O_2: under the conditions prevailing in the system, hydrogen and oxygen do not react to form water, so they are not in equilibrium and regarded as independent constituents.

- -

Self-test 8.1 Give the number of components in the following systems: (a) water, allowing for its autoprotolysis; (b) aqueous acetic acid; (c) magnesium carbonate in equilibrium with its decomposition products.

[(a) 1; (b) 2; (c) 2]

The **variance**, F, of a system is the number of intensive variables that can be changed independently without disturbing the number of phases in equilibrium. In a single-component, single-phase system ($C = 1$, $P = 1$), the pressure and temperature may be changed independently without changing the number of phases, so $F = 2$. We say that such a system is *bivariant*, or that it has two *degrees of freedom*. On the other hand, if two phases are in equilibrium (a liquid and its vapour, for instance) in a single-component system ($C = 1$, $P = 2$), the temperature (or the pressure) can be changed at will, but the change in temperature (or pressure) demands an accompanying change in pressure (or temperature) to preserve the number of phases in equilibrium. That is, the variance of the system has fallen to 1.

8.2 The phase rule

In one of the most elegant calculations of the whole of chemical thermodynamics, J.W. Gibbs[1] deduced the **phase rule**, which is a general relation between the variance, F, the

---

1 Josiah Willard Gibbs spent most of his working life at Yale, and may justly be regarded as the originator of chemical thermodynamics. He reflected for years before publishing his conclusions, and then did so in precisely expressed papers in an obscure journal (*The Transactions of the Connecticut Academy of Arts and Sciences*). He needed interpreters before the power of his work was recognized and before it could be applied to industrial processes. He is regarded by many as the first great American theoretical scientist.

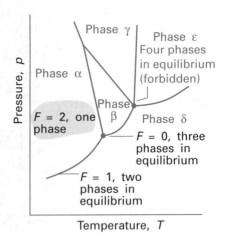

8.2 The typical regions of a one-component phase diagram. The lines represent conditions under which the two adjoining phases are in equilibrium. A point represents the unique set of conditions under which three phases coexist in equilibrium. Four phases cannot mutually coexist in equilibrium.

number of components, C, and the number of phases at equilibrium, P, for a system of any composition:

$$F = C - P + 2 \tag{1}$$

Justification 8.1

We begin by counting the total number of intensive variables (properties that do not depend on the size of the system). The pressure, p, and temperature, T, count as 2. We can specify the composition of a phase by giving the mole fractions of $C - 1$ components. We need specify only $C - 1$ and not all C mole fractions because $x_1 + x_2 + \cdots + x_C = 1$, and all mole fractions are known if all except one are specified. Because there are P phases, the total number of composition variables is $P(C - 1)$. At this stage, the total number of intensive variables is $P(C - 1) + 2$.

At equilibrium, the chemical potential of a component J must be the same in every phase (Section 6.4):

$$\mu_{J,\alpha} = \mu_{J,\beta} = \cdots \qquad \text{for } P \text{ phases}$$

That is, there are $P - 1$ equations of this kind to be satisfied for each component J. As there are C components, the total number of equations is $C(P - 1)$. Each equation reduces our freedom to vary one of the $P(C - 1) + 2$ intensive variables. It follows that the total number of degrees of freedom is

$$F = P(C - 1) + 2 - C(P - 1) = C - P + 2$$

which is eqn 1.

We shall now go on to see how the phase rule summarizes what we already know about one-component systems and then apply it to more complex cases.

(a) One-component systems

For a one-component system, such as pure water, $F = 3 - P$. When only one phase is present, $F = 2$ and both p and T can be varied independently without changing the number of phases. In other words, a single phase is represented by an *area* on a phase diagram. When two phases are in equilibrium $F = 1$, which implies that pressure is not freely variable if the temperature is set; indeed, at a given temperature, a liquid has a characteristic vapour pressure. It follows that the equilibrium of two phases is represented by a *line* in the phase diagram. Instead of selecting the temperature, we could select the pressure, but having done so the two phases would be in equilibrium at a single definite temperature. Therefore, freezing (or any other phase transition) occurs at a definite temperature at a given pressure.

When three phases are in equilibrium, $F = 0$ and the system is invariant. This special condition can be established only at a definite temperature and pressure that is characteristic of the substance and outside our control. The equilibrium of three phases is therefore represented by a *point*, the triple point, on the phase diagram. Four phases cannot be in equilibrium in a one-component system because F cannot be negative. These features are summarized in Fig. 8.2.

The features summarized in the illustration can be identified in the experimentally determined phase diagram for water (Fig. 8.3). This diagram summarizes the changes that take place as a sample, such as that at a, is cooled at constant pressure. The sample remains entirely gaseous until the temperature reaches b, when liquid appears. Two phases are now in equilibrium and $F = 1$. Because we have decided to specify the pressure, which uses up the single degree of freedom, the temperature at which this equilibrium occurs is not under our control. Lowering the temperature takes the system to c in the one-phase, liquid region.

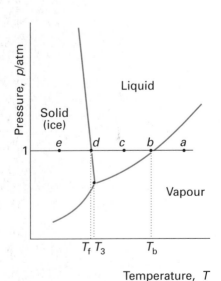

8.3 The phase diagram for water, a simplified version of Fig. 6.5. The label T_3 marks the temperature of the triple point, T_b the normal boiling point, and T_f the normal freezing point.

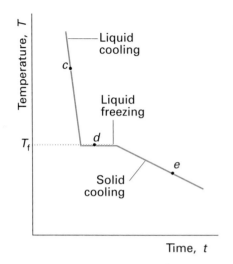

8.4 The cooling curve for the isobar *cde* in Fig. 8.3. The halt marked *d* corresponds to the pause in the fall of temperature while the first-order exothermic transition (freezing) occurs. This pause enables T_f to be located even if the transition cannot be observed visually.

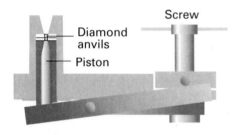

8.5 Ultra-high pressures (up to about 2 Mbar) can be achieved using a diamond anvil. The sample, together with a ruby for pressure measurement and a drop of liquid for pressure transmission, are placed between two gem-quality diamonds. The principle of its action is like a nutcracker: the pressure is exerted by turning the screw by hand.

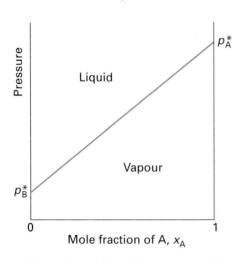

8.6 The variation of the total vapour pressure of a binary mixture with the mole fraction of A in the liquid when Raoult's law is obeyed.

The temperature can now be varied around the point *c* at will, and only when ice appears at *d* does the variance become 1 again.

(b) Experimental procedures

Detecting a phase change is not always as simple as seeing a kettle boil, so special techniques have been developed. One technique is **thermal analysis**, which takes advantage of the effect of the enthalpy change during a first-order transition (Section 6.7). In this method, a sample is allowed to cool and its temperature is monitored. At a first-order transition, heat is evolved and the cooling stops until the transition is complete. The cooling curve along the isobar *cde* in Fig. 8.3 therefore has the shape shown in Fig. 8.4. The transition temperature is obvious, and is used to mark point *d* on the phase diagram. This technique is useful for solid-solid transitions, where simple visual inspection of the sample may be inadequate.

Modern work on phase transitions often deals with systems at very high pressures, and more sophisticated detection procedures must be adopted. Some of the highest pressures currently attainable are produced in a *diamond-anvil cell* like that illustrated in Fig. 8.5. The sample is placed in a minute cavity between two gem-quality diamonds, and then pressure is exerted simply by turning the screw. The advance in design this represents is quite remarkable for, with a turn of the screw, pressures of up to about 1 Mbar can be reached which a few years ago could not be reached with equipment weighing tons.

The pressure is monitored spectroscopically by observing the shift of spectral lines in small pieces of ruby added to the sample, and the properties of the sample itself are observed optically through the diamond anvils. One application of the technique is to study the transition of covalent solids to metallic solids. Iodine, I_2, for instance, becomes metallic at around 200 kbar and makes a transition to a monatomic metallic solid at around 210 kbar. Studies such as these are relevant to the structure of material deep inside the Earth (at the centre of the Earth the pressure is around 5 Mbar) and in the interiors of the giant planets, where even hydrogen may be metallic.

Two-component systems

When two components are present in a system, $C = 2$ and $F = 4 - P$. If the temperature is constant, the remaining variance is $F' = 3 - P$, which has a maximum value of 2. (The prime on F indicates that one of the degrees of freedom has been discarded, in this case the temperature.) One of these two remaining degrees of freedom is the pressure and the other is the composition (as expressed by the mole fraction of one component). Hence, one form of the phase diagram is a map of pressures and compositions at which each phase is stable. Alternatively, the pressure could be held constant and the phase diagram depicted in terms of temperature and composition. We shall introduce both types of diagram.

8.3 Vapour pressure diagrams

The partial vapour pressures of the components of an ideal solution of two volatile liquids are related to the composition of the liquid mixture by Raoult's law (Section 7.3a):

$$p_A = x_A p_A^* \qquad p_B = x_B p_B^* \qquad (2)°$$

where p_A^* is the vapour pressure of pure A and p_B^* that of pure B. The total vapour pressure p of the mixture is therefore

$$p = p_A + p_B = x_A p_A^* + x_B p_B^* = p_B^* + (p_A^* - p_B^*)x_A \qquad (3)°$$

This expression shows that the total vapour pressure (at some fixed temperature) changes linearly with the composition from p_B^* to p_A^* as x_A changes from 0 to 1 (Fig. 8.6).

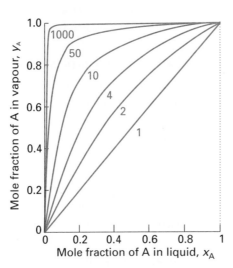

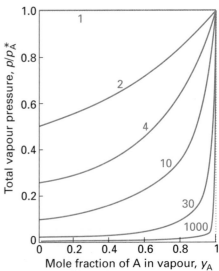

8.7 The mole fraction of A in the vapour of a binary ideal solution expressed in terms of its mole fraction in the liquid, calculated using eqn 5 for various values of p_A^*/p_B^* (the label on each curve) with A more volatile than B. In all cases the vapour is richer than the liquid in A.

8.8 The dependence of the vapour pressure of the same system as in Fig. 8.7, but expressed in terms of the mole fraction of A in the vapour by using eqn 6. Individual curves are labelled with the value of p_A^*/p_B^*.

(a) The composition of the vapour

The compositions of the liquid and vapour that are in mutual equilibrium are not necessarily the same, and common sense suggests that the vapour should be richer in the more volatile component. This expectation can be confirmed as follows. The partial pressures of the components are given by eqn 2. It follows from Dalton's law that the mole fractions in the gas, y_A and y_B, are

$$y_A = \frac{p_A}{p} \qquad y_B = \frac{p_B}{p} \tag{4}$$

Provided the mixture is ideal, the partial pressures and the total pressure may be expressed in terms of the mole fractions in the liquid by using eqn 2 for p_J and eqn 3 for the total vapour pressure p, which gives

$$y_A = \frac{x_A p_A^*}{p_B^* + (p_A^* - p_B^*)x_A} \qquad y_B = 1 - y_A \tag{5}°$$

Figure 8.7 shows the composition of the vapour plotted against the composition of the liquid for various values of $p_A^*/p_B^* > 1$. We see that in all cases $y_A > x_A$, that is, the vapour is richer than the liquid in the more volatile component. Note that if B is non-volatile, so that $p_B^* = 0$ at the temperature of interest, then it makes no contribution to the vapour ($y_B = 0$).

Equation 3 shows how the total vapour pressure of the mixture varies with the composition of the liquid. Because we can relate the composition of the liquid to the composition of the vapour through eqn 5, we can now also relate the total vapour pressure to the composition of the vapour:

$$p = \frac{p_A^* p_B^*}{p_A^* + (p_B^* - p_A^*)y_A} \tag{6}°$$

This expression is plotted in Fig. 8.8.

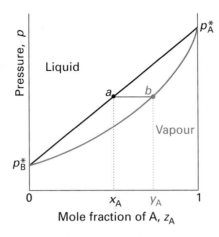

8.9 The dependence of the total vapour pressure of an ideal solution on the mole fraction of A in the entire system. A point between the two lines corresponds to both liquid and vapour being present; outside that region there is only one phase present. The mole fraction of A is denoted z_A, as explained in the text.

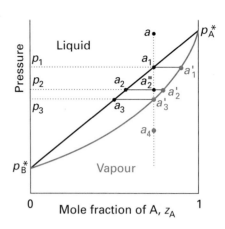

8.10 The points of the pressure–composition diagram discussed in the text. The vertical line through a is an isopleth, a line of constant composition of the entire system.

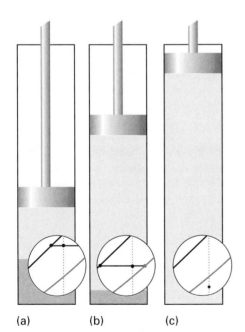

8.11 (a) A liquid in a container exists in equilibrium with its vapour. The superimposed fragment of the phase diagram shows the compositions of the two phases and their abundances (by the lever rule). (b) When the pressure is changed by drawing out a piston, the compositions of the phases adjust as shown by the tie line in the phase diagram. (c) When the piston is pulled so far out that all the liquid has vaporized and only the vapour is present, the pressure falls as the piston is withdrawn and the point on the phase diagram moves into the one-phase region.

(b) The interpretation of the diagrams

If we are interested in distillation, both the vapour and the liquid compositions are of equal interest. It is therefore sensible to combine the two preceding diagrams into one. The result is summarized in Fig. 8.9. The point a indicates the vapour pressure of a mixture of composition x_A, and the point b indicates the composition of the vapour that is in equilibrium with the liquid at that pressure. Note that, when two phases are in equilibrium, $P = 2$ so $F' = 1$ (as usual, the prime indicating that one degree of freedom, the temperature, has already been discarded). That is, if the composition is specified (so using up the only remaining degree of freedom), the pressure at which the two phases are in equilibrium is fixed.

A richer interpretation of the phase diagram is obtained if we interpret the horizontal axis as showing the *overall* composition, z_A, of the system. If the horizontal axis of the vapour pressure diagram is labelled with z_A, then all the points down to the solid diagonal line in the graph correspond to a system which is under such high pressure that it contains only a liquid phase (the applied pressure is higher than the vapour pressure), so $z_A = x_A$, the composition of the liquid. On the other hand, all points below the lower curve correspond to a system which is under such low pressure that it contains only a vapour phase (the applied pressure is lower than the vapour pressure), so $z_A = y_A$.

Points that lie *between* the two lines correspond to a system in which there are two phases present, one a liquid and the other a vapour. To see this interpretation, consider lowering the pressure on a liquid mixture of overall composition a in Fig. 8.10. The lowering of pressure can be achieved by drawing out a piston (Fig. 8.11). This degree of freedom is permitted by the phase rule because $F' = 2$ when $P = 1$, and even if the composition is selected one degree of freedom remains. The changes to the system do not affect the overall composition, so the state of the system moves down the vertical line that passes through a. This vertical line is called an **isopleth**, from the Greek words for 'equal abundance'. Until the point a_1 is reached (when the pressure has been reduced to p_1), the sample consists of a single liquid phase. At a_1 the liquid can exist in equilibrium with its vapour. As we have seen, the composition of the vapour phase is given by point a'_1. The horizontal line joining the two points is called a **tie line**. The composition of the liquid is the same as initially (a_1 lies on the

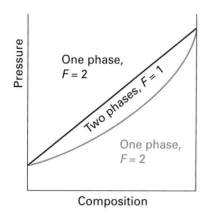

8.12 The general scheme of interpretation of a pressure–composition diagram (a vapour pressure diagram).

isopleth through a), so we have to conclude that at this pressure there is virtually no vapour present; however, the tiny amount of vapour that is present has the composition a_1'.

Now consider the effect of lowering the pressure to p_2, so taking the system to a pressure and overall composition represented by the point a_2''. This new pressure is below the vapour pressure of the original liquid, so it vaporizes until the vapour pressure of the remaining liquid falls to p_2. Now we know that the composition of such a liquid must be a_2. Moreover, the composition of the vapour in equilibrium with that liquid must be given by the point a_2' at the other end of the tie line. Note that two phases are now in equilibrium, so $F' = 1$ for all points between the two lines; hence, for a given pressure (such as at p_2) the variance is zero, and the vapour and liquid phases have fixed compositions (Fig. 8.12). If the pressure is reduced to p_3, a similar readjustment in composition takes place, and now the compositions of the liquid and vapour are represented by the points a_3 and a_3', respectively. The latter point corresponds to a system in which the composition of the vapour is the same as the overall composition, so we have to conclude that the amount of liquid present is now virtually zero, but the tiny amount of liquid that is present has the composition a_3. A further decrease in pressure takes the system to the point a_4; at this stage, only vapour is present and its composition is the same as the initial overall composition of the system (the composition of the original liquid).

(c) The lever rule

A point in the two-phase region of a phase diagram indicates not only qualitatively that both liquid and vapour are present, but represents quantitatively the relative amounts of each. To find the relative amounts of two phases α and β that are in equilibrium, we measure the distances l_α and l_β along the horizontal tie line, and then use the lever rule (Fig. 8.13):

$$n_\alpha l_\alpha = n_\beta l_\beta \tag{7}$$

where n_α is the amount of phase α and n_β the amount of phase β. In the case illustrated in Fig. 8.13, because $l_\beta \approx 2l_\alpha$, the amount of phase α is about twice the amount of phase β.

Justification 8.2

To prove the lever rule we write $n = n_\alpha + n_\beta$ and the overall amount of A as nz_A. The overall amount of A is also the sum of its amounts in the two phases:

$$nz_A = n_\alpha x_A + n_\beta y_A$$

Since also

$$nz_A = n_\alpha z_A + n_\beta z_A$$

by equating these two expressions it follows that

$$n_\alpha(x_A - z_A) = n_\beta(z_A - y_A)$$

which corresponds to eqn 7.

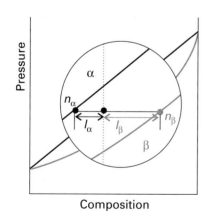

8.13 The lever rule. The distances l_α and l_β are used to find the proportions of the amounts of phases α (such as vapour) and β (for example, liquid) present at equilibrium. The lever rule is so called because a similar rule relates the masses at two ends of a lever to their distances from a pivot ($m_\alpha l_\alpha = m_\beta l_\beta$ for balance).

Illustration
. . . .
At p_1 in Fig. 8.10, the ratio l_{vap}/l_{liq} is almost infinite for this tie line, so n_{liq}/n_{vap} is also almost infinite, and there is only a trace of vapour present. When the pressure is reduced to p_2, the value of l_{vap}/l_{liq} is about 0.3, so $n_{liq}/n_{vap} \approx 0.3$ and the amount of liquid is about 0.3 times the amount of vapour. When the pressure has been reduced to p_3, the sample is almost completely gaseous and, because $l_{vap}/l_{liq} \approx 0$, we conclude that there is only a trace of liquid present.
. . . .

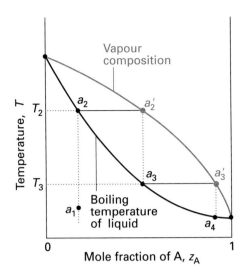

8.14 The temperature–composition diagram corresponding to an ideal mixture with the component A more volatile than component B. Successive boilings and condensations of a liquid originally of composition a_1 lead to a condensate that is pure A. The separation technique is called fractional distillation.

8.4 Temperature–composition diagrams

To discuss distillation we need a **temperature–composition diagram**, a phase diagram in which the boundaries show the compositions of the phases that are in equilibrium at various temperatures (and a given pressure, typically 1 atm). An example is shown in Fig. 8.14. Note that the liquid phase now lies in the lower part of the diagram.

(a) The distillation of mixtures

The region between the lines in Fig. 8.14 is a two-phase region where $F' = 1$ (as usual, the prime indicates that one degree of freedom has been discarded; in this case, the pressure is being kept fixed), and hence at a given temperature the compositions of the phases in equilibrium are fixed. The regions outside the phase lines correspond to a single phase, so $F' = 2$, and the temperature and composition are both independently variable.

Consider what happens when a liquid of composition a_1 is heated. It boils when the temperature reaches T_2. Then the liquid has composition a_2 (the same as a_1) and the vapour (which is present only as a trace) has composition a_2'. The vapour is richer in the more volatile component A (the component with the lower boiling point). From the location of a_2, we can state the vapour's composition at the boiling point, and from the location of the tie line joining a_2 and a_2' we can read off the boiling temperature (T_2) of the original liquid mixture.

In a simple distillation, the vapour is withdrawn and condensed. If the vapour in this example is drawn off and completely condensed, the first drop gives a liquid of composition a_3, which is richer in the more volatile component, A, than the original liquid. In **fractional distillation**, the boiling and condensation cycle is repeated successively. We can follow the changes that occur by seeing what happens when the condensate of composition a_3 is reheated. The phase diagram shows that this mixture boils at T_3 and yields a vapour of composition a_3', which is even richer in the more volatile component. That vapour is drawn off, and the first drop condenses to a liquid of composition a_4. The cycle can then be repeated until in due course almost pure A is obtained.

The efficiency of a fractionating column is expressed in terms of the number of **theoretical plates**, the number of effective vaporization and condensation steps that are required to achieve a condensate of given composition from a given distillate. Thus, to achieve the degree of separation shown by the dotted lines in Fig. 8.15a, the fractionating column must correspond to 3 theoretical plates. To achieve the same separation for the system shown in

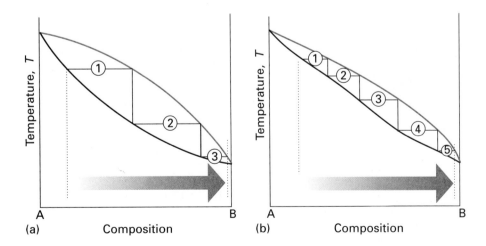

8.15 The number of theoretical plates is the number of steps needed to bring about a specified degree of separation of two components in a mixture. The two systems shown correspond to (a) 3, (b) 5 theoretical plates.

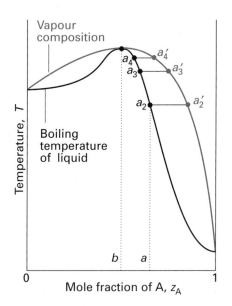

8.16 A high-boiling azeotrope. When the liquid of composition a is distilled, the composition of the remaining liquid changes towards b but no further.

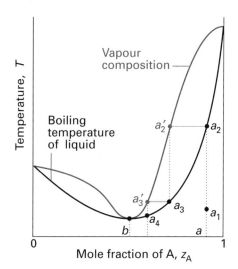

8.17 A low-boiling azeotrope. When the mixture at a is fractionally distilled, the vapour in equilibrium in the fractionating column moves towards b and then remains unchanged.

Fig 8.15b, in which the components have more similar partial vapour pressures, the fractionating column must be designed to correspond to five theoretical plates.

(b) Azeotropes

Although many liquids have temperature–composition phase diagrams resembling the ideal version in Fig. 8.14, in a number of important cases there are marked deviations. A maximum in the phase diagram (Fig. 8.16) may occur when the favourable interactions between A and B molecules reduce the vapour pressure of the mixture below the ideal value: in effect, the A–B interactions stabilize the liquid. In such cases the excess Gibbs energy, G^E (Section 7.4), is negative (more favourable to mixing than ideal). Examples of this behaviour include trichloromethane/propanone and nitric acid/water mixtures. Phase diagrams showing a minimum (Fig. 8.17) indicate that the mixture is destabilized relative to the ideal solution, the A–B interactions then being unfavourable. For such mixtures G^E is positive (less favourable to mixing than ideal), and there may be contributions from both enthalpy and entropy effects. Examples include dioxane/water and ethanol/water mixtures.

Deviations from ideality are not always so strong as to lead to a maximum or minimum in the phase diagram, but when they do there are important consequences for distillation. Consider a liquid of composition a on the right of the maximum in Fig. 8.16. The vapour (at a_2') of the boiling mixture (at a_2) is richer in A. If that vapour is removed (and condensed elsewhere), then the remaining liquid will move to a composition that is richer in B, such as that represented by a_3, and the vapour in equilibrium with this mixture will have composition a_3'. As that vapour is removed, the composition of the boiling liquid shifts to a point such as a_4, and the composition of the vapour shifts to a_4'. Hence, as evaporation proceeds, the composition of the remaining liquid shifts towards B as A is drawn off. The boiling point of the liquid rises, and the vapour becomes richer in B. When so much A has been evaporated that the liquid has reached the composition b, the vapour has the same composition as the liquid. Evaporation then occurs without change of composition. The mixture is said to form an **azeotrope** (which comes from the Greek words for 'boiling without changing'). When the azeotropic composition has been reached, distillation cannot separate the two liquids because the condensate has the same composition as the azeotropic liquid. One example of azeotrope formation is hydrochloric acid/water, which is azeotropic at 80 per cent by mass of water and boils unchanged at 108.6 °C.

The system shown in Fig. 8.17 is also azeotropic, but shows its azeotropy in a different way. Suppose we start with a mixture of composition a_1, and follow the changes in the composition of the vapour that rises through a fractionating column (essentially a vertical glass tube packed with glass rings to give a large surface area). The mixture boils at a_2 to give a vapour of composition a_2'. This vapour condenses in the column to a liquid of the same composition (now marked a_3). That liquid reaches equilibrium with its vapour at a_3', which condenses higher up the tube to give a liquid of the same composition, which we now call a_4. The fractionation therefore shifts the vapour towards the azeotropic composition at b, but not beyond, and the azeotropic vapour emerges from the top of the column. An example is ethanol/water, which boils unchanged when the water content is 4 per cent by mass and the temperature is 78 °C.

(c) Immiscible liquids

Finally we consider the distillation of two immiscible liquids, such as octane and water. At equilibrium, there is a tiny amount of A dissolved in B, and similarly a tiny amount of B dissolved in A: both liquids are saturated with the other component (Fig. 8.18a). As a result, the total vapour pressure of the mixture is close to $p = p_A^* + p_B^*$. If the temperature is raised to the value at which this total vapour pressure is equal to the atmospheric pressure, boiling

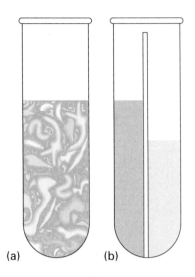

8.18 The distillation of (a) two immiscible liquids can be regarded as (b) the joint distillation of the separated components, and boiling occurs when the sum of the partial pressures equals the external pressure.

commences and the dissolved substances are purged from their solution. However, this boiling results in a vigorous agitation of the mixture, so each component is kept saturated in the other component, and the purging continues as the very dilute solutions are replenished. This intimate contact is essential: two immiscible liquids heated in a container like that shown in Fig. 8.18b would not boil at the same temperature. The presence of the saturated solutions means that the 'mixture' boils at a lower temperature than either component would alone because boiling begins when the total vapour pressure reaches 1 atm, not when either vapour pressure reaches 1 atm. This distinction is the basis of **steam distillation**, which enables some heat-sensitive, water-insoluble organic compounds to be distilled at a lower temperature than their normal boiling point. The only snag is that the composition of the condensate is in proportion to the vapour pressures of the components, so oils of low volatility distil in low abundance.

8.5 Liquid–liquid phase diagrams

Now we consider temperature-composition diagrams for systems that consist of pairs of **partially miscible liquids**, which are liquids that do not mix in all proportions at all temperatures. An example is hexane and nitrobenzene. The same principles of interpretation apply as to liquid-vapour diagrams. When $P = 2, F' = 1$ (the prime denoting the adoption of constant pressure), and the selection of a temperature implies that the compositions of the immiscible liquid phases are fixed. When $P = 1$ (corresponding to a system in which the two liquids are fully mixed), both the temperature and the composition may be adjusted.

(a) Phase separation

Suppose a small amount of a liquid B is added to a sample of another liquid A at a temperature T'. It dissolves completely, and the binary system remains a single phase. As more B is added, a stage comes at which no more dissolves. The sample now consists of two phases in equilibrium with each other ($P = 2$), the most abundant one consisting of A saturated with B, the minor one a trace of B saturated with A. In the temperature-composition diagram drawn in Fig. 8.19, the composition of the former is represented by the point a' and that of the latter by the point a''. The relative abundances of the two phases are given by the lever rule.

When more B is added, A dissolves in it slightly. The compositions of the two phases in equilibrium remain a' and a'' (because $P = 2$ implies that $F' = 0$, and hence that the compositions of the phases are invariant at a fixed temperature and pressure), but the amount of one phase increases at the expense of the other. A stage is reached when so much B is present that it can dissolve all the A, and the system reverts to a single phase. The addition of more B now simply dilutes the solution, and from then on it remains a single phase.

The compositions of the two phases at equilibrium vary with the temperature. For hexane and nitrobenzene, raising the temperature increases their miscibility. The two-phase system therefore becomes less extensive, because each phase in equilibrium is richer in its minor component: the A-rich phase is richer in B and the B-rich phase is richer in A. The entire phase diagram can be constructed by repeating the observations at different temperatures and drawing the envelope of the two-phase region.

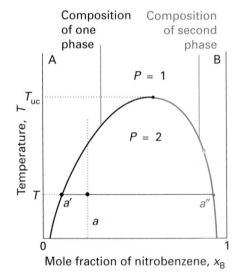

8.19 The temperature–composition diagram for hexane and nitrobenzene at 1 atm. The region below the curve corresponds to the compositions and temperatures at which the liquids form two phases. The upper critical temperature, T_{uc}, is the temperature above which the two liquids are miscible in all proportions.

Example 8.2 Interpreting a liquid–liquid phase diagram

A mixture of 50 g of hexane (0.59 mol C_6H_{14}) and 50 g nitrobenzene (0.41 mol $C_6H_5NO_2$) was prepared at 290 K. What are the compositions of the phases, and in what proportions

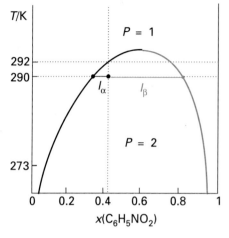

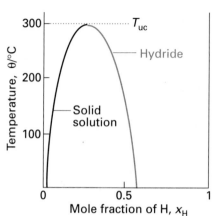

8.20 The temperature–composition diagram for hexane and nitrobenzene at 1 atm again, with the points and lengths discussed in the text.

8.21 The phase diagram for palladium and palladium hydride, which has an upper critical temperature at 300 °C.

do they occur? To what temperature must the sample be heated in order to obtain a single phase?

Method The compositions of phases in equilibrium are given by the points where the tie line through the point representing the temperature and overall composition of the system intersects the phase boundary. Their proportions are given by the lever rule (eqn 7). The temperature at which the components are completely miscible is found by following the isopleth upwards and noting the temperature at which it enters the one-phase region of the phase diagram.

Answer We denote hexane by H and nitrobenzene by N; refer to Fig. 8.20, which is a simplified version of Fig. 8.19. The point $x_N = 0.41$, $T = 290$ K occurs in the two-phase region of the phase diagram. The horizontal tie line cuts the phase boundary at $x_N = 0.35$ and $x_N = 0.83$, so those are the compositions of the two phases. The ratio of the amounts of each phase is equal to the ratio of the distances l_α and l_β:

$$\frac{n_\alpha}{n_\beta} = \frac{l_\beta}{l_\alpha} = \frac{0.83 - 0.41}{0.41 - 0.35} = \frac{0.42}{0.06} = 7$$

That is, there is about 7 times more nitrobenzene-rich phase than hexane-rich phase. Heating the sample to 292 K takes it into the single-phase region.

Comment Because the phase diagram has been constructed experimentally, these conclusions are not based on any assumptions about ideality. They would be modified if the system were subjected to a different pressure.

- -

Self-test 8.2 Repeat the problem for 50 g of hexane and 100 g of nitrobenzene at 273 K.
$$[x_N = 0.09 \text{ and } 0.95 \text{ in ratio } 1:1.3;\ 294\text{ K}]$$

(b) Critical solution temperatures

The **upper critical solution temperature**, T_{uc}, is the highest temperature at which phase separation occurs.[2] Above the upper critical temperature the two components are fully miscible. This temperature exists because the greater thermal motion overcomes any

- -

2   The upper critical solution temperature is also called the 'upper consolute temperature'.

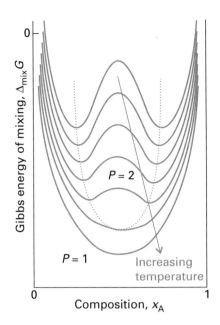

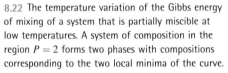

8.22 The temperature variation of the Gibbs energy of mixing of a system that is partially miscible at low temperatures. A system of composition in the region $P = 2$ forms two phases with compositions corresponding to the two local minima of the curve.

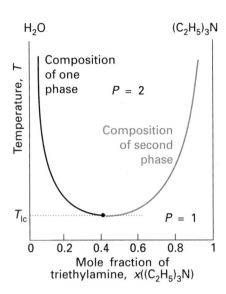

8.23 The temperature–composition diagram for water and triethylamine. This system shows a lower critical temperature at 292 K. The labels indicate the interpretation of the boundaries.

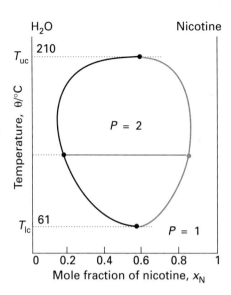

8.24 The temperature–composition diagram for water and nicotine, which has both upper and lower critical temperatures. Note the high temperatures for the liquid (especially the water): the diagram corresponds to a sample under pressure.

potential energy advantage in molecules of one type being close together. One example is the nitrobenzene/hexane system shown in Fig. 8.19; an example of a solid solution is the palladium/hydrogen system, which shows two phases, one a solid solution of hydrogen in palladium and the other a palladium hydride, up to 300 °C but which forms a single phase at higher temperatures (Fig. 8.21).

The thermodynamic interpretation of the upper critical solution temperature focuses on the Gibbs energy of mixing and its variation with temperature. The Gibbs energy of mixing of a partially miscible system behaves as shown in Fig. 8.22. The double minima in the curves indicate the compositions of the partially miscible phases. As the temperature rises, the two minima blend together and merge into a single minimum at the upper critical solution temperature.

Some systems show a **lower critical solution temperature**, T_{lc}, below which they mix in all proportions and above which they form two phases.[3] An example is water and triethylamine (Fig. 8.23). In this case, at low temperatures the two components are more miscible because they form a weak complex; at higher temperatures the complexes break up and the two components are less miscible.

Some systems have both upper and lower critical solution temperatures. They occur because, after the weak complexes have been disrupted, leading to partial miscibility, the thermal motion at higher temperatures homogenizes the mixture again, just as in the case of ordinary partially miscible liquids. The most famous example is nicotine and water, which are partially miscible between 61 °C and 210 °C (Fig. 8.24).

(c) The distillation of partially miscible liquids

Consider a pair of liquids that are partially miscible and form a low-boiling azeotrope. This combination is quite common because both properties reflect the tendency of the two kinds

---

3   The lower critical solution temperature is also called the 'lower consolute temperature'.

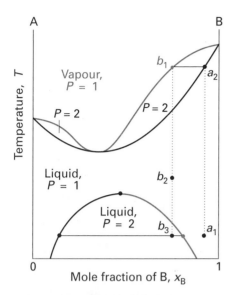

8.25 The temperature–composition diagram for a binary system in which the upper critical temperature is less than the boiling point at all compositions. The mixture forms a low-boiling azeotrope.

of molecule to avoid each other. There are two possibilities: one in which the liquids become fully miscible before they boil, the other in which boiling occurs before mixing is complete.

Figure 8.25 shows the phase diagram for two components that become fully miscible before they boil. Distillation of a mixture of composition a_1 leads to a vapour of composition b_1, which condenses to the completely miscible single-phase solution at b_2. Phase separation occurs only when this distillate is cooled to a point in the two-phase liquid region, such as b_3. This description applies only to the first drop of distillate. If distillation continues, the composition of the remaining liquid changes. In the end, when the whole sample has evaporated and condensed, the composition is back to a_1.

Figure 8.26 shows the second possibility, in which there is no upper critical solution temperature. The distillate obtained from a liquid initially of composition a_1 has composition b_3 and is a two-phase mixture. One phase has composition b_3' and the other has composition b_3''.

The behaviour of a system of composition represented by the isopleth e in Fig. 8.26 is interesting. A system at e_1 forms two phases, which persist (but with changing proportions) up to the boiling point at e_2. The vapour of this mixture has the same composition as the liquid (the liquid is an azeotrope). Similarly, condensing a vapour of composition e_3 gives a two-phase liquid of the same overall composition. At a fixed temperature, the mixture vaporizes and condenses like a single substance.

Example 8.3 Interpreting a phase diagram

State the changes that occur when a mixture of composition $x_B = 0.95$ (a_1) in Fig. 8.27 is boiled and the vapour condensed.

Method The area in which the point lies gives the number of phases; the compositions of the phases are given by the points at the intersections of the horizontal tie line with the phase boundaries; the relative abundances are given by the lever rule (eqn 7).

Answer The initial point is in the one-phase region. When heated it boils at 350 K (a_2) giving a vapour of composition $x_B = 0.66$ (b_1). The liquid gets richer in B, and the last drop (of pure B) evaporates at 392 K. The boiling range of the liquid is therefore 350 to 392 K. If the initial vapour is drawn off, it has a composition $x_B = 0.66$. This composition would be maintained if the sample were very large, but for a finite sample it shifts to higher values and ultimately to $x_B = 0.95$. Cooling the distillate corresponds to moving down the $x_B = 0.66$ isopleth. At 330 K, for instance, the liquid phase has composition $x_B = 0.87$, the vapour $x_B = 0.49$; their relative proportions are 1:3. At 320 K the sample is entirely liquid, and consists of three phases: the vapour and two liquids. One liquid phase has composition $x_B = 0.30$; the other has composition $x_B = 0.80$ in the ratio 0.62:1. Further cooling moves the system into the two-phase region, and at 298 K the compositions are 0.20 and 0.90 in the ratio 0.82:1. As further distillate boils over, the overall composition of the distillate becomes richer in B. When the last drop has been condensed, the phase composition is the same as at the beginning.

- -

Self-test 8.3 Repeat the discussion, beginning at the point $x_B = 0.4$, $T = 298$ K.

8.6 Liquid–solid phase diagrams

Solid and liquid phases may both be present in a system at temperatures below the boiling point. An example is a pair of metals that are almost completely immiscible right up to their melting points (such as antimony and bismuth).

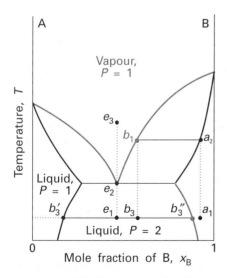

8.26 The temperature–composition diagram for a binary system in which boiling occurs before the two liquids are fully miscible.

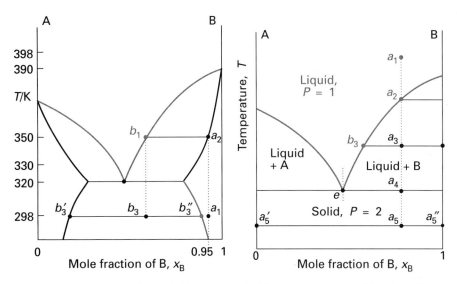

8.27 The points of the phase diagram in Fig. 8.26 that are discussed in Example 8.3.

8.28 The temperature–composition phase diagram for two almost immiscible solids and their completely miscible liquids. Note the similarity to Fig. 8.26. The isopleth through e corresponds to the eutectic composition, the mixture with lowest melting point.

Consider the two-component liquid of composition a_1 in Fig. 8.28. The changes that occur may be expressed as follows.

(1) $a_1 \rightarrow a_2$. The system enters the two-phase region labelled 'Liquid + B'. Pure solid B begins to come out of solution and the remaining liquid becomes richer in A.

(2) $a_2 \rightarrow a_3$. More of the solid forms, and the relative amounts of the solid and liquid (which are in equilibrium) are given by the lever rule. At this stage there are roughly equal amounts of each. The liquid phase is richer in A than before (its composition is given by b_3) because some B has been deposited.

(3) $a_3 \rightarrow a_4$. At the end of this step, there is less liquid than at a_3, and its composition is given by e. This liquid now freezes to give a two-phase system of pure B and pure A.

(a) Eutectics

The isopleth at e in Fig. 8.28 corresponds to the **eutectic composition**, the name coming from the Greek words for 'easily melted'. A liquid with the eutectic composition freezes at a single temperature, without previously depositing solid A or B. A solid with the eutectic composition melts, without change of composition, at the lowest temperature of any mixture. Solutions of composition to the right of e deposit B as they cool, and solutions to the left deposit A: only the eutectic mixture (apart from pure A or pure B) solidifies at a single definite temperature ($F' = 0$ when $C = 2$ and $P = 3$) without gradually unloading one or other of the components from the liquid.

One technologically important eutectic is solder, which has mass composition of about 67 per cent tin and 33 per cent lead and melts at $183\,°C$. The eutectic formed by 23 per cent NaCl and 77 per cent H_2O by mass melts at $-21.1\,°C$. When salt is added to ice under isothermal conditions (for example, when spread on an icy road) the mixture melts if the temperature is above $-21.1\,°C$ (and the eutectic composition has been achieved). When salt is added to ice under adiabatic conditions (for example, when added to ice in a vacuum flask) the ice melts, but in doing so it absorbs heat from the rest of the mixture. The temperature of the system falls and, if enough salt is added, cooling continues down to the eutectic temperature. Eutectic formation occurs in the great majority of binary alloy systems, and is

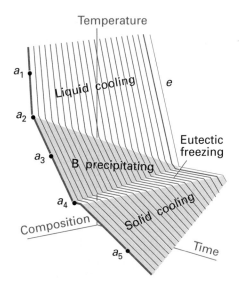

8.29 The cooling curves for the system shown in Fig. 8.28. For isopleth a, the rate of cooling slows at a_2 because solid B deposits from solution. There is a complete halt at a_4 while the eutectic solidifies. This halt is longest for the eutectic isopleth, e. The eutectic halt shortens again for compositions beyond e (richer in A). Cooling curves are used to construct the phase diagram.

of great importance for the microstructure of solid materials. Although a eutectic solid is a two-phase system, it crystallizes out in a nearly homogeneous mixture of microcrystals. The two microcrystalline phases can be distinguished by microscopy and structural techniques such as X-ray diffraction.

Thermal analysis is a very useful practical way of detecting eutectics. We can see how it is used by considering the rate of cooling down the isopleth through a_1 in Fig. 8.28. The liquid cools steadily until it reaches a_2, when B begins to be deposited (Fig. 8.29). Cooling is now slower because the solidification of B is exothermic and retards the cooling. When the remaining liquid reaches the eutectic composition, the temperature remains constant ($F' = 0$) until the whole sample has solidified: this region of constant temperature is the **eutectic halt**. If the liquid has the eutectic composition e initially, the liquid cools steadily down to the freezing temperature of the eutectic, when there is a long eutectic halt as the entire sample solidifies (like the freezing of a pure liquid).

Monitoring the cooling curves at different overall compositions gives a clear indication of the structure of the phase diagram. The solid–liquid boundary is given by the points at which the rate of cooling changes. The longest eutectic halt gives the location of the eutectic composition and its melting temperature.

(b) Reacting systems

Many binary mixtures react to produce compounds, and technologically important examples of this behaviour include the III/V semiconductors, such as the gallium arsenide system, which forms the compound GaAs. Although three constituents are present, there are only two components because GaAs is formed from the reaction $Ga + As \rightleftharpoons GaAs$. We shall illustrate some of the principles involved with a system that forms a compound C which also forms eutectic mixtures with the species A and B (Fig. 8.30).

A system prepared by mixing an excess of B with A consists of C and unreacted B. This is a binary B, C system, which we suppose forms a eutectic. The principal change from the eutectic phase diagram in Fig. 8.28 is that the whole of the phase diagram is squeezed into the range of compositions lying between equal amounts of A and B ($x_B = 0.5$, marked C in Fig. 8.30) and pure B. The interpretation of the information in the diagram is obtained in the same way as for Fig. 8.28. The solid deposited on cooling along the isopleth a is the compound C. At temperatures below a_4 there are two solid phases, one consisting of C and the other of B.

(c) Incongruent melting

In some cases the compound C is not stable as a liquid. An example is the alloy Na_2K, which survives only as a solid (Fig. 8.31). Consider what happens as a liquid at a_1 is cooled:

(1) $a_1 \rightarrow a_2$. Some solid Na is deposited, and the remaining liquid is richer in K.
(2) $a_2 \rightarrow$ just below a_3. The sample is now entirely solid, and consists of solid Na and solid Na_2K.

Now consider the isopleth through b_1:

(1) $b_1 \rightarrow b_2$. No obvious change occurs until the phase boundary is reached at b_2 when solid Na begins to deposit.
(2) $b_2 \rightarrow b_3$. Solid Na deposits, but at b_3 a reaction occurs to form Na_2K: this compound is formed by the K atoms diffusing into the solid Na.

At this stage the liquid Na/K mixture is in equilibrium with a little solid Na_2K, but there is still no liquid compound.

(3) $b_3 \rightarrow b_4$. As cooling continues, the amount of solid compound increases until at b_4 the liquid reaches its eutectic composition. It then solidifies to give a two-phase solid consisting of solid K and solid Na_2K.

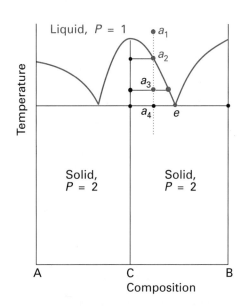

8.30 The phase diagram for a system in which A and B react to form a compound C = AB. This diagram resembles two versions of Fig. 8.28 in each half of the diagram. The constituent C is a true compound, not just an equimolar mixture.

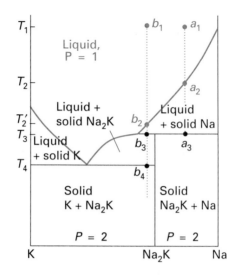

8.31 The phase diagram for an actual system (sodium and potassium) like that shown in Fig. 8.30, but with two differences. One is that the compound is Na₂K, corresponding to A₂B and not AB as in that illustration. The second is that the compound exists only as the solid, not as the liquid. The transformation of the compound at its melting point is an example of incongruent melting.

If the solid is reheated, the sequence of events is reversed. No liquid Na_2K forms at any stage because it is too unstable to exist as a liquid. This behaviour is an example of **incongruent melting**, in which a compound melts into its components and does not itself form a liquid phase.

8.7 Ultrapurity and controlled impurity

Advances in technology have called for materials of extreme purity. For example, semiconductor devices consist of almost perfectly pure silicon or germanium doped to a precisely controlled extent. For these materials to operate successfully, the impurity level must be kept down to less than 1 ppb (1 part in 10^9, which corresponds to a small grain of salt in 5 tons of sugar).

In the technique of **zone refining** the sample is in the form of a narrow cylinder. This cylinder is heated in a thin disk-like zone which is swept from one end of the sample to the other. The advancing liquid zone accumulates the impurities as it passes. In practice, a train of hot and cold zones is swept repeatedly from one end to the other (Fig. 8.32). The zone at the end of the sample is the impurity dump: when the heater has gone by, it cools to a dirty solid which can be discarded.

The technique makes use of the non-equilibrium properties of the system. It relies on the impurities being more soluble in the molten sample than in the solid, and sweeps them up by passing a molten zone repeatedly from one end to the other along a sample. The phase diagram in Fig. 8.33 gives some insight into the process. Consider a liquid (this represents the molten zone) on the isopleth through a_1, and let it cool without the entire sample coming to overall equilibrium. If the temperature falls to a_2 a solid of composition b_2 is deposited and the remaining liquid (the zone where the heater has moved on) is at a_2'. Cooling that liquid down an isopleth passing through a_2' deposits solid of composition b_3 and leaves liquid at a_3'. The process continues until the last drop of liquid to solidify is heavily contaminated with B. There is plenty of everyday evidence that impure liquids freeze in this way. For example, an ice cube is clear near the surface but misty in the core: the water used to make ice normally contains dissolved air; freezing proceeds from the outside, and air is accumulated in the

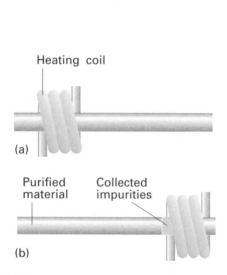

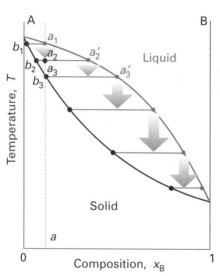

8.32 The procedure for zone refining. (a) Initially, impurities are distributed uniformly along the sample. (b) After a molten zone is passed along the rod, the impurities are more concentrated at the right. In practice, a series of molten zones is passed along the rod from left to right.

8.33 A binary temperature–composition diagram can be used to discuss zone refining, as explained in the text.

retreating liquid phase. It cannot escape from the interior of the cube, and so when that freezes it occludes the air in a mist of tiny bubbles.

A modification of zone refining is **zone levelling**. It is used to introduce controlled amounts of impurity (for example, of indium into germanium). A sample rich in the required dopant is put at the head of the main sample, and made molten. The zone is then dragged repeatedly in alternate directions through the sample, where it deposits a uniform distribution of the impurity.

Checklist of key ideas

Phases, components, and degrees of freedom

8.1 Definitions
- [ ] phase
- [ ] constituent
- [ ] component
- [ ] variance
- [ ] degree of freedom

8.2 The phase rule
- [ ] phase rule (1)
- [ ] application of the phase rule to one-component systems
- [ ] thermal analysis
- [ ] high-pressure studies

Two-component systems

8.3 Vapour pressure diagrams
- [ ] the vapour pressure of a mixture (2,3)
- [ ] the composition of the vapour (5)
- [ ] the total vapour pressure of a mixture (6)
- [ ] the overall composition and the interpretation of phase diagrams
- [ ] isopleth
- [ ] tie line
- [ ] the lever rule (7)

8.4 Temperature–composition diagrams
- [ ] temperature–composition diagram

- [ ] fractional distillation
- [ ] theoretical plate
- [ ] azeotropes and the effect on distillation
- [ ] the behaviour of immiscible liquids
- [ ] steam distillation

8.5 Liquid–liquid phase diagrams
- [ ] partially miscible liquids
- [ ] phase separation and the interpretation of a liquid–liquid phase diagram
- [ ] upper critical solution temperature
- [ ] lower critical solution temperature
- [ ] the interpretation of distillation of partially

miscible liquids in terms of phase diagrams

8.6 Liquid–solid phase diagrams
- [ ] the interpretation of solid–liquid phase diagrams
- [ ] eutectic composition
- [ ] eutectic halt
- [ ] cooling curve
- [ ] the effect of reactions on the appearance of phase diagrams
- [ ] incongruent melting

8.7 Ultrapurity and controlled impurity
- [ ] zone refining
- [ ] zone levelling

Further reading

Articles of general interest

M. Zhao, Z. Wang, and L. Xiao, Determining the number of independent components by Brinkley's method. *J. Chem. Educ.* **69**, 539 (1992).

R.J. Stead and K. Stead, Phase diagrams for ternary liquid systems. *J. Chem. Educ.* **67**, 385 (1990).

J. Kochansky, Liquid systems with more than two immiscible phases. *J. Chem. Educ.* **68**, 655 (1991).

N.K. Kildahl, Journey around a phase diagram. *J. Chem. Educ.* **71**, 1052 (1994).

R. Battino, The critical point and the number of degrees of freedom. *J. Chem. Educ.* **68**, 276 (1991).

A. Jayaraman, The diamond-anvil high-pressure cell. *Scientific American* **250**(4), 42 (1984).

W.B. Daniels, High-pressure techniques. In *Encyclopedia of applied physics* (ed. G.L. Trigg), **7**, 495. VCH, New York (1995).

A.D. Pelton, Phase equilibria. In *Encyclopedia of applied physics* (ed. G.L. Trigg), **13**, 297. VCH, New York (1995).

J. Zhou, W.S.W. Ho, and N.N. Li, Separation processes. In *Encyclopedia of applied physics* (ed. G.L. Trigg), **17**, 599. VCH, New York (1996).

Texts and sources of data and information

S.I. Sandler, *Chemical and engineering thermodynamics*. Wiley, New York (1989).

A. Alper, *Phase diagrams*, Vols 1, 2, and 3. Academic Press, New York (1970).

A. Reisman, *Phase equilibria*. Academic Press, New York (1970).

J. Wisniak, *Phase diagrams: a literature source book*. Elsevier, Amsterdam (1981–86).

S.I. Sandler, *Models for thermodynamic and phase equilibria*. Marcel Dekker, New York (1994).

Exercises

8.1 (a) At 90°C, the vapour pressure of methylbenzene is 400 Torr and that of 1,2-dimethylbenzene is 150 Torr. What is the composition of a liquid mixture that boils at 90°C when the pressure is 0.50 atm? What is the composition of the vapour produced?

8.1 (b) At 90°C, the vapour pressure of 1,2-dimethylbenzene is 20 kPa and that of 1,3-dimethylbenzene is 18 kPa. What is the composition of a liquid mixture that boils at 90°C when the pressure is 19 kPa? What is the composition of the vapour produced?

8.2 (a) The vapour pressure of pure liquid A at 300 K is 575 Torr and that of pure liquid B is 390 Torr. These two compounds form ideal liquid and gaseous mixtures. Consider the equilibrium composition of a mixture in which the mole fraction of A in the vapour is 0.350. Calculate the total pressure of the vapour and the composition of the liquid mixture.

8.2 (b) The vapour pressure of pure liquid A at 293 K is 68.8 kPa and that of pure liquid B is 82.1 kPa. These two compounds form ideal liquid and gaseous mixtures. Consider the equilibrium composition of a mixture in which the mole fraction of A in the vapour is 0.612. Calculate the total pressure of the vapour and the composition of the liquid mixture.

8.3 (a) It is found that the boiling point of a binary solution of A and B with $x_A = 0.6589$ is 88°C. At this temperature the vapour pressures of pure A and B are 957.0 Torr and 379.5 Torr, respectively. (a) Is this solution ideal? (b) What is the initial composition of the vapour above the solution?

8.3 (b) It is found that the boiling point of a binary solution of A and B with $x_A = 0.4217$ is 96°C. At this temperature the vapour pressures of pure A and B are 110.1 kPa and 94.93 kPa, respectively. (a) Is this solution ideal? (b) What is the initial composition of the vapour above the solution?

8.4 (a) Dibromoethene (DE, $p_{DE}^* = 172$ Torr at 358 K) and dibromopropene (DP, $p_{DP}^* = 128$ Torr at 358 K) form a nearly ideal solution. If $z_{DE} = 0.60$, what is (a) p_{total} when the system is all liquid, (b) the composition of the vapour when the system is still almost all liquid?

8.4 (b) Benzene and toluene form nearly ideal solutions. At 20°C the vapour pressures of pure benzene and toluene are 74 Torr and 22 Torr, respectively. A solution consisting of 1.00 mol of each component is boiled by reducing the external pressure below the vapour pressure. Calculate (a) the pressure when boiling begins, (b) the composition of each component in the vapour, and (c) the vapour pressure when only a few drops of liquid remain. Assume that the rate of vaporization is low enough for the temperature to remain constant at 20°C.

8.5 (a) The following temperature/composition data were obtained for a mixture of octane (O) and methylbenzene (M) at 760 Torr, where x is the mole fraction in the liquid and y the mole fraction in the vapour at equilibrium.

$\theta/°C$	110.9	112.0	114.0	115.8	117.3	119.0	121.1	123.0
x_M	0.908	0.795	0.615	0.527	0.408	0.300	0.203	0.097
y_M	0.923	0.836	0.698	0.624	0.527	0.410	0.297	0.164

The boiling points are 110.6°C and 125.6°C for M and O, respectively. Plot the temperature/composition diagram for the mixture. What is the composition of the vapour in equilibrium with the liquid of composition (a) $x_M = 0.250$ and (b) $x_O = 0.250$?

8.5 (b) The following temperature/composition data were obtained for a mixture of two liquids A and B at 1.00 atm, where x is the mole fraction in the liquid and y the mole fraction in the vapour at equilibrium.

$\theta/°C$	125	130	135	140	145	150
x_A	0.91	0.65	0.45	0.30	0.18	0.098
y_A	0.99	0.91	0.77	0.61	0.45	0.25

The boiling points are 124°C for A and 155°C for B. Plot the temperature/composition diagram for the mixture. What is the composition of the vapour in equilibrium with the liquid of composition (a) $x_A = 0.50$ and (b) $x_B = 0.33$?

8.6 (a) State the number of components in the following systems. (a) NaH_2PO_4 in water at equilibrium with water vapour but disregarding the fact that the salt is ionized. (b) The same, but taking into account the ionization of the salt.

8.6 (b) State the number of components for a system in which $AlCl_3$ is dissolved in water, noting that hydrolysis and precipitation of $Al(OH)_3$ occur.

8.7 (a) Blue $CuSO_4 \cdot 5H_2O$ crystals release their water of hydration when heated. How many phases and components are present in an otherwise empty heated container?

8.7 (b) Ammonium chloride, NH_4Cl, decomposes when it is heated. (a) How many components and phases are present when the salt is heated in an otherwise empty container? (b) Now suppose that additional ammonia is also present. How many components and phases are present?

8.8 (a) A saturated solution of Na_2SO_4, with excess of the solid, is present at equilibrium with its vapour in a closed vessel. (a) How many phases and components are present? (b) What is the variance (the number of degrees of freedom) of the system? Identify the independent variables.

8.8 (b) Suppose that the solution referred to in Exercise 8.8a is not saturated. (a) How many phases and components are present? (b) What is the variance (the number of degrees of freedom) of the system? Identify the independent variables.

8.9 (a) Draw phase diagrams for the following types of systems. Label the regions and intersections of the diagrams, stating what materials (possibly compounds or azeotropes) are present and whether they are solid, liquid, or gas. (a) One-component, pressure-temperature diagram; liquid density greater than that of solid. (b) Two-component, temperature-composition, solid-liquid diagram; one compound AB formed that melts congruently; negligible solid-solid solubility.

8.9 (b) Draw phase diagrams for the following types of systems. Label the regions and intersections of the diagrams, stating what materials (possibly compounds or azeotropes) are present and whether they are solid, liquid, or gas. (a) Two-component, temperature-composition, solid-liquid diagram; one compound of formula AB_2 that melts incongruently; negligible solid-solid solubility. (b) Two-component, temperature-composition, liquid-vapour diagram; formation of an azeotrope at $x_B = 0.333$; complete miscibility.

8.10 (a) Label the regions of the phase diagram in Fig. 8.34. State what substances (if compounds, give their formulas) exist in each region. Label each substance in each region as solid, liquid, or gas.

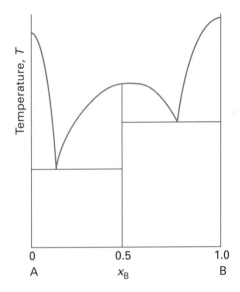

Fig. 8.34

8.10 (b) Label the regions of the phase diagram in Fig. 8.35. State what substances (if compounds, give their formulas) exist in each region. Label each substance in each region as solid, liquid, or gas.

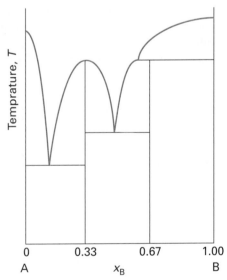

Fig. 8.35

8.11 (a) Methylethyl ether (A) and diborane, B_2H_6 (B), form a compound which melts congruently at 133 K. The system exhibits two eutectics, one at 25 mole per cent B and 123 K and a second at

90 mole per cent B and 104 K. The melting points of pure A and B are 131 K and 110 K, respectively. Sketch the phase diagram for this system. Assume negligible solid-solid solubility.

8.11 (b) Sketch the phase diagram of the system NH_3/N_2H_4 given that the two substances do not form a compound with each other, that NH_3 freezes at $-78\,^{\circ}C$ and N_2H_4 freezes at $+2\,^{\circ}C$, and that a eutectic is formed when the mole fraction of N_2H_4 is 0.07 and that the eutectic melts at $-80\,^{\circ}C$.

8.12 (a) Figure 8.36 shows the phase diagram for two partially miscible liquids, which can be taken to be that for water (A) and 2-methyl-1-propanol (B). Describe what will be observed when a mixture of composition $x_B = 0.8$ is heated, at each stage giving the number, composition, and relative amounts of the phases present.

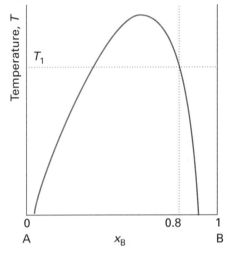

Fig. 8.36

8.12 (b) Figure 8.37 is the phase diagram for silver and tin. Label the regions, and describe what will be observed when liquids of compositions a and b are cooled to 200 K.

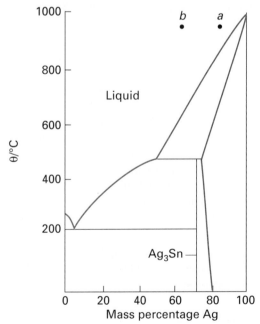

Fig. 8.37

8.13 (a) Indicate on the phase diagram in Fig. 8.38 the feature that denotes incongruent melting. What is the composition of the eutectic mixture and at what temperature does it melt?

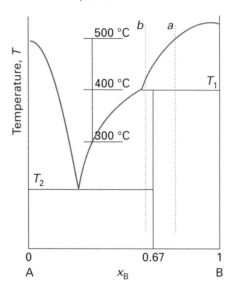

Fig. 8.38

8.13 (b) Indicate on the phase diagram in Fig. 8.39 the feature that denotes incongruent melting. What is the composition of the eutectic mixture and at what temperature does it melt?

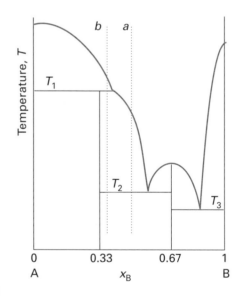

Fig. 8.39

8.14 (a) Sketch the cooling curves for the isopleths a and b in Fig. 8.38.

8.14 (b) Sketch the cooling curves for the isopleths a and b in Fig. 8.39.

8.15 (a) Use the phase diagram in Fig. 8.37 to state (a) the solubility of Ag in Sn at 800 °C, (b) the solubility of Ag_3Sn in Ag at 460 °C, (c) the solubility of Ag_3Sn in Ag at 300 °C.

8.15 (b) Use the phase diagram in Fig. 8.38 to state (a) the solubility of B in A at 500 °C, (b) the solubility of AB_2 in A at 390 °C, (c) the solubility of AB_2 in B at 300 °C.

8.16 (a) Figure 8.40 shows the experimentally determined phase diagrams for the nearly ideal solution of hexane and heptane. (a) Label the regions of the diagrams to which phases are present. (b) For a solution containing 1 mol each of hexane and heptane, estimate the vapour pressure at 70 °C when vaporization on reduction of the external pressure just begins. (c) What is the vapour pressure of the solution at 70 °C when just one drop of liquid remains? (d) Estimate from the figures the mole fraction of hexane in the liquid and vapour phases for the conditions of part b. (e) What are the mole fractions for the conditions of part c? (f) At 85 °C and 760 Torr, what are the amounts of substance in the liquid and vapour phases when $z_{Heptane} = 0.40$?

8.16 (b) Uranium tetrafluoride and zirconium tetrafluoride melt at 1035 °C and 912 °C, respectively. They form a continuous series of solid solutions with a minimum melting temperature of 765 °C and composition $x(ZrF_4) = 0.77$. At 900 °C, the liquid solution of composition $x(ZrF_4) = 0.28$ is in equilibrium with a solid solution of composition $x(ZrF_4) = 0.14$. At 850 °C the two compositions are 0.87 and 0.90, respectively. Sketch the phase diagram for this system and state what is observed when a liquid of composition $x(ZrF_4) = 0.40$ is cooled slowly from 900 °C to 500 °C.

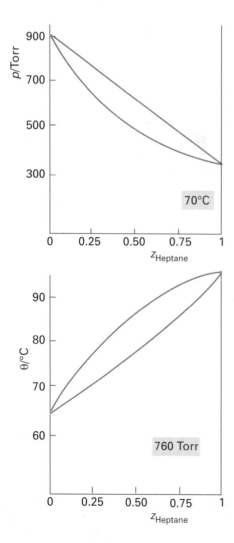

Fig. 8.40

8.17 (a) Methane (melting point 91 K) and tetrafluoromethane (melting point 89 K) do not form solid solutions with each other, and as liquids they are only partially miscible. The upper critical temperature of the liquid mixture is 94 K at $x(CF_4) = 0.43$ and the eutectic temperature is 84 K at $x(CF_4) = 0.88$. At 86 K, the phase in equilibrium with the tetrafluoromethane-rich solution changes from solid methane to a methane-rich liquid. At that temperature, the two liquid solutions that are in mutual equilibrium have the compositions $x(CF_4) = 0.10$ and $x(CF_4) = 0.80$. Sketch the phase diagram.

8.17 (b) Describe the phase changes that take place when a liquid mixture of 4.0 mol B_2H_6 (melting point 131 K) and 1.0 mol CH_3OCH_3 (melting point 135 K) is cooled from 140 K to 90 K. These substances form a compound $(CH_3)_2OB_2H_6$ that melts congruently at 133 K. The system exhibits one eutectic at $x(B_2H_6) = 0.25$ and 123 K and another at $x(B_2H_6) = 0.90$ and 104 K.

8.18 (a) Refer to the information in Exercise 8.17b and sketch the cooling curves for liquid mixtures in which $x(B_2H_6)$ is (a) 0.10, (b) 0.30, (c) 0.50, (d) 0.80, and (e) 0.95.

8.18 (b) Refer to the information in Exercise 8.17a and sketch the cooling curves for liquid mixtures in which $x(CF_4)$ is (a) 0.10, (b) 0.30, (c) 0.50, (d) 0.80, and (e) 0.95.

8.19 (a) Hexane and perfluorohexane show partial miscibility below 22.70°C. The critical concentration at the upper critical temperature is $x = 0.355$, where x is the mole fraction of C_6F_{14}. At 22.0°C the two solutions in equilibrium have $x = 0.24$ and $x = 0.48$, respectively, and at 21.5°C the mole fractions are 0.22 and 0.51. Sketch the phase diagram. Describe the phase changes that occur when perfluorohexane is added to a fixed amount of hexane at (a) 23°C, (b) 22°C.

8.19 (b) Two liquids, A and B, show partial miscibility below 52.4°C. The critical concentration at the upper critical temperature is $x = 0.459$, where x is the mole fraction of A. At 40.0°C the two solutions in equilibrium have $x = 0.22$ and $x = 0.60$, respectively, and at 42.5°C the mole fractions are 0.24 and 0.48. Sketch the phase diagram. Describe the phase changes that occur when B is added to a fixed amount of A at (a) 48°C, (b) 52.4°C.

Problems

Numerical problems

8.1 The compound p-azoxyanisole forms a liquid crystal. 5.0 g of the solid was placed in a tube, which was then evacuated and sealed. Use the phase rule to prove that the solid will melt at a definite temperature and that the liquid crystal phase will make a transition to a normal liquid phase at a definite temperature.

8.2 Magnesium oxide and nickel oxide withstand high temperatures. However, they do melt when the temperature is high enough and the behaviour of mixtures of the two is of considerable interest to the ceramics industry. Draw the temperature–composition diagram for the system using the data below, where x is the mole fraction of MgO in the solid and y its mole fraction in the liquid.

$\theta/°C$	1960	2200	2400	2600	2800
x	0	0.35	0.60	0.83	1.00
y	0	0.18	0.38	0.65	1.00

State (a) the melting point of a mixture with $x = 0.30$, (b) the composition and proportion of the phases present when a solid of composition $x = 0.30$ is heated to 2200°C, (c) the temperature at which a liquid of composition $y = 0.70$ will begin to solidify.

8.3 The bismuth–cadmium phase diagram is of interest in metallurgy, and its general form can be estimated from expressions for the depression of freezing point. Construct the diagram using the following data: $T_f(Bi) = 544.5$ K, $T_f(Cd) = 594$ K, $\Delta_{fus}H(Bi) = 10.88$ kJ mol^{-1}, $\Delta_{fus}H(Cd) = 6.07$ kJ mol^{-1}. The metals are mutually insoluble as solids. Use the phase diagram to state what would be observed when a liquid of composition $x(Bi) = 0.70$ is cooled slowly from 550 K. What are the relative abundances of the liquid and solid at (a) 460 K and (b) 350 K? Sketch the cooling curve for the mixture.

8.4 Phosphorus and sulfur form a series of binary compounds. The best characterized are P_4S_3, P_4S_7, and P_4S_{10}, all of which melt congruently. Assuming that only these three binary compounds of the two elements exist, (a) draw schematically the P/S phase diagram. Label each region of the diagram with the substance that exists in that region and indicate its phase. Label the horizontal axis as x_S and give the numerical values of x_S that correspond to the compounds. The melting point of pure phosphorus is 44°C and that of pure sulfur is 119°C. (b) Draw, schematically, the cooling curve for a mixture of composition $x_S = 0.28$. Assume that a eutectic occurs at $x_S = 0.2$ and assume negligible solid–solid solubility.

8.5 The table below gives the break and halt temperatures found in the cooling curves of two metals A and B. Construct a phase diagram consistent with the data of these curves. Label the regions of the diagram, stating what phases and substances are present. Give the probable formulas of any compounds that form.

$100x_B$	$\theta_{break}/°C$	$\theta_{halt,1}/°C$	$\theta_{halt,2}/°C$
0		1100	
10.0	1060	700	
20.0	1000	700	
30.0	940	700	400
40.0	850	700	400
50.0	750	700	400
60.0	670	400	
70.0	550	400	
80.0		400	
90.0	450	400	
100.0		500	

8.6 Consider the phase diagram in Fig. 8.41, which represents a solid–liquid equilibrium. Label all regions of the diagram according to the chemical species that exist in that region and their phases. Indicate the number of species and phases present at the points labelled b, d, e, f, g, and k. Sketch cooling curves for compositions $x_B = 0.16, 0.23, 0.57, 0.67$, and 0.84.

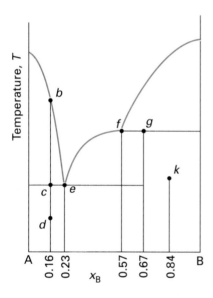

Fig. 8.41

8.7 Solutions of 3-methylphenylamine (MP) were made up in glycerol and then warmed from room temperature. The mixture became turbid at θ_1 and then cleared at θ_2. Plot the phase diagram using the data below, and find the upper and lower critical temperatures.

Mass%	18	20	40	60	80	85
$\theta_1/°C$	48	18	8	10	19	25
$\theta_2/°C$	53	90	120	118	83	53

Mass% denotes the mass percentage composition of MP. State what happens as MP is added dropwise to glycerol at 60°C. State the number of phases present at each composition and their relative amounts.

8.8 Sketch the phase diagram for the Mg/Cu system using the following information: $\theta_f(Mg) = 648°C$, $\theta_f(Cu) = 1085°C$; two intermetallic compounds are formed with $\theta_f(MgCu_2) = 800°C$ and $\theta_f(Mg_2Cu) = 580°C$; eutectics of mass percentage Mg composition and melting points 10 per cent (690°C), 33 per cent (560°C), and 65 per cent (380°C). A sample of Mg/Cu alloy containing 25 per cent Mg by mass was prepared in a crucible heated to 800°C in an inert atmosphere. Describe what will be observed if the melt is cooled slowly to room temperature. Specify the composition and relative abundances of the phases and sketch the cooling curve.

8.9 Iron(II) chloride (melting point 677°C) and potassium chloride (melting point 776°C) form the compounds $KFeCl_3$ and K_2FeCl_4 at elevated temperatures. $KFeCl_3$ melts congruently at 380°C and K_2FeCl_4 melts incongruently at 399°C. Eutectics are formed with compositions $x = 0.38$ (melting point 351°C) and $x = 0.54$ (melting point 393°C), where x is the mole fraction of $FeCl_2$. The KCl solubility curve intersects the K_2FeCl_4 curve at $x = 0.34$. Sketch the phase diagram. State the phases that are in equilibrium when a mixture of composition $x = 0.36$ is cooled from 400°C to 300°C.

Theoretical problems

8.10 Show that two phases are in thermal equilibrium only if their temperatures are the same.

8.11 Show that two phases are in mechanical equilibrium only if their pressures are equal.

Additional problems supplied by Carmen Giunta and Charles Trapp

8.12 1-Butanol and chlorobenzene form a minimum-boiling azeotropic system. The mole fraction of 1-butanol in the liquid (x) and vapour (y) phases at 1.000 atm is given below for a variety of boiling temperatures (H. Artigas, C. Lafuente, P. Cea, F.M. Royo, and J.S. Urieta, *J. Chem. Eng. Data* **42**, 132 (1997)).

T/K	396.57	393.94	391.60	390.15	389.03	388.66	388.57
x	0.1065	0.1700	0.2646	0.3687	0.5017	0.6091	0.7171
y	0.2859	0.3691	0.4505	0.5138	0.5840	0.6409	0.7070

Pure chlorobenzene boils at 404.86 K. (a) Construct the chlorobenzene-rich portion of the phase diagram from the data. (b) Estimate the temperature at which a solution whose mole fraction of 1-butanol is 0.300 begins to boil. (c) State the compositions and relative proportions of the two phases present after a solution initially 0.300 1-butanol is heated to 393.94 K.

8.13 Carbon dioxide at high pressure is used to separate various compounds in citrus oil. The mole fraction of CO_2 in the liquid (x) and vapour (y) at 323.2 K is given below for a variety of pressures (Y. Iwai, T. Morotomi, K. Sakamoto, Y. Koga, and Y. Arai, *J. Chem. Eng. Data* **41**, 951 (1996)).

p/MPa	3.94	6.02	7.97	8.94	9.27
x	0.2873	0.4541	0.6650	0.7744	0.8338
y	0.9982	0.9980	0.9973	0.9958	0.9922

(a) Plot the portion of the phase diagram represented by these data. (b) State the compositions and relative proportions of the two phases present after an equimolar gas mixture is compressed to 6.02 MPa at 323.2 K.

8.14 An *et al.* investigated the liquid–liquid coexistence curve of N,N-dimethylacetamide and heptane (X. An, H. Zhao, F. Fuguo, and W. Shen, *J. Chem. Thermodynamics* **28**, 1221 (1996)). Mole fractions of N,N-dimethylacetamide in the upper (x_1) and lower (x_2) phases of a two-phase region are given below as a function of temperature.

T/K	309.820	309.422	309.031	308.006	306.686
x_1	0.473	0.400	0.371	0.326	0.293
x_2	0.529	0.601	0.625	0.657	0.690
T/K	304.553	301.803	299.097	296.000	294.534
x_1	0.255	0.218	0.193	0.168	0.157
x_2	0.724	0.758	0.783	0.804	0.814

(a) Plot the phase diagram. (b) State the proportions and compositions of the two phases that form from mixing 0.750 mol of N,N-dimethylacetamide with 0.250 mol of heptane at 296.0 K. To what temperature must the mixture be heated to form a single-phase mixture?

8.15 The following data have been obtained for the liquid–vapour equilibrium compositions of mixtures of nitrogen and oxygen at 100 kPa.

T/K	77.3	78	80	82	84	86	88	90.2
$x(O_2)$	0	10	34	54	70	82	92	100
$y(O_2)$	0	2	11	22	35	52	73	100
$p^*(O_2)/\text{Torr}$	154	171	225	294	377	479	601	760

Plot the data on a temperature–composition diagram and determine the extent to which it fits the predictions for an ideal solution by calculating the activity coefficients of O_2 at each composition.

9

Chemical equilibrium

This chapter develops the concept of chemical potential and shows how it can be used to account for the equilibrium composition of chemical reactions. The equilibrium composition corresponds to a minimum in the Gibbs energy plotted against the extent of reaction, and by locating this minimum we establish the relation between the equilibrium constant and the standard Gibbs energy of reaction. The thermodynamic formulation of equilibrium enables us to establish the quantitative effects of changes in pressure and temperature. The final section of the chapter applies the information to three important types of equilibria.

Chemical reactions move towards a dynamic equilibrium in which both reactants and products are present but have no further tendency to undergo net change. In some cases, the concentration of products in the equilibrium mixture is so much greater than the concentration of unchanged reactants that for all practical purposes the reaction is 'complete'. However, in many important cases the equilibrium mixture has significant concentrations of both reactants and products. In this chapter we see how to use thermodynamics to predict the equilibrium composition under any reaction conditions.

Spontaneous chemical reactions

We have seen that the direction of spontaneous change at constant temperature and pressure is towards lower values of the Gibbs energy, G. The idea is entirely general, and in this chapter we apply it to the discussion of reactions.

9.1 The Gibbs energy minimum

We locate the equilibrium composition of a reaction mixture by calculating the Gibbs energy of the reaction mixture and identifying the composition that corresponds to minimum G.

(a) The reaction Gibbs energy

We begin with the simplest possible chemical equilibrium: $A \rightleftharpoons B$. Even though this reaction looks trivial, there are many examples of it, such as the isomerization of pentane to 2-methylbutane and the conversion of L-alanine to D-alanine. Suppose an infinitesimal amount $d\xi$ of A turns into B, then the change in the amount of A present is $dn_A = -d\xi$ and the change in the amount of B present is $dn_B = +d\xi$. The quantity ξ is called the **extent of reaction**; it has the dimensions of amount of substance, and is reported in moles. When the extent of reaction changes by a finite amount $\Delta\xi$, the amount of A present changes from $n_{A,0}$ to $n_{A,0} - \Delta\xi$ and the amount of B changes from $n_{B,0}$ to $n_{B,0} + \Delta\xi$. So, if initially 2.0 mol A is present and we wait until $\Delta\xi = 1.5$ mol, the amount of A remaining will be 0.5 mol.

The **reaction Gibbs energy**, $\Delta_r G$, is defined as the slope of the graph of the Gibbs energy plotted against the extent of reaction:

$$\Delta_r G = \left(\frac{\partial G}{\partial \xi}\right)_{p,T} \tag{1}$$

Although Δ normally signifies a *difference* in values, Δ_r signifies a *derivative*, the slope of G with respect to ξ. However, to see that there is a close relationship with the normal usage, suppose the reaction advances by $d\xi$. The corresponding change in Gibbs energy is

$$dG = \mu_A\, dn_A + \mu_B\, dn_B = -\mu_A\, d\xi + \mu_B\, d\xi = (\mu_B - \mu_A)\, d\xi$$

This equation can be reorganized into

$$\left(\frac{\partial G}{\partial \xi}\right)_{p,T} = \mu_B - \mu_A$$

That is,

$$\Delta_r G = \mu_B - \mu_A \tag{2}$$

We see that $\Delta_r G$ can be interpreted as the difference between the chemical potentials of the reactants and products *at the composition of the reaction mixture*.

Because the chemical potentials vary with composition, the slope of the plot of Gibbs energy against extent of reaction changes as the reaction proceeds. Moreover, because the reaction runs in the direction of decreasing G (that is, down the slope of G versus ξ), we see from eqn 2 that the reaction $A \rightarrow B$ is spontaneous when $\mu_A > \mu_B$, whereas the reverse reaction is spontaneous when $\mu_B > \mu_A$. The slope is zero, and the reaction is spontaneous in neither direction, when

$$\Delta_r G = 0 \tag{3}$$

This condition occurs when $\mu_B = \mu_A$ (Fig. 9.1). It follows that, if we can find the composition of the reaction mixture that ensures $\mu_B = \mu_A$, then we can identify the composition of the reaction mixture at equilibrium.

(b) Exergonic and endergonic reactions

The spontaneity of a reaction at constant temperature and pressure is expressed by the reaction Gibbs energy:

If $\Delta_r G < 0$, the forward reaction is spontaneous.
If $\Delta_r G > 0$, the reverse reaction is spontaneous.
If $\Delta_r G = 0$, the reaction is at equilibrium.

Reactions for which $\Delta_r G < 0$ are called **exergonic** (from the Greek words for work-producing). The name signifies that, because they are spontaneous, they can be used to drive other processes, such as other reactions, or used to do non-expansion work. Reactions for which $\Delta_r G > 0$ are called **endergonic** (signifying work-consuming). They can be made to

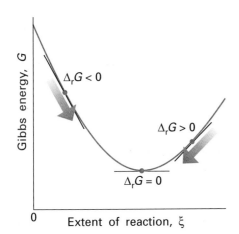

9.1 As the reaction advances (represented by motion from left to right along the horizontal axis) the slope of the Gibbs energy changes. Equilibrium corresponds to zero slope, at the foot of the valley.

occur only by doing work on them (like electrolysing water to reverse its spontaneous formation reaction). Reactions at equilibrium are spontaneous in neither direction: they are neither exergonic nor endergonic.

(c) Perfect gas equilibria

When A and B are perfect gases we can use eqn 5.20 ($\mu = \mu^{\ominus} + RT \ln p$, with p interpreted as $p/p^{\ominus}$) to write

$$
\begin{aligned}
\Delta_{\rm r} G &= \mu_{\rm B} - \mu_{\rm A} \\
&= \left(\mu_{\rm B}^{\ominus} + RT \ln p_{\rm B}\right) - \left(\mu_{\rm A}^{\ominus} + RT \ln p_{\rm A}\right) \\
&= \Delta_{\rm r} G^{\ominus} + RT \ln \left(\frac{p_{\rm B}}{p_{\rm A}}\right)
\end{aligned}
\tag{4}
$$

If we denote the ratio of partial pressures by Q, we obtain

$$
\Delta_{\rm r} G = \Delta_{\rm r} G^{\ominus} + RT \ln Q \qquad Q = \frac{p_{\rm B}}{p_{\rm A}}
\tag{5}
$$

The ratio Q is an example of a **reaction quotient**. It ranges from 0 (pure A) to infinity (pure B). The **standard reaction Gibbs energy**, $\Delta_{\rm r} G^{\ominus}$, is defined (like the standard reaction enthalpy) as the difference in the standard molar Gibbs energies of the reactants and products. For our reaction

$$
\Delta_{\rm r} G^{\ominus} = G_{\rm B,m}^{\ominus} - G_{\rm A,m}^{\ominus}
\tag{6}
$$

In Section 4.7 we saw that the difference in standard molar Gibbs energies of the products and reactants is equal to the difference in their standard Gibbs energies of formation, so in practice we calculate $\Delta_{\rm r} G^{\ominus}$ from

$$
\Delta_{\rm r} G^{\ominus} = \Delta_{\rm f} G^{\ominus}({\rm B}) - \Delta_{\rm f} G^{\ominus}({\rm A})
\tag{7}°
$$

At equilibrium $\Delta_{\rm r} G = 0$. The ratio of partial pressures at equilibrium is denoted K, and eqn 5 becomes

$$
0 = \Delta_{\rm r} G^{\ominus} + RT \ln K
$$

which rearranges to

$$
RT \ln K = -\Delta_{\rm r} G^{\ominus} \qquad K = \left(\frac{p_{\rm B}}{p_{\rm A}}\right)_{\rm equilibrium}
\tag{8}°
$$

This relation is a special case of one of the most important equations in chemical thermodynamics: it is the link between tables of thermodynamic data, such as those in the *Data section* at the end of this volume, and the chemically important equilibrium constant, K.

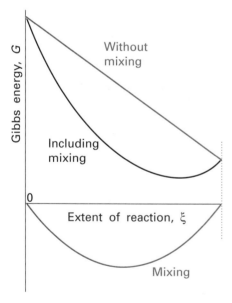

9.2 If the mixing of reactants and products is ignored, the Gibbs energy changes linearly from its initial value (pure reactants) to its final value (pure products) and the slope of the line is $\Delta_{\rm r} G^{\ominus}$. However, as products are produced, there is a further contribution to the Gibbs energy arising from their mixing (lowest curve). The sum of the two contributions has a minimum. That minimum corresponds to the equilibrium composition of the system.

Molecular interpretation 9.1 In molecular terms, the minimum in the Gibbs energy, which corresponds to $\Delta_{\rm r} G = 0$, stems from the Gibbs energy of mixing of the two gases. Hence, an important contribution to the position of chemical equilibrium is the mixing of the products with the reactants as the products are formed.

Consider the hypothetical reaction in which the A molecules change to B molecules without mingling together. Then the Gibbs energy of the system would change from $G^{\ominus}({\rm A})$ to $G^{\ominus}({\rm B})$ in proportion to the amount of B that had been formed, and the slope of the plot of G against the extent of reaction would be constant and equal to $\Delta_{\rm r} G^{\ominus}$ at all stages of the reaction (Fig. 9.2). However, in fact the newly produced B molecules mix with the surviving

A molecules. We have seen that the contribution of a mixing process to the change in Gibbs energy (eqn 7.17) is

$$\Delta_{mix}G = nRT(x_A \ln x_A + x_B \ln x_B)$$

This expression makes a U-shaped contribution to the total change in Gibbs energy. As can be seen from the illustration, there is now a minimum in the Gibbs energy, and its position corresponds to the equilibrium composition of the reaction mixture.

We see from eqn 8 that, when $\Delta_r G^{\ominus} > 0$, $K < 1$. Therefore, at equilibrium the partial pressure of A exceeds that of B, which means that the reactant A is favoured in the equilibrium. When $\Delta_r G^{\ominus} < 0$, $K > 1$, so at equilibrium the partial pressure of B exceeds that of A. Now the product B is favoured in the equilibrium.

(d) The general case of a reaction

The argument that led to eqn 8 can easily be extended to a general reaction. First, we need to generalize the concept of extent of reaction. We define ξ so that, if the change in it is $\Delta\xi$, then the change in the amount of any species J is $\nu_J \Delta\xi$, where ν_J is the stoichiometric number of J in the chemical equation.[1]

Illustration

Consider the reaction

$$N_2(g) + 3H_2(g) \longrightarrow 2NH_3(g) \tag{9}$$

The stoichiometric numbers are $\nu_{N_2} = -1$, $\nu_{H_2} = -3$, and $\nu_{NH_3} = +2$. Therefore, if initially there is 10 mol N_2 present, then, when the extent of reaction changes from $\xi = 0$ to $\xi = 1$ mol so that $\Delta\xi = +1$ mol, the amount of N_2 changes from 10 mol to 9 mol. All the N_2 has been consumed when $\xi = 10$ mol. When $\Delta\xi = 1$ mol, the amount of H_2 changes by $-3 \times (1 \text{ mol}) = -3$ mol and the amount of NH_3 changes by $+2 \times (1 \text{ mol}) = +2$ mol.

The reaction Gibbs energy, $\Delta_r G$, is defined in the same way as before, eqn 1. In the *Justification* below, we show that the Gibbs energy of reaction can always be written

$$\Delta_r G = \Delta_r G^{\ominus} + RT \ln Q \tag{10}$$

with the standard reaction Gibbs energy calculated from

$$\Delta_r G^{\ominus} = \sum_{Products} \nu\Delta_f G^{\ominus} - \sum_{Reactants} \nu\Delta_f G^{\ominus} \tag{11}$$

or, more formally,

$$\Delta_r G^{\ominus} = \sum_J \nu_J \Delta_f G^{\ominus}(J) \tag{12}$$

The reaction quotient, Q, has the form

$$Q = \frac{\text{activities of products}}{\text{activities of reactants}} \tag{13}$$

with each species raised to the power given by its stoichiometric coefficient. More formally, to write the general expression for Q we introduce the symbol $\prod$ to denote the product of what follows it (just as $\sum$ denotes the sum), and write

$$Q = \prod_J a_J^{\nu_J} \tag{[14]}$$

1   Recall that stoichiometric numbers are positive for products and negative for reactants.

Because reactants have negative stoichiometric numbers, they automatically appear as the denominator when the product is written out explicitly. Recall from Table 7.3 that, for pure solids and liquids, the activity is 1, so such substances make no contribution to Q even though they may appear in the chemical equation. For a gas, $a_J = f_J/p^{\ominus}$, where f_J is its fugacity. For a perfect gas, $f_J = p_J$, the partial pressure of J.

Justification 9.1

Consider the reaction

$$2A + 3B \longrightarrow C + 2D$$

When the reaction advances by $d\xi$, the amounts of reactants and products change as follows:

$$dn_A = -2d\xi \qquad dn_B = -3d\xi \qquad dn_C = +d\xi \qquad dn_D = +2d\xi$$

and in general $dn_J = \nu_J\, d\xi$. The resulting infinitesimal change in the Gibbs energy at constant temperature and pressure is

$$dG = \mu_C\, dn_C + \mu_D\, dn_D + \mu_A\, dn_A + \mu_B\, dn_B$$
$$= (\mu_C + 2\mu_D - 2\mu_A - 3\mu_B)\, d\xi$$

The general form of this expression is

$$dG = \left(\sum_J \nu_J\mu_J\right) d\xi \tag{15}$$

It follows that

$$\Delta_r G = \left(\frac{\partial G}{\partial \xi}\right)_{p,T} = -2\mu_A - 3\mu_B + \mu_C + 2\mu_D \tag{16}$$

To make further progress, we note that the chemical potential of a species J is related to its activity by

$$\mu_J = \mu_J^{\ominus} + RT \ln a_J$$

When this expression is substituted into eqn 16 we obtain eqn 10 with

$$Q = \frac{a_C a_D^2}{a_A^2 a_B^3}$$

which is a special case of eqn 14.

Now we conclude the argument based on eqn 10. At equilibrium, the slope of G is zero: $\Delta_r G = 0$. The activities then have their equilibrium values, and we can write

$$K = \left(\frac{a_C a_D^2}{a_A^2 a_B^3}\right)_{\text{equilibrium}}$$

and in general

$$K = \left(\prod_J a_J^{\nu_J}\right)_{\text{equilibrium}} \tag{[17]}$$

These expressions have the same form as Q, eqn 14, but are evaluated using equilibrium activities. From now on, we shall not write the 'equilibrium' subscript explicitly, and will rely on the context to make it clear that for K we use equilibrium values and for Q we use the values at the specified stage of the reaction.

An equilibrium constant K expressed in terms of activities (or fugacities) is called a **thermodynamic equilibrium constant**. Note that, because activities are dimensionless numbers, the thermodynamic equilibrium constant is also dimensionless. In elementary applications, the activities that occur in eqn 17 are often replaced by the numerical values of molalities or molar concentrations, and fugacities are replaced by partial pressures. In either case, the resulting expressions are only approximations. The approximation is particularly severe for electrolyte solutions, for in them activity coefficients differ from 1 even in very dilute solutions.

At this point we set $\Delta_r G = 0$ in eqn 10 and replace Q by K. We immediately obtain

$$RT \ln K = -\Delta_r G^{\ominus} \tag{18}$$

This is an exact and highly important thermodynamic relation, for it enables us to predict the equilibrium constant of any reaction from tables of thermodynamic data, and hence to predict the equilibrium composition of the reaction mixture.[2]

Example 9.1 Calculating an equilibrium constant

Calculate the equilibrium constant for the ammonia synthesis reaction, eqn 9, at 298 K and show how K is related to the partial pressures of the species at equilibrium when the overall pressure is low enough for the gases to be treated as perfect.

Method Calculate the standard reaction Gibbs energy from eqn 11 and convert it to the value of the equilibrium constant by using eqn 18. The expression for the equilibrium constant is obtained from eqn 17 (or via eqn 13) and, because the gases are taken to be perfect, we replace each fugacity by a partial pressure.

Answer The standard Gibbs energy of the reaction is

$$\Delta_r G^{\ominus} = 2\Delta_f G^{\ominus}(NH_3, g) - \{\Delta_f G^{\ominus}(N_2, g) + 3\Delta_f G^{\ominus}(H_2, g)\}$$
$$= 2\Delta_f G^{\ominus}(NH_3, g) = 2 \times (-16.5 \text{ kJ mol}^{-1})$$

Then, because $RT = 2.48 \text{ kJ mol}^{-1}$,

$$\ln K = -\frac{2 \times (-16.5 \text{ kJ mol}^{-1})}{2.48 \text{ kJ mol}^{-1}} = 13.3$$

Hence, $K = 6.0 \times 10^5$. This result is thermodynamically exact. The thermodynamic equilibrium constant for the reaction is

$$K = \frac{a_{NH_3}^2}{a_{N_2} a_{H_2}^3} = \frac{f_{NH_3}^2 p^{\ominus 2}}{f_{N_2} f_{H_2}^3}$$

and this ratio has exactly the value we have just calculated. However, at low overall pressures, when fugacities can be replaced by partial pressures, an approximate form of the equilibrium constant is

$$K = \frac{p_{NH_3}^2 p^{\ominus 2}}{p_{N_2} p_{H_2}^3}$$

- -

Self-test 9.1 Evaluate the equilibrium constant for $N_2O_4(g) \rightleftharpoons 2NO_2(g)$ at 298 K.

$$[K = 0.15]$$

---

2   In Chapter 20 we shall see that the right-hand side may be expressed in terms of spectroscopic data; so this expression also provides a link between spectroscopy and equilibrium composition.

Example 9.2 Estimating the degree of dissociation at equilibrium

The standard Gibbs energy of reaction for the decomposition $H_2O(g) \rightarrow H_2(g) + \frac{1}{2}O_2(g)$ is $+118.08$ kJ mol^{-1} at 2300 K. What is the degree of dissociation of H_2O at 2300 K and 1.00 bar?

Method The equilibrium constant is obtained from the standard Gibbs energy of reaction by using eqn 18, so the task is to relate the degree of dissociation, α, to K and then to find its numerical value. Proceed by expressing the equilibrium compositions in terms of α, and solve for α in terms of K. Because the standard Gibbs energy of reaction is large and positive, we can anticipate that K will be small, and hence that $\alpha \ll 1$, which opens the way to making approximations to obtain its numerical value.

Answer The equilibrium constant is obtained from eqn 18 in the form

$$\ln K = -\frac{\Delta_r G^{\ominus}}{RT} = -\frac{118.08 \times 10^3 \text{ J mol}^{-1}}{(8.3145 \text{ J K}^{-1} \text{ mol}^{-1}) \times (2300 \text{ K})} = -6.175$$

It follows that $K = 2.08 \times 10^{-3}$. The equilibrium composition can be expressed in terms of α by drawing up the following table:

	H_2O	H_2	O_2
Initial amount	n	0	0
Change to reach equilibrium	$-\alpha n$	$+\alpha n$	$+\frac{1}{2}\alpha n$
Amount at equilibrium	$(1-\alpha)n$	αn	$\frac{1}{2}\alpha n$
Mole fraction	$\dfrac{1-\alpha}{1+\frac{1}{2}\alpha}$	$\dfrac{\alpha}{1+\frac{1}{2}\alpha}$	$\dfrac{\frac{1}{2}\alpha}{1+\frac{1}{2}\alpha}$
Partial pressure	$\dfrac{(1-\alpha)p}{1+\frac{1}{2}\alpha}$	$\dfrac{\alpha p}{1+\frac{1}{2}\alpha}$	$\dfrac{\frac{1}{2}\alpha p}{1+\frac{1}{2}\alpha}$

The equilibrium constant is therefore

$$K = \frac{p_{H_2} p_{O_2}^{1/2}}{p_{H_2O}} = \frac{\alpha^{3/2} p^{1/2}}{(1-\alpha)(2+\alpha)^{1/2}}$$

In this expression, we have written p in place of $p/p^{\ominus}$, to keep the notation simple. Now make the approximation that $\alpha \ll 1$, and hence obtain

$$K \approx \frac{\alpha^{3/2} p^{1/2}}{\sqrt{2}}$$

Under the stated conditions, $p = 1.00$ (that is, $p/p^{\ominus} = 1.00$), so

$$\alpha \approx (\sqrt{2}K)^{2/3} = 0.0205$$

That is, about 2 per cent of the water has decomposed.

Comment Always check that the approximation is consistent with the final answer. In this case $\alpha \ll 1$, in accord with the original assumption.

- -

Self-test 9.2 Given that the standard Gibbs energy of reaction at 2000 K is $+135.2$ kJ mol^{-1} for the same reaction, suppose that steam at 200 kPa is passed through a furnace tube at that temperature. Calculate the mole fraction of O_2 present in the output gas stream.

[0.00221]

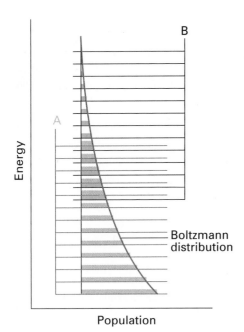

9.3 The Boltzmann distribution of populations over the energy levels of two species A and B with similar densities of energy levels; the reaction A → B is endothermic in this example. The bulk of the population is associated with the species A, so that species is dominant at equilibrium.

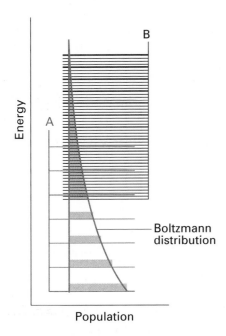

9.4 Even though the reaction A → B is endothermic, the density of energy levels in B is so much greater than that in A that the population associated with B is greater than that associated with A, so B is dominant at equilibrium.

(e) The relation between equilibrium constants

The only remaining problem is to express the thermodynamic equilibrium constant in terms of the mole fractions, x_J, or molalities, b_J, of the species. To do so, we need to know the activity coefficients, and then to use $a_J = \gamma_J x_J$ or $a_J = \gamma_J b_J / b^{\ominus}$ (recalling that the activity coefficients depend on the choice). For example, in the latter case, for an equilibrium of the form $A + B \rightleftharpoons C + D$, where all four species are solutes, we write

$$K = \frac{a_C a_D}{a_A a_B} = \frac{\gamma_C \gamma_D}{\gamma_A \gamma_B} \times \frac{b_C b_D}{b_A b_B} = K_{\gamma} K_b \tag{19}$$

The activity coefficients must be evaluated at the equilibrium composition of the mixture, which may involve a complicated calculation, because the latter is known only if the equilibrium composition is already known. In elementary applications, and to begin the iterative calculation of the concentrations in a real example, the assumption is often made that the activity coefficients are all so close to 1 or they cancel so that $K_{\gamma} = 1$. Then we obtain the result widely used in elementary chemistry that $K \approx K_b$, and equilibria are discussed in terms of molalities (or molar concentrations) themselves. In Chapter 10 we shall see a way of making better estimates of activity coefficients for equilibria involving ions.

Molecular interpretation 9.2 A deeper insight into the origin and significance of the equilibrium constant can be obtained by considering the Boltzmann distribution of molecules over the available states of a system composed of reactants and products. When atoms can exchange partners, as in a reaction, the available states of the system include arrangements in which the atoms are present in the form of reactants and in the form of products: these arrangements have their characteristic sets of energy levels, but the Boltzmann distribution does not distinguish between their identities, only their energies. The atoms distribute themselves over both sets of energy levels in accord with the Boltzmann distribution (Fig. 9.3). At a given temperature, there will be a specific distribution of populations, and hence a specific composition of the reaction mixture.

It can be appreciated from the illustration that, if the reactants and products both have similar arrays of molecular energy levels, then the dominant species in a reaction mixture at equilibrium will be the species with the lower set of energy levels. However, the fact that the Gibbs energy occurs in the expression is a signal that entropy plays a role as well as energy. Its role can be appreciated by referring to Fig. 9.4. We see that, although the B energy levels lie higher than the A energy levels, in this instance they are much more closely spaced. As a result, their total population may be considerable and B could even dominate in the reaction mixture at equilibrium. Closely spaced energy levels correlate with a high entropy, so in this case we see that entropy effects dominate adverse energy effects. This competition is mirrored in eqn 18, as can be seen most clearly by using $\Delta_r G^{\ominus} = \Delta_r H^{\ominus} - T \Delta_r S^{\ominus}$ and writing it in the form

$$K = e^{-\Delta_r H^{\ominus}/RT} e^{\Delta_r S^{\ominus}/R} \tag{20}$$

Note that a positive reaction enthalpy results in a lowering of the equilibrium constant (that is, an endothermic reaction can be expected to have an equilibrium composition that favours the reactants). However, if there is positive reaction entropy, then the equilibrium composition may favour products, despite the endothermic character of the reaction.

The response of equilibria to the conditions

There is one type of response that can be dismissed quickly: the equilibrium constant for a reaction is unaffected by the presence of a catalyst or an enzyme (a biological catalyst).

Catalysts increase the rate at which equilibrium is attained but do not affect its position. However, it is important to note that in industry reactions rarely reach equilibrium, partly on account of the rates at which reactants mix. Under these non-equilibrium conditions, catalysts can have some unexpected effects and may change the composition of the reaction mixture.

9.2 How equilibria respond to pressure

The equilibrium constant depends on the value of $\Delta_r G^{\ominus}$, which is defined at a single, standard pressure. The value of $\Delta_r G^{\ominus}$, and hence of K, is therefore independent of the pressure at which the equilibrium is actually established. Formally we may express this independence as

$$\left(\frac{\partial K}{\partial p}\right)_T = 0 \tag{21}$$

The conclusion that K is independent of pressure does not necessarily mean that the equilibrium composition is independent of the pressure. However, before considering the consequences of pressure, we need to distinguish between the two ways in which pressure may be applied. The pressure within a reaction vessel can be increased by injecting an inert gas into it. Provided the gases are perfect, this addition of gas leaves all the partial pressures of the reacting gases unchanged: the partial pressure of a perfect gas is the pressure it would exert if it were alone in the container, so the presence of another gas has no effect. Put another way, the addition of an inert gas leaves the molar concentrations of the original gases unchanged, as they continue to occupy the same volume. It follows that pressurization by the addition of an inert gas has no effect on the equilibrium composition of the system (provided the gases are perfect). Alternatively, the pressure of the system may be increased by confining the gases to a smaller volume (that is, by compression). Now the partial pressures are changed. Put another way, their molar concentrations are modified because the volume the gases occupy is reduced.

We need to consider the role of compression and understand how changes in partial pressures can be consistent with the general result expressed in eqn 21 that the equilibrium constant itself is independent of the pressure. We shall see that compression can adjust the individual partial pressures of the reactants and products in such a way that, although each one changes, their ratio (as it appears in the equilibrium constant) remains the same. Consider, for instance, the perfect gas equilibrium $A \rightleftharpoons 2B$, for which the equilibrium constant is

$$K = \frac{p_B^2}{p_A p^{\ominus}} \tag{22}$$

The right-hand side of this expression remains constant only if an increase in p_A cancels an increase in the *square* of p_B. This relatively steep increase of p_A compared to p_B will occur if the equilibrium composition shifts in favour of A at the expense of B. Then the number of A molecules will increase as the volume of the container is decreased and its partial pressure will rise more rapidly than can be ascribed to a simple change in volume alone (Fig. 9.5).

The increase in the number of A molecules and the corresponding decrease in the number of B molecules brought about by compression is a special case of a principle proposed by the French chemist (and inventor of oxyacetylene welding) Henri Le Chatelier. Le Chatelier's principle states that:

A system at equilibrium, when subjected to a disturbance, responds in a way that tends to minimize the effect of the disturbance.

The principle implies that, if a system at equilibrium is compressed, then the reaction will adjust so as to minimize the increase in pressure. This it can do by reducing the number of

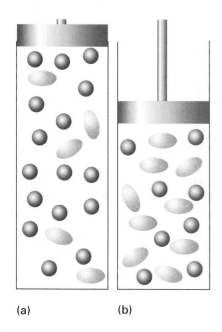

(a)   (b)

9.5 When a reaction at equilibrium is compressed (from a to b), the reaction responds by reducing the number of molecules in the gas phase (in this case by producing the dimers represented by the ellipses).

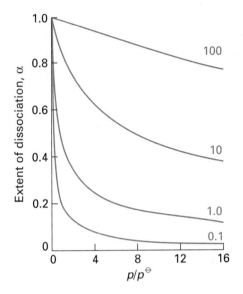

9.6 The pressure dependence of the degree of dissociation, α, at equilibrium for an $A(g) \rightleftharpoons 2B(g)$ reaction for different values of the equilibrium constant K. The value $\alpha = 0$ corresponds to pure A; $\alpha = 1$ corresponds to pure B.

particles in the gas phase, which implies a shift $A \leftarrow 2B$. The quantitative treatment of the effect of compression set out in the *Justification* below leads to the conclusion that the extent of dissociation, α, of A into 2B is

$$\alpha = \frac{1}{\left(1 + 4p/K\right)^{1/2}} \qquad \{23\}$$

where p is to be understood as $p/p^{\ominus}$. This formula shows that, even though K is independent of pressure, the amounts of A and B do depend on pressure (Fig. 9.6). It also shows that, as p is increased, α decreases, in accord with Le Chatelier's principle.

Justification 9.2

Suppose that there is an amount n of A present initially (and no B). At equilibrium the amount of A is $(1 - \alpha)n$ and the amount of B is $2\alpha n$. It follows that the mole fractions present at equilibrium are

$$x_A = \frac{(1 - \alpha)n}{(1 - \alpha)n + 2\alpha n} = \frac{1 - \alpha}{1 + \alpha} \qquad x_B = \frac{2\alpha}{1 + \alpha}$$

The equilibrium constant for the reaction (with p understood as $p/p^{\ominus}$) is

$$K = \frac{p_B^2}{p_A} = \frac{x_B^2 p^2}{x_A p} = \frac{4\alpha^2 p}{1 - \alpha^2}$$

This expression rearranges into eqn 23.

Illustration

To predict the effect of an increase in pressure on the composition of the ammonia synthesis at equilibrium, eqn 9, we note that the number of gas molecules decreases (from 4 to 2). So, Le Chatelier's principle predicts that an increase in pressure will favour the product. The equilibrium constant is

$$K = \frac{p_{NH_3}^2}{p_{N_2} p_{H_2}^3} = \frac{x_{NH_3}^2 p^2}{x_{N_2} x_{H_2}^3 p^4} = \frac{K_x}{p^2}$$

Therefore, doubling the pressure must increase K_x by a factor of 4 to preserve the value of K.

Self-test 9.3 Predict the effect of a 10-fold pressure increase on the equilibrium composition of the reaction $3N_2(g) + H_2(g) \rightarrow 2N_3H(g)$.

[100-fold increase in K_x]

9.3 The response of equilibria to temperature

Le Chatelier's principle predicts that a system at equilibrium will tend to shift in the endothermic direction if the temperature is raised, for then energy is absorbed as heat. Conversely, an equilibrium can be expected to shift in the exothermic direction if the temperature is lowered, for then the reduction in temperature is opposed. These conclusions can be summarized as follows:

Exothermic reactions: increased temperature favours the reactants.
Endothermic reactions: increased temperature favours the products.

We shall now justify these remarks and see how to express the changes quantitatively.

(a) The van't Hoff equation

The van't Hoff equation, which is derived in the *Justification* below, is an expression for the slope of a plot of the equilibrium constant (specifically, ln K) as a function of temperature. It may be expressed in either of two ways:

$$\text{(a)} \quad \frac{d \ln K}{dT} = \frac{\Delta_r H^{\ominus}}{RT^2} \qquad \text{(b)} \quad \frac{d \ln K}{d(1/T)} = -\frac{\Delta_r H^{\ominus}}{R} \tag{24}$$

Justification 9.3

From eqn 18, we know that

$$\ln K = -\frac{\Delta_r G^{\ominus}}{RT}$$

Differentiation of ln K with respect to temperature then gives

$$\frac{d \ln K}{dT} = -\frac{1}{R} \frac{d(\Delta_r G^{\ominus}/T)}{dT}$$

The differentials are complete because K and $\Delta_r G^{\ominus}$ depend only on temperature, not on pressure. To develop this equation we use the Gibbs–Helmholtz equation (eqn 5.13) in the form

$$\frac{d(\Delta_r G^{\ominus}/T)}{dT} = -\frac{\Delta_r H^{\ominus}}{T^2}$$

where $\Delta_r H^{\ominus}$ is the standard reaction enthalpy at the temperature T. Combining the two equations gives the van't Hoff equation, eqn 24a. The second form of the equation is obtained by noting that

$$\frac{d(1/T)}{dT} = -\frac{1}{T^2}, \text{ so } dT = -T^2 d(1/T)$$

Equation 24a shows that d ln $K/dT < 0$ (and therefore that $dK/dT < 0$) for a reaction that is exothermic under standard conditions ($\Delta_r H^{\ominus} < 0$). A negative slope means that ln K, and therefore K itself, decreases as the temperature rises. Therefore, as asserted above, in the case of an exothermic reaction the equilibrium shifts away from products. The opposite occurs in the case of endothermic reactions.

Some insight into the thermodynamic basis of this behaviour can be found in the expression $\Delta_r G = \Delta_r H - T\Delta_r S$ written in the form $-\Delta_r G/T = -\Delta_r H/T + \Delta_r S$. When the reaction is exothermic, $-\Delta_r H/T$ corresponds to a positive change of entropy of the surroundings, and favours the formation of products. When the temperature is raised, $-\Delta_r H/T$ decreases, and the increasing entropy of the surroundings has a less potent role. As a result, the equilibrium lies less to the right. When the reaction is endothermic, the principal factor is the increasing entropy of the reaction system. The importance of the unfavourable change of entropy of the surroundings is reduced if the temperature is raised (because then $\Delta_r H/T$ is smaller), and the reaction is able to shift towards products.

Molecular interpretation 9.3 The typical arrangement of energy levels for an endothermic reaction is shown in Fig. 9.7a. When the temperature is increased, the Boltzmann distribution adjusts and the populations change as shown. The change corresponds to an increased population of the higher energy states at the expense of the population of the lower energy states. We see that the states that arise from the B molecules become more populated at the expense of the A molecules. Therefore, the total

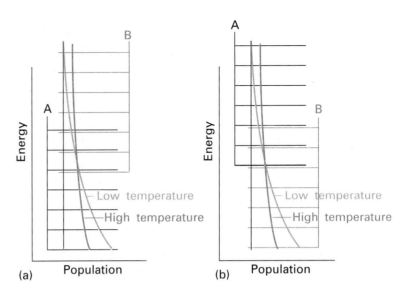

9.7 The effect of temperature on a chemical equilibrium can be interpreted in terms of the change in the Boltzmann distribution with temperature and the effect of that change in the population of the species. (a) In an endothermic reaction, the population of B increases at the expense of A as the temperature is raised. (b) In an exothermic reaction, the opposite happens.

population of B states increases, and B becomes more abundant in the equilibrium mixture. Conversely, if the reaction is exothermic (Fig. 9.7b), then an increase in temperature increases the population of the A states (which start at higher energy) at the expense of the B states, so the reactants become more abundant.

Example 9.3 Measuring a reaction enthalpy

The data below show the temperature variation of the equilibrium constant of the reaction

$$Ag_2CO_3(s) \rightleftharpoons Ag_2O(s) + CO_2(g)$$

Calculate the standard reaction enthalpy of the decomposition.

T/K	350	400	450	500
K	3.98×10^{-4}	1.41×10^{-2}	1.86×10^{-1}	1.48

Method It follows from eqn 24b that, provided the reaction enthalpy can be assumed to be independent of temperature, a plot of $-\ln K$ against $1/T$ should be a straight line of slope $\Delta_r H^\ominus / R$.

Answer We draw up the following table:

T/K	350	400	450	500
$(10^3 \text{ K})/T$	2.86	2.50	2.22	2.00
$-\ln K$	7.83	4.26	1.68	-0.39

These points are plotted in Fig. 9.8. The slope of the graph is $+9.5 \times 10^3$, so

$$\Delta_r H^\ominus = (+9.5 \times 10^3 \text{ K}) \times R = +79 \text{ kJ mol}^{-1}$$

Comment This is a non-calorimetric method of determining $\Delta_r H^\ominus$. A drawback is that the reaction enthalpy is actually temperature dependent, so the plot is not expected to be perfectly linear. However, the temperature dependence is weak in many cases, so the plot is reasonably straight. In practice, the method is not very accurate, but it is often the only method available.

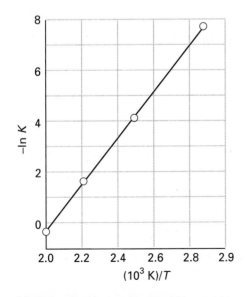

9.8 When $-\ln K$ is plotted against $1/T$, a straight line is expected with slope equal to $\Delta_r H^\ominus / R$. This is a non-calorimetric method for the measurement of reaction enthalpies.

Self-test 9.4 The equilibrium constant of the reaction $2SO_2(g) + O_2(g) \rightleftharpoons 2SO_3(g)$ is 4.0×10^{24} at 300 K, 2.5×10^{10} at 500 K, and 3.0×10^4 at 700 K. Estimate the reaction enthalpy at 500 K.

[-200 kJ mol^{-1}]

(b) The value of K at different temperatures

To find the value of the equilibrium constant at a temperature T_2 in terms of its value K_1 at another temperature T_1, we integrate eqn 24b between these two temperatures:

$$\ln K_2 - \ln K_1 = -\frac{1}{R} \int_{1/T_1}^{1/T_2} \Delta_r H^{\ominus} \, d(1/T) \tag{25}$$

If we suppose that $\Delta_r H^{\ominus}$ varies only slightly with temperature over the temperature range of interest, then we may take it outside the integral. It follows that

$$\ln K_2 - \ln K_1 = -\frac{\Delta_r H^{\ominus}}{R} \left(\frac{1}{T_2} - \frac{1}{T_1} \right) \tag{26}$$

Illustration

To estimate the equilibrium constant for the synthesis of ammonia at 500 K from its value at 298 K (6.0×10^5 for the reaction as written in eqn 9) we use the standard reaction enthalpy, which can be obtained from Table 2.6 in the *Data section* because $\Delta_r H^{\ominus} = 2\Delta_f H^{\ominus}(NH_3, g)$, and assume that its value is constant over the range of temperatures. Then, with $\Delta_r H^{\ominus} = -92.2$ kJ mol^{-1}, from eqn 26 we find

$$\ln K_2 = \ln (6.0 \times 10^5) - \frac{(-92.2 \text{ kJ mol}^{-1})}{8.3145 \text{ J K}^{-1} \text{mol}^{-1}} \left(\frac{1}{500 \text{ K}} - \frac{1}{298 \text{ K}} \right)$$

$$= -1.73$$

It follows that $K_2 = 0.18$.

Self-test 9.5 The equilibrium constant for $N_2O_4(g) \rightleftharpoons 2NO_2(g)$ was calculated in Self-test 9.1. Estimate its value at 100°C.

[16]

Applications to selected systems

In this section we look at some of the conclusions that can be drawn from the existence of equilibrium constants and from the equation $\Delta_r G^{\ominus} = -RT \ln K$. Note that $K > 1$ when $\Delta_r G^{\ominus} < 0$, and products then dominate reactants.

9.4 The extraction of metals from their oxides

Metals can be obtained from their oxides by reduction with carbon or carbon monoxide if any of the equilibria

$$MO(s) + C(s) \rightleftharpoons M(s) + CO(g)$$
$$MO(s) + \tfrac{1}{2}C(s) \rightleftharpoons M(s) + \tfrac{1}{2}CO_2(g)$$
$$MO(s) + CO \rightleftharpoons M(s) + CO_2(g)$$

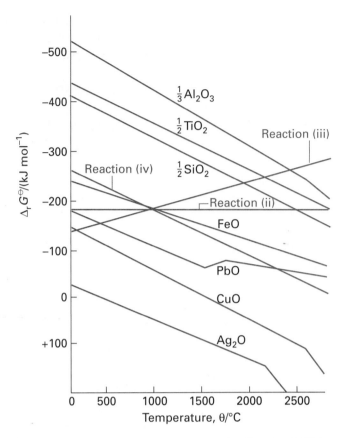

9.9 An Ellingham diagram for the discussion of metal ore reduction. Note that $\Delta_r G^{\ominus}$ is most negative at the top of the diagram.

lie to the right (that is, have $K > 1$). As we shall see, these equilibria can be discussed in terms of the thermodynamic functions for the reactions

(i) $M(s) + \frac{1}{2}O_2(g) \longrightarrow MO(s)$
(ii) $\frac{1}{2}C(s) + \frac{1}{2}O_2(g) \longrightarrow \frac{1}{2}CO_2(g)$
(iii) $C(s) + \frac{1}{2}O_2(g) \longrightarrow CO(g)$
(iv) $CO(g) + \frac{1}{2}O_2(g) \longrightarrow CO_2(g)$

The temperature dependences of the standard Gibbs energies of reactions (i)–(iv) depend on the reaction entropy through $d\Delta_r G^{\ominus}/dT = -\Delta_r S^{\ominus}$. Because in reaction (iii) there is a net *increase* in the amount of gas, the standard reaction entropy is large and positive; therefore, its $\Delta_r G^{\ominus}$ decreases sharply with increasing temperature. In reaction (iv), there is a similar net *decrease* in the amount of gas, so $\Delta_r G^{\ominus}$ increases sharply with increasing temperature. In reaction (ii), the amount of gas is constant, so the entropy change is small and $\Delta_r G^{\ominus}$ changes only slightly with temperature. These remarks are summarized in Fig. 9.9, which is called an **Ellingham diagram**. Note that $\Delta_r G^{\ominus}$ decreases upwards!

At room temperature, $\Delta_r G^{\ominus}$ is dominated by the contribution of the reaction enthalpy ($T\Delta_r S^{\ominus}$ being relatively small), so the order of increasing $\Delta_r G^{\ominus}$ is the same as the order of increasing $\Delta_r H^{\ominus}$ (Al_2O_3 is most exothermic, Ag_2O is least). The standard reaction entropy is similar for all metals because in each case gaseous oxygen is eliminated and a compact, solid oxide is formed. As a result, the temperature dependence of the standard Gibbs energy of oxidation should be similar for all metals, as is shown by the similar slopes of the lines in the diagram. The kinks at high temperatures correspond to the evaporation of the metals; less pronounced kinks occur at the melting temperatures of the metals and the oxides.

Successful reduction of the oxide depends on the outcome of the competition of the carbon for the oxygen bound to the metal. The standard Gibbs energies for the reductions can be expressed in terms of the standard Gibbs energies for the reactions above:

$$MO(s) + C(s) \longrightarrow M(s) + CO(g) \qquad \Delta_r G^\ominus = \Delta_r G^\ominus(\text{iii}) - \Delta_r G^\ominus(\text{i})$$
$$MO(s) + \tfrac{1}{2}C(s) \longrightarrow M(s) + \tfrac{1}{2}CO_2(g) \qquad \Delta_r G^\ominus = \Delta_r G^\ominus(\text{ii}) - \Delta_r G^\ominus(\text{i})$$
$$MO(s) + CO(g) \longrightarrow M(s) + CO_2(g) \qquad \Delta_r G^\ominus = \Delta_r G^\ominus(\text{iv}) - \Delta_r G^\ominus(\text{i})$$

The equilibrium lies to the right if $\Delta_r G^\ominus < 0$. This is the case when the line for reaction (i) lies below (is more positive than) the line for one of the reactions (ii) to (iv).

The spontaneity of a reduction at any temperature can be predicted simply by looking at the diagram: a metal oxide is reduced by any carbon reaction lying above it, because the overall reaction then has $\Delta_r G^\ominus < 0$. For example, CuO can be reduced to Cu at any temperature above room temperature. Even in the absence of carbon, Ag_2O decomposes when heated above 200 °C because then the standard Gibbs energy for reaction (i) becomes positive (and the reverse reaction is then spontaneous). On the other hand, Al_2O_3 is not reduced by carbon until the temperature has been raised to above 2300 °C.

9.5 Acids and bases

One of the most important examples of chemical equilibrium is the one that exists when acids and bases are present in solution. According to the **Brønsted–Lowry classification**:

An acid is a proton donor; a base is a proton acceptor.

These definitions make no mention of the solvent (and apply even if no solvent is present); however, by far the most important medium is aqueous solution, and we confine our attention to that. One of the properties of central interest in aqueous solutions of acids and bases is the **pH**, which is defined as

$$pH = -\log a_{H_3O^+} \tag{27}$$

where H_3O^+ is the hydronium ion, a representation of the state of the proton in aqueous solution. At low concentrations, the activity of hydronium ions is approximately equal to their molality and molar concentration, so a determination of pH is an indication of hydronium ion concentration. However, many thermodynamic observables depend on pH itself, and there is no need to make this approximation and interpretation.

(a) Acid–base equilibria in water

An acid HA takes part in the following proton transfer equilibrium in water:

$$HA(aq) + H_2O(l) \rightleftharpoons H_3O^+(aq) + A^-(aq) \qquad K = \frac{a_{H_3O^+} a_{A^-}}{a_{HA} a_{H_2O}} \tag{28}$$

In this expression, A^- is the **conjugate base** of the acid. If we confine attention to dilute solutions, the activity of water is close to 1 (the value for pure water), and the equilibrium can be expressed in terms of the **acidity constant**, K_a:

$$K_a = \frac{a_{H_3O^+} a_{A^-}}{a_{HA}} \tag{29}$$

When it simplifies the discussion, we shall make the approximations of replacing the activities in acidity constants by the numerical values of the molar concentrations and writing

$$K_a \approx \frac{[H_3O^+][A^-]}{[HA]} \tag{30}$$

where (as indicated by the equation label $\{\cdots\}$) $[J]$ should be understood as $[J]/(\text{mol L}^{-1})$. The approximation $a_J \approx [J]$ is legitimate only if *all* the ions are present at low concentration, not merely the ions of interest, because (as we shall see quantitatively in Section 10.2) all ions contribute to the departures from ideality. Concentrations must be very low for this approximation to be permissible; even for a 10^{-3} M solution of a 1:1 electrolyte in water at $25\,^{\circ}\text{C}$, activity coefficients are about 0.96, and their neglect introduces an error approaching 10 per cent into the interpretation of equilibrium constants. When concentrations are not low enough for concentrations to be used, activity coefficients can be found from tables (or estimated from the equations derived in Section 10.2).

It is common to report values of K_a in terms of its negative logarithm, pK_a:

$$pK_a = -\log K_a \tag{31}$$

A high value of pK_a signifies a very small value of K_a (because $K_a = 10^{-pK_a}$) and hence a very weak acid. We shall see that the use of pK_a in place of K_a simplifies the appearance of a number of equations. This simplification stems from the fact that the K_a of the proton transfer equilibrium is related to the standard Gibbs energy of the reaction by

$$\Delta_r G^{\ominus} = -RT \ln K_a = (RT \ln 10) \times pK_a \tag{32}$$

Hence, manipulations of pK_a are in fact manipulations of $\Delta_r G^{\ominus}$ in disguise.

For a base B in water, the characteristic proton transfer equilibrium is

$$B(aq) + H_2O(l) \rightleftharpoons HB^+(aq) + OH^-(aq) \qquad K = \frac{a_{HB^+} a_{OH^-}}{a_B a_{H_2O}}$$

In this expression, HB^+ is the **conjugate acid** of the base B. In dilute solutions, the activity of water is 1, and we can express this equilibrium in terms of the **basicity constant**, K_b:

$$K_b = \frac{a_{HB^+} a_{OH^-}}{a_B} \tag{33}$$

Although the basicity constant can be used to assess the strength of a base, it is common to express proton transfer equilibria involving a base in terms of its conjugate acid:

$$HB^+(aq) + H_2O(l) \rightleftharpoons H_3O^+(aq) + B(aq) \qquad K_a = \frac{a_{H_3O^+} a_B}{a_{HB^+}}$$

As may be verified by multiplying the expressions for K_a and K_b together, the acidity constant of the conjugate acid HB^+ is related to the basicity constant of the base B by

$$K_a K_b = K_w \tag{34}$$

where K_w is the **autoprotolysis constant** of water:

$$2H_2O(l) \rightleftharpoons H_3O^+(aq) + OH^-(aq) \qquad K_w = a_{H_3O^+} a_{OH^-} \tag{35}$$

At $25\,^{\circ}\text{C}$, $K_w = 1.008 \times 10^{-14}$ ($pK_w = 14.00$), showing that only a few of the water molecules are ionized. If we introduce $pOH = -\log a_{OH^-}$ by analogy with pH, then it follows that

$$pK_w = pH + pOH \tag{36}$$

Because the molar concentrations of H_3O^+ and OH^- are equal in pure water, it further follows that, for pure water at $25\,^{\circ}\text{C}$,

$$pH = \tfrac{1}{2}pK_w \approx 7.00 \tag{37}$$

(b) pH calculations

The calculations of the pH of solutions of acids and bases are fully treated in introductory chemistry courses, and this section is a review of that material. For a strong acid (one that is

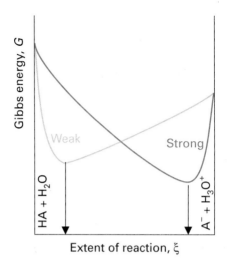

9.10 The Gibbs energy for a solution of a weak acid has a minimum that lies close to HA, and little deprotonation occurs at equilibrium. For a strong acid, the minimum lies close to products and deprotonation is almost complete.

fully ionized in solution), the molar concentration of hydronium ions is the same as the nominal molar concentration[3] of the acid, as each HA molecule generates one H_3O^+ ion. Similarly, for a Group 1 hydroxide of formula MOH in solution, the molar concentration of hydroxide ions is the same as the nominal molar concentration of the base, as each B molecule generates one OH^- ion. For Group 2 hydroxides, of formula $M(OH)_2$, the OH^- concentration is twice the nominal molar concentration of the base. It follows that we can estimate pOH from the nominal molar concentration of the base, and convert the value to pH by using eqn 36.

For a weak acid or base, we need to take the partial ionization into account by considering the proton transfer equilibrium. The distinction between weak and strong acids and bases is an illustration of the different types of behaviour shown in Fig. 9.10: strong acids and bases are those for which the minimum Gibbs energy of the solution lies close to the (ionized) products; weak acids and bases are those for which the minimum lies close to the (non-ionized) reactants. However, because the extent of ionization is so small (for typical solutions), an approximation is that the molar concentration of HA or B is unchanged from its nominal value. Moreover, because the molar concentrations of the species produced by the proton transfer are equal (to a good approximation), the expression for K_a in eqn 30 simplifies to

$$K_a \approx \frac{[H_3O^+]^2}{[HA]} \qquad \text{or} \quad [H_3O^+] \approx (K_a[HA])^{1/2} \qquad \{38\}$$

It then follows, by taking the negative logarithms of both sides, that

$$pH \approx \tfrac{1}{2}pK_a - \tfrac{1}{2}\log[HA] \qquad \{39\}$$

Illustration

The pK_a of hydrocyanic acid, HCN(aq), is given in Table 9.1 as 9.31. Therefore, the pH of 0.20 M HCN(aq) is

$$pH \approx \tfrac{1}{2}pK_a - \tfrac{1}{2}\log[HCN]$$
$$\approx \tfrac{1}{2} \times 9.31 - \tfrac{1}{2}\log 0.20 = 5.0$$

Self-test 9.6 Calculate the pH of 0.10 M NH_3(aq).

[11.1]

(c) Acid–base titrations

At the stoichiometric point[4] of a titration of a weak acid (such as CH_3COOH) and strong base (NaOH) the analyte (the solution being titrated) has become an aqueous solution of the weak acid–strong base salt (sodium acetate). At this point, the solution contains $CH_3CO_2^-$ and Na^+ ions together with any ions stemming from autoprotolysis. The presence of the Brønsted base $CH_3CO_2^-$ means that we can expect a pH of greater than 7. At the stoichiometric point of a titration of a weak base (such as NH_3) and a strong acid (HCl), the analyte is a solution of a strong acid–weak base salt (ammonium chloride) and contains NH_4^+ and Cl^- ions. Because Cl^- is a negligibly weak Brønsted base and NH_4^+ is a weak Brønsted acid, the solution is acidic and its pH will be less than 7.

Table 9.1* Acidity constants in water at 298 K

	pK_{a1}	pK_{a2}	pK_{a3}
Acetic acid, CH_3COOH	4.75		
Ammonium ion, NH_4^+	9.25		
Carbonic acid, H_2CO_3	6.37	10.25	
Hydrocyanic acid	9.31		
Phosphoric acid, H_3PO_4	2.12	7.21	12.67

* More values are given in the *Data section* at the end of this volume.

3   By nominal molar concentration is meant the molar concentration of the solute as prepared, ignoring any proton transfer.

4   The stoichiometric point of a titration is still widely called the equivalence point.

The next few paragraphs develop this argument and show how to predict the pH at any stage of an acid–base titration. We shall suppose that we are titrating a volume V_A of a solution of a weak acid of nominal molar concentration A_0 (the analyte) with a solution of a strong base MOH of molar concentration B (the titrant). When a volume V_B of the titrant has been added to the analyte the total volume of the analyte is $V = V_A + V_B$.

The approximations we shall make are based on the fact that the acid is weak, and therefore that HA is more abundant than any A^- ions in the solution. Furthermore, when HA is present, it provides hydronium ions that greatly outnumber any that stem from the autoprotolysis of water. Finally, when excess base is present, the OH^- ions it provides dominate any that come from the water autoprotolysis. At the start of the titration, the analyte is a solution of a weak acid, so we can use eqn 39 to estimate its pH:

$$pH = \tfrac{1}{2}pK_a - \tfrac{1}{2}\log A_0 \qquad \{40\}$$

The corresponding expression for a solution of a weak base is

$$pH = \tfrac{1}{2}pK_w + \tfrac{1}{2}pK_a + \tfrac{1}{2}\log B_0 \qquad \{41\}$$

Illustration

The pH of 0.10 M HClO(aq) is

$$pH = \tfrac{1}{2} \times 7.43 - \tfrac{1}{2}\log 0.010 = 4.7$$

We have used the information in Table 9.1 in the *Data section*.

Self-test 9.7 Estimate the pH of 0.010 M NH_3(aq).

[10.6]

After the addition of some base (but before the stoichiometric point is reached), the concentration of A^- ions stems almost entirely from the salt that is present, for the weak acid present provides only a few A^- ions. Therefore, $[A^-] = S$, the molar concentration of the salt (the base). The amount of HA molecules that remains is the original amount, $A_0 V_A$ less the amount of HA molecules that have been converted to salt by the addition of base, so the molar concentration of HA is $A' = A - S$. This calculation ignores the small additional loss of HA as a result of its ionization in solution. Hence

$$K_a = \frac{a_{H_3O^+} a_{A^-}}{a_{HA}} \approx \frac{a_{H_3O^+} S}{A'} \qquad (42)$$

The derivation has made the doubtful approximation that the activity coefficient of the A^- ions is close to 1. It follows that

$$pH = pK_a - \log\left(\frac{A'}{S}\right) \qquad (43)$$

This expression is called the **Henderson–Hasselbalch equation**. The general form of this equation, once we recognize that A' is the molar concentration of acid in the solution and S is the molar concentration of base, is

$$pH = pK_a - \log\frac{[\text{acid}]}{[\text{base}]} \qquad (44)$$

When the molar concentrations of acid and salt (acting as the base) are equal,

$$pH = pK_a \qquad (45)$$

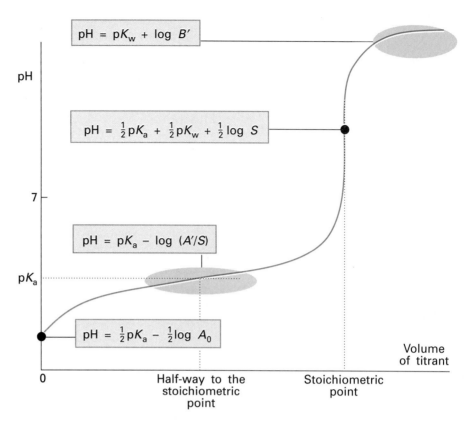

9.11 A summary of the regions of the pH curve of the titration of a weak acid with a strong base, and the equations used in different regions.

Hence the pK_a of the acid can be measured directly from the pH of the mixture. In practice this is done by recording the pH during a titration and then examining the record for the pH halfway to the stoichiometric point (Fig. 9.11).

At the stoichiometric point, the H_3O^+ ions in the solution stem from the influence of the OH^- ions on the autoprotolysis equilibrium, and the OH^- ions are produced by the proton transfer equilibrium from H_2O to A^-. Because only a small amount of HA is formed in this way, the concentration of A^- ions is almost exactly that of the salt, and we can write $[A^-] \approx S$. The number of OH^- ions that arise from the proton transfer equilibrium greatly outnumber those produced by the water autoprotolysis, so we can set $[HA] \approx [OH^-]$:

$$K_b = \frac{a_{HA}a_{OH^-}}{a_{A^-}} \approx \frac{[OH^-]^2}{S}$$

The equilibrium constant for the base protolysis can be expressed in terms of the acidity constant of the conjugate acid HA by eqn 34 (that $K_a K_b = K_w$), so

$$[OH^-] \approx \left(\frac{SK_w}{K_a}\right)^{1/2} \qquad \{46\}$$

Therefore, after taking negative logarithms of both sides,

$$pOH = \tfrac{1}{2}pK_w - \tfrac{1}{2}pK_a - \tfrac{1}{2}\log S$$

Finally, because $pH = pK_w - pOH$, it follows that

$$pH = \tfrac{1}{2}pK_a + \tfrac{1}{2}pK_w + \tfrac{1}{2}\log S \qquad \{47\}$$

The corresponding expression for the pH at the stoichiometric point of a titration of a strong acid with a weak base is

$$pH = \tfrac{1}{2}pK_a - \tfrac{1}{2}\log S \qquad \{48\}$$

Illustration

The stoichiometric point of a titration of 25.00 mL of 0.100 M HClO(aq) with 0.100 M NaOH(aq) occurs when the molar concentration of NaClO is 0.050 mol L^{-1} (because the volume of the solution has increased from 25.00 mL to 50.00 mL). Therefore, the pH is

$$pH = \tfrac{1}{2} \times 7.43 + \tfrac{1}{2} \times 14.00 + \tfrac{1}{2}\log 0.050 = 10.1$$

Self-test 9.8 Estimate the pH at the stoichiometric point of a titration of 25.00 mL of 0.200 M NH$_3$(aq) with 0.300 M HCl(aq).

[5.1]

When so much strong base has been added that the titration has been carried well past the stoichiometric point, the pH is determined by the excess base present. Then, $[H_3O^+] \approx K_w/[OH^-]$ so, if we write the molar concentration of excess base as B', this expression can be written

$$pH = pK_w + \log B' \qquad \{49\}$$

In this expression, as in all the preceding ones, the molar concentrations must take into account the change of volume that occurs as the titrant is added to the analyte.

The general form of the pH curve throughout a titration is illustrated in Fig. 9.11. The pH rises slowly from the value given by the 'weak acid alone' formula (eqn 40) following the values given by the Henderson–Hasselbalch equation (eqn 44) until the stoichiometric point is approached. It then changes rapidly to and through the value given by the 'salt alone' formula (eqn 47 or 48). It then climbs less rapidly towards the value given by the 'base in excess' formula (eqn 49). The stoichiometric point can be detected easily by observing where the pH changes rapidly through the value given by the 'salt alone formula' (eqn 47 or 48).

(d) Buffers and indicators

The slow variation of the pH in the vicinity of $S = A'$, when the molar concentrations of the salt and acid are equal, is the basis of **buffer action**, the ability of a solution to oppose changes in pH when small amounts of strong acids and bases are added to the solution. The mathematical basis of buffer action is the logarithmic dependence given by the Henderson–Hasselbalch equation (eqn 44), which is quite flat near pH = pK_a. The physical basis of buffer action is that the existence of an abundant supply of A$^-$ ions (because a salt is present) can remove most of the H$_3$O$^+$ ions brought by additional strong acid; moreover, the numerous HA molecules can supply H$_3$O$^+$ ions to react with any strong base that is added.

Example 9.4 Estimating the pH of a buffer solution

Estimate the pH of an aqueous buffer solution that contains 0.200 mol L^{-1} KH$_2$PO$_4$ and 0.100 mol L^{-1} K$_2$HPO$_4$.

Method The pH of a solution of a weak acid and its salt can be estimated from the Henderson–Hasselbalch equation. To do so, we must identify the acid HA and its conjugate base A$^-$.

Answer In this example, the acid is the anion $H_2PO_4^-$ and its conjugate base is the anion HPO_4^{2-}:

$$H_2PO_4^-(aq) + H_2O(l) \rightleftharpoons H_3O^+(aq) + HPO_4^{2-}(aq)$$

The acidity constant we require is therefore pK_{a2} for H_3PO_4, which from Table 9.1 is 7.21. Then, with $A' = 0.200 \text{ mol L}^{-1}$ and $S = 0.100 \text{ mol L}^{-1}$, eqn 44 gives the pH of the solution as

$$pH = 7.21 - \log\left(\frac{0.200}{0.100}\right) = 6.91$$

Hence, the solution should buffer close to $pH = 7$.

Self-test 9.9 Calculate the pH of an aqueous buffer solution which contains 0.100 mol L^{-1} NH_3 and 0.200 mol L^{-1} NH_4Cl.

[8.95; more realistically: 9]

(e) Acid–base indicators

The rapid change of pH near the stoichiometric point in a titration is the basis of indicator detection. An **acid–base indicator** is normally some large, water-soluble, weakly acidic organic molecule which can exist as acid (HIn) or conjugate base (In⁻) forms that differ in colour. The two forms are in equilibrium in solution:

$$HIn(aq) + H_2O(l) \rightleftharpoons In^-(aq) + H_3O^+(aq)$$

and, if we make the usual assumption that the solution is so dilute that the activity of water is 1, then the equilibrium is described by the constant

$$K_{In} = \frac{a_{H_3O^+} a_{In^-}}{a_{HIn}} \tag{50}$$

The ratio of acid and base forms at a given pH is found by rearranging this expression to

$$\log\frac{[HIn]}{[In^-]} \approx pK_{In} - pH \tag{51}$$

Therefore, when the pH is less than pK_{In}, the indicator is predominantly in its acidic form and has the corresponding colour; when the pH is greater than pK_{In}, the indicator is mainly in its basic form. The **end point** is the pH of the solution when both forms are present in equal abundance, which occurs when $pH = pK_{In}$.

At the stoichiometric point of an acid–base titration, the pH changes sharply through several units and, if the pH passes through pK_{In}, there is a pronounced colour change. With a well-chosen indicator, the end point coincides with the stoichiometric point of the titration.

Care must be taken to use an indicator that changes colour at the pH appropriate to the type of titration. Thus, in a weak acid–strong base titration, the stoichiometric point lies at the pH given by eqn 47, so an indicator that changes at that pH must be selected. Broadly speaking, an indicator with $pK_{In} > 7$ is required because the stoichiometric point lies at $pH > 7$. Similarly, in a strong acid–weak base titration, an indicator changing near the pH given by eqn 48 should be used. The stoichiometric points of such titrations lie at $pH < 7$, so an indicator with $pK_{In} < 7$ is required.

9.6 Biological activity: the thermodynamics of ATP

An important biochemical is adenosine triphosphate, ATP (**1**). Its function is to store the energy made available when food is metabolized and then to supply it on demand to a wide

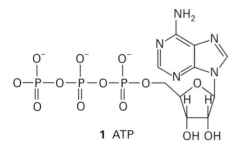

1 ATP

variety of biological processes. The essence of ATP's action is its ability to lose its terminal phosphate group by hydrolysis and to form adenosine diphosphate (ADP):

$$ATP(aq) + H_2O(l) \longrightarrow ADP(aq) + P_i^-(aq) + H_3O^+(aq)$$

(P_i^- denotes an inorganic phosphate group, such as $H_2PO_4^-$.) This reaction is exergonic and can drive an endergonic reaction if suitable enzymes are available.

(a) Biological standard states

The conventional standard state of hydrogen ions (unit activity, pH = 0) is not appropriate to normal biological conditions. Therefore, in biochemistry it is common to adopt the **biological standard state**, in which pH = 7 (an activity of 10^{-7}, neutral solution). We shall adopt this convention in this section, and label the corresponding standard thermodynamic functions as $G^\oplus$, $H^\oplus$, and $S^\oplus$ (some texts use $X^{\ominus\prime}$). The relation between the thermodynamic and biological standard Gibbs energies of reaction for a reaction of the form

$$A + \nu H^+(aq) \longrightarrow P \tag{52}$$

is

$$\Delta_r G^\oplus = \Delta_r G^\ominus + 7\nu RT \ln 10 \tag{53}$$

Note that there is no difference between the two standard values if hydrogen ions are not involved in the reaction ($\nu = 0$).

Justification 9.4

The reaction Gibbs energy is

$$\Delta_r G = \mu_P - \mu_A - \nu\mu_{H^+}$$

If all the species other than H^+ are in their standard states, this expression becomes

$$\Delta_r G = \mu_P^\ominus - \mu_A^\ominus - \nu\mu_{H^+}$$

Then, because

$$\mu_{H^+} = \mu_{H^+}^\ominus + RT \ln a_{H^+} = \mu_{H^+}^\ominus - (RT \ln 10) \times pH$$

(because $\ln x = \ln 10 \log x$, with $\ln 10 = 2.303$), this equation becomes

$$\Delta_r G = \mu_P^\ominus - \mu_A^\ominus - \nu\mu_{H^+}^\ominus + (RT\nu \ln 10) \times pH$$

Equation 53 then follows when we set pH = 7.

Illustration

Consider the reaction

$$NADH(aq) + H^+(aq) \longrightarrow NAD(aq) + H_2(g)$$

at 37°C, for which $\Delta_r G^\ominus = -21.8 \text{ kJ mol}^{-1}$. NADH is the reduced form of nicotinamide adenine dinucleotide and NAD^+ is its oxidized form; the molecules play an important role in the later stages of the respiratory process. It follows that because $\nu = 1$ and $7 \ln 10 = 16.1$,

$$\Delta_r G^\oplus = -21.8 \text{ kJ mol}^{-1} + 16.1 \times (8.314 \text{ J K}^{-1} \text{mol}^{-1}) \times (310 \text{ K})$$
$$= +19.7 \text{ kJ mol}^{-1}$$

Note that the biological standard value is opposite in sign (in this example) to the thermodynamic standard value.

Self-test 9.10 For a particular reaction of the form $A \rightarrow B + 2H^+$ in aqueous solution, it was found that $\Delta_r G^{\ominus} = +20 \text{ kJ mol}^{-1}$ at 28 °C. Estimate the value of $\Delta_r G^{\oplus}$.

[-61 kJ mol^{-1}]

The standard values for ATP hydrolysis at 37 °C (310 K, blood temperature) are $\Delta_r G^{\oplus} = -30 \text{ kJ mol}^{-1}$, $\Delta_r H^{\oplus} = -20 \text{ kJ mol}^{-1}$, and $\Delta_r S^{\oplus} = +34 \text{ J K}^{-1} \text{mol}^{-1}$. The hydrolysis is therefore exergonic ($\Delta_r G^{\oplus} < 0$) under these conditions, and 30 kJ mol^{-1} is available for driving other reactions. Moreover, because the reaction entropy is large, the reaction Gibbs energy is sensitive to temperature. In view of its exergonicity the ADP-phosphate bond has been called a 'high-energy phosphate bond'. The name is intended to signify a high tendency to undergo reaction, and should not be confused with 'strong' bond. In fact, even in the biological sense it is not of very 'high energy'. The action of ATP depends on it being intermediate in activity. Thus it acts as a phosphate donor to a number of acceptors (for example, glucose), but is recharged by more powerful phosphate donors in the respiration cycle.

(b) Anaerobic and aerobic metabolism

The efficiency of some biological processes can be gauged in terms of the value of $\Delta_r G^{\oplus}$ given above, as we shall consider by considering aerobic and anaerobic metabolism. Aerobic metabolism is a series of reactions in which inhaled oxygen plays a role; anaerobic metabolism is a form of metabolism in which inhaled oxygen plays no role. The energy source of anaerobic cells is glycolysis, the partial oxidation of glucose to lactic acid, and at blood temperature $\Delta_r G^{\oplus} = -218 \text{ kJ mol}^{-1}$. The standard reaction enthalpy is -120 kJ mol^{-1}, the exergonicity exceeding the exothermicity on account of the large increase of entropy accompanying the fracture of the glucose molecule. The glycolysis is coupled to a reaction in which two ADP molecules are converted into two ATP molecules:

$$\text{Glucose} + 2P_i^- + 2\text{ADP} \longrightarrow 2\text{Lactate}^- + 2\text{ATP} + 2H_2O$$

The standard reaction Gibbs energy is $(-218) - 2(-30) \text{ kJ mol}^{-1} = -158 \text{ kJ mol}^{-1}$. The reaction is exergonic, and therefore spontaneous: the metabolism of the food has been used to 'recharge' the ATP.

Metabolism by aerobic respiration is much more efficient. The standard Gibbs energy of combustion of glucose is $-2880 \text{ kJ mol}^{-1}$, and so terminating its oxidation at lactic acid is a poor use of resources. In aerobic respiration the oxidation is carried out to completion, and an extremely complex set of reactions preserves as much of the energy released as possible. In the overall reaction, 38 ATP molecules are generated for each glucose molecule consumed. Each mole of ATP extracts 30 kJ from the 2880 kJ supplied by 1 mol $C_6H_{12}O_6$ (180 g of glucose), and so 1140 kJ has been stored for later use.

Each ATP molecule can be used to drive an endergonic reaction for which $\Delta_r G^{\oplus}$ does not exceed $+30 \text{ kJ mol}^{-1}$. For example, the biosynthesis of sucrose from glucose and fructose can be driven (if a suitable enzyme system is available) because the reaction is endergonic to the extent $\Delta_r G^{\oplus} = +23 \text{ kJ mol}^{-1}$. The biosynthesis of proteins is strongly endergonic, not only on account of the enthalpy change but also on account of the large decrease in entropy that occurs when many amino acids are assembled into a precisely determined sequence. For instance, the formation of a peptide link is endergonic, with $\Delta_r G^{\oplus} = +17 \text{ kJ mol}^{-1}$, but the biosynthesis occurs indirectly and is equivalent to the consumption of three ATP molecules for each link. In a moderately small protein like myoglobin, with about 150 peptide links, the construction alone requires 450 ATP molecules, and therefore about 12 mol of glucose molecules for 1 mol of protein molecules.

Checklist of key ideas

Spontaneous chemical reactions

9.1 The Gibbs energy minimum
- [ ] extent of reaction (ξ)
- [ ] reaction Gibbs energy ($\Delta_r G$, 1)
- [ ] relation of $\Delta_r G$ to the chemical potentials of the species (2)
- [ ] the condition of equilibrium (3)
- [ ] exergonic reaction
- [ ] endergonic reaction
- [ ] reaction quotient (Q)
- [ ] standard reaction Gibbs energy ($\Delta_r G^\ominus$, 6)
- [ ] the equilibrium constant (K) in terms of $\Delta_r G^\ominus$ (8)
- [ ] the general expression for $\Delta_r G$ at an arbitrary stage of the reaction (10)
- [ ] $\Delta_r G^\ominus$ in terms of Gibbs energies of formation (11,12)
- [ ] the reaction quotient (13,14)
- [ ] thermodynamic equilibrium constant (18)

- [ ] the relation between equilibrium constants (19)

The response of equilibria to the conditions

9.2 How equilibria respond to pressure
- [ ] the response of K to catalysts and to changes in pressure (21)
- [ ] the effect of pressure on composition
- [ ] Le Chatelier's principle for the effect of pressure

9.3 The response of equilibria to temperature
- [ ] Le Chatelier's principle for the effect of temperature
- [ ] the van't Hoff equation for the effect of temperature on K (24)
- [ ] the equilibrium constant at one temperature in terms of its value at another temperature (26)

Applications to selected systems

9.4 The extraction of metals from their oxides
- [ ] the construction and interpretation of an Ellingham diagram

9.5 Acids and bases
- [ ] the Brønsted–Lowry classification of acids and bases
- [ ] the pH of a solution (27)
- [ ] conjugate base
- [ ] acidity constant (29)
- [ ] the significance of pK_a (31)
- [ ] conjugate acid
- [ ] basicity constant (33)
- [ ] the relation between K_a and K_b (34)
- [ ] autoprotolysis constant (35)
- [ ] the relation between pH and pOH (36)
- [ ] the pH of a solution of a weak acid (39)
- [ ] the changes in pH in the course of a titration

- [ ] the pH of a solution of a weak base (41)
- [ ] the Henderson–Hasselbalch equation for the pH of a mixed solution (43)
- [ ] the pH at the stoichiometric point (47,48)
- [ ] the pH after the stoichiometric point (49)
- [ ] buffer action
- [ ] the selection of buffers
- [ ] acid–base indicator
- [ ] indicator constant (50)
- [ ] the variation of composition with pH (51)
- [ ] selection of indicators

9.6 Biological activity: the thermodynamics of ATP
- [ ] the hydrolysis of ATP
- [ ] biological standard state
- [ ] the relation between thermodynamic and biological standard states (53)
- [ ] the efficiency of metabolism

Further reading

Articles of general interest

N.C. Craig, The chemists' delta. *J. Chem. Educ.* **64**, 668 (1987).

J.J. MacDonald, Equilibria and $\Delta G^\ominus$. *J. Chem. Educ.* **67**, 745 (1990).

H.R. Kemp, The effect of temperature and pressure on equilibria: a derivation of the van't Hoff rules. *J. Chem. Educ.* **64**, 482 (1987).

M. Deumiè, B. Boulil, and O. Henri-Rousseau, On the minimum of the Gibbs free energy involved in chemical equilibrium: some further comments on the stability of the system. *J. Chem. Educ.* **64**, 201 (1987).

J.J. MacDonald, Equilibrium, free energy, and entropy: rates and differences. *J. Chem. Educ.* **67**, 380 (1990).

H. Xijun and Y. Xiuping, Influences of temperature and pressure on chemical equilibrium in non-ideal systems. *J. Chem. Educ.* **68**, 295 (1991).

A. Dumon, A. Lichanot, and E. Poquet, Describing chemical transformation: from the extent of reaction ξ to the reaction advancement ratio χ. *J. Chem. Educ.* **70**, 29 (1993).

R. deLevie, Explicit expression of the general form of the titration curve in terms of concentration: writing a single closed-form expression for the titration curve for a variety of titrations without using approximations or segmentation. *J. Chem. Educ.* **70**, 209 (1993).

J. Gold and V. Gold, Le Chatelier's principle and the laws of van't Hoff. *Educ. in Chem.* **22**, 82 (1985).

R.T. Allsop and N.H. George, Le Chatelier—a redundant principle? *Educ. in Chem.* **21**, 54 (1984).

S.R. Logan, Entropy of mixing and homogeneous equilibria. *Educ. in Chem.* **25**, 44 (1988).

R.S. Treptow, Free energy versus extent of reaction: understanding the difference between ΔG and $\partial G/\partial \xi$. *J. Chem. Educ.* **73**, 51 (1996).

G. Schmitz, The uncertainty of pH. *J. Chem. Educ.* **71**, 117 (1994).

K. Anderson, Practical calculation of the equilibrium constant and the enthalpy of reaction at different temperatures. *J. Chem. Educ.* **71**, 474 (1994).

R. Holub and P. Vónka, *The chemical equilibrium of gaseous system.* Reidel, Boston (1976).

W.R. Smith and R.W. Missen, *Chemical reaction equilibrium analysis.* Wiley, New York (1982).

J.T. Edsall and H. Guttfreund, *Biothermodynamics.* Wiley, New York (1983).

W.E. Dasent, *Inorganic energetics.* Cambridge University Press (1982).

Texts and sources of data and information

M.J. Blandamer, *Chemical equilibria in solution.* Ellis Horwood/Prentice-Hall, Hemel Hempstead (1992).

K. Denbigh, *The principles of chemical equilibrium.* Cambridge University Press (1981).

D.A. Johnson, *Some thermodynamic aspects of inorganic chemistry.* Cambridge University Press (1982).

F.M. Harold, *The vital force: a study of bioenergetics.* W.H. Freeman & Co, New York (1986).

Exercises

9.1 (a) The equilibrium constant for the isomerization of *cis*-2-butene to *trans*-2-butene is $K = 2.07$ at 400 K. Calculate the standard reaction Gibbs energy.

9.1 (b) The equilibrium constant for the dissociation of Br_2 at 1600 K is $K = 0.255$. Calculate the standard reaction Gibbs energy.

9.2 (a) The standard reaction Gibbs energy of the isomerization of *cis*-2-pentene to *trans*-2-pentene at 400 K is $-3.67 \text{ kJ mol}^{-1}$. Calculate the equilibrium constant of the isomerization.

9.2 (b) The standard reaction Gibbs energy of the decomposition of $CaCO_3$ to CaO and CO_2 at 1173 K is $+0.178 \text{ kJ mol}^{-1}$. Calculate the equilibrium constant of the decomposition.

9.3 (a) At 2257 K and 1.00 atm total pressure, water is 1.77 per cent dissociated at equilibrium by way of the reaction $2H_2O(g) \rightleftharpoons 2H_2(g) + O_2(g)$. Calculate (a) K, (b) $\Delta_r G^{\ominus}$, and (c) $\Delta_r G$ at this temperature.

9.3 (b) For the equilibrium, $N_2O_4(g) \rightleftharpoons 2NO_2(g)$, the degree of dissociation, α_e, at 298 K is 0.201 at 1.00 bar total pressure. Calculate (a) $\Delta_r G$, (b) K, and (c) $\Delta_r G^{\ominus}$ at 298 K.

9.4 (a) Dinitrogen tetroxide is 18.46 per cent dissociated at 25°C and 1.00 bar in the equilibrium $N_2O_4(g) \rightleftharpoons 2NO_2(g)$. Calculate (a) K, (b) $\Delta_r G^{\ominus}$, (c) K at 100°C given that $\Delta_r H^{\ominus} = +57.2 \text{ kJ mol}^{-1}$ over the temperature range.

9.4 (b) Molecular bromine is 24 per cent dissociated at 1600 K and 1.00 bar in the equilibrium $Br_2(g) \rightleftharpoons 2Br(g)$. Calculate (a) K, (b) $\Delta_r G^{\ominus}$, (c) K at 2000°C given that $\Delta_r H^{\ominus} = +112 \text{ kJ mol}^{-1}$ over the temperature range.

9.5 (a) From information in the *Data section*, calculate the standard Gibbs energy and the equilibrium constant at (a) 298 K and (b) 400 K for the reaction $PbO(s) + CO(g) \rightleftharpoons Pb(s) + CO_2(g)$. Assume that the reaction enthalpy is independent of temperature.

9.5 (b) From information in the *Data section*, calculate the standard Gibbs energy and the equilibrium constant at (a) 25°C and (b) 50°C

for the reaction $CH_4(g) + 3Cl_2(g) \rightleftharpoons CHCl_3(l) + 3HCl(g)$. Assume that the reaction enthalpy is independent of temperature.

9.6 (a) In the gas-phase reaction $2A + B \rightleftharpoons 3C + 2D$, it was found that, when 1.00 mol A, 2.00 mol B, and 1.00 mol D were mixed and allowed to come to equilibrium at 25°C, the resulting mixture contained 0.90 mol C at a total pressure of 1.00 bar. Calculate (a) the mole fractions of each species at equilibrium, (b) K_x, (c) K, and (d) $\Delta_r G^{\ominus}$.

9.6 (b) In the gas-phase reaction $A + B \rightleftharpoons C + 2D$, it was found that, when 2.00 mol A, 1.00 mol B, and 3.00 mol D were mixed and allowed to come to equilibrium at 25°C, the resulting mixture contained 0.79 mol C at a total pressure of 1.00 bar. Calculate (a) the mole fractions of each species at equilibrium, (b) K_x, (c) K, and (d) $\Delta_r G^{\ominus}$.

9.7 (a) The standard reaction enthalpy of $Zn(s) + H_2O(g) \rightarrow ZnO(s) + H_2(g)$ is approximately constant at $+224 \text{ kJ mol}^{-1}$ from 920 K up to 1280 K. The standard reaction Gibbs energy is $+33 \text{ kJ mol}^{-1}$ at 1280 K. Estimate the temperature at which the equilibrium constant becomes greater than 1.

9.7 (b) The standard enthalpy of a certain reaction is approximately constant at $+125 \text{ kJ mol}^{-1}$ from 800 K up to 1500 K. The standard reaction Gibbs energy is $+22 \text{ kJ mol}^{-1}$ at 1120 K. Estimate the temperature at which the equilibrium constant becomes greater than 1.

9.8 (a) The equilibrium constant of the reaction $2C_3H_6(g) \rightleftharpoons C_2H_4(g) + C_4H_8(g)$ is found to fit the expression

$$\ln K = -1.04 - \frac{1088}{(T/K)} + \frac{1.51 \times 10^5}{(T/K)^2}$$

between 300 K and 600 K. Calculate the standard reaction enthalpy and standard reaction entropy at 400 K.

9.8 (b) The equilibrium constant of a reaction is found to fit the expression

$$\ln K = -2.04 - \frac{1176}{(T/K)} + \frac{2.1 \times 10^7}{(T/K)^3}$$

between 400 K and 500 K. Calculate the standard reaction enthalpy and standard reaction entropy at 450 K.

9.9 (a) The standard reaction Gibbs energy of the isomerization of borneol ($C_{10}H_{17}OH$) to isoborneol in the gas phase at 503 K is $+9.4$ kJ mol^{-1}. Calculate the reaction Gibbs energy in a mixture consisting of 0.15 mol of borneol and 0.30 mol of isoborneol when the total pressure is 600 Torr.

9.9 (b) The equilibrium pressure of H_2 over solid uranium and uranium hydride, UH_3, at 500 K is 1.04 Torr. Calculate the standard Gibbs energy of formation of $UH_3(s)$ at 500 K.

9.10 (a) Calculate the percentage change in the equilibrium constant K_x of the reaction $H_2CO(g) \rightleftharpoons CO(g) + H_2(g)$ when the total pressure is increased from 1.0 bar to 2.0 bar at constant temperature.

9.10 (b) Calculate the percentage change in the equilibrium constant K_x of the reaction $CH_3OH(g) + NOCl(g) \rightleftharpoons HCl(g) + CH_3NO_2(g)$ when the total pressure is increased from 1.0 bar to 2.0 bar at constant temperature.

9.11 (a) The equilibrium constant for the gas-phase isomerization of borneol ($C_{10}H_{17}OH$) to isoborneol at 503 K is 0.106. A mixture consisting of 7.50 g of borneol and 14.0 g of isoborneol in a container of volume 5.0 L is heated to 503 K and allowed to come to equilibrium. Calculate the mole fractions of the two substances at equilibrium.

9.11 (b) The equilibrium constant for the reaction $N_2(g) + O_2(g) \rightleftharpoons 2NO(g)$ is 1.69×10^{-3} at 2300 K. A mixture consisting of 5.0 g of nitrogen and 2.0 g of oxygen in a container of volume 1.0 L is heated to 2300 K and allowed to come to equilibrium. Calculate the mole fraction of NO at equilibrium.

9.12 (a) Use the data in Table 2.6 of the *Data section* to decide which of the following reactions have $K > 1$ at 298 K.

(a) $HCl(g) + NH_3(g) \rightleftharpoons NH_4Cl(s)$

(b) $2Al_2O_3(s) + 3Si(s) \rightleftharpoons 3SiO_2(s) + 4Al(s)$

(c) $Fe(s) + H_2S(g) \rightleftharpoons FeS(s) + H_2(g)$

9.12 (b) Use the data in Table 2.6 of the *Data section* to decide which of the following reactions have $K > 1$ at 298 K.

(a) $FeS_2(s) + 2H_2(g) \rightleftharpoons Fe(s) + 2H_2S(g)$

(b) $2H_2O_2(l) + H_2S(g) \rightleftharpoons H_2SO_4(l) + 2H_2(g)$

9.13 (a) Which of the equilibria in Exercise 9.12a are favoured (in the sense of K increasing) by a rise in temperature at constant pressure?

9.13 (b) Which of the equilibria in Exercise 9.12b are favoured (in the sense of K increasing) by a reduction in temperature at constant pressure?

9.14 (a) What is the standard enthalpy of a reaction for which the equilibrium constant is (a) doubled, (b) halved when the temperature is increased by 10 K at 298 K?

9.14 (b) What is the standard enthalpy of a reaction for which the equilibrium constant is (a) doubled, (b) halved when the temperature is increased by 15 K at 310 K?

9.15 (a) The standard Gibbs energy of formation of $NH_3(g)$ is -16.5 kJ mol^{-1} at 298 K. What is the reaction Gibbs energy when the partial pressures of the N_2, H_2, and NH_3 (treated as perfect gases) are 3.0 bar, 1.0 bar, and 4.0 bar, respectively? What is the spontaneous direction of the reaction in this case?

9.15 (b) The dissociation vapour pressure of NH_4Cl at 427 °C is 608 kPa but at 459 °C it has risen to 1115 kPa. Calculate (a) the equilibrium constant, (b) the standard reaction Gibbs energy, (c) the standard enthalpy, (d) the standard entropy of dissociation, all at 427 °C. Assume that the vapour behaves as a perfect gas and that $\Delta H^\ominus$ and $\Delta S^\ominus$ are independent of temperature in the range given.

9.16 (a) Estimate the temperature at which $CaCO_3$(calcite) decomposes.

9.16 (b) Estimate the temperature at which $CuSO_4 \cdot 5H_2O$ undergoes dehydration.

9.17 (a) At the half-way point in the titration of a weak acid with a strong base the pH was measured as 5.40. (a) What is the acidity constant and the pK_a of the acid? (b) What is the pH of the solution that is 0.015 M in the acid?

9.17 (b) At the half-way point in the titration of a weak acid with a strong base the pH was measured as 4.82. (a) What is the acidity constant and the pK_a of the acid? (b) What is the pH of the solution that is 0.025 M in the acid?

9.18 (a) Calculate the pH of (a) 0.10 M NH_4Cl(aq), (b) 0.10 M $NaCH_3CO_2$, (c) 0.100 M CH_3COOH(aq).

9.18 (b) Calculate the pH of (a) 0.10 M $NaHCO_2$(aq), (b) 0.20 M $NaC_6H_5CO_2$, (c) 0.150 M HCN(aq).

9.19 (a) Calculate the pH at the stoichiometric point of the titration of 25.00 mL of 0.100 M lactic acid with 0.175 M NaOH(aq).

9.19 (b) Calculate the pH at the stoichiometric point of the titration of 25.00 mL of 0.100 M chlorous acid with 0.175 M NaOH(aq). The pK_a of chlorous acid is 1.96.

9.20 (a) Sketch the pH curve of a solution containing 0.10 M $NaCH_3CO_2$(aq) and a variable amount of acetic acid.

9.20 (b) Sketch the pH curve of a solution containing 0.15 M $NaC_6H_5CO_2$(aq) and a variable amount of benzoic acid.

9.21 (a) From the information in Table 9.1, select suitable buffers for (a) pH = 2.2 and (b) pH = 7.0.

9.21 (b) From the information in Table 9.1, select suitable buffers for (a) pH = 4.6 and (b) pH = 10.8.

Problems

Numerical problems

9.1 The equilibrium constant for the reaction, $I_2(s) + Br_2(g) \rightleftharpoons 2IBr(g)$ is 0.164 at 25°C. (a) Calculate $\Delta_r G^\ominus$ for this reaction. (b) Bromine gas is introduced into a container with excess solid iodine. The pressure and temperature are held at 0.164 atm and 25°C. Find the partial pressure of IBr(g) at equilibrium. Assume that all the bromine is in the liquid form and that the vapour pressure of iodine is negligible. (c) In fact, solid iodine has a measurable vapour pressure at 25°C. In this case, how would the calculation have to be modified?

9.2 Consider the dissociation of methane, $CH_4(g)$, into the elements $H_2(g)$ and C(s, graphite). (a) Given that $\Delta_f H^\ominus(CH_4, g) = -74.85 \text{ kJ mol}^{-1}$ and that $\Delta_f S^\ominus(CH_4, g) = -80.67 \text{ J K}^{-1} \text{ mol}^{-1}$ at 298 K, calculate the value of the equilibrium constant at 298 K. (b) Assuming that $\Delta_f H^\ominus$ is independent of temperature, calculate K at 50°C. (c) Calculate the degree of dissociation, α_e, of methane at 25°C and a total pressure of 0.010 bar. (d) Without doing any numerical calculations, explain how the degree of dissociation for this reaction will change as the pressure and temperature are varied.

9.3 The equilibrium pressure of H_2 over U(s) and $UH_3(s)$ between 450 K and 715 K fits the expression

$$\ln(p/\text{Pa}) = 69.32 - \frac{1.464 \times 10^4}{T/\text{K}} - 5.65 \ln(T/\text{K})$$

Find an expression for the standard enthalpy of formation of $UH_3(s)$ and from it calculate $\Delta_r C_p^\ominus$.

9.4 The degree of dissociation, α_e, of $CO_2(g)$ into CO(g) and $O_2(g)$ at high temperatures was found to vary with temperature as follows:

T/K	1395	1443	1498
$\alpha_e/10^{-4}$	1.44	2.50	4.71

Assuming $\Delta_r H^\ominus$ to be constant over this temperature range, calculate K, $\Delta_r G^\ominus$, $\Delta_r H^\ominus$, and $\Delta_r S^\ominus$. Make any justifiable approximations.

9.5 The standard reaction enthalpy of the decomposition of $CaCl_2 \cdot NH_3(s)$ into $CaCl_2(s)$ and $NH_3(g)$ is nearly constant at $+78 \text{ kJ mol}^{-1}$ between 350 K and 470 K. The equilibrium pressure of NH_3 in the presence of $CaCl_2 \cdot NH_3$ is 12.8 Torr at 400 K. Find an expression for the temperature dependence of $\Delta_r G^\ominus$ in the same range.

9.6 Calculate the equilibrium constant of the reaction $CO(g) + H_2(g) \rightleftharpoons H_2CO(g)$ given that, for the production of liquid formaldehyde, $\Delta_r G^\ominus = +28.95 \text{ kJ mol}^{-1}$ at 298 K and that the vapour pressure of formaldehyde is 1500 Torr at that temperature.

9.7 Acetic acid was evaporated in a container of volume 21.45 cm³ at 437 K and at an external pressure of 764.3 Torr, and the container was then sealed. The mass of acid present in the sealed container was 0.0519 g. The experiment was repeated with the same container but at 471 K, and it was found that 0.0380 g of acetic acid was present.

Calculate the equilibrium constant for the dimerization of the acid in the vapour and the enthalpy of vaporization.

9.8 Hydrogen and carbon monoxide have been investigated for use in fuel cells, so their solubilities in molten salts are of interest. Their solubilities in a molten $NaNO_3/KNO_3$ mixture were examined (E. Desimoni and P.G. Zambonin, *J. Chem. Soc. Faraday Trans. I*, 2014 (1973)) with the following results:

$$\log s(H_2) = -5.39 - \frac{768}{T/\text{K}} \qquad \log s(CO) = -5.98 - \frac{980}{T/\text{K}}$$

where s is the solubility in $\text{mol cm}^{-3} \text{bar}^{-1}$. Calculate the standard molar enthalpies of solution of the two gases at 570 K.

9.9 The dissociation of I_2 can be monitored by measuring the total pressure, and three sets of results are as follows:

T/K	973	1073	1173
$100p/\text{atm}$	6.244	7.500	9.181
$10^4 n_I$	2.4709	2.4555	2.4366

where n_I is the amount of I atoms per mole of I_2 molecules in the mixture, which occupied 342.68 cm³. Calculate the equilibrium constants of the dissociation and the standard enthalpy of dissociation at the mean temperature.

Theoretical problems

9.10 Show that if K_p increases with pressure, then K_ϕ must decrease, where $K = K_p K_\phi$, ϕ denoting the fugacity coefficient.

9.11 Express the equilibrium constant of a gas-phase reaction $A + 3B \rightleftharpoons 2C$ in terms of the equilibrium value of the extent of reaction, ξ, given that initially A and B were present in stoichiometric proportions. Find an expression for ξ as a function of the total pressure, p, of the reaction mixture and sketch a graph of the expression obtained.

9.12 When light passes through a cell of length l containing an absorbing gas at a pressure p, the absorption is proportional to pl. Consider the equilibrium $2NO_2 \rightleftharpoons N_2O_4$, with NO_2 the absorbing species. Show that when two cells of lengths l_1 and l_2 are used, and the pressures needed to obtain equal absorptions are p_1 and p_2, respectively, then the equilibrium constant is given by

$$K = \frac{(p_1 \rho^2 - p_2)^2}{\rho(\rho - 1)(p_2 - p_1 \rho)p^\ominus}$$

with $\rho = l_1/l_2$. The following data were obtained (R.J. Nordstrum and W.H. Chan, *J. Phys. Chem.* **80**, 847 (1976)):

Absorbance	p_1/Torr	p_2/Torr
0.05	1.00	5.47
0.10	2.10	12.00
0.15	3.15	18.65

with $l_1 = 395$ mm and $l_2 = 75$ mm. Determine the equilibrium constant of the reaction.

9.13 Find an expression for the standard reaction Gibbs energy at a temperature T' in terms of its value at another temperature T and the coefficients a, b, and c in the expresson for the molar heat capacity listed in Table 2.2. Evaluate the standard Gibbs energy of formation of $H_2O(l)$ at 372 K from its value at 298 K.

Additional problems supplied by Carmen Giunta and Charles Trapp

9.14 Thorn *et al.* recently carried out a study of $Cl_2O(g)$ by photoelectron ionization (R.P. Thorn, L.J. Stief, S.-C. Kuo, and R.B. Klemm, *J. Phys. Chem.* **100**, 14178 (1996)). From their measurements, they report $\Delta_f H^{\ominus}(Cl_2O) = +77.2$ kJ mol^{-1}. They combined this measurement with literature data on the reaction $Cl_2O(g) + H_2O(g) \rightarrow 2HOCl(g)$, for which $K = 8.2 \times 10^{-2}$ and $\Delta_r S^{\ominus} = +16.38$ J K^{-1} mol^{-1}, and with readily available thermodynamic data on water vapour to report a value for $\Delta_f H^{\ominus}(HOCl)$. Calculate that value. All quantities refer to 298 K.

9.15 The dimerization of ClO in the Antarctic winter stratosphere is believed to play an important part in that region's severe seasonal depletion of ozone. The following equilibrium constants are based on measurements by Cox and Hayman (R.A. Cox and G.D. Hayman, *Nature* **332**, 796 (1988)) on the reaction $2ClO(g) \rightarrow (ClO)_2(g)$.

T/K	233	248	258	268	273
K	4.13×10^8	5.00×10^7	1.45×10^7	5.37×10^6	3.20×10^6
T/K	280	288	295	303	
K	9.62×10^5	4.28×10^5	1.67×10^5	7.02×10^4	

(a) Derive the values of $\Delta_r H^{\ominus}$ and $\Delta_r S^{\ominus}$ for this reaction. (b) Compute the standard enthalpy of formation and the standard molar entropy of $(ClO)_2$ given $\Delta_f H^{\ominus}(ClO) = +101.8$ kJ mol^{-1} and $S_m^{\ominus}(ClO) = 226.6$ J K^{-1} mol^{-1} (*CRC Handbook* (1995)).

9.16 Acid rain is an environmental concern in many parts of Western Europe and eastern North America. In assessing the acidity of rainfall, it is important to have an idea of the acidity of natural rainwater. Assuming that natural rainwater (that is, rainwater uncontaminated with nitric or sulfuric acids) is in equilibrium with 3.6×10^{-4} atm CO_2 (the Henry's law constant is 1.25×10^6 Torr), what is the pH of natural rainwater? What would the pH of natural rainwater have been in pre-industrial times, when the partial pressure of CO_2 was about 2.8×10^{-4} atm? State your approximations explicitly.

9.17 The 1980s saw reports of $\Delta_f H^{\ominus}(SiH_2)$ ranging from 243 to 289 kJ mol^{-1}. For example, the lower value was cited in the review article by Walsh (R. Walsh, *Acc. Chem. Res.* **14**, 246 (1981)); Walsh now leans toward the upper end of the range (H.M. Frey, R. Walsh, and I.M. Watts, *J. Chem. Soc., Chem. Commun.* 1189 (1986)). The higher value was reported by S.-K. Shin and J.L. Beauchamp (*J. Phys. Chem.* **90**, 1507 (1986)). If the standard enthalpy of formation is uncertain by this amount, by what factor is the equilibrium constant for the formation of SiH_2 from its elements uncertain at (a) 298 K, (b) 700 K?

9.18 Nitric acid hydrates have received much attention as possible catalysts for heterogeneous reactions which bring about the Antarctic ozone hole. Worsnop *et al.* investigated the thermodynamic stability of these hydrates under conditions typical of the polar winter stratosphere (D.R. Worsnop, L.E. Fox, M.S. Zahniser, and S.C. Wofsy, *Science* **259**, 71 (1993)). Standard reaction Gibbs energies can be computed for the following reactions at 190 K from their data.

(1) $H_2O(g) \longrightarrow H_2O(s)$ $\Delta_r G^{\ominus} = -23.6$ kJ mol^{-1}
(2) $H_2O(g) + HNO_3(g) \longrightarrow HNO_3 \cdot H_2O(s)$
  $\Delta_r G^{\ominus} = -57.2$ kJ mol^{-1}
(3) $2H_2O(g) + HNO_3(g) \longrightarrow HNO_3 \cdot 2H_2O(s)$
  $\Delta_r G^{\ominus} = -85.6$ kJ mol^{-1}
(4) $3H_2O(g) + HNO_3(g) \longrightarrow HNO_3 \cdot 3H_2O(s)$
  $\Delta_r G^{\ominus} = -112.8$ kJ mol^{-1}

Which solid is thermodynamically most stable at 190 K if $p_{H_2O} = 1.3 \times 10^{-7}$ bar and $p_{HNO_3} = 4.1 \times 10^{-10}$ bar? (*Hint*: Try computing $\Delta_r G^{\ominus}$ for each reaction under the prevailing conditions; if more than one solid forms spontaneously, check on $\Delta_r G^{\ominus}$ for conversion of one solid to another.)

9.19 Alberty *et al.* report thermochemical properties of polycyclic aromatic hydrocarbons, many estimated by group additivity methods (R.A. Alberty, M.B. Chung, and A.K. Reif, *J. Phys. Chem. Ref. Data* **18**, 77 (1989)). Some of their data are shown below for three isomeric pyrenes of formula $C_{20}H_{12}$ at 1000 K.

Species	$\Delta_f H^{\ominus}/(\text{kJ mol}^{-1})$	$S_m^{\ominus}/(\text{J K}^{-1}\text{mol}^{-1})$
Perylene	+253.2	987.9
Benzo(e)pyrene	+253.2	993.7
Benzo(a)pyrene	+262.4	999.4

What are the equilibrium mole fractions of these three isomers at 1000 K?

10

Equilibrium electrochemistry

The principles of thermodynamics established in the preceding chapters can be applied to solutions of electrolytes. One important extension of the previous material, however, is the need to take into account activity coefficients, for they can differ significantly from 1 on account of the strong ionic interactions in electrolyte solutions. These coefficients are best treated as empirical quantities, but it is possible to estimate them in very dilute solutions.

The bulk of the chapter is concerned with the description of the thermodynamic properties of reactions that take place in electrochemical cells, in which, as the reaction proceeds, it drives electrons through an external circuit. Thermodynamic arguments can be used to derive an expression for the electric potential of such cells and the potential can be related to their composition. There are two major topics developed in this connection. One is the definition and tabulation of standard potentials; the second is the use of these standard potentials to predict the equilibrium constants of chemical reactions.

Although the thermodynamic properties of electrolyte solutions can be discussed in terms of chemical potentials and activities in much the same way as solutions of non-electrolytes, they have a number of distinctive features. One is the presence of strong interactions between ions in solution, which means that deviations from ideality are marked even in quite dilute systems. Therefore, we must equip ourselves with means of dealing with activity coefficients that differ significantly from 1. A second feature is that, because many reactions of ions involve the transfer of electrons, they can be studied (and utilized) by allowing them to take place in an electrochemical cell. Measurements like those described in this chapter provide data that are very useful for discussing the characteristics of electrolyte solutions and of ionic equilibria in solution.

The thermodynamic properties of ions in solution

Many of the concepts described in previous chapters carry over without change into the discussion of electrolyte solutions. In this section we focus on departures from ideality.

10.1 Thermodynamic functions of formation

The standard enthalpy and Gibbs energy of a reaction involving ions in solution are expressed in terms of standard enthalpies and Gibbs energies of formation listed in Table 2.6 in the *Data section*. These properties are used in exactly the same way as those for neutral compounds. The values of $\Delta_f H^{\ominus}$ and $\Delta_f G^{\ominus}$ refer to the formation of solutions of ions from the reference states of the parent elements. However, solutions of cations cannot be prepared without their accompanying anions. Therefore, although the standard enthalpy of an overall reaction such as

$$Ag(s) + \tfrac{1}{2}Cl_2(g) \longrightarrow Ag^+(aq) + Cl^-(aq)$$

for which

$$\Delta_r H^{\ominus} = \Delta_f H^{\ominus}(Ag^+, aq) + \Delta_f H^{\ominus}(Cl^-, aq)$$

is meaningful and measurable (and found to be $-61.58 \text{ kJ mol}^{-1}$), the enthalpies of the individual formation reactions are not measurable.

(a) Standard functions of formation of ions

The problem is solved by defining one ion, conventionally the hydrogen ion, to have zero standard enthalpy and Gibbs energy of formation:

$$\Delta_f H^{\ominus}(H^+, aq) = 0 \qquad \Delta_f G^{\ominus}(H^+, aq) = 0 \tag{1}$$

at all temperatures. In essence, this definition adjusts the actual values of the enthalpies and Gibbs energies of formation of ions by a fixed amount, which is chosen so that the standard value for one of them, $H^+(aq)$, has the value zero. Then in the reaction

$$\tfrac{1}{2}H_2(g) + \tfrac{1}{2}Cl_2(g) \longrightarrow H^+(aq) + Cl^-(aq) \qquad \Delta_r G^{\ominus} = -131.23 \text{ kJ mol}^{-1}$$

we can write

$$\Delta_r G^{\ominus} = \Delta_f G^{\ominus}(H^+, aq) + \Delta_f G^{\ominus}(Cl^-, aq) = \Delta_f G^{\ominus}(Cl^-, aq)$$

and hence identify $\Delta_f G^{\ominus}(Cl^-, aq)$ as $-131.23 \text{ kJ mol}^{-1}$. All the Gibbs energies and enthalpies of formation of ions in Tables 10.1 and 2.6 were calculated in the same way.

Illustration

With the value of $\Delta_f G^{\ominus}(Cl^-, aq)$ established, we can find the value of $\Delta_f G^{\ominus}(Ag^+, aq)$ from

$$Ag(s) + \tfrac{1}{2}Cl_2(g) \longrightarrow Ag^+(aq) + Cl^-(aq) \qquad \Delta_r G^{\ominus} = -54.12 \text{ kJ mol}^{-1}$$

which leads to $\Delta_f G^{\ominus}(Ag^+, aq) = +77.11 \text{ kJ mol}^{-1}$.

Table 10.1* Standard thermodynamic functions of formation of ions in aqueous solution at 298 K

Ion	$\Delta_f H^{\ominus}/(\text{kJ mol}^{-1})$	$\Delta_f G^{\ominus}/(\text{kJ mol}^{-1})$
Cl^-	-167.2	-131.2
Cu^{2+}	$+64.8$	$+65.5$
H^+	0	0
K^+	-252.4	-283.3
Na^+	-240.1	-261.9
PO_4^{3-}	-1277.0	-1019.0

* More values are given in the *Data section* at the end of this volume; see Table 2.6.

Self-test 10.1 The standard enthalpy of formation of $AgNO_3(aq)$ is $-99.4 \text{ kJ mol}^{-1}$ at 298 K. Calculate the standard enthalpy of formation of the nitrate ion in water by using the standard enthalpy of formation of $Ag^+(aq)$ in Table 2.6.

$$[-205.0 \text{ kJ mol}^{-1}]$$

(b) Contributions to the Gibbs energy of formation

The factors responsible for the magnitude of the Gibbs energy of formation of an ion in solution can be identified by analysing it in terms of a thermodynamic cycle. As an illustration, we consider the reasons for the difference between the standard Gibbs energies

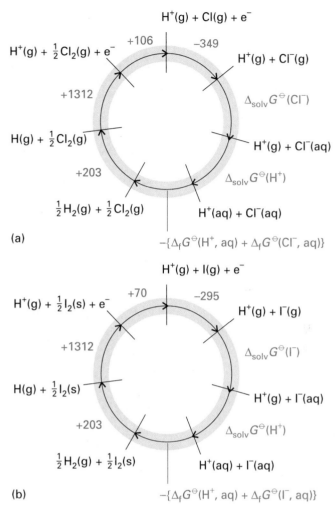

10.1 The thermodynamic cycles for the discussion of the Gibbs energies of solvation (hydration) and formation of (a) chloride ions, (b) iodide ions in aqueous solution. The changes in Gibbs energies around the cycle sum to zero because G is a state function.

of formation of Cl^- and I^- in water, which are -131 kJ mol^{-1} and -52 kJ mol^{-1}, respectively. We do so by treating their formation in the reaction

$$\tfrac{1}{2}H_2(g) + \tfrac{1}{2}X_2(g) \longrightarrow H^+(aq) + X^-(aq)$$

as the outcome of the sequence of steps shown in Fig. 10.1 (with values taken from the *Data section*, particularly Tables 2.6, 13.4, and 13.5).[1] In each case we use the fact that the sum of the Gibbs energies for all the steps around a closed cycle is zero. It follows that

$$\Delta_f G^{\ominus}(Cl^-, aq) = 1272 \text{ kJ mol}^{-1} + \Delta_{solv} G^{\ominus}(H^+) + \Delta_{solv} G^{\ominus}(Cl^-)$$
$$\Delta_f G^{\ominus}(I^-, aq) = 1290 \text{ kJ mol}^{-1} + \Delta_{solv} G^{\ominus}(H^+) + \Delta_{solv} G^{\ominus}(I^-)$$

---

1   The standard Gibbs energies of formation of the gas-phase ions are unknown. We have therefore used ionization energies and electron affinities (Section 13.4f), and have assumed that any differences from the Gibbs energies arising from conversion to enthalpy and the inclusion of entropies to obtain Gibbs energies in the formation of H^+ are cancelled by the corresponding terms in the electron gain of X. The conclusions from the cycles are therefore only approximate.

Table 10.2* Relative permittivities (dielectric constants) at 298 K

	ε_r
Ammonia	16.9
	22.4 (at $-33°C$)
Benzene	2.274
Ethanol	24.30
Water	78.54

* More values are given in the *Data section*.

An important point to note is that the value of $\Delta_f G^{\ominus}$ of an ion is not determined by the properties of the ion alone but includes contributions from the dissociation, ionization, and hydration of hydrogen. The difference between the two values is

$$\Delta_f G^{\ominus}(\text{Cl}^-, \text{aq}) - \Delta_f G^{\ominus}(\text{I}^-, \text{aq})$$
$$= \Delta_{\text{solv}} G^{\ominus}(\text{Cl}^-) - \Delta_{\text{solv}} G^{\ominus}(\text{I}^-) - 18 \text{ kJ mol}^{-1} \tag{2}$$

The two unknown quantities are the standard Gibbs energies of solvation $\Delta_{\text{solv}} G^{\ominus}$, the standard reaction Gibbs energy for the processes

$$\text{M}^+(\text{g}) \longrightarrow \text{M}^+(\text{solution}) \qquad \text{X}^-(\text{g}) \longrightarrow \text{A}^-(\text{solution}) \tag{3}$$

Experimentally, we know that

$$\Delta_f G^{\ominus}(\text{Cl}^-, \text{aq}) - \Delta_f G^{\ominus}(\text{I}^-, \text{aq}) = -79 \text{ kJ mol}^{-1}$$

so we can conclude that

$$\Delta_{\text{solv}} G^{\ominus}(\text{Cl}^-) - \Delta_{\text{solv}} G^{\ominus}(\text{I}^-) = -61 \text{ kJ mol}^{-1}$$

Gibbs energies of solvation of individual ions may be estimated from an equation derived by Max Born, who identified $\Delta_{\text{solv}} G^{\ominus}$ with the electrical work of transferring an ion from a vacuum into the solvent treated as a continuous dielectric of relative permittivity ε_r (Table 10.2). The resulting **Born equation**, which is derived in the *Justification* below, is

$$\Delta_{\text{solv}} G^{\ominus} = -\frac{z_i^2 e^2 N_A}{8\pi\varepsilon_0 r_i}\left(1 - \frac{1}{\varepsilon_r}\right) \tag{4}$$

where z_i is the charge number of the ion and r_i its radius[2] (N_A is the Avogadro constant). Note that $\Delta_{\text{solv}} G^{\ominus} < 0$, and that it is strongly negative for small, highly charged ions in media of high relative permittivity. For water at 25°C,

$$\Delta_{\text{solv}} G^{\ominus} = -\frac{z_i^2}{(r_i/\text{pm})} \times (6.86 \times 10^4 \text{ kJ mol}^{-1}) \tag{5}$$

Justification 10.1

The electrical concepts required in this derivation are reviewed in *Further information 5*. The strategy of the calculation is to identify the Gibbs energy of solvation with the work of transferring an ion from a vacuum into the solvent. That work is calculated by taking the difference of the work of charging an ion when it is in the solution and the work of charging the same ion when it is in a vacuum. We model an ion as a sphere of radius r_i immersed in a medium of permittivity ε. When the charge of the sphere is q, the electric potential, ϕ, at its surface is

$$\phi = \frac{q}{4\pi\varepsilon r_i}$$

The work of bringing up a charge dq to the sphere is $\phi \, dq$. Therefore, the total work of charging the sphere from 0 to $z_i e$ is

$$w = \int_0^{z_i e} \phi \, dq = \frac{1}{4\pi\varepsilon r_i} \int_0^{z_i e} q \, dq = \frac{z_i^2 e^2}{8\pi\varepsilon r_i}$$

This electrical work of charging, when multiplied by the Avogadro constant, is the molar Gibbs energy for charging the ions.

The work of charging an ion in a vacuum is obtained by setting $\varepsilon = \varepsilon_0$, the vacuum permittivity. The corresponding value for charging the ion in a medium is obtained by

2   Ionic radii are given in Table 21.3.

setting $\varepsilon = \varepsilon_r\varepsilon_0$, where ε_r is the relative permittivity of the medium. It follows that the change in molar Gibbs energy that accompanies the transfer of ions from a vacuum to a solvent is the difference of these two quantities:

$$\Delta_{solv}G^{\ominus} = \frac{z_i^2 e^2 N_A}{8\pi\varepsilon_0\varepsilon_r r_i} - \frac{z_i^2 e^2 N_A}{8\pi\varepsilon_0 r_i}$$

which can easily be rearranged into eqn 4.

Illustration

To see how closely the Born equation reproduces the experimental data, we calculate that the difference in the values of $\Delta_f G^{\ominus}$ for Cl^- and I^- in water, for which $\varepsilon_r = 78.54$ at $25\,°C$, given their radii as 181 pm and 220 pm (Table 21.3), respectively, is

$$\Delta_{solv}G^{\ominus}(Cl^-) - \Delta_{solv}G^{\ominus}(I^-) = -\left(\frac{1}{181} - \frac{1}{220}\right) \times (6.86 \times 10^4 \text{ kJ mol}^{-1})$$

$$= -67 \text{ kJ mol}^{-1}$$

This estimated difference is in good agreement with the experimental difference.

Self-test 10.2 Estimate the value of $\Delta_{solv}G^{\ominus}(Cl^-, aq) - \Delta_{solv}G^{\ominus}(Br^-, aq)$ from experimental data and from the Born equatioin.

[-26 kJ mol^{-1} experimental; -29 kJ mol^{-1} calculated]

(c) Standard entropies of ions in solution

Although the partial molar entropy of the solute in an electrolyte solution can be measured, there is no experimental way of ascribing a part of that entropy to the cations and a part to the anions. Therefore, yet again we are forced to define the partial molar entropy of one species and set up a table of values for other ions on that basis. The entropies of ions in solution are reported on a scale in which the standard entropy of the H^+ ions in water is taken as zero at all temperatures:

$$S^{\ominus}(H^+, aq) = 0 \qquad [6]$$

The values based on this choice are listed in Table 10.3 and more fully in Table 2.6 in the *Data section*.

Because the entropies of ions in water are values relative to the hydrogen ion in water, they may be either positive or negative. A positive entropy means that an ion has a higher partial molar entropy than H^+ in water and a negative entropy means that the ion has a lower partial molar entropy than H^+ in water. For instance, the entropy of $Cl^-(aq)$ is $+57$ J K^{-1} mol^{-1} and that of $Mg^{2+}(aq)$ is -128 J K^{-1} mol^{-1}. Partial molar ion entropies vary as expected on the basis that they are related to the degree to which the ions order the water molecules around them in the solution. Small, highly charged ions induce local structure in the surrounding water, and the disorder of the solution is decreased more than in the case of large, singly charged ions. The absolute, Third-Law standard partial molar entropy of the proton in water can be estimated by proposing a model of the structure it induces, and there is some agreement on the value -21 J K^{-1} mol^{-1}. The negative value indicates that the proton induces order in the solvent.

Table 10.3* Standard entropies of ions in aqueous solution at 298 K

	$S_m^{\ominus}/(\text{J K}^{-1}\text{ mol}^{-1})$	
Cl^-	$+56.5$	
Cu^{2+}	-99.6	
H^+	0	by definition
K^+	-102.5	
Na^+	-59.0	
PO_4^{3-}	-221.8	

* More values are given in the *Data section*; see Table 2.6.

10.2 Ion activities

Interactions between ions are so strong that the approximation of replacing activities by molalities is valid only in very dilute solutions (less than 10^{-3} mol kg^{-1} in total ion concentration) and in precise work activities themselves must be used.

(a) The definition of activity

We saw in Section 7.7c that the chemical potential of a solute in a real solution is related to its activity a by

$$\mu = \mu^{\ominus} + RT \ln a \tag{7}$$

where the standard state is a hypothetical solution with molality $b^{\ominus} = 1$ mol kg^{-1} in which the ions are behaving ideally. The activity is related to the molality,[3] b, by

$$a = \gamma b/b^{\ominus} \tag{8}$$

where the activity coefficient, γ, depends on the composition, molality, and temperature of the solution. As the solution approaches ideality (in the sense of obeying Henry's law) at low molalities, the activity coefficient tends towards 1:

$$\gamma \to 1 \quad \text{and} \quad a \to b/b^{\ominus} \quad \text{as } b \to 0 \tag{9}$$

Because all deviations from ideality are carried in the activity coefficient, the chemical potential can be written[4]

$$\mu = \mu^{\ominus} + RT \ln b + RT \ln \gamma = \mu^{\text{ideal}} + RT \ln \gamma \tag{10}$$

where μ^{ideal} is the chemical potential of the ideal-dilute solution of the same molality.

(b) Mean activity coefficients

If the chemical potential of a univalent cation M^+ is denoted μ_+ and that of a univalent anion X^- is denoted μ_-, the total Gibbs energy of the ions in the electrically neutral solution is the sum of these partial molar quantities. The molar Gibbs energy of an ideal solution is

$$G_m^{\text{ideal}} = \mu_+^{\text{ideal}} + \mu_-^{\text{ideal}} \tag{11}°$$

However, for a real solution of M^+ and X^- of the same molality,

$$\begin{aligned} G_m &= \mu_+ + \mu_- = \mu_+^{\text{ideal}} + \mu_-^{\text{ideal}} + RT \ln \gamma_+ + RT \ln \gamma_- \\ &= G_m^{\text{ideal}} + RT \ln \gamma_+\gamma_- \end{aligned} \tag{12}$$

All the deviations from ideality are contained in the last term.

There is no experimental way of separating the product $\gamma_+\gamma_-$ into contributions from the cations and the anions. The best we can do experimentally is to assign responsibility for the nonideality equally to both kinds of ion. Therefore, for a 1,1-electrolyte, we introduce the **mean activity coefficient** as the geometric mean[5] of the individual coefficients:

$$\gamma_{\pm} = \left(\gamma_+\gamma_-\right)^{1/2} \tag{13}$$

and express the individual chemical potentials of the ions as

$$\mu_+ = \mu_+^{\text{ideal}} + RT \ln \gamma_{\pm} \qquad \mu_- = \mu_-^{\text{ideal}} + RT \ln \gamma_{\pm} \tag{14}$$

---

3   The IUPAC recommendation for the symbol is either m or b; whereas the former is more commonly used, b has the advantage that it is less likely to be confused with mass, m.

4   In accord with the convention adopted in earlier chapters, the notation has been simplified by interpreting b as $b/b^{\ominus}$ and labelling the equation $\{\cdots\}$.

5   The geometric mean of x and y is $(xy)^{1/2}$.

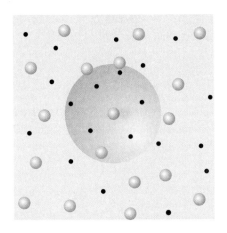

10.2 The picture underlying the Debye-Hückel theory is of a tendency for anions to be found around cations, and of cations to be found around anions (one such local clustering region is shown by the sphere). The ions are in ceaseless motion, and the diagram represents the time average of their motion.

The sum of these two chemical potentials is the same as before, eqn 12, but now the nonideality is shared equally.

This approach can be generalized to the case of a compound M_pX_q that dissolves to give a solution of p cations and q anions from each formula unit. The molar Gibbs energy of the ions is the sum of their partial molar Gibbs energies:

$$G_m = p\mu_+ + q\mu_- = G_m^{ideal} + pRT \ln \gamma_+ + qRT \ln \gamma_- \qquad (15)$$

If we introduce the mean activity coefficient

$$\gamma_{\pm} = (\gamma_+^p \gamma_-^q)^{1/s} \qquad s = p + q \qquad [16]$$

and write the chemical potential of each ion as

$$\mu_i = \mu_i^{ideal} + RT \ln \gamma_{\pm} \qquad (17)$$

we get the same expression as in eqn 15 for G_m when we write

$$G = p\mu_+ + q\mu_- \qquad (18)$$

However, both types of ion now share equal responsibility for the nonideality.

(c) The Debye-Hückel limiting law

The long range and strength of the Coulombic interaction between ions means that it is likely to be primarily responsible for the departures from ideality in ionic solutions and to dominate all the other contributions to nonideality. This domination is the basis of the **Debye-Hückel theory** of ionic solutions, which was devised by Peter Debye and Erich Hückel in 1923. We give here a qualitative account of the theory and its principal conclusions. The calculation itself, which is a profound example of how a seemingly intractable problem can be formulated and then resolved by drawing on physical insight, is outlined in the *Justification* on p. 250.

Oppositely charged ions attract one another. As a result, anions are more likely to be found near cations in solution, and vice versa (Fig. 10.2). Overall the solution is electrically neutral, but near any given ion there is an excess of counter-ions (ions of opposite charge). Averaged over time, counter-ions are more likely to be found near any given ion. This time-averaged, spherical haze, in which counter-ions outnumber ions of the same charge as the central ion, has a net charge equal in magnitude but opposite in sign to that on the central ion, and is called its **ionic atmosphere**. The energy, and therefore the chemical potential, of any given central ion is lowered as a result of its electrostatic interaction with its ionic atmosphere. This lowering of energy appears as the difference between the molar Gibbs energy G_m and the ideal value G_m^{ideal} of the solute, and hence can be identified with $sRT \ln \gamma_{\pm}$. The stabilization of ions by their interaction with their ionic atmospheres is part of the explanation why chemists commonly use dilute solutions, in which the stabilization is less important, to achieve precipitation of ions from electrolyte solutions.

The model leads to the result that at very low concentrations the activity coefficient can be calculated from the **Debye-Hückel limiting law**

$$\log \gamma_{\pm} = -|z_+ z_-| A I^{1/2} \qquad (19)$$

where $A = 0.509$ for an aqueous solution at $25\,^{\circ}C$ and I is the dimensionless **ionic strength** of the solution:

$$I = \tfrac{1}{2} \sum_i z_i^2 (b_i/b^{\ominus}) \qquad [20]$$

In this expression z_i is the charge number of an ion i (positive for cations and negative for anions) and b_i is its molality. The ionic strength occurs widely wherever ionic solutions are

Table 10.4 Ionic strength and molality, $I = k \times b/b^{\ominus}$

k	X^-	X^{2-}	X^{3-}	X^{4-}
M^+	1	3	6	10
M^{2+}	3	4	15	12
M^{3+}	6	15	9	42
M^{4+}	10	12	42	16

For example, the ionic strength of an M_2X_3 solution of molality b, which is understood to give M^{3+} and X^{2-} ions in solution, is $15b/b^{\ominus}$.

discussed, as we shall see. The sum extends over all the ions present in the solution. For solutions consisting of two types of ion at molalities b_+ and b_-,

$$I = \tfrac{1}{2}(b_+ z_+^2 + b_- z_-^2)/b^{\ominus} \tag{21}$$

The ionic strength emphasizes the charges of the ions because the charge numbers occur as their squares. Table 10.4 summarizes the relation of ionic strength and molality in an easily usable form.

Justification 10.2

Imagine a solution in which all the ions have their actual positions, but in which their Coulombic interactions have been turned off. The difference in molar Gibbs energy between the ideal and real solutions is equal to w_e, the electrical work of charging the system in this arrangement. Therefore, for a salt M_pX_q,

$$\ln \gamma_{\pm} = \frac{w_e}{sRT} \qquad s = p + q \tag{22}$$

It follows that we must first find the final distribution of the ions and then the work of charging them in that distribution.

The Coulomb potential at a distance r from an isolated ion of charge $z_i e$ in a medium of permittivity ε is

$$\phi_i = \frac{Z_i}{r} \qquad Z_i = \frac{z_i e}{4\pi\varepsilon} \tag{23}$$

The ionic atmosphere causes the potential to decay with distance more sharply than this expression implies. Such shielding is a familiar problem in electrostatics, and its effect is taken into account by replacing the Coulomb potential by the **shielded Coulomb potential**, an expression of the form

$$\phi_i = \frac{Z_i}{r} e^{-r/r_D} \tag{24}$$

where r_D is called the **Debye length**. When r_D is large, the shielded potential is virtually the same as the unshielded potential. When it is small, the shielded potential is much smaller than the unshielded potential, even for short distances (Fig. 10.3).

To calculate r_D, we need to know how the charge density, ρ_i, of the ionic atmosphere, the charge in a small region divided by the volume of the region, varies with distance from the ion. This step draws on another standard result of electrostatics, in which charge density and potential are related by **Poisson's equation**. Because we are considering only a spherical ionic atmosphere, we can use a simplified form of this equation in which the charge density varies only with distance from the central ion:

$$\frac{1}{r^2}\frac{d}{dr}\left(r^2 \frac{d\phi_i}{dr}\right) = -\frac{\rho_i}{\varepsilon} \tag{25}$$

Substitution of the expression for the shielded potential, eqn 24, results in

$$r_D^2 = -\frac{\varepsilon\phi_i}{\rho_i} \tag{26}$$

To solve this equation we need to relate ρ_i and ϕ_i.

For the next step we draw on the fact that the energy of an ion depends on its closeness to the central ion, and then use the Boltzmann distribution to work out the probability that an ion will be found at each distance. The energy of an ion of charge $z_j e$ at a distance

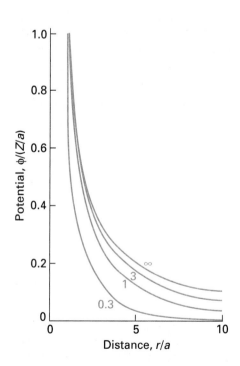

10.3 The variation of the shielded Coulomb potential with distance for different values of the Debye length, r/a. The smaller the Debye length, the more sharply the potential decays to zero. In each case, a is an arbitrary unit of length.

where it experiences the potential ϕ_i of the central ion i relative to its energy when it is far away in the bulk solution is its charge times the potential:

$$E = z_j e \phi_i \tag{27}$$

Therefore, according to the Boltzmann distribution (see the *Introduction*), the ratio of the molar concentration, c_j, of ions at a distance r and the molar concentration in the bulk, c_j°, where the energy is zero, is:

$$\frac{c_j}{c_j^\circ} = e^{-E/kT} \tag{28}$$

The charge density, ρ_i, at a distance r from the ion i is the molar concentration of each type of ion multiplied by the charge per mole of ions, $z_i e N_A$. The quantity $e N_A$, the magnitude of the charge per mole of electrons, occurs widely throughout electrochemistry, and is called the **Faraday constant**, F:

$$F = e N_A = 96.485 \ \text{kC} \, \text{mol}^{-1} \tag{29}$$

It follows that

$$\rho_i = c_+ z_+ F + c_- z_- F = c_+^\circ z_+ F e^{-z_+ e \phi_i / kT} + c_-^\circ z_- F e^{-z_- e \phi_i / kT} \tag{30}$$

At this stage we need to simplify the expression to avoid the awkward exponential terms. Because the average electrostatic interaction energy is small compared with kT, we may write eqn 30 as

$$\rho_i = (c_+^\circ z_+ + c_-^\circ z_-)F - (c_+^\circ z_+^2 + c_-^\circ z_-^2)\left(\frac{F^2 \phi_i}{RT}\right) + \cdots \tag{31}$$

(To obtain this expression, we have replaced an e by F/N_A and recognized that $N_A k = R$.) The first term in the expansion is zero because it is the charge density in the bulk, uniform solution, and the solution is electrically neutral. The unwritten terms are assumed to be too small to be significant. The one remaining term can be expressed in terms of the ionic strength, eqn 20, by noting that in the dilute aqueous solutions we are considering there is little difference between molality and molar concentration, and $c \approx b\rho$, where ρ is the mass density of the solvent

$$c_+^\circ z_+^2 + c_-^\circ z_-^2 \approx (b_+^\circ z_+^2 + b_-^\circ z_-^2)\rho = 2Ib^{\ominus}\rho$$

With these approximations, eqn 31 becomes

$$\rho_i = -\frac{2\rho F^2 I b^{\ominus} \phi_i}{RT} \tag{32}$$

We can now solve eqn 26 for r_D:

$$r_D = \left(\frac{\varepsilon RT}{2\rho F^2 I b^{\ominus}}\right)^{1/2} \tag{33}$$

To calculate the activity coefficient we need to find the electrical work of charging the central ion when it is surrounded by its atmosphere. To do so, we need to know the potential at the ion due to its atmosphere, ϕ_{atmos}. This potential is the difference between the total potential, given by eqn 24, and the potential due to the central ion itself:

$$\phi_{\text{atmos}} = \phi - \phi_{\text{central ion}} = Z_i \left(\frac{e^{-r/r_D}}{r} - \frac{1}{r}\right) \tag{34}$$

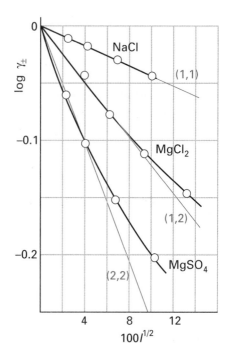

10.4 An experimental test of the Debye–Hückel limiting law. Although there are marked deviations for moderate ionic strengths, the limiting slopes as $I \to 0$ are in good agreement with the theory, so the law can be used for extrapolating data to very low molalities.

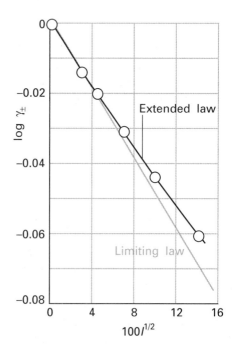

10.5 The extended Debye–Hückel law gives agreement with experiment over a wider range of molalities (as shown here for a 1,1-electrolyte), but it fails at higher molalities.

The potential at the central ion (at $r = 0$) is obtained by taking the limit of this expression as $r \to 0$, and is

$$\phi_{atmos}(0) = -\frac{z_i}{r_D} \tag{35}$$

This expression shows us that the potential of the ionic atmosphere is equivalent to the potential arising from a single charge of equal magnitude but opposite sign to that of the central ion and located at a distance r_D from the ion. If the charge of the central ion were q and not $z_i e$, the potential due to its atmosphere would be

$$\phi_{atmos}(0) = -\frac{q}{4\pi\varepsilon r_D}$$

The work of adding a charge dq to a region where the electrical potential is $\phi_{atmos}(0)$ is

$$dw_e = \phi_{atmos}(0)\, dq$$

Therefore, the total molar work of fully charging the ions is

$$w_e = N_A \int_0^{z_i e} \phi_{atmos}(0)\, dq = -\frac{N_A}{4\pi\varepsilon r_D}\int_0^{z_i e} q\, dq$$

$$= -\frac{N_A z_i^2 e^2}{8\pi\varepsilon r_D} = -\frac{z_i^2 F^2}{8\pi\varepsilon N_A r_D} \tag{36}$$

It follows from eqn 22 that the mean activity coefficient of the ions is

$$\ln \gamma_\pm = \frac{p w_{e,+} + q w_{e,-}}{sRT} = -\frac{(pz_+^2 + qz_-^2)F^2}{8\pi\varepsilon s N_A RT r_D}$$

However, for neutrality $pz_+ + qz_- = 0$, so[6]

$$\ln \gamma_\pm = -\frac{|z_+ z_-|F^2}{8\pi\varepsilon N_A RT r_D} \tag{37}$$

When we replace r_D by using eqn 33, and convert to common logarithms, this expression turns into eqn 19 with

$$A = \frac{F^3}{4\pi N_A \ln 10}\left(\frac{\rho b^{\ominus}}{2\varepsilon^3 R^3 T^3}\right)^{1/2} \tag{38}$$

Illustration

The mean activity coefficient of 5.0×10^{-3} mol kg^{-1} KCl(aq) at 25 °C is calculated by writing

$$I = \tfrac{1}{2}(b_+ + b_-)/b^{\ominus} = b/b^{\ominus}$$

where b is the molality of the solution (and $b_+ = b_- = b$). Then, from eqn 19,

$$\log \gamma_\pm = -0.509 \times (5.0 \times 10^{-3})^{1/2} = -0.036$$

Hence, $\gamma_\pm = 0.92$. The experimental value is 0.927.

6   For this step, multiply $pz_+ + qz_- = 0$ by p and also, separately, by q; add the two expressions and rearrange the result by using $p + q = s$ and $z_+ z_- = -|z_+ z_-|$.

Table 10.5* Mean activity coefficients in water at 298 K

$b/(mol\,kg^{-1})$	KCl	CaCl$_2$
0.001	0.966	0.888
0.01	0.902	0.732
0.1	0.770	0.524
1.0	0.607	0.725

* More values are given in the *Data section*.

Self-test 10.3 Calculate the ionic strength and the mean activity coefficient of 1.00×10^{-3} mol kg^{-1} CaCl$_2$(aq) at 25 °C.

$$[3.00 \times 10^{-3} \text{ mol kg}^{-1}, 0.880]$$

The name 'limiting law' is applied to eqn 19 because ionic solutions of moderate molalities may have activity coefficients that differ from the values given by this expression, yet all solutions are expected to conform in the limit of arbitrarily low molalities. Some experimental values of activity coefficients for salts of various valence types are listed in Table 10.5. Figure 10.4 shows some of these values plotted against $I^{1/2}$, and compares them with the theoretical straight lines calculated from eqn 19. The agreement at very low molalities (less than about 1 mmol kg^{-1}, depending on charge type) is impressive and convincing evidence in support of the model. Nevertheless, the departures from the theoretical curves above these molalities are large and show that the approximations are valid only at very low concentrations.

(d) The extended Debye–Hückel law

When the ionic strength of the solution is too high for the limiting law to be valid, it is found that the activity coefficient may be estimated from the **extended Debye–Hückel Law**:

$$\log \gamma_{\pm} = -\frac{A|z_+ z_-|I^{1/2}}{1 + BI^{1/2}} \tag{39}$$

where B is another dimensionless constant. Although B can be interpreted as a measure of the closest approach of the ions, it is best regarded as an adjustable empirical parameter. A curve drawn in this way is shown in Fig. 10.5. It is clear that eqn 39 accounts for some activity coefficients over a moderate range of dilute solutions (up to about 0.1 mol kg^{-1}); nevertheless it remains very poor near 1 mol kg^{-1}.

Current theories of activity coefficients for ionic solutes take an indirect route. They set up a theory for the dependence of the activity coefficient of the *solvent* on the concentration of the solute, and then use the Gibbs–Duhem equation (eqn 7.12) to estimate the activity coefficient of the solute. The results are reasonably reliable for solutions with molalities greater than about 0.1 mol kg^{-1} and are valuable for the discussion of mixed salt solutions, such as sea water.

Electrochemical cells

An **electrochemical cell** consists of two **electrodes**, or metallic conductors, in contact with an **electrolyte**, an ionic conductor (which may be a solution, a liquid, or a solid). An electrode and its electrolyte comprise an **electrode compartment**. The two electrodes may share the same compartment. The various kinds of electrode are summarized in Table 10.6 and illustrated in Fig. 10.6. When an 'inert metal' is part of the specification, it is present to act as a source or sink of electrons, but takes no other part in the reaction other than acting as a catalyst for it. If the electrolytes are different, the two compartments may be joined by a **salt bridge**, which is a concentrated electrolyte solution in agar jelly that completes the electrical circuit and enables the cell to function.

A **galvanic cell** is an electrochemical cell that produces electricity as a result of the spontaneous reaction occurring inside it. An **electrolytic cell** is an electrochemical cell in which a non-spontaneous reaction is driven by an external source of current.

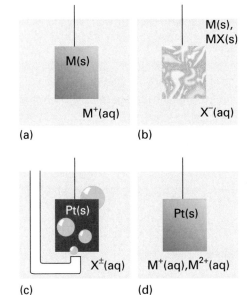

10.6 Typical electrode types: (a) metal/metal ion; (b) metal/insoluble salt; (c) gas; (d) redox electrodes.

Table 10.6 Varieties of electrode

Electrode type	Designation	Redox couple	Half-reaction
Metal/metal-ion	$M(s)|M^+(aq)$	M^+/M	$M^+(aq) + e^- \rightarrow M(s)$
Gas electrode	$Pt(s)|X_2(g)|X^+(aq)$	X^+/X_2	$X^+(aq) + e^- \rightarrow \frac{1}{2}X_2(g)$
	$Pt(s)|X_2(g)|X^-(aq)$	X_2/X^-	$\frac{1}{2}X_2(g) + e^- \rightarrow X^-(aq)$
Metal/insoluble-salt	$M(s)|MX(s)|X^-(aq)$	$MX/M,X^-$	$MX(s) + e^- \rightarrow M(s) + X^-(aq)$
Redox	$Pt(s)|M^+(aq),M^{2+}(aq)$	M^{2+}/M^+	$M^{2+}(aq) + e^- \rightarrow M^+(aq)$

10.3 Half-reactions and electrodes

It will be familiar from introductory chemistry courses that **oxidation** is the removal of electrons from a species and **reduction** is the addition of electrons to a species. A **redox reaction** is a reaction in which there is a transfer of electrons from one species to another. The electron transfer may be accompanied by other events, such as atom or ion transfer, but the net effect is electron transfer and hence a change in oxidation number of an element. The **reducing agent** (or 'reductant') is the electron donor; the **oxidizing agent** (or 'oxidant') is the electron acceptor.

(a) Half-reactions

Any redox reaction may be expressed as the difference of two reduction **half-reactions**, which are conceptual reactions showing the gain of electrons. For example, the reduction of Cu^{2+} ions by zinc can be expressed as the difference of the following two half-reactions:

$$Cu^{2+}(aq) + 2e^- \longrightarrow Cu(s) \qquad Zn^{2+}(aq) + 2e^- \longrightarrow Zn(s)$$

The difference of the two (copper − zinc) is

$$Cu^{2+}(aq) + Zn(s) \longrightarrow Cu(s) + Zn^{2+}(aq) \tag{40}$$

Even reactions that are not redox reactions may be expressed as the difference of two reduction half-reactions.

Example 10.1 Expressing a reaction in terms of half-reactions

Express the dissolution of silver chloride in water as the difference of two reduction half-reactions.

Method First, write the overall chemical equation. Then select one of the reactants, and write a half-reaction in which it is reduced to one of the products. Next, subtract that half-reaction from the overall reaction to identify the second half-reaction. Finally, write the second half-reaction as a reduction.

Answer The chemical equation of the overall reaction is

$$AgCl(s) \longrightarrow Ag^+(aq) + Cl^-(aq)$$

We select as one half-reaction the reduction of AgCl (more precisely, the reduction of the Ag(I) in AgCl to Ag(0)):

$$AgCl(s) + e^- \longrightarrow Ag(s) + Cl^-(aq)$$

Subtraction of this equation from the overall reaction leaves

$$-e^- \longrightarrow Ag^+(aq) - Ag(s)$$

which rearranges to

$$Ag^+(aq) + e^- \longrightarrow Ag(s)$$

Comment There is no net change of oxidation number in dissolution of AgCl, so it is not a redox reaction.

- -

Self-test 10.4 Express the formation of H_2O from H_2 and O_2 in acidic solution (a redox reaction) as the difference of two reduction half-reactions.

$$[4H^+(aq) + 4e^- \rightarrow 2H_2(g), \; O_2(g) + 4H^+(aq) + 4e^- \rightarrow 2H_2O(l)]$$

The oxidized and reduced substances in a half-reaction form a **redox couple**, denoted Ox/Red. Thus, the redox couples mentioned so far are Cu^{2+}/Cu and Zn^{2+}/Zn. In general we write a couple as Ox/Red and the corresponding reduction half-reaction as

$$Ox + \nu e^- \longrightarrow Red \tag{41}$$

We shall often find it useful to express the composition of an electrode compartment in terms of the reaction quotient, Q, for the half-reaction. This quotient is defined like the reaction quotient for the overall reaction, but the electrons are ignored. Thus, for the copper half-reaction $Cu^{2+}(aq) + 2e^- \rightarrow Cu(s)$, we write

$$Q = \frac{1}{a_{Cu^{2+}}}$$

We have used the fact that the pure metal (the standard state of the element) has unit activity (recall Table 7.3).

================

Example 10.2 Writing the half-reaction and reaction quotient for an electrode

Write the half-reaction and the reaction quotient for the reduction of oxygen to water in dilute acidic solution.

Method The first step is a simple balancing exercise: use H^+ ions to balance the H atoms and electrons to balance the charge. For the reaction quotient, include activities of products in the numerator and reactants (other than electrons) in the denominator.

Answer The reduction of O_2 in acidic solution produces H_2O according to the half-reaction

$$O_2(g) + 4H^+(aq) + 4e^- \longrightarrow 2H_2O(l)$$

The reaction quotient for the half-reaction is therefore

$$Q = \frac{a_{H_2O}^2}{a_{H^+}^4 (f_{O_2}/p^{\ominus})} \approx \frac{p^{\ominus}}{a_{H^+}^4 p_{O_2}}$$

The approximations used in the second step are that the activity of water is 1 (because the solution is dilute and the water almost pure) and the oxygen behaves like a perfect gas.

- -

Self-test 10.5 Write the half-reaction and the reaction quotient for a chlorine gas electrode.

$$[Cl_2(g) + 2e^- \rightarrow 2Cl^-(aq), \; Q = a_{Cl^-}^2 p^{\ominus}/p_{Cl_2}]$$

================

(b) Reactions at electrodes

In an electrochemical cell, the reduction and oxidation processes responsible for the overall reaction are separated in space: oxidation takes place in one electrode compartment and

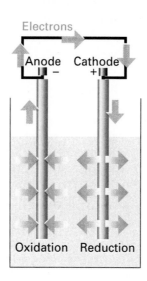

10.7 When a spontaneous reaction takes place in a galvanic cell, electrons are deposited in one electrode (the site of oxidation, the anode) and collected from another (the site of reduction, the cathode), and so there is a net flow of current which can be used to do work. Note that the + sign of the cathode can be interpreted as indicating the electrode at which electrons enter the cell, and the − sign of the anode is where the electrons leave the cell.

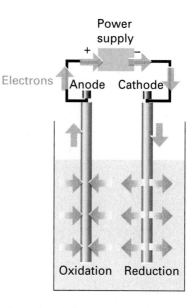

10.8 In an electrolytic cell, electrons are forced through the circuit by an external source. Although the cathode is still the site of reduction, it is now the negative electrode whereas the anode, the site of oxidation, is positive.

reduction takes place in the other compartment. As the reaction proceeds, the electrons released in the oxidation

$$\text{Red}_1 \longrightarrow \text{Ox}_1 + \nu\,e^-$$

at one electrode travel through the external circuit and re-enter the cell through the other electrode. There they bring about reduction:

$$\text{Ox}_2 + \nu\,e^- \longrightarrow \text{Red}_2$$

The electrode at which oxidation occurs is called the **anode**; the electrode at which reduction occurs is called the **cathode**.

In a *galvanic cell*, the cathode has a higher potential than the anode: the species undergoing reduction, Ox_2, withdraws electrons from its electrode (the cathode, Fig. 10.7), so leaving a relative positive charge on it (corresponding to a high potential). At the anode, oxidation results in the transfer of electrons to the electrode, so giving it a relative negative charge (corresponding to a low potential). In an *electrolytic cell*, the anode is still the location of oxidation (by definition), but now electrons must be withdrawn from the species in that compartment because oxidation does not occur spontaneously, and at the cathode there must be a supply of electrons to drive the reduction. Therefore, in an electrolytic cell the anode must be made relatively positive to the cathode (Fig. 10.8).

10.4 Varieties of cells

The simplest type of cell has a single electrolyte common to both electrodes (as in Fig. 10.7). In some cases it is necessary to immerse the electrodes in different electrolytes, as in the 'Daniell cell' in which the redox couple at one electrode is Cu^{2+}/Cu and at the other is Zn^{2+}/Zn (Fig. 10.9). In an **electrolyte concentration cell**, the electrode compartments are

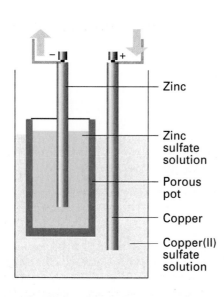

Zinc

Zinc
sulfate
solution

Porous
pot

Copper

Copper(II)
sulfate
solution

10.9 One version of the Daniell cell. The copper electrode is the cathode and the zinc electrode is the anode. Electrons leave the cell from the zinc and enter it again through the copper electrode.

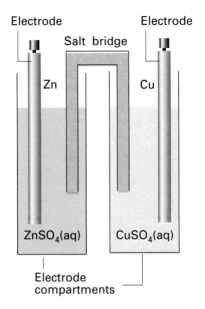

10.10 The salt bridge, essentially an inverted U-tube full of concentrated salt solution in a jelly, has two opposing liquid junction potentials which almost cancel.

identical except for the concentrations of the electrolytes. In an **electrode concentration cell** the electrodes themselves have different concentrations, either because they are gas electrodes operating at different pressures or because they are amalgams (solutions in mercury) with different concentrations.

(a) Liquid junction potentials

In a cell with two different electrolyte solutions in contact, as in the Daniell cell, there is an additional source of potential difference, the **liquid junction potential**, E_{lj}, across the interface of the two electrolytes. Another example of a junction potential is that between different concentrations of hydrochloric acid. At the junction, the mobile H^+ ions diffuse into the more dilute solution. The bulkier Cl^- ions follow, but initially do so more slowly, which results in a potential difference at the junction. The potential then settles down to a value such that, after that brief initial period, the ions diffuse at the same rates. Electrolyte concentration cells always have a liquid junction; electrode concentration cells do not.

The contribution of the liquid junction to the potential can be reduced (to about 1 to 2 mV) by joining the electrolyte compartments through a salt bridge (Fig. 10.10). The reason for the success of the salt bridge is that the liquid junction potentials at either end are largely independent of the concentrations of the two dilute solutions, and so nearly cancel.

(b) Notation

In the notation for cells, phase boundaries are denoted by a vertical bar. For example, the cell in Fig. 10.11 is denoted

$$Pt|H_2(g)|HCl(aq)|AgCl(s)|Ag(s)$$

A liquid junction is denoted by $\vdots$, so the cell in Fig. 10.9 is denoted

$$Zn(s)|ZnSO_4(aq)\vdots CuSO_4(aq)|Cu(s)$$

A double vertical line, ∥, denotes an interface for which it is assumed that the junction potential has been eliminated. Thus the cell in Fig. 10.10 is denoted

$$Zn(s)|ZnSO_4(aq)\|CuSO_4(aq)|Cu(s)$$

An electrolyte concentration cell in which the liquid junction potential is assumed to be eliminated is denoted

$$Pt|H_2(g)|HCl(aq, b_1)\|HCl(aq, b_2)|H_2(g)|Pt$$

(c) The cell reaction

The current produced by a galvanic cell arises from the spontaneous chemical reaction taking place inside it. The cell reaction is the reaction in the cell written on the assumption that the right-hand electrode is the cathode, and hence that the spontaneous reaction is one in which reduction is taking place in the right-hand compartment. Later we see how to predict if the right-hand electrode is in fact the cathode; if it is, then the cell reaction is spontaneous as written. If the left-hand electrode turns out to be the cathode, then the reverse of the cell reaction is spontaneous.

To write the cell reaction corresponding to a cell diagram, we first write the right-hand half-reaction as a reduction (because we have assumed that to be spontaneous). Then we subtract from it the left-hand reduction half-reaction (for, by implication, that electrode is the site of oxidation). Thus, in the cell

$$Zn(s)|ZnSO_4(aq)\|CuSO_4(aq)|Cu(s)$$

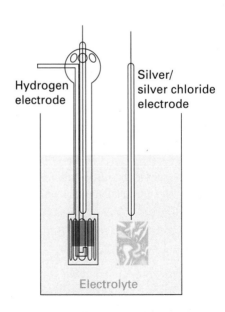

10.11 A typical cell for measuring a standard potential consists of a hydrogen electrode (on the left) and the electrode for the couple of interest (on the right).

the two electrodes and their reduction half-reactions are

$$\text{Right}: \quad Cu^{2+}(aq) + 2e^- \longrightarrow Cu(s)$$
$$\text{Left}: \quad Zn^{2+}(aq) + 2e^- \longrightarrow Zn(s)$$

Hence, the overall cell reaction is the difference:

$$Cu^{2+}(aq) + Zn(s) \longrightarrow Cu(s) + Zn^{2+}(aq)$$

(d) The cell potential

A cell in which the overall cell reaction has not reached chemical equilibrium can do electrical work as the reaction drives electrons through an external circuit. The work that a given transfer of electrons can accomplish depends on the potential difference between the two electrodes. This potential difference is called the **cell potential** and is measured in volts, V. When the cell potential is large, a given number of electrons travelling between the electrodes can do a large amount of electrical work. When the cell potential is small, the same number of electrons can do only a small amount of work. A cell in which the overall reaction is at equilibrium can do no work, and then the cell potential is zero.

According to the discussion in Section 4.6d, we know that the maximum electrical work that a system (the cell) can do is given by the value of ΔG, and in particular that, for a spontaneous process (in which both ΔG and w are negative) at constant temperature and pressure,

$$w_{e,max} = \Delta G \tag{42}$$

Therefore, to make thermodynamic measurements on the cell by measuring the work it can do, we must ensure that it is operating reversibly. Only then is it producing maximum work and only then can eqn 42 be used to relate that work to ΔG. Moreover, we saw in Section 9.1a that the reaction Gibbs energy, $\Delta_r G$, is actually a derivative evaluated at a specified composition of the reaction mixture. Therefore, to measure $\Delta_r G$ we must ensure that the cell is operating reversibly at a specific, constant composition. Both these conditions are achieved by measuring the cell potential when it is balanced by an exactly opposing source of potential so that the cell reaction occurs reversibly and the composition is constant: in effect, the cell reaction is poised for change, but not actually changing. The resulting potential difference is called the **zero-current cell potential**, E (formerly, and still commonly, the 'electromotive force', or emf, of the cell).

(e) The relation between E and $\Delta_r G$

The relation between the reaction Gibbs energy and the zero-current cell potential is

$$-\nu F E = \Delta_r G \tag{43}$$

where F is the Faraday constant. This equation, which is derived in the *Justification* below, is the key connection between electrical measurements on the one hand and thermodynamic properties on the other. It will be the basis of all that follows.

Justification 10.3

We consider the change in G when the cell reaction advances by an infinitesimal amount $d\xi$ at some composition. We saw in *Justification 9.1* that, at constant temperature and pressure, G changes by

$$dG = \sum_J \mu_J \, dn_J = \sum_J \nu_J \mu_J \, d\xi$$

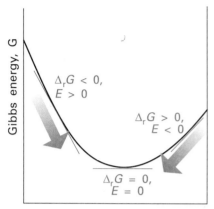

Gibbs energy, G

$\Delta_r G < 0,$
$E > 0$

$\Delta_r G > 0,$
$E < 0$

$\Delta_r G = 0,$
$E = 0$

Extent of reaction, ξ

10.12 As explained in Chapter 9, a spontaneous reaction occurs in the direction of decreasing Gibbs energy. When expressed in terms of a cell potential, the spontaneous direction of change can be expressed in terms of the cell potential E. The reaction is spontaneous as written (from left to right on the illustration) when $E > 0$. The reverse reaction is spontaneous when $E < 0$. When the cell reaction is at equilibrium, the cell potential is zero.

The reaction Gibbs energy, $\Delta_r G$, at the specified composition is

$$\Delta_r G = \left(\frac{\partial G}{\partial \xi}\right)_{p,T} = \sum_J \nu_J \mu_J$$

So we can write

$$dG = \Delta_r G \, d\xi$$

The maximum non-expansion work that the reaction can do as it advances by $d\xi$ at constant temperature and pressure is therefore

$$dw_e = \Delta_r G \, d\xi$$

This work is infinitesimal, and the composition of the system is virtually constant when it occurs.

Suppose that the reaction advances by $d\xi$, then $\nu \, d\xi$ electrons must travel from the anode to the cathode. The total charge transported between the electrodes when this change occurs is $-\nu e N_A \, d\xi$ (because $\nu \, d\xi$ is the amount of electrons and the charge per mole of electrons is $-e N_A$). Hence, the total charge transported is $-\nu F \, d\xi$ because $e N_A = F$.

The work done when an infinitesimal charge $-\nu F \, d\xi$ travels from the anode to the cathode is equal to the product of the charge and the potential difference E (see Table 2.1 and *Further information 5*):

$$dw_e = -\nu F E \, d\xi$$

When this relation is equated to the one above, the advancement $d\xi$ cancels, and eqn 43 is obtained.

It follows from eqn 43 that, by knowing the reaction Gibbs energy at a specified composition, we can state the zero-current cell potential at that composition. Note that a negative reaction Gibbs energy, corresponding to a spontaneous cell reaction, corresponds to a positive zero-current cell potential. Another way of looking at the content of eqn 43 is that it shows that the driving power of a cell (that is, the cell potential), is proportional to the slope of the Gibbs energy with respect to the extent of reaction. It is plausible that a reaction that is far from equilibrium (when the slope is steep) has a strong tendency to drive electrons through an external circuit (Fig. 10.12). When the slope is close to zero (when the cell reaction is close to equilibrium), the cell potential is small.

Illustration
....

To estimate the potential that can be expected for a typical cell we set $\Delta_r G \approx -100 \text{ kJ mol}^{-1}$ and $\nu = 1$; we find

$$E \approx -\frac{(-100 \times 10^3 \text{ J mol}^{-1})}{1 \times (96 \times 10^3 \text{ C mol}^{-1})} \approx 1 \text{ V}$$

We have used $1 \text{ J} = 1 \text{ C V}$.

....

(f) The Nernst equation

We can go on to relate the zero-current cell potential to the activities of the participants in the cell reaction. We know from eqn 9.10 that the reaction Gibbs energy is related to the composition of the reaction mixture by

$$\Delta_r G = \Delta_r G^{\ominus} + RT \ln Q$$

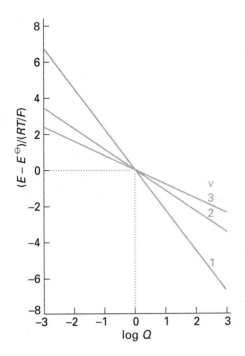

10.13 The variation of cell potential with the value of the reaction quotient for the cell reaction for different values of ν (the number of electrons transferred). At 298 K, $RT/F = 25.69$ mV, so the vertical scale refers to multiples of this value.

with Q the reaction quotient. It follows, on division of both sides by $-\nu F$, that

$$E = -\frac{\Delta_r G^\ominus}{\nu F} - \frac{RT}{\nu F} \ln Q$$

The first term on the right is written

$$E^\ominus = -\frac{\Delta_r G^\ominus}{\nu F} \qquad [44]$$

and called the **standard cell potential**. That is, the standard cell potential, $E^\ominus$, is the standard reaction Gibbs energy expressed as a potential (in volts). It follows that

$$E = E^\ominus - \frac{RT}{\nu F} \ln Q \qquad (45)$$

This equation for the cell potential in terms of the composition is called the **Nernst equation**; the dependence of cell potential on composition that it predicts is summarized in Fig. 10.13.

We see from eqn 45 that the standard cell potential (which will shortly move to centre stage of the exposition) can be interpreted as the zero-current cell potential when all the reactants and products are in their standard states, for then all activities are 1, so $Q = 1$ and $\ln Q = 0$. However, the fact that it is merely a disguised form of the standard reaction Gibbs energy should always be kept in mind and underlies all its applications.

Illustration

Because $RT/F = 25.7$ mV at 25 °C, a practical form of the Nernst equation is

$$E = E^\ominus - \frac{25.7 \text{ mV}}{\nu} \ln Q$$

It then follows that, for a reaction in which $\nu = 1$, if Q is increased by a factor of 10, then the cell potential decreases by 59.2 mV.

(g) Concentration cells

The Nernst equation can be used to derive an expression for the potential of an electrolyte concentration cell. Consider the cell

$$M|M^+(aq, L)\|M^+(aq, R)|M$$

where the solutions L and R have different molalities. The cell reaction is

$$M^+(aq, R) \longrightarrow M^+(aq, L) \qquad Q = \frac{a_L}{a_R} \quad \nu = 1$$

The standard cell potential is zero, because a cell cannot drive a current through a circuit when the two electrode compartments are identical (and, specifically, $\Delta_r G^\ominus = 0$ for the cell reaction). Therefore, the cell potential when the compartments have different concentrations is

$$E = -\frac{RT}{F} \ln \frac{a_L}{a_R} \approx -\frac{RT}{F} \ln \frac{b_L}{b_R} \qquad (46)$$

If R is the more concentrated solution, $E > 0$. Physically, the positive potential arises because positive ions tend to be reduced, so withdrawing electrons from the electrode. This process is dominant in the more concentrated right-hand electrode compartment.

One important example of a system that resembles this description is the biological cell wall, which is more permeable to K^+ ions than to either Na^+ or Cl^- ions. The concentration of K^+ inside the cell is about 20 to 30 times that on the outside, and is maintained at that level by a specific pumping operation fuelled by ATP and governed by enzymes. It follows from eqn 46 that the potential difference between the two sides is predicted to be about 77 mV. This estimate accords quite well with the measured value.

The transmembrane potential difference plays a particularly interesting role in the transmission of nerve impulses. Potassium and sodium ion pumps occur throughout the nervous system, and when the nerve is inactive there is a high K^+ concentration inside the cells and a high Na^+ concentration outside. The potential difference across the cell wall is about 70 mV. When the cell wall is subjected to a pulse of about 20 mV, the structure of the membrane adjusts and it becomes permeable to Na^+. This adjustment causes a decrease in membrane potential as the Na^+ flood into the interior of the cell. The change in potential difference triggers the adjacent part of the cell wall, and the pulse of collapsing potential passes along the nerve. Behind the pulse, the sodium and potassium pumps restore the concentration difference ready for the next pulse.

(h) Cells at equilibrium

A special case of the Nernst equation has great importance in electrochemistry. Suppose the reaction has reached equilibrium; then $Q = K$, where K is the equilibrium constant of the cell reaction. However, a chemical reaction at equilibrium cannot do work, and hence it generates zero potential difference between the electrodes of a galvanic cell. Therefore, setting $E = 0$ and $Q = K$ in the Nernst equation gives

$$\ln K = \frac{\nu F E^{\ominus}}{RT} \tag{47}$$

This very important equation lets us predict equilibrium constants from measured standard cell potentials.

Illustration

....

Because the standard potential of the Daniell cell is $+1.10$ V, the equilibrium constant for the cell reaction (eqn 40, for which $\nu = 2$) is $K = 1.5 \times 10^{37}$. We conclude that the displacement of copper by zinc goes virtually to completion.

....

10.5 Standard potentials

A galvanic cell is a combination of two electrodes, and each one can be considered as making a characteristic contribution to the overall cell potential. Although it is not possible to measure the contribution of a single electrode, we can define the potential of one of the electrodes as having a zero potential and then assign values to others on that basis. The specially selected electrode is the **standard hydrogen electrode** (SHE):

$$\text{Pt}|H_2(g)|H^+(aq) \qquad E^{\ominus} = 0 \tag{48}$$

at all temperatures. The **standard potential**, $E^{\ominus}$, of another couple is then assigned by constructing a cell in which it is the right-hand electrode and the standard hydrogen electrode is the left-hand electrode. For example, the standard potential of the Ag^+/Ag couple is the standard potential of the following cell:

$$\text{Pt}|H_2(g)|H^+(aq)\|Ag^+(aq)|Ag(s) \qquad E^{\ominus}(Ag^+/Ag) = E^{\ominus} = +0.80 \text{ V}$$

Likewise, the standard potential of the $AgCl/Ag, Cl^-$ couple is the standard potential of the following cell:

$$\text{Pt}|H_2(g)|H^+(aq)Cl^-(aq)|AgCl(s)|Ag(s)$$

$$E^{\ominus}(AgCl/Ag, Cl^-) = E^{\ominus} = +0.22 \text{ V}$$

Table 10.7* Standard potentials at 298 K

Couple	$E^{\ominus}/V$
$Ce^{4+}(aq) + e^- \rightarrow Ce^{3+}(aq)$	+1.61
$Cu^{2+}(aq) + 2e^- \rightarrow Cu(s)$	+0.34
$AgCl(s) + e^- \rightarrow Ag(s) + Cl^-(aq)$	+0.22
$2H^+(aq) + 2e^- \rightarrow H_2(g)$	0
$Zn^{2+}(aq) + 2e^- \rightarrow Zn(s)$	−0.76
$Na^+(aq) + e^- \rightarrow Na(s)$	−2.71

* More values are given in the *Data section*.

(Here and from now on, all values refer to 298 K.) Although a standard potential is written as though it refers to a half-reaction such as

$$AgCl(s) + e^- \longrightarrow Ag(s) + Cl^-(aq) \qquad E^{\ominus}(AgCl/Ag, Cl^-) = +0.22 \text{ V}$$

it should be understood that these equations are only shorthand for writing

$$AgCl(s) + \tfrac{1}{2}H_2(g) \longrightarrow Ag(s) + H^+(aq) + Cl^-(aq) \qquad E^{\ominus} = +0.22 \text{ V}$$

and that the standard potential is determined by properties of the hydrogen electrode as well as the couple to which the potential refers. Table 10.7 lists standard potentials at 298 K.

An important feature of standard cell potentials and standard potentials is that they are unchanged if the chemical equation for the cell reaction or a half-reaction is multiplied by a numerical factor. A numerical factor increases the value of the standard Gibbs energy for the reaction, but it also increases the number of electrons transferred by the same factor, and by eqn 44 the value of $E^{\ominus}$ remains unchanged.

(a) The standard potential of a cell in terms of individual standard potentials

The standard potential of a cell formed from any two electrodes can be calculated by taking the difference of their standard potentials. This rule follows from the fact that a cell such as

$$Ag(s)|Ag^+(aq)\|Cl^-(aq)|AgCl(s)|Ag(s)$$

is equivalent to two cells joined back-to-back:

$$Ag(s)|Ag^+(aq)\|H^+(aq)|H_2(g)|Pt\text{---}Pt|H_2(g)|H^+(aq)\|Cl^-(aq)|AgCl(s)|Ag(s)$$

The overall potential of this composite cell, and therefore of the cell of interest, is

$$E^{\ominus} = E^{\ominus}(AgCl/Ag, Cl^-) - E^{\ominus}(Ag^+/Ag) = -0.58 \text{ V}$$

The standard potentials in Table 10.7 can all be used in the same way, and the standard cell potential is the difference right − left of the corresponding standard potentials. Because $\Delta G^{\ominus} = -\nu F E^{\ominus}$, it then follows that, if the result gives $E^{\ominus} > 0$, then the corresponding cell reaction has $K > 1$.

Example 10.3 Identifying the spontaneous direction of a reaction

One of the reactions important in corrosion in an acidic environment is

$$Fe(s) + 2H^+(aq) + \tfrac{1}{2}O_2(g) \longrightarrow Fe^{2+}(aq) + H_2O(l)$$

Does the equilibrium constant favour the formation of $Fe^{2+}(aq)$?

Method We need to decide whether the standard potential for the reaction as written is positive, for a positive value would imply that $\Delta G^{\ominus} < 0$ and hence that $K > 1$. The sign of the cell potential is found by identifying the half-reactions that make up the overall reaction, and then taking their standard potentials from Table 10.7 in the *Data section* at the end of the book.

Answer The two reduction half-reactions are

(a) $Fe^{2+}(aq) + 2e^- \longrightarrow Fe(s) \qquad E^{\ominus} = -0.44 \text{ V}$
(b) $2H^+(aq) + \tfrac{1}{2}O_2(g) + 2e^- \longrightarrow H_2O(l) \qquad E^{\ominus} = +1.23 \text{ V}$

The difference (b) − (a) is

$$Fe(s) + 2H^+(aq) + \tfrac{1}{2}O_2(g) \longrightarrow Fe^{2+}(aq) + H_2O(l) \qquad E^{\ominus} = +1.67 \text{ V}$$

Therefore, because $E^\ominus > 0$, the reaction has $K > 1$, favouring products.

Comment Recall the point made earlier that the chemical equation for a half-reaction can be multiplied by any common factor without affecting its standard potential. Therefore, if either half-reaction needs to be multiplied by a factor before forming the difference (to ensure that the electrons in the equations cancel), the standard potentials are not affected.

Self-test 10.6 Does the displacement of copper by iron (that is, the reduction of Cu^{2+} by iron metal) have $K > 1$?

[Yes]

Example 10.4 Calculating an equilibrium constant

Calculate the equilibrium constant for the disproportionation $2Cu^+(aq) \rightarrow Cu(s) + Cu^{2+}(aq)$ at 298 K.

Method The strategy is to calculate the standard potential for the cell in which the reaction of interest is the cell reaction, and then to use eqn 47. To proceed, express the overall reaction as the difference of two reduction half-reactions and then find the corresponding standard potentials by referring to Table 10.7 in the *Data section*. Use $RT/F = 0.025693$ V at 298.15 K.

Answer The half-reactions and standard potentials we require are

$$R: \ Cu(s)|Cu^+(aq) \qquad Cu^+(aq) + e^- \longrightarrow Cu(s) \qquad E^\ominus = +0.52 \text{ V}$$
$$L: \ Pt|Cu^{2+}(aq), Cu^+(aq) \qquad Cu^{2+}(aq) + e^- \longrightarrow Cu^+(aq) \qquad E^\ominus = +0.16 \text{ V}$$

The standard cell potential is therefore

$$E^\ominus = +0.52 \text{ V} - 0.16 \text{ V} = +0.36 \text{ V}$$

Then, because $\nu = 1$,

$$\ln K = \frac{0.36 \text{ V}}{0.025693 \text{ V}} = 14$$

Hence, $K = 1.2 \times 10^6$.

Comment The equilibrium lies strongly towards the right of the reaction as written, and so Cu^+ disproportionates almost totally in solution.

Self-test 10.7 Calculate the equilibrium constant for the reaction $Sn^{2+}(aq) + Pb(s) \rightarrow Sn(s) + Pb^{2+}(aq)$ at 298 K.

[0.5]

(b) The measurement of standard potentials

The procedure for measuring a standard potential can be illustrated by considering a specific case, the silver chloride electrode. The measurement is made on the 'Harned cell':

$$Pt|H_2(g)|HCl(aq)|AgCl(s)|Ag(s) \qquad \tfrac{1}{2}H_2(g) + AgCl(s) \longrightarrow HCl(aq) + Ag(s)$$

for which

$$E = E^\ominus(AgCl/Ag, Cl^-) - \frac{RT}{F} \ln\left(\frac{a_{H^+} a_{Cl^-}}{f_{H_2}/p^\ominus}\right)$$

We shall set $f = p^{\ominus}$ from now on. The activities can be expressed in terms of the molality, b, and the mean activity coefficient, $\gamma_{\pm}$, through eqn 13:

$$E = E^{\ominus}(\text{AgCl/Ag, Cl}^-) - \frac{RT}{F}\ln b^2 - \frac{RT}{F}\ln \gamma_{\pm}^2 \qquad \{49\}$$

This expression rearranges to

$$E + \frac{2RT}{F}\ln b = E^{\ominus}(\text{AgCl/Ag, Cl}^-) - \frac{2RT}{F}\ln \gamma_{\pm} \qquad \{50\}$$

From the Debye–Hückel limiting law for a 1,1-electrolyte,

$$\ln \gamma_{\pm} \propto -b^{1/2}$$

(the natural logarithm used here is proportional to the common logarithm that appears in eqn 19). Therefore, with the constant of proportionality in this relation written A':

$$E + \frac{2RT}{F}\ln b = E^{\ominus}(\text{AgCl/Ag, Cl}^-) + \frac{2RTA'}{F}b^{1/2} \qquad \{51\}$$

(In precise work, the $b^{1/2}$ term is brought to the left, and a higher order correction term from the extended Debye–Hückel law is used on the right.) The expression on the left is evaluated at a range of molalities, plotted against $b^{1/2}$, and extrapolated to $b = 0$. The intercept at $b^{1/2} = 0$ is the value of $E^{\ominus}(\text{Ag/AgCl, Cl}^-)$.

Example 10.5 Determining the standard potential of a cell

The potential of the cell $\text{Zn}|\text{ZnCl}_2(\text{aq},b)|\text{AgCl}(s)|\text{Ag}$ at $25\,°\text{C}$ has the following values:

$b/(10^{-3}b^{\ominus})$	0.772	1.253	1.453	3.112	6.022
E/V	1.2475	1.2289	1.2235	1.1953	1.1742

Determine the standard potential of the cell.

Method　Proceed as described above, but adjusted to the cell reaction. Start by writing the Nernst equation for the cell, and then express the activities that occur in Q in terms of the mean activity coefficient. The latter can be written as proportional to $b^{1/2}$ by using the Debye–Hückel limiting law. However, there is no need to write all the constants because the standard cell potential is obtained by extrapolation, as explained in the text.

Answer　The ionic equation for the cell reaction is

$$\text{Zn}(s) + 2\text{AgCl}(s) \longrightarrow 2\text{Ag}(s) + \text{Zn}^{2+}(\text{aq}) + 2\text{Cl}^-(\text{aq})$$

$$Q = a_{\text{Zn}^{2+}}a_{\text{Cl}^-}^2 \qquad \nu = 2$$

because all the solids are at unit activity. The Nernst equation is therefore

$$E = E^{\ominus} - \frac{RT}{2F}\ln a_{\text{Zn}^{2+}}a_{\text{Cl}^-}^2$$

The activities are related to the molality, b, of ZnCl_2 by

$$a_{\text{Zn}^{2+}}a_{\text{Cl}^-}^2 = \gamma_{\pm}^3 b_{\text{Zn}^{2+}}b_{\text{Cl}^-}^2 = 4\gamma_{\pm}^3 b^3$$

because $b_{\text{Zn}^{2+}} = b$ and $b_{\text{Cl}^-} = 2b$ for a fully dissociated salt. Only a little work is then needed to convert the Nernst equation into

$$E + \frac{3RT}{2F}\ln b + \frac{RT}{2F}\ln 4 = E^{\ominus} + Cb^{1/2}$$

where C is a collection of constants that come from the limiting law. We now draw up the

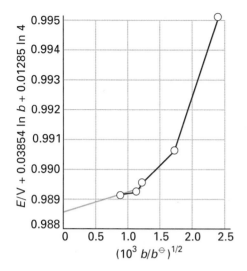

10.14 The plot, and the extrapolation used for the experimental measurement of a standard cell potential. The intercept at $b^{1/2} = 0$ is $E^{\ominus}$.

following table, using $RT/2F = 0.01285$ V:

$b/(10^{-3}\,b^{\ominus})$	0.772	1.253	1.453	3.112	6.022
$(b/(10^{-3}\,b^{\ominus}))^{1/2}$	0.879	1.119	1.205	1.764	2.454
$E/\mathrm{V} + 0.03854\ln b$ $+ 0.01285\ln 4$	0.9891	0.9892	0.9895	0.9906	0.9950

The data are plotted in Fig. 10.14; as can be seen, they extrapolate to $E^{\ominus} = +0.9886$ V.

Self-test 10.8 The data below are for the cell $Pt|H_2(g, p^{\ominus})|HBr(aq, b)|AgBr(s)|Ag$ at 25 °C. Determine the standard cell potential.

$b/(10^{-4}\,b^{\ominus})$	4.042	8.444	37.19
E/V	0.47381	0.43636	0.36173

[0.071 V]

(c) The measurement of activity coefficients

Once the standard potential of an electrode in a cell is known, the activities of the ions with respect to which it is reversible can be determined simply by measuring the cell potential with the ions at the concentration of interest. For example, the mean activity coefficient of the ions in hydrochloric acid of molality b is obtained from eqn 50 in the form

$$\ln \gamma_{\pm} = \frac{E^{\ominus}(\mathrm{AgCl/Ag, Cl^-}) - E}{2RT/F} - \ln b \qquad \{52\}$$

once E has been measured.

Applications of standard potentials

Zero-current cell potentials are a convenient source of data on the Gibbs energies, enthalpies, and entropies of reactions. In practice the standard values of these quantities are the ones normally determined.

10.6 The electrochemical series

We have seen that for two redox couples, Ox_1/Red_1 and Ox_2/Red_2, and the cell

$$Red_1, Ox_1 \| Red_2, Ox_2 \qquad E^{\ominus} = E_2^{\ominus} - E_1^{\ominus} \qquad (53a)$$

that the cell reaction

$$Red_1 + Ox_2 \longrightarrow Ox_1 + Red_2 \qquad (53b)$$

is spontaneous as written if $E^{\ominus} > 0$, and therefore if $E_2^{\ominus} > E_1^{\ominus}$. Because in the cell reaction Red_1 reduces Ox_2, we can conclude that Red_1 has a thermodynamic tendency to reduce Ox_2 if $E_1^{\ominus}$ is lower than $E_2^{\ominus}$. More briefly: low reduces high.

Illustration

Because $E^{\ominus}(Zn^{2+}, Zn) = -0.76$ V $< E^{\ominus}(Cu^{2+}, Cu) = +0.34$ V, zinc has a thermodynamic tendency to reduce Cu^{2+} ions in aqueous solution.

Table 10.8 shows a part of the **electrochemical series**, the metallic elements (and hydrogen) arranged in the order of their reducing power as measured by their standard potential in aqueous solution. A metal low in the series (with a lower standard potential) can

Table 10.8* The electrochemical series of the metals

Least strongly reducing

Gold
Platinum
Silver
Mercury
Copper
(Hydrogen)
Lead
Tin
Nickel
Iron
Zinc
Chromium
Aluminium
Magnesium
Sodium
Calcium
Potassium

Most strongly reducing

* The complete series can be inferred from Table 10.7

reduce the ions of metals with higher standard potentials. This conclusion is qualitative. The quantitative value of K is obtained by doing the calculations we have described previously. For example, to determine whether zinc can displace magnesium from aqueous solutions at 298 K, we note that zinc lies above magnesium in the electrochemical series, so zinc cannot reduce magnesium ions in aqueous solution. Zinc can reduce hydrogen ions, because hydrogen lies higher in the series. However, even for reactions that are thermodynamically favourable, there may be kinetic factors that result in very slow rates of reaction.

10.7 Solubility constants

We can discuss the solubility S (the molality of the saturated solution) of a sparingly soluble salt MX in terms of the equilibrium

$$\text{MX(s)} \rightleftharpoons \text{M}^+(\text{aq}) + \text{X}^-(\text{aq}) \qquad K_s = a_{\text{M}^+} a_{\text{X}^-} \qquad [54]$$

where the activities are those at equilibrium (that is, in the saturated solution) and we have used $a = 1$ for a pure solid. The equilibrium constant K_s is called the **solubility constant** (formerly, and still commonly, the *solubility product*) of the salt. When the solubility is so low that $\gamma_\pm \approx 1$ even in the saturated solution, we can write $a = b/b^\ominus$; moreover, because both molalities are equal to S in the saturated solution, we can conclude that

$$K_s \approx S^2$$

and hence that

$$S \approx K_s^{1/2} \qquad \{55\}$$

It follows that we can estimate S from the standard potential of a cell with a reaction corresponding to the solubility equilibrium.

Example 10.6 Evaluating a solubility from electrochemical data

Evaluate the solubility of AgCl(s) from cell potential data at 298 K.

Method We need to find an electrode combination that reproduces the solubility equilibrium, and then identify the solubility constant with the equilibrium constant of the cell reaction. The solubility itself is obtained from an equation like eqn 55 in conjunction with eqn 47 ($\ln K = \nu F E^\ominus / RT$).

Answer The solubility equilibrium is

$$\text{AgCl(s)} \rightleftharpoons \text{Ag}^+(\text{aq}) + \text{Cl}^-(\text{aq}) \qquad K_s = a_{\text{Ag}^+} a_{\text{Cl}^-}$$

and we saw in Example 10.1 that this equation can be expressed as the difference of the following half-reactions:

$$\text{AgCl(s)} + \text{e}^- \longrightarrow \text{Ag(s)} + \text{Cl}^-(\text{aq}) \qquad E^\ominus = +0.22 \text{ V}$$
$$\text{Ag}^+(\text{aq}) + \text{e}^- \longrightarrow \text{Ag(s)} \qquad E^\ominus = +0.80 \text{ V}$$

The cell potential is therefore -0.58 V. Then, because $\nu = 1$,

$$\ln K_s = \frac{\nu E^\ominus}{RT/F} = \frac{1 \times (-0.58 \text{ V})}{2.5693 \times 10^{-2} \text{ V}} = -23$$

Therefore, $K_s = 1.0 \times 10^{-10}$ and $S = 1.0 \times 10^{-5} \text{ mol kg}^{-1}$.

Self-test 10.9 Calculate the solubility constant and the solubility of mercury(I) chloride at 298 K. (*Hint*: The mercury(I) ion is the diatomic species Hg_2^{2+}.)

$$[2.6 \times 10^{-18}, 8.7 \times 10^{-7} \; mol\,kg^{-1}]$$

10.8 The measurement of pH and pK

The potential of a hydrogen electrode in which the half-reaction is

$$H^+(aq) + e^- \longrightarrow \tfrac{1}{2}H_2(g) \qquad Q = \frac{(f_{H_2}/p^{\ominus})^{1/2}}{a_{H^+}} \qquad \nu = 1$$

with $f_{H_2} = p^{\ominus}$ is

$$E(H^+/H_2) = \frac{RT}{F} \ln a_{H^+} = -\frac{RT \ln 10}{F} pH \tag{56}$$

(because $E^{\ominus}(H^+/H_2) = 0$). This expression makes sense physically. Increasing the pH above 0 (lowering the hydrogen ion activity below 1) decreases the tendency of the positive ions to discharge at the electrode, so we should expect its potential to become negative. At 25 °C, when $RT/F = 25.69$ mV, this relation becomes

$$E(H^+/H_2) = -59.16 \; mV \times pH \tag{57}$$

Each unit increase in pH decreases the electrode potential by 59 mV.

(a) The determination of pH

The measurement of the pH of a solution is simple in principle, for it is based on the measurement of the potential of a hydrogen electrode immersed in the solution. The left-hand electrode of the cell is typically a saturated calomel ($Hg_2Cl_2(s)$) reference electrode with potential $E(cal)$; the right-hand electrode is the hydrogen electrode with potential given by eqn 56. The pH of the cell is therefore

$$pH = \frac{E + E(cal)}{(-59.16 \; mV)} \tag{58}$$

The practical definition of the pH of a solution X is

$$pH = pH(S) - \frac{FE}{RT \ln 10} \tag{[59]}$$

where E is the potential of the cell

$$Pt|H_2(g)|S(aq)\|3.5 \; M \; KCl(aq)\|X(aq)|H_2(g)|Pt$$

and S is a solution of standard pH. The currently recommended primary standards include a saturated aqueous solution of potassium hydrogen tartrate, which has pH = 3.557 at 25 °C and 0.0100 $mol\,kg^{-1}$ disodium tetraborate, which has pH = 9.180 at that temperature.

In practice, indirect methods are much more convenient, and the hydrogen electrode is replaced by the glass electrode (Fig. 10.15). This electrode is sensitive to hydrogen ion activity, and has a potential proportional to pH. It is filled with a phosphate buffer containing Cl^- ions, and conveniently has $E = 0$ when the external medium is at pH = 7. The glass electrode is much more convenient to handle than the gas electrode itself, and can be calibrated using solutions of known pH.

The responsiveness of a glass electrode to the hydronium ion activity is a result of complex processes at the interface between the glass membrane and the solutions on either side of it. The membrane itself is permeable to Na^+ and Li^+ ions but not to H^+ ions. Therefore, the potential difference across the glass membrane must arise by a mechanism that is different from that responsible for biological transmembrane potentials. A clue to the

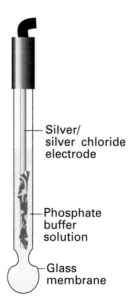

— Silver/
silver chloride
electrode

— Phosphate
buffer
solution

— Glass
membrane

10.15 The glass electrode. It is usually used in conjunction with a calomel electrode that makes contact with the test solution through a salt bridge.

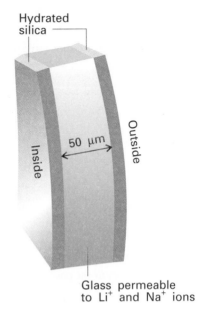

10.16 A section through the wall of a glass electrode.

mechanism comes from a detailed inspection of the glass membrane, for each face is coated with a thin layer of hydrated silica (Fig. 10.16). The hydrogen ions in the test solution modify this layer to an extent that depends on their activity in the solution, and the charge modification of the outside layer is transmitted to the inner layer by the Na^+ and Li^+ ions in the glass. The hydronium ion activity gives rise to a membrane potential by this indirect mechanism.

The electrochemical determination of pH opens up a route to the electrochemical determination of pK_a, for we saw in Section 9.5c that the pK_a of an acid is equal to the pH of a solution containing equal amounts of the acid and its conjugate base.

(b) Species-selective electrodes

A suitably adapted glass electrode can be used to detect the presence of certain gases. A simple form of a gas-sensing electrode consists of a glass electrode contained in an outer sleeve filled with an aqueous solution and separated from the test solution by a membrane that is permeable to gas. When a gas such as sulfur dioxide or ammonia diffuses into the aqueous solution, it modifies its pH, which in turn affects the potential of the glass electrode.

Somewhat more sophisticated devices are used as ion-selective electrodes that give potentials according to the presence of specific ions present in a test solution. In one arrangement, a porous lipophilic (hydrocarbon-attracting) membrane is attached to a small reservoir of a hydrophobic (water-repelling) liquid, such as dioctylphenylphosphonate, that saturates it (Fig. 10.17). The liquid contains a chelating agent, such as $(RO)_2PO_2^-$ with R a C_8 to C_{18} chain, that acts as a kind of solubilizing agent for the ions with which it can form a complex. The chelated ions are able to migrate through the lipophilic membrane, and hence give rise to a transmembrane potential, which is detected by a silver/silver chloride electrode in the interior of the assembly. Electrodes of this construction can be designed to be sensitive to a variety of ionic species, including calcium, zinc, iron, lead, and copper ions.

10.9 Thermodynamic functions from cell potential measurements

The standard cell potential is related to the standard reaction Gibbs energy through eqn 44 ($\Delta_r G^\ominus = -\nu F E^\ominus$). Therefore, by measuring $E^\ominus$ we can obtain this important thermodynamic quantity. Its value can then be used to calculate the Gibbs energy of formation of ions using the convention explained in Section 10.1a.

Illustration

The cell reaction of

$$H_2|H^+(aq)\|Ag^+(aq)|Ag \qquad E^\ominus = +0.7996\ V$$

is

$$Ag^+(aq) + \tfrac{1}{2}H_2(g) \longrightarrow H^+(aq) + Ag(s) \qquad \Delta_r G^\ominus = -\Delta_f G^\ominus(Ag^+, aq)$$

Therefore, with $\nu = 1$, we find

$$\Delta_f G^\ominus(Ag^+, aq) = -(-FE^\ominus) = +77.10\ kJ\,mol^{-1}$$

as in Table 2.6 in the *Data section*.

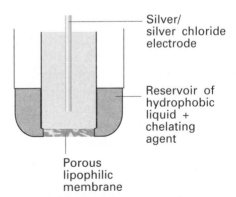

10.17 The structure of an ion-selective electrode. Chelated ions are able to migrate through the lipophilic membrane.

Example 10.7 Evaluating a standard potential from two others

Given that the standard potentials of the Cu^{2+}/Cu and Cu^+/Cu couples are $+0.340$ V and $+0.522$ V, respectively, evaluate $E^{\ominus}(Cu^{2+}, Cu^+)$.

Method First, we note that reaction Gibbs energies may be added (as in Hess's law analyses of reaction enthalpies). Therefore, we should convert the $E^{\ominus}$ values to $\Delta G^{\ominus}$ values by using eqn 44, add them appropriately, and then convert the overall $\Delta G^{\ominus}$ to the required $E^{\ominus}$ by using eqn 44 again. This roundabout procedure is necessary because, as we shall see, although the factor F cancels, the factor ν in general does not.

Answer The electrode reactions are as follows:

(a) $Cu^{2+}(aq) + 2e^- \longrightarrow Cu(s)$

$$E^{\ominus} = +0.340 \text{ V, so } \Delta_r G^{\ominus} = -2(0.340 \text{ V})F$$

(b) $Cu^+(aq) + e^- \longrightarrow Cu(s)$

$$E^{\ominus} = +0.522 \text{ V, so } \Delta_r G^{\ominus} = -(0.522 \text{ V})F$$

The required reaction is

(c) $Cu^{2+}(aq) + e^- \longrightarrow Cu^+(aq)$ $E^{\ominus} = -\Delta_r G^{\ominus}/F$

Because (c) = (a) − (b), the standard Gibbs energy of reaction (c) is

$$\Delta_r G^{\ominus} = \Delta_r G^{\ominus}(a) - \Delta_r G^{\ominus}(b) = -(0.160 \text{ V}) \times F$$

Therefore, $E^{\ominus} = +0.160$ V.

Comment Note that we cannot combine the $E^{\ominus}$ values directly and we must always work via $\Delta G^{\ominus}$. Alternatively, note that the generalization of the calculation illustrated above is

$$\nu_c E^{\ominus}(c) = \nu_a E^{\ominus}(a) - \nu_b E^{\ominus}(b)$$

- -

Self-test 10.10 Calculate the standard potential of the Fe^{3+}/Fe couple from the values for the Fe^{3+}/Fe^{2+} and Fe^{2+}/Fe couples.

$$[-0.037 \text{ V}]$$

The temperature coefficient of the cell potential gives the entropy of the cell reaction. This conclusion follows from the thermodynamic relation $(\partial G/\partial T)_p = -S$ and eqn 44, which combine to give

$$\frac{dE^{\ominus}}{dT} = \frac{\Delta_r S^{\ominus}}{\nu F} \tag{60}$$

(The derivative is complete because $E^{\ominus}$, like $\Delta_r G^{\ominus}$, is independent of the pressure.) Hence we have an electrochemical technique for obtaining standard reaction entropies and through them the entropies of ions in solution.

Finally, we can combine the results obtained so far and use them to obtain the standard reaction enthalpy:

$$\Delta_r H^{\ominus} = \Delta_r G^{\ominus} + T\Delta_r S^{\ominus} = -\nu F\left(E^{\ominus} - T\frac{dE^{\ominus}}{dT}\right) \tag{61}$$

This expression provides a noncalorimetric method for measuring $\Delta_r H^{\ominus}$ and, through the convention $\Delta_f H^{\ominus}(H^+, aq) = 0$, the standard enthalpies of formation of ions in solution. Thus, electrical measurements can be used to calculate all the thermodynamic properties with which this chapter began.

Example 10.8 Using the temperature coefficient of the cell potential

The standard cell potential of

$$Pt|H_2(g)|HBr(aq)|AgBr(s)|Ag(s)$$

was measured over a range of temperatures, and the data were fitted to the following polynomial:

$$E^{\ominus}/V = 0.07131 - 4.99 \times 10^{-4}(T/K - 298) - 3.45 \times 10^{-6}(T/K - 298)^2$$

Evaluate the standard reaction Gibbs energy, enthalpy, and entropy at 298 K.

Method The standard Gibbs energy of reaction is obtained by using eqn 44 after evaluating $E^{\ominus}$ at 298 K. The standard entropy of reaction is obtained by using eqn 60, which involves differentiating the polynomial with respect to T and then setting $T = 298$ K. The reaction enthalpy is obtained by combining the values of the standard Gibbs energy and entropy.

Answer At $T = 298$ K, $E^{\ominus} = +0.07131$ V, so

$$\Delta_r G^{\ominus} = -\nu F E^{\ominus} = -(1) \times (96.485 \text{ kC mol}^{-1}) \times (+0.07131 \text{ V})$$
$$= -6.880 \text{ kJ mol}^{-1}$$

The temperature coefficient of the cell potential is

$$\frac{dE^{\ominus}}{dT} = -4.99 \times 10^{-4} \text{ V K}^{-1} - 2(3.45 \times 10^{-6})(T/K - 298) \text{ V K}^{-1}$$

At $T = 298$ K this expression evaluates to

$$\frac{dE}{dT} = -4.99 \times 10^{-4} \text{ V K}^{-1}$$

So, from eqn 60, the reaction entropy is

$$\Delta_r S^{\ominus} = 1 \times (9.6485 \times 10^4 \text{ C mol}^{-1}) \times (-4.99 \times 10^{-4} \text{ V K}^{-1})$$
$$= -48.2 \text{ J K}^{-1} \text{ mol}^{-1}$$

It then follows that

$$\Delta_r H^{\ominus} = \Delta_r G^{\ominus} + T \Delta_r S^{\ominus}$$
$$= -6.880 \text{ kJ mol}^{-1} + (298 \text{ K}) \times (-0.0482 \text{ kJ K}^{-1} \text{ mol}^{-1})$$
$$= -21.2 \text{ kJ mol}^{-1}$$

Comment One difficulty with this procedure lies in the accurate measurement of small temperature coefficients of cell potential. Nevertheless, it is another example of the striking ability of thermodynamics to relate the apparently unrelated, in this case to relate electrical measurements to thermal properties.

- -

Self-test 10.11 Predict the standard potential of the Harned cell at 303 K from tables of thermodynamic data.

[+0.2190 V]

Checklist of key ideas

The thermodynamic properties of ions in solution

10.1 Thermodynamic functions of formation
- [ ] the standard enthalpy of formation of ions in solution
- [ ] the definition of standard enthalpy and Gibbs energy of formation of ions (1)
- [ ] the analysis of contributions to $\Delta_f G^\ominus$
- [ ] the use of a thermodynamic cycle
- [ ] the Born equation (4)
- [ ] the standard entropy of a hydrogen ion in water (6)
- [ ] the standard (partial molar) entropies of other ions

10.2 Ion activities
- [ ] the activity of an ion in solution (7)
- [ ] the activity coefficient (8)
- [ ] the chemical potential of an ion (10)
- [ ] mean activity coefficient (13,16)
- [ ] total Gibbs energy of ions in solution (18)
- [ ] Debye–Hückel theory
- [ ] ionic atmosphere
- [ ] Debye–Hückel limiting law (19)

- [ ] ionic strength (20)
- [ ] shielded Coulomb potential (24)
- [ ] Debye length (24,33)
- [ ] Poisson's equation (25)
- [ ] Faraday constant (29)
- [ ] extended Debye–Hückel law (39)

Electrochemical cells
- [ ] electrochemical cell
- [ ] electrode
- [ ] electrolyte
- [ ] electrode compartment
- [ ] salt bridge
- [ ] galvanic cell
- [ ] electrolytic cell

10.3 Half-reactions and electrodes
- [ ] oxidation
- [ ] reduction
- [ ] redox reaction
- [ ] reducing agent
- [ ] oxidizing agent
- [ ] half-reaction
- [ ] redox couple
- [ ] reaction quotient for a half-reaction
- [ ] anode
- [ ] cathode

10.4 Varieties of cells
- [ ] electrolyte concentration cell

- [ ] electrode concentration cell
- [ ] liquid junction potential
- [ ] cell notation
- [ ] the convention for writing the cell reaction
- [ ] cell potential
- [ ] zero-current cell potential (emf)
- [ ] the cell potential in terms of the reaction Gibbs energy (43)
- [ ] standard cell potential (44)
- [ ] Nernst equation for the cell potential (45)
- [ ] the potential of an electrolyte concentration cell (46) and transmembrane potentials
- [ ] the relation between the equilibrium constant for the cell reaction and the standard cell potential (47)

10.5 Standard potentials
- [ ] standard hydrogen electrode (SHE)
- [ ] standard potential ($E^\ominus$)
- [ ] the standard cell potential in terms of the standard potentials of the electrodes
- [ ] the graphical determination of standard potential
- [ ] the electrochemical determination of activity coefficients (52)

Applications of standard potentials

10.6 The electrochemical series
- [ ] the interpretation of the relative sizes of standard potentials
- [ ] the electrochemical series

10.7 Solubility constants
- [ ] solubility constant (54)
- [ ] the relation between solubility constant and solubility (55)

10.8 The measurement of pH and pK
- [ ] the variation of cell potential with pH (57)
- [ ] practical definition of pH
- [ ] glass electrode
- [ ] the determination of pK_a
- [ ] the action of ion-selective electrodes

10.9 Thermodynamic functions from cell potential measurements
- [ ] the temperature coefficient of cell potential (60)
- [ ] the determination of standard reaction entropy and enthalpy (60,61)

Further reading

Articles of general interest

P.J. Morgan and E. Gileadi, Alleviating the common confusion caused by polarity in electrochemistry. *J. Chem. Educ.* **66**, 912 (1989).

J.J. MacDonald, Cathodes, terminals, and signs. *Educ. in Chem.* **25**, 52 (1988).

Y. Marcus, Ionic radii in aqueous solutions. *Chem. Rev.* **88**, 1475 (1988).

A.K. Covington, R.G. Bates, and D.A. Durst, Definition of pH scales, standard reference values, and related terminology. *Pure Appl. Chem.* **57**, 531 (1985).

R.L. DeKock, Tendency of reaction, electrochemistry, and units. *J. Chem. Educ.* **73**, 955 (1996).

P. Millet, Electric potential distribution in an electrochemical cell. *J. Chem. Educ.* **73**, 956, (1996).

A.-S. Feiner and A.J. McEvoy, The Nernst equation. *J. Chem. Educ.* **71**, 493, (1994).

Texts and sources of data and information

D.R. Crow, *Principles and applications of electrochemistry.* Blackie, London (1994).

J. Koryta, *Ions, electrodes, and membranes*. Wiley, New York (1992).

V.M.M. Lobo, *Handbook of electrolyte solutions*. Elsevier, Amsterdam (1989).

A.J. Bard and L.R. Faulkner, *Electrochemical methods*. Wiley, New York (1980).

A.J. Bard, R. Parsons, and J. Jordan, *Standard potentials in aqueous solution*. Marcel Dekker, New York (1985).

A.J. Bard (ed.), *Encyclopedia of electrochemistry of the elements*, 1-15. Marcel Dekker (1973–1984).

M.S. Antelman, *The encyclopedia of chemical electrode potentials*. Plenum, New York (1982).

J. Goodisman, *Electrochemistry: theoretical foundations*. Wiley, New York (1987).

D. Pletcher and F.C. Walsh, *Industrial electrochemistry*. Chapman & Hall, London (1989).

P. Reiger, *Electrochemistry*. Prentice-Hall, Englewood Cliffs (1987).

J.O'M. Bockris, B.E. Conway, *et al.* (ed.), *Comprehensive treatise on electrochemistry*, 1–10. Plenum, New York (1980–85).

R.R. Adžić and E.B. Yeager, Electrochemistry. In *Encyclopedia of applied physics* (ed. G.L. Trigg), 5, 223. VCH, New York (1993).

Exercises

10.1 (a) Calculate $\Delta_r H^\ominus$ for the reaction $Zn(s) + CuSO_4(aq) \rightarrow ZnSO_4(aq) + Cu(s)$ from the information in Table 2.6 in the *Data section*.

10.1 (b) Calculate $\Delta_r H^\ominus$ for the reaction $NaCl(aq) + AgNO_3(aq) \rightarrow AgCl(s) + NaNO_3(aq)$ from the information in Table 2.6 in the *Data section*.

10.2 (a) Calculate the molar solubility of mercury(II) chloride at $25\,°C$ from standard Gibbs energies of formation.

10.2 (b) Calculate the molar solubility of lead(II) sulfide at $25\,°C$ where the standard Gibbs energy of formation is $-98.7\ kJ\,mol^{-1}$.

10.3 (a) Estimate the standard Gibbs energy of formation of $F^-(aq)$ from the value for $Cl^-(aq)$, taking the radius of F^- as 131 pm.

10.3 (b) Estimate the standard Gibbs energy of formation of $NO_3^-(aq)$ from the value for $Cl^-(aq)$, taking the radius of the nitrate ion as 189 pm (its thermochemical radius).

10.4 (a) Relate the ionic strengths of (a) KCl, (b) $FeCl_3$, and (c) $CuSO_4$ solutions to their molalities, b.

10.4 (b) Relate the ionic strengths of (a) $MgCl_2$, (b) $Al_2(SO_4)_3$, and (c) $Fe_2(SO_4)_3$ solutions to their molalities, b.

10.5 (a) Calculate the ionic strength of a solution that is $0.10\ mol\,kg^{-1}$ in KCl(aq) and $0.20\ mol\,kg^{-1}$ in $CuSO_4$(aq).

10.5 (b) Calculate the ionic strength of a solution that is $0.040\ mol\,kg^{-1}$ in $K_3[Fe(CN)_6]$(aq), $0.030\ mol\,kg^{-1}$ in KCl(aq), and $0.050\ mol\,kg^{-1}$ in NaBr(aq).

10.6 (a) Calculate the masses of (a) $Ca(NO_3)_2$ and, separately, (b) NaCl to add to a $0.150\ mol\,kg^{-1}$ solution of KNO_3(aq) containing 500 g of solvent to raise its ionic strength to 0.250.

10.6 (b) Calculate the masses of (a) KNO_3 and, separately, (b) $Ba(NO_3)_2$ to add to a $0.110\ mol\,kg^{-1}$ solution of KNO_3(aq) containing 500 g of solvent to raise its ionic strength to 1.00.

10.7 (a) What molality of $CuSO_4$ has the same ionic strength as $1.00\ mol\,kg^{-1}$ KCl(aq)?

10.7 (b) What molality of $Al_2(SO_4)_3$ has the same ionic strength as $0.500\ mol\,kg^{-1}\ Ca(NO_3)_2$(aq)?

10.8 (a) Express the mean activity coefficient of the ions in a solution of $CaCl_2$ in terms of the activity coefficients of the individual ions.

10.8 (b) Express the mean activity coefficient of the ions in a solution of $Al_2(SO_4)_3$ in terms of the activity coefficients of the individual ions.

10.9 (a) Estimate the mean ionic activity coefficient for $CaCl_2$ in a solution that is $0.010\ mol\,kg^{-1}\ CaCl_2$(aq) and $0.030\ mol\,kg^{-1}$ NaF(aq).

10.9 (b) Estimate the mean ionic activity coefficient for NaCl in a solution that is $0.020\ mol\,kg^{-1}$ NaCl(aq) and $0.035\ mol\,kg^{-1}\ Ca(NO_3)_2$(aq).

10.10 (a) The mean activity coefficient in a $0.500\ mol\,kg^{-1}\ LaCl_3$(aq) solution is 0.303 at $25\,°C$. What is the percentage error in the value predicted by the Debye–Hückel limiting law?

10.10 (b) The mean activity coefficient in a $0.100\ mol\,kg^{-1}\ CaCl_2$(aq) solution is 0.524 at $25\,°C$. What is the percentage error in the value predicted by the Debye–Hückel limiting law?

10.11 (a) The mean activity coefficients of HBr in three dilute aqueous solutions at $25\,°C$ are 0.930 (at $5.0\ mmol\,kg^{-1}$), 0.907 (at $10.0\ mmol\,kg^{-1}$), and 0.879 (at $20.0\ mmol\,kg^{-1}$). Estimate the value of B in the extended Debye–Hückel law.

10.11 (b) The mean activity coefficients of KCl in three dilute aqueous solutions at $25\,°C$ are 0.927 (at $5.0\ mmol\,kg^{-1}$), 0.902 (at $10.0\ mmol\,kg^{-1}$), and 0.816 (at $50.0\ mmol\,kg^{-1}$). Estimate the value of B in the extended Debye–Hückel law.

10.12 (a) For CaF_2, $K_s = 3.9 \times 10^{-11}$ at $25\,°C$ and the standard Gibbs energy of formation of CaF_2(s) is $-1167\ kJ\,mol^{-1}$. Calculate the standard Gibbs energy of formation of CaF_2(aq).

10.12 (b) For PbI_2, $K_s = 1.4 \times 10^{-8}$ at $25\,°C$ and the standard Gibbs energy of formation of $PbI_2(s)$ is -173.64 kJ mol^{-1}. Calculate the standard Gibbs energy of formation of $PbI_2(aq)$.

10.13 (a) Consider a hydrogen electrode in aqueous HBr solution at $25\,°C$ operating at 1.15 atm. Calculate the change in the electrode potential when the molality of the acid is changed from 5.0 mmol kg^{-1} to 20.0 mol kg^{-1}. Activity coefficients are given in Exercise 10.11a.

10.13 (b) Consider a hydrogen electrode in aqueous HCl solution at $25\,°C$ operating at 105 kPa. Calculate the change in the electrode potential when the molality of the acid is changed from 5.0 mmol kg^{-1} to 50 mol kg^{-1}. Activity coefficients are given in Table 10.5.

10.14 (a) Devise a cell in which the cell reaction is $Mn(s) + Cl_2(g) \to MnCl_2(aq)$. Give the half-reactions for the electrodes and from the standard cell potential of 2.54 V deduce the standard potential of the Mn^{2+}/Mn couple.

10.14 (b) Devise a cell in which the cell reaction is $Cd(s) + Ni(OH)_3(s) \to Cd(OH)_2(s) + Ni(OH)_2(s)$. Give the half-reactions for the electrodes.

10.15 (a) Write the cell reactions and electrode half-reactions for the following cells:

(a) $Zn|ZnSO_4(aq)\|AgNO_3(aq)|Ag$

(b) $Cd|CdCl_2(aq)\|HNO_3(aq)|H_2(g)|Pt$

(c) $Pt|K_3[(CN)_6](aq), K_4[(CN)_6](aq)\|CrCl_3(aq)|Cr$

10.15 (b) Write the cell reactions and electrode half-reactions for the following cells:

(a) $Pt|Cl_2(g)|HCl(aq)\|K_2CrO_4(aq)|Ag_2CrO_4(s)|Ag$

(b) $Pt|Fe^{3+}(aq), Fe^{2+}(aq)\|Sn^{4+}(aq), Sn^{2+}(aq)|Pt$

(c) $Cu|Cu^{2+}(aq)\|Mn^{2+}(aq), H^+(aq)|MnO_2(s)|Pt$

10.16 (a) Devise cells in which the following are the reactions:

(a) $Zn(s) + CuSO_4(aq) \longrightarrow ZnSO_4(aq) + Cu(s)$

(b) $2AgCl(s) + H_2(g) \longrightarrow 2HCl(aq) + 2Ag(s)$

(c) $2H_2(g) + O_2(g) \longrightarrow 2H_2O(l)$

10.16 (b) Devise cells in which the following are the reactions:

(a) $2Na(s) + 2H_2O(l) \longrightarrow 2NaOH(aq) + H_2(g)$

(b) $H_2(g) + I_2(g) \longrightarrow 2HI(aq)$

(c) $H_3O^+(aq) + OH^-(aq) \longrightarrow 2H_2O(l)$

10.17 (a) Use standard potentials to calculate the standard potentials of the cells in Exercise 10.15a.

10.17 (b) Use standard potentials to calculate the standard potentials of the cells in Exercise 10.15b.

10.18 (a) Use standard potentials to calculate the standard potentials of the cells in Exercise 10.16a.

10.18 (b) Use standard potentials to calculate the standard potentials of the cells in Exercise 10.16b.

10.19 (a) Calculate the standard Gibbs energies at $25\,°C$ of the following reactions from the standard potential data in Table 10.7:

(a) $2Na(s) + 2H_2O(l) \longrightarrow 2NaOH(aq) + H_2(g)$

(b) $2K(s) + 2H_2O(l) \longrightarrow 2KOH(aq) + H_2(g)$

10.19 (b) Calculate the standard Gibbs energies at $25\,°C$ of the following reactions from the standard potential data in Table 10.7:

(a) $K_2S_2O_8(aq) + 2KI(aq) \longrightarrow I_2(s) + 2K_2SO_4(aq)$

(b) $Pb(s) + Zn(NO_3)_2(aq) \longrightarrow Pb(NO_3)_2(aq) + Zn(s)$

10.20 (a) The standard reaction Gibbs energy for

$$K_2CrO_4(aq) + 2Ag(s) + 2FeCl_3(aq) \longrightarrow$$
$$Ag_2CrO_4(s) + 2FeCl_2(aq) + 2KCl(aq)$$

is -62.5 kJ mol^{-1} at 298 K. (a) Calculate the standard potential of the corresponding galvanic cell and (b) the standard potential of the $Ag_2CrO_4/Ag, CrO_4^{2-}$ couple.

10.20 (b) Two half-cell reactions may be combined in such a way as to form (a) a new half-cell reaction or (b) a complete cell reaction. Illustrate both (a) and (b) by using the half-cell reactions listed below and calculate $E^\ominus$ for both the new half-cell and complete cell reaction.

(i) $2H_2O(l) + 2e^- \longrightarrow H_2(g) + 2OH^-(aq) \qquad E_1^\ominus = -0.828$ V

(ii) $Ag^+(aq) + e^- \longrightarrow Ag(s) \qquad E_2^\ominus = +0.799$ V

10.21 (a) Calculate the standard potential of the couple $Ag_2S, H_2O/Ag, S^{2-}, O_2, H^+$ from the following data:

$$Ag_2S(s) + 2e^- \longrightarrow 2Ag(s) + S^{2-}(aq) \qquad E^\ominus = -0.69 \text{ V}$$
$$O_2(g) + 4H^+(aq) + 4e^- \longrightarrow 2H_2O(l) \qquad E^\ominus = +1.23 \text{ V}$$

10.21 (b) Consider the cell $Pt|H_2(g,p^\ominus)|HCl(aq)|AgCl(s)|Ag$, for which the cell reaction is $2AgCl(s) + H_2(g) \to 2Ag(s) + 2HCl(aq)$. At $25\,°C$ and a molality of HCl of 0.010 mol kg^{-1}, $E = +0.4658$ V. (a) Write the Nernst equation for the cell reaction. (b) Calculate $\Delta_r G$ for the cell reaction. (c) Assuming that the Debye–Hückel limiting law holds at this concentration, calculate $E^\ominus(AgCl, Ag)$.

10.22 (a) Use the Debye–Hückel limiting law and the Nernst equation to estimate the potential of the cell $Ag|AgBr(s)|KBr(aq, 0.050 \text{ mol kg}^{-1})\|Cd(NO_3)_2(aq, 0.010 \text{ mol kg}^{-1})|Cd$ at $25\,°C$.

10.22 (b) Use the information in Table 10.7 to calculate the standard potential of the cell $Ag|AgNO_3(aq)\|Fe(NO_3)_2(aq)|Fe$ and the standard Gibbs energy and enthalpy of the cell reaction at $25\,°C$. Estimate the value of $\Delta_r G^\ominus$ at $35\,°C$.

10.23 (a) Calculate the equilibrium constants of the following reactions at $25\,°C$ from standard potential data:

(a) $Sn(s) + Sn^{4+}(aq) \rightleftharpoons 2Sn^{2+}(aq)$

(b) $Sn(s) + 2AgCl(s) \rightleftharpoons SnCl_2(aq) + 2Ag(s)$

10.23 (b) Calculate the equilibrium constants of the following reactions at $25\,°C$ from standard potential data:

(a) $Sn(s) + CuSO_4(aq) \rightleftharpoons Cu(s) + SnSO_4(aq)$

(b) $Cu^{2+}(aq) + Cu(s) \rightleftharpoons 2Cu^+(aq)$

10.24 (a) Use the standard potentials of the couples Au^+/Au $(+1.69$ V), Au^{3+}/Au $(+1.40$ V), and Fe^{3+}/Fe^{2+} $(+0.77$ V) to calculate $E^\ominus$ and the equilibrium constant for the reaction $2Fe^{2+}(aq) + Au^{3+}(aq) \rightleftharpoons 2Fe^{3+}(aq) + Au^+(aq)$.

10.24 (b) Determine the standard potential of a cell in which the reaction is $Co^{3+}(aq) + 3Cl^-(aq) + 3Ag(s) \to 3AgCl(s) + Co(s)$ from the standard potentials of the couples $AgCl/Ag, Cl^-$ $(+0.22$ V), Co^{3+}/Co^{2+} $(+1.81$ V), and Co^{2+}/Co $(-0.28$ V).

10.25 (a) The solubilities of AgCl and $BaSO_4$ in water are 1.34×10^{-5} mol kg^{-1} and 9.51×10^{-4} mol kg^{-1} respectively at $25\,°C$. Calculate their solubility constants. Is there any significant difference when activity coefficients are ignored?

10.25 (b) The solubilities of AgI and Bi_2S_3 in water are 1.2×10^{-8} mol kg^{-1} and 1.6×10^{-20} mol kg^{-1}, respectively, at 25°C. Calculate their solubility constants. Is there any significant difference when activity coefficients are ignored?

10.26 (a) Derive an expression for the potential of an electrode for which the half-reaction is the reduction of $Cr_2O_7^{2-}$ ions to Cr^{3+} ions in acidic solution.

10.26 (b) Derive an expression for the potential of an electrode for which the half-reaction is the reduction of MnO_4^- ions to Mn^{2+} ions in acidic solution.

10.27 (a) The zero-current potential of the cell $Pt|H_2(g)|HCl(aq)|AgCl(s)|Ag$ was $+0.322$ V at 25°C. What is the pH of the electrolyte solution?

10.27 (b) The zero-current potential of the cell $Pt|H_2(g)|HI(aq)|AgI(s)|Ag$ was 1.00 V at 25°C. What is the pH of the electrolyte solution?

10.28 (a) The solubility of AgBr is 2.6 μmol kg^{-1} at 25°C. What is the zero-current potential of the cell $Ag|AgBr(aq)|AgBr(s)|Ag$ at that temperature?

10.28 (b) The solubility of AgI is 12 nmol kg^{-1} at 25°C. What is the zero-current potential of the cell $Ag|AgI(aq)|AgI(s)|Ag$ at that temperature?

10.29 (a) The standard potential of the cell $Ag|AgI(s)|AgI(aq)|Ag$ is $+0.9509$ V at 25°C. Calculate (a) the solubility of AgI and (b) its solubility constant.

10.29 (b) The standard potential of the cell $Bi|Bi_2S_3(s)|Bi_2S_3(aq)|Bi$ is $+0.96$ V at 25°C. Calculate (a) the solubility of Bi_2S_3 and (b) its solubility constant.

Problems

Numerical problems

10.1 Devise a cell in which the overall reaction is $Pb(s) + Hg_2SO_4(s) \rightarrow PbSO_4(s) + 2Hg(l)$. What is its potential when the electrolyte is saturated with both salts at 25°C? (The solubility constants of Hg_2SO_4 and $PbSO_4$ are 6.6×10^{-7} and 1.6×10^{-8}, respectively.)

10.2 Given that $\Delta_r G^{\ominus} = -212.7$ kJ mol^{-1} for the reaction in the Daniell cell at 25°C, and $b(CuSO_4) = 1.0 \times 10^{-3}$ mol kg^{-1} and $b(ZnSO_4) = 3.0 \times 10^{-3}$ mol kg^{-1}, calculate (a) the ionic strengths of the solutions, (b) the mean ionic activity coefficients in the compartments, (c) the reaction quotient, (d) the standard cell potential, and (e) the cell potential. (Take $\gamma_+ = \gamma_- = \gamma_\pm$ in the respective compartments.)

10.3 Although the hydrogen electrode may be conceptually the simplest electrode and is the basis for our reference state of electrical potential in electrochemical systems, it is cumbersome to use. Therefore, several substitutes for it have been devised. One of these alternatives is the quinhydrone electrode (quinhydrone, $Q \cdot QH_2$, is a complex of quinone, $C_6H_4O_2 = Q$, and hydroquinone, $C_6H_4O_2H_2 = QH_2$). The electrode half-reaction is $Q(aq) + 2H^+(aq) + 2e^- \rightarrow QH_2(aq)$, $E^{\ominus} = +0.6994$ V. If the cell $Hg|Hg_2Cl_2(s)|HCl(aq)|Q \cdot QH_2|Au$ is prepared, and the measured cell potential is $+0.190$ V, what is the pH of the HCl solution? Assume that the Debye–Hückel limiting law is applicable.

10.4 A fuel cell develops an electric potential from the chemical reaction between reagents supplied from an outside source. What is the zero-current potential of a cell fuelled by (a) hydrogen and oxygen, (b) the combustion of butane at 1.0 atm and 298 K?

10.5 Consider the cell, $Zn(s)|ZnCl_2(0.0050$ mol kg$^{-1})$ $|Hg_2Cl_2(s)|Hg(l)$, for which the cell reaction is $Hg_2Cl_2(s) + Zn(s) \rightarrow 2Hg(l) + 2Cl^-(aq) + Zn^{2+}(aq)$. Given that $E^{\ominus}(Zn^{2+}, Zn) = -0.7628$ V, $E^{\ominus}(Hg_2Cl_2, Hg) = +0.2676$ V, and that the measured

value of the cell potential is $+1.2272$ V, (a) write the Nernst equation for the cell. Determine (b) the standard cell potential, (c) $\Delta_r G$, $\Delta_r G^{\ominus}$, and K for the cell reaction, (d) the mean ionic activity and activity coefficient of $ZnCl_2$ from the measured cell potential, and (e) the mean ionic activity coefficient of $ZnCl_2$ from the Debye–Hückel limiting law. (f) Given that $(\partial E/\partial T)_p = -4.52 \times 10^{-4}$ V K^{-1}, calculate ΔS and ΔH.

10.6 The zero-current potential of the cell $Pt|H_2(g, p^{\ominus})|HCl(aq, b)|Hg_2Cl_2(s)|Hg(l)$ has been measured with high precision (G.J. Hills and D.J.G. Ives, *J. Chem. Soc.*, 311 (1951)) with the following results at 25°C:

$b/(\text{mmol kg}^{-1})$	1.6077	3.0769	5.0403	7.6938	10.9474
E/V	0.60080	0.56825	0.54366	0.52267	0.50532

Determine the standard potential of the cell and the mean activity coefficient of HCl at these molalities. (Make a least-squares fit of the data to the best straight line.)

10.7 Careful measurements of the potential of the cell $Pt|H_2(g, p^{\ominus})$ $|NaOH(aq, 0.0100$ mol kg$^{-1}), NaCl(aq, 0.01125$ mol kg$^{-1})$ $|AgCl(s)|Ag$ have been reported (C.P. Bezboruah, M.F.G.F.C. Camoes, A.K. Covington, and J.V. Dobson, *J. Chem. Soc. Faraday Trans. I*, **69**, 949 (1973)). Among the data is the following information:

$\theta/°C$	20.0	25.0	30.0
E/V	1.04774	1.04864	1.04942

Calculate pK_w at these temperatures and the standard enthalpy and entropy of the autoprotolysis of water at 25.0°C.

10.8 Measurements of the potentials of cells of the type $Ag|AgX(s)MX(b_1)|M_xHg|MX(b_2)|AgX(s)|Ag$, where M_xHg denotes an amalgam and the electrolyte is an alkali metal halide dissolved in ethylene glycol, have been reported (U. Sen, *J. Chem. Soc. Faraday Trans. I* **69**, 2006 (1973)) and some values for LiCl are given below.

Estimate the activity coefficient at the concentration marked * and then use this value to calculate activity coefficients from the measured cell potential at the other concentrations. Base your answer on the following version of the extended Debye–Hückel law:

$$\log \gamma_\pm = -\frac{AI^{1/2}}{1 - BI^{1/2}} + kI$$

with $A = 1.461$, $B = 1.70$, $k = 0.20$, and $I = b/b^{\ominus}$. For $b_2 = 0.09141 \text{ mol kg}^{-1}$:

$b_1/(\text{mol kg}^{-1})$	0.0555	0.09141*	0.1652	0.2171	1.040	1.350
E/V	−0.0220	0.0000	0.0263	0.0379	0.1156	0.1336

10.9 Suppose the extended Debye–Hückel law for a 1,1-electrolyte is written in the simplified form $\log \gamma_\pm = -0.509 I^{1/2} + kI$, where k is a constant and $I = b/b^{\ominus}$. Show that a plot of y against I, where

$$y = E + 0.1183 \log I - 0.0602\, I^{1/2}$$

should give a straight line with intercept $E^{\ominus}$ and slope $-0.1183k$. Apply the technique to the following data (at 25 °C) on the cell $\text{Pt}|\text{H}_2(g,p^{\ominus})|\text{HCl}(aq,b)|\text{AgCl}(s)|\text{Ag}$:

$b/(\text{mmol kg}^{-1})$	123.8	25.63	9.138	5.619	3.215
E/mV	341.99	418.24	468.60	492.57	520.53

(a) Find the standard cell potential and the standard potential of the AgCl/Ag, Cl⁻ couple. (b) The zero-current cell potential was measured as 352.4 mV when $b = 100.0 \text{ mol kg}^{-1}$. What is the pH and the mean ionic activity coefficient?

10.10 The mean activity coefficients for aqueous solutions of NaCl at 25 °C are given below. Confirm that they support the Debye–Hückel limiting law and that an improved fit is obtained with the extended law.

$b/(\text{mmol kg}^{-1})$	1.0	2.0	5.0	10.0	20.0
$\gamma_\pm$	0.9649	0.9519	0.9275	0.9024	0.8712

10.11 The standard potential of the AgCl/Ag, Cl⁻ couple has been measured very carefully over a range of temperature (R.G. Bates and V.E. Bowers, *J. Res. Nat. Bur. Stand.* **53**, 283 (1954)) and the results were found to fit the expression

$$E^{\ominus}/\text{V} = 0.23659 - 4.8564 \times 10^{-4}(\theta/°\text{C})$$
$$- 3.4205 \times 10^{-6}(\theta/°\text{C})^2 + 5.869 \times 10^{-9}(\theta/°\text{C})^3$$

Calculate the standard Gibbs energy and enthalpy of formation of Cl⁻(aq) and its entropy at 298 K.

10.12 Use the data below to confirm that the Debye–Hückel limiting law correctly predicts the limiting values of the mean activity coefficient of acetic acid by demonstrating that pK'_a, where $K_a = K'_a K_\gamma$, plotted against $(\alpha b)^{1/2}$, where b is the molality of the acid and α its degree of ionization, should be a straight line.

$b/(\text{mmol kg}^{-1})$	0.0280	0.1114	0.2184	1.0283	2.414	5.9115
α	0.5393	0.3277	0.2477	0.1238	0.0829	0.0540

10.13 The $\text{Sb}|\text{Sb}_2\text{O}_3(s)|\text{OH}^-(aq)$ electrode is reversible with respect to OH⁻ ions. Derive an expression for its potential in terms of (a) the pOH and (b) the pH of the solution. (c) By how much does the potential change when the molality of NaOH(aq) in the electrode compartment is increased from 0.010 mol kg⁻¹ to 0.050 mol kg⁻¹ at 25 °C? Use the Debye–Hückel limiting law to estimate any activity coefficients required.

10.14 Superheavy elements are now of considerable interest. Shortly before it was (falsely) believed that the first had been discovered, an attempt was made to predict the chemical properties of ununpentium (Uup, element 115, O.L. Keller, C.W. Nestor, and B. Fricke, *J. Phys. Chem.* **78**, 1945 (1974)). In one part of the paper the standard enthalpy and entropy of the reaction $\text{Uup}^+(aq) + \frac{1}{2}\text{H}_2(g) \rightarrow \text{Uup}(s) + \text{H}^+(aq)$ were estimated from the following data: $\Delta_{sub}H^{\ominus}(\text{Uup}) = +1.5 \text{ eV}$, $I(\text{Uup}) = 5.52 \text{ eV}$, $\Delta_{hyd}H^{\ominus}(\text{Uup}^+, aq) = -3.22 \text{ eV}$, $S^{\ominus}(\text{Uup}^+, aq) = +1.34 \text{ meV/K}^{-1}$, $S^{\ominus}(\text{Uup}, s) = 0.69 \text{ meV K}^{-1}$. Estimate the expected standard potential of the Uup⁺/Uup couple.

Theoretical problems

10.15 Show that the solubility, S, of a sparingly soluble 1 : 1 salt is related to its solubility constant by $S = K_s^{1/2} e^{1.172 S^{1/2}}$.

10.16 Suppose that a sparingly soluble salt MX has solubility constant K_s and solubility S. Show that, in an ideal solution that is of concentration C of a freely soluble salt NX, the solubility of MX is changed to

$$S' = \frac{1}{2}(C^2 + 4K_s^2)^{1/2} - \frac{1}{2}C$$

and that $S' \approx K_s/C$ when K_s is small (in a sense to be specified).

10.17 Show that, if the ionic strength of a solution of the sparingly soluble salt MX and the freely soluble salt NX is dominated by the concentration C of the latter, and if it is valid to use the Debye–Hückel limiting law, the solubility S' in the mixed solution is given by

$$S' = \frac{K_s e^{4.606 AC^{1/2}}}{C}$$

when K_s is small (in a sense to be specified).

10.18 Show that the freezing-point depression of a real solution in which the solvent of molar mass M has activity a_A obeys

$$\frac{d \ln a_A}{d(\Delta T)} = -\frac{M}{K_f}$$

and use the Gibbs–Duhem equation to show that

$$\frac{d \ln a_B}{d(\Delta T)} = -\frac{1}{b_B K_f}$$

where a_B is the solute activity and b_B is its molality. Use the Debye–Hückel limiting law to show that the osmotic coefficient (ϕ, Problem 7.12) is given by $\phi = 1 - \frac{1}{3}A'I$ with $A' = 2.303A$ and $I = b/b^{\ominus}$.

Additional problems supplied by Carmen Giunta and Charles Trapp

10.19 The list of standard potentials compiled by Bratsch shows that europium has the least negative reduction potential for the M^{3+}/M couple in strong acid (pH = 0) of all the lanthanides (S.G. Bratsch, *J. Phys. Chem. Ref. Data* **18**, 1 (1989)). $E^{\ominus}(Eu^{3+}/Eu) = -1.991$ V, while the next easiest to reduce is ytterbium, with $E^{\ominus}(Yb^{3+}/Yb) = -2.19$ V. (a) Describe the criteria for a reducing agent that would deposit Eu metal from an acidic solution of M^{3+} lanthanide ions while leaving the other lanthanides in solution under standard conditions (unit activities and fugacities; pH = 0). Is there such a reagent in Table 10.7? (b) Now suppose the conditions are not standard. What is the maximum value of the ratio $a_{Yb^{3+}}/a_{Eu^{3+}}$ such that Eu still deposits spontaneously (just barely) but Yb does not?

10.20 Larson studied the thermochemistry of vanadium (V) ions in aqueous solution (J.W. Larson, *J. Chem. Eng. Data* **40**, 1276 (1995)). At neutral pH, the dominant species are $H_2VO_4^-$ and $V_4O_{12}^{4-}$. The equilibrium constant for the reaction $4H_2VO_4^-(aq) \rightarrow V_4O_{12}^{4-}(aq) + 4H_2O(l)$ was determined to be 106.4 at 50°C. How is vanadium distributed between these two species at 50°C and unit ionic strength if the total vanadium concentration is 1.0×10^{-2} mol kg^{-1}? Take the ionic strength to be 1.000 and use the following extended Debye–Hückel equation to estimate activity coefficients where $A = 0.5373$ at 50°C.

$$\log \gamma = -\frac{Az^2 I^{1/2}}{1 + I^{1/2}}$$

10.21 (a) Derive a general relation for $(\partial E/\partial p)_{T,n}$ for electrochemical cells employing reactants in any state of matter. (b) E. Cohen and K. Piepenbroek (*Z. physik. Chem.* **167A**, 365 (1933)) calculated the change in volume for the reaction $TlCl(s) + CNS^-(aq) \rightarrow TlCNS(s) + Cl^-(aq)$ at 30°C from density data and obtained $\Delta_r V = -2.666 \pm 0.080$ cm^3 mol^{-1}. They also measured the potential of the cell $Tl(Hg)|TlCNS(s)|KCNS\!:\!KCl|TlCl(s)|Tl(Hg)$ at temperatures up to 1500 atm. Their results are given in the following table.

p/atm	1.00	250	500	750	1000	1250	1500
E/mV	8.56	9.27	9.98	10.69	11.39	12.11	12.82

From this information, obtain $(\partial E/\partial p)_{T,n}$ at 30°C and compare to the value obtained from $\Delta_r V$. (c) Fit the data to a polynomial for E against p. How constant is $(\partial E/\partial p)_{T,n}$? (d) From the polynomial, estimate an effective isothermal compressibility for the cell as a whole.

10.22 Because the hydrogen electrode is the fundamental reference electrode, it is important to study the effect of pressure on cells containing it. W.R. Hainsworth, H.J. Rowley, and D.A. MacInnes (*J. Am. Chem. Soc.* **46**, 1437 (1924)) investigated the influence of pressure on the hydrogen/calomel cell, $Pt|H_2(g, p)|HCl(0.1 \text{ M})|Hg_2Cl_2(s)|Hg(l)$, up to pressures of 1000 atm at 25°C with the following results, where $\Delta E = E(p) - E(1 \text{ atm})$.

p/atm	1.00	10	38	51	108	210
ΔE/mV	0	29.5	47	51	61	69
p/atm	380	430	560	720	900	1020
ΔE/mV	79	82	86	91	95	98

(a) Derive an expression for ΔE using the perfect gas law. Compare the expression with the experimental results. (b) Fit the values of ΔE and $(\partial E/\partial p)_T$ to a polynomial expression in p. (c) Obtain a theoretical equation for these quantities using both the virial form of the van der Waals equation and the empirical virial equation $Z = 1 + 0.000537(p/\text{atm}) + 3.5 \times 10^{-8}(p/\text{atm})^2$ used by Hainsworth *et al.* Plot the experimental and theoretical data for this cell on the same graph. How well do they fit? (d) Calculate the fugacity of H_2 from the cell potential data.

Equilibrium

1.6 Low and high temperature heat capacity of Ce$_2$Si$_2$O$_7$

It is found that the molar heat capacity of β-Ce$_2$Si$_2$O$_7$ at 1 bar in the range of temperatures 0–30 K is described by

$$C_{p,m}(T) = a(T/K)^3\left(1 - e^{-b/T^2}\right)$$

where $a = 5.597 \times 10^{-4}\,\mathrm{J\,mol^{-1}\,K^{-4}}$ and $b = 1701.8\,\mathrm{K^2}$ are empirical parameters. In the high-temperature range, 500–900 K, the molar heat capacity is described by

$$C_{p,m}(T)/(\mathrm{J\,mol^{-1}\,K^{-1}}) = 222.3716$$
$$+\,0.076\,790\,42(T/\mathrm{K}) - 3\,631\,886(T/\mathrm{K})^{-2}$$

No phase transition has been detected below 900 K.

(a) Draw graphs of $S_m(T)$, $H_m(T) - H_m(0)$, and $G_m(T) - G_m(0)$ in the range $0 \leq T \leq 30$ K.

(b) According to the law of Dulong and Petit, the high-temperature heat capacity per mole of atoms of a crystalline material is $3R$. How does the classical prediction compare with the experimental?

1.7 The destruction of chlorofluorocarbon stockpiles

Protocols have outlawed the use of certain CFC refrigerants. It has been suggested that sodium oxalate, Na$_2$(CO$_2$)$_2$, at elevated temperatures can be used to destroy existing CFC stockpiles. The mineralization of freon-12, for example, is

$$\mathrm{CF_2Cl_2(g) + 2Na_2(CO_2)_2(s) \longrightarrow}$$
$$\mathrm{2NaF(s) + 2NaCl(s) + C(s) + 4CO_2(g)}$$

(a) Without performing a calculation, what is your expectation concerning the spontaneity of the reaction? Why is the high temperature required? Why should reaction mixtures not be heated to temperatures much higher than 270°C?

(b) Use library, internet, or PC database resources to find thermodynamic data which will make possible the calculation of $\Delta_r G^{\ominus}(T)$. Estimate the value of any thermodynamic property that cannot be found. For general information about internet chemistry resources try these URLs: http://www.acs.org/; http://www.indiana.edu/cheminfo/. For specific thermodynamic data try the NIST Chemistry Webbook: http://webbook.nist.gov/.

1.8 Calorimetric determination of hydrogen bond strengths

The excess enthalpy of mixing data for 1-propanol and tripropylamine at 25°C is found to fit the expression

$$H^{\mathrm{E}}/(\mathrm{J\,mol^{-1}}) = x_1(1 - x_1)\sum_{r=0}^{3} A_r(1 - 2x_1)^r$$

where $A_r/(\mathrm{J\,mol^{-1}}) = -1131,\ 1355,\ -855,\ 1127$ for $r = 0, 1, 2, 3$.

(a) Draw a graph of H^{E} against x_1 and compare the graph shape to that expected for a solution in which $H^{\mathrm{E}} = \omega x_1 x_2$. What do the graphs imply about intermolecular interaction energies?

(b) At 25°C and infinite dilution the excess enthalpy of mixing 1-propanol in heptane is $H^{\mathrm{E}} = +25.3\,\mathrm{kJ\,mol^{-1}}$, whereas for the mixing of tripropylamine in heptane $H^{\mathrm{E}} = +121\,\mathrm{J\,mol^{-1}}$. Deduce the strength of the alcohol–amine hydrogen bond. Assume that intermolecular structural reorganization enthalpies cancel for the cycle.

1.9 Vapour–liquid equilibrium

The compound *tert*-amyl methyl ether (TAME) has potential use as a gasoline antiknock agent. The Redlich–Kister empirical equation for the excess Gibbs energy of mixing of methanol (1) and TAME (2) is:

$$G^{\mathrm{E}}/RT = x_1 x_2 \sum_{n=0}^{4} A_n(1 - 2x_1)^n$$

where $A_n = 1.4170,\ -0.05245,\ 0.1599,\ 0.1061,\ -0.09159$ for $n = 0, 1, \cdots, 4$. The activity coefficients are given by

$$\ln \gamma_1 = x_2^2 \sum_{n=0}^{4} A_n f_n(x_1) \qquad \ln \gamma_2 = x_1^2 \sum_{n=0}^{4} (-1)^n A_n f_n(x_2)$$

where

$$f_n(x) = \{1 - 2(n+1)x\}(1 - 2x)^{n-1}$$

The vapour pressure and molar volume of pure TAME are 6.09 kPa and 131.78 cm^3 mol^{-1} at 288.15 K and those of methanol are 9.86 kPa and 40.254 cm^3 mol^{-1}.

(a) What is the thermodynamic criterion for solution–vapour equilibrium? What law relates vapour pressure to ideal solution composition? How is this law modified so as to describe the vapour pressure of real solutions?

(b) Draw a graph of $\ln \gamma$ against x_1. Do the graphs reflect correct values at infinite dilutions of either TAME or methanol?

(c) Draw a graph of $G^{\mathrm{E}}(x_1)$. What does this graph reveal about the spontaneity of mixing? Does the curve have the shape expected for a regular solution?

(d) Draw a graph of total vapour pressure, $p = p_1 + p_2$, and also show the curves for $p(x_1)$ and $p(y_1)$, where y is the mole fraction in the vapour. What is the vapour pressure and composition of the azeotrope? What is the equilibrium pressure when $x_1 = 0.2$?

(e) Draw a graph of total vapour and partial vapour pressures. The curves would be $p(x_1)$, $p_1(x_1)$, and $p_2(x_1)$. What do these curves reveal about deviations from ideality? Calculate the Henry's law constant for both methanol and TAME.

(f) Estimate the vapour pressure of pure TAME at 288.15 K when subjected to an applied pressure of 2.0 bar.

1.10 Solubility and miscibility of Cu/Pb mixtures

The illustration below shows the $\Delta_{\text{mix}}G(x_{\text{Pb}}, T)$ for a mixture of copper and lead.

(a) What does the graph reveal about the miscibility of copper and lead and the spontaneity of solution formation? What is the variance (F) at (i) 1500 K, (ii) 1100 K?

(b) Suppose that at 1500 K a mixture of composition (i) $x_{\text{Pb}} = 0.1$, (ii) $x_{\text{Pb}} = 0.7$, is slowly cooled to 1100 K. What is the equilibrium composition of the final mixture? Include an estimate of the relative amounts of each phase.

(c) What is the solubility of (i) lead in copper, (ii) copper in lead at 1100 K?

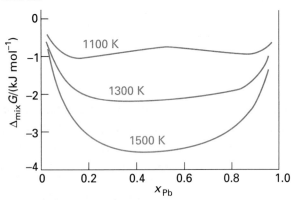

1.11 The temperature–composition diagram for the Ca/Si binary system

(a) Identify eutectics, congruent melting compounds, and incongruent melting compounds that are shown in the calcium/silicon phase diagram shown below.

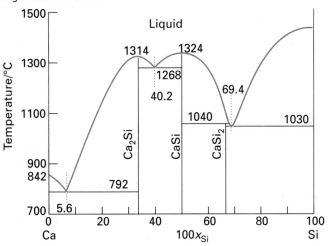

(b) If a 20 per cent by atom composition melt of silicon at 1500°C is cooled to 1000°C, what phases (and phase composition) would be at equilibrium? Estimate the relative amounts of each phase.

(c) If $CaSi_2$ melts completely at 1040°C, what phases (and phase composition and relative amounts) would be at equilibrium?

(d) Describe the equilibrium phases observed when an 80 per cent by atom composition Si melt is cooled to 1030°C. What phases, and relative amounts, would be at equilibrium at a temperature (i) slightly higher than 1030°C, (ii) slightly lower than 1030°C? Draw a graph of the mole percentages of both Si(s) and $CaSi_2$(s) as a function of mole percentage of melt that is freezing at 1030°C.

1.12 Aspects of ammonia production

Suppose that an iron catalyst at a particular plant produces ammonia in the most cost-effective manner at 450°C when the pressure is such that Δ_rG for the reaction $\frac{1}{2}N_2(g) + \frac{3}{2}H_2(g) \rightarrow NH_3(g)$ is equal to $-500\ \text{J mol}^{-1}$. What pressure is needed? Now suppose that a new catalyst is developed that is most cost-effective at 400°C when the pressure gives the same value of Δ_rG. What pressure is needed when the new catalyst is used? What are the advantages of the new catalyst? Assume that (a) all gases are perfect gases or that (b) all gases are van der Waals gases. Isotherms of $\Delta_rG(T,p)$ in the pressure range $100\ \text{atm} \leq p \leq 400\ \text{atm}$ are needed to derive the answer. Do these graphs confirm Le Chatelier's principle concerning the response of equilibrium changes in temperature and pressure?

1.13 Silica reduction with graphite

Large quantities of silicon are made by reducing SiO_2 with carbon in arc furnaces at elevated temperatures. Many chemical species may be present in these high-temperature SiO_2/graphite mixtures, including C(s), SiO_2(s), SiO_2(l), Si(s), Si(l), SiC(s), SiO(g), and CO(g). The table below contains information about the temperature dependence of $G(T) - H_{\text{SER}} = a + bT$ for each of these species, and several others, in the temperature range 1500–2500 K and at 1 bar, where $G(T) - H_{\text{SER}}$ is the Gibbs energy at temperature T relative to the enthalpies of the pure constituent elements in their reference states at the conventional temperature. Use these data alone to answer the following questions. Assume perfect gas behaviour and ideal solutions when necessary.

	$a/(\text{kJ mol}^{-1})$	$b/(\text{kJ K}^{-1}\text{mol}^{-1})$
graphite	34.38	−0.04054
Si(s)	44.281	−0.06671
SiO_2(l)	−757.039	−0.18456
Si(g)	484.667	−0.20808
SiO(g)	−41.075	−0.28123
O_2(g)	57.858	−0.26856
SiO_2(s)	−818.448	−0.15281
SiC(s)	2.521	−0.09838
Si(l)	96.980	−0.09743
SiO_2(g)	−209.039	−0.34203
CO(g)	−54.581	−0.26312
CO_2(g)	−302.873	−0.32021

(a) Calculate the normal melting point of silica and silicon. Make graphs of the vapour pressures of each in the temperature range 1500–2500 K and at a pressure of 1 bar.

(b) Draw an Ellingham diagram for the smelting reduction of silica with graphite in the temperature range 1500–2500 K and at a pressure of 1 bar. What does the diagram reveal about the evolution of carbon dioxide during smelting?

(c) Determine the equilibrium purity of silicon made by reducing silica with graphite at 2000 K and 1 bar. Assume that graphite is soluble in the silicon but that SiO_2 and SiC are not. Also determine the equilibrium partial pressures of $SiO(g)$ and $CO(g)$. Equilibria to consider include $SiO_2(l) + C(graphite, s) \rightleftharpoons SiO(g) + CO(g)$, $SiO_2(l) + 2C(graphite, s) \rightleftharpoons Si(l) + 2CO(g)$, and $SiO_2(l) + 3C(graphite, s) \rightleftharpoons SiC(s) + 2CO(g)$.

1.14 pH–concentration curves for carbon dioxide

The ocean's ability to dissolve carbon is an important sink when modelling the carbon cycle. Draw a graph of α_J against pH for carbon dioxide solutions in the pH range 0–14 where the fraction α_J of species J is defined as

$$\alpha_J = \frac{[J]}{[\text{total carbon}]}$$
$$= \frac{[J]}{[CO_2] + [H_2CO_3] + [HCO_3^-] + [CO_3^{2-}]}$$

Assume that solutions are ideal.

(a) What are the major chemical species at the physiological pH of 7.4 and equilibrium?

(b) What is the maximum mass of ocean carbon dissolved as CO_2, H_2CO_3, HCO_3^-, and CO_3^{2-} if the ocean has an average pH of 8, an average temperature of 298 K, and a volume of 1.37×10^{18} m^3? The partial pressure of atmospheric carbon dioxide is 3.3×10^{-4} atm.

1.15 Sulfuric acid solutions and lead electrochemistry

A cell of the lead–acid battery may be represented as

$$Pb(s)|PbSO_4(s)|H_2SO_4(aq)|PbO_2(s)|PbSO_4(s)|Pb(s)$$

where the aqueous solution is about 35 per cent by mass sulfuric acid. The sulfuric acid molality, b, is related to the solution density, d, at 25°C by the empirical equation:

$$b(H_2SO_4)/(\text{mol kg}^{-1}) = a(d - d_{25})/(\text{g cm}^{-3})$$
$$+ c(d - d_{25})^2/(\text{g cm}^{-3})^2$$

where d_{25} is the density of pure water at 25°C, $a = 14.523$, and $c = 25.031$.

(a) Draw graphs of molality, mass percentage, and molar concentration functions of density for densities in the range 1.0–1.4 g cm^{-3}.

(b) Write balanced equations for the half-reactions and the cell reaction for the lead–acid battery. Determine $E^\ominus$, $\Delta_rG^\ominus$, $\Delta_rS^\ominus$, and $\Delta_rH^\ominus$ for the cell reaction at 25°C. Estimate $E^\ominus(15°C)$. Calculate the cell potential at 25°C when $Q = 6.0 \times 10^{-5}$.

(c) The illustration below shows a simplified 'Pourbaix diagram' for the redox chemistry of $PbO_2/PbSO_4/Pb$ systems. The horizontal and slanted lines of the diagram are equilibrium half-cell potentials when all chemical species have unit activity. Vertical lines are non-redox equilibria. Prove that the redox potential lines at 25°C are given by $E = E^\ominus - (0.0592 \text{ V})\nu pH$, where ν is the number of hydrogen ions used in the half-reaction. Calculate $E(PbO_2/PbSO_4)$ at pH values of 5 and 8. What is $E(PbSO_4/Pb)$ at any pH?

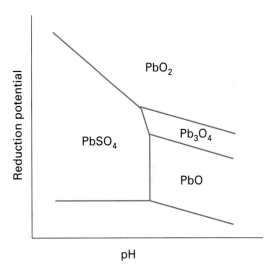

1.16 Weak acidity of 2-aminopyridinium chloride

The table below summarizes the zero current potentials observed for the cell $Pd|H_2(g, 1 \text{ bar})|BH(aq, b), B(aq, b)|AgCl(s)|Ag$. Each measurement is made at equimolal concentrations of 2-aminopyridinium chloride (BH) and 2-aminopyridine (B). The data are for 25°C and it is found that $E^\ominus = 0.22251$ V. Use the data to determine pK_a for the acid at 25°C and the mean activity coefficient $(\gamma_\pm)$ of BH as a function of molality (b) and ionic strength (I). Use the extended Debye–Hückel equation for the mean activity coefficient in the form:

$$-\log \gamma_\pm = \frac{AI^{1/2}}{1 + BI^{1/2}} - kb$$

where $A = 0.5091$ and B and k are parameters that depend upon the ions. Draw a graph of the mean activity coefficient with $b = 0.04$ mol kg^{-1} and $0 \leq I \leq 0.1$.

$b/(\text{mol kg}^{-1})$	0.01	0.02	0.03	0.04	0.05
$E^\ominus(25°C)/V$	0.744 52	0.728 53	0.719 28	0.713 14	0.708 09
$b/(\text{mol kg}^{-1})$	0.06	0.07	0.08	0.09	0.10
$E^\ominus(25°C)/V$	0.703 80	0.700 59	0.697 90	0.695 71	0.693 38

The table below summarizes p$K_a(T)$ for BH; add your value of p$K_a(25°C)$ to those of the table. Determine the coefficients a_0, a_1, and a_2 such that the regression equation $pK_a(T) = a_0 + a_1/(T/K) + a_2 \ln(T/K)$ gives the best least-squares fit of the data. Draw graphs of $\Delta_rG^\ominus(T)$, $\Delta_rH^\ominus(T)$, and $\Delta_rS^\ominus(T)$ for the proton transfer reaction.

$\theta/°C$	5	10	15	20	25	30	35	40
pK_a	7.177	7.060	6.949	6.841	?	6.640	6.543	6.449

Part 2

Structure

In Part 1 we examined the properties of bulk matter from the viewpoint of thermodynamics. In Part 2 we turn to the study of individual atoms and molecules from the viewpoint of quantum mechanics. The two viewpoints merge in Chapter 19.

11 Quantum theory: introduction and principles

This chapter introduces some of the basic principles of quantum mechanics. First, it reviews the experimental results that overthrew the concepts of classical physics. These experiments led to the conclusion that particles may not have an arbitrary energy and that the classical concepts of 'particle' and 'wave' blend together. The overthrow of classical mechanics inspired the formulation of a new set of concepts and the formulation of quantum mechanics. In quantum mechanics, all the properties of a system are expressed in terms of a wavefunction which is obtained by solving the Schrödinger equation. We see how to interpret wavefunctions. Finally, we introduce some of the techniques of quantum mechanics in terms of operators, and see that they lead to the uncertainty principle, one of the most profound departures from classical mechanics.

To understand the structures of individual atoms and molecules, we need to know how subatomic particles move in response to the forces they experience. It was once thought that the motion of atoms and subatomic particles could be expressed using the laws of **classical mechanics** introduced in the seventeenth century by Isaac Newton, for these laws were very successful at explaining the motion of everyday objects and planets. However, towards the end of the nineteenth century, experimental evidence accumulated showing that classical mechanics failed when it was applied to very small particles, and it took until the 1920s to discover the appropriate concepts and equations for describing them. We describe the concepts of this new mechanics, which is called **quantum mechanics**, in this chapter, and apply them throughout the remainder of the text.

The origins of quantum mechanics

The basic principles of classical mechanics are reviewed in *Further information 4*. In brief, they show that classical physics (1) predicts a precise trajectory for particles, with precisely specified locations and momenta at each instant, and (2) allows the translational, rotational, and vibrational modes of motion to be excited to any energy simply by controlling the forces

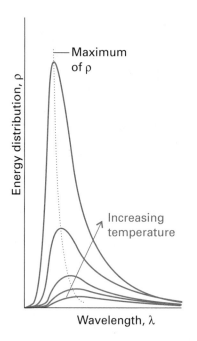

11.1 The energy distribution in a black-body cavity at several temperatures. Note how the energy density increases in the visible region as the temperature is raised, and how the peak shifts to shorter wavelengths. The total energy density (the area under the curve) increases as the temperature is increased (as T^4).

that are applied. These conclusions agree with everyday experience. Everyday experience, however, does not extend to individual atoms, and careful experiments of the type described below have shown that classical mechanics fails when applied to the transfers of very small quantities of energy and to objects of very small mass.

11.1 The failures of classical physics

In this section we review some of the experimental evidence which showed that several concepts of classical mechanics are untenable. In particular, we shall see that observations of black-body radiation, heat capacities, and atomic and molecular spectra indicate that systems can take up energy only in discrete amounts.

(a) Black-body radiation

A hot object emits electromagnetic radiation. An appreciable proportion of the radiation is in the visible region of the spectrum at high temperatures, and a higher proportion of short-wavelength blue light is generated as the temperature is raised. This behaviour is seen when a heated iron bar glowing red hot becomes white hot when heated further. The dependence is illustrated in Fig. 11.1, which shows how the energy output varies with wavelength at several temperatures. The curves are those of an ideal emitter called a **black body**, which is an object capable of emitting and absorbing all frequencies of radiation uniformly. A good approximation to a black body is a pinhole in an empty container maintained at a constant temperature, because any radiation leaking out of the hole has been absorbed and re-emitted inside so many times that it has come to thermal equilibrium with the walls (Fig. 11.2).

Figure 11.1 shows that the peak in the energy output shifts to shorter wavelengths as the temperature is raised. As a result, the short-wavelength tail of the energy distribution strengthens in the visible region and the perceived colour shifts towards the blue, as already mentioned. An analysis of the data led Wilhelm Wien (in 1893) to formulate the **Wien displacement law**:

$$T\lambda_{max} = \tfrac{1}{5}c_2 \qquad c_2 = 1.44 \text{ cm K} \tag{1}$$

where λ_{max} is the wavelength corresponding to the maximum of the distribution at a temperature T. The constant c_2 is called the **second radiation constant**. Using its value, we can predict that $\lambda_{max} \approx 2900$ nm at 1000 K.

A second feature of black-body radiation had been noticed in 1879 by Josef Stefan, who considered the **total energy density**, $\mathcal{E}$, the total electromagnetic energy in a region divided by the volume of the region ($\mathcal{E} = E/V$). The energy density of the electromagnetic field inside the container in Fig. 11.2 increases as the temperature is increased, and specifically the **Stefan–Boltzmann law** states that

$$\mathcal{E} = aT^4 \tag{2a}$$

Ludwig Boltzmann's name is attached to this law because he explained it theoretically. An alternative form of the law is in terms of the **excitance**, M, the power[1] emitted by a region of surface divided by the area of the surface: the excitance is a measure of the brightness of the emission. Because the excitance is proportional to the energy density in the container, M is also proportional to T^4, and we can write

$$M = \sigma T^4 \qquad \sigma = 5.67 \times 10^{-8} \text{ W m}^{-2} \text{ K}^{-4} \tag{2b}$$

The constant σ is called the **Stefan–Boltzmann constant**. The Stefan–Boltzmann law implies that 1 cm^2 of the surface of a black body at 1000 K radiates about 6 W when all

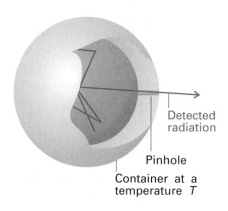

11.2 An experimental representation of a black body is a pinhole in an otherwise closed container. The radiation is reflected many times within the container and comes to thermal equilibrium with the walls at a temperature T. Radiation leaking out through the pinhole is characteristic of the radiation within the container.

---

1 Power is the rate of supply of energy. Its SI units are watts, W: 1 W = 1 J s^{-1}.

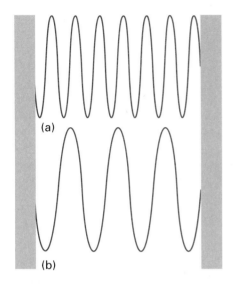

11.3 The electromagnetic vacuum can be regarded as able to support oscillations of the electromagnetic field. When a high-frequency short-wavelength oscillator (a) is excited, that frequency of radiation is present. The presence of low-frequency long-wavelength radiation (b) signifies that an oscillator of the corresponding frequency has been excited.

wavelengths of the emitted radiation are taken into account. The explanation of black-body radiation was a major challenge for nineteenth-century scientists, and in due course it was found to be beyond the capabilities of classical physics. The physicist Lord Rayleigh studied it theoretically from a classical viewpoint, and thought of the electromagnetic field as a collection of oscillators of all possible frequencies. He regarded the presence of radiation of frequency ν (and therefore of wavelength $\lambda = c/\nu$, where c is the speed of light) as signifying that the electromagnetic oscillator of that frequency had been excited (Fig. 11.3). Rayleigh used the equipartition principle (see the *Introduction*) to calculate the average energy of each oscillator as kT. Then, with minor help from James Jeans, he arrived at the **Rayleigh–Jeans law**:

$$d\mathcal{E} = \rho \, d\lambda \qquad \rho = \frac{8\pi kT}{\lambda^4} \tag{3}$$

where ρ is the proportionality constant between $d\lambda$ and the energy density in that range of wavelengths; k is the Boltzmann constant ($k = 1.381 \times 10^{-23}$ J K^{-1}).

Unfortunately (for Rayleigh, Jeans, and classical physics), although the Rayleigh–Jeans law is quite successful at long wavelengths (low frequencies), it fails badly at short wavelengths (high frequencies). Thus, as λ decreases, ρ increases without going through a maximum (Fig. 11.4). The equation therefore predicts that oscillators of very short wavelength (corresponding to ultraviolet light, X-rays, and even γ-rays) are strongly excited even at room temperature. This absurd result, which implies that a large amount of energy is radiated in the high-frequency region of the electromagnetic spectrum, is called the **ultraviolet catastrophe**. According to classical physics, even cool objects should radiate in the visible and ultraviolet regions: according to classical physics, objects should glow in the dark; there should in fact be no darkness.

(b) The Planck distribution

The German physicist Max Planck studied black-body radiation from the viewpoint of thermodynamics. In 1900 he found that he could account for the experimental observations by proposing that *the energy of each electromagnetic oscillator is limited to discrete values and cannot be varied arbitrarily*. This proposal is quite contrary to the viewpoint of classical physics (on which the equipartition principle used by Rayleigh is based), in which all possible energies are allowed. The limitation of energies to discrete values is called the **quantization of energy**. In particular, Planck found that he could account for the observed distribution of energy if he supposed that the permitted energies of an electromagnetic oscillator of frequency ν are integer multiples of $h\nu$:

$$E = nh\nu \qquad n = 0, 1, 2, \ldots \tag{4}$$

where h is a fundamental constant now known as the **Planck constant**.

On the basis of this assumption, Planck was able to derive the **Planck distribution**:

$$d\mathcal{E} = \rho d\lambda \qquad \rho = \frac{8\pi hc}{\lambda^5} \left(\frac{1}{e^{hc/\lambda kT} - 1} \right) \tag{5}$$

This expression fits the experimental curve very well at all wavelengths (Fig. 11.5), and the value of h, which is an undetermined parameter in the theory, may be obtained by varying its value until a best fit is obtained. The currently accepted value for h is 6.62608×10^{-34} J s.

The Planck distribution resembles the Rayleigh–Jeans law (eqn 3) apart from the all-important exponential factor in the denominator. For short wavelengths, $hc/\lambda kT$ is large and $e^{hc/\lambda kT} \rightarrow \infty$ faster than $\lambda^5 \rightarrow 0$; therefore $\rho \rightarrow 0$ as $\lambda \rightarrow 0$ or $\nu \rightarrow \infty$. Hence, the energy density approaches zero at high frequencies, in agreement with observation. For long

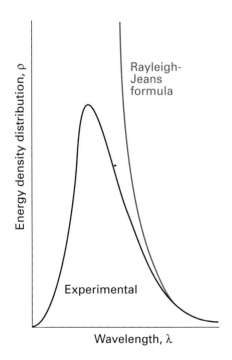

11.4 The Rayleigh–Jeans law (eqn 3) predicts an infinite energy density at short wavelengths. This prediction is called the ultraviolet catastrophe.

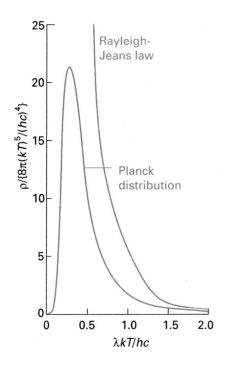

11.5 The Planck distribution (eqn 5) accounts very well for the experimentally determined distribution of radiation. Planck's quantization hypothesis essentially quenches the contributions of high-frequency, short-wavelength oscillators. The distribution coincides with the Rayleigh–Jeans distribution at long wavelengths.

wavelengths, $hc/\lambda kT \ll 1$, and the denominator in the Planck distribution can be replaced by

$$
e^{hc/\lambda kT} - 1 = \left(1 + \frac{hc}{\lambda kT} + \cdots\right) - 1 \approx \frac{hc}{\lambda kT}
$$

When this approximation is substituted into eqn 5, we find that the Planck distribution reduces to the Rayleigh–Jeans law.

The Planck distribution also accounts for the Stefan–Boltzmann and Wien laws. The former is obtained by integrating the energy density over all wavelengths from $\lambda = 0$ to $\lambda = \infty$, which gives

$$
\mathcal{E} = \int_0^\infty \rho \, d\lambda = aT^4 \qquad a = \frac{4\sigma}{c}, \; \sigma = \frac{2\pi^5 k^4}{15c^2 h^3} \tag{6}
$$

Substitution of the values of the fundamental constants gives $\sigma = 56.704 \text{ nW m}^{-2} \text{ K}^{-4}$, in accord with the experimental value. The Wien law is obtained by looking for the wavelength at which $d\rho/d\lambda = 0$, the condition for the maximum in the distribution. When we take the derivative, set it equal to zero, and make the approximation that the wavelength is so short that $hc/\lambda \gg kT$, we obtain

$$
T\lambda_{\text{max}} = \frac{hc}{5k} \tag{7}
$$

This result lets us identify the second radiation constant as $c_2 = hc/k = 1.439 \text{ cm K}$, which is also in good agreement with experiment.

It is quite easy to see why Rayleigh's approach was unsuccessful and Planck's hypothesis was successful. The thermal motion of the atoms in the walls of the black body excites the oscillators of the electromagnetic field. According to classical mechanics, all the oscillators of the field share equally in the energy supplied by the walls, so even the highest frequencies are excited. The excitation of very high frequency oscillators results in the ultraviolet catastrophe. According to Planck's hypothesis, however, oscillators are excited only if they can acquire an energy of at least $h\nu$. This energy is too large for the walls to supply in the case of the very high frequency oscillators, so the latter remain unexcited. The effect of quantization is to reduce the contribution from the high frequency oscillators, for they cannot be significantly excited with the energy available.

(c) Heat capacities

In the early nineteenth century, the French scientists Pierre-Louis Dulong and Alexis-Thérèse Petit determined the heat capacities of a number of monatomic solids.[2] On the basis of some somewhat slender experimental evidence, they proposed that the molar heat capacities of all monatomic solids are the same, and close to $25 \text{ J K}^{-1} \text{ mol}^{-1}$ (in modern units).

Dulong and Petit's law is easy to justify in terms of classical physics. If classical physics were valid, the equipartition principle could be used to calculate the heat capacity of a solid. According to this principle, the mean energy of an atom as it oscillates about its mean position in a solid is kT for each direction of displacement. As each atom can oscillate in three dimensions, the average energy of each atom is $3kT$; for N atoms the total energy is $3NkT$. The contribution of this motion to the molar internal energy is therefore

$$
U_{\text{m}} = 3N_{\text{A}}kT = 3RT
$$

---

2 As explained in Section 2.4b, the constant-volume heat capacity, C_V, is defined as $C_V = (\partial U/\partial T)_V$. A small heat capacity indicates that a large rise in temperature results from a given transfer of energy.

because $N_A k = R$, the gas constant. The molar constant-volume heat capacity (eqn 2.19) is then predicted to be

$$C_{V,m} = \left(\frac{\partial U_m}{\partial T}\right)_V = 3R \tag{8}$$

This result, with $3R = 24.9 \text{ J K}^{-1} \text{ mol}^{-1}$, is in striking accord with Dulong and Petit's value.

Significant deviations from Dulong and Petit's law were observed when technological advances made it possible to measure heat capacities at low temperatures. It was found that the molar heat capacities of all metals are lower than $3R$ at low temperatures, and that the values approach zero as $T \to 0$. To account for these observations, Einstein (in 1905) assumed that each atom oscillated about its equilibrium position with a single frequency ν. He then invoked Planck's hypothesis to assert that the energy of oscillation is confined to discrete values, and specifically to $nh\nu$, where n is an integer. Einstein first calculated the contribution of the oscillations of the atoms to the total molar energy of the metal (by a method described in Section 20.4) and obtained

$$U_m = \frac{3N_A h\nu}{e^{h\nu/kT} - 1}$$

in place of the classical expression $3RT$. Then he found the heat capacity by differentiating U_m with respect to T. The resulting expression is now known as the **Einstein formula**:

$$C_{V,m} = 3Rf^2 \qquad f = \frac{\theta_E}{T}\left(\frac{e^{\theta_E/2T}}{e^{\theta_E/T} - 1}\right) \tag{9}$$

where the **Einstein temperature**, $\theta_E = h\nu/k$, is a way of expressing the frequency of oscillation of the atoms as a temperature: a high frequency corresponds to a high Einstein temperature.

At high temperatures (when $T \gg \theta_E$) the exponentials in f can be expanded as $1 + \theta_E/T + \cdots$ and higher terms ignored. The result is

$$f = \frac{\theta_E}{T}\left\{\frac{1 + \theta_E/2T + \cdots}{(1 + \theta_E/T + \cdots) - 1}\right\} \approx 1 \tag{10a}$$

Consequently, the classical result ($C_{V,m} = 3R$) is obtained at high temperatures. At low temperatures, when $T \ll \theta_E$,

$$f \approx \frac{\theta_E}{T}\left(\frac{e^{\theta_E/2T}}{e^{\theta_E/T}}\right) = \frac{\theta_E}{T} e^{-\theta_E/2T} \tag{10b}$$

The strongly decaying exponential function goes to zero more rapidly than $1/T$ goes to infinity; so $f \to 0$ as $T \to 0$, and the heat capacity therefore approaches zero too. We see that Einstein's formula accounts for the decrease of heat capacity at low temperatures. The physical reason for this success is that at low temperatures only a few oscillators possess enough energy to oscillate significantly. At higher temperatures, there is enough energy available for all the oscillators to become active: all $3N$ oscillators contribute, and the heat capacity approaches its classical value.

The temperature dependence of the heat capacity predicted by the Einstein formula is plotted in Fig. 11.6. The general shape of the curve is satisfactory, but the numerical agreement is in fact quite poor. The poor fit arises from Einstein's assumption that all the atoms oscillate with the same frequency, whereas in fact they oscillate over a range of frequencies from zero up to a maximum value, ν_D. This complication is taken into account by averaging over all the frequencies present, the final result being the **Debye formula**:

$$C_{V,m} = 3Rf \qquad f = 3\left(\frac{T}{\theta_D}\right)^3 \int_0^{\theta_D/T} \frac{x^4 e^x}{(e^x - 1)^2}\, dx \tag{11}$$

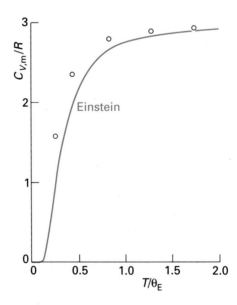

11.6 Experimental low-temperature molar heat capacities and the temperature dependence predicted on the basis of Einstein's theory. His equation (eqn 9) accounts for the dependence fairly well, but is everywhere too low.

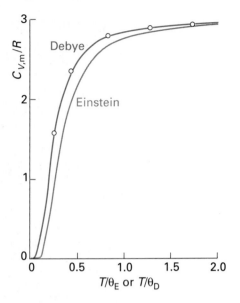

11.7 Debye's modification of Einstein's calculation (eqn 11) gives very good agreement with experiment. For copper, $T/\theta_D = 2$ corresponds to about 670 K.

where $\theta_D = h\nu_D/k$ is the **Debye temperature**. The integral in eqn 11 has to be evaluated numerically, but that is simple with mathematical software. The details of this modification, which, as Fig. 11.7 shows, gives improved agreement with experiment, need not distract us at this stage from the main conclusion, which is that quantization must be introduced in order to explain the thermal properties of solids.

(d) Atomic and molecular spectra

The most compelling evidence for the quantization of energy comes from the observation of the frequencies of radiation absorbed and emitted by atoms and molecules.

A typical atomic spectrum is shown in Fig. 11.8, and a typical molecular spectrum is shown in Fig. 11.9. The obvious feature of both is that radiation is emitted or absorbed at a series of discrete frequencies. This observation can be understood if the energy of the atoms or molecules is also confined to discrete values, for then energy can be discarded or absorbed only in discrete amounts (Fig. 11.10). Then, if the energy of an atom decreases by ΔE, the energy is carried away as radiation of frequency $\nu = \Delta E/h$, and a line appears in the spectrum.

11.2 Wave–particle duality

At this stage we have established that the energies of the electromagnetic field and of oscillating atoms are quantized. In this section we shall see the experimental evidence that led to the revision of two other basic concepts concerning the nature of the world. One experiment shows that electromagnetic radiation—which classical physics treats as wave-like—actually also displays the characteristics of particles. Another experiment shows that electrons—which classical physics treats as particles—also display the characteristics of waves.

(a) The particle character of electromagnetic radiation

The observation that electromagnetic radiation of frequency ν can possess only the energies $0, h\nu, 2h\nu, \ldots$ suggests that it can be thought of as consisting of $0, 1, 2, \ldots$ particles, each particle having an energy $h\nu$. Then, if one of these particles is present, the energy is $h\nu$, if two are present the energy is $2h\nu$, and so on. These particles of electromagnetic radiation are now called **photons**. The observation of discrete spectra from atoms and molecules can be pictured as the atom or molecule generating a photon of energy $h\nu$ when it discards an energy of magnitude ΔE, with $\Delta E = h\nu$.

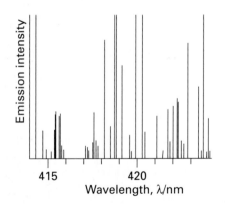

11.8 A region of the spectrum of radiation emitted by excited iron atoms consists of radiation at a series of discrete wavelengths (or frequencies).

Example 11.1 Calculating the number of photons

Calculate the number of photons emitted by a 100 W yellow lamp in 1.0 s. Take the wavelength of yellow light as 560 nm and assume 100 per cent efficiency.

Method Each photon has an energy $h\nu$, so the total number of photons needed to produce an energy E is $E/h\nu$. To use this equation, we need to know the frequency of the radiation (from $\nu = c/\lambda$) and the total energy emitted by the lamp. The latter is given by the product of the power (P, in watts) and the time ($E = Pt$). In general, to avoid rounding and other numerical errors, it is best to carry out algebraic calculations first, and to substitute numerical values into a single, final formula.

Answer The number of photons is

$$N = \frac{E}{h\nu} = \frac{Pt}{h(c/\lambda)} = \frac{\lambda Pt}{hc}$$

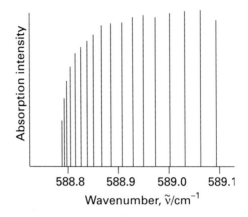

11.9 When a molecule changes its state, it does so by absorbing radiation at definite frequencies. This suggests that it can possess only discrete energies, not an arbitrary energy. This spectrum is part of that due to the vibrations and rotations of dinitrogen oxide (N_2O) molecules.

Substitution of the data gives

$$N = \frac{(5.60 \times 10^{-7} \text{ m}) \times (100 \text{ J s}^{-1}) \times (1.0 \text{ s})}{(6.626 \times 10^{-34} \text{ J s}) \times (2.998 \times 10^8 \text{ m s}^{-1})} = 2.8 \times 10^{20}$$

Comment Note that it would take nearly 40 min to produce 1 mol of these photons.

- -

Self-test 11.1 How many photons does a monochromatic (single-frequency) infrared rangefinder of power 1 mW and wavelength 1000 nm emit in 0.1 s?

[5×10^{14}]

Further evidence for the particle-like character of radiation comes from the measurement of the energies of electrons produced in the **photoelectric effect**. This effect is the ejection of electrons from metals when they are exposed to ultraviolet radiation. The experimental characteristics of the photoelectric effect are as follows:

1. No electrons are ejected, regardless of the intensity of the radiation, unless the frequency of the radiation exceeds a threshold value characteristic of the metal.
2. The kinetic energy of the ejected electrons increases linearly with the frequency of the incident radiation but is independent of the intensity of the radiation.
3. Even at low light intensities, electrons are ejected immediately if the frequency is above threshold.

The second characteristic is illustrated by the experimental data in Fig. 11.11.

These observations strongly suggest that the photoelectric effect depends on the ejection of an electron when it is involved in a collision with a particle-like projectile that carries

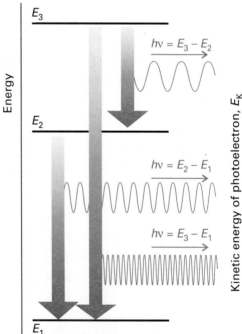

11.10 Spectral lines can be accounted for if we assume that a molecule emits a photon as it changes between discrete energy levels. Note that high-frequency radiation is emitted when the energy change is large.

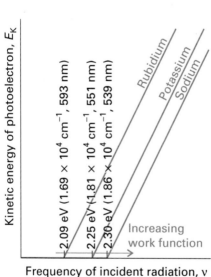

11.11 In the photoelectric effect, it is found that no electrons are ejected when the incident radiation has a frequency below a value characteristic of the metal and, above that value, the kinetic energy of the photoelectrons varies linearly with the frequency of the incident radiation.

enough energy to eject the electron from the metal. If we suppose that the projectile is a photon of energy $h\nu$, where ν is the frequency of the radiation, then the conservation of energy requires that the kinetic energy of the ejected electron should obey

$$\tfrac{1}{2}m_e v^2 = h\nu - \Phi \tag{12}$$

In this expression Φ is a characteristic of the metal called its **work function**, the energy required to remove an electron from the metal to infinity (Fig. 11.12). Photoejection cannot occur if $h\nu < \Phi$ because the photon brings insufficient energy: this conclusion accounts for observation (1). Equation 12 predicts that the kinetic energy of an ejected electron should increase linearly with frequency, in agreement with observation (2). When a photon collides with an electron, it gives up all its energy, so we should expect electrons to appear as soon as the collisions begin, provided the photons have sufficient energy: this conclusion agrees with observation (3).

(b) The wave character of particles

Although contrary to the long-established wave theory of light, the view that light consists of particles had been held before, but discarded. No significant scientist, however, had taken the view that matter is wave-like. Nevertheless, experiments carried out in 1925 forced people to even that conclusion. The crucial experiment was performed by the American physicists Clinton Davisson and Lester Germer, who observed the diffraction of electrons by a crystal (Fig. 11.13). Diffraction is a characteristic property of waves because it occurs when there is interference between their peaks and troughs. Depending on whether the interference is constructive or destructive, the result is a region of enhanced or diminished intensity. Davisson and Germer's success was a lucky accident, because a chance rise of temperature caused their polycrystalline sample to anneal, and the ordered planes of atoms then acted as a diffraction grating. At almost the same time, G.P. Thomson, working in Scotland, showed that a beam of electrons was diffracted when passed through a thin gold foil.

The Davisson–Germer experiment, which has since been repeated with other particles (including molecular hydrogen), shows clearly that particles have wave-like properties. We have also seen that waves of electromagnetic radiation have particle-like properties. Thus we

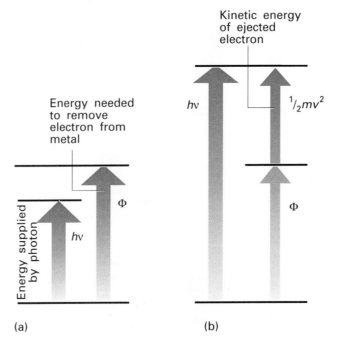

11.12 The photoelectric effect can be explained if it is supposed that the incident radiation is composed of photons that have energy proportional to the frequency of the radiation. (a) The energy of the photon is insufficient to drive an electron out of the metal. (b) The energy of the photon is more than enough to eject an electron, and the excess energy is carried away as the kinetic energy of the photoelectron (the ejected electron).

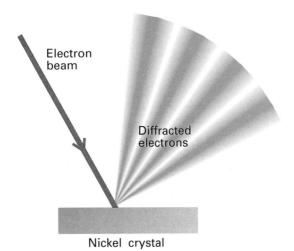

11.13 The Davisson–Germer experiment. The scattering of an electron beam from a nickel crystal shows a variation of intensity characteristic of a diffraction experiment in which waves interfere constructively and destructively in different directions.

Electron beam

Diffracted electrons

Nickel crystal

are brought to the heart of modern physics. When examined on an atomic scale, the classical concepts of particle and wave melt together, particles taking on the characteristics of waves, and waves the characteristics of particles.

Some progress towards coordinating these properties had already been made by the French physicist Louis de Broglie when, in 1924, he suggested that any particle, not only photons, travelling with a linear momentum p should have (in some sense) a wavelength given by the **de Broglie relation**:

$$\lambda = \frac{h}{p} \tag{13}$$

That is, a particle with a high linear momentum has a short wavelength (Fig. 11.14). Macroscopic bodies have such high momenta (even when they are moving slowly) that their wavelengths are undetectably small, and the wave-like properties cannot be observed.

Example 11.2 Estimating the de Broglie wavelength

Estimate the wavelength of electrons that have been accelerated from rest through a potential difference of 40 kV.

Method To use the de Broglie relation, we need to know the linear momentum, p, of the electrons. To calculate the linear momentum, we note that the energy acquired by an electron accelerated through a potential difference $\mathcal{V}$ is $e\mathcal{V}$, where e is the magnitude of its charge. At the end of the period of acceleration, all the acquired energy is in the form of kinetic energy, $p^2/2m_e$, so we can determine p by setting $p^2/2m_e$ equal to $e\mathcal{V}$. As before, carry through the calculation algebraically before substituting the data.

Answer The expression

$$\frac{p^2}{2m_e} = e\mathcal{V}$$

solves to

$$p = (2m_e e \mathcal{V})^{1/2}$$

Then, from the de Broglie relation,

$$\lambda = \frac{h}{(2m_e e \mathcal{V})^{1/2}}$$

Short wavelength, high momentum

Long wavelength, low momentum

11.14 An illustration of the de Broglie relation between momentum and wavelength. The wave is associated with a particle (shortly this wave will be seen to be the wavefunction of the particle). A particle with high momentum has a wavefunction with a short wavelength, and vice versa.

Substitution of the data and the fundamental constants (from inside the front cover) gives

$$\lambda = \frac{6.626 \times 10^{-34} \text{ J s}}{\{2 \times (9.109 \times 10^{-31} \text{ kg}) \times (1.609 \times 10^{-19} \text{ C}) \times (4.0 \times 10^4 \text{ V})\}^{1/2}}$$
$$= 6.1 \times 10^{-12} \text{ m}$$

Comment The wavelength of 6.1 pm is shorter than typical bond lengths in molecules (about 100 pm). Electrons accelerated in this way are used in the technique of electron diffraction (Section 21.10) for the determination of molecular structure.

- -

Self-test 11.2 Calculate the wavelength of a neutron with a translational kinetic energy equal to kT at 300 K.

[178 pm]

We now have to conclude that, not only has electromagnetic radiation the character classically ascribed to particles, but electrons (and all other particles) have the characteristics classically ascribed to waves. This joint particle and wave character of matter and radiation is called **wave–particle duality**. Duality strikes at the heart of classical physics, where particles and waves are treated as entirely separate entities. We have also seen that the energies of electromagnetic radiation and of matter cannot be varied continuously, and that for small objects the discreteness of energy is highly significant. In classical mechanics, in contrast, energies could be varied continuously. Such total failure of classical physics for small objects implied that its basic concepts were false. A new mechanics had to be devised to take its place.

The dynamics of microscopic systems

Quantum mechanics acknowledges the wave–particle duality of matter by supposing that, rather than travelling along a definite path, a particle is distributed through space like a wave. This remark may seem mysterious at this stage: it will be interpreted more fully shortly. The wave that in quantum mechanics replaces the classical concept of trajectory is called a **wavefunction**, ψ (psi).

11.3 The Schrödinger equation

In 1926, the Austrian physicist Erwin Schrödinger proposed an equation for finding the wavefunction of any system. The time-independent **Schrödinger equation** for a particle of mass m moving in one dimension with energy E is

$$-\frac{\hbar^2}{2m}\frac{d^2\psi}{dx^2} + V(x)\psi = E\psi \tag{14}$$

The factor $V(x)$ is the potential energy of the particle at the point x; $\hbar$ (which is read h-cross or h-bar) is a convenient modification of the Planck constant:

$$\hbar = \frac{h}{2\pi} = 1.054\,57 \times 10^{-34} \text{ J s} \tag{15}$$

Various ways of expressing the Schrödinger equation, of incorporating the time dependence of the wavefunction, and of extending it to more dimensions, are collected in Table 11.1. In Chapter 12 we shall solve the equation for a number of important cases; in this chapter we are mainly concerned with its significance, the interpretation of its solutions, and seeing how it implies that energy is quantized.

Table 11.1 The Schrödinger equation

For one-dimensional systems:

$$-\frac{\hbar^2}{2m}\frac{d^2\psi}{dx^2} + V(x)\psi = E\psi$$

where $V(x)$ is the potential energy of the particle and E is its total energy. For three-dimensional systems

$$-\frac{\hbar^2}{2m}\nabla^2\psi + V\psi = E\psi$$

where V may depend on position and ∇^2 ('del squared') is

$$\nabla^2 = \frac{\partial^2}{\partial x^2} + \frac{\partial^2}{\partial y^2} + \frac{\partial^2}{\partial z^2}$$

In systems with spherical symmetry:

$$\nabla^2 = \frac{\partial^2}{\partial r^2} + \frac{2}{r}\frac{\partial}{\partial r} + \frac{1}{r^2}\Lambda^2$$

where

$$\Lambda^2 = \frac{1}{\sin^2\theta}\frac{\partial^2}{\partial\phi^2} + \frac{1}{\sin\theta}\frac{\partial}{\partial\theta}\sin\theta\frac{\partial}{\partial\theta}$$

In the general case the Schrödinger equation is written

$$H\psi = E\psi$$

where H is the hamiltonian operator for the system:

$$H = -\frac{\hbar^2}{2m}\nabla^2 + V$$

For the evolution of a system with time, it is necessary to solve the time-dependent Schrödinger equation

$$H\Psi = i\hbar\frac{\partial\Psi}{\partial t}$$

Justification 11.1

Although the Schrödinger equation should be regarded as a postulate, like Newton's equations of motion, it can be seen to be plausible by noting that it implies the de Broglie relation for a freely moving particle. First, eqn 14 can be rearranged into

$$\frac{d^2\psi}{dx^2} = -\frac{2m}{\hbar^2}\{E - V(x)\}\psi$$

If the potential has a constant value V, a solution of this equation is

$$\psi = e^{ikx} = \cos kx + i\sin kx \qquad k = \left\{\frac{2m(E-V)}{\hbar^2}\right\}^{1/2}$$

For this result, we have used the mathematical relation $e^{ix} = \cos x + i\sin x$, where $i = (-1)^{1/2}$. Now we recognize that $\cos kx$ (or $\sin kx$) is a wave of wavelength $\lambda = 2\pi/k$, as can be seen by comparing $\cos kx$ with the standard form of a harmonic wave, $\cos(2\pi x/\lambda)$. The quantity $E - V$ is equal to the kinetic energy of the particle, E_K, so $k = (2mE_K/\hbar^2)^{1/2}$, which implies that $E_K = k^2\hbar^2/2m$. Because $E_K = p^2/2m$, it follows that

$$p = k\hbar$$

Therefore, the linear momentum is related to the wavelength of the wavefunction by

$$p = \frac{2\pi}{\lambda} \times \frac{h}{2\pi} = \frac{h}{\lambda}$$

which is the de Broglie relation.

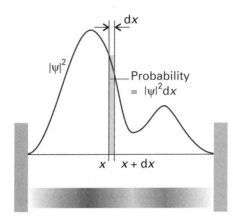

11.15 The wavefunction ψ is a probability amplitude in the sense that its square modulus ($\psi^*\psi$ or $|\psi|^2$) is a probability density. The probability of finding a particle in the region dx located at x is proportional to $|\psi|^2$ dx.

11.4 The Born interpretation of the wavefunction

It is a principal tenet of quantum mechanics that *the wavefunction contains all the dynamical information about the system it describes*. Here we shall concentrate on the information it carries about the location of the particle.

The interpretation of the wavefunction in terms of the location of the particle is based on a suggestion made by Max Born. He made use of an analogy with the wave theory of light, in which the square of the amplitude of an electromagnetic wave in a region is interpreted as its intensity and therefore (in quantum terms) as a measure of the probability of finding a photon present in the region. The **Born interpretation** of the wavefunction focuses on the square of the wavefunction (or the square modulus, $|\psi|^2 = \psi^*\psi$, if ψ is complex).[3] It states that the value of $|\psi|^2$ at a point is proportional to the probability of finding the particle at that point. Specifically, for a one-dimensional system (Fig. 11.15):

> If the wavefunction of a particle has the value ψ at some point x, the probability of finding the particle between x and $x+$dx is proportional to $|\psi|^2$ dx.

Thus, $|\psi|^2$ is the **probability density**, and to obtain the probability it must be multiplied by the length of the infinitesimal region dx. The wavefunction ψ itself is called the **probability amplitude**. For a particle free to move in three dimensions (for example, an electron near a nucleus in an atom), the wavefunction depends on the point r with coordinates x, y, and z, and the interpretation of $\psi(r)$ is as follows (Fig. 11.16):

> If the wavefunction of a particle has the value ψ at some point r, the probability of finding the particle in an infinitesimal volume d$\tau=$dx dy dz at that point is proportional to $|\psi|^2$ dτ.

The Born interpretation does away with any worry about the significance of a negative (and, in general, complex) value of ψ because $|\psi|^2$ is real and never negative. There is no *direct* significance in the negative (or complex) value of a wavefunction: only the square modulus, a positive quantity, is directly physically significant, and both negative and positive regions of a wavefunction may correspond to a high probability of finding a particle in a region (Fig. 11.17). However, later we shall see that the presence of positive and negative regions of a wavefunction is of great *indirect* significance, because it gives rise to the possibility of constructive and destructive interference between different wavefunctions.

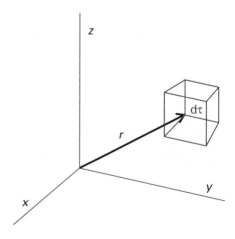

11.16 The Born interpretation of the wavefunction in three-dimensional space implies that the probability of finding the particle in the volume element d$\tau = dx$ dy dz at some location r is proportional to the product of dτ and the value of $|\psi|^2$ at that location.

Example 11.3 Interpreting a wavefunction

We shall see in Chapter 12 that the wavefunction of an electron in the lowest energy state of a hydrogen atom is proportional to e^{-r/a_0}, with a_0 a constant and r the distance from the nucleus. (Notice that this wavefunction depends only on this distance, not the angular position relative to the nucleus.) Calculate the relative probabilities of finding the electron inside a region of volume 1.0 pm^3, which is small even on the scale of the atom, located at (a) the nucleus, (b) a distance a_0 from the nucleus.

Method The region of interest is so small on the scale of the atom that we can ignore the variation of ψ within it and write the probability, P, as proportional to the probability density (ψ^2; note that ψ is real) evaluated at the point of interest multiplied by the volume of interest, δV. That is, $P \propto \psi^2 \delta V$.

---

3   To form the complex conjugate, ψ^*, of a complex function, replace i wherever it occurs by $-$i. For instance, the complex conjugate of e^{ikx} is e^{-ikx}. If the wavefunction is real, $|\psi|^2 = \psi^2$.

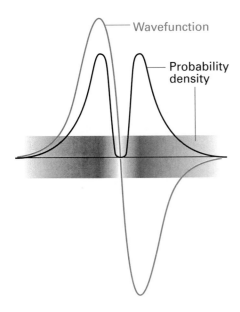

Wavefunction

Probability density

11.17 The sign of a wavefunction has no direct physical significance: the positive and negative regions of this wavefunction both correspond to the same probability distribution (as given by the square modulus of ψ and depicted by the density of shading).

Answer In each case $\delta V = 1.0\ \text{pm}^3$. (a) At the nucleus, $r = 0$, so

$$P \propto e^0 \times (1.0\ \text{pm}^3) = (1.0) \times (1.0\ \text{pm}^3)$$

(b) At a distance $r = a_0$ in an arbitrary direction,

$$P \propto e^{-2} \times (1.0\ \text{pm}^3) = (0.14) \times (1.0\ \text{pm}^3)$$

Therefore, the ratio of probabilities is $1.0/0.14 = 7.1$.

Comment Note that it is more probable (by a factor of 7.1) that the electron will be found at the nucleus than in the same volume element located at a distance a_0 from the nucleus. The negatively charged electron is attracted to the positively charged nucleus, and is likely to be found close to it.

- -

Self-test 11.3 The wavefunction for the lowest energy wavefunction in the ion He$^+$ is proportional to e^{-2r/a_0}. Repeat the calculation for this ion. Any comment?

[55; more compact wavefunction]

(a) Normalization

A mathematical feature of the Schrödinger equation is that, if ψ is a solution, then so is $N\psi$, where N is any constant. This feature is confirmed by noting that ψ occurs in every term in eqn 14, so any constant factor can be cancelled. This freedom to vary the wavefunction by a constant factor means that it is always possible to find a **normalization constant**, N, such that the proportionality of the Born interpretation becomes an equality.

We find the normalization constant by noting that, for a normalized wavefunction $N\psi$, the probability that a particle is in the region dx is equal to $(N\psi^*)(N\psi)\,dx$ (we are taking N to be real). Furthermore, the sum over all space of these individual probabilities must be 1 (the probability of the particle being somewhere is 1). Expressed mathematically, the latter requirement is

$$N^2 \int \psi^* \psi \, dx = 1 \tag{16}$$

where the integral is over all the space accessible to the particle (for instance, from $-\infty$ to $+\infty$ if the particle can be anywhere in an infinite range). It follows that

$$N = \frac{1}{\left(\int \psi^* \psi \, dx \right)^{1/2}} \tag{17}$$

Therefore, by evaluating the integral, we can find the value of N and hence 'normalize' the wavefunction. From now on, unless we state otherwise, we always use wavefunctions that have been normalized to 1; that is, from now on we assume that ψ already includes a factor which ensures that (in one dimension)

$$\int \psi^* \psi \, dx = 1 \tag{18}$$

In three dimensions, the wavefunction is normalized if

$$\int \psi^* \psi \, dx\,dy\,dz = 1 \tag{19}$$

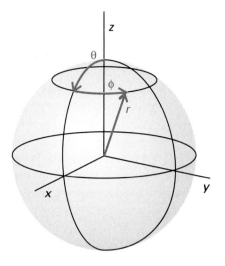

11.18 The spherical coordinates used for discussing systems with spherical symmetry.

or, more succinctly, if

$$\int \psi^* \psi \, d\tau = 1 \tag{20}$$

where $d\tau = dx\,dy\,dz$. In all such integrals, the integration is over all the space accessible to the particle. For systems with spherical symmetry, it is best to work in **spherical polar coordinates** r, θ, ϕ (Fig. 11.18):

$$x = r \sin\theta \cos\phi \qquad y = r \sin\theta \sin\phi \qquad z = r \cos\theta \tag{21a}$$

The volume element in spherical polar coordinates is

$$d\tau = r^2 \sin\theta \, dr \, d\theta \, d\phi \tag{21b}$$

To cover all space, the radius r ranges from 0 to ∞, the colatitude, θ, ranges from 0 to π, and the azimuth, ϕ, ranges from 0 to 2π (Fig. 11.19).

Example 11.4 Normalizing a wavefunction

Normalize the wavefunction used for the hydrogen atom in Example 11.3.

Method We need to find the factor N that guarantees that the integral in eqn 20 is equal to 1. Because the wavefunction is spherically symmetrical, it is sensible to work in spherical polar coordinates.

Answer The integration we require is

$$\int \psi^* \psi \, d\tau = N^2 \left(\int_0^\infty r^2 e^{-2r/a_0} \, dr \right) \left(\int_0^\pi \sin\theta \, d\theta \right) \left(\int_0^{2\pi} d\phi \right)$$
$$= N^2 \times \tfrac{1}{4} a_0^3 \times 2 \times 2\pi = \pi a_0^3 N^2$$

Therefore, for this integral to equal 1,

$$N = \left(\frac{1}{\pi a_0^3} \right)^{1/2}$$

and the normalized wavefunction is

$$\psi = \left(\frac{1}{\pi a_0^3} \right)^{1/2} e^{-r/a_0}$$

Comment If Example 11.3 is now repeated, we can obtain the actual probabilities of finding the electron in the volume element at each location, not just their relative values. Given (from the end-papers) that $a_0 = 52.9$ pm, the results are (a) 2.2×10^{-6}, corresponding to 1 chance in about 500 000 inspections of finding the electron in the test volume, and (b) 2.9×10^{-7}, corresponding to 1 chance in 3.4 million.

Self-test 11.4 Normalize the wavefunction given in Self-test 11.3.

$$[N = (8/\pi a_0^3)^{1/2}]$$

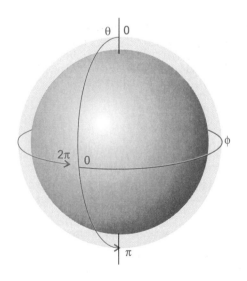

11.19 The surface of a sphere is covered by allowing θ to range from 0 to π, and then sweeping that arc around a complete circle by allowing ϕ to range from 0 to 2π.

The quantity $|\psi|^2 \, d\tau$ is a dimensionless probability and $d\tau$ has the dimensions of volume, $(\text{length})^d$, where d is the number of spatial dimensions. Therefore, the dimensions of a normalized wavefunction are $1/(\text{length})^{d/2}$. Thus, in one spatial dimension, $d = 1$ and a normalized wavefunction has the dimensions of $1/(\text{length})^{1/2}$. For a three-dimensional system, the wavefunction has the dimensions of $1/(\text{length})^{3/2}$, as we saw in Example 11.4.

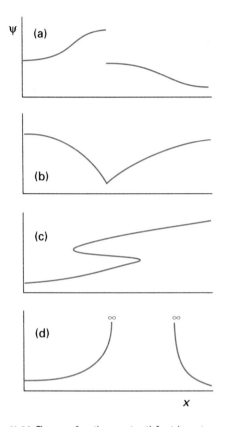

11.20 The wavefunction must satisfy stringent conditions for it to be acceptable. (a) Unacceptable because it is not continuous; (b) unacceptable because its slope is discontinuous; (c) unacceptable because it is not single-valued; (d) unacceptable because it is infinite over a finite region.

(b) Quantization

The Born interpretation puts severe restrictions on the acceptability of wavefunctions. The principal constraint is that ψ must not be infinite anywhere.[4] If it were, the integral in eqn 20 would be infinite and the normalization constant would be zero. The normalized function would then be zero everywhere, except where it is infinite, which would be unacceptable. The requirement that ψ is finite everywhere rules out many possible solutions of the Schrödinger equation, because many mathematically acceptable solutions rise to infinity and are therefore physically unacceptable. We shall meet several examples shortly.

The requirement that ψ is finite everywhere is not the only restriction implied by the Born interpretation. We could imagine (and in Section 12.6a will meet) a solution of the Schrödinger equation that gives rise to more than one value of $|\psi|^2$ at a single point. The Born interpretation implies that such solutions are unacceptable, because it would be absurd to have more than one probability that a particle is at some point. This restriction is expressed by saying that the wavefunction must be *single-valued*, that is, have only one value at each point of space.

The Schrödinger equation itself also implies some mathematical restrictions on the type of functions that will occur. Because it is a second-order differential equation, the second derivative of ψ must be well-defined if the equation is to be applicable everywhere. We can take the second derivative of a function only if it is continuous (so there are no sharp steps in it, Fig. 11.20) and if its first derivative, its slope, is continuous (so there are no kinks).[5]

At this stage we see that ψ must be continuous, have a continuous slope, be single-valued, and be finite everywhere. An acceptable wavefunction cannot be zero everywhere, because the particle it describes must be somewhere. These are such severe restrictions that acceptable solutions of the Schrödinger equation do not in general exist for arbitrary values of the energy E. In other words, *a particle may possess only certain energies, for otherwise its wavefunction would be physically unacceptable*. That is, the energy of a particle is quantized. We can find the acceptable energies by solving the Schrödinger equation for motion of various kinds, and selecting the solutions that conform to the restrictions listed above. That is the task of the next chapter.

Quantum mechanical principles

We have claimed that a wavefunction contains all the information it is possible to obtain about the dynamical properties (for example, its location and momentum) of the particle. We have seen that the Born interpretation tells us as much as we can know about location, but how do we find any additional information?

11.5 The information in a wavefunction

The Schrödinger equation for a particle of mass m free to move parallel to the x-axis with zero potential energy ($V = 0$ everywhere) is

$$-\frac{\hbar^2}{2m}\frac{d^2\psi}{dx^2} = E\psi \tag{22}$$

---

4   Infinitely sharp spikes are acceptable provided they have zero width. The true constraint is that the wavefunction must not be infinite over any finite region. In elementary quantum mechanics the simpler restriction, to finite ψ, is sufficient.

5   There are cases, and we shall meet them, where acceptable wavefunctions have kinks. These cases arise when the potential energy has peculiar properties, such as rising abruptly to infinity. When the potential energy is smoothly well-behaved and finite, the slope of the wavefunction must be continuous; if the potential energy becomes infinite, then the slope of the wavefunction need not be continuous. There are only two cases of this behaviour in elementary quantum mechanics, and the peculiarity will be mentioned when we meet them.

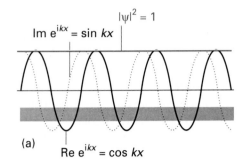

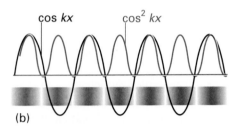

11.21 (a) The square modulus of a wavefunction corresponding to a definite state of linear momentum is a constant, so it corresponds to a uniform probability of finding the particle anywhere. (b) The probability distribution corresponding to the superposition of states of equal magnitude of linear momentum but opposite direction of travel.

The solutions of this equation have the form

$$\psi = Ae^{ikx} + Be^{-ikx} \qquad E = \frac{k^2\hbar^2}{2m} \tag{23}$$

where A and B are constants. To verify that ψ is a solution of eqn 22, we simply substitute it into the left-hand side of the equation and confirm that we obtain $E\psi$.

(a) The probability density

Suppose that $B = 0$ in eqn 23;[6] then the wavefunction is simply

$$\psi = Ae^{ikx} \tag{24}$$

Where is the particle? We form the square modulus to find the probability density of the particle:

$$|\psi|^2 = (Ae^{ikx})^*(Ae^{ikx}) = (A^*e^{-ikx})(Ae^{ikx}) = |A|^2 \tag{25}$$

This probability is *independent* of x; so, wherever we look along the x-axis, there is an equal probability of finding the particle (Fig. 11.21a). In other words, if the wavefunction of the particle is given by eqn 24, we cannot predict where we will find the particle. The same would be true if the wavefunction in eqn 23 had $A = 0$; then the probability density would be $|B|^2$, a constant.[7]

Now suppose that in the wavefunction $A = B$. Then eqn 23 becomes

$$\psi = A(e^{ikx} + e^{-ikx}) = 2A\cos kx \tag{26}$$

The probability density now has the form

$$|\psi|^2 = (2A\cos kx)^*(2A\cos kx) = 4|A|^2\cos^2 kx \tag{27}$$

This function is illustrated in Fig. 11.21b. As we see, the probability density periodically varies between 0 and $4|A|^2$. The locations where the probability density is zero correspond to nodes in the wavefunction: particles will never be found at the nodes. Specifically, a **node** is a point where a wavefunction passes through zero.

(b) Eigenvalues and eigenfunctions

Because the total energy of the particle is its kinetic energy, $p^2/2m$, it follows from eqn 23 that

$$p = k\hbar \tag{28}$$

This value is independent of the values of A and B.

To find a *systematic* way of extracting information from the wavefunction, we first note that any Schrödinger equation (such as those in eqn 14 and eqn 22) may be written in the succinct form

$$H\psi = E\psi \tag{29}$$

with (in one dimension)

$$H = -\frac{\hbar^2}{2m}\frac{d^2}{dx^2} + V(x) \tag{30}$$

The quantity H is an **operator**, something that carries out a mathematical operation on the function ψ. In this case, the operation is to take the second derivative of ψ and (after

---

6   We shall see later what determines the values of A and B; for the time being we can treat them as arbitrary constants.

7   It follows that if x is allowed to range from $-\infty$ to $+\infty$, the normalization constants, A or B, are 0. To avoid this embarrassing problem, x is allowed to range from $-L$ to $+L$, and L is allowed to go to infinity at the end of all calculations. We shall ignore this complication here.

multiplication by $-\hbar^2/2m$) to add the result to the outcome of multiplying ψ by V. The operator H plays a special role in quantum mechanics, and is called the **hamiltonian operator** after the nineteenth century mathematician William Hamilton. Hamilton developed a form of classical mechanics that, it subsequently turned out, is well suited to the formulation of quantum mechanics and which shows very clearly the relation between the two theories. The hamiltonian operator is the operator corresponding to the total energy of the system, the sum of the kinetic and potential energies. Consequently, we can infer that the first term in eqn 30 (the term proportional to the second derivative) must be the operator for the kinetic energy.

When the Schrödinger equation is written as in eqn 29, it is seen to be an **eigenvalue equation**, an equation of the form

$$(\text{operator})(\text{function}) = (\text{constant factor}) \times (\text{same function})$$

If we denote a general operator by $\hat{\Omega}$ and a constant factor by ω, this statement is

$$\hat{\Omega}\psi = \omega\psi \tag{31}$$

The factor ω is called the **eigenvalue** of the operator $\hat{\Omega}$. In eqn 29, the eigenvalue is the energy. The function ψ is called an **eigenfunction** and is different for each eigenvalue. In eqn 29, the eigenfunction is the wavefunction corresponding to the energy E. It follows that another way of saying 'solve the Schrödinger equation' is 'find the eigenvalues and eigenfunctions of the hamiltonian operator for the system'. The wavefunctions are the eigenfunctions of the hamiltonian operator, and the corresponding eigenvalues are the allowed energies.

Example 11.5 Identifying an eigenfunction

Show that e^{ax} is an eigenfunction of the operator d/dx, and find the corresponding eigenvalue. Show that e^{ax^2} is not an eigenfunction of d/dx.

Method We need to operate on the function with the operator and check whether the result is a constant factor times the original function.

Answer For $\hat{\Omega} = d/dx$ and $\psi = e^{ax}$:

$$\hat{\Omega}\psi = \frac{d}{dx}e^{ax} = ae^{ax} = a\psi$$

Therefore e^{ax} is indeed an eigenfunction of d/dx, and its eigenvalue is a. For $\psi = e^{ax^2}$,

$$\hat{\Omega}\psi = \frac{d}{dx}e^{ax^2} = 2axe^{ax^2} = 2ax \times \psi$$

which is not an eigenvalue equation even though the same function ψ occurs on the right, because ψ is now multiplied by a variable factor $(2ax)$, not a constant factor. Alternatively, if the right-hand side is written $2a(xe^{ax^2})$, we see that it is a constant times a *different* function.

Comment Much of quantum mechanics involves looking for functions that are eigenfunctions of a given operator, especially of the hamiltonian operator for the energy.

- -

Self-test 11.5 Is the function $\cos ax$ an eigenfunction of (a) d/dx, (b) d^2/dx^2?

$$[(a) \text{ No, (b) yes}]$$

The importance of eigenvalue equations is that the pattern

$$(\text{energy operator})\psi = (\text{energy})\psi$$

exemplified by the Schrödinger equation is repeated for other **observables**, or measurable properties of a system, such as the momentum or the electric dipole moment. Thus, it is often the case that we can write

$$(\text{operator corresponding to an observable})\psi = (\text{value of observable}) \times \psi$$

The symbol $\hat{\Omega}$ in eqn 31 is then interpreted as an operator (for example, the hamiltonian, H) corresponding to an observable (for example, the energy), and the eigenvalue ω is the value of that observable (for example, the value of the energy, E). Therefore, if we know both the wavefunction ψ and the operator $\hat{\Omega}$ corresponding to the observable Ω of interest, and the wavefunction is an eigenfunction of the operator $\hat{\Omega}$, we can predict the outcome of an observation of the property Ω (for example, an atom's energy) by picking out the factor ω in the eigenvalue equation, eqn 31.

(c) Operators

To make these abstract procedures concrete, we need to set up and use the operator corresponding to a given observable. The procedure is summarized by the following rule:

Observables, Ω, are represented by operators, $\hat{\Omega}$, built from the following position and momentum operators:

$$\hat{x} = x \times \qquad \hat{p}_x = \frac{\hbar}{i}\frac{d}{dx} \qquad [32]$$

That is, the operator for location along the x-axis is multiplication (of the wavefunction) by x and the operator for linear momentum parallel to the x-axis is proportional to taking the derivative (of the wavefunction) with respect to x.

For example, to deduce the value of the linear momentum given a specific wavefunction, we set up the eigenvalue equation

$$\hat{p}_x\psi = p_x\psi \qquad (33)$$

in the form

$$\frac{\hbar}{i}\frac{d\psi}{dx} = p_x\psi \qquad (34)$$

If the wavefunction is the one given in eqn 23 with $B = 0$,

$$\frac{\hbar}{i}\frac{d\psi}{dx} = \frac{\hbar}{i}A\frac{de^{ikx}}{dx} = \frac{\hbar}{i}A \times ike^{ikx} = k\hbar Ae^{ikx} = k\hbar\psi \qquad (35)$$

This is an eigenvalue equation, and by comparing it with eqn 33 we find that $p_x = +k\hbar$. The positive value implies that the linear momentum is directed towards positive x. Now suppose instead that the wavefunction is the one in eqn 23 with $A = 0$; then the same kind of calculation gives $p_x = -k\hbar$. It follows that a particle described by the second wavefunction has the same magnitude of momentum (and the same kinetic energy) as before, but its motion is towards $-x$.

The definitions in eqn 32 are used to construct operators for other observables. For example, suppose we wanted the operator for a potential energy of the form

$$V = \tfrac{1}{2}kx^2 \qquad (36)$$

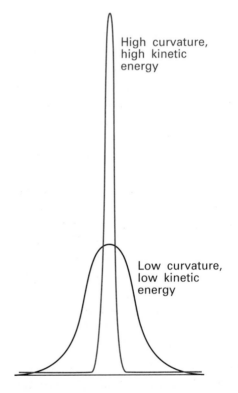

High curvature, high kinetic energy

Low curvature, low kinetic energy

11.22 Even if a wavefunction does not have the form of a periodic wave, it is still possible to infer from it the average kinetic energy of a particle by noting its average curvature. This illustration shows two wavefunctions: the sharply curved function corresponds to a higher kinetic energy than that of the less sharply curved function.

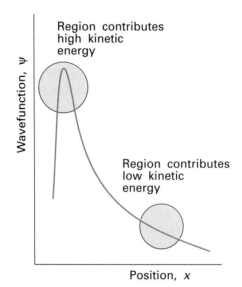

11.23 The observed kinetic energy of a particle is an average of contributions from the entire space covered by the wavefunction. Sharply curved regions contribute a high kinetic energy to the average; slightly curved regions contribute only a small kinetic energy.

with k a constant (later, we shall see that this potential describes the vibrations of atoms in molecules). Then it follows from eqn 32 that the operator corresponding to V is multiplication by x^2:

$$\hat{V} = \tfrac{1}{2}kx^2 \times \tag{37}$$

In normal practice, the multiplication sign is omitted. To construct the operator for kinetic energy, we make use of the classical relation between kinetic energy and linear momentum, which in one dimension is

$$E_K = \frac{p_x^2}{2m} \tag{38a}$$

Then, using the operator for p_x in eqn 32 we find:

$$\hat{E}_K = \frac{1}{2m}\left(\frac{\hbar}{i}\frac{d}{dx}\right)\left(\frac{\hbar}{i}\frac{d}{dx}\right) = -\frac{\hbar^2}{2m}\frac{d^2}{dx^2} \tag{38b}$$

It follows that the operator for the total energy, the hamiltonian operator, is

$$H = \hat{E}_K + \hat{V} = -\frac{\hbar^2}{2m}\frac{d^2}{dx^2} + \hat{V} \tag{39}$$

The expression for the kinetic energy operator, eqn 38b, gives another clue to the qualitative interpretation of a wavefunction. In mathematics, the second derivative of a function is a measure of its curvature: a large second derivative indicates a sharply curved function (Fig. 11.22). It follows that a sharply curved wavefunction is associated with a high kinetic energy, and one with a low curvature is associated with a low kinetic energy. This interpretation is consistent with the de Broglie relation, which predicts a short wavelength (a sharply curved wavefunction) when the linear momentum (and hence the kinetic energy) is high. However, it extends the interpretation to wavefunctions that do not spread through space and resemble those shown in Fig. 11.22. The curvature of a wavefunction in general varies from place to place. Wherever a wavefunction is sharply curved, its contribution to the total kinetic energy is large (Fig. 11.23). Wherever the wavefunction is not sharply curved, its contribution to the overall kinetic energy is low. As we shall shortly see, the observed kinetic energy of the particle is an integral of all the contributions of the kinetic energy from each region. Hence, we can expect a particle to have a high kinetic energy if the average curvature of its wavefunction is high.

The association of high curvature with high kinetic energy will turn out to be a valuable guide to the interpretation of wavefunctions and the prediction of their shapes. For example, suppose we need to know the wavefunction of a particle with a given total energy and a potential energy that decreases with increasing x (Fig. 11.24). Because the difference $E - V = E_K$ increases from left to right, the wavefunction must become more sharply curved as x increases: its wavelength decreases as the local contributions to its kinetic energy increase. We can therefore guess that the wavefunction will look like the function sketched in the illustration, and more detailed calculation confirms this to be so.

(d) Superpositions and expectation values

Suppose now that the wavefunction is the one given in eqn 26 (with $A = B$). What is the linear momentum of the particle it describes? We quickly run into trouble if we use the operator technique. When we operate with $\hat{p}_x$, we find

$$\frac{\hbar}{i}\frac{d\psi}{dx} = \frac{2\hbar}{i}A\frac{d\cos kx}{dx} = -\frac{2k\hbar}{i}A\sin kx \tag{40}$$

This expression is not an eigenvalue equation, because the function on the right is different from that on the left.

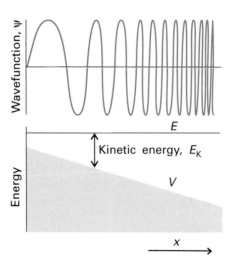

11.24 The wavefunction of a particle in a potential decreasing towards the right and hence subjected to a constant force to the right. Only the real part of the wavefunction is shown; the imaginary part is similar, but displaced to the right.

When the wavefunction of a particle is not an eigenfunction of an operator, the property to which the operator corresponds does not have a definite value. However, in the current example the momentum is not completely indefinite because the cosine wavefunction is a **linear combination**, or sum, of e^{ikx} and e^{-ikx}, and these two functions, as we have seen, individually correspond to definite momentum states. We say that the total wavefunction is a **superposition** of more than one wavefunction. Symbolically we can write the superposition as

$$\psi = \underset{\substack{\text{Particle with} \\ \text{linear momentum} \\ +k\hbar}}{\psi_{\rightarrow}} \quad + \quad \underset{\substack{\text{Particle with} \\ \text{linear momentum} \\ -k\hbar}}{\psi_{\leftarrow}}$$

The interpretation of this composite wavefunction is that, if the momentum of the particle is repeatedly measured in a long series of observations, then its *magnitude* will found to be $k\hbar$ in all the measurements (because that is the value for each component of the wavefunction). However, because the two component wavefunctions occur equally in the superposition, half the measurements will show that the particle is moving to the right ($p_x = +k\hbar$), and half the measurements will show that it is moving to the left ($p_x = -k\hbar$). According to quantum mechanics, we cannot predict in which direction the particle will in fact be found to be travelling; all we can say is that, in a long series of observations, there are equal probabilities of finding the particle travelling to the right and to the left.

The same interpretation applies to any wavefunction written as a linear combination of eigenfunctions of an operator. Thus, suppose the wavefunction is known to be a superposition of many different linear momentum eigenfunctions and is written as the linear combination

$$\psi = c_1\psi_1 + c_2\psi_2 + \cdots = \sum_k c_k\psi_k \tag{41}$$

where the c_k are numerical coefficients and the ψ_k correspond to different momentum states. Then, according to quantum mechanics,

1. When the momentum is measured, in a single observation *one* of the eigenvalues corresponding to the ψ_k that contribute to the superposition will be found.
2. The probability of measuring a particular eigenvalue in a series of observations is proportional to the square modulus ($|c_k|^2$) of the corresponding coefficient in the linear combination.
3. The average value of a large number of observations is given by the expectation value $\langle \Omega \rangle$ of the operator $\hat{\Omega}$ corresponding to the observable of interest.

The **expectation value** of an operator $\hat{\Omega}$ is defined as

$$\langle \Omega \rangle = \int \psi^* \hat{\Omega} \psi \, d\tau \tag{42}$$

This formula is valid only for normalized wavefunctions. As we see in the *Justification* below, an expectation value is the weighted average of a large number of observations of a property.

Justification 11.2

If ψ is an eigenfunction of $\hat{\Omega}$ with eigenvalue ω, the expectation value of Ω is

$$\langle \Omega \rangle = \int \psi^* \hat{\Omega} \psi \, d\tau = \int \psi^* \omega \psi \, d\tau = \omega \int \psi^* \psi \, d\tau = \omega$$

because ω is a constant and may be taken outside the integral, and the resulting integral is equal to 1 for a normalized wavefunction. The interpretation of this expression is that, because every observation of the property Ω results in the value ω (because the wavefunction is an eigenfunction of $\hat{\Omega}$), the mean value of all the observations is also ω.

A wavefunction that is not an eigenfunction of the operator of interest can be written as a linear combination of eigenfunctions. For simplicity, suppose the wavefunction is the sum of two eigenfunctions (the general case, eqn 41, can easily be developed). Then

$$\langle\Omega\rangle = \int (c_1\psi_1 + c_2\psi_2)^* \hat{\Omega}(c_1\psi_1 + c_2\psi_2)\,d\tau$$

$$= \int (c_1\psi_1 + c_2\psi_2)^* (c_1\omega_1\psi_1 + c_2\omega_2\psi_2)\,d\tau$$

$$= c_1^*c_1\omega_1 \int \psi_1^*\psi_1\,d\tau + c_2^*c_2\omega_2 \int \psi_2^*\psi_2\,d\tau$$

$$+ c_1^*c_2\omega_2 \int \psi_1^*\psi_2\,d\tau + c_2^*c_1\omega_1 \int \psi_2^*\psi_1\,d\tau$$

The first two integrals on the right are both equal to 1 because the wavefunctions are normalized. To deal with the remaining two integrals we need to make use of another property of eigenfunctions, called 'orthogonality': to say that two functions are orthogonal means that

$$\int \psi_i^*\psi_j\,d\tau = 0 \tag{43}$$

A very general rule in quantum mechanics is that *eigenfunctions corresponding to different eigenvalues of the same operator are orthogonal.*[8] For example, if ψ_1 corresponds to one energy, and ψ_2 corresponds to a different energy, then we know at once that the two functions are orthogonal and that the integral of their product is zero. Because ψ_1 and ψ_2 do correspond to different eigenvalues in the current example, they are orthogonal, so we can conclude that

$$\langle\Omega\rangle = |c_1|^2\omega_1 + |c_2|^2\omega_2 \tag{44}$$

This expression shows that the expectation value is the sum of the two eigenvalues weighted by the probabilities that each one will be found in a series of measurements. Hence, the expectation value is the weighted mean of a series of observations.

Example 11.6 Calculating an expectation value

Calculate the average value of the distance of an electron from the nucleus in the hydrogen atom in its state of lowest energy.

Method The average radius is the expectation value of the operator corresponding to the distance from the nucleus, which is multiplication by r. To evaluate $\langle r\rangle$, we need to know the normalized wavefunction (from Example 11.4) and then evaluate the integral in eqn 42. A useful integral for calculations on atomic wavefunctions is

$$\int_0^\infty x^n e^{-ax}\,dx = \frac{n!}{a^{n+1}}$$

where $n!$ denotes factorial n: $n! = n(n-1)(n-2)\cdots 1$.

---

8 Strictly speaking, this rule applies only to 'Hermitian operators', which are operators for which $\int \psi_i^*\hat{\Omega}\psi_j\,d\tau = \left(\int \psi_j^*\hat{\Omega}\psi_i\,d\tau\right)^*$. We shall be dealing only with Hermitian operators.

Answer The average value is given by the expectation value

$$\langle r \rangle = \int \psi^* \hat{r} \psi \, d\tau$$

which we evaluate by using spherical polar coordinates. Using the normalized function in Example 11.4 gives

$$\langle r \rangle = \frac{1}{\pi a_0^3} \left(\int_0^\infty r^3 e^{-2r/a_0} \, dr \right) \left(\int_0^\pi \sin\theta \, d\theta \right) \left(\int_0^{2\pi} d\phi \right)$$

$$= \frac{1}{\pi a_0^3} \times \frac{3! a_0^4}{2^4} \times 2 \times 2\pi = \tfrac{3}{2} a_0$$

Because $a_0 = 52.9$ pm (see end-papers), $\langle r \rangle = 79.4$ pm.

Comment The result means that, if a very large number of measurements of the distance of the electron from the nucleus are made, their mean value will be 79.4 pm. However, each different observation will give a different and unpredictable individual result, because the wavefunction is not an eigenfunction of the operator corresponding to r.

Self-test 11.6 Evaluate the root mean square distance, $\langle r^2 \rangle^{1/2}$, of the electron from the nucleus in the hydrogen atom.

$$[3^{1/2} a_0]$$

The mean kinetic energy of a particle in one dimension is the expectation value of the operator given in eqn 38b. Therefore, we can write

$$\langle E_K \rangle = \int \psi^* \hat{E}_K \psi \, d\tau = -\frac{\hbar^2}{2m} \int \psi^* \frac{d^2\psi}{dx^2} \, d\tau \tag{45}$$

We see that the kinetic energy is a kind of average over the curvature of the wavefunction: we get a large contribution to the observed value from regions where the wavefunction is sharply curved (so $d^2\psi/dx^2$ is large) and the wavefunction itself is large (so that ψ^* is large too).

11.6 The uncertainty principle

We have seen that, if the wavefunction is Ae^{ikx}, then the particle it describes has a definite state of linear momentum, namely travelling to the right with momentum $p_x = +k\hbar$. However, we have also seen that the position of the particle described by this wavefunction is completely unpredictable. In other words, *if the momentum is specified precisely, it is impossible to predict the location of the particle*. This statement is one-half of a special case of the **Heisenberg uncertainty principle**, one of the most celebrated results of quantum mechanics:

> **It is impossible to specify simultaneously, with arbitrary precision, both the momentum and the position of a particle.**

Before discussing the principle further, we must establish its other half: that if we know the position of a particle exactly, then we can say nothing about its momentum. The argument draws on the idea of regarding a wavefunction as a superposition of eigenfunctions, and runs as follows.

If we know that the particle is at a definite location, its wavefunction must be large there and zero everywhere else (Fig. 11.25). Such a wavefunction can be created by superimposing a large number of harmonic (sine and cosine) functions, or, what is equivalent, a number of

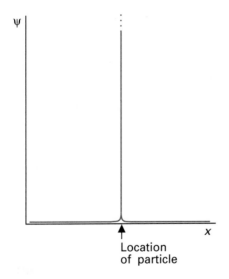

Location
of particle

11.25 The wavefunction for a particle at a well-defined location is a sharply spiked function which has zero amplitude everywhere except at the particle's position.

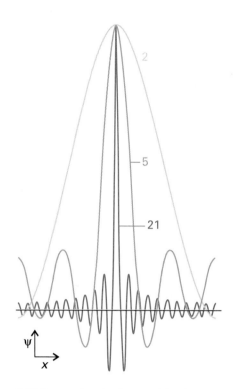

e^{ikx} functions.[9] In other words, we can create a sharply localized wavefunction by forming a linear combination of wavefunctions that correspond to many different linear momenta. The superposition of a few harmonic functions gives a wavefunction that spreads over a range of locations (Fig. 11.26). However, as the number of wavefunctions in the superposition increases, the wavefunction becomes sharper on account of the more complete interference between the positive and negative regions of the individual waves. When an infinite number of components is used, the wavefunction is a sharp, infinitely narrow spike, which corresponds to perfect localization of the particle. Now the particle is perfectly localized. However, we have lost all information about its momentum because, as we saw above, a measurement of the momentum will give a result corresponding to any one of the infinite number of waves in the superposition, and which one it will give is unpredictable. Hence, if we know the location of the particle precisely (implying that its wavefunction is a superposition of an infinite number of momentum eigenfunctions), its momentum is completely unpredictable.

A quantitative version of this result is

$$\Delta p \Delta q \geq \tfrac{1}{2}\hbar \tag{46}$$

In this expression Δp is the 'uncertainty' in the linear momentum parallel to the axis q, and Δq is the uncertainty in position along that axis. These 'uncertainties' are precisely defined, for they are the root mean square deviations of the properties from their mean values:

$$\Delta p = \{\langle p^2 \rangle - \langle p \rangle^2\}^{1/2} \qquad \Delta q = \{\langle q^2 \rangle - \langle q \rangle^2\}^{1/2} \tag{47}$$

If there is complete certainty about the position of the particle ($\Delta q = 0$), the only way that eqn 46 can be satisfied is for $\Delta p = \infty$, which implies complete uncertainty about the momentum. Conversely, if the momentum is known exactly ($\Delta p = 0$), then the position must be completely uncertain ($\Delta q = \infty$).

The p and q that appear in eqn 46 refer to the same direction in space. Therefore, whereas position on the x-axis and momentum parallel to the x-axis are restricted by the uncertainty relation, simultaneous location of position on x and motion parallel to y or z is not restricted.

11.26 The wavefunction for a particle with an ill-defined location can be regarded as the superposition of several wavefunctions of definite wavelength which interfere constructively in one place but destructively elsewhere. As more waves are used in the superposition (as given by the numbers attached to the curves), the location becomes more precise at the expense of uncertainty in the particle's momentum. An infinite number of waves is needed to construct the wavefunction of a perfectly localized particle.

Example 11.7 Using the uncertainty principle

The speed of a projectile of mass 1.0 g is known to within 1×10^{-6} m s^{-1}. Calculate the minimum uncertainty in its position.

Method Estimate Δp from $m\Delta v$ where Δv is the uncertainty in the speed; then use eqn 46 to estimate the minimum uncertainty in position, Δq.

Answer The minimum uncertainty in position is

$$\Delta q = \frac{\hbar}{2m\Delta v}$$

$$= \frac{1.055 \times 10^{-34} \text{ J s}}{2 \times (1.0 \times 10^{-3} \text{ kg}) \times (1 \times 10^{-6} \text{ m s}^{-1})} = 5 \times 10^{-26} \text{ m}$$

Comment The uncertainty is completely negligible for all practical purposes concerning macroscopic objects. However, if the mass is that of an electron, the same uncertainty in speed implies an uncertainty in position far larger than the diameter of an atom, so the

9   These sums are equivalent, because $e^{ikx} = \cos kx + i \sin kx$.

concept of a trajectory, the simultaneous possession of a precise position and momentum, is untenable.

Self-test 11.7 Estimate the minimum uncertainty in the speed of an electron in a one-dimensional region of length $2a_0$.

$$[500 \text{ km s}^{-1}]$$

==

The Heisenberg uncertainty principle is more general than eqn 46 implies. It applies to any pair of observables called **complementary observables**, which are defined in terms of the properties of their operators. Specifically, two observables Ω_1 and Ω_2 are complementary if

$$\hat{\Omega}_1\hat{\Omega}_2 \neq \hat{\Omega}_2\hat{\Omega}_1 \tag{48}$$

When the effect of two operators depends on their order (as this equation implies), we say that they do not **commute**.

Illustration

To show that the operators for position and momentum do not commute (and hence are complementary observables) we consider the effect of $\hat{x}\hat{p}_x$ on a wavefunction ψ:

$$\hat{x}\hat{p}_x\psi = x \times \frac{\hbar}{\mathrm{i}}\frac{\mathrm{d}\psi}{\mathrm{d}x}$$

Next, we consider the effect of $\hat{p}_x\hat{x}$ on the same function:

$$\hat{p}_x\hat{x}\psi = \frac{\hbar}{\mathrm{i}}\frac{\mathrm{d}}{\mathrm{d}x}x\psi = \frac{\hbar}{\mathrm{i}}\left(\psi + x\frac{\mathrm{d}\psi}{\mathrm{d}x}\right)$$

For this step we have used the standard rule about differentiating a product of functions. The second expression is clearly different from the first, so the two operators do not commute.

With the discovery that some pairs of observables are complementary (we meet more examples in the next chapter), we are at the heart of the difference between classical and quantum mechanics. Classical mechanics supposed, falsely as we now know, that the position and momentum of a particle could be specified simultaneously with arbitrary precision. However, quantum mechanics shows that position and momentum are complementary, and that we have to make a choice: we can specify position at the expense of momentum, or momentum at the expense of position.

The realization that some observables are complementary allows us to make considerable progress with the calculation of atomic and molecular properties, but it does away with some of classical physics' most cherished concepts.

Checklist of key ideas

- ☐ classical mechanics
- ☐ quantum mechanics

The origins of quantum mechanics

11.1 The failures of classical physics
- ☐ black body
- ☐ Wien displacement law (1)
- ☐ second radiation constant
- ☐ total energy density
- ☐ Stefan–Boltzmann law (2a)

- ☐ excitance
- ☐ Stefan–Boltzmann constant
- ☐ Rayleigh–Jeans law (3)
- ☐ ultraviolet catastrophe
- ☐ quantization of energy
- ☐ Planck constant
- ☐ Planck distribution (5)

- ☐ Einstein formula (9)
- ☐ Einstein temperature
- ☐ Debye formula (11)
- ☐ Debye temperature

11.2 Wave–particle duality
- ☐ photon

- ☐ photoelectric effect
- ☐ work function
- ☐ de Broglie relation (13)
- ☐ wave–particle duality

The dynamics of microscopic systems

- ☐ wavefunction

11.3 The Schrödinger equation
- ☐ Schrödinger equation (14, Table 11.1)

11.4 The Born interpretation of the wavefunction
- ☐ Born interpretation
- ☐ probability density
- ☐ probability amplitude
- ☐ normalization constant (17)
- ☐ spherical polar coordinates (21)

Quantum mechanical principles

11.5 The information in a wavefunction
- ☐ node
- ☐ operator
- ☐ hamiltonian operator (30)
- ☐ eigenvalue equation (31)
- ☐ eigenvalue
- ☐ eigenfunction

- ☐ observables
- ☐ linear combination (41)
- ☐ superposition
- ☐ expectation value (42)

11.6 The uncertainty princple
- ☐ Heisenberg uncertainty principle (46)
- ☐ complementary observables
- ☐ commute

Further reading

Texts and sources of data and information

P.W. Atkins, *Quanta: a handbook of concepts*. Oxford University Press (1991).

P.W. Atkins and R.S. Friedman, *Molecular quantum mechanics*. Oxford University Press (1997).

B. Cagnac and J.C. Pebay-Peyroula, *Modern atomic physics: fundamental principles*. Macmillan, London (1975).

T.S. Kuhn, *Black-body theory and the quantum discontinuity 1894–1912*. Oxford University Press (1978).

M. Jammer, *The conceptual development of quantum mechanics*. McGraw-Hill, New York (1966).

D.C. Cassidy, *Uncertainty: the life and science of Werner Heisenberg*. W.H. Freeman & Co., New York (1992).

A. Pais, *Niels Bohr's times: in physics, philosophy, and polity*. Clarendon Press, Oxford (1991).

W.J. Moore, *Schrödinger: life and thought*. Cambridge University Press (1989).

Exercises

11.1 (a) Calculate the power radiated by a 2.0 m × 3.0 m section of the surface of a hot body at 1500 K.

11.1 (b) Calculate the power radiated by the surface of a cylindrical wire of length 5.0 cm and radius 0.12 mm that is heated to 3300 K by an electric current.

11.2 (a) Calculate the average power output of a photodetector that collects 8.0×10^7 photons in 3.8 ms from monochromatic light of wavelength 325 nm.

11.2 (b) Calculate the average power output of a photosensitive plate that collects 1.20×10^8 photons in 5.9 ms from monochromatic light of wavelength 297 nm.

11.3 (a) Determine the wavelength of the radiation of the most intense electromagnetic radiation emitted from the surface of the star Sirius, which has a surface temperature of 11 000 K.

11.3 (b) Determine the wavelength of the radiation of the most intense electromagnetic radiation emitted from a furnace at 2500°C.

11.4 (a) Calculate the speed of an electron of wavelength 3.0 cm.

11.4 (b) Calculate the speed of a neutron of wavelength 3.0 cm.

11.5 (a) The fine-structure constant, α, plays a special role in the structure of matter; its approximate value is 1/137. What is the wavelength of an electron travelling at a speed αc, where c is

the speed of light? (Note that the circumference of the first Bohr orbit in the hydrogen atom is 331 pm.)

11.5 (b) A certain diffraction experiment requires the use of electrons of wavelength 0.45 nm. Calculate the speed of the electrons.

11.6 (a) Calculate the linear momentum of photons of wavelength 750 nm. What speed does an electron need to travel to have the same linear momentum?

11.6 (b) Calculate the linear momentum of photons of wavelength 350 nm. What speed does a hydrogen molecule need to travel to have the same linear momentum?

11.7 (a) The energy required for the ionization of a certain atom is 3.44×10^{-18} J. The absorption of a photon of unknown wavelength ionizes the atom and ejects an electron with velocity 1.03×10^6 m s^{-1}. Calculate the wavelength of the incident radiation.

11.7 (b) The energy required for the ionization of a certain atom is 5.12 aJ. The absorption of a photon of unknown wavelength ionizes the atom and ejects an electron with velocity 345 km s^{-1}. Calculate the wavelength of the incident radiation.

11.8 (a) The speed of a certain proton is 4.5×10^5 m s^{-1}. If the uncertainty in its momentum is to be reduced to 0.0100 per cent, what uncertainty in its location must be tolerated?

11.8 (b) The speed of a certain electron is 995 km s^{-1}. If the uncertainty in its momentum is to be reduced to 0.0010 per cent, what uncertainty in its location must be tolerated?

11.9 (a) Calculate the energy per photon and the energy per mole of photons for radiation of wavelength (a) 600 nm (red), (b) 550 nm (yellow), (c) 400 nm (blue).

11.9 (b) Calculate the energy per photon and the energy per mole of photons for radiation of wavelength (a) 200 nm (ultraviolet), (b) 150 pm (X-ray), (c) 1.00 cm (microwave).

11.10 (a) Calculate the speed to which a stationary H atom would be accelerated if it absorbed each of the photons used in Exercise 11.9a.

11.10 (b) Calculate the speed to which a stationary ^{4}He atom (mass 4.0026 u) would be accelerated if it absorbed each of the photons used in Exercise 11.9b.

11.11 (a) A glow-worm of mass 5.0 g emits red light (650 nm) with a power of 0.10 W entirely in the backward direction. To what speed will it have accelerated after 10 y if released into free space and assumed to live?

11.11 (b) A photon-powered spacecraft of mass 10.0 kg emits radiation of wavelength 225 nm with a power of 1.50 kW entirely in the backward direction. To what speed will it have accelerated after 10.0 y if released into free space?

11.12 (a) A sodium lamp emits yellow light (550 nm). How many photons does it emit each second if its power is (a) 1.0 W, (b) 100 W?

11.12 (b) A laser used to read CDs emits red light of wavelength 700 nm. How many photons does it emit each second if its power is (a) 0.10 W, (b) 1.0 W?

11.13 (a) The peak of the Sun's emission occurs at about 480 nm; estimate the temperature of its surface.

11.13 (b) The peak of the emission from the hot iron in a steel furnace occurs at about 1600 nm; estimate the temperature of the steel.

11.14 (a) The work function for metallic caesium is 2.14 eV. Calculate the kinetic energy and the speed of the electrons ejected by light of wavelength (a) 700 nm, (b) 300 nm.

11.14 (b) The work function for metallic rubidium is 2.09 eV. Calculate the kinetic energy and the speed of the electrons ejected by light of wavelength (a) 650 nm, (b) 195 nm.

11.15 (a) Calculate the size of the quantum involved in the excitation of (a) an electronic oscillation of period 1.0 fs, (b) a molecular vibration of period 10 fs, (c) a pendulum of period 1.0 s. Express the results in joules and kilojoules per mole.

11.15 (b) Calculate the size of the quantum involved in the excitation of (a) an electronic oscillation of period 2.50 fs, (b) a molecular vibration of period 2.21 fs, (c) a balance wheel of period 1.0 ms. Express the results in joules and kilojoules per mole.

11.16 (a) Calculate the de Broglie wavelength of (a) a mass of 1.0 g travelling at 1.0 cm s^{-1}, (b) the same, travelling at 100 km s^{-1}, (c) an He atom travelling at 1000 m s^{-1} (a typical speed at room temperature).

11.16 (b) Calculate the de Broglie wavelength of an electron accelerated from rest through a potential difference of (a) 100 V, (b) 1.0 kV, (c) 100 kV.

11.17 (a) Calculate the minimum uncertainty in the speed of a ball of mass 500 g that is known to be within 1.0 μm of a certain point on a bat. What is the minimum uncertainty in the position of a bullet of mass 5.0 g that is known to have a speed somewhere between 350.00001 m s^{-1} and 350.00000 m s^{-1}?

11.17 (b) An electron is confined to a linear region with a length of the same order as the diameter of an atom (about 100 pm). Calculate the minimum uncertainties in its position and speed.

11.18 (a) In an X-ray photoelectron experiment, a photon of wavelength 150 pm ejects an electron from the inner shell of an atom and it emerges with a speed of 2.14×10^7 m s^{-1}. Calculate the binding energy of the electron.

11.18 (b) In an X-ray photoelectron experiment, a photon of wavelength 121 pm ejects an electron from the inner shell of an atom and it emerges with a speed of 5.69×10^7 m s^{-1}. Calculate the binding energy of the electron.

Problems

Numerical problems

11.1 The Planck distribution gives the energy in the wavelength range dλ at the wavelength λ. Calculate the energy density in the range 650 nm to 655 nm inside a cavity of volume 100 cm^3 when its temperature is (a) 25°C, (b) 3000°C.

11.2 The wavelength of the emission maximum from a small pinhole in an electrically heated container was determined at a series of temperatures, and the results are given below. Deduce a value for the Planck constant.

$\theta/$°C	1000	1500	2000	2500	3000	3500
$\lambda_{max}/$nm	2181	1600	1240	1035	878	763

11.3 Write a computer program (or use mathematical software) to evaluate the Planck distribution at any temperature and wavelength or frequency, and add to it a routine for evaluating integrals for the energy density of the radiation between any two wavelengths. Use it to calculate the total energy density in the visible region (600 nm to

350 nm) for a black body at (a) $100°C$, (b) $500°C$, (c) 700 K. What are the classical values at these temperatures?

11.4 The Einstein frequency is often expressed in terms of an equivalent temperature θ_E, where $\theta_E = h\nu/k$. Confirm that θ_E has the dimensions of temperature, and express the criterion for the validity of the high-temperature form of the Einstein equation in terms of it. Evaluate θ_E for (a) diamond, for which $\nu = 4.65 \times 10^{13}$ Hz and (b) for copper, for which $\nu = 7.15 \times 10^{12}$ Hz. What fraction of the Dulong and Petit value of the heat capacity does each substance reach at $25°C$?

11.5 The ground-state wavefunction for a particle confined to a one-dimensional box of length L is

$$\psi = \left(\frac{2}{L}\right)^{1/2} \sin\left(\frac{\pi x}{L}\right)$$

Suppose the box is 10.0 nm long. Calculate the probability that the particle is (a) between $x = 4.95$ nm and 5.05 nm, (b) between $x = 1.95$ nm and 2.05 nm, (c) between $x = 9.90$ nm and 10.00 nm, (d) in the right half of the box, (e) in the central third of the box.

11.6 The ground-state wavefunction of a hydrogen atom is

$$\psi = \left(\frac{1}{\pi a_0^3}\right)^{1/2} e^{-r/a_0}$$

where $a_0 = 53$ pm (the Bohr radius). (a) Calculate the probability that the electron will be found somewhere within a small sphere of radius 1.0 pm centred on the nucleus. (b) Now suppose that the same sphere is located at $r = a_0$. What is the probability that the electron is inside it?

Theoretical problems

11.7 Derive Wien's law, that $\lambda_{max}T$ is a constant, from the Planck distribution, and deduce an expression for the constant.

11.8 Normalize the following wavefunctions: (a) $\sin(n\pi x/L)$ in the range $0 \leq x \leq L$, (b) a constant in the range $-L \leq x \leq L$, (c) $e^{-r/a}$ in three-dimensional space, (d) $xe^{-r/2a}$ in three-dimensional space. *Hint:* The volume element in three dimensions is $d\tau = r^2 \, dr \sin\theta \, d\theta \, d\phi$, with $0 \leq r < \infty, 0 \leq \theta \leq \pi, 0 \leq \phi \leq 2\pi$. A useful integral was given in Example 11.6.

11.9 Two (unnormalized) excited state wavefunctions of the H atom are (a) $\psi = (2 - r/a_0)e^{-r/2a_0}$, (b) $\psi = r\sin\theta\cos\phi\, e^{-r/2a_0}$. Normalize both functions to 1.

11.10 Identify which of the following functions are eigenfunctions of the operator d/dx: (a) e^{ikx}, (b) $\cos kx$, (c) k, (d) kx, (e) $e^{-\alpha x^2}$. Give the corresponding eigenvalue where appropriate.

11.11 Determine which of the following functions are eigenfunctions of the inversion operator i (which has the effect of making the replacement $x \rightarrow -x$): (a) $x^3 - kx$, (b) $\cos kx$, (c) $x^2 + 3x - 1$. State the eigenvalue of i when relevant.

11.12 Which of the functions in Problem 11.10 are (a) also eigenfunctions of d^2/dx^2 and (b) only eigenfunctions of d^2/dx^2? Give the eigenvalues where appropriate.

11.13 A particle is in a state described by the wavefunction

$$\psi = (\cos\chi)e^{ikx} + (\sin\chi)e^{-ikx}$$

where χ is a parameter. What is the probability that the particle will be found with a linear momentum (a) $+k\hbar$, (b) $-k\hbar$? (c) What form would the wavefunction have if it were 90 per cent certain that the particle had linear momentum $+k\hbar$?

11.14 Evaluate the kinetic energy of the particle with wavefunction given in Problem 11.13.

11.15 Calculate the average linear momentum of a particle described by the following wavefunctions: (a) e^{ikx}, (b) $\cos kx$, (c) $e^{-\alpha x^2}$, where in each one x ranges from $-\infty$ to $+\infty$.

11.16 Evaluate the expectation values of r and r^2 for a hydrogen atom with wavefunctions given in Problem 11.9.

11.17 Calculate (a) the mean potential energy and (b) the mean kinetic energy of an electron in the ground state of a hydrogenic atom.

11.18 Write a computer program, or use mathematical software, for constructing superpositions of cosine functions and explore how the wavefunction becomes more localized as more components are included. Include routines that determine the probability that a given momentum will be observed. If you plot the superposition (which you should), set $x = 0$ at the centre of the screen and build the superposition there. Include a routine that includes the evaluation of the root mean square location of the packet, $\langle x^2 \rangle^{1/2}$.

11.19 Determine the commutators (that is, the value of $\hat{\Omega}_1\hat{\Omega}_2 - \hat{\Omega}_2\hat{\Omega}_1$) of the operators (a) d/dx and x, (b) d/dx and x^2, (c) a and $a^\dagger$, where $a = (x + ip)/2^{1/2}$ and $a^\dagger = (x - ip)/2^{1/2}$.

Additional problems supplied by Carmen Giunta and Charles Trapp

11.20 Demonstrate explicitly that the Planck distribution reduces to the Rayleigh–Jeans law at long wavelengths.

11.21 The temperature of the Sun's surface is approximately 5800 K. On the assumption that the human eye evolved to be most sensitive at the wavelength of light corresponding to the maximum in the Sun's radiant energy distribution, determine the colour of light to which the eye is the most sensitive.

11.22 Solar energy strikes the top of the Earth's atmosphere at a rate of 343 W m^{-2}. About 30 per cent of this energy is reflected directly back into space by the Earth or the atmosphere. The Earth–atmosphere system absorbs the remaining energy and re-radiates it into space as black-body radiation. What is the average black-body temperature of the Earth? What is the wavelength of the most plentiful of the Earth's black-body radiation?

11.23 A star too small and cold to shine has been found by B.R. Oppenheimer, S. Kulkarni, K. Matthews, and T. Nakajima (*Science* 270, 1478, (1995)). The spectrum of the object shows the presence of methane which, according to the authors, would not exist at temperatures much above 1000 K. The mass of the object, as determined from its gravitational effect upon a companion star, is roughly 20 times the mass of Jupiter. With this mass, it is very unlikely that the object formed as a planet; hence it is considered a brown dwarf star, the coolest ever found. (a) From available thermodynamic data test the stability of methane at temperatures above 1000 K. (b) What is λ_{max} for this star? (c) What are the energy density and excitance of this star relative to that of the Sun (6000 K)? (d) To determine whether the star will shine, estimate the fraction of the energy density of the star that appears in the visible portion of the spectrum.

11.24 Max Planck was the first to determine the Boltzmann constant, k, and the value of the constant now known by his name from the experimental data on black-body radiation. Calculate values for k and h from the following data. The excitance, $\mathcal{M}$, from a surface of area 1.000 m^2 at 2000 K is 904.48 kW; at this temperature $\lambda_{max} = 1.451 \times 10^{-6}$ m. *Hint.* Obtain λ_{max} from the Planck distribution by differentiation with respect to λ.

12 Quantum theory: techniques and applications

To find the properties of systems according to quantum mechanics we need to solve the appropriate Schrödinger equation. This chapter presents the essentials of the solutions for three basic types of motion: translation, vibration, and rotation. We shall see that only certain wavefunctions and their corresponding energies are acceptable. Hence, quantization emerges as a natural consequence of the equation and the conditions imposed on it. The solutions bring to light a number of highly nonclassical, and therefore surprising, features of particles, particularly their ability to tunnel into and through regions where classical physics would forbid them to be found. We shall also encounter a property of the electron, its spin, that has no classical counterpart.

The three basic modes of motion—translation (motion through space), vibration, and rotation—all play an important role in chemistry because they are ways in which molecules store energy. Gas-phase molecules, for instance, undergo translational motion, and their kinetic energy is a contribution to the total internal energy of a sample. Molecules can also store energy as rotational kinetic energy, and transitions between their rotational energy states can be observed spectroscopically. Energy is also stored as molecular vibration, and transitions between vibrational states are responsible for the appearance of infrared spectra.

Translational motion

The quantum mechanical description of free motion in one dimension was introduced in Section 11.5. We saw there that the Schrödinger equation is

$$-\frac{\hbar^2}{2m}\frac{\mathrm{d}^2\psi}{\mathrm{d}x^2} = E\psi \tag{1a}$$

or more succinctly

$$H\psi = E\psi \qquad H = -\frac{\hbar^2}{2m}\frac{\mathrm{d}^2}{\mathrm{d}x^2} \tag{1b}$$

The general solutions are

$$\psi_k = A\mathrm{e}^{ikx} + B\mathrm{e}^{-ikx} \qquad E_k = \frac{k^2\hbar^2}{2m} \qquad (2)$$

Note that we are now labelling both the wavefunctions and the energies (that is, the eigenfunctions and eigenvalues of H) with the index k. That these functions are solutions can be verified by substituting ψ_k into the left-hand side of eqn 1a and showing that the result is equal to $E_k\psi_k$. In this case, all values of k, and therefore all values of the energy, are permitted. It follows that the translational energy of a free particle is not quantized.

We saw in Section 11.5c that a wavefunction of the form e^{ikx} describes a particle with linear momentum $p_x = +k\hbar$, corresponding to motion towards positive x (to the right), and that a wavefunction of the form e^{-ikx} describes a particle with the same magnitude of linear momentum but travelling towards negative x (to the left). That is, e^{ikx} is an eigenfunction of the operator $\hat{p}_x$ with eigenvalue $+k\hbar$, and e^{-ikx} is an eigenfunction with eigenvalue $-k\hbar$. In either state, $|\psi|^2$ is independent of x, which implies that the position of the particle is completely unpredictable. This conclusion is consistent with the uncertainty principle because, if the momentum is certain, then the position cannot be specified (the operators $\hat{x}$ and $\hat{p}_x$ do not commute, Section 11.5e).

12.1 A particle in a box

In this section, we consider the problem of a **particle in a box**, in which a particle of mass m is confined between two walls at $x = 0$ and $x = L$. In an **infinite square well**, the potential energy is zero inside the box but rises abruptly to infinity at the walls (Fig. 12.1). This potential energy is an idealization of the potential energy of a gas-phase molecule that is free to move in a one-dimensional container.

(a) The Schrödinger equation

The Schrödinger equation for the region between the walls (where $V = 0$) is the same as for a free particle (eqn 1), so the general solutions given in eqn 2 are also the same. It is convenient to write them as[1]

$$\psi_k(x) = C\sin kx + D\cos kx \qquad E_k = \frac{k^2\hbar^2}{2m} \qquad (3)$$

(b) The acceptable solutions

For a free particle, any value of E_k corresponds to an acceptable solution. However, when the particle is confined within a region, the acceptable wavefunctions must satisfy certain **boundary conditions**, or constraints on the function at certain locations. It is physically impossible for the particle to be found with an infinite potential energy, so the wavefunction must be zero where V is infinite, at $x < 0$ and $x > L$. The continuity of the wavefunction then requires it to vanish just inside the well at $x = 0$ and $x = L$. That is, the boundary conditions are $\psi_k(0) = 0$ and $\psi_k(L) = 0$.

Consider the wall at $x = 0$. According to eqn 3, $\psi(0) = D$ (because $\sin 0 = 0$ and $\cos 0 = 1$). But because $\psi(0) = 0$ we must have $D = 0$. It follows that the wavefunction must be of the form $\psi_k(x) = C\sin kx$. The value of ψ at the other wall (at $x = L$) is $\psi_k(L) = C\sin kL$, which must also be zero. Taking $C = 0$ would give $\psi_k(x) = 0$ for all x,

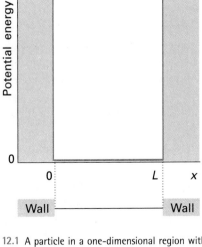

12.1 A particle in a one-dimensional region with impenetrable walls. Its potential energy is zero between $x = 0$ and $x = L$, and rises abruptly to infinity as soon as it touches the walls.

1   We use $\mathrm{e}^{\pm ix} = \cos x \pm i\sin x$ and absorb all numerical factors into the coefficients C and D.

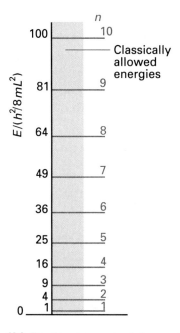

12.2 The allowed energy levels for a particle in a box. Note that the energy levels increase as n^2, and that their separation increases as the quantum number increases.

which would conflict with the Born interpretation (the particle must be somewhere). Therefore, kL must be chosen so that $\sin kL = 0$, which is satisfied by

$$kL = n\pi \qquad n = 1, 2, \ldots \tag{4}$$

The value $n = 0$ is ruled out, because it implies $k = 0$ and $\psi_k(x) = 0$ everywhere (because $\sin 0 = 0$), which is again unacceptable, and negative values of n merely change the sign of $\sin kL$ (because $\sin(-x) = -\sin x$). The wavefunctions are therefore

$$\psi_n(x) = C \sin \frac{n\pi x}{L} \qquad n = 1, 2, \ldots \tag{5}$$

(At this point we have started to label the solutions with the index n instead of k.) Because k and E_k are related by eqn 3, it follows that the energy of the particle is limited to the values

$$E_n = \frac{(n\pi/L)^2 \hbar^2}{2m} = \frac{n^2 h^2}{8mL^2} \qquad n = 1, 2, \ldots \tag{6}$$

We see that the energy of the particle is quantized, and that the quantization arises from the boundary conditions that ψ must satisfy if it is to be an acceptable wavefunction. This is a general conclusion: *the need to satisfy boundary conditions implies that only certain wavefunctions are acceptable, and hence restricts observables to discrete values*. So far, only energy has been quantized; shortly we shall see that other physical observables may also be quantized.

(c) Normalization

Before discussing the solution in more detail, we shall complete the derivation of the wavefunctions (which are real, that is, do not contain i) by finding the normalization constant (here written C). To do so, we look for the value of C that ensures that the integral of ψ^2 over all the space available to the particle (that is, from $x = 0$ to $x = L$) is equal to 1:[2]

$$\int_0^L \psi^2 \, dx = C^2 \int_0^L \sin^2 \frac{n\pi x}{L} \, dx = C^2 \times \tfrac{1}{2}L = 1, \qquad \text{so } C = \left(\frac{2}{L}\right)^{1/2}$$

for all n. Therefore, the complete solution to the problem is

$$E_n = \frac{n^2 h^2}{8mL^2} \qquad n = 1, 2, \ldots$$

$$\psi_n(x) = \left(\frac{2}{L}\right)^{1/2} \sin\left(\frac{n\pi x}{L}\right) \qquad \text{for } 0 \le x \le L \tag{7}$$

The energies and wavefunctions are labelled with the 'quantum number' n. A **quantum number** is an integer (in some cases, as we shall see, a half-integer) that labels the state of the system. For a particle in a box there is an infinite number of acceptable solutions, and the quantum number n specifies the one of interest (Fig. 12.2). As well as acting as a label, a quantum number can often be used to calculate the energy corresponding to the state and to write down the wavefunction explicitly (in the present example, by using the information in eqn 7).

(d) The properties of the solutions

Figure 12.3 shows some of the wavefunctions of a particle in a box: they are all sine functions with the same amplitude but different wavelengths. With these images in mind, it is easy to see the origin of the quantization: each wavefunction is a standing wave and, to fit

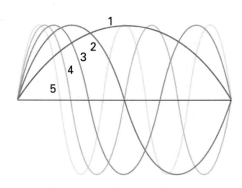

12.3 The first five normalized wavefunctions of a particle in a box. Each wavefunction is a standing wave, and successive functions possess one more half wave and a correspondingly shorter wavelength.

2   To evaluate the integral, we use the standard form

$$\int \sin^2 ax \, dx = \tfrac{1}{2}x - \tfrac{1}{4a}\sin 2ax + \text{constant}$$

into the cavity, successive wavefunctions must possess one more half-wavelength. Shortening the wavelength results in a sharper average curvature of the wavefunction and therefore an increase in the kinetic energy of the particle. Note that the number of nodes (points where the wavefunction *passes through* zero) also increases as n increases, and that the wavefunction ψ_n has $n-1$ nodes. Increasing the number of nodes between walls of a given separation increases the average curvature of the wavefunction and hence the kinetic energy of the particle.

Example 12.1 Deriving the energies of a particle in a box

Derive the energy levels of a particle in a box from the de Broglie relation and the boundary conditions on the wavefunction.

Method We see from Fig. 12.3 that, to fit into the box, successive wavefunctions possess one more half-wavelength. Therefore, the first thing to do is to find an expression for the permitted wavelengths. To convert the permitted wavelengths into energies, we use the de Broglie relation to express wavelength as linear momentum and then use the expression for the kinetic energy in terms of the momentum to find the permitted energies.

Answer The permitted wavelengths satisfy

$$L = n \times \tfrac{1}{2}\lambda \qquad n = 1, 2, \ldots$$

and therefore

$$\lambda = \frac{2L}{n}, \qquad n = 1, 2, \ldots$$

According to the de Broglie relation, these wavelengths correspond to the momenta

$$p = \frac{h}{\lambda} = \frac{nh}{2L}$$

The particle has only kinetic energy inside the box (where $V = 0$), so the permitted energies are

$$E_n = \frac{p^2}{2m} = \frac{n^2 h^2}{8mL^2}$$

as obtained more formally earlier.

- -

Self-test 12.1 What is the average value of the linear momentum of a particle in a box with quantum number n?

$$[\langle p \rangle = 0]$$

The linear momentum of a particle in a box is not well defined because the wavefunction $\sin kx$ is a standing wave and, like the example of $\cos kx$ treated in Section 11.5d, not an eigenfunction of the linear momentum operator. However, each wavefunction is a superposition of momentum eigenfunctions:

$$\psi_n = \left(\frac{2}{L}\right)^{1/2} \sin\left(\frac{n\pi x}{L}\right) = \frac{1}{2i}\left(\frac{2}{L}\right)^{1/2}\left(e^{ikx} - e^{-ikx}\right) \qquad k = \frac{n\pi}{L} \qquad (8)$$

It follows that measurement of the linear momentum will give the value $+k\hbar$ for half the measurements of momentum and $-k\hbar$ for the other half. This detection of opposite directions of travel with equal probability is the quantum mechanical version of the classical picture that a particle in a box rattles from wall to wall, and in any given period spends half its time travelling to the left and half travelling to the right.

Because n cannot be zero, the lowest energy that the particle may possess is not zero (as would be allowed by classical mechanics, corresponding to a stationary particle) but

$$E_1 = \frac{h^2}{8mL^2} \tag{9}$$

This lowest, irremovable energy is called the **zero-point energy**.

The physical origin of the zero-point energy can be explained in two ways. First, the uncertainty principle requires a particle to possess kinetic energy if it is confined to a finite region: the particle's location is not completely indefinite, so its momentum cannot be precisely zero. Hence it has nonzero kinetic energy. Second, if the wavefunction is to be zero at the walls, but smooth, continuous, and not zero everywhere, then it must be curved, and curvature in a wavefunction implies the possession of kinetic energy.

The separation between adjacent energy levels with quantum numbers n and $n+1$ is

$$E_{n+1} - E_n = \frac{(n+1)^2 h^2}{8mL^2} - \frac{n^2 h^2}{8mL^2} = (2n+1)\frac{h^2}{8mL^2} \tag{10}$$

This separation decreases as the length of the container increases, and is very small when the container has macroscopic dimensions. The separation of adjacent levels becomes zero when the walls are infinitely far apart. Atoms and molecules free to move in normal laboratory-sized vessels may therefore be treated as though their translational energy is not quantized. The translational energy of completely free particles (those not confined by walls) is not quantized.

Illustration

For a container of length $L = 1.0$ nm, $h^2/8m_eL^2 = 6.0 \times 10^{-20}$ J. Therefore, the zero-point energy is $E_1 = 6.0 \times 10^{-20}$ J (corresponding to 0.37 eV). The minimum excitation energy is given by eqn 10 with $n = 1$, and is $3E_1$, or 1.8×10^{-19} J, which corresponds to 1.1 eV.

Self-test 12.2 Estimate a typical nuclear excitation energy by calculating the first excitation energy of a proton confined to a one-dimensional infinite square well with a length roughly equal to the diameter of a nucleus (1 fm).

[0.6 GeV]

The probability density for a particle in a box is

$$\psi^2(x) = \frac{2}{L}\sin^2\left(\frac{n\pi x}{L}\right) \tag{11}$$

This probability density varies with position within the box. The nonuniformity is pronounced when n is small (Fig. 12.4), but $\psi^2(x)$ becomes more uniform as n increases. The distribution at high quantum numbers reflects the classical result that a particle bouncing between the walls spends, on the average, equal times at all points. That the quantum result corresponds to the classical prediction at high quantum numbers is an illustration of the **correspondence principle**, which states that classical mechanics emerges from quantum mechanics as high quantum numbers are reached.

Example 12.2 Using the particle in a box solutions

What is the probability, P, of locating the electron between $x = 0$ (the left-hand edge) and $x = 0.2$ nm in its lowest energy state in a box of length 1.0 nm?

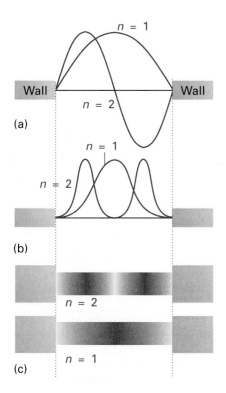

12.4 (a) The first two wavefunctions, (b) the corresponding probability distributions, and (c) a representation of the probability distribution in terms of the darkness of shading.

Method The value of $\psi^2\,dx$ is the probability of finding the particle in the small region dx located at x; therefore, the total probability of finding the electron in the specified region is the integral of $\psi^2\,dx$ over that region. The wavefunction of the electron is given in eqn 7 with $n = 1$.

Answer The probability of finding the particle in a region between $x = 0$ and $x = l$ is

$$P = \frac{2}{L}\int_0^l \sin^2\left(\frac{n\pi x}{L}\right)\,dx = \frac{l}{L} - \frac{1}{2n\pi}\sin\left(\frac{2n\pi l}{L}\right)$$

We then set $n = 1$ and $l = 0.2$ nm, which gives $P = 0.05$.

Comment The result corresponds to a chance of 1 in 20 of finding the electron in the region. As n becomes infinite, the sine term, which is multiplied by $1/n$, makes no contribution to P and the classical result, $P = l/L$, is obtained.

- -

Self-test 12.3 Calculate the probability that a particle in the state with $n = 1$ will be found between $x = 0.25L$ and $x = 0.75L$ in a box of length L (with $x = 0$ at the left-hand wall).

[0.82]

(e) Orthogonality and the bracket notation

A property of wavefunctions first mentioned in *Justification 11.2* can now be illustrated more fully. Two wavefunctions are **orthogonal** if the integral of their product vanishes. Specifically, the functions ψ_n and $\psi_{n'}$ are orthogonal if

$$\int \psi_n^* \psi_{n'}\,d\tau = 0 \tag{12a}$$

where the integration is over all space. A general feature of quantum mechanics is that wavefunctions corresponding to different energies are orthogonal; therefore, we can be confident that all the wavefunctions of a particle in a box are mutually orthogonal.

Illustration
....
We can verify the orthogonality of wavefunctions of a particle in a box with $n = 1$ and $n = 3$ (Fig. 12.5):

$$\int_0^L \psi_1^* \psi_3\,dx = \frac{2}{L}\int_0^L \sin\left(\frac{\pi x}{L}\right)\sin\left(\frac{3\pi x}{L}\right)\,dx = 0$$

from the general properties of integrals over trigonometric functions.
....

The integral in eqn 12a is often written

$$\langle n|n'\rangle = 0 \qquad (n' \neq n) \tag{12b}$$

This **Dirac bracket notation** is much more succinct than writing out the integral in full. It also introduces the words 'bra' and 'ket' into the language of quantum mechanics. Thus, the bra $\langle n|$ corresponds to the complex conjugate of the wavefunction ψ_n and the ket $|n'\rangle$ corresponds to the wavefunction $\psi_{n'}$. When the bra and ket are put together as in eqn 12b, the integration over all space is understood. Similarly, the normalization condition in eqn 11.20 becomes simply

$$\langle n|n\rangle = 1 \tag{13}$$

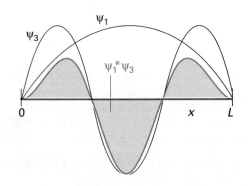

12.5 Two functions are orthogonal if the integral of their product is zero. Here the calculation of the integral is illustrated graphically for two wavefunctions of a particle in a square well. The integral is equal to the total area beneath the graph of the product, and is zero.

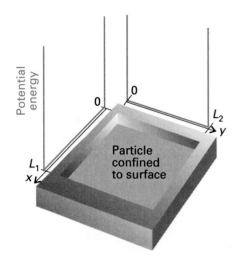

12.6 A two-dimensional square well. The particle is confined to the plane bounded by impenetrable walls. As soon as it touches the walls, its potential energy rises to infinity.

in bracket notation. These two expressions can be combined into a single expression:

$$\langle n | n' \rangle = \delta_{nn'} \tag{14}$$

where $\delta_{nn'}$, which is called the **Kronecker delta**, is 1 when $n' = n$ and 0 when $n' \neq n$. We shall see more of this notation later.

The property of orthogonality is of great importance in quantum mechanics because it enables us to eliminate a large number of integrals from calculations. Orthogonality plays a central role in the theory of chemical bonding (Chapter 14) and spectroscopy (Chapter 17).

12.2 Motion in two dimensions

Next, we consider a two-dimensional version of the particle in a box. Now the particle is confined to a rectangular surface of length L_1 in the x-direction and L_2 in the y-direction; the potential energy is zero everywhere except at the walls, where it is infinite (Fig. 12.6). The wavefunction is now a function of both x and y and the Schrödinger equation is

$$-\frac{\hbar^2}{2m}\left(\frac{\partial^2 \psi}{\partial x^2} + \frac{\partial^2 \psi}{\partial y^2}\right) = E\psi \tag{15}$$

We need to see how to solve this *partial* differential equation, an equation in more than one variable.

(a) Separation of variables

Some partial differential equations can be simplified by the **separation of variables technique**, which divides the equation into two or more ordinary differential equations, one for each variable. The method works in this case, as can be seen by testing whether a solution of eqn 15 can be found by writing the wavefunction as a product of functions, one depending only on x and the other only on y:

$$\psi(x, y) = X(x)Y(y) \tag{16}$$

The notation $X(x)Y(y)$ reminds us that the two functions into which the wavefunction is factored depend only on x and only on y for X and Y, respectively. We show in the *Justification* below that with this substitution, eqn 15 separates into two ordinary differential equations, one for each coordinate:

$$-\frac{\hbar^2}{2m}\frac{d^2 X}{dx^2} = E_X X \qquad -\frac{\hbar^2}{2m}\frac{d^2 Y}{dy^2} = E_Y Y \qquad E = E_X + E_Y \tag{17}$$

The quantity E_X is the energy associated with the motion of the particle parallel to the x-axis, and likewise for E_Y and motion parallel to the y-axis.

Justification 12.1

The first step in the justification of the separability of the wavefunction into the product of two functions X and Y is to note that, because X is independent of y and Y is independent of x, we can write

$$\frac{\partial^2 \psi}{\partial x^2} = \frac{\partial^2 XY}{\partial x^2} = Y\frac{d^2 X}{dx^2} \qquad \frac{\partial^2 \psi}{\partial y^2} = \frac{\partial^2 XY}{\partial y^2} = X\frac{d^2 Y}{dy^2}$$

Then eqn 15 becomes

$$-\frac{\hbar^2}{2m}\left(Y\frac{d^2 X}{dx^2} + X\frac{d^2 Y}{dy^2}\right) = EXY$$

When both sides are divided by XY, the resulting equation can be rearranged into

$$\frac{1}{X}\frac{d^2X}{dx^2} + \frac{1}{Y}\frac{d^2Y}{dy^2} = -\frac{2mE}{\hbar^2}$$

The first term on the left is independent of y, so if y is varied only the second term can change. But the sum of these two terms is a constant given by the right-hand side of the equation; therefore, even the second term cannot change when y is changed. In other words, the second term is a constant, which we write $-2mE_Y/\hbar^2$. By a similar argument, the first term is a constant when x changes, and we write it $-2mE_X/\hbar^2$, and $E = E_X + E_Y$. Therefore, we can write

$$\frac{1}{X}\frac{d^2X}{dx^2} = -\frac{2mE_X}{\hbar^2} \qquad \frac{1}{Y}\frac{d^2Y}{dy^2} = -\frac{2mE_Y}{\hbar^2}$$

which rearrange into the two ordinary (that is, single variable) differential equations in eqn 17.

Each of the two ordinary differential equations in eqn 17 is the same as the one-dimensional square-well Schrödinger equation; hence we can adapt the results in eqn 7 without further calculation:

$$X_{n_1}(x) = \left(\frac{2}{L_1}\right)^{1/2} \sin\left(\frac{n_1\pi x}{L_1}\right) \qquad Y_{n_2}(y) = \left(\frac{2}{L_2}\right)^{1/2} \sin\left(\frac{n_2\pi y}{L_2}\right)$$

Then, because $\psi = XY$ and $E = E_X + E_Y$, we obtain

$$\psi_{n_1 n_2}(x,y) = \frac{2}{(L_1 L_2)^{1/2}} \sin\left(\frac{n_1\pi x}{L_1}\right) \sin\left(\frac{n_2\pi y}{L_2}\right) \qquad 0 \le x \le L_1, 0 \le y \le L_2$$

$$E_{n_1 n_2} = \left(\frac{n_1^2}{L_1^2} + \frac{n_2^2}{L_2^2}\right)\frac{h^2}{8m} \tag{18}$$

with the quantum numbers taking the values $n_1 = 1, 2, \ldots$ and $n_2 = 1, 2, \ldots$ independently. Some of these functions are plotted in Fig. 12.7. They are the two-dimensional versions of the wavefunctions shown in Fig. 12.3. Note that two quantum numbers are needed in this

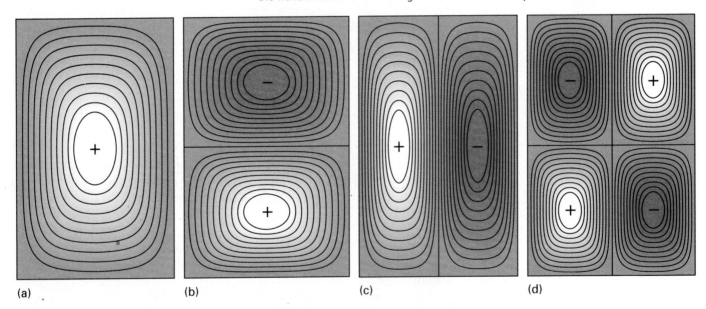

(a)   (b)   (c)   (d)

12.7 The wavefunctions for a particle confined to a rectangular surface depicted as contours of equal amplitude. (a) $n_1 = 1$, $n_2 = 1$, the state of lowest energy, (b) $n_1 = 1$, $n_2 = 2$, (c) $n_1 = 2$, $n_2 = 1$, and (d) $n_1 = 2$, $n_2 = 2$.

two-dimensional problem, and in the Dirac bracket notation we denote the states by the ket $|n_1, n_2\rangle$.

A particle in a three-dimensional box can be treated in the same way. The wavefunctions have another factor (for the z-dependence), and the energy has an additional term in n_3^2/L_3^2.

(b) Degeneracy

An interesting feature of the solutions is obtained when the plane surface is square, when $L_1 = L$ and $L_2 = L$. Then eqn 18 becomes

$$\psi_{n_1 n_2}(x, y) = \frac{2}{L} \sin\left(\frac{n_1 \pi x}{L}\right) \sin\left(\frac{n_2 \pi y}{L}\right) \qquad E_{n_1 n_2} = (n_1^2 + n_2^2)\frac{h^2}{8mL^2} \qquad (19)$$

Consider the cases $n_1 = 1, n_2 = 2$ and $n_1 = 2, n_2 = 1$:

$$\psi_{1,2}(x, y) = \frac{2}{L} \sin\left(\frac{\pi x}{L}\right) \sin\left(\frac{2\pi y}{L}\right) \qquad E_{1,2} = \frac{5h^2}{8mL^2}$$

$$\psi_{2,1}(x, y) = \frac{2}{L} \sin\left(\frac{2\pi x}{L}\right) \sin\left(\frac{\pi y}{L}\right) \qquad E_{2,1} = \frac{5h^2}{8mL^2}$$

We see that different wavefunctions correspond to the same energy, the condition called **degeneracy**. In this case, in which there are two degenerate wavefunctions, we say that the energy level $5(h^2/8mL^2)$ is 'doubly degenerate'. Alternatively, we say that the states $|1, 2\rangle$ and $|2, 1\rangle$ are degenerate.

The occurrence of degeneracy is related to the symmetry of the system. Fig. 12.8 shows contour diagrams of the two degenerate functions $\psi_{1,2}$ and $\psi_{2,1}$. Because the box is square, we can convert one wavefunction into the other simply by rotating the plane by 90°. Interconversion by rotation through 90° is not possible when the plane is not square, and $\psi_{1,2}$ and $\psi_{2,1}$ are then not degenerate. We shall see many examples of degeneracy in the pages that follow (for example, in the hydrogen atom), and all of them can be traced to the symmetry properties of the system (see Section 15.4b).

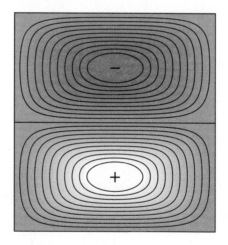 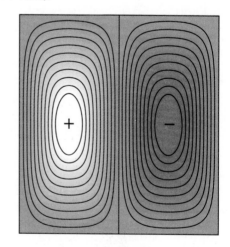

12.8 The wavefunctions for a particle confined to a square surface. Note that one wavefunction can be converted into the other by a rotation of the box by 90°. The two functions correspond to the same energy. Degeneracy and symmetry are closely related.

12.3 Tunnelling

If the potential energy of a particle does not rise to infinity when it is in the walls of the container, and $E < V$, the wavefunction does not decay abruptly to zero. If the walls are thin (so that the potential energy falls to zero again after a finite distance), the wavefunction oscillates inside the box, varies smoothly inside the region representing the wall, and oscillates again on the other side of the wall outside the box (Fig. 12.9). Hence the particle might be found on the outside of a container even though according to classical mechanics

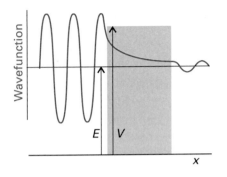

12.9 A particle incident on a barrier from the left has an oscillating wavefunction, but inside the barrier there are no oscillations (for $E < V$). If the barrier is not too thick, the wavefunction is nonzero at its opposite face, and so oscillations begin again there. (Only the real component of the wavefunction is shown.)

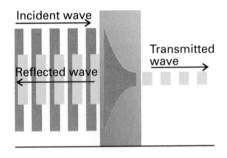

12.10 When a particle is incident on a barrier from the left, the wavefunction consists of a wave representing linear momentum to the right, a reflected component representing momentum to the left, a varying but not oscillating component inside the barrier, and a (weak) wave representing motion to the right on the far side of the barrier.

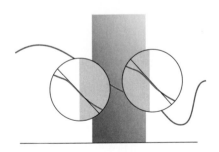

12.11 The wavefunction and its slope must be continuous at the edges of the barrier. The conditions for continuity enable us to connect the wavefunctions in the three zones and hence to obtain relations between the coefficients that appear in the solutions of the Schrödinger equation.

it has insufficient energy to escape. Such leakage by penetration through classically forbidden zones is called **tunnelling**.

The Schrödinger equation can be used to calculate the probability of tunnelling of a particle of mass m incident on a finite barrier from the left. On the left of the barrier (for $x < 0$) the wavefunctions are those of a particle with $V = 0$, so from eqn 2 we can write

$$\psi = Ae^{ikx} + Be^{-ikx} \qquad k\hbar = (2mE)^{1/2} \tag{20}$$

The Schrödinger equation for the region representing the barrier (for $0 \leq x \leq L$), where the potential energy is the constant V, is

$$-\frac{\hbar^2}{2m}\frac{d^2\psi}{dx^2} + V\psi = E\psi \tag{21}$$

We shall consider particles that have $E < V$, so $V - E$ is positive. The general solutions of this equation are then

$$\psi = Ce^{\kappa x} + De^{-\kappa x} \qquad \kappa\hbar = \{2m(V - E)\}^{1/2} \tag{22}$$

as can readily be verified by differentiating ψ twice with respect to x. The important feature to note is that the two exponentials are now *real* functions (as distinct from the complex, oscillating functions for the region where $V = 0$).[3] To the right of the barrier ($x > L$), where $V = 0$ again, the wavefunctions are

$$\psi = A'e^{ikx} + B'e^{-ikx} \qquad k\hbar = (2mE)^{1/2} \tag{23}$$

The complete wavefunction for a particle incident from the left consists of an incident wave, a wave reflected from the barrier, the exponentially changing amplitudes inside the barrier, and an oscillating wave representing the propagation of the particle to the right after tunnelling through the barrier successfully (Fig. 12.10). The *acceptable* wavefunctions have to obey the conditions set out in Section 11.4b. In particular, they must be continuous at the edges of the barrier (at $x = 0$ and $x = L$, remembering that $e^0 = 1$):

$$A + B = C + D \qquad Ce^{\kappa L} + De^{-\kappa L} = A'e^{ikL} + B'e^{-ikL} \tag{24}$$

Their slopes (their first derivatives) must also be continuous there (Fig. 12.11):

$$ikA - ikB = \kappa C - \kappa D \qquad \kappa Ce^{\kappa L} - \kappa De^{-\kappa L} = ikA'e^{ikL} - ikB'e^{-ikL} \tag{25}$$

At this stage, we have four equations for the six unknown coefficients. If the particles are shot towards the barrier from the left, there can be no particles travelling to the left on the *right* of the barrier. Therefore, we can set $B' = 0$, which removes one more unknown. We cannot set $B = 0$ because some particles may be reflected back from the barrier toward negative x.

The probability that a particle is travelling towards positive x (to the right) on the left of the barrier is proportional to $|A|^2$, and the probability that it is travelling to the right on the right of the barrier is proportional to $|A'|^2$. The ratio of these two probabilities is called the **transmission probability**, T. After some algebra we find

$$T = \left\{1 + \frac{(e^{\kappa L} - e^{-\kappa L})^2}{16\varepsilon(1 - \varepsilon)}\right\}^{-1} \tag{26}$$

where $\varepsilon = E/V$. This function is plotted in Fig. 12.12; the transmission coefficient for $E > V$ is shown there too. For high, wide barriers (in the sense that $\kappa L \gg 1$), eqn 26 simplifies to

$$T \approx 16\varepsilon(1 - \varepsilon)e^{-2\kappa L} \tag{27}$$

3   Oscillating functions would be obtained if $E > V$.

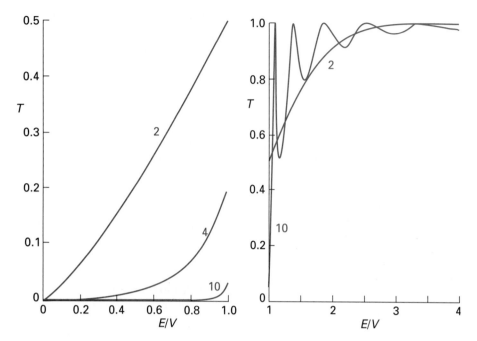

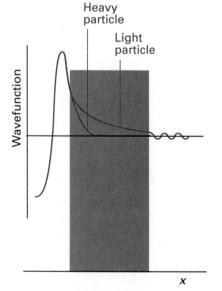

12.12 The transition probabilities for passage through a barrier. The horizontal axis is the energy of the incident particle expressed as a multiple of the barrier height. The curves are labelled with the value of $L(2mV)^{1/2}/\hbar$. The graph on the left is for $E<V$ and that on the right for $E>V$. Note that $T>0$ for $E<V$, whereas classically T would be zero. However, $T<1$ for $E>V$, whereas classically T would be 1.

The transmission probability decreases exponentially with the thickness of the barrier and with $m^{1/2}$. It follows that particles of low mass are more able to tunnel through barriers than heavy ones (Fig. 12.13). Tunnelling is very important for electrons and muons, and moderately important for protons; for heavier particles it is less important. A number of effects in chemistry (for example, the isotope-dependence of some reaction rates) depend on the ability of the proton to tunnel more readily than the deuteron. The very rapid equilibration of proton transfer reactions (which were discussed in Chapter 9) is also a manifestation of the ability of protons to tunnel through barriers and transfer quickly from an acid to a base. The important technique of 'scanning tunnelling microscopy' (STM), which is described in more detail in Section 28.2f, relies on the exponential dependence of electron tunnelling on the thickness of the region between a point and a surface.

Illustration

To estimate the relative probabilities that a proton and a deuteron can tunnel through the same barrier of height 1.00 eV (1.60×10^{-19} J) and length 100 pm when their energy is 0.9 eV, so $E - V = 0.10$ eV, we first evaluate

$$\kappa = \left\{ \frac{2(m/u) \times (1.67 \times 10^{-27} \text{ kg}) \times (1.6 \times 10^{-20} \text{ J})}{(1.055 \times 10^{-34} \text{ J s})^2} \right\}^{1/2}$$

$$= \frac{(m/u)^{1/2}}{14 \text{ pm}}$$

The values of κ for a proton ($m = 1.0$ u) and a deuteron ($m = 2.0$ u) are $1/(14$ pm) and $1/(9.9$ pm), respectively, so $\kappa L \gg 1$ and eqn 27 can be used. The ratio of transmission probabilities is then

$$\frac{T_{\text{H}}}{T_{\text{D}}} = e^{-2(\kappa_{\text{H}} - \kappa_{\text{D}})L} = 3.7 \times 10^2$$

(The ratio is very sensitive to rounding errors.) The result shows that the tunnelling probability of a proton (in the system specified) is about 370 times greater than that of a deuteron.

12.13 The wavefunction of a heavy particle decays more rapidly inside a barrier than that of a light particle. Consequently, a light particle has a greater probability of tunnelling through the barrier.

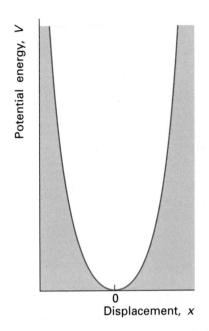

12.14 The parabolic potential energy $V = \frac{1}{2}kx^2$ of a harmonic oscillator, where x is the displacement from equilibrium. The narrowness of the curve depends on the force constant k: the larger the value of k, the narrower the well.

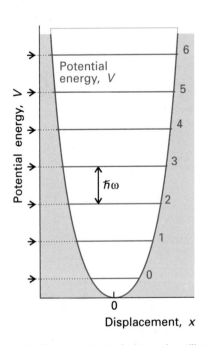

12.15 The energy levels of a harmonic oscillator are evenly spaced with separation $\hbar\omega$, with $\omega = (k/m)^{1/2}$. Even in its lowest state, an oscillator has an energy greater than zero.

Self-test 12.4 Calculate the relative tunnelling probabilities when the barrier is twice as long, the other conditions being unchanged.

$$[1.4 \times 10^5]$$

Vibrational motion

A particle undergoes **harmonic motion** if it experiences a restoring force proportional to its displacement:

$$F = -kx \qquad (28)$$

where k is the **force constant**: the stiffer the 'spring', the greater the value of k. Because force is related to potential energy by $F = -dV/dx$ (see *Further information 4*), the force in eqn 28 corresponds to a potential energy

$$V = \frac{1}{2}kx^2 \qquad (29)$$

This expression, which is the equation of a parabola (Fig. 12.14), is the origin of the term 'parabolic potential energy' for the potential energy characteristic of a harmonic oscillator. The Schrödinger equation for the particle is therefore

$$-\frac{\hbar^2}{2m}\frac{d^2\psi}{dx^2} + \frac{1}{2}kx^2\psi = E\psi \qquad (30)$$

12.4 The energy levels

Equation 30 is a standard equation in the theory of differential equations and its solutions are well known to mathematicians (see below).[4] When the boundary conditions, that the oscillator will not be found with infinitely large compressions or extensions, are applied, it is found that the permitted energy levels are

$$E_v = (v + \tfrac{1}{2})\hbar\omega \qquad \omega = \left(\frac{k}{m}\right)^{1/2} \qquad v = 0, 1, 2, \ldots \qquad (31)$$

Note that ω increases with increasing force constant and decreasing mass. It follows that the separation between adjacent levels is

$$E_{v+1} - E_v = \hbar\omega \qquad (32)$$

which is the same for all v. Therefore, the energy levels form a uniform ladder of spacing $\hbar\omega$ (Fig. 12.15). The energy separation $\hbar\omega$ is negligibly small for macroscopic objects (with large mass), but is of great importance for objects with mass similar to that of atoms.

Illustration

The force constant of a typical X–H chemical bond is around 500 N m^{-1}. Because the mass of a proton is about 1.7×10^{-27} kg, $\omega \approx 5 \times 10^{14}$ s^{-1} and the separation of adjacent levels is $\hbar\omega \approx 6 \times 10^{-20}$ J (about 0.4 eV). This energy separation corresponds to 30 kJ mol^{-1}, which is chemically significant. The excitation of the vibration of the bond from one level to the level immediately above requires 6×10^{-20} J. Therefore, if it is caused by a photon, the excitation requires radiation of frequency $\nu = \Delta E/h = 9 \times 10^{13}$ Hz and wavelength $\lambda = c/\nu = 3$ μm. It follows that transitions between adjacent vibrational energy levels of molecules are stimulated by or emit infrared radiation (Chapter 16).

4 For the details of the solution, see *Further reading*.

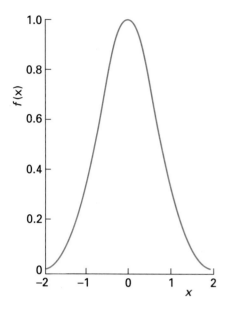

12.16 The graph of the Gaussian function, $f(x) = e^{-x^2}$.

Because the smallest permitted value of v is 0, it follows from eqn 31 that a harmonic oscillator has a zero-point energy

$$E_0 = \tfrac{1}{2}\hbar\omega \tag{33}$$

For the typical molecular oscillator specified in the *Illustration*, the zero-point energy is about 3×10^{-20} J, which corresponds to 0.2 eV, or 15 kJ mol^{-1}. The mathematical reason for the zero-point energy is that v cannot take negative values, for if it did the wavefunction would be ill-behaved. The physical reason is the same as for the particle in a square well: the particle is confined, its position is not completely uncertain, and therefore its momentum, and hence its kinetic energy, cannot be exactly zero. We can picture this zero-point state as one in which the particle fluctuates incessantly around its equilibrium position; classical mechanics would allow the particle to be perfectly still.

12.5 The wavefunctions

It is helpful at the outset to identify the similarities between the harmonic oscillator and the particle in a box, for then we shall be able to anticipate the form of the oscillator wavefunctions without detailed calculation. Like the particle in a box, a particle undergoing harmonic motion is trapped in a symmetrical well in which the potential energy rises to large values (and ultimately to infinity) for sufficiently large displacements (compare Figs 12.1 and 12.14). However, there are two important differences. First, because the potential energy climbs towards infinity only as x^2 and not abruptly, the wavefunction approaches zero more slowly at large displacements than for the particle in a box. Second, as the kinetic energy of the oscillator depends on the displacement in a more complex way (on account of the variation of the potential energy), the curvature of the wavefunction also varies in a more complex way.

(a) The form of the wavefunctions

The detailed solution of eqn 30 shows that the wavefunction for a harmonic oscillator has the form

$$\psi(x) = N \times (\text{polynomial in } x) \times (\text{bell-shaped Gaussian function})$$

where N is a normalization constant. A **Gaussian function** is a function of the form e^{-x^2} (Fig. 12.16). The precise form of the wavefunctions is

$$\psi_v(x) = N_v H_v(y) e^{-y^2/2} \qquad y = \frac{x}{\alpha} \qquad \alpha = \left(\frac{\hbar^2}{mk}\right)^{1/4} \tag{34}$$

The factor $H_v(y)$ is a **Hermite polynomial** (Table 12.1). For instance, because $H_0(y) = 1$, the wavefunction for the ground state (the lowest energy state, with $v = 0$) of the harmonic oscillator is

$$\psi_0(x) = N_0 e^{-y^2/2} = N_0 e^{-x^2/2\alpha^2} \tag{35}$$

It follows that the probability density is the bell-shaped Gaussian function

$$\psi_0^2(x) = N_0^2 e^{-x^2/\alpha^2} \tag{36}$$

The wavefunction and the probability distribution are shown in Fig. 12.17. Both curves have their largest values at zero displacement (at $x = 0$), so they capture the classical picture of the zero-point energy as arising from the ceaseless fluctuation of the particle about its equilibrium position.

Table 12.1 The Hermite polynomials $H_v(y)$

v	H_v
0	1
1	$2y$
2	$4y^2 - 2$
3	$8y^3 - 12y$
4	$16y^4 - 48y^2 + 12$
5	$32y^5 - 160y^3 + 120y$
6	$64y^6 - 480y^4 + 720y^2 - 120$

The Hermite polynomials (which continue up to infinite v) satisfy the equation

$$H_v'' - 2yH_v' + 2vH_v = 0$$

and the recursion relation

$$H_{v+1} = 2yH_v - 2vH_{v-1}$$

An important integral is

$$\int_{-\infty}^{\infty} H_{v'}H_v e^{-y^2}\,dy = \begin{cases} 0 & \text{if } v' \neq v \\ \pi^{1/2}2^v v! & \text{if } v' = v \end{cases}$$

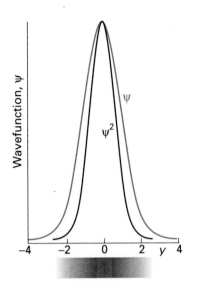

12.17 The normalized wavefunction and probability distribution (shown also by shading) for the lowest energy state of a harmonic oscillator.

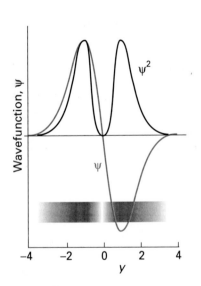

12.18 The normalized wavefunction and probability distribution (shown also by shading) for the first excited state of a harmonic oscillator.

The wavefunction for the first excited state of the oscillator, the state with $v = 1$, is obtained by noting that $H_1(y) = 2y$ (note that some of the Hermite polynomials are very simple functions!):

$$\psi_1(x) = N_1 \times 2ye^{-y^2/2} = \frac{2N_1}{\alpha}xe^{-x^2/2\alpha^2} \tag{37}$$

This function has a node at zero displacement ($x = 0$), and the probability density has maxima at $x = \pm\alpha$, corresponding to $y = \pm 1$ (Fig. 12.18).

The shapes of several wavefunctions are shown in Fig. 12.19. The shading in Fig. 12.20 that represents the probability density is based on the squares of these functions. At high quantum numbers, harmonic oscillator wavefunctions have their largest amplitudes near the turning points of the classical motion (the locations at which $V = E$, so the kinetic energy is zero). We see classical properties emerging in the correspondence limit of high quantum numbers, for a classical particle is most likely to be found at the turning points (where it travels most slowly) and is least likely to be found at zero displacement (where it travels most rapidly).

Example 12.3 Normalizing a harmonic oscillator wavefunction

Find the normalization constant for the harmonic oscillator wavefunctions.

Method Normalization is always carried out by evaluating the integral of $|\psi|^2$ over all space and then finding the normalization factor from eqn 11.17. The normalized wavefunction is then equal to $N\psi$. In this one-dimensional problem, the volume element is dx and the integration is from $-\infty$ to $+\infty$. The wavefunctions are expressed in terms of the dimensionless variable $y = x/\alpha$, so begin by expressing the integral in terms of y by using $dx = \alpha\,dy$. The integrals required are given in Table 12.1.

Answer The unnormalized wavefunction is

$$\psi_v(x) = H_v(y)e^{-y^2/2}$$

It follows from the integrals given in Table 12.1 that

$$\int_{-\infty}^{\infty} \psi_v^* \psi_v\, dx = \alpha \int_{-\infty}^{\infty} \psi_v^* \psi_v\, dy = \alpha \int_{-\infty}^{\infty} H_v^2(y)e^{-y^2}\, dy$$
$$= \alpha\pi^{1/2}2^v v!$$

where $v! = v(v-1)(v-2)\cdots 1$. Therefore,

$$N_v = \frac{1}{(\alpha\pi^{1/2}2^v v!)^{1/2}}$$

Note that for a harmonic oscillator N_v is different for each value of v.

Comment The Hermite polynomials are members of a class of functions called *orthogonal polynomials*. These polynomials have a wide range of important properties which allow a number of quantum mechanical calculations to be done with relative ease. See *Further reading* for a reference to their properties.

Self-test 12.5 Confirm, by explicit evaluation of the integral, that ψ_0 and ψ_1 are orthogonal.

[Evaluate the integral $\int_{-\infty}^{\infty} \psi_0^* \psi_1\, dx$ by using the information in Table 12.1.]

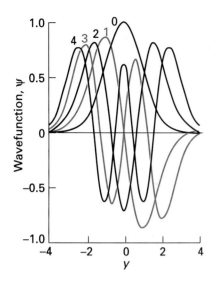

12.19 The normalized wavefunctions for the first five states of a harmonic oscillator. Even values of v are black; odd values are green. Note that the number of nodes is equal to v and that alternate wavefunctions are symmetrical or antisymmetrical about $y = 0$ (zero displacement).

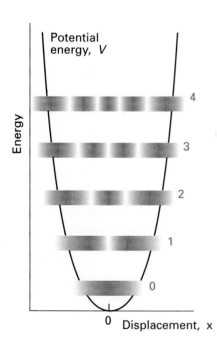

12.20 The probability distributions for the first five states of a harmonic oscillator represented by the density of shading. Note how the regions of highest probability (the regions of densest shading) move towards the turning points of the classical motion as v increases.

(b) The properties of oscillators

With the wavefunctions that are available, we can start calculating the properties of a harmonic oscillator. For instance, we can calculate the expectation values of an observable Ω by evaluating integrals of the type

$$\langle \Omega \rangle = \int_{-\infty}^{\infty} \psi_v^* \hat{\Omega} \psi_v \, \mathrm{d}x \tag{38}$$

(Here and henceforth, the wavefunctions are all taken to be normalized.) A tidier expression is obtained by using Dirac bracket notation, in which an integral is replaced by a bracket labelled with the quantum numbers of the states:

$$\langle v' | \hat{\Omega} | v \rangle = \int_{-\infty}^{\infty} \psi_{v'}^* \hat{\Omega} \psi_v \, \mathrm{d}x \tag{39}$$

This bracket is also called a **matrix element** of the operator $\hat{\Omega}$. Note how the operator stands between the bra and the ket (which may denote different states), in the place of the c in $\langle \text{bra}|c|\text{ket} \rangle$. An integration is implied whenever a complete bracket is written. In this notation, the expectation value is

$$\langle \Omega \rangle = \langle v | \hat{\Omega} | v \rangle \tag{40}$$

with the bra and the ket corresponding to the same state (with quantum number v and wavefunction ψ_v).

When the explicit wavefunctions are substituted, the integrals look fearsome, but the Hermite polynomials have many simplifying features. For instance, we show in the following example that the mean displacement, $\langle x \rangle$, and the mean square displacement, $\langle x^2 \rangle$, of the oscillator when it is in the state with quantum number v are

$$\langle x \rangle = 0 \qquad \langle x^2 \rangle = (v + \tfrac{1}{2}) \frac{\hbar}{(mk)^{1/2}} \tag{41}$$

The result for $\langle x \rangle$ shows that the oscillator is equally likely to be found on either side of $x = 0$ (like a classical oscillator). The result for $\langle x^2 \rangle$ shows that the mean square displacement increases with v. This increase is apparent from the probability densities in Fig. 12.20, and corresponds to the classical amplitude of swing increasing as the oscillator becomes more highly excited.

Example 12.4 Calculating properties of a harmonic oscillator

Calculate the mean displacement of the oscillator when it is in a quantum state v.

Method Normalized wavefunctions must be used to calculate the expectation value. The operator for position along x is multiplication by the value of x (Section 11.5c). The resulting integral can be evaluated either by inspection (the integrand is the product of an odd and an even function), or by explicit evaluation using the formulas in Table 12.1. To give practice in this type of calculation, we illustrate the latter procedure. We shall need the relation $x = \alpha y$, which implies that $\mathrm{d}x = \alpha \, \mathrm{d}y$.

Answer The integral we require is

$$\langle x \rangle = \int_{-\infty}^{\infty} \psi_v^* x \psi_v \, \mathrm{d}x = N_v^2 \int_{-\infty}^{\infty} (H_v \mathrm{e}^{-y^2/2}) x (H_v \mathrm{e}^{-y^2/2}) \, \mathrm{d}x$$

$$= \alpha^2 N_v^2 \int_{-\infty}^{\infty} (H_v \mathrm{e}^{-y^2/2}) y (H_v \mathrm{e}^{-y^2/2}) \, \mathrm{d}y$$

$$= \alpha^2 N_v^2 \int_{-\infty}^{\infty} H_v y H_v \mathrm{e}^{-y^2} \, \mathrm{d}y$$

Now use the recursion relation in Table 12.1 to form

$$yH_v = vH_{v-1} + \tfrac{1}{2}H_{v+1}$$

which turns the integral into

$$\int_{-\infty}^{\infty} H_v y H_v e^{-y^2}\,dy = v\int_{-\infty}^{\infty} H_{v-1}H_v e^{-y^2}\,dy + \tfrac{1}{2}\int_{-\infty}^{\infty} H_{v+1}H_v e^{-y^2}\,dy$$

Both integrals are zero (Table 12.1), so $\langle x \rangle = 0$.

Self-test 12.6 Calculate the mean square displacement $\langle x^2 \rangle$ of the particle from its equilibrium position. (Use the recursion relation twice.)

[eqn 41]

The mean potential energy of an oscillator, the expectation value of $V = \tfrac{1}{2}kx^2$, can now be calculated very easily:

$$\langle V \rangle = \langle \tfrac{1}{2}kx^2 \rangle = \tfrac{1}{2}(v + \tfrac{1}{2})\hbar\left(\frac{k}{m}\right)^{1/2} = \tfrac{1}{2}(v + \tfrac{1}{2})\hbar\omega \tag{42}$$

Because the total energy in the state with quantum number v is $(v + \tfrac{1}{2})\hbar\omega$, it follows that

$$\langle V \rangle = \tfrac{1}{2}E_v \tag{43}$$

The total energy is the sum of the potential and kinetic energies, so it follows at once that the mean kinetic energy of the oscillator is

$$\langle E_K \rangle = \tfrac{1}{2}E_v \tag{44}$$

The result that the mean potential and kinetic energies of a harmonic oscillator are equal (and therefore that both are equal to half the total energy) is a special case of the **virial theorem**:

> If the potential energy of a particle has the form $V = ax^b$, then its mean potential and kinetic energies are related by

$$2\langle E_K \rangle = b\langle V \rangle \tag{45}$$

For a harmonic oscillator $b = 2$, so $\langle E_K \rangle = \langle V \rangle$, as we have found. The virial theorem is a short cut to the establishment of a number of useful results, and we shall use it again.

An oscillator may be found at extensions with $V > E$ that are forbidden by classical physics, for they correspond to negative kinetic energy. For example, it follows from the shape of the wavefunction (see the *Justification* below) that in its lowest energy state there is about an 8 per cent chance of finding an oscillator stretched beyond its classical limit and an 8 per cent chance of finding it with a classically forbidden compression. These tunnelling probabilities are independent of the force constant and mass of the oscillator. The probability of being found in classically forbidden regions decreases quickly with increasing v, and vanishes entirely as v approaches infinity, as we would expect from the correspondence principle. Macroscopic oscillators (such as pendulums) are in states with very high quantum numbers, so the probability that they will be found in a classically forbidden region is wholly negligible. Molecules, however, are normally in their vibrational ground states, and for them the probability is very significant.

Justification 12.2

According to classical mechanics, the turning point, x_{tp}, of an oscillator occurs when its kinetic energy is zero, which is when its potential energy $\frac{1}{2}kx^2$ is equal to its total energy E. This equality occurs when

$$x_{tp}^2 = \frac{2E}{k} \quad \text{or} \quad x_{tp} = \pm\left(\frac{2E}{k}\right)^{1/2}$$

with E given by eqn 31. The probability of finding the oscillator stretched beyond a displacement x_{tp} is the sum of the probabilities $\psi^2\,dx$ of finding it in any of the intervals dx lying between x_{tp} and infinity:

$$P = \int_{x_{tp}}^{\infty} \psi_v^2\,dx$$

The variable of integration is best expressed in terms of $y = x/\alpha$ with α specified in eqn 34, and then the turning point on the right lies at

$$y_{tp} = x_{tp}/\alpha = \left(\frac{2(v+\frac{1}{2})\hbar\omega}{\alpha^2 k}\right)^{1/2} = (2v+1)^{1/2}$$

For the state of lowest energy ($v = 0$), $y_{tp} = 1$ and the probability is

$$P = \int_{x_{tp}}^{\infty} \psi_0^2\,dx = \alpha N_0^2 \int_{1}^{\infty} e^{-y^2}\,dy$$

The integral is a special case of the **error function**, erf z, which is defined as follows:

$$\text{erf } z = 1 - \frac{2}{\pi^{1/2}} \int_{z}^{\infty} e^{-y^2}\,dy \qquad [46]$$

The values of this function are tabulated (just like sine and cosine functions), and a small selection of values is given in Table 12.2. In the present case

$$P = \tfrac{1}{2}(1 - \text{erf } 1) = \tfrac{1}{2}(1 - 0.843) = 0.079$$

It follows that in 7.9 per cent of a large number of observations, any oscillator in the state $v = 0$ will be found stretched to a classically forbidden extent. There is the same probability of finding the oscillator with a classically forbidden compression. The total probability of finding the oscillator tunnelled into a classically forbidden region (stretched or compressed) is about 16 per cent.

Rotational motion

The treatment of rotational motion can be broken down into two parts. The first deals with motion in two dimensions and the second with rotation in three dimensions. It may be helpful to review the classical description of rotational motion given in *Further information 4*, particularly the concepts of moment of inertia and angular momentum.

12.6 Rotation in two dimensions

We consider a particle of mass m constrained to move in a circular path of radius r in the xy-plane (Fig. 12.21). The total energy is equal to the kinetic energy, because $V = 0$ everywhere. We can therefore write $E = p^2/2m$. According to classical mechanics, the angular momentum, J_z, around the z-axis (which lies perpendicular to the xy-plane) is $J_z = \pm pr$,

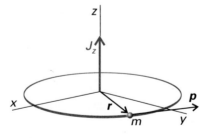

12.21 The angular momentum of a particle of mass m on a circular path of radius r in the xy-plane is represented by a vector of magnitude pr perpendicular to the plane.

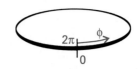

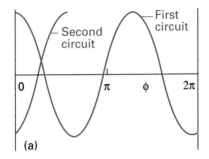

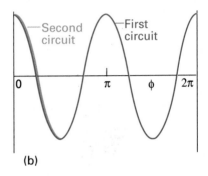

12.22 Two solutions of the Schrödinger equation for a particle on a ring. The circumference has been opened out into a straight line; the points at $\phi = 0$ and 2π are identical. The solution in (a) is unacceptable because it is not single-valued. Moreover, on successive circuits it interferes destructively with itself, and does not survive. The solution in (b) is acceptable: it is single-valued, and on successive circuits it reproduces itself.

so the energy can be expressed as $J_z^2/2mr^2$. Because mr^2 is the moment of inertia, I, of the mass on its path, it follows that

$$E = \frac{J_z^2}{2I} \qquad (47)$$

We shall now see that not all the values of the angular momentum are permitted in quantum mechanics, and therefore that both angular momentum and rotational energy are quantized.

(a) The qualitative origin of quantized rotation

Because $J_z = \pm pr$ and, from the de Broglie relation, $p = h/\lambda$, the angular momentum about the z-axis is

$$J_z = \pm \frac{hr}{\lambda}$$

Opposite signs correspond to opposite directions of travel. This equation shows that, the shorter the wavelength of the particle on a circular path of given radius, the greater the angular momentum of the particle. It follows that, if we can see why the wavelength is restricted to discrete values, then we shall understand why the angular momentum is quantized.

Suppose for the moment that λ can take an arbitrary value. In that case, the wavefunction depends on the azimuthal angle ϕ as shown in Fig. 12.22a. When ϕ increases beyond 2π, the wavefunction continues to change, but for an arbitrary wavelength it gives rise to a different value at each point, which is unacceptable (Section 11.4b). An acceptable solution is obtained only if the wavefunction reproduces itself on successive circuits, as in Fig. 12.22b. Because only some wavefunctions have this property, it follows that only some angular momenta are acceptable, and therefore that only certain rotational energies exist. Hence, the energy of the particle is quantized. Specifically, the only allowed wavelengths are

$$\lambda = \frac{2\pi r}{m_l}$$

with m_l, the conventional notation for this quantum number, taking integral values including 0.[5] The angular momentum is therefore limited to the values

$$J_z = \pm \frac{hr}{\lambda} = \frac{m_l hr}{2\pi r} = \frac{m_l h}{2\pi}$$

where we have allowed m_l to have positive or negative values. That is,

$$J_z = m_l \hbar \qquad m_l = 0, \pm 1, \pm 2, \ldots \qquad (48)$$

Positive values of m_l correspond to rotation in a clockwise sense around the z-axis (as viewed in the direction of z, Fig. 12.23) and negative values of m_l correspond to counter-clockwise rotation around z. It then follows from eqn 47 that the energy is limited to the values

$$E = \frac{J_z^2}{2I} = \frac{m_l^2 \hbar^2}{2I} \qquad (49)$$

We shall see shortly that the corresponding normalized wavefunctions are

$$\psi_{m_l}(\phi) = \frac{e^{im_l\phi}}{(2\pi)^{1/2}} \qquad (50)$$

The wavefunction with $m_l = 0$ is $\psi_0(\phi) = 1/(2\pi)^{1/2}$, and has the same value at all points on the circle.

We have arrived at a number of conclusions about rotational motion by cobbling together some classical notions and the de Broglie relation. Such a procedure can be very

5   The value $m_l = 0$ corresponds to $\lambda = \infty$; a 'wave' of infinite wavelength has a constant height at all values of ϕ.

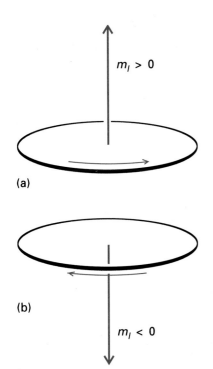

(a)

(b)

$m_l > 0$

$m_l < 0$

12.23 The angular momentum of a particle confined to a plane can be represented by a vector of length $|m_l|$ units along the z-axis and with an orientation that indicates the direction of motion of the particle. The direction is given by the right-hand screw rule.

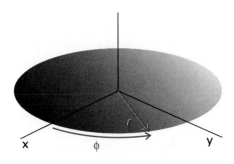

12.24 The cylindrical coordinates r and ϕ for discussing systems with axial (cylindrical) symmetry.

useful for establishing the general form (and, as in this case, the exact energies) for a quantum mechanical system. However, to be sure that the correct solutions have been obtained, and to obtain practice for more complex problems where this less formal approach is inadequate, we need to solve the Schrödinger equation explicitly. The formal solution is described in the *Justification* that follows.

Justification 12.3

The hamiltonian for a particle of mass m in a plane (with $V = 0$) is the same as that given in eqn 15:

$$H = -\frac{\hbar^2}{2m}\left(\frac{\partial^2}{\partial x^2} + \frac{\partial^2}{\partial y^2}\right)$$

and the Schrödinger equation is $H\psi = E\psi$, with the wavefunction a function of the angle ϕ. It is always a good idea to use coordinates that reflect the full symmetry of the system, so we introduce the coordinates r and ϕ (Fig. 12.24), where $x = r\cos\phi$ and $y = r\sin\phi$. By standard manipulations we can write

$$\frac{\partial^2}{\partial x^2} + \frac{\partial^2}{\partial y^2} = \frac{\partial^2}{\partial r^2} + \frac{1}{r}\frac{\partial}{\partial r} + \frac{1}{r^2}\frac{\partial^2}{\partial \phi^2} \tag{51}$$

However, because the radius of the path is fixed, the derivative with respect to r can be discarded; the hamiltonian then becomes

$$H = -\frac{\hbar^2}{2mr^2}\frac{d^2}{d\phi^2}$$

The moment of inertia $I = mr^2$ has appeared automatically, so H may be written

$$H = -\frac{\hbar^2}{2I}\frac{d^2}{d\phi^2} \tag{52}$$

and the Schrödinger equation is

$$\frac{d^2\psi}{d\phi^2} = -\frac{2IE}{\hbar^2}\psi \tag{53}$$

The normalized general solutions of the equation are

$$\psi_{m_l}(\phi) = \frac{e^{im_l\phi}}{(2\pi)^{1/2}} \qquad m_l = \pm\frac{(2IE)^{1/2}}{\hbar}$$

The quantity m_l is just a dimensionless number at this stage.

We now select the acceptable solutions from among these general solutions by imposing the condition that the wavefunction should be single-valued. That is, the wavefunction ψ must satisfy a **cyclic boundary condition**, and match at points separated by a complete revolution: $\psi(\phi + 2\pi) = \psi(\phi)$. On substituting the general wavefunction into this condition, we find

$$\psi_{m_l}(\phi + 2\pi) = \frac{e^{im_l(\phi + 2\pi)}}{(2\pi)^{1/2}} = \frac{e^{im_l\phi}e^{2im_l\pi}}{(2\pi)^{1/2}} = \psi(\phi)e^{2im_l\pi}$$

As $e^{i\pi} = -1$, this relation is equivalent to

$$\psi_{m_l}(\phi + 2\pi) = (-1)^{2m_l}\psi(\phi) \tag{54}$$

Because we require $(-1)^{2m_l} = 1$, $2m_l$ must be a positive or negative even integer (including 0), and therefore m_l must be an integer: $m_l = 0, \pm 1, \pm 2, \ldots$.

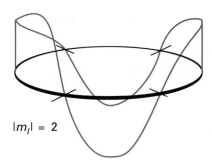

$|m_l| = 2$

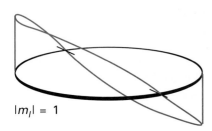

$|m_l| = 1$

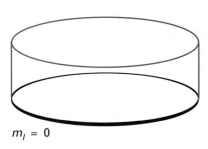

$m_l = 0$

12.25 The real parts of the wavefunctions of a particle on a ring. As shorter wavelengths are achieved, the magnitude of the angular momentum around the z-axis grows in steps of $\hbar$

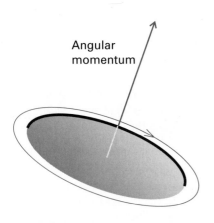

Angular momentum

12.26 The basic ideas of the vector representation of angular momentum: the magnitude of the angular momentum is represented by the length of the vector, and the orientation of the motion in space by the orientation of the vector (using the right-hand screw rule).

(b) Quantization of rotation

We can summarize the conclusions so far as follows. The energy is quantized and restricted to the values given in eqn 49 ($E = m_l^2 \hbar^2 / 2I$). The occurrence of m_l as its square means that the energy of rotation is independent of the sense of rotation (the sign of m_l), as we expect physically. In other words, states with a given value of $|m_l|$ are doubly degenerate, except for $m_l = 0$, which is non-degenerate. Although the result has been derived for the rotation of a single mass point, it also applies to any body of moment of inertia I constrained to rotate about one axis.

We have also seen that the angular momentum is quantized and confined to the values given in eqn 48 ($J_z = m_l \hbar$). The increasing angular momentum is associated with the increasing number of nodes in the real and imaginary parts of the wavefunction:[6] the wavelength decreases stepwise as $|m_l|$ increases, so the momentum with which the particle travels round the ring increases (Fig. 12.25). As shown in the following *Justification*, the same conclusion can be obtained formally by using the argument about the relations between eigenvalues and the values of observables that were established in Section 11.5.

Justification 12.4

In the discussion of translational motion in one dimension, we saw that the opposite signs in the wavefunctions e^{ikx} and e^{-ikx} correspond to opposite directions of travel, and that the linear momentum is given by the eigenvalue of the linear momentum operator. The same conclusions can be drawn here, but now we need the eigenvalues of the *angular* momentum operator. In classical mechanics the orbital angular momentum l_z about the z-axis is defined as[7]

$$l_z = xp_y - yp_x \tag{55}$$

where p_x is the component of linear motion parallel to the x-axis and p_y is the component parallel to the y-axis. The operators for the two linear momentum components are given in eqn 11.32, so the operator for angular momentum about the z-axis, which we denote $\hat{l}_z$, is

$$\hat{l}_z = \frac{\hbar}{i} \left(x \frac{\partial}{\partial y} - y \frac{\partial}{\partial x} \right) \tag{56}$$

When expressed in terms of the coordinates r and ϕ, this equation becomes

$$\hat{l}_z = \frac{\hbar}{i} \frac{\partial}{\partial \phi} \tag{57}$$

With the angular momentum operator available, we can test the wavefunction in eqn 50. Disregarding the normalization constant, we find

$$\hat{l}_z \psi_{m_l} = \frac{\hbar}{i} \frac{d\psi_{m_l}}{d\phi} = im_l \frac{\hbar}{i} e^{im_l \phi} = m_l \hbar \psi_{m_l} \tag{58}$$

That is, ψ_{m_l} is an eigenfunction of $\hat{l}_z$, and corresponds to an angular momentum $m_l \hbar$. When m_l is positive, the angular momentum is positive (clockwise when seen from below); when m_l is negative, the angular momentum is negative (counter-clockwise when seen from below). These features are the origin of the **vector representation** of angular momentum, in which the magnitude is represented by the length of a vector and the direction of motion by its orientation (Fig. 12.26).

6   The complex function $e^{im_l \phi}$ does not have nodes; however, it may be written as $\cos m_l \phi + i \sin m_l \phi$, and the real ($\cos m_l \phi$) and imaginary ($\sin m_l \phi$) components do have nodes.

7   The angular momentum in three dimensions is defined as $l = r \times p$; to obtain eqn 55, expand the vector product and identify the z-component.

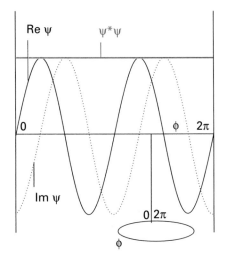

12.27 The probability density for a particle in a definite state of angular momentum is uniform, so there is an equal probability of finding the particle anywhere on the ring.

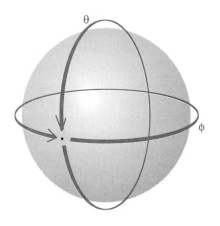

12.28 The wavefunction of a particle on the surface of a sphere must satisfy two cyclic boundary conditions; this requirement leads to two quantum numbers for its state of angular momentum.

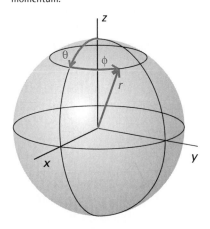

12.29 Spherical polar coordinates. For a particle confined to the surface of a sphere, only the colatitude, θ, and the azimuth, ϕ, can change.

To locate the particle given its wavefunction in eqn 50, we form the probability density:

$$\psi_{m_l}^* \psi_{m_l} = \left(\frac{e^{im_l\phi}}{(2\pi)^{1/2}}\right)^* \left(\frac{e^{im_l\phi}}{(2\pi)^{1/2}}\right) = \left(\frac{e^{-im_l\phi}}{(2\pi)^{1/2}}\right) \left(\frac{e^{im_l\phi}}{(2\pi)^{1/2}}\right)$$

$$= \frac{1}{2\pi} \tag{59}$$

Because this probability density is independent of ϕ, the probability of locating the particle somewhere on the ring is also independent of ϕ (Fig. 12.27). Hence the location of the particle is completely indefinite, and knowing the angular momentum precisely eliminates the possibility of specifying the particle's location. Angular momentum and angle are a pair of complementary observables (in the sense defined in Section 11.5e), and the inability to specify them simultaneously with arbitrary precision is another example of the uncertainty principle.

12.7 Rotation in three dimensions

We now consider a particle of mass m that is free to move anywhere on the surface of a sphere of radius r. We shall need the results of this calculation when we come to describe the states of electrons in atoms (Chapter 13) and of rotating molecules (Chapter 16). The latter application arises from the fact that the rotation of a solid body of moment of inertia I can be represented by a single point of mass m rotating at a radius r, which is defined so that $I = mr^2$. The requirement that the wavefunction should match as a path is traced over the poles as well as round the equator of the sphere surrounding the central point introduces a second cyclic boundary condition and therefore a second quantum number (Fig. 12.28).

(a) The Schrödinger equation

The hamiltonian for motion in three dimensions (Table 11.1) is

$$H = -\frac{\hbar^2}{2m}\nabla^2 + V \qquad \nabla^2 = \frac{\partial^2}{\partial x^2} + \frac{\partial^2}{\partial y^2} + \frac{\partial^2}{\partial z^2} \tag{60}$$

The symbol ∇^2 is a convenient abbreviation for the sum of the three second derivatives; it is called the **laplacian**, and read either 'del squared' or 'nabla squared'. For the particle confined to a spherical surface, $V = 0$ wherever it is free to travel, and the radius r is a constant. The wavefunction is therefore a function of the **colatitude**, θ, and the **azimuth**, ϕ (Fig. 12.29), and we write it $\psi(\theta, \phi)$. The Schrödinger equation is

$$-\frac{\hbar^2}{2m}\nabla^2\psi = E\psi \tag{61}$$

This partial differential equation can be simplified by the separation of variables procedure by expressing the wavefunction (for constant r) as the product

$$\psi(\theta, \phi) = \Theta(\theta)\Phi(\phi) \tag{62}$$

where Θ is a function only of θ and Φ is a function only of ϕ.

Justification 12.5

The laplacian in spherical polar coordinates is

$$\nabla^2 = \frac{\partial^2}{\partial r^2} + \frac{2}{r}\frac{\partial}{\partial r} + \frac{1}{r^2}\Lambda^2 \tag{63}$$

where the **legendrian**, Λ^2, is

$$\Lambda^2 = \frac{1}{\sin^2\theta}\frac{\partial^2}{\partial\phi^2} + \frac{1}{\sin\theta}\frac{\partial}{\partial\theta}\sin\theta\frac{\partial}{\partial\theta} \tag{64}$$

Because r is constant, we can discard the part of the laplacian that involves differentiation with respect to r, and so write the Schrödinger equation as

$$\frac{1}{r^2}\Lambda^2\psi = -\frac{2mE}{\hbar^2}\psi$$

or, because $I = mr^2$, as

$$\Lambda^2\psi = -\varepsilon\psi \qquad \varepsilon = \frac{2IE}{\hbar^2}$$

To verify that this expression is separable, we substitute $\psi = \Theta\Phi$:

$$\frac{\Theta}{\sin^2\theta}\frac{d^2\Phi}{d\phi^2} + \frac{\Phi}{\sin\theta}\frac{d}{d\theta}\sin\theta\frac{d\Theta}{d\theta} = -\varepsilon\Theta\Phi$$

We have made use of the fact that Θ and Φ are each functions of one variable, so the partial derivatives become complete derivatives. Division through by $\Theta\Phi$ and multiplication by $\sin^2\theta$ give

$$\frac{1}{\Phi}\frac{d^2\Phi}{d\phi^2} + \frac{\sin\theta}{\Theta}\frac{d}{d\theta}\sin\theta\frac{d\Theta}{d\theta} = -\varepsilon\sin^2\theta$$

The first term on the left depends only on ϕ and the remaining two terms depend only on θ. We met a similar situation when discussing a particle on a rectangular surface (*Justification 12.1*) and, by the same argument, the complete equation can be separated. Thus, if we set the first term equal to the numerical constant $-m_l^2$ (a constant clearly chosen with an eye to the future), the separated equations are

$$\frac{1}{\Phi}\frac{d^2\Phi}{d\phi^2} = -m_l^2$$

$$\frac{\sin\theta}{\Theta}\frac{d}{d\theta}\sin\theta\frac{d\Theta}{d\theta} + \varepsilon\sin^2\theta = m_l^2$$

The first of these two equations is the same as in *Justification 12.3*, so it has the same solutions (eqn 50). The second is much more complicated to solve, but the solutions are tabulated as the **associated Legendre functions**. The cyclic boundary conditions on θ result in the introduction of a second quantum number, l, which identifies the acceptable solutions. The presence of the quantum number m_l in the second equation implies, as we see below, that the range of acceptable values of m_l is restricted by the value of l.

As indicated in the *Justification*, solution of the Schrödinger equation shows that the acceptable wavefunctions are specified by two quantum numbers l and m_l which are restricted to the values

$$l = 0, 1, 2, \ldots \qquad m_l = l, l-1, \ldots, -l \tag{65}$$

Note that the quantum number l is non-negative and that, for a given value of l, there are $2l + 1$ permitted values of m_l. The normalized wavefunctions are usually denoted $Y_{l,m_l}(\theta,\phi)$ and are called the **spherical harmonics**. Some of the spherical harmonics are listed in Table 12.3, and their amplitudes at different points on the spherical surface are illustrated in Fig. 12.30.

Table 12.3 The spherical harmonics $Y_{l,m_l}(\theta,\phi)$

l	m_l	Y_{l,m_l}
0	0	$\left(\dfrac{1}{4\pi}\right)^{1/2}$
1	0	$\left(\dfrac{3}{4\pi}\right)^{1/2}\cos\theta$
	± 1	$\mp\left(\dfrac{3}{8\pi}\right)^{1/2}\sin\theta\, e^{\pm i\phi}$
2	0	$\left(\dfrac{5}{16\pi}\right)^{1/2}(3\cos^2\theta - 1)$
	± 1	$\mp\left(\dfrac{15}{8\pi}\right)^{1/2}\cos\theta\sin\theta\, e^{\pm i\phi}$
	± 2	$\left(\dfrac{15}{32\pi}\right)^{1/2}\sin^2\theta\, e^{\pm 2i\phi}$
3	0	$\left(\dfrac{7}{16\pi}\right)^{1/2}(5\cos^3\theta - 3\cos\theta)$
	± 1	$\mp\left(\dfrac{21}{64\pi}\right)^{1/2}(5\cos^2\theta - 1)\sin\theta\, e^{\pm i\phi}$
	± 2	$\left(\dfrac{105}{32\pi}\right)^{1/2}\sin^2\theta\cos\theta\, e^{\pm 2i\phi}$
	± 3	$\mp\left(\dfrac{35}{64\pi}\right)^{1/2}\sin^3\theta\, e^{\pm 3i\phi}$

Normalization and orthogonality:

$$\int_0^\pi\int_0^{2\pi} Y_{l',m_l'}^* Y_{l,m_l}\sin\theta\, d\theta\, d\phi = \delta_{l'l}\delta_{m_l'm_l}$$

Triple integral

$$\int_0^\pi\int_0^{2\pi} Y_{l'',m_l''}^* Y_{l',m_l'} Y_{l,m_l}\sin\theta\, d\theta\, d\phi = 0$$

unless $m_l'' = m_l + m_l'$ and l'', l', l can form a triangle.

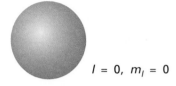

$l = 0$, $m_l = 0$

$l = 1$, $m_l = 0$

$l = 2$, $m_l = 0$

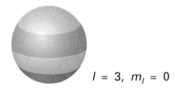

$l = 3$, $m_l = 0$

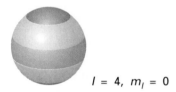

$l = 4$, $m_l = 0$

12.30 A representation of the wavefunctions of a particle on the surface of a sphere. Note that the number of nodes increases as the value of l increases. All these wavefunctions correspond to $m_l = 0$; a path around the vertical z-axis of the sphere does not cut through any nodes.

It also follows from the solution of the Schrödinger equation that the energy E of the particle is restricted to the values

$$E = l(l+1)\frac{\hbar^2}{2I} \qquad l = 0, 1, 2, \dots \tag{66}$$

We see that the energy is quantized, and that it is independent of m_l. Because there are $2l + 1$ different wavefunctions (one for each value of m_l) that correspond to the same energy, it follows that a level with quantum number l is $(2l + 1)$-fold degenerate.

(b) Angular momentum

The energy of a rotating particle is related classically to its angular momentum J by $E = J^2/2I$ (see *Further information 4*). Therefore, by comparing this equation with eqn 66, we can deduce that, because the energy is quantized, then so too is the magnitude of the angular momentum, and confined to the values

$$\text{magnitude of angular momentum} = \{l(l+1)\}^{1/2}\hbar \qquad l = 0, 1, 2, \dots \tag{67a}$$

We have already seen (in the context of rotation in a plane) that the angular momentum about the z-axis is quantized, and that it has the values

$$z\text{-component of angular momentum} = m_l\hbar \qquad m_l = l, l-1, \dots, -l \tag{67b}$$

A feature of the (real or imaginary parts of the) wavefunction $\psi_{l,m_l}(\theta, \phi)$ is that, the higher the value of l, the larger the number of nodal lines in the wavefunction (the positions at which ψ passes through 0). This feature reflects the fact that higher angular momentum implies higher kinetic energy, and therefore a more sharply buckled wavefunction. We can also see that the states corresponding to high angular momentum around the z-axis are those in which most nodal lines cut the equator: a high kinetic energy now arises from motion parallel to the equator because the curvature is greatest in that direction.

Illustration

The moment of inertia of H_2 is 4.603×10^{-48} kg m^2. It follows that

$$\frac{\hbar^2}{2I} = \frac{(1.05457 \times 10^{-34} \text{ J s})^2}{2 \times (4.603 \times 10^{-48} \text{ kg m}^2)} = 1.208 \times 10^{-21} \text{ J}$$

or 1.208 zJ (where z is the little-used but useful SI prefix zepto, denoting 10^{-21}). This energy corresponds to 0.727 kJ mol^{-1}. The first few rotational energy levels are therefore 0 ($l = 0$), 2.416 zJ ($l = 1$), 7.248 zJ ($l = 2$), and 14.496 zJ ($l = 3$). The degeneracies of these levels are 1, 3, 5, and 7, respectively (from $2l + 1$) and the magnitudes of the angular momentum of the molecule are 0, $2^{1/2}\hbar$, $6^{1/2}\hbar$, and $(12)^{1/2}\hbar$ (from eqn 67a).

Self-test 12.7 Repeat the calculation for a deuterium molecule (same bond length, approximately twice the mass).

[Energies smaller by a factor of two; same angular momenta and numbers of components]

(c) Space quantization

The result that m_l is confined to the values $l, l-1, \dots, -l$ for a given value of l means that the component of angular momentum about the z-axis may take only $2l + 1$ values. If the angular momentum is represented by a vector of length proportional to its magnitude (that

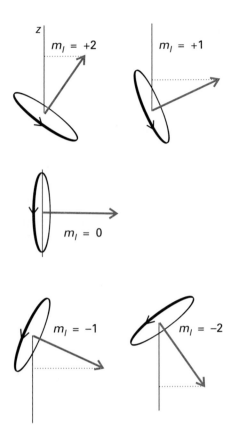

12.31 The permitted orientations of angular momentum when $l = 2$. We shall see soon that this representation is too specific because the azimuthal orientation of the vector (its angle around z) is indeterminate.

is, of length $\{l(l+1)\}^{1/2}$ units), then to represent correctly the value of the component of angular momentum, the vector must be oriented so that its projection on the z-axis is of length m_l units. In classical terms, this means that the plane of rotation of the particle can take only a discrete range of orientations (Fig. 12.31). The remarkable implication is that the orientation of a rotating body is quantized.

The quantum mechanical result that a rotating body may not take up an arbitrary orientation with respect to some specified axis (for example, an axis defined by the direction of an externally applied electric or magnetic field) is called **space quantization**. It was confirmed by an experiment first performed by Otto Stern and Walther Gerlach in 1921, who shot a beam of silver atoms through an inhomogeneous magnetic field (Fig. 12.32). The idea behind the experiment was that a rotating charged body behaves like a magnet and interacts with the applied field. According to classical mechanics, because the orientation of the angular momentum can take any value, the associated magnet can take any orientation. Because the direction in which the magnet is driven by the inhomogeneous field depends on the magnet's orientation, it follows that a broad band of atoms is expected to emerge from the region where the magnetic field acts. According to quantum mechanics, however, because the angular momentum is quantized, the associated magnet lies in a number of discrete orientations, and so several sharp bands of atoms are expected.

In their first experiment, Stern and Gerlach appeared to confirm the classical prediction. However, the experiment is difficult because collisions between the atoms in the beam blur the bands. When the experiment was repeated with a beam of very low intensity (so that collisions were less frequent), Stern and Gerlach observed discrete bands, and so confirmed the quantum prediction.

(d) The vector model

Throughout the preceding discussion, we have referred to the z-component of angular momentum (the component about an arbitrary axis, which is conventionally denoted z), and have made no reference to the x- and y-components (the components about the two axes perpendicular to z). The reason for this omission is that, because the operators for the three components do not commute with one another (Section 11.5e), the uncertainty principle forbids the simultaneous, exact specification of more than one component (unless $l = 0$). Therefore, if l_z is known, it is impossible to ascribe values to the other two components. It follows that the illustration in Fig. 12.31, which is summarized in Fig. 12.33a, gives a false impression of the state of the system, because it suggests definite values for the x- and y-components. A better picture must reflect the impossibility of specifying l_x and l_y if l_z is known.

The vector model of angular momentum uses pictures like that in Fig. 12.33b. The cones are drawn with side $\{l(l+1)\}^{1/2}$ units, and represent the magnitude of the angular momentum. Each cone has a definite projection (of m_l units) on the z-axis, representing the system's precise value of l_z. The l_x and l_y projections, however, are indefinite. The vector

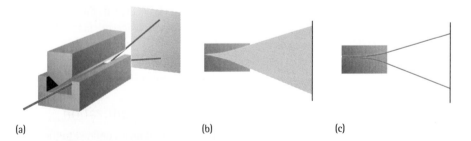

(a)   (b)   (c)

12.32 (a) The experimental arrangement for the Stern–Gerlach experiment: the magnet provides an inhomogeneous field. (b) The classically expected result. (c) The observed outcome using silver atoms.

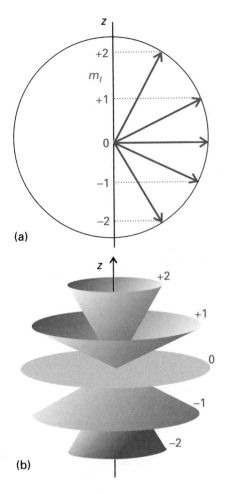

(a)

(b)

12.33 (a) A summary of Fig. 12.31. However, because the azimuthal angle of the vector around the z-axis is indeterminate, a better representation is as in (b), where each vector lies at an unspecified azimuthal angle on its cone.

representing the state of angular momentum can be thought of as lying with its tip on any point on the mouth of the cone. At this stage it should not be thought of as sweeping round the cone; that aspect of the model will be added later when we allow the picture to convey more information.

12.8 Spin

Stern and Gerlach observed *two* bands of Ag atoms in their experiment. This observation seems to conflict with one of the predictions of quantum mechanics, because an angular momentum l gives rise to $2l + 1$ orientations, which is equal to 2 only if $l = \frac{1}{2}$, contrary to the conclusion that l must be an integer. The conflict was resolved by the suggestion that the angular momentum they were observing was not due to orbital angular momentum (the motion of an electron around the atomic nucleus) but arose instead from the motion of the electron about its own axis. This intrinsic angular momentum of the electron is called its **spin**.

The spin of an electron does not have to satisfy the same boundary conditions as those for a particle circulating around a central point, so the quantum number for spin angular momentum is subject to different restrictions. To distinguish this spin angular momentum from orbital angular momentum we use the quantum number s (in place of l; like l, s is a non-negative number) and m_s for the projection on the z-axis. The magnitude of the spin angular momentum is $\{s(s + 1)\}^{1/2}\hbar$ and the component $m_s\hbar$ is restricted to the $2s + 1$ values

$$m_s = s, s - 1, \ldots, -s \tag{68}$$

The detailed analysis of the spin of a particle is sophisticated (it is rooted in special relativity), and shows that the property should not be taken to be an actual spinning motion. However, that picture can be very useful when used with care. For an electron it turns out that only one value of s is allowed, namely $s = \frac{1}{2}$, corresponding to an angular momentum of magnitude $\frac{1}{2}\sqrt{3}\hbar = 0.866\hbar$. This spin angular momentum is an intrinsic property of the electron, like its rest mass and its charge, and every electron has exactly the same value: the magnitude of the spin angular momentum of an electron cannot be changed. The spin may lie in $2s + 1 = 2$ different orientations (Fig. 12.34). One orientation corresponds to $m_s = +\frac{1}{2}$ (this state is often denoted α or $\uparrow$); the other orientation corresponds to $m_s = -\frac{1}{2}$ (this state is denoted β or $\downarrow$).

The outcome of the Stern–Gerlach experiment can now be explained if we suppose that each Ag atom possesses an angular momentum due to the spin of a single electron, because the two bands of atoms then correspond to the two spin orientations. Why the atoms behave like this will be explained in Chapter 13.[8]

Like the electron, other elementary particles have characteristic, constant spin angular momenta. For example, protons and neutrons are spin-$\frac{1}{2}$ particles (that is, $s = \frac{1}{2}$) and invariably spin with angular momentum $\left(\frac{3}{4}\right)^{1/2}\hbar$. Because the masses of a proton and a neutron are so much greater than the mass of an electron, yet they all have the same spin angular momentum, the classical picture would be of these two particles spinning much more slowly than an electron. Some elementary particles have $s = 1$, and so have an intrinsic angular momentum of magnitude $2^{1/2}\hbar$. Some mesons are spin-1 particles (as are some atomic nuclei), but for our purposes the most important spin-1 particle is the photon.[9] We shall see the importance of photon spin in the next chapter.

8 It is already probably familiar from introductory chemistry that the ground-state configuration of a silver atom is $[Kr]4d^{10}5s^1$, a single unpaired electron outside a closed shell.

9 A photon has zero rest mass, zero charge, an energy $h\nu$, a linear momentum h/λ or $h\nu/c$, an intrinsic angular momentum of $2^{1/2}\hbar$, and travels at the speed c.

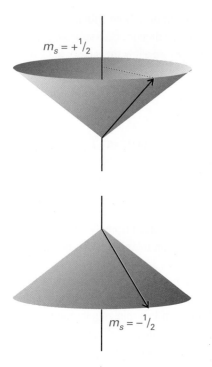

12.34 An electron spin ($s = \frac{1}{2}$) can take only two orientations with respect to a specified axis. An α electron (top) is an electron with $m_s = +\frac{1}{2}$; a β electron (bottom) is an electron with $m_s = -\frac{1}{2}$. The vector representing the magnitude of the spin angular momentum lies at an angle of 55° to the z-axis (more precisely, at $\arccos(1/3^{1/2})$).

Particles with half-integral spin are called **fermions** and those with integral spin (including 0) are called **bosons**. Thus, electrons and protons are fermions and photons are bosons. It is a very deep feature of nature that all the elementary particles that constitute matter are fermions whereas the fundamental particles that are responsible for the forces that bind fermions together are all bosons. (Photons, for example, transmit the electromagnetic force that binds together electrically charged particles.) Matter, therefore, is an assembly of fermions held together by forces conveyed by bosons.

The properties of angular momentum that we have developed are set out in Table 12.4. As mentioned there, when we use the quantum numbers l and m_l we shall mean orbital angular momentum (circulation in space); when we use s and m_s we shall mean spin angular momentum (intrinsic angular momentum); and when we use j and m_j we shall mean either (or, in some contexts to be described in Chapter 13, a combination of orbital and spin momenta).

Table 12.4 Angular momentum

The quantum numbers:
    Orbital angular momentum quantum number: $l = 0, 1, 2, \ldots$
    Orbital magnetic quantum number: $m_l = 0, \pm 1, \ldots, \pm l$
    Spin angular momentum quantum number: $s = \frac{1}{2}$
    Spin magnetic quantum number: $m_s = \pm \frac{1}{2}$

In general:
    Angular momentum quantum number: j
    Magnetic quantum number: m_j

The magnitude of the angular momentum is equal to $\{j(j+1)\}^{1/2}\hbar$ and the z-component of angular momentum is equal to $m_j\hbar$ with the $2j + 1$ values $j, j - 1, \ldots, -j$.

For the total angular momentum of a composite system see Section 13.8.

Checklist of key ideas

Translational motion

12.1 A particle in a box
- [ ] particle in a box
- [ ] infinite square well
- [ ] boundary conditions
- [ ] quantum number
- [ ] zero-point energy (9)
- [ ] correspondence principle
- [ ] orthogonal (12a)
- [ ] Dirac bracket notation (12b, 39)
- [ ] Kronecker delta
- [ ] energy levels (7)
- [ ] wavefunctions (7)

12.2 Motion in two dimensions
- [ ] separation of variables technique
- [ ] degeneracy

12.3 Tunnelling
- [ ] tunnelling
- [ ] transmission probability

Vibrational motion

- [ ] harmonic motion
- [ ] force constant
- [ ] parabolic potential

12.4 The energy levels
- [ ] energy levels (31)
- [ ] wavefunctions (34)

12.5 The wavefunctions
- [ ] matrix element
- [ ] Gaussian function
- [ ] Hermite polynomial
- [ ] virial theorem (45)
- [ ] error function (46)

Rotational motion

12.6 Rotation in two dimensions
- [ ] cyclic boundary condition
- [ ] vector representation
- [ ] energy levels (48)
- [ ] wavefunctions (50)

12.7 Rotation in three dimensions
- [ ] laplacian (60)
- [ ] colatitude
- [ ] azimuth
- [ ] legendrian (64)
- [ ] associated Legendre function
- [ ] spherical harmonics
- [ ] space quantization
- [ ] energy levels (66)

12.8 Spin
- [ ] spin
- [ ] fermion
- [ ] boson

Further reading

Articles of general interest

G.L. Breneman, The two-dimensional particle in a box. *J. Chem. Educ.* **67**, 866 (1990).

H.F. Blanck, Introduction to a quantum mechanical harmonic oscillator using a modified particle-in-a-box problem. *J. Chem. Educ.* **69**, 98 (1992).

K. Volkamer and M.W. Lerom, More about the particle-in-a-box system: the confinement of matter and the wave–particle dualism. *J. Chem. Educ.* **69**, 100 (1992).

C.A. Hollingsworth, Accidental degeneracies of the particle in a box. *J. Chem. Educ.* **67**, 999 (1990).

W.-K. Li and S.M. Blinder, Particle in an equilateral triangle: exact solution of a nonseparable problem. *J. Chem. Educ.* **64**, 130 (1987).

T. McDermott and G. Henderson, Spherical harmonics in a cartesian frame. *J. Chem. Educ.* **67**, 915 (1990).

K.V. Mikkelsen and M.A. Ratner, Electron tunneling in solid-state electron-transfer reactions. *Chem. Rev.* **87**, 113 (1987).

Y.Q. Liang, H. Zhang, and Y.X. Dardenne, Momentum distributions for a particle in a box. *J. Chem. Educ.* **72**, 148 (1995).

G.I. Gellene, Resonant states of a one-dimensional piecewise constant potential. *J. Chem. Educ.* **72**, 1015 (1995).

Texts and sources of data and information

P.W. Atkins, *Quanta: a handbook of concepts.* Oxford University Press (1991).

P.W. Atkins and R.S. Friedman, *Molecular quantum mechanics.* Oxford University Press (1997).

P.J.E. Peebles, *Quantum mechanics.* Princeton University Press, Princeton (1992).

G.C. Schatz and M.A. Ratner, *Quantum mechanics in chemistry.* Ellis Horwood/Prentice-Hall, Hemel Hempstead (1993).

J.P. Lowe, *Quantum chemistry.* Academic Press, San Diego (1993).

R.E. Christofferson, *Basic principles and techniques of molecular quantum mechanics.* Springer, New York (1989).

D.A. McQuarrie, *Quantum chemistry.* University Science Books, Mill Valley (1983).

L. Pauling and E.B. Wilson, *Introduction to quantum mechanics.* McGraw-Hill, New York (1935).

Exercises

12.1 (a) Calculate the energy separations in joules, kilojoules per mole, electronvolts, and reciprocal centimetres between the levels (a) $n = 2$ and $n = 1$, (b) $n = 6$ and $n = 5$ of an electron in a box of length 1.0 nm.

12.1 (b) Calculate the energy separations in joules, kilojoules per mole, electronvolts, and reciprocal centimetres between the levels (a) $n = 3$ and $n = 1$, (b) $n = 7$ and $n = 6$ of an electron in a box of length 1.50 nm.

12.2 (a) Calculate the probability that a particle will be found between $0.49L$ and $0.51L$ in a box of length L when it has (a) $n = 1$, (b) $n = 2$. Take the wavefunction to be a constant in this range.

12.2 (b) Calculate the probability that a particle will be found between $0.65L$ and $0.67L$ in a box of length L when it has (a) $n = 1$, (b) $n = 2$. Take the wavefunction to be a constant in this range.

12.3 (a) Calculate the expectation values of p and p^2 for a particle in the state $n = 1$ in a square-well potential.

12.3 (b) Calculate the expectation values of p and p^2 for a particle in the state $n = 2$ in a square-well potential.

12.4 (a) What are the most likely locations of a particle in a box of length L in the state $n = 3$?

12.4 (b) What are the most likely locations of a particle in a box of length L in the state $n = 5$?

12.5 (a) Consider a particle in a cubic box. What is the degeneracy of the level that has an energy three times that of the lowest level?

12.5 (b) Consider a particle in a cubic box. What is the degeneracy of the level that has an energy $\frac{14}{3}$ times that of the lowest level?

12.6 (a) Calculate the percentage change in a given energy level of a particle in a cubic box when the length of the edge of the cube is decreased by 10 per cent in each direction.

12.6 (b) A nitrogen molecule is confined in a cubic box of volume 1.00 m^3. Assuming that the molecule has an energy equal to $\frac{3}{2}kT$ at $T = 300$ K, what is the value of $n = (n_x^2 + n_y^2 + n_z^2)^{1/2}$ for this particle? What is the energy separation between the levels n and $n + 1$? What is its de Broglie wavelength? Would it be appropriate to describe this particle as classical?

12.7 (a) Calculate the zero-point energy of a harmonic oscillator consisting of a particle of mass 2.33×10^{-26} kg and force constant $155 \ N\,m^{-1}$.

12.7 (b) Calculate the zero-point energy of a harmonic oscillator consisting of a particle of mass 5.16×10^{-26} kg and force constant $285 \ N\,m^{-1}$.

12.8 (a) For a harmonic oscillator consisting of a particle of mass 1.33×10^{-25} kg, the difference in adjacent energy levels is 4.82×10^{-21} J. Calculate the force constant of the oscillator.

12.8 (b) For a harmonic oscillator consisting of a particle of mass 2.88×10^{-25} kg, the difference in adjacent energy levels is 3.17 zJ. Calculate the force constant of the oscillator.

12.9 (a) Calculate the wavelength of a photon needed to excite a transition between neighbouring energy levels of a harmonic oscillator of mass equal to that of a proton (1.0078 u) and force constant $855 \ N\,m^{-1}$.

12.9 (b) Calculate the wavelength of a photon needed to excite a transition between neighbouring energy levels of a harmonic oscillator of mass equal to that of an oxygen atom (15.9949 u) and force constant $544 \ N\,m^{-1}$.

12.10 (a) Refer to Exercise 12.9a and calculate the wavelength that would result from doubling the mass of the particle.

12.10 (b) Refer to Exercise 12.9b and calculate the wavelength that would result from doubling the mass of the particle.

12.11 (a) Calculate the minimum excitation energies of (a) a pendulum of length 1.0 m on the surface of the Earth, (b) the balance-wheel of a clockwork watch ($\nu = 5$ Hz).

12.11 (b) Calculate the minimum excitation energies of (a) the 33 kHz quartz crystal of a watch, (b) the bond between two O atoms in O_2, for which $k = 1177 \ N\,m^{-1}$.

12.12 (a) Confirm that the wavefunction for the ground state of a one-dimensional linear harmonic oscillator given in Table 12.1 is a solution of the Schrödinger equation for the oscillator and that its energy is $\frac{1}{2}\hbar\omega$.

12.12 (b) Confirm that the wavefunction for the first excited state of a one-dimensional linear harmonic oscillator given in Table 12.1 is a solution of the Schrödinger equation for the oscillator and that its energy is $\frac{3}{2}\hbar\omega$.

12.13 (a) Assuming that the vibrations of a $^{35}Cl_2$ molecule are equivalent to those of a harmonic oscillator with a force constant $k = 329 \ N\,m^{-1}$, what is the zero-point energy of vibration of this molecule? The mass of a ^{35}Cl atom is 34.9688 u.

12.13 (b) Assuming that the vibrations of a $^{14}N_2$ molecule are equivalent to those of a harmonic oscillator with a force constant $k = 2293.8 \ N\,m^{-1}$, what is the zero-point energy of vibration of this molecule? The mass of a ^{14}N atom is 14.0031 u.

12.14 (a) The wavefunction, $\psi(\phi)$, for the motion of a particle in a ring is of the form $\psi = Ne^{im_l\phi}$. Determine the normalization constant, N.

12.14 (b) Confirm that wavefunctions for a particle in a ring with different values of the quantum number m_l are orthogonal.

12.15 (a) A point mass rotates in a circle with $l = 1$. Calculate the magnitude of its angular momentum and the possible projections of the angular momentum on an arbitrary axis.

12.15 (b) A point mass rotates in a circle with $l = 2$. Calculate the magnitude of its angular momentum and the possible projections of the angular momentum on an arbitrary axis.

12.16 (a) Draw scale vector diagrams to represent the states (a) $s = \frac{1}{2}$, $m_s = +\frac{1}{2}$, (b) $l = 1$, $m_l = +1$, (c) $l = 2$, $m_l = 0$.

12.16 (b) Draw the vector diagram for all the permitted states of a particle with $l = 6$.

Problems

Numerical problems

12.1 Calculate the separation between the two lowest levels for an O_2 molecule in a one-dimensional container of length 5.0 cm. At what value of n does the energy of the molecule reach $\frac{1}{2}kT$ at 300 K, and what is the separation of this level from the one immediately below?

12.2 To a crude first approximation, a π electron in a linear polyene may be considered to be a particle in a one-dimensional box. The polyene β-carotene contains 22 conjugated C atoms, and the average internuclear distance is 140 pm. Each state up to $n = 11$ is occupied by two electrons. Calculate (a) the separation in energy between the ground state and the first excited state in which one electron occupies the state with $n = 12$, (b) the frequency of the radiation required to produce a transition between these two states, and (c) the total probability of finding an electron between C atoms 11 and 12 in the ground state of the 22-electron molecule.

12.3 The mass to use in the expression for the vibrational frequency of a diatomic molecule is the effective mass $\mu = m_A m_B/(m_A + m_B)$, where m_A and m_B are the masses of the individual atoms. The following data on the infrared absorption wavenumbers (in cm^{-1}) of molecules is taken from *Spectra of diatomic molecules*, G. Herzberg, van Nostrand (1950):

$H^{35}Cl$	$H^{81}Br$	HI	CO	NO
2990	2650	2310	2170	1904

Calculate the force constants of the bonds and arrange them in order of increasing stiffness.

12.4 The rotation of an $^1H^{127}I$ molecule can be pictured as the orbital motion of an H atom at a distance 160 pm from a stationary I atom. (This is quite a good picture; to be precise, both atoms rotate around their common centre of mass, which is very close to the I nucleus.) Suppose that the molecule rotates only in a plane. Calculate the energy needed to excite the molecule into rotation. What, apart from 0, is the minimum angular momentum of the molecule?

12.5 Calculate the energies of the first four rotational levels of $^1H^{127}I$ free to rotate in three dimensions, using for its moment of inertia $I = \mu R^2$, with $\mu = m_H m_I/(m_H + m_I)$ and $R = 160$ pm.

Theoretical problems

12.6 Set up the Schrödinger equation for a particle of mass m in a three-dimensional square well with sides L_1, L_2, and L_3. Show that the wavefunction is defined by three quantum numbers and that the Schrödinger equation is separable. Find the energy levels, and specialize the result to a cubic box of side L.

12.7 The wavefunction inside a long barrier of height V is $\psi = Ne^{-\kappa x}$. Calculate (a) the probability that the particle is inside the barrier and (b) the average penetration depth of the particle into the barrier.

12.8 Confirm that a function of the form e^{-gx^2} is a solution of the Schrödinger equation for the ground state of a harmonic oscillator and find an expression for g in terms of the mass and force constant of the oscillator.

12.9 Calculate the mean kinetic energy of a harmonic oscillator using the relations in Table 12.1.

12.10 Calculate the values of $\langle x^3 \rangle$ and $\langle x^4 \rangle$ for a harmonic oscillator using the relations in Table 12.1.

12.11 Determine the values of $\Delta x = \{\langle x^2 \rangle - \langle x \rangle^2\}^{1/2}$ and $\Delta p = \{\langle p^2 \rangle - \langle p \rangle^2\}^{1/2}$ for (a) a particle in a box of length L and (b) a harmonic oscillator. Discuss these quantities with reference to the uncertainty principle.

12.12 We shall see in Chapter 16 that the intensities of spectroscopic transitions between the vibrational states of a molecule are proportional to the square of the integral $\int \psi_{v'} x \psi_v \, dx$ over all space. Use the relations between Hermite polynomials given in Table 12.1 to show that the only permitted transitions are those for which $v' = v \pm 1$ and evaluate the integral in these cases.

12.13 Use the virial theorem to obtain an expression for the relation between the mean kinetic and potential energies of an electron in a hydrogen atom.

12.14 Evaluate the z-component of the angular momentum and the kinetic energy of a particle on a ring that is described by the

(unnormalized) wavefunctions (a) $e^{i\phi}$, (b) $e^{-2i\phi}$, (c) $\cos \phi$, and (d) $(\cos \chi)e^{i\phi} + (\sin \chi)e^{-i\phi}$.

12.15 Confirm that the spherical harmonics (a) $Y_{0,0}$, (b) $Y_{2,-1}$, and (c) $Y_{3,+3}$ satisfy the Schrödinger equation for a particle free to rotate in three dimensions, and find its energy and angular momentum in each case.

12.16 Confirm that $Y_{3,+3}$ is normalized to 1. (The integration required is over the surface of a sphere.)

12.17 Derive an expression in terms of l and m_l for the half-angle of the apex of the cone used to represent an angular momentum according to the vector model. Evaluate the expression for an α spin. Show that the minimum possible angle approaches 0 as $l \rightarrow \infty$.

12.18 Show that the function $f = \cos ax \cos by \cos cz$ is an eigenfunction of ∇^2, and determine its eigenvalue.

12.19 Derive (in Cartesian coordinates) the quantum mechanical operators for the three components of angular momentum starting from the classical definition of angular momentum, $l = r \times p$. Show that any two of the components do not mutually commute, and find their commutator.

12.20 Starting from the definition $l_z = xp_y - yp_x$, prove that in spherical polar coordinates $\hat{l}_z = -i\hbar \partial/\partial \phi$.

Additional problems supplied by Carmen Giunta and Charles Trapp

12.21 Scanning tunnelling microscopy is an imaging technique based on detecting electrons tunnelling across the vacuum between a conducting sample and a conducting probe tip. The tunnelling current is very sensitive to the distance between the tip and the sample, so sensitive that imaging of atoms has been accomplished through this technique. To get an idea of the distance dependence of this tunnelling current, suppose that the wavefunction of the electron in the gap between sample and tip is given by $\psi = Be^{-\kappa x}$, where $\kappa = \{2m_e(V - E)/\hbar^2\}^{1/2}$; take $V - E$ to be 2.0 eV. By what factor would the current drop if the probe is moved from 0.50 nm to 0.60 nm from the surface?

12.22 A particle is confined to move in a one-dimensional box of length L. (a) If the particle is classical, show that the average value of x is equal to $\frac{1}{2}L$, and that the root mean square value is $L/3^{1/2}$. (b) Show that, for large values of n, a quantum particle approaches the classical values. This result is an example of a very general principle called the *correspondence principle*, which states that, for very large values of the quantum numbers, quantum mechanics approaches classical mechanics.

13 Atomic structure and atomic spectra

The principles of quantum mechanics introduced in the preceding two chapters are now used to describe the internal structures of atoms. We see what experimental information is available from a study of the spectrum of atomic hydrogen. Then we set up the Schrödinger equation for an electron in an atom and separate it into angular and radial parts. The wavefunctions obtained are the 'atomic orbitals' of hydrogenic atoms. Next, we use these hydrogenic atomic orbitals to describe the structures of many-electron atoms. In conjunction with the Pauli exclusion principle, we account for the periodicity of atomic properties. The spectra of many-electron atoms are more complicated than those of hydrogen, but the same principles apply. We see in the closing sections of the chapter how such spectra are described in terms of term symbols, the origin of the finer details of their appearance, and the effects on them of an applied magnetic field.

In this chapter we see how to use quantum mechanics to describe the **electronic structure** of an atom, the arrangement of electrons around a nucleus. The concepts we meet are of central importance for understanding the structures and reactions of atoms and molecules, and hence have extensive chemical applications. We need to distinguish between two types of atoms. A **hydrogenic atom** is a one-electron atom or ion of general atomic number Z; examples of hydrogenic atoms are H, He^+, Li^{2+}, and U^{91+}. A **many-electron atom** is an atom or ion with more than one electron; examples include all neutral atoms other than H. So even He, with only two electrons, is a many-electron atom. Hydrogenic atoms are important because their Schrödinger equations can be solved exactly. They also provide a set of concepts that are used to describe the structures of many-electron atoms and, as we shall see in the next chapter, the structures of molecules too.

One of the principal experimental techniques for determining the electronic structures of atoms is **spectroscopy**, the detection and analysis of the electromagnetic radiation absorbed or emitted by a species. The record of spectral intensity as a function of frequency (ν), wavelength (λ), or wavenumber ($\tilde{\nu}$)[1] of the radiation emitted or absorbed by an atom or a

---

1   The relation between these quantities was described in the *Introduction*: $\nu = c/\lambda$, $\tilde{\nu} = 1/\lambda = \nu/c$.

molecule is called its **spectrum** (from the Greek word for appearance; plural 'spectra'). The spectrum of an atom consists of a series of 'lines', or sharply defined emission or absorption peaks.

The structure and spectra of hydrogenic atoms

When an electric discharge is passed through gaseous hydrogen, the H_2 molecules are dissociated and the energetically excited H atoms that are produced emit light of discrete frequencies (Fig. 13.1). The first important contribution to the interpretation of this spectrum was made by the Swiss schoolteacher Johann Balmer, who pointed out in 1885 that (in modern terms) the wavenumbers of the lines in the visible region fit the expression

$$\tilde{\nu} \propto \frac{1}{2^2} - \frac{1}{n^2} \qquad n = 3, 4, \ldots$$

The transitions this formula describes are now called the **Balmer series**. When further lines were discovered in the ultraviolet, giving the **Lyman series**, and in the infrared, the **Paschen series**, the Swedish spectroscopist Johannes Rydberg noted (in 1890) that all of them could be fitted to the expression

$$\tilde{\nu} = \mathcal{R}_H \left(\frac{1}{n_1^2} - \frac{1}{n_2^2} \right) \qquad \mathcal{R}_H = 109\,677 \text{ cm}^{-1} \tag{1}$$

with $n_1 = 1$ (the Lyman series), 2 (the Balmer series), and 3 (the Paschen series), and that in each case $n_2 = n_1 + 1, n_1 + 2, \ldots$. The constant $\mathcal{R}_H$ is now called the **Rydberg constant** for the hydrogen atom.

Illustration
....

The transition with the longest wavelength (lowest wavenumber) in the Lyman series ($n_1 = 1$) is the one with $n_2 = 2$; its wavenumber is

$$\tilde{\nu} = \mathcal{R}_H \left(\frac{1}{1^2} - \frac{1}{2^2} \right) = (109\,677 \text{ cm}^{-1}) \times \tfrac{3}{4} = 82\,258 \text{ cm}^{-1}$$

Its wavelength is therefore

$$\lambda = \frac{1}{\tilde{\nu}} = \frac{1}{8.2258 \times 10^6 \text{ m}^{-1}} = 1.2157 \times 10^{-7} \text{ m}$$

or 121.57 nm, in the vacuum ultraviolet region of the spectrum.
....

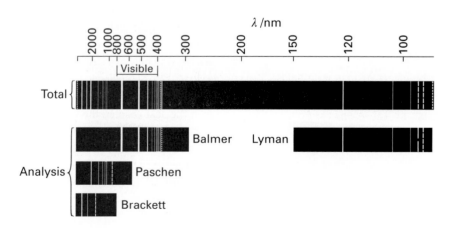

13.1 The spectrum of atomic hydrogen. Both the observed spectrum and its resolution into overlapping series are shown. Note that the Balmer series lies in the visible region.

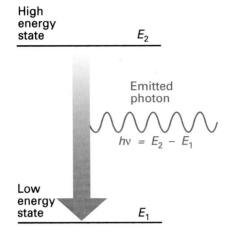

13.2 Energy is conserved when a photon is emitted, so the difference in energy of the atom before and after the emission event must be equal to the energy of the photon emitted.

Self-test 13.1 Calculate the shortest wavelength transition in the Paschen series.

[821 nm]

The form of eqn 1 strongly suggests that the wavenumber of each spectral line can be written as the difference of two **terms**, each of the form

$$T_n = \frac{\mathcal{R}_H}{n^2} \tag{2}$$

The **Ritz combination principle** states that *the wavenumber of any spectral line is the difference between two terms*. We say that two terms T_1 and T_2 'combine' to produce a spectral line of wavenumber

$$\tilde{\nu} = T_1 - T_2 \tag{3}$$

The Ritz combination principle applies to all types of atoms and molecules, but only for hydrogenic atoms do the terms have the simple form (constant)/n^2. The Ritz combination principle is readily explained in terms of photons and the conservation of energy. Thus, a spectroscopic line arises from the transition of an atom from one energy level (a term) to another (another term) with the emission of the difference in energy as a photon (Fig. 13.2). This interpretation leads to the **Bohr frequency condition**, which states that, when an atom changes its energy by ΔE, the difference is carried away as a photon of frequency ν, where

$$\Delta E = h\nu \tag{4}$$

Thus, if each spectroscopic term represents an energy hcT, the difference in energy when the atom undergoes a transition between two terms is $\Delta E = hcT_1 - hcT_2$, and the frequency of the light emitted is given by $\nu = cT_1 - cT_2$. This expression rearranges into the Ritz formula when expressed in terms of wavenumbers (on division by c).

Because spectroscopic observations show that electromagnetic radiation is absorbed and emitted by atoms only at certain wavenumbers, it follows that only certain energy states of atoms are permitted. Our tasks in the first part of this chapter are to determine the origin of this energy quantization, to find the permitted energy levels, and to account for the value of $\mathcal{R}_H$.

13.1 The structure of hydrogenic atoms

The Coulomb potential energy of an electron in a hydrogenic atom of atomic number Z (and nuclear charge Ze) is

$$V = -\frac{Ze^2}{4\pi\varepsilon_0 r} \tag{5}$$

where r is the distance of the electron from the nucleus and ε_0 is the vacuum permittivity. The hamiltonian for the electron and a nucleus of mass m_N is therefore

$$H = \hat{E}_{K,\text{electron}} + \hat{E}_{K,\text{nucleus}} + \hat{V}$$
$$= -\frac{\hbar^2}{2m_e}\nabla_e^2 - \frac{\hbar^2}{2m_N}\nabla_N^2 - \frac{Ze^2}{4\pi\varepsilon_0 r} \tag{6}$$

The subscripts on ∇^2 indicate differentiation with respect to the electron or nuclear coordinates.

(a) The separation of internal motion

Physical intuition suggests that the full Schrödinger equation ought to separate into two equations, one for the motion of the atom as a whole through space and the other for the motion of the electron relative to the nucleus. We have already solved the first of these equations, because it corresponds to the free translational motion of a particle of mass $m = m_e + m_N$ (Section 11.5). The initial strategy of the calculation is therefore to separate the relative motion of the electron and the nucleus from the motion of the atom as a whole. As we show in the following *Justification*, the resulting expression for the hamiltonian for the internal motion of the electron relative to the nucleus is

$$H = -\frac{\hbar^2}{2\mu}\nabla^2 - \frac{Ze^2}{4\pi\varepsilon_0 r} \qquad \frac{1}{\mu} = \frac{1}{m_e} + \frac{1}{m_N} \tag{7}$$

The quantity μ is called the **reduced mass**. The reduced mass is very similar to the electron mass because m_N, the mass of the nucleus, is much larger than the mass of an electron, so $1/\mu \approx 1/m_e$. In all except the most precise work, the reduced mass can be replaced by m_e.

Justification 13.1

Consider a one-dimensional system in which the potential energy depends only on the separation of the two particles; the total energy is

$$E = \frac{p_1^2}{2m_1} + \frac{p_2^2}{2m_2} + V$$

where $p_1 = m_1\dot{x}_1$ and $p_2 = m_2\dot{x}_2$, the dot signifying differentiation with respect to time. The centre of mass (Fig. 13.3) is located at

$$X = \frac{m_1}{m}x_1 + \frac{m_2}{m}x_2 \qquad m = m_1 + m_2$$

and the separation of the particles is $x = x_1 - x_2$. It follows that

$$x_1 = X + \left(\frac{m_2}{m}\right)x \qquad x_2 = X - \left(\frac{m_1}{m}\right)x$$

The linear momenta of the particles can be expressed in terms of x and X:

$$p_1 = m_1\dot{x}_1 = m_1\dot{X} + \left(\frac{m_1 m_2}{m}\right)\dot{x}$$

$$p_2 = m_2\dot{x}_2 = m_2\dot{X} - \left(\frac{m_1 m_2}{m}\right)\dot{x}$$

Then it follows that

$$\frac{p_1^2}{2m_1} + \frac{p_2^2}{2m_2} = \tfrac{1}{2}m\dot{X}^2 + \tfrac{1}{2}\mu\dot{x}^2$$

where μ is given in eqn 7. By writing $P = m\dot{X}$ for the linear momentum of the system as a whole and defining p as $\mu\dot{x}$, we find

$$E = \frac{P^2}{2m} + \frac{p^2}{2\mu} + V$$

The corresponding hamiltonian (generalized to three dimensions) is therefore

$$H = -\frac{\hbar^2}{2m}\nabla_{c.m.}^2 - \frac{\hbar^2}{2\mu}\nabla^2 + V$$

where the first term differentiates with respect to the centre of mass coordinates and the second with respect to the relative coordinates.

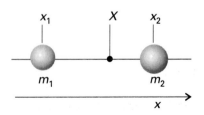

13.3 The coordinates used for discussing the separation of the relative motion of two particles from the motion of the centre of mass.

Now we write the overall wavefunction as the product $\psi_{total} = \psi_{c.m.}\psi$, where the first factor is a function of only the centre of mass coordinates and the second is a function of only the relative coordinates. The overall Schrödinger equation, $H\psi_{total} = E_{total}\psi_{total}$, then separates by the argument that we have used twice already:

$$-\frac{\hbar^2}{2m}\nabla_{c.m.}^2\psi_{c.m.} = E_{c.m.}\psi_{c.m.}$$

$$-\frac{\hbar^2}{2\mu}\nabla^2\psi + V\psi = E\psi$$

with $E_{total} = E_{c.m.} + E$.

From now on we consider only the internal, relative coordinates. The Schrödinger equation, $H\psi = E\psi$, is

$$-\frac{\hbar^2}{2\mu}\nabla^2\psi - \frac{Ze^2}{4\pi\varepsilon_0 r}\psi = E\psi \tag{8}$$

Because the potential energy is centrosymmetric (independent of angle), we can suspect that the equation is separable into radial and angular components. Therefore, we write

$$\psi(r,\theta,\phi) = R(r)Y(\theta,\phi) \tag{9}$$

and examine whether the Schrödinger equation can be separated into two equations, one for R and the other for Y. As shown in the *Justification* below, the equation does separate, and the equations we have to solve are

$$\Lambda^2 Y = -l(l+1)Y \tag{10}$$

$$-\frac{\hbar^2}{2\mu}\left(\frac{d^2R}{dr^2} + \frac{2}{r}\frac{dR}{dr}\right) + V_{eff}R = ER \tag{11}$$

where

$$V_{eff} = -\frac{Ze^2}{4\pi\varepsilon_0 r} + \frac{l(l+1)\hbar^2}{2\mu r^2} \tag{12}$$

Justification 13.2

The laplacian in three dimensions is given in eqn 12.63. It follows that the Schrödinger equation in eqn 8 is

$$-\frac{\hbar^2}{2\mu}\left(\frac{\partial^2}{\partial r^2} + \frac{2}{r}\frac{d}{dr} + \frac{1}{r^2}\Lambda^2\right)RY + VRY = ERY$$

Because R depends only on r and Y depends only on the angular coordinates, this equation becomes

$$-\frac{\hbar^2}{2\mu}\left(Y\frac{d^2R}{dr^2} + \frac{2Y}{r}\frac{dR}{dr} + \frac{R}{r^2}\Lambda^2 Y\right) + VRY = ERY$$

If we multiply through by r^2/RY, we obtain

$$-\frac{\hbar^2}{2\mu R}\left(r^2\frac{d^2R}{dr^2} + 2r\frac{dR}{dr}\right) + Vr^2 - \frac{\hbar^2}{2\mu Y}\Lambda^2 Y = Er^2$$

At this point we employ the usual argument. The term in Y is the only one that depends on the angular variables, so it must be a constant. When we write this constant as $\hbar^2 l(l+1)/2\mu$, eqn 12 follows immediately.

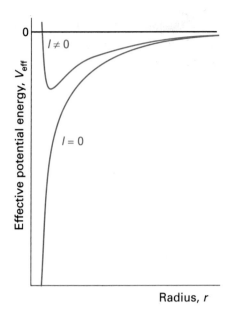

Radius, r

13.4 The effective potential energy of an electron in the hydrogen atom. When the electron has zero orbital angular momentum, the effective potential energy is the Coulombic potential energy. When the electron has nonzero orbital angular momentum, the centrifugal effect gives rise to a positive contribution which is very large close to the nucleus. We can expect the $l = 0$ and $l \neq 0$ wavefunctions to be very different near the nucleus.

Equation 10 is the same as the Schrödinger equation for a particle free to move round a central point, and we considered it in Section 12.7. The solutions are the spherical harmonics (Table 12.3), and are specified by the quantum numbers l and m_l. We consider them in more detail shortly. Equation 11 is called the **radial wave equation**. The radial wave equation is the description of the motion of a particle of mass μ in a one-dimensional region where the potential energy is V_{eff}.

(b) The radial solutions

We can anticipate some features of the shapes of the radial wavefunctions by analysing the shape of V_{eff}. The first term in eqn 12 is the Coulomb potential energy of the electron in the field of the nucleus. The second term stems from the centrifugal force that arises from the angular momentum of the electron around the nucleus. When $l = 0$, the electron has no angular momentum, and the effective potential energy is purely Coulombic and attractive at all radii (Fig. 13.4). When $l \neq 0$, the centrifugal term gives a positive contribution to the effective potential energy. When the electron is close to the nucleus ($r \approx 0$), this repulsive term, which is proportional to $1/r^2$, dominates the attractive Coulombic component, which is proportional to $1/r$, and the net effect is an effective repulsion of the electron from the nucleus. The two effective potential energies, the one for $l = 0$ and the one for $l \neq 0$, are qualitatively very different close to the nucleus; however, they are similar at large distances because the centrifugal contribution tends to zero more rapidly than the Coulombic contribution. Therefore, we can expect the solutions with $l = 0$ and $l \neq 0$ to be quite different near the nucleus but similar far away from it.

We shall not go through the technical steps of solving the radial equation (see *Further reading*). It is sufficient to know that acceptable solutions can be found only for integral values of a quantum number n, and that the allowed energies are

$$E_n = -\frac{Z^2 \mu e^4}{32\pi^2 \varepsilon_0^2 \hbar^2 n^2} \tag{13}$$

with $n = 1, 2, \ldots$.

The radial wave equation depends on l, and the radial wavefunctions, which depend on the values of both n and l (but not on m_l), all have the form

$$R(r) = (\text{polynomial in } r) \times (\text{decaying exponential in } r) \tag{14}$$

These functions are most simply written in terms of the dimensionless quantity ρ (rho), where

$$\rho = \frac{2Zr}{a_0} \qquad a_0 = \frac{4\pi\varepsilon_0 \hbar^2}{m_e e^2} \tag{15}$$

The **Bohr radius**, a_0, has the value 52.9177 pm; it is so called because the same quantity appeared in Bohr's early model of the hydrogen atom as the radius of the electron orbit of lowest energy. Specifically, the radial wavefunctions for an electron with quantum numbers n and l are the (real) functions

$$R_{n,l}(r) = N_{n,l} \left(\frac{\rho}{n}\right)^l L_{n,l} e^{-\rho/2n} \tag{16}$$

where L is a polynomial in ρ called an **associated Laguerre polynomial**. Expressions for some radial wavefunctions are given in Table 13.1 and their appearance is illustrated in Fig. 13.5. Note that, because R is proportional to ρ^l, all radial wavefunctions are zero at the nucleus unless $l = 0$.

Table 13.1 Hydrogénic radial wavefunctions

Orbital	n	l	$R_{n,l}$
$1s$	1	0	$2\left(\dfrac{Z}{a_0}\right)^{3/2} e^{-\rho/2}$
$2s$	2	0	$\dfrac{1}{2(2)^{1/2}}\left(\dfrac{Z}{a_0}\right)^{3/2}(2 - \tfrac{1}{2}\rho)e^{-\rho/4}$
$2p$	2	1	$\dfrac{1}{4(6)^{1/2}}\left(\dfrac{Z}{a_0}\right)^{3/2}\rho e^{-\rho/4}$
$3s$	3	0	$\dfrac{1}{9(3)^{1/2}}\left(\dfrac{Z}{a_0}\right)^{3/2}(6 - 2\rho + \tfrac{1}{9}\rho^2)e^{-\rho/6}$
$3p$	3	1	$\dfrac{1}{27(6)^{1/2}}\left(\dfrac{Z}{a_0}\right)^{3/2}(4 - \tfrac{1}{3}\rho)\rho e^{-\rho/6}$
$3d$	3	2	$\dfrac{1}{81(30)^{1/2}}\left(\dfrac{Z}{a_0}\right)^{3/2}\rho^2 e^{-\rho/6}$

The full wavefunction is $\psi = RY$, where Y is given in Table 12.3. In the table, $\rho = 2Zr/a_0$.

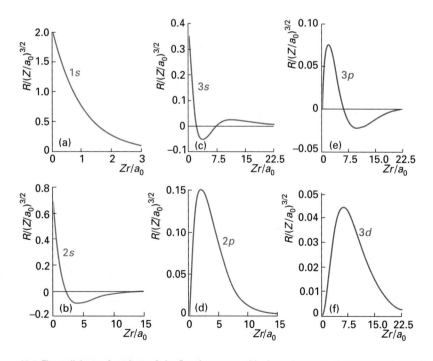

13.5 The radial wavefunctions of the first few states of hydrogenic atoms of atomic number Z. Note that the s orbitals have a nonzero and finite value at the nucleus. The horizontal scales are different in each case: orbitals with high principal quantum numbers are relatively distant from the nucleus.

Illustration

To calculate the probability density for a 1s-electron at the nucleus, we set $n = 1$, $l = 0$, $m_l = 0$ and evaluate ψ at $r = 0$:

$$\psi_{1,0,0}(0, \theta, \phi) = R_{1,0}(0)Y_{0,0}(\theta, \phi) = 2\left(\frac{Z}{a_0}\right)^{3/2}\left(\frac{1}{4\pi}\right)^{1/2}$$

The probability density is therefore

$$\psi^2_{1,0,0}(0, \theta, \phi) = \frac{Z^3}{\pi a_0^3}$$

which evaluates to 2.15×10^{-6} pm^{-3} when $Z = 1$.

Self-test 13.2 Evaluate the probability density of the electron at the nucleus for a 2s-electron.

$$[(Z/a_0)^3/8\pi]$$

13.2 Atomic orbitals and their energies

An **atomic orbital** is a one-electron wavefunction for an electron in an atom. Each hydrogenic atomic orbital is defined by three quantum numbers (Table 13.2), designated n, l, and m_l. When an electron is described by one of these wavefunctions, we say that it 'occupies' that orbital. We could go on to say that the electron is in the state $|n, l, m_l\rangle$. For instance, an electron described by the wavefunction $\psi_{1,0,0}$ and in the state $|1, 0, 0\rangle$ is said to occupy the orbital with $n = 1$, $l = 0$, and $m_l = 0$.

One quantum number, n, is called the **principal quantum number**; it can take the values $n = 1, 2, 3, \ldots$ and determines the energy of the electron:

An electron in an orbital with quantum number n has an energy given by eqn 13.

The two other quantum numbers, l and m_l, come from the angular solutions, and specify the angular momentum of the electron around the nucleus:

An electron in an orbital with quantum number l has an angular momentum of magnitude $\{l(l+1)\}^{1/2}\hbar$, with $l = 0, 1, 2, \ldots, n-1$.
An electron in an orbital with quantum number m_l has a z-component of angular momentum $m_l\hbar$, with $m_l = 0, \pm 1, \pm 2, \ldots \pm l$.

Note how the value of the principal quantum number, n, controls the maximum value of l, and how l in turn controls the range of values of m_l.

To define the state of an electron in a hydrogenic atom fully we need to specify not only the orbital it occupies but also its spin state. We saw in Section 12.8 that an electron possesses an intrinsic angular momentum that is described by the two quantum numbers s and m_s (the analogues of l and m_l). The value of s is fixed at $\frac{1}{2}$ for an electron, so we do not need to consider it further at this stage. However, m_s may be either $+\frac{1}{2}$ or $-\frac{1}{2}$, and to specify the electron's state in a hydrogenic atom we need to specify which of these values describes it. It follows that, to specify the state of an electron in a hydrogenic atom, we need to give the values of four quantum numbers, namely n, l, m_l, and m_s.

Table 13.2 Hydrogenic atoms

The wavefunctions of hydrogenic atoms depend on three quantum numbers:

Principal quantum number: $n = 1, 2, 3, \ldots$

Angular momentum quantum number: $l = 0, 1, 2, \ldots, n - 1$

Magnetic quantum number: $m_l = l, l - 1, l - 2, \ldots, -l$

The energy is related to n by

$$E_n = -\frac{hc\mathcal{R}_\text{atom}}{n^2} \qquad hc\mathcal{R}_\text{atom} = \frac{Z^2 \mu e^4}{32\pi^2 \varepsilon_0^2 \hbar^2}$$

The magnitude of the orbital angular momentum of the electron $\{l(l + 1)\}^{1/2}\hbar$ and its component on an arbitrary axis is $m_l \hbar$. Each energy level is n^2-fold degenerate.

The wavefunctions are products of radial and angular components:

$$\psi = R(r) Y(\theta, \phi)$$

The angular wavefunctions Y are the spherical harmonics (Table 12.3) and the radial wavefunctions R are the normalized associated Laguerre polynomials multiplied by an exponential factor (Table 13.1).

The selection rules for spectroscopic transitions are

$$\Delta m_l = 0, \pm 1 \quad \Delta n \text{ unrestricted} \quad \Delta l = \pm 1$$

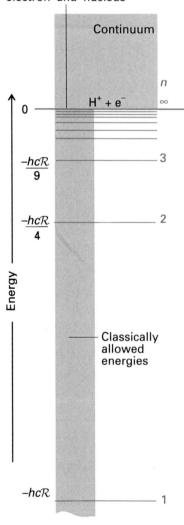

Energy of widely separated stationary electron and nucleus

13.6 The energy levels of a hydrogen atom. The values are relative to an infinitely separated, stationary electron and a proton.

(a) The energy levels

The energy levels predicted by eqn 13 are depicted in Fig. 13.6. The energies, and also the separation of neighbouring levels, are proportional to Z^2, so the levels are four times as widely apart (and the ground state four times deeper in energy) in He$^+$ ($Z = 2$) than in H ($Z = 1$). All the energies given by eqn 13 are negative. They refer to the **bound states** of the atom, in which the energy of the atom is lower than that of the infinitely separated, stationary electron and nucleus (which corresponds to the zero of energy). There are also solutions of the Schrödinger equation with positive energies. These solutions correspond to **unbound states** of the electron, the states to which an electron is raised when it is ejected from the atom by a high-energy collision or photon. The energies of the unbound electron are not quantized and form the **continuum states** of the atom.

Equation 13 is consistent with the spectroscopic result summarized by eqn 1, and we can identify the Rydberg constant for hydrogen ($Z = 1$) by writing

$$hc\mathcal{R}_\text{H} = \frac{\mu_\text{H} e^4}{32\pi^2 \varepsilon_0^2 \hbar^2} \qquad [\mathbf{17}]$$

where μ_H is the reduced mass for hydrogen. The **Rydberg constant** itself, $\mathcal{R}$, is defined by the same expression except for the replacement of μ by the mass of an electron, m_e:

$$\mathcal{R}_\text{H} = \frac{\mu_\text{H}}{m_\text{e}} \mathcal{R} \qquad \mathcal{R} = \frac{m_\text{e} e^4}{8\varepsilon_0^2 h^3 c} \qquad [\mathbf{18}]$$

Insertion of the values of the fundamental constants into the expression for $\mathcal{R}_\text{H}$ gives almost exact agreement with the experimental value. The only discrepancies arise from the neglect of relativistic corrections, which the non-relativistic Schrödinger equation ignores.

(b) Ionization energies

The ionization energy, I, of an element is the minimum energy required to remove an electron from the **ground state**, the state of lowest energy, of one of its atoms. The ground state of hydrogen is the state with $n = 1$, which has energy

$$E_1 = -hc\mathcal{R}_\text{H}$$

The atom is ionized when the electron has been excited to the level corresponding to $n = \infty$ (see Fig. 13.6). Therefore, the energy that must be supplied is

$$I = hc\mathcal{R}_H \tag{19}$$

The value of I is 2.179 aJ (a, for atto, is the prefix that denotes 10^{-18}), which corresponds to 13.60 eV.

Example 13.1 Measuring an ionization energy spectroscopically

The spectrum of atomic hydrogen shows lines at 82 259, 97 492, 102 824, 105 292, 106 632, 107 440 cm^{-1}. Determine (a) the ionization energy of the lower state, (b) the value of the Rydberg constant.

Method The spectroscopic determination of ionization energies depends on the determination of the series limit, the wavenumber at which the series terminates and becomes a continuum. If the upper state lies at an energy $-hc\mathcal{R}_H/n^2$, then, when the atom makes a transition to E_{lower}, a photon of wavenumber

$$\tilde{\nu} = -\frac{\mathcal{R}_H}{n^2} - E_{lower}/hc$$

is emitted. However, because $I = -E_{lower}$, it follows that

$$\tilde{\nu} = I/hc - \frac{\mathcal{R}_H}{n^2}$$

A plot of the wavenumbers against $1/n^2$ should give a straight line of slope $-\mathcal{R}_H$ and intercept I/hc. Use a computer (or a calculator) to make a least-squares fit of the data to get a result that reflects the precision of the data.

Answer The wavenumbers are plotted against $1/n^2$ in Fig. 13.7. The (least-squares) intercept lies at 109 679 cm^{-1}, so the ionization energy is 2.1788 aJ (1312.1 kJ mol^{-1}). The slope is, in this instance, numerically the same, so $\mathcal{R}_H = 109\,679$ cm^{-1}.

Comment A similar extrapolation procedure can be used for many-electron atoms (see Section 13.6).

Self-test 13.3 The spectrum of atomic deuterium shows lines at 15 238, 20 571, 23 039, 24 380 cm^{-1}. Determine (a) the ionization energy of the lower state, (b) the ionization energy of the ground state, (c) the mass of the deuteron (by expressing the Rydberg constant in terms of the reduced mass of the electron and the deuteron, and solving for the mass of the deuteron).

[(a) 328.1 kJ mol^{-1}, (b) 1312.4 kJ mol^{-1}, (c) 2.8×10^{-27} kg, a result very sensitive to $\mathcal{R}_D$.]

(c) Shells and subshells

All the orbitals of a given value of n are said to form a single **shell** of the atom. In a hydrogenic atom, all orbitals of given n, and therefore belonging to the same shell, have the same energy. It is common to refer to successive shells by letters:

$$n = \quad 1 \quad 2 \quad 3 \quad 4 \quad \cdots$$
$$\quad\quad K \quad L \quad M \quad N \quad \cdots$$

Thus, all the orbitals of the shell with $n = 2$ form the L shell of the atom, and so on.

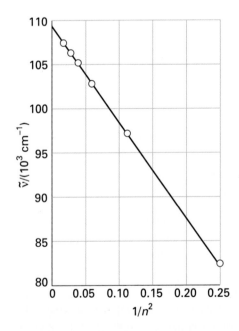

13.7 The plot of the data in Example 13.1 used to determine the ionization energy of an atom (in this case, of H).

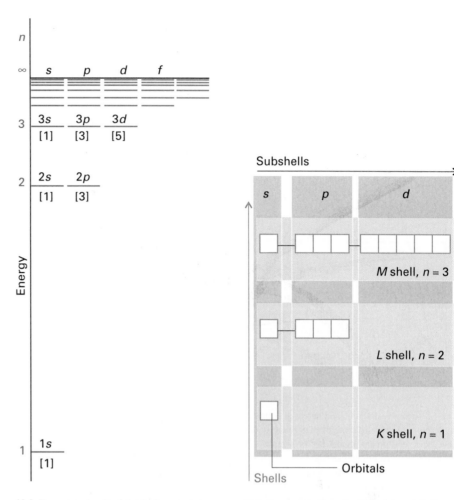

13.8 The energy levels of the hydrogen atom showing the subshells and (in square brackets) the numbers of orbitals in each subshell. In hydrogenic atoms, all orbitals of a given shell have the same energy.

13.9 The organization of orbitals into subshells (characterized by l) and shells (characterized by n).

The orbitals with the same value of n but different values of l are said to form a **subshell** of a given shell. These subshells are generally referred to by letters:

$$l = \quad 0 \quad 1 \quad 2 \quad 3 \quad 4 \quad 5 \quad 6 \quad \dots$$
$$\quad\quad s \quad p \quad d \quad f \quad g \quad h \quad i \quad \dots$$

The letters then run alphabetically (j is not used). Figure 13.8 is a version of Fig. 13.6 which shows the subshells explicitly. Because l can range from 0 to $n - 1$, giving n values in all, it follows that there are n subshells of a shell with principal quantum number n. Thus, when $n = 1$, there is only one subshell, the one with $l = 0$.

When $n = 2$, there are two subshells, the $2s$ subshell (with $l = 0$) and the $2p$ subshell (with $l = 1$). When $n = 1$ there is only one subshell, that with $l = 0$, and that subshell contains only one orbital, with $m_l = 0$ (the only value of m_l permitted). When $n = 2$, there are four orbitals, one in the s subshell with $l = 0$ and $m_l = 0$, and three in the $l = 1$ subshell with $m_l = +1, 0, -1$. When $n = 3$ there are nine orbitals (one with $l = 0$, three with $l = 1$, and five with $l = 2$). The organization of orbitals in the shells is summarized in Fig. 13.9. In general, the number of orbitals in a shell of principal quantum number n is n^2, so in a hydrogenic atom each shell is n^2-fold degenerate.

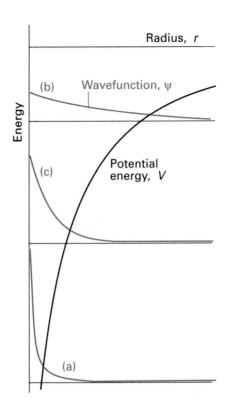

13.10 The balance of kinetic and potential energies that accounts for the structure of the ground state of hydrogen (and similar atoms). (a) The sharply curved but localized orbital has high mean kinetic energy, but low mean potential energy; (b) the mean kinetic energy is low, but the potential energy is not very favourable; (c) the compromise of moderate kinetic energy and moderately favourable potential energy.

(d) *s* orbitals

The orbital occupied in the ground state is the one with $n = 1$ (and therefore with $l = 0$ and $m_l = 0$, the only possible values of these quantum numbers when $n = 1$). From Table 13.1 we can write (for $Z = 1$):

$$\psi = \frac{1}{(\pi a_0^3)^{1/2}} e^{-r/a_0} \tag{20}$$

This wavefunction is independent of angle and has the same value at all points of constant radius, that is, the 1*s* orbital is spherically symmetrical. The wavefunction decays exponentially from a maximum value of $1/(\pi a_0^3)^{1/2}$ at the nucleus (at $r = 0$). It follows that the most probable point at which the electron will be found is at the nucleus itself.

We can understand the general form of the ground-state wavefunction by considering the contributions of the potential and kinetic energies to the total energy of the atom. The closer the electron is to the nucleus on average, the lower its average potential energy. This dependence suggests that the lowest potential energy should be obtained with a sharply peaked wavefunction that has a large amplitude at the nucleus and is zero everywhere else (Fig. 13.10). However, this shape implies a high kinetic energy, because such a wavefunction has a very high average curvature. The electron would have very low kinetic energy if its wavefunction had only a very low average curvature. However, such a wavefunction spreads to great distances from the nucleus and the average potential energy of the electron will be high. The actual ground-state wavefunction is a compromise between these two extremes: the wavefunction spreads away from the nucleus (so the expectation value of the potential energy is not as low as in the first example, but nor is it very high) and has a reasonably low average curvature (so the expectation of the kinetic energy is not very low, but nor is it as high as in the first example).

One way of depicting the probability density of the electron is to represent $|\psi|^2$ by density of shading (Fig. 13.11). A simpler procedure is to show only the **boundary surface**, the surface that captures about 90 per cent of the electron probability. For the 1*s* orbital, the boundary surface is a sphere centred on the nucleus (Fig. 13.12).

All *s* orbitals are spherically symmetric, but differ in the number of radial nodes. For instance, the 2*s* orbital has radial nodes where the polynomial factor (Table 13.1) is equal to zero:

$$2 - \frac{\rho}{2} = 0 \text{ at } \rho = 4, \text{ which implies that } r = \frac{2a_0}{Z}$$

(remember that $\rho = 2Zr/a_0$). Hence, the 2*s* orbital of a hydrogenic atom with atomic number Z has a radial node at $2a_0/Z$ (see Fig. 13.5). Similarly, the 3*s* orbital has two nodes which are found by solving

$$6 - 2\rho + (\tfrac{1}{3}\rho)^2 = 0$$

One radial node is at $1.90a_0/Z$ and the other is at $7.10a_0/Z$ (see Fig. 13.5).

The energies of the *s* orbitals increase (the electron becomes less tightly bound) as n increases because the average distance of the electron from the nucleus increases. By the virial theorem with $b = -1$ (Section 12.5b, eqn 12.45), $\langle E_K \rangle = -\frac{1}{2}\langle V \rangle$, so, even though the average kinetic energy decreases as n increases, the total energy is equal to $\frac{1}{2}\langle V \rangle$, which becomes less negative as n increases.

Example 13.2 Calculating the mean radius of an orbital

Use hydrogenic orbitals to calculate the mean radius of a 1*s* orbital.

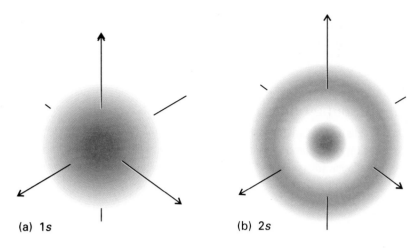

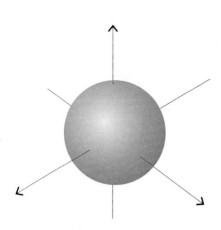

(a) 1s   (b) 2s

13.11 Representations of the 1s and 2s hydrogenic atomic orbitals in terms of their electron densities (as represented by the density of shading).

13.12 The boundary surface of an s orbital, within which there is a 90 per cent probability of finding the electron.

Method The mean radius is the expectation value

$$\langle r \rangle = \int \psi^* \hat{r} \psi \, d\tau = \int r |\psi|^2 \, d\tau$$

We therefore need to evaluate the integral using the wavefunctions given in Table 13.1 and $d\tau = r^2 \, dr \sin\theta \, d\theta \, d\phi$. The angular parts of the wavefunction are normalized in the sense that

$$\int_0^\pi \int_0^{2\pi} |Y_{l,m_l}|^2 \sin\theta \, d\theta \, d\phi = 1$$

The integral over r required is given in Example 11.6.

Answer With the wavefunction written in the form $\psi = RY$, the integration is

$$\langle r \rangle = \int_0^\infty \int_0^\pi \int_0^{2\pi} r R_{n,l}^2 |Y_{l,m_l}|^2 r^2 \, dr \sin\theta \, d\theta \, d\phi = \int_0^\infty r^3 R_{n,l}^2 \, dr$$

For a 1s orbital,

$$R_{1,0} = 2\left(\frac{Z}{a_0^3}\right)^{1/2} e^{-Zr/a_0}$$

Hence

$$\langle r \rangle = \left(\frac{4Z}{a_0^3}\right) \int_0^\infty r^3 e^{-2Zr/a_0} \, dr = \frac{3a_0}{2Z}$$

Comment The general expression for the mean radius of an orbital with quantum numbers l and n is

$$\langle r \rangle_{n,l} = n^2 \left\{ 1 + \tfrac{1}{2}\left(1 - \frac{l(l+1)}{n^2}\right) \right\} \frac{a_0}{Z}$$

The variation with n and l is shown in Fig. 13.13. Note that, for a given principal quantum number, the mean radius decreases as l increases.

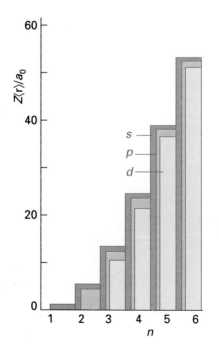

13.13 The variation of the mean radius of a hydrogenic atom with the principal and orbital angular momentum quantum numbers. Note that the mean radius lies in the order $d < p < s$ for a given value of n.

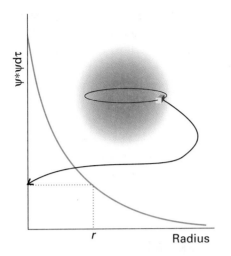

13.14 A constant-volume electron-sensitive detector (the small cube) gives its greatest reading at the nucleus, and a smaller reading elsewhere. The same reading is obtained anywhere on a circle of given radius: the *s* orbital is spherically symmetrical.

Self-test 13.4 Evaluate the mean radius (a) of a $3s$ orbital by integration, and (b) of a $3p$ orbital by using the general formula.

[(a) $27a_0/2Z$, (b) $25a_0/2Z$]

(e) Radial distribution functions

The wavefunction tells us, through the value of $|\psi|^2$, the probability of finding an electron in any region. We can imagine a probe with a volume $d\tau$ and sensitive to electrons, and which we can move around near the nucleus of a hydrogen atom. Because the probability density in the ground state of the atom is

$$|\psi|^2 \propto e^{-2Zr/a_0}$$

the reading from the detector decreases exponentially as the probe is moved out along any radius but is constant if the probe is moved on a circle of constant radius (Fig. 13.14).

Now consider the probability of finding the electron *anywhere* on a spherical shell of thickness dr at a radius r. The sensitive volume of the probe is now the volume of the shell (Fig. 13.15), which is $4\pi r^2\,dr$. The probability that the electron will be found between the inner and outer surfaces of this shell is the probability density at the radius r multiplied by the volume of the probe, or $|\psi|^2 \times 4\pi r^2\,dr$. This expression has the form $P(r)\,dr$, where

$$P(r) = 4\pi r^2\psi^2 \tag{21}$$

This expression is valid only for spherically symmetric orbitals. For all other orbitals we have to use the more general expression

$$P(r) = r^2 R(r)^2 \tag{22}$$

where $R(r)$ is the radial wavefunction for the orbital in question.

Justification 13.3

The probability of finding an electron in a volume element $d\tau$ when its wavefunction is $\psi = RY$ is $|RY|^2\,d\tau$ with $d\tau = r^2\,dr\,\sin\theta\,d\theta\,d\phi$. The total probability of finding the electron at any angle at a constant radius is the integral of this probability over the surface of a sphere of radius r, and is written $P(r)dr$, so

$$P(r)dr = \int_0^{2\pi}\int_0^{\pi} R(r)^2|Y(\theta,\phi)|^2 r^2\,dr\,\sin\theta\,d\theta\,d\phi$$

$$= r^2 R(r)^2\,dr\int_0^{2\pi}\int_0^{\pi}|Y(\theta,\phi)|^2\sin\theta\,d\theta\,d\phi = r^2 R(r)^2\,dr$$

The last equality follows from the fact that the spherical harmonics are normalized (see Example 13.2). It follows that $P(r) = r^2 R(r)^2$, as stated in the text.

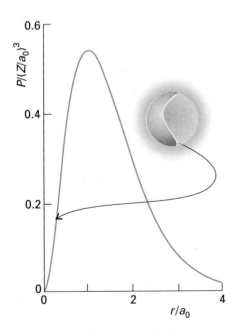

13.15 The radial distribution function P gives the probability that the electron will be found anywhere in a shell of radius r. For a $1s$ electron in hydrogen, P is a maximum when r is equal to the Bohr radius a_0. The value of P is equivalent to the reading that a detector shaped like a spherical shell would give as its radius was varied.

The **radial distribution function**, $P(r)$, is a probability density in the sense that, when it is multiplied by dr, it gives the probability of finding the electron anywhere in a shell of thickness dr at the radius r. For a $1s$ orbital,

$$P(r) = \frac{4Z^3}{a_0^3} r^2 e^{-2Zr/a_0} \tag{23}$$

Because r^2 increases with radius from zero at the nucleus, and the exponential term decreases towards zero at infinity, $P(0) = 0$ and $P(r) \rightarrow 0$ as $r \rightarrow \infty$ and passes through a maximum at an intermediate radius (see Fig. 13.15). The maximum of $P(r)$, which can be found by differentiation, marks the **most probable radius** at which the electron will be found, and for a 1s orbital in hydrogen occurs at $r = a_0$, the Bohr radius. When we carry through the same calculation for the radial distribution function of the 2s orbital in hydrogen, we find that the most probable radius is $5.2a_0 = 275$ pm. This larger value reflects the expansion of the atom as its energy increases.

Example 13.3 Calculating the most probable radius

Calculate the most probable radius, r^*, at which an electron will be found when it occupies a 1s orbital of a hydrogenic atom of atomic number Z, and tabulate the values for the one-electron species from H to Ne^{9+}.

Method We find the radius at which the radial distribution function of the hydrogenic 1s orbital has a maximum value by solving $dP/dr = 0$.

Answer The radial distribution function is given in eqn 23. It follows that

$$\frac{dP}{dr} = \frac{4Z^3}{a_0^3}\left(2r - \frac{2Zr^2}{a_0}\right)e^{-2Zr/a_0} = 0$$

at $r = r^*$. Therefore,

$$r^* = \frac{a_0}{Z}$$

Then, with $a_0 = 52.9$ pm,

	H	He^+	Li^{2+}	Be^{3+}	B^{4+}	C^{5+}	N^{6+}	O^{7+}	F^{8+}	Ne^{9+}
r^*/pm	52.9	26.5	17.6	13.2	10.6	8.82	7.56	6.614	5.88	5.29

Comment Notice how the 1s orbital is drawn towards the nucleus as the nuclear charge increases. At uranium the most probable radius is only 0.58 pm, almost 100 times closer than for hydrogen. (On a scale where $r^* = 10$ cm for H, $r^* = 1$ mm for U.) The electron then experiences strong accelerations, and relativistic effects are important.

Self-test 13.5 Find the most probable distance of a 2s electron from the nucleus in a hydrogenic atom.

$$[(3 + \sqrt{5})a_0/Z]$$

(f) p orbitals

A p electron has nonzero angular momentum (its actual magnitude is $2^{1/2}\hbar$). This momentum has a profound effect on the shape of the wavefunction close to the nucleus, for p orbitals have zero amplitude at $r = 0$. This difference from s orbitals can be understood classically in terms of the centrifugal effect of the angular momentum, which tends to fling the electron away from the nucleus. It is also what we expect from the form of the effective potential energy shown in Fig. 13.4, which rises to infinity as $r \to 0$ and excludes the wavefunction from the nucleus. The same centrifugal effect appears in all orbitals with $l > 0$ (such as the d orbitals and the f orbitals). We see from eqn 16, in fact, that close to the nucleus a wavefunction is proportional to r^l, so p wavefunctions are proportional to r, d wavefunctions to r^2, and so on (Fig. 13.16). The increasingly strong dependence on r as l increases can be regarded classically as the outcome of increasing centrifugal effects arising from the angular momentum. As remarked previously, all orbitals with $l > 0$ have zero amplitude at the nucleus, and consequently zero probability of finding the electron there.

The three 2p orbitals are distinguished by the three different values that m_l can take when $l = 1$. Because the quantum number m_l tells us the angular momentum around an

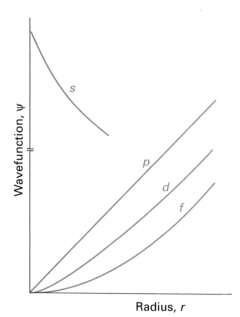

13.16 Close to the nucleus, p orbitals are proportional to r, d orbitals are proportional to r^2, and f orbitals are proportional to r^3. Electrons are progressively excluded from the neighbourhood of the nucleus as l increases. An s orbital has a finite, nonzero value at the nucleus.

axis, these different values of m_l denote orbitals in which the electron has different angular momenta around an arbitrary z-axis but the same magnitude of momentum (because l is the same for all three). The orbital with $m_l = 0$, for instance, has zero angular momentum around the z-axis. Its angular variation is proportional to $\cos\theta$, so the probability density, which is proportional to $\cos^2\theta$, has its maximum value on either side of the nucleus along the z-axis (at $\theta = 0$ and $180°$).

The wavefunction of a $2p$ orbital with $m_l = 0$ is

$$p_0 = R_{2,1}(r)Y_{1,0}(\theta, \phi) = \frac{1}{4(2\pi)^{1/2}}\left(\frac{Z}{a_0}\right)^{5/2} r\cos\theta\, e^{-Zr/2a_0}$$

$$= r\cos\theta f(r)$$

where $f(r)$ is a function only of r. Because in spherical polar coordinates $z = r\cos\theta$, this wavefunction may also be written

$$p_z = zf(r) \tag{24}$$

All p-orbitals with $m_l = 0$ have wavefunctions of this form regardless of the value of n. This way of writing the orbital is the origin of the name 'p_z-orbital': its boundary surface is shown in Fig. 13.17. The wavefunction is zero everywhere in the xy-plane, where $z = 0$, so the xy-plane is a **nodal plane** of the orbital: the wavefunction changes sign on going from one side of the plane to the other.

The wavefunctions of $2p$ orbitals with $m_l = \pm 1$ have the following form:

$$p_{\pm 1} = R_{2,1}(r)Y_{1,\pm 1}(\theta, \phi) = \mp\frac{1}{8\pi^{1/2}}\left(\frac{Z}{a_0}\right)^{5/2} re^{-Zr/2a_0}\sin\theta\, e^{\pm i\phi}$$

$$= \mp\frac{1}{2^{1/2}}r\sin\theta\, e^{\pm i\phi}f(r)$$

These functions do have angular momentum about the z-axis: as we have seen (in Section 12.6b), wavefunctions with this ϕ dependence correspond to a particle with angular momentum either clockwise or counter-clockwise around the z-axis: $e^{+i\phi}$ corresponds to clockwise rotation when viewed from below, and $e^{-i\phi}$ corresponds to counter-clockwise rotation (from the same viewpoint). They have zero amplitude where $\theta = 0$ and $180°$ (along the z-axis) and maximum amplitude at $90°$, which is in the xy-plane. To draw the functions it is usual to take the real linear combinations

$$p_x = -\frac{1}{2^{1/2}}(p_{+1} - p_{-1}) = r\sin\theta\cos\phi f(r) = xf(r)$$

$$p_y = \frac{i}{2^{1/2}}(p_{+1} + p_{-1}) = r\sin\theta\sin\phi f(r) = yf(r) \tag{25}$$

These linear combinations are standing waves with no net angular momentum around the z-axis, as they are composed of equal and opposite values of m_l. The p_x orbital has the same

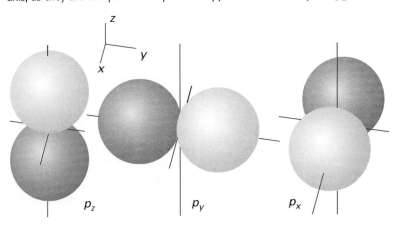

13.17 The boundary surfaces of p orbitals. A nodal plane passes through the nucleus and separates the two lobes of each orbital. The dark and light areas denote regions of opposite sign of the wavefunction.

shape as a p_z orbital, but it is directed along the x-axis (see Fig. 13.17); the p_y orbital is similarly directed along the y-axis. The wavefunction of any p orbital of a given shell can be written as a product of x, y, or z and the same radial function (which depends on the value of n).

Justification 13.4

In this remark, we justify the step of taking linear combinations of degenerate orbitals when we want to indicate a particular point. The freedom to do so rests on the fact that, whenever two or more wavefunctions correspond to the same energy, any linear combination of them is an equally valid solution of the Schrödinger equation.

Suppose ψ_1 and ψ_2 are both solutions of the Schrödinger equation with energy E; then we know that

$$H\psi_1 = E\psi_1 \qquad H\psi_2 = E\psi_2$$

Now consider the linear combination

$$\psi = c_1\psi_1 + c_2\psi_2$$

where c_1 and c_2 are arbitrary coefficients. Then it follows that

$$H\psi = H(c_1\psi_1 + c_2\psi_2) = c_1H\psi_1 + c_2H\psi_2$$
$$= c_1E\psi_1 + c_2E\psi_2 = E\psi$$

Hence, the linear combination is also a solution corresponding to the same energy E.

(g) d orbitals

When $n = 3$, l can be 0, 1, or 2. As a result, this shell consists of one $3s$ orbital, three $3p$ orbitals, and five $3d$ orbitals. The five d orbitals have $m_l = +2, +1, 0, -1, -2$ and correspond to five different angular momenta around the z-axis (but the same magnitude of angular momentum, because $l = 2$ in each case). As for the p orbitals, d orbitals with opposite values of m_l (and hence opposite senses of motion around the z-axis) may be combined in pairs to give real standing waves, and the boundary surfaces of the resulting shapes are shown in Fig. 13.18. The real combinations have the following forms:

$$d_{xy} = xyf(r) \qquad d_{yz} = yzf(r) \qquad d_{zx} = zxf(r)$$
$$d_{x^2-y^2} = \tfrac{1}{2}(x^2 - y^2)f(r) \qquad d_{z^2} = \tfrac{1}{2\sqrt{3}}(3z^2 - r^2)f(r) \tag{26}$$

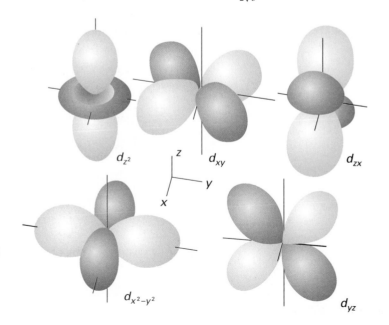

13.18 The boundary surfaces of d orbitals. Two nodal planes in each orbital intersect at the nucleus and separate the lobes of each orbital. The dark and light areas denote regions of opposite sign of the wavefunction.

13.3 Spectroscopic transitions and selection rules

The energies of the hydrogenic atoms are given by eqn 13. When the electron undergoes a **transition**, a change of state, from an orbital with quantum numbers n_1, l_1, m_{l1} to another (lower energy) orbital with quantum numbers n_2, l_2, m_{l2}, it undergoes a change of energy ΔE and discards the excess energy as a photon of electromagnetic radiation with a frequency ν given by the Bohr frequency condition (eqn 4).

It is tempting to think that all possible transitions are permissible, and that a spectrum arises from the transition of an electron from any initial orbital to any other orbital. However, this is not so, because a photon has an intrinsic spin angular momentum corresponding to $s = 1$ (Section 12.8). The change in angular momentum of the electron must compensate for the angular momentum carried away by the photon. Thus, an electron in a d orbital ($l = 2$) cannot make a transition into an s orbital ($l = 0$) because the photon cannot carry away enough angular momentum. Similarly, an s electron cannot make a transition to another s orbital, because there would then be no change in the electron's angular momentum to make up for the angular momentum carried away by the photon. It follows that some spectroscopic transitions are **allowed**, meaning that they can occur, whereas others are **forbidden**, meaning that they cannot occur.

A **selection rule** is a statement about which transitions are allowed. They are derived (for atoms) by identifying the transitions that conserve angular momentum when a photon is emitted or absorbed. The selection rules for hydrogenic atoms are

$$\Delta l = \pm 1 \qquad \Delta m_l = 0, \pm 1 \tag{27}$$

The principal quantum number n can change by any amount consistent with the Δl for the transition, because it does not relate directly to the angular momentum.

Justification 13.5

The formal derivation of a selection rule is based on the evaluation of a **transition dipole moment**, $\boldsymbol{\mu}_{\mathrm{fi}}$, between the initial and final states, where

$$\boldsymbol{\mu}_{\mathrm{fi}} = \langle f|\boldsymbol{\mu}|i \rangle \tag{28}$$

and $\boldsymbol{\mu}$ is the electric dipole moment operator. For a one-electron atom it is identified with multiplication by $-e\boldsymbol{r}$ with components $\mu_x = -ex$, $\mu_y = -ey$, and $\mu_z = -ez$. If the transition dipole moment is zero, the transition is forbidden. If it is nonzero, the transition is allowed and its intensity is proportional to the square modulus of the transition dipole moment. Physically, the transition dipole moment is a measure of the dipolar 'kick' that the electron gives to or receives from the electromagnetic field. To evaluate a transition dipole moment, we consider each component in turn. For example, for the z-component,

$$\mu_{z,\mathrm{fi}} = -e\langle f|z|i \rangle = -e \int \psi_f^* z \psi_i \, d\tau \tag{29}$$

To evaluate the integral, we note from Table 12.3 that $z = (4\pi/3)^{1/2} r Y_{1,0}$, so

$$\int \psi_f^* z \psi_i \, d\tau = \left(\frac{4\pi}{3}\right)^{1/2} \int_0^\infty R_{n_f,l_f}^* r R_{n_i,l_i} r^2 \, dr$$

$$\times \int_0^\pi \int_0^{2\pi} Y_{l_f,m_{l,f}}^*(\theta,\phi) Y_{1,0}(\theta,\phi) Y_{l_i,m_{l,i}}(\theta,\phi) \sin\theta \, d\theta \, d\phi$$

It follows from the properties of the spherical harmonics (Table 12.3) that the integral

$$\int_0^\pi \int_0^{2\pi} Y_{l_f,m_{l,f}}^*(\theta,\phi) Y_{1,m}(\theta,\phi) Y_{l_i,m_{l,i}}(\theta,\phi) \sin\theta \, d\theta \, d\phi$$

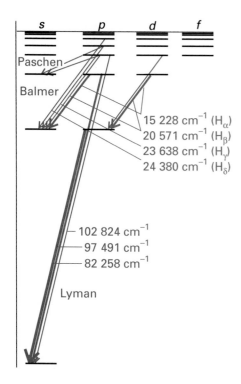

13.19 A Grotrian diagram that summarizes the appearance and analysis of the spectrum of atomic hydrogen. The thicker the line, the more intense the transition.

is zero unless $l_f = l_i \pm 1$ and $m_{l,f} = m_{l,i} + m$. Because $m = 0$ in the present case, the angular integral, and hence the z-component of the transition dipole moment, is zero unless $\Delta l = \pm 1$ and $\Delta m_l = 0$, which is a part of the set of selection rules. The same procedure, but considering the x- and y-components, results in the complete set of rules.

Illustration

To identify the orbitals to which a $4d$ electron may make radiative transitions, we first identify the value of l and then apply the selection rule for this quantum number. Because $l = 2$, the final orbital must have $l = 1$ or 3. Thus, an electron may make a transition from a $4d$ orbital to any np orbital (subject to $\Delta m_l = 0, \pm 1$) and to any nf orbital (subject to the same rule). However, it cannot undergo a transition to any other orbital, so a transition to any ns orbital or to another nd orbital is forbidden.

Self-test 13.6 To what orbitals may a $4s$-electron make radiative transitions?

[To np orbitals only]

The selection rules account for the structure of a **Grotrian diagram** (Fig. 13.19), which summarizes the energies of the states and the transitions between them. The thicknesses of the transition lines in the diagram denote their relative intensities in the spectrum.

The structures of many-electron atoms

The Schrödinger equation for a many-electron atom is highly complicated because all the electrons interact with one another. Even for a helium atom, with its two electrons, no analytical expression for the orbitals and energies can be given, and we are forced to make approximations. We shall adopt a simple approach based on what we already know about the structure of hydrogenic atoms. Later we shall see the kind of numerical computations that are currently used to obtain accurate wavefunctions and energies.

13.4 The orbital approximation

The wavefunction of a many-electron atom is a very complicated function of the coordinates of all the electrons, and we should write it $\Psi(\boldsymbol{r}_1, \boldsymbol{r}_2, \ldots)$, where $\boldsymbol{r}_i$ is the vector from the nucleus to electron i. However, in the **orbital approximation** we suppose that a reasonable first approximation to this exact wavefunction is obtained by thinking of each electron as occupying its 'own' orbital, and writing

$$\Psi(\boldsymbol{r}_1, \boldsymbol{r}_2, \ldots) = \psi(\boldsymbol{r}_1)\psi(\boldsymbol{r}_2) \ldots \tag{30}$$

We can think of the individual orbitals as resembling the hydrogenic orbitals, but with nuclear charges that are modified by the presence of all the other electrons in the atom. This description is only approximate, but it is a useful model for discussing the chemical properties of atoms, and is the starting point for more sophisticated descriptions of atomic structure.

Justification 13.6

The orbital approximation would be exact if there were no interactions between electrons. To demonstrate the validity of this remark, we need to consider a system in which the hamiltonian for the energy is the sum of two contributions, one for electron 1 and the other for electron 2:

$$H = H_1 + H_2$$

In an actual atom (such as helium atom), there is an additional term corresponding to the interaction of the two electrons, but we are ignoring that term. We shall now show that, if $\psi(r_1)$ is an eigenfunction of H_1 with energy E_1, and $\psi(r_2)$ is an eigenfunction of H_2 with energy E_2, then the product $\Psi(r_1, r_2) = \psi(r_1)\psi(r_2)$ is an eigenfunction of the combined hamiltonian H. To do so we write

$$\begin{aligned}
H\Psi(r_1, r_2) &= (H_1 + H_2)\psi(r_1)\psi(r_2) \\
&= \{H_1\psi(r_1)\}\psi(r_2) + \psi(r_1)\{H_2\psi(r_2)\} \\
&= \{E_1\psi(r_1)\}\psi(r_2) + \psi(r_1)\{E_2\psi(r_2)\} \\
&= (E_1 + E_2)\psi(r_1)\psi(r_2) = E\Psi(r_1, r_2)
\end{aligned}$$

where $E = E_1 + E_2$. This is the result we need to prove. However, if the electrons interact (as they do in fact), then the proof fails.

(a) The helium atom

The orbital approximation allows us to express the electronic structure of an atom by reporting its **configuration**, the list of occupied orbitals (usually, but not necessarily, in its ground state). Thus, as the ground state of a hydrogenic atom consists of the single electron in a $1s$ orbital, we report its configuration as $1s^1$.

The He atom has two electrons. We can imagine forming the atom by adding the electrons in succession to the orbitals of the bare nucleus (of charge $2e$). The first electron occupies a $1s$ hydrogenic orbital, but because $Z = 2$ that orbital is more compact than in H itself. The second electron joins the first in the $1s$ orbital, so the electron configuration of the ground state of He is $1s^2$.

(b) The Pauli principle

Lithium, with $Z = 3$, has three electrons. The first two occupy a $1s$ orbital drawn even more closely than in He around the more highly charged nucleus. The third electron, however, does not join the first two in the $1s$ orbital because that configuration is forbidden by the Pauli exclusion principle:

> **No more than two electrons may occupy any given orbital and, if two do occupy one orbital, then their spins must be paired.**

Electrons with paired spins, which we denote ↑↓, have zero net spin angular momentum because the spin of one electron is cancelled by the spin of the other. Specifically, one electron has $m_s = +\frac{1}{2}$, the other has $m_s = -\frac{1}{2}$, and they are orientated on their respective cones so that the resultant spin is zero (Fig. 13.20). The exclusion principle is the key to the structure of complex atoms, to chemical periodicity, and to molecular structure. It was proposed by Wolfgang Pauli in 1924 when he was trying to account for the absence of some lines in the spectrum of helium. Later he was able to derive a very general form of the principle from theoretical considerations.

Justification 13.7

The Pauli exclusion principle in fact applies to any pair of identical fermions (particles with half integral spin). Thus it applies to protons, neutrons, and ^{13}C nuclei (all of which have spin $\frac{1}{2}$) and to ^{35}Cl nuclei (which have spin $\frac{3}{2}$). It does not apply to identical bosons (particles

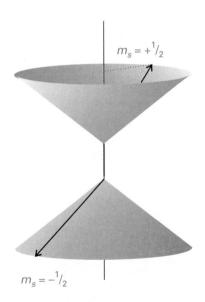

$m_s = +\frac{1}{2}$

$m_s = -\frac{1}{2}$

13.20 Electrons with paired spins have zero resultant spin angular momentum. They can be represented by two vectors that lie at an indeterminate position on the cones shown here but, wherever one lies on its cone, the other points in the opposite direction; their resultant is zero.

with integral spin), which include photons (spin 1), ^{12}C nuclei (spin 0). Any number of identical bosons may occupy the same orbital.

The Pauli exclusion principle is a special case of a general statement called the **Pauli principle**:

> **When the labels of any two identical fermions are exchanged, the total wavefunction changes sign. When the labels of any two identical bosons are exchanged, the total wavefunction retains the same sign.**

By 'total wavefunction' is meant the entire wavefunction, including the spin of the particles.

Consider the wavefunction for two electrons $\Psi(1, 2)$. The Pauli principle implies that it is a fact of nature (which has its roots in the theory of relativity) that the wavefunction must change sign if we interchange the labels 1 and 2 wherever they occur in the function:

$$\Psi(2, 1) = -\Psi(1, 2) \tag{31}$$

Suppose the two electrons in an atom occupy an orbital ψ, then in the orbital approximation the overall wavefunction is $\psi(1)\psi(2)$. To apply the Pauli principle, we must deal with the *total* wavefunction, the wavefunction including spin. There are several possibilities for two spins: both α, denoted $\alpha(1)\alpha(2)$, both β, denoted $\beta(1)\beta(2)$, and one α the other β, denoted either $\alpha(1)\beta(2)$ or $\alpha(2)\beta(1)$. Because we cannot tell which electron is α and which is β, in the last case it is appropriate to express the spin states as the (normalized) linear combinations

$$\sigma_+(1, 2) = \frac{1}{2^{1/2}}\{\alpha(1)\beta(2) + \beta(1)\alpha(2)\} \qquad \sigma_-(1, 2) = \frac{1}{2^{1/2}}\{\alpha(1)\beta(2) - \beta(1)\alpha(2)\} \tag{32}$$

because these allow one spin to be α and the other β with equal probability. The total wavefunction of the system is therefore the product of the orbital part and one of the four spin states:

$$\psi(1)\psi(2)\alpha(1)\alpha(2) \quad \psi(1)\psi(2)\beta(1)\beta(2) \quad \psi(1)\psi(2)\sigma_+(1, 2) \quad \psi(1)\psi(2)\sigma_-(1, 2)$$

The Pauli principle says that, for a wavefunction to be acceptable (for electrons), it must change sign when the electrons are exchanged. In each case, exchanging the labels 1 and 2 converts the factor $\psi(1)\psi(2)$ into $\psi(2)\psi(1)$, which is the same, because the order of multiplying the functions does not change the value of the product. The same is true of $\alpha(1)\alpha(2)$ and $\beta(1)\beta(2)$. Therefore, the first two overall products are not allowed, because they do not change sign. The combination $\sigma_+(1, 2)$ changes to

$$\sigma_+(2, 1) = \frac{1}{2^{1/2}}\{\alpha(2)\beta(1) + \beta(2)\alpha(1)\} = \sigma_+(1, 2)$$

because it is simply the original function written in a different order. The third overall product is therefore also disallowed. Finally, consider $\sigma_-(1, 2)$:

$$\sigma_-(2, 1) = \frac{1}{2^{1/2}}\{\alpha(2)\beta(1) - \beta(2)\alpha(1)\} = -\frac{1}{2^{1/2}}\{\alpha(1)\beta(2) - \beta(1)\alpha(2)\}$$
$$= -\sigma_-(1, 2)$$

This combination does change sign (it is 'antisymmetric'). The product $\psi(1)\psi(2)\sigma_-(1, 2)$ also changes sign under particle exchange, and therefore it is acceptable.

Now we see that only one of the four possible states is allowed by the Pauli principle, and the one that survives has paired α and β spins.[2] This is the content of the Pauli exclusion principle. The exclusion principle is irrelevant when the orbitals occupied by the

---

2   The distinction between σ_+ and σ_-, which both have one α and one β spin, is explained in Section 13.7.

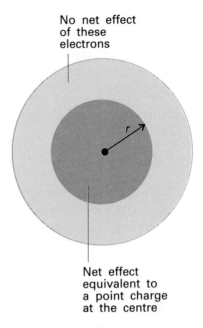

No net effect
of these
electrons

Net effect
equivalent to
a point charge
at the centre

13.21 An electron at a distance r from the nucleus experiences a Coulombic repulsion from all the electrons within a sphere of radius r and which is equivalent to a point negative charge located on the nucleus. The negative charge reduces the effective nuclear charge of the nucleus from Ze to $Z_{eff}e$.

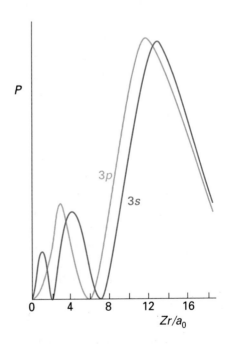

13.22 An electron in an s orbital (here a $3s$ orbital) is more likely to be found close to the nucleus than an electron in a p orbital of the same shell (note the closeness of the innermost peak of the $3s$ orbital to the nucleus at $r = 0$). Hence an s electron experiences less shielding and is more tightly bound than a p electron.

electrons are different, and both electrons may then have (but need not have) the same spin state. Nevertheless, even then the overall wavefunction must still be antisymmetric overall, and must still satisfy the Pauli principle itself.

In Li ($Z = 3$), the third electron cannot enter the $1s$ orbital because that orbital is already full: we say the K shell is **complete** and that the two electrons form a **closed shell**. Because a similar closed shell is characteristic of the He atom, we denote it [He]. The third electron is excluded from the K shell and must occupy the next available orbital, which is one with $n = 2$ and hence belonging to the L shell. However, we now have to decide whether the next available orbital is the $2s$ orbital or a $2p$ orbital, and therefore whether the lowest energy configuration of the atom is [He]$2s^1$ or [He]$2p^1$.

(c) Penetration and shielding

Unlike in hydrogenic atoms, the $2s$ and $2p$ orbitals (and, in general, all subshells of a given shell) are not degenerate in many-electron atoms. For reasons we shall now explain, s orbitals generally lie lower in energy than p orbitals of a given shell, and p orbitals lie lower than d orbitals.

An electron in a many-electron atom experiences a Coulombic repulsion from all the other electrons present. If it is at a distance r from the nucleus, it experiences a repulsion that can be represented by a point negative charge located at the nucleus and equal in magnitude to the total charge of the electrons within a sphere of radius r (Fig. 13.21). The effect of this point negative charge, when averaged over all the locations of the electron, is to reduce the full charge of the nucleus from Ze to $Z_{eff}e$, the **effective nuclear charge**. We say that the electron experiences a **shielded** nuclear charge, and the difference between Z and Z_{eff} is called the **shielding constant**, σ:

$$Z_{eff} = Z - \sigma \qquad [33]$$

The electrons do not actually 'block' the full Coulombic attraction of the nucleus: the shielding constant is simply a way of expressing the net outcome of the nuclear attraction and the electronic repulsions in terms of a single equivalent charge at the centre of the atom.

The shielding constant is different for s and p electrons because they have different radial distributions (Fig. 13.22). An s electron has a greater **penetration** through inner shells than a p electron, in the sense that it is more likely to be found close to the nucleus than a p electron of the same shell (the wavefunction of a p orbital, remember, is zero at the nucleus). Because only electrons inside the sphere defined by the location of the electron (in effect, the core electrons) contribute to shielding, an s electron experiences less shielding than a p electron. Consequently, by the combined effects of penetration and shielding, an s electron is more tightly bound than a p electron of the same shell. Similarly, a d electron penetrates less than a p electron of the same shell (recall that the wavefunction of a d orbital varies as r^2 close to the nucleus, whereas a p orbital varies as r), and therefore experiences more shielding.

Shielding constants for different types of electrons in atoms have been calculated from their wavefunctions obtained by numerical solution of the Schrödinger equation for the atom (Table 13.3). We see that, in general, valence-shell s electrons do experience higher effective nuclear charges than p electrons, although there are some discrepancies. We return to this point shortly.

The consequence of penetration and shielding is that the energies of subshells in a many-electron atom in general lie in the order

$$s < p < d < f$$

Table 13.3* Screening constants for atoms

Element	Z	Orbital	σ
He	2	$1s$	0.3125
C	6	$1s$	0.3273
		$2s$	2.7834
		$2p$	2.8642

* More values are given in the *Data section* at the end of this volume.

The individual orbitals of a given subshell remain degenerate because they all have the same radial characteristics and so experience the same effective nuclear charge.

We can now complete the Li story. Because the shell with $n = 2$ consists of two non-degenerate subshells, with the $2s$ orbital lower in energy than the three $2p$ orbitals, the third electron occupies the $2s$ orbital. This occupation results in the ground-state configuration $1s^2 2s^1$, with the central nucleus surrounded by a complete helium-like shell of two $1s$ electrons, and around that a more diffuse $2s$ electron. The electrons in the outermost shell of an atom in its ground state are called the **valence electrons** because they are largely responsible for the chemical bonds that the atom forms. Thus, the valence electron in Li is a $2s$ electron and its other two electrons belong to its core.

(d) The building-up principle

The extension of the procedure used for H, He, and Li to other atoms is called the **building-up principle**, or the '*Aufbau* principle', from the German word for building up. The building-up principle proposes an order of occupation of the hydrogenic orbitals that accounts for the experimentally determined ground-state configurations of neutral atoms.[3]

We imagine the bare nucleus of atomic number Z, and then feed into the orbitals Z electrons in succession. The order of occupation is

$$1s \quad 2s \quad 2p \quad 3s \quad 3p \quad 4s \quad 3d \quad 4p \quad 5s \quad 4d \quad 5p \quad 6s \ldots$$

and each orbital may accommodate up to two electrons. This order of occupation is approximately the order of energies of the individual orbitals, because, in general, the lower the energy of the orbital, the lower the total energy of the atom as a whole when that orbital is occupied. However, there are complicating effects arising from electron–electron repulsions that are important when the orbitals have very similar energies (such as the $4s$ and $3d$ orbitals near Ca and Sc), and we must take special care then.

We feed the Z electrons in succession into the orbitals subject to the demand of the exclusion principle that no more than two electrons can occupy any one orbital. Because an s subshell consists of only one orbital, up to two electrons may occupy it. A p subshell consists of three orbitals, so it can accommodate up to six electrons; a d subshell consists of five orbitals and can accommodate up to ten electrons.

As an example, consider the carbon atom, for which $Z = 6$ and there are six electrons to accommodate. Two electrons enter and fill the $1s$ orbital, two enter and fill the $2s$ orbital, leaving two electrons to occupy the orbitals of the $2p$ subshell. Hence the ground-state configuration of C is $1s^2 2s^2 2p^2$, or more succinctly [He]$2s^2 2p^2$, with [He] the helium-like $1s^2$ core. However, we can be more precise: we can expect the last two electrons to occupy different $2p$ orbitals because they will then be further apart on average and repel each other less than if they were in the same orbital. Thus one electron can be thought of as occupying the $2p_x$ orbital and the other the $2p_y$ orbital (the x, y, z designation is arbitrary, and it would be equally valid to use the complex forms of these orbitals), and the lowest energy configuration of the atom is [He]$2s^2 2p_x^1 2p_y^1$. The same rule applies whenever degenerate orbitals of a subshell are available for occupation. Thus, another rule of the building-up principle is:

Electrons occupy different orbitals of a given subshell before doubly occupying any one of them.

Thus nitrogen ($Z = 7$) has the configuration [He]$2s^2 2p_x^1 2p_y^1 2p_z^1$, and only when we get to oxygen ($Z = 8$) is a $2p$ orbital doubly occupied, giving [He]$2s^2 2p_x^2 2p_y^1 2p_z^1$.

---

3   Electron configurations are determined either spectroscopically or by measurements of magnetic properties.

An additional point arises when electrons occupy orbitals singly, for there is then no requirement that their spins should be paired. We need to know whether the lowest energy is achieved when the electron spins are the same (both α, for instance, denoted $\uparrow\uparrow$, if there are two electrons in question, as in C) or when they are paired ($\uparrow\downarrow$). This question is resolved by an empirical observation known as **Hund's rule**:

> **An atom in its ground state adopts a configuration with the greatest number of unpaired electrons.**

The explanation of Hund's rule is subtle, but it reflects the quantum mechanical property of **spin correlation**, that, as demonstrated in the *Justification* below, electrons with parallel spins behave as if they have a tendency to stay well apart, and hence repel each other less.[4] We can now conclude that, in the ground state of the carbon atom, the two $2p$ electrons have the same spin, that all three $2p$ electrons in the N atoms have the same spin, and that the two $2p$ electrons in different orbitals in the O atom have the same spin (the two in the $2p_x$ orbital are necessarily paired).

Justification 13.8

Suppose electron 1 is described by a wavefunction $\psi_a(r_1)$ and electron 2 is described by a wavefunction $\psi_b(r_2)$; then, in the orbital approximation, the joint wavefunction of the electrons is the product $\Psi = \psi_a(r_1)\psi_b(r_2)$. However, this wavefunction is not acceptable, because it suggests that we know which electron is in which orbital, whereas we cannot keep track of electrons. According to quantum mechanics, the correct description is either of the two following wavefunctions:

$$\Psi_\pm = \frac{1}{2^{1/2}}\{\psi_a(r_1)\psi_b(r_2) \pm \psi_b(r_1)\psi_a(r_2)\}$$

According to the Pauli principle (*Justification 13.7*), because Ψ_+ is symmetrical under particle interchange, it must be multiplied by an antisymmetric spin function (the one denoted σ_- in *Justification 13.7*). That combination corresponds to a spin-paired state. Conversely, Ψ_- is antisymmetric, so it must be multiplied by one of the three symmetric spin states. These three symmetric states correspond to electrons with parallel spins.[5]

Now consider the values of the two combinations when one electron approaches another, and $r_1 = r_2$. We see that Ψ_- vanishes, which means that there is zero probability of finding the two electrons at the same point in space when they have parallel spins. The other combination does not vanish when the two electrons are at the same point in space. Because the two electrons have different relative spatial distributions depending on whether their spins are parallel or not, it follows that their Coulombic interaction is different, and hence that the two states have different energies.

Neon, with $Z = 10$, has the configuration [He]$2s^2 2p^6$, which completes the L shell. This closed-shell configuration is denoted [Ne], and acts as a core for subsequent elements. The next electron must enter the $3s$ orbital and begin a new shell, so an Na atom, with $Z = 11$, has the configuration [Ne]$3s^1$. Like lithium with the configuration [He]$2s^1$, sodium has a single s electron outside a complete core.

This analysis has brought us to the origin of chemical periodicity. The L shell is completed by eight electrons, so the element with $Z = 3$ (Li) should have similar properties to the

---

4   The effect of spin correlation is to allow the atom to shrink slightly, so the electron–nucleus interaction is improved when the spins are parallel.

5   See Section 13.7 for an explanation of this point.

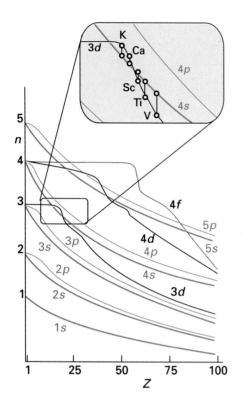

13.23 The orbital energies of the elements. Note the relative energies of the 3*d* and 4*s* orbitals close to potassium (see inset).

element with $Z = 11$ (Na). Likewise, Be ($Z = 4$) should be similar to $Z = 12$ (Mg), and so on, up to the noble gases He ($Z = 2$), Ne ($Z = 10$), and Ar ($Z = 18$).

Argon has complete 3*s* and 3*p* subshells, and as the 3*d* orbitals are high in energy it counts as having a closed-shell configuration. Indeed, the 3*d* orbitals are so high in energy that the next electron (for K) occupies the 4*s* orbital, and the configuration of a K atom is analogous to that of an Na atom. The same is true of a Ca atom, which has the configuration [Ar]4s^2. However, at this point, the 3*d* orbitals become comparable in energy to the 4*s* orbitals (Fig. 13.23), and they commence to be filled.

Ten electrons can be accommodated in the five 3*d* orbitals, which accounts for the electron configurations of scandium to zinc. However, the building-up principle has less clear-cut predictions about the ground-state configurations of these elements because electron–electron repulsions are comparable to the energy difference between the 4*s* and 3*d* orbitals, and a simple analysis no longer works. At gallium, the energy of the 3*d* orbitals has fallen so far below those of the 4*s* and 4*p* orbitals that the 3*d* orbitals can be largely ignored, and the building-up principle can be used in the same way as in preceding periods. Now the 4*s* and 4*p* subshells constitute the valence shell, and the period terminates with krypton. Because 18 electrons have intervened since argon, this period is the first 'long period' of the periodic table. The existence of the *d*-block elements (the 'transition metals') reflects the stepwise occupation of the 3*d* orbitals, and the subtle shades of energy differences along this series give rise to the rich complexity of inorganic *d*-metal chemistry. A similar intrusion of the *f* orbitals in Periods 6 and 7 accounts for the existence of the *f* block of the periodic table (the lanthanides and actinides).

(e) The configurations of ions

We derive the configurations of cations of elements in the *s*, *p*, and *d* blocks of the periodic table by removing electrons from the ground-state configuration of the neutral atom in a specific order. First, we remove valence *p* electrons, then valence *s* electrons, and then as many *d* electrons as are necessary to achieve the stated charge. For instance, because the configuration of Fe is [Ar]3$d^6$4s^2, the Fe^{3+} cation has the configuration [Ar]3d^5.

The configurations of anions are derived by continuing the building-up procedure and adding electrons to the neutral atom until the configuration of the next noble gas has been reached. Thus, the configuration of the O^{2-} ion is achieved by adding two electrons to [He]2$s^2$2p^4, giving [He]2$s^2$2p^6, the same as the configuration of neon.

(f) Ionization energies and electron affinities

The minimum energy necessary to remove an electron from a many-electron atom is the **first ionization energy**, I_1, of the element.[6] The **second ionization energy**, I_2, is the minimum energy needed to remove a second electron (from the singly charged cation). The variation of the first ionization energy through the periodic table is shown in Fig. 13.24 and some numerical values are given in Table 13.4. In thermodynamic considerations we often need the **standard enthalpy of ionization**, $\Delta_{ion}H^{\ominus}$. As shown in the *Justification* below, the two are related by

$$\Delta_{ion}H^{\ominus}(T) = I + \tfrac{5}{2}RT \tag{34}$$

At 298 K, the difference between the ionization enthalpy and the corresponding ionization energy is 6.20 kJ mol^{-1}.

Table 13.4* First and second ionization energies, I_1/(kJ mol^{-1}) and I_2/(kJ mol^{-1})

H	1312	
He	2372	5251
Mg	738	1451
Na	496	4562

*More values are given in the *Data section*.

6   The symbol recommended by IUPAC for ionization energy is E_i; but this notation is open to confusion.

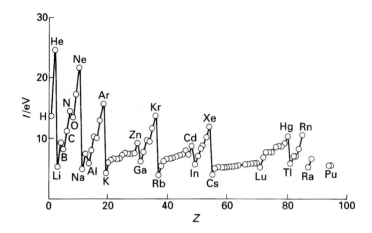

13.24 The first ionization energies of the elements plotted against atomic number.

Justification 13.9

It follows from Kirchhoff's law (Section 2.9 and eqn 2.45) that the reaction enthalpy for

$$M(g) \longrightarrow M^+(g) + e^-(g)$$

at a temperature T is related to the value at $T = 0$ by

$$\Delta_r H^{\ominus}(T) = \Delta_r H^{\ominus}(0) + \int_0^T \Delta_r C_p^{\ominus}\, dT$$

The molar constant-pressure heat capacity of each species in the reaction is $\frac{5}{2}R$, so $\Delta_r C_p^{\ominus} = +\frac{5}{2}R$. The integral in this expression therefore evaluates to $+\frac{5}{2}RT$. The reaction enthalpy at $T = 0$ is the same as the (molar) ionization energy, I. Equation 34 then follows. The same expression applies to each successive ionization step, so the *overall* ionization enthalpy for the formation of M^{2+} is

$$\Delta_r H^{\ominus}(T) = I_1 + I_2 + 5RT$$

Table 13.5* Electron affinities, $E_{ea}/(\text{kJ mol}^{-1})$

Cl	349		
F	322		
H	73		
O	141	O^-	−844

*More values are given in the *Data section*.

The **electron affinity**, E_{ea}, is the energy released when an electron attaches to a gas-phase atom (Table 13.5). In a common, logical, but not universal convention (which we adopt), the electron affinity is positive if energy is released when the electron attaches to the atom (that is, $E_{ea} > 0$ implies that electron attachment is exothermic). It follows from a similar argument to that given in the *Justification* above that the **standard enthalpy of electron gain**, $\Delta_{eg}H^{\ominus}$, at a temperature T is related to the electron affinity by

$$\Delta_{eg}H^{\ominus}(T) = -E_{ea} - \frac{5}{2}RT \qquad (35)$$

Note the change of sign. In typical thermodynamic cycles the $\frac{5}{2}RT$ that appears in eqn 35 cancels that in eqn 34, so ionization energies and electron affinities can be used directly. A final preliminary point is that the electron-gain enthalpy of a species X is the negative of the ionization enthalpy of its negative ion:

$$\Delta_{eg}H^{\ominus}(X) = -\Delta_{ion}H^{\ominus}(X^-) \qquad (36)$$

As ionization energy is often easier to measure than electron affinity, this relation can be used to determine numerical values of the latter.

Ionization energies and electron affinities show periodicities, but the former is more regular and we concentrate on it. Lithium has a low first ionization energy: its outermost electron is well-shielded from the nucleus by the core ($Z_{eff} = 1.3$, compared with $Z = 3$) and it is easily removed. Beryllium has a higher nuclear charge than lithium, and its outermost electron (one of the two $2s$ electrons) is more difficult to remove: its ionization energy is higher. The ionization energy decreases between beryllium and boron because in the latter the outermost electron occupies a 2p orbital and is less strongly bound than if it had been a 2s electron. The ionization energy increases between boron and carbon because the latter's outermost electron is also 2p and the nuclear charge has increased. Nitrogen has a still higher ionization energy because of the further increase in nuclear charge.

There is now a kink in the curve which reduces the ionization energy of oxygen below what would be expected by simple extrapolation. The explanation is that at oxygen a 2p orbital must become doubly occupied, and the electron–electron repulsions are increased above what would be expected by simple extrapolation along the row. In addition, the loss of a 2p electron results in a configuration with a half-filled subshell (like that of N), which is an arrangement of low energy, so the energy of $O^+ + e^-$ is lower than might be expected, and the ionization energy is correspondingly low too. (The kink is less pronounced in the next row, between phosphorus and sulfur because their orbitals are more diffuse.) The values for oxygen, fluorine, and neon fall roughly on the same line, the increase of their ionization energies reflecting the increasing attraction of the more highly charged nuclei for the outermost electrons.

The outermost electron in sodium is 3s. It is far from the nucleus, and the latter's charge is shielded by the compact, complete neon-like core. As a result, the ionization energy of sodium is substantially lower than that of neon. The periodic cycle starts again along this row, and the variation of the ionization energy can be traced to similar reasons.

Electron affinities are greatest close to fluorine, for the incoming electron enters a vacancy in a compact valence shell and can interact strongly with the nucleus. The attachment of an electron to an anion (as in the formation of O^{2-} from O^-) is invariably endothermic, so E_{ea} is negative. The incoming electron is repelled by the charge already present. Electron affinities are also small, and may be negative, when an electron enters an orbital that is far from the nucleus (as in the heavier alkali metal atoms) or is forced by the Pauli principle to occupy a new shell (as in the noble gas atoms).

13.5 Self-consistent field orbitals

The central difficulty of the Schrödinger equation is the presence of the electron–electron interaction terms. The potential energy of the electrons is

$$V = -\sum_i \frac{Ze^2}{4\pi\varepsilon_0 r_i} + \tfrac{1}{2}\sum_{i,j}' \frac{e^2}{4\pi\varepsilon_0 r_{ij}} \tag{37}$$

The prime on the second sum indicates that $i \neq j$, and the factor of one-half prevents double-counting of electron pair repulsions (1 with 2 is the same as 2 with 1). The first term is the total attractive interaction between the electrons and the nucleus. The second term is the total repulsive interaction; r_{ij} is the distance between electrons i and j. It is hopeless to expect to find analytical solutions of a Schrödinger equation with such a complicated potential energy term, but computational techniques are available that give very detailed and reliable numerical solutions for the wavefunctions and energies. The techniques were originally introduced by D.R. Hartree (before computers were available) and then modified by V. Fock to take into account the Pauli principle correctly. In broad outline, the **Hartree–Fock self-consistent field** (SCF) procedure is as follows.

Imagine that we have a rough idea of the structure of the atom. In the Ne atom, for instance, the orbital approximation suggests the configuration $1s^2 2s^2 2p^6$ with the orbitals approximated by hydrogenic atomic orbitals. Now consider one of the 2p electrons. A Schrödinger equation can be written for this electron by ascribing to it a potential energy due to the nuclear attraction and the repulsion from the other electrons. This equation has the form

$$-\frac{\hbar^2}{2m_e}\nabla_1^2\psi_{2p}(\boldsymbol{r}_1) - \frac{Ze^2}{4\pi\varepsilon_0 r_1}\psi_{2p}(\boldsymbol{r}_1)$$
$$+ 2\sum_i\left\{\int\frac{\psi_i^*(\boldsymbol{r}_2)\psi_i(\boldsymbol{r}_2)e^2}{4\pi\varepsilon_0 r_{12}}\,d\tau_2\right\}\psi_{2p}(\boldsymbol{r}_1)$$
$$-\sum_i\left\{\int\frac{\psi_i^*(\boldsymbol{r}_2)\psi_{2p}(\boldsymbol{r}_2)e^2}{4\pi\varepsilon_0 r_{12}}\,d\tau_2\right\}\psi_i(\boldsymbol{r}_1)$$
$$= E_{2p}\psi_{2p}(\boldsymbol{r}_1) \tag{38}$$

The orbitals are labelled i, and the sums on the left are over all the occupied orbitals. A similar equation can be written for the $1s$ and $2s$ orbitals in the atom.

Equation 38 is fearsome, but it can be interpreted by examining each term. The first term on the left is the usual kinetic energy contribution. The second term is the potential energy of attraction of the electron to the nucleus. The third term is the potential energy of the electron of interest due to the charge density $-e|\psi_i(\boldsymbol{r}_2)|^2$ of the electrons in the other occupied orbitals. The fourth term takes into account the spin correlation effects discussed earlier. Note that, although the equation is for the $2p$ orbital in neon, it depends on the wavefunctions of all the other occupied orbitals in the atom.

There is no hope of solving eqn 38 analytically. However, it can be solved numerically if we guess an approximate form of the wavefunctions of all the orbitals except $2p$. The procedure is then repeated for the other orbitals in the atom, the $1s$ and $2s$ orbitals. This sequence of calculations gives the form of the $2p$, $2s$, and $1s$ orbitals, and in general they will differ from the set used initially to start the calculation.[7] These improved orbitals can be used in another cycle of calculation, and a second improved set of orbitals is obtained. The recycling continues until the orbitals and energies obtained are insignificantly different from those used at the start of the current cycle. The solutions are then **self-consistent** and accepted as solutions of the problem.

Plots of some of the self-consistent field (SCF) Hartree–Fock radial distribution functions for sodium are shown in Fig. 13.25. They show the grouping of electron density into shells, as was anticipated by the early chemists, and the differences of penetration as discussed above. These SCF calculations therefore support the qualitative discussions that are used to explain chemical periodicity. They also considerably extend that discussion by providing detailed wavefunctions and precise energies.

The spectra of complex atoms

The spectra of atoms rapidly become very complicated as the number of electrons increases, but there are some important and moderately simple features. The general idea is straightforward: lines in the spectrum (in either emission or absorption) occur when the atom undergoes a change of state with a change of energy $|\Delta E|$, and emits or absorbs a photon of frequency $\nu = |\Delta E|/h$ and wavenumber $\tilde{\nu} = |\Delta E|/hc$. Hence, we can expect the spectrum to give information about the energies of electrons in atoms. However, the actual energy levels are not given solely by the energies of the orbitals, because the electrons interact with one another in various ways, and there are contributions to the energy in addition to those we have already considered.

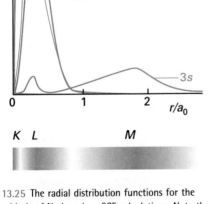

13.25 The radial distribution functions for the orbitals of Na based on SCF calculations. Note the shell-like structure, with the $3s$ orbital outside the inner K and L shells.

---

7 In practice, much more efficient procedures are used, and the equations for the wavefunctions are solved simultaneously.

13.6 Quantum defects and ionization limits

One application of atomic spectroscopy is to the determination of ionization energies. However, we cannot use the procedure illustrated in Example 13.1 indiscriminately because the energy levels of a many-electron atom do not in general vary as $1/n^2$. If we confine attention to the outermost electrons, then we know that, as a result of penetration and shielding, they experience a nuclear charge of slightly more than $1e$ because in a neutral atom the other $Z - 1$ electrons cancel all but about one unit of nuclear charge. Typical values of Z_{eff} are a little more than 1, so we expect binding energies to be given by a term of the form $-hc\mathcal{R}/n^2$, but lying slightly lower in energy than this formula predicts. We therefore introduce a **quantum defect**, δ, and write the energy as $-hc\mathcal{R}/(n - \delta)^2$. The quantum defect is best regarded as a purely empirical quantity.

There are some states that are so diffuse that the $1/n^2$ variation is valid: these states are called **Rydberg states**. In such cases we can write

$$\tilde{\nu} = \frac{I}{hc} - \frac{\mathcal{R}}{n^2} \tag{39}$$

and a plot of wavenumber against $1/n^2$ can be used to obtain I by extrapolation; in practice, one would use a linear regression fit using a computer. If the lower state is not the ground state (a possibility if we wish to generalize the concept of ionization energy), the ionization energy of the ground state can be determined by adding the appropriate energy difference to the ionization energy obtained as described here.

13.7 Singlet and triplet states

Suppose we were interested in the energy levels of a He atom, with its two electrons. We know that the ground-state configuration is $1s^2$, and can anticipate that an excited configuration will be one in which one of the electrons has been promoted into a $2s$ orbital, giving the configuration $1s^1 2s^1$. The two electrons need not be paired because they occupy different orbitals. According to Hund's rule, the state of the atom with the spins parallel lies lower in energy than the state in which they are paired. Both states are permissible, and can contribute to the spectrum of the atom.

Parallel and antiparallel (paired) spins differ in their overall spin angular momentum. In the paired case, the two spin momenta cancel each other, and there is zero net spin (as was depicted in Fig. 13.20). The paired-spin arrangement is called a **singlet**. Its spin state is the one we denoted σ_- in the discussion of the Pauli principle:

$$\sigma_-(1,2) = \frac{1}{2^{1/2}}\{\alpha(1)\beta(2) - \beta(1)\alpha(2)\} \tag{40a}$$

The angular momenta of two parallel spins add together to give a nonzero total spin, and the resulting state is called a **triplet**. As illustrated in Fig. 13.26, there are three ways of achieving a nonzero total spin, but only one way to achieve zero spin. The three spin states are the symmetric combinations introduced earlier:

$$\alpha(1)\alpha(2) \qquad \sigma_+(1,2) = \frac{1}{2^{1/2}}\{\alpha(1)\beta(2) + \beta(1)\alpha(2)\} \qquad \beta(1)\beta(2) \tag{40b}$$

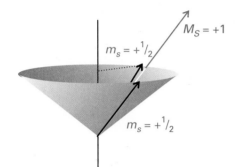

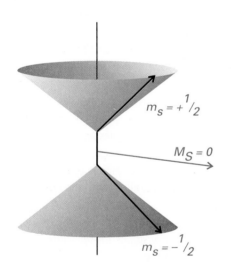

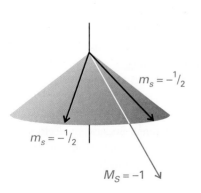

13.26 When two electrons have parallel spins, they have a nonzero total spin angular momentum. There are three ways of achieving this resultant, which are shown by these vector representations. Note that, although we cannot know the orientation of the spin vectors on the cones, the angle between the vectors is the same in all three cases, for all three arrangements have the same total spin angular momentum (that is, the resultant of the two vectors has the same length in each case, but points in different directions). Compare this diagram with Fig. 13.20, which shows the antiparallel case. Note that, whereas two paired spins are precisely antiparallel, two 'parallel' spins are not strictly parallel.

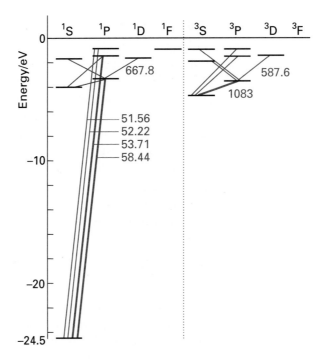

13.27 Part of the Grotrian diagram for a helium atom. Note that there are no transitions between the singlet and triplet levels. Transition wavelengths are given in nanometers.

The fact that the parallel arrangement of spins in the $1s^12s^1$ configuration of the He atom lies lower in energy than the antiparallel arrangement can now be expressed by saying that the triplet state of the $1s^12s^1$ configuration of He lies lower in energy than the singlet state. This is a general conclusion that applies to other atoms (and molecules), and, *for states arising from the same configuration, the triplet state generally lies lower than the singlet state*. The origin of the energy difference lies in the effect of spin correlation on the Coulombic interactions between electrons, as we saw in the case of Hund's rule for ground-state configurations. Because the Coulombic interaction between electrons in an atom is strong, the difference in energies between singlet and triplet states of the same configuration can be large. The two states of $1s^12s^1$ He, for instance, differ by 6421 cm^{-1} (corresponding to 0.7961 eV).

The spectrum of atomic helium is more complicated than that of atomic hydrogen, but there are two simplifying features. One is that the only excited configurations it is necessary to consider are of the form $1s^1nl^1$: that is, only one electron is excited. Excitation of two electrons requires an energy greater than the ionization energy of the atom, so the He$^+$ ion is formed instead of the doubly excited atom. Second, *no transitions take place between singlet and triplet states* because the relative orientation of the two electron spins cannot change during a transition. Thus, there is a spectrum arising from transitions between singlet states (including the ground state) and between triplet states, but not between the two. Spectroscopically, helium behaves like two distinct species, and the early spectroscopists actually thought of helium as consisting of 'parahelium' and 'orthohelium'. The Grotrian diagram for helium in Fig. 13.27 shows the two sets of transitions.

13.8 Spin–orbit coupling

Electron spin has a further implication for the energies of atoms. Because an electron has spin angular momentum, and because moving charges generate magnetic fields, an electron has a magnetic moment that arises from its spin (Fig. 13.28). Similarly, an electron with orbital angular momentum (that is, an electron in an orbital with $l > 0$) is in effect a

13.28 Angular momentum gives rise to a magnetic moment (μ). For an electron, the magnetic moment is antiparallel to the orbital angular momentum, but proportional to it. For spin angular momentum, there is a factor 2, which increases the magnetic moment to twice its expected value (see Section 13.10).

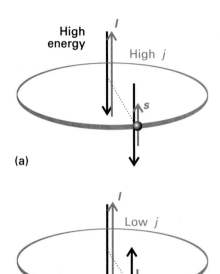

(a)

(b)

13.29 Spin–orbit coupling is a magnetic interaction between spin and orbital magnetic moments. When the angular momenta are parallel, as in (a), the magnetic moments are aligned unfavourably; when they are opposed, as in (b), the interaction is favourable. This magnetic coupling is the cause of the splitting of a configuration into levels.

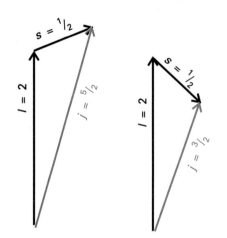

13.30 The coupling of the spin and orbital angular momenta of a d electron ($l = 2$) gives two possible values of j depending on the relative orientations of the spin and orbital angular momenta of the electron.

circulating current, and possesses a magnetic moment that arises from its orbital momentum. The interaction of the spin and orbital magnetic moments is called **spin–orbit coupling**. The strength of the coupling, and its effect on the energy levels of the atom, depends on the relative orientations of the spin and orbital magnetic moments, and therefore on the relative orientations of the two angular momenta (Fig. 13.29).

(a) The total angular momentum

One way of expressing the dependence of the spin–orbit interaction on the relative orientation of the spin and orbital momenta is to say that it depends on the **total angular momentum** of the electron, the vector sum of its spin and orbital momenta. Thus, when the spin and orbital angular momenta are nearly parallel, the total angular momentum is high; when the two angular momenta are opposed, the total angular momentum is low.

The total angular momentum of an electron is described by the quantum numbers j and m_j, with $j = l + \frac{1}{2}$ (when the two angular momenta are in the same direction) or $j = l - \frac{1}{2}$ (when they are opposed, Fig. 13.30). The different values of j that can arise for a given value of l label **levels** of a term. For $l = 0$, the only permitted value is $j = \frac{1}{2}$ (the total angular momentum is the same as the spin angular momentum because there is no other source of angular momentum in the atom). When $l = 1$, j may be either $\frac{3}{2}$ (the spin and orbital angular momenta are in the same sense) or $\frac{1}{2}$ (the spin and angular momenta are in opposite senses).

Example 13.4 Identifying the levels of a configuration

Identify the levels that may arise from the configurations (a) d^1, (b) s^1.

Method In each case, identify the value of l and then the possible values of j. For these one-electron systems, the total angular momentum is the sum or difference of the orbital and spin momenta.

Answer (a) For a d electron, $l = 2$ and there are two levels in the configuration, one with $j = 2 + \frac{1}{2} = \frac{5}{2}$ and the other with $j = 2 - \frac{1}{2} = \frac{3}{2}$. (b) For an s electron $l = 0$, so only one level is possible, and $j = \frac{1}{2}$.

- -

Self-test 13.7 Identify the levels of the configurations (a) p^1 and (b) f^1.

$$[(a)\ \tfrac{3}{2}, \tfrac{1}{2};\ (b)\ \tfrac{7}{2}, \tfrac{5}{2}]$$

The dependence of the spin–orbit interaction on the value of j is expressed in terms of the **spin–orbit coupling constant**, A (which is typically expressed as a wavenumber). A quantum mechanical calculation leads to the result that the energies of the levels with quantum numbers s, l, and j are given by

$$E_{l,s,j} = \tfrac{1}{2}hcA\{j(j+1) - l(l+1) - s(s+1)\} \qquad (41)$$

Justification 13.10

The energy of a magnetic moment $\boldsymbol{\mu}$ in a magnetic field $\boldsymbol{B}$ is equal to their scalar product $-\boldsymbol{\mu} \cdot \boldsymbol{B}$. If the magnetic field arises from the orbital angular momentum of the electron, it is proportional to $\boldsymbol{l}$; if the magnetic moment $\boldsymbol{\mu}$ is that of the electron spin, then it is proportional to $\boldsymbol{s}$. It then follows that the energy of interaction is proportional to the scalar product $\boldsymbol{s} \cdot \boldsymbol{l}$:

$$\text{energy of interaction} = -\boldsymbol{\mu} \cdot \boldsymbol{B} \propto \boldsymbol{s} \cdot \boldsymbol{l}$$

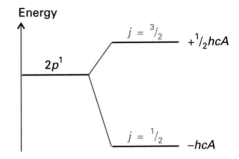

13.31 The levels of a ^2P term arising from spin–orbit coupling. Note that the low-j level lies below the high-j level.

Next, we note that the total angular momentum is the vector sum of the spin and orbital momenta: $\boldsymbol{j} = \boldsymbol{l} + \boldsymbol{s}$. The magnitude of the vector $\boldsymbol{j}$ is calculated by evaluating

$$\boldsymbol{j} \cdot \boldsymbol{j} = (\boldsymbol{l} + \boldsymbol{s}) \cdot (\boldsymbol{l} + \boldsymbol{s}) = \boldsymbol{l} \cdot \boldsymbol{l} + \boldsymbol{s} \cdot \boldsymbol{s} + 2\boldsymbol{s} \cdot \boldsymbol{l}$$

That is,

$$\boldsymbol{s} \cdot \boldsymbol{l} = \tfrac{1}{2}\{j^2 - l^2 - s^2\}$$

This is a classical result. To make the transition to quantum mechanics, we treat all the quantities as operators, and write

$$\hat{\boldsymbol{s}} \cdot \hat{\boldsymbol{l}} = \tfrac{1}{2}\{\hat{j}^2 - \hat{l}^2 - \hat{s}^2\} \tag{42}$$

At this point, we evaluate the expectation value:

$$\langle j, l, s | \hat{\boldsymbol{s}} \cdot \hat{\boldsymbol{l}} | j, l, s \rangle = \tfrac{1}{2}\langle j, l, s | \hat{j}^2 - \hat{l}^2 - \hat{s}^2 | j, l, s \rangle$$
$$= \tfrac{1}{2}\{j(j+1) - l(l+1) - s(s+1)\}\hbar^2 \tag{43}$$

Then, by inserting this expression into the formula for the energy, and writing the constant of proportionality as $hcA/\hbar^2$, we obtain eqn 41. The calculation of A is much more complicated: see *Further reading*.

Illustration

The unpaired electron in the ground state of an alkali metal atom has $l = 0$, so $j = \tfrac{1}{2}$. Because the orbital angular momentum is zero in this state, the spin–orbit coupling energy is zero (as is confirmed by setting $j = s$ and $l = 0$ in eqn 41). When the electron is excited to an orbital with $l = 1$, it has orbital angular momentum and can give rise to a magnetic field that interacts with its spin. In this configuration the electron can have $j = \tfrac{3}{2}$ or $j = \tfrac{1}{2}$, and the energies of these levels are

$$E_{3/2} = \tfrac{1}{2}hcA\{\tfrac{3}{2} \times \tfrac{5}{2} - 1 \times 2 - \tfrac{1}{2} \times \tfrac{3}{2}\} = \tfrac{1}{2}hcA$$
$$E_{1/2} = \tfrac{1}{2}hcA\{\tfrac{1}{2} \times \tfrac{3}{2} - 1 \times 2 - \tfrac{1}{2} \times \tfrac{3}{2}\} = -hcA$$

The corresponding energies are shown in Fig. 13.31. Note that the 'centre of gravity' of the levels is unchanged, because there are four states of energy $\tfrac{1}{2}hcA$ and two of energy $-hcA$.

The strength of the spin–orbit coupling depends on the nuclear charge. To understand why this is so, imagine riding on the orbiting electron and seeing a charged nucleus apparently orbiting around us (like the sun rising and setting). As a result, we find ourselves at the centre of a ring of current. The greater the nuclear charge, the greater this current, and therefore the stronger the magnetic field we detect. Because the spin magnetic moment of the electron interacts with this orbital magnetic field, it follows that, the greater the nuclear charge, the stronger the spin–orbit interaction. The coupling increases sharply with atomic number (as Z^4 in a hydrogenic atom). Whereas it is only small in H (giving rise to shifts of energy levels of no more than about 0.4 cm^{-1}), in heavy atoms like Pb it is very large (giving shifts of the order of thousands of reciprocal centimetres).

(b) Fine structure

Two spectral lines are observed when the p electron of an electronically excited alkali metal atom undergoes a transition and falls into a lower s orbital. One line is due to a transition starting in a $j = \tfrac{3}{2}$ level and the other line is due to a transition starting in the $j = \tfrac{1}{2}$ level of the same configuration. The two lines are an example of **fine structure**. Fine structure can be clearly seen in the emission spectrum from sodium vapour excited by an electric discharge

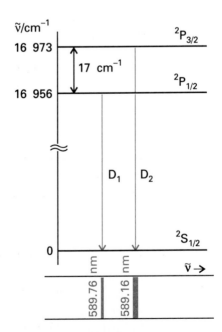

13.32 The energy-level diagram for the formation of the sodium D lines. The splitting of the spectral lines (by 17 cm^{-1}) reflects the splitting of the levels of the ^2P term.

(for example, in one kind of street lighting). The yellow line at 589 nm (close to 17 000 cm^{-1}) is actually a doublet composed of one line at 589.76 nm (16 956.2 cm^{-1}) and another at 589.16 nm (16 973.4 cm^{-1}); the components of this doublet are the 'D lines' of the spectrum (Fig. 13.32). Therefore, in Na, the spin–orbit coupling affects the energies by about 17 cm^{-1}.

Example 13.5 Analysing a spectrum for the spin-orbit coupling constant

The origin of the D lines in the spectrum of atomic sodium is shown in Fig. 13.32. Calculate the spin–orbit coupling constant for the upper configuration of the Na atom.

Method We see from Fig. 13.32 that the splitting of the lines is equal to the energy separation of the $j = \frac{3}{2}$ and $\frac{1}{2}$ levels of the excited configuration. This separation can be expressed in terms of A by using eqn 41. Therefore, set the observed splitting equal to the energy separation calculated from eqn 41 and solve the equation for A.

Answer The two levels are split by

$$\Delta\tilde{\nu} = A\tfrac{1}{2}\{\tfrac{3}{2}(\tfrac{3}{2}+1) - \tfrac{1}{2}(\tfrac{1}{2}+1)\} = \tfrac{3}{2}A$$

The experimental value is 17.2 cm^{-1}; therefore

$$A = \tfrac{2}{3} \times (17.2 \text{ cm}^{-1}) = 11.5 \text{ cm}^{-1}$$

Comment The same calculation repeated for the other alkali metal atoms gives Li: 0.23 cm^{-1}; K: 38.5 cm^{-1}; Rb: 158 cm^{-1}; Cs: 370 cm^{-1}. Note the increase of A with atomic number (but more slowly than Z^4 for these many-electron atoms).

- -

Self-test 13.8 The configuration . . . $4p^6 5d^1$ of rubidium has two levels at 25 700.56 cm^{-1} and 25 703.52 cm^{-1} above the ground state. What is the spin–orbit coupling constant in this excited state?

[1.18 cm^{-1}]

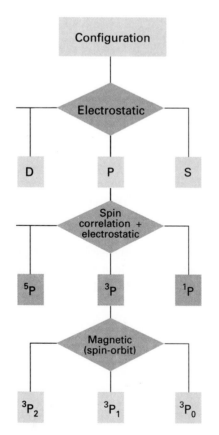

13.33 A summary of the types of interaction that are responsible for the various kinds of splitting of energy levels in atoms. For light atoms, magnetic interactions are small, but in heavy atoms they may dominate the electrostatic (charge–charge) interactions.

13.9 Term symbols and selection rules

We have used expressions such as 'the $j = \frac{3}{2}$ level of a configuration'. A **term symbol**, which is a symbol looking like $^2P_{3/2}$ or $^3D_2$, conveys this information much more succinctly. The convention of using lower-case letters to label orbitals and upper-case letters to label overall states applies throughout spectroscopy, not just to atoms.

A term symbol gives three pieces of information:

1. The letter (for example, P or D in the examples) indicates the total orbital angular momentum quantum number, L.
2. The left superscript in the term symbol (for example, the 2 in $^2P_{3/2}$) gives the multiplicity of the term.
3. The right subscript on the term symbol (for example, the $\frac{3}{2}$ in $^2P_{3/2}$) is the value of the total angular momentum quantum number, J.

We shall now say what each of these statements means; the contributions to the energies which we are about to discuss are summarized in Fig. 13.33.

(a) The total orbital angular momentum

When several electrons are present, it is necessary to judge how their individual orbital angular momenta add together or oppose each other. The **total orbital angular momentum quantum number**, L, tells us the magnitude of the angular momentum through $\{L(L+1)\}^{1/2}\hbar$. It has $2L+1$ orientations distinguished by the quantum number M_L, which can take the values $L, L-1, \ldots, -L$. Similar remarks apply to the **total spin quantum number**, S, and the quantum number M_S, and the **total angular momentum quantum number**, J, and the quantum number M_J. The value of L (a non-negative integer) is obtained by coupling the individual orbital angular momenta by using the **Clebsch–Gordan series**:

$$L = l_1 + l_2, l_1 + l_2 - 1, \ldots, |l_1 - l_2| \tag{44}$$

The modulus signs are attached to $l_1 - l_2$ because L is non-negative. The maximum value, $L = l_1 + l_2$, is obtained when the two orbital angular momenta are in the same direction; the lowest value, $|l_1 - l_2|$, is obtained when they are in opposite directions. The intermediate values represent possible intermediate relative orientations of the two momenta (Fig. 13.34). For two p electrons (for which $l_1 = l_2 = 1$), $L = 2, 1, 0$. The code for converting the value of L into a letter is the same as for the $s, p, d, f, \ldots$ designation of orbitals, but uses upper-case Roman letters:

L:	0	1	2	3	4	5	6 ...
	S	P	D	F	G	H	I ...

Thus, a p^2 configuration can give rise to D, P, and S terms. The terms differ in energy on account of the different spatial distribution of the electrons and the consequent differences in repulsion between them.

A closed shell has zero orbital angular momentum because all the individual orbital angular momenta sum to zero. Therefore, when working out term symbols, we need consider only the electrons of the unfilled shell. In the case of a single electron outside a closed shell, the value of L is the same as the value of l; so the configuration [Ne]$3s^1$ has only an S term.

Example 13.6 Deriving the total angular momentum of a configuration

Find the terms that can arise from the configurations (a) d^2, (b) p^3.

Method Use the Clebsch–Gordan series and begin by finding the minimum value of L (so that we know where the series terminates). When there are more than two electrons to couple together, use two series in succession: first couple two electrons, and then couple the third to each combined state, and so on.

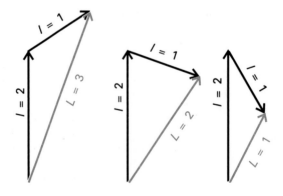

13.34 The total angular orbital momenta of a p electron and a d electron correspond to $L = 3, 2$, and 1 and reflect the different relative orientations of the two momenta.

Answer (a) Minimum value: $|l_1 - l_2| = |2 - 2| = 0$. Therefore,

$$L = 2 + 2, 2 + 2 - 1, \ldots, 0 = 4, 3, 2, 1, 0$$

corresponding to G, F, D, P, S terms, respectively. (b) First coupling: minimum value: $|1 - 1| = 0$. Therefore,

$$L' = 1 + 1, 1 + 1 - 1, \ldots, 0 = 2, 1, 0$$

Now couple l_3 with $L' = 2$, to give $L = 3, 2, 1$; with $L' = 1$, to give $L = 2, 1, 0$; and with $L' = 0$, to give $L = 1$. The overall result is

$$L = 3, 2, 2, 1, 1, 1, 0$$

giving one F, two D, three P, and one S term.

- -

Self-test 13.9 Repeat the question for the configurations (a) $f^1 d^1$ and (b) d^3.

[(a) H, G, F, D, P; (b) I, 2H, 3G, 4F, 5D, 3P, S]

(b) The multiplicity

When there are several electrons to be taken into account, we must assess their total spin angular momentum quantum number, S (a non-negative integer or half integer).[8] Once again, we use the Clebsch–Gordan series in the form

$$S = s_1 + s_2, s_1 + s_2 - 1, \ldots, |s_1 - s_2| \tag{45}$$

to decide on the value of S, noting that each electron has $s = \frac{1}{2}$, which gives $S = 1, 0$ (Fig. 13.35). If there are three electrons, the total spin angular momentum is obtained by coupling the third spin to each of the values of S for the first two spins, which results in $S = \frac{3}{2}, \frac{1}{2}$ and $S = \frac{1}{2}$.

The **multiplicity** of a term is the value of $2S + 1$. When $S = 0$ (as for a closed shell) the electrons are all paired and there is no net spin: this arrangement gives a singlet term, such as ^1S. A single electron has $S = s = \frac{1}{2}$, so a configuration such as [Ne]$3s^1$ can give rise to a doublet term, ^2S. The configuration [Ne]$3p^1$ likewise is a doublet, ^2P. When there are two unpaired electrons $S = 1$, so $2S + 1 = 3$, giving a triplet term, such as ^3D. We discussed the relative energies of singlets and triplets in Section 13.7 and saw that their energies differ on account of the different effects of spin correlation.

(c) The total angular momentum

As we have seen, the quantum number j tells us the relative orientation of the spin and orbital angular momenta of a single electron. The total angular momentum quantum number, J (a non-negative integer or half integer), does the same for several electrons. If there is a single electron outside a closed shell, $J = j$, with j either $l + \frac{1}{2}$ or $|l - \frac{1}{2}|$. The [Ne]$3s^1$ configuration has $j = \frac{1}{2}$ (because $l = 0$ and $s = \frac{1}{2}$), so the ^2S term has a single level, which we denote $^2S_{1/2}$. The [Ne]$3p^1$ configuration has $l = 1$; therefore $j = \frac{3}{2}$ and $\frac{1}{2}$; the ^2P term therefore has two levels, $^2P_{3/2}$ and $^2P_{1/2}$. These levels lie at different energies on account of the magnetic spin–orbit interaction.

If there are several electrons outside a closed shell we have to consider the coupling of all the spins and all the orbital angular momenta. This complicated problem can be simplified when the spin–orbit coupling is weak (for atoms of low atomic number), for then we can use the **Russell–Saunders coupling** scheme. This scheme is based on the view that, if spin–orbit coupling is weak, then it is effective only when all the orbital momenta are operating

13.35 For two electrons (which have $s = \frac{1}{2}$), only two total spin states are permitted ($S = 0, 1$). The state with $S = 0$ can have only one value of M_S ($M_S = 0$) and is a singlet; the state with $S = 1$ can have any of three values of M_S ($+1, 0, -1$) and is a triplet. The vector representations of the singlet and triplet states are shown in Figs. 13.20 and 13.26, respectively.

8 Distinguish italic S, the total spin quantum number, from Roman S, the term label.

cooperatively. We therefore imagine that all the orbital angular momenta of the electrons couple to give a total L, and that all the spins are similarly coupled to give a total S. Only at this stage do we imagine the two kinds of momenta coupling through the spin–orbit interaction to give a total J. The permitted values of J are given by the Clebsch–Gordan series

$$J = L + S, L + S - 1, \ldots, |L - S| \tag{46}$$

For example, in the case of the ^3D term of the configuration $[Ne]2p^13p^1$, the permitted values of J are 3, 2, 1 (because ^3D has $L = 2$ and $S = 1$), so the term has three levels, $^3D_3$, $^3D_2$, and $^3D_1$.

When $L > S$, the multiplicity is equal to the number of levels. For example, a ^2P term has the two levels $^2P_{3/2}$ and $^2P_{1/2}$, and ^3D has the three levels $^3D_3$, $^3D_2$, and $^3D_1$. However, this is not the case when $L < S$: the term ^2S, for example, has only the one level $^2S_{1/2}$.

Example 13.7 Deriving term symbols

Write the term symbols arising from the ground-state configurations of (a) Na and (b) F, and (c) the excited configurations $1s^22s^22p^13p^1$ of C.

Method Begin by writing the configurations, but ignore inner closed shells. Then couple the orbital momenta to find L and the spins to find S. Next, couple L and S to find J. Finally, express the term as $^{2S+1}\{L\}_J$, where $\{L\}$ is the appropriate letter. For F, for which the valence configuration is $2p^5$, treat the single gap in the closed-shell $2p^6$ configuration as a single particle.

Answer (a) For Na, the configuration is $[Ne]3s^1$, and we consider the single $3s$ electron. Because $L = l = 0$ and $S = s = \frac{1}{2}$, it is possible for $J = j = s = \frac{1}{2}$ only. Hence the term symbol is $^2S_{1/2}$.

(b) For F, the configuration is $[He]2s^22p^5$, which we can treat as $[Ne]2p^{-1}$ (where the notation $2p^{-1}$ signifies the absence of a $2p$ electron). Hence $L = 1$, and $S = s = \frac{1}{2}$. Two values of $J = j$ are allowed: $J = \frac{3}{2}, \frac{1}{2}$. Hence, the term symbols for the two levels are $^2P_{3/2}$, $^2P_{1/2}$.

(c) For C, the configuration is effectively $2p^13p^1$. This is a two-electron problem, and $l_1 = l_2 = 1, s_1 = s_2 = \frac{1}{2}$. It follows that $L = 2, 1, 0$ and $S = 1, 0$. The terms are therefore ^3D and ^1D, ^3P and ^1P, and ^3S and ^1S. For ^3D, $L = 2$ and $S = 1$; hence $J = 3, 2, 1$ and the levels are $^3D_3$, $^3D_2$, and $^3D_1$. For ^1D, $L = 2$ and $S = 0$, so the single level is $^1D_2$. The triplet of levels of ^3P is $^3P_2$, $^3P_1$, and $^3P_0$, and the singlet is $^1P_1$. For the ^3S term there is only one level, $^3S_1$ (because $J = 1$ only), and the singlet term is $^1S_0$.

Comment The reason why we have treated an excited configuration of carbon is that in the ground configuration, $2p^2$, the Pauli principle forbids some terms, and deciding which survive (^1D, ^3P, ^1S, in fact) is quite complicated. That is, there is a distinction between 'equivalent electrons', which are electrons that occupy the same orbitals, and 'inequivalent electrons', which are electrons that occupy different orbitals.

- -

Self-test 13.10 Write down the terms arising from the configurations (a) $2s^12p^1$, (b) $2p^13d^1$.

[(a) $^3P_2$, $^3P_1$, $^3P_0$, $^1P_1$;
(b) $^3F_4$, $^3F_3$, $^3F_2$, $^1F_3$, $^3D_3$, $^3D_2$, $^3D_1$, $^1D_2$,
$^3P_2$, $^3P_1$, $^3P_0$, $^1P_1$]

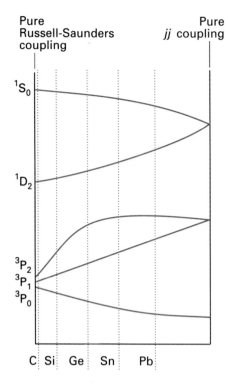

Pure
Russell-Saunders
coupling

Pure
jj coupling

$^1S_0$

$^1D_2$

$^3P_2$
$^3P_1$
$^3P_0$

C Si Ge Sn Pb

13.36 The correlation diagram for some of the states of a two-electron system. All atoms lie between the two extremes but, the heavier the atom, the closer it lies to the pure jj-coupling case.

Russell–Saunders coupling fails when the spin–orbit coupling is large (in heavy atoms). In that case, the individual spin and orbital momenta of the electrons are coupled into individual j values; then these momenta are combined into a grand total, J. This scheme is called **jj-coupling**. For example, in a p^2 configuration, the individual values of j are $\frac{3}{2}$ and $\frac{1}{2}$ for each electron. If the spin and the orbital angular momentum of each electron are coupled together strongly, it is best to consider each electron as a particle with angular momentum $j = \frac{3}{2}$ or $\frac{1}{2}$. These individual total momenta then couple as follows:

$$j_1 = \tfrac{3}{2} \text{ and } j_2 = \tfrac{3}{2} \quad J = 3, 2, 1, 0$$
$$j_1 = \tfrac{3}{2} \text{ and } j_2 = \tfrac{1}{2} \quad J = 2, 1$$
$$j_1 = \tfrac{1}{2} \text{ and } j_2 = \tfrac{3}{2} \quad J = 2, 1$$
$$j_1 = \tfrac{1}{2} \text{ and } j_2 = \tfrac{1}{2} \quad J = 1, 0$$

For heavy atoms, in which jj-coupling is appropriate, it is best to discuss their energies using these quantum numbers.

Although jj-coupling should be used for assessing the energies of heavy atoms, the term symbols derived from Russell–Saunders coupling can still be used as labels. To see why this procedure is valid, we need to examine how the energies of the atomic states change as the spin–orbit coupling increases in strength. Such a **correlation diagram** is shown in Fig. 13.36. It shows that there is a correspondence between the low spin–orbit coupling (Russell–Saunders coupling) and high spin–orbit coupling (jj-coupling) schemes, so the labels derived by using the Russell–Saunders scheme can be used to label the states of the jj-coupling scheme.

(d) Selection rules

Any state of the atom, and any spectral transition, can be specified by using term symbols. For example, the transitions giving rise to the yellow sodium doublet (which were shown in Fig. 13.32) are

$$3p^1 \, {}^2P_{3/2} \rightarrow 3s^1 \, {}^2S_{1/2} \qquad 3p^1 \, {}^2P_{1/2} \rightarrow 3s^1 \, {}^2S_{1/2}$$

By convention, the upper term precedes the lower. The corresponding absorptions are therefore denoted

$$^2P_{3/2} \leftarrow {}^2S_{1/2} \qquad {}^2P_{1/2} \leftarrow {}^2S_{1/2}$$

(The configurations have been omitted.)

We have seen that selection rules arise from the conservation of angular momentum during a transition and from the fact that a photon has a spin of 1. They can therefore be expressed in terms of the term symbols, because the latter carry information about angular momentum. A detailed analysis leads to the following rules:

$$\Delta S = 0 \qquad \Delta L = 0, \pm 1 \qquad \Delta l = \pm 1$$
$$\Delta J = 0, \pm 1, \text{ but } J = 0 \not\leftrightarrow J = 0 \tag{47}$$

The rule about ΔS (no change of overall spin) stems from the fact that the light does not affect the spin directly. The rules about ΔL and Δl express the fact that the orbital angular momentum of an individual electron must change (so $\Delta l = \pm 1$), but whether or not this results in an overall change of orbital momentum depends on the coupling.

The selection rules given above apply when Russell–Saunders coupling is valid (in light atoms). If we insist on labelling the terms of heavy atoms with symbols like ^3D, then we shall find that the selection rules progressively fail as the atomic number increases because the quantum numbers S and L become ill defined as jj-coupling becomes more appropriate. As explained above, Russell–Saunders term symbols are only a convenient way of labelling the terms of heavy atoms: they do not bear any direct relation to the actual angular momenta of the electrons in a heavy atom. For this reason, transitions between singlet and triplet states (for which $\Delta S = \pm 1$), while forbidden in light atoms, are allowed in heavy atoms.

13.10 The effect of magnetic fields

Orbital and spin angular momenta give rise to magnetic moments (recall the evidence provided for electron spin by the Stern–Gerlach experiment, Section 12.8). It can be expected that the application of a magnetic field should modify an atom's spectrum. We shall first establish how the energies of an atom depend on the strength of an external field and then see how the spectrum is affected.

(a) The magnetic moment of an electron

The orbital angular momentum of an electron around the z-axis (which we now take as the direction of the applied field) is $m_l\hbar$. Because the component of magnetic moment on the z-axis, μ_z, is proportional to the angular momentum around that axis, we can write

$$\mu_z = \gamma_e m_l \hbar \tag{48}$$

where γ_e is a constant called the **magnetogyric ratio** of the electron. If the magnetic moment is treated as arising from the circulation of an electron of charge $-e$, standard electromagnetic theory gives

$$\gamma_e = -\frac{e}{2m_e} \tag{49}$$

The negative sign (arising from the sign of the electron's charge) shows that the orbital magnetic moment of the electron is antiparallel to its orbital angular momentum (as was depicted in Fig. 13.28). It follows that the possible values of μ_z are

$$\mu_z = -\frac{e}{2m_e} \times m_l\hbar = -\mu_B m_l \tag{50}$$

where the **Bohr magneton**, μ_B, is

$$\mu_B = \frac{e\hbar}{2m_e} \tag{51}$$

Its numerical value is $9.274 \times 10^{-24}\ \mathrm{J\,T^{-1}}$. The Bohr magneton is often regarded as the fundamental quantum of magnetic moment.

The energy of a magnetic moment in a magnetic field of magnitude B in the z-direction is[9]

$$E = -\mu_z B \tag{52}$$

Therefore, in the presence of a magnetic field, an electron in a state with quantum number m_l has an additional contribution to its energy given by

$$E_{m_l} = \mu_B m_l B \tag{53}$$

The same expression, but with m_l replaced by M_L, applies when the orbital magnetic moment arises from several electrons.

A p electron has $l = 1$ and $m_l = 0, \pm 1$. In the absence of a magnetic field, these three states are degenerate. When a field is present, the degeneracy is removed: the state with

13.37 The different energies of the m_l states in a magnetic field are represented by different rates of precession of the vectors representing the angular momentum.

9   This is a result from standard magnetic theory. B is actually the magnetic induction, and is measured in tesla, T; $1\ \mathrm{T} = 1\ \mathrm{kg\,s^{-2}\,A^{-1}}$. The unit gauss, G, is also occasionally used: $1\ \mathrm{T} = 10^4\ \mathrm{G}$.

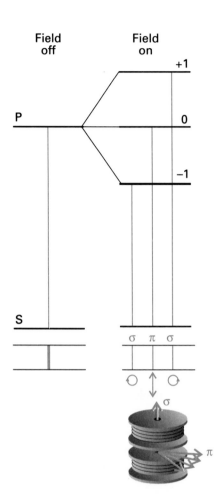

13.38 The normal Zeeman effect. On the left, when the field is off, a single spectral line is observed. When the field is on, the line splits into three, with different polarizations. The circularly polarized lines are called the σ-lines; the plane-polarized lines are called π-lines. Which line is observed depends on the orientation of the observer.

$m_l = +1$ moves up in energy by $\mu_B\mathcal{B}$, the state with $m_l = 0$ is unchanged, and the state with $m_l = -1$ moves down by $\mu_B\mathcal{B}$:

$$E_{+1} = +\mu_B\mathcal{B} \qquad E_0 = 0 \qquad E_{-1} = -\mu_B\mathcal{B}$$

The different energies arising from an interaction with an external field are sometimes represented on the vector model by picturing the vectors as **precessing**, or sweeping round their cones, with the rate of precession proportional to the energy of the state (Fig. 13.37).

The spin magnetic moment of an electron is also proportional to its angular momentum. However, it is not given by $\gamma_e m_s \hbar$ but by about twice this value:

$$\mu_z = g_e\gamma_e m_s\hbar \qquad g_e = 2.002\,319\dots \qquad (54)$$

The extra factor g_e is called the **g-value** of the electron. The factor 2 (as distinct from 2.0023) is derived from the Dirac equation; the additional 0.0023 arises from interactions of the electron with the electromagnetic fluctuations in the vacuum that surrounds the electron. The energy of an electron in a state m_s in a magnetic field of magnitude $\mathcal{B}$ in the z-direction is

$$E_{m_s} = -g_e\gamma_e m_s\hbar\mathcal{B} = g_e\mu_B m_s\mathcal{B} \qquad (55)$$

The same expression, but with m_s replaced by M_S, applies to the total magnetic moment arising from the spin of several electrons.

(b) The Zeeman effect

The **Zeeman effect** is the modification of an atomic spectrum by the application of a strong magnetic field. In particular, the **normal Zeeman effect** is the observation of three lines in the spectrum where, in the absence of the field, there is only one (Fig. 13.38). The splitting is in fact very small: a field of 2 T (20 kG) is needed to produce a splitting of about 1 cm^{-1}, which should be compared with typical optical transition wavenumbers of 20 000 cm^{-1} and more.

Much more common than the normal Zeeman effect is the **anomalous Zeeman effect**, in which the original line splits into more than three components. The origin of this complexity is the anomalous magnetic moment of electron spin, which results in a more complicated splitting pattern.

Checklist of key ideas

- [ ] electronic structure
- [ ] hydrogenic atom
- [ ] many–electron atom
- [ ] spectroscopy
- [ ] spectrum

The structure and spectra of hydrogenic atoms

- [ ] Balmer series
- [ ] Lyman series
- [ ] Paschen series
- [ ] Rydberg constant
- [ ] Ritz combination principle
- [ ] Bohr frequency condition (4)

13.1 The structure of hydrogenic atoms
- [ ] reduced mass (7)
- [ ] radial wave equation (11)
- [ ] Bohr radius (15)
- [ ] associated Laguerre polynomial (16)

13.2 Atomic orbitals and their energies
- [ ] atomic orbital
- [ ] principal quantum number
- [ ] bound states
- [ ] unbound states
- [ ] continuum states

- [ ] ground state
- [ ] shell
- [ ] subshell
- [ ] boundary surface
- [ ] radial distribution function
- [ ] most probable radius
- [ ] nodal plane

13.3 Spectroscopic transitions and selection rules
- [ ] transition
- [ ] allowed transition
- [ ] forbidden transition
- [ ] selection rule (27)

- [ ] transition dipole moment (28)
- [ ] Grotrian diagram

The structures of many-electron atoms

13.4 The orbital approximation
- [ ] orbital approximation
- [ ] configuration
- [ ] Pauli exclusion principle
- [ ] Pauli principle
- [ ] complete shell
- [ ] closed shell
- [ ] effective nuclear charge

- ☐ shielded nuclear charge
- ☐ shielding constant (33)
- ☐ penetration
- ☐ valence electrons
- ☐ building-up principle
- ☐ Hund's rule
- ☐ spin correlation
- ☐ first ionization energy
- ☐ second ionization energy
- ☐ standard enthalpy of ionization (34)
- ☐ electron affinity
- ☐ standard enthalpy of electron gain (35)

13.5 Self-consistent field orbitals
- ☐ Hartree–Fock self-consistent field (SCF)
- ☐ self-consistent

The spectra of complex atoms

13.6 Quantum defects and ionization limits
- ☐ quantum defect
- ☐ Rydberg state

13.7 Singlet and triplet states
- ☐ singlet
- ☐ triplet

13.8 Spin–orbit coupling
- ☐ spin–orbit coupling
- ☐ total angular momentum
- ☐ level
- ☐ spin–orbit coupling constant
- ☐ fine structure

13.9 Term symbols and selection rules
- ☐ term symbol
- ☐ total orbital angular momentum quantum number
- ☐ total spin quantum number
- ☐ total angular momentum quantum number

- ☐ Clebsch–Gordan series (44)
- ☐ multiplicity
- ☐ Russell–Saunders coupling
- ☐ jj-coupling
- ☐ correlation diagram

13.10 The effect of magnetic fields
- ☐ magnetogyric ratio (49)
- ☐ Bohr magneton (51)
- ☐ precessing
- ☐ g-value (54)
- ☐ Zeeman effect
- ☐ normal Zeeman effect
- ☐ anomalous Zeeman effect

Further reading

Articles of general interest

R.D. Allendoerfer, Teaching the shapes of the hydrogenlike and hybrid atomic orbitals. *J. Chem. Educ.* **67**, 37 (1990).

G.L. Breneman, Order out of chaos: shapes of hydrogen orbitals. *J. Chem. Educ.* **65**, 31 (1988).

J. Mason, Periodic contractions among the elements: or, on being the right size. *J. Chem. Educ.* **65**, 17 (1988).

N. Agmon, Ionization potentials for isoelectronic series. *J. Chem. Educ.* **65**, 42 (1988).

E.M.R. Kiremire, A numerical algorithm technique for deriving Russell–Saunders (R–S) terms. *J. Chem. Educ.* **64**, 951 (1987).

L. Guofan and M.L. Elizey Jr, Finding the terms of configurations of equivalent electrons by partitioning total spins. *J. Chem. Educ.* **64**, 771 (1987).

E.R. Scerri, Transition metal configurations and limitations of the orbital approximation. *J. Chem. Educ.* **66**, 481 (1989).

R.T. Myers, The periodicity of electron affinity. *J. Chem. Educ.* **67**, 307 (1990).

N. Shenkuan, The physical basis of Hund's rule: orbital contraction effects. *J. Chem. Educ.* **69**, 800 (1992).

K.D. Sen, M. Slamet, and V. Sahni, Atomic shell structure in Hartree–Fock theory. *Chem. Phys. Letts.* **205**, 313 (1993).

N.C. Pyper and M. Berry, Ionization energies revisited. *Educ. in Chem.* **27**, 135 (1990).

P.G. Nelson, Relative energies of $3d$ and $4s$ orbitals. *Educ. in Chem.* **29**, 84 (1992).

K. D. Sen, Electronegativity. In *Structure and bonding*, Vol. 6. Springer, New York (1987).

T.L. Meek, Electronegativities of the noble gases. *J. Chem. Educ.* **72**, 17 (1995).

N.E. Robertson, S.O. Boggioni, and F.E. Henriquez, A simple mathematical approach to experimental ionization energies of atoms: a new method for estimating the extent of shielding. *J. Chem. Educ.* **71**, 101 (1994).

L. Lessinger, Birge–Sponer extrapolation, and the electronic absorption spectrum of I_2. *J. Chem. Educ.* **71**, 388 (1994).

J.C. Wheeler, Electron affinities of the alkaline earth metals and the sign convention for electron affinity. *J. Chem. Educ.* **74**, 123 (1997).

D.F. Fitts, Ladder operator treatment of the radial equation for the hydrogenlike atom. *J. Chem. Educ.* **72**, 1066 (1995).

H. Kleindienst, A. Lüchow, and R. Benois, Pauli principle and permutation symmetry. *J. Chem. Educ.* **72**, 1019 (1995).

C.W. Haigh, The theory of atomic spectroscopy: jj coupling, intermediate coupling, and configuration interaction. *J. Chem. Educ.* **72**, 206 (1995).

J.C.A. Boeyens, Understanding electron spin. *J. Chem. Educ.* **72**, 412 (1995).

M. Melrose and E.R. Scerri, Why the $4s$ orbital is occupied before the $3d$. *J. Chem. Educ.* **74**, 498 (1996).

L.G. Vanquickenborne, K. Pierloot, and D. Devoghel, Transition metals and the *Aufbau* principle. *J. Chem. Educ.* **71**, 469 (1994).

S.J. Gauerke and M.L. Campbell, A simple, systematic method for determining J levels for jj coupling. *J. Chem. Educ.* **71**, 457 (1994).

K. Bonin and W. Happer, Atomic spectroscopy. In *Encyclopedia of applied physics* (ed. G.L. Trigg), **2**, 245. VCH, New York (1991).

B. Bederson, Atoms. In *Encyclopedia of applied physics* (ed. G.L. Trigg), **2**, 285. VCH, New York (1991).

J.C. Morrison, A.W. Weiss, K. Kirby, and D. Cooper, Electronic structure of atoms and molecules. In *Encyclopedia of applied physics* (ed. G.L. Trigg), **6**, 45. VCH, New York (1993).

Texts and sources of data and information

P.W. Atkins, *Quanta: a handbook of concepts.* Oxford University Press (1991).

P.W. Atkins and R.S. Friedman, *Molecular quantum mechanics.* Oxford University Press (1997).

T.P. Softley, *Atomic spectra*, Oxford Chemistry Primers. Oxford University Press (1994).

P.R. Scott and W.G. Richards, *Energy levels in atoms and molecules*, Oxford Chemistry Primers, Oxford University Press (1994).

P.A. Cox, *Introduction to quantum theory and atomic structure*, Oxford Chemistry Primers. Oxford University Press (1996).

C.F. Fischer, *The Hartree–Fock method for atoms.* Wiley, New York (1977).

E.U. Condon and H. Odabaşi, *Atomic structure.* Cambridge University Press (1980).

S. Bashkin and J.O. Stonor, Jr, *Atomic energy levels and Grotrian diagrams.* North-Holland, Amsterdam (1975–1982).

Exercises

13.1 (a) When ultraviolet radiation of wavelength 58.4 nm from a helium lamp is directed on to a sample of krypton, electrons are ejected with a speed of 1.59×10^6 m s^{-1}. Calculate the ionization energy of krypton.

13.1 (b) When ultraviolet radiation of wavelength 58.4 nm from a helium lamp is directed on to a sample of xenon, electrons are ejected with a speed of 1.79×10^6 m s^{-1}. Calculate the ionization energy of xenon.

13.2 (a) Consider the $2s$ radial wavefunction. Show that it has two extrema in its amplitude, and locate them.

13.2 (b) Consider the $3s$ radial wavefunction. Show that it has three extrema in its amplitude, and locate them.

13.3 (a) Locate the radial nodes in the $3s$ orbital of an H atom.

13.3 (b) Locate the radial nodes in the $3p$ orbital of an H atom.

13.4 (a) The wavefunction for the ground state of a hydrogen atom is Ne^{-r/a_0}. Determine the normalization constant N.

13.4 (b) The wavefunction for the $2s$ orbital of a hydrogen atom is $N(2 - r/a_0)e^{-r/2a_0}$. Determine the normalization constant N.

13.5 (a) Calculate the average kinetic and potential energies of an electron in the ground state of a hydrogen atom.

13.5 (b) Calculate the average kinetic and potential energies of a $2s$ electron in a hydrogenic atom of atomic number Z.

13.6 (a) Write down the expression for the radial distribution function of a $2s$ electron in a hydrogenic atom and determine the radius at which the electron is most likely to be found.

13.6 (b) Write down the expression for the radial distribution function of a $3s$ electron in a hydrogenic atom and determine the radius at which the electron is most likely to be found.

13.7 (a) What is the orbital angular momentum of an electron in the orbitals (a) $1s$, (b) $3s$, (c) $3d$? Give the numbers of angular and radial nodes in each case.

13.7 (b) What is the orbital angular momentum of an electron in the orbitals (a) $4d$, (b) $2p$, (c) $3p$? Give the numbers of angular and radial nodes in each case.

13.8 (a) Calculate the permitted values of j for (a) a d electron, (b) an f electron.

13.8 (b) Calculate the permitted values of j for (a) a p electron, (b) an h electron.

13.9 (a) An electron in two different states of an atom is known to have $j = \frac{3}{2}$ and $\frac{1}{2}$. What is its orbital angular momentum quantum number in each case?

13.9 (b) What are the allowed total angular momentum quantum numbers of a composite system in which $j_1 = 5$ and $j_2 = 3$?

13.10 (a) State the orbital degeneracy of the levels in a hydrogen atom that have energy (a) $-hc\mathcal{R}_H$; (b) $-\frac{1}{9}hc\mathcal{R}_H$; (c) $-\frac{1}{25}hc\mathcal{R}_H$.

13.10 (b) State the orbital degeneracy of the levels in a hydrogenic atom (Z in parentheses) that have energy (a) $-4hc\mathcal{R}_{atom}$ (2), (b) $-\frac{1}{4}hc\mathcal{R}_{atom}$ (4), and (c) $-hc\mathcal{R}_{atom}$ (5).

13.11 (a) What information does the term symbol $^1D_2$ provide about the angular momentum of an atom?

13.11 (b) What information does the term symbol $^3F_4$ provide about the angular momentum of an atom?

13.12 (a) At what radius does the probability of finding an electron at a point in the H atom fall to 50 per cent of its maximum value?

13.12 (b) At what radius in the H atom does the radial distribution function of the ground state have (a) 50 per cent, (b) 75 per cent of its maximum value?

13.13 (a) Which of the following transitions are allowed in the normal electronic emission spectrum of an atom: (a) $2s \rightarrow 1s$, (b) $2p \rightarrow 1s$, (c) $3d \rightarrow 2p$?

13.13 (b) Which of the following transitions are allowed in the normal electronic emission spectrum of an atom: (a) $5d \rightarrow 2s$, (b) $5p \rightarrow 3s$, (c) $5p \rightarrow 3f$?

13.14 (a) How many electrons can occupy the following subshells: (a) $1s$, (b) $3p$, (c) $3d$, and (d) $6g$?

13.14 (b) How many electrons can occupy the following subshells: (a) $2s$, (b) $4d$, (c) $6f$, and (d) $6h$?

13.15 (a) (a) Write the electronic configuration of the Ni^{2+} ion. (b) What are the possible values of the total spin quantum numbers S and M_S for this ion?

13.15 (b) (a) Write the electronic configuration of the V^{2+} ion. (b) What are the possible values of the total spin quantum numbers S and M_S for this ion?

13.16 (a) Suppose that an atom has (a) 2, (b) 3 electrons in different orbitals. What are the possible values of the total spin quantum number S? What is the multiplicity in each case?

13.16 (b) Suppose that an atom has (a) 4, (b) 5, electrons in different orbitals. What are the possible values of the total spin quantum number S? What is the multiplicity in each case?

13.17 (a) What atomic terms are possible for the electron configuration $ns^1 nd^1$? Which term is likely to lie lowest in energy?

13.17 (b) What atomic terms are possible for the electron configuration $np^1 nd^1$? Which term is likely to lie lowest in energy?

13.18 (a) What values of J may occur in the terms (a) ^1S, (b) ^2P, (c) ^3P? How many states (distinguished by the quantum number M_J) belong to each level?

13.18 (b) What values of J may occur in the terms (a) ^3D, (b) ^4D, (c) ^2G? How many states (distinguished by the quantum number M_J) belong to each level?

13.19 (a) Give the possible term symbols for (a) Li $[He]2s^1$, (b) Na $[Ne]3p^1$.

13.19 (b) Give the possible term symbols for (a) Sc $[Ar]3d^1 4s^2$, (b) Br $[Ar]3d^{10} 4s^2 4p^5$.

13.20 (a) Calculate the magnetic induction, B, required to produce a splitting of 1.0 cm^{-1} between the states of a ^1P term.

13.20 (b) Calculate the magnetic induction, B, required to produce a splitting of 0.784 cm^{-1} between the states of a ^1D term.

Problems

Numerical problems

13.1 The Humphreys series is another group of lines in the spectrum of atomic hydrogen. It begins at $12\,368$ nm and has been traced to 3281.4 nm. What are the transitions involved? What are the wavelengths of the intermediate transitions?

13.2 A series of lines in the spectrum of atomic hydrogen lie at 656.46 nm, 486.27 nm, 434.17 nm, and 410.29 nm. What is the wavelength of the next line in the series? What is the ionization energy of the atom when it is in the lower state of the transitions?

13.3 The Li^{2+} ion is hydrogenic and has a Lyman series at $740\,747$ cm^{-1}, $877\,924$ cm^{-1}, $925\,933$ cm^{-1}, and beyond. Show that the energy levels are of the form $-hc\mathcal{R}/n^2$ and find the value of $\mathcal{R}$ for this ion. Go on to predict the wavenumbers of the two longest-wavelength transitions of the Balmer series of the ion and find the ionization energy of the ion.

13.4 A series of lines in the spectrum of neutral Li atoms rise from combinations of $1s^2 2p^1\ ^2P$ with $1s^2 nd^1\ ^2D$ and occur at 610.36 nm, 460.29 nm, and 413.23 nm. The d orbitals are hydrogenic. It is known that the ^2P term lies at 670.78 nm above the ground state, which is $1s^2 2s^1\ ^2S$. Calculate the ionization energy of the ground-state atom.

13.5 The characteristic emission from K atoms when heated is purple and lies at 770 nm. On close inspection, the line is found to have two closely spaced components, one at 766.70 nm and the other at 770.11 nm. Account for this observation, and deduce what information you can.

13.6 Calculate the mass of the deuteron given that the first line in the Lyman series of H lies at $82\,259.098$ cm^{-1} whereas that of D lies at $82\,281.476$ cm^{-1}. Calculate the ratio of the ionization energies of H and D.

13.7 Positronium consists of an electron and a positron (same mass, opposite charge) orbiting round their common centre of mass. The broad features of the spectrum are therefore expected to be hydrogen-like, the differences arising largely from the mass differences. Predict the wavenumbers of the first three lines of the Balmer series of positronium. What is the binding energy of the ground state of positronium?

13.8 In 1976 it was mistakenly believed that the first of the 'superheavy' elements had been discovered in a sample of mica. Its atomic number was believed to be 126. What is the most probable distance of the innermost electrons from the nucleus of an atom of this element? (In such elements, relativistic effects are very important, but ignore them here.)

Theoretical problems

13.9 Is an electron further from the nucleus on average when it is in a $2s$ orbital or a $2p$ orbital?

13.10 What is the most probable point (not radius) at which a $2p$ electron will be found in the hydrogen atom?

13.11 Show by explicit integration that (a) hydrogenic $1s$ and $2s$ orbitals, (b) $2p_x$ and $2p_y$ orbitals are mutually orthogonal.

13.12 Determine whether the p_x and p_y orbitals are eigenfunctions of l_z. If not, does a linear combination exist that is an eigenfunction of l_z?

13.13 Show that l_z and l^2 both commute with the hamiltonian for a hydrogen atom. What is the significance of this result?

13.14 The 'size' of an atom is sometimes considered to be measured by the radius of a sphere that contains 90 per cent of the charge density of the electrons in the outermost occupied orbital. Calculate the 'size' of a hydrogen atom in its ground state according to this definition.

13.15 One of the most famous of the obsolete theories of the hydrogen atom was proposed by Bohr. It has been replaced by quantum mechanics but, by a remarkable coincidence (not the only one where the Coulomb potential is concerned), the energies it predicts agree exactly with those obtained from the Schrödinger equation. In the Bohr atom, an electron travels in a circle around the nucleus. The Coulombic force of attraction $(Ze^2/4\pi\varepsilon_0 r^2)$ is balanced by the centrifugal effect of the orbital motion. Bohr proposed that the angular momentum is limited to integral values of $\hbar$. When the two forces are balanced, the atom remains in a stationary state until it makes a spectral transition. Calculate the energies of a hydrogenic atom by using the Bohr model.

13.16 The Bohr model of the atom is specified in Problem 13.15. What features of it are untenable according to quantum mechanics? How does the Bohr ground state differ from the actual ground state? Is there an experimental distinction between the Bohr and quantum mechanical models of the ground state?

13.17 Atomic units of length and energy may be based on the properties of a particular atom. The usual choice is that of a hydrogen atom, with the unit of length being the Bohr radius, a_0, and the unit of energy being the energy of the $1s$ orbital. If the positronium atom (e^+, e^-) were used instead, with analogous definitions of units of length and energy, what would be the relation between these two sets of atomic units?

Additional problems supplied by Carmen Giunta and Charles Trapp

13.18 The diameters of atoms can be estimated from their densities in a condensed state. Calculate the diameters of hydrogen and uranium atoms in this manner from information in the *Data section*. One finds that all atoms are roughly the same size, with $r \approx 0.3 \pm 0.1$ nm. Why is that? In a plot of atomic radius against atomic number some periodicity is evident, but not to the extent seen in a plot of first ionization energies against atomic number. Explain this observation.

13.19 In the Bohr model of the hydrogen atom, the electron orbits the nucleus at a distance of 52.9 pm. Calculate the speed of the electron in the first Bohr orbit. Considering the electron and proton to be classical charges, calculate the electrical field strength at the electron and the magnetic field strength at the proton. Compare to field strengths commonly available in the laboratory.

13.20 Use the radial wave equation for the hydrogen atom to demonstrate that the energies of the $2s$ and $2p$ orbitals are identical.

13.21 Dimensionless ratios that occur in the physical sciences are thought to be of fundamental significance. These ratios tend to be clustered around $(10^{20})^n$, where $n = 0, 1, 2, 3$, and 4. One such ratio in the $n = 0$ group is the mass ratio of the two fundamental particles, the proton and the electron. Scientists are puzzled as to why this ratio should be close to 2000. The precise value of the ratio is determined by comparison of the atomic spectral lines in H and He$^+$. (a) Derive the following relations for the first line in any of the series (Lyman, Balmer, etc.) for H and He$^+$:

$$\gamma = \frac{\frac{1}{4}\tilde{\nu}_{He} - \tilde{\nu}_H}{\tilde{\nu}_H} = \frac{\mu_{He} - \mu_H}{\mu_H}$$

where μ is a reduced mass. (b) Calculate m_H/m_e from the following data.

	$\lambda(n_2 \rightarrow n_1)$/nm	$\mathcal{R}_J$/cm^{-1}
H	121.5664	109 677.7
He$^+$	30.3779	109 722.4

First do the calculation from the wavelength data; then derive a formula for the mass ratio in terms of the Rydberg constants $\mathcal{R}_J$ of the species J and repeat the calculation of the mass ratio from that data.

13.22 Highly excited atoms are said to be in a 'high Rydberg state' and have electrons with large principal quantum numbers. Such 'Rydberg atoms' have several unusual properties and have attracted much attention in recent years, for example, in astrophysics and radioastronomy. For hydrogen atoms with large n, derive a relation for the separation of energy levels. Calculate this separation for $n = 100$; also calculate the average radius, the geometric cross-section, and the ionization energy. Could a thermal collision with another hydrogen atom ionize this Rydberg atom? What minimum velocity of the second atom is required? Could a normal-sized neutral H atom simply pass through the Rydberg atom leaving it undisturbed? What might the radial wavefunction for a $100s$ orbital be like?

13.23 W.P. Wijesundera, S.H. Vosko, and F.A. Parpia (*Phys. Rev. A* **51**, 278 (1995)) attempted to determine the electron configuration of the ground state of lawrencium, element 103. The two contending configurations are $[Rn]5f^{14}7s^27p^1$ and $[Rn]5f^{14}6d^7s^2$. Write down the term symbols for each of these configurations, and identify the lowest level within each configuration. Which level would be lowest according to a simple estimate of spin–orbit coupling?

13.24 Stern–Gerlach splittings of atomic beams are small and require either large magnetic field gradients or long magnets for their observation. For a beam of atoms with zero orbital angular momentum, such as H or Ag, the deflection is given by $x = \pm(\mu_B L^2/4E_K)d\mathcal{B}/dz$, where L is the length of the magnet, E_k is the average kinetic energy of the atoms in the beam, and $d\mathcal{B}/dz$ is the magnetic field gradient. (a) Use the Maxwell–Boltzmann velocity distribution to show that the average translational kinetic energy of atoms emerging as a beam from a pinhole in an oven at temperature T is $2kT$. (b) Calculate the magnetic field gradient required to produce a splitting of 1.00 mm in a beam of Ag atoms from an oven at 1000 K with a magnet of length 50 cm.

14

Molecular structure

The concepts developed in Chapter 13, particularly those of orbitals, can be extended to a description of the electronic structures of molecules. There are two principal quantum mechanical theories of molecular electronic structure. In valence-bond theory the starting point is the concept of the shared electron pair. We see how to write the wavefunction for such a pair, and how it may be extended to account for the structures of a wide variety of molecules. The theory introduces the concepts of σ and π bonds, promotion, and hybridization that are used widely in chemistry. In molecular orbital theory (with which the bulk of the chapter is concerned), the concept of atomic orbital is extended to that of molecular orbital, which is a wavefunction that spreads over all the atoms in a molecule. This theory may be extended to the description of the electronic properties of solids, and used to account for electrical conduction and semiconduction.

The Born–Oppenheimer approximation

All theories of molecular structure make the same simplification at the outset. Whereas the Schrödinger equation for a hydrogen atom can be solved exactly, an exact solution is not possible for any molecule because the simplest molecule consists of three particles (two nuclei and one electron). The **Born–Oppenheimer approximation** is therefore adopted, in which it is supposed that the nuclei, being so much heavier than an electron, move relatively slowly, and may be treated as stationary while the electrons move relative to them. We can therefore think of the nuclei as being fixed at an arbitrary separation R, and then solve the Schrödinger equation for the wavefunction of the electrons alone.

The approximation is quite good for ground-state molecules, for calculations suggest that the nuclei in H_2 move through only about 1 pm while the electron speeds through 1000 pm, so the error of assuming that the nuclei are stationary is small. Exceptions to the approximation's validity include certain excited states of polyatomic molecules and the

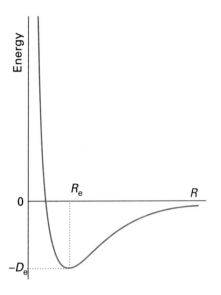

14.1 A molecular potential energy curve. The equilibrium bond length corresponds to the energy minimum.

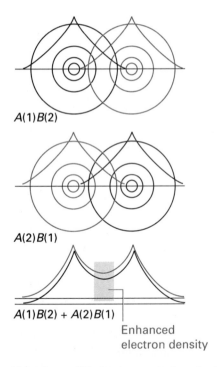

$A(1)B(2)$

$A(2)B(1)$

$A(1)B(2) + A(2)B(1)$

Enhanced electron density

14.2 It is very difficult to represent valence-bond wavefunctions because they refer to two electrons simultaneously. However, this illustration is an attempt. The atomic orbital for electron 1 is represented by the black contours, and that of electron 2 is represented by the green contours. The top illustration represents $A(1)B(2)$, and the middle illustration represents the contribution $A(2)B(1)$. When the two contributions are superimposed, there is interference between the black contributions and between the green contributions, resulting in an enhanced (two-electron) density in the internuclear region.

ground states of cations; both types of species are important when considering photoelectron spectroscopy (Section 17.8) and mass spectrometry.

The Born–Oppenheimer approximation allows us to select an internuclear separation, and (in principle) to solve the Schrödinger equation for the electrons at that nuclear separation. Then we choose a different separation and repeat the calculation, and so on. In this way we can explore how the energy of the molecule varies with bond length (and, in more complex molecules, with angles too), and obtain a **molecular potential energy curve** (Fig. 14.1).[1] It is called a *potential* energy curve because the kinetic energy of the stationary nuclei is zero. Once the curve has been calculated or determined experimentally (by using the spectroscopic techniques described in Chapters 16 and 17), we can identify the equilibrium bond length (the internuclear separation at the minimum of the curve) and the **bond dissociation energy**, D_0, which is closely related to the depth of the minimum below the energy of the infinitely widely separated atoms.[2]

Valence-bond theory

The **valence-bond theory** (VB theory) of bonding was the first to be developed. The language it introduced, which includes concepts such as spin-pairing, σ and π bonds, and hybridization, is widely used throughout chemistry. It is particularly widespread in the description of the properties and reactions of organic compounds.

14.1 The hydrogen molecule

The simplest molecule with an electron pair bond is H_2. We shall use this molecule to introduce the basic concepts of the theory.

(a) The spatial wavefunction

The wavefunction for an electron on each of two widely separated H atoms is

$$\psi = \psi_{H1sA}(\boldsymbol{r}_1)\psi_{H1sB}(\boldsymbol{r}_2)$$

if electron 1 is on atom A and electron 2 is on atom B. For simplicity, we shall write this wavefunction as $\psi = A(1)B(2)$. When the atoms are close, it is not possible to know whether it is electron 1 that is on A or electron 2. An equally valid description is therefore $\psi = A(2)B(1)$, in which electron 2 is on A and electron 1 is on B. When two outcomes are equally probable, quantum mechanics instructs us to describe the true state of the system as a superposition of the wavefunctions for each possibility (Section 11.5d), so a better description of the molecule than either wavefunction alone is

$$\psi = A(1)B(2) \pm A(2)B(1) \tag{1}$$

(These linear combinations are not normalized.) It turns out (as shown in the *Justification* below) that the combination with lower energy is the one with a $+$ sign, so the valence-bond wavefunction of the H_2 molecule is

$$\psi = A(1)B(2) + A(2)B(1) \tag{2}$$

---

1   When more than one molecular parameter is changed in a polyatomic molecule, we obtain a potential energy surface.

2   The dissociation energy differs from the depth of the well by an energy equal to the zero-point vibrational energy of the bonded atoms. If the depth of the well is denoted D_e, then $D_0 = D_e - \frac{1}{2}\hbar\omega$, where ω is the vibrational frequency of the bond.

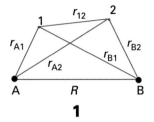

1

Justification 14.1

The VB wavefunction for H_2 is an approximate solution of the Schrödinger equation in which the potential energy of the two electrons is

$$V = -\frac{e^2}{4\pi\varepsilon_0}\left(\frac{1}{r_{A1}} + \frac{1}{r_{A2}} + \frac{1}{r_{B1}} + \frac{1}{r_{B2}}\right) + \frac{e^2}{4\pi\varepsilon_0 r_{12}}$$

The coordinates are specified in (1). The four terms in parentheses are the attractive contribution from the interaction between the electrons and the nuclei. The remaining term is the repulsive interaction between the two electrons. The energy of the molecule is calculated by evaluating the expectation value of the hamiltonian

$$H = -\frac{\hbar^2}{2m_e}\nabla_1^2 - \frac{\hbar^2}{2m_e}\nabla_2^2 + V + \frac{e^2}{4\pi\varepsilon_0 R}$$

with the expression for V given above; the final term is the potential energy of the nucleus–nucleus repulsion. When the wavefunctions in eqn 1 are used, the expectation value turns out to be

$$E_\pm = 2E_H + \frac{J \pm K}{1 \pm S^2} + \frac{e^2}{4\pi\varepsilon_0 R} \tag{3}$$

where E_H is the energy of a hydrogen atom and J and K are complicated collections of integrals over the wavefunctions. These integrals represent the interaction of the electrons with the nuclei and the mutual repulsion of the electrons. The integral S is the overlap integral, which is discussed in more detail shortly. The integrals J and K are both negative and the lower energy is achieved with the $+$ sign in eqn 1.

The formation of the bond in H_2 can be pictured as due to the high probability that the two electrons will be found between the two nuclei and hence will bind them together. More formally, the wave pattern represented by the term $A(1)B(2)$ interferes constructively with the wave pattern represented by the contribution $A(2)B(1)$, and there is an enhancement in the value of the wavefunction in the internuclear region (Fig. 14.2).

The electron distribution described by the wavefunction in eqn 2 is called a **σ bond**. A σ bond has cylindrical symmetry around the internuclear axis, and is so called because, when viewed along the internuclear axis, it resembles a pair of electrons in an s orbital (and σ is the Greek equivalent of s). More precisely, the electrons in a σ bond have zero orbital angular momentum about the internuclear axis.[3]

The molecular potential energy curve for H_2 is calculated by changing the internuclear separation R and evaluating the expectation value of the energy at each selected separation. The resulting graph is shown in Fig. 14.3. The energy falls below that of two separated H atoms as the two atoms are brought within bonding distance and each electron is free to migrate to the other atom. However, the energy reduction that follows from this process is counteracted by an increase in energy from the Coulombic repulsion between the two positively charged nuclei. This positive contribution to the energy becomes large as R becomes small. Consequently, the total potential energy curve passes through a minimum and then climbs to a strongly positive value at small internuclear separations.

(b) The role of electron spin

So far, the electron spin has not played a role in the argument, yet a chemist's picture of a covalent bond is one in which the spins of two electrons pair as the atomic orbitals overlap.

14.3 The molecular potential energy curve for the hydrogen molecule showing the variation of the energy of the molecule as the bond length is changed. The calculated curve refers to the valence-bond model.

---

[3] Recall from Section 12.6 that the orbital angular momentum of an electron is related to the number of angular nodes in its wavefunction, but there are no angular nodes in the wavefunction of a σ bond, so it has zero orbital angular momentum.

14.4 The orbital overlap and spin-pairing between electrons in two collinear p orbitals that result in the formation of a σ bond.

The origin of the role of spin is that the wavefunction given in eqn 2 can be formed only by a pair of electrons with opposed spins. Thus, spin-pairing is not an end in itself: it is a means of achieving a spatial wavefunction (and the probability distribution it implies) that corresponds to a low energy.

Justification 14.2

The Pauli principle requires the wavefunction of two electrons to change sign when the labels of the electrons are interchanged (see *Justification 13.7*). The total VB wavefunction for two electrons is

$$\Psi(1,2) = \{A(1)B(2) + A(2)B(1)\}\sigma(1,2)$$

where σ represents the spin component of the wavefunction. When the labels 1 and 2 are interchanged, this wavefunction becomes

$$\Psi(2,1) = \{A(2)B(1) + A(1)B(2)\}\sigma(2,1)$$
$$= \{A(1)B(2) + A(2)B(1)\}\sigma(2,1)$$

The Pauli principle requires that $\Psi(2,1) = -\Psi(1,2)$, which is satisfied only if $\sigma(2,1) = -\sigma(1,2)$. The combination of two spins that has this property is

$$\sigma_-(1,2) = \frac{1}{2^{1/2}}\{\alpha(1)\beta(2) - \alpha(2)\beta(1)\}$$

which corresponds to paired electron spins (Section 13.7). Therefore, we conclude that the state of lower energy (and hence the formation of a chemical bond) is achieved if the electron spins are paired.

14.2 Homonuclear diatomic molecules

The essential features of valence-bond theory are the pairing of the electrons and the accumulation of electron density in the internuclear region that stems from that pairing. The same description can be applied to more complex molecules, such as **homonuclear diatomic molecules**, which are diatomic molecules in which both atoms belong to the same element. Nitrogen, N_2, is an example. To construct the valence-bond description of N_2, we consider the valence electron configuration of each atom:

$$\text{N} \qquad 2s^2 2p_x^1 2p_y^1 2p_z^1$$

It is conventional to take the z-axis to be the internuclear axis, so we can imagine each atom as having a $2p_z$ orbital pointing towards a $2p_z$ orbital on the other atom (Fig. 14.4), with the $2p_x$ and $2p_y$ orbitals perpendicular to the axis. A σ bond is then formed by spin-pairing between the two electrons in the opposing $2p_z$ orbitals. Its spatial wavefunction is given by eqn 2, but now A and B stand for the two $2p_z$ orbitals.

The remaining $2p$ orbitals cannot merge to give σ bonds as they do not have cylindrical symmetry around the internuclear axis. Instead, the electrons in them merge to form two **π bonds** (Fig. 14.5). A π bond arises from the spin-pairing of electrons in two p orbitals that approach side-by-side. It is so called because, viewed along the internuclear axis, a π bond resembles a pair of electrons in a p orbital (and π is the Greek equivalent of p). More precisely, an electron in a π bond has one unit of orbital angular momentum about the internuclear axis, for the wavefunction has one angular node.

There are two π bonds in N_2, one formed by spin-pairing in two neighbouring $2p_x$ orbitals and the other by spin-pairing in two neighbouring $2p_y$ orbitals. The overall bonding pattern

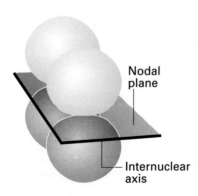

Nodal plane

Internuclear axis

14.5 A π bond results from spin pairing and orbital overlap of p orbitals that approach side by side.

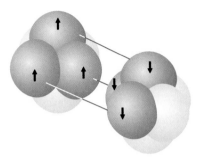

14.6 The structure of bonds in a nitrogen molecule which consists of one σ bond and two π bonds. The electron density has cylindrical symmetry around the internuclear axis.

in N_2 is therefore a σ bond plus two π bonds (Fig. 14.6), which is consistent with the Lewis structure : N≡N : for nitrogen.

Illustration

To obtain the VB description of Cl_2, note that the ground-state electron configuration of a Cl atom is $[Ar]3s^2 3p_x^2 3p_y^2 3p_z^1$. A σ bond can be formed between two atoms by spin-pairing of the electrons in the $3p_z$ orbitals. This description is consistent with the Lewis structure : Ċl—Ċl : for chlorine. The VB wavefunction for the bonding pair is the same as in eqn 2 but with A and B now standing for the two $Cl3p_z$ orbitals.

Self-test 14.1 Describe the ground state of HCl in valence-bond terms.

[eqn 2 with $A = \psi_{H1s}$, $B = \psi_{Cl2p_z}$]

14.3 Polyatomic molecules

Each σ bond in a polyatomic molecule is formed by the spin-pairing of electrons in any atomic orbitals with cylindrical symmetry about the relevant internuclear axis. Likewise, π bonds are formed by pairing electrons that occupy atomic orbitals of the appropriate symmetry.

The valence-bond description of H_2O will make this clear. The valence electron configuration of an O atom is $2s^2 2p_x^2 2p_y^1 2p_z^1$. The two unpaired electrons in the $O2p$ orbitals can each pair with an electron in an $H1s$ orbital, and each combination results in the formation of a σ bond (each bond has cylindrical symmetry about the respective O–H internuclear axis). Because the $2p_y$ and $2p_z$ orbitals lie at 90° to each other, the two σ bonds also lie at 90° to each other (Fig. 14.7). We can predict, therefore, that H_2O should be an angular molecule, which it is. However, the theory predicts a bond angle of 90°, whereas the actual bond angle is 104.5°.

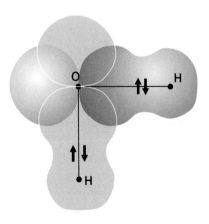

14.7 A first approximation to the valence-bond description of bonding in an H_2O molecule. Each σ bond arises from the overlap of an $H1s$ orbital with one of the $O2p$ orbitals. This model suggests that the bond angle should be 90°, which is significantly different from the experimental value.

Example 14.1 Predicting the shape of a molecule by using valence-bond theory

Describe the valence-bond structure of NH_3, and predict the bond angle of the molecule on the basis of this description.

Method Write down the ground-state configuration of an N atom, and decide which electrons and orbitals can be used to form bonds. Then, from the spatial arrangement of those atomic orbitals, infer the shape of the resulting molecule.

Answer The valence electron configuration of an N atom is N $2s^2 2p_x^1 2p_y^1 2p_z^1$. This configuration suggests that three H atoms can form bonds by spin-pairing with the electrons in the three half-filled $2p$ orbitals. The latter are perpendicular to each other, so we predict a trigonal pyramidal molecule with a bond angle of 90°.

Comment The molecule is trigonal pyramidal, but the experimental bond angle is 107°. The origin of this discrepancy is discussed below.

Self-test 14.2 Use valence-bond theory to suggest a shape for the hydrogen peroxide molecule, H_2O_2.

[Each H–O–O bond 90°]

(a) Promotion

An apparent deficiency of valence-bond theory is its inability to account for carbon's tetravalence (its ability to form four bonds). The ground-state configuration of C is $2s^2 2p_x^1 2p_y^1$, which suggests that a carbon atom should be capable of forming only two bonds, not four. This deficiency is overcome by allowing for **promotion**, the excitation of an electron to an orbital of higher energy. Although electron promotion requires an investment of energy, it is worthwhile if that energy can be more than recovered from the greater strength or number of bonds that it allows to be formed. Promotion is not a 'real' process in which an atom somehow becomes excited and then forms bonds: it is a contribution to the overall energy change that occurs when bonds form.

In carbon, for example, the promotion of a $2s$ electron to a $2p$ orbital can be thought of as leading to the configuration $2s^1 2p_x^1 2p_y^1 2p_z^1$, with four unpaired electrons in separate orbitals. These electrons may pair with four electrons in orbitals provided by four other atoms (such as four H1s orbitals if the molecule is CH_4), and hence form four σ bonds. Although energy was required to promote the electron, it is more than recovered by the atom's ability to form four bonds in place of the two bonds of the unpromoted atom. Promotion, and the formation of four bonds, is a characteristic feature of carbon because the promotion energy is quite small: the promoted electron leaves a doubly occupied $2s$ orbital and enters a vacant $2p$ orbital, hence significantly relieving the electron–electron repulsion it experiences in the former.

(b) Hybridization

The description of the bonding in CH_4 (and other alkanes) is still incomplete because it appears to imply the presence of three σ bonds of one type (formed from H1s and C2p orbitals) and a fourth σ bond of a distinctly different character (formed from H1s and C2s). This problem is overcome by realizing that the electron density distribution in the promoted atom is equivalent to the electron density in which each electron occupies a **hybrid orbital** formed by interference between the C2s and C2p orbitals. The origin of the hybridization can be appreciated by thinking of the four atomic orbitals, which are waves centred on a nucleus, as being like ripples spreading from a single point on the surface of a lake:[4] the waves interfere destructively and constructively in different regions, and give rise to four new shapes.

The specific linear combinations that give rise to four equivalent hybrid orbitals are

$$h_1 = s + p_x + p_y + p_z \qquad h_2 = s - p_x - p_y + p_z$$
$$h_3 = s - p_x + p_y - p_z \qquad h_4 = s + p_x - p_y - p_z \tag{4}$$

As a result of the interference between the component orbitals, each hybrid orbital consists of a large lobe pointing in the direction of one corner of a regular tetrahedron (Fig. 14.8). The angle between the axes of the hybrid orbitals is the tetrahedral angle, $\cos(-1/3) = 109.47°$. Because each hybrid is built from one s orbital and three p orbitals, it is called an sp^3 **hybrid orbital**.

It is now easy to see how the valence-bond description of the CH_4 molecule leads to a tetrahedral molecule containing four equivalent C–H bonds. Each hybrid orbital of the promoted C atom contains a single unpaired electron; an H1s electron can pair with each one, giving rise to a σ bond pointing in a tetrahedral direction. For example, the (un-normalized) wavefunction for the bond formed by the hybrid orbital h_1 and the 1s_A orbital (with wavefunction that we shall denote A) is

$$\psi = h_1(1)A(2) + h_1(2)A(1)$$

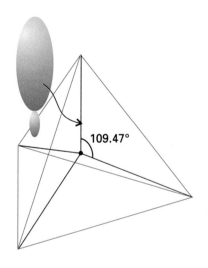

14.8 An sp^3 hybrid orbital formed from the superposition of s and p orbitals on the same atom. There are four such hybrids: each one points towards the corner of a regular tetrahedron. The overall electron density remains spherically symmetrical.

109.47°

---

4   It is admittedly difficult to imagine how a ripple resembling a p orbital could be contrived, but the general idea should be clear.

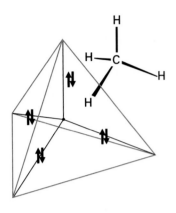

14.9 Each sp^3 hybrid orbital forms a σ bond by overlap with an H1s orbital located at the corner of the tetrahedron. This model accounts for the equivalence of the four bonds in CH_4.

Because each sp^3 hybrid orbital has the same composition, all four σ bonds are identical apart from their orientation in space (Fig. 14.9).

A further feature of hybridization is that a hybrid orbital has pronounced directional character, in the sense that it has an enhanced amplitude in the internuclear region. This directional character arises from the constructive interference between the s orbital and the positive lobes of the p orbitals (Fig. 14.10). As a result of the enhanced amplitude in the internuclear region, the bond strength is greater than for an s or p orbital alone. This increased bond strength is another factor that helps to repay the promotion energy.

Hybridization can also be used to describe the structure of an ethene molecule, $H_2C{=}CH_2$, and the torsional rigidity of double bonds. An ethene molecule is planar, with HCH and HCC bond angles close to $120°$. To reproduce the σ bonding structure, we promote each C atom to a $2s^1 2p^3$ configuration. However, instead of using all four orbitals to form hybrids, we form sp^2 **hybrid orbitals** by the superposition of an s orbital and two p orbitals. As shown in Fig. 14.11, the three hybrid orbitals

$$h_1 = s + 2^{1/2} p_y$$
$$h_2 = s + \left(\tfrac{3}{2}\right)^{1/2} p_x - \left(\tfrac{1}{2}\right)^{1/2} p_y \tag{5}$$
$$h_3 = s - \left(\tfrac{3}{2}\right)^{1/2} p_x - \left(\tfrac{1}{2}\right)^{1/2} p_y$$

lie in a plane and point towards the corners of an equilateral triangle. The third $2p$ orbital ($2p_z$) is not included in the hybridization, and its axis is perpendicular to the plane in which the hybrids lie.

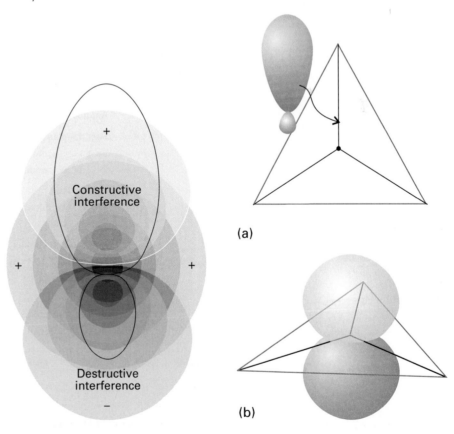

Constructive interference

Destructive interference

(a)

(b)

14.10 A more detailed representation of the formation of an sp^3 hybrid by interference between wavefunctions centred on the same atomic nucleus. (To simplify the representation, we have ignored the radial node of the $2s$ orbital.)

14.11 (a) An s orbital and two p orbitals can be hybridized to form three equivalent orbitals that point towards the corners of an equilateral triangle. (b) The remaining, unhybridized p orbital is perpendicular to the plane.

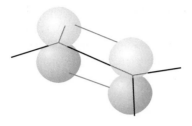

14.12 A representation of the structure of a double bond in ethene; only the π bond is shown explicitly.

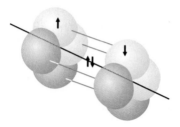

14.13 A representation of the structure of a triple bond in ethyne; only the π bonds are shown explicitly. The overall electron density has cylindrical symmetry around the axis of the molecule.

The structure of $CH_2=CH_2$ can now be described as follows. The sp^2-hybridized C atoms each form three σ bonds by spin-pairing with either the h_1 hybrid of the other C atom or with $H1s$ orbitals. The σ framework therefore consists of C–H and C–C σ bonds at 120° to each other. When the two CH_2 groups lie in the same plane, the two electrons in the unhybridized p orbitals can pair and form a π bond (Fig. 14.12). The formation of this π bond locks the framework into the planar arrangement, for any rotation of one CH_2 group relative to the other leads to a weakening of the π bond (and consequently an increase in energy of the molecule).

A similar description applies to ethyne, $HC\equiv CH$, a linear molecule. Now the C atoms are **sp hybridized**, and the σ bonds are formed using hybrid atomic orbitals of the form

$$h_1 = s + p_z \qquad h_2 = s - p_z \tag{6}$$

These two orbitals lie along the internuclear axis. The electrons in them pair either with an electron in the corresponding hybrid orbital on the other C atom or with an electron in one of the $H1s$ orbitals. Electrons in the two remaining p orbitals on each atom, which are perpendicular to the molecular axis, pair to form two perpendicular π bonds (Fig. 14.13).

Other hybridization schemes, particularly those involving d orbitals, are often invoked to be consistent with other molecular geometries (Table 14.1). The hybridization of N atomic orbitals always results in the formation of N hybrid orbitals. For example, sp^3d^2 hybridization results in six equivalent hybrid orbitals pointing towards the corners of a regular octahedron. This octahedral hybridization scheme is sometimes invoked to account for the structure of octahedral molecules, such as SF_6.

Table 14.1* Some hybridization schemes

Coordination number	Arrangement	Composition
2	Linear	sp, pd, sd
	Angular	sd
3	Trigonal planar	sp^2, p^2d
	Unsymmetrical planar	spd
	Trigonal pyramidal	pd^2
4	Tetrahedral	sp^3, sd^3
	Irregular tetrahedral	spd^2, p^3d, pd^3
	Square planar	p^2d^2, sp^2d
5	Trigonal bipyramidal	sp^3d, spd^3
	Tetragonal pyramidal	$sp^2d^2, sd^4, pd^4, p^3d^2$
	Pentagonal planar	p^2d^3
6	Octahedral	sp^3d^2
	Trigonal prismatic	spd^4, pd^5
	Trigonal antiprismatic	p^3d^3

*Source: H. Eyring, J. Walter, and G.E. Kimball, *Quantum chemistry*. Wiley (1944).

Molecular orbital theory

In **molecular orbital theory** (MO theory), it is accepted that electrons should not be regarded as belonging to particular bonds but should be treated as spreading throughout the entire molecule. This theory has been more fully developed than VB theory and provides the language that is widely used in modern discussions of bonding. To introduce it, we follow the same strategy as in Chapter 13, where the one-electron H atom was taken as the fundamental species for discussing atomic structure and then this discussion was developed into a description of many-electron atoms. In this chapter we use the simplest molecular species of all, the hydrogen molecule-ion, H_2^+, to introduce the essential features of bonding, and then use it as a guide to the structures of more complex systems.

14.4 The hydrogen molecule–ion

The hamiltonian for the single electron in H_2^+ is

$$H = -\frac{\hbar^2}{2m_e}\nabla_1^2 + V \qquad V = -\frac{e^2}{4\pi\varepsilon_0}\left(\frac{1}{r_{A1}} + \frac{1}{r_{B1}} - \frac{1}{R}\right) \tag{7}$$

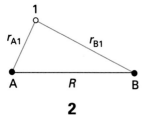

2

where r_{A1} and r_{B1} are the distances of the electron from the two nuclei (**2**). The one-electron wavefunctions obtained by solving the Schrödinger equation $H\psi = E\psi$ are called **molecular orbitals** (MO). A molecular orbital ψ gives, through the value of $|\psi|^2$, the distribution of the electron in the molecule. A molecular orbital is like an atomic orbital, but spreads throughout the molecule.

The Schrödinger equation can be solved for H_2^+ (within the Born–Oppenheimer approximation), but the wavefunctions are very complicated functions; moreover, the solution cannot be extended to polyatomic systems. Therefore, we shall adopt a simpler procedure that, while more approximate, can be extended readily to other molecules.

(a) Linear combinations of atomic orbitals

If an electron can be found in an atomic orbital belonging to atom A and also in an atomic orbital belonging to atom B, the overall wavefunction is a superposition of the two atomic orbitals:

$$\psi_\pm = N(A \pm B) \tag{8}$$

where, for H_2^+, A denotes ψ_{H1sA} and B denotes ψ_{H1sB} and N is a normalization factor. The technical term for the superposition in eqn 8 is a **linear combination of atomic orbitals** (LCAO). An approximate molecular orbital formed from a linear combination of atomic orbitals is called an **LCAO-MO**. A molecular orbital that has cylindrical symmetry around the internuclear axis, such as the one we are discussing, is called a **σ orbital** because it resembles an s orbital when viewed along the axis and, more precisely, because it has zero orbital angular momentum around the internuclear axis.

Example 14.2 Normalizing a molecular orbital

Normalize the molecular orbital ψ_+ in eqn 8.

Method We need to find the factor N such that

$$\int \psi^* \psi \, d\tau = 1$$

To proceed, substitute the LCAO into this integral, and make use of the fact that the atomic orbitals are individually normalized.

Answer When we substitute the wavefunction, we find

$$\int \psi^* \psi \, d\tau = N^2\left\{\int A^2 \, d\tau + \int B^2 \, d\tau + 2\int AB \, d\tau\right\}$$
$$= N^2(1 + 1 + 2S)$$

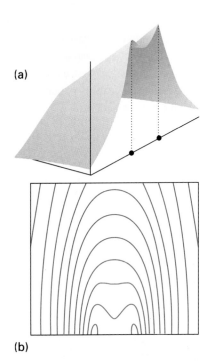

(a)

(b)

14.14 (a) The amplitude of the bonding molecular orbital in a hydrogen molecule-ion in a plane containing the two nuclei and (b) a contour representation of the amplitude.

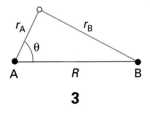

3

where $S = \int AB \, d\tau$. For the integral to be equal to 1, we require

$$N = \frac{1}{\{2(1+S)\}^{1/2}}$$

Comment In H_2^+, $S \approx 0.59$, so $N = 0.56$.

- -

Self-test 14.3 Normalize the orbital ψ_- in eqn 8.

$$[N = 1/\{2(1-S)\}^{1/2}, \text{ so } N = 1.10]$$

Figure 14.14 shows the contours of constant amplitude for the molecular orbital ψ_+ in eqn 8, and Fig. 14.15 shows its boundary surface. Plots like these are readily obtained using commercially available software. The calculation is quite straightforward, because all we need do is feed in the mathematical forms of the two atomic orbitals, and let the program do the rest. In this case, we use

$$A = \frac{e^{-r_A/a_0}}{(\pi a_0^3)^{1/2}} \qquad B = \frac{e^{-r_B/a_0}}{(\pi a_0^3)^{1/2}} \tag{9}$$

and note that r_A and r_B are not independent (3):

$$r_B = \{r_A^2 + R^2 - 2r_A R \cos\theta\}^{1/2} \tag{10}$$

To make this plot, we have taken $N^2 = 0.31$ (Example 14.2).

(b) Bonding orbitals

According to the Born interpretation, the probability density of the electron in H_2^+ is proportional to the square modulus of its wavefunction. The probability density corresponding to the (real) wavefunction ψ_+ in eqn 8 is

$$\psi_+^2 = N^2(A^2 + B^2 + 2AB) \tag{11}$$

This probability density is plotted in Fig. 14.16.

An important feature of the probability density becomes apparent when we examine the internuclear region, where both atomic orbitals have similar amplitudes. According to

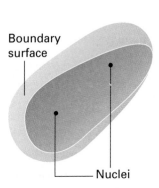

Boundary surface

Nuclei

14.15 The boundary surface of a σ orbital encloses the region where the electrons that occupy the orbital are most likely to be found. Note that the orbital has cylindrical symmetry.

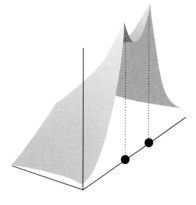

14.16 The electron density calculated by forming the square of the wavefunction used to construct Fig. 14.14. Note the accumulation of electron density in the internuclear region.

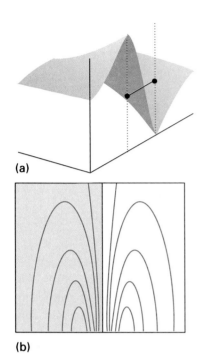

(a)

(b)

14.20 (a) The amplitude of the antibonding molecular orbital in a hydrogen molecule-ion in a plane containing the two nuclei and (b) a contour representation of the amplitude. Note the internuclear node.

each individual orbital before doubly occupying any one orbital (because that minimizes electron–electron repulsions). We also take note of Hund's rule (Section 13.4d) that, if electrons do occupy different degenerate orbitals, then a lower energy is obtained if they do so with parallel spins.

(a) The hydrogen and helium molecules

Consider H_2, the simplest many-electron diatomic molecule. Each H atom contributes a $1s$ orbital (as in H_2^+), so we can form the 1σ and $2\sigma^*$ orbitals from them, as we have seen already. At the experimental internuclear separation these orbitals will have the energies shown in Fig. 14.23, which is called a **molecular orbital energy level diagram.** Note that from two atomic orbitals we can build two molecular orbitals. In general, from N atomic orbitals we can build N molecular orbitals.

There are two electrons to accommodate, and both can enter 1σ by pairing their spins. The ground-state configuration is therefore $1\sigma^2$ and the atoms are joined by a bond consisting of an electron pair in a bonding σ orbital. This approach shows that an electron pair, which was the focus of Lewis's account of chemical bonding, represents the maximum number of electrons that can enter a bonding molecular orbital.

The same argument shows why He does not form diatomic molecules.[6] Each He atom contributes a $1s$ orbital, so 1σ and $2\sigma^*$ molecular orbitals can be constructed. Although these orbitals differ in detail from those in H_2, the general shape is the same, and we can use the same qualitative energy level diagram in the discussion. There are four electrons to accommodate. Two can enter the 1σ orbital, but then it is full, and the next two must enter the $2\sigma^*$ orbital (Fig. 14.24). The ground electronic configuration of He_2 is therefore $1\sigma^2 2\sigma^{*2}$. We see that there is one bond and one antibond. Because an antibond is slightly more antibonding than a bond is bonding, an He_2 molecule has a higher energy than the separated atoms, so it is unstable relative to the individual atoms.

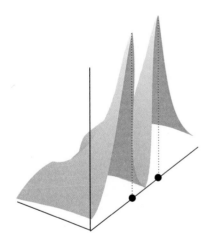

14.21 The electron density calculated by forming the square of the wavefunction used to construct Fig. 14.20. Note the elimination of electron density from the internuclear region.

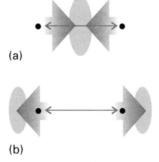

(a)

(b)

14.22 A partial explanation of the origin of bonding and antibonding effects. (a) In a bonding orbital, the nuclei are attracted to the accumulation of electron density in the internuclear region. (b) In an antibonding orbital, the nuclei are attracted to an accumulation of electron density outside the internuclear region.

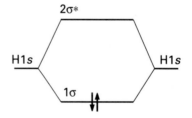

14.23 A molecular orbital energy level diagram for orbitals constructed from the overlap of H$1s$ orbitals; the separation of the levels corresponds to that found at the equilibrium bond length. The ground electronic configuration of H_2 is obtained by accommodating the two electrons in the lowest available orbital (the bonding orbital).

6   Diatomic helium 'molecules' have been prepared quite recently: they consist of pairs of atoms held together by weak van der Waals forces of the type described in Chapter 22.

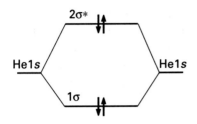

14.24 The ground electronic configuration of the hypothetical four-electron molecule He₂ has two bonding electrons and two antibonding electrons. It has a higher energy than the separated atoms, and so is unstable.

Table 14.2* Bond lengths

Bond	Order	R_e/pm
HH	1	74.14
NN	3	109.76
HCl	1	127.45
CH	1	*114*
CC	1	*154*
CC	2	*134*
CC	3	*120*

*More values will be found in the *Data section* at the end of this volume. Numbers in italics are mean values for polyatomic molecules.

Table 14.3* Bond dissociation energies

Bond	Order	D_e/(kJ mol⁻¹)
HH	1	432.1
NN	3	941.7
HCl	1	427.7
CH	1	*435*
CC	1	*368*
CC	2	*720*
CC	3	*962*

*More values will be found in the *Data section*. Numbers in italics are mean values for polyatomic molecules.

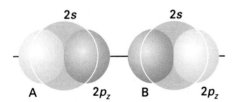

14.25 According to molecular orbital theory, σ orbitals are built from all orbitals that have the appropriate symmetry. In homonuclear diatomic molecules of Period 2, that means that two 2s and two 2pz orbitals should be used. From these four orbitals, four molecular orbitals can be built.

(b) Bond order

A measure of the net bonding in a diatomic molecule is its **bond order**, b:

$$b = \tfrac{1}{2}(n - n^*) \qquad [15]$$

where n is the number of electrons in bonding orbitals and n^* is the number of electrons in antibonding orbitals. Thus each electron pair in a bonding orbital increases the bond order by 1 and each pair in an antibonding orbital decreases b by 1. For H_2, $b = 1$, corresponding to a single bond, H–H, between the two atoms. In He_2, $b = 0$, and there is no bond.

As we shall see, the bond order is a useful parameter for discussing the characteristics of bonds, because it correlates with bond length and bond strength:

The greater the bond order between atoms of a given pair of elements, the shorter the bond.
The greater the bond order, the greater the bond strength.

Table 14.2 lists some typical bond lengths in diatomic and polyatomic molecules. The strength of a bond is measured by its **bond dissociation energy**, D_e, the energy required to separate the atoms to infinity.[7] Table 14.3 lists some experimental values of dissociation energies.

(c) Period 2 diatomic molecules

We now see how the concepts we have introduced apply to homonuclear diatomic molecules in general. In elementary treatments, only the orbitals of the valence shell are used to form molecular orbitals.

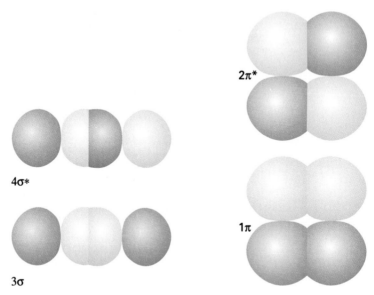

14.26 A representation of the composition of bonding and antibonding σ orbitals built from the overlap of p orbitals. These illustrations are schematic.

14.27 A schematic representation of the structure of π bonding and antibonding molecular orbitals.

7   Bond dissociation energies are commonly used in thermodynamic cycles, where bond enthalpies, $\Delta_{bond}H^{\ominus}$, should be used instead. It follows from the same kind of argument used in *Justification 13.9* concerning ionization enthalpies that

$$X_2(g) \longrightarrow 2\,X(g) \qquad \Delta_{bond}H^{\ominus}(T) = D_e + \tfrac{3}{2}RT$$

To derive this relation, we have supposed that the molar constant-pressure heat capacity of X_2 is $\tfrac{7}{2}R$, for there is a contribution from two rotational modes as well as three translational modes.

14.28 In a linear molecule, the electron density in a π orbital has cylindrical symmetry around the internuclear axis.

In Period 2, the valence orbitals are $2s$ and $2p$. A general principle of molecular orbital theory is that *all orbitals of the appropriate symmetry contribute to a molecular orbital*. Thus, to build σ orbitals, we form linear combinations of all atomic orbitals that have cylindrical symmetry about the internuclear axis. These orbitals include the $2s$ orbitals on each atom and the $2p_z$ orbitals on the two atoms (Fig. 14.25). Thus, the general form of the σ orbitals that may be formed is

$$\psi = c_{A2s}\psi_{A2s} + c_{B2s}\psi_{B2s} + c_{A2p_z}\psi_{A2p_z} + c_{B2p_z}\psi_{B2p_z} \tag{16}$$

From these four atomic orbitals we can form four molecular orbitals of σ symmetry by an appropriate choice of the coefficients c.

The procedure for calculating the coefficients will be described in Section 14.7. At this stage we adopt a simpler route, and suppose that, because the $2s$ and $2p_z$ orbitals have distinctly different energies, they may be treated separately. That is, the four σ orbitals fall approximately into two sets, one consisting of two molecular orbitals of the form

$$\psi = c_{A2s}\psi_{A2s} + c_{B2s}\psi_{B2s} \tag{17a}$$

and another consisting of two orbitals of the form

$$\psi = c_{A2p_z}\psi_{A2p_z} + c_{B2p_z}\psi_{B2p_z} \tag{17b}$$

Because atoms A and B are identical, the energies of their $2s$ orbitals are the same, so the coefficients are equal (apart from a possible difference in sign); the same is true of the $2p_z$ orbitals. Therefore, the two sets of orbitals have the form $\psi_{A2s} \pm \psi_{B2s}$ and $\psi_{A2p_z} \pm \psi_{B2p_z}$.

The $2s$ orbitals on the two atoms overlap to give a bonding and an antibonding σ orbital (1σ and 2σ, respectively) in exactly the same way as we have already seen for $1s$ orbitals. The two $2p_z$ orbitals directed along the internuclear axis overlap strongly. They may interfere either constructively or destructively, and give a bonding or antibonding σ orbital, respectively (Fig. 14.26). These two σ orbitals are labelled 3σ and $4\sigma^*$, respectively. In general, note how the numbering follows the order of increasing energy.[8]

(d) π orbitals

Now consider the $2p_x$ and $2p_y$ orbitals of each atom. These orbitals are perpendicular to the internuclear axis and may overlap broadside-on. This overlap may be constructive or destructive, and results in a bonding or an antibonding π **orbital** (Fig. 14.27). The notation π is the analogue of p in atoms for, when viewed along the axis of the molecule, a π orbital looks like a p orbital, and has one unit of orbital angular momentum around the internuclear axis. The two $2p_x$ orbitals overlap to give a bonding and antibonding π_x orbital, and the two $2p_y$ orbitals overlap to give two π_y orbitals. The π_x and π_y bonding orbitals are degenerate; so too are their antibonding partners. Strictly, because we are dealing with molecules with cylindrical symmetry, we should consider the complex forms of the p orbitals, one corresponding to circulation about the internuclear axis clockwise and the other anticlockwise. That is, we form $\pi_\pm \propto p_x \pm ip_y$, corresponding to angular momenta $\lambda\hbar$ with $\lambda = \pm 1$. Each complex orbital is like a cylindrical torus (Fig. 14.28). Although it is conventional to draw the real forms, it should not be forgotten that each π orbital in a linear molecule corresponds to a cylindrical distribution of charge.

In some cases, π orbitals are less strongly bonding than σ orbitals because their maximum overlap occurs off-axis. This relative weakness suggests that the molecular orbital energy level diagram ought to be as shown in Fig. 14.29. However, we must remember that we have constructed the diagram on the assumption that the $2s$ and $2p_z$ orbitals contribute to different sets of molecular orbitals, whereas in fact all four atomic orbitals contribute jointly

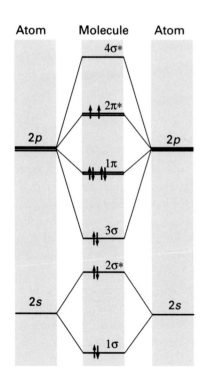

14.29 The molecular orbital energy level diagram for homonuclear diatomic molecules. As remarked in the text, this diagram should be used for O_2 and F_2.

---

8   In an alternative system of notation, 1σ and 2σ are used to designate the molecular orbitals formed from the core $1s$ orbitals of the atoms; the orbitals we are considering would then be labelled from 3 to 6. We are ignoring orbitals formed from core orbitals.

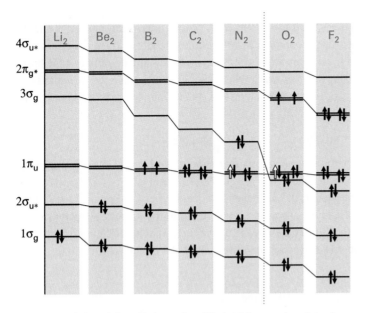

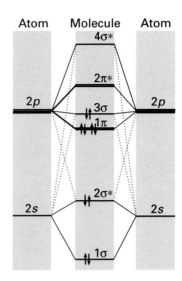

14.30 The variation of the orbital energies of Period 2 homonuclear diatomics. The g and u labels are explained later (Section 14.6a).

14.31 An alternative molecular orbital energy level diagram for homonuclear diatomic molecules. As remarked in the text, this diagram should be used for diatomics up to and including N_2.

to the four σ orbitals. Hence, there is no guarantee that this order of energies should prevail, and it is found experimentally (by spectroscopy) and by detailed calculation that the order varies along Period 2 (Fig. 14.30). The order shown in Fig. 14.31 is appropriate as far as N_2, and Fig. 14.29 applies for O_2 and F_2. The relative order is controlled by the separation of the $2s$ and $2p$ orbitals in the atoms, which increases across the group. The consequent switch in order occurs at about N_2.

(e) The overlap integral

The extent to which two atomic orbitals on different atoms overlap is measured by the **overlap integral**, S:

$$S = \int \psi_A^* \psi_B \, d\tau \qquad [18]$$

If the atomic orbital ψ_A on A is small wherever the orbital ψ_B on B is large, or vice versa, then the product of their amplitudes is everywhere small and the integral—the sum of these products—is small (Fig. 14.32). If ψ_A and ψ_B are simultaneously large in some region of space, then S may be large. If the two normalized atomic orbitals are identical (for example, $1s$ orbitals on the same nucleus), then $S = 1$. In some cases, simple formulas can be given for overlap integrals and the variation of S with bond length plotted (Fig. 14.33). It follows that $S = 0.59$ for two H$1s$ orbitals at the equilibrium bond length in H_2^+, which is an unusually large value. Typical values for orbitals with $n = 2$ are in the range 0.2 to 0.3.

Now consider the arrangement in which an s orbital is superimposed on a p_x orbital of a different atom (Fig. 14.34). The integral over the region where the product of orbitals is positive exactly cancels the integral over the region where the product of orbitals is negative, so overall $S = 0$ exactly. Therefore, there is no net overlap between the s and p orbitals in this arrangement.

(f) The structures of homonuclear diatomic molecules

We show the general layout of the valence-shell atomic orbitals of Period 2 atoms on the left and right of the molecular orbital energy level diagrams in Figs. 14.29 and 14.31. The lines in

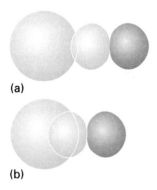

14.32 (a) When two orbitals are on atoms that are far apart, the wavefunctions are small where they overlap, so S is small. (b) When the atoms are closer, both orbitals have significant amplitudes where they overlap, and S may approach 1. Note that S will decrease again as the two atoms approach more closely than shown here, because the region of negative amplitude of the p orbital starts to overlap the positive overlap of the s orbital. When the centres of the atoms coincide, $S = 0$.

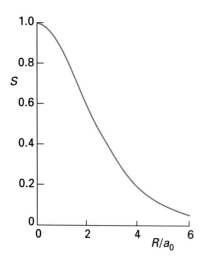

14.33 The overlap integral, S, between two H1s orbitals as a function of their separation R.

the middle are an indication of the energies of the molecular orbitals that can be formed by overlap of atomic orbitals: from the eight valence shell orbitals (four from each atom), we can form eight molecular orbitals. With the orbitals established, we can deduce the ground configurations of the molecules by adding the appropriate number of electrons to the orbitals and following the building-up rules. Anionic species (such as the peroxide ion, O_2^{2-}) need more electrons than the parent neutral molecules; cationic species (such as O_2^+) need fewer.

Consider N_2, which has 10 valence electrons. For this molecule, we use Fig. 14.31. Two electrons pair, occupy, and fill the 1σ orbital; the next two occupy and fill the $2\sigma^*$ orbital. Six electrons remain. There are two 1π orbitals, so four electrons can be accommodated in them. The last two enter the 3σ orbital. The ground-state configuration of N_2 is therefore $1\sigma^2 2\sigma^{*2} 1\pi^4 3\sigma^2$ and the bond order is $\frac{1}{2}(8-2) = 3$. This bond order accords with the Lewis structure of the molecule (: N≡N :) and is consistent with its high dissociation energy (942 kJ mol^{-1}).

The ground-state electron configuration of O_2, with 12 valence electrons, is based on Fig. 14.29, and is $1\sigma^2 2\sigma^{*2} 3\sigma^2 1\pi^4 2\pi^{*2}$. Its bond order is 2. According to the building-up principle, however, the two $2\pi^*$ electrons occupy different orbitals: one will enter $2\pi_x^*$ and the other will enter $2\pi_y^*$. Because the electrons are in different orbitals, they will have parallel spins. Therefore, we can predict that an O_2 molecule will have a net spin angular momentum $S = 1$ and, in the language introduced in Section 13.7, be in a triplet state. Because electron spin is the source of a magnetic moment, we can go on to predict that oxygen should be paramagnetic.[9] This prediction, which valence-bond theory does not make, is confirmed by experiment.

An F_2 molecule has two more electrons than an O_2 molecule. Its configuration is therefore $1\sigma^2 2\sigma^{*2} 3\sigma^2 1\pi^4 2\pi^{*4}$ and $b = 1$. We conclude that F_2 is a singly bonded molecule, in agreement with its Lewis structure. The low bond order is consistent with its low dissociation energy (154 kJ mol^{-1}). The hypothetical molecule dineon, Ne_2, has two further electrons: its configuration is $1\sigma^2 2\sigma^{*2} 3\sigma^2 1\pi^4 2\pi^{*4} 4\sigma^{*2}$ and $b = 0$. The zero bond order is consistent with the monatomic nature of Ne.

Example 14.3 Judging the relative bond strengths of molecules and ions

Judge whether N_2^+ is likely to have a larger or smaller dissociation energy than N_2.

Method Because the molecule with the larger bond order is likely to have the larger dissociation energy, compare their electronic configurations and assess their bond orders.

Answer From Fig. 14.31, the electron configurations and bond orders are

$$N_2 \quad 1\sigma^2 2\sigma^{*2} 1\pi^4 3\sigma^2 \quad b = 3$$
$$N_2^+ \quad 1\sigma^2 2\sigma^{*2} 1\pi^4 3\sigma^1 \quad b = 2\tfrac{1}{2}$$

Because the cation has the smaller bond order, we expect it to have the smaller dissociation energy.

Comment The experimental dissociation energies are 945 kJ mol^{-1} for N_2 and 842 kJ mol^{-1} for N_2^+.

- -

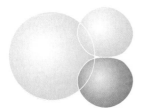

14.34 A p orbital in the orientation shown here has zero net overlap ($S = 0$) with the s orbital at all internuclear separations.

9   A paramagnetic substance tends to move into a magnetic field; a diamagnetic substance tends to move out of one. Paramagnetism, the rarer property, arises when the molecules have unpaired electron spins. Both properties are discussed in more detail in Section 22.6.

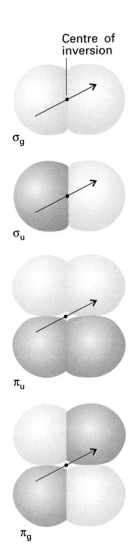

Centre of inversion

σ_g

σ_u

π_u

π_g

14.35 The parity of an orbital is even (g) if its wavefunction is unchanged under inversion in the centre of symmetry of the molecule, but odd (u) if the wavefunction changes sign. Heteronuclear diatomic molecules do not have a centre of inversion, so for them the g, u classification is irrelevant.

Self-test 14.4 Which can be expected to have the higher dissociation energy, F_2 or F_2^+?

[F_2^+]

14.6 More about notation

We have seen how to label molecular orbitals by taking note of their symmetries with respect to rotation around the internuclear axis. Certain other features of their symmetry can also be used. As we shall see in later chapters, these symmetry designations are used to formulate selection rules in molecular spectroscopy. Symmetry designations are described in detail in Chapter 15, and the following remarks are expanded there.

(a) Parity

The molecular orbitals of homonuclear diatomic molecules are labelled with a subscript g or u which specifies their **parity**, their behaviour under inversion. To decide on the parity, consider any point in a homonuclear diatomic molecule, and note the sign of the orbital. Then imagine travelling on a straight line through the centre of the molecule to a point the same distance out on the other side; this process is called **inversion** and the central point is the **centre of inversion** (Fig. 14.35). If the orbital has the same sign, it has even parity and is denoted g (from *gerade*, the German word for even). If the orbital has opposite sign, then it has odd parity and is denoted u (from *ungerade*, uneven). The parity designation applies only to homonuclear diatomic molecules, because heteronuclear diatomic molecules (such as HCl) do not have a centre of inversion.

We see from Fig. 14.35 that a bonding σ orbital has even parity; so we write it σ_g; an antibonding σ orbital has odd parity and is written σ_u. A bonding π orbital has odd parity and is denoted π_u and an antibonding π orbital has even parity, denoted π_g.[10]

(b) Term symbols

The term symbols of linear molecules (the analogues of the symbols ^2P, etc. for atoms) are constructed in a similar way to those for atoms, but now we must pay attention to the component of total orbital angular momentum about the internuclear axis, $\Lambda\hbar$. The value of $|\Lambda|$ is denoted by the symbols $\Sigma, \Pi, \Delta, \ldots$ for $|\Lambda| = 0, 1, 2\ldots$, respectively. These labels are the analogues of $S, P, D, \ldots$ for atoms.

The value of Λ is the sum of the values of λ for the individual electrons in a molecule.[11] A single electron in a σ orbital has $\lambda = 0$: the orbital is cylindrically symmetrical and has no angular nodes when viewed along the internuclear axis. Therefore, if that is the only electron present, $\Lambda = 0$. The term symbol for H_2^+ is therefore Σ. As in atoms, we use a superscript with the value of $2S + 1$ to denote the multiplicity of the term. In this case, because there is only one electron, $S = s = \frac{1}{2}$ and the term symbol is $^2\Sigma$, a doublet term. The overall parity of the term is added as a right subscript, and (if there are several electrons) is calculated by using

$$g \times g = g \qquad u \times u = g \qquad u \times g = u \tag{19}$$

(The rules can be generated by interpreting g as +1 and u as −1.) For H_2^+, the parity of the only occupied orbital is g, so the term itself is also g, and in full dress is $^2\Sigma_g$. The term symbol for any closed-shell homonuclear diatomic molecule is $^1\Sigma_g$ because the spin is zero (all electrons paired), there is no orbital angular momentum from a closed shell, and the overall parity is g.

---

10 For simplicity in comparing homonuclear and heteronuclear molecules, we ignore the parity subscripts when numbering orbitals; however, a more formal convention is to number the g and u orbitals separately.

11 Recall from Section 14.5d that $\lambda\hbar$ is the component of orbital angular momentum on the internuclear axis.

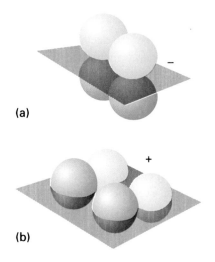

(a)

(b)

14.36 The $\pm$ in a term symbol refers to the symmetry of an orbital when it is reflected in a plane containing the two nuclei.

A π electron in a diatomic molecule has one unit of orbital angular momentum about the internuclear axis ($\lambda = \pm 1$) and, if it is the only electron outside a closed shell, gives rise to a Π term. If there are two π electrons (as in O_2) then the term symbol may be either Σ (if the electrons are travelling in opposite directions, which is the case if they occupy different π orbitals, one with $\lambda = +1$ and the other with $\lambda = -1$) or Δ (if they are travelling in the same direction, which is the case if they occupy the same π orbital, both $\lambda = +1$, for instance). For O_2, the two π electrons occupy different orbitals with parallel spins, so the ground term is $^3\Sigma$. The overall parity of the molecule is

$$(\text{closed shell}) \times \text{g} \times \text{g} = \text{g}$$

The term symbol is therefore $^3\Sigma_g$.

For Σ terms, a $\pm$ superscript denotes the behaviour of the molecular wavefunction under reflection in a plane containing the nuclei (Fig. 14.36). If, for convenience, we think of O_2 as having one electron in $2\pi_x$, which changes sign under reflection in the xz-plane, and the other electron in $2\pi_y$, which does not change sign under reflection in the same plane, the overall reflection symmetry is

$$(\text{closed shell}) \times (+) \times (-) = (-)$$

and the full term symbol is $^3\Sigma_g^-$. The need for all this dressing of a basic symbol will become apparent when we deal with the spectroscopic selection rules in Chapter 17.

14.7 Heteronuclear diatomic molecules

A heteronuclear diatomic molecule is a diatomic molecule formed from atoms of two different elements, such as CO and HCl. The electron distribution in the covalent bond between the atoms is not evenly shared because it is energetically favourable for the electron pair to be found closer to one atom than the other. This imbalance results in a **polar bond**, a covalent bond in which the electron pair is shared unequally by the two atoms. The bond in HF, for instance, is polar, with the electron pair closer to the F atom. The accumulation of the electron pair near the F atom results in that atom having a net negative charge, which is called a **partial negative charge** and denoted $\delta-$. There is a matching **partial positive charge**, $\delta+$, on the H atom.

(a) Polar bonds

A polar bond consists of two electrons in an orbital of the form

$$\psi = c_A A + c_B B \qquad (20)$$

with unequal coefficients. The proportion of the atomic orbital A in the bond is $|c_A|^2$ and that of B is $|c_B|^2$. A nonpolar bond has $|c_A|^2 = |c_B|^2$ and a pure ionic bond has one coefficient zero (so the species A^+B^- would have $c_A = 0$ and $c_B = 1$). The atomic orbital with the lower energy makes the larger contribution to the bonding molecular orbital. The opposite is true of the antibonding orbital, for which the dominant component comes from the atomic orbital with higher energy.

These points can be illustrated by considering HF, and judging the energies of the atomic orbitals from the ionization energies of the atoms. The general form of the molecular orbitals is

$$\psi = c_H \psi_H + c_F \psi_F \qquad (21)$$

where ψ_H is an H1s orbital and ψ_F is an F2p orbital. The H1s orbital lies at 13.6 eV below the zero of energy (the separated proton and electron) and the F2p orbital lies at 18.6 eV below the zero of energy (Fig. 14.37). Hence, the bonding σ orbital in HF is mainly F2p and the antibonding σ orbital is mainly H1s orbital in character. The two electrons in the bonding orbital are most likely to be found in the F2p orbital, so there is a partial negative charge on the F atom and a partial positive charge on the H atom.

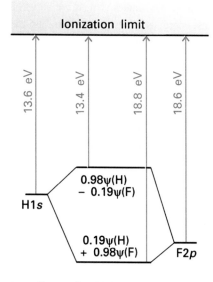

14.37 The atomic orbital energy levels of H and F atoms and the molecular orbitals they form.

Table 14.4* Pauling electronegativities

Element	χ_P
H	2.2
C	2.6
N	3.0
O	3.4
F	4.0
Cl	3.2
Cs	0.79

* More values will be found in the *Data section*.

(b) Electronegativity

The charge distribution in bonds is commonly discussed in terms of the **electronegativity**, χ (chi), of the elements involved. The electronegativity is a parameter introduced by Linus Pauling as a measure of the power of an atom to attract electrons to itself when it is part of a compound. Pauling used valence-bond arguments to suggest that an appropriate numerical scale of electronegativities could be defined in terms of bond dissociation energies, D, and proposed that the difference in electronegativities could be expressed as

$$|\chi_A - \chi_B| = 0.102\{D(A{-}B) - \tfrac{1}{2}[D(A{-}A) + D(B{-}B)]\}^{1/2} \qquad [22]$$

Electronegativities based on this definition are called **Pauling electronegativities**. A list of Pauling electronegativities is given in Table 14.4. The most electronegative elements are those close to fluorine; the least are those close to caesium. It is found that, the greater the difference in electronegativities, the greater the polar character of the bond. The difference for HF, for instance, is 1.78; a C–H bond, which is commonly regarded as almost nonpolar, has an electronegativity difference of 0.51.

The American spectroscopist R.S. Mulliken proposed an alternative definition of electronegativity. He argued that an element is likely to be highly electronegative if it has a high ionization energy (so it will not release electrons readily) and a high electron affinity (so it is energetically favorable to acquire electrons). The **Mulliken electronegativity scale** is therefore based on the definition

$$\chi_M = \tfrac{1}{2}(I + E_{ea}) \qquad [23]$$

where I is the ionization energy of the element and E_{ea} is its electron affinity, both in electronvolts (Section 13.4f).[12] The Mulliken and Pauling scales are approximately in line with one another.[13]

(c) The variation principle

A more systematic way of discussing bond polarity and finding the coefficients in the linear combinations used to build molecular orbitals is provided by the **variation principle**:

> **If an arbitrary wavefunction is used to calculate the energy, the value calculated is never less than the true energy.**

This principle is the basis of all modern molecular structure calculations. The arbitrary wavefunction is called the **trial wavefunction**. The principle implies that, if we vary the coefficients in the trial wavefunction until the lowest energy is achieved (by evaluating the expectation value of the hamiltonian for each wavefunction), then those coefficients will be the best. We might get a lower energy if we use a more complicated wavefunction (for example, by taking a linear combination of several atomic orbitals on each atom), but we shall have the optimum (minimum energy) molecular orbital that can be built from the chosen **basis set**, the given set of atomic orbitals.

The method can be illustrated by the trial wavefunction in eqn 20. We show in the *Justification* below that the coefficients are given by the solutions of the two **secular equations**[14]

$$
\begin{aligned}
(\alpha_A - E)c_A + (\beta - ES)c_B &= 0 \\
(\beta - ES)c_A + (\alpha_B - E)c_B &= 0
\end{aligned}
\qquad (24)
$$

12 There are certain technical difficulties with this definition in connection with the electronic state chosen to represent the state of the atom in a compound.

13 A reasonably reliable conversion between the two is $\chi_P = 1.35\chi_M^{1/2} - 1.37$.

14 The name 'secular' is derived from the Latin word for age or generation. The term comes via astronomy, where the same equations appear in connection with slowly accumulating modifications of planetary orbits.

The parameter α is called a **Coulomb integral**. It can be interpreted as the energy of the electron when it occupies A (for α_A) or B (for α_B), and is negative. In a homonuclear diatomic molecule, $\alpha_A = \alpha_B$. The parameter β is called a **resonance integral** (for classical reasons). It vanishes when the orbitals do not overlap, and at equilibrium bond lengths it is normally negative.

Justification 14.4

The trial wavefunction in eqn 20 is real but not normalized because at this stage the coefficients can take arbitrary values. Therefore, we can write $\psi^* = \psi$ but do not assume that $\int \psi^2 \, d\tau = 1$. The energy of the trial wavefunction is the expectation value of the energy operator (the hamiltonian, H, Section 11.5):

$$E = \frac{\int \psi^* H \psi \, d\tau}{\int \psi^* \psi \, d\tau} \tag{25}$$

We must search for values of the coefficients in the trial function that minimize the value of E. This is a standard problem in calculus, and is solved by finding the coefficients for which

$$\frac{\partial E}{\partial c_A} = 0 \qquad \frac{\partial E}{\partial c_B} = 0$$

The first step is to express the two integrals in terms of the coefficients. The denominator is

$$\int \psi^2 \, d\tau = \int (c_A A + c_B B)^2 \, d\tau$$
$$= c_A^2 \int A^2 \, d\tau + c_B^2 \int B^2 \, d\tau + 2c_A c_B \int AB \, d\tau$$
$$= c_A^2 + c_B^2 + 2c_A c_B S$$

because the individual atomic orbitals are normalized and the third integral is the overlap integral S (eqn 18). The numerator is

$$\int \psi H \psi \, d\tau = \int (c_A A + c_B B) H (c_A A + c_B B) \, d\tau$$
$$= c_A^2 \int AHA \, d\tau + c_B^2 \int BHB \, d\tau + c_A c_B \int AHB \, d\tau$$
$$+ c_A c_B \int BHA \, d\tau$$

There are some complicated integrals in this expression, but we can combine them all into the parameters

$$\alpha_A = \int AHA \, d\tau \qquad \alpha_B = \int BHB \, d\tau \qquad \beta = \int AHB \, d\tau = \int BHA \, d\tau$$

$$[26]$$

Then

$$\int \psi H \psi \, d\tau = c_A^2 \alpha_A + c_B^2 \alpha_B + 2c_A c_B \beta$$

The complete expression for E is

$$E = \frac{c_A^2 \alpha_A + c_B^2 \alpha_B + 2c_A c_B \beta}{c_A^2 + c_B^2 + 2c_A c_B S} \tag{27}$$

Its minimum is found by differentiation with respect to the two coefficients and setting the results equal to 0. This involves elementary but slightly tedious work, and the end result is eqn 24.

To solve the secular equations for the coefficients we need to know the energy E of the orbital. As for any set of simultaneous equations, the secular equations have a solution if the **secular determinant**, the determinant of the coefficients, is zero, that is, if

$$\begin{vmatrix} \alpha_A - E & \beta - ES \\ \beta - ES & \alpha_B - E \end{vmatrix} = 0 \tag{28}$$

This determinant expands to a quadratic equation in E (see Example 14.4). Its two roots give the energies of the bonding and antibonding molecular orbitals formed from the atomic orbitals and, according to the variation principle, these roots are the best energies for the given basis set.

Example 14.4 Finding the roots of a secular determinant

Find the energies E of the bonding and antibonding orbitals of a homonuclear diatomic molecule by solving eqn 28.

Method We need to know that a 2×2 determinant expands as follows:

$$\begin{vmatrix} a & b \\ c & d \end{vmatrix} = ad - bc$$

Answer When we apply the determinant expansion rule to eqn 28 with $\alpha_A = \alpha_B = \alpha$, we get

$$\begin{vmatrix} \alpha - E & \beta - ES \\ \beta - ES & \alpha - E \end{vmatrix} = (\alpha - E)^2 - (\beta - ES)^2 = 0$$

The solutions of this equation are

$$E_\pm = \frac{\alpha \pm \beta}{1 \pm S}$$

- -

Self-test 14.5 Find the coefficients corresponding to these two energies.

[See below; eqn 30]

The values of the coefficients in the linear combination are obtained by solving the secular equations using the two energies obtained from the secular determinant. The lower energy gives the coefficients for the bonding molecular orbital, the upper energy the coefficients for the antibonding molecular orbital. The secular equations give expressions for the ratio of the coefficients in each case, so we need a further equation in order to find their individual values. This equation is obtained by demanding that the best wavefunction should also be normalized. This condition means that, at this final stage, we must also ensure that

$$\int \psi^2 \, d\tau = c_A^2 + c_B^2 + 2c_A c_B S = 1 \tag{29}$$

(d) Two simple cases

The complete solutions of the secular equations are very cumbersome, even for 2×2 determinants, but there are two cases where the roots can be written down very simply.

We saw in Example 14.4 and its Self-test that, when the two atoms are the same, and we can write $\alpha_A = \alpha_B = \alpha$, the solutions are

$$E_+ = \frac{\alpha + \beta}{1 + S} \qquad c_A = \frac{1}{\{2(1 + S)\}^{1/2}} \qquad c_B = c_A$$

$$E_- = \frac{\alpha - \beta}{1 - S} \qquad c_A = \frac{1}{\{2(1 - S)\}^{1/2}} \qquad c_B = -c_A \tag{30}$$

In this case, the bonding orbital has the form

$$\psi_+ = \frac{A + B}{\{2(1 + S)\}^{1/2}} \tag{31a}$$

and the corresponding antibonding orbital is

$$\psi_- = \frac{A - B}{\{2(1 - S)\}^{1/2}} \tag{31b}$$

in agreement with the discussion of homonuclear diatomics we have already given, but now with the normalization constant in place.

The second simple case is for a heteronuclear diatomic molecule but with $S = 0$ (a common approximation in elementary work). The secular determinant is then

$$\begin{vmatrix} \alpha_A - E & \beta \\ \beta & \alpha_B - E \end{vmatrix} = (\alpha_A - E)(\alpha_B - E) - \beta^2 = 0$$

The solutions can be expressed in terms of the parameter ζ (zeta), with[15]

$$\zeta = \tfrac{1}{2} \arctan \frac{2|\beta|}{\alpha_B - \alpha_A} \tag{32}$$

and are

$$E_- = \alpha_A - \beta \cot \zeta \qquad \psi_- = -A \sin \zeta + B \cos \zeta$$

$$E_+ = \alpha_B + \beta \cot \zeta \qquad \psi_+ = A \cos \zeta + B \sin \zeta \tag{33}$$

An important feature revealed by these solutions is that, as the difference $|\alpha_A - \alpha_B|$ increases, the value of ζ decreases.[16] When the energy difference is large the energies of the molecular orbitals differ only slightly from those of the atomic orbitals, which implies in turn that the bonding and antibonding effects are small. That is, *the strongest bonding and antibonding effects are obtained when the two contributing orbitals have closely similar energies.* The difference in energy between core and valence orbitals is the justification for neglecting the contribution of core orbitals to bonding. The core orbitals of one atom have a similar energy to the core orbitals of the other atom; but core–core interaction is largely negligible because the overlap between them (and hence the value of β) is so small.

Example 14.5 Calculating the molecular orbitals of HF

Calculate the wavefunctions and energies of the σ orbitals in the HF molecule, taking $\beta = -1.0$ eV and the following ionization energies: H1s: 13.6 eV, F2s: 40.2 eV, F2p: 18.6 eV.

---

15 arctan x is the same as tan^{-1} x.

16 Because tan $x \approx x$ and cot $x \approx 1/x$ when $x \ll 1$, when $|\alpha_A - \alpha_B| \gg 2|\beta|$ we can write $\zeta \approx |\beta|/(\alpha_B - \alpha_A)$, which implies that tan $\zeta \approx |\beta|/(\alpha_B - \alpha_A)$, and hence that cot $\zeta \approx (\alpha_B - \alpha_A)/|\beta|$. Then (noting that $\beta/|\beta| = -1$) the energies of the two molecular orbitals are

$$E_- \approx \alpha_B \qquad E_+ \approx \alpha_A$$

Because sin $\zeta \approx \zeta$ and cos $\zeta \approx 1$ when $\zeta \ll 1$, the orbitals are respectively almost pure B and almost pure A.

Method Because the F2p and H1s orbitals are much closer in energy than the F2s and H1s orbitals, to a first approximation neglect the contribution of the F2s orbital. To use eqn 33, we need to know the values of the Coulomb integrals α_H and α_F. Because these integrals represent the energies of the H1s and F2p electrons, respectively, they are approximately equal to (the negative of) the ionization energies of the atoms. Calculate ζ from eqn 32 (with A identified as F and B as H), and then write the wavefunctions by using eqn 33.

Answer Refer to Fig. 14.37. Setting $\alpha_H = -13.6\,eV$ and $\alpha_F = -18.6\,eV$ gives $\tan 2\zeta = 0.40$; so $\zeta = 10.9°$. Then

$$E_- = -13.4\ eV \qquad \psi = 0.98\psi_H - 0.19\psi_F$$
$$E_+ = -18.8\ eV \qquad \psi_+ = 0.19\psi_H + 0.98\psi_F$$

Comment Notice how the lower energy orbital (the one with energy -18.8 eV) has a composition that is more F2p orbital than H1s, and that the opposite is true of the higher energy, antibonding orbital.

- -

Self-test 14.6 The ionization energy of Cl is 13.1 eV; find the form and energies of the σ orbitals in the HCl molecule using $\beta = -1.0$ eV.

$$[E_- = -12.3\ eV,\ \psi_- = -0.62\psi_H + 0.79\psi_{Cl};$$
$$E_+ = -14.4\ eV,\ \psi_+ = 0.79\psi_H + 0.62\psi_{Cl}]$$

Molecular orbitals for polyatomic systems

The molecular orbitals of polyatomic molecules are built in the same way as in diatomic molecules, the only difference being that we use more atomic orbitals to construct the molecular orbitals. As for diatomic molecules, polyatomic molecular orbitals spread over the entire molecule. A molecular orbital has the general form

$$\psi = \sum_i c_i \psi_i \tag{34}$$

where ψ_i is an atomic orbital and the sum extends over all the valence orbitals of all the atoms in the molecule. To find the coefficients, we set up the secular equations and the secular determinant, just as for diatomic molecules, solve the latter for the energies, and then use these energies in the secular equations to find the coefficients of the atomic orbitals for each molecular orbital.

The principal difference between diatomic and polyatomic molecules lies in the greater range of shapes that are possible: a diatomic molecule is necessarily linear, but a triatomic molecule, for instance, may be either linear or angular with a characteristic bond angle. The shape of a polyatomic molecule—the specification of its bond lengths and its bond angles—can be predicted by calculating the total energy of the molecule for a variety of nuclear positions, and then identifying the conformation that corresponds to the lowest energy. However, more insight into the features that control molecular geometry can be obtained by analysing the orbitals and their energies in a more pictorial fashion. We shall illustrate what is involved by considering H_2O, which has an experimental bond angle of 104°.

14.8 Walsh diagrams

The molecular orbitals of H_2O (and of H_2X molecules in general) have the form

$$\psi = c_1\psi_{H_A1s} + c_2\psi_{H_B1s} + c_3\psi_{O2s} + c_4\psi_{O2p_x} + c_5\psi_{O2p_y} + c_6\psi_{O2p_z}$$

There are six such orbitals (because they are built from six atomic orbitals) and eight valence electrons to accommodate in them. We shall consider two hypothetical conformations of the molecule, the linear 180° molecule and the angular 90° molecule, and then decide how the molecular orbitals of one shape turn into the molecular orbitals of the other as the bond angle changes from 180° to 90°. The procedure results in the construction of a **Walsh diagram**, a diagram showing the variation of orbital energy with molecular geometry.

(a) The Walsh diagram for H₂X molecules

The molecular orbitals of a hypothetical linear HOH molecule are classified as either σ or π (Fig. 14.38):

$$\psi_{\sigma_g} = c_1\psi_{O2s} + c_2(\psi_{H_A1s} + \psi_{H_B1s}) \qquad \text{(two orbitals)}$$
$$\psi_{\pi_u} = \psi_{O2p_x}, \ \psi_{O2p_z} \qquad \text{(one orbital each)} \qquad (35)$$
$$\psi_{\sigma_u} = c_3\psi_{O2p_y} + c_4(\psi_{H_A1s} - \psi_{H_B1s}) \qquad \text{(two orbitals)}$$

We have added the parity labels, but they no longer tell us which is bonding and antibonding. Thus, there are two σ_g orbitals, one bonding (with the two coefficients the same sign) and the other antibonding (with the coefficients of opposite sign). There are no orbitals of π symmetry on the H atoms, so the $O2p_x$ and $O2p_z$ orbitals do not form bonding and antibonding molecular orbitals. They are examples of **nonbonding orbitals**, orbitals that do not contribute directly to the bonding between atoms. The coefficients in the molecular orbitals may be found in the normal way, by setting up and solving the secular determinants using estimates of the Coulomb and resonance integrals, and the energies of the orbitals are shown on the left of the diagram in Fig. 14.39.

The molecular orbitals of a hypothetical 90° molecule are formed from the following groupings of atomic orbitals (Fig. 14.40):

$$\psi_{a_1} = c_1\psi_{O2s} + c_2\psi_{O2p_z} + c_3(\psi_{H_A1s} + \psi_{H_B1s}) \qquad \text{(three orbitals)}$$
$$\psi_{b_1} = \psi_{O2p_x} \qquad (36)$$
$$\psi_{b_2} = c_4\psi_{O2p_y} + c_5(\psi_{H_A1s} - \psi_{H_B1s}) \qquad \text{(two orbitals)}$$

(The coefficients are different from those in eqn 35.) We can no longer classify the orbitals as σ and π because those labels apply only when there is an axis of symmetry; the labels used here will be explained in Chapter 15 (as will be the choice of the orbitals from which each molecular orbital is built).[17]

The lowest energy orbital in 90° H₂O is the one labelled $1a_1$, which is built from the overlap of the $O2s$ and $O2p_z$ orbitals with the $\psi_{H_A1s} + \psi_{H_B1s}$ combination of H1s orbitals. The energy of the $1a_1$ orbital rises as the bond angle increases, in part because the weakly bonding H–H overlap decreases and in part because the loss of p_z character diminishes the overlap with the H–H combination. The energy of the $1b_2$ orbital is lowered because the H1s orbitals move into a better position for overlap with the $O2p_y$ orbital; their weakly *anti*bonding H–H overlap is also reduced. The biggest change occurs for the $2a_1$ orbital. This molecular orbital is principally an $O2s$ orbital in the 90° molecule, but correlates with a pure $O2p_z$ orbital in the 180° molecule. Hence, it shows a steep rise in energy as the bond angle increases. The $1b_1$ orbital is a nonbonding $O2p$ orbital perpendicular to the molecular plane in the 90° molecule and remains nonbonding in the linear molecule. Hence, its energy barely changes with angle.

---

17 As remarked earlier, the central feature of molecular orbital theory is the formation of molecular orbitals from all the atomic orbitals available that have the appropriate symmetry, and the linear combinations listed above can be regarded as a grouping of the atomic orbitals into different symmetry classes. This grouping is the subject of Chapter 15.

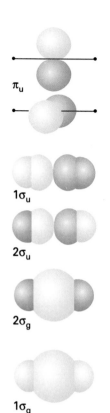

π_u

$1\sigma_u$

$2\sigma_u$

$2\sigma_g$

$1\sigma_g$

14.38 The molecular orbitals that can be constructed from the H1s, O2s, and O2p atomic orbitals in a hypothetical linear H₂O molecule.

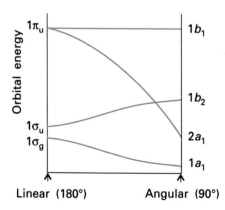

Orbital energy

$1\pi_u$ $1b_1$

$1b_2$

$1\sigma_u$
$1\sigma_g$ $2a_1$

$1a_1$

Linear (180°)   Angular (90°)

14.39 The Walsh diagram for H₂O. The energies of the linear molecule are shown on the left (see Fig. 14.38 for their compositions) and those of the 90° molecule are shown on the right (see Fig. 14.40). The actual molecule has a bond angle of 104°.

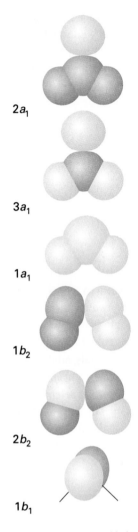

$2a_1$

$3a_1$

$1a_1$

$1b_2$

$2b_2$

$1b_1$

14.40 The molecular orbitals that can be constructed from the H1s, O2s, and O2p atomic orbitals in a hypothetical 90° H_2O molecule.

4

The principal feature that determines whether or not the H_2O molecule is bent is whether the $2a_1$ orbital is occupied. This orbital has considerable O2s character in the bent molecule but not in the linear molecule. Therefore, a lower total energy is achieved if, when it is occupied, the molecule is bent. The shape adopted by an H_2O molecule therefore depends on the number of electrons that occupy the orbitals.

Example 14.6 Using a Walsh diagram to predict a shape

Predict the shape of the H_2O molecule from the Walsh diagram.

Method Choose an intermediate bond angle along the horizontal axis of the H_2O diagram in Fig. 14.39, and accommodate eight electrons. Then consider whether the energy can be reduced by a modification of the bond angle. To do so, look at the effect on the energies of the occupied orbitals of a change in bond angle.

Answer The resulting configuration is $1a_1^2 2a_1^2 1b_2^2 1b_1^2$. The $2a_1$ orbital is occupied, so we expect the nonlinear molecule to have a lower energy than the linear molecule.

- -

Self-test 14.7 Predict the shape of the BeH_2 molecule.

[Linear]

14.9 The Hückel approximation

Molecular orbital theory takes large molecules and extended aggregates of atoms, such as solid materials, into its stride. First we shall consider **conjugated molecules**, in which there is an alternation of single and double bonds along a chain of carbon atoms.

Although the classification of an orbital as σ or π is strictly valid only in linear molecules, it is also used to denote the *local* symmetry with respect to a given A–B bond axis. Moreover, in nonlinear molecules, there is no orbital angular momentum around the bond axis: the π orbital is a (real) standing wave with electron density on each side of the local molecular plane.

The π molecular orbital energy level diagrams of conjugated molecules can be constructed using a set of approximations suggested by Erich Hückel in 1931. In his approach, the π orbitals are treated separately from the σ orbitals, and the latter form a rigid framework that determines the general shape of the molecule. All the C atoms are treated identically, so all the Coulomb integrals α for the atomic orbitals that contribute to the π orbitals are set equal. For example, in ethene, we take the σ bonds as fixed, and concentrate on finding the energies of the single π bond and its companion antibond. In butadiene (**4**), the σ framework is taken as fixed, and we concentrate on finding the π orbitals spreading across the four C atoms.

(a) The secular determinant

We express the π orbitals as LCAOs of the C2p orbitals that lie perpendicular to the molecular plane. In ethene we would write

$$\psi = c_A A + c_B B \tag{37}$$

and in butadiene

$$\psi = c_A A + c_B B + c_C C + c_D D \tag{38}$$

where the A is a C2p orbital on atom A, and so on. Next, the optimum coefficients and energies are found by the variation principle as explained in Section 14.7c. That is, we have to solve the secular determinant, which in the case of ethene is eqn 28 with $\alpha_A = \alpha_B = \alpha$.

The determinant for butadiene is similar, but more atoms contribute and, being at various distances from each other, they have different overlap and resonance integrals:

Ethene:

$$\begin{vmatrix} \alpha - E & \beta - ES \\ \beta - ES & \alpha - E \end{vmatrix} = 0 \tag{39}$$

Butadiene:

$$\begin{vmatrix} \alpha - E & \beta_{AB} - ES_{AB} & \beta_{AC} - ES_{AC} & \beta_{AD} - ES_{AD} \\ \beta_{BA} - ES_{BA} & \alpha - E & \beta_{BC} - ES_{BC} & \beta_{BD} - ES_{BD} \\ \beta_{CA} - ES_{CA} & \beta_{CB} - ES_{CB} & \alpha - E & \beta_{CD} - ES_{CD} \\ \beta_{DA} - ES_{DA} & \beta_{DB} - ES_{DB} & \beta_{DC} - ES_{DC} & \alpha - E \end{vmatrix} = 0 \tag{40}$$

The roots of the ethene determinant can be found very easily (they are the same as those in Example 14.4). However, for elementary calculations, the roots of the butadiene determinant are obviously going to prove difficult to find. In a modern computation all the resonance integrals and overlap integrals would be included, but an indication of the molecular orbital energy level diagram can be obtained very readily if we make the following additional **Hückel approximations**:

1. All overlap integrals are set equal to zero.
2. All resonance integrals between non-neighbours are set equal to zero.
3. All remaining resonance integrals are set equal (to β).

These approximations are obviously very severe, but they let us calculate at least a general picture of the molecular orbital energy levels with very little work. The assumptions result in the following structure of the secular determinant:

1. All diagonal elements: $\alpha - E$.
2. Off-diagonal elements between neighbouring atoms: β.
3. All other elements: 0.

(b) Ethene and frontier orbitals

For ethene, the Hückel approximations lead to

$$\begin{vmatrix} \alpha - E & \beta \\ \beta & \alpha - E \end{vmatrix} = (\alpha - E)^2 - \beta^2 = 0 \tag{41}$$

The roots of the equation are

$$E_\pm = \alpha \pm \beta \tag{42}$$

The + sign corresponds to the bonding combination (β is negative) and the − sign corresponds to the antibonding combination (Fig. 14.41).[18] The building-up principle then leads to the configuration $1\pi^2$, because each carbon atom supplies one electron to the π system. We can also estimate the $\pi^* \leftarrow \pi$ excitation energy ($2|\beta|$). The constant β is often left as an adjustable parameter; an approximate value for (C2p,C2p)-overlap π-bonds is about -75 kJ mol^{-1}, corresponding to -0.8 eV.

The **highest occupied molecular orbital** in ethene, its HOMO, is the 1π orbital; the **lowest unfilled molecular orbital**, its LUMO, is the $2\pi^*$ orbital. These two orbitals jointly form the **frontier orbitals** of the molecule. The frontier orbitals are important because they are largely responsible for many of the chemical and spectroscopic properties of the molecule.

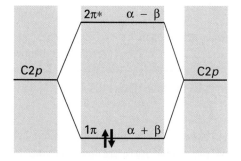

14.41 The Hückel molecular orbital energy levels of ethene. Two electrons occupy the lower π orbital.

18 To see the effect of neglecting overlap, compare the result obtained here with eqn 30.

(c) Butadiene and π-electron binding energy

For butadiene, the approximations result in the determinant

$$
\begin{vmatrix}
\alpha - E & \beta & 0 & 0 \\
\beta & \alpha - E & \beta & 0 \\
0 & \beta & \alpha - E & \beta \\
0 & 0 & \beta & \alpha - E
\end{vmatrix} = 0 \tag{43}
$$

Example 14.7 Finding the roots of a determinant

Find the roots of the butadiene secular determinant.

Method A 4×4 determinant is expanded in a series of steps like the 2×2 determinant treated in Example 14.4. After expansion, the terms are grouped to give a polynomial in E, which is set equal to 0 and then solved for E. A 4×4 determinant expands into a quartic equation, but we shall see that it may be expressed as a quadratic equation that can be solved by elementary methods.

Answer

$$
\begin{vmatrix}
\alpha - E & \beta & 0 & 0 \\
\beta & \alpha - E & \beta & 0 \\
0 & \beta & \alpha - E & \beta \\
0 & 0 & \beta & \alpha - E
\end{vmatrix}
$$

$$
= (\alpha - E)
\begin{vmatrix}
\alpha - E & \beta & 0 \\
\beta & \alpha - E & \beta \\
0 & \beta & \alpha - E
\end{vmatrix}
- \beta
\begin{vmatrix}
\beta & \beta & 0 \\
0 & \alpha - E & \beta \\
0 & \beta & \alpha - E
\end{vmatrix}
$$

$$
= (\alpha - E)^2
\begin{vmatrix}
\alpha - E & \beta \\
\beta & \alpha - E
\end{vmatrix}
- \beta(\alpha - E)
\begin{vmatrix}
\beta & \beta \\
0 & \alpha - E
\end{vmatrix}
$$

$$
- \beta^2
\begin{vmatrix}
\alpha - E & \beta \\
\beta & \alpha - E
\end{vmatrix}
+ \beta^2
\begin{vmatrix}
0 & \beta \\
0 & \alpha - E
\end{vmatrix}
$$

$$
= (\alpha - E)^4 - (\alpha - E)^2 \beta^2 - (\alpha - E)^2 \beta^2 - (\alpha - E)^2 \beta^2 + \beta^4
$$

$$
= (\alpha - E)^4 - 3(\alpha - E)^2 \beta^2 + \beta^4 = 0
$$

With $x = (\alpha - E)^2/\beta^2$, the expanded determinant has the form of a quadratic equation

$$
x^2 - 3x + 1 = 0
$$

The roots are $x = 2.62$ and 0.38. Therefore, the energies of the four LCAO-MOs are

$$
E = \alpha \pm 1.62\beta, \qquad \alpha \pm 0.62\beta
$$

- -

Self-test 14.8 Write down and expand the secular determinant for cyclobutadiene.

[See Example 14.8, below]

We have seen in Example 14.7 that the energies of the four LCAO-MOs are

$$
E = \alpha \pm 1.62\beta, \qquad \alpha \pm 0.62\beta \tag{44}
$$

These orbitals and their energies are drawn in Fig. 14.42. Note that, the greater the number of internuclear nodes, the higher the energy of the orbital. There are four electrons to accommodate, so the ground-state configuration is $1\pi^2 2\pi^2$. The frontier orbitals of butadiene are the 2π orbital (the HOMO, which is largely bonding) and the 3π orbital (the

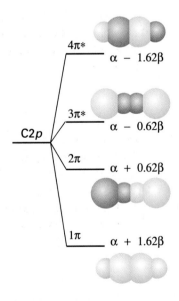

14.42 The Hückel molecular orbital energy levels of butadiene and the top view of the corresponding π orbitals. The four p electrons (one supplied by each C) occupy the two lower π orbitals. Note that the orbitals are delocalized.

C2p

$4\pi*$ $\alpha - 1.62\beta$

$3\pi*$ $\alpha - 0.62\beta$

2π $\alpha + 0.62\beta$

1π $\alpha + 1.62\beta$

LUMO, which is largely antibonding). 'Largely' bonding means that an orbital has both bonding and antibonding interactions between various neighbours, but the bonding effects dominate. 'Largely antibonding' indicates that the antibonding effects dominate.

An important point emerges when we calculate the total **π-electron binding energy**, E_π, the sum of the energies of each π electron, and compare it with what we find in ethene. In ethene the total energy is

$$E_\pi = 2(\alpha + \beta) = 2\alpha + 2\beta$$

In butadiene it is

$$E_\pi = 2(\alpha + 1.62\beta) + 2(\alpha + 0.62\beta) = 4\alpha + 4.48\beta$$

Therefore, the energy of the butadiene molecule lies lower by 0.48β (about -36 kJ mol^{-1}) than the sum of two individual π bonds. This extra stabilization of a conjugated system is called the **delocalization energy**.

Example 14.8 Estimating the delocalization energy

Use the Hückel approximation to find the energies of the π orbitals of cyclobutadiene, and estimate the delocalization energy.

Method Set up the secular determinant using the same basis as for butadiene, but note that atoms A and D are also now neighbours. Then solve for the roots of the secular equation and assess the total π-bond energy. For the delocalization energy, subtract from the total π-bond energy the energy of two π bonds.

Answer The secular determinant is

$$\begin{vmatrix} \alpha - E & \beta & 0 & \beta \\ \beta & \alpha - E & \beta & 0 \\ 0 & \beta & \alpha - E & \beta \\ \beta & 0 & \beta & \alpha - E \end{vmatrix} = 0$$

This determinant expands to

$$x(x-4) = 0 \qquad x = \left(\frac{\alpha - E}{\beta}\right)^2$$

The solutions are $x = 0$ and $x = 4$, so the energies of the orbitals are

$$E = \alpha + 2\beta, \qquad \alpha, \qquad \alpha, \qquad \alpha - 2\beta$$

Four electrons must be accommodated. Two occupy the lowest orbital (of energy $\alpha + 2\beta$), and two occupy the doubly degenerate orbitals (of energy α). The total energy is therefore $4\alpha + 4\beta$. Two isolated π bonds would have an energy $4\alpha + 4\beta$; therefore, in this case, the delocalization energy is zero.

- -

Self-test 14.9 Repeat the calculation for benzene.

[Next subsection]

(d) Benzene and aromatic stability

The most notable example of delocalization conferring extra stability is benzene and the aromatic molecules based on its structure. Benzene is often expressed in a mixture of valence-bond and molecular orbital terms with, typically, valence-bond language used for its σ framework and molecular orbital language used to describe its π electrons.

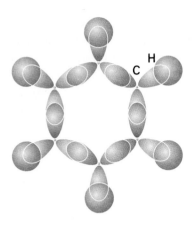

14.43 The σ framework of benzene is formed by the overlap of Csp^2 hybrids, which fit without strain into a hexagonal arrangement.

First, the valence-bond component. The six C atoms are regarded as sp^2 hybridized, with a single unhybridized perpendicular $2p$ orbital. One H atom is bonded by (Csp^2,H1s) overlap to each C carbon, and the remaining hybrids overlap to give a regular hexagon of atoms (Fig. 14.43). The internal angle of a regular hexagon is 120°, so sp^2 hybridization is ideally suited for forming σ bonds. We see that benzene's hexagonal shape permits strain-free σ bonding.

Now consider the molecular orbital component of the description. The six C$2p$ orbitals overlap to give six π orbitals that spread all round the ring. Their energies are calculated within the Hückel approximation by solving the secular determinant

$$
\begin{vmatrix}
\alpha - E & \beta & 0 & 0 & 0 & \beta \\
\beta & \alpha - E & \beta & 0 & 0 & 0 \\
0 & \beta & \alpha - E & \beta & 0 & 0 \\
0 & 0 & \beta & \alpha - E & \beta & 0 \\
0 & 0 & 0 & \beta & \alpha - E & \beta \\
\beta & 0 & 0 & 0 & \beta & \alpha - E
\end{vmatrix} = 0
\tag{45}
$$

When this determinant is expanded in the same way as in Example 14.7, the roots are found to be simply

$$
E = \alpha \pm 2\beta, \qquad \alpha \pm \beta, \qquad \alpha \pm \beta
\tag{46}
$$

as shown in Fig. 14.44. The orbitals there have been given symmetry labels which we explain in Chapter 15. Note that the lowest energy orbital is bonding between all neighbouring atoms, the highest energy orbital is antibonding between each pair of neighbours, and the intermediate orbitals are a mixture of bonding, nonbonding, and antibonding character between adjacent atoms.

We now apply the building-up principle to the π system. There are six electrons to accommodate (one from each C atom), so the three lowest orbitals (a_{2u} and the doubly degenerate pair e_{1g}) are fully occupied, giving the ground-state configuration $a_{2u}^2 e_{1g}^4$. A significant point is that the only molecular orbitals occupied are those with net bonding character.

The π-electron energy of benzene is

$$
E_\pi = 2(\alpha + 2\beta) + 4(\alpha + \beta) = 6\alpha + 8\beta
$$

If we ignored delocalization and thought of the molecule as having three isolated π bonds, it would be ascribed a π-electron energy of only $3(2\alpha + 2\beta) = 6\alpha + 6\beta$. The delocalization energy is therefore $2\beta \approx -150\ \text{kJ mol}^{-1}$, which is considerably more than for butadiene.

This discussion suggests that aromatic stability can be traced to two main contributions. First, the shape of the regular hexagon is ideal for the formation of strong σ bonds: the σ framework is relaxed and without strain. Second, the π orbitals are such as to be able to accommodate all the electrons in bonding orbitals, and the delocalization energy is large.

(e) Semi-empirical and ab initio methods

Modern techniques of molecular electronic structure calculation have moved on considerably from the techniques we have been describing, but they are clear descendants of these more elementary methods. They still involve expressing molecular orbitals as linear combinations of atomic orbitals, setting up secular determinants in which various integrals appear, finding their roots, and then solving secular equations for the coefficients. However, the principal difference is the inclusion of electron–electron repulsion into the energy calculation and looking for self-consistent solutions, in much the same way as for atoms

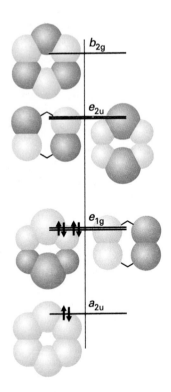

14.44 The Hückel orbitals of benzene and the corresponding energy levels. The symmetry labels are explained in Chapter 15. The bonding and antibonding character of the delocalized orbitals reflects the numbers of nodes between the atoms. In the ground state, only the net bonding orbitals are occupied.

(Section 13.5). There are two main strategies of calculation. In **semi-empirical methods** many of the integrals are estimated by appealing to spectroscopic data or physical properties such as ionization energies, and using a series of rules to set certain integrals equal to zero. In the *ab initio methods*, an attempt is made to calculate all the integrals that appear in the secular determinant. Both procedures employ a great deal of computational effort and, along with cryptanalysts and meteorologists, theoretical chemists are among the heaviest users of the fastest computers.

The Hückel method is a primitive example of a semi-empirical procedure: all the properties of the π system are expressed in terms of the two parameters α and β and all overlap integrals are set equal to zero. In a more sophisticated procedure, we write the π orbitals as linear combination of atomic orbitals, but use the full hamiltonian, including the electron–electron repulsions proportional to $1/r_{ij}$. Moreover, we also make sure that the many-electron wavefunction (the product of the individual occupied molecular orbitals) satisfies the Pauli principle. When all this is worked through, it turns out that the secular determinant includes integrals of the form

$$(AB|CD) = \int A(1)B(1)\left(\frac{e^2}{4\pi\varepsilon_0 r_{12}}\right)C(2)D(2)\,\mathrm{d}\tau_1\mathrm{d}\tau_2 \tag{47}$$

where A, B, C, and D are atomic orbitals which in general may be centred on different nuclei. It can be appreciated that, if there are several dozen atomic orbitals used to build the molecular orbitals, then there will be tens of thousands of integrals of this form to evaluate.

One severe approximation is called **complete neglect of differential overlap** (CNDO), in which all integrals are set to zero unless A and B are the same orbitals centred on the same nucleus, and likewise for C and D. The surviving integrals are then adjusted until the energy levels are in good agreement with experiment. The more recent semi-empirical methods make less draconian decisions about which integrals are to be ignored, but they are all descendants of the early CNDO technique. These procedures are now readily available in commercial software packages and can be used with very little detailed knowledge of their mode of calculation. The packages also have sophisticated graphical output procedures, which enable one to analyse the shapes of orbitals and the distribution of electric charge in molecules. The latter is important when assessing, for instance, the likelihood that a given molecule will bind to an active site in an enzyme. Such studies can greatly reduce the time and cost of screening compounds for potential pharmacological activity.

Commercial packages are also available for *ab initio* calculations. Here the problem is to evaluate as efficiently as possible thousands of integrals. This task is greatly facilitated by expressing the atomic orbitals used in the LCAOs as linear combinations of Gaussian orbitals. A **Gaussian type orbital** (GTO) is a function of the form $e^{-\alpha r^2}$. The advantage of GTOs over the correct orbitals (which are proportional to e^{-r}) is that the product of two Gaussian functions is itself a Gaussian function that lies between the centres of the two contributing functions (Fig. 14.45). In this way, the four-centre integrals like that in eqn 47 become two-centre integrals of the form

$$(AB|CD) = \int X(1)\left(\frac{e^2}{4\pi\varepsilon_0 r_{12}}\right)Y(2)\,\mathrm{d}\tau_1\mathrm{d}\tau_2 \tag{48}$$

where X is the Gaussian corresponding to the product AB and Y is the corresponding Gaussian from BD. Integrals of this form are much easier and faster to evaluate numerically than the original four-centre integrals. Although more GTOs have to be used to simulate the atomic orbitals, there is an overall increase in speed of computation.

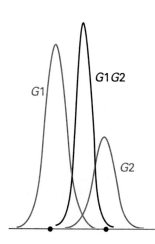

14.45 The product of two Gaussian functions (the green curves) is itself a Gaussian function located between the two contributing Gaussians.

14.10 The band theory of solids

The extreme case of delocalization is a solid, in which atom after atom lies in a three-dimensional array and takes part in bonding spreading throughout the crystal. Two types of solid are distinguished by the temperature dependence of their electrical conductivity:

A **metallic conductor** is a substance with a conductivity that decreases as the temperature is raised.

A **semiconductor** is a substance with a conductivity that increases as the temperature is raised.

A semiconductor generally has a lower conductivity than that typical of metals, but the magnitude of the conductivity is not the criterion of the distinction. It is conventional to classify semiconductors with very low electrical conductivities as **insulators**. We shall use this term, but it should be appreciated that it is one of convenience rather than one of fundamental significance.

We shall consider a one-dimensional solid, which consists of a single, infinitely long line of atoms, each one having one s orbital available for forming molecular orbitals. We can construct the LCAO-MOs of the solid by adding N atoms in succession to a line, and then find the electronic structure using the building-up principle.

(a) The formation of bands

One atom contributes one s orbital at a certain energy (Fig. 14.46). When a second atom is brought up it overlaps the first, and forms a bonding and antibonding orbital. The third atom overlaps its nearest neighbour (and only slightly the next-nearest), and from these three atomic orbitals three molecular orbitals are formed: one is fully bonding, one fully antibonding, and the intermediate orbital is nonbonding between neighbours. The fourth atom leads to the formation of a fourth molecular orbital. At this stage, we can begin to see that the general effect of bringing up successive atoms is to spread the range of energies covered by the molecular orbitals, and also to fill in the range of energies with more and more orbitals (one more for each atom). When N atoms have been added to the line, there

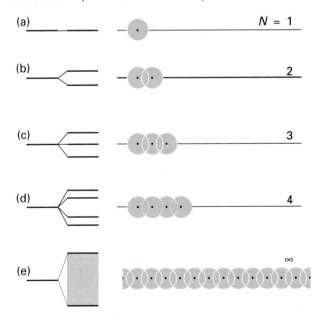

14.46 The formation of a band of N molecular orbitals by successive addition of N atoms to a line. Note that the band remains of finite width as $N \rightarrow \infty$ and, although it looks continuous, it consists of N different orbitals.

are N molecular orbitals covering a band of energies of finite width, and the Hückel secular determinant is

$$\begin{vmatrix} \alpha - E & \beta & 0 & 0 & 0 & \cdots & 0 \\ \beta & \alpha - E & \beta & 0 & 0 & \cdots & 0 \\ 0 & \beta & \alpha - E & \beta & 0 & \cdots & 0 \\ 0 & 0 & \beta & \alpha - E & \beta & \cdots & 0 \\ 0 & 0 & 0 & \beta & \alpha - E & \cdots & 0 \\ \vdots & \vdots & \vdots & \vdots & \vdots & \cdots & \vdots \\ 0 & 0 & 0 & 0 & 0 & \cdots & \alpha - E \end{vmatrix} = 0 \tag{49}$$

where β is now the (s, s) resonance integral. The theory of determinants applied to such a symmetrical example as this (technically a 'tridiagonal determinant') leads to the following expression for the roots:

$$E_k = \alpha + 2\beta \cos\left(\frac{k\pi}{N + 1}\right) \qquad k = 1, 2, \ldots, N \tag{50}$$

When N is infinitely large, the difference between neighbouring energy levels (the energies corresponding to k and $k + 1$) is infinitely small, but the band still has finite width overall:

$$E_N - E_1 \to 4\beta \text{ as } N \to \infty \tag{51}$$

We can think of this band as consisting of N different molecular orbitals, the lowest-energy orbital ($k = 1$) being fully bonding, and the highest-energy orbital ($k = N$) being fully antibonding between adjacent atoms (Fig. 14.47). Similar bands form in three-dimensional solids.

The band formed from overlap of s orbitals is called the **s band**. If the atoms have p orbitals available, the same procedure leads to a **p band** (as shown in the upper half of Fig. 14.47). If the atomic p orbitals lie higher in energy than the s orbitals, then the p band

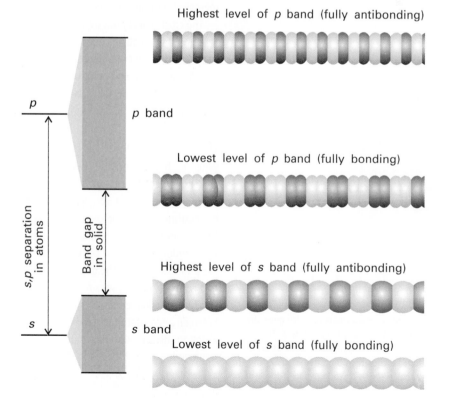

14.47 The overlap of s orbitals gives rise to an s band, and the overlap of p orbitals gives rise to a p band. In this case, the s and p orbitals of the atoms are so widely spaced that there is a band gap. In many cases the separation is less, and the bands overlap.

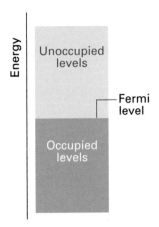

14.48 When N electrons occupy a band of N orbitals, it is only half full and the electrons near the Fermi level (the top of the filled levels) are mobile.

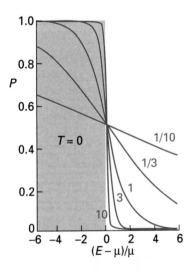

14.49 The Fermi–Dirac distribution, which gives the population of the levels at a temperature T. The high-energy tail decays exponentially towards zero. The curves are labelled with the value of μ/kT. The tinted grey region shows the occupation of levels at $T = 0$.

lies higher than the s band, and there may be a **band gap**, a range of energies to which no orbital corresponds.

(b) The occupation of orbitals at $T = 0$

Now consider the electronic structure of a solid formed from atoms each able to contribute one electron (for example, the alkali metals). There are N atomic orbitals and therefore N molecular orbitals squashed into an apparently continuous band. There are N electrons to accommodate.

At $T = 0$, only the lowest $\frac{1}{2}N$ molecular orbitals are occupied (Fig. 14.48), and the HOMO is called the **Fermi level**. However, unlike in the discrete molecules we have considered so far, there are empty orbitals very close in energy to the Fermi level, so it requires hardly any energy to excite the uppermost electrons. Some of the electrons are therefore very mobile, and give rise to electrical conductivity.

(c) The occupation of orbitals at $T > 0$

At temperatures above absolute zero, electrons can be excited by the thermal motion of the atoms. The population, P, of the orbitals is given by the **Fermi–Dirac distribution**, a version of the Boltzmann distribution that takes into account the effect of the Pauli principle:

$$P = \frac{1}{e^{(E-\mu)/kT} + 1} \tag{52}$$

The quantity μ is the **chemical potential**,[19] which in this context is the energy of the level for which $P = \frac{1}{2}$ (note that the chemical potential changes as the temperature changes). The shape of the Fermi–Dirac distribution is shown in Fig. 14.49. For energies well above μ, the 1 in the denominator can be neglected, and then

$$P \approx e^{-(E-\mu)/kT} \tag{53}$$

The population now resembles a Boltzmann distribution, decaying exponentially with increasing energy. The higher the temperature, the longer the exponential tail.

The electrical conductivity of a metallic solid decreases with increasing temperature even though more electrons are excited into empty orbitals. This apparent paradox is resolved by noting that the increase in temperature causes more vigorous thermal motion of the atoms, so collisions between the moving electrons and an atom are more likely. That is, the electrons are scattered out of their paths through the solid, and are less efficient at transporting charge.

(d) Insulators and semiconductors

When each atom provides two electrons, the $2N$ electrons fill the N orbitals of the s band. The Fermi level now lies at the top of the band (at $T = 0$), and there is a gap before the next band begins (Fig. 14.50). As the temperature is increased, the tail of the Fermi–Dirac distribution extends across the gap, and electrons populate the empty orbitals of the upper band. They are now mobile, and the solid is an electric conductor. In fact, the solid is a semiconductor, because the electrical conductivity depends on the number of electrons that are promoted across the gap, and that number increases as the temperature is raised. If the gap is large, though, very few electrons will be promoted at ordinary temperatures, and the conductivity will remain close to zero, giving an insulator. Thus, the conventional distinction between an insulator and a semiconductor is related to the size of the band gap and is not an absolute distinction like that between a metal (incomplete bands at $T = 0$) and a semiconductor (full bands at $T = 0$).

19 Note that the 'chemical potential' in eqn 52 is an energy, not a molar Gibbs energy, as in the thermodynamic use of the term.

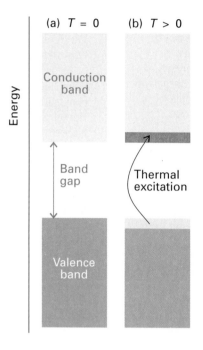

14.50 (a) When $2N$ electrons are present, the band is full and the material is an insulator at $T = 0$. (b) At temperatures above $T = 0$, electrons populate the levels of the upper 'conduction' band at the expense of the filled 'valence' band and the solid is a semiconductor.

14.51 (a) A dopant with fewer electrons than its host can form a narrow band that accepts electrons from the valence band. The holes in the band are mobile, and the substance is a p-type semiconductor. (b) A dopant with more electrons than its host forms a narrow band that can supply electrons to the conduction band. The electrons it supplies are mobile, and the substance is an n-type semiconductor.

Another method of increasing the number of charge carriers and enhancing the semiconductivity of a solid is to implant foreign atoms into an otherwise pure material. If these **dopants** can trap electrons, they withdraw electrons from the filled band, leaving holes which allow the remaining electrons to move (Fig. 14.51). This procedure gives rise to **p-type semiconductivity**, the p indicating that the holes are positive relative to the electrons in the band. Alternatively, a dopant might carry excess electrons (for example, phosphorus atoms introduced into germanium), and these additional electrons occupy otherwise empty bands, giving **n-type semiconductivity**, where n denotes the negative charge of the carriers. The preparation of doped but otherwise ultrapure materials was described in Section 8.7.

Checklist of key ideas

The Born–Oppenheimer approximation

☐ Born–Oppenheimer approximation
☐ molecular potential energy curve
☐ bond dissociation energy

Valence-bond theory

☐ valence-bond theory (VB theory)

14.1 The hydrogen molecule

☐ σ bond

14.2 Homonuclear diatomic molecules

☐ homonuclear diatomic molecules
☐ π bond

14.3 Polyatomic molecules

☐ promotion
☐ hybrid orbital
☐ sp^3 hybrid orbital
☐ sp^2 hybrid orbital
☐ sp hybrid orbital

Molecular orbital theory

☐ molecular orbital theory (MO theory)

14.4 The hydrogen molecule-ion

☐ molecular orbital (MO)
☐ linear combination of atomic orbitals (LCAO)
☐ LCAO-MO
☐ σ orbital
☐ overlap density
☐ bonding orbital
☐ σ electron
☐ antibonding orbital

14.5 The structures of diatomic molecules

- [ ] molecular orbital energy level diagram
- [ ] bond order (15)
- [ ] bond dissociation energy
- [ ] π orbital
- [ ] overlap integral (18)

14.6 More about notation

- [ ] parity
- [ ] centre of inversion

14.7 Heteronuclear diatomic molecules

- [ ] polar bond

- [ ] partial negative charge
- [ ] partial positive charge
- [ ] electronegativity
- [ ] Pauling electronegativity
- [ ] Mulliken electronegativity
- [ ] variation principle
- [ ] trial wavefunction
- [ ] basis set
- [ ] secular equations
- [ ] Coulomb integral
- [ ] resonance integral
- [ ] secular determinant

Molecular orbitals for polyatomic systems

14.8 Walsh diagrams

- [ ] Walsh diagram
- [ ] nonbonding orbital

14.9 The Hückel approximation

- [ ] conjugated molecule
- [ ] Hückel approximations
- [ ] highest occupied molecular orbital (HOMO)
- [ ] lowest unoccupied molecular orbital (LUMO)
- [ ] frontier orbital
- [ ] π-electron binding energy
- [ ] delocalization energy
- [ ] semi-empirical methods
- [ ] *ab initio* methods

- [ ] complete neglect of differential overlap (CNDO)
- [ ] Gaussian type orbital (GTO)

14.10 The band theory of solids

- [ ] metallic conductor
- [ ] semiconductor
- [ ] insulator
- [ ] *s* band
- [ ] *p* band
- [ ] band gap
- [ ] Fermi level
- [ ] Fermi–Dirac distribution
- [ ] chemical potential
- [ ] dopant
- [ ] p-type semiconductivity
- [ ] n-type semiconductivity

Further reading

Articles of general interest

G.A. Gallup, The Lewis electron-pair model, spectroscopy, and the role of the orbital picture in describing the electronic structure of molecules. *J. Chem. Educ.* **65**, 671 (1988).

A.B. Sannigrahi and T. Kar, Molecular orbital theory of bond order and valency. *J. Chem. Educ.* **65**, 674 (1988).

S. Nordholm, Delocalization—the key concept of covalent bonding. *J. Chem. Educ.* **65**, 581 (1988).

A.A. Woolf, Oxidation numbers and their limitations. *J. Chem. Educ.* **65**, 45 (1988).

A.D. Buckingham and T.W. Rowlands, Can addition of a bonding electron weaken a bond? *J. Chem. Educ.* **68**, 282 (1991).

E.I. von Nagy-Felsobuki, Hückel theory and photoelectron spectroscopy. *J. Chem. Educ.* **66**, 821 (1989).

J.R. Dias, A facile Hückel molecular orbital solution of buckminsterfullerene using chemical graph theory. *J. Chem. Educ.* **66**, 1012 (1989).

D.J. Klein and N.Trinajstić, Valence-bond theory and chemical structure. *J. Chem. Educ.* **67**, 633 (1990).

A. Pisanty, The electronic structure of graphite: a chemist's introduction to band theory. *J. Chem. Educ.* **68**, 804 (1991).

R. Parson, Visualizing the variation principle: an intuitive approach to interpreting the theorem in geometric terms. *J. Chem. Educ.* **70**, 115 (1993).

J.B. Holbrook, R. Salry-Grant, B.C. Smith, and T.V. Tandel, Lattice enthalpies of ionic halides, hydrides, oxides, and sulfides: second-electron affinities of atomic oxygen and sulfur. *J. Chem. Educ.* **67**, 304 (1990).

A.F. Gaines and F.M. Page, The strengths of σ and π bonds in gaseous compounds. *J. Chem. Res. (S)* 200 (1980).

J.C. Morrison, A.W. Weiss, K. Kirby, and D. Cooper, Electronic structure of atoms and molecules. In *Encyclopedia of applied physics* (ed. G.L. Trigg), **6**, 45. VCH, New York (1993).

P. Engelking, Molecules. In *Encyclopedia of applied physics* (ed. G.L. Trigg), **10**, 525. VCH. New York (1994).

R.M. Metzger, Semiconductors to superconductors: organic lower-dimensional systems. In *Encyclopedia of applied physics* (ed. G.L. Trigg), **17**, 215. VCH, New York (1996).

S. Tiwari, Semiconductors, compound—electronic properties. In *Encyclopedia of applied physics* (ed. G.L. Trigg), **17**, 303. VCH, New York (1996).

Texts and sources of data and information

M.J. Winter, *Chemical bonding*, Oxford Chemistry Primers. Oxford University Press (1993).

J.N. Murrell, S.F.A. Kettle, and J.M. Tedder, *The chemical bond*. Wiley, New York (1985).

M.F.C. Ladd, *Chemical bonding in solids and fluids*. Ellis Horwood/Prentice-Hall, Hemel Hempstead (1993).

C.A. Coulson, *The shape and structure of molecules* (revised by R. McWeeny). Oxford University Press (1982).

J.K. Burdett, *Chemical bonding: a dialogue*. Wiley, New York (1996).

Y. Jean, F. Volatron, and J.K. Burdett, *An introduction to chemical bonding*. Oxford University Press, New York (1993).

P.R. Scott and W.G. Richards, *Energy levels in atoms and molecules*, Oxford Chemistry Primers. Oxford University Press (1994).

R. McWeeny, *Coulson's Valence*. Oxford University Press (1979).

D.F. Shriver, P.W. Atkins, and C.H. Langford, *Inorganic chemistry*. Oxford University Press, Oxford and W.H. Freeman & Co, New York (1994).

L. Pauling, *The nature of the chemical bond*. Cornell University Press, Ithaca (1960).

A. Hinchcliffe, *Computational quantum chemistry*. Wiley, New York (1988).

R.L. deKock and H.B. Gray, *Chemical structure and bonding*. University Science Books, Mill Valley (1989).

G.H. Grant and W.G. Richards, *Computational chemistry*, Oxford Chemistry Primers. Oxford University Press (1995).

A.R. Leach, *Molecular modelling: principles and applications*. Longman, Harlow (1996).

B. Webster, *Chemical bonding theory*. Blackwell Scientific, Oxford (1990).

A. Zewail (ed.), *The chemical bond: structure and dynamics*. Academic Press, San Diego (1992).

P.A. Cox, *The electronic structure and chemistry of solids*. Oxford University Press (1987).

Z.B. Maksić (ed.), *Theoretical models of chemical bonding*, Vols 1–4. Springer-Verlag, Berlin (1990–1991).

Exercises

14.1 (a) Give the ground-state electron configurations and bond orders of (a) Li_2, (b) Be_2, and (c) C_2.

14.1 (b) Give the ground-state electron configurations of (a) H_2^-, (b) N_2, and (c) O_2.

14.2 (a) Give the ground-state electron configurations of (a) CO, (b) NO, and (c) CN^-.

14.2 (b) Give the ground-state electron configurations of (a) ClF, (b) CS, and (c) O_2^-.

14.3 (a) From the ground-state electron configurations of B_2 and C_2, predict which molecule should have the greater bond dissociation energy.

14.3 (b) Which of the molecules N_2, NO, O_2, C_2, F_2, and CN would you expect to be stabilized by (a) the addition of an electron to form AB^-, (b) the removal of an electron to form AB^+?

14.4 (a) Sketch the molecular orbital energy level diagram for XeF and deduce its ground-state electron configurations. Is XeF likely to have a shorter bond length than XeF^+?

14.4 (b) Sketch the molecular orbital energy level diagrams for BrCl and deduce its ground-state electron configurations. Is BrCl likely to have a shorter bond length than $BrCl^-$?

14.5 (a) Where appropriate, give the parity of (a) π^* in F_2, (b) σ^* in NO, (c) δ in Tl_2, (d) δ^* in Fe_2.

14.5 (b) Give the parities of the six π molecular orbitals of benzene.

14.6 (a) The term symbol for the ground state of N_2^+ is $^2\Sigma_g^+$. What is the total spin and total orbital angular momentum of the molecule? Show that the term symbol agrees with the electron configuration that would be predicted using the building-up principle.

14.6 (b) One of the excited states of the C_2 molecule has the valence electron configuration $1\sigma_g^2 2\sigma_u^{*2} 1\pi_u^3 2\pi_g^1$. Give the multiplicity and parity of the term.

14.7 (a) Use the electron configurations of NO and N_2 to predict which is likely to have the shorter bond length.

14.7 (b) Arrange the species O_2^+, O_2, O_2^-, O_2^{2-} in order of increasing bond length.

14.8 (a) Show that the sp^2 hybrid orbital $(s + 2^{1/2}p)/3^{1/2}$ is normalized to 1 if the s and p orbitals are normalized to 1.

14.8 (b) Normalize the molecular orbital $\psi_s(A) + \lambda\psi_s(B)$ in terms of the parameter λ and the overlap integral S.

14.9 (a) Confirm that the bonding and antibonding combinations $\psi_s(A) \pm \psi_s(B)$ are mutually orthogonal in the sense that their mutual overlap is zero.

14.9 (b) Suppose that a molecular orbital has the form $N(0.145A + 0.844B)$. Find a linear combination of the orbitals A and B that is orthogonal to this combination.

14.10 (a) Which of the following triatomic molecules and ions are expected to be linear: (a) CO_2, (b) NO_2, (c) NO_2^+? Give reasons in each case.

14.10 (b) Which of the following triatomic molecules and ions are expected to be linear: (a) NO_2^-, (b) SO_2, (c) H_2O, (d) H_2O^{2+}? Give reasons in each case.

14.11 (a) Construct the molecular orbital energy level diagrams of ethene (ethylene) on the basis that the molecule is formed from the appropriately hybridized CH_2 or CH fragments.

14.11 (b) Construct the molecular orbital energy level diagrams of ethyne (acetylene) on the basis that the molecule is formed from the appropriately hybridized CH_2 or CH fragments.

14.12 (a) Write down the secular determinants for (a) linear H_3, (b) cyclic H_3 within the Hückel approximation.

14.12 (b) Predict the electronic configurations of (a) the benzene anion, (b) the benzene cation. Estimate the π-bond energy in each case.

Problems

Numerical problems

14.1 Show that, if a wave $\cos kx$ centred on A (so that x is measured from A) interferes with a similar wave $\cos k'x$ centred on B (with x measured from B) a distance R away, then constructive interference occurs in the intermediate region when $k = k' = \pi/2R$ and destructive interference if $kR = \frac{3}{2}\pi$ and $k'R = \frac{3}{2}\pi$.

14.2 The overlap integral between two H1s orbitals on nuclei separated by a distance R is $S = \{1 + (R/a_0) + \frac{1}{3}(R/a_0)^2\}e^{-R/a_0}$. Plot this function for $0 \geq R < \infty$.

14.3 Before doing the calculation below, sketch how the overlap between an s orbital and a σp orbital can be expected to depend on their separation. The overlap integral between an H1s orbital and an H2p orbital on nuclei separated by a distance R is $S = (R/a_0)\{1 + (R/a_0) + \frac{1}{3}(R/a_0)^2\}e^{-R/a_0}$. Plot this function, and find the separation for which the overlap is a maximum.

14.4 Calculate the total amplitude of the normalized bonding and antibonding LCAO-MOs that may be formed from two H1s orbitals at a separation of 106 pm. Plot the two amplitudes for positions along the molecular axis both inside and outside the internuclear region.

14.5 Repeat the calculation in Problem 14.4, but plot the probability densities of the two orbitals. Then form the difference density, the difference between ψ^2 and $\frac{1}{2}\{\psi_s(A)^2 + \psi_s(B)^2\}$.

14.6 Imagine a small electron-sensitive probe of volume $1.00\ \text{pm}^3$ inserted into an H_2^+ molecule ion in its ground state. Calculate the probability that it will register the presence of an electron at the following positions: (a) at nucleus A, (b) at nucleus B, (c) halfway between A and B, (d) at a point 20 pm along the bond from A and 10 pm perpendicularly. Do the same for the molecule-ion the instant after the electron has been excited into the antibonding LCAO-MO.

14.7 The energy of H_2^+ with internuclear separation R is given by the expression

$$E = E_H - \frac{V_1 + V_2}{1 + S} + \frac{e^2}{4\pi\varepsilon_0 R}$$

where E_H is the energy of an isolated H atom, V_1 is the attractive potential energy between the electron centred on one nucleus and the charge of the other nucleus, V_2 is the attraction between the overlap density and one of the nuclei, S is the overlap integral. The values are given below. Plot the molecular potential energy curve and find the bond dissociation energy (in electronvolts) and the equilibrium bond length.

R/a_0	0	1	2	3	4
V_1/E_h	1.000	0.729	0.473	0.330	0.250
V_2/E_h	1.000	0.736	0.406	0.199	0.092
S		1.000	0.858	0.587	0.349	0.189

where $E_h = 27.3$ eV, $a_0 = 52.9$ pm, and $E_H = -\frac{1}{2}E_h$.

14.8 The same data as in Problem 14.7 may be used to calculate the molecular potential energy curve for the antibonding orbital, which is given by

$$E = E_H - \frac{V_1 - V_2}{1 - S} + \frac{e^2}{4\pi\varepsilon_0 R}$$

Plot the curve.

14.9 In the 'free electron molecular orbital' (FEMO) theory, the electrons in a conjugated molecule are treated as independent particles in a box of length L. Sketch the form of the two occupied orbitals in butadiene predicted by this model and predict the minium excitation energy of the molecule. The tetraene $CH_2=CHCH=CHCH=CHCH=CH_2$ can be treated as a box of length $8R$, where $R \approx 140$ pm (as in this case, an extra half bond-length is often added at each end of the box). Calculate the minimum excitation energy of the molecule and sketch the HOMO and LUMO. Estimate the colour a sample of the compound is likely to appear in white light.

Theoretical problems

14.10 An sp^2 hybrid orbital that lies in the xy plane and makes an angle of $120°$ to the x-axis has the form

$$\psi = \frac{1}{3^{1/2}}\left(s - \frac{1}{2^{1/2}}p_x + \frac{3^{1/2}}{2^{1/2}}p_y\right)$$

Use hydrogenic atomic orbitals to write the explicit form of the hybrid orbital. Show that it has its maximum amplitude in the direction specified.

14.11 Use the expressions in Problems 14.7 and 14.8 to show that the antibonding orbital is more antibonding than the bonding orbital is bonding at most internuclear separations.

14.12 Derive the expressions used in Problems 14.7 and 14.8 using the normalized LCAO-MOs for the H_2^+ molecule-ion. Proceed by evaluating the expectation value of the hamiltonian for the ion. Make use of the fact that $\psi_s(A)$ and $\psi_s(B)$ each individually satisfy the Schrödinger equation for an isolated H atom.

14.13 Construct the Walsh diagram for an AH_3 molecule, and use it to predict the shapes of (a) NH_3, (b) CH_3^+.

14.14 Take as a trial function for the ground state of the hydrogen atom (a) e^{-kr}, (b) e^{-kr^2} and use the variation principle to find the optimum value of k in each case. Identify the better wavefunction. The only part of the laplacian that need be considered is the part that involves radial derivatives (eqn 12.63).

Additional problems supplied by Carmen Giunta and Charles Trapp

14.15 J.G. Dojahn, E.C.M. Chen, and W.E. Wentworth (*J. Phys. Chem.* **100**, 9649 (1996)) characterized the potential energy curves of homonuclear diatomic halogen molecules and molecular anions. Among the properties they report are the the the equilibrium internuclear distance R_e, the vibrational wavenumber, $\tilde{\nu}$, and the dissociation energy, D_e.

Species	r_e/pm	$\tilde{\nu}$/cm^{-1}	D_e/eV
F_2	1.411	916.6	1.60
F_2^-	1.900	450.0	1.31

Rationalize these data by using qualitative molecular orbital configurations.

14.16 Rydberg molecules can be thought of as molecular analogues of Rydberg atoms. However Rydberg molecules do not involve atomic orbitals with analogously large quantum numbers ($n = 100$), but rather atomic orbitals with n one higher than the n values of the valence shells of the constituent atoms. Nevertheless speculate about the existence of Rydberg H_2 as formed from two H atoms with $100s$ electrons. Make reasonable guesses about the binding energy, the equilibrium internuclear separation, the vibrational force constant, and the rotational constant. Is such a molecule likely to exist under any circumstances?

14.17 Set up and solve the Hückel secular equations for the π electrons in NO_3^-. Express the energies in terms of the Coulomb integrals, α_O and α_N, and the resonance integral β. Determine the delocalization energy of the nitrate ion.

14.18 In Exercise 14.12a, you were invited to set up the Hückel secular determinants for linear and cyclic H_3. The same secular determinant applies to the molecular ions H_3^+ and D_3^+. The molecular ion H_3^+ was discovered as long ago as 1912 by J.J. Thomson, but only more recently has the equilateral-triangular structure been confirmed by M.J. Gaillard, *et al.* (*Phys. Rev.* **A17**, 1797 (1978)). The molecular ion H_3^+ is the simplest polyatomic species with a confirmed existence and plays an important role in interstellar chemistry. (a) Solve the Hückel secular equations for the energies of the H_3 system in terms of the parameters α and β, draw an energy level diagram for the orbitals, and determine the binding energies of the molecules H_3^{2+}, H_3^+, H_3, and H_3^-. (b) Accurate quantum mechanical calculations by G.D. Carney and R.N. Porter (*J. Chem. Phys.* **65**, 3547 (1976)) give the dissociation energy for the process $H_3^+(g) \rightarrow 2H(g) + H^+(g)$ as 849 kJ mol^{-1}. From this information and data in Table 2.6, calculate the enthalpy of the reaction $H^+(g) + H_2(g) \rightarrow H_3^+(g)$. Compare to the binding energy of $H_2(g)$. (c) From your equations and the information given, calculate a value for the the resonance integral, β, in H_3^+; then go on to calculate the binding energies for the other H_3 species in (a).

14.19 There is some indication that other hydrogen ring compounds and ions in addition to H_3 and D_3 species may play a role in interstellar chemistry. According to J.S. Wright and G.A. DiLabio (*J. Phys. Chem.* **96**, 10793 (1992)), H_5^-, H_6, and H_7^+ are particularly stable, whereas H_4 and H_5^+ are not. Confirm these statements by Hückel calculations.

15 Molecular symmetry

In this chapter we sharpen the concept of 'shape' into a precise definition of 'symmetry', and show that symmetry may be discussed systematically. We see how to classify any molecule according to its symmetry, and how to use this classification to discuss molecular properties. After describing the symmetry properties of molecules themselves, we turn to a consideration of the effect of symmetry transformations on orbitals, and see that their transformation properties can be used to set up a labelling scheme. These symmetry labels are used to identify what integrals necessarily vanish. One important integral is the overlap integral between two orbitals. By knowing which atomic orbitals may have nonzero overlap, we can decide which ones can contribute to the formation of molecular orbitals. We also see how to select linear combinations of atomic orbitals that match the symmetry of the nuclear framework. Finally, by considering the symmetry properties of integrals, we see that it is possible to derive the selection rules that govern spectroscopic transitions.

The systematic discussion of symmetry is called **group theory**. Much of group theory is a summary of common sense about the symmetries of objects. However, because group theory is systematic, its rules can be applied in a straightforward, mechanical way, and in some cases it gives unexpected results. In most cases the theory gives a simple, direct method for arriving at useful conclusions with the minimum of calculation, and this is the aspect we stress here.

The symmetry elements of objects

Some objects are 'more symmetrical' than others. A sphere is more symmetrical than a cube because it looks the same after it has been rotated through any angle about any diameter. A cube looks the same only if it is rotated through certain angles about specific axes, such as 90°, 180°, or 270° about an axis passing through the centres of any of its opposite faces (Fig. 15.1), or by 120° or 240° about an axis passing through any of its opposite corners. Similarly, an NH_3 molecule is 'more symmetrical' than an H_2O molecule because NH_3 looks the same after rotations of 120° or 240° about the axis shown in Fig. 15.2, whereas H_2O looks the same only after a rotation of 180°.

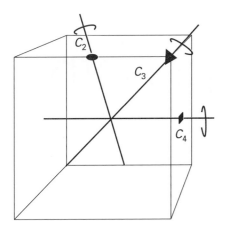

15.1 Some of the symmetry elements of a cube. The twofold, threefold, and fourfold axes are labelled with the conventional symbols.

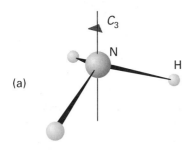

(a)

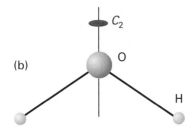

(b)

15.2 (a) An NH_3 molecule has a threefold (C_3) axis and (b) an H_2O molecule has a twofold (C_2) axis. Both have other symmetry elements too.

An action that leaves an object looking the same after it has been carried out is called a **symmetry operation**. Typical symmetry operations include rotations, reflections, and inversions. There is a corresponding **symmetry element** for each symmetry operation, which is the point, line, or plane with respect to which the symmetry operation is performed. For instance, a rotation (a symmetry operation) is carried out around an axis (the corresponding symmetry element). We shall see that we can classify molecules by identifying all their symmetry elements, and grouping together molecules that possess the same set of symmetry elements. This procedure, for example, puts the trigonal pyramidal species NH_3 and SO_3^{2-} into one group and the angular species H_2O and SO_2 into another group.

15.1 Operations and symmetry elements

The classification of objects according to symmetry elements corresponding to operations that leave at least one common point unchanged gives rise to the **point groups**. There are five kinds of symmetry operation (and five kinds of symmetry element) of this kind. When we consider crystals (Chapter 21), we shall meet symmetries arising from translation through space. These more extensive groups are called **space groups**.

The **identity**, E, consists of doing nothing; the corresponding symmetry element is the entire object. Because every object is indistinguishable from itself if nothing is done to it, every object possesses at least the identity element. One reason for including the identity is that some molecules have only this symmetry element (**1**); another reason is technical and connected with the detailed formulation of group theory.

An **n-fold rotation** (the operation) about an **n-fold axis of symmetry**, C_n (the corresponding element), is a rotation through $360°/n$. The operation C_1 is a rotation through $360°$, and is equivalent to the identity operation E. An H_2O molecule has one twofold axis, C_2. An NH_3 molecule has one threefold axis, C_3, with which is associated two symmetry operations, one being $120°$ rotation in a clockwise sense and the other $120°$ rotation in a counter-clockwise sense.[1] A pentagon has a C_5 axis, with two (clockwise and counter-clockwise) rotations through $72°$ associated with it. It also has an axis denoted C_5^2, corresponding to two successive C_5 rotations; there are two such operations, one through $144°$ in a clockwise sense and the other through $144°$ in a counter-clockwise sense. A cube has three C_4 axes, four C_3 axes, and six C_2 axes. However, even this high symmetry is exceeded by a sphere, which possesses an infinite number of symmetry axes (along any diameter) of all possible integral values of n. If a molecule possesses several rotation axes, then the one (or more) with the greatest value of n is called the **principal axis**. The principal axis of a benzene molecule is the sixfold axis perpendicular to the hexagonal ring (**2**).

A **reflection** (the operation) in a **mirror plane**, σ (the element), may contain the principal axis of a molecule or be perpendicular to it. If the plane is parallel to the principal axis, it is called 'vertical' and denoted σ_v. An H_2O molecule has two vertical planes of symmetry

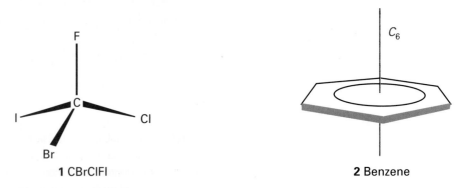

1 CBrClFI

2 Benzene

---

1   There is only one twofold rotation associated with a C_2 axis because clockwise and counter-clockwise $180°$ rotations are identical.

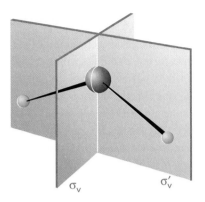

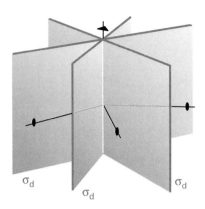

15.3 An H_2O molecule has two mirror planes. They are both vertical (that is, contain the principal axis) and so are denoted σ_v and σ'_v.

15.4 Dihedral mirror planes (σ_d) bisect the C_2 axes perpendicular to the principal axis.

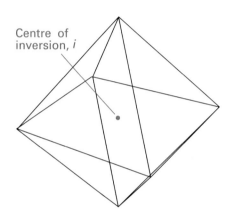

15.5 A regular octahedron has a centre of inversion (i).

(Fig. 15.3) and an NH_3 molecule has three. A vertical mirror plane that bisects the angle between two C_2 axes is called a 'dihedral plane' and is denoted σ_d (Fig. 15.4). When the plane of symmetry is perpendicular to the principal axis it is called 'horizontal' and denoted σ_h. A C_6H_6 molecule has a C_6 principal axis and a horizontal mirror plane (as well as several other symmetry elements).

In an **inversion** (the operation) through a **centre of symmetry**, i (the element), we imagine taking each point in a molecule, moving it to the centre of the molecule, and then moving it out the same distance on the other side,[2] that is, the point (x, y, z) is taken into the point $(-x, -y, -z)$. Neither an H_2O molecule nor an NH_3 molecule has a centre of inversion, but a sphere and a cube do have one. A C_6H_6 molecule does have a centre of inversion, as does a regular octahedron (Fig. 15.5); a regular tetrahedron and a CH_4 molecule do not.

An **n-fold improper rotation** (the operation) about an n-fold axis of improper rotation or an **n-fold improper rotation axis**, S_n (the symmetry element), is composed of two successive transformations. The first component is a rotation through $360°/n$, and the second is a reflection through a plane perpendicular to the axis of that rotation; neither operation alone needs to be a symmetry operation. A CH_4 molecule has three S_4 axes (Fig. 15.6).

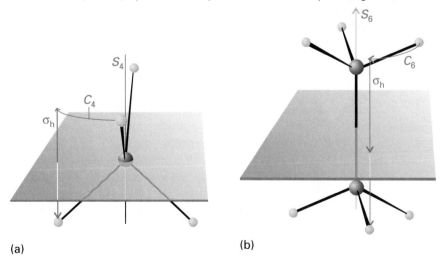

(a)   (b)

15.6 (a). A CH_4 molecule has a fourfold improper rotation axis (S_4): the molecule is indistinguishable after a $90°$ rotation followed by a reflection across the horizontal plane, but neither operation alone is a symmetry operation. (b) The staggered form of ethane has an S_6 axis composed of a $60°$ rotation followed by a reflection.

2   This operation was first encountered in Section 14.6a in connection with the parity classification of orbitals.

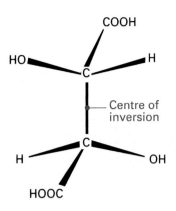

3 Meso-tartaric acid

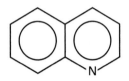

4 Quinoline

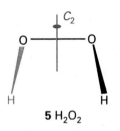

5 H_2O_2

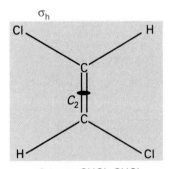

6 *trans*-CHCl=CHCl

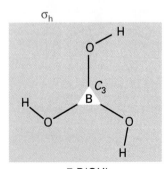

7 $B(OH)_3$

15.2 The symmetry classification of molecules

To classify molecules according to their symmetries, we list their symmetry elements and collect together molecules with the same list of elements. This procedure puts CH_4 and CCl_4, which both possess the same symmetry elements as a regular tetrahedron, into the same group, and H_2O into another group.

The name of the group to which a molecule belongs is determined by the symmetry elements it possesses. There are two systems of notation (Table 15.1). The **Schoenflies system** is more common for the discussion of individual molecules, and the **Hermann–Mauguin system**, or **International system**, is used almost exclusively in the discussion of crystal symmetry.

Table 15.1 The notation for point groups*

C_i	$\bar{1}$										
C_s	m										
C_1	1	C_2	2	C_3	3	C_4	4	C_6	6
		C_{2v}	$2mm$	C_{3v}	$3m$	C_{4v}	$4mm$	C_{6v}	$6mm$
		C_{2h}	$2/m$	C_{3h}	$\bar{6}$	C_{4h}	$4/m$	C_{6h}	$6/m$
		D_2	222	D_3	32	D_4	422	D_6	622
		D_{2h}	mmm	D_{3h}	$\bar{6}2m$	D_{4h}	$4/mmm$	D_{6h}	$6/mmm$
		D_{2d}	$\bar{4}2m$	D_{3d}	$\bar{3}m$	S_4	$\bar{4}/m$	S_6	$\bar{3}$
T	23	T_d	$\bar{4}3m$	T_h	$m3$				
O	432	O_h	$m3m$						

*In the International system (or Hermann–Mauguin system) for point groups, a number n denotes the presence of an n-fold axis and m denotes a mirror plane. A diagonal line / indicates that the mirror plane is perpendicular to the symmetry axis. It is important to distinguish symmetry elements of the same type but of different classes, as in $4/mmm$, in which there are three classes of mirror plane (σ_v, σ_h, and σ_d). A bar over a number indicates that the element is combined with an inversion. The only groups listed in this table are the so-called crystallographic point groups (Section 21.1).

(a) The groups C_1, C_i, and C_s

A molecule belongs to the group C_1 if it has no element other than the identity (as in (**1**)). It belongs to C_i if it has the identity and the inversion alone (**3**), and to C_s if it has the identity and a mirror plane alone (**4**).

(b) The groups C_n, C_{nv}, and C_{nh}

A molecule belongs to the group C_n if it possesses an n-fold axis.[3] An H_2O_2 molecule has the elements E and C_2 (**5**), so it belongs to the group C_2.

If in addition to the identity and a C_n axis a molecule has n vertical mirror planes σ_v, then it belongs to the group C_{nv}. An H_2O molecule, for example, has the symmetry elements E, C_2, and $2\sigma_v$, so it belongs to the group C_{2v}. An NH_3 molecule has the elements E, C_3, and $3\sigma_v$, so it belongs to the group C_{3v}. A heteronuclear diatomic molecule such as HCl belongs to the group $C_{\infty v}$ because all rotations around the axis and reflections across the axis are symmetry operations. Other members of the group $C_{\infty v}$ include the linear OCS molecule and a cone.

Objects that in addition to the identity and an n-fold principal axis also have a horizontal mirror plane σ_h belong to the groups C_{nh}. An example is *trans*-CHCl=CHCl (**6**), which has the elements E, C_2, and σ_h, and so belongs to the group C_{2h}; the molecule $B(OH)_3$ in the conformation shown in (**7**) belongs to the group C_{3h}. The presence of certain symmetry

3 Note that symbol C_n is now playing a triple role: as the label of a symmetry element, a symmetry operation, and a group.

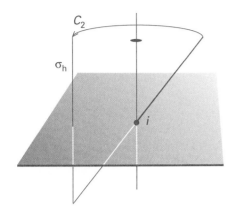

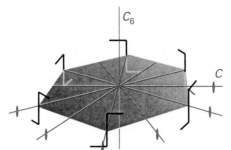

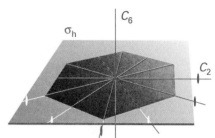

15.7 The presence of a twofold axis and a horizontal mirror plane jointly imply the presence of a centre of inversion in the molecule.

15.8 A molecule with n twofold rotation axes perpendicular to an n-fold rotation axis belongs to the group D_n.

15.9 A molecule with a mirror plane perpendicular to a C_n axis, and with n twofold axes in the plane, belongs to the group D_{nh}.

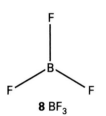

8 BF$_3$

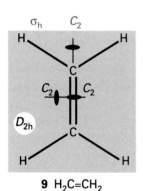

9 H$_2$C=CH$_2$

elements may be implied by the presence of others: thus, in C_{2h} the operations C_2 and σ_h jointly imply the presence of a centre of inversion (Fig. 15.7).

(c) The groups D_n, D_{nh}, and D_{nd}

A molecule that has an n-fold principal axis and n twofold axes perpendicular to C_n belongs to the group D_n (Fig. 15.8). A molecule belongs to D_{nh} if it also possesses a horizontal mirror plane (Fig. 15.9). The planar trigonal BF$_3$ molecule has the elements E, C_3, $3C_2$, and σ_h (with one C_2 axis along each B–F bond), and so belongs to D_{3h} (**8**). The C$_6$H$_6$ molecule has the elements E, C_6, $3C_2$, $3C_2'$, and σ_h together with some others that these elements imply,[4] so it belongs to D_{6h}. All homonuclear diatomic molecules, such as N$_2$, belong to the group $D_{\infty h}$ because all rotations around the axis are symmetry operations, as are end-to-end rotation and end-to-end reflection; $D_{\infty h}$ is also the group of the linear OCO and HCCH molecules and of a uniform cylinder. Other examples of D_{nh} molecules are shown in (**9**), (**10**), and (**11**).

A molecule belongs to the group D_{nd} if in addition to the elements of D_n it possesses n dihedral mirror planes σ_d. The twisted, 90° allene (**12**) belongs to D_{2d}, and the staggered conformation of ethane (**13**) belongs to D_{3d}.

(d) The groups S_n

Molecules that have not been classified into one of the groups mentioned so far, but which possess one S_n axis, belong to the group S_n. An example is tetraphenylmethane, which

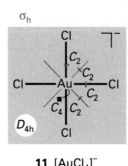

10 PCl$_5$

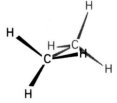

11 [AuCl$_4$]$^-$

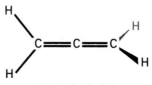

12 H$_2$C=C=CH$_2$

13 C$_2$H$_6$

---

4   The prime on $3C_2'$ indicates that the three C_2 axes are different from the other three C_2 axes.

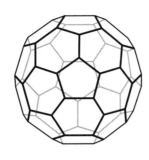

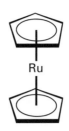

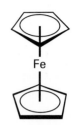

14 C(C$_6$H$_5$)$_4$ **15** Buckminsterfullerene, C$_{60}$ **16** Ruthenocene, Ru(C$_6$H$_5$)$_2$ **17** Excited ferrocene, Fe(C$_6$H$_5$)$_2^*$

belongs to the point group S_4 (**14**). Molecules belonging to S_n with $n > 4$ are rare. Note that the group S_2 is the same as C_i, so such a molecule will already have been classified as C_i.

(e) The cubic groups

A number of very important molecules (for example, CH$_4$ and SF$_6$) possess more than one principal axis. Most belong to the **cubic groups**, and in particular to the tetrahedral groups T, T_d, and T_h or to the octahedral groups O, O_h (Fig. 15.10). A few icosahedral (20-faced) molecules, belonging to the **icosahedral group**, I (Fig. 15.11), are also known: they include some of the boranes and buckminsterfullerene, C$_{60}$ (**15**). The groups T_d and O_h are the groups of the regular tetrahedron (for example, CH$_4$) and the regular octahedron (for example, SF$_6$), respectively. If the object possesses the rotational symmetry of the tetrahedron or the octahedron, but none of their planes of reflection, then it belongs to the simpler groups T or O (Fig. 15.12). The group T_h is based on T but also contains a centre of inversion (Fig. 15.13).

(f) The full rotation group

The **full rotation group**, R_3 (the 3 refers to rotation in three dimensions), consists of an infinite number of rotation axes with all possible values of n. A sphere and an atom belong to R_3, but no molecule does. Exploring the consequences of R_3 is a very important way of applying symmetry arguments to atoms, and is an alternative approach to the theory of orbital angular momentum.

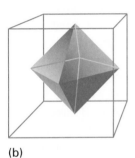

(a)

(b)

15.10 (a) Tetrahedral and (b) octahedral molecules are drawn in a way that shows their relation to a cube: they belong to the cubic groups T_d and O_h, respectively.

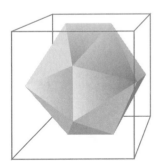

15.11 The relation of an icosahedron to a cube. The buckminsterfullerene molecule (**15**) is related to this object by cutting off each apex to form a regular pentagon.

Example 15.1 Identifying a point group of a molecule

Identify the point group to which a ruthenocene molecule (**16**) belongs.

Method The identification of a molecule's point group is simplified by referring to the flow diagram in Fig. 15.14 and the shapes shown in Fig. 15.15.

Answer The path to trace through the flow diagram in Fig. 15.14 is shown by a green line; it ends at D_{nh}. Because the molecule has a fivefold axis, it belongs to the group D_{5h}.

Comment If the rings were staggered, as they are in an excited state of ferrocene that lies 4 kJ mol^{-1} above the ground state (**17**), the horizontal reflection plane would be absent, but dihedral planes would be present.

- -

Self-test 15.1 Classify the pentagonal antiprismatic excited state of ferrocene.

$[D_{5d}]$

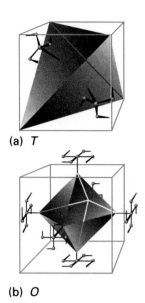

(a) T

(b) O

15.12 Shapes corresponding to the point groups (a) T and (b) O. The presence of the windmill-like structures reduces the symmetry of the object from T_d and O_h, respectively.

15.13 The shape of an object belonging to the group T_h.

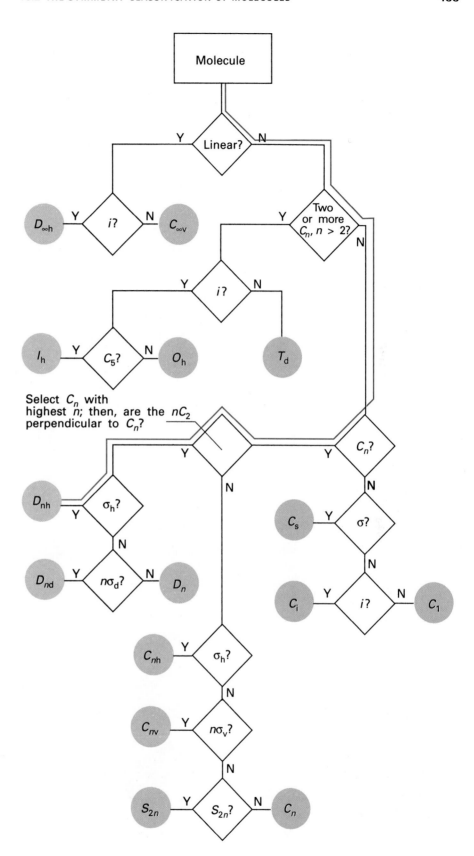

15.14 A flow diagram for determining the point group of a molecule. Start at the top and answer the question posed in each diamond (Y = yes, N = no).

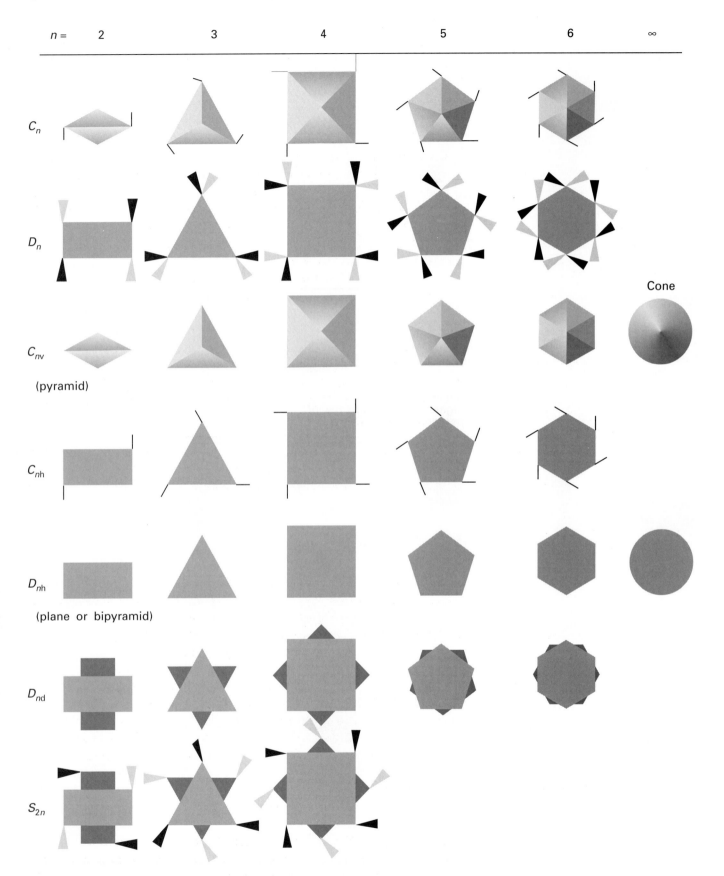

15.15 A summary of the shapes corresponding to different point groups. The group to which a molecule belongs can often be identified from this diagram without going through the formal procedure in Fig. 15.14.

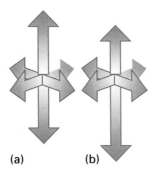

15.16 (a) A molecule with a C_n axis cannot have a dipole perpendicular to the axis, but (b) it may have one parallel to the axis. The arrows represent local contributions to the overall electric dipole, such as may arise from bonds between pairs of neighbouring atoms with different electronegativities.

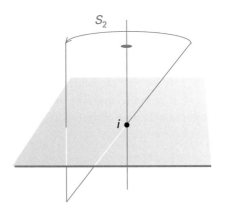

15.17 Some symmetry elements are implied by the other symmetry elements in a group. Any molecule containing an inversion also possesses at least an S_2 element because i and S_2 are equivalent.

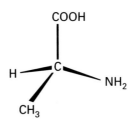

18 L-Alanine, $NH_2CH(CH_3)COOH$

19 Glycine, NH_2CH_2COOH

15.3 Some immediate consequences of symmetry

Some statements about the properties of a molecule can be made as soon as its point group has been identified.

(a) Polarity

A **polar molecule** is one with a permanent electric dipole moment (HCl, O_3, and NH_3 are examples). If the molecule belongs to the group C_n with $n > 1$, it cannot possess a charge distribution with a dipole moment perpendicular to the symmetry axis because the symmetry of the molecule implies that any dipole that exists in one direction perpendicular to the axis is cancelled by an opposing dipole (Fig. 15.16a). For example, the perpendicular component of the dipole associated with one OH bond in H_2O is cancelled by an equal but opposite component of the dipole of the second OH bond, so any dipole that the molecule has must be parallel to the twofold symmetry axis. However, as the group makes no reference to operations relating the two ends of the molecule, a charge distribution may exist that results in a dipole along the axis (Fig. 15.16b), and H_2O has a dipole moment parallel to its twofold symmetry axis. The same remarks apply to the group C_{nv}, so molecules belonging to any of the C_{nv} groups may be polar. In all the other groups, such as C_{3h}, D, etc., there are symmetry operations that take one end of the molecule into the other. Therefore, as well as having no dipole perpendicular to the axis, such molecules can have none along the axis, for otherwise these additional operations would not be symmetry operations.

We can conclude that only molecules belonging to the groups C_n, C_{nv}, and C_s may have a permanent electric dipole moment. For C_n and C_{nv}, that dipole moment must lie along the symmetry axis. Thus ozone, O_3, which is angular and belongs to the group C_{2v}, may be polar (and is), but carbon dioxide, CO_2, which is linear and belongs to the group $D_{\infty h}$, is not.

(b) Chirality

A **chiral molecule** (from the Greek word for 'hand') is a molecule that cannot be superimposed on its mirror image. Chiral molecules are optically active in the sense that they rotate the plane of polarized light (a property discussed in more detail in Section 22.2). A chiral molecule and its mirror-image partner constitute an **enantiomeric pair** of isomers and rotate the plane of polarization in equal but opposite directions.

It follows from the theory of optical activity that a molecule may be chiral only if it does not possess an axis of improper rotation, S_n. However, we need to be aware that such an axis may be present under a different name, and be implied by other symmetry elements that are present. For example, molecules belonging to the groups C_{nh} possess an S_n axis implicitly because they possess both C_n and σ_h, which are the two components of an improper rotation axis. Any molecule containing a centre of inversion, i, also possesses an S_2 axis, because i is equivalent to C_2 in conjunction with σ_h, and that combination of elements is S_2 (Fig. 15.17). It follows that all molecules with centres of inversion are achiral (that is, not chiral) and hence optically inactive. Similarly, because $S_1 = \sigma$, it follows that any molecule with a mirror plane is achiral.

A molecule may be chiral if it does not have a centre of inversion or a mirror plane, which is the case with the amino acid alanine (**18**), but not with glycine (**19**). However, a molecule may be achiral even though it does not have a centre of inversion. For example, the S_4 species (**20**) is achiral and optically inactive: though it lacks i, it does have an S_4 axis.

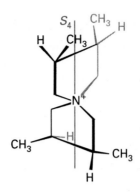

20 $N(CH_2CH(CH_3)CH(CH_3)CH_2)_2^+$

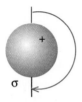

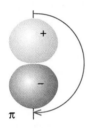

15.18 A rotation through 180° about the internuclear axis leaves the sign of a σ orbital unchanged but the sign of a π orbital is changed. In the language introduced in this chapter, the characters of the C_2 rotation are +1 and −1 for the σ and π orbitals, respectively.

C_2	(i.e. rotation by 180°)
σ +1	(i.e. no change of sign)
π −1	(i.e. change of sign)

Character tables

We shall now turn our attention away from the symmetries of molecules themselves and direct it towards the symmetry characteristics of orbitals that belong to the various atoms in a molecule. This material will enable us to discuss the formulation and labelling of molecular orbitals and selection rules in spectroscopy.

15.4 Character tables and symmetry labels

We saw in Chapter 14 that molecular orbitals of diatomic and linear polyatomic molecules are labelled σ, π, etc. These labels refer to the symmetries of the orbitals with respect to rotations around the principal symmetry axis of the molecule. Thus, a σ orbital does not change sign under a rotation through any angle, a π orbital changes sign when rotated by 180°, and so on (Fig. 15.18). The symmetry classifications σ and π can also be assigned to individual atomic orbitals in a linear molecule. For example, we can speak of an individual p_z orbital as having σ symmetry if the z-axis lies along the bond, because p_z is cylindrically symmetrical about the bond. This labelling of orbitals according to their behaviour under rotations can be generalized and extended to nonlinear polyatomic molecules, where there may be reflections and inversions to take into account as well as rotations.

Labels analogous to σ and π are also used to denote the symmetries of orbitals in polyatomic molecules. These labels look like a, a_1, e, e_g, and we first encountered them in Section 14.8a in connection with H_2O and in Fig. 14.44 in connection with the molecular orbitals of benzene. As we shall see, these labels indicate the behaviour of the orbitals under the symmetry operations of the relevant point group of the molecule.

(a) The structure of character tables

A label is assigned to an orbital by referring to the **character table** of the group, a table that characterizes the different symmetry types possible in the point group. Thus, to assign the labels σ and π, we use the table shown in the margin. This table is a fragment of the full character table for a linear molecule. The entry +1 shows that the orbital remains the same and the entry −1 shows that the orbital changes sign under the operation C_2 at the head of the column (as illustrated in Fig. 15.18). So, to assign the label σ or π to a particular orbital, we compare the orbital's behaviour with the information in the character table.

The entries in a complete character table are derived using the formal techniques of group theory, and are called **characters**, χ. These numbers characterize the essential features of each symmetry type in a way that we can illustrate by using the C_{3v} character table (Table 15.2). Character tables for other groups are given at the end of the *Data section* and are used in exactly the same way.

Table 15.2* The C_{3v} character table

$C_{3v}, 3m$	E	$2C_3$	$3\sigma_v$	$h = 6$	
A_1	1	1	1	z	$z^2, x^2 + y^2$
A_2	1	1	−1		
E	2	−1	0	(x, y)	$(xy, x^2 - y^2), (xz, yz)$

*More character tables are given at the end of the *Data section* at the end of this volume.

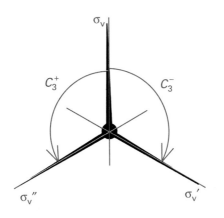

15.19 Symmetry operations in the same class are related to one another by the symmetry operations of the group. Thus, the three mirror planes shown here are related by threefold rotations, and the two rotations shown here are related by reflection in σ_v.

The columns in a character table are labelled with the symmetry operations of the group, which in C_{3v} are E, C_3, and σ_v. The numbers multiplying each operation are the numbers of members of each **class**. Symmetry operations fall into the same class if they are of the same type (for example, rotations) *and* can be transformed into one another by a symmetry operation of the group. From the C_{3v} character table we see that the two threefold rotations (clockwise and counter-clockwise rotations by 120°) belong to the same class: they are related by a reflection (Fig. 15.19). The three reflections (one through each of the three vertical mirror planes) also lie in the same class: they are related by the threefold rotations. The two reflections of the group C_{2v} fall into *different* classes: although they are both reflections, one cannot be transformed into the other by any symmetry operation of the group.

The total number of operations in a group is called the **order**, h, of the group. The order of C_{3v}, for instance, is 6.

The rows under the labels for the operations summarize the symmetry properties of the orbitals. They are labelled with the **symmetry species** (the analogues of the labels σ and π). More formally, the symmetry species label the **irreducible representations** of the group, which (as explained in the *Justification* below) are the basic types of behaviour that orbitals may show when subjected to the symmetry operations of the group. They are the analogues of the -1 and $+1$ given earlier, and which showed, respectively, whether an orbital changed sign or did not change sign when the molecule was subjected to a rotation of 180° about its internuclear axis. By convention, irreducible representations are labelled with upper-case Roman letters (such as A_1 and E) but the orbitals to which they apply are labelled with the lower-case italic equivalents (so an orbital of symmetry species A_1 is called an a_1 orbital). [5] Examples of each type of orbital are shown in Fig. 15.20.

Justification 15.1

Character tables are derived from the representation of the effects of symmetry operations by matrices. As an illustration, consider the C_{2v} molecule SO_2 and the valence p_x orbitals on each atom, which we shall denote p_S, p_A, and p_B (Fig. 15.21). Under σ_v, the change

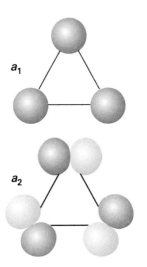

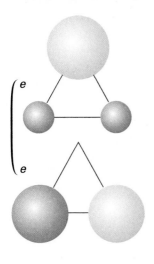

15.20 Typical symmetry-adapted linear combinations of orbitals in a C_{3v} molecule.

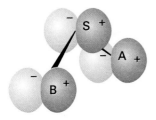

15.21 The three p_x orbitals that are used to illustrate the construction of a matrix representation in a C_{2v} molecule (SO_2).

---

5 Note that care must be taken to distinguish the identity element E (italic, a column heading) from the symmetry label E (Roman, a row label).

$(p_S, p_B, p_A) \leftarrow (p_S, p_A, p_B)$ takes place. We can express this transformation by using matrix multiplication:

$$(p_S, p_B, p_A) = (p_S, p_A, p_B) \begin{pmatrix} 1 & 0 & 0 \\ 0 & 0 & 1 \\ 0 & 1 & 0 \end{pmatrix} \tag{1a}$$

This relation can be expressed more succinctly as

$$(p_S, p_B, p_A) = (p_S, p_A, p_B)\mathbf{D}(\sigma_v) \tag{1b}$$

The matrix $\mathbf{D}(\sigma_v)$ is called a **representative** of the operation σ_v. Representatives take different forms according to the **basis**, the set of orbitals, that has been adopted.

We can use the same technique to find matrices that reproduce the other symmetry operations. For instance, C_2 has the effect $(-p_S, -p_B, -p_A) \leftarrow (p_S, p_A, p_B)$, and its representative is

$$\mathbf{D}(C_2) = \begin{pmatrix} -1 & 0 & 0 \\ 0 & 0 & -1 \\ 0 & -1 & 0 \end{pmatrix} \tag{2}$$

The effect of σ_v' is $(-p_S, -p_A, -p_B) \leftarrow (p_S, p_A, p_B)$, and its representative is

$$\mathbf{D}(\sigma_v') = \begin{pmatrix} -1 & 0 & 0 \\ 0 & -1 & 0 \\ 0 & 0 & -1 \end{pmatrix} \tag{3}$$

The identity operation has no effect on the basis, so its representative is the unit matrix:

$$\mathbf{D}(E) = \begin{pmatrix} 1 & 0 & 0 \\ 0 & 1 & 0 \\ 0 & 0 & 1 \end{pmatrix} \tag{4}$$

The set of matrices that represents all the operations of the group is called a **matrix representation**, Γ, of the group for the particular basis we have chosen. We denote this three-dimensional representation by the symbol $\Gamma^{(3)}$. The discovery of a matrix representation of the group means that we have found a link between the symbolic manipulations of the operations and algebraic manipulations involving numbers.

The character of an operation in a particular matrix representation is the sum of the diagonal elements of the representative of that operation. Thus, in the basis we are illustrating, the characters of the representatives are

$\mathbf{D}(E)$	$\mathbf{D}(C_2)$	$\mathbf{D}(\sigma_v)$	$\mathbf{D}(\sigma_v')$
3	−1	1	−3

The character of an operation depends on the basis.

Inspection of the representatives shows that they are all of **block-diagonal form**:

$$\mathbf{D} = \begin{pmatrix} \blacksquare & 0 & 0 \\ 0 & & \\ 0 & & \blacksquare \end{pmatrix}$$

The symmetry operations of C_{2v} never mix p_S with the other two functions. Consequently, the basis can be cut into two parts, one consisting of p_S alone and the other of (p_A, p_B). It is readily verified that the p_S orbital itself is a basis for the one-dimensional representation

$$\mathbf{D}(E) = 1 \qquad \mathbf{D}(C_2) = -1 \qquad \mathbf{D}(\sigma_v) = 1 \qquad \mathbf{D}(\sigma_v') = -1$$

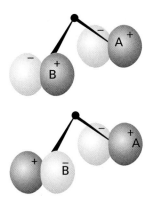

15.22 Two symmetry-adapted linear combinations of the basis orbitals shown in Fig. 15.21. The two combinations each span a one-dimensional irreducible representation, and their symmetry species are different.

which we shall call $\Gamma^{(1)}$. The remaining two basis functions are a basis for the two-dimensional representation $\Gamma^{(2)}$:

$$\mathbf{D}(E) = \begin{pmatrix} 1 & 0 \\ 0 & 1 \end{pmatrix} \quad \mathbf{D}(C_2) = \begin{pmatrix} 0 & -1 \\ -1 & 0 \end{pmatrix}$$

$$\mathbf{D}(\sigma_v) = \begin{pmatrix} 0 & 1 \\ 1 & 0 \end{pmatrix} \quad \mathbf{D}(\sigma_v') = \begin{pmatrix} -1 & 0 \\ 0 & -1 \end{pmatrix}$$

These matrices are the same as those of the original three-dimensional representation, except for the loss of the first row and column. We say that the original three-dimensional representation has been **reduced** to the **direct sum** of a one-dimensional representation **spanned** by p_S, and a two-dimensional representation spanned by (p_A, p_B). This reduction is consistent with the common sense view that the central orbital plays a role different from the other two. The reduction is denoted symbolically by writing

$$\Gamma^{(3)} = \Gamma^{(1)} + \Gamma^{(2)} \tag{5}$$

The one-dimensional representation cannot be reduced any further, and is called an **irreducible representation** of the group (an 'irrep'). We can demonstrate that the two-dimensional representation is reducible (for this basis in this group) by switching attention to the linear combinations $p_1 = p_A + p_B$ and $p_2 = p_A - p_B$. These combinations are sketched in Fig. 15.22. The representatives in the new basis can be constructed from the old by noting, for example, that under σ_v, $(p_B, p_A) \leftarrow (p_A, p_B)$. In this way we find the following representation in the new basis:

$$\mathbf{D}(E) = \begin{pmatrix} 1 & 0 \\ 0 & 1 \end{pmatrix} \quad \mathbf{D}(C_2) = \begin{pmatrix} -1 & 0 \\ 0 & 1 \end{pmatrix}$$

$$\mathbf{D}(\sigma_v) = \begin{pmatrix} 1 & 0 \\ 0 & -1 \end{pmatrix} \quad \mathbf{D}(\sigma_v') = \begin{pmatrix} -1 & 0 \\ 0 & -1 \end{pmatrix}$$

The new representatives are all in block-diagonal form, and the two combinations are not mixed with each other by any operation of the group. We have therefore achieved the reduction of $\Gamma^{(2)}$ to the sum of two one-dimensional representations. Thus, p_1 spans

$$\mathbf{D}(E) = 1 \qquad \mathbf{D}(C_2) = -1 \qquad \mathbf{D}(\sigma_v) = 1 \qquad \mathbf{D}(\sigma_v') = -1$$

which is the same one-dimensional representation as that spanned by p_S, and p_2 spans

$$\mathbf{D}(E) = 1 \qquad \mathbf{D}(C_2) = 1 \qquad \mathbf{D}(\sigma_v) = -1 \qquad \mathbf{D}(\sigma_v') = -1$$

which is a different one-dimensional representation; we shall denote it $\Gamma^{(1)'}$.

Now we can make the final link to the material in the text. *The character table of a group is the list of the characters of all its irreducible representations.* At this point we have found two irreducible representations of the group C_{2v} (Table 15.3). The two

Table 15.3* The C_{2v} character table

$C_{2v}, 2mm$	E	C_2	σ_v	σ_v'	$h = 4$	
A_1	1	1	1	1	z	z^2, y^2, x^2
A_2	1	1	-1	-1		xy
B_1	1	-1	1	-1	x	xz
B_2	1	-1	-1	1	y	yx

*More character tables are given at the end of the *Data section.*

irreducible representations are normally labelled B_1 and A_2, respectively. An A or a B is used to denote a one-dimensional representation; A is used if the character under the principal rotation is $+1$, and B is used if the character is -1. Subscripts are used to distinguish the irreducible representations if there is more than one of the same type: A_1 is reserved for the representation with character 1 for all operations. When higher dimensional irreducible representations are permitted, E denotes a two-dimensional irreducible representation and T a three-dimensional irreducible representation; all the irreducible representations of C_{2v} are one-dimensional.

There are in fact only two more species of irreducible representations of this group, for a surprising theorem of group theory states that

$$\text{Number of symmetry species} = \text{number of classes} \qquad (6)$$

In C_{2v} (Table 15.3), for instance, there are four classes (four columns in the character table), so there are only four species of irreducible representation. The character table in Table 15.3 therefore shows the characters of *all* the irreducible representations of this group.

(b) Character tables and orbital degeneracy

The characters of the identity operation, E, reveal the degeneracy of the orbitals. Thus, in a C_{3v} molecule, any orbital with a symmetry label a_1 or a_2 is non-degenerate. Any doubly degenerate pair of orbitals in C_{3v} must be labelled e because only E symmetry species have characters greater than 1.

Because there are no characters greater than 2 in the column headed E in C_{3v}, we know that there can be no triply degenerate orbitals in a C_{3v} molecule. This last point is a powerful result of group theory, for it means that, with a glance at the character table of a molecule, we can state the maximum possible degeneracy of its orbitals.

Example 15.2 Using a character table to judge degeneracy

Can a trigonal planar molecule such as BF_3 have triply degenerate orbitals? What is the minimum number of atoms from which a molecule can be built that does display triple degeneracy?

Method First, identify the point group, and then refer to the corresponding character table in the *Data section*. The maximum number in the column headed by the identity E is the maximum orbital degeneracy possible in a molecule of that point group. For the second part, consider the shapes that can be built from two, three, etc. atoms, and decide which number can be used to form a molecule that can have orbitals of symmetry species T.

Answer Trigonal planar molecules belong to the point group D_{3h}. Reference to the character table for this group shows that the maximum degeneracy is 2, as no character exceeds 2 in the column headed E. Therefore, the orbitals cannot be triply degenerate. A tetrahedral molecule (symmetry group T) has an irreducible representation with a T symmetry species. The minimum number of atoms needed to build such a molecule is four (as in P_4, for instance).

- -

Self-test 15.2 A buckminsterfullerene molecule, C_{60}, belongs to the icosahedral point group. What is the maximum possible degree of degeneracy of its orbitals?

[5]

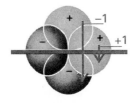

15.23 The two orbitals shown here have different properties under reflection through the mirror plane: one changes sign (character −1), the other does not (character +1).

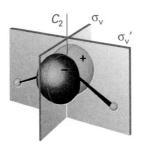

15.24 A p_x orbital on the central atom of a C_{2v} molecule and the symmetry elements of the group.

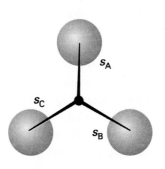

15.25 The three H1s orbitals used to construct symmetry-adapted linear combinations in a C_{3v} molecule such as NH_3.

(c) Characters and operations

The characters in the rows labelled A and B and in the columns headed by symmetry operations other than the identity E indicate the behaviour of an orbital under the corresponding operations: a +1 indicates that an orbital is unchanged, and a −1 indicates that it changes sign. It follows that we can identify the symmetry label of the orbital by comparing the changes that occur to an orbital under each operation, and then comparing the resulting +1 or −1 with the entries in a row of the character table for the point group concerned.

For the rows labelled E or T (which refer to the behaviour of sets of doubly and triply degenerate orbitals, respectively), the characters in a row of the table are the *sums* of the characters summarizing the behaviour of the individual orbitals in the basis. Thus, if one member of a doubly degenerate pair remains unchanged under a symmetry operation but the other changes sign (Fig. 15.23), then the entry is reported as $\chi = 1 - 1 = 0$. Care must be exercised with these characters because the transformations of orbitals can be quite complicated; nevertheless, the sums of the individual characters are commonly integers.

As an example, consider the O2p_x orbital in H_2O. Because H_2O belongs to the point group C_{2v}, we know by referring to the C_{2v} character table (Table 15.3) that the labels available for the orbitals are A_1, A_2, B_1, and B_2. We can decide the appropriate label for O2p_x by noting that under a 180° rotation (C_2) the orbital changes sign (Fig. 15.24), so it must be either B_1 or B_2, as only these two symmetry types have character −1 under C_2. The O2p_x orbital also changes sign under the reflection σ'_v, which identifies it as B_1. As we shall see, any molecular orbital built from this atomic orbital will also be a b_1 orbital. Similarly, O2p_y changes sign under C_2 but not under σ'_v; therefore, it can contribute to b_2 orbitals.

The behaviour of s, p, and d orbitals on a central atom under the symmetry operations of the molecule is so important that the symmetry species of these orbitals are generally indicated in a character table. To make these allocations, we look at the symmetry species of x, y, and z, which appear on the right-hand side of the character table. Thus, the position of z in Table 15.2 shows that p_z (which is proportional to $zf(r)$), has symmetry species A_1 in C_{3v}, whereas p_x and p_y (which are proportional to $xf(r)$ and $yf(r)$, respectively) are jointly of E symmetry. In technical terms, we say that p_x and p_y jointly span an irreducible representation of symmetry species E. An s orbital on the central atom always spans the fully symmetrical irreducible representation (typically labelled A_1) of a group as it is unchanged under all symmetry operations.

The five d orbitals of a shell are represented by xy for d_{xy}, etc., and are also listed on the right of the character table. We can see at a glance that in C_{3v}, d_{xy} and $d_{x^2-y^2}$ on a central atom jointly belong to E and hence form a doubly degenerate pair.

(d) The classification of linear combinations of orbitals

So far, we have dealt with the symmetry classification of individual orbitals. The same technique may be applied to linear combinations of orbitals on atoms that are related by symmetry transformations of the molecule, such as the combination $\psi_1 = \psi_A + \psi_B + \psi_C$ of the three H1s orbitals in the C_{3v} molecule NH_3 (Fig. 15.25). This combination remains unchanged under a C_3 rotation and under any of the three vertical reflections of the group, so its characters are

$$\chi(E) = 1 \qquad \chi(C_3) = 1 \qquad \chi(\sigma_v) = 1$$

Comparison with the C_{3v} character table shows that ψ_1 is of symmetry species A_1, and therefore that it contributes to a_1 molecular orbitals in NH_3.

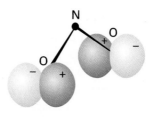

15.26 One symmetry-adapted linear combination of $O2p_x$ orbitals in the C_{2v} NO_2 molecule.

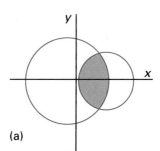

21

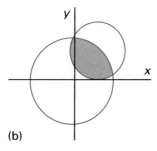

(a)

(b)

15.27 The value of an integral I (for example, an area) is independent of the coordinate system used to evaluate it. That is, I is a basis of a representation of symmetry species A_1 (or its equivalent).

Example 15.3 Identifying the symmetry species of orbitals

Identify the symmetry species of the orbital $\psi = \psi_A - \psi_B$ in a C_{2v} NO_2 molecule, where ψ_A is an $O2p_x$ orbital on one O atom and ψ_B that on the other O atom.

Method The negative sign in ψ indicates that the sign of ψ_B is opposite to that of ψ_A. We need to consider how the combination changes under each operation of the group, and then write the character as $+1$, -1, or 0 as specified above. Then we compare the resulting characters with each row in the character table for the point group, and hence identify the symmetry species.

Answer The combination is shown in Fig. 15.26. Under C_2, ψ changes into itself, implying a character of $+1$. Under the reflection σ_v, both orbitals change sign, so $\psi \rightarrow -\psi$, implying a character of -1. Under σ'_v, $\psi \rightarrow -\psi$, so the character for this operation is also -1. The characters are therefore

$$\chi(E) = 1 \qquad \chi(C_2) = 1 \qquad \chi(\sigma_v) = -1 \qquad \chi(\sigma'_v) = -1$$

These values match the characters of the A_2 symmetry species, so ψ can contribute to an a_2 orbital.

- -

Self-test 15.3 Identify the symmetry type of the combination $\psi_A - \psi_B + \psi_C - \psi_D$ in a square planar array of H atoms of point group D_{4h} (**21**).

$[B_{2g}]$

15.5 Vanishing integrals and orbital overlap

Suppose we had to evaluate the integral

$$I = \int f_1 f_2 \, d\tau \tag{7}$$

where f_1 and f_2 are functions. For example, f_1 might be an atomic orbital A on one atom and f_2 an atomic orbital B on another atom, in which case I would be their overlap integral. If we knew that the integral was zero, we could say at once that a molecular orbital does not result from (A, B) overlap in that molecule. We shall now see that character tables provide a quick way of judging whether an integral is necessarily zero.

(a) The criteria for vanishing integrals

The key point in dealing with the integral I is that the value of any integral, and of an overlap integral in particular, is independent of the orientation of the molecule (Fig. 15.27). In group theoretical language we express this by saying that I is invariant under any symmetry operation of the molecule, and that each operation brings about the trivial transformation $I \rightarrow I$. Because the volume element $d\tau$ is invariant under any symmetry operation, it follows that the integral is nonzero only if the integrand itself, the product $f_1 f_2$, is unchanged by any symmetry operation of the molecular point group. If the integrand changed sign under a symmetry operation, the integral would be the sum of equal and opposite contributions, and hence would be zero. It follows that the only contribution to a nonzero integral comes from functions for which, under any symmetry operation of the molecular point group, $f_1 f_2 \rightarrow f_1 f_2$ and hence for which the characters of the operations are all equal to $+1$. Therefore, for I not to be zero, the integrand $f_1 f_2$ must have symmetry species A_1 (or its equivalent in the specific molecular point group).

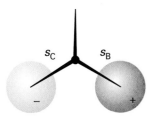

15.28 A symmetry-adapted linear combination that belongs to the symmetry species E in a C_{3v} molecule such as NH_3. This combination can form a molecular orbital by overlapping with the p_x orbital on the central atom (the orbital with its axis parallel to the width of the page; see Fig. 15.31c).

We use the following procedure to deduce the symmetry species spanned by the product $f_1 f_2$ and hence to see whether it does indeed span A_1.

1. Decide on the symmetry species of the individual functions f_1 and f_2 by reference to the character table, and write their characters in two rows in the same order as in the table.
2. Multiply the numbers in each column, writing the results in the same order.
3. Inspect the row so produced, and see if it can be expressed as a sum of characters from each column of the group. The integral must be zero if this sum does not contain A_1.

For example, if f_1 is the s_N orbital in NH_3 and f_2 is the linear combination $s_3 = s_B - s_C$ (Fig. 15.28) then, because s_N spans A_1 and s_3 is a member of the basis spanning E, we write

$$
\begin{array}{llll}
f_1: & 1 & 1 & 1 \\
f_2: & 2 & -1 & 0 \\
f_1 f_2: & 2 & -1 & 0
\end{array}
$$

The characters $2, -1, 0$ are those of E alone, so the integrand does not span A_1. It follows that the integral must be zero. Inspection of the form of the functions (see Fig. 15.28) shows why this is so: s_3 has a node running through s_N. Had we taken $f_1 = s_N$ and $f_2 = s_1$ instead, where $s_1 = s_A + s_B + s_C$, then because each spans A_1 with characters $1, 1, 1$:

$$
\begin{array}{llll}
f_1: & 1 & 1 & 1 \\
f_2: & 1 & 1 & 1 \\
f_1 f_2: & 1 & 1 & 1
\end{array}
$$

The characters of the product are those of A_1 itself. Therefore, s_1 and s_N may have nonzero overlap. A short cut that works when f_1 and f_2 are bases for irreducible representations of a group is to note their symmetry species: if they are different, the integral of their product must vanish; if they are the same, the integral may be nonzero.

It is important to note that group theory is specific about when an integral *must* be zero, but integrals that it allows to be nonzero *may* be zero for reasons unrelated to symmetry. For example, the N–H distance in ammonia may be so great that the (s_1, s_N) overlap integral is zero simply because the orbitals are so far apart.

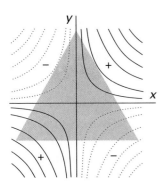

15.29 The integral of the function $f = xy$ over the tinted region is zero. In this case, the result is obvious by inspection, but group theory can be used to establish similar results in less obvious cases.

Example 15.4 Deciding if an integral must be zero (1)

May the integral of the function $f = xy$ be nonzero when evaluated over a region the shape of an equilateral triangle centred on the origin (Fig. 15.29)?

Method First, note that an integral over a single function f is included in the previous discussion if we take $f_1 = f$ and $f_2 = 1$ in eqn 7. Therefore, we need to judge whether f alone belongs to the symmetry species A_1 (or its equivalent) in the point group of the system. To decide that, we identify the point group and then examine the character table to see whether f belongs to A_1 (or its equivalent).

Answer An equilateral triangle has the point-group symmetry D_{3h}. If we refer to the character table of the group, we see that xy is a member of a basis that spans the irreducible representation E′. Therefore, its integral must be zero, because the integrand has no component that spans A_1'.

- -

Self-test 15.4 Can the function $x^2 + y^2$ have a nonzero integral when integrated over a regular pentagon centred on the origin?

[Yes, Fig. 15.30]

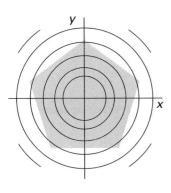

15.30 The integration of a function over a pentagonal region.

(a)

(b)

(c)

15.31 Orbitals of the same symmetry species may have non-vanishing overlap. This diagram illustrates the three bonding orbitals that may be constructed from (N2s, H1s) and (N2p, H1s) overlap in a C_{3v} molecule. (a) a_1; (b) and (c) the two components of the doubly degenerate e orbitals. (There are also three antibonding orbitals of the same species.)

(b) Orbitals with nonzero overlap

The rules just given let us decide which atomic orbitals may have nonzero overlap in a molecule. We have seen that s_N may have nonzero overlap with s_1 (the combination $1s_A + 1s_B + 1s_C$), so bonding and antibonding molecular orbitals can form from (s_N, s_1) overlap (Fig. 15.31). The general rule is that *only orbitals of the same symmetry species may have nonzero overlap*, so *only orbitals of the same symmetry species form bonding and antibonding combinations*. It should be recalled from Chapter 14 that the selection of atomic orbitals that have mutual nonzero overlap is the central and initial step in the construction of molecular orbitals by the LCAO procedure. We are therefore at the point of contact between group theory and the material introduced in that chapter. The molecular orbitals formed from a particular set of atomic orbitals with nonzero overlap are labelled with the lower-case letter corresponding to the symmetry species. Thus, the (s_N, s_1)-overlap orbitals are called a_1 orbitals (or a_1^*, if we wish to emphasize that they are antibonding).

The s_2 and s_3 linear combinations have symmetry species E. Does the N atom have orbitals that have nonzero overlap with them (and give rise to e molecular orbitals)? Intuition (as supported by Figs. 15.31b and c) suggests that N2p_x and N2p_y should be suitable. We can confirm this conclusion by noting that the character table shows that in C_{3v}, the functions x and y jointly belong to the symmetry species E. Therefore, N2p_x and N2p_y also belong to E, so may have nonzero overlap with s_2 and s_3. This conclusion can be verified by multiplying the characters and finding that the product of characters can be expressed as $E \times E = A_1 + A_2 + E$. The two e orbitals that result are shown in Fig. 15.31 (there are also two antibonding e orbitals).

The power of the method can be illustrated by exploring whether any d orbitals on the central atom can take part in bonding. As explained earlier, reference to the C_{3v} character table shows that d_{z^2} has A_1 symmetry and that the pairs $(d_{x^2-y^2}, d_{xy})$ and (d_{yz}, d_{zx}) each transform as E. It follows that molecular orbitals may be formed by (s_1, d_{z^2}) overlap and by overlap of the s_2, s_3 combinations with the E d orbitals. Whether or not the d orbitals are in fact important is a question group theory cannot answer because the extent of their involvement depends on energy considerations, not symmetry.

Example 15.5 Determining which orbitals can contribute to bonding

The four H1s orbitals of methane span $A_1 + T_2$. With which of the C atom orbitals can they overlap? What bonding pattern would be possible if the C atom had d orbitals available?

Method Refer to the T_d character table (in the *Data section*) and look for s, p, and d orbitals spanning A_1 or T_2.

Answer An s orbital spans A_1, so it may have nonzero overlap with the A_1 combination of H1s orbitals. The C2p orbitals span T_2, so they may have nonzero overlap with the T_2 combination. The d_{xy}, d_{yz}, and d_{zx} orbitals span T_2, so they may overlap the same combination. Neither of the other two d orbitals span A_1 (they span E), so they remain nonbonding orbitals.

Comment It follows that in methane there are (C2s,H1s)-overlap a_1 orbitals and (C2p,H1s)-overlap t_2 orbitals. The C3d orbitals might contribute to the latter. The lowest energy configuration is probably $a_1^2 t_2^6$, with all bonding orbitals occupied.

- -

Self-test 15.5 Consider the octahedral SF_6 molecule, with the bonding arising from overlap of S orbitals and a 2p orbital on each F directed towards the central S atom. The latter span $A_{1g} + E_g + T_{1u}$. What S orbitals have nonzero overlap? Suggest what the ground-state configuration is likely to be.

$$[3s(A_{1g}), 3p(T_{1u}), 3d(E_g); a_{1g}^2 t_{1u}^6 e_g^4]$$

(c) Symmetry-adapted linear combinations

So far, we have only asserted the forms of the linear combinations (such as s_1, etc.) that have a particular symmetry. Group theory also provides machinery that takes an arbitrary basis, or set of atomic orbitals (s_A, etc.), as input and generates combinations of the specified symmetry. Because these combinations are adapted to the symmetry of the molecule, they are called **symmetry-adapted linear combinations** (SALCs). Symmetry-adapted linear combinations are the building blocks of LCAO molecular orbitals, for they include combinations such as $\psi_{1s_A} \pm \psi_{1s_B}$ used to construct molecular orbitals in H_2O (Section 14.8a) and some of the more complex examples that we have seen since then. The construction of SALCs is the first step in any molecular orbital treatment of molecules.

The technique for building SALCs is derived by using the full power of group theory. We shall not show the derivation, which is very lengthy, but present the main conclusions as a set of rules:

1. Construct a table showing the effect of each operation on each orbital of the original basis.
2. To generate the combination of a specified symmetry species, take each column in turn and:

    (i)   Multiply each member of the column by the character of the corresponding operation.
    (ii) Add together all the orbitals in each column with the factors as determined in (i).
    (iii) Divide the sum by the order of the group.

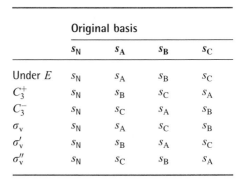

	Original basis		
	s_N	s_A	s_B	s_C
Under E	s_N	s_A	s_B	s_C
C_3^+	s_N	s_B	s_C	s_A
C_3^-	s_N	s_C	s_A	s_B
σ_v	s_N	s_A	s_C	s_B
σ_v'	s_N	s_B	s_A	s_C
σ_v''	s_N	s_C	s_B	s_A

For example, from the (s_N, s_A, s_B, s_C) basis in NH_3 we form the table shown in the margin. To generate the A_1 combination, we take the characters for A_1 $(1, 1, 1, 1, 1, 1)$; then rules (i) and (ii) lead to

$$\psi \propto s_N + s_N + \cdots = 6s_N$$

The order of the group (the number of elements) is 6, so the combination of A_1 symmetry that can be generated from s_N is s_N itself. Applying the same technique to the column under s_A gives

$$\psi = \tfrac{1}{6}(s_A + s_B + s_C + s_A + s_B + s_C) = \tfrac{1}{3}(s_A + s_B + s_C)$$

The same combination is built from the other two columns, so they give no further information. The combination we have just formed is the s_1 combination we used before (apart from the numerical factor).

We now form the overall molecular orbital by forming a linear combination of all the SALCs of the specified symmetry species. In this case, therefore, the a_1 molecular orbital is

$$\psi = c_N s_N + c_1 s_1$$

This is as far as group theory can take us. The coefficients must be found by solving the Schrödinger equation; they do not come directly from the symmetry of the system.

We run into a problem when we try to generate an SALC of symmetry species E, because, for representations of dimension 2 or more, the rules generate sums of SALCs. This problem can be illustrated as follows. In C_{3v}, the E characters are $2, -1, -1, 0, 0, 0$, so the column under s_N gives

$$\psi = \tfrac{1}{6}(2s_N - s_N - s_N + 0 + 0 + 0) = 0$$

The other columns give

$$\tfrac{1}{6}(2s_A - s_B - s_C) \qquad \tfrac{1}{6}(2s_B - s_A - s_C) \qquad \tfrac{1}{6}(2s_C - s_B - s_A)$$

However, any one of these three expressions can be written as a sum of the other two (they are not 'linearly independent'). The difference of the second and third gives $\tfrac{1}{2}(s_B - s_C)$, and this combination and the first, $\tfrac{1}{6}(2s_A - s_B - s_C)$, are the two (now linearly independent) SALCs we have used in the discussion of e orbitals.

15.6 Vanishing integrals and selection rules

Integrals of the form

$$I = \int f_1 f_2 f_3 \, d\tau \tag{8}$$

are also common in quantum mechanics for they include matrix elements of operators (Section 11.5d), and it is important to know when they are necessarily zero. For the integral to be nonzero, the product $f_1 f_2 f_3$ must span A_1 (or its equivalent). To test whether this is so, the characters of all three functions are multiplied together in the same way as in the rules set out above.

Example 15.6 Deciding if an integral must be zero (2)

Does the integral $\int (3d_{z^2}) x (3d_{xy}) \, d\tau$ vanish in a C_{2v} molecule?

Method We must refer to the C_{2v} character table (Table 15.3) and the characters of the irreducible representations spanned by $3z^2 - r^2$ (the form of the d_{z^2} orbital), x, and xy; then we can use the procedure set out above (with one more row of multiplication). Note that $3z^2 - r^2 = 2z^2 - x^2 - y^2$.

Answer We draw up the following table:

	E	C_2	σ_v	σ'_v	
$f_3 = d_{xy}$	1	1	-1	-1	A_2
$f_2 = x$	1	-1	1	-1	B_1
$f_1 = d_{z^2}$	1	1	1	1	A_1
$f_1 f_2 f_3$	1	-1	-1	1	

The characters are those of B_2. Therefore, the integral is necessarily zero.

- -

Self-test 15.6 Does the integral $\int (2p_x)(2p_y)(2p_z) \, d\tau$ necessarily vanish in an octahedral environment?

[No]

In Chapters 16 and 17 we shall see that the intensity of a spectral line arising from a molecular transition between some initial state $|i\rangle$ with wavefunction ψ_i and a final state $|f\rangle$

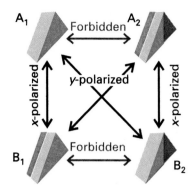

15.32 The polarizations of the allowed transitions in a C_{2v} molecule. The shading indicates the structure of the orbitals of the specified symmetry species. The perspective view of the molecule makes it look rather like a door-stop; however, from the side, each 'door-stop' is in fact an isosceles triangle.

with wavefunction ψ_f depends on the (electric) transition dipole moment $\boldsymbol{\mu}_{fi}$. The z-component of this vector is defined through

$$\mu_{z,fi} = \langle f|\mu_z|i\rangle = -e \int \psi_f^* z \psi_i \, d\tau \qquad [9]$$

where $-e$ is the charge of the electron. Stating the conditions for this quantity (and the x- and y-components) to be zero amounts to specifying the **selection rules** for the transition, the rules that specify which transitions are allowed. The transition moment has the form of the integral in eqn 8, so, once we know the symmetry species of the states, we can use group theory to decide which transitions have zero transition dipole moment and therefore are forbidden.

As an example, we investigate whether an electron in an a_1 orbital in H_2O (which belongs to the group C_{2v}) can make an electric dipole transition to a b_1 orbital (Fig. 15.32). We must examine all three components of the transition dipole moment, and take f_2 in eqn 8 as x, y, and z in turn. Reference to the C_{2v} character table shows that these components transform as B_1, B_2, and A_1, respectively. The three calculations run as follows:

	x-component				y-component				z-component				
	E	C_2	σ_v	σ_v'	E	C_2	σ_v	σ_v'	E	C_2	σ_v	σ_v'	
f_3	1	-1	1	-1	1	-1	1	-1	1	-1	1	-1	B_1
f_2	1	-1	1	-1	1	-1	-1	1	1	1	1	1	
f_1	1	1	1	1	1	1	1	1	1	1	1	1	A_1
$f_1 f_2 f_3$	1	1	1	1	1	1	-1	-1	1	-1	1	-1	

Only the first product (with $f_2 = x$) spans A_1, so only the x-component of the transition dipole moment may be nonzero. Therefore, we conclude that the electric dipole transitions between a_1 and b_1 are allowed. We can go on to state that the radiation emitted (or absorbed) is x-polarized and has its electric field vector in the x-direction, because that form of radiation couples with the x-component of a transition dipole.

Example 15.7 Deducing a selection rule

Is $p_x \rightarrow p_y$ an allowed transition in a tetrahedral environment?

Method We must decide whether the product $p_y q p_x$, with $q = x$, y, or z, spans A_1 by using the T_d character table.

Answer The procedure works out as follows:

	E	$8C_3$	$3C_2$	$6S_4$	$6\sigma_d$	
$f_3(p_y)$	3	0	-1	-1	1	T_2
$f_2(q)$	3	0	-1	-1	1	T_2
$f_1(p_x)$	3	0	-1	-1	1	T_2
$f_1 f_2 f_3$	27	0	-1	-1	1	

A_1 occurs (once) in this set of characters, so $p_x \rightarrow p_y$ is allowed.

Comment A more detailed analysis (using the matrix representatives rather than the characters) shows that only $q = z$ gives an integral that may be nonzero, so the transition is z-polarized. That is, the electromagnetic radiation involved in the transition has its electric vector aligned in the z-direction.

Self-test 15.7 What are the allowed transitions, and their polarizations, of a b_1 electron in a C_{4v} molecule?

$$[b_1 \rightarrow b_1(z); \, b_1 \rightarrow e(x,y)]$$

The following chapters will show many more examples of how the systematic use of symmetry using the techniques of group theory can greatly simplify the analysis of molecular structure and spectra.

Checklist of key ideas

☐ group theory

The symmetry elements of objects

☐ symmetry operation
☐ symmetry element

15.1 Operations and symmetry elements

☐ point groups
☐ space groups
☐ identity
☐ n-fold rotation
☐ principal axis
☐ reflection
☐ mirror plane
☐ inversion

☐ centre of symmetry
☐ n-fold improper rotation
☐ n-fold improper rotation axis

15.2 The symmetry classification of molecules

☐ Schoenflies system
☐ Hermann–Mauguin system
☐ International system
☐ cubic group
☐ icosahedral group

15.3 Some immediate consequences of symmetry

☐ polar molecule
☐ chiral molecule
☐ enantiomeric pair

Character tables

15.4 Character tables and symmetry labels

☐ character table
☐ character
☐ class
☐ order
☐ symmetry species
☐ irreducible representation
☐ representative
☐ basis
☐ matrix representation
☐ block-diagonal form
☐ reduced (representation)
☐ direct sum
☐ spanned (representation)

15.5 Vanishing integrals and orbital overlap

☐ symmetry-adapted linear combinations (SALCs)

15.6 Vanishing integrals and selection rules

☐ criteria for vanishing
☐ overlap integrals
☐ constructing SALCs
☐ selection rule

Further reading

Articles of general interest

F. Rioux, Quantum mechanics, group theory, and C_{60}. *J. Chem. Educ.* **71**, 464 (1994).

G.L. Breneman, Crystallographic symmetry point group notation flow chart. *J. Chem. Educ.* **64**, 216 (1987).

C. Contreras-Ortega, L. Vera, and E. Quiroz-Reyes, How great is the Great Orthogonality Theorem? *J. Chem. Educ.* **68**, 200 (1991).

C. Contreras-Ortega, L. Vera, and E. Quiroz-Reyes, More than one character table: a warning on the use of the rules of the irreducible representations and their characters. *J. Chem. Educ.* **72**, 821 (1995).

M. Hamermesh, Group theory. In *Encyclopedia of applied physics* (ed. G.L. Trigg), **7**, 365. VCH, New York (1993).

Texts and sources of data and information

F.A. Cotton, *Chemical applications of group theory*. Wiley, New York (1990).

S.F.A. Kettle, *Symmetry and structure: readable group theory for chemists*. Wiley, New York (1995).

E. Heilbronner and J.D. Dunitz, *Reflections on symmetry: in chemistry, and elsewhere*. VCH, Weinheim (1993).

P.W. Atkins and R.S. Friedman, *Molecular quantum mechanics*. Oxford University Press (1997).

D.C. Harris and M.D. Bertolucci, *Symmetry and spectroscopy*. Oxford University Press, New York (1978).

P.W. Atkins, M.S. Child, and C.S.G. Phillips, *Tables for group theory*. Oxford University Press (1970).

B.E. Douglas and C. Hollingsworth, *Symmetry in bonding and structure*. Academic Press, New York (1985).

Exercises

15.1 (a) The CH_3Cl molecule belongs to the point group C_{3v}. List the symmetry elements of the group and locate them in the molecule.

15.1 (b) The CCl_4 molecule belongs to the point group T_d. List the symmetry elements of the group and locate them in the molecule.

15.2 (a) Which of the following molecules may be polar: (a) pyridine (C_{2v}), (b) nitroethane (C_s), (c) gas-phase $HgBr_2$ ($D_{\infty h}$), (d) $B_3N_3H_6$ (D_{3h})?

15.2 (b) Which of the following molecules may be polar: (a) CH_3Cl (C_{3v}), (b) $HW_2(CO)_{10}$ (D_{4h}), (c) $SnCl_4$ (T_d)?

15.3 (a) Use symmetry properties to determine whether or not the integral $\int p_x z p_z \, d\tau$ is necessarily zero in a molecule with symmetry C_{4v}.

15.3 (b) Use symmetry properties to determine whether or not the integral $\int p_x z p_z \, d\tau$ is necessarily zero in a molecule with symmetry D_{6h}.

15.4 (a) Show that the transition $A_1 \rightarrow A_2$ is forbidden for electric dipole transitions in a C_{3v} molecule.

15.4 (b) Is the transition $A_{1g} \rightarrow E_{2u}$ forbidden for electric dipole transitions in a D_{6h} molecule?

15.5 (a) Show that the function xy has symmetry species B_2 in the group C_{4v}.

15.5 (b) Show that the function xyz has symmetry species A_1 in the group D_2.

15.6 (a) Molecules belonging to the point groups D_{2h} or C_{3h} cannot be chiral. Which elements of these groups rule out chirality?

15.6 (b) Molecules belonging to the point groups T_h or T_d cannot be chiral. Which elements of these groups rule out chirality?

15.7 (a) The group D_2 consists of the elements E, C_2, C_2', and C_2'', where the three twofold rotations are around mutually perpendicular axes. Construct the group multiplication table.

15.7 (b) The group C_{4v} consists of the elements E, $2C_4$, C_2, and $2\sigma_v$, $2\sigma_d$. Construct the group multiplication table.

15.8 (a) Identify the point groups to which the following objects belong: (a) a sphere, (b) an isosceles triangle, (c) an equilateral triangle, (d) an unsharpened cylindrical pencil.

15.8 (b) Identify the point groups to which the following objects belong: (a) a sharpened cylindrical pencil, (b) a three-bladed propellor, (c) a four-legged table, (d) yourself (approximately).

15.9 (a) List the symmetry elements of the following molecules and name the point groups to which they belong: (a) NO_2, (b) N_2O, (c) $CHCl_3$, (d) $CH_2{=}CH_2$, (e) *cis*-CHBr${=}$CHBr, (f) *trans*-CHCl${=}$CHCl.

15.9 (b) List the symmetry elements of the following molecules and name the point groups to which they belong: (a) naphthalene, (b) anthracene, (c) the three dichlorobenzenes.

15.10 (a) Assign (a) *cis*-dichloroethene and (b) *trans*-dichloroethene to point groups.

15.10 (b) Assign the following molecules to point groups: (a) HF, (b) IF_7 (pentagonal bipyramid), (c) XeO_2F_2 (see-saw), (d) $Fe_2(CO)_9$ **(22)**, (e) cubane, C_8H_8, (f) tetrafluorocubane, $C_8H_4F_4$ **(23)**.

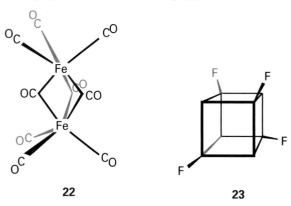

22   **23**

15.11 (a) Which of the molecules in Exercises 15.9a and 15.10a can be (a) polar, (b) chiral?

15.11 (b) Which of the molecules in Exercises 15.9b and 15.10b can be (a) polar, (b) chiral?

15.12 (a) Consider the C_{2v} molecule NO_2. The combination $p_x(A) - p_x(B)$ of the two O atoms (with x perpendicular to the plane) spans A_2. Is there any orbital of the central N atom that can have a nonzero overlap with that combination of O orbitals? What would be the case in SO_2, where $3d$ orbitals might be available?

15.12 (b) Consider the C_{3v} ion NO_3^-. Is there any orbital of the central N atom that can have a nonzero overlap with the combination $2p_z(A) - p_z(B) - p_z(C)$ of the three O atoms (with z perpendicular to the plane)? What would be the case in SO_3, where $3d$ orbitals might be available?

15.13 (a) The ground state of NO_2 is A_1 in the group C_{2v}. To what excited states may it be excited by electric dipole transitions, and what polarization of light is it necessary to use?

15.13 (b) The ClO_2 molecule (which belongs to the group C_{2v}) was trapped in a solid. Its ground state is known to be B_1. Light polarized parallel to the y-axis (parallel to the OO separation) excited the molecule to an upper state. What is the symmetry of that state?

15.14 (a) What states of (a) benzene, (b) naphthalene may be reached by electric dipole transitions from their (totally symmetrical) ground states?

15.14 (b) What states of (a) anthracene, (b) coronene **(24)** may be reached by electric dipole transitions from their (totally symmetrical) ground states?

24 Coronene

15.15 (a) Write $f_1 = \sin\theta$ and $f_2 = \cos\theta$, and show by symmetry arguments using the group C_s that the integral of their product over a symmetrical range around $\theta = 0$ is zero.

15.15 (b) Determine whether the integral over f_1 and f_2 in Exercise 15.15(a) is zero over a symmetric range about $\theta = 0$ in the group C_{3v}.

Problems

Numerical problems

15.1 List the symmetry elements of the following molecules and name the point groups to which they belong: (a) staggered CH_3CH_3, (b) chair and boat cyclohexane, (c) B_2H_6, (d) $[Co(en)_3]^{3+}$, where en is ethylenediamine (ignore its detailed structure), (e) crown-shaped S_8. Which of these molecules can be (i) polar, (ii) chiral?

15.2 The group C_{2h} consists of the elements E, C_2, σ_h, i. Construct the group multiplication table (the outcome of all multiplications R_iR_j, where R_i and R_j are operations of the group) and find an example of a molecule that belongs to the group.

15.3 The group D_{2h} has a C_2 axis perpendicular to the principal axis and a horizontal mirror plane. Show that the group must therefore have a centre of inversion.

15.4 Consider the H_2O molecule, which belongs to the group C_{2v}. Take as a basis the two H1s orbitals and the four valence orbitals of the O atom and set up the 6×6 matrices that represent the group in this basis. Confirm by explicit matrix multiplication that the group multiplications (a) $C_{2}\sigma_v = \sigma_v'$ and (b) $\sigma_v\sigma_v' = C_2$. Confirm, by calculating the traces of the matrices, (a) that symmetry elements in the same class have the same character, (b) that the representation is reducible, and (c) that the basis spans $3A_1 + B_1 + 2B_2$.

15.5 Confirm that the z-component of orbital angular momentum is a basis for an irreducible representation of A_2 symmetry in C_{3v}.

15.6 The (one-dimensional) matrices $D(C_3) = 1$ and $D(C_2) = 1$, and $D(C_3) = 1$ and $D(C_2) = -1$ both represent the group multiplication $C_3C_2 = C_6$ in the group C_{6v} with $D(C_6) = +1$ and -1, respectively. Use the character table to confirm these remarks. What are the representatives of σ_v and σ_d in each case?

15.7 Construct the multiplication table of the Pauli spin matrices, σ, and the 2×2 unit matrix:

$$\sigma_x = \begin{pmatrix} 0 & 1 \\ 1 & 0 \end{pmatrix} \qquad \sigma_y = \begin{pmatrix} 0 & -i \\ i & 0 \end{pmatrix}$$

$$\sigma_z = \begin{pmatrix} 1 & 0 \\ 0 & -1 \end{pmatrix} \qquad \sigma_0 = \begin{pmatrix} 1 & 0 \\ 0 & 1 \end{pmatrix}$$

Do the four matrices form a group under multiplication (in the sense that $\sigma_i\sigma_j \propto \sigma_k$ for all the matrices)?

15.8 What irreducible representations do the four H1s orbitals of CH_4 span? Are there s and p orbitals of the central C atom that may form molecular orbitals with them? Could d orbitals, even if they were present on the C atom, play a role in orbital formation in CH_4?

15.9 Suppose that a methane molecule became distorted to (a) C_{3v} symmetry by the lengthening of one bond, (b) C_{2v} symmetry, by a

kind of scissors action in which one bond angle opened and another closed slightly. Would more d orbitals become available for bonding?

15.10 The algebraic forms of the f orbitals are a radial function multiplied by one of the factors: (a) $z(5z^2 - 3r^2)$, (b) $y(5y^2 - 3r^2)$, (c) $x(5x^2 - 3r^2)$, (d) $z(x^2 - y^2)$, (e) $y(x^2 - z^2)$, (f) $x(z^2 - y^2)$, (g) xyz. Identify the irreducible representations spanned by these orbitals in (a) C_{2v}, (b) C_{3v}, (c) T_d, (d) O_h. Consider a lanthanide ion at the centre of (a) a tetrahedral complex, (b) an octahedral complex. What sets of orbitals do the seven f orbitals split into?

15.11 Does the product xyz necessarily vanish when integrated over (a) a cube, (b) a tetrahedron, (c) a hexagonal prism, each centred on the origin?

15.12 Treat the naphthalene molecule as belonging to the group C_{2v} with the C_2 axis perpendicular to the plane. Classify the irreducible representations spanned by the carbon $2p_z$ orbitals and find their symmetry-adapted linear combinations.

15.13 The NO_2 molecule belongs to the group C_{2v}, with the C_2 axis bisecting the ONO angle. Taking as a basis the N2s, N2p, and O2p orbitals, identify the irreducible representations they span, and construct the symmetry-adapted linear combinations.

15.14 Construct the symmetry-adapted linear combinations of C2p_z orbitals for benzene, and use them to calculate the Hückel secular determinant. This procedure leads to equations that are much easier to solve than those obtained by using the original orbitals. Show that the Hückel orbitals are those specified in Section 14.9d.

Additional problems supplied by Carmen Giunta and Charles Trapp

15.15 B.A. Bovenzi and G.A. Pearse, Jr (*J. Chem. Soc. Dalton Trans.* accepted, 1997) synthesized coordination compounds of the tridentate ligand pyridine-2,6-diamidoxime ($C_7H_9N_5O_2$, **25**). Reaction with $NiSO_4$ produced a complex in which two of the essentially planar ligands are bonded at right angles to a single Ni atom. Name the point group and the symmetry operations of the resulting $[Ni(C_7H_9N_5O_2)_2]^{2+}$ complex cation.

15.16 R. Eujen, B. Hoge, and D.J. Brauer (*Inorg. Chem.* **36**, 1464 (1997)) prepared and characterized several square-planar Ag(III) complex anions. In the complex anion [*trans*-Ag(CF$_3$)$_2$(CN)$_2$]$^-$, the Ag–CN groups are collinear. (a) Assuming free rotation of the CF$_3$ groups (that is, disregarding the AgCF and AgCH angles), name the point group of this complex anion. (b) Now suppose the CF$_3$ groups cannot rotate freely (because the ion was in a solid, for example). Structure (**26**) shows a plane which bisects the NC–Ag–CN axis and is perpendicular to it. Name the point group of the complex if each CF$_3$ group has a CF bond in that plane (so the CF$_3$ groups do not point to either CN group preferentially) and the CF$_3$ groups are (i) staggered, (ii) eclipsed.

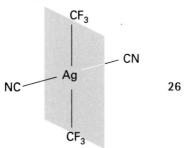

26

15.17 A computational study by C.J. Marsden (*Chem. Phys. Letts.* **245**, 475 (1995)) of AM$_x$ compounds, where A is in Group 14 of the periodic table and M is an alkali metal, shows several deviations from the most symmetric structures for each formula. (a) For example, most of the AM$_4$ structures were not tetrahedral but had two distinct values for MAM bond angles. They could be derived from a tetrahedron by a distortion shown in (**27**). What is the point group of the distorted tetrahedron? What is the symmetry species of the distortion considered as a vibration in the new, less symmetric group? (b) Some AM$_6$ structures are not octahedral, but could be derived from an octahedron by translating a C–M–C axis as in (**28**). What is

27

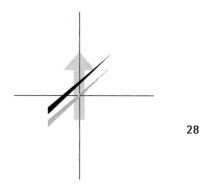

28

the point group of the distorted octahedron? What is the symmetry species of the distortion considered as a vibration in the new, less symmetric group?

15.18 In a spectroscopic study of C$_{60}$, F. Negri, G. Orlandi, and F. Zerbetto (*J. Phys. Chem.* **100**, 10849 (1996)) assigned peaks in the fluorescence spectrum. The molecule has icosahedral symmetry (I_h). The ground electronic state is A$_{1g}$, and the lowest-lying excited states are T$_{1g}$ and G$_g$. Are photon-induced transitions allowed from the ground state to either of these excited states? Explain your answer. What if the transition is accompanied by a vibration that breaks the parity?

15.19 The H$_3^+$ molecule, which plays an important role in chemical reactions occurring in interstellar clouds, is known to be equilateral triangular. (a) Identify the symmetry elements and determine the point group of this molecule. (b) Take as a basis for a representation of this molecule the three H1s orbitals and set up the matrices that represent the group in this basis. (c) Obtain the group multiplication table by explicit multiplication of the matrices. (d) Determine if the representation is reducible and, if so, give the irreducible representations obtained.

15.20 The H$_3^+$ ion has recently been found in the interstellar medium and in the atmospheres of Jupiter, Saturn, and Uranus. The H$_4$ analogues have not yet been found, and the square planar structure is thought to be unstable with respect to vibration. Take as a basis for a representation of the point group of this molecule the four H1s orbitals and determine if this representation is reducible.

16

Spectroscopy 1: rotational and vibrational spectra

The general strategy we adopt in the chapter is to set up expressions for the energy levels of molecules, and then to apply selection rules and considerations of populations to infer the form of spectra. Rotational energy levels are considered first, and we see how to derive expressions for their values and then how to interpret rotational spectra in terms of molecular dimensions. Not all molecules can occupy all rotational states: we see the experimental evidence for this restriction and its explanation in terms of nuclear spin and the Pauli principle. Next, we consider the vibrational energy levels of diatomic molecules, and see that we can use the properties of harmonic oscillators developed in Chapter 12. Then we consider polyatomic molecules and find that their vibrations may be discussed as though they consisted of a set of independent harmonic oscillators, so the same approach as that employed for diatomic molecules may be used. We also see that the symmetry properties of the vibrations of polyatomic molecules are helpful for deciding which modes can be studied spectroscopically.

The origin of spectral lines in molecular spectroscopy is the emission or absorption of a photon when the energy of a molecule changes. The difference from atomic spectroscopy is that the energy of a molecule can change not only as a result of electronic transitions but also because it can undergo changes of rotational and vibrational state. Molecular spectra are therefore more complex than atomic spectra. However, they also contain information relating to more properties, and their analysis leads to values of bond strengths, lengths, and angles. They also provide a way of determining a variety of molecular properties, particularly molecular dimensions, shapes, and dipole moments.

Pure rotational spectra, in which only the rotational state of a molecule changes, can be observed in the gas phase. Vibrational spectra of gaseous samples show features that arise from rotational transitions that accompany the excitation of vibration. Electronic spectra, which are described in Chapter 17, show features arising from simultaneous vibrational and rotational transitions. The simplest way of dealing with these complexities is to tackle each type of transition in turn, and then to see how simultaneous changes affect the appearance of spectra.

General features of spectroscopy

All types of spectra have some features in common, and we examine these first. We shall often need to use the relations between the frequency, ν, wavelength, λ, and wavenumber, $\tilde{\nu}$, of electromagnetic radiation that were first mentioned in the *Introduction*:

$$\lambda = \frac{c}{\nu} \qquad \tilde{\nu} = \frac{\nu}{c} \tag{1}$$

The units of wavenumber are almost always chosen as reciprocal centimetres (cm^{-1}).

Figure 16.1 summarizes the frequencies, wavelengths, and wavenumbers of the various regions of the electromagnetic spectrum and anticipates the type of molecular excitation that is characteristic of each region.

16.1 Experimental techniques

In **emission spectroscopy**, a molecule undergoes a transition from a state of high energy E_1 to a state of lower energy E_2 and emits the excess energy as a photon. In **absorption spectroscopy**, the net absorption[1] of nearly monochromatic (single-frequency) incident radiation is monitored as the radiation is swept over a range of frequencies. The energy, $h\nu$, of the photon emitted or absorbed, and therefore the frequency, ν, of the radiation emitted or absorbed, is given by the Bohr frequency condition

$$h\nu = E_1 - E_2 \tag{2}$$

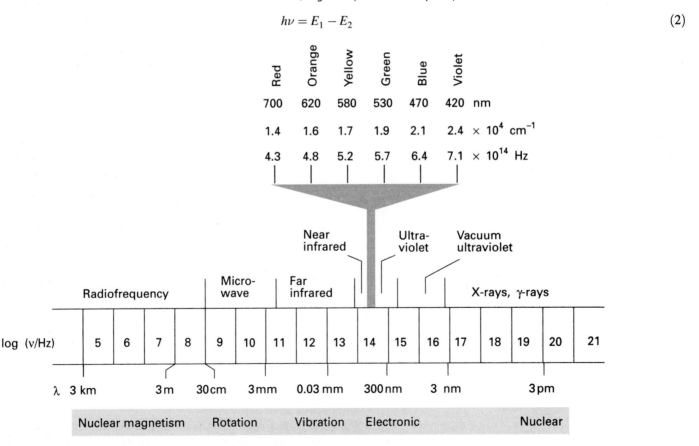

16.1 The electromagnetic spectrum and the classification of the spectral regions. The band at the bottom of the illustration indicates the types of transitions that absorb or emit in the various regions. ('Nuclear magnetism' refers to the types of transitions discussed in Chapter 18; 'nuclear' on the right refers to transitions within the nucleus.)

---

1   We say *net* absorption, because it will become clear that, when a sample is irradiated, both absorption and emission at a given frequency are stimulated, and the detector measures the difference, the net absorption.

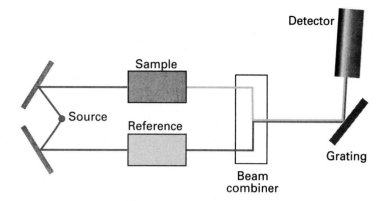

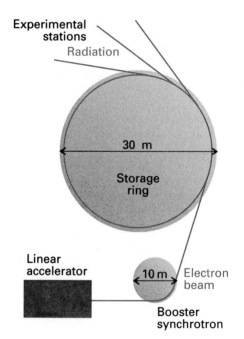

16.2 The layout of a typical absorption spectrometer. The beams pass alternately through the sample and reference cells, and the detector is synchronized with them so that the relative absorption can be determined.

16.3 A synchrotron storage ring. The electrons injected into the ring from the linear accelerator and booster synchrotron are accelerated to high speed in the main ring. An electron in a curved path is subject to constant acceleration, and an accelerated charge radiates electromagnetic energy.

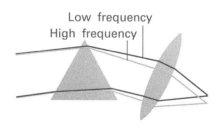

16.4 One simple dispersing element is a prism, which separates frequencies spatially by making use of the higher refractive index of matter for high-frequency radiation. The shortest wavelength for which a glass prism can be used is about 400 nm, but quartz can be used down to 180 nm.

Emission and absorption spectroscopy give the same information about energy level separations, but practical considerations generally determine which technique is employed. Emission spectroscopy, if it is used at all, is normally used only for visible and ultraviolet spectroscopy; absorption spectroscopy is much more widely employed, and we shall concentrate on it. Absorption spectra are also often easier to interpret than emission spectra.

(a) Sources of radiation

The general layout of a spectrometer is summarized in Fig. 16.2. The source generally produces radiation spanning a range of frequencies. For the far infrared, the source is a mercury arc inside a quartz envelope, most of the radiation being generated by the hot quartz. A *Nernst filament* is used to generate radiation in the near infrared. This device consists of a heated ceramic filament containing rare-earth (lanthanide) oxides, which emits radiation closely resembling that of a true black body. For the visible region of the spectrum, a tungsten/iodine lamp is used, which gives out intense white light. A discharge through deuterium gas or xenon in quartz is still widely used for the near ultraviolet. In a few cases, the source generates monochromatic radiation which can be swept over a range of values. One such generator is the *klystron*, an electronic device used to generate microwaves. Lasers, which are discussed in more detail in Chapter 17, generate monochromatic electromagnetic radiation that can often be tuned over a range of frequencies; different types of laser are used to cover different regions of the electromagnetic spectrum.

For certain applications, synchrotron radiation from a *synchrotron storage ring* is appropriate. A synchrotron storage ring consists of an electron beam (actually a series of closely spaced packets of electrons) travelling in a circular path of several metres in diameter. As electrons travelling in a circle are constantly accelerated by the forces that constrain them to their path, they generate radiation (Fig. 16.3). Synchrotron radiation spans a wide range of frequencies, including the far ultraviolet and beyond to X-rays, and in all except the microwave region is much more intense than can be obtained by most conventional sources. The disadvantage of the source is that it is so large and costly that it is essentially a national facility, and not a laboratory commonplace.

(b) The dispersing element

In all but specialized techniques using monochromatic microwave radiation and lasers, absorption spectrometers include a component for separating the frequencies of the radiation so that the variation of the absorption with frequency can be monitored. In conventional spectrometers, this component is a **dispersing element** that separates radiation with different frequencies into different spatial directions.

The simplest dispersing element is a glass or quartz prism, which utilizes the variation of refractive index with the frequency of the incident radiation (Fig. 16.4). Materials generally

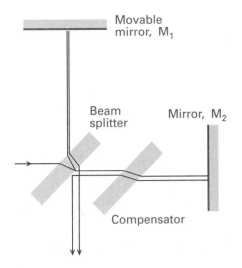

16.5 A Michelson interferometer. The beam-splitting element divides the incident beam into two beams with a path difference that depends on the location of the mirror M_1. The compensator ensures that both beams pass through the same thickness of material.

have a higher refractive index for high-frequency radiation than low-frequency radiation, and therefore high-frequency radiation undergoes a greater deflection when passing through a prism. Problems of absorption by the prism can be avoided by replacing it by a *diffraction grating*. A diffraction grating consists of a glass or ceramic plate into which fine grooves have been cut about 1000 nm apart (a separation comparable to the wavelength of visible light) and covered with a reflective aluminium coating. The grating causes interference between waves reflected from its surface, and constructive interference occurs at specific angles that depend on the wavelength of the radiation.

(c) Fourier transform techniques

Modern spectrometers, particularly those operating in the infrared, now almost always use **Fourier transform** techniques of spectral detection and analysis. The heart of a Fourier transform spectrometer is a *Michelson interferometer*, a device for analysing the frequencies present in a composite signal. The total signal from a sample is like a chord played on a piano, and the Fourier transform of the signal is equivalent to the separation of the chord into its individual notes, its spectrum.

A Michelson interferometer works by splitting the beam from the sample into two and introducing a varying path difference, p, into one of them (Fig. 16.5). When the two components recombine, there is a phase difference between them, and they interfere either constructively or destructively depending on the difference in path lengths. The detected signal oscillates as the two components alternately come into and out of phase as the path difference is changed (Fig. 16.6). If the radiation has wavenumber $\tilde{\nu}$, the intensity of the detected signal due to radiation in the range of wavenumbers $\tilde{\nu}$ to $\tilde{\nu} + d\tilde{\nu}$, which we denote $\mathcal{I}(p, \tilde{\nu})\,d\tilde{\nu}$, varies with p as

$$\mathcal{I}(p, \tilde{\nu})\,d\tilde{\nu} = \mathcal{I}(\tilde{\nu})(1 + \cos 2\pi\tilde{\nu}p)\,d\tilde{\nu} \tag{3}$$

Hence, the interferometer converts the presence of a particular wavenumber component in the signal into a variation in intensity of the radiation reaching the detector. An actual signal consists of radiation spanning a large number of wavenumbers, and the total intensity at the detector, which we write $\mathcal{I}(p)$, is the sum of contributions from all the wavenumbers present in the signal (Fig. 16.7):

$$\mathcal{I}(p) = \int_0^\infty \mathcal{I}(p, \tilde{\nu})\,d\tilde{\nu} = \int_0^\infty \mathcal{I}(\tilde{\nu})(1 + \cos 2\pi\tilde{\nu}p)\,d\tilde{\nu} \tag{4}$$

The problem is to find $\mathcal{I}(\tilde{\nu})$, the variation of intensity with wavenumber, which is the spectrum we require, from the record of values of $\mathcal{I}(p)$. This step is a standard technique of mathematics, and is the 'Fourier transformation' step from which this form of spectroscopy takes its name. Specifically:

$$\mathcal{I}(\tilde{\nu}) = 4 \int_0^\infty \{\mathcal{I}(p) - \tfrac{1}{2}\mathcal{I}(0)\} \cos 2\pi\tilde{\nu}p \, dp \tag{5}$$

Where $\mathcal{I}(0)$ is given by eqn 4 with $p = 0$. This integration is carried out in a computer that is interfaced to the spectrometer, and the output, $\mathcal{I}(\tilde{\nu})$, is the absorption spectrum of the sample (Fig. 16.8).[2]

A major advantage of the Fourier transform procedure is that all the radiation emitted by the source is monitored continuously. This is in contrast to a spectrometer in which a monochromator discards most of the generated radiation. As a result, Fourier transform spectrometers have a higher sensitivity than conventional spectrometers. The resolution

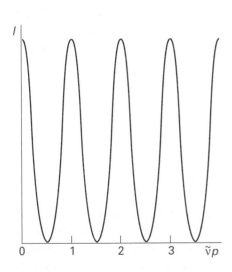

16.6 An interferogram produced as the path length p is changed in the interferometer shown in Fig. 16.5. Only a single frequency component is present in the radiation.

2　More precisely, it is the *transmission spectrum*, for the signal depends on the transmitted intensity. However, the absorption and transmission spectra carry the same information and the former term is normally employed.

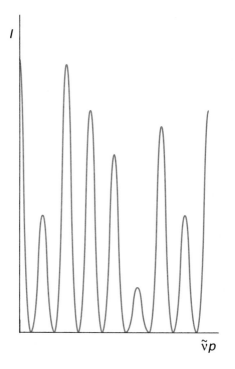

16.7 An interferogram obtained when several (in this case, three) frequencies are present in the radiation.

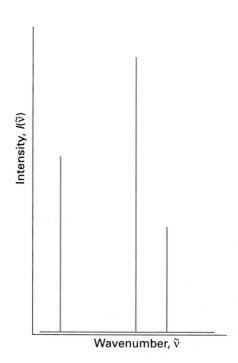

16.8 The three frequency components and their intensities that account for the appearance of the interferogram in Fig. 16.7. This spectrum is the Fourier transform of the interferogram, and is a depiction of the contributing frequencies.

they can achieve is determined by the maximum path length difference, p_{max}, of the interferometer:

$$\Delta \tilde{\nu} = \frac{1}{2p_{max}} \tag{6}$$

To achieve a resolution of 0.1 cm^{-1} requires a maximum path length difference of 5 cm.

(d) Detectors

The third component of a spectrometer is the **detector**, the device that converts incident radiation into an electric current for the appropriate signal processing or plotting. Radiation-sensitive semiconductor devices, such as a *charge-coupled device* (CCD), are increasingly dominating this role in the spectrometer. A microwave detector is typically a *crystal diode* consisting of a tungsten tip in contact with a semiconductor, such as germanium, silicon, or gallium arsenide.

The intensity of the radiation arriving at the detector is usually modulated, because alternating signals are easier to amplify than a steady signal. In some cases the beam is chopped by a rotating shutter. In other cases, the absorption characteristics of the sample itself are modulated. Ways of achieving the latter kind of modulation are described later in the chapter and in Chapter 18.

(e) The sample

The highest resolution is obtained when the sample is gaseous and at such low pressure that collisions between the molecules are infrequent. Gaseous samples are essential for rotational (microwave) spectroscopy, for only then can molecules rotate freely. To achieve sufficient absorption, the path lengths through gaseous samples must be very long, of the order of metres; long path lengths are achieved by multiple passage of the beam between parallel mirrors at each end of the sample cavity.

The most common range for infrared spectroscopy is from 4000 cm^{-1} to 625 cm^{-1}. Ordinary glass and quartz absorb over most of this range, so some other material must be used as windows. Thus, the sample is typically a liquid held between windows of sodium chloride (which is transparent down to 625 cm^{-1}) or potassium bromide (which is transparent down to 400 cm^{-1}). Other ways of preparing the sample include grinding it into a paste with 'Nujol', a hydrocarbon oil, or pressing it into a solid disk (with powdered potassium bromide, for example).

(f) Raman spectroscopy

In **Raman spectroscopy**, the energy levels of molecules are explored by examining the frequencies present in the radiation scattered by molecules. In a typical experiment, a monochromatic incident beam is passed through the sample and the radiation scattered perpendicular to the beam is monitored (Fig. 16.9). About 1 in 10^7 of the incident photons collide with the molecules, give up some of their energy, and emerge with a lower energy. These scattered photons constitute the lower-frequency **Stokes radiation** from the sample. Other incident photons may collect energy from the molecules (if they are already excited), and emerge as higher-frequency **anti-Stokes radiation**. The component of radiation scattered into the forward direction without change of frequency is called **Rayleigh radiation**.

The shifts in frequency of the scattered radiation from the incident radiation are quite small, and the latter must be very monochromatic if the shifts are to be observed. Moreover,

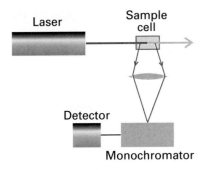

16.9 The arrangement adopted in Raman spectroscopy. The scattered radiation is monitored at right angles to the incident radiation.

the intensity of scattered radiation is low, so intense incident beams are needed. Lasers are ideal in both respects, and have entirely displaced the mercury arcs used originally. Although laser Raman spectra were originally examined using visible and ultraviolet incident radiation, radiation in the near infrared is now commonly used because its use avoids complications arising from the stimulation of fluorescence (Section 17.3). Detection is usually with a semiconductor device. Raman spectroscopy is often complementary to infrared spectroscopy because, as we shall see, different selection rules are obeyed.

16.2 The intensities of spectral lines

The ratio of the transmitted intensity, $\mathcal{I}$, to the incident intensity, $\mathcal{I}_0$, at a given frequency is called the **transmittance**, T, of the sample at that frequency:

$$T = \frac{\mathcal{I}}{\mathcal{I}_0} \tag{7}$$

It is found empirically that the transmitted intensity varies with the length, l, of the sample and the molar concentration, [J], of the absorbing species J in accord with the **Beer–Lambert law**:

$$\mathcal{I} = \mathcal{I}_0 10^{-\varepsilon[\text{J}]l} \tag{8}$$

The quantity ε is called the **molar absorption coefficient** (formerly, and still widely, the 'extinction coefficient'). The molar absorption coefficient depends on the frequency of the incident radiation and is greatest where the absorption is most intense. Its dimensions are $1/(\text{concentration} \times \text{length})$, and it is normally convenient to express it in litres per mole per centimetre ($\text{L mol}^{-1} \text{cm}^{-1}$).[3] The form of eqn 7 suggests that it is sensible to introduce the **absorbance**, A, of the sample at a given wavenumber as

$$A = \log \frac{\mathcal{I}_0}{\mathcal{I}} \qquad \text{or } A = -\log T \tag{9}$$

Then the Beer–Lambert law becomes

$$A = \varepsilon[\text{J}]l \tag{10}$$

The product $\varepsilon[\text{J}]l$ was known formerly as the *optical density* of the sample.

Justification 16.1

The Beer–Lambert law is an empirical result. However, it is simple to account for its form. The reduction in intensity, $d\mathcal{I}$, that occurs when light passes through a layer of thickness dl containing an absorbing species J at a molar concentration [J] is proportional to the thickness of the layer, the concentration of J, and the intensity, $\mathcal{I}$, incident on the layer (because the rate of absorption is proportional to the intensity, see below). We can therefore write

$$d\mathcal{I} = -\kappa[\text{J}]\mathcal{I}\, dl$$

where κ (kappa) is the proportionality coefficient, or equivalently

$$\frac{d\mathcal{I}}{\mathcal{I}} = -\kappa[\text{J}]\, dl$$

---

3   Alternative units are $\text{cm}^2 \text{mol}^{-1}$. This change of units emphasizes the point that ε is a molar cross-section for absorption and, the greater the cross-section of the molecule for absorption, the greater its ability to block the passage of the incident radiation.

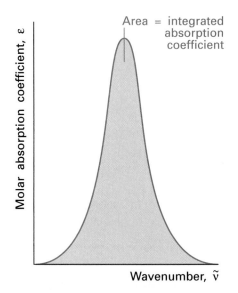

16.10 The intensity of a transition is the area under a plot of the molar absorption coefficient against the wavenumber of the incident radiation.

This expression applies to each successive layer into which the sample can be regarded as being divided. Therefore, to obtain the intensity that emerges from a sample of thickness l when the intensity incident on one face of the sample is $\mathcal{I}_0$, we sum all the successive changes:

$$\int_{\mathcal{I}_0}^{\mathcal{I}} \frac{d\mathcal{I}}{\mathcal{I}} = -\kappa \int_0^l [\text{J}]\, dl$$

If the concentration is uniform, [J] is independent of location, and the expression integrates to

$$\ln \frac{\mathcal{I}}{\mathcal{I}_0} = -\kappa[\text{J}]l$$

This expression gives the Beer–Lambert law when the logarithm is converted to base 10 by using $\ln x = (\ln 10) \log x$ and replacing κ by $\varepsilon \ln 10$.

Illustration

The Beer–Lambert law implies that the intensity of electromagnetic radiation transmitted through a sample at a given wavenumber decreases exponentially with the sample thickness and the molar concentration. If the transmittance is 0.1 for a path length of 1 cm (corresponding to a 90 per cent reduction in intensity), then it would be $(0.1)^2 = 0.01$ for a path of double the length (corresponding to a 99 per cent reduction in intensity overall).

The maximum value of the molar absorption coefficient, ε_{max}, is an indication of the intensity of a transition. However, as absorption bands generally spread over a range of wavenumbers, quoting the absorption coefficient at a single wavenumber might not give a true indication of the intensity of a transition. The **integrated absorption coefficient**, $\mathcal{A}$, is the sum of the absorption coefficients over the entire band (Fig. 16.10), and corresponds to the area under the plot of the molar absorption coefficient against wavenumber:

$$\mathcal{A} = \int_{\text{band}} \varepsilon(\tilde{\nu})\, d\tilde{\nu} \tag{11}$$

For lines of similar widths, the integrated absorption coefficients are proportional to the heights of the lines.

(a) Absorption intensities

Einstein identified three contributions to the transitions between states. **Stimulated absorption** is the transition from a low energy state to one of higher energy that is driven by the electromagnetic field oscillating at the transition frequency. The more intense the electromagnetic field (the more intense the incident radiation), the greater the rate at which transitions are induced and hence the stronger the absorption by the sample (Fig. 16.11). Einstein wrote the transition rate, w, from the lower to the upper state as[4]

$$w = B\rho \tag{12}$$

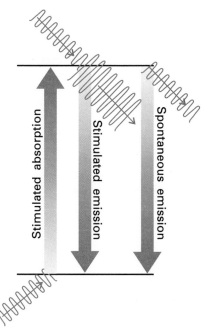

16.11 The processes that account for absorption and emission of radiation and the attainment of thermal equilibrium. The excited state can return to the lower state spontaneously as well as by a process stimulated by radiation already present at the transition frequency.

The constant B is the **Einstein coefficient of stimulated absorption** and $\rho\, d\nu$ is the energy density of radiation in the frequency range ν to $\nu + d\nu$, where ν is the frequency of the

4   Specifically, w is the rate of change of probability of the molecule being found in the upper state: $w = dP/dt$.

transition. When the molecule is exposed to black-body radiation from a source of temperature T, ρ is given by the Planck distribution (eqn 11.5):[5]

$$\rho = \frac{8\pi h \nu^3 / c^3}{e^{h\nu/kT} - 1} \tag{13}$$

For the time being, we can treat B as an empirical parameter that characterizes the transition: if B is large, then a given intensity of incident radiation will induce transitions strongly and the sample will be strongly absorbing. The **total rate of absorption**, W, the number of molecules excited during an interval divided by the duration of the interval, is the transition rate of a single molecule multiplied by the number of molecules N in the lower state: $W = Nw$.

Einstein considered that the radiation was also able to induce the molecule in the upper state to undergo a transition to the lower state, and hence to generate a photon of frequency ν. Thus, he wrote the rate of this stimulated emission as

$$w' = B'\rho \tag{14}$$

where B' is the **Einstein coefficient of stimulated emission**. Note that only radiation of the same frequency as the transition can stimulate an excited state to fall to a lower state. However, Einstein realized that stimulated emission was not the only means by which the excited state could generate radiation and return to the lower state, and suggested that an excited state could undergo **spontaneous emission** at a rate that was independent of the intensity of the radiation (of any frequency) that is already present. He therefore wrote the total rate of transition from the upper to the lower state as

$$w' = A + B'\rho \tag{15}$$

The constant A is the **Einstein coefficient of spontaneous emission**. The overall rate of emission is

$$W' = N'(A + B'\rho) \tag{16}$$

where N' is the population of the upper state.

As demonstrated in the *Justification* below, Einstein was able to show that the two coefficients of stimulated absorption and emission are equal, and that the coefficient of spontaneous emission is related to them by

$$A = \left(\frac{8\pi h\nu^3}{c^3}\right)B \tag{17}$$

Justification 16.2

At thermal equilibrium, the total rates of emission and absorption are equal, so

$$NB\rho = N'(A + B'\rho)$$

This expression rearranges into

$$\rho = \frac{N'A}{NB - N'B'} = \frac{A/B}{N/N' - B'/B} = \frac{A/B}{e^{h\nu/kT} - B'/B}$$

We have used the Boltzmann expression (see the *Introduction*) for the ratio of populations of states of energies E and E' in the last step:

$$\frac{N'}{N} = e^{-h\nu/kT} \qquad h\nu = E' - E$$

---

5   The slightly different form of the distribution stems from the fact that the ρ in eqn 11.5 is for the energy density written as $\rho \, d\lambda$, whereas here it is written as $\rho \, d\nu$, and $|d\lambda| = (c/\nu^2) \, d\nu$.

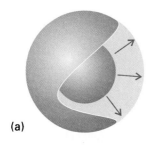

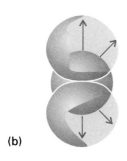

(a)

(b)

16.12 (a) When a $1s$ electron becomes a $2s$ electron, there is a spherical migration of charge; there is no dipole moment associated with this migration of charge; this transition is electric-dipole forbidden. (b) In contrast, when a $1s$ electron becomes a $2p$ electron, there is a dipole associated with the charge migration; this transition is allowed. (There are subtle effects arising from the sign of the wavefunction that give the charge migration a dipolar character, which this diagram does not attempt to convey.)

This result has the same form as the Planck distribution (eqn 13), which describes the radiation density at thermal equilibrium. Indeed, when we compare the two expressions for ρ, we can conclude that $B' = B$ and that A is related to B by eqn 17.

The growth of the importance of spontaneous emission with increasing frequency is a very important conclusion, as we shall see when we consider the operation of lasers (Section 17.5). The equality of the coefficients of stimulated emission and absorption implies that, if two states happen to have equal populations, then the rate of stimulated emission is equal to the rate of stimulated absorption, and there is then no net absorption.

Spontaneous emission can be largely ignored at the relatively low frequencies of rotational and vibrational transitions, and the intensities of these transitions can be discussed in terms of stimulated emission and absorption. Then the net rate of absorption is given by

$$W_{\text{net}} = NB\rho - N'B'\rho = (N - N')B\rho \tag{18}$$

and is proportional to the population difference of the two states involved in the transition.

(b) Selection rules and transition moments

We met the concept of a 'selection rule' in Sections 13.3 and 15.6 as a statement about whether a transition is forbidden or allowed. Selection rules also apply to molecular spectra, and the form they take depends on the type of transition. The underlying classical idea is that, for the molecule to be able to interact with the electromagnetic field and absorb or create a photon of frequency ν, it must possess, at least transiently, a dipole oscillating at that frequency. This transient dipole is expressed quantum mechanically in terms of the transition dipole moment, $\boldsymbol{\mu}_{\text{fi}}$, between states $|i\rangle$ and $|f\rangle$:

$$\boldsymbol{\mu}_{\text{fi}} = \langle f|\boldsymbol{\mu}|fi\rangle = \int \psi_f^* \boldsymbol{\mu} \psi_i \, d\tau \tag{19}$$

where $\boldsymbol{\mu}$ is the electric dipole moment operator. The size of the transition dipole can be regarded as a measure of the charge redistribution that accompanies a transition: a transition will be active (and generate or absorb photons) only if the accompanying charge redistribution is dipolar (Fig. 16.12).

The coefficient of stimulated absorption (and emission), and therefore the intensity of the transition, is proportional to the square of the transition dipole moment, and a detailed analysis gives

$$B = \frac{|\boldsymbol{\mu}_{\text{fi}}|^2}{6\varepsilon_0\hbar^2} \tag{20}$$

Only if the transition moment is nonzero does the transition contribute to the spectrum. We see that, to identify the selection rules, we must establish the conditions for which $\boldsymbol{\mu}_{\text{fi}} \neq 0$.

A **gross selection rule** specifies the general features a molecule must have if it is to have a spectrum of a given kind. For instance, we shall see that a molecule gives a rotational spectrum only if it has a permanent electric dipole moment. This rule, and others like it for other types of transition, will be explained in the relevant sections of the chapter.

A detailed study of the transition moment leads to the **specific selection rules** that express the allowed transitions in terms of the changes in quantum numbers. We have already encountered examples of specific selection rules when discussing atomic spectra (Section 13.3), such as the rule $\Delta l = \pm 1$ for the angular momentum quantum number. Specific selection rules can often be interpreted in terms of the changes of angular momentum when a photon (with its intrinsic spin angular momentum $s = 1$) enters or

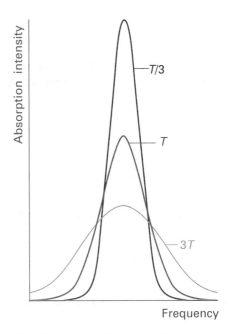

16.13 The shape of a Doppler-broadened spectral line reflects the Maxwell distribution of speeds in the sample at the temperature of the experiment. Notice that the line broadens as the temperature is increased.

leaves a molecule, and we shall discuss them once we have set up the quantum numbers needed to describe rotation and vibration.

16.3 Linewidths

A number of effects contribute to the widths of spectroscopic lines. Some contributions to linewidth can be modified by changing the conditions, and to achieve high resolutions we need to know how to mimimize these contributions. Other contributions cannot be changed, and represent an inherent limitation on resolution.

(a) Doppler broadening

One important broadening process in gaseous samples is the **Doppler effect**, in which radiation is shifted in frequency when the source is moving towards or away from the observer. When a source emitting electromagnetic radiation of frequency ν moves with a speed s relative to an observer, the observer detects radiation of frequency

$$\nu_{\text{receding}} = \nu\left(\frac{1-s/c}{1+s/c}\right)^{1/2} \qquad \nu_{\text{approaching}} = \nu\left(\frac{1+s/c}{1-s/c}\right)^{1/2} \tag{21}$$

where c is the speed of light. For nonrelativistic speeds ($s \ll c$), these expressions simplify to

$$\nu_{\text{receding}} \approx \frac{\nu}{1+s/c} \qquad \nu_{\text{approaching}} \approx \frac{\nu}{1-s/c} \tag{22}$$

Molecules reach high speeds in all directions in a gas, and a stationary observer detects the corresponding Doppler-shifted range of frequencies. Some molecules approach the observer, some move away; some move quickly, others slowly. The detected spectral 'line' is the absorption or emission profile arising from all the resulting Doppler shifts. The profile reflects the distribution of molecular velocities parallel to the line of sight (Section 1.3), which is a bell-shaped Gaussian curve (of the form e^{-x^2}). The Doppler line shape is therefore also a Gaussian (Fig. 16.13), and calculation shows that, when the temperature is T and the mass of the molecule is m, the width of the line at half-height (in terms of frequency or wavelength) is

$$\delta\nu = \frac{2\nu}{c}\left(\frac{2kT\ln 2}{m}\right)^{1/2} \qquad \delta\lambda = \frac{2\lambda}{c}\left(\frac{2kT\ln 2}{m}\right)^{1/2} \tag{23}$$

For a molecule like N_2 at room temperature ($T \approx 300$ K), $\delta\nu/\nu \approx 2.3 \times 10^{-6}$. For a typical rotational transition wavenumber of 1 cm^{-1} (corresponding to a frequency of 30 GHz), the linewidth is about 70 kHz.

Doppler broadening increases with temperature because the molecules acquire a wider range of speeds. Therefore, to obtain spectra of maximum sharpness, it is best to work with cold samples.

(b) Lifetime broadening

It is found that spectroscopic lines from gas-phase samples are not infinitely sharp even when Doppler broadening has been largely eliminated by working at low temperatures. The same is true of the spectra of samples in condensed phases and solution. This residual broadening is due to quantum mechanical effects. Specifically, when the Schrödinger equation is solved for a system that is changing with time, it is found that it is impossible to specify the energy levels exactly. If on average a system survives in a state for a time τ, the lifetime of the state, then its energy levels are blurred to an extent of order δE, where

$$\delta E \approx \frac{\hbar}{\tau} \tag{24}$$

This expression is reminiscent of the Heisenberg uncertainty principle (eqn 11.46), and consequently this **lifetime broadening** is often called 'uncertainty broadening'. When the energy spread is expressed as a wavenumber through $\delta E = hc\delta\tilde{\nu}$, and the values of the fundamental constants are introduced, this relation becomes

$$\delta\tilde{\nu} \approx \frac{5.3 \text{ cm}^{-1}}{\tau/\text{ps}} \tag{25}$$

No excited state has an infinite lifetime; therefore, all states are subject to some lifetime broadening and, the shorter the lifetimes of the states involved in a transition, the broader the corresponding spectral lines.

Two processes are responsible for the finite lifetimes of excited states. The dominant one for low-frequency transitions is **collisional deactivation**, which arises from collisions between molecules or with the walls of the container. If the **collisional lifetime**, the mean time between collisions, is τ_{col}, the resulting collisional linewidth is $\delta E_{\text{col}} \approx \hbar/\tau_{\text{col}}$. Because $\tau_{\text{col}} = 1/z$, where z is the collision frequency, and from the kinetic model of gases (Section 1.3) we know that z is proportional to the pressure, p, we see that the collisional linewidth is proportional to the pressure. The collisional linewidth can therefore be minimized by working at low pressures.

The rate of spontaneous emission cannot be changed. Hence it is a natural limit to the lifetime of an excited state, and the resulting lifetime broadening is the **natural linewidth** of the transition. The natural linewidth is an intrinsic property of the transition, and cannot be changed by modifying the conditions. Natural linewidths depend strongly on the transition frequency (they increase with the coefficient of spontaneous emission A and therefore as ν^3), so low-frequency transitions (such as the microwave transitions of rotational spectroscopy) have very small natural linewidths, and collisional and Doppler line-broadening processes are dominant. The natural lifetimes of electronic transitions are very much shorter than for vibrational and rotational transitions, so the natural linewidths of electronic transitions are much greater than those of vibrational and rotational transitions. For example, a typical electronic excited state natural lifetime is about 10^{-8} s (10 ns), corresponding to a natural width of about 5×10^{-4} cm^{-1} (15 MHz). A typical rotational state natural lifetime is about 10^3 s, corresponding to a natural linewidth of only 5×10^{-15} cm^{-1} (of the order of 10^{-4} Hz).

Pure rotation spectra

The general strategy we adopt for discussing molecular spectra and the information they contain is to find expressions for the energy levels of molecules and then to calculate the transition frequencies by applying the selection rules. We then predict the appearance of the spectrum by taking into account the transition moments and the populations of the states. In this section we illustrate the strategy by considering the rotational states of molecules.

16.4 Moments of inertia

The key molecular parameter we shall need is the **moment of inertia**, I, of the molecule (Section 12.6). The moment of inertia of a molecule is defined as the mass of each atom multiplied by the square of its distance from the rotational axis through the centre of mass of the molecule (Fig. 16.14):

$$I = \sum_i m_i r_i^2 \tag{26}$$

where r_i is the perpendicular distance of the atom i from the axis of rotation. The moment of inertia depends on the masses of the atoms present and the molecular geometry, so we can

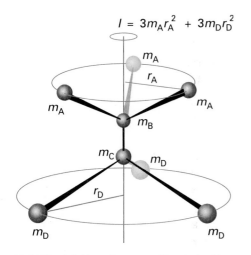

$$I = 3m_A r_A^2 + 3m_D r_D^2$$

16.14 The definition of moment of inertia. In this molecule there are three identical atoms attached to the B atom and three different but mutually identical atoms attached to the C atom. In this example, the centre of mass lies on the C_3 axis, and the perpendicular distances are measured from the axis passing through the B and C atoms.

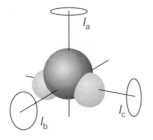

16.15 An asymmetric rotor has three different moments of inertia; all three rotation axes coincide at the centre of mass of the molecule.

suspect (and later shall see explicitly) that rotational spectroscopy will give information about bond lengths and bond angles.

In general, the rotational properties of any molecule can be expressed in terms of the moments of inertia about three perpendicular axes set in the molecule (Fig. 16.15). The convention is to label the moments of inertia I_a, I_b, and I_c, with the axes chosen so that $I_c \geq I_b \geq I_a$. For linear molecules, the moment of inertia around the internuclear axis is zero. The explicit expressions for the moments of inertia of some symmetrical molecules are given in Table 16.1.

Table 16.1 Moments of inertia†

1. Diatomics

$$I = \frac{m_A m_B}{m} R^2 = \mu R^2$$

2. Linear rotors

$$I = m_A R^2 + m_C R'^2$$
$$- \frac{(m_A R - m_C R')^2}{m}$$

$$I = 2 m_A R^2$$

3. Symmetric rotors

$$I_\parallel = 2 m_A R^2 (1 - \cos\theta)$$
$$I_\perp = m_A R^2 (1 - \cos\theta)$$
$$+ \frac{m_A}{m}(m_B + m_C) R^2 (1 + 2\cos\theta)$$
$$+ \frac{m_C R'}{m}\{(3m_A + m_B)R'$$
$$+ 6 m_A R[\tfrac{1}{3}(1 + 2\cos\theta)]^{1/2}\}$$

$$I_\parallel = 2 m_A R^2 (1 - \cos\theta)$$
$$I_\perp = m_A R^2 (1 - \cos\theta)$$
$$+ \frac{m_A m_B}{m} R^2 (1 + 2\cos\theta)$$

$$I_\parallel = 4 m_A R^2$$
$$I_\perp = 2 m_A R^2 + 2 m_C R'^2$$

4. Spherical rotors

$$I = \tfrac{8}{3} m_A R^2$$

$$I = 4 m_A R^2$$

† In each case m is the total mass of the molecule.

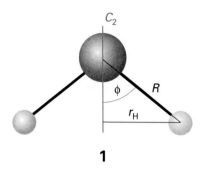

1

Example 16.1 Calculating the moment of inertia of a molecule

Calculate the moment of inertia of an H_2O molecule around its twofold axis (the bisector of the HOH angle (1)). The HOH bond angle is $104.5°$ and the bond length is 95.7 pm.

Method According to eqn 26, the moment of inertia is the sum of the masses multiplied by the squares of their distances from the axis of rotation. The latter can be expressed by using trigonometry and the bond angle and bond length.

Answer From eqn 26,

$$I = \sum_i m_i r_i^2 = m_H r_H^2 + 0 + m_H r_H^2 = 2m_H r_H^2$$

If the bond angle of the molecule is denoted 2ϕ and the bond length is R, trigonometry gives $r_H = R \sin \phi$. It follows that

$$I = 2m_H R^2 \sin^2 \phi$$

Substitution of the data gives

$$I = 2 \times (1.67 \times 10^{-27} \text{ kg}) \times (9.57 \times 10^{-11} \text{ m})^2 \times \sin^2 52.3°$$
$$= 1.91 \times 10^{-47} \text{ kg m}^2$$

Comment The mass of the O atom makes no contribution to the moment of inertia for this mode of rotation as the atom is immobile while the H atoms circulate around it.

- -

Self-test 16.1 Calculate the moment of inertia of a $CH^{35}Cl_3$ molecule around its threefold axis. The C–Cl bond length is 177 pm and the HCCl angle is $107°$; $m(^{35}Cl) = 34.97$ u.

[4.99×10^{-45} kg m^2]

We shall suppose initially that molecules are **rigid rotors**, bodies that do not distort under the stress of rotation. Rigid rotors can be classified into four types (Fig. 16.16):

Spherical rotors have three equal moments of inertia (examples: CH_4, SiH_4, and SF_6).
Symmetric rotors have two equal moments of inertia (examples: NH_3, CH_3Cl, and CH_3CN).
Linear rotors have one moment of inertia (the one about the axis) equal to zero (examples: CO_2, HCl, OCS, and $HC\equiv CH$).
Asymmetric rotors have three different moments of inertia (H_2O, H_2CO, and CH_3OH are examples).

In group theoretical language, a spherical rotor is a molecule that belongs to a cubic or icosahedral point group; a symmetric rotor is a molecule with at least a threefold axis of symmetry. All diatomic molecules are linear rotors. An asymmetric rotor is a molecule without a threefold (or higher) axis: it may have other elements of symmetry, such as a twofold axis or mirror planes. The energy levels of asymmetric rotors are complicated and we shall not consider them.

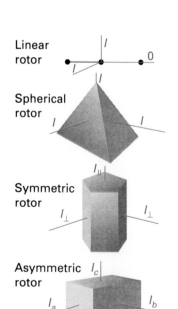

Linear rotor

Spherical rotor

Symmetric rotor

Asymmetric rotor

16.16 A schematic illustration of the classification of rigid rotors.

16.5 The rotational energy levels

The rotational energy levels of a rigid rotor may be obtained by solving the appropriate Schrödinger equation. Fortunately, however, there is a much less onerous short cut to the exact expressions that depends on noting the classical expression for the energy of a rotating body, expressing it in terms of the angular momentum, and then importing the quantum mechanical properties of angular momentum into the equations.

The classical expression for the energy of a body rotating about an axis a is

$$E_a = \tfrac{1}{2} I_a \omega_a^2 \tag{27}$$

where ω_a is the angular velocity (in radians per second, rad s^{-1}) about that axis and I_a is the corresponding moment of inertia. A body free to rotate about three axes has an energy

$$E = \tfrac{1}{2} I_a \omega_a^2 + \tfrac{1}{2} I_b \omega_b^2 + \tfrac{1}{2} I_c \omega_c^2$$

Because the classical angular momentum about the axis a is $J_a = I_a \omega_a$, with similar expressions for the other axes, it follows that

$$E = \frac{J_a^2}{2I_a} + \frac{J_b^2}{2I_b} + \frac{J_c^2}{2I_c} \tag{28}$$

This is the key equation. We described the quantum mechanical properties of angular momentum in Section 12.7b, and can now make use of them in conjunction with this equation to obtain the rotational energy levels.

(a) Spherical rotors

When all three momenta of inertia are equal to some value I, as in CH_4 and SF_6, the classical expression for the energy is

$$E = \frac{J_a^2 + J_b^2 + J_c^2}{2I} = \frac{\mathcal{J}^2}{2I}$$

where $\mathcal{J}$ is the magnitude of the angular momentum. We can immediately find the quantum expression by making the replacement

$$\mathcal{J}^2 \to J(J+1)\hbar^2 \qquad J = 0, 1, 2, \ldots$$

Therefore, the energy of a spherical rotor is confined to the values

$$E_J = J(J+1)\frac{\hbar^2}{2I} \qquad J = 0, 1, 2, \ldots \tag{29}$$

The resulting ladder of energy levels is illustrated in Fig. 16.17. The energy is normally expressed in terms of the **rotational constant**, B, of the molecule, where

$$hcB = \frac{\hbar^2}{2I} \qquad \text{so } B = \frac{\hbar}{4\pi cI} \tag{30}$$

The expression for the energy is then

$$E_J = hcBJ(J+1) \qquad J = 0, 1, 2, \ldots \tag{31}$$

The rotational constant as defined by eqn 31 is a wavenumber.[6] The energy of a rotational state is normally reported as the **rotational term**, $F(J)$, a wavenumber, by division by hc:

$$F(J) = BJ(J+1) \tag{32}$$

The separation of adjacent levels is

$$F(J) - F(J-1) = 2BJ \tag{33}$$

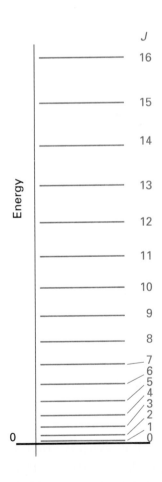

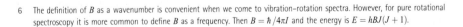

16.17 The rotational energy levels of a linear or spherical rotor. Note that the energy separation between neighbouring levels increases as J increases.

6 The definition of B as a wavenumber is convenient when we come to vibration–rotation spectra. However, for pure rotational spectroscopy it is more common to define B as a frequency. Then $B = \hbar/4\pi I$ and the energy is $E = hBJ(J+1)$.

Because the rotational constant decreases as I increases, we see that large molecules have closely spaced rotational energy levels. We can estimate the magnitude of the separation by considering CCl_4: from the bond lengths and masses of the atoms we find $I = 4.85 \times 10^{-45} \text{ kg m}^2$, and hence $B = 0.0577 \text{ cm}^{-1}$.

(b) Symmetric rotors

In symmetric rotors, two moments of inertia are equal but different from the third (as in CH_3Cl, NH_3, and C_6H_6); the unique axis of the molecule is its **principal axis** (or *figure axis*). We shall write the unique moment of inertia (that about the principal axis) as $I_\parallel$ and the other two as $I_\perp$. If $I_\parallel > I_\perp$, the rotor is classified as **oblate** (like a pancake, and C_6H_6); if $I_\parallel < I_\perp$ it is classified as **prolate** (like a cigar, and CH_3Cl). The classical expression for the energy, eqn 28, becomes

$$E = \frac{J_b^2 + J_c^2}{2I_\perp} + \frac{J_a^2}{2I_\parallel}$$

This expression can be written in terms of $\mathcal{J}^2 = J_a^2 + J_b^2 + J_c^2$:

$$E = \frac{\mathcal{J}^2 - J_a^2}{2I_\perp} + \frac{J_a^2}{2I_\parallel} = \frac{\mathcal{J}^2}{2I_\perp} + \left(\frac{1}{2I_\parallel} - \frac{1}{2I_\perp}\right)J_a^2 \tag{34}$$

Now we generate the quantum expression by replacing $\mathcal{J}^2$ by $J(J+1)\hbar^2$, where J is the angular momentum quantum number. We also know from the quantum theory of angular momentum (Section 12.7b) that the component of angular momentum about any axis is restricted to the values $K\hbar$, with $K = 0, \pm 1, \ldots, \pm J$. ($K$ is the quantum number used to signify a component on the principal axis; M_J is reserved for a component on an externally defined axis.) Therefore, we also replace J_a^2 by $K^2\hbar^2$. It follows that the rotational terms are

$$F(J,K) = BJ(J+1) + (A-B)K^2 \quad J = 0, 1, 2, \ldots \quad K = 0, \pm 1, \ldots, \pm J \tag{35}$$

with

$$A = \frac{\hbar}{4\pi c I_\parallel} \qquad B = \frac{\hbar}{4\pi c I_\perp} \tag{36}$$

Equation 35 matches what we should expect for the dependence of the energy levels on the two distinct moments of inertia of the molecule. When $K = 0$, there is no component of angular momentum about the principal axis, and the energy levels depend only on $I_\perp$ (Fig. 16.18). When $K = \pm J$, almost all the angular momentum arises from rotation around the principal axis, and the energy levels are determined largely by $I_\parallel$. The sign of K does not affect the energy because opposite values of K correspond to opposite senses of rotation, and the energy does not depend on the sense of rotation.

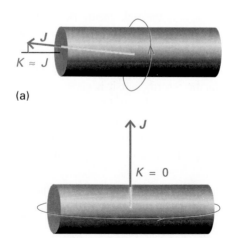

$K \approx J$

(a)

$K = 0$

(b)

16.18 The significance of the quantum number K. (a) When $|K|$ is close to its maximum value, J, most of the molecular rotation is around the principal axis. (b) When $K = 0$ the molecule has no angular momentum about its principal axis: it is undergoing end-over-end rotation.

Example 16.2 Calculating the rotational energy levels of a molecule

An $^{14}NH_3$ molecule is a symmetric rotor with bond length 101.2 pm and HNH bond angle 106.7°. Calculate its rotational terms.

Method Begin by calculating the rotational constants A and B by using the expressions for moments of inertia given in Table 16.1. Then use eqn 35 to find the rotational terms.

Answer Substitution of $m_A = 1.0078$ u, $m_B = 14.0031$ u, $R = 101.2$ pm, and $\theta = 106.7°$ into the second of the symmetric rotor expressions in Table 16.1 gives $I_\parallel = 4.4128 \times 10^{-47}$ kg m^2 and $I_\perp = 2.8059 \times 10^{-47}$ kg m^2. Hence, $A = 6.344$ cm^{-1} and $B = 9.977$ cm^{-1}. It follows from eqn 35 that

$$F(J,K)/\text{cm}^{-1} = 9.977J(J+1) - 3.633K^2$$

Comment For $J = 1$, the energy needed for the molecule to rotate mainly about its figure axis $(K = \pm J)$ is equivalent to 16.32 cm^{-1}, but end-over-end rotation $(K = 0)$ corresponds to 19.95 cm^{-1}.

- -

Self-test 16.2 A $CH_3{}^{35}Cl$ molecule has a C–Cl bond length of 178 pm, a C–H bond length of 111 pm, and an HCH angle of 110.5°. Calculate its rotational energy terms.

$$[F(J,K)/\text{cm}^{-1} = 0.444\,J(J+1) + 4.58K^2]$$

(c) Linear rotors

For a linear rotor (such as CO_2, HCl, and C_2H_2), in which the nuclei are regarded as mass points, the rotation occurs only about an axis perpendicular to the line of atoms and there is zero angular momentum around the line. Therefore, the component of angular momentum around the figure axis of a linear rotor is identically zero, and $K \equiv 0$ in eqn 35. The rotational terms of a linear molecule are therefore

$$F(J) = BJ(J+1) \qquad J = 0, 1, 2, \dots \tag{37}$$

This expression is the same as eqn 32 but we have arrived at it in a significantly different way: here $K \equiv 0$ but for a spherical rotor $A = B$.

(d) Degeneracies and the Stark effect

The energy of a symmetric rotor depends on J and K, and each level except those with $K = 0$ is doubly degenerate: the states with K and $-K$ have the same energy. However, we must not forget that the angular momentum of the molecule has a component on an external, laboratory-fixed axis. This component is quantized, and its permitted values are $M_J\hbar$, with $M_J = 0, \pm 1, \dots, \pm J$, giving $2J + 1$ values in all (Fig. 16.19). The quantum number M_J does not appear in the expression for the energy, but it is necessary for a complete specification of the state of the rotor. Consequently, all $2J + 1$ orientations of the rotating molecule have the same energy. It follows that a symmetric rotor level is $2(2J + 1)$-fold degenerate for $K \neq 0$ and $(2J + 1)$-fold degenerate for $K = 0$. A linear rotor has K fixed at 0, but the angular momentum may still have $2J + 1$ components on the laboratory axis, so its degeneracy is $2J + 1$.

A spherical rotor can be regarded as a version of a symmetric rotor in which A is equal to B. The quantum number K may still take any one of $2J + 1$ values, but the energy is independent of which value it takes. Therefore, as well as having a $(2J + 1)$-fold degeneracy arising from its orientation in space, the rotor also has a $(2J + 1)$-fold degeneracy arising from its orientation with respect to an arbitrary axis in the molecule. The overall degeneracy of a symmetric rotor with quantum number J is therefore $(2J + 1)^2$. This degeneracy increases very rapidly: when $J = 10$, for instance, there are 441 states of the same energy.

The degeneracy associated with the quantum number M_J (the orientation of the rotation in space) is partly removed when an electric field is applied to a polar molecule (for example, HCl or NH_3), as illustrated in Fig. 16.20. The splitting of states by an electric field is called the

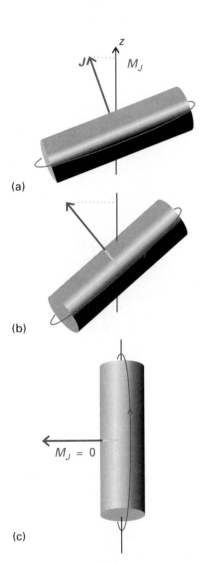

16.19 The significance of the quantum number M_J. (a) When M_J is close to its maximum value, J, most of the molecular rotation is around the laboratory z-axis. (b) An intermediate value of M_J. (c) When $M_J = 0$ the molecule has no angular momentum about the z-axis. All three diagrams correspond to a state with $K = 0$; there are corresponding diagrams for different values of K, in which the angular momentum makes a different angle to the molecule's principal axis.

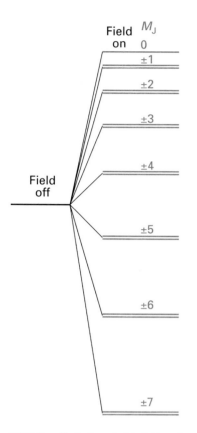

16.20 The effect of an electric field on the energy levels of a polar linear rotor. All levels are doubly degenerate except that with $M_J = 0$.

Stark effect. For a linear rotor in an electric field $\mathcal{E}$, the energy of the state $|J, M_J\rangle$ is given by

$$E(J, M_J) = hcBJ(J + 1) + a(J, M_J)\mu^2\mathcal{E}^2 \tag{38a}$$

where

$$a(J, M_J) = \frac{\{J(J + 1) - 3M_J^2\}}{2hcBJ(J + 1)(2J - 1)(2J + 3)} \tag{38b}$$

Note that the energy of a state with quantum number M_J depends on the square of the permanent electric dipole moment, μ. The observation of the Stark effect can therefore be used to measure this property, but the technique is limited to molecules that are sufficiently volatile to be studied by microwave spectroscopy. However, as spectra can be recorded for samples at pressures of only about 1 Pa, even some quite nonvolatile substances may be studied. Sodium chloride, for example, can be studied as diatomic NaCl molecules at high temperatures.

(e) Centrifugal distortion

We have treated molecules as rigid rotors. However, the atoms of rotating molecules are subject to centrifugal forces that tend to distort the molecular geometry and change the moments of inertia (Fig. 16.21). The effect of centrifugal distortion on a diatomic molecule is to stretch the bond and hence to increase the moment of inertia. As a result, centrifugal distortion reduces the rotational constant and consequently the energy levels are slightly closer than the rigid-rotor expressions predict. The effect is usually taken into account largely empirically by subtracting a term from the energy and writing

$$F(J) = BJ(J + 1) - D_J J^2(J + 1)^2 \tag{39}$$

The parameter D_J is the **centrifugal distortion constant**. It is large when the bond is easily stretched. The centrifugal distortion constant of a diatomic molecule is related to the vibrational wavenumber of the bond, $\tilde{\nu}$ (which, as we shall see later, is a measure of its stiffness), through the approximate relation

$$D_J = \frac{4B^3}{\tilde{\nu}^2} \tag{40}$$

Hence the observation of the convergence of the rotational levels as J increases can be interpreted in terms of the rigidity of the bond.

16.6 Rotational transitions

Typical values of B for small molecules are in the region of 0.1 to 10 cm^{-1} (for example, 0.356 cm^{-1} for NF$_3$ and 10.59 cm^{-1} for HCl), so rotational transitions lie in the microwave region of the spectrum. The transitions are detected by monitoring the net absorption of microwave radiation. Modulation of the transmitted intensity can be achieved by varying the energy levels with an oscillating electric field. In this **Stark modulation**, an electric field of about 10^5 V m^{-1} and a frequency of between 10 and 100 kHz is applied to the sample.

(a) Rotational selection rules

We have already remarked (Section 16.2) that the gross selection rule for the observation of a pure rotational spectrum is that a molecule must have a permanent electric dipole moment. That is, *for a molecule to give a pure rotational spectrum, it must be polar*. The classical basis of this rule is that a polar molecule appears to possess a fluctuating dipole when rotating, but a nonpolar molecule does not (Fig. 16.22). The permanent dipole can be regarded as a handle with which the molecule stirs the electromagnetic field into oscillation

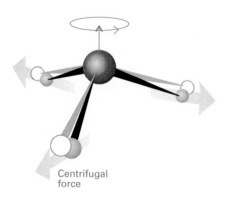

16.21 The effect of rotation on a molecule. The centrifugal force arising from rotation distorts the molecule, opening out bond angles and stretching bonds slightly. The effect is to increase the moment of inertia of the molecule and hence to decrease its rotational constant.

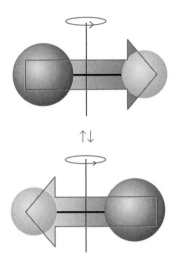

16.22 To a stationary observer, a rotating polar molecule looks like an oscillating dipole which can stir the electromagnetic field into oscillation. This picture is the classical origin of the gross selection rule for rotational transitions.

(and vice versa for absorption). Homonuclear diatomic molecules and symmetrical ($D_{\infty h}$) linear molecules such as CO_2 are rotationally inactive. Spherical rotors cannot have electric dipole moments unless they become distorted by rotation, so they are also inactive except in special cases. An example of a spherical rotor that does become sufficiently distorted for it to acquire a dipole moment is SiH_4, which has a dipole moment of about 8.3 μD by virtue of its rotation when $J \approx 10$ (for comparison, HCl has a permanent dipole moment of 1.1 D; molecular dipole moments and their units are discussed in Section 22.1). The pure rotational spectrum of SiH_4 has been detected by using long path lengths (10 m) through high-pressure (4 atm) samples.

Illustration

Of the molecules N_2, CO_2, OCS, H_2O, $CH_2{=}CH_2$, C_6H_6, only OCS and H_2O are polar, so only these two molecules have microwave spectra.

Self-test 16.3 Which of the molecules H_2, NO, N_2O, CH_4 can have a pure rotational spectrum?

[NO, N_2O]

The specific rotational selection rules are found by evaluating the transition dipole moment between rotational states. For a linear molecule, the transition moment vanishes unless the following conditions are fulfilled:

$$\Delta J = \pm 1 \qquad \Delta M_J = 0, \pm 1 \qquad (41)$$

The transition $\Delta J = +1$ corresponds to absorption and the transition $\Delta J = -1$ corresponds to emission. The allowed change in J in each case arises from the conservation of angular momentum when a photon, a spin-1 particle, is emitted or absorbed (Fig. 16.23). The change in M_J is also a consequence of the conservation of angular momentum, and takes into account the direction in which the photon leaves or enters the molecule.

When the transition moment is evaluated for all possible relative orientations of the molecule to the line of flight of the photon, it is found that the total $J + 1 \leftrightarrow J$ transition intensity is proportional to

$$|\mu_{J+1,J}|^2 = \left(\frac{J+1}{2J+1}\right)\mu^2 \rightarrow \tfrac{1}{2}\mu^2 \text{ for } J \gg 1 \qquad (42)$$

where μ is the permanent electric dipole moment of the molecule. Although the intensity of the absorption varies with J, the dependence is weak and the dominant effect on intensities is the population of the states. It should be noted that the intensity is proportional to the square of the permanent electric dipole moment, so strongly polar molecules give rise to much more intense rotational lines than less polar molecules.

A selection rule for K is needed for symmetric rotors. Any electric dipole moment possessed by a symmetric rotor must lie parallel to the principal axis, as in NF_3 (recall Fig. 15.16). Such a molecule cannot be accelerated into different states of rotation around the figure axis by the absorption of radiation, so $\Delta K = 0$ for a symmetric rotor.

(b) The appearance of rotational spectra

When these selection rules are applied to the expressions for the energy levels of a rigid rotor, it follows that the wavenumbers of the allowed $J + 1 \leftarrow J$ absorptions are

$$\tilde{\nu} = 2B(J + 1) \qquad J = 0, 1, 2, \ldots \qquad (43)$$

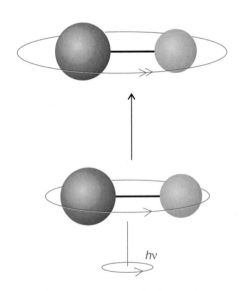

16.23 When a photon is absorbed by a molecule, the angular momentum of the combined system is conserved. If the molecule is rotating in the same sense as the spin of the incoming photon, then J increases by 1.

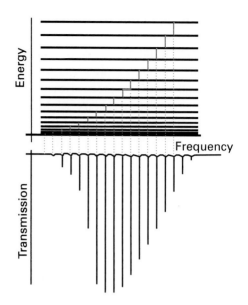

16.24 The rotational energy levels of a linear rotor, the transitions allowed by the selection rule $\Delta J = \pm 1$, and a typical pure rotational absorption spectrum (displayed here in terms of the radiation transmitted through the sample). The intensities reflect the populations of the initial level in each case and the strengths of the transition dipole moments.

When centrifugal distortion is taken into account, the corresponding expression is

$$\tilde{\nu} = 2B(J+1) - 4D_J(J+1)^3 \tag{44}$$

However, because the second term is typically very small compared with the first, the appearance of the spectrum closely resembles that predicted from eqn 43.

Example 16.3 Predicting the appearance of a rotational spectrum

Predict the form of the rotational spectrum of NH_3.

Method We calculated the energy levels in Example 16.2. The NH_3 molecule is a polar symmetric rotor, so the selection rules $\Delta J = \pm 1$ and $\Delta K = 0$ apply. For absorption, $\Delta J = +1$ and we can use eqn 43. Because $B = 9.977$ cm^{-1}, we can draw up the following table for the $J+1 \leftarrow J$ transitions.

J	0	1	2	3	$\cdots$
$\tilde{\nu}/\text{cm}^{-1}$	19.95	39.91	59.86	79.82	$\cdots$

The line spacing is 19.95 cm^{-1}.

- -

Self-test 16.4 Repeat the problem for $C^{35}ClH_3$ (see Self-test 16.2 for details).

[Lines of separation 0.888 cm^{-1}]

The form of the spectrum predicted by eqn 43 is shown in Fig. 16.24. The most significant feature is that it consists of a series of lines with wavenumbers $2B, 4B, 6B, \ldots$ and of separation $2B$. The intensities increase with increasing J and pass through a maximum before tailing off as J becomes large. It should be recalled from Section 16.2 that the observed absorption is the net outcome of the stimulated absorption less the stimulated emission, and that the intensity of each transition depends on the value of J. Hence the value of J corresponding to the most intense line is not quite the same as the value of J for the most highly populated level. The value of J for the most highly populated rotational energy level in a linear molecule is

$$J_{\max} \approx \left(\frac{kT}{2hcB}\right)^{1/2} - \tfrac{1}{2} \tag{45}$$

For a typical molecule (for example, OCS, with $B = 0.2$ cm^{-1}) at room temperature, $kT \approx 1000hcB$, so $J_{\max} \approx 30$.

Justification 16.3

There is a maximum in population because the Boltzmann distribution decays exponentially with increasing J, but the degeneracy of the levels, the number of states with a given energy, increases. Specifically, the population of a rotational energy level J is given by the Boltzmann expression

$$N_J \propto Ng_J e^{-E_J/kT}$$

where N is the total number of molecules and g_J is the degeneracy of the level J. The value of J corresponding to a maximum of this expression is found by treating J as a continuous variable, differentiating with respect to J, and then setting the result equal to zero. The result is eqn 45.

The measurement of the line spacing gives B, and hence the moment of inertia perpendicular to the principal axis of the molecule. Because the masses of the atoms are known, it is a simple matter to deduce the bond length of a diatomic molecule. However, in the case of a polyatomic molecule such as OCS or NH_3, the analysis gives only a single quantity, $I_\perp$, and it is not possible to infer both bond lengths (in OCS) or the bond length and bond angle (in NH_3). This difficulty can be overcome by using isotopically substituted molecules, such as ABC and A'BC; then, by assuming that $R(A—B) = R(A'—B)$, both A–B and B–C bond lengths can be extracted from the two moments of inertia. A famous example of this procedure is the study of OCS; the actual calculation is worked through in Problem 16.12. The assumption that bond lengths are unchanged by isotopic substitution is only an approximation, but it is a good approximation in most cases.

16.7 Rotational Raman spectra

The gross selection rule for rotational Raman transitions is that *the molecule must be anisotropically polarizable*. We begin by explaining what this means.

The distortion of a molecule in an electric field is determined by its polarizability, α (Section 22.1c). More precisely, if the strength of the field is $\mathcal{E}$, then the molecule acquires an induced dipole moment of magnitude

$$\mu = \alpha\mathcal{E} \tag{46}$$

in addition to any permanent dipole moment it may have. An atom is isotropically polarizable. That is, the same distortion is induced whatever the direction of the applied field. The polarizability of a spherical rotor is also isotropic. However, nonspherical rotors have polarizabilities that do depend on the direction of the field relative to the molecule, so these molecules are anisotropically polarizable (Fig. 16.25). The electron distribution in H_2, for example, is more distorted when the field is applied parallel to the bond than when it is applied perpendicular to it, and we write $\alpha_\parallel > \alpha_\perp$.

All linear molecules and diatomics (whether homonuclear or heteronuclear) have anisotropic polarizabilities, and so are rotationally Raman active. This activity is one reason for the importance of rotational Raman spectroscopy, for the technique can be used to study many of the molecules that are inaccessible to microwave spectroscopy. Spherical rotors such as CH_4 and SF_6, however, are rotationally Raman inactive as well as microwave inactive.[7]

The specific rotational Raman selection rules are

$$\text{Linear rotors: } \Delta J = 0, \pm 2$$
$$\text{Symmetric rotors: } \Delta J = 0, \pm 1, \pm 2; \Delta K = 0 \tag{47}$$

The classical origin of the ± 2 in these selection rules is outlined in the *Justification* below. The $\Delta J = 0$ transitions do not lead to a shift of the scattered photon's frequency in pure rotational Raman spectroscopy, and contribute to the unshifted Rayleigh radiation observed in the forward direction.[8]

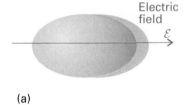

Electric field

$\mathcal{E}$

(a)

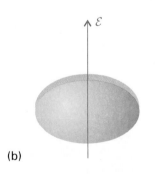

$\mathcal{E}$

(b)

16.25 An electric field applied to a molecule results in its distortion, and the distorted molecule acquires a contribution to its dipole moment (even if it is nonpolar initially). The polarizability may be different when the field is applied (a) parallel or (b) perpendicular to the molecular axis (or, in general, in different directions relative to the molecule); if that is so, then the molecule has an anisotropic polarizability.

---

7 This inactivity does not mean that such molecules are never found in rotationally excited states. Molecular collisions do not have to obey such restrictive selection rules, and hence collisions between molecules can result in the population of any rotational state.

8 See Section 23.5 for the information present in this component under different circumstances.

Justification 16.4

If the incident electric field is that of a light wave of frequency ω_i, the induced dipole moment of a molecule is

$$\mu = \alpha \mathcal{E}(t) = \alpha \mathcal{E} \cos \omega_i t$$

If the molecule is rotating at a circular frequency ω_R, to an external observer its polarizability is also time-dependent (if it is anisotropic), and we can write

$$\alpha = \alpha_0 + \Delta\alpha \cos 2\omega_R t$$

where $\Delta\alpha = \alpha_\parallel - \alpha_\perp$ and α ranges from $\alpha_0 + \Delta\alpha$ to $\alpha_0 - \Delta\alpha$ as the molecule rotates. The 2 appears because the polarizability returns to its initial value twice each revolution (Fig. 16.26). Substituting this expression into the expression for the induced dipole moment gives

$$\begin{aligned}
\mu &= (\alpha_0 + \Delta\alpha \cos 2\omega_R t) \times (\mathcal{E} \cos \omega_i t) \\
&= \alpha_0 \mathcal{E} \cos \omega_i t + \mathcal{E}\Delta\alpha \cos 2\omega_R t \cos \omega_i t \\
&= \alpha_0 \mathcal{E} \cos \omega_i t + \tfrac{1}{2}\mathcal{E}\Delta\alpha\{\cos(\omega_i + 2\omega_R)t + \cos(\omega_i - 2\omega_R)t\}
\end{aligned}$$

This calculation shows that the induced dipole has a component oscillating at the incident frequency (which generates Rayleigh radiation), and that it also has two components at $\omega_i \pm 2\omega_R$, which give rise to the shifted Raman lines. Note that these lines appear only if $\Delta\alpha \neq 0$; hence the polarizability must be anisotropic for there to be Raman lines.

The selection rules can also be explained in terms of the conservation of angular momentum, but the details are tricky because the incoming and scattered photons travel at right angles to each other. However, it should be clear that, because two photons are involved, and each one is a spin-1 particle, a maximum change in angular momentum quantum number of ± 2 is possible.

We can predict the form of the Raman spectrum of a linear rotor by applying the selection rule $\Delta J = \pm 2$ to the rotational energy levels (Fig. 16.27). When the molecule makes a transition with $\Delta J = +2$, the scattered radiation leaves it in a higher rotational state, so the wavenumber of the incident radiation, initially $\tilde{\nu}_i$, is decreased. These transitions account for the Stokes lines in the spectrum:

$$\tilde{\nu}(J+2 \leftarrow J) = \tilde{\nu}_i - \{F(J+2) - F(J)\} = \tilde{\nu}_i - 2B(2J+3) \tag{48a}$$

The Stokes lines appear to low frequency of the incident radiation and at displacements $6B, 10B, 14B, \ldots$ from $\tilde{\nu}_i$ for $J = 0, 1, 2, \ldots$. When the molecule makes a transition with $\Delta J = -2$, the scattered photon emerges with increased energy. These transitions account for the anti-Stokes lines of the spectrum:

$$\tilde{\nu}(J-2 \leftarrow J) = \tilde{\nu}_i + \{F(J) - F(J-2)\} = \tilde{\nu}_i + 2B(2J-1) \tag{48b}$$

The anti-Stokes lines occur at displacements of $6B, 10B, 14B, \ldots$ (for $J = 2, 3, 4, \ldots; J = 2$ is the lowest state that can contribute under the selection rule $\Delta J = -2$) to high frequency of the incident radiation. The separation of adjacent lines in both the Stokes and the anti-Stokes regions is $4B$, so from its measurement $I_\perp$ can be determined and then used to find the bond lengths exactly as in the case of microwave spectroscopy.

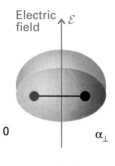

Electric field $\mathcal{E}$

0     $\alpha_\perp$

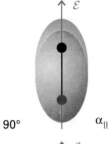

90° $\alpha_\parallel$

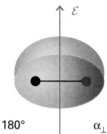

180° $\alpha_\perp$

270° $\alpha_\parallel$

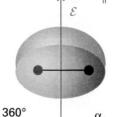

360° $\alpha_\perp$

16.26 The distortion induced in a molecule by an applied electric field returns to its initial value after a rotation of only 180° (that is, twice a revolution). This is the origin of the $\Delta J = \pm 2$ selection rule in rotational Raman spectroscopy.

Example 16.4 Predicting the form of a Raman spectrum

Predict the form of the rotational Raman spectrum of $^{14}N_2$, for which $B = 1.99$ cm^{-1}, when it is exposed to monochromatic 336.732 nm laser radiation.

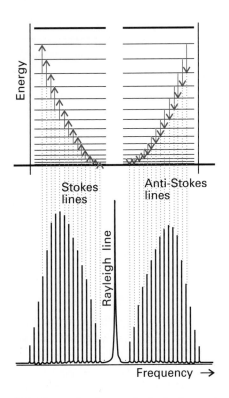

16.27 The rotational energy levels of a linear rotor and the transitions allowed by the $\Delta J = \pm 2$ Raman selection rules. The form of a typical rotational Raman spectrum is also shown.

Method The molecule is rotationally Raman active because end-over-end rotation modulates its polarizability as viewed by a stationary observer. The Stokes and anti-Stokes lines are given by eqn 48.

Answer Because $\lambda_i = 336.732$ nm corresponds to $\tilde{\nu}_i = 29\,697.2$ cm^{-1}, eqns 48a and 48b give the following line positions:

J	0	1	2	3
Stokes lines				
$\tilde{\nu}$/cm^{-1}	29 685.3	29 677.3	29 669.3	29 661.4
λ/nm	336.868	336.958	337.048	337.139
Anti-Stokes lines				
$\tilde{\nu}$/cm^{-1}			29 709.1	29 717.1
λ/nm			336.597	336.507

Comment There will be a strong central line at 336.732 nm accompanied on either side by lines of increasing and then decreasing intensity (as a result of transition moment and population effects). The spread of the entire spectrum is very small, so the incident light must be highly monochromatic.

- -

Self-test 16.5 Repeat the calculation for the rotational Raman spectrum of NH$_3$ ($B = 9.977$ cm^{-1}).

[Stokes lines at 29 637.3, 29 597.4, 29 557.5, 29 517.6 cm^{-1},
anti-Stokes lines at 29 757.1, 29 797.0 cm^{-1}]

16.8 Nuclear statistics and rotational states

If eqn 48 is used in conjunction with the rotational Raman spectrum of CO$_2$, the rotational constant is inconsistent with other measurements of C–O bond lengths. The results are consistent only if it is supposed that the molecule can exist in states with even values of J, so the Stokes lines are 2←0, 4←2, etc. and not 5←3, 3←1, etc.

The explanation of the missing lines is the Pauli principle and the fact that O nuclei are spin-0 bosons: just as the Pauli principle excludes certain electronic states, so too does it exclude certain molecular rotational states. The form of the Pauli principle given in *Justification 13.7* states that, when two identical bosons are exchanged, the overall wavefunction must remain unchanged in every respect, including sign. In particular, when a CO$_2$ molecule rotates through 180°, two identical O nuclei are interchanged, so the overall wavefunction of the molecule must remain unchanged. However, inspection of the form of the rotational wavefunctions (which have the same form as the s, p, etc. orbitals of atoms) shows that they change sign by $(-1)^J$ under such a rotation (Fig. 16.28). Therefore, only even values of J are permissible for CO$_2$, and hence the Raman spectrum shows only alternate lines.

The selective occupation of rotational states that stems from the Pauli principle is termed **nuclear statistics**. Nuclear statistics must be taken into account whenever a rotation interchanges equivalent nuclei. However, the consequences are not always as simple as for CO$_2$ because there are complicating features when the nuclei have nonzero spin: there may be several different relative nuclear spin orientations consistent with even values of J and a different number of spin orientations consistent with odd values of J. For molecular hydrogen and fluorine, for instance, with their two identical spin-$\frac{1}{2}$ nuclei, we show in the

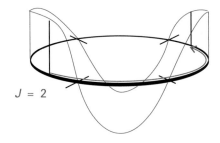

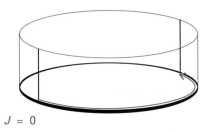

16.28 The symmetries of rotational wavefunctions (shown here, for simplicity as a two-dimensional rotor) under a rotation through 180°. Wavefunctions with J even do not change sign; those with J odd do change sign.

Justification below that there are three times as many ways of achieving a state with odd J than with even J, and there is a corresponding 3 : 1 alternation in intensity in their rotational Raman spectra (Fig. 16.29). In general, for a homonuclear diatomic molecule with nuclei of spin I, the numbers of ways of achieving states of odd and even J are in the ratio

$$\frac{\text{Number of ways of achieving odd } J}{\text{Number of ways of achieving even } J} = \begin{cases} (I+1)/I & \text{for half-integral spin nuclei} \\ I/(I+1) & \text{for integral spin nuclei} \end{cases}$$

(49)

For hydrogen, $I = \frac{1}{2}$, and the ratio is 3 : 1. For N_2, with $I = 1$, the ratio is 1 : 2.

Justification 16.5

Hydrogen nuclei are fermions, so the Pauli principle requires the overall wavefunction to change sign under particle interchange. However, the rotation of an H_2 molecule through 180° has a more complicated effect than merely relabelling the nuclei, because it interchanges their spin states too if the nuclear spins are paired ($\uparrow\downarrow$) but not if they are parallel ($\uparrow\uparrow$).

For the overall wavefunction of the molecule to change sign when the spins are parallel, the rotational wavefunction must change sign. Hence, only odd values of J are allowed. In contrast, if the nuclear spins are paired, their wavefunction is $\alpha(A)\beta(B) - \alpha(B)\beta(A)$, which changes sign when α and β are exchanged in order to bring about a simple $A \leftrightarrow B$ interchange overall (Fig. 16.30). Therefore, for the overall wavefunction to change sign in this case requires the rotational wavefunction *not* to change sign. Hence, only even values of J are allowed if the nuclear spins are paired.

As there are three nuclear spin states with parallel spins (just like the triplet state of two parallel electrons, as in Fig. 13.26), but only one state with paired spins (the analogue of the singlet state of two electrons, see Fig. 13.20), it follows that the populations of the odd J and even J states should be in the ratio of 3 : 1, and hence the intensities of transitions originating in these levels will be in the same ratio.

Different relative nuclear spin orientations change into one another only very slowly, so an H_2 molecule with parallel nuclear spins remains distinct from one with paired nuclear spins for long periods. The two forms of hydrogen can be separated by physical techniques, and stored. The form with parallel nuclear spins is called ***ortho*-hydrogen** and the form with paired nuclear spins is called ***para*-hydrogen**. Because *ortho*-hydrogen cannot exist in a state with $J = 0$, it continues to rotate at very low temperatures and has an effective rotational zero-point energy (Fig. 16.31). This energy is of some concern to manufacturers of liquid hydrogen, for the slow conversion of *ortho*-hydrogen into *para*-hydrogen (which can exist with $J = 0$) as nuclear spins slowly realign releases rotational energy, which vaporizes the liquid. Techniques are used to accelerate the conversion of *ortho*-hydrogen to *para*-hydrogen to avoid this problem. One such technique is to pass hydrogen over a metal surface: the molecules adsorb on the surface as atoms, which then recombine in the lower energy *para*-hydrogen form.

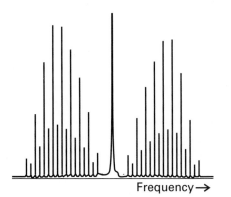

16.29 The rotational Raman spectrum of a diatomic molecule with two identical spin-$\frac{1}{2}$ nuclei shows an alternation in intensity as a result of nuclear statistics.

The vibrations of diatomic molecules

In this section, we adopt the same strategy of finding expressions for the energy levels, establishing the selection rules, and then discussing the form of the spectrum. We shall also see how the simultaneous excitation of rotation modifies the appearance of a vibrational spectrum.

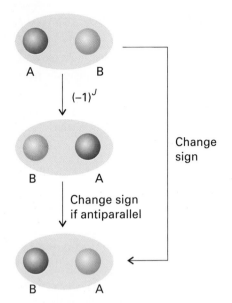

16.30 The interchange of two identical fermion nuclei results in the change in sign of the overall wavefunction. The relabelling can be thought of as occurring in two steps: the first is a rotation of the molecule; the second is the interchange of unlike spins (represented by the different colours of the nuclei). The wavefunction changes sign in the second step if the nuclei have antiparallel spins.

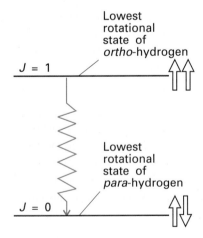

16.31 When hydrogen is cooled, the molecules with parallel nuclear spins accumulate in their lowest available rotational state, the one with $J = 0$. They can enter the lowest rotational state ($J = 0$) only if the spins change their relative orientation and become antiparallel. This is a slow process under normal circumstances, so energy is slowly released.

16.9 Molecular vibrations

We base our discussion on Fig. 16.32, which shows a typical potential energy curve (as in Fig. 14.1) of a diatomic molecule. In regions close to R_e (at the minimum of the curve) the potential energy can be approximated by a parabola, so we can write

$$V = \tfrac{1}{2}kx^2 \qquad x = R - R_e \tag{50}$$

where k is the **force constant** of the bond. The steeper the walls of the potential (the stiffer the bond), the greater the force constant.

To see the connection between the shape of the molecular potential energy curve and the value of k, note that we can expand the potential energy around its minimum by using a Taylor expansion:

$$V(x) = V(0) + \left(\frac{dV}{dx}\right)_0 x + \tfrac{1}{2}\left(\frac{d^2V}{dx^2}\right)_0 x^2 + \cdots \tag{51}$$

The term $V(0)$ can be set arbitrarily to zero. The first derivative of V is 0 at the minimum. Therefore, the first surviving term is proportional to the square of the displacement. For small displacements we can ignore all the higher terms, and so write

$$V(x) \approx \tfrac{1}{2}\left(\frac{d^2V}{dx^2}\right)_0 x^2 \tag{52}$$

Therefore, the first approximation to a molecular potential energy curve is a parabolic potential, and we can identify the force constant as

$$k = \left(\frac{d^2V}{dx^2}\right)_0 \tag{53}$$

We see that, if the potential energy curve is sharply curved close to its minimum, then k will be large. Conversely, if the potential energy curve is wide and shallow, then k will be small (Fig. 16.33).

The Schrödinger equation for the relative motion of two atoms of masses m_1 and m_2 with a parabolic potential energy is

$$-\frac{\hbar^2}{2m_{\text{eff}}}\frac{d^2\psi}{dx^2} + \tfrac{1}{2}kx^2\psi = E\psi \tag{54}$$

where m_{eff} is the **effective mass**:

$$m_{\text{eff}} = \frac{m_1 m_2}{m_1 + m_2} \tag{55}$$

These equations are derived in the same way as in *Justification 13.1*, but here the separation of variables procedure is used to separate the relative motion of the atoms from the motion of the molecule as a whole.[9]

The Schrödinger equation in eqn 54 is the same as eqn 12.30 for a particle of mass m_{eff} undergoing harmonic motion. Therefore, we can use the results of Section 12.4 to write down the permitted vibrational energy levels:

$$E_v = (v + \tfrac{1}{2})\hbar\omega \qquad \omega = \left(\frac{k}{m_{\text{eff}}}\right)^{1/2} \qquad v = 0, 1, 2, \ldots \tag{56}$$

9   In that context, the effective mass is called the 'reduced mass', and the name is widely used in this context too.

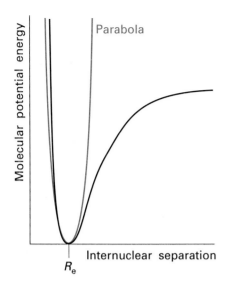

16.32 A molecular potential energy curve can be approximated by a parabola near the bottom of the well. The parabolic potential leads to harmonic oscillations. At high excitation energies the parabolic approximation is poor (the true potential is less confining), and is totally wrong near the dissociation limit.

The **vibrational terms** of a molecule, the energies of its vibrational states expressed in wavenumbers, are denoted $G(v)$, with $E_v = hcG(v)$, so

$$G(v) = (v + \tfrac{1}{2})\tilde{\nu} \qquad \tilde{\nu} = \frac{1}{2\pi c}\left(\frac{k}{m_{\text{eff}}}\right)^{1/2} \tag{57}$$

The vibrational wavefunctions are the same as those discussed in Section 12.5.

It is important to note that the vibrational terms depend on the effective mass of the molecule, not directly on its total mass. This dependence is physically reasonable for, if atom 1 were as heavy as a brick wall, then we would find $m_{\text{eff}} \approx m_2$, the mass of the lighter atom. The vibration would then be that of a light atom relative to that of a stationary wall (this is approximately the case in HI, for example, where the I atom barely moves and $m_{\text{eff}} \approx m_{\text{H}}$). For a homonuclear diatomic molecule $m_1 = m_2$, and the effective mass is half the total mass: $m_{\text{eff}} = \tfrac{1}{2}m$.

Illustration

An HCl molecule has a force constant of 516 N m^{-1}, a reasonably typical value. The effective mass of $^1\text{H}^{35}\text{Cl}$ is 1.63×10^{-27} kg (note that this mass is very close to the mass of the hydrogen atom, 1.67×10^{-27} kg, so the Cl atom is acting like a brick wall). These values imply $\omega = 5.63 \times 10^{14}$ s^{-1}, $\nu = 89.5$ THz (1 THz = 10^{12} Hz), $\tilde{\nu} = 2990$ cm^{-1}, $\lambda = 3.35$ μm. These characteristics correspond to electromagnetic radiation in the infrared region.

16.10 Selection rules

The gross selection rule for a molecular vibration is that *the electric dipole moment of the molecule must change when the atoms are displaced relative to one another*. Such vibrations are said to be **infrared active**. The classical basis of this rule is that the molecule can shake the electromagnetic field into oscillation if its dipole changes as it vibrates, and vice versa (Fig. 16.34); its formal basis is given in the *Justification* below. Note that the molecule need not have a permanent dipole: the rule requires only a change in dipole moment, possibly from zero. Some vibrations do not affect the molecule's dipole moment (for example, the stretching motion of a homonuclear diatomic molecule), so they neither absorb nor generate radiation: such vibrations are said to be **infrared inactive**. Homonuclear diatomic molecules are infrared inactive because their dipole moments remain zero however long the bond; heteronuclear diatomic molecules are infrared active.

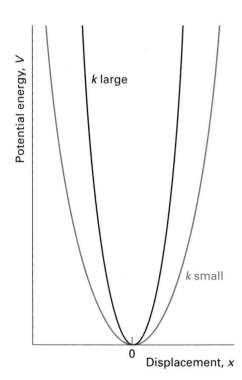

16.33 The force constant is a measure of the curvature of the potential energy close to the equilibrium extension of the bond. A strongly confining well (one with steep sides, a stiff bond) corresponds to high values of k.

Justification 16.6

The gross selection rule is based on an analysis of the transition dipole moment $\langle v_f|\boldsymbol{\mu}|v_i\rangle$. For simplicity, we shall consider a one-dimensional oscillator (like a diatomic molecule). The electric dipole moment operator depends on the location of all the electrons and all the nuclei in the molecule, so it varies as the internuclear separation changes (Fig. 16.35). If we think of the dipole moment as arising from two partial charges $\pm\delta q$ separated by a distance $R = R_e + x$, we can write its variation with displacement from the equilibrium separation, x, as

$$\mu = R\delta q = R_e\delta q + x\delta q = \mu_0 + x\delta q$$

where μ_0 is the electric dipole moment operator when the nuclei have their equilibrium separation. It then follows that, with f $\neq$ i,

$$\langle v_f|\mu|v_i\rangle = \mu_0\langle v_f|v_i\rangle + \delta q\langle v_f|x|v_i\rangle$$

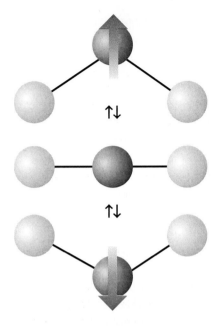

16.34 The oscillation of a molecule, even if it is nonpolar, may result in an oscillating dipole that can interact with the electromagnetic field.

The term proportional to μ_0 is zero because the states with different values of v are orthogonal. It follows that the transition dipole moment is

$$\langle v_f|\mu|v_i\rangle = \langle v_f|x|v_i\rangle \delta q$$

Because

$$\delta q = \frac{d\mu}{dx}$$

we can write the transition dipole moment more generally as

$$\langle v_f|\mu|v_i\rangle = \langle v_f|x|v_i\rangle \left(\frac{d\mu}{dx}\right) \tag{58}$$

and we see that the right-hand side is zero unless the dipole moment varies with displacement. We consider the matrix element of x in the next *Justification*.

Illustration

Of the molecules N_2, CO_2, OCS, H_2O, $CH_2{=}CH_2$, and C_6H_6, all except N_2 possess at least one vibrational mode that results in a change of dipole moment, so all except N_2 can show a vibrational absorption spectrum. Not all the modes of complex molecules are vibrationally active. For example, the symmetric stretch of CO_2, in which the O–C–O bonds stretch and contract symmetrically, is inactive because it leaves the dipole moment unchanged (at zero).

Self-test 16.6 Which of the molecules H_2, NO, N_2O, and CH_4 have infrared active vibrations?

[NO, N_2O, CH_4]

The specific vibrational selection rule, which is obtained from an analysis of the expression for the transition moment and the properties of integrals over harmonic oscillator wavefunctions (as shown in the *Justification* below), is

$$\Delta v = \pm 1 \tag{59}$$

Transitions for which $\Delta v = +1$ correspond to absorption and those with $\Delta v = -1$ correspond to emission.

Justification 16.7

The specific selection rule is determined by considering the value of the matrix element of x in eqn 58. We need to write out the wavefunctions in terms of the Hermite polynomials given in Section 12.5 and then to use their properties (Example 12.4 should be reviewed, for it gives further details of the calculation). We note that $x = \alpha y$ with $\alpha = (\hbar^2/m_{eff}k)^{1/4}$ (eqn 12.34), and write

$$\langle v_f|x|v_i\rangle = N_{v_f}N_{v_i}\int_{-\infty}^{\infty} H_{v_f}xH_{v_i}e^{-y^2}\,dx$$

$$= \alpha^2 N_{v_f}N_{v_i}\int_{-\infty}^{\infty} H_{v_f}yH_{v_i}e^{-y^2}\,dy$$

To evaluate the integral we use the recursion relation in Table 12.1:

$$yH_v = vH_{v-1} + \tfrac{1}{2}H_{v+1}$$

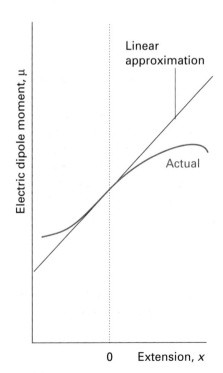

16.35 The electric dipole moment of a heteronuclear diatomic molecule varies as shown by the green curve. For small displacements the change in dipole moment is proportional to the displacement.

Linear approximation

Actual

Electric dipole moment, μ

0   Extension, x

This relation turns the matrix element into

$$\langle v_f | x | v_i \rangle$$
$$= \alpha^2 N_{v_f} N_{v_i} \left\{ v_i \int_{-\infty}^{\infty} H_{v_f} H_{v_i-1} e^{-y^2} \, dy + \tfrac{1}{2} \int_{-\infty}^{\infty} H_{v_f} H_{v_i+1} e^{-y^2} \, dy \right\}$$

We see from Table 12.1 that the first integral is zero unless $v_f = v_i - 1$ and that the second is zero unless $v_f = v_i + 1$. It follows that the transition dipole moment is zero unless $\Delta v = \pm 1$.

It follows from the specific selection rules that the wavenumbers of allowed vibrational transitions, which are denoted $\Delta G_{v+\frac{1}{2}}$ for the transition $v + 1 \leftarrow v$, are

$$\Delta G_{v+\frac{1}{2}} = G(v + 1) - G(v) = \tilde{\nu} \tag{60}$$

As we have seen, $\tilde{\nu}$ lies in the infrared region of the electromagnetic spectrum, so vibrational transitions absorb and generate infrared radiation.

At room temperature $kT/hc \approx 200$ cm^{-1}, and most vibrational wavenumbers are significantly greater than 200 cm^{-1}. It follows from the Boltzmann distribution that almost all the molecules will be in their vibrational ground states initially. Hence, the dominant spectral transition will be the **fundamental transition**, $1 \leftarrow 0$. As a result, the spectrum is expected to consist of a single absorption line. If the molecules are formed in a vibrationally excited state, such as when vibrationally excited HF molecules are formed in the reaction $H_2 + F_2 \rightarrow 2HF^*$, the transitions $5 \rightarrow 4$, $4 \rightarrow 3$, etc. may also appear (in emission). In the harmonic approximation, all these lines lie at the same frequency, and the spectrum is also a single line. However, as we shall now show, the breakdown of the harmonic approximation causes the transitions to lie at slightly different frequencies, so several lines are observed.

16.11 Anharmonicity

The vibrational terms in eqn 60 are only approximate because they are based on a parabolic approximation to the actual potential energy curve. A parabola cannot be correct at all extensions because it does not allow a bond to dissociate. At high vibrational excitations the swing of the atoms (more precisely, the spread of the vibrational wavefunction) allows the molecule to explore regions of the potential energy curve where the parabolic approximation is poor and additional terms in the Taylor expansion of V (eqn 51) must be retained. The motion then becomes **anharmonic**, in the sense that the restoring force is no longer proportional to the displacement. Because the actual curve is less confining than a parabola, we can anticipate that the energy levels become less widely spaced at high excitations.

(a) The convergence of energy levels

One approach to the calculation of the energy levels in the presence of anharmonicity is to use a function that resembles the true potential energy more closely. The **Morse potential energy** is

$$V = hcD_e \left\{ 1 - e^{-a(R-R_e)} \right\}^2 \qquad a = \left(\frac{m_{eff} \omega^2}{2hcD_e} \right)^{1/2} \tag{61}$$

where D_e is the depth of the potential minimum (Fig. 16.36). Near the well minimum the variation of V with displacement resembles a parabola (as can be checked by expanding the exponential as far as the first term) but, unlike a parabola, eqn 61 allows for dissociation at

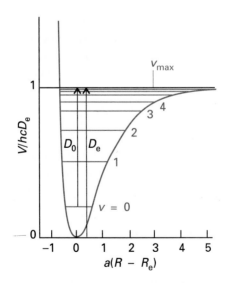

16.36 The Morse potential energy curve reproduces the general shape of a molecular potential energy curve. The corresponding Schrödinger equation can be solved, and the values of the energies obtained. The number of bound levels is finite. The illustration also shows the relation between the dissociation energy, D_0, and the minimum energy, D_e, of a molecular potential energy curve.

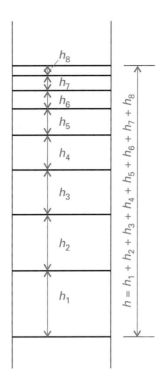

16.37 The dissociation energy is the sum of the separations of the vibrational energy levels up to the dissociation limit just as the length of a ladder is the sum of the separations of its rungs.

large displacements. The Schrödinger equation can be solved for the Morse potential and the permitted energy levels are

$$G(v) = (v + \tfrac{1}{2})\tilde{\nu} - (v + \tfrac{1}{2})^2 x_e \tilde{\nu} \qquad x_e = \frac{a^2 \hbar}{2\mu\omega} = \frac{\tilde{\nu}}{4D_e} \qquad (62)$$

The parameter x_e is called the **anharmonicity constant**. The number of vibrational levels of a Morse oscillator is finite, and $v = 0, 1, 2, \ldots, v_{max}$, as shown in Fig. 16.36. The second term in the expression for G subtracts from the first with increasing effect as v increases, and hence gives rise to the convergence of the levels at high quantum numbers.

Although the Morse oscillator is quite useful theoretically, in practice the more general expression

$$G(v) = (v + \tfrac{1}{2})\tilde{\nu} - (v + \tfrac{1}{2})^2 x_e \tilde{\nu} + (v + \tfrac{1}{2})^3 y_e \tilde{\nu} + \cdots \qquad (63)$$

where $x_e, y_e, \ldots$ are empirical constants characteristic of the molecule, is used to fit the experimental data and to find the dissociation energy of the molecule. When anharmonicities are present, the wavenumbers of transitions with $\Delta v = +1$ are

$$\Delta G_{v+\frac{1}{2}} = \tilde{\nu} - 2(v + 1)x_e \tilde{\nu} + \cdots \qquad (64)$$

The latter equation shows that when $x_e \neq 0$ the transitions move to lower wavenumbers as v increases.

Anharmonicity also accounts for the appearance of additional weak absorption lines corresponding to the transitions $2 \leftarrow 0$, $3 \leftarrow 0$, etc., even though these first, second,... **overtones** are forbidden by the selection rule $\Delta v = \pm 1$. The first overtone, for example, gives rise to an absorption at

$$G(v + 2) - G(v) = 2\tilde{\nu} - 2(2v + 3)x_e \tilde{\nu} + \cdots \qquad (65)$$

The reason for the appearance of overtones is that the selection rule is derived from the properties of harmonic oscillator wavefunctions, which are only approximately valid when anharmonicity is present. Therefore, the selection rule is also only an approximation. For an anharmonic oscillator, all values of Δv are allowed, but transitions with $\Delta v > 1$ are allowed only weakly if the anharmonicity is slight.

(b) The Birge–Sponer plot

When several vibrational transitions are detectable, a graphical technique called a **Birge–Sponer plot** may be used to determine the dissociation energy, D_0, of the bond. The basis of the Birge–Sponer plot is that the sum of successive intervals $\Delta G_{v+\frac{1}{2}}$ from the zero-point level to the dissociation limit is the dissociation energy:

$$D_0 = \Delta G_{1/2} + \Delta G_{3/2} + \cdots = \sum_v \Delta G_{v+\frac{1}{2}} \qquad (66)$$

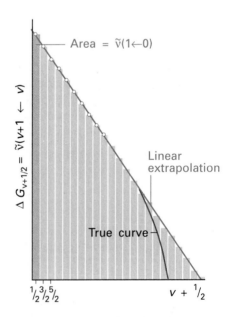

16.38 The area under a plot of transition wavenumber against vibrational quantum number is equal to the dissociation energy of the molecule. The assumption that the differences approach zero linearly is the basis of the Birge–Sponer extrapolation.

just as the height of the ladder is the sum of the separations of its rungs (Fig. 16.37). The construction in Fig. 16.38 shows that the area under the plot of $\Delta G_{v+\frac{1}{2}}$ against $v + \frac{1}{2}$ is equal to the sum, and therefore to D_0. The successive terms decrease linearly when only the x_e anharmonicity constant is taken into account and the inaccessible part of the spectrum can be estimated by linear extrapolation. Most actual plots differ from the linear plot as shown in the illustration, so the value of D_0 obtained in this way is usually an overestimate of the true value.

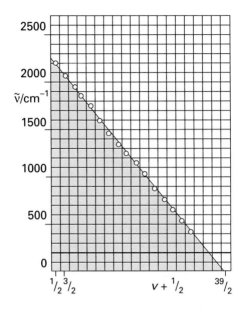

16.39 The Birge–Sponer plot used in Example 16.5. The area is obtained simply by counting the squares beneath the line or using the formula for the area of a right triangle.

Example 16.5 Using a Birge–Sponer plot

The observed vibrational intervals of H_2^+ lie at the following values for $1 \leftarrow 0$, $2 \leftarrow 1, \ldots$ respectively (in cm^{-1}): 2191, 2064, 1941, 1821, 1705, 1591, 1479, 1368, 1257, 1145, 1033, 918, 800, 677, 548, 411. Determine the dissociation energy of the molecule.

Method Plot the separations against $v + \tfrac{1}{2}$, extrapolate linearly to the point cutting the horizontal axis, and then measure the area under the curve.

Answer The points are plotted in Fig. 16.39, and a linear extrapolation is shown as a dotted line. The area under the curve (use the formula for the area of a triangle or count the squares) is 214. Each square corresponds to 100 cm^{-1} (refer to the scale of the vertical axis); hence the dissociation energy is 21 400 cm^{-1} (corresponding to 256 kJ mol^{-1}).

- -

Self-test 16.7 The vibrational levels of HgH converge rapidly, and successive intervals are 1203.7, 965.6, 632.4, and 172 cm^{-1}. Estimate the dissociation energy.

[35.6 kJ mol^{-1}]

16.12 Vibration–rotation spectra

Each line of the high-resolution vibrational spectrum of a gas-phase heteronuclear diatomic molecule is found to consist of a large number of closely spaced components (Fig. 16.40). Hence, molecular spectra are often called **band spectra**. The separation between the components is of the order of 10 cm^{-1}, which suggests that the structure is due to rotational transitions accompanying the vibrational transition. A rotational change should be expected because classically we can think of the transition as leading to a sudden increase or decrease in the instantaneous bond length. Just as ice-skaters rotate more rapidly when they bring their arms in, and more slowly when they throw them out, so the molecular rotation is either accelerated or retarded by a vibrational transition.

(a) Spectral branches

A detailed analysis of the quantum mechanics of simultaneous vibrational and rotational changes shows that the rotational quantum number J changes by ± 1 during the vibrational transition of a diatomic molecule. If the molecule also possesses angular momentum about its axis, as in the case of the electronic orbital angular momentum of the $^2\Pi$ molecule NO, then the selection rules also allow $\Delta J = 0$.

The appearance of the vibration–rotation spectrum of a diatomic molecule can be discussed in terms of the combined vibration–rotation terms, S:

$$S(v, J) = G(v) + F(J) \tag{67}$$

16.40 A high-resolution vibration–rotation spectrum of HCl. The lines appear in pairs because $H^{35}Cl$ and $H^{37}Cl$ both contribute (their abundance ratio is 3 : 1). There is no Q branch, because $\Delta J = 0$ is forbidden for this molecule.

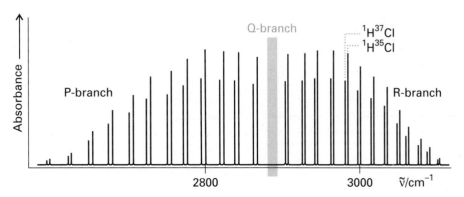

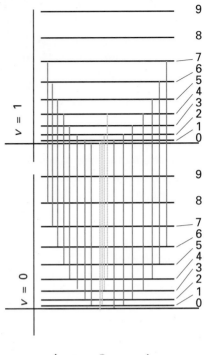

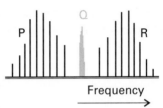

16.41 The formation of P, Q, and R branches in a vibration–rotation spectrum. The intensities reflect the populations of the initial rotational levels.

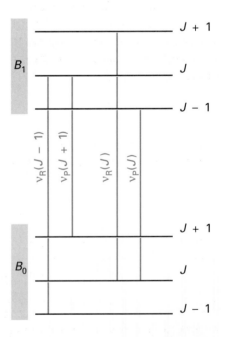

16.42 The method of combination differences makes use of the fact that some transitions share a common level.

If we ignore anharmonicity and centrifugal distortion,

$$S(v, J) = (v + \tfrac{1}{2})\tilde{v} + BJ(J + 1) \tag{68}$$

In a more detailed treatment, B is allowed to depend on the vibrational state because as v increases the molecule swells slightly and the moment of inertia changes. We shall continue with the simple expression initially.

When the vibrational transition $v + 1 \leftarrow v$ occurs, J changes by ± 1 and in some cases by 0 (when $\Delta J = 0$ is allowed). The absorptions then fall into three groups called **branches** of the spectrum. The **P branch** consists of all transitions with $\Delta J = -1$:

$$\tilde{v}_{\mathrm{P}}(J) = S(v + 1, J - 1) - S(v, J) = \tilde{v} - 2BJ \tag{69a}$$

This branch consists of lines at $\tilde{v} - 2B, \tilde{v} - 4B, \ldots$ with an intensity distribution reflecting both the populations of the rotational levels and the magnitude of the $J - 1 \leftarrow J$ transition moment (Fig. 16.41). The **Q branch** consists of all lines with $\Delta J = 0$, and its wavenumbers are all

$$\tilde{v}_{\mathrm{Q}}(J) = S(v + 1, J) - S(v, J) = \tilde{v} \tag{69b}$$

for all values of J. This branch, when it is allowed (as in NO), forms a single line at the vibrational transition wavenumber. In practice, because the rotational constants of the two vibrational levels are slightly different, the Q branch appears as a cluster of closely spaced lines. In Fig. 16.41 there is a gap at the expected location of the Q branch because it is forbidden in HCl. The **R branch** consists of lines with $\Delta J = +1$:

$$\tilde{v}_{\mathrm{R}}(J) = S(v + 1, J + 1) - S(v, J) = \tilde{v} + 2B(J + 1) \tag{69c}$$

This branch consists of lines displaced from $\tilde{v}$ to high wavenumber by $2B, 4B, \ldots$.

The separation between the lines in the P and R branches of a vibrational transition gives the value of B. Therefore, the bond length can be deduced without needing to take a pure rotational microwave spectrum. However, the latter is more precise.

(b) Combination differences

The rotational constant of the vibrationally excited state, B_1 (in general, B_v), is in fact slightly smaller than that of the ground vibrational state, B_0, because the anharmonicity of the vibration results in a slightly extended bond in the upper state. As a result, the Q branch (if it exists) consists of a series of closely spaced lines, the lines of the R branch converge slightly as J increases, and those of the P branch diverge:

$$\begin{aligned}
\tilde{v}_{\mathrm{P}}(J) &= \tilde{v} - (B_1 + B_0)J + (B_1 - B_0)J^2 \\
\tilde{v}_{\mathrm{Q}}(J) &= \tilde{v} + (B_1 - B_0)J(J + 1) \\
\tilde{v}_{\mathrm{R}}(J) &= \tilde{v} + (B_1 + B_0)(J + 1) + (B_1 - B_0)(J + 1)^2
\end{aligned} \tag{70}$$

To determine the two rotational constants individually, we use the method of **combination differences**. This procedure is used widely in spectroscopy to extract information about a particular state. It involves setting up expressions for the difference in the wavenumbers of transitions to a common state; the resulting expression then depends solely on properties of the other state.

As can be seen from Fig. 16.42, the transitions $\tilde{v}_{\mathrm{R}}(J - 1)$ and $\tilde{v}_{\mathrm{P}}(J + 1)$ have a common upper state, and hence can be anticipated to depend on B_1. Indeed, it is easy to show from eqn 70 that

$$\tilde{v}_{\mathrm{R}}(J - 1) - \tilde{v}_{\mathrm{P}}(J + 1) = 4B_0(J + \tfrac{1}{2}) \tag{71a}$$

Therefore, a plot of the combination difference against $J + \tfrac{1}{2}$ should be a straight line of slope $4B_0$, so the rotational constant of the molecule in the state $v = 0$ can be determined.

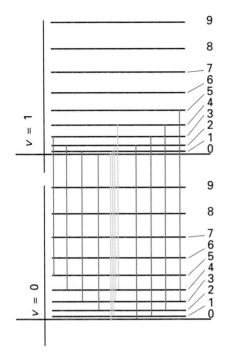

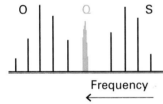

16.43 The formation of O, Q, and S branches in a vibration–rotation Raman spectrum of a linear rotor. Note that the frequency scale runs in the opposite direction to that in Fig. 16.41, because the higher energy transitions (on the right) extract more energy from the incident beam and leave it at lower frequency.

(Any deviation from a straight line is a consequence of centrifugal distortion, so that effect can be investigated too.) Similarly, $\tilde{\nu}_R(J)$ and $\tilde{\nu}_P(J)$ have a common lower state, and hence their combination difference gives information about the upper state:

$$\tilde{\nu}_R(J) - \tilde{\nu}_P(J) = 4B_1(J + \tfrac{1}{2}) \tag{71b}$$

The two rotational constants of $^1H^{35}Cl$ found in this way are $B_0 = 10.440$ cm^{-1} and $B_1 = 10.136$ cm^{-1}.

16.13 Vibrational Raman spectra of diatomic molecules

The gross selection rule for vibrational Raman transitions is that *the polarizability should change as the molecule vibrates*. As homonuclear and heteronuclear diatomic molecules swell and contract during a vibration, the control of the nuclei over the electrons varies, and hence the molecular polarizability changes. Both types of diatomic molecule are therefore vibrationally Raman active.

The specific selection rule for vibrational Raman transitions in the harmonic approximation is $\Delta v = \pm 1$. The lines to high frequency of the incident light, the anti-Stokes lines, are those for which $\Delta v = -1$. They are usually weak because very few molecules are in an excited vibrational state initially. The lines to low frequency, the Stokes lines, correspond to $\Delta v = +1$. In gas-phase spectra, these lines have a branch structure arising from the simultaneous rotational transitions that accompany the vibrational excitation (Fig. 16.43). The selection rules are $\Delta J = 0, \pm 2$ (as in pure rotational Raman spectroscopy), and give rise to the **O branch** ($\Delta J = -2$), the **Q branch** ($\Delta J = 0$), and the **S branch** ($\Delta J = +2$):

$$\begin{aligned} \tilde{\nu}_O(J) &= \tilde{\nu}_i - \tilde{\nu} - 2B + 4BJ \\ \tilde{\nu}_Q(J) &= \tilde{\nu}_i - \tilde{\nu} \\ \tilde{\nu}_S(J) &= \tilde{\nu}_i - \tilde{\nu} - 6B - 4BJ \end{aligned} \tag{72}$$

Note that, unlike in infrared spectroscopy, a Q branch is obtained for all linear molecules. The spectrum of CO, for instance, is shown in Fig. 16.44: the structure of the Q branch arises from the differences in rotational constants of the upper and lower vibrational states.

The information available from vibrational Raman spectra adds to that from infrared spectroscopy because homonuclear diatomics can also be studied. The spectra can be interpreted in terms of the force constants, dissociation energies, and bond lengths, and some of the information obtained is included in Table 16.2.

The vibrations of polyatomic molecules

There is only one mode of vibration for a diatomic molecule, the bond stretch. In polyatomic molecules there are several modes of vibration because all the bond lengths and angles may change.

16.14 Normal modes

We begin by calculating the total number of vibrational modes of a polyatomic molecule. We then see that we can choose combinations of these atomic displacements that give the simplest description of the vibrations.

(a) The number of vibrational modes

As shown in the *Justification* below, for a nonlinear molecule that consists of N atoms, there are $3N - 6$ independent modes of vibration. If the molecule is linear, there are $3N - 5$ independent vibrational modes.

Table 16.2* Properties of diatomic molecules

	$\tilde{\nu}/$cm^{-1}	$B/$cm^{-1}	$k/$(N m^{-1})
$^1H_2$	4400	60.86	575
$^1H^{35}Cl$	2991	10.59	516
$^1H^{127}I$	2309	6.61	313
$^{35}Cl_2$	560	0.244	323

*More values are given in the *Data section* at the end of this volume. See Tables 14.2 and 14.3 for related information (on D_e and R_e).

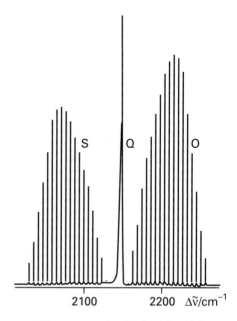

16.44 The structure of a vibrational line in the vibrational Raman spectrum of carbon monoxide, showing the O, Q, and S branches.

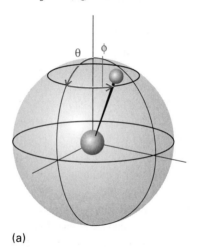

(a)

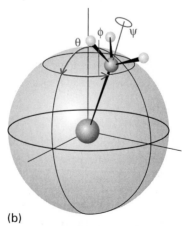

(b)

16.45 (a) The orientation of a linear molecule requires the specification of two angles. (b) The orientation of a nonlinear molecule requires the specification of three angles.

Justification 16.8

The total number of coordinates needed to specify the locations of N atoms is $3N$. Each atom may change its location by varying one of its three coordinates (x, y, and z), so the total number of displacements available is $3N$. These displacements can be grouped together in a physically sensible way. For example, three coordinates are needed to specify the location of the centre of mass of the molecule, so three of the $3N$ displacements correspond to the translational motion of the molecule as a whole. The remaining $3N - 3$ are non-translational 'internal' modes of the molecule.

Two angles are needed to specify the orientation of a linear molecule in space: in effect, we need to give only the latitude and longitude of the direction in which the molecular axis is pointing (Fig. 16.45a). However, three angles are needed for a nonlinear molecule because we also need to specify the orientation of the molecule around the direction defined by the latitude and longitude (Fig. 16.45b). Therefore, two (linear) or three (nonlinear) of the $3N - 3$ internal displacements are rotational. This leaves $3N - 5$ (linear) or $3N - 6$ (nonlinear) displacements of the atoms relative to one another: these are the vibrational modes. It follows that the number of modes of vibration N_{vib} is $3N - 5$ for linear molecules and $3N - 6$ for nonlinear molecules.

Illustration

Water, H_2O, is a nonlinear triatomic molecule, and has three modes of vibration (and three modes of rotation); CO_2 is a linear triatomic molecule, and has four modes of vibration (and only two modes of rotation). Even a middle-sized molecule such as naphthalene ($C_{10}H_8$) has 48 distinct modes of vibration.

(b) Combinations of displacements

The next step is to find the best description of the modes. One choice for the four modes of CO_2, for example, might be the ones in Fig. 16.46a. This illustration shows the stretching of one bond (the mode ν_L), the stretching of the other (ν_R), and the two perpendicular bending modes (ν_2). The description, while permissible, has a disadvantage: when one CO bond vibration is excited, the motion of the C atom sets the other CO bond in motion, so energy flows backwards and forwards between ν_L and ν_R. Moreover, the position of the centre of mass of the molecule varies in the course of either vibration.

The description of the vibrational motion is much simpler if linear combinations of ν_L and ν_R are taken. For example, one combination is ν_1 in Fig. 16.46b: this mode is the **symmetric stretch**. In this mode, the C atom is buffeted simultaneously from each side and the motion continues indefinitely. Another mode is ν_3, the **antisymmetric stretch**, in which the two O atoms always move in the same direction and opposite to that of the C atom. Both modes are independent in the sense that, if one is excited, then it does not excite the other. They are two of the 'normal modes' of the molecule, its independent, collective vibrational displacements. The two other normal modes are the bending modes ν_2. In general, a **normal mode** is an independent, synchronous motion of atoms or groups of atoms that may be excited without leading to the excitation of any other normal mode.

The four normal modes of CO_2, and the N_{vib} normal modes of polyatomics in general, are the key to the description of molecular vibrations. Each normal mode, q, behaves like an independent harmonic oscillator (if anharmonicities are neglected), so each has a series of terms

$$G_q(v) = (v + \tfrac{1}{2})\tilde{\nu}_q \qquad \tilde{\nu}_q = \frac{1}{2\pi c}\left(\frac{k_q}{m_q}\right)^{1/2} \tag{73}$$

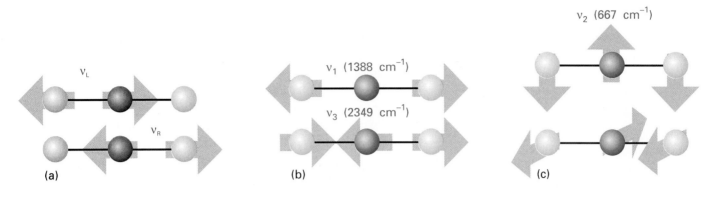

v_2 (667 cm^{-1})

v_1 (1388 cm^{-1})

v_3 (2349 cm^{-1})

v_L

v_R

(a)   (b)   (c)

16.46 Alternative descriptions of the vibrations of CO_2. (a) The stretching modes are not independent, and if one CO group is excited the other begins to vibrate. (b) The symmetric and antisymmetric stretches are independent, and one can be excited without affecting the other: they are normal modes. (c) The two perpendicular bending motions are also normal modes.

where $\tilde{\nu}_q$ is the wavenumber of mode q and depends on the force constant k_q for the mode and on the effective mass m_q of the mode. The effective mass of the mode is a measure of the mass that is swung about by the vibration and in general is a complicated function of the masses of the atoms. For example, in the symmetric stretch of CO_2, the C atom is stationary, and the effective mass depends on the masses of only the O atoms. In the antisymmetric stretch and in the bends, all three atoms move, so all contribute to the effective mass. The three normal modes of H_2O are shown in Fig. 16.47: note that the predominantly bending mode (ν_2) has a lower frequency than the others, which are predominantly stretching modes. It is generally the case that the frequencies of bending motions are lower than those of stretching modes. One point that must be appreciated is that only in special cases (such as the CO_2 molecule) are the normal modes purely stretches or purely bends. In general, a normal mode is a composite motion of simultaneous stretching and bending of bonds. Another point in this connection is that heavy atoms generally move less than light atoms in normal modes.

(c) The symmetry species of normal modes

One of the most powerful ways of dealing with normal modes, especially of complex molecules, is to classify them according to their symmetries. Each normal mode must belong to one of the symmetry species of the molecular point group, as discussed in Chapter 15.

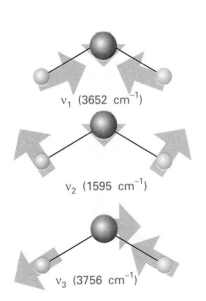

v_1 (3652 cm^{-1})

v_2 (1595 cm^{-1})

v_3 (3756 cm^{-1})

16.47 The three normal modes of H_2O. The mode ν_2 is predominantly bending, and occurs at lower wavenumber than the other two.

Example 16.6 Identifying the symmetry species of a normal mode

Establish the symmetry species of the normal mode vibrations of CH_4, which belongs to the group T_d.

Method The first step in the procedure is to identify the symmetry species of the irreducible representations spanned by all the $3N$ displacements of the atoms, using the characters of the molecular point group. Find these characters by counting 1 if the displacement is unchanged under a symmetry operation, -1 if it changes sign, and 0 if it is changed into some other displacement. Next, subtract the symmetry species of the translations. Translational displacements span the same symmetry species as x, y, and z, so they can be obtained from the right-most column of the character table. Finally, subtract the symmetry species of the rotations, which are also given in the character table (and denoted there by R_x, R_y, or R_z).

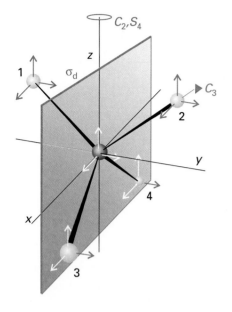

16.48 The atomic displacements of CH_4 and the symmetry elements used to calculate the characters.

Answer There are $3 \times 5 = 15$ degrees of freedom, of which $3 \times 5 - 6 = 9$ are vibrations. Refer to Fig. 16.48. Under E, no displacement coordinates are changed, so the character is 15. Under C_3, no displacements are left unchanged, so the character is 0. Under the C_2 indicated, the z-displacement of the central atom is left unchanged, whereas its x- and y-components both change sign. Therefore $\chi(C_2) = 1 - 1 - 1 + 0 + 0 + \cdots = -1$. Under the S_4 indicated, the z-displacement of the central atom is reversed, so $\chi(S_4) = -1$. Under σ_d, the x- and z-displacements of C, H_3, and H_4 are left unchanged and the y-displacements are reversed; hence $\chi(\sigma_d) = 3 + 3 - 3 = 3$. The characters are therefore $15, 0, -1, -1, 3$, corresponding to $A_1 + E + T_1 + 3T_2$. The translations span T_2; the rotations span T_1. Hence, the nine vibrations span $A_1 + E + 2T_2$.

Comment The modes themselves are shown in Fig. 16.49. We shall see that symmetry analysis gives a quick way of deciding which modes are active.

- -

Self-test 16.8 Establish the symmetry species of the normal modes of H_2O.

$$[2A_1 + B_2]$$

16.15 The vibrational spectra of polyatomic molecules

The gross selection rule for infrared activity is that *the motion corresponding to a normal mode should be accompanied by a change of dipole moment*. Deciding whether this is so can sometimes be done by inspection. For example, the symmetric stretch of CO_2 leaves the dipole moment unchanged (at zero, see Fig. 16.46), so this mode is infrared inactive. The antisymmetric stretch, however, changes the dipole moment because the molecule becomes unsymmetrical as it vibrates, so this mode is infrared active. Because the dipole moment change is parallel to the principal axis, the transitions arising from this mode are classified as **parallel bands** in the spectrum. Both bending modes are infrared active: they are accompanied by a changing dipole perpendicular to the principal axis, so transitions involving them lead to a **perpendicular band** in the spectrum. The latter bands eliminate the linearity of the molecule, and as a result a Q branch is observed; a parallel band does not have a Q branch.

(a) Symmetry and normal mode activity

It is best to use group theory to judge the activities of more complex modes of vibration. This is easily done by checking the character table of the molecular point group for the symmetry

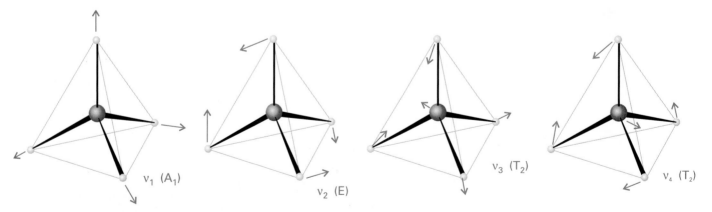

16.49 Typical normal modes of vibration of a tetrahedral molecule. There are in fact two modes of symmetry species E and three modes of each T_2 symmetry species.

species of the irreducible representations spanned by x, y, and z, for their species are also the symmetry species of the components of the electric dipole moment. Then apply the following rule:

If the symmetry species of a normal mode is the same as any of the symmetry species of x, y, or z, then the mode is infrared active.

Justification 16.9

The rule hinges on the form of the transition dipole moment between the ground-state vibrational wavefunction, ψ_0, and that of the first excited state, ψ_1. The x-component is

$$\mu_{x,10} = \langle 1|\mu_x|0 \rangle = -e \int \psi_1^* x \psi_0 \, d\tau \tag{74}$$

for the x-component, with similar expressions for the two other components of the transition moment. The ground-state vibrational wavefunction is a Gaussian function of the form e^{-x^2}, so it is symmetrical in x. The wavefunction for the first excited state gives a nonvanishing integral only if it is proportional to x, for then the integrand is proportional to x^2 rather than to xy or xz. Consequently, the excited state wavefunction must have the same symmetry as the displacement x.

Example 16.7 Identifying infrared active modes

Which modes of CH_4 are infrared active?

Method Refer to the T_d character table to establish the symmetry species of x, y, and z for this molecule, and then use the rule given above.

Answer The functions x, y, and z span T_2. We found in Example 16.6 that the symmetry species of the normal modes are $A_1 + E + 2T_2$. Therefore, only the T_2 modes are infrared active.

Comment The distortions accompanying these modes lead to a changing dipole moment. The A_1 mode, which is inactive, is the symmetrical 'breathing' mode of the molecule.

- -

Self-test 16.9 Which of the normal modes of H_2O are infrared active?

[All three]

(b) The appearance of the spectrum

The active modes are subject to the specific selection rule $\Delta v_q = \pm 1$ in the harmonic approximation, so the wavenumber of the fundamental transition (the 'first harmonic') of each active mode is $\tilde{v}_q$. From the analysis of the spectrum, a picture may be constructed of the stiffness of various parts of the molecule: that is, we can establish its **force field**, the set of force constants corresponding to all the displacements of the atoms. Superimposed on this simple scheme are the complications arising from anharmonicities and the effects of molecular rotation. Very often the sample is a liquid or a solid, and the molecules are unable to rotate freely. In a liquid, for example, a molecule may be able to rotate through only a few degrees before it is struck by another, so it changes its rotational state frequently. This random changing of orientation is called **tumbling**.

The lifetimes of rotational states in liquids are very short, so in most cases the rotational energies are ill-defined. Collisions occur at a rate of about 10^{13} s^{-1} and, even allowing for only a 10 per cent success rate in knocking the molecule into another rotational state, a lifetime broadening (eqn 24) of more than 1 cm^{-1} can easily result. The rotational structure of the vibrational spectrum is blurred by this effect, so the infrared spectra of molecules in condensed phases usually consist of broad lines spanning the entire range of the resolved gas-phase spectrum, and showing no branch structure.

One very important application of infrared spectroscopy to condensed phase samples, and for which the blurring of the rotational structure by random collisions is a welcome simplification, is to chemical analysis. The vibrational spectra of different groups in a molecule give rise to absorptions at characteristic frequencies. Their intensities are also transferable between molecules. Consequently, the molecules in a sample can often be identified by examining its infrared spectrum and referring to a table of characteristic frequencies and intensities (Table 16.3 and Fig. 16.50).

Table 16.3* Typical vibrational wavenumbers, $\tilde{\nu}/\text{cm}^{-1}$

C—H stretch	2850–2960
C—H bend	1340–1465
C—C stretch	700–1250
C≡C stretch	1620–1680

*More values are given in the *Data section*.

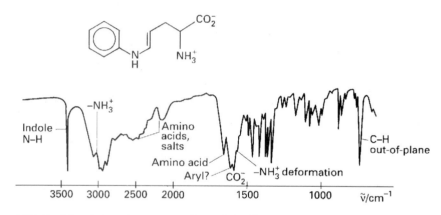

16.50 The infrared absorption spectrum of an amino acid, and a partial assignment.

16.16 Vibrational Raman spectra of polyatomic molecules

The normal modes of vibration of molecules are Raman active if they are accompanied by a changing polarizability. It is sometimes quite difficult to judge by inspection when this is so. The symmetric stretch of CO_2, for example, alternately swells and contracts the molecule: this motion changes the polarizability of the molecule, so the mode is Raman active. The other modes of CO_2 leave the polarizability unchanged, so they are Raman inactive.

(a) Symmetry aspects of Raman transitions

Group theory provides an explicit recipe for judging the Raman activity of a normal mode. In this case, the symmetry species of the quadratic forms (x^2, xy, etc.) listed in the character table are noted (they transform in the same way as the polarizability), and then we use the following rule:

> **If the symmetry species of a normal mode is the same as the symmetry species of a quadratic form, then the mode may be Raman active.**

Illustration
. . . .

To decide which of the vibrations of CH_4 are Raman active, refer to the T_d character table. It was established in Example 16.6 that the symmetry species of the normal modes are

$A_1 + E + 2T_2$. Because the quadratic forms span $A_1 + E + T_2$, all the normal modes are Raman active. All totally symmetric vibrations, whatever the point group of the molecule, are Raman active (and polarized; see below).

....

Self-test 16.10 Which of the vibrational modes of H_2O are Raman active?

[All three]

The **exclusion rule** also helps us to decide which modes are active:

If the molecule has a centre of symmetry, then no modes can be both infrared and Raman active.

(A mode may be inactive in both.) Because it is often possible to judge intuitively if a mode changes the molecular dipole moment, we can use this rule to identify modes that are not Raman active. The rule applies to CO_2 but to neither H_2O nor CH_4 because they have no centre of symmetry.

(b) Depolarization

The assignment of Raman lines to particular vibrational modes is aided by noting the state of polarization of the scattered light. The **depolarization ratio**, ρ, of a line is the ratio of the intensities, $\mathcal{I}$, of the scattered light with polarizations perpendicular and parallel to the plane of polarization of the incident radiation:

$$\rho = \frac{\mathcal{I}_\perp}{\mathcal{I}_\parallel} \qquad [75]$$

To measure ρ, the intensity of a Raman line is measured with a polarizing filter (a 'half-wave plate') first parallel and then perpendicular to the polarization of the incident beam. If the emergent light is not polarized, then both intensities are the same and ρ is close to 1; if the light retains its initial polarization, then $\mathcal{I}_\perp = 0$, so $\rho = 0$ (Fig. 16.51). A line is classified as **depolarized** if it has ρ close to or greater than 0.75 and as **polarized** if $\rho < 0.75$. Only totally symmetrical vibrations give rise to polarized lines in which the incident polarization is

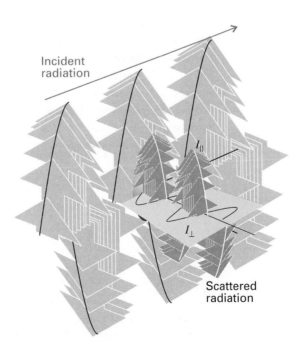

16.51 The definition of the planes used for the specification of the depolarization ratio, ρ, in Raman scattering. The fat arrows represent the electric vector of the incident (green) and scattered (grey) radiation. There is also a perpendicular scattered component, as indicated by the simple wave-like line.

largely preserved. Vibrations that are not totally symmetrical give rise to depolarized lines because the incident radiation can give rise to radiation in the perpendicular direction too.

(c) Resonance Raman spectra

A modification of the basic Raman effect involves using incident radiation that nearly coincides with the frequency of an electronic transition of the sample (Fig. 16.52). The technique is then called **resonance Raman spectroscopy**. It is characterized by a much greater intensity in the scattered radiation. Furthermore, because it is often the case that

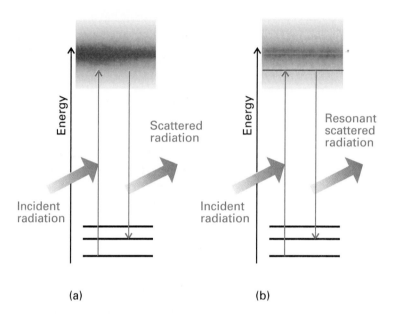

(a)   (b)

16.52 (a) In conventional Raman spectroscopy, the incident radiation does not match an absorption frequency of the molecule, and there is only a 'virtual' transition to an excited state. (b) However, in the resonance Raman effect, the incident radiation has a frequency that coincides with a molecular transition.

only a few vibrational modes contribute to the more intense scattering, the spectrum is greatly simplified. The resonance Raman spectrum shown in Fig. 16.53, for example, is of solid potassium chromate. The nine peaks that are identified are the Stokes lines that correspond to the excitation of the symmetric breathing mode of the tetrahedral CrO_4^{2-} ion and the transfer of up to nine vibrational quanta during the photon–ion collision. The high intensity of the resonance Raman transitions is employed to examine the metal ions in biological macromolecules (such as the iron in haemoglobin and cytochromes or the cobalt in vitamin B), which are present in such low abundances that conventional Raman spectroscopy cannot detect them. An additional advantage is that resonance picks out the

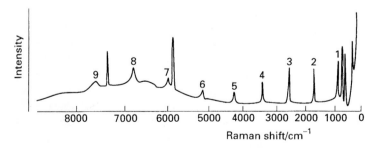

16.53 The resonance Raman spectrum of solid K_2CrO_4. The peaks are due to the totally symmetric stretching mode of the CrO_4^{2-} anion. (W. Kiefer and H.J. Bernstein, *Molec. Phys.* **23**, 835 (1972).)

fragment of a molecule that in conventional Raman spectroscopy would have a spectrum too complex to interpret.

(d) Coherent anti-Stokes Raman spectroscopy

The intensity of Raman transitions may be enhanced by **coherent anti-Stokes Raman spectroscopy** (CARS, Fig. 16.54). The technique relies on the fact that, if two laser beams of frequencies ν_1 and ν_2 pass through a sample, then they may mix together and give rise to coherent radiation of several different frequencies, one of which is

$$\nu' = 2\nu_1 - \nu_2 \tag{76}$$

Suppose that ν_2 is varied until it matches any Stokes line from the sample, such as the one with frequency $\nu_1 - \Delta\nu$; then the coherent emission will have frequency

$$\nu' = 2\nu_1 - (\nu_1 - \Delta\nu) = \nu_1 + \Delta\nu \tag{77}$$

which is the frequency of the corresponding anti-Stokes line. This coherent radiation forms a narrow beam of high intensity.

An advantage of CARS is that it can be used to study Raman transitions in the presence of competing incoherent background radiation, and so can be used to observe the Raman spectra of species in flames. The intensities of the transitions can then be interpreted in terms of the temperatures of different regions of the flame.

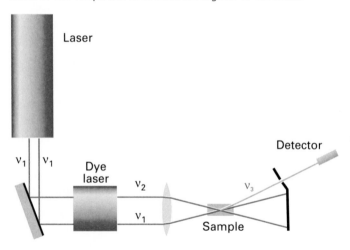

16.54 The experimental arrangement for the CARS experiment.

Checklist of key ideas

General features of spectroscopy

16.1 Experimental techniques
- [ ] emission spectroscopy
- [ ] absorption spectroscopy
- [ ] dispersing element
- [ ] Fourier transform
- [ ] detector
- [ ] Raman spectroscopy
- [ ] Stokes radiation
- [ ] anti-Stokes radiation
- [ ] Rayleigh radiation

16.2 The intensities of spectral lines
- [ ] transmittance (7)
- [ ] Beer–Lambert law (8, 10)
- [ ] molar absorption coefficient
- [ ] absorbance (9)
- [ ] integrated absorption coefficient (11)
- [ ] stimulated absorption
- [ ] Einstein coefficient of stimulated absorption (12)
- [ ] total rate of absorption

- [ ] Einstein coefficient of stimulated emission (14)
- [ ] spontaneous emission
- [ ] Einstein coefficient of spontaneous emission (15)
- [ ] gross selection rule
- [ ] specific selection rule

16.3 Linewidths
- [ ] Doppler effect
- [ ] lifetime
- [ ] lifetime broadening (24)
- [ ] collisional deactivation

- [ ] collisional lifetime
- [ ] natural linewidth

Pure rotation spectra

16.4 Moments of inertia
- [ ] moment of inertia (26)
- [ ] rigid rotor
- [ ] spherical rotor
- [ ] symmetric rotor
- [ ] linear rotor
- [ ] asymmetric rotor

16.5 The rotational energy levels

- [ ] rotational constant (30, 36)
- [ ] rotational term (32, 35)
- [ ] principal axis
- [ ] oblate
- [ ] prolate
- [ ] Stark effect
- [ ] centrifugal distortion constant

16.6 Rotational transitions

- [ ] Stark modulation
- [ ] energies (43, 44)

16.7 Rotational Raman spectra

- [ ] energies (48)

16.8 Nuclear statistics and rotational states

- [ ] nuclear statistics

- [ ] *ortho*-hydrogen
- [ ] *para*-hydrogen

The vibrations of diatomic molecules

16.9 Molecular vibrations

- [ ] force constant (53)
- [ ] effective mass (55)
- [ ] vibrational term (57)

16.10 Selection rules

- [ ] infrared active
- [ ] infrared inactive
- [ ] fundamental transition

16.11 Anharmonicity

- [ ] anharmonic
- [ ] Morse potential energy (61)
- [ ] anharmonicity constant (62)
- [ ] overtones
- [ ] Birge–Sponer plot

16.12 Vibration–rotation spectra

- [ ] band spectra
- [ ] branches (68)
- [ ] P-branch
- [ ] Q-branch
- [ ] R-branch
- [ ] combination difference

16.13 Vibrational Raman spectra of diatomic molecules

- [ ] O-branch
- [ ] S-branch

The vibrations of polyatomic molecules

16.14 Normal modes

- [ ] symmetric stretch
- [ ] antisymmetric stretch
- [ ] normal mode

16.15 The vibrational spectra of polyatomic molecules

- [ ] parallel band
- [ ] perpendicular band
- [ ] force field
- [ ] tumbling

16.16 Vibrational Raman spectra of polyatomic molecules

- [ ] exclusion rule
- [ ] depolarization ratio (75)
- [ ] polarized
- [ ] depolarized
- [ ] resonance Raman spectroscopy
- [ ] coherent anti-Stokes spectroscopy (CARS)

Further reading

Articles of general interest

N.C. Thomas, The early history of spectroscopy. *J. Chem. Educ.* **68**, 631 (1991).

R. Woods and G. Henderson, FTIR rotational spectroscopy. *J. Chem. Educ.* **64**, 921 (1987).

L. Glasser, Fourier transforms for chemists. Part I. Introduction to the Fourier transform. *J. Chem. Educ.* **64**, A228 (1987); Part II. Fourier transforms in chemistry and spectroscopy. *J. Chem. Educ.* **64**, A260 (1987).

W.D. Perkins, Fourier transform infrared spectroscopy: Part II. Advantages of FT-IR. *J. Chem. Educ.* **64**, A269 (1987).

P.L. Goodfriend, Diatomic vibrations revisited. *J. Chem. Educ.* **64**, 753 (1987).

A.R. Lacey, A student introduction to molecular vibrations. *J. Chem. Educ.* **64**, 756 (1987).

J.G. Verkade, A novel pictorial approach to teaching molecular motions in polyatomic molecules. *J. Chem. Educ.* **64**, 411 (1987).

F.A. Miller and G.B. Kauffman, C.V. Raman and the discovery of the Raman effect. *J. Chem. Educ.* **66**, 795 (1989).

M.-K. Ahn, A comparison of FTNMR and FTIR techniques. *J. Chem. Educ.* **66**, 802 (1989).

D.P. Strommen, Specific values of the depolarization ratio in Raman spectroscopy: their origins and significance. *J. Chem. Educ.* **69**, 803 (1992).

B.J. Bozlee, J.H. Luther, and M. Buraczewski, The infrared overtone intensity of a simple diatomic molecule: nitric oxide. *J. Chem. Educ.* **69**, 370 (1992).

G. Henderson and B. Logsdon, Stark effects on rigid-rotor wavefunctions: a quantum description of dipolar rotors trapped in electric fields. *J. Chem. Educ.* **72**, 1021 (1995).

E. Grunwald, J. Herzog, and C. Steel, Using Fourier transforms to understand spectral lineshapes. *J. Chem. Educ.* **72**, 210 (1995).

D.K. Graff, Fourier and Hadamard: transforms in chemistry. *J. Chem. Educ.* **72**, 304 (1995).

V.B.E. Thomsen, Why do spectral lines have a linewidth? *J. Chem. Educ.* **72**, 616 (1995).

R.S. Treptow, Bond energies and enthalpies: an often neglected difference. *J. Chem. Educ.* **72**, 497 (1995).

N.K. Kildahl, Bond energy data summarized. *J. Chem. Educ.* **72**, 423 (1995).

C.W. David, IR vibration–rotation spectrum of the ammonia molecule. *J. Chem. Educ.* **73**, 46 (1996).

D.A. Ramsay, Molecular spectroscopy. In *Encyclopedia of applied physics* (ed. G.L. Trigg), **10**, 491. VCH, New York (1994).

J.R. Lombardi, Radiation interaction with molecules. In *Encyclopedia of applied physics* (ed. G.L. Trigg), **15**, 509. VCH, New York (1996).

I.P. Herman, Raman scattering. In *Encyclopedia of applied physics* (ed. G.L. Trigg), **15**, 587. VCH, New York (1996).

H.G.M. Edwards, Raman spectroscopy instrumentation. In *Encyclopedia of applied physics* (ed. G.L. Trigg), **16**, 1. VCH, New York (1996).

Texts and sources of data and information

J.M. Hollas, *Modern spectroscopy*. Wiley, New York (1996).

J.M. Hollas, *High resolution spectroscopy*. Butterworth, London (1982).

P.F. Bernath, *Spectra of atoms and molecules*. Oxford University Press, New York (1995).

W. Gordy and R.L. Cook, *Microwave molecular spectra*. Wiley, New York (1974).

S. Wilson (ed.), *Methods in computational chemistry*, Vol 4. Molecular vibrations. Plenum, New York (1992).

K.P. Huber and G. Herzberg, *Molecular spectra and molecular structure IV. Constants of diatomic molecules*. Van Nostrand-Reinhold, New York (1979).

L. Bellamy, *The infrared spectra of complex molecules*. Chapman & Hall, London (1980).

E.A.V. Ebsworth, D.W.H. Rankin, and S. Cradock, *Structural methods in inorganic chemistry*. Blackwell Scientific, Oxford (1991).

R. Drago, *Physical methods for chemists*. Saunders, Philadelphia (1992).

G. Herzberg, *Infrared and Raman spectra of polyatomic molecules*. Van Nostrand, New York (1945).

E.B. Wilson, J.C. Decius, and P.C. Cross, *Molecular vibrations*. McGraw-Hill, New York (1955).

Exercises

16.1 (a) Calculate the ratio of the Einstein coefficients of spontaneous and stimulated emission, A and B, for transitions with the following characteristics: (a) 70.8 pm X-rays, (b) 500 nm visible light, (c) 3000 cm^{-1} IR radiation.

16.1 (b) Calculate the ratio of the Einstein coefficients of spontaneous and stimulated emission, A and B, for transitions with the following characteristics: (a) 500 MHz radiofrequency radiation, (b) 3.0 cm microwave radiation.

16.2 (a) Calculate the frequency of the $J = 4 \leftarrow 3$ transition in the pure rotational spectrum of $^{14}N^{16}O$. The equilibrium bond length is 115 pm.

16.2 (b) Calculate the frequency of the $J = 3 \leftarrow 2$ transition in the pure rotational spectrum of $^{12}C^{16}O$. The equilibrium bond length is 112.81 pm.

16.3 (a) If the wavenumber of the $J = 3 \leftarrow 2$ rotational transition of $^1H^{35}Cl$ considered as a rigid rotator is 63.56 cm^{-1}, what is (a) the moment of inertia of the molecule, (b) the bond length?

16.3 (b) If the wavenumber of the $J = 1 \leftarrow 0$ rotational transition of $^1H^{81}Br$ considered as a rigid rotator is 16.93 cm^{-1}, what is (a) the moment of inertia of the molecule, (b) the bond length?

16.4 (a) Given that the spacing of lines in the microwave spectrum of $^{27}Al^1H$ is constant at 12.604 cm^{-1}, calculate the moment of inertia and bond length of the molecule ($m(^{27}Al) = 26.9815$ u).

16.4 (b) Given that the spacing of lines in the microwave spectrum of $^{35}Cl^{19}F$ is constant at 1.033 cm^{-1}, calculate the moment of inertia and bond length of the molecule ($m(^{35}Cl) = 34.9688$ u, $m(^{19}F) = 18.9984$ u).

16.5 (a) The rotational constant of $^{127}I^{35}Cl$ is 0.1142 cm^{-1}. Calculate the ICl bond length ($m(^{35}Cl) = 34.9688$ u, $m(^{127}I) = 126.9045$ u).

16.5 (b) The rotational constant of $^{12}C^{16}O_2$ is 0.39021 cm^{-1}. Calculate the bond length of the molecule ($m(^{12}C) = 12$ u exactly, $m(^{16}O) = 15.9949$ u).

16.6 (a) Determine the HC and CN bond lengths in HCN from the rotational constants $B(^1H^{12}C^{14}N) = 44.316$ GHz, $B(^2H^{12}C^{14}N) = 36.208$ GHz.

16.6 (b) Determine the CO and CS bond lengths in OCS from the rotational constants $B(^{16}O^{12}C^{32}S) = 6081.5$ MHz, $B(^{16}O^{12}C^{34}S) = 5932.8$ MHz.

16.7 (a) The wavenumber of the incident radiation in a Raman spectrometer is 20 487 cm^{-1}. What is the wavenumber of the scattered Stokes radiation for the $J = 2 \leftarrow 0$ transition of $^{14}N_2$?

16.7 (b) The wavenumber of the incident radiation in a Raman spectrometer is 20 623 cm^{-1}. What is the wavenumber of the scattered Stokes radiation for the $J = 4 \leftarrow 2$ transition of $^{16}O_2$?

16.8 (a) Infrared absorption by $^1H^{81}Br$ gives rise to an R branch from $v = 0$. What is the wavenumber of the line originating from the rotational state with $J = 2$? Use the information in Table 16.2.

16.8 (b) Infrared absorption by $^1H^{127}I$ gives rise to an R branch from $v = 0$. What is the wavenumber of the line originating from the rotational state with $J = 2$? Use the information in Table 16.2.

16.9 (a) An object of mass 1.0 kg suspended from the end of a rubber band has a vibrational frequency of 2.0 Hz. Calculate the force constant of the rubber band.

16.9 (b) An object of mass 2.0 g suspended from the end of a spring has a vibrational frequency of 3.0 Hz. Calculate the force constant of the spring.

16.10 (a) Calculate the percentage difference in the fundamental vibration wavenumber of $^{23}Na^{35}Cl$ and $^{23}Na^{37}Cl$ on the assumption that their force constants are the same.

16.10 (b) Calculate the percentage difference in the fundamental vibration wavenumber of $^{1}H^{35}Cl$ and $^{2}H^{37}Cl$ on the assumption that their force constants are the same.

16.11 (a) The wavenumber of the fundamental vibrational transition of $^{35}Cl_2$ is 564.9 cm^{-1}. Calculate the force constant of the bond $(m(^{35}Cl) = 34.9688$ u).

16.11 (b) The wavenumber of the fundamental vibrational transition of $^{79}Br\ ^{81}Br$ is 323.2 cm^{-1}. Calculate the force constant of the bond $(m(^{79}Br) = 78.9183$ u, $m(^{81}Br) = 80.9163$ u).

16.12 (a) The molecule CH_2Cl_2 belongs to the point group C_{2v}. The displacements of the atoms span $5A_1 + 2A_2 + 4B_1 + 4B_2$. What are the symmetries of the normal modes of vibration?

16.12 (b) A carbon disulfide molecule belongs to the point group $D_{\infty h}$. The nine displacements of the three atoms span $A_{1g} + A_{1u} + A_{2g} + 2E_{1u} + E_{1g}$. What are the symmetries of the normal modes of vibration?

16.13 (a) Which of the following molecules may show a pure rotational microwave absorption spectrum: (a) H_2, (b) HCl, (c) CH_4, (d) CH_3Cl, (e) CH_2Cl_2?

16.13 (b) Which of the following molecules may show a pure rotational microwave absorption spectrum: (a) H_2O, (b) H_2O_2, (c) NH_3, (d) N_2O?

16.14 (a) Which of the following molecules may show infrared absorption spectra: (a) H_2, (b) HCl, (c) CO_2, (d) H_2O?

16.14 (b) Which of the following molecules may show infrared absorption spectra: (a) CH_3CH_3, (b) CH_4, (c) CH_3Cl, (d) N_2?

16.15 (a) Which of the following molecules may show a pure rotational Raman spectrum: (a) H_2, (b) HCl, (c) CH_4, (d) CH_3Cl?

16.15 (b) Which of the following molecules may show a pure rotational Raman spectrum: (a) CH_2Cl_2, (b) CH_3CH_3, (c) SF_6, (d) N_2O?

16.16 (a) What is the Doppler-shifted wavelength of a red (660 nm) traffic light approached at 80 km h^{-1}?

16.16 (b) At what speed of approach would a red (660 nm) traffic light appear green (520 nm)?

16.17 (a) A spectral line of $^{48}Ti^{8+}$ (of mass 47.95 u) in a distant star was found to be shifted from 654.2 nm to 706.5 nm and to be broadened to 61.8 pm. What is the speed of recession and the surface temperature of the star?

16.17 (b) A spectral line of $^{31}P^{3+}$ (of mass 30.97 u) in a distant star was found to be shifted from 326 nm to 365 nm and to be broadened to 45.8 pm. What is the speed of recession and the surface temperature of the star?

16.18 (a) Estimate the lifetime of a state that gives rise to a line of width (a) 0.10 cm^{-1}, (b) 1.0 cm^{-1}.

16.18 (b) Estimate the lifetime of a state that gives rise to a line of width (a) 100 MHz, (b) 2.14 cm^{-1}.

16.19 (a) A molecule in a liquid undergoes about 1.0×10^{13} collisions in each second. Suppose that (a) every collision is effective in deactivating the molecule vibrationally and (b) that one collision in 100 is effective. Calculate the width (in cm^{-1}) of vibrational transitions in the molecule.

16.19 (b) A molecule in a gas undergoes about 1.0×10^9 collisions in each second. Suppose that (a) every collision is effective in deactivating the molecule rotationally and (b) that one collision in 10 is effective. Calculate the width (in hertz) of rotational transitions in the molecule.

16.20 (a) Calculate the relative numbers of Cl_2 molecules $(\tilde{\nu} = 559.7$ cm$^{-1})$ in the ground and first excited vibrational states at (a) 298 K, (b) 500 K.

16.20 (b) Calculate the relative numbers of Br_2 molecules $(\tilde{\nu} = 321$ cm$^{-1})$ in the second and first excited vibrational states at (a) 298 K, (b) 800 K.

16.21 (a) The hydrogen halides have the following fundamental vibrational wavenumbers: 4141.3 cm^{-1} (HF); 2988.9 cm^{-1} (H^{35}Cl); 2649.7 cm^{-1} (H^{81}Br); 2309.5 cm^{-1} (H^{127}I). Calculate the force constants of the hydrogen–halogen bonds.

16.21 (b) From the data in Exercise 16.21a, predict the fundamental vibrational wavenumbers of the deuterium halides.

16.22 (a) For $^{16}O_2$, ΔG values for the transitions $v = 1\leftarrow 0$, $2\leftarrow 0$, and $3\leftarrow 0$ are, respectively, 1556.22, 3088.28, and 4596.21 cm^{-1}. Calculate $\tilde{\nu}$ and x_e. Assume y_e to be zero.

16.22 (b) For $^{14}N_2$, ΔG values for the transitions $v = 1\leftarrow 0$, $2\leftarrow 0$, and $3\leftarrow 0$ are, respectively, 2345.15, 4661.40, and 6983.73 cm^{-1}. Calculate $\tilde{\nu}$ and x_e. Assume y_e to be zero.

16.23 (a) The first five vibrational energy levels of HCl are at 1481.86, 4367.50, 7149.04, 9826.48, and 12 399.8 cm^{-1}. Calculate the dissociation energy of the molecule in reciprocal centimetres and electronvolts.

16.23 (b) The first five vibrational energy levels of HI are at 1144.83, 3374.90, 5525.51, 7596.66, and 9588.35 cm^{-1}. Calculate the dissociation energy of the molecule in reciprocal centimetres and electronvolts.

16.24 (a) The rotational Raman spectrum of $^{35}Cl_2$ $(m(^{35}Cl) = 34.9688$ u) shows a series of Stokes lines separated by 0.9752 cm^{-1} and a similar series of anti-Stokes lines. Calculate the bond length of the molecule.

16.24 (b) The rotational Raman spectrum of $^{19}F_2$ $(m(^{19}F) = 18.9984$ u) shows a series of Stokes lines separated by 3.5312 cm^{-1} and a similar series of anti-Stokes lines. Calculate the bond length of the molecule.

16.25 (a) How many normal modes of vibration are there for the following molecules: (a) H_2O, (b) H_2O_2, (c) C_2H_4?

16.25 (b) How many normal modes of vibration are there for the following molecules: (a) C_6H_6, (b) $C_6H_5CH_3$, (c) $HC{\equiv}C{-}C{\equiv}CH$?

16.26 (a) Which of the three vibrations of an AB_2 molecule are infrared or Raman active when it is (a) angular, (b) linear?

16.26 (b) Which of the vibrations of an AB_3 molecule are infrared or Raman active when it is (a) trigonal planar, (b) trigonal pyramidal?

16.27 (a) Consider the vibrational mode that corresponds to the uniform expansion of the benzene ring. Is it (a) Raman, (b) infrared active?

16.27 (b) Consider the vibrational mode that corresponds to the boat-like bending of a benzene ring. Is it (a) Raman, (b) infrared active?

Problems

Numerical problems

16.1 Calculate the Doppler width (as a fraction of the transition wavelength) for any kind of transition in (a) HCl, (b) ICl at 25°C. What would be the widths of the rotational and vibrational transitions in these molecules (in MHz and cm^{-1}, respectively), given $B(ICl) = 0.1142$ cm^{-1} and $\tilde{\nu}(ICl) = 384$ cm^{-1} and additional information in Table 16.2?

16.2 The collision frequency of a molecule of mass m in a gas of pressure p is $z = 4\sigma(kT/\pi m)^{1/2}p/kT$, where σ is the collision cross-section. Find an expression for the collision-limited lifetime of an excited state assuming that every collision is effective. Estimate the width of rotational transition in HCl ($\sigma = 0.30$ nm^2) at 25°C and 1.0 atm. To what value must the pressure of the gas be reduced in order to ensure that collision broadening is less important than Doppler broadening?

16.3 The rotational constant of NH_3 is equivalent to 298 GHz. Compute the separation of the pure rotational spectrum lines in GHz, cm^{-1}, and mm, and show that the value of B is consistent with an N–H bond length of 101.4 pm and a bond angle of 106.78°.

16.4 The rotational constant for CO is 1.9314 cm^{-1} and 1.6116 cm^{-1} in the ground and first excited vibrational states, respectively. By how much does the internuclear distance change as a result of this transition?

16.5 Pure rotational Raman spectra of gaseous C_6H_6 and C_6D_6 yield the following rotational constants: $B(C_6H_6) = 0.18960$ cm^{-1}, $B(C_6D_6) = 0.15681$ cm^{-1}. The moments of inertia of the molecules about any axis perpendicular to the C_6 axis were calculated from these data as $I(C_6H_6) = 147.59 \times 10^{-47}$ kg m^2, $I(C_6D_6) = 178.45 \times 10^{-47}$ kg m^2. Calculate the CC, CH, and CD bond lengths.

16.6 The vibrational energy levels of NaI lie at the wavenumbers 142.81, 427.31, 710.31, and 991.81 cm^{-1}. Show that they fit the expression $(v + \frac{1}{2})\tilde{\nu} - (v + \frac{1}{2})^2 x_e\tilde{\nu}$, and deduce the force constant, zero-point energy, and dissociation energy of the molecule.

16.7 Predict the shape of the nitronium ion, NO_2^+, from its Lewis structure and the VSEPR model. It has one Raman active vibrational mode at 1400 cm^{-1}, two strong IR active modes at 2360 and 540 cm^{-1}, and one weak IR mode at 3735 cm^{-1}. Are these data consistent with the predicted shape of the molecule? Assign the vibrational wavenumbers to the modes from which they arise.

16.8 Rotational absorption lines from $^1H^{35}Cl$ gas were found at the following wavenumbers (R.L. Hausler and R.A. Oetjen, *J. Chem. Phys.* **21**, 1340 (1953)): 83.32, 104.13, 124.73, 145.37, 165.89, 186.23, 206.60, 226.86 cm^{-1}. Calculate the moment of inertia and the bond length of the molecule. Predict the positions of the corresponding lines in $^2H^{35}Cl$.

16.9 Is the bond length in HCl the same as that in DCl? The wavenumbers of the $J = 1 \leftarrow 0$ rotational transitions for $H^{35}Cl$ and $^2H^{35}Cl$ are 20.8784 and 10.7840 cm^{-1}, respectively. Accurate atomic masses are 1.007825 u and 2.0140 u for ^1H and ^2H, respectively. The mass of ^{35}Cl is 34.96885 u. Based on this information alone, can you conclude that the bond lengths are the same or different in the two molecules?

16.10 The microwave spectrum of $^{16}O^{12}CS$ (C.H. Townes, A.N. Holden, and F.R. Merritt, *Phys. Rev.* **74**, 1113 (1948)) gave absorption lines (in GHz) as follows:

J	1	2	3	4
^{32}S	24.32592	36.48882	48.65164	60.81408
^{34}S	23.73233		47.46240	

Use the expressions for moments of inertia in Table 16.1 and assume that the bond lengths are unchanged by substitution; calculate the CO and CS bond lengths in OCS.

16.11 The HCl molecule is quite well described by the Morse potential with $D_e = 5.33$ eV, $\tilde{\nu} = 2989.7$ cm^{-1}, and $x_e\tilde{\nu} = 52.05$ cm^{-1}. Assuming that the potential is unchanged on deuteration, predict the dissociation energies (D_0) of (a) HCl, (b) DCl.

16.12 The Morse potential (eqn 61) is very useful as a simple representation of the actual molecular potential energy. When RbH was studied, it was found that $\tilde{\nu} = 936.8$ cm^{-1} and $x_e\tilde{\nu} = 14.15$ cm^{-1}. Plot the potential energy curve from 50 pm to 800 pm around $R_e = 236.7$ pm. Then go on to explore how the rotation of a molecule may weaken its bond by allowing for the kinetic energy of rotation of a molecule and plotting $V^* = V + hcBJ(J + 1)$ with $B = \hbar/4\pi c\mu R^2$. Plot these curves on the same diagram for $J = 40$, 80, and 100, and observe how the dissociation energy is affected by the rotation. (Taking $B = 3.020$ cm^{-1} at the equilibrium bond length will greatly simplify the calculation.)

Theoretical problems

16.13 Show that the moment of inertia of a diatomic molecule composed of atoms of masses m_A and m_B and bond length R is equal to $m_{eff}R^2$, where $m_{eff} = m_A m_B/(m_A + m_B)$.

16.14 Derive an expression for the value of J corresponding to the most highly populated rotational energy level of a diatomic rotor at a temperature T remembering that the degeneracy of each level is $2J + 1$. Evaluate the expression for ICl (for which $B = 0.1142$ cm^{-1}) at 25°C. Repeat the problem for the most highly populated level of a spherical rotor, taking note of the fact that each level is $(2J + 1)^2$-fold degenerate. Evaluate the expression for CH$_4$ (for which $B = 5.24$ cm^{-1}) at 25°C.

16.15 The moments of inertia of the linear mercury(II) halides are very large, so the O and S branches of their vibrational Raman spectra show little rotational structure. Nevertheless, the peaks of both branches can be identified and have been used to measure the rotational constants of the molecules (R.J.H. Clark and D.M. Rippon, *J. Chem. Soc. Faraday Soc. II* **69**, 1496 (1973)). Show, from a knowledge of the value of J corresponding to the intensity maximum, that the separation of the peaks of the O and S branches is given by the *Placzek–Teller relation* $\delta\tilde{\nu} = (32BkT/hc)^{1/2}$. The following widths were obtained at the temperatures stated:

	HgCl$_2$	HgBr$_2$	HgI$_2$
$\theta/°C$	282	292	292
$\delta\tilde{\nu}/cm^{-1}$	23.8	15.2	11.4

Calculate the bond lengths in the three molecules.

Additional problems supplied by Carmen Giunta and Charles Trapp

16.16 The noble gases and their complexes are favourite objects for the study of weak interatomic and intermolecular forces. J.-U. Grabow, A.S. Pine, G.T. Fraser, F.J. Lovas, R.D. Suenram, T. Emilsson, E. Arunan, and H.S. Gutowsky (*J. Chem. Phys.* **102**, 1181 (1995)) measured the pure rotational spectrum of ^{20}Ne^{40}Ar and reported its rotational constant (actually cB) as 2914.9 MHz. What is the internuclear separation of the complex? They also reported the centrifugal distortion constant (actually cD_J) as 231.01 kHz. Estimate the fundamental vibrational wavenumber and the force constant of the weak bond. $(m(^{40}Ar) = 39.963$ u; $m(^{20}Ne) = 19.992$ u.)

16.17 B.D. Shizgal (*J. Molec. Structure (Theochem)* **391**, 131 (1997)) reports the wavenumbers of vibrational transitions (based on quantum mechanical computations) for NeAr to be 1909.3, 1060.3, and 386.6 m^{-1} for 1–0, 2–1, and 3–2, respectively. Determine the Morse potential parameters D_e and a for this complex.

16.18 F. Luo, G.C. McBane, G. Kim, C.F. Giese, and W.R. Gentry (*J. Chem. Phys.* **98**, 3564 (1993)) reported experimental observation of the He$_2$ complex, a species which had escaped detection for a long time. The fact that the observation required temperatures in the

neighbourhood of 1 mK is consistent with computational studies which suggest that hcD_e for He$_2$ is about 1.51×10^{-23} J, hcD_0 about 2×10^{-26} J, and R_e about 297 pm. (a) Estimate the fundamental vibrational wavenumber, force constant, moment of inertia, and rotational constant based on the harmonic-oscillator and rigid-rotor approximations. (b) Such a weakly bound complex is hardly likely to be rigid. Estimate the vibrational wavenumber and anharmonicity constant based on the Morse potential.

16.19 In a study of the rotational spectrum of the linear FeCO radical, K. Tanaka, M. Shirasaka, and T. Tanaka (*J. Chem. Phys.* **106**, 6820 (1997)) report the following $J + 1 \leftarrow J$ transitions.

J	24	25	26
$\tilde{\nu}/m^{-1}$	214 777.7	223 379.0	231 981.2
J	27	28	29
$\tilde{\nu}/m^{-1}$	240 584.4	249 188.5	257 793.5

Evaluate the rotational constant of the molecule. Also, estimate the value of J for the most highly populated rotational energy level at 298 K and at 100 K.

16.20 In the constellation Ophiuchus, there is a gaseous interstellar cloud illuminated from behind by the star ζ-Ophiuci. Analysis of the Fraunhofer electronic–vibrational–rotational absorption lines obtained by H.S. Uhler and R.A. Patterson (*Astrophys. J.* **42**, 434 (1915)) shows the presence of CN molecules in the interstellar medium. A strong absorption line in the ultraviolet region at $\lambda = 387.5$ nm was observed corresponding to the transition $J = 0$–1. Unexpectedly, a second strong absorption line with 25 per cent of the intensity of the first was found at a slightly longer wavelength $(\Delta\lambda = 0.061$ nm) corresponding to the transition $J = 1$–1 (here allowed). Calculate the temperature of the CN molecules. Gerhard Herzberg, who was later to receive the Nobel Prize for his contributions to spectroscopy, calculated the temperature as 2.3 K. Although puzzled by this result, he did not realize its full significance. If he had, his prize might have been for the discovery of the cosmic background radiation.

16.21 The rotational energy levels for the H$_3^+$ molecular ion, an oblate symmetric rotor, are given by eqn 35, with C replacing A, when centrifugal distortion and other complications are ignored. Experimental values of the vibrational–rotational constants are $\tilde{\nu}_2(E') = 2521.6$ cm^{-1}, $B = 43.55$ cm^{-1}, and $C = 20.71$ cm^{-1}. (a) Show that for a nonlinear planar molecule such as H$_3^+$ that $I_C = 2I_B$. The rather large discrepancy with the experimental values is due to the factors ignored in eqn 35. (b) Calculate an approximate value of the internuclear distance in H$_3^+$. (c) The value for R_e obtained from the best quantum mechanical calculations by J.B. Anderson (*J. Chem. Phys.* **96**, 3702 (1991)) is 87.32 pm. Use this result to calculate values of the rotational constants B and C. (d) Assuming that the geometry and force constants are the same in D$_3^+$ and H$_3^+$, calculate the spectroscopic constants of D$_3^+$. The molecular ion D$_3^+$ was first produced by J.T. Shy, J.W. Farley, W.E. Lamb, Jr, and W.H. Wing (*Phys. Rev. Lett.* **45**, 535 (1980)) who observed the $\nu_2(E')$ band in the infrared.

17

Spectroscopy 2: electronic transitions

Simple analytical expressions for electronic energy levels cannot be given, so this chapter concentrates on the qualitative features of electronic transitions. A common theme throughout the chapter is that electronic transitions occur within a stationary nuclear framework. We pay particular attention to spontaneous radiative decay processes, which include fluorescence and phosphorescence. A specially important example of stimulated radiative decay is that responsible for the action of lasers, and we see how this stimulated emission may be achieved and employed.

An extreme case of electron excitation is photoionization, in which an electron is expelled completely from a molecule. Photoionization spectroscopy is used to build up detailed pictures of orbital energies and the role of particular electrons in bonding. Hence it is an experimental technique for exploring the concepts encountered in Chapter 14.

The energies needed to change the electron distributions of molecules are of the order of several electronvolts (1 eV is equivalent to about $8000\ cm^{-1}$ or $100\ kJ\ mol^{-1}$). Consequently, the photons emitted or absorbed when such changes occur lie in the visible and ultraviolet regions of the spectrum (Table 17.1). In some cases the relocation of electrons may be so extensive that it results in dissociation of the molecule.

Table 17.1* Colour, frequency, and energy of light

Colour	λ/nm	$\nu/(10^{14}\ Hz)$	$E/(kJ\ mol^{-1})$
Infrared	>1000	<3.0	<120
Red	700	4.3	170
Yellow	580	5.2	210
Blue	470	6.4	250
Ultraviolet	<300	>10	>400

*More values are given in the *Data section* at the end of this volume.

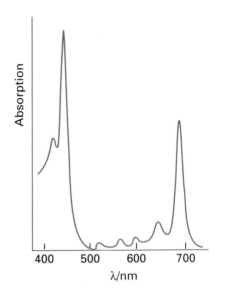

17.1 The absorption spectrum of chlorophyll in the visible region. Note that it absorbs in the red and blue regions, and that green light is not absorbed.

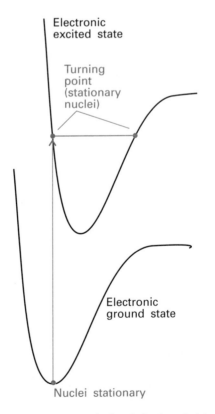

17.2 According to the Franck–Condon principle, the most intense vibronic transition is from the ground vibrational state to the vibrational state lying vertically above it. Transitions to other vibrational levels also occur, but with lower intensity.

The characteristics of electronic transitions

The nuclei in a molecule are subjected to different forces after an electronic transition has occurred, and the molecule may respond by starting to vibrate. The resulting vibrational structure of electronic transitions can be resolved for gaseous samples, but in a liquid or solid the lines usually merge together and result in a broad, almost featureless band (Fig. 17.1). Superimposed on the vibrational transitions that accompany the electronic transition of a molecule in the gas phase is an additional branch structure that arises from rotational transitions. The electronic spectra of gaseous samples are therefore very complicated, but rich in information.

17.1 The vibrational structure

The widths of electronic absorption bands in liquid samples can be traced to their vibrational structure, which is usually unresolved in solution. This structure, which can be resolved in gases and weakly interacting solvents, arises from the vibrational transitions that accompany electronic excitation.

(a) The Franck–Condon principle

The vibrational structure of an electronic transition is explained by the **Franck–Condon principle**:

> **Because the nuclei are so much more massive than the electrons, an electronic transition takes place very much faster than the nuclei can respond.**

As a result of the transition, electron density is rapidly built up in new regions of the molecule and removed from others, and the initially stationary nuclei suddenly experience a new force field. They respond to the new force by beginning to vibrate, and (in classical terms) swing backwards and forwards from their original separation, which was maintained during the rapid electronic excitation. The stationary equilibrium separation of the nuclei in the initial electronic state therefore becomes a stationary turning point in the final electronic state (Fig. 17.2).

The quantum mechanical version of the Franck–Condon principle refines this picture. Before the absorption, the molecule is in the lowest vibrational state of its lowest electronic state (Fig. 17.3); the most probable location of the nuclei is at their equilibrium separation, R_e. The electronic transition is most likely to take place when the nuclei have this separation. When the transition occurs, the molecule is excited to the state represented by the upper curve. According to the Franck–Condon principle, the nuclear framework remains constant during this excitation, so we may imagine the transition as being up the vertical line in Fig. 17.3. The vertical line is the origin of the expression **vertical transition**, which is used to denote an electronic transition that occurs without change of nuclear geometry.

The vertical transition cuts through several vibrational levels of the upper electronic state. The level marked * is the one in which the nuclei are most probably at the same initial separation R_e (because the vibrational wavefunction has maximum amplitude there), so this vibrational state is the most probable state for the termination of the transition. However, it is not the only accessible vibrational state because several nearby states have an appreciable probability of the nuclei being at the separation R_e. Therefore, transitions occur to all the vibrational states in this region, but most intensely to the state with a vibrational wavefunction that peaks most strongly near R_e.

The vibrational structure of the spectrum depends on the relative horizontal position of the two potential energy curves, and a long **vibrational progression**, a lot of vibrational structure, is stimulated if the upper potential energy curve is appreciably displaced horizontally from the lower. The upper curve is usually displaced to greater equilibrium bond lengths because electronically excited states usually have more antibonding character than electronic ground states.

The separation of the vibrational lines of an electronic absorption spectrum depends on the vibrational energies of the *upper* electronic state. Hence, electronic absorption spectra may be used to assess the force fields and dissociation energies of electronically excited molecules (for example, via a Birge–Sponer plot, Section 16.11b).

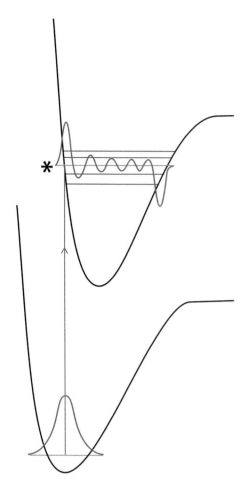

17.3 In the quantum mechanical version of the Franck–Condon principle, the molecule undergoes a transition to the upper vibrational state that most closely resembles the vibrational wavefunction of the vibrational ground state of the lower electronic state. The two wavefunctions shown here have the greatest overlap integral of all the vibrational states of the upper electronic state and hence are most closely similar.

(b) Franck–Condon factors

The quantitative form of the Franck–Condon principle is derived from the expression for the transition dipole moment, $\boldsymbol{\mu}_{fi} = \langle f|\boldsymbol{\mu}|i\rangle$. The dipole moment operator is a sum over all nuclei and electrons in the molecule:

$$\boldsymbol{\mu} = -e\sum_i \mathbf{r}_i + e\sum_I Z_I \mathbf{R}_I \tag{1}$$

where the vectors are the distances from the centre of charge of the molecule. The intensity of the transition is proportional to the square modulus, $|\boldsymbol{\mu}_{fi}|^2$, of the magnitude of the transition dipole moment (eqn 16.20), and we show in the *Justification* below that this intensity is proportional to the square modulus of the overlap integral, $S(v_f, v_i)$, between the vibrational states of the initial and final electronic states. This overlap integral is a measure of the match between the vibrational wavefunctions in the upper and lower electronic states: $S = 1$ for a perfect match and $S = 0$ when there is no similarity.

Justification 17.1

The overall state of the molecule consists of an electronic part, $|\varepsilon\rangle$, and a vibrational part, $|v\rangle$. Therefore, within the Born–Oppenheimer approximation, the transition dipole moment factorizes as follows:

$$\boldsymbol{\mu}_{fi} = \langle \varepsilon_f v_f| \left\{ -e\sum_i \mathbf{r}_i + e\sum_I Z_I \mathbf{R}_I \right\} |\varepsilon_i v_i\rangle$$

$$= -e\sum_i \langle \varepsilon_f|\mathbf{r}_i|\varepsilon_i\rangle\langle v_f|v_i\rangle + e\sum_I Z_I \langle \varepsilon_f|\varepsilon_i\rangle\langle v_f|\mathbf{R}_I|v_i\rangle$$

The second term on the right of the last row is zero, because $\langle \varepsilon_f|\varepsilon_i\rangle = 0$ for two different electronic states (they are orthogonal). Therefore,

$$\boldsymbol{\mu}_{fi} = -e\sum_i \langle \varepsilon_f|\mathbf{r}_i|\varepsilon_i\rangle\langle v_f|v_i\rangle = \boldsymbol{\mu}_{\varepsilon_f,\varepsilon_i} S(v_f, v_i) \tag{2}$$

where

$$\boldsymbol{\mu}_{\varepsilon_f,\varepsilon_i} = -e\sum_i \langle \varepsilon_f|\mathbf{r}_i|\varepsilon_i\rangle \qquad S(v_f, v_i) = \langle v_f|v_i\rangle \tag{3}$$

The matrix element $\boldsymbol{\mu}_{\varepsilon_f,\varepsilon_i}$ is the transition dipole moment arising from the redistribution of electrons (and a measure of the 'kick' this redistribution gives to the electromagnetic field, and vice versa for absorption). The factor $S(v_f, v_i)$ is the overlap integral between the vibrational state $|v_i\rangle$ in the initial electronic state of the molecule, and the vibrational state $|v_f\rangle$ in the final electronic state of the molecule.

Because the transition intensity is proportional to the square of the magnitude of the transition dipole moment, the intensity of an absorption is proportional to $|S(v_f, v_i)|^2$, which is known as the **Franck–Condon factor** for the transition. It follows that, the greater the

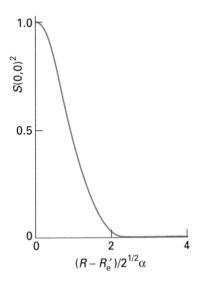

17.4 The Franck–Condon factor for the arrangement discussed in Example 17.1.

overlap of the vibrational state wavefunction in the upper electronic state with the vibrational wavefunction in the lower electronic state, the greater the absorption intensity of that particular simultaneous electronic and vibrational transition. This conclusion is the basis of the illustration in Fig. 17.3, where we see that the vibrational wavefunction of the ground state has the greatest overlap with the vibrational states that have peaks at similar bond lengths in the upper electronic state.

Example 17.1 Calculating a Franck–Condon factor

Consider the transition from one electronic state to another, their bond lengths being R_e and R'_e and their force constants equal. Calculate the Franck–Condon factor for the 0–0 transition and show that the transition is most intense when the bond lengths are equal.

Method We need to calculate $S(0,0)$, the overlap integral of the two ground-state vibrational wavefunctions, and then take its square. The difference between harmonic and anharmonic vibrational wavefunctions is negligible for $v = 0$, so harmonic oscillator wavefunctions can be used (Table 12.1).

Answer We use the (real) wavefunctions

$$\psi_0 = \left(\frac{1}{\alpha\pi^{1/2}}\right)^{1/2} e^{-y^2/2} \qquad \psi'_0 = \left(\frac{1}{\alpha\pi^{1/2}}\right)^{1/2} e^{-y'^2/2}$$

where $y = (R - R_e)/\alpha$ and $y' = (R - R'_e)/\alpha$, with $\alpha = (\hbar^2/mk)^{1/4}$ (Section 12.5a). The overlap integral is

$$S(0,0) = \langle 0|0 \rangle = \int_{-\infty}^{\infty} \psi'_0 \psi_0 \, dR = \frac{1}{\pi^{1/2}} \int_{-\infty}^{\infty} e^{-(y^2+y'^2)/2} \, dy$$

We now write $\alpha z = R - \frac{1}{2}(R_e + R'_e)$, and manipulate this expression into

$$S(0,0) = \frac{1}{\pi^{1/2}} e^{-(R_e - R'_e)^2/4\alpha^2} \int_{-\infty}^{\infty} e^{-z^2} \, dz$$

The value of the integral is $\pi^{1/2}$. Therefore, the overlap integral is

$$S(0,0) = e^{-(R_e - R'_e)^2/4\alpha^2}$$

and the Franck–Condon factor is

$$S(0,0)^2 = e^{-(R_e - R'_e)^2/2\alpha^2}$$

This factor is equal to 1 when $R'_e = R_e$ and decreases as the equilibrium bond lengths diverge from each other (Fig. 17.4).

Comment For Br_2, $R_e = 228$ pm and there is an upper state with $R'_e = 266$ pm. Taking the vibrational wavenumber as 250 cm^{-1}, gives $S(0,0)^2 = 5.1 \times 10^{-10}$, so the intensity of the 0–0 transition is only 5.1×10^{-10} what it would have been if the potential curves had been directly above each other.

- -

Self-test 17.1 Suppose the vibrational wavefunctions can be approximated by rectangular functions of width W and W', centred on the equilibrium bond lengths (Fig. 17.5). Find the corresponding Franck–Condon factors when the centres are coincident and $W' < W$.

$$[S^2 = W'/W]$$

17.5 The model wavefunctions used in Self-test 17.1.

17.2 Specific types of transitions

The absorption of a photon can often be traced to the excitation of specific types of electrons or to electrons that belong to a small group of atoms. For example, when a carbonyl group ($>C=O$) is present, an absorption at about 290 nm is normally observed, although its precise location depends on the nature of the rest of the molecule. Groups with characteristic optical absorptions are called **chromophores** (from the Greek for 'colour bringer'), and their presence often accounts for the colours of substances (Table 17.2).

Table 17.2* Absorption characteristics of some groups and molecules

Group	$\tilde{\nu}_{max}/cm^{-1}$	λ_{max}/nm	$\varepsilon_{max}/(L\,mol^{-1}\,cm^{-1})$
$C=O(\pi^* \leftarrow \pi)$	61 000	163	15 000
	57 300	174	5 500
$C=O(\pi^* \leftarrow n)$	37–35 000	270–290	10–20
$H_2O(\pi^* \leftarrow n)$	60 000	167	7 000

*More values are given in the *Data section*.

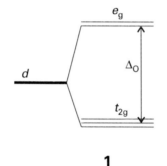

1

(a) d–d transitions

All five d orbitals of a given shell are degenerate in a free atom. In a d-metal complex, where the immediate environment of the atom is no longer spherical, the d orbitals are not all degenerate, and electrons can absorb energy by making transitions between them. In an octahedral complex, such as $[Ti(OH_2)_6]^{3+}$, the five d orbitals of the central atom are split into two sets (**1**), a triply degenerate set labelled t_{2g} and a doubly degenerate set labelled e_g. The three t_{2g} orbitals lie below the two e_g orbitals; the difference in energy is denoted Δ_O and called the **ligand-field splitting parameter** (the O denoting octahedral symmetry). The d orbitals also divide into two sets in a tetrahedral complex, but in this case the e orbitals lie below the t_2 orbitals and their separation is written Δ_T. Neither separation is large, so transitions between the two sets of orbitals typically occur in the visible region of the spectrum even though they are electronic. The transitions are responsible for many of the colours that are so characteristic of d-metal complexes. As an example, the spectrum of $[Ti(OH_2)_6]^{3+}$ near 20 000 cm^{-1} (500 nm) is shown in Fig. 17.6, and can be ascribed to the promotion of its single d electron from a t_{2g} orbital to an e_g orbital. The wavenumber of the absorption maximum suggests that $\Delta_O \approx 20\,000$ cm^{-1} for this complex, which corresponds to about 2.5 eV.

(b) Vibronic transitions

A major problem with the interpretation of visible spectra of octahedral complexes is that their d–d transitions are forbidden. The **Laporte selection rule** for centrosymmetric complexes (those with a centre of inversion) and atoms states that:

> **The only allowed transitions are transitions that are accompanied by a change of parity.**

That is, u → g and g → u transitions are allowed, but g → g and u → u transitions are forbidden.

Justification 17.2

The transition dipole moment in eqn 16.19 vanishes unless the integrand is invariant under all symmetry operations of the molecule. Hence, in a centrosymmetric complex it must

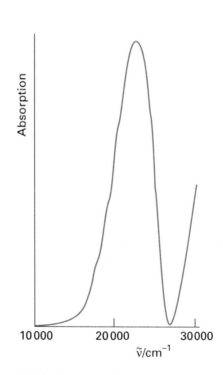

17.6 The electronic absorption spectrum of $[Ti(OH_2)_6]^{3+}$ in aqueous solution.

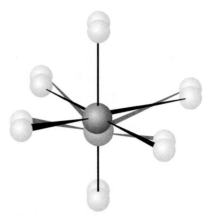

17.7 A d–d transition is parity-forbidden because it corresponds to a g–g transition. However, a vibration of the molecule can destroy the inversion symmetry of the molecule and the g, u classification no longer applies. The removal of the centre of symmetry gives rise to a vibronically allowed transition.

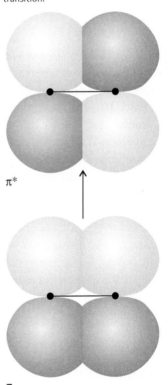

π^*

π

17.8 A C=C double bond acts as a chromophore. One of its important transitions is the $\pi^* \leftarrow \pi$ transition illustrated here, in which an electron is promoted from a π orbital to the corresponding antibonding orbital.

2

have even (g) parity (in O_h it needs to be A_{1g}, but only the g symmetry is important for this argument). The three components of the dipole moment operator transform like x, y, and z, and are all u. Therefore, for a d–d transition (a g $\rightarrow$ g transition), the overall parity of the transition dipole is g × u × g = u, so it must be zero. Likewise, for a u $\rightarrow$ u transition, the overall parity is u × u × u = u, so it must also vanish. Hence, transitions without a change of parity are forbidden.

A forbidden g $\rightarrow$ g transition can become allowed if the centre of symmetry is eliminated by an asymmetrical vibration, such as the one shown in Fig. 17.7. When the centre of symmetry is lost, d–d transitions are no longer parity-forbidden, so the $e_g \leftarrow t_{2g}$ transition becomes weakly allowed. A transition that derives its intensity from an asymmetrical vibration of a molecule is called a **vibronic transition**.

(c) Charge-transfer transitions

A complex may absorb radiation as a result of the transfer of an electron from the ligands into the d orbitals of the central atom, or vice versa. In such **charge-transfer transitions** the electron moves through a considerable distance, which means that the transition dipole moment may be large and the absorption is correspondingly intense. This mode of chromophore activity is shown by the permanganate ion, MnO_4^-, and accounts for its intense violet colour (which arises from strong absorption within the range 420–700 nm). In this oxoanion, the electron migrates from an orbital that is largely confined to the O atom ligands to an orbital that is largely confined to the Mn atom. It is therefore an example of a **ligand-to-metal charge-transfer transition** (LMCT). The reverse migration, a **metal-to-ligand charge-transfer transition** (MLCT), can also occur. An example is the transfer of a d electron into the antibonding π orbitals of an aromatic ligand. The resulting excited state may have a very long lifetime if the electron is extensively delocalized over several aromatic rings, and such species can participate in photochemically induced redox reactions.

(d) $\pi^* \leftarrow \pi$ and $\pi^* \leftarrow n$ transitions

Absorption by a C=C double bond excites a π electron into an antibonding π^* orbital (Fig. 17.8). The chromophore activity is therefore due to a $\pi^* \leftarrow \pi$ transition (which is normally read 'π to π-star transition'). Its energy is about 7 eV for an unconjugated double bond, which corresponds to an absorption at 180 nm (in the ultraviolet). When the double bond is part of a conjugated chain, the energies of the molecular orbitals lie closer together and the $\pi^* \leftarrow \pi$ transition moves to longer wavelength; it may even lie in the visible region if the conjugated system is long enough.

An important example of a $\pi^* \leftarrow \pi$ transition is provided by the photochemical mechanism of vision. The retina of the eye contains 'visual purple', which is a protein in combination with 11-*cis*-retinal (2). The 11-*cis*-retinal acts as a chromophore, and is the primary receptor for photons entering the eye. A solution of 11-*cis*-retinal absorbs at about 380 nm, but in combination with the protein (a link which might involve the elimination of the terminal carbonyl) the absorption maximum shifts to about 500 nm and tails into the blue. The conjugated double bonds are responsible for the ability of the molecule to absorb over the entire visible region, but they also play another important role. In its electronically excited state the conjugated chain can isomerize, one half of the chain being able to twist about an excited C=C bond and forming 11-*trans*-retinal (3). The primary step in vision therefore appears to be photon absorption followed by isomerization: the uncoiling of the molecule then triggers a nerve impulse to the brain.

3

The transition responsible for absorption in carbonyl compounds can be traced to the lone pairs of electrons on the O atom. The Lewis concept of a 'lone pair' of electrons is represented in molecular orbital theory by a pair of electrons in an orbital confined largely to one atom and not appreciably involved in bond formation. One of these electrons may be excited into an empty π^* orbital of the carbonyl group (Fig. 17.9), which gives rise to a $\pi^* \leftarrow n$ transition (an 'n to π-star transition'). Typical absorption energies are about 4 eV (290 nm). Because $\pi^* \leftarrow n$ transitions in carbonyls are symmetry forbidden, the absorptions are weak.

The fates of electronically excited states

A **radiative decay process** is a process in which a molecule discards its excitation energy as a photon. A more common fate is **nonradiative decay**, in which the excess energy is transferred into the vibration, rotation, and translation of the surrounding molecules. This thermal degradation converts the excitation energy into thermal motion of the environment (that is, to 'heat'). An excited molecule may also take part in a chemical reaction, as we discuss in Part 3.

17.3 Fluorescence and phosphorescence

In **fluorescence**, the spontaneously emitted radiation ceases immediately after the exciting radiation is extinguished (Fig. 17.10). In **phosphorescence**, the spontaneous emission may persist for long periods (even hours, but characteristically seconds or fractions of seconds). The difference suggests that fluorescence is an immediate conversion of absorbed radiation into re-emitted energy, and that phosphorescence involves the storage of energy in a reservoir from which it slowly leaks.

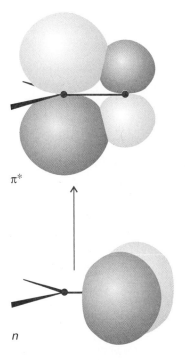

π^*

n

17.9 A carbonyl (CO) group acts as a chromophore primarily on account of the excitation of a nonbonding O lone-pair electron to an antibonding CO π orbital.

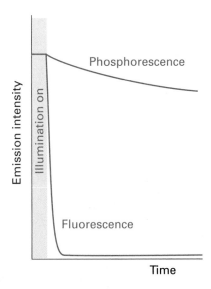

17.10 The empirical (observation-based) distinction between fluorescence and phosphorescence is that the former is extinguished immediately the exciting source is removed, whereas the latter continues with relatively slowly diminishing intensity.

(a) Fluorescence

Figure 17.11 shows the sequence of steps involved in fluorescence. The initial absorption takes the molecule to an excited electronic state, and if the absorption spectrum were monitored it would look like the one shown in Fig. 17.12a. The excited molecule is subjected to collisions with the surrounding molecules, and as it gives up energy nonradiatively it steps down the ladder of vibrational levels to the lowest vibrational level of the electronically excited molecular state. The surrounding molecules, however, might now be unable to accept the larger energy difference needed to lower the molecule to the ground electronic state. It might therefore survive long enough to undergo spontaneous emission, and emit the remaining excess energy as radiation. The downward electronic transition is vertical (in accord with the Franck–Condon principle) and the fluorescence spectrum has a vibrational structure characteristic of the *lower* electronic state (Fig. 17.12b).

Provided they can be seen, the 0–0 absorption and fluorescence transitions can be expected to be coincident. The absorption spectrum arises from 0–0, 1–0, 2–0, etc. transitions and the peaks occur at progressively higher wavenumber and with intensities governed by the Franck–Condon principle. The fluorescence spectrum arises from 0–0, 0–1, 0–2, etc. *downward* transitions, and hence the peaks occur with decreasing wavenumbers. The 0–0 absorption and fluorescence peaks are not always exactly coincident because the solvent may interact differently with the solute in the ground and excited states (for instance, the hydrogen bonding pattern might differ). Because the solvent molecules do not have time to rearrange during the transition, the absorption occurs in an environment characteristic of the solvated ground state; however, the fluorescence occurs in an environment characteristic of the solvated excited state (Fig. 17.13).

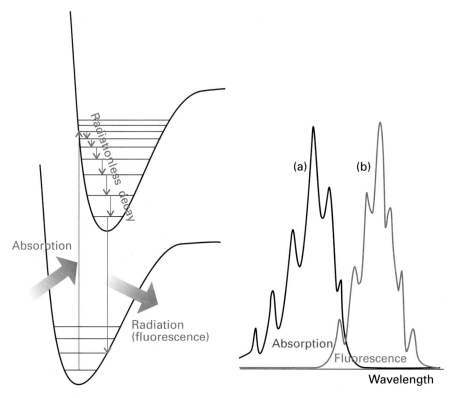

17.11 The sequence of steps leading to fluorescence. After the initial absorption, the upper vibrational states undergo radiationless decay by giving up energy to the surroundings. A radiative transition then occurs from the vibrational ground state of the upper electronic state.

17.12 An absorption spectrum (a) shows a vibrational structure characteristic of the upper state. A fluorescence spectrum (b) shows a structure characteristic of the lower state; it is also displaced to lower frequencies (but the 0–0 transitions are coincident) and resembles a mirror image of the absorption.

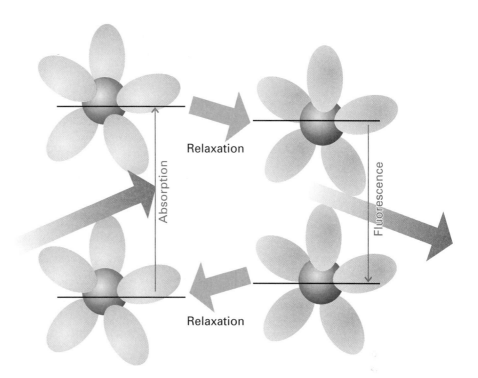

17.13 The solvent can shift the fluorescence spectrum relative to the absorption spectrum. On the left we see that the absorption occurs with the solvent (the ellipses) in the arrangement characteristic of the ground electronic state of the molecule (the sphere). However, before fluorescence occurs, the solvent molecules relax into a new arrangement, and that arrangement is preserved during the subsequent radiative transition.

Fluorescence occurs at a lower frequency than the incident radiation because the emissive transition occurs after some vibrational energy has been discarded into the surroundings. The vivid oranges and greens of fluorescent dyes are an everyday manifestation of this effect: they absorb in the ultraviolet and blue, and fluoresce in the visible. The mechanism also suggests that the intensity of the fluorescence ought to depend on the ability of the solvent molecules to accept the electronic and vibrational quanta. It is indeed found that a solvent composed of molecules with widely spaced vibrational levels (such as water) can in some cases accept the large quantum of electronic energy and so extinguish, or 'quench', the fluorescence.

(b) Phosphorescence

Figure 17.14 shows the sequence of events leading to phosphorescence for a molecule with a singlet ground state. The first steps are the same as in fluorescence, but the presence of a triplet excited state plays a decisive role.[1] The singlet and triplet excited states share a common geometry at the point where their potential energy curves intersect. Hence, if there is a mechanism for unpairing two electron spins (and achieving the conversion of $\uparrow\downarrow$ to $\uparrow\uparrow$), the molecule may undergo **intersystem crossing** and become a triplet state. We saw in the discussion of atomic spectra (Section 13.9d) that singlet–triplet transitions may occur in the presence of spin–orbit coupling, and the same is true in molecules. We can expect intersystem crossing to be important when a molecule contains a moderately heavy atom (such as S), because then the spin–orbit coupling is large.

If an excited molecule crosses into a triplet state, it continues to deposit energy into the surroundings. However, it is now stepping down the triplet's vibrational ladder, and at the lowest energy level it is trapped because the triplet state is at a lower energy than the corresponding singlet (recall Hund's rule, Section 13.7). The solvent cannot absorb the final, large quantum of electronic excitation energy, and the molecule cannot radiate its energy because return to the ground state is spin-forbidden. The radiative transition, however, is not totally forbidden because the spin–orbit coupling that was responsible for the

17.14 The sequence of steps leading to phosphorescence. The important step is the intersystem crossing, the switch from a singlet state to a triplet state brought about by spin–orbit coupling. The triplet state acts as a slowly radiating reservoir because the return to the ground state is spin-forbidden.

---

1   We first encountered triplet states in Section 13.7: they are states in which two electrons have parallel spins.

intersystem crossing also breaks the selection rule. The molecules are therefore able to emit weakly, and the emission may continue long after the original excited state was formed.

The mechanism accounts for the observation that the excitation energy seems to get trapped in a slowly leaking reservoir. It also suggests (as is confirmed experimentally) that phosphorescence should be most intense from solid samples: energy transfer is then less efficient and intersystem crossing has time to occur as the singlet excited state steps slowly past the intersection point. The mechanism also suggests that the phosphorescence efficiency should depend on the presence of a moderately heavy atom (with strong spin-orbit coupling), which is in fact the case. The confirmation of the mechanism is the experimental observation (using the sensitive resonance techniques described in Chapter 18) that the sample is paramagnetic while the reservoir state, with its unpaired electron spins, is populated.

The various types of nonradiative and radiative transitions that can occur in molecules are often represented on a schematic **Jablonski diagram** of the type shown in Fig. 17.15.

17.4 Dissociation and predissociation

Another fate for an electronically excited molecule is **dissociation**, the breaking of bonds (Fig. 17.16). The onset of dissociation can be detected in an absorption spectrum by seeing

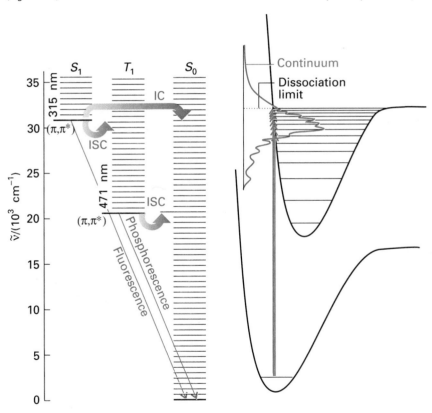

17.15 A Jablonski diagram (here, for naphthalene) is a simplified portrayal of the relative positions of the electronic energy levels of a molecule. Vibrational levels of states of a given electronic state lie above each other, but the relative horizontal locations of the columns bear no relation to the nuclear separations in the states. The ground vibrational states of each electronic state are correctly located vertically but the other vibrational states are shown only schematically. (IC: internal conversion; ISC: intersystem crossing.)

17.16 When absorption occurs to unbound states of the upper electronic state, the molecule dissociates and the absorption is a continuum. Below the dissociation limit the electronic spectrum shows a normal vibrational structure.

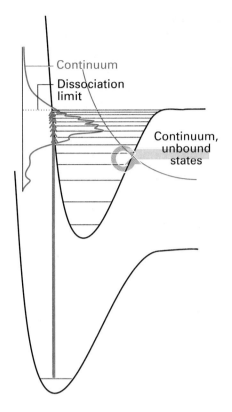

17.17 When a dissociative state crosses a bound state, as in the upper part of the illustration, molecules excited to levels near the crossing may dissociate. This process is called predissociation, and is detected in the spectrum as a loss of vibrational structure that resumes at higher frequencies.

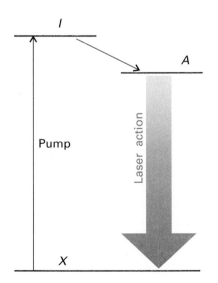

17.18 The transitions involved in one kind of three-level laser. The pumping pulse populates the intermediate state I, which in turn populates the laser state A. The laser transition is the stimulated emission $A \rightarrow X$.

that the vibrational structure of a band terminates at a certain energy. Absorption occurs in a continuous band above this **dissociation limit** because the final state is an unquantized translational motion of the fragments. Locating the dissociation limit is a valuable way of determining the bond dissociation energy.

In some cases, the vibrational structure disappears but resumes at higher photon energies. This **predissociation** can be interpreted in terms of the molecular potential energy curves shown in Fig. 17.17. When a molecule is excited to a vibrational level, its electrons may undergo a reorganization that results in it undergoing an **internal conversion**, a radiationless conversion to another state of the same multiplicity. An internal conversion occurs most readily at the point of intersection of the two molecular potential energy curves, because there the nuclear geometries of the two states are the same. The state into which the molecule converts may be dissociative, so the states near the intersection have a finite lifetime, and hence their energies are imprecisely defined. As a result, the absorption spectrum is blurred in the vicinity of the intersection. When the incoming photon brings enough energy to excite the molecule to a vibrational level high above the intersection, the internal conversion does not occur (the nuclei are unlikely to have the same geometry). Consequently, the levels resume their well-defined, vibrational character with correspondingly well-defined energies, and the line structure resumes on the high-frequency side of the blurred region.

Lasers

Lasers have transformed chemistry as much as they have transformed the everyday world. In this section, we see some of the principles of their operation, and then explore their applications in chemistry. Lasers lie very much on the frontier of physics and chemistry, for their operation depends on details of optics and, in some cases, of solid-state processes. We shall concentrate on the more chemical aspects of their operation, particularly the materials from which they are made and the events taking place within them.

17.5 General principles of laser action

The word laser is an acronym formed from light amplification by stimulated emission of radiation. In stimulated emission, an excited state is stimulated to emit a photon by radiation of the same frequency; the more photons that are present, the greater the probability of the emission. The essential feature of laser action is positive-feedback: the more photons present of the appropriate frequency, the more photons of that frequency that will be stimulated to form.

(a) Population inversion

One requirement of laser action is the existence of a **metastable excited state**, an excited state with a long enough lifetime for it to participate in stimulated emission. Another requirement is the existence of a greater population in the metastable state than in the lower state where the transition terminates, for then there will be a net emission of radiation. Because at thermal equilibrium the opposite is true, it is necessary to achieve a **population inversion** in which there are more molecules in the upper state than in the lower.

One way of achieving population inversion is illustrated in Fig. 17.18. The molecule is excited to an intermediate state I, which then gives up some of its energy nonradiatively and changes into a lower state A; the laser transition is the return of A to the ground state X. Because three energy levels are involved overall, this arrangement leads to a **three-level laser**. In practice, I consists of many states, all of which can convert to the upper of the two

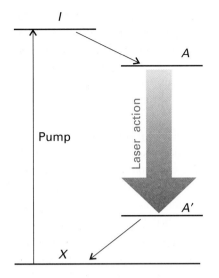

17.19 The transitions involved in a four-level laser. Because the laser transition terminates in an excited state (A'), the population inversion betweeen A and A' is much easier to achieve.

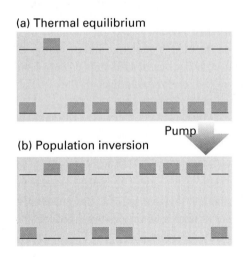

(a) Thermal equilibrium

(b) Population inversion

Pump

(c) Laser action

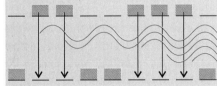

17.20 A schematic illustration of the steps leading to laser action. (a) The Boltzmann population of states, with more atoms in the ground state. (b) When the initial state absorbs, the populations are inverted (the atoms are pumped to the excited state). (c) A cascade of radiation then occurs, as one emitted photon stimulates another atom to emit, and so on. The radiation is coherent (phases in step).

laser states A. The $I \leftarrow X$ transition is stimulated with an intense flash of light in the process called **pumping**. The pumping is often achieved with an electric discharge through xenon or with the light of another laser. The conversion of I to A should be rapid, and the laser transitions from A to X should be relatively slow.

The disadvantage of this three-level arrangement is that it is difficult to achieve population inversion, because so many ground-state molecules must be converted to the excited state by the pumping action. The arrangement adopted in a **four-level laser** simplifies this task by having the laser transition terminate in a state A' other than the ground state (Fig. 17.19). Because A' is unpopulated initially, any population in A corresponds to a population inversion, and we can expect laser action if A is sufficiently metastable. Moreover, this population inversion can be maintained if the $A' \rightarrow X$ transitions are rapid, for these transitions will deplete any population in A' that stems from the laser transition, and keep the state A' relatively empty.

(b) Cavity and mode characteristics

The laser medium is confined to a cavity that ensures that only certain photons of a particular frequency, direction of travel, and state of polarization are generated abundantly. The cavity is essentially a region between two mirrors, which reflect the light back and forth. This arrangement can be regarded as a version of the particle in a box, with the particle now being a photon. As in the treatment of a particle in a box (Section 12.1), the only wavelengths that can be sustained satisfy

$$n \times \tfrac{1}{2}\lambda = L \tag{4}$$

where n is an integer and L is the length of the cavity. That is, only an integral number of half-wavelengths fit into the cavity; all other waves undergo destructive interference with themselves. In addition, not all wavelengths that can be sustained by the cavity are amplified by the laser medium (many fall outside the range of frequencies of the laser transitions), so only a few contribute to the laser radiation. These wavelengths are the **resonant modes** of the laser.

Photons with the correct wavelength for the resonant modes of the cavity and the correct frequency to stimulate the laser transition are highly amplified. One photon might be generated spontaneously, and travel through the medium. It stimulates the emission of another photon, which in turn stimulates more (Fig. 17.20). The cascade of energy builds up rapidly, and soon the cavity is an intense reservoir of radiation at all the resonant modes it can sustain. Some of this radiation can be withdrawn if one of the mirrors is partially transmitting.

The resonant modes of the cavity have various natural characteristics, and to some extent may be selected. Only photons that are travelling strictly parallel to the axis of the cavity undergo more than a couple of reflections, so only they are amplified, all others simply vanishing into the surroundings. Hence, laser light generally forms a beam with very low divergence. It may also be polarized, with its electric vector in a particular plane (or in some other state of polarization), by including a polarizing filter into the cavity or by making use of polarized transitions in a solid medium.

Laser radiation is **coherent** in the sense that the electromagnetic waves are all in step. In **spatial coherence** the waves are in step across the cross-section of the beam emerging from the cavity. In **temporal coherence** the waves remain in step along the beam. The latter is normally expressed in terms of a **coherence length**, l_C, and is related to the range of wavelengths, $\Delta\lambda$, present in the beam:

$$l_C = \frac{\lambda^2}{2\Delta\lambda} \tag{5}$$

$Y_3Al_5O_{12}$), and is then known as a Nd-YAG laser. A cheaper medium is glass, but glass is a poorer thermal conductor than YAG and the laser must be pulsed. A neodymium laser operates at a number of wavelengths in the infrared, the band at 1064 nm being most common. The transition at 1064 nm is very efficient and the laser is capable of substantial power output. The power is great enough for frequency doubling to be used efficiently. **Frequency doubling** is a technique in which the laser beam is converted to radiation with twice (and in general a multiple) of its initial frequency as it passes through a suitable transparent material. A frequency-doubled Nd-YAG laser produces green light at 532 nm.

Example 17.3 Accounting for multiphoton phenomena

Show that if a substance responds nonlinearly to incident radiation of frequency ω, then it may act as the source of radiation of twice the incident frequency.

Method Radiation of a particular frequency arises from oscillations of an electric dipole at that frequency. Therefore, express the induced electric dipole moment of the system in terms of powers of the applied electric field, and write the powers of harmonic (cosine) terms as sums and differences of cosine terms. Inspect the sum to see if $\cos 2\omega t$ is present.

Answer The incident electric field $\mathcal{E}$ induces an electric dipole of magnitude μ and, allowing for nonlinear response, we can write

$$\mu = \alpha\mathcal{E} + \beta\mathcal{E}^2 + \cdots$$

The nonlinear terms can be expanded as follows if we suppose that the incident electric field is $\mathcal{E}_0 \cos \omega t$:

$$\beta\mathcal{E}^2 = \beta\mathcal{E}_0^2 \cos^2 \omega t = \tfrac{1}{2}\beta\mathcal{E}_0^2(1 + \cos 2\omega t)$$

Hence, the nonlinear term contributes an induced electric dipole that oscillates at the frequency 2ω and which can act as a source of radiation of that frequency.

Self-test 17.3 Show that, if a substance responds nonlinearly to two sources of radiation, one of frequency ω_1 and the other of frequency ω_2, then it may give rise to radiation of the sum and difference of the two frequencies.
$$[\beta\mathcal{E}^2 \propto \cos(\omega_1 + \omega_2)t + \cos(\omega_1 - \omega_2)t]$$

(b) Gas lasers

Because gas lasers can be cooled by a rapid flow of the gas through the cavity, they can be used to generate high powers. The pumping is normally achieved using a gas that is different from the gas responsible for the laser emission itself.

In the **helium-neon laser** the active medium is a mixture of helium and neon in a mole ratio of about 5 : 1 (Fig. 17.27). The initial step is the excitation of an He atom to the metastable $1s^1 2s^1$ configuration by using an electric discharge (the collisions of electrons and ions cause transitions that are not restricted by electric-dipole selection rules). The excitation energy of this transition happens to match an excitation energy of neon, and during an He-Ne collision efficient transfer of energy may occur, leading to the production of highly excited, metastable Ne atoms with unpopulated intermediate states. Laser action generating 633 nm radiation (among about 100 other lines) then occurs.

The **argon-ion laser** (Fig. 17.28), one of a number of 'ion lasers', consists of argon at about 1 Torr, through which is passed an electric discharge. The discharge results in the formation of Ar^+ and Ar^{2+} ions in excited states, which undergo a laser transition to a lower

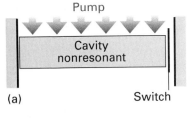

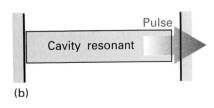

(a)

Switch

(b)

17.21 The principle of Q-switching. The excited state is populated while the cavity is nonresonant. Then the resonance characteristics are suddenly restored, and the stimulated emission emerges in a giant pulse.

If the beam were perfectly monochromatic, with strictly one wavelength present, then $\Delta\lambda$ would be zero, and the waves would remain in step for an infinite distance. When many wavelengths are present, the waves get out of step in a short distance and the coherence length is small. A typical light bulb gives out light with a coherence length of only about 400 nm; a He-Ne laser with $\Delta\lambda \approx 2$ pm has a coherence length of about 10 cm.

(c) Q-switching

A laser can generate radiation for as long as the population inversion is maintained. A laser can operate continuously when heat is easily dissipated, for then the population of the upper level can be replenished by pumping. Practical considerations govern whether or not continuous pumping is feasible, as we shall see when we consider some particular lasers. When overheating is a problem, the laser can be operated only in pulses, perhaps of microsecond or millisecond duration, so that the medium has a chance to cool or the lower state discard its population. However, it is sometimes desirable to have pulses of radiation rather than a continuous output, with a lot of power concentrated into a brief pulse. One way of achieving pulses is by **Q-switching**, the modification of the resonance characteristics of the laser cavity.[2]

Example 17.2 Relating the power and energy of a laser

A laser rated at 0.10 J can generate radiation in 3.0 ns pulses. What is the average power output per pulse?

Method The power output, P, is the energy released in an interval divided by the duration of the interval, and is expressed in watts (1 W = 1 J s^{-1}). So, to calculate the power, divide the energy output by the time over which the pulse is generated.

Answer From the data,

$$P = \frac{0.10 \text{ J}}{3.0 \times 10^{-9} \text{ s}} = 3.3 \times 10^7 \text{ J s}^{-1}$$

That is, the pulses deliver 33 MW of power.

Comment The answer gives the average power; the peak power will be larger if the pulse is not rectangular.

Self-test 17.2 Calculate the average power output of a laser in which a 2.0 J pulse can be delivered in 1.0 ns.
[2.0 GW]

The aim of Q-switching is to achieve a healthy population inversion in the absence of the resonant cavity, then to plunge the population-inverted medium into a cavity, and hence to obtain a sudden pulse of radiation. The switching may be achieved by impairing the resonance characteristics of the cavity in some way while the pumping pulse is active, and then suddenly to improve them (Fig. 17.21). One technique is to use a **saturable dye**, a dye that loses its power to absorb when many of its molecules have been excited by intense radiation. It then suddenly becomes transparent, and the cavity becomes resonant. In practice, Q-switching can give pulses of about 10 ns duration.

2 The name comes from the 'Q-factor' used as a measure of the quality of a resonance cavity in microwave engineering.

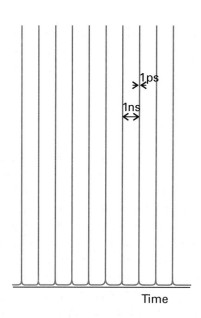

17.22 The output of a mode-locked laser consists of a stream of very narrow pulses separated by an interval equal to the time it takes for light to make a round trip inside the cavity.

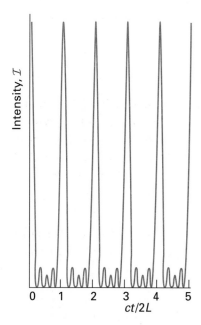

17.23 The function derived in *Justification 17.3* showing in more detail the structure of the pulses generated by a mode-locked laser.

(d) Mode locking

The technique of **mode locking** can produce pulses of picosecond duration and less. A laser radiates at a number of different frequencies, depending on the precise details of the resonance characteristics of the cavity and in particular on the number of half-wavelengths of radiation that can be trapped between the mirrors (the cavity modes). The resonant modes differ in frequency by multiples of $c/2L$ (as can be inferred from eqn 4 with $\nu = c/\lambda$). Normally, these modes have random phases relative to each other. However, it is possible to lock their phases together. Then interference occurs to give a series of sharp peaks, and the energy of the laser is obtained in picosecond bursts (Fig. 17.22). The sharpness of the peaks depends on the range of modes superimposed and, the wider the range, the narrower the pulses. In a laser with a cavity of length 30 cm, the peaks will be separated by 2 ns. If 1000 modes contribute, the width of the pulses will be 4 ps.

Justification 17.3

The general expression for a (complex) wave of amplitude $\mathcal{E}_0$ and frequency ω is $\mathcal{E}_0 e^{i\omega t}$. Therefore, each wave that can be supported by a cavity of length L has the form

$$\mathcal{E}_n(t) = \mathcal{E}_0 e^{2\pi i(\nu + nc/2L)t}$$

where ν is the lowest frequency. A wave formed by superimposing N modes with $n = 0, 1, \cdots N - 1$ has the form

$$\mathcal{E}(t) = \sum_n \mathcal{E}_n(t) = \mathcal{E}_0 e^{2\pi i \nu t} \sum_{n=0}^{N-1} e^{i\pi nct/L}$$

The sum is a geometrical progression:

$$\sum_{n=0}^{N-1} e^{i\pi nct/L} = 1 + e^{i\pi ct/L} + e^{2i\pi ct/L} + \cdots$$

$$= \frac{\sin(N\pi ct/2L)}{\sin(\pi ct/2L)} \times e^{(N-1)i\pi ct/2L}$$

The intensity, $\mathcal{I}$, of the radiation is proportional to the square modulus of the total amplitude, so

$$\mathcal{I} \propto \mathcal{E}^* \mathcal{E} = \mathcal{E}_0^2 \frac{\sin^2(N\pi ct/2L)}{\sin^2(\pi ct/2L)}$$

This function is shown in Fig. 17.23. We see that it is a series of peaks with maxima separated by $t = 2L/c$, the round-trip transit time of the light in the cavity, and that the peaks become sharper as N is increased.

Mode locking is achieved by varying the Q factor of the cavity periodically at the frequency $c/2L$. The modulation can be pictured as the opening of a shutter in synchrony with the round-trip travel time of the photons in the cavity, so only photons making the journey in that time are amplified. The modulation can be achieved by linking a prism in the cavity to a transducer driven by a radiofrequency source at a frequency $c/2L$. The transducer sets up standing-wave vibrations in the prism and modulates the loss it introduces into the cavity. Mode locking may also be accomplished passively by including a saturable dye. This procedure makes use of the fact that the **gain**, the growth in intensity of a component, is very sensitive to amplification and, once a particular frequency begins to grow, it can quickly dominate. If a saturable dye is included in the cavity, a spontaneous fluctuation in intensity—a bunching of photons—may result in its becoming transparent, and the bunch can pass through and travel to the far end of the cavity, amplifying as it goes. The dye immediately shuts down again (if it is well chosen), but opens when the intense pulse returns

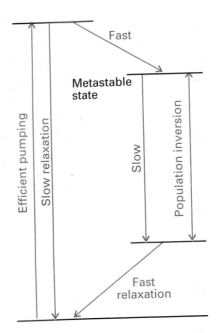

17.24 A summary of the features needed for efficient laser action.

from the mirror at the far end and saturates it. In this way, that par~ may grow to considerable intensity because it alone is stimulating

17.6 Practical lasers

Figure 17.24 summarizes the requirements for an efficient laser. requirements can be satisfied by using a variety of different systems, reviews some that are commonly available. For completeness, we include operate by using other than electronic transitions.

(a) Solid-state lasers

A **solid-state laser** is one in which the active medium is in the form of a single glass. The first successful laser, the ruby laser built by Theodore Maiman in \ example (Fig. 17.25). Ruby is Al_2O_3 containing a small proportion of Cr^{3+} ions.[3] three-level system, and the ground state, which is also the lower level of the laser tr is $^4A_2$ with three unpaired spins on each Cr^{3+} ion. The population inversion resul pumping a majority of the Cr^{3+} ions into an excited state by using an intense flas another source, followed by a radiationless transition to another excited state. The pur flash need not be monochromatic because the upper level actually consists of several s\ spanning a band of frequencies. The transition from the lower of the two excited state the ground state ($^2E \rightarrow {}^4A_2$) is the laser transition, and gives rise to red 694 nm radiati The population inversion is very difficult to sustain continuously, and in practice the rub laser is pulsed. Typical pulses from a Q-switched ruby laser might consist of 2 J pulses persisting for 10 ns, corresponding to an average power of 0.2 GW.

The **neodymium laser** is an example of a four-level laser (Fig. 17.26). In one form it consists of Nd^{3+} ions at low concentration in yttrium aluminium garnet (YAG, specifically

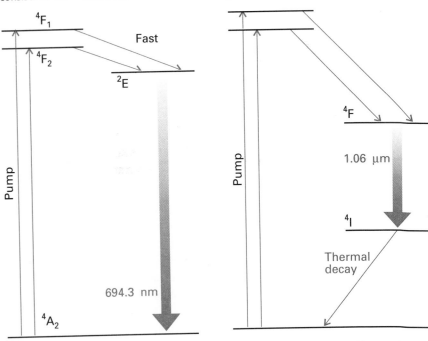

17.25 The transitions involved in a ruby laser. The laser medium, ruby, consists of Al_2O_3 doped with Cr^{3+} ions.

17.26 The transitions involved in a neodymium laser. The laser action takes place between two excited states, and the population inversion is easier to achieve than in the ruby laser.

3   The normal green of Cr^{3+} is modified to red by the distortion of the local crystal field stemming from the replacement of an Al^{3+} ion by a slightly larger Cr^{3+} ion.

If the beam were perfectly monochromatic, with strictly one wavelength present, then $\Delta\lambda$ would be zero, and the waves would remain in step for an infinite distance. When many wavelengths are present, the waves get out of step in a short distance and the coherence length is small. A typical light bulb gives out light with a coherence length of only about 400 nm; a He–Ne laser with $\Delta\lambda \approx 2$ pm has a coherence length of about 10 cm.

(c) Q-switching

A laser can generate radiation for as long as the population inversion is maintained. A laser can operate continuously when heat is easily dissipated, for then the population of the upper level can be replenished by pumping. Practical considerations govern whether or not continuous pumping is feasible, as we shall see when we consider some particular lasers. When overheating is a problem, the laser can be operated only in pulses, perhaps of microsecond or millisecond duration, so that the medium has a chance to cool or the lower state discard its population. However, it is sometimes desirable to have pulses of radiation rather than a continuous output, with a lot of power concentrated into a brief pulse. One way of achieving pulses is by **Q-switching**, the modification of the resonance characteristics of the laser cavity.[2]

Example 17.2 Relating the power and energy of a laser

A laser rated at 0.10 J can generate radiation in 3.0 ns pulses. What is the average power output per pulse?

Method The power output, P, is the energy released in an interval divided by the duration of the interval, and is expressed in watts ($1\ \text{W} = 1\ \text{J s}^{-1}$). So, to calculate the power, divide the energy output by the time over which the pulse is generated.

Answer From the data,

$$P = \frac{0.10\ \text{J}}{3.0 \times 10^{-9}\ \text{s}} = 3.3 \times 10^{7}\ \text{J s}^{-1}$$

That is, the pulses deliver 33 MW of power.

Comment The answer gives the average power; the peak power will be larger if the pulse is not rectangular.

Self-test 17.2 Calculate the average power output of a laser in which a 2.0 J pulse can be delivered in 1.0 ns.

[2.0 GW]

The aim of Q-switching is to achieve a healthy population inversion in the absence of the resonant cavity, then to plunge the population-inverted medium into a cavity, and hence to obtain a sudden pulse of radiation. The switching may be achieved by impairing the resonance characteristics of the cavity in some way while the pumping pulse is active, and then suddenly to improve them (Fig. 17.21). One technique is to use a **saturable dye**, a dye that loses its power to absorb when many of its molecules have been excited by intense radiation. It then suddenly becomes transparent, and the cavity becomes resonant. In practice, Q-switching can give pulses of about 10 ns duration.

2   The name comes from the 'Q-factor' used as a measure of the quality of a resonance cavity in microwave engineering.

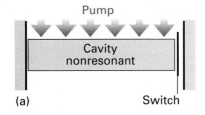

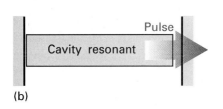

17.21 The principle of Q-switching. The excited state is populated while the cavity is nonresonant. Then the resonance characteristics are suddenly restored, and the stimulated emission emerges in a giant pulse.

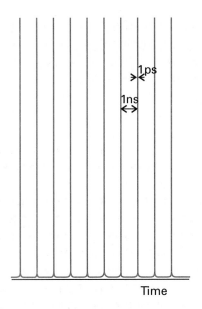

17.22 The output of a mode-locked laser consists of a stream of very narrow pulses separated by an interval equal to the time it takes for light to make a round trip inside the cavity.

(d) Mode locking

The technique of **mode locking** can produce pulses of picosecond duration and less. A laser radiates at a number of different frequencies, depending on the precise details of the resonance characteristics of the cavity and in particular on the number of half-wavelengths of radiation that can be trapped between the mirrors (the cavity modes). The resonant modes differ in frequency by multiples of $c/2L$ (as can be inferred from eqn 4 with $\nu = c/\lambda$). Normally, these modes have random phases relative to each other. However, it is possible to lock their phases together. Then interference occurs to give a series of sharp peaks, and the energy of the laser is obtained in picosecond bursts (Fig. 17.22). The sharpness of the peaks depends on the range of modes superimposed and, the wider the range, the narrower the pulses. In a laser with a cavity of length 30 cm, the peaks will be separated by 2 ns. If 1000 modes contribute, the width of the pulses will be 4 ps.

Justification 17.3

The general expression for a (complex) wave of amplitude $\mathcal{E}_0$ and frequency ω is $\mathcal{E}_0 e^{i\omega t}$. Therefore, each wave that can be supported by a cavity of length L has the form

$$\mathcal{E}_n(t) = \mathcal{E}_0 e^{2\pi i(\nu + nc/2L)t}$$

where ν is the lowest frequency. A wave formed by superimposing N modes with $n = 0, 1, \cdots N - 1$ has the form

$$\mathcal{E}(t) = \sum_n \mathcal{E}_n(t) = \mathcal{E}_0 e^{2\pi i \nu t} \sum_{n=0}^{N-1} e^{i\pi nct/L}$$

The sum is a geometrical progression:

$$\sum_{n=0}^{N-1} e^{i\pi nct/L} = 1 + e^{i\pi ct/L} + e^{2i\pi ct/L} + \cdots$$

$$= \frac{\sin(N\pi ct/2L)}{\sin(\pi ct/2L)} \times e^{(N-1)i\pi ct/2L}$$

The intensity, $\mathcal{I}$, of the radiation is proportional to the square modulus of the total amplitude, so

$$\mathcal{I} \propto \mathcal{E}^* \mathcal{E} = \mathcal{E}_0^2 \frac{\sin^2(N\pi ct/2L)}{\sin^2(\pi ct/2L)}$$

This function is shown in Fig. 17.23. We see that it is a series of peaks with maxima separated by $t = 2L/c$, the round-trip transit time of the light in the cavity, and that the peaks become sharper as N is increased.

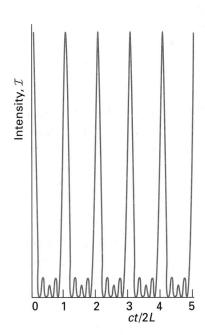

17.23 The function derived in *Justification 17.3* showing in more detail the structure of the pulses generated by a mode-locked laser.

Mode locking is achieved by varying the Q factor of the cavity periodically at the frequency $c/2L$. The modulation can be pictured as the opening of a shutter in synchrony with the round-trip travel time of the photons in the cavity, so only photons making the journey in that time are amplified. The modulation can be achieved by linking a prism in the cavity to a transducer driven by a radiofrequency source at a frequency $c/2L$. The transducer sets up standing-wave vibrations in the prism and modulates the loss it introduces into the cavity. Mode locking may also be accomplished passively by including a saturable dye. This procedure makes use of the fact that the **gain**, the growth in intensity of a component, is very sensitive to amplification and, once a particular frequency begins to grow, it can quickly dominate. If a saturable dye is included in the cavity, a spontaneous fluctuation in intensity—a bunching of photons—may result in its becoming transparent, and the bunch can pass through and travel to the far end of the cavity, amplifying as it goes. The dye immediately shuts down again (if it is well chosen), but opens when the intense pulse returns

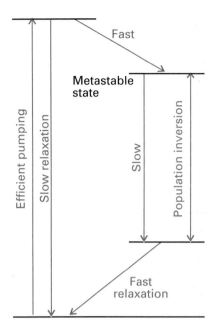

17.24 A summary of the features needed for efficient laser action.

from the mirror at the far end and saturates it. In this way, that particular bunch of photons may grow to considerable intensity because it alone is stimulating emission in the cavity.

17.6 Practical lasers

Figure 17.24 summarizes the requirements for an efficient laser. In practice, the requirements can be satisfied by using a variety of different systems, and this section reviews some that are commonly available. For completeness, we include some lasers that operate by using other than electronic transitions.

(a) Solid-state lasers

A **solid-state laser** is one in which the active medium is in the form of a single crystal or a glass. The first successful laser, the ruby laser built by Theodore Maiman in 1960, is an example (Fig. 17.25). Ruby is Al_2O_3 containing a small proportion of Cr^{3+} ions.[3] Ruby is a three-level system, and the ground state, which is also the lower level of the laser transition, is $^4A_2$ with three unpaired spins on each Cr^{3+} ion. The population inversion results from pumping a majority of the Cr^{3+} ions into an excited state by using an intense flash from another source, followed by a radiationless transition to another excited state. The pumping flash need not be monochromatic because the upper level actually consists of several states spanning a band of frequencies. The transition from the lower of the two excited states to the ground state ($^2E \rightarrow ^4A_2$) is the laser transition, and gives rise to red 694 nm radiation. The population inversion is very difficult to sustain continuously, and in practice the ruby laser is pulsed. Typical pulses from a Q-switched ruby laser might consist of 2 J pulses persisting for 10 ns, corresponding to an average power of 0.2 GW.

The **neodymium laser** is an example of a four-level laser (Fig. 17.26). In one form it consists of Nd^{3+} ions at low concentration in yttrium aluminium garnet (YAG, specifically

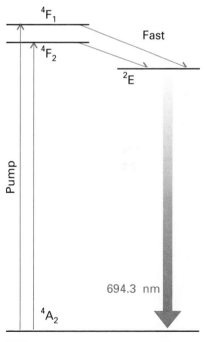

17.25 The transitions involved in a ruby laser. The laser medium, ruby, consists of Al_2O_3 doped with Cr^{3+} ions.

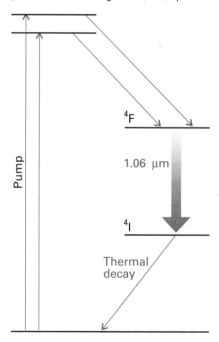

17.26 The transitions involved in a neodymium laser. The laser action takes place between two excited states, and the population inversion is easier to achieve than in the ruby laser.

---

3   The normal green of Cr^{3+} is modified to red by the distortion of the local crystal field stemming from the replacement of an Al^{3+} ion by a slightly larger Cr^{3+} ion.

$Y_3Al_5O_{12}$), and is then known as a Nd-YAG laser. A cheaper medium is glass, but glass is a poorer thermal conductor than YAG and the laser must be pulsed. A neodymium laser operates at a number of wavelengths in the infrared, the band at 1064 nm being most common. The transition at 1064 nm is very efficient and the laser is capable of substantial power output. The power is great enough for frequency doubling to be used efficiently. **Frequency doubling** is a technique in which the laser beam is converted to radiation with twice (and in general a multiple) of its initial frequency as it passes through a suitable transparent material. A frequency-doubled Nd-YAG laser produces green light at 532 nm.

Example 17.3 Accounting for multiphoton phenomena

Show that if a substance responds nonlinearly to incident radiation of frequency ω, then it may act as the source of radiation of twice the incident frequency.

Method Radiation of a particular frequency arises from oscillations of an electric dipole at that frequency. Therefore, express the induced electric dipole moment of the system in terms of powers of the applied electric field, and write the powers of harmonic (cosine) terms as sums and differences of cosine terms. Inspect the sum to see if $\cos 2\omega t$ is present.

Answer The incident electric field $\mathcal{E}$ induces an electric dipole of magnitude μ and, allowing for nonlinear response, we can write

$$\mu = \alpha\mathcal{E} + \beta\mathcal{E}^2 + \cdots$$

The nonlinear terms can be expanded as follows if we suppose that the incident electric field is $\mathcal{E}_0 \cos \omega t$:

$$\beta\mathcal{E}^2 = \beta\mathcal{E}_0^2 \cos^2 \omega t = \tfrac{1}{2}\beta\mathcal{E}_0^2(1 + \cos 2\omega t)$$

Hence, the nonlinear term contributes an induced electric dipole that oscillates at the frequency 2ω and which can act as a source of radiation of that frequency.

- -

Self-test 17.3 Show that, if a substance responds nonlinearly to two sources of radiation, one of frequency ω_1 and the other of frequency ω_2, then it may give rise to radiation of the sum and difference of the two frequencies.

$$[\beta\mathcal{E}^2 \propto \cos(\omega_1 + \omega_2)t + \cos(\omega_1 - \omega_2)t]$$

(b) Gas lasers

Because gas lasers can be cooled by a rapid flow of the gas through the cavity, they can be used to generate high powers. The pumping is normally achieved using a gas that is different from the gas responsible for the laser emission itself.

In the **helium–neon laser** the active medium is a mixture of helium and neon in a mole ratio of about 5 : 1 (Fig. 17.27). The initial step is the excitation of an He atom to the metastable $1s^1 2s^1$ configuration by using an electric discharge (the collisions of electrons and ions cause transitions that are not restricted by electric-dipole selection rules). The excitation energy of this transition happens to match an excitation energy of neon, and during an He–Ne collision efficient transfer of energy may occur, leading to the production of highly excited, metastable Ne atoms with unpopulated intermediate states. Laser action generating 633 nm radiation (among about 100 other lines) then occurs.

The **argon-ion laser** (Fig. 17.28), one of a number of 'ion lasers', consists of argon at about 1 Torr, through which is passed an electric discharge. The discharge results in the formation of Ar^+ and Ar^{2+} ions in excited states, which undergo a laser transition to a lower

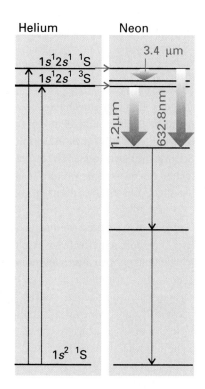

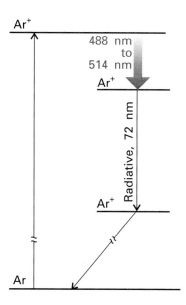

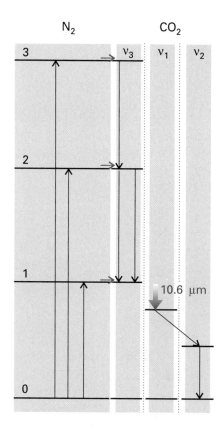

17.27 The transitions involved in a helium–neon laser. The pumping (of the neon) depends on a coincidental matching of the helium and neon energy separations, so excited He atoms can transfer their excess energy to Ne atoms during a collision.

17.28 The transitions involved in an argon-ion laser.

17.29 The transitions involved in a carbon dioxide laser. The pumping also depends on the coincidental matching of energy separations; in this case the vibrationally excited N_2 molecules have excess energies that correspond to a vibrational excitation of the antisymmetric stretch of CO_2. The laser transition is from $\nu_3 = 1$ to $\nu_1 = 1$.

state. These ions then revert to their ground states by emitting hard ultraviolet radiation (at 72 nm), and are then neutralized by a series of electrodes in the laser cavity. One of the design problems is to find materials that can withstand this damaging residual radiation. There are many lines in the laser transition because the excited ions may make transitions to many lower states, but two strong emissions from Ar^+ are at 488 nm (blue) and 514 nm (green); other transitions occur elsewhere in the visible region, in the infrared, and in the ultraviolet. The **krypton-ion laser** works similarly. It is less efficient, but gives a wider range of wavelengths, the most intense being at 647 nm (red), but it can also generate a yellow line. Both lasers are widely used in laser light shows (for this application argon and krypton are often used simultaneously in the same cavity) as well as being used as laboratory sources of high-power radiation.

The **carbon dioxide laser** works on a slightly different principle (Fig. 17.29), for its radiation (between 9.2 μm and 10.8 μm, with the strongest emission at 10.6 μm, in the infrared) arises from vibrational transitions. Most of the working gas is nitrogen, which becomes vibrationally excited by electronic and ionic collisions in an electric discharge. The vibrational levels happen to coincide with the ladder of antisymmetric stretch (ν_3, see Fig. 16.46) energy levels of CO_2, which pick up the energy during a collision. Laser action then occurs from the lowest excited level of ν_3 to the lowest excited level of the symmetric stretch (ν_1), which has remained unpopulated during the collisions. This transition is allowed by anharmonicities in the molecular potential energy. Some helium is included in the gas to help remove energy from this state and maintain the population inversion.

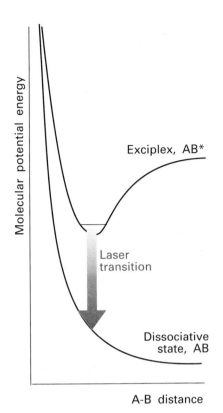

17.30 The molecular potential energy curves for an exciplex. The species can survive only as an excited state, because on discarding its energy it enters the lower, dissociative state. Because only the upper state can exist, there is never any population in the lower state.

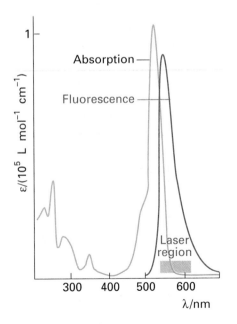

17.31 The optical absorption spectrum of the dye Rhodamine G and the region used for laser action.

In the **nitrogen laser**, the efficiency of the stimulated transition (at 337 nm, in the ultraviolet, the transition $C^3\Pi_u \rightarrow B^3\Pi_g$) is so great that a single passage of a pulse of radiation is enough to generate laser radiation and mirrors are unnecessary: such lasers are said to be **superradiant**.

(c) Chemical and exciplex lasers

Chemical reactions may also be used to generate molecules with non-equilibrium, inverted populations. For example, the photolysis of Cl_2 leads to the formation of Cl atoms which attack H_2 molecules in the mixture and produce HCl and H. The latter then attacks Cl_2 to produce vibrationally excited ('hot') HCl molecules. Because the newly formed HCl molecules have non-equilibrium vibrational populations, laser action can result as they return to lower states. Such processes are remarkable examples of the direct conversion of chemical energy into coherent electromagnetic radiation.

The population inversion needed for laser action is achieved in a more underhand way in **exciplex lasers**,[4] for in these (as we shall see) the lower state does not effectively exist. This odd situation is achieved by forming an **exciplex**, a combination of two atoms that survives only in an excited state and which dissociates as soon as the excitation energy has been discarded. An example is a mixture of xenon, chlorine, and neon (which acts as a buffer gas). An electric discharge through the mixture produces excited Cl atoms, which attach to the Xe atoms to give the exciplex XeCl*. The exciplex survives for about 10 ns, which is time enough for it to participate in laser action at 308 nm (in the ultraviolet). As soon as XeCl* has discarded a photon, the atoms separate because the molecular potential energy curve of the ground state is dissociative, and the ground state of the exciplex cannot become populated (Fig. 17.30). The KrF* exciplex laser is another example: it produces radiation at 249 nm.

(d) Dye lasers

A solid-state laser and a gas laser operate at discrete frequencies and, although the frequency required may be selected by suitable optics, the laser cannot be tuned continuously. The tuning problem is overcome by using a **dye laser**, which has broad spectral characteristics because the solvent broadens the vibrational structure of the transitions into bands. Hence, it is possible to scan the wavelength continuously (by rotating the diffraction grating in the cavity) and achieve laser action at any chosen wavelength. A commonly used dye is Rhodamine 6G in methanol (Fig. 17.31). As the gain is very high, only a short length of the optical path need be through the dye. The excited states of the active medium, the dye, are sustained by another laser or a flash lamp, and the dye solution is flowed through the laser cavity to avoid thermal degradation (Fig. 17.32).

(e) Light-emitting diodes and semiconductor lasers

We have seen (in Section 14.10d) that a semiconductor is classified as 'n-type' if its conduction band is partly populated and as 'p-type' if its valence band has a small number of holes. In this section we need to consider the properties of a **p–n junction**, the interface of the two types of semiconductor.

The band structure at the junction is shown in Fig. 17.33. When a 'forward bias' is applied to the junction, in the sense that electrons are supplied through an external circuit to the n side of the junction, the electrons in the conduction band of the n-type semiconductor fall into the holes in the valence band of the p-type semiconductor. As they fall, they emit energy. In silicon semiconductors this energy is largely in the form of heat because the wavefunctions of the relevant states of the bands differ in linear momentum, so the

---

4   The term 'excimer laser' is also widely encountered and used loosely when 'exciplex laser' is more appropriate. An exciplex has the form AB^* whereas an excimer, an excited dimer, is AA^*.

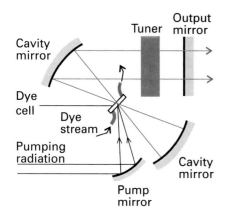

17.32 The configuration used for a dye laser. The dye is flowed through the cell inside the laser cavity. The flow helps to keep it cool and prevents degradation.

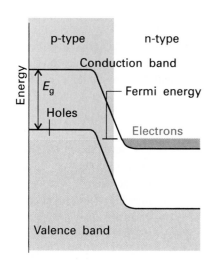

(a)

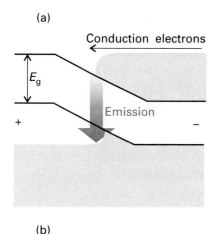

(b)

17.33 The structure of a diode junction (a) without bias and (b) with bias.

transition can occur only if the electron transfers linear momentum to the lattice, and the device becomes warm. However, in some materials, most notably gallium arsenide, GaAs, the wavefunctions of the states involved correspond to the same linear momentum, so transition can occur without the lattice needing to be involved and the energy is emitted as light. Practical **light-emitting diodes** of this kind are widely used in electronic displays. Gallium arsenide itself emits infrared light, but the band gap is widened by incorporating phosphorus, and a material of composition approximately $GaAs_{0.6}P_{0.4}$ emits light in the red region of the spectrum.

A light-emitting diode is not a laser, because no resonance cavity and stimulated emission are involved. However, it is easy (in principle) to employ the light emission of electron-hole recombination as the basis of laser action. The population inversion can be sustained by sweeping away the electrons that fall into the holes of the p-type semiconductor, and a resonant cavity can be formed by using the high refractive index of the semiconducting material and cleaving single crystals so that the light is trapped by the sudden variation of refractive index. One widely used material is $Ga_{1-x}Al_xAs$, which produces infrared laser radiation and is widely used in compact-disc (CD) players.

17.7 Applications of lasers in chemistry

Laser radiation has five striking characteristics (Table 17.3). Each of them (sometimes in combination with the others) opens up interesting opportunities in spectroscopy, giving rise to 'laser spectroscopy' and, in photochemistry, giving rise to 'laser photochemistry'.

(a) Spectroscopy at high photon fluxes

The high spectral power density of a laser—the high intensity of the radiation it produces at well-defined frequencies—is an aid to conventional spectroscopy. Thus, it reduces the problem of detector noise and the interfering effects of background radiation. The high intensity is particularly advantageous in Raman spectroscopy, which until the introduction of lasers was plagued by the low intensity of the scattered radiation (which could be overcome only by using long exposures) and by interference from background scattering (which obscured the signal).

Table **17.3** Characteristics of laser radiation and their chemical applications

Characteristic	Advantages	Applications
High power	Multiphoton process	Non-linear spectroscopy
		Saturation spectroscopy
	Low detector noise	Improved sensitivity
	High scattering intensity	Raman spectroscopy
Monochromatic	High resolution	Spectroscopy
	State selection	Isotope separation
		Photochemically precise
		State-to-state reaction dynamics
Collimated beam	Long path lengths	Sensitivity
	Forward-scattering observable	Non-linear Raman spectroscopy
Coherent	Interference between separate beams	CARS
Pulsed	Precise timing of excitation	Fast reactions
		Relaxation
		Energy transfer

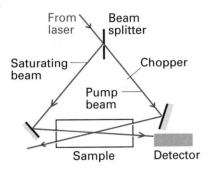

17.34 The configuration of laser radiation used for saturation spectroscopy.

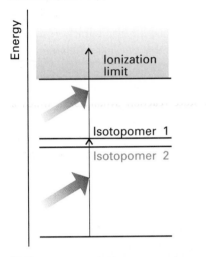

17.35 In one method of isotope separation, one photon excites an isotopomer to an excited state, and then a second photon achieves photoionization. The success of the first step depends on the nuclear mass.

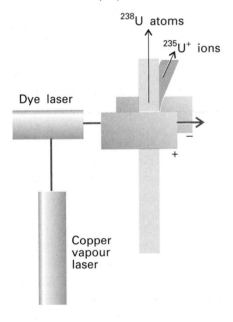

17.36 An experimental arrangement for isotope separation. The dye laser, which is pumped by a copper-vapour laser, photoionizes the U atoms selectively according to their mass, and the ions are deflected by the electric field applied between the plates.

The large number of photons in an incident beam generated by a laser gives rise to a qualitatively different branch of spectroscopy, for the photon density is so high that more than one photon may be absorbed by a single molecule and give rise to **multiphoton processes**. One application of multiphoton processes is that states inaccessible by conventional one-photon spectroscopy become observable because the overall transition occurs with no change of parity. For example, in one-photon spectroscopy, only g ↔ u transitions are observable; in two-photon spectroscopy, however, the overall outcome of absorbing two photons is a g → g or a u → u transition.

High powers and monochromatic beams make possible the technique of **saturation spectroscopy**, which permits the very precise location of absorption maxima. As illustrated in Fig. 17.34, the output of a tunable laser is divided into an intense saturating beam and a less intense probe beam that pass through the sample cavity in nearly opposite directions. The chopped saturating beam periodically excites molecules that are Doppler-shifted to its frequency. The probe beam gives a modulated signal at the detector, but only if it is interacting with the same Doppler-shifted molecules despite the fact that it is coming from an opposite direction. Because those molecules must be ones that are not moving parallel to the beams, the technique selects molecules that have essentially zero Doppler shift and hence gives very high resolution.

(b) Collimated beams

The collimated beams generated by most kinds of lasers permit the use of very long path lengths through spectroscopic samples. A well-defined beam also implies that the detector can be designed to collect only the radiation that has passed through a sample, and can be screened much more effectively against stray scattered light. Moreover, with a collimated beam, the interaction zone in Raman spectroscopy is much more well-defined than in conventional spectroscopy, so the optics of the spectrometer can be optimized.

The availability of nondivergent beams makes possible a qualitatively different kind of spectroscopy. The beam is so well-defined that it is possible to observe Raman transitions very close to the direction of propagation of the incident beam (rather than perpendicular to it). This configuration is employed in the technique called **stimulated Raman spectroscopy**. In this form of spectroscopy, the Stokes and anti-Stokes radiation in the forward direction are powerful enough to undergo more scattering and hence give up or acquire more quanta of energy from the molecules in the sample. This multiple scattering results in lines of frequency $\nu_i \pm 2\nu_M$, $\nu_i \pm 3\nu_M$, and so on, where ν_i is the frequency of the incident radiation and ν_M the frequency of a molecular excitation.

Raman spectroscopy was revitalized by the introduction of lasers. We have already commented on the enhancements of the technique that stem from the high powers and collimation of the incident beam. Its monochromaticity is also a great advantage, for it is now possible to observe scattered light that differs by only fractions of reciprocal centimetres from the incident radiation. Such high resolution is particularly useful for observing the rotational structure of Raman lines because rotational transitions are of the order of a few reciprocal centimetres. Monochromaticity also allows observations to be made very close to absorption frequencies, giving rise to the technique of resonance Raman spectroscopy (Section 16.16c). Modern, small, and efficient semiconductor lasers have also allowed the development of Fourier transform Raman spectrometers.

(c) Precision-specified transitions

The monochromatic character of laser radiation is a very powerful characteristic because it allows us to excite specific states with very high precision. One consequence of state-specificity for photochemistry is that the illumination of a sample may be photochemically

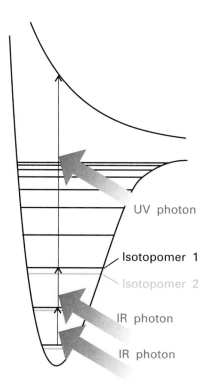

17.37 Isotopomers may be separated by making use of their selective absorption of infrared photons followed by photodissociation with an ultraviolet photon.

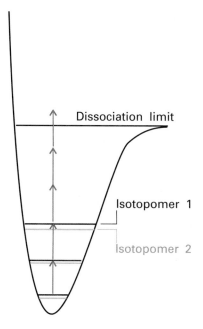

17.38 In an alternative scheme for separating isotopomers, multiphoton absorption of infrared photons is used to reach the dissociation limit of a ground electronic state.

precise and hence efficient in stimulating a reaction, because its frequency can be tuned exactly to an absorption. The specific excitation of a particular excited state of a molecule may greatly enhance the rate of a reaction even at low temperatures. The rate of a reaction is generally increased by raising the temperature because the energies of the various modes of motion of the molecule are enhanced. However, this enhancement increases the energy of all the modes, even those that do not contribute appreciably to the reaction rate. With a laser we can excite the kinetically significant mode, so rate enhancement is achieved most efficiently. An example is the reaction

$$BCl_3 + C_6H_6 \longrightarrow C_6H_5\text{---}BCl_2 + HCl$$

which normally proceeds only above 600°C in the presence of a catalyst; exposure to 10.6 μm CO_2 laser radiation results in the formation of products at room temperature without a catalyst. The commercial potential of this procedure is considerable (provided laser photons can be produced sufficiently cheaply), because heat-sensitive compounds, such as pharmaceuticals, may perhaps be made at lower temperatures than in conventional reactions.

A related application is the study of **state-to-state reaction dynamics**, in which a specific state of a reactant molecule is excited and we monitor not only the rate at which it forms products but also the states in which they are produced. Studies such as these give highly detailed information about the deployment of energy in chemical reactions (Chapter 27).

(d) Isotope separation

The precision state-selectivity of lasers is also of considerable potential for laser isotope separation. Isotope separation is possible because two **isotopomers**, or species that differ only in their isotopic composition, have slightly different energy levels and hence slightly different absorption frequencies.

One approach is to use **photoionization**, the ejection of an electron by the absorption of electromagnetic radiation. Direct photoionization by the absorption of a single photon does not distinguish between isotopomers because the upper level belongs to a continuum; to distinguish isotopomers it is necessary to deal with discrete states. At least two absorption processes are required. In the first step, a photon excites an atom to a higher state; in the second step, a photon achieves photoionization from that state (Fig. 17.35). The energy separation between the two states involved in the first step depends on the nuclear mass. Therefore, if the laser radiation is tuned to the appropriate frequency, only one of the isotopomers will undergo excitation and hence be available for photoionization in the second step. An example of this procedure is the photoionization of uranium vapour, in which the incident laser is tuned to excite ^{235}U but not ^{238}U. The ^{235}U atoms in the atomic beam are ionized in the two-step process; they are then attracted to a negative electrode, and may be collected (Fig. 17.36). This procedure is being used in the latest generation of uranium separation plants.

Molecular isotopomers are used in techniques based on **photodissociation**, the fragmentation of a molecule following absorption of electromagnetic radiation. The key problem is to achieve both mass selectivity (which requires excitation to take place between discrete states) and dissociation (which requires excitation to continuum states). In one approach, two lasers are used: an infrared photon excites one isotopomer selectively to a higher vibrational level, and then an ultraviolet photon completes the process of photodissociation (Fig. 17.37). An alternative procedure is to make use of multiphoton absorption within the ground electronic state (Fig. 17.38); the efficiency of absorption of the first few photons depends on the match of their frequency to the energy level separations, so it is sensitive to nuclear mass. The absorbed photons open the door to a subsequent influx of enough photons to complete the dissociation process. The isotopomers $^{32}SF_6$ and $^{34}SF_6$ have been separated in this way.

In a third approach, a selectively vibrationally excited species may react with another species and give rise to products that can be separated chemically. This procedure has been employed successfully to separate isotopes of B, N, O, and, most efficiently, H. A variation on this procedure is to achieve selective **photoisomerization**, the conversion of a species to one of its isomers (particularly a geometrical isomer) on absorption of electromagnetic radiation. Once again, the initial absorption, which is isotope selective, opens the way to subsequent further absorption and the formation of a geometrical isomer that can be separated chemically. The approach has been used with the photoisomerization of CH_3NC to CH_3CN.

A different, more physical approach, that of **photodeflection**, is based on the recoil that occurs when a photon is absorbed by an atom and the linear momentum of the photon (which is equal to h/λ) is transferred to the atom. The atom is deflected from its original path only if the absorption actually occurs, and the incident radiation can be tuned to a particular isotope. The deflection is very small, so an atom must absorb dozens of photons before its path is changed sufficiently to allow collection. For instance, if a Ba atom absorbs about 50 photons of 550 nm light, it will be deflected by only about 1 mm after a flight of 1 m.

(e) Pulse techniques

The ability of lasers to produce pulses of very short duration is particularly useful in chemistry when we want to monitor processes in time. Q-switched lasers produce nanosecond pulses, which are generally fast enough to study reactions with rates controlled by the speed with which reactants can move through a fluid medium. However, when we want to study the rates at which energy is converted from one mode to another within a molecule, we need the shorter timescale of picosecond pulses. These timescales are available from mode-locked lasers, and modern techniques have reduced timescales of pulses to the femtosecond region (1 fs = 10^{-15} s), the shortest pulse currently reported being about 6 fs, corresponding to a packet of electromagnetic radiation of only a few wavelengths long. We shall see some of the information obtained from this femtosecond spectroscopy in Section 27.5f. Pulse techniques are used to study ultrafast dynamical processes such as energy transfer and conversion from one mode of motion to another. They are used to study relaxation of a disturbed set of level populations to thermal equilibrium, and, of particular importance in chemistry, to study the rates of fast reactions.

Photoelectron spectroscopy

The technique of **photoelectron spectroscopy** (PES) measures the ionization energies of molecules when electrons are ejected from different orbitals, and uses the information to infer the orbital energies. The technique is also used to study solids, and in Chapter 28 we shall see the important information that it gives about species at or on surfaces.

17.8 The technique

Because energy is conserved when a photon ionizes a sample, the energy of the incident photon $h\nu$ must be equal to the sum of the ionization energy, I, of the sample and the kinetic energy of the **photoelectron**, the ejected electron (Fig. 17.39):

$$h\nu = \tfrac{1}{2}m_ev^2 + I \tag{6}$$

This equation (which is like the one used for the photoelectric effect, Section 11.2a) can be refined in two ways. First, photoelectrons may originate from one of a number of different

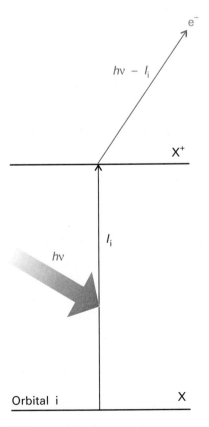

17.39 An incoming photon carries an energy $h\nu$; an energy I_i is needed to remove an electron from an orbital i, and the difference appears as the kinetic energy of the electron.

orbitals, and each one has a different ionization energy. Hence, a series of different kinetic energies of the photoelectrons will be obtained, each one satisfying

$$h\nu = \tfrac{1}{2}m_e v^2 + I_i \tag{7}$$

where I_i is the ionization energy for ejection of an electron from an orbital i. Therefore, by measuring the kinetic energies of the photoelectrons, and knowing ν, these ionization energies can be determined. Photoelectron spectra are interpreted in terms of an approximation called **Koopmans' theorem**, which states that the ionization energy I_i is equal to the orbital energy of the ejected electron (formally: $I_i = -\varepsilon_i$). That is, we can identify the ionization energy with the energy of the orbital from which it is ejected. The theorem is only an approximation because it ignores the fact that the remaining electrons adjust their distributions when ionization occurs.

The ejection of an electron may leave an ion in a vibrationally excited state. Then not all the excess energy of the photon appears as kinetic energy of the photoelectron, and we should write

$$h\nu = \tfrac{1}{2}m_e v^2 + I_i + E_{vib}^+ \tag{8}$$

where E_{vib}^+ is the energy used to excite the ion into vibration. Each vibrational quantum that is excited leads to a different kinetic energy of the photoelectron, and gives rise to the vibrational structure in the photoelectron spectrum.

The ionization energies of molecules are several electronvolts even for valence electrons, so it is essential to work in at least the ultraviolet region of the spectrum and with wavelengths of less than about 200 nm. Much work has been done with radiation generated by a discharge through helium: the He(I) line ($1s^1 2p^1 \rightarrow 1s^2$) lies at 58.43 nm, corresponding to a photon energy of 21.22 eV. Its use gives rise to the technique of **ultraviolet photoelectron spectroscopy** (UPS). When core electrons are being studied, photons of even higher energy are needed to expel them: X-rays are used, and the technique is denoted XPS. A modern version of PES makes use of synchrotron radiation (Section 16.1) which may be continuously tuned between UV and X-ray energies. The additional information that stems from the variation of the photoejection probability with wavelength is a valuable guide to the identity of the molecule and the orbital from which photoionization occurs.

Illustration

....

Photoelectrons ejected from N_2 with He(I) radiation had kinetic energies of 5.63 eV (1 eV = 8065.5 cm^{-1}). Helium(I) radiation of wavelength 58.43 nm has wavenumber 1.711×10^5 cm^{-1} and therefore corresponds to an energy of 21.22 eV. Then, from eqn 7, 21.22 eV = 5.63 eV + I_i, so $I_i = 15.59$ eV. This ionization energy is the energy needed to remove an electron from the HOMO of the N_2 molecule, the $3\sigma_g$ bonding orbital (see Fig. 14.29).

....

Self-test 17.4 Under the same circumstances, photoelectrons are also detected at 4.53 eV. To what ionization energy does that correspond? Suggest an origin.

[16.7 eV, $1\pi_u$]

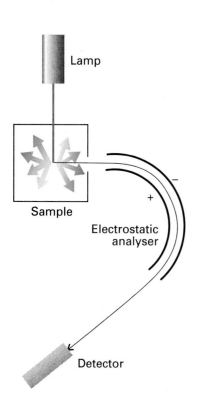

17.40 A photoelectron spectrometer consists of a source of ionizing radiation (such as a helium discharge lamp for UPS and an X-ray source for XPS), an electrostatic analyser, and an electron detector. The deflection of the electron path caused by the analyser depends on the speed at which they are ejected from the sample.

The kinetic energies of the photoelectrons are measured using an electrostatic deflector which produces different deflections in the paths of the photoelectrons as they pass between charged plates (Fig. 17.40). As the field strength is increased, electrons of different speeds, and therefore kinetic energies, reach the detector. The electron flux can be recorded and plotted against kinetic energy to obtain the photoelectron spectrum.

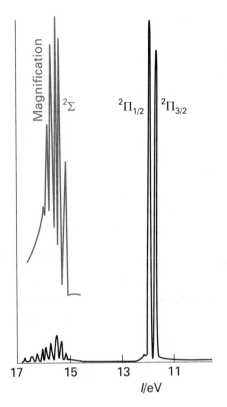

17.41 The photoelectron spectrum of HBr. The lowest ionization energy bands (Π) correspond to the ionization of a Br lone-pair electron. The higher ionization energy band (Σ) corresponds to the ionization of a bonding electron. The structure on the latter is due to the vibrational excitation of HBr$^+$ that results from the ionization.

17.9 Ultraviolet photoelectron spectroscopy

A typical photoelectron spectrum (of HBr) is shown in Fig. 17.41. If we disregard the fine structure, we see that the HBr lines fall into two main groups. The least tightly bound electrons (with the lowest ionization energies and hence highest kinetic energies when ejected) are those in the nonbonding lone pairs of Br (with $I = 11.8$ eV). The next ionization energy lies at 15.2 eV, and corresponds to the removal of an electron from the H–Br σ bond.

The HBr spectrum shows that ejection of a σ electron is accompanied by a long vibrational progression. The Franck–Condon principle would account for this progression if ejection were accompanied by an appreciable change of equilibrium bond length between HBr and HBr$^+$ because the ion is formed in a bond-compressed state, which is consistent with the important bonding effect of the σ electrons. The lack of much vibrational structure in the two bands labelled $^2\Pi$ is consistent with the nonbonding role of the Br$2p\pi$ lone pair of electrons, for the equilibrium bond length is little changed when one is removed.

Example 17.4 Interpreting a UV photoelectron spectrum

The highest kinetic-energy electrons in the spectrum of H_2O using 21.22 eV He radiation are at about 9 eV and show a large vibrational spacing of 0.41 eV. The symmetric stretching mode of the neutral H_2O molecule lies at 3652 cm^{-1}. What conclusions can be drawn from the nature of the orbital from which the electron is ejected?

Method We need to interpret the vibrational fine structure, which indicates the vibrational characteristics of the ion, in relation to the information about the vibrational characteristics of the neutral molecule.

Answer Because 0.41 eV corresponds to 3310 cm^{-1}, which is similar to the 3652 cm^{-1} of the nonionized molecule, we can suspect that the electron is ejected from an orbital that has little influence on the bonding in the molecule. That is, photoejection is from a largely nonbonding orbital.

- -

Self-test 17.5 In the same spectrum of H_2O, the band near 7.0 eV shows a long vibrational series with spacing 0.125 eV. The bending mode of H_2O lies at 1596 cm^{-1}. What conclusions can you draw about the characteristics of the orbital occupied by the photoelectron?

[The electron contributed to non-neighbour H–H bonding]

17.10 X-ray photoelectron spectroscopy

In XPS, the energy of the incident photon is so great that electrons are ejected from inner cores of atoms. As a first approximation, core ionization energies are insensitive to the bonds between atoms because they are too tightly bound to be greatly affected by the changes that accompany bond formation, so core ionization energies are characteristic of the individual atom rather than the overall molecule. Consequently, XPS gives lines characteristic of the elements present in a compound or alloy. For instance, the K-shell ionization energies of the second row elements are

Li	Be	B	C	N	O	F
50	110	190	280	400	530	690 eV

Detection of one of these values (and values corresponding to ejection from other inner shells) indicates the presence of the corresponding element. This application is responsible

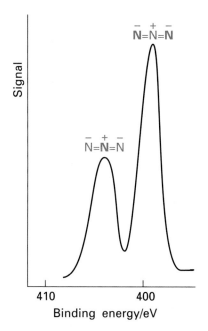

17.42 The photoelectron spectrum of solid NaN_3 excited by Al K-radiation, showing the region of N core ionization and the assignment. (K. Siegbahn, *et al.*, *Science*, **176**, **245** (1972).)

for the alternative name **electron spectroscopy for chemical analysis (ESCA)**. The technique is mainly limited to the study of surface layers (as we explore in Chapter 28) because, even though X-rays may penetrate into the bulk sample, the ejected electrons cannot escape except from within a few nanometres of the surface. Despite (or because of) this limitation, the technique is very useful for studying the surface state of heterogeneous catalysts, the differences between surface and bulk structures, and the processes that can cause damage to high-temperature superconductors and semiconductor wafers.

Whereas it is largely true that core ionization energies are unaffected by bond formation, it is not entirely true, and small but detectable shifts can be detected and interpreted in terms of the environments of the atoms. For example, the azide ion, N_3^-, gives the spectrum shown in Fig. 17.42. Although the spectrum lies in the region of 400 eV (and hence is typical of N1s electrons), it has a doublet structure with splitting 6 eV. This splitting can be understood by noting that the structure of the ion is $\bar{N}=N=\bar{N}$, with more negative charge on the outer two N atoms than on the inner (the formal charges are $(-1, +1, -1)$). The presence of the negative charges on the terminal atoms lowers the core ionization energies, whereas the positive charge on the central atom raises it. This inequivalence of the atoms results in two lines in the spectrum with intensities in the ratio 2 : 1. Observations like this can be used to obtain valuable information about the presence of chemically inequivalent atoms of the same element.

Checklist of key ideas

The characteristics of electronic transitions

17.1 The vibrational structure
- [ ] Franck–Condon principle
- [ ] vertical transition
- [ ] vibrational progression
- [ ] Franck–Condon factor

17.2 Specific types of transitions
- [ ] chromophore
- [ ] ligand-field splitting parameter
- [ ] Laporte selection rule
- [ ] vibronic transition
- [ ] charge-transfer transition
- [ ] ligand-to-metal charge-transfer transition (LMCT)
- [ ] metal-to-ligand charge-transfer transition (MLCT)

The fates of electronically excited states

- [ ] radiative decay process
- [ ] nonradiative decay

17.3 Fluorescence and phosphorescence
- [ ] fluorescence
- [ ] phosphorescence
- [ ] intersystem crossing
- [ ] Jablonski diagram

17.4 Dissociation and predissociation
- [ ] dissociation
- [ ] dissociation limit
- [ ] predissociation
- [ ] internal conversion

Lasers

17.5 General principles of laser action
- [ ] metastable excited state
- [ ] population inversion
- [ ] three-level laser
- [ ] pumping
- [ ] four-level laser
- [ ] resonant modes
- [ ] coherent radiation
- [ ] spatial coherence
- [ ] temporal coherence

- [ ] coherence length
- [ ] Q-switching
- [ ] saturable dye
- [ ] mode locking
- [ ] gain

17.6 Practical lasers
- [ ] solid-state laser
- [ ] neodymium laser
- [ ] frequency doubling
- [ ] helium–neon laser
- [ ] argon-ion laser
- [ ] krypton-ion laser
- [ ] carbon dioxide laser
- [ ] nitrogen laser
- [ ] superradiant
- [ ] exciplex laser
- [ ] exciplex
- [ ] dye laser
- [ ] p–n junction
- [ ] light-emitting diode

17.7 Applications of lasers in chemistry
- [ ] multiphoton processes
- [ ] saturation spectroscopy

- [ ] stimulated Raman spectroscopy
- [ ] state-to-state reaction dynamics
- [ ] photoionization
- [ ] photodissociation
- [ ] photoisomerization
- [ ] photodeflection

Photoelectron spectroscopy

- [ ] photoelectron spectroscopy (PES)

17.8 The technique
- [ ] photoelectron
- [ ] Koopmans' theorem
- [ ] ultraviolet photoelectron spectroscopy (UPS)

17.9 Ultraviolet photoelectron spectroscopy

17.10 X-ray photoelectron spectroscopy
- [ ] electron spectroscopy for chemical analysis (ESCA)

Further reading

Articles of general interest

R.B. Snadden, The iodine spectrum revisited. *J. Chem. Educ.* **64**, 919 (1987).

M. Allan, Electron spectroscopic techniques in teaching. *J. Chem. Educ.* **64**, 418 (1987).

F. Ahmed, A good example of the Franck–Condon principle. *J. Chem. Educ.* **64**, 427 (1987).

M.G.D. Baumann, J.C. Wright, A.B. Ellis, T. Kuech, and G.C. Lisesnky, Diode lasers. *J. Chem. Educ.* **69**, 89 (1992).

B.J. Duke and B. O'Leary, Non-Koopmans' molecules. *J. Chem. Educ.* **72**, 501 (1995).

P. Engelking, Laser photochemistry. In *Encyclopedia of applied physics* (ed. G.L. Trigg), **8**, 283. VCH, New York (1994).

P.L. Kelley and J.J. Zayhowski, Laser physics. In *Encyclopedia of applied physics* (ed. G.L. Trigg), **8**, 299. VCH, New York (1994).

H. Takuma, Laser technology. In *Encyclopedia of applied physics* (ed. G.L. Trigg), **8**, 321. VCH, New York (1994).

D.A. Ramsay, Molecular spectroscopy. In *Encyclopedia of applied physics* (ed. G.L. Trigg), **10**, 491. VCH, New York (1994).

F.J. Himpsell and I. Lindau, Photoemission and photoelectron spectra. In *Encyclopedia of applied physics* (ed. G.L. Trigg), **13**, 477. VCH, New York (1995).

Texts and sources of data and information

J.M. Hollas, *Modern spectroscopy*. Wiley, New York (1996).

J.M. Hollas, *High resolution spectroscopy*. Butterworth, London (1982).

E.A.V. Ebsworth, D.W.H. Rankin, and S. Cradock, *Structural methods in inorganic chemistry*. Blackwell Scientific, Oxford (1991).

R. Drago, *Physical methods for chemists*. Saunders, Philadelphia (1992).

G. Herzberg, *Spectra of diatomic molecules*. Van Nostrand, New York (1950).

G. Herzberg, *Electronic spectra and electronic structure of polyatomic molecules*. Van Nostrand, New York (1966).

A.G. Gaydon, *Dissociation energies*. Chapman & Hall, London (1968).

R.P. Wayne, *Principles and applications of photochemistry*. Oxford University Press (1988).

D.L. Andrews, *Lasers in chemistry*. Springer-Verlag, Berlin (1990).

D.L. Andrews, *An introduction to laser spectroscopy*. Plenum, New York (1995).

A.E. Siegman, *Lasers*. University Science Books, Mill Valley (1988).

W. Demtröder, *Laser spectroscopy*. Springer, Berlin (1988).

A.B. Myers and T.R. Rizzo (ed.), *Techniques in chemistry XXIII: laser techniques in chemistry*. Wiley, New York (1995).

G.R. Fleming, *Chemical applications of ultrafast spectroscopy*. Oxford University Press (1986).

J.H.D. Eland, *Photoelectron spectra*. Butterworth, London (1984).

T.L. Barr, *Modern ESCA: the principles and practice of x-ray photoelectron spectroscopy*. CRC Press, Boca Raton (1994).

C.R. Brundle and A.D. Baker (ed.), *Electron spectroscopy: theory, techniques, and applications*, Vols 1–4. Academic Press, London (1977–81).

Exercises

17.1 (a) The molar absorption coefficient of a substance dissolved in hexane is known to be 855 L mol^{-1} cm^{-1} at 270 nm. Calculate the percentage reduction in intensity when light of that wavelength passes through 2.5 mm of a solution of concentration 3.25 mmol L^{-1}.

17.1 (b) The molar absorption coefficient of a substance dissolved in hexane is known to be 327 L mol^{-1} cm^{-1} at 300 nm. Calculate the percentage reduction in intensity when light of that wavelength passes through 1.50 mm of a solution of concentration 2.22 mmol L^{-1}.

17.2 (a) A solution of an unknown component of a biological sample when placed in an absorption cell of path length 1.00 cm transmits 20.1 per cent of light of 340 nm incident upon it. If the concentration of the component is 1.11×10^{-4} mol L^{-1}, what is the molar absorption coefficient?

17.2 (b) When light of wavelength 400 nm passes through 3.5 mm of a solution of an absorbing substance at a concentration 6.67×10^{-4} mol L^{-1}, the transmission is 65.5 per cent. Calculate the molar absorption coefficient of the solute at this wavelength and express the answer in cm^2 mol^{-1}.

17.3 (a) The molar absorption coefficient of a solute at 540 nm is 286 L mol^{-1} cm^{-1}. When light of that wavelength passes through a 6.5 mm cell containing a solution of the solute, 46.5 per cent of the light was absorbed. What is the concentration of the solution?

17.3 (b) The molar absorption coefficient of a solute at 440 nm is 323 L mol^{-1} cm^{-1}. When light of that wavelength passes through a 7.50 mm cell containing a solution of the solute, 52.3 per cent of the light was absorbed. What is the concentration of the solution?

17.4 (a) The absorption associated with a particular transition begins at 230 nm, peaks sharply at 260 nm, and ends at 290 nm. The maximum value of the molar absorption coefficient is 1.21×10^4 L mol^{-1} cm^{-1}. Estimate the integrated absorption coefficient of the transition assuming a triangular lineshape (see eqn 16.11).

17.4 (b) The absorption associated with a certain transition begins at 199 nm, peaks sharply at 220 nm, and ends at 275 nm. The maximum value of the molar absorption coefficient is 2.25×10^4 L mol^{-1} cm^{-1}. Estimate the integrated absorption coefficient of the transition assuming an inverted parabolic lineshape (Fig. 17.43; use eqn 16.11).

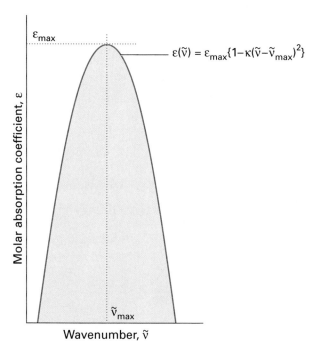

$$\varepsilon(\tilde{v}) = \varepsilon_{max}\{1 - \kappa(\tilde{v} - \tilde{v}_{max})^2\}$$

Fig. 17.43

17.5 (a) The two compounds 2,3-dimethyl-2-butene and 2,5-dimethyl-2,4-hexadiene are to be distinguished by their ultraviolet absorption spectra. The maximum absorption in one compound occurs at 192 nm and in the other at 243 nm. Match the maxima to the compounds and justify the assignment.

17.5 (b) 1,3,5-hexatriene (a kind of 'linear' benzene) was converted into benzene itself. On the basis of a free-electron molecular orbital model (in which hexatriene is treated as a linear box and benzene as a

ring), would you expect the lowest energy absorption to rise or fall in energy?

17.6 (a) The following data were obtained for the absorption by Br_2 in carbon tetrachloride using a 2.0 mm cell. Calculate the molar absorption coefficient of bromine at the wavelength employed:

$[Br_2]/(mol\,L^{-1})$	0.0010	0.0050	0.0100	0.0500
$T/(per\ cent)$	81.4	35.6	12.7	3.0×10^{-3}

17.6 (b) The following data were obtained for the absorption by a dye dissolved in methylbenzene using a 2.50 mm cell. Calculate the molar absorption coefficient of the dye at the wavelength employed:

$[dye]/(mol\,L^{-1})$	0.0010	0.0050	0.0100	0.0500
$T/(per\ cent)$	73	21	4.2	1.33×10^{-5}

17.7 (a) A 2.0 mm cell was filled with a solution of benzene in a non-absorbing solvent. The concentration of the benzene was 0.010 mol L^{-1} and the wavelength of the radiation was 256 nm (where there is a maximum in the absorption). Calculate the molar absorption coefficient of benzene at this wavelength given that the transmission was 48 per cent. What will the transmittance be in a 4.0 mm cell at the same wavelength?

17.7 (b) A 2.50 mm cell was filled with a solution of a dye. The concentration of the dye was 0.0155 mol L^{-1}. Calculate the molar absorption coefficient of dye at this wavelength given that the transmission was 32 per cent. What will the transmittance be in a 4.50 mm cell at the same wavelength?

17.8 (a) A swimmer enters a gloomier world (in one sense) on diving to greater depths. Given that the mean molar absorption coefficient of sea water in the visible region is 6.2×10^{-5} L mol^{-1} cm^{-1}, calculate the depth at which a diver will experience (a) half the surface intensity of light, (b) one-tenth the surface intensity.

17.8 (b) Given that the maximum molar absorption coefficient of a molecule containing a carbonyl group at a concentration of 1.00 mol L^{-1} is 30 L mol^{-1} cm^{-1} near 280 nm, calculate the thickness of a sample that will result in (a) half the initial intensity of radiation, (b) one-tenth the initial intensity.

17.9 (a) The electronic absorption bands of many molecules in solution have half-widths at half-height of about 5000 cm^{-1}. Estimate the integrated absorption coefficients of bands for which (a) $\varepsilon_{max} \approx 1 \times 10^4$ L mol^{-1} cm^{-1}, (b) $\varepsilon_{max} \approx 5 \times 10^2$ L mol^{-1} cm^{-1}.

17.9 (b) The electronic absorption band of a compound in solution had a Gaussian lineshape and a half-width at half-height of 4233 cm^{-1} and $\varepsilon_{max} = 1.54 \times 10^4$ L mol^{-1} cm^{-1}. Estimate the integrated absorption coefficient.

17.10 (a) The photoionization of H_2 by 21 eV photons produces H_2^+. Explain why the intensity of the $v = 2 \leftarrow 0$ transition is stronger than that of the $0 \leftarrow 0$ transition.

17.10 (b) The photoionization of F_2 by 21 eV photons produces F_2^+. Would you expect the $2 \leftarrow 0$ transition to be weaker or stronger than that of the $0 \leftarrow 0$ transition? Justify your answer.

Problems

Numerical problems

17.1 The vibrational wavenumber of the oxygen molecule in its electronic ground state is 1580 cm^{-1}, whereas that in the first excited state $(B\,^3\Sigma_u^-)$ to which there is an allowed electronic transition is 700 cm^{-1}. If the separation in energy between the minima in their respective potential energy curves of these two electronic states is 6.175 eV, what is the wavenumber of the lowest energy transition in the band of transitions originating from the $v = 0$ vibrational state of the electronic ground state to this excited state? Ignore any rotational structure or anharmonicity.

17.2 A Birge–Sponer extrapolation yields 7760 cm^{-1} as the area under the curve for the B state of the oxygen molecule described in Problem 17.1. Given that the B state dissociates to ground-state atoms (at zero energy, ^{3}P) and $15\,870$ cm^{-1} (^{1}D) and the lowest vibrational state of the B state is $49\,363$ cm^{-1} above the lowest vibrational state of the ground electronic state, calculate the dissociation energy of the molecular ground state to the ground-state atoms.

17.3 The electronic spectrum of the IBr molecule shows two low-lying, well defined convergence limits at $14\,660$ and $18\,345$ cm^{-1}. Energy levels for the iodine and bromine atoms occur at $0, 7598$ cm^{-1}; and $0, 3685$ cm^{-1}, respectively. Other atomic levels are at much higher energies. What possibilities exist for the numerical value of the dissociation energy of IBr? Decide which is the correct possibility by calculating this quantity from $\Delta_f H^{\ominus}(\text{IBr, g}) = +40.79$ kJ mol^{-1} and the dissociation energies of $I_2(\text{g})$ and $Br_2(\text{g})$ which are 146 and 190 kJ mol^{-1}, respectively.

17.4 In many cases it is possible to assume that an absorption band has a Gaussian lineshape (one proportional to e^{-x^2}) centred on the band maximum. Assume such a lineshape, and show that $\mathcal{A} \approx 1.0645\varepsilon_{\max}\Delta\tilde{\nu}_{1/2}$, where $\Delta\tilde{\nu}_{1/2}$ is the width at half-height. The absorption spectrum of azoethane ($CH_3CH_2N_2$) between $24\,000$ cm^{-1} and $34\,000$ cm^{-1} is shown in Fig. 17.44. First, estimate $\mathcal{A}$ for the band

by assuming that it is Gaussian. Then integrate the absorption band graphically. The latter can be done either by ruling and counting squares, or by tracing the lineshape on to paper and weighing. A more sophisticated procedure would be to use mathematical software to fit a polynomial to the absorption band (or a Gaussian), and then to integrate the result analytically.

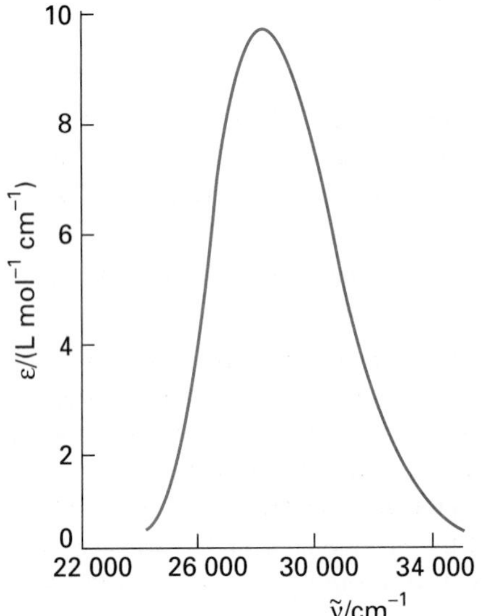

17.5 A lot of information about the energy levels and wavefunctions of small inorganic molecules can be obtained from their ultraviolet spectra. An example of a spectrum with considerable vibrational structure, that of gaseous SO_2 at $25°C$, is shown in Fig. 17.45. Estimate the integrated absorption coefficient for the transition. What electronic states are accessible from the A_1 ground state of this C_{2v} molecule by electric dipole transitions?

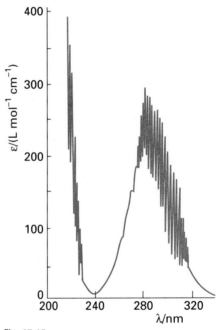

Fig. 17.45

17.6 A certain molecule fluoresces at a wavelength of 400 nm with a half-life of 1.0 ns. It phosphoresces at 500 nm. If the ratio of the transition probabilities for stimulated emission for the $S^* \to S$ to the $T \to S$ transitions is 1.0×10^5, what is the half-life of the phosphorescent state?

17.7 The photoelectron spectra of N_2 and CO are shown in Fig. 17.46. Ascribe the lines to the ionization processes involved and classify the orbitals from which the electrons are ejected as bonding, nonbonding, or antibonding in the light of the extent of vibrational structure in the band. Analyse the bands near 4 eV in terms of the vibrational energy levels of the ions.

Fig. 17.44

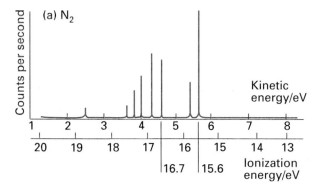

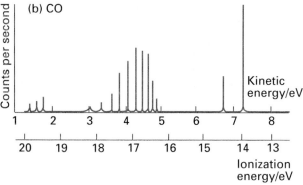

Fig. 17.46

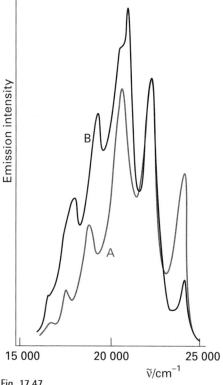

Fig. 17.47

17.8 The photoelectron spectrum of NO can be described as follows (D.W. Turner, in *Physical methods in advanced inorganic chemistry* (ed. H.A.O. Hill and P. Day), Wiley, Chichester (1968)). Using He 58.4 pm (21.21 eV) radiation there is a single strong peak at kinetic energy 4.69 eV and a long series of 24 lines starting at 5.56 eV and ending at 2.2 eV. A shorter series of six lines begins at 12.0 eV and ends at 10.7 eV. Account for this spectrum.

Theoretical problems

17.9 Assume that the electronic states of the π electrons of a conjugated molecule can be approximated by the wavefunctions of a particle in a one-dimensional box, and that the dipole moment can be related to the displacement along this length by $\mu = -ex$. Show that the transition probability connecting states 1 and 2 is nonzero, whereas that connecting states 1 and 3 is zero.

17.10 Use a group theoretical argument to decide which of the following transitions are electric-dipole allowed: (a) the $\pi^* \leftarrow \pi$ transition in ethene, (b) the $\pi^* \leftarrow n$ transition in a carbonyl group in a C_{2v} environment.

17.11 The line marked A in Fig. 17.47 is the fluorescence spectrum of benzophenone in solid solution in ethanol at low temperatures observed when the sample is illuminated with 360 nm light. What can be said about the vibrational energy levels of the carbonyl group in (a) its ground electronic state and (b) its excited electronic state? When naphthalene is illuminated with 360 nm light it does not absorb, but the line marked B in the illustration is the phosphorescence spectrum of a solid solution of a mixture of naphthalene and benzophenone in ethanol. Now a component of fluorescence from naphthalene can be detected. Account for this observation.

17.12 The fluorescence spectrum of anthracene vapour shows a series of peaks of increasing intensity with individual maxima at 440 nm, 410 nm, 390 nm, and 370 nm followed by a sharp cut-off at shorter wavelengths. The absorption spectrum rises sharply from zero to a maximum at 360 nm with a trail of peaks of lessening intensity at 345 nm, 330 nm, and 305 nm. Account for these observations.

17.13 Suppose that you are a colour chemist and had been asked to intensify the colour of a dye without changing the type of compound, and that the dye in question was a polyene. Would you choose to lengthen or to shorten the chain? Would the modification to the length shift the apparent colour of the dye towards the red or the blue?

17.14 One measure of the intensity of a transition of frequency ν is the oscillator strength, f, which is defined as

$$f = \frac{8\pi^2 m_e \nu |\mu_{fi}|^2}{3he^2}$$

Consider an electron in an atom to be oscillating harmonically in one dimension (the three-dimensional version of this model was used in early attempts to describe atomic structure). The wavefunctions for such an electron are those in Table 12.1. Show that the oscillator strength for the transition of this electron from its ground state is exactly $\frac{1}{3}$.

17.15 Estimate the oscillator strength (see Problem 17.14) of a charge-transfer transition modelled as the migration of an electron from a hydrogen $1s$ orbital on one atom to another hydrogen $1s$ orbital on an atom a distance R away. Approximate the transition moment by $-eRS$ where S is the overlap integral of the two orbitals. Sketch the oscillator strength as a function of R using the curve for S given in Fig. 14.31. Why does the intensity fall to zero as R approaches 0 and infinity?

Additional problems supplied by Carmen Giunta and Charles Trapp

17.16 Refer to Fig. 17.25 and estimate the maximum laser power that can be delivered from a ruby crystal of length 5.0 cm and diameter 0.50 cm, in a pulse of duration 100 ns. 'Pink ruby' consists of about 0.050 per cent by mass Cr^{3+} and the mass density of Al_2O_3 is 3.97 $g\,cm^{-3}$. Assume that the pumping radiation is of sufficient intensity to pump all the chromium ions out of their ground state at a rate faster than they decay back to the ground state.

17.17 J.G. Dojahn, E.C.M. Chen, and W.E. Wentworth (*J. Phys. Chem.* **100**, 9649 (1996)) characterized the potential energy curves of the ground and electronic states of homonuclear diatomic halogen anions. These anions have a $^2\Sigma_u^+$ ground state and $^2\Pi_g$, $^2\Pi_u$, and $^2\Sigma_g^+$ excited states. To which of the excited states are transitions by absorption of photons allowed? Explain.

17.18 M. Schwell, H.-W. Jochims, B. Wassermann, U. Rockland, R. Flesch, and E. Rühl (*J. Phys. Chem.* **100**, 10070 (1996)) measured the ionization energies of Cl_2O_2 by photoelectron spectroscopy in which the ionized fragments were detected using a mass spectrometer. From their data, we can infer that the ionization enthalpy of Cl_2O_2 is 11.05 eV and the enthalpy of the dissociative ionization $Cl_2O_2 \rightarrow Cl + OClO^+ + e^-$ is 10.95 eV. They used this information to make some inferences about the structure of Cl_2O_2. Computational studies had suggested that the lowest energy isomer is ClOOCl, but that ClClO$_2$ (C_{2v}) and ClOClO are not very much higher in energy. The Cl_2O_2 in the photoionization step is the lowest energy isomer, whatever its structure may be, and its enthalpy of formation had previously been reported as +133 kJ mol^{-1}. The Cl_2O_2 in the dissociative ionization step is unlikely to be ClOOCl, for the product can be derived from it only with substantial rearrangement. Given $\Delta_f H^{\ominus}(OClO^+) = +1096$ kJ mol^{-1} and $\Delta_f H^{\ominus}(e^-) = 0$, determine whether the Cl_2O_2 in the dissociative ionization is the same as that in the photoionization. If different, how much greater is its $\Delta_f H^{\ominus}$? Are these results consistent with or contradictory to the computational studies?

17.19 G.C.G. Wachewsky, R. Horansky, and V. Vaida (*J. Phys. Chem.* **100**, 11559 (1996)) examined the UV absorption spectrum of CH_3I, a species of interest in connection with stratospheric ozone chemistry. They found the integrated absorption coefficient to be dependent on temperature and pressure to an extent inconsistent with internal structural changes in isolated CH_3I molecules; they explained the changes as due to dimerization of a substantial fraction of the CH_3I, a process which would naturally be pressure- and temperature-dependent. (a) Compute the integrated absorption coefficient over a triangular lineshape in the range 31 250 to 34 483 cm^{-1} and a maximal molar absorption coefficient of 150 $L\,mol^{-1}\,cm^{-1}$ at 31 250 cm^{-1}. (b) Suppose 1 per cent of the CH_3I units in a sample at 2.4 Torr and 373 K exists as dimers. Compute the absorbance expected at 31 250 cm^{-1} in a sample cell of length 12.0 cm. (c) Suppose 18 per cent of the CH_3I units in a sample at 100 Torr and

373 K exists as dimers. Compute the absorbance expected at 31 250 cm^{-1} in a sample cell of length 12.0 cm; compute the molar absorption coefficient that would be inferred from this absorbance if dimerization were not considered.

17.20 The abundance of ozone is typically inferred from measurements of UV absorption and is often expressed in terms of Dobson units (DU): 1 DU is equivalent to a layer of pure ozone 10^{-3} cm thick at 1 atm and 0°C. Compute the absorbance of UV radiation at 300 nm expected for an ozone abundance of 300 DU (a typical value) and 100 DU (a value reached during seasonal Antarctic ozone depletions) given a molar absorption coefficient of 476 L mol^{-1} cm^{-1}.

17.21 Ozone absorbs ultraviolet radiation in a part of the electromagnetic spectrum that is energetic enough to disrupt DNA in biological organisms and that is absorbed by no other abundant atmospheric constituent. This spectral range, denoted UV-B, spans the wavelengths of about 290 nm to 320 nm. The molar extinction coefficient of ozone over this range is given in the table below (W.B. DeMore, S.P. Sander, D.M. Golden, R.F. Hampson, M.J. Kurylo, C.J. Howard, A.R. Ravishankara, C.E. Kolb, and M.J. Molina, *Chemical kinetics and photochemical data for use in stratospheric modeling: Evaluation Number 11*, JPL Publication 94–26 (1994)).

λ/nm	292.0	296.3	300.8	305.4	310.1	315.0	320.0
ε/(L mol^{-1} cm^{-1})	1512	865	477	257	135.9	69.5	34.5

Compute the integrated absorption coefficient of ozone over the wavelength range 290–320 nm. (*Hint:* $\varepsilon(\tilde{\nu})$ can be fitted to an exponential function quite well.)

17.22 One of the principal methods of obtaining the electronic spectra of unstable radicals is to study the spectra of comets, which consist almost entirely of radical spectra. Many radical spectra have been found in comets including that due to CN. These radicals are produced in comets by the absorption of far ultraviolet solar radiation by their parent compounds. Subsequently, their fluorescence is excited by sunlight of longer wavelength. The spectra of comet Hale–Bopp (C/1995 O1) have been the subject of many recent studies. One such study is that of the fluorescence spectrum of CN in the coma of Hale–Bopp at large heliocentric distances by R.M. Wagner and D.G. Schleicher (*Science* **275**, 1918 (1997)), in which the authors determine the spatial distribution and rate of production of CN in the coma. The (0–0) vibrational band is centred on 387.6 nm and the weaker (1–1) band with relative intensity 0.1 is centred on 386.4 nm. The band heads for (0–0) and (0–1) are known to be 388.3 and 421.6 nm, respectively. From these data, calculate the energy of the excited S_1 state relative to the ground S_0 state, the vibrational wavenumbers and the difference in the vibrational wavenumbers of the two states, and the relative populations of the $v = 0$ and $v = 1$ vibrational levels of the S_1 state. Also estimate the effective temperature of the molecule in the excited S_1 state. Only eight rotational levels of the S_1 state are said to be populated. Is that statement consistent with the effective temperature of the S_1 state?

18 Spectroscopy 3: magnetic resonance

One of the most widely used spectroscopic procedures in chemistry makes use of the classical concept of resonance. The chapter begins with an account of conventional nuclear magnetic resonance which shows how the resonance frequency of a magnetic nucleus is affected by its electronic environment and the presence of magnetic nuclei in its vicinity. Then we turn to the modern versions of NMR, which are based on the use of pulses of electromagnetic radiation and the processing of the resulting signal by Fourier transform techniques. The experimental techniques for electron spin resonance resemble those used in the early days of NMR. The information obtained, however, is very useful for the determination of the properties of radicals and d-metal complexes.

When two pendulums share a slightly flexible support and one is set in motion, the other is forced into oscillation by the motion of the common axle. As a result, energy flows between the two pendulums. The energy transfer occurs most efficiently when the frequencies of the two pendulums are identical. The condition of strong effective coupling when the frequencies of two oscillators are identical is called **resonance**.

Resonance is the basis of a number of everyday phenomena, including the response of radios to the weak oscillations of the electromagnetic field generated by a distant transmitter. In this chapter we explore some spectroscopic applications that, as originally developed (and in some cases still), depend on matching a set of energy levels to a source of monochromatic radiation and observing the strong absorption that occurs at resonance.

Nuclear magnetic resonance

The basic nuclear magnetic resonance (NMR) experiment is the resonant absorption of radiofrequency radiation by nuclei exposed to a magnetic field. Although simple in concept, NMR spectra can be highly complex; yet they have proved invaluable in chemistry, for they reveal so much structural information. A magnetic nucleus is a very sensitive, noninvasive probe of the surrounding electronic structure.

18.1 Nuclear magnetic moments

Many nuclei possess spin angular momentum. A nucleus with spin quantum number I (which is a fixed characteristic property of a nucleus and may be an integer or a half-integer but is never negative) has the following properties:

1. An angular momentum of magnitude $\{I(I+1)\}^{1/2}\hbar$.
2. A component of angular momentum $m_I\hbar$ on an arbitrary axis, where $m_I = I, I-1, \ldots, -I$.
3. If $I > 0$, a magnetic moment with a constant magnitude and an orientation that is determined by the value of m_I.

To say that a nucleus has a magnetic moment means that, to some extent, it behaves like a small bar magnet.

According to the second property, the spin, and hence the magnetic moment, of the nucleus may lie in $2I + 1$ different orientations relative to an axis. A proton has $I = \frac{1}{2}$ and its spin may adopt either of two orientations; a ^{14}N nucleus has $I = 1$ and its spin may adopt any of three orientations. For much of this chapter we shall consider **spin-$\frac{1}{2}$ nuclei**, which are nuclei with $I = \frac{1}{2}$, but NMR is applicable to nuclei with any nonzero spin. As well as protons, which are the most common nuclei studied by NMR, spin-$\frac{1}{2}$ nuclei include ^{13}C, ^{19}F, and ^{31}P nuclei. The state with $m_I = +\frac{1}{2}$ ($\uparrow$) is denoted α and the state with $m_I = -\frac{1}{2}$ ($\downarrow$) is denoted β. It is worth bearing in mind that two very common nuclei, ^{12}C and ^{16}O, have zero spin, and hence zero magnetic moment, and so are invisible in magnetic resonance.

18.2 The energies of nuclei in magnetic fields

The nuclear magnetic moment of a nucleus is denoted $\boldsymbol{\mu}$. The component of the nuclear magnetic moment on the z-axis, μ_z, is proportional to the component of spin angular momentum on that axis, $m_I\hbar$, and we write

$$\mu_z = \gamma \hbar m_I \tag{1}$$

The coefficient of proportionality γ is called the **magnetogyric ratio** of the nucleus, and is an experimentally determined quantity (Table 18.1). The magnetic moment is sometimes expressed in terms of the **nuclear g-factor**, g_I, and the **nuclear magneton**, μ_N, by using

$$\gamma \hbar = g_I \mu_N \qquad \mu_N = \frac{e\hbar}{2m_p} = 5.051 \times 10^{-27} \text{ J T}^{-1} \tag{2}$$

where m_p is the mass of the proton. Nuclear g-factors are numbers of the order of 1 (Table 18.1): positive values of g_I and γ denote a magnetic moment that is parallel to the spin; negative values indicate that the magnetic moment and spin are antiparallel. The nuclear magneton is about 2000 times smaller than the Bohr magneton, so nuclear magnetic moments are about 2000 times weaker than the electron spin magnetic moment.

Table 18.1* Nuclear spin properties

Nuclide	Natural abundance/%	Spin I	g-value, g_I	$\gamma/(10^7 \text{ T}^{-1}\text{s}^{-1})$
^{1}n		$\frac{1}{2}$	-3.826	-18.32
^{1}H	99.98	$\frac{1}{2}$	5.586	26.75
^{2}H	0.02	1	0.857	4.10
^{13}C	1.11	$\frac{1}{2}$	1.405	6.73
^{14}N	99.64	1	0.404	1.93

*More values are given in the *Data section* at the end of this volume. Note that $\gamma = g_I\mu_N/\hbar$.

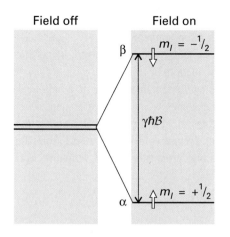

Field off   Field on

18.1 The nuclear spin energy levels of a spin-$\frac{1}{2}$ nucleus (for example, ^{1}H or ^{13}C) in a magnetic field. Resonance occurs when the energy separation of the levels matches the energy of the photons in the electromagnetic field.

(a) The basic resonance experiment

Each value of m_I corresponds to a different orientation of the nuclear spin and therefore of the nuclear magnetic moment too. In a magnetic field $\mathcal{B}$ in the z-direction, the $2I + 1$ orientations of the nucleus have different energies, which are given by

$$E_{m_I} = -\mu_z\mathcal{B} = -\gamma\hbar\mathcal{B}m_I \tag{3}$$

These energies are often expressed in terms of the **Larmor frequency**, ν_L:

$$E_{m_I} = -m_I h\nu_L \qquad \nu_L = \frac{\gamma\mathcal{B}}{2\pi} \tag{4}$$

The stronger the magnetic field, the higher the Larmor frequency. A field of 12 T corresponds to a Larmor frequency of about 500 MHz for protons.

The energy separation of the two states of spin-$\frac{1}{2}$ nuclei is

$$\Delta E = E_\beta - E_\alpha = \tfrac{1}{2}\gamma\hbar\mathcal{B} - (-\tfrac{1}{2}\gamma\hbar\mathcal{B}) = \gamma\hbar\mathcal{B} = h\nu_L \tag{5}$$

For most nuclei γ is positive. In such cases, the β state lies above the α state, and there are slightly more α spins than β spins. When the sample is exposed to radiation of frequency ν, the energy separations come into resonance with the radiation when the frequency satisfies the **resonance condition** (Fig. 18.1):

$$h\nu = \gamma\hbar\mathcal{B} = h\nu_L \tag{6}$$

That is, there is resonance when $\nu = \nu_L$. At resonance there is strong coupling between the nuclear spins and the radiation, and strong absorption occurs as the spins make the transition $\alpha \rightarrow \beta$. At 12 T, protons come into resonance at about 500 MHz (the Larmor frequency at that magnetic field).

(b) The technique

In its simplest form, **nuclear magnetic resonance** (NMR) is the study of the properties of molecules containing magnetic nuclei by applying a magnetic field and observing the frequency of the resonant electromagnetic field. Larmor frequencies of nuclei at the fields normally employed typically lie in the radiofrequency region of the electromagnetic spectrum, so NMR is a radiofrequency technique.

An NMR spectrometer consists of a magnet that can produce a uniform, intense field and the appropriate sources of radiofrequency electromagnetic radiation. In simple instruments, the magnetic field is provided by a permanent magnet. For serious work, a superconducting magnet capable of producing fields of the order of 2 T and more is used (Fig. 18.2). The sample is placed in the cylindrically wound magnet. In some cases the sample is rotated rapidly to remove magnetic inhomogeneities. However, sample spinning is a source of noise,

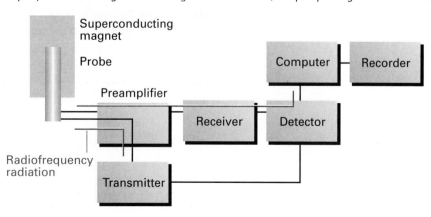

18.2 The layout of a typical NMR spectrometer. The link from the transmitter to the detector indicates that the high frequency of the transmitter is subtracted from the high-frequency received signal to give a low-frequency signal for processing.

and is often avoided. Although a superconducting magnet operates at the temperature of liquid helium (4 K), the sample itself is normally at room temperature.

The use of high magnetic fields has several advantages, the most important being that they simplify the appearance of spectra and so allow them to be interpreted more readily. A further advantage is that the rate of energy uptake by the sample is greater in a high field. There are two contributions to this increase. One comes from the greater population difference between the upper and lower spin states at high fields, for the population difference is approximately proportional to $\mathcal{B}$. The second contribution stems from the greater energy of each absorbed photon, which is also proportional to $\mathcal{B}$. It follows that overall the signal is proportional to $\mathcal{B}^2$.

Justification 18.1

According to the Boltzmann distribution, the ratio of populations is

$$\frac{N_\beta}{N_\alpha} = e^{-\Delta E/kT} \approx 1 - \frac{\Delta E}{kT}$$

It follows that

$$\frac{N_\alpha - N_\beta}{N_\alpha + N_\beta} \approx \frac{\Delta E}{2kT} = \frac{\gamma \hbar \mathcal{B}}{2kT}$$

and the population difference is proportional to $\mathcal{B}$. The energy of the photon absorbed when a nucleus makes a transition from its lower state to its higher state is $h\nu$; at resonance ν is equal to ν_L, and ν_L is proportional to $\mathcal{B}$. Hence, at resonance, each photon has an energy that is proportional to $\mathcal{B}$. The net rate of energy absorption is proportional to the population difference multiplied by the energy of each absorption event (the photon energy), so overall the net rate is proportional to $\mathcal{B}^2$.

18.3 The chemical shift

Nuclear magnetic moments interact with the *local* magnetic field. The local field may differ from the applied field because the latter induces electronic orbital angular momentum (that is, the circulation of electronic currents) which gives rise to a small additional magnetic field $\delta\mathcal{B}$ at the nuclei. This additional field is proportional to the applied field, and it is conventional to write

$$\delta\mathcal{B} = -\sigma\mathcal{B} \tag{7}$$

where the dimensionless quantity σ is called the **shielding constant** of the nucleus (σ is usually positive but may be negative). The ability of the applied field to induce an electronic current in the molecule, and the strength of the resulting local magnetic field experienced by the nucleus, depend on the details of the electronic structure near the magnetic nucleus of interest, so nuclei in different chemical groups have different shielding constants. The calculation of reliable values of the shielding constant is very difficult, but trends in it are quite well understood, and we concentrate on them.

(a) The δ scale of chemical shifts

Because the total local field is

$$\mathcal{B}_{loc} = \mathcal{B} + \delta\mathcal{B} = (1 - \sigma)\mathcal{B} \tag{8}$$

the Larmor frequency is

$$\nu_L = \frac{\gamma \mathcal{B}_{loc}}{2\pi} = (1 - \sigma)\frac{\gamma \mathcal{B}}{2\pi} \tag{9}$$

This frequency is different for nuclei in different environments. Hence, different nuclei, even of the same element, come into resonance at different frequencies.

It is conventional to express the resonance frequencies in terms of an empirical quantity called the **chemical shift**, which is related to the difference between the resonance frequency, ν, of the nucleus in question and that of a reference standard, ν°:

$$\delta = \frac{\nu - \nu^\circ}{\nu^\circ} \times 10^6 \tag{10}$$

The standard for protons is the proton resonance in tetramethylsilane ($Si(CH_3)_4$, commonly referred to as TMS), which bristles with protons and dissolves without reaction in many liquids. Other references are used for other nuclei. For ^{13}C, the reference frequency is the ^{13}C resonance in TMS; for ^{31}P it is the ^{31}P resonance in 85 per cent $H_3PO_4(aq)$. The advantage of the δ-scale is that shifts reported on it are independent of the applied field (because both numerator and denominator are proportional to the applied field).

Illustration

A nucleus with $\delta = 1.00$ (which is often, but unnecessarily, expressed as 1.00 ppm on account of the 10^6 in the definition of δ) in a spectrometer operating at 500 MHz will have a shift relative to the reference equal to

$$\nu - \nu^\circ = (500\ \text{MHz}) \times (1.00) \times 10^{-6} = 500\ \text{Hz}$$

In a spectrometer operating at 100 MHz, the shift relative to the reference would be only 100 Hz.

The relation between δ and σ is obtained by substituting eqn 8 into eqn 10:

$$\delta = \frac{(1-\sigma)\mathcal{B} - (1-\sigma^\circ)\mathcal{B}}{(1-\sigma^\circ)\mathcal{B}} \times 10^6 = \frac{\sigma^\circ - \sigma}{1 - \sigma^\circ} \times 10^6$$
$$\approx (\sigma^\circ - \sigma) \times 10^6 \tag{11}$$

As the shielding, σ, gets smaller, δ *increases*. Therefore, we speak of nuclei with large chemical shift as being strongly **deshielded**. Some typical chemical shifts are given in Fig. 18.3. As can be seen from the illustration, the nuclei of different elements have very different ranges of chemical shifts. The ranges exhibit the variety of electronic environments of the nuclei in molecules.

By convention, NMR spectra are plotted with δ increasing from right to left. Consequently, in a given applied magnetic field the Larmor frequency also increases from right to left. In a continuous wave (CW) spectrometer, in which the radiofrequency is held constant and the magnetic field is varied (a 'field sweep experiment'), the spectrum is displayed with the applied magnetic field increasing from left to right: a nucleus with a small chemical shift experiences a relatively low local magnetic field, so it needs a higher applied magnetic field to bring it into resonance with the radiofrequency field. Consequently, the right-hand end (low chemical shift) end of the spectrum is commonly referred to as the 'high field end' of the spectrum.

(b) Resonance of different groups of nuclei

The existence of a chemical shift explains the general features of the spectrum of ethanol shown in Fig. 18.4. The CH_3 protons form one group of nuclei with $\delta \approx 1$. The two CH_2

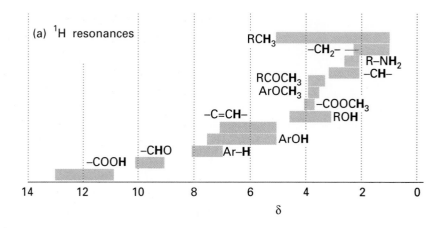

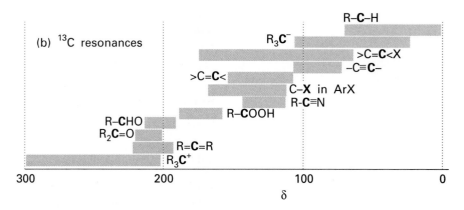

18.3 The range of typical chemical shifts for (a) ^{1}H resonances and (b) ^{13}C resonances.

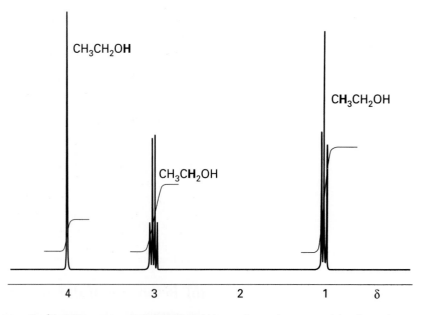

18.4 The ^{1}H-NMR spectrum of ethanol. The bold letters denote the protons giving rise to the resonance peak, and the step-like curve is the integrated signal.

protons are in a different part of the molecule, experience a different local magnetic field, and resonate at $\delta \approx 3$. Finally, the OH proton is in another environment, and has a chemical shift of $\delta \approx 4$. The increasing value of δ (that is, the increase in deshielding) is consistent with the electron-withdrawing power of the O atom: it reduces the electron density of the OH proton most, and that proton is strongly deshielded. It reduces the electron density of the distant methyl protons least, and those nuclei are least deshielded.

The relative intensities of the signal (the areas under the absorption lines) can be used to help distinguish which group of lines corresponds to which chemical group. The determination of the area under an absorption line is referred to as the **integration** of the signal (just as any area under a curve may be determined by mathematical integration). Spectrometers can integrate the absorption automatically (as indicated in Fig. 18.4). In ethanol the group intensities are in the ratio 3 : 2 : 1 because there are three CH_3 protons, two CH_2 protons, and one OH proton in each molecule. Counting the number of magnetic nuclei as well as noting their chemical shifts helps to identify a compound present in a sample.

(c) The origin of shielding constants

The calculation of shielding constants is very difficult, even for small molecules, for it requires detailed information about the distribution of electron density in the ground and excited states and the excitation energies of the molecule. Some success has been achieved with the calculation for diatomic molecules and small polyatomic molecules such as H_2O and CH_4, but large molecules are much more difficult. Nevertheless, a considerable body of useful empirical information about a variety of contributions to chemical shifts in large molecules has been compiled, and has been used to understand and interpret observations reasonably systematically.

The empirical approach supposes that the observed shielding constant is the sum of three contributions:

$$\sigma = \sigma(\text{local}) + \sigma(\text{neighbour}) + \sigma(\text{solvent}) \tag{12}$$

The **local contribution**, $\sigma(\text{local})$, is essentially the contribution of the electrons of the atom that contains the nucleus in question. The **neighbouring group contribution**, $\sigma(\text{neighbour})$, is the contribution from the groups of atoms that form the rest of the molecule. The **solvent contribution**, $\sigma(\text{solvent})$, is the contribution from the solvent molecules.

(d) The local contribution

It is convenient to regard the local contribution to the shielding constant as the sum of a positive **diamagnetic contribution**, σ_d, and a negative **paramagnetic contribution**, σ_p:

$$\sigma(\text{local}) = \sigma_d + \sigma_p \tag{13}$$

The total local contribution is positive if the diamagnetic contribution dominates, and is negative if the paramagnetic contribution dominates.

The diamagnetic contribution arises from the ability of the applied field to generate a circulation of charge in the ground-state electron distribution of the atom. The circulation generates a magnetic field that opposes the applied field and hence shields the nucleus. The magnitude of σ_d depends on the electron density close to the nucleus and can be calculated from the **Lamb formula**:

$$\sigma_d = \frac{e^2 \mu_0}{12\pi m_e} \left\langle \frac{1}{r} \right\rangle \tag{14}$$

where μ_0 is the vacuum permeability (a fundamental constant, see inside the front cover) and r is the electron–nucleus distance.

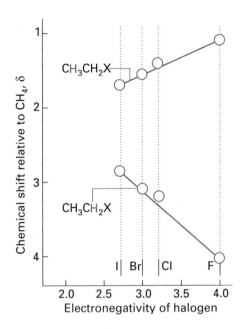

18.5 The variation of chemical shielding with electronegativity. The shifts for the methylene protons agree with the trend expected with increasing electronegativity. However, to emphasize that chemical shifts are subtle phenomena, notice that the trend for the methyl protons is opposite to that expected. For these protons another contribution (the magnetic anisotropy of C–H and C–X bonds) is dominant.

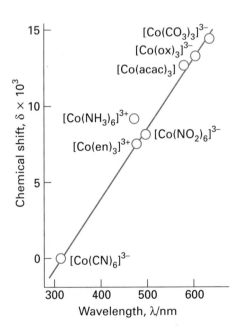

18.6 The correlation of chemical shifts relative to $[Co(CN)_6]^{3-}$ and the ligand field parameter for complexes of cobalt. The wavelength is that of the lowest energy transition (and $\Delta \propto 1/\lambda$, approximately). Data from R. Freeman, G.R. Murray, and R.E. Richards, *Proc. Roy. Soc.* A242, 455 (1957).

Example 18.1 Using the Lamb formula

Calculate the shielding constant for the proton in a free H atom.

Method To use the Lamb formula, we need to calculate the expectation value of $1/r$ for a hydrogen $1s$ orbital. Wavefunctions are given in Table 13.1, and a useful integral is given in Example 11.6.

Answer Because $d\tau = r^2 dr \sin\theta \, d\theta \, d\phi$, we can write

$$\left\langle \frac{1}{r} \right\rangle = \int \frac{\psi^*\psi}{r} \, d\tau = \frac{1}{\pi a_0^3} \int_0^{2\pi} d\phi \int_0^\pi \sin\theta \, d\theta \int_0^\infty r e^{-2r/a_0} \, dr$$

$$= \frac{4}{a_0^3} \int_0^\infty r e^{-2r/a_0} \, dr = \frac{1}{a_0}$$

Therefore,

$$\sigma_d = \frac{e^2 \mu_0}{12\pi m_e a_0}$$

With the values of the fundamental constants inside the front cover, this expression evaluates to 1.78×10^{-5}.

Comment The shielding constant is inversely proportional to the Bohr radius. This distance dependence can be understood as arising from the classical result that the magnetic moment of a current loop is proportional to its area (which for a hydrogen atom is of the order of a_0^2) and the magnetic field that it generates at the nucleus is inversely proportional to the cube of the latter's distance (a_0^3). Hence, the local field is proportional to $a_0^2 \times 1/a_0^3 = 1/a_0$.

Self-test 18.1 Calculate σ_d for a hydrogenic atom with atomic number Z.

$$[\sigma_d = Z\sigma_d(H)]$$

The diamagnetic contribution is the only contribution in closed-shell free atoms. It is also the only contribution to the local shielding for distributions of charge that have spherical or cylindrical symmetry. Thus, it is the only contribution to the local shielding from inner cores of atoms, for cores remain spherical even though the atom may be a component of a molecule and its valence electron distribution highly distorted. The diamagnetic contribution is broadly proportional to the electron density of the atom containing the nucleus of interest. It follows that the shielding is decreased if the electron density on the atom is reduced by the influence of an electronegative atom nearby. That reduction in shielding translates into an increase in deshielding, and hence to an increase in the chemical shift δ as the electronegativity of a neighbouring atom increases (Fig. 18.5). That is, as the electronegativity increases, δ decreases.

The local paramagnetic contribution, σ_p, arises from the ability of the applied field to force the electrons to circulate through the molecule by making use of orbitals that are unoccupied in the ground state. It is zero in free atoms and around the axes of linear molecules (such as ethyne, HC≡CH) where the electrons can circulate freely and a field applied along the internuclear axis is unable to force them into other orbitals.

The magnitude of the paramagnetic contribution depends on the ease with which the applied field can promote electrons into unoccupied orbitals. Hence, it is inversely proportional to the energy separation of the highest filled (HOMO) and lowest unfilled (LUMO) orbitals of the molecule, Δ. The strength of the magnetic field generated by the magnetic moment of the resulting circulation of charge is inversely proportional to the cube

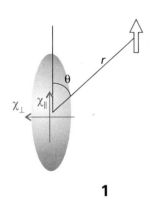

18.7 The variation of the function $1 - 3\cos^2\theta$ with the angle θ.

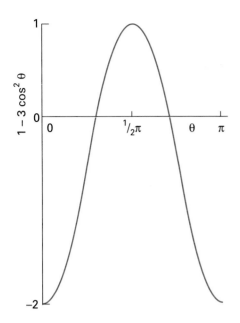

1

of the distance of the nucleus from the circulating current, so the field at the nucleus is proportional to $\langle r^{-3} \rangle$. Overall, therefore,

$$\sigma_p \propto -\frac{\langle r^{-3} \rangle}{\Delta} \tag{15}$$

We can therefore expect large paramagnetic contributions from small atoms in molecules with low-lying excited states. In fact, the paramagnetic contribution is the dominant local contribution for atoms other than hydrogen. The shielding constants of the nuclei of d-metal ions in complexes correlate quite well with spectroscopic data if Δ is identified with the ligand-field splitting parameter (Fig. 18.6).

(e) Neighbouring group contributions

The neighbouring group contribution arises from the currents induced in nearby groups of atoms. The effect of either kind of current (diamagnetic or paramagnetic) is to shield or deshield the nucleus depending on the relative location of the nucleus to the neighbouring group.

The applied field generates currents in the electron distribution of the neighbouring group and gives rise to a magnetic moment proportional to the applied field; the constant of proportionality is the magnetic susceptibility, χ, of the group.[1] This induced magnetic moment gives rise to a magnetic field at the nucleus with a strength that is inversely proportional to the cube of the distance of the nucleus from the group of atoms. The field varies with the orientation of the molecule, but it does not average to zero because the magnetic susceptibility also changes as the molecule presents different orientations to the applied field. The result is that the shielding constant depends on three quantities: the difference in the magnetic susceptibilities parallel and perpendicular to the group (we are assuming that the group has cylindrical symmetry), the angle θ that the vector to the magnetic nucleus makes to the axis of symmetry of the group, and the distance r of the nucleus from the group (1):

$$\sigma(\text{neighbour}) \propto (\chi_\parallel - \chi_\perp)\left(\frac{1 - 3\cos^2\theta}{r^3}\right) \tag{16}$$

This expression shows that the neighbouring group contribution may be positive or negative according to the relative magnitudes of the two magnetic susceptibilities and the relative orientation of the nucleus. The latter effect is easy to anticipate: if $54.7° < \theta < 125.3°$, then $1 - 3\cos^2\theta$ is positive, but it is negative otherwise (Fig. 18.7).

A $-C{\equiv}C-$ group is linear, and an applied field cannot induce a paramagnetic current when it is parallel to the group's axis.[2] The pattern of shielding and deshielding resulting from the diamagnetic current is shown in Fig. 18.8. Protons lying on the axis of the group (as

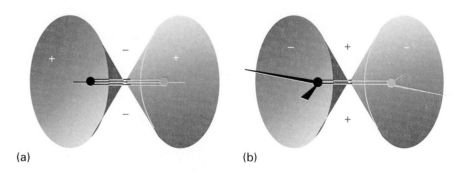

18.8 The neighbouring group effect in NMR. (a) The protons in HC≡CH are shielded by the currents induced in the triple bond, but a proton perpendicular to the bond is deshielded. (b) The opposite is true for protons near a C=C double bond because the applied field can induce a paramagnetic current parallel to the axis of a double bond.

(a)   (b)

---

1   Magnetic susceptibilities are discussed in Section 22.6.

2   The electrons are in orbitals that are eigenfunctions of the angular momentum operator, l_z, for circulation about the axis of the molecule. When the field is applied along that axis, it gives rise to a perturbation proportional to $\mathcal{B}l_z$, which cannot mix excited states into its own eigenfunctions.

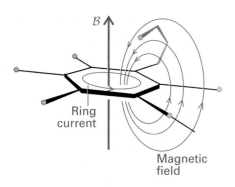

18.9 The shielding and deshielding effects of the ring current induced in the benzene ring by the applied field. Protons attached to the ring are deshielded but a proton attached to a substituent that projects above the ring is shielded.

in ethyne itself) are shielded, but a proton that lies perpendicular to the bond (as part of a larger molecule) is deshielded. The opposite is true for protons near a C=C double bond because in this nonlinear group the applied field can induce a paramagnetic current when it lies parallel to the axis.

A special case of a neighbouring group effect is found in aromatic compounds. The strong anisotropy of the magnetic susceptibility of the benzene ring is ascribed to the ability of the field to induce a **ring current**, a circulation of electrons around the ring, when it is applied perpendicular to the molecular plane. Protons in the plane are deshielded (Fig. 18.9), but any that happen to lie above or below the plane (as members of substituents of the ring) are shielded.

(f) The solvent contribution

A solvent can influence the local magnetic field experienced by a nucleus in a variety of ways. Some of these effects arise from specific interactions between the solute and the solvent (such as hydrogen-bond formation and other forms of Lewis acid–base complex formation). The magnetic susceptibility of the solvent molecules, especially if they are aromatic, can also be the source of a local magnetic field. Moreover, if there are steric interactions that result in a loose but specific interaction between a solute molecule and a solvent molecule, then protons in the solute molecule may experience shielding or deshielding effects according to their location relative to the solvent molecule (Fig. 18.10). We shall see that the NMR spectra of species that contain protons with widely different chemical shifts are easier to interpret than those in which the shifts are similar, so the appropriate choice of solvent may help to simplify the appearance and interpretation of a spectrum.

18.4 The fine structure

The splitting of resonances into individual lines in Fig. 18.4 is called the **fine structure** of the spectrum. It arises because each magnetic nucleus may contribute to the local field experienced by the other nuclei and so modify their resonance frequencies. The strength of the interaction is expressed in terms of the **scalar coupling constant**, J, and reported in hertz (Hz).[3] Spin coupling constants are independent of the strength of the applied field because they do not depend on the latter's ability to generate local fields. If the resonance line of a particular nucleus is split by a certain amount by a second nucleus, then the resonance line of the second nucleus is split by the first to the same extent.

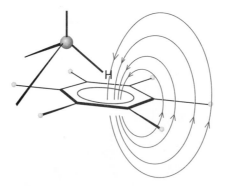

18.10 An aromatic solvent (benzene here) can give rise to local currents that shield or deshield a proton in a solvent molecule. In this relative orientation of the solvent and solute, the proton on the solute molecule is shielded.

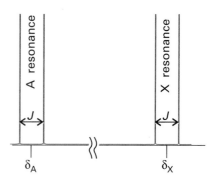

18.11 The effect of spin–spin coupling on an AX spectrum. Each resonance is split into two lines separated by J. The pairs of resonances are centred on the chemical shifts of the protons in the absence of spin–spin coupling.

---

3   The scalar coupling constant is so called because the interaction it describes is proportional to the scalar product of the two interacting spins: $E \propto I_1 \cdot I_2$. The constant of proportionality in this expression is J. (More precisely, it is $hJ/\hbar^2$, because each angular momentum is proportional to $\hbar$.)

(a) Patterns of coupling

In NMR, letters far apart in the alphabet (typically A and X) are used to indicate nuclei with very different chemical shifts; letters close together (such as A and B) are used for nuclei with similar chemical shifts. We shall consider first an AX system, a molecule that contains two spin-$\frac{1}{2}$ nuclei A and X with very different chemical shifts in the sense that the difference in chemical shifts is large compared with their spin–spin coupling.

Suppose the spin of X is α; then the spin of A will have a Larmor frequency as a result of the combined effect of the external field, the shielding constant, and the spin–spin interaction of A with X. The spin–spin coupling will result in one line in the spectrum of A being shifted by $-\frac{1}{2}J$ from the frequency it would have in the absence of coupling. If the spin of X is β, the spin of A will have a Larmor frequency shifted by $+\frac{1}{2}J$. Therefore, instead of a single line from A, we get a doublet of lines separated by J and centred on the chemical shift characteristic of A (Fig. 18.11). The same splitting occurs in the X resonance: instead of a single line, the resonance is a doublet with splitting J (the same value as for the splitting of A) centred on the chemical shift characteristic of X.

A subtle point is that the X resonance in an AX_n species (such as an AX_2 or AX_3 species) is also a doublet with splitting J. As we shall explain below, *a group of equivalent nuclei resonates like a single nucleus*. The only difference for the X resonance of an AX_n species is that the intensity is n times as great as that of an AX species (Fig. 18.12). The A resonance in an AX_n species, though, is quite different from the A resonance in an AX species. For example, consider an AX_2 species with two equivalent X nuclei. The resonance of A is split into a doublet of separation J by one X, and each line of that doublet is split again by the same amount by the second X (Fig. 18.13). This splitting results in three lines in the intensity ratio 1 : 2 : 1 (because the central frequency can be obtained in two ways). The A resonance

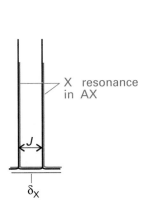

X resonance in AX

δ_X

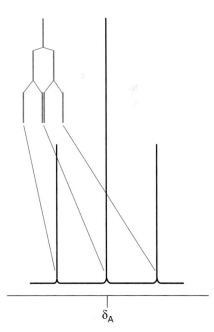

δ_A

18.12 The X resonance of an AX_2 species is also a doublet, because the two equivalent X nuclei behave like a single nucleus; however, the overall absorption is twice as intense as that of an AX species.

18.13 The origin of the 1 : 2 : 1 triplet in the A resonance of an AX_2 species. The resonance of A is split into two by coupling with one X nucleus (as shown in the inset), and then each of those two lines is split into two by coupling to the second X nucleus. Because each X nucleus causes the same splitting, the two central transitions are coincident and give rise to an absorption line of double the intensity of the outer lines.

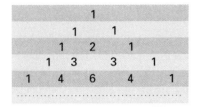

2

of an A_nX_2 species would also be a $1:2:1$ triplet of splitting J, the only difference being that the intensity of the A resonance would be n times as great as that of AX_2.

Three equivalent X nuclei (an AX_3 species) split the resonance of A into four lines of intensity ratio $1:3:3:1$ and separation J (Fig. 18.14). The X resonance, though, is still a doublet of separation J. In general, n equivalent spin-$\frac{1}{2}$ nuclei split the resonance of a nearby spin or group of equivalent spins into $n+1$ lines with an intensity distribution given by Pascal's triangle (**2**). The easiest way of constructing the pattern of fine structure is to draw a diagram in which successive rows show the splitting of a subsequent proton. The procedure is illustrated in Fig. 18.15 and was used in Figs. 18.13 and 18.14. It is easily extended to molecules containing nuclei with $I > \frac{1}{2}$ (Fig. 18.16).

Example 18.2 Accounting for the fine structure in a spectrum

Account for the fine structure in the NMR spectrum of the C–H protons of ethanol.

Method Consider how each group of equivalent protons (for example, three methyl protons) split the resonance of the other groups of protons. There is no splitting within groups of equivalent protons. Each splitting pattern can be decided by referring to Pascal's triangle.

Answer The three protons of the CH_3 group split the resonance of the CH_2 protons into a $1:3:3:1$ quartet with a splitting J. Likewise, the two protons of the CH_2 group split the resonance of the CH_3 protons into a $1:2:1$ triplet with the same splitting J. All the lines mentioned so far are split into doublets by the OH proton, but the splitting cannot be detected because the OH protons migrate rapidly from molecule to molecule and their effect averages to zero.

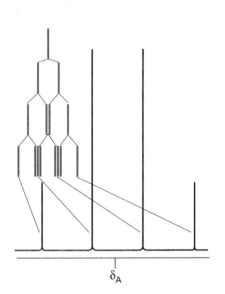

18.14 The origin of the $1:3:3:1$ quartet in the A resonance of an AX_3 species. The third X nucleus splits each of the lines shown in Fig. 18.13 for an AX_2 species into a doublet, and the intensity distribution reflects the number of transitions that have the same energy.

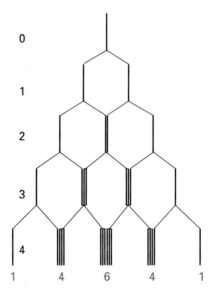

18.15 The intensity distribution of the A resonance of an AX_n resonance can be constructed by considering the splitting caused by $1, 2, \ldots n$ protons, as in Figs. 18.13 and 18.14. The resulting intensity distribution has a binomial distribution and is given by the integers in the corresponding row of Pascal's triangle. Note that, although the lines have been drawn side-by-side for clarity, the members of each group are coincident. Four protons, in AX_4, split the A resonance into a $1:4:6:4:1$ quintet.

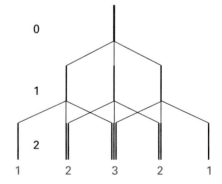

18.16 The intensity distribution arising from spin-spin interaction with nuclei with $I = 1$ can be constructed similarly, but each successive nucleus splits the lines into three equal intensity components. Two equivalent spin-1 nuclei give rise to a $1:2:3:2:1$ quintet.

Self-test 18.2 What fine structure can be expected for the protons in $^{14}NH_4^+$? The spin quantum number of nitrogen is 1.

[1 : 1 : 1 triplet from N]

(b) The energy levels of coupled systems

It will be useful for later discussions to consider an NMR spectrum in terms of the energy levels of the nuclei and the transitions between them. The energy level diagram for a single spin-$\frac{1}{2}$ nucleus and its single transition were shown in Fig. 18.1, and nothing more needs to be said. For a spin-$\frac{1}{2}$ AX system there are four spin states:

$$\alpha_A\alpha_X \qquad \alpha_A\beta_X \qquad \beta_A\alpha_X \qquad \beta_A\beta_X$$

The energy depends on the orientation of the spins in the external magnetic field, and, if spin–spin coupling is neglected,

$$E = -\gamma\hbar(1 - \sigma_A)\mathcal{B}m_A - \gamma\hbar(1 - \sigma_X)\mathcal{B}m_X = -h\nu_A m_A - h\nu_X m_X \qquad (17)$$

where ν_A and ν_X are the Larmor frequencies of A and X and m_A and m_X are their quantum numbers. This expression gives the four lines on the left of Fig. 18.17. The spin–spin coupling depends on the relative orientation of the two nuclear spins, so it is proportional to the product $m_A m_X$; the constant of proportionality is hJ. Therefore, the energy including spin–spin coupling is

$$E = -h\nu_A m_A - h\nu_X m_X + hJm_A m_X \qquad (18)$$

If $J > 0$, a lower energy is obtained when $m_A m_X < 0$, which is the case if one spin is α and the other is β. A higher energy is obtained if both spins are α or both spins are β. The opposite is true if $J < 0$. The resulting energy level diagram (for $J > 0$) is shown on the right of Fig. 18.17. We see that the $\alpha\alpha$ and $\beta\beta$ states are both raised by $\frac{1}{4}hJ$ and that the $\alpha\beta$ and $\beta\alpha$ states are both lowered by $\frac{1}{4}hJ$.

When a transition of nucleus A occurs, nucleus X remains unchanged. Therefore, the A resonance is a transition for which $\Delta m_A = +1$ and $\Delta m_X = 0$. There are two such transitions, one in which $\beta_A \leftarrow \alpha_A$ occurs when the X nucleus is α_X, and the other in which $\beta_A \leftarrow \alpha_A$ occurs when the X nucleus is β_X. They are shown in Fig. 18.17 and in a slightly different

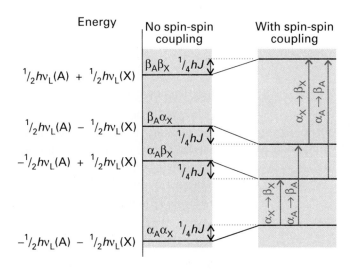

18.17 The energy levels of an AX system. The four levels on the left are those of the two spins in the absence of spin–spin coupling. The four levels on the right show how a positive spin–spin coupling constant affects the energies. The transitions shown are for $\beta \leftarrow \alpha$ of A or X, the other nucleus (X or A, respectively) remaining unchanged.

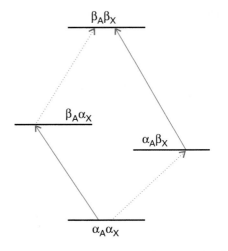

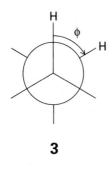

3

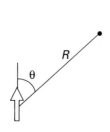

4

18.18 An alternative depiction of the energy levels and transitions shown in Fig. 18.17.

form in Fig. 18.18. The energies of the transitions are

$$\Delta E = h\nu_A \pm \tfrac{1}{2}hJ \tag{19}$$

Therefore, the A resonance consists of a doublet of separation J centred on the chemical shift of A (as in Fig. 18.11).

Similar remarks apply to the X resonance, which consists of two transitions according to whether the A nucleus is α or β (Fig. 18.18). The transition energies are

$$\Delta E = h\nu_X \pm \tfrac{1}{2}hJ \tag{20}$$

It follows that the X resonance also consists of two lines of separation J, but they are centred on the chemical shift of X (as shown in Fig. 18.11).

(c) The magnitudes of coupling constants

The scalar coupling constant of two nuclei separated by N bonds is denoted $^N J$, with subscripts for the types of nuclei involved. Thus, $^1 J_{CH}$ is the coupling constant for a proton joined directly to a ^{13}C atom, and $^2 J_{CH}$ is the coupling constant when the same two nuclei are separated by two bonds (as in $^{13}C-C-H$). A typical value of $^1 J_{CH}$ is in the range 120 to 250 Hz; $^2 J_{CH}$ is between 0 and 10 Hz. Both $^3 J$ and $^4 J$ give detectable effects in a spectrum, but couplings over larger numbers of bonds can generally be ignored. One of the longest couplings that has been detected is $^9 J_{HH} = 0.4$ Hz for CH_3 and CH_2 protons in $CH_3C\equiv CC\equiv CC\equiv CCH_2OH$.

The sign of J_{XY} indicates whether the energy of two spins is lower when they are parallel ($J < 0$) or when they are antiparallel ($J > 0$). It is found that $^1 J_{CH}$ is often positive, $^2 J_{HH}$ is often negative, $^3 J_{HH}$ is often positive, and so on. An additional point is that J varies with the angle between the bonds (Fig. 18.19). Thus, a $^3 J_{HH}$ coupling constant is often found to depend on the angle ϕ (3) according to the **Karplus equation**:

$$J = A + B \cos\phi + C \cos 2\phi \tag{21}$$

with A, B, and C empirical constants with values close to $+7$ Hz, -1 Hz, and $+5$ Hz, respectively. It follows that the measurement of $^3 J_{HH}$ in a series of related compounds can be used to determine their conformations. The coupling constant $^1 J_{CH}$ also depends on the hybridization of the C atom, as the following values indicate:

	sp	sp^2	sp^3
$^1 J_{CH}$/Hz:	250	160	125

(d) The origin of spin–spin coupling

Spin–spin coupling is a very subtle phenomenon, and it is better to treat J as an empirical parameter than to use calculated values. However, we can get some insight into its origins, if not its precise magnitude—or always reliably its sign—by considering the magnetic interactions within molecules.

A nucleus with spin projection m_I gives rise to a magnetic field with z component $\mathcal{B}_{nuc}$ at a distance R, where

$$\mathcal{B}_{nuc} = -\frac{\gamma\hbar\mu_0}{4\pi R^3}(1 - 3\cos^2\theta)m_I \tag{22}$$

The angle θ is defined in (4). The magnitude of this field is about 0.1 mT when $R = 0.3$ nm, corresponding to a splitting of resonance signal of about 10^4 Hz, and is of the order of magnitude of the splitting observed in solid samples (see Section 18.9a).

In a liquid, the angle θ sweeps over all values as the molecule tumbles, and $1 - 3\cos^2\theta$

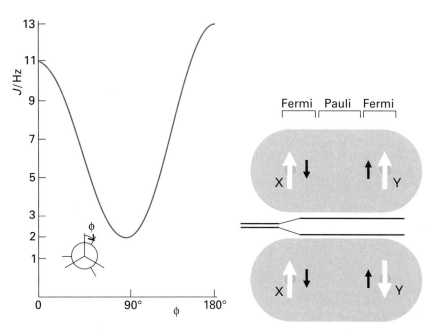

18.19 The variation of the spin–spin coupling constant with angle predicted by the Karplus equation.

18.20 The polarization mechanism for spin–spin coupling ($^1J_{HH}$). The two arrangements have slightly different energies. In this case, J is positive, corresponding to a lower energy when the nuclear spins are antiparallel.

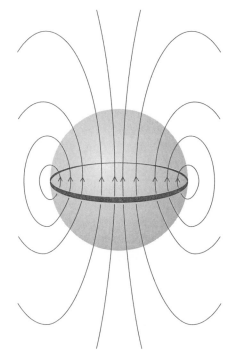

18.21 The origin of the Fermi contact interaction. From far away, the magnetic field pattern arising from a ring of current (representing the rotating charge of the nucleus, the pale grey sphere) is that of a point dipole. However, if an electron can sample the field close to the region indicated by the sphere, the field distribution differs significantly from that of a point dipole. For example, if the electron can penetrate the sphere, then the spherical average of the field it experiences is not zero.

averages to zero.[4] Hence the direct dipolar interaction between spins cannot account for the fine structure of the spectra of rapidly tumbling molecules. The direct interaction does make an important contribution to the spectra of solid samples and to the spectra of molecules that tumble only slowly in solution, such as biological and synthetic macromolecules.

Spin–spin coupling in molecules in solution can be explained in terms of the **polarization mechanism**, in which the interaction is transmitted through the bonds. The simplest case to consider is that of $^1J_{XY}$ where X and Y are spin-$\frac{1}{2}$ nuclei joined by an electron-pair bond (Fig. 18.20). The coupling mechanism depends on the fact that in some atoms it is favourable for the nucleus and a nearby electron spin to be parallel (both α or both β), but in others it is favourable for them to be antiparallel (one α and the other β). The electron–nucleus coupling is magnetic in origin, and may be either a dipolar interaction between the magnetic moments of the electron and nuclear spins or a **Fermi contact interaction**. As shown in the *Justification* below, the latter depends on the very close approach of an electron to the nucleus and hence can occur only if the electron occupies an *s* orbital. We shall suppose that it is energetically favourable for an electron spin and a nuclear spin to be antiparallel (as is the case for a proton and an electron in a hydrogen atom).

Justification 18.2

A pictorial description of the Fermi contact interaction is as follows. First, we regard the magnetic moment of the nucleus as arising from the circulation of a current in a tiny loop with a radius similar to that of the nucleus (Fig. 18.21). Far from the nucleus the field generated by this loop is indistinguishable from the field generated by a point magnetic dipole. Close to the loop, however, the field differs from that of a point dipole. The

---

4 The volume element in polar coordinates is proportional to $\sin\theta\,d\theta$, and θ ranges from 0 to π. Therefore the average value of $\mathcal{B}_{nuc}$ for a tumbling molecule is proportional to

$$\int_0^\pi (1 - 3\cos^2\theta)\sin\theta\,d\theta = 0$$

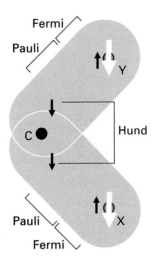

18.22 The polarization mechanism for $^2J_{HH}$ spin–spin coupling. The spin information is transmitted from one bond to the next by a version of the mechanism that accounts for the lower energy of electrons with parallel spins in different atomic orbitals (Hund's rule of maximum multiplicity). In this case, J is negative, corresponding to a lower energy when the nuclear spins are parallel.

magnetic interaction between this non-dipolar field and the electron's magnetic moment is the contact interaction. The lines of force depicted in Fig. 18.21 correspond to those for a proton with α spin. The lower energy state of an electron spin in such a field is the β state.

If the X nucleus is α, a β electron of the bonding pair will tend to be found nearby (since that is energetically favourable for it). The second electron in the bond, which must have α spin if the other is β, will be found mainly at the far end of the bond (because electrons tend to stay apart to reduce their mutual repulsion). Because it is energetically favourable for the spin of Y to be antiparallel to an electron spin, a Y nucleus with β spin has a lower energy, and hence a lower Larmor frequency, than a Y nucleus with α spin. The opposite is true when X is β, for now the α spin of Y has the lower energy. In other words, the antiparallel arrangement of nuclear spins lies lower in energy than the parallel arrangement as a result of their magnetic coupling with the bond electrons. That is, $^1J_{HH}$ is positive.

To account for the value of $^2J_{XY}$, as in H–C–H, we need a mechanism that can transmit the spin alignments through the central C atom (which may be ^{12}C, with no nuclear spin of its own). In this case (Fig. 18.22), an X nucleus with α spin polarizes the electrons in its bond, and the α electron is likely to be found closer to the C nucleus. The more favourable arrangement of two electrons on the same atom is with their spins parallel (Hund's rule, Section 13.4d), so the more favourable arrangement is for the α electron of the neighbouring bond to be close to the C nucleus. Consequently, the β electron of that bond is more likely to be found close to the Y nucleus, and therefore that nucleus will have a lower energy if it is α. Hence, according to this mechanism, the lower Larmor frequency of Y will be obtained if its spin is parallel to that of X. That is, $^2J_{HH}$ is negative.

The coupling of nuclear spin to electron spin by the Fermi contact interaction is most important for proton spins, but it is not necessarily the most important mechanism for other nuclei. These nuclei may also interact by a dipolar mechanism with the electron magnetic moments and with their orbital motion, and there is no simple way of specifying whether J will be positive or negative.

(e) Equivalent nuclei

A group of nuclei are **chemically equivalent** if they are related by a symmetry operation of the molecule and have the same chemical shifts. Chemically equivalent nuclei are nuclei that would be regarded as 'equivalent' according to ordinary chemical criteria. Nuclei are **magnetically equivalent** if, as well as being chemically equivalent, they also have identical spin–spin interactions with any other magnetic nuclei in the molecule.

The difference between chemical and magnetic equivalence is illustrated by CH_2F_2 and $H_2C{=}CF_2$, in both of which the protons are chemically equivalent: they are related by symmetry and undergo the same chemical reactions. However, although the protons in CH_2F_2 are magnetically equivalent, those in $CH_2{=}CF_2$ are not. One proton in the latter has spin-coupling interactions with a *cis* F nucleus which might be α whereas the other proton has a *trans* interaction with it. In CH_2F_2 both protons are equally distant from the two F nuclei, so there is no distinction between them. Strictly speaking, the CH_3 protons in ethanol (and other compounds) are magnetically inequivalent on account of their different interactions with the CH_2 protons in the next group. However, they are in practice made magnetically equivalent by the rapid rotation of the CH_3 group, which averages out any differences. Magnetically inequivalent species can give very complicated spectra (for instance, the spectrum of $H_2C{=}CF_2$ consists of ten lines), and we shall not consider them further.

An important feature of chemically equivalent magnetic nuclei is that, although they do couple together, the coupling has no effect on the appearance of the spectrum. The reason for the invisibility of the coupling is set out in the *Justification* below, but qualitatively it is

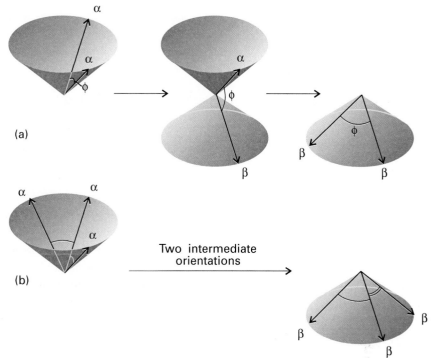

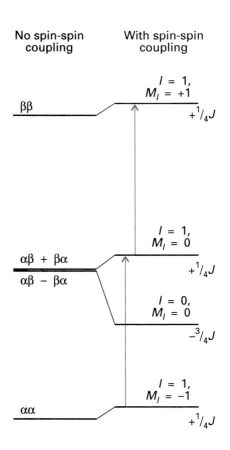

18.24 The energy levels of an A_2 system in the absence of spin–spin coupling are shown on the left. When spin–spin coupling is taken into account, the energy levels on the right are obtained. Note that the three states with total nuclear spin $I = 1$ correspond to parallel spins and give rise to the same increase in energy (J is positive); the one state with $I = 0$ (antiparallel nuclear spins) has a lower energy in the presence of spin–spin coupling. The only allowed transitions are those that preserve the angle between the spins, and so take place between the three states with $I = 1$. They occur at the same resonance frequency as they would have in the absence of spin–spin coupling.

18.23 (a) A group of two equivalent nuclei realigns as a group, without change of angle between the spins, when a resonant absorption occurs. Hence it behaves like a single nucleus and the spin–spin coupling between the individual spins of the group is undetectable. (b) Three equivalent nuclei also realign as a group without change of their relative orientations.

that all allowed nuclear spin transitions are collective reorientations of groups of equivalent nuclear spins that do not change the relative orientations of the spins within the group (Fig. 18.23). Then, because the relative orientations of nuclear spins are not changed in any transition, the magnitude of the coupling between them is undetectable. Hence, an isolated CH_3 group gives a single, unsplit line because all the allowed transitions of the group of three protons occur without change of their relative orientations.

Justification 18.3

Consider an A_2 system of two chemically equivalent spin-$\frac{1}{2}$ nuclei. First, consider the energy levels in the absence of spin–spin coupling. There are four spin states which (just as for two electrons) can be classified according to their total spin I (the analogue of S for two electrons) and their total projection M_I on the z-axis. The states are analogous to those we developed for two electrons in singlet and triplet states:[5]

Spins parallel, $I = 1$:	$M_I = +1$	$\alpha\alpha$
	$M_I = 0$	$(1/2^{1/2})\{\alpha\beta + \beta\alpha\}$
	$M_I = -1$	$\beta\beta$
Spins paired, $I = 0$:	$M_I = 0$	$(1/2^{1/2})\{\alpha\beta - \beta\alpha\}$

The effect of a magnetic field on these four states is shown on the left in Fig. 18.24: the energies of the two states with $M_I = 0$ are unchanged by the field because they are composed of equal proportions of α and β spins.

---

5   As in Section 13.7, the states we have selected are those with a definite resultant, and hence a well defined value of I. The $+$ sign in $\alpha\beta + \beta\alpha$ signifies an in-phase alignment of spins and $I = 1$; the $-$ sign in $\alpha\beta - \beta\alpha$ signifies an alignment out of phase by π, and hence $I = 0$. See Fig. 13.26.

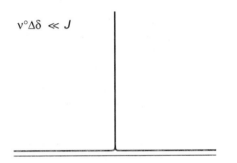

$\nu°\Delta\delta \ll J$

$\nu°\Delta\delta \approx J$

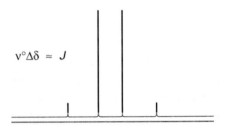

$\nu°\Delta\delta \approx 5J$

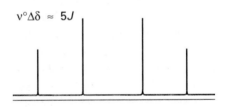

$\nu°\Delta\delta \gg J$

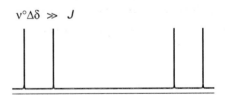

18.25 The NMR spectrum of an A_2 system (top) and an AX system (bottom) are simple, and give rise to 'first-order spectra'. At intermediate relative values of the chemical shift difference and the spin–spin coupling, complex 'strongly coupled' spectra are obtained. Note how the inner two lines of the bottom spectrum move together, grow in intensity, and form the single central line of the top spectrum. The two outer lines diminish in intensity and are absent in the top spectrum.

The spin–spin coupling energy is proportional to the scalar product of the vectors representing the spins, and we write $E = (hJ/\hbar^2)\boldsymbol{I}_1 \cdot \boldsymbol{I}_2$. The scalar product can be expressed in terms of the total nuclear spin by noting that

$$I^2 = (\boldsymbol{I}_1 + \boldsymbol{I}_2) \cdot (\boldsymbol{I}_1 + \boldsymbol{I}_2) = I_1^2 + I_2^2 + 2\boldsymbol{I}_1 \cdot \boldsymbol{I}_2$$

and replacing the magnitudes by their quantum mechanical values:

$$\boldsymbol{I}_1 \cdot \boldsymbol{I}_2 = \tfrac{1}{2}\{I(I+1) - I_1(I_1 + 1) - I_2(I_2 + 1)\}\hbar^2 \tag{23}$$

Then, because $I_1 = I_2 = \tfrac{1}{2}$, it follows that

$$E = \tfrac{1}{2}hJ\{I(I+1) - \tfrac{3}{2}\} \tag{24}$$

For parallel spins, $I = 1$ and $E = +\tfrac{1}{4}hJ$; for antiparallel spins $I = 0$ and $E = -\tfrac{3}{4}hJ$, as in the illustration. We see that three of the states move in energy in one direction and the fourth (the one with antiparallel spins) moves three times as much in the opposite direction. The resulting energy levels are shown in Fig. 18.24.

The NMR spectrum of the A_2 species arises from transitions between the levels. However, the radiofrequency field affects the two equivalent protons equally, so it cannot change the orientation of one proton relative to the other; therefore, the transitions take place within the set of states that correspond to parallel spin (those labelled $I = 1$), and no spin-parallel state can change to a spin-antiparallel state (the state with $I = 0$). Put another way, the allowed transitions are subject to the selection rule $\Delta I = 0$. This selection rule is in addition to the rule $\Delta M_I = \pm 1$ that arises from the conservation of angular momentum and the unit spin of the photon. The allowed transitions are shown in Fig. 18.24: we see that there are only two transitions, and that they occur at the same resonance frequency that the nuclei would have in the absence of spin–spin coupling. Hence, the spin–spin coupling interaction does not affect the appearance of the spectrum.

(f) Strongly coupled nuclei

NMR spectra are usually much more complex than the foregoing simple analysis suggests. We have described the extreme case in which the differences in chemical shifts are much greater than the spin–spin coupling constants. In such cases it is simple to identify groups of magnetically equivalent nuclei and to think of the groups of nuclear spins as reorientating relative to each other. The spectra that result are called **first-order spectra**.

Transitions cannot be allocated to definite groups when the differences in their chemical shifts are comparable to their spin–spin coupling interactions. The complicated spectra that are then obtained are called **strongly coupled** (or 'second-order spectra') and are much more difficult to analyse (Fig. 18.25). Because the difference in resonance frequencies increases with field, but spin–spin coupling constants are independent of it, a second-order spectrum may become simpler (and first-order) at high fields because individual groups of nuclei become identifiable again.

A clue to the type of analysis that is appropriate is given by the notation for the types of spins involved. Thus, an AX spin system (which consists of two nuclei with a large chemical shift difference) has a first-order spectrum. An AB system, on the other hand (with two nuclei of similar chemical shifts), gives a spectrum typical of a strongly coupled system. An AX system may have widely different chemical shifts because A and X are nuclei of different elements (such as ^{13}C and ^{1}H), in which case they form a **heteronuclear spin system**. AX may also denote a **homonuclear spin system** in which the nuclei are of the same element but in markedly different environments.

(g) Dilute and abundant spins: spin decoupling

Carbon-13 is a **dilute-spin species** in the sense that it is unlikely that more than one ^{13}C nucleus will be found in any given small molecule (provided the sample has not been enriched with that isotope; the natural abundance of ^{13}C is only 1.1 per cent). Even in large molecules, although more than one ^{13}C nucleus may be present, it is unlikely that they will be close enough to give an observable splitting. Hence, it is not normally necessary to take into account ^{13}C—^{13}C spin–spin coupling within a molecule.

Protons are **abundant-spin species** in the sense that a molecule is likely to contain many of them. If we were observing a ^{13}C-NMR spectrum, we would obtain a very complex spectrum on account of the coupling of the one ^{13}C nucleus with all the protons that are present. To avoid this difficulty, ^{13}C-NMR spectra are normally observed using the technique of **proton decoupling**. Thus, if the CH_3 protons of ethanol are irradiated with a second, strong, resonant radiofrequency source, they undergo rapid spin reorientations and the ^{13}C nucleus senses an average orientation. As a result, its resonance is a single line and not a $1:3:3:1$ quartet. Proton decoupling has the additional advantage of enhancing sensitivity, because the intensity is concentrated into a single transition frequency instead of being spread over several transition frequencies. If care is taken to ensure that the other parameters on which the strength of the signal depends are kept constant, the intensities of proton-decoupled spectra are proportional to the number of ^{13}C nuclei present. The technique is widely used to characterize synthetic polymers.

Pulse techniques in NMR

Modern methods of detecting the energy separation between nuclear spin states are more sophisticated than simply looking for the frequency at which resonance occurs. One of the best analogies that has been suggested to illustrate the difference between the old and new ways of observing an NMR spectrum is that of detecting the spectrum of vibrations of a bell. We could stimulate the bell with a gentle vibration at a gradually increasing frequency, and note the frequencies at which it resonated with the stimulation. A lot of time would be spent getting zero response when the stimulating frequency was between the bell's vibrational modes. However, if we were simply to hit the bell with a hammer, we would immediately obtain a clang composed of all the frequencies that the bell can produce. The equivalent in NMR is to monitor the radiation nuclear spins emit as they return to equilibrium after the appropriate stimulation. The resulting **Fourier-transform NMR** gives greatly increased sensitivity, so opening up the entire periodic table to the technique. Moreover, multiple-pulse FT-NMR gives chemists unparalleled control over the information content and display of spectra. We need to understand how the equivalent of the hammer blow is delivered and how the signal is monitored and interpreted. These features are generally expressed in terms of the vector model of angular momentum introduced in Section 12.7d.

18.5 The magnetization vector

Consider a sample composed of many identical spin-$\frac{1}{2}$ nuclei. As we saw in Section 12.7d, an angular momentum can be represented by a vector of length $\{I(I+1)\}^{1/2}$ units with a component of length m_I units along the z-axis. As the uncertainty principle does not allow us to specify the x- and y-components of the angular momentum, all we know is that the vector lies somewhere on a cone around the z-axis. For $I = \frac{1}{2}$, the length of the vector is $\frac{1}{2}\sqrt{3}$ and it makes an angle of $55°$ to the z-axis (Fig. 18.26).

In the absence of a magnetic field, the sample consists of equal numbers of α and β nuclear spins with their vectors lying at random angles on the cones. These angles are

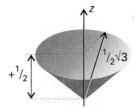

18.26 The vector model of angular momentum for a single spin-$\frac{1}{2}$ nucleus. The angle around the z-axis is indeterminate.

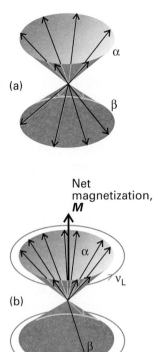

(a)

α

β

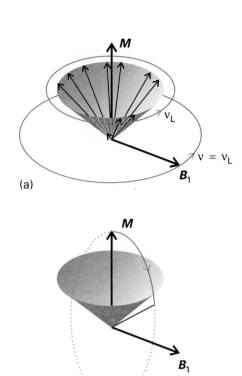

M

ν_L

ν = ν_L

B_1

(a)

M

B_1

(b)

Net
magnetization,
M

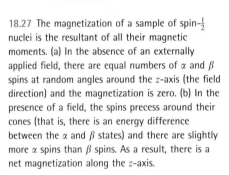

α

ν_L

(b)

β

Precession

18.27 The magnetization of a sample of spin-$\frac{1}{2}$ nuclei is the resultant of all their magnetic moments. (a) In the absence of an externally applied field, there are equal numbers of α and β spins at random angles around the z-axis (the field direction) and the magnetization is zero. (b) In the presence of a field, the spins precess around their cones (that is, there is an energy difference between the α and β states) and there are slightly more α spins than β spins. As a result, there is a net magnetization along the z-axis.

18.28 (a) In a resonance experiment, a circularly polarized radiofrequency magnetic field $\mathcal{B}_1$ is applied in the xy-plane (the magnetization vector lies along the z-axis). (b) If we step into a frame rotating at the Larmor frequency, the radiofrequency field appears to be stationary if its frequency is the same as the Larmor frequency. When the two frequencies coincide, the magnetization vector of the sample begins to rotate around the direction of the $\mathcal{B}_1$ field.

unpredictable, and at this stage we picture the spin vectors as stationary. The **magnetization, M**, of the sample, its net nuclear magnetic moment, is zero (Fig. 18.27a).

(a) The effect of the static field

Two changes occur in the magnetization when a magnetic field is present. First, the energies of the two orientations change, the α spins moving to low energy and the β spins to high energy (provided $\gamma > 0$). At 10 T, the Larmor frequency for protons is 427 MHz, and in the vector model the individual vectors are pictured as **precessing**, or sweeping round their cones, at this rate. This motion is a pictorial representation of the change in energy of the spin states (it is not an actual representation of reality). As the field is increased, the Larmor frequency increases and the precession becomes faster. Secondly, the populations of the two spin states (the numbers of α and β spins) change, and there will be more α spins than β spins. Because $h\nu_L/kT \approx 7 \times 10^{-5}$ for protons at 300 K and 10 T, there is only a tiny imbalance of populations, and it is even smaller for other nuclei with their smaller magnetogyric ratios. However, despite its smallness, the imbalance means that there is a net magnetization that we can represent by a vector **M** pointing in the z-direction and with a length proportional to the population difference (Fig.18.27b).

(b) The effect of the radiofrequency field

We now consider the effect of a circularly polarized radiofrequency field in the xy-plane. We have considered the electric component of this field in the other forms of spectroscopy that we have treated. However, in this chapter we consider only the magnetic component, for it is this component that interacts with the nuclear magnetic moment. The strength of the oscillating magnetic field is $\mathcal{B}_1$.

Suppose we choose the frequency of the oscillating field to be equal to the Larmor frequency of the spins. This choice is equivalent to selecting the resonance condition in the conventional experiment. The nuclei now experience a steady $\mathcal{B}_1$ field because the rotating magnetic field is in step with the precessing spins (Fig. 18.28). Under the influence of this effectively steady field, the magnetization vector begins to precess around the direction of $\mathcal{B}_1$ at a rate that is proportional to $\mathcal{B}_1$. If we apply the $\mathcal{B}_1$ field in a pulse of a certain duration, the magnetization precesses into the xy-plane, and we say that we have applied a **90° pulse** (or a '$\pi/2$ pulse'). The duration of the pulse depends on the strength of the $\mathcal{B}_1$ field, but is typically of the order of microseconds. To a stationary external observer (a radiofrequency coil, Fig. 18.29), the magnetization vector is now rotating in the xy-plane at the Larmor frequency (at about 430 MHz). The rotating magnetization induces a 430 MHz signal in the coil, which can be amplified and processed. In practice, the processing takes place after subtraction of a constant high-frequency component, so that all the signal manipulation takes place at frequencies of a few kilohertz.

As time passes, the individual spins move out of step (partly because they are precessing at slightly different rates, as we shall explain later), so the magnetization vector shrinks exponentially with a time constant T_2 and induces an ever weaker signal in the detector coil. The form of the signal that we can expect is therefore the oscillating-decaying **free-induction decay** (FID) shown in Fig. 18.30, and the y-component of the magnetization

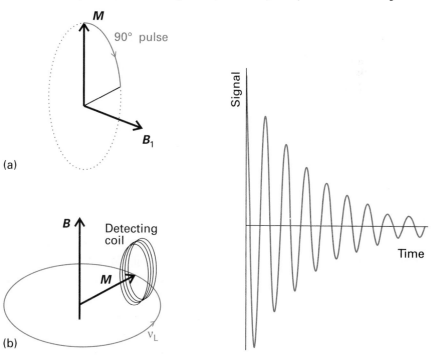

(a)

(b)

18.29 (a) If the radiofrequency field is applied for a certain time, the magnetization vector is rotated into the xy-plane. (b) To an external stationary observer (the coil), the magnetization vector is rotating at the Larmor frequency, and can induce a signal in the coil.

18.30 A simple free induction decay of a sample of spins with a single resonance frequency.

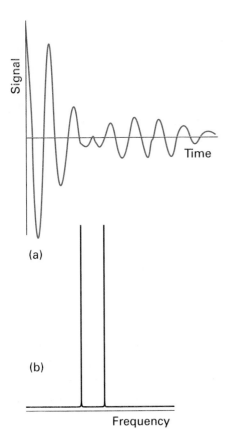

18.31 (a) A free induction decay signal of a sample of AX species and (b) its analysis into its frequency components.

varies as

$$M_y(t) = M_0 \cos(2\pi\nu_L t)e^{-t/T_2} \tag{25}$$

We have considered the effect of a pulse applied at exactly the Larmor frequency. However, virtually the same effect is obtained off resonance, provided that the pulse is applied close to ν_L. If the difference in frequency is small compared to the inverse of the duration of the $90°$ pulse, the magnetization will end up in the xy-plane. Note that we do not need to know the Larmor frequency beforehand: the short pulse is the analogue of the hammer blow on the bell, exciting a range of frequencies. The detected signal shows that a particular resonant frequency is present.

(c) Time- and frequency-domain spectra

We can think of the magnetization vector of a homonuclear AX spin system with $J = 0$ as consisting of two parts, one formed by the A spins and the other by the X spins. When the $90°$ pulse is applied, both magnetization vectors are rotated into the xy-plane. However, because the A and X nuclei precess at different frequencies, they induce two signals in the detector coils, and the overall FID curve may resemble that in Fig. 18.31a. The composite FID curve is the analogue of the struck bell emitting a rich tone composed of all the frequencies at which it can vibrate.

The problem we must address is how to recover the resonance frequencies present in a free-induction decay. We encountered a similar problem when discussing Fourier-transform infrared spectra in Section 16.1c, where all the vibrational frequencies were detected at once. The same technique is used here. We know that the FID curve is a sum of oscillating functions, so the problem is to analyse it into its harmonic components.

The analysis of the FID curve is achieved by the standard mathematical technique of Fourier transformation. We start by noting that the signal $S(t)$ in the time domain, the total FID curve, is the sum (more precisely, the integral) over all the contributing frequencies[6]

$$S(t) = \int_{-\infty}^{\infty} \mathcal{I}(\nu)e^{-2\pi i\nu t}\,d\nu \tag{26}$$

We need $\mathcal{I}(\nu)$, the spectrum in the frequency domain; it is obtained by evaluating the integral

$$\mathcal{I}(\nu) = 2\mathrm{Re}\int_0^{\infty} S(t)e^{2i\pi\nu t}\,dt \tag{27}$$

where Re means take the real part of the following expression. This integral is very much like an overlap integral: it gives a nonzero value if $S(t)$ contains a component that matches the oscillating function $e^{2i\pi\nu t}$. The integration is carried out at a series of frequencies ν on a computer that is built into the spectrometer. When the signal in Fig. 18.31a is transformed in this way, we get the frequency-domain spectrum shown in Fig. 18.31b. One line represents the Larmor frequency of the A nuclei and the other that of the X nuclei.

The FID curve in Fig. 18.32 is obtained from a sample of ethanol. The frequency-domain spectrum obtained from it by Fourier transformation is the one that we have already discussed (Fig. 18.4). We can now see why the FID curve in Fig. 18.32 is so complex: it arises from the precession of a magnetization vector that is composed of eight components, each with a characteristic frequency.

6   Because $e^{2\pi i\nu t} = \cos(2\pi\nu t) + i\sin(2\pi\nu t)$, this expression is a sum over harmonically oscillating functions, with each one weighted by the intensity $\mathcal{I}(\nu)$.

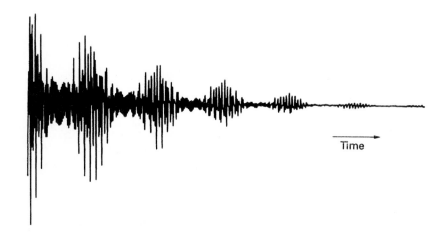

18.32 A free induction decay signal of a sample of ethanol. Its Fourier transform is the frequency-domain spectrum shown in Fig. 18.4.

18.6 Linewidths and rate processes

The linewidths of NMR spectra, in common with other spectroscopic techniques, provide information about the rates of processes relating to the molecules in the sample. We have seen that the FID signal decreases with time, which implies that the component of the magnetization vector in the xy-plane must be shrinking. In this section we see some of the processes involved.

(a) Spin relaxation

There are two reasons why the component of the magnetization vector in the xy-plane shrinks. Both reflect the fact that the nuclear spins are not in thermal equilibrium with their surroundings (for then M lies parallel to z). The return to equilibrium is the process called **spin relaxation**.

At thermal equilibrium the spins have a Boltzmann distribution, with more α spins than β spins; however, a magnetization vector in the xy-plane immediately after a 90° pulse has equal numbers of α and β spins. The populations revert to their thermal equilibrium values exponentially. As they do so, the z-component of magnetization reverts to its equilibrium value M_0 with a time constant called the **longitudinal relaxation time**, T_1 (Fig. 18.33):

$$M_z(t) - M_0 \propto e^{-t/T_1} \tag{28}$$

Because this relaxation process involves giving up energy to the surroundings (the 'lattice') as β spins revert to α spins, the time constant T_1 is also called the **spin–lattice relaxation time**. Spin–lattice relaxation is caused by fluctuating local magnetic fields arising from the motion of the molecules. These fluctuations can stimulate the spins to change from β to α, and vice versa, and hence to relax towards the thermal equilibrium population.

A second aspect of spin relaxation is the fanning-out of the spins in the xy-plane if they precess at different rates (Fig. 18.34). The magnetization vector is large when all the spins are bunched together immediately after a 90° pulse. However, this orderly bunching of spins is not at equilibrium and, even if there were no spin–lattice relaxation, we would expect the individual spins to spread out until they were uniformly distributed with all possible angles around the z-axis. At that stage, the component of magnetization vector in the plane would be zero. The randomization of the spin directions occurs exponentially with a time constant called the **transverse relaxation time**, T_2:

$$M_y(t) \propto e^{-t/T_2} \tag{29}$$

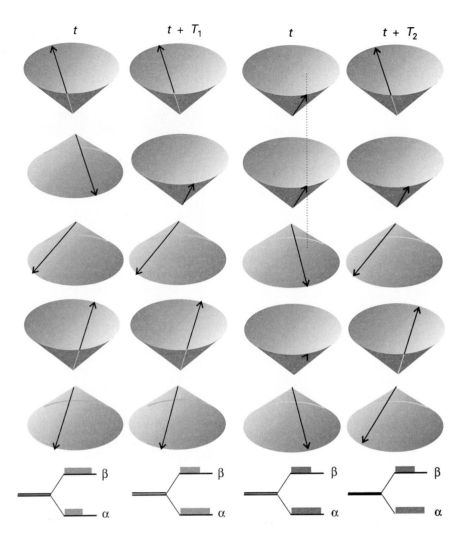

t $t + T_1$ t $t + T_2$

β   β   β   β

α   α   α   α

18.33 In longitudinal relaxation the spins relax back towards their thermal equilibrium populations. On the left we see the precessional cones representing spin-$\frac{1}{2}$ angular momenta, and they do not have their thermal equilibrium populations (there are more β-spins than α-spins). On the right, which represents the sample after a time T_1, the populations are those characteristic of a Boltzmann distribution.

18.34 The transverse relaxation time, T_2, is the time it takes for the phases of the spins to become randomized (another condition for equilibrium) and to change from the orderly arrangement shown on the left to the disorderly arrangement on the right. Note that the populations of the states remain the same; only the relative phase of the spins relaxes.

Because the relaxation involves the relative orientation of the spins, T_2 is also known as the **spin–spin relaxation time**.

If the y-component of magnetization decays with a time constant T_2, the spectral line is broadened (Fig. 18.35), and its width at half-height becomes

$$\Delta\nu_{1/2} = \frac{1}{\pi T_2} \tag{30}$$

Typical values of T_2 in proton NMR are of the order of seconds, so linewidths of around 0.1 Hz can be anticipated, in broad agreement with observation. In mobile liquids, $T_2 \approx T_1$.

So far, we have assumed that the equipment, and in particular the magnet, are perfect, and that the differences in Larmor frequencies arise solely from interactions within the sample. In practice, the magnet is not perfect, and the field is different at different locations in the sample. The inhomogeneity broadens the resonance, and in most cases this

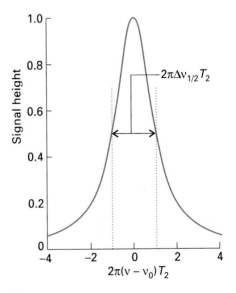

18.35 A Lorentzian absorption line. The width at half-height is inversely proportional to the parameter T_2 (so $\Delta\nu_{1/2}T_2$ is a constant) and, the longer the transverse relaxation time, the narrower the line.

inhomogeneous broadening dominates the broadening we have discussed so far. It is common to express the extent of inhomogeneous broadening in terms of an **effective transverse relaxation time**, T_2^*, by using a relation like eqn 30, but writing

$$T_2^* = \frac{1}{\pi \Delta \nu_{1/2}} \tag{31}$$

where $\Delta \nu_{1/2}$ is the observed width at half-height.[7]

Illustration

If a line in a spectrum has a width of 10 Hz, the effective transverse relaxation time is

$$T_2^* = \frac{1}{\pi \times (10 \text{ s}^{-1})} = 32 \text{ ms}$$

(b) Conformational conversion and exchange processes

The appearance of an NMR spectrum is changed if magnetic nuclei can jump rapidly between different environments. Consider a fluxional molecule, such as N,N-dimethylformamide, that can jump between conformations; in its case, the methyl shifts depend on whether they are *cis* or *trans* to the carbonyl group (Fig. 18.36). When the jumping rate is low, the spectrum shows two sets of lines, one each from from molecules in each conformation. When the inversion is fast, the spectrum shows a single line at the mean of the two chemical shifts. At intermediate inversion rates, the line is very broad. This maximum broadening occurs when the lifetime, τ, of a conformation gives rise to a linewidth that is comparable to the difference of resonance frequencies, $\delta \nu$, and both broadened lines blend together into a very broad line. Coalescence of the two lines occurs when

$$\tau = \frac{\sqrt{2}}{\pi \delta \nu} \tag{32}$$

For example, if the chemical shifts differ by 100 Hz, the spectrum collapses into a single line when the conformation lifetime is less than about 5 ms.

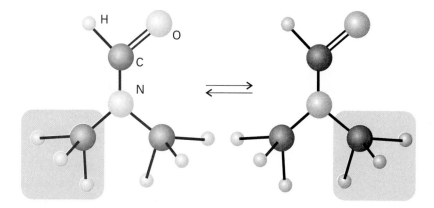

18.36 When a molecule changes from one conformation to another, the positions of its protons are interchanged and jump between magnetically distinct environments.

---

7   This formula assumes that the lineshape is Lorentzian; that is, of the form $y = 1/(1 + x^2)$.

Example 18.3 Interpreting line broadening

The NO group in N,N-dimethylnitrosamine, $(CH_3)_2N$—NO, rotates and, as a result, the magnetic environments of the two CH_3 groups are interchanged. In a 600 MHz spectrometer the two CH_3 resonances are separated by 390 Hz. At what rate of interconversion will the resonance collapse to a single line?

Method Use eqn 32 for the average lifetimes of the conformations. The rate of interconversion is the inverse of their lifetime.

Answer With $\delta\nu = 390$ Hz,

$$\tau = \frac{\sqrt{2}}{\pi \times (390 \text{ s}^{-1})} = 1.2 \text{ ms}$$

It follows that the signal will collapse to a single line when the interconversion rate exceeds about 830 s^{-1}.

Comment The dependence of the rate of collapse on the temperature is used to determine the energy barrier to interconversion.

- -

Self-test 18.3 What would you deduce from the observation of a single line from the same molecule in a 300 MHz spectrometer?

[Conformation lifetime less than 2.3 ms]

A similar explanation accounts for the loss of structure in solvents able to exchange protons with the sample. For example, hydroxyl protons are able to exchange with water protons. When this **chemical exchange** occurs, a molecule ROH with an α-spin proton (we write this ROH_α) rapidly converts to ROH_β and then perhaps to ROH_α again because the protons provided by the solvent molecules in successive exchanges have random spin orientations. Therefore, instead of seeing a spectrum composed of contributions from both ROH_α and ROH_β molecules (that is, a spectrum showing a doublet structure due to the OH proton), we see a single, unsplit line at the mean position (as in Fig. 18.4). The effect is observed when the lifetime of a molecule due to this chemical exchange is so short that the lifetime broadening is greater than the doublet splitting. Because this splitting is often very small (about 1 Hz), a proton must remain attached to the same molecule for longer than about 0.1 s for the splitting to be observable. In water, the exchange rate is much faster than that, so alcohols show no splitting from the OH protons. In dry dimethylsulfoxide (DMSO), the exchange rate may be slow enough for the splitting to be detected.

(c) The measurement of T_1

The longitudinal relaxation time can be measured by the **inversion recovery technique**. The first step is to apply a 180° pulse to the sample. A 180° pulse is achieved by applying the B_1 field for twice as long as for a 90° pulse, so the magnetization vector precesses through 180° and points in the $-z$ direction (Fig. 18.37). No signal can be seen at this stage because there is no component of magnetization in the xy-plane (where the detection coils are sensitive). The β spins begin to relax back into α spins, and the magnetization vector shrinks exponentially back towards its thermal equilibrium value, M_z. After an interval τ, a 90° pulse is applied that rotates the magnetization into the xy-plane, where it starts to generate an FID signal. The frequency-domain spectrum is then obtained by Fourier transformation.

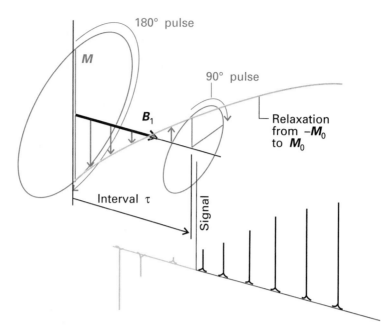

18.37 The result of applying a 180° pulse to the magnetization in the rotating frame and the effect of a subsequent 90° pulse. The amplitude of the frequency-domain spectrum varies with the interval between the two pulses because spin–lattice relaxation has time to occur.

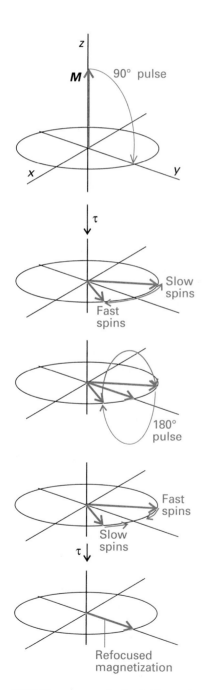

18.38 The sequence of pulses leading to the observation of a spin echo (see text).

The intensity of the spectrum obtained in this way depends on the length of the magnetization vector that is rotated into the xy-plane. The length of that vector returns exponentially to its thermal equilibrium value as the interval between the two pulses is increased, so the intensity of the spectrum also returns exponentially to its equilibrium intensity with increasing τ. We can therefore measure T_1 by fitting an exponential curve to the series of spectra obtained after different values of τ.

(d) Spin echoes

The measurement of T_2 (as distinct from T_2^*) depends on being able to eliminate the effects of inhomogeneous broadening. The cunning required is at the root of some of the most important advances that have been made in NMR since its introduction.

A **spin echo** is the magnetic analogue of an audible echo: transverse magnetization is created by a radiofrequency pulse, decays away, is reflected by a second pulse, and grows back to form an echo. The sequence of events is illustrated in Fig. 18.38. We can consider the overall magnetization as being made up of a number of different magnetizations, each of which arises from a **spin packet** of nuclei with very similar precession frequencies. The spread in these frequencies arises because the applied field $\mathcal{B}_0$ is inhomogeneous, so different parts of the sample experience different fields. The precession frequencies also differ if there is more than one chemical shift present. As will be seen, the importance of a spin echo is that it can suppress the effects of both field inhomogeneities and chemical shifts.

First, a 90° pulse is applied to the sample. The frame of reference is rotating at the same rate as the radiofrequency magnetic field of the pulse, with $\mathcal{B}_1$ applied along the x-axis, so the magnetization is rotated down into the xy-plane. The spin packets now begin to fan out because they have different Larmor frequencies, with some above the radiofrequency and some below. The detected signal depends on the resultant of the spin-packet magnetization vectors, and decays with a time constant T_2^* because of the combined effects of field inhomogeneity and spin–spin relaxation.

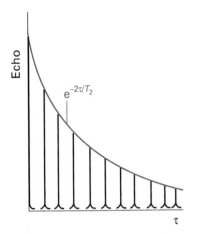

18.39 The exponential decay of the spin echoes can be used to determine the transverse relaxation time.

After an interval τ, a 180° pulse is applied to the sample; this time, about the y-axis of the rotating frame.[8] The pulse rotates the magnetization vectors of the faster spin packets into the positions previously occupied by the slower spin packets, and vice versa. Thus, as the vectors continue to precess, the fast vectors are now behind the slow; the fan begins to close up again, and the resultant signal begins to grow back into an echo. At time 2τ, all the vectors will once more be aligned along the y-axis, and the fanning out caused by the field inhomogeneity is said to have been **refocused**; the spin echo has reached its maximum. Because the effects of field inhomogeneities have been suppressed by the refocusing, the echo signal will have been attenuated by the factor $e^{-2\tau/T_2}$ caused by spin–spin relaxation alone. After the time 2τ, the magnetization will continue to precess, fanning out once again, giving a resultant that decays with time constant T_2^*.

The important feature of the technique is that the size of the echo is independent of any local fields that remain constant during the two τ intervals. If a spin packet is 'fast' because it happens to be composed of spins in a region of the sample that experiences higher than average fields, then it remains fast throughout both intervals, and what it gains on the first interval it makes up on the second interval. Hence, the size of the echo is independent of inhomogeneities in the magnetic field, for these remain constant. The true transverse relaxation arises from fields that fluctuate on a molecular timescale, and there is no guarantee that an individual 'fast' spin will remain 'fast' in the refocusing phase: the spins within the packets therefore spread with a time constant T_2. Hence, the effects of the true relaxation are not refocused, and the size of the echo decays with the time constant T_2 (Fig. 18.39).

18.7 The nuclear Overhauser effect

Spin relaxation can be used constructively to enhance the intensities of resonance lines. The enhancement is brought about by the **nuclear Overhauser effect** (NOE) which we shall explain by considering a simple AX system in which A is a ^{13}C nucleus and X is a proton.

We have seen already that one advantage of protons in NMR is their high magnetogyric ratio, which results in relatively large Boltzmann population differences and hence appreciable resonance intensities. In the nuclear Overhauser effect, relaxation processes involving internuclear dipole–dipole interactions are used to transfer this population advantage to another nucleus (to ^{13}C in the case we are considering), so that the latter's resonances are enhanced. A detailed calculation (which we do not reproduce here) shows that, if the relaxation of a nucleus A is dominated by its dipolar interaction with a nucleus X, and X is saturated by strong irradiation at its resonance frequency, then the signal enhancement is

$$\frac{\mathcal{I}_A}{\mathcal{I}_A^{\circ}} = 1 + \frac{\gamma_X}{2\gamma_A} \tag{33}$$

where $\mathcal{I}_J$ is the signal intensity of nucleus J. For ^{13}C coupled to a saturated proton, the ratio evaluates to 2.99, which shows that an enhancement of about a factor of 3 can be achieved.

The NOE is also used to determine interproton distances. The Overhauser enhancement of a proton A generated by saturating a spin X depends on the fraction of A's spin–lattice relaxation that is caused by its dipolar interaction with X. Because the dipolar field is proportional to r^{-3}, where r is the internuclear distance, and the relaxation effect is proportional to the square of the field, and therefore to r^{-6}, the NOE may be used to determine the geometries of molecules in solution. The determination of the structure of a small protein in solution involves the use of several hundred NOE measurements, effectively casting a net over the protons present.

---

8   The axis of the pulse is changed from x to y by a 90° phase shift of the radiofrequency radiation.

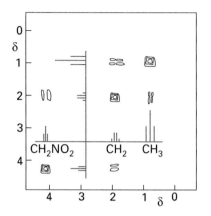

18.40 A typical two-dimensional ^{13}C-NMR spectrum obtained by correlation spectroscopy. The sample is I-nitropropane. Diagonal peaks show the normal one-dimensional spectrum, and cross-peaks at (δ_A, δ_X) appear where A and X are coupled. (Spectrum provided by Dr G. Morris.)

18.8 Two-dimensional NMR

An NMR spectrum contains a great deal of information and, if many protons are present, is very complex. Even a first-order spectrum is complex, for the fine structure of different groups of lines can overlap. The complexity would be reduced if we could use two axes to display the data, with resonances belonging to different groups lying at different locations on the second axis. This separation is essentially what is achieved in **two-dimensional NMR**.

We have seen that a spin-echo experiment refocuses spins that are in a constant environment. Hence, if two spins are in environments with different chemical shifts, they will be refocused and a single line will be obtained. That is, we can eliminate chemical shifts from a spectrum. Since we saw earlier that we can also remove the effects of spin–spin coupling by decoupling techniques, we can separate the two contributions to the spectrum. In practice, a clever choice of pulses and Fourier transformation techniques makes it possible to display spin coupling in one dimension and the chemical shifts in another, and so greatly simplify the appearance of a spectrum.

Much modern NMR work makes use of **correlation spectroscopy** (COSY) in which the basic pulse sequence is $90_x^\circ - t_1 - 90_x^\circ -$ acquire (t_2). A series of acquisitions is taken with a variable delay t_1, much as in a spin-echo experiment. The double Fourier transform is then performed on the real time-domain variable t_2 and then on the interferograms arising from the time delay t_1. A typical outcome for an AX system is shown in Fig. 18.40: the diagram shows contours of equal signal intensity.

The detailed analysis of the appearance of the contour plot is quite difficult, and a simple vector diagram of the processes involved cannot be given. However, the general rules of interpretation are quite straightforward (in simple cases, at least). The peaks across the diagonal constitute the normal four peaks of a one-dimensional NMR spectrum of an AX system, so they add nothing new. The interesting information is in the off-diagonal peaks, for they indicate that the protons to which they correlate by vertical and horizontal lines are spin–spin coupled. Although this information is trivial in this AX system, it can be of enormous help in the interpretation of more complex spectra. A complex spectrum that would be impossible to interpret in one-dimensional NMR can be interpreted reasonably rapidly by two-dimensional NMR. The techniques themselves are described in the books listed in *Further reading* at the end of the chapter.

18.9 Solid-state NMR

The principal difficulty with the application of NMR to solids is the low resolution that is characteristic of solid samples. Nevertheless, there are good reasons for seeking to overcome these difficulties. They include the possibility that a compound of interest is unstable in solution or that it is insoluble, so conventional solution NMR cannot be employed. Moreover, many species are intrinsically interesting as solids, and it is important to determine their structures and dynamics. Synthetic polymers are particularly interesting in this regard, and information can be obtained about the arrangement of molecules, their conformations, and the motion of different parts of the chain. This kind of information is crucial to an interpretation of the bulk properties of the polymer in terms of its molecular characteristics. Similarly, inorganic substances, such as the zeolites that are used as molecular sieves and shape-selective catalysts, can be studied using solid-state NMR, and structural problems can be resolved that cannot be tackled by X-ray diffraction.

Problems of resolution and linewidth are not the only features that plague NMR studies of solids. Because molecular rotation has almost ceased (except in special cases, including 'plastic crystals' in which the molecules continue to tumble), spin–lattice relaxation times are very long but spin–spin relaxation times are very short. Hence, in a pulse experiment, there need to be lengthy delays—of several seconds—between successive pulses so that the spin system has time to revert to equilibrium. Even gathering the murky information may therefore be a lengthy process. Moreover, because lines are so broad, very high powers of radiofrequency radiation may be required to achieve saturation. Whereas solution pulse NMR uses transmitters of a few tens of watts, solid-state NMR may require transmitters rated at several hundreds of watts.

(a) The origins of linewidths in solids

There are two principal contributions to the linewidths of solids. One is the direct magnetic dipolar interaction between nuclear spins. As we saw in the discussion of spin–spin coupling, a nuclear magnetic moment will give rise to a local magnetic field

$$\mathcal{B}_{\text{loc}} = -\frac{\gamma\hbar\mu_0 m_I}{4\pi R^3}(1 - 3\cos^2\theta) \tag{34}$$

Unlike in solution, this field is not motionally averaged to zero. Many nuclei may contribute to the total local field experienced by a nucleus of interest, and different nuclei in a sample may experience a wide range of fields. Typical dipole–dipole fields are of the order of 10^{-3} T, which corresponds to splittings and linewidths of the order of 10^4 Hz.

A second source of linewidth is the anisotropy of the chemical shift. We have seen that chemical shifts arise from the ability of the applied field to generate electron currents in molecules. In general, this ability depends on the orientation of the molecule relative to the applied field. In solution, when the molecule is tumbling rapidly, only the average value of the chemical shift is relevant. However, the anisotropy is not averaged to zero for stationary molecules in a solid, and molecules in different orientations have resonances at different frequencies. The chemical shift anisotropy also varies with the angle between the applied field and the principal axis of the molecule as $1 - 3\cos^2\theta$.

(b) The reduction of linewidths

Fortunately, there are techniques available for reducing the linewidths of solid samples. One technique, **magic-angle spinning** (MAS), takes note of the $1 - 3\cos^2\theta$ dependence of both the dipole–dipole interaction and the chemical shift anisotropy. The 'magic angle' is the angle at which $1 - 3\cos^2\theta = 0$, and corresponds to 54.74°. In the technique, the sample is spun at high speed at the magic angle to the applied field (Fig. 18.41). All the dipolar interactions and the anisotropies average to the value they would have at the magic angle, but at that angle they are zero. The difficulty with MAS is that the spinning frequency must not be less than the width of the spectrum, which is of the order of kilohertz. However, gas-driven sample spinners that can be rotated at up to 25 kHz are now routinely available, and a considerable body of work has been done.

The saturation and pulse techniques that we have described earlier in this section may also be used to reduce linewidths. The dipolar field of protons, for instance, may be reduced by a decoupling procedure. However, because the range of coupling strengths is so large, radiofrequency power of the order of 1 kW is required. Elaborate pulse sequences have also been devised that reduce linewidths by averaging procedures that make use of twisting the magnetization vector through an elaborate series of angles.

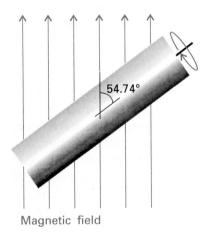

Magnetic field

18.41 In magic angle spinning, the sample spins at 54.74° (that is, arccos $1/3^{1/2}$) to the applied magnetic field. Rapid motion at this angle averages dipole–dipole interactions and chemical shift anisotropies to zero.

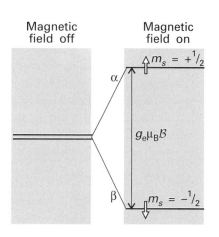

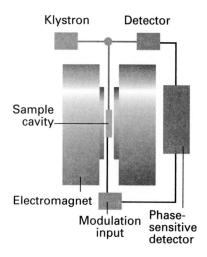

18.42 Electron spin levels in a magnetic field. Note that the β state is lower in energy than the α state (because the magnetogyric ratio of an electron is negative). Resonance is achieved when the frequency of the incident radiation matches the frequency corresponding to the energy separation.

18.43 The layout of an ESR spectrometer. A typical magnetic field is 0.3 T, which requires 9 GHz (3 cm) microwaves for resonance.

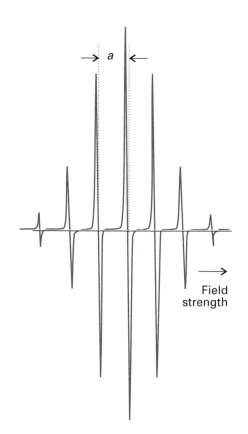

18.44 The ESR spectrum of the benzene radical anion, $C_6H_6^-$, in fluid solution. a is the hyperfine splitting of the spectrum; the centre of the spectrum is determined by the g-value of the radical.

Electron spin resonance

The energy levels of an electron spin in a magnetic field $\mathcal{B}$ (Fig. 18.42) are

$$E_{m_s} = g_e\mu_B\mathcal{B}m_s \qquad m_s = \pm\tfrac{1}{2} \tag{35}$$

where μ_B is the Bohr magneton and $g_e = 2.0023$ (Section 13.10a). This equation shows that the energy of an α electron ($m_s = +\tfrac{1}{2}$) increases and the energy of a β electron ($m_s = -\tfrac{1}{2}$) decreases as the field is increased, and that the separation of the levels is

$$\Delta E = E_\beta - E_\alpha = g_e\mu_B\mathcal{B} \tag{36}$$

When the sample is exposed to electromagnetic radiation of frequency ν, resonant absorption occurs when the resonance condition

$$h\nu = g_e\mu_B\mathcal{B} \tag{37}$$

is fulfilled. **Electron spin resonance (ESR)**, or **electron paramagnetic resonance (EPR)**, is the study of molecules and ions containing unpaired electrons by observing the magnetic fields at which they come into resonance with monochromatic radiation. Magnetic fields of about 0.3 T (the value used in most commercial ESR spectrometers) correspond to resonance with an electromagnetic field of frequency 10 GHz (10^{10} Hz) and wavelength 3 cm. Because 3 cm radiation falls in the X-band of the microwave region of the electromagnetic spectrum, ESR is a microwave technique.

The layout of an ESR spectrometer is shown in Fig. 18.43. It consists of a microwave source (a klystron), a cavity in which the sample is inserted in a glass or quartz container, a microwave detector, and an electromagnet with a field that can be varied in the region of 0.3 T. The ESR spectrum is obtained by monitoring the microwave absorption as the field is changed, and a typical spectrum (of the benzene radical anion, $C_6H_6^-$) is shown in Fig. 18.44. The peculiar appearance of the spectrum, which is in fact the first derivative of the absorption, arises from the detection technique, which is sensitive to the slope of the absorption curve (Fig. 18.45).

The sample must possess unpaired electron spins, so ESR is less widely applicable than NMR. It is used to study radicals formed during chemical reactions or by radiation, many d-metal complexes, and molecules in triplet states (such as those involved in phosphorescence, Section 17.3b). It is insensitive to normal, spin-paired molecules. The sample may be a gas, a liquid, or a solid, but the free rotation of molecules in the gas phase gives rise to complications.

18.10 The g-value

As in NMR, the spin magnetic moment interacts with the local magnetic field, and the resonance condition is normally written

$$h\nu = g\mu_B \mathcal{B} \tag{38}$$

where g is the **g-value** of the radical or complex.

Illustration

The centre of the ESR spectrum of the methyl radical occurred at 329.40 mT in a spectrometer operating at 9.2330 GHz. Its g-value is therefore

$$g = \frac{h\nu}{\mu_B \mathcal{B}} = \frac{(6.626\,08 \times 10^{-34}\ \text{J s}) \times (9.2330 \times 10^9\ \text{s}^{-1})}{(9.2740 \times 10^{-24}\ \text{J T}^{-1}) \times (0.32940\ \text{T})} = 2.0027$$

Comment Many organic radicals have g-values close to 2.0027; inorganic radicals have g-values typically in the range 1.9 to 2.1; d-metal complexes have g-values in a wider range (for example, 0 to 4).

Self-test 18.4 At what magnetic field would the methyl radical come into resonance in a spectrometer operating at 9.468 GHz?

[337.8 mT]

The deviation of g from $g_e = 2.0023$ depends on the ability of the applied field to induce local electron currents in the radical, and therefore its value gives some information about electronic structure. However, because g-values differ very little from g_e in many radicals (for example, 2.003 for H, 1.999 for NO_2, 2.01 for ClO_2), its main use in chemical applications is to aid the identification of the species present in a sample.

18.11 Hyperfine structure

The most important feature of ESR spectra is their **hyperfine structure**, the splitting of individual resonance lines into components. In general in spectroscopy, the term 'hyperfine structure' means the structure of a spectrum that can be traced to interactions of the electrons with nuclei other than as a result of the latter's point electric charge. The source of the hyperfine structure in ESR is the magnetic interaction between the electron spin and the magnetic dipole moments of the nuclei present in the radical.

(a) The effects of nuclear spin

Consider the effect on the ESR spectrum of a single H nucleus located somewhere in a radical. The proton spin is a source of magnetic field and, depending on the orientation of

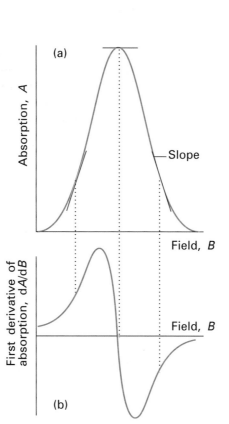

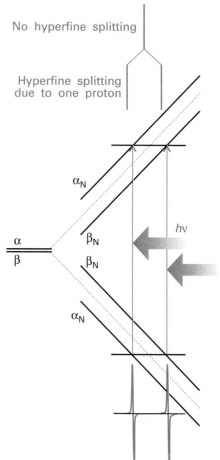

18.45 When phase-sensitive detection is used, the signal is the first derivative of the absorption intensity. (a) The absorption; (b) the signal, the slope of the absorption signal at each point. Note that the peak of the absorption corresponds to the point where the derivative passes through zero.

18.46 The hyperfine interaction between an electron and a spin-$\frac{1}{2}$ nucleus results in four energy levels in place of the original two. As a result, the spectrum consists of two lines (of equal intensity) instead of one. The intensity distribution can be summarized by a simple stick diagram. The diagonal lines show the energies of the states as the applied field is increased, and resonance occurs when the separation of states matches the fixed energy of the microwave photon.

the nuclear spin, the field it generates adds to or subtracts from the applied field. The total local field is therefore

$$\mathcal{B}_{\mathrm{loc}} = \mathcal{B} + a m_I \qquad m_I = \pm\tfrac{1}{2} \tag{39}$$

where a is the **hyperfine coupling constant**. Half the radicals in a sample have $m_I = +\frac{1}{2}$, so half resonate when the applied field satisfies the condition

$$h\nu = g\mu_B(\mathcal{B} + \tfrac{1}{2}a), \qquad \text{or } \mathcal{B} = \frac{h\nu}{g\mu_B} - \tfrac{1}{2}a \tag{40a}$$

The other half (which have $m_I = -\frac{1}{2}$) resonate when

$$h\nu = g\mu_B(\mathcal{B} - \tfrac{1}{2}a), \qquad \text{or } \mathcal{B} = \frac{h\nu}{g\mu_B} + \tfrac{1}{2}a \tag{40b}$$

Therefore, instead of a single line, the spectrum shows two lines of half the original intensity separated by a and centred on the field determined by g (Fig. 18.46).

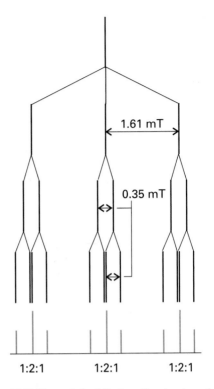

1.61 mT

0.35 mT

1:2:1   1:2:1   1:2:1

18.47 The analysis of the hyperfine structure of radicals containing one ^{14}N nucleus ($I = 1$) and two equivalent protons.

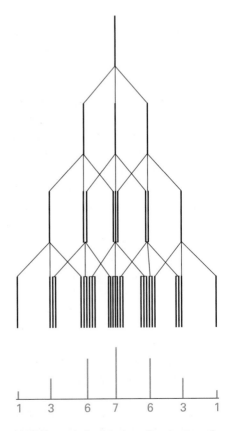

1   3   6   7   6   3   1

18.48 The analysis of the hyperfine structure of radicals containing three equivalent ^{14}N nuclei.

If the radical contains an ^{14}N atom ($I = 1$), its ESR spectrum consists of three lines of equal intensity, because the ^{14}N nucleus has three possible spin orientations, and each spin orientation is possessed by one-third of all the radicals in the sample. In general, a spin-I nucleus splits the spectrum into $2I + 1$ hyperfine lines of equal intensity.

When there are several magnetic nuclei present in the radical, each one contributes to the hyperfine structure. In the case of equivalent protons (for example, the two CH_2 protons in the radical CH_3CH_2) some of the hyperfine lines are coincident. It is not hard to show that, if the radical contains N equivalent protons, then there are $N + 1$ hyperfine lines with a binomial intensity distribution (that is, the intensity distribution given by Pascal's triangle). The spectrum of the benzene radical anion in Fig. 18.44, which has seven lines with intensity ratio $1 : 6 : 15 : 20 : 15 : 6 : 1$, is consistent with a radical containing six equivalent protons.

Example 18.4 Predicting the hyperfine structure of an ESR spectrum

A radical contains one ^{14}N nucleus ($I = 1$) with hyperfine constant 1.61 mT and two equivalent protons ($I = \frac{1}{2}$) with hyperfine constant 0.35 mT. Predict the form of the ESR spectrum.

Method We should consider the hyperfine structure that arises from each type of nucleus or group of equivalent nuclei in succession. So, split a line with one nucleus; then each of those lines is split by a second nucleus (or group of nuclei), and so on. It is best to start with the nucleus with the largest hyperfine splitting; however, any choice could be made, and the order in which nuclei are considered does not affect the conclusion.

Answer The ^{14}N nucleus gives three hyperfine lines of equal intensity separated by 1.61 mT. Each line is split into doublets of spacing 0.35 mT by the first proton, and each line of these doublets is split into doublets with the same 0.35 mT splitting (Fig. 18.47). The central lines of each split doublet coincide, so the proton splitting gives $1 : 2 : 1$ triplets of internal splitting 0.35 mT. Therefore, the spectrum consists of three equivalent $1 : 2 : 1$ triplets.

Comment Often it is quicker to realize that a group of equivalent protons gives a characteristic hyperfine pattern (two giving a $1 : 2 : 1$ triplet, in this case), and to superimpose the patterns directly.

- -

Self-test 18.5 Predict the form of the ESR spectrum of a radical containing three equivalent ^{14}N nuclei.

[Fig. 18.48]

The hyperfine structure of an ESR spectrum is a kind of fingerprint that helps to identify the radicals present in a sample. Moreover, because the magnitude of the splitting depends on the distribution of the unpaired electron near the magnetic nuclei present, the spectrum can be used to map the molecular orbital occupied by the unpaired electron. For example, because the hyperfine splitting in $C_6H_6^-$ is 0.375 mT, and one proton is close to a C atom with one-sixth the unpaired electron spin density (because the electron is spread uniformly around the ring), the hyperfine splitting caused by a proton in the electron spin entirely confined to a single adjacent C atom should be 6×0.375 mT $= 2.25$ mT. If in another aromatic radical we find a hyperfine splitting constant a, then the **spin density**, ρ, the probability that an unpaired electron is on the atom, can be calculated from the **McConnell equation**:

$$a = Q\rho \qquad (41)$$

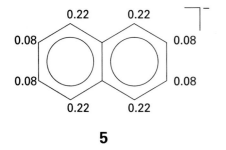

5

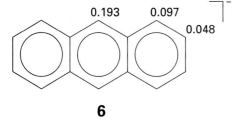

6

Table 18.2* Hyperfine coupling constants for atoms, a/mT

Nuclide	Isotropic coupling	Anisotropic coupling
^1H	50.8($1s$)	
^2H	7.8($1s$)	
^{14}N	55.2($2s$)	3.4($2p$)
^{19}F	1720($2s$)	108.4($2p$)

*More values are given in the *Data section*.

with $Q = 2.25$ mT. In this equation, ρ is the spin density on a C atom and a is the hyperfine splitting observed for the H atom to which it is attached.

Illustration
· ·
The hyperfine structure of the ESR spectrum of the radical anion (naphthalene)⁻ can be interpreted as arising from two groups of four equivalent protons. Those at the 1, 4, 5, and 8 positions in the ring have $a = 0.490$ mT and those in the 2, 3, 6, and 7 positions have $a = 0.183$ mT. The densities obtained by using the McConnell equation are 0.22 and 0.08, respectively (**5**).
· ·

Self-test 18.6 The spin density in (anthracene)⁻ is shown in (**6**). Predict the form of its ESR spectrum.

[A 1 : 2 : 1 triplet of splitting 0.43 mT split into a 1 : 4 : 6 : 4 : 1 quintet of splitting 0.22 mT, split into a 1 : 4 : 6 : 4 : 1 quintet of splitting 0.11 mT, $3 \times 5 \times 5 = 75$ lines in all]

(b) The origin of the hyperfine interaction

The hyperfine interaction is an interaction between the magnetic moments of the unpaired electron and the nuclei. There are two contributions to the interaction.

An electron in a p orbital does not approach the nucleus very closely, so it experiences a field that appears to arise from a point magnetic dipole. The resulting interaction is called the **dipole–dipole interaction**. The contribution of a magnetic nucleus to the local field experienced by the unpaired electron is given by an expression like that in eqn 34. A characteristic of this type of interaction is that it is anisotropic: that is, its magnitude (and sign) depends on the orientation of the radical with respect to the applied field. Furthermore, just as in the case of NMR, the dipole–dipole interaction averages to zero when the radical is free to tumble. Therefore, hyperfine structure due to the dipole–dipole interaction is observed only for radicals trapped in solids.

An s electron is spherically distributed around a nucleus and so has zero average dipole–dipole interaction with the nucleus even in a solid sample. However, because an s electron has a nonzero probability of being at the nucleus, it is incorrect to treat the interaction as one between two point dipoles. An s electron has a Fermi contact interaction with the nucleus, which as we saw in Section 18.4d is a magnetic interaction that occurs when the point dipole approximation fails. The contact interaction is isotropic (that is, independent of the radical's orientation), and consequently is shown even by rapidly tumbling molecules in fluids (provided the spin density has some s-character).

The dipole–dipole interactions of p electrons and the Fermi contact interaction of s electrons can be quite large. For example, a $2p$ electron in a nitrogen atom experiences an average field of about 3.4 mT from the ^{14}N nucleus. A $1s$ electron in a hydrogen atom experiences a field of about 50 mT as a result of its Fermi contact interaction with the central proton. More values are listed in Table 18.2. The magnitudes of the contact interactions in radicals can be interpreted in terms of the s orbital character of the molecular orbital occupied by the unpaired electron, and the dipole–dipole interaction can be interpreted in terms of the p character. The analysis of hyperfine structure therefore gives information about the composition of the orbital, and especially the hybridization of the atomic orbitals (see Problem 18.6).

We still have the source of the hyperfine structure of the $C_6H_6^-$ anion and other aromatic radical anions to explain. The sample is fluid, and as the radicals are tumbling the hyperfine structure cannot be due to the dipole–dipole interaction. Moreover, the protons lie in the

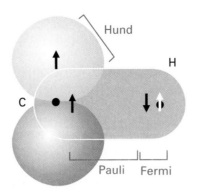
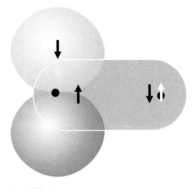

(a) Low energy

(b) High energy

18.49 The polarization mechanism for the hyperfine interaction in π-electron radicals. The arrangement in (a) is lower in energy than that in (b), so there is an effective coupling between the unpaired electron and the proton.

nodal plane of the π orbital occupied by the unpaired electron, so the structure cannot be due to a Fermi contact interaction. The explanation lies in a **polarization mechanism** similar to the one responsible for spin–spin coupling in NMR. There is a magnetic interaction between a proton and the σ electrons which results in one of the electrons tending to be found with a greater probability nearby (Fig. 18.49). The electron with opposite spin is therefore more likely to be close to the C atom at the other end of the bond. The unpaired electron on the C atom has a lower energy if it is parallel to that electron (Hund's rule favours parallel electrons on atoms), so the unpaired electron can detect the spin of the proton indirectly. Calculation using this model leads to a hyperfine interaction in agreement with the observed value of 2.25 mT.

Checklist of key ideas

Nuclear magnetic resonance

18.1 Nuclear magnetic moments
- [ ] spin-$\frac{1}{2}$ nuclei

18.2 The energies of nuclei in magnetic fields
- [ ] magnetogyric ratio
- [ ] nuclear g-factor
- [ ] nuclear magneton
- [ ] Larmor frequency
- [ ] resonance condition (6)
- [ ] nuclear magnetic resonance (NMR)

18.3 The chemical shift
- [ ] shielding constant
- [ ] chemical shift (10)
- [ ] deshielded nuclei
- [ ] integration
- [ ] local contribution
- [ ] neighbouring group contribution
- [ ] solvent contribution

- [ ] diamagnetic contribution
- [ ] paramagnetic contribution
- [ ] Lamb formula (14)
- [ ] ring current

18.4 The fine structure
- [ ] fine structure
- [ ] scalar coupling constant
- [ ] Karplus equation (21)
- [ ] polarization mechanism
- [ ] Fermi contact interaction
- [ ] chemically equivalent nuclei
- [ ] magnetically equivalent nuclei
- [ ] first-order spectra
- [ ] strongly coupled spectra
- [ ] heteronuclear spin system
- [ ] homonuclear spin system
- [ ] dilute-spin species
- [ ] abundant-spin species
- [ ] proton decoupling

Pulse techniques in NMR
- [ ] Fourier-transform NMR

18.5 The magnetization vector
- [ ] magnetization
- [ ] precessing
- [ ] 90° pulse
- [ ] free-induction decay (FID)

18.6 Linewidths and rate processes
- [ ] spin relaxation
- [ ] longitudinal relaxation time
- [ ] spin–lattice relaxation time
- [ ] transverse relaxation time
- [ ] spin–spin relaxation time
- [ ] inhomogeneous broadening
- [ ] effective transverse relaxation time (31)
- [ ] chemical exchange
- [ ] inversion recovery technique
- [ ] spin echo
- [ ] spin packet
- [ ] refocused

18.7 The nuclear Overhauser effect
- [ ] nuclear Overhauser effect (NOE)

18.8 Two-dimensional NMR
- [ ] two-dimensional NMR
- [ ] correlation spectroscopy (COSY)

18.9 Solid-state NMR
- [ ] magic-angle spinning (MAS)

Electron spin resonance
- [ ] electron spin resonance (ESR)
- [ ] electron paramagnetic resonance (EPR)
- [ ] resonance condition (38)

18.10 The g-value
- [ ] g-value

18.11 Hyperfine structure
- [ ] hyperfine structure
- [ ] hyperfine coupling constant
- [ ] spin density
- [ ] McConnell equation (41)
- [ ] dipole–dipole interaction
- [ ] polarization mechanism

Further reading

Articles of general interest

R.H. Orcutt, A straightforward way to determine relative intensities of spin–spin splitting lines of equivalent nuclei in NMR spectra. *J. Chem. Educ.* **64**, 763 (1987).

T.D. Lash and S.S. Lash, The use of Pascal-like triangles in describing first-order coupling patterns. *J. Chem. Educ.* **64**, 315 (1987).

T.A. Shaler, Generalization of Pascal's triangle to nuclei of any spin. *J. Chem. Educ.* **68**, 853 (1991).

L.J. Schwartz, A step-by-step picture of pulsed (time-domain) NMR. *J. Chem. Educ.* **65**, 959 (1988); idem, 752.

M.-K. Ahn, A comparison of FTNMR and FTIR techniques. *J. Chem. Educ.* **66**, 802 (1989).

D.J. Wink, Spin–lattice relaxation times in ^1H NMR spectroscopy. *J. Chem. Educ.* **66**, 810 (1989).

R.W. King and K.R. Williams, The Fourier transform in chemistry. Part 1. Nuclear magnetic resonance: introduction. *J. Chem. Educ.* **66**, A213 (1989); Part 2. Nuclear magnetic resonance: the single pulse experiment. *Ibid.* A243; Part 3. Multiple-pulse experiments. *J. Chem. Educ.* **67**, A9 (1990); Part 4. NMR: two-dimensional methods. *Ibid.* A125; A glossary of NMR terms. *Ibid.* A100.

R. Freeman, Selective excitation in high-resolution NMR. *Chem. Rev.* **91**, 1397 (1991).

D.H. Grant, Paramagnetic susceptibility by NMR: the "solvent correction" reexamined. *J. Chem. Educ.* **72**, 39 (1995).

B.E. Mann, The analysis of first-order coupling patterns in NMR. *J. Chem. Educ.* **72**, 614 (1995).

J.C. Paniagua and A. Moyano, On the way of introducing some basic NMR aspects: from the classical and naive quantum models to the density-operator approach. *J. Chem. Educ.* **73**, 310 (1996).

L.R. Dalton, A. Bain, and C.K. Westbrook, Recent advances in electron paramagnetic resonance. *Ann. Rev. Phys. Chem.* **41**, 389 (1990).

B.H. Suits, Magnetic resonance spectrometers. In *Encyclopedia of applied physics* (ed. G.L. Trigg), **9**, 71. VCH, New York (1994).

R.G. Barnes, Electron paramagnetic resonance. In *Encyclopedia of applied physics* (ed. G.L. Trigg), **5**, 475. VCH, New York (1993).

Texts and sources of data and information

P.J. Hore, *Nuclear magnetic resonance*, Oxford Chemistry Primers. Oxford University Press (1995).

R.J. Abraham, J. Fisher, and P. Lofthus, *Introduction to NMR spectroscopy*. Wiley, New York (1991).

R.K. Harris, *Nuclear magnetic resonance spectroscopy*. Longman, London (1986).

A.E. Derome, *Modern NMR techniques for chemistry research*. Pergamon, Oxford (1987).

J.K.M. Sanders and B.K. Hunter, *Modern NMR spectroscopy*. Oxford University Press (1993).

J.K.M. Sanders, E.C. Constable, and B.K. Hunter, *Modern NMR spectroscopy: a workbook of chemical problems*. Oxford University Press (1993).

R. Freeman, *A handbook of nuclear magnetic resonance spectrocopy*. Longman, London (1997).

R. Freeman, *Spin choreography: basic steps in high resolution NMR*. Spektrum, Oxford (1997).

H. Günther, *NMR spectroscopy: basic principles, concepts, and applications in chemistry*. Wiley, New York (1995).

D. Canet, *Nuclear magnetic resonance: concepts and methods*. Wiley, New York (1996).

J.N.S. Evans, *Biomolecular NMR spectroscopy*. Oxford University Press (1995).

E.A.V. Ebsworth, D.W.H. Rankin, and S. Cradock, *Structural methods in inorganic chemistry*. Blackwell Scientific, Oxford (1991).

R. Drago, *Physical methods for chemists*. Saunders, Philadelphia (1992).

D.M. Grant and R.K. Harris (ed.), *Encyclopedia of nuclear magnetic resonance*, Vols 1–8. Wiley, New York (1996).

N.M. Atherton, *Principles of electron spin resonance*. Ellis Horwood/Prentice-Hall, Hemel Hempstead (1993).

Exercises

18.1 (a) What is the resonance frequency of a proton in a magnetic field of 14.1 T?

18.1 (b) What is the resonance frequency of a ^{19}F nucleus in a magnetic field of 16.2 T?

18.2 (a) ^{32}S has a nuclear spin of $\frac{3}{2}$ and a nuclear g factor of 0.4289. Calculate the energies of the nuclear spin states in a magnetic field of 7.500 T.

18.2 (b) ^{14}N has a nuclear spin of 1 and a nuclear g factor of 0.404. Calculate the energies of the nuclear spin states in a magnetic field of 11.50 T.

18.3 (a) Calculate the frequency separation of the nuclear spin levels of a ^{13}C nucleus in a magnetic field of 14.4 T given that the magnetogyric ratio is $6.73 \times 10^7\ T^{-1}\ s^{-1}$.

18.3 (b) Calculate the frequency separation of the nuclear spin levels of a ^{14}N nucleus in a magnetic field of 15.4 T given that the magnetogyric ratio is $1.93 \times 10^7\ T^{-1}\ s^{-1}$.

18.4 (a) In which of the following systems is the energy level separation the largest: (a) a proton in a 600 MHz NMR spectrometer, (b) a deuteron in the same spectrometer?

18.4 (b) In which of the following systems is the energy level separation the largest: (a) a ^{14}N nucleus in a 600 MHz NMR spectrometer, (b) an electron in a radical in a field of 0.300 T?

18.5 (a) Calculate the energy difference between the lowest and highest nuclear spin states of a ^{14}N nucleus in a 15.00 T magnetic field.

18.5 (b) Calculate the magnetic field needed to satisfy the resonance condition for unshielded protons in a 150.0 MHz radiofrequency field.

18.6 (a) Use Table 18.1 to predict the magnetic fields at which (a) ^1H, (b) ^2H, (c) ^{13}C come into resonance at (i) 250 MHz, (ii) 500 MHz.

18.6 (b) Use Table 18.1 to predict the magnetic fields at which (a) ^{14}N, (b) ^{19}F, and (c) ^{31}P come into resonance at (i) 300 MHz, (ii) 750 MHz.

18.7 (a) Calculate the relative population differences ($\delta N/N$) for protons in fields of (a) 0.30 T, (b) 1.5 T, and (c) 10 T at 25°C.

18.7 (b) Calculate the relative population differences ($\delta N/N$) for ^{13}C nuclei in fields of (a) 0.50 T, (b) 2.5 T, and (c) 15.5 T at 25°C.

18.8 (a) The first generally available NMR spectrometers operated at a frequency of 60 MHz; today it is not uncommon to use a spectrometer that operates at 600 MHz. What are the relative population differences of ^{13}C spin states in these two spectrometers at 25°C?

18.8 (b) What are the relative values of the chemical shifts observed for nuclei in the spectrometers mentioned in Exercise 18.8a in terms of (a) δ values, (b) frequencies?

18.9 (a) The chemical shift of the CH_3 protons in acetaldehyde (ethanal) is $\delta = 2.20$ and that of the CHO proton is 9.80. What is the difference in local magnetic field between the two regions of the molecule when the applied field is (a) 1.5 T, (b) 15 T?

18.9 (b) The chemical shift of the CH_3 protons in diethyl ether is $\delta = 1.16$ and that of the CH_2 protons is $\delta = 3.36$. What is the difference in local magnetic field between the two regions of the molecule when the applied field is (a) 1.9 T, (b) 16.5 T?

18.10 (a) Sketch the appearance of the ^1H-NMR spectrum of acetaldehyde (ethanal) using $J = 2.90$ and the data in Exercise 18.9a in a spectrometer operating at (a) 250 MHz, (b) 500 MHz.

18.10 (b) Sketch the appearance of the ^1H-NMR spectrum of diethyl ether using $J = 6.97$ Hz and the data in Exercise 18.9b in a spectrometer operating at (a) 350 MHz, (b) 650 MHz.

18.11 (a) Two groups of protons are made equivalent by the isomerization of a fluxional molecule. At low temperatures, where the interconversion is slow, one group has $\delta = 4.0$ and the other has $\delta = 5.2$. At what rate of interconversion will the two signals merge in a spectrometer operating at 250 MHz?

18.11 (b) Two groups of protons are made equivalent by the isomerization of a fluxional molecule. At low temperatures, where the interconversion is slow, one group has $\delta = 5.5$ and the other has $\delta = 6.8$. At what rate of interconversion will the two signals merge in a spectrometer operating at 350 MHz?

18.12 (a) Sketch the form of the ^{19}F-NMR spectra of a natural sample of tetrafluoroborate ions, BF_4^-, allowing for the relative abundances of $^{10}BF_4^-$ and $^{11}BF_4^-$.

18.12 (b) From the data in Table 18.1, predict the frequency needed for ^{31}P-NMR in an NMR spectrometer designed to observe proton resonance at 500 MHz. Sketch the proton and ^{31}P resonances in the NMR spectrum of PH_4^+.

18.13 (a) Sketch the form of an $A_3M_2X_4$ spectrum, where A, M, and X are protons with distinctly different chemical shifts and $J_{AM} > J_{AX} > J_{MX}$.

18.13 (b) Sketch the form of an $A_2M_2X_5$ spectrum, where A, M, and X are protons with distinctly different chemical shifts and $J_{AM} > J_{AX} > J_{MX}$.

18.14 (a) Which of the following molecules have sets of nuclei that are chemically but not magnetically equivalent: (a) CH_3CH_3, (b) $CH_2=CH_2$?

18.14 (b) Which of the following molecules have sets of nuclei that are chemically but not magnetically equivalent: (a) $CH_2=C=CF_2$, (b) cis- and $trans$-$[Mo(CO)_4(PH_3)_2]$?

18.15 (a) The duration of a 90° or 180° pulse depends on the strength of the B_1 field. If a 90° pulse requires 10 μs, what is the strength of the B_1 field? How long would the corresponding 180° pulse require?

18.15 (b) The duration of a 90° or 180° pulse depends on the strength of the B_1 field. If a 180° pulse requires 12.5 μs, what is the strength of the B_1 field? How long would the corresponding 90° pulse require?

18.16 (a) What magnetic field would be required in order to use an ESR X-band spectrometer (9 GHz) to observe ^{1}H-NMR and a 300 MHz spectrometer to observe ESR?

18.16 (b) Some commercial ESR spectrometers use 8 mm microwave radiation (the Q band). What magnetic field is needed to satisfy the resonance condition?

18.17 (a) The centre of the ESR spectrum of atomic hydrogen lies at 329.12 mT in a spectrometer operating at 9.2231 GHz. What is the g-value of the atom?

18.17 (b) The centre of the ESR spectrum of atomic deuterium lies at 330.02 mT in a spectrometer operating at 9.2482 GHz. What is the g-value of the atom?

18.18 (a) A radical containing two equivalent protons shows a three-line spectrum with an intensity distribution 1 : 2 : 1. The lines occur at 330.2 mT, 332.5 mT, and 334.8 mT. What is the hyperfine coupling constant for each proton? What is the g-value of the radical given that the spectrometer is operating at 9.319 GHz?

18.18 (b) A radical containing three equivalent protons shows a four-line spectrum with an intensity distribution 1 : 3 : 3 : 1. The lines occur at 331.4 mT, 333.6 mT, 335.8 mT, and 338.0 mT. What is the hyperfine coupling constant for each proton? What is the g-value of the radical given that the spectrometer is operating at 9.332 GHz?

18.19 (a) A radical containing two inequivalent protons with hyperfine constants 2.0 mT and 2.6 mT gives a spectrum centred on 332.5 mT. At what fields do the hyperfine lines occur and what are their relative intensities?

18.19 (b) A radical containing three inequivalent protons with hyperfine constants 2.11 mT, 2.87 mT, and 2.89 mT gives a spectrum centred on 332.8 mT. At what fields do the hyperfine lines occur and what are their relative intensities?

18.20 (a) Predict the intensity distribution in the hyperfine lines of the ESR spectra of (a) $\cdot CH_3$, (b) $\cdot CD_3$.

18.20 (b) Predict the intensity distribution in the hyperfine lines of the ESR spectra of (a) $\cdot CH_2CH_3$, (b) $\cdot CD_2CD_3$.

18.21 (a) The benzene radical anion has $g = 2.0025$. At what field should you search for resonance in a spectrometer operating at (a) 9.302 GHz, (b) 33.67 GHz?

18.21 (b) The naphthalene radical anion has $g = 2.0024$. At what field should you search for resonance in a spectrometer operating at (a) 9.312 GHz, (b) 33.88 GHz?

18.22 (a) The ESR spectrum of a radical with a single magnetic nucleus is split into four lines of equal intensity. What is the nuclear spin of the nucleus?

18.22 (b) The ESR spectrum of a radical with two equivalent nuclei of a particular kind is split into five lines of intensity ratio 1 : 2 : 3 : 2 : 1. What is the spin of the nuclei?

18.23 (a) Sketch the form of the hyperfine structures of radicals XH_2 and XD_2, where the nucleus X has $I = \frac{5}{2}$.

18.23 (b) Sketch the form of the hyperfine structures of radicals XH_3 and XD_3, where the nucleus X has $I = \frac{3}{2}$.

Problems

Numerical problems

18.1 A scientist investigates the possibility of neutron spin resonance, and has available a commercial NMR spectrometer operating at 300 MHz. What field is required for resonance? What is the relative population difference at room temperature? Which is the lower energy spin state of the neutron?

18.2 Two groups of protons have $\delta = 4.0$ and $\delta = 5.2$ and are interconverted by a conformational change of a fluxional molecule. In a 60 MHz spectrometer the spectrum collapsed into a single line at 280 K but at 300 MHz the collapse did not occur until the temperature had been raised to 300 K. What is the activation energy of the interconversion?

18.3 The angular NO_2 molecule has a single unpaired electron and can be trapped in a solid matrix or prepared inside a nitrite crystal by radiation damage of NO_2^- ions. When the applied field is parallel to the OO direction the centre of the spectrum lies at 333.64 mT in a spectrometer operating at 9.302 GHz. When the field lies along the bisector of the ONO angle, the resonance lies at 331.94 mT. What are the g-values in the two orientations?

18.4 The hyperfine coupling constant in $\cdot CH_3$ is 2.3 mT. Use the information in Table 18.1 to predict the splitting between the

hyperfine lines of the spectrum of $\cdot CD_3$. What are the overall widths of the hyperfine spectra in each case?

18.5 The p-dinitrobenzene radical anion can be prepared by reduction of p-dinitrobenzene. The radical anion has two equivalent N nuclei ($I = 1$) and four equivalent protons. Predict the form of the ESR spectrum using $a(N) = 0.148$ mT and $a(H) = 0.112$ mT.

18.6 The hyperfine coupling constants observed in the radical anions (7), (8), and (9) are shown (in mT). Use the value for the benzene radical anion to map the probability of finding the unpaired electron in the π orbital on each C atom.

7

8

9

Theoretical problems

18.7 The z-component of the magnetic field at a distance R from a magnetic moment parallel to the z-axis is given by eqn 22. In a solid, a proton at a distance R from another can experience such a field and the measurement of the splitting it causes in the spectrum can be used to calculate R. In gypsum, for instance, the splitting in the H_2O resonance can be interpreted in terms of a magnetic field of 0.715 mT generated by one proton and experienced by the other. What is the separation of the protons in the H_2O molecule?

18.8 In a liquid crystal, a molecule might not rotate freely in all directions and the dipolar interaction might not average to zero. Suppose a molecule is trapped so that, although the vector separating two protons may rotate freely around the z-axis, the colatitude may vary only between 0 and θ'. Average the dipolar field over this restricted range of orientations and confirm that the average vanishes when $\theta' = \pi$ (corresponding to rotation over an entire sphere). What is the average value of the local dipolar field for the H_2O molecule in Problem 18.7 if it is dissolved in a liquid crystal that enables it to rotate up to $\theta' = 30°$?

18.9 The shape of a spectral line, $\mathcal{I}(\omega)$, is related to the free induction decay signal $G(t)$ by

$$\mathcal{I}(\omega) = a\mathrm{Re}\int_0^\infty G(t)\mathrm{e}^{\mathrm{i}\omega t}\,\mathrm{d}t$$

where a is a constant and 'Re' means take the real part of what follows. Calculate the lineshape corresponding to an oscillating, decaying function $G(t) = \cos\omega_0 t\,\mathrm{e}^{-t/\tau}$.

18.10 In the language of Problem 18.9, show that, if $G(t) = (a\cos\omega_1 t + b\cos\omega_2 t)\mathrm{e}^{-t/\tau}$, then the spectrum consists of two lines with intensities proportional to a and b and located at $\omega = \omega_1$ and ω_2, respectively.

Additional problems supplied by Carmen Giunta and Charles Trapp

18.11 Suppose that the FID in Fig. 18.30 was recorded in a 300 MHz spectrometer, and that the interval between maxima in the oscillations in the FID is 0.10 s. What is the Larmor frequency of the nuclei and the spin–spin relaxation time?

18.12 In a classical study of the application of NMR to the measurement of rotational barriers in molecules, P.M. Nair and J.D. Roberts (*J. Am. Chem. Soc.* **79**, 4565 (1957)) obtained the 40 MHz ^{19}F-NMR spectrum of $F_2BrCCBrCl_2$. Their spectra are reproduced in Fig. 18.50. At 193 K, the spectrum shows five resonance peaks. Peaks I and III are separated by 160 Hz, as are IV and V. The ratio of the integrated intensities of peak II to peaks I, III, IV,

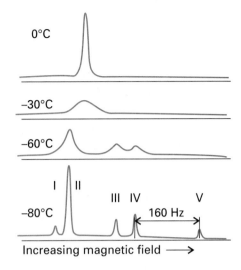

Fig. 18.50

and V is approximately 10 to 1. At 273 K, the five peaks have collapsed into one. Explain the spectrum and its change with temperature. At what rate of interconversion will the spectrum collapse to a single line? Calculate the rotational energy barrier between the rotational isomers on the assumption that it is related to the rate of interconversion between the isomers.

18.13 Various versions of the Karplus equation (eqn 21) have been used to correlate data on vicinal proton coupling constants in systems of the type $R_1R_2CHCHR_3R_4$. The original version, (M. Karplus, *J. Am. Chem. Soc.* **85**, 2870 (1963)), is $^3J_{HH} = A\cos^2\phi_{HH} + B$. When $R_3 = R_4 = H$, $^3J_{HH} = 7.3$ Hz; when $R_3 = CH_3$ and $R_4 = H$, $^3J_{HH} = 8.0$ Hz; when $R_3 = R_4 = CH_3$, $^3J_{HH} = 11.2$ Hz. Assume that only staggered conformations are important and determine which version of the Karplus equation fits the data better.

18.14 It might be unexpected that the Karplus equation, which was first derived for $^3J_{HH}$ coupling constants, should also apply to vicinal coupling between the nuclei of metals such as tin. T.N. Mitchell and B. Kowall (*Magn. Reson. Chem.* **33**, 325 (1995)) have studied the relation between $^3J_{HH}$ and $^3J_{SnSn}$ in compounds of the type $Me_3SnCH_2CHRSnMe_3$ and find that $^3J_{SnSn} = 78.86\,^3J_{HH} + 27.84$ Hz. (a) Does this result support a Karplus type equation for tin? Explain your reasoning. (b) Obtain the Karplus equation for $^3J_{SnSn}$ and plot it as a function of the dihedral angle. (c) Draw the preferred conformation.

18.15 The relative sensitivity of NMR lines for equal numbers of different nuclei at constant temperature for a given frequency is $R_\nu \propto (I+1)\mu\omega_0^2$ whereas for a given field it is $R_B \propto \{(I+1)/I^2\}\mu^3$. (a) From the data in Table 18.1, calculate these sensitivities for the deuteron, ^{13}C, ^{14}N, ^{19}F, and ^{31}P relative to the proton. (b) Derive the equation for R_B from the equation for R_ν.

19

Statistical thermodynamics: the concepts

Statistical thermodynamics provides the link between the microscopic properties of matter and its bulk properties. Two key ideas are introduced in this chapter. The first is the Boltzmann distribution. This enormously important result was described in the Introduction, where we saw that it can be used to predict the populations of states. In this chapter we see its derivation in terms of the distribution of particles over available states. The derivation leads naturally to the introduction of the partition function, which is the central mathematical concept of these two chapters. We see how to interpret the partition function and how to calculate it in a number of simple cases. The next part of the chapter shows how to extract thermodynamic information from the partition function.

In the final part of the chapter, we generalize the discussion to include systems that are composed of assemblies of interacting particles. Very similar equations are developed to those in the first part of the chapter, but they are much more widely applicable.

The preceding chapters of this part of the text have shown how the energy levels of molecules can be calculated, determined spectroscopically, and related to their structures. The next major step is to see how a knowledge of these energy levels can be used to account for the properties of matter in bulk. To do so, we now introduce the concepts of **statistical thermodynamics**, the link between molecular properties and bulk thermodynamic properties.

The crucial step in going from the quantum mechanics of individual molecules to the thermodynamics of bulk samples is to recognize that the latter deals with the *average* behaviour of large numbers of molecules. For example, the pressure of a gas depends on the average force exerted by its molecules, and there is no need to specify which molecules happen to be striking the wall at any instant. Nor is it necessary to consider the fluctuations in the pressure as different numbers of molecules collide with the wall at different moments. The fluctuations in pressure are very small compared with the steady pressure: it is highly improbable that there will be a sudden lull in the number of collisions, or a sudden surge.

This chapter introduces statistical thermodynamics in two stages. The first, the derivation of the Boltzmann distribution for individual particles, is of restricted applicability, but it has the advantage of taking us directly to a result of central importance in a straightforward and elementary way. We can *use* statistical thermodynamics once we have deduced the Boltzmann distribution. Then (in Section 19.5) we extend the arguments to systems composed of interacting particles.

The distribution of molecular states

We consider a closed system composed of N molecules. Although the total energy is constant at E, it is not possible to be definite about how that energy is shared between the molecules. Collisions result in the ceaseless redistribution of energy not only between the molecules but also among their different modes. The closest we can come to a description of the distribution of energy is to report the **population** of a state, the average number of molecules that occupy it, and to say that on average there are n_i molecules in a state of energy ε_i. The populations of the states remain almost constant, but the precise identities of the molecules in each state may change at every collision.

The problem we address in this section is the calculation of the populations of states for any type of molecule in any mode of motion at any temperature. The only restriction is that the molecules should be independent, in the sense that the total energy of the system is a sum of their individual energies. We are discounting (at this stage) the possibility that in a real system a contribution to the total energy may arise from interactions between molecules. We also adopt the **principle of equal *a priori* probabilities**,[1] the assumption that all possibilities for the distribution of energy are equally probable. That is, we assume that vibrational states of a certain energy, for instance, are as likely to be populated as rotational states of the same energy.

19.1 Configurations and weights

Any individual molecule may exist in states with energies $\varepsilon_0, \varepsilon_1, \ldots$. We shall always take ε_0, the lowest state, as the zero of energy ($\varepsilon_0 = 0$), and measure all other energies relative to that state. To obtain the actual internal energy, U, we may have to add a constant to the calculated energy of the system. For example, if we are considering the vibrational contribution to the internal energy, we must add the total zero-point energy of any oscillators in the sample.

(a) Instantaneous configurations

At any instant there will be n_0 molecules in the state with energy ε_0, n_1 with ε_1, and so on. The specification of the set of populations $n_0, n_1, \ldots$ in the form $\{n_0, n_1, \ldots\}$ is a statement of the instantaneous **configuration** of the system. The instantaneous configuration fluctuates with time because the populations change. We can picture a large number of different instantaneous configurations. One, for example, might be $\{N, 0, 0, \ldots\}$, corresponding to every molecule being in its ground state. Another might be $\{N-2, 2, 0, 0, \ldots\}$, in which two molecules are in the first excited state. The latter configuration is intrinsically more likely to be found than the former because it can be

---

1   *A priori* means in this context loosely 'as far as one knows'. We have no reason to presume otherwise than that all states are equally likely to be occupied whatever their nature.

achieved in more ways: $\{N, 0, 0, \ldots\}$ can be achieved in only one way, but $\{N-2, 2, 0, \ldots\}$ can be achieved in $\frac{1}{2}N(N-1)$ different ways (Fig. 19.1).[2]

Justification 19.1

One candidate for promotion to an upper state can be selected in N ways. There are $N-1$ candidates for the second choice, so the total number of choices is $N(N-1)$. However, we should not distinguish the choice (Jack, Jill) from the choice (Jill, Jack) because they lead to the same configurations. Therefore, only half the choices lead to distinguishable configurations, and the total number of distinguishable choices is $\frac{1}{2}N(N-1)$.

If, as a result of collisions, the system were to fluctuate between the configurations $\{N, 0, 0, \ldots\}$ and $\{N-2, 2, 0, \ldots\}$, it would almost always be found in the second, more likely state (especially if N were large). In other words, a system free to switch between the two configurations would show properties characteristic almost exclusively of the second configuration.

A general configuration $\{n_0, n_1, \ldots\}$ can be achieved in W different ways, where W is called the **weight** of the configuration. The weight of the configuration $\{n_0, n_1, \ldots\}$ is given by the expression

$$W = \frac{N!}{n_0! n_1! n_2! \cdots} \tag{1}$$

where $x!$, x factorial, denotes $x(x-1)(x-2)\cdots 1$, and by definition $0! = 1$. This expression is a generalization of the formula $W = \frac{1}{2}N(N-1)$, and reduces to it for the configuration $\{N-2, 2, 0, \ldots\}$.

Justification 19.2

Consider the number of ways of distributing N balls into bins. The first ball can be selected in N different ways, the next ball in $N-1$ different ways for the balls remaining, and so on. Therefore, there are $N(N-1)\cdots 1 = N!$ ways of selecting the balls for distribution over the bins. However, if there are n_0 balls in the bin labelled ε_0, there would be $n_0!$ different ways in which the same balls could have been chosen (Fig. 19.2). Similarly, there are $n_1!$ ways in which the n_1 balls in the bin labelled ε_1 can be chosen, and so on. Therefore, the total number of distinguishable ways of distributing the balls so that there are n_0 in bin ε_0, n_1 in bin ε_1, etc. regardless of the order in which the balls were chosen is $N!/n_0!n_1!\cdots$, which is the content of eqn 1.

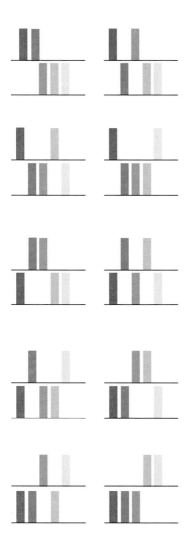

19.1 Whereas a configuration $\{5, 0, 0, \cdots\}$ can be achieved in only one way, a configuration $\{3, 2, 0, \cdots\}$ can be achieved in the ten different ways shown here, where the tinted blocks represent different molecules.

19.2 The 18 molecules shown here can be distributed into four receptacles (distinguished by the three vertical lines) in 18! different ways. However, 3! of the selections that put three molecules in the first receptacle are equivalent, 6! that put six molecules into the second receptacle are equivalent, and so on. Hence the number of distinguishable arrangements is 18!/3!6!5!4!.

$N = 18$

3!   6!   5!   4!

2   At this stage in the argument, we are ignoring the requirement that the total energy of the system should be constant (the second configuration has a higher energy than the first). The constraint of total energy is imposed later in this section.

Illustration

To calculate the number of ways of distributing 20 identical objects with the arrangement $1, 0, 3, 5, 10, 1$, we note that the configuration is $\{1, 0, 3, 5, 10, 1\}$ with $N = 20$; therefore the weight is

$$W = \frac{20!}{1!0!3!5!10!1!} = 9.31 \times 10^8$$

Self-test 19.1 Calculate the weight of the configuration in which 20 objects are distributed in the arrangement $0, 1, 5, 0, 8, 0, 3, 2, 0, 1$.

$$[4.19 \times 10^{10}]$$

It will turn out to be more convenient to deal with the natural logarithm of the weight, $\ln W$, rather than with the weight itself. We shall therefore need the expression

$$\ln W = \ln\left(\frac{N!}{n_0! n_1! n_2! \cdots}\right) = \ln N! - \ln(n_0! n_1! n_2! \cdots)$$
$$= \ln N! - (\ln n_0! + \ln n_1! + \ln n_2! + \cdots)$$
$$= \ln N! - \sum_i \ln n_i!$$

where in the first line we have used $\ln(x/y) = \ln x - \ln y$ and in the second $\ln xy = \ln x + \ln y$. One reason for introducing $\ln W$ is that it is easier to make approximations. In particular, we can simplify the factorials by using **Stirling's approximation** in the form[3]

$$\ln x! \approx x \ln x - x \tag{2}$$

Then the approximate expression for the weight is

$$\ln W = (N \ln N - N) - \sum_i (n_i \ln n_i - n_i)$$
$$= N \ln N - \sum_i n_i \ln n_i \tag{3}$$

The second line is derived by noting that the sum of n_i is equal to N, so the second and fourth terms on the right in the first line of eqn 3 cancel.

(b) The dominating configuration

We have seen that the configuration $\{N - 2, 2, 0, \ldots\}$ dominates $\{N, 0, 0, \ldots\}$, and it should be easy to believe that there may be other configurations that greatly dominate both. We shall see, in fact, that there is a configuration with so great a weight that it overwhelms all the rest in importance to such an extent that the system will almost always be found in it. The properties of the system will therefore be characteristic of that particular dominating configuration. This dominating configuration can be found by looking for the values of n_i that lead to a maximum value of W. Because W is a function of all the n_i, we can do this search by varying the n_i and looking for the values that correspond to $dW = 0$ (just as in the

---

3   The precise form of Stirling's approximation is

$$x! \approx (2\pi)^{1/2} x^{x+\frac{1}{2}} e^{-x}$$

and it is reliable when x is greater than about 10. We deal with far larger values of x, and the simplified version in eqn 2 is adequate.

search for the maximum of any function), or equivalently a maximum value of $\ln W$. However, there are two difficulties with this procedure.

The first difficulty is that the only permitted configurations are those corresponding to the specified, constant, total energy of the system. This requirement rules out many configurations; $\{N, 0, 0, \ldots\}$ and $\{N - 2, 2, 0, \ldots\}$, for instance, have different energies, so both cannot occur in the same isolated system. It follows that, in looking for the configuration with the greatest weight, we must ensure that the configuration also satisfies the condition

$$\text{Constant total energy:} \quad \sum_i n_i \varepsilon_i = E \tag{4}$$

where E is the total energy of the system.

The second constraint is that, because the total number of molecules present is also fixed (at N), we cannot arbitrarily vary all the populations simultaneously. Thus, increasing the population of one state by 1 demands that the population of another state must be reduced by 1. Therefore, the search for the maximum value of W is also subject to the condition

$$\text{Constant total number of molecules:} \quad \sum_i n_i = N \tag{5}$$

(c) The Boltzmann distribution

We are looking for the set of numbers $n_0, n_1, \ldots$ for which W has its maximum value. We show in the following *Justification* that the populations in the configuration of greatest weight depend on the energy of the state according to the **Boltzmann distribution**:

$$\frac{n_i}{N} = \frac{e^{-\beta \varepsilon_i}}{\sum_j e^{-\beta \varepsilon_j}} \qquad \beta = \frac{1}{kT} \tag{6}$$

where T is the thermodynamic temperature and k is the Boltzmann constant.

Justification 19.3

We have already remarked that it turns out to be simpler to find the condition for $\ln W$ being a maximum rather than dealing directly with W. Because $\ln W$ depends on all the n_i, when a configuration changes and the n_i change to $n_i + dn_i$, the function $\ln W$ changes to $\ln W + d \ln W$, where

$$d \ln W = \sum_i \left(\frac{\partial \ln W}{\partial n_i} \right) dn_i$$

At a maximum, $d \ln W = 0$. However, when the n_i change, they do so subject to the two constraints

$$\sum_i \varepsilon_i dn_i = 0 \qquad \sum_i dn_i = 0 \tag{7}$$

The first constraint recognizes that the total energy must not change, and the second recognizes that the total number of molecules must not change. These two constraints prevent us from solving $d \ln W = 0$ simply by setting all $(\partial \ln W / \partial n_i) = 0$ because the dn_i are not all independent.

The way to take constraints into account was devised by the French mathematician Lagrange, and is called the **method of undetermined multipliers**. The technique is described in *Further information 3*. All we need here is the rule that *a constraint should be multiplied by a constant and then added to the main variation equation*. The variables are then treated as though they were all independent, and the constants are evaluated at the end of the calculation.

We employ the technique as follows. The two constraints are multiplied by the constants $-\beta$ and α, respectively (the minus sign in $-\beta$ has been included for future convenience), and then added to the expression for $\mathrm{d}\ln W$:

$$\mathrm{d}\ln W = \sum_i \left(\frac{\partial \ln W}{\partial n_i}\right)\mathrm{d}n_i + \alpha \sum_i \mathrm{d}n_i - \beta \sum_i \varepsilon_i \mathrm{d}n_i$$

$$= \sum_i \left\{ \left(\frac{\partial \ln W}{\partial n_i}\right) + \alpha - \beta\varepsilon_i \right\}\mathrm{d}n_i$$

All the $\mathrm{d}n_i$ are now treated as independent. Hence the only way of satisfying $\mathrm{d}\ln W = 0$ is to require that, for each i,

$$\left(\frac{\partial \ln W}{\partial n_i}\right) + \alpha - \beta\varepsilon_i = 0 \tag{8}$$

when the n_i have their most probable values.

The expression for $\ln W$ is given in eqn 3. Differentiation of it with respect to n_i gives

$$\left(\frac{\partial \ln W}{\partial n_i}\right) = \frac{\partial(N \ln N)}{\partial n_i} - \sum_j \frac{\partial(n_j \ln n_j)}{\partial n_i}$$

The derivative of the first term is obtained as follows:

$$\frac{\partial(N \ln N)}{\partial n_i} = \left(\frac{\partial N}{\partial n_i}\right)\ln N + \frac{\partial N}{\partial n_i} = \ln N + 1$$

because $N = n_1 + n_2 + \cdots$ and its derivative with respect to any of the ns is 1. The derivative of the second term is[4]

$$\sum_j \frac{\partial(n_j \ln n_j)}{\partial n_i} = \sum_j \left\{ \left(\frac{\partial n_j}{\partial n_i}\right)\ln n_j + n_j \left(\frac{\partial \ln n_j}{\partial n_i}\right)\right\}$$

$$= \sum_j \left(\frac{\partial n_j}{\partial n_i}\right)(\ln n_j + 1) = \ln n_i + 1$$

and therefore

$$\frac{\partial \ln W}{\partial n_i} = -(\ln n_i + 1) + (\ln N + 1) = -\ln\left(\frac{n_i}{N}\right)$$

It follows from eqn 8 that

$$-\ln\left(\frac{n_i}{N}\right) + \alpha - \beta\varepsilon_i = 0$$

and therefore that

$$\frac{n_i}{N} = e^{\alpha - \beta\varepsilon_i}$$

---

4   We use

$$\frac{\partial \ln n_j}{\partial n_i} = \frac{1}{n_j}\left(\frac{\partial n_j}{\partial n_i}\right)$$

Then, if $i \neq j$, n_j is independent of n_i, so $\partial n_j/\partial n_i = 0$. However, if $i = j$,

$$\frac{\partial n_j}{\partial n_i} = \frac{\partial n_j}{\partial n_j} = 1$$

At this stage we note that

$$N = \sum_j n_j = Ne^\alpha \sum_j e^{-\beta\varepsilon_j}$$

(We are free to label the states with j instead of i.) Because the N cancels on each side of this equality, it follows that

$$e^\alpha = \frac{1}{\sum_j e^{-\beta\varepsilon_j}} \tag{9}$$

and the Boltzmann distribution in eqn 6 follows immediately. We justify the relation $\beta = 1/kT$ shortly (Section 19.3b).

19.2 The molecular partition function

From now on we write the Boltzmann distribution as

$$p_i = \frac{e^{-\beta\varepsilon_i}}{q} \tag{10}$$

where p_i is the fraction of molecules in the state i, $p_i = n_i/N$, and q is the **molecular partition function**:

$$q = \sum_j e^{-\beta\varepsilon_j} \tag{[11]}$$

The sum in q is sometimes expressed slightly differently. It may happen that several states have the same energy, and so give the same contribution to the sum. If, for example, g_j states have the same energy ε_j (so the level is g_j-fold degenerate), we could write

$$q = \sum_{\text{levels } j} g_j e^{-\beta\varepsilon_j} \tag{12}$$

where the sum is now over energy levels (sets of states with the same energy), not individual states.

Example 19.1 Writing a partition function

Write an expression for the partition function of a linear molecule (such as HCl) treated as a rigid rotor.

Method To use eqn 12 we need to know (a) the energies of the levels, (b) the degeneracies, the number of states that belong to each level. Whenever calculating a partition function, the energies of the levels are expressed relative to 0 for the state of lowest energy. The energy levels of a rigid linear rotor were derived in Section 16.5c.

Answer From eqn 16.37, the energy levels of a linear rotor are $hcBJ(J+1)$, with $J = 0, 1, 2, \ldots$. The state of lowest energy has zero energy, so no adjustment need be made to the energies given by this expression. Each level consists of $2J + 1$ degenerate states. Therefore,

$$q = \sum_{J=0}^{\infty} (2J + 1)e^{-\beta hcBJ(J+1)}$$

Comment The sum can be evaluated numerically by supplying the value of B (from spectroscopy or calculation) and the temperature. For reasons explained in Section 20.2b, this expression applies only to unsymmetrical linear rotors (for example, HCl, not CO_2; in general, to $C_{\infty v}$ and not $D_{\infty h}$ species).

Self-test 19.2 Write the partition function for a two-level system, the lower state (at energy 0) being non-degenerate, and the upper state (at an energy ε) doubly degenerate.

$$[q = 1 + 2e^{-\beta\varepsilon}]$$

(a) An interpretation of the partition function

Some insight into the significance of a partition function can be obtained by considering how q depends on the temperature. When T is close to zero, the parameter $\beta = 1/kT$ is close to infinity. Then every term except one in the sum defining q is zero because each one has the form e^{-x} with $x \to \infty$. The exception is the term with $\varepsilon_0 \equiv 0$ (or the g_0 terms at zero energy if the ground state is g_0-fold degenerate), because then $\varepsilon_0/kT \equiv 0$ whatever the temperature, including zero. As there is only one surviving term when $T = 0$, and its value is g_0, it follows that

$$\lim_{T \to 0} q = g_0 \tag{13}$$

That is, at $T = 0$, the partition function is equal to the degeneracy of the ground state.

Now consider the case when T is so high that for each term in the sum $\varepsilon_j/kT \approx 0$. Because $e^{-x} = 1$ when $x = 0$, each term in the sum now contributes 1. It follows that the sum is equal to the number of molecular states, which in general is infinite:

$$\lim_{T \to \infty} q \to \infty \tag{14}$$

In some idealized cases, the molecule may have only a finite number of states; then the upper limit of q is equal to the number of states. For example, if we were considering only the spin energy levels of a radical in a magnetic field, then there would be only two states ($m_s = \pm\frac{1}{2}$). The partition function for such a system can therefore be expected to rise towards 2 as T is increased towards infinity.

We see that *the molecular partition function gives an indication of the average number of states that are thermally accessible to a molecule at the temperature of the system*. At $T = 0$, only the ground level is accessible and $q = g_0$. At very high temperatures, virtually all states are accessible, and q is correspondingly large.

Example 19.2 Evaluating the partition function for a uniform ladder of energy levels

Evaluate the partition function for a molecule with an infinite number of equally spaced non-degenerate energy levels (Fig. 19.3). These levels can be thought of as the vibrational energy levels of a diatomic molecule in the harmonic approximation.

Method We expect the partition function to increase from 1 at $T = 0$ and approach infinity as $T \to \infty$. To evaluate eqn 11 explicitly, note that[5]

$$1 + x + x^2 + \cdots = \frac{1}{1-x}$$

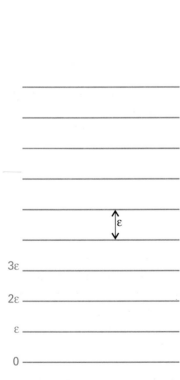

3ε

2ε

ε

0

19.3 The equally spaced infinite array of energy levels used in the calculation of the partition function. A harmonic oscillator has the same spectrum of levels.

---

5   The sum of the infinite series $S = 1 + x + x^2 + \cdots$ is obtained by multiplying both sides by x, which gives $xS = x + x^2 + x^3 + \cdots = S - 1$. This relation reorganizes into

$$S = \frac{1}{1-x}$$

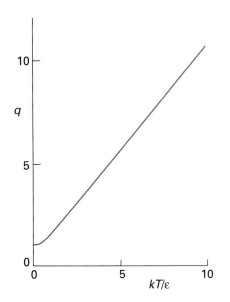

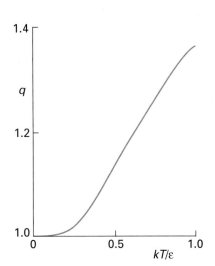

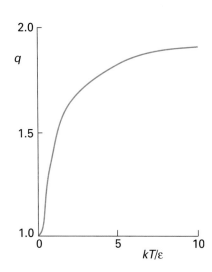

19.4 The partition function for the system shown in Fig. 19.3 (a harmonic oscillator) as a function of temperature.

19.5 The partition function for a two-level system as a function of temperature. The two graphs differ in the scale of the temperature axis to show the approach to 1 as $T \to 0$ and the slow approach to 2 as $T \to \infty$.

Answer If the separation of neighbouring levels is ε, the partition function is

$$q = 1 + e^{-\beta\varepsilon} + e^{-2\beta\varepsilon} + \cdots = 1 + e^{-\beta\varepsilon} + (e^{-\beta\varepsilon})^2 + \cdots$$
$$= \frac{1}{1 - e^{-\beta\varepsilon}}$$

This expression is plotted in Fig. 19.4: notice that it rises from 1 to infinity as the temperature is raised, as anticipated.

- -

Self-test 19.3 Find and plot an expression for the partition function of a system with one state at zero energy and another state at the energy ε.

$$[q = 1 + e^{-\beta\varepsilon}, \text{ Fig. 19.5}]$$

It follows from eqn 10 and the expression for q derived in Example 19.2 for a uniform ladder of states of spacing ε,

$$q = \frac{1}{1 - e^{-\beta\varepsilon}} \qquad (15)$$

that the fraction of molecules in the state with energy ε_i is

$$p_i = (1 - e^{-\beta\varepsilon})e^{-\beta\varepsilon_i} \qquad (16)$$

The variation of p_i with temperature is illustrated in Fig. 19.6. We see that at very low temperatures, where q is close to 1, only the lowest state is significantly populated. As the temperature is raised, the population breaks out of the lowest state, and the upper states become progressively more highly populated. At the same time the partition function rises from 1, and its value gives an indication of the range of states populated. The name 'partition function' reflects the sense in which q measures how the total number of molecules is distributed—partitioned—over the available states.

The corresponding expressions for a two-level system derived in Self-test 19.3 are

$$p_0 = \frac{1}{1 + e^{-\beta\varepsilon}} \qquad p_1 = \frac{e^{-\beta\varepsilon}}{1 + e^{-\beta\varepsilon}} \qquad (17)$$

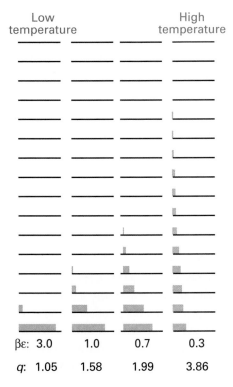

Low temperature		High temperature	
$\beta\varepsilon$: 3.0	1.0	0.7	0.3
q: 1.05	1.58	1.99	3.86

19.6 The populations of the energy levels of the system shown in Fig. 19.3 at different temperatures, and the corresponding values of the partition function calculated in Self-test 19.3. Note that $\beta = 1/kT$.

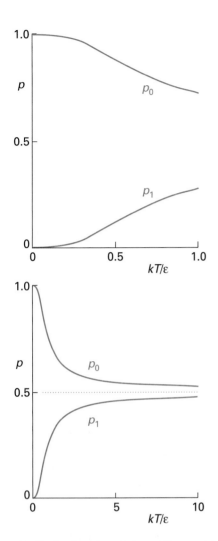

19.7 The fraction of populations of the two states of a two-level system as a function of temperature (eqn 17). Note that, as the temperature approaches infinity, the populations of the two states become equal (and the fractions both approach 0.5).

These functions are plotted in Fig. 19.7. Notice how the populations tend towards equality ($p_0 = 0.5, p_1 = 0.5$) as $T \to \infty$. A common error is to suppose that all the molecules in the system will be found in the upper energy state when $T = \infty$; however, we see from eqn 17 that, as $T \to \infty$, the populations of states become equal. The same conclusion is true of multi-level systems too: as $T \to \infty$, *all states become equally populated*.

Example 19.3 Using the partition function to calculate a population

Calculate the proportion of I_2 molecules in their ground, first excited, and second excited vibrational states at 25 °C. The vibrational wavenumber is 214.6 cm^{-1}.

Method Vibrational energy levels have a constant separation (in the harmonic approximation, Section 16.9), so the partition function is given by eqn 15 and the populations by eqn 16. To use the latter equation, we identify the index i with the quantum number v, and calculate p_v for $v = 0, 1$, and 2. At 298.15 K, $kT/hc = 207.226$ cm^{-1}.

Answer First, we note that

$$\beta\varepsilon = \frac{hc\tilde{\nu}}{kT} = \frac{214.6\ \text{cm}^{-1}}{207.226\ \text{cm}^{-1}} = 1.036$$

Then it follows from eqn 16 that the populations are

$$p_v = (1 - e^{-\beta\varepsilon})e^{-v\beta\varepsilon} = 0.645\, e^{-1.036v}$$

Therefore, $p_0 = 0.645$, $p_1 = 0.229$, $p_2 = 0.081$.

Comment The I–I bond is not stiff and the atoms are heavy: as a result, the vibrational energy separations are small and at room temperature several vibrational levels are significantly populated. The value of the partition function, $q = 1.55$, reflects this small but significant spread of populations.

- -

Self-test 19.4 At what temperature would the $v = 1$ level of I_2 have (a) half the population of the ground state, (b) the same population as the ground state?

[(a) 445 K, (b) infinite]

(b) Approximations and factorizations

In general, exact analytical expressions for partition functions cannot be obtained. However, closed approximate expressions can often be found and prove to be very important in a number of chemical applications. For instance, the expression for the partition function for a particle of mass m free to move in a one-dimensional container of length X can be evaluated by making use of the fact that the separation of energy levels is very small and that large numbers of states are accessible at normal temperatures. As shown in the *Justification* below, in this case

$$q_X = \left(\frac{2\pi m}{h^2\beta}\right)^{1/2} X \tag{18}$$

Justification 19.4

The energy levels of a molecule of mass m in a container of length X are given by eqn 12.7 with $L = X$:

$$E_n = \frac{n^2 h^2}{8mX^2} \qquad n = 1, 2, \ldots$$

The lowest level ($n = 1$) has energy $h^2/8mX^2$, so the energies relative to that level are

$$\varepsilon_n = (n^2 - 1)\varepsilon \qquad \varepsilon = \frac{h^2}{8mX^2}$$

The sum to evaluate is therefore

$$q_X = \sum_{n=1}^{\infty} e^{-(n^2-1)\beta\varepsilon}$$

The translational energy levels are very close together in a container the size of a typical laboratory vessel; therefore, the sum can be approximated by an integral:

$$q_X = \int_1^{\infty} e^{-(n^2-1)\beta\varepsilon} \, dn$$

The extension of the lower limit to $n = 0$ and the replacement of $n^2 - 1$ by n^2 introduces negligible error but turns the integral into standard form. We make the substitution $x^2 = n^2\beta\varepsilon$, implying $dn = dx/(\beta\varepsilon)^{\frac{1}{2}}$, and therefore that

$$q_X = \left(\frac{1}{\beta\varepsilon}\right)^{1/2} \int_0^{\infty} e^{-x^2} \, dx = \left(\frac{1}{\beta\varepsilon}\right)^{1/2} \left(\frac{\pi^{1/2}}{2}\right) = \left(\frac{2\pi m}{h^2\beta}\right)^{1/2} X$$

Another useful feature of partition functions is used to derive expressions when the energy of a molecule arises from several different, independent sources: *if the energy is a sum of independent contributions, the partition function is a product of partition functions for each mode of motion*. For instance, suppose the molecule we are considering is free to move in three dimensions. We take the length of the container in the y-direction to be Y and that in the z-direction to be Z. The total energy of a molecule ε is the sum of its translational energies in all three directions:

$$\varepsilon_{n_1 n_2 n_3} = \varepsilon_{n_1}^{(X)} + \varepsilon_{n_2}^{(Y)} + \varepsilon_{n_3}^{(Z)} \tag{19}$$

where n_1, n_2, and n_3 are the quantum numbers for motion in the x-, y-, and z-directions, respectively. Therefore, because $e^{a+b+c} = e^a e^b e^c$, the partition function factorizes as follows:

$$\begin{aligned}
q &= \sum_{\text{all } n} e^{-\beta\varepsilon_{n_1}^{(X)} - \beta\varepsilon_{n_2}^{(Y)} - \beta\varepsilon_{n_3}^{(Z)}} = \sum_{\text{all } n} e^{-\beta\varepsilon_{n_1}^{(X)}} e^{-\beta\varepsilon_{n_2}^{(Y)}} e^{-\beta\varepsilon_{n_3}^{(Z)}} \\
&= \left(\sum_{n_1} e^{-\beta\varepsilon_{n_1}^{(X)}}\right) \left(\sum_{n_2} e^{-\beta\varepsilon_{n_2}^{(Y)}}\right) \left(\sum_{n_3} e^{-\beta\varepsilon_{n_3}^{(Z)}}\right) = q_X q_Y q_Z
\end{aligned} \tag{20}$$

It is generally true that, *if the energy of a molecule can be written as the sum of independent terms, the partition function is the corresponding product of individual contributions*.

Equation 18 gives the partition function for translational motion in the x-direction. The only change for the other two directions is to replace the length X by the lengths Y or Z. Hence the partition function for motion in three dimensions is

$$q = \left(\frac{2\pi m}{h^2\beta}\right)^{3/2} XYZ \tag{21}$$

The product of lengths XYZ is the volume, V, of the container, so we can write

$$q = \frac{V}{\Lambda^3} \qquad \Lambda = h\left(\frac{\beta}{2\pi m}\right)^{1/2} = \frac{h}{(2\pi m k T)^{1/2}} \tag{22}$$

The quantity Λ has the dimensions of length and is called the **thermal wavelength** of the molecule.

Illustration
....

To calculate the translational partition function of an H_2 molecule confined to a 100 cm^3 vessel at 25°C we use $m = 2.016$ u; then

$$\Lambda = \frac{6.626 \times 10^{-34} \text{ J}}{\{2\pi \times (2.016 \times 1.6605 \times 10^{-27} \text{ kg}) \times (1.38 \times 10^{-23} \text{ J K}^{-1}) \times (298 \text{ K})\}^{1/2}}$$
$$= 7.12 \times 10^{-11} \text{ m}$$

Therefore,

$$q = \frac{1.00 \times 10^{-4} \text{ m}^3}{(7.12 \times 10^{-11} \text{ m})^3} = 2.77 \times 10^{26}$$

About 10^{26} quantum states are thermally accessible, even at room temperature and for this light molecule. Many states are occupied if the thermal wavelength (which in this case is 71.2 pm) is small compared with the linear dimensions of the container.
....

Self-test 19.5 Calculate the translational partition function for a D_2 molecule under the same conditions.

[$q = 7.8 \times 10^{26}$, $2^{3/2}$ times larger]

The internal energy and the entropy

The importance of the molecular partition function is that it contains all the information needed to calculate the thermodynamic properties of a system of independent particles. In this respect, q plays a role in statistical thermodynamics very similar to that played by the wavefunction in quantum mechanics: q is a kind of thermal wavefunction.

19.3 The internal energy

We shall begin to unfold the importance of q by showing how to derive an expression for the internal energy of the system.

(a) The relation between U and q

The total energy of the system is

$$E = \sum_i n_i \varepsilon_i \tag{23}$$

Because the most probable configuration is so strongly dominating, we can use the Boltzmann distribution for the populations and write

$$E = \frac{N}{q} \sum_i \varepsilon_i e^{-\beta \varepsilon_i} \tag{24}$$

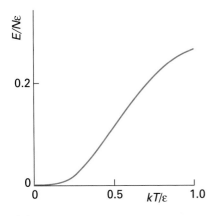

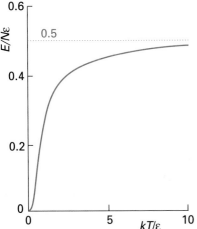

19.8 The total energy of a two-level system (expressed as a multiple of $N\varepsilon$) as a function of temperature, on two temperature scales. The graph at the top shows the slow rise away from zero energy at low temperatures; the slope of the graph at $T = 0$ is 0 (that is, the heat capacity is zero at $T = 0$). The graph below shows the slow rise to 0.5 as $T \to \infty$ as both states become equally populated (see Fig. 19.7).

To manipulate this expression into a form involving only q we note that

$$\frac{\mathrm{d}}{\mathrm{d}\beta}\mathrm{e}^{-\beta\varepsilon_i} = -\varepsilon_i\mathrm{e}^{-\beta\varepsilon_i}$$

It follows that

$$E = -\frac{N}{q}\sum_i\frac{\mathrm{d}}{\mathrm{d}\beta}\mathrm{e}^{-\beta\varepsilon_i} = -\frac{N}{q}\frac{\mathrm{d}}{\mathrm{d}\beta}\sum_i\mathrm{e}^{-\beta\varepsilon_i} = -\frac{N}{q}\frac{\mathrm{d}q}{\mathrm{d}\beta} \tag{25}$$

Illustration

From the two-level partition function $q = 1 + \mathrm{e}^{-\beta\varepsilon}$, we can deduce that the total energy of N two-level systems is

$$E = -\left(\frac{N}{1+\mathrm{e}^{-\beta\varepsilon}}\right)\frac{\mathrm{d}}{\mathrm{d}\beta}(1+\mathrm{e}^{-\beta\varepsilon}) = \frac{N\varepsilon\mathrm{e}^{-\beta\varepsilon}}{1+\mathrm{e}^{-\beta\varepsilon}}$$

$$= \frac{N\varepsilon}{1+\mathrm{e}^{\beta\varepsilon}}$$

This function is plotted in Fig. 19.8. Notice how the energy is zero at $T = 0$, when only the lower state (at the zero of energy) is occupied, and rises to $\frac{1}{2}N\varepsilon$ as $T \to \infty$, when the two levels become equally populated.

There are several points in relation to eqn 25 that need to be made. Because $\varepsilon_0 = 0$, (remember that we measure all energies from the lowest available level), E should be interpreted as the value of the internal energy relative to its value at $T = 0$, $U(0)$. Therefore, to obtain the conventional internal energy U, we must add the internal energy at $T = 0$:

$$U = U(0) + E \tag{26}$$

Secondly, because the partition function may depend on variables other than the temperature (for example, the volume), the derivative with respect to β in eqn 25 is actually a *partial* derivative with these other variables held constant. The complete expression relating the molecular partition function to the thermodynamic internal energy of a system of independent molecules is therefore

$$U = U(0) - \frac{N}{q}\left(\frac{\partial q}{\partial\beta}\right)_V \tag{27a}$$

An equivalent form is obtained by noting that $\mathrm{d}x/x = \mathrm{d}\ln x$:

$$U = U(0) - N\left(\frac{\partial\ln q}{\partial\beta}\right)_V \tag{27b}$$

These two equations confirm that we need know only the partition function (as a function of temperature) to calculate the internal energy relative to its value at $T = 0$.

(b) The value of β

We now confirm that the parameter β, which we have anticipated is equal to $1/kT$, does indeed have that value. To do so, we shall compare the equipartition expression for the internal energy of a monatomic perfect gas, which from *Molecular interpretation 2.2* we know to be

$$U = U(0) + \tfrac{3}{2}nRT \tag{28a}$$

with the value calculated from the translational partition function (see the following *Justification*), which is

$$U = U(0) + \frac{3N}{2\beta}$$ (28b)

It follows by comparing these two expressions that

$$\beta = \frac{N}{nRT} = \frac{nN_A}{nN_A kT} = \frac{1}{kT}$$ (29)

as was to be proved. (We have used $N = nN_A$, where n is the amount of gas molecules, N_A is the Avogadro constant, and $R = N_A k$.)

Justification 19.5

To use eqn 27, we introduce the translational partition function from eqn 22:

$$\left(\frac{\partial q}{\partial \beta}\right)_V = \left(\frac{\partial}{\partial \beta}\frac{V}{\Lambda^3}\right)_V = V\frac{d}{d\beta}\frac{1}{\Lambda^3} = -3\frac{V}{\Lambda^4}\frac{d\Lambda}{d\beta}$$

Then we note from the formula for Λ in eqn 22 that

$$\frac{d\Lambda}{d\beta} = \frac{d}{d\beta}\left\{\frac{h\beta^{1/2}}{(2\pi m)^{1/2}}\right\} = \frac{1}{2\beta^{1/2}} \times \frac{h}{(2\pi m)^{1/2}} = \frac{\Lambda}{2\beta}$$

and so obtain

$$\left(\frac{\partial q}{\partial \beta}\right)_V = -\frac{3V}{2\beta\Lambda^3}$$

Then, by eqn 27a,

$$U = U(0) - N\left(\frac{\Lambda^3}{V}\right)\left(-\frac{3V}{2\beta\Lambda^3}\right) = U(0) + \frac{3N}{2\beta}$$

19.4 The statistical entropy

If it is true that the partition function contains all thermodynamic information, then it must be possible to use it to calculate the entropy as well as the internal energy. Because we know (from Section 4.2) that entropy is related to the dispersal of energy and that the partition function is a measure of the number of states that are thermally accessible, we can be confident that the two are indeed related.

We shall develop the relation between the entropy and the partition function in two stages. In the first stage, we justify one of the most celebrated equations in statistical thermodynamics, the **Boltzmann formula** for the entropy:

$$S = k \ln W$$ [30]

In this expression, W is the weight of the most probable configuration of the system. In the second stage, we express W in terms of the partition function.

Justification 19.6

A change in the internal energy

$$U = U(0) + \sum_i n_i \varepsilon_i$$ (31)

may arise from either a modification of the energy levels of a system (when ε_i changes to $\varepsilon_i + \mathrm{d}\varepsilon_i$) or from a modification of the populations (when n_i changes to $n_i + \mathrm{d}n_i$). The most general change is therefore

$$\mathrm{d}U = \mathrm{d}U(0) + \sum_i n_i \mathrm{d}\varepsilon_i + \sum_i \varepsilon_i \mathrm{d}n_i \tag{32}$$

Because the energy levels do not change when a system is heated at constant volume (Fig. 19.9), in the absence of all changes other than heating

$$\mathrm{d}U = \sum_i \varepsilon_i \mathrm{d}n_i$$

We know from thermodynamics (and specifically from eqn 5.2) that under the same conditions

$$\mathrm{d}U = \mathrm{d}q_{\text{rev}} = T\mathrm{d}S$$

Therefore,

$$\mathrm{d}S = \frac{\mathrm{d}U}{T} = k\beta \sum_i \varepsilon_i \mathrm{d}n_i \tag{33}$$

We also know that for changes in the most probable configuration (the only one we need consider)

$$\left(\frac{\partial \ln W}{\partial n_i}\right) + \alpha - \beta\varepsilon_i = 0$$

(this is eqn 8). After rearranging this expression to

$$\beta\varepsilon_i = \left(\frac{\partial \ln W}{\partial n_i}\right) + \alpha$$

we find that

$$\mathrm{d}S = k \sum_i \left(\frac{\partial \ln W}{\partial n_i}\right)\mathrm{d}n_i + k\alpha \sum_i \mathrm{d}n_i$$

But the sum over the $\mathrm{d}n_i$ is zero, because the number of molecules is constant. Hence

$$\mathrm{d}S = k \sum_i \left(\frac{\partial \ln W}{\partial n_i}\right)\mathrm{d}n_i = k(\mathrm{d}\ln W) \tag{34}$$

This relation strongly suggests the definition $S = k \ln W$, as in eqn 30.

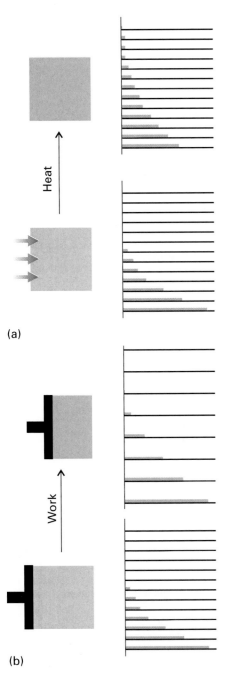

(a)

(b)

19.9 (a) When a system is heated, the energy levels are unchanged but their populations are changed. (b) When work is done on a system, the energy levels themselves are changed. The levels in this case are the one-dimensional particle-in-a-box energy levels of Chapter 12: they depend on the size of the container and move apart as its length is decreased. For simplicity in making the essential point, we have shown the energy levels as equally spaced; in fact, their separation increases with energy.

The statistical entropy behaves in exactly the same way as the thermodynamic entropy. Thus, as the temperature is lowered, the value of W, and hence of S, decreases because fewer configurations are compatible with the total energy. In the limit $T \to 0$, $W = 1$, so $\ln W = 0$, because only one configuration (every molecule in the lowest level) is compatible with $E = 0$. It follows that $S \to 0$ as $T \to 0$, which is compatible with the Third Law of thermodynamics, that the entropies of all perfect crystals approach the same value as $T \to 0$ (Section 4.4a).

Now we relate the Boltzmann formula for the entropy to the partition function. To do so, we substitute the expression for $\ln W$ given in eqn 3 into eqn 30 and, as shown in the *Justification* below, obtain

$$S = \frac{U - U(0)}{T} + Nk \ln q \tag{35}$$

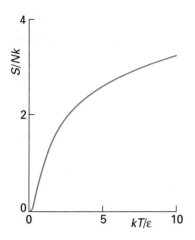

19.10 The temperature variation of the entropy of the system shown in Fig. 19.3 (expressed here as a multiple of Nk). The entropy approaches zero as $T \to 0$, and increases without limit as $T \to \infty$.

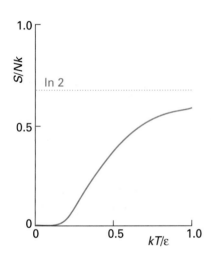

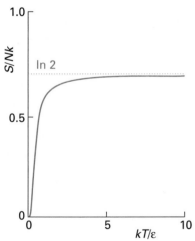

19.11 The temperature variation of the entropy of a two-level system (expressed as a multiple of Nk). As $T \to \infty$, the two states become equally populated and S approaches $Nk \ln 2$.

Justification 19.7

The first stage is to write

$$S = k \sum_i (n_i \ln N - n_i \ln n_i) = -k \sum_i n_i \ln\left(\frac{n_i}{N}\right) = -Nk \sum_i p_i \ln p_i \qquad (36)$$

where $p_i = n_i/N$, the fraction of molecules in state i. It follows from eqn 10 that

$$\ln p_i = -\beta\varepsilon_i - \ln q$$

and therefore that

$$S = -Nk\left(-\beta \sum_i p_i \varepsilon_i - \sum_i p_i \ln q\right) = k\beta\{U - U(0)\} + Nk \ln q$$

We have used the fact that the sum over the p_i is equal to 1 and the sum over $Np_i\varepsilon_i$ is equal to $U - U(0)$ (see eqn 31). We have already established that $\beta = 1/kT$, so eqn 35 immediately follows.

Example 19.4 Calculating the entropy of a collection of oscillators

Calculate the entropy of a collection of N independent harmonic oscillators, and evaluate it using vibrational data for I_2 at 25°C (Example 19.3).

Method To use eqn 31, we use the partition function for a molecule with evenly spaced vibrational energy levels, eqn 15. With the partition function available, the internal energy can be found by differentiation (as in eqn 27a), and the two expressions then combined to give S.

Answer The molecular partition function as given in eqn 15 is

$$q = \frac{1}{1 - e^{-\beta\varepsilon}}$$

The internal energy is obtained by using eqn 27a:

$$U - U(0) = -\frac{N}{q}\left(\frac{\partial q}{\partial \beta}\right)_V = \frac{N\varepsilon e^{-\beta\varepsilon}}{1 - e^{-\beta\varepsilon}} = \frac{N\varepsilon}{e^{\beta\varepsilon} - 1}$$

The entropy is therefore

$$S = Nk\left\{\frac{\beta\varepsilon}{e^{\beta\varepsilon} - 1} - \ln(1 - e^{-\beta\varepsilon})\right\}$$

This function is plotted in Fig. 19.10. For I_2 at 25°C, $\beta\varepsilon = 1.036$ (Example 19.3), so $S_m = 8.38 \text{ J K}^{-1}\text{ mol}^{-1}$.

Self-test 19.6 Evaluate the molar entropy of N two-level systems and plot the resulting expression. What is the entropy when the two states are equally thermally accessible?
$$[S/Nk = \beta\varepsilon/(1 + e^{\beta\varepsilon}) + \ln(1 + e^{-\beta\varepsilon}); \text{ see Fig. 19.11}; S = Nk \ln 2]$$

The canonical partition function

In this section we see how to generalize our conclusions to include systems composed of interacting molecules. We shall also see how to obtain the molecular partition function from the more general form of the partition function developed here.

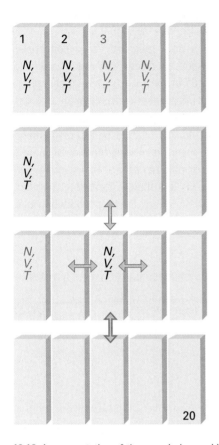

19.12 A representation of the canonical ensemble, in this case for $\tilde{N} = 20$. The individual replications of the actual system all have the same composition and volume. They are all in mutual thermal contact, and so all have the same temperature. Energy may be transferred between them as heat, and so they do not all have the same energy. The total energy ($\tilde{E}$) of all 20 replications is a constant because the ensemble is isolated overall.

19.5 The canonical ensemble

The crucial new concept we need when treating systems of interacting particles is the **ensemble**. Like so many scientific terms, the term has basically its normal meaning of 'collection', but it has been sharpened and refined into a precise significance.

(a) The concept of ensemble

To set up an ensemble, we take a closed system of specified volume, composition, and temperature, and think of it as replicated $\tilde{N}$ times (Fig. 19.12). All the identical closed systems are regarded as being in thermal contact with one another, so they can exchange energy. The total energy of all the systems is $\tilde{E}$ and, because they are in thermal equilibrium with one another, they all have the same temperature, T. This imaginary collection of replications of the actual system with a common temperature is called the **canonical ensemble**.[6]

The important point about an ensemble is that it is a collection of *imaginary* replications of the system, so we are free to let the number of members be as large as we like; when appropriate, we can let $\tilde{N}$ become infinite.[7] The number of members of the ensemble in a state with energy E_i is denoted $\tilde{n}_i$, and we can speak of the configuration of the ensemble (by analogy with the configuration of the system used in Section 19.1) and its weight, $\tilde{W}$.

(b) Dominating configurations

Just as in Section 19.1, some of the configurations of the ensemble will be very much more probable than others. For instance, it is very unlikely that the whole of the total energy, $\tilde{E}$, will accumulate in one system. By analogy with the earlier discussion, we can anticipate that there will be a dominating configuration, and that we can evaluate the thermodynamic properties by taking the average over the ensemble using that single, most probable, configuration. In the **thermodynamic limit** of $\tilde{N} \to \infty$, this dominating configuration is overwhelmingly the most probable, and it dominates the properties of the system virtually completely.

The quantitative discussion follows the argument in Section 19.1 with the modification that N and n_i are replaced by $\tilde{N}$ and $\tilde{n}_i$. The weight of a configuration $\{\tilde{n}_0, \tilde{n}_1, \ldots\}$ is

$$\tilde{W} = \frac{\tilde{N}!}{\tilde{n}_0! \tilde{n}_1! \cdots} \tag{37}$$

The configuration of greatest weight, subject to the constraints that the total energy of the ensemble is constant at $\tilde{E}$ and that the total number of members is fixed at $\tilde{N}$, is given by the **canonical distribution**:

$$\frac{\tilde{n}_i}{\tilde{N}} = \frac{e^{-\beta E_i}}{Q} \qquad Q = \sum_i e^{-\beta E_i} \tag{38}$$

The quantity Q, which is a function of the temperature, is called the **canonical partition function**.

---

6 The word 'canon' means 'according to a rule'. There are two other important ensembles. In the *microcanonical ensemble* the condition of constant temperature is replaced by the requirement that all the systems should have exactly the same energy: each system is individually isolated. In the *grand canonical ensemble* the volume and temperature of each system are the same, but they are open, which means that matter can be imagined as able to pass between the systems; the composition of each one may fluctuate, but now the chemical potential is the same in each system:

   Microcanonical ensemble: N, V, E common
   Canonical ensemble: N, V, T common
   Grand canonical ensemble: μ, V, T common

7 Note that $\tilde{N}$ is unrelated to N, the number of molecules in the actual system; $\tilde{N}$ is the number of imaginary replications of that system.

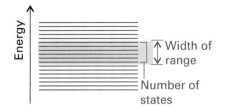

19.13 The energy density of states is the number of states in an energy range divided by the width of the range.

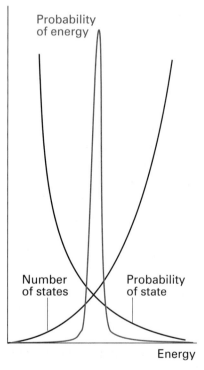

19.14 To construct the form of the distribution of members of the canonical ensemble in terms of their energies, we multiply the probability that any one is in a state of given energy, eqn 38, by the number of states corresponding to that energy (a steeply rising function). The product is a sharply peaked function at the mean energy, which shows that almost all the members of the ensemble have that energy.

(c) Fluctuations from the most probable distribution

The shape of the canonical distribution in eqn 38 is only *apparently* an exponentially decreasing function of the energy of the system. We must appreciate that eqn 38 gives the probability of occurrence of members in a single state i of the entire system of energy E_i. There may in fact be numerous states with almost identical energies. For example, in a gas the identities of the molecules moving slowly or quickly can change without necessarily affecting the total energy. The **density of states**, the number of states in an energy range divided by the width of the range (Fig. 19.13), is a very sharply increasing function of energy. It follows that the probability of a member of an ensemble having a specified energy (as distinct from being in a specified state) is given by eqn 38, a sharply decreasing function, multiplied by a sharply increasing function (Fig. 19.14). Therefore, the overall distribution is a sharply peaked function. We conclude that most members of the ensemble have an energy very close to the mean value.

19.6 The thermodynamic information in the partition function

Like the molecular partition function, the canonical partition function carries all the thermodynamic information about a system. However, Q is more general than q because it does not assume that the molecules are independent. We can therefore use Q to discuss the properties of condensed phases and real gases where molecular interactions are important.

(a) The internal energy

If the total energy of the ensemble is $\tilde{E}$, and there are $\tilde{N}$ members, the average energy of a member is $E = \tilde{E}/\tilde{N}$. We use this quantity to calculate the internal energy of the system in the limit of $\tilde{N}$ (and $\tilde{E}$) approaching infinity:

$$U = U(0) + E = U(0) + \frac{\tilde{E}}{\tilde{N}} \qquad \text{as } \tilde{N} \to \infty \tag{39}$$

The fraction, $\tilde{p}_i$, of members of the ensemble in a state i with energy E_i is given by the analogue of eqn 10 as

$$\tilde{p}_i = \frac{e^{-\beta E_i}}{Q} \tag{40}$$

It follows that the internal energy is given by

$$U = U(0) + \sum_i \tilde{p}_i E_i = U(0) + \frac{1}{Q} \sum_i E_i e^{-\beta E_i} \tag{41}$$

By the same argument that led to eqn 27,

$$U = U(0) - \frac{1}{Q} \left(\frac{\partial Q}{\partial \beta} \right)_V = U(0) - \left(\frac{\partial \ln Q}{\partial \beta} \right)_V \tag{42}$$

(b) The entropy

The total weight, $\tilde{W}$, of a configuration of the ensemble is the product of the average weight W of each member of the ensemble, $\tilde{W} = W^{\tilde{N}}$. Hence, we can calculate S from

$$S = k \ln W = k \ln \tilde{W}^{1/\tilde{N}} = \frac{k}{\tilde{N}} \ln \tilde{W} \tag{43}$$

It follows, by the same argument used in Section 19.4, that

$$S = \frac{U - U(0)}{T} + k \ln Q \tag{44}$$

19.7 Independent molecules

We shall now see how to recover the molecular partition function from the more general canonical partition function when the molecules are independent. When the molecules are independent and distinguishable (in the sense to be described), the relation between Q and q is

$$Q = q^N \tag{45}$$

Justification 19.8

The total energy of a collection of N independent molecules is the sum of the energies of the molecules. Therefore, we can write the total energy of a state i of the system as

$$E_i = \varepsilon_i(1) + \varepsilon_i(2) + \cdots + \varepsilon_i(N)$$

In this expression, $\varepsilon_i(1)$ is the energy of molecule 1 when the system is in the state i, $\varepsilon_i(2)$ the energy of molecule 2 when the system is in the same state i, and so on. The canonical partition function is then

$$Q = \sum_i e^{-\beta \varepsilon_i(1) - \beta \varepsilon_i(2) - \cdots - \beta \varepsilon_i(N)}$$

The sum over the states of the system can be reproduced by letting each molecule enter all its own individual states (although we meet an important proviso shortly). Therefore, instead of summing over the states i of the system, we can sum over all the individual states i of molecule 1, all the states i of molecule 2, and so on. This rewriting of the original expression leads to

$$Q = \left(\sum_i e^{-\beta \varepsilon_i} \right) \left(\sum_i e^{-\beta \varepsilon_i} \right) \cdots \left(\sum_i e^{-\beta \varepsilon_i} \right) = \left(\sum_i e^{-\beta \varepsilon_i} \right)^N = q^N$$

(a) Distinguishable and indistinguishable molecules

If all the molecules are identical and free to move through space, we cannot distinguish them and the relation $Q = q^N$ is not valid. Suppose that molecule 1 is in some state a, molecule 2 is in b, and molecule 3 is in c, then one member of the ensemble has an energy $E = \varepsilon_a + \varepsilon_b + \varepsilon_c$. This member, however, is indistinguishable from one formed by putting molecule 1 in state b, molecule 2 in state c, and molecule 3 in state a, or some other permutation. There are six such permutations in all, and $N!$ in general. In the case of indistinguishable molecules, it follows that we have counted too many states in going from the sum over system states to the sum over molecular states, so writing $Q = q^N$ overestimates the value of Q. The detailed argument is quite involved, but at all except very low temperatures it turns out that the correction factor is $1/N!$. Therefore:

(a) For distinguishable independent molecules: $Q = q^N$

(b) For indistinguishable independent molecules: $Q = \dfrac{q^N}{N!}$ $\tag{46}$

For molecules to be indistinguishable, they must be of the same kind: an Ar atom is never indistinguishable from a Ne atom. Their identity, however, is not the only criterion. Each identical molecule in a crystal lattice, for instance, can be 'named' with a set of coordinates. Identical molecules in a lattice are therefore distinguishable because their sites are distinguishable, and we use eqn 46*a* for any of their modes that may be considered independent of their neighbours. Equation 46*a* is also applicable to a collection of N molecules, each one of which is in its own box. On the other hand, identical molecules in a gas are free to move to different locations, and there is no way of keeping track of the identity of a given molecule; we therefore use eqn 46*b*.

(b) The entropy of a monatomic gas

An important application of the previous material is the derivation (as shown in the *Justification* below) of the **Sackur–Tetrode equation** for the entropy of a monatomic gas:

$$S = nR \ln\left(\frac{e^{5/2}V}{nN_A\Lambda^3}\right) \qquad \Lambda = \frac{h}{(2\pi mkT)^{1/2}} \qquad (47a)$$

Because the gas is perfect, we can use the relation $V = nRT/p$ to express the entropy in terms of the pressure as

$$S = nR \ln\left(\frac{e^{5/2}kT}{p\Lambda^3}\right) \qquad (47b)$$

Justification 19.9

For a gas of independent molecules, Q may be replaced by $q^N/N!$, with the result that eqn 44 becomes

$$S = \frac{U - U(0)}{T} + Nk \ln q - k \ln N!$$

Because the number of molecules ($N = nN_A$) in a typical sample is large, we can use Stirling's approximation (eqn 2) to write

$$S = \frac{U - U(0)}{T} + Nk \ln q - k(N \ln N - N)$$

The only mode of motion for a gas of atoms is translation, and the partition function is $q = V/\Lambda^3$ (eqn 22), where Λ is the thermal wavelength. The internal energy is given by eqn 28, so the entropy is

$$S = \tfrac{3}{2}nR + nR\left(\ln\frac{V}{\Lambda^3} - \ln nN_A + 1\right)$$
$$= nR\left(\ln e^{3/2} + \ln\frac{V}{\Lambda^3} - \ln nN_A + \ln e\right)$$

which rearranges into eqn 47.

Example 19.5 Using the Sackur–Tetrode equation

Calculate the standard molar entropy of gaseous argon at 25°C.

Method To calculate the *molar* entropy, S_m, from eqn 47b, divide both sides by n. To calculate the *standard* molar entropy, $S_m^\ominus$, set $p = p^\ominus$ in the expression for S_m:

$$S_m^\ominus = R \ln\left(\frac{e^{5/2}kT}{p^\ominus\Lambda^3}\right)$$

Answer The mass of an Ar atom is $m = 39.95$ u. At 25°C, its thermal wavelength is 16.0 pm (by the same kind of calculation as in the *Illustration* following eqn 22). Therefore,

$$S_m^\ominus = R \ln\left\{\frac{e^{5/2} \times (4.12 \times 10^{-21}\text{ J})}{(10^5\text{ N m}^{-2}) \times (1.60 \times 10^{-11}\text{ m})^3}\right\}$$
$$= 18.6R = 155\text{ J K}^{-1}\text{ mol}^{-1}$$

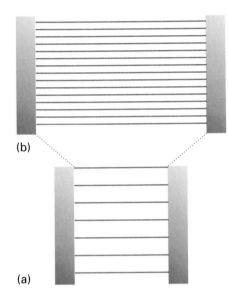

19.15 As the width of a container is increased (going from (a) to (b)), the energy levels become closer together (as $1/L^2$), and as a result more are thermally accessible at a given temperature. Consequently, the entropy of the system rises as the container expands. As before, this diagram is schematic: the separation of levels increases with energy.

(b)

(a)

Comment We can anticipate, on the basis of the number of accessible states for a lighter molecule, that the standard molar entropy of Ne is likely to be smaller than for Ar; its actual value is $17.60R$ at 298 K.

Self-test 19.7 Calculate the translational contribution to the standard molar entropy of H_2 at 25°C.

[14.2R]

The Sackur–Tetrode equation implies that, when a monatomic perfect gas expands isothermally from V_i to V_f, its entropy changes by

$$\Delta S = nR \ln(aV_f) - nR \ln(aV_i) = nR \ln\left(\frac{V_f}{V_i}\right) \tag{48}$$

where aV is the collection of quantities inside the logarithm of eqn 47a. This is exactly the expression we obtained by using classical thermodynamics (Example 4.1). Now, though, we see that that classical expression is in fact a consequence of the increase in the number of accessible translational states when the volume of the container is increased (Fig. 19.15).

Table 19.1 Key equations, with $\beta = 1/kT$

Definition of molecular partition function:

$$q = \sum_i e^{-\beta \varepsilon_i} \qquad q = \sum_{\text{levels, } i} g_i e^{-\beta \varepsilon_i}$$

Definition of canonical partition function:

$$Q = \sum_i e^{-\beta E_i} = \begin{cases} q^N & \text{distinguishable independent particles} \\ q^N/N! & \text{indistinguishable independent particles} \end{cases}$$

Two-level system, energies $0, \varepsilon$:

$$q = 1 + e^{-\beta \varepsilon}$$

Evenly spaced, infinite level system, energies $0, \varepsilon, 2\varepsilon, \cdots$:

$$q = (1 - e^{-\beta \varepsilon})^{-1}$$

Translational motion of particle of mass m in volume V:

$$q = \frac{V}{\Lambda^3} \qquad \Lambda = \left(\frac{h^2 \beta}{2\pi m}\right)^{1/2} = \frac{h}{(2\pi mkT)^{1/2}}$$

Boltzmann distribution:

$$p_i = \frac{e^{-\beta \varepsilon_i}}{q} \qquad p_i = \frac{n_i}{N}$$

Boltzmann formula:

$$S = k \ln W$$

Internal energy (independent particles):

$$U = U(0) - \frac{N}{q}\left(\frac{\partial q}{\partial \beta}\right)_V = U(0) - N\left(\frac{\partial \ln q}{\partial \beta}\right)_V$$

Internal energy (general):

$$U = U(0) - \frac{1}{Q}\left(\frac{\partial Q}{\partial \beta}\right)_V = U(0) - \left(\frac{\partial \ln Q}{\partial \beta}\right)_V$$

Entropy (independent particles):

$$S = \frac{U - U(0)}{T} + Nk \ln q$$

Entropy (general):

$$S = \frac{U - U(0)}{T} + k \ln Q$$

Checklist of key ideas

☐ statistical thermodynamics

The distribution of molecular states

☐ population
☐ principle of equal *a priori* probabilities

19.1 Configurations and weights
☐ configuration
☐ weight (1)
☐ Stirling's approximation (2)
☐ Boltzmann distribution (6)
☐ method of undetermined multipliers

19.2 The molecular partition function
☐ molecular partition function (11, 12)
☐ thermal wavelength

The internal energy and the entropy

☐ *q* for uniform array (15)
☐ *q* for translation (18, 22)

19.3 The internal energy
☐ *U* in terms of *q* (27)

19.4 The statistical entropy
☐ Boltzmann formula (30)
☐ *S* in terms of *q* (35)

The canonical partition function

19.5 The canonical ensemble
☐ ensemble
☐ canonical ensemble
☐ thermodynamic limit
☐ canonical distribution (38)
☐ canonical partition function
☐ density of states

19.6 The thermodynamic information in the partition function
☐ *U* in terms of *Q* (41)
☐ *S* in terms of *Q* (44)

19.7 Independent molecules
☐ distinguishable and indistinguishable molecules (46)
☐ Sackur–Tetrode equation (47)

Further reading

Articles of general interest

C.W. David, On the Legendre transformation and the Sackur–Tetrode equation. *J. Chem. Educ.* **65**, 876 (1988).

P.G. Nelson, Statistical mechanical interpretation of entropy. *J. Chem. Educ.* **71**, 103 (1994).

Texts and sources of data and information

P.W. Atkins, *The second law*. Scientific American Books, New York (1984, revised 1994).

R.P.H. Gasser and W.G. Richards, *Introduction to statistical thermodynamics*. World Scientific, Singapore (1995).

T.L. Hill, *An introduction to statistical mechanics*. Dover, New York (1986).

D. Chandler, *Introduction to statistical mechanics*. Oxford University Press (1987).

C.E. Hecht, *Statistical mechanics and kinetic theory*. W.H. Freeman & Co, New York (1990).

Exercises

19.1 (a) What are the relative populations of the states of a two-level system when the temperature is infinite?

19.1 (b) What is the temperature of a two-level system of energy separation equivalent to 300 cm^{-1} when the population of the upper state is one-half that of the lower state?

19.2 (a) Calculate the translational partition function at (a) 300 K and (b) 600 K of a molecule of molar mass 120 g mol^{-1} in a container of volume 2.00 cm^3.

19.2 (b) Calculate (a) the thermal wavelength, (b) the translational partition function of an Ar atom in a cubic box of side 1.00 cm at (i) 300 K and (ii) 3000 K.

19.3 (a) Calculate the ratio of the translational partition functions of D_2 and H_2 at the same temperature and volume.

19.3 (b) Calculate the ratio of the translational partition functions of xenon and helium at the same temperature and volume.

19.4 (a) A certain atom has a threefold degenerate ground level, a non-degenerate electronically excited level at 3500 cm^{-1}, and a threefold degenerate level at 4700 cm^{-1}. Calculate the partition function of these electronic states at 1900 K.

19.4 (b) A certain atom has a doubly degenerate ground level, a triply degenerate electronically excited level at 1250 cm^{-1}, and a doubly degenerate level at 1300 cm^{-1}. Calculate the partition function of these electronic states at 2000 K.

19.5 (a) Calculate the electronic contribution to the molar internal energy at 1900 K for a sample composed of the atoms specified in Exercise 19.4a.

19.5 (b) Calculate the electronic contribution to the molar internal energy at 2000 K for a sample composed of the atoms specified in Exercise 19.4b.

19.6 (a) A certain molecule has a non-degenerate excited state lying at 540 cm^{-1} above the non-degenerate ground state. At what temperature will 10 per cent of the molecules be in the upper state?

19.6 (b) A certain molecule has a doubly degenerate excited state lying at 360 cm^{-1} above the non-degenerate ground state. At what temperature will 15 per cent of the molecules be in the upper state?

19.7 (a) An electron spin can adopt either of two orientations in a magnetic field, and its energies are $\pm \mu_B B$, where μ_B is the Bohr magneton. Deduce an expression for the partition function and mean energy of the electron and sketch the variation of the functions with B. Calculate the relative populations of the spin states at (a) 4.0 K, (b) 298 K when $B = 1.0$ T.

19.7 (b) The spin of a nitrogen nucleus can adopt any of three orientations in a magnetic field, and its energies are $0, \pm \gamma_N \hbar B$, where γ_N is the magnetogyric ratio of the nucleus. Deduce an expression for the partition function and mean energy of the nucleus and sketch the variation of the functions with B. Calculate the relative populations of the spin states at (a) 1.0 K, (b) 298 K when $B = 20.0$ T.

19.8 (a) Consider a system of distinguishable particles having only two non-degenerate energy levels separated by an energy which is equal to the value of kT at 10 K. Calculate (a) the ratio of populations in the two states at (1) 1.0 K, (2) 10 K, and (3) 100 K, (b) the molecular partition function at 10 K, (c) the molar energy at 10 K, (d) the molar heat capacity at 10 K, (e) the molar entropy at 10 K.

19.8 (b) Consider a system of distinguishable particles having only three non-degenerate energy levels separated by an energy which is equal to the value of kT at 25.0 K. Calculate (a) the ratio of populations in the states at (1) 1.00 K, (2) 25.0 K, and (3) 100 K, (b)

the molecular partition function at 25.0 K, (c) the molar energy at 25.0 K, (d) the molar heat capacity at 25.0 K, (e) the molar entropy at 25.0 K.

19.9 (a) At what temperature would the population of the first excited vibrational state of HCl be $1/e$ times its population of the ground state?

19.9 (b) At what temperature would the population of the first rotational level of HCl be $1/e$ times the population of the ground state?

19.10 (a) Calculate the standard molar entropy of neon gas at (a) 200 K, (b) 298.15 K.

19.10 (b) Calculate the standard molar entropy of xenon gas at (a) 100 K, (b) 298.15 K.

19.11 (a) Calculate the vibrational contribution to the entropy of Cl_2 at 500 K given that the wavenumber of the vibration is 560 cm^{-1}.

19.11 (b) Calculate the vibrational contribution to the entropy of Br_2 at 600 K given that the wavenumber of the vibration is 321 cm^{-1}.

19.12 (a) Identify the systems for which it is essential to include a factor of $1/N!$ on going from Q to q: (a) a sample of helium gas, (b) a sample of carbon monoxide gas, (c) a solid sample of carbon monoxide, (d) water vapour.

19.12 (b) Identify the systems for which it is essential to include a factor of $1/N!$ on going from Q to q: (a) a sample of carbon dioxide gas, (b) a sample of graphite, (c) a sample of diamond, (d) ice.

Problems

Numerical problems

19.1 A certain atom has a doubly degenerate ground level pair and an upper level of four degenerate states at 450 cm^{-1} above the ground level. In an atomic beam study of the atoms it was observed that 30 per cent of the atoms were in the upper level, and the translational temperature of the beam was 300 K. Are the electronic states of the atoms in thermal equilibrium with the translational states?

19.2 Explore the conditions under which the 'integral' approximation for the translational partition function is not valid by considering the translational partition function of an Ar atom in a cubic box of side 1.00 cm. Estimate the temperature at which, according to the integral approximation, $q = 10$ and evaluate the exact partition function at that temperature.

19.3 (a) Calculate the electronic partition function of a tellurium atom at (i) 298 K, (ii) 5000 K by direct summation using the following data:

Term	Degeneracy	Wavenumber/cm^{-1}
Ground	5	0
1	1	4707
2	3	4751
3	5	10 559

(b) What proportion of the Te atoms are in the ground term and in the term labelled 2 at the two temperatures? (c) Calculate the electronic contribution to the standard molar entropy of gaseous Te atoms.

19.4 The four lowest electronic levels of a Ti atom are: $^3F_2, ^3F_3, ^3F_4,$ and $^5F_1$, at 0, 170, 387, and 6557 cm^{-1}, respectively. There are many other electronic states at higher energies. The boiling point of titanium is 3287°C. What are the relative populations of these levels at the boiling point? (*Hint*: The degeneracies of the levels are $2J + 1$.)

19.5 The NO molecule has a doubly degenerate excited level 121.1 cm^{-1} above the doubly degenerate ground term. Calculate and plot the electronic partition function of NO from $T = 0$ to 1000 K. Evaluate (a) the term populations and (b) the electronic contribution to the molar internal energy at 300 K. Calculate the electronic contribution to the molar entropy of the NO molecule at 300 K and 500 K.

19.6 Calculate, by explicit summation, the vibrational partition function and the vibrational contribution to the molar internal energy of I_2 molecules at (a) 100 K, (b) 298 K given that its vibrational energy levels lie at the following wavenumbers above the zero-point energy level: 0, 213.30, 425.39, 636.27, 845.93 cm^{-1}. What proportion of I_2 molecules are in the ground and first two excited levels at the two temperatures? Calculate the vibrational contribution to the molar entropy of I_2 at the two temperatures.

Theoretical problems

19.7 A sample consisting of five molecules has a total energy 5ε. Each molecule is able to occupy states of energy $j\varepsilon$, with $j = 0, 1, 2, \ldots$. (a) Calculate the weight of the configuration in which the molecules are distributed evenly over the available states. (b) Draw up a table with columns headed by the energy of the states and write beneath them all configurations that are consistent with the total energy. Calculate the weights of each configuration and identify the most probable configurations.

19.8 A sample of nine molecules is numerically tractable but on the verge of being thermodynamically significant. Draw up a table of configurations for $N = 9$, total energy 9ε in a system with energy levels $j\varepsilon$ (as in Problem 19.7). Before evaluating the weights of the configurations, guess (by looking for the most 'exponential' distribution of populations) which of the configurations will turn out to be the most probable. Go on to calculate the weights and identify the most probable configuration.

19.9 The most probable configuration is characterized by a parameter we know as the 'temperature'. The temperatures of the system specified in Problems 19.7 and 19.8 must be such as to give a mean value of ε for the energy of each molecule and a total energy $N\varepsilon$ for the system. (a) Show that the temperature can be obtained by plotting p_j against j, where p_j is the (most probable) fraction of molecules in the state with energy $j\varepsilon$. Apply the procedure to the system in Problem 19.8. What is the temperature of the system when ε corresponds to 50 cm^{-1}? (b) Choose configurations other than the most probable, and show that the same procedure gives a worse straight line, indicating that a temperature is not well-defined for them.

19.10 A certain molecule can exist in either a non-degenerate singlet state or a triplet state (with degeneracy 3). The energy of the triplet exceeds that of the singlet by ε. Assuming that the molecules are distinguishable (localized) and independent, (a) obtain the expression for the molecular partition function, (b) find expressions in terms of ε for the molar energy, molar heat capacity, and molar entropy of such molecules and calculate their values at $T = \varepsilon/k$.

19.11 Consider a system with energy levels $\varepsilon_j = j\varepsilon$ and N molecules. (a) Show that, if the mean energy per molecule is $a\varepsilon$, then the

temperature is given by

$$\beta = \frac{1}{\varepsilon} \ln \left(1 + \frac{1}{a} \right)$$

Evaluate the temperature for a system in which the mean energy is ε, taking ε equivalent to 50 cm^{-1}. (b) Calculate the molecular partition function q for the system when its mean energy is $a\varepsilon$. (c) Show that the entropy of the system is

$$S/k = (1 + a)\ln(1 + a) - a \ln a$$

and evaluate this expression for a mean energy ε.

19.12 Suppose that by some means we contrive to invert the population of a two-level system, in the sense that the upper and lower levels have the populations of the lower and upper levels, respectively, in the system in thermal equilibrium at a temperature T. Show that the relative populations are still given by a Boltzmann-like expression but with a temperature $-T$. Under what circumstances is it possible to speak of negative temperatures of an evenly spaced three-level system?

19.13 Consider Stirling's approximation for $\ln N!$ in the derivation of the Boltzmann distribution. What difference would it make if (a) a cruder approximation, $N! = N^N$, (b) the better approximation in footnote 3 of Section 19.1a were used instead?

Additional problems supplied by Carmen Giunta and Charles Trapp

19.14 Consider a system A consisting of subsystems A_1 and A_2, for which $W_1 = 1 \times 10^{20}$ and $W_2 = 2 \times 10^{20}$. What is the number of configurations available to the combined system? Also, compute the entropies S, S_1, and S_2. What is the significance of this result?

19.15 Consider 1.00×10^{22} ^{4}He atoms in a box of dimensions 1.0 cm $\times$ 1.0 cm $\times$ 1.0 cm. Calculate the occupancy of the first excited level at 1.0 mK, 2.0 K, and 4.0 K. Do the same for ^{3}He. What conclusions might you draw from the results of your calculations?

19.16 Given that for gases the canonical partition function, Q, is related to the molecular partition function q by $Q = q^N/N!$, prove, using the expression for q and general thermodynamic relations, the perfect gas law $pV = nRT$.

19.17 By what factor does the number of available configurations increase when 100 J of energy is added to a system containing 1.00 mol of particles at constant volume at 298 K?

19.18 By what factor does the number of available configurations increase when 20 m^3 of air at 1.00 atm and 300 K is allowed to expand by 0.0010 per cent at constant temperature?

19.19 (a) The standard molar entropy of graphite at 298, 410, and 498 K is 5.69, 9.03, and 11.63 J K^{-1} mol^{-1}, respectively. If 1.00 mol C(graphite) at 298 K is surrounded by thermal insulation and placed next to 1.00 mol C(graphite) at 498 K, also insulated, how many configurations are there altogether for the combined but independent systems? (b) If the same two samples are now placed in thermal contact and brought to thermal equilibrium, the final temperature will be 410 K. (Why might the final temperature not be the average? It isn't.) How many configurations are there now in the combined system? Neglect any volume changes. (c) Demonstrate that this process is spontaneous.

19.20 Obtain the barometric formula (Problem 1.35) from the Boltzmann distribution. Recall that the potential energy of a particle at height h above the surface of the Earth is mgh. Convert the barometric formula from pressure to number density, $\mathcal{N}$. Compare the relative number densities, $\mathcal{N}(h)/\mathcal{N}(0)$, for O_2 and H_2O at $h = 8.0$ km, a typical cruising altitude for commercial aircraft.

19.21 Over time planets lose their atmospheres unless they are replenished. A complete analysis of the overall process is very complicated and depends upon the radius of the planet, temperature, atmospheric composition, and other factors. Prove that the atmosphere of planets cannot be in an equilibrium state by demonstrating that the Boltzmann distribution leads to a uniform finite number density as $r \to \infty$. *Hint.* Recall that in a gravitational field the potential energy is $V(r) = -GMm/r$, where G is the gravitational constant, M is the planet's mass, and m the mass of the particle.

19.22 Consider the distribution of particles in a fluid (liquid) as a function of height in the fluid. Assume the particles have a slightly greater density than the fluid. (a) Show that the potential energy of such a particle is given by $V(h) = v(\rho - \rho_0)gh$, where ρ is the mass density of the particle, and ρ_0 is the mass density of the fluid. (b) Derive a formula for the number density of particles, $\mathcal{N}$, in the fluid as a function of height. (c) Perrin (1906) found for gamboge gum grains in water with density 1.21×10^3 kg m^{-3} and volume 1.03×10^{-19} m^3 at 4°C that the number density decreased to half its value at $h = 1.23 \times 10^{-5}$ m. From this result determine the Boltzmann constant and the Avogadro constant.

19.23 J. Sugar and A. Musgrove (*J. Phys. Chem. Ref. Data* **22**, 1213 (1993)) have published tables of energy levels for germanium atoms and cations from Ge^{+1} to Ge^{+31}. The lowest-lying energy levels in neutral Ge are as follows.

	${}^3P_0$	${}^3P_1$	${}^3P_2$	${}^1D_2$	${}^1S_0$
E/cm^{-1}	0.0	557.1	1410.0	7125.3	16 367.3

Calculate the electronic partition function at 298 K and 1000 K by direct summation. *Hint.* The degeneracy of a level is $2J + 1$.

20 Statistical thermodynamics: the machinery

In this chapter we apply the concepts of statistical thermodynamics to the calculation of chemically significant quantities. First, we establish the relations between thermodynamic functions and partition functions. Next, we show that the molecular partition function can be factorized into contributions from each mode of motion, and establish the formulas for the partition functions for translational, rotational, vibrational, and electronic modes of motion. These contributions can be calculated from spectroscopic data. Finally, we turn to specific calculations. These applications include the mean energies of modes of motion, the heat capacities of substances, and residual entropies. In the final section, we see how to calculate the equilibrium constant of a reaction and through that calculation understand some of the molecular features that determine the magnitudes of equilibrium constants and their temperature dependence.

A partition function is the bridge between thermodynamics, spectroscopy, and quantum mechanics. Once it is known, it can be used to calculate thermodynamic functions, heat capacities, entropies, and equilibrium constants. It also sheds light on the significance of these properties.

Fundamental relations

In this section we see how all the thermodynamic functions can be obtained once we know the partition function. Then we see how to calculate the molecular partition function, and through that the thermodynamic functions, from spectroscopic data.

20.1 The thermodynamic functions

We have already derived (in Chapter 19) the two expressions for calculating the internal energy and the entropy of a system from its canonical partition function, Q:

$$U - U(0) = -\left(\frac{\partial \ln Q}{\partial \beta}\right)_V \qquad S = \frac{U - U(0)}{T} + k \ln Q \tag{1}$$

Table 20.1 Statistical thermodynamic relations

In terms of the canonical partition function Q

$$U - U(0) = -\left(\frac{\partial \ln Q}{\partial \beta}\right)_V$$

$$S = \frac{U - U(0)}{T} + k \ln Q$$

$$A - A(0) = -kT \ln Q$$

$$p = kT\left(\frac{\partial \ln Q}{\partial V}\right)_T$$

$$H - H(0) = -\left(\frac{\partial \ln Q}{\partial \beta}\right)_V$$
$$+ kTV\left(\frac{\partial \ln Q}{\partial V}\right)_T$$

$$G - G(0) = -kT \ln Q + kTV\left(\frac{\partial \ln Q}{\partial V}\right)_T$$

For indistinguishable, independent particles $Q = q^N/N!$

$$U - U(0) = -N\left(\frac{\partial \ln q}{\partial \beta}\right)_V$$

$$S = \frac{U - U(0)}{T} + nR(\ln q - \ln N + 1)$$

$$G - G(0) = -nRT \ln\left(\frac{q_m}{N_A}\right)$$

where q_m is the molar partition function. For distinguishable, independent particles $Q = q^N$

$$U - U(0) = -N\left(\frac{\partial \ln q}{\partial \beta}\right)_V$$

$$S = \frac{U - U(0)}{T} + nR \ln q$$

$$G - G(0) = -nRT \ln q$$

In general, for indistinguishable independent particles,

$$Q = \frac{(q_{external}q_{internal})^N}{N!}$$
$$= \frac{(q_{external})^N}{N!} \times (q_{internal})^N$$

The thermodynamic functions are then the sums of internal and external (translational) contributions.

---

where $\beta = 1/kT$. If the molecules are independent, we can go on to make the substitutions $Q = q^N$ (if the molecules are also distinguishable, as in a solid) or $Q = q^N/N!$ (if they are indistinguishable, as in a gas). All the thermodynamic functions introduced in Part 1 are related to U and S, so we have a route to their calculation from Q. For later convenience, the expressions we derive here are collected in Table 20.1.

(a) The Helmholtz energy

The Helmholtz energy, A, is defined as $A = U - TS$. This relation implies that $A(0) = U(0)$, so substitution for U and S by using eqn 1 leads to the very simple expression

$$A - A(0) = -kT \ln Q \tag{2}$$

(b) The pressure

It follows from classical thermodynamics, using an argument like that leading to eqn 5.10,[1] that the pressure, p, and the Helmholtz energy are related by

$$p = -\left(\frac{\partial A}{\partial V}\right)_T \tag{3}$$

Therefore,

$$p = kT\left(\frac{\partial \ln Q}{\partial V}\right)_T \tag{4}$$

This relation is entirely general, and may be used for any type of substance, including perfect gases, real gases, and liquids.

Example 20.1 Deriving an equation of state

Derive an expression for the pressure of a gas of independent particles.

Method We can suspect that the pressure is that given by the perfect gas law. To proceed systematically, substitute the explicit formula for Q for a gas of independent, indistinguishable molecules (see Table 19.1) into eqn 4.

Answer For a gas of independent molecules, $Q = q^N/N!$ with $q = V/\Lambda^3$:

$$p = kT\left(\frac{\partial \ln Q}{\partial V}\right)_T = \frac{kT}{Q}\left(\frac{\partial Q}{\partial V}\right)_T = \frac{NkT}{q}\left(\frac{\partial q}{\partial V}\right)_T$$
$$= \frac{NkT\Lambda^3}{V} \times \frac{1}{\Lambda^3} = \frac{NkT}{V} = \frac{nRT}{V}$$

To derive this relation, we have used

$$\left(\frac{\partial q}{\partial V}\right)_T = \left(\frac{\partial(V/\Lambda^3)}{\partial V}\right)_T = \frac{1}{\Lambda^3}$$

and $NkT = nN_A kT = nRT$.

Comment The calculation shows that the equation of state of a gas of independent particles is indeed the perfect gas law. This calculation can be regarded as yet another way of deducing that $\beta = 1/kT$.

---

1 Specifically, from $A = U - TS$, it follows that $dA = -p\,dV - S\,dT$, so $(\partial A/\partial V)_T = -p$.

Self-test 20.1 Derive the equation of state of a sample for which $Q = q^N f / N!$, with $q = V / \Lambda^3$, where f depends on the volume.

$$[p = nRT/V + kT(\partial \ln f / \partial V)_T]$$

(c) The enthalpy

At this stage we can use the expressions for U and p in the definition $H = U + pV$ to obtain an expression for the enthalpy, H, of any substance:

$$H - H(0) = -\left(\frac{\partial \ln Q}{\partial \beta}\right)_V + kTV\left(\frac{\partial \ln Q}{\partial V}\right)_T \tag{5}$$

We have already seen that $U - U(0) = \frac{3}{2}nRT$ for a gas of independent particles (eqn 19.28a), and have just shown that $pV = nRT$. Therefore, for such a gas,[2]

$$H - H(0) = \frac{5}{2}nRT \tag{6}^\circ$$

(d) The Gibbs energy

One of the most important thermodynamic functions for chemistry is the Gibbs energy, $G = H - TS = A + pV$. We can now express this function in terms of the partition function by combining the expressions for A and p:

$$G - G(0) = -kT \ln Q + kTV\left(\frac{\partial \ln Q}{\partial V}\right)_T \tag{7}$$

This expression takes a simple form for a gas of independent molecules because pV in the expression $G = A + pV$ can be replaced by nRT:

$$G - G(0) = -kT \ln Q + nRT \tag{8}^\circ$$

Furthermore, because $Q = q^N / N!$, and therefore $\ln Q = N \ln q - \ln N!$, it follows that by using Stirling's approximation ($\ln N! \approx N \ln N - N$) we can write

$$
\begin{aligned}
G - G(0) &= -NkT \ln q + kT \ln N! + nRT \\
&= -nRT \ln q + kT(N \ln N - N) + nRT \\
&= -nRT \ln\left(\frac{q}{N}\right)
\end{aligned} \tag{9}^\circ
$$

with $N = nN_A$. Now we see another interpretation of the Gibbs energy: it is proportional to the logarithm of the average number of thermally accessible states per molecule.

It will turn out to be convenient to define the **molar partition function**, $q_m = q/n$ (with units mol^{-1}), for then

$$G - G(0) = -nRT \ln\left(\frac{q_m}{N_A}\right) \tag{10}^\circ$$

20.2 The molecular partition function

The energy of a molecule is the sum of contributions from its different modes of motion:

$$\varepsilon_i = \varepsilon_i^T + \varepsilon_i^R + \varepsilon_i^V + \varepsilon_i^E \tag{11}$$

where T denotes translation, R rotation, V vibration, and E the electronic contribution. This separation is only approximate (except for translation) because the modes are not completely independent, but in most cases it is satisfactory. The separation of the electronic and vibrational motions, for example, is justified by the Born–Oppenheimer approximation

2   Recall from Part 1 that we use a superscript $\circ$ on an equation number to denote a result valid only for a perfect gas.

(Chapter 14), and the separation of the vibrational and rotational modes is valid to the extent that a molecule can be treated as a rigid rotor.

Given that the energy is a sum of independent contributions, the partition function factorizes into a product of contributions (recall Section 19.2b):

$$q = \sum_i e^{-\beta\varepsilon_i} = \sum_i e^{-\beta\varepsilon_i^T - \beta\varepsilon_i^R - \beta\varepsilon_i^V - \beta\varepsilon_i^E}$$

$$= \left(\sum_i e^{-\beta\varepsilon_i^T}\right)\left(\sum_i e^{-\beta\varepsilon_i^R}\right)\left(\sum_i e^{-\beta\varepsilon_i^V}\right)\left(\sum_i e^{-\beta\varepsilon_i^E}\right) = q^T q^R q^V q^E \qquad (12)$$

This factorization means that we can investigate each contribution separately.

(a) The translational contribution

The translational partition function of a molecule of mass m in a container of volume V was derived in Section 19.2:

$$q^T = \frac{V}{\Lambda^3} \qquad \Lambda = h\left(\frac{\beta}{2\pi m}\right)^{1/2} = \frac{h}{(2\pi m k T)^{1/2}} \qquad (13)$$

Notice that $q^T \rightarrow \infty$ as $T \rightarrow \infty$ because an infinite number of states becomes accessible as the temperature is raised. Even at room temperature $q^T \approx 2 \times 10^{28}$ for an O_2 molecule in a vessel of volume 100 cm³.

The thermal wavelength, Λ, lets us judge whether the approximations that led to the expression for q^T are valid. The approximations are valid if many states are occupied, which requires V/Λ^3 to be large. That will be so if Λ is small compared with the linear dimensions of the container. For H_2 at 25°C, $\Lambda = 71$ pm, which is far smaller than any conventional container is likely to be (but comparable to pores in zeolites or cavities in clathrates). For O_2, a heavier molecule, $\Lambda = 18$ pm.

(b) The rotational contribution

As demonstrated in Example 19.1, the partition function of a nonsymmetrical (AB) linear rotor is

$$q^R = \sum_J (2J+1)e^{-\beta hcBJ(J+1)} \qquad (14)$$

The direct method of calculating q^R is to substitute the experimental values of the rotational energy levels into this expression and to sum the series numerically.

Example 20.2 Evaluating the rotational partition function explicitly

Evaluate the rotational partition function of $^1H^{35}Cl$ at 25°C, given that $B = 10.591$ cm⁻¹.

Method We use eqn 14, and evaluate it term by term. A useful relation is $kT/hc = 207.22$ cm⁻¹ at 298.15 K. The sum is readily evaluated using mathematical software.

Answer To show how successive terms contribute, we draw up the following table using $hcB/kT = 0.05111$ (Fig. 20.1).

J	0	1	2	3	4	$\cdots$	10
$(2J+1)e^{-0.05111J(J+1)}$	1	2.71	3.68	3.79	3.24	$\cdots$	0.08

The sum required by eqn 14 (the sum of the numbers in the last row of the table) is 19.9; hence $q^R = 19.9$ at this temperature. Taking J up to 50 gives $q^R = 19.902$.

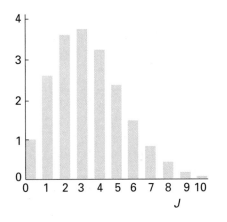

20.1 The contributions to the rotational partition function of an HCl molecule at 25°C. The vertical axis is the value of $(2J+1)e^{-\beta hcBJ(J+1)}$. Successive terms (which are proportional to the populations of the levels) pass through a maximum because the population of individual states decreases exponentially, but the degeneracy of the levels increases with J.

Comment Notice that about ten J-levels are significantly populated but the number of populated *states* is larger on account of the $(2J+1)$-fold degeneracy of each level. We shall shortly encounter the approximation that $q^R \approx kT/hcB$, which in the present case gives $q^R = 19.6$, in good agreement with the exact value, and with much less work.

Self-test 20.2 Evaluate the rotational partition function for HCl at 0°C.

[18.26]

At room temperature $kT/hc \approx 200 \text{ cm}^{-1}$. The rotational constants of many molecules are close to 1 cm^{-1} (Table 16.2) and often smaller. (The very light H_2 molecule, for which $B = 60.9 \text{ cm}^{-1}$, is one exception.) It follows that many rotational levels are populated at normal temperatures. When this is the case, the partition function may be approximated by

$$\text{Linear rotors: } q^R = \frac{kT}{hcB}$$

$$\text{Nonlinear rotors: } q^R = \left(\frac{kT}{hc}\right)^{3/2} \left(\frac{\pi}{ABC}\right)^{1/2}$$

(15)

where A, B, and C are the rotational constants of the molecule. However, before using these expressions, read on (to eqns 17 and 20).

Justification 20.1

When many rotational states are occupied and kT is much larger than the separation between neighbouring states, the sum in the partition function can be approximated by an integral, much as we illustrated for translational motion in *Justification 19.4*:

$$q^R = \int_0^\infty (2J+1)e^{-\beta hcBJ(J+1)} \, dJ$$

Although this integral looks complicated, it can be evaluated without much effort by noticing that it can also be written as

$$q^R = -\frac{1}{\beta hcB} \int_0^\infty \left(\frac{d}{dJ} e^{-\beta hcBJ(J+1)}\right) dJ$$

Then, because the integral of a derivative of a function is the function itself,

$$q^R = -\frac{1}{\beta hcB} e^{-\beta hcBJ(J+1)} \Big|_0^\infty = \frac{1}{\beta hcB}$$

The calculation for a nonlinear molecule is along the same lines, but slightly trickier (see *Further reading*).

Table 20.2* Rotational and vibrational temperatures

Molecule	Mode	θ_V/K	θ_R/K
H_2		6330	88
HCl		4300	9.4
I_2		309	0.053
CO_2	ν_1	1997	0.561
	ν_2	3380	
	ν_3	960	

*For more values, see Table 16.2 in the *Data section* at the end of this volume, and use $hc/k = 1.439 \text{ K cm}$.

A useful way of expressing the temperature above which the approximation is valid is to introduce the **rotational temperature**, $\theta_R = hcB/k$. Then 'high temperature' means $T \gg \theta_R$. Some typical values are shown in Table 20.2. The value for H_2 is abnormally high and we must be careful with the approximation for this molecule.

The general conclusion at this stage is that molecules with large moments of inertia (and hence small rotational constants and low rotational temperatures) have large rotational partition functions. The large value of q^R reflects the closeness in energy (compared with kT) of the rotational levels in large, heavy molecules, and the large number of them that are accessible at normal temperatures.

We must take care, however, not to include too many rotational states in the sum. For a homonuclear diatomic molecule or a symmetrical linear molecule (such as CO_2 or HC≡CH), a

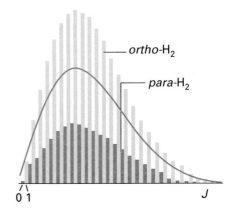

20.2 The values of the individual terms $(2J+1)e^{-\beta hcBJ(J+1)}$ contributing to the mean partition function of a $3:1$ mixture of *ortho*- and *para*-H_2. The partition function is the sum of all these terms. At high temperatures, the sum is approximately equal to the sum of the terms over all values of J, each with a weight of $\frac{1}{2}$. This is the sum of the contributions indicated by the curve.

rotation through $180°$ results in an indistinguishable state of the molecule. Hence, **the number of thermally accessible states is only half the number that can be occupied by a heteronuclear diatomic molecule**, where rotation through $180°$ does result in a distinguishable state. Therefore, for a symmetrical linear molecule,

$$q^R = \frac{kT}{2hcB} \tag{16}$$

The equations for symmetrical and nonsymmetrical molecules can be combined into a single expression by introducing the **symmetry number**, σ, which is the number of indistinguishable orientations of the molecule. Then

$$q^R = \frac{kT}{\sigma hcB} \tag{17}$$

For a heteronuclear diatomic molecule $\sigma = 1$; for a homonuclear diatomic molecule or a symmetrical ($D_{\infty h}$) linear molecule, $\sigma = 2$.

Justification 20.2

The quantum mechanical origin of the symmetry factor is the Pauli principle, which forbids the occupation of certain states. We saw in Section 16.8, for example, that H_2 may occupy rotational states with even J only if its nuclear spins are paired (*para*-hydrogen), and odd J states only if its nuclear spins are parallel (*ortho*-hydrogen). There are three states of *ortho*-H_2 to each value of J (because there are three parallel spin states of the two nuclei).

To set up the rotational partition function we note that 'ordinary' molecular hydrogen is a mixture of one part *para*-H_2 (with only its even-J rotational states occupied) and three parts *ortho*-H_2 (with only its odd-J rotational states occupied). Therefore, the average partition function per molecule is

$$q^R = \frac{1}{4}\left\{ \sum_{\text{even } J} (2J+1)e^{-\beta hcBJ(J+1)} + 3 \sum_{\text{odd } J} (2J+1)e^{-\beta hcBJ(J+1)} \right\} \tag{18}$$

The odd-J states are more heavily weighted than the even-J states (Fig. 20.2). From the illustration we see that we would obtain approximately the same answer for the partition function (the sum of all the populations) if each J term contributed half its normal value to the sum. That is, the last equation can be approximated as

$$q^R = \frac{1}{2} \sum_{J} (2J+1)e^{-\beta hcBJ(J+1)} \tag{19}$$

and this approximation is very good when many terms contribute (at high temperatures).

The same type of argument may be used for linear symmetrical molecules in which identical bosons are interchanged by rotation (such as CO_2). As pointed out in Section 16.8, if the nuclear spin of the bosons is 0, then only even-J states are admissible. Because only half the rotational states are occupied, the rotational partition function is only half the value of the sum obtained by allowing all values of J to contribute (Fig. 20.3).

The same care must be exercised for other types of symmetrical molecule, and for a nonlinear molecule we write

$$q^R = \frac{1}{\sigma}\left(\frac{kT}{hc}\right)^{3/2}\left(\frac{\pi}{ABC}\right)^{1/2} \tag{20}$$

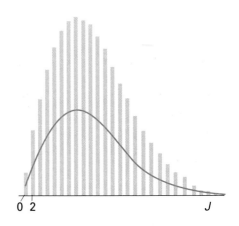

20.3 The relative populations of the rotational energy levels of CO_2. Only states with even J values are occupied. The full line shows the smoothed, averaged population of levels.

Some typical values of the symmetry numbers required are given in Table 20.2. The value $\sigma(H_2O) = 2$ reflects the fact that a $180°$ rotation about its C_2 axis interchanges two indistinguishable atoms. In NH_3, there are three indistinguishable orientations around its C_3

axis. For CH_4, any of three $120°$ rotations about any of its four C–H bonds leaves the molecule in an indistinguishable state, so the symmetry number is $3 \times 4 = 12$. For benzene, any of six orientations around its C_6 axis leaves it apparently unchanged, as does a rotation of $180°$ around any of six C_2 axes in the plane of the molecule.

A more formal way of arriving at the value of the symmetry number is to note that σ is the order (the number of elements) of the **rotational subgroup** of the molecule, the point group of the molecule with all but the identity and the rotations removed. The rotational subgroup of H_2O is $\{E, C_2\}$, so $\sigma = 2$. The rotational subgroup of NH_3 is $\{E, 2C_3\}$, so $\sigma = 3$. This recipe makes it easy to find the symmetry numbers for more complicated molecules. The rotational subgroup of CH_4 is obtained from the T character table as $\{E, 8C_3, 3C_2\}$, so $\sigma = 12$. For benzene, the rotational subgroup of D_{6h} is $\{E, 2C_6, 2C_3, C_2, 3C_2', 3C_2''\}$, so $\sigma = 12$.

Example 20.3 Estimating a rotational partition function

Estimate the rotational partition function of ethene at $25°C$ given that $A = 4.828$ cm^{-1}, $B = 1.0012$ cm^{-1}, and $C = 0.8282$ cm^{-1}.

Method Use eqn 20 with $kT/hc = 207.226$ cm^{-1}. Next, identify the molecular point group (for example, by using the chart in Fig. 15.14 or the shapes in Fig. 15.15). The symmetry number is obtained by deciding on the rotational subgroup of the molecular point group, and counting the number of elements in the group.

Answer The point group of the molecule is D_{2h}. From that group's character table (see the *Data section*), the rotational subgroup of D_{2h} consists of the elements $\{E, C_{2x}, C_{2y}, C_{2z}\}$. The order of this subgroup is 4; therefore $\sigma = 4$. Then, because $ABC = 4.0033$ cm^{-3}, it follows that $q^R = 661$.

Comment Ethene is quite a big molecule, the energy levels are close together (compared with kT at room temperature), and many are significantly populated at room temperature.

- -

Self-test 20.3 Evaluate the rotational partition function of pyridine, C_5H_5N, at room temperature ($A = 0.2014$ cm^{-1}, $B = 0.1936$ cm^{-1}, $C = 0.0987$ cm^{-1}).

$$[4.3 \times 10^4]$$

(c) The vibrational contribution

The vibrational partition function of a molecule is calculated by substituting the measured vibrational energy levels into the exponentials appearing in the definition of q^V, and summing them numerically. In a polyatomic molecule each normal mode (Section 16.14) has its own partition function (provided the anharmonicities are so small that the modes are independent). The overall vibrational partition function is the product of the individual partition functions, and we can write $q^V = q^V(1)q^V(2) \ldots$, where $q^V(K)$ is the partition function for the Kth normal mode and is calculated by direct summation of the observed spectroscopic levels.

Illustration

Given that a typical value of the vibrational partition function of one normal mode is about 1.14, and that a nonlinear molecule containing 10 atoms has $3N - 6 = 24$ normal modes (Section 16.14a), the overall vibrational partition function is approximately

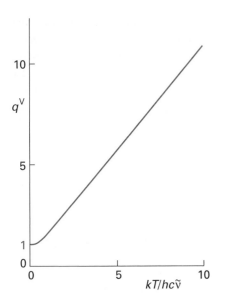

20.4 The vibrational partition function of a molecule in the harmonic approximation. Note that the partition function is linearly proportional to the temperature when the temperature is high $(T \gg \theta_V)$.

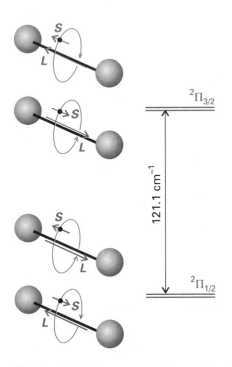

20.5 The doubly degenerate ground electronic level of NO (with the spin and orbital angular momentum around the axis in opposite directions) and the doubly degenerate first excited level (with the spin and orbital momenta parallel). The upper level is thermally accessible at room temperature.

$q^V \approx (1.1)^{24} = 9.8$. Even though each vibrational mode is not appreciably excited, there may be so many modes in a molecule that overall their excitation is significant.

If the vibrational excitation is not too great, the harmonic approximation may be made, and the vibrational energy levels written as

$$E_v = (v + \tfrac{1}{2})hc\tilde{\nu} \qquad v = 0, 1, 2, \ldots \tag{21}$$

If we measure energies from the zero-point level (our general rule), then the permitted values are $\varepsilon_v = vhc\tilde{\nu}$ and the partition function is

$$q^V = \sum_v e^{-\beta vhc\tilde{\nu}} = \sum_v (e^{-\beta hc\tilde{\nu}})^v \tag{22}$$

(because $e^{ax} = (e^x)^a$). We met this sum in Example 19.2 (which is no accident: the ladder-like array of levels in Fig. 19.3 is exactly the same as that of a harmonic oscillator). The series can be summed in the same way, and gives

$$q^V = \frac{1}{1 - e^{-\beta hc\tilde{\nu}}} \tag{23}$$

This function is plotted in Fig. 20.4. In a polyatomic molecule, each normal mode gives rise to a partition function of this form.

Example 20.4 Calculating a vibrational partition function

The wavenumbers of the three normal modes of H_2O are 3656.7 cm^{-1}, 1594.8 cm^{-1}, and 3755.8 cm^{-1}. Evaluate the vibrational partition function at 1500 K.

Method Use eqn 23 for each mode, and then form the product of the three contributions. At 1500 K, $kT/hc = 1042.6$ cm^{-1}.

Answer We can draw up the following table displaying the contributions of each mode:

Mode:	1	2	3
$\tilde{\nu}/\text{cm}^{-1}$	3656.7	1594.8	3755.8
$hc\tilde{\nu}/kT$	3.507	1.530	3.602
q^V	1.031	1.276	1.028

The overall vibrational partition function is therefore

$$q^V = 1.031 \times 1.276 \times 1.028 = 1.353$$

Comment The vibrations of H_2O are at such high wavenumbers that even at 1500 K most of the molecules are in their vibrational ground state.

Self-test 20.4 Repeat the calculation for CO_2, where the vibrational wavenumbers are 1388 cm^{-1}, 667.4 cm^{-1}, and 2349 cm^{-1}, the second being the doubly degenerate bending mode.

[6.79]

In many molecules the vibrational wavenumbers are so great that $\beta hc\tilde{\nu} > 1$. For example, the lowest vibrational wavenumber of CH_4 is 1306 cm^{-1}, so $\beta hc\tilde{\nu} = 6.3$ at room temperature. C–H stretches normally lie in the range 2850 to 2960 cm^{-1}, so for them $\beta hc\tilde{\nu} \approx 14$. In these cases, $e^{-\beta hc\tilde{\nu}}$ in the denominator of q^V is very close to zero (for example, $e^{-6.3} = 0.002$), and the vibrational partition function for a single mode is very close to 1

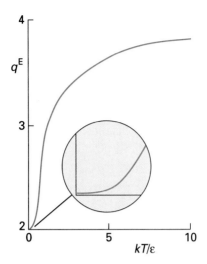

20.6 The variation with temperature of the electronic partition function of an NO molecule. Note that the curve resembles that for a two-level system (Fig. 19.5), but rises from 2 (the degeneracy of the lower level) and approaches 4 (the total number of states) at high temperatures.

($q^V = 1.002$ when $\beta hc\tilde{\nu} = 6.3$), implying that only the zero-point level is significantly occupied.

Now consider the case of bonds so weak that $\beta hc\tilde{\nu} \ll kT$. When this condition is satisfied, the partition function may be approximated by expanding the exponential ($e^x = 1 + x + \cdots$):

$$q^V = \frac{1}{1 - (1 - \beta hc\tilde{\nu} + \cdots)} \tag{24}$$

That is, for weak bonds at high temperatures,

$$q^V = \frac{1}{\beta hc\tilde{\nu}} = \frac{kT}{hc\tilde{\nu}} \tag{25}$$

The temperatures for which eqn 25 is valid can be expressed in terms of the **vibrational temperature**, $\theta_V = hc\tilde{\nu}/k$ (Table 20.2). In terms of the vibrational temperature, 'high temperature' means $T \gg \theta_V$. The value for H_2 is abnormally high because the atoms are so light and the vibrational frequency consequently high.

(d) The electronic contribution

Electronic energy separations from the ground state are usually very large, so for most cases $q^E = 1$. An important exception arises in the case of atoms and molecules having electronically degenerate ground states, in which case $q^E = g_0$, where g_0 is the degeneracy of the electronic ground state. Alkali metal atoms, for example, have doubly degenerate ground states (corresponding to the two orientations of their electron spin), so $q^E = 2$.

Some atoms and molecules have low-lying electronically excited states. (At high enough temperatures, all atoms and molecules have thermally accessible excited states.) An example is NO, which has a configuration of the form $\cdots \pi^1$ (the molecule has one electron more than N_2). The orbital angular momentum may take two orientations with respect to the molecular axis (corresponding to circulation clockwise or counter-clockwise around the axis), and the spin angular momentum may also take two, giving four states in all (Fig. 20.5). The energy of the two states in which the orbital and spin momenta are parallel (giving the $^2\Pi_{3/2}$ term) is slightly greater than that of the two other states in which they are antiparallel (giving the $^2\Pi_{1/2}$ term). The separation, which arises from spin–orbit coupling (Section 13.8), is only 121 cm^{-1}. Hence, at normal temperatures, all four states are thermally accessible. If we denote the energies of the two levels as $E_{1/2} = 0$ and $E_{3/2} = \varepsilon$, the partition function is

$$q^E = \sum_{\text{levels } j} g_j e^{-\beta\varepsilon_j} = 2 + 2e^{-\beta\varepsilon} \tag{26}$$

The variation of this function with temperature is shown in Fig. 20.6. At $T = 0$, $q^E = 2$, because only the doubly degenerate ground state is accessible. At high temperatures, q^E approaches 4 because all four states are accessible. At 25°C, $q^E = 3.1$.

(e) The overall partition function

The partition functions for each mode of motion of a molecule are collected in Table 20.3. The overall partition function is the product of each contribution. For a diatomic molecule with no low-lying electronically excited states and $T \gg \theta_R$,

$$q = g^E \left(\frac{V}{\Lambda^3} \right) \left(\frac{kT}{\sigma hcB} \right) \left(\frac{1}{1 - e^{-\beta hc\tilde{\nu}}} \right) \tag{27}$$

Overall partition functions obtained in this way are approximate because they assume that the rotational levels are very close and that the vibrational levels are harmonic. These approximations are avoided by using the energy levels identified spectroscopically and evaluating the sums explicitly.

Table 20.3* Symmetry numbers

Molecule	σ
H_2O	2
NH_3	3
CH_4	12
C_6H_6	12

*For more values, see Table 16.2 in the *Data section*.

Example 20.5 Calculating a thermodynamic function from spectroscopic data

Calculate the value of $G_m^{\ominus} - G_m^{\ominus}(0)$ for $H_2O(g)$ at 1500 K given that $A = 27.8778$ cm^{-1}, $B = 14.5092$ cm^{-1}, and $C = 9.2869$ cm^{-1} and the information in Example 20.4.

Method The starting point is eqn 10. For the standard value, we evaluate the translational partition function at $p^{\ominus}$ (that is, at 10^5 Pa exactly). The vibrational partition function was calculated in Example 20.4. Use the expressions in Table 20.4 for the other contributions.

Answer For this C_{2v} molecule, $\sigma = 2$. Because $m = 18.015$ u, $q_m^{T\,\ominus} = 1.706 \times 10^8$ mol^{-1}. For the vibrational contribution we have already found that $q^V = 1.352$. For the rotational contribution, $q^R = 486.7$. Therefore,

$$G_m^{\ominus} - G_m^{\ominus}(0) = -(8.3145 \text{ J K}^{-1}\text{ mol}^{-1}) \times (1500 \text{ K})$$
$$\times \ln\left(\frac{(1.706 \times 10^8 \text{ mol}^{-1}) \times 486.7 \times 1.352}{6.022\,14 \times 10^{23} \text{ mol}^{-1}}\right)$$
$$= -365.6 \text{ kJ mol}^{-1}$$

- -

Self-test 20.5 Repeat the calculation for CO_2. The vibrational data are given in Self-test 20.4; $B = 0.3902$ cm^{-1}.

$$[-366.6 \text{ kJ mol}^{-1}]$$

Table 20.4 Contributions to the molecular partition function*

Translation

$$q = \frac{V}{\Lambda^3} \qquad \Lambda = \left(\frac{h^2\beta}{2\pi m}\right)^{1/2}$$

$$\Lambda/\text{pm} = \frac{1749}{(T/K)^{1/2}(M/\text{g mol}^{-1})^{1/2}}$$

$$\frac{q_m^{\ominus}}{N_A} = \frac{kT}{p^{\ominus}\Lambda^3} = 2.561 \times 10^{-2}(T/K)^{5/2}(M/\text{g mol}^{-1})^{3/2}$$

Rotation

(a) Linear molecules

$$q = \frac{1}{\sigma hcB\beta} = \frac{0.6950}{\sigma} \times \frac{T/K}{(B/\text{cm}^{-1})}$$

(b) Non-linear molecules

$$q = \frac{1}{\sigma}(1/hc\beta)^{2/3}\left(\frac{\pi}{ABC}\right)^{1/2} = \frac{1.0270}{\sigma} \times \frac{(T/K)^{3/2}}{(ABC/\text{cm}^{-3})^{1/2}}$$

Vibration

$$q = \frac{1}{1 - e^{-hc\tilde{\nu}\beta}} = \frac{1}{1 - e^{-a}} \qquad a = \frac{1.4388(\tilde{\nu}/\text{cm}^{-1})}{T/K}$$

Electronic

$$q = g_0$$

where g_0 is the degeneracy of the electronic ground state (when that is the only accessible level); at high temperatures, evaluate q explicitly.

*$\beta = 1/kT$. It is often useful to note that

$$\frac{hc}{k} = 1.438\,79 \text{ cm K}$$

See also inside front cover for further information.

Using statistical thermodynamics

Any thermodynamic quantity can now be calculated from a knowledge of the energy levels of molecules: we have merged thermodynamics and spectroscopy. In this section, we indicate how to do the calculations for a number of important properties.

20.3 Mean energies

It is often useful to know the mean energy, $\langle \varepsilon \rangle$, of various modes of motion. When the molecular partition function can be factorized into contributions from each mode, the mean energy of each mode M is

$$\langle \varepsilon^{M} \rangle = -\frac{1}{q^{M}} \left(\frac{\partial q^{M}}{\partial \beta} \right)_{V} \qquad M = T, R, V, \text{ or } E \tag{28}$$

(a) The mean translational energy

To see a pattern emerging, we consider first a one-dimensional system of length X, for which $q^{T} = X/\Lambda$, with $\Lambda = h(\beta/2\pi m)^{1/2}$. Then, if we note that Λ is a constant times $\beta^{1/2}$,

$$\langle \varepsilon^{T} \rangle = -\frac{\Lambda}{X} \left(\frac{\partial}{\partial \beta} \frac{X}{\Lambda} \right)_{V} = -\beta^{1/2} \frac{d}{d\beta} \left(\frac{1}{\beta^{1/2}} \right) = \frac{1}{2\beta} = \tfrac{1}{2} kT \tag{29}$$

For a molecule free to move in three dimensions, the analogous calculation leads to

$$\langle \varepsilon^{T} \rangle = \tfrac{3}{2} kT \tag{30}$$

Both conclusions are in agreement with the classical equipartition theorem (see the *Introduction*), that the mean energy of each quadratic contribution to the energy is $\frac{1}{2}kT$. Furthermore, the fact that the mean energy is independent of the size of the container is consistent with the thermodynamic result that internal energy of a perfect gas is independent of its volume (Section 5.1b).

(b) The mean rotational energy

The mean rotational energy of a linear molecule is obtained from the partition function given in eqn 14. When the temperature is low $(T < \theta_{R})$, the series must be summed term by term, which gives

$$q^{R} = 1 + 3e^{-2\beta hcB} + 5e^{-6\beta hcB} + \cdots$$

Hence

$$\langle \varepsilon^{R} \rangle = \frac{hcB\left(6e^{-2\beta hcB} + 30e^{-6\beta hcB} + \cdots \right)}{1 + 3e^{-2\beta hcB} + 5e^{-6\beta hcB} + \cdots} \tag{31}$$

This function is plotted in Fig. 20.7. At high temperatures $(T \gg \theta_{R})$, q^{R} is given by eqn 17, and

$$\langle \varepsilon^{R} \rangle = -\frac{1}{q^{R}} \frac{dq^{R}}{d\beta} = -\sigma hcB\beta \frac{d}{d\beta} \frac{1}{\sigma hcB\beta} = \frac{1}{\beta} = kT \tag{32}$$

(q^{R} is independent of V, so the partial derivatives have been replaced by complete derivatives.) The high-temperature result is also in agreement with the equipartition theorem, for the classical expression for the energy of a linear rotor is $E_{K} = \frac{1}{2}I_{\perp}\omega_{a}^{2} + \frac{1}{2}I_{\perp}\omega_{b}^{2}$. (There is no rotation around the line of atoms.) It follows from the equipartition theorem that the mean rotational energy is $2 \times \frac{1}{2}kT = kT$.

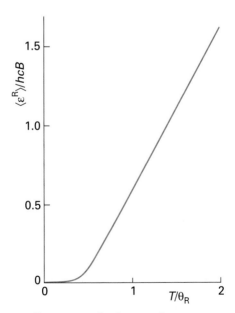

20.7 The mean rotational energy of a nonsymmetrical linear rotor as a function of temperature. At high temperatures $(T \gg \theta_{R})$, the energy is linearly proportional to the temperature, in accord with the equipartition theorem.

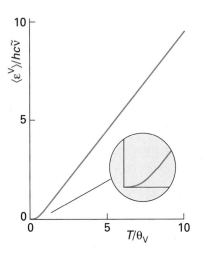

20.8 The mean vibrational energy of a molecule in the harmonic approximation as a function of temperature. At high temperatures ($T \gg \theta_V$), the energy is linearly proportional to the temperature, in accord with the equipartition theorem.

(c) The mean vibrational energy

The vibrational partition function in the harmonic approximation is given in eqn 23. Because q^V is independent of the volume, it follows that

$$\frac{dq^V}{d\beta} = \frac{d}{d\beta}\left(\frac{1}{1-e^{-\beta hc\tilde{\nu}}}\right) = -\frac{hc\tilde{\nu}e^{-\beta hc\tilde{\nu}}}{(1-e^{-\beta hc\tilde{\nu}})^2} \tag{33}$$

and hence that

$$\langle \varepsilon^V \rangle = \frac{hc\tilde{\nu}}{e^{\beta hc\tilde{\nu}} - 1} \tag{34}$$

The zero-point energy, $\frac{1}{2}hc\tilde{\nu}$, can be added to the right-hand side if the mean energy is to be measured from 0 rather than the lowest attainable level (the zero-point level). The variation of the mean energy with temperature is illustrated in Fig. 20.8.

At high temperatures, when $T \gg \theta_V$, or $\beta hc\tilde{\nu} \ll 1$, the exponential functions can be expanded ($e^x = 1 + x + \cdots$) and all but the leading terms discarded. This approximation leads to

$$\langle \varepsilon^V \rangle = \frac{hc\tilde{\nu}}{(1 + \beta hc\tilde{\nu} + \cdots) - 1} \approx \frac{1}{\beta} = kT \tag{35}$$

This result is in agreement with the value predicted by the classical equipartition theorem, because the energy of a one-dimensional oscillator is $E = \frac{1}{2}mv_x^2 + \frac{1}{2}kx^2$ and the mean value of each quadratic term is $\frac{1}{2}kT$.

20.4 Heat capacities

The constant-volume heat capacity is defined as $C_V = (\partial U/\partial T)_V$. The derivative with respect to T is converted into a derivative with respect to β by using

$$\frac{d}{dT} = \left(\frac{d\beta}{dT}\right)\frac{d}{d\beta} = -\frac{1}{kT^2}\frac{d}{d\beta} = -k\beta^2\frac{d}{d\beta} \tag{36}$$

It follows that

$$C_V = -k\beta^2\left(\frac{\partial U}{\partial \beta}\right)_V \tag{37}$$

Because the internal energy of a perfect gas is a sum of contributions, the heat capacity is also a sum of contributions from each mode. The contribution of mode M is

$$C_V^M = N\left(\frac{\partial\langle \varepsilon^M \rangle}{\partial T}\right)_V = -Nk\beta^2\left(\frac{\partial\langle \varepsilon^M \rangle}{\partial \beta}\right)_V \tag{38}$$

(a) The individual contributions

The temperature is always high enough (provided the gas is above its condensation temperature) for the mean translational energy to be $\frac{3}{2}kT$, the equipartition value. Therefore, the molar constant-volume heat capacity is

$$C_{V,m}^T = N_A\frac{d(\frac{3}{2}kT)}{dT} = \frac{3}{2}R \tag{39}°$$

Translation is the only mode of motion for a monatomic gas, so for such a gas $C_{V,m} = \frac{3}{2}R = 12.47\ \text{J K}^{-1}\ \text{mol}^{-1}$. This result is very reliable: helium, for example, has this

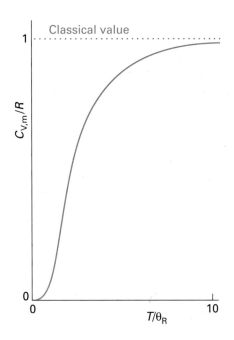

20.9 The temperature dependence of the rotational contribution to the heat capacity of a linear molecule.

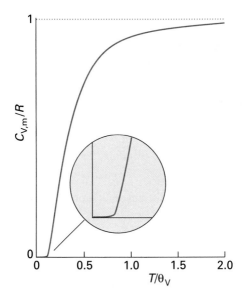

$C_{V,m}/R$

20.10 The temperature dependence of the vibrational heat capacity of a molecule in the harmonic approximation calculated by using eqn 41. Note that the heat capacity is within 10 per cent of its classical value for temperatures greater than θ_V.

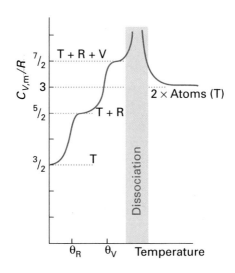

$C_{V,m}/R$

Dissociation

20.11 The general features of the temperature dependence of the heat capacity of diatomic molecules are as shown here. Each mode becomes active when its characteristic temperature is exceeded. The heat capacity becomes very large when the molecule dissociates because the energy is used to cause dissociation and not to raise the temperature. Then it falls back to the translation-only value of the atoms.

value over a range of 2000 K. We saw in Section 3.3a that $C_{p,m} - C_{V,m} = R$, so for a monatomic perfect gas $C_{p,m} = \frac{5}{2}R$, and therefore

$$\gamma = \frac{C_p}{C_V} = \frac{5}{3} \tag{40}°$$

When the temperature is high enough for the rotations of the molecules to be highly excited (when $T \gg \theta_R$), we can use the equipartition value kT for the mean rotational energy (for a linear rotor) to obtain $C_{V,m} = R$. For nonlinear molecules, the mean rotational energy rises to $\frac{3}{2}kT$, so the molar rotational heat capacity rises to $\frac{3}{2}R$ when $T \gg \theta_R$. Only the lowest rotational state is occupied when the temperature is very low, and then rotation does not contribute to the heat capacity. We can calculate the rotational heat capacity at intermediate temperatures by differentiating the equation for the mean rotational energy (eqn 31). The resulting (untidy) expression, which is plotted in Fig. 20.9, shows that the contribution rises from zero (when $T = 0$) to the equipartition value (when $T \gg \theta_R$). Because the translational contribution is always present, we can expect the molar heat capacity of a gas of diatomic molecules ($C_{V,m}^T + C_{V,m}^R$) to rise from $\frac{3}{2}R$ to $\frac{5}{2}R$ as the temperature is increased above θ_R.

Molecular vibrations contribute to the heat capacity, but only when the temperature is high enough for them to be significantly excited. The equipartition mean energy is kT for each mode, so the maximum contribution to the molar heat capacity is R. However, it is very unusual for the vibrations to be so highly excited that equipartition is valid, and it is more appropriate to use the full expression for the vibrational heat capacity, which is obtained by differentiating eqn 34:

$$C_{V,m}^V = Rf^2 \qquad f = \frac{\theta_V}{T}\left(\frac{e^{-\theta_V/2T}}{1 - e^{-\theta_V/T}}\right) \tag{41}$$

where $\theta_V = hc\tilde{\nu}/k$ is the vibrational temperature. The curve in Fig. 20.10 shows how the vibrational heat capacity depends on temperature. Note that even when the temperature is only slightly above the vibrational temperature the heat capacity is close to its equipartition value.[3]

(b) The overall heat capacity

The total heat capacity of a molecular substance is the sum of each contribution (Fig. 20.11). When equipartition is valid (when the temperature is well above the characteristic temperature of the mode, $T \gg \theta_M$) we can estimate the heat capacity by counting the numbers of modes that are active. In gases, all three translational modes are always active, and contribute $\frac{3}{2}R$ to the molar heat capacity. If we denote the number of active rotational modes by ν_R^* (so for most molecules at normal temperatures $\nu_R^* = 2$ for linear molecules, and 3 for nonlinear molecules), then the rotational contribution is $\frac{1}{2}\nu_R^*R$. If the temperature is high enough for ν_V^* vibrational modes to be active, the vibrational contribution to the molar heat capacity is ν_V^*R. In most cases $\nu_V^* \approx 0$. It follows that the total molar heat capacity is

$$C_{V,m} = \frac{1}{2}(3 + \nu_R^* + 2\nu_V^*)R \tag{42}$$

Example 20.6 Estimating the molar heat capacity of a gas

Estimate the molar constant-volume heat capacity of water vapour at 100°C. Vibrational wavenumbers are given in Example 20.4; the rotational constants of an H_2O molecule are 27.9, 14.5, and 9.3 cm^{-1}.

---

3 Equation 41 is essentially the same as the Einstein formula for the heat capacity of a solid (eqn 11.9) with θ_V the Einstein temperature, θ_E. The only difference is that vibrations can take place in three dimensions in a solid.

Method We need to assess whether the rotational and vibrational modes are active by computing their characteristic temperatures from the data (to do so, use $hc/k = 1.439\ cm\,K$).

Answer The characteristic temperatures (in round numbers) of the vibrations are 5300 K, 2300 K, and 5400 K; the vibrations are therefore not excited at 373 K. The three rotational modes have characteristic temperatures 40 K, 21 K, and 13 K, so they are fully excited, like the three translational modes. The translational contribution is $\frac{3}{2}R = 12.5\ J\,K^{-1}\,mol^{-1}$. Fully excited rotations contribute a further 12.5 $J\,K^{-1}\,mol^{-1}$. Therefore, a value close to 25 $J\,K^{-1}\,mol^{-1}$ is predicted.

Comment The experimental value is 26.1 $J\,K^{-1}\,mol^{-1}$. The discrepancy is probably due to deviations from perfect gas behaviour.

- -

Self-test 20.6 Estimate the molar constant-volume heat capacity of gaseous I_2 at 25°C ($B = 0.037\ cm^{-1}$; see Table 16.2 for more data).

$$[21\ J\,K^{-1}\,mol^{-1}]$$

20.5 Equations of state

The canonical partition function, Q, is a function of the volume and the temperature of the system and the number of molecules it contains. Therefore, eqn 4 for the pressure in terms of the partition function has the form $p = f(n, V, T)$. That is, eqn 4 is an equation of state. The relation between p and Q is a very important route to the equations of state of real gases in terms of intermolecular forces, for the latter can be built into Q.

We have already seen (Example 20.1) that the partition function for a gas of independent particles leads to the perfect gas equation of state, $pV = nRT$. Real gases differ from perfect gases in their equations of state and we saw in Section 1.4b that their equations of state may be written

$$\frac{pV_m}{RT} = 1 + \frac{B}{V_m} + \frac{C}{V_m^2} + \cdots \tag{43}$$

where B is the second virial coefficient and C is the third virial coefficient.

The total kinetic energy of a gas is the sum of the kinetic energies of the individual molecules. Therefore, even in a real gas the canonical partition function factorizes into a part arising from the kinetic energy, which is the same as for the perfect gas, and a factor called the **configuration integral**, Z, which depends on the intermolecular potentials. We therefore write

$$Q = \frac{Z}{\Lambda^{3N}} \tag{44}$$

By comparing this equation with eqn 19.46b ($Q = q^N/N!$, with $q = V/\Lambda^3$), we see that for a perfect gas

$$Z = \frac{V^N}{N!} \tag{45}$$

For a real gas, Z is related to the total potential energy $\mathcal{V}$ of interaction of all the particles by

$$Z = \frac{1}{N!} \int e^{-\beta \mathcal{V}}\, d\boldsymbol{r}_1 d\boldsymbol{r}_2 \cdots d\boldsymbol{r}_N \tag{46}$$

Illustration

When the molecules do not interact with one another, $\mathcal{V} = 0$, and hence $e^{-\beta \mathcal{V}} = 1$. Then

$$Z = \frac{1}{N!} \int d\boldsymbol{r}_1 d\boldsymbol{r}_2 \cdots d\boldsymbol{r}_N = \frac{V^N}{N!}$$

because $\int d\boldsymbol{r} = V$, where V is the volume of the container. This result coincides with eqn 45.

When we consider only interactions between pairs of particles the configuration integral simplifies to

$$Z = \frac{1}{2} \int e^{-\beta \mathcal{V}} d\boldsymbol{r}_1 d\boldsymbol{r}_2 \tag{47}$$

The second virial coefficient then turns out to be

$$B = -\frac{1}{2} \frac{N_A}{V} \int f \, d\boldsymbol{r}_1 d\boldsymbol{r}_2 \qquad f = e^{-\beta \mathcal{V}} - 1 \tag{48}$$

The quantity f is the **Mayer f-function**: it goes to zero when the two particles are so far apart that $\mathcal{V} = 0$. When the intermolecular interaction depends only on the separation r of the particles and not on their relative orientation, as in the interaction of closed-shell atoms and tetrahedral and octahedral molecules, eqn 48 simplifies to

$$B = -2\pi N_A \int_0^\infty f r^2 \, dr \tag{49}$$

The integral can be evaluated (usually numerically) by substituting an expression for the intermolecular potential energy.

Intermolecular potential energies are discussed in more detail in Chapter 22, where several expressions are developed for them. At this stage, we can illustrate how eqn 49 is used by considering the **hard-sphere potential**, which is infinite when the separation of the two molecules, r, is less than or equal to a certain value σ, and is zero for greater separations. Then

$$
\begin{aligned}
e^{-\beta \mathcal{V}} = 0 \qquad f = -1 \qquad &\text{when } r \le \sigma \ (\text{and } \mathcal{V} = \infty) \\
e^{-\beta \mathcal{V}} = 1 \qquad f = 0 \qquad &\text{when } r > \sigma \ (\text{and } \mathcal{V} = 0)
\end{aligned}
\tag{50}
$$

It follows from eqn 49 that the second virial coefficient is

$$B = 2\pi N_A \int_0^\sigma r^2 \, dr = \frac{2}{3}\pi N_A \sigma^3 \tag{51}$$

This calculation of B raises the question as to whether a potential can be found which, when the virial coefficients are evaluated, gives the van der Waals equation of state. Such a potential can be found: it consists of a hard-sphere repulsive core and a long-range, shallow attractive region. A further point is that, once a second virial coefficient has been calculated for a given intermolecular potential, it is possible to calculate other thermodynamic properties that depend on the form of the potential. For example, it is possible to calculate the isothermal Joule–Thomson coefficient, μ_T (Section 3.2c), from the thermodynamic relation

$$\lim_{p \to 0} \mu_T = B - T \frac{dB}{dT} \tag{52}$$

and from the result calculate the Joule–Thomson coefficient itself by using eqn 3.19.

20.6 Residual entropies

Entropies may be calculated from spectroscopic data; they may also be measured experimentally (Section 4.3d). In many cases there is good agreement, but in some the experimental entropy is less than the calculated value. One possibility is that the experimental determination failed to take a phase transition into account (and a term of the form $\Delta_{trs}H/T_{trs}$ incorrectly omitted from the sum). Another possibility is that some disorder is present in the solid even at $T = 0$. The entropy at $T = 0$ is then greater than zero, and is called the **residual entropy**.

The origin and magnitude of the residual entropy can be explained by considering a crystal composed of AB molecules, where A and B are similar atoms (such as CO, with its very small electric dipole moment). There may be so little energy difference between ...AB AB AB AB..., ...AB BA BA AB..., and other random arrangements that the molecules adopt either orientation at random in the solid. We can readily calculate the entropy arising from residual disorder by using the Boltzmann formula $S = k \ln W$. To do so, we suppose that two orientations are equally probable, and that the sample consists of N molecules. Because the same energy can be achieved in 2^N different ways (because each molecule can take either of two orientations), the total number of ways of achieving the same energy is $W = 2^N$. It follows that

$$S = k \ln 2^N = Nk \ln 2 = nR \ln 2 \tag{53}$$

We can therefore expect a residual molar entropy of $R \ln 2 = 5.8 \ \text{J K}^{-1} \text{mol}^{-1}$ for solids composed of molecules that can adopt either of two orientations at $T = 0$. If s orientations are possible, the residual molar entropy will be

$$S_m = R \ln s \tag{54}$$

An $FClO_3$ molecule, for example, can adopt four orientations with about the same energy, and the calculated residual molar entropy of $R \ln 4 = 11.5 \ \text{J K}^{-1} \text{mol}^{-1}$ is in good agreement with the experimental value ($10.1 \ \text{J K}^{-1} \text{mol}^{-1}$). For CO, the measured residual entropy is $5 \ \text{J K}^{-1} \text{mol}^{-1}$, which is close to $R \ln 2$, the value expected for a random structure of the form ...CO CO OC CO OC OC....

The residual entropy of ice is $3.4 \ \text{J K}^{-1} \text{mol}^{-1}$. This value can be explained in terms of the hydrogen-bonded structure of the solid. Each O atom is surrounded tetrahedrally by four H atoms, two of which are attached by short σ bonds, the other two being attached by long hydrogen bonds (Fig. 20.12). The randomness lies in which two of the four bonds are short, and an approximate analysis (see the *Justification* below) leads to $S_m(0) \approx R \ln \frac{3}{2} = 3.4 \ \text{J K}^{-1} \text{mol}^{-1}$, in good agreement with the experimental value.

Justification 20.3

Consider a sample of ice that consists of N H_2O molecules. Each of the $2N$ H atoms can be in one of two positions: either close to or far from an O atom (Fig. 20.13). There are therefore 2^{2N} possible arrangements. However, not all these arrangements are acceptable. Indeed, of the $2^4 = 16$ ways of arranging four H atoms around one O atom, only six have two short and two long OH distances and hence are acceptable. Therefore, the number of permitted arrangements is

$$W = 2^{2N} \left(\tfrac{6}{16}\right)^N = \left(\tfrac{3}{2}\right)^N$$

It then follows that the residual molar entropy is $R \ln(3/2)$, as stated in the text.

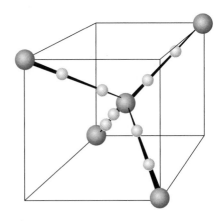

20.12 The possible locations of H atoms around a central O atom in an ice crystal are shown by the spheres. Only one of the locations on each bond may be occupied by an atom, and two H atoms must be close to the O atom and two H atoms must be distant from it.

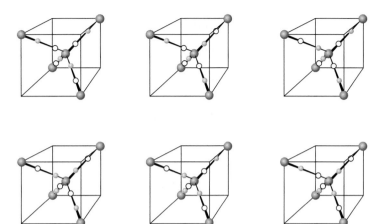

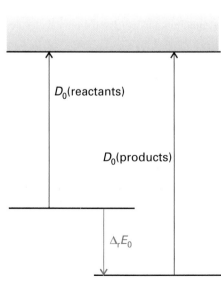

20.13 The six possible arrangements of H atoms in the locations identified in Fig. 20.12.

20.7 Equilibrium constants

The Gibbs energy of a gas of independent molecules is given by eqn 10 in terms of the molar partition function, $q_m = q/n$. The equilibrium constant K of a reaction is related to the standard Gibbs energy of reaction by $\Delta_r G^\ominus = -RT \ln K$. To calculate the equilibrium constant, we must combine these two equations. We shall consider gas-phase reactions in which the equilibrium constant is expressed in terms of the partial pressures of the reactants and products.

(a) The relation between K and the partition function

To find an expression for the standard reaction Gibbs energy we need expressions for the standard molar Gibbs energies, $G^\ominus/n$, of each species. For these expressions, we need the value of the molar partition function when $p = p^\ominus$ (where $p^\ominus = 1$ bar): we denote this **standard molar partition function** $q_m^\ominus$. Because only the translational component depends on the pressure, we can find $q_m^\ominus$ by evaluating the partition function with V replaced by $V_m^\ominus$, where $V_m^\ominus = RT/p^\ominus$. For a species J it follows that

$$G_{J,m}^\ominus = G_{J,m}^\ominus(0) - RT \ln\left(\frac{q_{J,m}^\ominus}{N_A}\right) \tag{55}°$$

where $q_{J,m}^\ominus$ is the standard molar partition function of J. By combining expressions like this one (as shown in the *Justification* below), it turns out that the equilibrium constant for the reaction

$$a\text{A} + b\text{B} \longrightarrow c\text{C} + d\text{D}$$

is given by the expression

$$K = \frac{(q_{C,m}^\ominus/N_A)^c (q_{D,m}^\ominus/N_A)^d}{(q_{A,m}^\ominus/N_A)^a (q_{B,m}^\ominus/N_A)^b} e^{-\Delta_r E_0/RT} \tag{56}$$

where $\Delta_r E_0$ is the difference in molar energies of the ground states of the products and reactants (this term is defined more precisely in the *Justification*), and is calculated from the bond dissociation energies of the species (Fig. 20.14).

D_0(reactants)

D_0(products)

$\Delta_r E_0$

20.14 The definition of $\Delta_r E_0$ for the calculation of equilibrium constants.

Justification 20.4

The standard molar reaction Gibbs energy for the reaction is

$$\Delta_r G^{\ominus} = cG_{C,m}^{\ominus} + dG_{D,m}^{\ominus} - aG_{A,m}^{\ominus} - bG_{B,m}^{\ominus}$$

$$= cG_{C,m}^{\ominus}(0) + dG_{D,m}^{\ominus}(0) - aG_{A,m}^{\ominus}(0) - bG_{B,m}^{\ominus}(0)$$

$$- RT\left\{ c\ln\left(\frac{q_{C,m}^{\ominus}}{N_A}\right) + d\ln\left(\frac{q_{D,m}^{\ominus}}{N_A}\right) - a\ln\left(\frac{q_{A,m}^{\ominus}}{N_A}\right) - b\ln\left(\frac{q_{B,m}^{\ominus}}{N_A}\right) \right\}$$

Because $G(0) = U(0)$, the first term on the right is

$$\Delta_r E_0 = cU_{C,m}^{\ominus}(0) + dU_{D,m}^{\ominus}(0) - aU_{A,m}^{\ominus}(0) - bU_{B,m}^{\ominus}(0) \tag{57}$$

the reaction internal energy at $T = 0$ (a molar quantity).

Now we can write

$$\Delta_r G^{\ominus} = \Delta_r E_0 - RT\left\{ \ln\left(\frac{q_{C,m}^{\ominus}}{N_A}\right)^c + \ln\left(\frac{q_{D,m}^{\ominus}}{N_A}\right)^d \right.$$

$$\left. - \ln\left(\frac{q_{A,m}^{\ominus}}{N_A}\right)^a - \ln\left(\frac{q_{B,m}^{\ominus}}{N_A}\right)^b \right\}$$

$$= \Delta_r E_0 - RT\ln\left\{ \frac{(q_{C,m}^{\ominus}/N_A)^c (q_{D,m}^{\ominus}/N_A)^d}{(q_{A,m}^{\ominus}/N_A)^a (q_{B,m}^{\ominus}/N_A)^b} \right\}$$

$$= -RT\left\{ -\frac{\Delta_r E_0}{RT} + \ln\left\{ \frac{(q_{C,m}^{\ominus}/N_A)^c (q_{D,m}^{\ominus}/N_A)^d}{(q_{A,m}^{\ominus}/N_A)^a (q_{B,m}^{\ominus}/N_A)^b} \right\} \right\}$$

At this stage we can pick out an expression for K by comparing this equation with $\Delta_r G^{\ominus} = -RT\ln K$, which gives

$$\ln K = -\frac{\Delta_r E_0}{RT} + \ln\left\{ \frac{(q_{C,m}^{\ominus}/N_A)^c (q_{D,m}^{\ominus}/N_A)^d}{(q_{A,m}^{\ominus}/N_A)^a (q_{B,m}^{\ominus}/N_A)^b} \right\}$$

This expression is easily rearranged into eqn 56 by taking antilogarithms of both sides.[4]

(b) A dissociation equilibrium

We shall illustrate the application of eqn 56 to an equilibrium in which a diatomic molecule X_2 dissociates into its atoms:

$$X_2(g) \rightleftharpoons 2X(g) \qquad K = \frac{p_X^2}{p_{X_2}p^{\ominus}} \tag{58}$$

According to eqn 56 (with $a = 1$, $b = 0$, $c = 2$, and $d = 0$):

$$K = \frac{(q_{X,m}^{\ominus}/N_A)^2}{q_{X_2,m}^{\ominus}/N_A}e^{-\Delta_r E_0/RT} = \frac{(q_{X,m}^{\ominus})^2}{q_{X_2,m}^{\ominus}N_A}e^{-\Delta_r E_0/RT} \tag{59}$$

with

$$\Delta_r E_0 = 2U_{X,m}^{\ominus}(0) - U_{X_2,m}^{\ominus}(0) = D_0(X—X) \tag{60}$$

where $D_0(X—X)$ is the dissociation energy of the X—X bond.

---

4 In terms of the general chemical equation for a reaction, eqn 2.40, we would write

$$K = \left\{ \prod_J \left(\frac{q_{J,m}^{\ominus}}{N_A}\right)^{\nu_J} \right\} e^{-\Delta_r E_0/RT}$$

The standard molar partition functions of the atoms X are

$$q^{\ominus}_{X,m} = g_X \left(\frac{V^{\ominus}_m}{\Lambda^3_X} \right) = \frac{RTg_X}{p^{\ominus} \Lambda^3_X}$$

because $V^{\ominus}_m = RT/p^{\ominus}$; g_X is the degeneracy of the electronic ground state of X. The diatomic molecule X_2 also has rotational and vibrational degrees of freedom, so its standard molar partition function is

$$q^{\ominus}_{X_2,m} = g_{X_2} \left(\frac{V^{\ominus}_m}{\Lambda^3_{X_2}} \right) q^R_{X_2} q^V_{X_2} = \frac{RTg_{X_2} q^R_{X_2} q^V_{X_2}}{p^{\ominus} \Lambda^3_{X_2}}$$

where g_{X_2} is the degeneracy of the electronic ground state of X_2. It follows that the equilibrium constant is

$$K = \frac{kTg^2_X \Lambda^3_{X_2}}{p^{\ominus} g_{X_2} q^R_{X_2} q^V_{X_2} \Lambda^6_X} e^{-D_0/RT} \qquad (61)$$

where we have used $R/N_A = k$, the Boltzmann constant. All the quantities in this expression can be calculated from spectroscopic data. The Λs are defined in Table 20.4 and depend on the masses of the species and the temperature; the expressions for the rotational and vibrational partition functions are also available in Table 20.4 and depend on the rotational constant and vibrational wavenumber of the molecule.

Example 20.7 Evaluating an equilibrium constant

Evaluate the equilibrium constant for the dissociation $Na_2(g) \rightleftharpoons 2Na(g)$ at 1000 K from the following data: $B = 0.1547$ cm^{-1}, $\tilde{\nu} = 159.2$ cm^{-1}, $D_0 = 70.4$ kJ mol^{-1}. The Na atoms have doublet ground terms.

Method The partition functions required are specified in eqn 61. They are evaluated by using the expressions in Table 20.4. For a homonuclear diatomic molecule, $\sigma = 2$. In the evaluation of $kT/p^{\ominus}$ use $p^{\ominus} = 10^5$ Pa and 1 Pa m$^3 = 1$ J.

Answer The partition functions and other quantities required are as follows:

$$\Lambda(Na_2) = 8.14 \text{ pm} \qquad \Lambda(Na) = 11.5 \text{ pm}$$
$$q^R(Na_2) = 2246 \qquad q^V(Na_2) = 4.885$$
$$g(Na) = 2 \qquad g(Na_2) = 1$$

Then, from eqn 61

$$K = \frac{(1.38 \times 10^{-23} \text{ J K}^{-1}) \times (1000 \text{ K}) \times 4 \times (8.14 \times 10^{-12} \text{ m})^3}{(10^5 \text{ Pa}) \times 2246 \times 4.885 \times (1.15 \times 10^{-11} \text{ m})^6} \times e^{-8.47}$$

$$= 2.42$$

Comment For conversion to an equilibrium constant in terms of molar concentrations, use $[J] = p_J/RT$.

- -

Self-test 20.7 Evaluate K at 1500 K.

[52]

(c) Contributions to the equilibrium constant

We are now in a position to appreciate the physical basis of equilibrium constants. To see what is involved, consider a simple $R \rightleftharpoons P$ gas-phase equilibrium (R for reactants, P for products).

Figure 20.15 shows two sets of energy levels; one set of states belongs to R, and the other belongs to P. The populations of the states are given by the Boltzmann distribution, and are independent of whether any given state happens to belong to R or to P. We can therefore imagine a single Boltzmann distribution spreading, without distinction, over the two sets of states. If the spacings of R and P are similar (as in Fig. 20.15), and P lies above R, the diagram indicates that R will dominate in the equilibrium mixture. However, if P has a high density of states (a large number of states in a given energy range, as in Fig. 20.16) then, even though its zero-point energy lies above that of R, the species P might still dominate at equilibrium.

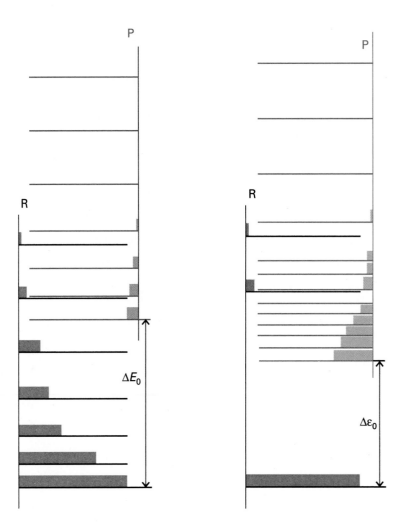

20.15 The array of R(eactants) and P(roducts) energy levels. At equilibrium all are accessible (to differing extents, depending on the temperature), and the equilibrium composition of the system reflects the overall Boltzmann distribution of populations. As ΔE_0 increases, R becomes dominant.

20.16 It is important to take into account the densities of states of the molecules. Even though P might lie well above R in energy (that is, ΔE_0 is large and positive), P might have so many states that its total population dominates in the mixture. In classical thermodynamic terms, we have to take entropies into account as well as enthalpies when considering equilibria.

It is quite easy to show (see the *Justification* below) that the ratio of numbers of R and P molecules at equilibrium is given by

$$\frac{N_P}{N_R} = \frac{q_P}{q_R} e^{-\Delta_r E_0/RT} \tag{62}$$

and therefore that the equilibrium constant for the reaction is

$$K = \frac{q_P}{q_R} e^{-\Delta_r E_0/RT} \tag{63}$$

just as would be obtained from eqn 56.[5]

Justification 20.5

The population in a state i of the composite (R, P) system is

$$n_i = \frac{Ne^{-\beta\varepsilon_i}}{q}$$

where N is the total number of molecules. The total number of R molecules is the sum of these populations taken over the states belonging to R; these states we label r with energies ε_r. The total number of P molecules is the sum over the states belonging to P; these states we label p with energies ε_p' (the prime is explained in a moment):

$$N_R = \sum_r n_r = \frac{N}{q}\sum_r e^{-\beta\varepsilon_r} \qquad N_P = \sum_p n_p = \frac{N}{q}\sum_p e^{-\beta\varepsilon_p'}$$

The sum over the states of R is its partition function, q_R, so

$$N_R = \frac{Nq_R}{q}$$

The sum over the states of P is also a partition function, but the energies are measured from the ground state of the *combined* system, which is the ground state of R. However, because $\varepsilon_p' = \varepsilon_p + \Delta\varepsilon_0$, where $\Delta\varepsilon_0$ is the separation of zero-point energies (as in Fig. 20.16),

$$N_P = \frac{N}{q}\sum_p e^{-\beta(\varepsilon_p + \Delta\varepsilon_0)} = \frac{N}{q}\left(\sum_p e^{-\beta\varepsilon_p}\right)e^{-\beta\Delta\varepsilon_0} = \frac{Nq_P}{q}e^{-\Delta_r E_0/RT}$$

The switch from $\Delta\varepsilon_0/k$ to $\Delta_r E_0/R$ in the last step is the conversion of molecular energies to molar energies.

The equilibrium constant of the $R \rightleftharpoons P$ reaction is proportional to the ratio of the numbers of the two types of molecule. Therefore,

$$K = \frac{N_P}{N_R} = \frac{q_P}{q_R} e^{-\Delta_r E_0/RT}$$

as in eqn 63.

The content of eqn 63 can be seen most clearly by exaggerating the molecular features that contribute to it. We shall suppose that R has only a single accessible level, which implies that $q_R = 1$. We also suppose that P has a large number of evenly, closely spaced levels (Fig. 20.17). The partition function of P is then $q_P = kT/\varepsilon$. In this model system, the

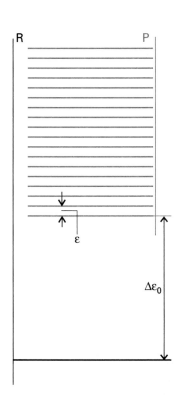

20.17 The model used in the text for exploring the effects of energy separations and densities of states on equilibria. The products P can dominate provided ΔE_0 is not too large and that P has an appreciable density of states.

5   For an $R \rightleftharpoons P$ equilibrium, the V factors in the partition functions cancel, so the appearance of q in place of $q^{\ominus}$ has no effect. In the case of a more general reaction, the conversion from q to $q^{\ominus}$ comes about at the stage of converting the pressures that occur in K to numbers of molecules.

equilibrium constant is

$$K = \frac{kT}{\varepsilon} e^{-\Delta_r E_0 / RT} \tag{64}$$

When $\Delta_r E_0$ is very large, the exponential term dominates and $K \ll 1$, which implies that very little P is present at equilibrium. When $\Delta_r E_0$ is small but still positive, K can exceed 1 because the factor kT/ε may be large enough to overcome the small size of the exponential term. The size of K then reflects the predominance of P at equilibrium on account of its high density of states. At low temperatures $K \ll 1$, and the system consists entirely of R. At high temperatures the exponential function approaches 1 and the pre-exponential factor is large. Hence P becomes dominant. We see that, in this endothermic reaction (endothermic because P lies above R), a rise in temperature favours P, because its states become accessible. This behaviour is what we saw, from the outside, in Chapter 9.

The model also shows why the Gibbs energy, G, and not just the enthalpy, determines the position of equilibrium. It shows that the density of states (and hence the entropy) of each species as well as their relative energies controls the distribution of populations and hence the value of the equilibrium constant.

Checklist of key ideas

Fundamental relations

20.1 The thermodynamic functions
- [ ] see Table 20.1
- [ ] molar partition function (10)

20.2 The molecular partition function
- [ ] see Table 20.4

- [ ] rotational temperature
- [ ] symmetry number
- [ ] rotational subgroup
- [ ] vibrational temperature

Using statistical thermodynamics

20.3 Mean energies
- [ ] mean energy of a mode (28)

20.4 Heat capacities
- [ ] heat capacity of a mode (38)
- [ ] overall heat capacity (42)

20.5 Equations of state
- [ ] configuration integral
- [ ] Mayer f-function (48)
- [ ] hard-sphere potential (50)

20.6 residual entropies
- [ ] residual entropy (54)

20.7 Equilibrium constants
- [ ] standard molar partition function
- [ ] equilibrium constant (56)

Further reading

Articles of general interest

C.W. David, A tractable model for studying solution thermodynamics. *J. Chem. Educ.* **64**, 484 (1987).

R.A. Alberty, The effect of a catalyst on the thermodynamic properties and partition functions of a group of isomers. *J. Chem. Educ.* **65**, 409 (1988).

M. Ross, Equations of state. In *Encyclopedia of applied physics* (ed. G.L. Trigg), **6**, 291. VCH New York (1993).

Texts and sources of data and information

A. Ben-Naim, *Statistical thermodynamics for chemists and biologists.* Plenum, New York (1992).

T.L. Hill, *An introduction to statistical mechanics.* Dover, New York (1986).

N. Davidson, *Statistical thermodynamics.* McGraw-Hill, New York (1962).

K. Lucas, *Applied statistical thermodynamics.* Springer-Verlag, New York (1991).

D. Chandler, *Introduction to statistical mechanics.* Oxford University Press (1987).

C.E. Hecht, *Statistical mechanics and kinetic theory.* W.H. Freeman & Co, New York (1990).

Exercises

20.1 (a) Use the equipartition theorem to estimate the constant-volume molar heat capacity of (a) I_2, (b) CH_4, (c) C_6H_6 in the gas phase at 25°C.

20.1 (b) Use the equipartition theorem to estimate the constant-volume molar heat capacity of (a) O_3, (b) C_2H_6, (c) CO_2 in the gas phase at 25°C.

20.2 (a) Estimate the value of $\gamma = C_p/C_V$ for gaseous ammonia and methane. Do this calculation with and without the vibrational contribution to the energy. Which is closer to the expected experimental value at 25°C?

20.2 (b) Estimate the value of $\gamma = C_p/C_V$ for carbon dioxide. Do this calculation with and without the vibrational contribution to the energy. Which is closer to the expected experimental value at 25°C?

20.3 (a) Estimate the rotational partition function of HCl at (a) 25°C and (b) 250°C.

20.3 (b) Estimate the rotational partition function of O_2 at (a) 25°C and (b) 250°C.

20.4 (a) Give the symmetry number for each of the following molecules: (a) CO, (b) O_2, (c) H_2S, (d) SiH_4, and (e) $CHCl_3$.

20.4 (b) Give the symmetry number for each of the following molecules: (a) CO_2, (b) O_3, (c) SO_3, (d) SF_6, and (e) Al_2Cl_6.

20.5 (a) Calculate the rotational partition function of H_2O at 298 K from its rotational constants 27.878 cm^{-1}, 14.509 cm^{-1}, and 9.287 cm^{-1}. Above what temperature is the high-temperature approximation valid?

20.5 (b) Calculate the rotational partition function of SO_2 at 298 K from its rotational constants 2.02736 cm^{-1}, 0.34417 cm^{-1}, and 0.293535 cm^{-1}. Above what temperature is the high-temperature approximation valid?

20.6 (a) From the results of Exercise 20.5a, calculate the rotational contribution to the molar entropy of gaseous water at 25°C.

20.6 (b) From the results of Exercise 20.5b, calculate the rotational contribution to the molar entropy of sulfur dioxide at 25°C.

20.7 (a) Calculate the rotational partition function of CH_4 (a) by direct summation of the energy levels at 298 K and 500 K, and (b) by the high-temperature approximation. Take $B = 5.2412$ cm^{-1}.

20.7 (b) Calculate the rotational partition function of CH_3CN (a) by direct summation of the energy levels at 298 K and 500 K, and (b) by the high-temperature approximation. Take $A = 5.28$ cm^{-1} and $B = 0.307$ cm^{-1}.

20.8 (a) The bond length of O_2 is 120.75 pm. Use the high-temperature approximation to calculate the rotational partition function of the molecule at 300 K.

20.8 (b) The NOF molecule is an asymmetric rotor with rotational constants 3.1752 cm^{-1}, 0.3951 cm^{-1}, and 0.3505 cm^{-1}. Calculate the rotational partition function of the molecule at (a) 25°C, (b) 100°C.

20.9 (a) Plot the molar heat capacity of a collection of harmonic oscillators as a function of T/θ_V, and predict the vibrational heat capacity of ethyne at (a) 298 K, (b) 500 K. The normal modes (and

their degeneracies in parentheses) occur at wavenumbers 612(2), 729(2), 1974, 3287, and 3374 cm^{-1}.

20.9 (b) Plot the molar entropy of a collection of harmonic oscillators as a function of T/θ_V, and predict the standard molar entropy of ethyne at (a) 298 K, (b) 500 K. For data, see the preceding exercise.

20.10 (a) A CO_2 molecule is linear, and its vibrational wavenumbers are 1388.2 cm^{-1}, 667.4 cm^{-1}, and 2349.2 cm^{-1}, the last being doubly degenerate and the others non-degenerate. The rotational constant of the molecule is 0.3902 cm^{-1}. Calculate the rotational and vibrational contributions to the molar Gibbs energy at 298 K.

20.10 (b) An O_3 molecule is angular, and its vibrational wavenumbers are 1110 cm^{-1}, 705 cm^{-1}, and 1042 cm^{-1}. The rotational constants of the molecule are 3.553 cm^{-1}, 0.4452 cm^{-1}, and 0.3948 cm^{-1}. Calculate the rotational and vibrational contributions to the molar Gibbs energy at 298 K.

20.11 (a) The ground level of Cl is $^2P_{3/2}$ and a $^2P_{1/2}$ level lies 881 cm^{-1} above it. Calculate the electronic contribution to the heat capacity of Cl atoms at (a) 500 K and (b) 900 K.

20.11 (b) The first electronically excited state of O_2 is $^1\Delta_g$ and lies 7918.1 cm^{-1} above the ground state, which is $^3\Sigma_g^-$. Calculate the electronic contribution to the molar Gibbs energy of O_2 at 400 K.

20.12 (a) The ground state of the Co^{2+} ion in $CoSO_4 \cdot 7H_2O$ may be regarded as $^4T_{9/2}$. The entropy of the solid at temperatures below 1 K is derived almost entirely from the electron spin. Estimate the molar entropy of the solid at these temperatures.

20.12 (b) Estimate the contribution of the spin to the molar entropy of a solid sample of a d-metal complex with $S = \frac{5}{2}$.

20.13 (a) Calculate the residual molar entropy of a solid in which the molecules can adopt (a) three, (b) five, (c) six orientations of equal energy at $T = 0$.

20.13 (b) Suppose that the hexagonal molecule $C_6H_nF_{6-n}$ has a residual entropy on account of the similarity of the H and F atoms. Calculate the residual for each value of n.

20.14 (a) An average human DNA molecule has 5×10^8 binucleotides (rungs on the DNA ladder) of four different kinds. If each rung were a random choice of one of these four possibilities, what would be the residual entropy associated with this typical DNA molecule?

20.14 (b) Calculate the standard molar entropy of $N_2(g)$ at 298 K from its rotational constant $B = 1.9987$ cm^{-1} and its vibrational wavenumber $\tilde{\nu} = 2358$ cm^{-1}. The thermochemical value is 192.1 J K^{-1} mol^{-1}. What does this suggest about the solid at $T = 0$?

20.15 (a) Calculate the equilibrium constant of the reaction $I_2(g) \rightleftharpoons 2I(g)$ at 1000 K from the following data for I_2: $\tilde{\nu} = 214.36$ cm^{-1}, $B = 0.0373$ cm^{-1}, $D_e = 1.5422$ eV. The ground state of the I atoms is $^2P_{3/2}$, implying fourfold degeneracy.

20.15 (b) Calculate the value of K at 298 K for the gas-phase isotopic exchange reaction $2\,^{79}Br^{81}Br \rightleftharpoons \,^{79}Br^{79}Br + \,^{81}Br^{81}Br$. The Br_2 molecule has a non-degenerate ground state, with no other electronic states nearby. Base the calculation on the wavenumber of the vibration of $^{79}Br^{81}Br$, which is 323.33 cm^{-1}.

Problems

Numerical problems

20.1 The NO molecule has a doubly degenerate electronic ground state and a doubly degenerate excited state at 121.1 cm^{-1}. Calculate the electronic contribution to the molar heat capacity of the molecule at (a) 50 K, (b) 298 K, and (c) 500 K.

20.2 Explore whether a magnetic field can influence the heat capacity of a paramagnetic molecule by calculating the electronic contribution to the heat capacity of an NO_2 molecule in a magnetic field. Estimate the total constant-volume heat capacity using equipartition, and calculate the percentage change in heat capacity brought about by a 5.0 T magnetic field at (a) 50 K, (b) 298 K.

20.3 The energy levels of a CH_3 group attached to a larger fragment are given by the expression for a particle on a ring, provided the group is rotating freely. What is the high-temperature contribution to the heat capacity and the entropy of such a freely rotating group at 25°C? The moment of inertia of CH_3 about its C_3 axis is 5.341×10^{-47} kg m^2.

20.4 Calculate the temperature dependence of the heat capacity of p-H_2 (in which only rotational states with even values of J are populated) at low temperatures on the basis that its rotational levels $J = 0$ and $J = 2$ constitute a system that resembles a two-level system except for the degeneracy of the upper level. Use $B = 60.864$ cm^{-1} and sketch the heat capacity curve. The experimental heat capacity of p-H_2 does in fact show a peak at low temperatures.

20.5 The pure rotational microwave spectrum of HCl has absorption lines at the following wavenumbers (in cm^{-1}): 21.19, 42.37, 63.56, 84.75, 105.93, 127.12, 148.31, 169.49, 190.68, 211.87, 233.06, 254.24, 275.43, 296.62, 317.80, 338.99, 360.18, 381.36, 402.55, 423.74, 444.92, 466.11, 487.30, 508.48. Calculate the rotational partition function at 25°C by direct summation.

20.6 Calculate and plot as a function of temperature, in the range 300 K to 1000 K, the equilibrium constant for the reaction $CD_4(g) + HCl(g) \rightleftharpoons CHD_3(g) + DCl(g)$ using the following data (numbers in parentheses are degeneracies): $\tilde{\nu}(CHD_3)/$cm$^{-1} =$ 2993(1), 2142(1), 1003(3), 1291(2), 1036(2); $\tilde{\nu}(CD_4)/$cm$^{-1} =$ 2109(1), 1092(2), 2259(3), 996(3); $\tilde{\nu}(HCl)/$cm$^{-1} = 2991$; $\tilde{\nu}(DCl)/$cm$^{-1} = 2145$; $B(HCl)/$cm$^{-1} = 10.59$; $B(DCl)/$cm$^{-1} =$ 5.445; $B(CHD_3)/$cm$^{-1} = 3.28$; $A(CHD_3)/$cm$^{-1} = 2.63$; $B(CO_9)/$cm$^{-1} = 2.63$.

20.7 The exchange of deuterium between acid and water is an important type of equilibrium, and we can examine it by using spectroscopic data on the molecules. Calculate the equilibrium constant at (a) 298 K and (b) 800 K for the gas-phase exchange reaction $H_2O + DCl \rightleftharpoons HDO + HCl$ from the following data: $\tilde{\nu}(H_2O)/$cm$^{-1} = 3656.7$, 1594.8, 3755.8; $\tilde{\nu}(HDO)/$cm$^{-1} = 2726.7$, 1402.2, 3707.5; $A(H_2O)/$cm$^{-1} = 27.88$, $B(H_2O)/$cm$^{-1} = 14.51$, $C(H_2O)/$cm$^{-1} = 9.29$; $A(HDO)/$cm$^{-1} = 23.38$, $B(HDO)/$cm$^{-1} =$ 9.102, $C(HDO)/$cm$^{-1} = 6.417$; $B(HCl)/$cm$^{-1} = 10.59$; $B(DCl)/$cm$^{-1} = 5.449$; $\tilde{\nu}(HCl)/$cm$^{-1} = 2991$; $\tilde{\nu}(OCl)/$cm$^{-1} = 2145$.

Theoretical problems

20.8 Derive the Sackur–Tetrode equation for a monatomic gas confined to a two-dimensional surface, and hence derive an expression for the standard molar entropy of condensation to form a mobile surface film.

20.9 Derive expressions for the internal energy, heat capacity, entropy, Helmholtz energy, and Gibbs energy of a harmonic oscillator. Express the results in terms of the vibrational temperature, θ_V and plot graphs of each property against T/θ_V.

20.10 Although expressions like $<\varepsilon> = -\mathrm{d}\ln q/\mathrm{d}\beta$ are useful for formal manipulations in statistical thermodynamics, and for expressing thermodynamic functions in neat formulas, they are sometimes more trouble than they are worth in practical applications. When presented with a table of energy levels, it is often much more convenient to evaluate the following sums directly:

$$q = \sum_j e^{-\beta \varepsilon_j} \qquad \dot{q} = \sum_j \beta \varepsilon_j e^{-\beta \varepsilon_j} \qquad \ddot{q} = \sum_j (\beta \varepsilon_j)^2 e^{-\beta \varepsilon_j}$$

(a) Derive expressions for the internal energy, heat capacity, and entropy in terms of these three functions. (b) Apply the technique to the calculation of the electronic contribution to the constant volume molar heat capacity of magnesium vapour at 5000 K using the following data:

Term	^1S	$^3P_0$	$^3P_1$	$^3P_2$	$^1P_1$	^3S
Degeneracy	1	1	3	5	3	3
$\tilde{\nu}/$cm^{-1}	0	21 850	21 870	21 911	35 051	41 197

20.11 Determine whether a magnetic field can influence the value of an equilibrium constant. Consider the equilibrium $I_2(g) \rightleftharpoons 2I(g)$ at 1000 K, and calculate the ratio of equilibrium constants $K(\mathcal{B})/K$, where $K(\mathcal{B})$ is the equilibrium constant when a magnetic field $\mathcal{B}$ is present and removes the degeneracy of the four states of the $^2P_{3/2}$ level. Data on the species are given in Exercise 20.15a. The electronic g-value of the atoms is $\frac{4}{3}$. Calculate the field required to change the equilibrium constant by 1 per cent.

20.12 The heat capacity ratio of a gas determines the speed of sound in it through the formula

$$c_s = \left(\frac{\gamma RT}{M} \right)^{1/2}$$

where $\gamma = C_p/C_V$ and M is the molar mass of the gas. Deduce an expression for the the speed of sound in a perfect gas of (a) diatomic, (b) linear triatomic, (c) nonlinear triatomic molecules at high temperatures (with translation and rotation active). Estimate the speed of sound in air at 25°C.

Additional problems supplied by Carmen Giunta and Charles Trapp

20.13 For H_2, at very low temperatures, only the translational contribution to the heat capacity is observed. At temperatures above $\theta_R = hcB/k$, the rotational contribution to the heat capacity becomes significant. At still higher temperatures, above $\theta_V = h\nu/k$, the vibrations contribute. But at this latter temperature, dissociation of the molecule into the atoms must be considered. (a) Explain the origin of the expressions for θ_R and θ_V, and calculate their values for hydrogen. (b) Obtain an expression for the molar constant-pressure heat capacity of hydrogen at all temperatures taking into account the dissociation of hydrogen. (c) Make a plot of the molar constant-pressure heat capacity as a function of temperature in the high-temperature region where dissociation of the molecule is significant.

20.14 J.G. Dojahn, E.C.M. Chen, and W.E. Wentworth (*J. Phys. Chem.* **100**, 9649 (1996)) characterized the potential energy curves of the ground and electronic states of homonuclear diatomic halogen anions. The ground state of F_2^- is $^2\Sigma_u^+$ with a fundamental vibrational wavenumber of 450.0 cm^{-1} and equilibrium internuclear distance of 190.0 pm. The first two excited states are at 1.609 and 1.702 eV above the ground state. Compute the standard molar entropy of F_2^- at 298 K.

20.15 R. Viswanathan, R.W. Schmude, Jr, and K.A. Gingerich (*J. Phys. Chem.* **100**, 10784 (1996)) studied thermodynamic properties of several boron–silicon gas-phase species experimentally and theoretically. These species can occur in the high-temperature chemical vapour deposition of silicon-based semiconductors. Among the computations they reported was computation of the Gibbs function of BSi(g) at several temperatures based on a $^4\Sigma^-$ ground state with equilibrium internuclear distance of 190.5 pm and fundamental vibrational wavenumber of 772 cm^{-1} and a $^2P_0$ first excited level 8000 cm^{-1} above the ground level. Compute the standard molar Gibbs function $G_m^{\ominus}(2000\ \text{K}) - G_m^{\ominus}(0)$.

20.16 In a spectroscopic study of the fullerene C_{60}, F. Negri, G. Orlandi, and F. Zerbetto (*J. Phys. Chem.* **100**, 10849 (1996)) reviewed the wavenumbers of all the vibrational modes of the molecule. The wavenumber for the single A_u mode is 976 cm^{-1}; wavenumbers for the four threefold degenerate T_{1u} modes are 525, 578, 1180, and 1430 cm^{-1}; wavenumbers for the five threefold degenerate T_{2u} modes are 354, 715, 1037, 1190, and 1540 cm^{-1}; wavenumbers for six fourfold degenerate G_u modes are 345, 757, 776, 963, 1315, and 1410 cm^{-1}; and wavenumbers for the seven fivefold degenerate H_u modes are 403, 525, 667, 738, 1215, 1342, and 1566 cm^{-1}. How many modes have a vibrational temperature θ_V below 1000 K? Estimate the molar constant-volume heat capacity of C_{60} at 1000 K, counting as active all modes with θ_V below this temperature.

20.17 J. Hutter, H.P. Lüthi, and F. Diederich (*J. Amer. Chem. Soc.* **116**, 750 (1994)) examined the geometric and vibrational structure of several carbon molecules of formula C_n. Given that the ground state of C_3, a molecule found in interstellar space and in flames, is a bent singlet with moments of inertia 39.340, 39.032, and 0.3082 u Å^2 and with vibrational wavenumbers of 63.4, 1224.5, and 2040 cm^{-1}, compute $G_m^{\ominus}(10.00\ \text{K}) - G_m^{\ominus}(0)$ and $G_m^{\ominus}(1000\ \text{K}) - G_m^{\ominus}(0)$ for C_3.

20.18 The molecule Cl_2O_2, which is believed to participate in the seasonal depletion of ozone over Antarctica, has been studied by several means. M. Birk, R.R. Friedl, E.A. Cohen, H.M. Pickett, and S.P. Sander (*J. Chem. Phys.* **91**, 6588 (1989)) report its rotational constants (actually cB) as 13 109.4, 2409.8, and 2139.7 MHz. They also report that its rotational spectrum indicates a molecule with a symmetry number of 2.19. J. Jacobs, M. Kronberg, H.S.P. Möller, and H. Willner (*J. Amer. Chem. Soc.* **116**, 1106 (1994)) report its vibrational wavenumbers as 753, 542, 310, 127, 646, and 419 cm^{-1}. Compute $G_m^{\ominus}(200\ \text{K}) - G_m^{\ominus}(0)$ of Cl_2O_2.

20.19 (a) Show that the number of molecules in any given rotational state of a linear molecule is given by $N_J = C(2J+1)e^{-hcBJ(J+1)/kT}$, where C is a constant. (b) Use this result to prove eqn 16.45 for the J value of the most highly populated rotational energy level. (c) Estimate the temperature at which the spectrum of HCl shown in Fig. 16.40 was taken.

21

Diffraction techniques

In this chapter we return to the techniques that are used to determine structure, but now the emphasis is on the geometrical arrangement of atoms and the distribution of electrons rather than energy levels. All the techniques described in this chapter make use of the property of diffraction of waves by objects with similar dimensions to the wavelength of the waves.

First, we see how to describe the regular arrangement of atoms in crystals and the symmetry of their arrangement. Then we consider the basic principles of X-ray diffraction, and show how the dimensions of unit cells and their symmetries can be inferred from experiments on powdered samples. After that, we turn to the most valuable technique, single-crystal X-ray diffraction, and show how the diffraction pattern can be interpreted in terms of the distribution of electron density in a unit cell. We then explore some of the principles that govern the crystal structures that X-ray diffraction reveals.

In the concluding sections of the chapter we see how neutron diffraction and electron diffraction bear close similarities to X-ray diffraction, but provide complementary information.

A characteristic property of waves is that they interfere with one another, giving a greater displacement where peaks or troughs coincide and a smaller displacement where peaks coincide with troughs (Fig. 21.1). According to classical electromagnetic theory, the intensity of electromagnetic radiation is proportional to the square of the amplitude of the waves. Therefore, the regions of constructive or destructive interference show up as regions of enhanced or diminished intensities. The phenomenon of **diffraction** is the interference caused by an object in the path of waves, and the pattern of varying intensity that results is called the **diffraction pattern**. Diffraction occurs when the dimensions of the diffracting object are comparable to the wavelength of the radiation.

X-rays have wavelengths comparable to bond lengths in molecules and the spacing of atoms in crystals (about 100 pm). By analysing an X-ray diffraction pattern, it is possible to draw up a detailed picture of the location of atoms even in such complex molecules as

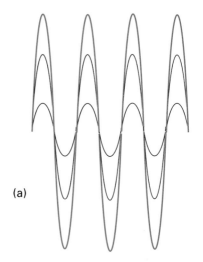

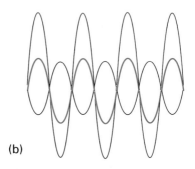

(a)

(b)

21.1 When two waves are in the same region of space they interfere. Depending on their relative phase, they may interfere (a) constructively, to give an enhanced amplitude, or (b) destructively, to give a smaller amplitude. The component waves are shown in black and the resultant in green.

proteins. Electrons moving at about $20\,000\ \mathrm{km\,s^{-1}}$ (after acceleration through about 4 kV) have wavelengths of 40 pm, and may also be diffracted by molecules, surfaces, and thin slices of solids. Neutrons generated in a nuclear reactor, and then slowed to thermal velocities, have similar wavelengths and may also be used for diffraction studies.

Crystal structure

Early in the history of modern science it was suggested that the regular external form of crystals implied an internal regularity of their constituents. In this section we see how to describe the arrangement of atoms inside crystals.

21.1 Lattices and unit cells

A crystal is built up from regularly repeating 'structural motifs', which may be atoms, molecules, or groups of atoms, molecules, or ions. A **space lattice** is the pattern formed by points representing the locations of these motifs (Fig. 21.2). The space lattice is, in effect, an abstract scaffolding for the crystal structure. More formally, the space lattice is a three-dimensional, infinite array of points, each of which is surrounded in an identical way by its neighbours, and which defines the basic structure of the crystal. In some cases there may be a structural motif centred on each lattice point, but that is not necessary. The **crystal structure** itself is obtained by associating with each lattice point an identical structural motif.

The **unit cell** is an imaginary parallelepiped (parallel-sided figure) that contains one unit of the translationally repeating pattern (Fig. 21.3). A unit cell can be thought of as the fundamental region from which the entire crystal may be constructed by purely translational displacements (like bricks in a wall). A unit cell is commonly formed by joining neighbouring lattice points by straight lines (Fig. 21.4). Such unit cells are called **primitive**. It is sometimes more convenient to draw non-primitive unit cells that also have lattice points at their centres or on pairs of opposite faces. An infinite number of different unit cells can describe the same lattice, but we normally choose the one with sides that have the shortest lengths and that are most nearly perpendicular to one another. The lengths of the sides of a

Lattice point

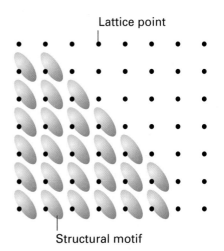

Structural motif

21.2 Each lattice point specifies the location of a structural motif (for example, a molecule or a group of molecules). The crystal lattice is the array of lattice points; the crystal structure is the collection of structural motifs arranged according to the lattice.

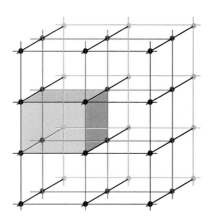

21.3 A unit cell is a parallel-sided (but not necessarily rectangular) figure from which the entire crystal structure can be constructed by using only translations (not reflections, rotations, or inversions).

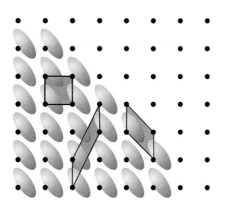

21.4 A unit cell can be chosen in a variety of ways, as shown here. It is conventional to choose the cell that represents the full symmetry of the lattice. In this rectangular lattice, the rectangular unit cell would normally be adopted.

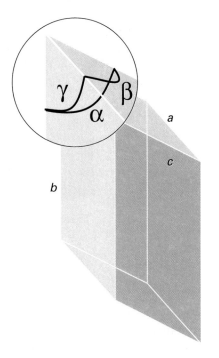

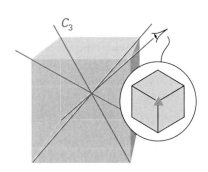

21.5 The notation for the sides and angles of a unit cell. Note that the angle α lies in the plane (b, c) and perpendicular to the axis a.

21.6 A unit cell belonging to the cubic system has four threefold axes arranged tetrahedrally. The insert shows the threefold symmetry.

Table 21.1 The seven crystal systems

System	Essential symmetries
Triclinic	None
Monoclinic	One C_2 axis
Orthorhombic	Three perpendicular C_2 axes
Rhombohedral	One C_3 axis
Tetragonal	One C_4 axis
Hexagonal	One C_6 axis
Cubic	Four C_3 axes in a tetrahedral arrangement

unit cell are denoted a, b, and c, and the angles between them are denoted α, β, and γ (Fig. 21.5).

Unit cells are classified into seven **crystal systems** by noting the rotational symmetry elements they possess. A cubic unit cell, for example, has four threefold axes in a tetrahedral array (Fig. 21.6). A monoclinic unit cell has one twofold axis; the unique axis is by convention the b-axis (Fig. 21.7). A triclinic unit cell has no rotational symmetry, and typically all three sides and angles are different (Fig. 21.8). The **essential symmetries**, the elements that must be present for the unit cell to belong to a particular crystal system, are listed in Table 21.1.

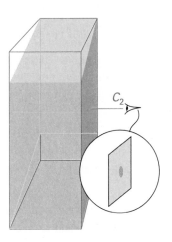

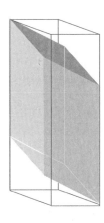

21.7 A unit belonging to the monoclinic system has a twofold axis (shown in more detail in the insert).

21.8 A triclinic unit cell has no axes of rotational symmetry.

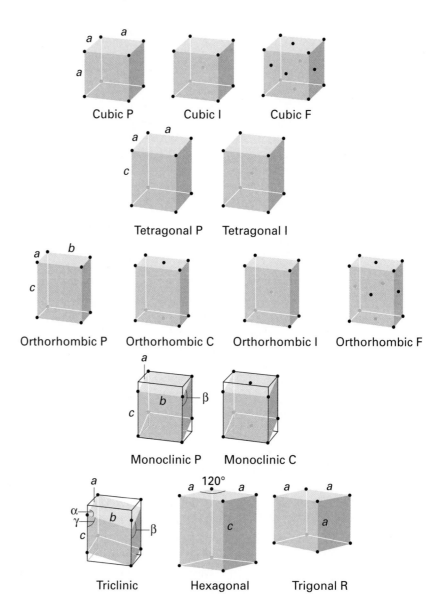

21.9 The fourteen Bravais lattices. The points are lattice points, and are not necessarily occupied by atoms. P denotes a primitive unit cell (R is used for a trigonal lattice), I a body-centred unit cell, F a face-centred unit cell, and C (or A or B) a cell with lattice points on two opposite faces.

There are only 14 distinct crystal lattices in three dimensions. These **Bravais lattices** are illustrated in Fig. 21.9. It is conventional to portray these lattices by primitive unit cells in some cases and by non-primitive unit cells in others. Primitive unit cells (with lattice points only at the corners) are denoted P. A **body-centred unit cell** (I) also has a lattice point at its centre. A **face-centred unit cell** (F) has lattice points at its corners and also at the centres of its six faces. A **side-centred unit cell** (A, B, or C) has lattice points at its corners and at the centres of two opposite faces. For simple structures, it is often convenient to choose an atom belonging to the structural motif, or the centre of a molecule, as the location of a lattice point or the vertex of a unit cell, but that is not a necessary requirement.

21.2 The identification of lattice planes

The spacing of the lattice points in a crystal is an important quantitative aspect of its structure and its investigation by diffraction techniques. However, there are many different sets of planes (Fig. 21.10), and we need to be able to label them. Two-dimensional lattices are easier to visualize than three-dimensional lattices, so we shall introduce the concepts involved by referring to two dimensions initially, and then extend the conclusions by analogy to three dimensions.

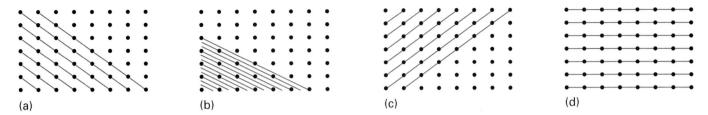

21.10 Some of the planes that can be drawn through the points of the space lattice and their corresponding Miller indices (*hkl*): (a) (110), (b) (230), (c) ($\bar{1}$10), (d) (010).

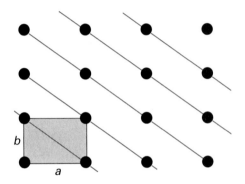

21.11 The dimensions of a unit cell and their relation to the plane passing through the lattice points.

(a) The Miller indices

Consider a two-dimensional rectangular lattice formed from a unit cell of sides *a*, *b* (as in Fig. 21.11). Each plane in the illustration (except the plane passing through the origin) can be distinguished by the distances at which it intersects the *a*- and *b*-axes. One way of labelling each set of parallel planes would therefore be to quote the smallest intersection distances. For example, we could denote the four sets in Fig. 21.10 as $(1a, 1b)$, $(\frac{1}{2}a, \frac{1}{3}b)$, $(-1a, 1b)$, and $(\infty a, 1b)$. However, if we agree to quote distances along the axes as multiples of the lengths of the unit cell, we can label the planes more simply as $(1, 1)$, $(\frac{1}{2}, \frac{1}{3})$, $(-1, 1)$, and $(\infty, 1)$. If the lattice in Fig. 21.10 is the top view of a three-dimensional orthorhombic lattice in which the unit cell has a length *c* in the *z*-direction, all four sets of planes intersect the *z*-axis at infinity. Therefore, the full labels are $(1, 1, \infty)$, $(\frac{1}{2}, \frac{1}{3}, \infty)$, $(-1, 1, \infty)$, and $(\infty, 1, \infty)$.

The presence of fractions and ∞ in the labels is inconvenient. They can be eliminated by taking the reciprocals of the labels. As we shall see, taking reciprocals turns out to have further advantages. The **Miller indices**, (*hkl*), are the reciprocals of intersection distances (with fractions cleared by multiplying through by an appropriate factor, if taking the reciprocal results in a fraction). For example, the $(1, 1, \infty)$ planes in Fig. 21.10a are the (110) planes in the Miller notation. Similarly, the $(\frac{1}{2}, \frac{1}{3}, \infty)$ planes are denoted (230). Negative indices are written with a bar over the number, and Fig. 21.10c shows the ($\bar{1}$10) planes. The Miller indices for the four sets of planes in Fig. 21.10 are therefore (110), (230), ($\bar{1}$10), and (010). A three-dimensional representation of a selection of planes, including one in a lattice with a non-orthogonal axes, is shown in Fig. 21.12.

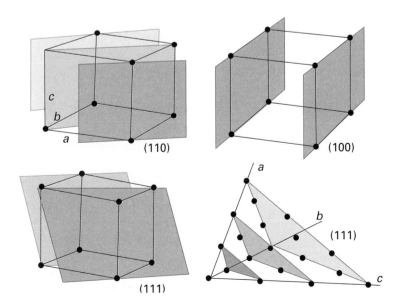

21.12 Some representative planes in three dimensions and their Miller indices. Note that a 0 indicates that a plane is parallel to the corresponding axis, and that the indexing may also be used for unit cells with non-orthogonal axes.

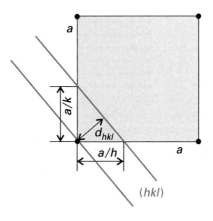

21.13 The calculation of the separation of the planes (*hkl*) in terms of the Miller indices for a square lattice.

A helpful feature to remember is that, the smaller the absolute value of *h* in (*hkl*), the more nearly parallel the plane is to the *a*-axis.[1] The same is true of *k* and the *b*-axis and *l* and the *c*-axis. When *h* = 0, the planes intersect the *a*-axis at infinity, so the (0*kl*) planes are parallel to the *a*-axis. Similarly, the (*h0l*) planes are parallel to *b* and the (*hk*0) planes are parallel to *c*.

(b) The separation of planes

The Miller indices are very useful for expressing the separation of planes. The separation of the (*hk*0) planes in the square lattice shown in Fig. 21.13 is given by

$$\frac{1}{d_{hk0}^2} = \frac{h^2 + k^2}{a^2} \qquad \text{or } d_{hk0} = \frac{a}{(h^2 + k^2)^{1/2}} \tag{1}$$

By extension to three dimensions, the separation of the (*hkl*) planes of a cubic lattice is given by

$$\frac{1}{d_{hkl}^2} = \frac{h^2 + k^2 + l^2}{a^2} \qquad \text{or } d_{hkl} = \frac{a}{(h^2 + k^2 + l^2)^{1/2}} \tag{2}$$

The corresponding expression for a general orthorhombic lattice is the generalization of this expression:

$$\frac{1}{d_{hkl}^2} = \frac{h^2}{a^2} + \frac{k^2}{b^2} + \frac{l^2}{c^2} \tag{3}$$

Example 21.1 Using the Miller indices

Calculate the separation of (a) the (123) planes and (b) the (246) planes of an orthorhombic cell with *a* = 0.82 nm, *b* = 0.94 nm, and *c* = 0.75 nm.

Method For the first part, simply substitute the information into eqn 3. For the second part, instead of repeating the calculation, note that, if all three Miller indices are multiplied by *n*, their separation is reduced by that factor (Fig. 21.14):

$$\frac{1}{d_{nh,nk,nl}^2} = \frac{(nh)^2}{a^2} + \frac{(nk)^2}{b^2} + \frac{(nl)^2}{c^2} = n^2 \left(\frac{h^2}{a^2} + \frac{k^2}{b^2} + \frac{l^2}{c^2} \right) = \frac{n^2}{d_{hkl}^2}$$

which implies that

$$d_{nh,nk,nl} = \frac{d_{hkl}}{n}$$

Answer Substituting the indices into eqn 3 gives

$$\frac{1}{d_{123}^2} = \frac{1^2}{(0.82 \text{ nm})^2} + \frac{2^2}{(0.94 \text{ nm})^2} + \frac{3^2}{(0.75 \text{ nm})^2} = 22 \text{ nm}^{-2}$$

Hence, $d_{123} = 0.21$ nm. It then follows immediately that d_{246} is one-half this value, or 0.11 nm.

- -

Self-test 21.1 Calculate the separation of (a) the (133) planes and (b) the (399) planes in the same lattice.

[0.19 nm, 0.063 nm]

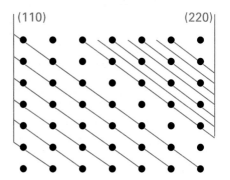

(110)   (220)

21.14 The separation of the (220) planes is half that of the (110) planes. In general, the separation of the planes (*nh, nk, nl*) is *n* times smaller than the separation of the (*hkl*) planes.

---

1 The (*h*00) planes are exceptions.

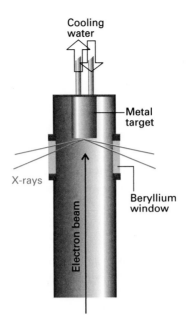

21.15 X-rays are generated by directing an electron beam on to a cooled metal target. Beryllium is transparent to X-rays (on account of the small number of electrons in each atom) and is used for the windows.

X-ray diffraction

X-rays, which are electromagnetic radiation with wavelengths of the order of 10^{-10} m, are typically generated by bombarding a metal with high-energy electrons (Fig. 21.15). The electrons decelerate as they plunge into the metal and generate radiation with a continuous range of wavelengths called **Bremsstrahlung**.[2] Superimposed on the continuum are a few high-intensity, sharp peaks (Fig. 21.16). These peaks arise from collisions of the incoming electrons with the electrons in the inner shells of the atoms. A collision expels an electron from an inner shell, and an electron of higher energy drops into the vacancy, emitting the excess energy as an X-ray photon (Fig. 21.17). If the electron falls into a K shell (that is, a shell with $n = 1$), the X-rays are classified as K-radiation, and similarly for transitions into the L ($n = 2$) and M ($n = 3$) shells. Strong, distinct lines are labelled K_α, K_β, and so on.

21.3 Bragg's law

Wilhelm Röntgen discovered X-rays in 1895. Seventeen years later, Max von Laue suggested that they might be diffracted when passed through a crystal, for by then he had realized that their wavelengths are comparable to the separations of lattice planes. von Laue's suggestion was confirmed almost immediately by Walter Friedrich and Paul Knipping, and has grown since then into a technique of extraordinary power.

An early approach to the analysis of diffraction patterns produced by crystals was to regard a lattice plane as a mirror, and to model a crystal as stacks of reflecting lattice planes of separation d (Fig. 21.18). The model makes it easy to calculate the angle the crystal must

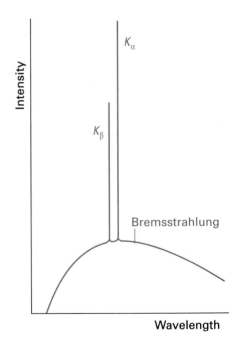

21.16 The X-ray emission from a metal consists of a broad, featureless Bremsstrahlung background, with sharp transitions superimposed on it. The label K indicates that the radiation comes from a transition in which an electron falls into a vacancy in the K shell of the atom.

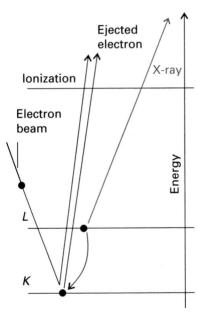

21.17 The processes that contribute to the generation of X-rays. An incoming electron collides with an electron (in the K shell), and ejects it. Another electron (from the L shell in this illustration) falls into the vacancy and emits its excess energy as an X-ray photon.

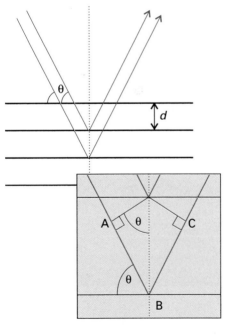

21.18 The conventional derivation of the Bragg law treats each lattice plane as a reflecting the incident radiation. The path lengths differ by $AB + BC$, which depends on the glancing angle, θ. Constructive interference (a 'reflection') occurs when $AB + BC$ is equal to an integer number of wavelengths.

---

2 *Bremse* is German for deceleration, *Strahlung* for ray.

make to the incoming beam of X-rays for constructive interference to occur. It has also given rise to the name **reflection** to denote an intense beam arising from constructive interference.

The path-length difference of the two rays shown in Fig. 21.18 is

$$AB + BC = 2d \sin \theta$$

where θ is the **glancing angle**. For many glancing angles the path-length difference is not an integer number of wavelengths, and the waves interfere largely destructively. However, when the path-length difference is an integer number of wavelengths ($AB + BC = n\lambda$), the reflected waves are in phase and interfere constructively. It follows that a bright reflection should be observed when the glancing angle satisfies **Bragg's law**:

$$n\lambda = 2d \sin \theta \qquad (4)$$

Reflections with $n = 2, 3, \ldots$ are called second-order, third-order, and so on; they correspond to path-length differences of $2, 3, \ldots$ wavelengths. In modern work it is normal to absorb the n into d, to write Bragg's law as

$$\lambda = 2d \sin \theta \qquad (5)$$

and to regard the nth-order reflection as arising from the (nh, nk, nl) planes (see Example 21.1).

The primary use of Bragg's law is in the determination of the spacing between the layers in the lattice for, once the angle θ corresponding to a reflection has been determined, d may readily be calculated.

Example 21.2 Using Bragg's law

A reflection from the (111) planes of a cubic crystal was observed at a glancing angle of 11.2° when Cu K_α X-rays of wavelength 154 pm were used. What is the length of the side of the unit cell?

Method The separation of the planes can be determined from Bragg's law. Because the crystal is cubic, the separation is related to the length of the side of the unit cell, a, by eqn 2, which may therefore be solved for a.

Answer According to eqn 5, the (111) planes responsible for the diffraction have separation

$$d_{111} = \frac{\lambda}{2 \sin \theta}$$

The separation of the (111) planes of a cubic lattice of side a is given by eqn 2 as

$$d_{111} = \frac{a}{3^{1/2}}$$

Therefore,

$$a = \frac{3^{1/2}\lambda}{2 \sin \theta} = \frac{3^{1/2} \times (154 \text{ pm})}{2 \sin 11.2°} = 687 \text{ pm}$$

- -

Self-test 21.2 Calculate the angle at which the same crystal will give a reflection from the (123) planes.

[24.8°]

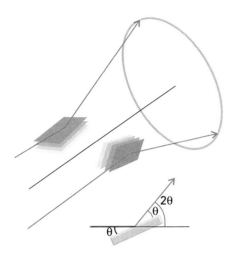

21.19 The same set of planes in two microcrystallites with different orientations around the direction of the incident beam gives diffracted rays that lie on a cone. The full powder diffraction pattern is formed by cones corresponding to reflections from all the sets of (*hkl*) planes that satisfy Bragg's law. (A reflection at a glancing angle θ gives rise to a reflection at an angle 2θ to the direction of the incident beam; see inset.)

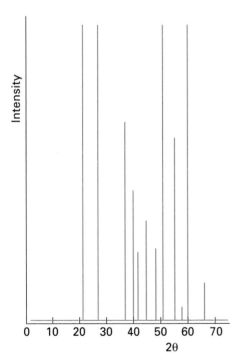

21.20 In the Debye–Scherrer method, a monochromatic X-ray beam is diffracted by a powder sample. The crystallites give rise to cones of intensity, which are detected electronically to give a pattern like that shown here.

21.4 The powder method

von Laue's original method consisted of passing a broad-band beam of X-rays into a single crystal, and recording the diffraction pattern photographically. The idea behind the approach was that a crystal might not be suitably orientated to act as a diffraction grating for a single wavelength but, whatever its orientation, Bragg's law would be satisfied for at least one of the wavelengths if a range of wavelengths was used. There is currently a resurgence of interest in this approach because synchrotron radiation spans a range of X-ray wavelengths (Section 16.1a).

(a) The Debye–Scherrer method

An alternative technique to von Laue's was developed by Peter Debye and Paul Scherrer and independently by Albert Hull. They used monochromatic radiation and a powdered sample. When the sample is a powder, at least some of the crystallites will be orientated so as to satisfy the Bragg condition for each set of planes (*hkl*). For example, some of the crystallites will be oriented so that their (111) planes, of spacing d_{111}, give rise to diffracted intensity at the glancing angle θ (Fig. 21.19). The crystallites with this glancing angle will lie at all possible angles around the incoming beam, so the diffracted beams lie on a cone around the incident beam of half-angle 2θ. Other crystallites will be oriented with different planes satisfying the Bragg condition. They give rise to a cone of diffracted intensity with a different half-angle. In principle, each set of (*hkl*) planes gives rise to a diffraction cone, because some of the randomly orientated crystallites will have the correct angle to diffract the incident beam. In modern powder diffractometers the intensities of the reflections are monitored electronically as the detector is rotated around the sample in a plane containing the incident ray (Fig. 21.20).

Powder diffraction techniques are used to identify a sample of a solid substance by comparison of the positions of the diffraction lines and their intensities with a large data bank (*The powder diffraction file*, which is maintained by the International Centre for Diffraction Data, ICDD, and contains information on about 50 000 crystalline phases). Powder diffraction data are also used to determine phase diagrams, for different solid phases result in different diffraction patterns, and to determine the relative amounts of each phase present in a mixture. The technique is also used for the initial determination of the dimensions and symmetries of unit cells, as the following section explains.

(b) Indexing the reflections

Bragg's law is used to interpret the angle θ of a reflection in terms of the separation of the lattice planes. If the values of *h*, *k*, and *l* for the planes responsible for that reflection are known, the dimensions of the unit cell can be deduced. The crux of the technique is therefore the **indexing** of the reflection, or ascribing the indices *hkl* to it.

Some types of unit cell give characteristic and easily recognizable patterns of lines. For example, in a cubic lattice of unit cell dimension *a* the spacing is given by eqn 2, so the angles at which the (*hkl*) planes give reflections are given by

$$\sin \theta = (h^2 + k^2 + l^2)^{1/2} \frac{\lambda}{2a} \tag{6}$$

The reflections are then predicted by substituting the values of *h*, *k*, and *l*:

(*hkl*)	(100)	(110)	(111)	(200)	(210)	(211)	(220)	(300)	(221)	(310)	$\cdots$
$h^2 + k^2 + l^2$	1	2	3	4	5	6	8	9	9	10	$\cdots$

Notice that 7 (and 15, . . .) is missing because the sum of the squares of three integers cannot equal 7 (or 15, . . .). Therefore the pattern has omissions that are characteristic of the cubic P lattice.

Example 21.3 Identifying the unit cell

A powder diffraction photograph of the element polonium gave lines at the following values of 2θ (in degrees) when 71.0 pm Mo X-rays were used: 12.1, 17.1, 21.0, 24.3, 27.2, 29.9, 34.7, 36.9, 38.9, 40.9, 42.8. Identify the unit cell and determine its dimensions.

Method From eqn 6 we write

$$\sin^2\theta = A(h^2 + k^2 + l^2) \qquad A = \left(\frac{\lambda}{2a}\right)^2$$

We need to determine the common factor A, and then find $h^2 + k^2 + l^2$.

Answer We draw up the following table:

$2\theta/°$	12.1	17.1	21.0	24.3	27.2	29.9	34.7	36.9	38.9	40.9	42.8
$\theta/°$	6.05	8.55	10.5	12.2	13.6	15.0	17.4	18.5	19.5	20.5	21.4
$100\sin^2\theta$	1.11	2.21	3.32	4.47	5.53	6.70	8.94	10.1	11.1	12.3	13.3

The common divisor is 1.11/100. Divide through to identify $h^2 + k^2 + l^2$:

$h^2 + k^2 + l^2$	1	2	3	4	5	6	8	9	10	11	12

The corresponding indices are

(100) (110) (111) (200) (210) (211) (220) (300) (310) (311) (222)

Note that indices like (120) and (012) are equivalent to (210) in this list. We have now indexed the lines. Note the absence of $h^2 + k^2 + l^2 = 7$, which indicates a primitive cubic (cubic P) cell. From $(\lambda/2a)^2 = 0.0111$, we find $a = 337$ pm.

Comment Later we shall see that additional information comes from the intensities of the lines.

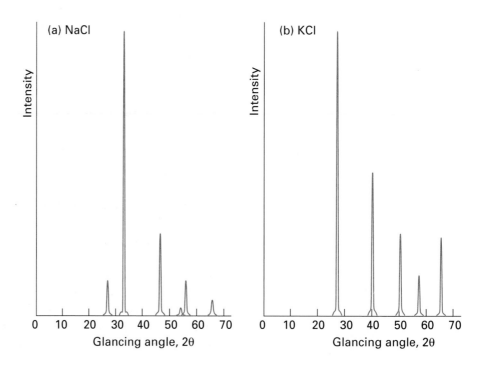

21.21 X-ray powder photographs of (a) NaCl, (b) KCl and the indexed reflections. The smaller number of lines in (b) is a consequence of the similarity of the K$^+$ and Cl$^-$ scattering factors, as discussed later in the chapter.

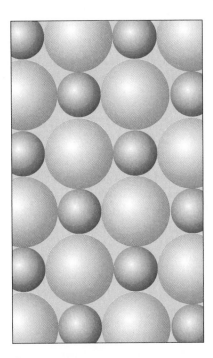

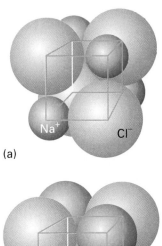

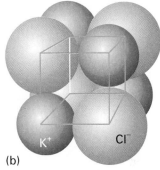

21.22 A fragment of the structure of NaCl and KCl: showing one plane of ions. Each cation (small spheres) has anions (large spheres) as its nearest neighbours. (For the three-dimensional structure, see Fig. 21.40.)

21.23 (a) One octant of the unit cell of NaCl; the Na$^+$ ions and Cl$^-$ ions have different numbers of electrons and hence have different scattering factors. (b) One octant of the unit cell of KCl; the K$^+$ ions and Cl$^-$ ions have the same numbers of electrons and hence have similar scattering factors. The diffraction pattern in this case is that of a primitive cubic lattice.

Self-test 21.3 In the same camera, another cubic crystal gave reflections at the following values of 2θ (in degrees): 10.4, 14.8, 18.2, 21.0, 23.6, 25.8, 27.7. Identify the unit cell and determine its dimension. See Fig. 21.26 for assistance.

[Cubic I; 550 pm]

(c) Systematic absences

The X-ray powder diffraction patterns for NaCl and KCl are remarkably different for two such similar structures (Fig. 21.21). Both crystals consist of two mutually interpenetrating face-centred cubic arrays of ions, one of Na$^+$ ions or K$^+$ ions and the other of Cl$^-$ ions (Fig. 21.22). The explanation of the difference is found in the scattering strengths of the ions and the interference between waves scattered by the cations and anions. Thus, some reflections from the Na$^+$ ions are in phase with the Cl$^-$ reflections, and the two reflections augment one another to give more intense maxima. For other orientations, the two sets of reflections may be out of phase, and tend to cancel, but, as the scattering strengths of the two ions are different, the cancellation is incomplete. For KCl, however, the scattering strengths of K$^+$ and Cl$^-$, which have the same numbers of electrons, are very similar, and cancellation is complete. The ions in KCl therefore all look very similar (Fig. 21.23) and, instead of appearing to be face-centred cubic, the powder diffraction pattern is that typical of a lattice with a primitive cubic unit cell.

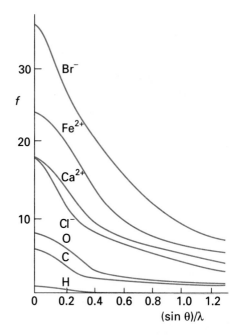

21.24 The variation of the scattering factor of atoms and ions with atomic number and angle. The scattering factor in the forward direction (at $\theta = 0$, and hence at $(\sin\theta)/\lambda = 0$) is equal to the number of electrons present in the species.

The general form of the diffraction pattern of a crystal composed of atoms and ions with different scattering strengths can be predicted by considering a crystal composed of A and B atoms with scattering strengths measured by their **scattering factors**, f_A and f_B. If the scattering factor is large, the atoms scatter X-rays strongly. The scattering factor of an atom is related to the electron density distribution in the atom, $\rho(r)$, by

$$f = 4\pi \int_0^\infty \rho(r) \frac{\sin kr}{kr} r^2 \, dr \qquad k = \frac{4\pi}{\lambda} \sin\theta \qquad (7)$$

The value of f is greatest in the forward direction and smaller for directions away from the forward direction (Fig. 21.24). The detailed analysis of the intensities of reflections must take this dependence on direction into account (in single-crystal studies as well as for powders). We show in the *Justification* below that, in the forward direction (for $\theta = 0$), f is equal to the total number of electrons in the atom.

Justification 21.1

As $\theta \to 0$, so $k \to 0$. Because $\sin x = x - \frac{1}{6}x^3 + \cdots$,

$$\lim_{x \to 0} \frac{\sin x}{x} = \lim_{x \to 0} \frac{x - \frac{1}{6}x^3 + \cdots}{x} = \lim_{x \to 0} \left(1 - \frac{1}{6}x^2 + \cdots\right) = 1$$

The factor $(\sin kr)/kr$ is therefore equal to 1 for forward scattering. It follows that in the forward direction

$$f = 4\pi \int_0^\infty \rho(r) r^2 \, dr$$

The integral over the electron density ρ (the number of electrons in an infinitesimal region divided by the volume of the region) multiplied by the volume element $4\pi r^2 \, dr$ is the total number of electrons, N_e, in the atom. Hence, in the forward direction, $f = N_e$. For example, the scattering factors of Na^+, K^+, and Cl^- are 8, 18, and 18, respectively.

The scattering factor is smaller in non-forward directions because $(\sin kr)/kr < 1$ for $\theta > 0$, so the integral is smaller than the value calculated above.

We shall now calculate the diffraction pattern to expect when two kinds of atom are present in a unit cell. We begin by showing in the following *Justification* that, if in the unit cell there is an A atom at the origin and a B atom at the coordinates (xa, yb, zc), where $x, y,$ and z lie in the range 0 to 1, then the phase difference between the hkl reflections of the A and B atoms is

$$\phi_{hkl} = 2\pi(hx + ky + lz) \qquad (8)$$

Justification 21.2

Consider the crystal shown schematically in Fig. 21.25. The reflection corresponds to two waves from adjacent A planes, the phase difference of the waves being 2π. If there is a B atom at a fraction x of the distance between the two A planes, then it gives rise to a wave with a phase difference $2\pi x$ relative to an A reflection. To see this conclusion, note that, if $x = 0$, there is no phase difference; if $x = \frac{1}{2}$ the phase difference is π; if $x = 1$, the B atom lies where the lower A atom is and the phase difference is 2π. Now consider a (200) reflection. There is now a $2 \times 2\pi$ difference between the waves from the two A layers, and if B were to lie at $x = 0.5$ it would give rise to a wave that differed in phase by 2π from the wave from the upper A layer. Thus, for a general fractional position x, the phase difference for a (200) reflection is $2 \times 2\pi x$. For a general (h00) reflection, the phase difference is therefore $h \times 2\pi x$. For three dimensions, this result generalizes to eqn 8.

The A and B reflections interfere destructively when the phase difference is π, and the total intensity is zero if the atoms have the same scattering power. For example, if the unit cells are cubic I with a B atom at $x = y = z = \frac{1}{2}$, then the A, B phase difference is $(h + k + l)\pi$. Therefore, all reflections for odd values of $h + k + l$ vanish because the waves are displaced in phase by π. Hence the diffraction pattern for a cubic I lattice can be constructed from that for the cubic P lattice (a cubic lattice without points at the centre of its unit cells) by striking out all reflections with odd values of $h + k + l$. Recognition of these **systematic absences** in a powder spectrum immediately indicates a cubic I lattice (Fig. 21.26).

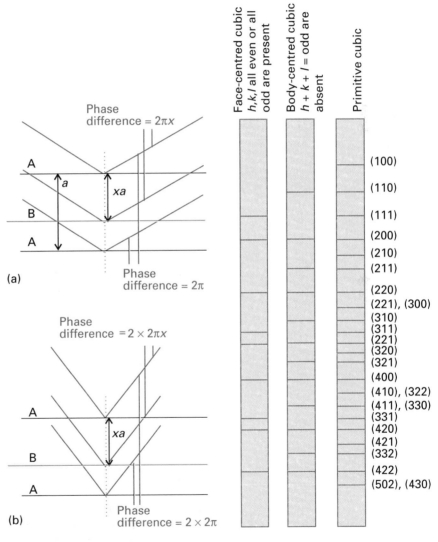

21.25 Diffraction from a crystal containing two kinds of atom. (a) For a (100) reflection from the A planes there is a phase difference of 2π between waves reflected by neighbouring planes, but (b) for a (200) reflection the phase difference is 4π. The reflection from a B plane at a fractional distance xa from an A plane has a phase that is x times these phase differences.

21.26 The powder diffraction patterns and the systematic absences of three versions of a cubic cell. Comparison of the observed pattern with patterns like these enables the unit cell to be identified. The locations of the lines give the cell dimensions.

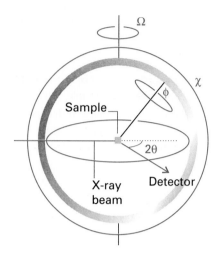

21.27 A four-circle diffractometer. The settings of the orientations (ϕ, χ, θ, and Ω) of the components are controlled by computer; each (hkl) reflection is monitored in turn, and their intensities are recorded.

If the amplitude of the waves scattered from A is f_A at the detector, that of the waves scattered from B is $f_B e^{i\phi_{hkl}}$, with ϕ_{hkl} the phase difference given in eqn 8. The total amplitude at the detector is therefore

$$F_{hkl} = f_A + f_B e^{i\phi_{hkl}} \tag{9}$$

Because the intensity is proportional to the square modulus of the amplitude of the wave, the intensity, I_{hkl}, at the detector is

$$I_{hkl} \propto F^*_{hkl} F_{hkl} = \{f_A + f_B e^{-i\phi_{hkl}}\}\{f_A + f_B e^{i\phi_{hkl}}\} \tag{10}$$

This expression expands to

$$I_{hkl} \propto f_A^2 + f_B^2 + f_A f_B (e^{i\phi_{hkl}} + e^{-i\phi_{hkl}}) = f_A^2 + f_B^2 + 2f_A f_B \cos\phi_{hkl} \tag{11}$$

The cosine term either adds to or subtracts from $f_A^2 + f_B^2$ depending on the value of ϕ_{hkl}, which in turn depends on h, k, and l and x, y, and z (through eqn 8). Hence, there is a variation in the intensities of the lines with different hkl, which is exactly what is observed for NaCl (Fig. 21.21a).

21.5 Single-crystal X-ray diffraction

The method developed by the Braggs (William and his son Lawrence, who later jointly won the Nobel Prize) is the foundation of almost all modern work in X-ray crystallography. They used a single crystal and a monochromatic beam of X-rays, and rotated the crystal until a reflection was detected. There are many different sets of planes in a crystal, so there are many angles at which a reflection occurs. The complete set of data consists of the list of angles at which reflections are observed and their intensities.

(a) The technique

The diffraction pattern produced by a single crystal is measured by using a **four-circle diffractometer** (Fig. 21.27). The computer linked to the diffractometer determines the unit cell dimensions and the angular settings of the diffractometer's four circles that are needed to observe any particular (hkl) reflection. The computer controls the settings, and moves the crystal and the detector for each one in turn. At each setting, the diffraction intensity is measured, and background intensities are assessed by making measurements at slightly different settings. Computing techniques are now available that lead not only to automatic indexing but also to the automated determination of the shape, symmetry, and size of the unit cell. Moreover, several techniques are now available for sampling large amounts of data, including area detectors and image plates, which sample whole regions of diffraction patterns simultaneously.

(b) Structure factors

The problem we now address is how to interpret the data from a diffractometer in terms of the structure of the crystal. To do so, we must go beyond Bragg's law.

Suppose the unit cell contains several atoms with scattering factors f_j and coordinates $(x_j a, y_j b, z_j c)$. The overall amplitude of a wave diffracted by the (hkl) planes is a generalization of the expression $F_{hkl} = f_A + f_B e^{i\phi_{hkl}}$ obtained earlier:

$$F_{hkl} = \sum_j f_j e^{i\phi_{hkl}(j)} \qquad \phi_{hkl}(j) = 2\pi(hx_j + ky_j + lz_j) \tag{12}$$

The sum is over all the atoms in the unit cell. The quantity F_{hkl} is called the **structure factor**, and the intensity of the (hkl) reflection is proportional to $|F_{hkl}|^2$.

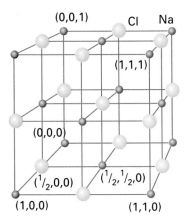

(0,0,1)   Cl   Na

(1,1,1)

(0,0,0)

($^1/_2$,0,0)   ($^1/_2$,$^1/_2$,0)

(1,0,0)   (1,1,0)

21.28 The location of the atoms for the structure factor calculation in Example 21.4. The dark spheres are Na$^+$, the light spheres are Cl$^-$.

Example 21.4 Calculating a structure factor

Calculate the structure factors for the unit cell in Fig. 21.28.

Method The structure factor is given by eqn 12. To use this equation, consider the ions at the locations specified in Fig. 21.28. Write f^+ for the Na$^+$ scattering factor and f^- for the Cl$^-$ scattering factor. Note that ions in the body of the cell contribute to the scattering with a strength f. However, ions on faces are shared between two cells (use $\frac{1}{2}f$), those on edges by four cells (use $\frac{1}{4}f$), and those at corners by eight cells (use $\frac{1}{8}f$). Two useful relations are

$$e^{i\pi} = -1 \qquad \cos\phi = \tfrac{1}{2}(e^{i\phi} + e^{-i\phi})$$

Answer From eqn 12, and summing over the coordinates of all 27 atoms in the illustration:

$$F_{hkl} = f^+\left(\tfrac{1}{8} + \tfrac{1}{8}e^{2\pi il} + \cdots + \tfrac{1}{2}e^{2\pi i(\frac{1}{2}h+\frac{1}{2}k+l)}\right)$$
$$+ f^-\left(e^{2\pi i(\frac{1}{2}h+\frac{1}{2}k+\frac{1}{2}l)} + \tfrac{1}{4}e^{2\pi i(\frac{1}{2}h)} + \cdots + \tfrac{1}{4}e^{2\pi i(\frac{1}{2}h+l)}\right)$$

To simplify this 27-term expression, we use

$$e^{2\pi ih} = e^{2\pi ik} = e^{2\pi il} = 1$$

because h, k, and l are all integers:

$$F_{hkl} = f^+\{1 + \cos(h+k)\pi + \cos(h+l)\pi + \cos(k+l)\pi\}$$
$$+ f^-\{(-1)^{h+k+l} + \cos k\pi + \cos l\pi + \cos h\pi\}$$

Then, because $\cos h\pi = (-1)^h$,

$$F_{hkl} = f^+\{1 + (-1)^{h+k} + (-1)^{h+l} + (-1)^{l+k}\}$$
$$+ f^-\{(-1)^{h+k+l} + (-1)^h + (-1)^k + (-1)^l\}$$

Now note that:
   if h, k, and l are all even, $k_{hkl} = f^+\{1+1+1+1\}$
                                                    $+ f^-\{1+1+1+1\} = 4(f^+ + f^-)$
   if h, k, and are all odd, $F_{hkl} = 4(f^+ - f^-)$
   if one index is odd and two are even, or vice versa, $F_{hkl} = 0$

Comment The hkl all-odd reflections are less intense than the hkl all-even.

Self-test 21.4 Deduce the rule for the systematic absences of a cubic I lattice.

[for $h + k + l$ odd, $F_{hkl} = 0$]

(c) The electron density

If we knew all the structure factors F_{hkl}, we could calculate the electron density distribution, $\rho(\mathbf{r})$, in the unit cell by using the expression

$$\rho(\mathbf{r}) = \frac{1}{V}\sum_{hkl} F_{hkl}e^{-2\pi i(hx+ky+lz)} \tag{13}$$

where V is the volume of the unit cell. Equation 13 is called a **Fourier synthesis** of the electron density.

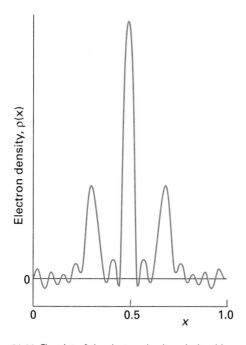

21.29 The plot of the electron density calculated in Example 21.5.

Example 21.5 Calculating an electron density by Fourier synthesis

Consider the $(h00)$ planes of a crystal extending indefinitely in the x-direction. In an X-ray analysis the structure factors were found as follows:

h:	0	1	2	3	4	5	6	7	8	9
F_h:	16	−10	2	−1	7	−10	8	−3	2	−3
h:	10	11	12	13	14	15				
F_h:	6	−5	3	−2	2	−3				

(and $F_{-h} = F_h$). Construct a plot of the electron density projected on to the x-axis of the unit cell.

Method Because $F_{-h} = F_h$, it follows from eqn 13 that

$$V\rho(x) = \sum_{h=-\infty}^{\infty} F_h e^{-2\pi i h x} = F_0 + \sum_{h=1}^{\infty}\left(F_h e^{-2\pi i h x} + F_{-h} e^{2\pi i h x}\right)$$

$$= F_0 + \sum_{h=1}^{\infty} F_h\left(e^{-2\pi i h x} + e^{2\pi i h x}\right) = F_0 + 2\sum_{h=1}^{\infty} F_h \cos(2\pi h x)$$

and we evaluate the sum (truncated at $h = 15$) for points $0 \leq x \leq 1$ using mathematical software.

Answer The results are plotted in Fig. 21.29.

Comment The positions of three atoms can be discerned very readily. The more terms that are included, the more accurate the density plot. Terms corresponding to high values of h (short-wavelength cosine terms in the sum) account for the finer details of the electron density; low values of h account for the broad features.

- -

Self-test 21.5 Use mathematical software to experiment with different structure factors (including changing signs as well as amplitudes). For example, use the same values of F_h as above, but with positive signs throughout.

[Fig. 21.30]

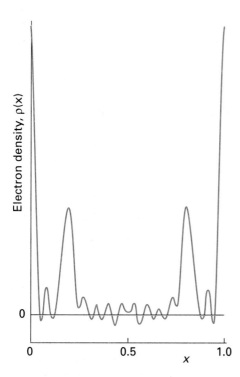

21.30 The plot of the electron density calculated in Self-test 21.5. Note how a different choice of phases for the structure factors leads to a markedly different structure.

(d) The phase problem

From the measured intensities, I_{hkl}, we get the structure factors, F_{hkl}, and then evaluate eqn 13 to find the electron density, $\rho(\mathbf{r})$, throughout the unit cell. Unfortunately, I_{hkl} is proportional to the square modulus $|F_{hkl}|^2$, so we cannot say whether we should use $+|F_{hkl}|$ or $-|F_{hkl}|$ in the sum. In fact, the difficulty is more severe for non-centrosymmetric unit cells because, if we write F_{hkl} as the complex number $|F_{hkl}|e^{i\alpha}$, where α is the phase of F_{hkl} and $|F_{hkl}|$ is its magnitude, the intensity lets us determine $|F_{hkl}|$ but tells us nothing of its phase, which may lie anywhere from 0 to 2π. This ambiguity is called the **phase problem**; its consequences are illustrated by comparing Figs. 21.29 and 21.30. Some way must be found to assign phases to the structure factors, for otherwise the sum for ρ could not be evaluated and the method would be useless.

The phase problem can be overcome to some extent by a variety of methods. One procedure that is widely used for inorganic materials with a reasonably small number of atoms in a unit cell and for organic molecules with a small number of heavy atoms, is the **Patterson synthesis**. Instead of the structure factors F_{hkl}, the values of $|F_{hkl}|^2$, which can be

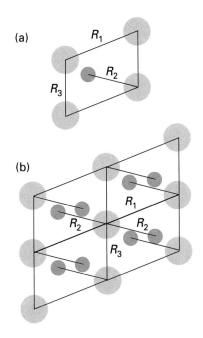

21.31 The Patterson synthesis corresponding to the pattern in (a) is the pattern in (b). The distance and orientation of each spot from the origin gives the orientation and separation of one atom–atom separation in (a). Some of the typical distances and their contribution to (b) are shown as R_1, etc.

obtained without ambiguity from the intensities, are used in an expression that resembles eqn 13:

$$P(\mathbf{r}) = \frac{1}{V} \sum_{hkl} |F_{hkl}|^2 e^{-2\pi i(hx+ky+lz)} \tag{14}$$

The outcome of a Patterson synthesis is a map of the *vector separations* of the atoms (the distances and directions between atoms) in the unit cell. Thus, if atom A is at the coordinates (x_A, y_A, z_A) and atom B is at (x_B, y_B, z_B), then there will be a peak at $(x_A - x_B, y_A - y_B, z_A - z_B)$ in the Patterson map. There will also be a peak at the negative of these coordinates, because there is a vector from B to A as well as a vector from A to B. The height of the peak in the map is proportional to the product of the atomic numbers of the two atoms, $Z_A Z_B$. For example, if the unit cell has the structure shown in Fig. 21.31a, the Patterson synthesis would be the map shown in Fig. 21.31b, where the location of each spot relative to the origin gives the separation and relative orientation of each pair of atoms in the original structure.

If some atoms are heavy, they dominate the scattering (because their scattering factors are large, of the order of their atomic number) and their locations may be deduced quite readily. The sign of F_{hkl} can now be calculated from the locations of the heavy atoms in the unit cell, and to a high probability the phase calculated for them will be the same as the phase for the entire unit cell. To see why this is so, we have to note that a structure factor of a centrosymmetric cell has the form

$$F = (\pm)f_{\text{heavy}} + (\pm)f_{\text{light}} + (\pm)f_{\text{light}} + \cdots \tag{15}$$

where f_{heavy} is the scattering factor of the heavy atom and f_{light} the scattering factors of the light atoms. (An expression of this form, but for atoms of similar atomic number, was derived in Example 21.4.) The f_{light} are all much smaller than f_{heavy}, and their phases are more or less random if the atoms are distributed throughout the unit cell. Therefore, the net effect of the f_{light} is to change F only slightly from f_{heavy}, and we can be reasonably confident that F will have the same sign as that calculated from the location of the heavy atom. This phase can then be combined with the observed $|F|$ (from the reflection intensity) to perform a Fourier synthesis of the full electron density in the unit cell, and hence to locate the light atoms as well as the heavy atoms.

Modern structural analyses make extensive use of **direct methods**. Direct methods are based on the possibility of treating the atoms in a unit cell as being virtually randomly distributed (from the radiation's point of view), and then to use statistical techniques to compute the probabilities that the phases have a particular value. It is possible to deduce relations between some structure factors and sums (and sums of squares) of others, which have the effect of constraining the phases to particular values (with high probability, so long as the structure factors are large). For example, the **Sayre probability relation** has the form

$$\text{sign of } F_{h+h',k+k',l+l'} \text{ is probably equal to } (\text{sign of } F_{hkl}) \times (\text{sign of } F_{h'k'l'}) \tag{16}$$

For example, if F_{122} and F_{232} are both large and negative, then it is highly likely that F_{354}, provided it is large, will be positive.

(e) Structure refinement

In the final stages of the determination of a crystal structure, the parameters describing the structure (atom positions, for instance) are adjusted systematically to give the best fit between the observed intensities and those calculated from the model of the structure deduced from the diffraction pattern. This process is called **structure refinement**. Not only does the procedure give accurate positions for all the atoms in the unit cell, but it also gives

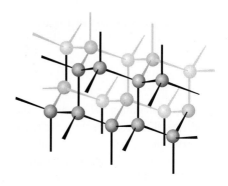

21.32 A fragment of the structure of diamond. Each C atom is tetrahedrally bonded to four neighbours. This framework-like structure results in a rigid crystal with a high thermal conductivity.

an estimate of the errors in those positions and in the bond lengths and angles derived from them. The procedure also provides information on the vibrational amplitudes of the atoms.

Information from X-ray analysis

The bonding within a solid may be of various kinds. Simplest of all (in principle) are **metals**, where electrons are delocalized over arrays of identical cations and bind them together into a rigid but ductile and malleable whole. In many cases the crystal structures of metals can be rationalized in terms of a model in which spherical metal cations pack together into an orderly array. In an **ionic solid**, both cations and anions are packed together.

In **covalent solids**, covalent bonds in a definite spatial orientation link the atoms in a network extending through the crystal. The demands of directional bonding, which have only a small effect on the structures of many metals, now override the geometrical problem of packing spheres together, and elaborate and extensive structures may be formed. A famous example of a covalent solid is diamond (Fig. 21.32), in which each sp^3-hybridized C atom is bonded tetrahedrally to its four neighbours.

Molecular solids, which are the subject of the overwhelming majority of modern structural determinations, are bonded together by van der Waals interactions (Chapter 22). The observed crystal structure is nature's solution to the problem of condensing objects of various shapes into an aggregate of minimum energy (actually, for temperatures above zero, of minimum Gibbs energy). The prediction of the structure is a very difficult task, and rarely possible. The problem is made more complicated by the role of hydrogen bonds, which in some cases dominate the crystal structure, as in ice (Fig. 21.33), but in others (for example, phenol) distort a structure that is determined largely by the van der Waals interactions. X-ray diffraction studies of molecular compounds reveal a huge amount of information, including interatomic distances, bond angles, the stereochemistry of the molecules, and vibrational parameters.

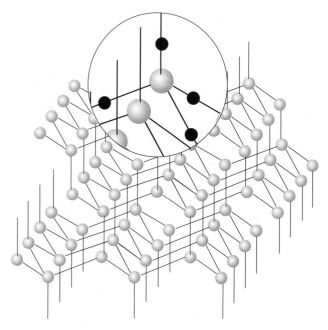

21.33 A fragment of the crystal structure of ice (ice-I). Each O atom is at the centre of a tetrahedron of four O atoms at a distance of 276 pm. The central O atom is attached by two short O–H bonds to two H atoms and by two long hydrogen bonds to the H atoms of two of the neighbouring molecules. Overall, the structure consists of planes of hexagonal puckered rings of H_2O molecules (like the chair form of cyclohexane).

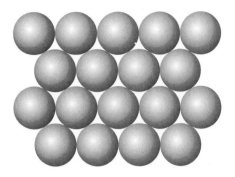

21.34 The first layer of close-packed spheres used to build a three-dimensional close-packed structure.

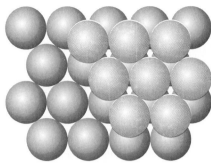

21.35 The second layer of close-packed spheres occupies the dips of the first layer. The two layers are the AB component of the close-packed structure.

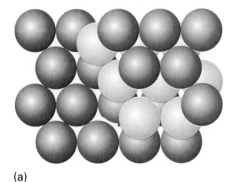

(a)

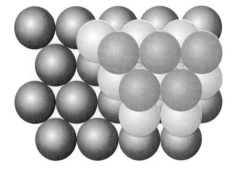

(b)

21.36 (a) The third layer of close-packed spheres might occupy the dips lying directly above the spheres in the first layer, resulting in an ABA structure, which corresponds to hexagonal close-packing. (b) Alternatively, the third layer might lie in the dips that are not above the spheres in the first layer, resulting in an ABC structure, which corresponds to cubic close-packing.

21.6 The packing of identical spheres: metal crystals

Most metallic elements crystallize in one of three simple forms, two of which can be explained in terms of rigid spheres packing together in the closest possible arrangement.

(a) Close packing

A **close-packed** layer of identical spheres, one with maximum utilization of space, is shown in Fig. 21.34. A close-packed three-dimensional structure can be envisaged as formed by stacking such close-packed layers on top of one another. However, this stacking can be done in different ways and can result in close-packed **polytypes**, or structures that are identical in two dimensions (the close-packed layers) but differ in the third dimension.

In all polytypes, the spheres of the second close-packed layer lie in the depressions of the first layer (Fig. 21.35). The third layer may be added in either of two ways. In one, the spheres are placed so that they reproduce the first layer (Fig. 21.36a), to give an ABA pattern of layers. Alternatively, the spheres may be placed over the gaps in the first layer (Fig. 21.36b), so giving an ABC pattern. Two polytypes are formed if the two stacking patterns are repeated in the vertical direction. If the ABA pattern is repeated, to give the sequence of layers ABABAB..., the spheres are **hexagonally close-packed** (hcp). Alternatively, if the ABC pattern is repeated, to give the sequence ABCABC..., the spheres are **cubic close-packed** (ccp). The origins of these names can be seen by referring to Fig. 21.37. The ccp structure gives rise to face-centred unit cells, so may also be denoted cubic F (or fcc, for

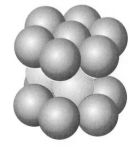

(a)

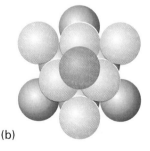

(b)

21.37 A fragment of the structure shown in Fig. 21.36 revealing the (a) hexagonal (b) cubic symmetry. The tints on the spheres are the same as for the layers in Fig. 21.36.

21.38 The calculation of the packing fraction of a ccp unit cell.

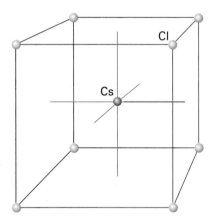

21.39 The caesium-chloride structure consists of two interpenetrating simple cubic arrays of ions, one of cations and the other of anions, so that each cube of ions of one kind has a counter-ion at its centre.

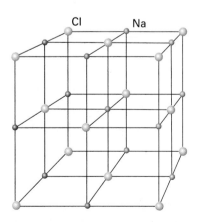

21.40 The rock-salt (NaCl) structure consists of two mutually interpenetrating slightly expanded face-centred cubic arrays of ions. The entire assembly shown here is the unit cell.

Table 21.2 The crystal structures of some elements

Structure	Element
hcp*	Be, Cd, Co, He, Mg, Sc, Ti, Zn
fcc* (ccp, cubic F)	Ag, Al, Ar, Au, Ca, Cu, Kr, Ne, Ni, Pd, Pb, Pt, Rh, Rn, Sr, Xe
bcc (cubic I)	Ba, Cs, Cr, Fe, K, Li, Mn, Mo, Rb, Na, Ta, W, V
cubic P	Po

* Close-packed structures.

face-centred cubic). It is also possible to have ABCABAB . . . structures and even random sequences; however, the hcp and ccp polytypes are the most important. Some elements possessing these structures are listed in Table 21.2.

The compactness of close-packed structures is indicated by their **coordination number**, the number of atoms immediately surrounding any selected atom, which is 12 in all cases. Another measure of their compactness is the **packing fraction**, the fraction of space occupied by the spheres, which is 0.740 (see the *Justification* below). That is, in a close-packed solid of identical spheres, only 26.0 per cent of the volume is empty space. The fact that many metals are close-packed accounts for their high densities.

Justification 21.3

To calculate a packing fraction of a ccp structure, we first calculate the volume of a unit cell, and then calculate the total volume of the spheres that fully or partially occupy it. The first part of the calculation is a straightforward exercise in geometry. The second part involves counting the fraction of spheres that occupy the cell.

Refer to Fig. 21.38. Because a diagonal of any face passes completely through one sphere and halfway through two other spheres, its length is $4R$. The length of a side is therefore $8^{1/2}R$ and the volume of the unit cell is $8^{3/2}R^3$. Because each cell contains the equivalent of $6 \times \frac{1}{2} + 8 \times \frac{1}{8} = 4$ spheres, and the volume of each sphere is $\frac{4}{3}\pi R^3$, the total occupied volume is $\frac{16}{3}\pi R^3$. The fraction of space occupied is therefore $\frac{16}{3}\pi R^3/8^{3/2}R^3 = 16\pi/8^{3/2}3$, or 0.740. Because an hcp structure has the same coordination number, its packing fraction is the same. The packing fractions of structures that are not close-packed are calculated similarly (see Problem 21.13).

(b) Less closely packed structures

As shown in Table 21.2, a number of common metals adopt structures that are less than close-packed. The departure from close packing suggests that specific covalent bonding between neighbouring atoms is beginning to influence the structure and impose a specific geometrical arrangement. One such arrangement results in a cubic I (bcc, for body-centred cubic) structure, with one sphere at the centre of a cube formed by eight others. The coordination number of a bcc structure is only 8, but there are six more atoms not much further away than the eight nearest neighbours. The packing fraction of 0.68 is not much smaller than the value for a close-packed structure (0.74), and shows that about two-thirds of the available space is actually occupied.

21.7 Ionic crystals

When crystals of compounds of monatomic ions are modelled by stacks of spheres it is essential to allow for the different ionic radii (typically with the cations smaller than the anions) and different charges. The **coordination number** of an ion is the number of nearest

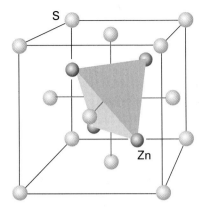

21.41 The structure of the sphalerite form of ZnS showing the location of the Zn atoms in the tetrahedral holes formed by the array of S atoms. (There is an S atom at the centre of the cube inside the tetrahedron of Zn atoms.)

Table 21.3* Ionic radii, R/pm

Na^+	102(6†), 116(8)
K^+	138(6), 151(8)
F^-	128(2), 131(4)
Cl^-	181 (close packing)

* More values are given in the *Data section* at the end of this volume.

† Coordination number.

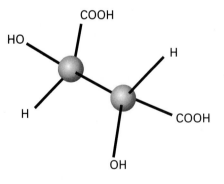

1 D(+)-Tartaric acid

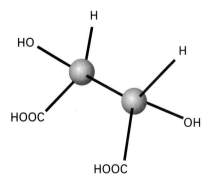

2 L(-)-Tartaric acid

neighbours of opposite charge; the structure itself is characterized as having (n_+, n_-) coordination, where n_+ is the coordination number of the cation and n_- that of the anion.

Even if, by chance, the ions have the same size, the problems of ensuring that the unit cells are electrically neutral make it impossible to achieve 12-coordinate close-packed structures. As a result, ionic solids are generally less dense than metals. The best packing that can be achieved is the (8, 8)-coordinate **caesium-chloride structure** in which each cation is surrounded by eight anions and each anion is surrounded by eight cations (Fig. 21.39). In this structure, an ion of one charge occupies the centre of a cubic unit cell with eight counter-ions at its corners. The structure is adopted by CsCl itself and also by CaS, CsCN (with some distortion), and CuZn.

When the radii of the ions differ more than in CsCl, even eight-coordinate packing cannot be achieved. One common structure adopted is the (6, 6)-coordinate **rock-salt structure** typified by NaCl (Fig. 21.40). In this structure, each cation is surrounded by six anions and each anion is surrounded by six cations. The rock-salt structure can be pictured as consisting of two interpenetrating slightly expanded cubic F (fcc) arrays, one of cations and the other of anions. This structure is adopted by NaCl itself and also by several other MX compounds, including KBr, AgCl, MgO, and ScN.

The switch from the caesium-chloride structure to the rock-salt structure occurs (sometimes) in accord with the **radius-ratio rule**, which is based on the value of the radius ratio, γ:

$$\gamma = \frac{r_{smaller}}{r_{larger}} \qquad [17]$$

The two radii are those of the larger and smaller ions in the crystal. The radius-ratio rule is derived by considering the geometrical problem of packing the maximum number of hard spheres of one radius around a hard sphere of a different radius. The rule states that the caesium-chloride structure should be expected when

$$\gamma > 3^{1/2} - 1 = 0.732$$

and that the rock-salt structure should be expected when

$$2^{1/2} - 1 = 0.414 < \gamma < 0.732$$

For $\gamma < 0.414$, the most efficient packing leads to four-coordination of the type exhibited by the sphalerite (or zinc blende) form of ZnS (Fig. 21.41). The deviation of a structure from that expected on the basis of the radius-ratio rule is often taken to be an indication of a shift from ionic towards covalent bonding; however, a major source of unreliability is the arbitrariness of ionic radii and their variation with coordination number.

Ionic radii are derived from the internuclear distance between adjacent ions in a crystal. However, we need to apportion the total distance between the two ions by defining the radius of one ion and then inferring the radius of the other ion. One scale that is widely used is based on the value 140 pm for the radius of the O^{2-} ion (Table 21.3). Other scales are also available (such as one based on F^- for discussing halides), and it is essential not to mix values from different scales. Because ionic radii are so arbitrary, predictions based on them must be viewed cautiously.

21.8 Absolute configurations

Although it has long been possible to separate enantiomers (mirror-image chiral isomers, Section 15.3b), it was not until X-ray diffraction techniques were developed that the absolute stereochemical configuration of an isomer could be determined. We now know, for example, that D-tartaric acid (**1**) is the isomer responsible for rotating light clockwise (that is, it is the (+) isomer), and that L-tartaric acid (**2**) is the (−) isomer. The X-ray method is not trivial, because enantiomers give almost identical diffraction patterns. The information about the absolute configuration is contained in small differences in diffraction intensities and is based on a technique developed by J.M. Bijvoet.

Consider first the diagrams in Fig. 21.42, which represent an idealized crystal and its mirror image. This model resembles the arrangement of Zn and S atoms in zinc blende, which was the first absolute configuration to be determined. The technique we are about to describe was used to show that the shiny (111) faces of the crystal have S atoms on the surface whereas the dull (111) faces have Zn atoms on the surface (Fig. 21.43). Each plane of atoms gives rise to a scattered wave, and their superpositions are shown on the left of Fig. 21.42. Note that the two superpositions have the same amplitude but differ in phase. The diffraction pattern therefore has the same intensity for each enantiomer and, at this stage, cannot be used to distinguish them.

The essence of the method is to use X-rays that are close to an absorption frequency of one species of atom in the sample. In the examination of ZnS, for instance, gold L_α radiation (127.6 pm) was used, which is close to the beginning of a zinc absorption band (which commences at 128.3 pm). In Bijvoet's development of this approach for a study of tartaric acid, a Rb atom is incorporated into the compound (he used sodium rubidium tartrate) with X-rays from a zirconium target. Atoms with absorptions close to the X-ray frequency introduce an extra phase shift in the scattered X-ray. A simple way to picture the additional phase shift is to imagine the X-rays as exciting the atom before being re-emitted, and being delayed in the process. The effect is called **anomalous scattering**.

Suppose the layer marked A in the crystal contains the anomalous scatterers; then the scattered waves are as shown on the right of Fig. 21.42. The essential point is that the superpositions now differ slightly in amplitude, not only phase, so the diffracted intensities are slightly different in each case. Therefore, the enantiomers can in fact be distinguished because the scattering intensities differ.

Modern diffractometers are so sensitive that the incorporation of a heavy atom is no longer strictly necessary. It is now possible to detect the small intensity variations arising from the light atoms normally present. However, the procedure is much easier and more

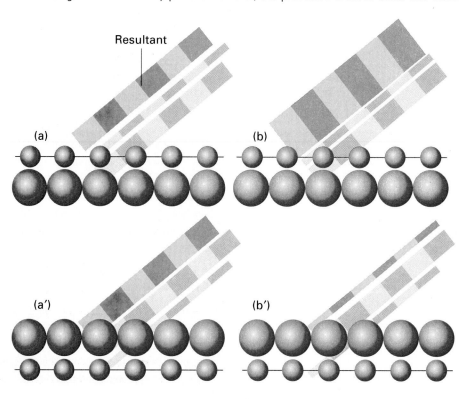

21.42 The two versions (top and bottom) of the two layers of atoms represent enantiomers. The interference between their scattered waves results in composite waves that differ in phase (a and a′), but the absolute phase cannot be determined, and the intensities of the reflections are identical. If, however, the atoms represented by the larger spheres modify the phase of the waves they scatter, then the resultant superpositions differ in amplitude as well as phase (b and b′), the reflections have different intensities, and the absolute configuration can be established. The green bands represent the waves scattered by the layers, with alternating positive and negative regions shown as light and dark. The width of a band indicates its intensity. The resultant in each instance is indicated by the grey band.

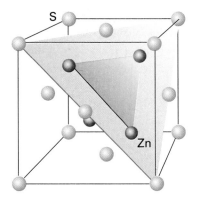

21.43 The (111) faces of the sphalerite crystal have either S atoms above Zn atoms or, as shown here, Zn atoms above S atoms.

reliable if some moderately heavy atoms (such as S or Cl) are present. Anomalous scattering depends strongly on the wavelength of the X-radiation. Thus, atoms lighter than S and Cl give little effect for Mo K_α radiation, and until recently Cu K_α radiation had to be used.

Neutron and electron diffraction

A neutron generated in a reactor and slowed to thermal velocities by repeated collisions with a moderator (such as graphite) until it is travelling at about 4 km s^{-1} has a wavelength of about 100 pm. Because 100 pm is comparable to X-ray wavelengths, similar diffraction phenomena can be expected. In practice, a range of wavelengths occurs in a neutron beam, but a monochromatic beam can be selected by diffraction from a germanium crystal.

Electrons can be accelerated to precisely controlled energies by a known potential difference. When accelerated from rest through 10 keV they acquire a wavelength of 12 pm, which makes them suitable for structural studies too.

Example 21.6 Calculating the typical wavelength of thermal neutrons

Calculate the typical wavelength of neutrons that have reached thermal equilibrium with their surroundings at 100°C.

Method We need to relate the wavelength to the temperature. There are two linking steps. First, the de Broglie relation expresses the wavelength in terms of the linear momentum. Then the linear momentum can be expressed in terms of the kinetic energy, the mean value of which is given in terms of the temperature by the equipartition theorem (see the *Introduction* and Section 20.3).

Answer The de Broglie relation states that $\lambda = h/p$. Then, from the equipartition theorem we know that the mean translational kinetic energy of a neutron at a temperature T travelling in the x-direction is $E_K = \frac{1}{2}kT$. The kinetic energy is also equal to $p^2/2m$, where p is the momentum of the neutron and m is its mass. Hence, $p = (mkT)^{1/2}$. It follows that the neutron's wavelength is

$$\lambda = \frac{h}{(mkT)^{1/2}}$$

Therefore, at 100°C,

$$\lambda = \frac{6.626 \times 10^{-34} \text{ J s}}{\{(1.675 \times 10^{-27} \text{ kg}) \times (1.381 \times 10^{-23} \text{ J K}^{-1}) \times (373 \text{ K})\}^{1/2}} \doteq 226 \text{ pm}$$

- -

Self-test 21.6 Calculate the temperature needed for the average wavelength of the neutrons to be 100 pm.

[1.6×10^3 °C]

21.9 Neutron diffraction

The scattering of X-rays is caused by the oscillations an incoming electromagnetic wave generates in the electrons of atoms. In contrast, the scattering of neutrons is a nuclear phenomenon. Neutrons pass through the electronic structures of atoms and interact with the nuclei through the strong nuclear force that is responsible for binding nucleons together. As a result, the intensity with which neutrons are scattered is independent of the number of electrons. Whereas X-ray scattering factors increase strongly with atomic

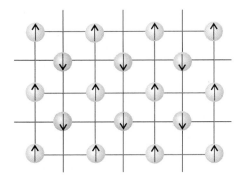

21.44 If the spins of atoms at lattice points are orderly, as in this antiferromagnetic material, where the spins of one set of atoms are aligned antiparallel to those of the other set, neutron diffraction detects two interpenetrating simple cubic lattices on account of the magnetic interaction of the neutron with the atoms, but X-ray diffraction would see only a single bcc lattice.

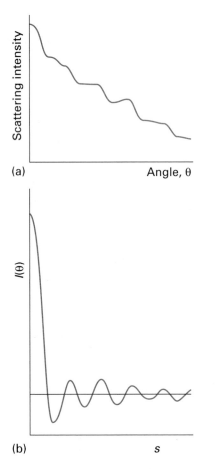

21.45 (a) The scattering intensity consists of a smoothly varying background with undulations superimposed. (b) The undulations are emphasized if a sector is rotated in front of the screen, and then the densitometer trace taken from the photograph plotted against $s = (4\pi/\lambda)\sin\frac{1}{2}\theta$.

number, neutron scattering factors vary much less strongly; nor do they vary with angle. As a result, in contrast to X-rays, neutron diffraction is not dominated by the heavy atoms present in a molecule. Neutron diffraction therefore shows up the positions of hydrogen nuclei much more clearly than X-rays do. Similarly, although neighbouring elements in the periodic table have almost identical X-ray scattering factors, and hence are almost indistinguishable in X-ray diffraction, their neutron scattering lengths may be significantly different. As a result, it is possible to distinguish atoms of elements such as Ni and Co that are present in the same compound and to study order–disorder phase transitions in FeCo.

The difference in sensitivity to hydrogen nuclei can have a pronounced effect on the measurement of C–H bond lengths. Because X-rays respond to accumulations of electrons, the weak peaks in an X-ray diffraction map represent the locations of the bulk of the electron density in the bonds, and this density may be shifted towards the C atom. For example, X-ray measurements on sucrose give $R(\text{C—H}) = 96$ pm; neutron measurements, which respond to the location of the nuclei, give $R(\text{C—H}) = 109.5$ pm. The O–H bond lengths in sucrose show similar differences, being 79 pm by X-rays but 97 pm by neutrons.

Another property of neutrons that distinguishes them from X-ray photons is their possession of a magnetic moment due to their spin. This magnetic moment can couple to the magnetic fields of ions in a crystal (if the ions have unpaired electrons) and modify the diffraction pattern. A simple example of this **magnetic scattering** is provided by metallic chromium. The lattice is cubic I (bcc), and the diffraction pattern using X-rays has systematic absences. These absences are not observed when neutrons are used, because the structure is such that atoms at the body-centre location have magnetic moments opposite to those at the corners, and the structure is better regarded as consisting of two interpenetrating arrays of magnetically different Cr atoms (Fig. 21.44). Therefore, although the atoms are identical as far as X-rays are concerned, they are different from the viewpoint of neutrons, and diffraction intensity is observed at the predicted systematic X-ray absences. Neutron diffraction is especially important for investigating these magnetically ordered lattices.

21.10 Electron diffraction

Electrons are scattered strongly by their interaction with the charges of electrons and nuclei, and so until recently could not be used to study the interiors of solid samples. However, they have been used for some time to study molecules in the gas phase, on surfaces, and in thin films. The application to surfaces, which is called 'low-energy electron diffraction' (LEED), is a major use of the technique and is discussed in Section 28.2e. Recent developments have extended electron diffraction techniques to solids, where they have certain advantages over X-ray diffraction. For instance, they are applicable to very small samples, and so may be used when single-crystal X-ray diffraction is impractical or powder diffraction too complex to interpret (see *Further reading*). A sample size of about 10^4 unit cells can be used for electron diffraction studies on solids, which is several million times smaller than for X-ray crystallography, even using synchrotron radiation.

In a typical gas-phase electron diffraction apparatus, electrons are emitted from a hot filament and are then accelerated through a potential gradient. They then pass through the stream of gas, and on to a fluorescent screen. The wavelength of electrons accelerated from rest through a potential difference $\mathcal{V}$ is

$$\lambda = \frac{h}{(2m_e e\mathcal{V})^{1/2}} \tag{18}$$

(see Example 11.2). For an accelerating potential difference of 40 kV, the wavelength is 6.1 pm.

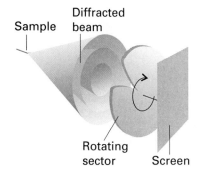

21.46 The layout of an electron diffraction apparatus. The diffraction pattern is photographed from the fluorescent screen. A rotating heart-shaped sector emphasizes the scattering from the nuclear positions and suppresses the smoothly varying background due to scattering from the continuous electron distribution in the molecules.

The gaseous sample presents all possible orientations of atom–atom separations to the electron beam. The resulting diffraction pattern consists of a series of concentric undulations on a background with an intensity that decreases steadily with increasing scattering angle (Fig. 21.45). The undulations are due to the **molecular scattering**, the sharply defined scattering from the nuclear positions. The background is due largely to the **atomic scattering**. One way of levelling the total intensity and hence to emphasize the undulations is to insert a rotating heart-shaped disk in front of the screen (Fig. 21.46).

The scattering from a pair of nuclei separated by a distance R_{ij} and orientated at a definite angle to the incident beam can be calculated. The overall diffraction pattern is then calculated by allowing for all possible orientations of this pair of atoms. When the molecule consists of a number of atoms, we sum over the contribution from all pairs, and find that the total intensity has an angular variation given by the **Wierl equation**:

$$I(\theta) = \sum_{i,j} f_i f_j \frac{\sin sR_{ij}}{sR_{ij}} \qquad s = \frac{4\pi}{\lambda} \sin \tfrac{1}{2}\theta \qquad (19)$$

where λ is the wavelength of the electrons in the beam and θ is the scattering angle. The **electron scattering factor**, f, is a measure of the intensity of the electron scattering powers of the atoms.

The electron diffraction pattern gives the distances between all possible pairs of atoms in the molecule (not just to those bonded together). When there are only a few atoms, the peaks can be analysed reasonably quickly, and the analysis proceeds by assuming a geometry and calculating the intensity pattern by using the Wierl equation. The best fit is then taken as the actual molecular geometry.

Checklist of key ideas

- ☐ diffraction
- ☐ diffraction pattern

Crystal structure

21.1 Lattices and unit cells
- ☐ space lattice
- ☐ crystal structure
- ☐ unit cell
- ☐ primitive unit cell
- ☐ crystal system
- ☐ essential symmetry
- ☐ Bravais lattice
- ☐ body-centred unit cell
- ☐ face-centred unit cell
- ☐ side-centred unit cell

21.2 The identification of lattice planes
- ☐ Miller indices

X-ray diffraction
- ☐ Bremsstrahlung

21.3 Bragg's law
- ☐ X-ray reflection
- ☐ glancing angle
- ☐ Bragg's law (5)

21.4 The powder method
- ☐ indexing
- ☐ scattering factor
- ☐ systematic absence

21.5 Single-crystal X-ray diffraction
- ☐ four-circle diffractometer
- ☐ structure factor
- ☐ Fourier synthesis (13)
- ☐ phase problem
- ☐ Patterson synthesis (14)
- ☐ direct methods

- ☐ Sayre probability relation (16)
- ☐ structure refinement

Information from X-ray analysis
- ☐ metal
- ☐ ionic solid
- ☐ covalent solid
- ☐ molecular solid

21.6 The packing of identical spheres: metal crystals
- ☐ close-packed
- ☐ polytype
- ☐ hexagonal close-packed (hcp)
- ☐ cubic close-packed (ccp)
- ☐ coordination number
- ☐ packing fraction

21.7 Ionic crystals
- ☐ coordination number (of ionic lattice)
- ☐ caesium-chloride structure
- ☐ rock-salt structure
- ☐ radius-ratio rule

21.8 Absolute configurations
- ☐ anomalous scattering

Neutron and electron diffraction

21.9 Neutron diffraction
- ☐ magnetic scattering

21.10 Electron diffraction
- ☐ molecular scattering
- ☐ atomic scattering
- ☐ Wierl equation (19)
- ☐ electron scattering factor

Further reading

Articles of general interest

J.P. Glusker, Teaching crystallography to noncrystallographers. *J. Chem. Educ.* **65**, 474 (1988).

J.H. Enemark, Introducing chemists to X-ray structure determination. *J. Chem. Educ.* **65**, 491 (1988).

Crystallographic resource list. *J. Chem. Educ.* **65**, 512 (1988).

C.G. Pope, X-ray diffraction and the Bragg equation. *J. Chem. Educ.* **74**, 129 (1997).

J.P. Chesick, Fourier transforms in chemistry. Part III. X-ray crystal structure analysis. *J. Chem. Educ.* **66**, 413 (1989).

T. Li and J.H. Worrell, Construction of the seven basic crystallographic units. *J. Chem. Educ.* **66**, 73 (1989).

S.F.A. Kettle and L.J. Norrby, Really, your lattices are all primitive, Mr. Bravais! *J. Chem. Educ.* **70**, 959 (1993).

J.P. Birk and P.R. Coffman, Finding the face-centred cube in the cubic close-packed structure. *J. Chem. Educ.* **69**, 953 (1992).

J. Toofan, A simple expression between critical radius ratio and coordination number. *J. Chem. Educ.* **71**, 147 (1994).

S.R. Elliott, Structure of amorphous materials. In *Encyclopedia of applied physics* (ed. G.L. Trigg), **1**, 559. VCH, New York (1991).

A.I. Goldman, Crystalline state. In *Encyclopedia of applied physics* (ed. G.L. Trigg), **4**, 365. VCH, New York (1992).

W.B. Pearson and C. Chieh, Crystallography. In *Encyclopedia of applied physics* (ed. G.L. Trigg), **4**, 385. VCH, New York (1992).

E. Schönherr, Crystal growth. In *Encyclopedia of applied physics* (ed. G.L. Trigg), **4**, 335. VCH, New York (1992).

T. Vogt, Neutron diffraction. In *Encyclopedia of applied physics* (ed. G.L. Trigg), **11**, 339. VCH, New York (1994).

F. Ebrahami and M.J. Kaufman, Structure of metals and alloys. In *Encyclopedia of applied physics* (ed. G.L. Trigg), **10**, 199. VCH, New York (1994).

R.M. Hanson and S.A. Bergman, Data-driven chemistry: building models of molecular structure (literally) from electron diffraction data. *J. Chem. Educ.* **71**, 150 (1994).

J.M. Cowley, Electron diffraction. In *Encyclopedia of applied physics* (ed. G.L. Trigg), **5**, 405. VCH, New York (1993).

Texts and sources of data and information

M.F.C. Ladd and R.A. Palmer, *Structure determination by X-ray crystallography*. Plenum, New York (1985).

J.P. Glusker and K.N. Trueblood, *Crystal structure analysis*. Oxford University Press (1985).

E.A.V. Ebsworth, D.W.H. Rankin, and S. Cradock, *Structural methods in inorganic chemistry*. Blackwell Scientific, Oxford (1991).

R. Drago, *Physical methods for chemists*. Saunders, Philadelphia (1992).

C. Giacovazzo (ed.), *Fundamentals of crystallography*. Oxford University Press (1992).

A.F. Wells, *Structural inorganic chemistry*. Clarendon Press, Oxford (1984).

R.W.G. Wycoff, *Crystal structure* (five sections, and supplements). Wiley-Interscience, New York (1959).

I. Hargittai and M. Hargittai (ed.), *Sterochemical applications of gas-phase electron diffraction*. VCH, Weinheim (1988).

J.M. Cowley, *Electron diffraction techniques*. Oxford University Press (1992).

C. Hammond, *The basics of crystallography and diffraction*. Oxford University Press (1997).

Exercises

21.1 (a) Equivalent lattice points within the unit cell of a Bravais lattice have identical surroundings. What points within a face-centred cubic unit cell are equivalent to the point $(\frac{1}{2},0,0)$?

21.1 (b) Equivalent lattice points within the unit cell of a Bravais lattice have identical surroundings. What points within a body-centred cubic unit cell are equivalent to the point $(\frac{1}{2},0,\frac{1}{2})$?

21.2 (a) Find the Miller indices of the planes that intersect the crystallographic axes at the distances $(2a, 3b, 2c)$ and $(2a, 2b, \infty c)$.

21.2 (b) Find the Miller indices of the planes that intersect the crystallographic axes at the distances $(1a, 3b, -c)$ and $(2a, 3b, 4c)$.

21.3 (a) Calculate the separations of the planes (111), (211), and (100) in a crystal in which the cubic unit cell has side 432 pm.

21.3 (b) Calculate the separations of the planes (121), (221), and (244) in a crystal in which the cubic unit cell has side 523 pm.

21.4 (a) The glancing angle of a Bragg reflection from a set of crystal planes separated by 99.3 pm is 20.85°. Calculate the wavelength of the X-rays.

21.4 (b) The glancing angle of a Bragg reflection from a set of crystal planes separated by 128.2 pm is 19.76°. Calculate the wavelength of the X-rays.

21.5 (a) What are the values of 2θ of the first three diffraction lines of bcc iron (atomic radius 126 pm) when the X-ray wavelength is 58 pm?

21.5 (b) What are the values of 2θ of the first three diffraction lines of fcc (gold atomic radius 144 pm) when the X-ray wavelength is 154 pm?

21.6 (a) Copper K_α radiation consists of two components of wavelengths 154.433 pm and 154.051 pm. Calculate the separation of the diffraction lines arising from the two components in a powder diffraction pattern recorded in a circular camera of radius 5.74 cm (with the sample at the centre) from planes of separation 77.8 pm.

21.6 (b) A synchrotron source produces X-radiation at a range of wavelengths. Consider two components of wavelengths 95.401 and 96.035 pm. Calculate the separation of the diffraction lines arising from the two components in a powder diffraction pattern recorded in a circular camera of radius 5.74 cm (with the sample at the centre) from planes of separation 82.3 pm.

21.7 (a) The compound Rb_3TlF_6 has a tetragonal unit cell with dimensions $a = 651$ pm and $c = 934$ pm. Calculate the volume of the unit cell.

21.7 (b) Calculate the volume of the hexagonal unit cell of sodium nitrate, for which the dimensions are $a = 1692.9$ pm and $c = 506.96$ pm.

21.8 (a) The orthorhombic unit cell of $NiSO_4$ has the dimensions $a = 634$ pm, $b = 784$ pm, and $c = 516$ pm, and the density of the solid is estimated as 3.9 g cm^{-3}. Determine the number of formula units per unit cell and calculate a more precise value of the density.

21.8 (b) An orthorhombic unit cell of a compound of molar mass 135.01 g mol^{-1} has the dimensions $a = 589$ pm, $b = 822$ pm, and $c = 798$ pm. The density of the solid is estimated as 2.9 g cm^{-3}. Determine the number of formula units per unit cell and calculate a more precise value of the density.

21.9 (a) The unit cells of $SbCl_3$ are orthorhombic with dimensions $a = 812$ pm, $b = 947$ pm, and $c = 637$ pm. Calculate the spacing of the (411) planes.

21.9 (b) An orthorhombic unit cell has dimensions $a = 679$ pm, $b = 879$ pm, and $c = 860$ pm. Calculate the spacing of the (322) planes.

21.10 (a) A substance known to have a cubic unit cell gives reflections with Cu K_α radiation (wavelength 154 pm) at glancing angles 19.4°, 22.5°, 32.6°, and 39.4°. The reflection at 32.6° is known to be due to the (220) planes. Index the other reflections.

21.10 (b) A substance known to have a cubic unit cell gives reflections with radiation of wavelength 137 pm at the glancing angles 10.7°, 13.6°, 17.7°, and 21.9°. The reflection at 17.7° is known to be due to the (111) planes. Index the other reflections.

21.11 (a) Potassium nitrate crystals have orthorhombic unit cells of dimensions $a = 542$ pm, $b = 917$ pm, and $c = 645$ pm. Calculate the glancing angles for the (100), (010), and (111) reflections using Cu K_α radiation (154 pm).

21.11 (b) Calcium carbonate crystals in the form of aragonite have orthorhombic unit cells of dimensions $a = 574.1$ pm, $b = 796.8$ pm, and $c = 495.9$ pm. Calculate the glancing angles for the (100), (010), and (111) reflections using radiation of wavelength 83.42 pm (from aluminium).

21.12 (a) Copper(I) chloride forms cubic crystals with four formula units per unit cell. The only reflections present in a powder photograph are those with either all even indices or all odd indices. What is the symmetry of the unit cell?

21.12 (b) A powder diffraction photograph from tungsten shows lines which index as (110), (200), (211), (220), (310), (222), (321), (400), ... Identify the symmetry of the unit cell.

21.13 (a) The coordinates, in units of a, of the atoms in a simple cubic lattice are $(0,0,0)$, $(0,1,0)$, $(0,0,1)$, $(0,1,1)$, $(1,0,0)$, $(1,1,0)$, $(1,0,1)$, and $(1,1,1)$. Calculate the structure factors F_{hkl} when all the atoms are identical.

21.13 (b) The coordinates, in units of a, of the atoms in a body-centred cubic lattice are $(0,0,0)$, $(0,1,0)$, $(0,0,1)$, $(0,1,1)$, $(1,0,0)$, $(1,1,0)$, $(1,0,1)$, $(1,1,1)$, and $(\frac{1}{2},\frac{1}{2},\frac{1}{2})$. Calculate the structure factors F_{hkl} when all the atoms are identical.

21.14 (a) Calculate the packing fraction for close-packed cylinders.

21.14 (b) Calculate the packing fraction for equilateral triangular rods stacked as shown in (3).

3

21.15 (a) Verify that the radius ratio for sixfold coordination is 0.414.

21.15 (b) Verify that the radius ratio for eightfold coordination is 0.732.

21.16 (a) From the data in Table 21.3 determine the radius of the smallest cation that can have (a) sixfold and (b) eightfold coordination with the O^{2-} ion.

21.16 (b) From the data in Table 21.3 determine the radius of the smallest cation that can have (a) sixfold and (b) eightfold coordination with the K^+ ion.

21.17 (a) Calculate the atomic packing factor for diamond.

21.17 (b) Calculate the atomic packing factor for a cubic C unit cell.

21.18 (a) The carbon–carbon bond length in diamond is 154.45 pm. If diamond were considered to be a close-packed structure of hard spheres with radii equal to half the bond length, what would be its expected density? The diamond lattice is face-centred cubic and its actual density is 3.516 g cm^{-3}. Can you explain the discrepancy?

21.18 (b) Although the crystallization of large biological molecules may not be as readily accomplished as that of small molecules, their crystal lattices are no different. Tobacco seed globulin forms face-centred cubic crystals with unit cell dimension of 12.3 nm and a density of 1.287 g cm^{-3}. Determine the globulin's molar mass.

21.19 (a) Is there an expansion or a contraction as titanium transforms from hcp to body-centred cubic? The atomic radius of titanium is 145.8 pm in hcp but 142.5 pm in bcc.

21.19 (b) Is there an expansion or a contraction as iron transforms from hcp to bcc? The atomic radius of iron is 126 pm in hcp but 122 pm in bcc.

21.20 (a) In a Patterson synthesis, the spots correspond to the lengths and directions of the vectors joining the atoms in a unit cell. Sketch the pattern that would be obtained for a planar, triangular isolated BF_3 molecule.

21.20 (b) In a Patterson synthesis, the spots correspond to the lengths and directions of the vectors joining the atoms in a unit cell. Sketch the pattern that would be obtained from the C atoms in an isolated benzene molecule.

21.21 (a) What velocity should neutrons have if they are to have wavelength 50 pm?

21.21 (b) Calculate the wavelength of neutrons that have reached thermal equilibrium by collision with a moderator at 300 K.

21.22 (a) What accelerating potential difference must be applied to electrons to generate a beam with wavelength 18 pm?

21.22 (b) Calculate the wavelengths of electrons that have been accelerated through (a) 1.0 kV, (b) 10 kV, (c) 40 kV.

21.23 (a) Predict from the Wierl equation the positions of the first maximum and first minimum in the neutron and electron diffraction patterns of the Cl_2 molecule obtained with neutrons of wavelength 80 pm and electrons of wavelength 4.0 pm.

21.23 (b) Predict from the Wierl equation the positions of the first maximum and first minimum in the neutron and electron diffraction patterns of a Br_2 molecule obtained with neutrons of wavelength 78 pm and electrons of wavelength 4.0 pm.

Problems

Numerical problems

21.1 In the early days of X-ray crystallography there was an urgent need to know the wavelengths of X-rays. One technique was to measure the diffraction angle from a mechanically ruled grating. Another method was to estimate the separation of lattice planes from the measured density of a crystal. The density of NaCl is 2.17 g cm^{-3} and the (100) reflection using Pd K_α radiation occurred at 6.0°. Calculate the wavelength of the X-rays.

21.2 The element polonium crystallizes in a cubic system. Bragg reflections, with X-rays of wavelength 154 pm, occur at $\sin\theta = 0.225$, 0.316, and 0.388 from the (100), (110), and (111) sets of planes. The separation between the sixth and seventh lines in the powder spectrum is larger than between the fifth and sixth lines. Is the unit cell simple, body-centred, or face-centred? Calculate the unit cell dimension.

21.3 The unit cell dimensions of NaCl, KCl, NaBr, and KBr, all of which crystallize in face-centred cubic lattices, are 562.8 pm, 627.7 pm, 596.2 pm, and 658.6 pm, respectively. In each case, anion and cation are in contact along an edge of the unit cell. Do the data support the contention that ionic radii are constants independent of the counter-ion?

21.4 The powder diffraction patterns of (a) tungsten, (b) copper obtained in a camera of radius 28.7 mm are shown in Fig. 21.47. Both were obtained with 154 pm X-rays and the scales are marked. Identify the unit cell in each case, and calculate the lattice spacing. Estimate the metallic radii of W and Cu.

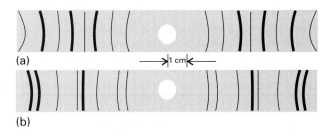

Fig. 21.47

21.5 Elemental silver reflects X-rays of wavelength 154.18 pm at angles of 19.076°, 22.171°, and 32.256°. However, there are no other reflections at angles of less than 33°. Assuming a cubic unit cell, determine its type and dimension. Calculate the density of silver.

21.6 Genuine pearls consist of concentric layers of calcite crystals ($CaCO_3$) in which the trigonal axes are oriented along the radii. The nucleus of a cultured pearl is a piece of mother-of-pearl that has been worked into a sphere on a lathe. The oyster then deposits concentric layers of calcite on the central seed. Suggest an X-ray method for distinguishing between real and cultured pearls.

21.7 In their book *X-rays and crystal structures* (which begins "It is now two years since Dr. Laue conceived the idea...") the Braggs give a number of simple examples of X-ray analysis. For instance, they report that the reflection from (100) planes in KCl occurs at 5° 23', but for NaCl it occurs at 6° 0' for X-rays of the same wavelength. If the side of the NaCl unit cell is 564 pm, what is the side of the KCl unit cell? The densities of KCl and NaCl are 1.99 g cm^{-3} and 2.17 g cm^{-3}, respectively. Do these values support the X-ray analysis?

21.8 The volume of a monoclinic unit cell is $abc \sin \beta$. Naphthalene has a monoclinic unit cell with two molecules per cell and sides in the ratio 1.377 : 1 : 1.436. The angle β is 122° 49' and the density of the solid is 1.152 g cm^{-3}. Calculate the dimensions of the cell.

21.9 Calculate the coefficient of thermal expansion of diamond given that the (111) reflection shifts from 22° 2' 25'' to 21° 57' 59'' on heating a crystal from 100 K to 300 K and 154.0562 pm X-rays are used.

21.10 Use the Wierl equation to predict the appearance of the electron diffraction pattern of CCl$_4$ with an (as yet) undetermined C–Cl bond length but of known tetrahedral symmetry. Take $f_{Cl} = 17f$ and $f_C = 6f$ and note that $R(Cl, Cl) = \left(\frac{8}{3}\right)^{1/2} R(C, Cl)$. Plot I/f^2 against $x = sR(C, Cl)$. In an actual experiment using 10.0 keV electrons the positions of the maxima occurred at 3° 10', 5° 22', and 7° 54' and minima occurred at 1° 46', 4° 6', 6° 40', and 9° 10'. What is the C–Cl bond length in CCl$_4$?

Theoretical problems

21.11 Show that the separation of the (hkl) planes in an orthorhombic crystal with sides a, b, and c is given by eqn 3.

21.12 Show that the volume of a triclinic unit cell of sides a, b, and c and angles α, β, and γ is

$$V = abc(1 - \cos^2 \alpha - \cos^2 \beta - \cos^2 \gamma + 2\cos \alpha \cos \beta \cos \gamma)^{1/2}$$

Use this expression to derive expressions for monoclinic and orthorhombic unit cells. For the derivation, it may be helpful to use the result from vector analysis that $V = \mathbf{a} \cdot \mathbf{b} \times \mathbf{c}$ and to calculate V^2 initially.

21.13 Calculate the packing fractions of (a) a primitive cubic lattice, (b) a bcc unit cell, (c) an fcc unit cell.

21.14 The coordinates of the four I atoms in the unit cell of KIO$_4$ are $(0, 0, 0)$, $(0, \frac{1}{2}, \frac{1}{2})$, $(\frac{1}{2}, \frac{1}{2}, \frac{1}{2})$, $(\frac{1}{2}, 0, \frac{3}{4})$. By calculating the phase of the I reflection in the structure factor, show that the I atoms contribute no net intensity to the (114) reflection.

21.15 The coordinates, in units of a, of the A atoms, with scattering factor f_A, in a cubic lattice are $(0, 0, 0)$, $(0, 1, 0)$, $(0, 0, 1)$, $(0, 1, 1)$, $(1, 0, 0)$, $(1, 1, 0)$, $(1, 0, 1)$, and $(1, 1, 1)$. There is also a B atom, with scattering factor f_B, at $(\frac{1}{2}, \frac{1}{2}, \frac{1}{2})$. Calculate the structure factors F_{hkl} and predict the form of the powder diffraction pattern when (a) $f_A = f$, $f_B = 0$, (b) $f_B = \frac{1}{2}f_A$, and (c) $f_A = f_B = f$.

Additional problems supplied by Carmen Giunta and Charles Trapp

21.16 B.A. Bovenzi and G.A. Pearse, Jr (*J. Chem. Soc. Dalton Trans.* (accepted, 1997)) synthesized coordination compounds of the tridentate ligand pyridine-2,6-diamidoxime (C$_7$H$_9$N$_5$O$_2$). The compound, which they isolated from the reaction of the ligand with CuSO$_4$(aq), did not contain a [Cu(C$_7$H$_9$N$_5$O$_2$)$_2$]$^{2+}$ complex cation as expected. Instead, X-ray diffraction analysis revealed a linear polymer of formula [Cu(Cu(C$_7$H$_9$N$_5$O$_2$)(SO$_4$) · 2H$_2$O]$_n$, which features bridging sulfate groups. The unit cell was primitive monoclinic with $a = 1.0427$ nm, $b = 0.8876$ nm, $c = 1.3777$ nm, and $\beta = 93.254°$. The mass density of the crystals is 2.024 g cm^{-3}. How many monomer units are there per unit cell?

21.17 D. Sellmann, M.W. Wemple, W. Donaubauer, and F.W. Heinemann (*Inorg. Chem.* **36**, 1397 (1997)) describe the synthesis and reactivity of the ruthenium nitrido compound [N(C$_4$H$_9$)$_4$][Ru(N)(S$_2$C$_6$H$_4$)$_2$]. The ruthenium complex anion has the two 1,2-benzenedithiolate ligands (4) at the base of a rectangular pyramid and the nitrido ligand at the apex. Compute the mass density of the compound given that it crystallizes into an orthorhombic unit cell with $a = 3.6881$ nm, $b = 0.9402$ nm, and $c = 1.7652$ nm and eight formula units per cell. Replacing the ruthenium with an osmium results in a compound with the same crystal structure and a unit cell with a volume less than 1 per cent larger. Estimate the mass density of the osmium analogue.

4

21.18 P.G. Radaelli, M. Marezio, M. Perroux, S. de Brion, J.L. Tholence, Q. Huang, and A. Santoro (*Science* **265**, 380 (1994)) report the synthesis and structure of a material that becomes superconducting at temperatures below 45 K. The compound is based on a layered compound Hg$_2$Ba$_2$YCu$_2$O$_{8-\delta}$, which has a tetragonal unit cell with $a = 0.38606$ nm and $c = 2.8915$ nm; each unit cell contains two formula units. The compound is made superconducting by partially replacing Y by Ca, accompanied by a change in unit cell volume by less than 1 per cent. Estimate the Ca content x in superconducting Hg$_2$Ba$_2$Y$_{1-x}$Ca$_x$Cu$_2$O$_{7.55}$ given that the mass density of the compound is 7.651 g cm^{-3}.

21.19 The scattering factor of an atom is given by eqn 7. In general this expression is difficult to evaluate as it requires knowledge of $\rho(r)$, which in turn requires knowledge of the wavefunction of the atom. The latter is not generally available in simple analytical form except for a hydrogenic atom. Derive an expression for the scattering factor of a hydrogenic atom of atomic number Z in its ground state.

21.20 Diamond forms a face-centred cubic lattice with eight atoms per unit cell. There are atoms at each lattice point and at points displaced by $\frac{1}{4}, \frac{1}{4}, \frac{1}{4}$ from each lattice point. Calculate the structure factor, F_{hkl}, for diamond. *Hint*. See Example 21.4.

21.21 Determine the relative intensities of the (100), (110), and (200) reflections in CsCl by calculating their structure factors. Assume that the atomic scattering factors are given by the number of electrons in the ions.

22

The electric and magnetic properties of molecules

In this chapter we examine some of the electric and magnetic properties of molecules and interpret them in terms of electronic structure. The properties we consider include the electric dipole moments and polarizabilities of molecules and some related properties that include refractive index, optical activity, and intermolecular forces. All these properties reflect the degree to which the nuclei of atoms exert control over the electrons in a molecule, either by causing electrons to accumulate in particular regions, or by permitting them to respond more or less strongly to the effects of external fields. We also discuss the analogous magnetic properties, particularly the magnetizabilities and magnetic susceptibilities of molecules, and see the origins of the distinction between paramagnetic and diamagnetic substances.

The electric properties, and to a smaller extent the magnetic properties, of molecules are responsible for many of the properties of bulk matter. The small imbalances of charge distributions in molecules allow them to interact with one another and with externally applied fields. One result of this interaction is the cohesion of molecules to form the bulk phases of matter. These interactions are also important for understanding the shapes adopted by biological and synthetic macromolecules, as we shall see in Chapter 23.

Electric properties

Many of the electrical properties of molecules can be traced to the competing influences of nuclei with different charges or the competion between the control exercised by a nucleus and the influence of an externally applied field. The former competition may result in an electric dipole moment. The latter may result in properties such as refractive index and optical activity.

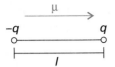

1 Electric dipole

22.1 Permanent and induced electric dipole moments

An **electric dipole** consists of two electric charges q and $-q$ separated by a distance R. This arrangement of charges is represented by a vector, the **electric dipole moment, μ,** that points from the negative charge to the positive charge (**1**).[1] The magnitude of μ is $\mu = qR$. The magnitudes of dipole moments are still commonly reported in the non-SI unit debye, D, where[2]

$$1\ \mathrm{D} = 3.335\,64 \times 10^{-30}\ \mathrm{C\,m} \tag{1}$$

The dipole moment of a pair of charges $+e$ and $-e$ separated by 100 pm is 1.6×10^{-29} C m, corresponding to 4.8 D. Dipole moments of small molecules are typically about 1 D.

(a) Classes of substance

A **polar molecule** is a molecule with a permanent electric dipole moment. The permanent dipole moment stems from the partial charges on the atoms in the molecule that arise from differences in electronegativity or other features of bonding (Section 14.7). Nonpolar molecules acquire an induced dipole moment in an electric field on account of the distortion the field causes in their electronic distributions and nuclear positions; however, this induced moment is only temporary, and disappears as soon as the perturbing field is removed. Polar molecules also have their existing dipole moments temporarily modified by an applied field.

The **polarization,** P, of a sample is the electric dipole moment density, the mean electric dipole moment of the molecules, $\langle \mu \rangle$, multiplied by the number density, $\mathcal{N}$:

$$P = \langle \mu \rangle \mathcal{N} \tag{2}$$

In the following pages we refer to the sample as a **dielectric**, by which is meant a polarizable, nonconducting medium.

The polarization of an isotropic fluid sample is zero in the absence of an applied field because the molecules adopt random orientations, so $\langle \mu \rangle = 0$. In the presence of a field, the dipoles become partially aligned because some orientations have lower energies than others. As a result, the electric dipole moment density is nonzero. Moreover, as we shall see, there is an additional contribution from the dipole moment induced by the field.

A **ferroelectric solid** is a solid that has a permanent polarization on account of a cooperative shift of some of its atoms in a given direction. For example, above 120°C, barium titanate, $BaTiO_3$, is electrically a normal solid, and the Ti ion is symmetrically placed inside an octahedron of O atoms. However, below 120°C, the Ti ion moves from the centre of the octahedron by about 10 pm and every unit cell over a large domain acquires an electric dipole moment that persists even in the absence of an applied field.

(b) Polar molecules

The Stark effect (Section 16.5) is used to measure the electric dipole moments of molecules for which a rotational spectrum can be observed. In many cases microwave spectroscopy cannot be used because the sample is not volatile, decomposes on vaporization, or consists of molecules that are so complex that their rotational spectra cannot be interpreted. In such cases the dipole moment may be obtained by measurements on a liquid or solid bulk sample using a method explained later.

---

1   In elementary chemistry, an electric dipole moment is represented by the arrow $\longmapsto$ added to the Lewis structure for the molecule, with the $+$ marking the positive end. Note that the direction of the arrow is opposite to that of μ.

2   The conversion factor stems from the original definition of the debye in terms of c.g.s. units: 1 D is the dipole moment of two equal and opposite charges of magnitude 1 e.s.u. separated by 1 Å. The debye is named after Peter Debye, a pioneer in the study of dipole moments of molecules.

Table 22.1* Dipole moments (μ) and polarizability volumes (α')

	μ/D	$\alpha'/(10^{-30}\,\text{m}^3)$
CCl_4	0	10.5
H_2	0	0.819
H_2O	1.85	1.48
HCl	1.08	2.63
HI	0.42	5.45

* More values are given in the *Data section* at the end of this volume.

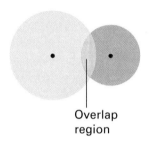

22.1 One of the contributions to the dipole moment of a molecule is the imbalance of charge arising from the overlap of orbitals of different radii. This diagram shows how the charge accumulation leads to a region of negative charge closer to the smaller atom.

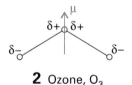

2 Ozone, O_3

All heteronuclear diatomic molecules are polar, and typical values of μ include 1.08 D for HCl and 0.42 D for HI (Table 22.1). A very approximate relation between the dipole moment and the difference in electronegativities of the two atoms, $\Delta\chi$, is

$$\mu/\text{D} \approx \Delta\chi \qquad (3)$$

The more electronegative atom is normally the negative end of the dipole, but there are exceptions, particularly when antibonding orbitals are occupied.[3] Thus the dipole moment of CO is very small (0.12 D), but the negative end of the dipole is on the C atom even though oxygen is more electronegative than carbon.

The interpretation and prediction of electric dipole moments is made even more complicated by the fact that a difference in atomic radii can result in an imbalance of electron density because the enhanced charge density associated with the overlap region lies closer to the nucleus of the smaller atom (Fig. 22.1). Such a **homopolar contribution** to the total dipole moment can arise even in the absence of a difference in electronegativity between the two atoms.

A polyatomic molecule is nonpolar if it fulfils certain symmetry criteria. We saw in Section 15.3a that a molecule is nonpolar if it belongs to a D point group or to one of the cubic or icosahedral point groups. We also saw that the dipole moment of a polar molecule with a symmetry axis cannot lie perpendicular to that axis (the dipole moment of NH_3, for instance, lies parallel to the molecular C_3 axis). The symmetry criterion is more important than the question of whether or not the atoms in the molecule are the same. Thus the homonuclear triatomic molecule O_3 (which is angular, with C_{2v} symmetry) is allowed to be polar by symmetry considerations, and in fact is polar because the electron density on the central O atom differs from that on the two outer O atoms. The dipole moment of the molecule lies parallel to the C_2 axis of the molecule (**2**). The heteronuclear triatomic molecule CO_2 (which is linear, with $D_{\infty h}$ symmetry) is strictly nonpolar by symmetry even though the C and O atoms have different electronegativities. The dipole moments associated with each CO bond point in opposite directions in CO_2, and cancel.

To a first approximation, the dipole moment of a polyatomic molecule can be resolved into contributions from various components (Fig. 22.2). Thus, 1,4-dichlorobenzene is nonpolar on account of the cancellation of the two equal but opposing moments associated with the presence of Cl atoms on opposite sides of the ring (the molecule has D_{2h} symmetry, so it is necessarily nonpolar). The isomer 1,2-dichlorobenzene (which has C_{2v} symmetry, with the C_2 axis lying along the bisector of the angle between the two CCl bonds) has a dipole moment that is approximately the resultant of two monochlorobenzene dipole moments arranged at 60°. The technique of vector addition can be applied with fair success

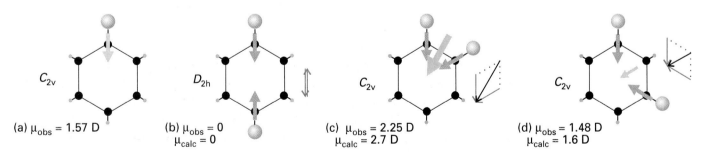

(a) $\mu_{obs} = 1.57$ D   (b) $\mu_{obs} = 0$ (c) $\mu_{obs} = 2.25$ D   (d) $\mu_{obs} = 1.48$ D
$\quad\qquad\qquad\qquad\qquad\quad\mu_{calc} = 0$ $\quad\mu_{calc} = 2.7$ D $\quad\mu_{calc} = 1.6$ D

22.2 The resultant dipole moments (grey) of the dichlorobenzene isomers (b to d) can be obtained approximately by vectorial addition of two chlorobenzene dipole moments (1.57 D).

3   We remarked in Section 14.7 that the major contribution to an antibonding orbital is made by the atomic orbitals of the less electronegative atom. Therefore, if an antibonding orbital is occupied there may be so much electron density on the less electronegative atom that that atom has a partial negative charge.

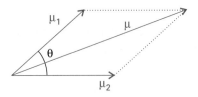

3 Addition of dipoles

to other series of related molecules, and the resultant of two dipole moments that make an angle θ to each other (**3**) is obtained from

$$\mu^2 = \mu_1^2 + \mu_2^2 + 2\mu_1\mu_2\cos\theta \tag{4a}$$

When the two dipole moments are equal, this equation simplifies to

$$\mu = 2\mu_1\cos\tfrac{1}{2}\theta \tag{4b}$$

The mean dipole moment of the molecules of a fluid sample is zero in the absence of an orientating electric field. In the presence of a field and at a temperature T the mean moment is nonzero, and we show in the *Justification* below that

$$\langle\mu_z\rangle = \frac{\mu^2\mathcal{E}}{3kT} \tag{5}$$

where z is the direction of the applied field. This nonzero value stems from the fact that some orientations of the dipole moment are energetically more favourable than others.

Justification 22.1

The probability $\mathrm{d}p$ that a dipole has an orientation in the range θ to $\theta + \mathrm{d}\theta$ is given by the Boltzmann distribution (Section 19.1c), which in this case is

$$\mathrm{d}p = \frac{\mathrm{e}^{-E(\theta)/kT}\sin\theta\,\mathrm{d}\theta}{\int_0^\pi \mathrm{e}^{-E(\theta)/kT}\sin\theta\,\mathrm{d}\theta}$$

where $E(\theta)$ is the energy of the dipole in the field: $E(\theta) = -\mu\mathcal{E}\cos\theta$, with $0 \le \theta \le \pi$. The average value of the component of the dipole moment parallel to the applied electric field is therefore

$$\langle\mu_z\rangle = \int \mu\cos\theta\,\mathrm{d}p = \mu\int\cos\theta\,\mathrm{d}p = \frac{\mu\int_0^\pi \mathrm{e}^{x\cos\theta}\cos\theta\sin\theta\,\mathrm{d}\theta}{\int_0^\pi \mathrm{e}^{x\cos\theta}\sin\theta\,\mathrm{d}\theta}$$

with $x = \mu\mathcal{E}/kT$. The integral takes on a simpler appearance when we write $y = \cos\theta$ and note that $\mathrm{d}y = -\sin\theta\,\mathrm{d}\theta$:

$$\langle\mu_z\rangle = \frac{\mu\int_{-1}^1 y\mathrm{e}^{xy}\,\mathrm{d}y}{\int_{-1}^1 \mathrm{e}^{xy}\,\mathrm{d}y}$$

At this point we use

$$\int_{-1}^1 \mathrm{e}^{xy}\,\mathrm{d}y = \frac{\mathrm{e}^x - \mathrm{e}^{-x}}{x} \qquad \int_{-1}^1 y\mathrm{e}^{xy}\,\mathrm{d}y = \frac{\mathrm{e}^x + \mathrm{e}^{-x}}{x} - \frac{\mathrm{e}^x - \mathrm{e}^{-x}}{x^2}$$

It is now straightforward algebra to combine these two results and to obtain

$$\langle\mu_z\rangle = \mu\mathcal{L}(x) \qquad \mathcal{L}(x) = \frac{\mathrm{e}^x + \mathrm{e}^{-x}}{\mathrm{e}^x - \mathrm{e}^{-x}} - \frac{1}{x} \qquad x = \frac{\mu\mathcal{E}}{kT} \tag{6}$$

The function $\mathcal{L}(x)$ is called the **Langevin function**.

Under most circumstances, x is very small (for example, if $\mu = 1\ \mathrm{D}$ and $T = 300\ \mathrm{K}$, then x exceeds 0.01 only if the field strength exceeds $100\ \mathrm{kV\,cm^{-1}}$, and most measurements are done at much lower strengths). When $x \ll 1$, the exponentials in the Langevin function can be expanded, and the largest term that survives is

$$\mathcal{L}(x) = \tfrac{1}{3}x + \cdots \tag{7}$$

Therefore, the average molecular dipole moment is given by eqn 5.

(c) Induced dipole moments

An applied electric field can distort a molecule as well as aligning its permanent electric dipole moment. The **induced dipole moment**, μ^*, is proportional to the field strength, $\mathcal{E}$, and we write[4]

$$\mu^* = \alpha\mathcal{E} \tag{8}$$

The constant of proportionality α is the **polarizability** of the molecule. The greater the polarizability, the larger is the induced dipole moment for a given applied field. When the applied field is very strong (as in laser beams), the induced moment is not strictly linear in the strength of the field, and we write

$$\mu^* = \alpha\mathcal{E} + \tfrac{1}{2}\beta\mathcal{E}^2 + \cdots \tag{9}$$

The coefficient β is the **hyperpolarizability** of the molecule.

Polarizability has the units (coulomb-metre)^2 per joule, $C^2\,m^2\,J^{-1}$. That collection of units is awkward, so α is often expressed as a **polarizability volume**, α', by using the relation

$$\alpha' = \frac{\alpha}{4\pi\varepsilon_0} \tag{10}$$

where ε_0 is the vacuum permittivity. Because the units of $4\pi\varepsilon_0$ are coulomb-squared per joule per metre ($C^2\,J^{-1}\,m^{-1}$), it follows that α' has the dimensions of volume (hence its name).[5] Polarizability volumes are similar in magnitude to actual molecular volumes (of the order of 10^{-30} m^3, 1 Å^3).

Some experimental polarizability volumes of molecules are given in Table 22.1. As shown in the *Justification* below, there is a correlation between the HOMO–LUMO separation in atoms and molecules. The electron distribution can be distorted readily if the LUMO lies close to the HOMO in energy, so the polarizability is then large. If the LUMO lies high above the HOMO, an applied field can perturb the electron distribution significantly, and the polarizability is low. Molecules with small HOMO–LUMO gaps are typically large, with numerous electrons.

Justification 22.2

The quantum mechanical expression for the mean polarizability is

$$\alpha = \tfrac{2}{3}\sum_n \frac{|\mu_{0n}|^2}{E_n - E_0} \tag{11}$$

where μ_{0n} is the magnitude of the transition dipole moment, the integral

$$\mu_{0n} = \int \psi_0^* \boldsymbol{\mu}\psi_n \, d\tau$$

with $\boldsymbol{\mu}$ the electric dipole moment operator. This integral is a measure of the extent to which electric charge is shifted when an electron migrates from a wavefunction ψ_0 to an excited-state wavefunction ψ_n. The sum is over the excited states, with energies E_n. The content of the expression for the polarizability can be appreciated by approximating the excitation energies by a mean value ΔE (an indication of the HOMO–LUMO separation),

---

4 We should use vector quantities and allow for the possibility that the induced dipole moment might not lie parallel to the applied field; for simplicity we discuss polarizabilities in terms of (scalar) magnitudes.

5 When using older compilations of data, it is useful to note that polarizability volumes have the same numerical values as the 'polarizabilities' reported using c.g.s. electrical units, so the tabulated values previously called 'polarizabilities' can be used directly.

and supposing that the most important transition dipole moment is approximately equal to the charge of an electron multiplied by the radius, R, of the molecule. Then

$$\alpha \approx \frac{2e^2R^2}{3\Delta E}$$

This expression shows that α increases with the size of the molecule and with the ease with which it can be excited (the smaller the value of ΔE).

If the excitation energy is approximated by the energy needed to remove an electron to infinity from a distance R from a single positive charge, we can write $\Delta E \approx e^2/4\pi\varepsilon_0 R$. When this expression is substituted into the equation above, both sides are divided by $4\pi\varepsilon_0$, and the factor of $\frac{2}{3}$ ignored in this approximation, we obtain $\alpha' \approx R^3$, which is of the same order of magnitude as the molecular volume.

For all molecules other than those belonging to one of the cubic or icosahedral groups, the polarizability depends on the orientation of the molecule relative to the field. The polarizability volume of benzene when the field is applied perpendicular to the ring is 12.3×10^{-30} m^3 and is 6.7×10^{-30} m^3 when the field is applied in the plane of the ring. The anisotropy of the polarizability determines whether a molecule is rotationally Raman active (Section 16.7).

(d) Polarization at high frequencies

When the applied field changes direction slowly, the permanent dipole moment has time to reorientate—the whole molecule rotates into a new direction—and follow the field. However, when the frequency of the field is high, a molecule cannot change direction fast enough to follow the change in direction of the applied field and the dipole moment then makes no contribution to the polarization of the sample. Because a molecule takes about 1 ps to turn through about 1 radian in a fluid, the loss of this contribution to the polarization occurs when measurements are made at frequencies greater than about 10^{11} Hz (in the microwave region). We say that the **orientation polarization**, the polarization arising from the permanent dipole moments, is lost at such high frequencies.

The next contribution to the polarization to be lost as the frequency is raised is the **distortion polarization**, the polarization that arises from the distortion of the positions of the nuclei by the applied field. The molecule is bent and stretched by the applied field, and the molecular dipole moment changes accordingly. The time taken for a molecule to bend is approximately the inverse of the molecular vibrational frequency, so the distortion polarization disappears when the frequency of the radiation is increased through the infrared. The disappearance of polarization occurs in stages: as shown in the *Justification* below, each successive stage occurs as the incident frequency rises above the frequency of a particular mode of vibration.

Justification 22.3

The quantum mechanical expression for the polarizability of a molecule in the presence of an electric field that is oscillating at a frequency ω is

$$\alpha(\omega) = \frac{2}{3\hbar}\sum_n \frac{\omega_{n0}|\mu_{0n}|^2}{\omega_{n0}^2 - \omega^2} \tag{12}$$

The quantities in this expression (which is valid provided that ω is not close to ω_{n0}) are the same as those in the previous *Justification*, with $\hbar\omega_{n0} = E_n - E_0$. As $\omega \to 0$, the equation reduces to eqn 11 for the static polarizability. As ω becomes very high (and much higher than any excitation frequency of the molecule), the polarizability becomes

$$\alpha(\omega) = -\frac{2}{3\hbar\omega^2} \sum_n \omega_{n0}|\mu_{0n}|^2 \to 0 \text{ as } \omega \to \infty$$

That is, when the incident frequency is higher than any excitation frequency, the polarizability becomes zero. The argument applies to each type of excitation, vibrational as well as electronic, and accounts for the successive decreases in polarizability as the frequency is increased.

At even higher frequencies, in the visible region, only the electrons are mobile enough to respond to the rapidly changing direction of the applied field. The polarization that remains is now due entirely to the distortion of the electron distribution, and the surviving contribution to the molecular polarizability is called the **electronic polarizability**.

(e) Relative permittivities

When two charges q_1 and q_2 are separated by a distance r in a vacuum, the potential energy of their interaction is

$$V = \frac{q_1 q_2}{4\pi\varepsilon_0 r} \tag{13a}$$

When the same two charges are immersed in a medium (such as air or a liquid), their potential energy is reduced to

$$V = \frac{q_1 q_2}{4\pi\varepsilon r} \tag{13b}$$

where ε is the **permittivity** of the medium. The permittivity is normally expressed in terms of the dimensionless **relative permittivity**, ε_r, (which is also called the dielectric constant) of the medium:[6]

$$\varepsilon_r = \frac{\varepsilon}{\varepsilon_0} \tag{14}$$

The relative permittivity can have a very significant effect on the strength of the interactions between ions in solution. For instance, water has a relative permittivity of 78 at 25°C, so the interionic Coulombic interaction energy is reduced by nearly two orders of magnitude from its vacuum value. Some of the consequences of this reduction for electrolyte solutions were explored in Chapter 10.

The relative permittivity of a substance is large if its molecules are polar or highly polarizable. The quantitative relation between the relative permittivity and the electric

---

6   The relative permittivity of a substance is measured by comparing the capacitance of a capacitor with and without the sample present (C and C_0, respectively) and using $\varepsilon_r = C/C_0$.

properties of the molecules is obtained by considering the polarization of a medium, and is expressed by the **Debye equation**:

$$\frac{\varepsilon_r - 1}{\varepsilon_r + 2} = \frac{\rho P_m}{M} \tag{15}$$

where ρ is the mass density of the sample, M is the molar mass of the molecules, and P_m is the **molar polarization**,[7] which is defined as

$$P_m = \frac{N_A}{3\varepsilon_0}\left(\alpha + \frac{\mu^2}{3kT}\right) \tag{16}$$

The term $\mu^2/3kT$ stems from the thermal averaging of the electric dipole moment in the presence of the applied field (eqn 5). The corresponding expression without the contribution from the permanent dipole moment is called the **Clausius–Mossotti equation**:

$$\frac{\varepsilon_r - 1}{\varepsilon_r + 2} = \frac{\rho N_A \alpha}{3 M \varepsilon_0} \tag{17}$$

The Clausius–Mossotti equation is used when there is no contribution from permanent electric dipole moments to the polarization, either because the molecules are nonpolar or because the frequency of the applied field is so high that the molecules cannot orientate quickly enough to follow the change in direction of the field.

Equation 16 implies that the polarizability and permanent dipole moment of the molecules in a sample can be determined by measuring ε_r at a series of temperatures, calculating P_m, and plotting it against $1/T$. The slope of the graph is $N_A\mu^2/9\varepsilon_0 k$ and its intercept at $1/T = 0$ is $N_A\alpha/3\varepsilon_0$.

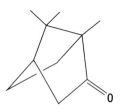

4 Camphor

Example 22.1 Determining dipole moment and polarizability

The relative permittivity of camphor (**4**) was measured at a series of temperatures with the results given below. Determine the dipole moment and the polarizability volume of the molecule.

$\theta/°C$	$\rho/(\mathrm{g\,cm^{-3}})$	ε_r
0	0.99	12.5
20	0.99	11.4
40	0.99	10.8
60	0.99	10.0
80	0.99	9.50
100	0.99	8.90
120	0.97	8.10
140	0.96	7.60
160	0.95	7.11
200	0.91	6.21

Method According to eqn 15, we need to calculate $(\varepsilon_r - 1)/(\varepsilon_r + 2)$ at each temperature, and then multiply by M/ρ to form P_m. Next, from eqn 16, we should plot P_m against $1/T$ and expect a straight line. The intercept at $1/T = 0$ is $N_A\alpha/3\varepsilon_0 = (4\pi N_A/3)\alpha'$ and the slope is $N_A\mu^2/9\varepsilon_0 k$.

---

7   Molar polarization is an unhappy but traditional name for P_m, which has the dimensions of volume per mole. H. Looyenga (*Mol. Phys.* 9, 501 (1965)) has argued that a better description is obtained if $(\varepsilon - 1)/(\varepsilon + 2)$ is replaced by $\varepsilon^{1/3} - 1$ in eqns 15 and 17.

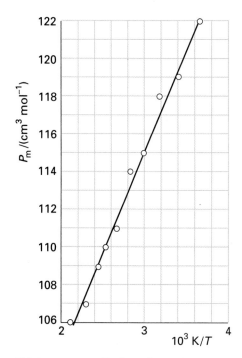

22.3 The plot of $P_m/(cm^3\,mol^{-1})$ against $(10^3\,K)/T$ used in Example 22.1 for the determination of the polarizability and dipole moment of camphor.

Answer For camphor, $M = 152.23\ g\,mol^{-1}$. We can therefore use the data to draw up the following table:

$\theta/^\circ C$	$(10^3\ K)/T$	ε_r	$(\varepsilon_r - 1)/(\varepsilon_r + 2)$	$P_m/(cm^3\,mol^{-1})$
0	3.66	12.5	0.793	122
20	3.41	11.4	0.776	119
40	3.19	10.8	0.766	118
60	3.00	10.0	0.750	115
80	2.83	9.50	0.739	114
100	2.68	8.90	0.725	111
120	2.54	8.10	0.703	110
140	2.42	7.60	0.688	109
160	2.31	7.11	0.670	107
200	2.11	6.21	0.634	106

The points are plotted in Fig. 22.3. The intercept lies at 82.7, so $\alpha' = 3.3 \times 10^{-23}\ cm^3$. The slope is 10.9, so $\mu = 4.46 \times 10^{-30}\ C\,m$, corresponding to 1.34 D.

Comment Because the Debye equation describes molecules that are free to rotate, the data show that camphor, which does not melt until 175°C, is rotating even in the solid. It is an approximately spherical molecule.

- -

Self-test 22.1 The relative permittivity of chlorobenzene is 5.71 at 20°C and 5.62 at 25°C. Assuming a constant density $(1.11\ g\,cm^{-3})$, estimate its polarizability volume and dipole moment.

$$[1.4 \times 10^{-23}\ cm^3,\ 1.2\ D]$$

22.2 Refractive index

One of the optical properties of matter that we are almost in a position to explain is the ability of a prism to separate light into its component colours. This effect depends on the **refractive index**, n_r, of the medium, the ratio of the speed of light in a vacuum, c, to its speed c' in the medium:

$$n_r = \frac{c}{c'} \tag{18}$$

It follows from the Maxwell equations[8] that the refractive index at a (visible or ultraviolet) specified frequency is related to the relative permittivity at that frequency by

$$n_r = \varepsilon_r^{1/2} \tag{19}$$

The molar polarization, P_m, and hence the molecular polarizability, α, can therefore be measured at frequencies typical of visible light (about 10^{15} to 10^{16} Hz) by measuring the refractive index of the sample (Table 22.2) and using the Clausius–Mossotti equation.

The refractive index is related to the molecular polarizability because the propagation of light through a medium can be imagined to occur by the incident light inducing an oscillating dipole moment, which then radiates light of the same frequency. The newly generated radiation is delayed slightly by this process, so it propagates more slowly through the medium than through a vacuum. Because photons of high-frequency light carry more energy than those of low-frequency light, they can distort the electronic distributions of the

Table 22.2* Refractive indices (at different wavelengths of light) relative to air at 20°C

	434 nm	589 nm	656 nm
$C_6H_6(l)$	1.524	1.501	1.497
$CS_2(l)$	1.675	1.628	1.618
$H_2O(l)$	1.340	1.333	1.331
$KI(s)$	1.704	1.666	1.658

*More values are given in the *Data section*.

8   The Maxwell equations describe the properties of electromagnetic radiation; they are not discussed in this text: see *Further reading*.

molecules in their path more effectively. Therefore, after allowing for the loss of contributions from low-frequency modes of motion, we can expect the electronic polarizabilities of molecules, and hence the refractive index, to increase as the incident frequency rises towards an absorption frequency. This dependence on frequency is the origin of the dispersion of white light by a prism: the refractive index is greater for blue light than for red, and therefore the blue rays are bent more than the red. The term **dispersion** is a term carried over from this phenomenon to mean the variation of the refractive index, or of any property, with frequency. Figure 22.4 shows the typical dispersion of the polarizability of a sample.

The concept of refractive index is closely related to the property of optical activity. An **optically active** substance is a substance that rotates the plane of polarization of plane-polarized light. As shown in the *Justification* below, the rotation arises from the difference in the refractive indices for right- and left-circularly polarized light, n_R and n_L, respectively. By convention, in right-handed circularly polarized light the electric vector rotates clockwise as seen by an observer facing the oncoming beam (Fig. 22.5). A sample in which these two refractive indices are different is said to be **circularly birefringent**.

Justification 22.4

Before entering the medium, the beam is plane-polarized (that is, the electric field oscillates in a plane containing the propagation direction). This beam may be regarded as a superposition of two oppositely rotating, circularly polarized components (Fig. 22.6). On entering the medium, one component propagates faster than the other if their refractive indices are different. If the sample is of length l, the difference in the times of passage is

$$\Delta t = \frac{l}{c_R} - \frac{l}{c_L}$$

where c_R and c_L are the speeds of the two components in the medium. In terms of the refractive indices, the difference is

$$\Delta t = (n_R - n_L)\frac{l}{c}$$

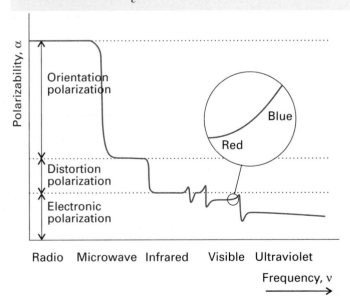

22.4 The general form of the variation of the polarizability with the frequency of the applied field. Note the considerable reduction in polarizability when the field is reversing direction so rapidly that the polar molecules cannot reorientate quickly enough to follow it. The inset shows the variation of the electronic polarizability in the visible region near an electronic excitation of the molecule.

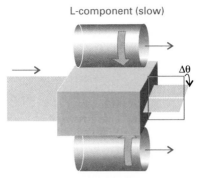

L-component (slow)

$\Delta\theta$

R-component (fast)

22.5 Linearly polarized light entering a sample (from the left) can be regarded as the superposition of two counter-rotating circularly polarized components (represented by the two cylindrical objects, which are actually superimposed inside the sample) with a definite phase relation. If one component propagates more rapidly than the other in the medium, when they emerge the phase relation is changed, and the resultant is plane-polarized light rotated through an angle $\Delta\theta$ to its original orientation.

The phase difference between the two components when they emerge from the sample is therefore

$$\Delta\theta = 2\pi\nu\Delta t = \frac{2\pi c \Delta t}{\lambda} = (n_{\mathrm{R}} - n_{\mathrm{L}}) \times \frac{2\pi l}{\lambda}$$

where λ is the wavelength of the light. The two rotating electric vectors have a different phase when they leave the sample from the value they had initially, so their superposition gives rise to a plane-polarized beam rotated through an angle $\Delta\theta$ relative to the plane of the incoming beam. It follows that the angle of optical rotation is proportional to the difference in refractive index, $n_{\mathrm{R}} - n_{\mathrm{L}}$.

To explain why the refractive indices depend on the handedness of the light, we must examine why the polarizabilities depend on the handedness. One interpretation is that, if a molecule has a helical structure (including, if the molecule is small, a structure that can be regarded as being a fragment of a helix), its polarizability depends on whether or not the electric field of the incident radiation rotates in the same sense as the helix. Molecules having a helical structure are chiral, which is the criterion for optical activity discussed in Section 15.3b.

The angle of optical rotation varies with the frequency of the radiation. This variation is called **optical rotatory dispersion** (ORD). It arises from the individual dispersions of the polarizabilities (and refractive indices) for left- and right-circularly polarized radiation. The effect can be used to investigate the stereochemistry of molecules.

Associated with the differences in the two refractive indices (the circular birefringence of the medium) is a difference in absorption intensities $\mathcal{I}_{\mathrm{R}}$ and $\mathcal{I}_{\mathrm{L}}$ for right- and left-circularly polarized radiation. This difference is known as **circular dichroism** (CD). The CD spectrum of a sample is a plot of the variation of $\mathcal{I}_{\mathrm{L}} - \mathcal{I}_{\mathrm{R}}$ with frequency of the radiation. Circular dichroism is particularly useful for determining the absolute configurations of d-metal complexes, because complexes with similar geometries have CD spectra with similar features.

Intermolecular forces

Van der Waals forces are the interactions between molecules that leave their chemical identities essentially unchanged. They include the interactions between the partial charges of polar molecules. There are also repulsive interactions that prevent the complete collapse of matter to nuclear densities. The repulsive interactions arise from Coulombic repulsions and, indirectly, from the Pauli principle and the exclusion of electrons from regions of space where the orbitals of neighbouring species overlap. In this section we consider the attractive forces between molecules, and see how they are related to the electrical properties treated in Section 22.1.

Right

Resultant

Left

Plane of polarization

22.6 The superposition shown in Fig. 22.5 as viewed by an observer facing the oncoming beam.

22.3 Interactions between dipoles

Most of the discussion in this section is based on the Coulombic potential energy of interaction between two charges (eqn 13a). It is easy to adapt this expression to obtain the potential energy of a charge and a dipole and to extend it to the interaction between two dipoles.

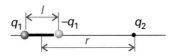

22.7 The potential energy of interaction between a dipole and a point charge is the sum of the repulsion of like charges and the attraction of opposite charges. For a point dipole, $l \ll r$.

(a) The potential energy of interaction

We show in the *Justification* below that the potential energy of interaction between a point dipole $\mu_1 = q_1 l$ and the point charge q_2 in the arrangement shown in Fig. 22.7 is

$$V = -\frac{\mu_1 q_2}{4\pi\varepsilon_0 r^2} \tag{20}$$

With μ in coulomb-metres, q_2 in coulombs, and r in metres, V is obtained in joules. A **point dipole** is a dipole in which the separation between the charges is much smaller than the distance at which the dipole is being observed, $l \ll r$. This expression should be multiplied by $\cos\theta$ when the point charge lies at an angle θ to the axis of the dipole. The potential energy rises towards zero (the value at infinite separation of the charge and the dipole) more rapidly (as $1/r^2$) than that between two point charges (which varies as $1/r$) because, from the viewpoint of the point charge, the partial charges of the dipole seem to merge and cancel as the distance r increases (Fig. 22.8).

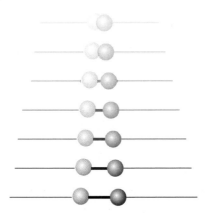

22.8 There are two contributions to the diminishing field of an electric dipole with distance (here seen from the side). The potential of the charges decreases (shown here by a fading intensity) and the two charges appear to merge, so their combined effect approaches zero more rapidly than by the distance effect alone.

Justification 22.5

The sum of the potential energies of repulsion between like charges and attraction between opposite charges in the orientation shown in Fig. 22.7 is

$$V = \frac{1}{4\pi\varepsilon_0}\left(-\frac{q_1 q_2}{r - \frac{1}{2}l} + \frac{q_1 q_2}{r + \frac{1}{2}l}\right)$$

Because $l \ll r$ for a point dipole, this expression can be simplified by writing

$$V = \frac{q_1 q_2}{4\pi\varepsilon_0 r}\left(-\frac{1}{1-x} + \frac{1}{1+x}\right)$$

where $x = l/2r$, and then expanding the terms in x by using

$$\frac{1}{1+x} = 1 - x + x^2 - \cdots \qquad \frac{1}{1-x} = 1 + x + x^2 + \cdots$$

and retaining only the leading term:

$$V = \frac{q_1 q_2}{4\pi\varepsilon_0 r}\{-(1 + x + \cdots) + (1 - x + \cdots)\}$$

$$\approx -\frac{2x q_1 q_2}{4\pi\varepsilon_0 r} = -\frac{q_1 q_2 l}{4\pi\varepsilon_0 r^2}$$

With $\mu_1 = q_1 l$, this expression becomes eqn 20.

22.9 The potential energy of interaction between two dipoles is the sum of the repulsions of like charges and the attractions of opposite charges. This illustration shows a collinear arrangement of dipoles.

Example 22.2 Calculating the interaction energy of two dipoles

Calculate the potential energy of interaction of two dipoles in the arrangement shown in Fig. 22.9 when their separation is r.

Method We proceed in exactly the same way as in the *Justification*, but now the total interaction energy is the sum of four pairwise terms: two attractions between opposite charges, which contribute negative terms to the potential energy, and two repulsions between like charges, which contribute positive terms.

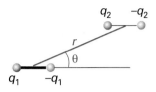

22.10 A parallel arrangement of electric dipoles.

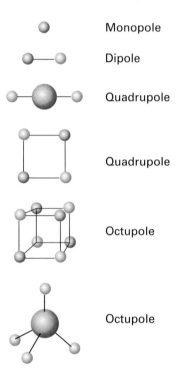

Monopole

Dipole

Quadrupole

Quadrupole

Octupole

Octupole

22.11 Typical charge arrays corresponding to electric multipoles. The field arising from an arbitrary finite charge distribution can be expressed as the superposition of the fields arising from a superposition of multipoles.

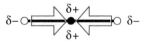

5 Carbon dioxide

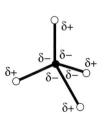

6 Methane

Answer The sum of the four contributions is

$$V = \frac{1}{4\pi\varepsilon_0}\left(-\frac{q_1 q_2}{r+l} + \frac{q_1 q_2}{r} + \frac{q_1 q_2}{r} - \frac{q_1 q_2}{r-l}\right)$$

$$= -\frac{q_1 q_2}{4\pi\varepsilon_0 r}\left(\frac{1}{1+x} - 2 + \frac{1}{1-x}\right)$$

with $x = l/r$. As before, we expand the two terms in x and retain only the first surviving term, which is equal to $2x^2$. This step results in the expression

$$V = -\frac{2x^2 q_1 q_2}{4\pi\varepsilon_0 r}$$

Therefore, because $\mu_1 = q_1 l$ and $\mu_2 = q_2 l$, the potential energy of interaction in the alignment shown in Fig. 22.9 is

$$V = -\frac{2\mu_1\mu_2}{4\pi\varepsilon_0 r^3}$$

Comment Notice that the interaction energy approaches zero more rapidly (as $1/r^3$) than for the previous case: now both interacting entities appear neutral to each other at large separations.

Self-test 22.2 Derive an expression for the potential energy when the dipoles are in the arrangement shown in Fig. 22.10.

$$[V = (\mu_1\mu_2/4\pi\varepsilon_0 r^3)(1 - 3\cos^2\theta)]$$

The various expressions for the interaction of charges and dipoles are summarized in Table 22.3. It is quite easy to extend the formulas given there to obtain expressions for the energy of interaction of higher **multipoles**, or arrays of point charges (Fig. 22.11). Specifically, an **n-pole** is an array of point charges with an n-pole moment but no lower moment. Thus, a monopole is a point charge, and the monopole moment is what we normally call the overall charge. A dipole, as we have seen, is an array of charges that has no monopole moment (no net charge). A quadrupole consists of an array of point charges that has neither net charge nor dipole moment (as for CO_2 molecules (**5**)). An octupole consists of an array of point charges that sum to zero and which has neither a dipole moment nor a quadrupole moment (as for CH_4 molecules (**6**)). The feature to remember is that the

Table 22.3 Multipole interaction potential energies

Interaction type	Distance dependence of potential energy	Typical energy/ ($\mathbf{kJ\,mol^{-1}}$)	Comment
Ion–ion	$1/r$	250	Only between ions
Ion–dipole	$1/r^2$	15	
Dipole–dipole	$1/r^3$	2	Between stationary polar molecules
	$1/r^6$	0.6	Between rotating polar molecules
London (dispersion)	$1/r^6$	2	Between all types of molecules

The energy of a hydrogen bond A—H$\cdots$B is typically 20 kJ mol^{-1} and occurs on contact for A, B = N, O, or F.

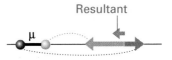

22.12 The electric field of a dipole is the sum of the opposing fields from the positive and negative charges, each of which is proportional to $1/r^2$. The difference, the net field, is proportional to $1/r^3$.

interaction energy falls off more rapidly the higher the order of the multipole. For the interaction of an n-pole with an m-pole, the potential energy varies with distance as

$$V \propto \frac{1}{r^{n+m-1}} \tag{21}$$

The reason for the even steeper decrease with distance is the same as before: the array of charges appears to blend together into neutrality more rapidly with distance the higher the number of individual charges that contribute to the multipole. Note that a given molecule may have a charge distribution that corresponds to a superposition of several different multipoles.

(b) The electric field

The same kind of argument as that used to derive expressions for the potential energy can be used to establish the distance dependence of the strength of the electric field generated by a dipole. We shall need this expression when we calculate the dipole moment induced in one molecule by another.

The starting point for the calculation is the strength of the electric field[9] generated by a point electric charge:

$$\mathcal{E} = \frac{q}{4\pi\varepsilon_0 r^2} \tag{22}$$

The field generated by a dipole is the sum of the fields generated by each partial charge. For the point-dipole arrangement shown in Fig. 22.12, the same procedure that was used to derive the potential energy gives

$$\mathcal{E} = \frac{2\mu}{4\pi\varepsilon_0 r^3} \tag{23}$$

The electric field of a multipole (in this case a dipole) decreases more rapidly with distance (as $1/r^3$ for a dipole) than that of a monopole (a point charge).

(c) Dipole–dipole interactions

The potential energy of interaction between two polar molecules is a complicated function of their relative orientation. When the two dipoles are parallel (as in Fig. 22.10), the potential energy is simply

$$V = \frac{\mu_1\mu_2 f(\theta)}{4\pi\varepsilon_0 r^3} \qquad f(\theta) = 1 - 3\cos^2\theta \tag{24}$$

This expression applies to polar molecules in a fixed, parallel orientation in a solid.

In a fluid of freely rotating molecules, the interaction between dipoles averages to zero because f changes sign as the orientation changes, and its average value is zero. Physically, the like partial charges of two freely rotating molecules are close together as much as the two opposite charges, and the repulsion of the former is cancelled by the attraction of the latter.

The interaction energy of two *freely* rotating dipoles is zero. However, because their mutual potential energy depends on their relative orientation, the molecules do not in fact rotate completely freely, even in a gas. In fact, the lower energy orientations are marginally favoured, so there is a nonzero average interaction between polar molecules. We show in the

---

9   The electric field is actually a vector, and we cannot simply add and subtract magnitudes without taking into account the directions of the fields. In the cases we consider, this will not be a complication because the two charges of the dipoles will be collinear and give rise to fields in the same direction. Be careful, though, with more general arrangements of charges.

following *Justification* that the average potential energy of two rotating molecules that are separated by a distance r is

$$\langle V \rangle = -\frac{C}{r^6} \qquad C = \frac{2\mu_1^2\mu_2^2}{3(4\pi\varepsilon_0)^2kT} \tag{25}$$

This expression describes the **Keesom interaction**.

Justification 22.6

The detailed calculation of the Keesom interaction energy is quite complicated, but the form of the final answer can be constructed quite simply. First, we note that the average interaction energy of two polar molecules rotating at a fixed separation r is given by

$$\langle V \rangle = \frac{\mu_1\mu_2\langle f \rangle}{4\pi\varepsilon_0 r^3}$$

where $\langle f \rangle$ now includes a weighting factor in the averaging that is equal to the probability that a particular orientation will be adopted. This probability is given by the Boltzmann distribution $p \propto e^{-E/kT}$, with E interpreted as the potential energy of interaction of the two dipoles in that orientation. That is,

$$p \propto e^{-V/kT} \qquad V = \frac{\mu_1\mu_2 f}{4\pi\varepsilon_0 r^3}$$

When the potential energy of interaction of the two dipoles is very small compared with the energy of thermal motion, we can use $V \ll kT$, expand the exponential function in p, and retain only the first two terms:

$$p \propto 1 - \frac{V}{kT} + \cdots$$

The weighted average of f is therefore

$$\langle f \rangle \propto \langle f \rangle_0 - \frac{\mu_1\mu_2}{4\pi\varepsilon_0 kTr^3}\langle f^2 \rangle_0 \cdots$$

where $\langle \cdots \rangle_0$ denotes an unweighted spherical average. The average value of f is zero, so the first term vanishes. However, the average value of f^2 is nonzero because f^2 is positive at all orientations, so we can write

$$\langle V \rangle = -\frac{\mu_1^2\mu_2^2\langle f^2 \rangle_0}{(4\pi\varepsilon_0)^2kTr^6}$$

The average value $\langle f^2 \rangle_0$ is a number that we can expect to be close to 1 (because f^2 ranges from 0 to 4) and in fact turns out to be $\frac{2}{3}$ when the calculation is carried through in detail. The final result is that quoted in eqn 25.

The important features of eqn 25 are its negative sign (the average interaction is attractive), the dependence of the average interaction energy on the inverse sixth power of the separation, and its inverse dependence on the temperature. The last feature reflects the way that the greater thermal motion overcomes the mutual orienting effects of the dipoles at higher temperatures. The inverse sixth power arises from the inverse third power of the interaction potential energy that is weighted by the energy in the Boltzmann term, which is also proportional to the inverse third power of the separation.

At 25°C the average interaction energy for pairs of molecules with $\mu = 1$ D is about $-0.07 \text{ kJ mol}^{-1}$ when the separation is 0.5 nm. This energy should be compared with the average molar kinetic energy of $\frac{3}{2}RT = 3.7 \text{ kJ mol}^{-1}$ at the same temperature. The interaction energy is much smaller than the energies involved in the making and breaking of chemical bonds.

(d) Dipole–induced-dipole interactions

A polar molecule with dipole moment μ_1 can induce a dipole μ_2^* in a neighbouring polarizable molecule. The induced dipole interacts with the permanent dipole of the first molecule, and the two are attracted together. It is shown in the *Justification* below that the average interaction energy when the separation of the molecules is r is[10]

$$V = -\frac{C}{r^6} \qquad C = \frac{\mu_1^2 \alpha_2'}{\pi \varepsilon_0} \tag{26}$$

where α_2' is the polarizability volume of molecule 2 and μ_1 is the permanent dipole moment of molecule 1.

Justification 22.7

The energy of interaction between a permanent dipole, μ_1, and an induced dipole, μ_2^*, is given in Example 22.2:

$$V = -\frac{2\mu_1 \mu_2^*}{4\pi \varepsilon_0 r^3}$$

The induced dipole moment depends on the field generated by the polar molecule, and hence on the separation of the two molecules. Because we can write $\mu_2^* = \alpha_2 \mathcal{E}$, where α_2 is the polarizability of molecule 2 and $\mathcal{E}$ is the field generated by molecule 1 (the polar molecule), the potential energy is:

$$V = -\frac{2\mu_1 \alpha_2 \mathcal{E}}{4\pi \varepsilon_0 r^3}$$

The electric field generated by the polar molecule is given by eqn 23, so:

$$V = -\left(\frac{2\mu_1 \alpha_2}{4\pi \varepsilon_0 r^3}\right)\left(\frac{2\mu_1}{4\pi \varepsilon_0 r^3}\right) = -\frac{4\mu_1^2 \alpha_2}{(4\pi \varepsilon_0)^2 r^6}$$

As the induced dipole follows the direction of the inducing dipole (Fig. 22.13), we do not need to take account of the effects of thermal motion: both dipoles remain aligned however fast the molecules tumble. Therefore, the interaction energy has approximately this value at all relative orientations. This expression rearranges into eqn 26 by noting that $\alpha_2' = \alpha_2/4\pi \varepsilon_0$.

The dipole–induced-dipole interaction energy is independent of the temperature because thermal motion has no effect on the averaging process. Moreover, like the dipole–dipole interaction, the potential energy depends on $1/r^6$: this distance dependence stems from the $1/r^3$ dependence of the field (and hence the magnitude of the induced dipole) and the $1/r^3$ dependence of the potential energy of interaction between the permanent and induced dipoles. For a molecule with $\mu = 1$ D (such as HCl) near a molecule of polarizability volume $\alpha' = 10 \times 10^{-30}$ m^3 (such as benzene, Table 22.1), the average interaction energy is about -0.8 kJ mol^{-1} when the separation is 0.3 nm.

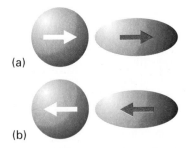

(a)

(b)

22.13 (a) A polar molecule (green arrow) can induce a dipole (white arrow) in a nonpolar molecule, and (b) the latter's orientation follows the former's, so the interaction does not average to zero.

10 Note that the C in this expression is different from the C in eqn 25 and other expressions below: we are using the same symbol in C/r^6 to emphasize the similarity of form of each expression.

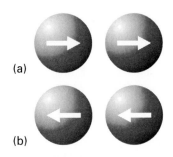

22.14 (a) In the dispersion interaction, an instantaneous dipole on one molecule induces a dipole on another molecule, and the two dipoles then interact to lower the energy. (b) The two instantaneous dipoles are correlated and, although they occur in different orientations at different instants, the interaction does not average to zero.

(e) Induced-dipole–induced-dipole interactions

Nonpolar molecules (including closed-shell atoms, such as Ar) attract one another even though neither has a permanent dipole moment. The abundant evidence for the existence of interactions between them is the formation of condensed phases of nonpolar substances, such as the condensation of hydrogen or argon to a liquid at low temperatures and the fact that benzene is a liquid at normal temperatures.

The interaction between nonpolar molecules arises from the transient dipoles that all molecules possess as a result of fluctuations in the instantaneous positions of electrons. To appreciate the origin of the interaction, suppose that the electrons in one molecule flicker into an arrangement that gives the molecule an instantaneous dipole moment μ_1^*. This dipole generates an electric field that polarizes the other molecule, and induces in that molecule an instantaneous dipole moment μ_2^*. The two dipoles attract each other and the potential energy of the pair is lowered. Although the first molecule will go on to change the size and direction of its instantaneous dipole, the electron distribution of the second molecule will follow, that is, the two dipoles are correlated in direction (Fig. 22.14). Because of this correlation, the attraction between the two instantaneous dipoles does not average to zero, and gives rise to an induced-dipole–induced-dipole interaction. This interaction is called either the **dispersion interaction** or the **London interaction** (for Fritz London, who first described it).

Polar molecules also interact by a dispersion interaction: such molecules also possess instantaneous dipoles, the only difference being that the time average of each fluctuating dipole does not vanish, but corresponds to the permanent dipole. Such molecules therefore interact both through their permanent dipoles and through the correlated, instantaneous fluctuations in these dipoles.

The strength of the dispersion interaction depends on the polarizability of the first molecule because the instantaneous dipole moment μ_1^* depends on the looseness of the control that the nuclear charge exercises over the outer electrons. The strength of the interaction also depends on the polarizability of the second molecule, for that polarizability determines how readily a dipole can be induced by another molecule. The actual calculation of the dispersion interaction is quite involved, but a reasonable approximation to the interaction energy is given by the **London formula**:

$$V = -\frac{C}{r^6} \qquad C = \tfrac{2}{3}\alpha_1'\alpha_2'\frac{I_1 I_2}{I_1 + I_2} \tag{27}$$

where I_1 and I_2 are the ionization energies of the two molecules (Table 13.4). This interaction energy is also proportional to the inverse sixth power of the separation of the molecules. The dispersion interaction generally dominates all the interactions between molecules other than hydrogen bonds.

Illustration

For two CH_4 molecules, we can substitute $\alpha' = 2.6 \times 10^{-30}$ m^3 and $I \approx 700$ kJ mol^{-1} to obtain $V = -2$ kJ mol^{-1} for $r = 0.3$ nm. A very rough check on this figure is the enthalpy of vaporization of methane, which is 8.2 kJ mol^{-1}. However, this comparison is insecure, partly because the enthalpy of vaporization is a many-body quantity and partly because the long-distance assumption breaks down.

(f) Hydrogen bonding

The interactions described so far are universal in the sense that they are possessed by all molecules independent of their specific identity. However, there is a type of interaction

Energy

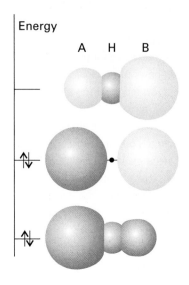

22.15 The molecular orbital interpretation of the formation of an A–H···B hydrogen bond. From the three A, H, and B orbitals, three molecular orbitals can be formed (their relative contributions are represented by the sizes of the spheres). Only the two lower energy orbitals are occupied, and there may therefore be a net lowering of energy compared with the separate AH and B species.

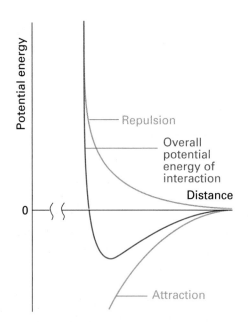

22.16 The general form of an intermolecular potential energy curve. At long range the interaction is attractive, but at close range the repulsions dominate.

possessed by molecules that have a particular constitution. A **hydrogen bond** is an attractive interaction between two species that arises from a link of the form A—H···B, where A and B are highly electronegative elements and B possesses a lone pair of electrons. Hydrogen bonding is conventionally regarded as being limited to N, O, and F but, if B is an anionic species (such as Cl^-), it may also participate in hydrogen bonding. There is no strict cut-off for an ability to participate in hydrogen bonding, but N, O, and F participate most effectively.

The formation of a hydrogen bond can be regarded as a particular example of delocalized molecular orbital formation in which A, H, and B each supply one atomic orbital from which three molecular orbitals are constructed (Fig. 22.15).[11] The A and H$1s$ orbitals are those used to form the A—H bond in the AH molecule and the B orbital originally accommodates the lone pair on B. In the combined species, there are four electrons to accommodate (two from the A—H bond, two from the lone pair of B), and they occupy the two lowest molecular orbitals of the AHB fragment. Because the uppermost (most antibonding) orbital is vacant, it is feasible for the net effect to be a lowering of energy, and hence the formation of a hydrogen bond.

In practice, the strength of the bond is found to be about 20 kJ mol^{-1}. Because the bonding depends on orbital overlap, it is virtually a contact-like interaction that is turned on when AH touches B and is zero as soon as the contact is broken. If hydrogen bonding is present, it dominates the other intermolecular interactions. The properties of liquid and solid water, for example, are dominated by the hydrogen bonding between H_2O molecules.

(g) The total attractive interaction

We shall consider molecules that are unable to participate in hydrogen bond formation. The total attractive interaction energy between rotating molecules is then the sum of the three van der Waals contributions discussed above. (Only the dispersion interaction contributes if both molecules are nonpolar.) In a fluid phase, all three contributions to the potential energy vary as the inverse sixth power of the separation of the molecules, so we may write

$$V = -\frac{C_6}{r^6} \qquad (28)$$

where C_6 is a coefficient that depends on the identity of the molecules.

Although attractive interactions between molecules are often expressed as in eqn 28, we must remember that this equation has only limited validity. First, we have taken into account only dipolar interactions of various kinds, for they have the longest range and are dominant if the average separation of the molecules is large. However, in a complete treatment we should also consider quadrupolar and higher-order multipole interactions, particularly if the molecules do not have permanent electric dipole moments. Secondly, the expressions have been derived by assuming that the molecules can rotate reasonably freely. That is not the case in most solids, and in rigid media the dipole–dipole interaction is proportional to $1/r^3$ because the Boltzmann averaging procedure is irrelevant when the molecules are trapped into a fixed orientation.

A different kind of limitation is that eqn 28 relates to the interactions of pairs of molecules. There is no reason to suppose that the energy of interaction of three (or more) molecules is the sum of the pairwise interaction energies alone. The total dispersion energy

---

11 A purely electrostatic description, in which the partial positive charge of H interacts Coulombically with the partial negative charge of B, is an alternative model of hydrogen bonding.

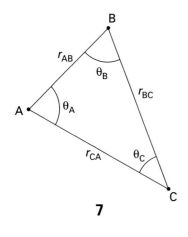

7

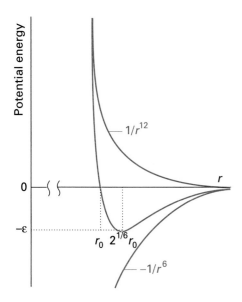

22.17 The Lennard-Jones potential, the relation of the parameters to the features of the curve, and the two contributions. Note that $2^{1/6} = 1.122\ldots$

Table 22.4* Lennard-Jones (12, 6) parameters

	$(\varepsilon/k)/\text{K}$	r_0/pm
Ar	111.84	362.3
CCl$_4$	376.86	624.1
N$_2$	91.85	391.9
Xe	213.96	426.0

*More values are given in the *Data section*.

† ε is expressed as an effective temperature on division by Boltzmann's constant k.

of three closed-shell atoms, for instance, is given approximately by the **Axilrod–Teller** formula:

$$V = -\frac{C_6}{r_{AB}^6} - \frac{C_6}{r_{BC}^6} - \frac{C_6}{r_{CA}^6} + \frac{C'}{(r_{AB}r_{BC}r_{CA})^3} \tag{29a}$$

where

$$C' = a(3\cos\theta_A\cos\theta_B\cos\theta_C + 1) \tag{29b}$$

The parameter a is approximately equal to $\frac{3}{4}\alpha'C_6$; the angles θ are the internal angles of the triangle formed by the three atoms (**7**). The term in C' (which represents the non-additivity of the pairwise interactions) is negative for a linear arrangement of atoms (so that arrangement is stabilized) and positive for an equilateral triangular cluster. It is found that the three-body term contributes about 10 per cent of the total interaction energy in liquid argon.

22.4 Repulsive and total interactions

When molecules are squeezed together, the nuclear and electronic repulsions and the rising electronic kinetic energy begin to dominate the attractive forces. The repulsions increase steeply with decreasing separation in a way that can be deduced only by very extensive, complicated molecular structure calculations of the kind described in Chapter 14 (Fig. 22.16).

In many cases, however, progress can be made by using a greatly simplified representation of the potential energy, where the details are ignored and the general features expressed by a few adjustable parameters. One such approximation is the **hard-sphere potential**, in which it is assumed that the potential energy rises abruptly to infinity as soon as the particles come within a separation d:

$$V = \infty \text{ for } r \leq d \qquad V = 0 \text{ for } r > d \tag{30}$$

This very simple potential is surprisingly useful for assessing a number of properties. Another widely used approximation is the **Mie potential**:

$$V = \frac{C_n}{r^n} - \frac{C_m}{r^m} \tag{31}$$

with $n > m$. The first term represents repulsions and the second term attractions. The **Lennard-Jones potential** is a special case of the Mie potential with $n = 12$ and $m = 6$ (Fig. 22.17); it is often written in the form

$$V = 4\varepsilon\left\{\left(\frac{r_0}{r}\right)^{12} - \left(\frac{r_0}{r}\right)^6\right\} \tag{32}$$

The two parameters are ε, the depth of the well, and r_0, the separation at which $V = 0$ (Table 22.4). The well minimum occurs at $r_e = 2^{1/6}r_0$. Although the Lennard-Jones potential has been used in many calculations, there is plenty of evidence to show that $1/r^{12}$ is a very poor representation of the repulsive potential, and that an exponential form, e^{-r/r_0}, is greatly superior. An exponential function is more faithful to the exponential decay of atomic wavefunctions at large distances, and hence to the overlap that is responsible for repulsion. The potential with an exponential repulsive term and a $1/r^6$ attractive term is known as an **exp-6 potential**. These potentials can be used to calculate the virial coefficients of gases, as explained in Section 20.5, and through them various properties of real gases, such as the Joule–Thompson coefficient. The potentials are also used to model the structures of condensed fluids.

Until recently, the potential energy of interaction of molecules was of primary interest. However, with the advent of **atomic force microscopy** (AFM), in which the force between a

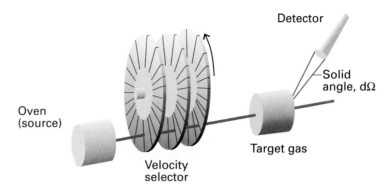

22.18 The basic arrangement of a molecular beam apparatus. The atoms or molecules emerge from a heated source, and pass through the velocity selector, a train of rotating disks. The scattering occurs from the target gas (which might take the form of another beam), and the flux of particles entering the detector set at some angle is recorded.

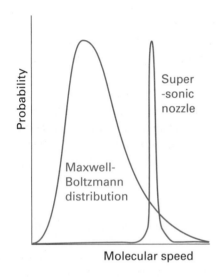

22.19 The shift in the mean speed and the width of the distribution brought about by use of a supersonic nozzle.

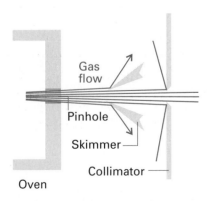

22.20 A supersonic nozzle skims off some of the molecules of the jet and leads to a beam with well-defined velocity.

molecular sized probe and a surface is monitored (see Section 28.2f), force is moving back into the centre of attention. As force, F, is the negative slope of potential, for a Lennard-Jones potential between individual molecules

$$F = -\frac{dV}{dr} = \frac{24\varepsilon}{r_0}\left\{2\left(\frac{r_0}{r}\right)^{13} - \left(\frac{r_0}{r}\right)^{7}\right\} \tag{33}$$

The net attractive force is greatest at $r = (26/7)^{1/6}r_0$, or $1.244r_0$, and at that distance is equal to $-144(7/26)^{7/6}\varepsilon/13r_0$, or $-2.397\varepsilon/r_0$. For typical parameters, the magnitude of this force is about 10 pN.

22.5 Molecular interactions in beams

Intermolecular forces can be studied in **molecular beams**, which consist of a collimated, narrow stream of molecules travelling though an evacuated vessel. The beam is directed towards other molecules, and the scattering that occurs on impact is related to the intermolecular interactions.

(a) The basic principles

The basic arrangement for a molecular beam experiment is shown in Fig. 22.18. If the pressure of vapour in the source is increased so that the mean free path of the molecules in the emerging beam is much shorter than the diameter of the pinhole, many collisions take place even outside the source. The net effect of these collisions, which give rise to **hydrodynamic flow**, is to transfer momentum into the direction of the beam. The molecules in the beam then travel with very similar speeds, so further downstream few collisions take place between them. This condition is called **molecular flow**. Because the spread in speeds is so small, the molecules are effectively in a state of very low translational temperature (Fig. 22.19). The translational temperature may reach as low as 1 K. Such jets are called **supersonic** because the average speed of the molecules in the jet is much greater than the speed of sound for the molecules that are not part of the jet.

A supersonic jet can be converted into a more parallel **supersonic beam** if it is 'skimmed' in the region of hydrodynamic flow and the excess gas pumped away. A skimmer consists of a conical nozzle shaped to avoid any supersonic shock waves spreading back into the gas and so increasing the translational temperature (Fig. 22.20). A jet or beam may also be formed by using helium or neon as the principal gas, and injecting molecules of interest into it in the hydrodynamic region of flow.

The low translational temperature of the molecules is reflected in the low rotational and vibrational temperatures of the molecules. In this context, a rotational or vibrational

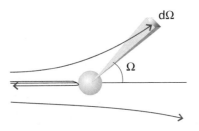

22.21 The definition of the solid angle, dΩ, for scattering.

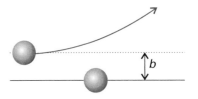

22.22 The definition of the impact parameter, b, as the perpendicular separation of the initial paths of the particles.

temperature means the temperature that should be used in the Boltzmann distribution to reproduce the observed populations of the states. However, as rotational modes equilibrate more slowly, and vibrational modes equilibrate even more slowly, the rotational and vibrational populations of the species correspond to somewhat higher temperatures, of the order of 10 K for rotation and 100 K for vibrations.

The target gas may be either a bulk sample or another molecular beam. The latter **crossed beam technique** gives a lot of information because the states of both the target and projectile molecules may be controlled. The intensity of the incident beam is measured by the **incident beam flux**, $\mathcal{I}$, which is the number of particles passing through a given area in a given interval divided by the area and the duration of the interval.

The detectors may consist of a chamber fitted with a sensitive pressure gauge, a bolometer, or an ionization detector, in which the incoming molecule is first ionized and then detected electronically. The state of the scattered molecules may also be determined spectroscopically, and is of interest when the collisions change their vibrational or rotational states.

(b) The experimental observations

The primary experimental information from a molecular beam experiment is the fraction of the molecules in the incident beam that are scattered into a particular direction. The fraction is normally expressed in terms of d$\mathcal{I}$, the rate at which molecules are scattered into a cone that represents the area covered by the 'eye' of the detector (Fig. 22.21). This rate is reported as the **differential scattering cross-section**, σ, the constant of proportionality between the value of d$\mathcal{I}$ and the intensity, $\mathcal{I}$, of the incident beam, the number density of target molecules, $\mathcal{N}$, and the infinitesimal path length dx through the sample:

$$d\mathcal{I} = \sigma \mathcal{I} \mathcal{N} \, dx \tag{34}$$

The value of σ (which has the dimensions of area) depends on the **impact parameter**, b, the initial perpendicular separation of the paths of the colliding molecules (Fig. 22.22), and the details of the intermolecular potential. The role of the impact parameter is most easily seen by considering the impact of two hard spheres (Fig. 22.23). If $b = 0$, the lighter projectile is on a trajectory that leads to a head-on collision, so the only scattering intensity is detected when the detector is at $\theta = \pi$. When the impact parameter is so great that the spheres do not make contact ($b > R_A + R_B$), there is no scattering and the scattering cross-section is zero at all angles except $\theta = 0$. Glancing blows, with $0 < b \le R_A + R_B$, lead to scattering intensity in cones around the forward direction.

(c) Scattering effects

The scattering pattern of real molecules, which are not hard spheres, depends on the details of the intermolecular potential, including the anisotropy that is present when the molecules are non-spherical. The scattering also depends on the relative speed of approach of the two

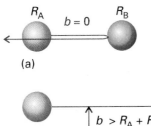

R_A $b = 0$ R_B

(a)

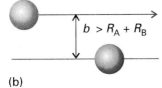

$b > R_A + R_B$

(b)

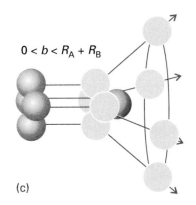

$0 < b < R_A + R_B$

(c)

22.23 Three typical cases for the collisions of two hard spheres: (a) $b = 0$, giving backward scattering; (b) $b > R_A + R_B$, giving forward scattering; (c) $0 < b < R_A + R_B$, leading to scattering into one direction on a ring of possibilities. (The target molecule is taken to be so heavy that it remains virtually stationary.)

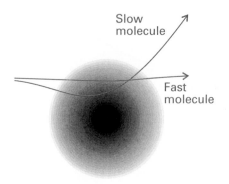

22.24 The extent of scattering may depend on the relative speed of approach as well as the impact parameter. The dark central zone represents the repulsive core; the fuzzy outer zone represents the long-range attractive potential.

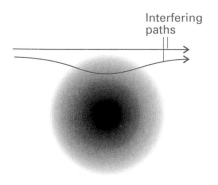

22.25 Two paths leading to the same destination will interfere quantum mechanically; in this case they give rise to quantum oscillations in the forward direction.

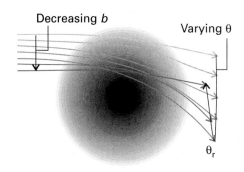

22.26 The interference of paths leading to rainbow scattering. The rainbow angle, θ_r, is the maximum scattering angle reached as b is decreased. Interference between the numerous paths at that angle modifies the scattering intensity markedly.

particles: a very fast particle might pass through the interaction region without much deflection, whereas a slower one on the same path might be temporarily captured and undergo considerable deflection (Fig. 22.24). The variation of the scattering cross-section with the relative speed of approach should therefore give information about the strength and range of the intermolecular potential.

A further point is that the outcome of collisions is determined by quantum, not classical, mechanics. The wave nature of the particles can be taken into account, at least to some extent, by drawing all classical trajectories that take the projectile particle from source to detector, and then considering the effects of interference between them.

Two quantum mechanical effects are of great importance. A particle with a certain impact parameter might approach the attractive region of the potential in such a way that the particle is deflected towards the repulsive core (Fig. 22.25), which then repels it out through the attractive region to continue its flight in the forward direction. Some molecules, however, also travel in the forward direction because they have impact parameters so large that they are undeflected. The wavefunctions of the particles that take the two types of path interfere, and the intensity in the forward direction is modified. The effect is called **quantum oscillation**. The same phenomenon accounts for the optical 'glory effect', in which a bright halo can sometimes be seen surrounding an illuminated object. (The coloured rings around the shadow of an aircraft cast on clouds by the sun, and often seen in flight, is an example of an optical glory.)

The second quantum effect we need consider is the observation of a strongly enhanced scattering in a nonforward direction. This effect is called **rainbow scattering** because the same mechanism accounts for the appearance of an optical rainbow. The origin of the phenomenon is illustrated in Fig. 22.26. As the impact parameter decreases, there comes a stage at which the scattering angle passes through a maximum and the interference between the paths results in a strongly scattered beam. The **rainbow angle**, θ_r, is the angle for which $d\theta/db = 0$ and the scattering is strong.

Another phenomenon that can occur in certain beams is the capturing of one species by another. The vibrational temperature in supersonic beams is so low that **van der Waals molecules** may be formed, which are complexes of the form AB in which A and B are held together by van der Waals forces or hydrogen bonds. Large numbers of such molecules have been studied spectroscopically, including ArHCl, $(HCl)_2$, $ArCO_2$, and $(H_2O)_2$. More recently, van der Waals clusters of water molecules have been pursued as far as $(H_2O)_6$. The study of their spectroscopic properties gives detailed information about the intermolecular potentials involved.

Magnetic properties

The magnetic and electric properties of molecules are analogous. For instance, some molecules possess permanent magnetic dipole moments, and an applied magnetic field can induce a magnetic moment.

22.6 Magnetic susceptibility

The analogue of the electric polarization, P, is the **magnetization**, M, the average molecular magnetic dipole moment multiplied by the number density of molecules in the sample. The magnetization induced by a field of strength $\mathcal{H}$ is proportional to $\mathcal{H}$, and we write

$$M = \chi \mathcal{H} \qquad [35]$$

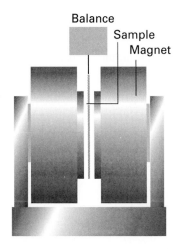

22.27 The arrangement of the Gouy balance for measuring magnetic susceptibilities. A paramagnetic sample appears to weigh more and a diamagnetic sample appears to weigh less, when the magnetic field is on. In a modern alternative version (not shown), a sensitive balance is used to measure the force exerted by the sample on a suspended permanent magnet.

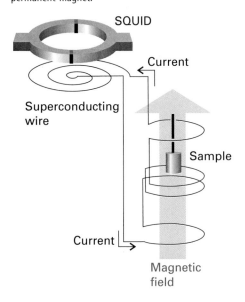

22.28 The arrangement used to measure magnetic susceptibility by using a SQUID. The sample is moved upwards in small increments and the potential difference across the SQUID is monitored.

Table 22.5* Magnetic susceptibilities at 289 K

	$\chi/10^{-6}$	$\chi_m/(10^{-5}\ cm^3\ mol^{-1})$
$H_2O(l)$	-90	-160
$NaCl(s)$	-13.9	-38
$Cu(s)$	-9.6	-6.8
$CuSO_4\cdot$ $5H_2O(s)$	$+176$	$+1930$

*More values are given in the *Data section*.

where χ is the dimensionless **volume magnetic susceptibility**. A closely related quantity is the **molar magnetic susceptibility**, χ_m:

$$\chi_m = \chi V_m \qquad [36]$$

where V_m is the molar volume of the substance (we shall soon see why it is sensible to introduce this quantity). The **magnetic flux density**, $\mathcal{B}$, is related to the applied field strength and the magnetization by

$$\mathcal{B} = \mu_0(\mathcal{H} + M) = \mu_0(1 + \chi)H \qquad [37]$$

where μ_0 is the **vacuum permeability**:

$$\mu_0 = 4\pi \times 10^{-7}\ J\,C^{-2}\,m^{-1}\,s^2 \qquad [38]$$

The magnetic flux density can be thought of as the density of magnetic lines of force permeating the medium. This density is increased if M adds to $\mathcal{H}$ (when $\chi > 0$), but the density is decreased if M opposes $\mathcal{H}$ (when $\chi < 0$). Materials for which χ is positive are called **paramagnetic**. Those for which χ is negative are called **diamagnetic**.

Just as polar molecules contribute a term proportional to $\mu^2/3kT$ to the electric polarization of a medium, so molecules with a permanent magnetic dipole moment of magnitude m contribute to the magnetization an amount proportional to $m^2/3kT$. An applied field can also induce a magnetic moment to an extent determined by the **magnetizability**, ξ (xi), of the molecules, and the magnetic analogue of eqn 16 is

$$\chi = \mathcal{N}\mu_0\left(\xi + \frac{m^2}{3kT}\right) \qquad (39)$$

We can now see why it is convenient to introduce χ_m, for the product of the number density $\mathcal{N}$ and the molar volume is the Avogadro constant, N_A:

$$\mathcal{N}V_m = \frac{NV_m}{V} = \frac{nN_A V_m}{nV_m} = N_A \qquad (40)$$

Hence

$$\chi_m = N_A\mu_0\left(\xi + \frac{m^2}{3kT}\right) \qquad (41)$$

and the density dependence of the susceptibility (which occurs in eqn 39 via $\mathcal{N} = N_A\rho/M$) has been eliminated. The expression for χ_m is in agreement with the empirical **Curie law**:

$$\chi_m = A + \frac{C}{T} \qquad (42)$$

with $A = N_A\mu_0\xi$ and $C = N_A\mu_0 m^2/3k$.

The magnetic susceptibility is traditionally measured with a **Gouy balance**. This instrument consists of a sensitive balance from which the sample hangs in the form of a narrow cylinder (Fig. 22.27) and lies between the poles of a magnet. If the sample is paramagnetic, it is drawn into the field, and its apparent weight is greater than when the field is off. A diamagnetic sample tends to be expelled from the field and appears to weigh less when the field is turned on. The balance is normally calibrated against a sample of known susceptibility. The modern version of the determination makes use of a **superconducting quantum interference device** (SQUID, Fig. 22.28).

Some experimental values are listed in Table 22.5; a typical paramagnetic volume susceptibility is about 10^{-3}, and a typical diamagnetic volume susceptibility is about $(-)10^{-5}$. The permanent magnetic moment can be extracted from susceptibility measurements by plotting χ against $1/T$.

22.7 The permanent magnetic moment

The permanent magnetic moment of a molecule arises from any unpaired electron spins in the molecule. We saw in Section 13.10a that the magnitude of the magnetic moment of an electron is proportional to the magnitude of the spin angular momentum, $\{s(s+1)\}^{1/2}\hbar$.

$$m = g_e\{s(s+1)\}^{1/2}\mu_B \qquad \mu_B = \frac{e\hbar}{2m_e} \tag{43}$$

where $g_e = 2.0023$. If there are several electron spins in each molecule, they combine to a total spin S, and then $s(s+1)$ should be replaced by $S(S+1)$. It follows that the spin contribution to the molar magnetic susceptibility is

$$\chi_m = \frac{N_A g_e^2 \mu_0 \mu_B^2 S(S+1)}{3kT} \tag{44}$$

This expression shows that the susceptibility is positive, so the spin magnetic moments contribute to the paramagnetic susceptibilities of materials. The contribution decreases with increasing temperature because the thermal motion randomizes the spin orientations. In practice, a contribution to the paramagnetism also arises from the orbital angular momenta of electrons: we have discussed the **spin-only contribution**.

Illustration

Consider a complex salt with three unpaired electrons per complex cation at 298 K, of mass density 3.24 g cm^{-3}, and molar mass 200 g mol^{-1}. First note that

$$\frac{N_A g_e^2 \mu_0 \mu_B^2}{3k} = 6.3001 \text{ cm}^3 \text{ K}^{-1} \text{ mol}^{-1}$$

Consequently,

$$\chi_m = 6.3001 \times \frac{S(S+1)}{T/K} \text{ cm}^3 \text{ mol}^{-1}$$

Substitution of the data with $S = \frac{3}{2}$ gives $\chi_m = 7.9 \times 10^{-2}$ cm^3 mol^{-1}. Note that the density is not needed at this stage. To obtain the volume magnetic susceptibility, the molar susceptibility is divided by the molar volume $V_m = M/\rho$, where ρ is the mass density. In this illustration, $V_m = 61.7$ cm^3 mol^{-1}, so $\chi = 1.3 \times 10^{-3}$.

At low temperatures, some paramagnetic solids make a phase transition to a state in which large domains of spins align with parallel orientations. This cooperative alignment gives rise to a very strong magnetization and is called **ferromagnetism** (Fig. 22.29). In other cases, the cooperative effect leads to alternating spin orientations: the spins are locked into a low-magnetization arrangement to give an **antiferromagnetic phase**. The ferromagnetic phase has a nonzero magnetization in the absence of an applied field, but the antiferromagnetic phase has a zero magnetization because the spin magnetic moments cancel. The ferromagnetic transition occurs at the **Curie temperature**, and the antiferromagnetic transition occurs at the **Néel temperature**.

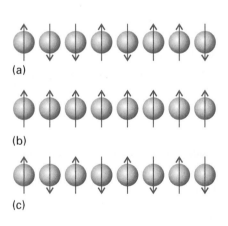

(a)

(b)

(c)

22.29 (a) In a paramagnetic material, the electron spins are aligned at random in the absence of an applied magnetic field. (b) In a ferromagnetic material, the electron spins are locked into a parallel alignment over large domains. (c) In an antiferromagnetic material, the electron spins are locked into an antiparallel arrangement. The latter two arrangements survive even in the absence of an applied field.

22.8 Induced magnetic moments

An applied magnetic field induces the circulation of electronic currents. These currents give rise to a magnetic field which usually opposes the applied field, so the substance is diamagnetic. In a few cases the induced field augments the applied field, and the substance is then paramagnetic.

The great majority of molecules with no unpaired electron spins are diamagnetic. In these cases, the induced electron currents occur within the orbitals of the molecule that are occupied in its ground state. In the few cases in which molecules are paramagnetic despite having no unpaired electrons, the induced electron currents flow in the opposite direction because they can make use of unoccupied orbitals that lie close to the HOMO in energy. This orbital paramagnetism can be distinguished from spin paramagnetism by the fact that it is temperature-independent: this is why the property is called **temperature-independent paramagnetism** (TIP).

We can summarize these remarks as follows. All molecules have a diamagnetic component to their susceptibility, but it is dominated by spin paramagnetism if the molecules have unpaired electrons. In a few cases (where there are low-lying excited states) TIP is strong enough to make the molecules paramagnetic even though their electrons are paired.

Checklist of key ideas

Electric properties

22.1 Permanent and induced electric dipole moments
- [ ] electric dipole
- [ ] electric dipole moment
- [ ] polar molecule
- [ ] polarization
- [ ] dielectric
- [ ] ferroelectric solid
- [ ] homopolar contribution
- [ ] Langevin function (6)
- [ ] induced dipole moment
- [ ] polarizability
- [ ] hyperpolarizability
- [ ] polarizability volume (10)
- [ ] orientation polarization
- [ ] distortion polarization
- [ ] electronic polarizability
- [ ] permittivity
- [ ] relative permittivity
- [ ] Debye equation (15)
- [ ] molar polarization (16)
- [ ] Clausius–Mossotti equation (17)

22.2 Refractive index
- [ ] refractive index (18)
- [ ] dispersion
- [ ] optically active
- [ ] circularly birefringent
- [ ] optical rotatory dispersion (ORD)
- [ ] circular dichroism (CD)

Intermolecular forces
- [ ] van der Waals forces

22.3 Interactions between dipoles
- [ ] point dipole
- [ ] multipole
- [ ] n-pole
- [ ] Keesom interaction (25)
- [ ] dispersion interaction
- [ ] London interaction
- [ ] London formula (27)
- [ ] hydrogen bond
- [ ] Axilrod–Teller formula (29)

22.4 Repulsive and total interactions
- [ ] hard-sphere potential (30)

- [ ] Mie potential (31)
- [ ] Lennard-Jones potential (32)
- [ ] exp-6 potential
- [ ] atomic force microscopy (AFM)

22.5 Molecular interactions in beams
- [ ] molecular beam
- [ ] hydrodynamic flow
- [ ] molecular flow
- [ ] supersonic jet
- [ ] supersonic beam
- [ ] crossed beam technique
- [ ] incident beam flux
- [ ] differential scattering cross-section
- [ ] impact parameter
- [ ] quantum oscillation
- [ ] rainbow scattering
- [ ] rainbow angle
- [ ] van der Waals molecule

Magnetic properties

22.6 Magnetic susceptibility
- [ ] magnetization

- [ ] volume magnetic susceptibility
- [ ] molar magnetic susceptibility
- [ ] magnetic flux density
- [ ] vacuum permeability (38)
- [ ] paramagnetic
- [ ] diamagnetic
- [ ] magnetizability
- [ ] Curie law (42)
- [ ] Gouy balance
- [ ] superconducting quantum interference device (SQUID)

22.7 The permanent magnetic moment
- [ ] spin-only contribution
- [ ] ferromagnetism
- [ ] antiferromagnetic phase
- [ ] Curie temperature
- [ ] Néel temperature

22.8 Induced magnetic moments
- [ ] temperature-independent paramagnetism (TIP)

Further reading

Articles of general interest

C.E. Dykstra, Electrical polarization in diatomic molecules. *J. Chem. Educ.* **65**, 198 (1988).

P. Hobza and R. Zahradnik, Intermolecular interactions between medium-sized systems. Nonempirical and empirical calculations of interaction energies: successes and failures. *Chem. Rev.* **88**, 871 (1988).

G. Chalasínski and M. Gutowski, Weak interactions between small systems. Models for studying the nature of intermolecular forces and challenging problems for *ab initio* calculations. *Chem. Rev.* **88**, 943 (1988).

A.D. Buckingham, P.W. Fowler, and J.M. Hutson, Theoretical studies of van der Waals molecules and intermolecular forces. *Chem. Rev.* **88**, 963 (1988).

F. Caudros, I. Cachadiña, and W. Ahamuda, Determination of Lennard-Jones interaction parameters using a new procedure. *Molec. Engineering* **6**, 319 (1996).

F. Cuardos, A. Mulero, and P. Rubio, The perturbative theories of fluids as a modern version of van der Waals theory. *J. Chem. Educ.* **71**, 956 (1994).

H. Guerin, Influence of the well width on the third virial coefficient of the square-well intermolecular potential. *J. Chem. Educ.* **69**, 203 (1992).

J.-L. Barrat and M.L. Klein, Molecular dynamics simulations of supercooled liquids near the glass transition. *Ann. Rev. Phys. Chem.* **42**, 23 (1991).

A. Gelessus, W. Thiel, and W. Weber, Multipoles and symmetry. *J. Chem. Educ.* **72**, 505 (1995).

L.N. Mulay and I.L. Mulay, Static magnetic techniques and applications. In *Techniques of chemistry* (ed. B.W. Rossiter and J.E. Hamilton), IIIB, 133 (1989).

W.E. Hatfield, Magnetic measurements. In *Solid-state chemistry: techniques* (ed. A.K. Cheetham and P. Day). Clarendon Press, Oxford (1987).

C.E. Dykstra, Intermolecular electrical interaction: a key ingredient in hydrogen bonding. *Acc. Chem. Res.* **21**, 355 (1988).

J. Israelachvili, Solvation forces and liquid structure, as probed by direct force measurements. *Acc. Chem. Res.* **20**, 415 (1987).

J.M. Hunter and M.F. Jarrold, Molecular and atomic clusters. In *Encyclopedia of applied physics* (ed. G.L. Trigg), **10**, 411. VCH, New York (1994).

C.D. Graham, Jr., Magnetic materials. In *Encyclopedia of applied physics* (ed. G.L. Trigg), **9**, 1. VCH, New York (1994).

L.M. Falicov, Diamagnetism. In *Encyclopedia of applied physics* (ed. G.L. Trigg), **4**, 557. VCH, New York (1992).

Y. Yafet, Paramagnetism. In *Encyclopedia of applied physics* (ed. G.L. Trigg), **13**, 101. VCH, New York (1995).

S. Foner, Measurement of magnetic properties and quantities. In *Encyclopedia of applied physics* (ed. G.L. Trigg), **9**, 463. VCH, New York (1994).

Texts and sources of data and information

J. Israelachvili, *Intermolecular and surface forces.* Academic Press, New York (1985).

C.P. Smyth, Determination of dipole moments. In *Techniques of chemistry* (ed. A. Weissberger and B.W. Rossiter), **4**, 351. Wiley-Interscience, New York (1972).

M. Rigby, E.B. Smith, W.A. Wakeham, and G.C. Maitland, *The forces between molecules.* Oxford University Press (1986).

G.C. Maitland, M. Rigby, E.B. Smith, and W.A. Wakeham, *Intermolecular forces: their origin and determination.* Clarendon Press, Oxford (1981).

M.A.D. Fluendy and K.P. Lawley, *Molecular beams in chemistry.* Chapman & Hall, London (1973).

E.A.V. Ebsworth, D.W.H. Rankin, and S. Cradock, *Structural methods in inorganic chemistry.* Blackwell Scientific, Oxford (1991).

R. Drago, *Physical methods for chemists.* Saunders, Philadelphia (1992).

Exercises

22.1 (a) Which of the following molecules may be polar: ClF_3, O_3, H_2O_2?

22.1 (b) Which of the following molecules may be polar: SO_3, XeF_4, SF_4?

22.2 (a) The molar polarization of fluorobenzene vapour varies linearly with T^{-1}, and is 70.62 cm³ mol⁻¹ at 351.0 K and 62.47 cm³ mol⁻¹ at 423.2 K. Calculate the polarizability and dipole moment of the molecule.

22.2 (b) The molar polarization of the vapour of a compound was found to vary linearly with T^{-1}, and is 75.74 cm³ mol⁻¹ at 320.0 K and 71.43 cm³ mol⁻¹ at 421.7 K. Calculate the polarizability and dipole moment of the molecule.

22.3 (a) At 0°C, the molar polarization of liquid chlorine trifluoride is 27.18 cm³ mol⁻¹ and its density is 1.89 g cm⁻³. Calculate the relative permittivity of the liquid.

22.3 (b) At 0°C, the molar polarization of a liquid is 32.16 cm³ mol⁻¹ and its density is 1.92 g cm⁻³. Calculate the relative permittivity of the liquid. Take $M = 85.0$ g mol⁻¹.

22.4 (a) The refractive index of CH_2I_2 is 1.732 for 656 nm light. Its density at 20°C is 3.32 g cm⁻³. Calculate the polarizability of the molecule at this wavelength.

22.4 (b) The refractive index of a compound is 1.622 for 643 nm light. Its density at 20°C is 2.99 g cm⁻³. Calculate the polarizability of the molecule at this wavelength. Take $M = 65.5$ g mol⁻¹.

22.5 (a) The dipole moments of the bonds C—O and C=O are 1.2 and 2.7 D, respectively. The bond lengths are 143 and 122 pm, respectively. Estimate the percentage ionic character of the bonds. How well do the results correlate with the electronegativity differences of the atoms in the bonds?

22.5 (b) The dipole moments of the bonds C–F and C–O are 1.4 and 1.2 D, respectively. The bond lengths are 141 and 143 pm, respectively. Estimate the percentage ionic character of the bonds. How well do the results correlate with the electronegativity differences of the atoms in the bonds?

22.6 (a) The electric dipole moment of toluene (methylbenzene) is 0.4 D. Estimate the dipole moments of the three xylenes (dimethylbenzene). Which answer can you be sure about?

22.6 (b) Calculate the resultant of two dipole moments of magnitude 1.5 D and 0.80 D that make an angle of 109.5° to each other.

22.7 (a) Calculate the magnitude and direction of the dipole moment of the following arrangement of charges in the xy-plane: $3e$ at $(0,0)$, $-e$ at $(0.32$ nm, $0)$, and $-2e$ at an angle of 20° from the x-axis and a distance of 0.23 nm from the origin.

22.7 (b) Calculate the magnitude and direction of the dipole moment of the following arrangement of charges in the xy-plane: $4e$ at $(0,0)$, $-2e$ at $(162$ pm, $0)$, and $-2e$ at an angle of 30° from the x-axis and a distance of 143 pm from the origin.

22.8 (a) The polarizability volume of H_2O is 1.48×10^{-24} cm³; calculate the dipole moment of the molecule (in addition to the permanent dipole moment) induced by an applied electric field of strength 1.0 kV cm^{-1}.

22.8 (b) The polarizability volume of NH_3 is 2.22×10^{-30} m³; calculate the dipole moment of the molecule (in addition to the permanent dipole moment) induced by an applied electric field of strength 15.0 kV m^{-1}.

22.9 (a) The polarizability volume of H_2O at optical frequencies is 1.5×10^{-24} cm³: estimate the refractive index of water. The experimental value is 1.33; what may be the origin of the discrepancy?

22.9 (b) The polarizability volume of a liquid of molar mass 72.3 g mol^{-1} and density 865 kg mol^{-1} at optical frequencies is 2.2×10^{-30} m³: estimate the refractive index of the liquid.

22.10 (a) The dipole moment of chlorobenzene is 1.57 D and its polarizability volume is 1.23×10^{-23} cm³. Estimate its relative permittivity at 25°C, when its density is 1.173 g cm^{-3}.

22.10 (b) The dipole moment of bromobenzene is 5.17×10^{-30} C m and its polarizability volume is approximately 1.5×10^{-29} m³. Estimate its relative permittivity at 25°C, when its density is 1491 kg m^{-3}.

22.11 (a) A solution of an optically active substance shows an optical rotation of 250° in a cell of length 10 cm at 500 nm. What is the difference of the refractive indices of left and right circularly polarized light through this substance?

22.11 (b) A solution of an optically active substance shows an optical rotation of 192° in a cell of length 15 cm at 450 nm. What is the difference of the refractive indices of left and right circularly polarized light through this substance?

22.12 (a) The magnetic moment of $CrCl_3$ is $3.81\mu_B$. How many unpaired electrons does the Cr possess?

22.12 (b) The magnetic moment of Mn^{2+} in its complexes is typically $5.3\mu_B$. How many unpaired electrons does the ion possess?

22.13 (a) Calculate the molar susceptibility of benzene given that its volume susceptibility is -7.2×10^{-7} and its density 0.879 g cm^{-3} at 25°C.

22.13 (b) Calculate the molar susceptibility of cyclohexane given that its volume susceptibility is -7.9×10^{-7} and its density 811 kg m^{-3} at 25°C.

22.14 (a) According to Lewis theory, an O_2 molecule should be diamagnetic. However, experimentally it is found that $\chi_m/(m^3\,mol^{-1}) = (1.22 \times 10^{-5}\,K)/T$. Determine the number of unpaired spins in O_2. How is the problem of the Lewis structure resolved?

22.14 (b) Predict the molar susceptibility of nitrogen dioxide at 298 K. Why does the molar susceptibility of a sample of nitrogen dioxide gas decrease as it is compressed?

22.15 (a) Data on a single crystal of MnF_2 give $\chi_m = 0.1463$ cm³ mol^{-1} at 294.53 K. Determine the effective number of unpaired electrons in this compound and compare your result with the theoretical value.

22.15 (b) Data on a single crystal of $NiSO_4\cdot7H_2O$ give $\chi_m = 6.00 \times 10^{-8}$ m³ mol^{-1} at 298 K. Determine the effective number of unpaired electrons in this compound and compare your result with the theoretical value.

22.16 (a) Estimate the spin-only molar susceptibility of $CuSO_4\cdot5H_2O$ at 25°C.

22.16 (b) Estimate the spin-only molar susceptibility of $MnSO_4\cdot4H_2O$ at 298 K.

22.17 (a) Approximately how large must the magnetic induction, B, be for the orientational energy of an $S = 1$ system to be comparable to kT at 298 K?

22.17 (b) Estimate the ratio of populations of the M_S states of a system with $S = 1$ in 15.0 T at 298 K.

Problems

Numerical problems

22.1 Suppose an H_2O molecule ($\mu = 1.85$ D) approaches an anion. What is the favourable orientation of the molecule? Calculate the electric field (in volts per metre) experienced by the anion when the water dipole is (a) 1.0 nm, (b) 0.3 nm, (c) 30 nm from the ion.

22.2 An H_2O molecule is aligned by an external electric field of strength 1.0 kV m^{-1} and an Ar atom ($\alpha' = 1.66 \times 10^{-24}$ cm³) is brought up slowly from one side. At what separation is it energetically favourable for the H_2O molecule to flip over and point towards the approaching Ar atom?

22.3 The relative permittivity of chloroform was measured over a range of temperatures with the following results:

$\theta/°C$	-80	-70	-60	-40	-20	0	20
ε_r	3.1	3.1	7.0	6.5	6.0	5.5	5.0
$\rho/(g\,cm^{-3})$	1.65	1.64	1.64	1.61	1.57	1.53	1.50

The freezing point of chloroform is $-64°C$. Account for these results and calculate the dipole moment and polarizability volume of the molecule.

22.4 The relative permittivities of methanol (m.p. $-95°C$) corrected for density variation are given below. What molecular information can be deduced from these values? Take $\rho = 0.791\,g\,cm^{-3}$ at 20°C.

$\theta/°C$	-185	-170	-150	-140	-110	-80	-50	-20	0	20
ε_r	3.2	3.6	4.0	5.1	67	57	49	43	38	34

22.5 In his classic book *Polar molecules*, Debye reports some early measurements of the polarizability of ammonia. From the selection below, determine the dipole moment and the polarizability volume of the molecule.

T/K	292.2	309.0	333.0	387.0	413.0	446.0
$P_m/(cm^3\,mol^{-1})$	57.57	55.01	51.22	44.99	42.51	39.59

The refractive index of ammonia at 273 K and 100 kPa is 1.000 379 (for yellow sodium light). Calculate the molar polarizability of the gas at this temperature and at 292.2 K. Combine the value calculated with the static molar polarizability at 292.2 K and deduce from this information alone the molecular dipole moment.

22.6 Values of the molar polarization of gaseous water at 100 kPa as determined from capacitance measurements are given below as a function of temperature.

T/K	384.3	420.1	444.7	484.1	522.0
$P_m/(cm^3\,mol^{-1})$	57.4	53.5	50.1	46.8	43.1

Calculate the dipole moment of H_2O and its polarizability volume.

Theoretical problems

22.7 Calculate the potential energy of the interaction between two linear quadrupoles when they are (a) collinear, (b) parallel and separated by a distance r.

22.8 Show that, in a gas (for which the refractive index is close to 1), the refractive index depends on the pressure as $n_r = 1 + \text{const} \times p$, and find the constant of proportionality. Go on to show how to deduce the polarizability volume of a molecule from measurements of the refractive index of a gaseous sample.

22.9 The refractive index of benzene is constant (at 1.51) from 0.4 GHz up to 0.55 GHz (in the microwave region of the spectrum), but then shows a series of oscillations between 1.47 and 1.54. Throughout the same frequency range, methylbenzene shows a higher refractive index (about 1.55), the same oscillations as in benzene, and additional oscillations between 1.52 and 1.56 near 0.4 GHz. Account for these observations.

22.10 Acetic acid vapour contains a proportion of planar, hydrogen-bonded dimers. The relative permittivity of pure liquid acetic acid is 7.14 at 290 K and increases with increasing temperature. Suggest an interpretation of the latter observation. What effect should

isothermal dilution have on the relative permittivity of solutions of acetic acid in benzene?

22.11 Show that the mean interaction energy of N atoms of diameter d interacting with a potential energy of the form C_6/R^6 is given by $U = -2N^2C_6/3Vd^3$, where V is the volume in which the molecules are confined and all effects of clustering are ignored. Hence, find a connection between the van der Waals parameter a and C_6 from $n^2a/V^2 = (\partial U/\partial V)_T$.

22.12 Suppose the repulsive term in a Lennard-Jones (12, 6) potential is replaced by an exponential function of the form $e^{-r/d}$. Sketch the form of the potential energy and locate the distance at which it is a minimum.

22.13 The *cohesive energy density*, $\mathcal{U}$, is defined as U/V, where U is the mean potential energy of attraction within the sample and V its volume. Show that $\mathcal{U} = -\frac{1}{2}\mathcal{N}^2 \int V(R)\,d\tau$, where $\mathcal{N}$ is the number density of the molecules and $V(R)$ is their attractive potential energy and where the integration ranges from d to infinity and over all angles. Go on to show that the cohesive energy density of a uniform distribution of molecules that interact by a van der Waals attraction of the form $-C_6/R^6$ is equal to $(2\pi/3)(N_A^2/d^3M^2)\rho^2C_6$, where ρ is the mass density of the solid sample and M is the molar mass of the molecules.

22.14 Consider the collision between a hard-sphere molecule of radius R_1 and mass m_1 and an infinitely massive impenetrable sphere of radius R_2. Plot the scattering angle θ as a function of the impact parameter b. Carry out the calculation using simple geometrical considerations.

22.15 The dependence of the scattering characteristics of atoms on the energy of the collision can be modelled as follows. We suppose that the two colliding atoms behave as impenetrable spheres, as in Problem 22.14, but that the effective radius of the heavy atoms depends on the speed v of the light atom. Suppose its effective radius depends on v as R_2e^{-v/v^*}, where v^* is a constant. Take $R_1 = \frac{1}{2}R_2$ for simplicity and an impact parameter $b = \frac{1}{2}R_2$, and plot the scattering angle as a function of (a) speed, (b) kinetic energy of approach.

22.16 The magnetizability, ξ, and the volume and molar magnetic susceptibilities can be calculated from the wavefunctions of molecules. For instance, the magnetizability of a hydrogenic atom is given by the expression $\xi = -(e^2/6m_e)\langle r^2\rangle$, where $\langle r^2\rangle$ is the (expectation) mean value of r^2 in the atom. Calculate ξ and χ_m for the ground state of a hydrogenic atom.

22.17 Nitrogen dioxide, a paramagnetic compound, is in equilibrium with its dimer, dinitrogen tetroxide, a diamagnetic compound. Derive an expression in terms of the equilibrium constant, K, for the dimerization to show how the molar susceptibility varies with the pressure of the sample. Suggest how the susceptibility might be expected to vary as the temperature is changed at constant pressure.

22.18 An NO molecule has thermally accessible electronically excited states. It also has an unpaired electron, and so may be expected to be paramagnetic. However, its ground state is not paramagnetic because the magnetic moment of the orbital motion of the unpaired electron almost exactly cancels the spin magnetic moment. The first excited state (at 121 cm^{-1}) is paramagnetic because the orbital magnetic moment adds to, rather than cancels, the spin magnetic moment. The upper state has a magnetic moment of $2\mu_B$. Because the upper state is thermally accessible, the paramagnetic susceptibility of NO shows a pronounced temperature dependence even near room temperature. Calculate the molar paramagnetic susceptibility of NO and plot it as a function of temperature.

Additional problems supplied by Carmen Giunta and Charles Trapp

22.19 The notion of group additivity of thermodynamic properties was introduced in Chapter 2. Results of a computational study of the polarizabilities of the homologous series of linear saturated silanes $Si_{2N}H_{4N+2}$ raise the possibility of group additivity of polarizability. B. Champagne, E.A. Perpète, and J.-M. André (*J. Molec. Structure (Theochem)* **391**, 67 (1997)) report the following values for the electronic contribution to the polarizability, α, of $Si_{2N}H_{4N+2}$ based on *ab initio* quantum mechanical calculations):

N	1	2	3	4	5
$\alpha/(10^{-40}\ J^{-1}\,C\,m^2)$	3.495	7.766	12.40	17.18	22.04
N	6	7	8	9	
$\alpha/(10^{-40}\ J^{-1}\,C\,m^2)$	26.92	31.82	36.74	41.63	

Determine the average contribution to this quantity per additional Si_2H_4 group, and report the root mean square deviation between the reported values and the best group-additivity fit.

22.20 F. Luo, G.C. McBane, G. Kim, C.F. Giese, and W.R. Gentry (*J. Chem. Phys.* **98**, 3564 (1993)) reported experimental observation of the He_2 complex, a species which had escaped detection for a long time. The fact that the observation required temperatures in the neighbourhood of 1 mK is consistent with computational studies which suggest that hcD_e for He_2 is about 1.51×10^{-23} J, hcD_0 about 2×10^{-26} J, and R_e about 297 pm. (a) Determine the Lennard-Jones parameters r_0 and ε and plot the Lennard-Jones potential for He–He

interactions. (b) Plot the Morse potential given that $a = 5.79 \times 10^{10}$ m^{-1}.

22.21 J.J. Dannenberg, D. Liotard, P. Halvick, and J.C. Rayez (*J. Phys. Chem.* **100**, 9631 (1996)) carried out theoretical studies of organic molecules consisting of chains of unsaturated four-membered rings. The calculations suggest that such compounds have large numbers of unpaired spins, and that they should therefore have unusual magnetic properties. For example, the lowest-energy state of the five-ring compound $C_{22}H_{14}$ (**8**) is computed to have $S = 3$, but the energies of $S = 2$ and $S = 4$ structures are each predicted to be 50 kJ mol^{-1} higher in energy. Compute the molar magnetic susceptibility of these three low-lying levels at 298 K. Estimate the molar susceptibility at 298 K if each level is present in proportion to its Boltzmann factor (effectively assuming that the degeneracy is the same for all three of these levels).

8

22.22 D.D. Nelson, G.T. Fraser, and W. Klemperer (*Science* **238**, 1670 (1987)) examined several weakly bound gas-phase complexes of ammonia in search of examples in which the H atoms in NH_3 formed hydrogen bonds, but found none. For example, they found that the complex of NH_3 and CO_2 has the carbon atom nearest the nitrogen (299 pm away): the CO_2 molecule is at right angles to the C–N 'bond', and the H atoms of NH_3 are pointing away from the CO_2. The permanent dipole moment of this complex is reported as 1.77 D. If the N and C atoms are the centres of the negative and positive charge distributions, respectively, what is the magnitude of those partial charges (as multiples of e)?

22.23 From data in Table 22.1 calculate the molar polarization, relative permittivity, and refractive index of methanol at 20°C. Its density at that temperature is 0.7914 g cm^{-3}.

23

Macromolecules and colloids

Macromolecules exhibit a range of properties and problems that illustrate a wide variety of physical chemical principles. They need to be characterized in terms of their molar mass, their size, and their shape. However, the molecules are so large that the solutions they form depart strongly from ideality, so techniques for accommodating these departures need to be developed. Although the shapes of large biomolecules can be determined by X-ray diffraction, synthetic polymers have less regular shapes in solution, and only their general shape can be inferred. Another major problem concerns the influences that determine the shapes of the molecules. We consider a range of influences in this chapter, beginning with the structureless random coil and ending with the structurally precise forces that operate in polypeptides.

Colloids, and other aggregates of molecules that are not chemically bonded together, exhibit some of the properties of molecules but have their own characteristic features arising from the very large surface-to-volume ratios of their constituent particles.

There are macromolecules everywhere, inside us and outside us. Some are natural: they include polysaccharides such as cellulose, polypeptides such as enzymes, and nucleic acids such as DNA. Others are synthetic: they include **polymers** such as nylon and polystyrene that are manufactured by stringing together and (in some cases) cross-linking smaller units known as **monomers**. Life in all its forms, from its intrinsic nature to its technological interaction with its environment, is the chemistry of macromolecules.

Macromolecules give rise to special problems that include the determination of their sizes, the shapes and the lengths of polymer chains, and the large deviations from ideality of their solutions. We concentrate on these special characteristics here.

Size and shape

X-ray diffraction (Chapter 21) can reveal the position of almost every atom even in highly complex molecules. However, there are several reasons why other techniques must also be used. In the first place, the sample might be a mixture of molecules with different chain lengths and extents of cross-linking, in which case sharp X-ray images are unobtainable. Even if all the molecules in the sample are identical, it might prove impossible to obtain a single crystal. Furthermore, although the work on enzymes, proteins, and DNA has shown how immensely stimulating the data can be, the information is incomplete. For instance, what can be said about the shape of the molecule in its natural environment, a biological cell? What can be said about the response of its shape to changes in its environment?

23.1 Mean molar masses

A pure protein is **monodisperse**, meaning that it has a single, definite molar mass. (There may be small variations, such as one amino acid replacing another, depending on the source of the sample.) A synthetic polymer, however, is **polydisperse** in the sense that a sample is a mixture of molecules with various chain lengths and molar masses. The various techniques that are used to measure molar masses result in different types of mean values of polydisperse systems. The mean obtained from the determination of molar mass by osmometry (Section 7.5e) gives the **number-average molar mass**, $\overline{M}_n$, which is the value obtained by weighting each molar mass by the number of molecules of that mass present in the sample:

$$\overline{M}_n = \frac{1}{N} \sum_i N_i M_i \tag{1}$$

where N_i is the number of molecules with molar mass M_i and there are N molecules in all. Viscosity measurements give the **viscosity-average molar mass**, $\overline{M}_v$, light-scattering experiments give the **weight-average molar mass**, $\overline{M}_w$, and sedimentation experiments give the **Z-average molar mass**, $\overline{M}_Z$. Although such averages are often best left as empirical quantities, some may be interpreted in terms of the composition of the sample. Thus, the weight-average molar mass is the average calculated by weighting the molar masses of the molecules by the mass of each one present in the sample:

$$\overline{M}_w = \frac{1}{m} \sum_i m_i M_i \tag{2}$$

In this expression, m_i is the total mass of molecules of molar mass M_i and m is the total mass of the sample. Because $m_i = N_i M_i / N_A$, we can also express this average as

$$\overline{M}_w = \frac{\sum_i N_i M_i^2}{\sum_i N_i M_i} \tag{3}$$

Hence, the weight-average molar mass is proportional to the mean square molar mass. Similarly, the Z-average molar mass can be interpreted in terms of the mean cubic molar mass:

$$\overline{M}_Z = \frac{\sum_i N_i M_i^3}{\sum_i N_i M_i^2} \tag{4}$$

Example 23.1 Calculating number and mass averages

Determine the number-average and the weight-average molar masses for a sample of poly(vinyl chloride) from the following data:

Molar mass interval/(kg mol^{-1})	Average molar mass within interval/(kg mol^{-1})	Mass of sample within interval/g
5–10	7.5	9.6
10–15	12.5	8.7
15–20	17.5	8.9
20–25	22.5	5.6
25–30	27.5	3.1
30–35	32.5	1.7

Method The relevant equations are eqns 1 and 2. The two averages are obtained by weighting the molar mass within each interval by the number and mass, respectively, of the molecule in each interval. The numbers in each interval are obtained by dividing the mass of the sample in each interval by the average molar mass for that interval. Because number of molecules is proportional to amount of substance (the number of moles), the number-weighted average can be obtained directly from the amounts in each interval.

Answer The amounts in each interval are as follows:

Interval	5–10	10–15	15–20	20–25	25–30	30–35
Molar mass/(kg mol^{-1})	7.50	12.5	17.5	22.5	27.5	32.5
Amount/mmol	1.30	0.70	0.51	0.25	0.11	0.052
					Total:	2.92

The number-average molar mass is therefore

$$\overline{M}_n/(\text{kg mol}^{-1}) = \frac{1}{2.92}(1.3 \times 7.5 + 0.70 \times 12.5 + 0.51 \times 17.5$$
$$+ 0.25 \times 22.5 + 0.11 \times 27.5 + 0.052 \times 32.5)$$
$$= 13$$

The weight-average molar mass is calculated directly from the data after noting that the total mass of the sample is 37.6 g:

$$\overline{M}_w/(\text{kg mol}^{-1}) = \frac{1}{37.6}(9.6 \times 7.5 + 8.7 \times 12.5$$
$$+ 8.9 \times 17.5 + 5.6 \times 22.5 + 3.1 \times 27.5 + 1.7 \times 32.5)$$
$$= 16$$

Comment Note the significantly different values of the two averages. In this instance, $\overline{M}_w/\overline{M}_n = 1.2$.

- -

Self-test 23.1 Evaluate the Z-average molar mass of the sample.

[19 kg mol^{-1}]

Whereas at first sight it might appear troublesome to have several types of average, the observation that they have different values gives additional information about the range of molar masses in the sample: the ratio $\overline{M}_w/\overline{M}_n$ is called the **heterogeneity index** (or 'polydispersity index'). In the determination of protein molar masses we expect the various averages to be the same because the sample is monodisperse (unless there has been degradation). In samples of synthetic polymers there is normally a range of molar masses and the different averages are expected to yield different values. Typical synthetic materials have $\overline{M}_w/\overline{M}_n \approx 4$. The term 'monodisperse' is conventionally applied to synthetic polymers in which this index is less than 1.1; commercial polyethylene samples might be much more heterogeneous, with a ratio close to 30. One consequence of a narrow molar mass distribution for synthetic polymers is often a higher crystallinity, and therefore density and melting point. The spread of values is controlled by the choice of catalyst and reaction conditions (Section 26.1).

The most direct method for measuring molar masses is by mass spectrometry. However, until recently this technique was thwarted by the difficulty of vaporizing the sample. That problem has been solved by embedding the macromolecules in a substrate; the substrate is vaporized, and carries the macromolecules into the vapour with it.

23.2 Colligative properties

The classical methods of determining molar mass utilize colligative properties (Section 7.5). For macromolecules, where the number of molecules in solution may be very small even though the mass of the solute may be appreciable, only osmometry is sufficiently sensitive; this procedure was described and illustrated in Section 7.5e. Osmometry can be used for molar masses up to about 100 kg mol^{-1}. Its lower limit of about 8 kg mol^{-1} is due to the use of membrane materials that allow these relatively small molecules to pass through.

Macromolecules give strongly nonideal solutions: being so large, they displace a large quantity of solvent instead of replacing individual solvent molecules with negligible disturbance. In thermodynamic terms, the displacement of solvent molecules implies that the entropy change is especially important when a macromolecule dissolves. Furthermore, its great bulk means that a macromolecule is unable to move freely through the solution because the molecule is excluded from the regions occupied by other solute molecules. There are also significant contributions to the Gibbs energy from the enthalpy of solution, largely because solvent–solvent interactions are more favourable than the macromolecule–solvent interactions that replace them.

The osmotic virial coefficient, B (see eqn 7.38), arises largely from the effect of excluded volume. If we imagine a solution of a macromolecule being built by the successive addition of macromolecules to the solvent, each one being excluded by the ones that preceded it, then the value of B turns out to be

$$B = \tfrac{1}{2}N_A v_P \tag{5}$$

where v_P is the excluded volume due to a single molecule.

Example 23.2 Estimating the volume of polymer molecules

Use the information in Example 7.5 to estimate the volume of the polymer molecules regarded as impenetrable spheres.

Method The excluded volume of spherical molecules of volume v is $v_P = 8v$ because the minimum separation of the centres of two spheres is the sum of their radii (Fig. 23.1). Estimate v_P from B by using eqn 5, and find B from the slope of the graph plotted in Fig. 7.25. To do so use (as in Example 7.5):

$$\frac{h}{c} = \frac{RT}{\rho g \overline{M}_n}\left(1 + \frac{B}{\overline{M}_n}c\right)$$

Answer The intercept $RT/\rho g\overline{M}_n$ was found in Example 7.5 to be 0.21 cm/(g L^{-1}). The slope of the straight line in Fig. 7.25, which is equal to $(RT/\rho g\overline{M}_n) \times B/\overline{M}_n$, is 0.073 (cm g^{-1} L)/(g L^{-1}). It follows that

$$\frac{\text{slope}}{\text{intercept}} = \frac{B}{\overline{M}_n} = \frac{0.073\ (\text{cm g}^{-1}\,\text{L})/(\text{g L}^{-1})}{0.21\ \text{cm g}^{-1}\,\text{L}} = \frac{0.35}{\text{g L}^{-1}}$$

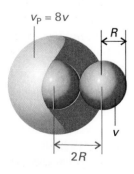

$v_P = 8v$

23.1 The excluded volume into which the centre of a molecule cannot penetrate is $8v$, where v is the volume of the molecule itself.

Therefore

$$B = (0.35/\text{g L}^{-1}) \times (123 \times 10^3 \text{ g mol}^{-1}) = 4.3 \times 10^4 \text{ L mol}^{-1}$$

Equation 5 then implies that

$$v_P = \frac{2B}{N_A} = 1.4 \times 10^{-22} \text{ m}^3$$

From this value of v_P it follows that the molecular volume is approximately $1.8 \times 10^4 \text{ nm}^3$.

Comment The radius of the molecule is approximately 16 nm.

- -

Self-test 23.2 Another sample in the same solvent resulted in the following heights of solution at the same temperature: 0.22, 0.53, 1.39, 3.32, 5.02 cm. Calculate the molar mass and the molecular volume of the solute.

$$[14 \text{ kg mol}^{-1}, 9 \times 10^{-23} \text{ m}^3]$$

In broad terms, the excluded volume contributes to the excess entropy of solution (the entropy change in excess of the ideal value, Section 7.4), and the attractions and repulsions between macromolecules contribute to the excess enthalpy. For most solute–solvent systems there is a unique temperature (which is not always experimentally attainable) at which these effects cancel and the solution is virtually ideal. This temperature (the analogue of the Boyle temperature for real gases) is called the **Flory theta temperature**, θ. At this temperature, the osmotic virial coefficient B is zero. As an example, for polystyrene in cyclohexane $\theta \approx 306$ K, the exact value depending on the average molar mass of the polymer. A solution at its Flory theta temperature is called a $\boldsymbol{\theta}$ **solution**. Because a θ solution behaves nearly ideally, its thermodynamic and structural properties are easier to describe even though the molar concentration is not low. In molecular terms, in a θ solution the molecules are in an unperturbed condition, whereas in other solutions expansion or contraction of the coiled molecule takes place as a result of interactions with the solvent.

(a) Vapour-phase osmometry

In the technique of **vapour-phase osmometry**, a droplet of solution is placed on one thermistor (a semiconductor temperature probe) and a droplet of pure solvent is placed on another thermistor (Fig. 23.2). The two droplets are surrounded by an atmosphere of solvent vapour. Because the vapour pressure of a solvent in a solution is lower than when it is pure, the net rate of condensation of solvent on to a droplet of solution is greater than the rate of condensation on to a droplet of pure solvent. It follows that more heat is liberated in the solution droplet, and the rise in temperature is greater there than in the droplet of pure solvent. The difference in temperature is measured for a series of concentrations and extrapolated to zero concentration. After calibration by using samples of known molar mass, the molar mass of the sample can be inferred from the temperature difference between the two thermistors. The method is not particularly sensitive, and is confined to samples of low molar mass.

(b) Polyelectrolytes and dialysis

Some polymers are strings of acid groups, as in poly(acrylic acid), $-(CH_2CHCOOH)_n-$, or strings of bases, as in nylon, $-[NH(CH_2)_6NHCO(CH_2)_4CO]_n-$; proteins have both acid and base groups. Macromolecules may therefore be **polyelectrolytes**, and, depending on their state of ionization, **polyanions** or **polycations**. A macromolecule with mixed cation and anion character is known as a **polyampholyte**.

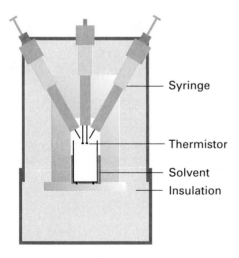

23.2 A vapour-phase osmometer. The syringes introduce droplets of solvent and solution on to the thermistors, and the difference in temperature (arising from the different rates of vaporization) is noted.

Syringe

Thermistor

Solvent

Insulation

One consequence of dealing with polyelectrolytes is that it is necessary to know the extent of ionization before osmotic data can be interpreted. For example, suppose the sodium salt of a polyelectrolyte is present in solution as ν Na^+ ions and a single polyanion $P^{\nu-}$; then if it is fully dissociated in solution it gives rise to $\nu + 1$ particles for each formula unit of salt that dissolves. If we guess that $\nu = 1$ when in fact $\nu = 10$, then the estimate of the molar mass will be wrong by an order of magnitude. We can find a way out of this difficulty by considering another feature of charged macromolecules.

Suppose the solution of the polyelectrolyte $Na_\nu P$ also contains added NaCl, and that it is in contact through a semipermeable membrane with another salt solution. Furthermore, suppose the membrane is permeable to the solvent and to the salt ions, but not to the polyanion. This arrangement is one that actually occurs in living systems, where osmosis is an important feature of cell operation. The presence of the salt affects the osmotic pressure because the anions and cations cannot migrate through the membrane to an arbitrary extent. Apart from small imbalances of charge close to the membrane and which give rise to transmembrane potentials, electrical neutrality must be preserved in the bulk on both sides of the membrane: if an anion migrates, a cation must accompany it.

The presence of a high concentration of added salt on each side of a semipermeable membrane ensures that the effective difference in concentrations is due solely to the presence of the polyanion P on one side of the membrane, for the number of cations the polymer provides is insignificant in comparison with the number supplied by the additional salt. Hence, under such circumstances we can expect the osmotic pressure to be given by $\Pi = RT[P]$, independent of the value of ν (a result confirmed in the *Justification* below). Therefore, if we measure the osmotic pressure in the presence of high concentrations of salt, the molar mass may be obtained unambiguously.

Justification 23.1

Suppose that $Na_\nu P$ is at a molar concentration $[P]$ on one side of the membrane, and that NaCl is added to each side. On the left (L) there are $P^{\nu-}$, Na^+, and Cl^- ions. On the right (R) there are Na^+ and Cl^- ions. The condition for equilibrium is that the Gibbs energy of NaCl in solution is the same on both sides of the membrane, so a net flow of Na^+ and Cl^- ions occurs until $G_m(NaCl, L) = G_m(NaCl, R)$. This equality occurs when

$$\mu^{\ominus}(NaCl) + RT \ln a_L(Na^+) + RT \ln a_L(Cl^-) =$$
$$\mu^{\ominus}(NaCl) + RT \ln a_R(Na^+) + RT \ln a_R(Cl^-)$$

or

$$RT \ln a_L(Na^+) a_L(Cl^-) = RT \ln a_R(Na^+) a_R(Cl^-)$$

If we ignore activity coefficients, the two expressions are equal when $[Na^+]_L[Cl^-]_L = [Na^+]_R[Cl^-]_R$. As the Na^+ ions are supplied by the polyelectrolyte as well as the added salt, the conditions for bulk electrical neutrality are $[Na^+]_L = [Cl^-]_L + \nu[P]$ and $[Na^+]_R = [Cl^-]_R$. We can now combine these three conditions to obtain expressions for the differences in ion concentrations across the membrane:

$$[Na^+]_L - [Na^+]_R = \frac{\nu[P][Na^+]_L}{[Na^+]_L + [Na^+]_R} = \frac{\nu[P][Na^+]_L}{2[Cl^-] + \nu[P]}$$

$$[Cl^-]_L - [Cl^-]_R = -\frac{\nu[P][Cl^-]_L}{[Cl^-]_L + [Cl^-]_R} = -\frac{\nu[P][Cl^-]_L}{2[Cl^-]} \qquad (6)$$

where $[Cl^-] = \frac{1}{2}([Cl^-]_L + [Cl^-]_R)$. The quantity $[Cl^-]$ is the average concentration of Cl^- ions on each side of the membrane.

The final step is to note that the osmotic pressure depends on the difference in the numbers of solute particles on each side of the membrane. That being so, the van't Hoff equation, $\Pi = RT[\text{Solute}]$, becomes

$$\Pi = RT\{([P] + [Na^+]_L + [Cl^-]_L) - ([Na^+]_R + [Cl^-]_R)\}$$

$$= RT[P](1 + B[P]) \qquad B = \frac{\nu^2 [Cl^-]_R}{4[Cl^-]^2 + 2\nu[Cl^-]_L[P]} \tag{7}$$

When the concentration of added salt is so great that $[Cl^-]_L$ and $[Cl^-]_R$ are both much larger than $[P]$, it follows that $B[P] \ll 1$ and this expression reduces to the one quoted in the text.

A second point arises from the effect of added salt. There is often interest in the extent to which ions are bound to macromolecules, especially when a membrane (such as a cell wall) separates two solutions. The equations

$$[Na^+]_L - [Na^+]_R = \frac{\nu[P][Na^+]_L}{2[Cl^-] + \nu[P]}$$

$$[Cl^-]_L - [Cl^-]_R = -\frac{\nu[P][Cl^-]_L}{2[Cl^-]} \tag{8}$$

which are derived in the *Justification*, show that cations will dominate the anions in the compartment containing the polyanion (because the concentration difference is positive for Na^+ and negative for Cl^-) as a result of the equilibrium and electroneutrality conditions. The equilibrium distribution of ions in two compartments in contact through a semipermeable membrane, in one of which there is a polyelectrolyte, is called a **Donnan equilibrium**.

Example 23.3 Analysing a Donnan equilibrium

Suppose that two equal volumes of 0.200 M NaCl(aq) are separated by a membrane and that a macromolecule of molar mass 55 kg mol^{-1}, which cannot pass through the membrane, is added as its sodium salt Na_6P to a concentration of 50 g L^{-1} to the left-hand compartment. Calculate the molar concentrations of Na^+ and Cl^- in each compartment.

Method The sums of the equilibrium concentrations of Na^+ and Cl^- in each compartment are

$$[Na^+]_L + [Na^+]_R = [Cl^-]_L + [Cl^-]_R + \nu[P] = 2[Cl^-] + \nu[P]$$

Then use $[Cl^-] = 0.200 \text{ mol L}^{-1}$.

Answer As $[P] = 9.1 \times 10^{-4} \text{ mol L}^{-1}$, we find

$$[Na^+]_L - [Na^+]_R = \frac{6 \times (9.1 \times 10^{-4} \text{ mol L}^{-1}) \times [Na^+]_L}{2 \times (0.200 \text{ mol L}^{-1}) + 6 \times (9.1 \times 10^{-4} \text{ mol L}^{-1})}$$

and the sum above gives

$$[Na^+]_L + [Na^+]_R = 2 \times (0.200 \text{ mol L}^{-1}) + 6 \times (9.1 \times 10^{-4} \text{ mol L}^{-1})$$
$$= 0.405 \text{ mol L}^{-1}$$

The solutions of these two equations are

$$[Na^+]_L = 0.204 \text{ mol L}^{-1} \qquad [Na^+]_R = 0.201 \text{ mol L}^{-1}$$

Then

$$[Cl^-]_R = [Na^+]_R = 0.201 \text{ mol L}^{-1}$$
$$[Cl^-]_L = [Na^+]_L - 6[P] = 0.199 \text{ mol L}^{-1}$$

Comment The Na^+ ions accumulate slightly in the compartment containing the macromolecule.

- -

Self-test 23.3 Repeat the calculation for 0.300 M NaCl(aq), a polyelectrolyte $Na_{10}P$ of molar mass 33 kg mol^{-1} at a mass concentration of 50.0 g L^{-1}.

[Na: 0.311 mol L^{-1}, 0.304 mol L^{-1}]

23.3 Sedimentation

In a gravitational field, heavy particles settle towards the foot of a column of solution by the process called **sedimentation**. The rate of sedimentation depends on the strength of the field and on the masses and shapes of the particles. Spherical molecules (and compact molecules in general) sediment faster than rod-like or extended molecules. For example, DNA helices sediment much faster when they are denatured to a random coil, so sedimentation rates can be used to study denaturation. When the sample is at equilibrium, the particles are dispersed over a range of heights in accord with the Boltzmann distribution (because the gravitational field competes with the stirring effect of thermal motion). The spread of heights depends on the masses of the molecules, so the equilibrium distribution is another way of determining molar mass.

Sedimentation is normally very slow, but it can be accelerated by replacing the gravitational field by a centrifugal field. The latter can be achieved in an ultracentrifuge, which is essentially a cylinder that can be rotated at high speed about its axis with a sample in a cell near its periphery (Fig. 23.3). Modern ultracentrifuges can produce accelerations equivalent to about 10^5 that of gravity ('10^5 g'). Initially the sample is uniform, but the 'top' (innermost) boundary of the solute moves outwards as sedimentation proceeds.

(a) The rate of sedimentation

A solute particle of mass m has an effective mass $m_{eff} = bm$ on account of the buoyancy of the medium, with

$$b = 1 - \rho v_s \tag{9}$$

where ρ is the solution density, v_s is the solute's partial specific volume ($v_s = (\partial V/\partial m_B)_T$, where m_B is the total mass of the solute), and ρv_s is the mass of solvent displaced per gram of solute. The solute particles at a distance r from the axis of a rotor spinning at an angular velocity ω experience a centrifugal force of magnitude $m_{eff}r\omega^2$. The acceleration outwards is countered by a frictional force proportional to the speed, s, of the particles through the medium. This force is written fs, where f is the **frictional coefficient**. The particles therefore adopt a **drift speed**, a steady speed through the medium, which is found by equating the two forces $m_{eff}r\omega^2$ and fs. The forces are equal when

$$s = \frac{m_{eff}r\omega^2}{f} = \frac{bmr\omega^2}{f} \tag{10}$$

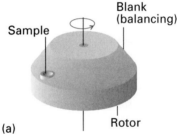

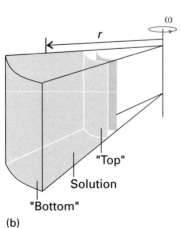

23.3 (a) An ultracentrifuge head. The sample on one side is balanced by a blank diametrically opposite. (b) Detail of the sample cavity: the 'top' surface is the inner surface, and the centrifugal force causes sedimentation towards the outer surface; a particle at a radius r experiences a force of magnitude $mr\omega^2$.

The drift speed depends on the angular velocity and the radius, and it is convenient to focus on the **sedimentation constant**, S, which is defined as

$$S = \frac{s}{r\omega^2} \qquad [11]$$

Then, because the average molecular mass is related to the average molar mass $\overline{M}_n$ through $m = \overline{M}_n/N_A$,

$$S = \frac{b\overline{M}_n}{f N_A} \qquad (12)$$

Example 23.4 Determining a sedimentation constant

The sedimentation of bovine serum albumin (BSA) was monitored at 25°C. The initial location of the solute surface was at 5.50 cm from the axis of rotation, and during centrifugation at 56 850 r.p.m. it receded as follows:

t/s	0	500	1000	2000	3000	4000	5000
r/cm	5.50	5.55	5.60	5.70	5.80	5.91	6.01

Calculate the sedimentation coefficient.

Method Equation 11 can be interpreted as a differential equation for $s = dr/dt$ in terms of r, so integrate it to obtain a formula for r in terms of t. The integrated expression, an expression for r as a function of t, will suggest how to plot the data and obtain from it the sedimentation constant.

Answer Equation 11 may be written

$$\frac{dr}{dt} = r\omega^2 S$$

This equation integrates to

$$\ln \frac{r}{r_0} = \omega^2 S t$$

It follows that a plot of $\ln (r/r_0)$ against t should be a straight line of slope $\omega^2 S$. Use $\omega = 2\pi\nu$, where ν is in cycles per second, and draw up the following table:

t/s	0	500	1000	2000	3000	4000	5000
$100 \ln (r/r_0)$	0	0.900	1.80	3.57	5.31	7.19	8.87

The straight-line graph (Fig. 23.4) has slope 1.78×10^{-5}, so $\omega^2 S = 1.79 \times 10^{-5} \text{ s}^{-1}$. Because $\omega = 2\pi \times (56\,850/60) \text{ s}^{-1} = 5.95 \times 10^3 \text{ s}^{-1}$, it follows that $S = 5.02 \times 10^{-13} \text{ s}$.

Comment We develop this result below. The unit 10^{-13} s is sometimes called a 'svedberg' and denoted Sv; in this case $S = 5.02$ Sv. Accurate results are obtained by extrapolating to zero concentration.

Self-test 23.4 Calculate the sedimentation constant given the following data (the other conditions being the same as above):

t/s	0	500	1000	2000	3000	4000	5000
r/cm	5.65	5.68	5.71	5.77	5.84	5.90	5.97

[3.11 Sv]

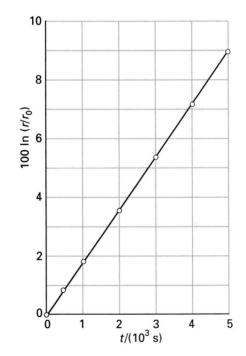

23.4 A plot of the data in Example 23.4.

Table 23.1* Frictional coefficients and molecular geometry†

a/b	Prolate	Oblate
2	1.04	1.04
4	1.18	1.17
6	1.31	1.28
8	1.43	1.37
10	1.54	1.46

* More values and analytical expressions are given in the *Data section* at the end of this volume.

† Entries are the ratio f/f_0 where $f_0 = 6\pi\eta c$ with $c = (ab^2)^{1/3}$ for prolate ellipsoids and $c = (a^2b)^{1/3}$ for oblate ellipsoids; $2a$ is the major axis and $2b$ is the minor axis.

Table 23.2* Diffusion coefficients in water at 20°C

	$M/(\text{kg mol}^{-1})$	$D/(\text{m}^2\,\text{s}^{-1})$
Sucrose	0.342	4.59×10^{-10}
Lysozyme	14.1	1.04×10^{-10}
Haemoglobin	68	6.9×10^{-11}
Collagen	345	6.9×10^{-12}

To make progress we need to know the frictional constant, f. For a spherical particle of radius a in a solvent of viscosity η, and for solute molecules that are not small compared with the solvent molecules, f is given by **Stokes' relation**

$$f = 6\pi a\eta \tag{13}$$

Therefore, for spherical molecules,

$$S = \frac{b\overline{M}_n}{6\pi a\eta N_A} \tag{14}$$

and S may be used to determine either $\overline{M}_n$ or a. If the molecules are not spherical, we use the appropriate value of f given in Table 23.1. As always when dealing with macromolecules, the measurements must be carried out at a series of concentrations and then extrapolated to zero concentration to avoid the complications that arise from the interference between bulky molecules.

At this stage, it appears that we need to know the molecular radius a (and in general the frictional coefficient f) to obtain the molar mass from the value of S. Fortunately, this requirement can be avoided by drawing on the **Stokes–Einstein relation** between f and the diffusion coefficient, D:

$$f = \frac{kT}{D} \tag{15}$$

The diffusion coefficient is a measure of the rate at which molecules spread down a concentration gradient (it is treated in detail in Section 24.11); this coefficient can be measured by observing the rate at which a concentration boundary spreads or the rate at which a more concentrated solution diffuses into a less concentrated one. Some typical values are given in Table 23.2. The diffusion coefficient can also be measured by using light scattering (Section 23.5). It follows from eqns 12 and 15 that

$$\overline{M}_n = \frac{SRT}{bD} \tag{16}$$

This result is independent of the shape of the solute molecules. It follows that we can find the molar mass by combining measurements of sedimentation and diffusion rates (for S and D, respectively).

(b) Sedimentation equilibria

The difficulty with using sedimentation rates to measure molar masses lies in the inaccuracies inherent in the determination of diffusion coefficients, such as the blurring of the boundary by convection currents. This problem can be avoided by allowing the system to reach equilibrium, for the transport property D is then no longer relevant. As we show in the *Justification* below, the weight-average molar mass can be obtained from the ratio of concentrations of the macromolecules at two different radii in a centrifuge operating at angular frequency ω:

$$\overline{M}_w = \frac{2RT}{(r_2^2 - r_1^2)b\omega^2} \ln \frac{c_2}{c_1} \tag{17}$$

An alternative treatment of the data leads to the Z-average molar mass. The centrifuge is run more slowly in this technique than in the sedimentation rate method to avoid having all the solute pressed in a thin film against the bottom of the cell. At these slower speeds, several days may be needed for equilibrium to be reached.

Justification 23.2

The distribution of particles is the outcome of the balance between the effect of the centrifugal force and the dispersing effect of diffusion down a concentration gradient. The kinetic energy of a particle of effective mass m at a radius r in a rotor spinning at a frequency ω is $\frac{1}{2}m\omega^2 r^2$, so the total chemical potential at a radius r is $\bar{\mu}(r) = \mu(r) - \frac{1}{2}M\omega^2 r^2$, where $\mu(r)$ is the contribution that depends on the concentration of solute. The condition for equilibrium is that the chemical potential is uniform, so

$$\left(\frac{\partial \bar{\mu}}{\partial r}\right)_T = \left(\frac{\partial \mu}{\partial r}\right)_T - M\omega^2 r = 0$$

To evaluate the partial derivative of μ, we write

$$\left(\frac{\partial \mu}{\partial r}\right)_T = \left(\frac{\partial \mu}{\partial p}\right)_{T,c}\left(\frac{\partial p}{\partial r}\right)_{T,c} + \left(\frac{\partial \mu}{\partial c}\right)_{T,p}\left(\frac{\partial c}{\partial r}\right)_{T,p}$$

$$= Mv\omega^2 r\rho + RT\left(\frac{\partial \ln c}{\partial r}\right)_{T,p}$$

The first result follows from the fact that $(\partial \mu/\partial p)_T = V_m$, the partial molar volume, and $V_m = Mv$. It also makes use of the fact that the hydrostatic pressure at r is $p(r) = p(r_0) + \frac{1}{2}\rho\omega^2(r^2 - r_0^2)$, where r_0 is the radius of the surface of the liquid in the sample holder (that is, the location of its meniscus), with ρ the density of the solution. The concentration term stems from the expression $\mu = \mu^{\ominus} + RT\ln c$. The condition for equilibrium is therefore

$$Mr\omega^2(1 - v\rho) - RT\left(\frac{\partial \ln c}{\partial r}\right)_{T,p} = 0$$

and therefore, at constant temperature,

$$d\ln c = \frac{Mr\omega^2(1 - v\rho)dr}{RT}$$

This expression integrates to eqn 17.

(c) Electrophoresis

Many macromolecules are charged and move in response to an electric field: this motion is called **electrophoresis**. In **gel electrophoresis** the migration takes place through a cross-linked polyacrylamide gel. The mobilities of macromolecules depend on their masses and their shapes, and a constant drift speed is reached when the driving force $ez\mathcal{E}$ (where z is the charge number and $\mathcal{E}$ is the field strength) is matched by the viscous retarding force fs.

One way to avoid the problem of knowing neither the hydrodynamic shape of the molecules nor their charge is to denature them in a controlled way. Sodium dodecylsulfate has been found to be very useful in this respect: it denatures proteins, whatever their initial shapes, into rods by forming a complex with them. Moreover, most proteins bind a constant amount of the anion to a given mass, so the charge per protein molecule is well regulated. The molar mass of the protein is determined by comparing its mobility in its rod-like complexed form with standard samples.

The charge on a protein depends on the pH, and hence the rate of migration varies with pH. This apparent difficulty can be used to distinguish proteins. For example, at a given pH the rate of migration of haemoglobin from people with sickle-cell anaemia is different from that of a sample taken from people without the disease. This difference is an indication that there are different charges on the protein molecule, which in turn is ascribed to the presence of a different amino acid residue in the polypeptide chain.

(d) Size-exclusion chromatography

All the techniques discussed so far have certain drawbacks, including the time needed to obtain data, and the often awkward interpretation of the data. Much of this difficulty has been swept away by a technique that makes use of beads of porous polymeric material about 0.1 mm in diameter that capture molecules selectively, according to their size. In the technique of **size-exclusion chromatography** (SEC), or **gel permeation chromatography** (GPC), which is now the most widely used technique for molar mass determinations of polymers, a solution of the polymer sample is filtered through a column. The small molecules, which can permeate into the porous structure of the gel, require a long elution time, or time to pass through a particular length of column, whereas the larger ones, which are not captured, pass through rapidly. The average molar mass of a macromolecule may therefore be determined by observing its elution time in a column calibrated against standard samples.

The range of molar masses that can be determined by SEC can be altered by selecting columns made from polymers with different degrees of cross-linking and different materials. The elution time depends on shape in a complicated way and the technique works best if the molecules are spherical. Polystyrene gels are used for investigations of nonpolar polymers in nonpolar solvents, and porous glass gels are used for more polar systems. Because elution is performed under pressure, molar mass determinations may be completed within a few minutes, in striking contrast to the time required for more classical techniques. Moreover, only a few milligrams of material are needed for highly reliable measurements.

23.4 Viscosity

The presence of a macromolecular solute increases the viscosity of a solution. The effect is large even at low concentrations, because big molecules affect the fluid flow over an extensive region surrounding them. At low concentrations the viscosity, η, of the solution is related to the viscosity of the pure solvent, η_0, by

$$\eta = \eta_0(1 + [\eta]c + \cdots)$$

(18)

The **intrinsic viscosity**, $[\eta]$, is the analogue of a virial coefficient (and has the dimensions of 1/concentration). It follows from eqn 18 that

$$[\eta] = \lim_{c \to 0}\left(\frac{\eta - \eta_0}{c\eta_0}\right) = \lim_{c \to 0}\left(\frac{\eta/\eta_0 - 1}{c}\right)$$

(19)

Viscosities are measured in several ways. In the **Ostwald viscometer** shown in Fig. 23.5, the time taken for a solution to flow through the capillary is noted, and compared with a standard sample. The method is well suited to the determination of $[\eta]$ because the ratio of the viscosities of the solution and the pure solvent is proportional to the drainage times t and t_0 after correcting for different densities ρ and ρ_0:

$$\frac{\eta}{\eta_0} = \frac{t}{t_0} \times \frac{\rho}{\rho_0}$$

(20)

(In practice, the two densities are only rarely significantly different.) This ratio can be used directly in eqn 19. Viscometers in the form of rotating concentric cylinders are also used (Fig. 23.6), and the torque on the inner cylinder is monitored while the outer one is rotated. Such **rotating drum viscometers** have the advantage over the Ostwald type that the shear gradient between the cylinders is simpler than in the capillary, and effects of the kind discussed shortly can be studied more easily.

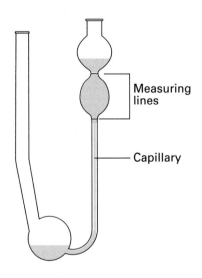

Measuring lines

Capillary

23.5 An Ostwald viscometer. The viscosity is measured by noting the time required for the liquid to drain between the two marks.

23.6 A rotating drum viscometer. The torque on the inner drum is observed when the outer container is rotated.

Table 23.3* Intrinsic viscosity

Macromolecule	Solvent	$\theta/°C$	$K/(cm^3\,g^{-1})$	a
Polystyrene	Benzene	25	9.5×10^{-3}	0.74
Polyisobutylene	Benzene	23	8.3×10^{-2}	0.50
Various proteins	Guanidine hydrochloride $+ HSCH_2CH_2OH$		7.2×10^{-3}	0.66

*More values are given in the *Data section*.

There are many complications in the interpretation of viscosity measurements. Much of the work is based on empirical observations, and the determination of molar mass is usually based on comparisons with standard, nearly monodisperse samples. Some regularities are observed that help in the determination. For example, it is found that θ solutions of macromolecules often fit the **Mark–Kuhn–Houwink–Sakurada equation**:

$$[\eta] = K\overline{M}_v^a \tag{21}$$

where K and a are constants that depend on the solvent and type of macromolecule (Table 23.3); the viscosity-average molar mass, $\overline{M}_v$, appears in this expression. As an example, solutions of poly(γ-benzyl-L-glutamate) in its rod-like form have an intrinsic viscosity about four times greater than when it is denatured and the rods collapse into random coils. Conversely, solutions of natural ribonuclease are less viscous than solutions of the denatured form: this observation indicates that the natural protein is more compact than the denatured form.

Example 23.5 Using intrinsic viscosity to measure molar mass

The viscosities of a series of solutions of polystyrene in toluene were measured at 25°C with the following results:

$c/(g\,L^{-1})$	0	2.0	4.0	6.0	8.0	10.0
$\eta/(10^{-4}\,kg\,m^{-1}\,s^{-1})$	5.58	6.15	6.74	7.35	7.98	8.64

Calculate the intrinsic viscosity and estimate the molar mass of the polymer by using eqn 21 with $K = 3.80 \times 10^{-5}\,L\,g^{-1}$ and $a = 0.63$.

Method The intrinsic viscosity is defined in eqn 19; therefore, form this ratio at the series of data points and extrapolate to $c = 0$. Interpret $\overline{M}_v$ as $\overline{M}_v/(g\,mol^{-1})$ in eqn 21.

Answer We draw up the following table:

$c/(g\,L^{-1})$	0	2.0	4.0	6.0	8.0	10.0
η/η_0	1	1.102	1.208	1.317	1.430	1.549
$100[(\eta/\eta_0) - 1]/(c/g\,L^{-1})$		5.11	5.20	5.28	5.38	5.49

The points are plotted in Fig. 23.7. The extrapolated intercept at $c = 0$ is 0.0504, so $[\eta] = 0.0504\,L\,g^{-1}$. Therefore,

$$\overline{M}_v = \left(\frac{[\eta]}{K}\right)^{1/a} = 9.0 \times 10^4\,g\,mol^{-1}$$

Comment When $\eta \approx \eta_0$,

$$\ln\frac{\eta}{\eta_0} = \ln\left(1 + \frac{\eta - \eta_0}{\eta_0}\right) \approx \frac{\eta - \eta_0}{\eta_0} = \frac{\eta}{\eta_0} - 1$$

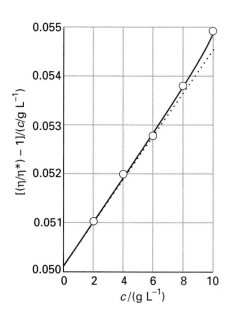

23.7 The plot used for the determination of intrinsic viscosity, which is taken from the intercept at $c = 0$; see Example 23.5.

This relation is exact in the limit that η coincides with η_0, which is true when $c = 0$. Hence, $[\eta]$ can also be defined as the limit of $(1/c) \ln (\eta/\eta_0)$ as $c \to 0$. The intercept is identified more precisely by plotting both functions.

Self-test 23.5 Evaluate the viscosity-average molar mass by using the second plotting technique.

$$[90 \text{ kg mol}^{-1}]$$

In some cases, it is found that the flow is non-Newtonian in the sense that the viscosity of the solution changes as the rate of flow increases. A decrease in viscosity with increasing rate of flow indicates the presence of long rod-like molecules that are orientated by the flow and hence slide past each other more freely. In some somewhat rare cases the stresses set up by the flow are so great that long molecules are broken up, with further consequences for the viscosity.

23.5 Light scattering

When electromagnetic radiation falls on an object, it forces the electron distribution in the object to oscillate and hence to radiate. If the medium is perfectly homogeneous (for example, a perfect crystal or a completely random collection of molecules that is homogeneous on the scale of the wavelength of the radiation, like a sample of water), the secondary waves interfere destructively except in the original propagation direction. Therefore, an observer sees the beam only when looking towards the source along the initial direction. If the medium is inhomogeneous (an imperfect crystal or a solution containing foreign bodies, such as macromolecules in a solvent or smoke in air), radiation is scattered into other directions too. A familiar example is light scattered by specks of dust in a sunbeam (and in advertisers' photographs of laser beams).

Scattering by particles with diameters much smaller than the wavelength of the incident radiation is called **Rayleigh scattering**. The intensity of Rayleigh scattered radiation depends on $1/\lambda^4$, so shorter wavelength radiation is scattered more intensely than longer wavelengths. The blue of the sky arises from the more intense scattering of the blue component of white sunlight by the molecules of the atmosphere. The intensity also depends on the scattering angle θ, and is proportional to $1 + \cos^2 \theta$ when the light is unpolarized and to $\cos^2 \theta$ when it is polarized (Fig. 23.8). In practice it turns out to be easier to make observations at an angle to the incident beam. The intensity also depends on the strength of

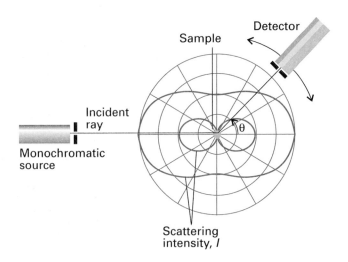

23.8 Rayleigh scattering from a sample of point-like particles follows a $1 + \cos^2 \theta$ dependence (outer trace on the polar plot) when unpolarized light is used, but a $\cos^2 \theta$ dependence (inner trace) when plane-polarized light is used.

the interaction of the light with the molecules: the interaction is strong when the polarizability of the molecules is large.

When these remarks are combined into a quantitative theory, it turns out that the scattering intensity, $\mathcal{I}(\theta)$, at the angle θ is

$$\mathcal{I} = A\mathcal{I}_0\overline{M}_\text{w}[\text{P}]g(\theta) \qquad g(\theta) = \begin{cases} 1 + \cos^2\theta \text{ for unpolarized light} \\ \cos^2\theta \text{ for polarized light} \end{cases} \qquad (22)$$

In this expression, $\mathcal{I}_0$ is the incident intensity, $[\text{P}]$ is the molar concentration of the solute, $\overline{M}_\text{w}$ its weight-average molar mass, and A is a constant that depends on the refractive index of the solution, the wavelength, and the distance of the detector from the sample. Equation 22 is an 'ideal' result in the sense that it ignores the complications that arise from the interactions between solute particles, and in an actual experiment it is important to extrapolate to zero concentration.

(a) Large-particle scattering

When the wavelength of the incident radiation is comparable to the size of the scattering particles, scattering may occur from different sites of the same molecule and the interference between different rays is important.[1] As a result, the scattering intensity is distorted from the form characteristic of small-particle, Rayleigh scattering of light given in eqn 22. A measure of the distortion is the ratio

$$P = \frac{\mathcal{I}_\text{observed}}{\mathcal{I}_\text{Rayleigh}} \qquad [23]$$

measured at several angles, where $\mathcal{I}_\text{observed}$ is the observed intensity at each angle and $\mathcal{I}_\text{Rayleigh}$ is the intensity predicted for Rayleigh scattering at that angle.

If the molecule is regarded as composed of a number of atoms i at distances R_i from a convenient point, interference occurs between the radiation scattered by each pair. The scattering from all the particles is then calculated by allowing for contributions from all possible orientations of each pair of atoms in each molecule. This description is very much like the one used in the discussion of electron diffraction (Section 21.10), so we can expect the intensity pattern to be described by a kind of Wierl equation. This turns out to be so and, if there are N atoms in the macromolecule, and if all are assumed to have the same scattering power, then

$$P = \frac{1}{N^2}\sum_{i,j}\frac{\sin sR_{ij}}{sR_{ij}} \qquad s = \frac{4\pi}{\lambda}\sin\tfrac{1}{2}\theta \qquad (24)$$

In this expression, R_{ij} is the separation of atoms i and j, and λ is the wavelength of the incident radiation. The observed intensity is equal to $P\mathcal{I}_\text{Rayleigh}$, with $\mathcal{I}_\text{Rayleigh}$ given by eqn 22.

(b) Small-particle scattering

When the molecule is much smaller than the wavelength of the incident radiation in the sense that $sR_{ij} \ll 1$ (for example, if $R = 5$ nm, and $\lambda = 500$ nm, all the sR_{ij} are about 0.1),

---

1   The effect accounts for the appearance of clouds, which, although we see them by scattered light, look white, not blue like the sky. The water molecules group together into droplets of a size comparable to the wavelength of light, and scatter cooperatively. Although blue light scatters more strongly, more molecules can contribute cooperatively when the wavelength is longer (as for red light), so the net result is uniform scattering for all wavelengths: white light scatters as white light. This paper looks white for the same reason. Cigarette smoke is blue before it is inhaled, but brownish after it is exhaled because the particles aggregate in the lungs.

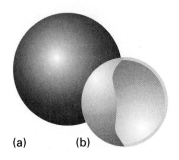

(a)   **(b)**

23.9 (a) A spherical molecule and (b) the hollow spherical shell that has the same rotational characteristics. The radius of the hollow shell is the radius of gyration of the molecule. The radius of gyration of a solid sphere of radius R is $0.77R$.

then we show in the *Justification* below that the deviation from Rayleigh scattering is proportional to the square of the **radius of gyration**, R_g, of the molecule:

$$P - 1 \propto R_g^2 \tag{25}$$

The radius of gyration is the radius of a thin hollow spherical shell of the same mass and moment of inertia as the molecule (Fig. 23.9), and is calculated formally from the expression:[2]

$$R_g = \frac{1}{N}\left(\tfrac{1}{2}\sum_{i,j} R_{ij}^2\right)^{1/2} \tag{26}$$

Justification 23.3

When $sR_{ij} \ll 1$ we can use the expansion $\sin x = x - \tfrac{1}{6}x^3 + \cdots$ to write

$$\sin sR_{ij} = sR_{ij} - \tfrac{1}{6}(sR_{ij})^3 + \cdots$$

and then

$$P(\theta) = \frac{1}{N^2}\sum_{i,j}\{1 - \tfrac{1}{6}(sR_{ij})^2 + \cdots\} = 1 - \frac{s^2}{6N^2}\sum_{i,j}R_{ij}^2 + \cdots$$

The sum over the squares of the separations gives the radius of gyration of the molecule (through eqn 26). Therefore

$$P(\theta) \approx 1 - \tfrac{1}{3}s^2R_g^2 = 1 - \frac{16\pi^2\sin^2\tfrac{1}{2}\theta}{3\lambda^2}R_g^2$$

shows that $P - 1$ is proportional to R_g^2.

Because the deviation from Rayleigh scattering depends on R_g, an analysis of the scattering intensity should give the value of R_g for the molecule in solution. This quantity in turn can be interpreted in terms of the size of the molecule. For example, a solid sphere of radius R has $R_g = (3/5)^{1/2}R$, and a long thin rod of length l has $R_g = l/2(3)^{1/2}$ for rotation about an axis perpendicular to the long axis. Once again, it must be emphasized that the analysis must be performed on data obtained by extrapolation to zero concentration. Some experimental values are listed in Table 23.4.

The use of laser light has led to further refinements in the application and interpretation of light scattering. There has been a shift of emphasis towards the investigation of the time dependence of the positions of atoms and the orientation of macromolecules in solution. These aspects of **polymer dynamics** can be studied by measuring the shift of frequency that occurs when monochromatic light is scattered by a moving target in the technique called **dynamic light scattering**. In particular, laser light scattering can be used for the direct determination of the diffusional characteristics of macromolecules, and provides a fast, direct, and reliable method for the measurement of diffusion coefficients, even of macromolecules of low stability. Polymer dynamics are also studied by inelastic neutron scattering, and are a target of computer simulation algorithms.

Table 23.4* Radius of gyration

	$M/(\mathrm{kg\,mol^{-1}})$	R_g/nm
Serum albumin	66	2.98
Polystyrene	3.2×10^3	50†
DNA	4×10^3	117

* More values are given in the *Data section*.
† In a poor solvent.

2   In Problem 23.25, this definition is shown to be equivalent to another and more easily visualized one in the case of a chain of identical atoms: the radius of gyration is the root mean square distance of the atoms from the centre of mass.

Conformation and configuration

The **primary structure** of a macromolecule is the sequence of small molecular residues making up the chain (or network if there is cross-linking). In the case of a synthetic polymer, virtually all the residues are identical, and it is sufficient to name the monomer used in the synthesis. Thus, the the repeating unit of polyethylene is —CH_2CH_2—, and the primary structure of the chain is specified by denoting it as —$(CH_2CH_2)_n$—.

The concept of primary structure ceases to be trivial in the case of synthetic copolymers and biological macromolecules, for in general these substances are chains formed from different molecules. Proteins, for example, are **polypeptides**, the name signifying chains formed from different amino acids (about twenty occur naturally) strung together by the peptide link, —CONH—. The determination of the primary structure is then a highly complex problem of chemical analysis called *sequencing*. The degradation of a polymer is a disruption of its primary structure, when the chain breaks into shorter components.

The **secondary structure** of macromolecules refers to the (often local) spatially well-characterized arrangement of the basic structural units. The secondary structure of an isolated molecule of polyethylene is a random coil, whereas that of a protein is a highly organized arrangement determined largely by hydrogen bonds, and taking the form of helices or sheets in various segments of the molecule. The loss of secondary structure is called **denaturation**. When the hydrogen bonds in a protein are destroyed (for instance, by heating, as when cooking an egg) the structure denatures into a random coil.

The difference between primary and secondary structure is closely related to the difference between the 'configuration' and the 'conformation' of a chain. The term **configuration** refers to the structural features that can be changed only by breaking chemical bonds and forming new ones. Thus, the chains —A—B—C— and —A—C—B— have different configurations. The term **conformation** refers to the spatial arrangement of the different parts of a chain, and one conformation can be changed into another by rotating one part of a chain around a bond.

By **tertiary structure** is meant the overall three-dimensional structure of the molecule. For instance, many proteins have a helical secondary structure, but in many proteins the helix is so bent and distorted that the molecule has a globular tertiary structure. The term **quaternary structure** refers to the manner in which some molecules are formed by the aggregation of others. Haemoglobin is a famous example: each molecule consists of four subunits of two types (the α and the β chains).

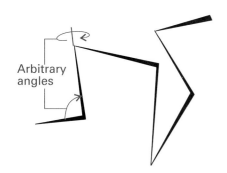

23.10 A freely jointed chain is like a three-dimensional random walk, each step being in an arbitrary direction but of the same length.

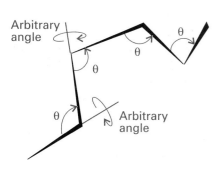

23.11 A better description is obtained by fixing the bond angle (for example, at the tetrahedral angle) and allowing free rotation about a bond direction.

23.6 Random coils

As the first step in unravelling the various aspects of structure, we consider the most likely conformation of a chain of identical units that are incapable of forming hydrogen bonds or any other type of specific bond. Polyethylene is a simple example, but the general idea applies to a denatured protein too. The simplest model is a **freely jointed chain**, in which any bond is free to make any angle with respect to the preceding one (Fig. 23.10); the residues are assumed to occupy zero volume, so different parts of the chain can occupy the same region of space. The model is obviously an oversimplification, because a bond is actually constrained to a cone of angles around a direction defined by its neighbour (Fig. 23.11).

The **random coil** is the least structured conformation of a polymer chain and corresponds to the state of greatest entropy. Any stretching of the coil introduces order and reduces the entropy. Conversely, the formation of a random coil from a more extended form is a spontaneous process (provided enthalpy contributions do not interfere). The random coil model is a helpful starting point for estimating the orders of magnitude of the hydrodynamic properties (such as sedimentation rates) of polymers and denatured proteins in solution.

The elasticity of a **perfect elastomer**, a flexible polymer in which the internal energy is independent of the extension, may also be discussed in terms of the properties of a random coil. The strategy is to set up an expression for the **conformational entropy**, the statistical entropy arising from the arrangement of bonds, and then to use various thermodynamic relations to establish an expression for the force needed to stretch the coil. The first part of the calculation leads to the result that the change in conformational entropy is

$$\Delta S = -\tfrac{1}{2}kN \ln\{(1+\nu)^{1+\nu}(1-\nu)^{1-\nu}\} \qquad \nu = n/N \tag{27}$$

when a polymer containing N bonds of length l is stretched or compressed by nl.

Justification 23.4

Consider a one-dimensional freely jointed polymer. The conformation of a molecule can be expressed in terms of the number of bonds pointing to the right (N_R) and the number pointing to the left (N_L). The distance between the ends of the chain is $(N_R - N_L)l$, where l is the length of an individual bond. We write $n = N_R - N_L$ and the total number of bonds as $N = N_R + N_L$.

The number of ways of forming a chain with a given end-to-end distance nl is the number of ways of having N_R right-pointing and N_L left-pointing bonds, and is given by the binomial coefficient

$$W = \frac{N!}{N_L! N_R!} = \frac{N!}{\{\tfrac{1}{2}(N+n)\}! \{\tfrac{1}{2}(N-n)\}!}$$

The conformational entropy of the chain, $S = k \ln W$, is therefore

$$S/k = \ln N! - \ln\{\tfrac{1}{2}(N+n)\}! - \ln\{\tfrac{1}{2}(N-n)\}!$$

Because the factorials are large (except for large extensions), we can use Stirling's approximation (Section 19.1a) in the form

$$\ln x! \approx \ln(2\pi)^{1/2} + (x+\tfrac{1}{2})\ln x - x$$

to obtain

$$S/k = -\ln(2\pi)^{1/2} + (N+1)\ln 2 + (N+\tfrac{1}{2})\ln N$$
$$-\tfrac{1}{2}\ln\{(N+n)^{N+n+1}(N-n)^{N-n+1}\}$$

The most probable conformation of the chain is the one with the ends close together ($n = 0$), as may be confirmed by differentiation. Therefore, the maximum entropy is

$$S/k = -\ln(2\pi)^{1/2} + (N+1)\ln 2 - \tfrac{1}{2}\ln N$$

The change in entropy when the chain is stretched or compressed by nl is therefore the difference of these two quantities, and the resulting expression is eqn 27 (Fig. 23.12).

Now for the thermodynamic part of the calculation. The work done on an elastomer when it is extended through a distance dx is $F\,dx$, where F is the restoring force. The change in internal energy is therefore

$$dU = T\,dS - p\,dV + F\,dx \tag{28}$$

It follows that

$$\left(\frac{\partial U}{\partial x}\right)_{T,V} = T\left(\frac{\partial S}{\partial x}\right)_{T,V} + F \tag{29}$$

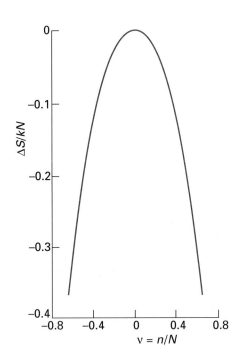

23.12 The change in molar entropy of a perfect rubber as its extension changes: $\nu = 1$ corresponds to complete extension; $\nu = 0$, the conformation of highest entropy, corresponds to the random coil.

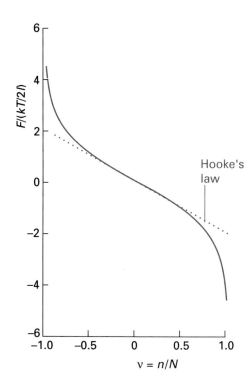

23.13 The restoring force, F, of a one-dimensional perfect rubber. For small deflections, F is linearly proportional to extension, corresponding to Hooke's law.

In a perfect elastomer, as in a perfect gas, the internal energy is independent of the dimensions (at constant temperature), so $(\partial U/\partial x)_{T,V} = 0$. The restoring force is therefore

$$F = -T\left(\frac{\partial S}{\partial x}\right)_{T,V} \tag{30}°$$

If eqn 27 is now substituted into this expression (we evade problems arising from the constraint of constant volume by supposing that the sample contracts laterally as it is stretched), we obtain

$$F = -\frac{T}{l}\left(\frac{\partial S}{\partial n}\right)_{T,V} = -\frac{T}{Nl}\left(\frac{\partial S}{\partial \nu}\right)_{T,V} = \frac{kT}{2l}\ln\left(\frac{1+\nu}{1-\nu}\right) \tag{31}°$$

This function is plotted in Fig. 23.13. At low extensions, when $\nu \ll 1$,

$$F \approx \frac{\nu kT}{l} = \frac{nkT}{Nl} \tag{32}°$$

and the sample obeys Hooke's law (that the restoring force is proportional to the displacement). For small displacements, therefore, the whole coil shakes with simple harmonic motion.

(a) The radial distribution

As shown in the *Justification* below, the same model used to discuss the elasticity of a polymer can be used to calculate the probability that the ends of a one-dimensional random coil are a distance nl apart:

$$P = \left(\frac{2}{\pi N}\right)^{1/2} e^{-\frac{1}{2}n^2/N} \tag{33}$$

This function is plotted in Fig. 23.14.

Justification 23.5

The probability of the separation being nl is

$$P = \frac{\text{number of polymers with } N_R \text{ bonds to the right}}{\text{total number of arrangements of bonds}}$$

$$= \frac{N!/N_R!(N-N_R)!}{2^N} = \frac{N!}{\{\frac{1}{2}(N+n)\}!\{\frac{1}{2}(N-n)\}!2^N}$$

At this point, the development follows the same route as in the previous *Justification*, and the use of Stirling's approximation gives (after quite a lot of algebra)

$$\ln P = \ln\left(\frac{2}{\pi N}\right)^{1/2} - \frac{1}{2}(N+n+1)\ln(1+\nu) - \frac{1}{2}(N-n+1)\ln(1-\nu)$$

For small extensions ($\nu \ll 1$) we can use the approximation $\ln(1\pm\nu) \approx \pm\nu - \frac{1}{2}\nu^2$, and so obtain

$$\ln P \approx \ln\left(\frac{2}{\pi N}\right)^{1/2} - \frac{1}{2}N\nu^2$$

which rearranges into eqn 33.

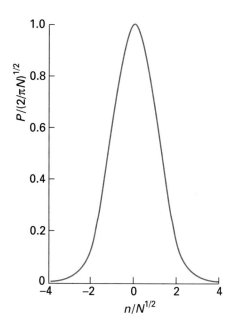

23.14 The probability distribution for the separation of the ends of a one-dimensional random coil. The separation of the ends is nl, where l is the bond length.

Equation 33 can be used to calculate the probability that in a three-dimensional coil the ends of the chain lie in the infinitesimal range R to $R + dR$. We write this probability as $f dR$, where

$$f = 4\pi \left(\frac{a}{\pi^{1/2}} \right)^3 R^2 e^{-a^2 R^2} \qquad a = \left(\frac{3}{2Nl^2} \right)^{1/2} \tag{34}$$

As usual, N is the number of bonds and l is the bond length.[3] Equation 34 shows that, in some coils (the proportion being given by the value of f with R large), the ends may be far apart, whereas in others their separation is small. An alternative interpretation is to regard each coil as ceaselessly writhing from one conformation to another; then $f dR$ is the probability that at any instant the chain will be found with the separation of its ends between R and $R + dR$.

(b) Measures of size

There are several measures of the geometrical size of a random coil. The **contour length**, R_c, is the length of the macromolecule measured along its backbone from atom to atom (the maximum distance that the random walker could walk). For a polymer of N monomer units each of length l, the contour length is

$$R_c = Nl \tag{35}$$

The **root mean square separation**, R_{rms}, is a measure of the average separation of the ends of a random coil: it is the square root of the mean value of R^2, calculated by weighting each possible value of R^2 with the probability that R occurs:

$$R_{rms} = \left(\int_0^\infty R^2 f \, dR \right)^{1/2} = N^{1/2} l \tag{36}$$

We see that, as the number of monomer units increases, the root mean square separation of its ends increases as $N^{1/2}$ (Fig. 23.15), and consequently its volume increases as $N^{3/2}$. Similarly, the radius of gyration of the coil is

$$R_g = \left(\frac{N}{6} \right)^{1/2} l \tag{37}$$

The radius of gyration also increases as $N^{1/2}$.

Example 23.6 Calculating the dimensions of a random coil

Calculate the mean separation of the ends of a freely jointed polymer chain of N bonds of length l.

Method The general expression for the mean nth power of the end-to-end separation is

$$\langle R^n \rangle = \int_0^\infty R^n f \, dR$$

which should be used with $n = 1$ and f from eqn 34.

Answer The mean separation is

$$\langle R \rangle = 4\pi \left(\frac{a}{\pi^{1/2}} \right)^3 \int_0^\infty R^3 e^{-a^2 R^2} \, dR = \frac{2}{a\pi^{1/2}} = \left(\frac{8N}{3\pi} \right)^{1/2} l$$

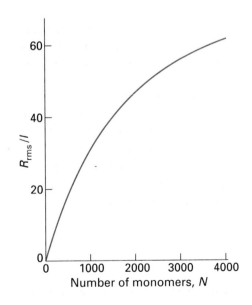

23.15 The variation of the root mean square separation of the ends of a three-dimensional random coil, R_{rms}, with the number of monomers.

---

3 Here and elsewhere we are ignoring the fact that the chain cannot be longer than Nl. Although eqn 34 gives a nonzero probability for $R > Nl$, the values are so small that the errors in pretending that R can range up to infinity are negligible.

The standard integral we have used is

$$\int_0^\infty x^3 e^{-a^2 x^2}\,\mathrm{d}x = \frac{1}{2a^4}$$

Comment The result must be multiplied by a factor when the chain is not freely jointed: see below.

- -

Self-test 23.6 Evaluate the root mean square separation of the ends of the chain.

$$[N^{1/2}l]$$

(c) Constrained chains

Before making use of these conclusions we must remove the freedom for bond angles to take any value. This adjustment is simple for long chains, for we can take groups of neighbouring bonds and consider the direction of their resultant. Although the individual bonds are constrained to a single cone of angle θ, the resultant of several bonds lies in a random direction. By concentrating on such groups rather than individuals, it turns out that for long chains the average values given above should be multiplied by

$$F = \left(\frac{1 - \cos\theta}{1 + \cos\theta}\right)^{1/2} \tag{38}$$

For tetrahedral bonds, for which $\cos\theta = -\frac{1}{3}$ (that is, $\theta = 109.5°$), $F = 2^{1/2}$. Therefore:

$$R_{\mathrm{rms}} = (2N)^{1/2}l \qquad R_{\mathrm{g}} = \left(\frac{N}{3}\right)^{1/2} l \tag{39}$$

Illustration
....

For a polyethylene chain with $M = 56\ \mathrm{kg\,mol^{-1}}$, corresponding to $N = 4000$, because $l = 154\ \mathrm{pm}$ for a C—C bond, we find $R_{\mathrm{rms}} = 14\ \mathrm{nm}$ and $R_{\mathrm{g}} = 5.6\ \mathrm{nm}$ (Fig. 23.16). This value of R_{g} means that, on average, the coils rotate like hollow spheres of radius 5.6 nm and mass equal to the molecular mass.
....

The model of a randomly coiled molecule is still an approximation, even after the bond angles have been restricted, because it does not take into account the impossibility of two or more atoms occupying the same place. Such self-avoidance tends to swell the coil, so it is better to regard R_{rms} and R_{g} as lower bounds to the actual values. The model also ignores the role of the solvent: a poor solvent will tend to cause the coil to tighten so that solute–solvent contacts are minimized; a good solvent does the opposite. A θ solvent leaves the coil in its natural state.

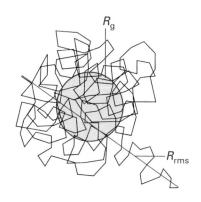

23.16 A random coil in three dimensions. This one contains about 200 units. The root mean square distance between the ends (R_{rms}) and the radius of gyration (R_{g}) are indicated.

23.7 Helices and sheets

Natural macromolecules need a precisely maintained conformation to function. The achievement of a specific conformation is the major remaining problem in protein synthesis for, although primary structures can be built, the product is inactive because the correct secondary structure is not adopted.

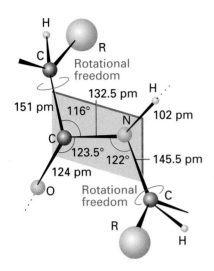

23.17 The dimensions that characterize the peptide link. The C—CO—NH—C atoms define a plane (the C—N bond has partial double-bond character), but there is rotational freedom around the C—CO and N—C bonds.

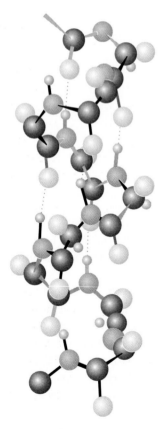

H C N O

23.18 The polypeptide α helix. There are 3.6 residues per turn, and a translation along the helix of 150 pm per residue, giving a pitch of 540 pm. The diameter (ignoring side chains) is about 600 pm.

(a) The Corey–Pauling rules

The origin of the secondary structures of proteins is found in the rules formulated by Linus Pauling and Robert Corey in 1951. The essential feature is the stabilization of structures by hydrogen bonds involving the peptide link. The latter can act both as a donor of the H atom (the NH part of the link) and as an acceptor (the CO part). The **Corey–Pauling rules** are as follows (Fig. 23.17):

1. The atoms of the peptide link lie in a plane.
2. The N, H, and O atoms of a hydrogen bond lie in a straight line (with displacements of H tolerated up to not more than 30° from the N—O vector).
3. All NH and CO groups are engaged in hydrogen bonding.

The rules are satisfied by two structures. One, in which hydrogen bonding occurs between peptide links of the same chain, is the **α helix**. The other, in which hydrogen bonding links different chains, is the **β-pleated sheet**; this form is the secondary structure of the protein fibroin, the constituent of silk.

The α helix is illustrated in Fig. 23.18. Each turn of the helix contains 3.6 amino acid residues, so the period of the helix corresponds to 5 turns (18 residues). The pitch of a single turn (the distance between points separated by 360°) is 544 pm. The N—H···O bonds lie parallel to the axis and residue i is linked to residues $i-4$ and $i+4$. There is freedom for the helix to be arranged as either a right- or a left-handed screw, but the overwhelming majority of natural polypeptides are right-handed on account of the preponderance of the L-configuration of the naturally occurring amino acids, as we explain below. The reason for their preponderance is uncertain, but it may be related to the symmetries of fundamental particles and the nonconservation of parity (the fact that this universe behaves differently from its hypothetical mirror image).

(b) Conformational energy

The stabilities of different polypeptide geometries can be investigated by calculating the total potential energy of all the interactions between nonbonded atoms, and looking for a minimum. It turns out, in agreement with experience, that a right-handed α helix of L-amino acids is marginally more stable than a left-handed helix of the same acids.

The geometry of the chain can be specified by two angles, ϕ (the torsional angle for the N—C bond) and ψ (the torsional angle for the C—C bond). The illustration in Fig. 23.19 defines these angles. The sign convention is that a positive angle means that the front atom must be rotated clockwise to bring it into an eclipsed position relative to the rear atom. The illustration shows the all-*trans* form of the chain, in which all ϕ and ψ are 180°. A helix is obtained when all the ϕ are equal and when all the ψ are equal. For a right-handed α helix, all $\phi = -57°$ and all $\psi = -47°$. For a left-handed α helix, both angles are positive. Because only two angles are needed to specify the conformation of a helix, and they range from $-180°$ to $+180°$, the potential energy of the entire molecule can be represented on a **Ramachandran plot**, a contour diagram in which one axis represents ϕ and the other represents ψ.

The potential energy of a given conformation (ϕ, ψ) can be calculated by using the expressions developed in Sections 22.3 and 22.4; the procedure is now widely automated in commercially available molecular modelling software. For example, the interaction energy of two atoms separated by a distance R (which we know once ϕ and ψ are specified) can be given the Lennard-Jones (12, 6) form. If the partial charges on the atoms (arising from ionic character in the bonds) are known, a Coulombic contribution of the form $1/R$ can be included. The inclusion of Coulombic interactions is sometimes accomplished by ascribing charges $-0.28e$ and $+0.28e$ to N and H, respectively, and $-0.39e$ and $+0.39e$ to O and C,

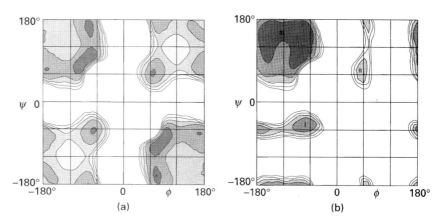

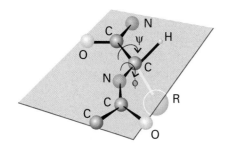

23.19 The definition of the torsional angles ψ and ϕ between two peptide units. In this case (an α-L-polypeptide) the chain has been drawn in its all-*trans* form, with $\psi = \phi = 180°$.

23.20 Ramachandran diagrams for (a) a glycyl residue of a polypeptide chain and (b) an alanyl residue. The darker the shading is, the lower the potential energy. The glycyl diagram is symmetrical, but regions I and II in the alanine diagram, which correspond to right- and left-handed helices, are unsymmetrical, and the minimum in region I lies lower than that in region II. (After D.A. Brant and P.J. Flory, *J. Mol. Biol.* **23**, 47 (1967).)

respectively. There is also a torsional contribution arising from the barrier to internal rotation of one bond relative to another (just like the barrier to internal rotation in ethane), and which is normally expressed as

$$V = A(1 + \cos 3\phi) + B(1 + \cos 3\psi) \tag{40}$$

in which A and B are constants of the order of 1 kJ mol^{-1}.

The potential energy contours for the helical form of polypeptide chains formed from the nonchiral amino acid glycine (R = H) and the chiral amino acid L-alanine are shown in Fig. 23.20. They were computed by summing all the contributions described above for each choice of angles, and then plotting contours of equal potential energy. The glycine map is symmetrical, with minima of equal depth at $\phi = -80°$, $\psi = +90°$ and at $\phi = +80°$, $\psi = -90°$. In contrast, the map for L-alanine is unsymmetrical, and there are three distinct low-energy conformations (marked I, II, III). The minima of regions I and II lie close to the angles typical of right- and left-handed α helices, but the former has a lower minimum, which is consistent with the formation of right-handed helices from the naturally occurring L-amino acids.

23.8 Higher-order structures

Helical polypeptide chains are folded into a tertiary structure if there are other bonding influences between the residues of the chain that are strong enough to overcome the interactions responsible for the secondary structure. The folding influences include disulfide (—S—S—) links, ionic interactions (which depend on the pH), and strong hydrogen bonds (such as O—H$\cdots$O—), and is illustrated by the structure of myoglobin (Fig. 23.21), the full structure (of 2600 atoms) having been determined by X-ray diffraction. The folding of the basic α helix caused by disulfide links can be seen in the structure: about 77 per cent of the structure is α helix, the rest being involved in the folds.

Proteins with $M > 50 \text{ kg mol}^{-1}$ are often found to be aggregates of two or more polypeptide chains. The possibility of such a quaternary structure often confuses the determination of their molar masses, for different techniques might give values differing by factors of 2 or more. Haemoglobin, which consists of four myoglobin-like chains, is an example.

Protein denaturation can be caused by several means, and different aspects of structure may be affected. The 'permanent waving' of hair, for example, is reorganization at the

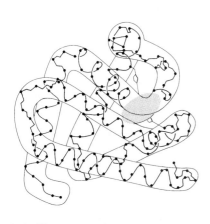

23.21 The structure of myoglobin. Only the α-carbon atom positions are shown. The haem group, the oxygen binding group, is shown as a shaded region. (Based on M.F. Perutz, copyright *Scientific American*, 1964; with permission.)

quaternary level. Hair is a form of the protein keratin, and its quaternary structure is thought to be a multiple helix, with the α helices bound together by disulfide links and hydrogen bonds (although there is some dispute about its precise structure). The process of permanent waving consists of disrupting these links, unravelling the keratin quaternary structure, and then reforming it into a more fashionable disposition. The 'permanence' is only temporary, however, because the structure of newly formed hair is genetically controlled. Incidentally, normal hair grows at a rate that requires at least 10 twists of the keratin helix to be produced each second, so very close inspection of the human scalp would show it to be literally writhing with activity.

Denaturation at the secondary level is brought about by agents that destroy hydrogen bonds. Thermal motion may be sufficient, in which case denaturation is a kind of intramolecular melting. When eggs are cooked, the albumin is denatured irreversibly, and the protein collapses into a structure resembling a random coil. This **helix–coil transition** is sharp, like ordinary melting, because it is a cooperative process: when one hydrogen bond has been broken it is easier to break the bonds to its neighbours, and then even easier to break their bonds, and so on. The disruption cascades down the helix, and the transition occurs sharply. Denaturation may also be brought about chemically. For instance, a solvent that forms stronger hydrogen bonds than those within the helix will compete successfully for the NH and CO groups. Acids and bases can cause denaturation by protonation or deprotonation of groups involved in hydrogen bonding.

Colloids and surfactants

Much of the material discussed in this chapter also applies to aggregates of particles, either in the form of small particles or of extended sheets, like those forming biological cell walls. However, these systems present a number of specific properties, and we concentrate on them.

23.9 The properties of colloids

A **colloid**, or **disperse phase**, is a dispersion of small particles of one material in another. In this context, 'small' means something less than about 500 nm in diameter (about the wavelength of visible light). In general, colloidal particles are aggregates of numerous atoms or molecules, but are too small to be seen with an ordinary optical microscope. They pass through most filter papers, but can be detected by light-scattering, sedimentation, and osmosis.

(a) Classification and preparation

The name given to the disperse phase depends on the two phases involved. A **sol** is a dispersion of a solid in a liquid (such as clusters of gold atoms in water) or of a solid in a solid (such as ruby glass, which is a gold-in-glass sol, and achieves its colour by scattering). An **aerosol** is a dispersion of a liquid in a gas (like fog and many sprays) or a solid in a gas (such as smoke): the particles are often large enough to be seen with a microscope. An **emulsion** is a dispersion of a liquid in a liquid (such as milk).

A further classification of colloids is as **lyophilic**, or solvent-attracting, and **lyophobic**, solvent-repelling. If the solvent is water, the terms **hydrophilic** and **hydrophobic**, respectively, are used instead. Lyophobic colloids include the metal sols. Lyophilic colloids generally have some chemical similarity to the solvent, such as –OH groups able to form hydrogen bonds. A **gel** is a semirigid mass of a lyophilic sol in which all the dispersion medium has penetrated into the sol particles.

The preparation of aerosols can be as simple as sneezing (which produces an imperfect aerosol). Laboratory and commercial methods make use of several techniques. Material (for example, quartz) may be ground in the presence of the dispersion medium. Passing a heavy electric current through a cell may lead to the sputtering (crumbling) of an electrode into colloidal particles. Arcing between electrodes immersed in the support medium also produces a colloid. Chemical precipitation sometimes results in a colloid. A precipitate (for example, silver iodide) already formed may be dispersed by the addition of a peptizing agent (for example, potassium iodide). Clays may be peptized by alkalis, the OH^- ion being the active agent.

Emulsions are normally prepared by shaking the two components together vigorously, although some kind of emulsifying agent usually has to be added to stabilize the product. This emulsifying agent may be a soap (the salt of a long-chain carboxylic acid) or other **surfactant** (surface active) species, or a lyophilic sol that forms a protective film around the dispersed phase. In milk, which is an emulsion of fats in water, the emulsifying agent is casein, a protein containing phosphate groups. It is clear from the formation of cream on the surface of milk that casein is not completely successful in stabilizing milk: the dispersed fats coalesce into oily droplets which float to the surface. This coagulation may be prevented by ensuring that the emulsion is dispersed very finely initially: intense agitation with ultrasonics brings this dispersion about, the product being 'homogenized' milk.

One way to form an aerosol is to tear apart a spray of liquid with a jet of gas. The dispersal is aided if a charge is applied to the liquid, for then electrostatic repulsions help to blast it apart into droplets. This procedure may also be used to produce emulsions, for the charged liquid phase may be directed into another liquid.

Colloids are often purified by dialysis. The aim is to remove much (but not all, for reasons explained later) of the ionic material that may have accompanied their formation. As in the discussion of the Donnan effect in Section 23.2b, a membrane (for example, cellulose) is selected that is permeable to solvent and ions, but not to the colloid particles. Dialysis is very slow, and is normally accelerated by applying an electric field and making use of the charges carried by many colloid particles; the technique is then called **electrodialysis**.

(b) Structure and stability

A disperse phase is thermodynamically unstable with respect to the bulk. This instability can be expressed thermodynamically by noting that, because the change in Gibbs energy, dG, when the surface area of the sample changes by $d\sigma$ at constant temperature and pressure is $dG = \gamma d\sigma$, where γ is the interfacial surface tension (Section 6.10), it follows that $dG < 0$ if $d\sigma < 0$. The survival of colloids must therefore be a consequence of the kinetics of collapse: colloids are thermodynamically unstable but kinetically nonlabile.

At first sight, even the kinetic argument seems to fail: colloidal particles attract each other over large distances, so there is a long-range force that tends to condense them into a single blob. The reasoning behind this remark is as follows. The energy of attraction between two individual atoms i and j separated by a distance R_{ij}, one in each colloidal particle, varies as their separation as $1/R_{ij}^6$ (Section 22.4). The sum of all these pairwise interactions, however, decreases only as approximately $1/R^2$ (the precise variation depending on the shape of the particles and their closeness), where R is the separation of the centres of the particles. The sum has a much longer range than the $1/R^6$ dependence characteristic of individual particles and small molecules.

Several factors oppose the long-range dispersion attraction. For example, there may be a protective film at the surface of the colloid particles that stabilizes the interface and cannot be penetrated when two particles touch. Thus the surface atoms of a platinum sol in water react chemically and are turned into $-Pt(OH)_3H_3$, and this layer encases the particle like a shell. A fat can be emulsified by a soap because the long hydrocarbon tails penetrate the oil droplet but the carboxylate head groups (or other hydrophilic groups in synthetic detergents) surround the surface, form hydrogen bonds with water, and give rise to a shell of negative charge that repels a possible approach from another similarly charged particle.

(c) Micelle formation and the hydrophobic interaction

Surfactant molecules or ions can cluster together as **micelles**, which are colloid-sized clusters of molecules, for their hydrophobic tails tend to congregate, and their hydrophilic heads provide protection (Fig. 23.22). Micelles form only above the **critical micelle concentration** (CMC) and above the **Krafft temperature**. The CMC is detected by noting a pronounced discontinuity in physical properties of the solution, particularly the molar conductivity (Fig. 23.23). The hydrocarbon interior of a micelle is like a droplet of oil. Nuclear magnetic resonance shows that the hydrocarbon tails are mobile, but slightly more restricted than in the bulk. Micelles are important in industry and biology on account of their solubilizing function: matter can be transported by water after it has been dissolved in their hydrocarbon interiors. For this reason, micellar systems are used as detergents and drug carriers, and for organic synthesis, froth flotation, and petroleum recovery.

Nonionic surfactant molecules may cluster together in clumps of 1000 or more, but ionic species tend to be disrupted by the electrostatic repulsions between head groups and are normally limited to groups of less than about 100. The micelle population is often polydisperse, and the shapes of the individual micelles vary with concentration. Spherical micelles do occur, but micelles are more commonly flattened spheres close to the CMC. Some micelles at concentrations well above the CMC form extended parallel sheets, called **lamellar micelles**, two molecules thick. The individual molecules lie perpendicular to the sheets, with hydrophilic groups on the outside in aqueous solution and on the inside in nonpolar media. Such lamellar micelles show a close resemblance to biological membranes, and are often a useful model on which to base investigations of biological structures. In concentrated solutions micelles formed from surfactant molecules may take the form of long cylinders and stack together in reasonably close-packed (hexagonal) arrays. These orderly arrangements of micelles are called **lyotropic mesomorphs**, and more colloquially 'liquid crystalline phases' (Section 24.5e).

Micelle formation in aqueous systems is commonly endothermic, with $\Delta H \approx 1$–2 kJ per mole of surfactant. That micelles do form above the CMC indicates that the entropy change accompanying their formation must then be positive, and measurements suggest a value of about $+140$ J K^{-1} mol^{-1} at room temperature. The fact that the entropy change is positive even though the molecules are clustering together shows that there must be a contribution to the entropy from the solvent, and that solvent molecules must be more free to move once the solute molecules have herded into small clusters. This interpretation is plausible, because each individual solute molecule is held in an organized solvent cage (Fig. 23.24), but once the micelle has formed the solvent molecules need form only a single (admittedly larger) cage. The increase in energy when hydrophobic groups cluster together and reduce their structural demands on the solvent is the origin of the **hydrophobic interaction**, which tends to stabilize groupings of hydrophobic groups in biological macromolecules. The hydrophobic interaction is an example of an ordering process that is stabilized by a tendency toward greater disorder of the solvent.

(d) The electrical double layer

A major source of kinetic stability of colloids is the existence of an electric charge on the surfaces of the particles. On account of this charge, ions of opposite charge tend to cluster nearby, and an ionic atmosphere is formed, just as for ions (Section 10.2c).

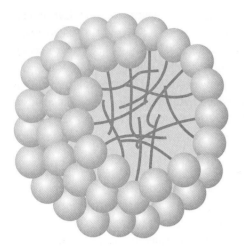

23.22 A schematic version of a spherical micelle. The hydrophilic groups are represented by spheres and the hydrophobic hydrocarbon chains are represented by the stalks; these stalks are mobile.

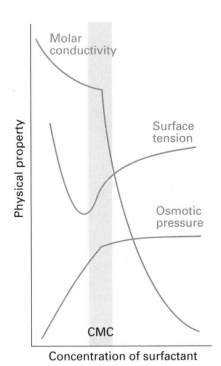

23.23 The typical variation of some physical properties of an aqueous solution of sodium dodecylsulfate close to the critical micelle concentration (CMC).

Two regions of charge must be distinguished. First, there is a fairly immobile layer of ions that adhere tightly to the surface of the colloidal particle, and which may include water molecules (if that is the support medium). The radius of the sphere that captures this rigid layer is called the **radius of shear**, and is the major factor determining the mobility of the particles. The electric potential at the radius of shear relative to its value in the distant, bulk medium is called the **zeta potential**, ζ, or the **electrokinetic potential**. Second, the charged unit attracts an oppositely charged atmosphere of mobile ions. The inner shell of charge and the outer ionic atmosphere is called the **electrical double layer**.

The theory of the stability of lyophobic dispersions was developed by B. Derjaguin and L. Landau and independently by E. Verwey and J.T.G. Overbeek, and is known as the **DLVO theory**. It assumes that there is a balance between the repulsive interaction between the charges of the electrical double layers on neighbouring particles and the attractive interactions arising from van der Waals interactions between the molecules in the particles. The potential energy arising from the repulsion of double layers on particles of radius a has the form

$$V_{\text{repulsion}} = +\frac{Aa^2\zeta^2}{R}\,e^{-s/r_D} \tag{41}$$

where A is a constant, ζ is the zeta potential,[4] R is the separation of centres, s is the separation of the surfaces of the two particles ($s = R - 2a$ for spherical particles of radius a), and r_D is the thickness of the double layer. This expression is valid for small particles with a thick double layer ($a \ll r_D$). When the double layer is thin ($a \gg r_D$), the expression is replaced by

$$V_{\text{repulsion}} = +\tfrac{1}{2}Aa\zeta^2 \ln(1 + e^{-s/r_D}) \tag{42}$$

In each case, the thickness of the double layer can be estimated from an expression like that derived for the thickness of the ionic atmosphere in the Debye–Hückel theory (eqn 10.33):

$$r_D = \left(\frac{\varepsilon RT}{2\rho F^2 I b^{\ominus}}\right)^{1/2} \tag{43}$$

where I is the ionic strength of the solution, ρ its mass density, and $b^{\ominus} = 1\ \text{mol kg}^{-1}$. The potential energy arising from the attractive interaction has the form

$$V_{\text{attraction}} = -\frac{B}{s} \tag{44}$$

where B is another constant. The variation of the total potential energy with separation is shown in Fig. 23.25.

At high ionic strengths, the ionic atmosphere is dense and the potential shows a secondary minimum at large separations. Aggregation of the particles arising from the stabilizing effect of this secondary minimum is called **flocculation**. The flocculated material can often be redispersed by agitation because the well is so shallow. **Coagulation**, the irreversible blending together of distinct particles into large particles, occurs when the separation of the particles is so small that they enter the primary minimum of the potential energy curve and van der Waals forces are dominant.

The ionic strength is increased by the addition of ions, particularly those of high charge type, so such ions act as flocculating agents. This increase is the basis of the empirical **Schulze–Hardy rule**, that hydrophobic colloids are flocculated most efficiently by ions of opposite charge type and high charge number. The Al^{3+} ions in alum are very effective, and are used to induce the congealing of blood. When river water containing colloidal clay flows

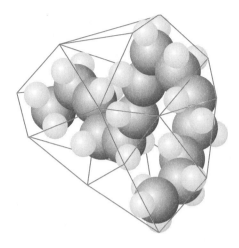

23.24 When a hydrocarbon molecule is surrounded by water, the H_2O molecules form a clathrate cage. As a result of this acquisition of structure, the entropy of the water decreases, so the dispersal of the hydrocarbon into the water is entropy-opposed; its coalescence is entropy-favoured.

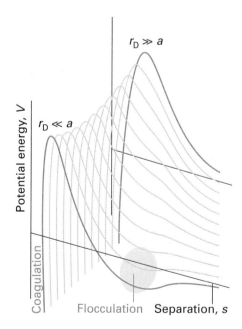

23.25 The potential energy of interaction as a function of the separation of the centres of the two particles and its variation with the ratio of the particle size to the thickness a of the electrical double layer r_D. The regions labelled coagulation and flocculation show the dips in the potential energy curves where these processes occur.

---

4   The actual potential is that of the surface of the particles; there is some danger in identifying it with the zeta potential. See the references in *Further reading*.

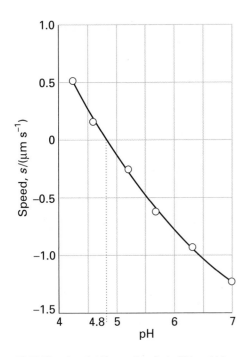

23.26 The plot of drift speed against pH by which the isoelectric point of a macromolecule can be determined: it corresponds to the pH at which the drift speed in the presence of an electric field is zero.

into the sea, the salt water induces flocculation and coagulation, and is a major cause of silting in estuaries.

Metal oxide sols tend to be positively charged whereas sulfur and the noble metals tend to be negatively charged. Naturally occurring macromolecules also acquire a charge when dispersed in water, and an important feature of proteins and other natural macromolecules is that their overall charge depends on the pH of the medium. For instance, in acidic environments protons attach to basic groups, and the net charge of the macromolecule is positive; in basic media the net charge is negative as a result of proton loss. At the **isoelectric point** the pH is such that there is no net charge on the macromolecule.

Example 23.7 Determining the isoelectric point

The drift speed of bovine serum albumin (BSA) under the influence of an electric field in aqueous solution was monitored at several values of pH, and the data are listed below (opposite signs indicate opposite directions of travel). What is the isoelectric point of the protein?

pH	4.20	4.56	5.20	5.65	6.30	7.00
Speed/$(\mu m\,s^{-1})$	+0.50	+0.18	−0.25	−0.65	−0.90	−1.25

Method The macromolecule has zero electrophoretic mobility when it is uncharged. Therefore, the isoelectric point is the pH at which it does not migrate in an electric field. We should therefore plot speed against pH and find by interpolation the pH of zero mobility.

Answer The data are plotted in Fig. 23.26. The drift speed is zero at pH = 4.8; hence pH = 4.8 is the isoelectric point.

Comment For some species, the isoelectric point must be obtained by extrapolation because the macromolecule might not be stable over the whole pH range.

- -

Self-test 23.7 The following data were obtained for another protein:

pH	4.5	5.0	5.5	6.0
Speed/$(\mu m\,s^{-1})$	−0.10	−0.20	−0.30	−0.35

Estimate the pH of the isoelectric point.

[4.3]

The primary role of the electrical double layer is to confer kinetic stability. Colliding colloidal particles break through the double layer and coalesce only if the collision is sufficiently energetic to disrupt the layers of ions and solvating molecules, or if thermal motion has stirred away the surface accumulation of charge. This disruption may occur at high temperatures, which is one reason why sols precipitate when they are heated. The protective role of the double layer is the reason why it is important not to remove all the ions when a colloid is being purified by dialysis, and why proteins coagulate most readily at their isoelectric point.

23.10 Surface films

The compositions of surface layers have been investigated by the simple (but technically elegant) procedure of slicing thin layers off the surfaces of solutions and analysing their compositions. The physical properties of surface films have also been investigated. Surface films one molecule thick, such as that formed by a surfactant, are called **monolayers**. When a

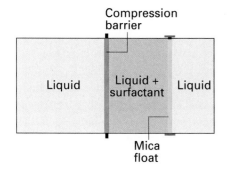

23.27 A schematic diagram of the apparatus used to measure the surface pressure and other characteristics of a surface film. The surfactant is spread on the surface of the liquid in the trough, and then compressed horizontally by moving the compression barrier towards the mica float. The latter is connected to a torsion wire, so the difference in force on either side of the float can be monitored.

monolayer has been transferred to a solid support, it is called a **Langmuir–Blodgett film**, after Irving Langmuir and Katherine Blodgett, who developed experimental techniques for studying them.

(a) Surface pressure

The principal apparatus used for the study of surface monolayers is a **surface film balance**, Fig. 23.27. This device consists of a shallow trough and a barrier that can be moved along the surface of the liquid in the trough, and hence compress any monolayer on the surface. The **surface pressure**, π, the difference between the surface tension of the pure solvent and the solution ($\pi = \gamma^* - \gamma$) is measured by using a torsion wire attached to a strip of mica that rests on the surface and against which one edge of the monolayer is pressed. The parts of the apparatus that are in touch with liquids are coated in polytetrafluoroethylene to eliminate effects arising from the liquid–solid interface. In an actual experiment, a small amount (about 0.01 mg) of the surfactant under investigation is dissolved in a volatile solvent and then poured on to the surface of the water; the compression barrier is then moved across the surface and the surface pressure exerted on the mica bar is monitored.

Some typical results are shown in Fig. 23.28. One parameter obtained from the isotherms is the area occupied by the molecules when the monolayer is closely packed. This quantity is obtained from the extrapolation of the steepest part of the isotherm to the horizontal axis. As can be seen from the illustration, even though stearic acid (**1**) and isostearic acid (**2**) are chemically very similar (they differ only in the location of a methyl group at the end of a long hydrocarbon chain), they occupy significantly different areas in the monolayer. Neither, though, occupies as much area as the tri-*p*-cresyl phosphate molecule (**3**), which is like a wide bush rather than a lanky tree.

The second feature to note from Fig. 23.28 is that the tri-*p*-cresyl phosphate isotherm is much less steep than the stearic acid isotherms. This difference indicates that the tri-*p*-cresyl phosphate film is more compressible than the stearic acid films, which is consistent with their different molecular structures.

A third feature of the isotherms is the **collapse pressure**, the highest surface pressure. When the monolayer is compressed beyond the point represented by the collapse pressure, the monolayer buckles and collapses into a film several molecules thick. As can be seen from the isotherms in Fig. 23.28, stearic acid has a high collapse pressure, but that of tri-*p*-cresyl phosphate is significantly smaller, indicating a much weaker film.

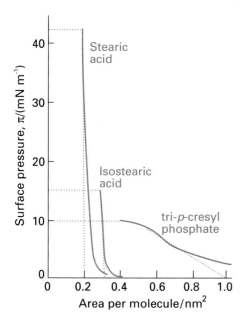

23.28 The variation of surface pressure with the area occupied by each surfactant molecule. The collapse pressures are indicated by the horizontal dotted lines.

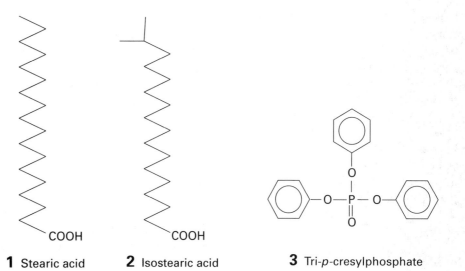

1 Stearic acid   **2** Isostearic acid   **3** Tri-*p*-cresylphosphate

(b) The thermodynamics of surface layers

A surfactant is active at the interface between two phases, such as at the interface between hydrophilic and hydrophobic phases. A surfactant accumulates at the interface, and modifies its surface tension and hence the surface pressure. To establish the relation between the concentration of surfactant at a surface and the change in surface tension it brings about, we consider two phases α and β in contact and suppose that the system consists of several components J, each one present in an overall amount n_J. If the components were distributed uniformly through the two phases right up to the interface, which is taken to be a plane of surface area σ, the total Gibbs energy, G, would be the sum of the Gibbs energies of both phases, $G = G(\alpha) + G(\beta)$. However, the components are not uniformly distributed because one may accumulate at the interface. As a result, the sum of the two Gibbs energies differs from G by an amount called the **surface Gibbs energy**, $G(\sigma)$:

$$G(\sigma) = G - \{G(\alpha) + G(\beta)\} \qquad [\mathbf{45}]$$

Similarly, if it is supposed that the concentration of a species J is uniform right up to the interface, then from its volume we would conclude that it contains an amount $n_J(\alpha)$ of J in phase α and an amount $n_J(\beta)$ in phase β. However, because a species may accumulate at the interface, the total amount of J differs from the sum of these two amounts by $n_J(\sigma) = n_J - \{n_J(\alpha) + n_J(\beta)\}$. This difference is expressed in terms of the **surface excess**, Γ_J:

$$\Gamma_J = \frac{n_J(\sigma)}{\sigma} \qquad [\mathbf{46}]$$

The surface excess may be either positive (an accumulation of J at the interface) or negative (a deficiency there).

The relation between the change in surface tension and the composition of a surface (as expressed by the surface excess) was derived by Gibbs. In the following *Justification* we derive the **Gibbs isotherm**, between the changes in the chemical potentials of the substances present in the interface and the change in surface tension:

$$d\gamma = -\sum_J \Gamma_J d\mu_J \qquad (47)$$

Justification 23.6

A general change in G is brought about by changes in T, p, and the n_J:

$$dG = -SdT + Vdp + \gamma d\sigma + \sum_J \mu_J \, dn_J$$

When this relation is applied to G, $G(\alpha)$, and $G(\beta)$ we find

$$dG(\sigma) = -S(\sigma)\, dT + \gamma \, d\sigma + \sum_J \mu_J \, dn_J(\sigma)$$

because at equilibrium the chemical potential of each component is the same in every phase, $\mu_J(\alpha) = \mu_J(\beta) = \mu_J(\sigma)$. Just as in the discussion of partial molar quantities (Section 7.1), the last equation integrates at constant temperature to

$$G(\sigma) = \gamma\sigma + \sum_J \mu_J n_J(\sigma)$$

We are seeking a connection between the change of surface tension $d\gamma$ and the change of composition at the interface. Therefore, we use the argument which in Section 7.1d led to the Gibbs–Duhem equation (eqn 7.12), but this time we compare the expression

$$dG(\sigma) = \gamma\, d\sigma + \sum_J \mu_J\, dn_J(\sigma)$$

(which is valid at constant temperature) with the expression for the same quantity but derived from the preceding equation:

$$dG(\sigma) = \gamma\, d\sigma + \sigma\, d\gamma + \sum_J \mu_J\, dn_J + \sum_J n_J(\sigma)\, d\mu_J$$

The comparison implies that, at constant temperature,

$$\sigma\, d\gamma + \sum_J n_J\, d\mu_J = 0$$

Division by σ then gives eqn 47.

Now consider a simplified model of the interface in which the 'oil' and 'water' phases are separated by a geometrically flat surface. This approximation implies that only the surfactant, S, accumulates at the surface, and hence that Γ_{Oil} and Γ_{Water} are both zero. Then the Gibbs equation becomes

$$d\gamma = -\Gamma_S\, d\mu_S \tag{48}$$

For dilute solutions,

$$d\mu_S = RT\, d\ln c \tag{49}°$$

where c is the molar concentration of the surfactant. It follows that

$$d\gamma = -RT\Gamma_S\, \frac{ds}{c}$$

at constant temperature, or

$$\left(\frac{\partial\gamma}{\partial c}\right)_T = -\frac{RT\Gamma_S}{c} \tag{50}°$$

If the surfactant accumulates at the interface, its surface excess is positive and eqn 50 implies that $(\partial\gamma/\partial c)_T < 0$. That is, the surface tension decreases when a solute accumulates at a surface. Conversely, if the concentration dependence of γ is known, the surface excess may be predicted and used to infer the area occupied by each surfactant molecule on the surface.

Checklist of key ideas

- [ ] polymers
- [ ] monomers

Size and shape

23.1 Mean molar masses
- [ ] monodisperse
- [ ] polydisperse
- [ ] number-average molar mass (1)

- [ ] viscosity-average molar mass (3)
- [ ] weight-average molar mass (2)
- [ ] Z-average molar mass (4)
- [ ] heterogeneity index

23.2 Colligative properties
- [ ] Florey theta temperature
- [ ] θ solution
- [ ] vapour-phase osmometry
- [ ] polyelectrolyte

- [ ] polyanion
- [ ] polycation
- [ ] polyampholyte
- [ ] Donnan equilibrium

23.3 Sedimentation
- [ ] sedimentation
- [ ] frictional coefficient
- [ ] drift speed
- [ ] sedimentation constant (11)
- [ ] Stokes' relation (13)
- [ ] Stokes–Einstein relation (15)

- [ ] electrophoresis
- [ ] gel electrophoresis
- [ ] size-exclusion chromatography (SEC)
- [ ] gel permeation chromatography (GPC)

23.4 Viscosity
- [ ] intrinsic viscosity (19)
- [ ] Ostwald viscometer
- [ ] rotating drum viscometer

- ☐ Mark–Kuhn–Houwink–Sakurada equation (21)

23.5 Light scattering
- ☐ Rayleigh scattering
- ☐ radius of gyration (26)
- ☐ polymer dynamics
- ☐ dynamic light scattering

Conformation and configuration
- ☐ primary structure
- ☐ polypeptides
- ☐ secondary structure
- ☐ denaturation
- ☐ configuration
- ☐ conformation
- ☐ tertiary structure
- ☐ quaternary structure

23.6 Random coils
- ☐ freely jointed chain
- ☐ random coil
- ☐ perfect elastomer
- ☐ conformational entropy (27)
- ☐ radial distribution (34)
- ☐ contour length (35)
- ☐ root mean square separation (36)

23.7 Helices and sheets
- ☐ Corey–Pauling rules
- ☐ Ramachandran plot
- ☐ α helix
- ☐ β-pleated sheet

23.8 Higher-order structures
- ☐ helix–coil transition

Colloids and surfactants

23.9 The properties of colloids
- ☐ colloid
- ☐ disperse phase
- ☐ sol
- ☐ aerosol
- ☐ emulsion
- ☐ lyophilic
- ☐ lyophobic
- ☐ hydrophilic
- ☐ hydrophobic
- ☐ gel
- ☐ surfactant
- ☐ electrodialysis
- ☐ micelle
- ☐ critical micelle concentration (CMC)
- ☐ Krafft temperature
- ☐ lamellar micelle

- ☐ lyotropic mesomorph
- ☐ hydrophobic interaction
- ☐ radius of shear
- ☐ zeta potential
- ☐ electrokinetic potential
- ☐ electrical double layer
- ☐ DLVO theory
- ☐ flocculation
- ☐ coagulation
- ☐ Schulze–Hardy rule
- ☐ isoelectric point

23.10 Surface films
- ☐ monolayer
- ☐ Langmuir–Blodgett film
- ☐ surface film balance
- ☐ surface pressure
- ☐ collapse pressure
- ☐ surface Gibbs energy (45)
- ☐ surface excess (46)
- ☐ Gibbs isotherm (47)

Further reading

Articles of general interest

J.P. Queslel and J.E. Mark, Advances in rubber elasticity and characterization of elastomeric networks. *J. Chem. Educ.* **64**, 491 (1987).

H. Bisswanger, Proteins and enzymes. In *Encyclopedia of applied physics* (ed. G.L. Trigg), **15**, 185. VCH, New York (1996).

C.M. Guttman and B. Fanconi, Molecular properties of polymers. In *Encyclopedia of applied physics* (ed. G.L. Trigg), **16**, 549. VCH, New York (1996).

R.H. Barth, Dialysis. In *Encyclopedia of applied physics* (ed. G.L. Trigg), **4**, 533. VCH, New York (1992).

B.Y.H. Liu and D.Y.H. Pui, Aerosols. In *Encyclopedia of applied physics* (ed. G.L. Trigg), **1**, 415. VCH, New York (1991).

R.J. Hunter, *Foundations of colloid science*, Vols 1 and 2. Oxford University Press (1987, 1989).

B. Dobiáš, *Coagulation and flocculation*. Marcel Dekker, New York (1993).

Y. Morai, *Micelles*. Plenum, New York (1982).

M. Takeo, Disperse systems. In *Encyclopedia of applied physics* (ed. G.L. Trigg), **5**, 87. VCH, New York (1993).

Texts and sources of data and information

F.W. Billmeyer, *Textbook of polymer science*. Wiley, New York (1984).

I.M. Ward and D.W. Hedley, *Mechanical properties of solid polymers*. Wiley, New York (1993).

H.R. Allcock and F.W. Lampe, *Contemporary polymer chemistry*. Prentice-Hall, Englewood Cliffs (1981).

E.G. Richards, *An introduction to physical properties of large molecules in solution*. Cambridge University Press (1980).

L.H. Sperling, *Physical polymer science*. Wiley-Interscience, New York (1986).

D. Freifelder, *Physical biochemistry*. W.H. Freeman & Co, New York (1982).

P. Flory, *Principles of polymer chemistry*. Cornell University Press, Ithaca (1953).

A.R. Leach, *Molecular modelling: principles and applications*. Longman, Harlow (1996).

D. Frenkel and B. Smit, *Understanding molecular simulation*. Academic Press, San Diego (1996).

Exercises

23.1 (a) Calculate the number-average molar mass and the mass-average molar mass of a mixture of equal amounts of two polymers, one having $M = 62$ kg mol^{-1} and the other $M = 78$ kg mol^{-1}.

23.1 (b) Calculate the number-average molar mass and the mass-average molar mass of a mixture of two polymers, one having $M = 62$ kg mol^{-1} and the other $M = 78$ kg mol^{-1}, with their amounts (numbers of moles) in the ratio 3 : 2.

23.2 (a) A polymer chain consists of 700 segments, each 0.90 nm long. If the chain were ideally flexible, what would be the r.m.s. separation of the ends of the chain?

23.2 (b) A polymer chain consists of 1200 segments, each 1.125 nm long. If the chain were ideally flexible, what would be the r.m.s. separation of the ends of the chain?

23.3 (a) The radius of gyration of a long chain molecule is found to be 7.3 nm. The chain consists of C–C links. Assume the chain is randomly coiled and estimate the number of links in the chain.

23.3 (b) The radius of gyration of a long chain molecule is found to be 18.9 nm. The chain consists of links of length 450 pm. Assume the chain is randomly coiled and estimate the number of links in the chain.

23.4 (a) Calculate the contour length (the length of the extended chain) and the root mean square separation (the end-to-end distance) for polyethylene with a molar mass of 280 kg mol^{-1}.

23.4 (b) Calculate the contour length (the length of the extended chain) and the root mean square separation (the end-to-end distance) for polypropylene of molar mass 174 kg mol^{-1}.

23.5 (a) What is the relative rate of sedimentation for two spherical particles of the same density, but which differ in radius by a factor of 10?

23.5 (b) What is the relative rate of sedimentation for two spherical particles with densities 1.10 g cm^{-3} and 1.18 g cm^{-3} and which differ in radius by a factor of 8.4, the former being the larger?

23.6 (a) Find the drift speed of a particle of radius 20 μm and density 1750 kg m^{-3} which is settling from suspension in water (density 1000 kg m^{-3}) under the influence of gravity alone. The viscosity of water is 8.9×10^{-4} kg m^{-1} s^{-1}.

23.6 (b) Find the drift speed of a particle of radius 15.5 μm and density 1250 kg m^{-3} which is settling from suspension in water (density 1000 kg m^{-3}) under the influence of gravity alone. The viscosity of water is 8.9×10^{-4} kg m^{-1} s^{-1}.

23.7 (a) Human haemoglobin has a specific volume of 0.749×10^{-3} m^3 kg^{-1}, a sedimentation constant of 4.48 Sv, and a diffusion coefficient of 6.9×10^{-11} m^2 s^{-1}. Determine its molar mass from this information.

23.7 (b) A synthetic polymer has a specific volume of 8.01×10^{-4} m^3 kg^{-1}, a sedimentation constant of 7.46 Sv, and a diffusion coefficient of 7.72×10^{-11} m^2 s^{-1}. Determine its molar mass from this information.

23.8 (a) At 20°C the diffusion coefficient of a macromolecule is found to be 8.3×10^{-11} m^2 s^{-1}. Its sedimentation constant is 3.2 Sv

in a solution of density 1.06 g cm^{-3}. The specific volume of the macromolecule is 0.656 cm^3 g^{-1}. Determine the molar mass of the macromolecule.

23.8 (b) At 20°C the diffusion coefficient of a macromolecule is found to be 7.9×10^{-11} m^2 s^{-1}. Its sedimentation constant is 5.1 Sv in a solution of density 997 kg m^{-3}. The specific volume of the macromolecule is 0.721 cm^3 g^{-1}. Determine the molar mass of the macromolecule.

23.9 (a) A solution consists of solvent, 30 per cent by mass of a dimer with $M = 30$ kg mol^{-1} and its monomer. What average molar mass would be obtained from measurement of: (a) osmotic pressure, (b) light scattering?

23.9 (b) A solution consists of 25 per cent by mass of a trimer with $M = 22$ kg mol^{-1} and its monomer. What average molar mass would be obtained by measurement of: (a) osmotic pressure, (b) light scattering?

23.10 (a) A polyelectrolyte Na$_{20}$P with $M = 100$ kg mol^{-1} at a concentration 1.00 g/(100 cm^3) was equilibrated in the presence of 0.0010 M NaCl(aq) (that is, $[Na^+]_R = 0.0010$ mol L^{-1}). What is the value of $[Na^+]_L$ at equilibrium?

23.10 (b) A polyelectrolyte K$_{15}$P with $M = 98.0$ kg mol^{-1} at a concentration 2.00 g/(100 cm^3) was equilibrated in the presence of 0.0015 M KCl(aq) (that is, $[K^+]_R = 0.0010$ mol L^{-1}). What is the value of $[K^+]_L$ at equilibrium?

23.11 (a) At the start of a membrane equilibrium experiment, the first compartment contains 1.00 L of solution with an NaX concentration of 0.100 mol L^{-1}, where X$^-$ cannot pass through the membrane. The second compartment has 2.00 L of 0.030 M NaCl(aq). Find the concentration of Cl$^-$ ions in the first compartment after equilibrium is established.

23.11 (b) At the start of a membrane equilibrium experiment, the first compartment contains 1.00 L of solution with a KX concentration of 0.150 mol L^{-1}, where X$^-$ cannot pass through the membrane. The second compartment has 2.00 L of 0.045 M KCl(aq). Find the concentration of Cl$^-$ ions in the first compartment after equilibrium is established.

23.12 (a) The data from a sedimentation equilibrium experiment performed at 300 K on a macromolecular solute in aqueous solution show that a graph of $\ln c$ against r^2 is a straight line with a slope of 729 cm^{-2}. The rotational rate of the centrifuge was 50 000 r.p.m. The specific volume of the solute is 0.61 cm^3 g^{-1}. Calculate the molar mass of the solute.

23.12 (b) The data from a sedimentation equilibrium experiment performed at 293 K on a macromolecular solute in aqueous solution show that a graph of $\ln c$ against $(r/\text{cm})^2$ is a straight line with a slope of 821. The rotation rate of the centrifuge was 1080 Hz. The specific volume of the solute is 7.2×10^{-4} m^3 kg^{-1}. Calculate the molar mass of the solute.

23.13 (a) Calculate the radial acceleration (as so many g) in a cell placed at 6.0 cm from the centre of rotation in an ultracentrifuge operating at 80 000 r.p.m.

23.13 (b) Calculate the radial acceleration (as so many g) in a cell placed at 5.50 cm from the centre of rotation in an ultracentrifuge operating at 1.32 kHz.

23.14 (a) Cotton consists of the polymer cellulose, which is a linear chain of glucose molecules. The chains are held together by hydrogen bonding. When a cotton shirt is ironed, it is first moistened, then heated under pressure. Explain this process.

23.14 (b) Sections of the solid fuel rocket boosters of the space shuttle *Challenger* were sealed together with O-ring rubber seals of circumference 11 m. These seals failed at 0°C, a temperature well above the crystallization temperature of the rubber. Speculate on why the failure occurred.

Problems

Numerical problems

23.1 The concentration dependence of the osmotic pressure of solutions of a macromolecule at 20°C was found to be as follows:

$c/(\text{g L}^{-1})$	1.21	2.72	5.08	6.60
Π/Pa	134	321	655	898

Determine the molar mass of the macromolecule and the osmotic virial coefficient.

23.2 The osmotic pressure of a fraction of poly(vinyl chloride) in a ketone solvent was measured at 25°C. The density of the solvent (which is virtually equal to the density of the solution) was 0.798 g cm^{-3}. Calculate the molar mass and the osmotic virial coefficient, B, of the fraction from the following data:

$c/(\text{g}/10^2 \text{ cm}^3)$	0.200	0.400	0.600	0.088	1.000
h/cm	0.48	1.2	1.86	2.76	3.88

23.3 The concentration dependence of the viscosity of a polymer solution is found to be as follows:

$c/(\text{g L}^{-1})$	1.32	2.89	5.73	9.17
$\eta/(\text{g m}^{-1}\text{s}^{-1})$	1.08	1.20	1.42	1.73

The viscosity of the solvent is 0.985 g m^{-1} s^{-1}. What is the intrinsic viscosity of the polymer?

23.4 In a sedimentation experiment the position of the boundary as a function of time was found to be as follows:

t/min	15.5	29.1	36.4	58.2
r/cm	5.05	5.09	5.12	5.19

The rotation rate of the centrifuge was 45 000 r.p.m. Calculate the sedimentation constant of the solute.

23.5 In an ultracentrifuge experiment at 20°C on bovine serum albumin the following data were obtained: $\rho = 1.001$ g cm^{-3}, $v_s = 1.112$ cm^3 g^{-1}, $\omega/2\pi = 322$ Hz,

r/cm	5.0	5.1	5.2	5.3	5.4
$c/(\text{mg cm}^{-3})$	0.536	0.284	0.148	0.077	0.039

Evaluate the molar mass of the sample.

23.6 Calculate the speed of operation (in r.p.m.) of an ultracentrifuge needed to obtain a readily measurable concentration gradient in a sedimentation equilibrium experiment. Take that gradient to be a concentration at the bottom of the cell about five times greater that at the top. Use $r_{\text{top}} = 5.0$ cm, $r_{\text{bottom}} = 7.0$ cm, $M \approx 10^5$ g mol^{-1}, $\rho v_s \approx 0.75$, $T = 298$ K.

23.7 At the start of a Donnan equilibrium experiment, the first compartment contains 2.00 L of solution which is 0.015 M in the polyelectrolyte $Na_2P(aq)$ and 0.010 M in NaCl(aq). The second compartment has 2.00 L of solution which is 0.0050 M in NaCl(aq). What is the potential difference across the membrane arising from the Na$^+$ ion concentration difference at 300 K?

23.8 Investigation of the composition of the solutions used to study the osmotic pressure due to a polyelectrolyte with $\nu = 20$ showed that at equilibrium the concentrations corresponded to $[\text{Cl}^-] \approx 0.020$ mol L^{-1}. Calculate the osmotic virial coefficient for $\nu = 20$. Does it dominate the effect of excluded volume?

23.9 Sedimentation studies on haemoglobin in water gave a sedimentation constant $S = 4.5$ Sv at 20°C. The diffusion coefficient is 6.3×10^{-11} m^2 s^{-1} at the same temperature. Calculate the molar mass of haemoglobin using $v_s = 0.75$ cm^3 g^{-1} for its partial specific volume and $\rho = 0.998$ g cm^{-3} for the density of the solution. Estimate the effective radius of the haemoglobin molecule given that the viscosity of the solution is 1.00×10^{-3} kg m^{-1} s^{-1}.

23.10 The times of flow of dilute solutions of polystyrene in benzene through a viscometer at 25°C are given in the table below. From these data, calculate the molar mass of the polystyrene samples. Since the solutions are dilute, assume that the densities of the solutions are the same as those of pure benzene. $\eta(\text{benzene}) = 0.601 \times 10^{-3}$ kg m^{-1} s^{-1} (0.601 cP) at 25°C.

$c/(\text{g L}^{-1})$	0.000	2.22	5.00	8.00	10.00
t/s	208.2	248.1	303.4	371.8	421.3

23.11 The rate of sedimentation of a recently isolated protein was monitored at 20°C and with a rotor speed of 50000 r.p.m. The boundary receded as follows:

t/s	0	300	600	900	1200	1500	1800
r/cm	6.127	6.153	6.179	6.206	6.232	6.258	6.284

Calculate the sedimentation constant and the molar mass of the protein on the basis that its partial specific volume is 0.728 cm^3 g^{-1} and its diffusion coefficient is 7.62×10^{-11} m^2 s^{-1} at 20°C, the density of the solution then being 0.9981 g cm^{-3}. Suggest a shape for the protein given that the viscosity of the solution is 1.00×10^{-3} kg m^{-1} s^{-1} at 20°C.

23.12 The viscosities of solutions of polyisobutylene in benzene were measured at 24°C (the θ temperature for the system) with the following results:

$c/(\text{g}/10^2\,\text{cm}^3)$	0	0.2	0.4	0.6	0.8	1.0
$\eta/(10^{-3}\,\text{kg m}^{-1}\,\text{s}^{-1})$	0.647	0.690	0.733	0.777	0.821	0.865

Use the information in Table 23.3 to deduce the molar mass of the polymer.

23.13 Evaluate the radius of gyration, R_g, of (a) a solid sphere of radius a, (b) a long straight rod of radius a and length l. Show that, in the case of a solid sphere of specific volume v_s,

$$R_g/\text{nm} \approx 0.056902 \times \{(v_s/\text{cm}^3\,\text{g}^{-1})(M/\text{g mol}^{-1})\}^{1/3}$$

Evaluate R_g for a species with $M = 100$ kg mol^{-1}, $v_s = 0.750$ cm^3 g^{-1}, and, in the case of the rod, of radius 0.50 nm.

23.14 Use the information below and the expression for R_g of a solid sphere quoted in the previous problem, to classify the species below as globular or rod-like.

	$M/(\text{g mol}^{-1})$	$v_s/(\text{cm}^3\,\text{g}^{-1})$	R_g/nm
Serum albumin	66×10^3	0.752	2.98
Bushy stunt virus	10.6×10^6	0.741	12.0
DNA	4×10^6	0.556	117.0

23.15 In formamide as solvent, poly(γ-benzyl-L-glutamate) is found by light scattering experiments to have a radius of gyration proportional to M; in contrast, polystyrene in butanone has R_g proportional to $M^{1/2}$. Present arguments to show that the first polymer is a rigid rod, while the second is a random coil.

23.16 The structures of crystalline macromolecules may be determined by X-ray diffraction techniques by methods similar to those for smaller molecules. Fully crystalline polyethylene has its chains aligned in an orthorhombic unit cell of dimensions 740 pm × 493 pm × 253 pm. There are two repeating CH_2CH_2 units per unit cell. Calculate the theoretical density of fully crystalline polyethylene. The actual density ranges from 0.92 to 0.95 g cm^{-3}.

Theoretical problems

23.17 A polymerization process produced a Gaussian distribution of polymers in the sense that the proportion of molecules having a molar mass in the range M to $M + dM$ was proportional to $e^{-(M-\bar{M})^2/2\gamma}dM$. What is the number-average molar mass when the distribution is narrow?

23.18 Consider the thermodynamic description of stretching rubber. The observables are the tension, t, and length, l (the analogues of p and V for gases). Because $dw = t\,dl$, the basic equation is $dU = T\,dS + t\,dl$. (The term pdV is supposed negligible throughout.) If $G = U - TS - tl$, find expressions for dG and dA, and deduce the Maxwell relations

$$\left(\frac{\partial S}{\partial l}\right)_T = -\left(\frac{\partial t}{\partial T}\right)_l \qquad \left(\frac{\partial S}{\partial t}\right)_T = \left(\frac{\partial l}{\partial T}\right)_t$$

Go on to deduce the equation of state for rubber,

$$\left(\frac{\partial U}{\partial l}\right)_T = t - \left(\frac{\partial t}{\partial T}\right)_l$$

23.19 On the assumption that the tension required to keep a sample at a constant length is proportional to the temperature ($t = aT$, the analogue of $p \propto T$), show that the tension can be ascribed to the dependence of the entropy on the length of the sample. Account for this result in terms of the molecular nature of the sample.

23.20 Radius of gyration is defined in eqn 26. Show that an equivalent definition is that R_g is the average root mean square distance of the atoms or groups (all assumed to be of the same mass), that is, that $R_g^2 = (1/N)\sum_j R_j^2$, where R_j is the distance of atom j from the centre of mass.

23.21 Use eqn 34 to deduce expressions for (a) the root mean square separation of the ends of the chain, (b) the mean separation of the ends, and (c) their most probable separation. Evaluate these three quantities for a fully flexible chain with $N = 4000$ and $l = 154$ pm.

Additional problems supplied by Carmen Giunta and Charles Trapp

23.22 Polystyrene in cyclohexane at 34.5°C forms a θ solution, with an intrinsic viscosity related to the molar mass by $[\eta] = KM^a$. The following data on polystyrene in cyclohexane are taken from L.J. Fetters, N. Hadjichristidis, J.S. Lindner, and J.W. Mays (*J. Phys. Chem. Ref. Data* **23**, 619 (1994)).

$M/(\text{kg mol}^{-1})$	10.0	19.8	106	249	359
$[\eta]/(\text{cm}^3\,\text{g}^{-1})$	8.90	11.9	28.1	44.0	51.2
$M/(\text{kg mol}^{-1})$	860	1800	5470	9720	56 800
$[\eta]/(\text{cm}^3\,\text{g}^{-1})$	77.6	113.9	195	275	667

Determine the parameters K and a. What is the molar mass of a polystyrene that forms a θ solution in cyclohexane with $[\eta] = 100$ cm^3 g^{-1}?

23.23 Polymer scientists often report their data in rather strange units. For example, in the determination of molar masses of polymers in solution by osmometry, osmotic pressures are often reported in grams per square centimetre (g cm^{-2}) and concentrations in grams per cubic centimetre (g cm^{-3}). (a) With these choices of units, what would be the units of R in the van't Hoff equation? (b) The data in the table below on the concentration dependence of the osmotic pressure of polyisobutene in chlorobenzene at 25°C have been adapted from J. Leonard and H. Daoust (*J. Polymer Sci.* **57**, 53 (1962)). From these data, determine the number average molar mass of polyisobutene by plotting Π/c against c. (c) Theta solvents are solvents for which the second osmotic virial coefficient is zero; for 'poor' solvents the plot is linear and for good solvents the plot is nonlinear. From your plot, how would you classify chlorobenzene as a solvent for polyisobutene? Rationalize the result in terms of the molecular structure of polymer and solvent. (d) Determine the second and third osmotic virial coefficients by fitting the curve to the virial form of the osmotic pressure equation. (e) Experimentally, it is often found that the virial expansion can be represented as

$$\frac{\Pi}{c} = \frac{RT}{M}\left(1 + B'c + gB'^2 c'^2 + \cdots\right)$$

and in good solvents, the parameter g is often about 0.25. With terms beyond the second power ignored, obtain an equation for $(\Pi/c)^{1/2}$ and plot this quantity against c. Determine the second and third virial coefficients from this plot and compare to the values from the first plot. Does this plot confirm the assumed value of g?

$10^{-2}(\Pi/c)/$ $(\text{g cm}^{-2}/\text{g cm}^{-3})$	2.6	2.9	3.6	4.3	6.0	12.0
$c/(\text{g cm}^{-3})$	0.0050	0.010	0.020	0.033	0.057	0.10

$10^{-2}(\Pi/c)/$ $(\text{g cm}^{-2}/\text{g cm}^{-3})$	19.0	31.0	38.0	52	63
$c/(\text{g cm}^{-3})$	0.145	0.195	0.245	0.27	0.29

23.24 A manufacturer of polystyrene beads claims that they have an average molar mass of 250 kg mol^{-1}. Solutions of these beads are studied by a physical chemistry student by dilute solution viscometry with an Ostwald viscometer in both the 'good' solvent toluene and the theta solvent cyclohexane. The drainage times, t_D, as a function of concentration for the two solvents are given in the table below. (a) Fit the data to the virial equation for viscosity,

$$\eta = \eta^*(1 + [\eta]c + k'[\eta]^2 c^2 + \cdots)$$

where k' is called the *Huggins constant* and is typically in the range 0.35–0.40. From the fit, determine the intrinsic viscosity and the Huggins constant. (b) Use the empirical Mark-Kuhn-Houwink-Sakurada equation (eqn 21) to determine the molar mass of polystyrene in the two solvents. For theta solvents, $a = 0.5$ and $K = 8.2 \times 10^{-5}$ L g^{-1} for cyclohexane; for the good solvent toluene $a = 0.72$ and $K = 1.15 \times 10^{-5}$ L g^{-1}. (c) According to a general theory proposed by Kirkwood and Riseman, the root mean square end-to-end distance of a polymer chain in solution is related to $[\eta]$ by $[\eta] = \Phi\langle r^2\rangle^{3/2}/M$, where Φ is a universal constant with the value 2.84×10^{26} when $[\eta]$ is expressed in litres per gram and the distance is in metres. Calculate this quantity for each solvent. (d) From the molar masses calculate the average number of styrene ($C_6H_5CH\!=\!CH_2$) monomer units, $\langle n\rangle$. (e) Calculate the length of a fully stretched, planar zigzag configuration, taking the C-C distance as 154 pm and the CCC bond angle to be 109°. (f) Use eqn 39 to calculate the radius of gyration, R_g. Also calculate $\langle r^2\rangle^{1/2} = n^{1/2}$. Compare this result with that predicted by the Kirkwood-Riseman theory: which gives the better fit? (g) Compare your values for M to the results of Problem 23.23. Is there any reason why they should or should not agree? Is the manufacturer's claim valid?

$c/(\text{g L}^{-1}\text{toluene})$	0	1.0	3.0	5.0
t_D/s	8.37	9.11	10.72	12.52
$c/(\text{g L}^{-1}\text{cyclohexane})$	0	1.0	1.5	2.0
t_D/s	8.32	8.67	8.85	9.03

23.25 K. Sato, F.R. Eirich, and J.E. Mark (*J. Polym. Sci., Polym. Phys.* **14**, 619 (1976)) have reported the data in the table below for the osmotic pressures of polychloroprene ($\rho = 1.25$ g cm^{-3}) in toluene ($\rho = 0.858$ g cm^{-3}) at 30°C. Determine the molar mass of polychloroprene and its second osmotic virial coefficient.

$c/(\text{mg cm}^{-3})$	1.33	2.10	4.52	7.18	9.87
$\Pi/(\text{N m}^{-2})$	30	51	132	246	390

23.26 Standard polystyrene solutions of known average molar masses continue to be used for the calibration of many methods of characterizing polymer solutions. M. Kolinsky and J. Janca (*J. Polym. Sci., Polym. Chem.* **12**, 1181 (1974)) studied polystyrene in tetrahydrofuran (THF) for use in calibrating a gel permeation chromatograph. Their results for the intrinsic viscosity, $[\eta]$, as a function of average molar mass at 25°C are given in the table below. (a) Obtain the Mark-Houwink constants that fit these data. (b) Compare your values to those in Table 23.3 and Example 23.5. How might you explain the differences?

$M_v/(10^3 \text{ g mol}^{-1})$	5.0	10.3	19.85	51	98.2	173	411	867
$[\eta]/(\text{cm}^3\text{ g}^{-1})$	5.2	8.8	14.0	27.6	43.6	67.0	125.0	206.7

23.27 There is much recent interest in electronically conducting polymers and the determination of their average molar masses is an important part of their characterization. S. Holdcroft (*J. Polym. Sci., Polym. Phys.* **29**, 1585 (1991)) has determined the molar mases and Mark-Houwink constants for the electronically conducting polymer, poly(3-hexylthiophene) (P3HT) in tetrahydrofuran (THF) at 25°C by methods similar to those used for nonconducting polymers. The values for molar mass and intrinsic viscosity in the table below are adapted from their data. Determine the constants in the Mark-Kuhn-Houwink-Sakurada equation from these results and compare to the values obtained in your solution to Problem 23.26.

$M_v/(10^3 \text{ g mol}^{-1})$	3.8	11.1	15.3	58.8
$[\eta]/(\text{cm}^3\text{ g}^{-1})$	6.23	17.44	23.73	85.28

23.28 A problem arises in the use of the Svedberg equation (eqn 16) for the determination of the molar masses of macromolecules due to the fact that values of S and D depend upon concentration. Consequently, accurate values for M_w must be obtained by extrapolation of the data to infinite dilution by using a virial expansion in the form

$$\frac{bD}{SRT} = \frac{1}{M_w}\left(1 + 2B'c + 3gB'^2 c^2 + \cdots\right)$$

where g is the parameter introduced in Problem 23.23. W.J. Closs, B.R. Jennings, and H.G. Gerrard (*Eur. Polymer J.* **4**, 639 (1968)) reported the data in the table below for polystyrene in cyclohexane at 35°C. The density of cyclohexane at this temperature, which can also be assumed to be that of the solution, is 0.765 g cm^{-3}, and the partial specific volume of polystyrene is 0.93 cm^3 g^{-1}. The dependence of the diffusion constant for these solutions has been determined empirically by T.A. King, A. Knox, W.I. Lee, and J.D.G. McAdam (*Polymer* **14**, 151 (1973)) to be given by the relation $D/(\text{cm}^2 \text{ s}^{-1}) = 1.3 \times 10^{-4}(M_w/(\text{g mol}^{-1}))^{-0.497}$. Determine the molar mass of polystyrene in cyclohexane and the second virial coefficient, B'. Compare the molar mass obtained here to that calculated in Problem 23.24. Is there any reason for them to be the same?

$c/(\text{mg cm}^{-3})$	2.0	3.0	4.0	5.0	6.0	7.0
$S/(10^{-13}\text{ s})$	14.8	13.9	13.1	12.4	11.8	11.2

MicroProjects Part 2:

Prepared by M. Cady and C. A. Trapp

2.1 Black–body radiation and the greenhouse effect

The experimentally observed average temperature of the Earth's surface is 288.16 K. This temperature is maintained in a steady state through an energy balance between solar radiation absorbed by the Earth and black-body radiation which is emitted by the Earth and lost to space. Energy balances of this type are often discussed in terms of energy flux, J, the energy passing through an area in an interval divided by the area and the duration of the interval and expressed in watts per square metre ($W\,m^{-2}$).

(a) Prove that $J = \frac{1}{4}c\mathcal{E}$, where $\mathcal{E}$ is the isotropic black-body energy density (eqn 11.5). *Hint*. Examine the radiation passing through area A in the time; use spherical coordinates centred on A and recognize that only volume elements within the hemisphere of radius c contribute to the flux through A. Also, determine $f(\tilde{\nu})$, where $=f(\tilde{\nu})\,d\tilde{\nu}$ and prove that the Stefan–Boltzmann constant is given by eqn 11.6. Use $f(\tilde{\nu})$ to demonstrate graphically that the Earth's black-body emissions are in the infrared.

(b) Consider an atmospheric model consisting of atmospheric nitrogen and oxygen only. Can these gases absorb any of Earth's black-body emissions? Why? Determine the value of the Earth's surface temperature that is predicted by this model. It is found experimentally that the solar energy flux at the edge of the Earth's atmosphere is $0.1353\ W\,cm^{-2}$ and that the fraction of the solar radiation scattered by gases and clouds of the atmosphere (the albedo) is 0.29. Consider that the magnitude of the solar radiation absorbed by the Earth equals the disk area of the Earth times the fraction of unscattered solar radiation times the solar radiation flux. The difference between the experimental value of the Earth's temperature and the temperature predicted by this model is due to the so-called greenhouse effect.

(c) Now, consider an atmospheric model consisting of nitrogen, oxygen, some water vapour, and some carbon dioxide. Why is it that water and carbon dioxide are able to absorb some of the Earth's black-body radiation? Which vibrational modes are responsible for this absorption? Water vapour shows strong absorption between $1300\ cm^{-1}$ and $1900\ cm^{-1}$ and also between $3550\ cm^{-1}$ and $3900\ cm^{-1}$. Carbon dioxide shows strong absorption between $500\ cm^{-1}$ and $725\ cm^{-1}$ and also between $2250\ cm^{-1}$ and $2400\ cm^{-1}$. Why are these bands so broad? Assume that these gases absorb all radiation falling within these bands and calculate the average surface temperature predicted by this atmospheric model. What percentage of the greenhouse effect is explained by the presence of atmospheric water and carbon dioxide?

2.2 One-dimensional tunnelling

Consider the one-dimensional space in which a particle can experience one of three potentials depending upon its position. They are: $V = 0$ for $-\infty < x \leq 0$, $V = V_2$ for $0 \leq x \leq L$, and $V = V_3$ for $L \leq x < \infty$.

The particle wavefunction is to have both a component $e^{ik_1 x}$ that is incident upon the barrier V_2 and a reflected component $e^{-ik_1 x}$ in region 1 $(-\infty < x \leq 0)$. In region 3 the wavefunction has only a forward component, $e^{ik_3 x}$, which represents a particle which has traversed the barrier. The energy of the particle, E, is somewhere in the range of the $V_2 > E > V_3$. The transmission probability, T, is the ratio of the square modulus of the region 3 amplitude to the square modulus of the incident amplitude.

(a) Base your calculation on the continuity of the amplitudes and the slope of the wavefunction at the locations of the zone boundaries and derive a general equation for T.

(b) Show that the general equation for T reduces to eqn 12.27 in the high, wide barrier limit when $V_1 = V_3 = 0$.

(c) Draw graphs of the probability of proton tunnelling when $V_3 = 0$, $L = 50$ pm, and $E = 10\ kJ\,mol^{-1}$ in the barrier range $E < V_2 < 2E$.

2.3 Hydrogenic orbitals

Explicit expressions for hydrogenic orbitals are given in Tables 13.1 and 13.2.

(a) Verify both that the $3p_x$ orbital is normalized (to 1) and that $3p_x$ and $3d_{xy}$ are mutually orthogonal.

(b) Determine the positions of both the radial nodes and nodal planes of the $3s$, $3p_x$, and $3d_{xy}$ orbitals.

(c) Determine the mean radius of the $3s$ orbital.

(d) Draw a graph of the radial distribution function for the three orbitals (of part (b)) and discuss the significance of the graphs for interpreting the properties of many-electron atoms.

(e) Create both xy-plane polar plots and boundary surface plots for these orbitals. Construct the boundary plots so that the distance from the origin to the surface is the absolute value of the angular part of the wavefunction. Compare the s, p, and d boundary surface plots with that of an f-orbital, for example, $\psi_f \propto x(5z^2 - r^2) \propto \sin\theta(5\cos^2\theta - 1)\cos\phi$.

2.4 A partition function paradox

Consider the electronic partition function of a perfect atomic hydrogen gas at a density of $1.99 \times 10^{-4}\ kg\,m^{-3}$ and 5780 K. These are the mean conditions within the Sun's photosphere, the surface layer of the Sun that is about 190 km thick.

(a) Show that this partition function, which involves a sum over an infinite number of quantum states that are solutions for the isolated atomic hydrogen atom, is infinite.

(b) Develop a theoretical argument for truncating the sum and estimate the maximum number of quantum states that contribute to the sum.

(c) Calculate the equilibrium probability that an atomic hydrogen electron is in each quantum state. Are there any general implications concerning electronic states that will be observed for other atoms and molecules? Is it wise to apply these calculations in the study of the Sun's photosphere?

2.5 Ammonia inversion

(a) Use quantum mechanical concepts to explain both the origin of ammonia microwave absorptions at 0.8 cm^{-1} and 36 cm^{-1} and an infrared absorption at 1000 cm^{-1}. All are associated with the umbrella-like inversion of ammonia (see the illustration); the absorption at 1000 cm^{-1} is the lowest-energy infrared transition associated with the inversion.

(b) Prove that

$$C = 2B^{1/2} \qquad D = \frac{s_e}{\text{arccosh } B^{1/4}}$$

for an assumed inversion potential of the form

$$V(s) = A\{1 + B \operatorname{sech}^4(s/D) - C \operatorname{sech}^2(s/D)\}$$

where s is the perpendicular distance of the nitrogen from the plane of the three hydrogens. At $s = s_e$ the potential has a minimum value equal to zero and the parameters A, B, C, and D are all positive.

(c) Use the absorption lines and $s_e = 0.381$ pm to determine both the potential parameters and the value of the inversion potential barrier height, ε. Simplify the computations by assuming that the inversion mode vibrational wavefunctions are adequately described by harmonic oscillator wavefunctions. Also assume that the distances between hydrogen atoms remain constant during inversion. This approximation simplifies the effective mass to: $m_{\text{eff}} = 3m_H m_N/(3m_H + m_N)$.

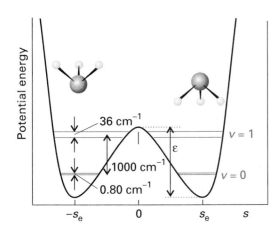

2.6 Vibration–rotation spectra and molecular constants

High-resolution absorption lines of three infrared vibrational bands of carbon monoxide are summarized in Table 1.

(a) Make the J and m assignments for each line, using J to represent the initial rotational state and defining $m = -J$ for P branches and $m = J + 1$ for R branches. Derive an equation for the dependence of the spectral lines upon m including terms for anharmonicity, centrifugal distortion, and rotation–vibration coupling. The rotation–vibration perturbation of the energy of the state (v,J) is equal to $-a(v + \frac{1}{2})J(J + 1)$, where a is the molecular rotation–vibration coupling constant.

(b) Determine all molecular constants associated with the absorption line by performing an appropriate regression analysis of Table 1 data.

Table 1 Wavenumbers ($\tilde{v}/\text{cm}^{-1}$) of the infrared absorption lines for vibrational bands of carbon monoxide. (From N. Mina-Camilde, C.I. Manzariares, and J.F. Caballero, *J. Chem. Educ.* **73**, 804 (1996).)

1←0		2←0		3←0	
2059.6	2146.9	4169.8	4263.2	6253.1	6353.8
2063.9	2150.7	4174.6	4267.2	6259.3	6357.2
2068.8	2154.1	4180.0	4270.6	6264.6	6361.0
2073.1	2158.0	4184.8	4274.4	6270.4	6364.4
2077.4	2161.8	4190.1	4277.8	6276.2	6367.8
2081.8	2165.2	4194.9	4281.2	6281.5	6370.7
2086.1	2169.0	4199.7	4284.6	6286.8	6374.1
2090.5	2172.4	4204.1	4287.9	6291.6	6377.0
2094.3	2175.8	4208.9	4290.8	6296.9	6379.9
2098.7	2179.2	4213.7	4294.2	6303.7	6382.7
2103.0	2183.0	4218.0	4297.1	6307.1	6385.2
2106.8	2186.4	4222.4	4300.5	6311.4	6388.0
2111.2	2189.8	4226.7	4303.4	6316.2	6390.5
2115.5	2193.1	4231.1	4306.6	6321.0	6392.9
2119.4	2196.5	4235.4	4308.7	6325.4	6394.8
2123.2	2199.4	4239.7	4311.6	6329.7	6397.2
2127.1	2202.8	4244.1	4314.0	6334.0	6399.1
2131.4	2206.2	4247.9	4316.9	6337.9	6401.1
2135.3	2209.1	4251.8	4319.3	6342.2	6403.0
2139.2	2212.4	4255.6	4321.7	6346.1	6404.6

(c) Determine each of the following: the moment of inertia and bond length defined by the harmonic oscillator and rigid rotor models, I_e and R_e; the moments of inertia and bond lengths for the $v = 0, 1, 2, 3$ vibration states (define these by analogy to the harmonic oscillator and rigid rotor models); the depth of the Morse potential; D_e (the 'spectroscopic dissociation energy'); and D_0 (the bond dissociation energy).

2.7 An IR absorption band of carbon dioxide

A mixture of carbon dioxide (2.1 per cent) and helium, at 1.00 bar and 298 K in a gas cell of length 10 cm has an IR absorption band centred at 2349 cm^{-1} with absorbances, $A(\tilde{\nu})$, described by:

$$A(\tilde{\nu}) = \frac{a_1}{1 + a_2(\tilde{\nu} - a_3)^2} + \frac{a_4}{1 + a_5(\tilde{\nu} - a_6)^2}$$

where the coefficients are $a_1 = 0.932$, $a_2 = 0.005050$ cm^2, $a_3 = 2333$ cm^{-1}, $a_4 = 1.504$, $a_5 = 0.01521$ cm^2, $a_6 = 2362$ cm^{-1}.

(a) Draw graphs of $A(\tilde{\nu})$ and $\varepsilon(\tilde{\nu})$. What is the origin of both the band and the band width? What are the allowed and forbidden transitions of this band?

(b) Calculate the transition wavenumbers and absorbances of the band with a simple harmonic oscillator-rigid rotor model and compare the result with the experimental spectra. The CO bond length is 116.2 pm.

(c) Within what height, h, is basically all the IR emission from the Earth in this band absorbed by atmospheric carbon dioxide? The mole fraction of CO_2 in the atmosphere is 3.3×10^{-4} and $T/K = 288 - 0.0065(h/m)$ below 10 km. Draw a surface plot of the atmospheric transmittance of the band as a function of both height and wavenumber.

2.8 σ and π bonding

Use the $2p_x$ and $2p_z$ hydrogenic atomic orbitals to construct simple LCAO descriptions of $2p\sigma$ and $2p\pi$ molecular orbitals.

(a) Make a probability density plot, and both surface and contour plots of the xz-plane amplitudes of the $2p_z\sigma$ and $2p_z\sigma^*$ molecular orbitals.

(b) Make surface and contour plots of the xz-plane amplitudes of the $2p_x\pi$ and $2p_x\pi^*$ molecular orbitals. Include plots for both of internuclear distances, R, of $10a_0$ and $3a_0$. Interpret the graphs, and describe why scientists are so interested in this graphical information.

2.9 Numerical analysis of the simple LCAO–MO description of H$_2^+$

The LCAO–MO constructed from normalized $1s$ hydrogenic wavefunctions centred on nuclei that are a distance R apart, and having g symmetry, does not describe the molecular hydrogen ion ground state accurately. It does provide insight into functional characteristics of wavefunctions, bonding, and numerical methods needed in quantum chemistry. The overlap, Coulomb, and resonance integrals of this LCAO–MO can be analytically evaluated to give the result in eqns 14.12 and 14.13. In this problem we evaluate the integrals numerically for S, j, and k and compare the results with the values determined from the analytically integrated forms of eqn 14.13.

(a) Use the LCAO-MO wavefunction and the H$_2^+$ hamiltonian to derive equations for the Coulomb and resonance integrals in terms of j and k; do not integrate j and k analytically. Evaluate the overlap, Coulomb, and resonance integrals numerically, and the total energy for the $1s\sigma_g$ MO in the range $a_0 < R < 4a_0$. Compare the results obtained through numerical integration with results obtained with the analytical equations.

(b) Use the results of the numerical integrations to draw a graph of the total energy, $E(R)$, and determine the minimum of total energy, the equilibrium internuclear distance, and the spectroscopic dissociation energy (D_e).

2.10 The variation method and H$_2^+$

A highly accurate description of the molecular hydrogen ion bond length, R_e, is provided by an MO constructed with the variation parameter η, within the LCAO-MO ground state consisting of two $1s$ hydrogenic orbitals centred upon nuclei A and B. With the nuclear separation R, the MO is

$$\psi = N\left(e^{-\eta r_A/a_0} + e^{-\eta r_B/a_0}\right) \qquad N = \left(\frac{\eta^3}{2\pi a_0^3(1+S)}\right)^{1/2}$$

(a) Use this wavefunction and the variation principle to determine η, R_e, the electronic energy (E_{el}), the minimum total energy (E), and D_e. The electronic Hamiltonian does not contain the nuclear repulsion term. Draw graphs of $\eta(R)$, $E_{el}(R)$, and $E(R)$. With this MO it is found that the electronic energy is the sum of the expectation value for kinetic energy ($\eta^2 F_1$) and the expectation value for electron potential energy (ηF_2):

$$E_{el} = \frac{e^2\{\eta^2 F_1(\omega) + \eta F_2(\omega)\}}{4\pi\varepsilon_0 a_0} \qquad \omega = \eta R/a_0$$

$$F_1(\omega) = \frac{1 + (1 + \omega - \frac{1}{3}\omega^2)e^{-\omega}}{2\{1 + S(\omega)\}}$$

$$F_2(\omega) = \frac{(1 + \omega)e^{-2\omega} - 1 - \omega - 2\omega(1 + \omega)e^{-\omega}}{\omega\{1 + S(\omega)\}}$$

$$S(\omega) = (1 + \omega + \frac{1}{3}\omega^2)e^{-\omega}$$

(b) Check numerically to determine whether or not the virial theorem is satisfied by this solution.

(c) Prove that the overlap integral, S, is correctly described by the above expression. The integration is facilitated with the ellipsoidal coordinates (μ, ν, ϕ) defined by the following relations:

$$R\mu = r_A + r_B \qquad r\nu = r_A - r_B \qquad d\tau = \frac{1}{8}R^3(\mu^2 - \nu^2)\,d\mu\,d\nu\,d\phi$$

with $1 \le \mu \le \infty$, $-1 \le \nu \le 1$, and $0 \le \phi \le 2\pi$.

2.11 Simple Hückel molecular orbitals

Solve the following in the context of simple Hückel theory.

(a) Prove that for an open chain of N conjugated carbons the characteristic polynomial of the secular determinant, $P_N(x)$, where $x = (\alpha - \beta)/\beta$, obeys the recurrence relation $P_N = xP_{N-1} - P_{N-2}$, with $P_1 = x$ and $P_0 = 1$.

(b) Determine a reasonable empirical estimate of the resonance integral for the homologous series consisting of ethene, butadiene, hexatriene, and octatetraene given that the $\pi^* \leftarrow \pi$ ultraviolet absorptions are at 61 500, 46 080, 39 750, and 32 900 cm^{-1}, respectively.

(c) Calculate the π-electron delocalization energy, E_{deloc}, of octatetraene where $E_{\mathrm{deloc}} = E_{\pi} - n(\alpha + \beta)$, where E_{π} is the total π-electron binding energy and n is the total number of π-electrons.

2.12 Equilibrium statistical thermodynamics

Treat carbon monoxide as a perfect gas and apply equilibrium statistical thermodynamics to the study of its properties, as specified below, in the temperature range 100–1000 K at 1 bar. $\nu = 2169.8$ cm^{-1},

$B = 1.931$ cm^{-1}, and $D_0 = 11.09$ eV; neglect anharmonicity and centrifugal distortion.

(a) Examine the probability distribution of molecules over available rotational and vibrational states.

(b) Explore numerically the differences, if any, between the rotational molecular partition function as calculated with the discrete energy distribution and that calculated with the classical, continuous energy distribution.

(c) Calculate the individual contributions to $U_{\mathrm{m}}(T) - U_{\mathrm{m}}(100 \text{ K})$, $C_{V,\mathrm{m}}(T)$, and $S_{\mathrm{m}}(T) - S_{\mathrm{m}}(100 \text{ K})$ made by the translational, rotational, and vibrational degrees of freedom.

Part 3 Change

Part 3 considers the processes by which change occurs. We prepare the ground for a discussion of the rates of reactions by considering the motion of molecules in gases and in liquids. Then we establish the precise meaning of reaction rate, and see how the overall rate, and the complex behaviour of some reactions, may be expressed in terms of elementary steps and the atomic events that take place when molecules meet. Characteristic physical and chemical events take place at surfaces, including catalysis, and we see how to describe them. A special type of surface is that of an electrode, and we shall see how to describe and understand the rate at which electrons are transferred between an electrode and species in solution.

24 Molecules in motion

One of the simplest types of molecular motion to describe is the random motion of molecules of a perfect gas. We see that the kinetic theory can be used to account for the rates at which molecules and energy migrate through gases and that simple expressions for the rates can be derived. Molecular mobility is particularly important in liquids, and we shall see a little of the structure of this phase and the motion of molecules in it. Another simple kind of motion is the largely uniform motion of ions in solution in the presence of an electric field. Molecular and ionic motion have common features and, by considering them from a more general viewpoint, we derive expressions that govern the migration of properties through matter. One of the most useful consequences of this general approach is the formulation of the diffusion equation, which is an equation that shows how matter and energy spread through media of various kinds. Finally, we build a simple model for all types of molecular motion, in which the molecules migrate in a series of small steps, and see that it accounts for many of the properties of migrating molecules in both gases and condensed phases.

The general approach we describe in this chapter provides techniques for discussing the motion of all kinds of particles in all kinds of fluids. We set the scene by considering a simple type of motion, that of molecules in a perfect gas, and go on to see that molecular motion in liquids shows a number of similarities.

Molecular motion in gases

In Section 1.3 we saw that the equilibrium properties of a gas can be understood in terms of the kinetic theory, which is based on a model of a gas in which the molecules are in ceaseless, random motion. Here we develop the kinetic theory to deal with gases that are not at internal equilibrium. In particular, we concentrate on the **transport properties** of a substance, its ability to transfer matter, energy, or some other property from one place to another. Four examples of transport properties are:

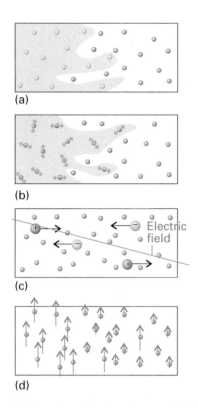

24.1 Four types of transport process: (a) diffusion, the spreading of one species into another; (b) thermal conduction, when molecules with different energies of thermal motion (represented by the arrows) spread into each others' regions; (c) electrical conduction, when ions migrate under the influence of an electric field; (d) viscosity, when molecules with different linear momenta (represented by the arrows) migrate.

Diffusion, the migration of matter down a concentration gradient.
Thermal conduction, the migration of energy down a temperature gradient.
Electric conduction, the migration of electric charge along a potential gradient.
Viscosity, the migration of linear momentum down a velocity gradient.

These processes are illustrated in Fig. 24.1. It is convenient to include **effusion**, the emergence of a gas from a container through a small hole, in the discussion.

We shall use two expressions derived in Chapter 1. One is for the mean free path, λ, of molecules in a gas:

$$\lambda = \frac{kT}{2^{1/2}\sigma p} \tag{1}°$$

where σ is the collision cross-section (this is eqn 1.33). The mean free path is independent of temperature in a container of constant volume because p is proportional to the temperature ($p = nRT/V$) and its variation cancels the T in the numerator. The second property is the mean speed, $\bar{c}$, of molecules of mass m and molar mass M:

$$\bar{c} = \left(\frac{8kT}{\pi m}\right)^{1/2} = \left(\frac{8RT}{\pi M}\right)^{1/2} \tag{2}°$$

This expression was derived in Example 1.6. The mean speed is proportional to $T^{1/2}$ and inversely proportional to $M^{1/2}$.

24.1 Collisions with walls and surfaces

The key to accounting for transport in the gas phase is the rate at which molecules strike an area (which may be an imaginary area embedded in the gas, or part of a real wall). The **collision flux**, Z_W, is the number of collisions with the area in a given time interval divided by the area and the duration of the interval. The **collision frequency**, the number of hits per second, is obtained by multiplication of the collision flux by the area of interest. We show in the *Justification* below that

$$Z_W = \frac{p}{(2\pi mkT)^{1/2}} \tag{3}°$$

When $p = 100$ kPa (1.00 bar) and $T = 300$ K, $Z_W \approx 3 \times 10^{23}$ cm^{-2} s^{-1}.

Justification 24.1

Consider a wall of area A perpendicular to the x-axis (Fig. 24.2). If a molecule has $v_x > 0$ (that is, it is travelling in the direction of positive x), then it will strike the wall within an interval Δt if it lies within a distance $v_x \Delta t$ of the wall. Therefore, all molecules in the volume $Av_x\Delta t$, and with positive x-component of velocities, will strike the wall in the interval Δt. The total number of collisions in this interval is therefore the volume $Av_x\Delta t$ multiplied by the number density, $\mathcal{N}$, of molecules. However, to take account of the presence of a range of velocities in the sample, we must sum the result over all the positive values of v_x weighted by the probability distribution of velocities (eqn 1.25):

$$\text{Number of collisions} = \mathcal{N}A\Delta t \int_0^\infty v_x f(v_x)\,dx$$

The collision flux is the number of collisions divided by A and Δt, so

$$Z_W = \mathcal{N}\int_0^\infty v_x f(v_x)\,dx$$

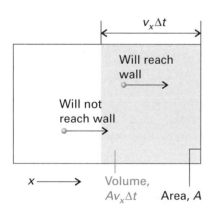

24.2 Only molecules within a distance $v_x\Delta t$ with $v_x > 0$ can reach the wall on the right in an interval Δt.

Then, using the velocity distribution in eqn 1.25,

$$\int_0^\infty v_x f(v_x)\, dv_x = \left(\frac{m}{2\pi kT}\right)^{1/2} \int_0^\infty v_x e^{-mv_x^2/2kT}\, dv_x = \left(\frac{kT}{2\pi m}\right)^{1/2}$$

Therefore,

$$Z_W = \mathcal{N}\left(\frac{kT}{2\pi m}\right)^{1/2} = \tfrac{1}{4}\bar{c}\mathcal{N} \qquad (4)°$$

Substitution of $\mathcal{N} = nN_A/V = p/kT$ gives eqn 3.

24.2 The rate of effusion

The essential empirical observations on effusion are summarized by **Graham's law of effusion**, which states that the rate of effusion is inversely proportional to the square root of the molar mass. The basis of this result is that, as remarked above, the mean speed of molecules is inversely proportional to $M^{1/2}$, so the rate at which they strike the area of the hole is similarly inversely proportional to $M^{1/2}$. However, by using the expression for the rate of collisions, we can obtain a more detailed expression for the rate of effusion and hence use effusion data more effectively.

When a gas at a pressure p and temperature T is separated from a vacuum by a small hole, the rate of escape of its molecules is equal to the rate at which they strike the area of the hole (which is given by eqn 3). Therefore, for a hole of area A_0,

$$\text{Rate of effusion} = Z_W A_0 = \frac{pA_0}{(2\pi mkT)^{1/2}} = \frac{pA_0 N_A}{(2\pi MRT)^{1/2}} \qquad (5)°$$

(In the last step we have used $R = N_A k$ and $M = mN_A$.) This rate is inversely proportional to $M^{1/2}$, in accord with Graham's law.

Example 24.1 Deducing the time dependence of the pressure inside an effusion oven

Derive an expression that shows how the pressure of a gas inside an effusion oven (a heated chamber with a small hole in one wall) varies with time if the oven is not replenished as the gas escapes.

Method The rate of effusion is proportional to the pressure of the gas in the container so, as gas effuses and the pressure falls, the rate of effusion will decrease. To find the explicit expression, set up a differential equation relating dp/dt to p, and then integrate it. The rate of effusion, as given by eqn 5, is the number of molecules that leave the container in a given interval divided by the duration of the interval. The first step is to relate the rate of change of pressure to the rate of change of number of molecules by using the perfect gas law in the form $pV = NkT$.

Answer The rate of change of pressure of a gas in a container at constant pressure and temperature is related to the rate of change of the number of molecules present by

$$\frac{dp}{dt} = \frac{kT}{V}\frac{dN}{dt}$$

The rate of change of the number of molecules is equal to the collision frequency with the hole, and that in turn is equal to the collision flux multiplied by the area of the hole:

$$\frac{dN}{dt} = -Z_W A_0 = -\frac{pA_0}{(2\pi mkT)^{1/2}}$$

Substitution of this expression into the one above gives

$$\frac{dp}{dt} = -\left(\frac{kT}{2\pi m}\right)^{1/2}\frac{pA_0}{V}$$

This expression integrates to

$$p = p_0 e^{-t/\tau} \qquad \tau = \left(\frac{2\pi m}{kT}\right)^{1/2}\frac{V}{A_0}$$

Comment The pressure falls exponentially towards zero; the decrease is faster the higher the temperature, the bigger the hole, and the lower the mass of the molecules.

- -

Self-test 24.1 Show that $t_{1/2}$, the time required for the pressure to decrease to half its initial value, is independent of the initial pressure.

$$[t_{1/2} = \tau \ln 2]$$

Equation 5 is the basis of the **Knudsen method** for the determination of the vapour pressures of liquids and solids, particularly of substances with very low vapour pressures. Thus, if the vapour pressure of a sample is p, and it is enclosed in a cavity with a small hole, then the rate of loss of mass from the container is proportional to p.

Example 24.2 Calculating the vapour pressure from a mass loss

Caesium (m.p. 29°C, b.p. 686°C) was introduced into a container and heated to 500°C. When a hole of diameter 0.50 mm was opened in the container for 100 s, a mass loss of 385 mg was measured. Calculate the vapour pressure of liquid caesium at 500°C.

Method The pressure of vapour is constant inside the container despite the effusion of atoms because the hot liquid metal replenishes the vapour. The rate of effusion is therefore constant, and given by eqn 5. To express the rate in terms of mass, the number of atoms that escape is multiplied by the mass of each atom.

Answer The mass loss Δm in an interval Δt is related to the collision flux by

$$\Delta m = Z_W A_0 m \Delta t$$

where A_0 is the area of the hole and m is the mass of one atom. It follows that

$$Z_W = \frac{\Delta m}{A_0 m \Delta t}$$

Because Z_W is related to the pressure by eqn 3, we can write

$$p = \left(\frac{2\pi RT}{M}\right)^{1/2}\frac{\Delta m}{A_0 \Delta t}$$

Because $M = 132.9\ \text{g mol}^{-1}$, substitution of the data gives $p = 11$ kPa (using $1\ \text{Pa} = 1\ \text{N m}^{-2} = 1\ \text{J m}^{-1}$), or 83 Torr.

- -

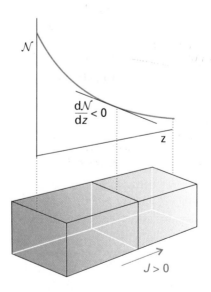

24.3 The flux of particles down a concentration gradient. Fick's first law states that the flux of matter (the number of particles passing through an imaginary window in a given interval divided by the area of the window and the length of the interval) is proportional to the density gradient at that point.

Self-test 24.2 How long would it take 1.0 g of Cs atoms to effuse out of the oven under the same conditions?

[260 s]

24.3 Migration down gradients

The rate of migration of a property is measured by its **flux**, J, the quantity of that property passing through a given area in a given time interval divided by the area and the duration of the interval. If matter is flowing (as in diffusion), we speak of a *matter flux* of so many molecules per square metre per second; if the property is energy (as in thermal conduction), then we speak of the *energy flux* and express it in joules per square metre per second, and so on.

Experimental observations on transport properties show that the flux of a property is usually proportional to the first derivative of some other related property. For example, the flux of matter diffusing parallel to the z-axis of a container is found to be proportional to the first derivative of the concentration:

$$J(\text{matter}) \propto \frac{d\mathcal{N}}{dz} \tag{6}$$

where $\mathcal{N}$ is the number density of particles with units number per metre cubed (m^{-3}). The SI units of J are number per metre squared per second ($\text{m}^{-2}\,\text{s}^{-1}$).The proportionality of the flux of matter to the concentration gradient is sometimes called **Fick's first law of diffusion**: the law implies that, if the concentration varies steeply with position, then diffusion will be fast. There is no net flux if the concentration is uniform ($d\mathcal{N}/dz = 0$). Similarly, the rate of thermal conduction (the flux of the energy associated with thermal motion) is found to be proportional to the temperature gradient:

$$J(\text{energy}) \propto \frac{dT}{dz} \tag{7}$$

The SI units of this flux are joules per metre squared per second ($\text{J}\,\text{m}^{-2}\,\text{s}^{-1}$).

A positive value of J signifies a flux towards positive z; a negative value of J signifies a flux towards negative z. Because matter flows down a concentration gradient, from high concentration to low concentration, J is positive if $d\mathcal{N}/dz$ is negative (Fig. 24.3). Therefore, the coefficient of proportionality in eqn 7 must be negative, and we write it $-D$:

$$J(\text{matter}) = -D\frac{d\mathcal{N}}{dz} \tag{8}$$

The constant D is called the **diffusion coefficient**; its SI units are metre squared per second ($\text{m}^2\,\text{s}^{-1}$). Energy migrates down a temperature gradient, and the same reasoning leads to

$$J(\text{energy}) = -\kappa\frac{dT}{dz} \tag{9}$$

where κ is the **coefficient of thermal conductivity**. The SI units of κ are joules per kelvin per metre per second ($\text{J}\,\text{K}^{-1}\,\text{m}^{-1}\,\text{s}^{-1}$). Some experimental values are given in Table 24.1.

To see the connection between the flux of momentum and the viscosity, consider a fluid in a state of **Newtonian flow**, which can be imagined as occurring by a series of layers moving past one another (Fig. 24.4). The layer next to the wall of the vessel is stationary, and the velocity of successive layers varies linearly with distance, z, from the wall. Molecules ceaselessly move between the layers and bring with them the x-component of linear momentum they possessed in their original layer. A layer is retarded by molecules arriving from a more slowly moving layer because they have a low momentum in the x-direction. A

Table 24.1* Transport properties of gases at 1 atm

	$\kappa/(\text{J}\,\text{K}^{-1}\,\text{m}^{-1}\,\text{s}^{-1})$	$\eta/\mu\text{P}$†	
	273 K	**273 K**	**293 K**
Ar	0.0163	210	223
CO_2	0.0145	136	147
He	0.1442	187	196
N_2	0.0240	166	176

*More values are given in the *Data section* at the end of this volume.

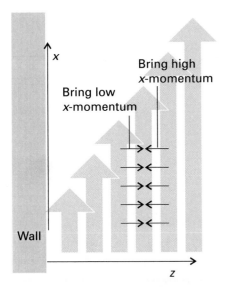

24.4 The viscosity of a gas arises from the transport of linear momentum. In this illustration the liquid is undergoing laminar flow, and particles bring their initial momentum when they enter a new layer. If they arrive with a high x-component of momentum they accelerate the layer; if they arrive with a low x-component of momentum they retard the layer.

layer is accelerated by molecules arriving from a more rapidly moving layer. We interpret the net retarding effect as the fluid's viscosity.

Because the retarding effect depends on the transfer of the x-component of linear momentum into the layer of interest, the viscosity depends on the flux of this x-component in the z-direction. The flux of the x-component of momentum is proportional to dv_x/dz because there is no net flux when all the layers move at the same velocity. We can therefore write

$$J(x\text{-component of momentum}) = -\eta\frac{dv_x}{dz} \tag{10}$$

The constant of proportionality, η, is the **coefficient of viscosity** (or simply 'the viscosity'). Its units are kilogram per metre per second ($\text{kg m}^{-1}\,\text{s}^{-1}$). Viscosities are often reported in poise (P), where $1\ \text{P} = 10^{-1}\ \text{kg m}^{-1}\,\text{s}^{-1}$. Some experimental values are given in Table 24.1.

24.4 Transport properties of a perfect gas

We shall now see how the kinetic theory can be used to justify Fick's law, and deduce the values of the transport coefficients of a perfect gas. These expressions show how transport properties vary with the conditions.

(a) Diffusion

As shown in the *Justification* below and summarized in Table 24.2, the kinetic theory leads to the result that, for a perfect gas,

$$D = \tfrac{1}{3}\lambda\bar{c} \tag{11}°$$

The mean free path, λ, decreases as the pressure is increased (Section 1.3c), so D decreases with increasing pressure and, as a result, the gas molecules diffuse more slowly. The mean speed, $\bar{c}$, increases with the temperature (Section 1.3a), so D also increases with temperature. As a result, molecules in a hot sample diffuse more quickly than those in a cool sample (for a given concentration gradient). Because the mean free path increases when the collision cross-section of the molecules decreases, the diffusion coefficient is greater for small molecules than for large molecules.

Justification 24.2

Consider the arrangement depicted in Fig. 24.5. On average, the molecules passing through the area A at $z = 0$ have travelled about one mean free path λ since their last collision. Therefore, the number density where they originated is $\mathcal{N}(z)$ evaluated at $z = -\lambda$. This number density is approximately[1]

$$\mathcal{N}(-\lambda) = \mathcal{N}(0) - \lambda\left(\frac{d\mathcal{N}}{dz}\right)_0 \tag{12}$$

where the subscript 0 indicates that the slope should be evaluated at $z = 0$. The average number of impacts on the imaginary window of area A_0 during an interval Δt is $Z_W A_0 \Delta t$, with $Z_W = \tfrac{1}{4}\mathcal{N}\bar{c}$ (eqn 4). Therefore, the flux from left to right, $J(L \rightarrow R)$, arising from the supply of molecules on the left, is

$$J(L \rightarrow R) = \frac{\tfrac{1}{4}A_0\mathcal{N}(-\lambda)\bar{c}\Delta t}{A_0\Delta t} = \tfrac{1}{4}\mathcal{N}(-\lambda)\bar{c} \tag{13a}$$

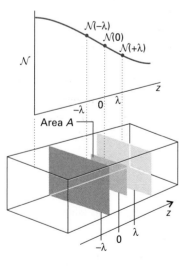

24.5 The calculation of the rate of diffusion of a gas considers the net flux of molecules through a plane of area A as a result of arrivals from on average a distance λ away in each direction, where λ is the mean free path.

---

1   This relation, and others like it that follow, is based on the Taylor expansion of a function, $f(x) = f(0) + (df/dx)_0 x + \cdots$, truncated after the second term.

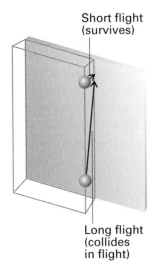

Short flight (survives)

Long flight (collides in flight)

24.6 One approximation ignored in the simple treatment is that some particles might make a long flight to the plane even though they are only a short perpendicular distance away, and therefore they have a higher chance of colliding during their journey.

Table 24.2 Transport properties of perfect gases

Property	Transported quantity	Simple kinetic theory	Units
Diffusion	Matter	$D = \frac{1}{3}\lambda\bar{c}$	$\mathrm{m^2\,s^{-1}}$
Thermal conductivity	Energy	$\kappa = \frac{1}{3}\lambda\bar{c}C_{V,m}[A]$ $= \dfrac{\bar{c}C_{V,m}}{(3\sqrt{2})\sigma N_A}$	$\mathrm{J\,K^{-1}\,m^{-1}\,s^{-1}}$
Viscosity	Momentum	$\eta = \frac{1}{3}\lambda\bar{c}m\mathcal{N}$ $= \dfrac{m\bar{c}}{(3\sqrt{2})\sigma}$	$\mathrm{kg\,m^{-1}\,s^{-1}}$

There is also a flux of molecules from right to left. On average, the molecules making the journey have originated from $z = +\lambda$ where the number density is $\mathcal{N}(\lambda)$. Therefore,

$$J(\mathrm{L} \leftarrow \mathrm{R}) = -\tfrac{1}{4}\mathcal{N}(\lambda)\bar{c} \tag{13b}$$

The average number density at $z = +\lambda$ is approximately

$$\mathcal{N}(\lambda) = \mathcal{N}(0) + \lambda\left(\frac{d\mathcal{N}}{dz}\right)_0 \tag{14}$$

The net flux is

$$\begin{aligned} J_z &= J(\mathrm{L} \to \mathrm{R}) + J(\mathrm{L} \leftarrow \mathrm{R}) \\ &= \tfrac{1}{4}\bar{c}\left\{ \left[\mathcal{N}(0) - \lambda\left(\frac{d\mathcal{N}}{dz}\right)_0\right] - \left[\mathcal{N}(0) + \lambda\left(\frac{d\mathcal{N}}{dz}\right)_0\right] \right\} \\ &= -\tfrac{1}{2}\bar{c}\lambda\left(\frac{d\mathcal{N}}{dz}\right)_0 \end{aligned} \tag{15}$$

This equation shows that the flux is proportional to the first derivative of the concentration, in agreement with Fick's law.

At this stage it looks as though we can pick out a value of the diffusion coefficient by comparing eqns 8 and 15, so obtaining $D = \frac{1}{2}\lambda\bar{c}$. It must be remembered, however, that the calculation is quite crude, and is little more than an assessment of the order of magnitude of D. One aspect that has not been taken into account is illustrated in Fig. 24.6, which shows that, although a molecule may have begun its journey very close to the window, it could have a long flight before it gets there. Because the path is long, the molecule is likely to collide before reaching the window, so it ought to be added to the graveyard of other molecules that have collided. To take this effect into account involves a lot of work, but the end result is the appearance of a factor of $\frac{2}{3}$ representing the lower flux. The modification results in eqn 11.

(b) Thermal conduction

According to the kinetic theory of gases, and as shown in the *Justification* below, the coefficient of thermal conductivity of a perfect gas A having molar concentration [A] is given by the expression

$$\kappa = \tfrac{1}{3}\lambda\bar{c}C_{V,m}[A] \tag{16}°$$

where $C_{V,m}$ is the molar heat capacity at constant volume.

Justification 24.3

According to the equipartition theorem (Section 20.3 and the *Introduction*), each molecule carries an average energy $\varepsilon = \nu kT$, where ν is a number of the order of 1. For monatomic particles, $\nu = \frac{3}{2}$. When one molecule passes through the imaginary window, it transports that energy on average. We suppose that the number density is uniform but that the temperature is not. On average, molecules arrive from the left after travelling a mean free path from their last collision in a hotter region, and therefore with a higher energy. Molecules also arrive from the right after travelling a mean free path from a cooler region. The two opposing energy fluxes are therefore

$$J(\mathrm{L} \to \mathrm{R}) = \tfrac{1}{4}\bar{c}\mathcal{N}\varepsilon(-\lambda) \qquad \varepsilon(-\lambda) = \nu k \left\{ T - \lambda \left(\frac{\mathrm{d}T}{\mathrm{d}z} \right)_0 \right\}$$

$$J(\mathrm{L} \gets \mathrm{R}) = -\tfrac{1}{4}\bar{c}\mathcal{N}\varepsilon(\lambda) \qquad \varepsilon(\lambda) = \nu k \left\{ T + \lambda \left(\frac{\mathrm{d}T}{\mathrm{d}z} \right)_0 \right\}$$

(17)

and the net flux is

$$J_z = J(\mathrm{L} \to \mathrm{R}) + J(\mathrm{L} \gets \mathrm{R}) = -\tfrac{1}{2}\nu k \lambda \bar{c} \mathcal{N} \left(\frac{\mathrm{d}T}{\mathrm{d}z} \right)_0 \tag{18}$$

As before, we multiply by $\frac{2}{3}$ to take long flight paths into account, and so arrive at

$$J_z = -\tfrac{1}{3}\nu k \lambda \bar{c} \mathcal{N} \left(\frac{\mathrm{d}T}{\mathrm{d}z} \right)_0 \tag{19}$$

The energy flux is proportional to the temperature gradient, as we wanted to show. Comparison of this equation with eqn 9 shows that

$$\kappa = \tfrac{1}{3}\nu k \lambda \bar{c} \mathcal{N} \tag{20}$$

Equation 16 then follows from $C_{V,\mathrm{m}} = \nu k N_{\mathrm{A}}$ for a perfect gas, where [A] is the molar concentration of A. For this step, we use $\mathcal{N} = N/V = n N_{\mathrm{A}}/V = N_{\mathrm{A}}[\mathrm{A}]$.

Because λ is inversely proportional to the pressure, and hence inversely proportional to the molar concentration of the gas, it follows from eqn 16 that the thermal conductivity is independent of the pressure. The physical reason for this independence is that the thermal conductivity can be expected to be large when many molecules are available to transport the energy, but the presence of so many molecules limits their mean free path and they cannot carry the energy over a great distance. These two effects balance. The thermal conductivity is indeed found experimentally to be independent of the pressure, except when the pressure is very low, when $\kappa \propto p$. At low pressures λ exceeds the dimensions of the apparatus, and the distance over which the energy is transported is determined by the size of the container and not by the other molecules present. The flux is still proportional to the number of carriers, but the length of the journey no longer depends on λ, so $\kappa \propto [\mathrm{A}]$, which implies that $\kappa \propto p$.

(c) The viscosity of a perfect gas

We have seen that viscosity is related to the flux of momentum. As shown in the *Justification* below, the expression obtained from the kinetic theory of gases is

$$\eta = \tfrac{1}{3}M \lambda \bar{c}[\mathrm{A}] \tag{21}°$$

where [A] is the molar concentration of the gas molecules and M is their molar mass.

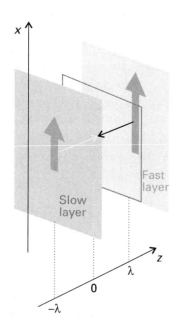

24.7 The calculation of the viscosity of a gas examines the net x-component of momentum brought to a plane from faster and slower layers on average a mean free path away in each direction.

Justification 24.4

Molecules travelling from the right in Fig. 24.7 (from a fast layer to a slower one) transport a momentum $mv_x(\lambda)$ to their new layer at $z = 0$; those travelling from the left transport $mv_x(-\lambda)$ to it. If it is assumed that the density is uniform, the collision flux is $\frac{1}{4}\mathcal{N}\bar{c}$. Those arriving from the right on average carry a momentum

$$mv_x(\lambda) = mv_x(0) + m\lambda\left(\frac{dv_x}{dz}\right)_0$$

Those arriving from the left bring a momentum

$$mv_x(-\lambda) = mv_x(0) - m\lambda\left(\frac{dv_x}{dz}\right)_0$$

The net flux of x-momentum in the z-direction is therefore

$$J = \frac{1}{4}\mathcal{N}\bar{c}\left\{\left[mv_x(0) - m\lambda\left(\frac{dv_x}{dz}\right)_0\right] - \left[mv_x(0) + m\lambda\left(\frac{dv_x}{dz}\right)_0\right]\right\}$$

$$= -\frac{1}{2}\mathcal{N}m\lambda\bar{c}\left(\frac{dv_x}{dz}\right)_0$$

The flux is proportional to the velocity gradient, as we wished to show. Comparison of this expression with eqn 10, and multiplication by $\frac{2}{3}$ in the normal way, leads to

$$\eta = \frac{1}{3}\mathcal{N}m\lambda\bar{c} \tag{22}$$

which can easily be converted into eqn 21.

The viscosity is independent of the pressure: $\lambda \propto 1/p$ and $[A] \propto p$, implying that $\eta \propto \bar{c}$, independent of p. The physical reason is the same as for the thermal conductivity: more molecules are available to transport the momentum, but they carry it less far on account of the decrease in mean free path. An unexpected result is that, because $\bar{c} \propto T^{1/2}$, the viscosity coefficient is proportional to $T^{1/2}$. That is, the viscosity of a gas increases with temperature. This conclusion is explained when we remember that at high temperatures the molecules travel more quickly, so the flux of momentum is greater.[2]

There are two main techniques for measuring viscosities of gases. One technique depends on the rate of damping of the torsional oscillations of a disk hanging in the gas. The half-life of the decay of the oscillation depends on the viscosity and the design of the apparatus, and the apparatus needs to be calibrated. The other method is based on **Poiseuille's formula** for the rate of flow of a fluid through a tube of radius r:

$$\frac{dV}{dt} = \frac{(p_1^2 - p_2^2)\pi r^4}{16l\eta p_0} \tag{23}$$

where V is the volume flowing, p_1 and p_2 are the pressures at each end of the tube of length l, and p_0 is the pressure at which the volume is measured.

Such measurements confirm that the viscosities of gases are independent of pressure over a wide range. For instance, the results for argon from 10^{-3} atm to 10^2 atm are shown in Fig. 24.8, and we see that η is constant from about 0.01 atm to 20 atm. The measurements also confirm (to a lesser extent) the $T^{1/2}$ dependence. The dotted line in the illustration shows the calculated values using $\sigma = 22 \times 10^{-20}$ m^2, implying a collision diameter of 260 pm, in contrast to the van der Waals diameter of 335 pm obtained from the density of

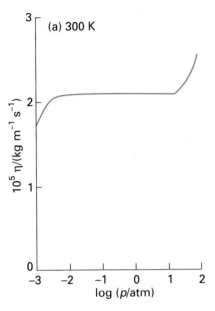

(a) 300 K

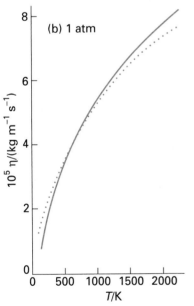

(b) 1 atm

24.8 The experimental results for (a) the pressure dependence of the viscosity of argon, and (b) its temperature dependence. The dotted line in the latter is the calculated value. Fitting the observed and calculated curves is one way of determining the collision cross-section.

---

2   As we shall see in Section 24.6, the viscosity of a liquid *decreases* with increase in temperature because intermolecular interactions must be overcome.

the solid. The agreement is not too bad, considering the simplicity of the model, especially the neglect of intermolecular forces.

Example 24.3 Using the Poiseuille formula to measure a viscosity

In a Poiseuille flow experiment to measure the viscosity of air at 298 K, the sample was allowed to flow through a tube of length 100 cm and internal diameter 1.00 mm. The high-pressure end was at 765 Torr and the low-pressure end was at 760 Torr. The volume was measured at the latter pressure. In 100 s, 90.2 cm^3 of air passed through the tube. What is the viscosity of air at 298 K?

Method The data can be used in the Poiseuille formula, eqn 23, reorganized into

$$\eta = \frac{(p_1^2 - p_2^2)\pi r^4}{16lp_0(dV/dt)}$$

To use this formula, convert the pressures to pascals by using 1 Torr = 133.3 Pa.

Answer The rate of flow is

$$\frac{dV}{dt} = \frac{9.02 \times 10^{-5} \text{ m}^3}{100 \text{ s}} = 9.02 \times 10^{-7} \text{ m}^3 \text{ s}^{-1}$$

Therefore, because

$$p_1^2 - p_2^2 = 1.355 \times 10^8 \text{ Pa}^2$$

and

$$\frac{\pi r^4}{16lp_0} = 1.94 \times 10^{-18} \text{ N}^{-1} \text{ m}^5$$

it follows that $\eta = 2.91 \times 10^{-4}$ kg m^{-1} s^{-1}.

Comment The kinetic theory expression gives $\eta = 1.4 \times 10^{-5}$ kg m^{-1} s^{-1}, so the agreement is quite good. Viscosities are commonly expressed in centipoise (cP) or (for gases) micropoise (μP), the conversion being 1 cP = 10^{-3} kg m^{-1} s^{-1}; the viscosity of air at 20°C is about 180 μP.

- -

Self-test 24.3 What volume would be collected if the pressure gradient were doubled, other conditions remaining constant?

[180 cm^3]

Motion in liquids

As a first step in dealing with the much more difficult problem of motion in liquids, we outline what is currently known about the structure of a simple liquid. Then we consider a particularly simple type of motion through a liquid, that of an ion, and see that the information that motion provides can be used to infer the behaviour of uncharged species too.

24.5 The structures of liquids

The starting point for the discussion of solids is the well-ordered structure of perfect crystals. The starting point for the discussion of gases is the completely disordered distribution of the molecules of a perfect gas. Liquids lie between these two extremes.

(a) The radial distribution function

The average relative locations of the particles of a liquid are expressed in terms of the **radial distribution function**, $g(r)$. This function is defined so that $g(r)r^2\,dr$ is the probability that a molecule will be found in the range dr at a distance r from another molecule. In a perfect crystal, $g(r)$ is a periodic array of sharp spikes, representing the certainty (in the absence of defects and thermal motion) that molecules (or ions) lie at definite locations. This regularity continues out to the edges of the crystal, so we say that crystals have **long-range order**. When the crystal melts, the long-range order is lost and, wherever we look at long distances from a given molecule, there is equal probability of finding a second molecule. Close to the first molecule, though, the nearest neighbours might still adopt approximately their original relative positions and, even if they are displaced by newcomers, the new particles might adopt their vacated positions. It is still possible to detect a sphere of nearest neighbours at a distance r_1, and perhaps beyond them a sphere of next-nearest neighbours at r_2. The existence of this **short-range order** means that the radial distribution function can be expected to oscillate at short distances, with a peak at r_1, a smaller peak at r_2, and perhaps some more structure beyond that.

The radial distribution function can be determined experimentally by X-ray diffraction, for $g(r)$ can be extracted from the diffuse diffraction pattern characteristic of liquid samples in much the same way as a crystal structure is obtained from X-ray diffraction of crystals. The shells of local structure shown in the example in Fig. 24.9 (for water) are unmistakable. Closer analysis shows that any given H_2O molecule is surrounded by other molecules at the corners of a tetrahedron, similar to the arrangement in ice (Fig. 21.33). The form of $g(r)$ at 100°C shows that the intermolecular forces (in this case, principally by hydrogen bonds) are strong enough to affect the local structure right up to the boiling point. Raman spectra and molecular dynamics calculations indicate that in liquid water most molecules participate in either three or four hydrogen bonds. Infrared spectra show that about 90 per cent of hydrogen bonds are intact at the melting point of ice, falling to about 20 per cent at the boiling point.

Despite its uniquely extensive hydrogen bonding, water has only a modest viscosity (much less than that of glycerol, for example). Indeed, water is a conundrum and a severe testing ground for theories of the liquid state. Water is a simple liquid when examined in the context of dynamic characteristics, and the challenge is to reconcile its extensive hydrogen-bonded structure with its dynamical simplicity.

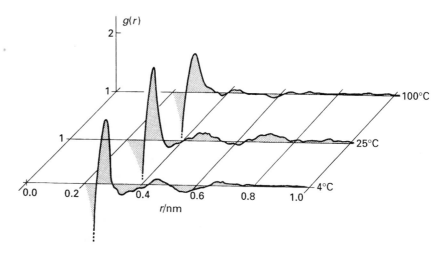

24.9 The radial distribution function of the oxygen atoms in liquid water at three temperatures. Note the expansion as the temperature is raised. (A.H. Narten, M.D. Danford, and H.A. Levy, *Discuss. Faraday. Soc.* **43**, 97 (1967).)

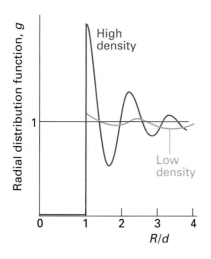

24.10 The radial distribution function for a simulation of a liquid using impenetrable hard spheres (ball-bearings).

(b) The calculation of $g(r)$

Because the radial distribution function can be calculated by making assumptions about the intermolecular forces, it can be used to test theories of liquid structure. However, even a fluid of hard spheres without attractive interactions (a collection of ball-bearings in a container) gives a function that oscillates near the origin (Fig. 24.10), and one of the factors influencing, and sometimes dominating, the structure of a liquid is the geometrical problem of stacking together reasonably hard spheres. Indeed, the radial distribution function of a liquid of hard spheres shows more pronounced oscillations at a given temperature than that of any other type of liquid. The attractive part of the potential modifies this basic structure, but sometimes only quite weakly. One of the reasons behind the difficulty of describing liquids theoretically is the similar importance of both the attractive and repulsive (hard core) components of the potential.

The formal expression for the radial distribution function for molecules 1 and 2 in a fluid consisting of N particles is the somewhat fearsome equation

$$g(r_{12}) = \frac{\int \int \cdots \int e^{-\beta V_N} \, d\mathbf{r}_3 d\mathbf{r}_4 \cdots d\mathbf{r}_N}{\mathcal{N}^2 \int \int \cdots \int e^{-\beta V_N} \, d\mathbf{r}_1 d\mathbf{r}_2 \cdots d\mathbf{r}_N} \tag{24}$$

where $\beta = 1/kT$ and V_N is the N-particle potential energy. There are several ways of building the intermolecular potential into the calculation of $g(r)$. Numerical methods take a box of about 10^3 particles (the number increases as computers grow more powerful), and the rest of the liquid is simulated by surrounding the box with replications of the original box (Fig. 24.11). Then, whenever a particle leaves the box through one of its faces, its image arrives through the opposite face. When calculating the interactions of a molecule in a box, it interacts with all the molecules in the box and all the periodic replications of those molecules and itself in the other boxes.

Once $g(r)$ is known it can be used to calculate the thermodynamic properties of liquids. For example, the contribution of the pairwise additive intermolecular potential, V_2, to the internal energy is given by the integral

$$U = \frac{2\pi N^2}{V} \int_0^\infty g V_2 r^2 \, dr \tag{25}$$

That is, U is essentially the average two-particle potential energy weighted by $g(r)r^2 \, dr$, which is the probability that the pair of particles have a separation between r and $r + dr$. Likewise, the contribution that pairwise interactions make to the pressure is

$$\frac{pV}{nRT} = 1 - \frac{2\pi N}{kTV} \int_0^\infty g v_2 r^2 \, dr \qquad v_2 = r\left(\frac{dV_2}{dr}\right) \tag{26}$$

The quantity v_2 is called the **virial** (hence the term 'virial equation of state').

(c) Monte Carlo methods

In the **Monte Carlo method**, the particles in the box are moved through small but otherwise random distances, and the change in total potential energy of the N particles in the box, ΔV_N, is calculated using one of the intermolecular potentials discussed in Chapter 22. Whether or not this new configuration is accepted is then judged from the following rules:

1. If the potential energy is not greater than before the change, then the configuration is accepted.

2. If the potential energy is greater than before the change, the factor $e^{-\Delta V_N/kT}$ is compared with a random number between 0 and 1: if the factor is larger than the random number, the configuration is accepted; if the factor is not larger, the configuration is rejected.

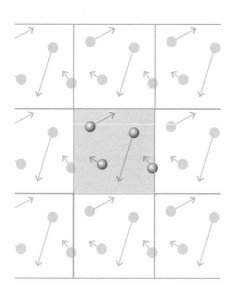

24.11 In a two-dimensional molecular dynamics simulation that uses periodic boundary conditions, when one particle leaves the cell its mirror image enters through the opposite face.

This procedure ensures that at equilibrium the probability of occurrence of any configuration is proportional to the Boltzmann factor $e^{-V_N/kT}$. The configurations generated in this way can then be used to construct $g(r)$ simply by counting the number of pairs of particles with a separation r and averaging the result over the whole collection of configurations.

(d) Molecular dynamics

In the **molecular dynamics** approach, the history of an initial arrangement is followed by calculating the trajectories of all the particles under the influence of the intermolecular potentials. Newton's laws are used to predict where each particle will be after a short time interval (about 10^{-15} s, which is shorter than the average time between collisions), and then the calculation is repeated for tens of thousands of such steps. The time-consuming part of the calculation is the evaluation of the net force on the molecule arising from all the other molecules present in the system.

A molecular dynamics calculation gives a series of snapshots of the liquid, and $g(r)$ can be calculated as before. The temperature of the system is inferred by computing the mean kinetic energy of the particles and using the equipartition result that

$$\langle \tfrac{1}{2}mv_q^2 \rangle = \tfrac{1}{2}kT \tag{27}$$

for each coordinate q.

(e) Liquid crystals

A feature that makes calculations even more difficult is the possibility that molecules have strongly anisotropic interactions. An important extreme case of anisotropy gives rise to a **mesophase**, a phase intermediate between solid and liquid. Mesophases are of great importance in biology, for they occur as lipid bilayers and in vesicular systems.

A mesophase may arise when molecules have highly anisotropic shapes, such as being long and thin, as in (**1**), or disk-like. When the solid melts, some aspects of the long-range order characteristic of the solid may be retained, and the new phase may be a **liquid crystal**, a substance having liquid-like imperfect long-range order in at least one direction in space but positional or orientational order in at least one other direction. One type of retained long-range order gives rise to a **smectic phase** (from the Greek word for soapy), in which the molecules align themselves in layers (Fig. 24.12a). Other materials, and some smectic liquid crystals at higher temperatures, lack the layered structure but retain a parallel alignment (Fig. 24.12b): this mesophase is called a **nematic phase** (from the Greek for thread, which refers to the observed defect structure of the phase). In the **cholesteric phase** (from the

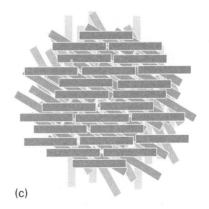

CN

1

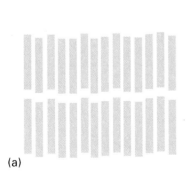

(a)

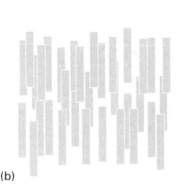

(b)

(c)

24.12 The arrangement of moleules in (a) the nematic phase, (b) the smectic phase, and (c) the cholesteric phase of liquid crystals. In the cholesteric phase, the stacking of layers continues to give a helical arrangement of molecules.

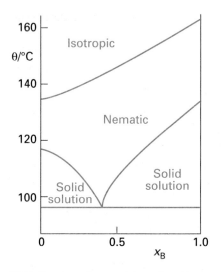

24.13 The phase diagram at 1 atm for a binary system of two liquid crystalline materials, 4, 4'-dimethoxyazoxybenzene (A) and 4, 4'-diethoxyazoxybenzene (B).

Table 24.3* Viscosities of liquids at 298 K, $\eta/(10^{-3}\,kg\,m^{-1}\,s^{-1})$

Benzene	0.601
Mercury	1.55
Pentane	0.224
Water†	0.891

* More values are given in the *Data section*.

† The viscosity of water corresponds to 0.891 cP.

Greek for bile solid) the molecules lie in sheets at angles that change slightly between each sheet (Fig. 24.12c). That is, they form helical structures with a pitch that depends on the temperature. As a result, cholesteric liquid crystals diffract light and have colours that depend on the temperature. The strongly anisotropic optical properties of nematic liquid crystals, and their response to electric fields, is the basis of their use as data displays (LCDs).

Although there are many liquid crystalline materials, some difficulty is often experienced in achieving a technologically useful temperature range for the existence of the mesophase. To overcome this difficulty, mixtures can be used. An example of the type of phase diagram that is then obtained is shown in Fig. 24.13. As can be seen, the mesophase exists over a wider range of temperatures than either liquid crystalline material alone.

24.6 Molecular motion in liquids

The motion of molecules in liquids can be studied experimentally by a variety of methods. Relaxation time measurements in NMR and ESR (Section 18.6b) can be interpreted in terms of the mobilities of the molecules, and have been used to show that big molecules in viscous fluids typically rotate in a series of small (about 5°) steps, whereas small molecules in nonviscous fluids typically jump through about 1 radian (57°) in each step. Another important technique is **inelastic neutron scattering**, in which the energy neutrons collect or discard as they pass through a sample is interpreted in terms of the motion of its particles. The same technique is used to examine the internal dynamics of macromolecules.

More mundane than these experiments are viscosity measurements (Table 24.3). For a molecule to move in a liquid, it must acquire at least a minimum energy to escape from its neighbours. The probability that a molecule has at least an energy E_a is proportional to $e^{-E_a/RT}$, so the mobility of the molecules in the liquid should follow this type of temperature

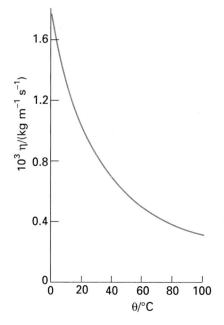

24.14 The experimental temperature dependence of the viscosity of water. As the temperature is increased, more molecules are able to escape from the potential wells provided by their neighbours, and so the liquid becomes more fluid. A plot of $\ln \eta$ against $1/T$ is a straight line (over a small range) with positive slope.

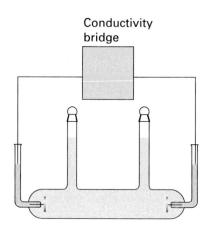

24.15 The conductivity of an electrolyte solution is measured by making a conductivity cell, like the one shown here, one arm of a resistance bridge.

dependence. Because the coefficient of viscosity, η, is inversely proportional to the mobility of the particles, we should expect that

$$\eta \propto e^{E_a/RT} \tag{28}$$

(Note the positive sign of the exponent.) This expression implies that the viscosity should decrease sharply with increasing temperature. Such a variation is found experimentally, at least over reasonably small temperature ranges (Fig. 24.14). The activation energy typical of viscosity is comparable to the mean potential energy of intermolecular interactions.

One problem with the interpretation of viscosity measurements is that the change in density of the liquid as it is heated makes a pronounced contribution to the temperature variation of the viscosity. Thus, the temperature dependence of viscosity at constant volume, when the density is constant, is much less than that at constant pressure. The intermolecular forces between the molecules of the liquid govern the magnitude of E_a, but the problem of calculating it is immensely difficult and still largely unsolved. At low temperatures, the viscosity of water decreases as the pressure is increased. This behaviour is consistent with the rupture of hydrogen bonds.

24.7 The conductivities of electrolyte solutions

Further insight into the nature of molecular motion can be obtained by studying the motion of ions in solution, for they can be dragged through the solution by the application of a potential difference between two electrodes immersed in the sample. By studying the transport of charge through electrolyte solutions it is possible to build up a picture of the events that occur in them and, in some cases, to extrapolate the conclusions to species that have zero charge, that is, to neutral molecules.

(a) Conductance and conductivity

The fundamental measurement used to study the motion of ions is that of the electrical resistance, $\mathcal{R}$, of the solution. The standard technique is to incorporate a conductivity cell into one arm of a resistance bridge and to search for the balance point, as explained in standard texts on electricity (Fig. 24.15). The main complication is that alternating current must be used because a direct current would lead to electrolysis and to **polarization**, which in this context means the modification of the composition of the layers of solution in contact with the electrodes. The use of alternating current (with a frequency of about 1 kHz) may avoid polarization because the charging that occurs on one half of the cycle is undone during the second half (if the reverse reaction is kinetically feasible).

The **conductance**, G, of a solution is the inverse of its resistance $\mathcal{R}$: $G = 1/\mathcal{R}$. As resistance is expressed in ohms, Ω, the conductance of a sample is expressed in Ω^{-1}. The reciprocal ohm used to be called the mho, but its official designation is now the siemens, S, and $1\ \mathrm{S} = 1\ \Omega^{-1}$. The conductance of a sample decreases with its length l and increases with its cross-sectional area A. We therefore write

$$G = \frac{\kappa A}{l} \tag{29}$$

where κ is the **conductivity**. With the conductance in siemens and the dimensions in metres, it follows that the SI units of κ are siemens per metre (S m^{-1}).

The conductivity of a solution depends on the number of ions present, and it is normal to introduce the **molar conductivity**, Λ_m, which is defined as

$$\Lambda_m = \frac{\kappa}{c} \tag{30}$$

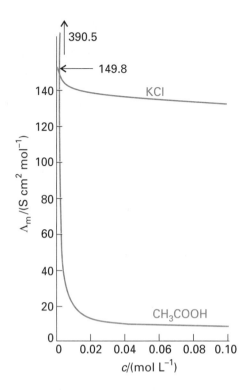

24.16 The concentration dependence of the molar conductivities of (a) a typical strong electrolyte (aqueous potassium chloride) and (b) a typical weak electrolyte (aqueous acetic acid).

where c is the molar concentration of the added electrolyte. The SI unit of molar conductivity is siemens metre-squared per mole ($S\,m^2\,mol^{-1}$), and typical values are about $10\,mS\,m^2\,mol^{-1}$ (where $1\,mS = 10^{-3}\,S$).

The molar conductivity of an electrolyte would be independent of concentration if κ were proportional to the concentration of the electrolyte. However, in practice, the molar conductivity is found to vary with the concentration. One reason for this variation is that the number of ions in the solution might not be proportional to the concentration of the electrolyte. For instance, the concentration of ions in a solution of a weak acid depends on the concentration of the acid in a complicated way, and doubling the concentration of the acid added does not double the number of ions. Secondly, because ions interact strongly with one another, the conductivity of a solution is not exactly proportional to the number of ions present.

The concentration dependence of molar conductivities indicates that there are two classes of electrolyte. The characteristic of a **strong electrolyte** is that its molar conductivity decreases only slightly as its concentration is increased (Fig. 24.16). The characteristic of a **weak electrolyte** is that its molar conductivity is normal at concentrations close to zero, but falls sharply to low values as the concentration increases. The classification depends on the solvent employed as well as the solute: lithium chloride, for example, is a strong electrolyte in water but a weak electrolyte in propanone.

(b) Strong electrolytes

Strong electrolytes are substances that are virtually fully ionized in solution, and include ionic solids and strong acids. As a result of their complete ionization, the concentration of ions in solution is proportional to the concentration of strong electrolyte added.

In an extensive series of measurements during the nineteenth century, Friedrich Kohlrausch showed that at low concentrations the molar conductivities of strong electrolytes vary linearly with the square root of the concentration:

$$\Lambda_m = \Lambda_m^\circ - \mathcal{K}c^{1/2} \tag{31}$$

This variation is called **Kohlrausch's law**. The constant Λ_m° is the **limiting molar conductivity**, the molar conductivity in the limit of zero concentration (when the ions are effectively infinitely far apart and do not interact with one another). The constant $\mathcal{K}$ is found to depend more on the stoichiometry of the electrolyte (that is, whether it is of the form MA, or M_2A, etc.) than on its specific identity.

Kohlrausch was also able to show that Λ_m° can be expressed as the sum of contributions from its individual ions. If the limiting molar conductivity of the cations is denoted λ_+ and that of the anions λ_-, then his **law of the independent migration of ions** states that

$$\Lambda_m^\circ = \nu_+\lambda_+ + \nu_-\lambda_- \tag{32}^\circ$$

where ν_+ and ν_- are the numbers of cations and anions per formula unit of electrolyte (for example, $\nu_+ = \nu_- = 1$ for HCl, NaCl, and $CuSO_4$, but $\nu_+ = 1$, $\nu_- = 2$ for $MgCl_2$). This simple result, which can be understood on the grounds that the ions migrate independently in the limit of zero concentration, lets us predict the limiting molar conductivity of any strong electrolyte from the data in Table 24.4.

Table 24.4* Limiting ionic conductivities in water at 298 K, $\lambda/(mS\,m^2\,mol^{-1})$

H^+	34.96	OH^-	19.91
Na^+	5.01	Cl^-	7.63
K^+	7.35	Br^-	7.81
Zn^{2+}	10.56	SO_4^{2-}	16.00

*More values are given in the *Data section*.

Illustration

The limiting molar conductivity of $BaCl_2$ in water at 298 K is

$$\Lambda_m^\circ = (12.72 + 2 \times 7.63)\ mS\,m^2\,mol^{-1} = 27.98\ mS\,m^2\,mol^{-1}$$

(c) Weak electrolytes

Weak electrolytes are not fully ionized in solution. They include weak Brønsted acids and bases, such as CH_3COOH and NH_3. The marked concentration dependence of their molar conductivities arises from the displacement of the equilibrium

$$HA(aq) + H_2O(l) \rightleftharpoons H_3O^+(aq) + A^-(aq) \qquad K_a = \frac{a(H_3O^+)a(A^-)}{a(HA)} \qquad (33)$$

towards products at low molar concentrations.

The conductivity depends on the number of ions in the solution, and therefore on the **degree of ionization**, α, of the electrolyte. The degree of ionization is defined so that, for the acid HA at a molar concentration c, at equilibrium

$$[H_3O^+] = \alpha c \qquad [A^-] = \alpha c \qquad [HA] = (1 - \alpha)c \qquad (34)$$

If we ignore activity coefficients, the acidity constant, K_a, is approximately

$$K_a = \frac{\alpha^2 c}{1 - \alpha} \qquad (35)°$$

from which it follows that

$$\alpha = \frac{K_a}{2c}\left\{\left(1 + \frac{4c}{K_a}\right)^{1/2} - 1\right\} \qquad (36)°$$

The electrolyte is fully ionized at infinite dilution, and its molar conductivity is then $\Lambda_m°$. Because only a fraction α is actually present as ions in the actual solution, the measured molar conductivity Λ_m is given by

$$\Lambda_m = \alpha \Lambda_m° \qquad (37)°$$

with α given by eqn 36.

Example 24.4 Using conductivity measurements to determine pK_a

The molar conductivity of $0.0100\,M\,CH_3COOH(aq)$ at 298 K was measured as $\Lambda_m = 1.65\,mS\,m^2\,mol^{-1}$. Determine the degree of ionization and the pK_a of the acid.

Method To calculate α, use eqn 37 with $\Lambda_m°$ calculated from the data in Table 24.4. Then calculate K_a by substitution of α into eqn 35 and form $pK_a = -\log K_a$.

Answer From the data in Table 24.4 we find $\Lambda_m° = 39.05\,mS\,m^2\,mol^{-1}$. Therefore, $\alpha = 0.0423$. It follows from eqn 35 that $K_a = 1.9 \times 10^{-5}$, implying that $pK_a = 4.72$.

Comment The thermodynamic value of pK_a is obtained by repeating the determination with different concentrations and extrapolating to zero concentration.

- -

Self-test 24.4 The molar conductivity of $0.0250\,M\,HCOOH(aq)$ was measured as $4.61\,mS\,m^2\,mol^{-1}$. Determine the pK_a of the acid.

[3.44]

Once we know K_a, we can use eqns 36 and 37 to predict the concentration dependence of the molar conductivity. The result agrees quite well with the experimental curve in Fig. 24.16. More usefully, we can use the concentration dependence of Λ_m in measurements of the limiting molar conductance. First, we rearrange eqn 36 into

$$\frac{1}{\alpha} = 1 + \frac{\alpha c}{K_a} \qquad (38)°$$

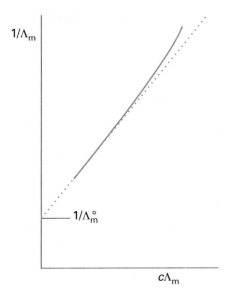

24.17 The graph used to determine the limiting value of the molar conductivity of a solution by extrapolation to zero concentration.

Then, by using eqn 37, we obtain **Ostwald's dilution law**:

$$\frac{1}{\Lambda_m} = \frac{1}{\Lambda_m^\circ} + \frac{\Lambda_m c}{K_a (\Lambda_m^\circ)^2} \tag{39}°$$

This equation implies that, if $1/\Lambda_m$ is plotted against $c\Lambda_m$, then the intercept at $c = 0$ will be $1/\Lambda_m^\circ$ (Fig. 24.17).

24.8 The mobilities of ions

To interpret conductivity measurements we need to know why ions move at different rates, why they have different molar conductivities, and why the molar conductivities of strong electrolytes decrease with the square root of the molar concentration. The central idea in this section is that, although the motion of an ion remains largely random, the presence of an electric field biases its motion, and the ion undergoes net migration through the solution.

(a) The drift speed

When the potential difference between two electrodes a distance l apart is $\Delta\phi$, the ions in the solution between them experience a uniform electric field of magnitude

$$\mathcal{E} = \frac{\Delta\phi}{l} \tag{40}$$

In such a field, an ion of charge[3] ze experiences a force of magnitude

$$\mathcal{F} = ze\mathcal{E} = \frac{ze\Delta\phi}{l} \tag{41}$$

A cation responds to the application of the field by accelerating towards the negative electrode and an anion responds by accelerating towards the positive electrode. However, this acceleration is short-lived. As the ion moves through the solvent it experiences a frictional retarding force $\mathcal{F}_{fric}$ proportional to its speed. If we assume that the Stokes formula (eqn 23.13) for a sphere of radius a and speed s applies even on a microscopic scale (and independent evidence from magnetic resonance suggests that it often gives at least the right order of magnitude), then we can write this retarding force as

$$\mathcal{F}_{fric} = fs \qquad f = 6\pi\eta a \tag{42}$$

The two forces act in opposite directions, and the ions quickly reach a terminal speed, the **drift speed**, when the accelerating force is balanced by the viscous drag. The net force is zero when

$$s = \frac{ze\mathcal{E}}{f} \tag{43}$$

Because the drift speed governs the rate at which charge is transported, we might expect the conductivity to decrease with increasing solution viscosity and ion size. Experiments confirm these predictions for bulky ions (such as R_4N^+ and RCO_2^-) but not for small ions. For example, the molar conductivities of the alkali metal ions increase from Li^+ to Cs^+ (Table 24.4) even though the ionic radii increase. The paradox is resolved when we realize that the radius a in the Stokes formula is the **hydrodynamic radius** (or 'Stokes' radius') of the ion, its effective radius in the solution taking into account all the H_2O molecules it carries in its hydration sphere. Small ions give rise to stronger electric fields than large ones,[4] so small ions are more extensively solvated than big ions. Thus, an ion of small ionic radius may have a large hydrodynamic radius because it drags many solvent molecules through the

---

3   In this chapter we disregard the sign of the charge number and so avoid notational complications.

4   The electric field at the surface of a sphere of radius r is proportional to ze/r^2, so the smaller the radius the stronger the field.

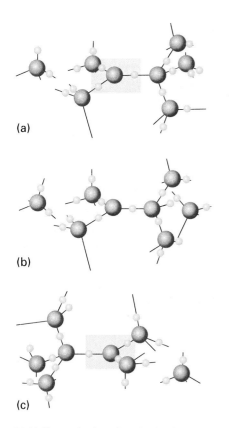

(a)

(b)

(c)

24.18 The mechanism of conduction by water as proposed by N. Agmon (*Chem. Phys. Letts.* **244**, 456 (1995)). Proton transfer between neighbouring molecules occurs when one molecule rotates into such a position that an O—H···O hydrogen bond can flip into being an O···H—O hydrogen bond. See text for a description of the steps.

solution as it migrates. The hydrating H_2O molecules are often very labile, however, and NMR and isotope studies have shown that the exchange between the coordination sphere of the ion and the bulk solvent is very rapid.

The proton, although it is very small, has a very high molar conductivity (Table 24.4)! Proton and ^{17}O-NMR show that the times characteristic of protons hopping from one molecule to the next are about 1.5 ps, which is comparable to the time that inelastic neutron scattering shows it takes a water molecule to reorientate through about 1 rad (1 to 2 ps). According to the **Grotthuss mechanism**,[5] there is an effective motion of a proton that involves the rearrangement of bonds in a group of water molecules. However, the actual mechanism is still highly contentious. Attention now focuses on the $H_9O_4^+$ unit, in which the nearly trigonal planar H_3O^+ ion[6] is linked to three strongly solvating H_2O molecules. This cluster of atoms is itself hydrated, but the hydrogen bonds in the secondary sphere are weaker than in the primary sphere. It is envisaged that the rate-determining step is the cleavage of one of the weaker hydrogen bonds of this secondary sphere (Fig. 24.18a). After this bond cleavage has taken place, and the released molecule has rotated through a few degrees (a process that takes about 1 ps), there is a rapid adjustment of bond lengths and angles in the remaining cluster, to form an $H_5O_2^+$ cation of structure $H_2O\cdots H^+\cdots OH_2$ (Fig. 24.18b). Shortly after this reorganization has occurred, a new $H_9O_4^+$ cluster forms as other molecules rotate into a position where they can become members of a secondary hydration sphere, but now the positive charge is located one molecule to the right of its initial location (Fig. 24.18c). According to this model, there is no coordinated motion of a proton along a chain of molecules, simply a very rapid hopping between neighbouring sites, with a low activation energy. The model is consistent with the observation that the molar conductivity of protons increases as the pressure is raised, for increasing pressure ruptures hydrogen bonds.

(b) Ion mobilities

According to eqn 43, the drift speed of an ion is proportional to the strength of the applied field. We write

$$s = u\mathcal{E} \qquad [44]$$

where u is called the **mobility** of the ion (Table 24.5). Comparison of eqns 43 and 44 and use of eqn 42 show that

$$u = \frac{ze}{f} = \frac{ze}{6\pi\eta a} \qquad (45)$$

Illustration

For an order of magnitude estimate we can take $z = 1$ and a the radius of an ion such as Cs^+ (which might be typical of a smaller ion plus its hydration sphere), which is 170 pm. For the viscosity, we use $\eta = 1.0$ cP (1.0×10^{-3} kg m^{-1} s^{-1}, Table 24.3). Then $u \approx 5 \times 10^{-8}$ m^2 V^{-1} s^{-1}. This value means that, when there is a potential difference of 1 V across a solution of length 1 cm (so $\mathcal{E} = 100$ V m^{-1}), the drift speed is typically about 5 μm s^{-1}. That speed might seem slow, but not when expressed on a molecular scale, for it corresponds to an ion passing about 10^4 solvent molecules per second.

Table 24.5* Ionic mobilities in water at 298 K, $u/(10^{-8}$ m^2 s^{-1} V^{-1})

H^+	36.23	OH^-	20.64
Na^+	5.19	Cl^-	7.91
K^+	7.62	Br^-	8.09
Zn^{2+}	5.47	SO_4^{2-}	8.29

*More values are given in the *Data section*.

---

5   The name of the mechanism is an allusion to an early view advanced by von Grotthuss, in which it was supposed that chains of dipoles were responsible for the transport of charge in water.

6   In the gas phase, H_3O^+ is trigonal pyramidal.

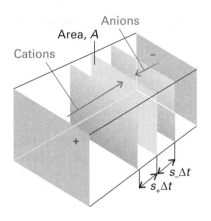

24.19 In the calculation of the current, all the cations within a distance $s_+\Delta t$ (that is, those in the volume $s_+A\Delta t$) will pass through the area A. The anions in the corresponding volume on the other side of the window will also contribute to the current similarly.

(c) Mobility and conductivity

The usefulness of ionic mobilities is that they provide a link between measurable and theoretical quantities. As a first step we establish in the *Justification* below the following relation between an ion's mobility and its molar conductivity:

$$\lambda = zuF \qquad (46)°$$

where F is the Faraday constant ($F = N_A e$).

Justification 24.5

To keep the calculation simple, we ignore signs in the following, and concentrate on the magnitudes of quantities: the direction of ion flux can always be decided by common sense.

Consider a solution of a fully dissociated strong electrolyte at a molar concentration c. Let each formula unit give rise to ν_+ cations of charge z_+e and ν_- anions of charge z_-e. The molar concentration of each type of ion is therefore νc (with $\nu = \nu_+$ or ν_-), and the number density of each type is νcN_A. The number of ions of one kind that pass through an imaginary window of area A during an interval Δt is equal to the number within the distance $s\Delta t$ (Fig. 24.19), and therefore to the number in the volume $s\Delta tA$. (The same sort of argument was used in Section 1.3 in the discussion of the pressure of a gas.) The number of ions of that kind in this volume is equal to $s\Delta tA\nu cN_A$. The flux through the window (the number of this type of ion passing through the window divided by the area of the window and the duration of the interval) is therefore

$$J(\text{ions}) = \frac{s\Delta tA\nu cN_A}{A\Delta t} = s\nu cN_A$$

Each ion carries a charge ze, so the flux of charge is

$$J(\text{charge}) = zs\nu ceN_A = zs\nu cF$$

Because $s = u\mathcal{E}$, the flux is

$$J(\text{charge}) = zu\nu cF\mathcal{E}$$

The current, I, through the window due to the ions we are considering is the charge flux times the area:

$$I = JA = zu\nu cF\mathcal{E}A$$

Because the electric field is the potential gradient, $\Delta\phi/l$, we can write

$$I = \frac{zu\nu cFA\Delta\phi}{l} \qquad (47)$$

Current and potential difference are related by Ohm's law, that

$$I = \frac{\Delta\phi}{\mathcal{R}} = G\Delta\phi = \frac{\kappa A\Delta\phi}{l}$$

where we have used eqn 29 in the form $\kappa = Gl/A$. Comparison of the last two expressions gives $\kappa = zu\nu cF$. Division by the molar concentration of ions, νc, then results in eqn 46.

Equation 46 applies to the cations and to the anions. Therefore, for the solution itself in the limit of zero concentration (when there are no interionic interactions),

$$\Lambda_m° = (z_+u_+\nu_+ + z_-u_-\nu_-)F \qquad (48)°$$

For a symmetrical $z : z$ electrolyte (for example, $CuSO_4$ with $z = 2$), this equation simplifies to

$$\Lambda_m^\circ = z(u_+ + u_-)F \qquad (49)^\circ$$

Illustration

Earlier, we estimated the typical ionic mobility as 5×10^{-8} m^2 V^{-1} s^{-1}; so, with $z = 1$ for both the cation and anion, we can estimate that a typical limiting molar conductivity should be about 10 mS m^2 mol^{-1}, in accord with experiment. The experimental value for KCl, for instance, is 15 mS m^2 mol^{-1}.

(d) Transport numbers

The **transport number**, $t_\pm$, is defined as the fraction of total current carried by the ions of a specified type. For a solution of two kinds of ion, the transport numbers of the cations (t_+) and anions (t_-) are

$$t_\pm = \frac{I_\pm}{I} \qquad [50]$$

where $I_\pm$ is the current carried by the cation (I_+) or anion (I_-) and I is the total current through the solution. Because the total current is the sum of the cation and anion currents, it follows that

$$t_+ + t_- = 1 \qquad (51)$$

The **limiting transport number**, $t_\pm^\circ$, is defined in the same way but for the limit of zero concentration of the electrolyte solution. We shall consider only these limiting values from now on, for that avoids the problem of ionic interactions.

The current that can be ascribed to each type of ion is related to the mobility of the ion by eqn 47. Hence the relation between $t_\pm^\circ$ and $u_\pm$ is

$$t_\pm^\circ = \frac{z_\pm \nu_\pm u_\pm}{z_+ \nu_+ u_+ + z_- \nu_- u_-} \qquad (52)^\circ$$

Because $z_+ \nu_+ = z_- \nu_-$ for all ionic species, this equation simplifies to

$$t_\pm^\circ = \frac{u_\pm}{u_+ + u_-} \qquad (53)^\circ$$

Moreover, because the ionic conductivities are related to the mobilities by eqn 46, it follows that

$$t_\pm^\circ = \frac{\nu_\pm \lambda_\pm}{\nu_+ \lambda_+ + \nu_- \lambda_-} = \frac{\nu_\pm \lambda_\pm}{\Lambda_m^\circ} \qquad (54)^\circ$$

and hence, for each type of ion,

$$\nu_\pm \lambda_\pm = t_\pm^\circ \Lambda_m^\circ \qquad (55)^\circ$$

Consequently, because there are independent ways of measuring transport numbers of ions, we can determine the individual ionic conductivities and (through eqn 46) the ionic mobilities.

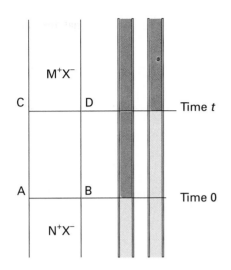

24.20 In the moving boundary method for the measurement of transport numbers, the distance moved by the boundary is observed as a current is passed. All the M ions in the volume between AB and CD must have passed through CD if the boundary moves from AB to CD.

(e) The measurement of transport numbers

One of the most accurate methods for measuring transport numbers is the **moving boundary method**, in which the motion of a boundary between two ionic solutions having a common ion is observed as a current flows.

Let MX be the salt of interest and NX a salt giving a denser solution. The solution of NX is called the **indicator solution**; it occupies the lower part of a vertical tube of cross-sectional area A (Fig. 24.20). The MX solution, which is called the **leading solution**, occupies the upper part of the tube. There is a sharp boundary between the two solutions. The indicator solution must be denser than the leading solution, and the mobility of the M ions must be greater than that of the N ions.[7] Thus, if any M ions diffuse into the lower solution, they will be pulled upwards more rapidly than the N ions around them, and the boundary will reform. The interpretation of the experiment makes use of the relation (see the *Justification* below) between the distance, l, moved by the boundary in the time Δt for which a current I is passed for a period Δt:

$$t_+ = \frac{z_+ c l A F}{I \Delta t} \tag{56}$$

Hence, by measuring the distance moved, the transport number and hence the conductivity and mobility of the ions can be determined.

Justification 24.6

When a current I is passed for a time Δt, the boundary moves from AB to CD, so all the M ions in the volume between AB and CD must have passed through CD. That number is $c l A N_A$, so the charge that the M ions transfer through the plane is $z_+ c l A e N_A$. However, the *total* charge transferred when a current I flows for an interval Δt is $I \Delta t$. Therefore, the fraction due to the motion of the M ions, which is their transport number, is given by eqn 56.

In the **Hittorf method**, an electrolytic cell is divided into three compartments and a charge $I \Delta t$ is passed. An amount $I \Delta t / z_+ F$ of cations is discharged at the cathode, but an amount $t_+ I \Delta t / z_+ F$ of cations migrates into the cathode compartment. The net change in the amount of cations in that compartment is therefore

$$\text{Net change} = (t_+ - 1) \frac{I \Delta t}{z_+ F} = -t_- \frac{I \Delta t}{z_+ F} \tag{57}$$

Hence, by measuring the change of composition in the cathode compartment, the anion transport number t_- can be deduced. Likewise, the change in composition of the anode compartment is $-t_+ I \Delta t / z_- F$, which gives the cation transport number, t_+.

Transport numbers may also be measured by using galvanic cells. In particular, the measurement is made on a **cell with transference**, which is a galvanic cell with a liquid boundary across which ions may pass from one electrode compartment to the other. An example is the cell

$$\text{Ag(s)}|\text{AgCl(s)}|\text{HCl}(m_1)|\text{HCl}(m_2)|\text{AgCl(s)}|\text{Ag(s)}$$

for which the zero-current cell potential is E_t and the electrodes are reversible with respect to anions (Cl$^-$). The corresponding cell without transference is

$$\text{Ag(s)}|\text{AgCl(s)}|\text{HCl}(m_1)|\text{H}_2\text{(g)}|\text{Pt(s)}|\text{H}_2\text{(g)}|\text{HCl}(m_2)|\text{AgCl(s)}|\text{Ag(s)}$$

---

7   One procedure is to add bromothymol blue indicator to a slightly alkaline solution of the ion of interest and to use a cadmium electrode at the lower end of the vertical tube. The electrode produces Cd^{2+} ions, which are slow moving and slightly acidic (the hydrated ion is a Brønsted acid), and the boundary is revealed by the colour change of the indicator.

and its zero-current potential is E. We show in the *Justification* below that the two potentials are related by

$$E_t = 2t_+E \qquad (58)$$

Therefore, comparison of the two cell potentials gives the transport number of the *counter-ion* of the ion with respect to which the electrodes are reversible (in this case, the transport number of H^+).

Justification 24.7

The argument is similar to that used to analyse the Hittorf method. Consider the consequences of passing 1 mol of electrons through the cell with transport specified above. In the right-hand electrode compartment, 1 mol Cl^- is formed but t_- mol Cl^- migrates out of it across the junction. The net change is $(1 - t_-)$ mol $= t_+$ mol. At the same time, t_+ mol H^+ migrates into the compartment. In the left-hand electrode compartment 1 mol Cl^- is removed from solution (to form 1 mol AgCl), but t_- mol Cl^- flows in across the junction. The net change is therefore $(-1 + t_-)$ mol $= -t_+$ mol Cl^-. At the same time, t_+ mol H^+ flows out. The reaction Gibbs energy is therefore

$$\Delta_r G = t_+\{\mu(Cl^-, m_2) - \mu(Cl^-, m_1) + \mu(H^+, m_2) - \mu(H^+, m_1)\}$$
$$= 2t_+RT \ln\frac{a_2}{a_1}$$

Because $\Delta_r G = -FE$, it follows that

$$E_t = -\frac{2t_+RT}{F}\ln\frac{a_2}{a_1}$$

For the same cell without transference, the Nernst equation gives

$$E = -\frac{RT}{F}\ln\frac{a_2}{a_1}$$

and the ratio of the two cell potentials is $2t_+$.

24.9 Conductivities and ion–ion interactions

The remaining problem is to account for the $c^{1/2}$ dependence of the Kohlrausch law (eqn 31). In Section 10.2c we saw something similar: the activity coefficients of ions at low concentrations also depend on $c^{1/2}$ and depend on their charge type rather than their specific identities. That $c^{1/2}$ dependence was explained in terms of the properties of the ionic atmosphere around each ion, and we can suspect that the same explanation applies here too.

To accommodate the effect of motion, we need to modify the picture of an ionic atmosphere as a spherical haze of charge. The ions forming the atmosphere do not adjust to the moving ion infinitely quickly, and the atmosphere is incompletely formed in front of the moving ion and incompletely decayed behind the ion (Fig. 24.21). The overall effect is the displacement of the centre of charge of the atmosphere a short distance behind the moving ion. Because the two charges are opposite, the result is a retardation of the moving ion. This reduction of the ions' mobility is called the **relaxation effect**. A confirmation of the picture is obtained by observing the conductivities of ions at high frequencies, which are greater than at low frequencies: the atmosphere does not have time to follow the rapidly changing direction of motion of the ion, and the effect of the field averages to zero.

The ionic atmosphere has another effect on the motion of the ions. We have seen that the moving ion experiences a viscous drag. When the ionic atmosphere is present this drag is enhanced because the ionic atmosphere moves in an opposite direction to the central ion.

(a)

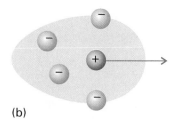

(b)

24.21 (a) In the absence of an applied field, the ionic atmosphere is spherically symmetric, but (b) when a field is present it is distorted and the centres of negative and positive charge no longer coincide. The attraction between the opposite charges retards the motion of the central ion.

Table 24.6* Debye–Hückel–Onsager coefficients for $(1,1)$-electrolytes at 298 K

Solvent	$A/(\text{mS m}^2 \text{ mol}^{-1}/ (\text{mol L}^{-1})^{1/2})$	$B/(\text{mol L}^{-1})^{-1/2}$
Methanol	15.61	0.923
Propanone	32.8	1.63
Water	6.02	0.229

*More values are given in the *Data section*.

The enhanced viscous drag, which is called the **electrophoretic effect**, reduces the mobility of the ions, and hence also reduces their conductivities.

The quantitative formulation of these effects is far from simple, but the **Debye–Hückel–Onsager theory** is an attempt to obtain quantitative expressions at about the same level of sophistication as the Debye–Hückel theory itself. The theory leads to a Kohlrausch-like expression in which

$$\mathcal{K} = A + B\Lambda_m^\circ \tag{59a}$$

with

$$A = \frac{z^2 e F^2}{3\pi\eta}\left(\frac{2}{\varepsilon RT}\right)^{1/2} \qquad B = \frac{qz^3 e F}{24\pi\varepsilon RT}\left(\frac{2}{\varepsilon RT}\right)^{1/2} \tag{59b}$$

where ε is the electric permittivity of the solvent (Section 22.1e) and $q = 0.586$ for a 1,1-electrolyte (Table 24.6). The slopes of the conductivity curves are predicted to depend on the charge type of the electrolyte, in accord with the Kohlrausch law, and some comparisons between theory and experiment are shown in Fig. 24.22. The agreement is quite good at very low molar concentrations (less than about 10^{-3} M, depending on the charge type).

Molecular dynamics calculations can also shed light on electric conduction. The key equation is a special case of a **Green–Kubo relation**, which expresses a transport property in terms of the fluctuations in microscopic properties of a system. The electrical conductivity is related to the fluctuations in the instantaneous electric current, j, in the sample that arises from variations in the velocities of the ions:

$$\kappa = \frac{1}{kTV}\int_0^\infty \langle j(0)j(t)\rangle\, dt \qquad j = \sum_i^N z_i e v_i \tag{60}$$

where v_i is the velocity of the ion i at a given instant and the angular brackets denote an average over the sample. If the ions are very mobile, there will be large fluctuations in the instantaneous currents in the sample, and the conductivity of the medium will be high. If the ions are locked into position, as in an ionic solid, there will be no instantaneous currents, and the ionic conductivity will be zero. The velocities of the ions are calculated explicitly in a molecular dynamics simulation, so the **correlation function**, the quantity $\langle j(0)j(t)\rangle$, can be evaluated reasonably simply.

Diffusion

We are now in a position to extend the discussion of ionic motion to cover the migration of neutral molecules and of ions in the absence of an applied electric field. We shall do this by expressing ion motion in a more general way than hitherto, and will then discover that the same equations apply even when the charge on the particles is zero.

24.10 The thermodynamic view

We saw in Part 1 that, at constant temperature and pressure, the maximum non-expansion work that can be done per mole when a substance moves from a location where its chemical potential is μ to a location where its chemical potential is $\mu + d\mu$ is $dw = d\mu$. In a system in which the chemical potential depends on the position x,

$$dw = d\mu = \left(\frac{\partial\mu}{\partial x}\right)_{p,T} dx \tag{61}$$

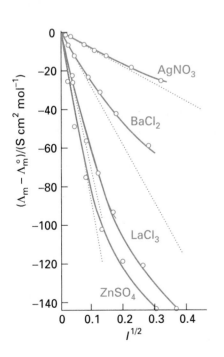

24.22 The dependence of molar conductivities on the square root of the ionic strength, and comparison (dotted lines) with the dependence predicted by the Debye–Hückel–Onsager theory.

We also saw in Chapter 2 (Table 2.1) that in general work can always be expressed in terms of an opposing force (which here we write $\mathcal{F}$), and that

$$dw = -\mathcal{F}\,dx \tag{62}$$

By comparing these two expressions, we see that the slope of the chemical potential can be interpreted as an effective force per mole of molecules. We write this **thermodynamic force** as

$$\mathcal{F} = -\left(\frac{\partial\mu}{\partial x}\right)_{p,T} \tag{63}$$

There is not necessarily a real force pushing the particles down the slope of the chemical potential. As we shall see, the force may represent the spontaneous tendency of the molecules to disperse as a consequence of the Second Law and the hunt for maximum entropy.

(a) The thermodynamic force of a concentration gradient

In a solution in which the activity of the solute is a, the chemical potential is

$$\mu = \mu^{\ominus} + RT \ln a$$

If the solution is not uniform the activity depends on the position and we can write

$$\mathcal{F} = -RT\left(\frac{\partial \ln a}{\partial x}\right)_{p,T} \tag{64}$$

If the solution is ideal, a may be replaced by the molar concentration c, and then

$$\mathcal{F} = -\frac{RT}{c}\left(\frac{\partial c}{\partial x}\right)_{p,T} \tag{65}°$$

because $(d \ln c/dx) = (1/c)dc/dx$.

Example 24.5 Calculating the thermodynamic force

Suppose the concentration of a solute decays exponentially along the length of a container. Calculate the thermodynamic force on the solute at 25°C given that the concentration falls to half its value in 10 cm.

Method According to eqn 65, the thermodynamic force is calculated by differentiating the concentration with respect to distance. Therefore, write an expression for the variation of the concentration with distance, and then differentiate it. Note that $1\,J = 1\,N\,m$.

Answer The concentration varies with position as

$$c = c_0 e^{-x/\lambda}$$

where λ is the decay constant. Therefore,

$$\frac{dc}{dx} = -\frac{c}{\lambda}$$

Equation 65 then implies that

$$\mathcal{F} = \frac{RT}{\lambda}$$

We know that the concentration falls to $\frac{1}{2}c_0$ at $x = 10$ cm, so we can find λ from $\frac{1}{2} = e^{-(10\ \text{cm})/\lambda}$. That is, $\lambda = (10\ \text{cm}/\ln 2)$. It follows that

$$\mathcal{F} = \frac{(8.31451\ \text{J K}^{-1}\,\text{mol}^{-1}) \times (298\ \text{K}) \times \ln 2}{1.0 \times 10^{-1}\ \text{m}} = 17\ \text{kN mol}^{-1}$$

- -

Self-test 24.5 Calculate the thermodynamic force on the molecules of molar mass M in a vertical tube in a gravitational field on the surface of the Earth, and evaluate $\mathcal{F}$ for molecules of molar mass $100\ \text{g mol}^{-1}$. Comment on its magnitude relative to that just calculated.

$[\mathcal{F} = -Mg,\ -0.98\ \text{N mol}^{-1}$; the force arising from
the concentration gradient greatly dominates that
arising from the gravitational gradient]

(b) Fick's first law of diffusion

In Section 24.4a it was shown that Fick's first law of diffusion (that the particle flux is proportional to the concentration gradient) could be deduced from the kinetic theory of gases. We shall now show that it can be deduced more generally and that it applies to the diffusion of species in condensed phases too.

We suppose that the flux of diffusing particles is motion in response to a thermodynamic force arising from a concentration gradient. The particles reach a steady drift speed, s, when the thermodynamic force, $\mathcal{F}$, is matched by the viscous drag. This drift speed is proportional to the thermodynamic force, and we write $s \propto \mathcal{F}$. However, the particle flux, J, is proportional to the drift speed, and the thermodynamic force is proportional to the concentration gradient, dc/dx. The chain of proportionalities ($J \propto s$, $s \propto \mathcal{F}$, and $\mathcal{F} \propto dc/dx$) implies that $J \propto dc/dx$, which is the content of Fick's law.

(c) The Einstein relation

Fick's law can be written[8]

$$J = -D\frac{dc}{dx} \tag{66}$$

In this expression, D is the diffusion coefficient and dc/dx is the slope of the molar concentration. The flux is related to the drift speed by

$$J = sc \tag{67}$$

This relation follows from the argument that we have used several times before. Thus, all particles within a distance $s\Delta t$, and therefore in a volume $s\Delta tA$, can pass through a window of area A in an interval Δt. Hence, the amount of substance that can pass through the window in that interval is $s\Delta tAc$. Therefore,

$$sc = -D\frac{dc}{dx}$$

If now we express dc/dx in terms of $\mathcal{F}$ by using eqn 65, we find

$$s = -\frac{D}{c}\frac{dc}{dx} = \frac{D\mathcal{F}}{RT} \tag{68}$$

8   This expression is derived from eqn 8 by dividing both sides by the Avogadro constant, which converts numbers into amounts (numbers of moles).

Therefore, once we know the effective force and the diffusion coefficient, D, we can calculate the drift speed of the particles (and vice versa) whatever the origin of the force.

There is one case where we already know the drift speed and the effective force acting on a particle: an ion in solution has a drift speed $s = u\mathcal{E}$ when it experiences a force $ez\mathcal{E}$ from an electric field of strength $\mathcal{E}$. Therefore, substituting these known values into eqn 68 gives

$$u\mathcal{E} = \frac{zF\mathcal{E}D}{RT}$$

and hence

$$u = \frac{zFD}{RT} \tag{69}$$

This equation rearranges into the very important result known as the **Einstein relation** between the diffusion coefficient and the ionic mobility:

$$D = \frac{uRT}{zF} \tag{70}°$$

On inserting the typical value $u = 5 \times 10^{-8}\ \mathrm{m^2\,s^{-1}\,V^{-1}}$, we find $D \approx 1 \times 10^{-9}\ \mathrm{m^2\,s^{-1}}$ at 25°C as a typical value of the diffusion coefficient of an ion in water.

(d) The Nernst–Einstein equation

The Einstein relation provides a link between the molar conductivity of an electrolyte and the diffusion coefficients of its ions. First, by using eqns 46 and 70 we write

$$\lambda = zuF = \frac{z^2DF^2}{RT} \tag{71}°$$

for each type of ion. Then, from $\Lambda_m° = \nu_+ \lambda_+ + \nu_- \lambda_-$, the limiting molar conductivity is

$$\Lambda_m° = (\nu_+ z_+^2 D_+ + \nu_- z_-^2 D_-)\frac{F^2}{RT} \tag{72}°$$

which is the **Nernst–Einstein equation**. One of its applications is to the determination of ionic diffusion coefficients from conductivity measurements; another is to the prediction of conductivities using models of ionic diffusion (see below).

(e) The Stokes–Einstein equation

Equations 45 ($u = ez/f$) and 70 relate the mobility of an ion to the frictional force and to the diffusion coefficient, respectively. The two expressions can be combined to give the **Stokes–Einstein equation**:

$$D = \frac{kT}{f} \tag{73}$$

If the frictional force is described by Stokes' law, then we also obtain a relation between the diffusion coefficient and the viscosity of the medium:

$$D = \frac{kT}{6\pi\eta a} \tag{74}$$

An important feature of eqn 73 (and of its special case, eqn 74) is that it makes no reference to the charge of the diffusing species. Therefore, the equation also applies in the limit of vanishingly small charge, that is, it also applies to neutral molecules. Consequently, we may use viscosity measurements to estimate the diffusion coefficients for electrically neutral molecules in solution (Table 24.7). It must not be forgotten, however, that both equations depend on the assumption that the viscous drag is proportional to the speed.

Table 24.7* Diffusion coefficients at 298 K, $D/(10^{-9}\ \mathrm{m^2\,s^{-1}})$

H$^+$ in water	9.31
I$_2$ in hexane	4.05
Na$^+$ in water	1.33
Sucrose in water	0.522

*More values are given in the *Data section*.

Example 24.6 Interpreting the mobility of an ion

Use the experimental value of the mobility to evaluate the diffusion coefficient, the limiting molar conductivity, and the hydrodynamic radius of a sulfate ion in aqueous solution.

Method The starting point is the mobility of the ion, which is given in Table 24.5. The diffusion coefficient can then be determined from the Einstein relation, eqn 70. The ionic conductivity is related to the mobility by eqn 46. To estimate the hydrodynamic radius, a, of the ion, use the Stokes–Einstein relation to find f and Stokes' law to relate f to a.

Answer From Table 24.5, the mobility of SO_4^{2-} is 8.29×10^{-8} m^2 s^{-1} V^{-1}. It follows from eqn 70 that

$$D = \frac{uRT}{zF} = 1.1 \times 10^{-9} \text{ m}^2 \text{ s}^{-1}$$

From eqn 46 it follows that

$$\lambda = zuF = 16 \text{ mS m}^2 \text{ mol}^{-1}$$

Finally, from $f = 6\pi\eta a$ using 1.00 cP (or 1.00×10^{-3} kg m^{-1} s^{-1}) for the viscosity of water (Table 24.3):

$$a = \frac{kT}{6\pi\eta D} = 200 \text{ pm}$$

Comment The bond length in SO_4^{2-} is 144 pm, so the radius calculated here is plausible and consistent with a small degree of solvation.

- -

Self-test 24.6 Repeat the calculation for the NH_4^+ ion.

$$[1.96 \times 10^{-9} \text{ m}^2 \text{ s}^{-1}, 7.4 \text{ mS m}^2 \text{ mol}^{-1}, 110 \text{ pm}]$$

Experimental support for the relations derived above comes from conductivity measurements. In particular, **Walden's rule** is the empirical observation that the product $\eta\Lambda_m$ is very approximately constant for the same ions in different solvents. Because $\Lambda_m \propto D$, and we have just seen that $D \propto 1/\eta$, we do indeed predict that $\Lambda_m \propto 1/\eta$, as Walden's rule implies. The usefulness of the rule, however, is muddied by the role of solvation: different solvents solvate the same ions to different extents, so both the hydrodynamic radius and the viscosity change with the solvent.

24.11 The diffusion equation

We now turn to the discussion of time-dependent diffusion processes, where we are interested in the spreading of inhomogeneities with time. One example is the temperature of a metal bar that has been heated at one end: if the source of heat is removed, then the bar gradually settles down into a state of uniform temperature. When the source of heat is maintained and the bar can radiate, it settles down into a steady state of nonuniform temperature. Another example (and one more relevant to chemistry) is the concentration distribution in a solvent to which a solute is added. We shall focus on the description of the diffusion of particles, but similar arguments apply to the diffusion of physical properties, such as temperature. Our aim is to obtain an equation for the rate of change of the concentration of particles in an inhomogeneous region.

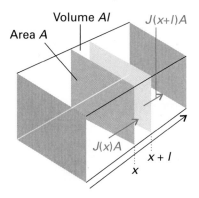

24.23 The net flux in a region is the difference between the flux entering from the region of high concentration (on the left) and the flux leaving to the region of low concentration (on the right).

The central equation of this section is the **diffusion equation**,[9] which relates the rate of change of concentration at a point to the spatial variation of the concentration at that point:

$$\frac{\partial c}{\partial t} = D \frac{\partial^2 c}{\partial x^2} \tag{75}$$

Justification 24.8

Consider a thin slab of cross-sectional area A that extends from x to $x + l$ (Fig. 24.23). Let the concentration at x be c at the time t. The amount (number of moles) of particles that enter the slab in the infinitesimal interval dt is $JAdt$, so the rate of increase in molar concentration inside the slab (which has volume Al) on account of the flux from the left is

$$\frac{\partial c}{\partial t} = \frac{JAdt}{Aldt} = \frac{J}{l}$$

There is also an outflow through the right-hand window. The flux through that window is J', and the rate of change of concentration that results is

$$\frac{\partial c}{dt} = -\frac{J'Adt}{Aldt} = -\frac{J'}{l}$$

The net rate of change of concentration is therefore

$$\frac{\partial c}{\partial t} = \frac{J - J'}{l}$$

Each flux is proportional to the concentration gradient at the window. So, by using Fick's first law, we can write

$$J - J' = -D \frac{\partial c}{\partial x} + D \frac{\partial c'}{\partial x} = -D \frac{\partial c}{\partial x} + D \frac{\partial}{\partial x} \left\{ c + \left(\frac{\partial c}{\partial x} \right) l \right\}$$

$$= Dl \frac{\partial^2 c}{\partial x^2}$$

When this relation is substituted into the expression for the rate of change of concentration in the slab, we get eqn 75.

(a) The significance of the diffusion equation

The diffusion equation shows that the rate of change of concentration is proportional to the curvature (more precisely, to the second derivative) of the concentration with respect to distance. If the concentration changes sharply from point to point (if the distribution is highly wrinkled), then the concentration changes rapidly with time. If the curvature is zero, then the concentration is constant in time. If the concentration decreases linearly with distance, then the concentration at any point is constant because the inflow of particles is exactly balanced by the outflow.

The diffusion equation can be regarded as a mathematical formulation of the intuitive notion that there is a natural tendency for the wrinkles in a distribution to disappear. More succinctly: Nature abhors a wrinkle.

(b) Diffusion with convection

The transport of particles arising from the motion of a streaming fluid is called **convection**. If for the moment we ignore diffusion, then the flux of particles through an area A in an

9   This equation used to be called 'Fick's second law of diffusion', but that name is now rarely used.

interval Δt when the fluid is flowing at a velocity v can be calculated in the way we have used several times before (by counting the particles within a distance $v\Delta t$), and is

$$J = \frac{cAv\Delta t}{A\Delta t} = cv \tag{76}$$

This J is called the **convective flux**. The rate of change of concentration in a slab of thickness l and area A is, by the same argument as before,

$$\frac{\partial c}{\partial t} = \frac{J - J'}{l} = \left\{ c - \left[c + \left(\frac{\partial c}{\partial x} \right) l \right] \right\} \frac{v}{l} = -v \frac{\partial c}{\partial x} \tag{77}$$

(We have assumed that the velocity does not depend on the position.)

When both diffusion and convection occur, the total change of concentration in a region is the sum of the two effects, and the **generalized diffusion equation** is

$$\frac{\partial c}{\partial t} = D \frac{\partial^2 c}{\partial x^2} - v \frac{\partial c}{\partial x} \tag{78}$$

A further refinement, which is important in chemistry, is the possibility that the concentrations of particles may change as a result of reaction. When reactions are included in eqn 78 (Section 27.3), we get a powerful differential equation for discussing the properties of reacting, diffusing, convecting systems and which is the basis of reactor design in chemical industry and of the utilization of resources in living cells.

(c) Solutions of the equation

The diffusion equation, eqn 75, is a second-order differential equation with respect to space and a first-order differential equation with respect to time. Therefore, we must specify two boundary conditions for the spatial dependence and a single initial condition for the time dependence.

As an illustration, consider a solvent in which the solute is initially coated on one surface of the container (for example, a layer of sugar on the bottom of a deep beaker of water). The single initial condition is that at $t = 0$ all N_0 particles are concentrated on the yz-plane (of area A) at $x = 0$. The two boundary conditions are derived from the requirements: (1) that the concentration must everywhere be finite and (2) that the total amount (number of moles) of particles present is n_0 (with $n_0 = N_0/N_A$) at all times. These requirements imply that the flux of particles is zero at the top and bottom surfaces of the system. Under these conditions it is found that

$$c(x, t) = \frac{n_0}{A(\pi Dt)^{1/2}} e^{-x^2/4Dt} \tag{79}$$

as may be verified by direct substitution. Figure 24.24 shows the shape of the concentration distribution at various times, and it is clear that the concentration spreads and tends to uniformity.

Another useful result is for a localized concentration of solute in a three-dimensional solvent (a sugar lump suspended in a large flask of water). The concentration of diffused solute is spherically symmetrical, and at a radius r is

$$c(r, t) = \frac{n_0}{8(\pi Dt)^{3/2}} e^{-r^2/4Dt} \tag{80}$$

Other chemically (and physically) interesting arrangements can be treated, but the solutions are more cumbersome.

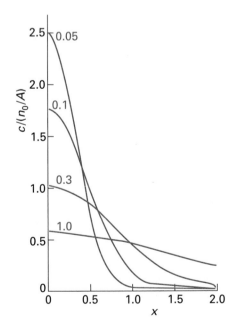

24.24 The concentration profiles above a plane from which a solute is diffusing. The curves are plots of eqn 79. The units of Dt and x are arbitrary, but are related so that Dt/x^2 is dimensionless. For example, if x is in metres, Dt would be in metres², so, for $D = 10^{-9}$ m² s⁻¹, $Dt = 0.1$ corresponds to $t = 10^4$ s.

(d) The measurement of diffusion coefficients

The solutions of the diffusion equation are useful for experimental determinations of diffusion coefficients. In the **capillary technique**, a capillary tube, open at one end and containing a solution, is immersed in a well stirred larger quantity of solvent, and the change of concentration in the tube is measured at a series of times. The solute diffuses from the open end of the capillary at a rate that can be calculated by solving the diffusion equation with the appropriate boundary conditions, so D may be determined. In the **diaphragm technique**, the diffusion occurs through the capillary pores of a sintered glass diaphragm separating the well-stirred solution and solvent. The concentrations are monitored and then related to the solutions of the diffusion equation corresponding to this arrangement.

24.12 Diffusion probabilities

The solutions of the diffusion equation can be used to predict the concentration of particles (or the value of some other physical quantity, such as the temperature in a nonuniform system) at any location. We can also use them to calculate the net distance through which the particles diffuse in a given time.

Example 24.7 Calculating the net distance of diffusion

Calculate the net distance travelled on average by particles in a time t if they are diffusing in a medium with diffusion constant D.

Method We need to calculate the probability that a particle will be found at a certain distance from the origin, and then calculate the average by weighting each distance by that probability.

Answer The number of particles in a slab of thickness dx and area A at x, where the molar concentration is c, is $cAN_A\,dx$. The probability that any of the $N_0 = n_0N_A$ particles is in the slab is therefore $cAN_A dx/N_0$. If the particle is in the slab, it has travelled a distance x from the origin. Therefore, the mean distance travelled by all the particles is the sum of each x weighted by the probability of its occurrence:

$$\langle x \rangle = \int_0^\infty \frac{xcAN_A}{N_0}\,dx = \frac{1}{(\pi Dt)^{1/2}}\int_0^\infty xe^{-x^2/4Dt}\,dx$$

$$= 2\left(\frac{Dt}{\pi}\right)^{1/2}$$

Comment The average distance of diffusion varies as the square root of the lapsed time. If we use the Stokes–Einstein relation for the diffusion coefficient, the mean distance travelled by particles of radius a in a solvent of viscosity η is

$$\langle x \rangle = \left(\frac{2kTt}{3\pi^2\eta a}\right)^{1/2}$$

Self-test 24.7 Derive an expression for the root mean square distance travelled by diffusing particles in a time t.

$$[\langle x^2 \rangle^{1/2} = (2Dt)^{1/2}]$$

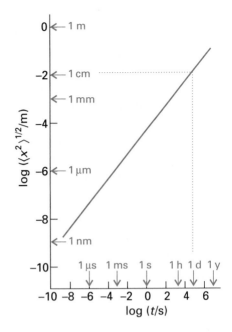

24.25 The root mean square distance covered by particles with $D = 5 \times 10^{-10} \, \text{m}^2 \, \text{s}^{-1}$. Note the great slowness of diffusion.

As shown in Example 24.7, the average distance travelled by diffusing particle in a time t (in an arrangement like that illustrated in Fig. 24.24) is

$$\langle x \rangle = 2 \left(\frac{Dt}{\pi} \right)^{1/2} \tag{81}$$

and the root mean square distance travelled in the same time is

$$\langle x^2 \rangle^{1/2} = (2Dt)^{1/2} \tag{82}$$

The latter is a valuable measure of the spread of particles when they can diffuse in both directions from the origin (for then $\langle x \rangle = 0$ at all times). The root mean square distance travelled by particles with a typical diffusion coefficient ($D = 5 \times 10^{-10} \, \text{m}^2 \, \text{s}^{-1}$) is illustrated in Fig. 24.25. The graph shows that diffusion is a very slow process (which is why solutions are stirred, to encourage mixing by convection).

24.13 The statistical view

An intuitive picture of diffusion is of the particles moving in a series of small steps and gradually migrating from their original positions. We shall explore this idea using a model in which the particles can jump through a distance λ in a time τ. The total distance travelled by a particle in a time t is therefore $t\lambda/\tau$. However, the particle will not necessarily be found at that distance from the origin. The direction of each step may be different, and the net distance travelled must take the changing directions into account.

If we simplify the discussion by allowing the particles to travel only along a straight line (the x-axis), and for each step (to the left or the right) to be through the same distance λ, then we obtain the **one-dimensional random walk**.[10] We show in the *Justification* below that the probability of a particle being at a distance x from the origin after a time t is

$$P = \left(\frac{2\tau}{\pi t} \right)^{1/2} e^{-x^2 \tau / 2t\lambda^2} \tag{83}$$

Justification 24.9

Consider a one-dimensional random walk in which each step is through a distance λ to the left or right. The net distance travelled after N steps is equal to the difference between the number of steps to the right (N_R) and to the left (N_L), and is $(N_R - N_L)\lambda$. We write $n = N_R - N_L$ and the total number of steps as $N = N_R + N_L$.

The number of ways of performing a walk with a given net distance of travel $n\lambda$ is the number of ways of making N_R steps to the right and N_L steps to the left, and is given by the binomial coefficient

$$W = \frac{N!}{N_L! N_R!} = \frac{N!}{\{\frac{1}{2}(N + n)\}! \{\frac{1}{2}(N - n)\}!}$$

The probability of the net distance walked being $n\lambda$ is

$$P = \frac{\text{number of paths with } N_R \text{ steps to the right}}{\text{total number of steps}}$$

$$= \frac{W}{2^N} = \frac{N!}{\{\frac{1}{2}(N + n)\}! \{\frac{1}{2}(N - n)\}! 2^N}$$

The use of Stirling's approximation (Section 19.1a) in the form

$$\ln x! \approx \ln (2\pi)^{1/2} + (x + \tfrac{1}{2}) \ln x - x$$

10　The same model was used in the discussion of a one-dimensional random coil in Section 23.6.

gives (after quite a lot of algebra)

$$\ln P = \ln \left(\frac{2}{\pi N}\right)^{1/2} - \tfrac{1}{2}(N + n + 1) \ln \left(1 + \frac{n}{N}\right) - \tfrac{1}{2}(N - n + 1) \ln \left(1 - \frac{n}{N}\right)$$

For small net distances ($n \ll N$) we can use the approximation $\ln (1 \pm x) \approx \pm x - \tfrac{1}{2}x^2$, and so obtain

$$\ln P \approx \ln \left(\frac{2}{\pi N}\right)^{1/2} - n^2/2N$$

At this point, we note that the number of steps taken in a time t is $N = t/\tau$ and the net distance travelled from the origin is $x = n\lambda$. Substitution of these quantities into the equation just derived gives an expression that rearranges into eqn 83.

The differences of detail between eqns 79 and 83 arise from the fact that in the present calculation the particles can migrate in either direction from the origin. Moreover, they can be found only at discrete points separated by λ instead of being anywhere on a continuous line. The fact that the two expressions are so similar suggests that diffusion can indeed be interpreted as the outcome of a large number of steps in random directions.

We can now relate the coefficient D to the step length λ and the rate at which the jumps occur. Thus, by comparing the two exponents in eqns 79 and 83 we can immediately write down the **Einstein–Smoluchowski equation**:

$$D = \frac{\lambda^2}{2\tau} \tag{84}$$

Illustration

Suppose that an SO_4^{2-} ion jumps through its own diameter each time it makes a move in an aqueous solution; then, because $D = 1.1 \times 10^{-9}$ $m^2 \, s^{-1}$ and $a = 210$ pm (as deduced from mobility measurements), it follows from $\lambda = 2a$ that $\tau = 80$ ps. Because τ is the time for one jump, the ion makes 1×10^{10} jumps per second.

The Einstein–Smoluchowski relation is the central connection between the microscopic details of particle motion and the macroscopic parameters relating to diffusion (for example, the diffusion coefficient and, through the Stokes–Einstein relation, the viscosity). It also brings us back full circle to the properties of the perfect gas. For, if we interpret λ/τ as $\bar{c}$, the mean speed of the molecules, and interpret λ as a mean free path, then we can recognize in the Einstein–Smoluchowski equation exactly the same expression as we obtained from the kinetic theory of gases, eqn 11. That is, the diffusion of a perfect gas is a random walk with an average step size equal to the mean free path.

Checklist of key ideas

Molecular motion in gases
- [ ] transport properties
- [ ] diffusion
- [ ] thermal conduction
- [ ] electric conduction
- [ ] viscosity
- [ ] effusion

24.1 Collisions with walls and surfaces
- [ ] collision flux (3)
- [ ] collision frequency

24.2 The rate of effusion
- [ ] Graham's law of effusion
- [ ] Knudsen method

24.3 Migration down gradients
- [ ] flux
- [ ] Fick's first law of diffusion (6)
- [ ] diffusion coefficient
- [ ] coefficient of thermal conductivity

- [ ] Newtonian flow
- [ ] coefficient of viscosity

24.4 Transport properties of a perfect gas
- [ ] diffusion coefficient of a perfect gas (11)

☐ coefficient of thermal conductivity of a perfect gas (16)
☐ viscosity of a perfect gas (21)
☐ Poiseuille's formula (23)

Motion in liquids

24.5 The structures of liquids
☐ radial distribution function
☐ long-range order
☐ short-range order
☐ virial
☐ Monte Carlo method
☐ molecular dynamics
☐ mesophase
☐ liquid crystal
☐ smectic phase
☐ nematic phase
☐ cholesteric phase

24.6 Molecular motion in liquids
☐ inelastic neutron scattering

24.7 The conductivities of electrolyte solutions
☐ polarization
☐ conductance
☐ conductivity (29)
☐ molar conductivity (30)
☐ strong electrolyte
☐ weak electrolyte
☐ Kohlrausch's law (31)
☐ limiting molar conductivity
☐ law of the independent migration of ions (32)
☐ degree of ionization (36)
☐ Ostwald's dilution law (39)

24.8 The mobilities of ions
☐ drift speed (43)
☐ hydrodynamic (Stokes) radius
☐ Grotthuss mechanism
☐ mobility of ion (44)
☐ mobility and conductivity (46)
☐ transport number (50)
☐ limiting transport number

☐ transport number and conductivity (55)
☐ moving boundary method
☐ indicator solution
☐ leading solution
☐ Hittorf method
☐ cell with transference

24.9 Conductivities and ion–ion interactions
☐ relaxation effect
☐ electrophoretic effect
☐ Debye–Hückel–Onsager theory (59)
☐ Green–Kubo relation (60)
☐ correlation function

Diffusion

24.10 The thermodynamic view
☐ thermodynamic force (63)
☐ thermodynamic force due to a concentration gradient (65)
☐ Einstein relation (70)

☐ Nernst–Einstein equation (72)
☐ Stokes–Einstein equation (73)
☐ Walden's rule

24.11 The diffusion equation
☐ diffusion equation (75)
☐ convection
☐ convective flux
☐ generalized diffusion equation (78)
☐ capillary technique
☐ diaphragm technique

24.12 Diffusion probabilities
☐ root mean square migration distance (82)

24.13 The statistical view
☐ one-dimensional random walk
☐ probability of location (83)
☐ Einstein–Smoluchowski equation (84)

Further reading

Articles of general interest

H.G. Herz, B.M. Braun, K.J. Müller, and R. Mauer, What is the physical significance of the pictures representing the Grotthuss H^+ conductance mechanism? *J. Chem. Educ.* **64**, 777 (1987).

N. Agmon, The Grotthuss mechanism. *Chem. Phys. Letts.* **244**, 456 (1995).

B.L. Earl, Confusion in the expressions for transport coefficients. *J. Chem. Educ.* **66**, 147 (1989).

T. Kenney, Graham's law: defining gas velocities. *J. Chem. Educ.* **67**, 871 (1990).

B.J. Alder and A.J. Ladd, Simulation by molecular dynamics. In *Encyclopedia of applied physics* (ed. G.L. Trigg), **18**, 281. VCH, New York (1997).

J.D. Lister and R. Shashidhar, Structure of liquid crystals. In *Encyclopedia of applied physics* (ed. G.L. Trigg), **8**, 515. VCH, New York (1994).

N.E. Cusack, Structure of simple liquids. In *Encyclopedia of applied physics* (ed. G.L. Trigg), **8**, 533. VCH, New York (1994).

D.G. Leaist, Diffusion and ionic conduction in liquids. In *Encyclopedia of applied physics* (ed. G.L. Trigg), **5**, 661. VCH, New York (1993).

Texts and sources of data and information

P.J. Collings, *Liquid crystals: Nature's delicate phase of matter.* Wiley, New York (1991).

G.A. Krestov, *et al., Ionic solvation.* Ellis Horwood/Prentice-Hall, Hemel Hempstead (1993).

W. Kauzmann, *Kinetic theory of gases.* Addison-Wesley, Reading (1966).

J.O. Hirschfelder, C.F. Curtiss, and R.B. Bird, *The molecular theory of gases and liquids.* Wiley, New York (1954).

D. Tabor, *Gases, liquids, and solids.* Cambridge University Press (1979).

A.J. Walton, *Three phases of matter.* Oxford University Press (1983).

R.B. Bird, W.E. Stewart, and E.N. Lightfoot, *Transport phenomena.* Wiley, New York (1960).

M. Spiro, Determination of transference numbers. In *Techniques of chemistry* (ed. A. Weissberger and B.W. Rossiter), **2A**, 205. Wiley-Interscience, New York (1971).

A.J. Bard and L.R. Faulkner, *Electrochemical methods: fundamentals and applications.* Wiley, New York (1980).

P.J. Dunlop, B.J. Steel, and J.E. Lane, Experimental methods for studying diffusion in liquids, gases, and solids. In *Techniques of chemistry* (ed. A. Weissberger and B.W. Rossiter), **4**, 205. Wiley-Interscience, New York (1972).

J. Crank, *The mathematics of diffusion*. Clarendon Press, Oxford (1975).

J.S. Rowlinson and F.L. Swinton, *Liquids and liquid mixtures*. Butterworth, London (1982).

J.N. Murrell and A.D. Jenkins, *Properties of liquids and solutions*. Wiley-Interscience, New York (1994).

A.R. Leach, *Molecular modelling: principles and applications*. Longman, Harlow (1996).

D. Frenkel and B. Smit, *Understanding molecular simulation*. Academic Press, San Diego (1996).

M.P. Allen and D. Tildesley, *Computer simulation of liquids*. Clarendon Press, Oxford (1986).

Exercises

24.1 (a) A solid surface with dimensions $2.5\ \text{mm} \times 3.0\ \text{mm}$ is exposed to argon gas at 90 Pa and 500 K. How many collisions do the Ar atoms make with this surface in 15 s?

24.1 (b) A solid surface with dimensions $3.5\ \text{mm} \times 4.0\ \text{cm}$ is exposed to helium gas at 111 Pa and 1500 K. How many collisions do the He atoms make with this surface in 10 s?

24.2 (a) An effusion cell has a circular hole of diameter 2.50 mm. If the molar mass of the solid in the cell is $260\ \text{g mol}^{-1}$ and its vapour pressure is 0.835 Pa at 400 K, by how much will the mass of the solid decrease in a period of 2.00 h?

24.2 (b) An effusion cell has a circular hole of diameter 3.00 mm. If the molar mass of the solid in the cell is $300\ \text{g mol}^{-1}$ and its vapour pressure is 0.224 Pa at 450 K, by how much will the mass of the solid decrease in a period of 24.00 h?

24.3 (a) Calculate the flux of energy arising from a temperature gradient of $2.5\ \text{K m}^{-1}$ in a sample of argon in which the mean temperature is 273 K.

24.3 (b) Calculate the flux of energy arising from a temperature gradient of $3.5\ \text{K m}^{-1}$ in a sample of hydrogen in which the mean temperature is 260 K.

24.4 (a) Use the experimental value of the thermal conductivity of neon (Table 24.1) to estimate the collision cross-section of Ne atoms at 273 K.

24.4 (b) Use the experimental value of the thermal conductivity of nitrogen (Table 24.1) to estimate the collision cross-section of N_2 molecules at 298 K.

24.5 (a) In a double-glazed window, the panes of glass are separated by 5.0 cm. What is the rate of transfer of heat by conduction from the warm room (25°C) to the cold exterior (−10°C) through a window of area $1.0\ \text{m}^2$? What power of heater is required to make good the loss of heat?

24.5 (b) Two sheets of copper of area $1.50\ \text{m}^2$ are separated by 10.0 cm. What is the rate of transfer of heat by conduction from the warm sheet (50°C) to the cold sheet (−10°C)? What is the rate of loss of heat?

24.6 (a) A manometer was connected to a bulb containing carbon dioxide under slight pressure. The gas was allowed to escape through a small pinhole, and the time for the manometer reading to drop from 75 cm to 50 cm was 52 s. When the experiment was repeated using

nitrogen (for which $M = 28.01\ \text{g mol}^{-1}$) the same fall took place in 42 s. Calculate the molar mass of carbon dioxide.

24.6 (b) A manometer was connected to a bulb containing nitrogen under slight pressure. The gas was allowed to escape through a small pinhole, and the time for the manometer reading to drop from 65.1 cm to 42.1 cm was 18.5 s. When the experiment was repeated using a fluorocarbon gas, the same fall took place in 82.3 s. Calculate the molar mass of the fluorocarbon.

24.7 (a) A space vehicle of internal volume $3.0\ \text{m}^3$ is struck by a meteor and a hole of radius 0.10 mm is formed. If the oxygen pressure within the vehicle is initially 80 kPa and its temperature 298 K, how long will the pressure take to fall to 70 kPa?

24.7 (b) A container of internal volume $22.0\ \text{m}^3$ was punctured, and a hole of radius 0.050 mm was formed. If the nitrogen pressure within the vehicle is initially 122 kPa and its temperature 293 K, how long will the pressure take to fall to 105 kPa?

24.8 (a) Use the experimental value of the coefficient of viscosity for neon (Table 24.1) to estimate the collision cross-section of Ne atoms at 273 K.

24.8 (b) Use the experimental value of the coefficient of viscosity for nitrogen (Table 24.1) to estimate the collision cross-section of the molecules at 273 K.

24.9 (a) Calculate the inlet pressure required to maintain a flow rate of $9.5 \times 10^5\ \text{L h}^{-1}$ of nitrogen at 293 K flowing through a pipe of length 8.50 m and diameter 1.00 cm. The pressure of gas as it leaves the tube is 1.00 bar. The volume of the gas is measured at that pressure.

24.9 (b) Calculate the inlet pressure required to maintain a flow rate of $8.70\ \text{cm}^3\ \text{s}^{-1}$ of nitrogen at 300 K flowing through a pipe of length 10.5 m and diameter 15 mm. The pressure of gas as it leaves the tube is 1.00 bar. The volume of the gas is measured at that pressure.

24.10 (a) Calculate the viscosity of air at (a) 273 K, (b) 298 K, (c) 1000 K. Take $\sigma \approx 0.40\ \text{nm}^2$. (The experimental values are 173 μP at 273 K, 182 μP at 20°C, and 394 μP at 600°C.)

24.10 (b) Calculate the viscosity of benzene vapour at (a) 273 K, (b) 298 K, (c) 1000 K. Take $\sigma \approx 0.88\ \text{nm}^2$.

24.11 (a) Calculate the thermal conductivities of (a) argon, (b) helium at 300 K and 1.0 mbar. Each gas is confined in a cubic vessel of side 10 cm, one wall being at 310 K and the one opposite at

295 K. What is the rate of flow of energy as heat from one wall to the other in each case?

24.11 (b) Calculate the thermal conductivities of (a) neon, (b) nitrogen at 300 K and 15 mbar. Each gas is confined in a cubic vessel of side 15 cm, one wall being at 305 K and the one opposite at 295 K. What is the rate of flow of energy as heat from one wall to the other in each case?

24.12 (a) The viscosity of carbon dioxide was measured by comparing its rate of flow through a long narrow tube (using Poiseuille's formula) with that of argon. For the same pressure differential, the same volume of carbon dioxide passed through the tube in 55 s as that of argon in 83 s. The viscosity of argon at 25°C is 208 μP; what is the viscosity of carbon dioxide? Estimate the molecular diameter of carbon dioxide.

24.12 (b) The viscosity of a chlorofluorocarbon (CFC) was measured by comparing its rate of flow through a long narrow tube (using Poiseuille's formula) with that of argon. For the same pressure differential, the same volume of the CFC passed through the tube in 72.0 s as that of argon in 18.0 s. The viscosity of argon at 25°C is 208 μP; what is the viscosity of the CFC? Estimate the molecular diameter of the CFC. Take $M = 200$ g mol^{-1}.

24.13 (a) Calculate the thermal conductivity of argon ($C_{V,m} = 12.5$ J K^{-1} mol^{-1}, $\sigma = 0.36$ nm^2) at room temperature (20°C).

24.13 (b) Calculate the thermal conductivity of nitrogen ($C_{V,m} = 20.8$ J K^{-1} mol^{-1}, $\sigma = 0.43$ nm^2) at room temperature (20°C).

24.14 (a) Calculate the diffusion constant of argon at 25°C and (a) 1.00 Pa, (b) 100 kPa, (c) 10.0 MPa. If a pressure gradient of 0.10 atm cm^{-1} is established in a pipe, what is the flow of gas due to diffusion?

24.14 (b) Calculate the diffusion constant of nitrogen at 25°C and (a) 10.0 Pa, (b) 100 kPa, (c) 15.0 MPa. If a pressure gradient of 0.20 bar m^{-1} is established in a pipe, what is the flow of gas due to diffusion?

24.15 (a) The mobility of a chloride ion in aqueous solution at 25°C is 7.91×10^{-8} m^2 s^{-1} V^{-1}. Calculate the molar ionic conductivity.

24.15 (b) The mobility of an acetate ion in aqueous solution at 25°C is 4.24×10^{-8} m^2 s^{-1} V^{-1}. Calculate the molar ionic conductivity.

24.16 (a) The mobility of a Rb$^+$ ion in aqueous solution is 7.92×10^{-8} m^2 s^{-1} V^{-1} at 25°C. The potential difference between two electrodes placed in the solution is 35.0 V. If the electrodes are 8.00 mm apart, what is the drift speed of the Rb$^+$ ion?

24.16 (b) The mobility of a Li$^+$ ion in aqueous solution is 4.01×10^{-8} m^2 s^{-1} V^{-1} at 25°C. The potential difference between two electrodes placed in the solution is 12.0 V. If the electrodes are 1.00 cm apart, what is the drift speed of the ion?

24.17 (a) What fraction of the total current is carried by Li$^+$ when current flows through an aqueous solution of LiBr at 25°C?

24.17 (b) What fraction of the total current is carried by Cl$^-$ when current flows through an aqueous solution of NaCl at 25°C?

24.18 (a) The limiting molar conductivities of KCl, KNO$_3$, and AgNO$_3$ are 14.99 mS m^2 mol^{-1}, 14.50 mS m^2 mol^{-1}, and 13.34 mS m^2 mol^{-1}, respectively (all at 25°C). What is the limiting molar conductivity of AgCl at this temperature?

24.18 (b) The limiting molar conductivities of NaI, NaCH$_3$CO$_2$, and Mg(CH$_3$CO$_2$)$_2$ are 12.69 mS m^2 mol^{-1}, 9.10 mS m^2 mol^{-1}, and 18.78 mS m^2 mol^{-1}, respectively (all at 25°C). What is the limiting molar conductivity of MgI$_2$ at this temperature?

24.19 (a) At 25°C the molar ionic conductivities of Li$^+$, Na$^+$, and K$^+$ are 3.87 mS m^2 mol^{-1}, 5.01 mS m^2 mol^{-1}, and 7.35 mS m^2 mol^{-1}, respectively. What are their mobilities?

24.19 (b) At 25°C the molar ionic conductivities of F$^-$, Cl$^-$, and Br$^-$ are 5.54 mS m^2 mol^{-1}, 7.635 mS m^2 mol^{-1}, and 7.81 mS m^2 mol^{-1}, respectively. What are their mobilities?

24.20 (a) The mobility of a NO$_3^-$ ion in aqueous solution at 25°C is 7.40×10^{-8} m^2 s^{-1} V^{-1}. Calculate its diffusion coefficient in water at 25°C.

24.20 (b) The mobility of a CH$_3$CO$_2^-$ ion in aqueous solution at 25°C is 4.24×10^{-8} m^2 s^{-1} V^{-1}. Calculate its diffusion coefficient in water at 25°C.

24.21 (a) The diffusion coefficient of CCl$_4$ in heptane at 25°C is 3.17×10^{-9} m^2 s^{-1}. Estimate the time required for a CCl$_4$ molecule to have a root mean square displacement of 5.0 mm.

24.21 (b) The diffusion coefficient of I$_2$ in hexane at 25°C is 4.05×10^{-9} m^2 s^{-1}. Estimate the time required for an iodine molecule to have a root mean square displacement of 1.0 cm.

24.22 (a) Estimate the effective radius of a sucrose molecule in water at 25°C given that its diffusion coefficient is 5.2×10^{-10} m^2 s^{-1} and that the viscosity of water is 1.00 cP.

24.22 (b) Estimate the effective radius of a glycine molecule in water at 25°C given that its diffusion coefficient is 1.055×10^{-9} m^2 s^{-1} and that the viscosity of water is 1.00 cP.

24.23 (a) The diffusion coefficient for molecular iodine in benzene is 2.13×10^{-9} m^2 s^{-1}. How long does a molecule take to jump through about one molecular diameter (approximately the fundamental jump length for translational motion)?

24.23 (b) The diffusion coefficient for CCl$_4$ in heptane is 3.17×10^{-9} m^2 s^{-1} and its viscosity is 0.387 kg m^{-1} s^{-1}. How long does a molecule take to jump through about one molecular diameter (approximately the fundamental jump length for translational motion)?

24.24 (a) What is the root mean square distance travelled by an iodine molecule in benzene at 25°C in 1.0 s?

24.24 (b) What is the root mean square distance travelled by a sucrose molecule in water at 25°C in 1.0 s?

24.25 (a) About how long, on average, does it take for the molecules in Exercise 24.24a to drift to a point (a) 1.0 mm, (b) 1.0 cm from their starting points?

24.25 (b) The diffusion coefficient of a particular kind of t-RNA molecule is $D = 1.0 \times 10^{-11}$ m^2 s^{-1} in the medium of a cell interior. How long does it take molecules produced in the cell nucleus to reach the walls of the cell at a distance 1.0 μm, corresponding to the radius of the cell?

Problems

Numerical problems

24.1 Enrico Fermi, the great Italian scientist, was a master at making good approximate calculations based on little or no actual data. Hence, such calculations are often called 'Fermi calculations'. Do a Fermi calculation on how long it would take for a gaseous air-borne cold virus of molar mass 100 kg mol^{-1} to travel the distance between two conversing people 1.0 m apart by diffusion in still air.

24.2 Calculate the ratio of the thermal conductivities of gaseous hydrogen at 300 K to gaseous hydrogen at 10 K. Be circumspect, and think about the modes of motion that are thermally active at the two temperatures.

24.3 In the Knudsen method for the determination of vapour pressure, a weighed amount of a sample is heated inside a container, in the wall of which there is a small hole. The mass loss over a period of time can be related to the vapour pressure at the temperature of the experiment. If Δw is the mass lost in an interval Δt through a circular hole of radius R, find an expression relating the vapour pressure, p, to Δw and Δt. A Knudsen cell was used to determine the vapour pressure of germanium at 1000°C. During an interval of 7200 s the mass loss through a hole of radius 0.50 mm amounted to 43 μg. What is the vapour pressure of germanium at 1000°C? Assume the gas to be monatomic.

24.4 In a study of the catalytic properties of a titanium surface it was necessary to maintain the surface free from contamination. Calculate the collision frequency per square centimetre of surface made by O$_2$ molecules at (a) 100 kPa, (b) 1.00 Pa and 300 K. Estimate the number of collisions made with a single surface atom in each second. The conclusions underline the importance of working at very low pressures (much lower than 1 Pa, in fact) in order to study the properties of uncontaminated surfaces. Take the nearest-neighbour distance as 291 pm.

24.5 The nuclide ^{244}Bk (berkelium) decays by producing α particles, which capture electrons and form He atoms. Its half-life is 4.4 h. A sample of mass 1.0 mg was placed in a container of volume 1.0 cm^3 that was impermeable to α radiation, but there was also a hole of radius 2.0 μm in the wall. What is the pressure of helium at 298 K, inside the container after (a) 1.0 h, (b) 10 h?

24.6 An atomic beam is designed to function with (a) cadmium, (b) mercury. The source is an oven maintained at 380 K, there being a small slit of dimensions 1.0 cm $\times$ 1.0 $\times$ 10^{-3} cm. The vapour pressure of cadmium is 0.13 Pa and that of mercury is 152 kPa at this temperature. What is the atomic current (the number of atoms per unit time) in the beams?

24.7 Conductivities are often measured by comparing the resistance of a cell filled with the sample to its resistance when filled with some standard solution, such as aqueous potassium chloride. The conductivity of water is 76 mS m^{-1} at 25°C and the conductivity of 0.100 mol L^{-1} KCl(aq) is 1.1639 S m^{-1}. A cell had a resistance of 33.21 Ω when filled with 0.100 mol L^{-1} KCl(aq) and 300.0 Ω when filled with 0.100 mol L^{-1} CH$_3$COOH. What is the molar conductivity of acetic acid at that concentration and temperature?

24.8 The resistances of a series of aqueous NaCl solutions, formed by successive dilution of a sample, were measured in a cell with cell constant (the constant C in the relation $\kappa = C/R$) 0.2063 cm^{-1}. The following values were found:

$c/(\text{mol L}^{-1})$	0.00050	0.0010	0.0050	0.010	0.020	0.050
R/Ω	3314	1669	342.1	174.1	89.08	37.14

Verify that the molar conductivity follows Kohlrausch's law and find the limiting molar conductivity. Determine the coefficient $\mathcal{K}$. Use the value of $\mathcal{K}$ (which should depend only on the nature, not the identity of the ions) and the information that $\lambda(\text{Na}^+) = 5.01$ mS m^2 mol^{-1} and $\lambda(\text{I}^-) = 7.68$ mS m^2 mol^{-1} to predict (a) the molar conductivity, (b) the conductivity, (c) the resistance it would show in the cell, of 0.010 mol L^{-1} NaI(aq) at 25°C.

24.9 After correction for the water conductivity, the conductivity of a saturated aqueous solution of AgCl at 25°C was found to be 0.1887 mS m^{-1}. What is the solubility of silver chloride at this temperature?

24.10 What are the drift speeds of Li$^+$, Na$^+$, and K$^+$ in water when a potential difference of 10 V is applied across a 1.00 cm conductivity cell? How long would it take an ion to move from one electrode to the other? In conductivity measurements it is normal to use alternating current: what are the displacements of the ions in (a) centimetres, (b) solvent diameters, about 300 pm, during a half cycle of 1.0 kHz applied potential?

24.11 The mobilities of H$^+$ and Cl$^-$ at 25°C in water are 3.623×10^{-7} m^2 s^{-1} V^{-1} and 7.91×10^{-8} m^2 s^{-1} V^{-1}, respectively. What proportion of the current is carried by the protons in 1.0 mM HCl(aq)? What fraction do they carry when the NaCl is added to the acid so that the solution is 1.0 mol L^{-1} in the salt? Note how concentration as well as mobility governs the transport of current.

24.12 In a moving boundary experiment on KCl the apparatus consisted of a tube of internal diameter 4.146 mm, and it contained aqueous KCl at a concentration of 0.021 mol L^{-1}. A steady current of 18.2 mA was passed, and the boundary advanced as follows:

$\Delta t/\text{s}$	200	400	600	800	1000
x/mm	64	128	192	254	318

Find the transport number of K$^+$, its mobility, and its ionic conductivity.

24.13 The proton possesses abnormal mobility in water, but does it behave normally in liquid ammonia? To investigate this question, a moving-boundary technique was used to determine the transport number of NH$_4^+$ in liquid ammonia (the analogue of H$_3$O$^+$ in liquid water) at −40°C (J. Baldwin, J. Evans, and J.B. Gill, *J. Chem. Soc. A*, 3389 (1971)). A steady current of 5.000 mA was passed for 2500 s, during which time the boundary formed between mercury(II) iodide

$0.013\,65 \text{ mol kg}^{-1}$ solution and 92.03 mm in a $0.042\,55 \text{ mol kg}^{-1}$ solution. Calculate the transport number of NH_4^+ at these concentrations, and comment on the mobility of the proton in liquid ammonia. The bore of the tube is 4.146 mm and the density of liquid ammonia is 0.682 g cm^{-3}.

24.14 A dilute solution of potassium permanganate in water at 25°C was prepared. The solution was in a horizontal tube of length 10 cm, and at first there was a linear gradation of intensity of the purple solution from the left (where the concentration was 0.100 mol L^{-1}) to the right (where the concentration was 0.050 mol L^{-1}). What is the magnitude and sign of the thermodynamic force acting on the solute (a) close to the left face of the container, (b) in the middle, (c) close to the right face? Give the force per mole and force per molecule in each case.

24.15 Estimate the diffusion coefficients and the effective hydrodynamic radii of the alkali metal cations in water from their mobilities at 25°C. Estimate the approximate number of water molecules that are dragged along by the cations. Ionic radii are given in Table 21.3.

24.16 Nuclear magnetic resonance can be used to determine the mobility of molecules in liquids. A set of measurements on methane in carbon tetrachloride showed that its diffusion coefficient is $2.05 \times 10^{-9} \text{ m}^2 \text{ s}^{-1}$ at 0°C and $2.89 \times 10^{-9} \text{ m}^2 \text{ s}^{-1}$ at 25°C. Deduce what information you can about the mobility of methane in carbon tetrachloride.

24.17 A concentrated sucrose solution is poured into a cylinder of diameter 5.0 cm. The solution consisted of 10 g of sugar in 5.0 cm³ of water. A further 1.0 L of water is then poured very carefully on top of the layer, without disturbing the layer. Ignore gravitational effects, and pay attention only to diffusional processes. Find the concentration at 5.0 cm above the lower layer after a lapse of (a) 10 s, (b) 1.0 y.

Theoretical problems

24.18 Show how the ratio of two transport numbers t' and t'' for two cations in a mixture depends on their concentrations c' and c'', and their mobilities u' and u''.

24.19 Confirm that eqn 79 is a solution of the diffusion equation with the correct initial value.

24.20 The diffusion equation is valid when many elementary steps are taken in the time interval of interest; but the random walk calculation lets us discuss distributions for short times as well as for long. Use eqn 83 to calculate the probability of being six paces from the origin (that is, at $x = 6\lambda$) after (a) four, (b) six, (c) twelve steps.

24.21 Write a program or use mathematical software to calculate P in a one-dimensional random walk, and evaluate the probability of

being at $x = n\lambda$ for $n = 6, 10, 14, \ldots, 60$. Compare the numerical value with the analytical value in the limit of a large number of steps. At what value of n is the discrepancy no more than 0.1 per cent?

Additional problems supplied by Carmen Giunta and Charles Trapp

24.22 A.K. Srivastava, R.A. Samant, and S.D. Patankar (*J. Chem. Eng. Data* **41**, 431 (1996)) measured the conductance of several salts in a binary solvent mixture of water and a dipolar aprotic solvent 1,3-dioxolan-2-one (ethylene carbonate). They report the following conductances at 25°C in a solvent 80 per cent 1,3-dioxolan-2-one by mass.

NaI

$c/(\text{mmol L}^{-1})$	32.02	20.28	12.06	8.64	2.85	1.24	0.83
$\Lambda_m/(\text{S cm}^2 \text{ mol}^{-1})$	50.26	51.99	54.01	55.75	57.99	58.44	58.67

KI

$c/(\text{mmol L}^{-1})$	17.68	10.88	7.19	2.67	1.28	0.83	0.19
$\Lambda_m/(\text{S cm}^2 \text{ mol}^{-1})$	42.45	45.91	47.53	51.81	54.09	55.78	57.42

Calculate Λ_m° for NaI and KI in this solvent and $\lambda^\circ(\text{Na}) - \lambda^\circ(\text{K})$. Compare your results to the analogous quantities in aqueous solution using Table 24.4 in the *Data section*.

24.23 A. Fenghour, W.A. Wakeham, V. Vesovic, J.T.R. Watson, J. Millat, and E. Vogel (*J. Phys. Chem. Ref. Data* **24**, 1649 (1995)) have compiled an extensive table of viscosity coefficients for ammonia in the liquid and vapour phases. Deduce the effective molecular diameter of NH_3 based on each of the following vapour-phase viscosity coefficients: (a) $\eta = 9.08 \times 10^{-6} \text{ kg m}^{-1} \text{ s}^{-1}$ at 270 K and 1.00 bar; (b) $\eta = 1.749 \times 10^{-5} \text{ kg m}^{-1} \text{ s}^{-1}$ at 490 K and 10.0 bar.

24.24 Interstellar space is quite a different medium from the gaseous environments we commonly encounter on Earth. For instance, a typical density of the medium is about 1 atom cm⁻³ and that atom is typically H; the effective temperature due to stellar background radiation is about 10 000 K. Estimate the diffusion coefficient and thermal conductivity of H under these conditions. (*Comment.* Energy is in fact transferred much more effectively by radiation.)

24.25 G. Bakale, K. Lacmann, and W.F. Schmidt (*J. Phys. Chem.* **100**, 12477 (1996)) measured the mobility of singly charged C_{60}^- ions in a variety of nonpolar solvents. In cyclohexane at 22°C, the mobility is $1.1 \text{ cm}^2 \text{ V}^{-1} \text{ s}^{-1}$. Estimate the effective radius of the C_{60}^- ion. The viscosity of the solvent is $0.93 \times 10^{-3} \text{ kg m}^{-1} \text{ s}^{-1}$. *Comment.* The researchers interpreted the substantial difference between this number and the van der Waals radius of neutral C_{60} in terms of a solvation layer around the ion.

25 The rates of chemical reactions

This chapter is the first of a sequence that explores the rates of chemical reactions. The chapter begins with a discussion of the definition of reaction rate and outlines the techniques for its measurement. The results of such measurements show that reaction rates depend on the concentration of reactants (and products) in characteristic ways that can be expressed in terms of differential equations known as rate laws. The solutions of these equations are used to predict the concentrations of species at any time after the start of the reaction. The form of the rate law also provides insight into the series of elementary steps by which a reaction takes place. The key task in this connection is the construction of a rate law from a proposed mechanism and its comparison with experiment. Simple elementary steps have simple rate laws, and these rate laws can be combined together by invoking one or more approximations. These approximations include the concept of the rate-determining stage of a reaction, the steady-state concentration of a reaction intermediate, and the existence of a pre-equilibrium.

This chapter introduces the principles of **chemical kinetics**, the study of reaction rates, by showing how the rates of reactions may be measured and interpreted. The remaining chapters of this part of the text then develop this material in more detail and apply it to more complicated or more specialized cases. The rate of a chemical reaction might depend on variables under our control, such as the pressure, the temperature, and the presence of a catalyst, and we may be able to optimize the rate by the appropriate choice of conditions. The study of reaction rates also leads to an understanding of the mechanisms of reactions, their analysis into a sequence of elementary steps. We saw in Chapter 9 that the Second Law accounts for the direction of spontaneous change. Here we explore why spontaneous chemical reactions occur at a finite rate and are not simply instantaneous.

Empirical chemical kinetics

The first step in the kinetic analysis of reactions is to establish the stoichiometry of the reaction and identify any side reactions. The basic data of chemical kinetics are then the concentrations of the reactants and products at different times after a reaction has been initiated. The rates of most chemical reactions are sensitive to the temperature, so in conventional experiments the temperature of the reaction mixture must be held constant throughout the course of the reaction. This requirement puts severe demands on the design of an experiment. Gas-phase reactions, for instance, are often carried out in a vessel held in contact with a substantial block of metal. Liquid-phase reactions, including flow reactions, must be carried out in an efficient thermostat. Special efforts have to be made to study reactions at low temperatures, as in the study of the kinds of reactions that take place in interstellar clouds. Thus, supersonic expansion of the reaction gas can be used to attain temperatures as low as 10 K. Non-isothermal conditions are sometimes employed. For instance, the shelf-life of an expensive pharmaceutical may be explored by slowly raising the temperature of a single sample.

25.1 Experimental techniques

The method used to monitor concentrations depends on the species involved and the rapidity with which their concentrations change. Many reactions reach equilibrium over periods of minutes or hours, and several techniques may then be used to follow the changing concentrations.

(a) Monitoring the progress of a reaction

A reaction in which at least one component is a gas might result in an overall change in pressure in a system of constant volume, so its progress may be followed by recording the variation of pressure with time.

Example 25.1 Monitoring the variation in pressure

Predict how the total pressure varies during the gas-phase decomposition $2N_2O_5(g) \rightarrow 4NO_2(g) + O_2(g)$.

Method The total pressure (at constant volume and temperature and assuming perfect gas behaviour) is proportional to the number of gas-phase molecules. Therefore, because each mole of N_2O_5 gives rise to $\frac{5}{2}$ mol of gas molecules, we can expect the pressure to rise to $\frac{5}{2}$ times its initial value. To confirm this conclusion, express the progress of the reaction in terms of the fraction, α, of N_2O_5 molecules that have reacted.

Answer Let the initial pressure be p_0 and the initial amount of N_2O_5 molecules present be n. When a fraction α of the N_2O_5 molecules has decomposed, the amounts of the components in the reaction mixture are:

	N_2O_5	NO_2	O_2	Total
Amount:	$n(1-\alpha)$	$2\alpha n$	$\frac{1}{2}\alpha n$	$n(1+\frac{3}{2}\alpha)$

When $\alpha = 0$ the pressure is p_0, so at any stage the total pressure is

$$p = (1 + \tfrac{3}{2}\alpha)p_0$$

When the reaction is complete, the pressure will have risen to $\frac{5}{2}$ times its initial value.

- -

Self-test 25.1 Repeat the calculation for $2NOBr(g) \rightarrow 2NO(g) + Br_2(g)$.

$$[p = (1 + \tfrac{1}{2}\alpha)p_0]$$

Spectrophotometry, the measurement of the intensity of absorption in a particular spectral region, is widely applicable, and is especially useful when one substance in the reaction mixture has a strong characteristic absorption in a conveniently accessible region of the electromagnetic spectrum. For example, the progress of the reaction

$$H_2(g) + Br_2(g) \longrightarrow 2HBr(g)$$

can be followed by measuring the absorption of visible light by bromine. If the reaction changes the number or type of ions present in a solution, then it may be followed by monitoring the electrical conductivity of the solution. The replacement of neutral molecules by ionic products can result in dramatic changes in the conductivity, as in the reaction

$$(CH_3)_3CCl(aq) + H_2O(l) \longrightarrow (CH_3)_3OH(aq) + H^+(aq) + Cl^-(aq)$$

If hydrogen ions are produced or consumed, the reaction may be followed by monitoring the pH of the solution.

Other methods of determining composition include mass spectrometry, gas chromatography, nuclear magnetic resonance, and electron spin resonance (for reactions involving radicals).

(b) Application of the techniques

In a **real-time analysis** the composition of the system is analysed while the reaction is in progress. Either a small sample is withdrawn or the bulk solution is monitored. In the **quenching method** the reaction is stopped after it has been allowed to proceed for a certain time, and the composition is analysed at leisure. The quenching (of the entire mixture or of a sample drawn from it) can be achieved either by cooling suddenly, by adding the mixture to a large volume of solvent, or by rapid neutralization of an acid reagent. This method is suitable only for reactions that are slow enough for there to be little reaction during the time it takes to quench the mixture. Many current investigations study fast reactions, which we shall take to be reactions complete in less than about 1 s (and often very much less), and the present thrust of chemical kinetics is to ever shorter timescales. With special laser techniques it is now possible to observe processes occurring in a few tens of femtoseconds.

In the **flow method** the reactants are mixed as they flow together in a chamber (Fig. 25.1). The reaction continues as the thoroughly mixed solutions flow through the outlet tube, and observation of the composition at different positions along the tube is equivalent to the observation of the reaction mixture at different times after mixing. The disadvantage of conventional flow techniques is that a large volume of reactant solution is necessary. This disadvantage is particularly important for fast reactions, because to spread the reaction over a length of tube the flow must be rapid. The **stopped-flow technique** avoids this disadvantage (Fig. 25.2). The suitability of the stopped-flow technique to the study of small samples means that it is appropriate for biochemical reactions, and it has been widely used to study the kinetics of enzyme action.

In **flash photolysis** the gaseous or liquid sample is exposed to a brief photolytic flash of light, and then the contents of the reaction chamber are monitored. Most work is now done with lasers with flashes of about 10 ns duration, but many studies are carried out at 1 ps, and some are done on a femtosecond timescale. Either emission or absorption spectroscopy may be used to monitor the reaction, and the spectra are recorded electronically at a series of times following the flash.

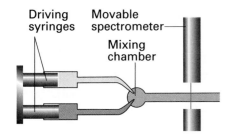

25.1 The arrangement used in the flow technique for studying reaction rates. The reactants are injected into the mixing chamber at a steady rate. The location of the spectrometer corresponds to different times after initiation.

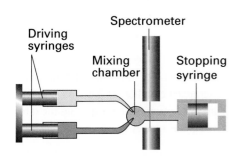

25.2 In the stopped-flow technique the reagents are driven quickly into the mixing chamber by the driving syringes and then the time dependence of the concentrations is monitored.

25.2 The rates of reactions

Reaction rates depend on the composition and the temperature of the reaction mixture. The next few sections look at these observations in more detail.

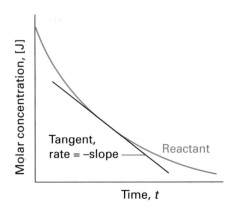

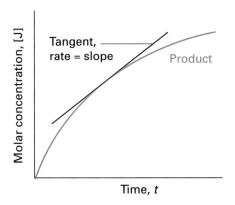

25.3 The definition of (instantaneous) rate as the slope of the tangent drawn to the curve showing the variation of concentration with time. For negative slopes, the sign is changed when reporting the rate, so all reaction rates are positive.

(a) The definition of rate

Consider a reaction of the form $A + 2B \rightarrow 3C + D$, in which at some instant the molar concentration of a participant J is [J]. The instantaneous **rate of consumption** of one of the reactants at a given time is $-d[R]/dt$, where R is A or B. This rate is a positive quantity (Fig. 25.3). The **rate of formation** of one of the products (C or D, which we denote P) is $d[P]/dt$ (note the difference in sign). This rate is also positive.

It follows from the stoichiometry for the reaction $A + 2B \rightarrow 3C + D$ that

$$\frac{d[D]}{dt} = \tfrac{1}{3}\frac{d[C]}{dt} = -\frac{d[A]}{dt} = -\tfrac{1}{2}\frac{d[B]}{dt}$$

so there are several rates connected with the reaction. The problem of having several possibly different rates to describe the same reaction is avoided by defining the unique **rate of reaction**, v, as

$$v = \frac{1}{\nu_J}\frac{d[J]}{dt} \qquad [1]$$

where ν_J is the stoichiometric number of substance J, with ν_J negative for reactants and positive for products (recall the notation introduced in Section 2.7b). Now there is a single rate for the entire reaction (for the chemical equation as written). With molar concentrations in moles per litre and time in seconds, reaction rates are reported in moles per litre per second ($mol\,L^{-1}\,s^{-1}$).

Illustration

If the rate of formation of NO in the reaction $2NOBr(g) \rightarrow 2NO(g) + Br_2(g)$ is reported as $1.6 \times 10^{-4}\,mol\,L^{-1}\,s^{-1}$, we use $\nu_{NO} = +2$ to report that $v = 8.0 \times 10^{-5}\,mol\,L^{-1}\,s^{-1}$. Because $\nu_{NOBr} = -2$ it follows that $d[NOBr]/dt = -1.6 \times 10^{-4}\,mol\,L^{-1}\,s^{-1}$. The rate of consumption of NOBr is therefore $1.6 \times 10^{-4}\,mol\,L^{-1}\,s^{-1}$.

Self-test 25.2 The rate of change of molar concentration of CH_3 radicals in the reaction $2CH_3(g) \rightarrow CH_3CH_3(g)$ was reported as $d[CH_3]/dt = -1.2\,mol\,L^{-1}\,s^{-1}$ under particular conditions. What are (a) the rate of reaction and (b) the rate of formation of CH_3CH_3?
[(a) $0.60\,mol\,L^{-1}\,s^{-1}$, (b) $0.60\,mol\,L^{-1}\,s^{-1}$]

(b) Rate laws and rate constants

The rate of reaction is often found to be proportional to the concentrations of the reactants raised to a power. For example, the rate of a reaction may be found to be proportional to the molar concentrations of two reactants A and B, in which case we write

$$v = k[A][B] \qquad (2)$$

where each concentration is raised to the first power. The coefficient k is called the **rate constant** for the reaction. The rate constant is independent of the concentrations but depends on the temperature. An experimentally determined equation of this kind is called the **rate law** of the reaction. More formally, a rate law is an equation that expresses the rate of reaction as a function of the concentrations of all the species present in the overall chemical equation for the reaction at some time:

$$v = f([A], [B], \ldots) \qquad [3]$$

The rate law of a reaction is determined experimentally, and cannot in general be inferred from the chemical equation for the reaction. The reaction of hydrogen and bromine, for example, has a very simple stoichiometry, $H_2(g) + Br_2(g) \rightarrow 2HBr(g)$, but its rate law is complicated:

$$v = \frac{k[H_2][Br_2]^{3/2}}{[Br_2] + k'[HBr]} \tag{4}$$

In certain cases the rate law does reflect the stoichiometry of the reaction, but that is either a coincidence or reflects a feature of the underlying reaction mechanism (see later).

A practical application of a rate law is that, once we know the law and the value of the rate constant, we can predict the rate of reaction from the composition of the mixture. Moreover, as we shall see later, by knowing the rate law, we can go on to predict the composition of the reaction mixture at a later stage of the reaction. Moreover, a rate law is a guide to the mechanism of the reaction, for any proposed mechanism must be consistent with the observed rate law.

(c) Reaction order

Many reactions are found to have rate laws of the form

$$v = k[A]^a[B]^b \cdots \tag{5}$$

The power to which the concentration of a species (a product or a reactant) is raised in a rate law of this kind is the **order** of the reaction with respect to that species. A reaction with the rate law in eqn 2 is **first-order** in A and first-order in B. The **overall order** of a reaction with a rate law like that in eqn 5 is the sum of the individual orders, $a + b + \cdots$. The rate law in eqn 2 is therefore second-order overall.

A reaction need not have an integral order, and many gas-phase reactions do not. For example, a reaction having the rate law

$$v = k[A]^{1/2}[B] \tag{6}$$

is half-order in A, first-order in B, and three-halves order overall. Some reactions obey a **zero-order rate law**, and therefore have a rate that is independent of the concentration of the reactant (so long as some is present). Thus, the catalytic decomposition of phosphine (PH_3) on hot tungsten at high pressures has the rate law

$$v = k \tag{7}$$

The PH_3 decomposes at a constant rate until it has almost entirely disappeared. Only heterogeneous reactions can have rate laws that are zero-order overall.

When a rate law is not of the form in eqn 5, the reaction does not have an overall order and may not even have definite orders with respect to each participant. Thus, although eqn 4 shows that the reaction of hydrogen and bromine is first-order in H_2, the reaction has an indefinite order with respect to both Br_2 and HBr and has no overall order.

These remarks point to three problems. First, we must see how to identify the rate law and obtain the rate constant from the experimental data. We concentrate on this aspect in this chapter. Second, we must see how to construct reaction mechanisms that are consistent with the rate law. We shall introduce the techniques of doing so in this chapter and develop them further in Chapter 26. Third, we must account for the values of the rate constants and explain their temperature dependence. We shall see a little of what is involved in this chapter, but leave the details until Chapter 27.

(d) The determination of the rate law

The determination of a rate law is simplified by the **isolation method** in which the concentrations of all the reactants except one are in large excess. If B is in large excess, for example, then to a good approximation its concentration is constant throughout the reaction. Although the true rate law might be $v = k[A][B]$, we can approximate $[B]$ by $[B]_0$ and write

$$v = k'[A] \qquad k' = k[B]_0 \tag{8}$$

which has the form of a first-order rate law. Because the true rate law has been forced into first-order form by assuming that the concentration of B is constant, it is called a **pseudofirst-order rate law**. The dependence of the rate on the concentration of each of the reactants may be found by isolating them in turn (by having all the other substances present in large excess), and so constructing a picture of the overall rate law.

In the **method of initial rates**, which is often used in conjunction with the isolation method, the rate is measured at the beginning of the reaction for several different initial concentrations of reactants. We shall suppose that the rate law for a reaction with A isolated is $v = k[A]^a$; then its initial rate, v_0, is given by the initial values of the concentration of A, and we write $v_0 = k[A]_0^a$. Taking logarithms gives:

$$\log v_0 = \log k + a \log [A]_0 \tag{9}$$

For a series of initial concentrations, a plot of the logarithms of the initial rates against the logarithms of the initial concentrations of A should be a straight line with slope a.

Example 25.2 Using the method of initial rates

The recombination of iodine atoms in the gas phase in the presence of argon was investigated and the order of the reaction was determined by the method of initial rates. The initial rates of reaction of $2I(g) + Ar(g) \rightarrow I_2(g) + Ar(g)$ were as follows:

$[I]_0/(10^{-5} \text{ mol L}^{-1})$	1.0	2.0	4.0	6.0
$v_0/(\text{mol L}^{-1}\text{s}^{-1})$ (a)	8.70×10^{-4}	3.48×10^{-3}	1.39×10^{-2}	3.13×10^{-2}
(b)	4.35×10^{-3}	1.74×10^{-2}	6.96×10^{-2}	1.57×10^{-1}
(c)	8.69×10^{-3}	3.47×10^{-2}	1.38×10^{-1}	3.13×10^{-1}

The Ar concentrations are (a) $1.0 \times 10^{-3} \text{ mol L}^{-1}$, (b) $5.0 \times 10^{-3} \text{ mol L}^{-1}$, and (c) $1.0 \times 10^{-2} \text{ mol L}^{-1}$. Determine the orders of reaction with respect to the I and Ar atom concentrations and the rate constant.

Method Plot the logarithm of the initial rate, $\log v_0$, against $\log [I]_0$ for a given concentration of Ar, and, separately, against $\log [Ar]_0$ for a given concentration of I. The slopes of the two lines are the orders of reaction with respect to I and Ar, respectively. The intercept with the vertical axis gives $\log k$.

Answer The plots are shown in Fig. 25.4. The slopes are 2 and 1 respectively, so the (initial) rate law is

$$v_0 = k[I]_0^2[Ar]_0$$

This rate law signifies that the reaction is second-order in $[I]$, first-order in $[Ar]$, and third-order overall. The intercept corresponds to $k = 9 \times 10^9 \text{ mol}^{-2}\text{L}^2\text{s}^{-1}$.

Comment The units of k come automatically from the calculation, and are always such as to convert the product of concentrations to concentration per unit time (for example, $\text{mol L}^{-1}\text{s}^{-1}$).

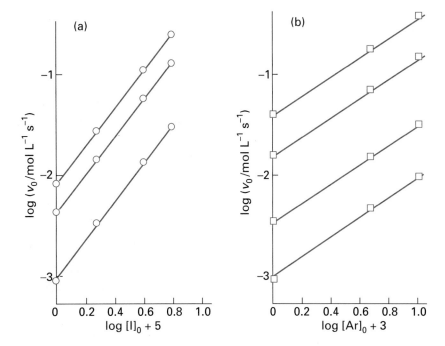

25.4 The plot of $\log v_0$ against (a) $\log[I]_0$ for a given $[Ar]_0$, and (b) $\log[Ar]_0$ for a given $[I]_0$.

Self-test 25.3 The initial rate of a reaction depended on concentration of a substance J as follows:

$[J]_0/(10^{-3}\ mol\ L^{-1})$	5.0	8.2	17	30
$v_0/(10^{-7}\ mol\ L^{-1}\ s^{-1})$	3.6	9.6	41	130

Determine the order of the reaction with respect to J and calculate the rate constant.

$$[2,\ 1.4 \times 10^{-2}\ L\ mol^{-1}\ s^{-1}]$$

The method of initial rates might not reveal the full rate law, for the products may participate in the reaction and affect the rate. For example, products participate in the synthesis of HBr, because eqn 4 shows that the full rate law depends on the concentration of HBr. To avoid this difficulty, the rate law should be fitted to the data throughout the reaction. The fitting may be done, in simple cases at least, by using a proposed rate law to predict the concentration of any component at any time, and comparing it with the data. A law should also be tested by observing whether the addition of products or, for gas-phase reactions, a change in the surface-to-volume ratio in the reaction chamber affects the rate.

25.3 Integrated rate laws

Rate laws are differential equations. Therefore, we must integrate them if we want to find the concentrations as a function of time. Even the most complex rate laws may be integrated numerically. However, in a number of simple cases analytical solutions are easily obtained, and prove to be very useful. We shall examine a few of these simple cases here, and illustrate the computational approach in Chapter 26.

(a) First-order reactions

As shown in the *Justification* below, the first-order rate law for the consumption of a reactant A

$$\frac{d[A]}{dt} = -k[A] \tag{10a}$$

has the solution

$$\ln\left(\frac{[A]}{[A]_0}\right) = -kt \qquad [A] = [A]_0 e^{-kt} \tag{10b}$$

These two equations are versions of an **integrated rate law**, the integrated form of the rate law.

Justification 25.1

Equation 10a rearranges to

$$\frac{d[A]}{[A]} = -k dt$$

which can be integrated directly because k is a constant independent of t. Initially (at $t = 0$) the concentration of A is $[A]_0$, and at a later time t it is $[A]$, so we make these values the limits of the integrals and write

$$\int_{[A]_0}^{[A]} \frac{d[A]}{[A]} = -k \int_0^t dt$$

Because the integral of $1/x$ is $\ln x$, eqn 10b is obtained immediately.

Equation 10b shows that, if $\ln\left([A]/[A]_0\right)$ is plotted against t, then a first-order reaction will give a straight line of slope $-k$. Some rate constants determined in this way are given in Table 25.1. The second expression in eqn 10b shows that in a first-order reaction the reactant concentration decreases exponentially with time with a rate determined by k (Fig. 25.5).

Example 25.3 Analysing a first-order reaction

The variation in the partial pressure of azomethane with time was followed at 600 K, with the results given below. Confirm that the decomposition

$$CH_3N_2CH_3(g) \longrightarrow CH_3CH_3(g) + N_2(g)$$

is first-order in azomethane, and find the rate constant at 600 K.

t/s	0	1000	2000	3000	4000
$p/(10^{-2}\ \text{Torr})$	8.20	5.72	3.99	2.78	1.94

Method As indicated in the text, to confirm that a reaction is first-order, plot $\ln\left([A]/[A]_0\right)$ against time and expect a straight line. Because the partial pressure of a gas is proportional to its concentration, it is equivalent to plot $\ln\left(p/p_0\right)$ against t. If a straight line is obtained, its slope can be identified with $-k$.

Table 25.1* Kinetic data for first-order reactions

Reaction	Phase	$\theta/^\circ C$	k/s^{-1}	$t_{1/2}$
$2N_2O_5 \rightarrow 4NO_2 + O_2$	g	25	3.38×10^{-5}	5.70 h
$2N_2O_5 \rightarrow 4NO_2 + O_2$	$Br_2(l)$	25	4.27×10^{-5}	4.51 h
$C_2H_6 \rightarrow 2CH_3$	g	700	5.36×10^{-4}	21.6 min

*More values are given in the *Data section* at the end of this volume.

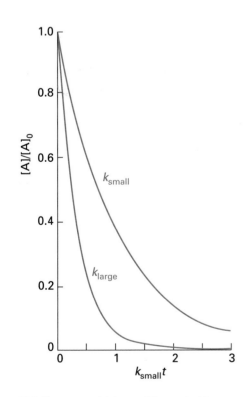

25.5 The exponential decay of the reactant in a first-order reaction. The larger the rate constant, the more rapid the decay: here $k_{large} = 3k_{small}$.

Table 25.3 Integrated rate laws

Order	Reaction	Rate law*	$t_{1/2}$
0	A→P	$v = k$	$\dfrac{[A]_0}{2k}$
		$kt = x$ for $0 \le x \le [A]_0$	
1	A→P	$v = k[A]$	$\dfrac{\ln 2}{k}$
		$kt = \ln\dfrac{[A]_0}{[A]_0 - x}$	
2	A→P	$v = k[A]^2$	$\dfrac{1}{k[A]_0}$
		$kt = \dfrac{x}{[A]_0([A]_0 - x)}$	
	A + B → P	$v = k[A][B]$	
		$kt = \dfrac{1}{[B]_0 - [A]_0}\ln\dfrac{[A]_0([B]_0 - x)}{([A]_0 - x)[B]_0}$	
	A + 2B → P	$v = k[A][B]$	
		$kt = \dfrac{1}{[B]_0 - 2[A]_0}\ln\dfrac{[A]_0([B]_0 - 2x)}{([A]_0 - x)[B]_0}$	
	A → P with autocatalysis	$v = k[A][P]$	
		$kt = \dfrac{1}{[A]_0 + [P]_0}\ln\dfrac{[A]_0([P]_0 + x)}{([A]_0 - x)[P]_0}$	
3	A + 2B→P	$v = k[A][B]^2$	
		$kt = \dfrac{2x}{(2[A]_0 - [B]_0)([B]_0 - 2x)[B]_0}$	
		$+ \dfrac{1}{(2[A]_0 - [B]_0)^2}\ln\dfrac{[A]_0([B]_0 - 2x)}{([A]_0 - x)[B]_0}$	
$n \ge 2$	A→P	$v = k[A]^n$	$\dfrac{2^{n-1} - 1}{(n-1)k[A]_0^{n-1}}$
		$kt = \dfrac{1}{n-1}\left\{\dfrac{1}{([A]_0 - x)^{n-1}} - \dfrac{1}{[A]_0^{n-1}}\right\}$	

*$x = [P]$, and $v = dx/dt$.

If the initial concentration of A is $[A]_0$, and no B is present initially, then at all times $[A] + [B] = [A]_0$. Therefore,

$$\frac{d[A]}{dt} = -k[A] + k'([A]_0 - [A]) = -(k + k')[A] + k'[A]_0 \tag{18}$$

The solution of this first-order differential equation (as may be checked by differentiation) is

$$[A] = \frac{k' + ke^{-(k+k')t}}{k + k'}[A]_0 \tag{19}$$

The time dependence predicted by this equation is drawn in Fig. 25.8.

As $t \to \infty$, the concentrations reach their equilibrium values, which are given by eqn 19 as:

$$[A]_{eq} = \frac{k'[A]_{eq}}{k + k'} \qquad [B]_{eq} = [A]_0 - [A]_\infty = \frac{k[A]_0}{k + k'} \tag{20}$$

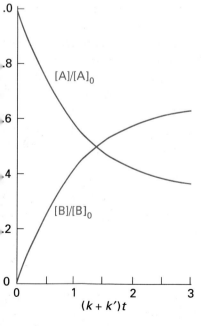

The approach of concentrations to their ...rium values as predicted by eqn 19 for a ...ion A ⇌ B that is first-order in each direction, ...r which $k = 2k'$.

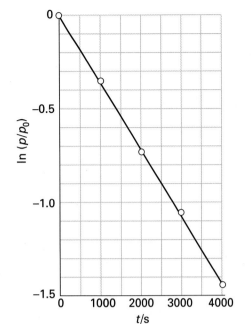

25.6 The determination of the rate constant of a first-order reaction: a straight line is obtained when $\ln[A]$ (or, as here, $\ln p$) is plotted against t; the slope gives k.

Answer We draw up the following table:

t/s	0	1000	2000	3000	4000
$\ln(p/p_0)$	1	−0.360	−0.720	−1.082	−1.441

Figure 25.6 shows the plot of $\ln(p/p_0)$ against t. The plot is straight, confirming a first-order reaction, and its slope is -3.6×10^{-4}. Therefore, $k = 3.6 \times 10^{-4}$ s^{-1}.

Self-test 25.4 In a particular experiment, it was found that the concentration of N_2O_5 in liquid bromine varied with time as follows:

t/s	0	200	400	600	1000
$[N_2O_5]/\text{mol L}^{-1}$	0.110	0.073	0.048	0.032	0.014

Confirm that the reaction is first-order in N_2O_5 and determine the rate constant.

$$[k = 2.1 \times 10^{-3}\ \text{s}^{-1}]$$

(b) Half-lives

A useful indication of the rate of a first-order chemical reaction is the **half-life**, $t_{1/2}$, of a substance, the time taken for the concentration of a reactant to fall to half its initial value. The time for $[A]$ to decrease from $[A]_0$ to $\frac{1}{2}[A]_0$ in a first-order reaction is given by eqn 10b as

$$kt_{1/2} = -\ln\left(\frac{\frac{1}{2}[A]_0}{[A]_0}\right) = -\ln\tfrac{1}{2} = \ln 2$$

Hence

$$t_{1/2} = \frac{\ln 2}{k} \tag{11}$$

($\ln 2 = 0.693$.) The main point to note about this result is that, for a first-order reaction, the half-life of a reactant is independent of its initial concentration. Hence, if the concentration of A at some *arbitrary* stage of the reaction is $[A]$, then it will have fallen to $\frac{1}{2}[A]$ after a further interval of $(\ln 2)/k$. Some half-lives are given in Table 25.1.

(c) Second-order reactions

We show in the *Justification* below that the integrated form of the second-order rate law

$$\frac{d[A]}{dt} = -k[A]^2 \tag{12a}$$

is

$$\frac{1}{[A]} - \frac{1}{[A]_0} = kt \qquad [A] = \frac{[A]_0}{1 + kt[A]_0} \tag{12b}$$

Justification 25.2

Equation 12a is integrated by rearranging it to

$$-\frac{d[A]}{[A]^2} = kdt$$

Table 25.2* Kinetic data for second-order reactions

Reaction	Phase	$\theta/^\circ\text{C}$	$k/(\text{L mol}^{-1}\,\text{s}^{-1})$
$2NOBr \rightarrow 2NO + Br_2$	g	10	0.80
$2I \rightarrow I_2$	g	23	7×10^9
$CH_3Cl + CH_3O^-$	$CH_3OH(l)$	20	2.29×10^{-6}

*More values are given in the *Data section*.

The concentration of A is $[A]_0$ at $t = 0$ and $[A]$ at a general time t later. Therefore, this expression integrates as follows:

$$-\int_{[A]_0}^{[A]} \frac{d[A]}{[A]^2} = k \int_0^t dt$$

Because the integral of $1/x^2$ is $-1/x$, we obtain eqn 12b by substitution of the limits.

The first expression in eqn 12b shows that to test for a second-order reaction we should plot $1/[A]$ against t and expect a straight line. The slope of the graph is k. Some rate constants determined in this way are given in Table 25.2. The second expression lets us predict the concentration of A at any time after the start of the reaction. It shows that the concentration of A approaches zero more slowly than in a first-order reaction with the same initial rate (Fig. 25.7).

It follows from eqn 12b by substituting $t = t_{1/2}$ and $[A] = \frac{1}{2}[A]_0$ that the half-life of a species A that is consumed in a second-order reaction is

$$t_{1/2} = \frac{1}{k[A]_0} \tag{13}$$

Therefore, unlike in a first-order reaction, the half-life of a substance in a second-order reaction varies with the initial concentration. A practical consequence is that species that decay by second-order reactions (which includes some environmentally harmful substances) may persist in low concentrations for long periods because their half-lives are long when their concentrations are low.

Another type of second-order reaction is one that is first-order in each of two reactants A and B:

$$\frac{d[A]}{dt} = -k[A][B] \tag{14}$$

Such a rate law cannot be integrated until we know how the concentration of B is related to that of A. For example, if the reaction is $A + B \rightarrow P$, where P denotes products, and the initial concentrations are $[A]_0$ and $[B]_0$, then it is shown in the *Justification* below that at a time t after the start of the reaction, the concentrations satisfy the relation

$$\ln\left(\frac{[B]/[B]_0}{[A]/[A]_0}\right) = ([B]_0 - [A]_0)kt \tag{15}$$

Therefore, a plot of the expression on the left against t should be a straight line from which k can be obtained. Note that, if $[A]_0 = [B]_0$, the solutions are those already given in eqn 12b (but this solution cannot be found simply by setting $[A]_0 = [B]_0$ in eqn 15).

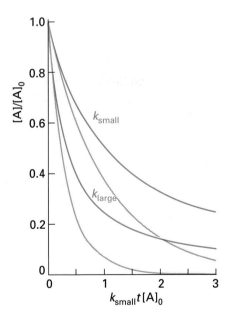

25.7 The variation with time of the concentration of a reactant in a second-order reaction. The grey line is the corresponding decay in a first-order reaction with the same initial rate. For this illustration, $k_{large} = 3k_{small}$.

Justification 25.3

It follows from the reaction stoichiometry that, when the concentration of A $[A]_0 - x$, the concentration of B will have fallen to $[B]_0 - x$ (because disappears entails the disappearance of one B). It follows that

$$\frac{d[A]}{dt} = -k([A]_0 - x)([B]_0 - x)$$

Then, because $d[A]/dt = -dx/dt$, the rate law is

$$\frac{dx}{dt} = k([A]_0 - x)([B]_0 - x)$$

The initial condition is that $x = 0$ when $t = 0$, so the integration required

$$\int_0^x \frac{dx}{([A]_0 - x)([B]_0 - x)} = k \int_0^t dt$$

The integral on the right is simply kt. It follows that

$$\begin{aligned}
kt &= \int_0^x \frac{dx}{([A]_0 - x)([B]_0 - x)} \\
&= \frac{1}{[B]_0 - [A]_0} \int_0^x \left\{ \frac{1}{[A]_0 - x} - \frac{1}{[B]_0 - x} \right\} dx \\
&= \frac{1}{[B]_0 - [A]_0} \left\{ \ln\left(\frac{[A]_0}{[A]_0 - x}\right) - \ln\left(\frac{[B]_0}{[B]_0 - x}\right) \right\}
\end{aligned}$$

This expression can be simplified and rearranged into eqn 15 by comb logarithms and noting that $[A] = [A]_0 - x$ and $[B] = [B]_0 - x$.

Similar calculations may be carried out to find the integrated rate laws and some are listed in Table 25.3.

25.4 Reactions approaching equilibrium

Because all the laws considered so far disregard the possibility that the re important, none of them describes the overall rate when the reaction is clos At that stage the products may be so abundant that the reverse reaction mu account. In practice, however, most kinetic studies are made on reactions t equilibrium, and the reverse reactions are unimportant.

(a) First-order reactions close to equilibrium

We can explore the variation of the composition with time close to chemic considering the reaction in which A forms B and both forward and reve first-order (as in some isomerizations). The scheme we consider is

$$\begin{aligned}
A &\longrightarrow B \qquad v = k[A] \\
B &\longrightarrow A \qquad v = k'[B]
\end{aligned}$$

The concentration of A is reduced by the forward reaction (at a rate $k[A]$) b by the reverse reaction (at a rate $k'[B]$). The net rate of change is therefor

$$\frac{d[A]}{dt} = -k[A] + k'[B]$$

It follows that the equilibrium constant of the reaction is

$$K = \frac{[\mathrm{B}]_{eq}}{[\mathrm{A}]_{eq}} = \frac{k}{k'} \tag{21}$$

Exactly the same conclusion can be reached—more simply, in fact—by noting that, at equilibrium, the forward and reverse rates must be the same, so

$$k[\mathrm{A}]_{eq} = k'[\mathrm{B}]_{eq} \tag{22}$$

This relation rearranges into eqn 21.

Equation 21 is very important, because it relates the thermodynamic quantity, the equilibrium constant, to quantities relating to rates. The practical importance of eqn 21 is that, if one of the rate constants can be measured, then the other may be obtained if the equilibrium constant is known.

For a more general reaction, the overall equilibrium constant can be expressed in terms of the rate constants for all the intermediate stages of the reaction mechanism:

$$K = \frac{k_a}{k'_a} \times \frac{k_b}{k'_b} \times \cdots$$

where the ks are the rate constants for the individual steps and the k's are for the corresponding reverse steps.

(b) Relaxation methods

The term **relaxation** denotes the return of a system to equilibrium. It is used in chemical kinetics to indicate that an externally applied influence has shifted the equilibrium position of a reaction, normally suddenly, and that the reaction is adjusting to the equilibrium composition characteristic of the new conditions (Fig. 25.9). We shall consider the response of reaction rates to a **temperature jump**, a sudden change in temperature. We know from Section 9.3a that the equilibrium composition of a reaction depends on the temperature (provided $\Delta_r H^{\ominus}$ is nonzero), so a shift in temperature acts as a perturbation on the system. One way of achieving a temperature jump is to discharge a capacitor through a sample made conducting by the addition of ions, but laser or microwave discharges can also be used. Temperature jumps of between 5 and 10 K can be achieved in about 1 μs. Some equilibria are also sensitive to pressure, and **pressure-jump techniques** may then also be used.

When a sudden temperature increase is applied to a simple $\mathrm{A} \rightleftharpoons \mathrm{B}$ equilibrium that is first-order in each direction, we show in the *Justification* below that the composition relaxes exponentially to the new equilibrium composition:

$$x = x_0 \mathrm{e}^{-t/\tau} \qquad \frac{1}{\tau} = k_a + k_b \tag{23}$$

where x is the departure from equilibrium at the new temperature and x_0 is the departure from equilibrium immediately after the temperature jump.

Justification 25.4

In the following analysis, we need to keep track of the fact that rate constants depend on temperature. At the initial temperature, when the rate constants are k'_a and k'_b, the net rate of change of $[\mathrm{A}]$ is

$$\frac{\mathrm{d}[\mathrm{A}]}{\mathrm{d}t} = -k'_a[\mathrm{A}] + k'_b[\mathrm{B}]$$

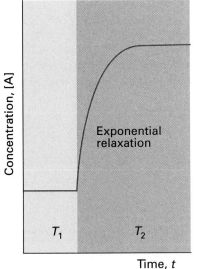

25.9 The relaxation to the new equilibrium composition when a reaction initially at equilibrium at a temperature T_1 is subjected to a sudden change of temperature, which takes it to T_2.

At equilibrium under these conditions, $d[A]/dt = 0$, and the concentrations are $[A]'_{eq}$ and $[B]'_{eq}$. Therefore,

$$k'_a[A]'_{eq} = k'_b[B]'_{eq}$$

When the temperature is increased suddenly, the rate constants change to k_a and k_b, but the concentrations of A and B remain for an instant at their old equilibrium values. As the system is no longer at equilibrium, it readjusts to the new equilibrium concentrations, which are now given by

$$k_a[A]_{eq} = k_b[B]_{eq}$$

and it does so at a rate that depends on the new rate constants.

We write the deviation of [A] from its new equilibrium value as x, so $[A] = x + [A]_{eq}$ and $[B] = [B]_{eq} - x$. The concentration of A then changes as follows:

$$\frac{d[A]}{dt} = -k_a(x + [A]_{eq}) + k_b(-x + [B]_{eq}) = -(k_a + k_b)x$$

because the two terms involving the equilibrium concentrations cancel. Because $d[A]/dt = dx/dt$, this equation is a first-order differential equation with the solution given in eqn 23.

Equation 23 shows that the concentrations of A and B relax into the new equilibrium at a rate determined by the sum of the two new rate constants. Because the equilibrium constant under the new conditions is $K = k_a/k_b$, its value may be combined with the relaxation time measurement to find the individual k_a and k_b.

Example 25.4 Analysing a temperature-jump experiment

The $H_2O(l) \rightarrow H^+(aq) + OH^-(aq)$ reaction relaxes to equilibrium with a time constant 37 μs at 298 K and pH $\approx$ 7, and $pK_w = 14.01$. Given that the forward reaction is first-order and the reverse is second-order overall, calculate the rate constants for the forward and reverse reactions.

Method We need to derive an expression for the relaxation time, τ, in terms of k_1 (forward, first-order reaction) and k_2 (reverse, second-order reaction). We can proceed as above, but it will be necessary to make the assumption that the deviation from equilibrium (x) is so small that terms in x^2 can be neglected. Relate k_1 and k_2 through the equilibrium constant, but be careful with units because K_w is dimensionless.

Answer The forward rate at the final temperature is $k_1[H_2O]$ and the reverse rate is $k_2[H^+][OH^-]$. The net rate of formation of H_2O is

$$\frac{d[H_2O]}{dt} = -k_1[H_2O] + k_2[H^+][OH^-]$$

We write $[H_2O] = [H_2O]_{eq} + x$, $[H^+] = [H^+]_{eq} - x$, and $[OH^-] = [OH^-]_{eq} - x$, and obtain

$$\frac{dx}{dt} = -\{k_1 + k_2([H^+]_{eq} + [OH^-]_{eq})\}x$$
$$- k_1[H_2O]_{eq} + k_2[H^+]_{eq}[OH^-]_{eq} + k_2x^2$$
$$\approx - \{k_1 + k_2([H^+]_{eq} + [OH^-]_{eq})\}x$$

where we have neglected the term in x^2 and used the equilibrium condition to eliminate the terms that are independent of x. It follows that

$$\frac{1}{\tau} = k_1 + k_2([\text{H}^+]_{\text{eq}} + [\text{OH}^-]_{\text{eq}})$$

The equilibrium condition is

$$k_1[\text{H}_2\text{O}]_{\text{eq}} = k_2[\text{H}^+]_{\text{eq}}[\text{OH}^-]_{\text{eq}}$$

From this expression it follows that

$$\frac{k_1}{k_2} = \frac{[\text{H}^+]_{\text{eq}}[\text{OH}^-]_{\text{eq}}}{[\text{H}_2\text{O}]_{\text{eq}}} = \frac{K_w \,(\text{mol L}^{-1})^2}{[\text{H}_2\text{O}]_{\text{eq}}} = \frac{K_w}{55.6} \,\text{mol L}^{-1}$$

because the molar concentration of pure water is $55.6\,\text{mol L}^{-1}$. If we write $K = K_w/55.6 = 1.8 \times 10^{-16}$, we obtain

$$\frac{1}{\tau} = k_2\{(K\,\text{mol L}^{-1}) + [\text{H}^+]_{\text{eq}} + [\text{OH}^-]_{\text{eq}}\}$$
$$= k_2\{K + K_w^{1/2} + K_w^{1/2}\}\,\text{mol L}^{-1} = (2.0 \times 10^{-7}) \times k_2\,\text{mol L}^{-1}$$

Hence,

$$k_2 = \frac{1}{(3.7 \times 10^{-5}\,\text{s}) \times (2.0 \times 10^{-7}\,\text{mol L}^{-1})} = 1.4 \times 10^{11}\,\text{L mol}^{-1}\,\text{s}^{-1}$$

It follows that

$$k_1 = k_2K\,\text{mol L}^{-1} = 2.4 \times 10^{-5}\,\text{s}^{-1}$$

Comment Notice how we keep track of units: K and K_w are dimensionless; k_2 is expressed in $\text{L mol}^{-1}\,\text{s}^{-1}$; and k_1 is expressed in s^{-1}. The reaction is faster in ice, where $k_2 = 8.6 \times 10^{12}\,\text{L mol}^{-1}\,\text{s}^{-1}$.

- -

Self-test 25.5 Derive an expression for the relaxation time of a concentration when the reaction $\text{A} + \text{B} \rightleftharpoons \text{C} + \text{D}$ is second-order in both directions.

$$[1/\tau = k([\text{A}] + [\text{B}])_{\text{eq}} + k'([\text{C}] + [\text{D}])_{\text{eq}}]$$

25.5 The temperature dependence of reaction rates

The rate constants of most reactions increase as the temperature is raised. Many reactions in solution fall somewhere in the range spanned by the hydrolysis of methyl ethanoate (where the rate constant at 35°C is 1.82 times that at 25°C) and the hydrolysis of sucrose (where the factor is 4.13).

(a) The Arrhenius parameters

It is found experimentally for many reactions that a plot of $\ln k$ against $1/T$ gives a straight line. This behaviour is normally expressed mathematically by introducing two parameters, one representing the intercept and the other the slope of the straight line, and writing the **Arrhenius equation**

$$\ln k = \ln A - \frac{E_a}{RT} \tag{24}$$

The parameter A, which is given by the intercept of the line at $1/T = 0$ (Fig. 25.10), is called the **pre-exponential factor** or the **frequency factor**. The parameter E_a, which is obtained

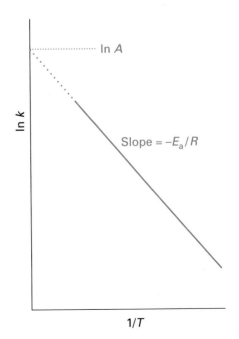

25.10 The Arrhenius plot of $\ln k$ against $1/T$ for the decomposition of CH_3CHO, and the best straight line. The slope gives $-E_a/R$ and the intercept at $1/T = 0$ gives $\ln A$.

Table 25.4* Arrhenius parameters

(1) First-order reactions	A/s^{-1}	$E_a/(\text{kJ mol}^{-1})$
$CH_3NC \rightarrow CH_3CN$	3.98×10^{13}	160
$2N_2O_5 \rightarrow 4NO_2+O_2$	4.94×10^{13}	103.4
(2) Second-order reactions	$A/(\text{L mol}^{-1}\,\text{s}^{-1})$	$E_a/(\text{kJ mol}^{-1})$
$OH + H_2 \rightarrow H_2O + H$	8.0×10^{10}	42
$NaC_2H_5O + CH_3I$ in ethanol	2.42×10^{11}	81.6

*More values are given in the *Data section*.

from the slope of the line $(-E_a/R)$, is called the **activation energy**. Collectively the two quantities are called the **Arrhenius parameters** (Table 25.4).

Example 25.5 Determining the Arrhenius parameters

The rate of the second-order decomposition of acetaldehyde (ethanal, CH_3CHO) was measured over the temperature range 700–1000 K, and the rate constants are reported below. Find E_a and A.

T/K	700	730	760	790	810	840	910	1000
$k/(\text{L mol}^{-1}\,\text{s}^{-1})$	0.011	0.035	0.105	0.343	0.789	2.17	20.0	145

Method According to eqn 24, the data can be analysed by plotting $\ln(k/\text{L mol}^{-1}\,\text{s}^{-1})$ against $1/(T/K)$ and getting a straight line. The slope of this line is $(-E_a/R)/K$ and the intercept at $1/T = 0$ is $\ln A$).

Answer We draw up the following table:

$10^3\ K/T$	1.43	1.37	1.32	1.27	1.23	1.19	1.10	1.00
$\ln(k/\text{L mol}^{-1}\,\text{s}^{-1})$	−4.51	−3.35	−2.25	−1.07	−0.24	0.77	3.00	4.98

Now plot $\ln k$ against $1/T$ (Fig. 25.11). The least-squares best fit of the line is with slope -2.27×10^4 and intercept 27.7. Therefore,

$$E_a = (2.21 \times 10^4\ \text{K}) \times (8.3145\ \text{J K}^{-1}\,\text{mol}^{-1}) = 188\ \text{kJ mol}^{-1}$$
$$A = e^{27.0}\ \text{L mol}^{-1}\,\text{s}^{-1} = 1.1 \times 10^{12}\ \text{L mol}^{-1}\,\text{s}^{-1}$$

Comment Note that A has the same units as k. The slopes and intercepts of graphs are always dimensionless, and care must be taken to relate the numerical value to the physical quantity by noting how the data have been plotted. In practice, A is obtained from one of the midrange data values rather than by using a lengthy extrapolation.

- -

Self-test 25.6 Determine A and E_a from the following data:

T/K	300	350	400	450	500
$k/(\text{L mol}^{-1}\,\text{s}^{-1})$	7.9×10^6	3.0×10^7	7.9×10^7	1.7×10^8	3.2×10^8

$$[8 \times 10^{10}\ \text{L mol}^{-1}\,\text{s}^{-1},\ 23\ \text{kJ mol}^{-1}]$$

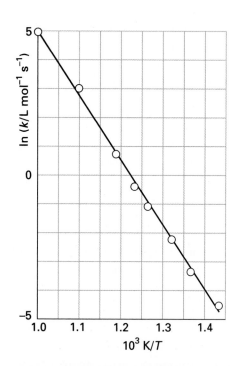

25.11 The Arrhenius plot using the data in Example 25.5.

The fact that E_a is given by the slope of the plot of $\ln k$ against $1/T$ means that the higher the activation energy, the stronger the temperature dependence of the rate constant (that is, the steeper the slope). *A high activation energy signifies that the rate constant depends strongly on temperature*. If a reaction has zero activation energy, its rate is independent of temperature. In some cases the activation energy is negative, which indicates that the rate decreases as the temperature is raised. We shall see that such behaviour is a signal that the reaction has a complex mechanism.

The temperature dependence of some reactions is not Arrhenius-like, in the sense that a straight line is not obtained when $\ln k$ is plotted against $1/T$. However, it is still possible to define an activation energy as

$$E_a = RT^2 \frac{\mathrm{d} \ln k}{\mathrm{d}T} \qquad [25]$$

This definition reduces to the earlier one (as the slope of a straight line) for a temperature-independent activation energy. However, the definition in eqn 25 is more general than eqn 24, because it allows E_a to be obtained from the slope (at the temperature of interest) of a plot of $\ln k$ against $1/T$ even if the Arrhenius plot is not a straight line. Non-Arrhenius behaviour is commonly a sign that quantum mechanical tunnelling is playing a significant role in the reaction.

(b) The interpretation of the parameters

For the present chapter we shall regard the Arrhenius parameters as purely empirical quantities that enable us to discuss the variation of rate constants with temperature. However, it is worth anticipating the interpretation of E_a in Section 27.1, which is motivated by writing eqn 24 as

$$k = Ae^{-E_a/RT} \qquad (26)$$

There we shall see that *the activation energy is the minimum kinetic energy that reactants must have in order to form products*. For example, in a gas-phase reaction there are numerous collisions each second, but only a tiny proportion are sufficiently energetic to lead to reaction. The fraction of collisions with a kinetic energy in excess of an energy E_a is given by the Boltzmann distribution as $e^{-E_a/RT}$. Hence, the exponential factor in eqn 26 can be interpreted as the fraction of collisions that have enough kinetic energy to lead to reaction.

The pre-exponential factor is a measure of the rate at which collisions occur irrespective of their energy. Hence, the product of A and the exponential factor, $e^{-E_a/RT}$, gives the rate of *successful* collisions. We shall develop these remarks in Chapter 27 and see that they have their analogues for reactions that take place in liquids.

Accounting for the rate laws

We now move on to the second stage of the analysis of kinetic data, their explanation in terms of a postulated reaction mechanism.

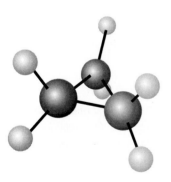

1 Cyclopropane

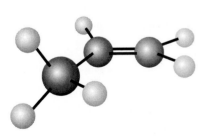

2 Propene

25.6 Elementary reactions

Most reactions occur in a sequence of steps called **elementary reactions**, each of which involves only a small number of molecules or ions. A typical elementary reaction is

$$H + Br_2 \longrightarrow HBr + Br$$

(We do not specify the phase of the species in the chemical equation for an elementary reaction.) This equation signifies that an H atom attacks a Br_2 molecule to produce an HBr molecule and a Br atom. The **molecularity** of an elementary reaction is the number of molecules coming together to react in an elementary reaction. In a **unimolecular reaction**, a single molecule shakes itself apart or its atoms into a new arrangement, as in the isomerization of cyclopropane (**1**) to propene (**2**). In a **bimolecular reaction**, a pair of molecules collide and exchange energy, atoms, or groups of atoms, or undergo some other kind of change. It is most important to distinguish molecularity from order:

Reaction order is an empirical quantity, and obtained from the experimental rate law.

The **molecularity** refers to an elementary reaction proposed as an individual step in a mechanism.

The rate law of a unimolecular elementary reaction is first-order in the reactant:

$$A \longrightarrow P \qquad \frac{d[A]}{dt} = -k[A] \tag{27}$$

where P denotes products (several different species may be formed). A unimolecular reaction is first-order because the number of A molecules that decay in a short interval is proportional to the number available to decay. (Ten times as many decay in the same interval when there are initially 1000 A molecules than when there are only 100 present.) Therefore, the rate of decomposition of A is proportional to its molar concentration.

An elementary bimolecular reaction has a second-order rate law:

$$A + B \longrightarrow P \qquad \frac{d[A]}{dt} = -k[A][B] \tag{28}$$

A bimolecular reaction is second-order because its rate is proportional to the rate at which the reactant species meet, which in turn is proportional to their concentrations. Therefore, if we believe that a reaction is a single-step, bimolecular process, we can write down the rate law (and then go on to test it). Bimolecular elementary reactions are believed to account for many homogeneous reactions, such as the dimerizations of alkenes and dienes and reactions such as

$$CH_3I(alc) + CH_3CH_2O^-(alc) \longrightarrow CH_3OCH_2CH_3(alc) + I^-(alc)$$

(where 'alc' signifies alcohol solution). The mechanism of this reaction is believed to be the single elementary step

$$CH_3I + CH_3CH_2O^- \longrightarrow CH_3OCH_2CH_3 + I^-$$

This mechanism is consistent with the observed rate law

$$v = k[CH_3I][CH_3CH_2O^-] \tag{29}$$

We shall see below how to string simple steps together into a mechanism and how to arrive at the corresponding rate law. For the present we emphasize that, *if the reaction is an elementary bimolecular process, then it has second-order kinetics but, if the kinetics are second-order, then the reaction might be complex.* The postulated mechanism can be explored only by detailed detective work on the system, and by investigating whether side products or intermediates appear during the course of the reaction. Detailed analysis of this kind was one of the ways, for example, in which the reaction $H_2(g) + I_2(g) \rightarrow 2HI(g)$ was shown to proceed by a complex reaction. For many years the reaction had been accepted on good, but insufficiently meticulous evidence, as a fine example of a simple bimolecular reaction in which atoms exchanged partners during a collision.

25.7 Consecutive elementary reactions

Some reactions proceed through the formation of an intermediate (I), as in the consecutive unimolecular reactions

$$A \xrightarrow{k_a} I \xrightarrow{k_b} P$$

An example is the decay of a radioactive family, such as

$$^{239}\text{U} \xrightarrow{\text{23.5 min}} {}^{239}\text{Np} \xrightarrow{\text{2.35 day}} {}^{239}\text{Pu}$$

(The times are half-lives.) We can discover the characteristics of this type of reaction by setting up the rate laws for the net rate of change of the concentration of each substance.

(a) The variation of concentrations with time

The rate of unimolecular decomposition of A is

$$\frac{d[A]}{dt} = -k_a[A] \tag{30}$$

and A is not replenished. The intermediate I is formed from A (at a rate $k_a[A]$) but decays to P (at a rate $k_b[I]$). The net rate of formation of I is therefore

$$\frac{d[I]}{dt} = k_a[A] - k_b[I] \tag{31}$$

The product P is formed by the unimolecular decay of I:

$$\frac{d[P]}{dt} = k_b[I] \tag{32}$$

We suppose that initially only A is present, and that its concentration is $[A]_0$.

The first of the rate laws, eqn 30, is an ordinary first-order decay, so we can write

$$[A] = [A]_0 e^{-k_a t} \tag{33}$$

When this equation is substituted into eqn 31, and we set $[I]_0 = 0$, the solution is

$$[I] = \frac{k_a}{k_b - k_a}\left(e^{-k_a t} - e^{-k_b t}\right)[A]_0 \tag{34}$$

At all times $[A] + [I] + [P] = [A]_0$, so it follows that

$$[P] = \left\{1 + \frac{k_a e^{-k_b t} - k_b e^{-k_a t}}{k_b - k_a}\right\}[A]_0 \tag{35}$$

The concentration of the intermediate I rises to a maximum, and then falls to zero (Fig. 25.12). The concentration of the product P rises from zero towards $[A]_0$.

Example 25.6 Analysing consecutive reactions

Suppose that in an industrial batch process a substance A produces the desired compound I which goes on to decay to a worthless product C, each step of the reaction being first-order. At what time will I be present in greatest concentration?

Method The time dependence of the concentration of I is given by eqn 34. We can find the time at which [I] passes through a maximum, t_{max}, by calculating $d[I]/dt$ and setting the resulting rate equal to zero.

Answer It follows from eqn 34 that

$$\frac{d[I]}{dt} = -\frac{k_a[A]_0\left(k_a e^{-k_a t} - k_b e^{-k_b t}\right)}{k_b - k_a}$$

This rate is equal to zero when

$$k_a e^{-k_a t} = k_b e^{-k_b t}$$

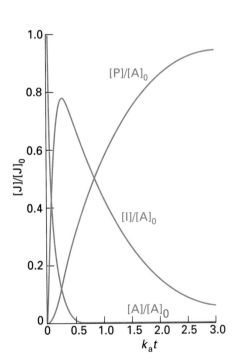

25.12 The concentrations of A, I, and P in the consecutive reaction scheme $A \rightarrow I \rightarrow P$. The curves are plots of eqns 33–35 with $k_a = 10k_b$. If the intermediate I is in fact the desired product, it is important to be able to predict when its concentration is greatest; see Example 25.6.

Therefore,

$$t_{max} = \frac{1}{k_a - k_b} \ln \frac{k_a}{k_b}$$

Comment For a given value of k_a, as k_b increases both the time at which $[I]$ is a maximum and the yield of I increase.

- -

Self-test 25.7 Calculate the maximum concentration of I and justify the last remark.

$$[[I]_{max}/[A]_0 = (k_a/k_b)^c, c = k_b/(k_b - k_a)]$$

(b) The rate-determining step

Suppose now that $k_b \gg k_a$; then, whenever an I molecule is formed, it decays rapidly into P. Because

$$e^{-k_b t} \ll e^{-k_a t} \qquad k_b - k_a \approx k_b$$

eqn 35 reduces to

$$[P] \approx (1 - e^{-k_a t})[A]_0 \tag{36}$$

which shows that the formation of the final product P depends on only the smaller of the two rate constants. That is, the rate of formation of P depends on the rate at which I is formed, not on the rate at which I changes into P. For this reason, the step $A \rightarrow I$ is called the **rate-determining step** of the reaction. Its existence has been likened to building a six-lane highway up to a single-lane bridge: the traffic flow is governed by the rate of crossing the bridge. Similar remarks apply to more complicated reaction mechanisms, and in general the rate-determining step is the one with the smallest rate constant.

(c) The steady-state approximation

One feature of the calculation so far has probably not gone unnoticed: there is a considerable increase in mathematical complexity as soon as the reaction mechanism has more than a couple of steps. A reaction scheme involving many steps is nearly always unsolvable analytically, and alternative methods of solution are necessary. One approach is to integrate the rate laws numerically. An alternative approach, which continues to be widely used because it leads to convenient expressions and more readily digestible results, is to make an approximation.

The **steady-state approximation** assumes that, after an initial **induction period**, an interval during which the concentrations of intermediates, I, rise from zero, and during the major part of the reaction, the rates of change of concentrations of all reaction intermediates are negligibly small (Fig. 25.13):

$$\frac{d[I]}{dt} \approx 0 \tag{37}$$

This approximation greatly simplifies the discussion of reaction schemes. For example, when we apply the approximation to the consecutive first-order mechanism, we set $d[I]/dt = 0$ in eqn 31, which then becomes

$$k_a[A] - k_b[I] \approx 0$$

Then

$$[I] \approx \frac{k_a}{k_b}[A] \tag{38}$$

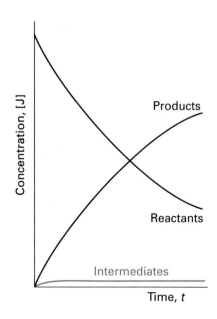

25.13 The basis of the steady-state approximation. It is supposed that the concentrations of intermediates remain small and hardly change during most of the course of the reaction.

On substituting this value of [I] into eqn 32, that equation becomes

$$\frac{d[P]}{dt} = k_b[I] \approx k_a[A] \tag{39}$$

and we see that P is formed by a first-order decay of A, with a rate constant k_a, the rate constant of the slower, rate-determining, step. We can write down the solution of this equation at once by substituting the solution for [A], eqn 33, and integrating:

$$[P] = k_a[A]_0 \int_0^t e^{-k_a t}\, dt = (1 - e^{-k_a t})[A]_0 \tag{40}$$

This is the same (approximate) result as before, eqn 36, but much more quickly obtained.

Example 25.7 Using the steady-state approximation

Devise the rate law for the decomposition of N_2O_5,

$$2N_2O_5(g) \longrightarrow 4NO_2(g) + O_2(g)$$

on the basis of the following mechanism:

$$
\begin{array}{ll}
N_2O_5 \longrightarrow NO_2 + NO_3 & k_a \\
NO_2 + NO_3 \longrightarrow N_2O_5 & k_a' \\
NO_2 + NO_3 \longrightarrow NO_2 + O_2 + NO & k_b \\
NO + N_2O_5 \longrightarrow 3NO_2 & k_c
\end{array}
$$

Method First identify the intermediates (the species that occur in the reaction steps but do not appear in the overall reaction) and write expressions for their net rates of formation. Then, all net rates of change of the concentrations of intermediates are set equal to zero and the resulting equations are solved algebraically.

Answer The intermediates are NO and NO_3; the net rates of change of their concentrations are

$$\frac{d[NO]}{dt} = k_b[NO_2][NO_3] - k_c[NO][N_2O_5] \approx 0$$

$$\frac{d[NO_3]}{dt} = k_a[N_2O_5] - k_a'[NO_2][NO_3] - k_b[NO_2][NO_3] \approx 0$$

The net rate of change of concentration of N_2O_5 is

$$\frac{d[N_2O_5]}{dt} = -k_a[N_2O_5] + k_a'[NO_2][NO_3] - k_c[NO][N_2O_5]$$

and replacing the concentrations of the intermediates by using the equations above gives

$$\frac{d[N_2O_5]}{dt} = -\frac{2k_a k_b[N_2O_5]}{k_a' + k_b}$$

Comment The decomposition of N_2O_5 is problematic because its rate decreases more quickly than expected at low pressures. It is believed that this decrease is due to changes in the rate constants themselves (particularly k_a').

Self-test 25.8 Derive the rate law for the decomposition of ozone in the reaction $2O_3(g) \rightarrow 3O_2(g)$ on the basis of the (incomplete) mechanism

$$O_3 \longrightarrow O_2 + O \qquad k_a$$
$$O_2 + O \longrightarrow O_3 \qquad k_a'$$
$$O + O_3 \longrightarrow 2O_2 \qquad k_b$$

$$[d[O_3]/dt = -k_a k_b [O_3]^2 / (k_a'[O_2] + k_b[O_3])]$$

(d) Pre-equilibria

From a simple sequence of consecutive reactions we now turn to a slightly more complicated mechanism in which an intermediate I reaches an equilibrium with the reactants A and B:

$$A + B \rightleftharpoons I \rightarrow P \qquad (41)$$

The rate constants are k_a and k_a' for the forward and reverse reactions of the equilibrium and k_b for the final step. This scheme involves a **pre-equilibrium**, in which an intermediate is in equilibrium with the reactants. A pre-equilibrium arises when the rates of formation of the intermediate and its decay back into reactants are much faster than its rate of formation of products; thus, the condition is possible when $k_a' \gg k_b$ but not when $k_b \gg k_a'$. Because we assume that A, B, and I are in equilibrium, we can write

$$K = \frac{[I]}{[A][B]} \qquad K = \frac{k_a}{k_a'} \qquad (42)$$

In writing these equations, we are presuming that the rate of reaction of I to form P is too slow to affect the maintenance of the pre-equilibrium (see the example below). The rate of formation of P may now be written:

$$\frac{d[P]}{dt} = k_b[I] = k_b K[A][B] \qquad (43)$$

This rate law has the form of a second-order rate law with a composite rate constant:

$$\frac{d[P]}{dt} = k[A][B] \qquad k = k_b K = \frac{k_a k_b}{k_a'} \qquad (44)$$

Example 25.8 Analysing a pre-equilibrium

Repeat the pre-equilibrium calculation but without ignoring the fact that I is slowly leaking away as it forms P.

Method Begin by writing the net rates of change of the concentrations of the substances and then invoke the steady-state approximation for the intermediate I. Use the resulting expression to obtain the rate of change of the concentration of P.

Answer The net rates of change of P and I are

$$\frac{d[P]}{dt} = k_b[I]$$

$$\frac{d[I]}{dt} = k_a[A][B] - k_a'[I] - k_b[I] \approx 0$$

The second equation solves to

$$[I] \approx \frac{k_a[A][B]}{k_a' + k_b}$$

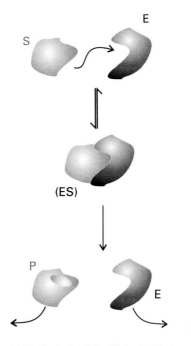

25.14 The basis of the Michaelis–Menten mechanism of enzyme action. Only a fragment of the large enzyme molecule E is shown.

When we substitute this into the expression for the rate of formation of P, we obtain

$$\frac{d[P]}{dt} \approx k[A][B] \qquad k = \frac{k_a k_b}{k_a' + k_b}$$

Comment This expression reduces to that in eqn 44 when the rate constant for the decay of I into products is much smaller than that for its decay into reactants, $k_b \ll k_a'$.

Self-test 25.9 Show that the pre-equilibrium mechanism in which $2A \rightleftharpoons I$ (K) followed by $I + B \rightarrow P$ (k_b) results in an overall third-order reaction.

$$[d[P]/dt = k_b K[A]^2[B]]$$

(e) The Michaelis–Menten mechanism

An example of a reaction in which an intermediate is formed is the **Michaelis–Menten mechanism** of enzyme action. The rate of an enzyme-catalysed reaction in which a substrate S is converted into products P is found to depend on the concentration of the enzyme E even though the enzyme undergoes no net change. The proposed mechanism, which is illustrated in Fig. 25.14, is

$$E + S \rightleftharpoons ES \rightarrow P + E \qquad k_a, k_a', k_b \tag{45}$$

In this mechanism, ES denotes a bound state of the enzyme and its substrate. This mechanism has the same form as that treated in Example 25.8, so we can conclude at once that

$$[ES] = \frac{k_a[E][S]}{k_a' + k_b} \tag{46}$$

[E] and [S] are the concentrations of the free enzyme and free substrate. If $[E]_0$ is the total concentration of enzyme, then

$$[E] + [ES] = [E]_0 \tag{47}$$

Because only a little enzyme is added, the free substrate concentration is almost the same as the total substrate concentration, and we can ignore the fact that [S] differs slightly from $[S]_{total}$. Therefore,

$$[ES] = \frac{k_a([E]_0 - [ES])[S]}{k_a' + k_b} \tag{48}$$

which rearranges to

$$[ES] = \frac{k_a[E]_0[S]}{k_a' + k_b + k_a[S]} \tag{49}$$

It follows that the rate of formation of product is

$$\frac{d[P]}{dt} = k[E]_0 \qquad k = \frac{k_b[S]}{K_M + [S]} \tag{50}$$

where the **Michaelis constant**, K_M, is

$$K_M = \frac{k_a' + k_b}{k_a} \tag{51}$$

According to eqn 50, the rate of enzymolysis varies linearly with the enzyme concentration, but in a more complicated manner with the concentration of substrate (Fig. 25.15). Thus, when $[S] \gg K_M$, the rate law in eqn 50 reduces to

$$\frac{d[P]}{dt} = k_b[E]_0 \tag{52}$$

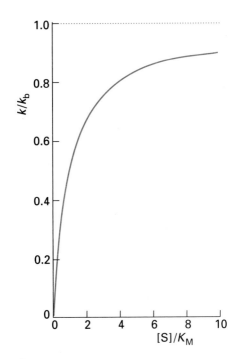

25.15 The variation of the effective rate constant k with substrate concentration according to the Michaelis–Menten mechanism.

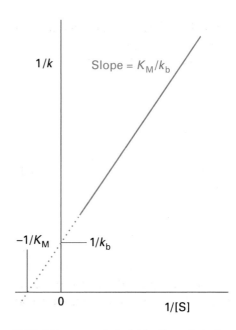

$1/k$ Slope = K_M/k_b

$-1/K_M$ $1/k_b$

0    $1/[S]$

25.16 A Lineweaver–Burk plot for the analysis of an enzymolysis that proceeds by a Michaelis–Menten mechanism, and the significance of the intercepts and the slope.

and is zero-order in S. This result means that under these conditions the rate is constant: there is so much S present that it remains at effectively the same concentration even though products are being formed. Moreover, the rate of formation of products is a maximum, and $k_b[E]_0$ is called the **maximum velocity** of the enzymolysis; k_b itself is called the **maximum turnover number**. When so little S is present that $[S] \ll K_M$, the rate of formation of products is

$$\frac{d[P]}{dt} = \frac{k_b}{K_M}[E]_0[S] \tag{53}$$

Now the rate is proportional to $[S]$ as well as to $[E]_0$.

It follows from eqn 50 that

$$\frac{1}{k} = \frac{1}{k_b} + \frac{K_M}{k_b[S]} \tag{54}$$

Hence, a **Lineweaver–Burk plot** of $1/k$ against $1/[S]$ will give k_b (from the intercept at $1/[S] = 0$) and K_M (from the slope, K_M/k_b, Fig. 25.16). However, the plot cannot give the individual rate constants k_a and k_a' that appear in K_M. The stopped-flow technique can give the additional data needed, because the rate of formation of the enzyme–substrate complex can be found by monitoring its concentration after mixing enzyme and substrate. This procedure gives k_a, and k_a' can then be found by combining this result with the value of K_M.

25.8 Unimolecular reactions

A number of gas-phase reactions follow first-order kinetics, as in the isomerization of cyclopropane mentioned earlier:

$$cyclo\text{-}C_3H_6 \longrightarrow CH_3CH{=}CH_2 \qquad v = k[cyclo\text{-}C_3H_6] \tag{55}$$

The problem with the interpretation of first-order rate laws is that presumably a molecule acquires enough energy to react as a result of its collisions with other molecules. However, collisions are simple bimolecular events, so how can they result in a first-order rate law? First-order gas-phase reactions are widely called 'unimolecular reactions' because they also involve an elementary unimolecular step in which the reactant molecule changes into the product. This term must be used with caution, though, because the overall mechanism has bimolecular as well as unimolecular steps.

(a) The Lindemann–Hinshelwood mechanism

The first successful explanation of unimolecular reactions was provided by Frederick Lindemann in 1921 and then elaborated by Cyril Hinshelwood. In the **Lindemann–Hinshelwood mechanism** it is supposed that a reactant molecule A becomes energetically excited by collision with another A molecule (Fig. 25.17):

$$A + A \longrightarrow A^* + A \qquad \frac{d[A^*]}{dt} = k_a[A]^2 \tag{56}$$

The energized molecule might lose its excess energy by collision with another molecule:

$$A + A^* \longrightarrow A + A \qquad \frac{d[A^*]}{dt} = -k_a'[A][A^*] \tag{57}$$

Alternatively, the excited molecule might shake itself apart and form products P. That is, it might undergo the unimolecular decay

$$A^* \longrightarrow P \qquad \frac{d[A^*]}{dt} = -k_b[A^*] \tag{58}$$

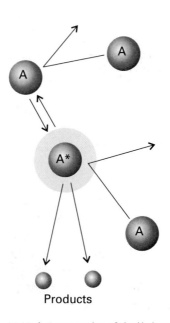

A

A

A*

A

Products

25.17 A representation of the Lindemann–Hinshelwood mechanism of unimolecular reactions. The species A is excited by collision with A, and the excited A molecule (A*) may either be deactivated by a collision with A or go on to decay by a unimolecular process to form products.

If the unimolecular step is slow enough to be the rate-determining step, the overall reaction will have first-order kinetics, as observed. This conclusion can be demonstrated explicitly by applying the steady-state approximation to the net rate of formation of A^*:

$$\frac{d[A^*]}{dt} = k_a[A]^2 - k_a'[A][A^*] - k_b[A^*] \approx 0 \tag{59}$$

This equation solves to

$$[A^*] = \frac{k_a[A]^2}{k_b + k_a'[A]} \tag{60}$$

so the rate law for the formation of P is

$$\frac{d[P]}{dt} = k_b[A^*] = \frac{k_a k_b[A]^2}{k_b + k_a'[A]} \tag{61}$$

At this stage the rate law is not first-order. However, if the rate of deactivation by (A^*, A) collisions is much greater than the rate of unimolecular decay, in the sense that

$$k_a'[A^*][A] \gg k_b[A^*] \qquad \text{or } k_a'[A] \gg k_b$$

then we can neglect k_b in the denominator and obtain

$$\frac{d[P]}{dt} \approx k[A] \qquad k = \frac{k_a k_b}{k_a'} \tag{62}$$

Equation 62 is a first-order rate law, as we set out to show.

The Lindemann–Hinshelwood mechanism can be tested because it predicts that, as the concentration (and therefore the partial pressure) of A is reduced, the reaction should switch to overall second-order kinetics. Thus, when $k_a'[A] \ll k_b$, the rate law in eqn 61 is

$$\frac{d[P]}{dt} \approx k_a[A]^2 \tag{63}$$

The physical reason for the change of order is that at low pressures the rate-determining step is the bimolecular formation of A^*. If we write the full rate law in eqn 61 as

$$\frac{d[P]}{dt} = k[A] \qquad k = \frac{k_a k_b[A]}{k_b + k_a'[A]} \tag{64}$$

then the expression for the effective rate-constant, k, can be rearranged to

$$\frac{1}{k} = \frac{k_a'}{k_a k_b} + \frac{1}{k_a[A]} \tag{65}$$

Hence, a test of the theory is to plot $1/k$ against $1/[A]$, and to expect a straight line.

Whereas the Lindemann–Hinshelwood mechanism agrees in general with the switch in order of unimolecular reactions, it does not agree in detail. A typical graph of $1/k$ against $1/[A]$ is shown in Fig. 25.18. The graph has a pronounced curvature, corresponding to a larger value of k (a smaller value of $1/k$) at high pressures (low $1/[A]$) than would be expected by extrapolation of the reasonably linear low pressure (high $1/[A]$) data.

(b) The activation energy of a composite reaction

Although the rate of each step of a complex mechanism might increase with temperature and show Arrhenius behaviour, is that true of a composite reaction? To answer this question, we consider the high-pressure limit of the Lindemann–Hinshelwood mechanism as

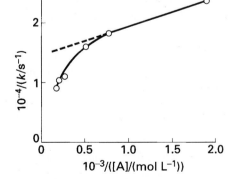

25.18 The pressure dependence of the unimolecular isomerization of *trans*-CHD=CHD showing a pronounced departure from the straight line predicted by eqn 65 based on the Lindemann–Hinshelwood mechanism.

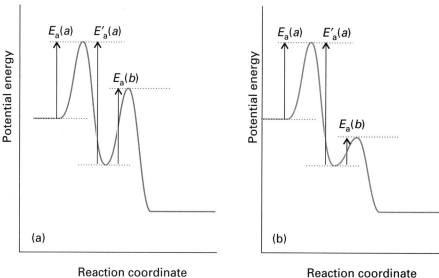

25.19 For a reaction with a pre-equilibrium, there are three activation energies to take into account, two referring to the reversible steps of the pre-equilibrium and one for the final step. The relative magnitudes of the activation energies determine whether the overall activation energy is (a) positive or (b) negative.

Reaction coordinate   Reaction coordinate

expressed in eqn 62. If each of the rate constants has an Arrhenius-like temperature dependence, we can use eqn 26 for each of them, and write

$$
k = \frac{k_a k_b}{k_a'} = \frac{\left(A(a)e^{-E_a(a)/RT}\right)\left(A(b)e^{-E_a(b)/RT}\right)}{\left(A'(a)e^{-E_a'(a)/RT}\right)}
$$
$$
= \frac{A(a)A(b)}{A'(a)} e^{-\{E_a(a)+E_a(b)-E_a'(a)\}/RT}
$$

(66)

That is, the composite rate constant k has an Arrhenius-like form with activation energy

$$
E_a = E_a(a) + E_a(b) - E_a'(a)
$$

(67)

Moreover, provided $E_a(a) + E_a(b) > E_a'(a)$, the activation energy is positive and the rate increases with temperature. However, it is conceivable that $E_a(a) + E_a(b) < E_a'(a)$ (Fig. 25.19), in which case the activation energy is negative and the rate will *decrease* as the temperature is raised. There is nothing remarkable about this behaviour: all it means is that the reverse reaction (corresponding to the deactivation of A*) is so sensitive to temperature that its rate increases sharply as the temperature is raised, and depletes the steady-state concentration of A*. The Lindemann–Hinshelwood mechanism is an unlikely candidate for this type of behaviour because the deactivation of A* has only a small activation energy, but there are reactions with analogous mechanisms in which a negative activation energy is observed.

When we examine the general rate law given in eqn 61, it is clear that the temperature dependence may be difficult to predict because each rate constant in the expression for k increases with temperature, and the outcome depends on whether the terms in the numerator dominate those in the denominator, or vice versa. The fact that so many reactions do show Arrhenius-like behaviour with positive activation energies suggests that their rate laws are in a 'simple' regime, like eqn 63 rather than eqn 61, and that the temperature dependence is dominated by the activation energy of the rate-determining stage. An enzyme reaction can show an even more complicated temperature dependence, because the enzyme may become denatured as the temperature is raised, and hence cease to function.

Checklist of key ideas

- [ ] chemical kinetics

Empirical chemical kinetics

25.1 Experimental techniques
- [ ] real-time analysis
- [ ] quenching method
- [ ] flow method
- [ ] stopped-flow technique
- [ ] flash photolysis

25.2 The rates of reactions
- [ ] rate of consumption
- [ ] rate of formation
- [ ] rate of reaction (1)
- [ ] rate constant
- [ ] rate law (3)
- [ ] reaction order (5)
- [ ] first-order reaction
- [ ] overall order
- [ ] zero-order rate law (7)
- [ ] isolation method

- [ ] pseudofirst-order rate law
- [ ] method of initial rates

25.3 Integrated rate laws
- [ ] integrated rate law
- [ ] first-order integrated rate law (10)
- [ ] half-life (11)
- [ ] second-order integrated rate law (12)
- [ ] half-life of second-order process (13)

25.4 Reactions approaching equilibrium
- [ ] integrated reversible rate laws
- [ ] equilibrium constants and rate constants (21)
- [ ] relaxation
- [ ] temperature jump
- [ ] pressure jump

25.5 The temperature dependence of reaction rates
- [ ] Arrhenius equation (24)
- [ ] pre-exponential factor
- [ ] frequency factor
- [ ] activation energy
- [ ] Arrhenius parameters
- [ ] formal definition of activation energy (25)

Accounting for the rate laws

25.6 Elementary reactions
- [ ] elementary reaction
- [ ] molecularity
- [ ] unimolecular reaction
- [ ] bimolecular reaction
- [ ] rate laws for elementary reactions (27, 28)

25.7 Consecutive elementary reactions
- [ ] integrated rate law for consecutive reactions
- [ ] rate-determining step
- [ ] steady-state approximation (37)
- [ ] induction period
- [ ] pre-equilibrium (41)
- [ ] Michaelis–Menten mechanism
- [ ] Michaelis constant
- [ ] maximum velocity (of enzymolysis) (52)
- [ ] maximum turnover number
- [ ] Lineweaver–Burk plot (54)

25.8 Unimolecular reactions
- [ ] Lindemann–Hinshelwood mechanism
- [ ] effective rate constant (65)
- [ ] composite activation energy (67)

Further reading

Articles of general interest

E. Levin and J.G. Eberhart, Simplified rate-law integration for reactions that are first-order in each of the two reactants. *J. Chem. Educ.* **66**, 705 (1989).

M.N. Berberan-Santos and J.M.G. Martinho, Integration of kinetic rate equations by matrix methods. *J. Chem. Educ.* **67**, 375 (1990).

J.C. Reeve, Some provocative opinions on the terminology of chemical kinetics. *J. Chem. Educ.* **68**, 728 (1991).

H. Maskill, The extent of reaction and chemical kinetics. *Educ. in Chem.* **21**, 122 (1984).

S.R. Logan, The meaning and significance of "the activation energy" of a chemical reaction. *Educ. in Chem.* **23**, 148 (1986).

J.G. Eberhardt and E. Levin, A simplified integration technique for reaction rate laws of integral order in several substances. *J. Chem. Educ.* **72**, 193 (1995).

G.I. Gellene, Application of kinetic approximations to the $A + B \rightleftarrows C$ reaction system. *J. Chem. Educ.* **72**, 196 (1995).

S. Bluestone and K.Y. Yan, A method to find the rate constants in chemical kinetics of a complex reaction. *J. Chem. Educ.* **72**, 884 (1995).

H. Bisswanger, Proteins and enzymes. In *Encyclopedia of applied physics* (ed. G.L. Trigg), **15**, 185. VCH, New York (1996).

R.W. Carr, Chemical kinetics. In *Encyclopedia of applied physics* (ed. G.L. Trigg), **3**, 345. VCH, New York (1992).

Texts and sources of data and information

M.J. Pilling and P.W. Seakins, *Reaction kinetics*. Oxford University Press (1995).

S.R. Logan, *Fundamentals of chemical kinetics*. Longman, Harlow (1996).

J.I. Steinfeld, J.S. Francisco, and W.L. Hase, *Chemical kinetics and dynamics*. Prentice-Hall, Englewood Cliffs (1989).

K.A. Connors, *Chemical kinetics*. VCH, Weinheim (1990).

K.J. Laidler, *Chemical kinetics*. Harper & Row, New York (1987).

I.H. Siegel, *Enzyme kinetics*. Wiley-Interscience, New York (1993).

C.H. Bamford and C.F. Tipper (ed.), *Comprehensive chemical kinetics*, Vols 1–26. Elsevier, Amsterdam (1969–86).

R.G. Compton (ed.), *Comprehensive chemical kinetics*, Vols 27–33. Elsevier, Amsterdam (1987–92).

B.B. Chance, Rapid flow methods. In *Techniques of chemistry* (ed. G.G. Hammes), **6B**, 5. Wiley-Interscience, New York (1974).

G.G. Hammes, Temperature-jump methods. In *Techniques of chemistry* (ed. G.G. Hammes), **6B**, 147. Wiley-Interscience, New York (1974).

W. Knoche, Pressure-jump methods. In *Techniques of chemistry* (ed. G.G. Hammes), **6B**, 187. Wiley-Interscience, New York (1974).

Exercises

25.1 (a) The rate of the reaction $A + 2B \rightarrow 3C + D$ was reported as $1.0 \text{ mol L}^{-1} \text{ s}^{-1}$. State the rates of formation and consumption of the participants.

25.1 (b) The rate of the reaction $A + 3B \rightarrow C + 2D$ was reported as $1.0 \text{ mol L}^{-1} \text{ s}^{-1}$. State the rates of formation and consumption of the participants.

25.2 (a) The rate of formation of C in the reaction $2A + B \rightarrow 2C + 3D$ is $1.0 \text{ mol L}^{-1} \text{ s}^{-1}$. State the reaction rate, and the rates of formation or consumption of A, B, and D.

25.2 (b) The rate of consumption of B in the reaction $A + 3B \rightarrow C + 2D$ is $1.0 \text{ mol L}^{-1} \text{ s}^{-1}$. State the reaction rate, and the rates of formation or consumption of A, C, and D.

25.3 (a) The rate law for the reaction in Exercise 25.1a was found to be $v = k[A][B]$. What are the units of k? Express the rate law in terms of the rates of formation and consumption of (a) A, (b) C.

25.3 (b) The rate law for the reaction in Exercise 25.1b was found to be $v = k[A][B]^2$. What are the units of k? Express the rate law in terms of the rates of formation and consumption of (a) A, (b) C.

25.4 (a) The rate law for the reaction in Exercise 25.2a was reported as $d[C]/dt = k[A][B][C]$. Express the rate law in terms of the reaction rate; what are the units for k in each case?

25.4 (b) The rate law for the reaction in Exercise 25.2b was reported as $d[C]/dt = k[A][B][C]^{-1}$. Express the rate law in terms of the reaction rate; what are the units for k in each case?

25.5 (a) At 518°C, the rate of decomposition of a sample of gaseous acetaldehyde, initially at a pressure of 363 Torr, was 1.07 Torr s^{-1} when 5.0 per cent had reacted and 0.76 Torr s^{-1} when 20.0 per cent had reacted. Determine the order of the reaction.

25.5 (b) At 400 K, the rate of decomposition of a gaseous compound initially at a pressure of 12.6 kPa, was 9.71 Pa s^{-1} when 10.0 per cent had reacted and 7.67 Pa s^{-1} when 20.0 per cent had reacted. Determine the order of the reaction.

25.6 (a) At 518°C, the half-life for the decomposition of a sample of gaseous acetaldehyde (ethanal) initially at 363 Torr was 410 s. When the pressure was 169 Torr, the half-life was 880 s. Determine the order of the reaction.

25.6 (b) At 400 K, the half-life for the decomposition of a sample of a gaseous compound initially at 55.5 kPa was 340 s. When the pressure was 28.9 kPa, the half-life was 178 s. Determine the order of the reaction.

25.7 (a) The rate constant for the first-order decomposition of N_2O_5 in the reaction $2N_2O_5(g) \rightarrow 4NO_2(g) + O_2(g)$ is $k = 3.38 \times 10^{-5} \text{ s}^{-1}$ at 25°C. What is the half-life of N_2O_5? What will be the pressure, initially 500 Torr, (a) 10 s, (b) 10 min after initiation of the reaction?

25.7 (b) The rate constant for the first-order decomposition of a compound A in the reaction $2A \rightarrow P$ is $k = 2.78 \times 10^{-7} \text{ s}^{-1}$ at 25°C. What is the half-life of A? What will be the pressure, initially 32.1 kPa, (a) 10 h, (b) 50 h after initiation of the reaction?

25.8 (a) A second-order reaction of the type $A + B \rightarrow P$ was carried out in a solution that was initially 0.050 mol L^{-1} in A and 0.080 mol L^{-1} in B. After 1.0 h the concentration of A had fallen to 0.020 mol L^{-1}. (a) Calculate the rate constant. (b) What is the half-life of the reactants?

25.8 (b) A second-order reaction of the type $A + 2B \rightarrow P$ was carried out in a solution that was initially 0.075 mol L^{-1} in A and 0.080 mol L^{-1} in B. After 1.0 h the concentration of A had fallen to 0.045 mol L^{-1}. (a) Calculate the rate constant. (b) What is the half-life of the reactants?

25.9 (a) If the rate laws are expressed with (a) concentrations in moles per litre, (b) pressures in kilopascals, what are the units of the second-order and third-order rate constants?

25.9 (b) If the rate laws are expressed with (a) concentrations in molecules per metre cubed, (b) pressures in newtons per metre squared, what are the units of the second-order and third-order rate constants?

25.10 (a) The half-life for the (first-order) radioactive decay of ^{14}C is 5730 y (it emits β rays with an energy of 0.16 MeV). An archaeological sample contained wood that had only 72 per cent of the ^{14}C found in living trees. What is its age?

25.10 (b) One of the hazards of nuclear explosions is the generation of ^{90}Sr and its subsequent incorporation in place of calcium in bones. This nuclide emits β rays of energy 0.55 MeV, and has a half-life of 28.1 y. Suppose $1.00 \mu g$ was absorbed by a newly born child. How much will remain after (a) 18 y, (b) 70 y if none is lost metabolically?

25.11 (a) The second-order rate constant for the reaction

$$CH_3COOC_2H_5(aq) + OH^-(aq) \longrightarrow$$
$$CH_3CO_2^-(aq) + CH_3CH_2OH(aq)$$

is 0.11 $L\,mol^{-1}\,s^{-1}$. What is the concentration of ester after (a) 10 s, (b) 10 min when ethyl acetate is added to sodium hydroxide so that the initial concentrations are $[NaOH] = 0.050\,mol\,L^{-1}$ and $[CH_3COOC_2H_5] = 0.100\,mol\,L^{-1}$?

25.11 (b) The second-order rate constant for the reaction $A + 2B \rightarrow C + D$ is $0.21\,L\,mol^{-1}\,s^{-1}$. What is the concentration of C after (a) 10 s, (b) 10 min when the reactants are mixed with initial concentrations of $[A] = 0.025\,mol\,L^{-1}$ and $[B] = 0.150\,mol\,L^{-1}$?

25.12 (a) A reaction $2A \rightarrow P$ has a second-order rate law with $k = 3.50 \times 10^{-4}\,L\,mol^{-1}\,s^{-1}$. Calculate the time required for the concentration of A to change from $0.260\,mol\,L^{-1}$ to $0.011\,mol\,L^{-1}$.

25.12 (b) A reaction $2A \rightarrow P$ has a third-order rate law with $k = 3.50 \times 10^{-4}\,L^2\,mol^{-2}\,s^{-1}$. Calculate the time required for the concentration of A to change from $0.077\,mol\,L^{-1}$ to $0.021\,mol\,L^{-1}$.

25.13 (a) The rate constant for the decomposition of a certain substance is $2.80 \times 10^{-3}\,L\,mol^{-1}\,s^{-1}$ at 30°C and $1.38 \times 10^{-2}\,L\,mol^{-1}\,s^{-1}$ at 50°C. Evaluate the Arrhenius parameters of the reaction.

25.13 (b) The rate constant for the decomposition of a certain substance is $1.70 \times 10^{-2}\,L\,mol^{-1}\,s^{-1}$ at 24°C and $2.01 \times 10^{-2}\,L\,mol^{-1}\,s^{-1}$ at 37°C. Evaluate the Arrhenius parameters of the reaction.

25.14 (a) The reaction mechanism

$$A_2 \rightleftharpoons 2A \quad \text{(fast)}$$
$$A + B \longrightarrow P \quad \text{(slow)}$$

involves an intermediate A. Deduce the rate law for the reaction.

25.14 (b) Consider the following mechanism for renaturation of a double helix from its strands A and B:

$$A + B \rightleftharpoons \text{unstable helix} \quad \text{(fast)}$$
$$\text{unstable helix} \longrightarrow \text{stable double helix} \quad \text{(slow)}$$

Derive the rate equation for the formation of the double helix and express the rate constant of the renaturation reaction in terms of the rate constants of the individual steps.

25.15 (a) Show that $t_{1/2} \propto 1/[A]^{n-1}$ for a reaction that is nth-order in A.

25.15 (b) Deduce an expression for the time it takes for the concentration of a substance to fall to one-third its initial value in an nth-order reaction.

25.16 (a) The enzyme-catalysed conversion of a substrate at 25°C has a Michaelis constant of $0.035\,mol\,L^{-1}$. The rate of the reaction is $1.15 \times 10^{-3}\,mol\,L^{-1}\,s^{-1}$ when the substrate concentration is $0.110\,mol\,L^{-1}$. What is the maximum velocity of this enzymolysis?

25.16 (b) The enzyme-catalysed conversion of a substrate at 25°C has a Michaelis constant of $0.042\,mol\,L^{-1}$. The rate of the reaction is $2.45 \times 10^{-4}\,mol\,L^{-1}\,s^{-1}$ when the substrate concentration is $0.890\,mol\,L^{-1}$. What is the maximum velocity of this enzymolysis?

25.17 (a) The effective rate constant for a gaseous reaction which has a Lindemann–Hinshelwood mechanism is $2.50 \times 10^{-4}\,s^{-1}$ at 1.30 kPa and $2.10 \times 10^{-5}\,s^{-1}$ at 12 Pa. Calculate the rate constant for the activation step in the mechanism.

25.17 (b) The effective rate constant for a gaseous reaction which has a Lindemann–Hinshelwood mechanism is $1.7 \times 10^{-3}\,s^{-1}$ at 1.09 kPa and $2.2 \times 10^{-4}\,s^{-1}$ at 25 Pa. Calculate the rate constant for the activation step in the mechanism.

25.18 (a) The pK_a of NH_4^+ is 9.25 at 25°C. The rate constant at 25°C for the reaction of NH_4^+ and OH^- to form aqueous NH_3 is $4.0 \times 10^{10}\,L\,mol^{-1}\,s^{-1}$. Calculate the rate constant for proton transfer to NH_3. What relaxation time would be observed if a temperature jump were applied to a solution of $0.15\,mol\,L^{-1}\,NH_3(aq)$ at 25°C?

25.18 (b) The equilibrium $A \rightleftharpoons B + C$ at 25°C is subjected to a temperature jump. The measured relaxation time is 3.0 μs. The equilibrium constant for the system is 2.0×10^{-16} at 25°C, and the equilibrium concentrations of B and C at 25°C are both $2.0 \times 10^{-4}\,mol\,L^{-1}$. Calculate the rate constants for the first-order forward and second-order reverse reactions.

Problems

Numerical problems

25.1 The data below apply to the formation of urea from ammonium cyanate, $NH_4CNO \rightarrow NH_2CONH_2$. Initially 22.9 g of ammonium cyanate was dissolved in enough water to prepare 1.00 L of solution. Determine the order of the reaction, the rate constant, and the mass of ammonium cyanate left after 300 min.

t/min	0	20.0	50.0	65.0	150
$m(\text{urea})$/g	0	7.0	12.1	13.8	17.7

25.2 The data below apply to the reaction, $(CH_3)_3CBr + H_2O \rightarrow (CH_3)_3COH + HBr$. Determine the order of the reaction, the rate constant, and the molar concentration of $(CH_3)_3CBr$ after 43.8 h.

t/h	0	3.15	6.20	10.00	18.30	30.80
$[(CH_3)_3CBr]/$ $(10^{-2}\,mol\,L^{-1})$	10.39	8.96	7.76	6.39	3.53	2.07

25.3 The thermal decomposition of an organic nitrile produced the following data:

$t/(10^3\,s)$	0	2.00	4.00	6.00	8.00	10.00	12.00	∞
$[\text{nitrile}]/(mol\,L^{-1})$	1.10	0.86	0.67	0.52	0.41	0.32	0.25	0

Determine the order of the reaction and the rate constant.

25.4 The following data have been obtained for the decomposition of $N_2O_5(g)$ at $67°C$ according to the reaction $2N_2O_5(g) \rightarrow 4NO_2(g) + O_2(g)$. Determine the order of the reaction, the rate constant, and the half-life. It is not necessary to obtain the result graphically; you may do a calculation using estimates of the rates of change of concentration.

t/min	0	1	2	3	4	5
$[N_2O_5]/(\text{mol L}^{-1})$	1.000	0.705	0.497	0.349	0.246	0.173

25.5 A first-order decomposition reaction is observed to have the following rate constants at the indicated temperatures. Estimate the activation energy.

$k/(10^{-3}\,\text{s}^{-1})$	2.46	45.1	576
$\theta/°C$	0	20.0	40.0

25.6 The gas-phase decomposition of acetic acid at 1189 K proceeds by way of two parallel reactions:

$$(1)\ CH_3COOH \longrightarrow CH_4 + CO_2 \qquad k_1 = 3.74\ s^{-1}$$

$$(2)\ CH_3COOH \longrightarrow H_2C=C=O + H_2O \qquad k_2 = 4.65\ s^{-1}$$

What is the maximum percentage yield of the ketene CH_2CO obtainable at this temperature?

25.7 The composition of a liquid-phase reaction $2A \rightarrow B$ was followed by a spectrophotometric method with the following results:

t/min	0	10	20	30	40	∞
$[B]/(\text{mol L}^{-1})$	0	0.089	0.153	0.200	0.230	0.312

Determine the order of the reaction and its rate constant.

25.8 Sucrose is readily hydrolysed to glucose and fructose in acidic solution. The hydrolysis is often monitored by measuring the angle of rotation of plane-polarized light passing through the solution. From the angle of rotation the concentration of sucrose can be determined. An experiment on the hydrolysis of sucrose in 0.50 M HCl(aq) produced the following data:

t/min	0	14	39	60	80	110	140	170	210
$[\text{sucrose}]/$ (mol L^{-1})	0.316	0.300	0.274	0.256	0.238	0.211	0.190	0.170	0.146

Determine the rate constant of the reaction and the average lifetime of a sucrose molecule.

25.9 The ClO radical decays rapidly by way of the reaction, $2ClO \rightarrow Cl_2 + O_2$. The following data have been obtained:

$t/(10^{-3}\,\text{s})$	0.12	0.62	0.96	1.60	3.20	4.00	5.75
$[ClO]/(10^{-6}\,\text{mol L}^{-1})$	8.49	8.09	7.10	5.79	5.20	4.77	3.95

Determine the rate constant of the reaction.

25.10 Cyclopropane isomerizes into propene when heated to $500°C$ in the gas phase. The extent of conversion for various initial pressures has been followed by gas chromatography by allowing the reaction to proceed for a time with various initial pressures:

p_0/Torr	200	200	400	400	600	600
t/s	100	200	100	200	100	200
p/Torr	186	173	373	347	559	520

where p_0 is the initial pressure and p is the final pressure of cyclopropane. What are the order and rate constant for the reaction under these conditions?

25.11 The addition of hydrogen halides to alkenes has played a fundamental role in the investigation of organic reaction mechanisms. In one study (M.J. Haugh and D.R. Dalton, *J. Amer. Chem. Soc.* **97**, 5674 (1975)), high pressures of hydrogen chloride (up to 25 atm) and propene (up to 5 atm) were examined over a range of temperatures and the amount of 2-chloropropane formed was determined by NMR. Show that if the reaction $A + B \rightarrow P$ proceeds for a short time δt, the concentration of product follows $[P]/[A] = k[A]^{m-1}[B]^n \delta t$ if the reaction is mth-order in A and nth-order in B. In a series of runs the ratio of [chloropropane] to [propene] was independent of [propene] but the ratio of [chloropropane] to [HCl] for constant amounts of propene depended on [HCl]. For $\delta t \approx 100$ h (which is short on the timescale of the reaction) the latter ratio rose from zero to 0.05, 0.03, 0.01 for $p(\text{HCl}) = 10$ atm, 7.5 atm, 5.0 atm, respectively. What are the orders of the reaction with respect to each reactant?

25.12 Show that the following mechanism can account for the rate law of the reaction in Problem 25.11:

$$2HCl \rightleftharpoons (HCl)_2 \qquad K_1$$
$$HCl + CH_3CH=CH_2 \rightleftharpoons \text{complex} \qquad K_2$$
$$(HCl)_2 + \text{complex} \longrightarrow CH_3CHClCH_3 + 2HCl \qquad k \quad (\text{slow})$$

What further tests could you apply to verify this mechanism?

25.13 In the experiments described in Problems 25.11 and 25.12 an inverse temperature dependence of the reaction rate was observed, the overall rate of reaction at $70°C$ being roughly one-third that at $19°C$. Estimate the apparent activation energy and the activation energy of the rate-determining step given that the enthalpies of the two equilibria are both of the order of -14 kJ mol^{-1}.

25.14 The second-order rate constants for the reaction of oxygen atoms with aromatic hydrocarbons have been measured (R. Atkinson and J.N. Pitts, *J. Phys. Chem.* **79**, 295 (1975)). In the reaction with benzene the rate constants are 1.44×10^7 L mol^{-1} s^{-1} at 300.3 K, 3.03×10^7 L mol^{-1} s^{-1} at 341.2 K, and 6.9×10^7 L mol^{-1} s^{-1} at 392.2 K. Find the pre-exponential factor and activation energy of the reaction.

25.15 In Problem 25.10 the isomerization of cyclopropane over a limited pressure range was examined. If the Lindemann mechanism of first-order reactions is to be tested we also need data at low pressures. These have been obtained (H.O. Pritchard, R.G. Sowden, and A.F. Trotman-Dickenson, *Proc. R. Soc.* **A217**, 563 (1953)):

p/Torr	84.1	11.0	2.89	0.569	0.120	0.067
$10^4\,k_{\text{eff}}/\text{s}^{-1}$	2.98	2.23	1.54	0.857	0.392	0.303

Test the Lindemann theory with these data.

25.16 The initial rate of O_2 production by the action of an enzyme on a substrate was measured for a range of substrate concentrations; the data are below. Evaluate the Michaelis constant for the reaction.

$[S]/(\text{mol L}^{-1})$	0.050	0.017	0.010	0.0050	0.0020
$v/(\text{mm}^3\,\text{min}^{-1})$	16.6	12.4	10.1	6.6	3.3

Theoretical problems

25.17 The equilibrium $A \rightleftharpoons B$ is first-order in both directions. Derive an expression for the concentration of A as a function of time when the initial molar concentrations of A and B are $[A]_0$ and $[B]_0$. What is the final composition of the system?

25.18 Derive an integrated expression for a second-order rate law $v = k[A][B]$ for a reaction of stoichiometry $2A + 3B \rightarrow P$.

25.19 Derive the integrated form of a third-order rate law $v = k[A]^2[B]$ in which the stoichiometry is $2A + B \rightarrow P$ and the reactants are initially present in (a) their stoichiometric proportions, (b) with B present initially in twice the amount.

25.20 Set up the rate equations for the reaction mechanism:

$$A \underset{k_a'}{\overset{k_a}{\rightleftharpoons}} B \underset{k_b'}{\overset{k_b}{\rightleftharpoons}} C$$

Show that the mechanism is equivalent to

$$A \underset{k_{eff}'}{\overset{k_{eff}}{\rightleftharpoons}} C$$

under specified circumstances.

25.21 Show that the ratio $t_{1/2}/t_{3/4}$, where $t_{1/2}$ is the half-life and $t_{3/4}$ is the time for the concentration of A to decrease to $\frac{3}{4}$ of its initial value (implying that $t_{3/4} < t_{1/2}$) can be written as a function of n alone, and can therefore be used as a rapid assessment of the order of a reaction.

25.22 Many enzyme-catalysed reactions are consistent with a modified version of the Michaelis–Menten mechanism in which the second step is also reversible. For this mechanism obtain an expression for the rate of formation of product and find its limiting behaviour for large and small concentrations of substrate.

25.23 Derive an equation for the steady-state rate of the sequence of reactions $A \rightleftharpoons B \rightleftharpoons C \rightleftharpoons D$, with [A] maintained at a fixed value and the product D removed as soon as it is formed.

Additional problems supplied by Carmen Giunta and Charles Trapp

25.24 *Prebiotic reactions* are reactions that might have occurred under the conditions prevalent on the Earth before the first living creatures emerged and that can lead to analogues of molecules necessary for life as we now know it. To qualify, a reaction must proceed with favourable rates and equilibria. M.P. Robertson and S.I. Miller (*Science* **268**, 702 (1995)) have studied the prebiotic synthesis of 5-substituted uracils, among them 5-hydroxymethyl-uracil (HMU). Amino acid analogues can be formed from HMU under prebiotic conditions by reaction with various nucleophiles, such as H_2S, HCN, indole, imidazole, etc. For the synthesis of HMU (the uracil

analogue of serine) from uracil and formaldehyde (HCHO), the rate of addition is given by $\log k/(\text{L mol}^{-1}\,\text{s}^{-1}) = 11.75 - 5488/(T/\text{K})$ (at pH = 7), and $\log K = -1.36 + 1794/(T/\text{K})$. For this reaction, calculate the rates and equilibrium constants over a range of temperatures corresponding to possible prebiotic conditions, such as 0–50°C, and plot them against temperature. Also, calculate the activation energy and the standard reaction Gibbs energy and enthalpy at 25°C. Prebiotic conditions are not likely to be standard conditions. Speculate about how the actual values of the reaction Gibbs energy and enthalpy might differ from the standard values. Do you expect that the reaction would still be favourable?

25.25 For the second-order reaction $A + B \rightarrow$ Products, the rate of reaction, v, may be written

$$v = \frac{dx}{dt} = k([A]_0 - x)([B]_0 + x)$$

where x is the decrease in concentration of A or B as a result of reaction. What are the conditions for the rate to be a maximum and a minimum? Draw a graph of v against x and, noting that v and x cannot be negative, identify the portion of the curve that corresponds to reality.

25.26 For the consecutive reaction $A \rightarrow I \rightarrow P$, Fig. 25.12 shows [I] plotted against time for $k_a = 10k_b$. For $[A]_0 = 1.0\,\text{mol L}^{-1}$ and $k_a = 1.0\,\text{min}^{-1}$, plot [I] against t for $k_a/k_b = 5$, 1, and 0.5. For each case determine the time at which [I] reaches a maximum.

25.27 For the only $A + B \rightarrow P$ reaction in Table 25.3, find an expression for x as a function of time.

25.28 T. Gierczak, R.K. Talukdar, S.C. Herndon, G.L. Vaghjiani, and A.R. Ravishankara (*J. Phys. Chem.* A **101**, 3125 (1997)) measured the rate constants for the elementary bimolecular gas-phase reaction of methane with the hydroxyl radical over a range of temperatures of importance to atmospheric chemistry. Deduce the Arrhenius parameters A and E_a from the following measurements.

T/K	295	295	223	218
$k/(10^6\,\text{L mol}^{-1}\,\text{s}^{-1})$	3.70	3.55	0.494	0.452
T/K	213	206	200	195
$k/(10^6\,\text{L mol}^{-1}\,\text{s}^{-1})$	0.379	0.295	0.241	0.217

25.29 The oxidation of HSO_3^- by O_2 in aqueous solution is a reaction of importance to the processes of acid rain formation and flue gas desulfurization. R.E. Connick, Y.-X. Zhang, S. Lee, R. Adamic, and P. Chieng (*Inorg. Chem.* **34**, 4543 (1995)) report that the reaction $2HSO_3^- + O_2 \rightarrow 2SO_4^{2-} + 2H^+$ follows the rate law $v = k[HSO_3^-]^2[H^+]^2$. Given a pH of 5.6 and an oxygen molar concentration of $2.4 \times 10^{-4}\,\text{mol L}^{-1}$ (both presumed constant), an initial HSO_3^- molar concentration of $5 \times 10^{-5}\,\text{mol L}^{-1}$, and a rate constant of $3.6 \times 10^6\,\text{L}^3\,\text{mol}^{-3}\,\text{s}^{-1}$, what is the initial rate of reaction? How long would it take for HSO_3^- to reach half its initial concentration?

25.30 Chlorine atoms react rapidly with ozone in the gas-phase bimolecular reaction $Cl + O_3 \rightarrow ClO + O_2$ with $k_2 = (1.7 \times 10^{10}\,\text{L mol}^{-1}\,\text{s}^{-1})e^{-260/(T/\text{K})}$ (W.B. DeMore, S.P. Sander, D.M. Golden, R.F. Hampson, M.J. Kurylo, C.J. Howard, A.R. Ravishankara, C.E. Kolb, and M.J. Molina, *Chemical kinetics and* *photochemical data for use in stratospheric modeling*: *Evaluation Number 11*, JPL Publication 94–26 (1994)). Estimate the rate of this reaction at (a) 20 km, where $[Cl] = 5 \times 10^{-17}$ mol L^{-1}, $[O_3] = 8 \times 10^{-9}$ mol L^{-1}, and $T = 220$ K; (b) 45 km, where $[Cl] = 3 \times 10^{-15}$ mol L^{-1}, $[O_3] = 8 \times 10^{-11}$ mol L^{-1}, and $T = 270$ K.

25.31 T. Gierczak, R.K. Talukdar, S.C. Herndon, G.L. Vaghjiani, and A.R. Ravishankara (*J. Phys. Chem.* **A 101**, 3125 (1997)) measured the rate constants for the bimolecular gas-phase reaction of methane with the hydroxyl radical $CH_4(g) + OH(g) \rightarrow CH_3(g) + H_2O(g)$ and found $A = 1.13 \times 10^9$ L mol^{-1} s^{-1} and $E_a = 14.1$ kJ mol^{-1} for the Arrhenius parameters. Reaction with OH is the main path by which CH_4 is removed from the lower atmosphere. (a) Estimate the rate of consumption of CH_4. Take the average OH concentration to be 1.5×10^{-21} mol L^{-1}, that of CH_4 to be 4.0×10^{-8} mol L^{-1}, and the temperature to be $-10°C$. (b) Estimate the global annual mass of CH_4 consumed by this reaction (which is slightly less than the amount introduced to the atmosphere) given an effective volume for the Earth's lower atmosphere of 4×10^{21} L.

25.32 P.W. Seakins, M.J. Pilling, L.T. Niiranen, D. Gutman, and L.N. Krasnoperov (*J. Phys. Chem.* **96**, 9847 (1992)) measured the forward and reverse rate constants for the gas-phase reaction $C_2H_5(g) + HBr(g) \rightarrow C_2H_6(g) + Br(g)$ and used their findings to compute thermodynamic parameters for C_2H_5. The reaction is bimolecular in both directions with Arrhenius parameters $A = 1.0 \times 10^9$ L mol^{-1} s^{-1}, $E_a = -4.2$ kJ mol^{-1} for the forward reaction and $k' = 1.4 \times 10^{11}$ L mol^{-1} s^{-1}, $E_a = 53.3$ kJ mol^{-1} for the reverse reaction. Compute $\Delta_f H^{\ominus}$, $S_m^{\ominus}$, and $\Delta_f G^{\ominus}$ of C_2H_5 at 298 K.

26 The kinetics of complex reactions

This chapter extends the material introduced in Chapter 25 by showing how complex reaction mechanisms are treated. In particular, we deal with chain reactions, and see that either complicated or simple rate laws can be obtained, depending on the conditions. Under certain circumstances, a chain reaction can become explosive, and we see some of the reasons for this behaviour. An important application of these more complicated techniques is to the kinetics of polymerization reactions. Here we shall see that there are two major classes of polymerization process, and the average molar mass of the product varies with time in distinctive ways. Finally, we consider reactions in which the concentrations of the intermediates and products oscillate in time: although their rate laws may be well known, under certain circumstances the composition of a system may be unpredictable.

Many reactions take place by mechanisms that involve several elementary steps, and some take place at a useful rate only if a catalyst is present. Certain other reactions take place by mechanisms in which there is positive or negative feedback, in which the products of the reaction influence the rate at which more products are produced. In this chapter we see how to develop the ideas introduced in Chapter 25 to deal with these special kinds of reactions.

Chain reactions

Many gas-phase reactions and liquid-phase polymerization reactions are **chain reactions**. In a chain reaction, an intermediate produced in one step generates an intermediate in a subsequent step; then that intermediate generates another intermediate, and so on.

26.1 The structure of chain reactions

The intermediates in a chain reaction are called **chain carriers**. In a **radical chain reaction** the chain carriers are radicals (species with unpaired electrons). Ions may also act as chain carriers. In nuclear fission the chain carriers are neutrons.

(a) The classification of reaction steps

Chain carriers are initially formed in the **initiation step** of the reaction. For example, Cl atoms are formed by the dissociation of Cl_2 molecules either as a result of vigorous intermolecular collisions in a **thermolysis**, a reaction initiated by heat, or as a result of absorption of a photon in a **photolysis**, a reaction stimulatated by the absorption of electromagnetic radiation. The chain carriers produced in the initiation step attack other reactant molecules in the **propagation steps**, and each attack gives rise to a new carrier. An example is the attack of a methyl radical on ethane:

$$\cdot CH_3 + CH_3CH_3 \longrightarrow CH_4 + \cdot CH_2CH_3$$

(The dot signifies the unpaired electron and marks the radical.) In some cases the attack results in the production of more than one chain carrier. An example of such a **branching step** is

$$\cdot O\cdot + H_2O \longrightarrow HO\cdot + HO\cdot$$

where the attack of one O atom on an H_2O molecule forms two $\cdot OH$ radicals (recall that an O atom has the ground-state configuration $[He]2s^2 2p^4$, with two unpaired electrons).

A chain carrier might attack a product molecule formed earlier in the reaction. Because this attack reduces the net rate of formation of product, it is called a **retardation step**. For example, in a photochemical reaction in which HBr is formed from H_2 and Br_2, an H atom might attack an HBr molecule, leading to H_2 and Br:

$$\cdot H + HBr \longrightarrow H_2 + \cdot Br$$

Retardation does not end the chain, because one radical ($\cdot H$) gives rise to another ($\cdot Br$), but it does deplete the concentration of the product. Elementary reactions in which radicals combine and end the chain are called **termination steps**, as in

$$CH_3CH_2\cdot + \cdot CH_2CH_3 \longrightarrow CH_3CH_2CH_2CH_3$$

In an **inhibition step**, radicals are removed other than by chain termination, such as by reaction with the walls of the vessel or with foreign radicals:

$$CH_3CH_2\cdot + \cdot R \longrightarrow CH_3CH_2R$$

An NO molecule has an unpaired electron and is a very efficient chain inhibitor. The observation that a reaction is quenched when NO is introduced is a good indication that a radical chain mechanism is in operation.

(b) The rate laws of chain reactions

A chain reaction can have a simple rate law. As a first example, consider the **pyrolysis**, or thermal decomposition in the absence of air, of acetaldehyde (ethanal, CH_3CHO), which is found to be three-halves order in CH_3CHO:

$$CH_3CHO(g) \xrightarrow{\Delta} CH_4(g) + CO(g) \qquad \frac{d[CH_4]}{dt} = k[CH_3CHO]^{3/2} \qquad (1)$$

Some ethane is also detected. The **Rice–Herzfeld mechanism** for this reaction is as follows:

(a) Initiation: $CH_3CHO \longrightarrow \cdot CH_3 + \cdot CHO \qquad v = k_a[CH_3CHO]$

(b) Propagation: $CH_3CHO + \cdot CH_3 \longrightarrow CH_4 + CH_3CO\cdot$

$$v = k_b[CH_3CHO][\cdot CH_3]$$

(c) Propagation: $CH_3CO\cdot \longrightarrow \cdot CH_3 + CO \qquad v = k_c[CH_3CO\cdot]$

(d) Termination: $\cdot CH_3 + \cdot CH_3 \longrightarrow CH_3CH_3 \qquad v = k_d[\cdot CH_3]^2$

As we shall see, this mechanism captures the principal features of the reaction; however, it does not accommodate the formation of various by-products, such as propanone (CH_3COCH_3) and propanal (CH_3CH_2CHO).

To test the proposed mechanism we need to show that it leads to the observed rate law. According to the steady-state approximation (Section 25.7c), the net rate of change of the intermediates ($\cdot CH_3$ and $CH_3CO\cdot$) may be set equal to zero:

$$\frac{d[\cdot CH_3]}{dt} = k_a[CH_3CHO] - k_b[\cdot CH_3][CH_3CHO] + k_c[CH_3CO\cdot]$$
$$- 2k_d[\cdot CH_3]^2 = 0$$

$$\frac{d[CH_3CO\cdot]}{dt} = k_b[\cdot CH_3][CH_3CHO] - k_c[CH_3CO\cdot] = 0$$

The sum of the two equations is

$$k_a[CH_3CHO] - 2k_d[\cdot CH_3]^2 = 0$$

which implies that the steady-state concentration of $\cdot CH_3$ radicals is

$$[\cdot CH_3] = \left(\frac{k_a}{2k_d}\right)^{1/2}[CH_3CHO]^{1/2} \tag{2}$$

It follows that the rate of formation of CH_4 is

$$\frac{d[CH_4]}{dt} = k_b[\cdot CH_3][CH_3CHO] = k_b\left(\frac{k_a}{2k_d}\right)^{1/2}[CH_3CHO]^{3/2} \tag{3}$$

which is in agreement with the three-halves order observed experimentally (eqn 1). However, as already indicated, the true mechanism must be more complicated because other products are formed in significant quantities.

In many cases, a chain reaction leads to a complicated rate law. An example is the reaction between H_2 and Br_2 for which the empirical rate law is

$$H_2(g) + Br_2(g) \longrightarrow 2HBr(g) \qquad \frac{d[HBr]}{dt} = \frac{k[H_2][Br_2]^{3/2}}{[Br_2] + k'[HBr]} \tag{4}$$

The following mechanism has been proposed to account for this rate law (Fig. 26.1):

(a) Initiation: $Br_2 + M \longrightarrow Br\cdot + Br\cdot + M \qquad v = k_a[Br_2][M]$

where M is either Br_2 or H_2.

(b) Propagation: $Br\cdot + H_2 \longrightarrow HBr + H\cdot \qquad v = k_b[Br\cdot][H_2]$
$\qquad\qquad\qquad H\cdot + Br_2 \longrightarrow HBr + Br\cdot \qquad v = k_b'[H\cdot][Br_2]$

(c) Retardation: $H\cdot + HBr \longrightarrow H_2 + Br\cdot \qquad v = k_c[H\cdot][HBr]$

(d) Termination: $Br\cdot + Br\cdot + M \longrightarrow Br_2 + M^* \qquad v = k_d[Br\cdot]^2[M]$

The third body M removes the energy of recombination. Other possible termination steps include the combination of H atoms to form H_2 and the combination of H and Br atoms; however, it turns out that only Br atom recombination is important. The net rate of formation of the product HBr is

$$\frac{d[HBr]}{dt} = k_b[Br\cdot][H_2] + k_b'[H\cdot][Br_2] - k_c[H\cdot][HBr]$$

We can now either analyse the rate equations numerically or look for approximate solutions and see if they agree with the empirical rate law. The latter approach is illustrated in the following example.

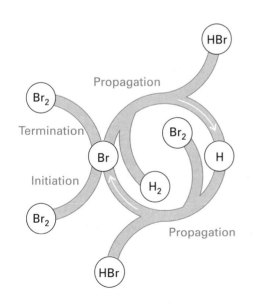

26.1 A schematic representation of the mechanism of the reaction between hydrogen and bromine. Note how the reactants and products are shown as arms to the circle, but the intermediates (H and Br) occur only within the circle. Similar diagrams are used to depict the action of catalysts.

Example 26.1 Deriving the rate equation of a chain reaction

Derive the rate law for the formation of HBr according to the mechanism given above.

Method Make the steady-state approximation for the concentrations of any intermediates (H· and Br· in the present case) and set the net rates of change of their concentrations equal to zero. Begin by writing down the net rates of formation of the intermediates. Set these expressions equal to zero, solve the resulting equations for the concentrations of the intermediates, and then use the resulting expressions in the equation for the net rate of formation of HBr.

Answer The net rates of formation of the two intermediates are

$$\frac{d[H\cdot]}{dt} = k_b[Br\cdot][H_2] - k_b'[H\cdot][Br_2] - k_c[H\cdot][HBr] = 0$$

$$\frac{d[Br\cdot]}{dt} = 2k_a[Br_2][M] - k_b[Br\cdot][H_2] + k_b'[H\cdot][Br_2] + k_c[H\cdot][HBr]$$
$$- 2k_d[Br\cdot]^2[M] = 0$$

The steady-state concentrations of the intermediates are obtained by solving these two simultaneous equations and are

$$[Br\cdot] = \left(\frac{k_a}{k_d}\right)^{1/2}[Br_2]^{1/2}$$

$$[H\cdot] = \frac{k_b(k_a/k_d)^{1/2}[H_2][Br_2]^{1/2}}{k_b'[Br_2] + k_c[HBr]}$$

Note that [M] has cancelled. When we substitute these concentrations into the expression for $d[HBr]/dt$, we obtain

$$\frac{d[HBr]}{dt} = \frac{2k_b(k_a/k_d)^{1/2}[H_2][Br_2]^{3/2}}{[Br_2] + (k_c/k_b')[HBr]}$$

This equation has the same form as the empirical rate law (eqn 4), so the two empirical rate constants can be identified as

$$k = 2k_b\left(\frac{k_a}{k_d}\right)^{1/2} \qquad k' = \frac{k_c}{k_b'}$$

Comment The presence of [HBr] in the denominator is a sign that HBr is acting as an inhibitor, and reducing the rate of formation of product. Likewise, the presence of [Br₂] stems from the role of Br₂ in the removal of the reactive ·H radicals from the chain.

- -

Self-test 26.1 Deduce the rate law for the production of HBr when the initiation step is a photochemical process in which $Br_2 \rightarrow Br\cdot + \cdot Br$ with rate $v = \mathcal{I}_{abs}$, where $\mathcal{I}_{abs}$ is the intensity of absorbed radiation.

[See eqn 6, below]

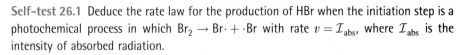

In the examples we have considered, the observed rate laws are reproduced by the mechanisms. That used to be essentially the end of the calculation (but not of the experimental investigation). Now, though, we can make use of computers to integrate the approximate rate law numerically, and hence predict the time dependence of the HBr concentration (Fig. 26.2). Mathematical software may also be used to integrate the original coupled rate laws without needing to invoke the steady-state approximation.

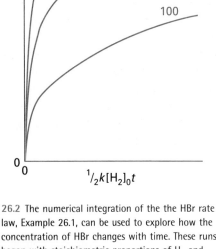

26.2 The numerical integration of the the HBr rate law, Example 26.1, can be used to explore how the concentration of HBr changes with time. These runs began with stoichiometric proportions of H_2 and Br_2; the curves are labelled with the value of $2k' - 1$.

26.2 Explosions

A **thermal explosion** is due to the rapid increase of reaction rate with increasing temperature. If the energy released by an exothermic reaction cannot escape, the temperature of the system rises and the reaction goes faster. The acceleration of the rate results in a faster rise of temperature, so the reaction goes even faster ... catastrophically fast. A **chain-branching explosion** may occur when there are chain-branching steps in a reaction, for then the number of chain centres grows exponentially and the rate of reaction may cascade into an explosion.

An example of both types of explosion is provided by the reaction between hydrogen and oxygen:

$$2H_2(g) + O_2(g) \longrightarrow 2H_2O(g)$$

Although the net reaction is very simple, the mechanism is very complex. A chain reaction is involved, and the chain carriers include $\cdot H$, $\cdot O\cdot$, $\cdot OH$, and $\cdot O_2H$. Some steps are:

Initiation: $\quad H_2 + O_2 \longrightarrow \cdot OH + \cdot OH$

Propagation: $\quad H_2 + \cdot OH \longrightarrow \cdot H + H_2O$

$\cdot (O_2)\cdot + \cdot H \longrightarrow \cdot O\cdot + \cdot OH$ (branching)

$\cdot O\cdot + H_2 \longrightarrow \cdot OH + \cdot H$ (branching)

$\cdot H + O_2 + M \longrightarrow HO_2\cdot + M^*$

The two branching steps can lead to a chain-branching explosion. The $HO_2\cdot$ radicals are eliminated by collisions with the walls.

The occurrence of an explosion depends on the temperature and pressure of the system, and the **explosion regions** for the reaction are shown in Fig. 26.3. At very low pressures the system is outside the explosion region and the mixture reacts smoothly. At these pressures the chain carriers produced in the branching steps can reach the walls of the container where they combine. Increasing the pressure (along a vertical line in the illustration) takes the system through the **first explosion limit** (if the temperature is greater than about 730 K). The mixture then explodes because the chain carriers react before reaching the walls and the branching reactions are explosively efficient. The reaction is smooth when the pressure is above the **second explosion limit**. The concentration of molecules in the gas is then so great that the radicals produced in the branching reaction combine in the body of the gas, and reactions such as $O_2 + \cdot H \rightarrow \cdot O_2H$ can occur. Recombination reactions like this one are facilitated by three-body collisions because the third body (M) can remove the excess energy. At low pressures three-particle collisions are unimportant and recombination is much slower. At higher pressures, when three-particle collisions are important, the explosive propagation of the chain by the radicals is partially quenched because the branching steps are diverted into simple propagation steps (see the following example). When the pressure is increased to above the **third explosion limit**, the reaction rate increases so much that a thermal explosion occurs. In this limit the reaction

$$HO_2\cdot + H_2 \longrightarrow H_2O_2 + H\cdot$$

dominates the elimination of $HO_2\cdot$ by the walls.

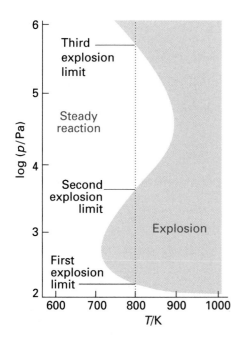

26.3 The explosion limits of the $H_2 + O_2$ reaction. In the explosive regions the reaction proceeds explosively when heated homogeneously.

Table 26.1 Photochemical processes†

Primary absorption $S \rightarrow S^*$
followed by vibrational and rotational relaxation

Physical processes
Fluorescence: $S^* \rightarrow S + h\nu$
Collision-induced emission:

$S^* + M \rightarrow S + M + h\nu$

Stimulated emission:

$S^* + h\nu \rightarrow S + 2h\nu$

Intersystem crossing (ISC): $S^* \rightarrow T^*$
Phosphorescence: $T^* \rightarrow S + h\nu$
Internal conversion (IC): $S^* \rightarrow S'$
Singlet electronic energy transfer:

$S^* + S \rightarrow S + S^*$

Energy pooling: $S^* + S^* \rightarrow S^{**} + S$
Triplet electronic energy transfer:

$T^* + S \rightarrow S + T^*$

Triplet–triplet absorption: $T^* + h\nu \rightarrow T^{**}$

Ionization
Penning ionization:

$A^* + B \rightarrow A + B^+ + e^-$

Dissociative ionization:

$A^* + B{-}C \rightarrow A{-}B^+ + C + e^-$

Collisional ionization:

$A^* + B \rightarrow A^+ + B + e^-$ (or B^-)

Associative ionization:

$A^* + B \rightarrow AB^+ + e^-$

Chemical processes
Dissociation: $A{-}B^* \rightarrow A + B$
Addition or insertion: $A^* + B \rightarrow AB$
Abstraction or fragmentation:

$A^* + B \rightarrow C + D$

Isomerization:

$A^* \rightarrow A'$

Dissociative excitation:

$A^* + C{-}D \rightarrow A + C^* + D$

† S denotes a singlet state and T a triplet state; A, B, and M are arbitrary.

Example 26.2 Examining the explosion behaviour of a chain reaction

Consider the following mechanism for the reaction of hydrogen and oxygen in the fuel-rich regime:

Initiation: $H_2 \longrightarrow H\cdot + H\cdot$ $v = $ constant (v_{init})
Propagation: $H_2 + \cdot OH \longrightarrow \cdot H + H_2O$ $v = k_1[H_2][\cdot OH]$
Branching: $\cdot O_2\cdot + \cdot H \longrightarrow \cdot O\cdot + \cdot OH$ $v = k_2[O_2][H\cdot]$
     $\cdot O\cdot + H_2 \longrightarrow \cdot OH + \cdot H$ $v = k_3[\cdot O\cdot][H_2]$
Termination: $H\cdot + $ wall $\longrightarrow \frac{1}{2}H_2$ $v = k_4[H\cdot]$
     $H\cdot + O_2 + M \longrightarrow HO_2\cdot + M$ $v = k_5[H\cdot][O_2][M]$

The generic form of this reaction mechanism is

Initiation: Reactant $\longrightarrow X$ $v = \mathcal{I}$
Propagation/branching: $X \longrightarrow \varepsilon X + $ product $v = k_a[X]$
Termination: $X \longrightarrow$ removed $v = k_b[X]$

where X is a radical. Show that an explosion occurs when the rate of chain branching exceeds that of chain termination.

Method Identify the onset of explosion with the rapid increase in the concentration of radicals. Set up the rate equation for the concentration of X, and integrate it. Be alert for different behaviour depending on the branching ratio, ε, and the propagation and termination rate constants.

Answer The rate of formation of radicals is

$$\frac{d[X]}{dt} = \mathcal{I} + \varepsilon k_a[X] - k_a[X] - k_b[X]$$
$$= \mathcal{I} + \phi[X]$$

where

$$\phi = (\varepsilon - 1)k_a - k_b$$

The solution of this simple differential equation is

$$[X](t) = \frac{\mathcal{I}}{\phi}(e^{\phi t} - 1)$$

When $\phi < 0$, corresponding to $k_b > (\varepsilon - 1)k_a$ (dominant termination), the concentration of X varies as

$$[X](t) = \frac{\mathcal{I}}{k_b - k_a(\varepsilon - 1)}(1 - e^{-\{k_b - k_a(\varepsilon - 1)\}t})$$

and settles into a steady state as $t \rightarrow \infty$:

$$[X](\infty) = \frac{\mathcal{I}}{k_b - k_a(\varepsilon - 1)}$$

See Fig. 26.4a. In contrast, when $\phi > 0$, corresponding to $k_b < (\varepsilon - 1)k_a$ (dominant propagation/branching), the concentration of X increases exponentially without limit. That is, there is an explosion:

$$[X](t) = \frac{\mathcal{I}}{k_a(\varepsilon - 1) - k_b}(e^{\{k_a(\varepsilon - 1) - k_b\}t} - 1)$$

This behaviour is shown in Fig. 26.4b.

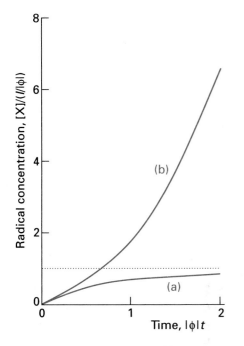

26.4 The concentration of radicals (a) when termination is dominant, (b) when propagation and branching are dominant.

Comment This calculation gives an indication of the basis for the transition from smooth combustion to explosive reaction conditions.

- -

Self-test 26.2 Calculate the variation in radical composition when the rates of propagation/branching and termination are equal.

$$[[X](t) = \mathcal{I}t]$$

26.3 Photochemical reactions

Many reactions can be initiated by the absorption of light by one of the mechanisms described in Section 17.2. The most important of all are the photochemical processes that capture the radiant energy of the Sun. Some of these reactions lead to the heating of the atmosphere during the daytime by absorption of ultraviolet radiation (Fig. 26.5). Others include the absorption of red and blue light by chlorophyll and the subsequent use of the energy to bring about the synthesis of carbohydrates from carbon dioxide and water. Without photochemical processes, the world would be simply a warm, sterile, rock. A summary of the processes that can occur following photochemical excitation is given in Table 26.1.

(a) Quantum yield

Even though a reactant molecule absorbs a photon, the excited molecule might not form products: there are many ways in which the excitation may be lost. We therefore speak of the **primary quantum yield**, ϕ, the number of reactant molecules producing specified

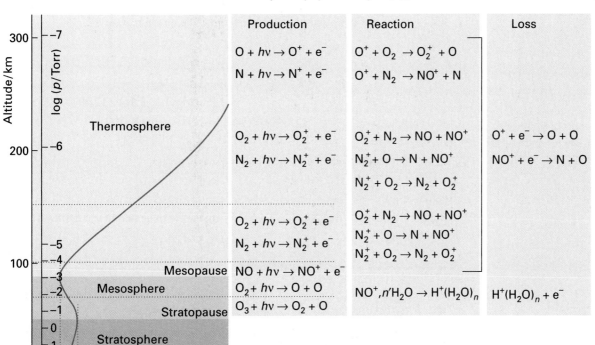

26.5 The temperature profile through the atmosphere and some of the reactions that occur. The temperature peak at about 50 km is due to the absorption of solar radiation by the O_2 and N_2 ionization reactions.

primary products (atoms or ions, for instance) for each photon absorbed. As a result of one successful initiation, many reactant molecules might be consumed. The **overall quantum yield**, Φ, is the number of reactant molecules that react for each photon absorbed. In the photolysis of HI, for example, the processes are

$$HI + h\nu \longrightarrow H\cdot + I\cdot$$
$$H\cdot + HI \longrightarrow H_2 + I\cdot$$
$$I\cdot + I\cdot + M \longrightarrow I_2 + M^*$$

The overall quantum yield is 2 because the absorption of one photon leads to the destruction of two HI molecules. In a chain reaction, Φ may be very large, and values of about 10^4 are common. In such cases the chain acts as a chemical amplifier of the initial dissociation step.

Example 26.3 Determining the quantum yield

When a sample of 4-heptanone was irradiated with 313 nm light with a power output of 50 W under conditions of total absorption for 100 s, it was found that 2.8×10^{-3} mol C_2H_4 was formed. What is the quantum yield for the formation of ethene?

Method First, calculate the amount of photons generated in an interval Δt: see Example 11.1. Then divide the amount of ethene molecules formed by the amount of photons absorbed.

Answer From Example 11.1, the amount (in moles) of photons absorbed is

$$n = \frac{P\Delta t}{(hc/\lambda)N_A}$$

If $n_{C_2H_4}$ is the amount of ethene molecule formed, the quantum yield is

$$\Phi = \frac{n_{C_2H_4}}{n} = \frac{n_{C_2H_4}N_A hc}{\lambda P \Delta t}$$

$$= \frac{(2.8 \times 10^{-3} \text{ mol}) \times (6.022 \times 10^{23} \text{ mol}^{-1}) \times}{(50 \text{ J s}^{-1}) \times (3.13 \times 10^{-7} \text{ m}) \times (100 \text{ s})}$$

$$= 0.21$$

Self-test 26.3 The overall quantum yield for another reaction at 290 nm is 0.30. For what length of time must irradiation with a 100 W source continue in order to destroy 1.0 mol of molecules?

[3.8 h]

(b) Photochemical rate laws

As an example of how to incorporate the photochemical activation step into a mechanism, consider the photochemical activation of the reaction

$$H_2(g) + Br_2(g) \longrightarrow 2HBr(g)$$

In place of the first step in the thermal reaction we have

$$Br_2 \xrightarrow{h\nu} Br\cdot + Br\cdot \qquad v = \mathcal{I}_{abs} \tag{5}$$

where $\mathcal{I}_{abs}$ is the rate at which photons of the appropriate frequency are absorbed divided by the volume in which absorption occurs. It follows that $\mathcal{I}_{abs}$ should take the place of $k_a[Br_2][M]$ in the thermal reaction scheme, so from Example 26.1 we can write

$$\frac{d[HBr]}{dt} = \frac{2k_b(1/k_d[M])^{1/2}[H_2][Br_2]\mathcal{I}_{abs}^{1/2}}{[Br_2] + (k_c/k_b')[HBr]} \tag{6}$$

Equation 6 predicts that the reaction rate should depend on the square root of the absorbed light intensity, which is confirmed experimentally.

(c) Photosensitization

The reactions of a molecule that does not absorb directly can be stimulated if another absorbing molecule is present, because the latter may be able to transfer its energy during a collision. An example of this **photosensitization** is the reaction often used to generate atomic hydrogen, the irradiation of hydrogen gas containing a trace of mercury vapour using radiation of wavelength 254 nm from a mercury discharge lamp. The Hg atoms are excited (to Hg*) by resonant absorption of the radiation, and then collide with H_2 molecules. Two reactions then take place:

$$Hg^* + H_2 \longrightarrow Hg + H\cdot + H\cdot$$
$$Hg^* + H_2 \longrightarrow HgH + H\cdot$$

The latter reaction is the initiation step for other mercury photosensitized reactions, such as the synthesis of formaldehyde from carbon monoxide and hydrogen:

$$H\cdot + CO \longrightarrow HCO\cdot$$
$$HCO\cdot + H_2 \longrightarrow HCHO + H\cdot$$
$$HCO\cdot + HCO\cdot \longrightarrow HCHO + CO$$

Note that the last step is termination by disproportionation rather than by combination. Photosensitization also plays an important role in solution kinetics, and molecules containing the carbonyl chromophore (such as benzophenone, $C_6H_5COC_6H_5$) are often used to trap the incident light and transfer it to some potentially reactive species.

(d) Quenching

Some photochemical reactions can be slowed down by the addition of a species that removes energy from the excited species, and its effect may be studied by monitoring the fluorescence from the excited state. Consider, for example, the scheme

$$S + h\nu_i \longrightarrow S^* \qquad v = \mathcal{I}$$
$$S^* \longrightarrow S + h\nu_f \qquad v = k_f[S^*]$$
$$S^* + Q \longrightarrow S + Q \qquad v = k_Q[S^*][Q]$$

in which S is an absorbing species, S^* an excited singlet state, Q a quenching agent, and $h\nu_i$ and $h\nu_f$ represent the incident and fluorescent photons, respectively. The steady-state approximation for the concentration of S^* implies that

$$\frac{d[S^*]}{dt} = \mathcal{I} - (k_f + k_Q[Q])[S^*] = 0 \tag{7}$$

and hence that

$$[S^*] = \frac{\mathcal{I}}{k_f + k_Q[Q]} \tag{8}$$

The fluorescence intensity is proportional to $k_f[S^*]$, so

$$\mathcal{I}_f \propto \frac{k_f\mathcal{I}}{k_f + k_Q[Q]} \tag{9}$$

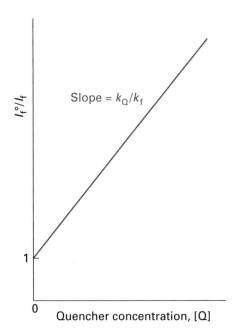

26.6 The format of a Stern–Volmer plot and the interpretation of the slope in terms of the rate constants for quenching and fluorescence.

If we denote the fluorescence intensity in the absence of Q as I_f°, it follows that

$$\frac{I_f^\circ}{I_f} = 1 + \frac{k_Q}{k_f}[Q] \tag{10}$$

Therefore, a plot of the left-hand side of this expression against [Q] should be a straight line with slope k_Q/k_f. Such a plot is called a **Stern–Volmer plot** (Fig. 26.6); it is used to determine the rate constant for quenching.

We need to disentangle k_Q/k_f to obtain the two rate constants. To do so, the time constant of the decay can be measured in a pulse experiment. The rate law immediately after a brief flash of radiation has formed an initial concentration of S* is

$$\frac{d[S^*]}{dt} = -(k_f + k_Q[Q])[S^*] \tag{11}$$

with the solution

$$[S^*]_t = [S^*]_0 e^{-t/\tau_f} \qquad \frac{1}{\tau_f} = k_f + k_Q[Q] \tag{12}$$

Therefore, a plot of $1/\tau_f$ (which can be obtained from the shape of the fluorescence intensity curve as a function of time) against [Q] gives k_f as the intercept at [Q] = 0 and k_Q as the slope.

Polymerization kinetics

In **chain polymerization** an activated monomer, M, attacks another monomer, links to it, then that unit attacks another monomer, and so on. The monomer is used up slowly through the reaction by linking to the growing chains (Fig. 26.7). High polymers are formed rapidly and, as we shall see in detail later, only the yield and not the average molar mass of the polymer is increased by allowing long reaction times. In **step polymerization** any two monomers present in the reaction mixture can link together at any time (Fig. 26.8), and growth is not confined to chains that are already forming. As a result, monomers are removed early in the reaction and (as we shall see) the average molar mass of the product grows with time.

26.4 Chain polymerization

Chain polymerization results in the rapid growth of an individual polymer chain for each activated monomer. It commonly occurs by addition, often by a radical chain process. Examples include the addition polymerizations of ethene, methyl methacrylate, and styrene, as in

$$—CH_2CHX\cdot + CH_2{=}CHX \longrightarrow —CH_2CHXCH_2CHX\cdot$$

and subsequent reactions. The central feature of the kinetic analysis (which is summarized in the *Justification* below), is that the rate of polymerization is proportional to the square root of the initiator concentration:

$$v = k[I]^{1/2}[M] \tag{13}$$

Justification 26.1

There are three basic types of reaction step in a chain polymerization process:

(a) Initiation:

$$I \longrightarrow R\cdot + R\cdot \qquad v = k_i[I]$$

$$M + \cdot R \longrightarrow \cdot M_1 \qquad \text{(fast)}$$

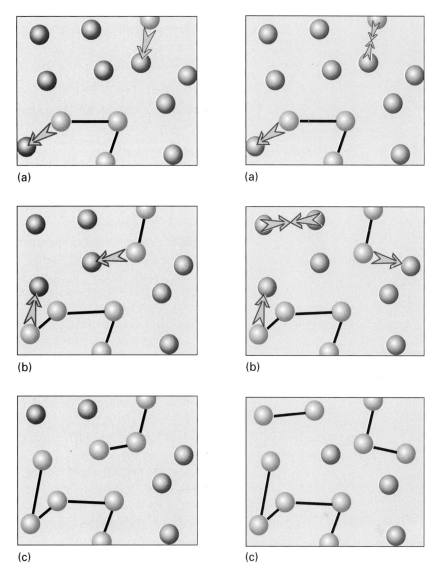

(a)

(b)

(c)

26.7 The process of chain polymerization. Chains grow as each chain acquires additional monomers.

(a)

(b)

(c)

26.8 In step polymerization, growth can start at any pair of monomers, and so new chains begin to form throughout the reaction.

where I is the initiator, $R\cdot$ the radicals I forms, and $\cdot M_1$ is a monomer radical. We have shown a reaction in which a radical is produced, but in some polymerizations the initiation step leads to the formation of an ionic chain carrier. The rate-determining step is the formation of the radicals $R\cdot$ by homolysis of the initiator, so the rate of initiation is equal to the v given above.

(b) Propagation:

$$M + \cdot M_1 \longrightarrow \cdot M_2$$
$$M + \cdot M_2 \longrightarrow \cdot M_3$$

$$\cdots\cdots\cdots\cdots\cdots\cdots\cdots\cdots\cdots$$

$$M + \cdot M_{n-1} \longrightarrow \cdot M_n \qquad v = k_p[M][\cdot M]$$

The rate at which the total concentration of radicals grows as a result of these steps is equal to the rate of the rate-determining initiation step. It follows that

$$\left(\frac{d[\cdot M]}{dt}\right)_{\text{production}} = 2\phi k_i[I] \tag{14}$$

where ϕ is the yield of the initiation step, the fraction of radicals $R\cdot$ that successfully initiate a chain.

(c) Termination:

$$\cdot M_n + \cdot M_m \longrightarrow M_{n+m}$$

If we suppose that the rate of termination is independent of the length of the chain, the rate law for termination is

$$v = k_t[\cdot M]^2 \tag{15}$$

and the rate of change of radical concentration by this process is

$$\left(\frac{d[\cdot M]}{dt}\right)_{\text{termination}} = -2k_t[\cdot M]^2 \tag{16}$$

In practice, other termination steps may intervene. Side reactions may also occur, such as **chain transfer**, in which the reaction

$$M + \cdot M_n \longrightarrow \cdot M + M_n$$

initiates a new chain at the expense of the one currently growing.

The total radical concentration is approximately constant throughout the main part of the polymerization, so the net rate of formation can be set equal to zero:

$$\frac{d[\cdot M]}{dt} = 2\phi k_i[I] - 2k_t[\cdot M]^2 = 0$$

The steady-state concentration of radical chains is therefore

$$[\cdot M] = \left(\frac{\phi k_i}{k_t}\right)^{1/2}[I]^{1/2} \tag{17}$$

Because the rate of propagation of the chains (the rate at which the monomer is consumed) is

$$\frac{d[\cdot M]}{dt} = -k_p[\cdot M][M] \tag{18}$$

it follows that

$$\frac{d[M]}{dt} = -k_p\left(\frac{\phi k_i}{k_t}\right)^{1/2}[I]^{1/2}[M] \tag{19}$$

which has the form of eqn 13.

The **kinetic chain length**, ν, is a measure of the efficiency of the chain propagation mechanism. It is defined as the ratio of the number of monomer units consumed per active centre produced in the initiation step:

$$\nu = \frac{\text{number of monomer units consumed}}{\text{number of active centres produced}} \tag{[20]}$$

The kinetic chain length is therefore equal to the ratio of the propagation and initiation rates:

$$\nu = \frac{\text{propagation rate}}{\text{initiation rate}} \tag{21}$$

Because the initiation rate is equal to the termination rate, we can write this expression (using the rate constants introduced in the *Justification*) as

$$\nu = \frac{k_p[\cdot M][M]}{2k_t[M\cdot]^2} = \frac{k_p[M]}{2k_t[\cdot M]} \tag{22}$$

When the steady-state expression, eqn 17, is substituted for the radical concentration, we obtain

$$\nu = k[\cdot M][I]^{-1/2} \qquad k = \tfrac{1}{2}k_p(\phi k_i k_t)^{-1/2} \tag{23}$$

We see that, the slower the initiation of the chain (the smaller the initiator concentration and the smaller the initiation rate constant), the greater the kinetic chain length.

Example 26.4 Using the kinetic chain length

Estimate the average number of units in a polymer produced by a chain mechanism in which termination occurs by combination of radicals.

Method Because termination occurs by the combination of two radicals, the average number of monomers in a polymer molecule, $\langle n \rangle$, produced by the reaction is the sum of the numbers in the two combining polymer chains. The latter are both, on average, equal to ν.

Answer The average number of monomers in a product molecule is

$$\langle n \rangle = 2\nu = 2k[M][I]^{-1/2}$$

with k given in eqn 23.

Comment We see that, the slower the initiation (as expressed by the rate constant k_i and the initiator concentration $[I]$), the longer the average chain length, and therefore the higher the average molar mass of the polymer. Some of the consequences of molar mass for polymers were explored in Chapter 23: now we see how we can exercise kinetic control over them.

- -

Self-test 26.4 Another termination mechanism is a disproportionation reaction of the form $M\cdot + \cdot M \to M + :M$. Calculate the average polymer length.

$$[\langle n \rangle = k[\cdot M][I]^{-1/2}]$$

26.5 Stepwise polymerization

Stepwise polymerization commonly proceeds by a condensation reaction, in which a small molecule (typically H_2O) is eliminated in each step. Stepwise polymerization is the mechanism of production of polyamides, as in the formation of nylon-66:

$$H_2N(CH_2)_6NH_2 + HOOC(CH_2)_4COOH$$
$$\longrightarrow H_2N(CH_2)_6NHCO(CH_2)_4COOH + H_2O$$
$$\longrightarrow H-[NH(CH_2)_6NHCO(CH_2)_4CO]_n-OH$$

Polyesters and polyurethanes are formed similarly (the latter without elimination). A polyester, for example, can be regarded as the outcome of the stepwise condensation of a hydroxyacid HO–M–COOH. We shall consider the formation of a polyester from such a monomer, and measure its progress in terms of the concentration of the –COOH groups in the sample (which we denote A), for these groups gradually disappear as the condensation proceeds. Because the condensation reaction can occur between molecules containing any number of monomer units, chains of many different lengths can grow in the reaction mixture. We shall show how to use the reaction scheme to predict molar mass distributions.

(a) The rate law of stepwise polymerization

The condensation can be expected to be overall second-order in the concentration of the –OH and –COOH (or A) groups:

$$\frac{d[A]}{dt} = -k[OH][A] \tag{24a}$$

However, because there is one –OH group for each –COOH group, this equation is the same as

$$\frac{d[A]}{dt} = -k[A]^2 \tag{24b}$$

If it is assumed that the rate constant for the condensation is independent of the chain length, then k remains constant throughout the reaction. The solution of this rate law is given by eqn 25.12, and is

$$[A] = \frac{[A]_0}{1 + kt[A]_0} \tag{25}$$

The fraction, p, of –COOH groups that have condensed at time t is

$$p = \frac{[A]_0 - [A]}{[A]_0} = \frac{kt[A]_0}{1 + kt[A]_0} \tag{26}$$

Example 26.5 Calculating the degree of polymerization

Find an expression for the growth with time of the degree of polymerization of a polymer formed by a stepwise process.

Method The number of monomers per polymer molecule is the ratio of the initial concentration of A, $[A]_0$, to the number of end groups, $[A]$, at the time of interest, because there is one A group per polymer molecule. For example, if there were initially 1000 A groups and there are now only 10, each polymer must be 100 units long on average. The value of $[A]$ can be expressed in terms of p, and hence in terms of the time by using eqn 26.

Answer The average number of monomers per polymer molecule is

$$\langle n \rangle = \frac{[A]_0}{[A]} = \frac{1}{1 - p}$$

This result is illustrated in Fig. 26.9. When we substitute the value for p given in eqn 26, we find

$$\langle n \rangle = 1 + kt[A]_0$$

Comment The average length grows linearly with time so, the longer a stepwise polymerization proceeds, the higher the average molar mass of the product.

- -

Self-test 26.5 Derive the expression for the time dependence of p for a stepwise polymerization in which the reaction is acid-catalysed by the –COOH acid functional group. The rate law is $d[A]/dt = -k[A]^2[OH]$.

$$[p = 1 - \alpha^{-1/2},\ \alpha = 1 + 2kt[A]_0^2]$$

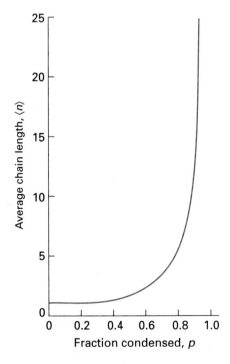

26.9 The average chain length of a polymer as a function of the fraction of reacted monomers, p. Note that p must be very close to 1 for the chains to be long.

Catalysis and oscillation

If the activation energy of a reaction is high, at normal temperatures only a small proportion of molecular encounters result in reaction. A **catalyst** is a substance that accelerates a reaction but undergoes no net chemical change. It functions by lowering the activation energy of the reaction by providing an alternative path that avoids the slow, rate-determining step of the uncatalysed reaction, and results in a higher reaction rate at the same temperature. Catalysts can be very effective; for instance, the activation energy for the decomposition of hydrogen peroxide in solution is 76 kJ mol^{-1}, and the reaction is slow at room temperature. When a little iodide is added, the activation energy falls to 57 kJ mol^{-1}, and the rate constant increases by a factor of 2000. Enzymes, which are biological catalysts, are very specific and can have a dramatic effect on the reaction they control. The activation energy for the acid hydrolysis of sucrose is 107 kJ mol^{-1}, but the enzyme saccharase reduces it to 36 kJ mol^{-1}, corresponding to an acceleration of the reaction by a factor of 10^{12} at body temperature (310 K).

A **homogeneous catalyst** is a catalyst that is in the same phase as the reaction mixture (for example, an acid added to an aqueous solution). A **heterogeneous catalyst** is in a different phase (for example, a solid catalyst for a gas-phase reaction). We examine heterogeneous catalysis in Chapter 28 and consider only homogeneous catalysis here.

26.6 Homogeneous catalysis

Some idea of the mode of action of homogeneous catalysts can be obtained by examining the kinetics of the bromide-catalysed decomposition of hydrogen peroxide:

$$2H_2O_2(aq) \longrightarrow 2H_2O(aq) + O_2(g)$$

The reaction is believed to proceed through the following pre-equilibrium:

$$H_3O^+ + H_2O_2 \rightleftharpoons H_3O_2^+ + H_2O \qquad K = \frac{[H_3O_2^+]}{[H_2O_2][H_3O^+]}$$

$$H_3O_2^+ + Br^- \longrightarrow HOBr + H_2O \qquad v = k[H_3O_2^+][Br^-]$$

$$HOBr + H_2O_2 \longrightarrow H_3O^+ + O_2 + Br^- \qquad \text{(fast)}$$

(In the equilibrium constant, the activity of H$_2$O has been set equal to 1.) Because the second step is rate-determining, the rate law of the overall reaction is obtained by setting the overall rate equal to the rate of the second step and using the equilibrium constant to express the concentration of H$_3$O$_2^+$ in terms of the reactants:

$$\frac{d[O_2]}{dt} = k_{eff}[H_2O_2][H_3O^+][Br^-] \qquad k_{eff} = kK \tag{27}$$

in agreement with the observed dependence of the rate on the Br$^-$ concentration and the pH of the solution. The observed activation energy is that of the effective rate coefficient kK. In the absence of Br$^-$ ions the reaction cannot proceed through the path set out above, and a different and much higher activation energy is observed.

In **acid catalysis** the central step is the transfer of a proton to the substrate:

$$X + HA \longrightarrow HX^+ + A^- \qquad HX^+ \longrightarrow \text{products}$$

and is the primary process in the solvolysis of esters and keto-enol tautomerism. In **base catalysis**, a hydrogen ion is transferred from the substrate to a base:

$$XH + B \longrightarrow X^- + BH^+ \qquad X^- \longrightarrow \text{products}$$

It is the primary step in the isomerization and halogenation of organic compounds, and of the Claisen and aldol reactions.

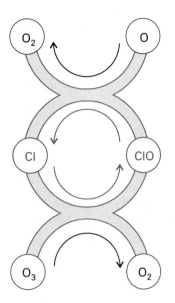

26.10 A catalytic cycle showing the propagation of ozone decomposition by chlorine atoms. Note how the Cl atoms are recycled.

Homogeneous catalytic cycles play a role in atmospheric chemistry. For instance, the presence of chlorine atoms resulting from the photolysis of CFCs (chlorofluorocarbons) can participate in the decomposition of ozone in the stratosphere (Fig. 26.10):

$$Cl + O_3 \longrightarrow ClO + O_2 \qquad ClO + O \longrightarrow Cl + O_2$$

26.7 Autocatalysis

The phenomenon of **autocatalysis** is the catalysis of a reaction by the products. For example, in a reaction A → P it may be found that the rate law is

$$v = k[A][P] \tag{28}$$

so the reaction rate increases as products are formed. The reaction gets started because there are usually other reaction routes for the formation of some P initially, which then takes part in the autocatalytic reaction proper. An example of autocatalysis is provided by two steps in the **Belousov–Zhabotinskii reaction** (BZ reaction), which will figure in discussions later in the chapter:

$$BrO_3^- + HBrO_2 + H_3O^+ \longrightarrow 2BrO_2 + 2H_2O$$
$$2BrO_2 + 2Ce(III) + 2H_3O^+ \longrightarrow 2HBrO_2 + 2Ce(IV) + 2H_2O$$

The product $HBrO_2$ is a reactant in the first step.

The industrial importance of autocatalysis (which occurs in a number of reactions, such as oxidations) is that the rate of the reaction can be maximized by ensuring that the optimum concentrations of reactant and product are always present.

Example 26.6 Calculating concentrations in an autocatalytic reaction

Integrate the rate equation for the A → P autocatalytic reaction in eqn 28.

Method The rate law is given in eqn 28, with $v = -d[A]/dt$. It is convenient to write $[A] = [A]_0 - x$, $[P] = [P]_0 + x$ and then to write down the expression for the rate of change of either species in terms of x.

Answer With the substitution indicated, the rate law becomes

$$\frac{dx}{dt} = k([A]_0 - x)([P]_0 + x)$$

Integration by partial fractions, using the relation

$$\frac{1}{([A]_0 - x)([P]_0 + x)} = \frac{1}{[A]_0 + [P]_0}\left(\frac{1}{[A]_0 - x} + \frac{1}{[P]_0 + x}\right)$$

gives

$$\frac{1}{[A]_0 + [P]_0} \ln\left(\frac{([P]_0 + x)[A]_0}{[P]_0([A]_0 - x)}\right) = kt$$

This expression can be rearranged into

$$\frac{x}{[P]_0} = \frac{e^{at} - 1}{1 + be^{at}} \qquad a = ([A]_0 + [P]_0)k \qquad b = \frac{[P]_0}{[A]_0}$$

Comment The solution is plotted in Fig. 26.11. The rate of reaction is slow initially (little P present), then fast (when P and A are both present), and finally slow again (when A has disappeared).

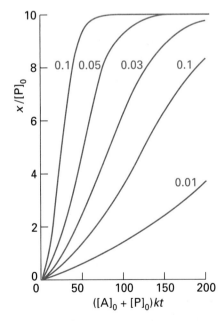

26.11 The concentration of product during the autocatalysed A → P reaction discussed in Example 26.6. Individual curves are labelled with the value of b.

Self-test 26.6 At what time is the reaction rate a maximum?

$$[t_{max} = -(1/ka)\ln b]$$

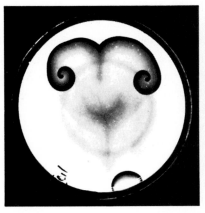

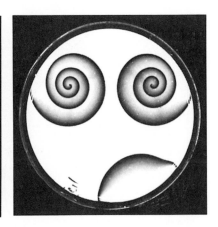

26.12 Some reactions show oscillations in time; some show spatially periodic variations. This sequence of photographs shows the emergence of a spatial pattern.

26.8 Oscillating reactions

One consequence of autocatalysis is the possibility that the concentrations of reactants, intermediates, and products will vary periodically either in space or in time (Fig. 26.12). Chemical oscillation is the analogue of electrical oscillation, with autocatalysis playing the role of positive feedback. Oscillating reactions are much more than a laboratory curiosity. While they are known to occur in only a few cases in industrial processes, there are many examples in biochemical systems where a cell plays the role of a chemical reactor. Oscillating reactions, for example, maintain the rhythm of the heartbeat. They are also known to occur in the glycolytic cycle, in which one molecule of glucose is used to produce (through enzyme-catalysed reactions involving ATP) two molecules of ATP. The concentrations of all the metabolites in the chain oscillate under some conditions, and do so with the same period but with different phases.

(a) The Lotka–Volterra mechanism

We shall use an autocatalytic reaction of a particularly simple form that illustrates how these oscillations may occur. The actual chemical examples that have been discovered so far have a different mechanism, as we shall see. The **Lotka–Volterra mechanism** is as follows:

$$
\begin{array}{lll}
\text{(a)} & \text{A} + \text{X} \longrightarrow \text{X} + \text{X} & \dfrac{d[\text{A}]}{dt} = -k_a[\text{A}][\text{X}] \\[2ex]
\text{(b)} & \text{X} + \text{Y} \longrightarrow \text{Y} + \text{Y} & \dfrac{d[\text{X}]}{dt} = -k_b[\text{X}][\text{Y}] \\[2ex]
\text{(c)} & \text{Y} \longrightarrow \text{B} & \dfrac{d[\text{B}]}{dt} = k_c[\text{Y}]
\end{array}
\tag{29}
$$

Steps (a) and (b) are autocatalytic. The concentration of A is held constant by supplying it to the reaction vessel as needed. (B plays no part in the reaction once it has been produced, so it is unnecessary to remove it; in practice, though, it would normally be removed.) These constraints leave [X] and [Y], the concentrations of the intermediates, as variables. Note that we are considering a **steady-state condition**, which is maintained by the flow of A into the reactor. This steady-state condition must not be confused with the steady-state approximation made earlier: in the present case we solve the rate equations exactly for the variable concentrations of X and Y, but hold [A] at an arbitrary but constant value.

The Lotka–Volterra equations can be solved numerically, and the results can be depicted in two ways. One way is to plot [X] and [Y] against time (Fig. 26.13). The same information

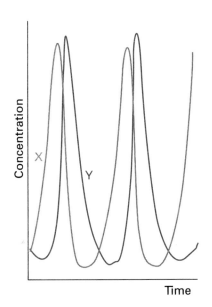

26.13 The periodic variation of the concentrations of the intermediates X and Y in a Lotka–Volterra reaction. The system is in a steady state, but not at equilibrium.

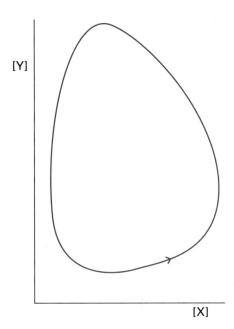

26.14 An alternative representation of the periodic variation of [X] and [Y] is to plot one against the other. Then the system describes closed orbits; different orbits are obtained with different starting conditions.

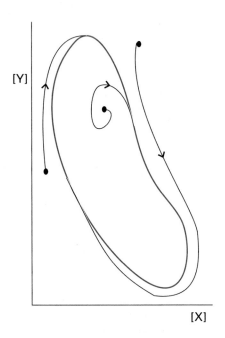

26.15 Some oscillating reactions approach a closed trajectory whatever their starting conditions. The closed trajectory (shown here in green) is called a limit cycle.

can be displayed more succinctly by plotting one concentration against the other (Fig. 26.14).

The periodic variation of the concentrations of the intermediates can be explained as follows. At some stage there may be only a little X present, but reaction (a) provides more, and the production of X autocatalyses the production of even more X. There is therefore a surge of X. However, as X is formed, reaction (b) can begin. It occurs slowly initially, because [Y] is small, but autocatalysis leads to a surge of Y. This surge, though, removes X, and so reaction (a) slows, and less X is produced. Because less X is now available, reaction (b) slows. As less Y becomes available to remove X, X has a chance to surge forward again, and so on.

(b) The brusselator

Another very interesting set of equations is called the **brusselator** (because it is an oscillator that originated with Ilya Prigogine's group in Brussels):

$$
\begin{array}{lll}
\text{(a)} & A \longrightarrow X & \dfrac{d[X]}{dt} = k_a[A] \\[2mm]
\text{(b)} & X + X + Y \longrightarrow Y + Y + Y & \dfrac{d[Y]}{dt} = k_b[X]^2[Y] \\[2mm]
\text{(c)} & B + X \longrightarrow Y + C & \dfrac{d[Y]}{dt} = k_c[B][X] \\[2mm]
\text{(d)} & X \longrightarrow D & \dfrac{d[X]}{dt} = -k_d[X]
\end{array}
\tag{30}
$$

Because the reactants (A and B) are maintained at constant concentration, the two variables are the concentrations of X and Y. These two concentrations may be calculated by solving the rate equations numerically, and the results are plotted in Fig. 26.15. The interesting feature is that, whatever the initial concentrations of X and Y, the system settles down into the same periodic variation of concentrations. The common trajectory to which the system migrates is called a **limit cycle**, and its period depends on the values of the rate constants. A limit cycle is an example of a structure called by mathematicians an **attractor** because it appears to attract to itself trajectories in its vicinity. In conventional chemistry, which deals with closed systems of reactants and products, the equilibrium state is an attractor determined by the minimum value of the Gibbs energy at a given temperature and pressure. In open systems maintained far from equilibrium, a limit cycle may act as an attractor.

(c) The oregonator

One of the first oscillating reactions to be reported and studied systematically was the BZ reaction mentioned earlier, which takes place in a mixture of potassium bromate, malonic acid, and a cerium(IV) salt in an acidic solution. The mechanism has been elucidated by Richard Noyes, and involves 18 elementary steps and 21 different chemical species. The main features of this awesomely complex mechanism can be reproduced by the following **oregonator** (so called because Noyes and his group work in Oregon), where A stands for BrO_3^-, C for HBrO, D for a product, X for $HBrO_2$, Y for Br^-, and Z for Ce^{4+}:

$$
\begin{array}{lll}
\text{(a)} & A + Y \longrightarrow X \\
\text{(b)} & X + Y \longrightarrow C \\
\text{(c)} & A + X \longrightarrow X + X + Z \\
\text{(d)} & X + X \longrightarrow D \\
\text{(e)} & Z \longrightarrow Y
\end{array}
\tag{31}
$$

A, B, C, and D are held constant (in the model) by maintaining an influx of reactants and removal of products. The oscillations arise in a similar way to the brusselator, and can be traced to the autocatalysis in step (c) and the linkage between the reactions provided by the other steps.

(d) Bistability

Attempts have been made to discover the underlying causes of oscillation more deeply than by simply recognizing the role of autocatalysis. It appears that in one class of systems three conditions must be fulfilled in order to obtain oscillations:

1. The reactions must be far from equilibrium.
2. The reactions must have autocatalytic steps.
3. The system must be able to exist in two steady states.

The last criterion is called **bistability**, and is a property that takes us well beyond what is familiar from the equilibrium properties of systems.

Consider a reaction in which there are two intermediates X and Y. If the concentration of Y is at some high value in a reactor, and X is added, then the concentration of Y might decrease as shown by the upper line in Fig. 26.16. If X is at some high value, then as Y is added the reaction might result in the slow increase of Y as shown by the lower line. However, in each case, a concentration may be reached at which the concentration will jump from one curve to the other (just as a supercooled liquid might suddenly solidify). The two curves represent the two stable states of the bistable system. Neither state is an equilibrium state in the thermodynamic sense: they occur in steady states that are well removed from equilibrium, and the concentrations of X and Y represent the consequences of reactants continuously flowing into and of products flowing out of the reactor.

Now consider what happens when a third type of intermediate, Z, is present. Suppose Z reacts with both X and Y. In the absence of Z the flows of material might correspond to the stable state on the upper curve of Fig. 26.17. However, as Z reacts with Y to produce X the state of the system moves along the curve (to the right, as Y decreases and X increases) until the sudden transition occurs to the lower curve. Then Z reacts with X and produces Y, which means that the composition moves to the left along the lower curve. There comes a point, however, when the concentration of X has been reduced so much, and that of Y has risen so

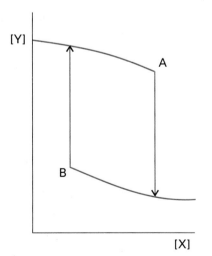

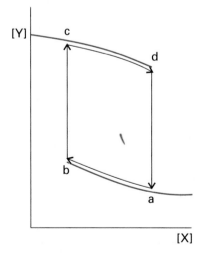

26.16 A system showing bistability. As the concentration of X is increased (by adding it to a reactor) the concentration of Y decreases along the upper curve but at A drops sharply to a low value. If X is then decreased, the concentration increases along the lower curve, but rises sharply to a high value at B.

26.17 Chemical oscillation in a bistable system occurs as a result of the effect of a third substance Z which can react with X to produce Y and with Y to produce X. As a result, the system switches periodically between the upper and lower curves.

much, that there is a sudden transition to the upper curve, when the process begins again. The leaping from one stable state to another is manifest as the sudden surge or depletion of the concentration of a species (Fig. 26.18).

By studying the regions of concentrations and the rate constants for the individual steps of a reaction it is now becoming possible to predict the occurrence of bistable chemical systems and to anticipate the occurrence of oscillations. We are still far, however, from being able to use these ideas to account for gene expression, the patterns on tigers and butterflies, and the oscillation of cool flames, in all of which it is thought that these processes play a part.

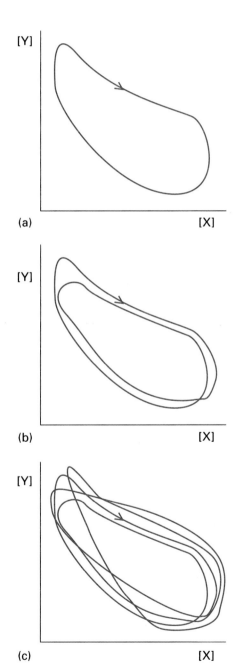

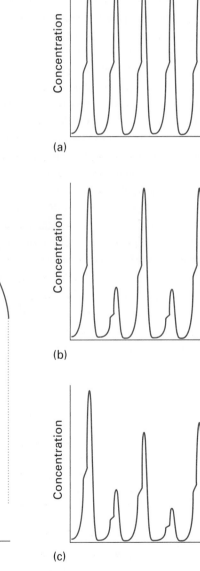

26.18 The leaping from one branch to another of a bistable system appears to the observer as a periodic surge and depletion of concentration.

26.19 The onset of chaos as a result of period-doubling. (a) Steady-state oscillation, (b) the oscillation after one period-doubling step, (c) the chaotic regime after many period-doubling steps.

26.20 Successive trajectories in concentration space (a) before period-doubling, (b) after one period-doubling step, (c) after several period-doubling steps.

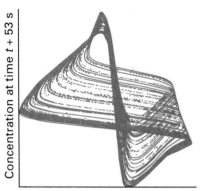

Concentration at time t + 53 s

Concentration at time t

26.21 A representation of a strange attractor for a system showing chemical chaos.

26.9 Chemical chaos

A remarkable development of chemical kinetics in recent years has been the discovery of rate laws for complex reactions that have innately chaotic solutions. Specifically, it is found that the solutions of certain sets of rate laws, although fully determined by the structure of the equations, cannot be used to predict the composition of the system from its initial composition. The kinetic equations that we have been considering are of such richness that it should be hardly surprising that they, or minor elaborations of them, can display such chaotic solutions. Instead of a reaction showing periodic oscillatory behaviour, the concentrations burst into chaotic oscillation, and the concentrations of the intermediates show unpredictable amplitudes or unpredictable frequencies (Fig. 26.19). Such behaviour can be literally a matter of life and death for, should the heartbeat become chaotic, death may result. The type of chaos that stems from the structure of well-defined (and often quite simple) nonlinear differential equations is called **deterministic chaos**, for the behaviour of the solutions is predictable but infinitely sensitive to the initial conditions.

There are several ways in which a reaction can be steered into a chaotic regime. For example, in certain systems the change in a parameter (such as a flow rate through the reaction vessel or the rate of stirring) may result in the successive doubling of the period of the limit cycle (Fig. 26.20). When this **period-doubling** occurs, the system must circulate twice round the cycle before an initial pair of concentrations is restored. Then, after a further modification of the parameter, the period doubles again, and the limit cycle circulates four times before repeating itself. As each period doubles, the periodicity of the motion becomes less apparent, and finally appears like random fluctuations. The trajectory of the composition of the system is now highly complex, and may take a path that never retraces its path or crosses itself. Such a path is called a **strange attractor** (Fig. 26.21).

It should be understood, however, that the term 'chaos' is in some respects misleading. First, the conditions under which certain systems of differential equations display period-doubling can be specified exactly, and the progression through period-doubling cycles shows considerable regularity that is common to many classes of systems. In so far as the composition can be specified exactly, then a later composition can be predicted. Our inability to predict the composition of a system that is in a chaotic regime stems from our inability experimentally to know the initial conditions exactly or make a later determination of the composition at an exact later instant. The unpredictability of a chaotic system lies not in the formulation or solution of the differential equations that describe the rates of processes, but in our ability to relate those solutions to the practical system of interest given the inherent imprecision of experimental observations.

Checklist of key ideas

Chain reactions

☐ chain reaction

26.1 The structure of chain reactions
☐ chain carrier
☐ radical chain reaction
☐ initiation step
☐ thermolysis
☐ photolysis
☐ propagation step
☐ branching step
☐ retardation step

☐ termination step
☐ inhibition step
☐ pyrolysis
☐ Rice–Herzfeld mechanism
☐ hydrogen–bromine reaction

26.2 Explosions
☐ thermal explosion
☐ chain-branching explosion
☐ explosion region
☐ first explosion limit
☐ second explosion limit
☐ third explosion limit

26.3 Photochemical reactions
☐ primary quantum yield
☐ overall quantum yield
☐ photochemical rate law
☐ photosensitization
☐ Stern–Volmer plot (10)

Polymerization kinetics

☐ chain polymerization
☐ step polymerization

26.4 Chain polymerization
☐ chain transfer
☐ kinetic chain length (20)

☐ rate law for chain polymerization (23)

26.5 Stepwise polymerization
☐ fraction polymerized (26)

Catalysis and oscillation

☐ catalyst
☐ homogeneous catalyst
☐ heterogeneous catalyst

26.6 Homogeneous catalysis
☐ acid catalysis
☐ base catalysis

26.7 Autocatalysis
☐ autocatalysis
☐ Belousov–Zhabotinskii (BZ) reaction

26.8 Oscillating reactions
☐ Lotka–Volterra mechanism
☐ steady-state condition
☐ brusselator

☐ limit cycle
☐ attractor
☐ oregonator
☐ bistability

26.9 Chemical chaos
☐ deterministic chaos
☐ period-doubling
☐ strange attractor

Further reading

Articles of general interest

K.J. Laidler, Rate-controlling step: a necessary or useful concept? *J. Chem. Educ.* **65**, 250 (1988).

A.F. Shaaban, The integrated rate equation of the nitric oxide–oxygen reaction. *J. Chem. Educ.* **67**, 869 (1990).

J.N. Spencer, Competitive and coupled reactions. *J. Chem. Educ.* **69**, 281 (1992).

R.T. Raines and D.E. Hansen, An intuitive approach to steady-state kinetics. *J. Chem. Educ.* **65**, 757 (1988).

P. Engelking, Laser photochemistry. In *Encyclopedia of applied physics* (ed. G.L. Trigg), **8**, 283. VCH, New York (1994).

G.F. Swiegers, Applying the principles of chemical kinetics to population growth problems. *J. Chem. Educ.* **70**, 364 (1993).

J.G. Eberhardt and E. Levin, A simplified integration technique for reaction rate laws of integral order in several substances. *J. Chem. Educ.* **72**, 193 (1995).

X. Tan, S. Lindenbaum, and N. Meltzer, A unified equation for chemical kinetics. *J. Chem. Educ.* **71**, 566 (1994).

G.I. Gellene, Application of kinetic approximations to the $A + B \rightleftarrows C$ reaction system. *J. Chem. Educ.* **72**, 196 (1995).

S. Bluestone and K.Y. Yan, A method to find the rate constants in chemical kinetics of a complex reaction. *J. Chem. Educ.* **72**, 884 (1995).

F. Mata-Perez and J.F. Perez-Benito, The kinetic rate law for autocatalytic reactions. *J. Chem. Educ.* **64**, 925 (1987).

L.J. Soltzberg, Self-organization in chemistry. *J. Chem. Educ.* **66**, 187 (1989).

R.J. Field, The language of dynamics. *J. Chem. Educ.* **66**, 188 (1989).

K.L. Stevenson and O. Horváth, Reactions induced by light. In *Encyclopedia of applied physics* (ed. G.L. Trigg), **16**, 117. VCH, New York (1996).

C.L. McCormick, Polymerization and polymer reactions. In *Encyclopedia of applied physics* (ed. G.L. Trigg), **14**, 445. VCH, New York (1996).

R.M. Noyes, Some models of chemical oscillators. *J. Chem. Educ.* **66**, 190 (1989).

I.R. Epstein, The role of flow systems in far-from-equilibrium dynamics. *J. Chem. Educ.* **66**, 191 (1989).

R.J. Field and F.W. Schneider, Oscillating chemical reactions and nonlinear dynamics. *J. Chem. Educ.* **66**, 195 (1989).

E. Hughes, Jr, Solving differential equations in kinetics by using power series. *J. Chem. Educ.* **66**, 46 (1989).

W. Jahnke and A.T. Winfree, Recipes for Belousov–Zhabotinsky reagents. *J. Chem. Educ.* **68**, 320 (1991).

S.K. Scott, Oscillations in simple models of chemical systems. *Acc. Chem. Res.* **20**, 186 (1987).

P. Brumer and M. Shapiro, Coherence chemistry: controlling chemical reactions with lasers. *Acc. Chem. Res.* **22**, 407 (1989).

H.G. Schuster, Chaotic phenomena. In *Encyclopedia of applied physics* (ed. G.L. Trigg), **3**, 189. VCH, New York (1992).

Texts and sources of data and information

M.J. Pilling and P.W. Seakins, *Reaction kinetics*. Oxford University Press (1995).

S.R. Logan, *Chemical kinetics*. Longman, Harlow (1996).

F.W. Billmeyer, *Textbook of polymer science*. Wiley-Interscience, New York (1984).

M. Kucera, *Comprehensive chemical kinetics: mechanism and kinetics of addition polymerizations*, Vol. 31. Elsevier, Amsterdam (1992).

R.P. Wayne, *Principles and applications of photochemistry*. Oxford University Press (1988).

R.P. Wayne, *Chemistry of atmospheres*. Clarendon Press, Oxford (1991).

C.H. Bamford and C.F. Tipper (ed.), *Comprehensive chemical kinetics*, Vols 1–26. Elsevier, Amsterdam (1969–86).

R.G. Compton (ed.), *Comprehensive chemical kinetics*, Vols 27–33, Elsevier, Amsterdam (1987–92).

S.K. Scott, *Oscillations, waves, and chaos in chemical kinetics*, Oxford Chemistry Primers. Oxford University Press (1994).

M. Boudart, *Kinetics of chemical processes*. Butterworth, London (1991).

P. Gray and S.K. Scott, *Chemical oscillations and instabilities*. Clarendon Press, Oxford (1990).

Exercises

In the following exercises and problems, it is recommended that rate constants be labelled with the number of the step in the proposed reaction mechanism, and any reverse steps be labelled similarly but with a prime.

26.1 (a) Derive the rate law for the decomposition of ozone in the reaction $2O_3(g) \rightarrow 3O_2(g)$ on the basis of the following proposed mechanism:

(1) $O_3 \rightleftharpoons O_2 + O$ (forward k_1, reverse k_1')

(2) $O + O_3 \longrightarrow 2O_2$

26.1 (b) On the basis of the following proposed mechanism, account for the experimental fact that the rate law for the decomposition $2N_2O_5(g) \rightarrow 4NO_2(g) + O_2(g)$ is $v = k[N_2O_5]$.

(1) $N_2O_5 \rightleftharpoons NO_2 + NO_3$ (forward k_1, reverse k_1')

(2) $NO_2 + NO_3 \longrightarrow NO_2 + O_2 + NO$

(3) $NO + N_2O_5 \longrightarrow 3NO_2$

26.2 (a) A slightly different mechanism for the decomposition of N_2O_5 from that in Exercise 26.1b has also been proposed. It differs only in the last step, which is replaced by

(3) $NO + NO_3 \longrightarrow 2NO_2$

Show that this mechanism leads to the same overall rate law.

26.2 (b) Consider the following mechanism for the thermal decomposition of R_2:

(1) $R_2 \longrightarrow 2R$

(2) $R + R_2 \longrightarrow P_B + R'$

(3) $R' \longrightarrow P_A + R$

(4) $2R \longrightarrow P_A + P_B$

where R_2, P_A, P_B are stable hydrocarbons and R and R' are radicals. Find the dependence of the rate of decomposition of R_2 on the concentration of R_2.

26.3 (a) Refer to Fig. 26.3 and determine the pressure range for a chain-branching explosion in the hydrogen–oxygen reaction at 800 K.

26.3 (b) Refer to Fig. 26.3 and determine the pressure range for a chain-branching explosion in the hydrogen–oxygen reaction at (a) 700 K, (b) 900 K.

26.4 (a) In a photochemical reaction $A \rightarrow 2B + C$, the quantum efficiency with 500 nm light is 2.1×10^2 mol einstein^{-1}. After exposure of 300 mmol of A to the light, 2.28 mmol of B is formed. How many photons were absorbed by A?

26.4 (b) In a photochemical reaction $A \rightarrow B + C$, the quantum efficiency with 500 nm light is 1.2×10^2 mol einstein^{-1}. After exposure of 200 mmol A to the light, 1.77 mmol B is formed. How many photons were absorbed by A?

26.5 (a) In an experiment to measure the quantum efficiency of a photochemical reaction, the absorbing substance was exposed to 490 nm light from a 100 W source for 45 min. The intensity of the transmitted light was 40 per cent of the intensity of the incident light. As a result of irradiation, 0.344 mol of the absorbing substance decomposed. Determine the quantum efficiency.

26.5 (b) In an experiment to measure the quantum efficiency of a photochemical reaction, the absorbing substance was exposed to 320 nm radiation from a 87.5 W source for 28.0 min. The intensity of the transmitted light was 0.257 that of the incident light. As a result of irradiation, 0.324 mol of the absorbing substance decomposed. Determine the quantum efficiency.

26.6 (a) The condensation reaction of propanone, $(CH_3)_2CO$, in aqueous solution is catalysed by bases, B, which react reversibly with propanone to form the carbanion $C_3H_5O^-$. The carbanion then reacts with a molecule of propanone to give the product. A simplified version of the mechanism is

(1) $AH + B \longrightarrow BH^+ + A^-$

(2) $A^- + BH^+ \longrightarrow AH + B$

(3) $A^- + AH \longrightarrow$ product

where AH stands for propanone and A^- denotes its carbanion. Use the steady-state approximation to find the concentration of the carbanion and derive the rate equation for the formation of the product.

26.6 (b) Consider the acid-catalysed reaction

(1) $HA + H^+ \rightleftharpoons HAH^+$

                 (k_1 forward, k_1' reverse; both fast)

(2) $HAH^+ + B \longrightarrow BH^+ + AH$ (k_2, slow)

Deduce the rate law and show that it can be made independent of the specific term $[H^+]$.

26.7 (a) Consider the following chain mechanism:

(1) $AH \longrightarrow A\cdot + H\cdot$

(2) $A\cdot \longrightarrow B\cdot + C$

(3) $AH + B \longrightarrow A\cdot + D$

(4) $A\cdot + B\cdot \longrightarrow P$

Identify the initiation, propagation, and termination steps, and use the steady-state approximation to deduce that the decomposition of AH is first-order in AH.

26.7 (b) Consider the following chain mechanism:

(1) $A_2 \longrightarrow 2A\cdot$

(2) $A\cdot \longrightarrow B\cdot + C$

(3) $A\cdot + P \longrightarrow B\cdot$

(4) $A\cdot + B\cdot \longrightarrow P$

Identify any initiation, propagation, retardation, inhibition, and termination steps, and use the steady-state approximation to deduce the rate law for the consumption of A_2.

Problems

Numerical problems

26.1 The number of photons falling on a sample can be determined by a variety of methods, of which the classical one is chemical actinometry. The decomposition of oxalic acid $(COOH)_2$, in the presence of uranyl sulfate, $(UO_2)SO_4$, proceeds according to the sequence

(1) $UO^{2+} + h\nu \longrightarrow (UO^{2+})^*$

(2) $(UO^{2+})^* + (COOH)_2 \longrightarrow UO^{2+} + H_2O + CO_2 + CO$

with a quantum efficiency of 0.53 at the wavelength used. The amount of oxalic acid remaining after exposure can be determined by titration (with $KMnO_4$) and the extent of decomposition used to find the number of incident photons. In a particular experiment, the actinometry solution consisted of 5.232 g anhydrous oxalic acid, 25.0 cm^3 water (together with the uranyl salt). After exposure for 300 s the remaining solution was titrated with 0.212 M $KMnO_4$(aq), and 17.0 cm^3 were required for complete oxidation of the remaining oxalic acid. What is the rate of incidence of photons at the wavelength of the experiment? Express the answer in photons/second and einstein/second.

26.2 When benzophenone is illuminated with ultraviolet light it is excited into a singlet state. This singlet changes rapidly into a triplet, which phosphoresces. Triethylamine acts as a quencher for the triplet. In an experiment in methanol as solvent, the phosphorescence intensity varied with amine concentration as shown below. A flash photolysis experiment had also shown that the half-life of the fluorescence in the absence of quencher is 29 μs. What is the value of k_q?

$[Q]/(mol\,L^{-1})$	0.0010	0.0050	0.0100
I_f/(arbitrary units)	0.41	0.25	0.16

26.3 Studies of combustion reactions depend on knowing the concentrations of H atoms and HO radicals. Measurements on a flow system using ESR for the detection of radicals gave information on the reactions

(1) $H + NO_2 \longrightarrow OH + NO$ $k_1 = 2.9 \times 10^{10}\ L\,mol^{-1}\,s^{-1}$

(2) $OH + OH \longrightarrow H_2O + O$ $k_2 = 1.55 \times 10^9\ L\,mol^{-1}\,s^{-1}$

(3) $O + OH \longrightarrow O_2 + H$ $k_3 = 1.1 \times 10^{10}\ L\,mol^{-1}\,s^{-1}$

(J.N. Bradley, W. Hack, K. Hoyermann, and H.G. Wagner, *J. Chem. Soc. Faraday Trans. I*, 1889 (1973)). Using initial H atom and NO_2 concentrations of 4.5×10^{-10} mol cm^{-3} and 5.6×10^{-10} mol cm^{-3}, respectively, compute and plot curves showing the O, O_2, and OH concentrations as a function of time in the range 0–10 ns.

26.4 In a flow study of the reaction between O atoms and Cl_2 (J.N. Bradley, D.A. Whytock, and T.A. Zaleski, *J. Chem. Soc. Faraday Trans. I*, 1251 (1973)) at high chlorine pressures, plots of $\ln([O]_0/[O])$ against distances l along the flow tube, where $[O]_0$ is the oxygen concentration at zero chlorine pressure, gave straight lines. Given the flow velocity as 6.66 m s^{-1} and the data below, find the rate coefficient for the reaction $O + Cl_2 \rightarrow ClO + Cl$.

l/cm	0	2	4	6	8	10	12	14	16	18
$\ln([O]_0/[O])$	0.27	0.31	0.34	0.38	0.45	0.46	0.50	0.55	0.56	0.60

with $[O]_0 = 3.3 \times 10^{-8}$ mol L^{-1}, $[Cl_2] = 2.54 \times 10^{-7}$ mol L^{-1}, and $p = 1.70$ Torr.

26.5 Models of population growth are analogous to chemical reaction rate equations. In the model due to Malthus (1798) the rate of change of the population N of the planet is assumed to be given by $dN/dt = $ births $-$ deaths. The numbers of births and deaths are proportional to the population, with proportionality constants b and d. Obtain the integrated rate law. How well does it fit the (very approximate) data below on the population of the planet as a function of time?

Year	1750	1825	1922	1960	1974	1987
$N/10^9$	0.5	1	2	3	4	5

Theoretical problems

26.6 An autocatalytic reaction $A \rightarrow P$ is observed to have the rate law $d[P]/dt = k[A]^2[P]$. Solve the rate law for initial concentrations $[A]_0$ and $[P]_0$. Calculate the time at which the rate reaches a maximum.

26.7 Another reaction with the stoichiometry $A \rightarrow P$ has the rate law $d[P]/dt = k[A][P]^2$; integrate the rate law for initial concentrations $[A]_0$ and $[P]_0$. Calculate the time at which the rate reaches a maximum.

26.8 The Rice–Herzfeld mechanism for the dehydrogenation of ethane is specified in Section 26.1b, and it is noted there that it led to first-order kinetics. Confirm this remark, and find the approximations that lead to the rate law quoted there. How may the conditions be changed so that the reaction shows different orders?

26.9 Photolysis of $Cr(CO)_6$ in the presence of certain molecules M, can give rise to the following reaction sequence:

(1) $Cr(CO)_6 + h\nu \longrightarrow Cr(CO)_5 + CO$

(2) $Cr(CO)_5 + CO \longrightarrow Cr(CO)_6$

(3) $Cr(CO)_5 + M \longrightarrow Cr(CO)_5M$

(4) $Cr(CO)_5M \longrightarrow Cr(CO)_5 + M$

Suppose that the absorbed light intensity is so weak that $I \ll k_4[Cr(CO)_5M]$. Find the factor f in the equation $d[Cr(CO)_5M]/dt = -f[Cr(CO)_5M]$. Show that a graph of $1/f$ against [M] should be a straight line.

26.10 The following mechanism has been proposed for the thermal decomposition of acetaldehyde (ethanal):

(1) $CH_3CHO \longrightarrow \cdot CH_3 + CHO$

(2) $\cdot CH_3 + CH_3CHO \longrightarrow CH_4 + \cdot CH_2CHO$

(3) $\cdot CH_2CHO \longrightarrow CO + \cdot CH_3$

(4) $\cdot CH_3 + \cdot CH_3 \longrightarrow CH_3CH_3$

Find an expression for the rate of formation of methane and the rate of disappearance of acetaldehyde.

26.11 Express the root mean square deviation, $\{\langle M^2 \rangle - \langle \overline{M} \rangle^2\}^{1/2}$, of the molar mass of a condensation polymer in terms of p, and deduce its time dependence.

26.12 Calculate the ratio of the mean cube molar mass to the mean square molar mass in terms of (a) the fraction p, (b) the chain length.

26.13 Derive an expression for the rate of disappearance of a species A in a photochemical reaction for which the mechanism is:

(1) initiation with light of intensity $\mathcal{I}$, $A \longrightarrow 2R\cdot$

(2) propagation, $A + R\cdot \longrightarrow R\cdot + B$

(3) termination, $R\cdot + R\cdot \longrightarrow R_2$

Hence, show that rate measurements will give only a combination of k_2 and k_3 if a steady state is reached, but that both may be obtained if a steady state is not reached.

26.14 The photochemical chlorination of chloroform in the gas phase has been found to follow the rate law $d[CCl_4]/dt = k[Cl_2]^{1/2}\mathcal{I}_a^{1/2}$. Devise a mechanism that leads to this rate law when the chlorine pressure is high.

26.15 Conventional equilibrium considerations do not apply when a reaction is being driven by light absorption. Thus the steady-state concentration of products and reactants might differ significantly from equilibrium values. For instance, suppose the reaction $A \rightarrow B$ is driven by light absorption, and that its rate is $\mathcal{I}_a$, but that the reverse reaction $B \rightarrow A$ is bimolecular and second-order with a rate $k[B]^2$. What is the stationary state concentration of B? Why does this 'photostationary state' differ from the equilibrium state?

26.16 Write a program for the integration of the Lotka–Volterra equations and arrange for it to plot the concentration of Y against that of X. Explore the consequences of varying the starting concentrations for the integration. Identify the conditions (the concentrations of X and Y) corresponding to the steady state of the Lotka–Volterra equations (the point at the centre of the orbits).

26.17 Set up the overall rate equations for the concentrations of X and Y in terms of the oregonator and explore the periodic properties of the solutions.

26.18 Many biological and biochemical processes involve auto-catalytic steps. In the SIR model of the spread and decline of infectious diseases the population is divided into three classes; the susceptibles, S, who can catch the disease, the infectives, I, who have the disease and can transmit it, and the removed class, R, who have either had the disease and recovered, are dead, or are immune or isolated. The model mechanism for this process implies the following rate laws:

$$\frac{dS}{dt} = -rSI \qquad \frac{dI}{dt} = rSI - aI \qquad \frac{dR}{dt} = aI$$

What are the autocatalytic steps of this mechanism? Find the conditions on the ratio a/r that decide whether the disease will spread (an epidemic) or die out. Show that a constant population is built into this system, namely that $S + I + R = N$, meaning that the timescales of births, deaths by other causes, and migration are assumed large compared to that of the spread of the disease.

Additional problems supplied by Carmen Giunta and Charles Trapp

26.19 J.D. Chapple-Sokol, C.J. Giunta, and R.G. Gordon (*J. Electrochem. Soc.* **136**, 2993 (1989)) proposed the following radical chain mechanism for the initial stages of the gas-phase oxidation of silane by nitrous oxide:

(1) $N_2O \longrightarrow N_2 + O$

(2) $O + SiH_4 \longrightarrow SiH_3 + OH$

(3) $OH + SiH_4 \longrightarrow SiH_3 + H_2O$

(4) $SiH_3 + N_2O \longrightarrow SiH_3O + N_2$

(5) $SiH_3O + SiH_4 \longrightarrow SiH_3OH + SiH_3$

(6) $SiH_3 + SiH_3O \longrightarrow (H_3Si)_2O$

Label each step with its role in the chain. Use the steady-state approximation to show that this mechanism predicts the following rate law for SiH_4 consumption (provided k_1 and k_6 are in some sense small):

$$\frac{d[SiH_4]}{dt} = -k[N_2O][SiH_4]^{1/2}$$

26.20 Ultraviolet radiation photolyses O_3 to O_2 and O. Determine the rate at which ozone is consumed by 305 nm radiation in a layer of the stratosphere of thickness 1 km. The quantum efficiency is 0.94 at 220 K, the concentration about 8×10^{-9} mol L^{-1}, the molar absorption coefficient 260 L mol^{-1} cm^{-1}, and the flux of 305 nm radiation about 1×10^{14} photons cm^{-2} s^{-1}. Data from W.B. DeMore, S.P. Sander, D.M. Golden, R.F. Hampson, M.J. Kurylo, C.J. Howard, A.R. Ravishankara, C.E. Kolb, and M.J. Molina, *Chemical kinetics and photochemical data for use in stratospheric modeling: Evaluation Number 11*, JPL Publication 94–26 (1994).

26.21 Arylmercury compounds have played an ever-increasing role in organic synthesis in recent years because of their ability to accommodate almost all important organic functional groups and their chemical and thermal stability. Y. Wang (*Int. J. Chem. Kinet.* **25**, 91 (1993)) has studied the kinetics and mechanism of the dimerization of arylmercurials (A), with formulas XC_6H_4HgCl (X = H, p-CH$_3$, m-CH$_3$, o-CH$_3$, p-Cl, and others) catalysed by $[ClRh(CO)_2]_2$ (C) in hexamethylphosphoramide (HMPA). The observed dimerization in each case is represented by the equation $2ArHgCl \rightarrow ArAr + HgCl_2 + Hg$, and plots of $x/([A]_0 - x)$, where x is the molar concentration of the arylmercury compound reacted and $[A]_0$ the initial concentration, was found to be a linear function of time. In each case, $[A]_0 = 0.100$ mol L^{-1} and $[C] = 5.00 \times 10^{-4}$ mol L^{-1}. Concentration data as a function of time at 60°C for the compounds with X = H and Cl are

X = H	$x/([A]_0 - x)$	0.14	0.48	0.63	0.84	1.12
	$t/$min	25	85	115	150	205
X = p-Cl	$x/([A]_0 - x)$	0.04	0.27	0.60	1.05	
	$t/$min	30	123	236	405	

Similar results were obtained for the other compounds in the series. (a) Plot these data and confirm that the reactions are second-order.

Would the first-order integrated rate law fit the data too? (b) The proposed mechanism for these reactions is

$$[ClRh(CO)_2]_2 + ArHgCl \rightleftharpoons [ArRh(CO)_2]_2 + HgCl_2$$
$$(1)\ k_1, k_{-1}, \text{fast}$$

$$[ArRh(CO)_2]_2 + ArHgCl \rightleftharpoons Ar_2Rh(CO)_2HgCl$$
$$(2)\ k_2, k_{-2}, \text{slow}$$

$$Ar_2Rh(CO)_2HgCl \longrightarrow [ClRh(CO)_2]_2 + ArAr + Hg$$
$$(3)\ k_3, \text{fast}$$

Derive the rate law for the rate of disappearence of ArHgCl and determine the conditions under which this mechanism supports the experimental results.

26.22 Because of its importance in atmospheric chemistry, the thermal decomposition of nitric oxide, $2NO(g) \rightarrow N_2(g) + O_2(g)$, has been among the most thoroughly studied of gas phase reactions. The commonly accepted mechanism has been that of H. Wise and M.F. Freech (*J. Chem. Phys.* **22**, 1724 (1952)):

(a) $NO \longrightarrow N_2O + O$ k_a

(b) $O + NO \longrightarrow O_2 + N$ k_b

(c) $N + NO \longrightarrow N_2 + O$ k_c

(d) $2O + M \longrightarrow O_2 + M$ k_d

(e) $O_2 + M \longrightarrow 2O + M$ k_{-d}

(a) Label the steps of this mechanism as initiation, propagation, etc. (b) Write down the full expression for the rate of disappearance of NO. What does this expression for the rate become on the basis of the assumptions that $v_b = v_c$ when [N] reaches its steady-state concentration, that the rate of the propagation step is more rapid than the rate of the initiation step, and that oxygen atoms are in equilibrium with oxygen molecules? (c) Find an expression for the effective activation energy, $E_{a,eff}$, for the overall reaction in terms of the activation energies of the individual steps of the reaction. (d) Estimate $E_{a,eff}$ from the bond energies of the species involved. (e) It has been pointed out by R.J. Wu and C.T. Yeh (*Int. J. Chem. Kinet.* **28**, 89 (1996)) that the reported experimental values of $E_{a,eff}$ obtained by different authors have varied from 253 to 357 kJ mol^{-1}. They suggest that the assumption of oxygen atoms and oxygen molecules being in equilibrium is unwarranted and that the steady-state approximation needs to be applied to the entire mechanism. Obtain the overall rate law based on the steady-state approximation and find the forms that it assumes for low NO conversion (low O_2 concentration). (f) When the

reaction conversion becomes significant, Wu and Yeh suggest that two additional elementary steps,

$$O_2 + M \longrightarrow 2O + M$$
$$NO + O_2 \longrightarrow O + NO_2$$

start to compete with step (a) as the initiation step. Obtain the rate laws based on these alternative mechanisms and again estimate the apparent activation energies. Is the range of these different theoretically estimated values for $E_{a,eff}$ consistent with the range of values obtained experimentally?

26.23 The water formation reaction has been studied many times and continues to be of interest. Despite the many studies there is not uniform agreement on the mechanism. But, as explosions are known to occur at certain critical values of the pressure, any proposed mechanism that is to be considered plausible must be consistent with the existence of these critical explosion limits. One such plausible mechanism is that of Example 26.2. Another is the following:

$$H_2 \longrightarrow 2H$$
$$H + O_2 \longrightarrow OH + O$$
$$O + H_2 \longrightarrow OH + H$$
$$H + O_2 \longrightarrow HO_2$$
$$HO_2 + H_2 \longrightarrow H_2O + OH$$
$$HO_2 + \text{wall} \longrightarrow \text{destruction}$$
$$H + M \longrightarrow \text{destruction}$$

In a manner similar to that in Example 26.2, determine whether or not this mechanism can lead to explosions under appropriate conditions.

26.24 For many years the reaction $H_2(g) + I_2(g) \rightarrow 2HI(g)$ and its reverse were assumed to be elementary bimolecular reactions. However, J.H. Sullivan (*J. Chem. Phys.* **46**, 73 (1967)) suggested that the following mechanism for the reaction, originally proposed by M. Bodenstein (*Z. physik. Chem.* **29**, 56 (1898)), provides a better explanation of the experimental results:

(a) $I_2 \longrightarrow I + I$ k_a, k_a'

(b) $I + I + H_2 \longrightarrow HI + HI$ k_b

Obtain the expression for the rate of formation of HI based on this mechanism. Under what conditions does this rate law reduce to the one for the originally accepted mechanism?

27 Molecular reaction dynamics

The simplest quantitative account of reaction rates is in terms of collision theory, which can be used only for the discussion of reactions between simple species in the gas phase. Reactions in solution can be classified into two types: diffusion-controlled and activation-controlled. The former can be expressed quantitatively in terms of the diffusion equation. In activated complex theory, it is assumed that the reactant molecules form a complex that can be discussed in terms of the population of its energy levels. Activated complex theory inspires a thermodynamic approach to reaction rates, in which the rate constant is expressed in terms of thermodynamic parameters. This approach is useful for parametrizing the rates of reactions in solution. The highest level of sophistication is in terms of potential energy surfaces and the motion of molecules through these surfaces. As we shall see, such an approach gives a very intimate picture of the events that occur when reactions occur, and is open to experimental study.

Now we are at the heart of chemistry. Here we examine the details of what happens to molecules at the climax of reactions. Extensive changes of structure are taking place and energies the size of dissociation energies are being redistributed among bonds: old bonds are being ripped apart and new bonds are being formed.

As may be imagined, the calculation of the rates of such processes from first principles is very difficult. Nevertheless, like so many intricate problems, the broad features can be established quite simply, and only when we enquire more deeply do the complications emerge. In this chapter we look at three levels of approach to the calculation of a rate constant for an elementary bimolecular reaction. Although a great deal of information can be obtained from gas-phase reactions, many reactions of interest take place in solution, and we shall also see to what extent their rates can be predicted.

Reactive encounters

In this section we consider two elementary approaches to the calculation of reaction rates, one relating to gas-phase reactions and the other to reactions in solution. Both approaches are based on the view that reactant molecules must meet, and that reaction takes place only if they have a certain minimum energy. In the **collision theory** of bimolecular gas-phase reactions, which we mentioned briefly in Section 25.5b, products are formed only if the collision is sufficiently energetic; otherwise the colliding reactant molecules separate again. In solution, the reactant molecules may simply diffuse together and then acquire energy from their immediate surroundings while they are in contact.

27.1 Collision theory

We shall consider the bimolecular elementary reaction

$$A + B \longrightarrow P \qquad v = k_2[A][B] \tag{1}$$

where P denotes products, and aim to calculate the second-order rate constant k_2.

We can anticipate the general form of the expression for k_2 by considering the physical requirements for reaction. We can expect the rate v to be proportional to the rate of collisions, and therefore to the mean speed of the molecules, $\bar{c} \propto (T/M)^{1/2}$ where M is the molar mass of the molecules, their collisional cross-section, σ, and the number densities of A and B:

$$v \propto \sigma \left(\frac{T}{M}\right)^{1/2} \mathcal{N}_A \mathcal{N}_B \propto \sigma \left(\frac{T}{M}\right)^{1/2} [A][B]$$

However, a collision will be successful only if the kinetic energy exceeds a minimum value, the activation energy, E_a, of the reaction. This requirement suggests that the rate constant should also be proportional to a Boltzmann factor of the form $e^{-E_a/RT}$. So we can anticipate, by writing the reaction rate in the form given in eqn 1, that

$$k_2 \propto \sigma \left(\frac{T}{M}\right)^{1/2} e^{-E_a/RT}$$

However, not every collision will lead to reaction even if the energy requirement is satisfied, because the reactants may need to collide in a certain relative orientation. This 'steric requirement' suggests that a further factor, P, should be introduced, and that

$$k_2 \propto P\sigma \left(\frac{T}{M}\right)^{1/2} e^{-E_a/RT} \tag{2}$$

As we shall see in detail below, this expression has the form predicted by collision theory. It reflects three aspects of a successful collision:

$$k_2 \propto \text{steric requirement} \times \text{encounter rate}$$
$$\times \text{minimum energy requirement}$$

(a) Collision rates in gases

We have anticipated that the reaction rate, and hence k_2, depends on the frequency with which molecules collide. The **collision density**, Z_{AB}, is the number of (A, B) collisions in a region of the sample in an interval of time divided by the volume of the region and the duration of the interval. The frequency of collisions of a single molecule in a gas was calculated in Section 1.3b. As shown in the *Justification* on the following page, that result can be adapted to deduce that

$$Z_{AB} = \sigma \left(\frac{8kT}{\pi\mu}\right)^{1/2} N_A^2 [A][B] \tag{3}$$

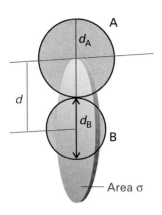

27.1 The collision cross-section for two molecules can be regarded to be the area within which the centre of the projectile molecule (A) must enter around the target molecule (B) in order for a collision to occur. If the diameters of the two molecules are d_A and d_B, the radius of the target area is $d = \frac{1}{2}(d_A + d_B)$ and the cross-section is πd^2.

where σ is the collision cross-section (Fig. 27.1):

$$\sigma = \pi d^2 \qquad d = \tfrac{1}{2}(d_A + d_B) \tag{4}$$

and μ is the reduced mass,

$$\mu = \frac{m_A m_B}{m_A + m_B} \tag{5}$$

Similarly, the collision density for like molecules at a molar concentration [A] is

$$Z_{AA} = \sigma \left(\frac{4kT}{\pi m_A}\right)^{1/2} N_A^2 [A]^2 \tag{6}$$

Collision densities may be very large. For example, in nitrogen at room temperature and pressure, with $d = 280$ pm, $Z = 5 \times 10^{34}$ m^{-3} s^{-1}.

Justification 27.1

It follows from eqn 1.30 that the collision frequency, z, for a single A molecule of mass m_A in a gas of other A molecules is

$$z = \sigma \bar{c}_{rel} \mathcal{N}_A \tag{7}$$

where $\mathcal{N}_A$ is the number density of A molecules and $\bar{c}_{rel}$ is their relative mean speed. As indicated in Section 1.3,

$$\bar{c}_{rel} = 2^{1/2} \bar{c} \qquad \bar{c} = \left(\frac{8kT}{\pi m}\right)^{1/2} \tag{8}$$

For future convenience, it is sensible to introduce $\mu = \frac{1}{2}m$ (for like molecules of mass m), and then to write

$$\bar{c}_{rel} = \left(\frac{8kT}{\pi \mu}\right)^{1/2} \tag{9}$$

This expression also applies to the mean relative speed of dissimilar molecules, provided that μ is interpreted as the reduced mass in eqn 5.

The total collision density is the collision frequency multiplied by the number density of A molecules:

$$Z_{AA} = \tfrac{1}{2} z \mathcal{N}_A = \tfrac{1}{2} \sigma \bar{c}_{rel} \mathcal{N}_A^2 \tag{10}$$

The factor of $\frac{1}{2}$ has been introduced to avoid double counting of the collisions (so one A molecule colliding with another A molecule is counted as one collision regardless of their actual identities). For collisions of A and B molecules present at number densities $\mathcal{N}_A$ and $\mathcal{N}_B$, the collision density is

$$Z_{AB} = \sigma \bar{c}_{rel} \mathcal{N}_A \mathcal{N}_B \tag{11}$$

Note that we have discarded the factor of $\frac{1}{2}$ because now we are considering an A molecule colliding with any of the B molecules as a collision.

The number density of a species J is $\mathcal{N}_J = N_A[J]$, where [J] is their molar concentration and N_A is the Avogadro constant. Equations 3 and 6 then follow.

(b) The energy requirement

According to collision theory, the rate of change in the molar concentration of A molecules is the product of the collision density and the probability that a collision occurs with sufficient energy. The latter condition can be incorporated by writing the collision cross-

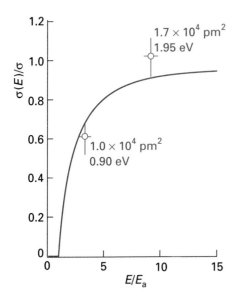

27.2 The variation of the reactive cross-section with energy as expressed by eqn 15. The data points are from experiments on the reaction H + D$_2$ → HD + D. (K. Tsukiyama, B. Katz, and R. Bersohn, *J. Chem. Phys.* **84**, 1934 (1986).)

section as a function of the kinetic energy of approach of the two colliding species, and setting the cross-section, $\sigma(\varepsilon)$, equal to zero if the kinetic energy of approach is below a certain threshold value, ε_a. Later, we shall identify $N_A\varepsilon_a$ as E_a, the (molar) activation energy of the reaction. Then, for a collision with a specific relative speed of approach v_{rel} (not, at this stage, a mean value),

$$\frac{d[A]}{dt} = -\sigma(\varepsilon)v_{rel}N_A[A][B] \tag{12}$$

The relative kinetic energy, ε, and the relative speed are related by $\varepsilon = \frac{1}{2}\mu v_{rel}^2$, so $v_{rel} = (2\varepsilon/\mu)^{1/2}$. At this point we recognize that a wide range of approach energies is present in a sample, so we should average the expression just derived over a Boltzmann distribution of energies, write

$$\frac{d[A]}{dt} = -\left\{\int_0^\infty \sigma(\varepsilon)v_{rel}f(\varepsilon)\,d\varepsilon\right\}N_A[A][B] \tag{13}$$

and hence recognize the rate constant as

$$k_2 = N_A\int_0^\infty \sigma(\varepsilon)v_{rel}f(\varepsilon)\,d\varepsilon \tag{14}$$

Now suppose that the reactive collision cross-section is zero below ε_a and that above that energy it varies as

$$\sigma(\varepsilon) = \left(1 - \frac{\varepsilon_a}{\varepsilon}\right)\sigma \tag{15}$$

This form of the energy dependence is broadly consistent with experimental determinations of the reaction between H and D$_2$ as determined by molecular beam measurements of the kind described later (Fig. 27.2). Then, in the *Justification* below, we show that

$$k_2 = N_A\sigma\bar{c}_{rel}e^{-E_a/RT} \tag{16}$$

Justification 27.2

The Maxwell–Boltzmann distribution of molecular speeds is given in Section 1.3a. It may be expressed in terms of the kinetic energy, ε, by writing $\varepsilon = \frac{1}{2}mv^2$, then $dv = d\varepsilon/(2m\varepsilon)^{1/2}$ and eqn 1.22 becomes

$$\begin{aligned}f(v)\,dv &= 4\pi\left(\frac{m}{2\pi kT}\right)^{3/2}\left(\frac{2\varepsilon}{m}\right)e^{-\varepsilon/kT}\frac{d\varepsilon}{(2m\varepsilon)^{1/2}} \\ &= 2\pi\left(\frac{1}{\pi kT}\right)^{3/2}\varepsilon^{1/2}e^{-\varepsilon/kT}\,d\varepsilon \\ &= f(\varepsilon)\,d\varepsilon\end{aligned} \tag{17}$$

The integral we need to evaluate is therefore

$$\begin{aligned}\int_0^\infty \sigma(\varepsilon)v_{rel}f(\varepsilon)\,d\varepsilon &= 2\pi\left(\frac{1}{\pi kT}\right)^{3/2}\int_0^\infty \sigma(\varepsilon)\left(\frac{2\varepsilon}{\mu}\right)^{1/2}\varepsilon^{1/2}e^{-\varepsilon/kT}\,d\varepsilon \\ &= \left(\frac{8}{\pi\mu kT}\right)^{1/2}\left(\frac{1}{kT}\right)\int_0^\infty \varepsilon\sigma(\varepsilon)e^{-\varepsilon/kT}\,d\varepsilon\end{aligned}$$

Now we introduce the approximation for $\sigma(\varepsilon)$ in eqn 15, and evaluate

$$\int_0^\infty \varepsilon\sigma(\varepsilon)e^{-\varepsilon/kT}\,d\varepsilon = \sigma\int_{\varepsilon_a}^\infty \varepsilon\left(1 - \frac{\varepsilon_a}{\varepsilon}\right)e^{-\varepsilon/kT}\,d\varepsilon = (kT)^2\sigma e^{-\varepsilon_a/kT}$$

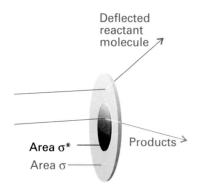

27.3 The collision cross-section is the target area that results in simple deflection of the projectile molecule; the reaction cross-section is the corresponding area for chemical change to occur on collision.

We have made use of the fact that $\sigma = 0$ for $\varepsilon < \varepsilon_a$. It follows that

$$\int_0^\infty \sigma(\varepsilon)v_{rel}f(\varepsilon)\,d\varepsilon = \sigma\left(\frac{8kT}{\pi\mu}\right)^{1/2}e^{-\varepsilon_a/kT}$$

as in eqn 16 (with $\varepsilon_a/kT = E_a/RT$).

Equation 16 has the Arrhenius form $k_2 = Ae^{-E_a/RT}$ provided the exponential temperature dependence dominates the weak square-root temperature dependence of the pre-exponential factor.[1] It follows that the activation energy, E_a, can be identified with the minimum kinetic energy along the line of approach that is needed for reaction, and that the pre-exponential factor is a measure of the rate at which collisions occur in the gas.

(c) The steric requirement

The simplest procedure for calculating k_2 is to use for σ the values obtained for non-reactive collisions (for example, typically those obtained from viscosity measurements) or from tables of molecular radii. Table 27.1 compares some values of the pre-exponential factor calculated in this way with values obtained from Arrhenius plots (Section 25.5a). One of the reactions shows fair agreement between theory and experiment, but for others there are major discrepancies. In some cases the experimental values are orders of magnitude smaller than those calculated, which suggests that the collision energy is not the only criterion for reaction and that some other feature, such as the relative orientation of the colliding species, is important. Moreover, one reaction in the table has a pre-exponential factor larger than theory, which seems to indicate that the reaction occurs more quickly than the particles collide!

We can express the disagreement between experiment and theory by introducing a **steric factor**, P, and expressing the **reactive cross-section**, σ^*, as a multiple of the collision cross-section, $\sigma^* = P\sigma$ (Fig. 27.3). Then the rate constant becomes

$$k_2 = P\sigma\left(\frac{8kT}{\pi\mu}\right)^{1/2}N_A e^{-E_a/RT} \tag{18}$$

This expression has the form we anticipated in eqn 2. The steric factor is normally found to be several orders of magnitude smaller than 1.

Table 27.1* Arrhenius parameters for gas-phase reactions

	$A/(\text{L mol}^{-1}\,\text{s}^{-1})$		$E_a/(\text{kJ mol}^{-1})$	P
	Experiment	Theory		
$2NOCl \rightarrow 2NO + 2Cl$	9.4×10^9	5.9×10^{10}	102.0	0.16
$2ClO \rightarrow Cl_2 + O_2$	6.3×10^7	2.5×10^{10}	0.0	2.5×10^{-3}
$H_2 + C_2H_4 \rightarrow C_2H_6$	1.24×10^6	7.3×10^{11}	180.0	1.7×10^{-6}
$K + Br_2 \rightarrow KBr + Br$	1.0×10^{12}	2.1×10^{11}	0.0	4.8

*More values are given in the *Data section*.

---

1 From the general definition of activation energy in eqn 25.25 and eqn 18:

$$RT^2\frac{d\ln k}{dT} = E_a + \tfrac{1}{2}RT$$

so the activation energy is weakly temperature-dependent. Commonly, $E_a \gg RT$, and the $\tfrac{1}{2}RT$ can be neglected.

Example 27.1 Estimating a steric factor (1)

Estimate the steric factor for the reaction $H_2 + C_2H_4 \rightarrow C_2H_6$ at 628 K given that the pre-exponential factor is $1.24 \times 10^6 \ L\,mol^{-1}\,s^{-1}$.

Method To calculate P, we need to calculate the pre-exponential factor, A, by using eqn 16 and then compare the answer with experiment: the ratio is P. Collision cross-sections for non-reactive encounters are listed in Table 1.3. The best way to estimate the collision cross-section for dissimilar spherical species is to calculate the collision diameter for each one (from $\sigma = \pi d^2$), to calculate the mean of the two diameters, and then to calculate the cross-section for that mean diameter. However, as neither species is spherical, a simpler but more approximate procedure is just to take the average of the two collision cross-sections.

Answer The reduced mass of the colliding pair is

$$\mu = \frac{m_1 m_2}{m_1 + m_2} = 3.15 \times 10^{-27} \ kg$$

because $m_1 = 2.016$ u for H_2 and $m_2 = 28.05$ u for C_2H_4 (the atomic mass unit, 1 u, is defined inside the front cover). Hence

$$\left(\frac{8kT}{\pi\mu}\right)^{1/2} = 2.65 \times 10^3 \ m\,s^{-1}$$

From Table 1.3, $\sigma(H_2) = 0.27 \ nm^2$ and $\sigma(C_2H_4) = 0.64 \ nm^2$, giving a mean collision cross-section of $\sigma = 0.46 \ nm^2$. Therefore,

$$A = \sigma \left(\frac{8kT}{\pi\mu}\right)^{1/2} N_A = 7.33 \times 10^{11} \ L\,mol^{-1}\,s^{-1}$$

Experimentally $A = 1.24 \times 10^6 \ L\,mol^{-1}\,s^{-1}$, so it follows that $P = 1.7 \times 10^{-6}$.

Comment The very small value of P is one reason why catalysts are needed to bring this reaction about at a reasonable rate. As a general guide, the more complex the molecules, the smaller the value of P.

Self-test 27.1 It is found for the reaction $NO + Cl_2 \rightarrow NOCl + Cl$ that $A = 4.0 \times 10^9 \ L\,mol^{-1}\,s^{-1}$ at 298 K. Use $\sigma(NO) = 0.42 \ nm^2$ and $\sigma(Cl_2) = 0.93 \ nm^2$ to estimate the P factor for the reaction.

[0.018]

An example of a reaction for which it is possible to estimate the steric factor is $K + Br_2 \rightarrow KBr + Br$, for which the experimental value of P is 4.8. In this reaction, the distance of approach at which reaction occurs appears to be considerably larger than the distance needed for deflection of the path of the approaching molecules in a non-reactive collision. It has been proposed that the reaction proceeds by a **harpoon mechanism**. This brilliant name is based on a model of the reaction which pictures the K atom as approaching a Br_2 molecule; when the two are close enough an electron (the harpoon) flips across from K to Br_2. In place of two neutral particles there are now two ions, and so there is a Coulombic attraction between them: this attraction is the line on the harpoon. Under its influence the ions move together (the line is wound in), the reaction takes place, and KBr + Br emerge. The harpoon extends the cross-section for the reactive encounter, and the reaction rate is greatly underestimated by taking for the collision cross-section the value for simple mechanical contact between $K + Br_2$.

Example 27.2 Estimating a steric factor (2)

Estimate the value of P for the harpoon mechanism by calculating the distance at which it is energetically favourable for the electron to leap from K to Br_2.

Method We should begin by identifying all the contributions to the energy of interaction between the colliding species. There are three contributions to the energy of the process $K + Br_2 \rightarrow K^+ + Br_2^-$. The first is the ionization energy, I, of K. The second is the electron affinity, E_{ea}, of Br_2. The third is the Coulombic interaction energy between the ions when they have been formed: when their separation is R this energy is $-e^2/4\pi\varepsilon_0 R$. The electron flips across when the sum of these three contributions changes from positive to negative (that is, when the sum is zero).

Answer The net change in energy when the transfer occurs at a separation R is

$$E = I - E_{ea} - \frac{e^2}{4\pi\varepsilon_0 R}$$

The ionization energy I is larger than E_{ea}, so E becomes negative only when R has decreased to less than some critical value R^* given by

$$\frac{e^2}{4\pi\varepsilon_0 R^*} = I - E_{ea}$$

When the particles are at this separation, the harpoon shoots across from K to Br_2, so we can identify the reactive cross-section as $\sigma^* = \pi R^{*2}$. This value of σ^* implies that the steric factor is

$$P = \frac{\sigma^*}{\sigma} = \frac{R^{*2}}{d^2} = \left\{ \frac{e^2}{4\pi\varepsilon_0 d(I - E_{ea})} \right\}^2$$

where $d = R(K) + R(Br_2)$. With $I = 420$ kJ mol^{-1} (corresponding to 7.0×10^{-19} J), $E_{ea} \approx 250$ kJ mol^{-1} (corresponding to 4.2×10^{-19} J), and $d = 400$ pm, we find $P = 4.2$, in good agreement with the experimental value (4.8).

- -

Self-test 27.2 Estimate the value of P for the harpoon reaction between Na and Cl_2 for which $d \approx 350$ pm; take $E_{ea} \approx 230$ kJ mol^{-1}.

[2.2]

Example 27.2 illustrates two points about steric factors. First, the concept of a steric factor is not wholly useless because in some cases its numerical value can be estimated. Second (and more pessimistically) most reactions are much more complex than $K + Br_2$, and we cannot expect to obtain P so easily. What we need is a more powerful theory that lets us calculate, and not merely guess, its value. We go some way to setting up that theory in Section 27.4 and the sections that follow.

27.2 Diffusion-controlled reactions

Encounters between reactants in solution occur in a very different manner from encounters in gases. Reactant molecules have to jostle their way through the solvent, so their encounter frequency is considerably less than in a gas. However, because a molecule also migrates only slowly away from a location, two reactant molecules that encounter each other stay near each other for much longer than in a gas. This lingering of one molecule near another on account of the hindering presence of solvent molecules is called the **cage effect**. Such an encounter pair may accumulate enough energy to react even though it does not have enough energy to do so when it is first formed. The activation energy of a reaction is a much more complicated quantity in solution than in a gas because the encounter pair is surrounded by solvent and the energy of the entire local assembly of reactant and solvent molecules must be considered.

(a) Classes of reaction

The complicated overall process can be divided into simpler parts by setting up a simple kinetic scheme. We suppose that the rate of formation of an encounter pair AB is first-order in each of the reactants A and B:

$$A + B \longrightarrow AB \qquad v = k_d[A][B] \tag{19}$$

As we shall see, k_d (where the d signifies diffusion) is determined by the diffusional characteristics of A and B. The encounter pair can break up without reaction or it can go on to form products P. If we suppose that both processes are pseudofirst-order reactions (with the solvent perhaps playing a role), then we can write

$$AB \longrightarrow A + B \qquad v = k_d'[AB] \tag{20}$$

and

$$AB \longrightarrow P \qquad v = k_a[AB] \tag{21}$$

The concentration of AB can now be found from the equation for the net rate of change of concentration of AB:

$$\frac{d[AB]}{dt} = k_d[A][B] - k_d'[AB] - k_a[AB] \approx 0 \tag{22}$$

This expression solves to

$$[AB] \approx \frac{k_d[A][B]}{k_a + k_d'} \tag{23}$$

The rate of formation of products is therefore

$$\frac{d[P]}{dt} \approx k_a[AB] = k_2[A][B] \qquad k_2 = \frac{k_a k_d}{k_a + k_d'} \tag{24}$$

Two limits can now be distinguished. If the rate of separation of the unreacted encounter pair is much slower than the rate at which it forms products, then $k_d' \ll k_a$ and the effective rate constant is

$$k_2 \approx \frac{k_a k_d}{k_a} = k_d \tag{25}$$

In this **diffusion-controlled limit**, the rate of reaction is governed by the rate at which the reactant molecules diffuse through the solvent. An indication that a reaction is diffusion-controlled is that its rate constant is of the order of 10^9 L mol^{-1} s^{-1} or greater. Because the combination of radicals involves very little activation energy, radical and atom recombination reactions are often diffusion-controlled.

An **activation-controlled reaction** arises when a substantial activation energy is involved in the reaction of AB. Then $k_a \ll k_d'$ and

$$k_2 \approx \frac{k_a k_d}{k_d'} = k_a K \tag{26}$$

where K is the equilibrium constant for $A + B \rightleftharpoons AB$. In this limit, the reaction proceeds at the rate at which energy accumulates in the encounter pair from the surrounding solvent. Some experimental data are given in Table 27.2.

Table 27.2* Arrhenius parameters for reactions in solution

	Solvent	$A/(\text{L mol}^{-1}\,\text{s}^{-1})$	$E_a/(\text{kJ mol}^{-1})$
$(CH_3)_3CCl$ solvolysis	Water	7.1×10^{16}	100
	Ethanol	3.0×10^{13}	112
	Chloroform	1.4×10^{4}	45
$CH_3CH_2Br + OH^-$	Ethanol	4.3×10^{11}	89.5

*More values are given in the Data section.

(b) Diffusion and reaction

The rate of a diffusion-controlled reaction is calculated by considering the rate at which the reactants diffuse together. As shown in the *Justification* below, the rate constant for a reaction in which the two reactant molecules react if they come within a distance R^* of one another is

$$k_d = 4\pi R^* D N_A \qquad (27)$$

where D is the sum of the diffusion coefficients of the two reactant species in the solution.

Justification 27.3

According to the diffusion equation ($D_B\nabla^2[B] = \partial[B]/\partial t$, Section 24.11), the concentration of B when the system has reached a steady state ($\partial[B]/\partial t = 0$) satisfies $\nabla^2[B]_r = 0$, where the subscript r signifies a quantity that varies with the distance r. For a spherically symmetrical system, ∇^2 can be replaced by radial derivatives alone (see Table 11.1), so the equation satisfied by $[B]_r$ is

$$\frac{d^2[B]_r}{dr^2} + \frac{2}{r}\frac{d[B]_r}{dr} = 0 \qquad (28)$$

The general solution of this equation is

$$[B]_r = a + \frac{b}{r} \qquad (29)$$

We need two boundary conditions to pin down the values of the two constants. One condition is that $[B]_r$ has its bulk value $[B]$ as $r \to \infty$. The second condition is that the concentration of B is zero at $r = R^*$, the distance at which reaction occurs. It follows that $a = [B]$ and $b = -R^*[B]$, and hence that (for $r \geq R^*$)

$$[B]_r = \left(1 - \frac{R^*}{r}\right)[B] \qquad (30)$$

The variation of concentration expressed by this equation is illustrated in Fig. 27.4.

The rate of reaction is the (molar) flux, J, of the reactant B towards A multiplied by the area of the spherical surface of radius R^*:

$$\text{rate of reaction} = 4\pi R^{*2}J \qquad (31)$$

From Fick's first law (eqn 24.8), the flux towards A is proportional to the concentration gradient, so at a radius R^*:

$$J = D_B\left(\frac{d[B]_r}{dr}\right)_{r=R^*} = \frac{D_B[B]}{R^*}$$

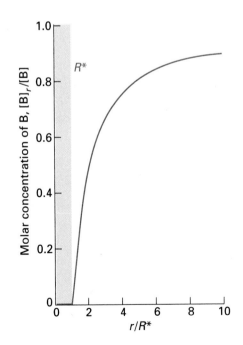

27.4 The concentration profile for reaction in solution when a molecule B diffuses towards another reactant molecule and reacts if it reaches R^*.

(A sign change has been introduced because we are interested in the flux towards decreasing values of r.) When this condition is substituted into the previous equation we obtain

$$\text{rate of reaction} = 4\pi R^* D_B [B] \tag{32}$$

The rate of the diffusion-controlled reaction is equal to the average flow of B molecules to all the A molecules in the sample. If the bulk concentration of A is [A], the number of A molecules in the sample of volume V is $N_A [A] V$; the global flow of all B to all A is therefore $4\pi R^* D_B N_A [A][B] V$. It is unrealistic to suppose that all A are stationary; this feature is removed by replacing D_B by the sum of the diffusion coefficients of the two species and writing $D = D_A + D_B$. Then the rate of change of concentration of AB is

$$\frac{d[AB]}{dt} = 4\pi R^* D N_A [A][B] \tag{33}$$

Hence, the diffusion-controlled rate constant is as given in eqn 27.

Equation 27 can be taken further by incorporating the Stokes–Einstein equation (eqn 24.73) relating the diffusion constant and the hydrodynamic radius R_A and R_B of each molecule in a medium of viscosity η:

$$D_A = \frac{kT}{6\pi\eta R_A} \qquad D_B = \frac{kT}{6\pi\eta R_B} \tag{34}$$

As these relations are approximate, little extra error is introduced if we write $R_A = R_B = \frac{1}{2}R^*$, which leads to

$$k_d = \frac{8RT}{3\eta} \tag{35}$$

(The R in this equation is the gas constant.) The radii have cancelled because, although the diffusion constants are smaller when the radii are large, the reactive collision radius is larger and the particles need travel a shorter distance to meet. In this approximation, the rate constant is independent of the identities of the reactants, and depends only on the temperature and the viscosity of the solvent.

Illustration

The rate constant for the recombination of I atoms in hexane at 298 K, when the viscosity of the solvent is 0.326 cP (with $1\,P = 10^{-1}\,kg\,m^{-1}\,s^{-1}$) is

$$k_d = \frac{8 \times (8.3145\,J\,K^{-1}\,mol^{-1}) \times (298\,K)}{3 \times (3.26 \times 10^{-4}\,kg\,m^{-1}\,s^{-1})} = 2.0 \times 10^7\,m^3\,mol^{-1}\,s^{-1}$$

Because $1\,m^3 = 10^3\,L$, this result corresponds to $2.0 \times 10^{10}\,L\,mol^{-1}\,s^{-1}$. The experimental value is $1.3 \times 10^{10}\,L\,mol^{-1}\,s^{-1}$, so the agreement is very good considering the approximations involved.

27.3 The material balance equation

The diffusion of reactants plays an important role in many chemical processes, such as the diffusion of O_2 molecules into red blood corpuscles and the diffusion of a gas towards a catalyst. We can have a glimpse of the kinds of calculations involved by considering the diffusion equation (Section 24.11) generalized to take into account the possibility that the diffusing, convecting molecules are also reacting.

(a) The formulation of the equation

Consider a small volume element in a chemical reactor (or a biological cell). The net rate at which J molecules enter the region by diffusion and convection is given by eqn 24.78:

$$\frac{\partial[J]}{\partial t} = D\frac{\partial^2[J]}{\partial x^2} - v\frac{\partial[J]}{\partial x} \tag{36}$$

The net rate of change of molar concentration due to chemical reaction is

$$\frac{\partial[J]}{\partial t} = -k[J] \tag{37}$$

if we suppose that J disappears by a pseudofirst-order reaction. Therefore, the overall rate of change of the concentration of J is

$$\frac{\partial[J]}{\partial t} = D\frac{\partial^2[J]}{\partial x^2} - v\frac{\partial[J]}{\partial x} - k[J] \tag{38}$$

Equation 38 is called the **material balance equation**. If the rate constant is large, then [J] will decline rapidly. However, if the diffusion constant is large, then the decline can be replenished as J diffuses rapidly into the region. The convection term, which may represent the effects of stirring, can sweep material either into or out of the region according to the sign of v.

(b) Solutions of the equation

The material balance equation is a second-order partial differential equation, and it is far from easy to solve in general. Some idea of how it is solved can be obtained by considering the special case in which there is no convective motion (as in an unstirred reaction vessel):

$$\frac{\partial[J]}{\partial t} = D\frac{\partial^2[J]}{\partial x^2} - k[J] \tag{39}$$

If the solution of this equation in the absence of reaction (that is, for $k = 0$) is [J], then the solution in the presence of reaction ($k > 0$) is

$$[J]^* = k\int_0^t [J]e^{-kt}\,dt + [J]e^{-kt} \tag{40}$$

We have already met one solution of the diffusion equation in the absence of reaction: eqn 24.79 is the solution for a system in which initially a layer of n_0N_A molecules is spread over a plane of area A:

$$[J] = \frac{n_0 e^{-x^2/4Dt}}{A(\pi Dt)^{1/2}} \tag{41}$$

When this expression is substituted into eqn 40 and the integral is evaluated, we obtain the concentration of J as it diffuses away from its initial surface layer and undergoes reaction in the solution above (Fig. 27.5).

Even this relatively simple example has led to an equation that is difficult to solve, and only in some special cases can the full material balance equation be solved analytically. Most modern work on reactor design and cell kinetics uses numerical methods to solve the equation, and detailed solutions for realistic environments can be obtained reasonably easily. Some of the interesting applications include the exploration of the spatial periodicity of autocatalytic reactions mentioned in Section 26.8.

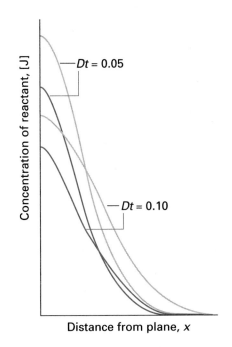

27.5 The concentration profiles for a diffusing, reacting system (for example, a column of solution) in which one reactant is initially in a layer at $x = 0$. In the absence of reaction (grey lines) the concentration profiles are the same as in Fig. 24.24. (Arbitrary values for D and k have been taken.)

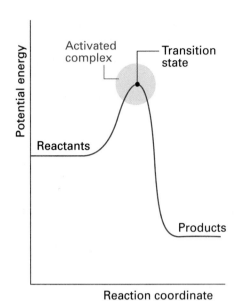

27.6 A reaction profile. The horizontal axis is the reaction coordinate, and the vertical axis is potential energy. The activated complex is the region near the potential maximum, and the transition state corresponds to the maximum itself.

Activated complex theory

We now consider a more detailed calculation of rate constants which uses the concepts of statistical thermodynamics developed in Chapter 20. The approach we describe, which is called **activated complex theory** (ACT), has the advantage that a quantity corresponding to the steric factor appears automatically, and P does not need to be grafted on to an equation as an afterthought. Activated complex theory is an attempt to identify the principal features governing the size of a rate constant in terms of a model of the events that take place during the reaction.

27.4 The reaction coordinate and the transition state

The general features of how the potential energy of the reactants A and B change in the course of a bimolecular elementary reaction are shown in Fig. 27.6. Initially, only the reactants A and B are present. As the reaction event proceeds, A and B come into contact, distort, and begin to exchange or discard atoms. The potential energy rises to a maximum, and the cluster of atoms that corresponds to the region close to the maximum is called the **activated complex**. After the maximum, the potential energy falls as the atoms rearrange in the cluster, and it reaches a value characteristic of the products. The climax of the reaction is at the peak of the potential energy. Here two reactant molecules have come to such a degree of closeness and distortion that a small further distortion will send them in the direction of products. This crucial configuration is called the **transition state** of the reaction. Although some molecules entering the transition state might revert to reactants, if they pass through this configuration then it is inevitable that products will emerge from the encounter.[2]

27.5 The Eyring equation

Activated complex theory pictures a reaction between A and B as proceeding through the formation of an activated complex, $C^{\ddagger}$, that falls apart by unimolecular decay into products, P, with a rate constant $k^{\ddagger}$:

$$C^{\ddagger} \longrightarrow P \qquad v = k^{\ddagger}[C^{\ddagger}] \tag{42}$$

The concentration of the activated complex is likely to be proportional to the concentrations of the reactants, and later we show explicitly that[3]

$$[C^{\ddagger}] = K^{\ddagger}[A][B] \tag{43}$$

where $K^{\ddagger}$ is a proportionality constant (with dimensions 1/concentration). It follows that

$$v = k_2[A][B] \qquad k_2 = k^{\ddagger}K^{\ddagger} \tag{44}$$

Our task is to calculate the unimolecular rate constant $k^{\ddagger}$ and the proportionality constant $K^{\ddagger}$.

(a) The rate of decay of the activated complex

An activated complex can form products if it passes through the transition state. If its vibration-like motion along the reaction coordinate occurs with a frequency ν, the frequency with which the cluster of atoms forming the complex approaches the transition state is also ν. However, it is possible that not every oscillation along the reaction coordinate

---

2   The terms activated complex and transition state are often used as synonyms; however, we shall preserve a distinction. Activated complex theory is also widely referred to as *transition state theory*.

3   This chapter inevitably puts heavy demands on the letter K; the various meanings are summarized in Table 27.3 at the end of the chapter.

takes the complex through the transition state. For instance, the centrifugal effect of rotations might also be an important contribution to the break-up of the complex, and in some cases the complex might be rotating too slowly, or rotating rapidly but about the wrong axis. Therefore, we suppose that the rate of passage of the complex through the transition state is proportional to the vibrational frequency along the reaction coordinate, and write

$$k^{\ddagger} = \kappa \nu \tag{45}$$

where κ is the **transmission coefficient**. In the absence of information to the contrary, κ is assumed to be about 1.

(b) The concentration of the activated complex

The simplest procedure for estimating the concentration of the activated complex is to assume that there is a pre-equilibrium between the reactants and the complex,[4] and to write

$$A + B \rightleftharpoons C^{\ddagger} \qquad K = \frac{(p_{C^{\ddagger}}/p^{\ominus})}{(p_A/p^{\ominus})(p_B/p^{\ominus})} = \frac{p_{C^{\ddagger}}p^{\ominus}}{p_A p_B}$$

The partial pressures, p_J, can be expressed in terms of the molar concentrations by using $p_J = RT[J]$, which gives

$$[C^{\ddagger}] = K \frac{RT}{p^{\ominus}}[A][B] \tag{46}$$

from which it follows that

$$K^{\ddagger} = K \frac{RT}{p^{\ominus}} \tag{47}$$

The remaining task is to calculate the equilibrium constant K.

We saw in Section 20.7 how to calculate equilibrium constants from structural data. Equation 20.56 of that section can be used directly, which in this case gives

$$K = \frac{N_A q_{C^{\ddagger}}^{\ominus}}{q_A^{\ominus} q_B^{\ominus}} e^{-\Delta E_0/RT} \tag{48}$$

where

$$\Delta E_0 = E_0(C^{\ddagger}) - E_0(A) - E_0(B) \tag{49}$$

and the $q_J^{\ominus}$ are the standard molar partition functions, as defined in Section 20.2. Note that the units of N_A and the q_J are mol^{-1}, so K is dimensionless (as is appropriate for an equilibrium constant).

In the final step of this part of the calculation, we focus attention on the partition function of the activated complex. We have already assumed that a vibration of the activated complex $C^{\ddagger}$ tips it through the transition state. The partition function for this vibration is

$$q = \frac{1}{1 - e^{-h\nu/kT}} \tag{50}$$

where ν is its frequency (the same frequency that determines $k^{\ddagger}$). This frequency is much lower than for an ordinary molecular vibration because the oscillation corresponds to the

4 In an earlier edition of this text a different line of argument was used: we recognized that almost nothing is known about the populations of the levels of the activated complex. Since nothing is known, the most honest approach is to assume that the populations depend on the energy and not on the identity of the level (that is, whether it belongs to A, B, or C^‡). This 'least-prejudiced' approach leads to the same results as here, but has the advantage that it avoids the presumption of equilibrium between the reactants and their activated complex. Its disadvantage is that it is slightly less direct. The second edition should be consulted for details.

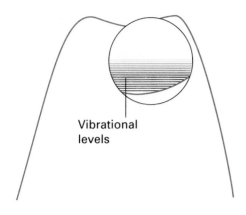

Vibrational
levels

Reaction coordinate

27.7 In an elementary depiction of the activated complex close to the transition state, there is a broad, shallow dip in the potential energy surface along the reaction coordinate. The complex vibrates harmonically and almost classically in this well. However, see footnote 5.

complex falling apart (Fig. 27.7), so the force constant is very low.[5] Therefore, because $h\nu/kT \ll 1$, the exponential may be expanded and the partition function reduces to

$$q = \frac{1}{1 - \left(1 - \frac{h\nu}{kT} + \cdots\right)} \approx \frac{kT}{h\nu} \tag{51}$$

We can therefore write

$$q_{C\ddagger} \approx \frac{kT}{h\nu}\bar{q}_{C\ddagger} \tag{52}$$

where $\bar{q}$ denotes the partition function for all the other modes of the complex. The constant $K^\ddagger$ is therefore

$$K^\ddagger = \frac{kT}{h\nu}\bar{K} \qquad \bar{K} = \left(\frac{RT}{p^\ominus}\right)\left(\frac{N_A\bar{q}_{C\ddagger}^\ominus}{q_A^\ominus q_B^\ominus}\right)e^{-\Delta E_0/RT} \tag{53}$$

with $(p^\ominus/RT)\bar{K}$ a kind of equilibrium constant, but with one vibrational mode of $C^\ddagger$ discarded.

(c) The rate constant

We can now combine all the parts of the calculation into

$$k_2 = k^\ddagger K^\ddagger = \kappa\nu\frac{kT}{h\nu}\bar{K} \tag{54}$$

At this stage the unknown frequencies ν cancel, and we obtain the **Eyring equation**:

$$k_2 = \kappa\frac{kT}{h}\bar{K} \tag{55}$$

The factor $\bar{K}$ is given by eqn 53 in terms of the partition functions of A, B, and $C^\ddagger$, so in principle we now have an explicit expression for calculating the second-order rate constant for a bimolecular reaction in terms of the molecular parameters for the reactants and the activated complex and the quantity κ.

The partition functions for the reactants can normally be calculated quite readily, using either spectroscopic information about their energy levels or the approximate expressions set out in Table 20.2. The difficulty with the Eyring equation, however, lies in the calculation of the partition function of the activated complex: $C^\ddagger$ cannot normally be investigated spectroscopically, and in general we need to make assumptions about the size, shape, and structure of the complex. We shall illustrate what is involved for two simple cases.

(d) The collision of structureless particles

As a first example, consider the case of two structureless particles, A and B, colliding to give an activated complex that resembles a diatomic molecule. Because the reactants $J = A, B$ are structureless 'atoms', the only contributions to their partition functions are the translational terms:

$$q_J^\ominus = \frac{V_m^\ominus}{\Lambda_J^3} \qquad \Lambda_J = \frac{h}{(2\pi m_J kT)^{1/2}} \qquad V_m^\ominus = \frac{RT}{p^\ominus} \tag{56}$$

The activated complex is a diatomic cluster of mass $m_C = m_A + m_B$ and moment of inertia I. It has one vibrational mode, but that mode corresponds to motion along the reaction

---

5   There is a real problem here. Figure 27.7 is probably an oversimplification, for in many cases there is no dip at the top of the barrier, and the curvature of the potential energy, and therefore the force constant, is negative. Formally, the vibrational frequency is then imaginary. We ignore this problem here, but see *Further reading*.

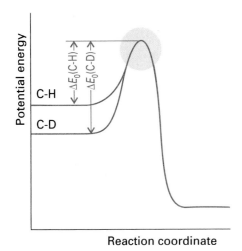

27.8 Changes in the reaction profile when a bond undergoing cleavage is deuterated. The only significant change is to the zero-point energy of the reactants, which is lower for C–D than for C–H. As a result, the activation energy is greater for C–D than for C–H.

coordinate and therefore does not appear in $\bar{q}_{C^{\ddagger}}$. It follows that the standard molar partition function of the activated complex is

$$
\bar{q}_{C^{\ddagger}} = \left(\frac{2IkT}{\hbar^2}\right)\frac{V_m^{\ominus}}{\Lambda_{C^{\ddagger}}^3} \tag{57}
$$

The moment of inertia of a diatomic molecule of bond length r is μr^2, where $\mu = m_A m_B/(m_A + m_B)$ is the effective mass, so the expression for the rate constant is

$$
\begin{aligned}
k_2 &= \kappa \frac{kT}{h}\frac{RT}{p^{\ominus}}\left(\frac{N_A \Lambda_A^3 \Lambda_B^3}{\Lambda_{C^{\ddagger}}^3 V_m^{\ominus}}\right)\left(\frac{2IkT}{\hbar^2}\right)e^{-\Delta E_0/RT} \\
&= \kappa \frac{kT}{h}N_A\left(\frac{\Lambda_A \Lambda_B}{\Lambda_{C^{\ddagger}}}\right)^3\left(\frac{2IkT}{\hbar^2}\right)e^{-\Delta E_0/RT} \\
&= \kappa N_A\left(\frac{8kT}{\pi\mu}\right)^{1/2}\pi r^2 e^{-\Delta E_0/RT}
\end{aligned} \tag{58}
$$

Finally, by identifying the reactive cross-section σ^* as $\kappa\pi r^2$, we arrive at precisely the same expression as that obtained from simple collision theory (eqn 16).

(e) The kinetic isotope effect

As a second example, consider the effect of deuteration on a reaction in which the rate-determining step is the scission of a C–H bond. The experimental observation we seek to explain is the **kinetic isotope effect**, the decrease in the rate of the reaction on deuteration. We shall demonstrate that the difference lies in the fact that C–H scission has a lower activation energy than C–D scission on account of the former's greater zero-point vibrational energy.

The reaction coordinate corresponds to the stretching of the C–H bond, and the potential energy profile is shown in Fig. 27.8. On deuteration, the dominant change is the reduction of the zero-point energy of the bond (because the deuterium atom is heavier). The whole reaction profile is not lowered, however, because the relevant vibration in the activated complex has a very low force constant, so there is little zero-point energy associated with the reaction coordinate in either the proton or the deuterium forms of the complex.

We assume that the deuteration affects only the reaction coordinate, and hence that the partition functions for all the other internal modes remain unchanged. The translational partition functions are changed by deuteration, but the mass of the rest of the molecule is normally so great that the change is insignificant. The value of ΔE_0 changes on account of the change of zero-point energy, and

$$
\begin{aligned}
\Delta E_0(\text{C—D}) - \Delta E_0(\text{C—H}) &= N_A\left\{\tfrac{1}{2}\hbar\omega(\text{C—H}) - \tfrac{1}{2}\hbar\omega(\text{C—D})\right\} \\
&= \tfrac{1}{2}N_A\hbar k_f^{1/2}\left(\frac{1}{\mu_{CH}^{1/2}} - \frac{1}{\mu_{CD}^{1/2}}\right)
\end{aligned} \tag{59}
$$

where k_f is the force constant of the bond and μ the relevant effective mass (Section 16.9). Because all the partition functions are the same (by assumption), the rate constants for the two species should be in the ratio

$$
\frac{k(\text{C—D})}{k(\text{C—H})} = e^{-\lambda} \qquad \lambda = \frac{\hbar k_f^{1/2}}{2kT}\left(\frac{1}{\mu_{CH}^{1/2}} - \frac{1}{\mu_{CD}^{1/2}}\right) \tag{60}
$$

Note that $\lambda > 0$ because $\mu_{CD} > \mu_{CH}$. This equation predicts that at room temperature C–H cleavage should be about seven times faster than C–D cleavage, other conditions being equal.

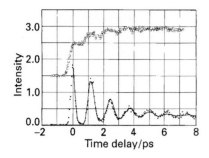

27.9 Femtosecond spectroscopic results for the reaction in which NaI separates into Na and I. The full circles are the absorption of the complex and the open circles the absorption of the free Na atoms. (A.H. Zewail, *Science* **242**, 1645 (1988).)

(f) The experimental observation of the activated complex

Until very recently there were no direct spectroscopic observations on activated complexes, for they have a very fleeting existence and often survive for only a picosecond or so. However, the development of femtosecond pulsed lasers and their application to chemistry in the form of **femtochemistry** has made it possible to make observations on species that have such short lifetimes that in a number of respects they resemble an activated complex.

In a typical experiment, a femtosecond pulse is used to excite a molecule to a dissociative state, and then a second femtosecond pulse is fired at an interval after the dissociating pulse. The frequency of the second pulse is set at an absorption of one of the free fragmentation products, so its absorption is a measure of the abundance of the dissociation product. For example, when ICN is dissociated by the first pulse, the emergence of CN from the photoactivated state can be monitored by watching the growth of the free CN absorption (or, more commonly, its laser-induced fluorescence). In this way it has been found that the CN signal remains zero until the fragments have separated by about 600 pm, which takes about 205 fs.

Some sense of the progress that has been made in the study of the intimate mechanism of chemical reactions can be obtained by considering an analogue of the harpoon reaction introduced in Section 27.1c. The decay of the ion pair Na^+X^-, where X is a halogen, has been studied by exciting it with a femtosecond pulse to an excited state that corresponds to a covalently bonded NaX molecule. The second probe pulse examines the system at an absorption frequency either of the free Na atom or at a frequency at which the atom absorbs when it is a part of the complex. The latter frequency depends on the Na–X distance, so an absorption (in practice, a laser-induced fluorescence) is obtained each time the vibration of the complex returns it to that separation.

A typical set of results for NaI is shown in Fig. 27.9. The bound Na absorption intensity shows up as a series of pulses that recur in about 1 ps, showing that the complex vibrates with about that period. The decline in intensity shows the rate at which the complex can dissociate as the two atoms swing away from each other. The complex does not dissociate on every outward-going swing because there is a chance that the I atom can be harpooned again, in which case it fails to make good its escape. The free Na absorption also grows in an oscillating manner, showing the periodicity of the vibration of the complex, each swing of which gives it a chance to dissociate. The precise period of the oscillation in NaI is 1.25 ps, corresponding to a vibrational wavenumber of 27 cm^{-1} (recall that activated complex theory assumes that such a vibration has a very low frequency). The complex survives for about ten oscillations. In contrast, although the oscillation frequency of NaBr is similar, it barely survives one oscillation.

Femtosecond spectroscopy has also been used to examine analogues of the activated complex involved in bimolecular reactions. Thus, a molecular beam can be used to produce a van der Waals molecule (Section 22.5c), such as IH $\cdots$ OCO. The HI bond can be dissociated by a femtosecond pulse, and the H atom is ejected towards the O atom of the neighbouring CO_2 molecule to form HOCO. Hence, the van der Waals molecule is a source of a species that resembles the activated complex of the reaction

$$H + CO_2 \longrightarrow [HOCO]^{\ddagger} \longrightarrow HO + CO$$

The probe pulse is tuned to the OH radical, which enables the evolution of $[HOCO]^{\ddagger}$ to be studied in real time.

27.6 Thermodynamic aspects

The statistical thermodynamic version of activated complex theory rapidly runs into difficulties because only rarely is anything known about the structure of the activated complex. However, the concepts that it introduces, principally that of an equilibrium between the reactants and the activated complex, have motivated a more general, empirical approach in which the activation process is expressed in terms of thermodynamic functions.

(a) Activation parameters

If we accept that $(p^{\ominus}/RT)K$ is an equilibrium constant (despite one mode of $C^{\ddagger}$ having been discarded), we can express it in terms of a **Gibbs energy of activation**, $\Delta^{\ddagger}G$, through the definition[6]

$$\Delta^{\ddagger}G = -RT\ln(p^{\ominus}/RT)K \qquad [61]$$

Then the rate constant becomes

$$k_2 = \kappa\frac{kT}{h}\frac{RT}{p^{\ominus}}e^{-\Delta^{\ddagger}G/RT} \qquad (62)$$

Because $G = H - TS$, the Gibbs energy of activation can be divided into an **entropy of activation**, $\Delta^{\ddagger}S$, and an **enthalpy of activation**, $\Delta^{\ddagger}H$, by writing

$$\Delta^{\ddagger}G = \Delta^{\ddagger}H - T\Delta^{\ddagger}S \qquad [63]$$

When eqn 63 is used in eqn 62 and κ is absorbed into the entropy term, we obtain

$$k_2 = Be^{\Delta^{\ddagger}S/R}e^{-\Delta^{\ddagger}H/RT} \qquad B = \frac{kT}{h}\frac{RT}{p^{\ominus}} \qquad (64)$$

The formal definition of activation energy, $E_a = RT^2(\partial\ln k/\partial T)$, then gives $E_a = \Delta^{\ddagger}H + 2RT$,[7] so

$$k_2 = e^2 Be^{\Delta^{\ddagger}S/R}e^{-E_a/RT} \qquad (65)$$

from which it follows that the Arrhenius factor A can be identified as

$$A = e^2 Be^{\Delta^{\ddagger}S/R} \qquad (66)$$

The entropy of activation is negative because two reactant species come together to form one species. However, if there is a reduction in entropy below what would be expected for the simple encounter of A and B, then A will be smaller than that expected on the basis of simple collision theory. Indeed, we can identify that additional reduction in entropy, $\Delta^{\ddagger}S_{steric}$, as the origin of the steric factor of collision theory, and write

$$P = e^{\Delta^{\ddagger}S_{steric}/R} \qquad (67)$$

Thus, the more complex the steric requirements of the encounter, the more negative the value of $\Delta^{\ddagger}S_{steric}$, and the smaller the value of P.

Gibbs energies, enthalpies, and entropies of activation are widely used to report experimental reaction rates, especially for organic reactions in solution. They are encountered when relationships between equilibrium constants and rates of reaction are explored using **correlation analysis**, in which $\ln K$ (which is equal to $-\Delta_r G^{\ominus}/RT$) is plotted against $\ln k$ (which is proportional to $-\Delta^{\ddagger}G/RT$). In many cases the correlation is linear, signifying that, as the reaction becomes thermodynamically more favourable, its rate constant increases (Fig. 27.10). This linear correlation is the origin of the alternative name **linear free energy relation** (LFER; see *Further reading*).

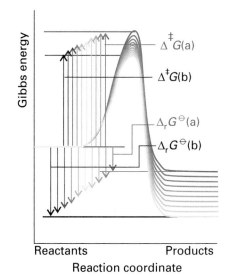

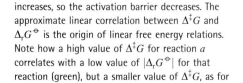

27.10 For a related series of reactions, as the magnitude of the standard reaction Gibbs energy increases, so the activation barrier decreases. The approximate linear correlation between $\Delta^{\ddagger}G$ and $\Delta_r G^{\ominus}$ is the origin of linear free energy relations. Note how a high value of $\Delta^{\ddagger}G$ for reaction a correlates with a low value of $|\Delta_r G^{\ominus}|$ for that reaction (green), but a smaller value of $\Delta^{\ddagger}G$, as for reaction b (black) correlates with a larger value of $|\Delta_r G^{\ominus}|$.

---

6 All the $\Delta^{\ddagger}X$ in this section are *standard* thermodynamic quantities, $\Delta^{\ddagger}X^{\ominus}$, but we shall omit the standard state sign to avoid overburdening the notation.

7 For reactions in solution, $E_a = \Delta^{\ddagger}H + RT$.

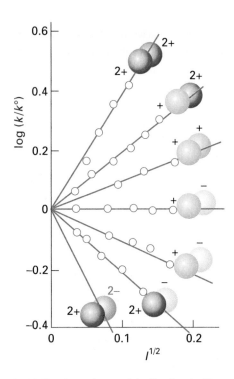

27.11 Experimental tests of the kinetic salt effect for reactions in water at 298 K. The ion types are shown as spheres, and the slopes of the lines are those given by the Debye–Hückel limiting law and eqn 73.

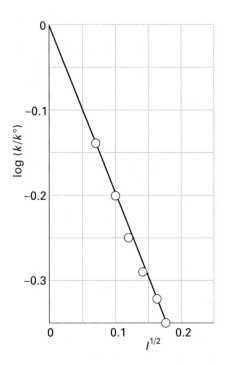

27.12 The experimental ionic strength dependence of the rate constant of a hydrolysis reaction: the slope gives information about the charge types involved in the activated complex of the rate-determining step. See Example 27.3.

(b) Reactions between ions

The thermodynamic version of activated complex theory simplifies the discussion of reactions in solution. The statistical thermodynamic theory is very complicated to apply because the solvent plays a role in the activated complex. In the thermodynamic approach we combine the rate law

$$\frac{d[P]}{dt} = k^{\ddagger}[C^{\ddagger}] \tag{68}$$

with the thermodynamic equilibrium constant

$$K = \frac{a_{C^{\ddagger}}}{a_A a_B} = K_{\gamma} \frac{[C^{\ddagger}]}{[A][B]} \qquad K_{\gamma} = \frac{\gamma_{C^{\ddagger}}}{\gamma_A \gamma_B} \tag{69}$$

Then

$$\frac{d[P]}{dt} = k_2[A][B] \qquad k_2 = \frac{k^{\ddagger}K}{K_{\gamma}} \tag{70}$$

If k_2° is the rate constant when the activity coefficients are 1 (that is, $k_2^{\circ} = k^{\ddagger}K$), we can write

$$k_2 = \frac{k_2^{\circ}}{K_{\gamma}} \tag{71}$$

At low concentrations the activity coefficients can be expressed in terms of the ionic strength, I, of the solution by using the Debye–Hückel limiting law (Section 10.2c, particularly eqn 10.19) in the form

$$\log \gamma_J = -A z_J^2 I^{1/2} \tag{72}$$

with $A = 0.509$ in aqueous solution at 298 K. Then

$$\log k_2 = \log k_2^{\circ} - A\{z_A^2 + z_B^2 - (z_A + z_B)^2\}I^{1/2}$$
$$= \log k_2^{\circ} + 2A z_A z_B I^{1/2} \tag{73}$$

The charge numbers of A and B are z_A and z_B, so the charge number of the activated complex is $z_A + z_B$; the z_J are positive for cations and negative for anions.

Equation 73 expresses the **kinetic salt effect**, the variation of the rate constant of a reaction between ions with the ionic strength of the solution (Fig. 27.11). If the reactant ions have the same sign (as in a reaction between cations or between anions), then increasing the ionic strength by the addition of inert ions increases the rate constant. The formation of a single, highly charged ionic complex from two less highly charged ions is favoured by a high ionic strength because the new ion has a denser ionic atmosphere and interacts with that atmosphere more strongly. Conversely, ions of opposite charge react more slowly in solutions of high ionic strength. Now the charges cancel and the complex has a less favourable interaction with its atmosphere than the separated ions.

Example 27.3 Analysing the kinetic salt effect

The rate constant for the base hydrolysis of $[CoBr(NH_3)_5]^{2+}$ varies with ionic strength as tabulated below. What can be deduced about the charge of the activated complex in the rate-determining stage?

I	0.0050	0.0100	0.0150	0.0200	0.0250	0.0300
k/k°	0.718	0.631	0.562	0.515	0.475	0.447

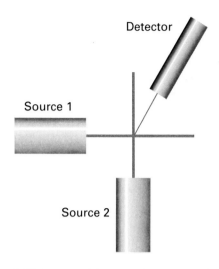

27.13 In a crossed-beam experiment, state-selected molecules are generated in two separate sources, and are directed perpendicular to one another. The detector responds to molecules (which may be product molecules if chemical reaction occurs) scattered into a chosen direction.

Method According to eqn 73, plot $\log(k/k^\circ)$ against $I^{1/2}$, when the slope will give $1.02z_A z_B$, from which we can infer the charges of the ions involved in the formation of the activated complex.

Answer Form the following table:

I	0.0050	0.0100	0.0150	0.0200	0.0250	0.0300
$I^{1/2}$	0.071	0.100	0.122	0.141	0.158	0.173
$\log(k/k^\circ)$	−0.14	−0.20	−0.25	−0.29	−0.32	−0.35

These points are plotted in Fig. 27.12. The slope of the (least squares) straight line is −2.04, indicating that $z_A z_B = -2$. Because $z_A = -1$ for the OH^- ion, if that ion is involved in the formation of the activated complex, then the charge number of the second ion is +2. This analysis suggests that the pentaamminebromocobalt(III) cation participates in the formation of the activated complex.

Comment The rate constant is also influenced by the relative permittivity of the medium.

- -

Self-test 27.3 An ion of charge number +1 is known to be involved in the activated complex of a reaction. Deduce the charge number of the other ion from the following data:

I	0.005	0.010	0.015	0.020	0.025	0.030
k/k°	0.930	0.902	0.884	0.867	0.853	0.841

[−1]

The dynamics of molecular collisions

We now come to the third and most detailed level of our examination of the factors that govern the rates of reactions. Molecular beams allow us to study collisions between molecules in preselected energy states, and can be used to determine the states of the products of a reactive collision. Information of this kind is essential if a full picture of the reaction is to be built, because the rate constant is an average over events in which reactants in different initial states evolve into products in their final states.

27.7 Reactive collisions

Detailed experimental information about the intimate processes that occur during reactive encounters comes from molecular beams, especially crossed molecular beams (Fig. 27.13). The detector for the products of the collision of two beams can be moved to different angles, so the angular distribution of the products can be determined. Because the molecules in the incoming beams can be prepared with different energies (for example, with different translational energies by using rotating sectors and supersonic nozzles, and with different vibrational energies by using selective excitation with lasers) and with different orientations (by using electric fields), it is possible to study the dependence of the success of collisions on these variables and to study how they affect the properties of the outcoming product molecules.

One method for examining the energy distribution in the products is **infrared chemiluminescence**, in which vibrationally excited molecules emit infrared radiation as they return to their ground states. By studying the intensities of the infrared emission spectrum, the populations of the vibrational states may be determined (Fig. 27.14). Another method makes use of **laser-induced fluorescence**. In this technique, a laser is used to excite a product molecule from a specific vibration–rotation level; the intensity of the fluorescence

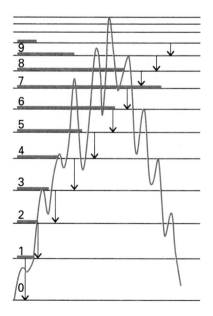

27.14 Infrared chemiluminescence from CO produced in the reaction $O + CS \rightarrow CO + S$ arises from the non-equilibrium populations of the vibrational states of CO and the radiative relaxation to equilibrium.

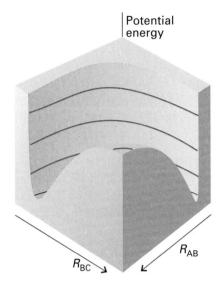

27.15 The potential energy surface for the $H + H_2 \rightarrow H_2 + H$ reaction when the atoms are constrained to be collinear.

from the upper state is monitored and interpreted in terms of the population of the initial vibration–rotation state.

The concept of collision cross-section was introduced in connection with collision theory in Section 27.1, where we saw that the second-order rate constant, k_2, can be expressed as a Boltzmann-weighted average of the reactive collision cross-section and the relative speed of approach. We shall write eqn 14 as

$$k_2 = \langle \sigma v_{\text{rel}} \rangle N_A \qquad (74)$$

where the angle brackets denote a Boltzmann average. Molecular beam studies provide a more sophisticated version of this quantity, for they provide the **state-to-state cross-section**, $\sigma_{nn'}$, and hence the **state-to-state rate constant**, $k_{nn'}$:

$$k_{nn'} = \langle \sigma_{nn'} v_{\text{rel}} \rangle N_A \qquad (75)$$

The rate constant k_2 is the sum of the state-to-state rate constant over all final states (because a reaction is successful whatever the final state of the products) and over a Boltzmann-weighted sum of initial states (because the reactants are initially present with a characteristic distribution of populations at a temperature T):

$$k_2 = \sum_{n,n'} k_{nn'}(T) f_n(T) \qquad (76)$$

where $f_n(T)$ is the Boltzmann factor at a temperature T.

It follows that, if we can determine or calculate the state-to-state cross-sections for a wide range of approach speeds and initial and final states, then we have a route to the calculation of the rate constant for the reaction.

27.8 Potential energy surfaces

One of the most important concepts for discussing beam results and calculating the state-to-state collision cross-section is the **potential energy surface** of a reaction, the potential energy as a function of the relative positions of all the atoms taking part in the reaction. For a collision between an H atom and an H_2 molecule, for instance, the potential energy surface is the plot of the potential energy for all relative locations of the three hydrogen nuclei. Detailed calculations show that the approach of an atom along the H–H axis requires less energy for reaction than any other approach, so initially we confine our attention to a collinear approach. Two parameters are required to define the nuclear separations: one is the H_A–H_B separation R_{AB}, and the other is the H_B–H_C separation R_{BC}.

At the start of the encounter R_{AB} is infinite and R_{BC} is the H_2 equilibrium bond length. At the end of a successful reactive encounter R_{AB} is equal to the equilibrium bond length and R_{BC} is infinite. The total energy of the three-atom system depends on their relative separations, and can be found by doing a molecular orbital calculation. The plot of the total energy of the system against R_{AB} and R_{BC} gives the potential energy surface of this collinear reaction (Fig. 27.15). This surface is normally depicted as a contour diagram (Fig. 27.16).

When R_{AB} is very large, the variation in potential energy represented by the surface as R_{BC} changes are those of an isolated H_2 molecule as its bond length is altered. A section through the surface at $R_{AB} = \infty$, for example, is the same as the H_2 bonding potential energy curve drawn in Fig. 14.18. At the edge of the diagram where R_{BC} is very large, a section through the surface is the molecular potential energy curve of an isolated $H_A H_B$ molecule.

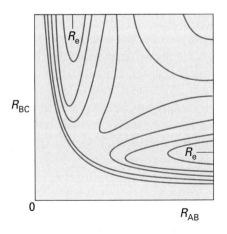

27.16 The contour diagram (with contours of equal potential energy) corresponding to the surface in Fig. 27.15. R_e marks the equilibrium bond length of an H_2 molecule (strictly, it relates to the arrangement when the third atom is at infinity).

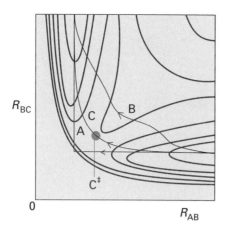

27.17 Various trajectories through the potential energy surface shown in Fig. 27.16. Path A corresponds to a route in which R_{BC} is held constant as H_A approaches; path B corresponds to a route in which R_{BC} lengthens at an early stage during the approach of H_A; path C is the route along the floor of the potential valley.

27.18 The transition state is a set of configurations (here, marked by the line across the saddle point) through which successful reactive trajectories must pass.

The actual path of the atoms in the course of the encounter depends on their total energy, the sum of their kinetic and potential energies. However, we can obtain an initial idea of the paths available to the system for paths that correspond to least potential energy. For example, consider the changes in potential energy as H_A approaches $H_B H_C$. If the H_B–H_C bond length is constant during the initial approach of H_A, then the potential energy of the H_3 cluster would rise along the path marked A in Fig. 27.17. We see that the potential energy rises to a high value as H_A is pushed into the molecule, and then decreases sharply as H_C breaks off and separates to a great distance. An alternative reaction path can be imagined (B) in which the H_B–H_C bond length increases while H_A is still far away. Both paths, although feasible if the molecules have sufficient initial kinetic energy, take the three atoms to regions of high potential energy in the course of the encounter.

The path of least potential energy is the one marked C, corresponding to R_{BC} lengthening as H_A approaches and begins to form a bond with H_B. The H_B–H_C bond relaxes at the demand of the incoming atom, and the potential energy climbs only as far as the saddle-shaped region of the surface, to the **saddle point** marked $C^{\ddagger}$. The encounter of least potential energy is one in which the atoms take route C up the floor of the valley, through the saddle point, and down the floor of the other valley as H_C recedes and the new H_A–H_B bond achieves its equilibrium length. This path is the reaction coordinate we met in Section 27.4.

We can now make contact with the activated complex theory of reaction rates. In terms of trajectories on potential surfaces, the transition state can be identified with a critical geometry such that every trajectory that goes through this geometry goes on to react (Fig. 27.18).

27.9 Some results from experiments and calculations

To travel successfully from reactants to products the incoming molecules must possess enough kinetic energy to be able to climb to the saddle point of the potential surface. Therefore, the shape of the surface can be explored experimentally by changing the relative speed of approach (by selecting the beam velocity) and the degree of vibrational excitation and observing whether reaction occurs and whether the products emerge in a vibrationally

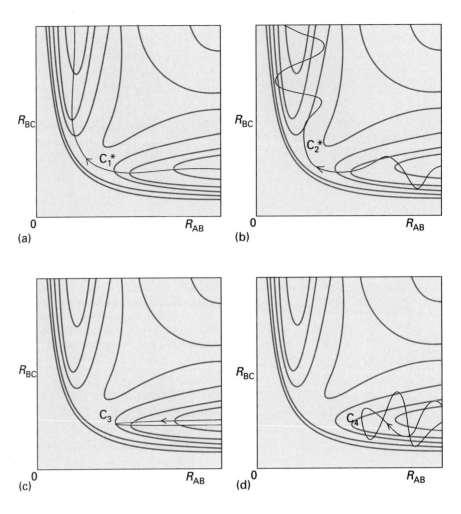

27.19 Some successful (*) and unsuccessful encounters. (a) C_1^* corresponds to the path along the foot of the valley; (b) C_2^* corresponds to an approach of H_A to a vibrating H_{BC} molecule, and the formation of a vibrating H_{AB} molecule as H_C departs. (c) C_3 corresponds to H_A approaching a non-vibrating H_{BC} molecule, but with insufficient translational kinetic energy; (d) C_4 corresponds to H_A approaching a vibrating H_{BC} molecule, but still the energy, and the phase of the vibration, is insufficient for reaction.

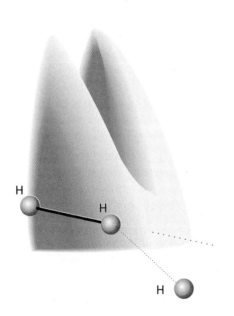

27.20 An indication of the anisotropy of the potential energy changes as H approaches H_2 with different angles of attack. The collinear attack has the lowest potential barrier to reaction. The surface indicates the potential energy profile along the reaction coordinate for each configuration.

excited state (Fig. 27.19). For example, one question that can be answered is whether it is better to smash the reactants together with a lot of translational kinetic energy or to ensure instead that they approach in highly excited vibrational states. Thus, is trajectory C_2^*, where the H_BH_C molecule is initially vibrationally excited, more efficient at leading to reaction than the trajectory C_1^*, in which the total energy is the same but has a high translational kinetic energy?

(a) The direction of attack and separation

Figure 27.20 shows the results of a calculation of the potential energy as an H atom approaches an H_2 molecule from different angles, the H_2 bond being allowed to relax to the optimum length in each case. The potential barrier is least for collinear attack, as we assumed earlier. (But we must be aware that other lines of attack are feasible and contribute to the overall rate.) In contrast, Fig. 27.21 shows the potential energy changes that occur as a Cl atom approaches an HI molecule. The lowest barrier occurs for approaches within a cone of half-angle 30° surrounding the H atom. The relevance of this result to the calculation of the steric factor of collision theory should be noted: not every collision is successful, because not every one lies within the reactive cone.

If the collision is sticky, so that when the reactants collide they orbit around each other, the products can be expected to emerge in random directions because all memory of the approach direction has been lost. A rotation takes about 1 ps, so if the collision is over in less than that time the complex will not have had time to rotate and the products will be thrown off in a specific direction. In the collision of K and I_2, for example, most of the products are

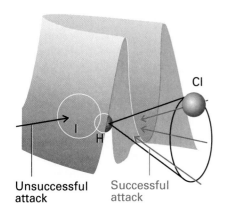

27.21 The potential energy barrier for the approach of Cl to HI. In this case, successful encounters occur only when Cl approaches within a cone surrounding the H atom.

thrown off in the forward direction.[8] This product distribution is consistent with the harpoon mechanism (Section 27.1c) because the transition takes place at long range. In contrast, the collision of K with CH_3I leads to reaction only if the molecules approach each other very closely. In this mechanism, K effectively bumps into a brick wall, and the KI product bounces out in the backward direction. The detection of this anisotropy in the angular distribution of products gives an indication of the distance and orientation of approach needed for reaction, as well as showing that the event is complete in less than 1 ps.

(b) Attractive and repulsive surfaces

Some reactions are very sensitive to whether the energy has been predigested into a vibrational mode or left as the relative translational kinetic energy of the colliding molecules. For example, if two HI molecules are hurled together with more than twice the activation energy of the reaction, then no reaction occurs if all the energy is translational. For $F + HCl \rightarrow Cl + HF$, for example, the reaction is about five times as efficient when the HCl is in its first vibrational excited state than when, although HCl has the same total energy, it is in its vibrational ground state.

The origin of these requirements can be found by examining the potential energy surface. Figure 27.22 shows an **attractive surface** in which the saddle point occurs early in the reaction coordinate. Figure 27.23 shows a **repulsive surface** in which the saddle point occurs late. A surface that is attractive in one direction is repulsive in the reverse direction.

Consider first the attractive surface. If the original molecule is vibrationally excited, then a collision with an incoming molecule takes the system along C. This path is bottled up in the region of the reactants, and does not take the system to the saddle point. If, however, the same amount of energy is present solely as translational kinetic energy, then the system moves along C* and travels smoothly over the saddle point into products. We can therefore

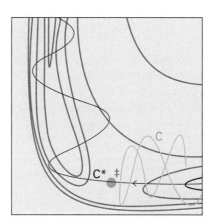

27.22 An attractive potential energy surface. A successful encounter (C*) involves high translational kinetic energy and results in a vibrationally excited product.

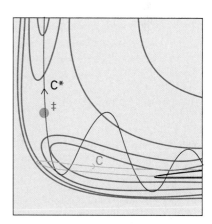

27.23 A repulsive potential energy surface. A successful encounter (C*) involves initial vibrational excitation and the products have high translational kinetic energy. A reaction that is attractive in one direction is repulsive in the reverse direction.

8   There is a subtlety here. In molecular beam work the remarks normally refer to directions in a centre-of-mass coordinate system. The origin of the coordinates is the centre of mass of the colliding reactants, and the collision takes place when the molecules are at the origin. The way in which centre-of-mass coordinates are constructed and the events in them interpreted involves too much detail for our present purposes, but we should bear in mind that 'forward' and 'backward' have unconventional meanings. The details are explained in the books in *Further reading*.

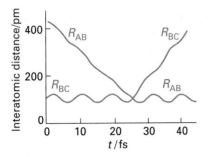

27.24 The calculated trajectories for a reactive encounter between H_A and a vibrating $H_B H_C$ molecule leading to the formation of a vibrating $H_A H_B$ molecule. This direct-mode reaction is between H and H_2. (M. Karplus, R.N. Porter, and R.D Sharma, *J. Chem. Phys.* **43**, 3258 (1965).)

conclude that reactions with attractive potential energy surfaces proceed more efficiently if the energy is in relative translational motion. Moreover, the potential surface shows that once past the saddle point the trajectory runs up the steep wall of the product valley, and then rolls from side to side as it falls to the foot of the valley as the products separate. In other words, the products emerge in a vibrationally excited state.

Now consider the repulsive surface (Fig. 27.23). On trajectory C the collisional energy is largely in translation. As the reactants approach, the potential energy rises. Their path takes them up the opposing face of the valley, and they are reflected back into the reactant region. This path corresponds to an unsuccessful encounter, even though the energy is sufficient for reaction. On C^* some of the energy is in the vibration of the reactant molecule and the motion causes the trajectory to weave from side to side up the valley as it approaches the saddle point. This motion may be sufficient to tip the system round the corner to the saddle point and then on to products. In this case, the product molecule is expected to be in an unexcited vibrational state. Reactions with repulsive potential surfaces can therefore be expected to proceed more efficiently if the excess energy is present as vibrations. This is the case with the $H + Cl_2 \rightarrow HCl + Cl$ reaction, for instance.

(c) Classical trajectories

A clear picture of the reaction event can be obtained using classical mechanics to calculate the trajectories of the atoms taking place in a reaction. Figure 27.24 shows the result of such a calculation of the positions of the three atoms in the reaction $H + H_2 \rightarrow H_2 + H$, the horizontal coordinate now being time and the vertical coordinate the separations. This illustration shows clearly the vibration of the original molecule and the approach of the attacking atom. The reaction itself, the switch of partners, takes place very rapidly and is an example of a **direct-mode process**. The newly formed molecule shakes, but quickly settles down to steady, harmonic vibration as the expelled atom departs. In contrast, Fig. 27.25 shows an example of a **complex-mode process**, in which the activated complex survives for an extended period. The reaction in the illustration is the exchange reaction $KCl + NaBr \rightarrow KBr + NaCl$. The tetraatomic activated complex survives for about 5 ps, during which time the atoms make about 15 oscillations before dissociating into products.

Although this kind of calculation gives a good sense of what happens during a reaction, its limitations must be kept in mind. In the first place, a real gas-phase reaction occurs with a wide variety of different speeds and angles of attack. In the second place, the motion of the atoms, electrons, and nuclei is governed by quantum mechanics. The concept of trajectory then fades and is replaced by the unfolding of a wavefunction that represents initially the

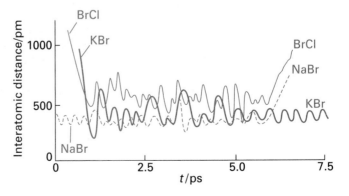

27.25 An example of the trajectories calculated for a complex-mode reaction, $KCl + NaBr \rightarrow KBr + NaCl$, in which the collision cluster has a long lifetime. (P. Brumer and M. Karplus, *Faraday Disc. Chem. Soc.* **55**, 80 (1973).)

reactants and finally the products. Nevertheless, recognition of these limitations should not be allowed to obscure the fact that recent advances in molecular reaction dynamics have given us a first glimpse of the processes going on at the core of reactions.

Table 27.3 Summary of uses of k

Symbol	Significance
k	Boltzmann constant
k_2	Second-order rate constant
k_2°	Rate constant at zero ionic strength
$k_a, k_b, \ldots$	Rate constants for individual steps
$k_a', k_b', \ldots$	Rate constants for individual reverse steps
$k^\ddagger$	Rate constant for unimolecular decay of activated complex
K	Equilibrium constant (dimensionless)
K_γ	Ratio of activity coefficients
$K^\ddagger$	Proportionality constant between $[C^\ddagger]$ and $[A][B]$ (dimensions, 1/concentration)
κ	Transmission coefficient
$\bar{K}$	When multiplied by $p^\ominus/RT$, a type of equilibrium constant but with one vibrational mode discarded (dimensions, 1/concentration)
k_f	Force constant

Checklist of key ideas

Reactive encounters

- [ ] collision theory

27.1 Collision theory
- [ ] collision density
- [ ] like–like collisions (10)
- [ ] dissimilar collisions (11)
- [ ] energy-dependent reactive collision cross-section (15)
- [ ] rate constant from collision theory (16)
- [ ] steric factor
- [ ] reactive cross-section
- [ ] harpoon mechanism

27.2 Diffusion-controlled reactions
- [ ] cage effect
- [ ] diffusion-controlled limit
- [ ] activation-controlled reaction

- [ ] rate constant and diffusion coefficient (27)
- [ ] rate constant and viscosity (35)

27.3 The material balance equation
- [ ] material balance equation (38)

Activated complex theory

- [ ] activated complex theory (ACT)

27.4 The reaction coordinate and the transition state
- [ ] activated complex
- [ ] transition state

27.5 The Eyring equation
- [ ] transmission coefficient (45)
- [ ] concentration of activated complex (46)
- [ ] Eyring equation (55)
- [ ] collision of structureless particles (58)
- [ ] kinetic isotope effect (60)
- [ ] femtochemistry

27.6 Thermodynamic aspects
- [ ] Gibbs energy of activation (61)
- [ ] entropy of activation (63)
- [ ] enthalpy of activation (63)
- [ ] steric factor and entropy (67)
- [ ] correlation analysis
- [ ] linear free energy relation (LFER)
- [ ] kinetic salt effect (73)

The dynamics of molecular collisions

27.7 Reactive collisions
- [ ] infrared chemiluminescence
- [ ] laser-induced fluorescence
- [ ] state-to-state cross-section
- [ ] state-to-state rate constant (75)

27.8 Potential energy surfaces
- [ ] potential energy surface
- [ ] saddle point

27.9 Some results from experiments and calculations
- [ ] attractive surface
- [ ] repulsive surface
- [ ] direct mode process
- [ ] complex mode process

Further reading

Articles of general interest

G.M. Fernandez, J.A. Sordo, and T.L. Sordo, Analysis of potential energy surfaces. *J. Chem. Educ.* **65**, 665 (1988).

K.J. Laidler, Just what is a transition state? *J. Chem. Educ.* **65**, 540 (1988).

M.A. Smith, The nature of distribution functions for colliding systems: calculation of averaged properties. *J. Chem. Educ.* **70**, 218 (1993).

H. Maskill, The Arrhenius equation. *Educ. in Chem.* **27**, 111 (1990).

S.R. Logan, The meaning and significance of "the activation energy" of a chemical reaction. *Educ. in Chem.* **23**, 148 (1986).

I. Powis, Energy redistribution in unimolecular ion dissociation. *Acc. Chem. Res.* **20**, 179 (1987).

C.E. Klots, The reaction coordinate and its limitations: an experimental perspective. *Acc. Chem. Res.* **21**, 16 (1988).

I.W.M. Smith, Vibrational adiabaticity in chemical reactions. *Acc. Chem. Res.* **23**, 101 (1990).

P.R. Brooks, Spectroscopy of transition region species. *Chem. Rev.* **88**, 407 (1988).

J. Keizer, Diffusion effects on rapid bimolecular chemical reactions. *Chem. Rev.* **87**, 167 (1987).

D.W. Lupo and M. Quack, IR-laser photochemistry. *Chem. Rev.* **87**, 181 (1987).

D.G. Trular, R. Steckler, and M.S. Gordon, Potential energy surfaces for polyatomic reaction dynamics. *Chem. Rev.* **87**, 181 (1987).

A.H. Zewail, Laser femtochemistry. *Science* **242**, 1645 (1988).

A.H. Zewail, Femtosecond transition-state dynamics. *Faraday Discuss. Chem. Soc.* **91**, 1 (1991).

M.S. Child, Molecular reaction dynamics. *Sci. Prog.* **70**, 73 (1986).

R.W. Carr, Chemical kinetics. In *Encyclopedia of applied physics* (ed. G.L. Trigg), 3, 345. VCH, New York (1992).

Texts and sources of data and information

K.J. Laidler, *Chemical kinetics.* Harper & Row, New York (1987).

G.D. Billing and K.V. Mikkelsen, *Molecular dynamics and chemical kinetics.* Wiley, New York (1996).

J.I. Steinfeld, J.S. Francisco, and W.L. Hase, *Chemical kinetics and dynamics.* Prentice-Hall, Englewood Cliffs (1989).

J. Simons, *Energetic principles of chemical reactions.* Jones & Bartlett, Portola Valley (1983).

R.A. Marcus, Activated complex theory: current status, extensions, and applications. In *Techniques of chemistry* (ed. E.S. Lewis), **6A**, 13. Wiley-Interscience, New York (1974).

M.J. Blandamer, *Chemical equilibria in solution.* Ellis Horwood/Prentice-Hall, Hemel Hempstead (1992).

P.C. Jordan, *Chemical kinetics and transport.* Plenum, New York (1979).

I.W.M. Smith, *Kinetics and dynamics of elementary gas reactions.* Butterworth, London (1980).

R.G. Gilbert and S.C. Smith, *Theory of unimolecular and recombination reactions.* Blackwell Scientific, Oxford (1990).

R.B. Bernstein, *Chemical dynamics via molecular beam and laser techniques.* Clarendon Press, Oxford (1982).

D.M. Hirst, *Potential energy surfaces.* Taylor & Francis, London (1985).

R.D. Levine and R.B. Bernstein, *Molecular reaction dynamics and chemical reactivity.* Clarendon Press, Oxford (1987).

R. van Eldik, T. Asano, and W.J. Le Noble, Activation and reaction volumes in solution. 2. *Chem. Rev.* **89**, 549 (1989).

Exercises

27.1 (a) Calculate the collision frequency, z, and the collision density, Z, in ammonia, $R = 190$ pm, at 25°C and 100 kPa. What is the percentage increase when the temperature is raised by 10 K at constant volume?

27.1 (b) Calculate the collision frequency, z, and the collision density, Z, in carbon monoxide, $R = 180$ pm at 25°C and 100 kPa. What is the percentage increase when the temperature is raised by 10 K at constant volume?

27.2 (a) Collision theory demands knowing the fraction of molecular collisions having at least the kinetic energy E_a along the line of flight. What is this fraction when (a) $E_a = 10$ kJ mol^{-1}, (b) $E_a = 100$ kJ mol^{-1} at (i) 300 K and (ii) 1000 K?

27.2 (b) Collision theory demands knowing the fraction of molecular collisions having at least the kinetic energy E_a along the line of flight. What is this fraction when (a) $E_a = 15$ kJ mol^{-1}, (b) $E_a = 150$ kJ mol^{-1} at (i) 300 K and (ii) 800 K?

27.3 (a) Calculate the percentage increase in the fractions in Exercise 27.2a when the temperature is raised by 10 K.

27.3 (b) Calculate the percentage increase in the fractions in Exercise 27.2b when the temperature is raised by 10 K.

27.4 (a) Use the collision theory of gas-phase reactions to calculate the theoretical value of the second-order rate constant for the reaction $H_2(g) + I_2(g) \rightarrow 2HI(g)$ at 650 K, assuming that it is elementary bimolecular. The collision cross-section is 0.36 nm^2, the reduced mass is 3.32×10^{-27} kg, and the activation energy is 171 kJ mol^{-1}.

27.4 (b) Use the collision theory of gas-phase reactions to calculate the theoretical value of the second-order rate constant for the reaction $D_2(g) + Br_2(g) \rightarrow 2DBr(g)$ at 450 K, assuming that it is elementary bimolecular. Take the collision cross-section as 0.30 nm^2, the reduced mass as 3.930 u, and the activation energy as 200 kJ mol^{-1}.

27.5 (a) A typical diffusion coefficient for small molecules in aqueous solution at 25°C is 5×10^{-9} m^2 s^{-1}. If the critical reaction distance is 0.4 nm, what value is expected for the second-order rate constant for a diffusion-controlled reaction?

27.5 (b) Suppose that the typical diffusion coefficient for a reactant in aqueous solution at 25°C is 4.2×10^{-9} m^2 s^{-1}. If the critical reaction distance is 0.50 nm, what value is expected for the second-order rate constant for the diffusion-controlled reaction?

27.6 (a) Calculate the magnitude of the diffusion-controlled rate constant at 298 K for a species in (a) water, (b) pentane. The viscosities are 1.00×10^{-3} kg m^{-1} s^{-1}, and 2.2×10^{-4} kg m^{-1} s^{-1}, respectively.

27.6 (b) Calculate the magnitude of the diffusion-controlled rate constant at 298 K for a species in (a) decylbenzene, (b) concentrated sulfuric acid. The viscosities are 3.36 cP and 27 cP, respectively.

27.7 (a) Calculate the magnitude of the diffusion-controlled rate constant at 298 K for the recombination of two atoms in water, for which $\eta = 0.89$ cP. Assuming the concentration of the reacting species is 1.0×10^{-3} mol L^{-1} initially, how long does it take for the concentration of the atoms to fall to half that value? Assume the reaction is elementary.

27.7 (b) Calculate the magnitude of the diffusion-controlled rate constant at 298 K for the recombination of two atoms in benzene, for which $\eta = 0.601$ cP. Assuming the concentration of the reacting species is 1.8×10^{-3} mol L^{-1} initially, how long does it take for the concentration of the atoms to fall to half that value? Assume the reaction is elementary.

27.8 (a) For the gaseous reaction $A + B \rightarrow P$, the reactive cross-section obtained from the experimental value of the pre-exponential factor is 9.2×10^{-22} m^2. The collision cross-sections of A and B estimated from the transport properties are 0.95 and 0.65 nm^2, respectively. Calculate the P-factor for the reaction.

27.8 (b) For the gaseous reaction $A + B \rightarrow P$, the reactive cross-section obtained from the experimental value of the pre-exponential factor is 8.7×10^{-22} m^2. The collision cross-sections of A and B

estimated from the transport properties are 0.88 and 0.40 nm^2, respectively. Calculate the P-factor for the reaction.

27.9 (a) Two neutral species, A and B, with diameters 588 pm and 1650 pm, respectively, undergo the diffusion-controlled reaction $A + B \rightarrow P$ in a solvent of viscosity 2.37×10^{-3} kg m^{-1} s^{-1} at 40°C. Calculate the initial rate $d[P]/dt$ if the initial concentrations of A and B are 0.150 mol L^{-1} and 0.330 mol L^{-1}, respectively.

27.9 (b) Two neutral species, A and B, with diameters 442 pm and 885 pm, respectively, undergo the diffusion-controlled reaction $A + B \rightarrow P$ in a solvent of viscosity 1.27 cP at 20°C. Calculate the initial rate $d[P]/dt$ if the initial concentrations of A and B are 0.200 mol L^{-1} and 0.150 mol L^{-1}, respectively.

27.10 (a) The reaction of propylxanthate ion in acetic acid buffer solutions has the mechanism $A^- + H^+ \rightarrow P$. Near 30°C the rate constant is given by the empirical expression $k_2 = (2.05 \times 10^{13})e^{-(8681\ K)/T}$ L mol^{-1} s^{-1}. Evaluate the energy and entropy of activation at 30°C.

27.10 (b) The reaction $A^- + H^+ \rightarrow P$ has a rate constant given by the empirical expression $k_2 = (8.72 \times 10^{12})e^{-(6134\ K)/T}$ L mol^{-1} s^{-1}. Evaluate the energy and entropy of activation at 25°C.

27.11 (a) When the reaction in Exercise 27.10a occurs in a dioxane/water mixture which is 30 per cent dioxane by mass, the rate constant fits $k_2 = (7.78 \times 10^{14})e^{-(9134\ K)/T}$ L mol^{-1} s^{-1} near 30°C. Calculate $\Delta^{\ddagger}G$ for the reaction at 30°C.

27.11 (b) A rate constant is found to fit the expression $k_2 = (6.45 \times 10^{13})e^{-(5375\ K)/T}$ L mol^{-1} s^{-1} near 25°C. Calculate $\Delta^{\ddagger}G$ for the reaction at 25°C.

27.12 (a) The gas-phase association reaction between F_2 and IF_5 is first-order in each of the reactants. The energy of activation for the reaction is 58.6 kJ mol^{-1}. At 65°C the rate constant is 7.84×10^{-3} kPa^{-1} s^{-1}. Calculate the entropy of activation at 65°C.

27.12 (b) A gas-phase recombination reaction is first-order in each of the reactants. The energy of activation for the reaction is 49.6 kJ mol^{-1}. At 55°C the rate constant is 0.23 m^3 s^{-1}. Calculate the entropy of activation at 55°C.

27.13 (a) Calculate the entropy of activation for a collision between two structureless particles at 300 K, taking $M = 50$ g mol^{-1} and $\sigma = 0.40$ nm^2.

27.13 (b) Calculate the entropy of activation for a collision between two structureless particles at 500 K, taking $M = 78$ g mol^{-1} and $\sigma = 0.62$ nm^2.

27.14 (a) The pre-exponential factor for the gas-phase decomposition of ozone at low pressures is 4.6×10^{12} L mol^{-1} s^{-1} and its activation energy is 10.0 kJ mol^{-1}. What are (a) the entropy of activation, (b) the enthalpy of activation, (c) the Gibbs energy of activation at 298 K?

27.14 (b) The pre-exponential factor for a gas-phase decomposition of ozone at low pressures is 2.3×10^{13} L mol^{-1} s^{-1} and its activation energy is 30.0 kJ mol^{-1}. What are (a) the entropy of activation, (b) the enthalpy of activation, (c) the Gibbs energy of activation at 298 K?

27.15 (a) The base-catalysed bromination of nitromethane-d_3 in water at room temperature (298 K) proceeds 4.3 times more slowly than the bromination of the undeuterated material. Account for this difference. Use $k_f(CH) = 450\ N\,m^{-1}$.

27.15 (b) Predict the order of magnitude of the isotope effect on the relative rates of displacement of (a) ^1H and ^3H, (b) ^{16}O and ^{18}O. Will raising the temperature enhance the difference? Take $k_f(CH) = 450\ N\,m^{-1}$, $k_f(CO) = 1750\ N\,m^{-1}$.

27.16 (a) The rate constant of the reaction $H_2O_2(aq) +$ $I^-(aq) + H^+(aq) \rightarrow H_2O(l) + HIO(aq)$ is sensitive to the ionic strength of the aqueous solution in which the reaction occurs. At 25°C, $k = 12.2\ L^2\,mol^{-2}\,min^{-1}$ at an ionic strength of 0.0525. Use the Debye–Hückel limiting law to estimate the rate constant at zero ionic strength.

27.16 (b) At 25°C, $k = 1.55\ L^2\,mol^{-3}\,min^{-1}$ at an ionic strength of 0.0241 for a reaction in which the rate-determining step involves the encounter of two singly charged cations. Use the Debye–Hückel limiting law to estimate the rate constant at zero ionic strength.

Problems

Numerical problems

27.1 In the dimerization of methyl radicals at 25°C, the experimental pre-exponential factor is $2.4 \times 10^{10}\ L\,mol^{-1}\,s^{-1}$. What is (a) the reactive cross-section, (b) the P-factor for the reaction if the C–H bond length is 154 pm?

27.2 Nitrogen dioxide reacts bimolecularly in the gas phase to give $2NO + O_2$. The temperature dependence of the second-order rate constant for the rate law $d[P]/dt = k[NO_2]^2$ is given below. What are the P-factor and the reactive cross-section for the reaction?

T/K	600	700	800	1000
$k/(cm^3\,mol^{-1}\,s^{-1})$	4.6×10^2	9.7×10^3	1.3×10^5	3.1×10^6

Take $\sigma = 0.60\ nm^2$.

27.3 The diameter of the methyl radical is about 308 pm. What is the maximum rate constant in the expression $d[C_2H_6]/dt = k[CH_3]^2$ for second-order recombination of radicals at room temperature? 10 per cent of a 1.0 L sample of ethane at 298 K and 100 kPa is dissociated into methyl radicals. What is the minimum time for 90 per cent recombination?

27.4 The rates of thermolysis of a variety of *cis*- and *trans*-azoalkanes have been measured over a range of temperatures in order to settle a controversy concerning the mechanism of the reaction. In ethanol an unstable *cis*-azoalkane decomposed at a rate that was followed by observing the N_2 evolution, and this led to the rate constants listed below (P.S. Engel and D.J. Bishop, *J. Amer. Chem. Soc.* **97**, 6754 (1975)). Calculate the enthalpy, entropy, energy, and Gibbs energy of activation at −20°C.

$\theta/°C$	−24.82	−20.73	−17.02	−13.00	−8.95
$10^4 \times k/s^{-1}$	1.22	2.31	4.39	8.50	14.3

27.5 In an experimental study of a bimolecular reaction in aqueous solution, the second-order rate constant was measured at 25°C and at a variety of ionic strengths and the results are tabulated below. It is known that a singly charged ion is involved in the rate-determining step. What is the charge on the other ion involved?

I	0.0025	0.0037	0.0045	0.0065	0.0085
$k/(L\,mol^{-1}\,s^{-1})$	1.05	1.12	1.16	1.18	1.26

27.6 The rate constant of the reaction $I^-(aq) + H_2O_2(aq) \rightarrow H_2O(l) + IO^-(aq)$ varies slowly with ionic strength, even though the Debye–Hückel limiting law predicts no effect. Use the following data from 25°C to find the dependence of $\log k_r$ on the ionic strength:

I	0.0207	0.0525	0.0925	0.1575
$k_r/(L\,mol^{-1}\,min^{-1})$	0.663	0.670	0.679	0.694

Evaluate the limiting value of k_r at zero ionic strength. What does the result suggest for the dependence of $\log \gamma$ on ionic strength for a neutral molecule in an electrolyte solution?

27.7 The total cross-sections for reactions between alkali metal atoms and halogen molecules are given in the table below (R.D. Levine and R.B. Bernstein, *Molecular reaction dynamics*, Clarendon Press, Oxford, 72 (1974)). Assess the data in terms of the harpoon mechanism.

σ^*/nm^2	Cl_2	Br_2	I_2
Na	1.24	1.16	0.97
K	1.54	1.51	1.27
Rb	1.90	1.97	1.67
Cs	1.96	2.04	1.95

Electron affinities are approximately 1.3 eV (Cl_2), 1.2 eV (Br_2), and 1.7 eV (I_2), and ionization energies are 5.1 eV (Na), 4.3 eV (K), 4.2 eV (Rb), and 3.9 eV (Cs).

Theoretical problems

27.8 Confirm that eqn 40 is a solution of eqn 39, where $[J]_t$ is a solution of the same equation but with $k = 0$ and for the same initial conditions.

27.9 Evaluate $[J]^*$ numerically using mathematical software for integration in eqn 40, and explore the effect of increasing reaction rate constant on the spatial distribution of J.

27.10 Estimate the orders of magnitude of the partition functions involved in a rate expression. State the order of magnitude of q_m^T/N_A, q^R, q^V, q^E for typical molecules. Check that in the collision of two structureless molecules the order of magnitude of the pre-exponential factor is of the same order as that predicted by collision theory. Go on to estimate the P-factor for a reaction in which $A + B \rightarrow P$, and A and B are nonlinear triatomic molecules.

27.11 Use the Debye–Hückel limiting law to show that changes in ionic strength can affect the rate of reaction catalysed by H^+ from the ionization of a weak acid. Consider the mechanism: $H^+(aq) + B(aq) \rightarrow P(aq)$, where H^+ comes from the ionization of the weak acid, HA. The weak acid has a fixed concentration. First show that $\log[H^+]$, derived from the ionization of HA, depends on the activity coefficients of ions and thus depends on the ionic strength. Then find the relationship between $\log(\text{rate})$ and $\log[H^+]$ to show that the rate also depends on the ionic strength.

27.12 The major difficulty in applying activated complex theory (and, it must be admitted, in devising straightforward problems to illustrate it) is to decide on the structure of the activated complex and to ascribe appropriate bond strengths and lengths to it. The following exercise gives some familiarity with the difficulties involved, yet leads to a numerical result for a reaction of some interest. Consider the attack of H on D_2, which is one step in the $H_2 + D_2$ reaction. Suppose that the H atom approaches D_2 from the side and forms a complex in the form of an isosceles triangle. Take the H–D distance as 30 per cent greater than in H_2 (74 pm) and the D–D distance as 20 per cent greater than in H_2. Let the critical coordinate be the antisymmetric stretching vibration in which one H–D bond stretches as the other shortens. Let all the vibrations be at about 1000 cm^{-1}. Estimate k_2 for this reaction at 400 K using the experimental activation energy of about 35 kJ mol^{-1}.

27.13 Now change the model of the activated complex in Problem 27.12 and make it linear. Use the same estimated molecular bond lengths and vibrational frequencies to calculate k_2 for this choice of model.

27.14 Clearly, there is much scope for modifying the parameters of the models of the activated complex in the last pair of problems. Write and run a program that allows you to vary the structure of the complex and the parameters in a plausible way, and look for a model (or more than one model) that gives a value of k close to the experimental value, 4×10^5 L mol^{-1} s^{-1}.

27.15 The Eyring equation can also be applied to physical processes. As an example, consider the rate of diffusion of an atom stuck to the surface of a solid. Suppose that in order to move from one site to another it has to reach the top of the barrier where it can vibrate classically in the vertical direction and in one horizontal direction, but vibration along the other horizontal direction takes it into the neighbouring site. Find an expression for the rate of diffusion, and evaluate it for W atoms on a tungsten surface ($E_a = 60$ kJ mol^{-1}). Suppose that the vibration frequencies at the transition state are (a) the same as, (b) one-half the value for the adsorbed atom. What is the value of the diffusion coefficient D at 500 K? (Take the site separation as 316 pm and $\nu = 1 \times 10^{11}$ Hz.)

27.16 Suppose now that the adsorbed, migrating species treated in Problem 27.15 is a spherical molecule, and that it can rotate classically as well as vibrate at the top of the barrier, but that at the adsorption site itself it can only vibrate. What effect does this have on the diffusion constant? Take the molecule to be methane, for which $B = 5.24$ cm^{-1}.

27.17 Show that the intensities of a molecular beam before and after passing through a chamber of length l containing inert scattering atoms are related by $I = I_0 e^{-\mathcal{N}\sigma l}$, where σ is the collision cross-section and $\mathcal{N}$ the number density of scattering atoms.

27.18 In a molecular beam experiment to measure collision cross-sections it was found that the intensity of a CsCl beam was reduced to 60 per cent of its intensity on passage through CH_2F_2 at 10 μTorr, but that when the target was Ar at the same pressure the intensity was reduced only by 10 per cent. What are the relative cross-sections of the two types of collision? Why is one much larger than the other?

Additional problems supplied by Carmen Giunta and Charles Trapp

27.19 T. Gierczak, R.K. Talukdar, S.C. Herndon, G.L. Vaghjiani, and A.R. Ravishankara (*J. Phys. Chem.* **A 101**, 3125 (1997)) measured the rate constants for the bimolecular gas-phase reaction of methane with the hydroxyl radical in several isotopic variations. From their data, the following Arrhenius parameters can be obtained.

	$A/(\text{L mol}^{-1}\,\text{s}^{-1})$	$E_a/(\text{kJ mol}^{-1})$
$CH_4 + OH \rightarrow CH_3 + H_2O$	1.13×10^9	14.1
$CD_4 + OH \rightarrow CD_3 + DOH$	6.0×10^8	17.5
$CH_4 + OD \rightarrow CH_3 + DOH$	1.01×10^9	13.6

Compute the rate constants at 298 K, and interpret the kinetic isotope effects.

27.20 R. Atkinson (*J. Phys. Chem. Ref. Data* **26**, 215 (1997)) has reviewed a large set of rate constants relevant to the atmospheric chemistry of volatile organic compounds. The recommended rate constant for the bimolecular association of O_2 with an alkyl radical R at 298 K is 4.7×10^9 L mol^{-1} s^{-1} for R = C_2H_5 and 8.4×10^9 L mol^{-1} s^{-1} for R = cyclohexyl. Assuming no energy barrier, compute the steric factor, P, for each reaction. *Hint.* Obtain collision diameters from collision cross-sections of similar molecules in the *Data section*.

27.21 M. Cyfert, B. Latko, and M. Wawrzeczyk (*Int. J. Chem. Kinet.* **28**, 103 (1996)) examined the oxidation of tris(1,10-phenanthroline)iron(II) by periodate in aqueous solution, a reaction that shows autocatalytic behaviour. To assess the kinetic salt effect, they measured rate constants at a variety of concentrations of Na_2SO_4 far in excess of reactant concentrations and reported the following data.

$[Na_2SO_4]/(\text{mol kg}^{-1})$	0.2	0.15	0.1	0.05
$k/(\text{L}^{1/2}\,\text{mol}^{-1/2}\,\text{s}^{-1})$	0.462	0.430	0.390	0.321

$[Na_2SO_4]/(\text{mol kg}^{-1})$	0.25	0.0125	0.005
$k/(\text{L}^{1/2}\,\text{mol}^{-1/2}\,\text{s}^{-1})$	0.283	0.252	0.224

What can be inferred about the charge of the activated complex of the rate-determining step?

27.22 R.H. Bisby and A.W. Parker (*J. Amer. Chem. Soc.* **117**, 5664 (1995)) studied the reaction of photochemically excited duroquinone with the antioxidant α-tocopherol in ethanol. Once the duroquinone was photochemically excited, a bimolecular reaction took place at a rate described as diffusion-limited. (a) Estimate the rate constant for a diffusion-limited reaction in ethanol. (b) The reported rate constant was 2.77×10^9 L mol^{-1} s^{-1}; estimate the critical reaction distance if the sum of diffusion constants is 1×10^{-9} m^2 s^{-1}.

27.23 For the thermal decomposition of F_2O by the reaction $2F_2O(g) \rightarrow 2F_2(g) + O_2(g)$, J. Czarnowski and H.J. Schuhmacher (*Chem. Phys. Lett.* **17**, 235 (1972)) have suggested the following mechanism:

(1) $F_2O + F_2O \longrightarrow F + OF + F_2O$ k_1
(2) $F + F_2O \longrightarrow F_2 + OF$ k_2
(3) $OF + OF \longrightarrow O_2 + F + F$ k_3
(4) $F + F + F_2O \longrightarrow F_2 + F_2O$ k_4

(a) Using the steady-state approximation, show that this mechanism is consistent with the experimental rate law $-d[F_2O]/dt = k[F_2O]^2 + k'[F_2O]^{3/2}$. (b) The experimentally determined Arrhenius parameters in the range 501–583 K are $A = 7.8 \times 10^{13}$ L mol^{-1} s^{-1}, $E_a/R = 1.935 \times 10^4$ K for k and $A = 2.3 \times 10^{10}$ L mol^{-1} s^{-1}, $E_a/R = 1.691 \times 10^4$ K for k'. At 540 K, $\Delta_f H^{\ominus}(F_2O) = +24.41$ kJ mol^{-1}, $D(F—F) = 160.6$ kJ mol^{-1}, and $D(O—O) = 498.2$ kJ mol^{-1}. Estimate the bond dissociation energies of the first and second F–O bonds and the Arrhenius activation energy of reaction 2.

27.24 Show that bimolecular reactions between nonlinear molecules are much slower than those between atoms even when the activation energies of both reactions are equal. Use activated complex theory and make the following assumptions. (1) All vibrational partition functions are close to unity; (2) all rotational partition functions are approximately $1 \times 10^{1.5}$, which is a reasonable order of magnitude number; (3) the translational partition function for each species is 1×10^{26}.

27.25 For the gas-phase reaction $A + A \rightarrow A_2$, the experimental rate constant, k_2, has been fitted to the Arrhenius equation with the preexponential factor $A = 4.07 \times 10^5$ L mol^{-1} s^{-1} at 300 K and an activation energy of 65.43 kJ mol^{-1}. Calculate $\Delta^{\ddagger}S$, $\Delta^{\ddagger}H$, $\Delta^{\ddagger}U$, and $\Delta^{\ddagger}G$ for the reaction.

27.26 One of the most historically significant studies of chemical reaction rates was that by M. Bodenstein (*Z. physik. Chem.* **29**, 295 (1899)) of the gas-phase reaction $2HI(g) \rightarrow H_2(g) + I_2(g)$ and its reverse, with rate constants k and k', respectively. The measured rate constants as a function of temperature are

T/K	647	666	683	700
$k/(22.4$ L mol^{-1} min$^{-1})$	0.230	0.588	1.37	3.10
$k'/(22.4$ L mol^{-1} min$^{-1})$	0.0140	0.0379	0.0659	0.172
T/K	716	781		
$k/(22.4$ L mol^{-1} min$^{-1})$	6.70	105.9		
$k'/(22.4$ L mol^{-1} min$^{-1})$	0.375	3.58		

Demonstrate that these data are consistent with the collision theory of bimolecular gas-phase reactions.

28 Processes at solid surfaces

In this chapter we see how solids grow at their surfaces and how the details of the structure and composition of solid surfaces can be determined experimentally. A major part of the material concerns the extent to which a surface is covered and the variation of the extent of coverage with the pressure and temperature. This material is used to discuss how surfaces affect the rate and course of chemical change by acting as the site of catalysis.

Processes at solid surfaces govern the viability of industry both constructively, as in catalysis, and destructively, as in corrosion. Chemical reactions at solid surfaces may differ sharply from reactions in the bulk, for reaction pathways of much lower activation energy may be provided, and hence result in catalysis. The concept of a solid surface has been extended in recent years with the availability of microporous aluminosilicates, the zeolites, in which the surface effectively extends deep inside the solid (Fig. 28.1). Most of the catalysts used in the petrochemical industry are now zeolites.

We shall see that acronyms are widely used in surface studies; for convenience, a list of the acronyms used in this text is given at the end of the chapter.

The growth and structure of solid surfaces

In this section we see how surfaces are extended and crystals grow. The attachment of particles to a surface is called **adsorption**. The substance that adsorbs is the **adsorbate** and the underlying material that we are concerned with in this section is the **adsorbent** or **substrate**. The reverse of adsorption is **desorption**.

28.1 Surface growth

A simple picture of a perfect crystal surface is as a tray of oranges in a grocery store (Fig. 28.2). A gas molecule that collides with the surface can be imagined as a ping-pong ball bouncing erratically over the oranges. The molecule loses energy as it bounces, but it is likely to escape from the surface before it has lost enough kinetic energy to be trapped. The same

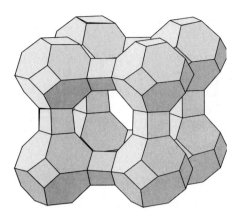

28.1 A framework representation showing the general layout of the atoms in a zeolite material. Note the truncated octahedra, the small cubic cages, and the large central cage.

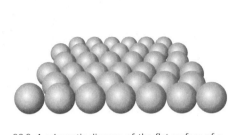

28.2 A schematic diagram of the flat surface of a solid. This primitive model is largely supported by surface-tunnelling microscope images (see Fig. 28.20).

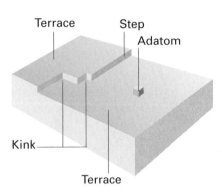

28.3 Some of the kinds of defects that may occur on otherwise perfect terraces. Defects play an important role in surface growth and catalysis.

is true, to some extent, of an ionic crystal in contact with a solution. There is little energy advantage for an ion in solution to discard some of its solvating molecules and stick at an exposed position on the surface.

(a) The role of defects

The picture changes when the surface has defects, for then there are ridges of incomplete layers of atoms or ions. A typical type of surface defect is a **step** between two otherwise flat layers of atoms called **terraces** (Fig. 28.3). A step defect might itself have defects for it might have kinks. When an atom settles on a terrace it bounces across it under the influence of the intermolecular potential, and might come to a step or a corner formed by a kink. Instead of interacting with a single terrace atom, the molecule now interacts with several, and the interaction may be strong enough to trap it. Likewise, when ions deposit from solution, the loss of the solvation interaction is offset by a strong Coulombic interaction between the arriving ions and several ions at the surface defect.

(b) Dislocations

Not all kinds of defect result in sustained surface growth. As the process of settling into ledges and kinks continues, there comes a stage when an entire lower terrace has been covered. At this stage the surface defects have been eliminated, and growth will cease. For continuing growth, a surface defect is needed that propagates as the crystal grows. We can see what form of defect this must be by considering the types of **dislocations**, or discontinuities in the regularity of the lattice, that exist in the bulk of a crystal. One reason for their formation may be that the crystal grows so quickly that its particles do not have time to settle into states of lowest potential energy before being trapped in position by the deposition of the next layer.

A special kind of dislocation is the **screw dislocation** shown in Fig. 28.4. Imagine a cut in the crystal, with the particles to the left of the cut pushed up through a distance of one unit cell. The unit cells now form a continuous spiral around the end of the cut, which is called the **screw axis**. A path encircling the screw axis spirals up to the top of the crystal, and where the dislocation breaks through to the surface it takes the form of a spiral ramp.

The surface defect formed by a screw dislocation is a step, possibly with kinks, where growth can occur. The incoming particles lie in ranks on the ramp, and successive ranks reform the step at an angle to its initial position. As deposition continues the step rotates around the screw axis, and is not eliminated. Growth may therefore continue indefinitely.

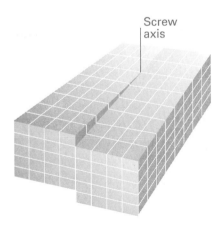

28.4 A screw dislocation occurs where one region of the crystal is pushed up through one or more unit cells relative to another region. The cut extends to the screw axis. As atoms lie along the step, the dislocation rotates round the screw axis, and is not annihilated.

28.5 The spiral growth arising from the propagation of a screw axis is clearly visible in this photograph of an alkane crystal. (B.R. Jennings and V.J. Morris, *Atoms in contact*, Clarendon Press, Oxford (1974); courtesy of Dr A.J. Forty.)

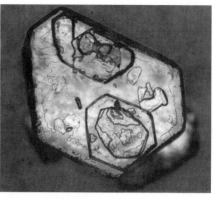

28.6 The spiral growth pattern is sometimes concealed because the terraces are subsequently completed by further deposition. This accounts for the appearance of this cadmium iodide crystal. (H.M. Rosenberg, *The solid state*, Clarendon Press, Oxford (1978).)

Several layers of deposition may occur, and the edges of the spirals might be cliffs several atoms high (Fig. 28.5).

Propagating spiral edges can also give rise to flat terraces (Fig. 28.6). Terraces are formed if growth occurs simultaneously at neighbouring left- and right-handed screw dislocations (Fig. 28.7). Successive tables of atoms may form as counter-rotating defects collide on successive circuits, and the terraces formed may then fill up by further deposition at their edges to give flat crystal planes.

The rapidity of growth depends on the crystal plane concerned, and the slowest growing faces dominate the appearance of the crystal. This feature is explained in Fig. 28.8, where we see that, although the horizontal face grows forward most rapidly, it grows itself out of existence, and the slower-growing faces survive.

28.2 Surface composition

Under normal conditions, a surface exposed to a gas is constantly bombarded with molecules and a freshly prepared surface is covered very quickly. Just how quickly can be estimated using the kinetic theory of gases and the expression (eqn 24.3) for the collision flux:

$$Z_W = \frac{p}{(2\pi mkT)^{1/2}} \qquad (1a)$$

A practical form of this equation is

$$Z_W = \frac{Z_0(p/\text{Pa})}{\{(T/\text{K})(M/(\text{g mol}^{-1}))\}^{1/2}} \qquad Z_0 = 2.63 \times 10^{24} \text{ m}^{-2}\text{ s}^{-1} \qquad (1b)$$

where M is the molar mass of the gas. For air ($M \approx 29 \text{ g mol}^{-1}$) at 1 atm and 25°C the collision flux is $3 \times 10^{27} \text{ m}^{-2}\text{ s}^{-1}$. Because 1 m^2 of metal surface consists of about 10^{19} atoms, each atom is struck about 10^8 times each second. Even if only a few collisions leave a molecule adsorbed to the surface, the time for which a freshly prepared surface remains clean is very short.

(a) High-vacuum techniques

The obvious way to retain cleanliness is to reduce the pressure. When it is reduced to 10^{-4} Pa (as in a simple vacuum system) the collision flux falls to about $10^{18} \text{ m}^{-2}\text{ s}^{-1}$, corresponding to one hit per surface atom in each 0.1 s. Even that is too brief in most experiments, and in **ultra-high vacuum** (UHV) techniques pressures as low as 10^{-7} Pa (when

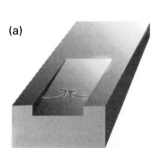

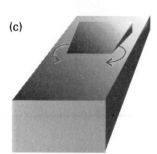

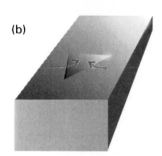

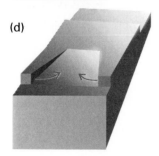

28.7 Counter-rotating screw dislocations on the same surface lead to the formation of terraces. Four stages of one cycle of growth are shown here. Subsequent deposition can complete each terrace.

(a)

(b)

(c)

(d)

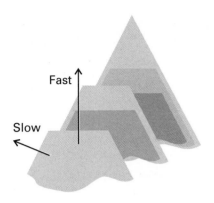

28.8 The slower-growing faces of a crystal dominate its final external appearance. Three successive stages of the growth are shown.

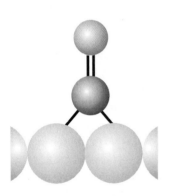

1 Bridge site

$Z_W = 10^{15}$ m^{-2} s^{-1}) are reached on a routine basis and 10^{-9} Pa (when $Z_W = 10^{13}$ m^{-2} s^{-1}) are reached with special care. These collision fluxes correspond to each surface atom being hit once every 10^5 to 10^6 s, or about once a day.

The layout of a typical UHV apparatus is such that the whole of the evacuated part can be heated to 150–250°C for several hours to drive gas molecules from the walls. All the taps and seals are usually of metal so as to avoid contamination from greases. The sample is usually in the form of a thin foil, a filament, or a sharp point. Where there is interest in the role of specific crystal planes the sample is a single crystal with a freshly cleaved face. Initial surface cleaning is achieved either by heating it electrically or by bombarding it with accelerated gaseous ions. The latter procedure demands care because ion bombardment can shatter the surface structure and leave it an amorphous jumble of atoms. High-temperature annealing is then required to return the surface to an ordered state.

(b) Ionization techniques

Surface composition can be determined by a variety of ionization techniques. The same techniques can be used to detect any remaining contamination after cleaning and to detect layers of material adsorbed later in the experiment. Their common feature is that the **escape depth** of the electrons, the maximum depth from which ejected electrons come, is in the range 0.1–1.0 nm, which ensures that only surface species contribute.

One technique that may be used is photoelectron spectroscopy (Section 17.8), which in surface studies is normally called **photoemission spectroscopy**. X-rays or hard ultraviolet ionizing radiation of energy in the range 5–40 eV may be used, giving rise to the techniques denoted XPS and UPS, respectively. XPS, which examines inner-shell binding energies, is able to fingerprint the materials present (Fig. 28.9). UPS, which examines electrons ejected from valence shells, is more suited to establishing the bonding characteristics and the details of valence shell electronic structures of substances on the surface. Its usefulness is its ability to reveal which orbitals of the adsorbate are involved in the bond to the substrate. For instance,

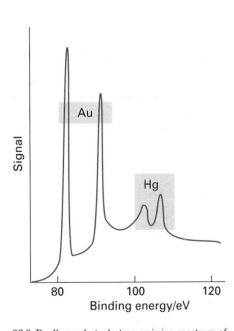

28.9 The X-ray photoelectron emission spectrum of a sample of gold contaminated with a surface layer of mercury. (M.W. Roberts and C.S. McKee, *Chemistry of the metal–gas interface*, Oxford (1978).)

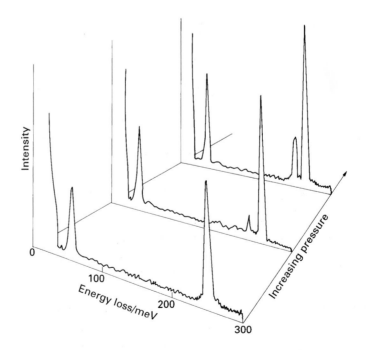

28.10 The electron energy-loss spectrum of CO adsorbed on Pt(111). The results for three different pressures are shown, and the growth of the additional peak at about 200 meV (1600 cm^{-1}) should be noted. (Spectra provided by Professor H. Ibach.)

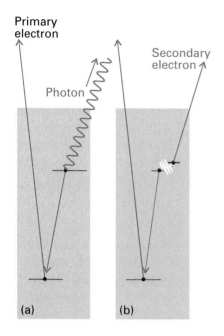

28.11 When an electron is expelled from a solid, (a) an electron of higher energy may fall into the vacated orbital and emit an X-ray photon to produce X-ray fluorescence. Alternatively, (b) the electron falling into the orbital may give up its energy to another electron, which is ejected in the Auger effect.

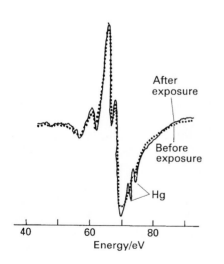

28.12 An Auger spectrum of the same sample used for Fig. 28.9 taken before and after deposition of mercury. (M.W. Roberts and C.S. McKee, *Chemistry of the metal–gas interface*, Oxford (1978).)

the principal difference between the photoemission results on free benzene and benzene adsorbed on palladium is in the energies of the π electrons. This difference is interpreted as meaning that the C_6H_6 molecules lie parallel to the surface and are attached to it by their π orbitals. In contrast, pyridine (C_6H_5N) stands more or less perpendicular to the surface, and is attached by a σ bond formed by the nitrogen lone pair.

(c) Energy-loss spectroscopy

Several kinds of vibrational spectroscopy have been developed to study adsorbates and to show whether dissociation has occurred. Infrared and Raman spectroscopy have been greatly improved by the development of laser and Fourier transform techniques, and have largely overcome the difficulties arising from low intensities on account of the low surface coverages normally encountered under laboratory conditions. **Surface-enhanced Raman scattering** (SERS) makes the Raman technique viable for surface studies despite its intrinsic weakness: the strong enhancement, which can reach a factor of 10^6 on roughened silver surfaces, is due in part to local accumulations of electron density at the features of the roughened surface and at regions where bonding occurs. Raman intensities remain a problem on flat single crystal surfaces, and SERS works for only certain metals.

A hybrid version of photoemission spectroscopy and vibrational spectroscopy is **electron energy-loss spectroscopy** (EELS, or HREELS, where HR denotes high resolution) in which the energy loss suffered by a beam of electrons is monitored when they are reflected from a surface. As in optical Raman spectroscopy, the spectrum of energy loss can be interpreted in terms of the vibrational spectrum of the adsorbate. High resolution and sensitivity are attainable, and the technique is sensitive to light elements (to which X-ray techniques are insensitive). Very tiny amounts of adsorbate can be detected, and one report estimated that about 48 atoms of phosphorus were detected in one sample. As an example, Fig. 28.10 shows the EELS result for CO on the (111) face of a platinum crystal as the extent of surface coverage increases. The main peak arises from CO attached perpendicular to the surface by a single Pt atom. As the coverage increases the neighbouring smaller peak increases in intensity. This peak is due to CO at a bridge site, attached to two Pt atoms, as in (1).

(d) Auger electron spectroscopy

A very important technique, which is widely used in the microelectronics industry, is **Auger electron spectroscopy** (AES). The **Auger effect** is the emission of a second electron after high-energy radiation has expelled another. The first electron to depart leaves a hole in a low-lying orbital, and an electron from a higher energy orbital falls into it. The energy this releases may result either in the generation of radiation, which is called **X-ray fluorescence** (Fig. 28.11a) or in the ejection of another electron (Fig. 28.11b). The latter is the secondary electron of the Auger effect. The energies of the secondary electrons are characteristic of the material present, and so the Auger effect effectively takes a fingerprint of the sample (Fig. 28.12). In practice, the Auger spectrum is normally obtained by irradiating the sample with an electron beam of energy in the range 1–5 keV rather than with electromagnetic radiation. In **scanning Auger electron microscopy** (SAM), the finely focused electron beam is scanned over the surface and a map of composition is compiled; the resolution can reach below about 50 nm.

The technique known as **surface-extended X-ray absorption fine structure spectroscopy** (SEXAFS) makes use of the intense X-radiation from synchrotron sources (Section 16.1a). The technique makes use of the oscillations in X-ray absorbance that are observed on the high-frequency side of the absorption edge (the start of an X-ray absorption band) of a substance. Oscillations arise from a quantum mechanical interference between the wavefunction of a photoejected electron and parts of that electron's

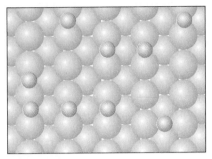

(a)

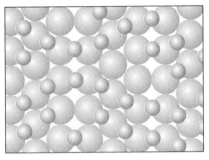

(b)

28.13 The restructuring of a surface that sometimes accompanies chemisorption. The diagrams illustrate the changes that occur when CO adsorbs on the (111) face of palladium. (a) Up to 0.3 monolayers, the surface structure is unchanged. (b) A metastable structure produced at low temperature when the surface is saturated.

wavefunction that are scattered by neighbouring atoms. If the waves interfere destructively, the photoelectron appears with lower probability and the X-ray absorption is correspondingly less. If the waves interfere constructively, then the photoelectron amplitude is higher, and the photoelectron has a higher probability of appearing; correspondingly, the X-ray absorption is greater. The oscillations therefore contain information about the number and distances of the neighbouring atoms. Such studies show that a solid's surface is much more plastic than had previously been thought, and that it undergoes **reconstruction**, or structural modification, in response to adsorbates that are present. An example is given in Fig. 28.13, which shows how a palladium surface is reconstructed in different ways and to different extents at different temperatures and with varying surface coverage of CO.

(e) Low-energy electron diffraction

One of the most informative techniques for determining the arrangement of the atoms close to the surface is **low-energy electron diffraction** (LEED). This technique is essentially electron diffraction (Section 21.10), but the sample is now the surface of a solid. The use of low-energy electrons (with energies in the range 10–200 eV, corresponding to wavelengths in the range 100–400 pm) ensures that the diffraction is caused only by atoms on and close to the surface. The experimental arrangement is shown in Fig. 28.14, and typical LEED patterns, obtained by photographing the fluorescent screen through the viewing port, are shown in Fig. 28.15.

The LEED pattern portrays the two-dimensional structure of the surface. By studying how the diffraction intensities depend on the energy of the electron beam it is also possible to infer some details about the vertical location of the atoms and to measure the thickness of the surface layer, but the interpretation of LEED data is much more complicated than the interpretation of bulk X-ray data. The pattern is sharp if the surface is well-ordered for distances long compared with the wavelength of the incident electrons. In practice, sharp patterns are obtained for surfaces ordered to depths of about 20 nm and more. Diffuse patterns indicate either a poorly ordered surface or the presence of impurities. If the LEED pattern does not correspond to the pattern expected by extrapolation of the bulk surface to the surface, then either a reconstruction of the surface has occurred or there is order in the arrangement of an adsorbed layer.

LEED experiments show that the surface of a crystal rarely has exactly the same form as a slice through the bulk. As a general rule, it is found that metal surfaces are simply truncations of the bulk lattice, but the distance between the top layer of atoms and the one

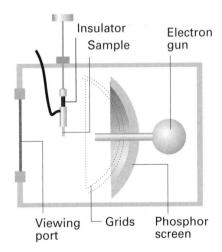

28.14 A schematic diagram of the apparatus used for a LEED experiment. The electrons diffracted by the surface layers are detected by the fluorescence they cause on the phosphor screen.

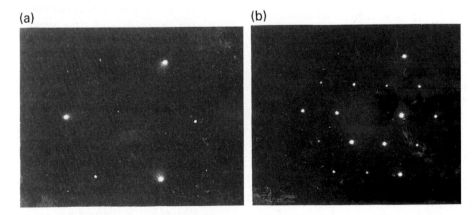

(a)   (b)

28.15 LEED photographs of (a) a clean platinum surface and (b) after its exposure to propyne, $CH_3C{\equiv}CH$. (Photographs provided by Professor G.A. Samorjai.)

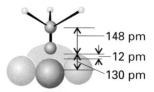

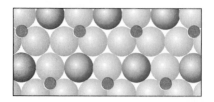

28.16 The structure of a surface close to the point of attachment of CH_3C- to the (111) surface of rhodium at 300 K and the changes in positions of the metal atoms that accompany chemisorption.

below is contracted by around 5 per cent. Semiconductors generally have surfaces reconstructed to a depth of several layers. Reconstruction occurs in ionic solids. For example, in lithium fluoride the Li^+ and F^- ions close to the surface apparently lie on slightly different planes. An actual example of the detail that can now be obtained from refined LEED techniques is shown in Fig. 28.16 for CH_3C- adsorbed on a (111) plane of rhodium.

A simple notation has been developed to describe the arrangement of atoms at a surface. An arrangement of atoms corresponding to the bulk unit cell is called the **substrate structure** and is designated (1×1). Thus, the substrate structure of the (111) face of a metal M would be designated $M(111)-(1 \times 1)$. An adsorbed species A that has the same structure would be denoted $(1 \times 1)-A$. Therefore, a monolayer of oxygen atoms on the (111) face of silicon would be denoted $Si(111)-(1 \times 1)-O$.

A surface structure with a unit cell side that is twice as large as the unit cell side of the substrate surface is denoted (2×2). In general, the relative size would be denoted $(n \times m)$, as illustrated for two common cases in Fig. 28.17. In a number of cases, the surface atoms form a lattice that is rotated with respect to the substrate. A surface structure rotated by 45°, for instance, would be denoted $R45°$, as in $(2^{1/2} \times 2^{1/2})R45°$.

The presence of terraces, steps, and kinks in a surface shows up in LEED patterns, and their surface density (the number of defects in a region divided by the area of the region) can be estimated. The importance of this type of measurement will emerge later. Three examples of how steps and kinks affect the pattern are shown in Fig. 28.18. The samples used were obtained by cleaving a crystal at different angles to a plane of atoms. Only terraces are produced when the cut is parallel to the plane, and the density of steps increases as the angle

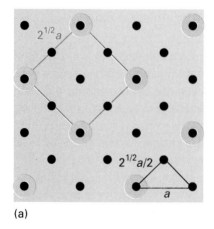

(a)

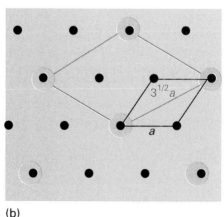

(b)

28.17 The designation of two adsorbed surface layers: (a) $(2^{1/2} \times 2^{1/2})R45°$, (b) $(3^{1/2} \times 3^{1/2})R30°$.

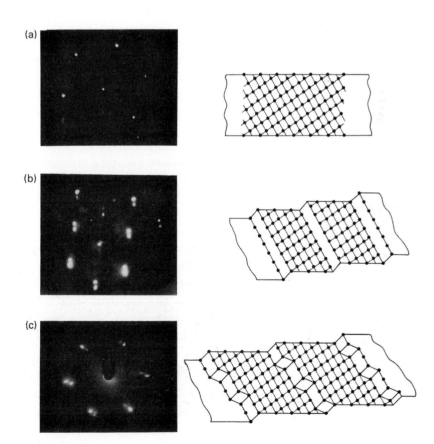

(a)

(b)

(c)

28.18 LEED patterns may be used to assess the defect density of a surface. The photographs correspond to a platinum surface with (a) low defect density, (b) regular steps separated by about six atoms, and (c) regular steps with kinks. (Photographs provided by Professor G.A. Samorjai.)

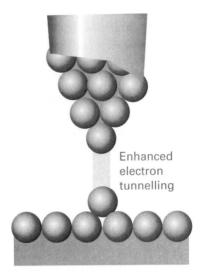

Enhanced electron tunnelling

28.19 A representation of the the tip of a scanning tunnelling microscope and the surface above which it lies. The tunnelling probability for electrons varies exponentially with the separation of the tip from the surface features, so the current (shown here by the green band) is representative of atomic-scale variations of the topography of the surface.

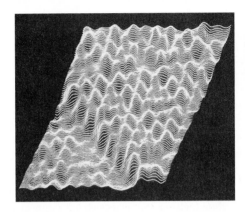

28.20 The type of image that can be obtained with scanning tunnelling microscopy. The sample is of a silicon surface and the cliff is one atom high. (Sang-il Park and C.F. Quaite.)

28.21 A scanning tunnelling microscope image of a liquid-crystal molecule (5-nonyl-2-nonoxylphenylpyrimidine) adsorbed on a graphite surface. (J.S. Foster, *et al.*, *Nature* **338**, 137 (1988).)

of the cut increases. The observation of additional structure in the LEED patterns, rather than blurring, shows that the steps are arrayed regularly.

(f) Scanning tunnelling and atomic force microscopy

The central component in **scanning tunnelling microscopy** (STM) is a platinum–rhodium or tungsten needle, which is scanned across the surface of a conducting solid. When the tip of the needle is brought very close to the surface, electrons tunnel across the intervening space (Fig. 28.19). In the constant-current mode of operation, the stylus moves up and down corresponding to the form of the surface, and the topography of the surface, including any adsorbates, can be mapped on an atomic scale. The vertical motion of the stylus is achieved by fixing it to a piezoelectric cylinder, which contracts or expands according to the potential difference it experiences. In the constant-z mode, the vertical position of the stylus is held constant and the current is monitored. Because the tunnelling probability is very sensitive to the size of the gap (Section 12.3), the microscope can detect tiny, atomic-scale variations in the height of the surface. An example of the kind of image obtained with a clean surface is shown in Fig. 28.20, where the cliff is only one atom high. A spectacular demonstration of the power of the technique for displaying the shape of adsorbed species is shown in Fig. 28.21, which shows a single liquid-crystal molecule adsorbed to a surface.

In **atomic force microscopy** (AFM) a sharpened stylus attached to a beam is scanned across the surface (Fig. 28.22). The force exerted by the surface and any adsorbate pushes or pulls on the stylus and deflects the beam. The deflection is monitored either by interferometry or by using a laser beam. Because no current is needed between the sample and the probe, the technique can be applied to nonconducting surfaces too.

(g) Molecular beam techniques

Whereas many important studies have been carried out simply by exposing a surface to a gas, modern work is increasingly making use of **molecular beam scattering** (MBS). One advantage is that the activity of specific crystal planes can be investigated by directing the beam on to an orientated surface with known step and kink densities (as measured by LEED). Furthermore, if the adsorbate reacts at the surface, the products (and their angular distributions) can be analysed as they are ejected from the surface and pass into a mass

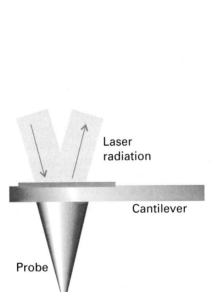

Laser radiation

Cantilever

Probe

Surface

28.22 A representation of the probe of an atomic force microscope. The probe tapers to a single atom; the deflection of the cantilever beam is monitored by using laser radiation reflected from a small mirror.

spectrometer. Another advantage is that the time of flight of a particle may be measured and interpreted in terms of its residence time on the surface. In this way a very detailed picture can be constructed of the events taking place during reactions at surfaces.

The extent of adsorption

The extent of surface coverage is normally expressed as the **fractional coverage**, θ:

$$\theta = \frac{\text{number of adsorption sites occupied}}{\text{number of adsorption sites available}} \qquad [2]$$

The fractional coverage is often expressed in terms of the volume of adsorbate adsorbed by $\theta = V/V_\infty$, where V_∞ is the volume of adsorbate corresponding to complete monolayer coverage. The **rate of adsorption**, $d\theta/dt$, is the rate of change of surface coverage, and can be determined by observing the change of fractional coverage with time.

Among the principal techniques for measuring $d\theta/dt$ are flow methods, in which the sample itself acts as a pump because adsorption removes particles from the gas. One commonly used technique is therefore to monitor the rates of flow of gas into and out of the system: the difference is the rate of gas uptake by the sample. Integration of this rate then gives the fractional coverage at any stage. In **flash desorption** the sample is suddenly heated (electrically) and the resulting rise of pressure is interpreted in terms of the amount of adsorbate originally on the sample. The interpretation may be confused by the desorption of a compound (for example, WO_3 from oxygen on tungsten). **Gravimetry**, in which the sample is weighed on a microbalance during the experiment, can also be used, as can radioactive tracers. In the latter case, the radioactivity of the sample is measured after exposure to an isotopically labelled gas.

28.3 Physisorption and chemisorption

Molecules and atoms can attach to surfaces in two ways. In **physisorption** (an abbreviation of 'physical adsorption'), there is a van der Waals interaction (for example, a dispersion or a dipolar interaction) between the adsorbate and the substrate. Van der Waals interactions have a long range but are weak, and the energy released when a particle is physisorbed is of the same order of magnitude as the enthalpy of condensation. Such small energies can be absorbed as vibrations of the lattice and dissipated as thermal motion, and a molecule bouncing across the surface will gradually lose its energy and finally adsorb to it in the process called **accommodation**. The enthalpy of physisorption can be measured by monitoring the rise in temperature of a sample of known heat capacity, and typical values are in the region of 20 kJ mol^{-1} (Table 28.1). This small enthalpy change is insufficient to lead to bond breaking, so a physisorbed molecule retains its identity, although it might be distorted by the presence of the surface.

(a) Chemisorption

In **chemisorption** (an abbreviation of 'chemical adsorption'), the molecules (or atoms) stick to the surface by forming a chemical (usually covalent) bond, and tend to find sites that maximize their coordination number with the substrate. The enthalpy of chemisorption is very much greater than that for physisorption, and typical values are in the region of 200 kJ mol^{-1} (Table 28.2). The distance between the surface and the closest adsorbate atom is also typically shorter for chemisorption than for physisorption. A chemisorbed molecule may be torn apart at the demand of the unsatisfied valencies of the surface atoms, and the existence of molecular fragments on the surface as a result of chemisorption is one reason why solid surfaces catalyse reactions.

Table 28.1* Maximum observed enthalpies of physisorption, $\Delta_{ad}H^{\ominus}/(\text{kJ mol}^{-1})$

CH_4	−21
H_2	−84
H_2O	−59
N_2	−21

*More values are given in the *Data section* at the end of this volume.

Table 28.2* Enthalpies of chemisorption, $\Delta_{ad}H^{\ominus}/(\text{kJ mol}^{-1})$

Adsorbate	Adsorbent (substrate)		
	Cr	Fe	Ni
C_2H_4	−427	−285	−243
CO		−192	
H_2	−188	−134	
NH_3		−188	−155

*More values are given in the *Data section*.

Except in special cases, chemisorption must be exothermic. A spontaneous process requires $\Delta G < 0$. Because the translational freedom of the adsorbate is reduced when it is adsorbed, ΔS is negative. Therefore, in order for $\Delta G = \Delta H - T\Delta S$ to be negative, ΔH must be negative (that is, the process is exothermic). Exceptions may occur if the adsorbate dissociates and has high translational mobility on the surface. For example, H_2 adsorbs endothermically on glass because there is a large increase of translational entropy accompanying the dissociation of the molecules into atoms that move quite freely over the surface. In its case, the entropy change in the process $H_2(g) \rightarrow 2H(glass)$ is sufficiently positive to overcome the small positive enthalpy change.

The principal test for distinguishing chemisorption from physisorption used to be the enthalpy of adsorption. Values less negative than $-25 \ kJ \ mol^{-1}$ were taken to signify physisorption, and values more negative than about $-40 \ kJ \ mol^{-1}$ were taken to signify chemisorption. However, this criterion is by no means foolproof, and spectroscopic techniques that identify the adsorbed species are now available.

The enthalpy of adsorption depends on the extent of surface coverage, mainly because the adsorbate particles interact. If the particles repel each other (as for CO on palladium) the adsorption becomes less exothermic (the enthalpy of adsorption less negative) as coverage increases. Moreover, LEED studies show that such species settle on the surface in a disordered way until packing requirements demand order. If the adsorbate particles attract one another (as for O_2 on tungsten), then they tend to cluster together in islands, and growth occurs at the borders. These adsorbates also show order–disorder transitions when they are heated enough for thermal motion to overcome the particle–particle interactions, but not so much that they are desorbed.

28.4 Adsorption isotherms

The free gas and the adsorbed gas are in dynamic equilibrium, and the fractional coverage of the surface depends on the pressure of the overlying gas. The variation of θ with pressure at a chosen temperature is called the **adsorption isotherm**.

(a) The Langmuir isotherm

The simplest physically plausible isotherm is based on three assumptions:

1. Adsorption cannot proceed beyond monolayer coverage.
2. All sites are equivalent and the surface is uniform (that is, the surface is perfectly flat on a microscopic scale).
3. The ability of a molecule to adsorb at a given site is independent of the occupation of neighbouring sites.

The dynamic equilibrium is

$$A(g) + M(surface) \rightleftharpoons AM(surface)$$

with rate constants k_a for adsorption and k_d for desorption. The rate of change of surface coverage due to adsorption is proportional to the partial pressure p of A and the number of vacant sites $N(1 - \theta)$, where N is the total number of sites:

$$\frac{d\theta}{dt} = k_a p N (1 - \theta) \tag{3}$$

The rate of change of θ due to desorption is proportional to the number of adsorbed species, $N\theta$:

$$\frac{d\theta}{dt} = -k_d N \theta \tag{4}$$

At equilibrium there is no net change (that is, the sum of these two rates is zero), and solving for θ gives the **Langmuir isotherm**:

$$\theta = \frac{Kp}{1 + Kp} \qquad K = \frac{k_a}{k_d} \tag{5}$$

Example 28.1 Using the Langmuir isotherm

The data given below are for the adsorption of CO on charcoal at 273 K. Confirm that they fit the Langmuir isotherm, and find the constant K and the volume corresponding to complete coverage. In each case V has been corrected to 1.00 atm.

p/Torr	100	200	300	400	500	600	700
V/cm^3	10.2	18.6	25.5	31.5	36.9	41.6	46.1

Method From eqn 5,

$$Kp\theta + \theta = Kp$$

With $\theta = V/V_\infty$, where V_∞ is the volume corresponding to complete coverage, this expression can be rearranged into

$$\frac{p}{V} = \frac{p}{V_\infty} + \frac{1}{KV_\infty}$$

Hence, a plot of p/V against p should give a straight line of slope $1/V_\infty$ and intercept $1/KV_\infty$.

Answer The data for the plot are as follows:

p/Torr	100	200	300	400	500	600	700
$(p/\text{Torr})/(V/\text{cm}^3)$	9.80	10.8	11.8	12.7	13.6	14.4	15.2

The points are plotted in Fig. 28.23. The (least squares) slope is 0.00904, so $V_\infty = 111 \text{ cm}^3$. The intercept at $p = 0$ is 8.99, so

$$K = \frac{1}{(111 \text{ cm}^3) \times (8.99 \text{ Torr cm}^{-3})} = 1.00 \times 10^{-3} \text{ Torr}^{-1}$$

Comment At high surface coverage, the data would deviate from a straight line. The dimensions of K are 1/pressure.

- -

Self-test 28.1 Repeat the calculation for the following data:

p/Torr	100	200	300	400	500	600	700
V/cm^3	10.3	19.3	27.3	34.1	40.0	45.5	48.0

$$[128 \text{ cm}^3, 8.89 \times 10^{-4} \text{ Torr}^{-1}]$$

For adsorption with dissociation, the rate of adsorption is proportional to the pressure and to the probability that both atoms will find sites, which is proportional to the square of the number of vacant sites,

$$\frac{d\theta}{dt} = k_a p \{N(1 - \theta)\}^2 \tag{6}$$

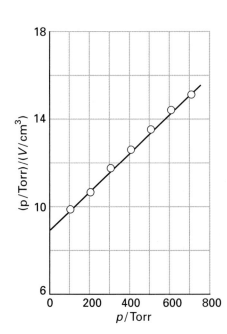

28.23 The plot of the data in Example 28.1. As illustrated here, the Langmuir isotherm predicts that a straight line should be obtained when p/V is plotted against p.

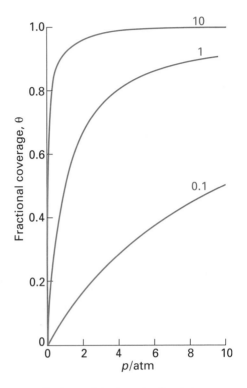

28.24 The Langmuir isotherm for dissociative adsorption ($X_2 \rightarrow 2X$) for different values of K.

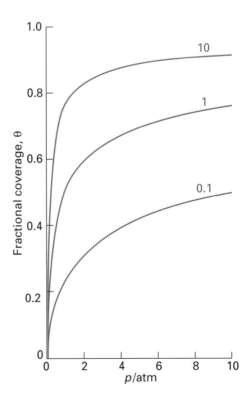

28.25 The Langmuir isotherm for non-dissociative adsorption for different values of K.

The rate of desorption is proportional to the frequency of encounters of atoms on the surface, and is therefore second-order in the number of atoms present:

$$\frac{d\theta}{dt} = -k_d(N\theta)^2 \tag{7}$$

The condition for no net change leads to the isotherm

$$\theta = \frac{(Kp)^{1/2}}{1 + (Kp)^{1/2}} \tag{8}$$

The surface coverage now depends more weakly on pressure than for non-dissociative adsorption.

The shapes of the Langmuir isotherms with and without dissociation are shown in Fig. 28.24 and Fig. 28.25. The fractional coverage increases with increasing pressure, and approaches 1 only at very high pressure, when the gas is forced on to every available site of the surface. Different curves (and therefore values of K) are obtained at different temperatures, and the temperature dependence of K can be used to determine the **isosteric enthalpy of adsorption**, $\Delta_{ad}H^{\ominus}$, the standard enthalpy of adsorption at a fixed surface coverage. To determine this quantity we recognize that K is essentially an equilibrium constant, and then use the van't Hoff equation (eqn 9.24) to write

$$\left(\frac{\partial \ln K}{\partial T}\right)_{\theta} = \frac{\Delta_{ad}H^{\ominus}}{RT^2} \tag{9}$$

Example 28.2 Measuring the isosteric enthalpy of adsorption

The data below show the pressures of CO needed for the volume of adsorption (corrected to 1.00 atm and 273 K) to be 10.0 cm³ using the same sample as in Example 28.1. Calculate the adsorption enthalpy at this surface coverage.

T/K	200	210	220	230	240	250
$p/$Torr	30.0	37.1	45.2	54.0	63.5	73.9

Method The Langmuir isotherm can be rearranged to

$$Kp = \frac{\theta}{1-\theta}$$

Therefore, when θ is constant,

$$\ln K + \ln p = \text{constant}$$

It follows from eqn 9 that

$$\left(\frac{\partial \ln p}{\partial T}\right)_{\theta} = -\left(\frac{\partial \ln K}{\partial T}\right)_{\theta} = -\frac{\Delta_{ad}H^{\ominus}}{RT^2}$$

With $d(1/T)/dT = -1/T^2$, this expression rearranges to

$$\left(\frac{\partial \ln p}{\partial(1/T)}\right)_{\theta} = \frac{\Delta_{ad}H^{\ominus}}{R}$$

Therefore, a plot of $\ln p$ against $1/T$ should be a straight line of slope $\Delta_{ad}H^{\ominus}/R$.

Answer We draw up the following table:

T/K	200	210	220	230	240	250
$10^3/(T/K)$	5.00	4.76	4.55	4.35	4.17	4.00
$\ln(p/$Torr$)$	3.40	3.61	3.81	3.99	4.15	4.30

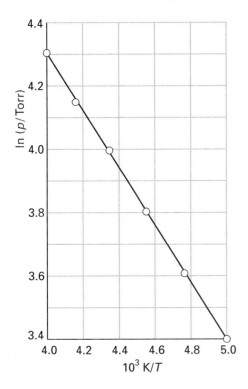

28.26 The isosteric enthalpy of adsorption can be obtained from the slope of the plot of $\ln p$ against $1/T$, where p is the pressure needed to achieve the specified surface coverage. The data used are from Example 28.2.

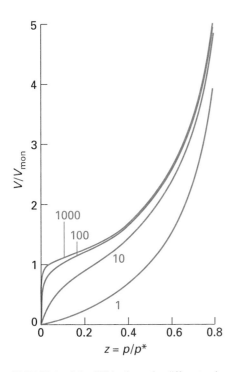

28.27 Plots of the BET isotherm for different values of c. The value of V/V_{mon} rises indefinitely because the adsorbate may condense on the covered substrate surface.

The points are plotted in Fig. 28.26. The slope (of the least squares fitted line) is -0.904, so $\Delta_{ad}H^{\ominus} = -(0.902 \times 10^3 \text{ K}) \times R = -7.50 \text{ kJ mol}^{-1}$.

Comment The value of K can be used to obtain a value of $\Delta_{ad}G^{\ominus}$, and then that value combined with $\Delta_{ad}H^{\ominus}$ to obtain the standard entropy of adsorption.

- -

Self-test 28.2 Repeat the calculation using the following data:

T/K	200	210	220	230	240	250
p/Torr	32.4	41.9	53.0	66.0	80.0	96.0

$$[-9.0 \text{ kJ mol}^{-1}]$$

===

(b) The BET isotherm

If the initial adsorbed layer can act as a substrate for further (for example, physical) adsorption then, instead of the isotherm levelling off to some saturated value at high pressures, it can be expected to rise indefinitely. The most widely used isotherm dealing with multilayer adsorption was derived by Stephen Brunauer, Paul Emmett, and Edward Teller, and is called the **BET isotherm**:

$$\frac{V}{V_{mon}} = \frac{cz}{(1-z)\{1-(1-c)z\}} \qquad z = \frac{p}{p^*} \qquad (10)$$

In this expression, p^* is the vapour pressure above a layer of adsorbate that is more than one molecule thick and which resembles a pure bulk liquid, V_{mon} is the volume corresponding to monolayer coverage, and c is a constant which is large when the enthalpy of desorption from a monolayer is large compared with the enthalpy of vaporization of the liquid adsorbate:

$$c = e^{(\Delta_{des}H^{\ominus} - \Delta_{vap}H^{\ominus})/RT} \qquad (11)$$

The shapes of BET isotherms are illustrated in Fig. 28.27. They rise indefinitely as the pressure is increased because there is no limit to the amount of material that may condense when multilayer coverage may occur. A BET isotherm is not accurate at all pressures, but it is widely used in industry to determine the surface areas of solids.

===

Example 28.3 Using the BET isotherm

The data below relate to the adsorption of N_2 on rutile (TiO_2) at 75 K. Confirm that they fit a BET isotherm in the range of pressures reported, and determine V_{mon} and c.

p/Torr	1.20	14.0	45.8	87.5	127.7	164.4	204.7
V/mm^3	601	720	822	935	1046	1146	1254

At 75 K, $p^* = 570$ Torr. The volumes have been corrected to 1.00 atm and 273 K and refer to 1.00 g of substrate.

Method Equation 10 can be reorganized into

$$\frac{z}{(1-z)V} = \frac{1}{cV_{mon}} + \frac{(c-1)z}{cV_{mon}}$$

It follows that $(c-1)/cV_{mon}$ can be obtained from the slope of a plot of the expression on the left against z, and cV_{mon} can be found from the intercept at $z = 0$. The results can then be combined to give c and V_{mon}.

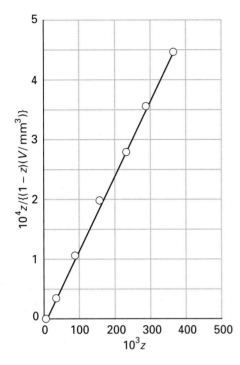

Answer We draw up the following table:

p/Torr	1.20	14.0	45.8	87.5	127.7	164.4	204.7
$10^3 z$	2.11	24.6	80.4	154	224	288	359
$10^4 z/(1-z)(V/\text{mm}^3)$	0.035	0.350	1.06	1.95	2.76	3.53	4.47

These points are plotted in Fig. 28.28. The least squares best line has an intercept at 0.0406, so

$$\frac{1}{cV_{\text{mon}}} = 4.06 \times 10^{-6} \text{ mm}^{-3}$$

The slope of the line is 1.23×10^{-2}, so

$$\frac{c-1}{cV_{\text{mon}}} = (1.23 \times 10^{-2}) \times 10^3 \times 10^{-4} \text{ mm}^{-3} = 1.23 \times 10^{-3} \text{ mm}^{-3}$$

The solutions of these equations are $c = 303$ and $V_{\text{mon}} = 814 \text{ mm}^3$.

Comment At 1.00 atm and 273 K, 810 mm^3 corresponds to 3.6×10^{-5} mol, or 2.2×10^{19} atoms. Because each atom occupies an area of about 0.16 nm^2, the surface area of the sample is about 3.5 m^2.

- -

Self-test 28.3 Repeat the calculation for the following data:

p/Torr	1.20	14.0	45.8	87.5	127.7	164.4	204.7
V/cm^3	235	559	649	719	790	860	950

$$[370, 615 \text{ cm}^3]$$

When $c \gg 1$, the BET isotherm takes the simpler form

$$\frac{V}{V_{\text{mon}}} = \frac{1}{1-z} \tag{12}$$

This expression is applicable to unreactive gases on polar surfaces, for which $c \approx 10^2$ because $\Delta_{\text{des}}H^{\ominus}$ is then significantly greater than $\Delta_{\text{vap}}H^{\ominus}$ (eqn 11). The BET isotherm fits experimental observations moderately well over restricted pressure ranges, but it errs by underestimating the extent of adsorption at low pressures and by overestimating it at high pressures.

(c) Other isotherms

An assumption of the Langmuir isotherm is the independence and equivalence of the adsorption sites. Deviations from the isotherm can often be traced to the failure of these assumptions. For example, the enthalpy of adsorption often becomes less negative as θ increases, which suggests that the energetically most favourable sites are occupied first. Various attempts have been made to take these variations into account. The **Temkin isotherm**,

$$\theta = c_1 \ln (c_2 p) \tag{13}$$

where c_1 and c_2 are constants, corresponds to supposing that the adsorption enthalpy changes linearly with pressure. The **Freundlich isotherm**

$$\theta = c_1 p^{1/c_2} \tag{14}$$

corresponds to a logarithmic change. Different isotherms agree with experiment more or less well over restricted ranges of pressure, but they remain largely empirical. Empirical, however, does not mean useless for, if the parameters of a reasonably reliable isotherm are known, reasonably reliable results can be obtained for the extent of surface coverage under various conditions. This kind of information is essential for any discussion of heterogeneous catalysis.

28.5 The rates of surface processes

Figure 28.29 shows how the potential energy of a molecule varies with its distance from the substrate surface. As the molecule approaches the surface its energy falls as it becomes physisorbed into the **precursor state** for chemisorption. Dissociation into fragments often takes place as a molecule moves into its chemisorbed state, and after an initial increase of energy as the bonds stretch there is a sharp decrease as the adsorbate–substrate bonds reach their full strength. Even if the molecule does not fragment, there is likely to be an initial increase of potential energy as the bonds adjust as the molecule approaches the surface.

In most cases, therefore, we can expect there to be a potential energy barrier separating the precursor and chemisorbed states. This barrier, though, might be low, and might not rise above the energy of a distant, stationary particle (as in Fig. 28.29a). In this case, chemisorption is not an activated process and can be expected to be rapid. Many gas adsorptions on clean metals appear to be non-activated. In some cases the barrier rises above the zero axis (as in Fig. 28.29b); such chemisorptions are activated and slower than the non-activated kind. An example is H_2 on copper, which has an activation energy in the region of 20–40 kJ mol^{-1}.

One point that emerges from this discussion is that rates are not good criteria for distinguishing between physisorption and chemisorption. Chemisorption can be fast if the activation energy is small or zero, but it may be slow if the activation energy is large. Physisorption is usually fast, but it can appear to be slow if adsorption is taking place on a porous medium.

(a) The rate of adsorption

The rate at which a surface is covered by adsorbate depends on the ability of the substrate to dissipate the energy of the incoming particle as thermal motion as it crashes on to the surface. If the energy is not dissipated quickly, the particle migrates over the surface until a vibration expels it into the overlying gas or it reaches an edge. The proportion of collisions with the surface that successfully lead to adsorption is called the **sticking probability**, s:

$$s = \frac{\text{rate of adsorption of particles by the surface}}{\text{rate of collision of particles with the surface}} \qquad [15]$$

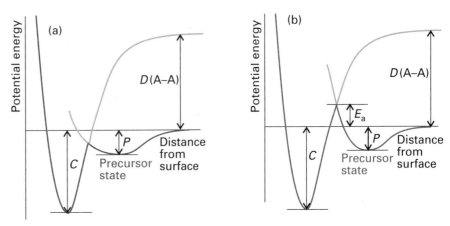

28.29 The potential energy profiles for the dissociative chemisorption of an A_2 molecule. In each case, P is the enthalpy of (non-dissociative) physisorption and C that for chemisorption (at $T = 0$). The relative locations of the curves determine whether the chemisorption is (a) not activated or (b) activated.

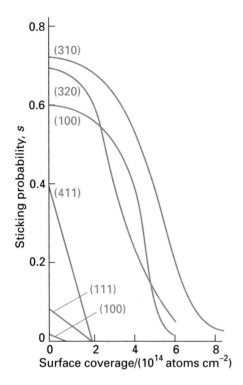

28.30 The sticking probability of N_2 on various faces of a tungsten crystal and its dependence on surface coverage. Note the very low sticking probability for the (110) and (111) faces. (Data provided by Professor D.A. King.)

The denominator can be calculated from kinetic theory, and the numerator can be measured by observing the rate of change of pressure.

Values of s vary widely. For example, at room temperature CO has s in the range 0.1–1.0 for several d-metal surfaces, but for N_2 on rhenium $s < 10^{-2}$, indicating that more than a hundred collisions are needed before one molecule sticks successfully. Beam studies on specific crystal planes show a pronounced specificity: for N_2 on tungsten, s ranges from 0.74 on the (320) faces down to less than 0.01 on the (110) faces at room temperature. The sticking probability decreases as the surface coverage increases (Fig. 28.30). A simple assumption is that s is proportional to $1 - \theta$, the fraction uncovered, and it is common to write

$$s = (1 - \theta)s_0 \tag{16}$$

where s_0 is the sticking probability on a perfectly clean surface. The results in the illustration do not fit this expression because they show that s remains close to s_0 until the coverage has risen to about 6×10^{13} molecules cm^{-2}, and then falls steeply. The explanation is probably that the colliding molecule does not enter the chemisorbed state at once, but moves over the surface until it encounters an empty site.

(b) The rate of desorption

Desorption is always activated because the particles have to be lifted from the foot of a potential well. A physisorbed particle vibrates in its shallow potential well, and might shake itself off the surface after a short time. The temperature dependence of the first-order rate of departure can be expected to be Arrhenius-like, with an activation energy for desorption, E_d, comparable to the enthalpy of physisorption:

$$k_d = Ae^{-E_d/RT} \tag{17}$$

Therefore, the half-life for remaining on the surface has a temperature dependence

$$t_{1/2} = \frac{\ln 2}{k_d} = \tau_0 e^{E_d/RT} \qquad \tau_0 = \frac{\ln 2}{A} \tag{18}$$

(Note the positive sign in the exponent.) If we suppose that $1/\tau_0$ is approximately the same as the vibrational frequency of the weak particle–surface bond (about 10^{12} Hz) and $E_d \approx 25$ kJ mol^{-1}, then residence half-lives of around 10 ns are predicted at room temperature. Lifetimes close to 1 s are obtained only by lowering the temperature to about 100 K. For chemisorption, with $E_d = 100$ kJ mol^{-1} and guessing that $\tau_0 = 10^{-14}$ s (because the adsorbate–substrate bond is quite stiff), we expect a residence half-life of about 3×10^3 s (about an hour) at room temperature, decreasing to 1 s at about 350 K.

The desorption activation energy can be measured in several ways. However, we must be guarded in its interpretation because it often depends on the fractional coverage, and so may change as desorption proceeds. Moreover, the transfer of concepts such as 'reaction order' and 'rate constant' from bulk studies to surfaces is hazardous, and there are few examples of strictly first-order or second-order desorption kinetics (just as there are few integral-order reactions in the gas phase).

If we disregard these complications, one way of measuring the desorption activation energy is to monitor the rate of increase in pressure when the sample is maintained at a series of temperatures, and to attempt to make an Arrhenius plot. A more sophisticated technique is **temperature programmed desorption** (TPD) or **thermal desorption spectroscopy** (TDS). The basic observation is a surge in desorption rate (as monitored by a mass spectrometer) when the temperature is raised linearly to the temperature at which desorption occurs rapidly; but once the desorption has occurred there is no more adsorbate to escape from the surface, so the desorption flux falls again as the temperature continues

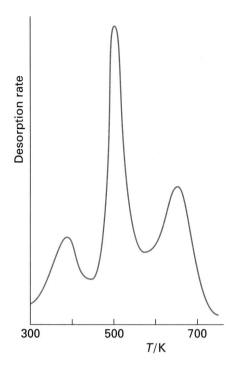

28.31 The flash desorption spectrum of H_2 on the (100) face of tungsten. The three peaks indicate the presence of three sites with different adsorption enthalpies and therefore different desorption activation energies. (P.W. Tamm and L.D. Schmidt, *J. Chem. Phys.* **51**, 5352 (1969).)

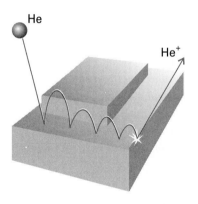

28.32 The events leading to an FIM image of a surface. The He atom migrates across the surface until it is ionized at an exposed atom, when it is pulled off by the externally applied potential. (The bouncing motion is due to the intermolecular potential, not gravity!)

28.33 FIM micrographs showing the migration of Re atoms on rhenium during 3 s intervals at 375 K. (Photographs provided by Professor G. Ehrlich.)

to rise. The TPD spectrum, the plot of desorption flux against temperature, therefore shows a peak, the location of which depends on the desorption activation energy. In the example shown in Fig. 28.31, three maxima are shown, indicating the presence of three sites with different activation energies.

In many cases only a single activation energy (and a single peak in the TPD spectrum) is observed. When several peaks are observed they might correspond to adsorption on different crystal planes or to multilayer adsorption. For instance, Cd atoms on tungsten show two activation energies, one of $18 \ kJ \ mol^{-1}$ and the other of $90 \ kJ \ mol^{-1}$. The explanation is that the more tightly bound Cd atoms are attached directly to the substrate, and the less strongly bound are in a layer (or layers) above the primary overlayer. Chemisorption cannot normally exceed monolayer coverage because a hydrocarbon gas, for instance, cannot chemisorb on to a surface already covered with its fragments; a metal adsorbate, on the other hand, as in this case, can provide a surface capable of further chemisorption.

Another example of a system showing two desorption activation energies is CO on tungsten, the values being $120 \ kJ \ mol^{-1}$ and $300 \ kJ \ mol^{-1}$. The explanation is believed to be the existence of two types of metal-adsorbate binding site, one involving a simple M–CO bond, the other adsorption with dissociation into individually adsorbed C and O atoms. In some cases two desorption peaks may occur even though there is only one type of site. This complication arises when there are significant interactions between the adsorbate particles with the result that at low surface coverages the enthalpy of adsorption is significantly different from its value at high coverages.

(c) Mobility on surfaces

A further aspect of the strength of the interactions between adsorbate and substrate is the mobility of the adsorbate. Mobility is often a vital feature of a catalyst's activity, because a catalyst might be impotent if the reactant molecules adsorb so strongly that they cannot migrate. The activation energy for diffusion over a surface need not be the same as for desorption because the particles may be able to move through valleys between potential peaks without leaving the surface completely. In general, the activation energy for migration is about 10–20 per cent of the energy of the surface-adsorbate bond, but the actual value depends on the extent of coverage. The defect structure of the sample (which depends on the temperature) may also play a dominant role because the adsorbed molecules might find it easier to skip across a terrace than to roll along the foot of a step, and these molecules might become trapped in vacancies in an otherwise flat terrace. Diffusion may also be easier across one crystal face than another, and so the surface mobility depends on which lattice planes are exposed.

Diffusion characteristics of an adsorbate can be examined by using STM to follow the change in surface characteristics or by **field-ionization microscopy** (FIM), which portrays the electrical characteristics of a surface by using the ionization of noble gas atoms to probe the surface (Fig. 28.32). An individual atom is imaged, the temperature is raised, and then lowered after a definite interval. A new image is then recorded, and the new position of the atom measured (Fig. 28.33). A sequence of images shows that the atom makes a random walk across the surface, and the diffusion coefficient, D, can be inferred from the mean

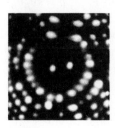
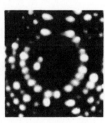
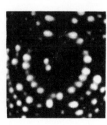
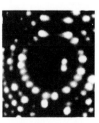

Table 28.3* Activation energies of catalysed reactions

Reaction	Catalyst	$E_a/(\text{kJ mol}^{-1})$
$2HI \rightarrow H_2 + I_2$	None	184
	Au	105
	Pt	59
$2NH_3 \rightarrow N_2 + 3H_2$	None	350
	W	162

*More values are given in the *Data section*.

distance, d, travelled in an interval τ by using the two-dimensional random walk expression $d = (D\tau)^{1/2}$. The value of D for different crystal planes at different temperatures can be determined directly in this way, and the activation energy for migration over each plane obtained from the Arrhenius-like expression

$$D = D_0 e^{-E_D/RT} \tag{19}$$

where E_D is the activation energy for diffusion. Typical values for W atoms on tungsten have E_D in the range 57–87 kJ mol^{-1} and $D_0 \approx 3.8 \times 10^{-11}$ m^2 s^{-1}. For CO on tungsten, the activation energy falls from 144 kJ mol^{-1} at low surface coverage to 88 kJ mol^{-1} when the coverage is high.

Catalytic activity at surfaces

A catalyst acts by providing an alternative reaction path with a lower activation energy (Table 28.3). It does not disturb the final equilibrium composition of the system, only the rate at which that equilibrium is approached. In this section we shall consider heterogeneous catalysis, in which (as mentioned in the introduction preceding Section 26.6) the catalyst and the reagents are in different phases. For simplicity, we shall consider only gas/solid systems.

Many catalysts depend on the **coadsorption** of species, the adsorption of two or more species. One consequence of the presence of a second species may be the modification of the electronic structure at the surface of a metal. For instance, partial coverage of d-metal surfaces by alkali metals has a pronounced effect on the electron distribution and reduces the work function of the metal. Such modifiers can act as promoters (to enhance the action of catalysts) or as poisons (to inhibit catalytic action).

28.6 Adsorption and catalysis

Heterogeneous catalysis normally depends on at least one reactant being adsorbed (usually chemisorbed) and modified to a form in which it readily undergoes reaction. Often this modification takes the form of a fragmentation of the reactant molecules. In practice, the active phase is dispersed as very small particles of linear dimension less than 2 nm on a porous oxide support. **Shape-selective catalysts**, such as the zeolites, which have a pore size that can distinguish shapes and sizes at a molecular scale (recall Fig. 28.1), have high internal specific surface areas, in the range of 100–500 m^2 g^{-1}.

(a) The Langmuir–Hinshelwood mechanism

In the **Langmuir–Hinshelwood mechanism (LH mechanism)** of surface-catalysed reactions, the reaction takes place by encounters between molecular fragments and atoms adsorbed on the surface. We therefore expect the rate law to be second-order in the extent of surface coverage:

$$A + B \longrightarrow P \qquad v = k\theta_A \theta_B \tag{20}$$

Insertion of the appropriate isotherms for A and B then gives the reaction rate in terms of the partial pressures of the reactants. For example, if A and B follow Langmuir isotherms, and adsorb without dissociation, so that

$$\theta_A = \frac{K_A p_A}{1 + K_A p_A + K_B p_B} \qquad \theta_B = \frac{K_B p_B}{1 + K_A p_A + K_B p_B} \tag{21}$$

then it follows that the rate law is

$$v = \frac{kK_A K_B p_A p_B}{(1 + K_A p_A + K_B p_B)^2} \tag{22}$$

The parameters K in the isotherms and the rate constant k are all temperature-dependent, so the overall temperature dependence of the rate may be strongly non-Arrhenius (in the sense that the reaction rate is unlikely to be proportional to $e^{-E_a/RT}$).

(b) The Eley–Rideal mechanism

In the **Eley–Rideal mechanism** (ER mechanism) of a surface-catalysed reaction, a gas-phase molecule collides with another molecule already adsorbed on the surface. The rate of formation of product is expected to be proportional to the partial pressure, p_B, of the non-adsorbed gas B and the extent of surface coverage, θ_A, of the adsorbed gas A. It follows that the rate law should be

$$A + B \longrightarrow P \qquad v = kp_B \theta_A \tag{23}$$

The rate constant, k, might be much larger than for the uncatalysed gas-phase reaction because the reaction on the surface has a low activation energy and the adsorption itself is often not activated.

If we know the adsorption isotherm for A, we can express the rate law in terms of its partial pressure, p_A. For example, if the adsorption of A follows a Langmuir isotherm in the pressure range of interest, then the rate law would be

$$v = \frac{kKp_A p_B}{1 + Kp_A} \tag{24}$$

If A were a diatomic molecule that adsorbed as atoms, we would substitute the isotherm given in eqn 8 instead.

According to eqn 24, when the partial pressure of A is high (in the sense $Kp_A \gg 1$) there is almost complete surface coverage, and the rate is equal to kp_B. Now the rate-determining step is the collision of B with the adsorbed fragments. When the pressure of A is low ($Kp_A \ll 1$), perhaps because of its reaction, the rate is equal to $kKp_A p_B$; now the extent of surface coverage is important in the determination of the rate.

Almost all thermal surface-catalysed reactions are thought to take place by the LH mechanism, but a number of reactions with an ER mechanism have also been identified from molecular beam investigations. For example, the reaction between H(g) and D(ad) to form HD(g) is thought to be by an ER mechanism involving the direct collision and pick-up of the adsorbed D atom by the incident H atom. However, the two mechanisms should really be thought of as ideal limits, and all reactions lie somewhere between the two and show features of each one.

Example 28.4 Interpreting the kinetics of a catalysed reaction

The decomposition of phosphine (PH_3) on tungsten is first-order at low pressures and zero-order at high pressures. Account for these observations.

Method Write down a plausible rate law in terms of an adsorption isotherm and explore its form in the limits of high and low pressure.

Answer If the rate is supposed to be proportional to the surface coverage, we can write

$$v = k\theta = \frac{kKp}{1 + Kp}$$

where p is the pressure of phosphine. When the pressure is so low that $Kp \ll 1$,

$$v = kKp$$

and the decomposition is first-order. When $Kp \gg 1$,

$$v = k$$

and the decomposition is zero-order.

Comment Many heterogeneous reactions are first-order, which indicates that the rate-determining stage is the adsorption process.

Self-test 28.4 Suggest the form of the rate law for the deuteration of NH_3 in which D_2 adsorbs dissociatively and extensively (that is, $Kp \gg 1$, with p the partial pressure of D_2), and NH_3 (with partial pressure p') adsorbs at different sites.

$$[v = k(Kp)^{1/2}K'p'/(1 + K'p')]$$

(c) Molecular beam studies

Molecular beam reactive scattering (MBRS) studies are able to provide detailed information about catalysed reactions. It has become possible to investigate how the catalytic activity of a surface depends on its structure as well as its composition. For instance, the cleavage of C–H and H–H bonds appears to depend on the presence of steps and kinks, and a terrace often has only minimal catalytic activity. The reaction $H_2 + D_2 \rightarrow 2HD$ has been studied in detail. For this reaction, terrace sites are inactive but one molecule in ten reacts when it strikes a step. Although the step itself might be the important feature, it may be that the presence of the step merely exposes a more reactive crystal face (the step face itself). Likewise, the dehydrogenation of hexane to hexene depends strongly on the kink density, and it appears that kinks are needed to cleave C–C bonds. These observations suggest a reason why even small amounts of impurities may poison a catalyst: they are likely to attach to step and kink sites, and so impair the activity of the catalyst entirely. A constructive outcome is that the extent of dehydrogenation may be controlled relative to other types of reactions by seeking impurities that adsorb at kinks and act as specific poisons.

Molecular beam studies can also be used to investigate the details of the reaction process, particularly by using **pulsed beams**, in which the beam is chopped into short slugs. The angular distribution of the products, for instance, can be used to assess the length of time that a species remains on the surface during the reaction, for a long residence time will result in a loss of memory of the incident beam direction.

28.7 Examples of catalysis

Almost the whole of modern chemical industry depends on the development, selection, and application of catalysts (Table 28.4). All we can hope to do in this section is to give a brief indication of some of the problems involved. Other than the ones we consider, these problems include the danger of the catalyst being poisoned by by-products or impurities, and economic considerations relating to cost and lifetime.

(a) Catalytic activity

The activity of a catalyst depends on the strength of chemisorption as indicated by the 'volcano' curve in Fig. 28.34 (which is so-called on account of its general shape). To be active, the catalyst should be extensively covered by adsorbate, which is the case if chemisorption is strong. On the other hand, if the strength of the substrate–adsorbate bond

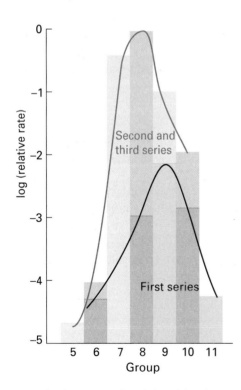

28.34 A volcano curve of catalytic activity arises because, although the reactants must adsorb reasonably strongly, they must not adsorb so strongly that they are immobilized. The lower curve refers to the first series of d-block metals, the upper curve to the second and third series d-block metals. The group numbers relate to the periodic table inside the back cover.

Table 28.4 Properties of catalysts

Catalyst	Function	Examples
Metals	Hydrogenation Dehydrogenation	Fe, Ni, Pt, Ag
Semiconducting oxides and sulfides	Oxidation Desulfurization	NiO, ZnO, MgO, Bi_2O_3/MoO_3, MoS_2
Insulating oxides	Dehydration	Al_2O_3, SiO_2, MgO
Acids	Polymerization Isomerization Cracking Alkylation	H_3PO_4, H_2SO_4, SiO_2/Al_2O_3, zeolites

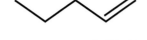

2

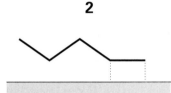

3

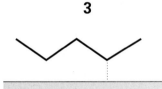

4

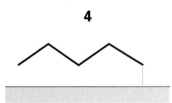

5

becomes too great, the activity declines either because the other reactant molecules cannot react with the adsorbate or because the adsorbate molecules are immobilized on the surface. This pattern of behaviour suggests that the activity of a catalyst should initially increase with strength of adsorption (as measured, for instance, by the enthalpy of adsorption) and then decline, and that the most active catalysts should be those lying near the summit of the volcano. Most active metals are those that lie close to the middle of the d-block.

Many metals are suitable for adsorbing gases, and the general order of adsorption strengths decreases along the series O_2, C_2H_2, C_2H_4, CO, H_2, CO_2, N_2. Some of these molecules adsorb dissociatively (for example, H_2). Elements from the d-block, such as iron, vanadium, and chromium, show a strong activity towards all these gases, but manganese and copper are unable to adsorb N_2 and CO_2. Metals towards the left of the periodic table (for example, magnesium and lithium) can adsorb (and, in fact, react with) only the most active gas (O_2). These trends are summarized in Table 28.5.

(b) Hydrogenation

An example of catalytic action is found in the hydrogenation of alkenes. The alkene (**2**) adsorbs by forming two bonds with the surface (**3**), and on the same surface there may be adsorbed H atoms. When an encounter occurs, one of the alkene–surface bonds is broken (forming (**4**) or (**5**)) and later an encounter with a second H atom releases the fully hydrogenated hydrocarbon, which is the thermodynamically more stable species.

The evidence for a two-stage reaction is the appearance of different isomeric alkenes in the mixture. The formation of isomers comes about because while the hydrocarbon chain is waving about over the surface of the metal, an atom in the chain might chemisorb again to

Table 28.5 Chemisorption abilities

	O_2	C_2H_2	C_2H_4	CO	H_2	CO_2	N_2
Ti, Cr, Mo, Fe	+	+	+	+	+	+	+
Ni, Co	+	+	+	+	+	+	−
Pd, Pt	+	+	+	+	+	−	−
Mn, Cu	+	+	+	+	±	−	−
Al, Au	+	+	+	+	−	−	−
Li, Na, K	+	+	−	−	−	−	−
Mg, Ag, Zn, Pb	+	−	−	−	−	−	−

+, Strong chemisorption; ±, chemisorption; −, no chemisorption. See G.C. Bond, *Heterogeneous catalysis*, Oxford University Press (1986) for further information.

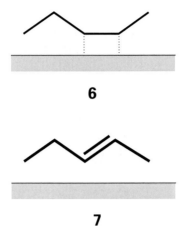

6

7

form (6) and then desorb to (7), an isomer of the original molecule. The new alkene would not be formed if the two hydrogen atoms attached simultaneously.

A major industrial application of catalytic hydrogenation is to the formation of edible fats from vegetable and animal oils. Raw oils obtained from sources such as the soya bean have the structure $CH_2(OOCR)CH(OOCR')CH_2(OOCR'')$, where R, R', and R'' are long-chain hydrocarbons with several double bonds. One disadvantage of the presence of many double bonds is that the oils are susceptible to atmospheric oxidation, and therefore are liable to become rancid. The geometrical configuration of the chains is responsible for the liquid nature of the oil, and in many applications a solid fat is at least much better and often necessary. Controlled partial hydrogenation of an oil with a catalyst, carefully selected so that hydrogenation is incomplete and so that the chains do not isomerize (finely divided nickel, in fact), is used on a wide scale to produce edible fats. The process, and the industry, is not made any easier by the seasonal variation of the number of double bonds in the oils.

(c) Oxidation

Catalytic oxidation is also widely used in industry and in pollution control. Although in some cases it is desirable to achieve complete oxidation (as in the production of nitric acid from ammonia), in others partial oxidation is the aim. For example, the complete oxidation of propene to carbon dioxide and water is wasteful, but its partial oxidation to propenal (acrolein, $CH_2{=}CHCHO$) is the start of important industrial processes. Likewise, the controlled oxidations of ethene to ethanol, acetaldehyde, and (in the presence of acetic acid or chlorine) to vinyl acetate or vinyl chloride, are the initial stages of very important chemical industries.

Some of these reactions are catalysed by d-metal oxides of various kinds. The physical chemistry of oxide surfaces is very complex, as can be appreciated by considering what happens during the oxidation of propene to acrolein on bismuth molybdate. The first stage is the adsorption of the propene molecule with loss of a hydrogen to form the allyl radical, $CH_2{=}CHCH_2$. An O atom in the surface can now transfer to this radical, leading to the formation of acrolein (propenal, $CH_2{=}CHCHO$) and its desorption from the surface. The H atom also escapes with a surface O atom, and goes on to form H_2O, which leaves the surface. The surface is left with vacancies and metal ions in lower oxidation states. These vacancies are attacked by O_2 molecules in the overlying gas, which then chemisorb as O^{2-} ions, so reforming the catalyst. This sequence of events, which is called the **Mars van Krevelen mechanism**, involves great upheavals of the surface, and some materials break up under the stress.

(d) Cracking and reforming

Many of the small organic molecules used in the preparation of all kinds of chemical products come from oil. These small building blocks of polymers, and petrochemicals in general, are usually cut from the long-chain hydrocarbons drawn from the Earth as petroleum. The catalytically induced fragmentation of the long-chain hydrocarbons is called **cracking**, and is often brought about on silica–alumina catalysts. These catalysts act by forming unstable carbocations, which dissociate and rearrange to more highly branched isomers. These branched isomers burn more smoothly and efficiently in internal combustion engines, and are used to produce higher octane fuels.

Catalytic **reforming** uses a dual-function catalyst, such as a dispersion of platinum and acidic alumina. The platinum provides the metal function, and brings about dehydrogenation and hydrogenation. The alumina provides the acidic function, being able to form carbocations from alkenes. The sequence of events in catalytic reforming shows up very clearly the complications that must be unravelled if a reaction as important as this is to be understood and improved. The first step is the attachment of the long-chain hydrocarbon by chemisorption to the platinum. In this process first one and then a second H atom is lost, and an alkene is formed. The alkene migrates to a Brønsted acid site, where it accepts a proton and attaches to the surface as a carbocation. This carbocation can undergo several different reactions. It can break into two, isomerize into a more highly branched form, or undergo varieties of ring-closure. Then the adsorbed molecule loses a proton, escapes from the surface, and migrates (possibly through the gas) as an alkene to a metal part of the catalyst where it is hydrogenated. We end up with a rich selection of smaller molecules which can be withdrawn, fractionated, and then used as raw materials for other products.

Table 28.6 Summary of acronyms

AES	Auger electron spectroscopy
AFM	Atomic force microscopy
BET isotherm	Brunauer, Emmett, Teller isotherm
EELS	Electron energy-loss spectroscopy
ER mechanism	Eley–Rideal mechanism
FIM	Field-ionization microscopy
HREELS	High-resolution electron energy-loss spectroscopy
LEED	Low-energy electron diffraction
LH mechanism	Langmuir–Hinshelwood mechanism
MBRS	Molecular beam reactive scattering
MBS	Molecular beam scattering
SAMS	Scanning Auger electron miscroscopy
SERS	Surface-enhanced Raman scattering
SEXAFS	Surface-extended X-ray absorption fine structure spectroscopy
STM	Scanning tunnelling microscopy
TDS	Thermal desorption spectroscopy
TPD	Temperature programmed desorption
UHV	Ultra-high vacuum technique
UPS	Ultraviolet photoemission spectroscopy
XPS	X-ray photoemission spectroscopy

Checklist of key ideas

The growth and structure of solid surfaces

- [ ] adsorption
- [ ] adsorbate
- [ ] adsorbent
- [ ] substrate
- [ ] desorption

28.1 Surface growth

- [ ] step defect
- [ ] terrace
- [ ] dislocation
- [ ] screw dislocation
- [ ] screw axis

28.2 Surface composition

- [ ] ultra-high vacuum (UHV) techniques
- [ ] escape depth
- [ ] photoemission spectroscopy (UPS, XPS)
- [ ] surface-enhanced Raman scattering (SERS)
- [ ] electron energy-loss spectroscopy (EELS, HREELS)
- [ ] Auger electron spectroscopy (AES)
- [ ] Auger effect
- [ ] X-ray fluorescence
- [ ] scanning Auger electron microscopy (SAM)
- [ ] surface-extended X-ray absorption fine structure spectroscopy (SEXAFS)
- [ ] surface reconstruction
- [ ] low-energy electron diffraction (LEED)
- [ ] substrate structure notation
- [ ] scanning tunnelling microscopy (STM)
- [ ] atomic force microscopy (AFM)
- [ ] molecular beam scattering (MBS)

The extent of adsorption

- [ ] fractional coverage (2)
- [ ] rate of adsorption
- [ ] flash desorption
- [ ] gravimetry

28.3 Physisorption and chemisorption

- [ ] physisorption
- [ ] accommodation
- [ ] chemisorption

28.4 Adsorption isotherms

- [ ] adsorption isotherm
- [ ] Langmuir isotherm (5)

☐ Langmuir isotherm with dissociation (8)
☐ isosteric enthalpy of adsorption (9)
☐ Brunauer, Emmett, Teller (BET) isotherm (10)
☐ Temkin isotherm (13)
☐ Freundlich isotherm (14)

28.5 The rates of surface processes
☐ precursor state
☐ sticking probability (15)
☐ half-life for desorption (18)
☐ temperature programmed desorption (TPD)
☐ thermal desorption spectroscopy (TDS)
☐ field-ionization microscopy (FIM)

☐ surface diffusion temperature dependence (19)

Catalytic activity at surfaces
☐ coadsorption

28.6 Adsorption and catalysis
☐ shape-selective catalyst
☐ Langmuir–Hinshelwood (LH) mechanism
☐ Eley–Rideal (ER) mechanism

☐ molecular beam reactive scattering (MBRS)
☐ pulsed beams

28.7 Examples of catalysis
☐ volcano curve
☐ hydrogenation
☐ oxidation
☐ Mars van Krevelen mechanism
☐ cracking
☐ reforming

Further reading

Articles of general interest

S.T. Oyama and G.A. Samorjai, Homogeneous, heterogeneous, and enzymatic catalysis. *J. Chem. Educ.* **65**, 765 (1988).

N.-F. Zhou, The availability of simple form of Gibbs adsorption equation for mixed surfactants. *J. Chem. Educ.* **66**, 137 (1989).

T.E. Mallouk and H. Lee, Designer solids and surfaces. *J. Chem. Educ.* **67**, 829 (1990).

L. Glasser, The BET isotherm in 3D. *Educ. in Chem.* **25**, 178 (1988).

E. Shustorovich, Chemisorption theory: in search of the elephant. *Acc. Chem. Res.* **21**, 183 (1988).

S.L. Bernasek, State-resolved dynamics of chemical reactions at surfaces. *Chem. Rev.* **87**, 91 (1987).

B.B. Laird and A.D.J. Haymet, The crystal/liquid interface: structure and properties from computer simulation. *Chem. Rev.* **92**, 1819 (1992).

R. Parsons, Electrical double layer: recent experimental and theoretical developments. *Chem. Rev.* **90**, 813 (1990).

E. Schönherr, Crystal growth. In *Encyclopedia of applied physics* (ed. G.L. Trigg), **4**, 335. VCH, New York (1992).

R.R. Corderman, Auger spectroscopy. In *Encyclopedia of applied physics* (ed. G.L. Trigg), **2**, 297. VCH, New York (1991).

F.J. Himpsell and I. Lindau, Photoemission and photoelectron spectra. In *Encyclopedia of applied physics* (ed. G.L. Trigg), **13**, 477. VCH, New York (1995).

W. Beall Fowler, R. Phillips, and A.E. Carlsson, Point and extended defects in crystals. In *Encyclopedia of applied physics* (ed. G.L. Trigg), **14**, 317. VCH, New York (1996).

J.M. Cowley, Electron diffraction. In *Encyclopedia of applied physics* (ed. G.L. Trigg), **5**, 405. VCH, New York (1993).

Texts and sources of data and information

B.C. Gates, *Catalytic chemistry*. Wiley, New York (1992).

J.M. Thomas and W.J. Thomas, *Principles and practice of heterogeneous catalysis*. VCH, Weinheim (1996).

G. Ertl, H. Knözinger, and J. Weitkamp, *Handbook of heterogeneous catalysis*. VCH, Weinheim (1997).

K.L. Mittal (ed.), *Particles on surfaces*, Vol. 3. Plenum, New York (1991).

V.P. Zhdanov, *Elementary physicochemical processes on solid surfaces*. Plenum, New York (1991).

S.R. Morrison, *The chemical physics of surfaces*. Plenum, New York (1990).

J. Israelachvili, *Intermolecular and surface forces*. Academic Press, New York (1985).

C.H. Bamford and C.F. Tipper (ed.), *Comprehensive chemical kinetics*, Vols 1–26. Elsevier, Amsterdam (1969–1986).

R.G. Compton (ed.), *Comprehensive chemical kinetics*, Vols 27–33. Elsevier, Amsterdam (1987–1992).

K. Tamaru, *Dynamic heterogeneous catalysis*. Academic Press, New York (1978).

A. Zangwill, *Physics at surfaces*. Cambridge University Press (1988).

J. Vickerman, *Surface analysis: techniques and applications*. Wiley, New York (1997).

R. Aveyard and D.A. Haydon, *An introduction to the principles of surface chemistry*. Cambridge University Press (1973).

A.W. Anderson, *Physical chemistry of surfaces*. Wiley-Interscience, New York (1990).

R.P.H. Gasser, *An introduction to chemisorption and catalysis by metals*. Clarendon Press, Oxford (1985).

M.W. Roberts and C.S. McKee, *Chemistry of the metal-gas interface*. Clarendon Press, Oxford (1978).

G.A. Somorjai, *Chemistry in two dimensions: surfaces.* Cornell University Press, Ithaca (1981).

G.A. Somorjai, *Surface chemistry and catalysis.* Wiley-Interscience, New York (1994).

G.C. Bond, *Heterogeneous catalysis: principles and applications.* Clarendon Press, Oxford (1986).

M. Boudart and G. Djéga-Mariadassou, *Kinetics of heterogeneous catalysis reactions.* Princeton University Press

---

Exercises

28.1 (a) Calculate the frequency of molecular collisions per square centimetre of surface in a vessel containing (a) hydrogen, (b) propane at 25°C when the pressure is (i) 100 Pa, (ii) 0.10 μTorr.

28.1 (b) Calculate the frequency of molecular collisions per square centimetre of surface in a vessel containing (a) nitrogen, (b) methane at 25°C when the pressure is (i) 10.0 Pa, (ii) 0.150 μTorr.

28.2 (a) What pressure of argon gas is required to produce a collision rate of 4.5×10^{20} s^{-1} at 425 K on a circular surface of diameter 1.5 mm?

28.2 (b) What pressure of nitrogen gas is required to produce a collision rate of 5.00×10^{19} s^{-1} at 525 K on a circular surface of diameter 2.0 mm?

28.3 (a) Calculate the average rate at which He atoms strike a Cu atom in a surface formed by exposing a (100) plane in metallic copper to helium gas at 80 K and a pressure of 35 Pa. Crystals of copper are face-centred cubic with a cell edge of 361 pm.

28.3 (b) Calculate the average rate at which He atoms strike an iron atom in a surface formed by exposing a (100) plane in metallic iron to helium gas at 100 K and a pressure of 24 Pa. Crystals of iron are body-centred cubic with a cell edge of 145 pm.

28.4 (a) A monolayer of N_2 molecules (effective area 0.165 nm^2) is adsorbed on the surface of 1.00 g of an Fe/Al$_2$O$_3$ catalyst at 77 K, the boiling point of liquid nitrogen. Upon warming, the nitrogen occupies 2.86 cm^3 at 0°C and 760 Torr. What is the surface area of the catalyst?

28.4 (b) A monolayer of CO molecules (effective area 0.165 nm^2) is adsorbed on the surface of 1.00 g of an Fe/Al$_2$O$_3$ catalyst at 77 K, the boiling point of liquid nitrogen. Upon warming, the carbon monoxide occupies 4.25 cm^3 at 0°C and 1.00 bar. What is the surface area of the catalyst?

28.5 (a) The volume of oxygen gas at 0°C and 101 kPa adsorbed on the surface of 1.00 g of a sample of silica at 0°C was 0.284 cm^3 at 142.4 Torr and 1.430 cm^3 at 760 Torr. What is the value of V_{mon}?

28.5 (b) The volume of gas at 20°C and 1.00 bar adsorbed on the surface of 1.50 g of a sample of silica at 0°C was 1.60 cm^3 at 52.4 kPa and 2.73 cm^3 at 104 kPa. What is the value of V_{mon}?

28.6 (a) The enthalpy of adsorption of CO on a surface is found to be -120 kJ mol^{-1}. Is the adsorption physisorption or chemisorption? Estimate the mean lifetime of a CO molecule on the surface at 400 K.

28.6 (b) The enthalpy of adsorption of ammonia on a nickel surface is found to be -155 kJ mol^{-1}. Estimate the mean lifetime of an NH$_3$ molecule on the surface at 500 K.

28.7 (a) The average time for which an oxygen atom remains adsorbed to a tungsten surface is 0.36 s at 2548 K and 3.49 s at 2362 K. Find the activation energy for desorption. What is the pre-exponential factor for these tightly chemisorbed atoms?

28.7 (b) The chemisorption of hydrogen on manganese is activated, but only weakly so. Careful measurements have shown that it proceeds 35 per cent faster at 1000 K than at 600 K. What is the activation energy for chemisorption?

28.8 (a) The adsorption of a gas is described by the Langmuir isotherm with $K = 0.85$ kPa^{-1} at 25°C. Calculate the pressure at which the fractional surface coverage is (a) 0.15, (b) 0.95.

28.8 (b) The adsorption of a gas is described by the Langmuir isotherm with $K = 0.777$ kPa^{-1} at 25°C. Calculate the pressure at which the fractional surface coverage is (a) 0.20, (b) 0.75.

28.9 (a) A certain solid sample adsorbs 0.44 mg of CO when the pressure of the gas is 26.0 kPa and the temperature is 300 K. The mass of gas adsorbed when the pressure is 3.0 kPa and the temperature is 300 K is 0.19 mg. The Langmuir isotherm is known to describe the adsorption. Find the fractional coverage of the surface at the two pressures.

28.9 (b) A certain solid sample adsorbs 0.63 mg of CO when the pressure of the gas is 36.0 kPa and the temperature is 300 K. The mass of gas adsorbed when the pressure is 4.0 kPa and the temperature is 300 K is 0.21 mg. The Langmuir isotherm is known to describe the adsorption. Find the fractional coverage of the surface at the two pressures.

28.10 (a) For how long on average would an H atom remain on a surface at 298 K if its desorption activation energy were (a) 15 kJ mol^{-1}, (b) 150 kJ mol^{-1}? Take $\tau_0 = 0.10$ ps. For how long on average would the same atoms remain at 1000 K?

28.10 (b) For how long on average would an atom remain on a surface at 400 K if its desorption activation energy were (a) 20 kJ mol^{-1}, (b) 200 kJ mol^{-1}? Take $\tau_0 = 0.12$ ps. For how long on average would the same atoms remain at 800 K?

28.11 (a) A solid in contact with a gas at 12 kPa and 25°C adsorbs 2.5 mg of the gas and obeys the Langmuir isotherm. The enthalpy change when 1.00 mmol of the adsorbed gas is desorbed is $+10.2$ kJ mol^{-1}. What is the equilibrium pressure for the adsorption of 2.5 mg of gas at 40°C?

28.11 (b) A solid in contact with a gas at 8.86 kPa and 25°C adsorbs 4.67 mg of the gas and obeys the Langmuir isotherm. The enthalpy change when 1.00 mmol of the adsorbed gas is desorbed is +12.2 J. What is the equilibrium pressure for the adsorption of the same mass of gas at 45°C?

28.12 (a) Hydrogen iodide is very strongly adsorbed on gold but only slightly adsorbed on platinum. Assume the adsorption follows the Langmuir isotherm and predict the order of the HI decomposition reaction on each of the two metal surfaces.

28.12 (b) Suppose it is known that ozone adsorbs on a particular surface in accord with a Langmuir isotherm. How could you use the pressure dependence of the fractional coverage to distinguish between adsorption (a) without dissociation, (b) with dissociation into $O + O_2$, (c) with dissociation into $O + O + O$?

28.13 (a) Nitrogen gas adsorbed on charcoal to the extent of 0.921 cm^3 g^{-1} at 490 kPa and 190 K, but at 250 K the same amount of adsorption was achieved only when the pressure was increased to 3.2 MPa. What is the enthalpy of adsorption of nitrogen on charcoal?

28.13 (b) Nitrogen gas adsorbed on a surface to the extent of 1.242 cm^3 g^{-1} at 350 kPa and 180 K, but at 240 K the same amount of adsorption was achieved only when the pressure was increased to 1.02 MPa. What is the enthalpy of adsorption of nitrogen on the surface?

28.14 (a) In an experiment on the adsorption of oxygen on tungsten it was found that the same volume of oxygen was desorbed in 27 min at 1856 K and 2.0 min at 1978 K. What is the activation energy of desorption? How long would it take for the same amount to desorb at (a) 298 K, (b) 3000 K?

28.14 (b) In an experiment on the adsorption of ethene on iron it was found that the same volume of the gas was desorbed in 1856 s at 873 K and 8.44 s at 1012 K. What is the activation energy of desorption? How long would it take for the same amount of ethene to desorb at (a) 298 K, (b) 1500 K?

Problems

Numerical problems

28.1 Nickel is face-centred cubic with a unit cell of side 352 pm. What is the number of atoms per square centimetre exposed on a surface formed by (a) (100), (b) (110), (c) (111) planes? Calculate the frequency of molecular collisions per surface atom in a vessel containing (a) hydrogen, (b) propane at 25°C when the pressure is (i) 100 Pa, (ii) 0.10 μTorr.

28.2 The data below are for the chemisorption of hydrogen on copper powder at 25°C. Confirm that they fit the Langmuir isotherm at low coverages. Then find the value of K for the adsorption equilibrium and the adsorption volume corresponding to complete coverage.

p/Torr	0.19	0.97	1.90	4.05	7.50	11.95
V/cm^3	0.042	0.163	0.221	0.321	0.411	0.471

28.3 The data for the adsorption of ammonia on barium fluoride are reported below. Confirm that they fit a BET isotherm and find values of c and V_{mon}.

(a) $\theta = 0$°C, $p^* = 3222$ Torr:

p/Torr	105	282	492	594	620	755	798
V/cm^3	11.1	13.5	14.9	16.0	15.5	17.3	16.5

(b) $\theta = 18.6$°C, $p^* = 6148$ Torr:

p/Torr	39.5	62.7	108	219	466	555	601	765
V/cm^3	9.2	9.8	10.3	11.3	12.9	13.1	13.4	14.1

28.4 The following data have been obtained for the adsorption of H_2 on the surface of 1.00 g of copper at 0°C. The volume of H_2 below is the volume that the gas would occupy at STP (0°C and 1 atm).

p/atm	0.050	0.100	0.150	0.200	0.250
V/mL	1.22	1.33	1.31	1.36	1.40

Determine the volume of H_2 necessary to form a monolayer and estimate the surface area of the copper sample. The density of liquid hydrogen is 0.0708 g cm^{-3}.

28.5 The designers of a new industrial plant wanted to use a catalyst code-named CR-1 in a step involving the fluorination of butadiene. As a first step in the investigation they determined the form of the adsorption isotherm. The volume of butadiene adsorbed per gram of CR-1 at 15°C varied with pressure as given below. Is the Langmuir isotherm suitable at this pressure?

p/Torr	100	200	300	400	500	600
V/cm^3	17.9	33.0	47.0	60.8	75.3	91.3

Investigate whether the BET isotherm gives a better description of the adsorption of butadiene on CR-1. At 15°C, p^*(butadiene) = 200 kPa. Find V_{mon} and c.

28.6 The adsorption of solutes on solids from liquids often follows a Freundlich isotherm. Check the applicability of this isotherm to the following data for the adsorption of acetic acid on charcoal at 25°C and find the values of the parameters c_1 and c_2.

[acid]/(mol L^{-1})	0.05	0.10	0.50	1.0	1.5
w_a/g	0.04	0.06	0.12	0.16	0.19

w_a is the mass adsorbed per unit mass of charcoal.

28.7 In some catalytic reactions the products may adsorb more strongly than the reacting gas. This is the case, for instance, in the catalytic decomposition of ammonia on platinum at 1000°C. As a first step in examining the kinetics of this type of process, show that the rate of ammonia decomposition should follow

$$\frac{dp_{NH_3}}{dt} = -k_c \frac{p_{NH_3}}{p_{H_2}}$$

in the limit of very strong adsorption of hydrogen. Start by showing that, when a gas J adsorbs very strongly, and its pressure is p_J, the fraction of uncovered sites is approximately $1/Kp_J$. Solve the rate equation for the catalytic decomposition of NH_3 on platinum and show that a plot of $F(t) = (1/t)\ln(p/p_0)$ against $G(t) = (p - p_0)/t$, where p is the pressure of ammonia, should give a straight line from which k_c can be determined. Check the rate law on the basis of the data below, and find k_c for the reaction.

t/s	0	30	60	100	160	200	250
p/Torr	100	88	84	80	77	74	72

Theoretical problems

28.8 The deposition of atoms and ions on a surface depends on their ability to stick, and therefore on the energy changes that occur. As an illustration, consider a two-dimensional square lattice of univalent positive and negative ions separated by 200 pm, and consider a cation approaching the upper terrace of this array from the top of the page. Calculate, by direct summation, its Coulombic interaction when it is in an empty lattice point directly above an anion. Now consider a high step in the same lattice, and let the approaching ion go into the corner formed by the step and the terrace. Calculate the Coulombic energy for this position, and decide on the likely settling point for a deposited cation.

28.9 Although the attractive van der Waals interaction between individual molecules varies as R^{-6}, the interaction of a molecule with a nearby solid (a homogeneous collection of molecules) varies as R^{-3}, where R is its vertical distance above the surface. Confirm this assertion. Calculate the interaction energy between an Ar atom and the surface of solid argon on the basis of a Lennard-Jones $(6, 12)$-potential. Estimate the equilibrium distance of an atom above the surface.

28.10 Use the Gibbs adsorption isotherm (another name for eqn 23.47), to show that the volume adsorbed per unit area of solid, V_a/σ, is related to the pressure of the gas by $V_a = -(\sigma/RT)(d\mu/d\ln p)$, where μ is the chemical potential of the adsorbed gas.

28.11 If the dependence of the chemical potential of the gas on the extent of surface coverage is known, the Gibbs adsorption isotherm, eqn 23.47, can be integrated to give a relation between V_a and p, as in a normal adsorption isotherm. For instance, suppose that the change in the chemical potential of a gas when it adsorbs is of the form $d\mu = -c_2(RT/\sigma)\,dV_a$, where c_2 is a constant of proportionality: show that the Gibbs isotherm leads to the Freundlich isotherm in this case.

28.12 Finally we come full circle and return to the Langmuir isotherm. Find the form of $d\mu$ that, when inserted in the Gibbs adsorption isotherm, leads to the Langmuir isotherm.

Additional problems supplied by Carmen Giunta and Charles Trapp

28.13 N.E. Shafer and R.N. Zare in the article 'Through a beer glass darkly' (*Phys. Today* **44**, 48 (1991)) observe that the radii of bubbles

ascending from the bottom of a glass of beer increase in size as they rise. The reason is that after the bottle of beer is opened the partial pressure of dissolved CO_2 in the beer is greater than the pressure of CO_2 in the bubble. Assuming that the pressure difference is approximately constant, the change in the number, N, of CO_2 molecules is proportional to the surface area of the bubble, that is, $dN/dt = 4\pi r^2 g$, where g is a constant and r is the time-dependent radius of the bubble. (a) Use the perfect gas law to show that the radius of the bubble increases linearly with time in accordance with $r = r_0 + vt$, where r_0 is the initial radius of the bubble and v is the rate of increase of its radius. Also find an expression for v. (b) Shafer and Zare observe that the velocity and spacing between bubbles in a linear stream of bubbles proceeding from the bottom of the glass increase with height. Their data on radius and height as a function of time, with typical uncertainties are reproduced below.

Time/s	Radius, r/cm	Height, z/cm
0.00	0.017 ± 0.004	0.0 ± 0.2
0.54 ± 0.04	0.020	1.2
1.08	0.026	3.4
2.16	0.025	5.2
2.70	0.030	9.6
3.24	0.031	12.4
3.78	0.034	15.6

Test the goodness of fit of the radius data to the equation. From the fit, determine v and the constant g. Also, find an empirical relation between height and time and difference in height and time.

28.14 A. Akgerman and M. Zardkoohi (*J. Chem. Eng. Data* **41**, 185 (1996)) examined the adsorption of phenol from aqueous solution on to fly ash at 20°C. They fitted their observations to a Freundlich isotherm of the form $c_{ads} = Kc_{sol}^{1/n}$, where c_{ads} is the concentration of adsorbed phenol and c_{sol} is the concentration of aqueous phenol. Among the data reported are the following.

$c_{sol}/(\text{mg g}^{-1})$	8.26	15.65	25.43	31.74	40.00
$c_{ads}/(\text{mg g}^{-1})$	4.4	19.2	35.2	52.0	67.2

Determine the constants K and n. What further information would be necessary in order to express the data in terms of fractional coverage, θ?

28.15 C. Huang and W.P. Cheng (*J. Colloid Interface Sci.* **188**, 270 (1997)) examined the adsorption of the hexacyanoferrate(III) ion, $[Fe(CN)_6]^{3-}$, on γ-Al_2O_3 from aqueous solution. They modelled the adsorption with a modified Langmuir isotherm, obtaining the following values of K at pH = 6.5.

T/K	283	298	308	318
$10^{-11} K$	2.642	2.078	1.286	1.085

Determine the isosteric enthalpy of adsorption, $\Delta_{ads}H^{\ominus}$, at this pH. The researchers also reported $\Delta_{ads}S^{\ominus} = +146 \text{ J K}^{-1} \text{ mol}^{-1}$ under these conditions. Determine $\Delta_{ads}G^{\ominus}$.

28.16 M.-G. Olivier and R. Jadot (*J. Chem. Eng. Data* **42**, 230 (1997)) studied the adsorption of butane on silica gel. They report the following amounts of absorption (in moles per kilogram of silica gel) at 303 K.

p/kPa	31.00	38.22	53.03	76.38	101.97
$n/(\text{mol kg}^{-1})$	1.00	1.17	1.54	2.04	2.49
p/kPa		130.47	165.06	182.41	205.75	219.91
$n/(\text{mol kg}^{-1})$		2.90	3.22	3.30	3.35	3.36

Fit these data to a Langmuir isotherm, and determine the value of n that corresponds to complete coverage and the constant K.

28.17 In a study relevant to automobile catalytic converters, C.E. Wartnaby, A. Stuck, Y.Y. Yeo, and D.A. King (*J. Phys. Chem.* **100**, 12483 (1996)) measured the enthalpy of adsorption of CO, NO, and O_2 on initially clean platinum 110 surfaces. They report $\Delta_{ads}H^{\ominus}$ for NO to be -160 kJ mol^{-1}. How much more strongly adsorbed is NO at 500°C than at 400°C?

28.18 The following data were obtained for the extent of adsorption, s, of acetone (propanone) on charcoal from an aqueous solution of concentration c at 18°C.

$c/(\text{mmol L}^{-1})$	15.0	23.0	42.0	84.0	165	390	800
$s/(\text{mmol acetone/g charcoal})$	0.60	0.75	1.05	1.50	2.15	3.50	5.10

Which isotherm fits this data best, Langmuir, Freundlich, or Temkin?

28.19 The removal or recovery of volatile organic compounds (VOCs) from exhaust gas streams is an important process in environmental engineering. Activated carbon has long been used as an adsorbent in this process, but the presence of moisture in the stream reduces its effectiveness. M.-S. Chou and J.-H. Chiou (*J. Envir. Engrg. ASCE* **123**, 437(1997)) have studied the effect of moisture content on the adsorption capacities of granular activated carbon (GAC) for normal hexane and cyclohexane in air streams. From their data for dry streams, shown in the table below, they conclude that GAC obeys a Langmuir type model in which $q_{VOC, RH=0} = abc_{VOC}/(1 + bc_{VOC})$, where $q_{VOC} = m_{VOC}/m_{GAC}$, RH is the relative humidity, a is the maximum adsorption capacity, b is the affinity parameter, and c is the concentration in parts per million (ppm).

	$q_{VOC,RH=0}$(cyclohexane)				
c/ppm	33.6°C	41.5°C	57.4°C	76.4°C	99°C
200	0.080	0.069	0.052	0.042	0.027
500	0.093	0.083	0.072	0.056	0.042
1000	0.101	0.088	0.076	0.063	0.045
2000	0.105	0.092	0.083	0.068	0.052
3000	0.112	0.102	0.087	0.072	0.058

(a) By linear regression of $1/q_{VOC,RH=0}$ against $1/c_{VOC}$, test the goodness of fit and determine values of a and b. (b) The parameters a and b can be related to $\Delta_{ad}H$, the enthalpy of adsorption, and $\Delta_b H$, the difference in activation energy for adsorption and desorption of the VOC molecules, through Arrhenius type equations: $a = k_a e^{-\Delta_{ad}H/RT}$; $b = k_b e^{-\Delta_b H/RT}$. Test the goodness of fit of the data to these equations and obtain values for k_a, k_b, $\Delta_{ad}H$, and $\Delta_b H$. (c) What interpretation might you give to k_a and k_b?

28.20 M.-S. Chou and J.-H. Chiou (*J. Envir. Engrg., ASCE* **123**, 437 (1997)) have studied the effect of moisture content on the adsorption capacities of granular activated carbon (GAC, Norit PK 1-3) for the volatile organic compounds (VOCs) normal hexane and cyclohexane in air streams. The following table shows the adsorption capacities ($q_{water} = m_{water}/m_{GAC}$) of GAC for pure water from moist air streams as a function of relative humidity (RH) in the absence of VOCs at 41.5°C.

RH	0.00	0.26	0.49	0.57	0.80	1.00
q_{water}	0.00	0.026	0.072	0.091	0.161	0.229

The authors conclude that the data at this and other temperatures obey a Freundlich type isotherm, $q_{water} = k(\text{RH})^{1/n}$. (a) Test this hypothesis for their data at 41.5°C and determine the constants k and n. (b) Why might VOCs obey the Langmuir model, but water the Freundlich model? (c) When both water vapour and cyclohexane were present in the stream the values given in the table below were determined for the ratio $r_{VOC} = q_{VOC}/q_{VOC,RH=0}$ at 41.5°C.

RH	0.00	0.10	0.25	0.40	0.53	0.76	0.81
r_{VOC}	1.00	0.98	0.91	0.84	0.79	0.67	0.61

The authors propose that these data fit the equation $r_{VOC} = 1 - q_{water}$. Test their proposal and determine values for k and n and compare to those obtained in part (b) for pure water. Suggest reasons for any differences.

28.21 The release of petroleum products by leaky underground storage tanks is a serious threat to clean ground water. BTEX compounds (benzene, toluene, ethylbenzene, and xylenes) are of primary concern due to their ability to cause health problems at low concentrations. D.S. Kershaw, B.C. Kulik, and S. Pamukcu (*J. Geotech. & Geoenvir. Engrg.* **123**, 324 (1997)) have studied the ability of ground tyre rubber to sorb (adsorb and absorb) benzene and o-xylene. Though sorption involves more than surface interactions, sorption data is usually found to fit one of the adsorption isotherms. In this study, the authors have tested how well their data fit the linear ($q = Kc_{eq}$), Freundlich ($q = K_F c_{eq}^{1/n}$), and Langmuir ($q = K_L M c_{eq}/(1 + K_L c_{eq})$ type isotherms, where q is the mass of solvent sorbed per gram of ground rubber (in milligrams per gram), the Ks and M are empirical constants, c_{eq} the equilibrium concentration of contaminant in solution (in milligrams per litre). (a) Determine the units of the empirical constants. (b) Determine which of the isotherms best fits the data in the table below for the sorption of benzene on ground rubber.

$c_{eq}/(\text{mg L}^{-1})$	97.10	36.10	10.40	6.51	6.21	2.48
$q/(\text{mg g}^{-1})$	7.13	4.60	1.80	1.10	0.55	0.31

(c) Compare the sorption efficiency of ground rubber to that of granulated activated charcoal which, for benzene has been shown to obey the Freundlich isotherm in the form $q = 1.0c_{eq}^{1.6}$ with coefficient of determination $R^2 = 0.94$.

29 Dynamic electrochemistry

In this final chapter we examine one more example of chemical change, that resulting from the transfer of electrons at electrodes. The approach we adopt is largely phenomenological and draws on the thermodynamic language inspired by activated complex theory. First, a model of a solution–electrode interface is constructed, and that model is used to derive a relation between the current density at the electrode and the overpotential, the difference between the electrode potential when a current is flowing and when it is not. The variation of current with overpotential can be used to infer details of the mechanism responsible for the redox process at the electrode. The Butler–Volmer equation can also be used to analyse the behaviour of working cells and to demonstrate how their electric potential when they are in use differs from the zero-current value. The same approach can be used to analyse the kinetics of reactions that are responsible for corrosion, and point the way to methods of decreasing its rate.

The economic consequences of electrochemistry are almost incalculable. Most of the modern methods of generating electricity are inefficient, and the development of fuel cells could revolutionize our production and deployment of energy. Today we produce energy inefficiently to produce goods that then decay by corrosion. Each step of this wasteful sequence could be improved by discovering more about the kinetics of electrochemical processes. Similarly, the techniques of organic and inorganic electrosynthesis, where an electrode is an active component of an industrial process, depend on a detailed knowledge of the factors affecting their rates.

Much of electrochemistry depends on processes that occur at the interface of an electrode and an ionic solution, and the kinetic problem examined in this chapter is the rate at which reducible or oxidizable species—in short, electroactive species—can gain or lose electrons. A measure of this rate is the **current density**, j, the charge flux through a region (the electric current divided by the area of the region). Most of the discussion that follows is concerned with the properties that control the current density and its consequences.

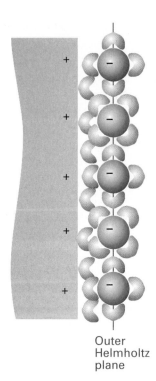

Outer
Helmholtz
plane

29.1 A simple model of the electrical double layer treats it as two rigid planes of charge, one plane, the outer Helmholtz plane (OHP), being due to the ions with their solvating molecules, the other being that on the electrode itself.

Processes at electrodes

When only equilibrium properties are of interest, there is no need to know the details of the charge separation responsible for the potential difference at an interface (just as we do not need to propose a reaction mechanism when discussing equilibria). However, a description of the interface is essential when we are interested in the rate of charge transfer.

29.1 The electrical double layer

An electrode becomes positively charged relative to the solution nearby if electrons leave the electrode and decrease the local cation concentration. The most primitive model of the interface is as an **electrical double layer**, which consists of a sheet of positive charge at the surface of the electrode and a sheet of negative charge next to it in the solution (or vice versa).

(a) The structure of the double layer

A more detailed picture of the interface can be constructed by speculating about the arrangement of ions and electric dipoles in the solution. In the **Helmholtz model** of the double layer the solvated ions range themselves along the surface of the electrode but are held away from it by their hydration spheres (Fig. 29.1). The location of the sheet of ionic charge, which is called the **outer Helmholtz plane** (OHP), is identified as the plane running through the solvated ions. In a refinement of this model, ions that have discarded their solvating molecules and have become attached to the electrode surface by chemical bonds are regarded as forming the **inner Helmholtz plane** (IHP). The Helmholtz model ignores the disrupting effect of thermal motion, which tends to break up and disperse the rigid outer plane of charge. In the **Gouy–Chapman model** of the **diffuse double layer**, the disordering effect of thermal motion is taken into account in much the same way as the Debye–Hückel model describes the ionic atmosphere of an ion (Section 10.2c) with the latter's single central ion replaced by an infinite, plane electrode.

Figure 29.2 shows how the local concentrations of cations and anions differ in the Gouy–Chapman model from their bulk concentrations. Ions of opposite charge cluster close to the electrode, and ions of the same charge are repelled from it. The modification of the local concentrations near an electrode implies that it might be misleading to use activity coefficients characteristic of the bulk to discuss the thermodynamic properties of ions near the interface. This is one of the reasons why measurements in dynamic electrochemistry are almost always done using a large excess of supporting electrolyte (for example, a 1 M solution of a salt, an acid, or a base). Under such conditions, the activity coefficients are almost constant because the inert ions dominate the effects of local changes caused by any reactions taking place.[1]

Neither the Helmholtz nor the Gouy–Chapman model is a very good representation of the structure of the double layer. The former overemphasizes the rigidity of the local solution; the latter underemphasizes its structure. The two are combined in the **Stern model**, in which the ions closest to the electrode are constrained into a rigid Helmholtz plane while outside that plane the ions are dispersed as in the Gouy–Chapman model.

(b) The electric potential at the double layer

The potential at the interface can be analysed by imagining the separation of the electrode from the solution, but with the charges of the metal and the solution frozen in position. A positive test charge at great distances from the isolated electrode experiences a Coulomb

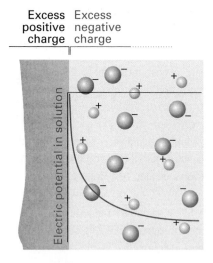

Excess positive charge | Excess negative charge

29.2 The Gouy–Chapman model of the electrical double layer treats the outer region as an atmosphere of counter-charge, similar to the Debye–Hückel theory of ion atmospheres.

---

1   The use of a concentrated solution also minimizes ion migration effects and ohmic losses.

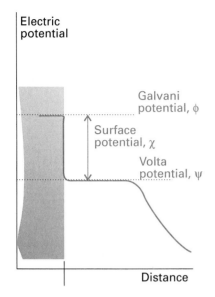

29.3 The variation of potential with distance from an electrode that has been separated from the electrolyte solution without there being an adjustment of charge. A similar diagram applies to the separated solution.

potential that varies inversely with distance (Fig. 29.3). As the test charge approaches the electrode it enters a region where the potential varies more slowly. This change in behaviour can be traced to the fact that the surface charge is not point-like but is spread over an area. At about 100 nm from the surface the potential varies only slightly with distance, and its value in this region is called the **Volta potential**, or the **outer potential**, ψ. As the test charge is taken through the skin of electrons on the surface of the electrode, the potential it experiences changes until the probe reaches the inner, bulk metal environment. where the potential is called the **Galvani potential**, ϕ. The difference between the Galvani and Volta potentials is called the **surface potential**, χ.

A similar sequence of changes of potential is observed as a positive test charge is brought up to and through the solution surface. The potential changes to its Volta value as the charge approaches the charged medium, then to its Galvani value as the probe is taken into the bulk.

Now consider bringing the electrode and solution back together again but without any change of charge distribution. The potential difference between points in the bulk metal and the bulk solution is the **Galvani potential difference**, $\Delta\phi$. Apart from a constant, this Galvani potential difference is the electrode potential that was at the centre of our discussion in Chapter 10. We shall ignore the constant, and identify changes in $\Delta\phi$ with changes in electrode potential.

Justification 29.1

To demonstrate the relation between $\Delta\phi$ and E, consider the cell $Pt|H_2(g)|H^+(g)\|M^+(aq)|M(s)$ and the half-reactions

$$M^+(aq) + e^- \longrightarrow M(s) \qquad H^+(aq) + e^- \longrightarrow \tfrac{1}{2}H_2(g)$$

The Gibbs energies of these two half-reactions can be expressed in terms of the chemical potentials, μ, of all the species. However, we must take into account the fact that the species are present in phases with different electric potentials. Thus, a cation in a region of positive potential has a higher chemical potential (is chemically more active in a thermodynamic sense) than in a region of zero potential.

The contribution of an electric potential to the chemical potential is calculated by noting that the electrical work of adding a charge ze to a region where the potential is ϕ is $ze\phi$, and therefore that the work per mole is $zF\phi$, where F is the Faraday constant. Because at constant temperature and pressure the maximum electrical work can be identified with the change in Gibbs energy (Section 10.2c), the difference in chemical potential of an ion with and without the electric potential present is $zF\phi$. The chemical potential of an ion in the presence of an electric potential is called its **electrochemical potential**, $\overline{\mu}$. It follows that

$$\overline{\mu} = \mu + zF\phi \qquad [1]$$

where μ is the chemical potential of the species when the potential is zero. When $z = 0$ (a neutral species), the electrochemical potential is equal to the chemical potential.

To express the Gibbs energy for the half-reactions in terms of the electrochemical potentials of the species we note that the cations M^+ are in the solution where the Galvani potential is ϕ_S and the electrons are in the electrode where it is ϕ_M. It follows that

$$\begin{aligned}
\Delta_r G_R &= \overline{\mu}(M) - \{\overline{\mu}(M^+) + \overline{\mu}(e^-)\} \\
&= \mu(M) - \{\mu(M^+) + F\phi_S + \mu(e^-) - F\phi_M\} \\
&= \mu(M) - \mu(M^+) - \mu(e^-) + F\Delta\phi_R
\end{aligned}$$

(a)   (b)

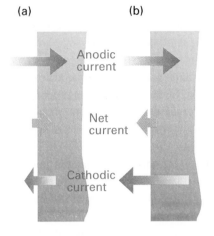

Anodic current

Net current

Cathodic current

29.4 The net current density is defined as the difference between the cathodic and anodic contributions. (a) When $j_a > j_c$, the net current is anodic, and there is a net oxidation of the species in solution. (b) When $j_c > j_a$, the net current is cathodic, and the net process is reduction.

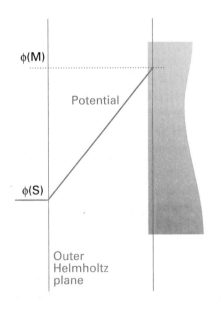

$\phi(M)$

Potential

$\phi(S)$

Outer Helmholtz plane

29.5 The potential, ϕ, varies linearly between two plane parallel sheets of charge, and its effect on the Gibbs energy of the transition state depends on the extent to which the latter resembles the species at the inner or outer planes.

where $\Delta\phi_R = \phi_M - \phi_S$ is the Galvani potential difference at the right-hand electrode. Likewise, in the hydrogen half-reaction, the electrons are in the platinum electrode at a potential ϕ_{Pt} and the H^+ ions are in the solution where the potential is ϕ_S:

$$\Delta_r G_L = \tfrac{1}{2}\overline{\mu}(H_2) - \{\overline{\mu}(H^+) + \overline{\mu}(e^-)\}$$
$$= \tfrac{1}{2}\mu(H_2) - \mu(H^+) - \mu(e^-) + F\Delta\phi_L$$

where $\Delta\phi_L = \phi_{Pt} - \phi_S$ is the Galvani potential difference at the left-hand electrode.

The overall reaction Gibbs energy is

$$\Delta_r G_R - \Delta_r G_L = \mu(M) + \mu(H^+) - \mu(M^+) - \tfrac{1}{2}\mu(H_2) + F(\Delta\phi_R - \Delta\phi_L)$$
$$= \Delta_r G + F(\Delta\phi_R - \Delta\phi_L)$$

where $\Delta_r G$ is the Gibbs energy of the cell reaction. When the cell is balanced against an external source of potential the entire system is at equilibrium. The overall reaction Gibbs energy is then zero,[2] and the last equation becomes

$$0 = \Delta_r G + F(\Delta\phi_R - \Delta\phi_L)$$

which rearranges to

$$\Delta_r G = -F(\Delta\phi_R - \Delta\phi_L) \tag{2}$$

If we compare this with the result established in Section 10.4e that

$$\Delta_r G = -FE \qquad E = E_R - E_L \tag{3}$$

we can conclude that

$$E_R - E_L = \Delta\phi_R - \Delta\phi_L \tag{4}$$

This is the result we wanted to show, for it implies that the Galvani potential difference at each electrode can differ from the electrode potential by a constant at most; that constant cancels when the difference is taken.

29.2 The rate of charge transfer

Because an electrode reaction is heterogeneous, it is natural to express its rate as the flux of products, the amount of material produced over a region of the electrode surface in an interval of time divided by the area of the region and the duration of the interval.

(a) The rate laws

A first-order heterogeneous rate law has the form

$$\text{Product flux} = k[\text{species}] \tag{5}$$

where [species] is the molar concentration of the relevant species in solution close to the electrode, just outside the double layer. The rate constant has dimensions of length/time (for example, $cm\,s^{-1}$). If the molar concentrations of the oxidized and reduced materials outside the double layer are [Ox] and [Red] respectively, the rate of reduction of Ox, v_{Ox}, is

$$v_{Ox} = k_c[\text{Ox}] \tag{6a}$$

and the rate of oxidation of Red, v_{Red}, is

$$v_{Red} = k_a[\text{Red}] \tag{6b}$$

(The notation k_c and k_a is justified below.)

---

2   The cell reaction itself is not necessarily at equilibrium: its tendency to change is balanced against the external source of potential and overall there is stalemate.

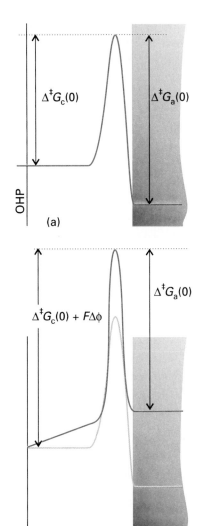

29.6 When the transition state resembles a species that has undergone reduction, the activation Gibbs energy for the anodic current is almost unchanged, but the full effect applies to the cathodic current. (a) Zero potential difference; (b) nonzero potential difference.

Now consider a reaction at the electrode in which an ion is reduced by the transfer of a single electron in the rate-determining step.[3] Then the net current density at the electrode is the difference between the current densities arising from the reduction of Ox and the oxidation of Red. Because the redox processes at the electrode involve the transfer of one electron per reaction event, the current densities, j, arising from the redox processes are the rates (as expressed above) multiplied by the charge transferred per mole of reaction, which is given by the Faraday constant. Therefore, there is a **cathodic current density** of magnitude

$$j_c = Fk_c[\text{Ox}] \qquad (7a)$$

arising from the reduction. There is also an opposing **anodic current density** of magnitude

$$j_a = Fk_a[\text{Red}] \qquad (7b)$$

arising from the oxidation.[4] The net current density at the electrode (Fig. 29.4) is the difference

$$j = j_a - j_c = Fk_a[\text{Red}] - Fk_c[\text{Ox}] \qquad (8)$$

Note that when $j_a > j_c$, so that $j > 0$, the current is anodic (Fig. 29.4a); when $j_c > j_a$, so that $j < 0$, the current is cathodic (Fig 29.4b).

(b) The activation Gibbs energy

If a species is to participate in reduction or oxidation at an electrode, it must discard any solvating molecules, migrate through the electrical double layer, and adjust its hydration sphere as it receives or discards electrons. Likewise, a species already at the inner plane must be detached and migrate into the bulk. Because both processes are activated, we can expect to write their rate constants in the form suggested by activated complex theory (Section 27.6) as

$$k = Be^{-\Delta^{\ddagger}G/RT} \qquad (9)$$

where $\Delta^{\ddagger}G$ is the activation Gibbs energy and B is a constant with the same dimensions as k.

When eqn 9 is inserted into eqn 8 we obtain

$$j = FB_a[\text{Red}]e^{-\Delta^{\ddagger}G_a/RT} - FB_c[\text{Ox}]e^{-\Delta^{\ddagger}G_c/RT} \qquad (10)$$

This expression allows the activation Gibbs energies to be different for the cathodic and anodic processes. That they are different is the central feature of the remaining discussion.

(c) The Butler–Volmer equation

Now we relate j to the Galvani potential difference, which varies across the double layer as shown schematically in Fig. 29.5.

Consider the reduction reaction, $\text{Ox} + e^- \rightarrow \text{Red}$, and the corresponding reaction profile. If the transition state of the activated complex is product-like (as represented by the peak of the reaction profile being close to the electrode in Fig. 29.6), the activation Gibbs energy is changed from $\Delta^{\ddagger}G_c(0)$, the value it has in the absence of a potential difference across the double layer, to

$$\Delta^{\ddagger}G_c = \Delta^{\ddagger}G_c(0) + F\Delta\phi \qquad (11)$$

Thus, if the electrode is more positive than the solution, $\Delta\phi > 0$, then more work has to be done to form an activated complex from Ox; in this case the activation Gibbs energy is

---

3 The last phrase is important: in the deposition of cadmium, for instance, only one electron is transferred in the rate-determining step even though overall the deposition involves the transfer of two electrons.

4 Recall from Chapter 10 that the cathode is the site of reduction; hence the cathodic current is the current arising from reduction. The anode is the site of oxidation, and the anodic current is the current arising from oxidation.

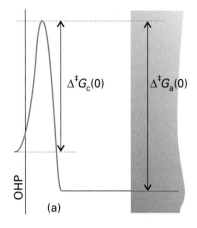

(a)

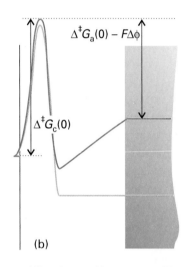

(b)

29.7 When the transition state resembles a species that has undergone oxidation, the cathodic current activation Gibbs energy is almost unchanged but the anodic current activation Gibbs energy is strongly affected. (a) Zero potential difference; (b) nonzero potential difference.

increased. If the transition state is reactant-like (represented by the peak of the reaction profile being close to the outer plane of the double layer in Fig. 29.7), then $\Delta^{\ddagger}G_c$ is independent of $\Delta\phi$. In a real system, the transition state has an intermediate resemblance to these extremes (Fig. 29.8) and the activation Gibbs energy for reduction may be written as

$$\Delta^{\ddagger}G_c = \Delta^{\ddagger}G_c(0) + \alpha F\Delta\phi \tag{12}$$

The parameter α is called the (cathodic) **transfer coefficient**, and lies in the range 0 to 1. Experimentally, α is often found to be about 0.5.

Now consider the oxidation reaction, $\text{Red} + e^- \rightarrow \text{Ox}$ and its reaction profile. Similar remarks apply. In this case, Red discards an electron to the electrode, so the extra work is zero if the transition state is reactant-like (represented by a peak close to the electrode, Fig. 29.6). The extra work is the full $-F\Delta\phi$ if it resembles the product (the peak close to the outer plane, Fig. 29.7. In general (Fig. 29.8), the activation Gibbs energy for this anodic process is

$$\Delta^{\ddagger}G_a = \Delta^{\ddagger}G_a(0) - (1-\alpha)F\Delta\phi \tag{13}$$

The two activation Gibbs energies can now be inserted in place of the values used in eqn 10 with the result that

$$j = FB_a[\text{Red}]e^{-\Delta^{\ddagger}G_a(0)/RT}e^{(1-\alpha)F\Delta\phi/RT} - FB_c[\text{Ox}]e^{-\Delta^{\ddagger}G_c(0)/RT}e^{-\alpha F\Delta\phi/RT} \tag{14}$$

This is an explicit, if complicated, expression for the net current density in terms of the potential difference.

The appearance of eqn 14 can be simplified. First, in a purely cosmetic step we write

$$f = \frac{F}{RT} \tag{15}$$

Next, we identify the individual cathodic and anodic current densities:

$$\left.\begin{array}{l} j_a = FB_a[\text{Red}]e^{-\Delta^{\ddagger}G_a(0)/RT}e^{(1-\alpha)f\Delta\phi} \\ j_c = FB_c[\text{Ox}]e^{-\Delta^{\ddagger}G_c(0)/RT}e^{-\alpha f\Delta\phi} \end{array}\right\} \quad j = j_a - j_c \tag{16}$$

Example 29.1 Calculating the current density

Calculate the change in cathodic current density at an electrode when the potential difference changes from 1.0 V to 2.0 V at 25°C.

Method Equation 16 can be used to express the ratio of cathodic current densities j_c' and j_c when the potential differences are $\Delta\phi'$ and $\Delta\phi$. Then insert a typical value of α (namely, $\frac{1}{2}$) and the data.

Answer The ratio is

$$\frac{j_c'}{j_c} = e^{-\alpha f(\Delta\phi' - \Delta\phi)}$$

So, with $\alpha = \frac{1}{2}$ and $\Delta\phi' - \Delta\phi = 1.0$ V,

$$\alpha f \times (\Delta\phi' - \Delta\phi) = \frac{\frac{1}{2} \times (9.6485 \times 10^4 \text{ C mol}^{-1}) \times (1.0 \text{ V})}{(8.3145 \text{ J K}^{-1} \text{ mol}^{-1}) \times (298 \text{ K})} = 19.47$$

(At this stage, we are preserving an unwarranted number of significant figures.) Hence,

$$\frac{j_c'}{j_c} = e^{-19.47} = 4 \times 10^{-9}$$

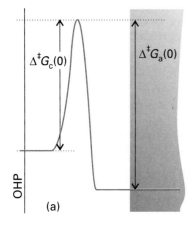

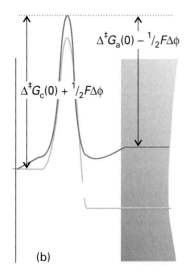

29.8 When the transition state is intermediate in its resemblance to reduced and oxidized species, as represented here by a peak located at an intermediate position as measured by α (with $0 < \alpha < 1$), both activation Gibbs energies are affected; here, $\alpha \approx 0.5$.

Comment This huge change in current density occurs for a very mild and easily applied change of conditions. We can appreciate why the change is so great by realizing that a change of potential difference by 1 V changes the activation Gibbs energy by $(1 \text{ V}) \times F$, or about 50 kJ mol^{-1}, which has an enormous effect on the rates.

Self-test 29.1 Calculate the change in anodic current density under the same circumstances.

$$[j'_a/j_a = 3 \times 10^8]$$

If the cell is balanced against an external source, the difference in Galvani potential, $\Delta\phi$, can be identified as the (zero-current) electrode potential, E, and we can write[5]

$$j_a = FB_a[\text{Red}]e^{-\Delta^{\ddagger}G_a(0)/RT}e^{(1-\alpha)fE}$$
$$j_c = FB_c[\text{Ox}]e^{-\Delta^{\ddagger}G_c(0)/RT}e^{-\alpha fE} \tag{17}$$

When these equations apply, there is no net current at the electrode (as the cell is balanced), and so the two current densities must be equal. From now on we denote them both as j_0, which is called the **exchange current density**.

When the cell is producing current (that is, when a load is connected between the electrode being studied and a second counter-electrode) the electrode potential changes from its zero-current value, E, to its working value, E', and the difference is the **overpotential**, η:

$$\eta = E' - E \tag{18}$$

Hence, $\Delta\phi$ changes to

$$\Delta\phi = E + \eta \tag{19}$$

It follows that the two current densities become

$$j_a = j_0 e^{(1-\alpha)f\eta} \qquad j_c = j_0 e^{-\alpha f\eta} \tag{20}$$

Then from eqn 14 we obtain the **Butler–Volmer equation**:

$$j = j_0\left\{e^{(1-\alpha)f\eta} - e^{-\alpha f\eta}\right\} \tag{21}$$

The Butler–Volmer equation is the basis of all that follows.

(d) The low overpotential limit

When the overpotential is so small that $f\eta \ll 1$ (in practice, η less than about 0.01 V) the exponentials in eqn 21 can be expanded by using $e^x = 1 + x + \cdots$ to give

$$j = j_0\{1 + (1-\alpha)f\eta + \cdots - (1 - \alpha f\eta + \cdots)\} \approx j_0 f\eta \tag{22}$$

This equation shows that the current density is proportional to the overpotential, so at low overpotentials the interface behaves like a conductor that obeys Ohm's law. When there is a small positive overpotential the current is anodic ($j > 0$ when $\eta > 0$), and when the overpotential is small and negative the current is cathodic ($j < 0$ when $\eta < 0$). The relation can also be reversed to calculate the potential difference that must exist if a current density j has been established by some external circuit:

$$\eta = \frac{RTj}{Fj_0} \tag{23}$$

The importance of this interpretation will become clear below.

------

5   We are assuming that we can identify the Galvani potential and the zero-current electrode potential. As explained earlier, they differ by a constant amount, which may be regarded as absorbed into the constant B.

Illustration

The exchange current density of a $Pt(s)|H_2(g)|H^+(aq)$ electrode at 298 K is 0.79 mA cm^{-2}. Therefore, the current density when the overpotential is +5.0 mV is obtained by using eqn 22 and $f = F/RT = 1/(25.69 \text{ mV})$:

$$j = j_0 f\eta = \frac{(0.79 \text{ mA cm}^{-2}) \times (5.0 \text{ mV})}{25.69 \text{ mV}} = 0.15 \text{ mA cm}^{-2}$$

The current through an electrode of total area 5.0 cm^2 is therefore 0.75 mA.

Self-test 29.2 What would be the current at pH = 2.0, the other conditions being the same?

[−18 mA (cathodic)]

(e) The high overpotential limit

When the overpotential is large and positive (in practice, $\eta \geq 0.12$ V), corresponding to the electrode being the anode in electrolysis, the second exponential in eqn 21 is much smaller than the first, and may be neglected. Then

$$j = j_0 e^{(1-\alpha)f\eta} \tag{24}$$

so

$$\ln j = \ln j_0 + (1-\alpha)f\eta \tag{25}$$

When the overpotential is large but negative (in practice, $\eta \leq -0.12$ V), corresponding to the cathode in electrolysis, the first exponential in eqn 21 may be neglected. Then

$$j = -j_0 e^{-\alpha f\eta} \tag{26}$$

so

$$\ln(-j) = \ln j_0 - \alpha f\eta \tag{27}$$

The plot of the logarithm of the current density against the overpotential is called a **Tafel plot**. The slope gives the value of α and the intercept at $\eta = 0$ gives the exchange current density.

The experimental arrangement used for a Tafel plot is shown in Fig. 29.9. The electrode of interest is called the **working electrode**, and the current flowing through it is controlled externally. If its area is A and the current is I, the current density across its surface is I/A. The potential difference across the interface cannot be measured directly, but the potential of the working electrode relative to a third electrode, the **reference electrode**, can be measured with a high impedance voltmeter, and no current flows in that half of the circuit. The reference electrode is in contact with the solution close to the working electrode through a 'Luggin capillary', which helps to eliminate any ohmic potential difference that might arise accidentally. Changing the current flowing through the working circuit causes a change of potential of the working electrode, and that change is measured with the voltmeter. The overpotential is then obtained by taking the difference between the potentials measured with and without a flow of current through the working circuit.

29.9 The general arrangement for electrochemical rate measurements. The external source establishes a current between the working electrodes, and its effect on the potential difference of either of them relative to the reference electrode is observed. No current flows in the reference circuit.

Example 29.2 Interpreting a Tafel plot

The data below refer to the anodic current through a platinum electrode of area 2.0 cm^2 in contact with an Fe^{3+}, Fe^{2+} aqueous solution at 298 K. Calculate the exchange current density and the transfer coefficient for the electrode process.

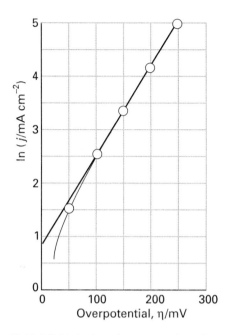

29.10 A Tafel plot is used to measure the exchange current density (given by the extrapolated intercept at $\eta = 0$) and the transfer coefficient (from the slope). The data are from Example 29.2.

η/mV	50	100	150	200	250
I/mA	8.8	25.0	58.0	131	298

Method The anodic process is the oxidation $Fe^{2+}(aq) \rightarrow Fe^{3+}(aq) + e^-$. To analyse the data, we make a Tafel plot (of $\ln j$ against η) using the anodic form (eqn 25). The intercept at $\eta = 0$ is $\ln j_0$ and the slope is $(1 - \alpha)f$.

Answer Draw up the following table:

η/mV	50	100	150	200	250
$j/(\text{mA cm}^{-2})$	4.4	12.5	29.0	56.6	149
$\ln(j/\text{mA cm}^{-2})$	1.48	2.53	3.37	4.18	5.00

The points are plotted in Fig. 29.10. The high overpotential region gives a straight line of intercept 0.92 and slope 0.0163. From the former it follows that $\ln(j_0/\text{mA cm}^{-2}) = 0.92$, so $j_0 = 2.5 \text{ mA cm}^{-2}$. From the latter,

$$(1 - \alpha)\frac{F}{RT} = 0.0163 \text{ mV}^{-1}$$

so $\alpha = 0.58$.

Comment Note that the Tafel plot is nonlinear for $\eta < 100$ mV; in this region $\alpha f \eta = 2.3$ and the approximation that $\alpha f \eta \gg 1$ fails.

Self-test 29.3 Repeat the analysis using the following cathodic current data:

η/mV	−50	−100	−150	−200	−250	−300
I/mA	−0.3	−1.5	−6.4	−27.6	−118.6	−510

$$[\alpha = 0.75, j_0 = 0.040 \text{ mA cm}^{-2}]$$

Some experimental values for the Butler–Volmer parameters are given in Table 29.1. From them we can see that exchange current densities vary over a very wide range. For example, the N_2, N_3^- couple on platinum has $j_0 = 10^{-76} \text{ A cm}^{-2}$, whereas the H^+, H_2 couple on platinum has $j_0 = 8 \times 10^{-4} \text{ A cm}^{-2}$, a difference of 73 orders of magnitude. Exchange currents are generally large when the redox process involves no bond breaking (as in the $[Fe(CN)_6]^{3-}$, $[Fe(CN)_6]^{4-}$ couple) or if only weak bonds are broken (as in Cl_2, Cl^-). They are generally small when more than one electron needs to be transferred, or when multiple or strong bonds are broken, as in the N_2, N_3^- couple and in redox reactions of organic compounds.

Table 29.1* Exchange current densities and transfer coefficients at 298 K

Reaction	Electrode	$j_0/(\text{A cm}^{-2})$	α
$2H^+ + 2e^-$	Pt	7.9×10^{-4}	
$\rightarrow H_2$	Ni	6.3×10^{-6}	0.58
	Pb	5.0×10^{-12}	
$Fe^{3+} + e^-$	Pt	2.5×10^{-3}	0.58
$\rightarrow Fe^{2+}$			

* More values are given in the *Data section* at the end of this volume.

29.3 Polarization

Electrodes with potentials that change only slightly when a current passes through them are classified as **non-polarizable**. Those with strongly current-dependent potentials are classified as **polarizable**. From the linearized equation (eqn 23) it is clear that the criterion for low polarizability is high exchange current density (so η may be small even though j is large). The calomel and H_2/Pt electrodes are both highly non-polarizable, which is one reason why they are so extensively used in equilibrium electrochemistry measurements.

(a) Concentration polarization

One of the assumptions in the derivation of the Butler–Volmer equation is the negligible conversion of the electroactive species at low current densities, resulting in uniformity of concentration near the electrode. This assumption fails at high current densities because the consumption of electroactive species close to the electrode results in a concentration gradient; diffusion of the species towards the electrode from the bulk is slow and may become rate-determining. A larger overpotential is then needed to produce a given current. This effect is called **concentration polarization** and its contribution to the total overpotential is called the **polarization overpotential**, η^c.

Consider a case for which the concentration polarization dominates all the rate processes and a redox couple of the type M^{z+}, M. Under zero-current conditions, when the net current density is zero, the electrode potential is related to the activity, a, of the ions in the solution by the Nernst equation (eqn 10.45):

$$E = E^\ominus + \frac{RT}{zF}\ln a \tag{28}$$

As remarked earlier, electrode kinetics are normally studied using a large excess of support electrolyte so as to keep the mean activity coefficients approximately constant. Therefore, the constant activity coefficient in $a = \gamma c$ may be absorbed into E, and we write the **formal potential**, E°, of the electrode as

$$E^\circ = E^\ominus + \frac{RT}{zF}\ln \gamma \tag{29}$$

Then the electrode potential is

$$E = E^\circ + \frac{RT}{zF}\ln c \tag{30}$$

When the cell is producing current, the active ion concentration at the OHP changes to c' and the electrode potential changes to

$$E' = E^\circ + \frac{RT}{zF}\ln c' \tag{31}$$

The concentration overpotential is therefore

$$\eta^c = E' - E = \frac{RT}{zF}\ln\left(\frac{c'}{c}\right) \tag{32}$$

We now suppose that the solution has its bulk concentration, c, up to a distance δ from the outer Helmholtz plane, and then falls linearly to c' at the plane itself. This **Nernst diffusion layer** is illustrated in Fig. 29.11. The thickness of the Nernst layer (which is typically 0.1 mm, and strongly dependent on the condition of hydrodynamic flow due to any stirring or convective effects) is quite different from that of the electrical double layer (which is typically less than 1 nm, and unaffected by stirring). The concentration gradient through the Nernst layer is

$$\frac{dc}{dx} = \frac{c' - c}{\delta} \tag{33}$$

This gradient gives rise to a flux of ions towards the electrode, which replenishes the cations as they are reduced. The (molar) flux, J, is proportional to the concentration gradient, and according to Fick's law (Section 24.3)

$$J = -D\left(\frac{\partial c}{\partial x}\right) \tag{34}$$

Therefore, the particle flux towards the electrode is

$$J = D\frac{c - c'}{\delta} \tag{35}$$

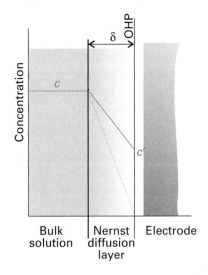

29.11 In a simple model of the Nernst diffusion layer there is a linear variation in concentration between the bulk and the outer Helmholtz plane; the thickness of the layer depends strongly on the state of flow of the fluid.

The cathodic current density towards the electrode is the product of the particle flux and the charge per mole of ions, zF:

$$j = zFJ = zFD\frac{c - c'}{\delta} \tag{36}$$

The maximum rate of diffusion across the Nernst layer occurs when the gradient is steepest, which is when $c' = 0$. This concentration occurs when an electron from an ion that diffuses across the layer is snapped over the activation barrier and on to the electrode. No flow of current can exceed the **limiting current density**, j_{lim}, which is given by

$$j_{lim} = zFJ_{lim} = \frac{zFDc}{\delta} \tag{37}$$

Example 29.3 Estimating the limiting current density

Estimate the limiting current density at 298 K for an electrode in a $0.10\,M\,Cu^{2+}(aq)$ unstirred solution in which the thickness of the diffusion layer is about 0.3 mm.

Method To use eqn 37, estimate D from the ionic conductivity $\lambda = 107\ S\,cm^2\,mol^{-1}$ (Table 24.4) and the Nernst–Einstein equation (eqn 24.72).

Answer Because $\lambda = z^2F^2D/RT$, eqn 37 may be written

$$j_{lim} = \frac{cRT\lambda}{zF\delta}$$

Therefore, with $\delta = 0.3$ mm, $c = 0.10\ mol\,L^{-1}$, $z = 2$, and $T = 298$ K, we obtain $j_{lim} = 5\ mA\,cm^{-2}$. The result implies that the current towards an electrode of area 1 cm^2 electrode cannot exceed 5 mA in this (unstirred) solution.

- -

Self-test 29.4 Evaluate the limiting current density for an $Ag(s)|Ag^+(aq)$ electrode in a $0.010\,M\,Ag^+(aq)$ solution at 298 K. Take $\delta = 0.03$ mm.

$$[5\ mA\,cm^{-2}]$$

It follows from eqn 36 that the concentration c' is related to the current density at the double layer by

$$c' = c - \frac{j\delta}{zFD} \tag{38}$$

Hence, as the current density is increased, the concentration falls below the bulk value. However, this decline in concentration is small when the diffusion constant is large, for then the ions are very mobile and can quickly replenish any ions that have been removed.

Finally, we substitute eqn 38 into eqn 32 and obtain the following expressions for the overpotential in terms of the current density, and vice versa:

$$\eta^c = \frac{RT}{zF}\ln\left(1 - \frac{j\delta}{zcFD}\right)$$
$$j = \frac{zcFD}{\delta}\left(1 - e^{zf\eta^c}\right) \tag{39}$$

(b) Voltammetry

The kinetics of electrode processes can be studied by **voltammetry**, in which the current is monitored as the potential of the electrode is changed, and by **chronopotentiometry**, in

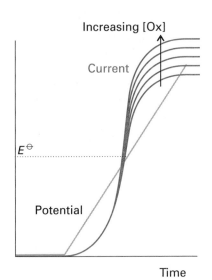

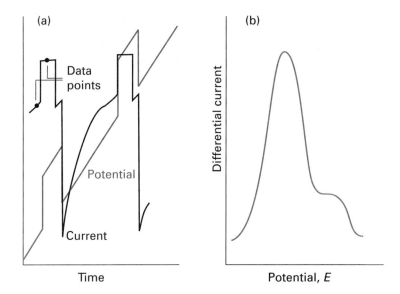

29.12 The change of potential with time and the resulting current/potential curve in a linear-sweep voltammetry experiment. The limiting value of the current density is proportional to the concentration of electroactive species (for instance, [Ox]) in solution.

29.13 A differential pulse polarography experiment. (a) The potential is swept linearly as a mercury droplet grows on the end of a capillary dipping into the sample, and then pulsed as shown by the green line. The resulting current is shown as the black line, and is sampled at the two points shown. (b) The data output is obtained as the difference of the currents at the two sampled points.

which the potential is monitored as the current flow is changed. Voltammetry may also be used to identify species present in solution and to determine their concentration.

The kind of output from **linear-sweep voltammetry** is illustrated in Fig. 29.12. Initially the absolute value of the potential is low, and the cathodic current is due to the migration of ions in the solution. However, as the potential approaches the reduction potential of the reducible solute, the cathodic current grows. Soon after the potential exceeds the reduction potential the current rises to a plateau at which it has its limiting value (as specified in eqn 37). This limiting current is proportional to the molar concentration of the species, so that concentration can be determined from the height of the plateau above the extrapolated baseline. In **differential pulse voltammetry** the current is monitored before and after a pulse of potential is applied, and the processed output is the slope of a curve like that obtained by linear-sweep voltammetry (Fig. 29.13). The area under the curve (in effect, the integral of the derivative displayed in the illustration) is proportional to the concentration of the species.

In **cyclic voltammetry** the potential is applied in a sawtooth manner to the working electrode and the current is monitored. A typical cyclic voltammogram is shown in Fig. 29.14. The shape of the curve is initially like that of a linear-sweep experiment, but after the absolute value of the potential begins to fall there is a rapid change in current on account of the high concentration of oxidizable species close to the electrode that were generated on the reductive sweep. When the potential is close to the value required to oxidize the reduced species, there is a substantial anodic current until all the oxidation is complete, and the current returns to zero.

When the reduction reaction at the electrode can be reversed, as in the case of the $[Fe(CN)_6]^{3-}/[Fe(CN)_6]^{4-}$ couple, the cyclic voltammogram is almost symmetric about the standard potential of the couple (as in Fig. 29.14b). The scan is initiated with $[Fe(CN)_6]^{3+}$ present in solution and, as the potential approaches $E^{\ominus}$ for the couple, the $[Fe(CN)_6]^{3-}$ near the electrode is reduced and current begins to flow. As the potential continues to change, the cathodic current begins to decline again because all the $[Fe(CN)_6]^{3-}$ near the electrode has been reduced and the current reaches its limiting value. The potential is now returned linearly to its initial value, and the reverse series of events occurs with the $[Fe(CN)_6]^{4-}$

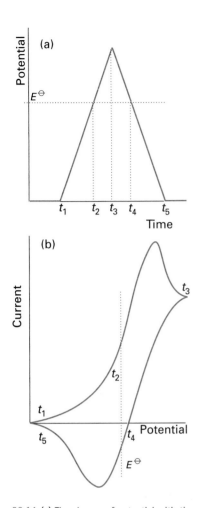

29.14 (a) The change of potential with time and (b) the resulting current/potential curve in a cyclic voltammetry experiment.

produced during the forward scan now undergoing oxidation. The peak of current lies on the other side of $E^{\ominus}$, so the species present and its standard potential can be identified, as indicated in the illustration, by noting the locations of the two peaks.

The overall shape of the curve gives details of the kinetics of the electrode process and the change in shape as the rate of change of potential is altered gives information on the rates of the processes involved. For example, the matching peak on the return phase of the sawtooth change of potential may be missing, which indicates that the oxidation (or reduction) is irreversible. The appearance of the curve may also depend on the timescale of the sweep for, if the sweep is too fast, some processes might not have time to occur. This style of analysis is illustrated in the following example.

Example 29.4 Analysing a cyclic voltammetry experiment

The electroreduction of p-bromonitrobenzene in liquid ammonia is believed to occur by the following mechanism:

$$BrC_6H_4NO_2 + e^- \longrightarrow BrC_6H_4NO_2^-$$
$$BrC_6H_4NO_2^- \longrightarrow \cdot C_6H_4NO_2 + Br^-$$
$$\cdot C_6H_4NO_2 + e^- \longrightarrow C_6H_4NO_2^-$$
$$C_6H_4NO_2^- + H^+ \longrightarrow C_6H_5NO_2$$

Suggest the likely form of the cyclic voltammogram expected on the basis of this mechanism.

Method Decide which steps are likely to be reversible on the timescale of the potential sweep: such processes will give symmetrical voltammograms. Irreversible processes will give unsymmetrical shapes as reduction (or oxidation) might not occur. However, at fast sweep rates, an intermediate might not have time to react, and a reversible shape will be observed.

Answer At slow sweep rates, the second reaction has time to occur, and a curve typical of a two-electron reduction will be observed, but there will be no oxidation peak on the second half of the cycle because the product, $C_6H_5NO_2$, cannot be oxidized (Fig. 29.15a). At fast sweep rates, the second reaction does not have time to take place before oxidation of the $BrC_6H_4NO_2^-$ intermediate starts to occur during the reverse scan, so the voltammogram will be typical of a reversible one-electron reduction (Fig. 29.15b).

29.15 (a) When a non-reversible step in a reaction mechanism has time to occur, the cyclic voltammogram may not show the reverse oxidation or reduction peak. (b) However, if the rate of sweep is increased, the return step may be caused to occur before the irreversible step has had time to intervene, and a typical 'reversible' voltammogram is obtained.

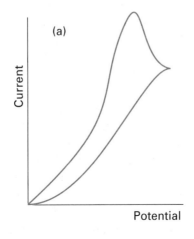

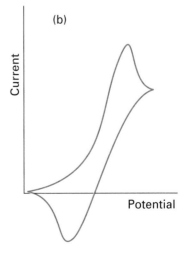

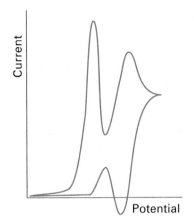

29.16 The cyclic voltammogram referred to in Self-test 29.5.

Self-test 29.5 Suggest an interpretation of the cyclic voltammogram shown in Fig. 29.16. The electroactive material is ClC_6H_4CN in acid solution; after reduction to $ClC_6H_4CN^-$, the radical anion may form C_6H_5CN irreversibly.

$$[ClC_6H_4CN + e^- \rightleftarrows ClC_6H_4CN^-,$$
$$ClC_6H_4CN^- + H^+ + e^- \rightarrow C_6H_5CN + Cl^-,$$
$$C_6H_5CN + e^- \rightleftarrows C_6H_5CN^-]$$

Electrochemical processes

To induce current to flow through an electrolytic cell and bring about a non-spontaneous cell reaction, the applied potential difference must exceed the zero-current potential by at least the **cell overpotential**. The cell overpotential is the sum of the overpotentials at the two electrodes and the ohmic drop (IR_s, where R_s is the internal resistance of the cell) due to the current through the electrolyte. The additional potential needed to achieve a detectable rate of reaction may need to be large when the exchange current density at the electrodes is small. For similar reasons, a working galvanic cell generates a smaller potential than under zero-current conditions. In this section we see how to cope with both aspects of the overpotential.

29.4 Electrolysis

The rate of gas evolution or metal deposition during electrolysis can be estimated from the Butler–Volmer equation and tables of exchange current densities. The exchange current density depends strongly on the nature of the electrode surface, and changes in the course of the electrodeposition of one metal on another. A very crude criterion is that significant evolution or deposition occurs only if the overpotential exceeds about 0.6 V.

Example 29.5 Estimating the relative rates of electrolysis

Derive an expression for the relative rates of electrodeposition and hydrogen evolution in a solution in which both may occur.

Method We can calculate the ratio of the cathodic currents by using eqn 26. For simplicity, assume equal transfer coefficients.

Answer From eqn 26, where j' is the current density for electrodeposition and j is that for hydrogen evolution, and j'_0 and j_0 are the corresponding exchange current densities.

$$\frac{j'}{j} = \frac{j'_0}{j_0} e^{(\eta - \eta')\alpha f}$$

Comment This equation shows that metal deposition is favoured by a large exchange current density and relatively high hydrogen evolution overvoltage (so $\eta - \eta'$ is positive and large). Note that $\eta < 0$ for a cathodic process, $-\eta' > 0$.

- -

Self-test 29.6 Deduce an expression for the ratio when the hydrogen evolution is limited by transport across a diffusion layer.

$$[j'/j = (\delta j'_0/cFD)e^{-\alpha\eta' f}]$$

A glance at Table 29.1 shows the wide range of exchange current densities for a metal/hydrogen electrode. The most sluggish exchange currents occur for lead and mercury, and the value of $1 \ \mathrm{pA \ cm^{-2}}$ corresponds to a monolayer of atoms being replaced in about 5 y. For such systems, a high overpotential is needed to induce significant hydrogen evolution. In contrast, the value for platinum ($1 \ \mathrm{mA \ cm^{-2}}$) corresponds to a monolayer being replaced in 0.1 s, so gas evolution occurs for a much lower overpotential.

The exchange current density also depends on the crystal face exposed. For the deposition of copper on copper, the (100) face has $j_0 = 1 \ \mathrm{mA \ cm^{-2}}$, so for the same overpotential the (100) face grows at 2.5 times the rate of the (111) face, for which $j_0 = 0.4 \ \mathrm{mA \ cm^{-2}}$.

29.5 The characteristics of working cells

We expect the cell potential to decrease as current is generated because it is then no longer working reversibly and can therefore do less than maximum work.

(a) The potentials of working cells

We shall consider the cell $M|M^+(aq)\|M'^+(aq)|M'$ and ignore all the complications arising from liquid junctions. The working potential of the cell is

$$E' = \Delta\phi_R - \Delta\phi_L \tag{40}$$

Because the working potential differences differ from their zero-current values by overpotentials, we can write

$$\Delta\phi_X = E_X + \eta_X \tag{41}$$

where X is L or R for the left or right electrode, respectively. The working cell potential is therefore

$$E' = E + \eta_R - \eta_L \tag{42}$$

with E the zero-current cell potential. We should subtract from this expression the ohmic potential difference IR_s, where R_s is the cell's internal resistance:

$$E' = E + \eta_R - \eta_L - IR_s \tag{43}$$

The ohmic term is a contribution to the cell's irreversibility—it is a thermal dissipation term—so the sign of IR_s is always such as to reduce the potential in the direction of zero.

The overpotentials in eqn 43 can be calculated from the Butler–Volmer equation for a given current, I, being drawn. We shall simplify the equations by supposing that the areas, A, of the electrodes are the same, that only one electron is transferred in the rate-determining steps at the electrodes, that the transfer coefficients are both $\frac{1}{2}$, and that the high-overpotential limit of the Butler–Volmer equation may be used. Then from eqns 26 and 43 we find

$$E' = E - IR_s - \frac{4RT}{F}\ln\left(\frac{I}{A\bar{j}}\right) \qquad \bar{j} = (j_{0L}j_{0R})^{1/2} \tag{44}$$

where j_{0L} and j_{0R} are the exchange current densities for the two electrodes.

The concentration overpotential also reduces the cell potential. If we use the Nernst diffusion layer model for each electrode, the total change of potential arising from

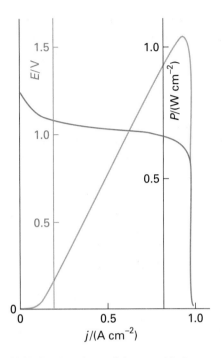

29.17 The dependence of the potential of a working cell on the current being drawn (green line) and the corresponding power output (grey line) calculated by using eqns 46 and 47, respectively. Notice the sharp decline in power just after the maximum.

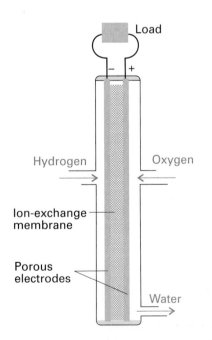

29.18 A single cell of a hydrogen/oxygen fuel cell. In practice, a battery of many cells is used.

concentration polarization is given by eqn 39 as

$$E' = E - \frac{RT}{zF} \ln\left\{\left(1 - \frac{I}{Aj_{\text{lim,L}}}\right)\left(1 - \frac{I}{Aj_{\text{lim,R}}}\right)\right\} \tag{45}$$

This contribution can be added to the one in eqn 44 to obtain a full (but still very approximate) expression for the cell potential when a current I is being drawn:

$$E' = E - IR_s - \frac{2RT}{zF}\ln g(I)$$

$$g(I) = \left(\frac{I}{A\bar{\bar{j}}}\right)^{2z}\left\{\left(1 - \frac{I}{Aj_{\text{lim,L}}}\right)\left(1 - \frac{I}{Aj_{\text{lim,R}}}\right)\right\}^{-1/2} \tag{46}$$

This equation depends on a lot of parameters, but an example of its general form is given in Fig. 29.17. Notice the very steep decline of working potential when the current is high and close to the limiting value for one of the electrodes.

(b) The power output of working cells

Because the power, P, supplied by a working cell is IE', from eqn 46 we can write

$$P = IE - I^2R_s - \frac{2IRT}{zF}\ln g(I) \tag{47}$$

The first term on the right is the power that would be produced if the cell retained its zero-current potential when delivering current. The second term is the power generated uselessly as heat as a result of the resistance of the electrolyte. The third term is the reduction of the potential at the electrodes as a result of drawing current.

The general dependence of power output on the current drawn is shown in Fig. 29.17 as the grey line. Notice how maximum power is achieved just before the concentration polarization quenches the cell's performance. Information of this kind is essential if the optimum conditions for operating electrochemical devices are to be found and their performance improved.

Power production and corrosion

The thermodynamics of galvanic cells were treated in Chapter 10. In this section we consider some kinetic considerations relating to power production and the processes responsible for corrosion.

29.6 Fuel cells and secondary cells

A fuel cell operates like a conventional galvanic cell with the exception that the reactants are supplied from outside rather than forming an integral part of its construction. A fundamental and important example of a fuel cell is the hydrogen/oxygen cell (Fig. 29.18). One of the electrolytes used is concentrated aqueous potassium hydroxide maintained at 200°C and 20-40 atm; the electrodes may be porous nickel in the form of sheets of compressed powder. The cathode reaction is the reduction

$$O_2(g) + 2H_2O(l) + 4e^- \longrightarrow 4OH^-(aq) \qquad E^{\ominus} = +0.40 \text{ V}$$

and the anode reaction is the oxidation

$$H_2(g) + 2OH^-(aq) \longrightarrow 2H_2O(l) + 2e^-$$

For the corresponding reduction, $E^{\ominus} = -0.83$ V. Because the overall reaction

$$2H_2(g) + O_2(g) \longrightarrow 2H_2O(l) \qquad E^{\ominus} = +1.23 \text{ V}$$

is exothermic as well as spontaneous, it is less favourable thermodynamically at 200°C than at 25°C, so the cell potential is lower at the higher temperature. However, the increased pressure compensates for the increased temperature and, at 200°C and 40 atm, $E \approx +1.2$ V.

One advantage of the hydrogen/oxygen system is the large exchange current density of the hydrogen reaction. Unfortunately, the oxygen reaction has an exchange current density of only about 0.1 nA cm^{-2}, which limits the current available from the cell. One way round the difficulty is to use a catalytic surface (to increase j_0) with a large surface area. One type of highly developed fuel cell has phosphoric acid as the electrolyte and operates with hydrogen and air at about 200°C; the hydrogen is obtained from a reforming reaction on natural gas. The power output of batteries of such cells has reached the order of 10 MW. Cells with molten carbonate electrolytes at about 600°C can make use of natural gas directly. Solid-state electrolytes are also used. They include one version in which the electrolyte is a solid polymeric ionic conductor at about 100°C, but in current versions it requires very pure hydrogen to operate successfully. Solid ionic conducting oxide cells operate at about 1000°C and can use hydrocarbons directly as fuel.

Electric storage cells operate as galvanic cells while they are producing electricity but as electrolytic cells while they are being charged by an external supply. The lead–acid battery is an old device, but one well suited to the job of starting cars (and the only one available). During charging the cathode reaction is the reduction of Pb^{2+} and its deposition as lead on the lead electrode. Deposition occurs instead of the reduction of the acid to hydrogen because the latter has a low exchange current density on lead. The anode reaction during charging is the oxidation of Pb(II) to Pb(IV), which is deposited as the oxide PbO$_2$. On discharge, the two reactions run in reverse. Because they have such high exchange current densities the discharge can occur rapidly, which is why the lead battery can produce large currents on demand.

29.7 Corrosion

A thermodynamic warning of the likelihood of corrosion is obtained by comparing the standard potentials of the metal reduction, such as

$$\text{Fe}^{2+}(\text{aq}) + 2\text{e}^- \longrightarrow \text{Fe}(\text{s}) \qquad E^{\ominus} = -0.44 \text{ V}$$

with the values for one of the following half-reactions:

In acidic solution:

(a) $2\text{H}^+(\text{aq}) + 2\text{e}^- \longrightarrow \text{H}_2(\text{g}) \qquad E^{\ominus} = 0$

(b) $4\text{H}^+(\text{aq}) + \text{O}_2(\text{g}) + 4\text{e}^- \longrightarrow 2\text{H}_2\text{O}(\text{l}) \qquad E^{\ominus} = +1.23 \text{ V}$

In basic solution:

(c) $2\text{H}_2\text{O}(\text{l}) + \text{O}_2(\text{g}) + 4\text{e}^- \longrightarrow 4\text{OH}^-(\text{aq}) \qquad E^{\ominus} = +0.40 \text{ V}$

Because all three redox couples have standard potentials more positive than $E^{\ominus}(\text{Fe}^{2+}/\text{Fe})$, all three can drive the oxidation of iron to iron(II). The electrode potentials we have quoted are standard values, and they change with the pH of the medium. For the first two:

$$E(\text{a}) = E^{\ominus}(\text{a}) + \frac{RT}{F} \ln a(\text{H}^+) = -(0.059 \text{ V})\text{pH}$$

$$E(\text{b}) = E^{\ominus}(\text{b}) + \frac{RT}{F} \ln a(\text{H}^+) = 1.23 \text{ V} - (0.059 \text{ V})\text{pH}$$

These expressions let us judge at what pH the iron will have a tendency to oxidize (see Chapter 10). A thermodynamic discussion of corrosion, however, only indicates whether a tendency to corrode exists. If there is a thermodynamic tendency, we must examine the

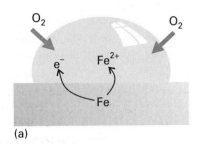

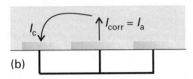

29.19 (a) A simple version of the corrosion process is that of a droplet of water, which is oxygen-rich near its boundary with air. The oxidation of the iron takes place in the region away from the oxygen because the electrons are transported through the metal. (b) The process may be modelled as a short-circuited electrochemical cell.

kinetics of the processes involved to see whether the process occurs at a significant rate.

(a) The rate of corrosion

A model of a corrosion system is shown in Fig. 29.19a. It can be taken to be a drop of slightly acidic (or basic) water containing some dissolved oxygen in contact with the metal. The oxygen at the edges of the droplet, where the O_2 concentration is higher, is reduced by electrons donated by the iron over an area A. Those electrons are replaced by others released elsewhere as $Fe \rightarrow Fe^{2+} + 2e^-$. This oxidative release occurs over an area A' under the oxygen-deficient inner region of the droplet. The droplet acts as a short-circuited galvanic cell (Fig. 29.19b).

The rate of corrosion is measured by the current of metal ions leaving the metal surface in the anodic region. This flux of ions gives rise to the **corrosion current**, I_{corr}, which can be identified with the anodic current, I_a. We show in the *Justification* below that the corrosion current is related to the cell potential of the corrosion couple by

$$I_{corr} = \bar{j}_0 \bar{A} e^{fE/4} \qquad \bar{j}_0 = (j_0 j'_0)^{1/2}, \quad \bar{A} = (AA')^{1/2} \qquad (48)$$

Justification 29.2

Because any current emerging from the anodic region must find its way to the cathodic region, the cathodic current, I_c, and the anodic current, I_a must both be equal to the corrosion current. In terms of the current densities at the oxidation and reduction sites, j and j', respectively, we can write

$$I_{corr} = jA = j'A' = (jj'AA')^{1/2} = \bar{j}\bar{A} \qquad \bar{j} = (jj')^{1/2}, \quad \bar{A} = (AA')^{1/2} \qquad (49)$$

The Butler–Volmer equation is now used to express the current densities in terms of overpotentials. For simplicity we assume that the overpotentials are large enough for the high-overpotential limit (eqn 26) to apply, that polarization overpotential can be neglected, that the rate-determining step is the transfer of a single electron, and that the transfer coefficients are $\frac{1}{2}$. We also assume that, since the droplet is so small, there is negligible potential difference between the cathode and anode regions of the solution. Moreover, because it is short-circuited by the metal, the potential of the metal is the same in both regions, and so the potential difference between the metal and the solution is the same in both regions too; it is denoted $\Delta\phi_{corr}$. The overpotentials in the two regions are therefore

$$\eta = \Delta\phi_{corr} - \Delta\phi \qquad \eta' = \Delta\phi_{corr} - \Delta\phi'$$

and the current densities are

$$j = j_0 e^{\eta f/2} = j_0 e^{f\Delta\phi_{corr}/2} e^{-f\Delta\phi/2}$$
$$j' = j'_0 e^{-\eta' f/2} = j'_0 e^{-f\Delta\phi_{corr}/2} e^{f\Delta\phi'/2}$$

These expressions can be substituted into the expression for I_{corr} and $\Delta\phi' - \Delta\phi$ replaced by the difference of electrode potentials E to give eqn 48.

Several conclusions can be drawn from eqn 48. First, the rate of corrosion depends on the surfaces exposed: if either A or A' is zero, then the corrosion current is zero. This interpretation points to a trivial, yet often effective, method of slowing corrosion: cover the surface with a coating, such as paint. Paint also increases the effective solution resistance between the cathode and anode patches on the surface. Second, for corrosion reactions with similar exchange current densities, the rate of corrosion is high when E is large. That is, rapid corrosion can be expected when the oxidizing and reducing couples have widely differing electrode potentials.

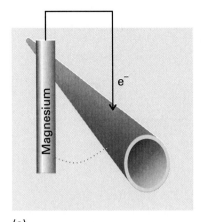

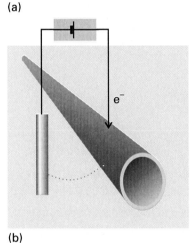

29.20 (a) In cathodic protection an anode of a more strongly reducing metal is sacrificed to maintain the integrity of the protected object (for example, a pipeline, bridge, or boat). (b) In impressed current cathodic protection electrons are supplied from an external cell so that the object itself is not oxidized. The dotted lines depict the completed circuit through the soil.

The effect of the exchange current density on the corrosion rate can be seen by considering the specific case of iron in contact with acidified water. Thermodynamically, either hydrogen or oxygen reduction reaction (a) or (b) on p. 875 is effective. However, the exchange current density of reaction (b) on iron is only about 10^{-14} A cm^{-2}, whereas for (a) it is 10^{-6} A cm^{-2}. The latter therefore dominates kinetically, and iron corrodes by hydrogen evolution in acidic solution.

(b) The inhibition of corrosion

Several techniques for inhibiting corrosion are available. Coating the surface with an impermeable layer, such as paint, may prevent the access of damp air. Unfortunately, this protection fails disastrously if the paint becomes porous. The oxygen then has access to the exposed metal and corrosion continues beneath the paintwork. Another form of surface coating is provided by **galvanizing**, the coating of an iron object with zinc. Because the latter's standard potential is -0.76 V, which is more negative than that of the iron couple, the corrosion of zinc is thermodynamically favoured and the iron survives (the zinc survives because it is protected by a hydrated oxide layer). In contrast, tin plating leads to a very rapid corrosion of the iron once its surface is scratched and the iron exposed because the tin couple ($E^{\ominus} = -0.14$ V) oxidizes the iron couple ($E^{\ominus} = -0.44$ V). Some oxides are inert kinetically in the sense that they adhere to the metal surface and form an impermeable layer over a fairly wide pH range. This passivation, or kinetic protection, can be seen as a way of decreasing the exchange currents by sealing the surface. Thus, aluminium is inert in air even though its standard potential is strongly negative (-1.66 V).

Another method of protection is to change the electric potential of the object by pumping in electrons that can be used to satisfy the demands of the oxygen reduction without involving the oxidation of the metal. In **cathodic protection**, the object is connected to a metal with a more negative standard potential (such as magnesium, -2.36 V). The magnesium acts as a **sacrificial anode**, supplying its own electrons to the iron and becoming oxidized to Mg^{2+} in the process (Fig. 29.20a). A block of magnesium replaced occasionally is much cheaper than the ship, building, or pipeline for which it is being sacrificed. In **impressed-current cathodic protection** (Fig. 29.20b) an external cell supplies the electrons and eliminates the need for iron to transfer its own.

Checklist of key ideas

☐ current density

Processes at electrodes

29.1 The electrical double layer
☐ electrical double layer
☐ Helmholtz model
☐ outer Helmholtz plane (OHP)
☐ inner Helmholtz plane (IHP)
☐ Gouy–Chapman model
☐ diffuse double layer
☐ Stern model
☐ Volta potential
☐ outer potential
☐ Galvani potential

☐ surface potential
☐ electrochemical potential (1)

29.2 The rate of charge transfer
☐ product flux
☐ rate of heterogeneous reaction (5)
☐ cathodic current density
☐ anodic current density
☐ transfer coefficient (12)
☐ exchange current density
☐ overpotential (18)
☐ Butler–Volmer equation (21)
☐ low overpotential limit

(22, 23)
☐ high overpotential limit (25, 27)
☐ Tafel plot
☐ working electrode
☐ reference electrode

29.3 Polarization
☐ non-polarizable electrode
☐ polarizable electrode
☐ concentration polarization
☐ polarization overpotential
☐ formal potential (29)
☐ Nernst diffusion layer
☐ limiting current density (37)

☐ voltammetry
☐ chronopotentiometry
☐ linear-sweep voltammetry
☐ differential pulse voltammetry
☐ cyclic voltammetry
☐ rate processes in voltammetry

Electrochemical processes

☐ cell overpotential

29.4 Electrolysis
☐ gas evolution and metal deposition

Further reading

Articles of general interest

R.M. Wightman and D.O. Wipf, High-speed cyclic voltammetry. *Acc. Chem. Res.* **23**, 64 (1990).

A.T. Hubbard, Electrochemistry at well-characterized surfaces. *Chem. Rev.* **88**, 633 (1988).

R. Parsons, Electrical double layer: recent experimental and theoretical developments. *Chem. Rev.* **90**, 813 (1990).

R. Adžić and E.B. Yeager, Electrochemistry. In *Encyclopedia of applied physics* (ed. G.L. Trigg), **5**, 223. VCH, New York (1993).

Texts and sources of data and information

J. Koryta, *Ions, electrodes, and membranes.* Wiley, New York (1991).

R.G. Compton (ed.), *Comprehensive chemical kinetics.* Elsevier, Amsterdam (1987–92).

J.O'M. Bockris, B.E. Conway, and R.E. White (ed.), *Modern aspects of electrochemistry*, Vol. 22. Plenum, New York (1992).

C.D.S. Tuck, *Modern battery construction.* Ellis Horwood, New York (1991).

D. Linden (ed.) *Handbook of batteries and cells.* McGraw-Hill, New York (1984).

D.R. Crow, *Principles and applications of electrochemistry.* Chapman & Hall, London (1988).

A.J. Bard and L.R. Faulkner, *Electrochemical techniques: fundamentals and applications.* Wiley-Interscience, New York (1979).

J.O'M. Bockris and S.U.M. Khan, *Surface electrochemistry: a molecular level approach.* Plenum, New York (1993).

M.G. Fontanna and R.W. Staehle (ed.), *Advances in corrosion science and technology.* Plenum, New York (1980).

Exercises

29.1 (a) The Helmholtz model of the electrical double layer is equivalent to a parallel plate capacitor. Hence the potential difference across the double layer is given by $\Delta\phi = \sigma d/\varepsilon$, where d is the distance between the plates and σ is the surface charge density. Assuming that this model holds for concentrated salt solutions calculate the magnitude of the electric field at the surface of silica in 5.0 M NaCl(aq) if the surface charge density is 0.10 C m^{-2}.

29.1 (b) Refer to the preceding exercise. Calculate the magnitude of the electric field at the surface of silica in 4.5 M NaCl(aq) if the surface charge density is 0.12 C m^{-2}.

29.2 (a) The transfer coefficient of a certain electrode in contact with M^{3+} and M^{4+} in aqueous solution at 25°C is 0.39. The current density is found to be 55.0 mA cm^{-2} when the overvoltage is 125 mV. What is the overvoltage required for a current density of 75 mA cm^{-2}?

29.2 (b) The transfer coefficient of a certain electrode in contact with M^{2+} and M^{3+} in aqueous solution at 25°C is 0.42. The current density is found to be 17.0 mA cm^{-2} when the overvoltage is 105 mV. What is the overvoltage required for a current density of 72 mA cm^{-2}?

29.3 (a) Determine the exchange current density from the information given in Exercise 29.2a.

29.3 (b) Determine the exchange current density from the information given in Exercise 29.2b.

29.4 (a) To a first approximation, significant evolution or deposition in occurs in electrolysis only if the overpotential exceeds about 0.6 V. To illustrate this criterion determine the effect that increasing the overpotential from 0.40 V to 0.60 V has on the current density in the electrolysis of 1.0 M NaOH(aq), which is 1.0 mA cm^{-2} at 0.4 V and 25°C. Take $\alpha = 0.5$.

29.4 (b) Determine the effect that increasing the overpotential from 0.50 V to 0.60 V has on the current density in the electrolysis of 1.0 M NaOH(aq), which is 1.22 mA cm^{-2} at 0.50 V for a particular electrode and 25°C. Take $\alpha = 0.50$.

29.5 (a) Use the data in Table 29.1 for the exchange current density and transfer coefficient for the reaction $2H^+ + 2e^- \rightarrow H_2$ on nickel at 25°C to determine what current density would be needed to obtain an overpotential of 0.20 V as calculated from (a) the Butler–Volmer equation, and (b) the Tafel equation. Is the validity of the Tafel approximation affected at higher overpotentials (of 0.4 V and more)?

29.5 (b) Use the data in Table 29.1 for the exchange current density and transfer coefficient for the reaction $Fe^{3+} + e^- \rightarrow Fe^{2+}$ on platinum at 25°C to determine what current density would be needed to obtain an overpotential of 0.30 V as calculated from (a) the Butler–Volmer equation, and (b) the Tafel equation. Is the validity of the Tafel approximation affected at higher overpotentials (of 0.4 V and more)?

29.6 (a) Estimate the limiting current density at an electrode in which the concentration of Ag^+ ions is 2.5 mmol L^{-1} at 25°C. The thickness of the Nernst diffusion layer is 0.40 mm. The ionic conductivity of Ag^+ at infinite dilution and 25°C is 6.19 mS m^2 mol^{-1}.

29.6 (b) Estimate the limiting current density at an electrode in which the concentration of Mg^{2+} ions is 1.5 mmol L^{-1} at 25°C. The thickness of the Nernst diffusion layer is 0.32 mm. The ionic conductivity of Mg^{2+} at infinite dilution and 25°C is 10.60 mS m^2 mol^{-1}.

29.7 (a) A 0.10 M $CdSO_4$(aq) solution is electrolysed between a cadmium cathode and a platinum anode with a current density of 1.00 mA cm^{-2}. The hydrogen overpotential is 0.60 V. What will be the concentration of Cd^{2+} ions when evolution of H_2 just begins at the cathode? Assume all activity coefficients are unity.

29.7 (b) A 0.10 M $FeSO_4$(aq) solution is electrolysed between a magnesium cathode and a platinum anode with a current density of 1.50 mA cm^{-2}. The hydrogen overpotential is 0.60 V. What will be the concentration of Fe^{2+} ions when evolution of H_2 just begins at the cathode? Assume all activity coefficients are unity.

29.8 (a) A typical exchange current density, that for H^+ discharge at platinum, is 0.79 mA cm^{-2} at 25°C. What is the current density at an electrode when its overpotential is (a) 10 mV, (b) 100 mV, (c) −0.50 V? Take $\alpha = 0.5$.

29.8 (b) The exchange current density for a $Pt|Fe^{3+}$, Fe^{2+} electrode is 2.5 mA cm^{-2}. The standard potential of the electrode is +0.77 V. Calculate the current flowing through an electrode of surface area 1.0 cm^2 as a function of the potential of the electrode. Take unit activity for both ions.

29.9 (a) Suppose that the electrode potential is set at 1.00 V. The exchange current density is 6.0×10^{-4} A cm^{-2} and $\alpha = 0.50$. Calculate the current density for the ratio of activities $a(Fe^{3+})/a(Fe^{2+})$ in the range 0.1 to 10.0 and at 25°C.

29.9 (b) Suppose that the electrode potential is set at 0.50 V. Calculate the current density for the ratio of activities $a(Cr^{3+})/a(Cr^{2+})$ in the range 0.1 to 10.0 and at 25°C.

29.10 (a) What overpotential is needed to sustain a current density of 20 mA cm^{-2} at a $Pt|Fe^{3+}$, Fe^{2+} electrode in which both ions are at a mean activity $a = 0.10$?

29.10 (b) What overpotential is needed to sustain a current density of 15 mA cm^{-2} at a $Pt|Ce^{4+}$, Ce^{3+} electrode in which both ions are at a mean activity $a = 0.010$? Take $j_0 = 6.0 \times 10^{-4}$ A cm^{-2}, $\alpha = 0.50$.

29.11 (a) How many electrons or protons are transported through the double layer in each second when the Pt, $H_2|H^+$, $Pt|Fe^{3+}$, Fe^{2+}, and Pb, $H_2|H^+$ electrodes are at equilibrium at 25°C? Take the area as 1.0 cm^2 in each case. Estimate the number of times each second a single atom on the surface takes part in a electron transfer event, assuming an electrode atom occupies about $(280 \text{ pm})^2$ of the surface.

29.11 (b) How many electrons or protons are transported through the double layer in each second when the Cu, $H_2|H^+$ and $Pt|Ce^{4+}$, Ce^{3+} electrodes are at equilibrium at 25°C? Take the area as 1.0 cm^2 in each case. Estimate the number of times each second a single atom on the surface takes part in a electron transfer event, assuming an electrode atom occupies about $(260 \text{ pm})^2$ of the surface.

29.12 (a) What is the effective resistance at 25°C of an electrode interface when the overpotential is small? Evaluate it for 1.0 cm^2 (a) Pt, $H_2|H^+$, (b) Hg, $H_2|H^+$ electrodes.

29.12 (b) Evaluate the effective resistance at 25°C of an electrode interface for 1.0 cm^2 (a) Pb, $H_2|H^+$, (b) $Pt|Fe^{2+}$, Fe^{3+} electrodes.

29.13 (a) State what happens when a platinum electrode in an aqueous solution containing both Cu^{2+} and Zn^{2+} ions at unit activity is made the cathode of an electrolysis cell.

29.13 (b) State what happens when a platinum electrode in an aqueous solution containing both Fe^{2+} and Ni^{2+} ions at unit activity is made the cathode of an electrolysis cell.

29.14 (a) What are the conditions that allow a metal to be deposited from aqueous acidic solution before hydrogen evolution occurs significantly at 293 K? Why may silver be deposited from aqueous silver nitrate?

29.14 (b) The overpotential for hydrogen evolution on cadmium is about 1 V at current densities of 1 mA cm^{-2}. Why may cadmium be deposited from aqueous cadmium sulfate?

29.15 (a) The exchange current density for H^+ discharge at zinc is about 50 pA cm^{-2}. Can zinc be deposited from a unit activity aqueous solution of a zinc salt?

29.15 (b) The standard potential of the $Zn^{2+}|Zn$ electrode is −0.76 V at 25°C. The exchange current density for H^+ discharge at platinum is 0.79 mA cm^{-2}. Can zinc be plated on to platinum at that temperature? (Take unit activities.)

29.16 (a) Can magnesium be deposited on a zinc electrode from a unit activity acid solution at 25°C?

29.16 (b) Can iron be deposited on a copper electrode from a unit activity acid solution at 25°C?

29.17 (a) Calculate the maximum (zero-current) potential difference of a nickel–cadmium cell, and the maximum possible power output when 100 mA is drawn at 25°C.

29.17 (b) Calculate the maximum (zero-current) potential difference of a lead–acid cell, and the maximum possible power output when 100 mA is drawn at 25°C.

29.18 (a) Calculate the thermodynamic limit to the zero-current potential of fuel cells operating on (a) hydrogen and oxygen, (b) methane and air. Use the Gibbs energy information in the *Data section*, and take the species to be in their standard states at 25°C.

29.18 (b) Calculate the thermodynamic limit to the zero-current potential of fuel cells operating on propane and air. Use the Gibbs energy information in the *Data section*, and take the species to be in their standard states at 25°C.

29.19 (a) Which of the following metals has a thermodynamic tendency to corrode in moist air at pH $= 7$: Fe, Cu, Pb, Al, Ag, Cr, Co? Take as a criterion of corrosion a metal ion concentration of at least 10^{-6} mol L^{-1}.

29.19 (b) Which of the following metals has a thermodynamic tendency to corrode in moist air at pH = 7: Ni, Cd, Mg, Ti, Mn? Take as a criterion of corrosion a metal ion concentration of at least 10^{-6} mol L^{-1}.

29.20 (a) The corrosion current density j_{corr} at an iron anode is 1.0 A m^{-2}. What is the corrosion rate in millimetres per year? Assume uniform corrosion.

29.20 (b) The corrosion current density j_{corr} at a zinc anode is 2.0 A m^{-2}. What is the corrosion rate in millimetres per year? Assume uniform corrosion.

Problems

Numerical problems

29.1 In an experiment on the Pt$|$H$_2|$H$^+$ electrode in dilute H$_2$SO$_4$ the following current densities were observed at 25°C. Evaluate α and j_0 for the electrode.

η/mV	50	100	150	200	250
j/(mA cm^{-2})	2.66	8.91	29.9	100	335

How would the current density at this electrode depend on the overpotential of the same set of magnitudes but of opposite sign?

29.2 The standard potentials of lead and tin are -126 mV and -136 mV respectively at 25°C, and the overvoltages for their deposition are close to zero. What should their relative activities be in order to ensure simultaneous deposition from a mixture?

29.3 The limiting current density for the reaction $I_3^- + 2e^- \rightarrow 3I^-$ at a platinum electrode is 28.9 μA cm^{-2} when the concentration of KI is 6.6×10^{-4} mol L^{-1} and the temperature 25°C. The diffusion coefficient of I_3^- is 1.14×10^{-9} m^2 s^{-1}. What is the thickness of the diffusion layer?

29.4 The maximum theoretical efficiency of a fuel cell may be defined as $\varepsilon = |\Delta_r G / \Delta_r H| = |\nu F E / \Delta_r H|$. However, as indicated in eqn 46, the potential of a working cell is reduced from the zero-current potential. In a hydrogen/oxygen fuel cell with platinum electrodes operating at 373 K, the exchange current densities are $j_a = 100$ mA m^{-2} and $j_c = 3.00$ mA m^{-2}. Calculate the efficiency of the cell when operated at a current density of 300 mA m^{-2} with internal resistance of 0.500 Ω m^{-2}. Assume $\alpha \approx 0.5$ and $j/j_{\lim} \approx 0.5$ for both electrodes and assume $E = E^\ominus$. Compare the resulting value to the theoretical maximum efficiency of the cell and to the theoretical maximum efficiency of a heat engine that uses the combustion of hydrogen and oxygen and operates between 373 K and 673 K.

29.5 Estimating the power output and potential of a cell under operating conditions is very difficult, but eqn 47 summarizes, in an approximate way, some of the parameters involved. As a first step in manipulating this expression, identify all the quantities that depend on the ionic concentrations. Express E in terms of the concentration and conductivities of the ions present in the cell. Estimate the parameters for Zn$|$ZnSO$_4$(aq)$||$CuSO$_4$(aq)$|$Cu. Take electrodes of area 5 cm^2 separated by 5 cm. Ignore both potential differences and resistance of the liquid junction. Take the concentration as 1 mol L^{-1}, the temperature 25°C, and neglect activity coefficients. Plot E as a function of the current drawn. On the same graph, plot the power output of the cell. What current corresponds to maximum power?

29.6 Consider a cell in which the current is activation-controlled. Show that the current for maximum power can be estimated by plotting $\log(I/I_0)$ and $c_1 - c_2 I$ against I (where $I_0 = A^2 j_0 j_0'$ and c_1 and c_2 are constants), and looking for the point of intersection of the curves. Carry through this analysis for the cell in Problem 29.5 ignoring all concentration overpotentials.

29.7 Estimate the magnitude of the corrosion current for a patch of zinc of area 0.25 cm^2 in contact with a similar area of iron in an aqueous environment at 25°C. Take the exchange current densities as 1 μA cm^{-2} and the local ion concentrations as 1 μmol L^{-1}.

29.8 The corrosion potential of iron immersed in a de-aerated acidic solution of pH = 3 is -0.720 V as measured at 25°C relative to the standard calomel electrode with potential 0.2802 V. A Tafel plot of cathodic current density against overpotential yields a slope of 18 V^{-1} and the hydrogen ion exchange current density $j_0 = 0.10$ μA cm^{-2}. Calculate the corrosion rate in milligrams of iron per square centimetre per day (mg cm^{-2} d^{-1}).

Theoretical problems

29.9 If $\alpha = \frac{1}{2}$, an electrode interface is unable to rectify alternating current because the current density curve is symmetrical about $\eta = 0$. When $\alpha \neq \frac{1}{2}$, the magnitude of the current density depends on the sign of the overpotential, and so some degree of 'faradaic rectification' may be obtained. Suppose that the overpotential varies as $\eta = \eta_0 \cos \omega t$. Derive an expression for the mean flow of current (averaged over a cycle) for general α, and confirm that the mean current is zero when $\alpha = \frac{1}{2}$. In each case work in the limit of small η_0 but to second-order in $\eta_0 F/RT$. Calculate the mean direct current at 25°C for a 1.0 cm^2 hydrogen–platinum electrode with $\alpha = 0.38$ when the overpotential varies between ± 10 mV at 50 Hz.

29.10 Now suppose that the overpotential is in the high overpotential region at all times even though it is oscillating. What waveform will the current across the interface show if it varies linearly and periodically (as a sawtooth waveform) between η_- and η_+ around η_0? Take $\alpha = \frac{1}{2}$.

29.11 Derive an expression for the current density at an electrode where the rate process is diffusion-controlled and η_c is known. Sketch the form of j/j_L as a function of η^c. What changes occur if anion currents are involved?

Additional problems supplied by Carmen Giunta and Charles Trapp

29.12 The rate of deposition of iron, v, on the surface of an iron electrode from an aqueous solution of Fe^{2+} has been studied as a function of potential, E, relative to the standard hydrogen electrode, by J. Konya (*J. Electroanal. Chem.* **84**, 83 (1977)). The values in the table below are based on the data obtained with an electrode of surface area 9.1 cm^2 in contact with a solution of concentration 1.70 μmol L^{-1} in Fe^{2+}. (a) Assuming unit activity coefficients, calculate the zero current potential of the Fe^{2+}/Fe cathode and the overpotential at each value of the working potential. (b) Calculate the cathodic current density, j_c, from the rate of deposition of Fe^{2+} for each value of E. (c) Examine the extent to which the data fit the Tafel equation and calculate the exchange current density.

$v/(\text{pmol s}^{-1})$	1.47	2.18	3.11	7.26
$-E/\text{mV}$	702	727	752	812

29.13 The thickness of the diffuse double layer according to the Gouy–Chapman model is given by eqn 23.43. Use this equation to calculate and plot the thickness as a function of concentration and electrolyte type at 25°C. For examples, choose aqueous solutions of NaCl and Na$_2$SO$_4$ ranging in concentration from 0.1 to 100 mmol L^{-1}.

29.14 V.V. Losev and A.P. Pchel'nikov (*Soviet Electrochem.* **6**, 34 (1970)) obtained the following current–voltage data for an indium anode relative to a standard hydrogen electrode at 293 K.

E/V	0.388	0.365	0.350	0.335
$j/(\text{A m}^{-2})$	0	0.590	1.438	3.507

Use these data to calculate the transfer coefficient and the exchange current density. What is the cathodic current density when the potential is 0.365 V?

29.15 The redox reactions of quinones have been the subject of many studies over the years and they continue to be of interest to electrochemists. A relatively recent study is that of E. Kariv, J. Hermolin, and E. Gileadi (*Electrochim. Acta* **16**, 1437 (1971)) on methone (1,1-dimethyl-3,5-cyclohexanedione). The current–voltage data for the reduction of this compound (M) in anhydrous butanol on a mercury electrode are:

$-E/\text{V}$	1.50	1.58	1.63	1.72	1.87	1.98	≥ 2.10
$j/(\text{A m}^{-2})$	10	30	50	100	200	250	290

(a) How well do these data fit the empirical Tafel equation? (b) The authors postulate that the reduction product is the dimer HMMH formed by the following mechanism:

$$M(\text{sol}) \rightleftharpoons M(\text{ads})$$
$$M(\text{ads}) + H^+ + e^- \rightleftharpoons MH(\text{ads})$$
$$2MH(\text{ads}) \rightleftharpoons HMMH$$

The affixes sol and ads refer to species in solution and on the surface of the electrode, respectively. Does this mechanism help to explain the current–voltage data?

29.16 The classic study of the hydrogen overpotential is that of H. Bowden and T. Rideal (*Proc. Roy. Soc.* **A120**, 59 (1928)), who measured the overpotential for H$_2$ evolution with a mercury electrode in dilute aqueous solutions of H$_2$SO$_4$ at 25°C. Determine the exchange current density and transfer coefficient, α, from their data:

$j/(\text{mA m}^{-2})$	2.9	6.3	28	100	250	630	1650	3300
η/V	0.60	0.65	0.73	0.79	0.84	0.89	0.93	0.96

Explain any deviations from the result expected from the Tafel equation.

MicroProjects Part 3:

Prepared by M. Cady and C. A. Trapp

3.1 Diffusion studies

(a) Show that, for the initial and boundary conditions $c(x,t) = c(x,0) = c_o$, $(0 < x \le \infty)$ and $c(0,t) = c_s$ $(0 \le t \le \infty$) where c_o and c_s are constants, the concentration, $c(x,t)$, of a species is given by

$$c(x,t) = c_o + (c_s - c_o)\{1 - \mathrm{erf}(\xi)\} \qquad \xi(x,t) = \frac{x}{(4Dt)^{1/2}}$$

where $\mathrm{erf}(\xi)$ is the error function (eqn 12.46) and the concentration $c(x,t)$ evolves by diffusion from the yz-plane of constant concentration, such as might occur if a condensed phase is absorbing a species from a gas phase. Draw graphs of concentration profiles at several different times of your choice for the diffusion of oxygen into water at 298 K (when $D = 2.10 \times 10^{-9}$ m^2 s^{-1}) on a spatial scale comparable to passage of oxygen from lungs through alveoli into the blood. Use $c_o = 0$ and set c_s equal to the solubility of oxygen in water.

(b) Draw graphs of vertical (x) iron concentration, $c(x,t)$, profiles for 5, 10, 15, and 20 h after filling a very wide iron container with molten aluminium at 1273 K, when $D = 2.000 \times 10^{-8}$ m^2 s^{-1}. Assume that the iron dissolves at a steady rate so that the mass flux of iron, $J(x,t)$ with $x \ge 0$, is constant at the iron–aluminium interface, with $J(0,t) = J_0$. In this case, the solution of the diffusion equation is

$$\frac{c(x,t)}{J_0} = \left(\frac{t}{\pi D}\right)^{1/2} e^{-x^2/4\pi t} - \frac{x}{2D}\mathrm{erfc}(\xi)$$

where $\mathrm{erfc}(\xi)$ is the complementary error function, $\mathrm{erfc}(\xi) = 1 - \mathrm{erf}(\xi)$. Determine J_0 when $c(3\,\mathrm{cm}, 24\,\mathrm{h}) = 1.2$ mmol L^{-1}.

(c) Is it possible that diffusion is the mechanism by which a volatile liquid, like a perfume, becomes distributed around the air of a room? List several situations for which diffusion is important.

3.2 Conductivities of weak (1,1)-electrolyte solutions

A dilute solution of a weak (1,1)-electrolyte contains both neutral ion pairs and ions in equilibrium (AB $\rightleftharpoons$ A$^+$ + B$^-$). At higher concentrations, or in media of low relative permittivity, the ion pairs may associate with simple ions to form triple ions of the form ABA$^+$ and BAB$^-$.

(a) Prove that, in the model for which only ion pairs and simple ions are present, molar conductivities are related to the degree of ionization by the equations

$$\frac{1}{\varLambda_m} = \frac{1}{\varLambda_\alpha} + \frac{(1-\alpha)\varLambda_m}{(\alpha\varLambda_\alpha)^2} \qquad \varLambda_\alpha = \lambda_+ + \lambda_- = \varLambda_m^\circ - \mathcal{K}(\alpha c)^{1/2}$$

where $\varLambda_m^\circ$ is the molar conductivity at infinite dilution and $\mathcal{K}$ is the constant in Kohlrausch's law (eqn 24.31).

(b) The following table contains molar conductivities at 298 K for aqueous iodic acid solutions (R.M. Fuoss and C.A. Kraus, *J. Am. Chem. Soc.* **55**, 476 (1933)).

$c/(\mathrm{mmol\,L^{-1}})$	0.068 517	0.096 291	0.144 572	0.168 284	0.233 308
$\varLambda_m/(\mathrm{S\,cm^2\,mol^{-1}})$	389.02	389.02	388.31	388.62	388.31
$c/(\mathrm{mmol\,L^{-1}})$	0.286 320	0.399 158	0.465 172	0.648 686	0.710 642
$\varLambda_m/(\mathrm{S\,cm^2\,mol^{-1}})$	388.04	387.46	387.00	386.12	385.76
$c/(\mathrm{mmol\,L^{-1}})$	0.986 010	1.023 35	1.436 15	1.494 29	1.528 09
$\varLambda_m/(\mathrm{S\,cm^2\,mol^{-1}})$	384.54	384.31	382.20	382.36	382.26
$c/(\mathrm{mmol\,L^{-1}})$	1.730 67	2.103 23	2.111 08	3.007 36	3.040 44
$\varLambda_m/(\mathrm{S\,cm^2\,mol^{-1}})$	380.92	380.09	380.06	376.58	376.77
$c/(\mathrm{mmol\,L^{-1}})$	5.384 53	5.479 88			
$\varLambda_m/(\mathrm{S\,cm^2\,mol^{-1}})$	370.33	370.13			

Develop an extrapolation method to evaluate $\varLambda_m^\circ$. Determine the degree of ionization and the mean activity coefficient at each concentration. Calculate the acidity constant. When evaluating the mean activity coefficient, assume the extended Debye–Hückel law of the form

$$- \log \gamma_\pm = \frac{A_\gamma(\alpha c)^{1/2}}{1 + B_\gamma(\alpha c)^{1/2}}$$

with $A_\gamma = 0.5044$ $(\mathrm{mol\,L^{-1}})^{-1/2}$ and $B_\gamma = 0.0153$ $(\mathrm{mol\,L^{-1}})^{-1/2}$. Assume that $\gamma_{AB} = 1$.

(c) The simple ion model in (a) is a basis for a description of the molar conductivities of (b). However, this is not always the case. As an example of the failure of the simple ion model, consider the following molar conductivities for solutions of tetrabutylammonium tetrafluoroborate in 15 mole per cent phenanthrene in anisole, a solvent with a low electric permittivity (A.P. Abott and D.J. Schiffrin, *J. Chem. Soc. Faraday Trans.*, **86**, 1453 (1990)).

$c^{1/2}/(\mathrm{mol\,L^{-1}})^{1/2}$	0.0115	0.0220	0.0278	0.0470	0.0992
$\varLambda_m/(\mathrm{S\,cm^2\,mol^{-1}})$	0.0433	0.0341	0.0333	0.0389	0.0705
$c^{1/2}/(\mathrm{mol\,L^{-1}})^{1/2}$	0.186	0.476	0.844	1.006	1.178
$\varLambda_m/(\mathrm{S\,cm^2\,mol^{-1}})$	0.173	0.964	1.51	1.42	1.20

Draw a graph of $\varLambda_m$ versus $c^{1/2}$. In what range of concentrations might the simple ion model describe conductivities adequately? At what approximate concentration does triple ion formation become important and how does triple ion formation affect conductivity? What happens at high concentrations? Why?

3.3 Destructive oxidation of arenes

The destructive oxidation of alkylarenes by aqueous permanganate may occur by attack upon the alkyl group or by attack upon the aromatic ring. In perchloric acid solution at pH ≥ 3 it is thought that permanganate anion attack upon the alkyl group predominates. However, at pH ≤ 0.3 permanganic acid attack upon the ring prevails. In this problem, data that concerns the ring oxidation alone will be presented and analysed. A proposed mechanism for the ring

Change

destruction envisions successive formation of a charge-transfer complex (CCT) and a σ-complex as shown below.

This mechanism suggests that reaction activation energies for a homologous series of alkylarenes might be multilinear (that is, $f(x, y) = c_1 x + c_2 y + c_3$) in both basicity constants (K_b) and ionization energies (I).

(a) Prove that, if the multilinear hypothesis is correct, the rate constant k is given by

$$\Delta pk(T) = \frac{aT_0 \Delta pK_b^\circ (cT - 1)}{T(cT_0 - 1)} - \frac{b\Delta I}{RT \ln 10}$$

where $\Delta pk = pk - pk_{ref}$, $\Delta pK_b = pK_b - pK_{b,ref}$, and $\Delta I = I - I_{ref}$; 'ref' values are those for a chosen reference compound within the series and K_b° is the basicity constant at T_0. The parameters a, b, and c are temperature-independent properties which may be treated as regression parameters and determined by fitting the equation to experimental rate constants. Assume that $\Delta(\Delta S_b) = c\Delta(\Delta H_b)$, where c is independent of temperature and $\Delta(\Delta)X = \Delta X - \Delta X_{ref}$. Specify any other assumption that you make.

(b) At 20, 30, 40, 50, 60, and 70°C, $k/(10^{-2} \text{ L mol}^{-1} \text{ s}^{-1}) = 0.86, 2.5, 5.4, 13, 47$, and 59, respectively, for 1,4-dimethylbenzene. Determine A and E_a. Derive equations for $\Delta^\ddagger H$ and $\Delta^\ddagger S$ that are appropriate for solution kinetics and calculate their values.

(c) The table below contains kinetic properties for a homologous series (E.S. Rudakov, V.L. Lobachev, and E.V. Zaichuk, *Kinetics and Catalysis* **37**, 500 (1996)). The Arrhenius parameters were determined with rate constants measured at the same temperatures as those of (b). Determine the values of a, b, and c using methylbenzene as the reference compound. Develop a graphical method for evaluating the validity of the multilinear hypothesis.

Arene	$\log(A/(\text{L mol}^{-1}\text{s}^{-1}))$	$E_a/(\text{kJ mol}^{-1})$	pK_b at 273 K	I/eV
benzene	12.5	104	9.2	9.25
methylbenzene	11.4	84	6.3	8.82
1,2-dimethylbenzene	11.2	76	5.3	8.56
1,3-dimethylbenzene	10.5	68	3.2	8.56
1,4-dimethylbenzene	Part b	Part b	5.7	8.44
mesitylene	10.0	56	0.4	8.40
pseudocumene	10.5	62	2.9	8.27
durene	10.6	56	2.2	8.02

3.4 Kinetics of a bromine substitution reaction

Here we consider the reaction $SiCl_3H(g) + Br_2(g) \rightarrow SiCl_3Br(g) + HBr(g)$. The temperature dependence of this reaction has been studied by measuring the degree of conversion, $\alpha(t)$, for initial mixtures of 4.0 volume percentage trichlorosilane and 10.0 volume percentage bromine in helium at 1 bar. The data at four different temperatures are as follows.

$\alpha(160°C, t) = 0.044$ at 6.07 min; 0.057 at 7.87 min; 0.076 at 10.00 min

$\alpha(180°C, t) = 0.137$ at 2.72 min; 0.188 at 3.91 min; 0.228 at 5.03 min; 0.273 at 6.67 min; 0.302 at 8.06 min

$\alpha(200°C, t) = 0.281$ at 1.39 min; 0.395 at 2.16 min; 0.516 at 3.36 min; 0.612 at 5.04 min

$\alpha(220°C, t) = 0.592$ at 0.67 min; 0.734 at 1.22 min; 0.866 at 2.16 min

For these reactants it is thought that the dissociation reaction $Br_2(g) \rightarrow 2Br(g)$ is at equilibrium and that, from 300–700 K, the equilibrium constant for the dissociation, can be approximated as $K = 10^{5.1}e^{-C/RT}$, where $C = 190.2$ kJ mol^{-1} and K is expressed in terms of concentrations in moles per cubic centimetre (mol cm^{-3}). No wall effects were observed. Reactions of HBr were discounted and no hexachlorodisilane was observed.

(a) Presume a rate law of the form $d[SiCl_3Br]/dt = k[SiCl_3H]^a[Br_2]^b$ and determine the reaction orders a and b.

(b) Determine the rate constant $k(T)$ for each of the four data sets. Evaluate the activation parameters.

(c) Show that the proposed mechanism

$$SiCl_3H(g) + \cdot Br(g) \longrightarrow \cdot SiCl_3(g) + HBr(g)$$
$$\cdot SiCl_3(g) + Br_2(g) \longrightarrow SiCl_3Br(g) + \cdot Br(g)$$

is compatible with the experimental rate law.

3.5 Photochemical ozone production

The photochemical formation of ozone is a very complex process in the Earth's atmosphere. To simplify the description, consider the experiment in which pure oxygen at 10 Torr and 298 K is exposed to measurable frequencies and intensities of UV radiation. According to the Chapman model, the following elementary reactions contribute to ozone production. (M is a collision partner that must be part of the elementary reaction so as to conserve angular momentum.)

$$O_2 + h\nu_1 \longrightarrow O + O \qquad k_1$$
$$O + O_2 + M \longrightarrow O_3 + M \qquad k_2$$
$$O_3 + h\nu_3 \longrightarrow O + O_2 \qquad k_3$$
$$O + O_3 \longrightarrow O_2 + O_2 \qquad k_4$$
$$O + O + M \longrightarrow O_2 + M \qquad k_5$$

(a) Look up the values of k_2, k_4, and k_5 in a source such as the *CRC Handbook of chemistry and physics*. The rate constants k_1 and k_3 depend upon the radiation conditions; assume values of 1.0×10^{-8} s^{-1} and 0.016 s^{-1}, respectively. Write the rate expressions for the concentration of each chemical species. Assume that the UV radiation is turned on at $t = 0$, and solve the rate expressions for the concentration of all species as a function of time over a period of 4 h. Examine relevant concentrations in the very early time period $t < 0.1$ s. You will need a software package for solving a set of differential equations and you will find that this set is 'stiff'. Stiff differential equations have at least two rate constants that have widely different values and relate to processes that contribute to the process on very different time scales. The solution to such equations usually requires that the total time period be broken into two or more and solved separately: one may be very short, another very long. State all assumptions. Is there any ozone present initially? Why must the pressure be low and the UV radiation intensities high for the production of ozone? Draw graphs of the time variations of both atomic oxygen and ozone on both the very short and the long time scales. What is the percentage of ozone after 4.0 h of irradiation?

(b) To study the effect of trace amounts of nitrogen dioxide of initial concentration 1.0×10^{12} molecules cm^{-3}, at least the following set of elementary reactions must be added to the Chapman model.

$$
\begin{array}{ll}
NO + O_3 \longrightarrow NO_2 + O_2 + M & k_6 \\
NO_2 + O \longrightarrow NO + O_2 & k_7 \\
NO_2 + h\nu_8 \longrightarrow NO + O & k_8 \\
NO_2 + O_3 \longrightarrow NO_3 + O_2 & k_9 \\
NO_3 + NO_2 + M \longrightarrow N_2O_5 + M & k_{10} \\
N_2O_5 + h\nu_{11} \longrightarrow NO_2 + NO_3 & k_{11}
\end{array}
$$

Assume that the values of k_8 and k_{11} are 1.0×10^{-4} s^{-1} and 1.0×10^{-7} s^{-1}, respectively, and look up the other rate constants in a reference source. Write the rate expressions for the concentration of all chemical species. Solve the set of equations for a time period that allows comparison with the experiment having no nitrogen oxides (a). State all assumptions. Is the ozone production affected in the very early, intermediate, and late time periods? Draw graphs of the time evolution of all chemical species on the relevant time scales. What are your comparative conclusions?

3.6 Autocatalytic mechanisms

(a) The Lotka–Volterra mechanism may be viewed as being a simple description of the predator–prey relationship with A representing the favourable environmental and nutritional conditions for the prey population, X and Y representing the prey and predator populations, respectively, and B representing predator death. Solve this predator-prey model for the normalized populations of both predator and prey for a 20 year period. Draw graphs for the evolution of the normalized populations. What is the approximate period of the observed oscillations? Why do the oscillations occur? Draw a graph that portrays the limit cycle. What obvious feature of population analysis is missing from this model? In your computations use $k_a = 1$ y, $k_b = 1$ y, $k_c = 2$ y and the initial normalized conditions $[\text{prey}]_0 = 1$,

$[\text{predator}]_0 = 0.5$. Assume that the environment is unaffected by either species and remains in the normalized steady-state [environ] $= 1$ throughout the period.

(b) Identify the attractor of Brusselator species when the normalized initial state is $[A]_0 = 1$, $[B]_0 = 1$, $[X]_0 = 30$, and $[Y]_0 = 0.5$. The rate constants are $k_a = 10$, $k_b = 0.01$, $k_c = 0.01$, and $k_d = 10$ in normalized (or 'reduced') units. Assume that A and B are systematically fed into the system so that their concentrations remain constant.

3.7 Adsorption isotherms

A series of experiments designed to measure the adsorption of methylene blue on activated carbon (J.H. Potgieter, *J. Chem. Ed.* **68**, 349 (1991)) uses 100.0 mL of 25.0 mg L^{-1} methylene blue solution. The respective amounts of activated carbon placed in solution were $m/\text{mg} = 1$, 5, 10, 12.5, 25, 30, and 100. After equilibration the corresponding methylene blue concentrations were found to be $c/(\text{mg L}^{-1}) = 19.9$, 10.8, 4.2, 3.5, 1.0, 0.70, and 0.20. Write the equations for the Langmuir and Freundlich adsorption isotherms in a form that is suitable for adsorption from solution. The equations should use concentration, c, and mass fraction adsorbed, x, where x is the mass of solute adsorbed divided by the mass of adsorbent. Perform suitable linear regression fits of the isotherms and critically evaluate the suitability of each fit; compare the statistics of the fits and provide evidence that one or the other is the better fit. Give a critique of the evidence. In addition to the linear relationship suggested in Section 28.4a, write the Langmuir isotherm equation in a form for which x^{-1} versus c^{-1} is expected to be linear and perform the regression analysis. The regression statistics should include the correlation coefficient, the standard deviation or the coefficient of variation, the standard deviation of the intercept, and the standard deviation of the slope. Determine the maximum value of the mass fraction, $x_{\max}$, for this activated carbon sample and use this property to estimate the specific surface area of the adsorbent. The model of methylene blue shown below has the C,N,S frame drawn to scale and can be used to estimate the area occupied by the cation.

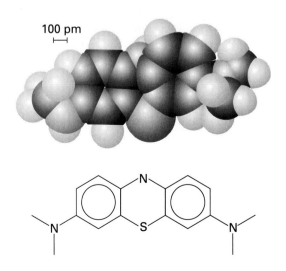

100 pm

3.8 The polarization curve and Tafel plots

Consider a one-electron transfer process for which the transfer is rate-determining.

(a) Draw a graph that shows the dependence of the relative current densities j_a/j_0, j_c/j_0, and j/j_0 on the overpotential. This is the polarization curve. How does variation of the transfer coefficient affect the polarization curve?

(b) Draw a Tafel plot that shows both anodic and cathodic current densities for the platinum electrode process $Ce^{4+} + e^- \rightarrow Ce^{3+}$. Assume that the electron transfer is rate-determining. Explain the graphical methodology for using a Tafel plot of experimental data for evaluation of both the transfer coefficient and the exchange current density. Draw a graph that illustrates the methodology.

3.9 Stationary electrode, reversible voltammetry

The analytic description of reversible redox kinetics at a stationary electrode is well established (R.S. Nicholson and I. Shain, *Anal. Chem.* **36**, 706 (1964)). For a linear-sweep voltammetry experiment with a constant potential scan rate, v, from $E_{initial}$, the equation for current density at a flat, non-reactive electrode with linear diffusion to and from the electrode is:

$$j(x) = zFc_{Ox}(\pi D_{Ox}zfv)^{1/2}\chi(x)$$

where $x = (E_{initial} - E)zf$, $f = RT/F$, and

$$\chi(x) = \frac{1}{\pi x^{1/2}(1 + \gamma\theta)} + \frac{1}{4\pi}\int_0^x \frac{1}{(x-y)^{1/2}\cosh^2\left(\frac{1}{2}\{\ln(\gamma\theta) - y\}\right)}$$

$$\gamma = \left(\frac{D_{Ox}}{D_{Red}}\right)^{1/2} \qquad \theta = e^{f(E_{initial} - E)}$$

Far from the electrode, the bulk concentrations are c_{Ox} and $c_{Red} = 0$; at the electrode $j_{Ox} = -j_{Red}$.

(a) Simulate the reversible electron transfer voltammogram for Ti^{3+}/Ti^{2+} reduction. Use $c_{Ox} = 1.0 \times 10^{-4}$ mol L^{-1} and $v = 10$ mV s^{-1}. Discuss the physical/chemical properties that cause the variations of the current density revealed by the simulation. How do the values of D_{Ox} and v affect the voltammogram?

(b) Let E_{pc} be the voltammogram peak cathodic potential. Show that both $E_{pc} \approx E^{\circ\prime} - (28.5\ \text{mV})/z$ and $E_{p/2} - E_{pc} \approx 56.5$ mV. List all the experimental criteria for reversibility that are suggested by this simulation. $E_{p/2}$ is the potential at which $j = \frac{1}{2}j(E_{pc})$ and $E^{\circ\prime}$ is the standard formal potential at the ionic strength of the solution.

Further information 1

Relations between partial derivatives

A partial derivative of a function of more than one variable, such as $f(x, y)$, is the slope of the function with respect to one of the variables, all the other variables being held constant (see Fig. 2.12). Although a partial derivative shows how a function changes when one variable changes, it may be used to determine how the function changes when more than one variable changes by an infinitesimal amount. Thus, if f is a function of x and y, then when x and y change by dx and dy, respectively, f changes by

$$df = \left(\frac{\partial f}{\partial x}\right)_y dx + \left(\frac{\partial f}{\partial y}\right)_x dy \tag{1}$$

For example, if $f = ax^3y + by^2$,

$$\left(\frac{\partial f}{\partial x}\right)_y = 3ax^2y \qquad \left(\frac{\partial f}{\partial y}\right)_x = ax^3 + 2by$$

Then, when x and y undergo infinitesimal changes, f changes by

$$df = 3ax^2y\, dx + (ax^3 + 2by)\, dy$$

Partial derivatives may be taken in any order:

$$\frac{\partial^2 f}{\partial x \partial y} = \frac{\partial^2 f}{\partial y \partial x} \tag{2}$$

For the function f given above, it is easy to verify that

$$\left(\frac{\partial}{\partial y}\left(\frac{\partial f}{\partial x}\right)_y\right)_x = 3ax^2 \qquad \left(\frac{\partial}{\partial x}\left(\frac{\partial f}{\partial y}\right)_x\right)_y = 3ax^2$$

In the following, z is a variable on which x and y depend (for example, x, y, and z might correspond to p, V, and T).

Relation no. 1. When x is changed at constant z:

$$\left(\frac{\partial f}{\partial x}\right)_z = \left(\frac{\partial f}{\partial x}\right)_y + \left(\frac{\partial f}{\partial y}\right)_x \left(\frac{\partial y}{\partial x}\right)_z \tag{3}$$

Relation no. 2 (the inverter).

$$\left(\frac{\partial x}{\partial y}\right)_z = \frac{1}{(\partial y/\partial x)_z} \tag{4}$$

Relation no. 3 (the permuter).

$$\left(\frac{\partial x}{\partial y}\right)_z = -\left(\frac{\partial x}{\partial z}\right)_y \left(\frac{\partial z}{\partial y}\right)_x \tag{5}$$

By combining this relation and Relation no. 2 we obtain **Euler's chain relation**:

$$\left(\frac{\partial x}{\partial y}\right)_z \left(\frac{\partial y}{\partial z}\right)_x \left(\frac{\partial z}{\partial x}\right)_y = -1 \tag{6}$$

Relation no. 4. This relation establishes whether or not $\mathrm{d}f$ is an exact differential.

$$\mathrm{d}f = g(x, y)\,\mathrm{d}x + h(x, y)\,\mathrm{d}y \text{ is exact if } \left(\frac{\partial g}{\partial y}\right)_x = \left(\frac{\partial h}{\partial x}\right)_y \tag{7}$$

If $\mathrm{d}f$ is exact, its integral between specified limits is independent of the path.

Differential equations

A differential equation is a relation between derivatives of a function and the function itself, as in

$$a\frac{\mathrm{d}^2 y}{\mathrm{d}x^2} + b\frac{\mathrm{d}y}{\mathrm{d}x} + cy = 0 \qquad (1)$$

The coefficients a, b, etc. may be functions of x. The **order** of the equation is the order of the highest derivative that occurs in it, so eqn 1 is a second-order equation. Only rarely in science is a differential equation of order higher than 2 encountered. A solution of a differential equation is an expression for y as a function of x. The process of solving a differential equation is commonly termed 'integration', and in simple cases simple integration can be employed to find $y(x)$. A **general solution** of a differential equation is the most general solution of the equation and is expressed in terms of a number of constants. When the constants are chosen to accord with certain specified **initial conditions** (if one variable is the time) or certain **boundary conditions** (to fulfil certain spatial restrictions on the solutions), we obtain the **particular solution** of the equation. A first-order differential equation requires the specification of *one* boundary (or initial) condition; a second-order differential equation requires the specification of *two* such conditions, and so on.

First-order differential equations may often be solved by direct integration. For example, the equation

$$\frac{\mathrm{d}y}{\mathrm{d}x} = axy$$

with a constant may be rearranged into

$$\frac{\mathrm{d}y}{y} = ax\,\mathrm{d}x$$

and then integrated to

$$\ln y = \tfrac{1}{2}ax^2 + A$$

where A is a constant. If we know that $y = y_0$ when $x = 0$ (for instance), then it follows that $A = \ln y_0$, and hence the particular solution of the equation is

$$\ln y = \tfrac{1}{2}ax^2 + \ln y_0$$

This expression rearranges to

$$y = y_0 e^{ax^2/2}$$

First-order equations of a more complex form can often be solved by the appropriate substitution. For example, it is sensible to try the substitution $y = sx$, and to change the variables from x and y to x and s. An alternative useful transformation is to write $x = u + a$ and $y = v + b$, and then to select a and b to simplify the form of the resulting expression.

Second-order differential equations are in general much more difficult to solve than first-order equations. The general solutions of many such equations are best found by referring to tables: the *Handbook of mathematical functions*, M. Abramowitz and I.A. Stegun, Dover, New York (1965), is a particularly helpful source of such information. Mathematical software is now capable of finding numerical and, in certain cases, analytical solutions of a wide variety of differential equations.

One powerful approach commonly used to lay siege to second-order differential equations is to express the solution as a power series:

$$y = \sum_{n=0}^{\infty} c_n x^n \tag{2}$$

and then to use the differential equation to find a relation between the coefficients. This approach results, for instance, in the Hermite polynomials that form part of the solution of the Schrödinger equation for the harmonic oscillator. All the second-order differential equations that occur in this text can be found tabulated in compilations of solutions, and the specialized techniques that are needed to establish the form of the solutions may be found in mathematical texts.

A **partial differential equation** is a differential in more than one variable. An example is

$$\frac{\partial^2 y}{\partial t^2} = a \frac{\partial^2 y}{\partial x^2} \tag{3}$$

with y a function of the two variables x and t. In certain cases, partial differential equations may be separated into ordinary differential equations. Thus, the Schrödinger equation for a particle in a two-dimensional square well (Section 12.2) may be separated by writing the wavefunction, $\psi(x, y)$, as the product $X(x)Y(y)$, which results in the separation of the second-order partial differential equation into two second-order differential equations in the variables x and y. A good guide to the likely success of such a **separation of variables** procedure is the symmetry of the system.

A common approach to the solution of awkward differential equations that appear to have no analytical solutions is to adopt a numerical procedure. Software packages are now readily available that can be used to solve almost any equation numerically. The general form of such programs to solve $df/dx = g(x)$, for instance, replaces the infinitesimal quantity $df = g(x)\,dx$ by the small quantity $\delta f = g(x)\,\delta x$, so that

$$f(x + \delta x) \approx f(x) + g(x)\,\delta x \tag{4}$$

and then proceeds to step along the x-axis numerically, generating $f(x)$ as it goes. The actual algorithms adopted are much more sophisticated than this primitive scheme, but stem from it.

Undetermined multipliers

Suppose we need to find the maximum (or minimum) value of some function f that depends on several variables $x_1, x_2, \ldots, x_n$. When the variables undergo a small change from x_i to $x_i + \delta x_i$ the function changes from f to $f + \delta f$, where

$$\delta f = \sum_i^n \left(\frac{\partial f}{\partial x_i} \right) \delta x_i \tag{1}$$

At a minimum or maximum, $\delta f = 0$, so then

$$\sum_i^n \left(\frac{\partial f}{\partial x_i} \right) \delta x_i = 0 \tag{2}$$

If the x_i were all independent, all the δx_i would be arbitrary, and this equation could be solved by setting each $(\partial f / \partial x_i) = 0$ individually. When the x_i are not all independent, the x_i are not all independent, and the simple solution is no longer valid. We proceed as follows.

Let the constraint connecting the variables be an equation of the form $g = 0$. For example, in Chapter 19, one constraint was $n_0 + n_1 + \cdots = N$, which can be written

$$g = 0, \quad \text{with } g = (n_0 + n_1 + \cdots) - N$$

The constraint $g = 0$ is always valid, so g remains unchanged when the x_i are varied:

$$\delta g = \sum_i \left(\frac{\partial g}{\partial x_i} \right) \delta x_i = 0 \tag{3}$$

Because δg is zero, we can multiply it by a parameter, λ, and add it to eqn 2:

$$\sum_i^n \left\{ \left(\frac{\partial f}{\partial x_i} \right) + \lambda \left(\frac{\partial g}{\partial x_i} \right) \right\} \delta x_i = 0 \tag{4}$$

This equation can be solved for one of the δx, δx_n for instance, in terms of all the other δx_i. All those other δx_i $(i = 1, 2, \ldots n - 1)$ are independent, because there is only one constraint on the system. But here is the trick: λ is arbitrary; therefore we can choose it so that the coefficient of δx_n in eqn 4 is zero. That is, we choose λ so that

$$\left(\frac{\partial f}{\partial x_n}\right) + \lambda\left(\frac{\partial g}{\partial x_n}\right) = 0 \tag{5}$$

Then eqn 4 becomes

$$\sum_i^{n-1}\left\{\left(\frac{\partial f}{\partial x_i}\right) + \lambda\left(\frac{\partial g}{\partial x_i}\right)\right\}\delta x_i = 0 \tag{6}$$

Now the $n-1$ variations δx_i *are* independent, so the solution of this equation is

$$\left(\frac{\partial f}{\partial x_i}\right) + \lambda\left(\frac{\partial g}{\partial x_i}\right) = 0 \qquad i = 1, 2, \ldots n-1 \tag{7}$$

But eqn 5 has exactly the same form as this equation, so the maximum or minimum of f can be found by solving

$$\left(\frac{\partial f}{\partial x_i}\right) + \lambda\left(\frac{\partial g}{\partial x_i}\right) = 0 \qquad i = 1, 2, \ldots n \tag{8}$$

The use of this approach was illustrated in the text for two constraints and therefore two undetermined multipliers λ_1 and λ_2 (α and $-\beta$).

The multipliers λ cannot always remain undetermined. One approach is to solve eqn 5 instead of incorporating it into the minimization scheme. In Chapter 19 we used the alternative procedure of keeping λ undetermined until a property was calculated for which the value was already known. Thus, we found that $\beta = 1/kT$ by calculating the internal energy of a perfect gas.

Classical mechanics

We shall see how classical mechanics describes the behaviour of objects in terms of two equations. One equation expresses the fact that the total energy is constant in the absence of external forces. The other equation expresses the response of particles to the forces acting on them.

1 The trajectory in terms of the energy

The total energy of a particle is the sum of the **kinetic energy**, E_K, the energy arising from the motion of the particle, and **potential energy**, $V(x)$, the energy arising from the position of the particle in a field of force:

$$E = E_K + V(x) \tag{1}$$

The **force**, F, is related to the potential energy by

$$F = -\frac{dV}{dx} \tag{2}$$

According to this expression, the direction of the force is towards decreasing potential energy (Fig. 1). The kinetic energy of a particle of mass m travelling with a speed v is

$$E_K = \tfrac{1}{2}mv^2 \tag{3}$$

It is often convenient to express kinetic energy in terms of the **linear momentum, p.** The linear momentum is a vector quantity (that is, it has direction as well as magnitude, like the velocity, v). The magnitude of the linear momentum, p, is related to the speed, v, of the particle by

$$p = mv \tag{4}$$

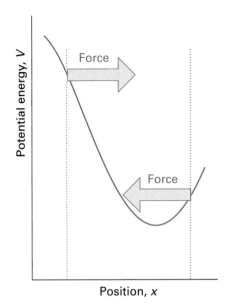

FI4.1 The force acting on a particle is determined by the slope of the potential energy at its location. The force points in the direction of decreasing potential energy.

A heavy particle moving slowly can have a higher momentum than a light particle moving rapidly. The linear momentum vector points in the direction of travel of the particle (Fig. 2). In terms of the linear momentum, the total energy of a particle is

$$E = \frac{p^2}{2m} + V(x) \tag{5}$$

These equations can be used in a number of ways. For example, it is easy to show that they predict that a particle will have a definite **trajectory**, or definite position and momentum at each instant. For example, consider a particle free to move in one direction (along the x-axis) in a region where $V = 0$ (so the energy is independent of position and there is no force acting). Because speed is the rate of change of position, $v = dx/dt$, it follows from eqns 4 and 5 that

$$\frac{dx}{dt} = \left(\frac{2E_K}{m}\right)^{1/2} \tag{6}$$

A solution of this equation is

$$x(t) = x(0) + \left(\frac{2E_K}{m}\right)^{1/2} t \tag{7}$$

The linear momentum is a constant:

$$p(t) = mv(t) = m\frac{dx}{dt} = (2mE_K)^{1/2} \tag{8}$$

Hence, if we know the initial position and momentum, we can predict all later positions and momenta exactly.

2 Newton's second law

The second basic equation of classical mechanics is **Newton's second law of motion**, that the rate of change of momentum is equal to the force acting on the particle. In one dimension:

$$\frac{dp}{dt} = F \tag{9}$$

Because $p = mdx/dt$ in one dimension, it is sometimes more convenient to write this equation as

$$m\frac{d^2x}{dt^2} = F \tag{10}$$

The second derivative, d^2x/dt^2, is the **acceleration** of the particle, its rate of change of velocity (along the x-axis). Newton's second law therefore states that the acceleration of a particle is proportional to the force it experiences. It follows that, if we know the force acting everywhere and at all times, then solving eqn 10 will also give the trajectory. This calculation is equivalent to the one based on E, but is more suitable in some applications. For example, it can be used to show that, if a particle of mass m is initially stationary and is subjected to a constant force F for a time τ, then its kinetic energy increases from zero to

$$E_K = \frac{F^2\tau^2}{2m} \tag{11}$$

and then remains at that energy after the forces ceases to act. Because the applied force, F, and the time, τ, for which it acts may be varied at will, the energy of the particle may be increased to any value.

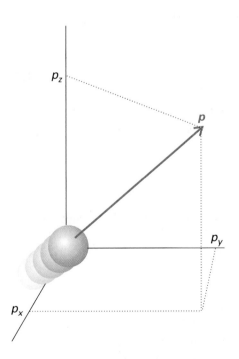

FI4.2 The linear momentum of a particle is a vector property, and points in the direction of motion.

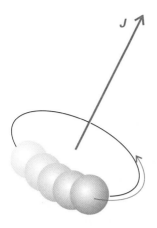

FI4.3 The angular momentum of a particle can be represented by a vector along the axis of rotation and perpendicular to the plane of rotation. The length of the vector denotes the magnitude of the angular momentum. The direction of motion is clockwise for an observer looking in the direction of the vector.

3 Rotational motion

The same type of calculation may be applied to free rotational motion, a form of motion that plays an important role in the electronic structures of atoms (where electrons are free to travel round the central nucleus) and in the spectroscopy of molecules (because free molecules rotate). Whereas the translational motion of a particle is expressed in terms of its linear momentum, the rotational motion of a particle about a central point is described by its **angular momentum, J**. The angular momentum has direction as well as magnitude: its magnitude gives the rate at which a particle circulates and its direction indicates the axis of rotation (Fig. 3). The magnitude of the angular momentum, J, is given by the expression

$$J = I\omega \qquad (12)$$

where ω is the **angular velocity** of the body, its rate of change of angular position (in radians per second), and I is the **moment of inertia**. The analogous roles of m and I, of v and ω, and of p and J in the translational and rotational cases, respectively, should be remembered, because they provide a ready way of constructing and recalling equations. For a point particle of mass m moving in a circle of radius r, the moment of inertia is given by the expression

$$I = mr^2 \qquad (13)$$

The angular momentum of a particle is large if it has a large moment of inertia (which is the case if it is heavy and moving in a circle of large radius) and its angular velocity is high (so that it is travelling rapidly in its circular path).

To accelerate a rotation it is necessary to apply a **torque**, T, a twisting force, and Newton's equation is then

$$\frac{dJ}{dt} = T \qquad (14)$$

If a constant torque is applied for a time τ, the rotational energy of an initially stationary body is increased to

$$E_K = \frac{T^2\tau^2}{2I} \qquad (15)$$

(The reasoning behind this expression is exactly the same as that leading to eqn 11.) The implication of this equation is that an appropriate torque and period for which it is applied can excite the rotation to an arbitrary energy.

4 The harmonic oscillator

A third fundamental type of motion is oscillatory motion, the kind of motion that atoms undergo in a vibrating molecule. A **harmonic oscillator** consists of a particle that experiences a restoring force proportional to its displacement from its equilibrium position:

$$F = -kx \qquad (16)$$

An example is a particle joined to a rigid support by a spring. The constant of proportionality k is called the **force constant**, and the stiffer the spring the greater the force constant. The negative sign in F signifies that the direction of the force is opposite to that of the displacement (force, like momentum, velocity, and angular momentum, has direction as well as magnitude). Thus, when x is positive (displacement to the right), the force is negative (pushing towards the left), and vice versa (Fig. 4). The motion of a particle that undergoes

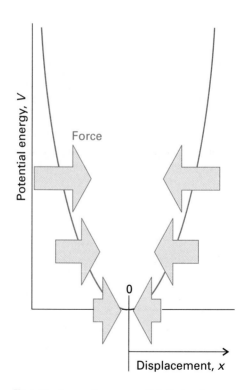

FI4.4 The force acting on a particle that undergoes simple harmonic motion. The force is directed towards zero displacement and is proportional to the displacement. The corresponding potential energy is parabolic ($V \propto x^2$).

harmonic motion is found by substituting the expression for the force, eqn 16, into Newton's equation, eqn 10. The resulting equation is

$$m\frac{d^2x}{dt^2} = -kx \tag{17}$$

and a solution is

$$x(t) = A\sin\omega t \qquad p(t) = m\omega A\cos\omega t \qquad \omega = \left(\frac{k}{m}\right)^{1/2} \tag{18}$$

(To verify the solution for x, substitute it into the differential equation; the expression for the momentum is obtained from $p = m\,dx/dt$.) These expressions show that the position of the particle varies **harmonically** (that is, as $\sin\omega t$) with a frequency $\nu = \omega/2\pi$. They also show that the particle is stationary ($p = 0$) when the displacement, x, has its maximum value, A, which is called the **amplitude** of the motion.

The total energy of a classical harmonic oscillator is proportional to the square of the amplitude of its motion. To confirm this remark we note that the kinetic energy is

$$E_K = \frac{p^2}{2m} = \frac{(m\omega A\cos\omega t)^2}{2m} = \tfrac{1}{2}m\omega^2 A^2\cos^2\omega t \tag{19a}$$

Then, because $\omega = (k/m)^{1/2}$, this expression may be written

$$E_K = \tfrac{1}{2}kA^2\cos^2\omega t \tag{19b}$$

The force on the oscillator is $F = -kx$, so it follows from the relation $F = -dV/dx$ that the potential energy of a harmonic oscillator is

$$V = \tfrac{1}{2}kx^2 = \tfrac{1}{2}kA^2\sin^2\omega t \tag{20}$$

The total energy is therefore

$$E = \tfrac{1}{2}kA^2\cos^2\omega t + \tfrac{1}{2}kA^2\sin^2\omega t = kA^2 \tag{21}$$

(We have used $\cos^2\omega t + \sin^2\omega t = 1$.) That is, the energy of the oscillator is constant and, for a given force constant, is determined by its maximum displacement. It follows that the energy of an oscillating particle can be raised to any value by stretching the spring to any desired amplitude A. It is important to note that the frequency of the motion depends only on the inherent properties of the oscillator (as represented by k and m) and is independent of the energy; the amplitude governs the energy, through $E = \tfrac{1}{2}kA^2$, and is independent of the frequency. In other words, the particle will oscillate at the same frequency regardless of the amplitude of its motion.

Electrical quantities

The fundamental expression in **electrostatics**, the interactions of stationary electric charges, is the **Coulomb potential energy** of one charge of magnitude q at a distance r from another charge q:

$$V = \frac{qq'}{4\pi\varepsilon_0 r} \tag{1}$$

That is, the potential energy is inversely proportional to the separation of the charges. The fundamental constant ε_0 is the **vacuum permittivity**; its value is $\varepsilon_0 = 8.854 \times 10^{-12}\ \mathrm{J^{-1}\,C^2\,m^{-1}}$. (Note that with r in metres, m, and the charges in coulombs, C, the potential energy is in joules, J.) The potential energy is equal to the work that must be done to bring up a charge q from infinity to a distance r from a charge q'. In a medium other than a vacuum, the potential energy of interaction between two charges is reduced, and the vacuum permittivity is replaced by the **permittivity**, ε, of the medium. It is common to express the permittivity as a multiple of the vacuum permittivity, and to write $\varepsilon = \varepsilon_r \varepsilon_0$, where ε_r is the **relative permittivity** (or dielectric constant) of the medium. For water at 25°C, $\varepsilon_r = 78.54$.

The potential energy of a charge q in the presence of another charge q' can be expressed in terms of the **Coulomb potential**, ϕ:

$$V = q\phi \qquad \phi = \frac{q'}{4\pi\varepsilon_0 r} \tag{2}$$

The units of potential are joules per coulomb, $\mathrm{J\,C^{-1}}$ so, when ϕ is multiplied by a charge in coulombs, the result is in joules. The combination joules per coulomb occurs widely in electrostatics, and is called a *volt*, V:

$$1\ \mathrm{V} = 1\ \mathrm{J\,C^{-1}}$$

(which implies that $1 \text{ V C} = 1 \text{ J}$). If there are several charges $q_1, q_2, \ldots$ present in the system, the total potential experienced by the charge q is the sum of the potential generated by each charge:

$$\phi = \phi_1 + \phi_2 + \cdots \tag{3}$$

The electrical force, F, exerted by a charge q on a second charge q' has magnitude

$$F = \frac{qq'}{4\pi\varepsilon_0 r^2} \tag{4}$$

The force is a vector quantity (that is, has direction), and is directed along the line joining the two charges. With charge in coulombs and distance in metres, the force is obtained in newtons.

The motion of charge gives rise to an electric **current**, I. Electric current is measured in *amperes*, A, where

$$1 \text{ A} = 1 \text{ C s}^{-1}$$

If the electric charge is that of electrons (as it is for conduction in metals and semiconductors), then a current of 1 A represents the flow of 6×10^{18} electrons per second. If the current flows from a region of potential ϕ_i to ϕ_f, through a **potential difference** $\Delta\phi = \phi_f - \phi_i$, the rate of doing work is the current (the rate of transfer of charge) multiplied by the potential difference, $I\Delta\phi$. The rate of doing work is called **power**, P, so

$$P = I\Delta\phi \tag{5}$$

With current in amperes and the potential difference in volts, the power works out in joules per second, or *watts*, W:

$$1 \text{ W} = 1 \text{ J s}^{-1}$$

The total energy, E, supplied in an interval Δt is the power (the energy per second) multiplied by the duration of the interval:

$$E = P\Delta t = I\Delta\phi\Delta t \tag{6}$$

The energy is obtained in joules with the current in amperes, the potential difference in volts, and the time in seconds.

Data section

The following tables reproduce and expand the data given in the short tables in the text, and follow their numbering. Standard states refer to a pressure of $p^{\ominus} = 1$ bar. The general references are as follows:

AIP: D.E. Gray (ed.), *American Institute of Physics handbook*. McGraw-Hill, New York (1972).

AS: M. Abramowitz and I.A. Stegun (ed.), *Handbook of mathematical functions*. Dover, New York (1963).

E: J. Emsley, *The elements*. Oxford University Press (1991).

HCP: R.C. Weast (ed.), *Handbook of chemistry and physics*. CRC Press, Boca Raton (1993).

JL: A.M. James and M.P. Lord, *Macmillan's chemical and physical data*. Macmillan, London (1992).

KL: G.W.C. Kaye and T.H. Laby (ed.), *Tables of physical and chemical constants*. Longman, London (1973).

LR: G.N. Lewis and M. Randall, revised by K.S. Pitzer and L. Brewer, *Thermodynamics*. McGraw-Hill, New York (1961).

NBS: *NBS tables of chemical thermodynamic properties*, published as *J. Phys. and Chem. Reference Data*, 11, Supplement 2 (1982).

RS: R.A. Robinson and R.H. Stokes, *Electrolyte solutions*. Butterworth, London (1959).

TDOC: J.B. Pedley, J.D. Naylor, and S.P. Kirby, *Thermochemical data of organic compounds*. Chapman & Hall, London (1986).

Contents

Physical properties of selected materials

	$\rho/(\text{g cm}^{-3})$ at 293 K†	T_f/K	T_b/K		$\rho/(\text{g cm}^{-3})$ at 293 K†	T_f/K	T_b/K
Elements				**Inorganic compounds**			
Aluminium(s)	2.698	933.5	2740	$CaCO_3$(s, calcite)	2.71	1612	1171d
Argon(g)	1.381	83.8	87.3	$CuSO_4 \cdot 5H_2O$(s)	2.284	383($-H_2O$)	423($-5H_2O$)
Boron(s)	2.340	2573	3931	HBr(g)	2.77	184.3	206.4
Bromine(l)	3.123	265.9	331.9	HCl(g)	1.187	159.0	191.1
Carbon(s, gr)	2.260	3700s		HI(g)	2.85	222.4	237.8
Carbon(s, d)	3.513			H_2O(l)	0.997	273.2	373.2
Chlorine(g)	1.507	172.2	239.2	D_2O(l)	1.104	277.0	374.6
Copper(s)	8.960	1357	2840	NH_3(g)	0.817	195.4	238.8
Fluorine(g)	1.108	53.5	85.0	KBr(s)	2.750	1003	1708
Gold(s)	19.320	1338	3080	KCl(s)	1.984	1049	1773s
Helium(g)	0.125		4.22	NaCl(s)	2.165	1074	1686
Hydrogen(g)	0.071	14.0	20.3	H_2SO_4(l)	1.841	283.5	611.2
Iodine(s)	4.930	386.7	457.5	**Organic compounds**			
Iron(s)	7.874	1808	3023	Acetaldehyde, CH_3CHO(l, g)	0.788	152	293
Krypton(g)	2.413	116.6	120.8	Acetic acid, CH_3COOH(l)	1.049	289.8	391
Lead(s)	11.350	600.6	2013	Acetone, $(CH_3)_2CO$(l)	0.787	178	329
Lithium(s)	0.534	453.7	1620	Aniline, $C_6H_5NH_2$(l)	1.026	267	457
Magnesium(s)	1.738	922.0	1363	Anthracene, $C_{14}H_{10}$(s)	1.243	490	615
Mercury(l)	13.546	234.3	629.7	Benzene, C_6H_6(l)	0.879	278.6	353.2
Neon(g)	1.207	24.5	27.1	Carbon tetrachloride, CCl_4(l)	1.63	250	349.9
Nitrogen(g)	0.880	63.3	77.4	Chloroform, $CHCl_3$(l)	1.499	209.6	334
Oxygen(g)	1.140	54.8	90.2	Ethanol, C_2H_5OH(l)	0.789	156	351.4
Phosphorus(s, wh)	1.820	317.3	553	Formaldehyde, HCHO(g)		181	254.0
Potassium(s)	0.862	336.8	1047	Glucose, $C_6H_{12}O_6$(s)	1.544	415	
Silver(s)	10.500	1235	2485	Methane, CH_4(g)		90.6	111.6
Sodium(s)	0.971	371.0	1156	Methanol, CH_3OH(l)	0.791	179.2	337.6
Sulfur(s, α)	2.070	386.0	717.8	Naphthalene, $C_{10}H_8$(s)	1.145	353.4	491
Uranium(s)	18.950	1406	4018	Octane, C_8H_{18}(l)	0.703	216.4	398.8
Xenon(g)	2.939	161.3	166.1	Phenol, C_6H_5OH(s)	1.073	314.1	455.0
Zinc(s)	7.133	692.7	1180	Sucrose, $C_{12}H_{22}O_{11}$(s)	1.588	457d	

d: decomposes;　s: sublimes;　Data: AIP, E, HCP, KL;　† For gases, at their boiling points.

Table 1.3 Collision cross-sections, σ/nm^2

Ar	0.36
C_2H_4	0.64
C_6H_6	0.88
CH_4	0.46
Cl_2	0.93
CO_2	0.52
H_2	0.27
He	0.21
N_2	0.43
Ne	0.24
O_2	0.40
SO_2	0.58

Data: KL

Table 1.4 Second virial coefficients, $B/(cm^3\,mol^{-1})$

	100 K	273 K	373 K	600 K
Air	−167.3	−13.5	3.4	19.0
Ar	−187.0	−21.7	−4.2	11.9
CH_4		−53.6	−21.2	8.1
CO_2		−142	−72.2	−12.4
H_2	−2.0	13.7	15.6	
He	11.4	12.0	11.3	10.4
Kr		−62.9	−28.7	1.7
N_2	−160.0	−10.5	6.2	21.7
Ne	−6.0	10.4	12.3	13.8
O_2	−197.5	−22.0	−3.7	12.9
Xe		−153.7	−81.7	−19.6

Data: AIP, JL. The values relate to the expansion in eqn 36 of Section 1.4b; convert to eqn 35 using $B' = B/RT$.

For Ar at 273 K, $C = 1200\,cm^6\,mol^{-1}$

Table 1.5 Critical constants

	p_c/atm	$V_c/(cm^3\,mol^{-1})$	T_c/K	Z_c	T_B/K		p_c/atm	$V_c/(cm^3\,mol^{-1})$	T_c/K	Z_c	T_B/K
Ar	48.00	75.25	150.72	0.292	411.5	HCl	81.5	81.0	324.7	0.248	
Br_2	102	135	584	0.287		He	2.26	57.76	5.21	0.305	22.64
C_2H_4	50.50	124	283.1	0.270		HI	80.8	423.2			
C_2H_6	48.20	148	305.4	0.285		Kr	54.27	92.24	209.39	0.291	575.0
C_6H_6	48.6	260	562.7	0.274		N_2	33.54	90.10	126.3	0.292	327.2
CH_4	45.6	98.7	190.6	0.288	510.0	Ne	26.86	41.74	44.44	0.307	122.1
Cl_2	76.1	124	417.2	0.276		NH_3	111.3	72.5	405.5	0.242	
CO_2	72.85	94.0	304.2	0.274	714.8	O_2	50.14	78.0	154.8	0.308	405.9
F_2	55	144				Xe	58.0	118.8	289.75	0.290	768.0
H_2	12.8	64.99	33.23	0.305	110.0						
H_2O	218.3	55.3	647.4	0.227							
HBr	84.0	363.0									

Data: AIP, KL

Table 1.6 van der Waals constants

	$a/(atm\,L^2\,mol^{-2})$	$b/(10^{-2}\,L\,mol^{-1})$		$a/(atm\,L^2\,mol^{-2})$	$b/(10^{-2}\,L\,mol^{-1})$
Ar	1.363	3.219	H_2S	4.490	4.287
C_2H_4	4.530	5.714	He	0.03457	2.370
C_2H_6	5.562	6.380	Kr	2.349	3.978
C_6H_6	18.24	11.54	N_2	1.408	3.913
CH_4	2.283	4.278	Ne	0.2135	1.709
Cl_2	6.579	5.622	NH_3	4.225	3.707
CO	1.505	3.985	O_2	1.378	3.183
CO_2	3.640	4.267	SO_2	6.803	5.636
H_2	0.2476	2.661	Xe	4.250	5.105
H_2O	5.536	3.049			

Data: HCP, JL

Table 2.2 Temperature variation of molar heat capacities†

	a	$b/(10^{-3}\,\text{K})$	$c/(10^5\,\text{K})$
Monatomic gases			
	20.78	0	0
Other gases			
Br_2	37.32	0.50	−1.26
Cl_2	37.03	0.67	−2.85
CO_2	44.22	8.79	−8.62
F_2	34.56	2.51	−3.51
H_2	27.28	3.26	0.50
I_2	37.40	0.59	−0.71
N_2	28.58	3.77	−0.50
NH_3	29.75	25.1	−1.55
O_2	29.96	4.18	−1.67
Liquids (from melting to boiling)			
$C_{10}H_8$, naphthalene	79.5	0.4075	0
I_2	80.33	0	0
H_2O	75.29	0	0
Solids			
Al	20.67	12.38	0
C (graphite)	16.86	4.77	−8.54
$C_{10}H_8$, naphthalene	−115.9	3.920×10^3	0
Cu	22.64	6.28	0
I_2	40.12	49.79	0
NcCl	45.94	16.32	0
Pb	22.13	11.72	0.96

† For $C_{p,m}/(\text{J K}^{-1}\,\text{mol}^{-1}) = a + bT + c/T^2$
Source: LR.

Table 2.3 Standard enthalpies of fusion and vaporization at the transition temperature, $\Delta_{\text{trs}}H^{\ominus}/(\text{kJ mol}^{-1})$

	T_f/K	Fusion	T_b/K	Vaporization		T_f/K	Fusion	T_b/K	Vaporization
Elements					CO_2	217.0	8.33	194.6	25.23 s
Ag	1234	11.30	2436	250.6	CS_2	161.2	4.39	319.4	26.74
Ar	83.81	1.188	87.29	6.506	H_2O	273.15	6.008	373.15	40.656
Br_2	265.9	10.57	332.4	29.45					44.016 at 298 K
Cl_2	172.1	6.41	239.1	20.41	H_2S	187.6	2.377	212.8	18.67
F_2	53.6	0.26	85.0	3.16	H_2SO_4	283.5	2.56		
H_2	13.96	0.117	20.38	0.916	NH_3	195.4	5.652	239.7	23.35
He	3.5	0.021	4.22	0.084	**Organic compounds**				
Hg	234.3	2.292	629.7	59.30	CH_4	90.68	0.941	111.7	8.18
I_2	386.8	15.52	458.4	41.80	CCl_4	250.3	2.5	350	30.0
N_2	63.15	0.719	77.35	5.586	C_2H_6	89.85	2.86	184.6	14.7
Na	371.0	2.601	1156	98.01	C_6H_6	278.61	10.59	353.2	30.8
O_2	54.36	0.444	90.18	6.820	C_6H_{14}	178	13.08	342.1	28.85
Xe	161	2.30	165	12.6	$C_{10}H_8$	354	18.80	490.9	51.51
K	336.4	2.35	1031	80.23	CH_3OH	175.2	3.16	337.2	35.27
Inorganic compounds									37.99 at 298 K
CCl_4	250.3	2.47		30.00	C_2H_5OH	158.7	4.60	352	43.5

Data: AIP; s denotes sublimation

Table 2.5 Thermodynamic data for organic compounds (all values are for 298 K)

	$M/(\text{g mol}^{-1})$	$\Delta_f H^{\ominus}/(\text{kJ mol}^{-1})$	$\Delta_f G^{\ominus}/(\text{kJ mol}^{-1})$	$S_m^{\ominus}/(\text{J K}^{-1}\text{ mol}^{-1})$	$C_{p,m}^{\ominus}/(\text{J K}^{-1}\text{ mol}^{-1})$	$\Delta_c H^{\ominus}/(\text{kJ mol}^{-1})$
C(s) (graphite)	12.011	0	0	5.740	8.527	−393.51
C(s) (diamond)	12.011	+1.895	+2.900	2.377	6.113	−395.40
CO_2(g)	44.010	−393.51	−394.36	213.74	37.11	
Hydrocarbons						
CH_4(g), methane	16.04	−74.81	−50.72	186.26	35.31	−890
CH_3(g), methyl	15.04	+145.69	+147.92	194.2	38.70	
C_2H_2(g), ethyne	26.04	+226.73	+209.20	200.94	43.93	−1300
C_2H_4(g), ethene	28.05	+52.26	+68.15	219.56	43.56	−1411
C_2H_6(g), ethane	30.07	−84.68	−32.82	229.60	52.63	−1560
C_3H_6(g), propene	42.08	+20.42	+62.78	267.05	63.89	−2058
C_3H_6(g), cyclopropane	42.08	+53.30	+104.45	237.55	55.94	−2091
C_3H_8(g), propane	44.10	−103.85	−23.49	269.91	73.5	−2220
C_4H_8(g), 1-butene	56.11	−0.13	+71.39	305.71	85.65	−2717
C_4H_8(g), cis-2-butene	56.11	−6.99	+65.95	300.94	78.91	−2710
C_4H_8(g), trans-2-butene	56.11	−11.17	+63.06	296.59	87.82	−2707
C_4H_{10}(g), butane	58.13	−126.15	−17.03	310.23	97.45	−2878
C_5H_{12}(g), pentane	72.15	−146.44	−8.20	348.40	120.2	−3537
C_5H_{12}(l)	72.15	−173.1				
C_6H_6(l), benzene	78.12	+49.0	+124.3	173.3	136.1	−3268
C_6H_6(g)	78.12	+82.93	+129.72	269.31	81.67	−3302
C_6H_{12}(l), cyclohexane	84.16	−156	+26.8		156.5	−3920
C_6H_{14}(l), hexane	86.18	−198.7		204.3		−4163
$C_6H_5CH_3$(g), methyl-benzene (toluene)	92.14	+50.0	+122.0	320.7	103.6	−3953
C_7H_{16}(l), heptane	100.21	−224.4	+1.0	328.6	224.3	
C_8H_{18}(l), octane	114.23	−249.9	+6.4	361.1		−5471
C_8H_{18}(l), iso-octane	114.23	−255.1				−5461
$C_{10}H_8$(s), naphthalene	128.18	+78.53				−5157
Alcohols and phenols						
CH_3OH(l), methanol	32.04	−238.66	−166.27	126.8	81.6	−726
CH_3OH(g)	32.04	−200.66	−161.96	239.81	43.89	−764
C_2H_5OH(l), ethanol	46.07	−277.69	−174.78	160.7	111.46	−1368
C_2H_5OH(g)	46.07	−235.10	−168.49	282.70	65.44	−1409
C_6H_5OH(s), phenol	94.12	−165.0	−50.9	146.0		−3054
Carboxylic acids, hydroxy acids, and esters						
$HCOOH$(l), formic	46.03	−424.72	−361.35	128.95	99.04	−255
CH_3COOH(l), acetic	60.05	−484.5	−389.9	159.8	124.3	−875
CH_3COOH(aq)	60.05	−485.76	−396.46	178.7		
$CH_3CO_2^-$(aq)	59.05	−486.01	−369.31	86.6	−6.3	
$(COOH)_2$(s), oxalic	90.04	−827.2			117	−254
C_6H_5COOH(s), benzoic	122.13	−385.1	−245.3	167.6	146.8	−3227
$CH_3CH(OH)COOH$(s), lactic	90.08	−694.0				−1344
$CH_3COOC_2H_5$(l), ethyl acetate	88.11	−479.0	−332.7	259.4	170.1	−2231
Alkanals and alkanones						
$HCHO$(g), methanal	30.03	−108.57	−102.53	218.77	35.40	−571
CH_3CHO(l), ethanal	44.05	−192.30	−128.12	160.2		−1166
CH_3CHO(g)	44.05	−166.19	−128.86	250.3	57.3	−1192
CH_3COCH_3(l), propanone	58.08	−248.1	−155.4	200.4	124.7	−1790

Table 2.5 (Continued)

	$M/(\text{g mol}^{-1})$	$\Delta_f H^{\ominus}/(\text{kJ mol}^{-1})$	$\Delta_f G^{\ominus}/(\text{kJ mol}^{-1})$	$S_m^{\ominus}/(\text{J K}^{-1}\,\text{mol}^{-1})$	$C_{p,m}^{\ominus}/(\text{J K}^{-1}\,\text{mol}^{-1})$	$\Delta_c H^{\ominus}/(\text{kJ mol}^{-1})$
Sugars						
$C_6H_{12}O_6(s)$, α-D-glucose	180.16	−1274				−2808
$C_6H_{12}O_6(s)$, β-D-glucose	180.16	−1268	−910	212		
$C_6H_{12}O_6(s)$, β-D-fructose	180.16	−1266				−2810
$C_{12}H_{22}O_{11}(s)$, sucrose	342.30	−2222	−1543	360.2		−5645
Nitrogen compounds						
$CO(NH_2)_2(s)$, urea	60.06	−333.51	−197.33	104.60	93.14	−632
$CH_3NH_2(g)$, methylamine	31.06	−22.97	+32.16	243.41	53.1	−1085
$C_6H_5NH_2(l)$, aniline	93.13	+31.1				−3393
$CH_2(NH_2)COOH(s)$, glycine	75.07	−532.9	−373.4	103.5	99.2	−969

Data: NBS, TDOC

Table 2.6 Thermodynamic data (all values relate to 298 K)

	$M/(\text{g mol}^{-1})$	$\Delta_f H^{\ominus}/(\text{kJ mol}^{-1})$	$\Delta_f G^{\ominus}/(\text{kJ mol}^{-1})$	$S_m^{\ominus}/(\text{J K}^{-1}\,\text{mol}^{-1})$	$C_{p,m}^{\ominus}/(\text{J K}^{-1}\,\text{mol}^{-1})$
Aluminium (aluminum)					
$Al(s)$	26.98	0	0	28.33	24.35
$Al(l)$	26.98	+10.56	+7.20	39.55	24.21
$Al(g)$	26.98	+326.4	+285.7	164.54	21.38
$Al^{3+}(g)$	26.98	+5483.17			
$Al^{3+}(aq)$	26.98	−531	−485	−321.7	
$Al_2O_3(s, \alpha)$	101.96	−1675.7	−1582.3	50.92	79.04
$AlCl_3(s)$	133.24	−704.2	−628.8	110.67	91.84
Argon					
$Ar(g)$	39.95	0	0	154.84	20.786
Antimony					
$Sb(s)$	121.75	0	0	45.69	25.23
$SbH_3(g)$	124.77	+145.11	+147.75	232.78	41.05
Arsenic					
$As(s, \alpha)$	74.92	0	0	35.1	24.64
$As(g)$	74.92	+302.5	+261.0	174.21	20.79
$As_4(g)$	299.69	+143.9	+92.4	314	
$AsH_3(g)$	77.95	+66.44	+68.93	222.78	38.07
Barium					
$Ba(s)$	137.34	0	0	62.8	28.07
$Ba(g)$	137.34	+180	+146	170.24	20.79
$Ba^{2+}(aq)$	137.34	−537.64	−560.77	9.6	
$BaO(s)$	153.34	−553.5	−525.1	70.43	47.78
$BaCl_2(s)$	208.25	−858.6	−810.4	123.68	75.14
Beryllium					
$Be(s)$	9.01	0	0	9.50	16.44
$Be(g)$	9.01	+324.3	+286.6	136.27	20.79

Table 2.6 (Continued)

	$M/(\text{g mol}^{-1})$	$\Delta_f H^\ominus/(\text{kJ mol}^{-1})$	$\Delta_f G^\ominus/(\text{kJ mol}^{-1})$	$S_m^\ominus/(\text{J K}^{-1}\text{mol}^{-1})$	$C_{p,m}^\ominus/(\text{J K}^{-1}\text{mol}^{-1})$
Bismuth					
Bi(s)	208.98	0	0	56.74	25.52
Bi(g)	208.98	+207.1	+168.2	187.00	20.79
Bromine					
Br_2(l)	159.82	0	0	152.23	75.689
Br_2(g)	159.82	+30.907	+3.110	245.46	36.02
Br(g)	79.91	+111.88	+82.396	175.02	20.786
Br^-(g)	79.91	−219.07			
Br^-(aq)	79.91	−121.55	−103.96	82.4	−141.8
HBr(g)	90.92	−36.40	−53.45	198.70	29.142
Cadmium					
Cd(s, γ)	112.40	0	0	51.76	25.98
Cd(g)	112.40	+112.01	+77.41	167.75	20.79
Cd^{2+}(aq)	112.40	−75.90	−77.612	−73.2	
CdO(s)	128.40	−258.2	−228.4	54.8	43.43
$CdCO_3$(s)	172.41	−750.6	−669.4	92.5	
Caesium (cesium)					
Cs(s)	132.91	0	0	85.23	32.17
Cs(g)	132.91	+76.06	+49.12	175.60	20.79
Cs^+(aq)	132.91	−258.28	−292.02	133.05	−10.5
Calcium					
Ca(s)	40.08	0	0	41.42	25.31
Ca(g)	40.08	+178.2	+144.3	154.88	20.786
Ca^{2+}(aq)	40.08	−542.83	−553.58	−53.1	
CaO(s)	56.08	−635.09	−604.03	39.75	42.80
$CaCO_3$(s) (calcite)	100.09	−1206.9	−1128.8	92.9	81.88
$CaCO_3$(s) (aragonite)	100.09	−1207.1	−1127.8	88.7	81.25
CaF_2(s)	78.08	−1219.6	−1167.3	68.87	67.03
$CaCl_2$(s)	110.99	−795.8	−748.1	104.6	72.59
$CaBr_2$(s)	199.90	−682.8	−663.6	130	

Carbon (for 'organic' compounds of carbon, see Table 2.5)

C(s) (graphite)	12.011	0	0	5.740	8.527
C(s) (diamond)	12.011	+1.895	+2.900	2.377	6.113
C(g)	12.011	+716.68	+671.26	158.10	20.838
C_2(g)	24.022	+831.90	+775.89	199.42	43.21
CO(g)	28.011	−110.53	−137.17	197.67	29.14
CO_2(g)	44.010	−393.51	−394.36	213.74	37.11
CO_2(aq)	44.010	−413.80	−385.98	117.6	
H_2CO_3(aq)	62.03	−699.65	−623.08	187.4	
HCO_3^-(aq)	61.02	−691.99	−586.77	91.2	
CO_3^{2-}(aq)	60.01	−677.14	−527.81	−56.9	
CCl_4(l)	153.82	−135.44	−65.21	216.40	131.75
CS_2(l)	76.14	+89.70	+65.27	151.34	75.7
HCN(g)	27.03	+135.1	+124.7	201.78	35.86
HCN(l)	27.03	+108.87	+124.97	112.84	70.63
CN^-(aq)	26.02	+150.6	+172.4	94.1	

Table 2.6 (Continued)

	$M/(\text{g mol}^{-1})$	$\Delta_f H^{\ominus}/(\text{kJ mol}^{-1})$	$\Delta_f G^{\ominus}/(\text{kJ mol}^{-1})$	$S_m^{\ominus}/(\text{J K}^{-1}\text{ mol}^{-1})$	$C_{p,m}^{\ominus}/(\text{J K}^{-1}\text{ mol}^{-1})$
Chlorine					
$Cl_2(g)$	70.91	0	0	223.07	33.91
$Cl(g)$	35.45	+121.68	+105.68	165.20	21.840
$Cl^-(g)$	35.45	−233.13			
$Cl^-(aq)$	35.45	−167.16	−131.23	56.5	−136.4
$HCl(g)$	36.46	−92.31	−95.30	186.91	29.12
$HCl(aq)$	36.46	−167.16	−131.23	56.5	−136.4
Chromium					
$Cr(s)$	52.00	0	0	23.77	23.35
$Cr(g)$	52.00	+396.6	+351.8	174.50	20.79
$CrO_4^{2-}(aq)$	115.99	−881.15	−727.75	50.21	
$Cr_2O_7^{2-}(aq)$	215.99	−1490.3	−1301.1	261.9	
Copper					
$Cu(s)$	63.54	0	0	33.150	24.44
$Cu(g)$	63.54	+338.32	+298.58	166.38	20.79
$Cu^+(aq)$	63.54	+71.67	+49.98	40.6	
$Cu^{2+}(aq)$	63.54	+64.77	+65.49	−99.6	
$Cu_2O(s)$	143.08	−168.6	−146.0	93.14	63.64
$CuO(s)$	79.54	−157.3	−129.7	42.63	42.30
$CuSO_4(s)$	159.60	−771.36	−661.8	109	100.0
$CuSO_4 \cdot H_2O(s)$	177.62	−1085.8	−918.11	146.0	134
$CuSO_4 \cdot 5H_2O(s)$	249.68	−2279.7	−1879.7	300.4	280
Deuterium					
$D_2(g)$	4.028	0	0	144.96	29.20
$HD(g)$	3.022	+0.318	−1.464	143.80	29.196
$D_2O(g)$	20.028	−249.20	−234.54	198.34	34.27
$D_2O(l)$	20.028	−294.60	−243.44	75.94	84.35
$HDO(g)$	19.022	−245.30	−233.11	199.51	33.81
$HDO(l)$	19.022	−289.89	−241.86	79.29	
Fluorine					
$F_2(g)$	38.00	0	0	202.78	31.30
$F(g)$	19.00	+78.99	+61.91	158.75	22.74
$F^-(aq)$	19.00	−332.63	−278.79	−13.8	−106.7
$HF(g)$	20.01	−271.1	−273.2	173.78	29.13
Gold					
$Au(s)$	196.97	0	0	47.40	25.42
$Au(g)$	196.97	+366.1	+326.3	180.50	20.79
Helium					
$He(g)$	4.003	0	0	126.15	20.786
Hydrogen (see also deuterium)					
$H_2(g)$	2.016	0	0	130.684	28.824
$H(g)$	1.008	+217.97	+203.25	114.71	20.784
$H^+(aq)$	1.008	0	0	0	0
$H^+(g)$	1.008	+1536.20			

Table 2.6 (Continued)

	$M/(\text{g mol}^{-1})$	$\Delta_f H^{\ominus}/(\text{kJ mol}^{-1})$	$\Delta_f G^{\ominus}/(\text{kJ mol}^{-1})$	$S_m^{\ominus}/(\text{J K}^{-1} \text{mol}^{-1})$	$C_{p,m}^{\ominus}/(\text{J K}^{-1} \text{mol}^{-1})$
Hydrogen (Continued)					
$H_2O(l)$	18.015	−285.83	−237.13	69.91	75.291
$H_2O(g)$	18.015	−241.82	−228.57	188.83	33.58
$H_2O_2(l)$	34.015	−187.78	−120.35	109.6	89.1
Iodine					
$I_2(s)$	253.81	0	0	116.135	54.44
$I_2(g)$	253.81	+62.44	+19.33	260.69	36.90
$I(g)$	126.90	+106.84	+70.25	180.79	20.786
$I^-(aq)$	126.90	−55.19	−51.57	111.3	−142.3
$HI(g)$	127.91	+26.48	+1.70	206.59	29.158
Iron					
$Fe(s)$	55.85	0	0	27.28	25.10
$Fe(g)$	55.85	+416.3	+370.7	180.49	25.68
$Fe^{2+}(aq)$	55.85	−89.1	−78.90	−137.7	
$Fe^{3+}(aq)$	55.85	−48.5	−4.7	−315.9	
$Fe_3O_4(s)$ (magnetite)	231.54	−1118.4	−1015.4	146.4	143.43
$Fe_2O_3(s)$ (haematite)	159.69	−824.2	−742.2	87.40	103.85
$FeS(s, \alpha)$	87.91	−100.0	−100.4	60.29	50.54
$FeS_2(s)$	119.98	−178.2	−166.9	52.93	62.17
Krypton					
$Kr(g)$	83.80	0	0	164.08	20.786
Lead					
$Pb(s)$	207.19	0	0	64.81	26.44
$Pb(g)$	207.19	+195.0	+161.9	175.37	20.79
$Pb^{2+}(aq)$	207.19	−1.7	−24.43	10.5	
$PbO(s, \text{yellow})$	223.19	−217.32	−187.89	68.70	45.77
$PbO(s, \text{red})$	223.19	−218.99	−188.93	66.5	45.81
$PbO_2(s)$	239.19	−277.4	−217.33	68.6	64.64
Lithium					
$Li(s)$	6.94	0	0	29.12	24.77
$Li(g)$	6.94	+159.37	+126.66	138.77	20.79
$Li^+(aq)$	6.94	−278.49	−293.31	13.4	68.6
Magnesium					
$Mg(s)$	24.31	0	0	32.68	24.89
$Mg(g)$	24.31	+147.70	+113.10	148.65	20.786
$Mg^{2+}(aq)$	24.31	−466.85	−454.8	−138.1	
$MgO(s)$	40.31	−601.70	−569.43	26.94	37.15
$MgCO_3(s)$	84.32	−1095.8	−1012.1	65.7	75.52
$MgCl_2(s)$	95.22	−641.32	−591.79	89.62	71.38
Mercury					
$Hg(l)$	200.59	0	0	76.02	27.983
$Hg(g)$	200.59	+61.32	+31.82	174.96	20.786
$Hg^{2+}(aq)$	200.59	+171.1	+164.40	−32.2	
$Hg_2^{2+}(aq)$	401.18	+172.4	+153.52	84.5	
$HgO(s)$	216.59	−90.83	−58.54	70.29	44.06

Table 2.6 (Continued)

	$M/(\text{g mol}^{-1})$	$\Delta_f H^{\ominus}/(\text{kJ mol}^{-1})$	$\Delta_f G^{\ominus}/(\text{kJ mol}^{-1})$	$S_m^{\ominus}/(\text{J K}^{-1}\text{mol}^{-1})$	$C_{p,m}^{\ominus}/(\text{J K}^{-1}\text{mol}^{-1})$
Mercury (Continued)					
$Hg_2Cl_2(s)$	472.09	−265.22	−210.75	192.5	102
$HgCl_2(s)$	271.50	−224.3	−178.6	146.0	
HgS(s, black)	232.65	−53.6	−47.7	88.3	
Neon					
Ne(g)	20.18	0	0	146.33	20.786
Nitrogen					
$N_2(g)$	28.013	0	0	191.61	29.125
N(g)	14.007	+472.70	+455.56	153.30	20.786
NO(g)	30.01	+90.25	+86.55	210.76	29.844
$N_2O(g)$	44.01	+82.05	+104.20	219.85	38.45
$NO_2(g)$	46.01	+33.18	+51.31	240.06	37.20
$N_2O_4(g)$	92.01	+9.16	+97.89	304.29	77.28
$N_2O_5(s)$	108.01	−43.1	+113.9	178.2	143.1
$N_2O_5(g)$	108.01	+11.3	+115.1	355.7	84.5
$HNO_3(l)$	63.01	−174.10	−80.71	155.60	109.87
$HNO_3(aq)$	63.01	−207.36	−111.25	146.4	−86.6
$NO_3^-(aq)$	62.01	−205.0	−108.74	146.4	−86.6
$NH_3(g)$	17.03	−46.11	−16.45	192.45	35.06
$NH_3(aq)$	17.03	−80.29	−26.50	111.3	
$NH_4^+(aq)$	18.04	−132.51	−79.31	113.4	79.9
$NH_2OH(s)$	33.03	−114.2			
$HN_3(l)$	43.03	+264.0	+327.3	140.6	43.68
$HN_3(g)$	43.03	+294.1	+328.1	238.97	98.87
$N_2H_4(l)$	32.05	+50.63	+149.43	121.21	139.3
$NH_4NO_3(s)$	80.04	−365.56	−183.87	151.08	84.1
$NH_4Cl(s)$	53.49	−314.43	−202.87	94.6	
Oxygen					
$O_2(g)$	31.999	0	0	205.138	29.355
O(g)	15.999	+249.17	+231.73	161.06	21.912
$O_3(g)$	47.998	+142.7	+163.2	238.93	39.20
$OH^-(aq)$	17.007	−229.99	−157.24	−10.75	−148.5
Phosphorus					
P(s, wh)	30.97	0	0	41.09	23.840
P(g)	30.97	+314.64	+278.25	163.19	20.786
$P_2(g)$	61.95	+144.3	+103.7	218.13	32.05
$P_4(g)$	123.90	+58.91	+24.44	279.98	67.15
$PH_3(g)$	34.00	+5.4	+13.4	210.23	37.11
$PCl_3(g)$	137.33	−287.0	−267.8	311.78	71.84
$PCl_3(l)$	137.33	−319.7	−272.3	217.1	
$PCl_5(g)$	208.24	−374.9	−305.0	364.6	112.8
$PCl_5(s)$	208.24	−443.5			
$H_3PO_3(s)$	82.00	−964.4			
$H_3PO_3(aq)$	82.00	−964.8			
$H_3PO_4(s)$	94.97	−1279.0	−1119.1	110.50	106.06
$H_3PO_4(l)$	94.97	−1266.9			
$H_3PO_4(aq)$	94.97	−1277.4	−1018.7	−222	

Table 2.6 (Continued)

	$M/(\text{g mol}^{-1})$	$\Delta_f H^{\ominus}/(\text{kJ mol}^{-1})$	$\Delta_f G^{\ominus}/(\text{kJ mol}^{-1})$	$S_m^{\ominus}/(\text{J K}^{-1}\,\text{mol}^{-1})$	$C_{p,m}^{\ominus}/(\text{J K}^{-1}\,\text{mol}^{-1})$
Phosphorus (Continued)					
PO_4^{3-}(aq)	94.97	−1277.4	−1018.7	−221.8	
P_4O_{10}(s)	283.89	−2984.0	−2697.0	228.86	211.71
P_4O_6(s)	219.89	−1640.1			
Potassium					
K(s)	39.10	0	0	64.18	29.58
K(g)	39.10	+89.24	+60.59	160.336	20.786
K^+(g)	39.10	+514.26			
K^+(aq)	39.10	−252.38	−283.27	102.5	21.8
KOH(s)	56.11	−424.76	−379.08	78.9	64.9
KF(s)	58.10	−576.27	−537.75	66.57	49.04
KCl(s)	74.56	−436.75	−409.14	82.59	51.30
KBr(s)	119.01	−393.80	−380.66	95.90	52.30
KI(s)	166.01	−327.90	−324.89	106.32	52.93
Silicon					
Si(s)	28.09	0	0	18.83	20.00
Si(g)	28.09	+455.6	+411.3	167.97	22.25
SiO_2(s, α)	60.09	−910.94	−856.64	41.84	44.43
Silver					
Ag(s)	107.87	0	0	42.55	25.351
Ag(g)	107.87	+284.55	+245.65	173.00	20.79
Ag^+(aq)	107.87	+105.58	+77.11	72.68	21.8
AgBr(s)	187.78	−100.37	−96.90	107.1	52.38
AgCl(s)	143.32	−127.07	−109.79	96.2	50.79
Ag_2O(s)	231.74	−31.05	−11.20	121.3	65.86
$AgNO_3$(s)	169.88	−129.39	−33.41	140.92	93.05
Sodium					
Na(s)	22.99	0	0	51.21	28.24
Na(g)	22.99	+107.32	+76.76	153.71	20.79
Na^+(aq)	22.99	−240.12	−261.91	59.0	46.4
NaOH(s)	40.00	−425.61	−379.49	64.46	59.54
NaCl(s)	58.44	−411.15	−384.14	72.13	50.50
NaBr(s)	102.90	−361.06	−348.98	86.82	51.38
NaI(s)	149.89	−287.78	−286.06	98.53	52.09
Sulfur					
S(s, α) (rhombic)	32.06	0	0	31.80	22.64
S(s, β) (monoclinic)	32.06	+0.33	+0.1	32.6	23.6
S(g)	32.06	+278.81	+238.25	167.82	23.673
S_2(g)	64.13	+128.37	+79.30	228.18	32.47
S^{2-}(aq)	32.06	+33.1	+85.8	−14.6	
SO_2(g)	64.06	−296.83	−300.19	248.22	39.87
SO_3(g)	80.06	−395.72	−371.06	256.76	50.67
H_2SO_4(l)	98.08	−813.99	−690.00	156.90	138.9
H_2SO_4(aq)	98.08	−909.27	−744.53	20.1	−293
SO_4^{2-}(aq)	96.06	−909.27	−744.53	20.1	−293
HSO_4^-(aq)	97.07	−887.34	−755.91	131.8	−84

Table 2.6 (Continued)

	$M/(\text{g mol}^{-1})$	$\Delta_f H^{\ominus}/(\text{kJ mol}^{-1})$	$\Delta_f G^{\ominus}/(\text{kJ mol}^{-1})$	$S_m^{\ominus}/(\text{J K}^{-1}\,\text{mol}^{-1})$	$C_{p,m}^{\ominus}/(\text{J K}^{-1}\,\text{mol}^{-1})$
Sulfur (Continued)					
$H_2S(g)$	34.08	−20.63	−33.56	205.79	34.23
$H_2S(aq)$	34.08	−39.7	−27.83	121	
$HS^-(aq)$	33.072	−17.6	+12.08	62.08	
$SF_6(g)$	146.05	−1209	−1105.3	291.82	97.28
Tin					
$Sn(s, \beta)$	118.69	0	0	51.55	26.99
$Sn(g)$	118.69	+302.1	+267.3	168.49	20.26
$Sn^{2+}(aq)$	118.69	−8.8	−27.2	−17	
$SnO(s)$	134.69	−285.8	−256.9	56.5	44.31
$SnO_2(s)$	150.69	−580.7	−519.6	52.3	52.59
Xenon					
$Xe(g)$	131.30	0	0	169.68	20.786
Zinc					
$Zn(s)$	65.37	0	0	41.63	25.40
$Zn(g)$	65.37	+130.73	+95.14	160.98	20.79
$Zn^{2+}(aq)$	65.37	−153.89	−147.06	−112.1	46
$ZnO(s)$	81.37	−348.28	−318.30	43.64	40.25

Source: NBS

Supplementary thermochemical information

Table 2.6a Lattice enthalpies, $\Delta H_L^{\ominus}/(\text{kJ mol}^{-1})$

	F	Cl	Br	I
Halides				
Li	1037	852	815	761
Na	926	787	752	705
K	821	717	689	649
Rb	789	695	668	632
Cs	750	676	654	620
Ag	969	912	900	886
Be		3017		
Mg		2524		
Ca		2255		
Sr		2153		

Oxides

MgO 3850   CaO 3461   SrO 3283   BaO 3114

Sulfides

MgS 3406   CaS 3119   SrS 2974   BaS 2832

Entries refer to $MX(s) \rightarrow M^+(g) + X^-(g)$.
Data: Principally D. Cubicciotti, *J. Chem. Phys.* **31**, 1646 (1959).

Table 2.6b Standard enthalpies of hydration at infinite dilution, $\Delta_{hyd} H^{\ominus}/(\text{kJ mol}^{-1})$

	Li^+	Na^+	K^+	Rb^+	Cs^+
F^-	−1026	−911	−828	−806	−782
Cl^-	−884	−783	−685	−664	−640
Br^-	−856	−742	−658	−637	−613
I^-	−815	−701	−617	−596	−572

Entries refer to $X^+(g) + Y^-(g) \rightarrow X^+(aq) + Y^-(aq)$.
Data: Principally J.O'M. Bockris and A.K.N. Reddy, *Modern electrochemistry*, Vol. 1. Plenum Press, New York (1970).

Table 2.6c Standard ion hydration enthalpies, $\Delta_{\text{hyd}}H^{\ominus}/(\text{kJ mol}^{-1})$ at 298 K

Cations								Anions							
H^+	(−1090)	Ag^+	−464	Mg^2	−1920			OH^-	−460						
Li^+	−520	NH_4^+	−301	Ca^{2+}	−1650			F^-	−506	Cl^-	−364	Br^-	−337	I^-	−296
Na^+	−405			Sr^{2+}	−1480										
K^+	−321			Ba^{2+}	−1360										
Rb^+	−300			Fe^{2+}	−1950										
Cs^+	−277			Cu^{2+}	−2100										
				Zn^{2+}	−2050										
				Al^{3+}	−4690										
				Fe^{3+}	−4430										

Entries refer to $X^{\pm}(g) \rightarrow X^{\pm}(aq)$ based on $H^+(g) \rightarrow H^+(aq)$; $\Delta H^{\ominus} = -1090 \text{ kJ mol}^{-1}$.
Data: Principally J.O'M. Bockris and A.K.N. Reddy, *Modern electrochemistry*, Vol. 1. Plenum Press, New York (1970).

Table 2.7 Benson thermochemical groups

Group	$\Delta_f H^{\ominus}/(\text{kJ mol}^{-1})$	$S_m^{\ominus}/(\text{J K}^{-1}\text{ mol}^{-1})$	$C_{p,m}^{\ominus}/(\text{J K}^{-1}\text{ mol}^{-1})$
$C(H)_3(C)$	−42.17	127.2	25.9
$C(H)_2(C)_2$	−20.7	39.4	22.8
$C(H)(C)_3$	−6.19	−50.50	18.7
$C(C)_4$	+8.16	−146.9	18.2
$C(Cl)(H)_2(C)$	−65.7	158	37
$C(Br)(H)_2(C)$	−22	169	38
$C(I)(H)_2(C)$	+37		
$C(Cl)(H)(C)_2$	−60.2	74.1	36
$C(Br)(H)(C)_2$	−9.6	84.9	
$C(Cl)(H)(C)_3$	−53.1	−32	36
$C(Br)(H)(C)_3$		−18	39
$C(Cl)_3(C)$		210.	66.1
$[O—(C)(H)] + [C—(O)(H)_3]$	−48.1	59.5	10.5
$[O—(C)(C)] + 2[C—(O)(H)_3]$	−45.3	69.4	15.7
$[C—(O)(C)(H)_2] + 2[C—(O)(H)_3]$	+2.0	−20.4	−1.1

Data: S.W. Benson, *Thermochemical kinetics*. McGraw-Hill, New York (1976).

Table 3.1 Expansion coefficients, α, and isothermal compressibilities, κ_T

	$a/(10^{-4}\text{ K}^{-1})$	$\kappa_T/(10^{-6}\text{ atm}^{-1})$
Liquids		
Benzene	12.4	92.1
Carbon tetrachloride	12.4	90.5
Ethanol	11.2	76.8
Mercury	1.82	38.7
Water	2.1	49.6
Solids		
Copper	0.501	0.735
Diamond	0.030	0.187
Iron	0.354	0.589
Lead	0.861	2.21

The values refer to 20 °C
Data: AIP(α), KL(κ_T).

Table 3.2 Inversion temperatures, normal freezing and boiling points, and Joule–Thomson coefficients at 1 atm and 298 K

	T_I/K	T_f/K	T_b/K	$\mu_{\text{JT}}/(\text{K atm}^{-1})$
Air	603			0.189 at 50°C
Argon	723	83.8	87.3	
Carbon dioxide	1500	194.7s		1.11 at 300 K
Helium	40		4.22	−0.062
Hydrogen	202	14.0	20.3	−0.03
Krypton	1090	116.6	120.8	
Methane	968	90.6	111.6	
Neon	231	24.5	27.1	
Nitrogen	621	63.3	77.4	0.27
Oxygen	764	54.8	90.2	0.31

s: sublimes
Data: AIP, JL, and M.W. Zemansky, *Heat and thermodynamics*. McGraw-Hill, New York (1957).

Table 4.1 Standard entropies (and temperatures) of phase transitions at 1 atm, $\Delta_{trs}S^{\ominus}/(J\,K^{-1}\,mol^{-1})$

	Fusion (at T_f)	Vaporization (at T_b)
Ar	14.17 (at 83.8 K)	74.53 (at 87.3 K)
Br_2	39.76 (at 265.9 K)	88.61 (at 332.4 K)
C_6H_6	38.00 (at 278.6 K)	87.19 (at 353.2 K)
CH_3COOH	40.4 (at 289.8 K)	61.9 (at 391.4 K)
CH_3OH	18.03 (at 175.2 K)	104.6 (at 337.2 K)
Cl_2	37.22 (at 172.1 K)	85.38 (at 239.0 K)
H_2	8.38 (at 14.0 K)	44.96 (at 20.38 K)
H_2O	22.00 (at 273.2 K)	109.0 (at 373.2 K)
H_2S	12.67 (at 187.6 K)	87.75 (at 212.0 K)
He	4.8 (at 1.8 K and 30 bar)	19.9 (at 4.22 K)
N_2	11.39 (at 63.2 K)	75.22 (at 77.4 K)
NH_3	28.93 (at 195.4 K)	97.41 (at 239.73 K)
O_2	8.17 (at 54.4 K)	75.63 (at 90.2 K)

Data: AIP

Table 4.2 Standard entropies of vaporization of liquids at their normal boiling point

	$\Delta_{vap}H^{\ominus}/(kJ\,mol^{-1})$	$\theta_b/^{\circ}C$	$\Delta_{vap}S^{\ominus}/(J\,K^{-1}\,mol^{-1})$
Benzene	30.8	80.1	+87.2
Carbon disulfide	26.74	46.25	+83.7
Carbon tetrachloride	30.00	76.7	+85.8
Cyclohexane	30.1	80.7	+85.1
Decane	38.75	174	+86.7
Dimethyl ether	21.51	−23	+86
Ethanol	38.6	78.3	+110.0
Hydrogen sulfide	18.7	−60.4	+87.9
Mercury	59.3	356.6	+94.2
Methane	8.18	−161.5	+73.2
Methanol	35.21	65.0	+104.1
Water	40.7	100.0	+109.1

Data: JL

Table 4.3 Standard Third-Law entropies at 298 K: see Tables 2.5 and 2.6

Table 4.4 Standard Gibbs energies of formation at 298 K: see Tables 2.5 and 2.6

Table 5.2 The fugacity coefficient of nitrogen at 273 K

p/atm	ϕ	p/atm	ϕ
1	0.99955	300	1.0055
10	0.9956	400	1.062
50	0.9812	600	1.239
100	0.9703	800	1.495
150	0.9672	1000	1.839
200	0.9721		

Data: LR

Table 6.1 Surface tensions of liquids at 293 K

	$\gamma/(\mathrm{mN\,m^{-1}})$
Benzene	28.88
Carbon tetrachloride	27.0
Ethanol	22.8
Hexane	18.4
Mercury	472
Methanol	22.6
Water	72.75
	72.0 at 25°C
	58.0 at 100°C

Data: KL

Table 7.1 Henry's law constants for gases at 298 K, $K/$Torr

	Water	Benzene
CH_4	3.14×10^5	4.27×10^5
CO_2	1.25×10^6	8.57×10^4
H_2	5.34×10^7	2.75×10^6
N_2	6.51×10^7	1.79×10^6
O_2	3.30×10^7	

Data: F. Daniels and R.A. Alberty, *Physical chemistry*. Wiley, New York (1980).

Table 7.2 Cryoscopic and ebullioscopic constants

	$K_f/(\mathrm{K\,kg\,mol^{-1}})$	$K_b/(\mathrm{K\,kg\,mol^{-1}})$
Acetic acid	3.90	3.07
Benzene	5.12	2.53
Camphor	40	
Carbon disulfide	3.8	2.37
Carbon tetrachloride	30	4.95
Naphthalene	6.94	5.8
Phenol	7.27	3.04
Water	1.86	0.51

Data: KL

Table 9.1 Acidity constants for aqueous solution at 298 K. (a) In order of acid strength

Acid	HA	A^-	K_a	pK_a
Hydroiodic	HI	I^-	10^{11}	-11
Perchloric	$HClO_4$	ClO_4^-	10^{10}	-10
Hydrobromic	HBr	Br^-	10^9	-9
Hydrochloric	HCl	Cl^-	10^7	-7
Sulfuric	H_2SO_4	HSO_4^-	10^2	-2
Hydronium ion	H_3O^+	H_2O	1	0.0
Oxalic	$(COOH)_2$	$HOOCCO_2^-$	5.9×10^{-2}	1.23
Sulfurous	H_2SO_3	HSO_3^-	1.5×10^{-2}	1.81
Hydrogensulfate ion	HSO_4^-	SO_4^{2-}	1.2×10^{-2}	1.92
Phosphoric	H_3PO_4	$H_2PO_4^-$	7.5×10^{-3}	2.12
Hydrofluoric	HF	F^-	3.5×10^{-4}	3.45
Formic	HCOOH	HCO_2^-	1.8×10^{-4}	3.75
Lactic	$CH_3CH(OH)COOH$	$CH_3CH(OH)CO_2^-$	1.4×10^{-4}	3.86
Hydrogenoxalate ion	$HOOCCO_2^-$	$(CO_2)_2^{2-}$	6.5×10^{-5}	4.19
Anilinium ion	$C_6H_5NH_3^+$	$C_6H_5NH_2$	2.3×10^{-5}	4.63
Acetic (ethanoic)	CH_3COOH	$CH_3CO_2^-$	1.8×10^{-5}	4.75
Butanoic	C_3H_7COOH	$C_3H_7CO_2^-$	1.5×10^{-5}	4.82
Propanoic	C_2H_5COOH	$C_2H_5CO_2^-$	1.4×10^{-5}	4.87
Pyridinium ion	$HC_5H_5N^+$	C_5H_5N	5.6×10^{-6}	5.25
Carbonic	H_2CO_3	HCO_3^-	4.3×10^{-7}	6.37
Hydrogen sulfide	H_2S	HS^-	9.1×10^{-8}	7.04
Dihydrogenphosphate ion	$H_2PO_4^-$	HPO_4^{2-}	6.2×10^{-8}	7.21
Hypochlorous	HClO	ClO^-	3.0×10^{-8}	7.53
Hydrazinium ion	$NH_2NH_3^+$	NH_2NH_2	5.9×10^{-9}	8.23
Hypobromous	HBrO	BrO^-	2.0×10^{-9}	8.69
Boric	$B(OH)_3$	$B(OH)_4^-$	7.2×10^{-10}	9.14
Ammonium ion	NH_4^+	NH_3	5.6×10^{-10}	9.25
Hydrogen cyanide	HCN	CN^-	4.9×10^{-10}	9.31
Glycinium ion	NH_2CH_2COOH	$NH_2CH_2CO_2^-$	1.7×10^{-10}	9.78
Trimethylammonium ion	$(CH_3)_3NH^+$	$(CH_3)_3N$	1.6×10^{-10}	9.81
Phenol	C_6H_5OH	$C_6H_5O^-$	1.3×10^{-10}	9.89
Hydrogencarbonate ion	HCO_3^-	CO_3^{2-}	5.6×10^{-11}	10.25
Hypoiodous	HIO	IO^-	2.3×10^{-11}	10.64
Methylammonium ion	$CH_3NH_3^+$	CH_3NH_2	2.2×10^{-11}	10.66
Dimethylammonium ion	$(CH_3)_2NH_2^+$	$(CH_3)_2NH$	1.9×10^{-11}	10.73
Triethylammonium ion	$(C_2H_5)_3NH^+$	$(C_2H_5)_3N$	1.7×10^{-11}	10.76
Ethylammonium ion	$C_2H_5NH_3^+$	$C_2H_5NH_2$	1.6×10^{-11}	10.81
Diethylammonium ion	$(C_2H_5)_2NH_2^+$	$(C_2H_5)_2NH$	1.0×10^{-11}	10.99
Hydrogenarsenate ion	$HAsO_4^{2-}$	AsO_4^{3-}	3.0×10^{-12}	11.53
Hydrogensulfide ion	HS^-	S^{2-}	1.1×10^{-12}	11.96
Hydrogenphosphate ion	HPO_4^{2-}	PO_4^{3-}	2.2×10^{-13}	12.67

Data: Principally HCP

Table 9.1 Acidity constants for aqueous solution at 298 K. (b) In alphabetical order of acid

Acid	HA	A^-	K_a	pK_a
Acetic	CH_3COOH	$CH_3CO_2^-$	1.8×10^{-5}	4.75
Ammonium ion	NH_4^+	NH_3	5.6×10^{-10}	9.25
Anilinium ion	$C_6H_5NH_3^+$	$C_6H_5NH_2$	2.3×10^{-5}	4.63
Boric	$B(OH)_3$	$B(OH)_4^-$	7.2×10^{-10}	9.14
Butanoic	C_3H_7COOH	$C_3H_7CO_2^-$	1.5×10^{-5}	4.82
Carbonic	H_2CO_3	HCO_3^-	4.3×10^{-7}	6.37
Diethylammonium ion	$(C_2H_5)_2NH_2^+$	$(C_2H_5)_2NH$	1.0×10^{-11}	10.99
Dihydrogenphosphate ion	$H_2PO_4^-$	HPO_4^{2-}	6.2×10^{-8}	7.21
Dimethylammonium ion	$(CH_3)_2NH_2^+$	$(CH_3)_2NH$	1.9×10^{-11}	10.73
Ethylammonium ion	$C_2H_5NH_3^+$	$C_2H_5NH_2$	1.6×10^{-11}	10.81
Formic	$HCOOH$	HCO_2^-	1.8×10^{-4}	3.75
Glycinium ion	NH_2CH_2COOH	$NH_2CH_2CO_2^-$	1.7×10^{-10}	9.78
Hydrazinium ion	$NH_2NH_3^+$	NH_2NH_2	5.9×10^{-9}	8.23
Hydroiodic	HI	I^-	10^{11}	-11
Hydrobromic	HBr	Br^-	10^9	-9
Hydrochloric	HCl	Cl^-	10^7	-7
Hydrofluoric	HF	F^-	3.5×10^{-4}	3.45
Hydrogenarsenate ion	$HAsO_4^{2-}$	AsO_4^{3-}	3.0×10^{-12}	11.53
Hydrogencarbonate ion	HCO_3^-	CO_3^{2-}	4.8×10^{-11}	10.32
Hydrogen cyanide	HCN	CN^-	4.9×10^{-10}	9.31
Hydrogenoxalate ion	$HOOCCO_2^-$	$(CO_2)_2^{2-}$	6.5×10^{-5}	4.19
Hydrogenphosphate ion	HPO_4^{2-}	PO_4^{3-}	2.2×10^{-13}	12.67
Hydrogensulfate ion	HSO_4^-	SO_4^{2-}	1.2×10^{-2}	1.92
Hydrogen sulfide	H_2S	HS^-	9.1×10^{-8}	7.04
Hydrogensulfide ion	HS^-	S^{2-}	1.1×10^{-12}	11.96
Hydronium ion	H_3O^+	H_2O	1	0.0
Hypobromous	$HBrO$	BrO^-	2.0×10^{-9}	8.69
Hypochlorous	$HClO$	ClO^-	3.0×10^{-8}	7.53
Hypoiodous	HIO	IO^-	2.3×10^{-11}	10.64
Lactic	$CH_3CH(OH)COOH$	$CH_3CH(OH)CO_2^-$	1.4×10^{-4}	3.86
Methylammonium ion	$CH_3NH_3^+$	CH_3NH_2	2.2×10^{-11}	10.66
Oxalic	$(COOH)_2$	$HOOCCO_2^-$	5.9×10^{-2}	1.23
Perchloric	$HClO_4$	ClO_4^-	10^{10}	-10
Phenol	C_6H_5OH	$C_6H_5O^-$	1.3×10^{-10}	9.89
Phosphoric	H_3PO_4	$H_2PO_4^-$	7.5×10^{-3}	2.12
Propanoic	C_2H_5COOH	$C_2H_5CO_2^-$	1.4×10^{-5}	4.87
Pyridinium ion	$HC_5H_5N^+$	C_5H_5N	5.6×10^{-6}	5.25
Sulfuric	H_2SO_4	HSO_4^-	10^2	-2
Sulfurous	H_2SO_3	HSO_3^-	1.5×10^{-2}	1.81
Triethylammonium ion	$(C_2H_5)_3NH^+$	$(C_2H_5)_3N$	1.7×10^{-11}	10.76
Trimethylammonium ion	$(CH_3)_3NH^+$	$(CH_3)_3N$	1.6×10^{-10}	9.81

Table 10.1 Standard thermodynamic functions of ions in solution at 298 K. See Table 2.6

Table 10.2 Relative permittivities (dielectric constants) at 298 K

Nonpolar molecules		Polar molecules	
Methane (at $-173°C$)	1.70	Water	78.54
			80.37 at 20°C
Carbon tetrachloride	2.228	Ammonia	16.9
			22.4 at $-33°C$
Cyclohexane	2.015	Hydrogen sulfide	9.26 at $-85°C$
Benzene	2.274	Methanol	32.63
		Ethanol	24.30
		Nitrobenzene	34.82

Data: HCP

Table 10.3 Standard molar entropies of ions in aqueous solution at 298 K. See Table 2.6

Table 10.5 Mean activity coefficients in water at 298 K

$b/b^{\ominus}$	HCl	KCl	$CaCl_2$	H_2SO_4	$LaCl_3$	$In_2(SO_4)_3$
0.001	0.966	0.966	0.888	0.830	0.790	
0.005	0.929	0.927	0.789	0.639	0.636	0.16
0.01	0.905	0.902	0.732	0.544	0.560	0.11
0.05	0.830	0.816	0.584	0.340	0.388	0.035
0.10	0.798	0.770	0.524	0.266	0.356	0.025
0.50	0.769	0.652	0.510	0.155	0.303	0.014
1.00	0.811	0.607	0.725	0.131	0.387	
2.00	1.011	0.577	1.554	0.125	0.954	

Data: RS, HCP, and S. Glasstone, *Introduction to electrochemistry*. Van Nostrand (1942).

Table 10.7 Standard potentials at 298 K. (a) In electrochemical order

Reduction half-reaction	$E^{\ominus}/V$	Reduction half-reaction	$E^{\ominus}/V$
Strongly oxidizing		$Cu^{2+} + e^- \rightarrow Cu^+$	+0.16
$H_4XeO_6 + 2H^+ + 2e^- \rightarrow XeO_3 + 3H_2O$	+3.0	$Sn^{4+} + 2e^- \rightarrow Sn^{2+}$	+0.15
$F_2 + 2e^- \rightarrow 2F^-$	+2.87	$AgBr + e^- \rightarrow Ag + Br^-$	+0.07
$O_3 + 2H^+ + 2e^- \rightarrow O_2 + H_2O$	+2.07	$Ti^{4+} + e^- \rightarrow Ti^{3+}$	0.00
$S_2O_8^{2-} + 2e^- \rightarrow 2SO_4^{2-}$	+2.05	$2H^+ + 2e^- \rightarrow H_2$	0, by definition
$Ag^{2+} + e^- \rightarrow Ag^+$	+1.98	$Fe^{3+} + 3e^- \rightarrow Fe$	−0.04
$Co^{3+} + e^- \rightarrow Co^{2+}$	+1.81	$O_2 + H_2O + 2e^- \rightarrow HO_2^- + OH^-$	−0.08
$H_2O_2 + 2H^+ + 2e^- \rightarrow 2H_2O$	+1.78	$Pb^{2+} + 2e^- \rightarrow Pb$	−0.13
$Au^+ + e^- \rightarrow Au$	+1.69	$In^+ + e^- \rightarrow In$	−0.14
$Pb^{4+} + 2e^- \rightarrow Pb^{2+}$	+1.67	$Sn^{2+} + 2e^- \rightarrow Sn$	−0.14
$2HClO + 2H^+ + 2e^- \rightarrow Cl_2 + 2H_2O$	+1.63	$AgI + e^- \rightarrow Ag + I^-$	−0.15
$Ce^{4+} + e^- \rightarrow Ce^{3+}$	+1.61	$Ni^{2+} + 2e^- \rightarrow Ni$	−0.23
$2HBrO + 2H^+ + 2e^- \rightarrow Br_2 + 2H_2O$	+1.60	$Co^{2+} + 2e^- \rightarrow Co$	−0.28
$MnO_4^- + 8H^+ + 5e^- \rightarrow Mn^{2+} + 4H_2O$	+1.51	$In^{3+} + 3e^- \rightarrow In$	−0.34
$Mn^{3+} + e^- \rightarrow Mn^{2+}$	+1.51	$Tl^+ + e^- \rightarrow Tl$	−0.34
$Au^{3+} + 3e^- \rightarrow Au$	+1.40	$PbSO_4 + 2e^- \rightarrow Pb + SO_4^{2-}$	−0.36
$Cl_2 + 2e^- \rightarrow 2Cl^-$	+1.36	$Ti^{3+} + e^- \rightarrow Ti^{2+}$	−0.37
$Cr_2O_7^{2-} + 14H^+ + 6e^- \rightarrow 2Cr^{3+} + 7H_2O$	+1.33	$Cd^{2+} + 2e^- \rightarrow Cd$	−0.40
$O_3 + H_2O + 2e^- \rightarrow O_2 + 2OH^-$	+1.24	$In^{2+} + e^- \rightarrow In^+$	−0.40
$O_2 + 4H^+ + 4e^- \rightarrow 2H_2O$	+1.23	$Cr^{3+} + e^- \rightarrow Cr^{2+}$	−0.41
$ClO_4^- + 2H^+ + 2e^- \rightarrow ClO_3^- + H_2O$	+1.23	$Fe^{2+} + 2e^- \rightarrow Fe$	−0.44
$MnO_2 + 4H^+ + 2e^- \rightarrow Mn^{2+} + 2H_2O$	+1.23	$In^{3+} + 2e^- \rightarrow In^+$	−0.44
$Br_2 + 2e^- \rightarrow 2Br^-$	+1.09	$S + 2e^- \rightarrow S^{2-}$	−0.48
$Pu^{4+} + e^- \rightarrow Pu^{3+}$	+0.97	$In^{3+} + e^- \rightarrow In^{2+}$	−0.49
$NO_3^- + 4H^+ + 3e^- \rightarrow NO + 2H_2O$	+0.96	$U^{4+} + e^- \rightarrow U^{3+}$	−0.61
$2Hg^{2+} + 2e^- \rightarrow Hg_2^{2+}$	+0.92	$Cr^{3+} + 3e^- \rightarrow Cr$	−0.74
$ClO^- + H_2O + 2e^- \rightarrow Cl^- + 2OH^-$	+0.89	$Zn^{2+} + 2e^- \rightarrow Zn$	−0.76
$Hg^{2+} + 2e^- \rightarrow Hg$	+0.86	$Cd(OH)_2 + 2e^- \rightarrow Cd + 2OH^-$	−0.81
$NO_3^- + 2H^+ + e^- \rightarrow NO_2 + H_2O$	+0.80	$2H_2O + 2e^- \rightarrow H_2 + 2OH^-$	−0.83
$Ag^+ + e^- \rightarrow Ag$	+0.80	$Cr^{2+} + 2e^- \rightarrow Cr$	−0.91
$Hg_2^{2+} + 2e^- \rightarrow 2Hg$	+0.79	$Mn^{2+} + 2e^- \rightarrow Mn$	−1.18
$Fe^{3+} + e^- \rightarrow Fe^{2+}$	+0.77	$V^{2+} + 2e^- \rightarrow V$	−1.19
$BrO^- + H_2O + 2e^- \rightarrow Br^- + 2OH^-$	+0.76	$Ti^{2+} + 2e^- \rightarrow Ti$	−1.63
$Hg_2SO_4 + 2e^- \rightarrow 2Hg + SO_4^{2-}$	+0.62	$Al^{3+} + 3e^- \rightarrow Al$	−1.66
$MnO_4^{2-} + 2H_2O + 2e^- \rightarrow MnO_2 + 4OH^-$	+0.60	$U^{3+} + 3e^- \rightarrow U$	−1.79
$MnO_4^- + e^- \rightarrow MnO_4^{2-}$	+0.56	$Sc^{3+} + 3e^- \rightarrow Sc$	−2.09
$I_2 + 2e^- \rightarrow 2I^-$	+0.54	$Mg^{2+} + 2e^- \rightarrow Mg$	−2.36
$Cu^+ + e^- \rightarrow Cu$	+0.52	$Ce^{3+} + 3e^- \rightarrow Ce$	−2.48
$I_3^- + 2e^- \rightarrow 3I^-$	+0.53	$La^{3+} + 3e^- \rightarrow La$	−2.52
$NiOOH + H_2O + e^- \rightarrow Ni(OH)_2 + OH^-$	+0.49	$Na^+ + e^- \rightarrow Na$	−2.71
$Ag_2CrO_4 + 2e^- \rightarrow 2Ag + CrO_4^{2-}$	+0.45	$Ca^{2+} + 2e^- \rightarrow Ca$	−2.87
$O_2 + 2H_2O + 4e^- \rightarrow 4OH^-$	+0.40	$Sr^{2+} + 2e^- \rightarrow Sr$	−2.89
$ClO_4^- + H_2O + 2e^- \rightarrow ClO_3^- + 2OH^-$	+0.36	$Ba^{2+} + 2e^- \rightarrow Ba$	−2.91
$[Fe(CN)_6]^{3-} + e^- \rightarrow [Fe(CN)_6]^{4-}$	+0.36	$Ra^{2+} + 2e^- \rightarrow Ra$	−2.92
$Cu^{2+} + 2e^- \rightarrow Cu$	+0.34	$Cs^+ + e^- \rightarrow Cs$	−2.92
$Hg_2Cl_2 + 2e^- \rightarrow 2Hg + 2Cl^-$	+0.27	$Rb^+ + e^- \rightarrow Rb$	−2.93
$AgCl + e^- \rightarrow Ag + Cl^-$	+0.22	$K^+ + e^- \rightarrow K$	−2.93
$Bi^{3+} + 3e^- \rightarrow Bi$	+0.20		

Table 10.7 Standard electrode potentials at 298 K. (b) In alphabetical order

Reduction half-reaction	$E^{\ominus}/V$	Reduction half-reaction	$E^{\ominus}/V$
$Li^+ + e^- \rightarrow Li$	-3.05	$I_2 + 2e^- \rightarrow 2I^-$	$+0.54$
$Ag^+ + e^- \rightarrow Ag$	$+0.80$	$I_3^- + 2e^- \rightarrow 3I^-$	$+0.53$
$Ag^{2+} + e^- \rightarrow Ag^+$	$+1.98$	$In^+ + e^- \rightarrow In$	-0.14
$AgBr + e^- \rightarrow Ag + Br^-$	$+0.0713$	$In^{2+} + e^- \rightarrow In^+$	-0.40
$AgCl + e^- \rightarrow Ag + Cl^-$	$+0.22$	$In^{3+} + 2e^- \rightarrow In^+$	-0.44
$Ag_2CrO_4 + 2e^- \rightarrow 2Ag + CrO_4^{2-}$	$+0.45$	$In^{3+} + 3e^- \rightarrow In$	-0.34
$AgF + e^- \rightarrow Ag + F^-$	$+0.78$	$In^{3+} + e^- \rightarrow In^{2+}$	-0.49
$AgI + e^- \rightarrow Ag + I^-$	-0.15	$K^+ + e^- \rightarrow K$	-2.93
$Al^{3+} + 3e^- \rightarrow Al$	-1.66	$La^{3+} + 3e^- \rightarrow La$	-2.52
$Au^+ + e^- \rightarrow Au$	$+1.69$	$Li^+ + e^- \rightarrow Li$	-3.05
$Au^{3+} + 3e^- \rightarrow Au$	$+1.40$	$Mg^{2+} + 2e^- \rightarrow Mg$	-2.36
$Ba^{2+} + 2e^- \rightarrow Ba$	-2.91	$Mn^{2+} + 2e^- \rightarrow Mn$	-1.18
$Be^{2+} + 2e^- \rightarrow Be$	-1.85	$Mn^{3+} + e^- \rightarrow Mn^{2+}$	$+1.51$
$Bi^{3+} + 3e^- \rightarrow Bi$	$+0.20$	$MnO_2 + 4H^+ + 2e^- \rightarrow Mn^{2+} + 2H_2O$	$+1.23$
$Br_2 + 2e^- \rightarrow 2Br^-$	$+1.09$	$MnO_4^- + 8H^+ + 5e^- \rightarrow Mn^{2+} + 4H_2O$	$+1.51$
$BrO^- + H_2O + 2e^- \rightarrow Br^- + 2OH^-$	$+0.76$	$MnO_4^- + e^- \rightarrow MnO_4^{2-}$	$+0.56$
$Ca^{2+} + 2e^- \rightarrow Ca$	-2.87	$MnO_4^{2-} + 2H_2O + 2e^- \rightarrow MnO_2 + 4OH^-$	$+0.60$
$Cd(OH)_2 + 2e^- \rightarrow Cd + 2OH^-$	-0.81	$Na^+ + e^- \rightarrow Na$	-2.71
$Cd^{2+} + 2e^- \rightarrow Cd$	-0.40	$Ni^{2+} + 2e^- \rightarrow Ni$	-0.23
$Ce^{3+} + 3e^- \rightarrow Ce$	-2.48	$NiOOH + H_2O + e^- \rightarrow Ni(OH)_2 + OH^-$	$+0.49$
$Ce^{4+} + e^- \rightarrow Ce^{3+}$	$+1.61$	$NO_3^- + 2H^+ + e^- \rightarrow NO_2 + H_2O$	-0.80
$Cl_2 + 2e^- \rightarrow 2Cl^-$	$+1.36$	$NO_3^- + 4H^+ + 3e^- \rightarrow NO + 2H_2O$	$+0.96$
$ClO^- + H_2O + 2e^- \rightarrow Cl^- + 2OH^-$	$+0.89$	$NO_3^- + H_2O + 2e^- \rightarrow NO_2^- + 2OH^-$	$+0.10$
$ClO_4^- + 2H^+ + 2e^- \rightarrow ClO_3^- + H_2O$	$+1.23$	$O_2 + 2H_2O + 4e^- \rightarrow 4OH^-$	$+0.40$
$ClO_4^- + H_2O + 2e^- \rightarrow ClO_3^- + 2OH^-$	$+0.36$	$O_2 + 4H^+ + 4e^- \rightarrow 2H_2O$	$+1.23$
$Co^{2+} + 2e^- \rightarrow Co$	-0.28	$O_2 + e^- \rightarrow O_2^-$	-0.56
$Co^{3+} + e^- \rightarrow Co^{2+}$	$+1.81$	$O_2 + H_2O + 2e^- \rightarrow HO_2^- + OH^-$	-0.08
$Cr^{2+} + 2e^- \rightarrow Cr$	-0.91	$O_3 + 2H^+ + 2e^- \rightarrow O_2 + H_2O$	$+2.07$
$Cr_2O_7^{2-} + 14H^+ + 6e^- \rightarrow 2Cr^{3+} + 7H_2O$	$+1.33$	$O_3 + H_2O + 2e^- \rightarrow O_2 + 2OH^-$	$+1.24$
$Cr^{3+} + 3e^- \rightarrow Cr$	-0.74	$Pb^{2+} + 2e^- \rightarrow Pb$	-0.13
$Cr^{3+} + e^- \rightarrow Cr^{2+}$	-0.41	$Pb^{4+} + 2e^- \rightarrow Pb^{2+}$	$+1.67$
$Cs^+ + e^- \rightarrow Cs$	-2.92	$PbSO_4 + 2e^- \rightarrow Pb + SO_4^{2-}$	-0.36
$Cu^+ + e^- \rightarrow Cu$	$+0.52$	$Pt^{2+} + 2e^- \rightarrow Pt$	$+1.20$
$Cu^{2+} + 2e^- \rightarrow Cu$	$+0.34$	$Pu^{4+} + e^- \rightarrow Pu^{3+}$	$+0.97$
$Cu^{2+} + e^- \rightarrow Cu^+$	$+0.16$	$Ra^{2+} + 2e^- \rightarrow Ra$	-2.92
$F_2 + 2e^- \rightarrow 2F^-$	$+2.87$	$Rb^+ + e^- \rightarrow Rb$	-2.93
$Fe^{2+} + 2e^- \rightarrow Fe$	-0.44	$S + 2e^- \rightarrow S^{2-}$	-0.48
$Fe^{3+} + 3e^- \rightarrow Fe$	-0.04	$S_2O_8^{2-} + 2e^- \rightarrow 2SO_4^{2-}$	$+2.05$
$Fe^{3+} + e^- \rightarrow Fe^{2+}$	$+0.77$	$Sc^{3+} + 3e^- \rightarrow Sc$	-2.09
$[Fe(CN)_6]^{3-} + e^- \rightarrow [Fe(CN)_6]^{4-}$	$+0.36$	$Sn^{2+} + 2e^- \rightarrow Sn$	-0.14
$2H^+ + 2e^- \rightarrow H_2$	0, by definition	$Sn^{4+} + 2e^- \rightarrow Sn^{2+}$	$+0.15$
$2H_2O + 2e^- \rightarrow H_2 + 2OH^-$	-0.83	$Sr^{2+} + 2e^- \rightarrow Sr$	-2.89
$2HBrO + 2H^+ + 2e^- \rightarrow Br_2 + 2H_2O$	$+1.60$	$Ti^{2+} + 2e^- \rightarrow Ti$	-1.63
$2HClO + 2H^+ + 2e^- \rightarrow Cl_2 + 2H_2O$	$+1.63$	$Ti^{3+} + e^- \rightarrow Ti^{2+}$	-0.37
$H_2O_2 + 2H^+ + 2e^- \rightarrow 2H_2O$	$+1.78$	$Ti^{4+} + e^- \rightarrow Ti^{3+}$	0.00
$H_4XeO_6 + 2H^+ + 2e^- \rightarrow XeO_3 + 3H_2O$	$+3.0$	$Tl^+ + e^- \rightarrow Tl$	-0.34
$Hg_2^{2+} + 2e^- \rightarrow 2Hg$	$+0.79$	$U^{3+} + 3e^- \rightarrow U$	-1.79
$Hg_2Cl_2 + 2e^- \rightarrow 2Hg + 2Cl^-$	$+0.27$	$U^{4+} + e^- \rightarrow U^{3+}$	-0.61
$Hg^{2+} + 2e^- \rightarrow Hg$	$+0.86$	$V^{2+} + 2e^- \rightarrow V$	-1.19
$2Hg^{2+} + 2e^- \rightarrow Hg_2^{2+}$	$+0.92$	$V^{3+} + e^- \rightarrow V^{2+}$	-0.26
$Hg_2SO_4 + 2e^- \rightarrow 2Hg + SO_4^{2-}$	$+0.62$	$Zn^{2+} + 2e^- \rightarrow Zn$	-0.76

Table 12.2 The error function

z	erf z	z	erf z
0	0	0.45	0.475 48
0.01	0.011 28	0.50	0.520 50
0.02	0.022 56	0.55	0.563 32
0.03	0.033 84	0.60	0.603 86
0.04	0.045 11	0.65	0.642 03
0.05	0.056 37	0.70	0.677 80
0.06	0.067 62	0.75	0.711 16
0.07	0.078 86	0.80	0.742 10
0.08	0.090 08	0.85	0.770 67
0.09	0.101 28	0.90	0.796 91
0.10	0.112 46	0.95	0.820 89
0.15	0.168 00	1.00	0.842 70
0.20	0.222 70	1.20	0.910 31
0.25	0.276 32	1.40	0.952 28
0.30	0.328 63	1.60	0.976 35
0.35	0.379 38	1.80	0.989 09
0.40	0.428 39	2.00	0.995 32

Data: AS

Table 13.3 Screening constants for atoms; values of $Z_{eff} = Z - \sigma$ for neutral ground-state atoms

	H							He
$1s$	1							1.6875
	Li	Be	B	C	N	O	F	Ne
$1s$	2.6906	3.6848	4.6795	5.6727	6.6651	7.6579	8.6501	9.6421
$2s$	1.2792	1.9120	2.5762	3.2166	3.8474	4.4916	5.1276	5.7584
$2p$			2.4214	3.1358	3.8340	4.4532	5.1000	5.7584
	Na	Mg	Al	Si	P	S	Cl	Ar
$1s$	10.6259	11.6089	11910	13.5754	14.5578	15.5409	16.5239	17.5075
$2s$	6.5714	7.3920	8.2136	9.0200	9.8250	10.6288	11.4304	12.2304
$2p$	6.8018	7.8258	8.9634	9.9450	10.9612	11.9770	12.9932	14.0082
$3s$	2.5074	3.3075	4.1172	4.9032	5.6418	6.3669	7.0683	7.7568
$3p$			4.0656	4.2852	4.8864	5.4819	6.1161	6.7641

Data: E. Clementi and D.L. Raimondi, *Atomic screening constants from SCF functions.*
IBM Res. Note NJ-27 (1963).

Table 13.4 Ionization energies, $I_i/(\text{kJ mol}^{-1})$

H							He
1312.0							2372.3
							5250.4
Li	**Be**	**B**	**C**	**N**	**O**	**F**	**Ne**
513.3	899.4	800.6	1086.2	1402.3	1313.9	1681	2080.6
7298.0	1757.1	2427	2352	2856.1	3388.2	3374	3952.2
Na	**Mg**	**Al**	**Si**	**P**	**S**	**Cl**	**Ar**
495.8	737.7	577.4	786.5	1011.7	999.6	1251.1	1520.4
4562.4	1450.7	1816.6	1577.1	1903.2	2251	2297	2665.2
		2744.6		2912			
K	**Ca**	**Ga**	**Ge**	**As**	**Se**	**Br**	**Kr**
418.8	589.7	578.8	762.1	947.0	940.9	1139.9	1350.7
3051.4	1145	1979	1537	1798	2044	2104	2350
		2963	2735				
Rb	**Sr**	**In**	**Sn**	**Sb**	**Te**	**I**	**Xe**
403.0	549.5	558.3	708.6	833.7	869.2	1008.4	1170.4
2632	1064.2	1820.6	1411.8	1794	1795	1845.9	2046
		2704	2943.0	2443			
Cs	**Ba**	**Tl**	**Pb**	**Bi**	**Po**	**At**	**Rn**
375.5	502.8	589.3	715.5	703.2	812	930	1037
2420	965.1	1971.0	1450.4	1610			
		2878	3081.5	2466			

Data: E

Table 13.5 Electron affinities, $E_{ea}/(\text{kJ mol}^{-1})$

H							He
72.8							−21
Li	**Be**	**B**	**C**	**N**	**O**	**F**	**Ne**
59.8	⩽0	23	122.5	−7	141	322	−29
					−844		
Na	**Mg**	**Al**	**Si**	**P**	**S**	**Cl**	**Ar**
52.9	⩽0	44	133.6	71.7	200.4	348.7	−35
					−532		
K	**Ca**	**Ga**	**Ge**	**As**	**Se**	**Br**	**Kr**
48.3	2.37	36	116	77	195.0	324.5	−39
Rb	**Sr**	**In**	**Sn**	**Sb**	**Te**	**I**	**Xe**
46.9	5.03	34	121	101	190.2	295.3	−41
Cs	**Ba**	**Tl**	**Pb**	**Bi**	**Po**	**At**	**Rn**
45.5	13.95	30	35.2	101	186	270	−41

Data: E

Table 14.2 Bond lengths, R_e/pm

(a) Bond lengths in specific molecules

Br_2	228.3
Cl_2	198.75
CO	112.81
F_2	141.78
H_2^+	106
H_2	74.138
HBr	141.44
HCl	127.45
HF	91.680
HI	160.92
N_2	109.76
O_2	120.75

(b) Mean bond lengths from covalent radii[*]

H	37						
C	77(1)	N	74(1)	O	66(1)	F	64
	67(2)		65(2)		57(2)		
	60(3)						
Si	118	P	110	S	104(1)	Cl	99
					95(2)		
Ge	122	As	121	Se	104	Br	114
		Sb	141	Te	137	I	133

[*]Values are for single bonds except where indicated otherwise (values in parentheses). The length of an A–B covalent bond (of given order) is the sum of the corresponding covalent radii.

Table 14.3a Bond dissociation enthalpies, $\Delta H^{\ominus}(A—B)/(kJ\,mol^{-1})$ at 298 K

Diatomic molecules

H—H	436	F—F	155	Cl—Cl	242	Br—Br	193	I—I	151
O=O	497	C=O	1076	N≡N	945				
H—O	428	H—F	565	H—Cl	431	H—Br	366	H—I	299

Polyatomic molecules

$H—CH_3$	435	$H—NH_2$	460	H—OH	492	$H—C_6H_5$	469
$H_3C—CH_3$	368	$H_2C=CH_2$	720	HC≡CH	962		
$HO—CH_3$	377	$Cl—CH_3$	352	$Br—CH_3$	293	$I—CH_3$	237
O=CO	531	HO—OH	213	$O_2N—NO_2$	54		

Data: HCP, KL

Table 14.3b Mean bond enthalpies, $\Delta H^{\ominus}(A—B)/(kJ\,mol^{-1})$

	H	C	N	O	F	Cl	Br	I	S	P	Si
H	436										
C	412	348(i)									
		612(ii)									
		838(iii)									
		518(a)									
N	388	305(i)	163(i)								
		613(ii)	409(ii)								
		890(iii)	946(iii)								
O	463	360(i)	157	146(i)							
		743(ii)		497(ii)							
F	565	484	270	185	155						
Cl	431	338	200	203	254	242					
Br	366	276				219	193				
I	299	238				210	178	151			
S	338	259			496	250	212		264		
P	322									201	
Si	318		374	466							226

(i) Single bond, (ii) double bond, (iii) triple bond, (a) aromatic.
Data: HCP and L. Pauling, *The nature of the chemical bond*. Cornell University Press (1960).

Table 14.4 Pauling (*italics*) and Mulliken electronegativities

H							He
2.20							
3.06							
Li	Be	B	C	N	O	F	Ne
0.98	*1.57*	*2.04*	*2.55*	*3.04*	*3.44*	*3.98*	
1.28	1.99	1.83	2.67	3.08	3.22	4.43	4.60
Na	Mg	Al	Si	P	S	Cl	Ar
0.93	*1.31*	*1.61*	*1.90*	*2.19*	*2.58*	*3.16*	
1.21	1.63	1.37	2.03	2.39	2.65	3.54	3.36
K	Ca	Ga	Ge	As	Se	Br	Kr
0.82	*1.00*	*1.81*	*2.01*	*2.18*	*2.55*	*2.96*	*3.0*
1.03	1.30	1.34	1.95	2.26	2.51	3.24	2.98
Rb	Sr	In	Sn	Sb	Te	I	Xe
0.82	*0.95*	*1.78*	*1.96*	*2.05*	*2.10*	*2.66*	*2.6*
0.99	1.21	1.30	1.83	2.06	2.34	2.88	2.59
Cs	Ba	Tl	Pb	Bi			
0.79	*0.89*	*2.04*	*2.33*	*2.02*			

Data: Pauling values: A.L. Allred, *J. Inorg. Nucl. Chem.* **17**, 215 (1961); L.C. Allen and J.E. Huheey, *ibid.*, **42**, 1523 (1980). Mulliken values: L.C. Allen, *J. Am. Chem. Soc.* **111**, 9003 (1989). The Mulliken values have been scaled to the range of the Pauling values.

Table 16.2 Properties of diatomic molecules

	$\tilde{\nu}_0/cm^{-1}$	θ_V/K	B/cm^{-1}	θ_R/K	r/pm	$k/(N\,m^{-1})$	$D/(kJ\,mol^{-1})$	σ
$^1H_2^+$	2321.8	3341	29.8	42.9	106	160	255.8	2
$^1H_2$	4400.39	6332	60.864	87.6	74.138	574.9	432.1	2
$^2H_2$	3118.46	4487	30.442	43.8	74.154	577.0	439.6	2
$^1H^{19}F$	4138.32	5955	20.956	30.2	91.680	965.7	564.4	1
$^1H^{35}Cl$	2990.95	4304	10.593	15.2	127.45	516.3	427.7	1
$^1H^{81}Br$	2648.98	3812	8.465	12.2	141.44	411.5	362.7	1
$^1H^{127}I$	2308.09	3321	6.511	9.37	160.92	313.8	294.9	1
$^{14}N_2$	2358.07	3393	1.9987	2.88	109.76	2293.8	941.7	2
$^{16}O_2$	1580.36	2274	1.4457	2.08	120.75	1176.8	493.5	2
$^{19}F_2$	891.8	1283	0.8828	1.27	141.78	445.1	154.4	2
$^{35}Cl_2$	559.71	805	0.2441	0.351	198.75	322.7	239.3	2
$^{12}C^{16}O$	2170.21	3122	1.9313	2.78	112.81	1903.17	1071.8	1
$^{79}Br^{81}Br$	323.2	465	0.0809	10.116	283.3	245.9	190.2	1

Data: AIP

Table 16.3 Typical vibrational wavenumbers, $\tilde{\nu}/cm^{-1}$

C—H stretch	2850–2960	C—F stretch	1000–1400
C—H bend	1340–1465	C—Cl stretch	600–800
C—C stretch, bend	700–1250	C—Br stretch	500–600
C=C stretch	1620–1680	C—I stretch	500
C≡C stretch	2100–2260	CO_3^{2-}	1410–1450
O—H stretch	3590–3650	NO_3^-	1350–1420
H-bonds	3200–3570	NO_2^-	1230–1250
C=O stretch	1640–1780	SO_4^{2-}	1080–1130
C≡N stretch	2215–2275	Silicates	900–1100
N—H stretch	3200–3500		

Data: L.J. Bellamy, *The infrared spectra of complex molecules* and *Advances in infrared group frequencies*. Chapman and Hall.

Table 17.1 Colour, frequency, and energy of light

Colour	λ/nm	$\nu/(10^{14}\,Hz)$	$\tilde{\nu}/(10^4\,cm^{-1})$	E/eV	$E/(kJ\,mol^{-1})$
Infrared	>1000	<3.00	<1.00	<1.24	<120
Red	700	4.28	1.43	1.77	171
Orange	620	4.84	1.61	2.00	193
Yellow	580	5.17	1.72	2.14	206
Green	530	5.66	1.89	2.34	226
Blue	470	6.38	2.13	2.64	254
Violet	420	7.14	2.38	2.95	285
Near ultraviolet	300	10.0	3.33	4.15	400
Far ultraviolet	<200	>15.0	>5.00	>6.20	>598

Data: J.G. Calvert and J.N. Pitts, *Photochemistry*. Wiley, New York (1966).

Table 17.2 Absorption characteristics of some groups and molecules

Group	$\tilde{\nu}_{max}/(10^4 \text{ cm}^{-1})$	λ_{max}/nm	$\varepsilon_{max}/(\text{L mol}^{-1}\text{cm}^{-1})$
$C{=}C(\pi^* \leftarrow \pi)$	6.10	163	1.5×10^4
	5.73	174	5.5×10^3
$C{=}O(\pi^* \leftarrow n)$	3.7–3.5	270–290	10–20
$-N{=}N-$	2.9	350	15
	>3.9	<260	Strong
$-NO_2$	3.6	280	10
	4.8	210	1.0×10^4
C_6H_5-	3.9	255	200
	5.0	200	6.3×10^3
	5.5	180	1.0×10^5
$[Cu(OH_2)_6]^{2+}(aq)$	1.2	810	10
$[Cu(NH_3)_4]^{2+}(aq)$	1.7	600	50
$H_2O(\pi^* \leftarrow n)$	6.0	167	7.0×10^3

Table 18.1 Nuclear spin properties

Nuclide	Natural abundance %	Spin I	Magnetic moment μ/μ_N	g-value	$\gamma/(10^7 \text{ T}^{-1}\text{s}^{-1})$	NMR frequency at 1 T, ν/MHz
$^1n^*$		$\frac{1}{2}$	-1.9130	-3.8260	-18.324	29.167
^1H	99.9844	$\frac{1}{2}$	2.792 85	5.5857	26.752	42.576
^2H	0.0156	1	0.857 45	0.857 45	4.1067	6.536
$^3H^*$		$\frac{1}{2}$	$-2.127 65$	-4.2553	-20.380	32.434
^{10}B	19.6	3	1.8005	0.6002	2.875	4.574
^{11}B	80.4	$\frac{3}{2}$	2.6884	1.7923	8.5841	13.660
^{13}C	1.108	$\frac{1}{2}$	0.7023	1.4046	6.7272	10.705
^{14}N	99.635	1	0.403 56	0.403 56	1.9328	3.076
^{17}O	0.037	$\frac{5}{2}$	-1.893	-0.7572	-3.627	5.772
^{19}F	100	$\frac{1}{2}$	2.628 35	5.2567	25.177	40.054
^{31}P	100	$\frac{1}{2}$	1.1317	2.2634	10.840	17.238
^{33}S	0.74	$\frac{3}{2}$	0.6434	0.4289	2.054	3.266
^{35}Cl	75.4	$\frac{3}{2}$	0.8218	0.5479	2.624	4.171
^{37}Cl	24.6	$\frac{3}{2}$	0.6841	0.4561	2.184	3.472

* Radioactive.

μ is the magnetic moment of the spin state with the largest value of m_I: $\mu = g_I\mu_N I$ and μ_N is the nuclear magneton (see inside front cover).

Data: KL

Table 18.2 Hyperfine coupling constants for atoms, a/mT

Nuclide	Spin	Isotropic coupling	Anisotropic coupling
^{1}H	$\frac{1}{2}$	50.8(1s)	
^{2}H	1	7.8(1s)	
^{13}C	$\frac{1}{2}$	113.0(2s)	6.6(2p)
^{14}N	1	55.2(2s)	4.8(2p)
^{19}F	$\frac{1}{2}$	1720(2s)	108.4(2p)
^{31}P	$\frac{1}{2}$	364(3s)	20.6(3p)
^{35}Cl	$\frac{3}{2}$	168(3s)	10.0(3p)
^{37}Cl	$\frac{3}{2}$	140(3s)	8.4(3p)

Data: P.W. Atkins and M.C.R. Symons, *The structure of inorganic radicals.* Elsevier, Amsterdam (1967).

Table 21.3 Ionic radii (r/pm)†

Li$^+$(4)	Be^{2+}(4)	B^{3+}(4)	N^{3-}	O^{2-}(6)	F$^-$(6)
59	27	12	171	140	133
Na$^+$(6)	Mg^{2+}(6)	Al^{3+}(6)	P^{3-}	S^{2-}(6)	Cl$^-$(6)
102	72	53	212	184	181
K$^+$(6)	Ca^{2+}(6)	Ga^{3+}(6)	As^{3-}(6)	Se^{2-}(6)	Br$^-$(6)
138	100	62	222	198	196
Rb$^+$(6)	Sr^{2+}(6)	In^{3+}(6)		Te^{2-}(6)	I$^-$(6)
149	116	79		221	220
Cs$^+$(6)	Ba^{2+}(6)	Tl^{3+}(6)			
167	136	88			

d-block elements (high-spin ions)

Sc^{3+}(6)	Ti^{4+}(6)	Cr^{3+}(6)	Mn^{3+}(6)	Fe^{2+}(6)	Co^{3+}(6)	Cu^{2+}(6)	Zn^{2+}(6)
73	60	61	65	63	61	73	75

† Numbers in parentheses are the coordination numbers of the ions. Values for ions without a coordination number stated are estimates.
Data: R.D. Shannon and C.T. Prewitt, *Acta Cryst.* B25, 925 (1969).

Table 22.1 Dipole moments, polarizabilities, and polarizability volumes

	$\mu/(10^{-30}\ \mathrm{C\,m})$	μ/D	$\alpha'/(10^{-30}\ \mathrm{m}^3)$	$\alpha'/(10^{-40}\ \mathrm{J}^{-1}\,\mathrm{C}^2\,\mathrm{m}^2)$
Ar	0	0	1.66	1.85
C_2H_5OH	5.64	1.69		
$C_6H_5CH_3$	1.20	0.36		
C_6H_6	0	0	10.4	11.6
CCl_4	0	0	10.5	11.7
CH_2Cl_2	5.24	1.57	6.80	7.57
CH_3Cl	6.24	1.87	4.53	5.04
CH_3OH	5.70	1.71	3.23	3.59
CH_4	0	0	2.60	2.89
$CHCl_3$	3.37	1.01	8.50	9.46
CO	0.390	0.117	1.98	2.20
CO_2	0	0	2.63	2.93
H_2	0	0	0.819	0.911
H_2O	6.17	1.85	1.48	1.65
HBr	2.67	0.80	3.61	4.01
HCl	3.60	1.08	2.63	2.93
He	0	0	0.20	0.22
HF	6.37	1.91	0.51	0.57
HI	1.40	0.42	5.45	6.06
N_2	0	0	1.77	1.97
NH_3	4.90	1.47	2.22	2.47
$1,2\text{-}C_6H_4(CH_3)_2$	2.07	0.62		

Data: HCP and C.J.F. Böttcher and P. Bordewijk, *Theory of electric polarization.* Elsevier, Amsterdam (1978).

Table 22.2 Refractive indices relative to air at 20°C

	434 nm	589 nm	656 nm
Benzene	1.5236	1.5012	1.4965
Carbon tetrachloride	1.4729	1.4676	1.4579
Carbon disulfide	1.6748	1.6276	1.6182
Ethanol	1.3700	1.3618	1.3605
KCl(s)	1.5050	1.4904	1.4973
KI(s)	1.7035	1.6664	1.6581
Methanol	1.3362	1.3290	1.3277
Methylbenzene	1.5170	1.4955	1.4911
Water	1.3404	1.3330	1.3312

Data: AIP

Table 22.4 Lennard-Jones (12,6)-potential parameters

	$(\varepsilon/k)/K$	r_0/pm
Ar	111.84	362.3
C_2H_2	209.11	463.5
C_2H_4	200.78	458.9
C_2H_6	216.12	478.2
C_6H_6	377.46	617.4
CCl_4	378.86	624.1
Cl_2	296.27	448.5
CO_2	201.71	444.4
F_2	104.29	357.1
Kr	154.87	389.5
N_2	91.85	391.9
O_2	113.27	365.4
Xe	213.96	426.0

Source: F. Cuadros, I. Cachadiña, and W. Ahamuda, *Molec. Engineering*, **6**, 319 (1996).

Table 22.5 Magnetic susceptibilities at 298 K

	$\chi/10^{-6}$	$\chi_m/(10^{-4}\,cm^3\,mol^{-1})$
Water	−90	−16.0
Benzene	−7.2	−6.4
Cyclohexane	−7.9	−8.5
Carbon tetrachloride	−8.9	−8.4
NaCl(s)	−13.9	−3.75
Cu(s)	−96	−6.8
S(s)	−12.9	−2.0
Hg(l)	−28.5	−4.2
$CuSO_4 \cdot 5H_2O$(s)	+176	+192
$MnSO_4 \cdot 4H_2O$(s)	+2640	$+2.79 \times 10^3$
$NiSO_4 \cdot 7H_2O$(s)	+416	+600
$FeSO_4(NH_4)_2SO_4 \cdot 6H_2O$(s)	+755	$+1.51 \times 10^3$
Al(s)	+22	+2.2
Pt(s)	+262	+22.8
Na(s)	+7.3	+1.7
K(s)	+5.6	+2.5

Data: KL and $\chi_m = \chi M/\rho$.

Table 23.1 Frictional coefficients and molecular geometry

Major axis/ Minor axis	Prolate	Oblate
2	1.04	1.04
3	1.11	1.10
4	1.18	1.17
5	1.25	1.22
6	1.31	1.28
7	1.38	1.33
8	1.43	1.37
9	1.49	1.42
10	1.54	1.46
50	2.95	2.38
100	4.07	2.97

Data: K.E. Van Holde, *Physical biochemistry*. Prentice-Hall, Englewood Cliffs (1971).

Sphere; radius $a, c = a f_0$

Prolate ellipsoid; major axis $2a$, minor axis $2b$, $c = (ab^2)^{1/3}$

$$\left\{ \frac{(1 - b^2/a^2)^{1/2}}{(b/a)^{2/3} \ln\{[1 + (1 - b^2/a^2)^{1/2}]/(b/a)\}} \right\} f_0$$

Oblate ellipsoid; major axis $2a$, minor axis $2b$, $c = (a^2 b)^{1/3}$

$$\left\{ \frac{(a^2/b^2 - 1)^{1/2}}{(a/b)^{2/3} \arctan[(a^2/b^2 - 1)^{1/2}]} \right\} f_0$$

Long rod; length l, radius $a, c = (3a^2/4)^{1/3}$

$$\left\{ \frac{(1/2a)^{2/3}}{(3/2)^{1/3}\{2 \ln (l/a) - 0.11\}} \right\} f_0$$

In each case $f_0 = 6\pi\eta c$ with the appropriate value of c.

Table 23.2 Diffusion coefficients of macromolecules in water at 20°C

	$M/(kg\,mol^{-1})$	$D/(10^{-10}\,m^2\,s^{-1})$
Sucrose	0.342	4.586
Ribonuclease	13.7	1.19
Lysozyme	14.1	1.04
Serum albumin	65	0.594
Haemoglobin	68	0.69
Urease	480	0.346
Collagen	345	0.069
Myosin	493	0.116

Data: C. Tanford, *Physical chemistry of macromolecules*. Wiley, New York (1961).

Table 23.3 Intrinsic viscosity

Macromolecule	Solvent	$\theta/^\circ C$	$K/(10^{-3}\ cm^3\ g^{-1})$	a
Polystyrene	Benzene	25	9.5	0.74
	Cyclohexane	34†	81	0.50
Polyisobutylene	Benzene	23†	83	0.50
	Cyclohexane	30	26	0.70
Amylose	0.33 M KCl(aq)	25†	113	0.50
Various protein‡	Guanidine hydrochloride + $HSCH_2CH_2OH$		7.16	0.66

† The θ temperature.
‡ Use $[\eta]=KN^a$; N is the number of amino acid residues.
Data: K.E. Van Holde, *Physical biochemistry*. Prentice-Hall, Englewood Cliffs (1971).

Table 23.4 Radius of gyration of some macromolecules

	$M/(kg\ mol^{-1})$	R_g/nm
Serum albumin	66	2.98
Myosin	493	46.8
Polystyrene	3.2×10^3	50 (in poor solvent)
DNA	4×10^3	117.0
Tobacco mosaic virus	3.9×10^4	92.4

Data: C. Tanford, *Physical chemistry of macromolecules*. Wiley, New York (1961).

Table 24.1 Transport properties of gases at 1 atm

	$\kappa/$ $(J\ K^{-1}\ m^{-1}\ s^{-1})$ 273 K	$\eta/\mu P$ 273 K	$\eta/\mu P$ 293 K
Air	0.0241	173	182
Ar	0.0163	210	223
C_2H_4	0.0164	97	103
CH_4	0.0302	103	110
Cl_2	0.079	123	132
CO_2	0.0145	136	147
H_2	0.1682	84	88
He	0.1442	187	196
Kr	0.0087	234	250
N_2	0.0240	166	176
Ne	0.0465	298	313
O_2	0.0245	195	204
Xe	0.0052	212	228

Data: KL

Table 24.3 Viscosities of liquids at 298 K, $\eta/(10^{-3}\ kg\ m^{-1}\ s^{-1})$

Benzene	0.601
Carbon tetrachloride	0.880
Ethanol	1.06
Mercury	1.55
Methanol	0.553
Pentane	0.224
Sulfuric acid	27
Water†	0.891

† The viscosity of water over its entire liquid range is represented with less than 1 per cent error by the expression

$$\log(\eta_{20}/\eta) = A/B,$$

$$A = 1.37023(t - 20) + 8.36 \times 10^{-4}\ (t - 20)^2$$

$$B = 109 + t \qquad t = \theta/^\circ C$$

Convert $kg\ m^{-1}\ s^{-1}$ to centipoise (cP) by multiplying by 10^3 (so $\eta \approx 1\ cP$ for water).
Data: AIP, KL

Table 24.4 Limiting ionic conductivities in water at 298 K, λ/(mS m^2 mol^{-1})

Cations		Anions	
Ba^{2+}	12.72	Br^-	7.81
Ca^{2+}	11.90	$CH_3CO_2^-$	4.09
Cs^+	7.72	Cl^-	7.635
Cu^{2+}	10.72	ClO_4^-	6.73
H^+	34.96	CO_3^{2-}	13.86
K^+	7.350	$(CO_2)_2^{2-}$	14.82
Li^+	3.87	F^-	5.54
Mg^{2+}	10.60	$[Fe(CN)_6]^{3-}$	30.27
Na^+	5.010	$[Fe(CN)_6]^{4-}$	44.20
$[N(C_2H_5)_4]^+$	3.26	HCO_2^-	5.46
$[N(CH_3)_4]^+$	4.49	I^-	7.68
NH_4^+	7.35	NO_3^-	7.146
Rb^+	7.78	OH^-	19.91
Sr^{2+}	11.89	SO_4^{2-}	16.00
Zn^{2+}	10.56		

Data: KL, RS

Table 24.5 Ionic mobilities in water at 298 K, u/(10^{-8} m^2 s^{-1} V^{-1})

Cations		Anions	
Ag^+	6.42	Br^-	8.09
Ca^{2+}	6.17	$CH_3CO_2^-$	4.24
Cu^{2+}	5.56	Cl^-	7.91
H^+	36.23	CO_3^{2-}	7.46
K^+	7.62	F^-	5.70
Li^+	4.01	$[Fe(CN)_6]^{3-}$	10.5
Na^+	5.19	$[Fe(CN)_6]^{4-}$	11.4
NH_4^+	7.63	I^-	7.96
$[N(CH_3)_4]^+$	4.65	NO_3^-	7.40
Rb^+	7.92	OH^-	20.64
Zn^{2+}	5.47	SO_4^{2-}	8.29

Data: Principally Table 24.4 and $u = \lambda/zF$

Table 24.6 Debye–Hückel–Onsager coefficients for (1,1)-electrolytes at 25 °C

Solvent	A/(mS m^2 mol^{-1}/ (mol L^{-1})$^{1/2}$)	B/(mol L^{-1})$^{-1/2}$
Acetone (propanone)	3.28	1.63
Acetonitrile	2.29	0.716
Ethanol	8.97	1.83
Methanol	15.61	0.923
Nitrobenzene	4.42	0.776
Nitromethane	111	0.708
Water	6.020	0.229

Data: J.O'M. Bockris and A.K.N. Reddy, *Modern electrochemistry*. Plenum, New York (1970).

Table 24.7 Diffusion coefficients at 25 °C, D/(10^{-9} m^2 s^{-1})

Molecules in liquids				Ions in water			
I_2 in hexane	4.05	H_2 in $CCl_4(l)$	9.75	K^+	1.96	Br^-	2.08
in benzene	2.13	N_2 in $CCl_4(l)$	3.42	H^+	9.31	Cl^-	2.03
CCl_4 in heptane	3.17	O_2 in $CCl_4(l)$	3.82	Li^+	1.03	F^-	1.46
Glycine in water	1.055	Ar in $CCl_4(l)$	3.63	Na^+	1.33	I^-	2.05
Dextrose in water	0.673	CH_4 in $CCl_4(l)$	2.89			OH^-	5.30
Sucrose in water	0.5216	H_2O in water	2.26				
		CH_3OH in water	1.58				
		C_2H_5OH in water	1.24				

Data: AIP and (for the ions) $\lambda = zuF$ in conjunction with Table 24.5.

Table 25.1 Kinetic data for first-order reactions

	Phase	$\theta/°C$	k/s^{-1}	$t_{1/2}$
$2N_2O_5 \rightarrow 4NO_2 + O_2$	g	25	3.38×10^{-5}	5.70 h
	$HNO_3(l)$	25	1.47×10^{-6}	131 h
	$Br_2(l)$	25	4.27×10^{-5}	4.51 h
$C_2H_6 \rightarrow 2CH_3$	g	700	5.36×10^{-4}	21.6 min
Cyclopropane $\rightarrow$ propene	g	500	6.71×10^{-4}	17.2 min
$CH_3N_2CH_3 \rightarrow C_2H_6 + N_2$	g	327	3.4×10^{-4}	34 min
Sucrose $\rightarrow$ glucose + fructose	$aq(H^+)$	25	6.0×10^{-5}	3.2 h

g: High pressure gas-phase limit.

Data: Principally K.J. Laidler, *Chemical kinetics*. Harper & Row, New York (1987); M.J. Pilling and P.W. Seakins, *Reaction kinetics*. Oxford University Press (1995); J. Nicholas, *Chemical kinetics*. Harper & Row, New York (1976). See also JL.

Table 25.2 Kinetic data for second-order reactions

	Phase	$\theta/°C$	$k/(L\,mol^{-1}\,s^{-1})$
$2NOBr \rightarrow 2NO + Br_2$	g	10	0.80
$2NO_2 \rightarrow 2NO + O_2$	g	300	0.54
$H_2 + I_2 \rightarrow 2HI$	g	400	2.42×10^{-2}
$D_2 + HCl \rightarrow DH + DCl$	g	600	0.141
$2I \rightarrow I_2$	g	23	7×10^9
	hexane	50	1.8×10^{10}
$CH_3Cl + CH_3O^-$	methanol	20	2.29×10^{-6}
$CH_3Br + CH_3O^-$	methanol	20	9.23×10^{-6}
$H^+ + OH^- \rightarrow H_2O$	water	25	1.35×10^{11}
	ice	−10	8.6×10^{12}

Data: Principally K.J. Laidler, *Chemical kinetics*. Harper & Row, New York (1987); M.J. Pilling and P.W. Seakins, *Reaction kinetics*. Oxford University Press (1995); J. Nicholas, *Chemical kinetics*. Harper & Row, New York (1976).

Table 25.4 Arrhenius parameters

First-order reactions	A/s^{-1}	$E_a/(kJ\,mol^{-1})$
Cyclopropane $\rightarrow$ propene	1.58×10^{15}	272
$CH_3NC \rightarrow CH_3CN$	3.98×10^{13}	160
cis-CHD=CHD $\rightarrow$ *trans*-CHD=CHD	3.16×10^{12}	256
Cyclobutane $\rightarrow$ $2C_2H_4$	3.98×10^{15}	261
$C_2H_5I \rightarrow C_2H_4 + HI$	2.51×10^{13}	209
$C_2H_6 \rightarrow 2CH_3$	2.51×10^{17}	384
$2N_2O_5 \rightarrow 4NO_2 + O_2$	4.94×10^{13}	103
$N_2O \rightarrow N_2 + O$	7.94×10^{11}	250
$C_2H_5 \rightarrow C_2H_4 + H$	1.0×10^{13}	167

Second-order, gas-phase	$A/(L\,mol^{-1}\,s^{-1})$	$E_a/(kJ\,mol^{-1})$
$O + N_2 \rightarrow NO + N$	1×10^{11}	315
$OH + H_2 \rightarrow H_2O + H$	8×10^{10}	42
$Cl + H_2 \rightarrow HCl + H$	8×10^{10}	23
$2CH_3 \rightarrow C_2H_6$	2×10^{10}	ca. 0
$NO + Cl_2 \rightarrow NOCl + Cl$	4.0×10^9	85
$SO + O_2 \rightarrow SO_2 + O$	3×10^8	27
$CH_3 + C_2H_6 \rightarrow CH_4 + C_2H_5$	2×10^8	44
$C_6H_5 + H_2 \rightarrow C_6H_6 + H$	1×10^8	ca. 25

Second-order, solution	$A/(L\,mol^{-1}\,s^{-1})$	$E_a/(kJ\,mol^{-1})$
$C_2H_5ONa + CH_3I$ in ethanol	2.42×10^{11}	81.6
$C_2H_5Br + OH^-$ in water	4.30×10^{11}	89.5
$C_2H_5I + C_2H_5O^-$ in ethanol	1.49×10^{11}	86.6
$CH_3I + C_2H_5O^-$ in ethanol	2.42×10^{11}	81.6
$C_2H_5Br + OH^-$ in ethanol	4.30×10^{11}	89.5
$CO_2 + OH^-$ in water	1.5×10^{10}	38
$CH_3I + S_2O_3^{2-}$ in water	2.19×10^{12}	78.7
Sucrose + H_2O in acidic water	1.50×10^{15}	107.9
$(CH_3)_3CCl$ solvolysis		
in water	7.1×10^{16}	100
in methanol	2.3×10^{13}	107
in ethanol	3.0×10^{13}	112
in acetic acid	4.3×10^{13}	111
in chloroform	1.4×10^4	45
$C_6H_5NH_2 + C_6H_5COCH_2Br$		
in benzene	91	34

Data: Principally J. Nicholas, *Chemical kinetics*. Harper & Row, New York (1976) and A.A. Frost and R.G. Pearson, *Kinetics and mechanism*. Wiley, New York (1961).

Table 27.1 Arrhenius parameters for gas-phase reactions

Reaction	$A/(\text{L mol}^{-1}\,\text{s}^{-1})$		$E_a/(\text{kJ mol}^{-1})$	P
	Experiment	Theory		
$2NOCl \rightarrow 2NO + Cl_2$	9.4×10^9	5.9×10^{10}	102.0	0.16
$2NO_2 \rightarrow 2NO + O_2$	2.0×10^9	4.0×10^{10}	111.0	5.0×10^{-2}
$2ClO \rightarrow Cl_2 + O_2$	6.3×10^7	2.5×10^{10}	0.0	2.5×10^{-3}
$H_2 + C_2H_4 \rightarrow C_2H_6$	1.24×10^6	7.3×10^{11}	180	1.7×10^{-6}
$K + Br_2 \rightarrow KBr + Br$	1.0×10^{12}	2.1×10^{11}	0.0	4.8

Data: Principally M.J. Pilling and P.W. Seakins, *Reaction kinetics*. Oxford University Press (1995).

Table 27.2 Arrhenius parameters for reactions in solution. See Table 25.4

Table 28.1 Maximum observed enthalpies of physisorption, $\Delta_{ad}H^{\ominus}/(\text{kJ mol}^{-1})$

C_2H_2	−38
C_2H_4	−34
CH_4	−21
Cl_2	−36
CO	−25
CO_2	−25
H_2	−84
H_2O	−59
N_2	−21
NH_3	−38
O_2	−21

Data: D.O. Haywood and B.M.W. Trapnell, *Chemisorption*. Butterworth (1964).

Table 28.2 Enthalpies of chemisorption, $\Delta_{ad}H^{\ominus}/(\text{kJ mol}^{-1})$

Adsorbate	Adsorbent (substrate)											
	Ti	Ta	Nb	W	Cr	Mo	Mn	Fe	Co	Ni	Rh	Pt
H_2		−188			−188	−167	−71	−134			−117	
N_2		−586						−293				
O_2						−720					−494	−293
CO	−640							−192	−176			
CO_2	−682	−703	−552	−456	−339	−372	−222	−225	−146	−184		
NH_3				−301				−188		−155		
C_2H_4		−577		−427	−427			−285		−243	−209	

Data: D.O. Haywood and B.M.W. Trapnell, *Chemisorption*. Butterworth (1964).

Table 28.3 Activation energies of catalysed reactions

Reaction	Catalyst	$E_a/(\text{kJ mol}^{-1})$
$2HI \rightarrow H_2 + I_2$	None	184
	Au(s)	105
	Pt(s)	59
$2NH_3 \rightarrow N_2 + 3H_2$	None	350
	W(s)	162
$2N_2O \rightarrow 2N_2 + O_2$	None	245
	Au(s)	121
	Pt(s)	134
$(C_2H_5)_2O$ pyrolysis	None	224
	$I_2(g)$	144

Data: G.C. Bond, *Heterogeneous catalysis*. Clarendon Press, Oxford (1986).

Table 29.1 Exchange current densities and transfer coefficients at 298 K

Reaction	Electrode	$j_0/(\text{A cm}^{-2})$	α
$2H^+ + 2e^- \rightarrow H_2$	Pt	7.9×10^{-4}	
	Cu	1×10^{-6}	
	Ni	6.3×10^{-6}	0.58
	Hg	7.9×10^{-13}	0.50
	Pb	5.0×10^{-12}	
$Fe^{3+} + e^- \rightarrow Fe^{2+}$	Pt	2.5×10^{-3}	0.58
$Ce^{4+} + e^- \rightarrow Ce^{3+}$	Pt	4.0×10^{-5}	0.75

Data: Principally J.O'M. Bockris and A.K.N. Reddy, *Modern electrochemistry*. Plenum, New York (1970).

Character tables

The groups C_1, C_s, C_i

C_1 (1)	E	$h = 1$
A	1	

$C_s = C_h$ (*m*)	E	σ_h	$h = 2$		
A$'$	1	1	x, y, R_z	x^2, y^2, z^2, xy
A$''$	1	-1	z, R_x, R_y	yz, xz

$C_i = S_2$ ($\bar{1}$)	E	i	$h = 2$		
A$_g$	1	1	R_x, R_y, R_z	$x^2, y^2, z^2, xy, xz, yz$
A$_u$	1	-1	x, y, z	

The groups C_{nv}

$C_{2v}, 2mm$	E	C_2	σ_v	σ_v'	$h = 4$	
A$_1$	1	1	1	1	z, z^2, x^2, y^2	
A$_2$	1	1	-1	-1	xy	R_z
B$_1$	1	-1	1	-1	x, xz	R_y
B$_2$	1	-1	-1	1	y, yz	R_x

$C_{3v}, 3m$	E	$2C_3$	$3\sigma_v$	$h = 6$	
A$_1$	1	1	1	$z, z^2, x^2 + y^2$	
A$_2$	1	1	-1		R_z
E	2	-1	0	$(x, y), (xy, x^2 - y^2)(xz, yz)$	(R_x, R_y)

$C_{4v}, 4mm$	E	C_2	$2C_4$	$2\sigma_v$	$2\sigma_d$	$h = 8$	
A$_1$	1	1	1	1	1	$z, z^2, x^2 + y^2$	
A$_2$	1	1	1	-1	-1		R_z
B$_1$	1	1	-1	1	-1	$x^2 - y^2$	
B$_2$	1	1	-1	-1	1	xy	
E	2	-2	0	0	0	$(x, y), (xz, yz)$	(R_x, R_y)

C_{5v}	E	$2C_5$	$2C_5^2$	$5\sigma_v$	$h = 10, \alpha = 72°$	
A_1	1	1	1	1	$z, z^2, x^2 + y^2$	
A_2	1	1	1	-1		R_z
E_1	2	$2\cos\alpha$	$2\cos 2\alpha$	0	$(x, y), (xz, yz)$	(R_x, R_y)
E_2	2	$2\cos 2\alpha$	$2\cos\alpha$	0	$(xy, x^2 - y^2)$	

$C_{6v}, 6mm$	E	C_2	$2C_3$	$2C_6$	$3\sigma_d$	$3\sigma_v$	$h = 12$	
A_1	1	1	1	1	1	1	$z, z^2, x^2 + y^2$	
A_2	1	1	1	1	-1	1		R_z
B_1	1	-1	1	-1	-1	1		
B_2	1	-1	1	-1	1	-1		
E_1	2	-2	-1	1	0	0	$(x, y), (xz, yz)$	(R_x, R_y)
E_2	2	2	-1	-1	0	0	$(xy, x^2 - y^2)$	

$C_{\infty v}$	E	$2C_\phi$†	$\infty\sigma_v$	$h = \infty$	
$A_1(\Sigma^+)$	1	1	1	$z, z^2, x^2 + y^2$	
$A_2(\Sigma^-)$	1	1	-1		R_z
$E_1(\Pi)$	2	$2\cos\phi$	0	$(x, y), (xz, yz)$	(R_x, R_y)
$E_2(\Delta)$	2	$2\cos 2\phi$	0	$(xy, x^2 - y^2)$	

† There is only one member of this class if $\phi = \pi$.

The groups D_n

$D_2, 222$	E	C_2^z	C_2^y	C_2^x	$h = 4$	
A_1	1	1	1	1	x^2, y^2, z^2	
B_1	1	1	-1	-1	z, xy	R_z
B_2	1	-1	1	-1	y, xz	R_y
B_3	1	-1	-1	1	x, yz	R_x

$D_3, 32$	E	$2C_3$	$3C_2'$	$h = 6$	
A_1	1	1	1	$z^2, x^2 + y^2$	
A_2	1	1	-1	z	R_z
E	2	-1	0	$(x, y), (xz, yz), (xy, x^2 - y)$	(R_x, R_y)

$D_4, 422$	E	C_2	$2C_4$	$2C_2'$	$2C_2''$	$h = 8$	
A_1	1	1	1	1	1	$z^2, x^2 + y^2$	
A_2	1	1	1	-1	-1	z	R_z
B_1	1	1	-1	1	-1	$x^2 - y^2$	
B_2	1	1	-1	-1	1	xy	
E	2	-2	0	0	0	$(x, y), (xz, yz)$	(R_x, R_y)

The groups D_{nh}

D_{3h}, $\bar{6}2m$	E	σ_h	$2C_3$	$2S_3$	$3C_2'$	$3\sigma_v$	$h = 12$	
A_1'	1	1	1	1	1	1	z^2, x^2+y^2	
A_2'	1	1	1	1	-1	-1		R_z
A_1''	1	-1	1	-1	1	-1		
A_2''	1	-1	1	-1	-1	1	z	
E'	2	2	-1	-1	0	0	$(x,y), (xy, x^2-y^2)$	
E''	2	-2	-1	1	0	0	(xz, yz)	(R_x, R_y)

$D_{\infty h}$	E	$2C_\phi$	$\infty C_2'$	i	$2iC_\phi$	iC_2'	$h = \infty$	
A_{1g} (Σ_g^+)	1	1	1	1	1	1	z^2, x^2+y^2	
A_{1u} (Σ_u^+)	1	1	1	-1	-1	-1	z	
A_{2g} (Σ_g^-)	1	1	-1	1	1	-1		R_z
A_{2u} (Σ_u^-)	1	1	-1	-1	1	1		
E_{1g} (Π_g)	2	$2\cos\phi$	0	2	$-2\cos\phi$	0	(xz, yz)	(R_x, R_y)
E_{1u} (Π_u)	2	$2\cos\phi$	0	-2	$2\cos\phi$	0	(x, y)	
E_{2g} (Δ_g)	2	$2\cos 2\phi$	0	2	$2\cos 2\phi$	0	(xy, x^2-y^2)	
E_{2u} (Δ_u)	2	$2\cos 2\phi$	0	-2	$-2\cos 2\phi$	0		
$\vdots$								

The cubic groups

T_d, $\bar{4}3m$	E	$8C_3$	$3C_2$	$6\sigma_d$	$6S_4$	$h = 24$	
A_1	1	1	1	1	1	$x^2+y^2+z^2$	
A_2	1	1	1	-1	-1		
E	2	-1	2	0	0	$(3z^2-r^2, x^2-y^2)$	
T_1	3	0	-1	-1	1		(R_x, R_y, R_z)
T_2	3	0	-1	1	-1	$(x,y,z), (xy,xz,yz)$	

O_h (*m3m*)	E	$8C_3$	$6C_2$	$6C_2$	$3C_2$ ($=C_4^2$)	i	$6S_4$	$8S_6$	$3\sigma_h$	$6\sigma_d$	$h = 48$	
A_{1g}	1	1	1	1	1	1	1	1	1	1	$x^2+y^2+z^2$	
A_{2g}	1	1	-1	-1	1	1	-1	1	1	-1		
E_g	2	-1	0	0	2	2	0	-1	2	0	$(2z^2-x^2-y^2, x^2-y^2)$	
T_{1g}	3	0	-1	1	-1	3	1	0	-1	-1	(R_x, R_y, R_z)	
T_{2g}	3	0	1	-1	-1	3	-1	0	-1	1	(xy, yz, xy)	
A_{1u}	1	1	1	1	1	-1	-1	-1	-1	-1		
A_{2u}	1	1	-1	-1	1	-1	1	-1	-1	1		
E_u	2	-1	0	0	2	-2	0	1	-2	0		
T_{1u}	3	0	-1	1	-1	-3	-1	0	1	1	(x,y,z)	
T_{2u}	3	0	1	-1	-1	-3	1	0	1	-1		

The icosahedral group

I	E	$12C_5$	$12C_5^2$	$20C_3$	$15C_2$	$h = 60$	
A	1	1	1	1	1		$z^2 + y^2 + z^2$
T_1	3	$\frac{1}{2}(1 + \sqrt{5})$	$\frac{1}{2}(1 - \sqrt{5})$	0	-1	(x, y, z) (R_x, R_y, R_z)	
T_2	3	$\frac{1}{2}(1 - \sqrt{5})$	$\frac{1}{2}(1 + \sqrt{5})$	0	-1		
G	4	-1	-1	1	0		
G	5	0	0	-1	1		$(2z^2 - x^2 - y^2, x^2 - y^2, xy, yz, zx)$

Further information: P.W. Atkins, M.S. Child, and C.S.G. Phillips, *Tables for group theory*. Oxford University Press (1970).

Answers to exercises

Detailed solutions to the (a) exercises can be found in the *Student's Solutions Manual for Physical Chemistry*, sixth edition, by P.W. Atkins, C.A. Trapp., M. Cady, and C. Giunta

Answers to 'a' exercises

1.1 10 atm.

1.2 (a) 24 atm; (b) 22 atm.

1.3 (a) 2.57 kTorr; (b) 3.38 atm.

1.4 30 K.

1.5 30 lb in^{-2}.

1.6 4.20×10^{-2} atm.

1.7 8.3147 J K^{-1} mol^{-1}.

1.8 S_8.

1.9 6.2 kg.

1.10 (a) 0.758, 0.242, 561 Torr, 179 Torr; (b) 0.751, 0.239, 0.010, 556 Torr, 177 Torr, 7.4 Torr.

1.11 169 g mol^{-1}.

1.12 $-272°C$.

1.13 (a) 9.975; (b) 1.

1.14 (a) 72 K; (b) 0.95 km s^{-1}; (c) 72 K.

1.15 81 mPa.

1.16 9.7×10^{-7} m.

1.17 (a) 5×10^{10} s^{-1}; (b) 5×10^9 s^{-1}; (c) 5×10^3 s^{-1}.

1.18 (a) 6.7 nm; (b) 67 nm; (c) 6.7 cm.

1.19 9.06×10^{-3}.

1.20 (a) 1.0 atm, 8.2×10^2 atm; (b) 1.0 atm, 1.7×10^3 atm.

1.21 67.8 mL mol^{-1}, 54.5 atm, 120 K.

1.22 (a) 0.88; (b) 1.2 L.

1.23 140 atm.

1.24 (a) 0.1353 L mol^{-1}, 0.6957; (b) 0.649.

1.25 (a) 50.7 atm; (b) 34.8 atm, 0.687.

1.26 (a) 0.67, 0.33; (b) 2.0 atm, 1.0 atm; (c) 3.0 atm.

1.27 32.9 cm^3 mol^{-1}, 1.33 L^2 atm mol^{-2}, 0.24 nm.

1.28 (a) 1.4 kK; (b) 0.28 nm.

1.29 (a) 3.64 kK, 8.7 atm; (b) 2.60 kK, 4.5 atm; (c) 46.7 K, 0.18 atm.

1.30 4.6×10^{-5} m^3 mol^{-1}, 0.66.

2.1 (a) 98 J; (b) 16 J.

2.2 2.6 kJ.

2.3 -1.0×10^2 J.

2.4 (a) $\Delta U = 0$, $\Delta H = 0$, $q = +1.57$ kJ, $w = -1.57$ kJ; (b) $\Delta U = 0$, $\Delta H = 0$, $q = +1.13$ kJ, $w = -1.13$ kJ; (c) all 0.

2.5 1.33 atm, $\Delta U = +1.25$ kJ, $w = 0$, $q = +1.25$ kJ.

2.6 (a) -88 J; (b) -167 J.

2.7 $+123$ J.

2.8 $\Delta U = -37.55$ kJ, $\Delta H = -40.656$ kJ, $q = -40.656$ kJ, $w = +3.10$ kJ.

2.9 -1.5 kJ.

2.10 85.0 MJ.

2.11 (a) $\Delta U = +26.8$ kJ, $\Delta H = +28.3$ kJ, $q = +28.3$ kJ, $w = -1.45$ kJ; (b) $\Delta U = +26.8$ kJ, $\Delta H = +28.3$ kJ, $q = +26.8$ kJ, $w = 0$.

2.12 131 K.

2.13 194 J.

2.14 22 kPa.

2.15 0.45 atm.

2.16 -125 kJ mol^{-1}.

2.17 $C_{p,m} = 30$ J K^{-1} mol^{-1}, $C_{V,m} = 22$ J K^{-1} mol^{-1}.

2.18 80 J K^{-1}.

2.19 $\Delta U = +1.6$ kJ, $\Delta H = +2.2$ kJ, $q = +2.2$ kJ.

2.20 $w = -3.2$ kJ, $q = 0$, $\Delta T = -38$ K, $\Delta U = -3.2$ kJ, $\Delta H = -4.5$ kJ.

2.21 $w = +4.1$ kJ, $q = 0$, $\Delta U = +4.1$ kJ, $\Delta H = +5.4$ kJ, $p_f = 5.2$ atm, $V_f = 11.8$ L.

2.22 9.4 L, 288 K, -0.46 kJ.

2.23 $+0.9$ mm^3.

2.24 $q = 0$, $w = \Delta U = -20$ J, $\Delta T = -0.35$ K, $\Delta H = -26$ J.

2.25 (a) 226 K; (b) 238 K.

2.26 $\Delta U = +12$ kJ, $\Delta H = +13$ kJ, $q = +13$ kJ, $w = -1.0$ kJ.

2.27 -4564.7 kJ mol^{-1}.

2.28 -126 kJ mol^{-1}.

2.29 $+53$ kJ mol^{-1}, -33 kJ mol^{-1}.

2.30 -432 kJ mol^{-1}.

2.31 641 J K^{-1}.

2.32 1.58 kJ K^{-1}, $+2.05$ K.

2.33 (a) -2.80 MJ mol^{-1}; (b) -2.80 MJ mol^{-1}; (c) -1.28 MJ mol^{-1}.

2.34 $+65.49$ kJ mol^{-1}.

2.35 -383 kJ mol^{-1}.

2.36 25 kJ, 9.8 m.

2.37 (a) -2205 kJ mol^{-1}; (b) -2200 kJ mol^{-1}.

2.38 (a) $\nu(CO_2) = +1$, $\nu(H_2O) = +2$, $\nu(CH_4) = -1$, $\nu(O_2) = -2$, exothermic; (b) $\nu(C_2H_2) = +1$, $\nu(C) = -2$, $\nu(H_2) = -1$, endothermic; (c) $\nu(Na^+) = +1$, $\nu(Cl^-) = +1$, $\nu(NaCl) = -1$, endothermic.

2.39 (a) -57.20 kJ mol^{-1}; (b) -176.01 kJ mol^{-1}.

2.40 (a) -114.40 kJ mol^{-1}, -109.44 kJ mol^{-1}; (b) -92.31 kJ mol^{-1}, -241.82 kJ mol^{-1}.

2.41 -1368 kJ mol^{-1}.

2.42 (a) -392.1 kJ mol^{-1}; (b) -946.6 kJ mol^{-1}; (c) $+52.5$ kJ mol^{-1}.

2.43 -56.98 kJ mol^{-1}.

2.44 (a) $+131.29$ kJ mol^{-1}, $+128.81$ kJ mol^{-1}; (b) $+132.56$ kJ mol^{-1}, $+129.42$ kJ mol^{-1}.

2.45 -1892.2 kJ mol^{-1}.

2.46 (a) -124.2 kJ mol^{-1}; (b) -222.46 kJ mol^{-1}.

3.1 (a) $\partial^2 f / \partial y \partial x = \partial / \partial y (2xy) = 2x$, $\partial^2 f / \partial x \partial y = \partial / \partial x (x^2 + 6y) = 2x$; (b) $\partial^2 f / \partial y \partial x = \partial / \partial y (\cos xy - xy \sin xy) = -x \sin xy - x \sin xy - x^2 y \cos xy = -2x \sin xy - x^2 y \cos xy$, $\partial^2 f / \partial x \partial y = \partial / \partial x (-x^2 \sin xy) = -2x \sin xy - x^2 y \cos xy$.

3.2 $dz = 2axy^3 \, dx + 3ax^2 y^2 \, dy$.

3.3 (a) $dz = (2x - 2y + 2) \, dx + (4y - 2x - 4) \, dy$.

3.4 $dz = (y + 1/x) \, dx + (x - 1) \, dy$.

3.5 $(\partial C_V / \partial V)_T = (\partial (\partial U / \partial V)_T \partial T)_V$.

3.6 $(\partial H / \partial U)_p = 1 + p(\partial V / \partial U)_p$.

3.7 $dV = (\partial V / \partial p)_T \, dp + (\partial V / \partial T)_p \, dT$; $d \ln V = -\kappa_T \, dp + \alpha \, dT$.

3.8 0, 0.

3.10 0.71 K atm^{-1}.

3.11 $\Delta U_m = +137$ J mol^{-1}, $q = +8.05 \times 10^3$ J mol^{-1}, $w = -7.91 \times 10^3$ J mol^{-1}.

3.12 1.31×10^{-3} K^{-1}.

3.13 1×10^3 atm.

3.14 -7.2 J atm^{-1} mol^{-1}, 8.1 kJ.

3.15 -4.2 atm.

4.1 (a) $+92$ J K^{-1}; (b) $+67$ J K^{-1}.

4.2 152.67 J K^{-1} mol^{-1}.

4.3 $+8.92$ J K^{-1}.

4.4 -22.1 J K^{-1}.

4.5 $w = +4.1$ kJ, $q = 0$, $\Delta U = +4.1$ kJ, $\Delta H = +5.4$ kJ, $\Delta S = 0$.

4.6 $+12.9$ J K^{-1}.

4.7 Not reversible.

4.8 (a) 54.9 kJ; (b) -195 J K^{-1}.

4.9 $+26$ J K^{-1}.

4.10 6.6 L.

4.11 $+2.8$ J K^{-1}.

4.12 ΔH(overall) $= 0$, ΔH(individual) $= \pm 1.9 \times 10^2$ kJ, ΔS(overall) $= +93.4$ J K^{-1}.

4.13 (a) $q = 0$; (b) $w = -20$ J; (c) $\Delta U = -20$ J; (d) $\Delta T = -0.35$ K; (e) $\Delta S = +0.60$ J K^{-1}.

4.14 (a) $+87.8$ J K^{-1} mol^{-1}; (b) -87.8 J K^{-1} mol^{-1}.

4.15 (a) -386.1 J K^{-1} mol^{-1}; (b) -49.0 J K^{-1} mol^{-1}; (c) -153.1 J K^{-1} mol^{-1}.

4.16 (a) -521.5 kJ mol^{-1}; (b) $+25.8$ kJ mol^{-1}; (c) -178.7 kJ mol^{-1}.

4.17 (a) -522.1 kJ mol^{-1}; (b) $+25.78$ kJ mol^{-1}; (c) -178.6 kJ mol^{-1}.

4.18 -93.05 kJ mol^{-1}.

4.19 -50 kJ mol^{-1}.

4.20 (a) $+2.9$ J K^{-1}, -2.9 J K^{-1}, 0; (b) $+2.9$ J K^{-1}, 0, $+2.9$ J K^{-1}; (c) 0, 0, 0.

4.21 $\Delta S = n(C_{V,m} - R) \ln 2$.

4.22 817.90 kJ mol^{-1}.

4.23 0.11, 0.38.

5.1 $(\partial S/\partial V)_T = \alpha/\kappa_T$.

5.2 -3.8 J.

5.3 -36.5 J K^{-1}.

5.4 $+10$ kJ.

5.5 (a) 15.7 atm; (b) $+8.25$ kJ.

5.6 $+7.3$ kJ mol^{-1}.

5.7 -0.55 kJ mol^{-1}.

5.8 -2.63×10^{-8} Pa^{-1}, 0.88.

5.9 $+10$ kJ.

5.10 $+11$ kJ mol^{-1}.

5.11 $p = RT/(V_m - b) - a/V_m^2$.

5.12 $(\partial S/\partial V)_T = nR/(V - nb)$, ΔS greater for van der Waals gas.

6.1 303 K (30°C).

6.2 $+16$ kJ mol^{-1}, $+45.2$ J K^{-1} mol^{-1}.

6.3 $+20.80$ kJ mol^{-1}.

6.4 (a) $+34.08$ kJ mol^{-1}; (b) 350.5 K.

6.5 281.8 K (8.7°C).

6.6 25 g s^{-1}.

6.7 (a) 1.7 kg; (b) 31 kg; (c) 1.4 g.

6.8 At 373 K, water vapour condenses to liquid. At 273 K, liquid water freezes. Ice remains at 260 K. There is a pause in the rate of cooling at 373 K and at 273 K.

6.9 (a) $+49$ kJ mol^{-1}; (b) 216°C; (c) $+99$ J K^{-1} mol^{-1}.

6.10 272.80 K.

6.11 0.07630 (7.6 per cent).

6.12 2.6 kPa.

6.13 72.8 mN m^{-1}.

6.14 728 kPa.

7.1 886.8 cm^3.

7.2 56.3 cm^3 mol^{-1}.

7.3 6.4 MPa.

7.4 0.13 MPa.

7.5 $K_f = 32$ K kg mol^{-1}, $K_b = 5.22$ K kg mol^{-1}.

7.6 82 g mol^{-1}.

7.7 381 g mol^{-1}.

7.8 -0.09°C.

7.9 $+1.2$ J K^{-1}, -0.35 kJ.

7.10 $+4.7$ J K^{-1} mol^{-1}.

7.11 (a) 1:1; (b) 0.8600.

7.12 (a) 3.4 mmol kg^{-1}; (b) 34 mmol kg^{-1}.

7.13 0.17 mol L^{-1}.

7.14 -0.16°C.

7.15 24 g kg^{-1}.

7.16 87 kg mol^{-1}.

7.17 Raoult's law basis: $a_A = 0.833$, $\gamma_A = 0.93$; Henry's law basis (concentration in mole fractions), $a_B = 0.125$, $\gamma_B = 1.25$; (concentration as molality) $a_B = 2.8$, $\gamma_B = 1.25$.

7.18 $p(CCl_4) = 32.2$ Torr, $p(Br_2) = 6.1$ Torr, p(Total) $= 38.3$ Torr, $y(CCl_4) = 0.841$, $y(Br_2) = 0.16$.

7.19 $a_A = 0.499$, $a_M = 0.668$, $\gamma_A = 1.25$, $\gamma_M = 1.11$.

8.1 $x_A = 0.920$, $y_A = 0.968$.

8.2 440 Torr, $x_A = 0.268$.

8.3 (a) yes; (b) $y_A = 0.830$.

8.4 (a) 154 Torr; (b) $y_{DE} = 0.67$.

8.5 (a) $y_M = 0.36$; (b) $y_M = 0.82$.

8.6 (a) 2; (b) 2.

8.7 2, 2.

8.8 (a) 3, 2; (b) 1.

8.12 At b_3, $C = 2$, $P = 2$, $F = 2$. The compositions of the phases are $x_A = 0.18$ and $x_A = 0.70$. At b_2, $C = 2$, $P = 1$, $F = 3$. Between the liquid line and b_1, $C = 2$, $P = 2$, $F = 2$. Above b_1, $C = 1$, $P = 1$, $F = 3$.

8.13 Incongruent melting occurs at 460°C. The composition of the eutectic is 4 per cent by mass of silver; it melts at $\theta_e = 215$°C.

8.15 (a) ≈ 80 per cent silver by mass; (b) compound decomposes; (c) ≈ 82 per cent silver by mass.

8.16 (b) 620 Torr; (c) 490 Torr; (d) $x_{Hexane} = 0.50$, $y_{Hexane} = 0.72$; (e) $y_{Hexane} = 0.50$, $x_{Hexane} = 0.30$; (f) $n_{vap} \approx 1.7$ mol, $n_{liq} \approx 0.3$ mol.

8.19 (a) The mixture has a single liquid phase at all compositions. (b) At $x(C_6F_{14}) = 0.24$, two liquid phases separate with compositions $x = 0.24$ and $x = 0.48$. At $x > 0.48$, a single phase forms.

9.1 -2.42 kJ mol^{-1}.

9.2 3.01.

9.3 (a) 2.85×10^{-6}; (b) $+240$ kJ mol^{-1}; (c) 0.

9.4 (a) 0.1411; (b) $+4.855$ kJ mol^{-1}; (c) 14.556.

9.5 (a) -68.26 kJ mol^{-1}, 9.2×10^{11}; (b) -69.7 kJ mol^{-1}, 1.3×10^9.

9.6 (a) 0.087 (A), 0.370 (B), 0.196 (C), 0.438 (D); (b) 0.33; (c) 0.33; (d) $+2.8$ kJ mol^{-1}.

9.7 1.5 kK.

9.8 $+2.77$ kJ mol^{-1}, -16.5 J K^{-1} mol^{-1}.

9.9 $+12.3$ kJ mol^{-1}.

9.10 50 per cent.

9.11 0.904, 0.096.

9.12 (a, c).

9.13 (b).

9.14 (a) $+53$ kJ mol^{-1}; (b) -53 kJ mol^{-1}.

9.15 -14.38 kJ mol^{-1}, product formation.

9.16 1110 K.

9.17 (a) 5.40; (b) 3.61.

9.18 (a) 5.13; (b) 8.88; (c) 2.88.

9.19 8.3.

9.21 (a) Na$_2$HPO$_4$/H$_3$PO$_4$; (b) H$_2$PO$_4^-$/HPO$_4^{2-}$.

10.1 -218.66 kJ mol^{-1}.

10.2 1.25×10^{-5} mol L^{-1}.

10.3 -291 kJ mol^{-1}.

10.4 (a) $I(KCl) = b/b^{\ominus}$; (b) $I(FeCl_3) = 6b/b^{\ominus}$; (c) $I(CuSO_4) = 4b/b^{\ominus}$.

10.5 0.90.

10.6 (a) 2.73 g; (b) 2.92 g.

10.7 0.25 mol kg^{-1}.

10.8 $\gamma_{\pm} = (\gamma_+\gamma_-^2)^{1/3}$.

10.9 0.56.

10.10 1×10^4 per cent.

10.11 2.01.

10.12 -1108 kJ mol^{-1}.

10.13 $+34.2$ mV.

10.14 -1.18 V.

10.15 (a) $Ag^+(aq) + e^- \longrightarrow Ag(s)$, $Zn^{2+}(aq) + 2e^- \longrightarrow Zn(s)$, $2Ag^+(aq) + Zn(s) \longrightarrow 2Ag(s) + Zn^{2+}(aq)$, $+1.56$ V; (b) $H^+(aq) + e^- \longrightarrow \frac{1}{2}H_2(g)$, $Cd^{2+}(aq) + 2e^- \longrightarrow Cd(s)$, $Cd(s) + 2H^+(aq) \longrightarrow Cd^{2+}(aq) + H_2(g)$, $+0.40$ V; (c) $Cr^{3+}(aq) + 3e^- \longrightarrow Cr(s)$, $[Fe(CN)_6]^{3-}(aq) + e^- \longrightarrow [Fe(CN)_6]^{4-}(aq)$, $Cr^{3+}(aq) + 3[Fe(CN)_6]^{4-}(aq) \longrightarrow Cr(s) + 3[Fe(CN)_6]^{3-}(aq)$, -1.10 V.

10.16 (a) $Zn(s)|ZnSO_4(aq) \| CuSO_4(aq)|Cu(s)$, $+1.10$ V; (b) $Pt|H_2(g)|HCl(aq)| AgCl(s)|Ag(s)$, $+0.22$ V; (c) $Pt|H_2(g)|H^+(aq), H_2O(l)|O_2(g)|Pt$, $+1.23$ V.

10.17 (a) $+1.56$ V; (b) $+0.40$ V; (c) -1.10 V.

10.18 (a) $+1.10$ V; (b) $+0.22$ V; (c) $+1.23$ V.

10.19 (a) -363 kJ mol^{-1}; (b) -405 kJ mol^{-1}.

10.20 (a) $+0.324$ V; (b) $+0.45$ V.

10.21 $+1.92$ V.

10.22 -0.62 V.

10.23 (a) 6.5×10^9; (b) 1.5×10^{12}.

10.24 $+0.49$ V; 4×10^{16}.

10.25 $1.80 \times 10^{-10} \longrightarrow 1.78 \times 10^{-10}$, $9.04 \times 10^{-7} \longrightarrow 5.1 \times 10^{-7}$.

10.26 $E = E^{\ominus} - (RT/6F) \ln[(a(Cr^{3+})^2)/(a(Cr_2O_7^{2-})a(H^+)^{14})]$.

10.27 0.86.

10.28 0.

10.29 (a) 1×10^{-8} mol kg^{-1}; (b) 1×10^{-16}.

11.1 1.7 MW.

11.2 1.3×10^{-2} W.

11.3 262 nm.

11.4 2.42 cm s^{-1}.

11.5 332 pm.

11.6 8.83×10^{-28} kg m s^{-1}, 0.969 km s^{-1}.

11.7 50.6 nm.

11.8 0.70 nm.

11.9 $E/(10^{-19}$ J$)$, $E/($kg mol$^{-1})$: (a) 3.31, 199; (b) 3.61, 218; (c) 4.97, 299.

11.10 (a) 0.66 m s^{-1}; (b) 0.72 m s^{-1}; (c) 0.99 m s^{-1}.

11.11 21 m s^{-1}.

11.12 (a) 2.8×10^{18}; (b) 2.8×10^{20}.

11.13 6 kK.

11.14 (a) no ejection; (b) 3.19×10^{-19} J, 837 km s^{-1}.

11.15 (a) 7×10^{-19} J, 400 kJ mol^{-1}; (b) 7×10^{-20} J, 40 kJ mol^{-1}; (c) 7×10^{-34} J, 4×10^{-13} kJ mol^{-1}.

11.16 (a) 6.6×10^{-29} m; (b) 6.6×10^{-36} m; (c) 99.7 pm.

11.17 1.1×10^{-28} m s^{-1}, 1×10^{-27} m.

11.18 1.12×10^{-15} J.

12.1 (a) 1.81×10^{-19} J, 110 kJ mol^{-1}, 1.1 eV, 9.1×10^3 cm^{-1}; (b) 6.6×10^{-19} J, 400 kJ mol^{-1}, 4.1 eV, 3.3×10^4 cm^{-1}.

12.2 (a) 0.04; (b) 0.

12.3 $-(\hbar^2/2m)(d^2\psi/dx^2) = E\psi$, 0, $h^2/4L^2$.

12.4 $L/6$, $L/2$, $5L/6$.

12.5 3.

12.6 23 per cent.

12.7 4.30×10^{-21} J.

12.8 278 N m^{-1}.

12.9 2.63 μm.

12.10 3.72 μm.

12.11 (a) 3.3×10^{-34} J; (b) 3.3×10^{-33} J.

12.13 5.61×10^{-21} J.

12.14 $N = 1/(2\pi)^{1/2}$.

12.15 1.49×10^{-34} J s; 0, $\pm 1.05 \times 10^{-34}$ J s.

13.1 14.0 eV.

13.2 $r = 4a_0$, 0.

13.3 101 pm, 376 pm.

13.4 $N = 2/a_0^{3/2}$.

13.5 $\langle V \rangle = 2E(1s)$, $\langle T \rangle = -E(1s)$.

13.6 $r^* = 5.24a_0/Z$.

13.7 (Angular momentum/$\hbar$, angular nodes, radial nodes) = (a) 0,0,0; (b) 0, 0, 2; (c) $6^{1/2}$, 2, 0.

13.8 (a) $\frac{5}{2}$, $\frac{3}{2}$; (b) $\frac{7}{2}$, $\frac{5}{2}$.

13.9 1, 1

13.10 (a) 1; (b) 9; (c) 25.

13.11 $L = 2$, $S = 0$, $J = 2$.

13.12 $r = 0.35a_0$.

13.13 (b, c).

13.14 (a) 2; (b) 6; (c) 10; (d) 18.

13.15 (a) [Ar]$3d^8$; (b) $S = 1$, $M_S = 0, \pm 1$, $S = 0$, $M_S = 0$.

13.16 (a) 1 (3), 0 (1); (b) $\frac{3}{2}$ (4), $\frac{1}{2}$ (2), $\frac{1}{2}$ (2).

13.17 $^3D_3$, $^3D_2$, $^3D_1$, $^1D_2$ with $^3D < {}^1D$.

13.18 (a) 0 (1); (b) $\frac{3}{2}$ (4), $\frac{1}{2}$ (2); (c) 2 (5), 1 (3), 2 (1).

13.19 (a) $^2S_{1/2}$; (b) $^2P_{3/2}$, $^2P_{1/2}$.

13.20 2.1 T.

14.1 (a) $1\sigma^2$ (1); (b) $1\sigma^2 2\sigma^{*2}$ (0); (c) $1\sigma^2 2\sigma^{*1} 1\pi^4$ (2).

14.2 (a) $1\sigma^2 2\sigma^{*2} 1\pi^4 3\sigma^2$; (b) $1\sigma^2 2\sigma^{*2} 1\pi^4 3\sigma^2 2\pi^{*1}$; (c) $1\sigma^2 2\sigma^{*2} 1\pi^4 3\sigma^2$.

14.3 C$_2$.

14.4 XeF$^+$ is shorter.

14.5 (a) g; (c) g; (d) u.

14.6 $\frac{1}{2}$, 0.

14.7 N$_2$ is shorter.

14.10 (a, c).

15.1 E, C_3, $3\sigma_v$.

15.2 (a, b).

15.3 Yes.

15.6 i, σ_h.

15.8 (a) R_3; (b) C_{2v}; (c) D_{3h}; (d) $D_{\infty h}$.

15.9 (a) C_{2v}; (b) $C_{\infty v}$; (c) C_{3v}; (d) D_{2h}; (e) C_{2v}; (f) C_{2h}.

15.10 (a) C_{2v}; (b) C_{2h}.

15.11 Polar: NO$_2$, N$_2$O, CHCl$_3$, and *cis*-CHBr=CHBr. Chiral: none.

15.12 d_{xy}.

15.13 B$_1(x)$, B$_2(y)$, A$_1(z)$.

15.14 (a) E$_{1u}$, A$_{2u}$; (b) B$_{3u}$, B$_{2u}$, B$_{1u}$.

16.1 (a) 0.0469 J m^{-3} s; (b) 1.33×10^{-13} J m^{-3} s; (c) 4.50×10^{-16} J m^{-3} s.

16.2 0.409 THz.

16.3 (a) 2.642×10^{-47} kg m^2; (b) 127.4 pm.

16.4 4.442×10^{-47} kg m^2, 165.9 pm.

16.5 232.1 pm.

16.6 106.5 pm, 115.6 pm.

16.7 20 475 cm^{-1}.

16.8 2699.77 cm^{-1}.

16.9 0.16 kN m^{-1}.

16.10 1.089 per cent.

16.11 328.7 N m^{-1}.

16.12 $4A_1 + A_2 + 2B_1 + 2B_2$.

16.13 b and d.

16.14 b, c, and d.

16.15 a, b, and d.

16.16 $0.999\,999\,925 \times 660$ nm.

16.17 2.4×10^7 m s^{-1}, 8.4×10^5 K.

16.18 (a) 5×10 ps; (b) 5 ps.

16.19 (a) 53 cm^{-1}; (b) 0.53 cm^{-1}.

16.20 (a) 0.067; (b) 0.20.

16.21 HF (967.0 N m^{-1}), HCl (515.6 N m^{-1}), HBr (411.8 N m^{-1}), HI (314.2 N m^{-1}).

16.22 1580.38 cm^{-1}, 7.644×10^{-3}.

16.23 5.15 eV.

16.24 198.9 pm.

16.25 (a) 3; (b) 6; (c) 12.

16.26 (a) All; (b) symmetric stretch: Raman, antisymmetric stretch and bends: IR.

16.27 Raman active.

17.1 80 per cent.

17.2 6.28×10^3 L mol^{-1} cm^{-1}.

17.3 1.5 mmol L^{-1}.

17.4 5.44×10^7 L mol^{-1} cm^{-2}.

17.5 Diene: 243 nm; butene: 192 nm.

17.6 450 L mol^{-1} cm^{-1}.

17.7 159 L mol^{-1} cm^{-1}, 23 per cent.

17.8 (a) 0.9 m; (b) 3 m.

17.9 (a) 5×10^7 L mol^{-1} cm^{-2}; (b) 2.5×10^6 L mol^{-1} cm^{-2}.

18.1 600 MHz.

18.2 $(-1.625 \times 10^{-26}$ J$) \times m_I$.

18.3 154 MHz.

18.4 (a) proton.

18.5 6.116×10^{-26} J.

18.6 (a) 5.87 T; (b) 38.3 T; (c) 23.4 T.

18.7 (a) 1×10^{-6}; (b) 5.1×10^{-6}; (c) 3.4×10^{-5}.

18.8 10.

18.9 (a) 11 μT; (b) 110 μT.

18.11 6.7×10^2 s^{-1}.

18.14 (b).

18.15 0.59 mT, 20 μs.

18.16 0.2 kT, 10 mT.

18.17 2.0022.

18.18 2.3 mT, 2.003.

18.19 330.2 mT, 332.2 mT, 332.8 mT, 334.8 mT, 1:1:1:1.

18.20 (a) 1:3:3:1; (b) 1:3:6:7:6:3:1.

18.21 (a) 331.9 mT; (b) 1.201 T.

18.22 $\frac{3}{2}$.

19.1 1.

19.2 (a) 2.57×10^{27}; (b) 7.26×10^{27}.

19.3 2.83.

19.4 3.156.

19.5 2.45 kJ mol^{-1}.

19.6 354 K.

19.7 (a) 0.71; (b) 0.996.

19.8 (a) 5×10^{-5}, 0.4, 0.905; (b) 1.4; (c) 22 J mol^{-1}; (d) 1.6 J K^{-1} mol^{-1}; (e) 4.8 J K^{-1} mol^{-1}.

19.9 4303 K.

19.10 (a) 138 J K^{-1} mol^{-1}; (b) 146 J K^{-1} mol^{-1}.

19.11 5.18 J K^{-1} mol^{-1}.

19.12 a, b, and d.

20.1 (a) $5R/2$; (b) $3R$; (c) $3R$.

20.2 NH$_3$: 1.33 and 1.11 (1.31); CH$_4$:1.33 and 1.08 (1.31).

20.3 (a) 19.6; (b) 34.3.

20.4 (a) 1; (b) 2; (c) 2; (d) 12; (e) 3.

20.5 43.1, 23.36 K.

20.6 43.76 J K^{-1} mol^{-1}.

20.7 (a) 36.95, 80.08; (b) 36.7, 79.7.

20.8 72.5.

20.9 (a) 14.93 J K^{-1} mol^{-1}; (b) 25.65 J K^{-1} mol^{-1}.

20.10 -13.8 kJ mol^{-1}, -0.20 kJ mol^{-1}.

20.11 (a) 0.236R; (b) 0.193R.

20.12 11.5 J K^{-1} mol^{-1}.

20.13 (a) 9 J K^{-1} mol^{-1}; (b) 13 J K^{-1} mol^{-1}; (c) 15 J K^{-1} mol^{-1}.

20.14 9.57×10^{-15} J K^{-1}.

20.15 3.70×10^{-3}.

21.1 $(1, \frac{1}{2}, 0)$, $(1, 0, \frac{1}{2})$, $(\frac{1}{2}, \frac{1}{2}, \frac{1}{2})$.

21.2 (323), (110).

21.3 249 pm, 176 pm, 432 pm.

21.4 70.7 pm.

21.5 16°, 23°, 28°.

21.6 0.215 cm.

21.7 3.96×10^{-28} m^3.

21.8 4, 4.01 g cm^{-3}.

21.9 190 pm.

21.10 (111), (200), (311).

21.11 8.17°, 4.82°, 11.75°.

21.12 fcc.

21.13 $F_{hkl} = f$.

21.14 0.9069.

21.16 (a) 58.0 pm; (b) 102 pm.

21.17 0.340.

21.18 7.654 g cm^{-3}.

21.19 $+1.6$ per cent.

21.21 7.9 km s^{-1}.

21.22 4.6 kV.

21.23 5.8°, 17°, 0.3°, 0.9°.

22.1 O$_3$ and H$_2$O$_2$, but rotation about the O–O bond in H$_2$O$_2$ averages out the polarity.

22.2 1.01×10^{-39} J^{-1} C^2 m^2 (9.1×10^{-24} cm^3), 1.7 D.

22.3 4.8.

22.4 1.42×10^{-39} J^{-1} C^2 m^2 (1.28×10^{-23} cm^3).

22.5 17 per cent, 23 per cent; no correlation.

22.6 (a) 0 (by symmetry); (b) 0.7 D; (c) 0.4 D.

22.7 37 D at 11.7° to x-axis.

22.8 4.9 μD.

22.9 1.34.

22.10 18.

22.11 6.9×10^{-6}.

22.12 3.

22.13 -6.4×10^{-5} cm^3 mol^{-1}.

22.14 2.

22.15 4.326.

22.16 $+0.016$ cm^3 mol^{-1}.

22.17 222 T.

23.1 70 kg mol^{-1}, 71 kg mol^{-1}.

23.2 24 nm.

23.3 1.37×10^4.

23.4 3.08 μm, 3.08 nm.

23.5 100.

23.6 0.73 mm s^{-1}.

23.7 63 kg mol^{-1}.

23.8 31 kg mol^{-1}.

23.9 (a) 18 kg mol^{-1}; (b) 20 kg mol^{-1}.

23.10 0.24 mmol L^{-1}.

23.11 6.7 mmol L^{-1}.

23.12 3.4 Mg mol^{-1}.

23.13 4.3×10^5 g.

23.14 24 ns, 14 ps.

24.1 1.9×10^{20}.

24.2 104 mg.

24.3 4.1×10^{-2} J m^{-2} s^{-1}.

24.4 0.056 nm^2.

24.5 17 W, 17 W.

24.6 43 g mol^{-1}.

24.7 30 h.

24.8 0.142 nm^2.

24.9 205 kPa.

24.10 (a) 130 μP; (b) 130 μP; (c) 240 μP.

24.11 (a) 5.4 mJ K^{-1} m^{-1} s^{-1}, 8.1 mW; (b) 29 mJ K^{-1} m^{-1} s^{-1}, 44 mW.

24.12 138 μP, 390 pm.

24.13 5.4×10^{-3} J K^{-1} m^{-1} s^{-1}.

24.14 (a) 11 m^2 s^{-1}, 4.4×10^2 mol m^{-2} s^{-1}; (b) 1.1×10^{-5} m^2 s^{-1}, 4.4×10^{-3} mol m^{-2} s^{-1}; (c) 1.1×10^{-7} m^2 s^{-1}, 4.4×10^{-5} mol m^{-2} s^{-1}.

24.15 7.63×10^{-3} S m^2 mol^{-1}.

24.16 347 μm s^{-1}.

24.17 0.331.

24.18 13.83 mS m^2 mol^{-1}.

24.19 4.01×10^{-8} m^2 V^{-1} s^{-1}, 5.19×10^{-8} m^2 V^{-1} s^{-1}, 7.62×10^{-8} m^2 V^{-1} s^{-1}.

24.20 1.90×10^{-9} m^2 s^{-1}.

24.21 1.3 ks.

24.22 420 pm.

24.23 27 ps.

24.24 113 μm.

24.25 (a) 78 s; (b) 7.8×10^3 s.

25.1 C: 3.0 mol L^{-1} s^{-1}, D: 1.0 mol L^{-1} s^{-1}, A: 1.0 mol L^{-1} s^{-1}, B: 2.0 mol L^{-1} s^{-1}.

25.2 v: 0.50 mol L^{-1} s^{-1}, D: 1.5 mol L^{-1} s^{-1}, A: 1.0 mol L^{-1} s^{-1}, B: 0.50 mol L^{-1} s^{-1}.

25.3 L mol^{-1} s^{-1}; (a) k[A][B]; (b) $3k$[A][B].

25.4 $v = \frac{1}{2}k$[A][B][C], L^2 mol^{-2} s^{-1}.

25.5 2.

25.6 2.

25.7 10.3 ks; (a) 499.7 Torr; (b) 480 Torr.

25.8 (a) 4.1×10^{-3} L mol^{-1} s^{-1}; (b) A: 2.6 ks, B: 7.4 ks.

25.9 (a) L mol^{-1} s^{-1}, L^2 mol^{-2} s^{-1}; (b) kPa^{-1} s^{-1}, kPa^{-2} s^{-1}.

25.10 2720 y.

25.11 (a) 45, 95 mmol L^{-1}; (b) 1, 51 mmol L^{-1}.

25.12 124 ks.

25.13 64.9 kJ mol^{-1}, 4.32×10^8 mol L^{-1} s^{-1}.

25.14 $v = k_2 K^{1/2}[A_2]^{1/2}[B]$.

25.16 1.52 mmol L^{-1} s^{-1}.

25.17 1.9 MPa^{-1} s^{-1}.

25.18 7.1×10^5 s^{-1}, 7.63 ns.

26.1 $v = k_1 k_2[O_3]^2/(k_1'[O_2] + k_2[O_3])$.

26.3 0.16 to 4.0 kPa.

26.4 3.3×10^{18}.

26.5 0.518.

26.6 $d[P]/dt = (k_1/k_3[AH]^2[B])/(k_2[BH]^+ + k_3[AH])$.

26.7 $d[AH]/dt = -k_{eff}[AH]$; k_{eff} is a complex combination of rate constants.

27.1 9.6×10^9 s^{-1}, 1.2×10^{35} m^{-3} s^{-1}.

27.2 (a) 0.018, 0.30; (b) 3.9×10^{-18}, 6.0×10^{-6}.

27.3 (a) 13 per cent, 1.2 per cent; (b) 130 per cent, 12 per cent.

27.4 1.7×10^{-2} L mol^{-1} s^{-1}.

27.5 3×10^{10} L mol^{-1} s^{-1}.

27.6 (a) 6.61×10^6 m^3 mol^{-1} s^{-1}; (b) 3.0×10^7 m^3 mol^{-1} s^{-1}.

27.7 7.4×10^9 L mol^{-1} s^{-1}; 0.14 μs.

27.8 1.2×10^{-3}.

27.9 1.9×10^8 mol L^{-1} s^{-1}.

27.10 69.7 kJ mol^{-1}, -25 J K^{-1} mol^{-1}.

27.11 $+71.9$ kJ mol^{-1}.

27.12 -96.6 J K^{-1} mol^{-1}.

27.13 -76 J K^{-1} mol^{-1}.

27.14 (a) -45.8 J K^{-1} mol^{-1}; (b) $+5.0$ kJ mol^{-1}; (c) $+18.7$ kJ mol^{-1}.

27.15 $k_D/k_H \approx 0.15$.

27.16 20.9 L^2 mol^{-2} min^{-1}.

28.1 (a) 1.07×10^{25} m^{-2} s^{-1}, 1.4×10^{18} m^{-2} s^{-1}; (b) 2.35×10^{24} m^{-2} s^{-1}, 3.1×10^{17} m^{-2} s^{-1}.

28.2 1.3×10^4 Pa.

28.3 3.4×10^5 s^{-1}.

28.4 12.7 m^2.

28.5 20.5 cm^3.

28.6 Chemisorption, 50 s.

28.7 $E_d = 610 \text{ kJ mol}^{-1}$,
$\tau_0 = 1.13 \times 10^{-13} \text{ s}, A = 6.15 \times 10^{12} \text{ s}^{-1}$.

28.8 (a) 0.21 kPa; (b) 22 kPa.

28.9 0.83, 0.36.

28.10 (a) 40 ps, 0.6 ps; (b) 20 Ts, 7 μs.

28.11 15 kPa.

28.12 0 on gold, 1 on platinum.

28.13 -13 kJ mol^{-1}.

28.14 700 kJ mol^{-1}; (a) 1.2×10^{97} min; (b) 2.8×10^{-6} min.

29.1 0.24 GV m^{-1}.

29.2 138 mV.

29.3 2.8 mA cm^{-2}.

29.4 Increase $\times 50$.

29.5 (a) 0.17 mA cm^{-2}; (b) 0.17 mA cm^{-2}.

29.6 0.99 A m^{-2}.

29.7 $0.2 \ \mu\text{mol L}^{-1}$.

29.8 (a) 0.31 mA cm^{-2}; (b) 5.41 mA cm^{-2}; (c) -2.19 mA cm^{-2}.

29.9 $a(\text{Fe}^{3+})/a(\text{Fe}^{2+}) = 0.1: 684 \text{ mA cm}^{-2}$;
$a(\text{Fe}^{3+})/a(\text{Fe}^{2+}) = 1: 215 \text{ mA cm}^{-2}$;
$a(\text{Fe}^{3+})/a(\text{Fe}^{2+}) = 10: 68 \text{ mA cm}^{-2}$.

29.10 108 mV.

29.11 (a) $4.9 \times 10^{15} \text{ cm}^{-2} \text{ s}^{-1}$, 3.8 s^{-1};
(b) $1.6 \times 10^{16} \text{ cm}^{-2} \text{ s}^{-1}$, 12 s^{-1};
(c) $3.1 \times 10^{7} \text{ cm}^{-2} \text{ s}^{-1}$, $2.4 \times 10^{-8} \text{ s}^{-1}$.

29.12 (a) 33 Ω; (b) 33 GΩ.

29.15 Yes.

29.16 No.

29.17 +1.30 V, 0.13 W.

29.18 (a) +1.23 V; (b) +1.06 V.

29.19 Fe, Al, Co, Cr if O_2 absent; all if O_2 present.

29.20 1.2 mm year^{-1}.

Answers to 'b' exercises

1.1 146 kPa.

1.2 (a) 10.5 bar; (b) 10.4 bar.

1.3 (a) 8.04×10^2 Torr; (b) 1.07 bar.

1.4 92.4 K.

1.5 119 kPa.

1.6 2.67×10^3 kg.

1.7 $8.206\,15 \times 10^{-2} \text{ L atm mol}^{-1} \text{ K}^{-1}$, $31.9987 \text{ g mol}^{-1}$.

1.8 P_4.

1.9 2.6 kg.

1.10 (a) 3.14 L; (b) 212 Torr.

1.11 16.4 g mol^{-1}.

1.12 $-270°C$.

1.13 (a) 7.079, (b) 1.

1.14 (a) 475 m s^{-1}; (b) 40 km; (c) 0.01 s^{-1}.

1.15 2.4×10^7 Pa.

1.16 4.1×10^{-7} m.

1.17 $9.9 \times 10^8 \text{ s}^{-1}$.

1.18 (a) 3.7×10^{-9} m; (b) 5.5×10^{-8} m; (c) 4.1×10^{-5} m.

1.19 9.6×10^{-2}.

1.20 (a) 1.0 atm, 270 atm; (b) 0.99 atm, 180 atm.

1.21 0.131 L mol^{-1}, 25.7 atm, 109 K.

1.22 (a) 1.12, repulsive; (b) 2.7 L mol^{-1}.

1.23 (a) 0.124 L mol^{-1}; (b) 0.108 L mol^{-1}.

1.24 (a) $31.728 \text{ L mol}^{-1}$, 0.996; (b) 0.996.

1.25 (a) 8.7 mL; (b) -0.15 L mol^{-1}.

1.26 (a) 0.63, 0.37; (b) $p(\text{N}_2) = 2.5$ atm, $p(\text{H}_2) = 1.5$ atm;
(c) 4.0 atm.

1.27 $a = 3.16 \text{ L}^2 \text{ atm mol}^{-2}$, $b = 0.493 \text{ L mol}^{-1}$,
$r = 1.94 \times 10^{-10}$ m.

1.28 (a) 1276 K; (b) 1.286×10^{-10} m.

1.29 (a) 2.6 atm, 881 K; (b) 2.2 atm, 718 K; (c) 1.4 atm, 356 K.

1.30 0.13 L mol^{-1}, 0.67.

2.1 (a) 4.9×10^3 J; (b) 1.9×10^3 J.

2.2 59 J.

2.3 -91 J.

2.4 $\Delta U = 0$, $\Delta H = 0$; (a) $w = -1.62$ kJ, $q = +1.62$ kJ;
(b) $w = -1.38$ kJ, $q = +1.38$ kJ; (c) $w = q = 0$.

2.5 $p_2 = 143$ kPa, $w = 0$, $q = \Delta U = +3.28$ kJ.

2.6 (a) -19 J; (b) -52.8 J.

2.7 6.01 J.

2.8 $q = \Delta H = -70.6$ kJ, $w = +5.60 \times 10^3$ J, $\Delta U = -65.0$ kJ.

2.9 -188 J.

2.10 3.07×10^4 kJ.

2.11 (a) $q = \Delta H = +14.9$ kJ, $w = -831$ J, $\Delta U = +14.1$ kJ;
(b) $q = \Delta U = +14.1$ kJ, $w = 0$, $\Delta H = +14.9$ kJ.

2.12 200 K.

2.13 -325 J.

2.14 8.5 Torr.

2.15 $p_i = 1.9$ atm, $p_f = 0.46$ atm.

2.16 -199 kJ mol^{-1}.

2.17 $C_{p,m} = 53 \text{ J K}^{-1} \text{ mol}^{-1}$, $C_{V,m} = 45 \text{ J K}^{-1} \text{ mol}^{-1}$.

2.18 $q_p = \Delta H = -2.3$ kJ, $C = 0.18 \text{ kJ K}^{-1}$.

2.19 $\Delta H = q = +2.0 \text{ kJ mol}^{-1}$, $\Delta U = +1.6 \text{ kJ mol}^{-1}$.

2.20 $q = 0$, $w = \Delta U = -3.5$ kJ, $\Delta T = -24$ K, $\Delta H = -4.5$ kJ.

2.21 $q = 0$, $w = \Delta U = +2.4$ kJ, $\Delta H = +3.1$ kJ, $V_f = 14$ L,
$p_f = 3.8 \times 10^5$ Pa.

2.22 20 L, 275 K, -0.75 kJ.

2.23 $1.8 \times 10^{-3} \text{ cm}^3$.

2.24 $q = 0$, $w = \Delta U = -36$ J, $\Delta T = -0.57$ K, $\Delta H = -50$ J.

2.25 (a) 164 K; (b) 171 K.

2.26 $q = \Delta H = +24$ kJ, $w = -1.6$ kJ, $\Delta H = +22.4$ kJ.

2.27 -3053.6 kJ mol^{-1}.

2.28 -126 kJ mol^{-1}.

2.29 -1152 kJ mol^{-1}.

2.30 -324.83 kJ mol^{-1}.

2.31 451 J K^{-1}.

2.32 69.3 J K^{-1}, 63.1 K.

2.33 (a) -2.81×10^3 kJ mol^{-1}; (b) -2.81×10^3 kJ mol^{-1}; (c) -1.27×10^3 kJ mol^{-1}.

2.34 84.40 kJ mol^{-1}.

2.35 1.90 kJ mol^{-1}.

2.36 39 kJ, 16 m.

2.37 (a) -2857 kJ mol^{-1}; (b) -2851 kJ mol^{-1}.

2.38 (a) $\nu(C(s, diamond)) = -1$, $\nu(C(s, graphite)) = +1$, exothermic; (b) $\nu(Fe_3O_4) = -1$, $\nu(CO) = -1$, $\nu(FeO) = +3$, $\nu(CO_2) = +1$, endothermic; (c) $\nu(FeO) = -3$, $\nu(CO_2) = -1$, $\nu(Fe_3O_4) = +1$, $\nu(CO) = +1$, exothermic.

2.39 (a) -32.88 kJ mol^{-1}; (b) -55.84 kJ mol^{-1}.

2.40 (a) -589.56 kJ mol^{-1}, -582.13 kJ mol^{-1}; (b) -26.48 kJ mol^{-1}, -241.82 kJ mol^{-1}.

2.41 -760.3 kJ mol^{-1}.

2.42 $+52.5$ kJ mol^{-1}.

2.43 -566.93 kJ mol^{-1}.

2.44 -1745 kJ mol^{-1}, -173 kJ mol^{-1} -176 kJ mol^{-1}.

2.45 -1587 kJ mol^{-1}.

2.46 (a) -229.6 kJ mol^{-1}; (b) -160.5 kJ mol^{-1}.

3.2 $dz = dx/[(1 + y)^2 - 2x dy/(1 + y)^3]$.

3.3 (a) $(3x^2 - 2y^2)dx - 4xy dy$.

3.4 $dz = (2xy + y^2)dx + (x^2 + 2xy)dy$.

3.5 $(\partial C_p/\partial p)_T = [\partial(\partial H/\partial p)_T/\partial T]_p$.

3.7 $dp = (\partial p/\partial V)_T dV + (\partial V/\partial T)_V dT$, $d \ln p = (1/p\kappa_T)(\alpha dT - dV/V)$.

3.8 $0, 0$.

3.9 $\kappa_T = 1/p$, $\alpha = 1/T$.

3.10 0.48 K atm^{-1}.

3.11 $\Delta U_m = +130$ J mol^{-1}, $q = +7.75$ kJ mol^{-1}, $w = -7.62$ kJ mol^{-1}.

3.12 1.27×10^{-3} K^{-1}.

3.13 3.6×10^2 atm.

3.14 -41.2 J atm^{-1} mol^{-1}, 27.2 kJ.

3.15 -340 kPa.

4.1 (a) 1.8×10^2 J K^{-1}; (b) 1.5×10^2 J K^{-1}.

4.2 152.65 J K^{-1} mol^{-1}.

4.3 9.08 J K^{-1}.

4.4 -7.3 J K^{-1}.

4.5 $q = 0$, $\Delta S = 0$, $\Delta U = w = +2.75$ kJ, $\Delta H = +3.58$ kJ.

4.6 76.9 J K^{-1}.

4.7 Not reversible.

4.8 (a) -58.2 kJ; (b) -193 J K^{-1}.

4.9 17 J K^{-1}.

4.10 6.00 L.

4.11 0.2 J K^{-1}.

4.12 $\Delta H_{total} = 0$, $\Delta S_{total} = +24$ J K^{-1}.

4.13 (a) 0; (b) -230 J; (c) -230 J; (d) -5.3 K; (e) $+3.21$ J K^{-1}.

4.14 (a) $+104.6$ J K^{-1}; (b) -104.6 J K^{-1}.

4.15 (a) -21.0 J K^{-1} mol^{-1}; (b) $+512$ J K^{-1} mol^{-1}.

4.16 (a) -212.40 kJ mol^{-1}; (b) -5798 kJ mol^{-1}.

4.17 (a) -212.55 kJ mol^{-1}; (b) -5798 kJ mol^{-1}.

4.18 -86.2 kJ mol^{-1}.

4.19 -197 kJ mol^{-1}.

4.20 (a) $+3.0$ J K^{-1}, -3.0 J K^{-1}, 0; (b) $+3.0$ J K^{-1}, 0, $+3.0$ J K^{-1}; (c) 0, 0, 0.

4.21 $\Delta S = \frac{3}{2}nR \ln 3$.

4.22 2108.11 kJ mol^{-1}.

4.23 (a) 0.500; (b) 0.50 kJ; (c) 0.5 kJ.

5.1 Both (Maxwell relation) $= -\alpha V$.

5.2 -2.0 J.

5.3 -42.8 J K^{-1}.

5.4 3.2 kJ.

5.5 (a) 274 kPa; (b) 3.45 kJ.

5.6 2.71 kJ mol^{-1}.

5.7 -0.93 kJ mol^{-1}.

5.8 -1.92×10^{-7} Pa^{-1}, 10^{-6}.

5.9 200 J.

5.10 $+2.88$ kJ mol^{-1}.

5.11 $V = (RT/p)(1 + B'p/RT + C'p^2/RT + D'p^3/RT)$.

6.1 $23°C$.

6.2 2.4 kJ mol^{-1}, 5.5 J K^{-1} mol^{-1}.

6.3 25.25 kJ mol^{-1}.

6.4 (a) 31.11 kJ mol^{-1}; (b) 276.9 K.

6.5 272 K.

6.6 3.6 kg s^{-1}.

6.7 Yes, $p \geq 3$ Torr.

6.8 (a) gas expands; (b) gas contracts; (c) gas freezes; (d) solid sublimes; (e) gas expands.

6.9 (a) 29 kJ mol^{-1}; (b) 0.22 atm, 0.76 atm.

6.10 272.40 K.

6.11 6.73×10^{-2}.

6.12 5.92 kPa.

6.13 7.12×10^{-2} N m^{-1}.

6.14 2.04×10^5 Pa.

7.1 843.5 cm^3.

7.2 18 cm^3.

7.3 $8.2 \times 10^3 \text{ kPa}$.

7.4 $1.5 \times 10^2 \text{ kPa}$.

7.5 $7.1 \text{ K kg mol}^{-1}$, $4.99 \text{ K kg mol}^{-1}$.

7.6 270 g mol^{-1}.

7.7 178 g mol^{-1}.

7.8 $-0.077°C$.

7.9 $6.34 \times 10^{-2} \text{ J K}^{-1}$, -17.3 J.

7.10 -3.43 kJ, $+11.5 \text{ J K}^{-1}$, 0.

7.11 (a) $n_B/n_E = 1$; (b) $m_B/m_E = 0.7358$.

7.12 N_2: $0.51 \text{ mmol kg}^{-1}$; O_2: $0.27 \text{ mmol kg}^{-1}$.

7.13 0.067 mol L^{-1}.

7.14 $-0.52°C$.

7.15 $11 \text{ kg Pb}/1 \text{ kg Bi}$.

7.16 14.0 kg mol^{-1}.

7.17 $a = 0.9701$, $\gamma = 0.980$.

7.18 $-3.54 \text{ kJ mol}^{-1}$, 212 Torr.

7.19 $a_A = 0.436$, $a_B = 0.755$, $\gamma_A = 1.98$, $\gamma_B = 0.968$.

8.1 $x_A = 0.5$, $y_A = 0.5$.

8.2 73.4 kPa, $x_A = 0.653$.

8.3   (a) yes; (b) $y_A = 0.458$, $y_B = 0.542$.

8.4   (a) 48 Torr; (b) $y_B = 0.77$, $y_T = 0.23$; (c) 34 Torr.

8.5   (a) $y_A = 0.81$; (b) $x_A = 0.67$, $y_A = 0.925$.

8.6 3.

8.7   (a) $C = 1$, $P = 2$; (b) $C = 2$, $P = 2$.

8.8   (a) $C = 2$, $P = 2$; (b) $F = 2$.

8.13 $x_B \approx 0.53$ at T_2 and $x_B = 0.82$ at T_3.

8.15 (a) $x_B \approx 0.75$; (b) $x_{AB_2} \approx 0.8$; (c) $x_{AB_2} \approx 0.6$.

8.16 A solid solution with $x(ZrF_4) = 0.24$ appears at $855°C$. The solid solution continues to form, and its ZrF_4 content increases until it reaches $x(ZrF_4) = 0.40$ at $820°C$. At that temperature, the entire sample is solid.

9.1 $+18.18 \text{ kJ mol}^{-1}$.

9.2 0.98.

9.3   (a) 0; (b) 0.168; (c) 4.41 kJ mol^{-1}.

9.4   (a) 0.24; (b) $+19 \text{ kJ mol}^{-1}$; (c) 2.96.

9.5   (a) $-308.84 \text{ kJ mol}^{-1}$, 1.3×10^{54}; (b) $-306.52 \text{ kJ mol}^{-1}$, 3.5×10^{49}.

9.6   (a) 0.178 (A), 0.31 (B), 0.116 (C), 0.674 (D); (b) 9.6; (c) 9.6; (d) -5.6 kJ mol^{-1}.

9.7 $1.4 \times 10^3 \text{ K}$.

9.8 $+7.2 \text{ kJ mol}^{-1}$, $-21 \text{ J K}^{-1} \text{ mol}^{-1}$.

9.9 $-41.0 \text{ kJ mol}^{-1}$.

9.10 No change.

9.11 1.6×10^{-2}.

9.12 (b).

9.13 (b).

9.14 (a) $+39 \text{ kJ mol}^{-1}$; (b) -39 kJ mol^{-1}.

9.15 (a) 9.24; (b) $-12.9 \text{ kJ mol}^{-1}$; (c) $+161 \text{ kJ mol}^{-1}$; (d) $+248 \text{ J K}^{-1} \text{ mol}^{-1}$.

9.16 397 K.

9.17 (a) 1.5×10^{-5}, 4.82; (b) 3.21.

9.18 (a) 8.37; (b) 8.74; (c) 5.07.

9.19 7.38.

9.21 (a) aniline and anilinium ion; (b) ethylammonium ion and ethylamine.

10.1 $-65.49 \text{ kJ mol}^{-1}$.

10.2 $9.3 \times 10^{-15} \text{ mol kg}^{-1}$.

10.3 -363 kJ mol^{-1}.

10.4 (a) $3b/b^{\ominus}$; (b) $15b/b^{\ominus}$; (c) $15b/b^{\ominus}$.

10.5 0.320.

10.6 (a) 45 g; (b) 38.8 g.

10.7 $0.100 \text{ mol kg}^{-1}$.

10.8 $\gamma_{\pm} = (\gamma_+^2 \gamma_-^3)^{1/5}$.

10.9 0.661.

10.10 47.1 per cent.

10.11 1.3.

10.12 $-128.8 \text{ kJ mol}^{-1}$.

10.13 $+56.3 \text{ mV}$.

10.14 $Cd(s)|CD(OH)_2(s)|OH^-(aq)|Ni(OH)_2(s)|Ni(OH)_3(s)|Pt$, L: $Cd(s) + 2OH^-(aq) \longrightarrow Cd(OH)_2(s) + 2e^-$, R: $Ni(OH)_3(s) + e^- \longrightarrow Ni(OH)_2(s) + OH^-$.

10.15 (a) $Ag_2CrO_4(s) + 2e^- \longrightarrow 2Ag(s) + CrO_4^{2-}(aq)$, $Cl_2(g) + 2e^- \longrightarrow 2Cl^-(aq)$, $Ag_2CrO_4(s) + 2Cl^-(aq) \longrightarrow 2Ag(s) + CrO_4^{2-}(aq) + Cl_2(g)$, -0.91 V; (b) $Sn^{4+}(aq) + 2e^- \longrightarrow Sn^{2+}(aq)$, $Fe^{3+}(aq) + e^- \longrightarrow Fe^{2+}(aq)$, $Sn^{4+}(aq) + 2Fe^{2+}(aq) \longrightarrow Sn^{2+}(aq) + 2Fe^{3+}(aq)$, -0.62 V; (c) $MnO_2(s) + 4H^+(aq) + 2e^- \longrightarrow Mn^{2+}(aq) + 2H_2O(l)$, $Cu^{2+}(aq) + 2e^- \longrightarrow Cu(s)$, $Cu(s) + MnO_2(s) + 4H^+(aq) \longrightarrow Cu^{2+}(aq) + Mn^{2+}(aq) + 2H_2O(l)$, $+0.89 \text{ V}$.

10.16 (a) $Na(s)|NaOH(aq)|H_2(g)|Pt$, $+1.88 \text{ V}$; (b) $Pt|H_2(g)|HI(aq)|I_2(s)|Pt$, $+0.54 \text{ V}$; (c) $Pt|H_2(g)|H_2O(l)|H_2(g)|Pt = 0.83 \text{ V}$.

10.17 (a) -0.91 V; (b) -0.62 V; (c) $+0.89 \text{ V}$.

10.18 (a) $+1.88 \text{ V}$; (b) $+0.54 \text{ V}$; (c) $+0.83 \text{ V}$.

10.19 (a) -291 kJ mol^{-1}; (b) $+122 \text{ kJ mol}^{-1}$.

10.20 (a) -2.455 V; (b) $+1.627 \text{ V}$.

10.21 (a) $E = E^{\ominus} - (2RT/F) \ln(\gamma_{\pm} b)$; (b) $-89.89 \text{ kJ mol}^{-1}$; (c) $+0.223 \text{ V}$.

10.22 -1.24 V, $+239 \text{ kJ mol}^{-1}$, $+300.3 \text{ kJ mol}^{-1}$, $+237 \text{ kJ mol}^{-1}$ at $35°C$.

10.23 1.7×10^{16}; (b) 8.2×10^{-7}.

10.24 $+0.20 \text{ V}$.

10.25 AgI: 1.4×10^{-16}; Bi_2S_3: 3.2×10^{-98}; no significant difference.

10.26 $E = E^{\ominus} - (RT/5F) \ln\{a(Mn^{2+})/[a(MnO_4^-)a(H^+)^8]\}$.

10.27 9.72.

10.28 0.

10.29 (a) 10^{-20} mol L^{-1}; (b) 10^{-98}.

11.1 2.5×10^2 W.

11.2 1.36×10^{-8} W.

11.3 $1.15 \ \mu m$.

11.4 1.3×10^{-5} m s^{-1}.

11.5 1.6×10^6 m s^{-1}.

11.6 1.89×10^{-27} kg m s^{-1}, 0.565 m s^{-1}.

11.7 38.4 nm.

11.8 5.8×10^{-6} m.

11.9 (a) 9.93×10^{-19} J, 5.98×10^5 J mol^{-1}; (b) 1.32×10^{-15} J, 7.98×10^8 J mol^{-1}; (c) 1.99×10^{-23} J, 12.0 J mol^{-1}.

11.10 (a) 0.499 m s^{-1}; (b) 665 m s^{-1}; (c) 9.9×10^{-6} m s^{-1}.

11.11 158 m s^{-1}.

11.12 (a) 3.52×10^{17} s^{-1}; (b) 3.52×10^{18} s^{-1}.

11.13 1800 K.

11.14 (a) 0; (b) 6.84×10^{-19} J, 1.23×10^6 m s^{-1}.

11.15 (a) 2.65×10^{-19} J, 160 kJ mol^{-1}; (b) 3.00×10^{-19} J, 181 kJ mol^{-1}; (c) 6.62×10^{-31} J, 4.0×10^{-10} kJ mol^{-1}.

11.16 (a) 1.23×10^{-10} m; (b) 3.9×10^{-11} m; (c) 3.88×10^{-12} m.

11.17 $\Delta x = 100$ pm, $\Delta v = 5.8 \times 10^5$ m s^{-1}.

11.18 1.67×10^{-16} J.

12.1 (a) 2.14×10^{-19} J, 1.29×10^2 kJ mol^{-1}, 1.34 eV, 1.08×10^4 cm^{-1}; (b) 3.48×10^{-19} J, 2.10×10^2 kJ mol^{-1}, 2.10 eV, 1.75×10^4 cm^{-1}.

12.2 (a) 0.03; (b) 0.03.

12.3 $\langle p \rangle = 0, \langle p^2 \rangle = h^2/L^2$.

12.4 $L/10, 3L/10, L/2, 7L/10, 9L/10$.

12.5 6.

12.6 $n = 7.26 \times 10^{10}$, $\Delta E = 1.76 \times 10^{-31}$ J, $\lambda = 27.5$ pm; may be treated classically.

12.7 3.92×10^{-21} J.

12.8 260 N m^{-1}.

12.9 13.2 μm.

12.10 18.7 μm.

12.11 (a) 2.2×10^{-29} J; (b) 3.14×10^{-20} J.

12.13 2.3421×10^{-20} J.

12.15 Magnitude: 2.58×10^{-34} J s^{-1}; projections: 0, $\pm 1.0546 \times 10^{-34}$ J s, and 2.11094×10^{-34} J s.

13.1 1.94×10^{-18} J.

13.2 $r = 11.5a_0/Z, 3.53a_0/Z, 0$.

13.3 $r = 0, 6a_0$.

13.4 $N = \frac{1}{4}(2\pi a_0^3)^{-1/2}$.

13.5 $\langle E_K \rangle = \hbar^2 Z^2/8ma_0^2, \langle V \rangle = -Z^2 e^2/16\pi\varepsilon_0 a_0$.

13.6 $P_{3s} = 4\pi r^2 (1/4\pi)(1/243)(Z/a_0)^3(6 - 6\rho + \rho^2)^2 e^{-\rho}$, $r = 0.74a_0/Z, 4.19a_0/Z$, and $13.08 \ a_0/Z$.

13.7 (a) 2.45×10^{-34} J s, 2, 1; (b) 1.49×10^{-34} J s, 1, 0; (c) 1.49×10^{-34} J s, 1, 1.

13.8 (a) $\frac{1}{2}, \frac{3}{2}$; (b) $\frac{9}{2}, \frac{11}{2}$.

13.9 8, 7, 6, 5, 4, 3, 2.

13.10 (a) 1; (b) 64; (c) 25.

13.11 $S = 1, L = 3, J = 4, (2S + 1) = 3$ is the multiplicity.

13.12 (a) 110 pm, 20.1 pm; (b) 86 pm, 29.4 pm.

13.13 (b).

13.14 (a) 2; (b) 10; (c) 14; (d) 22.

13.15 $1s^2 2s^2 2p^6 3s^2 3p^6 3d^3, S: \frac{1}{2}$ or $\frac{3}{2}, M_S: \pm\frac{1}{2}, \pm\frac{3}{2}$.

13.16 (a) 2 (5), 1 (3), 0 (1); (b) $\frac{5}{2}$ (6), $\frac{3}{2}$ (4), $\frac{1}{2}$ (2).

13.17 $^1F_3; ^3F_4, ^3F_3, ^3F_2; ^1D_2; ^3D_3, ^3D_2, ^3D_1; ^1P_1; ^3P_2, ^3P_1, ^3P_0; ^3F_2$ is lowest.

13.18 (a) $J = 3, 2, 1$ (7, 5, 3 states); (b) $J = \frac{7}{2}, \frac{5}{2}, \frac{3}{2}, \frac{1}{2}$ (8, 6, 4, 2 states); (c) $J = \frac{9}{2}, \frac{7}{2}$ (10, 8).

13.19 (a) $^2D_{5/2}, ^2D_{3/2}$; (b) $^2P_{3/2}, ^2P_{1/2}$.

13.20 1.68 T.

14.1 (a) $1\sigma^2 2\sigma^{*1}$; (b) $1\sigma^2 2\sigma^{*2} 1\pi^4 3\sigma^2$; (c) $1\sigma^2 2\sigma^{*2} 3\sigma^2 1\pi^4 2\pi^{*2}$.

14.2 (a) $1\sigma^2 2\sigma^{*2} 3\sigma^2 1\pi^4 2\pi^{*4}$; (b) $1\sigma^2 2\sigma^{*2} 1\pi^4 3\sigma^2$; (c) $1\sigma^2 2\sigma^{*2} 3\sigma^2 1\pi^4 2\pi^{*3}$.

14.3 (a) C_2, CN; (b) NO, O_2, F_2.

14.4 $1\sigma^2 1\pi^4 2\pi^{*2}$, BrCl is shorter.

14.5 $a_{2u}, e_{1g}, e_{2u}, b_{2g}$; hence parities are u, g, u, g.

14.6 3, u.

14.7 $O_2^+, O_2, O_2^-, O_2^{2-}$.

14.8 $N = 1/(1 + 2\lambda S + \lambda^2)^{1/2}$.

14.9 $N(0.844A - 0.145B)$.

14.10 none.

14.12 (a) $a_{2u}^2 e_{1e}^4 e_{2u}^1, 7\alpha + 7\beta$; (b) $a_{2u}^2 e_{1g}^3, 5\alpha + 7\beta$.

15.1 4 C_3 axes (each C–Cl axis), 4 C_2 axes (bisecting Cl–C–Cl angles), 3 S_4 axes (same as C_2 axes), and 6 dihedral mirror planes (each Cl–C–Cl plane).

15.2 (a).

15.3 necessarily 0.

15.4 forbidden.

15.6 T_d: (S_4 and S_1); T_h: (S_2).

15.8 (a) $C_{\infty v}$; (b) D_3; (c) C_{4v}; (d) C_s.

15.9 (a) 3 C_2 axes, 3 mirror planes, inversion centre; D_{2h}; (b) C_2 axis, $2\sigma_v$; D_{2h}; (c) *ortho* and *meta*: C_2 axis, $2\sigma_v$; C_{2v}. *para*: 3 C_2 axes, 3 mirror planes, inversion centre; D_{2h}.

15.10 (a) $C_{\infty v}$; (b) D_{5h}; (c) C_{2v}; (d) D_{3h}; (e) O_h; (f) T_d.

15.11 (a) *o*-dichlorobenzene, *m*-dichlorobenzene HF, XeO_2F_2; (b) none.

15.12 NO_3^-: p_x and p_y; SO_3: all *d*-orbitals except d_{z^2}.

15.13 A_2.

15.14 (a) B_{3u}, B_{2u}, B_{1u}; (b) A_{2u}, E_{1u}.

15.15 yes.

16.1 (a) 7.73×10^{-32} J m^{-3} s; (b) 6.2×10^{-28} J m^{-3} s.

16.2 3.4754×10^{-11} s^{-1}.

16.3 (a) 3.307×10^{-47} kg m^2; (b) 141.4 pm.

16.4 5.420×10^{-46} kg m^2, 162.8 pm.

16.5 116.21 pm.

16.6 116.1 pm, 155.9 pm.

16.7 20 603 cm^{-1}.

16.8 2347.16 cm^{-1}.

16.9 0.71 N m^{-1}.

16.10 28.4 per cent.

16.11 245.9 N m^{-1}.

16.12 $A_{1g} + A_{2g} + E_{1u}$.

16.13 all.

16.14 all but (d) N_2.

16.15 all but (c) SF_6.

16.16 6.36×10^7 m s^{-1}.

16.17 3.59×10^7 m s^{-1}, 1.19×10^6 K.

16.18 (a) 1.59 ns; (b) 2.48 ps.

16.19 (a) 1.6×10^2 MHz; (b) 16 MHz.

16.20 (a) 0.212; (b) 0.561.

16.21 3002.3 cm^{-1} (DF), 2143.7 cm^{-1} (DCl), 1885.8 cm^{-1} (DBr), 1640.1 cm^{-1} (DI).

16.22 2374.05 cm^{-1}, 6.087×10^{-3}.

16.23 3.235×10^4 cm^{-1}, 4.01 eV.

16.24 141.78 pm.

16.25 (a) 30; (b) 42; (c) 13.

16.26 (a) IR: A_2'', E', Raman: A_1', E'; (b) IR: A_1, E, Raman: A_1, E.

16.27 (a) inactive; (b) active.

17.1 22.2 per cent.

17.2 7.9×10^5 cm^2 mol^{-1}.

17.3 1.33×10^{-3} mol L^{-1}.

17.4 1.56×10^8 L mol^{-1} cm^{-2} (1.56×10^9 m mol^{-1}).

17.5 rise.

17.6 552 L mol^{-1} cm^{-1}.

17.7 128 L mol^{-1} cm^{-1}, 0.13.

17.8 (a) 0.020 cm; (b) 0.033 cm.

17.9 1.39×10^8 L mol^{-1} cm^{-2} (1.39×10^9 m mol^{-1}).

17.10 stronger.

18.1 649 MHz.

18.2 $E_{\pm 1} = \mp 2.35 \times 10^{-26}$ J, 0.

18.3 47.3 MHz.

18.4 (b)

18.5 3.523 T.

18.6 (a) 97.5 T, 244 T; (b) 7.49 T, 18.7 T; (c) 17.4 T, 43.5 T.

18.7 (a) 4.3×10^{-7}; (b) 2.2×10^{-6}; (c) 1.34×10^{-5}.

18.8 (a) 1; (b) 10.

18.9 (a) 4.2×10^{-6} T; (b) 3.63×10^{-5} T.

18.10 spectrum appears narrower at 650 MHz.

18.11 2.9×10^3 s^{-1}.

18.12 203 MHz.

18.14 neither.

18.15 9.40×10^{-4} T, 6.25 μs.

18.16 1.3 T.

18.17 2.0022.

18.18 2.2 mT, 1.992.

18.19 eight lines at $(332.8 \pm 1.055 \pm 1.435 \pm 1.445)$ mT, all of equal intensity (in a high resolution spectrometer).

18.20 a triplet (1:2:1) of quartets (1:3:3:1).

18.21 (a) 332.3 mT; (b) 1209 mT.

18.22 1.

19.1 623 K.

19.2 (a) 15.9 pm, 5.04 pm; (b) 2.47×10^{26}, 7.82×10^{27}.

19.3 187.9.

19.4 4.006.

19.5 7.605 kJ mol^{-1}.

19.6 213 K.

19.7 (a) 0.997, 0.994; (b) 0.99999, 0.99998.

19.8 (a) 1.00 K: $n_2/n_1 = 1.39 \times 10^{11}$, $n_3/n_1 = 1.93 \times 10^{-22}$; 25.0 K: $n_2/n_1 = 0.368$, $n_3/n_1 = 0.135$; 100 K: $n_2/n_1 = 0.779$, $n_3/n_1 = 0.607$; (b) 1.503; (c) 88.3 J mol^{-1}; (d) 3.53 J K^{-1} mol^{-1}; (e) 6.92 J K^{-1} mol^{-1}.

19.9 50.2 K.

19.10 (a) 147 J K^{-1} mol^{-1}; (b) 169.6 J K^{-1} mol^{-1}.

19.11 10.7 J K^{-1} mol^{-1}.

19.12 (a).

20.1 (a) 3*R*, 6*R*; (b) 3*R*, 21*R*; (c) 2.5*R*, 6.5*R*. (In each case the first value assumes no vibrational contribution; the second a full vibrational contribution.)

20.2 with: 1.15; without: 1.40; experimental: 1.29.

20.3 (a) 143; (b) 251.

20.4 (a) 2; (b) 2; (c) 6; (d) 24; (e) 4.

20.5 5840, 0.8479 K.

20.6 84.57 J K^{-1} mol^{-1}.

20.7 (a) 2.50×10^3, 5.43×10^3; (b) the same.

20.8 (a) 8.03×10^3; (b) 1.13×10^4.

20.9 (a) $5.70 \text{ J K}^{-1}\text{mol}^{-1}$; (b) $14.83 \text{ J K}^{-1} \text{ mol}^{-1}$.

20.10 $-20.1 \text{ kJ mol}^{-1}$, -110 J mol^{-1}.

20.11 $-3.65 \text{ kJ mol}^{-1}$.

20.12 $14.90 \text{ J K}^{-1} \text{ mol}^{-1}$.

20.14 $191.4 \text{ J K}^{-1} \text{ mol}^{-1}$, residual entropy is negligible.

20.15 ≈ 0.25.

21.1 $(\frac{1}{2}, \frac{1}{2}, 0)$, $(0, \frac{1}{2}, \frac{1}{2})$.

21.2 $(3, 1, \bar{3})$, $(6, 4, 3)$.

21.3 214 pm, 174 pm, 87.2 pm.

21.4 86.7 pm.

21.5 $38.2°$, $44.4°$, $64.6°$.

21.6 0.054 cm.

21.7 1.2582 nm^3.

21.8 $5, 2.90 \text{ g cm}^{-3}$.

21.9 182 pm.

21.10 (110), (110), (200).

21.11 $4.166°$, $3.000°$, $7.057°$.

21.12 body-centred cubic.

21.13 $2f$ for $h + k + l$ even; 0 for $h + k + l$ odd.

21.14 $\frac{2}{3}$.

21.16 (a) 57 pm; (b) 111 pm.

21.17 0.370.

21.18 $3.61 \times 10^5 \text{ g mol}^{-1}$.

21.19 contraction.

21.21 252 pm.

21.22 (a) 39 pm; (b) 12 pm; (c) 6.1 pm.

21.23 neutron: $0°$, $14.0°$; electron: $0°$, $0.72°$.

22.1 SF_4.

22.2 $\alpha = 2.55 \times 10^{-39} \text{ C}^2 \text{ m}^2 \text{ J}^{-1}$, $\mu = 3.23 \times 10^{-30} \text{ C m}$ (3.23 D).

22.3 5.57.

22.4 $3.40 \times 10^{-40} \text{ C}^2 \text{ m}^2 \text{ J}^{-1}$.

22.5 $\mu(\text{C-F}) > \mu(\text{C-O})$.

22.6 1.4 D.

22.7 $9.45 \times 10^{-29} \text{ C m}$, $194.0°$.

22.8 $3.71 \times 10^{-36} \text{ C m}$.

22.9 1.10.

22.10 16.

22.11 3.2×10^{-6}.

22.12 5.

22.13 $-8.2 \times 10^{-4} \text{ cm}^3 \text{ mol}^{-1}$.

22.14 $1.58 \times 10^{-8} \text{ m}^3 \text{ mol}^{-1}$, dimerization occurs.

22.15 2.52.

22.16 $1.85 \times 10^{-7} \text{ m}^3 \text{ mol}^{-1}$.

22.17 0.935.

23.1 $\overline{M}_n = 68 \text{ kg mol}^{-1}$, $\overline{M}_w = 69 \text{ kg mol}^{-1}$.

23.2 38.97 nm.

23.3 1.06×10^4.

23.4 $1.26 \times 10^{-6} \text{ m}$, $1.97 \times 10^{-8} \text{ m}$.

23.5 71.

23.6 $1.47 \times 10^{-4} \text{ m s}^{-1}$.

23.7 120 kg mol^{-1}.

23.8 56 kg mol^{-1}.

23.9 (a) 8.8 kg mol^{-1}; (b) 11 kg mol^{-1}.

23.10 $3.4 \times 10^{-3} \text{ mol L}^{-1}$.

23.11 $1.5 \times 10^{-2} \text{ mol L}^{-1}$.

23.12 $3.1 \times 10^3 \text{ kg mol}^{-1}$.

23.13 $3.9 \times 10^5 \text{ g}$.

24.1 1.1×10^{21}.

24.2 $4.89 \times 10^{-4} \text{ kg}$.

24.3 $0.17 \text{ J m}^{-2} \text{ s}^{-1}$.

24.4 $1.61 \times 10^{-19} \text{ m}^2$.

24.5 22 J s^{-1}

24.6 554 g mol^{-1}.

24.7 $1.5 \times 10^4 \text{ s}$.

24.8 $3.00 \times 10^{-19} \text{ m}^2$.

24.9 $1.00 \times 10^5 \text{ Pa}$.

24.10 (a) $0.95 \times 10^{-5} \text{ kg m}^{-1} \text{ s}^{-1}$; (b) $0.99 \times 10^{-5} \text{ kg m}^{-1} \text{ s}^{-1}$; (c) $1.81 \times 10^{-5} \text{ kg m}^{-1} \text{ s}^{-1}$.

24.11 (a) $0.0114 \text{ J m}^{-1} \text{ s}^{-1} \text{ K}^{-1}$, 0.017 J s^{-1}; (b) $9.0 \times 10^{-3} \text{ J m}^{-1} \text{ s}^{-1} \text{ K}^{-1}$, 0.014 J s^{-1}.

24.12 $52.0 \times 10^{-7} \text{ kg m}^{-1} \text{ s}^{-1}$, 923 pm.

24.13 $9.0 \times 10^{-3} \text{ J m}^{-1} \text{ s}^{-1} \text{ K}^{-1}$.

24.14 (a) $0.107 \text{ m}^2 \text{ s}^{-1}$, $0.87 \text{ mol m}^{-2} \text{ s}^{-1}$; (b) $1.07 \times 10^{-5} \text{ m}^2 \text{ s}^{-1}$, $8.7 \times 10^{-5} \text{ mol m}^{-2} \text{ s}^{-1}$; (c) $7.13 \times 10^{-8} \text{ m}^2 \text{ s}^{-1}$, $5.8 \times 10^{-7} \text{ mol m}^{-2} \text{ s}^{-1}$.

24.15 $4.09 \text{ mS m}^2 \text{ mol}^{-1}$.

24.16 $4.81 \times 10^{-5} \text{ m}^2 \text{ V}^{-1} \text{ s}^{-1}$.

24.17 0.604.

24.18 $25.96 \text{ mS m}^2 \text{ mol}^{-1}$.

24.19 $5.74 \times 10^{-8} \text{ m}^2 \text{ V}^{-1} \text{ s}^{-1}$, $7.913 \times 10^{-8} \text{ m}^2 \text{ V}^{-1} \text{ s}^{-1}$, $8.09 \times 10^{-8} \text{ m}^2 \text{ V}^{-1} \text{ s}^{-1}$.

24.20 $1.09 \times 10^{-9} \text{ m}^2 \text{ s}^{-1}$.

24.21 $4.1 \times 10^3 \text{ s}$.

24.22 207 pm.

24.23 20 ps.

24.24 $5.594 \times 10^{-5} \text{ m}$.

24.25 $1.7 \times 10^{-2} \text{ s}$.

25.1 A: $1.0 \text{ mol L}^{-1} \text{ s}^{-1}$, B: $3.0 \text{ mol L}^{-1} \text{ s}^{-1}$, C: $1.0 \text{ mol L}^{-1} \text{ s}^{-1}$, D: $2.0 \text{ mol L}^{-1} \text{ s}^{-1}$.

25.2 rate of: reaction, $0.33 \text{ mol L}^{-1} \text{ s}^{-1}$, formation of C, $0.33 \text{ mol L}^{-1} \text{ s}^{-1}$; formation of D, $0.66 \text{ mol L}^{-1} \text{ s}^{-1}$; consumption of A, $0.33 \text{ mol L}^{-1} \text{ s}^{-1}$.

25.3 k: $\text{L}^2 \text{ mol}^{-2} \text{ s}^{-1}$; (a) $d[A]/dt = -k[A][B]^2$; (b) $d[C]/dt = k[A][B]^2$.

25.4 k: s^{-1}, $k[A][B][C]^{-1}$.

25.5 2.

25.6 0.

25.7 1.80×10^6 s; (a) 31.5 kPa; (b) 29.0 kPa.

25.8 (a) $3.5 \times 10^{-3} \text{ L mol}^{-1} \text{ s}^{-1}$; (b) A: 2.4 h; B: 0.44 h.

25.9 $\text{m}^3 \text{ molecule}^{-1} \text{ s}^{-1}$, $\text{m}^6 \text{ molecule}^{-2} \text{ s}^{-1}$; $\text{Pa}^{-1} \text{ s}^{-1}$, $\text{Pa}^{-2} \text{ s}^{-1}$.

25.10 (a) $0.642 \ \mu\text{g}$; (b) $0.177 \ \mu\text{g}$.

25.11 (a) $6.5 \times 10^{-3} \text{ mol L}^{-1}$; (b) 0.025 mol L^{-1}.

25.12 1.5×10^6 s.

25.13 $E_a = 9.9 \text{ kJ mol}^{-1}$; $A = 0.94 \text{ L mol}^{-1} \text{ s}^{-1}$.

25.14 $v = k[A][B]$, $k = k_1 k_2 / k_2'$.

25.15 $\{(3^{n-1} - 1)/k(n-1)\}[A]_0^{(1-n)}$.

25.16 $2.57 \times 10^{-4} \text{ mol L}^{-1} \text{ s}^{-1}$.

25.17 $9.9 \times 10^{-6} \text{ s}^{-1} \text{ Pa}^{-1}$.

25.18 $k_r = 1.7 \times 10^{-7} \text{ s}^{-1}$, $k_f = 8.3 \times 10^8 \text{ L mol}^{-1} \text{ s}^{-1}$.

26.1 $k[N_2O_5] = k_1 k_2 [N_2O_5]/(k_1' + k_2)$.

26.2 $-k_1[R_2] - k_2(k_1/k_4)^{1/2}[R_2]^{3/2}$.

26.3 (a) does not occur; (b) 1.3×10^2 Pa to 3×10^4 Pa.

26.4 1.5×10^{-5} moles of photons.

26.5 1.11.

26.6 $(k_1 k_2 K_a^{1/2}/k_1')[HA]^{3/2}[B]$.

26.7 (1) initiation, (3) retardation, (4) termination; $k_1[A_2]$.

27.1 $6.64 \times 10^9 \text{ s}^{-1}$, $8.07 \times 10^{34} \text{ m}^3 \text{ s}^{-1}$, 1.6 per cent.

27.2 (a) 2.4×10^{-3}, 0.10; (b) 7.7×10^{-27}, 1.6×10^{-10}.

27.3 (a) 1.2, 1.03; (b) 7.4, 1.3.

27.4 $1.7 \times 10^{-12} \text{ L mol}^{-1} \text{ s}^{-1}$.

27.5 $3.2 \times 10^7 \text{ m}^3 \text{ mol}^{-1} \text{ s}^{-1}$.

27.6 (a) $1.97 \times 10^6 \text{ m}^3 \text{ mol}^{-1} \text{ s}^{-1}$; (b) $2.4 \times 10^5 \text{ m}^3 \text{ mol}^{-1} \text{ s}^{-1}$.

27.7 $1.10 \times 10^7 \text{ m}^3 \text{ mol}^{-1} \text{ s}^{-1}$, 5.05×10^{-8} s.

27.8 2.22×10^{-3}.

27.9 $1.54 \times 10^8 \text{ mol L}^{-1} \text{ s}^{-1}$.

27.10 $48.52 \text{ kJ mol}^{-1}$, $-32.2 \text{ J K}^{-1} \text{ mol}^{-1}$.

27.11 46.8 kJ mol^{-1}.

27.12 $-93 \text{ J K}^{-1} \text{ mol}^{-1}$.

27.13 $-80.0 \text{ J K}^{-1} \text{ mol}^{-1}$.

27.14 (a) $-24.1 \text{ J K}^{-1} \text{ mol}^{-1}$; (b) 27.5 kJ mol^{-1}; (c) 34.7 kJ mol^{-1}.

27.15 (a) $k_T/k_H \approx 0.06$; (b) $k_{18}/k_{16} \approx 0.89$.

27.16 $1.08 \text{ L}^2 \text{ mol}^{-2} \text{ min}^{-1}$.

28.1 (a) $2.88 \times 10^{23} \text{ m}^{-2} \text{ s}^{-1}$, $5.75 \times 10^{17} \text{ m}^{-2} \text{ s}^{-1}$; (b) $3.81 \times 10^{24} \text{ m}^{-2} \text{ s}^{-1}$, $7.60 \times 10^{17} \text{ m}^{-2} \text{ s}^{-1}$.

28.2 7.3×10^2 Pa.

28.3 $6.6 \times 10^4 \text{ s}^{-1}$.

28.4 18.8 m^2.

28.5 9.7 cm^3.

28.6 200 s.

28.7 3.7 kJ mol^{-1}.

28.8 (a) 0.32 kPa; (b) 3.9 kPa.

28.9 0.75, 0.25.

28.10 (a) 4.9×10^{-11} s, 2.4×10^{-12} s; (b) 1.6×10^{13} s, 1.4 s.

28.11 6.50 kPa.

28.13 $-6.40 \text{ kJ mol}^{-1}$.

28.14 $E_d = 2.85 \times 10^5 \text{ J mol}^{-1}$; (a) 1.48×10^{36} s; (b) 1.38×10^{-4} s.

29.1 $2.8 \times 10^8 \text{ V m}^{-1}$.

29.2 0.37 V.

29.3 1.6 mA cm^{-2}

29.4 8.5 mA cm^{-2}.

29.5 (a) 0.34 A cm^{-2}; (b) 0.34 A cm^{-2}.

29.6 1.3 A m^{-2}.

29.7 $4 \times 10^{-6} \text{ mol L}^{-1}$.

29.8 $(2.5 \text{ mA cm}^{-2})[e^{0.42E'/f}(3.41 \times 10^{-6}) - e^{-0.58E'/f}(3.55 \times 10^7)]$.

29.10 0.61 V.

29.11 $Cu, H_2|H^+$: $6.2 \times 10^{-12} \text{ s}^{-1} \text{ cm}^{-2}$, $4.2 \times 10^{-3} \text{ s}^{-1}$; $Pt|Ce^{4+}, Ce^{3+}$: $2.54 \times 10^{14} \text{ s}^{-1} \text{ cm}^{-2}$, 0.17 s^{-1}.

29.12 (a) $5.1 \text{ G}\Omega$; (b) $10 \ \Omega$.

29.15 no.

29.16 no.

29.17 1.80 V, 0.180 W.

29.18 0.97675 V.

29.19 all.

29.20 1.5 mm y^{-1}.

Answers to problems

Detailed solutions for selected problems (indicated with an asterisk) can be found in the *Student's Solutions Manual for Physical Chemistry*, sixth edition, by P.W. Atkins, C.A. Trapp, M. Cady, and C. Giunta.

1.1* 0.50 m^3.

1.2 1.5 kPa.

1.3* $-233°\text{N}$.

1.4 3.2×10^{-2} atm.

1.5* $p = \rho RT/M$, 46.0 g mol^{-1}.

1.6 $-272.95°\text{C}$.

1.7* (a) 4.6 kmol; (b) 130 kg; (c) 120 kg.

1.8 102 g mol^{-1}, CH_2FCF_3 or CHF_2CHF_2.

1.9* (a) 0.184 Torr; (b) 68.6 Torr; (c) 0.184 Torr.

1.10 0.33 atm (N_2), 0 (H_2), 1.33 atm (NH_3), 1.66 atm.

1.11*

1.12* (a) 2.8 km h^{-1} E; (b) 86 km h^{-1}; (c) 86 km h^{-1}.

1.13 (a) 1.89 m; (b) 1.89 m.

1.14* $v = (2gR)^{1/2}$.

1.15 (a) 12.5 L mol^{-1}; (b) 12.3 L mol^{-1}.

1.16* 0.927, 0.208 L.

1.17* (a) 0.939 L mol^{-1}; (b) 439 K.

1.18* (a) 0.1353 L mol^{-1}; (b) 0.6957; (c) 0.58 (from expansion in $1/V_m$), 0.71 (from expansion in p).

1.19 210 K, 0.28 nm.

1.20 5.649 L^2 atm mol^{-1}, 59.4 cm^3 mol^{-1}, 21 atm.

1.22* $c^* = (2kT/m)^{1/2}$.

1.23* $c_{\text{mean}} = (\pi kT/2m)^{1/2}$.

1.24 $v = 0.47v_{\text{initial}}$.

1.25*   (a) 0.39; (b) 0.61; (c) 0.47, 0.53.

1.26 3.02×10^{-3}, 4.9×10^{-6}.

1.27*

1.28 $B = b - a/RT$, $C = b^2$, 1.26 $\text{L}^2\,\text{atm}\,\text{mol}^{-2}$, 34.6 $\text{cm}^3\,\text{mol}^{-1}$.

1.29* $V_c = 3C/B$, $T_c = B^2/3RC$, $p_c = B^3/27C^2$, $\frac{1}{3}$.

1.30* $B' = B/RT$, $C' = (C - B^2)/R^2T^2$.

1.31 -0.18 atm^{-1}, -4.4 $\text{L}\,\text{mol}^{-1}$.

1.32 $(dV_{\text{m}}/dT)_p = (RV_{\text{m}} + b)/(2pV_{\text{m}} + RT)$.

1.33*   no.

1.34   1.11.

1.35* $(p - p_0)/p_0 =$ (a) 0.00; (b) 0.05.

1.36*   8.54, 15.1.

1.37   0.011.

1.38*   (a) $B = -1.32 \times 10^{-2}$ $\text{L}\,\text{mol}^{-1}$; (b) $B = -1.51 \times 10^{-2}$ $\text{L}\,\text{mol}^{-1}$, $C = 1.07 \times 10^{-3}$ $\text{L}^2\,\text{mol}^{-2}$.

1.39   (a) 1.12×10^{-3} mol, 2.8×10^{-9} $\text{mol}\,\text{L}^{-1}$; (b) 4.46×10^{-4} mol, 1.1×10^{-9} $\text{mol}\,\text{L}^{-1}$.

1.40*   (a) 1.1×10^{-11} $\text{mol}\,\text{L}^{-1}$, 2.2×10^{-11} $\text{mol}\,\text{L}^{-1}$; (b) 8.0×10^{-13} $\text{mol}\,\text{L}^{-1}$, 1.6×10^{-12} $\text{mol}\,\text{L}^{-1}$.

1.41*   (a) 7.1×10^{-14} $\text{cm}^3\,\text{mol}^{-1}$; on that basis alone, perfect gas law should apply, but other forces may be significant; (b) 1.6×10^7 K; (c) no, $T_{\text{perfect}} \approx T_{\text{vanderWaals}}$.

1.43*   51.5 km, 3.0×10^{-3} bar.

1.44 $Z = 0.611$ for all gases; the experimental value from Fig. 1.27 is close to 0.55.

2.1* $q = \Delta H = \Delta U = 2.6$ MJ.

2.2 $+37$ K, 4.09 kg.

2.3*   (a) -3.46 kJ; (b) 0; (c) -3.46 kJ; (d) $+24.0$ kJ; (e) $+27.5$ kJ.

2.4* $T_2 = 546$ K, $T_3 = 273$ K; Step 1→2: $w = -2.27$ kJ, $q = +5.67$ kJ, $\Delta U = +3.40$ kJ, $\Delta H = +5.67$ kJ; Step 2→3: $w = 0$, $q = -3.40$ kJ, $\Delta U = -3.40$ kJ, $\Delta H = -5.67$ kJ; Step 3→1: $w = +1.57$ kJ, $q = -1.57$ kJ, $\Delta U = 0$, $\Delta H = 0$; Cycle: $w = -0.70$ kJ, $q = +0.70$ kJ, $\Delta U = 0$, $\Delta H = 0$.

2.5   (a) -0.27 kJ; (b) -0.94 kJ.

2.6 -8.9 kJ, -8.9 kJ.

2.7* $+98.7$ $\text{kJ}\,\text{mol}^{-1}$, $+95.8$ $\text{kJ}\,\text{mol}^{-1}$.

2.8   36.5 L.

2.9* -87.33 $\text{kJ}\,\text{mol}^{-1}$.

2.10* -2.13 $\text{MJ}\,\text{mol}^{-1}$, -1.267 $\text{MJ}\,\text{mol}^{-1}$.

2.11 $+17.7$ $\text{kJ}\,\text{mol}^{-1}$, $+116.0$ $\text{kJ}\,\text{mol}^{-1}$.

2.12*   more exothermic by 5376 $\text{kJ}\,\text{mol}^{-1}$.

2.13   (a) 0.39 mol, 0.50 L, 0.50 L; (b) $+19$ kJ; (c) -3.0 kJ; (d) $\Delta U = 0$, for all 3 paths; path ACB: $q = -9.5 \times 10^2$ J; path ADB: $q = -1.9 \times 10^4$ J; path AB: $q = -3.0 \times 10^3$ J.

2.14*   (a) 60 kJ; (b) -70 J; (c) $+10$ J , $+50$ J.

2.15* $\Delta H = nC_{p,\text{m}}(T_{\text{f}} - T_{\text{i}})$.

2.16 $w_{\text{r}} = 3bw/a$; $w_{\text{r}} = -\frac{8}{9}nT_{\text{r}}\ln\{(V_{\text{r,2}} - \frac{1}{3})/(V_{\text{r,1}} - \frac{1}{3})\} - n(1/V_{\text{r,2}} - 1/V_{\text{r,1}})$; $w_{\text{r}}/n = -\frac{8}{9}\ln\{\frac{1}{2}(3x - 1)\} - 1/x + 1$.

2.17* $-25\,968$ $\text{kJ}\,\text{mol}^{-1}$, $+2357$ $\text{kJ}\,\text{mol}^{-1}$.

2.18 -994.30 $\text{kJ}\,\text{mol}^{-1}$.

2.19*   (a) $+16.2$ $\text{kJ}\,\text{mol}^{-1}$; (b) $+114.6$ $\text{kJ}\,\text{mol}^{-1}$; (c) $+122.0$ $\text{kJ}\,\text{mol}^{-1}$.

2.20   (a) 120.3 $\text{kJ}\,\text{mol}^{-1}$; (b) $+68.9$ $\text{kJ}\,\text{mol}^{-1}$; (c) $+48.1$ $\text{kJ}\,\text{mol}^{-1}$.

2.21*   (a) $+240$ $\text{kJ}\,\text{mol}^{-1}$; (b) $+228$ $\text{kJ}\,\text{mol}^{-1}$.

2.22   (a) -101.8 $\text{kJ}\,\text{mol}^{-1}$; (b) -344.2 $\text{kJ}\,\text{mol}^{-1}$; (c) $+44.0$ $\text{kJ}\,\text{mol}^{-1}$.

2.23*   (a) (1) isochoric step: $n = \infty$; isobaric step: $n = 0$; (2) adiabatic step: $n = \gamma$; isochoric step: $n = \infty$; (b) (1) isochoric step: $w = 0$, $\Delta U = q = 55.8$ kJ, $\Delta H = 78.1$ kJ; isobaric step: $w = 22.3$ kJ, $\Delta U = -55.8$ kJ, $\Delta H = q = -78.1$ kJ; overall: $w = -q = 22.3$ kJ, $\Delta U = \Delta H = 0$; (2) adiabatic step: $q = 0$, $\Delta U = w = 9.37$ kJ, $\Delta H = 13.1$ kJ; isochoric step: $w = 0$, $\Delta U = q = -9.37$ kJ, $\Delta H = -13.1$ kJ; overall: $w = -q = 9.37$ kJ, $\Delta U = \Delta H = 0$.

2.25* $k = -66.51$, $n = 0.9277$, $R = 0.999\,58$ (a good fit). Predicted $\Delta_c H^\ominus$ (decane) $= -6612.4$ $\text{kJ}\,\text{mol}^{-1}$; experimental value, -6772.5 $\text{kJ}\,\text{mol}^{-1}$; error, 2.36 per cent.

2.26*   (a) no, as a plot of $\ln p$ against $\ln V$ does not yield a straight line; (b) numerical integration yields $w = 685$ J; (c) fitting the data to the van der Waals equation yields $T = 350$ K.

3.1 2.18×10^{-11} Pa^{-1}, -0.220 cm^3, 997.2 cm^3.

3.2*   (a) $+0.75$ $\text{kJ}\,\text{mol}^{-1}$; (b) $+0.75$ $\text{kJ}\,\text{mol}^{-1}$.

3.3*   41.40 $\text{J}\,\text{K}^{-1}\,\text{mol}^{-1}$.

3.4* -30.5 $\text{J}\,\text{mol}^{-1}$.

3.5   1.67.

3.6*   not exact.

3.7*   not exact; dq/T is exact.

3.8 $dw = (y + z)\,dx + (x + z)\,dy + (x + y)\,dz$.

3.10 $(\partial H/\partial p)_T = -\mu C_p$.

3.11* $C_{p,\text{m}} - C_{V,\text{m}} = R$.

3.12*   **3.14***

3.15 $dp = \{R/(V_{\text{m}} - b)\}dT + \{2a/V_{\text{m}}^3 - RT/(V_{\text{m}} - b)^2\}\,dV_{\text{m}}$, $(\partial V/\partial T_p = RV^3(V - b)/(RTV^3 - 2a(V - b)^2)$.

3.16* $+3.80$ kJ.

3.17 $dp = \{a(V_{\text{m}} - 2b)/V_{\text{m}}^3 - p\}dV_{\text{m}}/(V_{\text{m}} - b) + (p + a/V_{\text{m}}^2)dT/T$.

3.18 $T = p(V - nb)/nR + na(V - nb)/RV^2$, $(\partial T/\partial p)_V = (V_{\text{m}} - b)/R$.

3.19* $\alpha = RT^2(V - nb)/\{RTV^3 - 2na(V - nb)^2\}$, $\kappa_T = V^2(V - nb)^2/\{RTV^3 - 2na(V - nb)^2\}$.

3.20 $\mu C_p = (1 - b\zeta/V_{\text{m}})V/(\zeta - 1)$, $\zeta = RTV_{\text{m}}^3/2a(V_{\text{m}} - b)^2$, 1.46 $\text{K}\,\text{atm}^{-1}$, $T_1 = (\frac{27}{4})T_c(1 - b/V_{\text{m}})^2$, 2021 K.

3.21*

3.22 $9.2 \text{ J K}^{-1} \text{mol}^{-1}$.

3.23* 322 m s^{-1}.

3.24* 0.80 m, 1.6 m, 2.8 m.

3.25 (a) 29.9 K MPa^{-1}; (b) -2.99 K.

3.26* (a) 23.5 K MPa^{-1}; (b) 14.0 K MPa^{-1}.

3.27* $T_1 = 842$ K, $T_2 = T_3 = 348$ K; $p_1 = p_2 = p_3 = 1.72$ bar; $V_1 = 40.7$ L, $V_2 = V_3 = 16.8$ L; $\Delta U_1 = 11.3$ kJ, $\Delta U_2 = \Delta U_3 = 1.04$ kJ; $\Delta U(\text{total}) = 13.4$ kJ.

3.28 Same answers as for problem 3.27. It does not matter whether the piston between chambers 2 and 3 is diathermic or adiabatic.

3.29* increase.

3.30 (a) $\mu = \alpha T^2/C_p$; (b) $C_V = C_p - R(1 + 2apT/R)^2$.

4.1 (a) $-21.3 \text{ J K}^{-1} \text{mol}^{-1}$, $+21.7 \text{ J K}^{-1} \text{mol}^{-1}$; (b) $-111.2 \text{ J K}^{-1} \text{mol}^{-1}$, $-1.5 \text{ J K}^{-1} \text{mol}^{-1}$.

4.2 $+11 \text{ J K}^{-1}$.

4.3* (a) $57.0°C$, -43.9 kJ, $+146 \text{ J K}^{-1}$, $+28 \text{ J K}^{-1}$; (b) $49.9°C$.

4.4* (a) $+50.74 \text{ J K}^{-1}$, -11.5 J K^{-1}; (b) ΔA_A is indeterminate, $\Delta A_B = +3.46$ kJ; (c) ΔG_A is indeterminate, $\Delta G_B = +3.46$ kJ; (d) $+39.2 \text{ J K}^{-1}$, -39.2 J K^{-1}.

4.5* Step 1→2: $w = -11.5$ kJ, $q = +11.5$ kJ, $\Delta U = 0$, $\Delta H = 0$, $\Delta S = +19.1 \text{ J K}^{-1}$, $\Delta S_{\text{tot}} = 0$; Step 2→3: $w = -3.74$ kJ, $q = 0$, $\Delta U = -3.74$ kJ, $\Delta H = -6.23$ kJ, $\Delta S = 0$, $\Delta S_{\text{sur}} = 0$; Step 3→4: $w = +5.74$ kJ, $q = -5.74$ kJ, $\Delta U = 0$, $\Delta H = 0$, $\Delta S = +19.1 \text{ J K}^{-1}$, $\Delta S_{\text{tot}} = 0$; Step 4→1: $w = +3.74$ kJ, $q = 0$, $\Delta U = +3.74$ kJ, $\Delta H = +6.23$ kJ, $\Delta S = 0$, $\Delta S_{\text{tot}} = 0$; Cycle: $w = -5.8$ kJ, $q = +5.8$ kJ, $\Delta U = 0$, $\Delta H = 0$, $\Delta S = 0$, $\Delta S_{\text{tot}} = 0$.

4.6* Path (a): $w = -2.74$ kJ, $q = +2.74$ kJ, $\Delta U = 0$, $\Delta H = 0$, $\Delta S = +9.13 \text{ J K}^{-1}$, $\Delta S_{\text{sur}} = -9.13 \text{ J K}^{-1}$, $\Delta S_{\text{tot}} = 0$; Path (b): $w = -1.66$ kJ, $q = +1.66$ kJ, $\Delta U = 0$, $\Delta H = 0$, $\Delta S = +9.13 \text{ J K}^{-1}$, $\Delta S_{\text{sur}} = -5.53 \text{ J K}^{-1}$, $\Delta S_{\text{tot}} = +3.60 \text{ J K}^{-1}$.

4.7* Path (a): $q = 0$, $w = -9.1 \times 10^2$ J, $\Delta H = -1.5$ kJ, $\Delta S = 0$, $\Delta S_{\text{sur}} = 0$, $\Delta S_{\text{tot}} = 0$; Path (b): $q = 0$, $w = -7.5$ kJ, $\Delta H = -1.2$ kJ, $\Delta S = +1.12 \text{ J K}^{-1}$, $\Delta S_{\text{sur}} = 0$, $\Delta S_{\text{tot}} = +1.12 \text{ J K}^{-1}$.

4.8 Process (a): $\Delta S = +5.8 \text{ J K}^{-1}$, $\Delta S_{\text{sur}} = -5.8 \text{ J K}^{-1}$, $\Delta H = 0$, $\Delta T = 0$, $\Delta A = -1.7$ kJ, $\Delta G = -1.7$ kJ; Process (b): $\Delta S = +5.8 \text{ J K}^{-1}$, $\Delta S_{\text{sur}} = -1.7 \text{ J K}^{-1}$, $\Delta H = 0$, $\Delta T = 0$, $\Delta A = -1.7$ kJ, $\Delta G = -1.7$ kJ; Process (c): $\Delta S = +3.9 \text{ J K}^{-1}$, $\Delta S_{\text{sur}} = 0$, $\Delta H = -0.84$ kJ, $\Delta T = -41$ K, ΔA and ΔG indeterminate.

4.9 (a) $200.7 \text{ J K}^{-1} \text{mol}^{-1}$; (b) $232.0 \text{ J K}^{-1} \text{mol}^{-1}$.

4.10 $+45.4 \text{ J K}^{-1}$, $+51.2 \text{ J K}^{-1}$.

4.11* $-160.07 \text{ kJ mol}^{-1}$.

4.12 (a) $+17.0 \text{ J K}^{-1}$; (b) $+36 \text{ J K}^{-1}$.

4.13* (a) 0.11 kJ mol^{-1}; (b) 0.11 kJ mol^{-1}.

4.14* (a) $63.88 \text{ J K}^{-1} \text{mol}^{-1}$; (b) $66.08 \text{ J K}^{-1} \text{mol}^{-1}$.

4.15 7.8 km.

4.16 at 298 K: $+41.16 \text{ kJ mol}^{-1}$, $+42.08 \text{ J K}^{-1} \text{mol}^{-1}$; at 398 K: $+40.84 \text{ kJ mol}^{-1}$, $+41.08 \text{ J K}^{-1} \text{mol}^{-1}$.

4.17* (a) $+76.9 \text{ J K}^{-1} \text{mol}^{-1}$; (b) $+96.864 \text{ J K}^{-1} \text{mol}^{-1}$.

4.18 at 200 K: $32.00 \text{ kJ mol}^{-1}$ $293.5 \text{ J K}^{-1} \text{mol}^{-1}$.

4.19* $+34.4 \text{ kJ mol}^{-1}$, $243 \text{ J K}^{-1} \text{mol}^{-1}$ at 298 K.

4.21*

4.23 $\Delta S = nC_{p,\text{m}} \ln(T_\text{f}/T_\text{h}) + nC_{p,\text{m}} \ln(T_\text{f}/T_\text{c})$, $+22.6 \text{ J K}^{-1}$.

4.24* -21 K, $+35.9 \text{ J K}^{-1} \text{mol}^{-1}$.

4.25* Step 1: $\Delta S = 0$, $\Delta S_{\text{sur}} = 0$; Step 2: $\Delta S = +33 \text{ J K}^{-1}$, $\Delta S_{\text{sur}} = -33 \text{ J K}^{-1}$; Step 3: $\Delta S = 0$, $\Delta S_{\text{sur}} = 0$; Step 4: $\Delta S = -33 \text{ J K}^{-1}$, $\Delta S_{\text{sur}} = +33 \text{ J K}^{-1}$.

4.26*

4.28* $+247.8 \text{ J K}^{-1} \text{mol}^{-1}$, $+336.6 \text{ J K}^{-1} \text{mol}^{-1}$, $+314.7 \text{ J K}^{-1} \text{mol}^{-1}$.

4.29 (a) $+0.8 \text{ kJ mol}^{-1}$; (b) $+11.9 \text{ kJ mol}^{-1}$; (c) $+15.0 \text{ kJ mol}^{-1}$.

4.30* $46.60 \text{ J K}^{-1} \text{mol}^{-1}$, $46.73 \text{ J K}^{-1} \text{mol}^{-1}$.

4.31 $\frac{4}{3}$

4.32*

5.1* -501 kJ mol^{-1}.

5.2 (a) $+7 \text{ kJ mol}^{-1}$; (b) $+107 \text{ kJ mol}^{-1}$.

5.3 -27 kJ mol^{-1}.

5.4* 73 atm.

5.6*

5.7* $(\partial S/\partial V)_T = \alpha/\kappa_T$ and $(\partial V/\partial S)_p = \alpha TV/C_p$.

5.8 $(\partial p/\partial S)_V = \alpha T/\kappa_T C_V$.

5.9*

5.10 (a) $(\partial H/\partial p)_T = 0$; (b) $(\partial H/\partial p)_T = \{nb - (2na/RT)\lambda^2\}/\{1 - (2na/RTV)\lambda^2\}$, $\lambda = 1 - b/V_\text{m} \approx -8.3 \text{ J atm}^{-1}$, -8 J.

5.11*

5.12 (a) 3.0×10^{-3} atm; (b) 0.30 atm.

5.13 $(\partial C_V/\partial V)_T = (RT/V_\text{m}^2)(\partial^2(BT)/\partial T^2)_V$.

5.14*

5.15 $\pi_T = ap/RTV_\text{m}$.

5.16 0.02 per cent.

5.17* (a) $\Delta_r G' = \tau\Delta_r G + (1 - \tau)\Delta_r H$; (b) $\Delta_r G' = \tau\Delta_r G + (1 - \tau)(\Delta_r H - T\Delta_r C_p) - T'\Delta C_p \ln\tau$, $\tau = T'/T$.

5.19* $q_{\text{rev}} = nRT \ln\{(V_\text{f} - nb)/(V_\text{i} - nb)\}$.

5.20 -0.50 kJ.

5.21* $G' = G + p^*V_0(1 - e^{-p/p^*})$, expansion.

5.22* $\ln\phi = Bp/RT + (C - B^2)p^2/2R^2T^2 + \cdots$, 0.999 atm.

5.23 $\phi = 2e^{\lambda^{1/2}-1}/(1 + \lambda^{1/2})$, $\lambda = 1 + 4pq/R$.

5.24* 13 per cent.

5.25 $+57.2 \text{ kJ mol}^{-1}$, $+85.6 \text{ kJ mol}^{-1}$, $+112.8 \text{ kJ mol}^{-1}$.

5.26* (a) 0.750; (b) 0.372.

5.28* The second option leads to $p = RT/(V - b) + \text{constant}$.

5.29* As pressure increases, entropy increases, remains constant, and decreases, respectively.

6.1* 196.0 K, 11.1 Torr.

6.2* 9 atm.

6.3 (a) $+5.56 \text{ kPa K}^{-1}$; (b) 2.5 per cent.

6.4* (a) $-22.0 \text{ J K}^{-1} \text{ mol}^{-1}$; (b) $-109.0 \text{ J K}^{-1} \text{ mol}^{-1}$; (c) $+110 \text{ J mol}^{-1}$.

6.5 (a) $-1.63 \text{ cm}^3 \text{ mol}^{-1}$; (b) $+30.1 \text{ L mol}^{-1}$, 0.6 kJ mol^{-1}.

6.6* 234.4 K.

6.7 22°C.

6.8* (a) 357 K (84°C); (b) $+37.8 \text{ kJ mol}^{-1}$.

6.9 (a) 227.5°C; (b) $+55 \text{ kJ mol}^{-1}$.

6.10* 6.12*

6.13 9.8 Torr.

6.14 $1/T_h = 1/T_b + Mgh/T\Delta_{\text{vap}}H$, 363 K (90°C).

6.15* $(\partial^2 \mu/\partial T^2)_p = -C_{p,m}/T$.

6.16*

6.18* (b) 112 K; (c) $+8.07 \text{ kJ mol}^{-1}$.

6.19 (b) 178.18 K; (c) 383.6 K; (d) $+33.0 \text{ kJ mol}^{-1}$.

6.20 $+31.6 \text{ kJ mol}^{-1}$.

6.21* $1.60 \times 10^4 \text{ bar}$.

7.1* $K_A = 15.58 \text{ kPa}$, $K_B = 47.03 \text{ kPa}$.

7.2* $17.5 \text{ cm}^3 \text{ mol}^{-1}$ (NaCl), $18.07 \text{ cm}^3 \text{ mol}^{-1}$ (H_2O).

7.3 $-1.4 \text{ cm}^3 \text{ mol}^{-1}$ ($MgSO_4$), $18.04 \text{ cm}^3 \text{ mol}^{-1}$ (H_2O).

7.4* $12.0 \text{ cm}^3 \text{ mol}^{-1}$.

7.5 57.9 mL ethanol, 45.8 mL water, 0.96 cm^3.

7.6* (b) $K_A = 450 \text{ Torr}$, $K_I = 465 \text{ Torr}$.

7.7 -4.6 kJ.

7.8* $\mu_A = \mu_A^* - RT \ln x_A + gRT x_B^2$.

7.10* $V_B(x_A, x_B) = V_B(0, 1) - \int_{V_A(0)}^{V_A(x_A)} \{x_A \, dV_A/(1 - x_A)\}$.

7.12*

7.14* $V_{\text{cyclo}} = 109.0 \text{ cm}^3 \text{ mol}^{-1}$, $V_{\text{polymer}} = 279.3 \text{ cm}^3 \text{ mol}^{-1}$.

7.15 (a) $V_1 = V_{m,1} + a_0 x_2^2 + a_1(3x_1 - x_2)x_2^2$, $V_2 = V_{m,2} + a_0 x_1^2 + a_1(x_1 - 3x_2)x_1^2$; (b) $V_1 = 75.63 \text{ cm}^3 \text{ mol}^{-1}$, $V_2 = 99.06 \text{ cm}^3 \text{ mol}^{-1}$.

7.16* For $x_T = 0.228, 0.511, 0.810$, $\gamma_T = 0.490, 0.723, 0.966$, and $\gamma_E = 1.031, 0.920, 0.497$.

7.17 $K_H = 371 \text{ bar}$; at $p = 60.0 \text{ bar}$, $\gamma_{CO_2} = 0.98$.

7.18* $S_0 = 19.89 \text{ mol L}^{-1}$, $\tau = 165 \text{ K}$; fits well, $R = 0.99978$.

8.2* (a) 2150°C; (b) $x(\text{MgO}) = 0.35$, $y(\text{MgO}) = 0.18$, ratio 0.4; (c) 2640°C.

8.3 (a) $n(\text{l})/n(\text{s}) = 5$; (b) no liquid.

8.4*

8.5* A compound with probable formula A_3B exists. It melts incongruently at 700°C. The proportions of A and B in the product are dependent upon the overall composition and the temperature. A eutectic exists at 400°C and $x_B \approx 0.83$.

8.6 The number of distinct chemical species (as opposed to components) and phases present at the indicated points are, respectively $b(3, 2)$, $d(2, 2)$, $e(4, 3)$, $f(4, 3)$, $g(4, 3)$, $k(2, 2)$.

8.7* $T_{\text{uc}} = 122°C$, $T_{\text{lc}} = 8°C$.

8.8 $MgCu_2$: 16 per cent Mg by mass; Mg_2Cu: 43 per cent Mg by mass.

8.9* No phases are in equilibrium as the system cools. The phases through which the system passes upon cooling are: liquid, liquid + solid K_2FeCl_4, solid K_2FeCl_4 + solid $KFeCl_3$.

8.11*

8.12* (b) 391.0 K; (c) $n_{\text{liq}}/n_{\text{vap}} = 0.532$.

8.13 (b) $n_{\text{liq}}/n_{\text{vap}} = 10.85$.

8.14 (b) $n_{\text{left}}/n_{\text{right}} = 0.093$; 302.5 K at $x = 0.750$.

8.15* Temperature (γ): 78 K (0.9); 80 K (1.08); 82 K (1.04); 84 K (1.00); 86 K (0.99); 88 K (0.99); 90.2 K (0.99). To within experimental uncertainties the solution appears ideal.

9.1* (a) $+4.48 \text{ kJ mol}^{-1}$; (b) 0.101 atm.

9.2 (a) 1.24×10^{-9}; (b) 1.29×10^{-8}; (c) 1.8×10^{-4}; (d) as pressure increases, α decreases; as temperature increases, α increases.

9.3 $\Delta_f H^\ominus = -(2.196 \times 10^4 \text{ K} - 8.84T)R$, $8.48R$.

9.4* $\Delta_r H^\ominus = +3.00 \times 10^5 \text{ J mol}^{-1}$ and $\Delta_r S^\ominus = +102 \text{ J K}^{-1} \text{ mol}^{-1}$ throughout this temperature range; at 1395 K: $K = 1.22 \times 10^{-6}$, $\Delta_r G^\ominus = +158 \text{ kJ mol}^{-1}$; at 1443 K: $K = 2.80 \times 10^{-6}$, $\Delta_r G^\ominus = +153 \text{ kJ mol}^{-1}$; at 1498 K: $K = 7.23 \times 10^{-6}$, $\Delta_r G^\ominus = +147 \text{ kJ mol}^{-1}$.

9.5 $\Delta_r G^\ominus(T)/(\text{kJ mol}^{-1}) = 78 - 0.161(T/\text{K})$.

9.6* 1.69×10^{-5}.

9.7 5.71, -103 kJ mol^{-1}.

9.8* 14.7 kJ mol^{-1}, $+18.8 \text{ kJ mol}^{-1}$.

9.9 1.800×10^{-3} (at 973 K), 1.109×10^{-2} (at 1073 K), 4.848×10^{-2} (at 1173 K), $+158 \text{ kJ mol}^{-1}$.

9.10*

9.11* $\xi = 1 - 1/(1 + ap/p^\ominus)^{1/2}$. 9.12 0.140.

9.13* $\Delta_r G' = \Delta_r G + (T - T')\Delta_r S + \alpha\Delta a + \beta\Delta b + \gamma\Delta c$, $\alpha = T' - T - T' \ln(T'/T)$, $\beta = \frac{1}{2}(T'^2 - T^2) - T'(T' - T)$, $\gamma = 1/T - 1/T' + \frac{1}{2}T'(1/T'^2 - 1/T^2)$; $-225.31 \text{ kJ mol}^{-1}$.

9.14* $+76.8 \text{ kJ mol}^{-1}$.

9.15 (a) $-72.4 \text{ kJ mol}^{-1}$, $-144 \text{ J K}^{-1} \text{ mol}^{-1}$; (b) $+131.2 \text{ kJ mol}^{-1}$, $+309.2 \text{ J K}^{-1} \text{ mol}^{-1}$.

9.16* now: 5.64; then: 5.70.

9.17 (a) 1.2×10^8; (b) 2.7×10^3.

9.18 the trihydrate.

9.19* 0.23, 0.46, 0.30.

10.1 $Pb(s)|PbSO_4(s)|PbSO_4(aq)|Hg_2SO_4(aq)|Hg_2SO_4(s)|Hg(l)$, +1.03 V.

10.2* (a) 4.0×10^{-3}, 1.2×10^{-2}; (b) 0.74, 0.60; (c) 5.9; (d) +1.102 V; (e) +1.079 V.

10.3* 2.0.

10.4 (a) +1.23 V; (b) +1.09 V.

10.5* (a) $E = E^{\ominus} - (38.54 \text{ mV}) \times \{\ln(4^{1/3}b) + \ln \gamma_{\pm}\}$; (b) 1.0304 V; (c) 6.84×10^{34}; (d) 0.763; (e) 0.75; (f) $-87.2 \text{ J K}^{-1} \text{ mol}^{-1}$, $-262.4 \text{ kJ mol}^{-1}$.

10.6* +0.268 38 V.

10.7 14.23 at 20.0°C, $+74.9 \text{ kJ mol}^{-1}$, $+80.0 \text{ kJ mol}^{-1}$, $-17.1 \text{ J K}^{-1} \text{ mol}^{-1}$.

10.8 0.533.

10.9* (a) +0.2223 V, +0.2223 V; (b) 1.10, 0.796.

10.10*

10.11 $-131.25 \text{ kJ mol}^{-1}$, $+56.7 \text{ J K}^{-1} \text{ mol}^{-1}$, $-167.10 \text{ kJ mol}^{-1}$.

10.12*

10.13 (a) $E = E^{\ominus} + (2.303RT/F)\text{pOH}$; (b) $E = E^{\ominus} + (2.303RT/F) \times (pK_w - \text{pH})$; (c) -37.6 mV.

10.14 -1.2 V.

10.15* **10.18***

10.19* (a) $-1.991 \text{ V} > E^{\ominus} > -2.19 \text{ V}$, scandium; (b) 1.4×10^{10}.

10.20 $b(H_2VO_4^-) = 0.0048 \text{ mol kg}^{-1}$, $b(V_4O_{12}^{4-}) = 0.0013 \text{ mol kg}^{-1}$.

10.21 (a) $(\partial E/\partial p)_{T,n} = -(\Delta_r V/\nu F)$; (b) $2.84 \times 10^{-6} \text{ V atm}^{-1}$ from fit, $2.98 \times 10^{-6} \text{ V atm}^{-1}$ from $\Delta_r V$; (c) $E_{30°C}/V = 0.00856 + 2.84 \times 10^{-6} p/(\text{atm})$. A second-order polynomial fit is slightly better; (d) $\kappa_T = -3.2 \times 10^{-7} \text{ atm}^{-1}$.

10.22* (a) $\Delta E = -(RT/\nu F) \ln p/(\text{atm})$; fits well below 100 atm, but deviates above that pressure; (b) $\Delta E = (RT/\nu F)\{\ln p/(\text{atm}) + Cp\}$, $(\partial E/\partial p)_T = (RT/\nu F)\{1/p + C\}$, $C = 6.665 \times 10^{-4} \text{ atm}^{-1}$, $R = 0.99940$; (c) From empirical virial equation: $\Delta E = (RT/\nu F) \ln p/(\text{atm}) + 0.000537(p-1)/(\text{atm}) + 1.75 \times 10^{-8}(p^2 - 1)/(\text{atm}^2)$; (d) p, $\phi/\phi_{1 \text{ atm}}$: 1.00 atm, 1.00; 10 atm, 0.994; 38 atm, 1.02; 51 atm, 1.04; 108 atm, 1.07; 210 atm, 1.03; 380 atm, 1.23; 430 atm, 1.38; 560 atm, 1.44; 720 atm, 1.64; 900 atm, 1.81; 1020 atm, 2.02.

11.1* (a) $1.6 \times 10^{-33} \text{ J m}^{-3}$; (b) $2.5 \times 10^{-4} \text{ J m}^{-3}$.

11.2* $6.29 \times 10^{-34} \text{ J s}$.

11.3 (a) $7.47 \times 10^{-29} \text{ J m}^{-3}$; (b) $4.59 \times 10^{-14} \text{ J m}^{-3}$; (c) $3.49 \times 10^{-11} \text{ J m}^{-3}$; classical values: (a) 0.807 J m^{-3}; (b) 1.67 J m^{-3}; (c) 2.10 J m^{-3}.

11.4 (a) 2231 K, $0.031R$; (b) 343 K, $0.897R$.

11.5* (a) 0.020; (b) 0.007; (c) 7×10^{-6}; (d) 0.5; (e) 0.61.

11.6 (a) 9.0×10^{-6}; (b) 1.2×10^{-6}.

11.7 $\lambda_{\text{max}} T = hc/5k$.

11.8* (a) $N = (2/L)^{1/2}$; (b) $1/c(2L)^{1/2}$; (c) $1/(\pi a^3)^{1/2}$; (d) $1/(32\pi a^5)^{1/2}$.

11.9 (a) $N = 1/(32\pi a_0^3)^{1/2}$; (b) $N = 1/(32\pi a_0^5)^{1/2}$.

11.10* (a) ik; (c) 0.

11.11 (a) -1; (b) -1.

11.12* (a) $-k^2$; (b) $-k^2$; (c) 0; (d) 0.

11.13* (a) $\cos^2 \chi$; (b) $\sin^2 \chi$; (c) $0.95e^{ikx} \pm 0.32e^{-ikx}$.

11.14 $\hbar^2 k^2/2m$.

11.15 (a) $k\hbar$; (b) 0; (c) 0.

11.16* (a) $6a_0$, $42a_0^2$; (b) $5a_0$, $30a_0^2$.

11.17* (a) $-e^2/4\pi\varepsilon_0 a_0$; (b) $\hbar^2/m_e a_0^2$.

11.19* (a) 1; (b) $2x$; (c) $\hbar$.

11.21* 500 nm, blue green.

11.22* 255 K, 11.3 μm.

11.23* (a) Assuming $\Delta_r C_p \approx$ constant, methane becomes unstable above ≈ 825 K; the authors' statement is confirmed; (b) $\lambda_{\text{max}}(1000 \text{ K}) = 2880$ nm; (c) $\rho(\text{brown dwarf})/\rho(\text{Sun}) = 8.8 \times 10^{-3}$ at $\lambda_{\text{max}}(\text{brown dwarf})$, $M(\text{brown dwarf})/M(\text{Sun}) = 7.7 \times 10^{-4}$; (d) 2.31×10^{-7}, it hardly shines.

11.24 $k = 1.382 \times 10^{-23} \text{ J K}^{-1}$, $h = 6.69 \times 10^{-34} \text{ J s}$.

12.1* 1.24×10^{-39} J, 2.2×10^9, 1.8×10^{-30} J.

12.2 (a) 1.60×10^{-19} J; (b) 2.42×10^{14} Hz; (c) total number of electrons = 1.12262.

12.3* CO $(1900 \text{ N m}^{-1}) > \text{NO } (1600 \text{ N m}^{-1}) > \text{HCl}$ $(516 \text{ N m}^{-1}) > \text{HBr } (412 \text{ N m}^{-1}) > \text{HI } (314 \text{ N m}^{-1})$.

12.4* 1.30×10^{-22} J, $\pm\hbar$.

12.5 $E/(10^{-22} \text{ J}) = 0, 2.62, 7.86, 15.72$.

12.6 $E = (h^2/8m) \times (n_1^2/L_1^2 + n_2^2/L_2^2 + n_3^2/L_3^2)$.

12.7* (a) $N^2/2\kappa$; (b) $N^2/4\kappa^2$.

12.8 $g = \frac{1}{2}(mk/\hbar^2)^{1/2}$.

12.9* $\langle T \rangle = \frac{1}{2}(v + \frac{1}{2})\hbar\omega$.

12.10 0, $\frac{3}{4}(2v^2 + 2v + 1)\alpha^4$.

12.11* (a) $L((1/12) - (1/2\pi^2 n^2))^{1/2}$, $nh/2L$; (b) $\{(v\frac{1}{2})\hbar/\omega m\}^{1/2}$, $\{(v + \frac{1}{2})\hbar\omega m\}^{1/2}$.

12.12* $\mu_{v+1,v} = \alpha\{(v+1)/2\}^{1/2}$, $\mu_{v-1,v} = \alpha(v/2)^{1/2}$.

12.13 $\langle T \rangle = -\frac{1}{2}\langle V \rangle$.

12.14 (a) $+\hbar$, $\hbar^2/2I$; (b) $-2\hbar$, $2\hbar^2/I$; (c) 0, $\hbar^2/2I$; (d) $\hbar \cos 2\chi$, $\hbar^2/2I$.

12.15* (a) 0, 0; (b) $3\hbar^2/I$, $6^{1/2}\hbar$; (c) $6\hbar^2/I$, $2(3^{1/2})\hbar$.

12.17* $\cos \theta = m_l/\{l(l+1)\}^{1/2}$, $54°44'$.

12.18 $-(a^2 + b^2 + c^2)$.

12.19* $l_x = (\hbar/i)(y\partial/\partial z - z\partial/\partial y)$ and cyclic permutations, $[l_x, l_y] = i\hbar l_z$.

12.20*

12.21* 0.49.

12.22 (b) $\langle x^2 \rangle_n^{1/2} = [L^2/3 - \frac{1}{4}(n\pi/L)^2]^{1/2}$; as $n \to \infty$, $\langle x^2 \rangle_n^{1/2} \to L/\sqrt{3}$.

13.1* $n_2 \rightarrow 6$, 7503, 5908, 5129, $\cdots$ 3908 nm.

13.2 397.13 nm, 3.40 eV.

13.3* 987 663 cm^{-1}, 137 175 cm^{-1}, 185 187 cm^{-1}, 122.5 eV.

13.4 5.39 eV.

13.5* $A = 38.50$ cm^{-1}.

13.6 3.3429×10^{-27} kg, $I_D/I_H = 1.000\,272$.

13.7* 7621 cm^{-1}, 10 228 cm^{-1}, 11 552 cm^{-1}, 6.80 eV.

13.8 0.420 pm.

13.9* $2s$.

13.10 106 pm.

13.12* $p_x \pm ip_y$.

13.13*

13.14* $2.66a_0$.

13.15 $E = -(Z^2 e^4 m_e/32\pi^2 \varepsilon_0^2 \hbar^2)(1/n^2)$.

13.17* $a_{Ps} = 2a_0$, $E_{1,Ps} = \frac{1}{2}E_{1,H}$.

13.18* $d_H \approx 0.314$ nm, $d_U \approx 0.301$ nm.

13.19* $v = 2.19 \times 10^6$ m s^{-1}, $\mathcal{E} = 5.14 \times 10^{11}$ V m^{-1}, $\mathcal{H} = 9.98 \times 10^6$ A m^{-1}.

13.20 $E_{2s} = E_{2p} = -(1/4)(Z^2\hbar^2/2\mu a_0^2)$.

13.21* (b) 1663 (from wavelength data), 1840 (from Rydberg constants).

13.22* $\Delta E = 4.29 \times 10^{-24}$ J $= 0.216$ cm^{-1}, $\langle r \rangle_{100} = 529$ nm, $I_{100} = 10.9677$ cm^{-1}; yes, thermal energy is 207 cm^{-1} and is sufficient, $v_{min} = 511$ m s^{-1}.

13.23 (1) $^2P_{1/2}$(lower) and $^2P_{3/2}$, (2) $^2D_{3/2}$(lower) and $^2D_{5/2}$; simple estimate suggests $^2D_{3/2}$ is the ground state.

13.24 (b) 23.8 T m^{-1}.

14.2*

14.3* $R_{max} = 2.1a_0$.

14.6 (a) 8.6×10^{-7}, 2.0×10^{-6}; (b) 8.6×10^{-7}, 2.0×10^{-6}; (c) 3.7×10^{-7}, 0; (d) 4.9×10^{-7}, 5.5×10^{-7}.

14.7* 1.9 eV, 130 pm.

14.8*

14.9* $\Delta E = 2.7$ eV, $\lambda = 460$ nm, orange.

14.11*

14.13* (a) nonplanar; (b) planar.

14.14 (a) $E = -hc\mathcal{R}_H$; (b) $-(8/3\pi)hc\mathcal{R}_H$.

14.16* Thermal motion would cause the molecule to break apart; it is not likely to exist for more than one vibrational period.

14.17 $E = \alpha_0$ (twice), $\frac{1}{2}\{(\alpha_0 + \alpha_N) \pm [(\alpha_0 - \alpha_N)^2 + 12\beta^2]^{1/2}\}$; E(delocalization) $= [(\alpha_0 - \alpha_N)^2 + 12\beta^2]^{1/2} - [(\alpha_0 - \alpha N)^3 + 4\beta^2]^{1/2}$.

14.18* (a) $E = (\alpha - \beta)$ (twice), $\alpha + 2\beta$; binding energies: $\alpha + 2\beta$, $2\alpha + 4\beta$, $3\alpha + 3\beta$, $4\alpha + 2\beta$, respectively; (b) $\Delta_r H = -413$ kJ mol^{-1}, slightly less than binding energy of H$_2$ (435.94 kJ mol^{-1}); (c) $\beta = (-849$ kJ mol$^{-1} - 2\alpha)/4$; H$_3^{2+}$: -425 kJ mol^{-1}; H$_3^+$: -849 kJ mol^{-1}; H$_3$: $3(\alpha/2 - 212$ kJ mol$^{-1})$; H$_3^-$: $3\alpha - 425$ kJ mol^{-1}.

14.19*

15.1* (a) D_{3d}; (b) D_{3d}, C_{2v}; (c) D_{2h}; (d) D_3; (e) D_{4d}.

15.2* *trans*-CHCl=CHCl.

15.3 $C_2\sigma_h = i$.

15.4* **15.5***

15.6 representation 1: $\mathbf{D}(\sigma_v) = \mathbf{D}(\sigma_d) = +1$ or -1; representation 2: $\mathbf{D}(\sigma_v) = -\mathbf{D}(\sigma_d) = +1$ or -1.

15.7* the matrices do not form a group.

15.8 $A_1 + T_2$, s and p, (d_{xy}, d_{yz}, d_{zx}) span T_2.

15.9 (a) all five d orbitals; (b) all except $d_{xy}(A_2)$.

15.10* (a) $2A_1 + A_2 + 2B_1 + 2B_2$; (b) $A_1 + 3E$; (c) $A_1 + T_1 + T_2$; (d) $A_{2u} + T_{1u} + T_{2u}$.

15.11 (a) yes; (b) no; (c) yes.

15.12* irreducible representations: $3A_1 + 2A_2 + 2B_1 + 3B_2$.

15.13 irreducible representations: $4A_1 + 2B_1 + 3B_2 + A_2$.

15.15 group: S_4; operations: S_4, C_2.

15.16* (a) D_{2h}; (b) (i) C_{2h}, (ii) C_{2v}.

15.17 (a) D_{2d}, A_1; (b) C_{4v}, A_1.

15.18* neither transition allowed; with vibration transition to T_{1g} becomes allowed, to G_g remains forbidden.

15.19* (a) symmetry elements: E, $2C_3$, $3C_2$, σ_h, $2S_3$, $3\sigma_v$; point group: D_{3h}; () reduces to: $A_1' + E'$.

15.20 reduces to $A_{1g} + B_{1g} + E_u$.

16.1* (a) 2.1×10^{-6}, 1.3 MHz, 0.0063 cm^{-1}; (b) 9.7×10^{-7}, 6.6 kHz, 0.0004 cm^{-1}.

16.2 700 MHz, 1 Torr.

16.3* 596 GHz, 19.9 cm^{-1}, 0.503 mm, $B = 9.941$ cm^{-1}.

16.4 From 112.83 pm to 123.52 pm.

16.5* $R(CC) = 139.6$ pm, $R(CH) = R(CD) = 108.5$ pm.

16.6 $k = 93.8$ N m^{-1}, 142.81 cm^{-1}, 3.36 eV.

16.7* linear, $\tilde{\nu}_1$: 1400 cm^{-1}; $\tilde{\nu}_2$ (bend): 540 cm^{-1}, $\tilde{\nu}_3$: 2360 cm^{-1} combination band, $\tilde{\nu}_1 + \tilde{\nu}_3$: 3735 cm^{-1}.

16.8* 2.728×10^{-47} kg m^2, 129.5 pm; DCl lines at 10.56, 21.11, 31.67,... cm^{-1}.

16.9 HCl: 128.393 pm, DCl: 128.13 pm.

16.10* $R(CO) = 116.28$ pm, $R(CS) = 155.97$ pm.

16.11 (a) 5.15 eV; (b) 5.20 eV.

16.12*

16.14* $J_{max} = (kT/2hcB)^{1/2} - \frac{1}{2}$, 30, $J_{max} = (kT/hcB)^{1/2} - \frac{1}{2}$, 6.

16.15 230 pm, 240 pm, 250 pm.

16.16* $R = 360.71$ pm $= 21.84$ cm^{-1}, $k = 0.3746$ N m^{-1}.

16.17 46.07 cm^{-1}, 1.769×10^8 cm^{-1}.

16.18* (a) 152 m^{-1}, 2.72×10^{-4} kg s^{-2}, 2.93×10^{-46} kg m^2, 95.5 m^{-1}; (b) 293 m^{-1}, 0.96.

16.19 14.35 m^{-1}, 26, 15.

16.20* 2.35 K.

16.21 (b) 87.61 pm (from B), 89.83 pm (from C), 88.7 pm (average); (c) $B = 43.84$ cm^{-1}, $C = 21.92$ cm^{-1}; (d) $\tilde{\nu}_2(D_3) = 1783.0$ cm^{-1}.

17.1* 49 364 cm^{-1}.

17.2* 5.1147 eV.

17.3 14 660 cm^{-1}.

17.4* 4.8×10^4 L mol^{-1} cm^{-2}.

17.5 1.1×10^6 L mol^{-1} cm^{-2}.

17.6* 0.2 ms.

17.8*

17.10* (a) allowed; (b) forbidden.

17.12*

17.13 lengthen; to blue.

17.15* $f = (RS/a_0)^2 f_0$.

17.16* 65 MW.

17.17* $^2\Sigma_g^+$, $^2\Pi_g$.

17.18 28 kJ mol^{-1} greater, consistent.

17.19* (a) 2.42×10^5 L mol^{-1} cm^{-2}; (b) 0.185; (c) 6.97, 135 L mol^{-1} cm^{-1}.

17.20 6.37, 2.12.

17.21 1.24×10^5 L mol^{-1} cm^{-2}.

17.22* $V_1 - V_0 = 3.1938$ eV, $\tilde{\nu}_1 - \tilde{\nu}_0 = 79.538$ cm^{-1}, $\tilde{\nu}_1 = 2113.8$ cm^{-1}, $\tilde{\nu}_0 = 2034.3$ cm^{-1}, relative populations $= 10$, $T_{eff} = 1.3 \times 10^3$ K.

18.1* 10.3 T, 2.42×10^{-5}, β.

18.2 57 kJ mol^{-1}.

18.3* 1.992, 2.002.

18.4 6.9 mT, 2.1 mT.

18.5* 1:2:3:2:1 quintet of 1:4:6:4:1 quintets.

18.6 (7) $\rho(3) = \rho(6) = 0.005$, $\rho(4) = \rho(5) = 0.076$; (8) $\rho(2) = \rho(4) = 0.200$, $\rho(3) = 0.048$, $\rho(6) = 0.121$; (9) all $\rho = 0.050$.

18.7* 158 pm.

18.8* -0.89 mT.

18.9 $\mathcal{I} = \frac{1}{2}A\tau/\{1 + (\omega_0 - \omega)^2\tau^2\}$.

18.10*

18.11* 300×10^6 Hz ± 10 Hz, 0.29 s.

18.12* 4×10^2 s^{-1}, 3.7 kJ mol^{-1}, 16 kJ mol^{-1}.

18.13 both fit data equally well.

18.14* (a) yes; (b) $^3J_{SnSn}/\text{Hz} = 580 - 79\cos\phi + 395\cos 2\phi$; (c) a staggered conformation with the two SnMe$_3$ groups at 180° to each other relative to the C—C bond.

18.15 (a) R_ν(nuclide)$/R_\nu$(proton): 0.409, 0.251, 0.193, 0.941, 0.405; R_B(nuclide)$/R_B$(proton) = 0.00965, 0.01590, 0.00101, 0.83350, 0.06654.

19.1 no.

19.2* 3.5 fK, 7.41.

19.3* (a) 5.00, 6.26; (b) 1.00 at 298 K, 0.80 at 5000 K; 6.5×10^{-11} at 298 K and 0.12 at 5000 K; (c) 13.38 J K^{-1} mol^{-1}, 18.07 J K^{-1} mol^{-1}.

19.4 0.257, 0.336, 0.396, 0.011.

19.5* (a) 0.64, 0.36; (b) 0.52 kJ mol^{-1}.

19.6 (a) 1.049, 123 J mol^{-1}, 1.65 J K^{-1} mol^{-1}; (b) 1.55, 1.348 kJ mol^{-1}, 8.17 J K^{-1} mol^{-1}; proportions are: $p_0 = $ (a) 0.953, (b) 0.645; $p_1 = $ (a) 0.044, (b) 0.230; $p_2 = $ (a) 0.002, (b) 0.083.

19.7* (a) 1; (b) $\{2, 2, 0, 1, 0, 0\}$ and $\{2, 1, 2, 0, 0, 0\}$.

19.8 $\{4, 2, 2, 1, 0, 0, 0, 0, 0, 0\}$.

19.9 (a) 160 K.

19.10* (a) $q = 1 + 3e^{-\varepsilon/kT}$, 2.104; (b) $0.5245RT$, 2.074 J K^{-1} mol^{-1}, 10.55 J K^{-1} mol^{-1}.

19.11* (a) 104 K; (b) $q = 1 + a$; (c) $2Nk \ln 2$.

19.12 if the separations $\varepsilon_2 - \varepsilon_1$ and $\varepsilon_1 - \varepsilon_0$ are equal.

19.13 (a) no difference; (b) no measurable difference.

19.14* $W = 2 \times 10^{40}$, $S = 1.282 \times 10^{21}$ J K^{-1}, $S_1 = 0.637 \times 10^{-21}$ J K^{-1}, $S_2 = 0.645 \times 10^{-21}$ J K^{-1}. Entropy is an extensive property.

19.15* ^{4}He: 7.69×10^5, 9, 3; ^{3}He: 1.18×10^6, 13, 5.

19.17* $\Delta W/W \approx 2.4 \times 10^{25}$.

19.18 $\Delta W/W \approx 4.8 \times 10^{21}$.

19.19* (a) $10^{5.44 \times 10^{23}}$; (b) $10^{5.69 \times 10^{23}}$, the heat capacity is temperature-dependent; (c) $\Delta S = +3.34$ J K^{-1}; therefore this is a spontaneous process, $\Delta S_{U,V} > 0$.

19.20 O_2: 0.36, H_2O: 0.57.

19.21*

19.22 (c) $k = 1.36 \times 10^{-23}$ J K^{-1}; $N_A = 6.11 \times 10^{23}$ mol^{-1}.

19.23 1.209, 3.004.

20.1* (a) $0.351R$; (b) $0.079R$; (c) $0.029R$.

20.2 (a) 0.1 per cent; (b) 4×10^{-3} per cent.

20.3* 4.2, 15 J K^{-1} mol^{-1}.

20.5* 19.89.

20.7* (a) 3.89; (b) 2.41.

20.8* $S_m = R \ln(2\pi e^2 m\sigma_m/h^2 N_A\beta)$, $\sigma_m = \sigma/n$; $\Delta S_m = R \ln\{(\sigma_m/V_m)(h^2\beta/2\pi me)^{1/2}\}$.

20.9 $U - U(0) = H - H(0) = (N\hbar\omega)/(e^x - 1)$, $C_V = kN\{x^2 e^x/(e^x - 1)^2\}$, $S = Nk\{x/(e^x - 1) - \ln(1 - e^{-x})\}$, $A - A(0) = G - G(0) = NkT \ln(1 - e^{-x})$.

20.10* (b) 5.41 J K^{-1} mol^{-1}.

20.11 100 T.

20.12* 350 m s^{-1}.

20.13 (a) $\theta_R = 87.55$ K, $\theta_V = 6330$ K; (b) $C_{V,m}$(equil.mixt.)$] = 2\alpha C_{V,m}(\text{H}) + (1 - \alpha)\,C_{V,m}(\text{H}_2)$,

$\alpha = [K/(K+4)]^{1/2}, C_{V,m}(H) = 3R/2,$
$C_{V,m}(H_2) = \frac{5}{2}R + \{(\theta_V/T)[e^{-(\theta_V/2T)}/(1 - e^{-(\theta_V/T)})]\}^2 R,$
$K = \{kT\Lambda^3(H_2)/pq^R(H_2)q^V(H_2)\Lambda^6(H)\}e^{-(D_0/RT)}.$

20.14* $199.4 \text{ J K}^{-1} \text{ mol}^{-1}$.

20.15 $513.5 \text{ kJ mol}^{-1}$.

20.16* $28, 258 \text{ J K}^{-1} \text{ mol}^{-1}$.

20.17 $0.6608 \text{ kJ mol}^{-1}, 241.5 \text{ kJ mol}^{-1}$.

20.18* $45.76 \text{ kJ mol}^{-1}$.

20.19 (c) $T \approx 374 \text{ K}$.

21.1 118 pm.

21.2* P, 342 pm.

21.3* yes.

21.4 (a) bcc, $315 \text{ pm}, 136 \text{ pm}$; (b) fcc, $364 \text{ pm}, 129 \text{ pm}$.

21.5* 10.51 g cm^{-3}.

21.7* 628 pm, yes.

21.8 $834 \text{ pm}, 606 \text{ pm}, 870 \text{ pm}$.

21.9* $\alpha_{volume} = 4.8 \times 10^{-5} \text{ K}^{-1}, \alpha_{linear} = 1.6 \times 10^{-5} \text{ K}^{-1}$.

21.10 177 pm.

21.11*

21.13* (a) 0.5236; (b) 0.6802; (c) 0.7405.

21.14*

21.15 (a) no absences; (b) alternation ($h + k + l$ odd or even); (c) $h + k + l$ odd missing.

21.16 4.

21.17* $1.385 \text{ g cm}^{-3}, 1.578 \text{ g cm}^{-3}$.

21.18 0.41.

21.19* $f = [1 + \frac{1}{4}(ka_0/Z)^2]^{-2}$.

21.20 $F_{hkl} = f\{1 + (-1)^{n+l} + (-1)^{k+l} + (-1)^{h+k} + [(-1)^{h+k+l}]^{1/2}[(-1)^l + (-1)^k + (-1)^h + (-1)^{h+k+l}]\}$.

21.21* $F_{100} = f(\text{Cs}^+) - f(\text{Cl}^-) = 54 - 18 = 36;$
$F_{110} = f(\text{Cs}^+) + f(\text{Cl}^-) = 54 + 18 = 72;$
$F_{200} = f(\text{Cs}^+) + f(\text{Cl}^-) = 54 + 18 = 72.$

22.1* (a) 0.11 GV m^{-1}; (b) 4 GV m^{-1}; (c) 4 kV m^{-1}.

22.2 2.4 nm.

22.3* $1.2 \times 10^{-23} \text{ cm}^3, 0.86 \text{ D}$.

22.4 $1.38 \times 10^{-23} \text{ cm}^3, 0.34 \text{ D}$.

22.5* $2.24 \times 10^{-24} \text{ cm}^3, 1.58 \text{ D}$.

22.6 $1.85 \text{ D}, 1.36 \times 10^{-24} \text{ cm}^3$.

22.7* (a) $6Q_1Q_2/\pi\varepsilon_0 r^5$; (b) $9Q_1Q_2/4\pi\varepsilon_0 r^5$.

22.8* $n_r = 1 + p(2\pi\alpha'/kT)$.

22.10* the relative permittivity should decrease.

22.11* $a = 2\pi N_A^2 C_6/3d^3$.

22.12 when $A = \sigma = 1$, a minimum occurs at $r = 1.63$.

22.14*

22.16* $\xi = -e^2 a_0^2/2m_e, \chi_m = -N_A\mu_0 e^2 a_0^2/2m_e$.

22.17 the susceptibility varies as $d = 1 - \{1/[(4p/K) + 1]\}^{1/2}$, where d = degree of dimerization.

22.18 $\chi_m = (25.2 \text{ cm}^3 \text{ mol}^{-1})/\{(T/K) \times [1 + e^{174(T/K)}]\}$.

22.19 $4.80 \times 10^{-40} \text{ J}^{-1} \text{ C m}^2, 0.26 \times 10^{-40} \text{ J}^{-1} \text{ C m}^2$.

22.20* (a) $1.51 \times 10^{-21} \text{ J}, 265 \text{ pm}$.

22.21* $0.127 \text{ cm}^3 \text{ mol}^{-1}$ ($S = 2$), $0.254 \text{ cm}^3 \text{ mol}^{-1}$ ($S = 3$), $0.423 \text{ cm}^3 \text{ mol}^{-1}$ ($S = 4$), $0.254 \text{ cm}^3 \text{ mol}^{-1}$.

22.22 0.123.

22.23 $8.14 \text{ cm}^3 \text{ mol}^{-1}, 1.76, 1.33$.

23.1* $23.1 \text{ kg mol}^{-1}, 1.02 \text{ m}^3 \text{ mol}^{-1}$.

23.2 $155 \text{ kg mol}^{-1}, 13.7 \text{ m}^3 \text{ mol}^{-1}$.

23.3* 0.0716 L g^{-1}.

23.4* 5.0 Sv.

23.5 65.6 kg mol^{-1}.

23.6 3500 r.p.m.

23.7* -29 mV.

23.8* $5 \text{ m}^3 \text{ mol}^{-1}$.

23.9 $69 \text{ kg mol}^{-1}, 3.4 \text{ nm}$.

23.10* 0.21 Mg mol^{-1}.

23.11 $5.14 \text{ Sv}, 60.1 \text{ kg mol}^{-1}$.

23.12* 158 kg mol^{-1}.

23.13 (a) $(\frac{3}{5})^{1/2}a$; (b) $l/2(3)^{1/2}, 2.40 \text{ nm}, 46 \text{ nm}$.

23.14* serum albumin and bushy stunt virus resemble spheres; DNA does not.

23.15 PBLG is rod-like; polystyrene is a random coil.

23.16* 1.01 g cm^{-3}.

23.17 $\bar{M}_n = \bar{M} + (2\gamma/\pi)^{1/2}$.

23.18* $dG = -S\,dT - l\,dt, dA = -S\,dT + t\,dl$.

23.19 $t = -T(\partial S/\partial l)_T$.

23.21* (a) $lN^{1/2}, 9.74 \text{ nm}$; (b) $(8N/3\pi)^{1/2}l, 8.97 \text{ nm}$; (c) $(2N/3)^{1/2}l, 7.95 \text{ nm}$.

23.22* $K = 2.73 \text{ cm}^3 \text{ g}^{-1} \text{ kg}^{-1/2} \text{ mol}^{1/2}, a = 0.500,$
$M = 1.34 \times 10^3 \text{ kg mol}^{-1}$.

23.23 (a) $\text{g cm}^3 \text{ mol}^{-1} \text{ K}^{-1}$, (b) $1.1 \times 10^5 \text{ g mol}^{-1}$, (c) 'good', (d) $B' = 21.4 \text{ cm}^3 \text{ g}^{-1}, C' = 211 \text{ cm}^6 \text{ g}^{-2}$, (e) $(\pi/c)^{1/2} = (RT/\bar{M}_n)^{1/2} \times (1 + \frac{1}{2}B'c); B' = 28.0 \text{ cm}^3\text{g}^{-1},$
$C' = 196 \text{ cm}^6 \text{ g}^{-2}$; yes.

23.24 (a) toluene: $0.086 \text{ L g}^{-1}, 0.37$; cyclohexane: 0.042 L g^{-1}, 0.35; (b) toluene: $2.4 \times 10^5 \text{ g mol}^{-1}$; cyclohexane: $2.6 \times 10^5 \text{ g mol}^{-1}$; (c) toluene: 42 nm; cyclohexane: 34 nm; (d) toluene: 2.3×10^3; cyclohexane: 2.5×10^3; (e) toluene: $5.8 \times 10^2 \text{ nm}$; cyclohexane: $6.2 \times 10^2 \text{ nm}$; (f) toluene: $2.1 \times 10^2 \text{ nm}, 7.4 \text{ nm}$; cyclohexane: $2.2 \times 10^2 \text{ nm}, 7.7 \text{ nm}$; (g) no reason for them to agree; the manufacturer's claim's valid.

23.25* $1.26 \times 10^5 \text{ g mol}^{-1}, 1.23 \times 10^4 \text{ L mol}^{-1}$.

23.26* (a) $K = 0.0117 \text{ cm}^3 \text{ g}^{-1}, a = 0.717$, (b) THF is polar. The constants depend upon both solute and solvent and their interactions.

23.27 $K = 2.38 \times 10^{-3}$ cm^3 g^{-1}, $a = 0.955$; the constants are considerably different and indicate that the conducting polymer exists in a linear chain form.

23.28 $M_W = 1.23 \times 10^6$ g mol^{-1}, $B' = 3.72 \times 10^{-2}$ cm^3 mg^{-1}. The average molar masses are different and there is no reason for them to be the same; different preparations of the same polymer can lead to much different average molar masses.

24.1* 100 Ms.

24.2 9.1.

24.3* 7.3 mPa.

24.4 (a) $Z_w = 2.69 \times 10^{23}$ cm^{-2} s^{-1}, $Z_{atom} = 2.0 \times 10^8$;
(b) 1.6×10^3 s^{-1}.

24.5* (a) 100 Pa; (b) 24 Pa.

24.6 (a) 2×10^{14} s^{-1}; (b) 1×10^{20} s^{-1}.

24.7* 5.3 S cm^2 mol^{-1}.

24.8* (a) 11.96 mS m^2 mol^{-1}; (b) 119.6 mS m^{-1}; (c) 172.5 Ω.

24.9 1.36×10^{-5} mol L^{-1}.

24.10* 40 μm s^{-1}, 52 μm s^{-1}, 76 μm s^{-1}; 250 s, 190 s, 130 s;
(a) 13 nm, 17 nm, 24 nm. (b) 43, 55, 81.

24.11 0.82, 0.0028.

24.12* 0.48, 7.5×10^{-8} m^2 s^{-1} V^{-1}, 72 S cm^2 mol^{-1}.

24.13 0.278, 0.278.

24.14* (a) 12 kN mol^{-1}, 2.1×10^{-20} N molecule^{-1};
(b) 17 kN mol^{-1}, 2.8×10^{-20} N molecule^{-1};
(c) 25 kN mol^{-1}, 4.1×10^{-20} N molecule^{-1}.

24.15* Li$^+$: 4 water molecules; Na$^+$: 1 to 2 water molecules.

24.16 $E_a = 9.3$ kJ mol^{-1}.

24.17* (a) ≈ 0; (b) 63 mmol L^{-1}.

24.18 $t'/t'' = c'u'z'/c''u''z''$.

24.20* (a) 0; (b) 0.016; (c) 0.054.

24.21 $n > 60$.

24.22* Λ_m°(NaI) = 60.7 S cm^2 mol^{-1},
Λ_m°(KI) = 58.9 S cm^2 mol^{-1}, λ°(Na$^+$) $- \lambda^\circ$(K$^+$) = 1.8 S cm^2 mol^{-1}; the analogous quantities in water are, respectively, 126.9, 150.3, and -23.4 S cm^2 mol^{-1}.

24.23 (a) 368 pm, (b) 307 pm.

24.24* 1.6×10^{16} m^2 s^{-1}, 0.34 J K^{-1} m^{-1} s^{-1}.

24.25 830 pm.

25.1* 2, 59 mL mol^{-1} min^{-1}, 2.94 g.

25.2 1, 1.51×10^{-5} s^{-1}, 9.82 mmol L^{-1}.

25.3* 1, 1.2×10^{-4} s^{-1}.

25.4 1, 5.84×10^{-3} s^{-1}, 1.98 min.

25.5* 97.0 kJ mol^{-1}.

25.6* 55.4 per cent.

25.7 1, 2.8×10^{-4} s^{-1}.

25.8* 3.65×10^{-3} min^{-1}, 274 min.

25.9 2.37×10^7 L mol^{-1} s^{-1}.

25.10* 1, 7.2×10^{-4} s^{-1}.

25.11* propene: 1; HCl: 3.

25.12* rate = kK_2K_1[HCl]^3[CH$_3$CH=CH$_2$].

25.13 -18 kJ mol^{-1}, $+10$ kJ mol^{-1}.

25.14* 1.14×10^{10} L mol^{-1} s^{-1}, 16.7 kJ mol^{-1}.

25.15 deviates from theory at low pressures.

25.16* 10 mmol L^{-1}.

25.17* $[B]_\infty/[A]_\infty = k/k'$.

25.19*

25.20 equivalent when B is a reactive intermediate.

25.21 $t_{1/2}/t_{3/4} = (2^{n-1} - 1)/\{(\frac{4}{3})^{n-1} - 1\}$.

25.22*

25.24* $E_a = 105$ kJ mol^{-1}, $\Delta G = -26.6$ kJ mol^{-1}, $\Delta H = -34.3$ kJ mol^{-1}; still favourable under prebiotic conditions.

25.25 $v_{max} = k(([A]_0 = [B]_0)/2)^2$ for $[B]_0 \leq [A]_0$, $(x/[A]_0) \leq 1$ corresponds to reality.

25.26* 2.01 min, 1 min, 0.693 min.

25.27 $x = \{[B]_0 - [A]_0 f(t)\}/(1 - f(t))$;
$f(t) = \exp\{-([A]_0 - [B]_0)kt\}$.

25.28* 1.03×10^9 L mol^{-1} s^{-1}, 13.9 kJ mol^{-1}.

25.29 9×10^{-10} mol L^{-1} s^{-1}, 3×10^5 s.

25.30* (a) 2.1×10^{-15} mol L^{-1} s^{-1}; (b) 1.6×10^{-15} mol L^{-1} s^{-1}.

25.31 (a) 1.1×10^{-16} mol L^{-1} s^{-1}; (b) 2.2×10^{11} kg or 220 Tg.

25.32* 121.2 kJ mol^{-1}, 247.0 J K^{-1} mol^{-1}.

26.1* 1.9×10^{20} s^{-1}, 3.1×10^{-4} einstein s^{-1}.

26.2 5.1×10^8 L mol^{-1} s^{-1}.

26.3*

26.4 5.0×10^7 L mol^{-1} s^{-1}.

26.5* $N(t) = N_0 e^{(b-d)t}$; fits data with $R^2 = 0.983$.

26.6* $([A]_0 + [P]_0)^2 kt_{max} = \frac{1}{2} - p - \ln 2p$.

26.7 $([A]_0 + [P]_0)^2 kt_{max} = (2 - p)/2p + \ln(2/p)$.

26.9* $f = k_2k_4$[CO]$/\{k_2$[CO] $+ k_3$[M]$\}$.

26.11* $\delta M = M\{kt[A]_0(1 + kt[A]_0)\}^{1/2}$.

26.12 (a) $M(1 + 4p + p^2)/(1 - p)^2$; (b) $(6\langle n \rangle^2 - 6\langle n \rangle + 1)\langle n \rangle$.

26.13* ratio = $M(1 + 4p + p^2)/(1 - p^2)$.

26.15 [B] $= (\mathcal{I}_a/k)^{1/2} \propto$ [A]$^{1/2}$.

26.16* [X] $= k_c/k_b$, [Y] $= k_a$[A]$/k_b$.

26.18* Step 1 is autocatalytic. $a/r < S_0$: infection spreads; $a/r > S_0$: infection dies out.

26.19*

26.20 5.9×10^{-13} mol L^{-1} s^{-1}.

26.21* (a) For both X = H and X = Cl, the fit of the data to the second-order integrated rate expression is superior to the fit to the first-order rate expressions as evidenced by comparing correlation coefficients and standard deviations of the slopes and intercepts.
(b) rate = $-k_2k_1$[(ClRh(CO)$_2$)$_2$][ArHgCl]^2/[HgCl$_2$].

26.22* (a) initiation, propagation, propagation, termination, initiation; (b) $d[NO]/dt = -2k_b(k_{-d}/k_d)^{1/2}[O_2]^{1/2}[NO]$; (c) $E_{a,eff} = E_b + \frac{1}{2}E_{-d} - \frac{1}{2}E_d$; (d) $E_{a,eff} = 381$ kJ mol^{-1}, consistent with high end of range; (e) $d[NO]/dt = -2k_b(k_a/2k_d[M])^{1/2}[NO]^2$; (f) $d[NO]/dt = -2k_b(k_e/2k_d[M])^{1/2}[O]^{1/2}[NO]^{3/2}$, where k_e is the rate constant for $NO + O_2 \rightarrow O + NO_2$; $E_{a,eff} = 253$ kJ mol^{-1}, consistent with low end of range.

26.24* When step (b) is rate-determining and step (a) is a rapid equilibrium so that [I] is in a steady state.

27.1 (a) 0.044 nm^2; (b) 0.15.

27.2* 0.007, 0.0040 nm^2.

27.3 1.7×10^{11} L mol^{-1} s^{-1}, 3.6 ns.

27.4* $+83.8$ kJ mol^{-1}, $+19.1$ J K^{-1} mol^{-1}, 85.9 kJ mol^{-1}, $+79.0$ kJ mol^{-1}.

27.5* 2$-$.

27.6 0.658 L mol^{-1} min^{-1}.

27.8*

27.10 $P = 5.2 \times 10^{-6}$.

27.11* $\log v \propto I^{1/2}$.

27.12* 1.4×10^6 L mol^{-1} s^{-1}.

27.13* 1.2×10^6 L mol^{-1} s^{-1}.

27.15* (a) 2.7×10^{-15} m^2 s^{-1}; (b)1.1×10^{-14} m^2 s^{-1}.

27.16 2×10^{-13} m^2 s^{-1} if $v^{\ddagger} = v$; 9×10^{-13} m^2 s^{-1}if $v^{\ddagger} = \frac{1}{2}v$.

27.17*

27.18 5.

27.19* $k_1 = 3.82 \times 10^6$ L mol^{-1} s^{-1}, $k_2 = 5.1 \times 10^5$ L mol^{-1} s^{-1}, $k_3 = 4.17 \times 10^6$ L mol^{-1} s^{-1}, $k_2/k_1 = 0.13$.

27.20 1.6×10^{-3}, 1.8×10^{-3}.

27.21* A complex of two univalent ions of the same sign.

27.22 6.23×10^9 L mol^{-1} s^{-1}, 0.37 nm.

27.23* $k = k_1k_2/k_4$, $k' = k_1$; (b) $D(FO-F) \approx E_{a1} = 140.6$ kJ mol^{-1}, $D(O-F) = 244.7$ kJ mol^{-1}, $E_{a2} = 20.3$ kJ mol^{-1}.

27.24 k_2 (atoms)$/k_2$(molecules) $\approx 3 \times 10^7$.

27.25 -148 J K^{-1} mol^{-1}, 60.44 kJ mol^{-1}, 62.9 kJ mol^{-1}, 104.8 kJ mol^{-1}, respectively.

27.26* ln k fits the Arrhenius equation with $A = 3.12 \times 10^{14}$ L mol^{-1} s^{-1}, $E_a = 193$ kJ mol^{-1}, $R = 0.99976$; ln k' fits with $A' = 7.29 \times 10^{11}$ L mol^{-1} s^{-1}, $E'_a = 175$ kJ mol^{-1}, $R = 0.99848$.

28.1 (a) 1.61×10^{15} cm^{-2}; (b) 1.14×10^{15} cm^{-2}; (c) 1.86×10^{15} cm^{-2}.

28.2* 0.37 Torr^{-1}.

28.3 (a) 164, 13.1 cm^3; (b) 264, 12.5 cm^3.

28.4* 1.4 mL, 5.9 m^2.

28.5 BET better; 75.4 cm^3, 3.98.

28.6* 2.4, 0.16.

28.7 0.02 Torr s^{-1}.

28.9* $U \propto C_6/R^3$, 294 pm.

28.10*

28.11*

28.12 $d\mu' = (RTV_\infty/\sigma)\, d \ln(1 - \theta)$.

28.13* (a) $v = gkT/p$; (b) $R = 0.959$ for the linear equation which yields $g = 1.0 \times 10^{21}$ m^{-3} s^{-1}. $z(t)$ is seen to be nonlinear; $z(t) = a(e^{bt} - 1)$, with $a = 5.71$ cm, $b = 0.35$ s^{-1}, $R = 0.994$.

28.14 $K = 0.138$ mg g^{-1}, $n = 0.58$, amount corresponding to monolayer coverage must be known.

28.15* -20.1 kJ mol^{-1}, -63.6 kJ mol^{-1}.

28.16 $n = 5.78$ mol kg^{-1}, $K = 7.02$ Pa^{-1}.

28.17* 40.4.

28.18 Regression analysis provides the following coefficients of correlation: R(Langmuir) $= 0.973$; R(Freundlich) $= 0.99994$; R(Temkin) $= 0.9590$. The fit to the Freundlich isotherm is best.

28.19* (a) R values range from 0.975 to 0.998; the fit is good at all temperatures. (b) $k_a = 3.68 \times 10^{-3}$, $k_b = 2.62 \times 10^{-5}$ ppm^{-1}, $\Delta_{ad}H = -8.67$ kJ mol^{-1}, $\Delta_b H = -15.7$ kJ mol^{-1}.

28.20* (a) $k = 0.2289$ and $n = 0.6180$; $R = 0.9995$. (c) $k = 0.5227$, $n = 0.7273$; $R = 0.996$.

28.21 (a) $K[(\text{mg L}^{-1})^{-1}]$, $K_F[(\text{mg L}^{-1})^{-1/n}]$, $K_L[(\text{mg L}^{-1})^{-1}]$; (b) R(linear) $= 0.9612$, R(Freundlich) $= 0.9682$, R(Langmuir) $= 0.9690$; on that basis alone the fits are equally satisfactory, but not good. The Langmuir isotherm can be eliminated, as it gives a negative value for K_L; the fit to the Freundlich isotherm has a large standard deviation. Hence, the linear isotherm seems best, but the Freundlich isotherm is preferred for this kind of system. (c) $q_{rubber}/q_{charcoal} = 0.164c_{eq}^{-0.46}$; hence much worse.

29.1* 0.38, 0.78 mA cm^{-2}.

29.2 $a(Sn^{2+}) \approx 2.2a(Pb^{2+})$.

29.3* 0.25 mm.

29.4 45 per cent, 45 per cent.

29.5* 87 mA.

29.6*

29.7 6 μA.

29.8* 0.28 mg cm^{-2} d^{-1}.

29.9* 7.2 μA.

29.11* $j/j_L = 1 - \exp(F\eta_c/RT)$.

29.12* (a) $E_0 = -0.618$ V, η/mV $= -84, -109, -134, -194$; (b) $j_c/(\mu\text{A cm}^{-2}) = 0.0324, 0.0469, 0.0663, 0.154$; (c) $j_0 = 0.00997$ μA cm^{-2}, $\alpha = 0.363$; excellent fit, $R = 0.99994$.

29.14 0.50, 0.150 A m^{-2}, -0.038 A m.

29.15* (a) No linear region exists; the Tafel equation cannot be used to calculate j_0 and α.

29.16* $j_0 = 2.00 \times 10^{-5}$ mA m^{-2}, $\alpha = 0.498$; $R = 0.9990$. There are no significant deviations.

Partial answers to microprojects

Part 1: Equilibrium

1.1 (c) 18.30 cm^3 mol^{-1}, 5789 cm^3 mol^{-1}.

1.2 (a) 501 K; (b) virial: 0.455 L mol^{-1}, 0.598 L mol^{-1}; perfect: 0.496 L mol^{-1}, 0.621 L mol^{-1}; (c) $\pi_T = 2.91$ kJ L^{-1}, $\Delta S_m = -40.4$ J K^{-1} mol^{-1}, $\Delta H_m = -0.99$ kJ mol^{-1}, $\Delta G_m = +14.1$ kJ mol^{-1}.

1.4 $\Delta_f H^{\ominus}(\text{Na}^+, \text{aq}) = -240.65$ kJ mol^{-1}.

1.5 (a) $\Delta_{sub}H(1\ \text{Torr}) = +80.84$ kJ mol^{-1}, $\Delta_{sub}S(1\ \text{Torr}) = +201.1$ J K^{-1} mol^{-1}; (b) $\Delta_{atom}H = +11\,525 \pm 5$ kJ mol^{-1}.

1.8 (b) -24.9 kJ mol^{-1}.

1.9 (e) 45.1 kPa for methanol in TAME, 25.3 kPa for TAME in methanol; (f) 6.16 kPa.

1.10 (a) $F(1500\ \text{K}) = 3$, $F(1100\ \text{K}) = 2$; (c) 0.67 g Pb per g Cu; 0.050 g Cu per g Pb.

1.12 (a) $p(450°\text{C}) = 156.5$ bar, $p(400°\text{C}) = 81.8$ bar; (b) $p(450°\text{C}) = 132.5$ bar, $p(400°\text{C}) = 210$ bar.

1.13 (a) mp(SiO$_2$) $= 1934.1$ K, mp(Si) $= 1715.5$ K; (c) 5.3 per cent carbon impurity.

1.14 (b) maximum carbon mass in ocean $\approx 2.67 \times 10^{12}$ tonne.

1.15 (b) $E^{\ominus} = 2.05$ V, $\Delta_r G^{\ominus} = -395.6$ kJ mol^{-1}, $\Delta_r H^{\ominus} = -586.21$ kJ mol^{-1}, $\Delta_r S^{\ominus} = -639.29$ J K^{-1} mol^{-1}; $E^{\ominus}(15°\text{C}) = 2.083$ V; $E(25°\text{C}, Q) = 2.185$ V; (c) pH $= 5$: 1.0997 V; pH $= 8$: 0.745 V; $E(\text{PbSO}_4/\text{Pb}) = -0.3588$ V.

1.16 $pK_a(25°\text{C}) = 6.736$; $a_0 = -14.102$, $a_1 = 2461$ K, $a_2 = 2.209$.

Part 2: Structure

2.1 (b) 255.1 K; (c) 281.9 K, 81.1 per cent.

2.3 (b) $3s$ radial nodes: $\rho = 3 \pm 3^{1/2}$, no nodal plane; $3p_x$ radial nodes: $\rho = 0$ and 4, yz nodal plane ($\phi = 90°$); $3d_{xy}$ radial node: $\rho = 0$ only, xz nodal plane ($\phi = 0$) and yz nodal plane ($\phi = 90°$); (c) $13.5a_0/Z$.

2.4 (b) $n_{max} \approx 29$.

2.5 (c) $A/hc = 46\,119$ cm^{-1}, $B = 1.4275$, $C = 2.3896$, $D = 88.99$ pm, $\varepsilon/hc = 1750$ cm^{-1}.

2.6 (b) $\nu = 2169.2$ cm^{-1}, $x\nu = 13.13$ cm^{-1}, $B = 1.94$ cm^{-1}, $\alpha = 0.0174$ cm^{-1}, $D_J = 6.19 \times 10^{-6}$ cm^{-1};
(c) $I_e = 1.444 \times 10^{-46}$ kg m^2, $R_e = 112.63$ pm, $D_e = 89\,595$ cm^{-1}, $D_0 = 88\,574$ cm^{-1}, $I_{v=0} = 1.451 \times 10^{-46}$ kg m^2, $I_{v=1} = 1.464 \times 10^{-46}$ kg m^2, $R_{v=0} = 112.88$ pm, $R_{v=1} = 113.39$ pm.

2.7 (c) $h \leq 30$ m.

2.9 (b) $E_{min} = -15.367$ eV, $R_e = 131.864$ pm, $D_e = 1.7612$ eV.

2.10 (a) $E_{min} = -15.96$ eV, $R_e = 106.009$ pm, $D_e = 2.354$ eV, $\eta = 1.238$.

2.11 (b) $\beta \approx -21\,135$ cm^{-1}; (c) $E_{deloc} = 1.5176\beta$.

Part 3: Change

3.1 (b) $J_0 = 7.55$ mmol m^{-2} h^{-1}.

3.2 (b) $\Lambda_m^0 = 391.347$ S cm^2 mol^{-1}, $K_a = 0.157$.

3.3 (b) $\log(A) = 11.09$, $E_a = 73.8$ kJ mol^{-1}; at 293.15 K: $\Delta^\ddagger H = 71.4$ kJ mol^{-1}, $\Delta^\ddagger S = -40.8$ J K^{-1} mol^{-1};
(c) $a = 0.411$, $b = -0.192$, $c = 1.369 \times 10^{-3}$ K^{-1}.

3.4 (a) $a = 1$, $b = 0.5$;
(b) $k(433.15$ K$) = 0.077$ cm$^{3/2}$ s^{-1} mol$^{-1/2}$, $A = 5.002 \times 10^{16}$ cm$^{3/2}$ s^{-1} mol$^{-1/2}$, $E_a = 147.7$ kJ mol^{-1}.

3.5 (a) 0.012 per cent ozone.

3.6 (a) period ≈ 5 years; (b) attractor: $[X] = 22.35$, $[Y] = 21.88$.

3.7 For x^{-1} versus c^{-1} regression: $r = 0.99836$, coefficient of variation $= 7.381$, $x_{max} = 0.380$, $K = 0.351$ L mg^{-1}, specific surface area $= 658$ m^2 g^{-1}.

Index

(T) after a page number refers to a Table in the text (and usually in the Data Section)

The Periodic Table

Period	1	2		3	4	5	6	7	8	9	10	11	12	13/III	14/IV	15/V	16/VI	17/VII	18/VIII
1	1 H 1.0079																		2 He 4.003
2	3 Li 6.941	4 Be 9.012												5 B 10.81	6 C 12.01	7 N 14.01	8 O 16.00	9 F 19.00	10 Ne 20.18
3	11 Na 22.99	12 Mg 24.30												13 Al 26.98	14 Si 28.09	15 P 30.97	16 S 32.07	17 Cl 35.45	18 Ar 39.95
4	19 K 39.10	20 Ca 40.08		21 Sc 44.96	22 Ti 47.87	23 V 50.94	24 Cr 52.00	25 Mn 54.94	26 Fe 55.85	27 Co 58.93	28 Ni 58.69	29 Cu 63.55	30 Zn 65.39	31 Ga 69.72	32 Ge 72.61	33 As 74.92	34 Se 78.96	35 Br 79.90	36 Kr 83.80
5	37 Rb 85.47	38 Sr 87.62		39 Y 88.91	40 Zr 91.22	41 Nb 92.91	42 Mo 95.94	43 Tc 98.91	44 Ru 101.1	45 Rh 102.9	46 Pd 106.4	47 Ag 107.9	48 Cd 112.4	49 In 114.8	50 Sn 118.7	51 Sb 121.8	52 Te 127.6	53 I 126.9	54 Xe 131.3
6	55 Cs 132.9	56 Ba 137.3		La–Lu	72 Hf 178.5	73 Ta 180.9	74 W 183.8	75 Re 186.2	76 Os 190.2	77 Ir 192.2	78 Pt 195.1	79 Au 197.0	80 Hg 200.6	81 Tl 204.4	82 Pb 207.2	83 Bi 209.0	84 Po 210.0	85 At 210.0	86 Rn 222.0
7	87 Fr 223.0	88 Ra 226.0		Ac–Lr	104 Rf	105 Db	106 Sg	107 Bh	108 Hs	109 Mt	110 Uun	111 Uuu	112 Uub	113 Uut					

s block   d block   p block

Lanthanides

57 La 138.9	58 Ce 140.1	59 Pr 140.9	60 Nd 144.2	61 Pm 146.9	62 Sm 150.4	63 Eu 152.0	64 Gd 157.2	65 Tb 158.9	66 Dy 162.5	67 Ho 164.9	68 Er 167.3	69 Tm 168.9	70 Yb 173.0	71 Lu 175.0

Actinides

89 Ac 227.0	90 Th 232.0	91 Pa 231.0	92 U 238.0	93 Np 237.0	94 Pu 239.1	95 Am 241.1	96 Cm 244.1	97 Bk 249.1	98 Cf 252.1	99 Es 252.1	100 Fm 257.1	101 Md 256.1	102 No 259.1	103 Lr 260.1

f block